# MOLECULAR CELL BIOLOGY

# ABOUT THE AUTHORS

**HARVEY LODISH** is Professor of Biology and Professor of Bioengineering at the Massachusetts Institute of Technology and a Founding Member of the Whitehead Institute for Biomedical Research. Dr. Lodish is also a member of the National Academy of Sciences and the American Academy of Arts and Sciences and was President (2004) of the American Society for Cell Biology. He is well known for his work on cell-membrane physiology, particularly the biosynthesis of many cell-surface proteins, and on the cloning and functional analysis of several cell-surface receptor proteins, such as the erythropoietin and TGF–β receptors. His laboratory also studies hematopoietic stem cells and has identified novel proteins that support their proliferation. Dr. Lodish teaches undergraduate and graduate courses in cell biology and biotechnology. **Photo credit:** John Soares/Whitehead Institute.

**ARNOLD BERK** holds the UCLA Presidential Chair in Molecular Cell Biology in the Department of Microbiology, Immunology, and Molecular Genetics and is a member of the Molecular Biology Institute at the University of California, Los Angeles. Dr. Berk is also a fellow of the American Academy of Arts and Sciences. He is one of the original discoverers of RNA splicing and of mechanisms for gene control in viruses. His laboratory studies the molecular interactions that regulate transcription initiation in mammalian cells, focusing in particular on adenovirus regulatory proteins. He teaches an advanced undergraduate course in cell biology of the nucleus and a graduate course in biochemistry.

**CHRIS A. KAISER** is Professor and Head of the Department of Biology at the Massachusetts Institute of Technology. His laboratory uses genetic and cell biological methods to understand the basic processes of how newly synthesized membrane and secretory proteins are folded and stored in the compartments of the secretory pathway. Dr. Kaiser is recognized as a top undergraduate educator at MIT, where he has taught genetics to undergraduates for many years.

**MONTY KRIEGER** is the Whitehead Professor in the Department of Biology at the Massachusetts Institute of Technology and a Senior Associate Member of the Broad Institute of MIT and Harvard. Dr. Krieger is also a member of the National Academy of Sciences. For his innovative teaching of undergraduate biology and human physiology as well as graduate cell-biology courses, he has received numerous awards. His laboratory has made contributions to our understanding of membrane trafficking through the Golgi apparatus and has cloned and characterized receptor proteins important for pathogen recognition and the movement of cholesterol into and out of cells, including the HDL receptor.

**ANTHONY BRETSCHER** is Professor of Cell Biology at Cornell University and a member of the Weill Institute for Cell and Molecular Biology. His laboratory is well known for identifying and characterizing new components of the actin cytoskeleton and elucidating the biological functions of those components in relation to cell polarity and membrane traffic. For this work, his laboratory exploits biochemical, genetic, and cell biological approaches in two model systems, vertebrate epithelial cells and the budding yeast. Dr. Bretscher teaches cell biology to undergraduates at Cornell University.

**HIDDE PLOEGH** is Professor of Biology at the Massachusetts Institute of Technology and a member of the Whitehead Institute for Biomedical Research. One of the world's leading researchers in immune system behavior, Dr. Ploegh studies the various tactics that viruses employ to evade our immune responses and the ways our immune system distinguishes friend from foe. Dr. Ploegh teaches immunology to undergraduate students at Harvard University and MIT.

**ANGELIKA AMON** is Professor of Biology at the Massachusetts Institute of Technology, a member of the Koch Institute for Integrative Cancer Research, and Investigator at the Howard Hughes Medical Institute. She is also a member of the National Academy of Sciences. Her laboratory studies the molecular mechanisms that govern chromosome segregation during mitosis and meiosis and the consequences—aneuploidy—when these mechanisms fail during normal cell proliferation and cancer development. Dr. Amon teaches undergraduate and graduate courses in cell biology and genetics.

# MOLECULAR CELL BIOLOGY

## SEVENTH EDITION

Harvey Lodish

Arnold Berk

Chris A. Kaiser

Monty Krieger

Anthony Bretscher

Hidde Ploegh

Angelika Amon

Matthew P. Scott

W. H. Freeman and Company

New York

PUBLISHER: Katherine Ahr Parker
ACQUISITIONS EDITOR: Beth McHenry
DEVELOPMENTAL EDITORS: Matthew Tontonoz, Erica Pantages Frost, Erica Champion
ASSISTANT EDITOR: Marni Rolfes
ASSOCIATE DIRECTOR OF MARKETING: Debbie Clare
SENIOR PROJECT EDITOR: Mary Louise Byrd
TEXT DESIGNER: Marsha Cohen
PAGE MAKEUP: Aptara®, Inc.
COVER DESIGN: Victoria Tomaselli
ILLUSTRATION COORDINATOR: Susan Timmins
ILLUSTRATIONS: Network Graphics, Erica Beade, H. Adam Steinberg
PHOTO EDITOR: Cecilia Varas
PHOTO RESEARCHER: Christina Micek
PRODUCTION MANAGER: Julia DeRosa
MEDIA AND SUPPLEMENTS EDITOR: Beth McHenry
MEDIA DEVELOPERS: Biostudio, Inc., Sumanas, Inc.
COMPOSITION: Aptara®, Inc.
MANUFACTURING: RR Donnelley & Sons Company

Library of Congress Control Number: 2012932495


ISBN-13: 978-1-4292-3413-9
ISBN-10: 1-4292-3413-X

Printed in the United States of America

Second printing

W. H. Freeman and Company
41 Madison Avenue, New York, NY 10010
Houndmills, Basingstoke
RG21 6XS, England

www.whfreeman.com

**To our students and to our teachers,
from whom we continue to learn, and to our families,
for their support, encouragement, and love**

In writing the seventh edition of *Molecular Cell Biology* we have incorporated many of the spectacular advances made over the past four years in biomedical science, driven in part by new experimental technologies that have revolutionized many fields. Fast techniques for sequencing DNA and RNA, for example, have uncovered many novel noncoding RNAs that regulate gene expression and identified hundreds of human genes that affect diseases such as diabetes, osteoporosis, and cancer. Genomics has also led to many novel insights into the evolution of life forms and the functions of individual members of multiprotein families. Exploring the most current developments in the field is always a priority in writing a new edition, but it is also important to us to communicate the basics of cell biology clearly. To this end, in addition to introducing new discoveries and technologies, we have streamlined and reorganized several chapters to clarify processes and concepts for students.

## New Co-Author, Angelika Amon

The new edition of *MCB* introduces a new member to our author team, respected researcher and teacher Angelika Amon of the Massachusetts Institute of Technology. Her laboratory uses the budding yeast *S. cerevisiae* and mouse and cell culture models to gain a detailed molecular understanding of the regulatory circuits that control chromosome segregation and the effects of aneuploidy on cell physiology. Dr. Amon also teaches undergraduate and graduate courses in Cell Biology and Genetics.

## Revised, Cutting Edge Content

The seventh edition of *Molecular Cell Biology* includes new and improved chapters:

• "Molecules, Cells and Evolution" (Chapter 1) now frames cell biology in the light of evolution: this perspective explains why scientists pick particular unicellular and multicellular "model" organisms to study specific genes and proteins that are important for cellular function.

• "Culturing, Visualizing, and Perturbing Cells" (Chapter 9) has been rewritten to include cutting edge methods including FRAP, FRET, siRNA, and chemical biology, making it a state-of-the-art methods chapter.

• "Signal Transduction and G Protein-Coupled Receptors" and "Signaling Pathways that Control Gene Expression" (Chapters 15 and 16) have been reorganized and illustrated with simplified overview figures, to help students navigate the complexity of signaling pathways.

• "The Eukaryotic Cell Cycle" (Chapter 19) now begins with the concepts of "START" (a cell's commitment to entering the cell cycle starting with DNA synthesis) and then progresses through the cycle stages. The chapter focuses on yeast and mammals and uses general names for cell cycle components when possible to improve student understanding.

• "Stem Cells, Cell Asymmetry, and Cell Death" (Chapter 21) now incorporates developmental topics, including new coverage of induced pluripotent stem (iPS) cells.

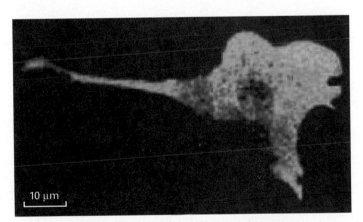

**FIGURE 9-22** In this mouse fibroblast, FRET has been used to reveal that the interaction between an active regulatory protein (Rac) and its binding partner is localized to the front of the migrating cell.

## Increased Clarity, Improved Pedagogy

As experienced teachers of both undergraduate and graduate students, we are always striving to improve student understanding. In this seventh edition, perennially confusing topics, such as cellular energetics, cell signaling, and immunology, have been streamlined and revised to improve student understanding. Each figure was reconsidered and, if possible, simplified to highlight key lessons. Heavily revised end-of-chapter materials include 30% new questions, including additional Analyze the Data problems to give students further practice at interpreting experimental evidence. The result is a balance of state-of-the-art currency and experimental focus with attention to clarity, organization, and pedagogy.

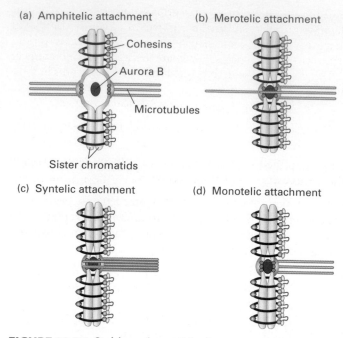

**FIGURE 19-25** Stable and unstable chromosome attachments.

(a) Amphitelic attachment
- Cohesins
- Aurora B
- Microtubules
- Sister chromatids

(b) Merotelic attachment

(c) Syntelic attachment

(d) Monotelic attachment

## New Discoveries, New Methodologies

- Covalent regulation of protein activity by ubiquitination/deubiquitination (Ch. 3)

- Molecular chaperones including the Hsp90 family of proteins (Ch. 3)

- Mammalian protein synthesis and the roles of polymerases delta (lagging strand) and epsilon (leading strand) in eukaryotic DNA synthesis (Ch. 4)

- Non-radioactive probes (for in-situ hybridization, for example) (Ch. 5)

- Quantitative PCR (and RT-PCR) and high-throughput DNA sequencing (Ch. 5)

- DNA fingerprinting using microsatellites and PCR (Ch. 6)

- Personal genome sequencing and the 1000 Genome Project (Ch. 6)

- Epigenetic mechanisms of transcriptional regulation (Ch. 7)

- Transcriptional regulation by non-coding RNAs (e.g., Xist in X-chromosome inactivation, siRNA-directed heterochromatin formation in fission yeast and DNA methylation in plants) (Ch. 7)

- Fluorescent mRNA labeling to follow mRNA localization in live cells (Ch. 8)

- Structure and function of the nuclear pore complex (Chs. 8 and 13)

- Additional coverage of FRAP, FRET, and siRNA techniques (Ch. 9)

- Lipid droplets and their formation (Ch. 10)

- Assembly of the multiprotein T-cell receptor complex (Ch. 10)

- Structure of the $Na^+/K^+$ ATPase (Ch. 11)

- Structure and mechanism of the multidrug transporter ABCB1 (MDR1) (Ch. 11)

- Structure and function of the cystic fibrosis transmembrane regulator (CFTR) (Ch. 11)

- The role of an anion antiporter in bone resorption (Ch. 11)

- Structures of complex I and II as well as the mechanism of electron flow and proton pumping in the electron transport chain (Ch. 12)

- Generation and inactivation of toxic reactive oxygen species (ROS) (Ch. 12)

- The mechanism of proton flow through the half-channels of ATP Synthase (Ch. 12)

- Tail-anchored membrane proteins (Ch. 13)

- How modifications of *N*-linked oligosaccharides are used to monitor protein folding and quality control (Ch. 13)

- The mechanism of formation of multivesicular endosomes involving ubiquitination and ESCRT (Ch. 14)

- Advances in our understanding of autophagy as a mechanism for recycling organelles and proteins (Ch. 14)

- Affinity purification techniques for studying signal transduction proteins (Ch. 15)

- Structure of the β-adrenergic receptor in the inactive and active states and with its associated trimeric G protein, $G_{\alpha s}$ (Ch. 15)

- Activation of EGF receptor by EGF via the formation of an asymmetric kinase domain dimer (Ch. 16)

- Hedgehog signaling in vertebrates involving primary cilia (Ch. 16)

- NF-κB signaling pathway and polyubiquitin scaffolds (Ch. 16)

- Integration of signals in fat cell differentiation via PPARγ (Ch. 16)

- Mechanism of Arp2/3 nucleation of actin filaments (Ch. 17)

- The dynamics of microfilaments during endocytosis and the role of endocytic membrane recycling during cell migration (Ch. 17)

- Intraflagellar transport and the function of primary cilia (Ch. 18)

- Plant mitosis and cytokinesis (Ch. 18)

- +TIPs as regulators of microtubule (+) end function (Ch. 18)

- Proteins involved in mitotic spindle formation and kinetochore attachment to microtubules (Ch. 19)

- Elastic fibers that permit many tissues to undergo repeated stretching and recoiling (Ch. 20)

- Extracellular matrix remodelling and degradation by matrix metalloproteinases (Ch. 20)

- Stem cells in the intestinal epithelium (Ch. 21)

- Regulation of gene expression in embryonic stem (ES) cells (Ch. 21)

- Generation of induced pluripotent stem (iPS) cells (Ch. 21)

- Advances in our understanding of regulated cell death (Ch. 21)

- Structure of the nicotinic acetylcholine receptor (Ch. 22)

- Molecular model of the MEC-4 touch receptor complex in *C. elegans* (Ch. 22)

- Synapse formation in neuromuscular junctions (Ch. 22)

- Toll-like receptors (TLRs) and the inflammasome (Ch. 23)

- Epigenetics and cancer (Ch. 24)

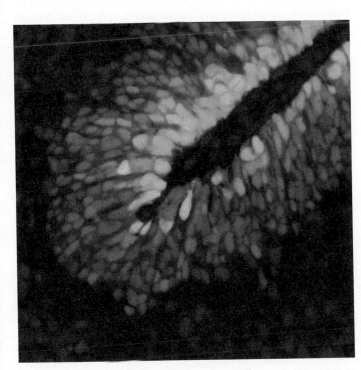

Cells being born in the developing cerebellum.

## Medical Relevance

Many advances in basic cellular and molecular biology have led to new treatments for cancer and other significant human diseases. These medical examples are woven throughout the chapters where appropriate to give students an appreciation for the clinical applications of the basic science they are learning. Many of these applications hinge on a detailed understanding of multiprotein complexes in cells—complexes that catalyze cell movements; regulate DNA transcription, replication, and repair; coordinate metabolism; and connect cells to other cells and to proteins and carbohydrates in their extracellular environment.

The following is a list of new medical examples.

- Cholesterol transport and atherosclerosis as an illustration of the hydrophobic effect (Ch. 2)

- Use of genetically engineered corn with high lysine content to promote the growth of livestock as an illustration of importance of essential amino acids (Ch. 2)

- Poliovirus and HIV-1 as examples of viruses that infect only certain cell types due to tissue-specific cell surface receptors (Ch. 4)

- HPV vaccine and its ability to protect against common types of HPV, and the development of cervical cancer (Ch. 4)

- Huntington's disease as an example of a microsatellite expansion disease (Ch. 6)

- Potential treatment of cystic fibrosis using small molecules that would allow the mutant protein to traffic normally to the cell surface (Ch. 11)

- Role of genetic defects in ClC-7, a chloride ion channel, in the hereditary bone disease osteopetrosis (Ch. 11)

- Mitochondrial diseases such as Charcot-Marie-Tooth disease and Miller syndrome (Ch. 12)

- Use of ligand-binding domains of cell-surface receptors as therapeutic drugs, such as the extracellular domain of TNFα receptor to treat arthritis and other inflammatory conditions (Ch. 15)

- Role of Hedgehog (Hh) signaling in human cancers including medulloblastomas and rhabdomyosarcomas (Ch. 16)

- Role of B-Raf kinase in melanoma and use of selective inhibitors of B-Raf in cancer treatment (Ch. 16)

- Defects in a regulator of dynein as a cause of lissencephaly (Ch. 18)

- Elastic fiber protein fibrillin 1 and Marfan's Syndrome (Ch. 20)

- Use of iPS cells in uncovering the molecular basis of ALS (Ch. 21)

- Variations in human sense of smell (Ch. 22)

- Microarray analysis of breast cancer tumors as a way to distinguish gene expression patterns and individualize treatment (Ch. 24)

## For Students

**\*NEW\* BioPortal for** *Molecular Cell Biology* A robust teaching and learning tool with all of the study and quizzing resources available through the Companion Web Site (listed below) as well as a fully-interactive eBook. BioPortal also includes **NEW LearningCurve**, a self-paced adaptive quizzing tool for students. With questions tailored to their target difficulty level and an engaging scoring system, LearningCurve encourages students to incorporate content from the text into their study routine and provides them with a study plan upon completion.

**Companion Web Site www.whfreeman.com/lodish7e**

- **Podcasts** narrated by the authors give students a deeper understanding of key figures in the text and a sense of the thrill of discovery.

- More than 125 **animations and research videos** show the dynamic nature of key cellular processes and important experimental techniques.

- **Classic Experiment** essays focus on classic groundbreaking experiments and explore the investigative process.

- **Online Quizzing** is provided, including multiple-choice and short answer questions.

**Student Solutions Manual** (ISBN:1-4641-0230-9), written by Brian Storrie of the University of Arkansas for Medical Sciences, Eric A. Wong, Richard Walker, Glenda Gillaspy, and Jill Sible of Virginia Polytechnic Institute and State University and updated by Tom Huxford of San Diego State University, Stephanie Bingham of Barry University, Brian Sato of University of California-Irvine, Steve Amato of Johns Hopkins University, Greg Kelly of University of Western Ontario, Tom Keller of Florida State University, and Elizabeth Good of University of Illinois-Urbana-Champaign, contains complete worked-out solutions to all the end-of-chapter problems in the textbook.

**eBook** (ISBN: 1-4641-0229-5) This customizable eBook fully integrates the complete contents of the text and its interactive media in a format that features a variety of helpful study tools, including full-text searching, note-taking, bookmarking, highlighting, and more. Easily accessible on any Internet-connected computer via a standard Web browser, the eBook enables students to take an active approach to their learning in an intuitive, easy-to-use format. Visit **http://ebooks.bfwpub.com** to learn more.

## For Instructors

**\*NEW\* BioPortal for** *Molecular Cell Biology* In addition to all student resources (including NEW LearningCurve quizzing tool) and a dynamic eBook, BioPortal also includes tools for instructors. Robust gradebook and assignment features allow instructors to assign any materials to their students and monitor their progress throughout the semester. Visit **http://courses.bfwpub.com** for more information.

**Companion Web Site www.whfreeman.com/lodish7e**

All the student resources, plus:

- All figures and tables from the book in **JPEG and Power-Point** formats, which instructors can edit and project section by section, allowing students to follow underlying concepts. Optimized for lecture-hall presentation, including enhanced colors, enlarged labels, and boldface type.

- **Test Bank** in editable Microsoft Word format now featuring *new and revised questions* for every chapter. The test bank is written by Brian Storrie of the University of Arkansas for Medical Sciences and Eric A. Wong, Richard Walker, Glenda Gillaspy, and Jill Sible of Virginia Polytechnic Institute and State University and revised by Cindy Klevickis of James Madison University and Greg M. Kelly of the University of Ontario.

- Additional **Analyze the Data** problems are available in PDF format.

- Lecture-ready **Personal Response System "clicker" questions** are available as Microsoft Word files and Microsoft PowerPoint slides.

**Instructor's Resource CD-ROM** (ISBN: 1-4292-0126-6) includes all the instructor's resources from the Web site, including all the illustrations from the text, animations, videos, test bank files, clicker questions, and the solutions manual files.

**Overhead Transparency Set** (ISBN: 1-4292-0477-X) contains 250 key illustrations from the text, optimized for lecture-hall presentation.

# ACKNOWLEDGMENTS

In updating, revising and rewriting this book, we were given invaluable help by many colleagues. We thank the following people who generously gave of their time and expertise by making contributions to specific chapters in their areas of interest, providing us with detailed information about their courses, or by reading and commenting on one or more chapters:

David Agard, *University of California, San Francisco*

Ravi Allada, *Northwestern University*

Stephen Amato, *Boston College*

James M. Anderson, *National Institutes of Health and University of North Carolina, Chapel Hill*

Kenneth Balazovich, *University of Michigan, Ann Arbor*

Amit Banerjee, *Wayne State University*

Amy Bejsovec, *Duke University*

Andrew Bendall, *University of Guelph, Ridgetown*

Stephanie Bingham, *Barry University, Dwayne O. Andreas School of Law*

Doug Black, *Howard Hughes Medical Institute and University of California, Los Angeles*

Heidi Blank, *Massachusetts Institute of Technology*

Jonathan Bogan, *Yale University School of Medicine*

Laurie Boyer, *Massachusetts Institute of Technology*

William J. Brown, *Cornell University*

Steve Burden, *New York University*

Monique Cadrin, *Université du Québec à Trois-Rivières*

Steven A. Carr, *Broad Institute of Harvard and Massachusetts Institute of Technology*

Paul Chang, *Massachusetts Institute of Technology*

Kuang Yu Chen, *Rutgers, The State University of New Jersey, Camden*

Orna Cohen-Fix, *National Institutes of Health*

Ronald Cooper, *University of California, Los Angeles*

David Daleke, *Indiana State University*

Elizabeth De Stasio, *Lawrence University*

Linda DeVeaux, *Idaho State University*

Richard Dickerson, *University of California, Los Angeles*

Patrick DiMario, *Louisiana State University*

Glenn Dorsam, *North Dakota State University*

William Dowhan, *University of Texas, Houston*

Janet Duerr, *Ohio University*

Fahd Z Eissa, King Abdul Aziz, *City for Science and Technology (KACST)*

Robert H. Fillingame, *University of Wisconsin Medical School*

Gerry Fink, *Massachusetts Institute of Technology*

David Foster, *City University of New York, Hunter College*

Gail Fraizer, *Kent State University, East Liverpool*

Margaret T. Fuller, *Stanford University School of Medicine*

Topher Gee, *University of North Carolina, Charlotte*

Mary Gehring, *Massachusetts Institute of Technology*

Elizabeth Good, *University of Illinois, Urbana-Champaign*

David Goodenough, *Harvard Medical School*

Mark Grimes, *University of Montana, Missoula*

Lawrence I. Grossman, *Wayne State University*

Michael Grunstein, *University of California, Los Angeles, School of Medicine*

Barry M. Gumbiner, *University of Virginia*

Yanlin Guo, *University of Southern Mississippi*

Leah Haimo, *University of California, Riverside*

Craig Hart, *Louisiana State University*

Michael Hemann, *Massachusetts Institute of Technology*

Chris Hill, *University of Utah*

H. Robert Horvitz, *Massachusetts Institute of Technology*

Tim C. Huffaker, *Cornell University*

Tom Huxford, *San Diego State University*

Richard Hynes, *Massachusetts Institute of Technology and Howard Hughes Medical Institute*

Naohiro Kato, *Louisiana State University*

Amy E. Keating, *Massachusetts Institute of Technology*

Thomas Keller, *Florida State University, Panama City*

Greg Kelly, *University of Western Ontario*

Leung Kim, *Florida International University, Biscayne Bay*

Gwendolyn M. Kinebrew, *John Carroll University*

Ashwini Kucknoor, *Lamar University*

Mark Lazzaro, *College of Charleston*

Maureen Leupold, *Genesee Community College, Batavia*

Robert Levine, *McGill University*

Fang Ju Lin, *Coastal Carolina University*

Susan Lindquist, *Massachusetts Institute of Techology*

Song-Tao Liu, *University of Toledo, Scott Park*

Elizabeth Lord, *University of California, Riverside*

Charles Mallery, *University of Miami*

C. William McCurdy, *University of California, Davis, and Lawrence Berkeley National Laboratory*

David McNabb, *University of Arkansas*

James McNew, *Rice University*

Raka Mitra, *Carleton College*

Ivona Mladenovic, *Simon Fraser University*

Vamsi K. Mootha, *Massachusetts General Hospital, Boston*

Roderick Morgan, *Grand Valley State University*

Dana Nayduch, *Georgia Southern University*

Brent Nielsen, *Brigham Young University*

Terry Orr-Weaver, *Massachusetts Institute of Technology*

Rekha Patel, *University of South Carolina, Lancaster*

David Paul, *Harvard Medical School*
Debra Pires, *University of California, Los Angeles*
Nicholas Quintyne, *Florida Atlantic University, Jupiter*
Alex Rich, *Massachusetts Institute of Technology*
Edmund Rucker, *University of Kentucky*
Brian Sato, *University of California, Irvine*
Robert Sauer, *Massachusetts Institute of Techology*
Thomas Schwartz, *Massachusetts Institute of Technology*
Gowri Selvan, *University of California, Irvine*
Jiahai Shi, *Whitehead Institute for Biomedical Research*
Daniel Simmons, *University of Delaware*
Stephen T. Smale, *University of California, Los Angeles*
Paul Teesdale-Spittle, *Victoria University of Wellington*
Fernando Tenjo, *Virginia Commonwealth University*
Andrei Tokmakoff, *Massachusetts Institute of Technology*
Harald Vaessin, *Ohio State University, Columbus*
Peter van der Geer, *San Diego State University*
Volker M. Vogt, *Cornell University*
Michael B. Yaffe, *Massachusetts Institute of Technology*
Jing Zhang, *University of Wisconsin*

We would also like to express our gratitude and appreciation to Leah Haimo of the University of California, Riverside, for her development of new Analyze the Data problems, to Cindy Klevickis of James Madison University and Greg M. Kelly of the University of Ontario for their authorship of excellent new Review the Concepts problems and Test Bank questions, and to Jill Sible of Virginia Polytechnic Institute and State University for her revision of the Online Quizzing problems. We are also grateful to Lisa Rezende of the University of Arizona for her development of the Classic Experiments and Podcasts.

This edition would not have been possible without the careful and committed collaboration of our publishing partners at W. H. Freeman and Company. We thank Kate Ahr Parker, Mary Louise Byrd, Debbie Clare, Marsha Cohen, Victoria Tomaselli, Christina Micek, Bill O'Neal, Marni Rolfes, Beth McHenry, Susan Timmins, Cecilia Varas, and Julia DeRosa for their labor and for their willingness to work overtime to produce a book that excels in every way.

In particular, we would like to acknowledge the talent and commitment of our text editors, Matthew Tontonoz, Erica Pantages Frost, and Erica Champion. They are remarkable editors. Thank you for all you've done in this edition.

We are also indebted to H. Adam Steinberg for his pedagogical insight and his development of beautiful molecular models and illustrations.

We would like to acknowledge those whose direct contributions to previous editions continue to influence in this edition; especially Ruth Steyn.

Thanks to our own staff: Sally Bittancourt, Diane Bush, Mary Anne Donovan, Carol Eng, James Evans, George Kokkinogenis, Julie Knight, Guicky Waller, Nicki Watson, and Rob Welsh.

Finally, special thanks to our families for inspiring us and for granting us the time it takes to work on such a book and to our mentors and advisers for encouraging us in our studies and teaching us much of what we know: *(Harvey Lodish)* my wife, Pamela; my children and grandchildren Heidi and Eric Steinert and Emma and Andrew Steinert; Martin Lodish, Kristin Schardt, and Sophia, Joshua, and Tobias Lodish; and Stephanie Lodish, Bruce Peabody, and Isaac and Violet Peabody; mentors Norton Zinder and Sydney Brenner; and also David Baltimore and Jim Darnell for collaborating on the first editions of this book; *(Arnold Berk)* my wife Sally, Jerry Berk, Shirley Berk, Angelina Smith, David Clayton, and Phil Sharp; *(Chris A. Kaiser)* my wife Kathy O'Neill; *(Monty Krieger)* my wife Nancy Krieger, parents I. Jay Krieger and Mildred Krieger, and children Jonathan Krieger and Joshua Krieger; my mentors Robert Stroud, Michael Brown, and Joseph Goldstein; *(Anthony Bretscher)* my wife Janice and daughters Heidi and Erika, and advisers A. Dale Kaiser and Klaus Weber; *(Hidde Ploegh)* my wife Anne Mahon; *(Angelika Amon)* my husband Johannes Weis, Theresa and Clara Weis, Gerry Fink and Frank Solomon.

# CONTENTS IN BRIEF

# CONTENTS

## Part IV  Cell Growth and Development

# Molecules, Cells, and Evolution

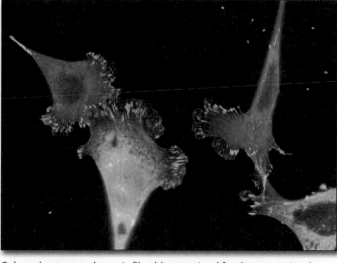

Cultured mouse embryonic fibroblasts stained for three proteins that form the cytoskeleton. [Courtesy of Ana M. Pasapera, Clare M. Waterman]

*Nothing in biology makes sense except in the light of evolution.*

—Theodosius Dobzhansky
(essay in *The American Biology Teacher* 35: 125–129, 1973)

Biology is a science fundamentally different from physics or chemistry, which deal with unchanging properties of matter that can be described by mathematical equations. Biological systems of course follow the rules of chemistry and physics, but biology is a historical science, as the forms and structures of the living world today are the results of billions of years of *evolution*. Through evolution, all organisms are related in a family tree extending from primitive single-celled organisms that lived in the distant past to the diverse plants, animals, and microorganisms of the present era (Figure 1-1, Table 1-1). The great insight of Charles Darwin (Figure 1-2) was the principle of natural selection: organisms vary randomly and compete within their environment for resources. Only those that survive to reproduce are able to pass down their genetic traits.

At first glance, the biological universe appears amazingly diverse—from tiny ferns to tall fir trees, from single-celled bacteria and protozoans visible only under a microscope to multicellular animals of all kinds. Yet the bewildering array of outward biological forms overlies a powerful uniformity: thanks to our common ancestry, all biological systems are composed of the same types of chemical molecules and employ similar principles of organization at the cellular level. Although the basic kinds of biological molecules have been conserved during the billions of years of evolution, the patterns in which they are assembled to form functioning cells and organisms have undergone considerable change.

We now know that **genes,** which chemically are composed of **deoxyribonucleic acid (DNA),** ultimately define biological structure and maintain the integration of cellular function. Many genes encode **proteins,** the primary molecules that make up cell structures and carry out cellular activities. Alterations in the structure and organization of genes, or **mutations,** provide the random variation that can alter biological structure and function. While the vast majority of random mutations have no observable effect on a gene's or protein's function, many are deleterious, and only a few confer an evolutionary advantage. In all organisms mutations in DNA are constantly occurring, allowing over time the small alterations

## OUTLINE

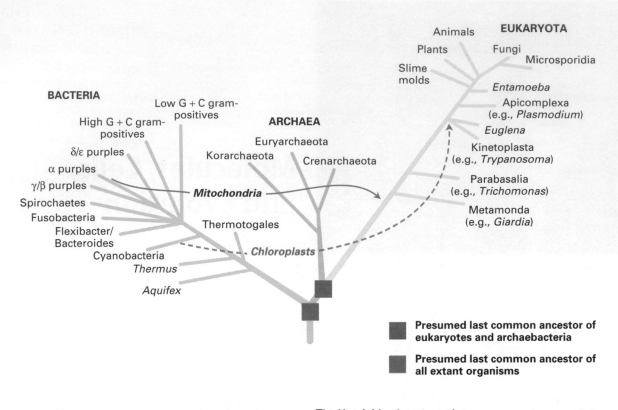

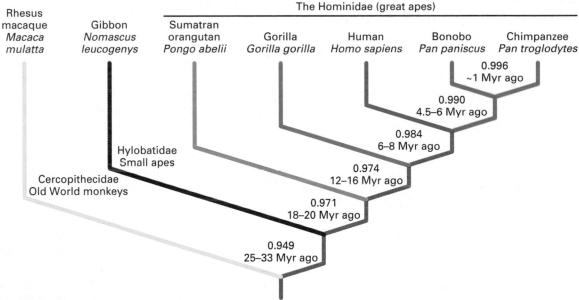

**FIGURE 1-1 All living organisms descended from a common ancestral cell.** (a) All organisms from simple bacteria to complex mammals probably evolved from a common single-celled ancestor. This family tree depicts the evolutionary relations among the three major lineages of organisms. The structure of the tree was initially ascertained from morphological criteria: creatures that look alike were put close together. More recently the sequences of DNA and proteins found in organisms have been examined as more information-rich criteria for assigning relationships. The greater the similarities in these macromolecular sequences, the more closely related organisms are thought to be. The trees based on morphological comparisons and the fossil record generally agree well with those based on molecular data. (b) Evolution of great apes, a small ape, and an Old World monkey with respect to humans, as estimated from the divergence among their genomic DNA sequences. Whole genome DNA sequences were aligned, and the average nucleotide divergence in unique DNA sequences was estimated. Estimates of the times different species diverged from each other, calculated in millions of years (Myr), are indicated at each node; ~1 Myr implies approximately 1 Myr or less. [Part (a) adapted from J. R. Brown, 2005, "Universal tree of life," in *Encyclopedia of Life Sciences,* Wiley InterScience (online). Part (b) adapted from D. P. Locke et al., 2011, *Nature* **469**:529.]

**TABLE 1-1 | Timeline for Evolution of Life on Earth, as Determined from the Fossil Record**

| | |
|---|---|
| 4600 million years ago | The planet Earth forms from material revolving around the young Sun. |
| ~3900–2500 million years ago | Cells resembling prokaryotes appear. These first organisms are chemoautotrophs: they use carbon dioxide as a carbon source and oxidize inorganic materials to extract energy. |
| 3500 million years ago | Lifetime of the last universal ancestor; the split between bacteria and archaea occurs. |
| 2700 million years ago | Photosynthesizing cyanobacteria evolve; they use water as a reducing agent, thereby producing oxygen as a waste product. |
| 1850 million years ago | Unicellular eukaryotic cells appear. |
| 1200 million years ago | Simple multicellular organisms evolve, mostly consisting of cell colonies of limited complexity. |
| 580–500 million years ago | Most modern phyla of animals begin to appear in the fossil record during the Cambrian explosion. |
| 535 million years ago | Major diversification of living things in the oceans: chordates, arthropods (e.g., trilobites, crustaceans), echinoderms, mollusks, brachiopods, foraminifers, radiolarians, etc. |
| 485 million years ago | First vertebrates with true bones (jawless fishes) evolve. |
| 434 million years ago | First primitive plants arise on land. |
| 225 million years ago | Earliest dinosaurs (prosauropods) and teleost fishes appear. |
| 220 million years ago | Gymnosperm forests dominate the land; herbivores grow to huge sizes. |
| 215 million years ago | First mammals evolve. |
| 65.5 million years ago | The Cretaceous-Tertiary extinction event eradicates about half of all animal species, including all of the dinosaurs. |
| 6.5 million years ago | First hominids evolve. |
| 2 million years ago | First members of the genus Homo appear. |
| 350 thousand years ago | Neanderthals appear. |
| 200 thousand years ago | Anatomically modern humans appear in Africa. |
| 30 thousand years ago | Extinction of Neanderthals. |

in cellular structures and functions that may prove to be advantageous. Entirely new structures rarely are created; more often, old structures are adapted to new circumstances. More rapid change is possible by rearranging or multiplying previously evolved components rather than by waiting for a wholly new approach to emerge. For instance, in a particular organism one gene may randomly become duplicated; one copy of the gene and its encoded protein may retain their original function while over time the second copy of the gene mutates such that its protein takes on a slightly different or even a totally new function. The cellular organization of organisms plays a fundamental role in this process because it allows these changes to come about by small alterations in previously evolved cells, giving them new abilities. The result is that closely related organisms have very similar genes, proteins, and cellular organization.

Living systems, including the human body, consist of such closely interrelated elements that no single element can be fully appreciated in isolation from the others. Organisms contain organs, organs are composed of tissues, tissues consist of cells, and cells are formed from molecules (Figure 1-3). The unity of living systems is coordinated by many levels of

**FIGURE 1-2 Charles Darwin (1809–1882).** Four years after his epic voyage on HMS Beagle, Darwin had already begun formulating in private notebooks his concept of natural selection, which would be published in *Origin of Species* (1859). [Walt Anderson/Visuals Unlimited, Inc.]

interrelationship: molecules carry messages from organ to organ and cell to cell and tissues are delineated and integrated with other tissues by molecules secreted by cells. Generally all the levels into which we fragment biological systems interconnect.

To learn about biological systems, however, we must examine one small portion of a living system at a time. The biology of cells is a logical starting point because an organism can be viewed as consisting of interacting cells, which are the closest thing to an autonomous biological unit that exists. The last common ancestor of all life on earth was a cell, and at the cellular level all life is remarkably similar. All cells use the same molecular building blocks, similar methods for the storage, maintenance, and expression of genetic information, and similar processes of energy metabolism, molecular transport, signaling, development, and structure.

In this chapter, we introduce the common features of cells. We begin with a brief discussion of the principal small molecules and macromolecules found in biological systems. Next we discuss the fundamental aspects of cell structure and function that are conserved in present-day organisms, and the use of prokaryotic organisms (single-celled organisms without a nucleus) to study the basic molecules of life.

In the third section we discuss the formation of tissues from individual cells, and the diverse types of unicellular and multicellular organisms used in investigations of molecular cell biology.

One focus of this chapter is DNA, as we now have the complete sequence of the genomes of over a hundred organisms and these have provided considerable insight into the evolution of genes and organisms. Recent studies, for instance, indicate that human and chimpanzee genomes are about 99 percent identical in sequence and that the ancestors of these species likely diverged from a common ape-like organism between 4.5 and 6 million years ago (see Figure 1-1b). This conclusion is consistent with the fossil record (see Table 1-1). Biologists use evolution as a research tool: if a gene and its protein have been conserved in, say, all **metazoans** (multicellular animals) but are not found in unicellular organisms, the protein likely has an important function in all metazoans and thus can be studied in whatever metazoan organism is most suitable for the investigation. Interwoven in the second and third sections of this chapter are discussions of the reasons scientists pick particular unicellular and multicellular "model" organisms to study specific genes and proteins that are important for cellular function.

## 1.1 The Molecules of Life

While large polymers are the focus of molecular cell biology, small molecules are the stage upon which all cellular processes are set. Water, inorganic ions, and a wide array of relatively small organic molecules (Figure 1-4) account for 75 to 80 percent of living matter by weight, and water accounts for about 75 percent of a cell's volume. These small molecules, including water, serve as substrates for many of the reactions that take place inside the cell, including energy metabolism and cell signaling. Cells acquire these small molecules in different ways. Ions, water, and many small organic molecules are imported into the cell (Chapter 11); other small molecules are synthesized within the cell, often by a series of chemical reactions.

Even in the structures of many small molecules, such as sugars, vitamins, and amino acids, we see the footprint of evolution. For example, all amino acids save glycine have an asymmetric carbon atom, yet only the L-stereoisomer, never the D-stereoisomer, is incorporated into proteins. Similarly, only the D-stereoisomer of glucose is invariably found in cells, never the mirror image L-stereoisomer (see Figure 1-4). At an early stage of biological evolution, our common cellular ancestor evolved the ability to catalyze reactions with one stereoisomer instead of the other. How these selections happened is unknown, but now these choices are locked in place.

An important and universally conserved small molecule is **adenosine triphosphate (ATP),** which stores readily available chemical energy in two of its chemical bonds (Figure 1-5). When one of these energy-rich bonds in ATP is broken, forming **ADP (adenosine diphosphate),** the released

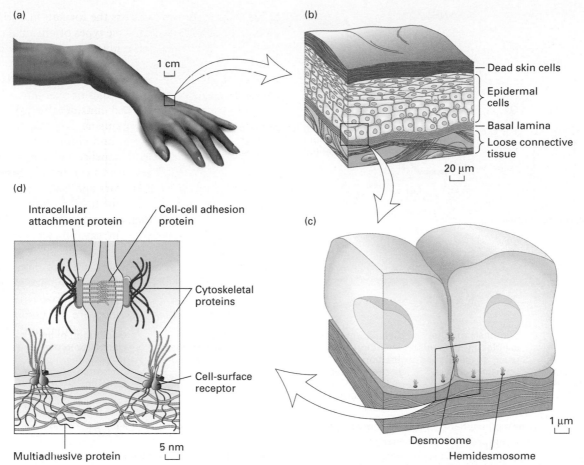

(a)

1 cm

(b)

Dead skin cells

Epidermal cells

Basal lamina

Loose connective tissue

20 μm

(d)

Intracellular attachment protein

Cell-cell adhesion protein

Cytoskeletal proteins

Cell-surface receptor

Multiadhesive protein

5 nm

(c)

Desmosome

Hemidesmosome

1 μm

**FIGURE 1-3 Living systems such as the human body consist of closely interrelated elements.** (a) The surface of our hand is covered by a living organ, skin, that is comprised of several layers of tissue. (b) An outer covering of hard, dead skin cells protects the body from injury, infection, and dehydration. This layer is constantly renewed by living epidermal cells, which also give rise to hair and fur in animals. Deeper layers of muscle and connective tissue give skin its tone and firmness.

(c) Tissues are formed through subcellular adhesion structures (desmosomes and hemidesmosomes) that join cells to each other and to an underlying layer of supporting fibers. (d) At the heart of cellular adhesion are its structural components: phospholipid molecules that make up the cell surface membrane, and large protein molecules. Protein molecules that traverse the cell membrane often form strong bonds with internal and external fibers made of multiple proteins.

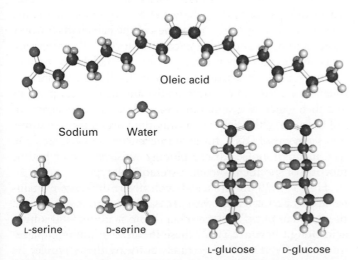

Oleic acid

Sodium     Water

L-serine     D-serine

L-glucose     D-glucose

**FIGURE 1-4 Some of the many small molecules found in cells.** Only the L-forms of amino acids such as serine are incorporated into proteins, not their D-mirror images; only the D-form of glucose, not its L-mirror image, can be metabolized to carbon dioxide and water.

energy can be harnessed to power an energy-requiring process such as muscle contraction or protein biosynthesis. To obtain energy for making ATP, all cells break down food molecules. For instance, when sugar is degraded to carbon dioxide and water, the energy stored in the sugar molecule's chemical bonds is released and much of it can be "captured" in the energy-rich bonds in ATP (Chapter 12). Bacterial, plant, and animal cells can all make ATP by this process. In addition, plants and a few other organisms can harvest energy from sunlight to form ATP in **photosynthesis.**

Other small molecules (e.g., hormones and growth factors) act as signals that direct the activities of cells (Chapters 15 and 16), and nerve cells communicate with one another by releasing and sensing certain small signaling molecules (Chapter 22). The powerful effect on our body of a frightening event comes from the instantaneous flooding of the body with the small-molecule hormone adrenaline, which mobilizes the "fight or flight" response.

Certain small molecules (**monomers**) can be joined to form **polymers,** also called **macromolecules,** through repetition of a

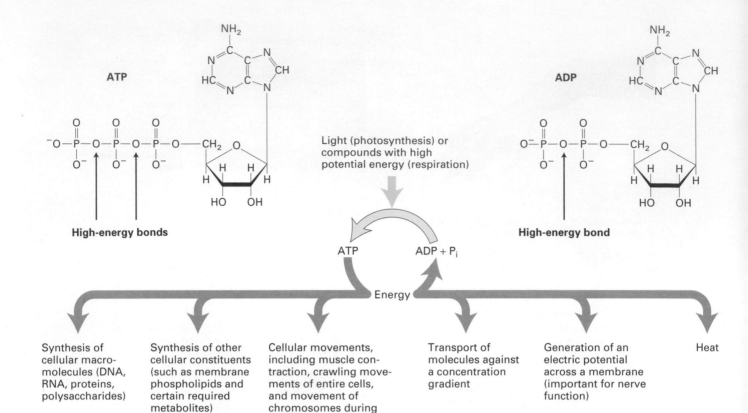

**FIGURE 1-5 Adenosine triphosphate (ATP) is the most common molecule used by cells to capture and transfer energy.** ATP is formed from adenosine diphosphate (ADP) and inorganic phosphate ($P_i$) by photosynthesis in plants, and by the breakdown of sugars and fats in most cells. The energy released by the splitting (hydrolysis) of $P_i$ from ATP drives many cellular processes.

single type of covalent chemical-linkage reaction (see Figure 2-1). Cells produce three types of large macromolecules: polysaccharides, proteins, and nucleic acids (Figure 1-6). Sugars, for example, are the monomers used to form polysaccharides. Different polymers of D-glucose form the cellulose component of plant cell walls and glycogen, a storage form of glucose found in liver and muscle. The cell is careful to provide the appropriate mix of small molecules needed as precursors for synthesis of macromolecules.

## Proteins Give Cells Structure and Perform Most Cellular Tasks

Proteins, the workhorses of the cell, are the most abundant and functionally versatile of the cellular macromolecules. Cells string together 20 different **amino acids** in a linear chain to form proteins (see Figure 2-14), which commonly range in length from 100 to 1000 amino acids. During its polymerization a linear chain of amino acids folds into a complex shape, conferring a distinctive three-dimensional structure and function on each protein (see Figure 1-6). Humans obtain amino acids either by synthesizing them from other molecules or by breaking down proteins that we eat.

Proteins have a variety of functions in the cell. Many proteins are **enzymes,** which accelerate (catalyze) chemical reactions involving small molecules or macromolecules (Chapter 3). Certain proteins catalyze steps in the synthesis of proteins; others catalyze synthesis of other macromolecules such as DNA and RNA. **Cytoskeletal proteins** serve as structural components of a cell, for example by forming an internal skeleton; others power the movement of subcellular structures such as chromosomes, and even of whole cells, by using energy stored in the chemical bonds of ATP (Chapters 17 and 18). Other proteins bind adjacent cells together or form parts of the extracellular matrix (see Figure 1-3). Proteins can be sensors that change shape as temperature, ion concentrations, or other properties of the cell change. Many proteins that are embedded in the cell surface (plasma) membrane import and export a variety of small molecules and ions (Chapter 11). Some proteins, such as insulin, are hormones; others are hormone receptors that bind their target proteins and then generate a signal that regulates a specific aspect of cell function. Other important classes of proteins bind to specific segments of DNA, turning genes on or off (Chapter 7). In fact, much of molecular cell biology consists of studying the function of specific proteins in specific cell types.

How can 20 amino acids form all the different proteins needed to perform these varied tasks? It seems impossible at first glance. But if a "typical" protein is about 400 amino acids long, there are $20^{400}$ possible different amino acid sequences. Even assuming that many of these would be functionally equivalent, unstable, or otherwise discountable, the number of possible proteins is astronomical.

Next we might ask how many protein molecules a cell needs to operate and maintain itself. To estimate this number,

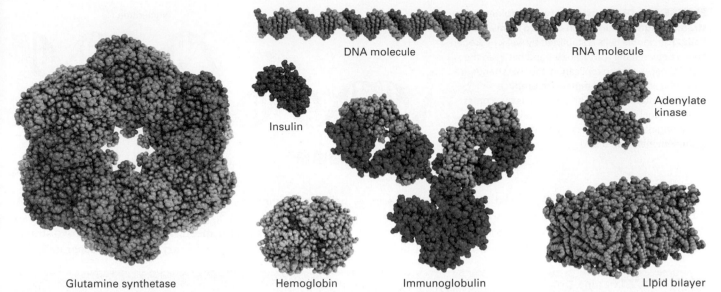

DNA molecule

RNA molecule

Insulin

Adenylate kinase

Glutamine synthetase

Hemoglobin

Immunoglobulin

Lipid bilayer

**FIGURE 1-6 Models of some representative proteins drawn to a common scale and compared with a small portion of a lipid bilayer sheet, a DNA molecule, and an RNA molecule.** Each protein has a defined three-dimensional shape held together by numerous chemical bonds. The illustrated proteins include enzymes (glutamine synthetase and adenylate kinase), an antibody (immunoglobulin), a hormone (insulin), and the blood's oxygen carrier (hemoglobin).

let's take a typical eukaryotic cell (a cell containing a nucleus), such as a hepatocyte (liver cell). This cell, roughly a cube 15 μm (0.0015 cm) on a side, has a volume of $3.4 \times 10^{-9}$ cm$^3$ (or milliliters, ml). Assuming a cell density of 1.03 g/ml, the cell would weigh $3.5 \times 10^{-9}$ g. Since protein accounts for approximately 20 percent of a cell's weight, the total weight of cellular protein is $7 \times 10^{-10}$ g. The average protein has a molecular weight of 52,700 g/mol; we can calculate the total number of protein molecules per liver cell as about $7.9 \times 10^9$ from the total protein weight and Avogadro's number, the number of molecules per mole of any chemical compound ($6.02 \times 10^{23}$). To carry this calculation one step further, consider that a liver cell contains about 10,000 different proteins; thus each cell would on average contain close to a million molecules of each type of protein. In fact, the abundance of different proteins varies widely, from the quite rare insulin-binding receptor protein (20,000 molecules per cell) to the abundant structural protein actin ($5 \times 10^8$ molecules per cell). Every cell closely regulates the level of each protein such that each is present in the appropriate quantity for its cellular functions, as we detail in Chapters 7 and 8.

## Nucleic Acids Carry Coded Information for Making Proteins at the Right Time and Place

The macromolecule that garners the most public attention is deoxyribonucleic acid (DNA), whose functional properties make it the cell's "master molecule." The three-dimensional structure of DNA, first proposed by James D. Watson and Francis H. C. Crick about 60 years ago (Figure 1-7), consists of two long helical strands that are coiled around a common axis to form a **double helix** (Figure 1-8). The double-helical structure of DNA, one of nature's most magnificent constructions, is critical to the phenomenon of heredity, the transfer of genetically determined characteristics from one generation to the next.

DNA strands are composed of monomers called **nucleotides;** these often are referred to as *bases* because their structures contain cyclic organic bases (Chapter 4). Four different

**FIGURE 1-7 James D. Watson (*left*) and Francis H. C. Crick (*right*) with the double-helical model of DNA they constructed in 1952–1953.** Their model ultimately proved correct in all its essential aspects. [A. Barrington Brown/Science Photo Researcher. From J. D. Watson, 1968, *The Double Helix,* Atheneum, Copyright 1968, p. 215; Courtesy of A. C. Barrington Brown.]

**FIGURE 1-8 DNA consists of two complementary strands wound around each other to form a double helix.** The double helix is stabilized by weak hydrogen bonds between the A and T bases and between the C and G bases. During replication, the two strands are unwound and used as templates to produce complementary strands. The outcome is two copies of the original double helix, each containing one of the original strands and one new daughter (complementary) strand.

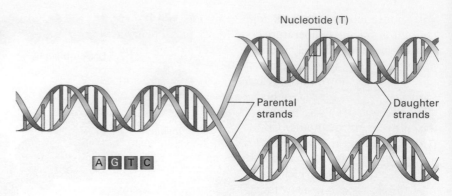

nucleotides, abbreviated A, T, C, and G, are joined to form a DNA strand, with the base parts projecting inward from the backbone of the strand. Two strands bind together via the bases, and twist to form a double helix. Each DNA double helix has a simple construction: wherever one strand has an A, the other strand has a T, and each C is matched with a G (see Figure 1-8). This **complementary** matching of the two strands is so strong that if complementary strands are separated, they

| TABLE 1-2 | Genome Sizes of Organisms Used in Molecular Cell Biology Research That Have Been Completely Sequenced | | | |
|---|---|---|---|---|
| **Bacteria** | **Base pairs (millions)** | **Encoded proteins** | **Chromosomes** | **Reference** |
| *Mycoplasma genitalum* | 0.58 | 482 | 1 | A |
| *Helicobacter pylori* | 1.67 | 1,587 | 1 | A |
| *Haemophilus influenza* | 1.83 | 1,737 | 1 | A |
| *Escherichia coli* | 4.64 | 4,289 | 1 | A |
| *Bacillus subtilis* | 4.22 | 4,245 | 1 | A |
| **Archaea** | | | | |
| *Methanococcus jannaschii* | 1.74 | 1,785 | 3 | A |
| *Sulfolobus solfataricus* | 2.99 | 2,960 | 1 | A |
| **Eukaryotes** | | | | |
| *Saccharomyces cerevisiae* | 12.16 | 5,885 | 16 | B |
| *Drosophila melanogaster* | 168 | 13,781 | 4 | C |
| *Caenorhabditis elegans* | 100 | 20,424 | 6 | D |
| *Danio rerio* | 1505 | 19,929 | 25 | C |
| *Gallus gallus* (chicken) | 1050 | 14,923 | 39 | C |
| *Mus musculus* | 3421 | 22,085 | 20 | C |
| *Homo sapiens* | 3279 | 21,077 | 23 | C |
| *Arabidopsis thaliana* | 135 | 27,416 | 5 | E |

Table courtesy of Dr. Fran Lewitter. SOURCES: A, http://cmr.jcvi.org/cgi-bin/CMR/shared/Genomes.cgi; B, http://www.yeastgenome.org/; C, http://uswest. ensembl.org/info/about/species.html; D, http://wiki.wormbase.org/index.php/WS222; E, http://www.arabidopsis.org/portals/genAnnotation/gene_structural_annotation/annotation_data.jsp.

will spontaneously zip back together under the right salt concentration and temperature conditions. Such **nucleic acid hybridization** is extremely useful for detecting one strand by using the other, as we learn in Chapter 5.

The genetic information carried by DNA resides in its sequence, the linear order of nucleotides along a strand. Specific segments of DNA, termed genes, carry instructions for making specific proteins. Commonly genes contain two parts: the *coding region* specifies the amino acid sequence of a protein; the *regulatory region* binds specific proteins and controls when and in which cells the protein is made.

Most bacteria have a few thousand genes; yeasts and other unicellular eukaryotes have about 5000. Humans and other metazoans have between 13,000 and 23,000, while many plants like *Arabidopsis* have more (Table 1-2). As we discuss later in this chapter, many bacterial genes encode proteins that are conserved throughout all living organisms. These catalyze reactions that occur universally, such as the metabolism of glucose and synthesis of nucleic acids and proteins. Studies on bacterial cells have yielded profound insights into these basic life processes. Similarly, many genes in unicellular eukaryotes such as yeasts encode proteins that are conserved throughout all eukaryotes; we will see how yeasts have been used to study processes such as cell division that have yielded profound insights into human diseases such as cancer.

Cells use two processes in series to convert the coded information in DNA into proteins (Figure 1-9). In the first, called **transcription,** the coding region of a gene is copied into a single-stranded **ribonucleic acid (RNA)** whose sequence is the same as one of the two in the double-stranded DNA. A large enzyme, **RNA polymerase,** catalyzes the linkage of nucleotides into an RNA chain using DNA as a template. In eukaryotic cells, the initial RNA product is processed into a smaller **messenger RNA (mRNA)** molecule, which moves out of the nucleus to the cytoplasm. Here the **ribosome,** an enormously complex molecular machine composed of both RNA and protein, carries out the second process, called **translation.** During translation, the ribosome assembles and links together amino acids in the precise order dictated by the mRNA sequence according to the nearly universal **genetic code.** We examine the cell components that carry out transcription and translation in detail in Chapter 4.

In addition to its role in transferring information from nucleus to cytoplasm, RNA can serve as a framework for building a molecular machine. For example, the ribosome is built of four RNA chains that bind to more than 50 proteins to make a remarkably precise and efficient mRNA reader and protein synthesizer. While most chemical reactions in cells are catalyzed by proteins, a few, such as the formation of the peptide bonds that connect amino acids in proteins, are catalyzed by RNA molecules.

Well before the entire human genome was sequenced it was apparent that only about 5 percent of human DNA codes for protein, and for many years most of the human genome was considered "junk DNA"! However, in recent years we've learned that much of the so-called junk DNA is actually copied into thousands of RNA molecules that,

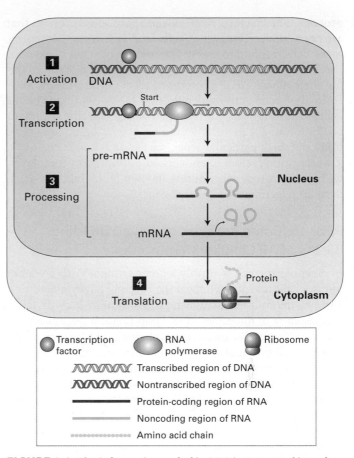

**FIGURE 1-9  The information coded in DNA is converted into the amino acid sequences of proteins by a multistep process.** Step **1**: Transcription factors bind to the regulatory regions of the specific genes they control and activate them. Step **2**: Following assembly of a multiprotein initiation complex bound to the DNA, RNA polymerase begins transcription of an activated gene at a specific location, the start site. The polymerase moves along the DNA linking nucleotides into a single-stranded pre-mRNA transcript using one of the DNA strands as a template. Step **3**: The transcript is processed to remove noncoding sequences. Step **4**: In a eukaryotic cell, the mature messenger RNA (mRNA) moves to the cytoplasm, where it is bound by ribosomes that read its sequence and assemble a protein by chemically linking amino acids into a linear chain.

though they do not code for proteins, serve equally important purposes in the cell (Chapter 6). Small *micro RNAs,* 20–25 nucleotides long, are abundant in metazoan cells and bind to and repress the activity of target mRNAs. By some estimates these small RNAs may indirectly regulate the activity of most or all genes, though the mechanisms and ubiquity of this type of regulation are still being explored (Chapter 8). Several long noncoding RNAs bind to DNA or chromosomal proteins and so affect chromosome structure and RNA synthesis, processing, and stability. However, we know the function of only very few of these abundant noncoding RNAs.

All organisms must control when and where their genes can be transcribed. Nearly all the cells in our bodies contain the full set of human genes, but in each cell type only some of these genes are active, or turned on, and used to make proteins. For instance, liver cells produce some proteins that are not produced by kidney cells, and vice versa. Moreover, many cells respond to external signals or changes in external conditions by turning specific genes on or off, thereby adapting their repertoire of proteins to meet current needs. Such control of gene activity depends on DNA-binding proteins called **transcription factors,** which bind to specific sequences of DNA and act as switches, either activating or repressing transcription of particular genes (see Figure 1-9 and Chapter 7). Transcription factors often work as multiprotein complexes, with each protein contributing its own DNA-binding specificity to selecting the regulated genes.

## Phospholipids Are the Conserved Building Blocks of All Cellular Membranes

In essence, any cell is simply a compartment with a watery interior that is separated from the external environment by a surface membrane, the plasma membrane, which prevents the free flow of molecules in and out. In addition, eukaryotic cells have extensive internal membranes that further subdivide the cell into multiple subcompartments, the organelles.

In all organisms cellular membranes are composed primarily of a bilayer (two layers) of phospholipid molecules. These bipartite molecules have a "water-loving" (hydrophilic) end and a "water-hating" (hydrophobic) end. The two phospholipid layers of a membrane are oriented with all the hydrophilic ends directed toward the inner and outer surfaces of the membrane and the hydrophobic ends buried within its interior (Figure 1-10). Smaller amounts of other lipids, such as cholesterol, are inserted into the phospholipid framework. Phospholipid membranes are impermeable to water, all ions, and virtually all hydrophilic small molecules. Thus each

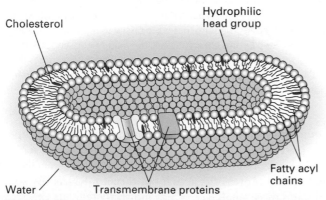

**FIGURE 1-10 The watery interior of cells is surrounded by the plasma membrane, a two-layered shell of phospholipids.** The phospholipid molecules are oriented with their hydrophobic fatty acyl chains (black squiggly lines) facing inward and their hydrophilic head groups (white spheres) facing outward. Thus both sides of the membrane are lined by head groups, mainly charged phosphates, adjacent to the watery spaces inside and outside the cell. All biological membranes have the same basic phospholipid bilayer structure. Cholesterol (red) and various proteins are embedded in the bilayer. The interior space is actually much larger relative to the volume of the plasma membrane than is depicted here.

membrane in each cell also contains groups of proteins that allow specific ions and small molecules to cross. Other membrane proteins serve to attach the cell to other cells or to polymers that surround it; still others give the cell its shape or allow its shape to change. We will learn more about membranes and how molecules cross them in Chapters 10 and 11.

New cells are always derived from parental cells by cell division. We've seen that the synthesis of new DNA molecules is templated by the two strands of the parental DNA such that each daughter DNA molecule has the same sequence as the parental one. In parallel, membranes are made by incorporation of lipids and proteins into existing membranes in the parental cell, and these are divided between daughter cells by fission. Thus membrane synthesis, similarly to DNA synthesis, is also templated by a parental structure.

## 1.2 Genomes, Cell Architecture, and Cell Function

The biological universe consists of two types of cells—prokaryotic and eukaryotic. Prokaryotic cells such as bacteria consist of a single closed compartment that is surrounded by the plasma membrane, lack a defined nucleus, and have a relatively simple internal organization (Figure 1-11). Eukaryotic cells, unlike prokaryotic cells, contain a defined membrane-bound nucleus and extensive internal membranes that enclose the **organelles** (Figure 1-12). The region of the cell lying between the plasma membrane and the nucleus is the **cytoplasm,** comprising the **cytosol** (water, dissolved ions, small molecules, and proteins) and the organelles. Eukaryotes include four kingdoms: the plants, animals, fungi, and protists. Prokaryotes comprise the fifth and sixth kingdoms: the eubacteria (true bacteria) and archaea.

Genome sequencing has provided profound insights into the function and evolution of both conserved and nonconserved genes and proteins found in multiple organisms. In the following section we describe some basic structural and functional features of prokaryotic and eukaryotic cells and relate these to insights provided from their genome sequences. We emphasize the conserved proteins found in multiple diverse species and explain why scientists have chosen several of these species as **model organisms,** systems in which the study of specific aspects of cellular function and development can serve as a model for other species (Figure 1-13).

## Prokaryotes Comprise True Bacteria and Archaea

In recent years, detailed analysis of the DNA sequences from a variety of prokaryotic organisms has revealed two distinct kingdoms: the eubacteria, often simply called "bacteria," and the archaea. Eubacteria, a numerous type of prokaryote, are single-celled organisms; included are the cyanobacteria, or blue-green algae, which can be unicellular or filamentous chains of cells. Figure 1-11 illustrates the general structure of a typical bacterial cell; archaeal cells have a similar structure.

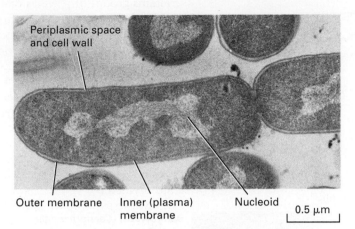

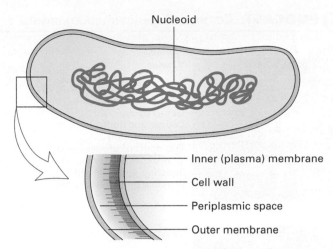

Nucleoid

Inner (plasma) membrane
Cell wall
Periplasmic space
Outer membrane

Periplasmic space and cell wall

Outer membrane    Inner (plasma) membrane    Nucleoid

0.5 μm

**FIGURE 1-11 Prokaryotic cells are have a relatively simple structure.** (Left) Electron micrograph of a thin section of *Escherichia coli*, a common intestinal bacterium. The nucleoid, consisting of the bacterial DNA, is not enclosed within a membrane. *E. coli* and other gram-negative bacteria are surrounded by two membranes separated by the periplasmic space. The thin cell wall is adjacent to the inner membrane. (Right) This artist's drawing shows the nucleoid (blue) and a magnification of the layers that surround the cytoplasm. Most of the cell is composed of water, proteins, ions, and other molecules that are too small to be depicted in the scale of this drawing. [Electron micrograph courtesy of I. D. J. Burdett and R. G. E. Murray. Illustration by D. Goodsell.]

Bacterial cells are commonly 1–2 μm in size and consist of a single closed compartment containing the cytoplasm and bounded by the plasma membrane. Although bacterial cells do not have a defined nucleus, the single circular DNA genome is extensively folded and condensed into the central region of the cell. In contrast, most ribosomes are found in the DNA-free region of the cell. Some bacteria also have an invagination of the cell membrane, called a *mesosome,* which is associated with synthesis of DNA and secretion of proteins. Many proteins are precisely localized within the cytosol or in the plasma membrane, indicating the presence of an elaborate internal organization.

Bacterial cells possess a cell wall, which lies adjacent to the external side of the plasma membrane. The cell wall is composed of layers of peptidoglycan, a complex of proteins and oligosaccharides; it helps protect the cell and maintain its shape. Some bacteria (e.g., *E. coli*) have a thin inner cell wall and an outer membrane separated from the inner cell

(a)

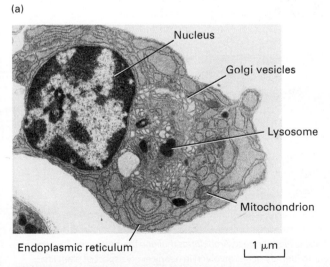

Nucleus

Golgi vesicles

Lysosome

Mitochondrion

Endoplasmic reticulum        1 μm

**FIGURE 1-12 Eukaryotic cells have a complex internal structure with many membrane-limited organelles.** (a) Electron micrograph and (b) diagram of a plasma cell, a type of white blood cell that secretes antibodies. A single membrane (the plasma membrane) surrounds the cell and the cell interior contains many membrane-limited compartments, or organelles. The defining characteristic of eukaryotic cells is segregation of the cellular DNA within a defined nucleus, which is bounded by a double membrane. The outer nuclear membrane is

(b)

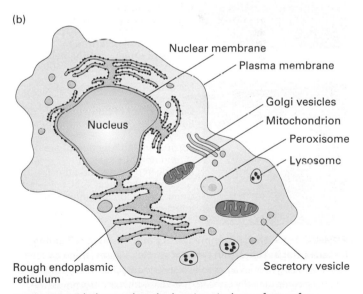

Nuclear membrane
Plasma membrane
Golgi vesicles
Mitochondrion
Peroxisome
Lysosome
Nucleus
Rough endoplasmic reticulum
Secretory vesicle

continuous with the rough endoplasmic reticulum, a factory for assembling secreted and membrane proteins. Golgi vesicles process and modify secreted and membrane proteins, mitochondria generate energy, lysosomes digest cell materials to recycle them, peroxisomes process molecules using oxygen, and secretory vesicles carry cell materials to the surface to release them. [From P. C. Cross and K. L. Mercer, 1993, *Cell and Tissue Ultrastructure: A Functional Perspective,* W. H. Freeman and Company.]

(a)

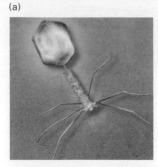

**Viruses**

Proteins involved in DNA, RNA, protein synthesis
Gene regulation
Cancer and control of cell proliferation
Transport of proteins and organelles inside cells
Infection and immunity
Possible gene therapy approaches

(b)

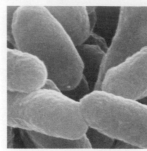

**Bacteria**

Proteins involved in DNA, RNA, protein synthesis, metabolism
Gene regulation
Targets for new antibiotics
Cell cycle
Signaling

(c)

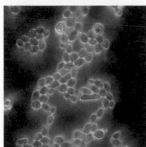

**Yeast** (*Saccharomyces cerevisiae*)

Control of cell cycle and cell division
Protein secretion and membrane biogenesis
Function of the cytoskeleton
Cell differentiation
Aging
Gene regulation and chromosome structure

(d)

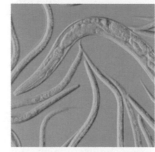

**Roundworm** (*Caenorhabditis elegans*)

Development of the body plan
Cell lineage
Formation and function of the nervous system
Control of programmed cell death
Cell proliferation and cancer genes
Aging
Behavior
Gene regulation and chromosome structure

(e)

**Fruit fly** (*Drosophila melanogaster*)

Development of the body plan
Generation of differentiated cell lineages
Formation of the nervous system, heart, and musculature
Programmed cell death
Genetic control of behavior
Cancer genes and control of cell proliferation
Control of cell polarization
Effects of drugs, alcohol, pesticides

(f)

**Zebrafish**

Development of vertebrate body tissues
Formation and function of brain and nervous system
Birth defects
Cancer

(g)

**Mice,** including cultured cells

Development of body tissues
Function of mammalian immune system
Formation and function of brain and nervous system
Models of cancers and other human diseases
Gene regulation and inheritance
Infectious disease

(h)

**Plant** (*Arabidopsis thaliana*)

Development and patterning of tissues
Genetics of cell biology
Agricultural applications
Physiology
Gene regulation
Immunity
Infectious disease

**FIGURE 1-13 Each experimental organism used in cell biology has advantages for certain types of studies.** Viruses (a) and bacteria (b) have small genomes amenable to genetic dissection. Many insights into gene control initially came from studies with these organisms. The yeast *Saccharomyces cerevisiae* (c) has the cellular organization of a eukaryote but is a relatively simple single-celled organism that is easy to grow and to manipulate genetically. In the nematode worm *Caenorhabditis elegans* (d), which has a small number of cells arranged in a nearly identical way in every worm, the formation of each individual cell can be traced. The fruit fly *Drosophila melanogaster* (e), first used to discover the properties of chromosomes, has been especially valuable in identifying genes that control embryonic development. Many of these genes are evolutionarily conserved in humans. The zebrafish *Danio rerio* (f) is used for rapid genetic screens to identify genes that control vertebrate development and organogenesis. Of the experimental animal systems, mice (*Mus musculus*) (g) are evolutionarily the closest to humans and have provided models for studying numerous human genetic and infectious diseases. The mustard-family weed *Arabidopsis thaliana* has been used for genetic screens to identify genes involved in nearly every aspect of plant life. Genome sequencing is completed for many viruses and bacterial species, the yeast *S. cerevisiae*, the roundworm *C. elegans*, the fruit fly *D. melanogaster*, humans, mice, zebrafish, and the plant *A. thaliana*. Other organisms, particularly frogs, sea urchins, chickens, and slime molds, have also had their genomes sequenced and continue to be immensely valuable for cell biology research. Increasingly, a wide variety of other species are used, especially for studies of evolution of cells and mechanisms. [Part (a) Visuals Unlimited, Inc. Part (b) Kari Lountmaa/Science Photo Library/Photo Researchers, Inc. Part (c) Scimat/Photo Researchers, Inc. Part (d) Photo Researchers, Inc. Part (e) Darwin Dale/Photo Researchers, Inc. Part (f) Inge Spence/Visuals Unlimited, Inc. Part (g) J. M. Labat/Jancana/Visuals Unlimited, Inc. Part (h) Darwin Dale/Photo Researchers, Inc.]

wall by the periplasmic space. Such bacteria are not stained by the Gram technique and thus are classified as gram-negative. Other bacteria (e.g., *Bacillus polymyxa*) that have a thicker cell wall and no outer membrane take the Gram stain and thus are classified as gram-positive.

Working on the assumption that similar organisms diverged more recently from a common ancestor than did dissimilar ones, researchers have developed the evolutionary lineage tree shown in Figure 1-1a. According to this tree, the archaea and the eukaryotes diverged from bacteria more than a billion years before they diverged from each other (Table 1-1). In addition to DNA sequence distinctions that define the three groups of organisms, archaeal cell membranes have chemical properties that differ dramatically from those of bacteria and eukaryotes.

Many archaeans grow in unusual, often extreme, environments that may resemble the ancient conditions that existed when life first appeared on earth. For instance, halophiles ("salt lovers") require high concentrations of salt to survive, and thermoacidophiles ("heat and acid lovers") grow in hot (80 °C) sulfur springs, where a pH of less than 2 is common. Still other archaeans live in oxygen-free milieus and generate methane ($CH_4$) by combining water with carbon dioxide.

## *Escherichia coli* Is Widely Used in Biological Research

The bacterial lineage includes *Escherichia coli,* a favorite experimental organism which in nature is common in soil and animal intestines. *E. coli* and several other bacteria have a number of advantages as experimental organisms. They grow rapidly in a simple and inexpensive medium containing glucose and salts, in which they can synthesize all necessary amino acids, lipids, vitamins, and other essential small molecules. Like all bacteria, *E. coli* possesses elegant mechanisms for controlling gene activity that are now well understood. Over time, workers have developed powerful systems for genetic analysis of this organism. These systems are facilitated by the small size of bacterial genomes, the ease of obtaining mutants, the availability of techniques for transferring genes into bacteria, an enormous wealth of knowledge about bacterial gene control and protein functions, and the relative simplicity of mapping genes relative to one another in the bacterial genome. In Chapter 5 we see how *E. coli* is used in recombinant DNA research.

Bacteria such as *E. coli* that grow in environments as diverse as the soil and the human gut have about 4000 genes encoding about the same number of proteins (see Table 1-2). Parasitic bacteria such as the *Mycoplasma* species acquire amino acids and other nutrients from their host cells, and lack the genes for enzymes that catalyze reactions in the synthesis of amino acids and certain lipids. Many bacterial genes encoding proteins essential for DNA, RNA, protein synthesis, and membrane function are conserved in all organisms, and much of our knowledge of these important cellular processes was uncovered first in *E. coli.* For example, certain *E. coli* cell membrane proteins that import amino acids across the plasma membrane are closely related in sequence, structure,

and function to membrane proteins in certain mammalian brain cells that import small nerve-to-nerve signaling molecules called **neurotransmitters** (Chapters 11 and 22).

## All Eukaryotic Cells Have Many of the Same Organelles and Other Subcellular Structures

**Eukaryotes** comprise all members of the plant and animal kingdoms, as well as fungi (e.g., yeasts, mushrooms, molds) and protozoans (*proto*, primitive; *zoan*, animal), which are exclusively unicellular. Eukaryotic cells are commonly about 10–100 μm across, generally much larger than bacteria. A typical human fibroblast, a connective tissue cell, is about 15 μm across with a volume and dry weight some thousands of times those of an *E. coli* cell. An amoeba, a single-celled protozoan, can have a cell diameter of approximately 0.5 mm, more than thirty times that of a fibroblast.

Eukaryotic cells, like prokaryotic cells, are surrounded by a plasma membrane. However, unlike prokaryotic cells, most eukaryotic cells (the human red blood cell is an exception) also contain extensive internal membranes that enclose specific subcellular compartments, the **organelles,** and separate them from the rest of the **cytoplasm,** the region of the cell lying outside the nucleus (see Figure 1-12). Many organelles are surrounded by a single phospholipid membrane, but the nucleus, mitochondrion, and chloroplast are enclosed by two membranes. Each type of organelle contains a collection of specific proteins, including enzymes that catalyze requisite chemical reactions. The membranes defining these subcellular compartments control their internal ionic composition so that it generally differs from that of the surrounding cytosol as well as that of the other organelles.

The largest organelle in a eukaryotic cell is generally the nucleus, which houses most of the cellular DNA. In animal and plant cells, most ATP is produced by large multiprotein "molecular machines" located in the organelles termed **mitochondria.** Plants carry out photosynthesis in **chloroplasts,** organelles that contain molecular machines for synthesizing ATP from ADP and phosphate, similar to those found in mitochondria. Similar molecular machines for generating ATP are located in the plasma membrane of bacterial cells. Both mitochondria and chloroplasts are thought to have originated as bacteria that took up residence inside eukaryotic cells and then became welcome collaborators (Chapter 12). Over time many of the bacterial genes "migrated" to the cell nucleus and became incorporated into the cell's nuclear genome. Both mitochondria and chloroplasts contain small genomes that encode a few of the essential organelle proteins; the sequences of these DNAs reveal their bacterial origins.

Cells need to break down worn-out or obsolete parts into small molecules that can be discarded or recycled. In animals this housekeeping task is assigned in part to **lysosomes,** organelles filled with degradative enzymes. The interior of a lysosome has a pH of about 5.0, much more acidic than that of the surrounding cytosol. This aids in the breakdown of materials by lysosomal enzymes, which can function at such a low pH. To create the low-pH environment,

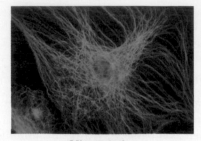
**Microtubules**

**Microfilaments**

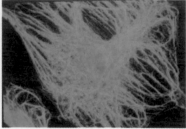
**Intermediate filaments**

**FIGURE 1-14 The three types of cytoskeletal filaments have characteristic distributions within mammalian cells.** Three views of the same cell. A cultured fibroblast was permeabilized and then treated with three different antibody preparations. Each antibody binds specifically to the protein monomers forming one type of filament and is chemically linked to a differently colored fluorescent dye (blue, red, or green). Visualization of the stained cell in a fluorescence microscope reveals the location of filaments bound to a particular dye-antibody preparation. In this case, microtubules are stained blue; microfilaments, red; and intermediate filaments, green. All three fiber systems contribute to the shape and movements of cells. [Courtesy of V. Small.]

proteins located in the lysosomal membrane pump hydrogen ions into the lysosome using energy supplied from ATP (Chapter 11). Plants and fungi contain a vacuole that also has a low-pH interior and stores certain salts and nutrients. **Peroxisomes** are another type of small organelle, found in virtually all eukaryotic cells, that is specialized for breaking down the lipid components of membranes.

The cytoplasm of eukaryotic cells contains an array of fibrous proteins collectively called the **cytoskeleton** (Chapters 17 and 18). Three classes of fibers compose the cytoskeleton: **microtubules** (20 nm in diameter), built of polymers of the protein tubulin; **microfilaments** (7 nm in diameter), built of the protein actin; and **intermediate filaments** (10 nm in diameter), built of one or more rod-shaped protein subunits (Figure 1-14). The cytoskeleton gives the cell strength and rigidity, thereby helping to maintain cell shape. Cytoskeletal fibers also control movement of structures within the cell; for example, some cytoskeletal fibers connect to organelles or provide tracks along which organelles and chromosomes move; other fibers play key roles in cell motility. Thus the cytoskeleton is important for "organizing" the cell.

The rigid cell wall, composed of cellulose and other polymers, that surrounds plant cells contributes to their strength and rigidity. Fungi are also surrounded by a cell wall, but its composition differs from that of bacterial or plant cell walls.

Each organelle membrane and each space in the interior of an organelle has a unique set of proteins that enable it to carry out its specific functions. For cells to work properly, the numerous proteins composing the various working compartments must be transported from where they are made to their proper locations (Chapters 17 and 18). Some proteins are made on ribosomes that are free in the cytoplasm; from there, some proteins are moved into the nucleus while others are directed into mitochondria, chloroplasts, or peroxisomes, depending on their specific functions. Proteins to be secreted from the cell and most membrane proteins, in contrast, are made on ribosomes associated with the **endoplasmic reticulum (ER)**. This organelle produces, processes, and ships out both proteins and lipids. Most protein chains produced on the ER move to the **Golgi complex**, where they are further

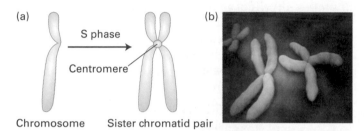

(a) Chromosome → S phase → Sister chromatid pair
Centromere

(b)

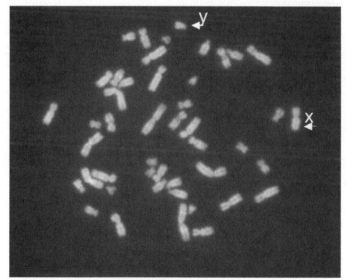

(c)

**FIGURE 1-15 Individual chromosomes can be seen in cells during cell division (mitosis).** (a) During the S phase of the cell cycle (see Figure 1-16) chromosomes are duplicated and the daughter "sister chromatids," each with a complete copy of the chromosomal DNA, remain attached at the centromere. (b) During the actual cell division process (mitosis) the chromosomal DNA becomes highly compacted and the pairs of sister chromatids can be seen in the electron microscope as depicted here. (c) Light microscopic image of a chromosomal spread from a cultured human male lymphoid cell arrested in the metaphase stage of mitosis by treatment with the microtubule-depolymerizing drug colcemid. There is a single copy of the duplicated X and Y chromosomes and two of each of the others. [Part (b) courtesy of Medical RF/The Medical File/Peter Arnold Inc. Part (c) courtesy of Tatyana Pyntikova.]

modified before being forwarded to their final destinations. Proteins that travel in this way contain short sequences of amino acids or attached sugar chains (oligosaccharides) that serve as addresses for directing them to their correct destinations. These addresses work because they are recognized and bound by other proteins that do the sorting and shipping in various cell compartments.

## Cellular DNA Is Packaged Within Chromosomes

In most prokaryotic cells, most or all of the genetic information resides in a single circular DNA molecule about a millimeter in length; this molecule lies, folded back on itself many times, in the central region of the micrometer-sized cell (Figure 1-11). In contrast, DNA in the nuclei of eukaryotic cells is distributed among multiple long linear structures called **chromosomes.** The length and number of chromosomes are the same in all cells of an organism, but vary among different types of organisms (see Table 1-2). Each chromosome comprises a single DNA molecule associated with numerous proteins, and the total DNA in the chromosomes of an organism is referred to as its **genome.** Chromosomes, which stain intensely with basic dyes, are visible in light and electron microscopes only during cell division, when the DNA becomes tightly compacted (Figure 1-15). Although the large genomic DNA molecule in prokaryotes is associated with proteins and often is referred to as a chromosome, the arrangement of DNA within a bacterial chromosome differs greatly from that within the chromosomes of eukaryotic cells.

## All Eukaryotic Cells Utilize a Similar Cycle to Regulate Their Division

Unicellular eukaryotes, animals, and plants use essentially the same **cell cycle,** a series of events that prepares a cell to divide, and the actual division process, called **mitosis.** The eukaryotic cell cycle commonly is represented as four stages

(Figure 1-16). The chromosomes and the DNA they carry are duplicated during the **S (synthesis) phase.** The replicated chromosomes separate during the **M (mitotic) phase,** with each daughter cell getting a copy of each chromosome during cell division. The M and S phases are separated by two gap stages, the $G_1$ **phase** and the $G_2$ **phase,** during which mRNAs and proteins are made and the cell increases in size.

In single-celled organisms, both daughter cells often (though not always) resemble the parent cell. In multicellular organisms, when many types of cells divide the daughters look a lot like the parent cell—liver cells, for instance, divide to generate two liver cells with the same characteristics and functions as their parent, as do insulin-producing cells in the pancreas. In contrast, **stem cells** and certain other undifferentiated cells can generate multiple types of differentiated descendant cells; these cells often divide such that the two daughter cells are different. Such **asymmetric cell division** is critical to the generation of different cell types in the body (Chapter 21). Often one daughter resembles its parent in that it remains undifferentiated and retains its ability to give rise to multiple types of differentiated cells. The other daughter divides many times and each of the daughter cells differentiates into a specific type of cell.

Under optimal conditions some bacteria, such as *E. coli*, can divide to form two daughter cells once every 30 minutes. Most eukaryotic cells take considerably longer to grow and divide, though cell divisions in the early *Drosophila* embryo require only 7 minutes. Moreover, the cell cycle in eukaryotes normally is highly regulated (Chapter 19). This tight control prevents imbalanced, excessive growth of cells and tissues if essential nutrients or certain hormonal signals are lacking. Some highly specialized cells in adult animals, such as nerve cells and striated muscle cells, divide rarely if at all. However, an organism usually replaces worn out cells or makes more cells in response to a new need, as exemplified by the growth of muscle in response to exercise or damage.

**⊙ OVERVIEW ANIMATION:** Life Cycle of a Cell

**FIGURE 1-16 During growth, all eukaryotic cells continually progress through the four stages of the cell cycle, generating new daughter cells.** In proliferating human cells, the four phases of the cell cycle proceed successively, taking from 10–20 hours depending on cell type and developmental state. Yeasts divide much faster. During interphase, which consists of the $G_1$, S, and $G_2$ phases, the cell roughly doubles its mass. Replication of DNA during the S phase leaves the cell with four copies of each type of chromosome. In the mitotic (M) phase, the chromosomes are evenly partitioned into two daughter cells, and the cytoplasm divides roughly in half in most cases. Under certain conditions, such as starvation or when a tissue has reached its final size, cells will stop cycling and remain in a waiting state called $G_0$. Most cells in $G_0$ can reenter the cycle if conditions change.

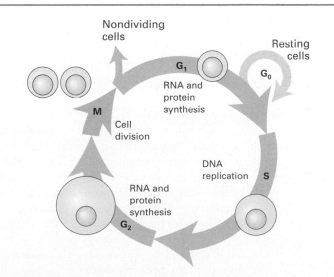

Another example is the formation of additional red blood cells when a person ascends to a higher altitude and needs more capacity to capture oxygen. The fundamental defect in cancer is loss of the ability to control the growth and division of cells. In Chapter 24 we examine the molecular and cellular events that lead to inappropriate, uncontrolled proliferation of cells.

## 1.3 Cells into Tissues: Unicellular and Metazoan Organisms Used for Molecular Cell Biology Investigations

Our current understanding of the molecular functioning of cells largely rests on studies of just a few types of organisms, termed *model organisms*. Because of the evolutionary conservation of genes, proteins, organelles, cell types, and so forth, discoveries about biological structures and functions obtained with one experimental organism often apply to others. Thus researchers generally conduct studies with the organism that is most suitable for rapidly and completely answering the question being posed, knowing that the results obtained in one organism are likely to be broadly applicable.

As we have seen, bacteria are excellent models for studies of several cellular functions, but they lack the organelles found in eukaryotes. Unicellular eukaryotes such as yeasts are used to study many fundamental aspects of eukaryotic cell structure and function. Multicellular, or **metazoan**, models are required to study more complex tissue and organ systems and development. As we will see in this section, several eukaryotic model organisms are in wide use to understand these complex cell systems and mechanisms.

### Single-Celled Eukaryotes Are Used to Study Fundamental Aspects of Eukaryotic Cell Structure and Function

One group of single-celled eukaryotes, the yeasts, has proven exceptionally useful in molecular and genetic analysis of eukaryotic cell formation and function. Yeasts and their multicellular cousins, the molds, which collectively constitute the fungi, have an important ecological role in breaking down plant and animal remains for reuse. They also make numerous antibiotics and are used in the manufacture of bread, beer, and wine.

The common yeast used to make bread and beer, *Saccharomyces cerevisiae*, appears frequently in this book because it has proven to be an extremely useful experimental organism. Homologs of many of the approximately 6000 different proteins expressed in an *S. cerevisiae* cell (Table 1-2) are found in most if not all eukaryotes and are important for cell division or for the functioning of individual eukaryotic organelles. Much of what we know of the proteins in the endoplasmic reticulum and Golgi apparatus that promote protein secretion was elucidated first in yeasts. Yeasts were also essential for the identification of many proteins that regulate the cell

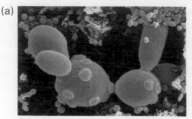

Budding (*S. cerevisiae*)

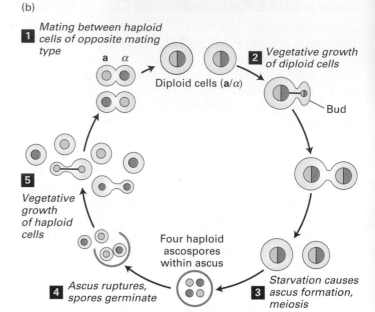

**FIGURE 1-17 The yeast *Saccharomyces cerevisiae* can grow as both haploids and diploids and can reproduce sexually and asexually.** (a) Scanning electron micrograph of the budding yeast *Saccharomyces cerevisiae*. These cells grow by an unusual type of mitosis termed mitotic budding. One daughter nucleus remains in the "mother" cell; the other daughter nucleus is transported into the bud, which grows in size and soon is released as a new cell. After each bud cell breaks free, a scar is left at the budding site, so the number of previous buds on the mother cell can be counted. The orange-colored cells are bacteria. (b) Haploid yeast cells can have different mating types, called **a** and **α**; both types contain a single copy of each yeast chromosome, half the usual number, and grow by mitotic budding. Two haploid cells that differ in mating type, one **a** and one **α,** can fuse together to form an **a/α** diploid cell that contains two copies of each chromosome; diploid cells also multiply by mitotic budding. Under starvation conditions, diploid cells undergo meiosis, a special type of cell division, to form haploid ascospores. Rupture of an ascus releases four haploid spores, which can germinate into haploid **a** and **α** cells. These also can multiply asexually. [Part (a) M. Abbey/VisualsUnlimited, Inc.]

cycle and catalyze DNA replication and transcription. *S. cerevisiae* (Figure 1-17a) and other yeasts offer many advantages to molecular and cellular biologists:

• Vast numbers of yeast cells can be grown easily and cheaply in culture from a single cell; such cell **clones** all have the same genes and the same biochemical properties. Individual proteins or multiprotein complexes can be purified from large amounts of cells and then studied in detail.

- Yeast cells can grow by mitosis both as haploids (containing one copy of each chromosome) and as diploids (containing two copies of each chromosome); this makes isolating and characterizing mutations in genes encoding essential cell proteins relatively straightforward.

- Yeasts, like many organisms, have a sexual cycle that allows exchange of genes between cells. Under starvation conditions, diploid cells undergo meiosis, a special type of cell division, to form haploid daughter cells, which are of two types, **a** and **α** cells. Haploid cells can also grow by mitosis. If haploid **a** and **α** cells encounter each other they can fuse, forming an **a/α** diploid cell that contains two copies of each chromosome (Figure 1-17b).

With the use of a single species like *S. cerevisiae* as a model organism, results from studies carried out by tens of thousands of scientists worldwide, using multiple experimental techniques, can be combined to yield a deeper level of understanding of a single type of cell. As we will see many times in this book, conclusions based on studies of *S. cerevisiae* are often generally true for all eukaryotes and form the basis for exploring the evolution of more complex processes in multicellular animals and plants.

## Mutations in Yeast Led to the Identification of Key Cell Cycle Proteins

Biochemical studies can tell us much about an individual protein, but they cannot prove that it is required for cell division or any other cell process. The importance of a protein is demonstrated most firmly if a mutation that prevents its synthesis or makes it nonfunctional adversely affects the process under study. A diploid organism generally carries two versions (alleles) of each gene, one derived from each parent. There are important exceptions, such as the genes on the X and Y chromosomes in males of some species, including our own.

In a classical genetics approach, scientists isolate and characterize mutants that lack the ability to do something a normal organism can do. Often large genetic "screens" are done to look for many different mutant individuals (e.g., fruit flies, yeast cells) that are unable to complete a certain process, such as cell division or muscle formation. Mutations usually are produced by treatment with a mutagen, a chemical or physical agent that promotes mutations in a largely random fashion. But how can we isolate and maintain mutant organisms or cells that are defective in some process, such as cell division, that is necessary for survival?

One way is to isolate organisms with a temperature-sensitive mutation. These mutants are able to grow at the permissive temperature, but not at another, usually higher temperature, the nonpermissive temperature. Normal cells can grow at either temperature. In most cases, a temperature-sensitive mutant produces an altered protein that works at the permissive temperature but unfolds and is nonfunctional at the nonpermissive temperature. Temperature-sensitive screens are most readily done with haploid organisms

like yeasts since there is only one copy of each gene, and a mutation in it will immediately have a consequence.

By analyzing the effects of numerous different temperature-sensitive mutations that altered division of haploid yeast cells, geneticists discovered most of the genes necessary for cell division without knowing anything, initially, about which proteins they encode or how these proteins participate in the process. The great power of genetics is to reveal the existence and relevance of proteins without prior knowledge of their biochemical identity or molecular function. Eventually these "mutation-defined" genes were isolated and replicated (cloned) with recombinant DNA techniques discussed in Chapter 5. With the isolated genes in hand, the encoded proteins could be produced in the test tube or in engineered bacteria or cultured cells. Then biochemists could investigate whether the proteins associate with other proteins or DNA or catalyze particular chemical reactions during cell division (Chapter 19).

Most of these yeast cell cycle genes are found in human cells as well, and the encoded proteins have similar amino acid sequences. Proteins from different organisms, but with similar amino acid sequences, are said to be **homologous**, and may have the same or similar functions. Remarkably, it has been shown that a human cell cycle protein, when expressed in a mutant yeast defective in the homologous yeast protein, is able to "rescue the defect" of the mutant yeast (that is, to allow the cell to grow normally), thus demonstrating the protein's ability to function in a very different type of eukaryotic cell. This experimental result, which garnered a Nobel Prize for Paul Nurse, was especially notable because the common ancestor cell of present day yeasts and humans is thought to have lived over a billion years ago. Clearly the eukaryotic cell cycle and the genes and proteins that catalyze it evolved early in biological evolution and have remained quite constant over a very long period of evolutionary time. Importantly, subsequent studies showed that mutations in many yeast cell cycle proteins that allow uncontrolled cell growth also frequently occur in human cancers (Chapter 24), again attesting to the important conserved functions of these proteins in all eukaryotes.

## Multicellularity Requires Cell-Cell and Cell Matrix Adhesions

The evolution of multicellular organisms most likely began when cells remained associated in small colonies after division instead of separating into individual cells. A few prokaryotes and several unicellular eukaryotes, such as many fungi and slime molds, exhibit such rudimentary social behavior. The full flowering of multicellularity, however, occurred in eukaryotic organisms whose cells became differentiated and organized into groups, or *tissues*, in which the cells performed a specialized, common function. Metazoans—be they invertebrates like the fruit fly *Drosophila melanogaster* and the roundworm *Caenorhabditis elegans*, or vertebrates such as mice and humans—contain between 13,000 and 23,000 protein-coding genes, about three to four times that of a yeast (Table 1-2). Many of these genes are conserved among the metazoans and essential for the formation and function of specific tissues and organs.

Animal cells are often "glued" together into a chain, a ball, or a sheet by **cell-adhesion proteins** (often called *c*ell *a*dhesion *m*olecules, or CAMs) on their surface (Figure 1-3). Some CAMs bind cells to one another; other types bind cells to the extracellular matrix, forming a cohesive unit. In animals, the matrix cushions cells and allows nutrients to diffuse toward them and waste products to diffuse away. A specialized, especially tough matrix called the **basal lamina,** comprised of multiple proteins such as collagen and polysaccharides, forms a supporting layer underlying cell sheets and prevents the cell aggregates from ripping apart. The cells of higher plants are encased in a network of chambers formed by the interlocking cell walls surrounding the cells, and are connected by cytoplasmic bridges called **plasmodesmata.**

## Tissues Are Organized into Organs

The specialized groups of differentiated cells form tissues, which are themselves the major components of organs. For example, the lumen of a blood vessel is lined with a sheetlike layer of endothelial cells, or **endothelium,** which prevents blood cells from leaking out (Figure 1-18). A layer of smooth muscle tissue encircles the endothelium and basal lamina and contracts to limit blood flow. During times of fright, constriction of smaller peripheral vessels forces more blood to the vital organs. The muscle layer of a blood vessel is wrapped in an outer layer of connective tissue, a network of fibers and cells that encase and protect the vessel walls from stretching and rupture. This hierarchy of tissues is copied in other blood vessels, which differ mainly in the thickness of the layers. The wall of a major artery must withstand much stress and is therefore thicker than a minor vessel. The strategy of grouping and layering different tissues is used to build other complex organs as well. In each case the function of the organ is determined by the specific functions of its component tissues, and each type of cell in a tissue produces the specific groups of proteins that enable the tissue to carry out its functions.

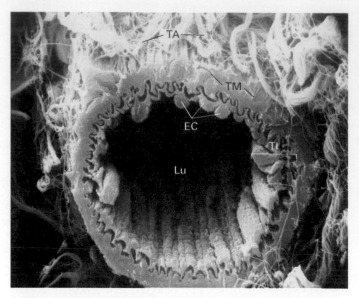

**FIGURE 1-18  All organs are organized arrangements of various tissues, as illustrated in this cross section of a small artery (arteriole).** Blood flows through the vessel lumen (Lu), which is lined by a thin sheet of endothelial cells (EC) forming the endothelium (TI) and by the underlying basal lamina. This tissue adheres to the overlying layer of smooth muscle tissue (TM); contraction of the muscle layer controls blood flow through the vessel. A fibrillar layer of connective tissue (TA) surrounds the vessel and connects it to other tissues. Dr. Richard Kessel & Dr. Randy Kardon/Visuals Unlimited, Inc.

## Body Plan and Rudimentary Tissues Form Early in Embryonic Development

The human body consists of some 100 trillion cells, yet it develops from a single cell, the zygote, resulting from fusion of a sperm and an egg. The early stages in the development of an embryo are characterized by rapid cell division (Figure 1-19) and the differentiation of cells into tissues. The embryonic *body plan,* the spatial pattern of cell types (tissues) and body parts, emerges from two influences: a program of genes that

**VIDEO:** Early Embryonic Development

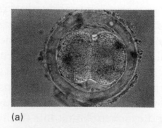

(a)

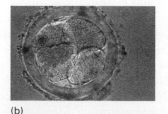

(b)

(c)

**FIGURE 1-19  The first few cell divisions of a fertilized egg set the stage for all subsequent development.** A developing mouse embryo is shown at the (a) two-cell, (b) four-cell, and (c) eight-cell stages. The embryo is surrounded by supporting membranes. The corresponding steps in human development occur during the first few days after fertilization. [Claude Edelmann/Photo Researchers, Inc.]

(a)

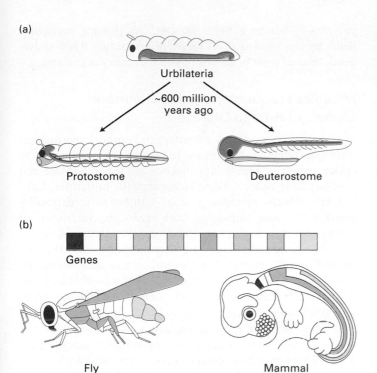

Urbilateria

~600 million
years ago

Protostome

Deuterostome

(b)

Genes

Fly
(protostome)

Mammal
(deuterostome)

**FIGURE 1-20 Similar genes, conserved during evolution, regulate early developmental processes in diverse animals.**
(a) Urbilateria is the presumed ancestor of all protostomes and deuterostomes that existed about 600 million years ago. The positions of the nerve chord (violet), surface ectoderm (mainly skin, white), and endoderm (mainly digestive tract and organs, light green) are shown. (b) Highly conserved proteins called Hox proteins are found in both protostomes and deuterostomes and determine the identity of body segments during embryonic development. *Hox* genes are found in clusters on the chromosomes of most or all animals, and encode related transcription factors that control the activities of other genes. In many animals *Hox* genes direct the development of different segments along the head-to-tail axis, as indicated by corresponding colors. Each gene is activated (transcriptionally) in a specific region along the head-to-tail axis and controls the growth of tissues there. For example, in the mouse, a deuterostome, the *Hox* genes are responsible for the distinctive shapes of vertebrae. Mutations affecting *Hox* genes in the fruit fly, a protostome, cause body parts to form in the wrong locations, such as legs in lieu of antennae on the head. In both organisms these genes provide a head-to-tail address and serve to direct the formation of structures in the appropriate places.

specify the pattern of the body, and local cell interactions that induce different parts of the program.

With only a few exceptions, most animals display axial symmetry; that is, their left and right sides mirror each other. This most basic of patterns is encoded in the genome. Developmental biologists have divided bilaterally symmetric animal phyla into two large groups depending on where the mouth and anus form in the early embryo. **Protostomes** develop a mouth close to a transient opening in the early embryo (the **blastopore**) and have a ventral nerve chord; protostomes include all worms, insects, and mollusks. The **deuterostomes** develop an anus close to this transient opening in the embryo and have a dorsal central nervous system; these include echinoderms (such as sea stars and sea urchins) and vertebrates. The bodies of both protostomes and deuterostomes are divided into discrete segments that form early in embryonic development. Protostomes and deuterostomes likely evolved from a common ancestor, termed **Urbilateria**, that lived approximately 600 million years ago (Figure 1-20a).

**Patterning genes** specify the general organization of an organism, beginning with the major body axes—anterior-posterior, dorsal-ventral, and left-right—and ending with body segments such as the head, chest, abdomen, and tail. The conservation of axial symmetry from the simplest worms to mammals is explained by the presence of conserved patterning genes in their genomes. Some patterning genes encode proteins that control expression of other genes; other patterning genes encode proteins that are important in cell adhesion or in cell signaling. This broad repertoire of patterning genes permits the integration and coordination of events in different parts of the developing embryo and gives each segment in the body its unique identity.

Remarkably, many patterning genes that are often called "master transcription factors," are highly conserved in both protostomes and deuterostomes (Figure 1-20b). This conservation of body plan reflects evolutionary pressure to preserve the commonalities in the molecular and cellular mechanisms controlling development in different organisms.

Fly eyes and human eyes are very different in structure, function, and nerve connections. Nonetheless, the so-called "master regulator genes" that initiate eye development—*eyeless* in the fly and *Pax6* in the human—encode highly related transcription factors that regulate the activities of other genes and are descended from the same ancestral gene. Mutations in the *eyeless* or *Pax6* genes cause major defects in eye formation (Figure 1-21).

## Invertebrates, Fish, and Other Organisms Serve as Experimental Systems for Study of Human Development

Studies of cells in specialized tissues make use of animal and plant model organisms. Nerve cells and muscle cells, for instance, traditionally were studied in mammals or in creatures with especially large or accessible cells, such as the giant neural cells of the squid and sea hare or the flight muscles of birds. More recently, muscle and nerve development have been extensively studied in fruit flies (*Drosophila melanogaster*), roundworms (*Caenorhabditis elegans*), and zebrafish (*Danio rerio*), in which mutants in muscle and nerve formation or function can be readily isolated (Figure 1-13).

Organisms with large-celled embryos that develop outside the mother (e.g., frogs, sea urchins, fish, and chickens) are extremely useful for tracing the fates of cells as they form

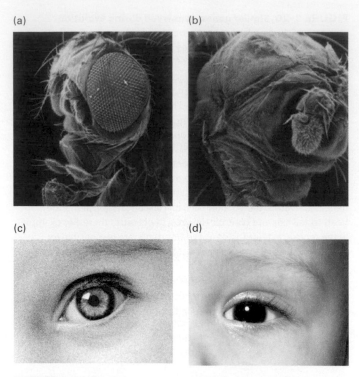

(a)

(b)

(c)

(d)

**FIGURE 1-21 Similar genes, conserved during evolution, regulate organ development in diverse animals.** (a) Development of the large compound eyes in fruit flies requires a gene called *eyeless* (named for the mutant phenotype). (b) Flies with inactivated *eyeless* genes lack eyes. (c) Normal human eyes require the gene *Pax6*, the homolog of *eyeless*. (d) People lacking adequate *Pax6* function have the genetic disease *aniridia*, a lack of irises in the eyes. *Pax6* and *eyeless* encode highly related transcription factors that regulate the activities of other genes and are descended from the same ancestral gene. [Parts (a) and (b) Andreas Hefti, Interdepartmental Electron Microscopy (IEM) Biocenter of the University of Basel. Part (c) © Simon Fraser/Photo Researchers, Inc. Part (d) Visuals Unlimited.]

different tissues and for making extracts for biochemical studies. For instance, a key protein in regulating mitosis was first identified in studies with frog and sea urchin embryos and subsequently purified from their extracts (Chapter 20).

Using recombinant DNA techniques, researchers can engineer specific genes to contain mutations that inactivate or increase production of their encoded proteins. Such genes can be introduced into the embryos of worms, flies, frogs, sea urchins, chickens, mice, a variety of plants, and other organisms, permitting the effects of these mutations to be assessed. This approach is being used extensively to produce mouse

versions of human genetic diseases. Inactivating particular genes by introducing short pieces of interfering RNA allows quick tests of gene functions possible in many organisms.

## Mice Are Frequently Used to Generate Models of Human Disease

Mice have one enormous advantage over other experimental organisms: they are the closest to humans of any animal for which powerful genetic approaches are feasible. Mice and humans have shared living structures for millennia, have similar immune systems, and are subject to infection by many of the same pathogens. Both organisms contain about the same number of genes, and about 99 percent of mouse protein-coding genes have homologs in the human, and vice versa. Over 90 percent of mouse and human genomes can be partitioned into regions of conserved synteny—that is, DNA segments that have the same order of unique DNA sequences and genes along a segment of a chromosome. This means that the gene order in the most recent common ancestor of humans and mice has been conserved in both species (Figure 1-22). This conserved synteny is consistent with archeological and other evidence that humans and mice descended from a common mammalian evolutionary ancestor that likely lived about 75 million years ago. Of course mice are not people; relative to humans, mice have expanded families of genes related to immunity, reproduction, and olfaction, likely reflecting the differences between the human and mouse lifestyle.

In Chapter 5 we will learn about the experimental utility of mouse embryonic stem (ES) cells, lines of cells derived from early mouse embryos that can be grown in culture in an undifferentiated state. Using techniques of recombinant DNA, scientists can introduce specific mutations into the mouse genome that mimic the corresponding mutations in human disease. For example, patients with a certain type of cancer accumulate inactivating mutations in a key cell cycle regulatory protein, and the analogous mutation can be introduced into the corresponding mouse gene. These gene-altered ES cells can be injected into an early mouse embryo, which is then implanted into a pseudopregnant female mouse (a mouse treated with hormones to trigger physiological changes needed for pregnancy). If the mice that develop from the injected ES cells exhibit a disease similar to the human cancer, then the link between the disease and mutations in a particular gene or genes is supported. Once mouse models of a human disease are available, further studies on

**FIGURE 1-22 Conservation of synteny between human and mouse.** Shown is a typical 510,000 base pair (bp) segment of mouse chromosome 12 that shares common ancestry with a 600,000 bp section of human chromosome 14. Blue lines connect the reciprocal unique DNA sequences in the two genomes. Mb, 1 million base pairs. [After Mouse Genome Sequencing Consortium, 2002, *Nature* **420**:520.]

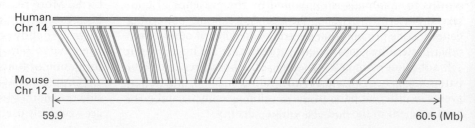

Human Chr 14

Mouse Chr 12

59.9

60.5 (Mb)

(a) T4 bacteriophage

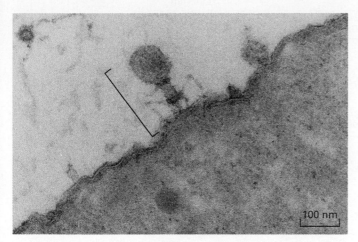

100 nm

(b) Tobacco mosaic virus

50 nm

(c) Adenovirus

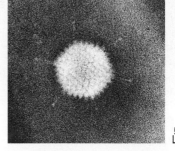

50 nm

**FIGURE 1-23 Viruses must infect a host cell to grow and reproduce.** These electron micrographs illustrate some of the structural variety exhibited by viruses. (a) T4 bacteriophage (bracket) attaches to an *E. coli* bacterial cell via a tail structure and injects its DNA, localized in the head, into the cell. Viruses that infect bacteria are called bacteriophages, or simply phages. (b) Tobacco mosaic virus causes a mottling of the leaves of infected tobacco plants and stunts their growth. (c) Adenovirus causes eye and respiratory tract infections in humans. This virus has an outer membranous envelope from which long glycoprotein spikes protrude. [Part (a) from A. Levine, 1991, *Viruses*, Scientific American Library, p. 20. Part (b) courtesy of R. C. Valentine. Part (c) courtesy of Robley C. Williams, University of California.]

the molecular defects causing the disease can be done and new treatments can be tested, thereby minimizing human exposure to untested treatments.

## Viruses Are Cellular Parasites That Are Widely Employed in Molecular Cell Biology Research

Virus-caused diseases are numerous and all too familiar, including chickenpox, influenza, some types of pneumonia, polio, measles, rabies, hepatitis, the common cold, and many others. Viral infections in plants (e.g., dwarf mosaic virus in corn) have a major economic impact on crop production. Almost all viruses have a rather limited host range, infecting only certain bacteria, plants, or animals (Figure 1-23). Viruses are much smaller than cells, on the order of 100 nanometers (nm) in diameter. A virus is typically composed of a protein coat that encloses a core containing the genetic material, which can be either DNA or RNA and carries the information for producing more viruses (Chapter 4). The coat protects a virus from the environment and allows it to stick to, or to enter, specific host cells. In some viruses, the protein coat is surrounded by an outer membrane-like envelope that is formed from the plasma membrane of the infected cell (Figure 14-34).

Because viruses cannot grow or reproduce on their own, a virus must infect a host cell and take over its internal machinery to synthesize viral proteins. All viruses use cellular ribosomes to synthesize viral proteins; most DNA viruses use cellular enzymes for replication of their DNA and for transcription of their DNA into mRNA. Thus studies of virus DNA replication and RNA synthesis are informative of the corresponding cellular processes. When newly made viruses are released by budding from the cell membrane or when the infected cell bursts, the cycle starts anew.

Consider the adenoviruses, which cause eye and respiratory tract infections in humans. Human adenoviruses have a genome of only approximately 35,000 base pairs—about 2 percent the size of a bacterial genome—and encode about 30 proteins, about half of which are conserved among adenoviruses that infect different species. These conserved proteins comprise structural proteins that form parts of the mature virus particle (virion) and proteins that catalyze steps in viral DNA replication. Late in adenovirus infection of human cells, the cell becomes a virtual factory for producing just a few viral proteins: about half of the non-ribosomal RNAs synthesized are viral mRNAs, and most of the proteins produced are viral. In the 1970s—before recombinant DNA techniques were developed—this permitted experiments on adenovirus mRNA synthesis that demonstrated that mature mRNAs had undergone *splicing*, or removal of noncoding sequences (see Figure 1-9). Only later was splicing shown to be a fundamental part of biogenesis of virtually all eukaryotic mRNAs.

A different type of virus, vesicular stomatitis virus, makes a single glycoprotein (a protein with attached carbohydrate chain) that is transported to the plasma membrane and then forms part of the membrane coat of this virus. Studies of this protein (Figures 14-2 and 14-3) elucidated many aspects of the biogenesis of membrane glycoproteins that were later shown to apply to all cellular glycoproteins.

Even today viruses are useful in many aspects of molecular cell biology. Many methods for genetically manipulating cells depend on using viruses to convey DNA molecules into cells. To do this, the portion of the viral genetic material that encodes proteins that are potentially harmful is replaced with other genetic material, including human genes; adenovirus is often employed for this purpose. The altered viruses, or vectors, still can enter cells toting the introduced genes

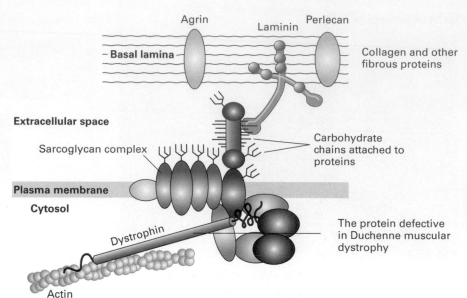

**FIGURE 1-24  The dystrophin glycoprotein complex (DGC) in skeletal muscle cells.**
Dystrophin—the protein defective in Duchenne muscular dystrophy—links the actin cytoskeleton to the multiprotein sarcoglycan complex in the plasma membrane. Other proteins in the complex bind to components of the basal lamina, such as laminin, that in turn bind to the collagen fibers that give the basal lamina strength and rigidity. Thus dystrophin is an important member of a group of proteins that links the muscle cell and its internal actin cytoskeleton with the surrounding basal lamina. [Adapted from S. J. Winder, 2001, *Trends Biochem. Sci.* **26**:118, and D. E. Michele and K. P. Campbell, 2003, *J. Biol. Chem.* **278**:15457.]

with them (Chapter 5). One day, diseases caused by defective genes may be treated by using viral vectors to introduce a normal copy of a defective gene into patients. Current research is dedicated to overcoming the considerable obstacles to this *gene therapy* approach, such as getting the introduced genes to work in the right cells at the correct times.

## Genetic Diseases Elucidate Important Aspects of Cell Function

Many genetic diseases are caused by mutations in a single protein; studies on humans with these diseases have shed light on the normal function of the protein.  As an example, consider Duchenne muscular dystrophy (DMD), the most common type of hereditary muscle-wasting diseases, collectively called muscular dystrophies. DMD is an X chromosome-linked disorder, affecting 1 in 3300 boys, that results in cardiac or respiratory failure, usually in the late teens or early twenties. The first clue to understanding the molecular basis of this disease came from the discovery that people with DMD carry mutations in the gene encoding a protein named dystrophin. This very large protein was later found to be a cytosolic adapter protein, binding to actin filaments that are part of the cytoskeleton (see Figure 1-14) and to a complex of muscle plasma membrane proteins termed the sarcoglycan complex (Figure 1-24). The resulting large multiprotein assemblage, the dystrophin glycoprotein complex (DGC), links the extracellular matrix protein laminin to the cytoskeleton within muscle and other types of cells. Mutations in dystrophin, other DGC components, or laminin can disrupt the DGC-mediated link between the exterior and interior of muscle cells and cause muscle weakness and eventual death. The first step in identifying the entire dystrophin glycoprotein complex involved cloning the dystrophin-encoding gene using DNA from normal individuals and patients with Duchenne muscular dystrophy.

## The Following Chapters Present Much Experimental Data That Explains How We Know What We Know About Cell Structure and Function

In subsequent chapters of this book we discuss cellular processes in much greater detail. We begin (Chapter 2) with a discussion of the chemical nature of the building blocks of cells and the basic chemical processes required to understand the macromolecular processes discussed in subsequent chapters. We go on to discuss the structure and function of proteins (Chapter 3) and how the information for their synthesis is encoded in DNA (Chapter 4). Chapter 5 describes many of the techniques used to study genes, gene expression, and protein function. Gene and chromosome structure and the regulation of gene expression are covered in Chapters 6, 7, and 8. Chapter 9 discusses many of the techniques biologists use to culture and fractionate cells and to visualize specific proteins and structures within cells. Biomembrane structure and transport of ions and small molecules across membranes are the topics of Chapters 10 and 11, and Chapter 12 discusses cellular energetics and the functions of mitochondria and chloroplasts. Membrane biogenesis, protein secretion, and protein trafficking—the sorting of proteins to their correct subcellular destinations—are the topics of Chapters 13 and 14. Chapters 15 and 16 discuss the many types of signals and signal receptors used by cells to communicate and regulate their activities. The cytoskeleton and cell movements are discussed in Chapters 17 and 18. Chapter 19 discusses the cell cycle and how cell division is regulated. The interactions among cells and between cells and the extracellular matrix that enable formation of tissues and organs are detailed in Chapter 20. Later chapters of the book discuss important types of specialized cells—stem cells (Chapter 21), nerve cells (Chapter 22), and cells of the immune system (Chapter 23). Chapter 24 discusses cancer and the multiple ways in which cell growth and differentiation can be altered by mutations.

# Chemical Foundations

Polarized light microscopic image of crystals of cholesterol. Cholesterol is a water-insoluble molecule that plays a critical structural role in many membranes of animal cells and is a precursor for the synthesis of steroid hormones, bile acids, and vitamin D. Excess deposition of cholesterol in artery walls is a key step in clogging of the arteries, a major cause of heart attacks and strokes. [Courtesy of National High Magnetic Field Laboratory/The Florida State University.]

The life of a cell depends on thousands of chemical interactions and reactions exquisitely coordinated with one another in time and space and under the influence of the cell's genetic instructions and its environment. By understanding at a molecular level these interactions and reactions, we can begin to answer fundamental questions about cellular life: How does a cell extract nutrients and information from its environment? How does a cell convert the energy stored in nutrients into the work of movement or metabolism? How does a cell transform nutrients into the cellular components required for its survival? How does a cell link itself to other cells to form a tissue? How do cells communicate with one another so that a complex, efficiently functioning organism can develop and thrive? One of the goals of *Molecular Cell Biology* is to answer these and other questions about the structure and function of cells and organisms in terms of the properties of individual molecules and ions.

For example, the properties of one such molecule, water, have controlled and continue to control the evolution, structure, and function of all cells. An understanding of biology is not possible without appreciating how the properties of water control the chemistry of life. Life first arose in a watery environment. Constituting 70–80 percent by weight of most cells, water is the most abundant molecule in biological systems. It is within this aqueous milieu that small molecules and ions, which make up about 7 percent of the weight of living matter, combine into the larger macromolecules and macromolecular assemblies that make up a cell's machinery and architecture and so the remaining mass of organisms. These small molecules include amino acids (the building blocks of proteins), nucleotides (the building blocks of DNA and RNA), lipids (the building blocks of biomembranes), and sugars (the building blocks of complex carbohydrates).

Many of the cell's biomolecules (e.g., sugars) readily dissolve in water; these molecules are called **hydrophilic** ("water liking"). Others (e.g., cholesterol) are oily, fatlike substances that shun water; these are said to be **hydrophobic** ("water fearing"). Still other biomolecules (e.g., phospholipids) contain both hydrophilic and hydrophobic regions; these molecules are said to be **amphipathic** ("both liking"). Phospholipids are used to build the flexible membranes that enclose cells and their internal organelles. The smooth functioning of cells, tissues, and organisms depends on all these molecules,

## OUTLINE

**(a) Molecular complementarity**

Protein A

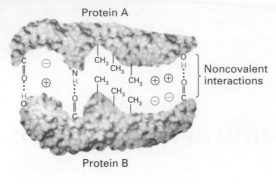

Noncovalent
interactions

Protein B

**(b) Chemical building blocks**

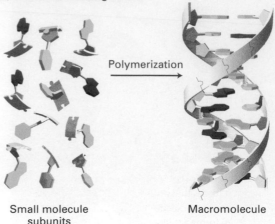

Polymerization

Small molecule
subunits

Macromolecule

**(c) Chemical equilibrium**

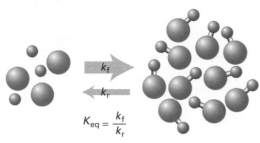

$$K_{eq} = \frac{k_f}{k_r}$$

**(d) Chemical bond energy**

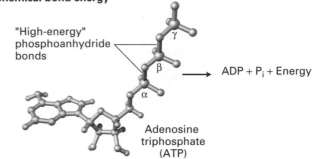

"High-energy"
phosphoanhydride
bonds

$\gamma$

$\beta$

$\alpha$

ADP + P$_i$ + Energy

Adenosine
triphosphate
(ATP)

**FIGURE 2-1 Chemistry of life: four key concepts.** (a) Molecular complementarity lies at the heart of all biomolecular interactions, as when two proteins with complementary shapes and chemical properties come together to form a tightly bound complex. (b) Small molecules serve as building blocks for larger structures. For example, to generate the information-carrying macromolecule DNA, four small nucleotide building blocks are covalently linked into long strings (polymers), which then wrap around each other to form the double helix. (c) Chemical reactions are reversible, and the distribution of the chemicals between starting reagents (*left*) and the products of the reactions (*right*) depends on the rate constants of the forward ($k_f$, upper arrow) and reverse ($k_r$, lower arrow) reactions. The ratio of these, $K_{eq}$, provides an informative measure of the relative amounts of products and reactants that will be present at equilibrium. (d) In many cases, the source of energy for chemical reactions in cells is the hydrolysis of the molecule ATP. This energy is released when a high-energy phosphoanhydride bond linking the $\beta$ and $\gamma$ phosphates in the ATP molecule (red) is broken by the addition of a water molecule, forming ADP and P$_i$.

from the smallest to the largest. Indeed, the chemistry of the simple proton (H$^+$) can be as important to the survival of a human cell as that of each gigantic DNA molecule (the mass of the DNA molecule in human chromosome 1 is $8.6 \times 10^{10}$ times that of a proton!). The chemical interactions of all of these molecules, large and small, with water and with one another, define the nature of life.

Luckily, although many types of biomolecules interact and react in numerous and complex pathways to form functional cells and organisms, a relatively small number of chemical principles are necessary to understand cellular processes at the molecular level (Figure 2-1). In this chapter we review these key principles, some of which you already know well. We begin with the covalent bonds that connect atoms into molecules and the noncovalent interactions that stabilize groups of atoms within and between molecules. We then consider the basic chemical building blocks of macromolecules and macromolecular assemblies. After reviewing those aspects of chemical equilibrium that are most relevant to

biological systems, we end the chapter with basic principles of biochemical energetics, including the central role of ATP (adenosine triphosphate) in capturing and transferring energy in cellular metabolism.

## 2.1 Covalent Bonds and Noncovalent Interactions

Strong and weak attractive forces between atoms are the "glue" that holds individual molecules together and permits interactions between different molecules. When two atoms share a single pair of electrons, the result is a **covalent bond**—a type of strong force that holds atoms together in molecules. Sharing of multiple pairs of electrons results in multiple covalent bonds (e.g., "double" or "triple" bonds). The weak attractive forces of **noncovalent interactions** are equally important in determining the properties and functions of biomolecules such as proteins, nucleic acids,

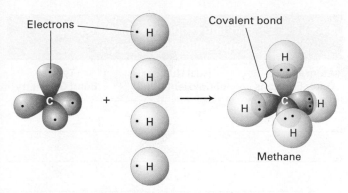

**FIGURE 2-2 Covalent bonds form by the sharing of electrons.** Covalent bonds, the strong forces that hold atoms together into molecules, form when atoms share electrons from their outermost electron orbitals. Each atom forms a defined number and geometry of covalent bonds.

carbohydrates, and lipids. In this section, we first review covalent bonds and then discuss the four major types of non-covalent interactions: ionic bonds, hydrogen bonds, Van der Waals interactions, and the hydrophobic effect.

## The Electronic Structure of an Atom Determines the Number and Geometry of Covalent Bonds It Can Make

Hydrogen, oxygen, carbon, nitrogen, phosphorus, and sulfur are the most abundant elements in biological molecules. These atoms, which rarely exist as isolated entities, readily form covalent bonds, using electrons in the outermost electron orbitals surrounding their nuclei (Figure 2-2). As a rule, each type of atom forms a characteristic number of covalent bonds with other atoms, with a well-defined geometry determined by the atom's size and by both the distribution of electrons around the nucleus and the number of electrons that it can share. In some cases, the number of stable covalent bonds an atom can make is fixed; carbon, for example, always forms four covalent bonds. In other cases, different numbers of stable covalent bonds are possible; for example, sulfur can form two, four, or six stable covalent bonds.

All the biological building blocks are organized around the carbon atom, which forms four covalent bonds. In these organic biomolecules, each carbon usually bonds to three or four other atoms. (Carbon can also bond to two other atoms, as in the linear molecule carbon dioxide, $CO_2$, which has two carbon-oxygen double bonds ($O=C=O$); however, such bond arrangements of carbon are not found in biological building blocks.) As illustrated in Figure 2-3a for formaldehyde, carbon can bond to three atoms, all in a common plane. The carbon atom forms two single bonds with two atoms and one double bond (two shared electron pairs) with the third atom. In the absence of other constraints, atoms joined by a single bond generally can rotate freely about the bond axis, whereas those connected by a double bond cannot. The rigid planarity imposed by double bonds has enormous

significance for the shapes and flexibility of biomolecules such as phospholipids, proteins, and nucleic acids.

Carbon can also bond to four rather than three atoms. As illustrated by methane ($CH_4$), when carbon is bonded to four other atoms, the angle between any two bonds is 109.5° and the positions of bonded atoms define the four points of a tetrahedron (Figure 2-3b). This geometry defines the structures of many biomolecules. A carbon (or any other) atom bonded to four dissimilar atoms or groups in a nonplanar configuration is said to be asymmetric. The tetrahedral orientation of bonds formed by an **asymmetric carbon atom** can be arranged in three-dimensional space in two different ways, producing molecules that are mirror images of each other, a property called *chirality* (from the Greek word *cheir*, meaning "hand") (Figure 2-4). Such molecules are called *optical isomers*, or **stereoisomers**. Many molecules in cells contain at least one asymmetric carbon atom, often called a *chiral carbon* atom. The different stereoisomers of a molecule usually have completely different biological activities because the arrangement of atoms within their structures, and thus their ability to interact with other molecules, differs.

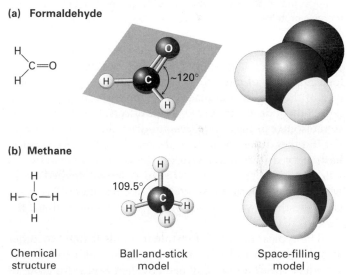

**(a) Formaldehyde**

**(b) Methane**

Chemical structure   Ball-and-stick model   Space-filling model

**FIGURE 2-3 Geometry of bonds when carbon is covalently linked to three or four other atoms.** (a) A carbon atom can be bonded to three atoms, as in formaldehyde ($CH_2O$). In this case, the carbon-bonding electrons participate in two single bonds and one double bond, which all lie in the same plane. Unlike atoms connected by a single bond, which usually can rotate freely about the bond axis, those connected by a double bond cannot. (b) When a carbon atom forms four single bonds, as in methane ($CH_4$), the bonded atoms (all H in this case) are oriented in space in the form of a tetrahedron. The letter representation on the left clearly indicates the atomic composition of the molecule and the bonding pattern. The ball-and-stick model in the center illustrates the geometric arrangement of the atoms and bonds, but the diameters of the balls representing the atoms and their nonbonding electrons are unrealistically small compared with the bond lengths. The sizes of the electron clouds in the space-filling model on the right more accurately represent the structure in three dimensions.

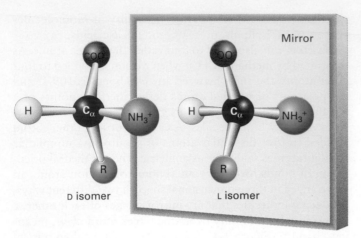

**FIGURE 2-4 Stereoisomers.** Many molecules in cells contain at least one asymmetric carbon atom. The tetrahedral orientation of bonds formed by an asymmetric carbon atom can be arranged in three-dimensional space in two different ways, producing molecules that are mirror images, or stereoisomers, of each other. Shown here is the common structure of an amino acid, with its central asymmetric carbon and four attached groups, including the R group, discussed in Section 2.2. Amino acids can exist in two mirror-image forms, designated L and D. Although the chemical properties of such stereoisomers are identical, their biological activities are distinct. Only L amino acids are found in proteins.

Some drugs are mixtures of the stereoisomers of small molecules in which only one stereoisomer has the biological activity of interest. The use of a pure single stereoisomer of the chemical in place of the mixture may result in a more potent drug with reduced side effects. For example, one stereoisomer of the antidepressant drug citalopram (Celexa) is 170 times more potent than the other. Some stereoisomers have very different activities. Darvon is a pain reliever, whereas its stereoisomer, Novrad (*Darvon* spelled backward), is a cough suppressant. One stereoisomer of ketamine is an anesthetic, whereas the other causes hallucinations. ∎

The typical number of covalent bonds formed by other atoms common to biomolecules is shown in Table 2-1. A hydrogen atom forms only one covalent bond. An atom of oxygen usually forms only two covalent bonds but has two additional pairs of electrons that can participate in noncovalent interactions. Sulfur forms two covalent bonds in hydrogen sulfide ($H_2S$) but also can accommodate six covalent bonds, as in sulfuric acid ($H_2SO_4$) and its sulfate derivatives. Nitrogen and phosphorus each have five electrons to share. In ammonia ($NH_3$), the nitrogen atom forms three covalent bonds; the pair of electrons around the atom not involved in a covalent bond can take part in noncovalent interactions. In the ammonium ion ($NH_4^+$), nitrogen forms four covalent bonds, which have a tetrahedral geometry. Phosphorus commonly forms five covalent bonds, as in phosphoric acid ($H_3PO_4$) and its phosphate derivatives, which form the backbone of nucleic acids. Phosphate groups covalently attached to proteins play a key role in regulating the activity of many proteins, and the central molecule in cellular energetics,

| TABLE 2-1 | Bonding Properties of Atoms Most Abundant in Biomolecules | |
|---|---|---|
| **Atom and Outer Electrons** | **Usual Number of Covalent Bonds** | **Typical Bond Geometry** |
| $\dot{H}$ | 1 | H |
| $\cdot\ddot{O}\cdot$ | 2 | O |
| $\cdot\ddot{S}\cdot$ | 2, 4, or 6 | S |
| $\cdot\ddot{N}\cdot$ | 3 or 4 | N |
| $\cdot\ddot{P}\cdot$ | 5 | P |
| $\cdot\dot{C}\cdot$ | 4 | C |

ATP, contains three phosphate groups (see Section 2.4). A summary of common covalent linkages and functional groups, which confer distinctive chemical properties to the molecules of which they are a part, is provided in Table 2-2.

## Electrons May Be Shared Equally or Unequally in Covalent Bonds

The extent of an atom's ability to attract an electron is called its *electronegativity*. In a bond between atoms with identical or similar electronegativities, the bonding electrons are essentially shared equally between the two atoms, as is the case for most carbon-carbon single bonds (C—C) and carbon-hydrogen single bonds (C—H). Such bonds are called **nonpolar**. In many molecules, the bonded atoms have different electronegativities, resulting in unequal sharing of electrons. The bond between them is said to be **polar**.

One end of a polar bond has a partial negative charge ($\delta^-$), and the other end has a partial positive charge ($\delta^+$). In an O—H bond, for example, the greater electronegativity of the oxygen atom relative to hydrogen results in the electrons spending more time around the oxygen atom than the hydrogen. Thus the O—H bond possesses an electric **dipole**, a positive charge separated from an equal but opposite negative charge. The amount of $\delta^-$ charge on the oxygen atom of a O—H dipole is approximately 25 percent of that of an electron, with an equivalent and opposite $\delta^+$ charge on the H atom. A common quantitative measure of the extent of charge separation, or strength, of a dipole is called the **dipole moment**, $\mu$, which for a chemical bond is the product of the partial charge on each atom and the distance between the two atoms. For a molecule with multiple dipoles, the amount of charge separation for the molecule as a whole depends in part on the dipole moments of all of its individual chemical bonds and in part on the geometry of the molecule (relative orientations of the individual dipole moments). Consider the example

**TABLE 2-2** | **Common Functional Groups and Linkages in Biomolecules**

**Functional Groups**

| —OH<br>Hydroxyl<br>(alcohol) | $\overset{\displaystyle O}{\underset{\displaystyle \parallel}{}}$<br>—C—R<br>Acyl<br>(triacylglycerol) | $\overset{\displaystyle O}{\underset{\displaystyle \parallel}{}}$<br>—C—<br>Carbonyl<br>(ketone) | $\overset{\displaystyle O}{\underset{\displaystyle \parallel}{}}$<br>—C—O⁻<br>Carboxyl<br>(carboxylic acid) |
|---|---|---|---|
| —SH<br>Sulfhydryl<br>(Thiol) | —NH₂ or —⁺NH₃<br>Amino<br>(amines) | $\overset{\displaystyle O}{\underset{\displaystyle O^-}{\overset{\displaystyle \parallel}{}}}$<br>—O—P—O⁻<br>Phosphate<br>(phosphorylated molecule) | $\overset{\displaystyle O \quad\quad O}{}$<br>—O—P—O—P—<br>O⁻    O⁻<br>Pyrophosphate<br>(diphosphate) |

**Linkages**

| $\overset{\displaystyle O}{}$<br>—C—O—C—<br>Ester | —C—O—C—<br>Ether | $\overset{\displaystyle O}{\underset{}{\parallel}}$<br>—N—C—<br>Amide |
|---|---|---|

of water ($H_2O$), which has two O—H bonds and thus two individual bond dipole moments. If water were a linear molecule with the two bonds on exact opposite sides of the O atom, the two dipoles on each end of the molecule would be identical in strength but would be oriented in opposite directions. The two dipole moments would cancel each other and the dipole moment of molecule as a whole would be zero. However, because water is a V-shaped molecule, with the individual dipoles of its two O—H bonds both pointing toward the oxygen, one end of the water molecule (the end with the oxygen atom) has a partial negative charge and the other end (the one with the two hydrogen atoms) has a partial positive charge. As a consequence, the molecule as a whole is a dipole with a well-defined dipole moment (Figure 2-5). This dipole

moment and the electronic properties of the oxygen and hydrogen atoms allow water to form electrostatic, noncovalent interactions with other waters and with other molecules. These interactions play a critical role in almost every biochemical interaction in cells and organisms and will be discussed shortly.

Another important example of polarity is the O=P double bond in $H_3PO_4$. In the structure of $H_3PO_4$ shown on the left below, lines represent single and double bonds and nonbonding electrons are shown as pairs of dots:

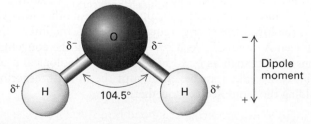

Because of the polarity of the O=P double bond, the $H_3PO_4$ can also be represented by the structure on the right, in which one of the electrons from the P=O double bond has accumulated around the O atom, giving it a negative charge and leaving the P atom with a positive charge. These charges are important in noncovalent interactions. Neither of these two models precisely describes the electronic state of $H_3PO_4$. The actual structure can be considered to be an intermediate, or hybrid, between these two representations, as indicated by the double-headed arrow between them. Such intermediate structures are called *resonance hybrids*.

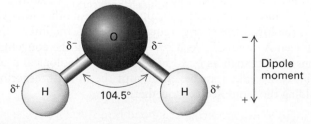

δ⁻    O    δ⁻

δ⁺ H    104.5°    H δ⁺

Dipole moment

**FIGURE 2-5 The dipole nature of a water molecule.** The symbol δ represents a partial charge (a weaker charge than the one on an electron or a proton). Because of the difference in the electronegativities of H and O, each of the polar H—O bonds in water is a dipole. The sizes and directions of the dipoles of each of the bonds determine the net distance and amount of charge separation, or dipole moment, of the molecule.

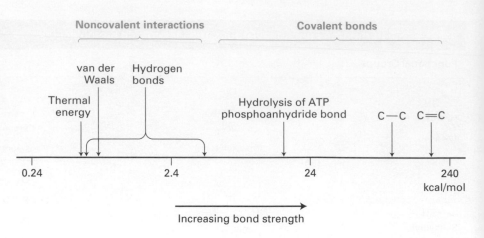

**FIGURE 2-6 Relative energies of covalent bonds and noncovalent interactions.** Bond energies are defined as the energy required to break a particular type of linkage. Shown here are the energies required to break a variety of linkages, arranged on a log scale. Covalent bonds, including those for single (C—C) and double (C=C) carbon-carbon bonds, are one to two powers of 10 stronger than noncovalent interactions. The latter are somewhat greater than the thermal energy of the environment at normal room temperature (25 °C). Many biological processes are coupled to the energy released during hydrolysis of a phosphoanhydride bond in ATP.

## Covalent Bonds Are Much Stronger and More Stable Than Noncovalent Interactions

Covalent bonds are considered to be strong because the energies required to break them are much greater than the thermal energy available at room temperature (25 °C) or body temperature (37 °C). As a consequence, they are stable at these temperatures. For example, the thermal energy at 25 °C is approximately 0.6 kilocalorie per mole (kcal/mol), whereas the energy required to break the C—C bond in ethane is about 140 times larger (Figure 2-6). Consequently, at room temperature (25 °C), fewer than 1 in $10^{12}$ ethane molecules is broken into a pair of $\cdot CH_3$ molecules, each containing an unpaired, nonbonding electron (called a radical).

Covalent single bonds in biological molecules have energies similar to the energy of the C—C bond in ethane. Because more electrons are shared between atoms in double bonds, they require more energy to break than single bonds. For instance, it takes 84 kcal/mol to break a single C—O bond but 170 kcal/mol to break a C=O double bond. The most common double bonds in biological molecules are C=O, C=N, C=C, and P=O.

In contrast, the energy required to break noncovalent interactions is only 1–5 kcal/mol, much less than the bond energies of covalent bonds (see Figure 2-6). Indeed, noncovalent interactions are weak enough that they are constantly being formed and broken at room temperature. Although these interactions are weak and have a transient existence at physiological temperatures (25–37 °C), multiple noncovalent interactions can, as we will see, act together to produce highly stable and specific associations between different parts of a large molecule or between different macromolecules. Protein-protein and protein-nucleic acid interactions are good examples of noncovalent interactions. Below, we review the four main types of noncovalent interactions and then consider their roles in the binding of biomolecules to one another and to other molecules.

## Ionic Interactions Are Attractions Between Oppositely Charged Ions

**Ionic interactions** result from the attraction of a positively charged ion—a **cation**—for a negatively charged ion—an **anion**. In sodium chloride (NaCl), for example, the bonding electron

contributed by the sodium atom is completely transferred to the chlorine atom. (Figure 2-7a). Unlike covalent bonds, ionic interactions do not have fixed or specific geometric orientations because the electrostatic field around an ion—its attraction for an opposite charge—is uniform in all directions. In solid NaCl, oppositely charged ions pack tightly together in an alternating pattern, forming the highly ordered crystalline array typical of salt crystals (Figure 2-7b). The energy required to break an ionic interaction depends on the distance between the ions and the electrical properties of the environment of the ions.

When solid salts dissolve in water, the ions separate from one another and are stabilized by their interactions with water molecules. In aqueous solutions, simple ions of biological significance, such as $Na^+$, $K^+$, $Ca^{2+}$, $Mg^{2+}$, and $Cl^-$, are hydrated, surrounded by a stable shell of water molecules held in place by ionic interactions between the central ion and the oppositely charged end of the water dipole (Figure 2-7c). Most ionic compounds dissolve readily in water because the energy of hydration, the energy released when ions tightly bind water molecules and spread out in an aqueous solution, is greater than the lattice energy that stabilizes the crystal structure. Parts or all of the aqueous *hydration shell* must be removed from ions when they directly interact with proteins. For example, water of hydration is lost when ions pass through protein pores in the cell membrane during nerve conduction.

The relative strength of the interaction between two oppositely charged ions, $A^-$ and $C^+$, depends on the concentration of other ions in a solution. The higher the concentration of other ions (e.g., $Na^+$ and $Cl^-$), the more opportunities $A^-$ and $C^+$ have to interact ionically with these other ions and thus the lower the energy required to break the interaction between $A^-$ and $C^+$. As a result, increasing the concentrations of salts such as NaCl in a solution of biological molecules can weaken and even disrupt the ionic interactions holding the biomolecules together.

## Hydrogen Bonds Are Noncovalent Interactions That Determine the Water Solubility of Uncharged Molecules

A **hydrogen bond** is the interaction of a partially positively charged hydrogen atom in a molecular dipole, such as water,

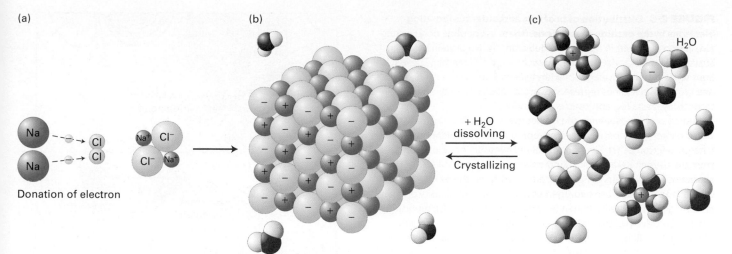

(a) (b) (c)

Na → Cl
Na → Cl

Na⁺ Cl⁻
Cl⁻ Na⁺

Donation of electron

+ H₂O
dissolving

Crystallizing

H₂O

**FIGURE 2-7 Electrostatic interactions of oppositely charged ions of salt (NaCl) in crystals and in aqueous solution.** (a) In crystalline table salt, sodium atoms are positively charged ions ($Na^+$) due to the loss of one electron each, whereas chloride atoms are correspondingly negatively charged ($Cl^-$) by gaining one electron each. (b) In solid form, ionic compounds form neatly ordered arrays, or crystals, of tightly packed ions in which the positive and negatively charged ions counter- balance each other. (c) When the crystals are dissolved in water, the ions separate and their charges, no longer balanced by immediately adjacent ions of opposite charge, are stabilized by interactions with polar water. Water molecules and the ions are held together by electrostatic interactions between the charges on the ion and the partial charges on the water's oxygen and hydrogen atoms. In aqueous solutions, all ions are surrounded by a hydration shell of water molecules.

with unpaired electrons from another atom, either in the same or a different molecule. Normally, a hydrogen atom forms a covalent bond with only one other atom. However, a hydrogen atom covalently bonded to an electronegative donor atom D may form an additional weak association, the hydrogen bond, with an acceptor atom A, which must have a nonbonding pair of electrons available for the interaction:

$$D^{\delta-}\!-\!H^{\delta+} + :A^{\delta-} \rightleftharpoons D^{\delta-}\!-\!H^{\delta+}\cdots:A^{\delta-}$$
**Hydrogen bond**

The length of the covalent D—H bond is a bit longer than it would be if there were no hydrogen bond because the acceptor "pulls" the hydrogen away from the donor. An important feature of all hydrogen bonds is directionality. In the strongest hydrogen bonds, the donor atom, the hydrogen atom, and the acceptor atom all lie in a straight line. Nonlinear hydrogen bonds are weaker than linear ones; still, multiple nonlinear hydrogen bonds help to stabilize the three-dimensional structures of many proteins.

Hydrogen bonds are both longer and weaker than covalent bonds between the same atoms. In water, for example, the distance between the nuclei of the hydrogen and oxygen atoms of adjacent, hydrogen-bonded molecules is about 0.27 nm, about twice the length of the covalent O—H bonds within a single water molecule (Figure 2-8a). The strength of a hydrogen bond between water molecules (approximately 5 kcal/mol) is much

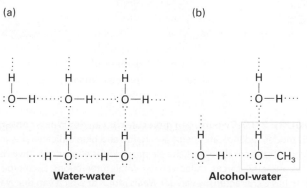

(a) (b)

Water-water    Alcohol-water

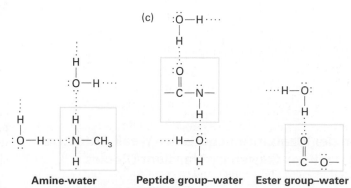

(c)

Amine-water    Peptide group–water    Ester group–water

**FIGURE 2-8 Hydrogen bonding of water with itself and with other compounds.** Each pair of nonbonding outer electrons in an oxygen or a nitrogen atom can accept a hydrogen atom in a hydrogen bond. The hydroxyl and the amino groups can also form hydrogen bonds with water. (a) In liquid water, each water molecule forms transient hydrogen bonds with several others, creating a dynamic network of hydrogen-bonded molecules. (b) Water also can form hydrogen bonds with alcohols and amines, accounting for the high solubility of these compounds. (c) The peptide group and ester group, which are present in many biomolecules, commonly participate in hydrogen bonds with water or polar groups in other molecules.

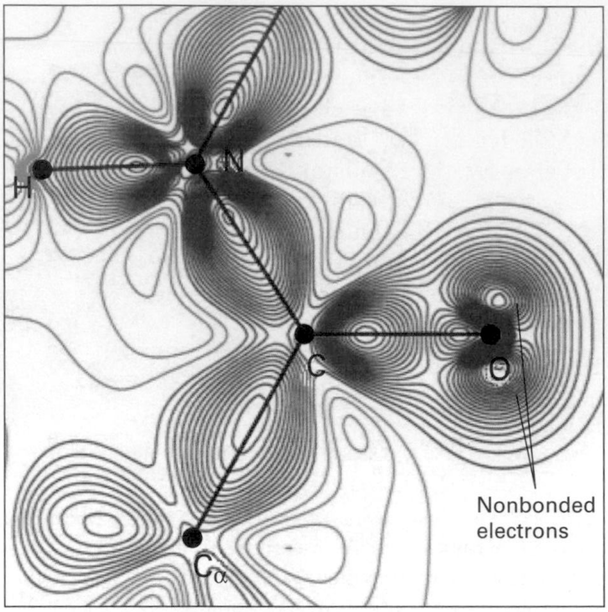

**FIGURE 2-9 Distribution of bonding and outer nonbonding electrons in the peptide group.** Shown here is a peptide bond linking two amino acids within a protein called crambin. No protein has been structurally characterized at higher resolution than crambin. The black lines represent the covalent bonds between atoms. The red (negative) and blue (positive) lines represent contours of charge determined using x-ray crystallography and computational methods. The greater the number of contour lines, the higher the charge. The high density of red contour lines between atoms represents the covalent bonds (shared electron pairs). The two sets of red contour lines emanating from the oxygen (O) and not falling on a covalent bond (black line) represent the two pairs of nonbonded electrons on the oxygen that are available to participate in hydrogen bonding. The high density of blue contour lines near the hydrogen (H) bonded to nitrogen (N) represents a partial positive charge, indicating that this H can act as a donor in hydrogen bonding. [From C. Jelsch et al., 2000, *Proc. Nat'l. Acad. Sci.* USA **97**:3171. Courtesy of M. M. Teeter.]

weaker than a covalent O—H bond (roughly 110 kcal/mol), although it is greater than that for many other hydrogen bonds in biological molecules (1–2 kcal/mol). Extensive intermolecular hydrogen bonding between water molecules accounts for many of its key properties, including its unusually high melting and boiling points and its ability to dissolve many other molecules.

The solubility of uncharged substances in an aqueous environment depends largely on their ability to form hydrogen bonds with water. For instance, the hydroxyl group (—OH) in an alcohol ($XCH_2OH$) and the amino group (—NH$_2$) in amines ($XCH_2NH_2$) can form several hydrogen bonds with water, allowing these molecules to dissolve in water to high concentrations (Figure 2-8b). In general, molecules with polar bonds that easily form hydrogen bonds with water, as well as charged molecules and ions that interact with the dipole in water, can readily dissolve in water; that is, they are hydrophilic. Many biological molecules contain, in addition to hydroxyl and amino groups, peptide and ester groups, which form hydrogen bonds with water via otherwise nonbonded electrons on their carbonyl oxygens (Figure 2-8c). X-ray crystallography combined with computational analysis permits an accurate depiction of the distribution of the outermost unbonded electrons of atoms as well as the electrons in covalent bonds, as illustrated in Figure 2-9.

## Van der Waals Interactions Are Weak Attractive Interactions Caused by Transient Dipoles

When any two atoms approach each other closely, they create a weak, nonspecific attractive force called a **van der Waals interaction**. These nonspecific interactions result from the momentary random fluctuations in the distribution of the electrons of any atom, which give rise to a transient unequal distribution of electrons. If two noncovalently bonded atoms are close enough, electrons of one atom will perturb the electrons of the other. This perturbation generates a transient dipole in the second atom, and the two dipoles will attract each other weakly (Figure 2-10). Similarly, a polar covalent bond in one molecule will attract an oppositely oriented dipole in another.

Van der Waals interactions, involving either transiently induced or permanent electric dipoles, occur in all types of molecules, both polar and nonpolar. In particular, van der Waals interactions are responsible for the cohesion between nonpolar molecules such as heptane, $CH_3$—$(CH_2)_5$—$CH_3$, that cannot form hydrogen bonds or ionic interactions with each other. The strength of van der Waals interactions

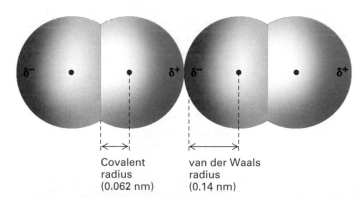

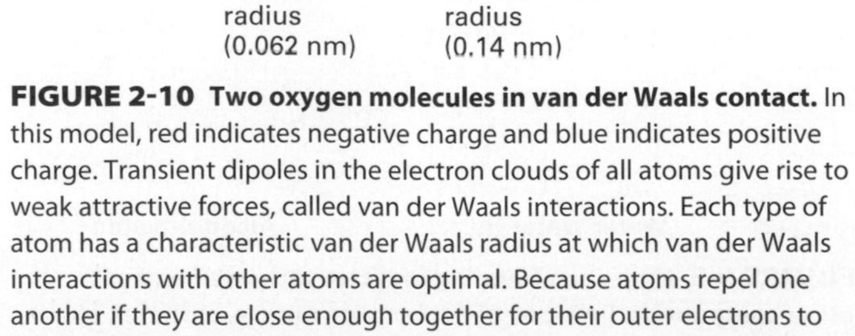

**FIGURE 2-10 Two oxygen molecules in van der Waals contact.** In this model, red indicates negative charge and blue indicates positive charge. Transient dipoles in the electron clouds of all atoms give rise to weak attractive forces, called van der Waals interactions. Each type of atom has a characteristic van der Waals radius at which van der Waals interactions with other atoms are optimal. Because atoms repel one another if they are close enough together for their outer electrons to overlap without being shared in a covalent bond, the van der Waals radius is a measure of the size of the electron cloud surrounding an atom. The covalent radius indicated here is for the double bond of O=O; the single-bond covalent radius of oxygen is slightly longer.

decreases rapidly with increasing distance; thus these noncovalent interactions can form only when atoms are quite close to one another. However, if atoms get too close together, the negative charges of their electrons create a repulsive force. When the van der Waals attraction between two atoms exactly balances the repulsion between their two electron clouds, the atoms are said to be in van der Waals contact. The strength of the van der Waals interaction is about 1 kcal/mol, weaker than typical hydrogen bonds and only slightly higher than the average thermal energy of molecules at 25 °C. Thus multiple van der Waals interactions, a van der Waals interaction together with other noncovalent interactions, or both are required to form stable attractions within and between molecules.

## The Hydrophobic Effect Causes Nonpolar Molecules to Adhere to One Another

Because nonpolar molecules do not contain charged groups, do not possess a dipole moment, or become hydrated, they are insoluble or almost insoluble in water; that is, they are hydrophobic. The covalent bonds between two carbon atoms and between carbon and hydrogen atoms are the most common nonpolar bonds in biological systems. **Hydrocarbons**—molecules made up only of carbon and hydrogen—are virtually insoluble in water. Large triacylglycerols (also known as triglycerides), which make up animal fats and vegetable oils, also are insoluble in water. As we will see later, the major part of these molecules consists of long hydrocarbon chains. After being shaken in water, triacylglycerols form a separate phase. A familiar example is the separation of oil from the water-based vinegar in an oil-and-vinegar salad dressing.

Nonpolar molecules or nonpolar parts of molecules tend to aggregate in water owing to a phenomenon called the **hydrophobic effect**. Because water molecules cannot form hydrogen bonds with nonpolar substances, they tend to form "cages" of relatively rigid hydrogen-bonded pentagons and hexagons around nonpolar molecules (Figure 2-11, *left*). This state is energetically unfavorable because it decreases the entropy, or randomness, of the population of water molecules. (The role of entropy in chemical systems is discussed in Section 2.4.) If nonpolar molecules in an aqueous environment aggregate with their hydrophobic surfaces facing each other, the net hydrophobic surface area exposed to water is reduced (Figure 2-11, *right*). As a consequence, less water is needed to form the cages surrounding the nonpolar molecules, entropy increases relative to the unaggregated state, and an energetically more favorable state is reached. In a sense, then, water squeezes the nonpolar molecules into spontaneously forming aggregates. Rather than constituting an attractive force such as in hydrogen bonds, the hydrophobic effect results from an avoidance of an unstable state—that is, extensive water cages around individual nonpolar molecules.

Nonpolar molecules can also associate, albeit weakly, through van der Waals interactions. The net result of the

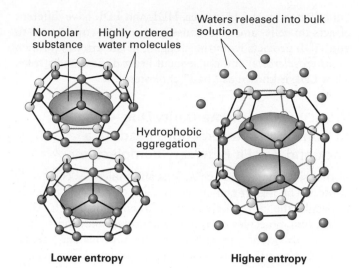

**FIGURE 2-11 Schematic depiction of the hydrophobic effect.** Cages of water molecules that form around nonpolar molecules in solution are more ordered than water molecules in the surrounding bulk liquid. Aggregation of nonpolar molecules reduces the number of water molecules involved in highly ordered cages, resulting in a higher-entropy, more energetically favorable state (*right*) compared with the unaggregated state (*left*).

hydrophobic effect and van der Waals interactions is a very powerful tendency for hydrophobic molecules to interact with one another, not with water. Simply put, *like dissolves like*. Polar molecules dissolve in polar solvents such as water; nonpolar molecules dissolve in nonpolar solvents such as hexane.

One well-known hydrophobic molecule is cholesterol (see the structure in Section 2.2). Cholesterol, as well as triglycerides and other poorly water-soluble molecules, is called a lipid. Unlike hydrophilic molecules such as glucose or amino acids, lipids cannot readily dissolve in the blood, which is an aqueous circulatory system for transporting molecules and cells throughout the body. Instead, lipids such as cholesterol must be packaged into hydrophilic carriers that can themselves dissolve in the blood and be transported throughout the body. There can be hundreds to thousands of lipids packed into the center, or core, of each carrier. The hydrophobic core is surrounded by amphipathic molecules that have hydrophilic parts that interact with water and hydrophobic parts that interact with each other and the core. The packaging of these lipids into special carriers, called lipoproteins (discussed in Chapter 14), permits their efficient transport in blood and is reminiscent of the containerization of cargo for efficient long-distance transport via cargo ships, trains and trucks. High-density lipoprotein (HDL) and low-density lipoprotein (LDL) are two such lipoprotein carriers that are associated with either reduced or increased heart disease, respectively, and are therefore often referred to as "good" and "bad" cholesterol. Actually, the cholesterol molecules and their derivatives that are carried by both HDL and LDL are essentially identical and in themselves neither "good" nor "bad." However, HDL and LDL have different effects on cells, and as a consequence LDL contributes to and HDL protects from the clogging of the arteries (known as *atherosclerosis*) and consequent heart disease and stroke. Thus LDL is known as "bad" cholesterol. ∎

## Molecular Complementarity Due to Noncovalent Interactions Leads to a Lock-and-Key Fit Between Biomolecules

Both inside and outside cells, ions and molecules constantly collide. The higher the concentration of any two types of molecule, the more likely they are to encounter each other. When two molecules encounter each other, they most likely will simply bounce apart because the noncovalent interactions that would bind them together are weak and have a transient existence at physiological temperatures. However, molecules that exhibit **molecular complementarity**, a lock-and-key kind of fit between their shapes, charges, or other physical properties, can form multiple noncovalent interactions at close range. When two such structurally complementary molecules bump into each other, these multiple interactions cause them to stick together, or bind.

Figure 2-12 illustrates how multiple, different weak interactions can cause two proteins to bind together tightly. Almost any other arrangement of the same groups on the two surfaces would not allow the molecules to bind so tightly. Such molecular complementarity between regions within a protein molecule allow it to fold into a unique three-dimensional shape (see Chapter 3); it is also what

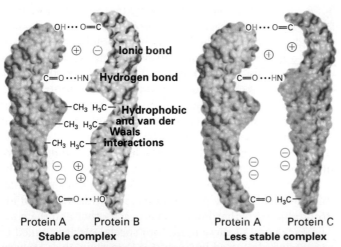

| Protein A | Protein B | Protein A | Protein C |
|---|---|---|---|
| **Stable complex** | | **Less stable complex** | |

**FIGURE 2-12 Molecular complementarity permits tight protein bonding via multiple noncovalent interactions.** The complementary shapes, charges, polarity, and hydrophobicity of two protein surfaces permit multiple weak interactions, which in combination produce a strong interaction and tight binding. Because deviations from molecular complementarity substantially weaken binding, a particular surface region of any given biomolecule usually can bind tightly to only one or a very limited number of other molecules. The complementarity of the two protein molecules on the left permits them to bind much more tightly than the two noncomplementary proteins on the right.

holds the two chains of DNA together in a double helix (see Chapter 4). Similar interactions underlie the association of groups of molecules into multimolecular assemblies, or complexes, leading, for example, to the formation of muscle fibers, to the gluelike associations between cells in solid tissues, and to numerous other cellular structures.

Depending on the number and strength of the noncovalent interactions between the two molecules and on their environment, their binding may be tight or loose and, as a consequence, either long lasting or transient. The higher the *affinity* of two molecules for each other, the better the molecular "fit" between them, the more noncovalent interactions can form, and the tighter they can bind together. An important quantitative measure of affinity is the binding dissociation constant $K_d$, described later.

As we discuss in Chapter 3, nearly all the chemical reactions that occur in cells also depend on the binding properties of enzymes. These proteins not only speed up, or catalyze, reactions but also do so with a high degree of *specificity*, a reflection of their ability to bind tightly to only one or a few related molecules. The specificity of intermolecular interactions and reactions, which depends on molecular complementarity, is essential for many processes critical to life.

## KEY CONCEPTS of Section 2.1

### Covalent Bonds and Noncovalent Interactions

- Covalent bonds consist of pairs of electrons shared by two atoms. Covalent bonds arrange the atoms of a molecule into a specific geometry.

- Covalent bonds are stable in biological systems because the relatively high energies required to break them (50–200 kcal/mol) are much larger than the thermal kinetic energy available at room (25 °C) or body (37 °C) temperatures.

- Many molecules in cells contain at least one asymmetric carbon atom, which is bonded to four dissimilar atoms. Such molecules can exist as optical isomers (mirror images), designated D and L (see Figure 2-4), which have different biological activities. In biological systems, nearly all sugars are D isomers, whereas nearly all amino acids are L isomers.

- Electrons may be shared equally or unequally in covalent bonds. Atoms that differ in electronegativity form polar covalent bonds, in which the bonding electrons are distributed unequally. One end of a polar bond has a partial positive charge and the other end has a partial negative charge (see Figure 2-5).

- Noncovalent interactions between atoms are considerably weaker than covalent bonds, with energies ranging from about 1–5 kcal/mol (see Figure 2-6).

- Four main types of noncovalent interactions occur in biological systems: ionic bonds, hydrogen bonds, van der Waals interactions, and interactions due to the hydrophobic effect.

- Ionic bonds result from the electrostatic attraction between the positive and negative charges of ions. In aqueous solutions, all cations and anions are surrounded by a shell of bound water molecules (see Figure 2-7c). Increasing the salt (e.g., NaCl) concentration of a solution can weaken the relative strength of and even break the ionic bonds between biomolecules.

- In a hydrogen bond, a hydrogen atom covalently bonded to an electronegative atom associates with an acceptor atom whose nonbonding electrons attract the hydrogen (see Figure 2-8).

- Weak and relatively nonspecific van der Waals interactions result from the attraction between transient dipoles associated with all molecules. They form when two atoms approach each other closely (see Figure 2-10).

- In an aqueous environment, nonpolar molecules or nonpolar parts of larger molecules are driven together by the hydrophobic effect, thereby reducing the extent of their direct contact with water molecules (see Figure 2-11).

- Molecular complementarity is the lock-and-key fit between molecules whose shapes, charges, and other physical properties are complementary. Multiple noncovalent interactions can form between complementary molecules, causing them to bind tightly (see Figure 2-12), but not between molecules that are not complementary.

- The high degree of binding specificity that results from molecular complementarity is one of the features that underlies intermolecular interactions in biology and thus is essential for many processes critical to life.

## 2.2 Chemical Building Blocks of Cells

A common theme in biology is the construction of large **macromolecules** and macromolecular structures out of smaller molecular subunits. Often these subunits are similar or identical. The three main types of biological macromolecules—**proteins, nucleic acids,** and **polysaccharides**—are all **polymers** composed of multiple covalently linked building block small molecules, or **monomers** (Figure 2-13). Proteins are linear polymers containing around 10 to several thousand amino acids linked by **peptide bonds.** Nucleic acids are linear polymers containing hundreds to millions of nucleotides linked by **phosphodiester bonds.** Polysaccharides are linear or branched polymers of monosaccharides (sugars) such as glucose linked by **glycosidic bonds.** Although the actual mechanisms of covalent bond formation between monomers are complex and will be discussed later, the formation of a covalent bond between two monomer molecules usually involves the net loss of a hydrogen (H) from one monomer and a hydroxyl (OH) from the other monomer—or the net loss of one water molecule—and can therefore be thought of as a *dehydration reaction*. The breakdown, or cleavage, of this bond in the polymer that releases a monomeric subunit involves the reverse reaction or the addition of water, called

*hydrolysis.* These bonds linking the monomers together are normally stable under normal biological conditions (e.g., 37 °C, neutral pH), and so these biopolymers are stable and can perform a wide variety of jobs in cells, such as store information, catalyze chemical reactions, serve as structural elements in defining cell shape and movement, and many others.

Macromolecular structures can also be assembled using noncovalent interactions. The two-ply, or "bilayer" structure of cellular membranes is built up by the noncovalent assembly of many thousands of small molecules called phospholipids (see Figure 2-13). In this chapter, we focus on the chemical building blocks making up cells—amino acids, nucleotides, sugars, and phospholipids. The structure, function, and assembly of proteins, nucleic acids, polysaccharides, and biomembranes are discussed in subsequent chapters.

### Amino Acids Differing Only in Their Side Chains Compose Proteins

The monomeric building blocks of proteins are 20 **amino acids,** which—when incorporated into a protein polymer—are sometimes called **residues.** All amino acids have a characteristic structure consisting of a central **alpha carbon atom ($C_\alpha$)** bonded to four different chemical groups: an amino ($NH_2$) group, a carboxyl or carboxylic acid (COOH) group (hence the name *amino acid*), a hydrogen (H) atom, and one variable group, called a **side chain** or **R group.** Because the α carbon in all amino acids except glycine is asymmetric, these molecules can exist in two mirror-image forms called by convention the D (dextro) and the L (levo) isomers (see Figure 2-4). The two isomers cannot be interconverted (one made identical to the other) without breaking and then re-forming a chemical bond in one of them. With rare exceptions, only the L forms of amino acids are found in proteins. However, D amino acids are prevalent in bacterial cell walls and other microbial products.

To understand the three-dimensional structures and functions of proteins, discussed in detail in Chapter 3, you must be familiar with some of the distinctive properties of amino acids, which are determined in part by their side chains. You need not memorize the detailed structure of each type of side chain to understand how proteins work because amino acids can be classified into several broad categories based on the size, shape, charge, hydrophobicity (a measure of water solubility), and chemical reactivity of the side chains (Figure 2-14). However, you should be familiar with the general properties of each category.

Amino acids with nonpolar side chains are called hydrophobic and are poorly soluble in water. The larger the nonpolar side chain, the more hydrophobic the amino acid. The side chains of *alanine, valine, leucine,* and *isoleucine* are linear or branched hydrocarbons that do not form a ring and are therefore called aliphatic amino acids. These amino acids are all nonpolar, as is *methionine*, which is similar except it contains one sulfur atom. *Phenylalanine, tyrosine,* and *tryptophan* have large, hydrophobic, aromatic rings in their side chains. In later chapters, we will see in detail how hydrophobic side chains under the influence of the hydrophobic effect

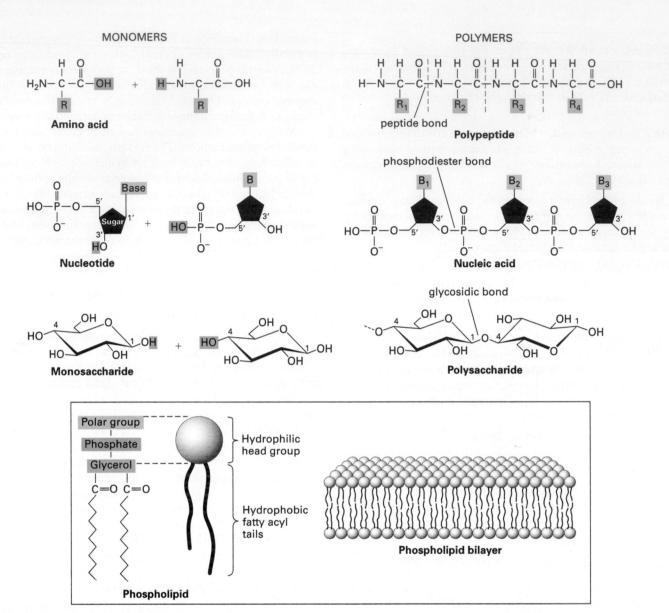

**FIGURE 2-13 Overview of the cell's principal chemical building blocks.** (*Top*) The three major types of biological macromolecules are each assembled by the polymerization of multiple small molecules (monomers) of a particular type: proteins from amino acids (see Chapter 3), nucleic acids from nucleotides (see Chapter 4), and polysaccharides from monosaccharides (sugars). Each monomer is covalently linked into the polymer by a reaction whose net result is loss of a water molecule (dehydration). (*Bottom*) In contrast, phospholipid monomers noncovalently assemble into a bilayer structure, which forms the basis of all cellular membranes (see Chapter 10).

often pack in the interior of proteins or line the surfaces of proteins that are embedded within hydrophobic regions of biomembranes.

Amino acids with polar side chains are called hydrophilic; the most hydrophilic of these amino acids is the subset with side chains that are charged (ionized) at the pH typical of biological fluids (~7), both inside and outside the cell (see Section 2.3). *Arginine* and *lysine* have positively charged side chains and are called basic amino acids; *aspartic acid* and *glutamic acid* have negatively charged side chains due to the carboxylic acid groups in their side chains (their charged forms are called *aspartate* and *glutamate*) and are called acidic. A fifth amino acid, *histidine*, has a side chain containing a ring with two nitrogens, called imidazole, which can

shift from being positively charged to uncharged depending on small changes in the acidity of its environment:

**pH 5.8**      **pH 7.8**

The activities of many proteins are modulated by shifts in environmental acidity (pH) through protonation or deprotonation of histidine side chains. *Asparagine* and *glutamine* are uncharged but have polar side chains containing amide

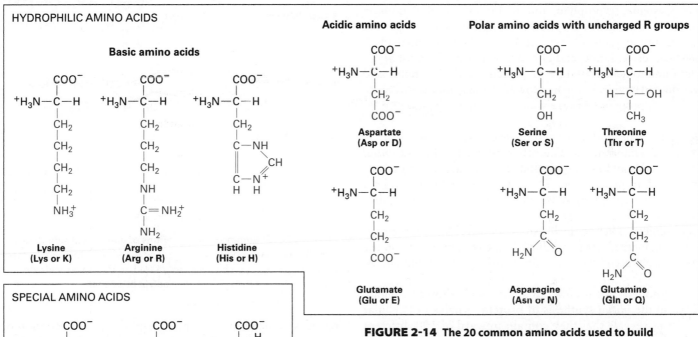

**FIGURE 2-14 The 20 common amino acids used to build proteins.** The side chain (R group; red) determines the characteristic properties of each amino acid and is the basis for grouping amino acids into three main categories: hydrophobic, hydrophilic, and special. Shown are the ionized forms that exist at the pH ($\approx$7) of the cytosol. In parentheses are the three-letter and one-letter abbreviations for each amino acid.

groups with extensive hydrogen-bonding capacities. Similarly, *serine* and *threonine* are uncharged but have polar hydroxyl groups, which also participate in hydrogen bonds with other polar molecules.

Lastly, cysteine, glycine, and proline exhibit special roles in proteins because of the unique properties of their side chains. The side chain of *cysteine* contains a reactive *sulfhydryl group* (—SH). On release of a proton (H$^+$), a sulfhydryl is converted into a thiolate anion (S$^-$). Thiolate anions can play important roles in catalysis, notably in certain enzymes that destroy proteins (proteases). In proteins, each of two adjacent sulfhydryl groups can be oxidized, each releasing a proton and an electron, to form a covalent *disulfide bond* (—S—S—):

Disulfide bonds serve to "cross-link" regions within a single polypeptide chain (intramolecular) or between two separate chains (intermolecular). Disulfide bonds stabilize the folded structure of some proteins. The smallest amino acid, *glycine,* has a single hydrogen atom as its R group. Its small size allows it to fit into tight spaces. Unlike the other common amino acids, the side chain of *proline* (pronounced pro-leen) bends around to form a ring by covalently bonding to the nitrogen atom in the amino group attached to the $C_\alpha$. As a result, proline is very rigid and the amino group is not available for typical hydrogen bonding. The presence of proline in a protein creates a fixed kink in the polymer chain, limiting how it can fold in the region of the proline residue.

Some amino acids are more abundant in proteins than others. Cysteine, tryptophan, and methionine are not common amino acids: together they constitute approximately 5 percent of the amino acids in a typical protein. Four amino acids—leucine, serine, lysine, and glutamic acid—are the most abundant amino acids, totaling 32 percent of all the amino acid residues in a typical protein. However, the amino acid compositions of proteins can vary widely from these values.

Humans and other mammals can synthesize 11 of the 20 amino acids. The other nine are called *essential amino acids* and must be included in the diet to permit normal protein production. These are phenylalanine, valine, threonine, tryptophan, isoleucine, methionine, leucine, lysine, and histidine. Adequate provision of these essential amino acids in feed is key to the livestock industry. Indeed, a genetically engineered corn with high lysine content is now in use as an "enhanced" feed to promote the growth of animals. ∎

Although cells use the 20 amino acids shown in Figure 2-14 in the *initial* synthesis of proteins, analysis of cellular proteins reveals that they contain upward of 100 different amino acids. The difference is due to the chemical modifications of some of amino acids after they are incorporated into protein by the addition of acetyl groups ($CH_3CO$) and a variety of other chemical groups (Figure 2-15). An important modification is the addition of a phosphate ($PO_4$) to hydroxyl groups in serine, threonine, and tyrosine residues, a process known as phosphorylation. We will encounter numerous examples of proteins whose activity is regulated by reversible phosphorylation and dephosphorylation. Phosphorylation of nitrogen in the side chain of histidine is well known in bacteria, fungi, and plants but less studied—perhaps because of the relative instability of phosphorylated histidine—and apparently rare in mammals. The side chains of asparagine, serine, and threonine are sites for glycosylation, the attachment of linear and branched carbohydrate chains. Many secreted proteins and membrane proteins contain glycosylated residues, and the reversible modification of hydroxyl groups on specific serines and threonines by a sugar called N-acetylglucosamine also regulates protein activities. Other amino acid modifications found in selected proteins include the hydroxylation of proline and lysine residues in collagen (see Chapter 19), the methylation of histidine

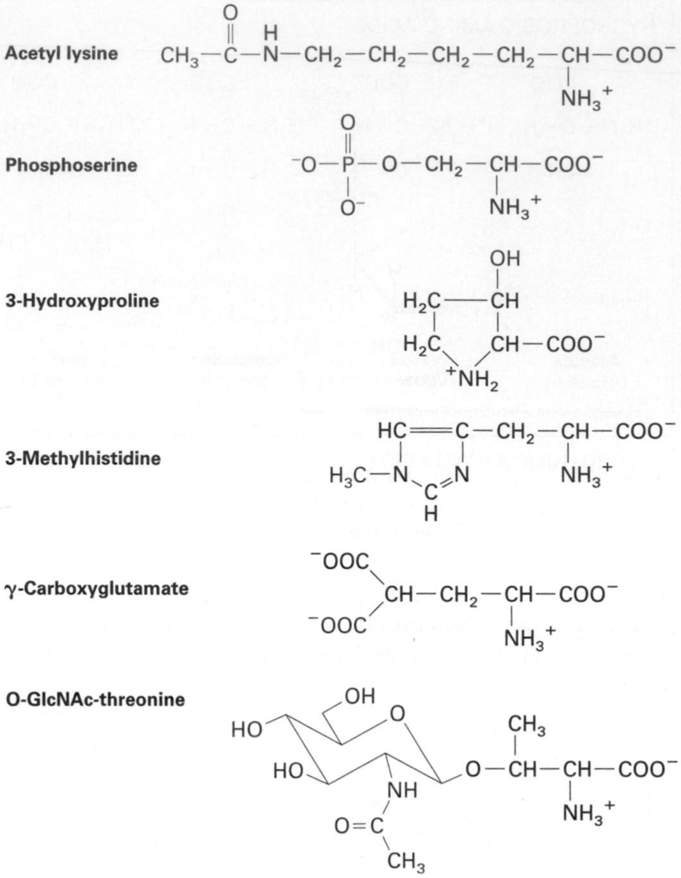

**FIGURE 2-15 Common modifications of amino acid side chains in proteins.** These modified residues and numerous others are formed by addition of various chemical groups (red) to the amino acid side chains during or after synthesis of a polypeptide chain.

residues in membrane receptors, and the $\gamma$ carboxylation of glutamate in blood-clotting factors such as prothrombin. Deamidation of Asn and Gln to the corresponding acids, Asp and Glu, is also a common occurrence.

Acetylation, addition of an acetyl group, to the amino group of the N-terminal residue, is the most common form of amino acid chemical modification, affecting an estimated 80 percent of all proteins:

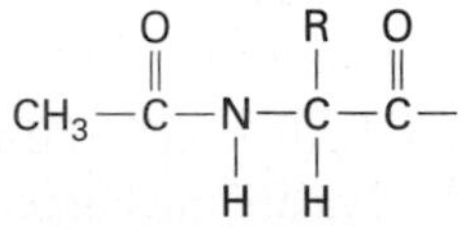

**Acetylated N-terminus**

This modification may play an important role in controlling the life span of proteins within cells because many nonacetylated proteins are rapidly degraded.

## Five Different Nucleotides Are Used to Build Nucleic Acids

Two types of chemically similar nucleic acids, **DNA (deoxyribonucleic acid)** and **RNA (ribonucleic acid)**, are the principal

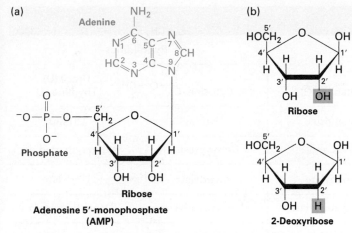

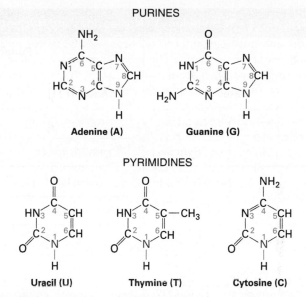

**FIGURE 2-16 Common structure of nucleotides.** (a) Adenosine 5′-monophosphate (AMP), a nucleotide present in RNA. By convention, the carbon atoms of the pentose sugar in nucleotides are numbered with primes. In natural nucleotides, the 1′ carbon is joined by a β linkage to the base (in this case adenine); both the base (blue) and the phosphate on the 5′ hydroxyl (red) extend above the plane of the sugar ring. (b) Ribose and deoxyribose, the pentoses in RNA and DNA, respectively.

**FIGURE 2-17 Chemical structures of the principal bases in nucleic acids.** In nucleic acids and nucleotides, nitrogen 9 of purines and nitrogen 1 of pyrimidines (red) are bonded to the 1′ carbon of ribose or deoxyribose. U is only found in RNA, and T is only found in DNA. Both RNA and DNA contain A, G, and C.

genetic-information-carrying molecules of the cell. The monomers from which DNA and RNA polymers are built, called **nucleotides**, all have a common structure: a phosphate group linked by a phosphoester bond to a pentose (five-carbon) sugar that in turn is linked to a nitrogen- and carbon-containing ring structure commonly referred to as a base (Figure 2-16a). In RNA, the pentose is ribose; in DNA, it is deoxyribose that at position 2′ has a proton rather than a hydroxyl group (Figure 2-16b). (We describe the structures of sugars in more detail below.) The bases *adenine, guanine,* and *cytosine* (Figure 2-17) are found in both DNA and RNA; *thymine* is found only in DNA, and *uracil* is found only in RNA.

Adenine and guanine are **purines**, which contain a pair of fused rings; cytosine, thymine, and uracil are **pyrimidines**, which contain a single ring (see Figure 2-17). The bases are often abbreviated A, G, C, T, and U, respectively; these same single-letter abbreviations are also commonly used to denote the entire nucleotides in nucleic acid polymers. In nucleotides, the 1′ carbon atom of the sugar (ribose or deoxyribose) is attached to the nitrogen at position 9 of a purine ($N_9$) or at position 1 of a pyrimidine ($N_1$). The acidic character of nucleotides is due to the phosphate group, which under normal intracellular conditions releases hydrogen ions ($H^+$), leaving the phosphate negatively charged (see Figure 2-16a). Most nucleic acids in cells are associated with proteins, which form ionic interactions with the negatively charged phosphates.

Cells and extracellular fluids in organisms contain small concentrations of **nucleosides**, combinations of a base and a sugar without a phosphate. Nucleotides are nucleosides that have one, two, or three phosphate groups esterified at the 5′ hydroxyl. Nucleoside monophosphates have a single esterified phosphate (see Figure 2-16a); nucleoside diphosphates contain a pyrophosphate group:

$$^-O-\overset{\overset{\displaystyle O}{\|}}{\underset{\underset{\displaystyle O^-}{|}}{P}}-O-\overset{\overset{\displaystyle O}{\|}}{\underset{\underset{\displaystyle O^-}{|}}{P}}-O-$$

**Pyrophosphate**

and nucleoside triphosphates have a third phosphate. Table 2-3 lists the names of the nucleosides and nucleotides in nucleic acids and the various forms of nucleoside phosphates. The nucleoside triphosphates are used in the synthesis of nucleic acids, which we cover in Chapter 4. Among their other functions in the cell, GTP participates in intracellular signaling and acts as an energy reservoir, particularly in protein synthesis, and ATP, discussed later in this chapter, is the most widely used biological energy carrier.

## Monosaccharides Covalently Assemble into Linear and Branched Polysaccharides

The building blocks of the polysaccharides are the simple sugars, or **monosaccharides**. Monosaccharides are **carbohydrates**, which are literally covalently bonded combinations of carbon and water in a one-to-one ratio $(CH_2O)_n$, where $n$ equals 3, 4, 5, 6, or 7. **Hexoses** ($n = 6$) and **pentoses** ($n = 5$) are the most common monosaccharides. All monosaccharides

TABLE 2-3 | Terminology of Nucleosides and Nucleotides

| Bases | | Purines | | Pyrimidines | |
|---|---|---|---|---|---|
| | | Adenine (A) | Guanine (G) | Cytosine (C) | Uracil (U) Thymine (T) |
| Nucleosides { | in RNA | Adenosine | Guanosine | Cytidine | Uridine |
| | in DNA | Deoxyadenosine | Deoxyguanosine | Deoxycytidine | Deoxythymidine |
| Nucleotides { | in RNA | Adenylate | Guanylate | Cytidylate | Uridylate |
| | in DNA | Deoxyadenylate | Deoxyguanylate | Deoxycytidylate | Deoxythymidylate |
| Nucleoside monophosphates | | AMP | GMP | CMP | UMP |
| Nucleoside diphosphates | | ADP | GDP | CDP | UDP |
| Nucleoside triphosphates | | ATP | GTP | CTP | UTP |
| Deoxynucleoside mono-, di-, and triphosphates | | dAMP, etc. | dGMP, etc. | dCMP, etc. | dTMP, etc. |

contain hydroxyl (—OH) groups and either an aldehyde or a keto group:

**Aldehyde**        **Keto**

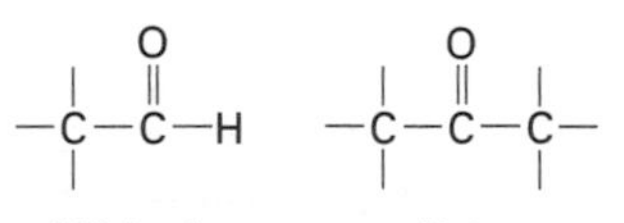

Many biologically important sugars are hexoses, including glucose, mannose, and galactose (Figure 2-18). Mannose is identical to glucose except that the orientation of the groups bonded to carbon 2 is reversed. Similarly, galactose, another hexose, differs from glucose only in the orientation of the groups attached to carbon 4. Interconversion of glucose and mannose or galactose requires the breaking and making of covalent bonds; such reactions are carried out by enzymes called *epimerases*.

D-Glucose ($C_6H_{12}O_6$) is the principal external source of energy for most cells in complex multicellular organisms and can exist in three different forms: a linear structure and two different hemiacetal ring structures (Figure 2-18a). If the aldehyde group on carbon 1 combines with the hydroxyl group on carbon 5, the resulting hemiacetal, D-glucopyranose, contains a six-member ring. In the α anomer of D-glucopyranose, the hydroxyl group attached to carbon 1 points "downward" from the ring as shown in Figure 2-18a; in the β anomer, this hydroxyl points "upward." In aqueous solution, the α and β anomers readily interconvert spontaneously; at equilibrium there is about one-third α anomer and two-thirds

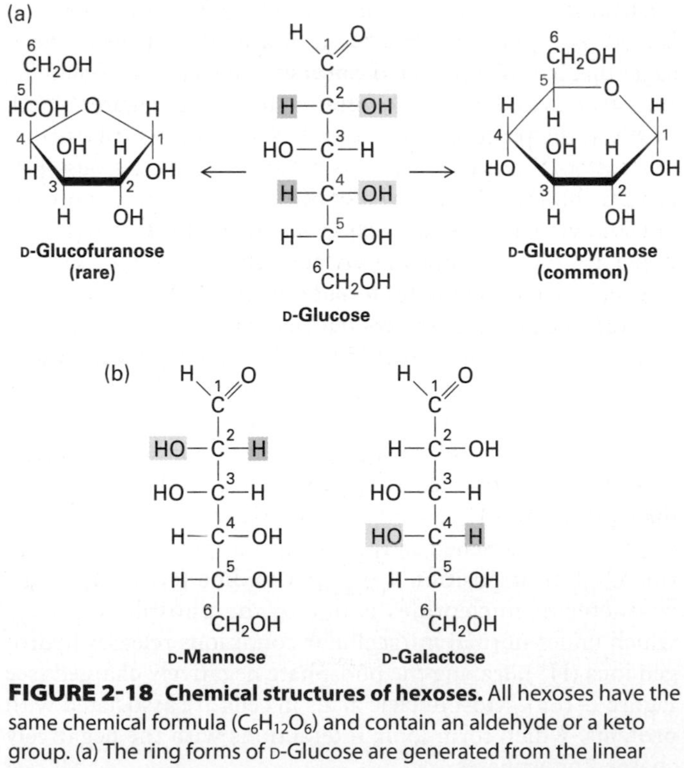

**FIGURE 2-18 Chemical structures of hexoses.** All hexoses have the same chemical formula ($C_6H_{12}O_6$) and contain an aldehyde or a keto group. (a) The ring forms of D-Glucose are generated from the linear molecule by reaction of the aldehyde at carbon 1 with the hydroxyl on carbon 5 or carbon 4. The three forms are readily interconvertible, although the pyranose form (*right*) predominates in biological systems. (b) In D-mannose and D-galactose, the configuration of the H (green) and OH (blue) bound to one carbon atom differs from that in glucose. These sugars, like glucose, exist primarily as pyranoses (six-member rings).

β, with very little of the open-chain form. Because enzymes can distinguish between the α and β anomers of D-glucose, these forms have distinct biological roles. Condensation of the hydroxyl group on carbon 4 of the linear glucose with its aldehyde group results in the formation of D-glucofuranose, a hemiacetal containing a five-member ring. Although all three forms of D-glucose exist in biological systems, the pyranose (six-member ring) form is by far the most abundant.

The pyranose ring in Figure 2-18a is depicted as planar. In fact, because of the tetrahedral geometry around carbon atoms, the most stable conformation of a pyranose ring has a nonplanar, chairlike shape. In this conformation, each bond from a ring carbon to a nonring atom (e.g., H or O) is either nearly perpendicular to the ring, referred to as axial (a), or nearly in the plane of the ring, referred to as equatorial (e):

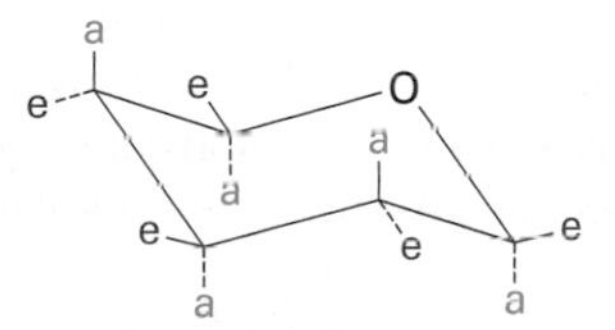

**Pyranoses** 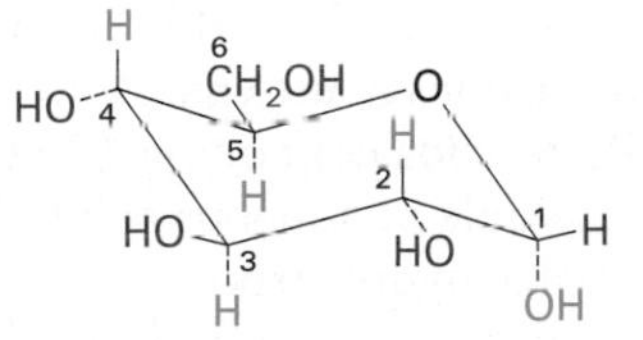　　　　**α-D-Glucopyranose**

**Disaccharides**, formed from two monosaccharides, are the simplest polysaccharides. The disaccharide lactose, composed of galactose and glucose, is the major sugar in milk; the disaccharide sucrose, composed of glucose and fructose, is a principal product of plant photosynthesis and is refined into common table sugar (Figure 2-19).

Larger polysaccharides, containing dozens to hundreds of monosaccharide units, can function as reservoirs for glucose, as structural components, or as adhesives that help hold cells together in tissues. The most common storage carbohydrate in animal cells is **glycogen**, a very long, highly branched polymer of glucose. As much as 10 percent by weight of the liver can be glycogen. The primary storage carbohydrate in plant cells, **starch**, also is a glucose polymer. It occurs in an unbranched form (amylose) and lightly branched form (amylopectin).

Both glycogen and starch are composed of the α anomer of glucose. In contrast, **cellulose**, the major constituent of plant cell walls that confers stiffness to many plant structures (see Chapter 19), is an unbranched polymer of the β anomer of glucose. Human digestive enzymes can hydrolyze the α glycosidic bonds in starch but not the β glycosidic bonds in cellulose. Many species of plants, bacteria, and molds produce cellulose-degrading enzymes. Cows and termites can break down cellulose because they harbor cellulose-degrading bacteria in their gut. Bacterial cell walls consist of **peptidoglycan**, a polysaccharide chain cross-linked by peptide cross-bridges, which confers rigidity and cell shape. Human tear and gastrointestinal fluids contains lysozyme, an enzyme capable of hydrolyzing peptidoglycan in the bacterial cell wall.

The enzymes that make the glycosidic bonds linking monosaccharides into polysaccharides are specific for the α or β anomer of one sugar and a particular hydroxyl group on the other. In principle, any two sugar molecules can be linked in a variety of ways because each monosaccharide has multiple hydroxyl groups that can participate in the formation of glycosidic bonds. Furthermore, any one monosaccharide has the potential of being linked to more than two other monosaccharides, thus generating a branch point and nonlinear polymers. Glycosidic bonds are usually formed between the growing polysaccharide chain and a covalently modified form of a monosaccharide. Such modifications include a phosphate (e.g., glucose 6-phosphate) or a nucleotide (e.g., UDP-galactose):

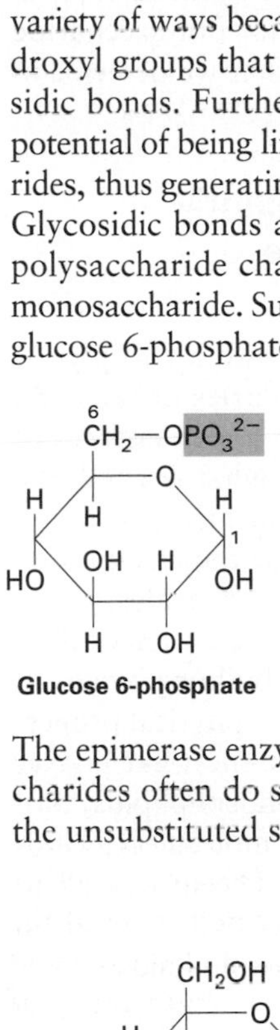

**Glucose 6-phosphate** 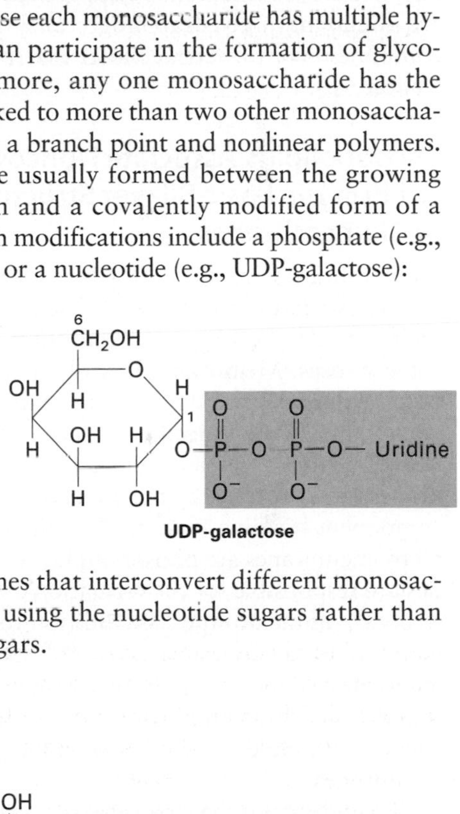　　　　**UDP-galactose**

The epimerase enzymes that interconvert different monosaccharides often do so using the nucleotide sugars rather than the unsubstituted sugars.

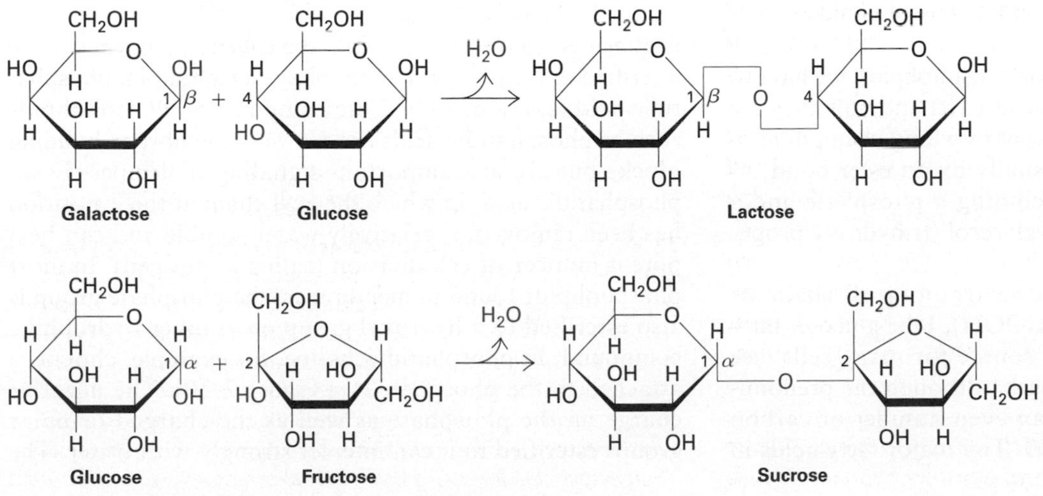

**Galactose**　　**Glucose**　　　　**Lactose**

**Glucose**　　**Fructose**　　　　**Sucrose**

**FIGURE 2-19 Formation of the disaccharides lactose and sucrose.** In any glycosidic linkage, the anomeric carbon of one sugar molecule (in either the α or β conformation) is linked to a hydroxyl oxygen on another sugar molecule. The linkages are named accordingly; thus lactose contains a β(1 → 4) bond, and sucrose contains an α(1 → 2) bond.

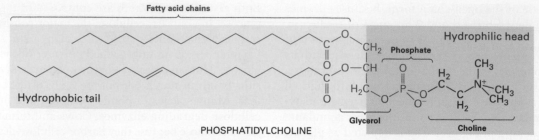

**FIGURE 2-20 Phosphatidylcholine, a typical phosphoglyceride.** All phosphoglycerides are amphipathic phospholipids, having a hydrophobic tail (yellow) and a hydrophilic head (blue) in which glycerol is linked via a phosphate group to an alcohol. Either or both of the fatty acyl side chains in a phosphoglyceride may be saturated or unsaturated. In phosphatidic acid (red), the simplest phospholipid, the phosphate is not linked to an alcohol.

Many complex polysaccharides contain modified sugars that are covalently linked to various small groups, particularly amino, sulfate, and acetyl groups. Such modifications are abundant in **glycosaminoglycans**, major polysaccharide components of the extracellular matrix that we describe in Chapter 19.

## Phospholipids Associate Noncovalently to Form the Basic Bilayer Structure of Biomembranes

Biomembranes are large, flexible sheets with a two-ply, or bilayer, structure that serve as the boundaries of cells and their intracellular organelles and form the outer surfaces of some viruses. Membranes literally define what is a cell (the outer membrane and the contents within the membrane) and what is not (the extracellular space outside the membrane). Unlike proteins, nucleic acids, and polysaccharides, membranes are assembled by the *noncovalent* association of their component building blocks. The primary building blocks of all biomembranes are **phospholipids**, whose physical properties are responsible for the formation of the sheet-like bilayer structure of membranes. In addition to phospholipids, biomembranes can contain a variety of other molecules, including cholesterol, glycolipids, and proteins. The structure and functions of biomembranes will be described in detail in Chapter 10. Here we will focus on the phospholipids in biomembranes.

To understand the structure of phospholipids, we have to understand each of their component parts and how they are assembled. Phospholipids consist of two long-chain, nonpolar fatty acid groups linked (usually by an ester bond) to small, highly polar groups, including a phosphate and a short organic molecule, such as glycerol (trihydroxy propanol) (Figure 2-20).

**Fatty acids** consist of a hydrocarbon (acyl) chain attached to a carboxyl group (—COOH). Like glucose, fatty acids are an important energy source for many cells (see Chapter 12). They differ in length, although the predominant fatty acids in cells have an even number of carbon atoms, usually 14, 16, 18, or 20. The major fatty acids in phospholipids are listed in Table 2-4. Fatty acids often are designated by the abbreviation Cx:y, where x is the number of carbons in the chain and y is the number of double bonds. Fatty acids containing 12 or more carbon atoms are nearly insoluble in aqueous solutions because of their long hydrophobic hydrocarbon chains.

Fatty acids in which all the carbon-carbon bonds are single bonds, that is, the fatty acids have no carbon-carbon double bonds, are said to be **saturated**; those with at least one carbon-carbon double bond are called **unsaturated**. Unsaturated fatty acids with more than one carbon-carbon double bond are referred to as **polyunsaturated**. Two "essential" polyunsaturated fatty acids, linoleic acid (C18:2) and linolenic acid (C18:3), cannot be synthesized by mammals and must be supplied in their diet. Mammals can synthesize other common fatty acids.

In phospholipids, fatty acids are covalently attached to another molecule by a type of dehydration reaction called *esterification,* in which the OH from the carboxyl group of the fatty acid and an H from a hydroxyl group on the other molecule are lost. In the combined molecule formed by this reaction, the part derived from the fatty acid is called an *acyl group,* or *fatty acyl group*. This is illustrated by the most common forms of phospholipids: **phosphoglycerides**, which contain two acyl groups attached to two of the hydroxyl groups of glycerol (see Figure 2-20).

In phosphoglycerides, one hydroxyl group of the glycerol is esterified to phosphate while the other two normally are esterified to fatty acids. The simplest phospholipid, phosphatidic acid, contains only these components. Phospholipids such as phosphatidic acids are not only membrane building blocks but are also important signaling molecules. Lysophosphatidic acid, in which the acyl chain at the 2 position has been removed, is relatively water soluble and can be a potent inducer of cell division (called a mitogen). In most phospholipids found in membranes, the phosphate group is also esterified to a hydroxyl group on another hydrophilic compound. In phosphatidylcholine, for example, choline is attached to the phosphate (see Figure 2-20). The negative charge on the phosphate as well as the charged or polar groups esterified to it can interact strongly with water. The

| TABLE 2-4 | Fatty Acids That Predominate in Phospholipids | | |
|---|---|---|---|
| **Common Name of Acid (ionized form in parentheses)** | | **Abbreviation** | **Chemical Formula** |
| **Saturated Fatty Acids** | | | |
| Myristic (myristate) | | C14:0 | $CH_3(CH_2)_{12}COOH$ |
| Palmitic (palmitate) | | C16:0 | $CH_3(CH_2)_{14}COOH$ |
| Stearic (stearate) | | C18:0 | $CH_3(CH_2)_{16}COOH$ |
| **Unsaturated Fatty Acids** | | | |
| Oleic (oleate) | | C18:1 | $CH_3(CH_2)_7CH{=}CH(CH_2)_7COOH$ |
| Linoleic (linoleate) | | C18:2 | $CH_3(CH_2)_4CH{=}CHCH_2CH{=}CH(CH_2)_7COOH$ |
| Arachidonic (arachidonate) | | C20:4 | $CH_3(CH_2)_4(CH{=}CHCH_2)_3CH{=}CH(CH_2)_3COOH$ |

phosphate and its associated esterified group, the "head" group of a phospholipid, is hydrophilic, whereas the fatty acyl chains, the "tails," are hydrophobic. Other common phosphoglycerides and associated head groups are shown in Table 2-5. Molecules such as phospholipids that have both hydrophobic and hydrophilic regions are called amphipathic. In Chapter 10, we will see how the amphipathic properties of phospholipids are responsible for the assembly of phospholipids into sheet-like bilayer biomembranes in which the fatty acyl tails point into the center of the sheet and the head groups point outward toward the aqueous environment (see Figure 2-13).

Fatty acyl groups also can be covalently linked into other fatty molecules, including **triacylglycerols**, or **triglycerides**, which contain three acyl groups esterfied to glycerol:

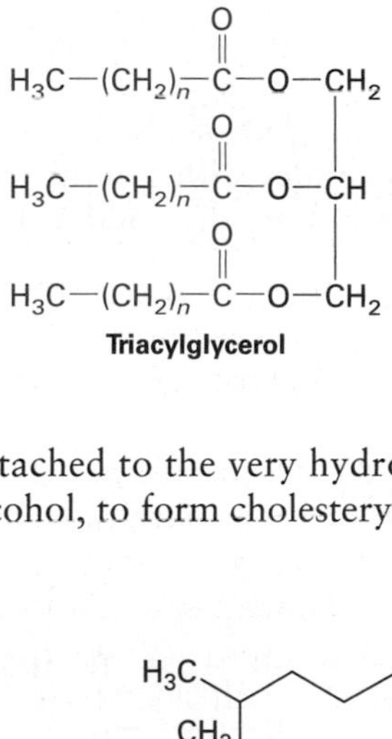

**Triacylglycerol**

and covalently attached to the very hydrophobic molecule cholesterol, an alcohol, to form cholesteryl esters:

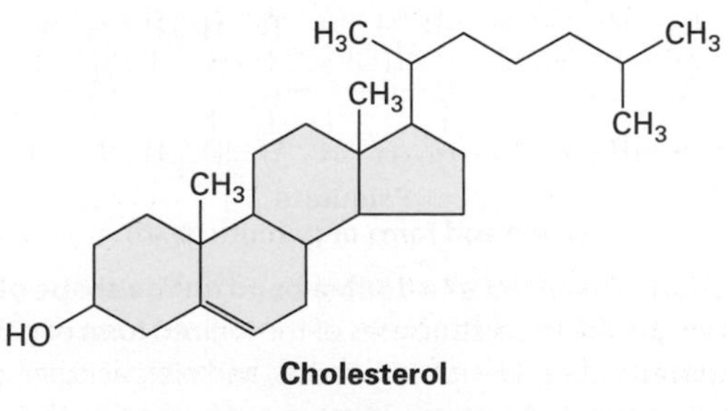

| TABLE 2-5 | Common Phosphoglycerides and Head Groups |
|---|---|
| **Common Phosphoglycerides** | **Head Group** |
| Phosphatidylcholine | **Choline** |
| Phosphatidylethanolamine | **Ethanolamine** |
| Phosphatidylserine | **Serine** |
| Phosphatidylinositol | **Inositol** |

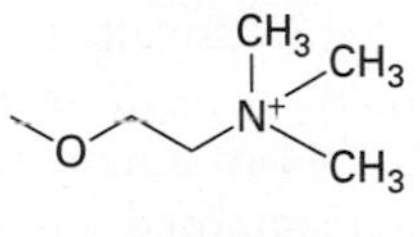

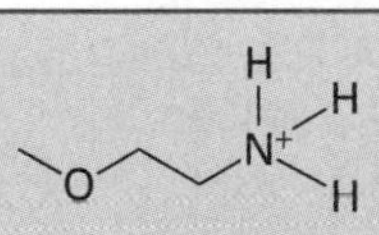

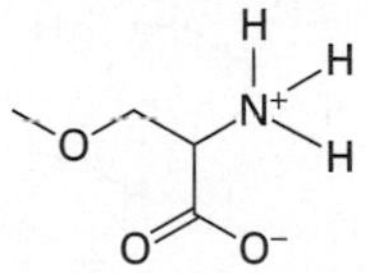

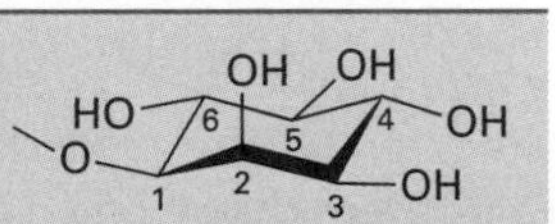

**Cholesterol**

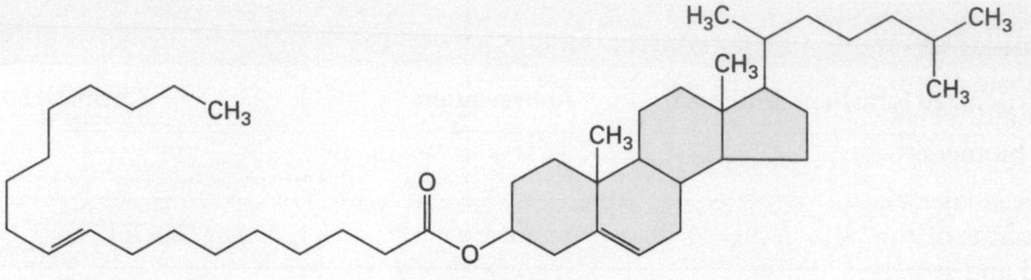

**Cholesteryl ester**

Triglycerides and cholesteryl esters are extremely water-insoluble molecules in which fatty acids and cholesterol are either stored or transported. Triglycerides are the storage form of fatty acids in the fat cells of adipose tissue and are the principal components of dietary fats. Cholesteryl esters and triglycerides are transported between tissues through the bloodstream in specialized carriers called lipoproteins (see Chapter 14).

We saw above that the fatty acids making up phospholipids (both phosphoglycerides and triglycerides) can be either saturated or unsaturated. An important consequence of the carbon-carbon double bond (C=C) in an unsaturated fatty acid is that two stereoisomeric configurations, cis and trans, are possible around each of these bonds:

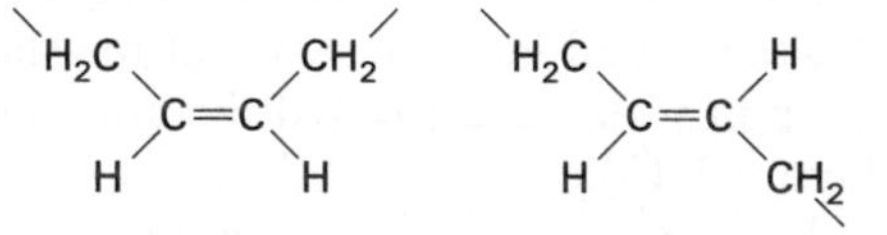

a cis double bond introduces a rigid kink in the otherwise flexible straight acyl chain of a saturated fatty acid (Figure 2-21). In general, the unsaturated fatty acids in biological systems contain only cis double bonds. Saturated fatty acids without the kink can pack together tightly and so have higher melting points than unsaturated fatty acids. The main fatty molecules in butter are triglycerides with saturated fatty acyl chains, which is why butter is usually solid at room temperature.

Unsaturated fatty acids or fatty acyl chains with the cis double bond kink cannot pack as closely together as saturated fatty acyl chains. Thus, vegetable oils, composed of triglycerides with unsaturated fatty acyl groups, usually are liquid at room temperature. Vegetable and similar oils are partially hydrogenated to convert some of their unsaturated fatty acyl chains to saturated fatty acyl chains. As a consequence, the hydrogenated vegetable oil can be molded into solid sticks of margarine. A by-product of the hydrogenation reaction is the conversion of some of the fatty acyl chains into trans fatty acids, popularly called "trans fats." The "trans fats," found in partially hydrogenated margarine and other food products, are not natural. Saturated and trans fatty acids have similar physical properties; for example, they tend to be solids at room temperature. Their consumption, relative to the consumption of unsaturated fats, is associated with increased plasma cholesterol levels and is discouraged by some nutritionists. ∎

## KEY CONCEPTS of Section 2.2

### Chemical Building Blocks of Cells

• Macromolecules are polymers of monomer subunits linked together by covalent bonds via dehydration reactions. Three major types of macromolecules are found in cells: proteins, composed of amino acids linked by peptide bonds; nucleic acids, composed of nucleotides linked by phosphodiester

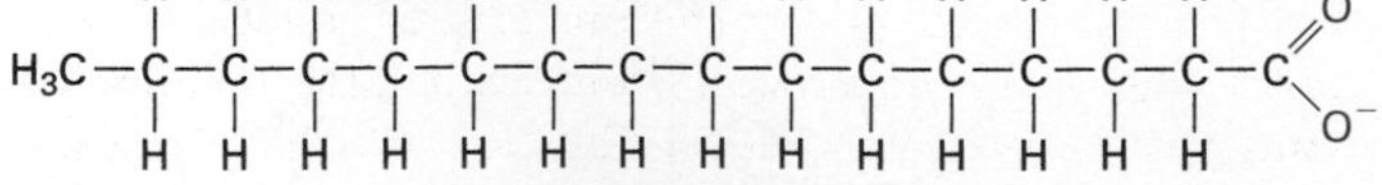

**Palmitate**
**(ionized form of palmitic acid)**

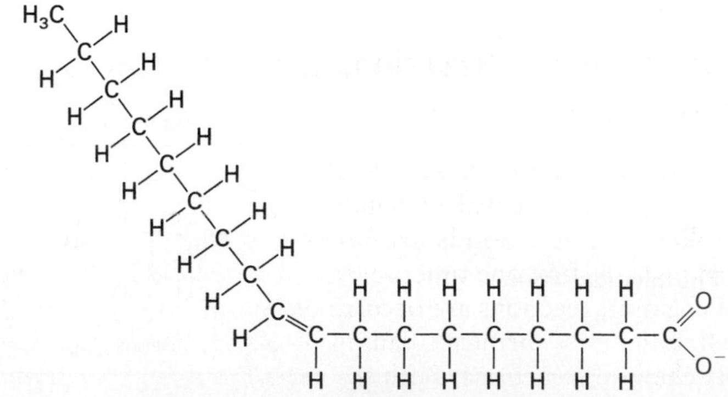

**Oleate**
**(ionized form of oleic acid)**

**FIGURE 2-21 The effect of a double bond on the shape of fatty acids.** Shown are chemical structures of the ionized form of palmitic acid, a saturated fatty acid with 16 C atoms, and oleic acid, an unsaturated one with 18 C atoms. In saturated fatty acids, the hydrocarbon chain is often linear; the cis double bond in oleate creates a rigid kink in the hydrocarbon chain. [After L. Stryer, 1994, *Biochemistry*, 4th ed., W. H. Freeman and Company, p. 265.]

bonds; and polysaccharides, composed of monosaccharides (sugars) linked by glycosidic bonds (see Figure 2-13). Phospholipids, the fourth major chemical building block, assemble noncovalently into biomembranes.

- Differences in the size, shape, charge, hydrophobicity, and reactivity of the side chains of the 20 common amino acids determine the chemical and structural properties of proteins (see Figure 2-14).

- The bases in the nucleotides composing DNA and RNA are carbon- and nitrogen-containing rings attached to a pentose sugar. They form two groups: the purines—adenine (A) and guanine (G)—and the pyrimidines—cytosine (C), thymine (T), and uracil (U) (see Figure 2-17). A, G, T, and C are found in DNA, and A, G, U, and C are found in RNA.

- Glucose and other hexoses can exist in three forms: an open-chain linear structure, a six-member (pyranose) ring, and a five-member (furanose) ring (see Figure 2-18). In biological systems, the pyranose form of D-glucose predominates.

- Glycosidic bonds are formed between either the α or the β anomer of one sugar and a hydroxyl group on another sugar, leading to formation of disaccharides and other polysaccharides (see Figure 2-19).

- Phospholipids are amphipathic molecules with a hydrophobic tail (often two fatty acyl chains) connected by a small organic molecule (often glycerol) to a hydrophilic head (see Figure 2-20).

- The long hydrocarbon chain of a fatty acid may be saturated (containing no carbon-carbon double bond) or unsaturated (containing one or more double bonds). Fatty substances such as butter that have primarily saturated fatty acyl chains tend to be solid at room temperature, whereas unsaturated fats with cis double bonds have kinked chains that cannot pack closely together and so tend to be liquids at room temperature.

## 2.3 Chemical Reactions and Chemical Equilibrium

We now shift our discussion to chemical reactions in which bonds, primarily covalent bonds in *reactant* chemicals, are broken and new bonds are formed to generate reaction *products*. At any one time, several hundred different kinds of chemical reactions are occurring simultaneously in every cell, and many chemicals can, in principle, undergo multiple chemical reactions. Both the *extent* to which reactions can proceed and the *rate* at which they take place determine the chemical composition of cells. In this section, we discuss the concepts of equilibrium and steady state as well as dissociation constants and pH. In Section 2.4, we discuss how energy influences the extents and rates of chemical reactions.

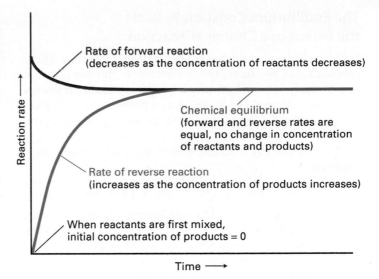

**FIGURE 2-22 Time dependence of the rates of a chemical reaction.** The forward and reverse rates of a reaction depend in part on the initial concentrations of reactants and products. The net forward reaction rate slows as the concentration of reactants decreases, whereas the net reverse reaction rate increases as the concentration of products increases. At equilibrium, the rates of the forward and reverse reactions are equal and the concentrations of reactants and products remain constant.

## A Chemical Reaction Is in Equilibrium When the Rates of the Forward and Reverse Reactions Are Equal

When reactants first mix together—before any products have been formed—the rate of the forward reaction to form products is determined in part by their initial concentrations, which determine the likelihood of reactants bumping into one another and reacting (Figure 2-22). As the reaction products accumulate, the concentration of each reactant decreases and so does the forward reaction rate. Meanwhile, some of the product molecules begin to participate in the reverse reaction, which re-forms the reactants. The ability of a reaction to go "backward" is called *microscopic reversibility*. The reverse reaction is slow at first but speeds up as the concentration of product increases. Eventually, the rates of the forward and reverse reactions become equal, so that the concentrations of reactants and products stop changing. The system is then said to be in **chemical equilibrium** (plural: *equilibria*).

The ratio of the concentrations of the products to reactants when they reach equilibrium, called the **equilibrium constant**, $K_{eq}$, is a fixed value. Thus $K_{eq}$ provides a measure of the extent to which a reaction occurs by the time it reaches equilibrium. The rate of a chemical reaction can be increased by a **catalyst**, but a catalyst does not change the equilibrium constant (see Section 2.4). A catalyst accelerates the making and breaking of covalent bonds but itself is not permanently changed during a reaction.

## The Equilibrium Constant Reflects the Extent of a Chemical Reaction

For any chemical reaction, $K_{eq}$ depends on the nature of the reactants and products, the temperature, and the pressure (particularly in reactions involving gases). Under standard physical conditions (25 °C and 1 atm pressure for biological systems), the $K_{eq}$ is always the same for a given reaction, whether or not a catalyst is present.

For the general reaction with three reactants and three products

$$aA + bB + cC \rightleftharpoons zZ + yY + xX \qquad (2\text{-}1)$$

where capital letters represent particular molecules or atoms and lowercase letters represent the number of each in the reaction formula; the equilibrium constant is given by

$$K_{eq} = \frac{[X]^x[Y]^y[Z]^z}{[A]^a[B]^b[C]^c} \qquad (2\text{-}2)$$

where brackets denote the concentrations of the molecules at equilibrium. The rate of the forward reaction (left to right in Equation 2-1) is

$$\text{Rate}_{forward} = k_f[A]^a[B]^b[C]^c$$

where $k_f$ is the **rate constant** for the forward reaction. Similarly, the rate of the reverse reaction (right to left in Equation 2-1) is

$$\text{Rate}_{reverse} = k_r[X]^x[Y]^y[Z]^z$$

where $k_r$ is the rate constant for the reverse reaction. It is important to remember that the forward and reverse rates of a reaction can change because of changes in reactant or product concentrations, yet at the same time the forward and reverse rate constants do not change; hence the name "constant." Confusing rates and rate constants is a common error. At equilibrium the forward and reverse rates are equal, so $\text{Rate}_{forward}/\text{Rate}_{reverse} = 1$. By rearranging these equations, we can express the equilibrium constant as the ratio of the rate constants:

$$K_{eq} = \frac{k_f}{k_r} \qquad (2\text{-}3)$$

The concept of $K_{eq}$ is particularly helpful when we want to think about the energy that is released or absorbed when a chemical reaction occurs. We will discuss this in considerable detail in Section 2.4.

## Chemical Reactions in Cells Are at Steady State

Under appropriate conditions and given sufficient time, a single biochemical reaction carried out in a test tube eventually will reach equilibrium and the concentration of reactants and products does not change with time because the rates of the forward and reverse reactions are equal. Within cells, however, many reactions are linked in pathways in which a product of one reaction has alternative fates to simply reconverting via a reverse reaction to the reactants and thus ultimately reaching equilibrium. For example, the product of one reaction might serve as a reactant in another, or it might be

(a) Test tube equilibrium concentrations

(b) Intracellular steady-state concentrations

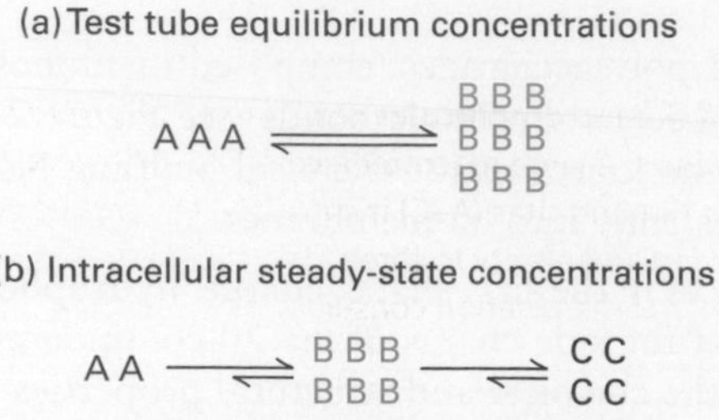

**FIGURE 2-23 Comparison of reactions at equilibrium and steady state.** (a) In the test tube, a biochemical reaction (A → B) eventually will reach equilibrium, in which the rates of the forward and reverse reactions are equal (as indicated by the reaction arrows of equal length). (b) In metabolic pathways within cells, the product B commonly would be consumed, in this example by conversion to C. A pathway of linked reactions is at steady state when the rate of formation of the intermediates (e.g., B) equals their rate of consumption. As indicated by the unequal length of the arrows, the individual reversible reactions constituting a metabolic pathway do not reach equilibrium. Moreover, the concentrations of the intermediates at steady state can differ from what they would be at equilibrium.

pumped out of the cell. In this more complex situation, the original reaction can never reach equilibrium because some of the products do not have a chance to be converted back to reactants. Nevertheless, in such non-equilibrium conditions the rate of formation of a substance can be equal to the rate of its consumption, and as a consequence the concentration of the substance remains constant over time. In such circumstances, the system of linked reactions for producing and consuming that substance is said to be in a **steady state** (Figure 2-23). One consequence of such linked reactions is that they prevent the accumulation of excess intermediates, protecting cells from the harmful effects of intermediates that are toxic at high concentrations. When the concentration of a product of an ongoing reaction is not changing over time, it might be a consequence of a state of equilibrium or it might be a consequence of a steady state. In biological systems when metabolite concentrations, such as blood glucose levels, are not changing with time—a condition called *homeostasis*—it is a consequence of a steady state rather than equilibrium.

## Dissociation Constants of Binding Reactions Reflect the Affinity of Interacting Molecules

The concept of equilibrium also applies to the binding of one molecule to another. Many important cellular processes depend on such binding "reactions," which involve the making and breaking of various noncovalent interactions rather than covalent bonds, as discussed above. A common example is the binding of a **ligand** (e.g., the hormone insulin or adrenaline) to its **receptor** on the surface of a cell, which triggers an intracellular signaling pathway (see Chapter 15). Another example is the binding of a protein to a specific sequence of base pairs in a molecule of DNA, which frequently causes the expression of a nearby gene to increase or decrease (see Chapter 7). If the equilibrium constant for a binding reaction is known, the stability of the resulting complex can be

**FIGURE 2-24 Macromolecules can have distinct binding sites for multiple ligands.** A large macromolecule (e.g., a protein, blue) with three distinct binding sites (A–C) is shown; each binding site exhibits molecular complementarity to three different binding partners (ligands A–C) with distinct dissociation constants ($K_{dA–C}$).

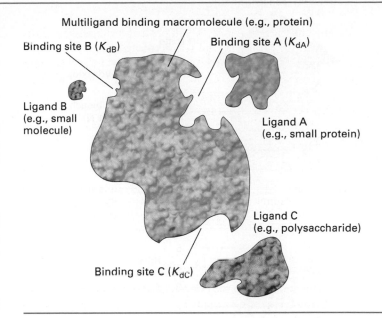

predicted. To illustrate the general approach for determining the concentration of noncovalently associated complexes, we will calculate the extent to which a protein (P) is bound to DNA (D), forming a protein-DNA complex (PD):

$$P + D \rightleftharpoons PD$$

Most commonly, binding reactions are described in terms of the **dissociation constant** $K_d$, which is the reciprocal of the equilibrium constant. For this binding reaction, the dissociation constant is given by

$$K_d = \frac{[P][D]}{[PD]} \qquad (2\text{-}4)$$

It is worth noting that in such a binding reaction, when half of the DNA is bound to the protein ([PD] = [D]), the concentration of P is equal to the $K_d$. The lower the $K_d$, the lower the concentration of P needed to bind to half of D. In other words, the lower the $K_d$, the tighter the binding (the higher the affinity) of P for D. Typical reactions in which a protein binds to a specific DNA sequence have a $K_d$ of $10^{-10}$ M, where M symbolizes molarity, or moles per liter (mol/L). To relate the magnitude of this dissociation constant to the intracellular ratio of bound to unbound DNA, let's consider the simple example of a bacterial cell having a volume of $1.5 \times 10^{-15}$ L and containing 1 molecule of DNA and 10 molecules of the DNA-binding protein P. In this case, given a $K_d$ of $10^{-10}$ M and the total concentration of the P in the cell ($\sim 111 \times 10^{-10}$ M, 100-fold higher than the $K_d$), 99 percent of the time this specific DNA sequence will have a molecule of protein bound to it and 1 percent of the time it will not, even though the cell contains only 10 molecules of the protein! Clearly, P and D have a high affinity for each other and bind tightly, as reflected by the low value of the dissociation constant for their binding reaction. For protein-protein and protein-DNA binding, $K_d$ values of $\leq 10^{-9}$ M (nanomolar) are considered to be tight, $\sim 10^{-6}$ M (micromolar) modestly tight, and $\sim 10^{-3}$ M (millimolar) relatively weak.

A large biological macromolecule, such as a protein, can have multiple binding surfaces for binding several molecules simultaneously (Figure 2-24). In some cases, these binding reactions are independent, with their own distinct $K_d$ values that are constant. In other cases, binding of a molecule at one site on a macromolecule can change the three-dimensional shape of a distant site, thus altering the binding interactions of that distant site with some other molecule. This is an important mechanism by which one molecule can alter, and thus regulate, the binding activity of another. We examine this regulatory mechanism in more detail in Chapter 3.

## Biological Fluids Have Characteristic pH Values

The solvent inside cells and in all extracellular fluids is water. An important characteristic of any aqueous solution is the concentration of positively charged hydrogen ions ($H^+$) and negatively charged hydroxyl ions ($OH^-$). Because these ions are the dissociation products of $H_2O$, they are constituents of all living systems, and they are liberated by many reactions that take place between organic molecules within cells. These ions also can be transported into or out of cells, as when highly acidic gastric juice is secreted by cells lining the walls of the stomach.

When a water molecule dissociates, one of its polar H—O bonds breaks. The resulting hydrogen ion, often referred to as a **proton**, has a short lifetime as a free ion and quickly combines with a water molecule to form a hydronium ion ($H_3O^+$). For convenience, we refer to the concentration of hydrogen ions in a solution, $[H^+]$, even though this really represents the concentration of hydronium ions, $[H_3O^+]$. Dissociation of $H_2O$ generates one $OH^-$ ion along with each $H^+$. The dissociation of water is a reversible reaction:

$$H_2O \rightleftharpoons H^+ + OH^-$$

At 25 °C, $[H^+][OH^-] = 10^{-14}$ M², so that in pure water, $[H^+] = [OH^-] = 10^{-7}$ M.

The concentration of hydrogen ions in a solution is expressed conventionally as its **pH**, defined as the negative log of the hydrogen ion concentration. The pH of pure water at 25 °C is 7:

$$pH = -\log[H^+] = \log\frac{1}{[H^+]} = \log\frac{1}{10^{-7}} = 7$$

It is important to keep in mind that a 1 unit difference in pH represents a tenfold difference in the concentration of protons. On the pH scale, 7.0 is considered neutral: pH values

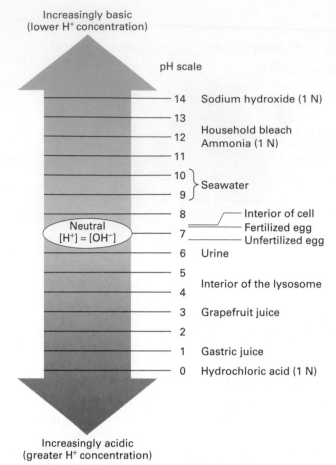

Increasingly basic
(lower H⁺ concentration)

pH scale

— 14  Sodium hydroxide (1 N)
— 13
— 12  Household bleach
        Ammonia (1 N)
— 11
— 10 ⎫ Seawater
— 9  ⎭
— 8  ——— Interior of cell
— 7  ——— Fertilized egg
        ——— Unfertilized egg

Neutral
[H⁺] = [OH⁻]

— 6   Urine
— 5
— 4   Interior of the lysosome
— 3   Grapefruit juice
— 2
— 1   Gastric juice
— 0   Hydrochloric acid (1 N)

Increasingly acidic
(greater H⁺ concentration)

**FIGURE 2-25  pH values of common solutions.** The pH of an aqueous solution is the negative log of the hydrogen ion concentration. The pH values for most intracellular and extracellular biological fluids are near 7 and are carefully regulated to permit the proper functioning of cells, organelles, and cellular secretions.

below 7.0 indicate acidic solutions (higher [H⁺]), and values above 7.0 indicate basic, or alkaline, solutions (Figure 2-25). For instance, gastric juice, which is rich in hydrochloric acid (HCl), has a pH of about 1. Its [H⁺] is roughly a millionfold greater than that of cytoplasm, with a pH of about 7.2.

Although the cytosol of cells normally has a pH of about 7.2, the interior of certain organelles in eukaryotic cells (see Chapter 9) can have a much lower pH. Lysosomes, for example, have a pH of about 4.5. The many degradative enzymes within lysosomes function optimally in an acidic environment, whereas their action is inhibited in the near neutral environment of the cytoplasm. As this example illustrates, maintenance of a specific pH is essential for proper functioning of some cellular structures. On the other hand, dramatic shifts in cellular pH may play an important role in controlling cellular activity. For example, the pH of the cytoplasm of an unfertilized egg of the sea urchin, an aquatic animal, is 6.6. Within 1 minute of fertilization, however, the pH rises to 7.2; that is, the [H⁺] decreases to about one-fourth its original value, a change necessary for subsequent growth and division of the egg.

## Hydrogen Ions Are Released by Acids and Taken Up by Bases

In general, an **acid** is any molecule, ion, or chemical group that tends to release a hydrogen ion ($H^+$), such as hydrochloric acid (HCl) or the carboxyl group (—COOH), which tends to dissociate to form the negatively charged carboxylate ion (—COO⁻). Likewise, a **base** is any molecule, ion, or chemical group that readily combines with a $H^+$, such as the hydroxyl ion (OH⁻); ammonia ($NH_3$), which forms an ammonium ion ($NH_4^+$); or the amino group (—NH₂).

When acid is added to an aqueous solution, the [H⁺] increases and the pH goes down. Conversely, when a base is added to a solution, the [H⁺] decreases and the pH goes up. Because $[H^+][OH^-] = 10^{-14} M^2$, any increase in [H⁺] is coupled with a commensurate decrease in [OH⁻] and vice versa.

Many biological molecules contain both acidic and basic groups. For example, in neutral solutions (pH = 7.0), many amino acids exist predominantly in the doubly ionized form, in which the carboxyl group has lost a proton and the amino group has accepted one:

$$\begin{array}{c} NH_3^+ \\ | \\ H-C-COO^- \\ | \\ R \end{array}$$

where R represents the uncharged side chain. Such a molecule, containing an equal number of positive and negative ions, is called a *zwitterion*. Zwitterions, having no net charge, are neutral. At extreme pH values, only one of these two ionizable groups of an amino acid will be charged.

The dissociation reaction for an acid (or acid group in a larger molecule) HA can be written as HA $\rightleftharpoons$ H⁺ + A⁻. The equilibrium constant for this reaction, denoted $K_a$ (the subscript a stands for "acid"), is defined as $K_a = [H^+][A^-]/[HA]$. Taking the logarithm of both sides and rearranging the result yields a very useful relation between the equilibrium constant and pH:

$$pH = pK_a + \log \frac{[A^-]}{[HA]} \qquad (2\text{-}5)$$

where $pK_a$ equals $-\log K_a$.

From this expression, commonly known as the *Henderson-Hasselbalch equation*, it can be seen that the $pK_a$ of any acid is equal to the pH at which half the molecules are dissociated and half are neutral (undissociated). This is because when [A⁻] = [HA], then log ([A⁻]/[HA]) = 0, and thus $pK_a$ = pH. The Henderson-Hasselbalch equation allows us to calculate the degree of dissociation of an acid if both the pH of the solution and the $pK_a$ of the acid are known. Experimentally, by measuring the [A⁻] and [HA] as a function of the solution's pH, one can calculate the $pK_a$ of the acid and thus the equilibrium constant $K_a$ for the dissociation reaction (Figure 2-26). Knowing the $pK_a$ of a molecule not only provides an important description of its properties but

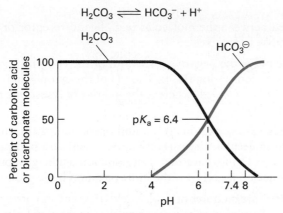

**FIGURE 2-26 The relationship between pH, p$K_a$, and the dissociation of an acid.** As the pH of a solution of carbonic acid rises from 0 to 8.5, the percentage of the compound in the undissociated, or non-ionized, form ($H_2CO_3$) decreases from 100 percent and that of the ionized form increases from 0 percent. When the pH (6.4) is equal to the acid's p$K_a$, half of the carbonic acid has ionized. When the pH rises to above 8, virtually all of the acid has ionized to the bicarbonate form ($HCO_3^-$).

also allows us to exploit these properties to manipulate the acidity of an aqueous solution and to understand how biological systems control this critical characteristic of their aqueous fluids.

## Buffers Maintain the pH of Intracellular and Extracellular Fluids

A living, actively metabolizing cell must maintain a constant pH in the cytoplasm of about 7.2–7.4 despite the metabolic production of many acids, such as lactic acid and carbon dioxide; the latter reacts with water to form carbonic acid ($H_2CO_3$). Cells have a reservoir of weak bases and weak acids, called **buffers**, which ensure that the cell's cytoplasmic pH remains relatively constant despite small fluctuations in the amounts of $H^+$ or $OH^-$ being generated by metabolism or by the uptake or secretion of molecules and ions by the cell. Buffers do this by "soaking up" excess $H^+$ or $OH^-$ when these ions are added to the cell or are produced by metabolism.

If additional acid (or base) is added to a buffered solution whose pH is equal to the p$K_a$ of the buffer ([HA] = [$A^-$]), the pH of the solution changes, but it changes less than it would if the buffer had not been present. This is because protons released by the added acid are taken up by the ionized form of the buffer ($A^-$); likewise, hydroxyl ions generated by the addition of base are neutralized by protons released by the undissociated buffer (HA). The capacity of a substance to release hydrogen ions or take them up depends partly on the extent to which the substance has already taken up or released protons, which in turn depends on the pH of the solution relative to the p$K_a$ of the substance. The ability of a buffer to minimize changes in pH, its *buffering capacity*,

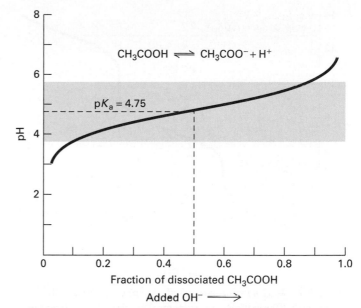

**FIGURE 2-27 The titration curve of the buffer acetic acid (CH$_3$COOH).** The p$K_a$ for the dissociation of acetic acid to hydrogen and acetate ions is 4.75. At this pH, half the acid molecules are dissociated. Because pH is measured on a logarithmic scale, the solution changes from 91 percent CH$_3$COOH at pH 3.75 to 9 percent CH$_3$COOH at pH 5.75. The acid has maximum buffering capacity in this pH range.

depends on the concentration of the buffer and the relationship between its p$K_a$ value and the pH, which is expressed by the Henderson-Hasselbalch equation.

The titration curve for acetic acid shown in Figure 2-27 illustrates the effect of pH on the fraction of molecules in the un-ionized (HA) and ionized forms ($A^-$). At one pH unit below the p$K_a$ of an acid, 91 percent of the molecules are in the HA form; at one pH unit above the p$K_a$, 91 percent are in the $A^-$ form. At pH values more than one unit above or below the p$K_a$, the buffering capacity of weak acids and bases declines rapidly. In other words, the addition of the same number of moles of acid to a solution containing a mixture of HA and $A^-$ that is at a pH near the p$K_a$ will cause less of a pH change than it would if the HA and $A^-$ were not present or if the pH were far from the p$K_a$ value.

All biological systems contain one or more buffers. Phosphate ions, the ionized forms of phosphoric acid, are present in considerable quantities in cells and are an important factor in maintaining, or buffering, the pH of the cytoplasm. Phosphoric acid ($H_3PO_4$) has three protons that are capable of dissociating, but they do not dissociate simultaneously. Loss of each proton can be described by a discrete dissociation reaction and p$K_a$, as shown in Figure 2-28. The titration curve for phosphoric acid shows that the p$K_a$ for the dissociation of the second proton is 7.2. Thus at pH 7.2, about 50 percent of cellular phosphate is $H_2PO_4^-$ and about 50 percent is $HPO_4^{2-}$ according to the Henderson-Hasselbalch equation. For this reason, phosphate is an excellent buffer at pH values around 7.2, the approximate pH of the cytoplasm of cells, and at pH 7.4, the pH of human blood.

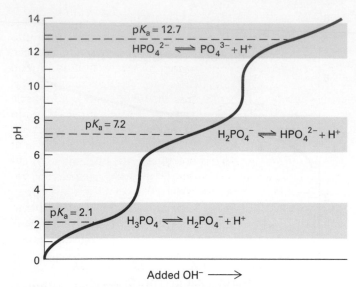

**FIGURE 2-28 The titration curve of phosphoric acid ($H_3PO_4$), a common buffer in biological systems.** This biologically ubiquitous molecule has three hydrogen atoms that dissociate at different pH values; thus phosphoric acid has three $pK_a$ values, as noted on the graph. The shaded areas denote the pH ranges—within one pH unit of the three $pK_a$ values—where the buffering capacity of phosphoric acid is high. In these regions, the addition of acid (or base) will cause relatively small changes in the pH.

## KEY CONCEPTS of Section 2.3

### Chemical Reactions and Chemical Equilibrium

• A chemical reaction is at equilibrium when the rate of the forward reaction is equal to the rate of the reverse reaction, and thus there is no net change in the concentration of the reactants or products.

• The equilibrium constant $K_{eq}$ of a reaction reflects the ratio of products to reactants at equilibrium and thus is a measure of the extent of the reaction and the relative stabilities of the reactants and products.

• The $K_{eq}$ depends on the temperature, pressure, and chemical properties of the reactants and products but is independent of the reaction rate and of the initial concentrations of reactants and products.

• For any reaction, the equilibrium constant $K_{eq}$ equals the ratio of the forward rate constant to the reverse rate constant ($k_f/k_r$). The rates of conversion of reactants to products and vice versa depend on the rate constants and the concentrations of the reactants or products.

• Within cells, the linked reactions in metabolic pathways generally are at steady state, not equilibrium, at which rate of formation of the intermediates equals their rate of consumption (see Figure 2-23) and thus the concentrations of the intermediates are not changing.

• The dissociation constant $K_d$ for the noncovalent binding of two molecules is a measure of the stability of the complex

formed between the molecules (e.g., ligand-receptor or protein-DNA complexes).

• The pH is the negative logarithm of the concentration of hydrogen ions ($-\log [H^+]$). The pH of the cytoplasm is normally about 7.2–7.4, whereas the interior of lysosomes has a pH of about 4.5.

• Acids release protons ($H^+$), and bases bind them. In biological molecules, the carboxyl (—COOH) and phosphoryl groups (—$H_2PO_4$) are the most common acidic groups; the amino group (—$NH_2$) is the most common basic group.

• Buffers are mixtures of a weak acid (HA) and its corresponding base form ($A^-$), which minimize the change in pH of a solution when acid or base is added. Biological systems use various buffers to maintain their pH within a very narrow range.

## 2.4 Biochemical Energetics

The transformation of energy, its storage, and its use are central to the economy of the cell. Energy may be defined as the ability to do work, a concept as applicable to cells as to automobile engines and electric power plants. The energy stored within chemical bonds can be harnessed to support chemical work and the physical movements of cells. In this section, we will review how energy influences the extents of chemical reactions, a discipline called chemical thermodynamics, and the rates of chemical reactions, a discipline called chemical kinetics.

### Several Forms of Energy Are Important in Biological Systems

There are two principal forms of energy: kinetic and potential. **Kinetic energy** is the energy of movement—the motion of molecules, for example. **Potential energy** is stored energy—the energy stored in covalent bonds, for example. Potential energy plays a particularly important role in the energy economy of cells.

*Thermal energy,* or heat, is a form of kinetic energy—the energy of the motion of molecules. For heat to do work, it must flow from a region of higher temperature—where the average speed of molecular motion is greater—to one of lower temperature. Although differences in temperature can exist between the internal and external environments of cells, these thermal gradients do not usually serve as the source of energy for cellular activities. The thermal energy in warm-blooded animals, which have evolved a mechanism for thermoregulation, is used chiefly to maintain constant organismic temperatures. This is an important homeostatic function because the rates of many cellular activities are temperature dependent. For example, cooling mammalian cells from their normal body temperature of 37 °C to 4 °C can virtually "freeze" or stop many cellular processes (e.g., intracellular membrane movements).

*Radiant energy* is the kinetic energy of photons, or waves of light, and is critical to biology. Radiant energy can be converted to thermal energy, for instance when light is absorbed by molecules and the energy is converted to molecular motion. Radiant energy absorbed by molecules can also change the electronic structure of the molecules, moving electrons into higher-energy orbitals, whence it can later be recovered to perform work. For example, during photosynthesis, light energy absorbed by pigment molecules such as chlorophyll is subsequently converted into the energy of chemical bonds (see Chapter 12).

*Mechanical energy,* a major form of kinetic energy in biology, usually results from the conversion of stored chemical energy. For example, changes in the lengths of cytoskeletal filaments generate forces that push or pull on membranes and organelles (see Chapters 17 and 18).

*Electric energy*—the energy of moving electrons or other charged particles—is yet another major form of kinetic energy, one with particular importance to membrane function, such as in electrically active neurons (see Chapter 22).

Several forms of potential energy are biologically significant. Central to biology is **chemical potential energy**, the energy stored in the bonds connecting atoms in molecules. Indeed, most of the biochemical reactions described in this book involve the making or breaking of at least one covalent chemical bond. We recognize this energy when chemicals undergo energy-releasing reactions. For example, the high potential energy in the covalent bonds of glucose can be released by controlled enzymatic combustion in cells (see Chapter 12). This energy is harnessed by the cell to do many kinds of work.

A second biologically important form of potential energy is the energy in a **concentration gradient**. When the concentration of a substance on one side of a barrier, such as a membrane, is different from that on the other side, a concentration gradient exists. All cells form concentration gradients between their interior and the external fluids by selectively exchanging nutrients, waste products, and ions with their surroundings. Also, organelles within cells (e.g., mitochondria, lysosomes) frequently contain different concentrations of ions and other molecules; the concentration of protons within a lysosome, as we saw in the last section, is about 500 times that of the cytoplasm.

A third form of potential energy in cells is an **electric potential**—the energy of charge separation. For instance, there is a gradient of electric charge of ~200,000 volts per cm across the plasma membrane of virtually all cells. We discuss how concentration gradients and the potential difference across cell membranes are generated and maintained in Chapter 11 and how they are converted to chemical potential energy in Chapter 12.

## Cells Can Transform One Type of Energy into Another

According to the first law of thermodynamics, energy is neither created nor destroyed but can be converted from one form to another. (In nuclear reactions, mass is converted to energy, but this is irrelevant to biological systems.) Energy conversions are very important in biology. In photosynthesis, for example, the radiant energy of light is transformed into the chemical potential energy of the covalent bonds between the atoms in a sucrose or starch molecule. In muscles and nerves, chemical potential energy stored in covalent bonds is transformed, respectively, into the kinetic energy of muscle contraction and the electric energy of nerve transmission. In all cells, potential energy—released by breaking certain chemical bonds—is used to generate potential energy in the form of concentration and electric potential gradients. Similarly, energy stored in chemical concentration gradients or electric potential gradients is used to synthesize chemical bonds or to transport molecules from one side of a membrane to another to generate a concentration gradient. The latter process occurs during the transport of nutrients such as glucose into certain cells and transport of many waste products out of cells.

Because all forms of energy are interconvertible, they can be expressed in the same units of measurement. Although the standard unit of energy is the joule, biochemists have traditionally used an alternative unit, the **calorie** (1 joule = 0.239 calorie). A calorie is the amount of energy required to raise the temperature of one gram of water by 1 °C. Throughout this book, we use the kilocalorie to measure energy changes (1 kcal = 1000 cal). When you read or hear about the "Calories" in food (note the capital C), the reference is almost always to kilocalories as defined here.

## The Change in Free Energy Determines If a Chemical Reaction Will Occur Spontaneously

Chemical reactions can be divided into two types depending on whether energy is absorbed or released in the process. In an **exergonic** ("energy-releasing") reaction, the products contain less energy than the reactants. Exergonic reactions take place spontaneously. The liberated energy is usually released as heat (the energy of molecular motion), and generally results in a rise in temperature, such as in the oxidation (burning) of wood. In an **endergonic** ("energy-absorbing") reaction, the products contain more energy than the reactants and energy is absorbed during the reaction. If there is no external source of energy to drive an endergonic reaction, it cannot take place. Endergonic reactions are responsible for the ability of instant cold packs often used to treat injuries to rapidly cool below room temperature. Crushing the pack mixes the reagents, initiating the reaction.

A fundamentally important concept in understanding if a reaction is exergonic or endergonic, and therefore if it occurs spontaneously or not, is **free energy, G,** named after J. W. Gibbs. Gibbs, who received the first PhD in engineering in America in 1863, showed that "all systems change in such a way that free energy [G] is minimized." In other words, a chemical reaction occurs spontaneously when the free energy of the products is lower than the free energy of the reactants. In the case of a chemical reaction, reactants $\rightleftharpoons$ products, the free-energy change, $\Delta G$, is given by

$$\Delta G = G_{products} - G_{reactants}$$

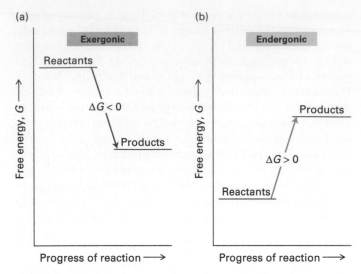

(a) Exergonic

Reactants

$\Delta G < 0$

Products

Free energy, $G$ →

Progress of reaction →

(b) Endergonic

Products

$\Delta G > 0$

Reactants

Free energy, $G$ →

Progress of reaction →

**FIGURE 2-29 Changes in the free energy ($\Delta G$) of exergonic and endergonic reactions.** (a) In exergonic reactions, the free energy of the products is lower than that of the reactants. Consequently, these reactions occur spontaneously and energy is released as the reactions proceed. (b) In endergonic reactions, the free energy of the products is greater than that of the reactants and these reactions do not occur spontaneously. An external source of energy must be supplied if the reactants are to be converted into products.

The relation of $\Delta G$ to the direction of any chemical reaction can be summarized in three statements:

• If $\Delta G$ is negative, the forward reaction will tend to occur spontaneously and energy usually will be released as the reaction takes place (exergonic reaction) (Figure 2-29). A reaction with a negative $\Delta G$ is called thermodynamically favorable.

• If $\Delta G$ is positive, the forward reaction will not occur spontaneously; energy will have to be added to the system in order to force the reactants to become products (endergonic reaction).

• If $\Delta G$ is zero, both forward and reverse reactions occur at equal rates and there will be no spontaneous net conversion of reactants to products, or vice versa; the system is at equilibrium.

By convention, the *standard free-energy change* of a reaction $\Delta G^{\circ\prime}$ is the value of the change in free energy under the conditions of 298 K (25 °C), 1 atm pressure, pH 7.0 (as in pure water), and initial concentrations of 1 M for all reactants and products except protons, which are kept at $10^{-7}$ M (pH 7.0). Most biological reactions differ from standard conditions, particularly in the concentrations of reactants, which are normally less than 1 M.

The free energy of a chemical system can be defined as $G = H - TS$, where $H$ is the bond energy, or **enthalpy**, of the system; $T$ is its temperature in degrees Kelvin (K); and $S$ is the **entropy**, a measure of its randomness or disorder. According to the second law of thermodynamics, the natural tendency of any system is to become more disordered—that is, for entropy to increase. A reaction can occur spontaneously only if the combined effects of changes in enthalpy and

entropy lead to a lower $\Delta G$. That is, if temperature remains constant, a reaction proceeds spontaneously only if the free-energy change, $\Delta G$, in the following equation is negative:

$$\Delta G = \Delta H - T\Delta S \qquad (2\text{-}6)$$

In an **exothermic** ("heat-releasing") chemical reaction, $\Delta H$ is negative. In an **endothermic** ("heat-absorbing") reaction, $\Delta H$ is positive. The combined effects of the changes in the enthalpy and entropy determine if the $\Delta G$ for a reaction is positive or negative and thus if the reaction occurs spontaneously. An exothermic reaction ($\Delta H < 0$), in which entropy increases ($\Delta S > 0$), occurs spontaneously ($\Delta G < 0$). An endothermic reaction ($\Delta H > 0$) will occur spontaneously if $\Delta S$ increases enough so that the $T\Delta S$ term can overcome the positive $\Delta H$.

Many biological reactions lead to an increase in order and thus a decrease in entropy ($\Delta S < 0$). An obvious example is the reaction that links amino acids to form a protein. A solution of protein molecules has a lower entropy than does a solution of the same amino acids unlinked because the free movement of any amino acid in a protein is more restricted (greater order) when it is bound into a long chain than when it is not. Thus when cells synthesize polymers such as proteins from their constituent monomers, the polymerizing reaction will only be spontaneous if the cells can efficiently transfer energy to both generate the bonds that hold the monomers together and overcome the loss in entropy that accompanies polymerization. Often cells accomplish this feat by "coupling" such synthetic, entropy-lowering reactions with independent reactions that have a very highly negative $\Delta G$ (see below). In this way, cells can convert sources of energy in their environment into the building of highly organized structures and metabolic pathways that are essential for life.

The actual change in free energy $\Delta G$ during a reaction is influenced by temperature, pressure, and the initial concentrations of reactants and products and usually differs from the standard free-energy change $\Delta G^{\circ\prime}$. Most biological reactions—like others that take place in aqueous solutions—also are affected by the pH of the solution. We can estimate free-energy changes for temperatures and initial concentrations that differ from the standard conditions by using the equation

$$\Delta G = \Delta G^{\circ\prime} + RT \ln Q = \Delta G^{\circ\prime} + RT \ln \frac{[\text{products}]}{[\text{reactants}]} \qquad (2\text{-}7)$$

where $R$ is the gas constant of 1.987 cal/(degree·mol), $T$ is the temperature (in degrees Kelvin), and $Q$ is the *initial* ratio of products to reactants. For a reaction $A + B \rightleftharpoons C$, in which two molecules combine to form a third, $Q$ in Equation 2-7 equals [C]/[A][B]. In this case, an increase in the initial concentration of either [A] or [B] will result in a larger negative value for $\Delta G$ and thus drive the reaction toward spontaneous formation of C.

Regardless of the $\Delta G^{\circ\prime}$ for a particular biochemical reaction, it will proceed spontaneously within cells only if $\Delta G$ is negative, given the intracellular concentrations of reactants and products. For example, the conversion of glyceraldehyde 3-phosphate (G3P) to dihydroxyacetone phosphate (DHAP), two intermediates in the breakdown of glucose,

$$G3P \rightleftharpoons DHAP$$

has a $\Delta G^{\circ\prime}$ of $-1840$ cal/mol. If the initial concentrations of G3P and DHAP are equal, then $\Delta G = \Delta G^{\circ\prime}$ because $RT \ln 1 = 0$; in this situation, the reversible reaction $G3P \rightleftharpoons DHAP$ will proceed spontaneously in the direction of DHAP formation until equilibrium is reached. However, if the initial [DHAP] is 0.1 M and the initial [G3P] is 0.001 M, with other conditions standard, then $Q$ in Equation 2-7 equals 0.1/0.001 = 100, giving a $\Delta G$ of $+887$ cal/mol. Under these conditions, the reaction will proceed in the direction of formation of G3P.

The $\Delta G$ for a reaction is independent of the reaction rate. Indeed, under usual physiological conditions, few if any of the biochemical reactions needed to sustain life would occur without some mechanism for increasing reaction rates. As we describe below and in more detail in Chapter 3, the rates of reactions in biological systems are usually determined by the activity of **enzymes**, the protein catalysts that accelerate the formation of products from reactants without altering the value of $\Delta G$.

## The $\Delta G^{\circ\prime}$ of a Reaction Can Be Calculated from Its $K_{eq}$

A chemical mixture at equilibrium is in a stable state of minimal free energy. For a system at equilibrium ($\Delta G = 0$, $Q = K_{eq}$), we can write

$$\Delta G^{\circ\prime} = -2.3RT \log K_{eq} = -1362 \log K_{eq} \quad (2\text{-}8)$$

under standard conditions (note the change to base 10 logarithms). Thus if we determine the concentrations of reactants and products at equilibrium (i.e., the $K_{eq}$), we can calculate the value of $\Delta G^{\circ\prime}$. For example, the $K_{eq}$ for the interconversion of glyceraldehyde 3-phosphate to dihydroxyacetone phosphate ($G3P \rightleftharpoons DHAP$) is 22.2 under standard conditions. Substituting this value into Equation 2-8, we can easily calculate the $\Delta G^{\circ\prime}$ for this reaction as $-1840$ cal/mol.

By rearranging Equation 2-8 and taking the antilogarithm, we obtain

$$K_{eq} = 10^{-(\Delta G^{\circ\prime}/2.3RT)} \quad (2\text{-}9)$$

From this expression, it is clear that if $\Delta G^{\circ\prime}$ is negative, the exponent will be positive and hence $K_{eq}$ will be greater than 1. Therefore, at equilibrium there will be more products than reactants; in other words, the formation of products from reactants is favored. Conversely, if $\Delta G^{\circ\prime}$ is positive, the exponent will be negative and $K_{eq}$ will be less than 1. The relationship between $K_{eq}$ and $\Delta G^{\circ\prime}$ further emphasizes the influence of relative free energies of reactants and

products on the extent to which a reaction will occur spontaneously.

## The Rate of a Reaction Depends on the Activation Energy Necessary to Energize the Reactants into a Transition State

As a chemical reaction proceeds, reactants approach each other; some bonds begin to form while others begin to break. One way to think of the state of the molecules during this transition is that there are strains in the electronic configurations of the atoms and their bonds. The collection of atoms moves from the relatively stable state of the reactants to this transient, intermediate, and higher-energy state during the course of the reaction (Figure 2-30). The state during a chemical reaction at which the system is at its highest energy level is called the **transition state**, and the collection of reactants in that state is called the **transition-state intermediate**. The energy needed to excite the reactants to this higher-energy state is called the **activation energy** of the reaction. The activation energy is usually represented by $\Delta G^{\ddagger}$, analogous to the representation of the change in Gibbs free energy ($\Delta G$) already discussed. From the transition state, the collection of atoms can either release energy as the reaction products are formed or release energy as the atoms go "backward" and re-form the original reactants. The velocity ($V$) at which products are generated from reactants during the reaction under a given set of conditions (temperature, pressure, reactant concentrations) will depend on the concentration of

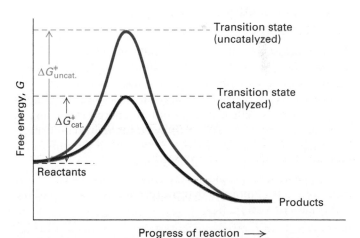

**FIGURE 2-30 Activation energy of uncatalyzed and catalyzed chemical reactions.** This hypothetical reaction pathway (blue) depicts the changes in free energy, $G$, as a reaction proceeds. A reaction will take place spontaneously if the free energy ($G$) of the products is less than that of the reactants ($\Delta G < 0$). However, all chemical reactions proceed through one (shown here) or more high-energy transition states, and the rate of a reaction is inversely proportional to the activation energy ($\Delta G^{\ddagger}$), which is the difference in free energy between the reactants and the transition state. In a catalyzed reaction (red), the free energies of the reactants and products are unchanged but the free energy of the transition state is lowered, thus increasing the velocity of the reaction.

material in the transition state, which in turn will depend on the activation energy and the characteristic rate constant ($v$) at which the transition state is converted to products. The higher the activation energy, the lower the fraction of reactants that reach the transition state and the slower the overall rate of the reaction. The relationship between the concentration of reactants, $v$, and $V$ is

$$V = v \text{ [reactants]} \times 10^{-(\Delta G^{\ddagger}/2.3RT)}$$

From this equation, we can see that lowering the activation energy—that is, decreasing the free energy of the transition state $\Delta G^{\ddagger}$—leads to an acceleration of the overall reaction rate $V$. A reduction in $\Delta G^{\ddagger}$ of 1.36 kcal/mol leads to a tenfold increase in the rate of the reaction, whereas a 2.72 kcal/mol reduction increases the rate 100-fold. Thus relatively small changes in $\Delta G^{\ddagger}$ can lead to large changes in the overall rate of the reaction.

Catalysts such as enzymes (discussed further in Chapter 3) accelerate reaction rates by lowering the relative energy of the transition state and so the activation energy required to reach it (see Figure 2-30). The relative energies of reactants and products will determine if a reaction is thermodynamically favorable (negative $\Delta G$), whereas the activation energy will determine how rapidly products form—that is, its reaction kinetics. Thermodynamically favorable reactions will not occur at appreciable rates if the activation energies are too high.

## Life Depends on the Coupling of Unfavorable Chemical Reactions with Energetically Favorable Ones

Many processes in cells are energetically unfavorable ($\Delta G > 0$) and will not proceed spontaneously. Examples include the synthesis of DNA from nucleotides and transport of a substance across the plasma membrane from a lower to a higher concentration. Cells can carry out an energy-requiring, or endergonic, reaction ($\Delta G_1 > 0$) by coupling it to an energy-releasing, or exergonic, reaction ($\Delta G_2 < 0$) if the sum of the two reactions has an overall net negative $\Delta G$.

Suppose, for example, that the reaction A $\rightleftharpoons$ B + X has a $\Delta G$ of +5 kcal/mol and that the reaction X $\rightleftharpoons$ Y + Z has a $\Delta G$ of −10 kcal/mol:

$$
\begin{array}{llll}
(1) & \text{A} \rightleftharpoons \text{B + X} & \Delta G & = +5 \text{ kcal/mol} \\
(2) & \underline{\text{X} \rightleftharpoons \text{Y + Z}} & \Delta G & = -10 \text{ kcal/mol} \\
Sum: & \text{A} \rightleftharpoons \text{B + Y + Z} & \Delta G^{\circ\prime} & = -5 \text{ kcal/mol}
\end{array}
$$

In the absence of the second reaction, there would be much more A than B at equilibrium. However, because the conversion of X to Y + Z is such a favorable reaction, it will pull the first process toward the formation of B and the consumption of A. Energetically unfavorable reactions in cells often are coupled to the energy-releasing hydrolysis of ATP, as we discuss next.

## Hydrolysis of ATP Releases Substantial Free Energy and Drives Many Cellular Processes

In almost all organisms, the nucleoside triphosphate adenosine triphosphate, or ATP (Figure 2-31), is the most important molecule for capturing, transiently storing, and subsequently transferring energy to perform work (e.g., biosynthesis, mechanical motion). Commonly referred to as a cell's energy "currency," ATP is a type of usable energy that cells can "spend" in order to power their activities. The storied history of ATP begins with its discovery in 1929 apparently simultaneously by Kurt Lohmann, who was working with the great biochemist Otto Meyerhof in Germany and who published first, and by Cyrus Fiske and Yellagaprada SubbaRow in the United States. Muscle contractions were shown to depend on ATP in the 1930s. The proposal that ATP is the main intermediary for the transfer of energy in cells is credited to Fritz Lipmann around 1941. Many Nobel Prizes have been awarded for the study of ATP and its role in cellular energy metabolism, and its importance in understanding molecular cell biology cannot be overstated.

The useful energy in an ATP molecule is contained in phosphoanhydride bonds, which are covalent bonds formed

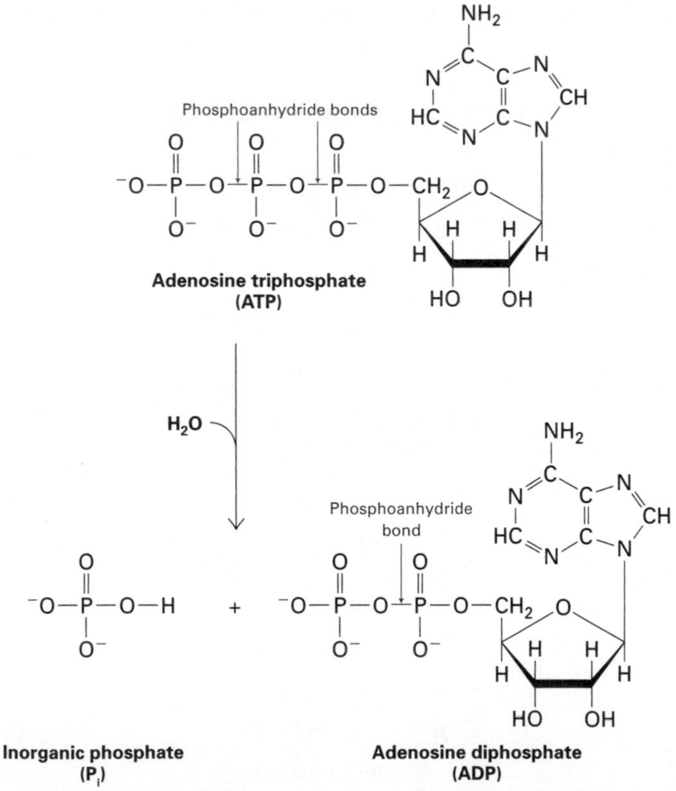

**FIGURE 2-31 Hydrolysis of adenosine triphosphate (ATP).** The two phosphoanhydride bonds (red) in ATP (*top*), which link the three phosphate groups, each have a $\Delta G^{\circ}$ of about −7.3 kcal/mol for hydrolysis. Hydrolysis of the terminal phosphoanhydride bond by the addition of water results in the release of phosphate and generation of ADP. Hydrolysis of the phosphoanhydride bonds of ATP, especially the terminal one, is the source of energy that drives many energy-requiring reactions in biological systems.

from the condensation of two molecules of phosphate by the loss of water:

$$O^- - \underset{\underset{O^-}{|}}{\overset{\overset{O}{\|}}{P}} - OH \;+\; HO - \underset{\underset{O^-}{|}}{\overset{\overset{O}{\|}}{P}} - O^- \;\rightleftharpoons$$

$$O^- - \underset{\underset{O^-}{|}}{\overset{\overset{O}{\|}}{P}} - O - \underset{\underset{O^-}{|}}{\overset{\overset{O}{\|}}{P}} - O^- \;+\; H_2O$$

As shown in Figure 2-31, an ATP molecule has two key phosphoanhydride (also called phosphodiester) bonds. Forming these bonds in ATP requires an input of energy. When these bonds are hydrolyzed, or broken by the addition of water, that energy is released. Hydrolysis of a phosphoanhydride bond (represented by the symbol ~) in each of the following reactions has a highly negative $\Delta G^{\circ\prime}$ of about $-7.3$ kcal/mol:

$$Ap{\sim}p{\sim}p + H_2O \rightarrow Ap{\sim}p + P_i + H^+$$
$$\text{(ATP)} \qquad\qquad \text{(ADP)}$$

$$Ap{\sim}p{\sim}p + H_2O \rightarrow Ap + PP_i + H^+$$
$$\text{(ATP)} \qquad\qquad \text{(AMP)}$$

$$Ap{\sim}p + H_2O \rightarrow Ap + P_i + H^+$$
$$\text{(ADP)} \qquad\qquad \text{(AMP)}$$

In these reactions that occur in biological systems, $P_i$ stands for inorganic phosphate ($PO_4^{3-}$) and $PP_i$ for inorganic pyrophosphate, two phosphate groups linked by a phosphoanhydride bond. As the top two reactions show, the removal of a phosphate or a pyrophosphate group from ATP leaves adenosine diphosphate (ADP) or adenosine monophosphate (AMP), respectively.

A phosphoanhydride bond or other "high-energy bond" (commonly denoted by ~) is not intrinsically different from other covalent bonds. High-energy bonds simply release substantial amounts of energy when hydrolyzed. For instance, the $\Delta G^{\circ\prime}$ for hydrolysis of a phosphoanhydride bond in ATP ($-7.3$ kcal/mol) is more than three times the $\Delta G^{\circ\prime}$ for hydrolysis of the phosphoester bond (red) in glycerol 3-phosphate ($-2.2$ kcal/mol):

$$HO - \underset{\underset{O^-}{|}}{\overset{\overset{O}{\|}}{P}} - O - CH_2 - \underset{\overset{|}{OH}}{CH} - CH_2OH$$

**Glycerol 3-phosphate**

A principal reason for this difference is that ATP and its hydrolysis products ADP and $P_i$ are highly charged at neutral pH. During synthesis of ATP, a large input of energy is required to force the negative charges in ADP and $P_i$ together. Conversely, this energy is released when ATP is hydrolyzed to ADP and $P_i$. In comparison, formation of the phosphoester bond between an uncharged hydroxyl in glycerol and $P_i$ requires less energy, and less energy is released when this bond is hydrolyzed.

Cells have evolved protein-mediated mechanisms for transferring the free energy released by hydrolysis of phosphoanhydride bonds to other molecules, thereby driving reactions that would otherwise be energetically unfavorable. For example, if the $\Delta G$ for the reaction $B + C \rightarrow D$ is positive but less than the $\Delta G$ for hydrolysis of ATP, the reaction can be driven to the right by coupling it to hydrolysis of the terminal phosphoanhydride bond in ATP. In one common mechanism of such *energy coupling*, some of the energy stored in this phosphoanhydride bond is transferred to one of the reactants by breaking the bond in ATP and forming a covalent bond between the released phosphate group and one of the reactants. The phosphorylated intermediate generated in this way can then react with C to form $D + P_i$ in a reaction that has a negative $\Delta G$:

$$B + Ap{\sim}p{\sim}p \rightarrow B{\sim}p + Ap{\sim}p$$
$$B{\sim}p + C \rightarrow D + P_i$$

The overall reaction

$$B + C + ATP \rightleftharpoons D + ADP + P_i$$

is energetically favorable ($\Delta G < 0$).

An alternative mechanism of energy coupling is to use the energy released by ATP hydrolysis to change the conformation of the molecule to an "energy-rich" stressed state. In turn, the energy stored as conformational stress can be released as the molecule "relaxes" back into its unstressed conformation. If this relaxation process can be mechanistically coupled to another reaction, the released energy can be harnessed to drive important cellular processes.

As with many biosynthetic reactions, transport of molecules into or out of the cell often has a positive $\Delta G$ and thus requires an input of energy to proceed. Such simple transport reactions do not *directly* involve the making or breaking of covalent bonds; thus the $\Delta G^{\circ\prime}$ is 0. In the case of a substance moving into a cell, Equation 2-7 becomes

$$\Delta G = RT \ln \frac{[C_{in}]}{[C_{out}]} \tag{2-10}$$

where $[C_{in}]$ is the initial concentration of the substance inside the cell and $[C_{out}]$ is its concentration outside the cell. We can see from Equation 2-10 that $\Delta G$ is positive for transport of a substance into a cell against its concentration gradient (when $[C_{in}] > [C_{out}]$); the energy to drive such "uphill" transport often is supplied by the hydrolysis of ATP. Conversely, when a substance moves down its concentration gradient ($[C_{out}] > [C_{in}]$), $\Delta G$ is negative. Such "downhill" transport releases energy that can be coupled to an energy-requiring reaction, say, the movement of another substance uphill across a membrane or the synthesis of ATP itself (see Chapters 11 and 12).

## ATP Is Generated During Photosynthesis and Respiration

ATP is continually being hydrolyzed to provide energy for many cellular activities. Some estimates suggest that humans daily hydrolyze a mass of ATP equal to their entire body weight. Clearly, to continue functioning, cells must constantly replenish their ATP supply. Constantly replenishing ATP requires that cells continually obtain energy from their environment. For nearly all cells, the ultimate source of energy used to make ATP is sunlight. Some organisms can use sunlight directly. Through the process of **photosynthesis**, plants, algae, and certain photosynthetic bacteria trap the energy of sunlight and use it to synthesize ATP from ADP and $P_i$. Much of the ATP produced in photosynthesis is hydrolyzed to provide energy for the conversion of carbon dioxide to six-carbon sugars, a process called **carbon fixation**:

$$6\ CO_2 + 6\ H_2O \longrightarrow C_6H_{12}O_6 + 6\ O_2 + energy$$

The sugars made during photosynthesis are a source of food, and thus energy, for the plants or other photosynthetic organisms making them and for the non-photosynthetic organisms, such as animals, that either consume the plants directly or indirectly by eating other animals that have eaten the plants. In this way sunlight is the direct or indirect source of energy for most organisms (see Chapter 12).

In plants, animals, and nearly all other organisms, the free energy in sugars and other molecules derived from food is released in the processes of **glycolysis** and **cellular respiration**. During cellular respiration, energy-rich molecules in food (e.g., glucose) are oxidized to carbon dioxide and water. The complete oxidation of glucose,

$$C_6H_{12}O_6 + 6\ O_2 \rightarrow 6\ CO_2 + 6\ H_2O$$

has a $\Delta G°'$ of $-686$ kcal/mol and is the reverse of photosynthetic carbon fixation. Cells employ an elaborate set of protein-mediated reactions to couple the oxidation of 1 molecule of glucose to the synthesis of as many as 30 molecules of ATP from 30 molecules of ADP. This oxygen-dependent (**aerobic**) degradation (**catabolism**) of glucose is the major pathway for generating ATP in all animal cells, non-photosynthetic plant cells, and many bacterial cells. Catabolism of fatty acids can also be an important source of ATP. We discuss the mechanisms of photosynthesis and cellular respiration in Chapter 12.

Although light energy captured in photosynthesis is the primary source of chemical energy for cells, it is not the only source. Certain microorganisms that live in or around deep ocean vents, where adequate sunlight is unavailable, derive the energy for converting ADP and $P_i$ into ATP from the oxidation of reduced inorganic compounds. These reduced compounds originate deep in the earth and are released at the vents.

## NAD$^+$ and FAD Couple Many Biological Oxidation and Reduction Reactions

In many chemical reactions, electrons are transferred from one atom or molecule to another; this transfer may or may not accompany the formation of new chemical bonds or the release of energy that can be coupled to other reactions. The loss of electrons from an atom or a molecule is called **oxidation**, and the gain of electrons by an atom or a molecule is called **reduction**. An example of oxidation is the removal of electrons from the sulfhydryl groups of two cysteines to form a disulfide bond, described above in Section 2.2. Because electrons are neither created nor destroyed in a chemical reaction, if one atom or molecule is oxidized, another must be reduced. For example, oxygen draws electrons from $Fe^{2+}$ (ferrous) ions to form $Fe^{3+}$ (ferric) ions, a reaction that occurs as part of the process by which carbohydrates are degraded in mitochondria. Each oxygen atom receives two electrons, one from each of two $Fe^{2+}$ ions:

$$2\ Fe^{2+} + ½\ O_2 \rightarrow 2\ Fe^{3+} + O^{2-}$$

Thus $Fe^{2+}$ is oxidized and $O_2$ is reduced. Such reactions in which one molecule is reduced and another oxidized often are referred to as **redox reactions**. Oxygen is an electron acceptor in many redox reactions in cells under aerobic conditions.

Many biologically important oxidation and reduction reactions involve the removal or the addition of hydrogen atoms (protons plus electrons) rather than the transfer of isolated electrons on their own. The oxidation of succinate to fumarate, which also occurs in mitochondria, is an example (Figure 2-32). Protons are soluble in aqueous solutions (as $H_3O^+$), but electrons are not and must be transferred directly from one atom or molecule to another without a water-dissolved intermediate. In this type of oxidation reaction, electrons often are transferred to small electron-carrying molecules, sometimes referred to as coenzymes. The most common of these electron carriers are **NAD$^+$ (nicotinamide adenine dinucleotide)**, which is reduced to NADH, and **FAD (flavin adenine dinucleotide)**, which is reduced to FADH$_2$ (Figure 2-33). The reduced forms of these coenzymes can transfer protons and electrons to other molecules, thereby reducing them.

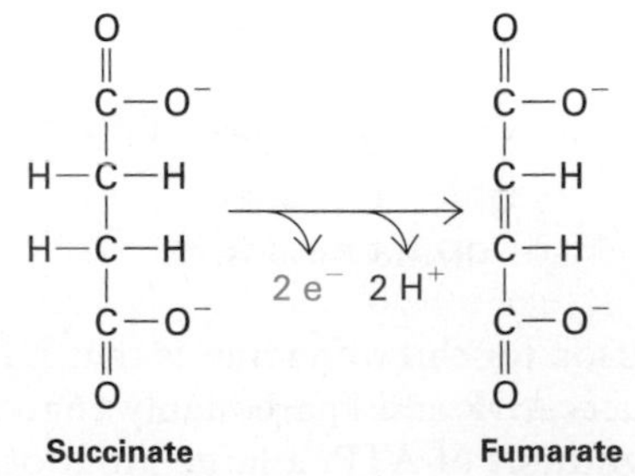

**FIGURE 2-32 Conversion of succinate to fumarate.** In this oxidation reaction, which occurs in mitochondria as part of the citric acid cycle, succinate loses two electrons and two protons. These are transferred to FAD, reducing it to FADH$_2$.

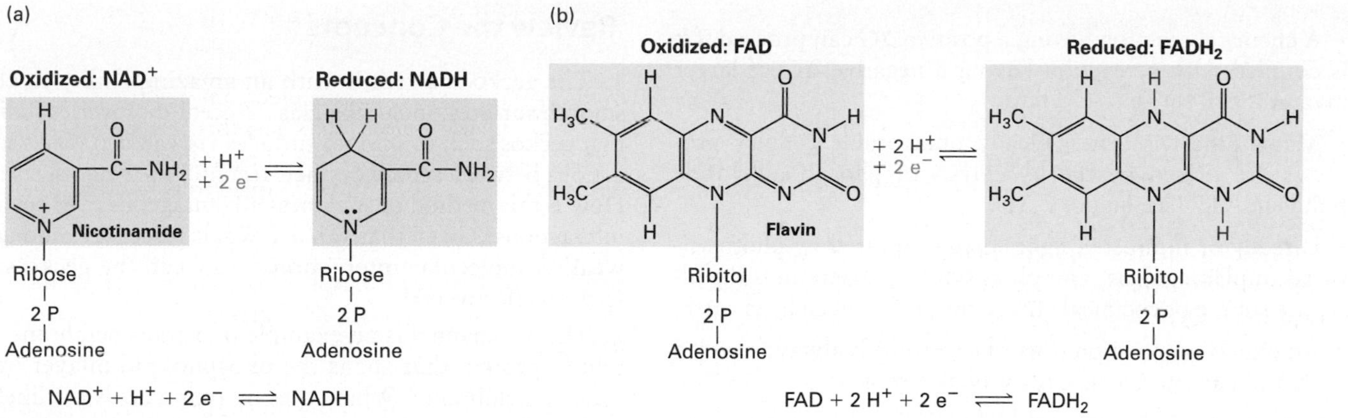

**(a)**

Oxidized: NAD$^+$      Reduced: NADH

Nicotinamide

Ribose
|
2 P
|
Adenosine

$$NAD^+ + H^+ + 2\,e^- \rightleftharpoons NADH$$

**(b)**

Oxidized: FAD      Reduced: FADH$_2$

Flavin

Ribitol
|
2 P
|
Adenosine

$$FAD + 2\,H^+ + 2\,e^- \rightleftharpoons FADH_2$$

**FIGURE 2-33 The electron-carrying coenzymes NAD$^+$ and FAD.**
(a) NAD$^+$ (nicotinamide adenine dinucleotide) is reduced to NADH by the addition of two electrons and one proton simultaneously. In many biological redox reactions, a pair of hydrogen atoms (two protons and two electrons) are removed from a molecule. In some cases, one of the protons and both electrons are transferred to NAD$^+$; the other proton is released into solution. (b) FAD (flavin adenine dinucleotide) is reduced to FADH$_2$ by the addition of two electrons and two protons, as occurs when succinate is converted to fumarate (see Figure 2-32). In this two-step reaction, addition of one electron together with one proton first generates a short-lived semiquinone intermediate (not shown), which then accepts a second electron and proton.

To describe redox reactions, such as the reaction of ferrous ion (Fe$^{2+}$) and oxygen (O$_2$), it is easiest to divide them into two half-reactions:

*Oxidation of* Fe$^{2+}$:    $2\,Fe^{2+} \rightarrow 2\,Fe^{3+} + 2\,e^-$
*Reduction of* O$_2$:    $2\,e^- + \frac{1}{2}\,O_2 \rightarrow O^{2-}$

In this case, the reduced oxygen (O$^{2-}$) readily reacts with two protons to form one water molecule (H$_2$O). The readiness with which an atom or a molecule *gains* an electron is its **reduction potential** $E$. The tendency to *lose* electrons, the **oxidation potential**, has the same magnitude but opposite sign as the reduction potential for the reverse reaction.

Reduction potentials are measured in volts (V) from an arbitrary zero point set at the reduction potential of the following half-reaction under standard conditions (25 °C, 1 atm, and reactants at 1 M):

$$H^+ + e^- \underset{\text{oxidation}}{\overset{\text{reduction}}{\rightleftharpoons}} \frac{1}{2}\,H_2$$

The value of $E$ for a molecule or an atom under standard conditions is its standard reduction potential, $E'_0$. A molecule or an ion with a positive $E'_0$ has a higher affinity for electrons than the H$^+$ ion does under standard conditions. Conversely, a molecule or ion with a negative $E'_0$ has a lower affinity for electrons than the H$^+$ ion does under standard conditions. Like the values of $\Delta G^{\circ\prime}$, standard reduction potentials may differ somewhat from those found under the conditions in a cell because the concentrations of reactants in a cell are not 1 M.

In a redox reaction, electrons move spontaneously toward atoms or molecules having *more positive* reduction potentials. In other words, a molecule having a more negative reduction potential can transfer electrons spontaneously to, or reduce, a molecule with a more positive reduction potential. In this type of reaction, the change in electric potential $\Delta E$ is the sum of the reduction and oxidation potentials for the two half-reactions. The $\Delta E$ for a redox reaction is related to the change in free energy $\Delta G$ by the following expression:

$$\Delta G \text{ (cal/mol)} = -n\,(23{,}064)\,\Delta E \text{ (volts)} \qquad (2\text{-}11)$$

where $n$ is the number of electrons transferred. Note that a redox reaction with a positive $\Delta E$ value will have a negative $\Delta G$ and thus will tend to proceed spontaneously from left to right.

## KEY CONCEPTS of Section 2.4

### Biochemical Energetics

• The change in free energy, $\Delta G$, is the most useful measure for predicting the potential of chemical reactions to occur spontaneously in biological systems. Chemical reactions tend to proceed spontaneously in the direction for which $\Delta G$ is negative. The magnitude of $\Delta G$ is independent of the reaction rate. A reaction with a negative $\Delta G$ is called thermodynamically favorable.

• The chemical free-energy change, $\Delta G^{\circ\prime}$, equals $-2.3\,RT \log K_{eq}$. Thus the value of $\Delta G^{\circ\prime}$ can be calculated from the experimentally determined concentrations of reactants and products at equilibrium.

• The rate of a reaction depends on the activation energy needed to energize reactants to a transition state. Catalysts such as enzymes speed up reactions by lowering the activation energy of the transition state.

• A chemical reaction having a positive $\Delta G$ can proceed if it is coupled with a reaction having a negative $\Delta G$ of larger magnitude.

• Many otherwise energetically unfavorable cellular processes are driven by the hydrolysis of phosphoanhydride bonds in ATP (see Figure 2-31).

• Directly or indirectly, light energy captured by photosynthesis in plants, algae, and photosynthetic bacteria is the ultimate source of chemical energy for nearly all cells on earth.

• An oxidation reaction (loss of electrons) is always coupled with a reduction reaction (gain of electrons).

• Biological oxidation and reduction reactions often are coupled by electron-carrying coenzymes such as $NAD^+$ and FAD (see Figure 2-33).

• Oxidation-reduction reactions with a positive $\Delta E$ have a negative $\Delta G$ and thus tend to proceed spontaneously.

## Key Terms

acid 46

adenosine triphosphate (ATP) 52

$\alpha$ carbon atom $(C_\alpha)$ 33

amino acid 33

amphipathic 23

base 46

buffer 47

catalyst 43

chemical potential energy 49

covalent bond 24

dehydration reaction 33

dipole 26

dissociation constant $(K_d)$ 45

disulfide bond 35

endergonic 49

endothermic 50

energy coupling 53

enthalpy $(H)$ 50

entropy $(S)$ 50

equilibrium constant $(K_{eq})$ 43

exergonic 49

exothermic 50

fatty acids 40

$\Delta G$ (free-energy change) 49

hydrogen bond 28

hydrophilic 23

hydrophobic 23

hydrophobic effect 31

ionic interactions 28

molecular complementarity 32

monomer 33

monosaccharide 37

noncovalent interactions 24

nucleoside 37

nucleotide 37

oxidation 54

pH 45

phosphoanhydride bond 52

phosphoglyceride 40

phospholipid bilayer 40

polar 26

polymer 33

redox reaction 54

reduction 54

saturated 40

steady state 44

stereoisomer 25

transition state 51

unsaturated 40

van der Waals interaction 30

## Review the Concepts

1. The gecko is a reptile with an amazing ability to climb smooth surfaces, including glass. Recent discoveries indicate that geckos stick to smooth surfaces via van der Waals interactions between septae on their feet and the smooth surface. How is this method of stickiness advantageous over covalent interactions? Given that van der Waals forces are among the weakest molecular interactions, how can the gecko's feet stick so effectively?

2. The $K^+$ channel is an example of a transmembrane protein (a protein that spans the phospholipid bilayer of the plasma membrane). What types of amino acids are likely to be found (a) lining the channel through which $K^+$ passes, (b) in contact with the hydrophobic core of the phospholipid bilayer containing fatty acyl groups, (c) in the cytosolic domain of the protein, and (d) in the extracellular domain of the protein?

3. V-M-Y-F-E-N: This is the single-letter amino acid abbreviation for a peptide. What is the net charge of this peptide at pH 7.0? An enzyme called a protein tyrosine kinase can attach phosphates to the hydroxyl groups of tyrosine. What is the net charge of the peptide at pH 7.0 after it has been phosphorylated by a tyrosine kinase? What is the likely source of phosphate utilized by the kinase for this reaction?

4. Disulfide bonds help to stabilize the three-dimensional structure of proteins. What amino acids are involved in the formation of disulfide bonds? Does the formation of a disulfide bond increase or decrease entropy $(\Delta S)$?

5. In the 1960s, the drug thalidomide was prescribed to pregnant women to treat morning sickness. However, thalidomide caused severe limb defects in the children of some women who took the drug, and its use for morning sickness was discontinued. It is now known that thalidomide was administered as a mixture of two stereoisomeric compounds, one of which relieved morning sickness and the other of which was responsible for the birth defects. What are stereoisomers? Why might two such closely related compounds have such different physiologic effects?

6. Name the compound shown below.

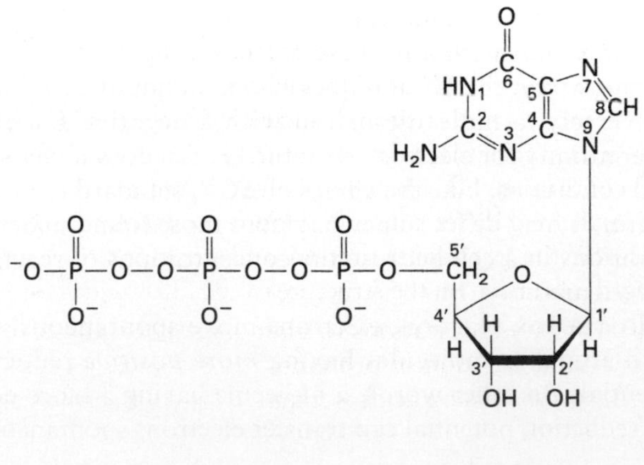

Is this nucleotide a component of DNA, RNA, or both? Name one other function of this compound.

7. The chemical basis of blood-group specificity resides in the carbohydrates displayed on the surface of red blood cells. Carbohydrates have the potential for great structural diversity. Indeed, the structural complexity of the oligosaccharides that can be formed from four sugars is greater than that for oligopeptides from four amino acids. What properties of carbohydrates make this great structural diversity possible?

8. Calculate the pH of 1 L of pure water at equilibrium. How will the pH change after 0.008 moles of the strong base NaOH are dissolved in the water? Now, calculate the pH of a 50 mM aqueous solution of the weak acid 3-(N-morpholino) propane-1-sulfonic acid (MOPS) in which 61% of the solute is in its weak acid form and 39% is in the form of MOPS conjugate base (the $pK_a$ for MOPS is 7.20). What is the final pH after 0.008 moles of NaOH are added to 1 L of this MOPS buffer?

9. Ammonia ($NH_3$) is a weak base that under acidic conditions becomes protonated to the ammonium ion in the following reaction:

$$NH_3 + H^+ \rightarrow NH_4^+$$

$NH_3$ freely permeates biological membranes, including those of lysosomes. The lysosome is a subcellular organelle with a pH of about 4.5–5.0; the pH of cytoplasm is ~7.0. What is the effect on the pH of the fluid content of lysosomes when cells are exposed to ammonia? *Note:* Protonated ammonia does not diffuse freely across membranes.

10. Consider the binding reaction L + R → LR, where L is a ligand and R is its receptor. When $1 \times 10^{-3}$ M L is added to a solution containing $5 \times 10^{-2}$ M R, 90 percent of the L binds to form LR. What is the $K_{eq}$ of this reaction? How will the $K_{eq}$ be affected by the addition of a protein that facilitates (catalyzes) this binding reaction? What is the dissociation equilibrium constant $K_d$?

11. What is the ionization state of phosphoric acid in the cytoplasm? Why is phosphoric acid such a physiologically important compound?

12. The $\Delta G^{\circ\prime}$ for the reaction X + Y → XY is −1000 cal/mol. What is the $\Delta G$ at 25 °C (298 Kelvin) starting with 0.01 M each X, Y, and XY? Suggest two ways one could make this reaction energetically favorable.

13. According to health experts, saturated fatty acids, which come from animal fats, are a major factor contributing to coronary heart disease. What distinguishes a saturated fatty acid from an unsaturated fatty acid, and to what does the term *saturated* refer? Recently, trans unsaturated fatty acids, or trans fats, which raise total cholesterol levels in the body, have also been implicated in heart disease. How does the cis stereoisomer differ from the trans configuration, and what effect does the cis configuration have on the structure of the fatty acid chain?

14. Chemical modifications to amino acids contribute to the diversity and function of proteins. For instance, γ-carboxylation of specific amino acids is required to make some proteins biologically active. What particular amino acid undergoes this modification, and what is the biological relevance? Warfarin, a derivative of coumarin, which is present in many plants, inhibits γ-carboxylation of this amino acid and was used in the past as a rat poison. At present, it is also used clinically in humans. What patients might be prescribed warfarin and why?

## Analyze the Data

1. During much of the "Age of Enlightenment" in eighteenth-century Europe, scientists toiled under the belief that living things and the inanimate world were fundamentally distinct forms of matter. Then in 1828, Friedrich Wöhler showed that he could synthesize urea, a well-known waste product of animals, from the minerals silver isocyanate and ammonium chloride. "I can make urea without kidneys!" he is said to have remarked. Of Wöhler's discovery the preeminent chemist Justus von Liebig wrote in 1837 that the "production of urea without the assistance of vital functions . . . must be considered one of the discoveries with which a new era in science has commenced." Slightly more than 100 years later, Stanley Miller discharged sparks into a mixture of $H_2O$, $CH_4$, $NH_3$, and $H_2$ in an effort to simulate the chemical conditions of an ancient reducing earth atmosphere (the sparks mimicked lightning striking a primordial sea or "soup") and identified many biomolecules in the resulting mixture, including amino acids and carbohydrates. What do these experiments suggest about the nature of biomolecules and the relationship between organic (living) and inorganic (nonliving) matter? What do they suggest about the evolution of life? What do they indicate about the value of chemistry in understanding living things?

2. The graph below illustrates the effect that the addition of a strong base such as sodium hydroxide has on the pH of an aqueous 0.1 M solution of an amino acid. Assume that prior to the addition of any $OH^-$, the entire dissolved amino acid

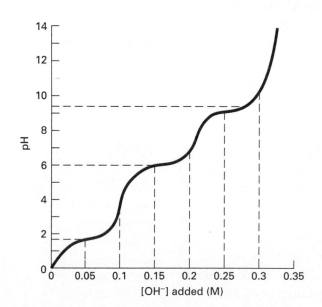

sample is in its fully protonated form. The addition of OH⁻ causes the expected steep increase in solution pH until, between roughly 0.03–0.07 M NaOH, the solution pH remains almost constant at a pH of approximately 1.8. What causes the resistance to change pH in this range? What are solutions that resist changes in pH called? What organic chemical group on the amino acid is most likely responsible for this phenomenon at pH 1.8? Additional base causes the pH to again increase rapidly until the base concentration reaches 0.15 M and 0.25 M, at which points the solution pH hovers around values of 6 and 9.3, respectively. What is the significance of these pH values? Which amino acid do you suspect is being titrated?

## References

Alberty, R. A., and R. J. Silbey. 2005. *Physical Chemistry,* 4th ed. Wiley.

Atkins, P., and J. de Paula. 2005. *The Elements of Physical Chemistry,* 4th ed. W. H. Freeman and Company.

Berg, J. M., J. L. Tymoczko, and L. Stryer. 2007. *Biochemistry,* 6th ed. W. H. Freeman and Company.

Cantor, P. R., and C. R. Schimmel. 1980. *Biophysical Chemistry.* W. H. Freeman and Company.

Davenport, H. W. 1974. *ABC of Acid-Base Chemistry,* 6th ed. University of Chicago Press.

Eisenberg, D., and D. Crothers. 1979. *Physical Chemistry with Applications to the Life Sciences.* Benjamin-Cummings.

Guyton, A. C., and J. E. Hall. 2000. *Textbook of Medical Physiology,* 10th ed. Saunders.

Hill, T. J. 1977. *Free Energy Transduction in Biology.* Academic Press.

Klotz, I. M. 1978. *Energy Changes in Biochemical Reactions.* Academic Press.

Murray, R. K., et al. 1999. *Harper's Biochemistry,* 25th ed. Lange.

Nicholls, D. G., and S. J. Ferguson. 1992. *Bioenergetics 2.* Academic Press.

Oxtoby, D., H. Gillis, and N. Nachtrieb. 2003. *Principles of Modern Chemistry,* 5th ed. Saunders.

Sharon, N. 1980. Carbohydrates. *Sci. Am.* **243**(5):90–116.

Tanford, C. 1980. *The Hydrophobic Effect: Formation of Micelles and Biological Membranes,* 2d ed. Wiley.

Tinoco, I., K. Sauer, and J. Wang. 2001. *Physical Chemistry—Principles and Applications in Biological Sciences,* 4th ed. Prentice Hall.

Van Holde, K., W. Johnson, and P. Ho. 1998. *Principles of Physical Biochemistry.* Prentice Hall.

Voet, D., and J. Voet. 2004. *Biochemistry,* 3d ed. Wiley.

Wood, W. B., et al. 1981. *Biochemistry: A Problems Approach,* 2d ed. Benjamin-Cummings.

# Protein Structure and Function

Molecular model of the proteasome from the heat- and acid-loving archaeon *T. acidophilium*, represented using both solvent-accessible surfaces (*bottom*) and ribbons (*top*). Proteasomes are protein-digesting molecular machines, comprising a middle catalytic core (red, beige, and gray), where degradation takes place, and two regulatory subunit caps (yellow and black), which recognize proteins that have been tagged for destruction by the addition of ubiquitin molecules. [Ramon Andrade 3Dciencia/Science Photo Library.]

Proteins, which are polymers of amino acids, come in many sizes and shapes. Their three-dimensional diversity principally reflects variations in their lengths and amino acid sequences. In general, the linear, unbranched polymer of amino acids composing any protein will fold into only one or a few closely related three-dimensional shapes—called **conformations**. The conformation of a protein together with the distinctive chemical properties of its amino acid side chains determines its function. Because of their many different shapes and chemical properties, proteins can perform a dazzling array of distinct functions inside and outside cells that either are essential for life or provide selective evolutionary advantage to the cell or organism that contains them. It is, therefore, not surprising that characterizing the structures and activities of proteins is a fundamental prerequisite for

understanding how cells work. Much of this textbook is devoted to examining how proteins act together to allow cells to live and function properly.

Although their structures are diverse, most individual proteins can be grouped into one of a few broad functional classes. *Structural proteins*, for example, determine the shapes of cells and their extracellular environments and serve as guide wires or rails to direct the intracellular movement of molecules and organelles. They usually are formed by the assembly of multiple protein subunits into very large, long structures. *Scaffold proteins* bring other proteins together into ordered arrays to perform specific functions more efficiently than if those proteins were not assembled together. *Enzymes* are proteins that catalyze chemical reactions. *Membrane transport proteins* permit the flow of ions

## OUTLINE

and molecules across cellular membranes. *Regulatory proteins* act as signals, sensors, and switches to control the activities of cells by altering the functions of other proteins and genes. Regulatory proteins include *signaling proteins*, such as hormones and cell-surface receptors that transmit extracellular signals to the cell interior. *Motor proteins* are responsible for moving other proteins, organelles, cells—even whole organisms. Any one protein can be a member of more than one protein class, as is the case with some cell-surface signaling receptors that are both enzymes and regulator proteins because they transmit signals from outside to inside cells by catalyzing chemical reactions. To accomplish efficiently their diverse missions, some proteins assemble into large complexes, often called *molecular machines*.

How do proteins perform so many diverse functions? They do this by exploiting a few simple activities. Most fundamentally, proteins *bind*—to one another, to other macromolecules such as DNA, and to small molecules and ions. In many cases such binding induces a conformational change in the protein and thus influences its activity. Binding is based on molecular complementarity between a protein and its binding partner, as described in Chapter 2. A second key activity is enzymatic *catalysis*. Appropriate folding of a protein will place some amino acid side chains and carboxyl and amino groups of its backbone into positions that permit the catalysis of covalent bond rearrangements. A third activity involves *folding* into a channel or pore within a membrane through which molecules and ions flow. Although these are especially crucial protein activities, they are not the only ones. For example, fish that live in frigid waters—the Antarctic borchs and Arctic cods—have antifreeze proteins in their circulatory systems to prevent water crystallization.

A complete understanding of how proteins permit cells to live and thrive requires the identification and characterization of all the proteins used by a cell. In a sense, molecular cell biologists want to compile a complete protein "parts list" and construct a "user's manual" that describes how these proteins work. Compiling a comprehensive inventory of proteins has become feasible in recent years with the sequencing of entire **genomes**—complete sets of genes—of more and more organisms. From a computer analysis of genome sequences, researchers can deduce the amino acid sequences and approximate number of the encoded proteins (see Chapter 5). The term **proteome** was coined to refer to the entire protein complement of an organism. The human genome contains some 20,000–23,000 genes that encode proteins. However, variations in mRNA production, such as alternative splicing (see Chapter 8), and more than 100 types of protein modifications may generate hundreds of thousands of distinct human proteins. By comparing the sequences and structures of proteins of unknown function to those of known function, scientists can often deduce much about what these proteins do. In the past, characterization of protein function by genetic, biochemical, or physiological methods often preceded the identification of particular proteins. In the modern genomic and proteomic era, a protein is usually identified prior to determining its function.

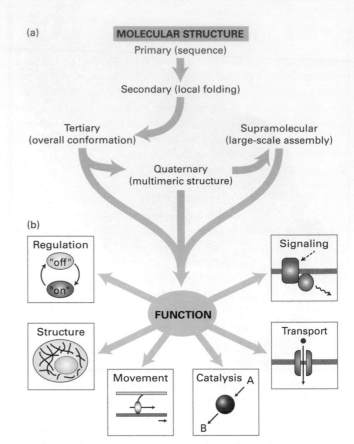

**FIGURE 3-1 Overview of protein structure and function.**
(a) Proteins have a hierarchical structure. A polypeptide's linear sequence of amino acids linked by peptide bonds (primary structure) folds into local helices or sheets (secondary structure) that pack into a complex three-dimensional shape (tertiary structure). Some individual polypeptides associate into multichain complexes (quaternary structure), which in some cases can be very large, consisting of tens to hundreds of subunits (supramolecular complexes). (b) Proteins perform numerous functions, including organizing in three-dimensional space the genome, organelles, the cytoplasm, protein complexes, and membranes (structure); controlling protein activity (regulation); monitoring the environment and transmitting information (signaling); moving small molecules and ions across membranes (transport); catalyzing chemical reactions (via enzymes); and generating force for movement (via motor proteins). These functions and others arise from specific binding interactions and conformational changes in the structure of a properly folded protein.

In this chapter, we begin our study of how the structure of a protein gives rise to its function, a theme that recurs throughout this book (Figure 3-1). The first section examines how linear chains of amino acid building blocks are arranged in a three-dimensional structural hierarchy. The next section discusses how proteins fold into these structures. We then turn to protein function, focusing on enzymes, the special class of proteins that catalyze chemical reactions. Various mechanisms that cells use to control the activities and life spans of proteins are covered next. The chapter concludes with a discussion of commonly used techniques for identifying, isolating, and characterizing proteins, including a discussion of the burgeoning field of proteomics.

## 3.1 Hierarchical Structure of Proteins

A protein chain folds into a distinct three-dimensional shape that is stabilized primarily by noncovalent interactions between regions in the linear sequence of amino acids. A key concept in understanding how proteins work is that *function is derived from three-dimensional structure, and three-dimensional structure is determined by both a protein's amino acid sequence and intramolecular noncovalent interactions.* Principles relating biological structure and function initially were formulated by the biologists Johann von Goethe (1749–1832), Ernst Haeckel (1834–1919), and D'Arcy Thompson (1860–1948), whose work has been widely influential in biology and beyond. Indeed, their ideas greatly influenced the school of "organic" architecture pioneered in the early twentieth century that is epitomized by the dicta "form follows function" (Louis Sullivan) and "form is function" (Frank Lloyd Wright). Here we consider the architecture of proteins at four levels of organization: primary, secondary, tertiary, and quaternary (Figure 3-2).

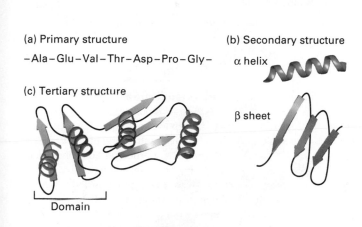

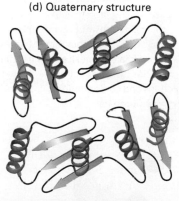

**FIGURE 3-2 Four levels of protein hierarchy.** (a) The linear sequence of amino acids linked together by peptide bonds is the primary structure. (b) Folding of the polypeptide chain into local α helices or β sheets represents secondary structure. (c) Secondary structural elements together with various loops and turns in a single polypeptide chain pack into a larger independently stable structure, which may include distinct domains; this is tertiary structure. (d) Some proteins consist of more than one polypeptide associated together in a quaternary structure.

## The Primary Structure of a Protein Is Its Linear Arrangement of Amino Acids

As discussed in Chapter 2, proteins are polymers constructed out of 20 different types of amino acids. Individual amino acids are linked together in linear, unbranched chains by covalent amide bonds, called **peptide bonds**. Peptide bond formation between the amino group of one amino acid and the carboxyl group of another results in the net release of a water molecule and thus is a form of dehydration reaction (Figure 3-3a). The repeated amide N, α carbon ($C_\alpha$), carbonyl C, and oxygen atoms of each amino acid residue form

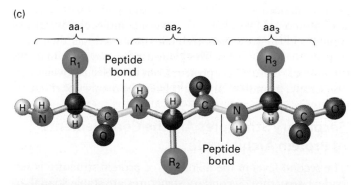

**FIGURE 3-3 Structure of a polypeptide.** (a) Individual amino acids are linked together by peptide bonds, which form via reactions that result in a loss of water (dehydration). $R_1$, $R_2$, etc., represent the side chains ("R groups") of amino acids. (b) Linear polymers of peptide-bond-linked amino acids are called *polypeptides*, which have a free amino end (N-terminus) and a free carboxyl end (C-terminus). (c) A ball-and-stick model shows peptide bonds (yellow) linking the amino nitrogen atom (blue) of one amino acid (aa) with the carbonyl carbon atom (gray) of an adjacent one in the chain. The R groups (green) extend from the α carbon atoms (black) of the amino acids. These side chains largely determine the distinct properties of individual proteins.

the backbone of a protein molecule from which the various side-chain groups project (Figure 3-3b, c). As a consequence of the peptide linkage, the backbone exhibits directionality, usually referred to as an N-to-C orientation, because all the amino groups are located on the same side of the C$_\alpha$ atoms. Thus one end of a protein has a free (unlinked) amino group (the *N-terminus*), and the other end has a free carboxyl group (the *C-terminus*). The sequence of a protein chain is conventionally written with its N-terminal amino acid on the left and its C-terminal amino acid on the right, and the amino acids are numbered sequentially starting from the amino terminus.

The **primary structure** of a protein is simply the linear covalent arrangement, or sequence, of the amino acid residues that compose it. The first primary structure of a protein determined was that of insulin in the early 1950s and today the number of known sequences exceeds 10 million and is growing daily. Many terms are used to denote the chains formed by the polymerization of amino acids. A short chain of amino acids linked by peptide bonds and having a defined sequence is called an **oligopeptide**, or just **peptide**; longer chains are referred to as **polypeptides**. Peptides generally contain fewer than 20–30 amino acid residues, whereas polypeptides are often 200–500 residues long. The longest protein described to date is the muscle protein titin with >35,000 residues. We generally reserve the term **protein** for a polypeptide (or complex of polypeptides) that has a well-defined three-dimensional structure.

The size of a protein or a polypeptide is expressed either as its mass in **daltons** (a dalton is 1 atomic mass unit) or as its molecular weight (MW), which is a dimensionless number equal to the mass in daltons. For example, a 10,000-MW protein has a mass of 10,000 daltons (Da), or 10 kilodaltons (kDa). Later in this chapter, we will consider different methods for measuring the sizes and other physical characteristics of proteins. The proteins encoded by the yeast genome have an average molecular weight of 52,728 and contain, on average, 466 amino acid residues. The average molecular weight of amino acids in proteins is 113, taking into account their average relative abundances. This value can be used to estimate the number of residues in a protein from its molecular weight or, conversely, its molecular weight from the number of residues.

## Secondary Structures Are the Core Elements of Protein Architecture

The second level in the hierarchy of protein structure is **secondary structure**. Secondary structures are stable spatial arrangements of segments of a polypeptide chain held together by hydrogen bonds between backbone amide and carbonyl groups and often involving repeating structural patterns. A single polypeptide may contain multiple types of secondary structure in various portions of the chain, depending on its sequence. The principal secondary structures are the **alpha (α) helix**, the **beta (β) sheet**, and a short U-shaped **beta (β) turn**. Parts of the polypeptide that don't form these structures but nevertheless have a well-defined, stable shape are said to have an *irregular* structure. The term *random coil* applies to

highly flexible parts of a polypeptide chain that have no fixed three-dimensional structure. In an average protein, 60 percent of the polypeptide chain exists as α helices and β sheets; the remainder of the molecule is in irregular structures, coils and turns. Thus α helices and β sheets are the major internal supportive elements in most proteins. In this section, we explore the shapes of secondary structures and the forces that favor their formation. In later sections, we examine how arrays of secondary structure fold together into larger, more complex arrangements called tertiary structure.

**The α Helix** In a polypeptide segment folded into an α helix, the backbone forms a spiral structure in which the carbonyl oxygen atom of each peptide bond is hydrogen-bonded to the amide hydrogen atom of the amino acid four residues farther along the chain in the direction of the C-terminus (Figure 3-4). Within an α helix, all the backbone amino and carboxyl groups are hydrogen-bonded to one another except at the very beginning and end of the helix. This periodic

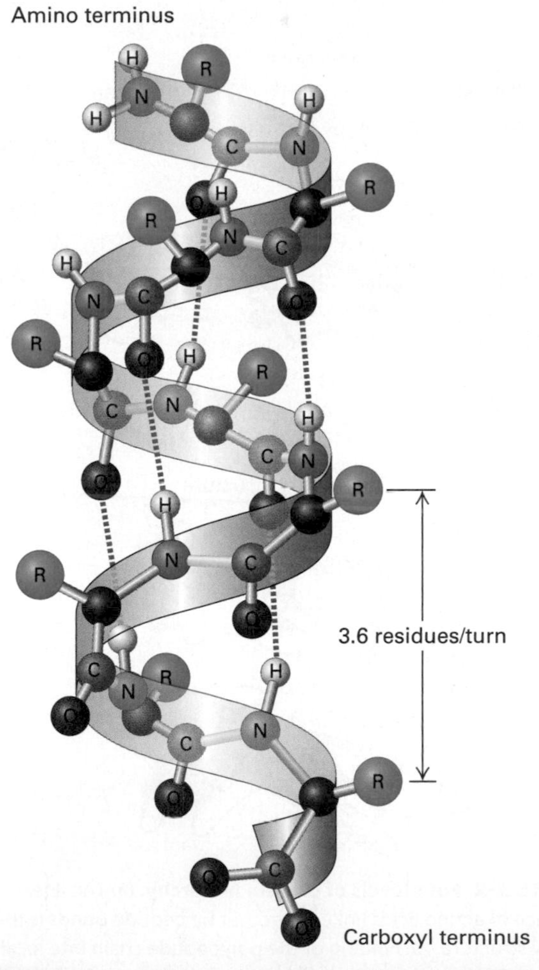

Amino terminus

3.6 residues/turn

Carboxyl terminus

**FIGURE 3-4 The α helix, a common secondary structure in proteins.** The polypeptide backbone (seen as a ribbon) is folded into a spiral that is held in place by hydrogen bonds between backbone oxygen and hydrogen atoms. Only hydrogens involved in bonding are shown. The outer surface of the helix is covered by the side-chain R groups (green).

arrangement of bonds confers an amino-to-carboxy-termi-nal directionality on the helix because all the hydrogen bond acceptors (i.e., the carbonyl groups) have the same orientation (pointing in the downward direction in Figure 3-4), resulting in a structure in which there is a complete turn of the spiral every 3.6 residues. An α helix 36 amino acids long has 10 turns of the helix and is 5.4 nm long (0.54 nm/turn).

The stable arrangement of hydrogen-bonded amino acids in the α helix holds the backbone in a straight, rodlike cylinder from which the side chains point outward. The relative hydrophobic or hydrophilic quality of a particular helix within a protein is determined entirely by the characteristics of the side chains. In water-soluble proteins, hydrophilic helices tend to be found on the outside surfaces, where they can interact with the aqueous environment, whereas hydrophobic helices tend to be buried within the core of the folded protein. The amino acid proline is usually not found in α helices because the covalent bonding of its amino group with a carbon in the side chain prevents its participation in stabilizing the backbone through normal hydrogen bonding. While the classic α helix is the most intrinsically stable and most common helical form in proteins, there are variations, such as more tightly or loosely twisted helices. For example, in a specialized helix called a coiled coil (described several sections farther on), the helix is more tightly wound (3.5 residues and 0.51 nm per turn).

**The β Sheet** Another type of secondary structure, the β sheet, consists of laterally packed β strands. Each β strand is a short (5- to 8-residue), nearly fully extended polypeptide segment. Unlike in the α helix, where hydrogen bonds occur between the amino and carboxyl groups in the backbone between nearly adjacent residues, hydrogen bonds in the β sheet occur between backbone atoms in separate, but adjacent, β strands and are oriented perpendicularly to the chains of backbone atoms (Figure 3-5a). These distinct β strands may be either within a single polypeptide chain, with short or long loops between the β strand segments, or on different polypeptide chains in a protein composed of multiple polypeptides. Figure 3-5b shows how two or more β strands align into adjacent rows, forming a nearly two-dimensional β pleated sheet (or simply *pleated sheet*), in which hydrogen bonds within the plane of the sheet hold the β strands together as the side chains stick out above and below the plane. Like α helices, β strands have a directionality defined by the orientation of the peptide bond. Therefore, in a pleated sheet, adjacent β strands can be oriented in the same (parallel) or alternating opposite (antiparallel) directions with respect to each other. In Figure 3-5a, you can see that the N-to-C orientations of the chains, indicated by arrows, alternate directions between adjacent chains, signifying an antiparallel sheet. In some proteins, β sheets form the floor of a binding pocket or a hydrophobic core; in proteins embedded in membranes the β sheets curve around and form a hydrophilic central pore through which ions and small molecules may flow (see Chapter 11).

**β Turns** Composed of four residues, β turns are located on the surface of a protein, forming sharp bends that reverse the

(a) Top view

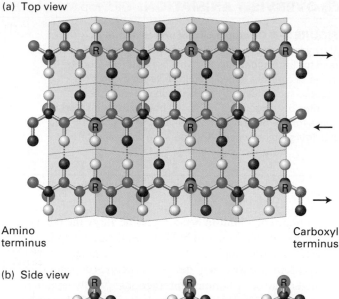

Amino terminus

Carboxyl terminus

(b) Side view

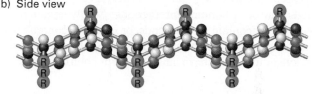

**FIGURE 3-5 The β sheet, another common secondary structure in proteins.** (a) Top view of a simple three-stranded β sheet with antiparallel β strands, as indicated by the arrows that represent the N-to-C orientations of the chains. The stabilizing hydrogen bonds between the β strands are indicated by green dashed lines. (b) Side view of a β sheet. The projection of the R groups (green) above and below the plane of the sheet is obvious in this view. The fixed bond angles in the polypeptide backbone produce a pleated contour.

direction of the polypeptide backbone, often toward the protein's interior. These short, U-shaped secondary structures are often stabilized by a hydrogen bond between their end residues (Figure 3-6). Glycine and proline are commonly present in turns. The lack of a large side chain in glycine and

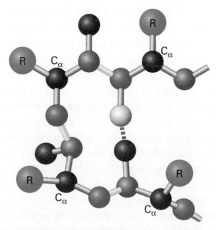

**FIGURE 3-6 Structure of a β turn.** Composed of four residues, β turns reverse the direction of a polypeptide chain (180° U-turn). The $C_\alpha$ carbons of the first and fourth residues are usually <0.7 nm apart, and those residues are often linked by a hydrogen bond. β turns facilitate the folding of long polypeptides into compact structures.

**FIGURE 3-7  Oil drop model of protein folding.**
The hydrophobic residues (blue) of a polypeptide chain tend to cluster together, somewhat like an oil drop, on the inside, or core, of a folded protein, driven away from the aqueous surroundings by the hydrophobic effect (see Chapter 2). Charged and uncharged polar side chains (red) appear on the protein's surface, where they can form stabilizing interactions with surrounding water and ions.

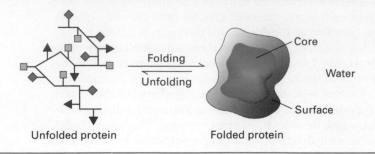

the presence of a built-in bend in proline allow the polypeptide backbone to fold into a tight U shape. β turns help large proteins to fold into highly compact structures. There are six types of well-defined turns, their detailed structures depending on the arrangement of H-bonding interactions. A polypeptide backbone also may contain longer bends, or loops. In contrast with tight β turns, which exhibit just a few well-defined conformations, longer loops can have many different conformations.

## Tertiary Structure Is the Overall Folding of a Polypeptide Chain

**Tertiary structure** refers to the overall conformation of a polypeptide chain—that is, the three-dimensional arrangement of all its amino acid residues. In contrast with secondary structures, which are stabilized only by hydrogen bonds, tertiary structure is primarily stabilized by hydrophobic interactions between nonpolar side chains, together with hydrogen bonds involving polar side chains and backbone amino and carboxyl groups. These stabilizing forces compactly hold together elements of secondary structure—α helices, β strands, turns, and coils. Because the stabilizing interactions are often weak, however, the tertiary structure of a protein is not rigidly fixed but undergoes continual, minute fluctuations, and some segments within the tertiary structure of a protein can be so very mobile they are considered to be disordered—that is, lacking well-defined, stable, three-dimensional structure. This variation in structure has important consequences for the function and regulation of proteins.

Chemical properties of amino acid side chains help define tertiary structure. **Disulfide bonds** between the side chains of cysteine residues in some proteins covalently link regions of proteins, thus restricting the proteins' flexibility and increasing the stability of their tertiary structures. Amino acids with charged hydrophilic polar side chains tend to be on the outer surfaces of proteins; by interacting with water, they help to make proteins soluble in aqueous solutions and can form noncovalent interactions with other water-soluble molecules, including other proteins. In contrast, amino acids with hydrophobic nonpolar side chains are usually sequestered away from the water-facing surfaces of a protein, in many cases forming a water-insoluble central core. This observation led to what's known as the "oil drop model" of

protein conformation because of the relatively hydrophobic, or "oily," core of a protein (Figure 3-7). Uncharged hydrophilic polar side chains are found on both the surface and inner core of proteins.

Proteins usually fall into one of three broad structural categories, based on their tertiary structure: globular proteins, fibrous proteins, and integral membrane proteins. *Globular proteins* are generally water-soluble, compactly folded structures, often but not exclusively spheroidal, that comprise a mixture of secondary structures (see the structure of myoglobin, below). *Fibrous proteins* are large, elongated, often stiff molecules. Some fibrous proteins are composed of a long polypeptide chain comprising many tandem copies of a short amino acid sequence that forms a single repeating secondary structure (see the structure of collagen, the most abundant protein in mammals, in Figure 20-24). Other fibrous proteins are composed of repeating globular protein subunits, such as the helical array of G-actin protein monomers that forms the F-actin microfilaments (see Chapter 17). Fibrous proteins, which often aggregate into large multiprotein fibers that do not readily dissolve in water, usually play a structural role or participate in cellular movements. *Integral membrane proteins* are embedded within the phospholipid bilayer of the membranes that enclose cells and organelles (see Chapter 10). The three broad categories of proteins noted here are not mutually exclusive—some proteins are made up of combinations of two or even all three categories.

## Different Ways of Depicting the Conformation of Proteins Convey Different Types of Information

The simplest way to represent three-dimensional protein structure is to trace the course of the backbone atoms, sometimes only the $C_\alpha$ atoms, with a solid line (called a $C_\alpha$ trace, Figure 3-8a); the most complex model shows every atom (Figure 3-8b). The former shows the overall fold of the polypeptide chain without consideration of the amino acid side chains; the latter, a ball-and-stick model (with balls representing atoms and sticks representing bonds), details the interactions between side-chain atoms, including those that stabilize the protein's conformation and interact with other

(a) C$_\alpha$ backbone trace

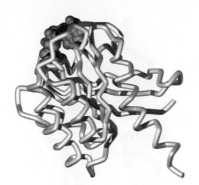

(b) Ball and stick

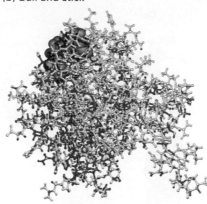

(c) Ribbons

(d) Solvent-accessible surface

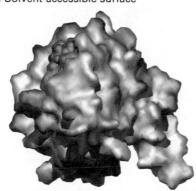

**FIGURE 3-8  Four ways to visualize protein structure.** Shown here are four distinct methods for representing the structure of a protein called ras, a monomeric (single polypeptide chain) protein that binds to guanosine diphosphate (GDP, depicted in blue). (a) The C$_\alpha$ backbone trace demonstrates how the polypeptide is tightly packed into a small volume. (b) A ball-and-stick representation reveals the location of all atoms. (c) A ribbon representation emphasizes how β strands (light blue) and α helices (red) are organized in the protein. Note the turns and loops connecting pairs of helices and strands. (d) A model of the water-accessible surface reveals the numerous lumps, bumps, and crevices on the protein surface. Regions of positive charge are shaded purple; regions of negative charge are shaded red.

molecules as well as the atoms of the backbone. Even though both views are useful, the elements of secondary structure are not always easily discerned in them. Another type of representation uses common shorthand symbols for depicting secondary structure—for example, coiled ribbons or solid cylinders for α helices, flat ribbons or arrows for β strands, and flexible thin strands for β turns, coils, and loops (Figure 3-8c). In a variation of the basic ribbon diagram, ball-and-stick or space-filling models of all or only a subset of side chains can be attached to the backbone ribbon. In this way, side chains that are of interest can be visualized in the context of the secondary structure that is especially clearly represented by the ribbons.

However, none of these three ways of representing protein structure conveys much information about the protein surface, which is of interest because it is where other molecules usually bind to a protein. Computer analysis can identify the surface atoms that are in contact with the watery environment. On this water-accessible surface, regions having a common chemical character, such as, hydrophobicity or hydrophilicity, and charge characteristics, such as positive (basic) or negative (acidic) side chains, can be indicated by coloring (Figure 3-8d). Such models reveal the topography of the protein surface and the distribution of charge, both important features of binding sites, as well as clefts in the surface where small molecules bind. This view represents a protein as it is "seen" by another molecule.

## Structural Motifs Are Regular Combinations of Secondary Structures

A particular combination of two or more secondary structures that form a distinct three-dimensional structure is called a **structural motif** when it appears in multiple proteins. A structural motif is often, but not always, associated with a specific function. Any particular structural motif will frequently perform a common function in different proteins, such as binding to a particular ion or small molecule, for example, calcium or ATP.

One common structural motif is the α helix–based **coiled coil**, or heptad repeat. Many proteins, including fibrous proteins and DNA-regulating proteins called transcription factors (see Chapter 7), assemble into dimers or trimers by using a coiled-coil motif, in which α helices from two, three, or even four separate polypeptide chains coil about one another—resulting in a coil of coils; hence the name (Figure 3-9a). The individual helices bind tightly to one another because each helix has a strip of aliphatic (hydrophobic, but not aromatic) side chains (leucine valine, etc.) running along one side of the helix that interacts with a similar strip in the adjacent helix, thus sequestering the hydrophobic groups away from water and stabilizing the assembly of multiple independent helices. These hydrophobic strips are generated along only one side of the helix because the primary structure of each helix is composed of repeating seven-amino-acid

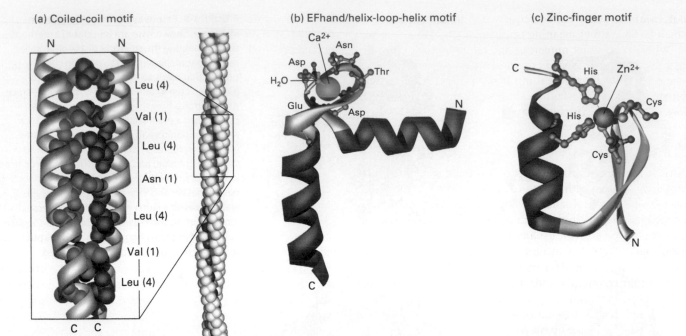

**(a) Coiled-coil motif**

N    N

Leu (4)

Val (1)

Leu (4)

Asn (1)

Leu (4)

Val (1)

Leu (4)

C    C

**(b) EFhand/helix-loop-helix motif**

Ca²⁺   Asn

Asp

H₂O   Thr

Glu

Asp

N

C

**(c) Zinc-finger motif**

C

His   Zn²⁺

Cys

His

Cys

N

**FIGURE 3-9 Motifs of protein secondary structure.** (a) The parallel two-stranded coiled-coil motif (*left*) is characterized by two α helices wound around each other. Helix packing is stabilized by interactions between hydrophobic side chains (red and blue) present at regular intervals along each strand and found along the seam of the intertwined helices. Each α helix exhibits a characteristic heptad repeat sequence with a hydrophobic residue often, but not always, at positions 1 and 4, as indicated. The coiled-coil nature of this structural motif is more apparent in long coiled coils containing many such motifs (*right*). (b) An EF hand, a type of helix-loop-helix motif, consists of two helices connected by a short loop in a specific conformation. This structural motif is common to many proteins, including many calcium-binding and DNA-binding regulatory proteins. In calcium-binding proteins such as calmodulin, oxygen atoms from five residues in the acidic glutamate- and aspartate-rich loop and one water molecule form ionic bonds with a Ca²⁺ ion. (c) The zinc-finger motif is present in many DNA-binding proteins that help regulate transcription. A Zn²⁺ ion is held between a pair of β strands (blue) and a single α helix (red) by a pair of cysteine residues and a pair of histidine residues. The two invariant cysteine residues are usually at positions 3 and 6, and the two invariant histidine residues are at positions 20 and 24 in this 25-residue motif. [See A. Lewit-Bentley and S. Rety, 2000, *Curr. Opin. Struc. Biol.* **10**:637–643; S. A. Wolfe, L. Nekludova, and C. O. Pabo, 2000, *Ann. Rev. Biophys. Biomol. Struc.* **29**:183–212.]

units, called heptads, in which the side chains of the first and fourth residues are aliphatic and the other side chains are often hydrophilic (Figure 3-9a). Because hydrophilic side chains extend from one side of the helix and hydrophobic side chains extend from the opposite side, the overall helical structure is **amphipathic**. Because leucine frequently appears in the fourth positions and the hydrophobic side chains merge together like the teeth of a zipper, these structural motifs are also called **leucine zippers**.

Many other structural motifs contain α helices. A common calcium-binding motif called the **EF hand** contains two short helices connected by a loop (Figure 3-9b). This structural motif, one of several **helix-turn-helix** structural motifs, is found in more than 100 proteins and is used for sensing the calcium levels in cells. The binding of a Ca²⁺ ion to oxygen atoms in conserved residues in the loop depends on the concentration of Ca²⁺ and often induces a conformational change in the protein, altering its activity. Thus calcium concentrations can directly control proteins' structures and functions. Somewhat different helix-turn-helix and **basic helix-loop-helix** (bHLH) structural motifs are used for protein binding to DNA and consequently the regulation of

gene activity (see Chapter 7). Yet another structural motif commonly found in proteins that bind RNA or DNA is the **zinc finger**, which contains three secondary structures—an α helix and two β strands with an antiparallel orientation—that form a fingerlike bundle held together by a zinc ion (Figure 3-9c).

The relationship between the primary structure of a polypeptide chain and the structural motifs into which it folds is not always straightforward. The amino acid sequences responsible for any given structural motif may be very similar to one another. In other words, a common *sequence motif* can result in a common structural motif. This is the case for the heptad repeats that form coiled coils. However, it is possible for seemingly unrelated amino acid sequences to fold into a common structural motif, so it is not always possible to predict which amino acids sequences will fold into a given structural motif. Conversely, it is possible that a commonly occurring sequence motif does not fold into a well-defined structural motif. Sometimes short sequence motifs that have an unusual abundance of a particular amino acid, for example, proline or aspartate or glutamate, are called "domains"; however, these and other short contiguous segments are more appropriately called sequence motifs than domains, which has a distinct meaning that is defined below.

We will encounter numerous additional motifs in later discussions of other proteins in this and other chapters. The presence of the same structural motif in different proteins with similar functions clearly indicates that these useful combinations of secondary structures have been conserved in evolution.

## Domains Are Modules of Tertiary Structure

Distinct regions of protein structure often are referred to as **domains**. There are three main classes of protein domains: functional, structural, and topological. A *functional domain* is a region of a protein that exhibits a particular activity characteristic of the protein, usually even when isolated from the rest of the protein. For instance, a particular region of a protein may be responsible for its catalytic activity (e.g., a kinase domain that covalently adds a phosphate group to another molecule) or binding ability (e.g., a DNA-binding domain or a membrane-binding domain). Functional domains are often identified experimentally by whittling down a protein to its smallest active fragment with the aid of **proteases**, enzymes that cleave one or more peptide bonds in a target polypeptide. Alternatively, the DNA encoding a protein can be modified so that when the modified DNA is used to generate a protein, only a particular region, or domain, of the full-length protein is made. Thus it is possible to determine if specific parts of a protein are responsible for particular activities exhibited by the protein. Indeed, functional domains are often also associated with corresponding structural domains.

A *structural domain* is a region ~40 or more amino acids in length, arranged in a single, stable, and distinct structure often comprising one or more secondary structures. Structural domains often can fold into their characteristic structures independently of the rest of the protein in which they are embedded. As a consequence, distinct structural domains can be linked together—sometimes by short or long spacers—to form a large, multidomain protein. Each of the polypeptide chains in the trimeric flu virus hemagglutinin, for example, contains a globular domain and a fibrous domain (Figure 3-10a). Like structural motifs (composed of secondary structures), structural domains are incorporated as modules into different proteins. The modular approach to protein architecture is particularly easy to recognize in large proteins, which tend to be mosaics of different domains that confer distinct activities and thus can perform different functions simultaneously. As many as 75 percent of the proteins in eukaryotes have multiple structural domains. Structural domains frequently are also functional domains in that they can have an activity independent of the rest of the protein.

The epidermal growth factor (EGF) domain is a structural domain present in several proteins (Figure 3-11). EGF is a small, soluble peptide hormone that binds to cells in the embryo and in skin and connective tissue in adults, causing them

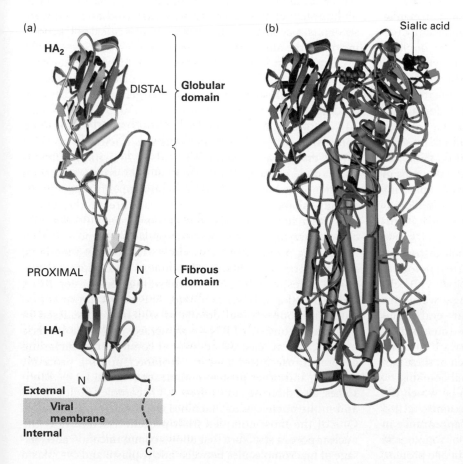

(a)

HA₂

DISTAL — Globular domain

PROXIMAL      N — Fibrous domain

HA₁

N

**External**
**Viral membrane**
**Internal**

C

(b)

Sialic acid

**FIGURE 3-10 Tertiary and quaternary levels of structure.** The protein pictured here, hemagglutinin (HA), is found on the surface of the influenza virus. This long, multimeric molecule has three identical subunits, each composed of two polypeptide chains, HA₁ and HA₂. (a) Tertiary structure of each HA subunit comprises the folding of its helices and strands into a compact structure that is 13.5 nm long and divided into two domains. The membrane-distal domain (silver) is folded into a globular conformation. The membrane-proximal domain (gold) has a fibrous, stemlike conformation owing to the alignment of two long α helices (cylinders) of HA₂ with β strands in HA₁. Short turns and longer loops, often at the surface of the molecule, connect the helices and strands in each chain. (b) Quaternary structure of HA is stabilized by lateral interactions between the long helices (cylinders) in the fibrous domains of the three subunits (gold, blue, and green), forming a triple-stranded coiled-coil stalk. Each of the distal globular domains in HA binds sialic acid (red) on the surface of target cells. Like many membrane proteins, HA contains several covalently linked carbohydrate chains (not shown).

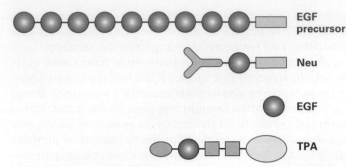

**FIGURE 3-11 Modular nature of protein domains.** Epidermal growth factor (EGF) is generated by proteolytic cleavage of a precursor protein containing multiple EGF domains (green) and a membrane-spanning domain (blue). The EGF domain is also present in the Neu protein and in tissue plasminogen activator (TPA). These proteins also contain other widely distributed domains, indicated by shape and color. [Adapted from I. D. Campbell and P. Bork, 1993, *Curr. Opin. Struc. Biol.* **3**:385.]

to divide. It is generated by proteolytic cleavage (breaking of a peptide bond) between repeated EGF domains in the EGF precursor protein, which is anchored in the cell membrane by a membrane-spanning domain. EGF domains with sequences similar to, but not identical to, those in the EGF peptide hormone are present in other proteins and can be liberated by proteolysis. These proteins include tissue plasminogen activator (TPA), a protease that is used to dissolve blood clots in heart attack victims; Neu protein, which takes part in embryonic differentiation; and Notch protein, a receptor protein in the plasma membrane that functions in developmentally important signaling (see Chapter 16). Besides the EGF domain, these proteins have other domains in common with other proteins. For example, TPA possesses a trypsin domain, a functional domain in some proteases. It is estimated that there are about 1000 different types of structural domains in all proteins. Some of these are not very common, whereas others are found in many different proteins. Indeed, by some estimates only nine major types of structural domains account for as much as a third of all the structural domains in all proteins. Structural domains can be recognized in proteins whose structures have been determined by x-ray crystallography or nuclear magnetic resonance (NMR) analysis or in images captured by electron microscopy.

Regions of proteins that are defined by their distinctive spatial relationships to the rest of the protein are *topological domains*. For example, some proteins associated with cell-surface membranes can have a part extending inward into the cytoplasm (cytoplasmic domain), a part embedded within the phospholipid bilayer membrane (membrane-spanning domain), and a part extending outward into the extracellular space (extracellular domain). Each of these can comprise one or more structural and functional domains.

In Chapter 6 we consider the mechanism by which the gene segments that correspond to domains became shuffled in the course of evolution, resulting in their appearance in many proteins. Once a functional, structural, or topological domain has been identified and characterized in one protein,

it is possible to use that information to search for similar domains in other proteins and to suggest potentially similar functions for those domains in those proteins.

## Multiple Polypeptides Assemble into Quaternary Structures and Supramolecular Complexes

**Multimeric** proteins consist of two or more polypeptide chains, which in this context are referred to as subunits. A fourth level of structural organization, **quaternary structure**, describes the number (stoichiometry) and relative positions of the subunits in multimeric proteins. Flu virus hemagglutinin, for example, is a trimer of three identical subunits (homotrimer) held together by noncovalent bonds (see Figure 3-10b). Other multimeric proteins can be composed of various numbers of identical (homomeric) or different (heteromeric) subunits. Hemoglobin, the oxygen-carrying molecule in blood, is an example of a heteromeric multimeric protein. It has two copies each of two different polypeptide chains (discussed below). Often, the individual monomer subunits of a multimeric protein cannot function normally unless they are assembled into the multimeric protein. In some cases, assembly into a multimeric protein permits proteins that act sequentially in a pathway to increase their efficiency of operation owing to their juxtaposition in space, a phenomenon referred to as "metabolic coupling." Classic examples of this coupling are fatty acid synthases, the enzymes in fungi that synthesize fatty acids, and the polyketide synthases, the large multiprotein complexes in bacteria that synthesize a diverse set of pharmacologically relevant molecules called polyketides, including the antibiotic erythromycin.

The highest level in the hierarchy of protein structure is the association of proteins into supramolecular complexes. Typically, such structures are very large, in some cases exceeding 1 MDa in mass, approaching 30–300 nm in size, and containing tens to hundreds of polypeptide chains and sometimes other biopolymers such as nucleic acids. The capsid that encases the nucleic acids of the viral genome is an example of a supramolecular complex with a structural function. The bundles of cytoskeletal filaments that support and give shape to the plasma membrane are another example. Other supramolecular complexes act as molecular machines, carrying out the most complex cellular processes by integrating multiple proteins, each with distinct functions, into one large assembly. For example, a transcriptional machine is responsible for synthesizing messenger RNA (mRNA) using a DNA template. This transcriptional machine, the operational details of which are discussed in Chapter 4, consists of RNA polymerase, itself a multimeric protein, and at least 50 additional components, including general transcription factors, promoter-binding proteins, helicase, and other protein complexes (Figure 3-12). Ribosomes, also discussed in Chapter 4, are complex multiprotein and multi-nucleic acid machines that synthesize proteins. One of the most complex multiprotein assemblies is the nuclear pore, a structure that allows communication and passage of macromolecules between nucleoplasm and cytoplasm

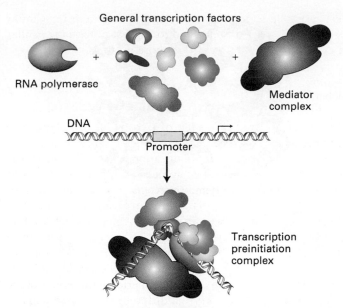

General transcription factors

RNA polymerase

+

+

Mediator complex

DNA

Promoter

Transcription preinitiation complex

**FIGURE 3-12 A macromolecular machine: the transcription-initiation complex.** The core RNA polymerase, general transcription factors, a mediator complex containing about 20 subunits, and other protein complexes not depicted here assemble at a promoter in DNA. The polymerase carries out transcription of DNA; the associated proteins are required for initial binding of polymerase to a specific promoter. The multiple components function together as a machine.

(see Chapter 14). It is composed of multiple copies of about 30 distinct proteins and forms an assembly with an estimated mass of around 50 megadaltons. The fatty acid synthases and polyketide synthases referred to above are also supramolecular machines.

## Members of Protein Families Have a Common Evolutionary Ancestor

Studies of myoglobin and hemoglobin, the oxygen-carrying proteins in muscle and red blood cells, respectively, provided early evidence that a protein's function derives from its three-dimensional structure, which in turn is specified by amino acid sequence. X-ray crystallographic analysis showed that the three-dimensional structures of myoglobin (a monomer) and the α and β subunits of hemoglobin (a $\alpha_2\beta_2$ tetramer) are remarkably similar. Sequencing of myoglobin and the hemoglobin subunits revealed that many identical or chemically similar residues are found in equivalent positions throughout the primary structures of both proteins. A mutation in the gene encoding the β chain that results in the substitution of a valine for a glutamic acid disturbs the folding and function of hemoglobin and causes sickle-cell anemia.

Similar comparisons between other proteins conclusively confirmed the relation between the amino acid sequence, three-dimensional structure, and function of proteins. Use of sequence comparisons to deduce protein function has expanded substantially in recent years as the genomes of more and more organisms have been sequenced. While this

comparative approach is very powerful, caution must always be exercised when attributing to one protein, or a part of a protein, a similar function or structure to another based only on amino acid sequence similarities. There are examples in which proteins with similar overall structures display different functions and cases in which functionally unrelated proteins with dissimilar amino acid sequences nevertheless have very similar folded tertiary structures, as will be explained below. Nevertheless, in many cases such comparisons provide important insights into protein structure and function.

The molecular revolution in biology during the last decades of the twentieth century created a new scheme of biological classification based on similarities and differences in the amino acid sequences of proteins. Proteins that have a common ancestor are referred to as **homologs**. The main evidence for **homology** among proteins, and hence for their common ancestry, is similarity in their sequences, which is often also reflected in similar structures. We can describe homologous proteins as belonging to a "family" and can trace their lineage from comparisons of their sequences. Generally, more closely related proteins will exhibit greater sequence similarity than more distantly related proteins because, over evolutionary time, mutations accumulate in the genes encoding these proteins. The folded three-dimensional structures of homologous proteins can be similar even if parts of their primary structure show little evidence of sequence homology. Initially, proteins with relatively high sequence similarities (>50 percent exact matches, or "identities") and related functions or structures were defined as an evolutionarily related *family*, while a *superfamily* encompassed two or more families in which the interfamily sequences matched less well (~30–40 percent identities) than within one family. It is generally thought that proteins with 30 percent sequence identity are likely to have similar three-dimensional structures; however, such high sequence identity is not required for proteins to share similar structures. Recently, revised definitions of *family* and *superfamily* have been proposed, in which a family comprises proteins with a clear evolutionary relationship (>30 percent identity or additional structural and functional information showing common descent but <30 percent identity), while a superfamily comprises proteins with only a probable common evolutionary origin—for example, lower percent sequence identities but one or more common motifs or domains.

The kinship among homologous proteins is most easily visualized by a tree diagram based on sequence analyses. For example, the amino acid sequences of globins—the proteins hemoglobin and myoglobin and their relatives from bacteria, plants, and animals—suggest that they evolved from an ancestral monomeric, oxygen-binding protein (Figure 3-13). With the passage of time, the gene for this ancestral protein slowly changed, initially diverging into lineages leading to animal and plant globins. Subsequent changes gave rise to myoglobin, the monomeric oxygen-storing protein in muscle, and to the α and β subunits of the tetrameric hemoglobin molecule ($\alpha_2\beta_2$) of the vertebrate circulatory system.

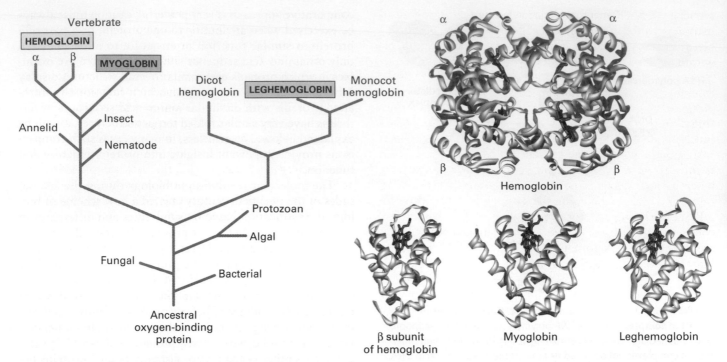

**FIGURE 3-13 Evolution of the globin protein family.** *Left:* A primitive monomeric oxygen-binding globin is thought to be the ancestor of modern-day blood hemoglobins, muscle myoglobins, and plant leghemoglobins. Sequence comparisons have revealed that evolution of the globin proteins parallels the evolution of animals and plants. Major junctions occurred with the divergence of plant globins from animal globins and of myoglobin from hemoglobin. Later gene duplication gave rise to the α and β subunits of hemoglobin. *Right:* Hemoglobin is a tetramer of two α and two β subunits. The structural similarity of these subunits with leghemoglobin and myoglobin, both of which are monomers, is evident. A heme molecule (red) noncovalently associated with each globin polypeptide is directly responsible for oxygen-binding in these proteins. [Adapted from R. C. Hardison, 1996, *Proc. Nat'l Acad. Sci. USA* **93:**5675.]

## KEY CONCEPTS of Section 3.1

### Hierarchical Structure of Proteins

• Proteins are linear polymers of amino acids linked together by peptide bonds. A protein can have a single polypeptide chain or multiple polypeptide chains. The primary structure of a polypeptide chain is the sequence of covalently linked amino acids that compose the chain. Various, mostly noncovalent interactions between amino acids in the linear sequence stabilize a protein's specific folded three-dimensional structure, or conformation.

• The α helix, β strand and sheet, and β turn are the most prevalent elements of protein secondary structure. Secondary structures are stabilized by hydrogen bonds between atoms of the peptide backbone (see Figures 3-4 through 3-6).

• Protein tertiary structure results from hydrophobic interactions between nonpolar side groups and hydrogen bonds and ionic interactions involving polar side groups and the polypeptide backbone. These interactions stabilize folding of the protein, including its secondary structural elements, into an overall three-dimensional arrangement.

• Certain combinations of secondary structures give rise to different structural motifs, which are found in a variety of proteins and are often associated with specific functions (see Figure 3-9).

• Proteins often contain distinct domains, independently folded regions with characteristic structural, functional, and topological properties (see Figure 3-10).

• The incorporation of domains as modules in different proteins in the course of evolution has generated diversity in protein structure and function.

• The number and organization of individual polypeptide subunits in multimeric proteins define their quaternary structure.

• Cells contain large supramolecular assemblies, sometimes called molecular machines, in which all the necessary participants in complex cellular processes (e.g., DNA, RNA, and protein synthesis; photosynthesis; signal transduction) are bound together.

• Homologous proteins are proteins that evolved from a common ancestor and thus have similar sequences, structures, and functions. They can be classified into families and superfamilies.

## 3.2 Protein Folding

As noted above, when it comes to the architecture of proteins, "form follows function." Thus it is essential that when a polypeptide is synthesized with its particular amino acid sequence, it folds into the proper three-dimensional conformation with the appropriate secondary, tertiary, and possibly quaternary structure if it is to fulfill its biological role within or outside cells. How is a protein with a proper sequence generated? A polypeptide chain is synthesized by a complex process called **translation,** which occurs in the cytoplasm on a large protein–nucleic acid complex called a **ribosome.** During translation, a sequence of **messenger RNA (mRNA)** serves as a template from which the assembly of a corresponding amino acid sequence is directed. The mRNA is initially generated by a process called **transcription,** whereby a nucleotide sequence in DNA is converted, by transcriptional machinery in the nucleus, into a sequence of mRNA. The intricacies of transcription and translation are considered in Chapter 4. Here we describe the key determinants of the proper folding of a newly formed or forming (nascent) polypeptide chain as it emerges from the ribosome.

## Planar Peptide Bonds Limit the Shapes into Which Proteins Can Fold

A critical structural feature of polypeptides that limits how the chain can fold is the planar peptide bond. Figure 3-3 illustrates the amide group in peptide bonds in a polypeptide chain. Because the peptide bond itself behaves partially like a double bond,

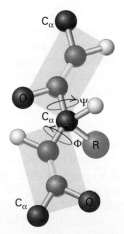

the carbonyl carbon and amide nitrogen and those atoms directly bonded to them must all lie in a fixed plane (Figure 3-14);

**FIGURE 3-14 Rotation between planar peptide groups in proteins.** Rotation about the $C_\alpha$–amino nitrogen bond (the $\Phi$ angle) and the $C_\alpha$–carbonyl carbon bond (the $\Psi$ angle) permits polypeptide backbones, in principle, to adopt a very large number of potential conformations. However, steric restraints due to the structure of the polypeptide backbone and the properties of the amino acid side chains dramatically restrict the potential conformations that any given protein can assume.

there is no rotation possible about the peptide bond itself. As a consequence, the only flexibility in a polypeptide chain backbone, allowing it to twist and turn—and thus fold into different three-dimensional shapes—is rotation of the fixed planes of adjacent peptide bonds with respect to one another about two bonds: the $C_\alpha$–amino nitrogen bond (rotational angle called $\Phi$) and the $C_\alpha$–carbonyl carbon bond (rotational angle called $\Psi$).

Yet a further constraint on the potential conformations that a polypeptide backbone chain can adopt is the fact that only a limited number of $\Phi$ and $\Psi$ angles are possible because for most $\Phi$ and $\Psi$ angles, the backbone or side chain atoms would come too close to one another and thus the associated conformation would be highly unstable or even physically impossible to achieve.

## The Amino Acid Sequence of a Protein Determines How It Will Fold

While the constraints of backbone bond angles seem very restrictive, any polypeptide chain containing only a few residues could, in principle, still fold into many conformations. For example, if the $\Phi$ and $\Psi$ angles were limited to only eight combinations, an $n$-residue-long peptide would potentially have $8^n$ conformations; for even a small polypeptide of only 10 residues, that's about 8.6 million possible conformations! In general, however, any particular protein adopts only one or just a few very closely related conformations called the *native state*; for the vast majority of proteins, the native state is the most stably folded form of the molecule and the one that permits it to function normally. In thermodynamic terms, the native state is usually the conformation with the lowest free energy ($G$) (see Chapter 2).

What features of proteins limit their folding from very many potential conformations to just one? The properties of the side chains (e.g., size, hydrophobicity, ability to form hydrogen and ionic bonds), together with their particular sequence along the polypeptide backbone, impose key restrictions. For example, a large side chain such as that of tryptophan might sterically block one region of the chain from packing closely against another region, whereas a side chain with a positive charge such as arginine might attract a segment of the polypeptide that has a complementary negatively charged side chain (e.g., aspartic acid). Another example we have already discussed is the effect of the aliphatic side chains in heptad repeats in promoting the association of helices and the consequent formation of coiled coils. Thus a polypeptide's primary structure determines its secondary, tertiary, and quaternary structures.

The initial evidence that the information necessary for a protein to fold properly is encoded in its amino acid sequence came from in vitro studies on the refolding of purified proteins, especially the Nobel Prize–winning studies in the 1960s by Christian Anfinsen of the refolding of ribonuclease A, an enzyme that cleaves RNA. Others had previously shown that various chemical and physical perturbations can disrupt the weak noncovalent interactions that stabilize the native conformation of a protein, leading to the loss of its normal tertiary structure. The process by which a protein's structure (and this can include secondary as well as tertiary structure) is disrupted is called **denaturation**. Denaturation can be induced by thermal energy from heat, extremes of pH that alter the charges on amino acid side chains, and exposure to *denaturants* such as urea or guanidine hydrochloride at concentrations of 6–8 M, all of which disrupt structure-stabilizing noncovalent interactions. Treatment with reducing agents, such as β-mercaptoethanol, that break disulfide bonds can further destabilize disulfide-containing proteins. Under such unfolding or denaturing conditions, a population of uniformly folded molecules is destabilized and converted into a collection of many unfolded, or denatured, molecules that have many different non-native and biologically inactive conformations. As we have seen, a large number of possible non-native conformations exist (e.g., $8^n - 1$).

The spontaneous unfolding of proteins under denaturing conditions is not surprising, given the substantial increase in entropy that accompanies the denatured protein assuming many non-native conformations. What is striking, however, is that when a pure sample of a single type of unfolded protein in a test tube is shifted back very carefully to normal conditions (body temperature, normal pH levels, reduction in the concentration of denaturants), some denatured polypeptides can spontaneously refold into their native, biologically active states as in Anfinsen's experiments. This kind of refolding experiment, as well as studies that show synthetic proteins made chemically can fold properly, showed that the information contained in a protein's primary structure can be sufficient to direct correct refolding. Newly synthesized proteins appear to fold into their proper conformations just as denatured proteins do. The observed similarity in the folded, three-dimensional structures of proteins with similar amino acid sequences, noted in Section 3.1, provided additional evidence that the primary sequence also determines protein folding in vivo. It appears that formation of secondary structures and structural motifs occurs early in the folding process, followed by assembly of more complex structural domains, which then associate into more complex tertiary and quaternary structures (Figure 3-15).

## Folding of Proteins in Vivo Is Promoted by Chaperones

The conditions of refolding of a purified, denatured protein in a test tube differ markedly from the conditions under which a newly synthesized polypeptide folds in a cell. The presence of other biomolecules, including many other proteins at very high concentration (~300 mg/ml in mammalian cells), some of which are themselves nascent and in the process of folding, can potentially interfere with the autonomous, spontaneous folding of a protein. Furthermore, although protein folding into the native state can occur in vitro, this does not happen for all unfolded molecules in a timely fashion. Given such impediments, cells require a faster, more efficient mechanism for folding proteins into their correct shapes than sequence alone provides. Without

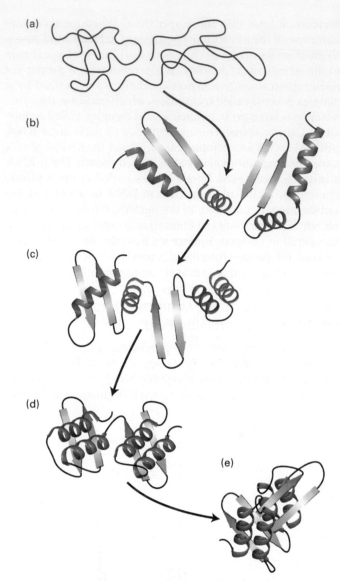

**FIGURE 3-15 Hypothetical protein-folding pathway.** Folding of a monomeric protein follows the structural hierarchy of primary (a) → secondary (b–d) → tertiary (e) structure. Formation of small structural motifs (c) appears to precede formation of domains (d) and the final tertiary structure (e).

such help, cells might waste much energy in the synthesis of improperly folded, nonfunctional proteins, which would have to be destroyed to prevent their disrupting cell function. Cells clearly have such mechanisms since more than 95 percent of the proteins present within cells have been shown to be in their native conformations. The explanation for the cell's remarkable efficiency in promoting proper protein folding is that cells make a set of proteins, called **chaperones**, that facilitate proper folding of nascent proteins.

The importance of chaperones is highlighted by the observations that many are evolutionarily conserved. Chaperones are found in all organisms from bacteria to humans, and some are homologs with high sequence similarity that use almost identical mechanisms to assist protein folding.

Chaperones use ATP binding, ATP hydrolysis to ADP, and exchange of a new ATP molecule for the ADP to induce a series of conformational changes essential for their function. Chaperones can fold newly made proteins into functional conformations, refold misfolded or unfolded proteins into functional conformations, disassemble potentially toxic protein aggregates that form due to protein misfolding, and assemble and dismantle large multiprotein complexes. Chaperones, which in eukaryotes are located in every cellular compartment and organelle, bind to the target proteins whose folding they will assist. There are several different classes of chaperones with distinct structures that all use ATP binding and hydrolysis in a variety of ways to facilitate folding. These include (1) enhancing the binding of protein substrates and (2) switching the conformation of the chaperones. The ATP-dependent conformational switch is used (1) to optimize folding after one substrate is folded, (2) to return the chaperone to its initial state so that it is available to help fold another molecule, and (3) to set the time permitted for refolding, which can be determined by the rate of ATP hydrolysis.

Two general families of chaperones have been identified:

• **Molecular chaperones,** which bind to a short segment of a protein substrate and stabilize unfolded or partly folded proteins, thereby preventing these proteins from aggregating and being degraded;

• **Chaperonins,** which form small folding chambers into which all or part of an unfolded protein can be sequestered, giving it time and an appropriate environment to fold properly.

One reason that chaperones are needed for intracellular protein folding is that they help prevent aggregation of unfolded proteins. Unfolded and partly folded proteins tend to aggregate into large, often water-insoluble masses, from which it is extremely difficult for a protein to dissociate and then fold into its proper conformation. In part this aggregation is due to the exposure of hydrophobic side chains that have not yet had a chance to be buried in the inner core of the folded protein. These exposed hydrophobic side chains on different molecules will stick to one another, owing to the hydrophobic effect (see Chapter 2) and thus promote aggregation. The risk for such aggregation is especially high for newly synthesized proteins that have not yet completed their proper folding. Chaperones prevent aggregation by binding to the target polypeptide or sequestering it from other partially or fully unfolded proteins, thus giving the nascent protein time to fold properly.

**Molecular Chaperones** The heat-shock protein Hsp70 in the cytosol and its homologs (Hsp70 in the mitochondrial matrix, BiP in the endoplasmic reticulum, and DnaK in bacteria) are molecular chaperones. They were first identified by their rapid appearance after a cell has been stressed by heat shock (*Hsp* stands for "*heat-shock protein*"). Hsp70 and its homologs are the major chaperones in all organisms. When bound to ATP, the monomeric Hsp70 protein assumes an open form in which an exposed hydrophobic substrate binding pocket

transiently binds to exposed hydrophobic regions of an incompletely folded or partially denatured target protein and then rapidly releases this substrate as long as ATP is bound (step **1** in Figure 3-16a). Hydrolysis of the bound ATP causes the molecular chaperone to assume a closed form that binds its substrate protein much more tightly and this tighter binding appears to facilitate the target protein's folding, in part by preventing it from aggregating with other unfolded proteins. The exchange of ATP for the protein-bound ADP causes a conformational change in the chaperone that releases the target protein. If the target is now properly folded, it cannot rebind to an Hsp70. If it remains at least partially unfolded, it can bind again to give it another chance to fold properly.

Additional proteins, such as the co-chaperone Hsp40 in eukaryotes (DnaJ in bacteria), help increase efficiency of Hsp70-mediated folding of many proteins by stimulating—together with the binding of substrate—the rate of hydrolysis of ATP by Hsp70/DnaK by 100- to 1000-fold (see step **2** in Figure 3-16a). Members of four different families of nucleotide exchange factors (e.g., GrpE in bacteria; BAG, HspBP, and Hsp110 families in eukaryotes) also interact with the Hsp70/DnaK, promoting the exchange of ATP for ADP. Multiple molecular chaperones are thought to bind all nascent polypeptide chains as they are being synthesized on ribosomes. In bacteria, 85 percent of the proteins are released from their chaperones and proceed to fold normally; an even higher percentage of proteins in eukaryotes follow this pathway.

The Hsp70 protein family is not the only class of molecular chaperones. Another distinct class of molecular chaperones is the Hsp90 family. Hsp90 family members are present in all organisms except archaea. In eukaryotes there are distinct Hsp90s located in different organelles and Hsp90 is one of the most abundant proteins in the cytosol (1–2 percent of total protein). Although the range of protein substrates of Hsp90 chaperones is not as broad as for some other chaperones, the Hsp90s are critically important in cells. They help cells cope with denatured proteins generated by stress (e.g., heat shock) and they ensure that some of their substrates, usually called "clients," can be converted from an inactive to an active state or otherwise held in a functional conformation. In some cases the Hsp90s form a relatively stable complex with their clients until an appropriate signal causes their dissociation from the client, freeing the client to perform some regulated function in the cells. These clients include transcription factors, such as the receptors for the steroid hormones estrogen or testosterone that regulate sexual development and function by controlling the activities of many genes (see Chapter 7). Another type of Hsp90 client is enzymes called kinases, which control the activities of many proteins by phosphorylation (see Chapters 15 and 16). It is estimated that as many as 20 percent of all proteins in yeast are directly or indirectly influenced by the activities of Hsp90.

Unlike monomeric Hsp70, Hsp90 functions as a dimer in a cycle in which ATP binding, hydrolysis, and ADP release are coupled to major conformational changes and to binding, activation, and release of clients (Figure 3-16b). Although much about the mechanism of Hsp90 remains to be learned,

**FIGURE 3-16 Molecular chaperone-mediated protein folding.** (a) Hsp70. Many proteins fold into their proper three-dimensional structures with the assistance of Hsp70-like proteins. These molecular chaperones transiently bind to a nascent polypeptide as it emerges from a ribosome or to proteins that have otherwise unfolded. In the Hsp70 cycle, an unfolded protein substrate binds in rapid equilibrium to the open conformation of the substrate-binding domain (SBD, orange) of the monomeric Hsp70, to which an ATP (red oval) is bound in the nucleotide-binding domain (NBD, blue) (step **1**). The substrate binding pocket is shown as a green patch on the substrate-binding domain. Co-chaperone accessory proteins (DnaJ/Hsp40) stimulate the hydrolysis of ATP to ADP (blue oval) and conformational change in Hsp70, resulting in the closed form, in which the substrate is locked into the SBD; here proper folding is facilitated (step **2**). Exchange of ATP for the bound ADP, stimulated by other accessory co-chaperone proteins (GrpE/BAG1), converts the Hsp70 back to the open form (step **3**), releasing the properly folded substrate (step **4**). (b) Hsp90. Hsp90 proteins are dimers, whose monomers contains an N-terminal NBD domain (blue), a central substrate (client) binding domain (SBD, orange), and a C-terminal dimerization domain (gray). The Hsp90 cycle begins when there is no nucleotide bound to the NBD and the dimer is in a very flexible, open (Y-shaped) configuration that can bind substrates (step **1**). Rapid ATP binding leads to a slow conformational change in which the NBDs dimerize and the SBDs move together into a closed conformation (step **2**). ATP hydrolysis results in folding of the client and client protein release (steps **3** and **4**). The ADP-bound form of Hsp90 can adopt several conformations, including a highly compact form. Release of ADP regenerates the initial state, which can then interact with additional clients (step **4**). [Part (b) modified from M. Taipale, D. F. Jarosz, and S. Lindquist, 2010, *Nat. Rev. Mol. Cell Biol.* **11**(7):515–528.]

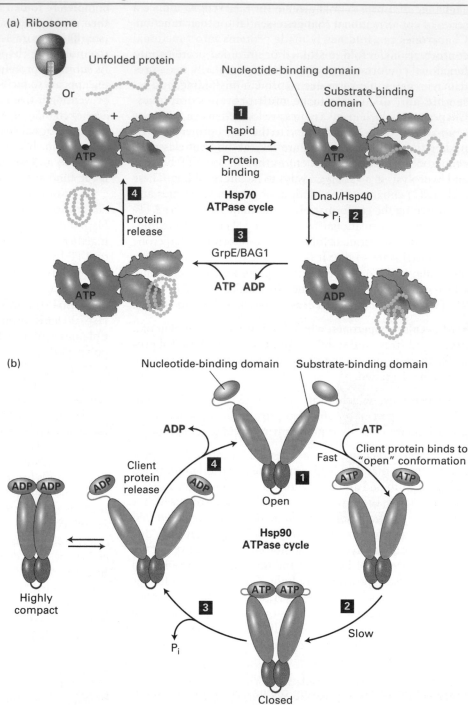

it is clear that clients bind to the "open" conformation, that ATP binding leads to interaction of the ATP binding domains and formation of a "closed" conformation, and that hydrolysis of ATP plays an important role in activating some client proteins and their subsequent release from the Hsp90. We also know that there are at least 20 co-chaperones that can have profound effects on the activity of Hsp90, including ATPase activity and determining which proteins will be clients (client specificity). Co-chaperones also can help coordinate the activities of Hsp90 and Hsp70. For example, Hsp70 can help begin the folding of a client that is then handed off by a co-chaperone to Hsp90 for additional processing. Hsp90 activity can also be influenced by its covalent modification by small molecules. Finally, Hsp90s can help cells recognize misfolded proteins that are unable to refold and facilitate their degradation by mechanisms discussed later in this chapter.

Thus, as part of the quality-control system in cells, chaperones can help properly fold proteins or facilitate the destruction of those that cannot fold properly.

**Chaperonins** The proper folding of a large variety of newly synthesized proteins also requires the assistance of another class of proteins, the chaperonins, also called Hsp60s. These huge cylindrical supramolecular assemblies are formed from two rings of oligomers. There are two distinct groups of chaperonins that differ somewhat in their structures, detailed molecular mechanisms, and locations. Group I chaperonins, found in prokaryotes, chloroplasts, and mitochondria, are composed of two rings, each having seven subunits that interact with a homoheptameric co-chaperone "lid." The bacteria group I chaperonin, known as GroEL/GroES, is shown in Figure 3-17a. In the bacterium *E. coli*, GroEL is thought to participate in the folding of about 10 percent of all proteins. Group II chaperonins, which are found in the cytosol of eukaryotic cells (e.g., TriC in mammals) and in archaea, can have eight to nine either homomeric or heteromeric subunits in each ring, and the "lid" function is incorporated in those subunits themselves—no separate lid is needed. It appears that ATP hydrolysis triggers the closing of the lid of group II chaperonins.

Figure 3-17b illustrates the GroEL/GroES cycle of protein folding. A partly folded or misfolded polypeptide <60 kD in

**FOCUS ANIMATION:** GroEL ATPase Cycle

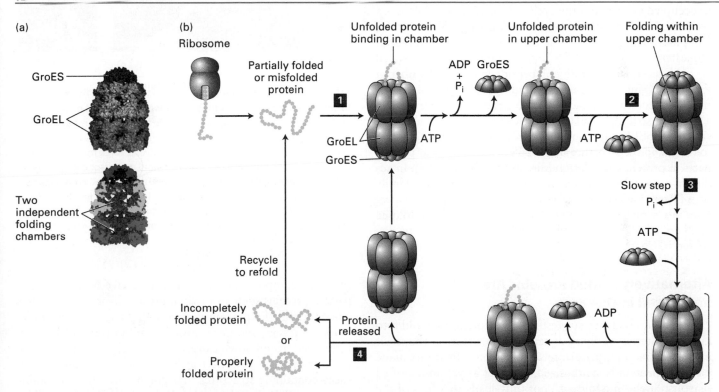

**FIGURE 3-17 Chaperonin-mediated protein folding.** Proper folding of some proteins depends on chaperonins such as the prokaryotic group I chaperonin GroEL. (a) GroEL is a barrel-shaped complex of 14 identical ~60,000-MW subunits arranged in two stacked rings (blue and red) of seven subunits each that form two distinct internal polypeptide folding chambers. Homoheptameric (10,000-MW subunits) lids, GroES (yellow), can bind to either end of the barrel and seal the chamber on that side. (b) The GroEL-GroES folding cycle. A partly folded or misfolded polypeptide enters one of the folding chambers (step **1**). The second chamber is blocked by a GroES lid. Each ring of seven GroEL subunits binds seven ATPs, hydrolyzes them, and releases the ADPs in a set order coordinated with GroES binding and release and polypeptide binding, folding, and release. The major conformational changes in the GroEL rings that take place control the binding of the GroES lid that seals the chamber (step **2**). The polypeptide remains encased in the chamber capped by the lid, where it can undergo folding until ATP hydrolysis, the slowest, rate-limiting step in the cycle ($t_{1/2} \sim 10$ s) (step **3**), induces binding of ATP and a different GroES to the other ring (transient intermediate shown in brackets). This then causes the GroES lid and ADP bound to the peptide-containing ring to be released, opening the chamber and permitting the folded protein to diffuse out of the chamber (step **4**). If the polypeptide folded properly, it can proceed to function in the cell. If it remains partially folded or misfolded, it can rebind to an unoccupied GroEL and the cycle can be repeated. [Part (a) modified from David L. Nelson and Michael M. 2000, Cox, *Lehninger: Principles of Biochemistry*, 3d ed., W. H. Freeman and Co.]

mass is captured by hydrophobic residues near the entrance of the GroEL chamber and enters one of the folding chambers (upper chamber in Figure 3-17b). The second chamber is blocked by a GroES lid. Each of the 14 subunits of GroEL can bind ATP, hydrolyze it, and subsequently release ADP. These reactions are concerted for each set of seven subunits in a single ring and lead to major conformational changes. These changes control both the binding of the GroES lid that seals the chamber and the environment of the chamber in which polypeptide folding takes place. The polypeptide remains encased in the chamber capped by the lid. There it can undergo folding until ATP hydrolysis in that chamber, the slowest, rate-limiting step in the cycle ($t_{1/2} \sim 10$ s), induces binding of ATP and a different GroES to the other ring. This then causes the GroES lid and ADP bound to the peptide containing ring to be released, opening the chamber and permitting the folded protein to diffuse out of the chamber. If the polypeptide folded properly, it can proceed to function in the cell. If it remains partially folded or misfolded, it can rebind to an unoccupied GroEL and the cycle can be repeated. There is a reciprocal relationship between the two rings in one GroEL complex. The capping of one chamber by GroES to permit sequestered substrate folding in that chamber is accompanied by the release of substrate polypeptide from the chamber of the second ring (simultaneous binding, folding, and release from the second chamber is not illustrated in Figure 3-17b). There is a striking similarity between the capped-barrel design of GroEL/GroES, in which proteins are sequestered for folding, and the structure of the 26S proteasome that participates in protein degradation (discussed in Section 3.4). In addition, a group of proteins that are part of the AAA$^+$ family of ATPases are composed of hexameric rings with a central pore into which substrates can enter for folding or unfolding or in some cases proteolysis; examples of these will be discussed in Chapter 13.

## Alternatively Folded Proteins Are Implicated in Diseases

Recent evidence suggests that a protein may fold into an alternative three-dimensional structure as the result of mutations, inappropriate covalent modifications made after the protein is synthesized, or other as yet unidentified reasons. Such "misfolding" not only leads to a loss of the normal function of the protein but often marks it for proteolytic degradation. However, when degradation isn't complete or doesn't keep pace with misfolding, the subsequent accumulation of the misfolded protein or its proteolytic fragments contributes to certain degenerative diseases characterized by the presence of insoluble, disordered aggregates of twisted-together protein, or *plaques,* in various organs, including the liver and brain.

Some neurodegenerative diseases, including Alzheimer's disease and Parkinson's disease in humans and transmissible spongiform encephalopathy ("mad cow" disease) in cows and sheep, are marked by the formation of tangled filamentous plaques in a deteriorating brain (Figure 3-18). The *amyloid*

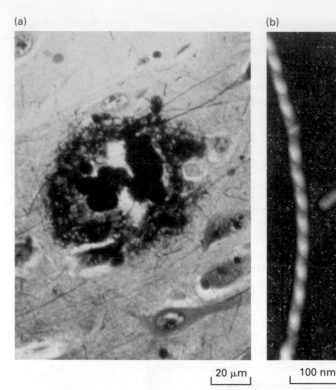

**FIGURE 3-18 Alzheimer's disease is characterized by the formation of insoluble plaques composed of amyloid protein.** (a) At low resolution, an amyloid plaque in the brain of an Alzheimer's patient appears as a tangle of filaments. (b) The regular structure of filaments from plaques is revealed in the atomic force microscope. Proteolysis of the naturally occurring amyloid precursor protein yields a short fragment, called β-amyloid protein, that for unknown reasons changes from an α-helical to a β-sheet conformation. This alternative structure aggregates into the highly stable filaments (amyloid) found in plaques. Similar pathologic changes in other proteins cause other degenerative diseases. [Courtesy of K. Kosik.]

*filaments* composing these structures derive from abundant natural proteins such as amyloid precursor protein, which is embedded in the plasma membrane; Tau, a microtubule-binding protein; and prion protein, an "infectious" protein. Influenced by unknown causes, these α helix–containing proteins or their proteolytic fragments fold into alternative β sheet–containing structures that polymerize into very stable filaments. Whether the extracellular deposits of these filaments or the soluble alternatively folded proteins are toxic to the cell is unclear. ■

## KEY CONCEPTS of section 3.2

### Protein Folding

• The primary structure (amino acid sequence) of a protein determines its three-dimensional structure, which determines its function. In short, function derives from structure; structure derives from sequence.

- Because protein function derives from protein structure, newly synthesized proteins must fold into the correct shape to function properly.

- The planar structure of the peptide bond limits the number of conformations a polypeptide can have (see Figure 3-14).

- The amino acid sequence of a protein dictates its folding into a specific three-dimensional conformation, the native state. Proteins will unfold, or denature, if treated under conditions that disrupt the noncovalent interactions stabilizing their three-dimensional structures.

- Protein folding in vivo occurs with assistance from ATP-dependent chaperones. Chaperones can influence proteins in several ways, including preventing misfolding and aggregation, facilitating proper folding, and maintaining an appropriate, stable structure required for subsequent protein activity (see Figure 3-16).

- There are two broad classes of chaperones: (1) molecular chaperones, which bind to a short segment of a protein substrate, and (2) chaperonins, which form folding chambers into which all or part of an unfolded protein can be sequestered, giving it time and an appropriate environment to fold properly.

- Some neurodegenerative diseases are caused by aggregates of proteins that are stably folded in an alternative conformation.

## 3.3 Protein Binding and Enzyme Catalysis

Proteins perform an extraordinarily diverse array of activities both inside and outside cells, yet most of these diverse functions are based on the ability of proteins to engage in a common activity: binding. Proteins bind to themselves, to other macromolecules, to small molecules, and to ions. In this section, we describe some key features of protein binding and then turn to look at one group of proteins, enzymes, in greater detail. The activities of the other functional classes of proteins (structural, scaffold, transport, regulatory, motor) will be described in later chapters.

### Specific Binding of Ligands Underlies the Functions of Most Proteins

The molecule to which a protein binds is called its **ligand**. In some cases, ligand binding causes a change in the shape of a protein. Such conformational changes are integral to the mechanism of action of many proteins and are important in regulating protein activity.

Two properties of a protein characterize how it binds ligands. *Specificity* refers to the ability of a protein to bind one molecule or a very small group of molecules in preference to all other molecules. *Affinity* refers to the tightness or strength of binding, usually expressed as the dissociation constant ($K_d$). The $K_d$ for a protein-ligand complex, which is the inverse of the equilibrium constant $K_{eq}$ for the binding reaction, is the most common quantitative measure of affinity (see Chapter 2). The stronger the interaction between a protein and ligand, the lower the value of $K_d$. Both the specificity and the affinity of a protein for a ligand depend on the structure of the *ligand-binding site*. For high-affinity and highly specific interactions to take place, the shape and chemical properties of the binding site must be complementary to that of the ligand molecule, a property termed **molecular complementarity**. As we saw in Chapter 2, molecular complementarity allows molecules to form multiple noncovalent interactions at close range and thus stick together.

One of the best-studied examples of protein-ligand binding, involving high affinity and exquisite specificity, is that of **antibodies** binding to **antigens**. Antibodies are proteins that circulate in the blood and are made by the immune system (see Chapter 23) in response to antigens, which are usually macromolecules present in infectious agents (e.g., a bacterium or a virus) or other foreign substances (e.g., proteins or polysaccharides in pollens). Different antibodies are generated in response to different antigens, and these antibodies have the remarkable characteristic of binding specifically to ("recognizing") a part of the antigen, called an **epitope**, which initially induced the production of the antibody, and not to other molecules. Antibodies act as specific sensors for antigens, forming antibody-antigen complexes that initiate a cascade of protective reactions in cells of the immune system.

All antibodies are Y-shaped molecules formed from two identical longer, or *heavy*, chains and two identical shorter, or *light*, chains (Figure 3-19a). Each arm of an antibody molecule contains a single light chain linked to a heavy chain by a disulfide bond. Near the end of each arm are six highly variable loops, called *complementarity-determining regions (CDRs)*, which form the antigen-binding sites. The sequences of the six loops are highly variable among antibodies, generating unique complementary ligand-binding sites that make them specific for different epitopes (Figure 3-19b). The intimate contact between these two surfaces, stabilized by numerous noncovalent interactions, is responsible for the extremely precise binding specificity exhibited by an antibody.

The specificity of antibodies is so precise that they can distinguish between the cells of individual members of a species and in some cases between proteins that differ by only a single amino acid, or even between proteins with identical sequences and only different post-translational modifications. Because of their specificity and the ease with which they can be produced (see Chapter 23), antibodies are highly useful reagents used in many of the experiments discussed in subsequent chapters.

We will see many examples of protein-ligand binding throughout this book, including hormones binding to receptors (see Chapter 15), regulatory molecules binding to DNA (see Chapter 7), and cell-adhesion molecules binding to extracellular matrix (see Chapter 20), to name just a few. Here we focus on how the binding of one class of proteins, enzymes,

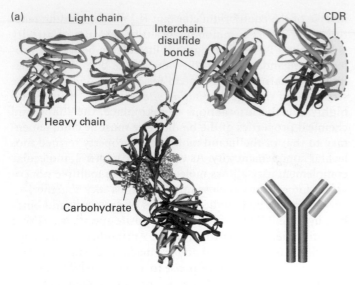

(a) Light chain
Interchain disulfide bonds
CDR
Heavy chain
Carbohydrate

(b)

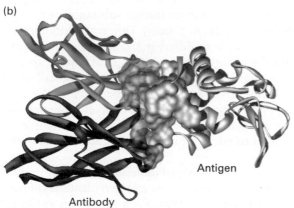

Antigen

Antibody

**FIGURE 3-19 Protein-ligand binding of antibodies.** (a) Ribbon model of an antibody. Every antibody molecule of the immunoglobulin IgG class consists of two identical heavy chains (light and dark red) and two identical light chains (blue) covalently linked by disulfide bonds. The inset shows a diagram of the overall structure containing the two heavy and two light chains. (b) The hand-in-glove fit between an antibody and the site to which it binds (epitope) on its target antigen—in this case, chicken egg-white lysozyme. Regions where the two molecules make contact are shown as surfaces. The antibody contacts the antigen with residues from all its complementarity-determining regions (CDRs). In this view, the molecular complementarity of the antigen and antibody is especially apparent where "fingers" extending from the antigen surface are opposed to "clefts" in the antibody surface.

to their ligands results in the catalysis of the chemical reactions essential for the survival and function of cells.

## Enzymes Are Highly Efficient and Specific Catalysts

The subset of proteins that catalyze chemical reactions, the making and breaking of covalent bonds, is called **enzymes**, and enzymes' ligands are called **substrates**. Enzymes make up a large and very important class of proteins—indeed, almost every chemical reaction in the cell is catalyzed by a

specific catalyst, usually an enzyme. Another form of catalytic macromolecule in cells is made from RNA. These RNAs are called **ribozymes** (see Chapter 4).

Thousands of different types of enzymes, each of which catalyzes a single chemical reaction or set of closely related reactions, have been identified. Certain enzymes are found in the majority of cells because they catalyze the synthesis of common cellular products (e.g., proteins, nucleic acids, and phospholipids) or take part in harvesting energy from nutrients (e.g., by the conversion of glucose and oxygen into carbon dioxide and water during cellular respiration). Other enzymes are present only in a particular type of cell because they catalyze chemical reactions unique to that cell type (e.g., the enzymes in nerve cells that convert tyrosine into dopamine, a neurotransmitter). Although most enzymes are located within cells, some are secreted and function at extracellular sites such as the blood, the digestive tract, or even outside the organism (e.g., toxic enzymes in the venom of poisonous snakes).

Like all **catalysts** (see Chapter 2), enzymes increase the rate of a reaction but do not affect the extent of a reaction, which is determined by the change in free energy $\Delta G$ between reactants and products and are not themselves permanently changed as a consequence of the reaction they catalyze. Enzymes increase the reaction rate by lowering the energy of the **transition state** and therefore the **activation energy** required to reach it (Figure 3-20). In the test tube, catalysts such as charcoal and platinum facilitate reactions but usually only at high temperatures or pressures, at extremes of high or low pH, or in organic solvents. Within cells, however, enzymes

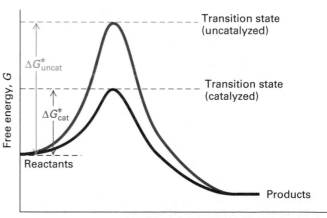

**FIGURE 3-20 Effect of an enzyme on the activation energy of a chemical reaction.** This hypothetical reaction pathway depicts the changes in free energy, *G*, as a reaction proceeds. A reaction will take place spontaneously only if the total *G* of the products is less than that of the reactants (negative $\Delta G$). However, all chemical reactions proceed through one or more high-energy transition states, and the rate of a reaction is inversely proportional to the activation energy ($\Delta G^{\ddagger}$), which is the difference in free energy between the reactants and the transition state (highest point along the pathway). Enzymes and other catalysts accelerate the rate of a reaction by reducing the free energy of the transition state and thus $\Delta G^{\ddagger}$.

must function effectively in an aqueous environment at 37 °C and 1 atmosphere pressure and at physiologic pH values, usually 6.5–7.5 but sometimes lower. Remarkably, enzymes exhibit immense catalytic power, accelerating the rates of reactions $10^6$–$10^{12}$ times that of the corresponding uncatalyzed reactions under otherwise similar conditions.

## An Enzyme's Active Site Binds Substrates and Carries Out Catalysis

Certain amino acids of an enzyme are particularly important in determining its specificity and catalytic power. In the native conformation of an enzyme, critically important amino acids (which usually come from different parts of the linear sequence of the polypeptide) are brought into proximity, forming a cleft in the surface called the **active site** (Figure 3-21). An active site usually makes up only a small part of the total protein, with the remaining part involved in folding of the polypeptide, regulation of the active site, and interactions with other molecules.

An active site consists of two functionally important regions: the *substrate-binding site,* which recognizes and binds the substrate or substrates, and the *catalytic site,* which carries out the chemical reaction once the substrate has bound. The catalytic groups in the catalytic site are amino acid side chains and backbone carbonyl and amino groups. In some enzymes, the catalytic and substrate-binding sites overlap; in others, the two regions are structurally distinct.

The substrate-binding site is responsible for the remarkable specificity of enzymes—their ability to act selectively on one substrate or a small number of chemically similar substrates. The alteration of the structure of an enzyme's substrate by only one or a few atoms, or a subtle change in the geometry (e.g., stereochemistry) of the substrate, can result in a variant molecule that is no longer a substrate of the enzyme. As with the specificity of antibodies for antigens described above, the specificity of enzymes for substrates is a consequence of the precise molecular complementarity between its substrate-binding site and the substrate. Usually only one or a few substrates can fit precisely into a binding site.

The idea that substrates might bind to enzymes in the manner of a key fitting into a lock was suggested first by Emil Fischer in 1894. In 1913 Leonor Michaelis and Maud Leonora Menten provided crucial evidence supporting this hypothesis. They showed that the rate of an enzymatic reaction was proportional to the substrate concentration at low substrate concentrations, but that as the substrate concentrations increased, the rate reached a **maximal velocity, $V_{max}$,** and became substrate concentration independent, with the value of $V_{max}$ being directly proportional to the amount of enzyme present in the reaction mixture (Figure 3-22).

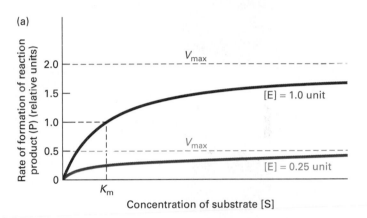

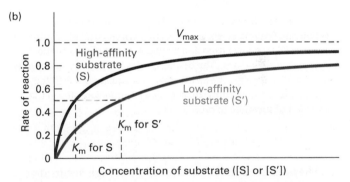

**FIGURE 3-22 $K_m$ and $V_{max}$ for an enzyme-catalyzed reaction.** $K_m$ and $V_{max}$ are determined from analysis of the dependence of the initial reaction velocity on substrate concentration. The shape of these hypothetical kinetic curves is characteristic of a simple enzyme-catalyzed reaction in which one substrate (S) is converted into product (P). The initial velocity is measured immediately after addition of enzyme to substrate before the substrate concentration changes appreciably. (a) Plots of the initial velocity at two different concentrations of enzyme [E] as a function of substrate concentration [S]. The [S] that yields a half-maximal reaction rate is the Michaelis constant $K_m$, a measure of the affinity of E for turning S into P. Quadrupling the enzyme concentration causes a proportional increase in the reaction rate, and so the maximal velocity $V_{max}$ is quadrupled; the $K_m$, however, is unaltered. (b) Plots of the initial velocity versus substrate concentration with a substrate S for which the enzyme has a high affinity and with a substrate S' for which the enzyme has a lower affinity. Note that the $V_{max}$ is the same with both substrates because [E] is the same but that $K_m$ is higher for S', the low-affinity substrate.

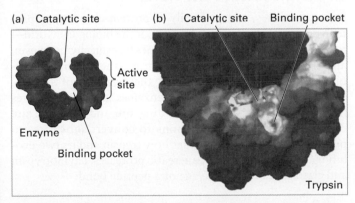

**FIGURE 3-21 Active site of the enzyme trypsin.** (a) An enzyme's active site is composed of a binding pocket, which binds specifically to a substrate, and a catalytic site, which carries out catalysis. (b) A surface representation of the serine protease trypsin. Active site clefts containing the catalytic site (side chains of the catalytic triad Ser-195, Asp-102, and His-57 shown as stick figures) and the substrate side chain specificity binding pocket are clearly visible. [Part (b) courtesy of P. Teesdale-Spittle.]

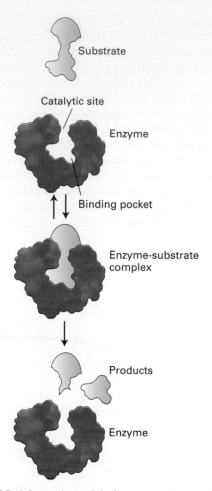

**FIGURE 3-23 Schematic model of an enzyme's reaction mechanism.** Enzyme kinetics suggest that enzymes (E) bind substrate molecules (S) through a fixed and limited number of sites on the enzymes (the active sites). The bound species is known as an enzyme-substrate (ES) complex. The ES complex is in equilibrium with the unbound enzyme and substrate and is an intermediate step in the conversion of substrate to products (P).

They deduced that this saturation at high substrate concentrations was due to the binding of substrate molecules (S) to a fixed and limited number of sites on the enzymes (E), and they called the bound species the enzyme-substrate (ES) complex. They proposed that the ES complex is in equilibrium with the unbound enzyme and substrate and is an intermediate step in the ultimately irreversible conversion of substrate to product (P) (Figure 3-23):

$$E + S \rightleftharpoons ES \rightarrow E + P$$

and that the rate $V_0$ of formation of product at a particular substrate concentration [S] is given by what is now called the *Michaelis-Menten equation*:

$$V_0 = V_{max} \frac{[S]}{[S] + K_m} \qquad (3\text{-}1)$$

where the **Michaelis constant**, $K_m$, a measure of the affinity of an enzyme for its substrate, is the substrate concentration that yields a half-maximal reaction rate (i.e., 1/2 $V_{max}$ in Figure 3-22). The $K_m$ is somewhat similar in nature, but not identical, to the dissociation constant, $K_d$ (see Chapter 2). The smaller the value of $K_m$, the more effective the enzyme is at making product from dilute solutions of substrate and the lower the substrate concentration needed to reach half-maximal velocity. The smaller the $K_d$, the lower the ligand concentration needed to reach 50 percent of binding. The concentrations of the various small molecules in a cell vary widely, as do the $K_m$ values for the different enzymes that act on them. A good rule of thumb is that the intracellular concentration of a substrate is approximately the same as, or somewhat greater than, the $K_m$ value of the enzyme to which it binds.

The rates of reaction at substrate saturation vary enormously among enzymes. The maximum number of substrate molecules converted to product at a single enzyme active site per second is called the *turnover number,* which can be less than 1 for very slow enzymes. The turnover number for carbonic anhydrase, one of the fastest enzymes, is $6 \times 10^5$ molecules/s.

Many enzymes catalyze the conversion of substrates to products by dividing the process into multiple, discrete chemical reactions that involve multiple, distinct enzyme substrate complexes (ES, ES′, ES″, etc.) generated prior to the final release of the products:

$$E + S \rightleftharpoons ES \rightleftharpoons ES' \rightleftharpoons ES'' \rightleftharpoons \dots E + P$$

The energy profiles for such multistep reactions involve multiple hills and valleys (Figure 3-24), and methods have been developed to trap the intermediates in such reactions to learn more about the details of how enzymes catalyze reactions.

## Serine Proteases Demonstrate How an Enzyme's Active Site Works

Serine proteases, a large family of protein-cleaving, or proteolytic, enzymes, are used throughout the biological world—to digest meals (the pancreatic enzymes trypsin, chymotrypsin, and elastase), to control blood clotting (the enzyme thrombin), even to help silk moths chew their way out of their cocoons (cocoonase). This class of enzymes usefully illustrates how an enzyme's substrate-binding site and catalytic site cooperate in multistep reactions to convert substrates to products. Here we will consider how trypsin and its two evolutionarily closely related pancreatic proteases, chymotrypsin and elastase, catalyze cleavage of a peptide bond:

$$P_1-\overset{\overset{\displaystyle O}{\|}}{C}\cdots\underset{\underset{\displaystyle H}{|}}{N}-P_2 + H_2O \rightleftharpoons P_1-\overset{\overset{\displaystyle O}{\|}}{C}-\overset{O}{\underset{O}{\cdots}} + NH_3^+-P_2$$

where in the polypeptide substrate $P_1$ is the part of the protein on the N-terminal side of the peptide bond and $P_2$ is the

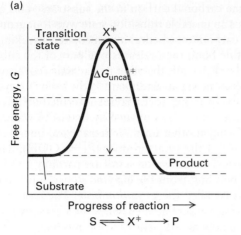

**(a)**

Transition state $X^‡$

Free energy, $G$

$\Delta G_{uncat}^‡$

Product

Substrate

Progress of reaction →

$S \rightleftharpoons X^‡ \longrightarrow P$

**(b)**

Enzyme–transition state complex

$EX^‡$

Free energy, $G$

Enzyme + substrate

Enzyme – substrate complex

$\Delta G_{cat}^‡$

Enzyme + product

$E + S$

ES

$E + P$

Progress of reaction →

$E + S \rightleftharpoons ES \rightleftharpoons EX^‡ \longrightarrow E + P$

**FIGURE 3-24 Free-energy reaction profiles of uncatalyzed and multistep enzyme-catalyzed reactions.** (a) The free-energy reaction profile of a hypothetical simple uncatalyzed reaction converting substrate (S) to product (P) via a single high-energy transition state. (b) Many enzymes catalyze such reactions by dividing the process into multiple discrete steps, in this case the initial formation of an ES complex followed by conversion via a single transition state ($EX^‡$) to the free enzyme (E) and P. The activation energy for each of these steps is significantly less than the activation energy for the uncatalyzed reaction; thus the enzyme dramatically enhances the reaction rate.

portion on the C-terminal side. We first consider how serine proteases bind specifically to their substrates and then show in detail how catalysis takes place.

Figure 3-25a shows how a substrate polypeptide binds to the substrate-binding site in the active site of trypsin. There are two key binding interactions. First, the substrate (black polypeptide backbone) and enzyme (blue polypeptide backbone) form hydrogen bonds that resemble a β sheet. Second, a key side chain of the substrate that determines which peptide in the substrate is to be cleaved extends into the enzyme's *side-chain specificity binding pocket*, at the bottom of which resides the negatively charged side chain of the enzyme's Asp-189. Trypsin has a marked preference for hydrolyzing substrates at the carboxyl (C=O) side of an amino acid with a long positively charged side chain (arginine or lysine) because the side chain is stabilized in the enzyme's specificity binding pocket by the negative Asp-189.

Slight differences in the structures of otherwise similar specificity pockets help explain the differing substrate specificities of two related serine proteases: chymotrypsin prefers large aromatic groups (as in Phe, Tyr, Trp), and elastase prefers the small side chains of Gly and Ala (Figure 3-25b). The uncharged Ser-189 in chymotrypsin allows large, uncharged, hydrophobic side chains to bind stably in the pocket. The

**(a)**

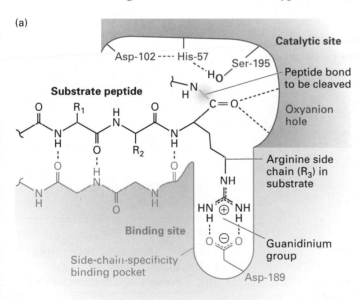

Catalytic site

Asp-102 ---- His-57    Ser-195

Substrate peptide

Peptide bond to be cleaved

Oxyanion hole

Arginine side chain (R₃) in substrate

Guanidinium group

Binding site

Side-chain-specificity binding pocket

Asp-189

**(b)**

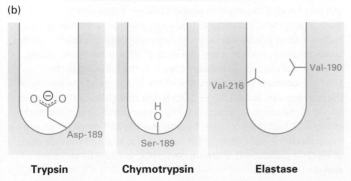

Trypsin    Chymotrypsin    Elastase

Asp-189    Ser-189    Val-216, Val-190

**FIGURE 3-25 Substrate binding in the active site of trypsin-like serine proteases.** (a) The active site of trypsin (purple and blue molecule) with a bound substrate (black molecule). The substrate forms a two-stranded β sheet with the binding site, and the side chain of an arginine (R₃) in the substrate is bound in the side-chain-specificity binding pocket. Its positively charged guanidinium group is stabilized by the negative charge on the side chain of the enzyme's Asp-189. This binding aligns the peptide bond of the arginine appropriately for hydrolysis catalyzed by the enzyme's active-site catalytic triad (side chains of Ser-195, His-57, and Asp-102). (b) The amino acids lining the side-chain-specificity binding pocket determine its shape and charge and thus its binding properties. Trypsin accommodates the positively charged side chains of arginine and lysine; chymotrypsin, large, hydrophobic side chains such as phenylalanine; and elastase, small side chains such as glycine and alanine. [Part (a) modified from J. J. Perona and C. S. Craik, 1997, *J. Biol. Chem.* **272**(48):29987–29990.]

specificity of elastase is influenced by the replacement of glycines in the sides of the pocket in trypsin with the branched aliphatic side chains of valines (Val-216 and Val-190) that obstruct the pocket (Figure 3-25b). As a consequence, large side chains in substrates are prevented from fitting into the pocket of elastase, whereas substrates with the short alanine or glycine side chains at this position can bind well and be subject to subsequent cleavage.

In the catalytic site, all three enzymes use the hydroxyl group on the side chain of a serine in position 195 to catalyze the hydrolysis of peptide bonds in protein substrates. A catalytic triad formed by the three side chains of Ser-195, His-57, and Asp-102 participates in what is essentially a two-step hydrolysis reaction. Figure 3-26 shows how the catalytic triad cooperates in breaking the peptide bond, with Asp-102 and His-57 supporting the attack of the hydroxyl oxygen of Ser-195

on the carbonyl carbon in the substrate. This attack initially forms an unstable transition state with four groups attached to this carbon (tetrahedral intermediate). Breaking of the C—N peptide bond then releases one part of the substrate protein ($NH_3$—$P_2$), while the other part remains covalently attached to the enzyme via an ester bond to the serine's oxygen, forming a relatively stable acyl enzyme intermediate. The subsequent replacement of this oxygen by one from water, in a reaction involving another unstable tetrahedral intermediate, leads to release of the final product ($P_1$—COOH). The tetrahedral intermediate transition states are partially stabilized by hydrogen bonding from the enzyme's backbone amino groups in what is called the *oxyanion hole*. The large family of serine proteases and related enzymes with an active site serine illustrates how an efficient reaction mechanism is used over and over by distinct enzymes to catalyze similar reactions.

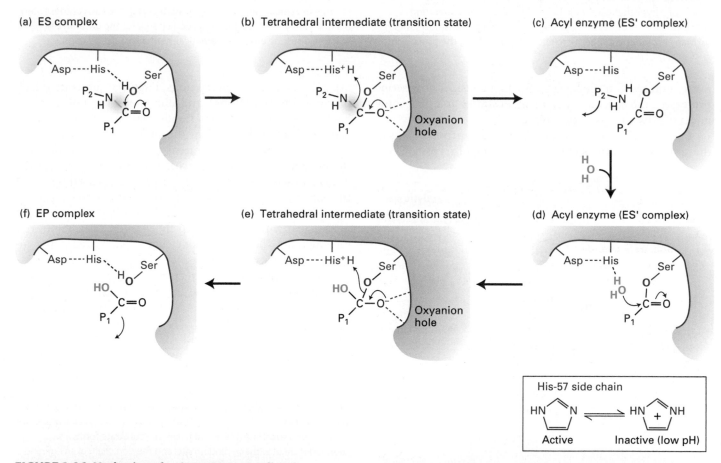

**FIGURE 3-26 Mechanism of serine-protease-mediated hydrolysis of peptide bonds.** The catalytic triad of Ser-195, His-57, and Asp-102 in the active sites of serine proteases employs a multistep mechanism to hydrolyze peptide bonds in target proteins. (a) After a polypeptide substrate binds to the active site (see Figure 3-25), forming an ES complex, the hydroxyl oxygen of Ser-195 attacks the carbonyl carbon of the substrate's targeted peptide bond (yellow). Movements of electrons are indicated by arrows. (b) This attack results in the formation of a transition state called the *tetrahedral intermediate*, in which the negative charge on the substrate's oxygen is stabilized by hydrogen bonds formed with the enzyme's *oxyanion hole*. (c) Additional electron movements result in the breaking of the peptide bond,

release of one of the reaction products ($NH_2$—$P_2$), and formation of the acyl enzyme (ES' complex). (d) An oxygen from a solvent water molecule then attacks the carbonyl carbon of the acyl enzyme. (e) This attack results in the formation of a second tetrahedral intermediate. (f) Additional electron movements result in the breaking of the Ser-195–substrate bond (formation of the EP complex) and release of the final reaction product ($P_1$—COOH). The side chain of His-57, which is held in the proper orientation by hydrogen bonding to the side chain of Asp-102, facilitates catalysis by withdrawing and donating protons throughout the reaction (*inset*). If the pH is too low and the side chain of His-57 is protonated, it cannot participate in catalysis and the enzyme is inactive.

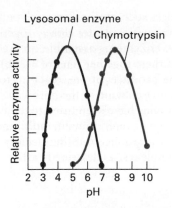

**FIGURE 3-27 pH dependence of enzyme activity.** Ionizable (pH-titratable) groups in the active sites or elsewhere in enzymes often must be either protonated or deprotonated to permit proper substrate binding or catalysis or to permit the enzyme to adopt the correct conformation. Measurement of enzyme activity as a function of pH can be used to identify the $pK_a$'s of these groups. The pancreatic serine proteases, such as chymotrypsin (*right curve*), exhibit maximum activity around pH 8 because of titration of the active site His-57 (required for catalysis, $pK_a \sim 6.8$) and of the amino terminus of the protein (required for proper conformation, $pK_a \sim 9$). Many lysosomal hydrolases have evolved to exhibit a lower pH optimum ($\sim 4.5$, *left curve*) to match the low internal pH in lysosomes in which they function. [Adapted from P. Lozano, T. De Diego, and J. L. Iborra, 1997, *Eur. J. Biochem.* **248**(1):80–85, and W. A. Judice et al., 2004, *Eur. J. Biochem.* **271**(5):1046–1053.]

The serine protease mechanism points out several general key features of enzymatic catalysis: (1) enzyme catalytic sites have evolved to stabilize the binding of a transition state, thus lowering the activation energy and accelerating the overall reaction; (2) multiple side chains, together with the polypeptide backbone, carefully organized in three dimensions, work together to chemically transform substrate into product, often by multistep reactions; and (3) acid-base catalysis mediated by one or more amino acid side chains is often used by enzymes, as when the imidazole group of His-57 in serine proteases acts as a base to remove the hydrogen from Ser-195's hydroxyl group. As a consequence, often only a particular ionization state (protonated or nonprotonated) of one or more amino acid side chains in the catalytic site is compatible with catalysis, and thus the enzyme's activity is pH dependent.

For example, the imidazole of His-57 in serine proteases, whose $pK_a$ is $\sim 6.8$, can help the Ser-195 hydroxyl attack the substrate only if it is not protonated. Thus the activity of the protease is low at pH $<6.8$, and the shape of the pH activity profile in the pH range 4–8 matches the titration of the His-57 side chain, which is governed by the Henderson-Hasselbalch equation, with an inflection near pH 6.8 (see chymotrypsin data in Figure 3-27 and see Chapter 2). The activity drops at higher pH values, generating a bell-shaped curve, because the proper folding of the protein is disrupted when the amino group at the protein's amino terminus ($pK_a \sim 9$) is deprotonated; the conformation near the active site changes as a consequence.

The pH sensitivity of an enzyme's activity can be due to changes in the ionization of catalytic groups, groups that participate directly in substrate binding, or groups that influence the conformation of the protein. Pancreatic proteases evolved to function in the neutral or slightly basic conditions in the intestines; hence their pH optima are $\sim 8$. Proteases and other hydrolytic enzymes that function in acidic conditions must employ a different catalytic mechanism. This is the case for enzymes within the stomach (pH $\sim 1$) such as the protease pepsin or those within lysosomes (pH $\sim 4.5$), which play a key role in degrading macromolecules within cells (see lysosomal enzyme data in Figure 3-27). Indeed, lysosomal hydrolases, which degrade a wide variety of biomolecules (proteins, lipids, etc.), are relatively inactive at the pH in the cytosol ($\sim 7$), helping protect a cell from self-digestion if these enzymes escape the confines of the membrane-bound lysosome.

One key feature of enzymatic catalysis not seen in serine proteases but found in many other enzymes is a *cofactor* or *prosthetic group*. This "helper" group is a nonpolypeptide small molecule or ion (e.g., iron, zinc, copper, manganese) that is bound in the active site and plays an essential role in the reaction mechanism. Small organic prosthetic groups in enzymes are also called *coenzymes*. Some of these are chemically modified during the reaction and thus need to be replaced or regenerated after each reaction; others are not. Examples include $NAD^+$ (nicotinamide adenine dinucleotide), FAD (flavin adenine dinucleotide) (see Figure 2-33), and heme groups that bind oxygen in hemoglobin or transfer electrons in some cytochromes (see Figure 12-14). Thus the chemistry catalyzed by enzymes is not restricted by the limited number of types of amino acids in polypeptide chains. Many vitamins—for example, the B vitamins, thiamine ($B_1$), riboflavin ($B_2$), niacin ($B_3$), and pyridoxine ($B_6$), and vitamin C—which cannot be synthesized in mammalian cells, function as or are used to generate coenzymes. That is why supplements of vitamins must be added to the liquid medium in which mammalian cells are grown in the laboratory (see Chapter 9).

Small molecules that can bind to active sites and disrupt catalytic reactions are called *enzyme inhibitors*. Such inhibitors are useful tools for studying the roles of enzymes in cells and whole organisms. Inhibitors that directly bind to an enzyme's binding site and thus compete directly with the normal substrates' ability to bind substrate are called competitive inhibitors. Alternatively, inhibitors can interfere with enzyme activity in other ways, for example by binding to some other site on the enzyme and changing its conformation; this is called noncompetitive inhibition. Inhibitors complement the use of mutations in genes and a technique called RNA interference (RNAi) for probing an enzyme's function in cells (see Chapter 5). In all three approaches the cellular consequences of disrupting an enzyme's activity can be used to deduce the normal function of the enzyme. The same approaches can be used to study the functions of non-enzymatic macromolecules. However, interpreting results of inhibitor studies can be complicated if, as is often the case, the inhibitors block the activity of more than one protein.

Small-molecule inhibition of protein activity is the basis for most drugs and also for chemical warfare agents. Aspirin inhibits enzymes called cyclooxygenases, whose products can cause pain. Sarin and other nerve gases react with the active serine hydroxyl groups of both serine proteases and a related enzyme, acetylcholine esterase, which is a key enzyme in regulating nerve conduction (see Chapter 22). ∎

## Enzymes in a Common Pathway Are Often Physically Associated with One Another

Enzymes taking part in a common metabolic process (e.g., the degradation of glucose to pyruvate during glycolysis; see Chapter 12) are generally located in the same cellular compartment, be it the cytosol, at a membrane, or within a particular organelle. Within this compartment, products from one reaction can move by diffusion to the next enzyme in the pathway. However, diffusion entails random movement and can be a slow, relatively inefficient process for moving molecules between enzymes (Figure 3-28a). To overcome this impediment, cells have evolved mechanisms for bringing enzymes in a common pathway into close proximity, a process called metabolic coupling.

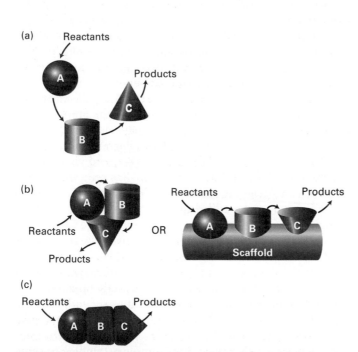

**FIGURE 3-28 Assembly of enzymes into efficient multienzyme complexes.** In the hypothetical reaction pathways illustrated here, the initial reactants are converted into final products by the sequential action of three enzymes: A, B, and C. (a) When the enzymes are free in solution or even constrained within the same cellular compartment, the intermediates in the reaction sequence must diffuse from one enzyme to the next, an inherently slow process. (b) Diffusion is greatly reduced or eliminated when the individual enzymes associate into multisubunit complexes, either by themselves or with the aid of a scaffold protein. (c) The closest integration of different catalytic activities occurs when the enzymes are fused at the genetic level, becoming domains in a single polypeptide chain.

In the simplest such mechanism, polypeptides with different catalytic activities cluster closely together as subunits of a multimeric enzyme or assemble on a common "scaffold" that holds them together (Figure 3-28b). This arrangement allows the products of one reaction to be channeled directly to the next enzyme in the pathway. In some cases, independent proteins have been fused together at the genetic level to create a single multidomain, multifunctional enzyme (Figure 3-28c). Metabolic coupling usually involves large multiprotein complexes, as described earlier in this chapter.

## KEY CONCEPTS of Section 3.3

### Protein Binding and Enzyme Catalysis

• A protein's function depends on its ability to bind other molecules known as ligands. For example, antibodies bind to a group of ligands known as antigens and enzymes bind to reactants called substrates that will be converted by chemical reactions into products.

• The specificity of a protein for a particular ligand refers to the preferential binding of one or a few closely related ligands. The affinity of a protein for a particular ligand refers to the strength of binding, usually expressed as the dissociation constant $K_d$.

• Proteins are able to bind to ligands because of molecular complementarity between the ligand-binding sites and the corresponding ligands.

• Enzymes are catalytic proteins that accelerate the rate of cellular reactions by lowering the activation energy and stabilizing transition-state intermediates (see Figure 3-20).

• An enzyme's active site, which is usually only a small part of the protein, comprises two functional parts: a substrate-binding site and a catalytic site. The substrate-binding site is responsible for the exquisite specificity of enzymes owing to its molecular complementarity with the substrate and the transition state.

• The initial binding of substrates (S) to enzymes (E) results in the formation of an enzyme-substrate complex (ES), which then undergoes one or more reactions catalyzed by the catalytic groups in the active site until the products (P) are formed.

• From plots of reaction rate versus substrate concentration, two characteristic parameters of an enzyme can be determined: the Michaelis constant, $K_m$, a rough measure of the enzyme's affinity for converting substrate into product, and the maximal velocity, $V_{max}$, a measure of its catalytic power (see Figure 3-22).

• The rates of enzyme-catalyzed reactions vary enormously, with the turnover numbers (number of substrate molecules converted to products at a single active site at substrate saturation) ranging between <1 to $6 \times 10^5$ molecules/s.

- Many enzymes catalyze the conversion of substrates to products by dividing the process into multiple discrete chemical reactions that involve multiple distinct enzyme substrate complexes (ES', ES'' etc.).

- Serine proteases hydrolyze peptide bonds in protein substrates using as catalytic groups the side chains of Ser-195, His-57, and Asp-102. Amino acids lining the specificity binding pocket in the binding site of serine proteases determine the residue in a protein substrate whose peptide bond will be hydrolyzed and account for differences in the specificity of trypsin, chymotrypsin, and elastase.

- Enzymes often use acid-base catalysis mediated by one or more amino acid side chains, such as the imidazole group of His-57 in serine proteases, to catalyze reactions. The pH dependence of protonation of catalytic groups ($pK_a$) is often reflected in the pH-rate profile of the enzyme's activity.

- In some enzymes, nonpolypeptide small molecules or ions, called *cofactors* or *prosthetic groups*, can bind to the active site and play an essential role in enzymatic catalysis. Small organic prosthetic groups in enzymes are also called *coenzymes*; vitamins, which cannot be synthesized in higher animal cells, function as or are used to generate coenzymes.

- Enzymes in a common pathway are located within specific cell compartments and may be further associated as domains of a monomeric protein, subunits of a multimeric protein, or components of a protein complex assembled on a common scaffold (see Figure 3-28).

## 3.4 Regulating Protein Function

Most processes in cells do not take place independently of one another or at a constant rate. The activities of all proteins and other biomolecules are regulated to integrate their functions for optimal performance for survival. For example, the catalytic activity of enzymes is regulated so that the amount of reaction product is just sufficient to meet the needs of the cell. As a result, the steady-state concentrations of substrates and products will vary depending on cellular conditions. Regulation of nonenzymatic proteins—the opening or closing of membrane channels or the assembly of a macromolecular complex, for example—is also essential.

In general, there are three ways to regulate protein activity. First, cells can increase or decrease the steady-state level of the protein by altering its rate of synthesis, its rate of degradation, or both. Second, cells can change the intrinsic activity, as distinct from the amount, of the protein. For example, through noncovalent and covalent interactions, cells can change the affinity of substrate binding, or the fraction of time the protein is in an active versus inactive conformation. Third, there can be a change in location or concentration within the cell of the protein itself, the target of the protein's activity (e.g., an enzyme's substrate), or some other molecule required for the protein's activity (e.g., an enzyme's cofactor).

All three types of regulation play essential roles in the lives and functions of cells. In this section, we first discuss mechanisms for regulating the amount of a protein and then turn to noncovalent and covalent interactions that regulate protein activity.

### Regulated Synthesis and Degradation of Proteins Is a Fundamental Property of Cells

The rate of synthesis of proteins is determined by the rate at which the DNA encoding the protein is converted to mRNA (transcription), the steady-state amount of the active mRNA in the cell, and the rate at which the mRNA is converted into newly synthesized protein (translation). These important pathways are described in detail in Chapter 4.

The life span of intracellular proteins varies from as short as a few minutes for mitotic cyclins, which help regulate passage through the mitotic stage of cell division (see Chapter 19), to as long as the age of an organism for proteins in the lens of the eye. Protein life span is controlled primarily by regulated protein degradation.

There are two especially important roles for protein degradation. First, degradation removes proteins that are potentially toxic, improperly folded or assembled, or damaged—including the products of mutated genes and proteins damaged by chemically active cell metabolites or stress (e.g., heat shock). Despite the existence of chaperone-mediated protein folding, some of the newly made proteins are rapidly degraded because they are misfolded. This might occur due to failure of timely engagement of the necessary chaperones to guide their folding or their defective assembly into complexes. Most other proteins are degraded more slowly, about 1–2 percent degradation per hour in mammalian cells. Second, the controlled destruction of otherwise normal proteins provides a powerful mechanism for maintaining the appropriate levels of the proteins and their activities and for permitting rapid changes in these levels to help the cells respond to changing conditions.

Eukaryotic cells have several pathways for degrading proteins. One major pathway is degradation by enzymes within lysosomes, membrane-limited organelles whose acidic interior (pH ~4.5) is filled with a host of hydrolytic enzymes. Lysosomal degradation is directed primarily toward aged or defective organelles of the cell—a process called autophagy (see Chapter 14)—and toward extracellular proteins taken up by the cell. Lysosomes will be discussed at length in later chapters. Here we will focus on another important pathway: cytoplasmic protein degradation by proteasomes, which can account for up to 90 percent of the protein degradation in mammalian cells.

### The Proteasome Is a Molecular Machine Used to Degrade Proteins

**Proteasomes** are very large, protein-degrading macromolecular machines that influence many different cellular functions, including the cell cycle (see Chapter 19), transcription,

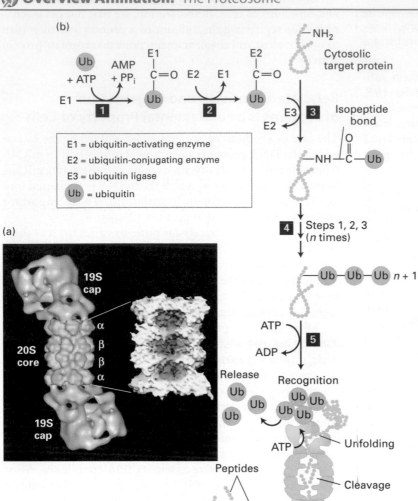

(b)

E1 = ubiquitin-activating enzyme
E2 = ubiquitin-conjugating enzyme
E3 = ubiquitin ligase
Ub = ubiquitin

(a)

**FIGURE 3-29 Ubiquitin- and proteasome-mediated proteolysis.** (a) *Left:* The 26S proteasome has a cylindrical structure with a 19S cap (blue) at each end of a 20S core consisting of four stacked heptameric rings (~110 Å diameter × 160 Å long) each containing either α (outer rings) or β (inner rings) subunits (yellow). *Right:* Cutaway view of 20S core showing inner chambers. Proteolysis of ubiquitin-tagged proteins occurs within the inner chamber of the core formed by the β rings. (b) Proteins are targeted for proteasomal degradation by polyubiquitination. Enzyme E1 is activated by attachment of a ubiquitin (Ub) molecule (step **1**) and then transfers this Ub molecule to a cysteine residue in E2 (step **2**). Ubiquitin ligase (E3) transfers the bound Ub molecule on E2 to the side-chain —$NH_2$ of a lysine residue in a target protein, forming an isopeptide bond (step **3**). Additional Ub molecules are added to the Ub-modified target protein via isopeptide bonds to the previously added Ub by repeating steps **1**–**3**, forming a polyubiquitin chain (step **4**). The polyubiquitinylated target is recognized by the proteasome cap, which uses deubiquitinase enzymes to remove the Ub groups and ATP hydrolysis to drive unfolding and transfer of the unfolded protein into the proteolysis chamber in the core, from which the short peptide digestion fragments are later released (step **5**) [Part (a) from W. Baumeister et al., 1998, *Cell* **92**:357, courtesy of W. Baumeister and modified from M. Bochtler et al.,1999, *Ann. Rev. Biophys. Biomol. Struct.* 28:295–317.]

DNA repair (see Chapter 4), programmed cell death or **apoptosis** (see Chapter 21), recognition and response to infection by foreign organisms (see Chapter 23), and removal of misfolded proteins. There are approximately 30,000 proteasomes in a typical mammalian cell.

Proteasomes consist of ~50 protein subunits and have a mass of 2–2.4 × 10⁶ Da. Proteasomes have a cylindrical, barrel-like catalytic core called the *20S proteasome* (where *S* is a Svedberg unit based on the sedimentation properties of the particle and is proportional to its size), which is approximately 14.8 nm tall and 11.3 nm in diameter. Bound to the ends of this core are either one or two 19S cap complexes (Figure 3-29a) that regulate the activity of the 20S catalytic core. When the core and one or two caps are combined, they are both referred to as the 26S complex, even though the two-cap-containing complex is larger (30S). A 19S cap has 16–18 protein subunits, six of which can hydrolyze ATP (i.e., they are AAA-type ATPases) to provide the energy needed to unfold protein substrates and selectively transfer them into the inner chamber of the proteasome's catalytic core. Genetic studies in yeast have shown that cells cannot survive without functional proteasomes, thus demonstrating their importance. Furthermore, proper proteasomal activity is so important that cells will expend as much as 30 percent of the energy needed to synthesize a protein to degrade it in a proteasome.

The 20S proteasomal catalytic core comprises two inner rings of seven β subunits each, with three proteolytic active sites per ring facing toward the inner chamber of the ~1.7-nm-diameter barrel and two outer rings of seven α subunits each that control substrate access (Figure 3-29a). Proteasomes can degrade most proteins thoroughly because the three active sites in each β subunit ring can cleave after hydrophobic residues, acidic residues, or basic residues. Polypeptide substrates must enter the chamber via a regulated ~1.3-nm-diameter aperture at the center of the outer α subunit rings. In the 26S proteasome, the opening of the aperture, which is narrow and often allows the entry of only unfolded proteins, is controlled by ATPases in the 19S cap that also participate in selectively binding and unfolding the substrate (Figure 3-29b, *bottom right*). The short peptide products of proteasomal digestion (2–24 residues long) exit the chamber and are further degraded rapidly by cytosolic peptidases, eventually being converted to individual ("free") amino acids. One researcher has quipped that a proteasome is a "cellular chamber of doom" in which proteins suffer a "death by a thousand cuts."

Inhibitors of proteasome function can be used therapeutically. Because of the global importance of proteasomes for cells, continuous, complete inhibition of proteasomes kills cells. However, partial proteasome inhibition for short intervals has been introduced as an approach to cancer chemotherapy. To survive and grow, cells normally require the robust activity of a regulatory protein called $NF_kB$ (see Chapter 16) as well as other, similar "pro-survival" proteins. In turn, $NF_kB$ can function fully and promote survival only when its inhibitor, $I_kB$, is disengaged and degraded by proteasomes (see Chapter 16). Partial inhibition of proteasomal activity by a small-molecule inhibitor drug results in increased levels of $I_kB$ and, consequently, reduced $NF_kB$ activity (that is, loss of pro-survival activity). Cells subsequently die by apoptosis. Because at least some types of tumor cells are more sensitive to being killed by proteasome inhibitors than normal cells are, *controlled* administration of proteasome inhibitors, at levels that kill the cancer cells but not normal cells, has proven to be an effective therapy for at least one type of lethal cancer, multiple myeloma. ∎

## Ubiquitin Marks Cytosolic Proteins for Degradation in Proteasomes

If proteasomes are to rapidly degrade only those proteins that are either defective or scheduled to be removed, they must be able to distinguish between those proteins that need to be degraded and those that don't. Cells mark proteins that should be degraded by covalently attaching to them a linear chain of multiple copies of a 76-residue polypeptide called **ubiquitin** that is highly conserved from yeast to humans. This "polyubiquitin tail" serves as a cellular "kiss of death," marking the protein for destruction in the proteasome. The ubiquitination process (Figure 3-29b) involves three distinct steps:

1. Activation of *ubiquitin-activating enzyme (E1)* by the addition of a ubiquitin molecule, a reaction that requires ATP;

2. Transfer of this ubiquitin molecule to a cysteine residue in *ubiquitin-conjugating enzyme (E2)*;

3. Formation of a covalent bond between the carboxyl group of the C-terminal glycine 76 of the ubiquitin bound to E2 and the amino group of the side chain of a lysine residue in the target protein, a reaction catalyzed by *ubiquitin-protein ligase (E3)*. This type of bond is called an *isopeptide bond* because it covalently links a side chain amino group, rather than the α amino group, to the carboxyl group.

Subsequent ligase reactions covalently attach the C-terminal glycine of an additional ubiquitin via an isopeptide bond to the side chain of lysine 48 of the previously added ubiquitin to generate a polyubiquitin chain covalently attached to the target protein.

Following attachment of four or more ubiquitins in the polyubiquitin chain, the 19S regulatory cap of the 26S proteasome (sometimes with the help of accessory proteins) recognizes the polyubiquitin-labeled proteins and unfolds and transports them into the proteasome for degradation (see Figure 3-29b). As a polyubiquitinated substrate is unfolded and passed into the core of the proteasome, enzymes called deubiquitinases (Dubs) hydrolyze the bonds between individual ubiquitins and between the targeted protein and ubiquitin, recycling the ubiquitins for additional rounds of protein modification (see Figure 3-29b). Analysis of the human genome sequence indicates the presence of ~90 Dubs, about 80 percent of which use cysteine in a catalytic triad, similar to serine proteases described earlier. In some Dubs, zinc is a key participant in the catalytic reactions.

**Specificity of Degradation** Targeting of specific proteins for proteasomal degradation is primarily achieved through the substrate specificity of the E3 ligase. As a testament to their importance, there are more than 600 predicted ubiquitin ligase genes in the human genome. The many E3 ligases in mammalian cells ensure that the wide variety of proteins to be polyubiquitinated can be modified when necessary. Some E3 ligases are associated with chaperones that recognize unfolded or misfolded proteins; for example, the E3 ligase CHIP is a co-chaperone for Hsp70. These and other proteins (co-chaperones, escort factors, adaptors) can mediate E3-ligase-catalyzed polyubiquitination of these dysfunctional proteins that cannot be readily refolded properly and their delivery to proteasomes for degradation. In such cases, the chaperone-ubiquitination-proteasome system works in concert for protein quality control.

In addition to quality control, the ubiquitin-proteasome system can be used to regulate the activity of important cellular proteins. An example is the regulated degradation of proteins called **cyclins**, which control the cell cycle (see Chapter 19). Cyclins contain the internal sequence Arg-X-X-Leu-Gly-X-Ile-Gly-Asp/Asn (X can be any amino acid), which is recognized by specific ubiquitinating enzyme complexes. At a specific time in the cell cycle, each cyclin is phosphorylated by a cyclin kinase. This phosphorylation is thought to cause a conformational change that exposes the recognition sequence to the ubiquitinating enzymes, leading to polyubiquitination and proteasomal degradation.

**Other Functions of Ubiquitin and Ubiquitin-Related Molecules** There are several close relatives of ubiquitin that employ similar E1-, E2-, and E3-dependent mechanisms of activation and transfer to acceptor substrates. These ubiquitin-like modifications control processes as diverse as nuclear import, regulated by the ubiquitin-like modifier Sumo, and autophagy, regulated by the ubiquitin-like modifier Atg8/LC3 (see Chapter 14). The attachment of ubiquitin to a target protein can be used for purposes other than to mark the protein for degradation, as we will discuss later in this chapter, and some of these involve polyubiquitin linkages other than those via Lys48.

As is the case for ubiquitination itself, deubiquitination is involved in processes other than proteasome-mediated protein degradation. Large-scale, mass-spectroscopy-based "proteomic" methods described later in this chapter together with sophisticated computational approaches have suggested that Dubs, which are often bound in multiprotein complexes, are involved in an extraordinarily wide range of cell processes. These vary from cell division and cell cycle control (see Chapter 19) to membrane trafficking (see Chapter 14) to cell signaling pathways (see Chapters 15 and 16).

## Noncovalent Binding Permits Allosteric, or Cooperative, Regulation of Proteins

In addition to regulating the amount of a protein, cells can also regulate the intrinsic activity of a protein. One of the most important mechanisms for regulating protein function is through allosteric interactions. Broadly speaking, **allostery** (from the Greek "other shape") refers to any change in a protein's tertiary or quaternary structure, or in both, induced by the noncovalent binding of a ligand. When a ligand binds to one site (A) in a protein and induces a conformational change and associated change in activity of a different site (B), the ligand is called an *allosteric effector* of the protein, while site A is called an *allosteric binding site*, and the protein is called an *allosteric protein*. By definition, allosteric proteins have multiple binding sites, at least one for the allosteric effector and at least one for other molecules with which the protein interacts. The allosteric change in activity can be positive or negative; that is, it can induce an increase or a decrease in protein activity. Negative allostery often involves the end product of a multistep biochemical pathway binding to and reducing the activity of an enzyme that catalyzes an early, rate-controlling step for that pathway. In this way excessive buildup of the product is prevented. This kind of regulation of a metabolic pathway is also called *end-product inhibition* or *feedback inhibition*. Allosteric regulation is particularly prevalent in multimeric enzymes and other proteins where conformational changes in one subunit are transmitted to an adjacent subunit.

*Cooperativity* is a term often used synonymously with allostery and usually refers to the influence (positive or negative) that the binding of a ligand at one site has on the binding of another molecule of the *same* type of ligand at a different site. Hemoglobin presents a classic example of positive cooperative binding in that the binding of a single ligand, molecular oxygen ($O_2$), increases the affinity of the binding of the next oxygen molecule. Each of the four subunits in hemoglobin contains one heme molecule. The heme groups are the oxygen-binding components of hemoglobin (see Figure 3-13). The binding of oxygen to the heme molecule in one of the four hemoglobin subunits induces a local

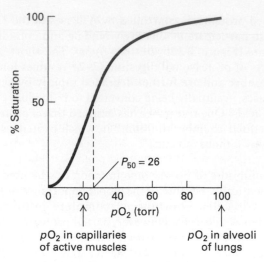

**EXPERIMENTAL FIGURE 3-30 Hemoglobin binds oxygen cooperatively.** Each tetrameric hemoglobin protein has four oxygen-binding sites; at saturation all the sites are loaded with oxygen. The oxygen concentration in tissues is commonly measured as the partial pressure ($pO_2$). $P_{50}$ is the $pO_2$ at which half the oxygen-binding sites at a given hemoglobin concentration are occupied; it is somewhat analogous to the $K_m$ for an enzymatic reaction. The large change in the amount of oxygen bound over a small range of $pO_2$ values permits efficient unloading of oxygen in peripheral tissues such as muscle. The sigmoidal shape of a plot of percent saturation versus ligand concentration is indicative of cooperative binding, in which the binding of one oxygen molecule allosterically influences the binding of subsequent oxygens. In the absence of cooperative binding, a binding curve is a hyperbola, similar to the curves in Figure 3-22. [Adapted from L. Stryer, 1995, *Biochemistry*, 4th ed., W. H. Freeman and Company.]

conformational change whose effect spreads to the other subunits, lowering the $K_d$ (increasing the affinity) for the binding of additional oxygen molecules to the remaining hemes and yielding a sigmoidal oxygen-binding curve (Figure 3-30). Because of the sigmoidal shape of the oxygen-saturation curve, it takes only a fourfold increase in oxygen concentration for the percent saturation of the oxygen-binding sites in hemoglobin to go from 10 to 90 percent. Conversely, if there were no cooperativity and the shape of the curve was typical of that for Michaelis-Menten-type, or noncooperative, binding, it would take an 81-fold increase in oxygen concentration to accomplish the same increase in loading of its binding sites in hemoglobin. This cooperativity permits hemoglobin to take up oxygen very efficiently in the lungs, where the oxygen concentration is high and unload it in tissues, where the concentration is low. Thus cooperativity amplifies the sensitivity of a system to concentration changes in its ligands, providing in many cases selective evolutionary advantage.

## Noncovalent Binding of Calcium and GTP Are Widely Used as Allosteric Switches to Control Protein Activity

Unlike oxygen, which causes graded allosteric changes in the activity of hemoglobin, some other allosteric effectors act as switches, turning the activity of many different proteins on or off by binding to them noncovalently. Two important allosteric switches that we will encounter many times throughout this book, especially in the context of cell signaling pathways (see Chapters 15 and 16), are $Ca^{2+}$ and GTP.

**$Ca^{2+}$/Calmodulin-Mediated Switching** The concentration of $Ca^{2+}$ free in the cytosol (not bound to molecules other than water) is kept very low ($\sim 10^{-7}$ M) by specialized membrane transport proteins that continually pump excess $Ca^{2+}$ out of the cytosol. However, as we learn in Chapter 11, the cytosolic $Ca^{2+}$ concentration can increase from 10- to 100-fold when $Ca^{2+}$-permeable channels in the cell-surface membranes open and allow extracellular $Ca^{2+}$ to flow into the cell. This rise in cytosolic $Ca^{2+}$ is sensed by specialized $Ca^{2+}$-binding proteins, which alter cellular behavior by turning the activities of other proteins on or off. The importance of extracellular $Ca^{2+}$ for cell activity was first documented by S. Ringer in 1883, when he discovered that isolated rat hearts suspended in an NaCl solution made with "hard" ($Ca^{2+}$-rich) London tap water contracted beautifully, whereas they beat poorly and stopped quickly if distilled water was used.

Many of the $Ca^{2+}$-binding proteins bind $Ca^{2+}$ using the EF hand/helix-loop-helix structural motif discussed earlier (see Figure 3-9b). A well-studied EF hand protein, **calmodulin**, is found in all eukaryotic cells and may exist as an individual monomeric protein or as a subunit of a multimeric protein. A dumbbell-shaped molecule, calmodulin contains four $Ca^{2+}$-binding EF hands with $K_d$'s of $\sim 10^{-6}$ M. The binding of $Ca^{2+}$ to calmodulin causes a conformational change that permits $Ca^{2+}$/calmodulin to bind to conserved sequences in various target proteins (Figure 3-31), thereby switching their activities on or off. Calmodulin and similar EF hand proteins thus function as *switch proteins*, acting in concert with changes in $Ca^{2+}$ levels to modulate the activity of other proteins.

**Switching Mediated by Guanine-Nucleotide-Binding Proteins** Another group of intracellular switch proteins constitutes the **GTPase superfamily**. As the name suggests, these proteins are enzymes, GTPases, that can hydrolyze GTP (guanosine triphosphate) to GDP (guanosine diphosphate). They include the monomeric Ras protein, whose structure is shown in Figure 3-8 with bound GDP shown in blue, and the $G_\alpha$ subunit of the trimeric G proteins, both discussed at length in Chapters 15 and 16. Both Ras and $G_\alpha$ can bind to the plasma membrane, function in cell signaling, and play a key role in cell proliferation and differentiation. Other members of the GTPase superfamily function in protein synthesis, the transport of proteins between the nucleus and the cytoplasm, the formation of coated vesicles and their fusion with target membranes, and rearrangements of the actin cytoskeleton. The Hsp70 chaperone protein we encountered earlier is an example of an ATP/ADP switch, similar in many respects to a GTP/GDP switch.

All the GTPase switch proteins exist in two forms, or conformations (Figure 3-32): (1) an active ("on") form with bound GTP that can influence the activity of specific target

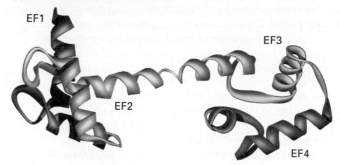

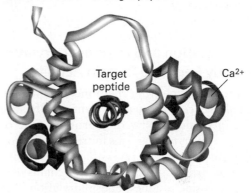

(b) $Ca^{2+}$/calmodulin bound to target peptide

**FIGURE 3-31 Conformational changes induced by $Ca^{2+}$ binding to calmodulin.** Calmodulin is a widely distributed cytosolic protein that contains four $Ca^{2+}$-binding sites, one in each of its EF hands. Each EF hand has a helix-loop-helix motif. At cytosolic $Ca^{2+}$ concentrations above about $5 \times 10^{-7}$ M, binding of $Ca^{2+}$ to calmodulin changes the protein's conformation from the dumbbell-shaped, unbound form (a) to one in which hydrophobic side chains become more exposed to solvent. The resulting $Ca^{2+}$-calmodulin can wrap around exposed helices of various target proteins (b), thereby altering their activity.

proteins to which they bind and (2) an inactive ("off") form with bound GDP. The switch is turned on, that is, the conformation of the protein changes from inactive to active, when a GTP molecule replaces a bound GDP in the inactive conformation. The switch is turned off when the relatively slow GTPase activity of the protein hydrolyzes bound GTP, converting it to GDP and leading the conformation to change to the inactive form. The amount of time any given GTPase switch remains in the active, GTP-bound form depends on how long the GTP remains bound before it is converted to GDP. This time depends on the rate of the GTPase activity. Thus the GTPase activity acts as a timer to control this switch. Cells contain a variety of proteins that can modulate the baseline (or intrinsic) rate of GTPase activity for any given GTPase switch and so can control how long the switch remains on. Cells also have specific proteins whose function is to regulate the conversion of inactive GTPases to active ones—that is, turn the switch on—by mediating the replacement of bound GDP with a GTP. These are called GTP exchange factors, or GEFs. We examine the role of various GTPase switch proteins in regulating intracellular signaling and other processes in several later chapters.

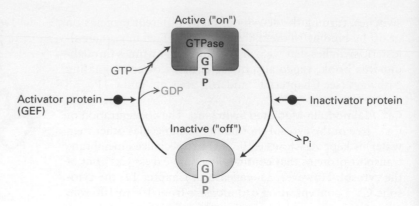

**FIGURE 3-32 The GTPase switch.** GTPases are enzymes that bind to and hydrolyze GTP to GDP. When bound to GTP, the GTPase protein adopts its active or "on" conformation and can interact with target proteins to regulate their activities; when the bound GTP is hydrolyzed to GDP by the intrinsic GTPase activity of the protein, the GTPase with GDP bound assumes an inactive or "off" conformation. The GTPase switch can be turned back on when another protein, called a GEF (guanine nucleotide exchange factor), mediates the replacement of the bound GDP with a GTP molecule from the surrounding fluid. The binding of the active form of the GTPases to its targets is a form of noncovalent regulation. Various proteins can influence the rates of GTP hydrolysis (i.e., inactivator proteins) and exchange of GDP for GTP (GEFs).

## Phosphorylation and Dephosphorylation Covalently Regulate Protein Activity

In addition to exploiting the noncovalent regulators described above, cells can use covalent modifications to regulate the intrinsic activity of a protein. One of the most common covalent mechanisms for regulating protein activity is **phosphorylation**, the reversible addition of phosphate groups to hydroxyl groups on the side chains of serine, threonine, or tyrosine residues. Phosphorylation is catalyzed by enzymes called protein **kinases**, while the removal of phosphates, known as dephosphorylation, is catalyzed by **phosphatases**. The counteracting activities of kinases and phosphatases provide cells with a "switch" that can turn on or turn off the function of various proteins (Figure 3-33). Sometimes phosphorylation sites are masked transiently by the reversible covalent modification with the sugar N-acetylglucosamine as an additional means of regulation. Phosphorylation changes a protein's charge and can lead to a

conformational change that can significantly alter ligand binding or other features of the protein, causing an increase or decrease in its activity. In addition, several conserved protein domains specifically bind to phosphorylated peptides. Thus phosphorylation can mediate the formation of protein complexes that can generate or extinguish a wide variety of cellular activities, discussed in many subsequent chapters.

Nearly 3 percent of all yeast proteins are protein kinases or phosphatases, indicating the importance of phosphorylation and dephosphorylation reactions even in simple cells. All classes of proteins—including structural proteins, scaffolds, enzymes, membrane channels, and signaling molecules—have members regulated by kinase/phosphatase modifications. Different protein kinases and phosphatases are specific for different target proteins and so can regulate distinct cellular pathways, as discussed in later chapters. Some kinases have many targets and so a single kinase can serve to integrate the activities of many targets simultaneously. Frequently, the target of the kinase (and phosphatase) is yet another kinase or phosphatase, creating a cascade effect. There are many examples of such kinase cascades, which permit amplification of a signal and many levels of fine-tuning control (see Chapters 15 and 16).

## Ubiquitination and Deubiquitination Covalently Regulate Protein Activity

Both ubiquitin and ubiquitin-like proteins (of which there are more than a dozen in humans) can be covalently linked to a target protein in a regulated fashion, analogously to phosphorylation. Deubiquitinases can reverse ubiquitination, analogously to the action of phosphatases. These ubiquitin modifications are structurally far more complex than relatively simple phosphorylation, however, and so can mediate many distinct interactions between the ubiquitinated protein and other cellular proteins. Ubiquitination can involve attachment of a single ubiquitin to a protein (**monoubiquitination**), addition of multiple, single ubiquitin molecules to different sites on one target protein (**multiubiquitination**), or addition of a polymeric chain of ubiquitins to a protein (**polyubiquitination**). An additional source of variation is that

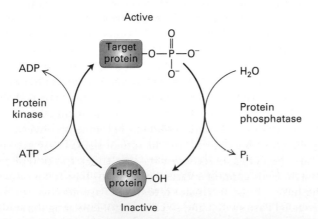

**FIGURE 3-33 Regulation of protein activity by phosphorylation and dephosphorylation.** The cyclic phosphorylation and dephosphorylation of a protein is a common cellular mechanism for regulating protein activity. In this example, the target protein is active (*top*) when phosphorylated and inactive (*bottom*) when dephosphorylated; some proteins have the opposite response to phosphorylation.

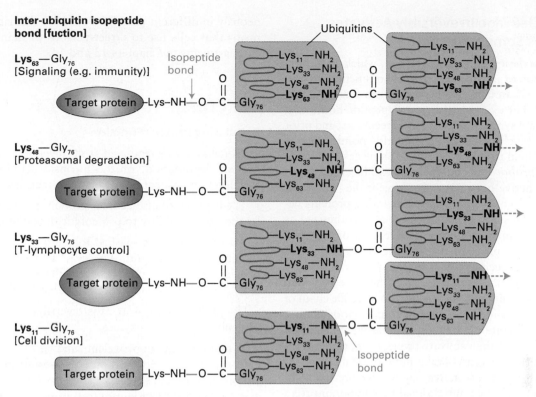

**Inter-ubiquitin isopeptide bond [fuction]**

Lys$_{63}$—Gly$_{76}$
[Signaling (e.g. immunity)]

Lys$_{48}$—Gly$_{76}$
[Proteasomal degradation]

Lys$_{33}$—Gly$_{76}$
[T-lymphocyte control]

Lys$_{11}$—Gly$_{76}$
[Cell division]

**FIGURE 3-34 Determination of polyubiquitin function by the lysine used for inter-ubiquitin isopeptide bonds.** Different ubiquitin ligases catalyze polyubiquitination of distinct target (substrate) proteins (colored ovals) using distinct lysine side chains of ubiquitin molecules (purple) to generate the inter-ubiquitin isopeptide linkages (blue) with Gly76 of the adjacent ubiquitin. Dotted blue arrows represent additional ubiquitins in the chain that are not shown.

The lysine used for the isopeptide bonds determines the function of the polyubiquitination. For example, polyubiquitins with Lys48:Gly76 isopeptide bonds direct the target to proteasomes for degradation. Those with Lys63, Lys33, and Lys11 influence signaling, T-lymphocyte control, and cell division, respectively. Isopeptide bonds involving ubiquitin's Lys6, Lys27, and Lys29 and bonds using its N-terminal amino group (not shown) can also be used to generate polyubiquitin chains.

different amino groups in the ubiquitin can be used to form an isopeptide bond with the carboxy terminal Gly76 in an adjacent ubiquitin in a polyubiquitin chain. All seven lysine residues in ubiquitin (Lys6, Lys11, Lys27, Lys29, Lys33, Lys48, and Lys63) and the N-terminal amino group of ubiquitin can participate in inter-ubiquitin linkages. Different ubiquitin ligases exhibit specificity for both the target (substrate) to be ubiquitinated and the lysine side chains on the ubiquitins that participate in the inter-ubiquitin isopeptide linkages (Lys63 or Lys 48, etc.) (Figure 3-34). These multiple forms of ubiquitination result in the generation of a wide variety of recognition surfaces that can participate in many protein-protein interactions with the hundreds of proteins (>200 in humans) that contain more than a dozen distinct ubiquitin-binding domains (UBD). In addition, any given polyubiquitin chain has the potential to bind simultaneously more than one UBD-containing protein, leading to the formation of ubiquitination-dependent multiprotein complexes. Some deubiquitinases can remove an intact polyubiquitin chain from a modified protein ("anchored" chain) and thus generate a polyubiquitin chain not covalently linked to another protein ("unanchored" chain). Even these unanchored chains may serve a regulatory

role. With this great structural diversity, it is not surprising that cells use ubiquitination and deubiquitination to control many different cellular functions.

We have already seen how polyubiquitination via Lys48 residues is used to tag proteins for proteasomal degradation. Ubiquitination unrelated to protein degradation also can control diverse cell functions, including repair of damaged DNA, metabolism, messenger RNA synthesis (transcription), defense against pathogens, cell division/cell cycle progression, cell signaling pathways, trafficking of proteins within a cell, and programmed cell death (apoptosis). The lysine used to form the inter-ubiquitin isopeptide bonds can vary depending on the cellular system that is regulated (see Figure 3-34). For example, polyubiquitination with Lys63 linkages is used in many cellular identification and signaling systems, such as recognition of the presence of intracellular viral RNA and the consequent induction of a protective immune response. Lys11-linked polyubiquitin chains regulate cell division. Lys33-linked chains help suppress the activity of receptors on specialized white blood cells, called T lymphocytes (see Chapter 23) and so control the activity and function of the lymphocytes that bear these receptors.

## Proteolytic Cleavage Irreversibly Activates or Inactivates Some Proteins

Unlike phosphorylation and ubiquitination, which are reversible, the activation or inactivation of protein function by proteolytic cleavage is an irreversible mechanism for regulating protein activity. For example, many polypeptide hormones, such as insulin, are synthesized as longer precursors, and prior to secretion from cells some of their peptide bonds must be hydrolyzed for them to fold properly. In some cases, a single long precursor *prohormone* polypeptide can be cleaved into several distinct active hormones. To prevent the pancreatic serine proteases from inappropriately digesting proteins before they reach the small intestines, they are synthesized as *zymogens*, inactive precursor enzymes. Cleavage of a peptide bond near the N-terminus of trypsinogen (the zymogen of trypsin) by a highly specific protease in the small intestine generates a new N-terminal residue (Ile-16), whose amino group can form an ionic bond with the carboxylic acid side chain of an internal aspartic acid. This causes a conformation change that opens the substrate-binding site, activating the enzyme. The active trypsin can then activate trypsinogen, chymotrypsinogen, and other zymogens. Similar but more elaborate protease cascades (one protease activating inactive precursors of others) that can amplify an initial signal play important roles in several systems, such as the blood-clotting cascade and the complement system (see Chapter 23). The importance of carefully regulating such systems is clear—inappropriate clotting could fatally clog the circulatory system, while insufficient clotting could lead to uncontrolled bleeding.

An unusual and rare type of proteolytic processing, termed *protein self-splicing*, takes place in bacteria and some eukaryotes. This process is analogous to editing film: an internal segment of a polypeptide is removed and the ends of the polypeptide are rejoined (ligated). Unlike other forms of proteolytic processing, protein self-splicing is an autocatalytic process, which proceeds by itself without the participation of other enzymes. The excised peptide appears to eliminate itself from the protein by a mechanism similar to that used in the processing of some RNA molecules (see Chapter 8). In vertebrate cells, the processing of some proteins includes self-cleavage, but the subsequent ligation step is absent. One such protein is Hedgehog, a membrane-bound signaling molecule that is critical to a number of developmental processes (see Chapter 16).

## Higher-Order Regulation Includes Control of Protein Location and Concentration

All the regulatory mechanisms heretofore described affect a protein locally at its site of action, turning its activity on or off. Normal functioning of a cell, however, also requires the segregation of proteins to particular compartments such as the mitochondria, nucleus, and lysosomes. In regard to enzymes, compartmentation not only provides an opportunity for controlling the delivery of substrate or the exit of product but also permits competing reactions to take place simulta-

neously in different parts of a cell. We describe the mechanisms that cells use to direct various proteins to different compartments in Chapters 13 and 14.

## KEY CONCEPTS of Section 3.4

### Regulating Protein Function

- Proteins may be regulated at the level of protein synthesis, protein degradation, or the intrinsic activity of proteins through noncovalent or covalent interactions.

- The life span of intracellular proteins is largely determined by their susceptibility to proteolytic degradation.

- Many proteins are marked for destruction with a polyubiquitin tag by ubiquitin ligases and then degraded within proteasomes, large cylindrical complexes with multiple protease active sites in their interior chambers (see Figure 3-29).

- Ubiquitination of proteins is reversible due to the activity of deubiquitinating enzymes.

- In allostery, the noncovalent binding of one ligand molecule, the allosteric effector, induces a conformational change that alters a protein's activity or affinity for other ligands. The allosteric effector can be identical in structure to or different from the other ligands, whose binding it affects. The allosteric effector can be an activator or an inhibitor.

- In multimeric proteins, such as hemoglobin, that bind multiple identical ligand molecules (e.g., oxygen), the binding of one ligand molecule may increase or decrease the binding affinity for subsequent ligand molecules. This type of allostery is known as cooperativity (see Figure 3-30).

- Several allosteric mechanisms act as switches, turning protein activity on and off in a reversible fashion.

- Two classes of intracellular switch proteins regulate a variety of cellular processes: (1) $Ca^{2+}$-binding proteins (e.g., calmodulin) and (2) members of the GTPase superfamily (e.g., Ras), which cycle between active GTP-bound and inactive GDP-bound forms (see Figure 3-32).

- The phosphorylation and dephosphorylation of hydroxyl groups on serine, threonine, or tyrosine residue side chains by protein kinases and phosphatases provide reversible on/off regulation of numerous proteins (see Figure 3-33).

- Variations in the nature of the covalent attachment of ubiquitin to proteins (mono-, multi-, and polyubiquitination involving a variety of linkages between the ubiquitin monomers) are involved in a wide variety of cellular functions other than proteasome-mediated degradation, such as changes in the location or activity of proteins (see Figure 3-34).

- Many types of covalent and noncovalent regulation are reversible, but some forms of regulation, such as proteolytic cleavage, are irreversible.

- Higher-order regulation includes compartmentation of proteins and control of protein concentration.

## 3.5 Purifying, Detecting, and Characterizing Proteins

A protein often must be purified before its structure and the mechanism of its action can be studied in detail. However, because proteins vary in size, shape, oligomerization state, charge, and water solubility, no single method can be used to isolate all proteins. To isolate one particular protein from the estimated 10,000 different proteins in a particular type of cell is a daunting task that requires methods both for separating proteins and for detecting the presence of specific proteins.

Any molecule, whether protein, carbohydrate, or nucleic acid, can be separated, or *resolved*, from other molecules on the basis of their differences in one or more physical or chemical characteristics. The larger and more numerous the differences between two proteins, the easier and more efficient their separation. The three most widely used characteristics for separating proteins are *size*, defined as either length or mass; net electrical charge; and *binding affinity* for specific ligands. In this section, we briefly outline several important techniques for separating proteins; these separation techniques are also useful for the separation of nucleic acids and other biomolecules. (Specialized methods for removing membrane proteins from membranes are described in Chapter 10 after the unique properties of these proteins are discussed.) We then consider the use of radioactive compounds for tracking biological activity. Finally, we consider several techniques for characterizing a protein's mass, sequence, and three-dimensional structure.

### Centrifugation Can Separate Particles and Molecules That Differ in Mass or Density

The first step in a typical protein purification scheme is centrifugation. The principle behind centrifugation is that two particles in suspension (cells, cell fragments, organelles, or molecules) with different masses or densities will settle to the bottom of a tube at different rates. Remember, mass is the weight of a sample (measured in daltons or molecular weight units), whereas density is the ratio of its mass to volume (often expressed as grams/liter because of the methods used to measure density). Proteins vary greatly in mass but not in density. Unless a protein has an attached lipid or carbohydrate, its density will not vary by more than 15 percent from 1.37 g/cm$^3$, the average protein density. Heavier or more dense molecules settle, or sediment, more quickly than lighter or less dense molecules.

A centrifuge speeds sedimentation by subjecting particles in suspension to centrifugal forces as great as 1 million times the force of gravity, *g*, which can sediment particles as small as 10 kDa. Modern ultracentrifuges achieve these forces by reaching speeds of 150,000 revolutions per minute (rpm) or greater. However, small particles with masses of 5 kDa or less will not sediment uniformly even at such remarkably high rotation rates. The extraordinary technical achievements of modern ultracentrifuges can be appreciated by considering that they can rotate a several-pound rotor (about the size of an American football) that holds the samples in tubes at rates as high as 2500 revolutions per second!

Centrifugation is used for two basic purposes: (1) as a preparative technique to separate one type of material from others and (2) as an analytical technique to measure physical properties (e.g., molecular weight, density, shape, and equilibrium binding constants) of macromolecules. The sedimentation constant, *s*, of a protein is a measure of its sedimentation rate. The sedimentation constant is commonly expressed in svedbergs (S), where a typical, large protein complex is about 3–5S, a proteasome is 26S, and a eukaryotic ribosome is 80S.

**Differential Centrifugation** The most common initial step in protein purification from cells or tissues is the separation of water-soluble proteins from insoluble cellular material by *differential centrifugation*. A starting mixture, commonly a cell homogenate (mechanically broken cells), is poured into a tube and spun at a rotor speed and for a period of time that forces cell organelles such as nuclei and large unbroken cells or large cell fragments to collect as a pellet at the bottom; the soluble proteins remain in the supernatant (Figure 3-35a). The supernatant fraction then is poured off, and either it or the pellet can be subjected to other purification methods to separate the many different proteins that they contain.

**Rate-Zonal Centrifugation** On the basis of differences in their masses, water-soluble proteins can be separated by centrifugation through a solution of increasing density called a *density gradient*. A concentrated sucrose solution is commonly used to form density gradients. When a protein mixture is layered on top of a sucrose gradient in a tube and subjected to centrifugation, each protein in the mixture migrates down the tube at a rate controlled by the factors that affect the sedimentation constant. All the proteins start from a thin zone at the top of the tube and separate into bands (actually, disks) of proteins of different masses. In this separation technique, called *rate-zonal centrifugation*, samples are centrifuged just long enough to separate the molecules of interest into discrete bands (Figure 3-35b). If a sample is centrifuged for too short a time, the different protein molecules will not separate sufficiently. If a sample is centrifuged much longer than necessary, all the proteins will end up in a pellet at the bottom of the tube.

Although the sedimentation rate is strongly influenced by particle mass, rate-zonal centrifugation is seldom effective in determining precise molecular weights because variations in shape also affect sedimentation rate. The exact effects of shape are hard to assess, especially for proteins or other molecules, such as single-stranded nucleic acid molecules, that can assume many complex shapes. Nevertheless, rate-zonal centrifugation has proved to be a practical method for separating many different types of polymers and particles. A second density-gradient technique, called *equilibrium density-gradient centrifugation*, is used mainly to separate DNA, lipoproteins that carry lipids through the circulatory system, or organelles (see Figure 9-35).

## EXPERIMENTAL FIGURE 3-35

**Centrifugation techniques separate particles that differ in mass or density.** (a) In differential centrifugation, a cell homogenate or other mixture is spun long enough to sediment the larger particles (e.g., cell organelles, cells), which collect as a pellet at the bottom of the tube (step **2**). The smaller particles (e.g., soluble proteins, nucleic acids) remain in the liquid supernatant, which can be transferred to another tube (step **3**). (b) In rate-zonal centrifugation, a mixture is spun (step **1**) just long enough to separate molecules that differ in mass but may be similar in shape and density (e.g., globular proteins, RNA molecules) into discrete zones within a density gradient commonly formed by a concentrated sucrose solution. Fractions are removed from the bottom of the tube and subjected to testing (assayed).

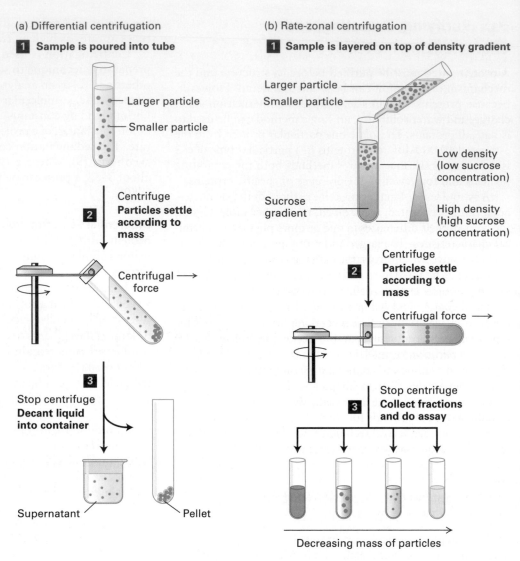

**(a) Differential centrifugation**

**1** Sample is poured into tube

Larger particle
Smaller particle

Centrifuge
**Particles settle according to mass**

**2**

Centrifugal force →

**3**

Stop centrifuge
**Decant liquid into container**

Supernatant
Pellet

**(b) Rate-zonal centrifugation**

**1** Sample is layered on top of density gradient

Larger particle
Smaller particle

Sucrose gradient

Low density (low sucrose concentration)
High density (high sucrose concentration)

Centrifuge
**2** **Particles settle according to mass**

Centrifugal force →

Stop centrifuge
**3** **Collect fractions and do assay**

Decreasing mass of particles

## Electrophoresis Separates Molecules on the Basis of Their Charge-to-Mass Ratio

**Electrophoresis** is a technique for separating molecules in a mixture under the influence of an applied electric field and is one of the most frequently used techniques to study proteins and nucleic acids. Dissolved molecules in an electric field move, or migrate, at a speed determined by their charge-to-mass (charge:mass) ratio and the physical properties of the medium through which they migrate. For example, if two molecules have the same mass and shape, the one with the greater net charge will move faster toward an electrode of the opposite polarity.

**SDS-Polyacrylamide Gel Electrophoresis** Because many proteins or nucleic acids that differ in size and shape have nearly identical charge:mass ratios, electrophoresis of these macromolecules in solution results in little or no separation of molecules of different lengths. However, successful separation of proteins and nucleic acids can be accomplished by electrophoresis in various gels (semisolid suspensions in water similar to the congealed gelatin found in desserts) rather than in a liquid solution. Electrophoretic separation of proteins is most commonly performed in polyacrylamide gels. When a mixture of proteins is placed in a gel and an electric current is applied, smaller proteins migrate faster through the gel than do larger proteins because the gel acts as a sieve, with smaller species able to maneuver more rapidly through the pores in the gel than larger species. The shape of a molecule can also influence its rate of migration (long asymmetric molecules migrate more slowly than spherical ones of the same mass).

Gels are cast into flat, relatively thin slabs between a pair of glass plates by polymerizing a solution of acrylamide monomers into polyacrylamide chains and simultaneously cross-linking the chains into a semisolid matrix. The pore size of a gel can be varied by adjusting the concentrations of polyacrylamide and the cross-linking reagent. The rate at which a protein moves through a gel is influenced by the gel's pore size and the strength of the electric field. By suitable adjustment of these parameters, proteins of widely varying sizes can be resolved (separated from one another) by polyacrylamide gel electrophoresis (PAGE).

**EXPERIMENTAL FIGURE 3-36 SDS-polyacrylamide gel electrophoresis (SDS-PAGE) separates proteins primarily on the basis of their masses.** (a) Initial treatment with SDS, a negatively charged detergent, dissociates multimeric proteins and denatures all the polypeptide chains (step **1**). During electrophoresis, the SDS-protein complexes migrate through the polyacrylamide gel (step **2**). Small complexes are able to move through the pores faster than larger ones. Thus the proteins separate into bands according to their sizes as they migrate. The separated protein bands are visualized by staining with a dye (step **3**). (b) Example of SDS-PAGE separation of all the proteins in a whole-cell lysate (detergent solubilized cells): (*left*) the many separate stained proteins, appearing almost as a continuum; (*right*) a protein purified from the lysate by a single step of antibody-affinity chromatography. The proteins were visualized by staining with a silver-based dye. [Part (b) modified from B. Liu and M. Krieger, 2002, *J. Biol. Chem.* **277**(37):34125–34135.]

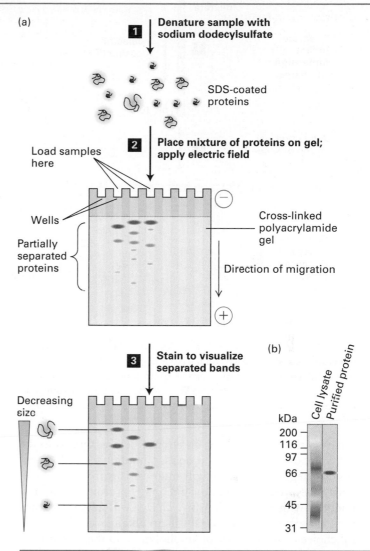

In the most powerful technique for resolving protein mixtures, proteins are exposed to the ionic detergent SDS (sodium dodecylsulfate) before and during gel electrophoresis (Figure 3-36). SDS denatures proteins, in part because it binds to hydrophobic side chains, destabilizing the hydrophobic interactions in the core of a protein that contribute to its stable conformation. (SDS treatment is usually combined with heating, often in the presence of reducing agents that break disulfide bonds.) Under these conditions, multimeric proteins dissociate into their subunits, and typically the amount of SDS that binds to the protein is proportional to the length of the polypeptide chain and relatively independent of the sequence. Two proteins of similar size will bind the same absolute quantity of SDS, whereas a protein twice that size will bind twice the amount of SDS. Denaturation of a complex protein mixture with SDS in combination with heat usually forces each polypeptide chain into an extended conformation and imparts on each of the proteins in the mixture a constant mass/charge ratio because the dodecylsulfate is negatively charged. As the SDS-bound proteins move through the polyacrylamide gel, they are separated according to size by the sieving action of the gel. SDS treatment thus eliminates the effect of differences in shape in native structures; therefore, chain length, which corresponds to mass, is the principal determinant of the migration rate of proteins in *SDS-polyacrylamide electrophoresis (SDS-PAGE)*. Even chains that differ in molecular weight by less than 10 percent can be resolved by this technique. Moreover, the molecular weight of a protein can be estimated by comparing the distance that it migrates through a gel with the distances that proteins of known molecular weight migrate (there is roughly a linear relationship between migration distance and the log of the molecular weight). Proteins within the gels can be extracted for further analysis (e.g., identification by methods described below).

**Two-Dimensional Gel Electrophoresis** Electrophoresis of all cellular proteins by SDS-PAGE can separate proteins having relatively large differences in mass but cannot readily resolve proteins having similar masses (e.g., a 41-kDa protein versus a 42-kDa protein). To separate proteins of similar masses, another physical characteristic must be exploited. Most commonly, this characteristic is electric charge, which is determined by the pH and the relative number of the protein's positively and negatively charged groups, which is in turn dependent on the $pK_a$'s of the ionizable groups (see Chapter 2). Two unrelated proteins having similar masses are unlikely to have identical net charges because their sequences, and thus the number of acidic and basic residues, are different.

In two-dimensional electrophoresis, proteins are separated sequentially, first by their charges and then by their masses (Figure 3-37a). In the first step, a cell or tissue extract is fully denatured by high concentrations (8 M) of urea (and sometimes SDS) and then layered on a gel strip that contains urea, which removes any bound SDS, and a continuous pH gradient. The pH gradient is formed by ampholytes, a mixture of polyanionic and polycationic small molecules, that

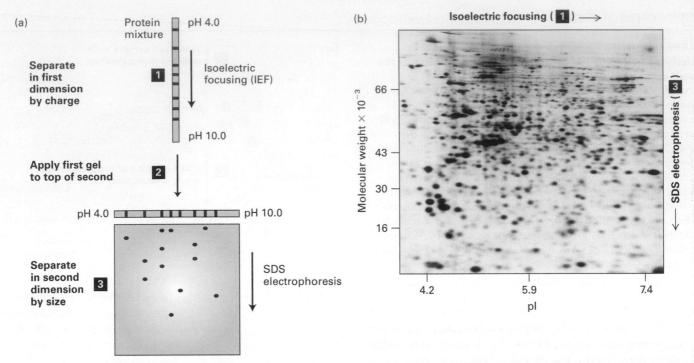

**EXPERIMENTAL FIGURE 3-37 Two-dimensional gel electrophoresis separates proteins on the basis of charge and mass.** (a) In this technique, proteins are first separated into bands on the basis of their charges by isoelectric focusing (step **1**). The resulting gel strip is applied to an SDS-polyacrylamide gel (step **2**), and the proteins are separated into spots by mass (step **3**). (b) In this two-dimensional gel of a protein extract from cultured cells, each spot represents a single polypeptide. Polypeptides can be detected by dyes, as here, or by other techniques such as autoradiography. Each polypeptide is characterized by its isoelectric point (pI) and molecular weight. [Part (b) courtesy of J. Celis.]

are cast into the gel. When an electric field is applied to the gel, ampholytes will migrate, so that ampholytes with an excess of negative charges will migrate toward the anode, where they establish an acidic pH (many protons), while ampholytes with an excess positive charge will migrate toward the cathode, where they establish an alkaline pH. The careful choice of the mixture of ampholytes and the preparation of the gel allows the construction of stable pH gradients anywhere from pH 3 to pH 10. A charged protein placed onto one end of such a gel will migrate through the gradient until it reaches its **isoelectric point** (**pI**), the pH at which the net charge of the protein is zero. With no net charge, the protein will not migrate under the influence of the electric field. This technique, called *isoelectric focusing* (IEF), can resolve proteins that differ by only one charge unit. Proteins that have been separated on an IEF gel can then be separated in a second dimension on the basis of their molecular weights. To accomplish this separation, the IEF gel strip is placed lengthwise on one outside edge of a sheet-like (two-dimensional, or slab) polyacrylamide gel, this time saturated with SDS to confer on each separated protein a more or less constant mass:charge ratio. When an electric field is imposed, the proteins will migrate from the IEF gel into the SDS slab gel and then separate according to their masses.

The sequential resolution of proteins by charge and mass can achieve excellent separation of cellular proteins (Figure 3-37b). For example, two-dimensional gels have been very useful in comparing the proteomes in undifferentiated and differentiated cells or in normal and cancer cells because as many as 1000 proteins can be resolved as individual spots simultaneously. Unfortunately, membrane proteins separate relatively poorly using this technique. Sophisticated methods have been developed to permit the comparison of complex patterns of proteins in two-dimensional gels from related, but distinct, samples (e.g., tissue from a normal versus a mutant individual) to permit identification of differences in the types or amounts of proteins in the samples (see Section 3.6, on proteomics, below). Sophisticated mass spectrometry methods, described below, are often used in place of two-dimensional gel electrophoresis to identify the protein components of a complex sample.

## Liquid Chromatography Resolves Proteins by Mass, Charge, or Binding Affinity

A third common technique for separating mixtures of proteins or fragments of proteins, as well as other molecules, is based on the principle that molecules dissolved in a solution can differentially interact (bind and dissociate) with a particular solid surface, depending on the physical and chemical properties of the molecule and the surface. If the solution is allowed to flow across the surface, then molecules that interact frequently with the surface will spend more time bound to the surface and thus flow past the surface more slowly than molecules that interact infrequently with it. In this technique, called **liquid chromatography** (LC), the sample is placed on top of a tightly packed column of spherical beads held within a glass or plastic cylinder. The sample then flows down the column, usually driven by gravitational or hydrostatic forces alone or with the assistance of a pump, and small aliquots of fluid flowing out of the column, called *fractions*, are collected sequentially for subsequent analysis for the presence of the proteins of interest. The nature of the beads in the column determines whether the separation of proteins depends on differences in mass, charge, or binding affinity.

**Gel Filtration Chromatography** Proteins that differ in mass can be separated on a column composed of porous beads made from polyacrylamide, dextran (a bacterial polysaccharide), or agarose (a seaweed derivative)—a technique called gel filtration chromatography. Although proteins flow around the spherical beads in gel filtration chromatography, they spend some time within the large depressions that cover a bead's surface. Because smaller proteins can penetrate into these depressions more readily than larger proteins can, they travel through a gel filtration column more slowly than larger proteins (Figure 3-38a). (In contrast, proteins migrate *through* the pores in an electrophoretic gel; thus smaller proteins move faster than larger ones.) The total volume of liquid required to elute (or separate and remove) a protein from a gel filtration column depends on its mass: the smaller the mass, the more time it is trapped on the beads, the greater the elution volume. By use of proteins of known mass as standards to calibrate the column, the elution volume can be used to estimate the mass of a protein in a mixture. A protein's shape as well as its mass can influence the elution volume.

**Ion-Exchange Chromatography** In ion-exchange chromatography, a second type of liquid chromatography, proteins are separated on the basis of differences in their charges. This technique makes use of specially modified beads whose surfaces are covered by amino groups or carboxyl groups and thus carry either a positive charge ($NH_3^+$) or a negative charge ($COO^-$) at neutral pH.

The proteins in a mixture carry various net charges at any given pH. When a solution of a protein mixture flows through a column of positively charged beads, only proteins with a net negative charge (acidic proteins) adhere to the beads; neutral and positively charged (basic) proteins flow unimpeded through the column (Figure 3-38b). The acidic proteins are then eluted selectively from the column by passing a solution of increasing concentrations of salt (a salt gradient) through the column. At low salt concentrations, protein molecules and beads are attracted by their opposite charges. At higher salt concentrations, negative salt ions bind to the positively charged beads, displacing the negatively charged proteins. In a gradient of increasing salt concentration, weakly bound proteins, those with relatively low charge, are eluted first and highly charged proteins are eluted last. Similarly, a negatively charged column can be used to retain and fractionate basic (positively charged) proteins.

**Affinity Chromatography** The ability of proteins to bind specifically to other molecules is the basis of affinity chroma-

tography. In this technique, ligand or other molecules that bind to the protein of interest are covalently attached to the beads used to form the column. Ligands can be enzyme substrates, inhibitors or their analogs, or other small molecules that bind to specific proteins. In a widely used form of this technique—*antibody-affinity*, or *immunoaffinity*, *chromatography*—the attached molecule is an antibody specific for the desired protein (Figure 3-38c). (We discuss antibodies as tools to study proteins next; see also Chapter 23, which describes how antibodies are made.)

An affinity column in principle will retain only those proteins that bind the molecule attached to the beads; the remaining proteins, regardless of their charges or masses, will pass through the column because they do not bind. However, if a retained protein is in turn bound to other molecules, forming a complex, then the entire complex is retained on the column. The proteins bound to the affinity column are then eluted by adding an excess of a soluble form of the ligand, by exposure of bound materials to detergents, or by changing the salt concentration or pH such that the binding to the molecule on the column is disrupted. The ability of this technique to separate particular proteins depends on the selection of appropriate binding partners that bind more tightly to the protein of interest than to other proteins.

## Highly Specific Enzyme and Antibody Assays Can Detect Individual Proteins

The purification of a protein, or any other molecule, requires a specific assay that can detect the presence of the molecule of interest as it is separated from other molecules (e.g., in column or density-gradient fractions or gel bands or spots). An assay capitalizes on some highly distinctive characteristic of a protein: the ability to bind a particular ligand, to catalyze a particular reaction, or to be recognized by a specific antibody. An assay must also be simple and fast to minimize errors and the possibility that the protein of interest becomes denatured or degraded while the assay is performed. The goal of any purification scheme is to isolate sufficient amounts of a given protein for study; thus a useful assay must also be sensitive enough that only a small proportion of the available material is consumed by it. Many common protein assays require just $10^{-9}$ to $10^{-12}$ g of material.

**Chromogenic and Light-Emitting Enzyme Reactions** Many assays are tailored to detect some functional aspect of a protein. For example, enzymatic activity assays are based on the ability to detect the loss of substrate or the formation of product. Some enzymatic assays utilize chromogenic substrates, which change color in the course of the reaction. (Some substrates are naturally chromogenic; if they are not, they can be linked to a chromogenic molecule.) Because of the specificity of an enzyme for its substrate, only samples that contain the enzyme will change color in the presence of a chromogenic substrate; the rate of the reaction provides a measure of the quantity of enzyme present. Enzymes that catalyze chromogenic reactions can also be fused or chemically linked to an

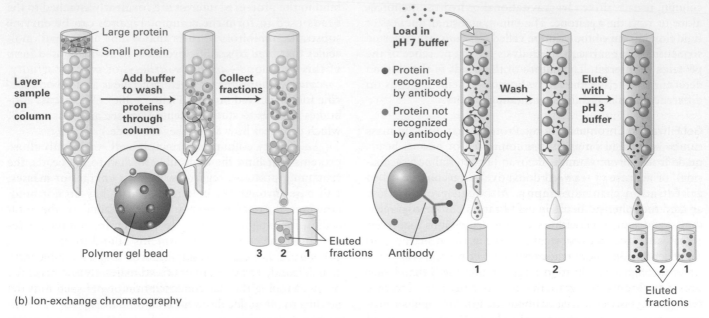

**(a) Gel filtration chromatography**

Large protein
Small protein

Layer sample on column

Add buffer to wash proteins through column

Collect fractions

Polymer gel bead

3  2  1  Eluted fractions

**(b) Ion-exchange chromatography**

Negatively charged protein ●
Positively charged protein ●

Layer sample on column

Cations elute out

Anions retained by beads

Elute negatively charged protein with salt solution (NaCl)

Positively charged gel bead

Na⁺  Cl⁻

Eluted fractions

3  2  1  Eluted fractions

4  3  2  1

**(c) Antibody-affinity chromatography**

Load in pH 7 buffer

● Protein recognized by antibody
● Protein not recognized by antibody

Wash

Elute with pH 3 buffer

Antibody

1  2  3  2  1

Eluted fractions

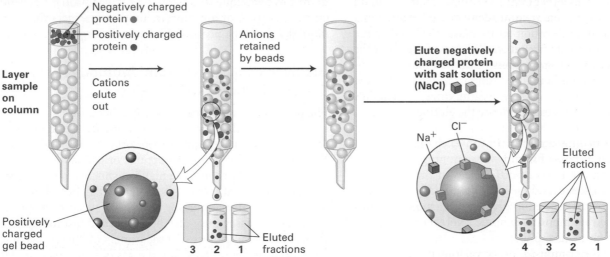

**EXPERIMENTAL FIGURE 3-38  Three commonly used liquid chromatographic techniques separate proteins on the basis of mass, charge, or affinity for a specific binding partner.** (a) Gel filtration chromatography separates proteins that differ in size. A mixture of proteins is carefully layered on the top of a cylinder packed with porous beads. Smaller proteins travel through the column more slowly than larger proteins. Thus different proteins emerging in the eluate flowing out of the bottom of the column at different times (different elution volumes) can be collected in separate tubes, called *fractions*. (b) Ion-exchange chromatography separates proteins that differ in net charge in columns packed with special beads that carry either a positive charge (shown here) or a negative charge. Proteins having the same net charge as the beads are repelled and flow through the column, whereas proteins having the opposite charge bind to the beads more or less tightly, depending on their structures. Bound proteins—in this case, negatively charged—are subsequently eluted by passing a salt gradient (usually of NaCl or KCl) through the column. As the ions bind to the beads, they displace the proteins (more tightly bound proteins require higher salt concentration in order to be released). (c) In antibody-affinity chromatography, a mixture of proteins is passed through a column packed with beads to which a specific antibody is covalently attached. Only protein with high affinity for the antibody is retained by the column; all the nonbinding proteins flow through. After the column is washed, the bound protein is eluted with an acidic solution or some other solution that disrupts the antigen-antibody complexes; the released protein then flows out of the column and is collected.

antibody and used to "report" the presence or location of an antigen to which the antibody binds (see below).

**Antibody Assays**  As noted earlier, antibodies have the distinctive characteristic of binding tightly and specifically to antigens. As a consequence, preparations of antibodies that recognize a protein antigen of interest can be generated and used to detect the presence of the protein, either in a complex mixture of other proteins (finding a needle in a haystack, as it were) or in a partially purified preparation of a particular protein. An antibody molecule will generally only bind tightly to a small part of a target molecule (antigen) that exhibits molecular complementarity with the antibody. This antibody-binding region of the target is called the antibody's cognate epitope, or simply epitope. Thus the presence of the antigen that contains an epitope can be visualized by labeling the antibody with an enzyme, a fluorescent molecule, or radioactive isotopes. For example, luciferase, an enzyme present in fireflies and some bacteria, can be linked to an antibody. In the presence of ATP and the substrate luciferin, luciferase catalyzes a light-emitting reaction. In either case, after the antibody binds to the protein of interest (the antigen) and unbound antibody is washed away, substrates of the linked enzyme are added and the appearance of color or emitted light is monitored. The intensity is proportional to the amount of enzyme-linked antibody, and thus antigen, in the sample. A variation of this technique, particularly useful in detecting specific proteins within living cells, makes use of *green fluorescent protein* (*GFP*), a naturally fluorescent protein found in jellyfish (see Figure 9-15). Alternatively, after the first antibody that is not otherwise labeled binds to its target protein, a second, labeled antibody that can recognize the first antibody is used to bind to the complex of the first antibody and its target. This combination of two antibodies permits very high sensitivity in the detection of a target protein because the labeled second antibody is often a mixture of antibodies that bind to multiple sites on the first antibody. It is important to remember that an antibody binds to or recognizes its cognate epitope on a target antigen. If that epitope is altered, for example by partial unfolding or post-translational modifications, or is blocked when the antigen protein is bound to some other molecule, the ability of the antibody to bind may be reduced or completely lost. Thus the absent of antibody binding does not necessarily mean the antigen is not present in a sample, only that the epitope is not present or accessible for binding.

To generate the antibodies, the intact protein or a fragment of the protein is injected into an animal (usually a rabbit, mouse, or goat). Sometimes a short synthetic peptide of 10–15 residues based on the sequence of the protein is used as the antigen to induce antibody formation. A synthetic peptide, when coupled to a large protein carrier, can induce an animal to produce antibodies that bind to that part (the epitope) of the full-size, natural protein. Biosynthetically or chemically attaching the epitope to an unrelated protein is called *epitope tagging*. As we'll see throughout this book, antibodies generated using either peptide epitopes or intact proteins are extremely versatile reagents for isolating, detecting, and characterizing proteins.

**Detecting Proteins in Gels** Proteins embedded within a one- or two-dimensional gel usually are not visible. The two general approaches for detecting proteins in gels are either to label or stain the proteins while they are still within the gel or to electrophoretically transfer the proteins to a membrane made of nitrocellulose or polyvinylidene difluoride and then detect them. Proteins within gels are usually stained with an organic dye or a silver-based stain, both detected with normal visible light, or with a fluorescent dye that requires specialized detection equipment. Coomassie blue is the most commonly used organic dye, typically used to detect ~1000 ng of protein, with a lower limit of detection of ~4–10 ng. Silver staining or fluorescence staining are more sensitive (lower limit of ~1 ng). Coomassie and other stains can also be used to visualize proteins after transfer to membranes; however, the most common method to visualize proteins in these membranes is immunoblotting, also called Western blotting.

**Immunoblotting** combines the resolving power of gel electrophoresis and the specificity of antibodies. This multistep procedure is commonly used to separate proteins and then identify a specific protein of interest. As shown in Figure 3-39, two different antibodies are used in this method, one that is specific for the desired protein and a second that binds to the first and is linked to an enzyme or other molecule that permits detection of the first antibody (and thus the protein of interest to which it binds). Enzymes to which the second antibody is attached can either generate a visible colored product or, by a process called *chemiluminescence*, produce light that can readily be recorded by film or a sensitive detector. The two different antibodies, sometimes called a "sandwich," are used to amplify the signals and improve sensitivity. If an antibody is not available but the gene encoding the protein is available and can be used to express the protein, recombinant DNA methods (see Chapter 5) can incorporate a small peptide epitope (epitope tagging) into the normal sequence of the protein that can be detected by a commercially available antibody to that epitope.

## Radioisotopes Are Indispensable Tools for Detecting Biological Molecules

A sensitive method for tracking a protein or other biological molecule is by detecting the radioactivity emitted from radioisotopes introduced into the molecule. At least one atom in a radiolabeled molecule is present in a radioactive form, called a **radioisotope**.

**Radioisotopes Useful in Biological Research** Hundreds of biological molecules (e.g., amino acids, nucleosides, and numerous metabolic intermediates) labeled with various radioisotopes are commercially available. These preparations vary considerably in their *specific activity*, which is the amount of radioactivity per unit of material, measured in disintegrations per minute (dpm) per millimole. The specific activity of a labeled compound depends on the probability of decay of the radioisotope, determined by its *half-life*, which is the time required for half the atoms to undergo radioactive decay. In general, the shorter the half-life of a radioisotope, the higher its specific activity (Table 3-1).

The specific activity of a labeled compound must be high enough that sufficient radioactivity is incorporated into molecules to be accurately detected. For example, methionine and cysteine labeled with sulfur-35 ($^{35}S$) are widely used to biosynthetically label cellular proteins because preparations of these amino acids with high specific activities ($>10^{15}$ dpm/mmol) are available. Likewise, commercial preparations of $^3H$-labeled nucleic acid precursors have much higher specific activities than those of the corresponding $^{14}C$-labeled preparations. In most experiments, the former are preferable

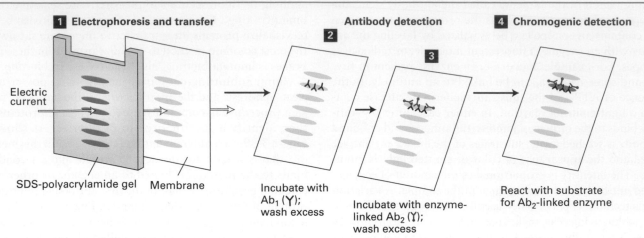

**① Electrophoresis and transfer**

Electric current

SDS-polyacrylamide gel — Membrane

**② Antibody detection**

Incubate with Ab₁ (Y); wash excess

Incubate with enzyme-linked Ab₂ (Y); wash excess

**④ Chromogenic detection**

React with substrate for Ab₂-linked enzyme

**EXPERIMENTAL FIGURE 3-39 Western blotting (immunoblotting) combines several techniques to resolve and detect a specific protein.** Step **①**: After a protein mixture has been electrophoresed through an SDS gel, the separated bands (or spots, for a two-dimensional gel) are transferred (blotted) from the gel onto a porous membrane from which it is not readily removed. Step **②**: The membrane is flooded with a solution of antibody (Ab₁) specific for the desired protein and allowed to incubate for a while. Only the membrane-bound band containing this protein binds the antibody, forming a layer of antibody molecules (whose position cannot be seen at this point). Then the membrane is washed to remove unbound Ab₁. Step **③**: The membrane is incubated with a second antibody (Ab₂) that specifically recognizes and binds to the first Ab₁. This second antibody is covalently linked to either an enzyme (e.g., alkaline phosphatase, which can catalyze a chromogenic reaction), a radioactive isotope, or some other substance whose presence can be detected with great sensitivity. Step **④**: Finally, the location and amount of bound Ab₂ are detected (e.g., by a deep-purple precipitate from chromogenic reaction), permitting the electrophoretic mobility (and therefore the mass) of the desired protein to be determined, as well as its quantity (based on band intensity).

because they allow RNA or DNA to be adequately labeled after a shorter time of incorporation or require a smaller cell sample. Various phosphate-containing compounds in which the phosphorus atom is the radioisotope phosphorus-32 are readily available. Because of their high specific activity, $^{32}$P-labeled nucleotides are routinely used to label nucleic acids in cell-free systems.

Labeled compounds in which a radioisotope replaces atoms normally present in the molecule have virtually the same chemical properties as the corresponding nonlabeled compounds. Enzymes, for instance, generally cannot distinguish between substrates labeled in this way and their nonla-beled substrates. The presence of such radioactive atoms is indicated with the isotope in brackets (no hyphen) as a prefix (e.g., [$^3$H]leucine). In contrast, labeling almost all biomolecules (e.g., protein or nucleic acid) with the radioisotope iodine-125 ($^{125}$I) requires the covalent addition of $^{125}$I to a molecule that normally does not have iodine as part of its structure. Because this labeling procedure modifies the chemical structure, the biological activity of the labeled molecule may differ somewhat from that of the nonlabeled form. The presence of such radioactive atoms is indicated with the isotope as a prefix with a hyphen (no bracket) (e.g., $^{125}$I-trypsin). Standard methods for labeling proteins with $^{125}$I result in covalent attachment of the $^{125}$I primarily to the aromatic rings of tyrosine side chains (mono- and diiodotyrosine). Nonradioactive isotopes find increasing use in cell biology, especially in nuclear magnetic resonance studies and in mass spectroscopy applications, as will be explained below.

**Labeling Experiments and Detection of Radiolabeled Molecules** Whether labeled compounds are detected by **autoradiography**, a semiquantitative visual assay, or their radioactivity is measured in an appropriate "counter," a highly quantitative assay that can determine the amount of a radiolabeled compound in a sample, depends on the nature of the experiment. In some experiments, both types of detection are used.

In one use of autoradiography, a tissue, cell, or cell constituent is labeled with a radioactive molecule, unassociated

| TABLE 3-1 | Radioisotopes Commonly Used in Biological Research |
| --- | --- |
| **Isotope** | **Half-Life** |
| Phosphorus-32 | 14.3 days |
| Iodine-125 | 60.4 days |
| Sulfur-35 | 87.5 days |
| Tritium (hydrogen-3) | 12.4 years |
| Carbon-14 | 5730.4 years |

radioactive material is washed away, and the structure of the sample is stabilized either by chemically cross-linking the macromolecules in the sample ("fixation") or by freezing it. The sample is then overlaid with a photographic emulsion sensitive to radiation. Development of the emulsion yields small silver grains whose distribution corresponds to that of the radioactive material and is usually detected by microscopy. Autoradiographic studies of whole cells were crucial in determining the intracellular sites where various macromolecules are synthesized and the subsequent movements of these macromolecules within cells. Various techniques employing fluorescent microscopy, which we describe in Chapter 9, have largely supplanted autoradiography for studies of this type. However, autoradiography is sometimes used in various assays for detecting specific isolated DNA or RNA sequences at specific tissue locations (see Chapter 5) in a technique referred to as in situ hybridization.

Quantitative measurements of the amount of radioactivity in a labeled material are performed with several different instruments. A Geiger counter measures ions produced in a gas by the β particles or γ rays emitted from a radioisotope. These instruments are mostly handheld devices used to monitor radioactivity in the laboratory to protect investigators from excess exposure. In a scintillation counter, a radiolabeled sample is mixed with a liquid containing a fluorescent compound that emits a flash of light when it absorbs the energy of the β particles or γ rays released in the decay of the radioisotope; a phototube in the instrument detects and counts these light flashes. Phosphorimagers are used to detect radioactivity using a two-dimensional array detector, storing digital data on the number of decays in disintegrations per minute per small pixel of surface area. These instruments, which can be thought of as a kind of reusable electronic film, are commonly used to quantitate radioactive molecules separated by gel electrophoresis and are replacing photographic emulsions for this purpose.

A combination of labeling and biochemical techniques and of visual and quantitative detection methods is often employed in labeling experiments. For instance, to identify the major proteins synthesized by a particular cell type, a sample of the cells is incubated with a radioactive amino acid (e.g., [$^{35}$S]methionine) for a few minutes, during which time the labeled amino acid enters the cells and mixes with the cellular pool of unlabeled amino acids and some of the labeled amino acid is biosynthetically incorporated into newly synthesized protein. Subsequently unincorporated radioactive amino acid is washed away from the cells. The cells are harvested; the mixture of cellular proteins is extracted from the cells (for example, by a detergent solution) and then separated by any of the commonly used methods to resolve complex protein mixtures into individual components. Gel electrophoresis in combination with autoradiography or phosphorimager analysis is often the method of choice. The radioactive bands in the gel correspond to newly synthesized proteins, which have incorporated the radiolabeled amino acid. To detect a specific protein of interest rather than the entire ensemble of biosynthetically radiolabeled proteins, a specific antibody to the protein of interest can be used to precipitate the protein away from the other proteins in the sample (immunoprecipitation). The precipitate is then solubilized under denaturing conditions, for example, in an SDS-containing buffer, to separate the antibody from the protein, and the sample is analyzed by SDS-PAGE followed by autoradiography. In this type of experiment, a fluorescent compound that is activated by the radiation ("scintillator") may be infused into the gel on completion of the electrophoretic separation so that the light emitted can be used to detect the presence of the labeled protein, either using film or a two-dimensional electronic detector. This method is particularly useful for weak β emitters such as $^3$H.

**Pulse-chase** experiments are particularly useful for tracing changes in the intracellular location of proteins or the modification of a protein or metabolite over time. In this experimental protocol, a cell sample is exposed to a radiolabeled compound that can be incorporated or otherwise attached to a cellular molecule of interest—the "pulse"—for a brief period. The pulse ends when the unincorporated radioactive molecules are washed away and the cells are exposed to a vast excess of the identical, but unlabeled compound to dilute the radioactivity of any remaining, but unincorporated radioactive compound. This procedure prevents further incorporation of significant amounts of radiolabel after the "pulse" period and initiates the "chase" period (Figure 3-40). Samples taken periodically during the chase period are assayed to determine the location or chemical form of the radiolabel as a function of time. Often, pulse-chase experiments, in which the protein is detected by autoradiography after immunoprecipitation and SDS-PAGE, are used to follow the rate of synthesis, modification, and degradation of proteins by adding radioactive amino acid precursors during the pulse and then detecting the amounts and characteristics of the radioactive protein during the chase. One can thus observe postsynthetic modifications of the protein that change its electrophoretic mobility and the rate of degradation of a specific protein, which is detected as the loss of signal with increasing time of chase. A classic use of the pulse-chase technique was in studies to elucidate the pathway traversed by secreted proteins from their site of synthesis in the endoplasmic reticulum to the cell surface (see Chapter 14).

## Mass Spectrometry Can Determine the Mass and Sequence of Proteins

Mass spectrometry (MS) is a powerful technique for characterizing proteins, especially for determining the mass of a protein or fragments of a protein. With such information in hand, it is also possible to determine part of or all of the protein's sequence. This method permits the very highly accurate direct determination of the ratio of the mass ($m$) of a charged molecule (molecular ion) to its charge ($z$), or $m/z$. Techniques are then used to deduce the absolute mass of the molecular ion. There are four key features of all mass spectrometers. The first is an ion source, from which charge, usually in the form of protons, is transferred to the peptide or

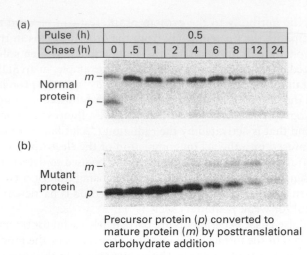

| Pulse (h) | 0.5 | | | | | | | | |
|---|---|---|---|---|---|---|---|---|---|
| Chase (h) | 0 | .5 | 1 | 2 | 4 | 6 | 8 | 12 | 24 |

(a)

Normal protein    *m* —    *p* —

(b)

Mutant protein    *m* —    *p* —

Precursor protein (*p*) converted to mature protein (*m*) by posttranslational carbohydrate addition

**EXPERIMENTAL FIGURE 3-40 Pulse-chase experiments can track the pathway of protein modification or movement within cells.** (a) To follow the fate of a specific newly synthesized protein in a cell, cells were incubated with [$^{35}$S]methionine for 0.5 h (the pulse) to label all newly synthesized proteins, and the radioactive amino acid not incorporated into the cells was then washed away. The cells were further incubated (the chase) for varying times up to 24 hours, and samples from each time of chase were subjected to immunoprecipitation to isolate one specific protein (here the low-density lipoprotein receptor). SDS-PAGE of the immunoprecipitates followed by autoradiography permitted visualization of the one specific protein, which is initially synthesized as a small precursor (*p*) and then rapidly modified to a larger mature form (*m*) by addition of carbohydrates. About half of the labeled protein was converted from *p* to *m* during the pulse; the rest was converted after 0.5 hour of chase. The protein remains stable for 6–8 hours before it begins to be degraded (indicated by reduced band intensity). (b) The same experiment was performed in cells in which a mutant form of the protein is made. The mutant *p* form cannot be properly converted to the *m* form, and it is more quickly degraded than the normal protein. [Adapted from K. F. Kozarsky, H. A. Brush, and M. Krieger, 1986, *J. Cell Biol.* **102**(5):1567–1575.]

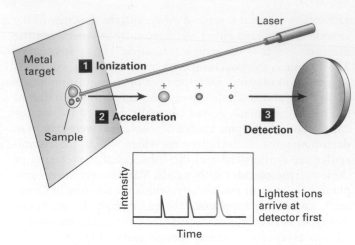

**EXPERIMENTAL FIGURE 3-41 Molecular mass can be determined by matrix-assisted laser desorption/ionization time-of-flight (MALDI-TOF) mass spectrometry.** In a MALDI-TOF mass spectrometer, pulses of light from a laser ionize a protein or peptide mixture that is absorbed on a metal target (step **1**). An electric field accelerates the ions in the sample toward the detector (steps **2** and **3**). The time to the detector is proportional to the square root of the mass-to-charge (*m/z*) ratio. For ions having the same charge, the smaller ions move faster (shorter time to the detector). The molecular weight is calculated using the time of flight of a standard.

protein molecules. The formation of these ions occurs in the presence of a high electric field that then directs the charged molecular ions into the second key component, the mass analyzer. The mass analyzer, which is always in a high vacuum chamber, physically separates the ions on the basis of their differing mass-to-charge (*m/z*) ratios. The mass-separated ions are subsequently directed to strike a detector, the third key component, which provides a measure of the relative abundances of each of the ions in the sample. The fourth essential component is a computerized data system that is used to calibrate the instrument; acquire, store, and process the resulting data; and often direct the instrument automatically to collect additional specific types of data from the sample, based on the initial observations. This type of automated feedback is used for the tandem MS (MS/MS) peptide-sequencing methods described below.

The two most frequently used methods of generating ions of proteins and protein fragments are (1) matrix-assisted laser desorption/ionization (MALDI) and (2) electrospray

(ES). In MALDI (Figure 3-41) the peptide or protein sample is mixed with a low-molecular-weight, UV-absorbing organic acid (the matrix) and then dried on a metal target. Energy from a laser ionizes and vaporizes the sample, producing singly charged molecular ions from the constituent molecules. In ES (Figure 3-42a), the sample of peptides or proteins in solution is converted into a fine mist of tiny droplets by spraying through a narrow capillary at atmospheric pressure. The droplets are formed in the presence of a high electric field, rendering them highly charged. The droplets evaporate in their short flight (mm) to the entrance of the mass spectrometer's analyzer, forming multiply charged ions from the peptides and proteins. The gaseous ions are sampled into the analyzer region of the MS, where they are then accelerated by electric fields and separated by the mass analyzer on the basis of their *m/z*.

The two most frequently used mass analyzers are time-of-flight (TOF) instruments and ion traps. TOF instruments exploit the fact that the time it takes an ion to pass through the length of the analyzer before reaching the detector is proportional to the square root of *m/z* (smaller ions move faster than larger ones with the same charge; see Figure 3-41). In ion-trap analyzers, tunable electric fields are used to capture, or "trap," ions with a specific *m/z* and to sequentially pass the trapped ions out of the analyzer onto the detector (see Figure 3-42a). By varying the electric fields, ions with a wide range of *m/z* values can be examined one by one, producing a mass spectrum, which is a graph of *m/z* (x axis) versus relative abundance (y axis) (Figure 3-42b, *top panel*).

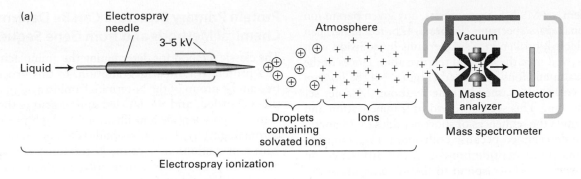

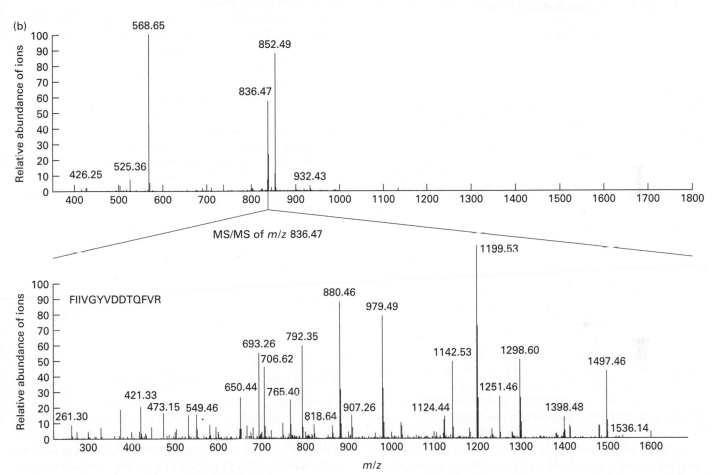

**EXPERIMENTAL FIGURE 3-42 Molecular mass of proteins and peptides can be determined by electrospray ionization ion-trap mass spectrometryz.** (a) Electrospray (ES) ionization converts proteins and peptides in a solution into highly charged gaseous ions by passing the solution through a needle (forming the droplets) that has a high voltage across it (charging the droplets). Evaporation of the solvent produces gaseous ions that enter a mass spectrometer. The ions are analyzed by an ion-trap mass analyzer that then directs ions to the detector. (b) *Top panel:* Mass spectrum of a mixture of three major and several minor peptides is presented as the relative abundance of the ions striking the detector (y axis) as a function of the mass-to-charge (*m/z*) ratio (x axis). *Bottom panel:* In an MS/MS instrument such as the ion trap shown in part (a), a specific peptide ion can be selected for fragmentation into smaller ions that are then analyzed and detected. The MS/MS spectrum (also called the product-ion spectrum) provides detailed structural information about the parent ion, including sequence information for peptides. Here the ion with an *m/z* of 836.47 was selected and fragmented and the *m/z* mass spectrum of the product ions measured. Note there is no longer an ion with an *m/z* of 836.47 present because it was fragmented. From the varying sizes of the product ions, the understanding that peptide bonds are often broken in such experiments, the known *m/z* values for individual amino acid fragments, and database information, the sequence of the peptide, FIIVGYVDDTQFVR, can be deduced. [Part (a) based on a figure from S. Carr; part (b), unpublished data from S. Carr.]

In tandem, or MS/MS, instruments, any given parent ion in the original mass spectrum (Figure 3-42b, *top panel*) can be mass-selected, broken into smaller ions by collision with an inert gas, and then the *m/z* and relative abundances of the resulting fragment ions measured (Figure 3-42b, *bottom panel*), all within the same machine in about 0.1 s per selected parent ion. This second round of fragmentation and analysis permits the sequences of short peptides (<25 amino acids) to be determined because collisional fragmentation occurs primarily at peptide bonds, so the differences in masses between ions correspond to the in-chain masses of the individual amino acids, permitting deduction of the sequence in conjunction with database sequence information (Figure 3-42b, *bottom panel*).

Mass spectrometry is highly sensitive, able to detect as little as $1 \times 10^{-16}$ mol (100 attomoles) of a peptide or $10 \times 10^{-15}$ mol (10 femtomoles) of a protein of 200,000 MW. Errors in mass measurement accuracy are dependent on the specific mass analyzer used but are typically ~0.01 percent for peptides and 0.05–0.1 percent for proteins. As described in Section 3.6, below, it is possible to use MS to analyze complex mixtures of proteins as well as purified proteins. Most commonly, protein samples are digested by proteases, and the peptide digestion products are subjected to analysis. An especially powerfully application of MS is to take a complex mixture of proteins from a biological specimen, digest it with trypsin or other proteases, partially separate the components using liquid chromatography (LC), and then transfer the solution flowing out of the chromatographic column directly into an ES tandem mass spectrometer. This technique, called *LC-MS/MS*, permits the nearly continuous analysis of a very complex mixture of proteins.

The abundances of ions determined by mass spectrometry in any given sample are relative, not absolute, values. Therefore, if one wants to use MS to compare the amounts of a particular protein in two different samples (e.g., from a normal versus a mutant organism), it is necessary to have an internal standard in the samples whose amounts do not differ between the two samples. One then determines the amounts of the protein of interest relative to that of the standard in each sample. This permits quantitatively accurate inter-sample comparisons of protein levels. An alternative approach involves simultaneously comparing the amounts of proteins from two different cell or tissue samples that are mixed together. To do this, investigators first incubate one of the samples with amino acids containing "heavy" isotope atoms. These are biosynthetically incorporated into all of the proteins of that sample. Proteins from the two samples are then mixed together and analyzed by mass spectrometry. Proteins and peptides derived from the "heavy" sample can be distinguished in the mass spectrometer from those from the other, "light," sample because of their higher masses. Thus a direct comparison of the relative amounts of each protein in each sample can be made. When the samples are cells grown in the laboratory, the method is called *s*table *i*sotope *l*abeling with *a*mino *a*cids in *c*ell *c*ulture (SILAC).

## Protein Primary Structure Can Be Determined by Chemical Methods and from Gene Sequences

The classic method for determining the amino acid sequence of a protein is Edman degradation. In this procedure, the free amino group of the N-terminal amino acid of a polypeptide is labeled, and the labeled amino acid is then cleaved from the polypeptide and identified by high-pressure liquid chromatography. The polypeptide is left one residue shorter, with a new amino acid at the N-terminus. The cycle is repeated on the ever-shortening polypeptide until all the residues have been identified.

Before about 1985, biologists commonly used the Edman chemical procedure for determining protein sequences. Now, however, complete protein sequences usually are determined primarily by analysis of genome sequences. The complete genomes of several organisms have already been sequenced, and the database of genome sequences from humans and numerous model organisms is expanding rapidly. As discussed in Chapter 5, the sequences of proteins can be deduced from DNA sequences that are predicted to encode proteins.

A powerful approach for determining the primary structure of an isolated protein combines MS and the use of sequence databases. First, the peptide "mass fingerprint" of the protein is obtained by MS. A *peptide mass fingerprint* is the list of the molecular weights of peptides that are generated from the protein by digestion with a specific protease, such as trypsin. The molecular weights of the parent protein and its proteolytic fragments are then used to search genome databases for any similar-size protein with identical or similar peptide mass maps. Mass spectrometry can also be used to directly sequence peptides using MS/MS, as described above.

## Protein Conformation Is Determined by Sophisticated Physical Methods

In this chapter, we have emphasized that protein function is dependent on protein structure. Thus, to figure out exactly how a protein works, its three-dimensional structure must be determined. Determining a protein's conformation requires sophisticated physical methods and complex analyses of the experimental data. We briefly describe three methods used to generate three-dimensional models of proteins.

**X-ray Crystallography** The use of **x-ray crystallography** to determine the three-dimensional structures of proteins was pioneered by Max Perutz and John Kendrew in the 1950s. In this technique, beams of x-rays are passed through a protein crystal in which millions of protein molecules are precisely aligned with one another in a rigid crystalline array. The wavelengths of x-rays are about 0.1–0.2 nm, short enough to determine the positions of individual atoms in the protein. The electrons in the atoms of the crystal scatter the x-rays, which produce a diffraction pattern of discrete spots when they are intercepted by photographic film or an electronic detector (Figure 3-43). Such patterns are extremely complex—composed of as many as 25,000 diffraction spots, or reflections,

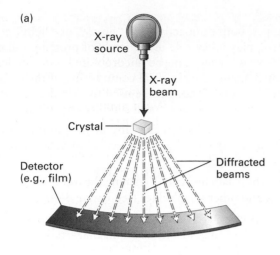

(a)

X-ray source

X-ray beam

Crystal

Detector (e.g., film)

Diffracted beams

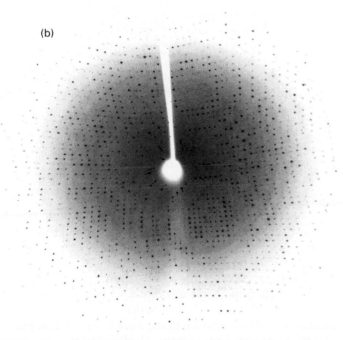

(b)

**EXPERIMENTAL FIGURE 3-43 X-ray crystallography provides diffraction data from which the three-dimensional structure of a protein can be determined.** (a) Basic components of an x-ray crystallographic determination. When a narrow beam of x-rays strikes a crystal, part of it passes straight through and the rest is scattered (diffracted) in various directions. The intensity of the diffracted waves, which form periodic arrangements of diffraction spots, is recorded on an x-ray film or with a solid-state electronic detector. (b) X-ray diffraction pattern for a protein crystal collected on a solid-state detector. From complex analyses of patterns of spots like this one, the location of the atoms in a protein can be determined. [Part (a) adapted from L. Stryer, 1995, *Biochemistry*, 4th ed., W. H. Freeman and Company, p. 64; part (b) courtesy of J. Berger.]

whose measured intensities vary depending on the distribution of the electrons, which is, in turn, determined by the atomic structure and three-dimensional conformation of the protein. Elaborate calculations and modifications of the protein (such as the binding of heavy metals) must be made to interpret the diffraction pattern and calculate the distribution

of electrons (called the *electron density map*). With the three-dimensional electron density map in hand, one then "fits" a molecular model of the protein to match the electron density, and it is these models that one sees in the various diagrams of proteins throughout this book (e.g., Figure 3-8). The process is analogous to reconstructing the precise shape of a rock from the ripples that it creates when thrown into a pond. Although sometimes the structures of parts of the protein cannot be clearly defined, using x-ray crystallography, researchers are systematically determining the structures of representative types of most proteins. To date, the detailed three-dimensional structures of more than 18,000 proteins have been established using x-ray crystallography. These structures can be found in the Research Collaboratory for Structural Bioinformatics Protein Data Bank (http://www.rcsb.org/pdb/home/home.do), each with its own PDB entry.

**Cryoelectron Microscopy** Although some proteins readily crystallize, obtaining crystals of others—particularly large multisubunit proteins and membrane-associated proteins—requires a time-consuming, often robot-assisted trial-and-error effort to find just the right conditions, if they can be found at all. (Growing crystals suitable for structural studies is as much an art as a science.) There are several ways to determine the structures of such difficult-to-crystallize proteins. One is cryoelectron microscopy. In this technique, a protein sample is rapidly frozen in liquid helium to preserve its structure and then examined in the frozen, hydrated state in a cryoelectron microscope. Pictures of the protein are taken at various angles and recorded on film using a low dose of electrons to prevent radiation-induced damage to the structure. Sophisticated computer programs analyze the images and reconstruct the protein's structure in three dimensions. Recent advances in cryoelectron microscopy permit researchers to generate molecular models that can help provide insight into how the protein functions. The use of cryoelectron microscopy and other types of electron microscopy for visualizing cell structures is discussed in Chapter 9.

**NMR Spectroscopy** The three-dimensional structures of small proteins containing as many as 200 amino acids can be studied routinely with nuclear magnetic resonance (NMR) spectroscopy. Specialized approaches can be used to extend the size range to somewhat larger proteins. In this technique, a concentrated protein solution is placed in a magnetic field, and the effects of different radio frequencies on the nuclear spin states of different atoms are measured. The spin state of any atom is influenced by neighboring atoms in adjacent residues, with closely spaced residues having a greater influence than distant residues. From the magnitude of the effect, the distances between residues can be calculated by a triangulation-like process; these distances are then used to generate a model of the three-dimensional structure of the protein. An important distinction between x-ray crystallography and NMR spectroscopy is that the former method directly determines the locations of the atoms while the later directly determines the distances between the atoms.

Although NMR does not require the crystallization of a protein, a definite advantage, this technique is limited to proteins smaller than about 20 kDa. However, NMR analysis can provide information about the ability of a protein to adopt a set of closely related, but not exactly identical, conformations and to move between these conformations (protein dynamics). This is a common feature of proteins, which are not absolutely rigid structures but can "breathe" or exhibit slight variations in the relative positions of their constituent atoms. In some cases these variations can have functional significance, for example in how proteins bind to one another. NMR structural analysis has been particularly useful in studying isolated protein domains, which can often be obtained as stable structures and tend to be small enough for this technique. To date, there are more than 5000 NMR-determined protein structures available in the Protein Data Bank (http://www.rcsb.org/pdb/home/home.do).

## KEY CONCEPTS of Section 3.5

### Purifying, Detecting, and Characterizing Proteins

• Proteins can be separated from other cell components and from one another on the basis of differences in their physical and chemical properties.

• Various assays are used to detect and quantify proteins. Some assays use a light-producing reaction to generate a readily detected signal. Other assays produce an amplified colored signal with enzymes and chromogenic substrates.

• Centrifugation separates proteins on the basis of their rates of sedimentation, which are influenced by their masses and shapes (see Figure 3-35).

• Electrophoresis separates proteins on the basis of their rates of movement in an applied electric field. SDS-polyacrylamide gel electrophoresis (SDS-PAGE) can resolve polypeptide chains differing in molecular weight by 10 percent or less (see Figure 3-36). Two-dimensional gel electrophoresis provides additional resolution by separating proteins first by charge (first dimension) and then by mass (second dimension).

• Liquid chromatography separates proteins on the basis of their rates of movement through a column packed with spherical beads. Proteins differing in mass are resolved on gel filtration columns; those differing in charge, on ion-exchange columns; and those differing in ligand-binding properties, on affinity columns, including antibody-based affinity chromatography (see Figure 3-38).

• Antibodies are powerful reagents used to detect, quantify, and isolate proteins.

• Immunoblotting, also called *Western blotting*, is a frequently used method to study specific proteins that exploits the high specificity and sensitivity of protein detection by antibodies and the high-resolution separation of proteins by SDS-PAGE (see Figure 3-39).

• Isotopes, both radioactive and "heavy" or "light" nonradioactive, play a key role in the study of proteins and other biomolecules. They can be incorporated into molecules without changing the chemical composition of the molecule or as add-on tags. They can be used to help detect the synthesis, location, processing, and stability of proteins.

• Autoradiography is a semiquantitative technique for detecting radioactively labeled molecules in cells, tissues, or electrophoretic gels.

• Pulse-chase labeling can determine the intracellular fate of proteins and other metabolites (see Figure 3-40).

• Mass spectrometry is a very sensitive and highly precise method of detecting, identifying, and characterizing proteins and peptides.

• Three-dimensional structures of proteins are obtained by x-ray crystallography, cryoelectron microscopy, and NMR spectroscopy. X-ray crystallography provides the most detailed structures but requires protein crystallization. Cryoelectron microscopy is most useful for large protein complexes, which are difficult to crystallize. Only relatively small proteins are amenable to NMR analysis.

## 3.6 Proteomics

For most of the twentieth century, the study of proteins was restricted primarily to the analysis of individual proteins. For example, one would study an enzyme by determining its enzymatic activity (substrates, products, rate of reaction, requirement for cofactors, pH, etc.), its structure, and its mechanism of action. In some cases, the relationships between a few enzymes that participate in a metabolic pathway might also be studied. On a broader scale, the localization and activity of an enzyme would be examined in the context of a cell or tissue. The effects of mutations, diseases, or drugs on the expression and activity of the enzyme might also be the subject of investigation. This multipronged approach provided deep insight into the function and mechanisms of action of individual proteins or relatively small numbers of interacting proteins. However, such a one-by-one approach to studying proteins does not readily provide a global picture of what is happening in the proteome of a cell, tissue, or entire organism.

### Proteomics Is the Study of All or a Large Subset of Proteins in a Biological System

The advent of genomics (sequencing of genomic DNA and its associated technologies, such as simultaneous analysis of the levels of all mRNAs in cells and tissues) clearly showed that a global, or systems, approach to biology could provide unique and highly valuable insights. Many scientists recognized that a global analysis of the proteins in biological systems had the potential for equally valuable contributions to our understanding. Thus a new field was born—**proteomics**.

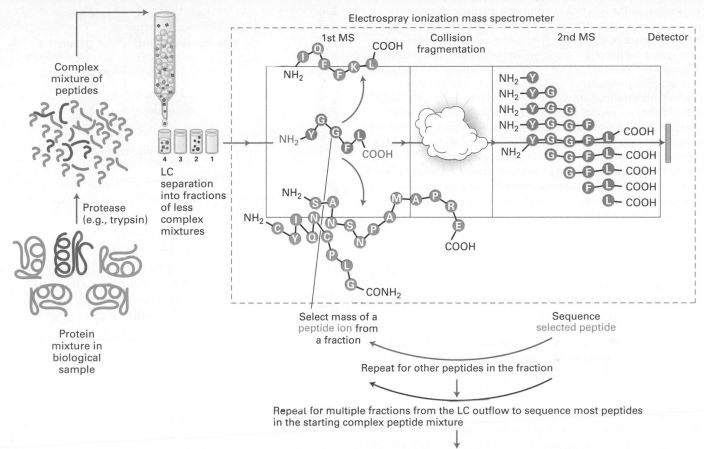

**EXPERIMENTAL FIGURE 3-44 LC-MS/MS is used to identify the proteins in a complex biological sample.** A complex mixture of proteins in a biological sample (e.g., isolated preparation of Golgi organelles) is digested with a protease; the mixture of resulting peptides is fractionated by liquid chromatography (LC) into multiple, less complex, fractions, which are slowly but continuously injected by electrospray ionization into a tandem mass spectrometer. The fractions are then sequentially subjected to multiple cycles of MS/MS until masses and sequences of many of the peptides are determined and used to identify the proteins in the original biological sample through comparison with protein databases. [Based on a figure provided by S. Carr.]

Proteomics is the systematic study of the amounts, modifications, interactions, localization, and functions of all or subsets of proteins at the whole-organism, tissue, cellular, and subcellular levels.

A number of broad questions are addressed in proteomic studies:

• In a given sample (whole organism, tissue, cell, subcellular compartment), what fraction of the whole proteome is expressed (i.e., which proteins are present)?

• Of those proteins present in the sample, what are their relative abundances?

• What are the relative amounts of the different splice forms and chemically modified forms (e.g., phosphorylated, methylated, fatty acylated) of the proteins?

• Which proteins are present in large multiprotein complexes, and which proteins are in each complex? What are the functions of these complexes, and how do they interact?

• When the state (e.g., growth rate, stage of cell cycle, differentiation, stress level) of a cell changes, do the proteins in the cell or secreted from the cell change in a characteristic (*fingerprint*-like) fashion? Which proteins change and how (relative amounts, modifications, splice forms, etc.)? (This is a form of *protein expression profiling* that complements the *transcriptional (mRNA) profiling* discussed in Chapter 7.)

• Can such fingerprint-like changes be used for diagnostic purposes? For example, do certain cancers or heart disease cause characteristic changes in blood proteins? Can the proteomic fingerprint help determine if a given cancer is resistant or sensitive to a particular chemotherapeutic drug? Proteomic

**EXPERIMENTAL FIGURE 3-45 Density-gradient centrifugation and LC-MS/MS can be used to identify many of the proteins in organelles.** (a) The cells in liver tissue were mechanically broken to release the organelles, and the organelles were partially separated by density-gradient centrifugation. The locations of the organelles—which were spread out through the gradient and somewhat overlapped with one another—were determined using immunoblotting with antibodies that recognize previously identified, organelle-specific proteins. Fractions from the gradient were subjected to proteolysis and LC-MS/MS to identify the peptides, and hence the proteins, in each fraction. Comparisons with the locations of the organelles in the gradient (called protein correlation profiling) permitted assignment of many individual proteins to one or more organelles (organelle proteome identification). (b) The hierarchical breakdown of data derived from the procedures in part (a). Note that not all proteins identified could be assigned to organelles and some proteins were assigned to more than one organelle. [From L. J. Foster et al., 2006, *Cell* **125**(1):187–199.]

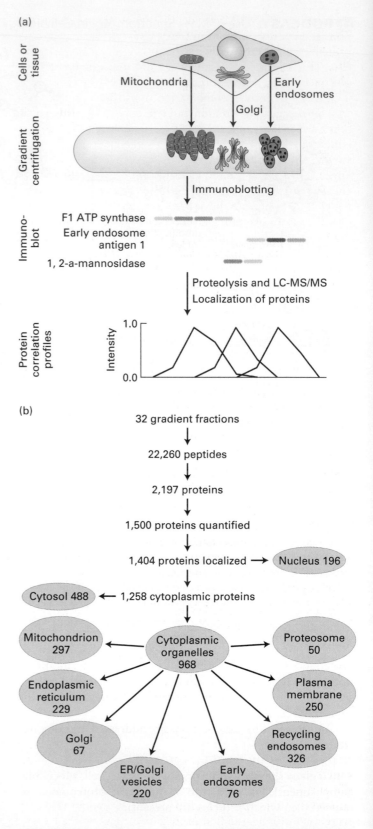

fingerprints can also be the starting point for studies of the mechanisms underlying the change of state. Proteins (and other biomolecules) that show changes that are diagnostic of a particular state are called *biomarkers*.

• Can changes in the proteome help define targets for drugs or suggest mechanisms by which that drug might induce toxic side effects? If so, it might be possible to engineer modified versions of the drug with fewer side effects.

These are just a few of the questions that can be addressed using proteomics. The methods used to answer these questions are as diverse as the questions themselves, and their numbers are growing rapidly.

## Advanced Techniques in Mass Spectrometry Are Critical to Proteomic Analysis

Advances in proteomics technologies (e.g., mass spectrometry) profoundly affect the types of questions that can be practically studied. For many years, two-dimensional gel electrophoresis has allowed researchers to separate, display, and characterize a complex mixture of proteins (see Figure 3-37). The spots on a two-dimensional gel can be excised, the protein fragmented by proteolysis (e.g., by trypsin digestion), and the fragments identified by MS. An alternative to this two-dimensional gel method is *high-throughput LC-MS/MS*. Figure 3-44 outlines the general LC-MS/MS approach, in which a complex mixture of proteins is digested with a protease; the myriad resulting peptides are fractionated by LC into multiple, less complex fractions, which are slowly but continuously injected by electrospray ionization into a tandem mass spectrometer. The fractions are then sequentially subjected to multiple cycles of MS/MS until sequences of many of the peptides are determined and used to identify from databases the proteins in the original biological sample.

An example of the use of LC-MS/MS to identify many of the proteins in each organelle is seen in Figure 3-45. Cells from murine (mouse) liver tissue were mechanically broken to

release the organelles, and the organelles were partially separated by density-gradient centrifugation. The locations of the organelles in the gradient were determined using immunoblotting with antibodies that recognize previously identified, organelle-specific proteins. Fractions from the gradient were subjected to LC-MS/MS to identify the proteins in each fraction, and the distributions in the gradient of many individual proteins were compared with the distributions of the organelles. This permitted assignment of many individual proteins to one or more organelles (organelle proteome proteome profiling). More recently, a combination of organelle purification, MS, biochemical localization and computational methods has been used to show that at least 1000 distinct proteins are localized in the mitochondria of humans and mice.

Proteomics combined with molecular genetics methods are currently being used to identify all protein complexes in a eukaryotic cell, the yeast *Saccharomyces cerevisiae*. Approximately 500 complexes have been identified, with an average of 4.9 distinct proteins per complex, and these in turn are involved in at least 400 complex-to-complex interactions. Such systematic proteomic studies are providing new insights into the organization of proteins within cells and how proteins work together to permit cells to live and function.

## KEY CONCEPTS of Section 3.6

### Proteomics

• Proteomics is the systematic study of the amounts (and changes in the amounts), modifications, interactions, localization, and functions of all or subsets of all proteins in biological systems at the whole-organism, tissue, cellular, and subcellular levels.

• Proteomics provides insights into the fundamental organization of proteins within cells and how this organization is influenced by the state of the cells (e.g., differentiation into distinct cell types; response to stress, disease, and drugs).

• A wide variety of methods are used for proteomic analyses, including two-dimensional gel electrophoresis, density-gradient centrifugation, and mass spectroscopy (MALDI-TOF and LC-MS/MS).

• Proteomics has helped begin to identify the proteomes of organelles ("organelle proteome profiling") and the organization of individual proteins into multiprotein complexes that interact in a complex network to support life and cellular function (see Figure 3-45).

## Perspectives for the Future

Impressive expansion of the computational power of computers is at the core of advances in determining the three-dimensional structures of proteins. For example, vacuum tube computers running on programs punched on cards were used to solve the first protein structures on the basis of x-ray crystallography, a process that at the time took years but can now be accomplished in a matter of days and in some cases hours. In the future, researchers aim to predict the structures of proteins using only amino acid sequences deduced from gene sequences. This computationally challenging problem requires supercomputers or large clusters of computers working in synchrony. Currently, only the structures of very small domains containing 100 residues or fewer can be predicted at a low resolution. However, continued developments in computing and models of protein folding, combined with large-scale efforts to solve the structures of all protein structural motifs by x-ray crystallography, will allow the prediction of the structures of larger proteins. With an exponentially expanding database of structurally defined motifs, domains, and proteins, scientists will be able to identify the motifs in an unknown protein, match the motif to the sequence, and use this to predict the three-dimensional structure of the entire protein.

New combined approaches will also help in determining high-resolution structures of molecular machines. Although these very large macromolecular assemblies usually are difficult to crystallize and thus to solve by x-ray crystallography, they can be imaged in a cryoelectron microscope at liquid helium temperatures and high electron energies. From millions of individual "particles," each representing a random view of the protein complex, the three-dimensional structure can be built. Because subunits of the complex may already be solved by crystallography, a composite structure consisting of the x-ray-derived subunit structures fit to the EM-derived model will be generated.

Methods for rapid structure determination combined with identification of novel substrates and inhibitors will help determine the structures of enzyme-substrate complexes and transition states and thus help provide detailed information regarding the mechanisms of enzyme catalysis. Membrane proteins, because of the specialized environment in which they reside and their solubility characteristics, remain challenging, although progress in this area is accelerating.

Although our understanding of chaperone structure and activity continues to grow exponentially, a number of critical questions remain a mystery. We do not understand precisely how cells make the distinction between unfolded and misfolded versus properly folded. Clearly, the exposure of hydrophobic side chains plays a role, but what are the other determinants of this key recognition process? How is the decision made to turn from trying to refold a protein to degrading it?

The rapid development of new technologies can be expected to help solve some of the still outstanding problems in proteomics. It is becoming possible to identify and characterize intact proteins as large as 30–70 kDa in complex mixtures using MS techniques without first digesting the samples into peptides—a method called the "top-down" approach, in contrast to starting with fragments of the protein ("bottom-up" approach). An ongoing problem in proteomic analysis of complex mixtures is that it is difficult to detect and identify protein fragments from samples whose concentrations in the sample differ by more than 1000-fold: some samples, such as blood plasma, contain proteins whose concentrations vary over a $10^{11}$-fold range. Routine analysis of specimens with such diverse concentrations should dramatically improve the mechanistic and diagnostic value of blood plasma proteomics.

## Key Terms

## Review the Concepts

**1.** The three-dimensional structure of a protein is determined by its primary, secondary, and tertiary structures. Define the *primary*, *secondary*, and *tertiary structures*. What are some of the common secondary structures? What are the forces that hold together the secondary and tertiary structures?

**2.** Proper folding of proteins is essential for their biological activity. In general, the functional conformation of a protein is the conformation with lowest energy. This means that if an unfolded protein is allowed to reach equilibrium, it should assemble automatically into its native, functioning folded state. Why then is there a need for molecular chaperones and chaperonins in cells? What different roles do molecular chaperones and chaperonins play in the folding of proteins?

**3.** Enzymes catalyze chemical reactions. What constitutes the active site of an enzyme? What are the turnover number ($k_{cat}$), the Michaelis constant ($K_m$), and the maximal velocity ($V_{max}$) of an enzyme? The $k_{cat}$ for carbonic anhydrase is $5 \times 10^5$ molecules/s. This is a "rate constant" but not a "rate." What is the difference? By what concentration would you multiply this rate constant in order to determine an actual rate of product formation ($V$)? Under what circumstances would this rate become equal to the maximal velocity ($V_{max}$) of the enzyme?

**4.** The following reaction coordinate diagram charts the energy of a substrate molecule (S) as it passes through a transition state ($X^{\ddagger}$) on its way to becoming a stable product (P) alone or in the presence of one of two different enzymes (E1 and E2). How does the addition of either enzyme affect the change in Gibb's free energy ($\Delta G$) for the reaction? Which of the two enzymes binds with greater affinity to the substrate?

Which enzyme better stabilizes the transition state? Which enzyme functions as a better catalyst?

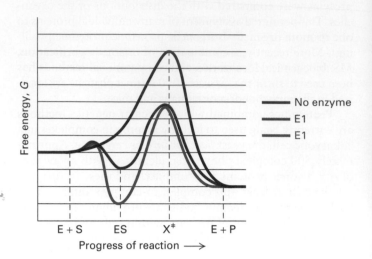

**5.** A healthy adaptive immune system can raise antibodies that recognize and bind with high affinity to almost any stable molecule. The molecule to which an antibody binds is known as "antigen." Antibodies have been exploited by enterprising scientists to generate valuable tools for research, diagnosis, and therapy. One clever application is the generation of antibodies that function like enzymes to catalyze complicated chemical reactions. If you wished to produce such a "catalytic" antibody, what would you suggest using as the antigen? Should it be the substrate of the reaction? The product? Something else?

**6.** Proteins are degraded in cells. What is ubiquitin, and what role does it play in tagging proteins for degradation? What is the role of proteasomes in protein degradation? How might proteasome inhibitors serve as chemotherapeutic (cancer-treating) agents?

**7.** The function of proteins can be regulated in a number of ways. What is cooperativity, and how does it influence protein function? Describe how protein phosphorylation and proteolytic cleavage can modulate protein function.

**8.** A number of techniques can separate proteins on the basis of their differences in mass. Describe the use of two of these techniques, centrifugation and gel electrophoresis. The blood proteins transferrin (MW 76 kDa) and lysozyme (MW 15 kDa) can be separated by rate-zonal centrifugation or SDS-polyacrylamide gel electrophoresis. Which of the two proteins will sediment faster during centrifugation? Which will migrate faster during electrophoresis?

**9.** Chromatography is an analytical method used to separate proteins. Describe the principles for separating proteins by gel filtration, ion-exchange, and affinity chromatography.

**10.** Various methods have been developed for detecting proteins. Describe how radioisotopes and autoradiography can be used for labeling and detecting proteins. How does Western blotting detect proteins?

**11.** Physical methods are often used to determine protein conformation. Describe how x-ray crystallography, cryoelectron microscopy, and NMR spectroscopy can be used to determine the shape of proteins. What are the advantages and disadvantages of each method? Which is better for small proteins? Large proteins? Huge macromolecular assemblies?

12. Mass spectrometry is a powerful tool in proteomics. What are the four key features of a mass spectrometer? Describe briefly how MALDI and two-dimensional polyacrylamide gel electrophoresis (2D-PAGE) could be used to identify a protein expressed in cancer cells but not in normal healthy cells.

## Analyze the Data

1. Beautiful models of macromolecules such as proteins and nucleic acids are generated from files of atomic coordinates obtained usually from x-ray diffraction of crystallized samples or NMR analysis of the molecules in solution. The Protein Data Bank (PDB) is a publicly accessible repository of macromolecular atomic coordinate files that can be accessed online at http://www.rcsb.org. Access the PDB and familiarize yourself with its homepage. How many molecular structures does it contain today? What is the "Molecule of the Month"? Download a coordinate file for the serine protease chymotrypsin by typing the accession code "1ACB" into the search window. This will take you to a page describing the x-ray crystal structure of a complex between bovine alpha-chymotrypsin and the small pseudosubstrate inhibitor protein eglin-c. When and in what journal was the study reporting this structural model published? Click on the "Download File" link, select "PDB File (Text)", and download the file "1ACB.pdb". This is an atomic coordinate (.pdb) file that specifies the relative positions for each atom in this protein complex as determined experimentally by x-ray crystallography. Open the file in a text viewer or word processor and look at its format. The first several hundred lines contain background information including the names of the molecules, their natural sources, how they were prepared for the experiment, statistical analysis of the model quality, and bibliographic information. Eventually, you will arrive at a long list of lines that each begin with "ATOM". These are the coordinates, listed by atom number, atom type, amino acid type, and chain number. Each "ATOM" line ends with five numbers representing the atomic position on an x, y, z axis, its "occupancy," and its "thermal factor." Close the file and download software for viewing the molecular model. There are many, such as RasMol, iMol, Swiss-PDB Viewer, and PyMol, that are available for download in a free format for educational purposes. Open the 1ACB.pdb file and twirl it around in the viewer. Can you identify the protease? The inhibitor protein? Can you find the enzyme's active site? What other observations can you make about serine proteases from the model of this inactivated complex?

2. Proteomics involves the global analysis of protein expression. In one approach, all the proteins in control cells and treated cells are extracted and subsequently separated using two-dimensional gel electrophoresis. Typically, hundreds or thousands of protein spots are resolved and the steady-state levels of each protein are compared between control and treated cells. In the following example, only a few protein spots are shown for simplicity. Proteins are separated in the first dimension on the basis of charge by isoelectric focusing (pH 4–10) and then separated by size by SDS-polyacrylamide gel electrophoresis. Proteins are detected with a stain such as Coomassie blue and assigned numbers for identification.

a. Cells are treated with a drug ("1 Drug") or left untreated ("Control"), and then proteins are extracted and separated by two-dimensional gel electrophoresis. The stained gels are shown below. What do you conclude about the effect of the drug on the steady-state levels of proteins 1–7?

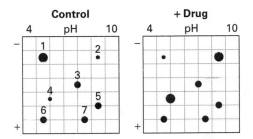

b. You suspect that the drug may be inducing a protein kinase and so repeat the experiment in part (a) in the presence of $^{32}$P-labeled inorganic phosphate. In this experiment the two-dimensional gels are exposed to x-ray film to detect the presence of $^{32}$P-labeled proteins. The x-ray films are shown below. What do you conclude from this experiment about the effect of the drug on proteins 1–7?

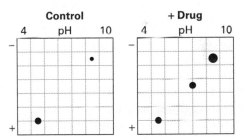

c. To determine the cellular localization of proteins 1–7, the cells from part (a) were separated into nuclear and cytoplasmic fractions by differential centrifugation. Two-dimensional gels were run, and the stained gels are shown below. What do you conclude about the cellular localization of proteins 1–7?

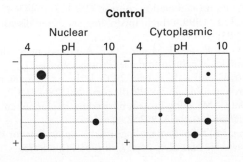

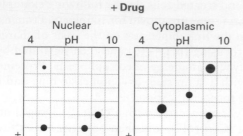

**+ Drug**

Nuclear | Cytoplasmic

d. Summarize the overall properties of proteins 1–7, combining the data from parts (a), (b), and (c). Describe how you could determine the identity of any one of the proteins.

# References

## General References

Berg, J. M., J. L. Tymoczko, and L. Stryer. 2007. *Biochemistry,* 6th ed. W. H. Freeman and Company.

Nelson, D. L., and M. M. Cox. 2005. *Lehninger Principles of Biochemistry,* 4th ed. W. H. Freeman and Company.

## Web Sites

Entry site into proteins, structures, genomes, and taxonomy: http://www.ncbi.nlm.nih.gov/Entrez/

The protein 3-D structure database: http://www.rcsb.org/

Structural classifications of proteins: http://scop.berkeley.edu/

Sites containing general information about proteins: http://www.expasy.ch/; http://www.proweb.org/; http://scop.berkeley.edu/intro.html

PROSITE database of protein families and domains: http://www.expasy.org/prosite/

Domain organization of proteins and large collection of multiple sequence alignments: http://www.sanger.ac.uk/Software/Pfam/; http://people.cryst.bbk.ac.uk/~ubcg16z/cpn/elmovies.html

MitoCarta: An Inventory of Mammalian Mitochondrial Genes:http://www.broadinstitute.org/pubs/MitoCarta/index.html

## Hierarchical Structure of Proteins and Protein Folding

Branden, C., and J. Tooze. 1999. *Introduction to Protein Structure.* Garland.

Broadley, S. A., and F. U. Hartl. 2009. The role of molecular chaperones in human misfolding diseases. *FEBS Lett.* 583(16):2647–2653.

Brodsky, J. L., and G. Chiosis. 2006. Hsp70 molecular chaperones: emerging roles in human disease and identification of small molecule modulators. *Curr. Top. Med. Chem.* 6(11):1215–1225.

Bukau, B, J. Weissman, and A. Horwich. 2006. Molecular chaperones and protein quality control. *Cell* 125(3):443–451.

Cohen, F. E. 1999. Protein misfolding and prion diseases. *J. Mol. Biol.* 293:313–320.

Coulson, A. F., and J. Moult. 2002. A unifold, mesofold, and superfold model of protein fold use. *Proteins* 46:61–71.

Daggett, V, and A. R. Fersht. 2003. Is there a unifying mechanism for protein folding? *Trends Biochem. Sci.* 28(1):18–25.

Dobson, C. M. 1999. Protein misfolding, evolution, and disease. *Trends Biochem. Sci.* 24:329–332.

Gimona, M. 2006. Protein linguistics—a grammar for modular protein assembly? *Nat. Rev. Mol. Cell Biol.* 7(1):68–73.

Gough, J. 2006. Genomic scale sub-family assignment of protein domains. *Nucl. Acids Res.* 34(13):3625–3633.

Koonin, E. V., Y. I. Wolf, and G. P. Karev. 2002. The structure of the protein universe and genome evolution. *Nature* 420:218–223.

Lesk, A. M. 2001. *Introduction to Protein Architecture.* Oxford.

Levitt M. 2009. Nature of the protein universe. *Proc. Natl. Acad. Sci. USA* 106(27):11079–11084.

Lin, Z., and H. S. Rye. 2006. GroEL-mediated protein folding: making the impossible, possible. *Crit. Rev. Biochem. Mol. Biol.* 41(4):211–239.

Orengo, C. A., D. T. Jones, and J. M. Thornton. 1994. Protein superfamilies and domain superfolds. *Nature* 372:631–634.

Patthy, L. 1999. *Protein Evolution.* Blackwell Science.

Rochet, J.-C., and P. T. Landsbury. 2000. Amyloid fibrillogenesis: themes and variations. *Curr. Opin. Struc. Biol.* 10:60–68.

Taipale, M., D. F. Jarosz, and S. Lindquist. 2010. HSP90 at the hub of protein homeostasis: emerging mechanistic insights. *Nat. Rev. Mol. Cell Biol.* 11(7):515–528.

Vogel, C., and C. Chothia. 2006. Protein family expansions and biological complexity. *PLoS Comput. Biol.* 2(5):e48.

Yaffe, M. B. 2006. "Bits" and pieces. *Sci STKE.* 2006(340):pe28.

Young, J. C., et al. 2004. Pathways of chaperone-mediated protein folding in the cytosol. *Nat. Rev. Mol. Cell Biol.* 5:781–791.

Wlodarski, T., and B. Zagrovic. 2009. Conformational selection and induced fit mechanism underlie specificity in noncovalent interactions with ubiquitin. *Proc. Natl. Acad. Sci. USA* 106(46):19346–19351.

## Protein Binding and Enzyme Catalysis

Dressler, D. H., and H. Potter. 1991. *Discovering Enzymes.* Scientific American Library.

Fersht, A. 1999. *Enzyme Structure and Mechanism,* 3d ed. W. H. Freeman and Company.

Jeffery, C. J. 2004. Molecular mechanisms for multitasking: recent crystal structures of moonlighting proteins. *Curr. Opin. Struc. Biol.* 14(6):663–668.

Marnett, A. B., and C. S. Craik. 2005. Papa's got a brand new tag: advances in identification of proteases and their substrates. *Trends Biotechnol.* 23(2):59–64.

Polgar, L. 2005. The catalytic triad of serine peptidases. *Cell Mol. Life Sci.* 62(19–20):2161–2172.

Radisky, E. S., et al. 2006. Insights into the serine protease mechanism from atomic resolution structures of trypsin reaction intermediates. *Proc. Nat'l Acad. Sci. USA* 103(18):6835–6840.

Schenone, M, B. C. Furie, and B. Furie. 2004. The blood coagulation cascade. *Curr. Opin. Hematol.* 11(4):272–277.

Schramm, V. L. 2005. Enzymatic transition states and transition state analogues. *Curr. Opin. Struc. Biol.* 15(6):604–613.

## Regulating Protein Function

Bellelli, A., et al. 2006. The allosteric properties of hemoglobin: insights from natural and site directed mutants. *Curr. Prot. Pep. Sci.* 7(1):17–45.

Bochtler, M., et al. 1999. The proteasome. *Ann. Rev. Biophys. Biomol. Struct.* 28:295–317.

Burack, W. R., and A. S. Shaw. 2000. Signal transduction: hanging on a scaffold. *Curr. Opin. Cell Biol.* 12:211–216.

Gallastegui, N., and M. Groll. 2010. The 26S proteasome: assembly and function of a destructive machine. *Trends Biochem. Sci.* 35(11):634–642.

Glen, R., et al. 2008. Regulatory monoubiquitination of phosphoenolpyruvate carboxylase in germinating castor oil seeds *J. Biol. Chem.* 283:29650–29657.

Glickman, M. H., and A. Ciechanover. 2002. The ubiquitin-proteasome proteolytic pathway: destruction for the sake of construction. *Physiol. Rev.* 82(2):373–428.

Goldberg, A. L. 2003. Protein degradation and protection against misfolded or damaged proteins. *Nature* 426:895–899.

Goldberg, A. L, S. J. Elledge, and J. W. Harper. 2001. The cellular chamber of doom. *Sci. Am.* 284(1):68–73.

Groll, M., and R. Huber. 2005. Purification, crystallization, and x-ray analysis of the yeast 20S proteasome. *Meth. Enzymol.* 398:329–336.

Halling, D. B, P. Aracena-Parks, and S. L. Hamilton. 2006. Regulation of voltage-gated $Ca^{21}$ channels by calmodulin. *Sci STKE* 2005 (315):re15.

Horovitz, A., et al. 2001. Review: allostery in chaperonins. *J. Struc. Biol.* 135:104–114.

Huang, H., et al. 2010. K33-linked polyubiquitination of T cell receptor-ζ regulates proteolysis-independent T cell signaling. *Immunity.* 33(1):60–70.

Katz, E. J., M. Isasa, and B. Crosas. 2010. A new map to understand deubiquitination. *Biochem. Soc. Trans.* 38(pt. 1):21–28.

Kern, D., and E. R. Zuiderweg. 2003. The role of dynamics in allosteric regulation. *Curr. Opin. Struc. Biol.* 13(6):748–757.

Kisselev, A. F., A. Caliard, and A. L. Goldberg. 2006. Importance of the different proteolytic sites of the proteasome and the efficacy of inhibitors varies with the protein substrate. *J. Biol. Chem.* 281(13):8582–8590.

Lane, K. T., and L. S. Beese. 2006. Thematic review series: lipid posttranslational modifications. Structural biology of protein farnesyltransferase and geranylgeranyltransferase type I. *J. Lipid Res.* 47(4):681–699.

Lim, W. A. 2002. The modular logic of signaling proteins: building allosteric switches from simple binding domains. *Curr. Opin. Struc. Biol.* 12:61–68.

Martin, C., and Y. Zhang. 2005. The diverse functions of histone lysine methylation. *Nat. Rev. Mol. Cell Biol.* 6(11):838–849.

Rabl, J., et al. 2008. Mechanism of gate opening in the 20S proteasome by the proteasomal ATPases. *Mol. Cell.* 30(3):360–368.

Rechsteiner, M., and C. P. Hill. 2005. Mobilizing the proteolytic machine: cell biological roles of proteasome activators and inhibitors. *Trends Cell Biol.* 15(1):27–33.

Sawyer, T. K., et al. 2005. Protein phosphorylation and signal transduction modulation: chemistry perspectives for small-molecule drug discovery. *Med. Chem.* 1(3):293–319.

Sowa, M. E., et al. 2009. Defining the human deubiquitinating enzyme interaction landscape. *Cell* 138(2):389–403.

Xia, Z., and D. R. Storm. 2005. The role of calmodulin as a signal integrator for synaptic plasticity. *Nat. Rev. Neurosci.* 6(4):267–276.

Yap, K. L., et al. 1999. Diversity of conformational states and changes within the EF-hand protein superfamily. *Proteins* 37:499–507.

Zeng, W., et al. 2010. Reconstitution of the RIG-I pathway reveals a signaling role of unanchored polyubiquitin chains in innate immunity. *Cell* 141(2):315–330.

Zhou, P. 2006. REGgamma: a shortcut to destruction. *Cell* 124(2):256–257.

Zolk, O., C. Schenke, and A. Sarikas. 2006. The ubiquitin-proteasome system: focus on the heart. *Cardiovasc. Res.* 70(3):410–421.

### Purifying, Detecting, and Characterizing Proteins

Domon, B., and R. Aebersold. 2006. Mass spectrometry and protein analysis. *Science* 312(5771):212–217.

Encarnacion, S., et al. 2005. Comparative proteomics using 2-D gel electrophoresis and mass spectrometry as tools to dissect stimulons and regulons in bacteria with sequenced or partially sequenced genomes. *Biol. Proc. Online* 7:117–135.

Hames, B. D. *A Practical Approach.* Oxford University Press. A methods series that describes protein purification methods and assays.

O'Connell, M. R., R. Gamsjaeger, and J. P. Mackay. 2009. The structural analysis of protein-protein interactions by NMR spectroscopy. *Proteomics* 9(23):5224–5232.

Patton, W. F. 2002. Detection technologies in proteome analysis. *J. Chromatogr. B. Analyt. Technol. Biomed. Life Sci.* 771(1–2):3–31.

White, I. R., et al. 2004. A statistical comparison of silver and SYPRO Ruby staining for proteomic analysis. *Electrophoresis* 25(17):3048–3054.

### Proteomics

Calvo, S. E., and V. K. Mootha. 2010. The mitochondrial proteome and human disease. *Annu. Rev. Genomics Hum. Genet.* 11:25–34.

Foster, L. J., et al. 2006. A mammalian organelle map by protein correlation profiling. *Cell* 125(1):187–199.

Fu, Q., and J. E. Van Eyk. 2006. Proteomics and heart disease: identifying biomarkers of clinical utility. *Expert Rev. Proteomics* 3(2):237–249.

Gavin, A. C., et al. 2006. Proteome survey reveals modularity of the yeast cell machinery. *Nature* 440(7084):631–636.

Kellie, J. F., et al. 2010. The emerging process of Top Down mass spectrometry for protein analysis: biomarkers, protein-therapeutics, and achieving high throughput. *Mol. Biosyst.* 6(9):1532–1539.

Kislinger, T., et al. 2006. Global survey of organ and organelle protein expression in mouse: combined proteomic and transcriptomic profiling. *Cell* 125(1):173–186.

Kolker, E., R. Higdon, and J. M. Hogan. 2006. Protein identification and expression analysis using mass spectrometry. *Trends Microbiol.* 14(5):229–235.

Krogan, N. J., et al. 2006. Global landscape of protein complexes in the yeast *Saccharomyces cerevisiae*. *Nature* 440(7084):637–643.

Ong, S. E., and M. Mann. 2005. Mass spectrometry-based proteomics turns quantitative. *Nat. Chem. Biol.* 1(5):252–262.

Rifai, N., M. A. Gillette, and S. A. Carr. 2006. Protein biomarker discovery and validation: the long and uncertain path to clinical utility. *Nat. Biotech.* 24(8):971–983.

Walther, T. C., and M. Mann. 2010. Mass spectrometry-based proteomics in cell biology. *J. Cell Biol.* 190(4):491–500.

Zhou, M., and C. V. Robinson. When proteomics meets structural biology. *Trends Biochem. Sci.* 35:522–539.

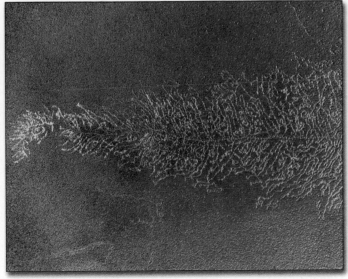

# Basic Molecular Genetic Mechanisms

**Colored transmission electron micrograph of one ribosomal RNA transcription unit from a *Xenopus* oocyte.** Transcription proceeds from left to right, with nascent ribosomal ribonucleoprotein complexes (rRNPs) growing in length as each successive RNA polymerase I molecule moves along the DNA template at the center. In this preparation each rRNP is oriented either above or below the central strand of DNA being transcribed, so that the overall shape is similar to a feather. In the nucleolus of a living cell, the nascent rRNPs extend in all directions, like a bottle brush. [Professor Oscar L. Miller/Science Photo Library.]

The extraordinary versatility of proteins as molecular machines and switches, cellular catalysts, and components of cellular structures was described in Chapter 3. In this chapter we consider how proteins are made, as well as other cellular processes that are critical for the survival of an organism and its descendants. Our focus will be on the vital molecules known as **nucleic acids**, and how they ultimately are responsible for governing all cellular function. As introduced in Chapter 2, nucleic acids are linear polymers of four types of nucleotides (see Figures 2-13, 2-16, and 2-17). These macromolecules (1) contain in the precise sequence of their nucleotides the information for determining the amino acid sequence and hence the structure and function of all the proteins of a cell, (2) are critical functional components of the cellular macromolecular factories that select and align amino acids in the correct order as a polypeptide chain is being synthesized, (3) catalyze a number of fundamental chemical reactions in cells, including formation of peptide bonds between amino acids during protein synthesis, and (4) regulate the expression of genes.

Deoxyribonucleic acid (DNA) is an informational molecule that contains in the sequence of its nucleotides the information required to build all the proteins of an organism, and hence the cells and tissues of that organism. It is ideally suited to perform this function on a molecular level. Chemically, it is extraordinarily stable under most terrestrial conditions, as exemplified by the ability to recover DNA sequence from bones and tissues that are tens of thousands of years old. Because of this, and because of repair mechanisms that operate in living cells, the long polymers that make up a DNA molecule can be up to $10^9$ nucleotides long. Virtually all the information required for the development of a fertilized human egg into an adult made of trillions of cells with specialized functions can be stored in the sequence of the four types of nucleotides that makes up the $\approx 3 \times 10^9$ base pairs in the human genome. Because of the principles of base pairing discussed in the following, the information is readily copied with an error rate of $<1$ in $10^9$ nucleotides per generation. The exact replication of this information in any

## OUTLINE

species assures its genetic continuity from generation to generation and is critical to the normal development of an individual. DNA fulfills these functions so well that it is the vessel for genetic information in all forms of life known (excluding RNA viruses, which are limited to extremely short genomes because of the relative instability of RNA compared to DNA, as we will see). The discovery that virtually all forms of life use DNA to encode their genetic information, and also use nearly the identical genetic code, implies that all forms of life descended from a common ancestor based on the storage of information in nucleic acid sequence. This information is accessed and replicated by specific base pairing between nucleotides. The information stored in DNA is arranged in hereditary units, known as genes, that control identifiable traits of an organism. In the process of transcription, the information stored in DNA is copied into **ribonucleic acid (RNA)**, which has three distinct roles in protein synthesis.

Portions of the DNA nucleotide sequence are copied into **messenger RNA (mRNA)** molecules that direct the synthesis of a specific protein. The nucleotide sequence of an mRNA molecule contains information that specifies the correct order of amino acids during the synthesis of a protein. The remarkably accurate, stepwise assembly of amino acids into proteins occurs by **translation** of mRNA. In this process, the nucleotide sequence of an mRNA molecule is "read" by a second type of RNA called **transfer RNA (tRNA)** with the aid of a third type of RNA, **ribosomal RNA (rRNA)**, and their associated proteins. As the correct amino acids are brought into sequence by tRNAs, they are linked by peptide bonds to make proteins. RNA synthesis is called **transcription** because the nucleotide sequence "language" of DNA is precisely copied, or *transcribed,* into the nucleotide sequence of an RNA molecule. Protein synthesis is referred to as **translation** because the nucleotide sequence "language" of DNA and RNA is *translated* into the amino acid sequence "language" of proteins.

Discovery of the structure of DNA in 1953 and subsequent elucidation of how DNA directs synthesis of RNA, which then directs assembly of proteins—the so-called *central dogma*—were monumental achievements marking the early days of molecular biology. However, the simplified representation of the central dogma as DNA → RNA → protein does not reflect the role of proteins in the synthesis of nucleic acids. Moreover, as discussed here for bacteria and in later chapters for eukaryotes, proteins are largely responsible for *regulating* gene expression, the entire process whereby the information encoded in DNA is decoded into proteins in the correct cells at the correct times in development. As a consequence, hemoglobin is expressed only in cells in the bone marrow (reticulocytes) destined to develop into circulating red blood cells (erythrocytes), and developing neurons make the proper synapses (connections) with $10^{11}$ other developing neurons in the human brain. The fundamental molecular genetic processes of DNA replication, transcription, and translation must be carried out with extraordinary fidelity, speed, and accurate regulation for the normal development of organisms as complex as prokaryotes and eukaryotes. This is achieved by chemical processes that operate with extraordinary accuracy coupled with multiple layers of checkpoint or surveillance mechanisms that test whether critical steps in these processes have occurred correctly before the next step is initiated. The highly regulated expression of genes necessary for the development of a multicellular organism requires integrating information from signals sent by distant cells in the developing organism, as well as from neighboring cells, and an intrinsic developmental program determined by earlier steps in embryogenesis taken by that cell's progenitors. All of this regulation is dependent on control sequences in the DNA that function with proteins called *transcription factors* to coordinate the expression of every gene. RNA sequences we discuss in Chapter 8 that regulate RNA processing and translation also are encoded in DNA originally. Nucleic acids function as the "brains and central nervous system" of the cell, while proteins carry out the functions they specify.

In this chapter, we first review the structures and properties of DNA and RNA, and explore how the different characteristics of each type of nucleic acid make them suited for their respective functions in the cell. In the next several sections we discuss the basic processes summarized in Figure 4-1: transcription of DNA into RNA precursors, processing of these precursors to make functional RNA molecules, translation of mRNAs into proteins, and the replication of DNA. Proteins regulate cell structure and most of the biochemical reactions in cells, so we first consider how the amino acid sequences of proteins, which determines their three-dimensional structures and hence their functions, is encoded in DNA and translated. After outlining functions of mRNA, tRNA, and rRNA in protein synthesis, we present a detailed description of the components and biochemical steps in translation. Understanding these processes gives us a deep appreciation of the need to copy the nucleotide sequence of DNA precisely. Consequently, we next consider the molecular problems involved in DNA replication and the complex cellular machinery for ensuring accurate copying of the genetic material. Along the way, we compare these processes in prokaryotes and eukaryotes. The next section describes how damage to DNA is repaired, and how regions of different DNA molecules are exchanged in the process of **recombination** to generate new combinations of traits in the individual organisms of a species. The final section of the chapter presents basic information about viruses, parasites that exploit the cellular machinery for DNA replication, transcription, and protein synthesis. In addition to being significant pathogens, viruses are important model organisms for studying these cellular mechanisms of macromolecular synthesis and other cellular processes. Viruses have relatively simple structures compared to cells, and small genomes that made them tractable for historic early studies of these basic cellular processes. Viruses continue to teach important lessons in molecular cell biology today and have been adapted as experimental tools for introducing any desired genes into cells, tools that are currently being tested for their effectiveness in human gene therapy.

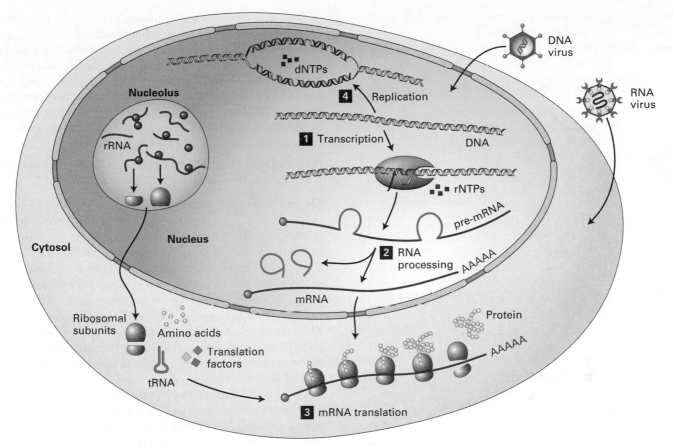

**FIGURE 4-1 Overview of four basic molecular genetic processes.** In this chapter we cover the three processes that lead to production of proteins (**1**–**3**) and the process for replicating DNA (**4**). Because viruses utilize host-cell machinery, they have been important models for studying these processes. During transcription of a protein-coding gene by RNA polymerase (**1**), the four-base DNA code specifying the amino acid sequence of a protein is copied, or *transcribed*, into a precursor messenger RNA (pre-mRNA) by the polymerization of ribonucleoside triphosphate monomers (rNTPs). Removal of noncoding sequences and other modifications to the pre-mRNA (**2**), collectively known as *RNA processing*, produce a functional mRNA, which is transported to the cytoplasm. During *translation* (**3**), the four-base code of the mRNA is decoded into the 20–amino acid language of proteins. Ribosomes, the macromolecular machines that translate the mRNA code, are composed of two subunits assembled in the nucleolus from ribosomal RNAs (rRNAs) and multiple proteins (*left*). After transport to the cytoplasm, ribosomal subunits associate with an mRNA and carry out protein synthesis with the help of transfer RNAs (tRNAs) and various translation factors. During DNA replication (**4**), which occurs only in cells preparing to divide, deoxyribonucleoside triphosphate monomers (dNTPs) are polymerized to yield two identical copies of each chromosomal DNA molecule. Each daughter cell receives one of the identical copies.

## 4.1 Structure of Nucleic Acids

DNA and RNA are chemically very similar. The primary structures of both are linear **polymers** composed of **monomers** called **nucleotides**. Both function primarily as informational molecules, carrying information in the exact sequence of their nucleotides. Cellular RNAs range in length from less than one hundred to many thousands of nucleotides. Cellular DNA molecules can be as long as several hundred million nucleotides. These large DNA units in association with proteins can be stained with dyes and visualized in the light microscope as chromosomes, so named because of their stainability. Though chemically similar, DNA and RNA exhibit some very important differences. For example, RNA can also function as a catalytic molecule. As we will see, it is the different and unique properties of DNA and RNA that makes them each suited for their specific roles in the cell.

## A Nucleic Acid Strand Is a Linear Polymer with End-to-End Directionality

In all organisms, DNA and RNA are each comprised of only four different nucleotides. Recall from Chapter 2 that all nucleotides consist of an organic base linked to a five-carbon sugar that has a phosphate group attached to carbon 5. In RNA, the sugar is ribose; in DNA, deoxyribose (see Figure 2-16). The nucleotides used in synthesis of DNA and RNA contain five different bases. The bases *adenine* (A) and *guanine* (G) are **purines**, which contain a pair of fused rings; the bases *cytosine* (C), *thymine* (T), and *uracil* (U) are **pyrimidines**, which contain a single ring (see Figure 2-17). Three of these bases—A, G, and C—are found in both DNA and RNA; however, T is found only in DNA, and U only in RNA. (Note that the single-letter abbreviations for these bases are also commonly used to denote the entire nucleotides in nucleic acid polymers.)

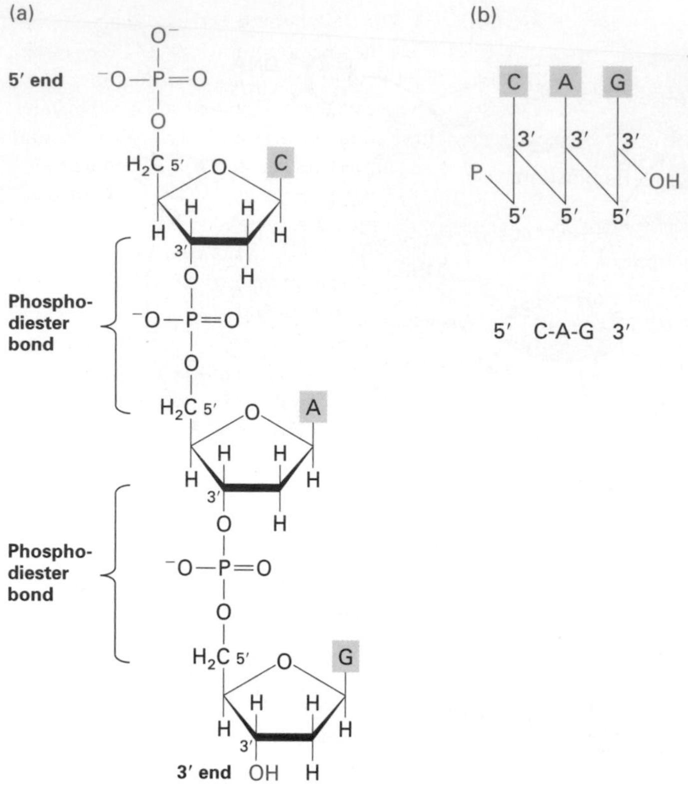

**FIGURE 4-2 Chemical directionality of a nucleic acid strand.**
Shown here are alternative representations of a single strand of DNA containing only three bases: cytosine (C), adenine (A), and guanine (G). (a) The chemical structure shows a hydroxyl group at the 3′ end and a phosphate group at the 5′ end. Note also that two phosphoester bonds link adjacent nucleotides; this two-bond linkage commonly is referred to as a *phosphodiester bond*. (b) In the "stick" diagram (*top*), the sugars are indicated as vertical lines and the phosphodiester bonds as slanting lines; the bases are denoted by their single-letter abbreviations. In the simplest representation (*bottom*), only the bases are indicated. By convention, a polynucleotide sequence is always written in the 5′→3′ direction (left to right) unless otherwise indicated.

A single nucleic acid strand has a *backbone* composed of repeating pentose-phosphate units from which the purine and pyrimidine bases extend as side groups. Like a polypeptide, a nucleic acid strand has an end-to-end chemical orientation: the *5′ end* has a hydroxyl or phosphate group on the 5′ carbon of its terminal sugar; the *3′ end* usually has a hydroxyl group on the 3′ carbon of its terminal sugar (Figure 4-2). This directionality, plus the fact that synthesis proceeds 5′ to 3′, has given rise to the convention that polynucleotide sequences are written and read in the 5′→3′ direction (from left to right); for example, the sequence AUG is assumed to be (5′)AUG(3′). As we will see, the 5′→3′ directionality of a nucleic acid strand is an important property of the molecule. The chemical linkage between adjacent nucleotides, commonly called a **phosphodiester bond**, actually consists of two phosphoester bonds, one on the 5′ side of the phosphate and another on the 3′ side.

The linear sequence of nucleotides linked by phosphodiester bonds constitutes the primary structure of nucleic acids. Like polypeptides, polynucleotides can twist and fold into three-dimensional conformations stabilized by noncovalent bonds. Although the primary structures of DNA and RNA are generally similar, their three-dimensional conformations are quite different. These structural differences are critical to the different functions of the two types of nucleic acids.

## Native DNA Is a Double Helix of Complementary Antiparallel Strands

The modern era of molecular biology began in 1953 when James D. Watson and Francis H. C. Crick proposed that DNA has a double-helical structure. Their proposal was based on analysis of x-ray diffraction patterns of DNA fibers generated by Rosalind Franklin and Maurice Wilkins, which showed that the structure was helical, and analyses of the base composition of DNA from multiple organisms by Erwin Chargaff and colleagues. Chargaff's studies revealed that while the base composition (percent of A, T, G, and C) varies greatly between distantly related organisms, in all organisms the percent of A always equals the percent of T, and the percent of G always equals the percent of C. Based on these discoveries and the structures of the four nucleotides, Watson and Crick performed careful molecular model building, proposing a **double helix,** with A always hydrogen-bonded to T and G always hydrogen-bonded to C at the axis of the double helix. The Watson and Crick model proved correct and paved the way for our modern understanding of how DNA functions as the genetic material. Today, our most accurate models for DNA structure come from high-resolution x-ray diffraction studies of crystals of DNA, made possible by the chemical synthesis of large amounts of short DNA molecules of uniform length and sequence that are amenable to crystallization (Figure 4-3a).

DNA consists of two associated polynucleotide strands that wind together to form a double helix. The two sugar-phosphate backbones are on the outside of the double helix, and the bases project into the interior. The adjoining bases in each strand stack on top of one another in parallel planes (Figure 4-3a). The orientation of the two strands is *antiparallel;* that is, their 5′→3′ directions are opposite. The strands are held in precise register by formation of **base pairs** between the two strands: A is paired with T through two hydrogen bonds; G is paired with C through three hydrogen bonds (Figure 4-3b). This base-pair complementarity is a consequence of the size, shape, and chemical composition of the bases. The presence of thousands of such hydrogen bonds in a DNA molecule contributes greatly to the stability of the double helix. Hydrophobic and van der Waals interactions between the stacked adjacent base pairs further stabilize the double-helical structure.

In natural DNA, A always hydrogen bonds with T, and G with C, forming A·T and G·C base pairs as shown in Figure 4-3b. These associations, always between a larger purine and a smaller pyrimidine, are often called *Watson-Crick*

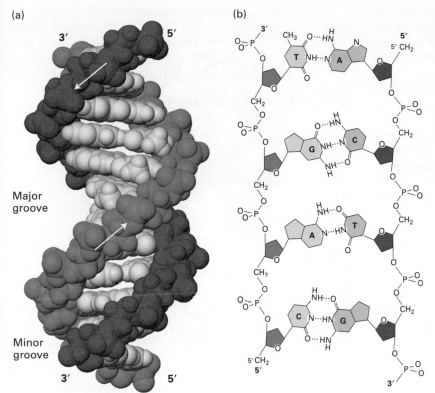

(a)

3′  5′

Major groove

Minor groove

3′  5′

(b)

**FIGURE 4-3 The DNA double helix.** (a) Space-filling model of B DNA, the most common form of DNA in cells. The bases (light shades) project inward from the sugar-phosphate backbones (dark red and blue) of each strand, but their edges are accessible through major and minor grooves. Arrows indicate the 5′→3′ direction of each strand. Hydrogen bonds between the bases are in the center of the structure. The major and minor grooves are lined by potential hydrogen bond donors and acceptors (highlighted in yellow). (b) Chemical structure of DNA double helix. This extended schematic shows the two sugar-phosphate backbones and hydrogen bonding between the Watson-Crick base pairs, A·T and G·C. [Part (a) adapted from R. Wing et al., 1980, *Nature* **287**:755. Part (b) adapted from R. E. Dickerson, 1983, *Sci. Am.* **249**:94.]

*base pairs.* Two polynucleotide strands, or regions thereof, in which all the nucleotides form such base pairs are said to be **complementary.** However, in theory and in synthetic DNAs, other base pairs can form. For example, guanine (a purine) could theoretically form hydrogen bonds with thymine (a pyrimidine), causing only a minor distortion in the helix. The space available in the helix also would allow pairing between the two pyrimidines cytosine and thymine. Although the nonstandard G·T and C·T base pairs are normally not found in DNA, G·U base pairs are quite common in double-helical regions that form within otherwise single-stranded RNA. Nonstandard base pairs do not occur naturally in duplex DNA because the DNA copying enzyme, which is described later in this chapter, does not permit them.

Most DNA in cells is a *right-handed* helix. The x-ray diffraction pattern of DNA indicates that the stacked bases are regularly spaced 0.34 nm apart along the helix axis. The helix makes a complete turn every 3.4 to 3.6 nm, depending on the sequence; thus there are about 10–10.5 base pairs per turn. This is referred to as the *B form* of DNA, the normal form present in most DNA stretches in cells. On the outside of B-form DNA, the spaces between the intertwined strands form two helical grooves of different widths described as the *major groove* and the *minor groove* (see Figure 4-3a). As a consequence, the atoms on the edges of each base within these grooves are accessible from outside the helix, forming two types of binding surfaces. DNA-binding proteins can "read" the sequence of bases in duplex DNA by contacting atoms in either the major or the minor grooves.

Under laboratory conditions in which most of the water is removed from DNA, the crystallographic structure of DNA changes to the *A form*, which is wider and shorter than B-form DNA, with a wider and deeper major groove and a more narrow and shallow minor groove (Figure 4-4). RNA-DNA and RNA-RNA helices exist in this form in cells and in vitro.

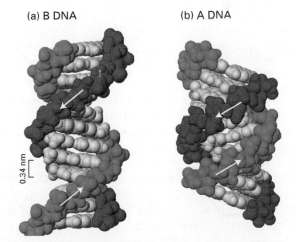

(a) B DNA    (b) A DNA

0.34 nm

**FIGURE 4-4 Comparison of A- and B-Form DNA.** The sugar-phosphate backbones of the two strands, which are on the outside of both structures, are shown in red and blue; the bases (lighter shades) are oriented inward. (a) The B form of DNA has ≈10.5 base pairs per helical turn. Adjacent stacked base pairs are 0.34 nm apart. (b) The more compact A form of DNA has 11 base pairs per turn with a much deeper major groove and much more shallow minor groove than B form DNA.

Important modifications in the structure of standard B-form DNA come about as a result of protein binding to specific DNA sequences. Although the multitude of hydrogen and hydrophobic bonds between the bases provides stability to DNA, the double helix is flexible about its long axis. Unlike the α helix in proteins (see Figure 3-4), there are no hydrogen bonds parallel to the axis of the DNA helix. This property allows DNA to bend when complexed with a DNA-binding protein (Figure 4-5). Bending of DNA is critical to the dense packing of DNA in chromatin, the protein-DNA complex in which nuclear DNA occurs in eukaryotic cells (Chapter 6).

Why did DNA evolve to be the carrier of genetic information in cells as opposed to RNA? The hydrogen at the 2′ position in the deoxyribose of DNA makes it a far more stable molecule than RNA, which instead has a hydroxyl group at the 2′ position of ribose (see Figure 2-16). The 2′-hydroxyl groups in RNA participate in the slow, $OH^-$-catalyzed hydrolysis of phosphodiester bonds at neutral pH (Figure 4-6). The absence of 2′-hydroxyl groups in DNA prevents this process. Therefore, the presence of deoxyribose in DNA makes it a more stable molecule—a characteristic that is critical to its function in the long-term storage of genetic information.

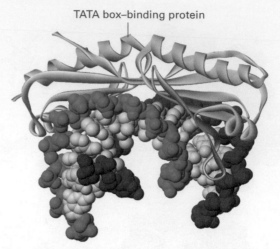

**FIGURE 4-5 Protein interaction can bend DNA.** The conserved C-terminal domain of the TATA box–binding protein (TBP) binds to the minor groove of specific DNA sequences rich in A and T, untwisting and sharply bending the double helix. Transcription of most eukaryotic genes requires participation of TBP. [Adapted from D. B. Nikolov and S. K. Burley, 1997, *Proc. Nat'l Acad. Sci. USA* **94**:15.]

## DNA Can Undergo Reversible Strand Separation

During replication and transcription of DNA, the strands of the double helix must separate to allow the internal edges of the bases to pair with the bases of the nucleotides being polymerized into new polynucleotide chains. In later sections, we describe the cellular mechanisms that separate and subsequently reassociate DNA strands during replication and transcription. Here we discuss fundamental factors that influence the separation and reassociation of DNA strands. These properties of DNA were elucidated by in vitro experiments.

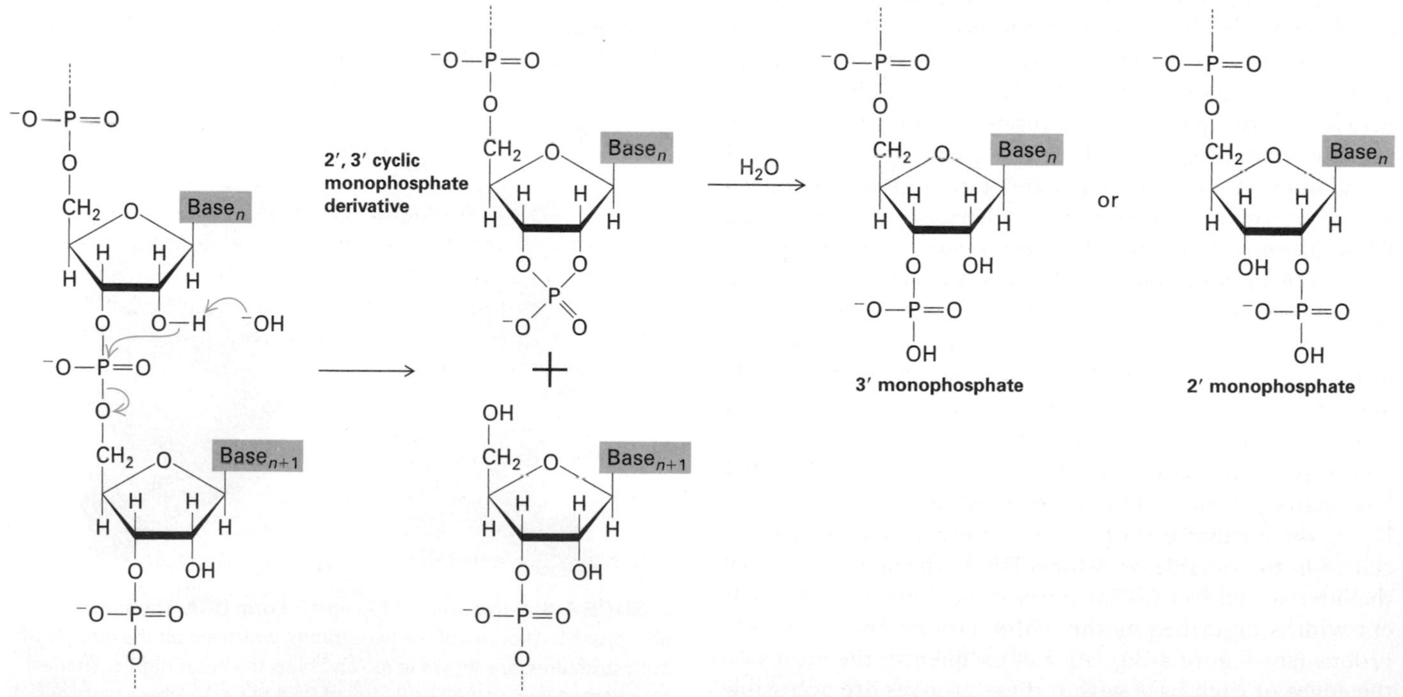

**FIGURE 4-6 Base-catalyzed hydrolysis of RNA.** The 2′-hydroxyl group in RNA can act as a nucleophile, attacking the phosphodiester bond. The 2′,3′ cyclic monophosphate derivative is further hydrolyzed to a mixture of 2′ and 3′ monophosphates. This mechanism of phosphodiester bond hydrolysis cannot occur in DNA, which lacks 2′-hydroxyl groups. [Adapted from Nelson et al., *Lehninger Principles of Biochemistry*, 4th ed., W. H. Freeman and Company.]

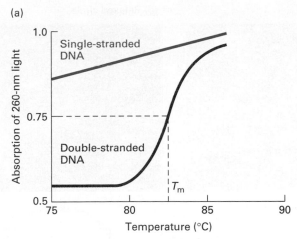

(a)

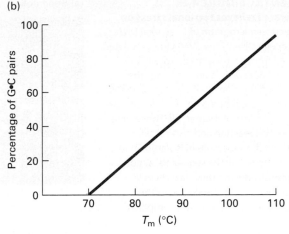

(b)

**EXPERIMENTAL FIGURE 4-7  G·C content of DNA affects melting temperature.** The temperature at which DNA denatures increases with the proportion of G·C pairs. (a) Melting of doubled-stranded DNA can be monitored by the absorption of ultraviolet light at 260 nm. As regions of double-stranded DNA unpair, the absorption of light by those regions increases almost twofold. The temperature at which half the bases in a double-stranded DNA sample have denatured is denoted $T_m$ (for "temperature of melting"). Light absorption by single-stranded DNA changes much less as the temperature is increased. (b) The $T_m$ is a function of the G·C content of the DNA; the higher the G+C percentage, the greater the $T_m$.

The unwinding and separation of DNA strands, referred to as **denaturation,** or "melting," can be induced experimentally by increasing the temperature of a solution of DNA. As the thermal energy increases, the resulting increase in molecular motion eventually breaks the hydrogen bonds and other forces that stabilize the double helix; the strands then separate, driven apart by the electrostatic repulsion of the negatively charged deoxyribose-phosphate backbone of each strand. Near the denaturation temperature, a small increase in temperature causes a rapid, nearly simultaneous loss of the multiple weak interactions holding the strands together along the entire length of the DNA molecules. Because the stacked base pairs in duplex DNA absorb less ultraviolet (UV) light than the unstacked bases in single-stranded DNA, this leads to an abrupt increase in the absorption of UV light, a phenomenon known as *hyperchromicity* (Figure 4-7a).

The *melting temperature ($T_m$)* at which DNA strands will separate depends on several factors. Molecules that contain a greater proportion of G·C pairs require higher temperatures to denature because the three hydrogen bonds in G·C pairs make these base pairs more stable than A·T pairs, which have only two hydrogen bonds. Indeed, the percentage of G·C base pairs in a DNA sample can be estimated from its $T_m$ (Figure 4-7b). The ion concentration also influences the $T_m$ because the negatively charged phosphate groups in the two strands are shielded by positively charged ions. When the ion concentration is low, this shielding is decreased, thus increasing the repulsive forces between the strands and reducing the $T_m$. Agents that destabilize hydrogen bonds, such as formamide or urea, also lower the $T_m$. Finally, extremes of pH denature DNA at low temperature. At low (acid) pH, the bases become protonated and thus positively charged, repelling each other. At high (alkaline) pH, the bases lose protons and become negatively charged,

again repelling each other because of the similar charge. In cells, pH and temperature are, for the most part, maintained. These features of DNA separation are most useful for manipulating DNA in a laboratory setting.

The single-stranded DNA molecules that result from denaturation form random coils without an organized structure. Lowering the temperature, increasing the ion concentration, or neutralizing the pH causes the two complementary strands to reassociate into a perfect double helix. The extent of such *renaturation* is dependent on time, the DNA concentration, and the ionic concentration. Two DNA strands that are not related in sequence will remain as random coils and will not renature; most importantly, they will not inhibit complementary DNA partner strands from finding each other and renaturing. Denaturation and renaturation of DNA are the basis of nucleic acid **hybridization,** a powerful technique used to study the relatedness of two DNA samples and to detect and isolate specific DNA molecules in a mixture containing numerous different DNA sequences (see Figure 5-16).

## Torsional Stress in DNA Is Relieved by Enzymes

Many bacterial genomic DNAs and many viral DNAs are circular molecules. Circular DNA molecules also occur in mitochondria, which are present in almost all eukaryotic cells, and in chloroplasts, which are present in plants and some unicellular eukaryotes.

Each of the two strands in a circular DNA molecule forms a closed structure without free ends. Localized unwinding of a circular DNA molecule, which occurs during DNA replication, induces torsional stress into the remaining portion of the molecule because the ends of the strands are not free to rotate. As a result, the DNA molecule twists back

**EXPERIMENTAL FIGURE 4-8**

**Topoisomerase I relieves torsional stress on DNA.** (a) Electron micrograph of SV40 viral DNA. When the circular DNA of the SV40 virus is isolated and separated from its associated protein, the DNA duplex is underwound and assumes the supercoiled configuration. (b) If a supercoiled DNA is nicked (i.e., one strand cleaved), the strands can rewind, leading to loss of a supercoil. Topoisomerase I catalyzes this reaction and also reseals the broken ends. All the supercoils in isolated SV40 DNA can be removed by the sequential action of this enzyme, producing the relaxed-circle conformation. For clarity, the shapes of the molecules at the bottom have been simplified.

(a) Supercoiled

(b) Relaxed circle

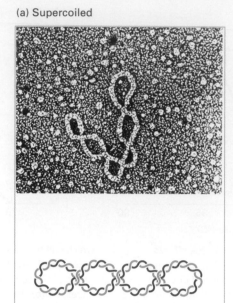
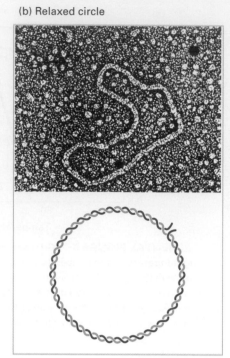

on itself, like a twisted rubber band, forming *supercoils* (Figure 4-8a). In other words, when part of the DNA helix is underwound, left-handed supercoils are introduced into the circular DNA molecule, as in Figure 4-8a. Bacterial and eukaryotic cells, however, contain *topoisomerase I,* which can relieve any torsional stress that develops in cellular DNA molecules during replication or other processes. This enzyme binds to DNA at random sites and breaks a phosphodiester bond in one strand. Such a one-strand break in DNA is called a *nick.* The broken end then winds around the uncut strand, leading to loss of supercoils (Figure 4-8b). Finally, the same enzyme joins (ligates) the two ends of the broken strand. Another type of enzyme, *topoisomerase II,* makes breaks in both strands of a double-stranded DNA and then religates them. As a result, topoisomerase II can both relieve torsional stress and link together two circular DNA molecules as in the links of a chain.

Although eukaryotic nuclear DNA is linear, long loops of DNA are fixed in place within chromosomes (Chapter 6). Thus torsional stress and the consequent formation of supercoils also could occur during replication of nuclear DNA. As in bacterial cells, abundant topoisomerase I in eukaryotic nuclei relieves any torsional stress in nuclear DNA that would develop in the absence of this enzyme.

## Different Types of RNA Exhibit Various Conformations Related to Their Functions

The primary structure of RNA is generally similar to that of DNA with two exceptions: the sugar component of RNA, ribose, has a hydroxyl group at the 2' position (see Figure 2-16b), and thymine in DNA is replaced by uracil in RNA.

The presence of thymine rather than uracil in DNA is important to the long-term stability of DNA because of its function in DNA repair (see Section 4.7). As noted earlier, the hydroxyl group on the 2' C of ribose makes RNA more chemically labile than DNA. As a result of this lability, RNA is cleaved into mononucleotides by alkaline solution (see Figure 4-6), whereas DNA is not. The 2'-C hydroxyl of RNA also provides a chemically reactive group that takes part in RNA-mediated catalysis. Like DNA, RNA is a long polynucleotide that can be double-stranded or single-stranded, linear or circular. It can also participate in a hybrid helix composed of one RNA strand and one DNA strand. As discussed above, RNA-RNA and RNA-DNA double helices have a compact conformation like the A form of DNA (see Figure 4-4b).

Unlike DNA, which exists primarily as a very long double helix, most cellular RNAs are single-stranded and exhibit a variety of conformations (Figure 4-9). Differences in the sizes and conformations of the various types of RNA permit them to carry out specific functions in a cell. The simplest secondary structures in single-stranded RNAs are formed by pairing of complementary bases. "Hairpins" are formed by pairing of bases within ≈5–10 nucleotides of each other, and "stem-loops" by pairing of bases that are separated by >10 to several hundred nucleotides. These simple folds can cooperate to form more complicated tertiary structures, one of which is termed a "pseudoknot."

As discussed in detail later, tRNA molecules adopt a well-defined three-dimensional architecture in solution that is crucial in protein synthesis. Larger rRNA molecules also have locally well-defined three-dimensional structures, with more flexible linkers in between. Secondary and tertiary structures

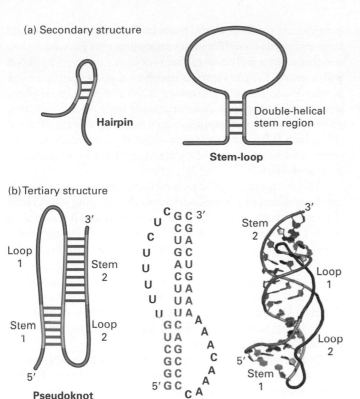

**(a) Secondary structure**

Hairpin

Loop

Double-helical stem region

**Stem-loop**

**(b) Tertiary structure**

Loop 1

Stem 2

Stem 1

Loop 2

5′

**Pseudoknot**

Stem 2

Loop 1

Loop 2

Stem 1

**FIGURE 4-9 RNA secondary and tertiary structures.** (a) Stem-loops, hairpins, and other secondary structures can form by base pairing between distant complementary segments of an RNA molecule. In stem-loops, the single-stranded loop between the base-paired helical stem may be hundreds or even thousands of nucleotides long, whereas in hairpins, the short turn may contain as few as four nucleotides. (b) Pseudoknots, one type of RNA tertiary structure, are formed by interaction of secondary loops through base pairing between complementary bases. The structure shown forms the core domain of the human telomerase RNA. *Left:* Secondary-structure diagram with base-paired nucleotides in green and blue and single-stranded regions in red. *Middle:* Sequence of the telomerase RNA core domain, colored to correspond to the secondary structure diagram at the left. *Right:* Diagram of the telomerase core domain structure determined by 2D-NMR, showing bases-paired bases only and a tube for the sugar phosphate backbone, colored to correspond to the diagrams at left. [Part (b) *middle and right* adapted from C. A. Theimer et al., 2005, *Mol. Cell* **17**:671.]

also have been recognized in mRNA, particularly near the ends of molecules. Clearly, then, RNA molecules are like proteins in that they have structured domains connected by less structured, flexible stretches.

The folded domains of RNA molecules not only are structurally analogous to the α helices and β strands found in proteins, but in some cases also have catalytic capacities. Such catalytic RNAs are called **ribozymes.** Although ribozymes usually are associated with proteins that stabilize the ribozyme structure, it is the RNA that acts as a catalyst. Some ribozymes can catalyze splicing, a remarkable process in which an internal RNA sequence is cut and removed, and the two resulting chains then ligated. This process occurs

during formation of the majority of functional mRNA molecules in multicellular eukaryotes, and also occurs in single-celled eukaryotes such as yeast, bacteria, and archaea. Remarkably, some RNAs carry out *self-splicing,* with the catalytic activity residing in the sequence that is removed. The mechanisms of splicing and self-splicing are discussed in detail in Chapter 8. As noted later in this chapter, rRNA plays a catalytic role in the formation of peptide bonds during protein synthesis.

In this chapter, we focus on the functions of mRNA, tRNA, and rRNA in gene expression. In later chapters we will encounter other RNAs, often associated with proteins, that participate in other cell functions.

## KEY CONCEPTS of Section 4.1

### Structure of Nucleic Acids

• Deoxyribonucleic acid (DNA), the genetic material, carries information to specify the amino acid sequences of proteins. It is transcribed into several types of ribonucleic acid (RNA), including messenger RNA (mRNA), transfer RNA (tRNA), and ribosomal RNA (rRNA), which function in protein synthesis (see Figure 4-1).

• All DNAs and most RNAs are long, unbranched polymers of nucleotides, which consist of a phosphorylated pentose linked to an organic base, either a purine or a pyrimidine.

• The purines adenine (A) and guanine (G) and the pyrimidine cytosine (C) are present in both DNA and RNA. The pyrimidine thymine (T) present in DNA is replaced by the pyrimidine uracil (U) in RNA.

• Adjacent nucleotides in a polynucleotide are linked by phosphodiester bonds. The entire strand has a chemical directionality with 5′ and 3′ ends (see Figure 4-2).

• Natural DNA (B DNA) contains two complementary antiparallel polynucleotide strands wound together into a regular right-handed double helix with the bases on the inside and the two sugar-phosphate backbones on the outside (see Figure 4-3). Base pairing between the strands and hydrophobic interactions between adjacent base pairs stacked perpendicular to the helix axis stabilize this native structure.

• The bases in nucleic acids can interact via hydrogen bonds. The standard Watson-Crick base pairs are G·C, A·T (in DNA), and G·C, A·U (in RNA). Base pairing stabilizes the native three-dimensional structures of DNA and RNA.

• Binding of protein to DNA can deform its helical structure, causing local bending or unwinding of the DNA molecule.

• Heat causes the DNA strands to separate (denature). The melting temperature $T_m$ of DNA increases with the percentage of G·C base pairs. Under suitable conditions, separated complementary nucleic acid strands will renature.

- Circular DNA molecules can be twisted on themselves, forming supercoils (see Figure 4-8). Enzymes called topoisomerases can relieve torsional stress and remove supercoils from circular DNA molecules. Long linear DNA can also experience torsional stress because long loops are fixed in place within chromosomes.

- Cellular RNAs are single-stranded polynucleotides, some of which form well-defined secondary and tertiary structures (see Figure 4-9). Some RNAs, called ribozymes, have catalytic activity.

## 4.2 Transcription of Protein-Coding Genes and Formation of Functional mRNA

The simplest definition of a gene is a "unit of DNA that contains the information to specify synthesis of a single polypeptide chain or functional RNA (such as a tRNA)." The DNA molecules of small viruses contain only a few genes, whereas the single DNA molecule in each of the chromosomes of higher animals and plants may contain several thousand genes. The vast majority of genes carry information to build protein molecules, and it is the RNA copies of such *protein-coding genes* that constitute the mRNA molecules of cells.

During synthesis of RNA, the four-base language of DNA containing A, G, C, and T is simply copied, or *transcribed*, into the four-base language of RNA, which is identical except that U replaces T. In contrast, during protein synthesis, the four-base language of DNA and RNA is *translated* into the 20–amino acid language of proteins. In this section, we focus on formation of functional mRNAs from protein-coding genes (see Figure 4-1, step **1**). A similar process yields the precursors of rRNAs and tRNAs encoded by rRNA and tRNA genes; these precursors are then further modified to yield functional rRNAs and tRNAs (Chapter 8). Similarly, thousands of **micro RNAs (miRNAs)** that regulate translation of specific target mRNAs are transcribed into precursors by RNA polymerases and processed into functional miRNAs (Chapter 8). Other non-protein-coding (or simply "noncoding") RNAs help to regulate transcription of specific protein-coding genes. Regulation of transcription allows distinct sets of genes to be expressed in the multiple different types of cells that make up a multicellular organism. It also allows different amounts of mRNA to be transcribed from different genes, resulting in differences in the amounts of the encoded proteins in a cell. Regulation of transcription is addressed in Chapter 7.

### A Template DNA Strand Is Transcribed into a Complementary RNA Chain by RNA Polymerase

During transcription of DNA, one DNA strand acts as a *template*, determining the order in which ribonucleoside triphosphate (rNTP) monomers are polymerized to form a complementary RNA chain. Bases in the template DNA strand base-pair with complementary incoming rNTPs, which then are joined in a polymerization reaction catalyzed by **RNA polymerase.** Polymerization involves a nucleophilic attack by the 3' oxygen in the growing RNA chain on the α phosphate of the next nucleotide precursor to be added, resulting in formation of a phosphodiester bond and release of pyrophosphate ($PP_i$). As a consequence of this mechanism, RNA molecules are always synthesized in the 5'→3' direction (Figure 4-10a).

The energetics of the polymerization reaction strongly favors addition of ribonucleotides to the growing RNA chain because the high-energy bond between the α and β phosphates of rNTP monomers is replaced by the lower-energy phosphodiester bond between nucleotides. The equilibrium for the reaction is driven further toward chain elongation by pyrophosphatase, an enzyme that catalyzes cleavage of the released $PP_i$ into two molecules of inorganic phosphate. Like the two strands in DNA, the template DNA strand and the growing RNA strand that is base-paired to it have opposite 5'→3' directionality.

By convention, the site on the DNA at which RNA polymerase begins transcription is numbered +1 (Figure 4-10b). *Downstream* denotes the direction in which a template DNA strand is transcribed; *upstream* denotes the opposite direction. Nucleotide positions in the DNA sequence downstream from a start site are indicated by a positive (+) sign; those upstream, by a negative (−) sign. Because RNA is synthesized 5'→3', RNA polymerase moves down the template DNA strand in a 3'→5' direction. The newly synthesized RNA is complementary to the template DNA strand; therefore it is identical to the nontemplate DNA strand, with uracil in place of thymine.

**Stages in Transcription** To carry out transcription, RNA polymerase performs several distinct functions, as depicted in Figure 4-11. During transcription *initiation*, RNA polymerase, with the help of initiation factors discussed later, recognizes and binds to a specific site, called a **promoter,** in double-stranded DNA (step **1**). After binding, RNA polymerase and the initiation factors separate the DNA strands to make the bases in the template strand available for base pairing with the bases of the ribonucleoside triphosphates that it will polymerize. RNA polymerases and initiation factors then melt 12–14 base pairs of DNA around the transcription start site, which is located on the template strand within the promoter region (step **2**). This allows the template strand to enter the active site of the enzyme that catalyzes phosphodiester bond formation between ribonucleotide triphosphates that are complementary to the promoter template strand at the start site of transcription. The 12- to 14-base-pair region of melted DNA in the polymerase is known as the *transcription bubble*. Transcription initiation is considered complete when the first two ribonucleotides of an RNA chain are linked by a phosphodiester bond (step **3**).

After several ribonucleotides have been polymerized, RNA polymerase dissociates from the promoter DNA and

(a)

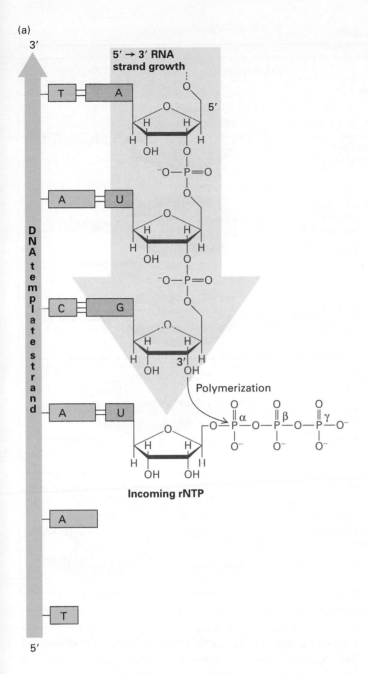

general transcription factors. During the stage of *strand elongation,* RNA polymerase moves along the template DNA one base at a time, opening the double-stranded DNA in front of its direction of movement and guiding the strands together so that they hybridize at the upstream end of the transcription bubble (Figure 4-11, step **4**). One ribonucleotide at a time is added to the 3' end of the growing (*nascent*) RNA chain during strand elongation by the polymerase. The enzyme maintains a melted region of approximately 14 base pairs in the transcription bubble. Approximately eight nucleotides at the 3' end of the growing RNA strand remain base-paired to the template DNA strand in the transcription bubble. The elongation complex, comprising RNA polymerase, template DNA, and the growing (nascent) RNA strand, is extraordinarily stable. For example, RNA polymerase transcribes the longest known mammalian gene, containing about 2 million base pairs, without dissociating from

**FIGURE 4-10 RNA is synthesized 5'→3'.** (a) Polymerization of ribonucleotides by RNA polymerase during transcription. The ribonucleotide to be added at the 3' end of a growing RNA strand is specified by base pairing between the next base in the template DNA strand and the complementary incoming ribonucleoside triphosphate (rNTP). A phosphodiester bond is formed when RNA polymerase catalyzes a reaction between the 3' O of the growing strand and the α phosphate of a correctly base-paired rNTP. RNA strands always are synthesized in the 5'→3' direction and are opposite in polarity to their template DNA strands. (b) Conventions for describing RNA transcription. *Top:* The DNA nucleotide where RNA polymerase begins transcription is designated +1. The direction the polymerase travels on the DNA is "downstream," and bases are marked with positive numbers. The opposite direction is "upstream," and bases are noted with negative numbers. Some important gene features lie upstream of the transcription start site, including the promoter sequence that localizes RNA polymerase to the gene. *(Bottom)* The DNA strand that is being transcribed is the template strand; its complement, the nontemplate strand. The RNA being synthesized is complementary to the template strand, and therefore identical with the nontemplate strand sequence, except with uracil in place of thymine. [Part (b) adapted from Griffiths et al., *Modern Genetic Analysis,* 2d ed., W. H. Freeman and Company.]

(b)

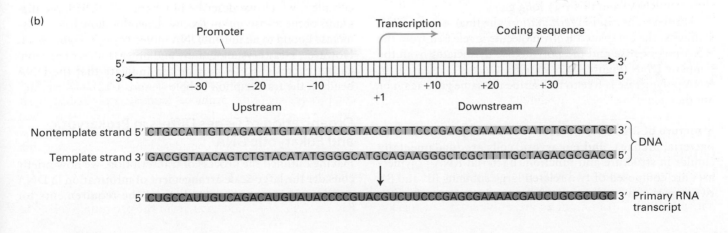

**FIGURE 4-11 Three stages in transcription.** During initiation of transcription, RNA polymerase forms a transcription bubble and begins polymerization of ribonucleotides (rNTPs) at the start site, which is located within the promoter region. Once a DNA region has been transcribed, the separated strands reassociate into a double helix. The nascent RNA is displaced from its template strand except at its 3′ end. The 5′ end of the RNA strand exits the RNA polymerase through a channel in the enzyme. Termination occurs when the polymerase encounters a specific termination sequence (stop site). See the text for details. For simplicity, the diagram depicts transcription of four turns of the DNA helix encoding ≈40 nucleotides of RNA. Most RNAs are considerably longer, requiring transcription of a longer region of DNA.

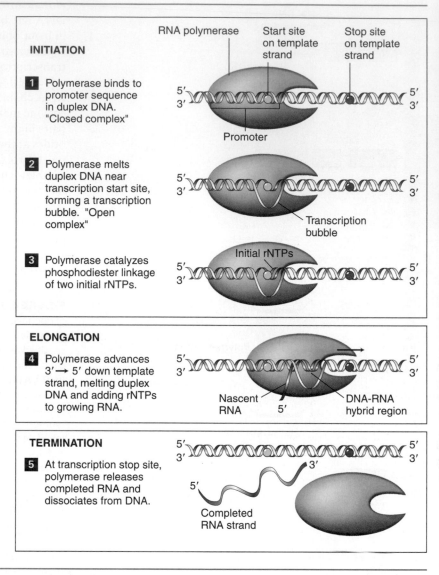

**INITIATION**

**1** Polymerase binds to promoter sequence in duplex DNA. "Closed complex"

**2** Polymerase melts duplex DNA near transcription start site, forming a transcription bubble. "Open complex"

**3** Polymerase catalyzes phosphodiester linkage of two initial rNTPs.

**ELONGATION**

**4** Polymerase advances 3′ ⟶ 5′ down template strand, melting duplex DNA and adding rNTPs to growing RNA.

**TERMINATION**

**5** At transcription stop site, polymerase releases completed RNA and dissociates from DNA.

the DNA template or releasing the nascent RNA. RNA synthesis occurs at a rate of about 1000 nucleotides per minute at 37 °C, so the elongation complex must remain intact for more than 24 hours to assure continuous RNA synthesis of the pre-mRNA from this very long gene.

During transcription *termination,* the final stage in RNA synthesis, the completed RNA molecule is released from the RNA polymerase and the polymerase dissociates from the template DNA (Figure 4-11, step **5**). Once it is released, an RNA polymerase is free to transcribe the same gene again or another gene.

**Structure of RNA Polymerases** The RNA polymerases of bacteria, archaea, and eukaryotic cells are fundamentally similar in structure and function. Bacterial RNA polymerases are composed of two related large subunits (β′ and β), two copies of a smaller subunit (α), and one copy of a fifth

subunit (ω) that is not essential for transcription or cell viability but that stabilizes the enzyme and assists in the assembly of its subunits. Archaeal and eukaryotic RNA polymerases have several additional small subunits associated with this core complex, which we describe in Chapter 7. Schematic diagrams of the transcription process generally show RNA polymerase bound to an unbent DNA molecule, as in Figure 4-11. However, x-ray crystallography and other studies of an elongating bacterial RNA polymerase indicate that the DNA bends at the transcription bubble (Figure 4-12).

## Organization of Genes Differs in Prokaryotic and Eukaryotic DNA

Having outlined the process of transcription, we now briefly consider the large-scale arrangement of information in DNA and how this arrangement dictates the requirements for

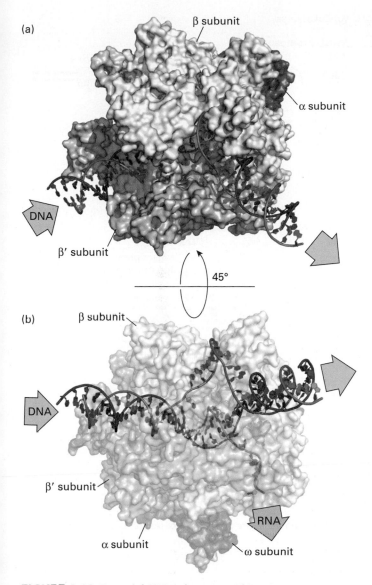

(a)

β subunit

α subunit

DNA

β' subunit

45°

(b)

β subunit

DNA

β' subunit

RNA

α subunit          ω subunit

**FIGURE 4-12 Bacterial RNA polymerase.** This structure corresponds to the polymerase molecule in the elongation phase (step **4**) of Figure 4-11. In these diagrams, transcription is proceeding in the leftward direction. Arrows indicate where downstream DNA enters the polymerase and upstream DNA exits at an angle from the downstream DNA; the coding strand is red, the noncoding strand blue; nascent RNA green. The RNA polymerase β' subunit is gold, β is light gray, and the α subunit visible from this angle brown. The upper diagram is a space-filling model of the elongation complex viewed from an angle that emphasizes the bend in the DNA as it passes through the polymerase. The elongation complex is rotated in the bottom diagram as shown, and proteins are made largely transparent to reveal the structure of the transcription bubble inside the polymerase that is not visible in the space-filling model. Nucleotides complementary to the template DNA are added to the 3' end of the nascent RNA strand (at the left). The newly synthesized nascent RNA exits the polymerase at the bottom through a channel formed between the β and β' subunits. The ω subunit and the other α subunit are visible from this angle. [Courtesy of Seth Darst; see N. Korzheva et al., 2000, *Science* **289**:619–625, and N. Opalka et al., 2003, *Cell* **114**:335–345.]

RNA synthesis so that information transfer goes smoothly. In recent years, sequencing of the entire **genomes** from multiple organisms has revealed not only large variations in the number of protein-coding genes but also differences in their organization in bacteria and eukaryotes.

The most common arrangement of protein-coding genes in all bacteria has a powerful and appealing logic: genes encoding proteins that function together, for example, the enzymes required to synthesize the amino acid tryptophan, are most often found in a contiguous array in the DNA. Such an arrangement of genes in a functional group is called an **operon,** because it operates as a unit from a single promoter. Transcription of an operon produces a continuous strand of mRNA that carries the message for a related series of proteins (Figure 4-13a). Each section of the mRNA represents the unit (or gene) that encodes one of the proteins in the series. This arrangement results in the *coordinate expression* of all the genes in the operon. Every time an RNA polymerase molecule initiates transcription at the promoter of the operon, all the genes of the operon are transcribed and translated. In prokaryotic DNA the genes are closely packed with very few noncoding gaps, and the DNA is transcribed directly into mRNA. Because DNA is not sequestered in a nucleus in prokaryotes, ribosomes have immediate access to the translation start sites in the mRNA as they emerge from the surface of the RNA polymerase. Consequently, translation of the mRNA begins even while the 3' end of the mRNA is still being synthesized at the active site of the RNA polymerase.

This economic clustering of genes devoted to a single metabolic function does not occur in eukaryotes, even simple ones such as yeasts, which can be metabolically similar to bacteria. Rather, eukaryotic genes encoding proteins that function together are most often physically separated in the DNA; indeed, such genes usually are located on different chromosomes. Each gene is transcribed from its own promoter, producing one mRNA, which generally is translated to yield a single polypeptide (Figure 4-13b).

Early research on the structure of eukaryotic genes involved studies of viruses that infect animals. When researchers analyzed the regions of a viral DNA molecule that encode viral mRNAs, they were surprised to observe that the sequence of a single viral mRNA was encoded in several regions of the viral DNA separated by DNA sequences that are not present in the mRNA. Later, the development of gene cloning and DNA sequencing (see Chapter 5) allowed researchers to compare the genomic DNA sequence of multicellular organisms with the sequences of their mRNAs. This revealed that cellular mRNAs also are usually encoded in several separate regions of genomic DNA, called exons, separated by sequences of DNA called introns. Further studies showed that a gene is first transcribed into a long primary transcript that includes both exon sequences and the intron sequences that separate them. Subsequently, exons are spliced together, precisely removing the introns (Chapter 8). Although introns are common in multicellular eukaryotes, they are extremely rare in bacteria and archaea and uncommon in many unicellular eukaryotes such as baker's yeast.

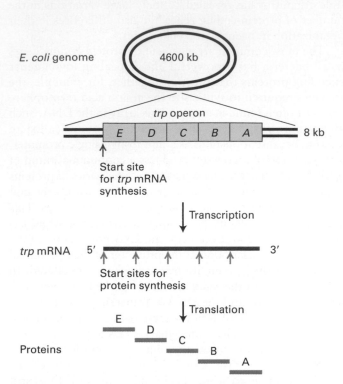

(a) Prokaryotes

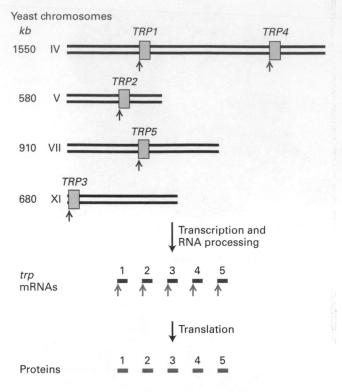

(b) Eukaryotes

**FIGURE 4-13 Gene organization in prokaryotes and eukaryotes.**
(a) The tryptophan (*trp*) operon is a continuous segment of the *E. coli* chromosome, containing five genes (blue) that encode the enzymes necessary for the stepwise synthesis of tryptophan. The entire operon is transcribed from one promoter into one long continuous *trp* mRNA (red). Translation of this mRNA begins at five different start sites, yielding five proteins (green). The order of the genes in the bacterial genome parallels the sequential function of the encoded proteins in the tryptophan pathway. (b) The five genes encoding the enzymes required for tryptophan synthesis in yeast (*Saccharomyces cerevisiae*) are carried on four different chromosomes. Each gene is transcribed from its own promoter to yield a primary transcript that is processed into a functional mRNA encoding a single protein. The lengths of the various chromosomes are given in kilobases ($10^3$ bases).

## Eukaryotic Precursor mRNAs Are Processed to Form Functional mRNAs

In bacterial cells, which have no nuclei, translation of an mRNA into protein can begin from the 5′ end of the mRNA even while the 3′ end is still being synthesized by RNA polymerase. In other words, transcription and translation occur concurrently in bacteria. In eukaryotic cells, however, not only is the site of RNA synthesis—the nucleus—separated from the site of translation—the cytoplasm—but also the primary transcripts of protein-coding genes are precursor mRNAs (**pre-mRNAs**) that must undergo several modifications, collectively termed *RNA processing*, to yield a functional mRNA (see Figure 4-1, step **2**). This mRNA then must be exported to the cytoplasm before it can be translated into protein. Thus transcription and translation cannot occur concurrently in eukaryotic cells.

All eukaryotic pre-mRNAs initially are modified at the two ends, and these modifications are retained in mRNAs. As the 5′ end of a nascent RNA chain emerges from the surface of RNA polymerase, it is immediately acted on by several enzymes that together synthesize the 5′ *cap*, a 7-methylguanylate

that is connected to the terminal nucleotide of the RNA by an unusual 5′,5′ triphosphate linkage (Figure 4-14). The cap protects an mRNA from enzymatic degradation and assists in its export to the cytoplasm. The cap also is bound by a protein factor required to begin translation in the cytoplasm.

Processing at the 3′ end of a pre-mRNA involves cleavage by an endonuclease to yield a free 3′-hydroxyl group to which a string of adenylic acid residues is added one at a time by an enzyme called *poly(A) polymerase*. The resulting *poly(A) tail* contains 100–250 bases, being shorter in yeasts and invertebrates than in vertebrates. Poly(A) polymerase is part of a complex of proteins that can locate and cleave a transcript at a specific site and then add the correct number of A residues, in a process that does not require a template.

Another step in the processing of many different eukaryotic mRNA molecules is **RNA splicing**: the internal cleavage of a transcript to excise the introns and stitch together the coding exons. Figure 4-15 summarizes the basic steps in eukaryotic mRNA processing, using the β-globin gene as an example. We examine the cellular machinery for carrying out processing of mRNA, as well as tRNA and rRNA, in Chapter 8.

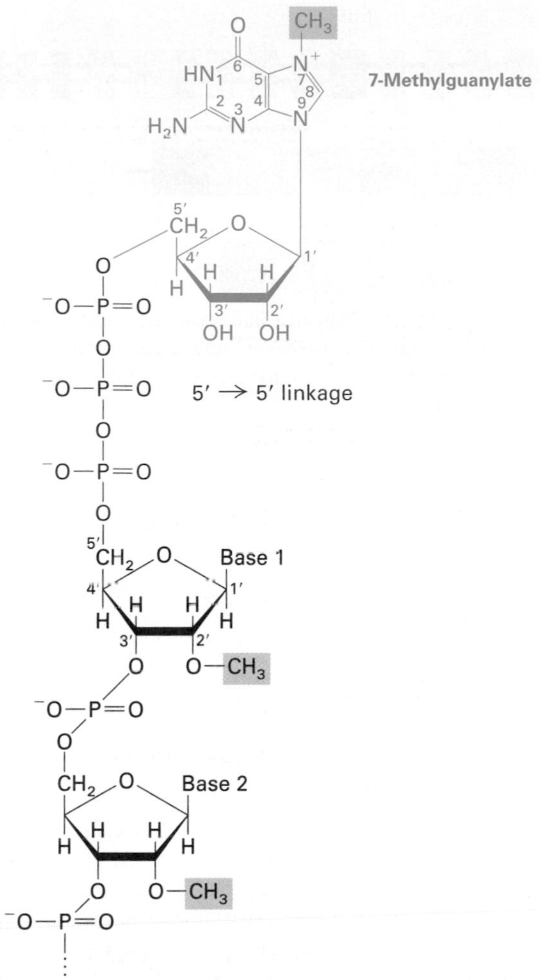

7-Methylguanylate

$5' \rightarrow 5'$ linkage

Base 1

$O$—$CH_3$

Base 2

$O$—$CH_3$

**FIGURE 4-14 Structure of the 5′ methylated cap.** The distinguishing chemical features of the 5′ methylated cap on eukaryotic mRNA are (1) the 5′→5′ linkage of 7-methylguanylate to the initial nucleotide of the mRNA molecule and (2) the methyl group on the 2′ hydroxyl of the ribose of the first nucleotide (base 1). Both these features occur in all animal cells and in cells of higher plants; yeasts lack the methyl group on nucleotide 1. The ribose of the second nucleotide (base 2) also is methylated in vertebrates. [See A. J. Shatkin, 1976, *Cell* **9**:645.]

The functional eukaryotic mRNAs produced by RNA processing retain noncoding regions, referred to as *5′ and 3′ untranslated regions (UTRs)*, at each end. In mammalian mRNAs, the 5′ UTR may be a hundred or more nucleotides long, and the 3′ UTR may be several kilobases in length. Bacterial mRNAs also usually have 5′ and 3′ UTRs, but these are much shorter than those in eukaryotic mRNAs, generally containing fewer than 10 nucleotides.

## Alternative RNA Splicing Increases the Number of Proteins Expressed from a Single Eukaryotic Gene

In contrast to bacterial and archaeal genes, the vast majority of genes in higher, multicellular eukaryotes contain multiple introns. As noted in Chapter 3, many proteins from higher eukaryotes have a multidomain tertiary structure (see Figure 3-11). Individual repeated protein domains often are

🔁 **OVERVIEW ANIMATION:** Life Cycle of an mRNA

**FIGURE 4-15 Overview of RNA processing.** RNA processing produces functional mRNA in eukaryotes. The β-globin gene contains three protein-coding exons (constituting the coding region, red) and two intervening noncoding introns (blue). The introns interrupt the protein-coding sequence between the codons for amino acids 31 and 32 and 105 and 106. Transcription of eukaryotic protein-coding genes starts before the sequence that encodes the first amino acid and extends beyond the sequence encoding the last amino acid, resulting in noncoding regions (gray) at the ends of the primary transcript. These untranslated regions (UTRs) are retained during processing. The 5′ cap ($m^7$Gppp) is added during formation of the primary RNA transcript, which extends beyond the poly(A) site. After cleavage at the poly(A) site and addition of multiple A residues to the 3′ end, splicing removes the introns and joins the exons. The small numbers refer to positions in the 147–amino acid sequence of β-globin.

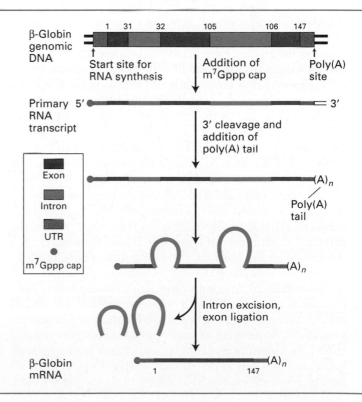

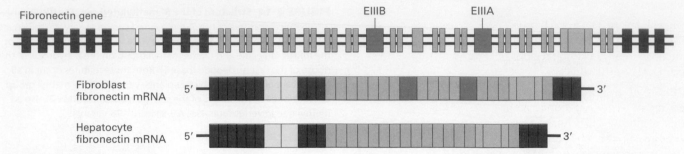

**FIGURE 4-16 Alternative splicing.** The ≈75-kb fibronectin gene *(top)* contains multiple exons; splicing of fibronectin varies by cell type. The EIIIB and EIIIA exons (green) encode binding domains for specific proteins on the surface of fibroblasts. The fibronectin mRNA produced in fibroblasts includes the EIIIA and EIIIB exons, whereas these exons are spliced out of fibronectin mRNA in hepatocytes. In this diagram, introns (black lines) are not drawn to scale; most of them are much longer than any of the exons.

encoded by one exon or a small number of exons that code for identical or nearly identical amino acid sequences. Such repeated exons are thought to have evolved by the accidental multiple duplication of a length of DNA lying between two sites in adjacent introns, resulting in insertion of a string of repeated exons, separated by introns, between the original two introns. The presence of multiple introns in many eukaryotic genes permits expression of multiple, related proteins from a single gene by means of **alternative splicing.** In higher eukaryotes, alternative splicing is an important mechanism for production of different forms of a protein, called **isoforms,** by different types of cells.

Fibronectin, a multidomain protein found in mammals, provides a good example of alternative splicing (Figure 4-16). Fibronectin is a long, adhesive protein secreted into the extracellular space that can bind other proteins together. What and where it binds depends on which domains are spliced together. The fibronectin gene contains numerous exons, grouped into several regions corresponding to specific domains of the protein. Fibroblasts produce fibronectin mRNAs that contain exons EIIIA and EIIIB; these exons encode amino acid sequences that bind tightly to proteins in the fibroblast plasma membrane. Consequently, this fibronectin isoform adheres fibroblasts to the extracellular matrix. Alternative splicing of the fibronectin primary transcript in hepatocytes, the major type of cell in the liver, yields mRNAs that lack the EIIIA and EIIIB exons. As a result, the fibronectin secreted by hepatocytes into the blood does not adhere tightly to fibroblasts or most other cell types, allowing it to circulate. During formation of blood clots, however, the fibrin-binding domains of hepatocyte fibronectin binds to fibrin, one of the principal constituents of clots. The bound fibronectin then interacts with integrins on the membranes of passing platelets, thereby expanding the clot by addition of platelets.

More than 20 different isoforms of fibronectin have been identified, each encoded by a different, alternatively spliced mRNA composed of a unique combination of fibronectin gene exons. Sequencing of large numbers of mRNAs isolated from various tissues and comparison of their sequences with genomic DNA has revealed that nearly 90 percent of all human genes are expressed as alternatively spliced mRNAs.

Clearly, alternative RNA splicing greatly expands the number of proteins encoded by the genomes of higher, multicellular organisms.

## KEY CONCEPTS of Section 4.2

### Transcription of Protein-Coding Genes and Formation of Functional mRNA

• Transcription of DNA is carried out by RNA polymerase, which adds one ribonucleotide at a time to the 3′ end of a growing RNA chain (see Figure 4-11). The sequence of the template DNA strand determines the order in which ribonucleotides are polymerized to form an RNA chain.

• During transcription initiation, RNA polymerase binds to a specific site in DNA (the promoter), locally melts the double-stranded DNA to reveal the unpaired template strand, and polymerizes the first two nucleotides complementary to the template strand. The melted region of 12–14 base pairs is known as the transcription bubble.

• During strand elongation, RNA polymerase moves down the DNA, melting the DNA ahead of the polymerase so that the template strand can enter the active site of the enzyme and allowing the complementary strands of the region just transcribed to reanneal behind it. The transcription bubble moves with the polymerase as the enzyme adds ribonucleotides complementary to the template strand to the 3′ end of the growing RNA chain.

• When RNA polymerase reaches a termination sequence in the DNA, the enzyme stops transcription, leading to release of the completed RNA and dissociation of the enzyme from the template DNA.

• In prokaryotic DNA, several protein-coding genes commonly are clustered into a functional region, an operon, which is transcribed from a single promoter into one mRNA encoding multiple proteins with related functions (see Figure 4-13a). Translation of a bacterial mRNA can begin before synthesis of the mRNA is complete.

- In eukaryotic DNA, each protein-coding gene is transcribed from its own promoter. The initial primary transcript very often contains noncoding regions (introns) interspersed among coding regions (exons).

- Eukaryotic primary transcripts must undergo RNA processing to yield functional RNAs. During processing, the ends of nearly all primary transcripts from protein-coding genes are modified by addition of a 5′ cap and 3′ poly(A) tail. Transcripts from genes containing introns undergo splicing, the removal of the introns and joining of the exons (see Figure 4-15).

- The individual domains of multidomain proteins found in higher eukaryotes are often encoded by individual exons or a small number of exons. Distinct isoforms of such proteins often are expressed in specific cell types as the result of alternative splicing of exons.

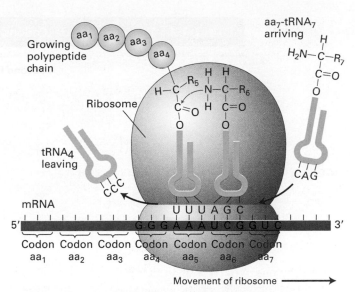

**FIGURE 4-17 The three roles of RNA in protein synthesis.** Messenger RNA (mRNA) is translated into protein by the joint action of transfer RNA (tRNA) and the ribosome, which is composed of numerous proteins and three (bacterial) or four (eukaryotic) ribosomal RNA (rRNA) molecules (not shown). Note the base pairing between tRNA anticodons and complementary codons in the mRNA. Formation of a peptide bond between the amino-group N on the incoming aa-tRNA and the carboxy-terminal C on the growing protein chain (green) is catalyzed by one of the rRNAs. aa = amino acid; R = side group. [Adapted from A. J. F. Griffiths et al., 1999, *Modern Genetic Analysis*, W. H. Freeman and Company.]

## 4.3 The Decoding of mRNA by tRNAs

Although DNA stores the information for protein synthesis and mRNA conveys the instructions encoded in DNA, most biological activities are carried out by proteins. As we saw in Chapter 3, the linear order of amino acids in each protein determines its three-dimensional structure and activity. For this reason, assembly of amino acids in their correct order, as encoded in DNA, is critical to production of functional proteins and hence the proper functioning of cells and organisms.

Translation is the whole process by which the nucleotide sequence of an mRNA is used as a template to join the amino acids in a polypeptide chain in the correct order (see Figure 4-1, step **3**). In eukaryotic cells, protein synthesis occurs in the cytoplasm, where three types of RNA molecules come together to perform different but cooperative functions (Figure 4-17):

1. **Messenger RNA (mRNA)** carries the genetic information transcribed from DNA in a linear form. The mRNA is read in sets of three-nucleotide sequences, called **codons,** each of which specifies a particular amino acid.

2. **Transfer RNA (tRNA)** is the key to deciphering the codons in mRNA. Each type of amino acid has its own subset of tRNAs, which bind the amino acid and carry it to the growing end of a polypeptide chain when the next codon in the mRNA calls for it. The correct tRNA with its attached amino acid is selected at each step because each specific tRNA molecule contains a three-nucleotide sequence, an **anticodon,** that can base-pair with its complementary codon in the mRNA.

3. **Ribosomal RNA (rRNA)** associates with a set of proteins to form **ribosomes.** These complex structures, which physically move along an mRNA molecule, catalyze the assembly of amino acids into polypeptide chains. They also bind

tRNAs and various accessory proteins necessary for protein synthesis. Ribosomes are composed of a large and a small subunit, each of which contains its own rRNA molecule or molecules.

These three types of RNA participate in the synthesis of proteins in all organisms. In this section, we focus on the decoding of mRNA by tRNA adaptors, and how the structure of each of these RNAs relates to its specific task. How they work together with rRNA, ribosomes, and other protein factors to synthesize proteins is detailed in the following section. Because translation is essential for protein synthesis, the two processes commonly are referred to interchangeably. However, the polypeptide chains resulting from translation undergo post-translational folding and often other changes (e.g., chemical modifications, association with other chains) that are required for production of mature, functional proteins (Chapter 3).

### Messenger RNA Carries Information from DNA in a Three-Letter Genetic Code

As noted above, the **genetic code** used by cells is a *triplet* code, with every three-nucleotide sequence, or codon, being "read" from a specified starting point in the mRNA. Of the 64 possible codons in the genetic code, 61 specify individual amino acids and three are stop codons. Table 4-1 shows that most amino acids are encoded by more than one codon.

**TABLE 4-1** | The Genetic Code (Codons to Amino Acids)*

**Second Position**

| First Position (5′ End) | | U | C | A | G | Third Position (3′ End) |
|---|---|---|---|---|---|---|
| U | | Phe | Ser | Tyr | Cys | U |
| | | Phe | Ser | Tyr | Cys | C |
| | | Leu | Ser | Stop | Stop | A |
| | | Leu | Ser | Stop | Trp | G |
| C | | Leu | Pro | His | Arg | U |
| | | Leu | Pro | His | Arg | C |
| | | Leu | Pro | Gln | Arg | A |
| | | Leu (Met)* | Pro | Gln | Arg | G |
| A | | Ile | Thr | Asn | Ser | U |
| | | Ile | Thr | Asn | Ser | C |
| | | Ile | Thr | Lys | Arg | A |
| | | Met (Start) | Thr | Lys | Arg | G |
| G | | Val | Ala | Asp | Gly | U |
| | | Val | Ala | Asp | Gly | C |
| | | Val | Ala | Glu | Gly | A |
| | | Val (Met)* | Ala | Glu | Gly | G |

*AUG is the most common initiator codon; GUG usually codes for valine and CUG for leucine, but, rarely, these codons can also code for methionine to initiate a protein chain.

Only two—methionine and tryptophan—have a single codon; at the other extreme, leucine, serine, and arginine are each specified by six different codons. The different codons for a given amino acid are said to be synonymous. The code itself is termed *degenerate,* meaning that a particular amino acid can be specified by multiple codons.

Synthesis of all polypeptide chains in prokaryotic and eukaryotic cells begins with the amino acid methionine. In bacteria, a specialized form of methionine is used with a formyl group linked to its amino group. In most mRNAs, the *start (initiator) codon* specifying this amino-terminal methionine is AUG. In a few bacterial mRNAs, GUG is used as the initiator codon, and CUG occasionally is used as an initiator codon for methionine in eukaryotes. The three codons UAA, UGA, and UAG do not specify amino acids but constitute *stop (termination) codons* that mark the carboxyl terminus of polypeptide chains in almost all cells. The sequence of codons that runs from a specific start codon to a stop codon is called a **reading frame.** This precise linear array of ribonucleotides in groups of three in mRNA specifies the precise linear sequence of amino acids in a polypeptide chain and also signals where synthesis of the chain starts and stops.

Because the genetic code is a non-overlapping triplet code without divisions between codons, a particular mRNA theoretically could be translated in three different reading frames. Indeed, some mRNAs have been shown to contain overlapping information that can be translated in different reading frames, yielding different polypeptides (Figure 4-18). The vast majority of mRNAs, however, can be read in only one frame because stop codons encountered in the other two possible reading frames terminate translation before a functional protein is produced. Very rarely, another unusual coding arrangement occurs because of *frame-shifting.* In this case the protein-synthesizing machinery may read four nucleotides as one amino acid and then continue reading triplets, or it may back up one base and read all succeeding triplets in the new frame until termination of the chain occurs. Only a few dozen such instances are known.

The meaning of each codon is the same in most known organisms—a strong argument that life on earth evolved

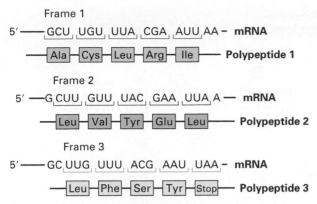

**FIGURE 4-18 Multiple reading frames in an mRNA sequence.** If translation of the mRNA sequence shown begins at three different upstream start sites (not shown), then three overlapping reading frames are possible. In this example, the codons are shifted one base to the right in the middle frame and two bases to the right in the third frame, which ends in a stop codon. As a result, the same nucleotide sequence specifies different amino acids during translation. Although regions of sequence that are translated in two of the three possible reading frames are rare, there are examples in both prokaryotes and eukaryotes, and especially in their viruses, where the same sequence is used in two alternative mRNAs expressed from the same region of DNA, and the sequence is read in one reading frame in one mRNA and in an alternative reading frame in the other mRNA. There are even a few instances where the same short sequence is read in all three possible reading frames.

only once. In fact, the genetic code shown in Table 4-1 is known as the *universal code*. However, the genetic code has been found to differ for a few codons in many mitochondria, in ciliated protozoans, and in *Acetabularia,* a single-celled plant. As shown in Table 4-2, most of these changes involve reading of normal stop codons as amino acids, not an exchange of one amino acid for another. These exceptions to the universal code probably were later evolutionary developments; that is, at no single time was the code immutably fixed, although massive changes were not tolerated once a general code began to function early in evolution.

## The Folded Structure of tRNA Promotes Its Decoding Functions

Translation, or decoding, of the four-nucleotide language of DNA and mRNA into the 20–amino acid language of proteins requires tRNAs and enzymes called *aminoacyl-tRNA synthetases*. To participate in protein synthesis, a tRNA molecule must become chemically linked to a particular amino acid via a high-energy bond, forming an **aminoacyl-tRNA** (Figure 4-19). The anticodon in the tRNA then base-pairs with a codon in mRNA so that the activated amino acid can be added to the growing polypeptide chain (see Figure 4-17).

Some 30–40 different tRNAs have been identified in bacterial cells and as many as 50–100 in animal and plant cells. Thus the number of tRNAs in most cells is more than the number of amino acids used in protein synthesis (20) and also differs from the number of amino acid codons in the genetic code (61). Consequently, many amino acids have more than one tRNA to which they can attach (explaining how there can be more tRNAs than amino acids); in addition, many tRNAs can pair with more than one codon (explaining how there can be more codons than tRNAs).

The function of tRNA molecules, which are 70–80 nucleotides long, depends on their precise three-dimensional structures. In solution, all tRNA molecules fold into a similar stem-loop arrangement that resembles a cloverleaf when drawn in two dimensions (Figure 4-20a). The four stems are short double helices stabilized by Watson-Crick base pairing; three of the four stems have loops containing seven or eight bases at their ends, while the remaining, unlooped stem contains the free 3′ and 5′ ends of the chain. The three nucleotides composing the anticodon are located at the center of the middle loop, in an accessible position that facilitates codon-anticodon base pairing. In all tRNAs, the 3′ end of the unlooped amino acid *acceptor stem* has the sequence CCA, which in most cases is added after synthesis and processing of the tRNA are complete. Several bases in most tRNAs also are modified after transcription, creating nonstandard nucleotides such as inosine, dihydrouridine, and pseudouridine. As we will see shortly, some of these modified bases are known to play an important role in protein

| TABLE 4-2 | Known Deviations from the Universal Genetic Code | | |
|---|---|---|---|
| Codon | Universal Code | Unusual Code* | Occurrence |
| UGA | Stop | Trp | *Mycoplasma, Spiroplasma,* mitochondria of many species |
| CUG | Leu | Thr | Mitochondria in yeasts |
| UAA, UAG | Stop | Gln | *Acetabularia, Tetrahymena, Paramecium.* etc. |
| UGA | Stop | Cys | *Euplotes* |

*Found in nuclear genes of the listed organisms and in mitochondrial genes as indicated.
SOURCE: S. Osawa et al., 1992, *Microbiol. Rev.* 56:229.

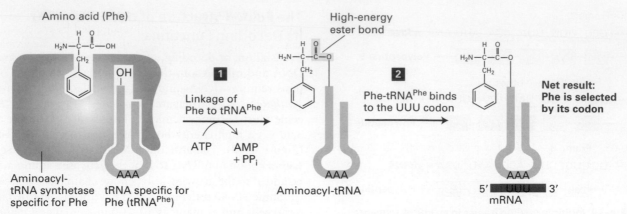

**FIGURE 4-19 Decoding nucleic acid sequence into amino acid sequence.** The process for translating nucleic acid sequences in mRNA into amino acid sequences in proteins involves two steps. Step **1**: An aminoacyl-tRNA synthetase first couples a specific amino acid, via a high-energy ester bond (yellow), to either the 2′ or 3′ hydroxyl of the terminal adenosine in the corresponding tRNA. Step **2**: A three-base sequence in the tRNA (the anticodon) then base-pairs with a codon in the mRNA specifying the attached amino acid. If an error occurs in either step, the wrong amino acid may be incorporated into a polypeptide chain. Phe = phenylalanine.

synthesis. Viewed in three dimensions, the folded tRNA molecule has an L shape with the anticodon loop and acceptor stem forming the ends of the two arms (Figure 4-20b).

## Nonstandard Base Pairing Often Occurs Between Codons and Anticodons

If perfect Watson-Crick base pairing were demanded between codons and anticodons, cells would have to contain at least 61 different types of tRNAs, one for each codon that specifies an amino acid. As noted above, however, many cells contain fewer than 61 tRNAs. The explanation for the smaller number lies in the capability of a single tRNA anticodon to recognize more than one, but not necessarily every, codon corresponding to a given amino acid. This broader recognition can occur because of nonstandard pairing between bases in the so-called *wobble* position: that is, the third (3′) base in an mRNA codon and the corresponding first (5′) base in its tRNA anticodon.

The first and second bases of a codon almost always form standard Watson-Crick base pairs with the third and second bases, respectively, of the corresponding anticodon, but four nonstandard interactions can occur between bases in the wobble position. Particularly important is the G·U base pair,

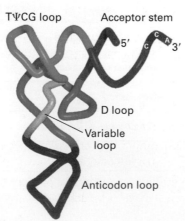

**FIGURE 4-20 Structure of tRNAs.** (a) Although the exact nucleotide sequence varies among tRNAs, they all fold into four base-paired stems and three loops. The CCA sequence at the 3′ end also is found in all tRNAs. Attachment of an amino acid to the 3′ A yields an aminoacyl-tRNA. Some of the A, C, G, and U residues are modified post-transcriptionally in most tRNAs (see key). Dihydrouridine (D) is nearly always present in the D loop; likewise, ribothymidine (T) and pseudouridine (Ψ) are almost always present in the TΨCG loop. Yeast alanine tRNA, represented here, also contains other modified bases. The triplet at the tip of the anticodon loop base-pairs with the corresponding codon in mRNA. (b) Three-dimensional model of the generalized backbone of all tRNAs. Note the L shape of the molecule. [Part (a) see R. W. Holly et al., 1965, *Science* **147**:1462. Part (b) adapted from J. G. Arnez and D. Moras, 1997, *Trends Biochem. Sci.* **22**:211.]

| 321 | C | A | G | U | I | |
|-----|---|---|---|---|---|---|
| 123 | G | U | C | A | C | then the tRNA may recognize codons in mRNA having these bases in third position |
| | | | U | G | A | |
| | | | | | U | |

If these bases are in first, or wobble, position of anticodon

If these bases are in third, or wobble, position of codon of an mRNA

| 123 | C | A | G | U | |
|-----|---|---|---|---|---|
| 321 | G | U | C | A | then the codon may be recognized by a tRNA having these bases in first position of anticodon |
| | I | I | U | G | |
| | | | | I | |

**FIGURE 4-21 Nonstandard base pairing at the wobble position.** The base in the third (or wobble) position of an mRNA codon often forms a nonstandard base pair with the base in the first (or wobble) position of a tRNA anticodon. Wobble pairing allows a tRNA to recognize more than one mRNA codon (*top*); conversely, it allows a codon to be recognized by more than one kind of tRNA (*bottom*), although each tRNA will bear the same amino acid. Note that a tRNA with I (inosine) in the wobble position can "read" (become paired with) three different codons, and a tRNA with G or U in the wobble position can read two codons. Although A is theoretically possible in the wobble position of the anticodon, it is almost never found in nature.

which structurally fits almost as well as the standard G·C pair. Thus, a given anticodon in tRNA with G in the first (wobble) position can base-pair with the two corresponding codons that have either pyrimidine (C or U) in the third position (Figure 4-21). For example, the phenylalanine codons UUU and UUC (5'→3') are both recognized by the tRNA that has GAA (5'→3') as the anticodon. In fact, any two codons of the type NNPyr (N = any base; Pyr = pyrimidine) encode a single amino acid and are decoded by a single tRNA with G in the first (wobble) position of the anticodon.

Although adenine rarely is found in the anticodon wobble position, many tRNAs in plants and animals contain inosine (I), a deaminated product of adenine, at this position. Inosine can form nonstandard base pairs with A, C, and U. A tRNA with inosine in the wobble position thus can recognize the corresponding mRNA codons with A, C, or U in the third (wobble) position (see Figure 4-21). For this reason, inosine-containing tRNAs are heavily employed in translation of the synonymous codons that specify a single amino

acid. For example, four of the six codons for leucine (CUA, CUC, CUU, and UUA) are all recognized by the same tRNA with the anticodon 3'-GAI-5'; the inosine in the wobble position forms nonstandard base pairs with the third base in the four codons. In the case of the UUA codon, a nonstandard G·U pair also forms between position 3 of the anticodon and position 1 of the codon.

## Amino Acids Become Activated When Covalently Linked to tRNAs

Recognition of the codon or codons specifying a given amino acid by a particular tRNA is actually the second step in decoding the genetic message. The first step, attachment of the appropriate amino acid to a tRNA, is catalyzed by a specific *aminoacyl-tRNA synthetase*. Each of the 20 different synthetases recognizes *one* amino acid and *all* its compatible, or *cognate*, tRNAs. These coupling enzymes link an amino acid to the free 2' or 3' hydroxyl of the adenosine at the 3' terminus of tRNA molecules by an ATP-requiring reaction. In this reaction, the amino acid is linked to the tRNA by a high-energy bond and thus is said to be *activated*. The energy of this bond subsequently drives formation of the peptide bonds linking adjacent amino acids in a growing polypeptide chain. The equilibrium of the aminoacylation reaction is driven further toward activation of the amino acid by hydrolysis of the high-energy phosphoanhydride bond in the released pyrophosphate (see Figure 4-19).

Aminoacyl-tRNA synthetases recognize their cognate tRNAs by interacting primarily with the anticodon loop and acceptor stem, although interactions with other regions of a tRNA also contribute to recognition in some cases. Also, specific bases in incorrect tRNAs that are structurally similar to a cognate tRNA will inhibit charging of the incorrect tRNA. Thus, recognition of the correct tRNA depends on both positive interactions and the absence of negative interactions. Still, because some amino acids are so similar structurally, aminoacyl-tRNA synthetases sometimes make mistakes. These are corrected, however, by the enzymes themselves, which have a *proofreading* activity that checks the fit in their amino acid–binding pocket. If the wrong amino acid becomes attached to a tRNA, the bound synthetase catalyzes removal of the amino acid from the tRNA. This crucial function helps guarantee that a tRNA delivers the correct amino acid to the protein-synthesizing machinery. The overall error rate for translation in *E. coli* is very low, approximately 1 per 50,000 codons, evidence of both the fidelity of tRNA recognition and the importance of proofreading by aminoacyl-tRNA synthetases.

## KEY CONCEPTS of Section 4.3

### The Decoding of mRNA by tRNAs

- Genetic information is transcribed from DNA into mRNA in the form of an overlapping, degenerate triplet code.

- Each amino acid is encoded by one or more three-nucleotide sequences (codons) in mRNA. Each codon specifies one amino acid, but most amino acids are encoded by multiple codons (see Table 4-1).

• The AUG codon for methionine is the most common start codon, specifying the amino acid at the $NH_2$-terminus of a protein chain. Three codons (UAA, UAG, UGA) function as stop codons and specify no amino acids.

• A reading frame, the uninterrupted sequence of codons in mRNA from a specific start codon to a stop codon, is translated into the linear sequence of amino acids in a polypeptide chain.

• Decoding of the nucleotide sequence in mRNA into the amino acid sequence of proteins depends on tRNAs and aminoacyl-tRNA synthetases.

• All tRNAs have a similar three-dimensional structure that includes an acceptor arm for attachment of a specific amino acid and a stem-loop with a three-base anticodon sequence at its ends (see Figure 4-20). The anticodon can base-pair with its corresponding codon in mRNA.

• Because of nonstandard interactions, a tRNA may base-pair with more than one mRNA codon; conversely, a particular codon may base-pair with multiple tRNAs. In each case, however, only the proper amino acid is inserted into a growing polypeptide chain.

• Each of the 20 aminoacyl-tRNA synthetases recognizes a single amino acid and covalently links it to a cognate tRNA, forming an aminoacyl-tRNA (see Figure 4-19). This reaction activates the amino acid, so it can participate in peptide bond formation.

## 4.4 Stepwise Synthesis of Proteins on Ribosomes

The previous sections have introduced two of the major participants in protein synthesis—mRNA and aminoacylated tRNA. Here we first describe the third key player in protein synthesis—the rRNA-containing ribosome—before taking a detailed look at how all three components are brought together to carry out the biochemical events leading to formation of polypeptide chains on ribosomes. Similar to transcription, the complex process of translation can be divided into three stages—initiation, elongation, and termination—which we consider in order. We focus our description on translation in eukaryotic cells, but the mechanism of translation is fundamentally the same in all cells.

### Ribosomes are Protein-Synthesizing Machines

If the many components that participate in translating mRNA had to interact in free solution, the likelihood of simultaneous collisions occurring would be so low that the rate of amino acid polymerization would be very slow. The efficiency of translation is greatly increased by the binding of the mRNA and the individual aminoacyl-tRNAs within a ribosome. The ribosome, the most abundant RNA-protein complex in the cell, directs elongation of a polypeptide at a rate of three to five amino acids added per second. Small proteins of 100–200 amino acids are therefore made in a minute or less. On the other hand, it takes 2–3 hours to make the largest known protein, titin, which is found in muscle and contains about 30,000 amino acid residues. The cellular machine that accomplishes this task must be precise and persistent.

With the aid of the electron microscope, ribosomes were first discovered as small, discrete, RNA-rich particles in cells that secrete large amounts of protein. However, their role in protein synthesis was not recognized until reasonably pure ribosome preparations were obtained. In vitro radiolabeling experiments with such preparations showed that radioactive amino acids were first incorporated into growing polypeptide chains that were associated with ribosomes before appearing in finished chains.

Though there are differences between the ribosomes of prokaryotes and eukaryotes, the great structural and functional similarities between ribosomes from all species reflects the common evolutionary origin of the most basic constituents of living cells. A ribosome is composed of three (in bacteria) or four (in eukaryotes) different rRNA molecules and as many as 83 proteins, organized into a large subunit and a small subunit (Figure 4-22). The ribosomal subunits and the rRNA molecules are commonly designated in Svedberg units (S), a measure of the sedimentation rate of macromolecules centrifuged under standard conditions—essentially, a measure of size. The small ribosomal subunit contains a single rRNA molecule, referred to as *small rRNA*. The large subunit contains a molecule of *large rRNA* and one molecule of 5S rRNA, plus an additional molecule of 5.8S rRNA in vertebrates. The lengths of the rRNA molecules, the quantity of proteins in each subunit, and consequently the sizes of the subunits differ between bacterial and eukaryotic cells. The assembled ribosome is 70S in bacteria and 80S in vertebrates.

The sequences of the small and large rRNAs from several thousand organisms are now known. Although the primary nucleotide sequences of these rRNAs vary considerably, the same parts of each type of rRNA theoretically can form base-paired stem-loops, which would generate a similar three-dimensional structure for each rRNA in all organisms. The three-dimensional structures of bacterial rRNAs have been determined by x-ray crystallography of isolated 50S and 30S subunits and entire 70S ribosomes (Figure 4-23). The multiple, much smaller ribosomal proteins for the most part are associated with the surface of the rRNAs. Although the number of protein molecules in ribosomes greatly exceeds the number of RNA molecules, RNA constitutes about 60 percent of the mass of a ribosome. The sites where tRNAs are bound by ribosomes are known as the *A site,* the *P site,* and the *E site.* As we will see shortly, tRNAs move between these sites as protein synthesis takes place. Crystal structures of the yeast 80S ribosome, a protozoan 40S subunit, as well

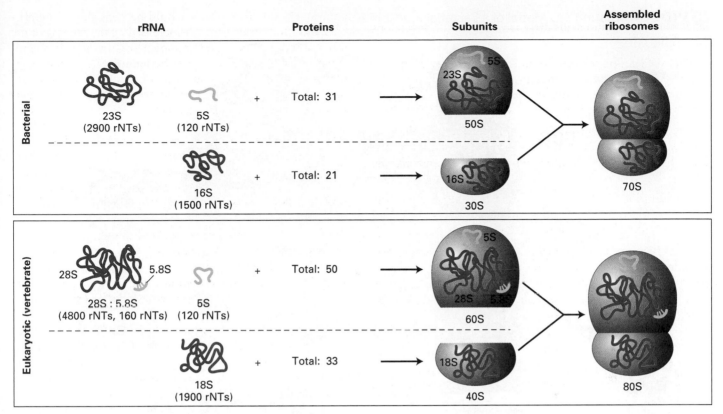

**FIGURE 4-22 Prokaryotic and eukaryotic ribosome components.** In all cells, each ribosome consists of a large and a small subunit. The two subunits contain rRNAs (red) of different lengths, as well as a different set of proteins. All ribosomes contain two major rRNA molecules (23S and 16S rRNA in bacteria; 28S and 18S rRNA in vertebrates) and a 5S rRNA. The large subunit of vertebrate ribosomes also contains a 5.8S rRNA base-paired to the 28S rRNA. The number of ribonucleotides (rNTs) in each rRNA type is indicated.

as cryo-electron microscopy of plant ribosomes have been reported recently. They are generally similar to bacterial ribosomes but are about 50 percent larger because of eukaryotic-specific insertions of RNA segments into regions of the bacterial rRNAs as well as the presence of a larger number of proteins (see Figure 4-22). Basic aspects of protein synthesis are thought to be similar, although initiation of translation in eukaryotes, discussed later, is more complex and subject to additional mechanisms of regulation.

During translation, a ribosome moves along an mRNA chain, interacting with various protein factors and tRNAs and undergoing large conformational changes. Despite the complexity of the ribosome, great progress has been made in determining the overall structure of bacterial ribosomes and how they function in protein synthesis. More than 50 years after the initial discovery of ribosomes, their overall structure and function during protein synthesis are finally becoming clear.

## Methionyl-tRNA$_i^{Met}$ Recognizes the AUG Start Codon

As noted earlier, the AUG codon for methionine functions as the start codon in the vast majority of mRNAs. A critical aspect of translation initiation is to begin protein synthesis at the start codon, thereby establishing the correct reading frame for the entire mRNA. Both prokaryotes and eukaryotes contain two different methionine tRNAs: tRNA$_i^{Met}$ can initiate protein synthesis, and tRNA$^{Met}$ can incorporate methionine only into a growing protein chain. The same aminoacyl-tRNA synthetase (MetRS) charges both tRNAs with methionine. But *only* Met-tRNA$_i^{Met}$ (i.e., activated methionine attached to tRNA$_i^{Met}$) can bind at the appropriate site on the small ribosomal subunit, the *P site,* to begin synthesis of a polypeptide chain. The regular Met-tRNA$^{Met}$ and all other charged tRNAs bind only to the *A site,* as described later. tRNAs enter the exit or *E* site after transferring their covalently bound amino acid to the growing polypeptide chain.

## Eukaryotic Translation Initiation Usually Occurs at the First AUG Closest to the 5′ End of an mRNA

During the first stage of translation, the small and large ribosomal subunits assemble around an mRNA that has an activated initiator tRNA correctly positioned at the start codon in the ribosomal P site. In eukaryotes, this process is mediated by a special set of proteins known as **eukaryotic** translation **initiation factors (eIFs)**. As each individual component joins the complex, it is guided by interactions with specific eIFs. Several of these initiation factors bind GTP, and the hydrolysis

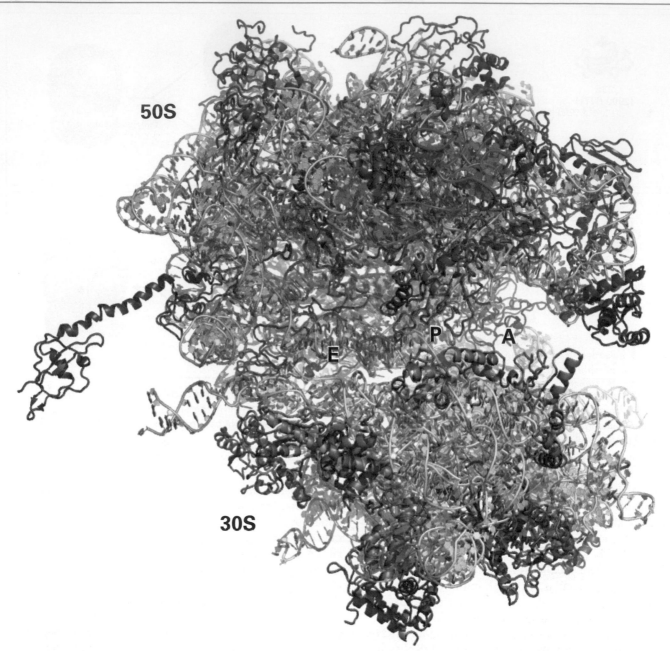

**FIGURE 4-23 Structure of *E. coli* 70S ribosome as determined by x-ray crystallography.** Model of the ribosome viewed along the interface between the large (50S) and small (30S) subunits. The 16S rRNA and proteins in the small subunit are colored light green and dark green, respectively; the 23S rRNA and proteins in the large subunit are colored light purple and dark purple, respectively; and the 5S rRNA is colored dark blue. The positions of the ribosomal A, P, and E sites are indicated. Note that the ribosomal proteins are located primarily on the surface of the ribosome, and the rRNAs on the inside, lining the A, P, and E sites. [From B. S. Schuwirth et al., 2005, *Science* **310:**827.]

of GTP to GDP functions as a proofreading switch that only allows subsequent steps to proceed if the preceding step has occurred correctly. Before GTP hydrolysis, the complex is unstable, allowing a second attempt at complex formation until the correct complex assembles, resulting in GTP hydrolysis and stabilization of the appropriate complex. Considerable

progress has been made in the past few years in understanding translation initiation in vertebrates.

The current model for initiation of translation in vertebrates is depicted in Figure 4-24. Large and small ribosomal subunits released from a previous round of translation are kept apart by the binding of eIFs 1, 1A, and 3 to the small

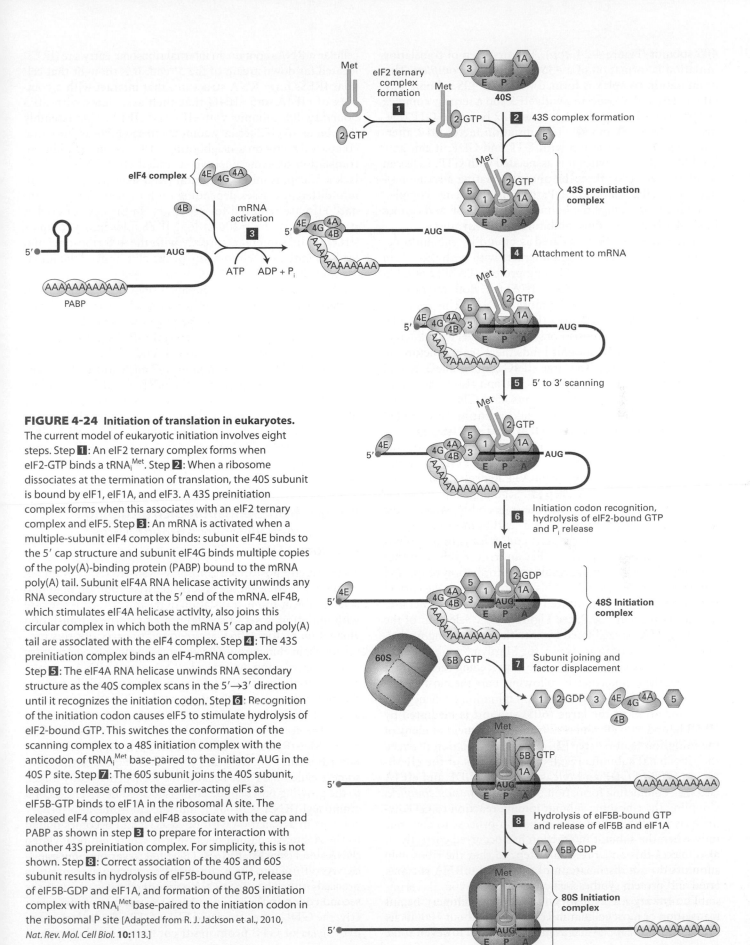

**FIGURE 4-24 Initiation of translation in eukaryotes.**
The current model of eukaryotic initiation involves eight steps. Step **1**: An eIF2 ternary complex forms when eIF2-GTP binds a tRNA$_i^{Met}$. Step **2**: When a ribosome dissociates at the termination of translation, the 40S subunit is bound by eIF1, eIF1A, and eIF3. A 43S preinitiation complex forms when this associates with an eIF2 ternary complex and eIF5. Step **3**: An mRNA is activated when a multiple-subunit eIF4 complex binds: subunit eIF4E binds to the 5' cap structure and subunit eIF4G binds multiple copies of the poly(A)-binding protein (PABP) bound to the mRNA poly(A) tail. Subunit eIF4A RNA helicase activity unwinds any RNA secondary structure at the 5' end of the mRNA. eIF4B, which stimulates eIF4A helicase activity, also joins this circular complex in which both the mRNA 5' cap and poly(A) tail are associated with the eIF4 complex. Step **4**: The 43S preinitiation complex binds an eIF4-mRNA complex. Step **5**: The eIF4A RNA helicase unwinds RNA secondary structure as the 40S complex scans in the 5'→3' direction until it recognizes the initiation codon. Step **6**: Recognition of the initiation codon causes eIF5 to stimulate hydrolysis of eIF2-bound GTP. This switches the conformation of the scanning complex to a 48S initiation complex with the anticodon of tRNA$_i^{Met}$ base-paired to the initiator AUG in the 40S P site. Step **7**: The 60S subunit joins the 40S subunit, leading to release of most the earlier-acting eIFs as eIF5B-GTP binds to eIF1A in the ribosomal A site. The released eIF4 complex and eIF4B associate with the cap and PABP as shown in step **3** to prepare for interaction with another 43S preinitiation complex. For simplicity, this is not shown. Step **8**: Correct association of the 40S and 60S subunit results in hydrolysis of eIF5B-bound GTP, release of eIF5B-GDP and eIF1A, and formation of the 80S initiation complex with tRNA$_i^{Met}$ base-paired to the initiation codon in the ribosomal P site [Adapted from R. J. Jackson et al., 2010, *Nat. Rev. Mol. Cell Biol.* **10**:113.]

40S subunit (Figure 4-24, *top*). The first step of translation initiation is formation of a *43S preinitiation complex*. This preinitiation complex is formed when the 40S subunit with eIFs 1, 1A, and 3 associates with eIF5 and a ternary complex consisting of the Met-tRNA$_i^{Met}$ and eIF2 bound to GTP (Figure 4-24, steps **1** and **2**). The initiation factor eIF2 alternates between association with GTP and GDP; it can only bind Met-tRNA$_i^{Met}$ when it is associated with GTP. Cells can regulate protein synthesis by phosphorylating a serine residue on the eIF2 bound to GDP; the phosphorylated complex is unable to exchange the bound GDP for GTP and cannot bind Met-tRNA$_i^{Met}$, thus inhibiting protein synthesis.

The mRNA to be translated is bound by the multiple-subunit eIF4 complex, which interacts with both the 5'-cap structure and the poly(A)-binding protein (PABP) bound in multiple copies to the mRNA poly(A) tail. Both interactions are required for translation of most mRNAs. This results in formation of a circular complex (Figure 4-24, step **3**). The eIF4 cap-binding complex consists of several subunits with different functions. The eIF4E subunit binds the 5'-cap structure on mRNAs (Figure 4-14). The large eIF4G subunit interacts with PABP bound to the mRNA poly(A) tail, and also forms a scaffold to which the other eIF4 subunits bind. The mRNA-eIF4 complex then associates with the preinitiation complex through an interaction between eIF4G and eIF3 (step **4**).

The initiation complex then slides along, or *scans,* the associated mRNA as the **helicase** activity of eIF4A, stimulated by eIF4B, uses energy from ATP hydrolysis to unwind RNA secondary structure (step **5**). Scanning stops when the tRNA$_i^{Met}$ anticodon recognizes the start codon, which is the first AUG downstream from the 5' end in most eukaryotic mRNAs. Recognition of the start codon leads to hydrolysis of the GTP associated with eIF2, an irreversible step that prevents further scanning, resulting in formation of the *48S initiation complex* (step **6**). This commitment to the correct initiation codon is facilitated by eIF5, an eIF2 GTPase-activating protein (GAP, see Figure 3-32). Selection of the initiating AUG is facilitated by specific surrounding nucleotides called the *Kozak sequence*, for Marilyn Kozak, who defined it: (5') ACCAUGG (3'). The A preceding the AUG (in bold) and the G immediately following it are the most important nucleotides affecting translation initiation efficiency.

Association of the large (60S) subunit is mediated by eIF5B bound to GTP, and results in displacement of many of the initiation factors (step **7**). Correct association between the ribosomal subunits results in hydrolysis of the eIF5B-bound GTP to GDP and release of eIF5B-GDP and eIF1A (step **8**), completing formation of an *80S initiation complex*. Coupling the ribosome subunit joining reaction to GTP hydrolysis by eIF5B allows the initiation process to continue only when the subunit interaction has occurred correctly. It also makes this an irreversible step, so that the ribosomal subunits do not dissociate until the entire mRNA is translated and protein synthesis is terminated.

The eukaryotic protein-synthesizing machinery begins translation of most cellular mRNAs within about 100 nucleotides of the 5'-capped end as just described. However, some cellular mRNAs contain an internal ribosome entry site (IRES) located far downstream of the 5' end. It is thought that cellular IRESs form RNA structures that interact with a complex of eIF4A and eIF4G that then associates with eIF3 bound to 40S subunits with eIF1 and eIF1A. This assembly then binds an eIF2 ternary complex to assemble an initiation complex directly on a neighboring AUG codon. In addition, translation of some positive-stranded viral RNAs, which lack a 5' cap, is initiated at viral IRES sequences. These fall into different classes depending on how many of the standard eIFs are required for initiation. In the case of cricket paralysis virus, the ≈200-nt-long IRES folds into a complex structure that interacts directly with the 40S ribosome and leads to initiation without any of the eIFs or even the initiator Met-tRNA$_i^{Met}$!

In bacteria, binding of the small ribosomal subunit to an initiation site occurs by a different mechanism that allows initiation at internal sites in the polycistronic mRNAs transcribed from operons. In bacterial mRNAs, an ≈6-base sequence complementary to the 3' end of the small rRNA precedes the AUG start codon by 4–7 nucleotides. Base pairing between this sequence in the mRNA, called the Shine-Dalgarno sequence after its discoverers, and the small rRNA places the small ribosomal subunit in the proper position for initiation. Initiation factors comparable to eIF1A, eIF2, eIF3, and f-Met-tRNA$_i^{Met}$ then associate with the small subunit, followed by association of the large subunit to form the complete bacterial ribosome.

## During Chain Elongation Each Incoming Aminoacyl-tRNA Moves Through Three Ribosomal Sites

The correctly positioned ribosome–Met-tRNA$_i^{Met}$ complex is now ready to begin the task of stepwise addition of amino acids by the in-frame translation of the mRNA. As is the case with initiation, a set of special proteins, termed translation **elongation factors (EFs)**, are required to carry out this process of chain elongation. The key steps in elongation are entry of each succeeding aminoacyl-tRNA with a tRNA complementary to the next codon, formation of a peptide bond, and the movement, or *translocation,* of the ribosome one codon at a time along the mRNA.

At the completion of translation initiation, as noted already, Met-tRNA$_i^{Met}$ is bound to the P site on the assembled 80S ribosome (Figure 4-25, *top*). This region of the ribosome is called the *P* site because the tRNA chemically linked to the growing poly*peptide* chain is located here. The second aminoacyl-tRNA is brought into the ribosome as a ternary complex in association with EF1α·GTP and becomes bound to the *A* site, so named because it is where *a*minoacylated tRNAs bind (step **1**). EF1α·GTP bound to various aminoacyl-tRNAs diffuse into the A site, but the next step in translation proceeds only when the tRNA anticodon base-pairs with the second codon in the coding region. When that occurs properly, the GTP in the associated EF1α·GTP is hydrolyzed. The hydrolysis of GTP promotes a conformational change in

**FIGURE 4-25 Peptidyl chain elongation in eukaryotes.** Once the 80S ribosome with Met-tRNA$_i^{Met}$ in the ribosome P site is assembled (*top*), a ternary complex bearing the second amino acid (aa$_2$) coded by the mRNA binds to the A site (step **1**). Following a conformational change in the ribosome induced by hydrolysis of GTP in EF1α·GTP (step **2**), the large rRNA catalyzes peptide bond formation between Met$_i$ and aa$_2$ (step **3**). Hydrolysis of GTP in EF2·GTP causes another conformational change in the ribosome that results in its translocation one codon along the mRNA and shifts the unacylated tRNA$_i^{Met}$ to the E site and the tRNA with the bound peptide to the P site (step **4**). The cycle can begin again with binding of a ternary complex bearing aa$_3$ to the now open A site. In the second and subsequent elongation cycles, the tRNA at the E site is ejected during step **2** as a result of the conformational change induced by hydrolysis of GTP in EF1α·GTP. [Adapted from K. H. Nierhaus et al., 2000, in R. A. Garrett et al., eds., *The Ribosome: Structure, Function, Antibiotics, and Cellular Interactions,* ASM Press, p. 319.]

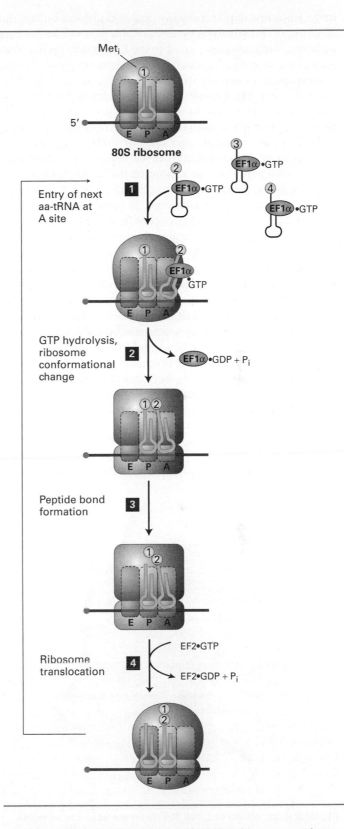

EF1α that leads to release of the resulting EF1α·GDP complex and tight binding of the aminoacyl-tRNA in the A site (step **2**). This conformational change also positions the aminoacylated 3′ end of the tRNA in the A site in close proximity to the 3′ end of the Met-tRNA$_i^{Met}$ in the P site. GTP hydrolysis, and hence tight binding, does not occur if the anticodon of the incoming aminoacyl-tRNA cannot base-pair with the codon at the A site. In this case, the ternary complex diffuses away, leaving an empty A site that can associate with other aminoacyl-tRNA–EF1α·GTP complexes until a correctly base-paired tRNA is bound. Thus, GTP hydrolysis by EF1α is another proofreading step that allows protein synthesis to proceed only when the correct aminoacylated tRNA is bound to the A site. This phenomenon contributes to the fidelity of protein synthesis.

With the initiating Met-tRNA$_i^{Met}$ at the P site and the second aminoacyl-tRNA tightly bound at the A site, the α-amino group of the second amino acid reacts with the "activated" (ester-linked) methionine on the initiator tRNA, forming a peptide bond (Figure 4-25, step **3**; see Figure 4-17). This *peptidyltransferase reaction* is catalyzed by the large rRNA, which precisely orients the interacting atoms, permitting the reaction to proceed. The 2′-hydroxyl of the terminal A of the peptidyl-tRNA in the P site also participates in catalysis. The catalytic ability of the large rRNA in bacteria has been demonstrated by carefully removing the vast majority of the protein from large ribosomal subunits. The nearly pure bacterial 23S rRNA can catalyze a peptidyltransferase reaction between analogs of aminoacylated-tRNA and peptidyl-tRNA. Further support for the catalytic role of large rRNA in protein synthesis came from high-resolution crystallographic studies showing that no proteins lie near the site of peptide bond synthesis in the crystal structure of the bacterial large subunit.

Following peptide bond synthesis, the ribosome translocates along the mRNA a distance equal to one codon. This translocation step is monitored by hydrolysis of the GTP in eukaryotic EF2·GTP. Once translocation has occurred correctly, the bound GTP is hydrolyzed, another irreversible process that prevents the ribosome from moving along the RNA in the wrong direction or from translocating an incorrect number of nucleotides. As a result of conformational changes in the ribosome that accompany proper translocation and the resulting GTP hydrolysis by EF2, tRNA$_i^{Met}$, now without its activated methionine, is moved to the *E* (*exit*) *site* on the ribosome; concurrently, the second tRNA, now covalently bound to a dipeptide (a peptidyl-tRNA), is moved to the P site (Figure 4-25, step **4**). Translocation thus returns the ribosome conformation to a state in which the A site is open and able to accept another aminoacylated tRNA complexed with EF1α·GTP, beginning another cycle of chain elongation.

Repetition of the elongation cycle depicted in Figure 4-25 adds amino acids one at a time to the C-terminus of the growing polypeptide as directed by the mRNA sequence, until a stop codon is encountered. In subsequent cycles, the conformational change that occurs in step **2** ejects the un-acylated tRNA from the E site. As the nascent polypeptide chain becomes longer, it threads through a channel in the large ribosomal subunit, exiting at a position opposite the side that interacts with the small subunit (Figure 4-26).

In the absence of the ribosome, the three-base-pair RNA-RNA hybrid between the tRNA anticodons and the mRNA codons in the A and P sites would not be stable; RNA-RNA duplexes between separate RNA molecules must be considerably longer to be stable under physiological conditions. However, multiple interactions between the large and small rRNAs and general domains of tRNAs (e.g., the D and TΨCG loops, see Figure 4-20) stabilize the tRNAs in the A and P sites, while other RNA-RNA interactions sense correct codon-anticodon base pairing, assuring that the genetic code is read properly. Then, interactions between rRNAs and the general domains of all tRNAs result in the movement of the tRNAs between the A, P, and E sites as the ribosome translocates along the mRNA one three-nucleotide codon at a time.

## Translation Is Terminated by Release Factors When a Stop Codon Is Reached

The final stages of translation, like initiation and elongation, require highly specific molecular signals that decide the fate of the mRNA–ribosome–tRNA-peptidyl complex. Two types of specific protein **release factors** (RFs) have been discovered. Eukaryotic eRF1, whose shape is similar to that of tRNAs, acts by binding to the ribosomal A site and recognizing stop codons directly. Like some of the initiation and elongation factors discussed previously, the second eukaryotic release factor, eRF3, is a GTP-binding protein. The eRF3·GTP acts in concert with eRF1 to promote cleavage of the peptidyl-tRNA, thus releasing the completed protein chain (Figure 4-27). Bacteria have two release factors (RF1 and RF2) that are functionally analogous to eRF1 and a GTP-binding factor (RF3) that is analogous to eRF3. Once again, the eRF3 GTPase monitors the correct recognition of a stop codon by eRF1. The peptidyl-tRNA bond of the tRNA in the P site is not cleaved, terminating translation, until one of the three stop codons is correctly recognized by eRF1, another example of a proofreading step in protein synthesis.

Release of the completed protein leaves a free tRNA in the P site and the mRNA still associated with the 80S ribosome with eRF1 and eRF3-GDP bound in the A site. In eukaryotes, ribosome recycling occurs when this post-termination complex is bound by a protein called ABCE1, which uses energy from ATP hydrolysis to separate the subunits and release the mRNA. Initiation factors eIF1, eIF1A, and eIF3 are also required and load onto the 40S subunit, making it ready for another round of initiation (Figure 4-24, *top*). In reality, a free mRNA is never released as diagrammed in Figure 4-27 for simplicity. Rather, the mRNA has other ribosomes associated with it in various stages of elongation, PABP bound to the polyA tail, and eIF4 complex associated with the 5'-cap, ready to associate with another 43S preinitiation complex (Figure 4-24).

## Polysomes and Rapid Ribosome Recycling Increase the Efficiency of Translation

Translation of a single eukaryotic mRNA molecule to yield a typical-sized protein takes one to two minutes. Two phenomena significantly increase the overall rate at which cells can synthesize a protein: the simultaneous translation of a single mRNA molecule by multiple ribosomes, and rapid recycling of ribosomal subunits after they disengage from the 3' end of an mRNA. Simultaneous translation of an mRNA by multiple ribosomes is readily observable in electron micrographs and by sedimentation analysis, revealing mRNA attached to multiple ribosomes bearing nascent growing polypeptide chains. These structures, referred to as **polyribosomes** or *polysomes*, were seen to be circular in electron micrographs of some tissues. Subsequent studies with purified initiation factors explained

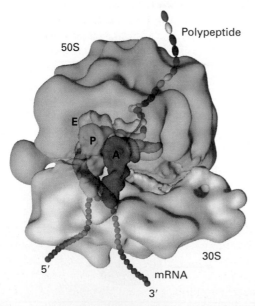

**FIGURE 4-26 Model of *E. coli* 70S ribosome bound to an mRNA with a nascent polypeptide chain in the exit tunnel.** The model is based on cryo-EM studies. Three tRNAs are superimposed on the A (pink), P (green), and E (yellow) sites. The nascent polypeptide chain is buried in a tunnel in the large ribosomal subunit that begins close to the acceptor stem of the tRNA in the P site. [See I. S. Gabashvili et al., 2000, *Cell* **100**:537; courtesy of J. Frank.]

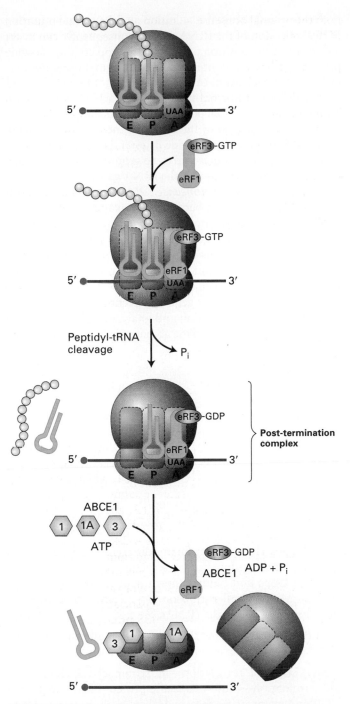

**FIGURE 4-27 Termination of translation in eukaryotes.** When a ribosome bearing a nascent protein chain reaches a stop codon (UAA, UGA, UAG), release factor eRF1 enters the A site together with eRF3·GTP. Hydrolysis of the bound GTP is accompanied by cleavage of the peptide chain from the tRNA in the P site and ejection of the tRNA in the E site, forming a post-termination complex. The ribosomal subunits are separated by the action of the ABCE1 ATPase together with eIF1, eIF1A, and eIF3. The 40S subunit is released bound to these eIFs, ready to initiate another cycle of translation (see Figure 4-24).

tail and the eIF4G subunit of eIF4. Since the eIF4E subunit of eIF4 binds to the cap structure on the 5′ end of an mRNA, the two ends of an mRNA molecule are bridged by the intervening proteins, forming a "circular" mRNA (Figure 4-28a). Because the two ends of a polysome are relatively close together, ribosomal subunits that disengage from the 3′ end are positioned near the 5′ end, facilitating re-initiation by the interaction of the 40S subunit and its associated initiation factors with eIF4 bound to the 5′ cap. The circular pathway depicted in Figure 4-28b is thought to enhance ribosome recycling and thus increase the efficiency of protein synthesis.

## GTPase-Superfamily Proteins Function in Several Quality Control Steps of Translation

We can now see that one or more GTP-binding proteins participate in each stage of translation. These proteins belong to the **GTPase superfamily** of switch proteins that cycle between a GTP-bound active form and GDP-bound inactive form (see Figure 3-32). Hydrolysis of the bound GTP causes a conformational change in the GTPase itself and other associated proteins that are critical to various complex molecular processes. In translation initiation, for instance, hydrolysis of eIF2·GTP to eIF2·GDP prevents further scanning of the mRNA once the start site is encountered and allows binding of the large ribosomal subunit to the small subunit (see Figure 4-24, step **6**). Similarly, hydrolysis of EF1α·GTP to EF1α·GDP during chain elongation occurs only when the A site is occupied by a charged tRNA with an anticodon that base-pairs with the codon in the A site. GTP hydrolysis causes a conformational change in EF1α resulting in release of its bound tRNA, allowing the aminoacylated 3′ end of the charged tRNA to move into the position required for peptide bond formation (Figure 4-26, step **2**). Hydrolysis of EF2·GTP to EF2·GDP leads to correct translocation of the ribosome along the mRNA, (see Figure 4-26, step **4**), and hydrolysis of eRF3·GTP to eRF3·GDP assures correct termination of translation (Figure 4-27). Since hydrolysis of the high energy β-γ phosphodiester bond of GTP is irreversible, coupling of these steps in protein synthesis to GTP hydrolysis prevents them from going in the reverse direction.

## Nonsense Mutations Cause Premature Termination of Protein Synthesis

One kind of mutation that can inactivate a gene in any organism is a base-pair change that converts a codon normally encoding an amino acid into a stop codon, e.g., UAC (encoding tyrosine) → UAG (stop). When this occurs early in the reading frame, the resulting truncated protein usually is nonfunctional. Such mutations are called *nonsense* mutations because when the genetic code equating each triplet codon sequence with a single amino acid was being deciphered by researchers, the three stop codons were found to not encode any amino acid—they did not "make sense."

In genetic studies with the bacterium *E. coli*, it was discovered that the effect of a nonsense mutation can be suppressed

the circular shape of polyribosomes and suggested the mechanism by which ribosomes recycle efficiently.

These studies revealed that multiple copies of the poly(A)-binding protein (PABP) interact with both an mRNA poly(A)

(a)

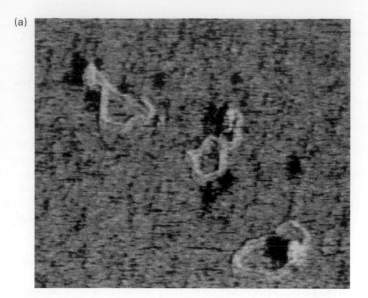

(b)

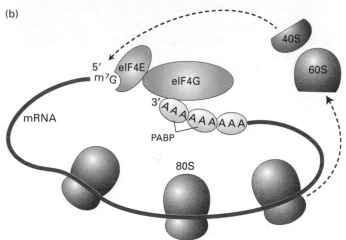

**EXPERIMENTAL FIGURE 4-28 Circular structure of mRNA increases translation efficiency.** Eukaryotic mRNA forms a circular structure owing to interactions of three proteins. (a) In the presence of purified poly(A)-binding protein (PABP), eIF4E, and eIF4G, eukaryotic mRNAs form circular structures, visible in this force-field electron micrograph. In these structures, protein-protein and protein-mRNA interactions form a bridge between the 5' and 3' ends of the mRNA. (b) Model of protein synthesis on circular polysomes and recycling of ribosomal subunits. Multiple individual ribosomes can simultaneously translate a eukaryotic mRNA, shown here in circular form stabilized by interactions between proteins bound at the 3' and 5' ends. When a ribosome completes translation and dissociates from the 3' end, the separated subunits can rapidly find the nearby 5' cap (m⁷G) and PABP-bound poly(A) tail and initiate another round of synthesis. [Part (a) courtesy of A. Sachs.]

by a second mutation in a tRNA gene. This occurs when the sequence encoding the anticodon in the tRNA gene is changed to a triplet that is complementary to the original stop codon, e.g., a mutation in tRNA$^{Tyr}$ that changes its anticodon from GUA to CUA, which can base-pair with the UAG stop codon. The mutant tRNA can still be recognized by the tyrosine amino acyl synthetase and coupled to tyrosine. Cells with

both the original nonsense mutation and the second mutation in the anticodon of the tRNA$^{Tyr}$ gene consequently can insert a tyrosine at the position of the mutant stop codon, allowing protein synthesis to continue past the original nonsense mutation. This mechanism of suppression is not highly efficient, so translation of normal mRNAs with a UAG stop codon terminates at the normal position in most instances. If enough of the protein encoded by the original gene with the nonsense mutation is produced to provide its essential functions, the effect of the first mutation is said to be *suppressed* by the second mutation in the anticodon of the tRNA gene.

This mechanism of *nonsense suppression* is a powerful tool in genetic studies in bacteria. For example, mutant bacterial viruses can be isolated that cannot grow in normal cells but can grow in cells expressing a nonsense-suppressing tRNA because the mutant virus has a nonsense mutation in an essential gene. Such mutant viruses grown on the nonsense-suppressing cells can then be used in experiments to analyze the function of the mutant gene by infecting normal cells that do not suppress the mutation and analyzing what step in the viral life cycle is defective in the absence of the mutant protein.

## KEY CONCEPTS of Section 4.4

### Stepwise Synthesis of Proteins on Ribosomes

• Both prokaryotic and eukaryotic ribosomes—the large ribonucleoprotein complexes on which translation occurs—consist of a small and a large subunit (see Figure 4-22). Each subunit contains numerous different proteins and one major rRNA molecule (small or large). The large subunit also contains one accessory 5S rRNA in bacteria and two accessory rRNAs in eukaryotes (5S and 5.8S in vertebrates).

• Analogous rRNAs from many different species fold into quite similar three-dimensional structures containing numerous stem-loops and binding sites for proteins, mRNA, and tRNAs. Much smaller ribosomal proteins are associated with the periphery of the rRNAs.

• Of the two methionine tRNAs found in all cells, only one (tRNA$_i^{Met}$) functions in initiation of translation.

• Each stage of translation—initiation, chain elongation, and termination—requires specific protein factors including GTP-binding proteins that hydrolyze their bound GTP to GDP when a step has been completed successfully.

• During initiation, the ribosomal subunits assemble near the translation start site in an mRNA molecule with the tRNA carrying the amino-terminal methionine (Met-tRNA$_i^{Met}$) base-paired with the start codon (Figure 4-24).

• Chain elongation entails a repetitive four-step cycle: loose binding of an incoming aminoacyl-tRNA to the A site on the ribosome; tight binding of the correct aminoacyl-tRNA to the A site accompanied by release of the previously used tRNA from the E site; transfer of the growing peptidyl chain to the incoming amino acid catalyzed by large rRNA; and translocation of the ribosome to the next codon, thereby moving the peptidyl-tRNA in the A site to the P site and the now unacylated tRNA in the P site to the E site (see Figure 4-25).

• In each cycle of chain elongation, the ribosome undergoes two conformational changes monitored by GTP-binding proteins. The first (EF1α) permits tight binding of the incoming aminoacyl-tRNA to the A site and ejection of a tRNA from the E site, and the second (EF2) leads to translocation.

• Termination of translation is carried out by two types of termination factors: those that recognize stop codons and those that promote hydrolysis of peptidyl-tRNA (see Figure 4-27). Once again, correct recognition of a stop codon is monitored by a GTPase (eRF3).

• The efficiency of protein synthesis is increased by the simultaneous translation of a single mRNA by multiple ribosomes, forming a polyribosome, or simply polysome. In eukaryotic cells, protein-mediated interactions bring the two ends of a polyribosome close together, thereby promoting the rapid recycling of ribosomal subunits, which further increases the efficiency of protein synthesis (see Figure 4-28b).

## 4.5 DNA Replication

Now that we have seen how genetic information encoded in the nucleotide sequence of DNA is translated into the proteins that perform most cell functions, we can appreciate the necessity for precisely copying DNA sequences during DNA replication, in preparation for cell division (see Figure 4-1, step **4**). The regular pairing of bases in the double-helical DNA structure suggested to Watson and Crick that new DNA strands are synthesized by using the existing (*parental*) strands as templates in the formation of new, *daughter* strands complementary to the parental strands.

This base-pairing template model theoretically could proceed either by a *conservative* or a *semiconservative* mechanism. In a conservative mechanism, the two daughter strands would form a new double-stranded (*duplex*) DNA molecule and the parental duplex would remain intact. In a semiconservative mechanism, the parental strands are permanently separated and each forms a duplex molecule with the daughter strand base-paired to it. Definitive evidence that duplex DNA is replicated by a semiconservative mechanism came from a now classic experiment conducted by M. Meselson and W. F. Stahl, outlined in Figure 4-29.

Copying of a DNA template strand into a complementary strand thus is a common feature of DNA replication, transcription of DNA into RNA, and, as we will see later in this chapter, DNA repair and recombination. In all cases, the information in the template in the form of the specific sequence of nucleotides is preserved. In some viruses, single-stranded RNA molecules function as templates for synthesis of complementary RNA or DNA strands. However, the vast preponderance of RNA and DNA in cells is synthesized from preexisting duplex DNA.

## DNA Polymerases Require a Primer to Initiate Replication

Analogous to RNA, DNA is synthesized from deoxynucleoside 5′-triphosphate precursors (dNTPs). Also like RNA synthesis, DNA synthesis always proceeds in the 5′→3′ direction because chain growth results from formation of a phosphoester bond between the 3′ oxygen of a growing strand and the α phosphate of a dNTP (see Figure 4-10a). As discussed earlier, an RNA polymerase can find an appropriate transcription start site on duplex DNA and initiate the synthesis of an RNA complementary to the template DNA strand (see Figure 4-11). In contrast, **DNA polymerases** cannot initiate chain synthesis de novo; instead, they require a short, preexisting RNA or DNA strand, called a **primer**, to begin chain growth. With a primer base-paired to the template strand, a DNA polymerase adds deoxynucleotides to the free hydroxyl group at the 3′ end of the primer as directed by the sequence of the template strand:

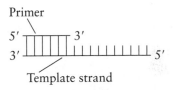

When RNA is the primer, the daughter strand that is formed is RNA at the 5′ end and DNA at the 3′ end.

## Duplex DNA Is Unwound, and Daughter Strands Are Formed at the DNA Replication Fork

In order for duplex DNA to function as a template during replication, the two intertwined strands must be unwound, or melted, to make the bases available for pairing with the bases of the dNTPs that are polymerized into the newly synthesized daughter strands. This unwinding of the parental DNA strands is by specific **helicases**, beginning at unique segments in a DNA molecule called *replication origins*, or simply *origins*. The nucleotide sequences of origins from different organisms vary greatly, although they usually contain A·T-rich sequences. Once helicases have unwound the parental DNA at an origin, a specialized RNA polymerase called **primase** forms a short RNA primer complementary to the unwound template strands. The primer, still base-paired to its complementary DNA strand, is then elongated by a DNA polymerase, thereby forming a new daughter strand.

The DNA region at which all these proteins come together to carry out synthesis of daughter strands is called the **replication fork**. As replication proceeds, the replication fork and associated proteins move away from the origin. As noted earlier, local unwinding of duplex DNA produces torsional stress, which is relieved by topoisomerase I. In order for DNA polymerases to move along and copy a duplex DNA, helicase must sequentially unwind the duplex and topoisomerase must remove the supercoils that form.

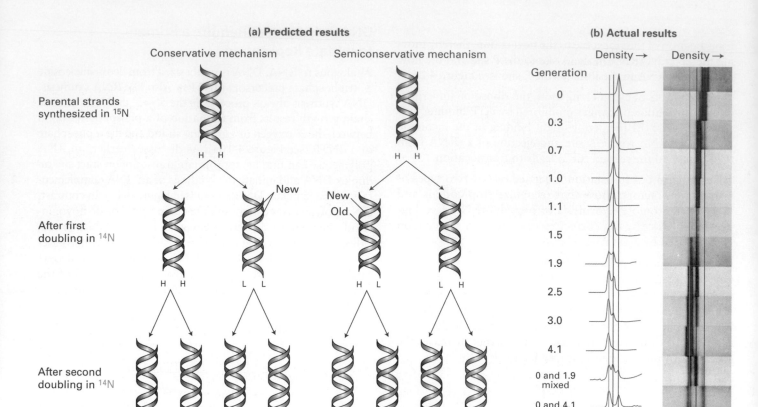

**(a) Predicted results**

Conservative mechanism     Semiconservative mechanism

**(b) Actual results**

Density →     Density →

Generation

Parental strands synthesized in $^{15}$N

After first doubling in $^{14}$N

After second doubling in $^{14}$N

0
0.3
0.7
1.0
1.1
1.5
1.9
2.5
3.0
4.1
0 and 1.9 mixed
0 and 4.1 mixed

L-L   H-L   H-H     L-L   H-L   H-H

**EXPERIMENTAL FIGURE 4-29  The Meselson-Stahl experiment.**
This experiment showed that DNA replicates by a semiconservative mechanism. *E. coli* cells initially were grown in a medium containing ammonium salts prepared with "heavy" nitrogen ($^{15}$N) until all the cellular DNA was labeled. After the cells were transferred to a medium containing the normal "light" isotope ($^{14}$N), samples were removed periodically from the cultures and the DNA in each sample was analyzed by equilibrium density-gradient centrifugation, a procedure that separates macromolecules on the basis of their density. This technique can separate heavy-heavy (H-H), light-light (L-L), and heavy-light (H-L) duplexes into distinct bands. (a) Expected composition of daughter duplex molecules synthesized from $^{15}$N-labeled DNA after *E. coli* cells are shifted to $^{14}$N-containing medium if DNA replication occurs by a conservative or semiconservative mechanism. Parental heavy (H) strands are in red; light (L) strands synthesized after shift to $^{14}$N-containing medium are in blue. Note that the conservative mechanism never generates H-L DNA and that the semiconservative mechanism never generates H-H DNA but does generate H-L DNA during the first and subsequent doublings. With additional replication cycles, the $^{15}$N-labeled (H) strands from the original DNA are diluted, so that the vast bulk of the DNA would consist of L-L duplexes

with either mechanism. (b) Actual banding patterns of DNA subjected to equilibrium density-gradient centrifugation before and after shifting $^{15}$N-labeled *E. coli* cells to $^{14}$N-containing medium. DNA bands were visualized under UV light and photographed. The traces on the left are a measure of the density of the photographic signal, and hence the DNA concentration, along the length of the centrifuge cells from left to right. The number of generations (*far left*) following the shift to $^{14}$N-containing medium was determined by counting the concentration of *E. coli* cells in the culture. This value corresponds to the number of DNA replication cycles that had occurred at the time each sample was taken. After one generation of growth, all the extracted DNA had the density of H-L DNA. After 1.9 generations, approximately half the DNA had the density of H-L DNA; the other half had the density of L-L DNA. With additional generations, a larger and larger fraction of the extracted DNA consisted of L-L duplexes; H-H duplexes never appeared. These results match the predicted pattern for the semiconservative replication mechanism depicted in (a). The bottom two centrifuge cells contained mixtures of H-H DNA and DNA isolated at 1.9 and 4.1 generations in order to clearly show the positions of H-H, H-L, and L-L DNA in the density gradient. [Part (b) from M. Meselson and F. W. Stahl, 1958, *Proc. Nat'l Acad. Sci. USA* **44:**671.]

A major complication in the operation of a DNA replication fork arises from two properties: the two strands of the parental DNA duplex are antiparallel, and DNA polymerases (like RNA polymerases) can add nucleotides to the growing new strands only in the 5'→3' direction. Synthesis of one daughter strand, called the **leading strand,** can proceed continuously from a single RNA primer in the 5'→3' direction, *the same direction as movement of the replication*

*fork* (Figure 4-30). The problem comes in synthesis of the other daughter strand, called the **lagging strand.**

Because growth of the lagging strand must occur in the 5'→3' direction, copying of its template strand must somehow occur in the *opposite* direction from the movement of the replication fork. A cell accomplishes this feat by synthesizing a new primer every few hundred bases or so on the second parental strand, as more of the strand is exposed by

**ANIMATION:** Nucleotide Polymerization by DNA Polymerase

**FIGURE 4-30 Leading-strand and lagging-strand DNA synthesis.** Nucleotides are added by a DNA polymerase to each growing daughter strand in the 5′→3′ direction (indicated by arrowheads). The leading strand is synthesized continuously from a single RNA primer (red) at its 5′ end. The lagging strand is synthesized discontinuously from multiple RNA primers that are formed periodically as each new region of the parental duplex is unwound. Elongation of these primers initially produces Okazaki fragments. As each growing fragment approaches the previous primer, the primer is removed and the fragments are ligated. Repetition of this process eventually results in synthesis of the entire lagging strand.

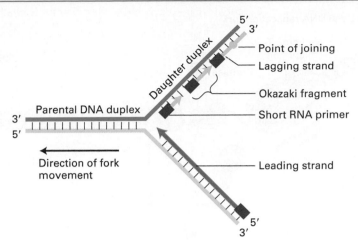

unwinding. Each of these primers, base-paired to their template strand, is elongated in the 5′→3′ direction, forming discontinuous segments called **Okazaki fragments** after their discoverer Reiji Okazaki (see Figure 4-30). The RNA primer of each Okazaki fragment is removed and replaced by DNA chain growth from the neighboring Okazaki fragment; finally, an enzyme called *DNA ligase* joins the adjacent fragments.

## Several Proteins Participate in DNA Replication

Detailed understanding of the eukaryotic proteins that participate in DNA replication initially came largely from studies with small viral DNAs, particularly SV40 DNA, the circular genome of a small virus that infects monkeys. Virus-infected cells replicate large numbers of the simple viral genome in a short period of time, making them ideal model systems for studying basic aspects of DNA replication. Because simple viruses like SV40 depend largely on the DNA replication machinery of their host cells (in this case monkey cells), they offer a unique opportunity to study DNA replication of multiple identical small DNA molecules by cellular proteins. Figure 4-31 depicts the multiple proteins that coordinate copying of SV40 DNA at a replication fork. The assembled proteins at a replication fork further illustrate the concept of molecular machines introduced in Chapter 3. These multicomponent complexes permit the cell to carry out an ordered sequence of events that accomplishes essential cell functions.

The molecular machine that replicates SV40 DNA contains only one viral protein. All other proteins involved in SV40 DNA replication are provided by the host cell. This viral protein, *large T-antigen,* forms a hexameric *replicative helicase,* a protein that uses energy from ATP hydrolysis to unwind the parental strands at a replication fork. Primers for leading and lagging daughter-strand DNA are synthesized by a complex of *primase,* which synthesizes a short RNA primer of ≈10 nucleotides, and *DNA polymerase α* (Pol α), which extends the RNA primer with deoxynucleotides for another ≈20 nucleotides, forming a mixed RNA-DNA primer.

The primer is extended into daughter-strand DNA by *DNA polymerase δ* (Pol δ) and *DNA polymerase ε* (Pol ε), which are less likely to make errors during copying of the template strand than is Pol α because of their proofreading mechanism (see Section 4.7). Recent results indicate that during the replication of cellular DNA, Pol δ synthesizes lagging-strand DNA, while Pol ε synthesizes most of the length of the leading strand. Pol δ and Pol ε each form a complex with *Rfc* (replication *f*actor C) and *PCNA* (proliferating *c*ell *n*uclear *a*ntigen), which displaces the primase–Pol α complex following primer synthesis. As illustrated in Figure 4-31b, PCNA is a homotrimeric protein that has a central hole through which the daughter duplex DNA passes, thereby preventing the PCNA-Rfc–Pol δ and PCNA-Rfc-Pol ε complexes from dissociating from the template. As such, PCNA is known as a *sliding clamp* that enables Pol δ and Pol ε to remain stably associated with a single template strand for thousands of nucleotides. Rfc functions to open the PCNA ring so that it can encircle the short region of double-stranded DNA synthesized by Pol α. Consequently, Rfc is often called a *clamp loader.*

After parental DNA is separated into single-stranded templates at the replication fork, it is bound by multiple copies of RPA (replication protein A), a heterotrimeric protein (Figure 4-31c). Binding of RPA maintains the template in a uniform conformation optimal for copying by DNA polymerases. Bound RPA proteins are dislodged from the parental strands by Pol α, Pol δ, and Pol ε as they synthesize the complementary strands base-paired with the parental strands.

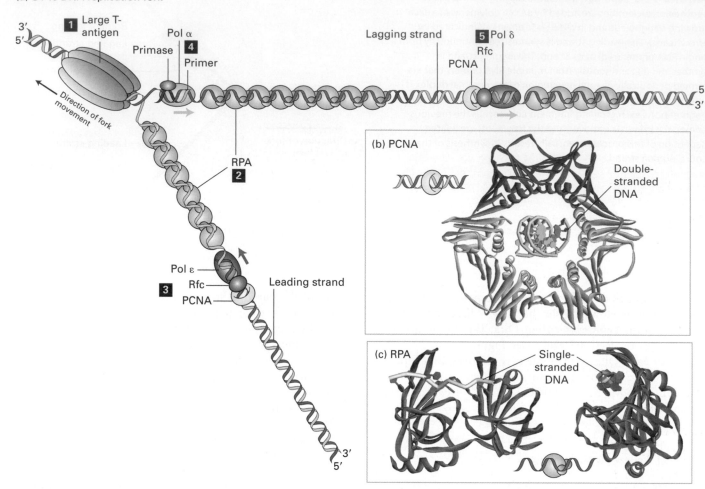

**FIGURE 4-31 Model of an SV40 DNA replication fork.** (a) A hexamer of large T-antigen (**1**), a viral protein, functions as a helicase to unwind the parental DNA strands. Single-strand regions of the parental template unwound by large T-antigen are bound by multiple copies of the heterotrimeric protein RPA (**2**). The leading strand is synthesized by a complex of DNA polymerase ε (Pol ε), PCNA, and Rfc (**3**). Primers for lagging-strand synthesis (red, RNA; light blue, DNA) are synthesized by a complex of DNA polymerase α (Pol α) and primase (**4**). The 3′ end of each primer synthesized by Pol α–primase is then bound by a PCNA-Rfc–Pol δ complex, which proceeds to extend the primer and synthesize most of each Okazaki fragment (**5**). (b) The three subunits of PCNA, shown in different colors, form a circular structure with a central hole through which double-stranded DNA passes. A diagram of DNA is shown in the center of a ribbon model of the PCNA trimer. The diagram at the upper left shows the icon representing PCNA bound to DNA in (a). (c) The large subunit of RPA contains two domains that bind single-stranded DNA. On the left, the structure determined for the two DNA-binding domains of the large subunit bound to single-stranded DNA is shown with the DNA backbone (white backbone with blue bases) parallel to the plane of the page. Note that the single DNA strand is extended with the bases exposed, an optimal conformation for replication by a DNA polymerase. On the right, the view is down the length of the single DNA strand, revealing how RPA β strands wrap around the DNA. The diagram at the bottom shows the icon representing RPA bound to DNA in part (a). [Part (a) adapted from S. J. Flint et al., 2000, *Virology: Molecular Biology, Pathogenesis, and Control,* ASM Press. Part (b) adapted from J. M. Gulbis et al., 1996, *Cell* **87:**297. Part (c) adapted from A. Bochkarev et al., 1997, *Nature* **385:**176.]

Several other eukaryotic proteins that function in DNA replication are not depicted in Figure 4-31. For example, topoisomerase I associates with the parental DNA ahead of the replicative helicase, i.e., to the left of T-antigen in Figure 4-31, to remove torsional stress introduced by the unwinding of the parental strands (see Figure 4-8a). Ribonuclease H and FEN I remove the ribonucleotides at the 5′ ends of Okazaki fragments; these are replaced by deoxynucleotides added by DNA polymerase δ as it extends the upstream Okazaki fragment. Successive Okazaki fragments are coupled by DNA

**EXPERIMENTAL FIGURE 4-32 Bidirectional replication in SV40 DNA.** Electron microscopy of replicating SV40 DNA indicates bidirectional growth of DNA strands from an origin. The replicating viral DNA from SV40-infected cells was cut by the restriction enzyme *Eco*RI, which recognizes one site in the circular DNA. This was done to provide a landmark for a specific sequence in the SV40 genome: the *Eco*RI recognition sequence is now easily recognized as the ends of linear DNA molecules visualized by electron microscopy. Electron micrographs of *Eco*RI-cut replicating SV40 DNA molecules showed a collection of cut molecules with increasingly longer replication "bubbles," whose centers are a constant distance from each end of the cut molecules. This finding is consistent with chain growth in two directions from a common origin located at the center of a bubble, as illustrated in the corresponding diagrams. [See G. C. Fareed et al., 1972, *J. Virol.* **10**:484; photographs courtesy of N. P. Salzman.]

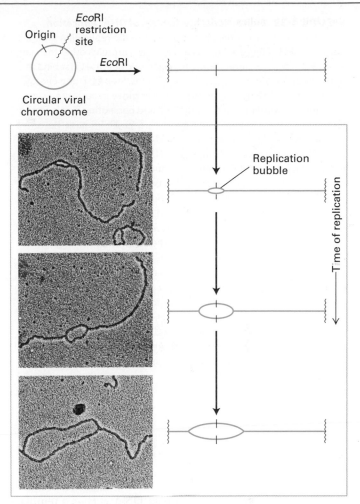

ligase through standard 5′→3′ phosphoester bonds. Other specialized DNA polymerases are involved in repair of mismatches and damaged lesions in DNA (see Section 4.7).

## DNA Replication Occurs Bidirectionally from Each Origin

As indicated in Figures 4-30 and 4-31, both parental DNA strands that are exposed by local unwinding at a replication fork are copied into a daughter strand. In theory, DNA replication from a single origin could involve one replication fork that moves in one direction. Alternatively, two replication forks might assemble at a single origin and then move in opposite directions, leading to *bidirectional growth* of both daughter strands. Several types of experiments, including the one shown in Figure 4-32, provided early evidence in support of bidirectional strand growth.

The general consensus is that all prokaryotic and eukaryotic cells employ a bidirectional mechanism of DNA replication. In the case of SV40 DNA, replication is initiated by binding of two large T-antigen hexameric helicases to the single SV40 origin and assembly of other proteins to form two replication forks. These then move away from the SV40 origin in opposite directions, with leading- and lagging-strand synthesis occurring at both forks. As shown in Figure 4-33, the left replication fork extends DNA synthesis in the leftward direction; similarly, the right replication fork extends DNA synthesis in the rightward direction.

Unlike SV40 DNA, eukaryotic chromosomal DNA molecules contain multiple replication origins separated by tens to hundreds of kilobases. A six-subunit protein called *ORC*, for *o*rigin *r*ecognition *c*omplex, binds to each origin and associates with other proteins required to load cellular hexameric helicases composed of six homologous *MCM* proteins. Two opposed MCM helicases separate the parental strands at an origin, with RPA proteins binding to the resulting single-stranded DNA. Synthesis of primers and subsequent steps in replication of cellular DNA are thought to be analogous to those in SV40 DNA replication (see Figures 4-31 and 4-33).

Replication of cellular DNA and other events leading to proliferation of cells are tightly regulated, so that the appropriate numbers of cells constituting each tissue are produced during development and throughout the life of an organism. Control of the initiation step is the primary mechanism for regulating cellular DNA replication. Activation of MCM helicase activity, which is required to initiate cellular DNA replication, is regulated by specific protein kinases called S-phase **cyclin-dependent kinases.** Other cyclin-dependent kinases regulate additional aspects of cell proliferation, including the complex process of mitosis by which a eukaryotic cell divides into two daughter cells. Mitosis and another specialized type of cell division called meiosis, which generates haploid sperm and egg cells, are discussed in Chapter 5. We discuss the various regulatory mechanisms that determine the rate of cell division in Chapter 20.

**FIGURE 4-33 Bidirectional mechanism of DNA replication.**
The left replication fork here is comparable to the replication fork diagrammed in Figure 4-31, which also shows proteins other than large T-antigen. *Top:* Two large T-antigen hexameric helicases first bind at the replication origin in opposite orientations. Step **1**: Using energy provided from ATP hydrolysis, the helicases move in opposite directions, unwinding the parental DNA and generating single-strand templates that are bound by RPA proteins. Step **2**: Primase–Pol α complexes synthesize short primers (red) base-paired to each of the separated parental strands. Step **3**: PCNA-Rfc–Pol δ/ε complexes replace the primase–Pol α complexes and extend the short primers, generating the leading strands (dark green) at each replication fork. Step **4**: The helicases further unwind the parental strands, and RPA proteins bind to the newly exposed single-strand regions. Step **5**: PCNA-Rfc–Pol δ complexes extend the leading strands further. Step **6**: Primase–Pol α complexes synthesize primers for lagging-strand synthesis at each replication fork. Step **7**: PCNA-Rfc–Pol δ complexes displace the primase–Pol α complexes and extend the lagging-strand Okazaki fragments (light green), which eventually are ligated to the 5′ ends of the leading strands. The position where ligation occurs is represented by a circle. Replication continues by further unwinding of the parental strands and synthesis of leading and lagging strands as in Steps **4**–**7**. Although depicted as individual steps for clarity, unwinding and synthesis of leading and lagging strands occur concurrently.

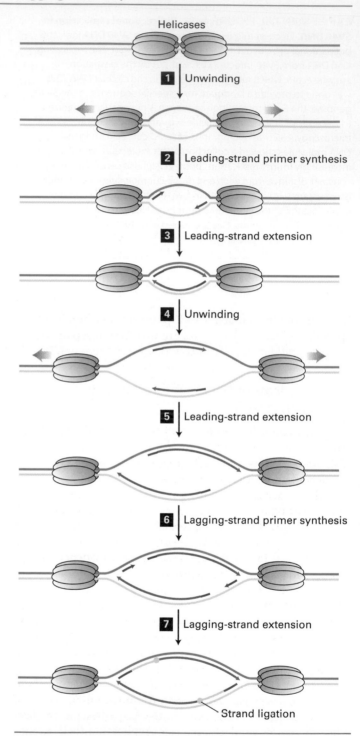

Helicases

**1** Unwinding

**2** Leading-strand primer synthesis

**3** Leading-strand extension

**4** Unwinding

**5** Leading-strand extension

**6** Lagging-strand primer synthesis

**7** Lagging-strand extension

Strand ligation

## KEY CONCEPTS of Section 4.5

### DNA Replication

• Each strand in a parental duplex DNA acts as a template for synthesis of a daughter strand and remains base-paired to the new strand, forming a daughter duplex (semiconservative mechanism). New strands are formed in the 5′→3′ direction.

• Replication begins at a sequence called an *origin*. Each eukaryotic chromosomal DNA molecule contains multiple replication origins.

• DNA polymerases, unlike RNA polymerases, cannot unwind the strands of duplex DNA and cannot initiate synthesis of new strands complementary to the template strands.

• At a replication fork, one daughter strand (the leading strand) is elongated continuously. The other daughter strand (the lagging strand) is formed as a series of discontinuous Okazaki fragments from primers synthesized every few hundred nucleotides (Figure 4-30).

• The ribonucleotides at the 5′ end of each Okazaki fragment are removed and replaced by elongation of the 3′ end of the next Okazaki fragment. Finally, adjacent Okazaki fragments are joined by DNA ligase.

• Helicases use energy from ATP hydrolysis to separate the parental (template) DNA strands which are initially bound

by multiple copies of a single-stranded DNA-binding protein, RPA. Primase synthesizes a short RNA primer, which remains base-paired to the template DNA. This initially is extended at the 3′ end by DNA polymerase α (Pol α), resulting in a short (5′)RNA-(3′)DNA daughter strand.

• Most of the DNA in eukaryotic cells is synthesized by Pol δ and Pol ε, which take over from Pol α and continue elongation of the daughter strand in the 5′→3′ direction. Pol δ synthesizes most of the length of the lagging strand, while Pol ε synthesizes the leading strand. Pol δ and Pol ε remain stably associated with the template by binding to Rfc protein, which in turn binds to PCNA, a trimeric protein that encircles the daughter duplex DNA, functioning as a sliding clamp (see Figure 4-31).

• DNA replication generally occurs by a bidirectional mechanism in which two replication forks form at an origin and move in opposite directions, with both template strands being copied at each fork (see Figure 4-33).

• Synthesis of eukaryotic DNA in vivo is regulated by controlling the activity of the MCM helicases that initiate DNA replication at multiple origins spaced along chromosomal DNA.

## 4.6 DNA Repair and Recombination

Damage to DNA is unavoidable and arises in many ways. DNA damage can be caused by spontaneous cleavage of chemical bonds in DNA, by environmental agents such as ultraviolet and ionizing radiation, and by reaction with genotoxic chemicals that are by-products of normal cellular metabolism or occur in the environment. A change in the normal DNA sequence, called a **mutation**, can occur during replication when a DNA polymerase inserts the wrong nucleotide as it reads a damaged template. Mutations also occur at a low frequency as the result of copying errors introduced by DNA polymerases when they replicate an undamaged template. If such mutations were left uncorrected, cells might accumulate so many mutations that they could no longer function properly. In addition, the DNA in germ cells might incur too many mutations for viable offspring to be formed. Thus the prevention of DNA sequence errors in all types of cells is important for survival, and several cellular mechanisms for repairing damaged DNA and correcting sequence errors have evolved. One of these mechanisms for repairing double-stranded DNA breaks, by the process of recombination, is also used by eukaryotic cells to generate new combinations of maternal and paternal genes on each chromosome through the exchange of segments of the chromosomes during the production of germ cells (e.g., sperm and eggs).

Significantly, defects in DNA repair mechanisms and cancer are closely related. When repair mechanisms are compromised, mutations accumulate in the cell's DNA. If these mutations affect genes that are normally involved in the careful regulation of cell division, cells can begin to divide uncontrollably, leading to tumor formation, and cancer. Chapter 25 outlines in detail how cancer arises from defects in DNA repair. We will encounter a few examples in this section as well, as we first consider the ways in which DNA

integrity can be compromised, and then discuss the repair mechanisms that cells have evolved to ensure the fidelity of this very important molecule.

### DNA Polymerases Introduce Copying Errors and Also Correct Them

The first line of defense in preventing mutations is DNA polymerase itself. Occasionally, when replicative DNA polymerases progress along the template DNA, an incorrect nucleotide is added to the growing 3′ end of the daughter strand. *E. coli* DNA polymerases, for instance, introduce about 1 incorrect nucleotide per $10^4$ (ten thousand) polymerized nucleotides. Yet the measured mutation rate in bacterial cells is much lower: about 1 mistake in $10^9$ (one billion) nucleotides incorporated into a growing strand. This remarkable accuracy is largely due to *proofreading* by *E. coli* DNA polymerases. Eukaryotic Pol δ and Pol ε employ a similar mechanism.

Proofreading depends on a *3′→5′ exonuclease activity* of some DNA polymerases. When an incorrect base is incorporated during DNA synthesis, base pairing between the 3′ nucleotide of the nascent strand and the template strand does not occur. As a result, the polymerase pauses, then transfers the 3′ end of the growing chain to the exonuclease site, where the incorrect mispaired base is removed (Figure 4-34). Then the 3′ end is transferred back to the polymerase site, where this region is copied correctly. All three *E. coli* DNA polymerases have proofreading activity, as do the two eukaryotic DNA polymerases, δ and ε, used for replication of most chromosomal DNA in animal cells. It seems likely that proofreading is indispensable for all cells to avoid excessive mutations.

### Chemical and Radiation Damage to DNA Can Lead to Mutations

DNA is continually subjected to a barrage of damaging chemical reactions; estimates of the number of DNA damage events in a single human cell range from $10^4$ to $10^6$ per day! Even if DNA were not exposed to damaging chemicals, certain aspects of DNA structure are inherently unstable. For example, the bond connecting a purine base to deoxyribose is prone to hydrolysis at a low rate under physiological conditions, leaving a sugar without an attached base. Thus coding information is lost, and this can lead to a mutation during DNA replication. Normal cellular reactions, including the movement of electrons along the electron-transport chain in mitochondria and lipid oxidation in peroxisomes, produce several chemicals that react with and damage DNA, including hydroxyl radicals and superoxide ($O_2^-$). These too can cause mutations, including those that lead to cancers.

Many spontaneous mutations are **point mutations**, which involve a change in a single base pair in the DNA sequence. This can introduce a stop codon, causing a *nonsense* mutation as discussed earlier, or a change in the amino acid sequence of an encoded protein, called a *missense* mutation. *Silent* mutations do not change the amino acid sequence (e.g., GAG to GAA; both encode glutamine). Point mutations can

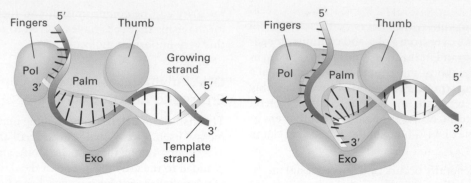

**FIGURE 4-34 Proofreading by DNA polymerase.** All DNA polymerases have a similar three-dimensional structure, which resembles a half-opened right hand. The "fingers" bind the single-stranded segment of the template strand, and the polymerase catalytic activity (Pol) lies in the junction between the fingers and palm. As long as the correct nucleotides are added to the 3′ end of the growing strand, it remains in the polymerase site. Incorporation of an incorrect base at the 3′ end causes melting of the newly formed end of the duplex. As a result, the polymerase pauses, and the 3′ end of the growing strand is transferred to the 3′→5′ exonuclease site (Exo) about 3 nm away, where the mispaired base and probably other bases are removed. Subsequently, the 3′ end flips back into the polymerase site and elongation resumes. [Adapted from C. M. Joyce and T. T. Steitz, 1995, *J. Bacteriol.* **177:**6321, and S. Bell and T. Baker, 1998, *Cell* **92:**295.]

also occur in a non-protein-coding DNA sequence that functions in the regulation of a gene's transcription, as discussed in Chapter 7. One of the most frequent point mutations comes from *deamination* of a cytosine (C) base, which converts it into a uracil (U) base. In addition, the common modified base 5-methyl cytosine forms thymine when it is deaminated. If these alterations are not corrected before the DNA is replicated, the cell will use the strand containing U or T as template to form a U·A or T·A base pair, thus creating a permanent change to the DNA sequence (Figure 4-35).

## High-Fidelity DNA Excision Repair Systems Recognize and Repair Damage

In addition to proofreading, cells have other repair systems for preventing mutations due to copying errors and exposure

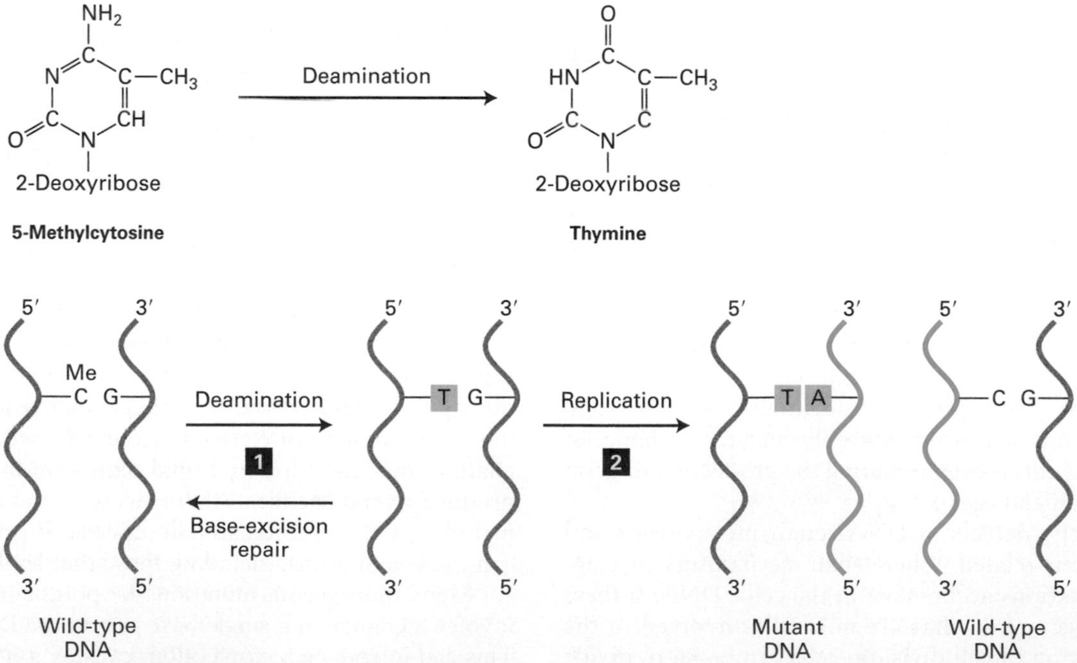

**FIGURE 4-35 Deamination leads to point mutations.** A spontaneous point mutation can form by deamination of 5-methylcytosine (C) to form thymine (T). If the resulting T·G base pair is not restored to the normal C·G base pair by base excision-repair mechanisms (step **1**), it will lead to a permanent change in sequence (i.e., a mutation) following DNA replication (step **2**). After one round of replication, one daughter DNA molecule will have the mutant T·A base pair and the other will have the wild-type C·G base pair.

to chemicals and radiation. Several DNA **excision-repair systems** that normally operate with a high degree of accuracy have been well studied. These systems were first elucidated through a combination of genetic and biochemical studies in *E. coli.* Homologs of the key bacterial proteins exist in eukaryotes from yeast to humans, indicating that these error-free mechanisms arose early in evolution to protect DNA integrity. Each of these systems functions in a similar manner—a segment of the damaged DNA strand is excised, and the gap is filled by DNA polymerase and ligase using the complementary DNA strand as template.

We will now turn to a closer look at some of the mechanisms of DNA repair, ranging from repair of single base mutations to repair of DNA broken across both strands. Some of these accomplish their repairs with great accuracy; others are less precise.

## Base Excision Repairs T-G Mismatches and Damaged Bases

In humans, the most common type of point mutation is a C to T, which is caused by deamination of 5-methyl C to T (see Figure 4-35). The conceptual problem with *base excision repair* in this case is determining which is the normal and which is the mutant DNA strand, and repairing the latter so that it is properly base-paired with the normal strand. But since a G·T mismatch is almost invariably caused by chemical conversion of C to U or 5-methyl C to T, the repair system evolved to remove the T and replace it with a C.

The G·T mismatch is recognized by a DNA glycosylase that flips the thymine base out of the helix and then hydrolyzes the bond that connects it to the sugar-phosphate DNA backbone. Following this initial incision, an apurinic (AP) endonuclease cuts the DNA strand near the abasic site. The deoxyribose phosphate lacking the base is then removed and replaced with a C by a specialized repair DNA polymerase that reads the G in the template strand (Figure 4-36).

As mentioned earlier, this repair must take place prior to DNA replication, because the incorrect base in this pair, T, occurs naturally in normal DNA. Consequently, it would be able to engage in normal Watson-Crick base pairing during replication, generating a stable point mutation that is now unable to be recognized by repair mechanisms (see Figure 4-35, step **2**).

Human cells contain a battery of glycosylases, each of which is specific for a different set of chemically modified DNA bases. For example, one removes 8-oxyguanine, an oxidized form of guanine, allowing its replacement by an undamaged G, and others remove bases modified by alkylating agents. The resulting nucleotide lacking a base is then replaced by the repair mechanism discussed above. A similar mechanism also functions in the repair of lesions resulting from *depurination,* the loss of a guanine or adenine base from DNA resulting from hydrolysis of the glycosylic bond between deoxyribose and the base. Depurination occurs spontaneously and is fairly common in mammals. The resulting abasic sites, if left unrepaired, generate mutations during DNA replication because they cannot specify the appropriate paired base.

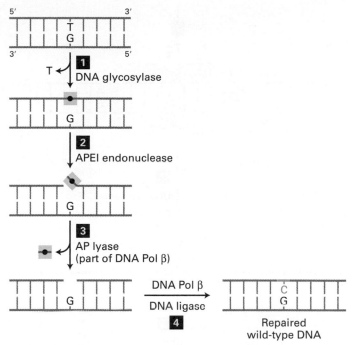

**FIGURE 4-36 Base excision repair of a T·G mismatch.** A DNA glycosylase specific for G·T mismatches, usually formed by deamination of 5-mC residues (see Figure 4-35), flips the thymine base out of the helix and then cuts it away from the sugar-phosphate DNA backbone (step **1**), leaving just the deoxyribose (black dot). An endonuclease specific for the resultant baseless site (apurinic endonuclease I, APE1) then cuts the DNA backbone (step **2**), and the deoxyribose phosphate is removed by an endonuclease, apurinic lyase (AP lyase), associated with DNA polymerase β, a specialized DNA polymerase used in repair (step **3**). The gap is then filled in by DNA Pol β and sealed by DNA ligase (step **4**), restoring the original G·C base pair. [Adapted from O. Schärer, 2003, *Angewandte Chemie* **42**:2946.]

## Mismatch Excision Repairs Other Mismatches and Small Insertions and Deletions

Another process, also conserved from bacteria to humans, principally eliminates base-pair mismatches and insertions or deletions of one or a few nucleotides that are accidentally introduced by DNA polymerases during replication. As with base excision repair of a T in a T·G mismatch, the conceptual problem with *mismatch excision repair* is determining which is the normal and which is the mutant DNA strand, and repairing the latter. How this happens in human cells is not known with certainty. It is thought that the proteins that bind to the mismatched segment of DNA distinguish the template and daughter strands; then the mispaired segment of the daughter strand—the one with the replication error—is excised and repaired to become an exact complement of the template strand (Figure 4-37). In contrast to base excision repair, mismatch excision repair occurs after DNA replication.

Predisposition to a colon cancer known as hereditary nonpolyposis colorectal cancer results from an inherited loss-of-function mutation in one copy of either the *MLH1* or the

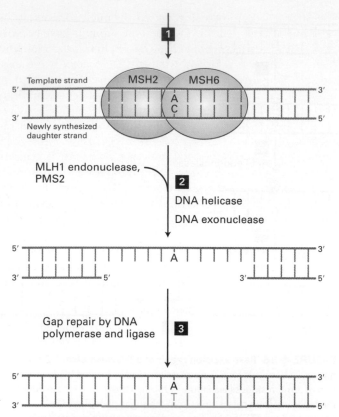

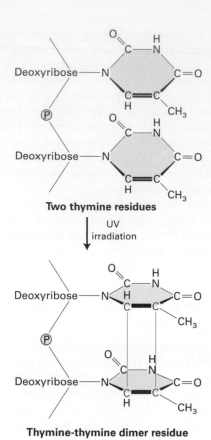

**FIGURE 4-37 Mismatch excision repair in human cells.** The mismatch excision-repair pathway corrects errors introduced during replication. A complex of the MSH2 and MSH6 proteins (bacterial *MutS* homologs 1 and 6) binds to a mispaired segment of DNA in such a way as to distinguish between the template and newly synthesized daughter strands (step **1**). This triggers binding of MLH1 and PMS2 (both homologs of bacterial MutL). The resulting DNA-protein complex then binds an endonuclease that cuts the newly synthesized daughter strand. Next a DNA helicase unwinds the helix, and an exonuclease removes several nucleotides from the cut end of the daughter strand, including the mismatched base (step **2**). Finally, as with base excision repair, the gap is then filled in by a DNA polymerase (Pol δ, in this case) and sealed by DNA ligase (step **3**).

**FIGURE 4-38 Formation of thymine-thymine dimers.** The most common type of DNA damage caused by UV irradiation, thymine-thymine dimers, can be repaired by an excision-repair mechanism.

*MSH2* gene. The MSH2 and MLH1 proteins are essential for DNA mismatch repair (see Figure 4-37). Cells with at least one functional copy of each of these genes exhibit normal mismatch repair. However, tumor cells frequently arise from those cells that have experienced a random mutation in the second copy; when both copies of one gene are not functional, the mismatch repair system is lost. Inactivating mutations in these genes are also common in noninherited forms of colon cancer. ∎

## Nucleotide Excision Repairs Chemical Adducts that Distort Normal DNA Shape

Cells use *nucleotide excision repair* to fix DNA regions containing chemically modified bases, often called chemical adducts, that distort the normal shape of DNA locally. A key to this type of repair is the ability of certain proteins to slide along the surface of a double-stranded DNA molecule looking

for bulges or other irregularities in the shape of the double helix. For example, this mechanism repairs *thymine-thymine dimers*, a common type of damage caused by UV light (Figure 4-38); these dimers interfere with both replication and transcription of DNA.

Figure 4-39 illustrates how the nucleotide excision-repair system repairs damaged DNA. Some 30 proteins are involved in this repair process, the first of which were identified through a study of the defects in DNA repair in cultured cells from individuals with xeroderma pigmentosum, a hereditary disease associated with a predisposition to cancer. Individuals with this disease frequently develop the skin cancers called melanomas and squamous cell carcinomas if their skin is exposed to the UV rays in sunlight. Cells of affected patients lack a functional nucleotide excision-repair system. Mutations in any of at least seven different genes, called *XP-A* through *XP-G*, lead to inactivation of this repair system and cause xeroderma pigmentosum; all produce the same phenotype and have the same consequences. The roles of most of these XP proteins in nucleotide excision repair are now well understood (see Figure 4-39). ∎

Remarkably, five polypeptide subunits of TFIIH, a general transcription factor required for transcription of all genes

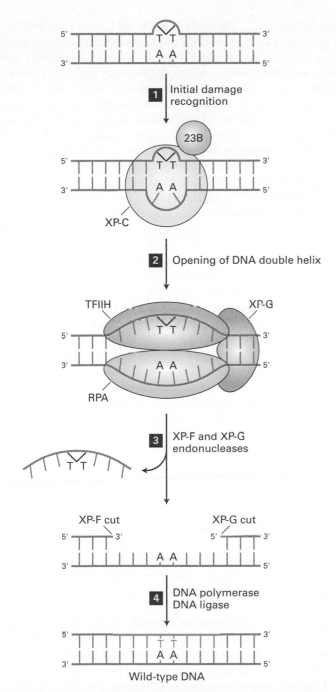

**1** Initial damage recognition

23B

XP-C

**2** Opening of DNA double helix

TFIIH                    XP-G

RPA

**3** XP-F and XP-G endonucleases

T T

XP-F cut            XP-G cut

A A

**4** DNA polymerase DNA ligase

A A

Wild-type DNA

**FIGURE 4-39 Nucleotide excision repair in human cells.** A DNA lesion that causes distortion of the double helix, such as a thymine dimer, is initially recognized by a complex of the XP-C (*xeroderma pigmentosum C* protein) and 23B proteins (step **1**). This complex then recruits transcription factor TFIIH, whose helicase subunits, powered by ATP hydrolysis, partially unwind the double helix. XP-G and RPA proteins then bind to the complex and further unwind and stabilize the helix until a bubble of ≈25 bases is formed (step **2**). Then XP-G (now acting as an endonuclease) and XP-F, a second endonuclease, cut the damaged strand at points 24–32 bases apart on each side of the lesion (step **3**). This releases the DNA fragment with the damaged bases, which is degraded to mononucleotides. Finally the gap is filled by DNA polymerase exactly as in DNA replication, and the remaining nick is sealed by DNA ligase (step **4**). [Adapted from J. Hoeijmakers, 2001, *Nature* **411:**366, and O. Schärer, 2003, *Angewandte Chemie* **42:**2946.]

(see Figure 7-16) are also required for nucleotide excision repair in eukaryotic cells. Two of these subunits have homology to helicases, as shown in Figure 4-39. In transcription, the helicase activity of TFIIH unwinds the DNA helix at the start site, allowing RNA polymerase to initiate (see Figure 7-16). It appears that nature has used a similar protein assembly in two different cellular processes that require helicase activity.

The use of shared subunits in transcription and DNA repair may help explain the observation that DNA damage in higher eukaryotes is repaired at a much faster rate in regions of the genome being actively transcribed than in nontranscribed regions—so-called transcription-coupled repair. Since only a small fraction of the genome is transcribed in any one cell in higher eukaryotes, transcription-coupled repair efficiently directs repair efforts to the most critical regions. In this system, if an RNA polymerase becomes stalled at a lesion on DNA (e.g., a thymine-thymine dimer), a small protein, CSB, is recruited to the RNA polymerase; this triggers opening of the DNA helix at that point, recruitment of TFIIH, and the reactions of steps **2** through **4** depicted in Figure 4-39.

## Two Systems Utilize Recombination to Repair Double-Strand Breaks in DNA

Ionizing radiation (e.g., x- and γ-radiation) and some anticancer drugs cause double-strand breaks in DNA. These are particularly severe lesions because incorrect rejoining of double strands of DNA can lead to gross chromosomal rearrangements that can affect the functioning of genes. For example, incorrect joining could create a "hybrid" gene that codes for the N-terminal portion of one amino acid sequence fused to the C-terminal portion of a completely different amino acid sequence; or a chromosomal rearrangement could bring the promoter of one gene into close proximity to the coding region of another gene, changing the level or cell type in which that gene is expressed.

Two systems have evolved to repair double-strand breaks: *homologous recombination*, discussed in the next section, and *nonhomologous end-joining (NHEJ)*, which is error-prone, since several nucleotides are invariably lost at the point of repair.

**Error-Prone Repair by Nonhomologous End-Joining** The predominant mechanism for repairing double-strand breaks in multicellular organisms involves rejoining the nonhomologous ends of two DNA molecules. Even if the joined DNA fragments come from the same chromosome, the repair process results in loss of several base pairs at the joining point (Figure 4-40). Formation of such a possibly mutagenic deletion is one example of how repair of DNA damage can introduce mutations.

Since movement of DNA within the protein-dense nucleus is fairly minimal, the correct ends are generally rejoined together, albeit with loss of base pairs. Occasionally, however, broken ends from different chromosomes are joined together,

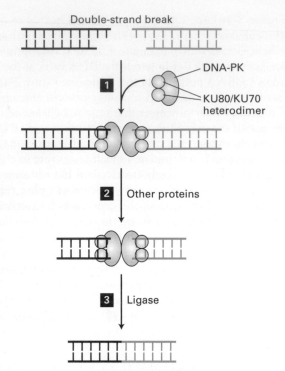

**FIGURE 4-40 Nonhomologous end-joining.** When sister chromatids are not available to help repair double-strand breaks, nucleotide sequences are butted together that were not apposed in the unbroken DNA. These DNA ends are usually from the same chromosome locus, and when linked together, several base pairs are lost. Occasionally, ends from different chromosomes are accidentally joined together. A complex of two proteins, Ku and DNA-dependent protein kinase (DNA-PK), binds to the ends of a double-strand break (step **1**). After formation of a synapse, the ends are further processed by nucleases, resulting in removal of a few bases (step **2**), and the two double-stranded molecules are ligated together (step **3**). As a result, the double-strand break is repaired, but several base pairs at the site of the break are removed. [Adapted from G. Chu, 1997, *J. Biol. Chem.* **272:**24097; M. Lieber et al., 1997, *Curr. Opin. Genet. Devel.* **7:**99; and D. van Gant et al., 2001, *Nature Rev. Genet.* **2:**196.]

leading to translocation of pieces of DNA from one chromosome to another. Such translocations may generate chimeric genes that can have drastic effects on normal cell function, such as uncontrollable cell growth, which is the hallmark of cancer (see Figure 6-42). The devastating effects of double-strand breaks make these the "most unkindest cuts of all," to borrow a phrase from Shakespeare's *Julius Caesar*.

## Homologous Recombination Can Repair DNA Damage and Generate Genetic Diversity

At one time homologous recombination was thought to be a minor repair process in human cells. This changed when it was realized that several human cancers are potentiated by inherited mutations in genes essential for homologous recombination repair (see Table 25-1). For example, some women with an inherited susceptibility to breast cancer have

a mutation in one allele of either the *BRCA-1* or the *BRCA-2* genes that encode proteins participating in this repair process. Loss or inactivation of the second allele inhibits the homologous recombination repair pathway and thus tends to induce cancer in mammary or ovarian epithelial cells. Yeasts can repair double-strand breaks induced by γ-irradiation. Isolation and analysis of radiation-sensitive (*RAD*) mutants that are deficient in this repair system facilitated study of the process. Virtually all the yeast Rad proteins have homologs in the human genome, and the human and yeast proteins function in an essentially identical fashion.

A variety of DNA lesions that are not repaired by mechanisms discussed earlier can be repaired by mechanisms in which the damaged sequence is copied from an undamaged copy of the same or highly homologous DNA sequence on the homologous chromosome of diploid organisms or the sister chromosome following DNA replication in haploid and diploid organisms. These mechanisms involve an exchange of strands between separate DNA molecules and hence are referred to as **DNA recombination**.

In addition to providing a mechanism for DNA repair, similar recombination mechanisms generate genetic diversity among the individuals of a species by causing the exchange of large regions of chromosomes between the maternal and paternal pair of homologous chromosomes during *meiosis*, the special type of cellular division that generates germ cells (sperm and eggs) (Figure 5-3). In fact, the exchange of regions of homologous chromosomes, called *crossing-over*, is required for proper segregation of chromosomes during the first meiotic cell division. Meiosis and the consequences of generating new combinations of maternal and paternal genes on one chromosome by recombination are discussed further in Chapter 5. The mechanisms leading to proper segregation of chromosomes during meiosis are discussed in Chapter 20. Here we will focus on the molecular mechanisms of DNA recombination, highlighting the exchange of DNA strands between two recombining DNA molecules.

**Repair of a Collapsed Replication Fork** An example of recombinational DNA repair is the repair of a "collapsed" replication fork. If a break in the phosphodiester backbone of one DNA strand is not repaired before a replication fork passes, the replicated portions of the daughter chromosomes become separated when the replication helicase reaches the break in the parental DNA strand because there are no covalent bonds between the two fragments of the parental strand on either side of the nick. This process is called *replication fork collapse* (Figure 4-41, step **1**). If it is not repaired, it is generally lethal to at least one daughter cell following cell division, because of the loss of genetic information between the break and the end of the chromosome. The recombination process that repairs the resulting double-stranded break and regenerates a replication fork involves multiple enzymes and other proteins, only some of which are mentioned here.

The first step in the repair of the double-strand break is exonucleolytic digestion of the strand with its 5′ end at the

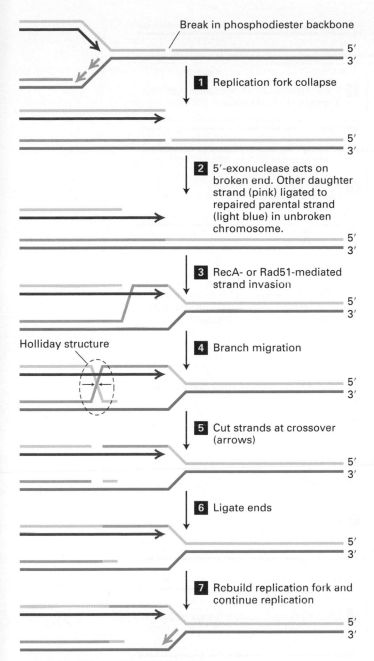

**Break in phosphodiester backbone**

5′
3′

**1** Replication fork collapse

5′
3′

**2** 5′-exonuclease acts on broken end. Other daughter strand (pink) ligated to repaired parental strand (light blue) in unbroken chromosome.

5′
3′

**3** RecA- or Rad51-mediated strand invasion

5′
3′

Holliday structure

**4** Branch migration

5′
3′

**5** Cut strands at crossover (arrows)

5′
3′

**6** Ligate ends

5′
3′

**7** Rebuild replication fork and continue replication

5′
3′

**FIGURE 4-41 Recombinational repair of a collapsed replication fork.** Parental strands are light and dark blue. The leading daughter strand is dark red, and the lagging daughter strand pink. Diagonal lines in step **3** and beyond represent a single phosphodiester bond from the DNA strand of the corresponding color. Small black arrows following step **4** represent cleavage of the phosphodiester bonds at the crossover of DNA strands in the Holliday structure. See http:// www.sheffield.ac.uk/mbb/ruva for an animation of branch migration catalyzed by *E. coli* proteins RuvA and RuvB. See the text for a discussion. [Adapted from D. L. Nelson and M. M. Cox, 2005, *Lehninger Principles of Biochemistry*, 4th ed., W. H. Freeman and Company.]

broken end of DNA, leaving the strand with a 3′ end at the break single-stranded (Figure 4-41, step **2**). The lagging nascent strand (pink) base paired to the unbroken parent strand (dark blue) is ligated to the unreplicated portion of the parent chromosome (light blue), as shown in Figure 4-41, step **2**. A critical protein required for the next step is RecA in bacteria, or the homologous Rad51 in *S. cerevisiae* and other eukaryotes. Multiple RecA/Rad51 molecules bind to the single-stranded DNA and catalyze its hybridization to a perfectly or nearly perfectly complementary sequence in another, homologous, double-stranded DNA molecule. The complementary strand of this target double-stranded DNA (dark blue) is displaced as a single-stranded loop of DNA over the region of hybridization to the invading strand (Figure 4-41, step **3**). This RecA/Rad51 catalyzed *invasion* of a duplex DNA by a single-stranded complement of one of the strands is key to the recombination process. Since no base pairs are lost or gained in this process, called **strand invasion**, it does not require an input of energy.

Next, the hybrid region between target DNA and the invading strand is extended in the direction away from the break by proteins that use energy from ATP hydrolysis. This process is called *branch migration* (Figure 4-41, step **4**) because the position where the target DNA strand crosses from one complementary strand (dark blue) to its complement in the broken DNA molecule (dark red), i.e. the pink diagonal line after step **3**, is called a *branch* in the DNA structure. In this diagram, the diagonal lines represent only one phosphodiester bond. Molecular modeling and other studies show that the first base on either side of the branch is base-paired to a complementary nucleotide. As this branch *migrates* to the left, the number of base pairs remains constant; one new base pair formed with the (red) invading strand is matched by the loss of one base pair with the parental (blue) strand.

When the region of hybrid extends beyond the 5′ end of the broken strand (light blue), the single-stranded parental DNA strand generated (light blue) base pairs with the complementary region of the other parental strand (dark blue) that becomes single-stranded as the branch migrates to the left (Figure 4-41, step **4**). The resulting structure is called a **Holliday structure**, after Robin Holliday, the geneticist who first proposed it as an intermediate in genetic recombination. Again, the diagonal lines in the diagram following step **4** represent single phosphodiester bonds, and all bases in the Holliday structure are base-paired to complementary bases in the parental strands. Cleavage of the phosphodiester bonds that *cross over* from one parental strand to the other (step **5**) and ligation of the 5′ and 3′ ends base paired to the same parental strands (step **6**) result in the generation of a structure similar to a replication fork. Rebinding of replication fork proteins results in extension of the leading strand past the point of the original strand break and re-initiation of lagging-strand synthesis (step **7**), thus regenerating a replication fork. The overall process allows the ligated upper strand in the lower molecule following step **2** to serve as template for extension of the leading strand in step **7**.

**Double-Stranded DNA Break Repair by Homologous Recombination** A similar mechanism called **homologous recombination** can repair a double-strand break in a chromosome and can also exchange large segments of two double-stranded

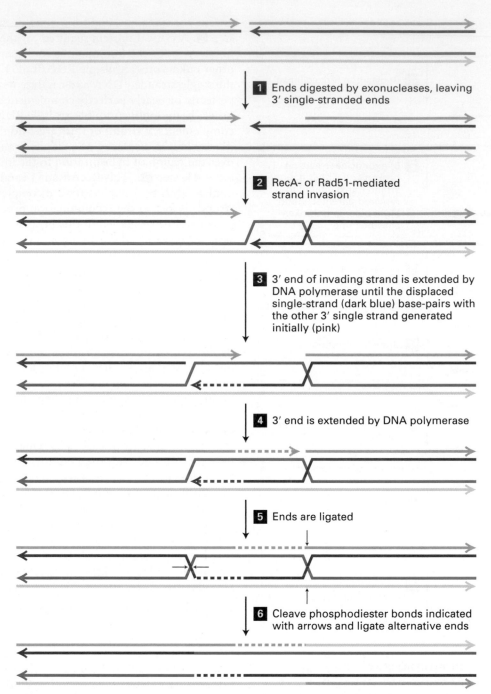

**FIGURE 4-42 Double-strand DNA break repair by homologous recombination.** For simplicity, each DNA double helix is represented by two parallel lines with the polarities of the strands indicated by arrowheads at their 3′ ends. The upper molecule has a double-strand break.

Note that in the diagram of the upper DNA molecule the strand with its 3′ end at the right is on the top, while in the diagram of the lower DNA molecule this strand is drawn on the bottom. See the text for discussion. [Adapted from T. L. Orr-Weaver and J. W. Szostak, 1985, *Microbiol. Rev.* **49:**33.]

DNA molecules (Figure 4-42). First, the broken ends of the DNA molecule are resected by exonucleases that leave a single-stranded region of DNA with a 3′ end (step **1**). RecA in bacteria and Rad51 in eukaryotes then catalyzes strand invasion of one of these 3′ ends into the homologous region of the homologous chromosome as discussed above for repair of a collapsed replication fork (step **2**). The 3′ end of the invading DNA strand is then extended by a DNA polymerase,

displacing the parental strand as an enlarging single-stranded loop of DNA (dark blue) (step **3**). When the loop extends to a sequence that is complementary to the other broken end of DNA (the fragment on the left following step **1**), the complementary sequences base-pair (diagram following step **3**). This 3′ end is then extended by a DNA polymerase using the displaced single-stranded loop of parental DNA (dark blue) as template (step **4**).

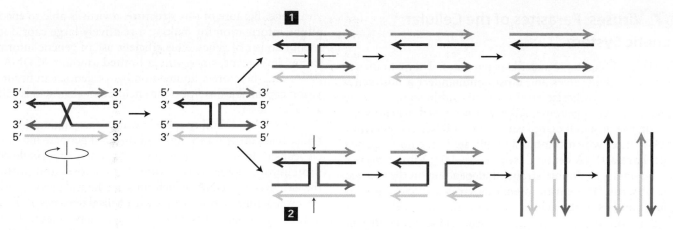

**FIGURE 4-43 Alternative resolution of a Holliday structure.** Diagonal and vertical lines represent a single phosphodiester bond. It is simplest to diagram the process by rotating the diagram of the bottom molecule 180° so that the top and bottom molecules have the same strand orientations. Cutting the bonds as shown in **1** and ligating the ends as indicated regenerates the original chromosomes. Cutting the strands as shown in **2** and religating as shown at the bottom generates recombinant chromosomes. See http://engels. genetics.wisc.edu/Holliday/holliday3D.html for a three-dimensional animation of the Holliday structure and its resolution.

Next, the new 3′ ends are ligated (step **5**) to the exonuclease-digested 5′ ends. This generates two Holliday structures in the paired molecules (step **5**). Branch migration of these Holliday structures can occur in either direction (not diagrammed). Finally, cleavage of the strands at the positions shown by the arrows, and ligation of the alternative 5′ and 3′ ends at each cleaved Holliday structure generates two *recombinant* chromosomes that contain the DNA of one *parental* DNA molecule on one side of the break point (pink and red strands), and the DNA of the other parental DNA molecule on the other side of the initial break point (light and dark blue) (step **6**). The region in the immediate vicinity of the initial break point forms a *heteroduplex,* in which one strand from one parent is base-paired to the complementary strand of the other parent (pink or red strand base-paired to the light or dark blue strand). Base-pair mismatches between the two parental strands are usually repaired by repair mechanisms discussed above to generate a complementary base pair. In the process, sequence differences between the two parents are lost, a process referred to as **gene conversion**.

Figure 4-43 diagrams how cleavage of one or the other pair of strands at the four-way strand junction in the Holliday structure generates parental or recombinant molecules. This process, called *resolution* of the Holiday structure, separates DNA molecules initially joined by RecA/Rad51-catalyzed strand invasion. Each Holliday structure in the intermediate following Figure 4-42, step **5**, can be cleaved and religated in the two possible ways shown by the two sets of small black arrows in Figure 4-43. Consequently, there are four possible products of the recombination process shown in Figure 4-42. Two of these regenerate the parental chromosomes [with the exception of the heteroduplex region at the break point that is repaired into the sequence of one parent or the other (gene conversion)]. The other two possible products generate recombinant chromsomes as shown in Figure 4-42.

## KEY CONCEPTS of Section 4.6

### DNA Repair and Recombination

- Changes in the DNA sequence result from copying errors and the effects of various physical and chemical agents.

- Many copying errors that occur during DNA replication are corrected by the proofreading function of DNA polymerases that can recognize incorrect (mispaired) bases at the 3′ end of the growing strand and then remove them by an inherent 3′→5′ exonuclease activity (see Figure 4-34).

- Eukaryotic cells have three excision-repair systems for correcting mispaired bases and for removing UV-induced thymine-thymine dimers or large chemical adducts from DNA. Base excision repair, mismatch repair, and nucleotide excision repair operate with high accuracy and generally do not introduce errors.

- Repair of double-strand breaks by the nonhomologous end-joining pathway can link segments of DNA from different chromosomes, possibly forming an oncogenic translocation. The repair mechanism also produces a small deletion, even when segments from the same chromosome are joined.

- Error-free repair of double-strand breaks in DNA is accomplished by homologous recombination using the undamaged sister chromatid as template.

- Inherited defects in the nucleotide excision-repair pathway, as in individuals with xeroderma pigmentosum, predispose them to skin cancer. Inherited colon cancer frequently is associated with mutant forms of proteins essential for the mismatch repair pathway. Defects in repair by homologous recombination are associated with inheritance of one mutant allele of the *BRCA-1* or *BRCA-2* gene and result in predisposition to breast and uterine cancer.

## 4.7 Viruses: Parasites of the Cellular Genetic System

**Viruses** are obligate, intracellular parasites. They cannot reproduce by themselves and must commandeer a host cell's machinery to synthesize viral proteins and in some cases to replicate the viral genome. RNA viruses, which usually replicate in the host-cell cytoplasm, have an RNA genome, and DNA viruses, which commonly replicate in the host-cell nucleus, have a DNA genome (see Figure 4-1). Viral genomes may be single- or double-stranded, depending on the specific type of virus. The entire infectious virus particle, called a **virion,** consists of the nucleic acid and an outer shell of protein that both protects the viral nucleic acid and functions in the process of host-cell infection. The simplest viruses contain only enough RNA or DNA to encode four proteins; the most complex can encode ≈200 proteins. In addition to their obvious importance as causes of disease, viruses are extremely useful as research tools in the study of basic biological processes, such as those discussed in this chapter.

### Most Viral Host Ranges Are Narrow

The surface of a virion contains many copies of one type of protein that binds specifically to multiple copies of a receptor protein on a host cell. This interaction determines the *host range*—the group of cell types that a virus can infect—and begins the infection process. Most viruses have a rather limited host range.

A virus that infects only bacteria is called a **bacteriophage,** or simply a *phage*. Viruses that infect animal or plant cells are referred to generally as animal viruses or plant viruses. A few viruses can grow in both plants or animals and the insects that feed on them. The highly mobile insects serve as vectors for transferring such viruses between susceptible animal or plant hosts. Wide host ranges are also characteristic of some strictly animal viruses, such as vesicular stomatitis virus, which grows in insect vectors and in many different types of mammals. Most animal viruses, however, do not cross phyla, and some (e.g., poliovirus) infect only closely related species such as primates. The host-cell range of some animal viruses is further restricted to a limited number of cell types because only these cells have appropriate surface receptors to which the virions can attach. One example is poliovirus, which only infects cells in the intestine and, unfortunately, motor neurons in the spinal chord, causing paralysis. Another is HIV-1, discussed further below, which infects cells essential for the immune response called CD4$^+$ T-lymphocytes, causing AIDS (see Chapter 23), and certain neurons and other cells of the central nervous system called glial cells.

### Viral Capsids Are Regular Arrays of One or a Few Types of Protein

The nucleic acid of a virion is enclosed within a protein coat, or **capsid,** composed of multiple copies of one protein or a few different proteins, each of which is encoded by a single viral gene. Because of this structure, a virus is able to encode all the information for making a relatively large capsid in a small number of genes. This efficient use of genetic information is important, since only a limited amount of DNA or RNA, and therefore a limited number of genes, can fit into a virion capsid. A capsid plus the enclosed nucleic acid is called a **nucleocapsid.**

Nature has found two basic ways of arranging the multiple capsid protein subunits and the viral genome into a nucleocapsid. In some viruses, multiple copies of a single coat protein form a *helical* structure that encloses and protects the viral RNA or DNA, which runs in a helical groove within the protein tube. Viruses with such a helical nucleocapsid, such as tobacco mosaic virus, have a rodlike shape (Figure 4-44a). The other major structural type is based on the *icosahedron,* a solid, approximately spherical object built of 20 identical faces, each of which is an equilateral triangle (Figure 4-44b). During infection, some icosahedral viruses interact with host cell-surface receptors via clefts in between the capsid subunits, others interact via long fiberlike proteins extending from the nucleocapsid.

In many DNA bacteriophages, the viral DNA is located within an icosahedral "head" that is attached to a rodlike "tail." During infection, viral proteins at the tip of the tail bind to host-cell receptors, and then the viral DNA passes down the tail into the cytoplasm of the host cell (Figure 4-44c).

In some viruses, the symmetrically arranged nucleocapsid is covered by an external membrane, or **envelope,** which consists mainly of a phospholipid bilayer but also contains one or two types of virus-encoded glycoproteins (Figure 4-44d). The phospholipids in the viral envelope are similar to those in the plasma membrane of an infected host cell. The viral envelope is, in fact, derived by budding from that membrane, but contains mainly viral glycoproteins, as we will discuss shortly.

### Viruses Can Be Cloned and Counted in Plaque Assays

The number of infectious viral particles in a sample can be quantified by a **plaque assay.** This assay is performed by culturing a dilute sample of viral particles on a plate covered with host cells and then counting the number of local lesions, called *plaques,* that develop (Figure 4-45). A plaque develops on the plate wherever a single virion initially infects a single cell. The virus replicates in this initial host cell and then lyses (ruptures) the cell, releasing many progeny virions that infect the neighboring cells on the plate. After a few such cycles of infection, enough cells are lysed to produce a visible clear area, or plaque, in the layer of remaining uninfected cells.

Since all the progeny virions in a plaque are derived from a single parental virus, they constitute a virus **clone.** This type of plaque assay is in standard use for bacterial and animal viruses. Plant viruses can be assayed similarly by counting local lesions on plant leaves inoculated with viruses. Analysis of viral mutants, which are commonly isolated by

(a)

50 nm

Tobacco mosaic virus

(b) 10 nm

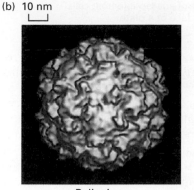

Poliovirus

(c)  50 nm

Bacteriophage T4

(d)  50 nm

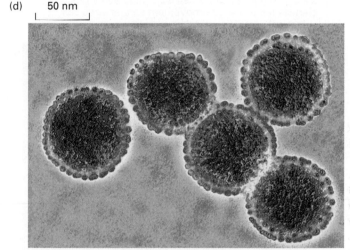

Avian influenza virus

**FIGURE 4-44 Virion structures.** (a) Helical tobacco mosaic virus. (b) Small icosahedral virus. An icosahedron is composed of 20 equilateral triangular faces. The example shown is poliovirus. In poliovirus, each face is built from three capsomeres, outlined in red. The numbers show how five capsomeres associate at the 12 vertices of the icosahedron. (c) Bacteriophage T4. (d) Influenza virus, an example of an enveloped virus. [Part (a): O. Bradfute, Peter Arnold/Science Photo Library; Part (b) courtesy of T. S. Baker; Part (c): Dept. of Microbiology, Biozentrum/Science Photo Library; Part (d) : James Cavallini/Photo Researchers, Inc.]

plaque assays, has contributed extensively to our current understanding of molecular cellular processes.

## Lytic Viral Growth Cycles Lead to Death of Host Cells

Although details vary among different types of viruses, those that exhibit a *lytic cycle* of growth proceed through the following general stages:

1. *Adsorption*—Virion interacts with a host cell by binding of multiple copies of capsid protein to specific receptors on the cell surface.

2. *Penetration*—Viral genome crosses the plasma membrane. For some viruses, viral proteins packaged inside the capsid also enter the host cell.

3. *Replication*—Viral mRNAs are produced with the aid of the host-cell transcription machinery (DNA viruses) or by viral enzymes (RNA viruses). For both types of viruses, viral mRNAs are translated by the host-cell translation machinery. Production of multiple copies of the viral genome is carried out either by viral proteins alone or with the help of host-cell proteins.

4. *Assembly*—Viral proteins and replicated genomes associate to form progeny virions.

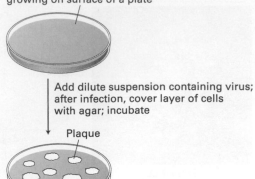

(a) Confluent layer of susceptible host cells growing on surface of a plate

Add dilute suspension containing virus; after infection, cover layer of cells with agar; incubate

Plaque

Each plaque represents cell lysis initiated by one viral particle (agar restricts movement so that virus can infect only contiguous cells)

(b)

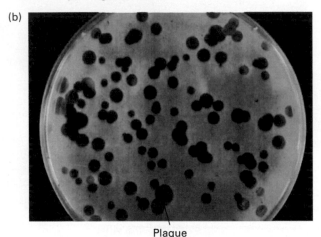

Plaque

**EXPERIMENTAL FIGURE 4-45 Plaque assay.** The plaque assay determines the number of infectious particles in a viral suspension. (a) Each lesion, or plaque, which develops where a single virion initially infected a single cell, constitutes a pure viral clone. (b) Plaques on a lawn of *Pseudomonas fluorescens* bacteria made by bacteriophage ΦS1. [Part (b) Courtesy of Dr. Pierre ROSSI, Ecole Polytechnique fédérale de Lausanne (LBE–EPFL).]

5. *Release*—Infected cell either ruptures suddenly (**lysis**), releasing all the newly formed virions at once, or disintegrates gradually, with slow release of virions. Both cases lead to the death of the infected cell.

Figure 4-46 illustrates the lytic cycle for T4 bacteriophage, a nonenveloped DNA virus that infects *E. coli*. Viral capsid proteins generally are made in large amounts because many copies of them are required for the assembly of each progeny virion. In each infected cell, about 100–200 T4 progeny virions are produced and released by lysis.

The lytic cycle is somewhat more complicated for DNA viruses that infect eukaryotic cells. In most such viruses, the DNA genome is transported (with some associated proteins) into the cell nucleus. Once inside the nucleus, the viral DNA is transcribed into RNA by the host's transcription machinery.

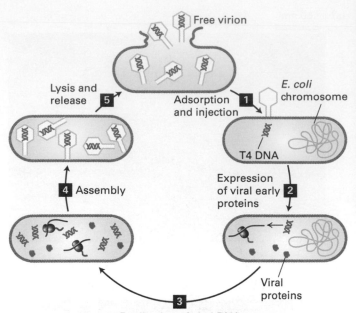

**FIGURE 4-46 Lytic replication cycle of a nonenveloped, bacterial virus.** *E. coli* bacteriophage T4 has a double-stranded DNA genome and lacks a membrane envelope. After viral coat proteins at the tip of the tail in T4 interact with specific receptor proteins on the exterior of the host cell, the viral genome is injected into the host (step **1**). Host-cell enzymes then transcribe viral "early" genes into mRNAs and subsequently translate these into viral "early" proteins (step **2**). The early proteins replicate the viral DNA and induce expression of viral "late" proteins by host-cell enzymes (step **3**). The viral late proteins include capsid and assembly proteins and enzymes that degrade the host-cell DNA, supplying nucleotides for synthesis of more viral DNA. Progeny virions are assembled in the cell (step **4**) and released (step **5**) when viral proteins lyse the cell. Newly liberated viruses initiate another cycle of infection in other host cells.

Processing of the viral RNA primary transcript by host-cell enzymes yields viral mRNA, which is transported to the cytoplasm and translated into viral proteins by host-cell ribosomes, tRNA, and translation factors. The viral proteins are then transported back into the nucleus, where some of them either replicate the viral DNA directly or direct cellular proteins to replicate the viral DNA, as in the case of SV40 discussed in an earlier section. Assembly of the capsid proteins with the newly replicated viral DNA occurs in the nucleus, yielding thousands to hundreds of thousands of progeny virions.

Most plant and animal viruses with an RNA genome do not require nuclear functions for lytic replication. In some of these viruses, a virus-encoded enzyme that enters the host during penetration transcribes the genomic RNA into mRNAs in the cell cytoplasm. The mRNA is directly translated into viral proteins by the host-cell translation machinery. One or more of these proteins then produces additional copies of the viral RNA genome. Finally, progeny genomes are assembled with newly synthesized capsid proteins into progeny virions in the cytoplasm.

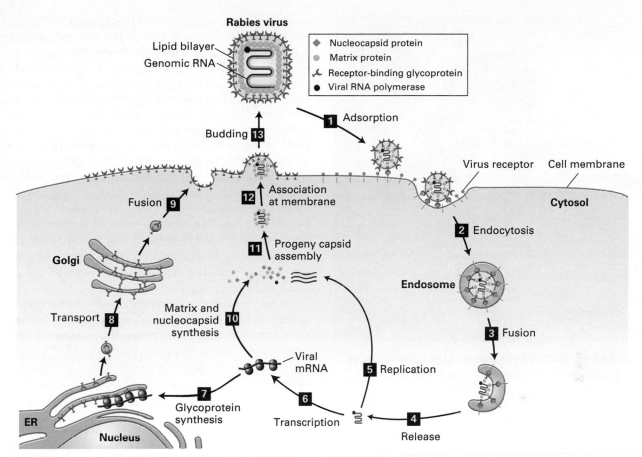

**FIGURE 4-47 Lytic replication cycle of an enveloped animal virus.** Rabies virus is an enveloped virus with a single-stranded RNA genome. The structural components of this virus are depicted at the top. After a virion adsorbs to multiple copies of a specific host membrane protein (step **1**), the cell engulfs it in an endosome (step **2**). A cellular protein in the endosome membrane pumps H$^+$ ions from the cytosol into the endosome interior. The resulting decrease in endosomal pH induces a conformational change in the viral glycoprotein, leading to fusion of the viral envelope with the endosomal lipid bilayer membrane and release of the nucleocapsid into the cytosol (steps **3** and **4**). Viral RNA polymerase uses ribonucleoside triphosphates in the cytosol to replicate the viral RNA genome (step **5**) and to synthesize viral mRNAs (step **6**). One of the viral mRNAs encodes the viral transmembrane glycoprotein, which is inserted into the membrane of the endoplasmic reticulum (ER) as it is synthesized on ER-bound ribosomes (step **7**). Carbohydrate is added to the large folded domain inside the ER lumen and is modified as the membrane and the associated glycoprotein pass through the Golgi apparatus (step **8**). Vesicles with mature glycoprotein fuse with the host plasma membrane, depositing viral glycoprotein on the cell surface with the large receptor-binding domain outside the cell (step **9**). Meanwhile, other viral mRNAs are translated on host-cell ribosomes into nucleocapsid protein, matrix protein, and viral RNA polymerase (step **10**). These proteins are assembled with replicated viral genomic RNA (bright red) into progeny nucleocapsids (step **11**), which then associate with the cytosolic domain of viral transmembrane glycoproteins in the plasma membrane (step **12**). The plasma membrane is folded around the nucleocapsid, forming a "bud" that eventually is released (step **13**).

After the synthesis of hundreds to hundreds of thousands of new virions has been completed, depending on the type of virus and host cell, most infected bacterial cells and some infected plant and animal cells are lysed, releasing all the virions at once. In many plant and animal viral infections, however, no discrete lytic event occurs; rather, the dead host cell releases the virions as it gradually disintegrates.

As noted previously, enveloped animal viruses are surrounded by an outer phospholipid bilayer derived from the plasma membrane of host cells and containing abundant viral glycoproteins. The processes of adsorption and release of enveloped viruses differ substantially from these processes for nonenveloped viruses. To illustrate lytic replication of enveloped viruses, we consider the rabies virus, whose nucleocapsid consists of a single-stranded RNA genome surrounded by multiple copies of nucleocapsid protein. Like other lytic RNA viruses, rabies virions are replicated in the cytoplasm and do not require host-cell nuclear enzymes. As shown in Figure 4-47, a rabies virion is adsorbed by endocytosis, and release of progeny virions occurs by *budding* from the host-cell plasma membrane. Budding virions are clearly visible in electron micrographs of infected cells, as illustrated in Figure 4-48. Many tens of thousands of progeny virions bud from an infected host cell before it dies.

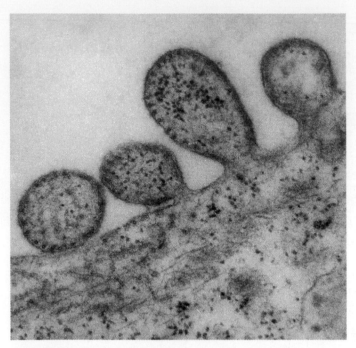

**EXPERIMENTAL FIGURE 4-48 Release of progeny virions by budding.** Progeny virions of enveloped viruses are released by budding from infected cells. In this transmission electron micrograph of a cell infected with measles virus, virion buds are clearly visible protruding from the cell surface. Measles virus is an enveloped RNA virus with a helical nucleocapsid, like rabies virus, and replicates as illustrated in Figure 4-47. [From A. Levine, 1991, *Viruses,* Scientific American Library, p. 22.]

## Viral DNA Is Integrated into the Host-Cell Genome in Some Nonlytic Viral Growth Cycles

Some bacterial viruses, called *temperate phages,* can establish a nonlytic association with their host cells that does not kill the cell. For example, when bacteriophage λ infects *E. coli,* most of the time it causes a lytic infection. Occasionally, however, the viral DNA is integrated into the host-cell chromosome rather than being replicated. The integrated viral DNA, called a *prophage,* is replicated as part of the cell's DNA from one host-cell generation to the next. This phenomenon is referred to as **lysogeny.** If the host cell suffers extensive damage to its DNA from ultraviolet light, the prophage DNA is activated, leading to its excision from the host-cell chromosome, entrance into the lytic cycle, and subsequent production and release of progeny virions before the host cell dies.

The genomes of a number of animal viruses also can integrate into the host-cell genome. Among the most important are the **retroviruses,** which are enveloped viruses with a genome consisting of two identical strands of RNA. These viruses are known as *retroviruses* because their RNA genome acts as a template for formation of a DNA molecule— the opposite flow of genetic information compared with the more common transcription of DNA into RNA. In the retroviral life cycle (Figure 4-49), a viral enzyme called **reverse transcriptase** initially copies the viral RNA genome into single-stranded DNA complementary to the virion RNA; the same enzyme then catalyzes synthesis of a complementary DNA strand. (This complex reaction is detailed in Chapter 6 when we consider closely related intracellular parasites called retrotransposons.) The resulting double-stranded DNA is integrated into the chromosomal DNA of the infected cell. Finally, the integrated DNA, called a *provirus,* is transcribed by the cell's own machinery into RNA, which either is translated into viral proteins or is packaged within virion coat proteins to form progeny virions that are released by budding from the host-cell membrane. Because most retroviruses do not kill their host cells, infected cells can replicate, producing daughter cells with integrated proviral DNA. These daughter cells continue to transcribe the proviral DNA and bud progeny virions.

Some retroviruses contain cancer-causing genes (**oncogenes**), and cells infected by such retroviruses are oncogenically transformed into tumor cells. Studies of oncogenic retroviruses (mostly viruses of birds and mice) have revealed a great deal about the processes that lead to transformation of a normal cell into a cancer cell (Chapter 24).

Among the known human retroviruses are human T-cell lymphotrophic virus (HTLV), which causes a form of leukemia, and human immunodeficiency virus (HIV-1), which causes acquired immune deficiency syndrome (AIDS). Both of these viruses can infect only specific cell types, primarily certain cells of the immune system and, in the case of HIV-1, some central nervous system neurons and glial cells. Only these cells have cell-surface receptors that interact with viral envelope proteins, accounting for the host-cell specificity of these viruses. Unlike most other retroviruses, HIV-1 eventually kills its host cells. The eventual death of large numbers of immune-system cells results in the defective immune response characteristic of AIDS.

Some DNA viruses also can integrate into a host-cell chromosome. One example is the human papillomaviruses (HPVs), which most commonly cause warts and other benign skin lesions. The genomes of certain HPV serotypes, however, occasionally integrate into the chromosomal DNA of infected cervical epithelial cells, initiating development of cervical cancer. Routine Pap smears can detect cells in the early stages of the transformation process initiated by HPV integration, permitting effective treatment before cancer develops. A vaccine for the types of HPV associated with cervical cancer has been developed and can protect against the initial infection by these viruses, and consequently, against development of cervical cancer. However, once an individual is infected with these HPVs, the "window of opportunity" is missed, and the vaccine does not protect against the development of cancer. Because the vaccine is not 100 percent effective, even vaccinated women should have regular Pap smears. ∎

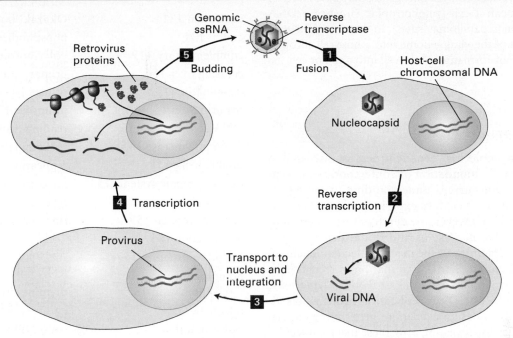

**FIGURE 4-49 Retroviral life cycle.** Retroviruses have a genome of two identical copies of single-stranded RNA and an outer envelope. Step **1**: After viral glycoproteins in the envelope interact with a specific host-cell membrane protein, the retroviral envelope fuses directly with the plasma membrane, allowing entry of the nucleocapsid into the cytoplasm of the cell. Step **2**: Viral reverse transcriptase and other proteins copy the viral ssRNA genome into a double-stranded DNA. Step **3**: The viral dsDNA is transported into the nucleus and integrated into one of many possible sites in the host-cell chromosomal DNA. For simplicity, only one host-cell chromosome is depicted. Step **4**: The integrated viral DNA (provirus) is transcribed by the host-cell RNA polymerase, generating mRNAs (dark red) and genomic RNA molecules (bright red). The host-cell machinery translates the viral mRNAs into glycoproteins and nucleocapsid proteins. Step **5**: Progeny virions then assemble and are released by budding as illustrated in Figure 4-47.

## KEY CONCEPTS of Section 4.7

### Viruses: Parasites of the Cellular Genetic System

• Viruses are small parasites that can replicate only in host cells. Viral genomes may be either DNA (DNA viruses) or RNA (RNA viruses) and either single- or double-stranded.

• The capsid, which surrounds the viral genome, is composed of multiple copies of one or a small number of virus-encoded proteins. Some viruses also have an outer envelope, which is similar to the plasma membrane but contains viral transmembrane proteins.

• Most animal and plant DNA viruses require host-cell nuclear enzymes to carry out transcription of the viral genome into mRNA and production of progeny genomes. In contrast, most RNA viruses encode enzymes that can transcribe the RNA genome into viral mRNA and produce new copies of the RNA genome.

• Host-cell ribosomes, tRNAs, and translation factors are used in the synthesis of all viral proteins in infected cells.

• Lytic viral infection entails adsorption, penetration, synthesis of viral proteins and progeny genomes (replication), assembly of progeny virions, and release of hundreds to thousands of virions, leading to death of the host cell (see Figure 4-46). Release of enveloped viruses occurs by budding through the host-cell plasma membrane (see Figure 4-47).

• Nonlytic infection occurs when the viral genome is integrated into the host-cell DNA and generally does not lead to cell death.

• Retroviruses are enveloped animal viruses containing a single-stranded RNA genome. After a host cell is penetrated, reverse transcriptase, a viral enzyme carried in the virion, converts the viral RNA genome into double-stranded DNA, which integrates into chromosomal DNA (see Figure 4-49).

• Unlike infection by other retroviruses, HIV infection eventually kills host cells, causing the defects in the immune response characteristic of AIDS.

• Tumor viruses, which contain oncogenes, may have an RNA genome (e.g., human T-cell lymphotrophic virus) or a DNA genome (e.g., human papillomaviruses). In the case of these viruses, integration of the viral genome into a host-cell chromosome can cause transformation of the cell into a tumor cell.

## Perspectives for the Future

The basic cellular molecular genetic processes discussed in this chapter form the foundation of contemporary molecular cell biology. Our current understanding of these processes is grounded in a wealth of experimental results and is not likely to change. However, the depth of our understanding will continue to increase as additional details of the structures and interactions of the macromolecular machines involved are uncovered. The determination in recent years of the three-dimensional structures of RNA polymerases, ribosomal subunits, and DNA replication proteins has allowed researchers to design ever more penetrating experimental approaches for revealing how these macromolecules operate at the molecular level. The detailed level of understanding currently being developed may allow the design of new and more effective drugs for treating illnesses of humans, crops, and livestock. For example, the recent high-resolution structures of ribosomes are providing insights into the mechanism by which antibiotics inhibit bacterial protein synthesis without affecting the function of mammalian ribosomes. This new knowledge may allow the design of even more effective antibiotics. Similarly, detailed understanding of the mechanisms regulating transcription of specific human genes may lead to therapeutic strategies that can reduce or prevent inappropriate immune responses that lead to multiple sclerosis and arthritis, the inappropriate cell division that is the hallmark of cancer, and other pathological processes.

Much of current biological research is focused on discovering how molecular interactions endow cells with decision-making capacity and their special properties. For this reason, several of the following chapters describe current knowledge about how such interactions regulate transcription and protein synthesis in multicellular organisms and how such regulation endows cells with the capacity to perform their specialized functions. Other chapters deal with how protein-protein interactions underlie the construction of specialized organelles in cells, and how they determine cell shape and movement. The rapid advances in molecular cell biology in recent years hold promise that in the not too distant future we will understand more deeply how the regulation of specialized cell function, shape, and mobility, coupled with regulated cell replication and cell death (apoptosis), lead to the growth of complex organisms such as flowering plants and human beings.

## Key Terms

| | |
|---|---|
| anticodon 131 | Okazaki fragments 147 |
| codon 131 | phosphodiester bond 118 |
| complementary 119 | polyribosome 142 |
| crossing over 156 | primary transcript 127 |
| deamination 152 | primer 145 |
| depurination 153 | promoter 124 |
| DNA end-joining 155 | reading frame 132 |
| DNA polymerase 145 | recombination 116 |
| double helix 118 | replication fork 145 |
| excision-repair systems 153 | retroviruses 164 |
| exon 127 | reverse transcriptase 164 |
| gene conversion 159 | ribosomal RNA (rRNA) 116 |
| genetic code 131 | ribosome 131 |
| Holliday structure 157 | RNA polymerase 124 |
| homologous recombination 157 | thymine-thymine dimer 154 |
| intron 127 | transcription 116 |
| isoform 130 | transfer RNA (tRNA) 116 |
| lagging strand 146 | translation 116 |
| leading strand 146 | viral envelope 160 |
| messenger RNA (mRNA) 116 | Watson-Crick base pairs 118 |
| mutation 151 | |

## Review the Concepts

1. What are Watson-Crick base pairs? Why are they important?

2. Preparing plasmid (double-stranded, circular) DNA for sequencing involves annealing a complementary, short, single-stranded oligonucleotide DNA primer to one strand of the plasmid template. This is routinely accomplished by heating the plasmid DNA and primer to 90 °C and then slowly bringing the temperature down to 25 °C. Why does this protocol work?

3. What difference between RNA and DNA helps to explain the greater stability of DNA? What implications does this have for the function of DNA?

4. What are the major differences in the synthesis and structure of prokaryotic and eukaryotic mRNAs?

5. While investigating the function of a specific growth factor receptor gene from humans, researchers found that two types of proteins are synthesized from this gene. A larger protein containing a membrane-spanning domain functions to recognize growth factors at the cell surface, stimulating a specific downstream signaling pathway. In contrast, a related,

smaller protein is secreted from the cell and functions to bind available growth factor circulating in the blood, thus inhibiting the downstream signaling pathway. Speculate on how the cell synthesizes these disparate proteins.

**6.** The transcription of many bacterial genes relies on functional groups called *operons,* such as the tryptophan operon (Figure 4-13a). What is an operon? What advantages are there to having genes arranged in an operon, compared with the arrangement in eukaryotes?

**7.** How would a mutation in the poly(A)-binding protein gene affect translation? How would an electron micrograph of polyribosomes from such a mutant differ from the normal pattern?

**8.** What characteristic of DNA results in the requirement that some DNA synthesis is discontinuous? How are Okazaki fragments and DNA ligase utilized by the cell?

**9.** Eukaryotes have repair systems that prevent mutations due to copying errors and exposure to mutagens. What are the three excision-repair systems found in eukaryotes, and which one is responsible for correcting thymine-thymine dimers that form as a result of UV light damage to DNA?

**10.** DNA-repair systems are responsible for maintaining genomic fidelity in normal cells despite the high frequency with which mutational events occur. What type of DNA mutation is generated by (a) UV irradiation and (b) ionizing radiation? Describe the system responsible for repairing each of these types of mutations in mammalian cells. Postulate why a loss of function in one or more DNA-repair systems typifies many cancers.

**11.** What is the name given to the process that can repair DNA damage *and* generate genetic diversity? Briefly describe the similarities and differences of the two processes.

**12.** The genome of a retrovirus can integrate into the host-cell genome. What gene is unique to retroviruses, and why is the protein encoded by this gene absolutely necessary for maintaining the retroviral life cycle? A number of retroviruses can infect certain human cells. List two of them, briefly describe the medical implications resulting from these infections, and describe why only certain cells are infected.

**13. a.** Which of the following DNA strands, the top or bottom, would serve as a template for RNA transcription if the DNA molecule were to unwind in the indicated direction?

5'ACGGACTGTACCGCTGAAGTCATGGACGCTCGA 3'
3'TGCCTGACATGGCGACTTCAGTACCTGCGAGCT 5'

———————→

Direction of DNA unwinding

  **b.** What would be the resulting RNA sequence (written 5' to 3')?

**14.** Contrast prokaryotic and eukaryotic gene characteristics.

**15.** You have learned about the events surrounding DNA replication and the central dogma. Identify the steps associated with these processes that will be adversely affected in the following scenarios.

  **a.** Helicases unwind the DNA, but stabilizing proteins are mutated and cannot bind to the DNA.

  **b.** The mRNA molecule forms a hairpin loop on itself via complementary base pairing in an area spanning the AUG start site.

  **c.** The cell is unable to produce functional tRNA$_i^{met}$.

**16.** Use the key provided below to determine the amino acid sequence of the polypeptide produced from the following DNA sequence. Intron sequences are highlighted. Note: Not all amino acids in the key will be used.

5' TTCTAAACGCATGAAGCACCGTCTCAGAGCCAGTGA 3'
3' AAGATTTGCGTACTTCGTGGCAGAGTCTCGGTCACT 5'

———————→

Direction of DNA unwinding

Asn = AAU  Cys = TCG  Gly = CAG  His = CAU  Lys = AAG
Met = AUG  Phe = UUC  Ser = AGC  Tyr = UAC  Val = GUC; GUA

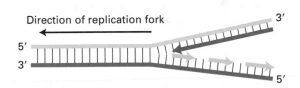

Direction of replication fork

**17. a.** Look at the figure above. Explain why it is necessary for Okazaki fragments to be formed as the lagging strand is produced (instead of a continuous strand).

  **b.** If the DNA polymerase in the figure above could only bind to the lower template strand, under what condition(s) would it be able to produce a leading strand?

**18.** The DNA repair systems preferentially target the newly synthesized strand. Why is this important?

**19.** Identify the specific types of point mutations below (you are viewing the direct DNA version of the RNA sequence).

Original sequence: 5' AUG TCA GGA CGT CAC TCA GCT 3'
    Mutation A: 5' AUG TCA GGA CGT CAC TGA GCT 3'
    Mutation B: 5' AUA TCA GGA CGT CAC TCA GCT 3'

**20. a.** Detail the key differences between lytic and nonlytic viral infection and provide an example of each.

  **b.** Which of the following processes occurs in both lytic and nonlytic viral infections?

    **(i)** Infected cell ruptures to release viral particles.

    **(ii)** Viral mRNAs are transcribed by the host-cell translation machinery.

    **(iii)** Viral proteins and nucleic acids are packaged to produce virions.

## Analyze the Data

Protein synthesis in eukaryotes normally begins at the first AUG codon in the mRNA. Sometimes, however, the ribosomes do not begin protein synthesis at this first AUG but scan past it (leaky scanning), and protein synthesis begins instead at an internal AUG. In order to understand what features of an mRNA affect efficiency of initiation at the first AUG, studies have been undertaken in which the synthesis of chloramphenicol acetyltransferase was examined. Translation of its message can give rise to a protein referred to as preCAT or give rise to a slightly smaller protein, CAT (see M. Kozak, 2005, *Gene* **361**:13). The two proteins differ in that CAT lacks several amino acids found at the N-terminus of preCAT. CAT is not derived by cleavage of preCAT but, instead, by initiation of translation of the mRNA at an internal AUG:

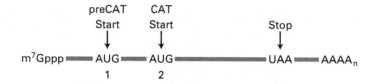

**a.** Results from a number of studies have given rise to the hypothesis that the sequence (−3)ACCAUGG(+4), in which the start codon AUG is shown in boldface, provides an optimal context for initiation of protein synthesis and ensures that ribosomes do not scan past this first AUG to begin initiation instead at a downstream AUG. In the numbering scheme used here, the A of the AUG initiation is designated (11); bases 5′ of this are given negative numbers [so that the first base of this sequence is (−3)], and bases 3′ to the (+1) A are given positive numbers [so that the last base of this sequence is (+4)]. To test the hypothesis that the start site sequence (−3)ACCAUGG(+4) prevents leaky scanning, the chloramphenicol acetyltransferase mRNA sequence was modified and the resulting effects on translation assessed. In the following figure, the sequence (red) surrounding the first AUG codon (black) of the mRNA that gives rise to the synthesis of preCAT is shown above lane 3. Modification of this message is shown above the other gel lanes (altered nucleotides are in blue), and the completed proteins generated from each modified message appear as bands on the SDS-polyacrylamide gel below. The intensity of each band is an indication of the amount of that protein synthesized. Analyze the alterations to the wild-type sequence, and describe how they affect translation. Are the positions of some nucleotides more important than others? Do the data shown in this figure provide support for the hypothesis that the context in which the first AUG is present affects efficiency of translation from this site? Is ACCAUGG an optimal context for initiation from the first AUG?

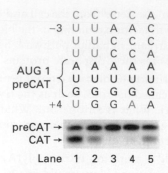

**b.** What are some additional alterations to this message, other than those shown in the figure, that would further elucidate the importance of the ACCAUGG sequence as an optimal context for synthesis of preCAT rather than CAT? How would you further examine whether A at the (−3) position and G at the (+4) position are the most important nucleotides to provide context for the AUG start?

**c.** A mutation causing a severe blood disease has been found in a single family (see T. Matthes et al., 2004, *Blood* **104**:2181). The mutation, shown in red in the figure below, has been mapped to the 5′-untranslated region of the gene encoding hepcidin and has been found to alter the gene's mRNA. The shaded regions indicate the coding sequence of the normal and mutant genes. No hepcidin is produced from the altered mRNA, and lack of hepcidin results in the disease. Can you provide a reasonable explanation for the lack of synthesis of hepcidin in the family members who have inherited this mutation? What can you deduce about the importance of the context in which the start site for initiation of protein synthesis occurs in this case?

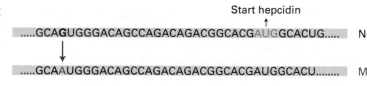

## References

### Structure of Nucleic Acids

Arnott, S. 2006. Historical article: DNA polymorphism and the early history of the double helix. *Trends Biochem. Sci.* **31**:349–354.

Berger, J. M., and J. C. Wang. 1996. Recent developments in DNA topoisomerase II structure and mechanism. *Curr. Opin. Struc. Biol.* **6**:84–90.

Cech, T. R. 2009. Evolution of biological catalysis: ribozyme to RNP enzyme. *Cold Spring Harbor Symp. Quant. Biol.* **74**:11–16.

Dickerson, R. E. 1992. DNA Structure from A to Z. *Methods Enzymol.* **211**:67–111.

Kornberg, A., and T. A. Baker. 2005. *DNA Replication.* University Science, chap. 1. A good summary of the principles of DNA structure.

Lilley, D. M. 2005. Structure, folding and mechanisms of ribozymes. *Curr. Opin. Struc. Biol.* **15**:313–323.

Vicens, Q., and T. R. Cech. 2005. Atomic level architecture of group I introns revealed. *Trends Biochem. Sci.* **31**:41–51.

Wang, J. C. 1980. Superhelical DNA. *Trends Biochem. Sci.* 5:219–221.

Wigley, D. B. 1995. Structure and mechanism of DNA topoisomerases. *Ann. Rev. Biophys. Biomol. Struc.* 24:185–208.

## Transcription of Protein-Coding Genes and Formation of Functional mRNA

Brenner, S., F. Jacob, and M. Meselson. 1961. An unstable intermediate carrying information from genes to ribosomes for protein synthesis. *Nature* 190:576–581.

Brueckner F., J. Ortiz, and P. Cramer 2009. A movie of the RNA polymerase nucleotide addition cycle. *Curr. Opin. Struc. Biol.* 19:294–299.

Murakami, K. S., and S. A. Darst. 2003. Bacterial RNA polymerases: the whole story. *Curr. Opin. Struc. Biol.* 13:31–39.

Okamoto K., Y. Sugino, and M. Nomura. 1962. Synthesis and turnover of phage messenger RNA in E. coli infected with bacteriophage T4 in the presence of chloromycetin. *J. Mol. Biol.* 5:527–534.

Steitz, T. A. 2006. Visualizing polynucleotide polymerase machines at work. *EMBO J.* 25:3458–3468.

## The Decoding of mRNA by tRNAs

Alexander, R. W., and P. Schimmel. 2001. Domain-domain communication in aminoacyl-tRNA synthetases. *Prog. Nucl. Acid Res. Mol. Biol.* 69:317–349.

Hatfield, D. L., and V. N. Gladyshev. 2002. How selenium has altered our understanding of the genetic code. *Mol. Cell Biol.* 22:3565–3576.

Hoagland, M. B., et al. 1958. A soluble ribonucleic acid intermediate in protein synthesis. *J. Biol. Chem.* 231:241–257.

Ibba, M., and D. Soll. 2004. Aminoacyl-tRNAs: setting the limits of the genetic code. *Genes Dev.* 18:731–738.

Khorana, G. H., et al. 1966. Polynucleotide synthesis and the genetic code. *Cold Spring Harbor Symp. Quant. Biol.* 31:39–49.

Nakanishi, K., and O. Nureki. 2005. Recent progress of structural biology of tRNA processing and modification. *Mol. Cells* 19:157–166.

Nirenberg, M., et al. 1966. The RNA code in protein synthesis. *Cold Spring Harbor Symp. Quant. Biol.* 31:11–24.

Rich, A., and S.-H. Kim. 1978. The three-dimensional structure of transfer RNA. *Sci. Am.* 240(1):52–62 (offprint 1377).

## Stepwise Synthesis of Proteins on Ribosomes

Belousoff, M. J., et al. 2010. Ancient machinery embedded in the contemporary ribosome. *Biochem. Soc. Trans.* 38:422–427.

Frank, J., and R. L. Gonzalez, Jr. 2010. Structure and dynamics of a processive Brownian motor: the translating ribosome. *Annu. Rev. Biochem.* 79:381–412.

Jackson, R.J., et al. 2010. The mechanism of eukaryotic translation initiation and principles of its regulation. *Nat. Rev. Mol. Cell Biol.* 11:113–127.

Korostelev, A., D. N. Ermolenko, and H. F. Noller. 2008. Structural dynamics of the ribosome. *Curr. Opin. Chem. Biol.* 12:674–683.

Livingstone, M., et al. 2010. Mechanisms governing the control of mRNA translation. *Phys. Biol.* 7(2):021001.

Sarnow, P., R. C. Cevallos, and E. Jan. 2005. Takeover of host ribosomes by divergent IRES elements. *Biochem. Soc. Trans.* 33:1479–1482.

Steitz, T. A. 2008. A structural understanding of the dynamic ribosome machine. *Nat. Rev. Mol. Cell Biol.* 9:242–253.

## DNA Replication

DePamphilis, M. L., ed. 2006. *DNA Replication and Human Disease.* Cold Spring Harbor Laboratory Press.

Gai, D., Y. P. Chang, and X. S. Chen. 2010. Origin DNA melting and unwinding in DNA replication. *Curr. Opin. Struc. Biol.* 20(6):756–762.

Kornberg, A., and T. A. Baker. 2005. *DNA Replication.* University Science.

Langston, L. D., C. Indiani, and M. O'Donnell. 2009. Whither the replisome: emerging perspectives on the dynamic nature of the DNA replication machinery. *Cell Cycle* 8:2686–2691.

Langston, L. D., and M. O'Donnell. 2006. DNA replication: keep moving and don't mind the gap. *Mol. Cells* 23:155–160.

Schoeffler, A. J., and J. M. Berger. 2008. DNA topoisomerases: harnessing and constraining energy to govern chromosome topology. *Quart. Rev. Biophys.* 41:41–101.

Stillman, B. 2008. DNA polymerases at the replication fork in eukaryotes. *Cell* 30:259–260.

## DNA Repair and Recombination

Andressoo, J. O., and J. H. Hoeijmakers. 2005. Transcription-coupled repair and premature aging. *Mutat. Res.* 577:179–194.

Barnes, D. E., and T. Lindahl. 2004. Repair and genetic consequences of endogenous DNA base damage in mammalian cells. *Ann. Rev. Genet.* 38:445–476.

Bell, C. E. 2005. Structure and mechanism of *Escherichia coli* RecA ATPase. *Mol. Microbiol.* 58:358–366.

Friedberg, E. C., et al. 2006. DNA repair: from molecular mechanism to human disease. *DNA Repair* 5:986–996.

Haber, J. E. 2000. Partners and pathways repairing a double-strand break. *Trends Genet.* 16:259–264.

Jiricny, J. 2006. The multifaceted mismatch-repair system. *Nat. Rev. Mol. Cell Biol.* 7:335–346.

Khuu, P. A., et al. 2006. The stacked-X DNA Holliday junction and protein recognition. *J. Mol. Recog.* 19:234–242.

Lilley, D. M., and R. M. Clegg. 1993. The structure of the four-way junction in DNA. *Annu. Rev. Biophys. Biomol. Struc.* 22:299–328.

Mirchandani, K. D., and A. D. D'Andrea. 2006. The Fanconi anemia/BRCA pathway: a coordinator of cross-link repair. *Exp. Cell Res.* 312:2647–2653.

Mitchell, J. R., J. H. Hoeijmakers, and L. J. Niedernhofer. 2003. Divide and conquer: nucleotide excision repair battles cancer and aging. *Curr. Opin. Cell Biol.* 15:232–240.

Orr-Weaver, T. L., and J. W. Szostak. 1985. Fungal recombination. *Microbiol. Rev.* 49:33–58.

Shin, D. S., et al. 2004. Structure and function of the double strand break repair machinery. *DNA Repair* 3:863–873.

Wood, R. D., M. Mitchell, and T. Lindahl. Human DNA repair genes. *Mutat. Res.* 577:275–283.

Yoshida, K., and Y. Miki. 2004. Role of BRCA1 and BRCA2 as regulators of DNA repair, transcription, and cell cycle in response to DNA damage. *Cancer Sci.* 95:866–871.

## Viruses: Parasites of the Cellular Genetic System

Flint, S. J., et al. 2000. *Principles of Virology: Molecular Biology, Pathogenesis, and Control.* ASM Press.

Hull, R. 2002. *Mathews' Plant Virology.* Academic Press.

Klug, A. 1999. The tobacco mosaic virus particle: structure and assembly. *Phil. Trans. R. Soc. Lond. B Biol. Sci.* 354:531–535.

Knipe, D. M., and P. M. Howley, eds. 2001. *Fields Virology.* Lippincott Williams & Wilkins.

Kornberg, A., and T. A. Baker. 1992. *DNA Replication,* 2d ed. W. H. Freeman and Company. Good summary of bacteriophage molecular biology.

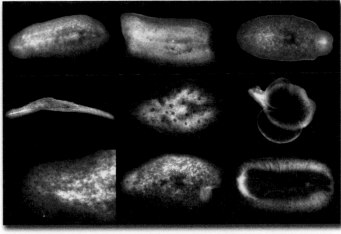

# Molecular Genetic Techniques

The Planarian is a freshwater flatworm with an amazing capacity for regeneration. If the head and tail of an adult Planarian are sliced off, the worm will readily regenerate these structures (as shown in the upper left frame). The role of specific genes in the regeneration process can be tested by shutting off gene expression by RNA interference (RNAi) before cutting off the head and tail. The remaining eight frames show the variety of regeneration defects that can be observed after RNAi of different genes responsible for regeneration. The genes inhibited by RNAi are from left to right starting at the upper center panel are: smad4, β-catenin-1, carcinoma antigen, POU2/3, rootletin, Novel, tolloid, and piwi. [Courtesy of Peter Reddien/MIT, Whitehead Institute.]

In the field of molecular cell biology, reduced to its most basic elements, we seek an understanding of the biological behavior of cells in terms of the underlying chemical and molecular mechanisms. Often the investigation of a new molecular process focuses on the function of a particular protein. There are three fundamental questions that cell biologists usually ask about a newly discovered protein: what is the function of the protein in the context of a living cell, what is the biochemical function of the purified protein, and where is the protein located? To answer these questions, investigators employ three molecular genetic tools: the gene that encodes the protein, a mutant cell line or organism that lacks the functional protein, and a source of the purified protein for biochemical studies. In this chapter, we consider various aspects of two basic experimental strategies for obtaining all three tools (Figure 5-1).

The first strategy, often referred to as *classical genetics*, begins with isolation of a mutant that appears to be defective in some process of interest. Genetic methods then are used to identify and isolate the affected gene. The isolated gene can be manipulated to produce large quantities of the protein for biochemical experiments and to design probes for studies of

where and when the encoded protein is expressed in an organism. The second strategy follows essentially the same steps as the classical approach but in reverse order, beginning with isolation of an interesting protein or its identification based on analysis of an organism's genomic sequence. Once the corresponding gene has been isolated, the gene can be altered and then reinserted into an organism. In both strategies, by examining the phenotypic consequences of mutations that inactivate a particular gene, geneticists are able to connect knowledge about the sequence, structure, and biochemical activity of the encoded protein to its function in the context of a living cell or multicellular organism.

An important component in both strategies for studying a protein and its biological function is isolation of the corresponding gene. Thus we discuss various techniques by which researchers can isolate, sequence, and manipulate specific regions of an organism's DNA. Next we introduce a variety of techniques that are commonly used to analyze where and when a particular gene is expressed and where in the cell its protein is localized. In some cases, knowledge of protein function can lead to significant medical advances, and the first step in developing treatments for an inherited disease is

## OUTLINE

**FIGURE 5-1 Overview of two strategies for relating the function, location, and structure of gene products.** A mutant organism is the starting point for the classical genetic strategy (green arrows). The reverse strategy (orange arrows) usually begins with identification of a protein-coding sequence by analysis of genome-sequence databases. In both strategies, the actual gene is isolated either from a DNA library or by specific amplification of the gene sequence from genomic DNA. Once a cloned gene is isolated, it can be used to produce the encoded protein in bacterial or eukaryotic expression systems. Alternatively, a cloned gene can be inactivated by one of various techniques and used to generate mutant cells or organisms.

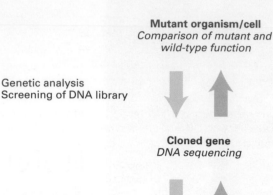

**Mutant organism/cell**
*Comparison of mutant and wild-type function*

Genetic analysis
Screening of DNA library

Gene inactivation

**Cloned gene**
*DNA sequencing*

Expression in cultured cells

Database search to identify protein-coding sequence
PCR isolation of corresponding gene

**Protein**
*Localization*
*Biochemical studies*
*Determination of structure*

to identify and isolate the affected gene, which we describe here. Finally, we discuss techniques that abolish normal protein function in order to analyze the role of the protein in the cell.

## 5.1 Genetic Analysis of Mutations to Identify and Study Genes

As described in Chapter 4, the information encoded in the DNA sequence of genes specifies the sequence—and therefore the structure and function—of every protein molecule in a cell. The power of genetics as a tool for studying cells and organisms lies in the ability of researchers to selectively alter every copy of just one type of protein in a cell by making a change in the gene for that protein. Genetic analyses of mutants defective in a particular process can reveal (a) new genes required for the process to occur, (b) the order in which gene products act in the process, and (c) whether the proteins encoded by different genes interact with one another. Before seeing how genetic studies of this type can provide insights into the mechanism of a complicated cellular or developmental process, we first explain some basic genetic terms used throughout our discussion.

The different forms, or variants, of a gene are referred to as **alleles.** Geneticists commonly refer to the numerous naturally occurring genetic variants that exist in populations, particularly human populations, as alleles. The term **mutation** usually is reserved for instances in which an allele is known to have been newly formed, such as after treatment of an experimental organism with a **mutagen,** an agent that causes a heritable change in the DNA sequence.

Strictly speaking, the particular set of alleles for all the genes carried by an individual is its **genotype.** However, this term also is used in a more restricted sense to denote just the alleles of the particular gene or genes under examination. For experimental organisms, the term **wild type** often is used

to designate a standard genotype for use as a reference in breeding experiments. Thus the normal, nonmutant allele will usually be designated as the wild type. Because of the enormous naturally occurring allelic variation that exists in human populations, the term *wild type* usually denotes an allele that is present at a much higher frequency than any of the other possible alternatives.

Geneticists draw an important distinction between the *genotype* and the **phenotype** of an organism. The phenotype refers to all the physical attributes or traits of an individual that are the consequence of a given genotype. In practice, however, the term *phenotype* often is used to denote the physical consequences that result from just the alleles that are under experimental study. Readily observable phenotypic characteristics are critical in the genetic analysis of mutations.

## Recessive and Dominant Mutant Alleles Generally Have Opposite Effects on Gene Function

A fundamental genetic difference between experimental organisms is whether their cells carry a single set of chromosomes or two copies of each chromosome. The former are referred to as haploid; the latter as diploid. Complex multicellular organisms (e.g., fruit flies, mice, humans) are diploid, whereas many simple unicellular organisms are haploid. Some organisms, notably the yeast *Saccharomyces cerevisiae*, can exist in either haploid or diploid states. The normal cells of some organisms, both plants and animals, carry more than two copies of each chromosome and are thus designated polyploid. Moreover, cancer cells begin as diploid cells but through the process of transformation into cancer cells can gain extra copies of one or more chromosomes and are thus designated as aneuploid. However, our discussion of genetic techniques and analysis relates to diploid organisms, including diploid yeasts.

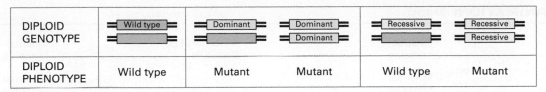

| DIPLOID GENOTYPE | Wild type | Dominant | Dominant / Dominant | Recessive | Recessive / Recessive |
|---|---|---|---|---|---|
| DIPLOID PHENOTYPE | Wild type | Mutant | Mutant | Wild type | Mutant |

**FIGURE 5-2 Effects of dominant and recessive mutant alleles on phenotype in diploid organisms.** A single copy of a dominant allele is sufficient to produce a mutant phenotype, whereas both copies of a recessive allele must be present to cause a mutant phenotype. Recessive mutations usually cause a loss of function; dominant mutations usually cause a gain of function or an altered function.

Although many different alleles of a gene might occur in different organisms in a population, any individual diploid organism will carry two copies of each gene and thus at most can have two different alleles. An individual with two different alleles is **heterozygous** for a gene, whereas an individual that carries two identical alleles is **homozygous** for a gene. A recessive mutant allele is defined as one in which both alleles must be mutant in order for the mutant phenotype to be observed; that is, the individual must be homozygous for the mutant allele to show the mutant phenotype. In contrast, the phenotypic consequences of a **dominant** mutant allele can be observed in a heterozygous individual carrying one mutant and one wild-type allele (Figure 5-2).

Whether a mutant allele is recessive or dominant provides valuable information about the function of the affected gene and the nature of the causative mutation. Recessive alleles usually result from a mutation that inactivates the affected gene, leading to a partial or complete *loss of function*. Such recessive mutations may remove part of the gene or the entire gene from the chromosome, disrupt expression of the gene, or alter the structure of the encoded protein, thereby altering its function. Conversely, dominant alleles are often the consequence of a mutation that causes some kind of *gain of function*. Such dominant mutations may increase the activity of the encoded protein, confer a new function on it, or lead to its inappropriate spatial or temporal pattern of expression.

Dominant mutations in certain genes, however, are associated with a loss of function. For instance, some genes are *haplo-insufficient*, in that removing or inactivating one of the two alleles of such a gene leads to a mutant phenotype because not enough gene product is made. In other rare instances, a dominant mutation in one allele may lead to a structural change in the protein that interferes with the function of the wild-type protein encoded by the other allele. This type of mutation, referred to as a *dominant-negative,* produces a phenotype similar to that obtained from a loss-of-function mutation.

Some alleles can exhibit both recessive and dominant properties. In such cases, statements about whether an allele is dominant or recessive must specify the phenotype. For example, the allele of the hemoglobin gene in humans designated $Hb^s$ has more than one phenotypic consequence. Individuals who are homozygous for this allele ($Hb^s/Hb^s$) have the debilitating disease sickle-cell anemia, but heterozygous individuals ($Hb^s/Hb^a$) do not have the disease. Therefore, $Hb^s$ is *recessive* for the trait of sickle-cell disease. On the other hand, heterozygous ($Hb^s/Hb^a$) individuals are more resistant to malaria than homozygous ($Hb^a/Hb^a$) individuals, revealing that $Hb^s$ is *dominant* for the trait of malaria resistance. ∎

A commonly used agent for inducing mutations (mutagenesis) in experimental organisms is ethylmethane sulfonate (EMS). Although this mutagen can alter DNA sequences in several ways, one of its most common effects is to chemically modify guanine bases in DNA, ultimately leading to the conversion of a G·C base pair into an A·T base pair. Such an alteration in the sequence of a gene, which involves only a single base pair, is known as a **point mutation**. A *silent* point mutation causes no change in the amino acid sequence or activity of a gene's encoded protein. However, observable phenotypic consequences due to changes in a protein's activity can arise from point mutations that result in substitution of one amino acid for another (*missense* mutation), introduction of a premature stop codon (*nonsense* mutation), or a change in the reading frame of a gene (*frameshift* mutation). Because alterations in the DNA sequence leading to a decrease in protein activity are much more likely than alterations leading to an increase or qualitative change in protein activity, mutagenesis usually produces many more recessive mutations than dominant mutations.

## Segregation of Mutations in Breeding Experiments Reveals Their Dominance or Recessivity

Geneticists exploit the normal life cycle of an organism to test for the dominance or recessivity of alleles. To see how this is done, we need first to review the type of cell division that gives rise to gametes (sperm and egg cells in higher plants and animals). Whereas the body (somatic) cells of most multicellular organisms divide by mitosis, the germ cells that give rise to gametes undergo meiosis. Like somatic cells, premeiotic germ cells are diploid, containing two homologs of each morphological type of chromosome. The two homologs constituting each pair of homologous chromosomes are descended from different parents, and thus their genes may exist in different allelic forms. Figure 5-3 depicts the major events in mitotic and meiotic cell division. In mitosis, DNA replication is always followed by cell division, yielding two diploid daughter cells. In meiosis, *one* round of DNA replication is followed by *two* separate cell divisions,

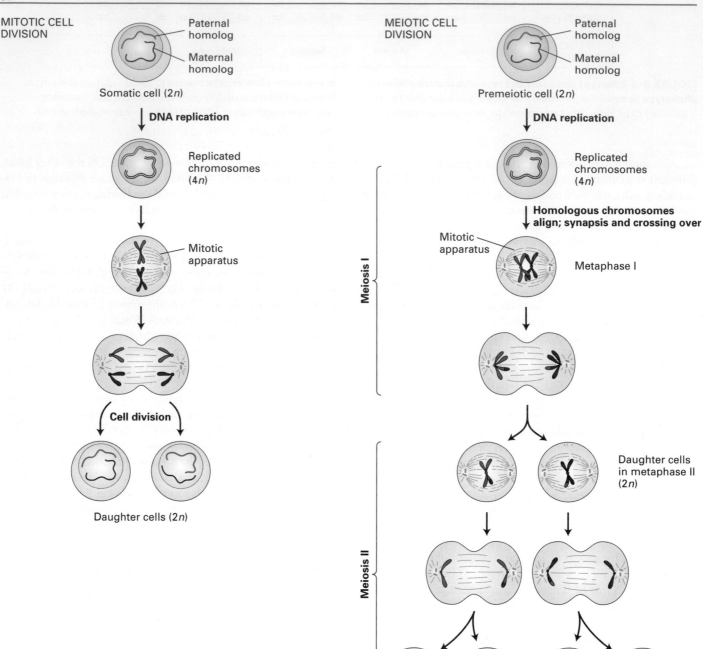

**MITOTIC CELL DIVISION**

Somatic cell (2*n*)

Paternal homolog
Maternal homolog

**DNA replication**

Replicated chromosomes (4*n*)

Mitotic apparatus

Cell division

Daughter cells (2*n*)

**MEIOTIC CELL DIVISION**

Premeiotic cell (2*n*)

Paternal homolog
Maternal homolog

**DNA replication**

Replicated chromosomes (4*n*)

**Homologous chromosomes align; synapsis and crossing over**

Mitotic apparatus

Metaphase I

Meiosis I

Daughter cells in metaphase II (2*n*)

Meiosis II

1*n*  1*n*  1*n*  1*n*

**FIGURE 5-3 Comparison of mitosis and meiosis.** Both somatic cells and premeiotic germ cells have two copies of each chromosome (2*n*), one maternal and one paternal. In mitosis, the replicated chromosomes, each composed of two sister chromatids, align at the cell center in such a way that both daughter cells receive a maternal and paternal homolog of each morphological type of chromosome. During the first *meiotic* division, however, each replicated chromosome pairs with its homologous partner at the cell center; this pairing off is referred to as *synapsis*, and crossing over between homologous chromosomes is evident at this stage. One replicated chromosome of each morphological type then goes into each daughter cell. The resulting cells undergo a second division without intervening DNA replication, with the sister chromatids of each morphological type being apportioned to the daughter cells. In the second meiotic division the alignment of chromatids and their equal segregation into daughter cells is the same as in mitotic division. The alignment of pairs of homologous chromosomes in metaphase I is random with respect to other chromosome pairs, resulting in a mix of paternally and maternally derived chromosomes in each daughter cell.

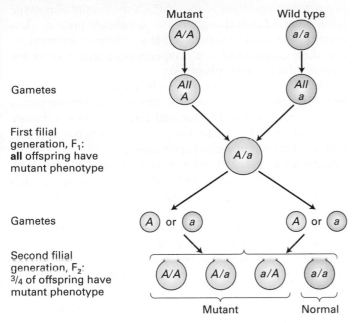

(a) Segregation of **dominant** mutation

Mutant
A/A

Wild type
a/a

Gametes

All
A

All
a

First filial
generation, F₁:
**all** offspring have
mutant phenotype

A/a

Gametes

A or a

A or a

Second filial
generation, F₂:
³/₄ of offspring have
mutant phenotype

A/A    A/a    a/A    a/a

Mutant    Normal

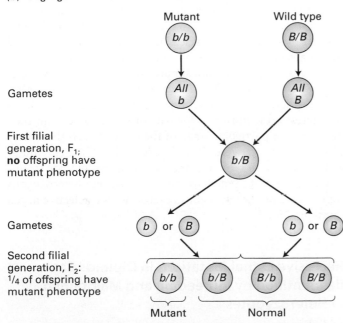

(b) Segregation of **recessive** mutation

Mutant
b/b

Wild type
B/B

Gametes

All
b

All
B

First filial
generation, F₁:
**no** offspring have
mutant phenotype

b/B

Gametes

b or B

b or B

Second filial
generation, F₂:
¹/₄ of offspring have
mutant phenotype

b/b    b/B    B/b    B/B

Mutant    Normal

**FIGURE 5-4 Segregation patterns of dominant and recessive mutations in crosses between true-breeding strains of diploid organisms.** All the offspring in the first (F₁) generation are heterozygous. If the mutant allele is dominant, the F₁ offspring will exhibit the mutant phenotype, as in part (a). If the mutant allele is recessive, the F₁ offspring will exhibit the wild-type phenotype, as in part (b). Crossing of the F₁ heterozygotes among themselves also produces different segregation ratios for dominant and recessive mutant alleles in the F₂ generation.

yielding four haploid (1n) cells that contain only one chromosome of each homologous pair. The apportionment, or **segregation**, of the replicated homologous chromosomes to daughter cells during the first meiotic division is random; that is, maternally and paternally derived homologs segregate independently, yielding daughter cells with different mixes of paternal and maternal chromosomes.

As a way to avoid unwanted complexity, geneticists usually strive to begin breeding experiments with strains that are homozygous for the genes under examination. In such *true-breeding* strains, every individual will receive the same allele from each parent and therefore the composition of alleles will not change from one generation to the next. When a true-breeding mutant strain is mated to a true-breeding wild-type strain, all the first filial (F₁) progeny will be heterozygous (Figure 5-4). If the F₁ progeny exhibit the mutant trait, then the mutant allele is dominant; if the F₁ progeny exhibit the wild type trait, then the mutant is recessive. Further crossing between F₁ individuals will also reveal different patterns of inheritance according to whether the mutation is dominant or recessive. When F₁ individuals that are heterozygous for a dominant allele are crossed among themselves, three-fourths of the resulting F₂ progeny will exhibit the mutant trait. In contrast, when F₁ individuals that are heterozygous for a recessive allele are crossed among themselves, only one-fourth of the resulting F₂ progeny will exhibit the mutant trait.

As noted earlier, the yeast *S. cerevisiae*, an important experimental organism, can exist in either a haploid or a diploid state. In these unicellular eukaryotes, crosses between haploid cells can determine whether a mutant allele is dominant or recessive. Haploid yeast cells, which carry one copy of each chromosome, can be of two different mating types known as **a** and α. Haploid cells of opposite mating type can mate to produce **a**/α diploids, which carry two copies of each chromosome. If a new mutation with an observable phenotype is isolated in a haploid strain, the mutant strain can be mated to a wild-type strain of the opposite mating type to produce **a**/α diploids that are heterozygous for the mutant allele. If these diploids exhibit the mutant trait, then the mutant allele is dominant, but if the diploids appear as wild type, then the mutant allele is recessive. When **a**/α diploids are placed under starvation conditions, the cells undergo meiosis, giving rise to a tetrad of four haploid spores, two of type **a** and two of type α. Sporulation of a heterozygous diploid cell yields two spores carrying the mutant allele and two carrying the wild-type allele (Figure 5-5). Under appropriate conditions, yeast spores will germinate, producing vegetative haploid strains of both mating types.

## Conditional Mutations Can Be Used to Study Essential Genes in Yeast

The procedures used to identify and isolate mutants, referred to as *genetic screens*, depend on whether the experimental organism is haploid or diploid and, if the latter, whether the mutation is recessive or dominant. Genes that encode pro-

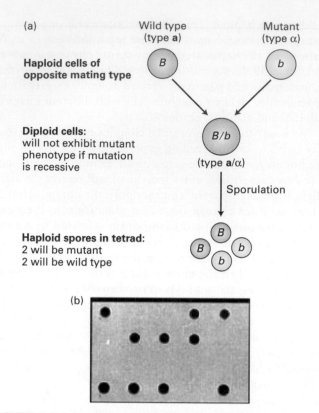

**(a)**

Haploid cells of opposite mating type

Wild type (type **a**)    Mutant (type α)

*B*    *b*

Diploid cells: will not exhibit mutant phenotype if mutation is recessive

*B/b*

(type **a**/α)

Sporulation

Haploid spores in tetrad:
2 will be mutant
2 will be wild type

*B*
*B*    *b*
*b*

**(b)**

**FIGURE 5-5 Segregation of alleles in yeast.** (a) Haploid *Saccharomyces* cells of opposite mating type (i.e., one of mating type α and one of mating type **a**) can mate to produce an **a**/α diploid. If one haploid carries a dominant wild-type allele and the other carries a recessive mutant allele of the same gene, the resulting heterozygous diploid will express the dominant trait. Under certain conditions, a diploid cell will form a tetrad of four haploid spores. Two of the spores in the tetrad will express the recessive trait and two will express the dominant trait. (b) If the mutant phenotype is not viable under restrictive growth conditions, each tetrad, shown here as four spores separated vertically and grown into colonies on restrictive media, consists of two viable and two nonviable spores. [Part (b) from B. Senger et al., 1998, *EMBO J* **17**:2196]

teins essential for life are among the most interesting and important ones to study. Since phenotypic expression of mutations in essential genes leads to death of the individual, clever genetic screens are needed to isolate and maintain organisms with a lethal mutation.

In haploid yeast cells, essential genes can be studied through the use of *conditional mutations*. Among the most common conditional mutations are **temperature-sensitive mutations**, which can be isolated in bacteria and lower eukaryotes but not in warm-blooded eukaryotes. For instance, a single missense mutation may cause the resulting mutant protein to have reduced thermal stability such that the protein is fully functional at one temperature (e.g., 23 °C) but begins to denature and is inactive at another temperature (e.g., 36 °C), whereas the normal protein would be fully stable and functional at both temperatures. A temperature at which the mu-

tant phenotype is observed is called *nonpermissive*; a *permissive* temperature is one at which the mutant phenotype is not observed even though the mutant allele is present. Thus mutant strains can be maintained at a permissive temperature and then subcultured at a nonpermissive temperature for analysis of the mutant phenotype.

An example of a particularly important screen for temperature-sensitive mutants in the yeast *S. cerevisiae* comes from the studies of L. H. Hartwell and colleagues in the late 1960s and early 1970s. They set out to identify genes important in regulation of the cell cycle during which a cell synthesizes proteins, replicates its DNA, and then undergoes mitotic cell division, with each daughter cell receiving a copy of each chromosome. Exponential growth of a single yeast cell for 20–30 cell divisions forms a visible yeast colony on solid agar medium. Since mutants with a complete block in the cell cycle would not be able to form a colony, conditional mutants were required to study mutations that affect this basic cell process. To screen for such mutants, the researchers first identified mutagenized yeast cells that could grow normally at 23 °C but that could not form a colony when placed at 36 °C (Figure 5-6a).

Once temperature-sensitive mutants were isolated, further analysis revealed that some indeed were defective in cell division. In *S. cerevisiae,* cell division occurs through a budding process, and the size of the bud, which is easily visualized by light microscopy, indicates a cell's position in the cell cycle. Each of the mutants that could not grow at 36 °C was examined by microscopy after several hours at the nonpermissive temperature. Examination of many different temperature-sensitive mutants revealed that about 1 percent exhibited a distinct block in the cell cycle. These mutants were therefore designated *cdc* (cell division cycle) mutants. Importantly, these yeast mutants did not simply fail to grow, as they might if they carried a mutation affecting general cellular metabolism. Rather, at the nonpermissive temperature, the mutants of interest grew normally for part of the cell cycle but then arrested at a particular stage of the cell cycle, so that many cells at this stage were seen (Figure 5-6b). Most *cdc* mutations in yeast are recessive; that is, when haploid *cdc* strains are mated to wild-type haploids, the resulting heterozygous diploids are neither temperature sensitive nor defective in cell division.

## Recessive Lethal Mutations in Diploids Can Be Identified by Inbreeding and Maintained in Heterozygotes

In diploid organisms, phenotypes resulting from recessive mutations can be observed only in individuals homozygous for the mutant alleles. Since mutagenesis in a diploid organism typically changes only one allele of a gene, yielding heterozygous mutants, genetic screens must include inbreeding steps to generate progeny that are homozygous for the mutant alleles. The geneticist H. Muller developed a general and efficient procedure for carrying out such inbreeding experiments in the

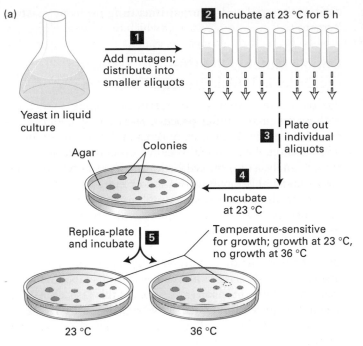

(a)

Yeast in liquid culture

**1** Add mutagen; distribute into smaller aliquots

**2** Incubate at 23 °C for 5 h

**3** Plate out individual aliquots

Agar    Colonies

**4** Incubate at 23 °C

**5** Replica-plate and incubate

Temperature-sensitive for growth; growth at 23 °C, no growth at 36 °C

23 °C        36 °C

**EXPERIMENTAL FIGURE 5-6  Haploid yeasts carrying temperature-sensitive lethal mutations are maintained at permissive temperature and analyzed at nonpermissive temperature.** (a) Genetic screen for temperature-sensitive cell-division cycle (*cdc*) mutants in yeast. Yeasts that grow and form colonies at 23 °C (permissive temperature) but not at 36 °C (nonpermissive temperature) may carry a lethal mutation that blocks cell division. (b) Assay of temperature-sensitive colonies for blocks at specific stages in the cell cycle. Shown here are micrographs of wild-type yeast and two different temperature-sensitive mutants after incubation at the nonpermissive temperature for 6 h. Wild-type cells, which continue to grow, can be seen with all different sizes of buds, reflecting different stages of the cell cycle. In contrast, cells in the lower two micrographs exhibit a block at a specific stage in the cell cycle. The *cdc28* mutants arrest at a point before emergence of a new bud and therefore appear as unbudded cells. The *cdc7* mutants, which arrest just before separation of the mother cell and bud (emerging daughter cell), appear as cells with large buds. [Part (a) see L. H. Hartwell, 1967, *J. Bacteriol.* **93**:1662; part (b) from L. M. Hereford and L. H. Hartwell, 1974, *J. Mol. Biol.* **84**:445.]

(b)
Wild type

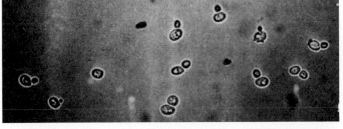

cdc28 mutants

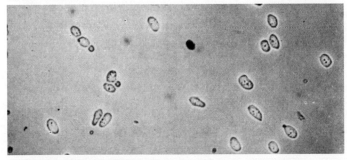

cdc7 mutants

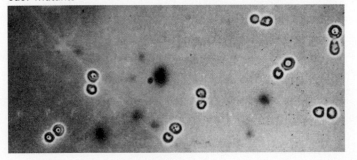

fruit fly *Drosophila*. Recessive lethal mutations in *Drosophila* and other diploid organisms can be maintained in heterozygous individuals and their phenotypic consequences analyzed in homozygotes.

The Muller approach was used to great effect by C. Nüsslein-Volhard and E. Wieschaus, who systematically screened for recessive lethal mutations affecting embryogenesis in *Drosophila*. Dead homozygous embryos carrying recessive lethal mutations identified by this screen were examined under the microscope for specific morphological defects in the embryos. Current understanding of the molecular mechanisms underlying development of multicellular organisms is based, in large part, on the detailed picture of embryonic development revealed by characterization of these *Drosophila* mutants.

## Complementation Tests Determine Whether Different Recessive Mutations Are in the Same Gene

In the genetic approach to studying a particular cellular process, researchers often isolate multiple recessive mutations that produce the same phenotype. A common test for determining whether these mutations are in the same gene or in different genes exploits the phenomenon of genetic complementation, that is, the restoration of the wild-type phenotype by mating of two different mutants. If two recessive mutations, *a* and *b*, are in the *same* gene, then a diploid organism heterozygous for both mutations (i.e., carrying one *a* allele and one *b* allele) will exhibit the mutant phenotype because neither allele provides a functional copy of the gene. In contrast, if mutations *a* and *b* are in *separate* genes, then heterozygotes carrying a single copy of each mutant allele will not exhibit the mutant phenotype because a wild-type allele of each gene will also be present.

In this case, the mutations are said to *complement* each other. Complementation analysis cannot be performed on dominant mutants because the phenotype conferred by the mutant allele is displayed even in the presence of a wild-type allele of the gene.

Complementation analysis of a set of mutants exhibiting the same phenotype can distinguish the individual genes in a set of functionally related genes, all of which must function to produce a given phenotypic trait. For example, the screen for *cdc* mutations in *Saccharomyces* described previously yielded many recessive temperature-sensitive mutants that appeared arrested at the same cell cycle stage. To determine how many genes were affected by these mutations, Hartwell and his colleagues performed complementation tests on all of the pair-wise combinations of *cdc* mutants following the general protocol outlined in Figure 5-7. These tests identified more than 20 different *CDC* genes. The subsequent molecular characterization of the *CDC* genes and their encoded proteins, as described in detail in Chapter 20, has provided a framework for understanding how cell division is regulated in organisms ranging from yeast to humans.

## Double Mutants Are Useful in Assessing the Order in Which Proteins Function

Based on careful analysis of mutant phenotypes associated with a particular cellular process, researchers often can deduce the order in which a set of genes and their protein products function. Two general types of processes are amenable to such analysis: (a) biosynthetic pathways in which a precursor material is converted via one or more intermediates to a final product and (b) signaling pathways that regulate other processes and involve the flow of information rather than chemical intermediates.

**Ordering of Biosynthetic Pathways** A simple example of the first type of process is the biosynthesis of a metabolite such as the amino acid tryptophan in bacteria. In this case, each of the enzymes required for synthesis of tryptophan catalyzes the

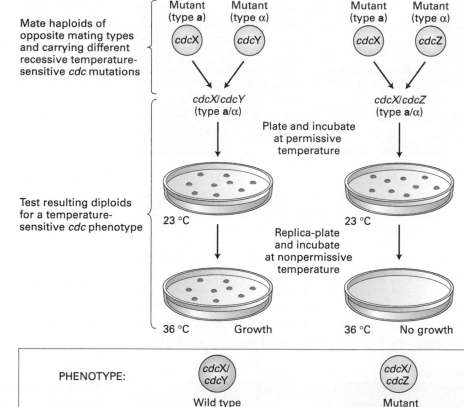

**EXPERIMENTAL FIGURE 5-7 Complementation analysis determines whether recessive mutations are in the same or different genes.** Complementation tests in yeast are performed by mating haploid **a** and α cells carrying different recessive mutations to produce diploid cells. In the analysis of *cdc* mutations, pairs of different haploid temperature-sensitive *cdc* strains were systematically mated and the resulting diploids tested for growth at the permissive and nonpermissive temperatures. In this hypothetical example, the *cdcX* and *cdcY* mutants complement each other and thus have mutations in different genes, whereas the *cdcX* and *cdcZ* mutants have mutations in the same gene.

conversion of one of the intermediates in the pathway to the next. In *E. coli*, the genes encoding these enzymes lie adjacent to one another in the genome, constituting the *trp* operon (see Figure 4-13a). The order of action of the different genes for these enzymes, hence the order of the biochemical reactions in the pathway, initially was deduced from the types of intermediate compounds that accumulated in each mutant. In the case of complex synthetic pathways, however, phenotypic analysis of mutants defective in a single step may give ambiguous results that do not permit conclusive ordering of the steps. Double mutants defective in two steps in the pathway are particularly useful in ordering such pathways (Figure 5-8a).

In Chapter 14, we discuss the classic use of the double-mutant strategy to help elucidate the secretory pathway. In this pathway, proteins to be secreted from the cell move from their site of synthesis on the rough endoplasmic reticulum (ER) to the Golgi complex, then to secretory vesicles, and finally to the cell surface.

**Ordering of Signaling Pathways** As we learn in later chapters, expression of many eukaryotic genes is regulated by signaling pathways that are initiated by extracellular hormones, growth factors, or other signals. Such signaling pathways may include numerous components, and double-mutant analysis often can provide insight into the functions and interactions of these components. The only prerequisite for obtaining useful information from this type of analysis is that the two mutations must have opposite effects on the output of the same regulated pathway. Most commonly, one mutation represses expression of a particular reporter gene even when the signal is present, while another mutation results in reporter gene expression even when the signal is absent (i.e., constitutive expression). As illustrated in Figure 5-8b, two simple regulatory mechanisms are consistent with such single mutants, but the double-mutant phenotype can distinguish between them. This general approach has enabled geneticists to delineate many of the key steps in a variety of different regulatory pathways, setting the stage for more specific biochemical assays.

Note that this technique differs from the complementation analysis just described in that both dominant and recessive mutants can be subjected to double mutant analysis. When two recessive mutations are tested, the double mutant created must be *homozygous* for both mutations. Furthermore, dominant mutants can be subjected to double-mutant analysis.

## Genetic Suppression and Synthetic Lethality Can Reveal Interacting or Redundant Proteins

Two other types of genetic analysis can provide additional clues about how proteins that function in the same cellular process may interact with one another in the living cell. Both of these methods, which are applicable in many experimental organisms, involve the use of double mutants in which the phenotypic effects of one mutation are changed by the presence of a second mutation.

**(a) Analysis of a biosynthetic pathway**

A mutation in **A** accumulates intermediate 1.
A mutation in **B** accumulates intermediate 2.

| | |
|---|---|
| PHENOTYPE OF DOUBLE MUTANT: | A double mutation in A and B accumulates intermediate 1. |
| INTERPRETATION: | *The reaction catalyzed by A precedes the reaction catalyzed by B.* |

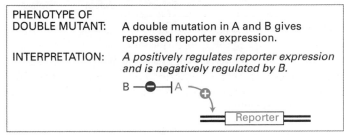

**(b) Analysis of a signaling pathway**

A mutation in **A** gives repressed reporter expression.
A mutation in **B** gives constitutive reporter expression.

| | |
|---|---|
| PHENOTYPE OF DOUBLE MUTANT: | A double mutation in A and B gives repressed reporter expression. |
| INTERPRETATION: | *A positively regulates reporter expression and is negatively regulated by B.* |

| | |
|---|---|
| PHENOTYPE OF DOUBLE MUTANT: | A double mutation in A and B gives constitutive reporter expression. |
| INTERPRETATION: | *B negatively regulates reporter expression and is negatively regulated by A.* |

**FIGURE 5-8 Analysis of double mutants often can order the steps in biosynthetic or signaling pathways.** When mutations in two different genes affect the same cellular process but have distinctly different phenotypes, the phenotype of the double mutant can often reveal the order in which the two genes must function. (a) In the case of mutations that affect the same biosynthetic pathway, a double mutant will accumulate the intermediate immediately preceding the step catalyzed by the protein that acts earlier in the wild-type organism. (b) Double-mutant analysis of a signaling pathway is possible if two mutations have opposite effects on expression of a reporter gene. In this case, the observed phenotype of the double mutant provides information about the order in which the proteins act and whether they are positive or negative regulators.

**Suppressor Mutations** The first type of analysis is based on *genetic suppression*. To understand this phenomenon, suppose that point mutations lead to structural changes in one protein (A) that disrupt its ability to associate with another protein (B) involved in the same cellular process. Similarly, mutations in protein B lead to small structural changes that inhibit its ability to interact with protein A. Assume, furthermore, that the normal functioning of proteins A and B depends on their interacting. In theory, a specific structural change in protein A might be suppressed by compensatory

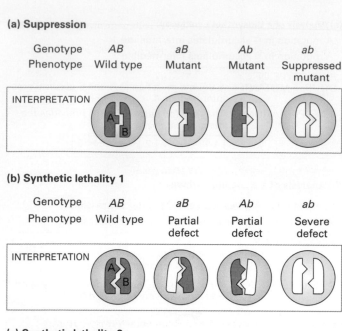

**(a) Suppression**

| Genotype | *AB* | *aB* | *Ab* | *ab* |
|---|---|---|---|---|
| Phenotype | Wild type | Mutant | Mutant | Suppressed mutant |

INTERPRETATION

**(b) Synthetic lethality 1**

| Genotype | *AB* | *aB* | *Ab* | *ab* |
|---|---|---|---|---|
| Phenotype | Wild type | Partial defect | Partial defect | Severe defect |

INTERPRETATION

**(c) Synthetic lethality 2**

| Genotype | *AB* | *aB* | *Ab* | *ab* |
|---|---|---|---|---|
| Phenotype | Wild type | Wild type | Wild type | Mutant |

INTERPRETATION

**FIGURE 5-9 Mutations that result in genetic suppression or synthetic lethality reveal interacting or redundant proteins.** (a) Observation that double mutants with two defective proteins (A and B) have a wild-type phenotype but that single mutants give a mutant phenotype indicates that the function of each protein depends on interaction with the other. (b) Observation that double mutants have a more severe phenotypic defect than single mutants also is evidence that two proteins (e.g., subunits of a heterodimer) must interact to function normally. (c) Observation that a double mutant is nonviable but that the corresponding single mutants have the wild-type phenotype indicates that two proteins function in redundant pathways to produce an essential product.

changes in protein B, allowing the mutant proteins to interact. In the rare cases in which such suppressor mutations occur, strains carrying both mutant alleles would be normal, whereas strains carrying only one or the other mutant allele would have a mutant phenotype (Figure 5-9a).

The observation of genetic suppression in yeast strains carrying a mutant actin allele (*act1-1*) and a second mutation (*sac6*) in another gene provided early evidence for a direct interaction in vivo between the proteins encoded by the two genes. Later biochemical studies showed that these two proteins—Act1 and Sac6—do indeed interact in the construction of functional actin structures within the cell.

**Synthetic Lethal Mutations** Another phenomenon, called *synthetic lethality,* produces a phenotypic effect opposite to that of suppression. In this case, the deleterious effect of one mutation is greatly exacerbated (rather than suppressed) by a second mutation in a related gene. One situation in which such synthetic lethal mutations can occur is illustrated in Figure 5-9b. In this example, a heterodimeric protein is partially, but not completely, inactivated by mutations in either one of the nonidentical subunits. However, in double mutants carrying specific mutations in the genes encoding both subunits, little interaction between subunits occurs, resulting in severe phenotypic effects. Synthetic lethal mutations also can reveal nonessential genes whose encoded proteins function in redundant pathways for producing an essential cell component. As depicted in Figure 5-9c, if either pathway alone is inactivated by a mutation, the other pathway will be able to supply the needed product. However, if both pathways are inactivated at the same time, the essential product cannot be synthesized and the double mutants will be nonviable.

## Genes Can Be Identified by Their Map Position on the Chromosome

The preceding discussion of genetic analysis illustrates how a geneticist can gain insight into gene function by observing the phenotypic effects produced by joining together different combinations of mutant alleles in the same cell or organism. For example, combinations of different alleles of the same gene in a diploid can be used to determine whether a mutation is dominant or recessive or whether two different recessive mutations are in the same gene. Furthermore, combinations of mutations in different genes can be used to determine the order of gene function in a pathway or to identify functional relationships between genes such as suppression and synthetic enhancement. Generally speaking, all these methods can be viewed as analytical tests based on *gene function*. We will now consider a fundamentally different type of genetic analysis based on *gene position*. Studies designed to determine the position of a gene on a chromosome, often referred to as genetic mapping studies, can be used to identify the gene affected by a particular mutation or to determine whether two mutations are in the same gene.

In many organisms, genetic mapping studies rely on exchanges of genetic information that occur during meiosis. As shown in Figure 5-10a, genetic **recombination** takes place before the first meiotic cell division in germ cells, when the replicated chromosomes of each homologous pair align with each other. At this time, homologous DNA sequences on maternally and paternally derived chromatids can exchange with each other, a process known as crossing over. We now know that the resulting crossovers between homologous chromosomes provide structural links that are important for the proper segregation of pairs of homologous chromatids to opposite poles during the first meiotic cell division (for discussion see Chapter 19).

Consider two different mutations, one inherited from each parent, that are located close to each other on the same chromosome. Two different types of gametes can be produced according to whether a crossover occurs between the

(a)

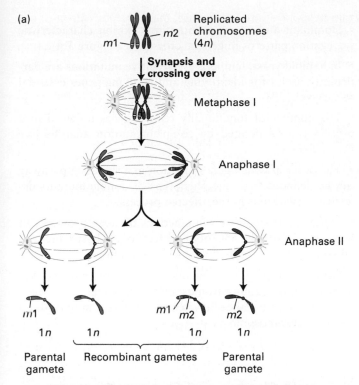

Replicated chromosomes (4n)

**Synapsis and crossing over**

Metaphase I

Anaphase I

Anaphase II

*m1* · *m2*

1n · 1n · 1n · 1n

Parental gamete · Recombinant gametes · Parental gamete

(b) Consider two linked genes *A* and *B* with recessive alleles *a* and *b*.

Cross of two mutants to construct a doubly heterozygous strain:

$$A/A\ b/b \times a/a\ B/B$$

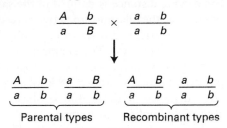

$$\frac{A\quad b}{a\quad B}$$

Cross of double heterozygote to test strain:

$$\frac{A\quad b}{a\quad B} \times \frac{a\quad b}{a\quad b}$$

$$\frac{A\quad b}{a\quad b}\quad \frac{a\quad B}{a\quad b} \qquad \frac{A\quad B}{a\quad b}\quad \frac{a\quad b}{a\quad b}$$

Parental types · Recombinant types

Genetic distance between *A* and *B* can be determined from frequency of parental and recombinant gametes:

$$\text{Genetic distance in cM} = 100 \times \frac{\text{recombinant gametes}}{\text{total gametes}}$$

**FIGURE 5-10 Recombination during meiosis can be used to map the position of genes.** (a) Shown is an individual that carries two mutations, designated *m1* (yellow) and *m2* (green), that are on the maternal and paternal versions of the same chromosome. If crossing over occurs at an interval between *m1* and *m2* before the first meiotic division, then two recombinant gametes are produced; one carries both *m1* and *m2*, whereas the other carries neither mutation. The longer the distance between two mutations on a chromatid, the more likely they are to be separated by recombination and the greater the proportion of recombinant gametes produced. (b) In a typical mapping experiment, a strain that is heterozygous for two different genes is constructed. The frequency of parental or recombinant gametes produced by this strain can be determined from the phenotypes of the progeny in a testcross to a homozygous recessive strain. The genetic map distance in centimorgans (cM) is given as the percent of the gametes that are recombinant.

mutations during meiosis. If no crossover occurs between them, gametes known as *parental types,* which contain either one or the other mutation, will be produced. In contrast, if a crossover occurs between the two mutations, gametes known as *recombinant types* will be produced. In this example, recombinant chromosomes would contain either both mutations or neither of them. The sites of recombination occur more or less at random along the length of chromosomes; thus the closer together two genes are, the less likely that recombination will occur between them during meiosis. In other words, *the less frequently recombination occurs between two genes on the same chromosome, the closer together they are.* Two genes that are on the same chromosome and that are sufficiently close together such that there are significantly fewer recombinant gametes produced than parental gametes are considered to be *genetically linked.*

The technique of recombinational mapping was devised in 1911 by A. Sturtevant while he was an undergraduate working in the laboratory of T. H. Morgan at Columbia University. Originally used in studies on *Drosophila,* this technique is still used today to assess the distance between two genetic loci on the same chromosome in many experimental organisms. A typical experiment designed to determine the map distance between two genetic positions would involve two

steps. In the first step, a strain is constructed that carries a different mutation at each position, or locus. In the second step, the progeny of this strain are assessed to determine the relative frequency of inheritance of parental or recombinant types. A typical way to determine the frequency of recombination between two genes is to cross one diploid parent heterozygous at each of the genetic loci to another parent homozygous for each gene. For such a cross, the proportion of recombinant progeny is readily determined because recombinant phenotypes will differ from the parental phenotypes. By convention, one *genetic map unit* is defined as the distance between two positions along a chromosome that results in one recombinant individual in 100 total progeny. The distance corresponding to this 1 percent recombination frequency is called a *centimorgan* (cM) in honor of Sturtevant's mentor, Morgan (Figure 5-10b).

A complete discussion of the methods of genetic mapping experiments is beyond the scope of this introductory discussion; however, two features of measuring distances by recombination mapping need particular emphasis. First, the frequency of genetic exchange between two loci is strictly proportional to the physical distance in base pairs separating them only for loci that are relatively close together (say, less than about 10 cM). For linked loci that are farther apart than this, a distance measured by the frequency of genetic exchange tends to underestimate the physical distance because of the possibility of two or more crossovers occurring within an interval. In the limiting case in which the number of recombinant types will equal the number of parental types, the two loci under consideration could be far apart on the same chromosome or they could be on different chromosomes, and in such cases the loci are considered to be *unlinked*.

A second important concept needed for interpretation of genetic mapping experiments in different types of organisms is that although genetic distance is defined in the same way for different organisms, the relationship between recombination frequency (i.e., genetic map distance) and physical distance varies between organisms. For example, a 1 percent recombination frequency (i.e., a genetic distance of 1 cM) represents a physical distance of about 2.8 kilobases in yeast compared with a distance of about 400 kilobases in *Drosophila* and about 780 kilobases in humans.

One of the chief uses of genetic mapping studies is to locate the gene that is affected by a mutation of interest. The presence of many different already mapped genetic traits, or genetic markers, distributed along the length of a chromosome permits the position of an unmapped mutation to be determined by assessing its segregation with respect to these marker genes during meiosis. Thus the more markers that are available, the more precisely a mutation can be mapped. In Section 5.4, we will see how the genes affected in inherited human diseases can be identified using such methods. A second general use of mapping experiments is to determine whether two different mutations are in the same gene. If two mutations are in the same gene, they will exhibit tight **linkage** in mapping experiments, but if they are in different genes, they will usually be unlinked or exhibit weak linkage.

## KEY CONCEPTS of Section 5.1

### Genetic Analysis of Mutations to Identify and Study Genes

- Diploid organisms carry two copies (alleles) of each gene, whereas haploid organisms carry only one copy.

- Recessive mutations lead to a loss of function, which is masked if a normal allele of the gene is present. For the mutant phenotype to occur, both alleles must carry the mutation.

- Dominant mutations lead to a mutant phenotype in the presence of a normal allele of the gene. The phenotypes associated with dominant mutations often represent a gain of function but in the case of some genes result from a loss of function.

- In meiosis, a diploid cell undergoes one DNA replication and two cell divisions, yielding four haploid cells in which maternal and paternal chromosomes and their associated alleles are randomly assorted (see Figure 5-3).

- Dominant and recessive mutations exhibit characteristic segregation patterns in genetic crosses (see Figure 5-4).

- In haploid yeast, temperature-sensitive mutations are particularly useful for identifying and studying genes essential to survival.

- The number of functionally related genes involved in a process can be defined by complementation analysis (see Figure 5-7).

- The order in which genes function in a signaling pathway can be deduced from the phenotype of double mutants defective in two steps in the affected process.

- Functionally significant interactions between proteins can be deduced from the phenotypic effects of allele-specific suppressor mutations or synthetic lethal mutations.

- Genetic mapping experiments make use of crossing over between homologous chromosomes during meiosis to measure the genetic distance between two different mutations on the same chromosome.

## 5.2 DNA Cloning and Characterization

Detailed studies of the structure and function of a gene at the molecular level require large quantities of the individual gene in pure form. A variety of techniques, often referred to as *recombinant DNA technology,* are used in **DNA cloning**, which permits researchers to prepare large numbers of identical DNA molecules. **Recombinant DNA** is simply any DNA molecule composed of sequences derived from different sources.

The key to cloning a DNA fragment of interest is to link it to a **vector** DNA molecule that can replicate within a host cell. After a single recombinant DNA molecule, composed of a vector plus an inserted DNA fragment, is introduced into a host cell, the inserted DNA is replicated along with the vector, generating a large number of identical DNA molecules. The basic scheme can be summarized as follows:

Vector + DNA fragment
↓
Recombinant DNA
↓
Replication of recombinant DNA within host cells
↓
Isolation, sequencing, and manipulation
of purified DNA fragment

Although investigators have devised numerous experimental variations, this flow diagram indicates the essential steps in DNA cloning. In this section, we first describe methods for isolating a specific sequence of DNA from a sea of other DNA sequences. This process often involves cutting the genome into fragments and then placing each fragment in a vector so that the entire collection can be propagated as recombinant molecules in separate host cells. While many different types of vectors exist, our discussion will mainly focus on plasmid vectors in *E. coli* host cells, which are commonly used. Various techniques can then be employed to identify the sequence of interest from this collection of DNA fragments, known as a DNA library. Once a specific DNA fragment is isolated, it is typically characterized by determining the exact sequence of nucleotides in the molecule. We end with a discussion of the polymerase chain reaction (PCR). This powerful and versatile technique can be used in many ways to generate large quantities of a specific sequence and otherwise manipulate DNA in the laboratory. The various uses of cloned DNA fragments are discussed in subsequent sections.

## Restriction Enzymes and DNA Ligases Allow Insertion of DNA Fragments into Cloning Vectors

A major objective of DNA cloning is to obtain discrete, small regions of an organism's DNA that constitute specific genes. In addition, only relatively small DNA molecules can be cloned in any of the available vectors. For these reasons, the very long DNA molecules that compose an organism's genome must be cleaved into fragments that can be inserted into the vector DNA. Two types of enzymes—**restriction enzymes** and DNA ligases—facilitate production of such recombinant DNA molecules.

**Cutting DNA Molecules into Small Fragments** Restriction enzymes are endonucleases produced by bacteria that typically recognize specific 4- to 8-bp sequences, called *restriction sites,* and then cleave both DNA strands at this site. Restriction sites commonly are short *palindromic* sequences; that is, the restriction-site sequence is the same on each DNA strand when read in the 5′ to 3′ direction (Figure 5-11).

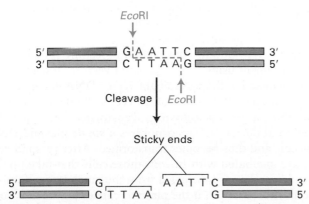

**FIGURE 5-11 Cleavage of DNA by the restriction enzyme EcoRI.** This restriction enzyme from *E. coli* makes staggered cuts at the specific 6-bp palindromic sequence shown, yielding fragments with single-stranded, complementary "sticky" ends. Many other restriction enzymes also produce fragments with sticky ends.

For each restriction enzyme, bacteria also produce a *modification enzyme,* which protects a host bacterium's own DNA from cleavage by modifying the host DNA at or near each potential cleavage site. The modification enzyme adds a methyl group to one or two bases, usually within the restriction site. When a methyl group is present there, the restriction endonuclease is prevented from cutting the DNA. Together with the restriction endonuclease, the methylating enzyme forms a restriction-modification system that protects the host DNA while it destroys incoming foreign DNA (e.g., bacteriophage DNA or DNA taken up during transformation) by cleaving it at all the restriction sites in the DNA.

Many restriction enzymes make staggered cuts in the two DNA strands at their recognition site, generating fragments that have a single-stranded "tail" at both ends, *sticky ends* (see Figure 5-11). The tails on the fragments generated at a given restriction site are complementary to those on all other fragments generated by the same restriction enzyme. At room temperature, these single-stranded regions can transiently base-pair with those on other DNA fragments generated with the same restriction enzyme. A few restriction enzymes, such as *Alu*I and *Sma*I, cleave both DNA strands at the same point within the restriction site, generating fragments with "blunt" (flush) ends in which all the nucleotides at the fragment ends are base-paired to nucleotides in the complementary strand.

The DNA isolated from an individual organism has a specific sequence, which purely by chance will contain a specific set of restriction sites. Thus a given restriction enzyme will cut the DNA from a particular source into a reproducible set of fragments called restriction fragments. The frequency with which a restriction enzyme cuts DNA, and thus the average size of the resulting restriction fragments, depends largely on the length of the recognition site. For example, a restriction enzyme that recognizes a 4-bp site will cleave DNA an average of once every $4^4$, or 256, base pairs, whereas an enzyme that recognizes an 8-bp sequence will cleave DNA an average of once every $4^8$ base pairs (~65 kbp). Restriction enzymes have been purified from several hundred different species of bacteria, allowing DNA molecules to be cut at a large number of different sequences corresponding to the recognition sites of these enzymes (Table 5-1).

**Inserting DNA Fragments into Vectors** DNA fragments with either sticky ends or blunt ends can be inserted into vector DNA with the aid of DNA ligases. During normal DNA replication, DNA ligase catalyzes the end-to-end joining (ligation) of short fragments of DNA called Okazaki fragments. For purposes of DNA cloning, purified DNA ligase is used to covalently join the ends of a restriction fragment and vector DNA that have complementary ends (Figure 5-12). The vector DNA and restriction fragment are covalently ligated together through the standard 3′ to 5′ phosphodiester bonds of DNA. In addition to ligating complementary sticky ends, the DNA ligase from bacteriophage T4 can ligate any two blunt DNA ends. However, blunt-end ligation is inherently inefficient and requires a higher concentration of both DNA and DNA ligase than does ligation of sticky ends.

## TABLE 5-1 Selected Restriction Enzymes and Their Recognition Sequences

| Enzyme | Source Microorganism | Recognition Site* | Ends Produced |
|--------|---------------------|-------------------|---------------|
| BamHI | Bacillus amyloliquefaciens | ↓<br>-G-G-A-T-C-C-<br>-C-C-T-A-G-G-<br>↑ | Sticky |
| Sau3A | Staphylococcus aureus | ↓<br>-G-A-T-C-<br>-C-T-A-G-<br>↑ | Sticky |
| EcoRI | Escherichia coli | ↓<br>-G-A-A-T-T-C-<br>-C-T-T-A-A-G<br>↑ | Sticky |
| HindIII | Haemophilus influenzae | ↓<br>-A-A-G-C-T-T-<br>-T-T-C-G-A-A-<br>↑ | Sticky |
| SmaI | Serratia marcescens | ↓<br>-C-C-C-G-G-G-<br>-G-G-G-C-C-C-<br>↑ | Blunt |
| NotI | Nocardia otitidis-caviarum | ↓<br>-G-C-G-G-C-C-G-C-<br>-C-G-C-C-G-G-C-G-<br>↑ | Sticky |

*Many of these recognition sequences are included in a common polylinker sequence (see Figure 5-13).

## E. coli Plasmid Vectors Are Suitable for Cloning Isolated DNA Fragments

**Plasmids** are circular, double-stranded DNA (dsDNA) molecules that are separate from a cell's chromosomal DNA. These extrachromosomal DNAs, which occur naturally in bacteria and in lower eukaryotic cells (e.g., yeast), exist in a parasitic or symbiotic relationship with their host cell. Like the host-cell chromosomal DNA, plasmid DNA is duplicated before every cell division. During cell division, copies of the plasmid DNA segregate to each daughter cell, ensuring continued propagation of the plasmid through successive generations of the host cell.

The plasmids most commonly used in recombinant DNA technology are those that replicate in E. coli. Investigators have engineered these plasmids to optimize their use as vectors in DNA cloning. For instance, removal of unneeded portions from naturally occurring E. coli plasmids yields plasmid vectors ~1.2–3 kb in circumferential length that contain three regions essential for DNA cloning: a replication origin; a marker that permits selection, usually a drug-resistance gene; and a region in which exogenous DNA fragments can be inserted (Figure 5-13). Host-cell enzymes replicate a plasmid beginning at the replication origin (ORI), a specific DNA sequence of 50–100 base pairs. Once DNA replication is initiated at the ORI, it continues around the circular plasmid regardless of its nucleotide sequence. Thus any DNA sequence inserted into such a plasmid is replicated along with the rest of the plasmid DNA.

Figure 5-14 outlines the general procedure for cloning a DNA fragment using E. coli plasmid vectors. When E. coli cells are mixed with recombinant vector DNA under certain conditions, a small fraction of the cells will take up the plasmid DNA, a process known as **transformation**. Typically, 1 cell in about 10,000 incorporates a *single* plasmid DNA molecule and thus becomes transformed. After plasmid vectors are incubated with E. coli, those cells that take up the plasmid can be easily selected from the much larger number of cells. For instance, if the plasmid carries a gene that confers resistance to the antibiotic ampicillin, transformed cells can be selected by growing them in an ampicillin-containing medium.

DNA fragments from a few base pairs up to ~10 kb can be inserted into plasmid vectors. When a recombinant plasmid

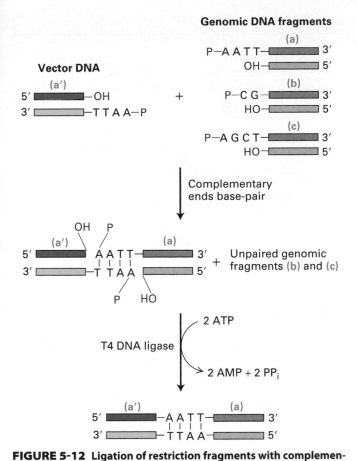

**Vector DNA**

(a')

5' ▬▬▬—OH     +

3' ▬▬▬—T T A A—P

**Genomic DNA fragments**

(a)

P—A A T T—▬▬▬ 3'

OH—▬▬▬ 5'

(b)

P—C G—▬▬▬ 3'

HO—▬▬▬ 5'

(c)

P—A G C T—▬▬▬ 3'

HO—▬▬▬ 5'

↓ Complementary ends base-pair

OH  P

(a')      (a)

5' ▬▬▬ A A T T—▬▬▬ 3'   +   Unpaired genomic
           | | | |                fragments (b) and (c)
3' ▬▬▬—T T A A ▬▬▬ 5'

P   HO

2 ATP ↘

T4 DNA ligase

2 AMP + 2 PP$_i$ ↗

(a')      (a)

5' ▬▬▬—A A T T—▬▬▬ 3'
          | | | |
3' ▬▬▬—T T A A—▬▬▬ 5'

**FIGURE 5-12 Ligation of restriction fragments with complementary sticky ends.** In this example, vector DNA cut with *Eco*RI is mixed with a sample containing restriction fragments produced by cleaving genomic DNA with several different restriction enzymes. The short base sequences composing the sticky ends of each fragment type are shown. The sticky end on the cut vector DNA (a') base-pairs only with the complementary sticky ends on the *Eco*RI fragment (a) in the genomic sample. The adjacent 3' hydroxyl and 5' phosphate groups (red) on the base-paired fragments then are covalently joined (ligated) by T4 DNA ligase.

with an inserted DNA fragment transforms an *E. coli* cell, all the antibiotic-resistant progeny cells that arise from the initial transformed cell will contain plasmids with the same inserted DNA. The inserted DNA is replicated along with the rest of the plasmid DNA and segregates to daughter cells as

the colony grows. In this way, the initial fragment of DNA is replicated in the colony of cells into a large number of identical copies. Since all the cells in a colony arise from a single transformed parental cell, they constitute a **clone** of cells, and the initial fragment of DNA inserted into the parental plasmid is referred to as *cloned DNA* or a *DNA clone*.

The versatility of an *E. coli* plasmid vector is increased by the addition of a *polylinker*, a synthetically generated sequence containing one copy of several different restriction sites that are not present elsewhere in the plasmid sequence (see Figure 5-13). When such a vector is treated with a restriction enzyme that recognizes a restriction site in the polylinker, the vector is cut only once within the polylinker. Subsequently, any DNA fragment of appropriate length produced with the same restriction enzyme can be inserted into the cut plasmid with DNA ligase. Plasmids containing a polylinker permit a researcher to use the same plasmid vector when cloning DNA fragments generated with different restriction enzymes, which simplifies experimental procedures.

For some purposes, such as the isolation and manipulation of large segments of the human genome, it is desirable to clone DNA segments as large as several megabases [1 megabase (Mb) = 1 million nucleotides]. For this purpose specialized plasmid vectors known as *BACs* (bacterial artificial chromosomes) have been developed. One type of BAC uses a replication origin derived from an endogenous plasmid of *E. coli* known as the *F factor*. The F factor and cloning vectors derived from it can be stably maintained at a single copy per *E. coli* cell even when they contain inserted sequences of up to about 2 Mb. Production of BAC libraries requires special methods for the isolation, ligation, and transformation of large segments of DNA because segments of DNA larger than about 20 kb are highly vulnerable to mechanical breakage by even standard manipulations such as pipetting.

## cDNA Libraries Represent the Sequences of Protein-Coding Genes

A collection of DNA molecules each cloned into a vector molecule is known as a **DNA library**. When genomic DNA from a particular organism is the source of the starting DNA, the set of clones that collectively represent all the DNA sequences in the genome is known as a *genomic library*. Such genomic

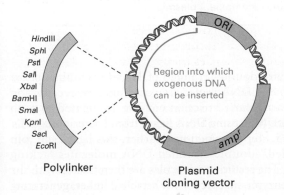

*Hind*III
*Sph*I
*Pst*I
*Sal*I
*Xba*I
*Bam*HI
*Sma*I
*Kpn*I
*Sac*I
*Eco*RI

Polylinker

ORI

Region into which
exogenous DNA
can be inserted

*amp*$^r$

Plasmid
cloning vector

**FIGURE 5-13 Basic components of a plasmid cloning vector that can replicate within an E. coli cell.** Plasmid vectors contain a selectable gene such as *amp*$^r$, which encodes the enzyme β-lactamase and confers resistance to ampicillin. Exogenous DNA can be inserted into the bracketed region without disturbing the ability of the plasmid to replicate or express the *amp*$^r$ gene. Plasmid vectors also contain a replication origin (ORI) sequence where DNA replication is initiated by host-cell enzymes. Inclusion of a synthetic polylinker containing the recognition sequences for several different restriction enzymes increases the versatility of a plasmid vector. The vector is designed so that each site in the polylinker is unique on the plasmid.

**EXPERIMENTAL FIGURE 5-14 DNA cloning in a plasmid vector permits amplification of a DNA fragment.** A fragment of DNA to be cloned is first inserted into a plasmid vector containing an ampicillin-resistance gene (*amp'*), such as that shown in Figure 5-13. Only the few cells transformed by incorporation of a plasmid molecule will survive on ampicillin-containing medium. In transformed cells, the plasmid DNA replicates and segregates into daughter cells, resulting in formation of an ampicillin-resistant colony.

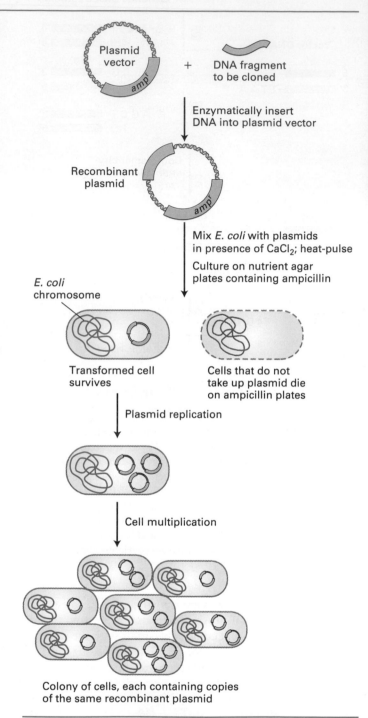

libraries are ideal for representing the genetic content of relatively simple organisms such as bacteria or yeast but present certain experimental difficulties for higher eukaryotes. First, the genes from such organisms usually contain extensive intron sequences and therefore can be too large to be inserted intact into plasmid vectors. As a result, the sequences of individual genes are broken apart and carried in more than one clone. Moreover, the presence of introns and long intergenic regions in genomic DNA often makes it difficult to identify the important parts of a gene that actually encode protein sequences. For example, only about 1.5 percent of the human genome actually represents protein-coding gene sequences. Thus for many studies, cellular mRNAs, which lack the noncoding regions present in genomic DNA, are a more useful starting material for generating a DNA library. In this approach, DNA copies of mRNAs, called **complementary DNAs (cDNAs)**, are synthesized and cloned into plasmid vectors. A large collection of the resulting cDNA clones, representing all the mRNAs expressed in a cell type, is called a *cDNA library*.

## cDNAs Prepared by Reverse Transcription of Cellular mRNAs Can Be Cloned to Generate cDNA Libraries

The first step in preparing a cDNA library is to isolate the total mRNA from the cell type or tissue of interest. Because of their poly(A) tails, mRNAs are easily separated from the much more prevalent rRNAs and tRNAs present in a cell extract by use of a column to which short strings of thymidylate (oligo-dTs) are linked to the matrix. The general procedure for preparing a cDNA library from a mixture of cellular mRNAs is outlined in Figure 5-15. The enzyme reverse transcriptase, which is found in retroviruses, is used to synthesize a strand of DNA complementary to each mRNA molecule, starting from an oligo-dT primer (steps **1** and **2**). The resulting cDNA-mRNA hybrid molecules are converted in several steps to double-stranded cDNA molecules corresponding to all the mRNA molecules in the original preparation (steps **3**–**5**). Each double-stranded cDNA contains an oligo-dC·oligo-dG double-stranded region at one end and an oligo-dT·oligo-dA double-stranded region at the other end. Methylation of the cDNA protects it from subsequent restriction enzyme cleavage (step **6**).

To prepare double-stranded cDNAs for cloning, short double-stranded DNA molecules containing the recognition site for a particular restriction enzyme are ligated to both ends of the cDNAs using DNA ligase from bacteriophage T4 (Figure 5-15, step **7**). As noted earlier, this ligase can join "blunt-ended" double-stranded DNA molecules lacking sticky ends. The resulting molecules are then treated with the restriction enzyme specific for the attached linker, generating

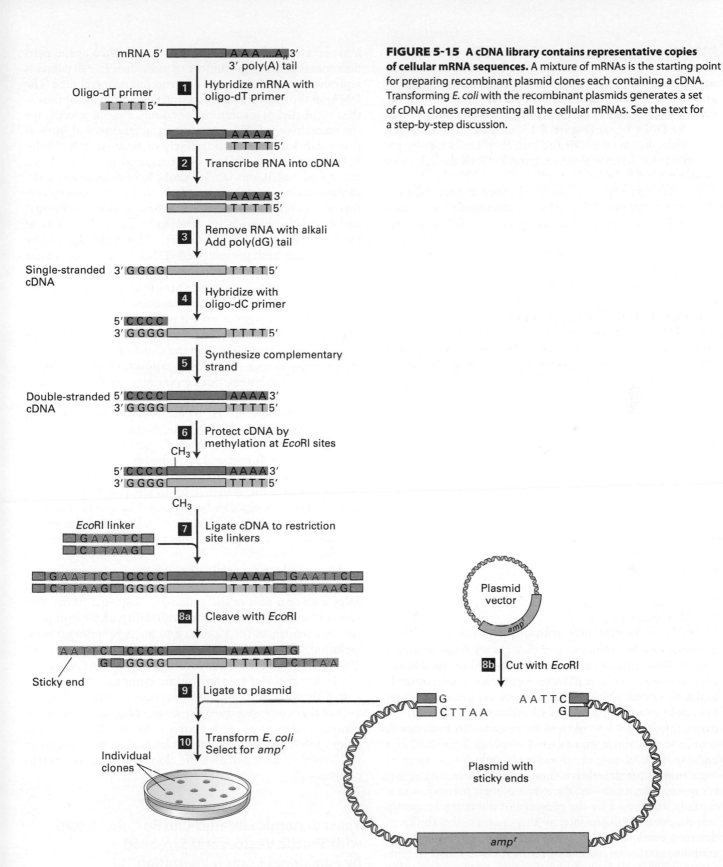

**FIGURE 5-15 A cDNA library contains representative copies of cellular mRNA sequences.** A mixture of mRNAs is the starting point for preparing recombinant plasmid clones each containing a cDNA. Transforming *E. coli* with the recombinant plasmids generates a set of cDNA clones representing all the cellular mRNAs. See the text for a step-by-step discussion.

cDNA molecules with sticky ends (step **8a**). In a separate procedure, plasmid DNA is treated with the same restriction enzyme to produce the appropriate sticky ends (step **8b**).

The vector and the collection of cDNAs, all containing complementary sticky ends, then are mixed and joined covalently by DNA ligase (Figure 5-15, step **9**). The resulting DNA molecules are transformed into *E. coli* cells to generate individual clones; each clone carrying a cDNA derived from a single mRNA.

Because different genes are transcribed at very different rates, cDNA clones corresponding to abundantly transcribed genes will be represented many times in a cDNA library, whereas cDNAs corresponding to infrequently transcribed genes will be extremely rare or not present at all. This property is advantageous if an investigator is interested in a gene that is transcribed at a high rate in a particular cell type. In this case, a cDNA library prepared from mRNAs expressed in that cell type will be enriched in the cDNA of interest, facilitating isolation of clones carrying that cDNA from the library. However, to have a reasonable chance of including clones corresponding to slowly transcribed genes, mammalian cDNA libraries must contain $10^6$–$10^7$ individual recombinant clones.

## DNA Libraries Can Be Screened by Hybridization to an Oligonucleotide Probe

Both genomic and cDNA libraries of various organisms contain hundreds of thousands to upward of a million individual clones in the case of higher eukaryotes. Two general approaches are available for screening libraries to identify clones carrying a gene or other DNA region of interest: (1) detection with oligonucleotide **probes** that bind to the clone of interest and (2) detection based on expression of the encoded protein. Here we describe the first method; an example of the second method is presented in the next section.

The basis for screening with oligonucleotide probes is **hybridization**, the ability of complementary single-stranded DNA or RNA molecules to associate (hybridize) specifically with each other via base pairing. As discussed in Chapter 4, double-stranded (duplex) DNA can be denatured (melted) into single strands by heating in a dilute salt solution. If the temperature then is lowered and the ion concentration raised, complementary single strands will reassociate (hybridize) into duplexes. In a mixture of nucleic acids, only complementary single strands (or strands containing complementary regions) will reassociate; moreover, the extent of their reassociation is virtually unaffected by the presence of noncomplementary strands. As we will see later in this chapter, the ability to identify a particular DNA or RNA sequence within a highly complex mixture of molecules through nucleic acid hybridization is the basis for many techniques employed to study gene expression.

The steps involved in screening an *E. coli* plasmid cDNA library are depicted in Figure 5-16. First, the DNA to be screened must be attached to a solid support. A *replica* of the petri dish containing a large number of individual *E. coli* clones is reproduced on the surface of a nitrocellulose membrane. The DNA on the membrane is denatured, and the membrane is then incubated in a solution containing a probe specific for the recombinant DNA containing the fragment of interest that is labeled either radioactively or fluorescently. Under hybridization conditions (near neutral pH, 40–65 °C, 0.3–0.6 M NaCl), this labeled probe hybridizes to any complementary nucleic acid strands bound to the membrane. Any excess probe that does not hybridize is washed away, and the labeled hybrids are detected by autoradiography or by fluorescent imaging of the filter. This technique can be used to screen both genomic and cDNA libraries but is most commonly used to isolate specific cDNAs.

Clearly, identification of specific clones by the membrane-hybridization technique depends on the availability of complementary radiolabeled probes. For an oligonucleotide to be useful as a probe, it must be long enough for its sequence to occur uniquely in the clone of interest and not in any other clones. For most purposes, this condition is satisfied by oligonucleotides containing about 20 nucleotides. This is because a specific 20-nucleotide sequence occurs once in every $4^{20}$ ($\sim 10^{12}$) nucleotides. Since all genomes are much smaller ($\sim 3 \times 10^9$ nucleotides for humans), a specific 20-nucleotide sequence in a genome usually occurs only once. With automated instruments now available, researchers can program the chemical synthesis of oligonucleotides of specific sequence up to about 100 nucleotides long. Longer probes can be prepared by the polymerase chain reaction (PCR), a widely used technique for amplifying specific DNA sequences that is described later in this chapter.

How might an investigator design an oligonucleotide probe to identify a clone encoding a particular protein? It helps if all or a part of the amino acid sequence of the protein is known. Thanks to the availability of the complete genomic sequences for humans and many other organisms, including all of the model organisms such as the mouse, *Drosophila,* and the roundworm *Caenorhabditis elegans,* a researcher can use an appropriate computer program to search the genomic sequence database for the coding sequence that corresponds to the amino acid sequence of the protein under study. If a match is found, then a single, unique DNA probe based on this known genomic sequence will hybridize perfectly with the clone encoding the protein of interest.

## Yeast Genomic Libraries Can Be Constructed with Shuttle Vectors and Screened by Functional Complementation

In some cases, a DNA library can be screened for the ability to express a functional protein that complements a **recessive** mutation. Such a screening strategy would be an efficient way to

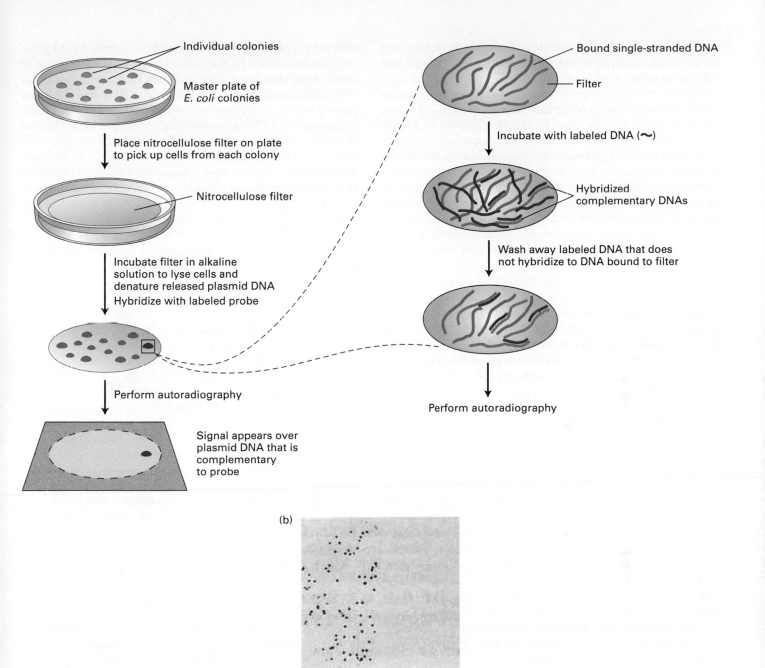

**EXPERIMENTAL FIGURE 5-16 cDNA libraries can be screened with a radiolabeled probe to identify a clone of interest.** The appearance of a spot on the autoradiogram indicates the presence of a recombinant clone containing DNA complementary to the probe. The position of the spot on the autoradiogram is the mirror image of the position of that particular clone on the original petri dish (although for ease of comparison, it is not shown reversed here). Aligning the autoradiogram with the original petri dish will locate the corresponding clone from which *E. coli* cells can be recovered. (b) This autoradiogram shows five colonies (arrows) of *E. coli* containing the desired cDNA. [Part (b) from H. Fromm and N.-H. Chua, 1992, *Plant. Mol. Biol. Rep.* **10**:199.]

isolate a cloned gene that corresponds to an interesting recessive mutation identified in an experimental organism. To illustrate this method, referred to as **functional complementation,** we describe how yeast genes cloned in special *E. coli* plasmids can be introduced into mutant yeast cells to identify the wild-type gene that is defective in the mutant strain.

Libraries constructed for the purpose of screening among yeast gene sequences usually are constructed from genomic DNA rather than cDNA. Because *Saccharomyces* genes do not contain multiple introns, they are sufficiently compact that the entire sequence of a gene can be included in a genomic DNA fragment inserted into a plasmid vector. To construct a plasmid genomic library that is to be screened by functional complementation in yeast cells, the plasmid vector must be capable of replication in both *E. coli* cells and yeast cells. This type of vector, capable of propagation in two different hosts, is called a shuttle vector. The structure of a typical yeast shuttle vector is shown in Figure 5-17a. This vector contains the basic elements that permit cloning of DNA fragments in *E. coli*. In addition, the shuttle vector contains an autonomously replicating sequence (ARS), which functions as an origin for DNA replication in yeast; a yeast centromere (called CEN), which allows faithful segregation of the plasmid during yeast cell division; and a yeast gene encoding an enzyme for uracil synthesis (*URA3*), which serves as a selectable marker in an appropriate yeast mutant.

To increase the probability that all regions of the yeast genome are successfully cloned and represented in the plasmid library, the genomic DNA usually is only partially digested to yield overlapping restriction fragments of ~10 kb. These fragments are then ligated into the shuttle vector in which the polylinker has been cleaved with a restriction enzyme that produces sticky ends complementary to those on the yeast DNA fragments (Figure 5-17b). Because the 10-kb restriction fragments of yeast DNA are incorporated into the shuttle vectors randomly, at least $10^5$ *E. coli* colonies, each containing a particular recombinant shuttle vector, are necessary to ensure that each region of yeast DNA has a high probability of being represented in the library at least once.

Figure 5-18 outlines how such a yeast genomic library can be screened to isolate the wild-type gene corresponding to one of the temperature-sensitive *cdc* mutations mentioned earlier in this chapter. The starting yeast strain is a double mutant that requires uracil for growth due to a *ura3* mutation and is temperature sensitive due to a *cdc28* mutation identified by its phenotype (see Figure 5-6). Recombinant plasmids isolated from the yeast genomic library are mixed with yeast cells under conditions that promote transformation of the cells with foreign DNA. Since transformed yeast cells carry a plasmid-borne copy of the wild-type *URA3* gene, they can be selected by their ability to grow in the absence of uracil. Typically, about 20 petri dishes, each containing about 500 yeast transformants, are sufficient to represent the entire yeast genome. This collection of yeast transformants can be maintained at 23 °C, a temperature permissive for growth of the *cdc28* mutant. The entire collection on 20 plates is then transferred to replica plates, which are placed at 36 °C, a nonpermissive temperature for *cdc* mutants. Yeast colonies that carry recombinant plasmids expressing a wild-type copy

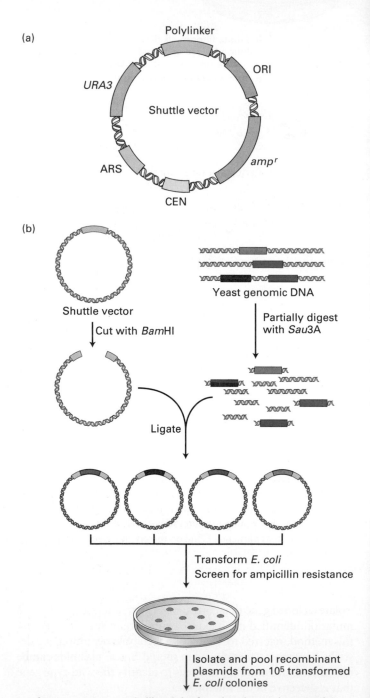

**EXPERIMENTAL FIGURE 5-17  A yeast genomic library can be constructed in a plasmid shuttle vector that can replicate in yeast and *E. coli*.** (a) Components of a typical plasmid shuttle vector for cloning *Saccharomyces* genes. The presence of a yeast origin of DNA replication (ARS) and a yeast centromere (CEN) allows stable replication and segregation in yeast. Also included is a yeast selectable marker such as *URA3*, which allows a *ura3* mutant to grow on medium lacking uracil. Finally, the vector contains sequences for replication and selection in *E. coli* (ORI and *amp*$^r$) and a polylinker for easy insertion of yeast DNA fragments. (b) Typical protocol for constructing a yeast genomic library. Partial digestion of total yeast genomic DNA with *Sau*3A is adjusted to generate fragments with an average size of about 10 kb. The vector is prepared to accept the genomic fragments by digestion with *Bam*HI, which produces the same sticky ends as *Sau*3A. Each transformed clone of *E. coli* that grows after selection for ampicillin resistance contains a single type of yeast DNA fragment.

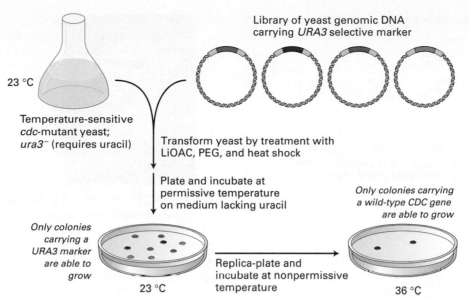

**EXPERIMENTAL FIGURE 5-18 Screening of a yeast genomic library by functional complementation can identify clones carrying the normal form of a mutant yeast gene.** In this example, a wild-type *CDC* gene is isolated by complementation of a *cdc* yeast mutant. The *Saccharomyces* strain used for screening the yeast library carries *ura3* and a temperature-sensitive *cdc* mutation. This mutant strain is grown and maintained at a permissive temperature (23 °C). Pooled recombinant plasmids prepared as shown in Figure 5-17 are incubated with the mutant yeast cells under conditions that promote transformation. The relatively few transformed yeast cells, which contain recombinant plasmid DNA, can grow in the absence of uracil at 23 °C. When transformed yeast colonies are replica-plated and placed at 36 °C (a nonpermissive temperature), only clones carrying a library plasmid that contains the wild-type copy of the *CDC* gene will survive. LiOAC = lithium acetate; PEG = polyethylene glycol.

of the *CDC28* gene will be able to grow at 36 °C. Once temperature-resistant yeast colonies have been identified, plasmid DNA can be extracted from the cultured yeast cells and analyzed by subcloning and DNA sequencing, topics we take up next.

## Gel Electrophoresis Allows Separation of Vector DNA from Cloned Fragments

In order to manipulate or sequence a cloned DNA fragment, it sometimes must first be separated from the vector DNA. This can be accomplished by cutting the recombinant DNA clone with the same restriction enzyme used to produce the recombinant vectors originally. The cloned DNA and vector DNA then are subjected to gel electrophoresis, a powerful method for separating DNA molecules of different size (Figure 5-19).

Near neutral pH, DNA molecules carry a large negative charge and therefore move toward the positive electrode during gel electrophoresis. Because the gel matrix restricts random diffusion of the molecules, molecules of the same length migrate together as a band whose width equals that of the well into which the original DNA mixture was placed at the start of the electrophoretic run. Smaller molecules move through the gel matrix more readily than larger molecules, so that molecules of different length migrate as distinct bands. Smaller DNA molecules from about 10 to 2000

nucleotides can be separated electrophoretically on *poly-acrylamide gels*, and larger molecules from about 200 nucleotides to more than 20 kb on *agarose gels*.

A common method for visualizing separated DNA bands on a gel is to incubate the gel in a solution containing the fluorescent dye ethidium bromide. This planar molecule binds to DNA by intercalating between the base pairs. Binding concentrates ethidium in the DNA and also increases its intrinsic fluorescence. As a result, when the gel is illuminated with ultraviolet light, the regions of the gel containing DNA fluoresce much more brightly than the regions of the gel without DNA.

Once a cloned DNA fragment, especially a long one, has been separated from vector DNA, it often is treated with various restriction enzymes to yield smaller fragments. After separation by gel electrophoresis, all or some of these smaller fragments can be ligated individually into a plasmid vector and cloned in *E. coli* by the usual procedure. This process, known as *subcloning*, is an important step in rearranging parts of genes into useful new configurations. For instance, an investigator who wants to change the conditions under which a gene is expressed might use subcloning to replace the normal promoter associated with a cloned gene with a DNA segment containing a different promoter. Subcloning also can be used to obtain cloned DNA fragments that are of an appropriate length for determining the nucleotide sequence.

**(a)** DNA restriction fragments

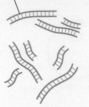

↓ Place mixture in the well of an agarose or polyacrylamide gel. Apply electric field

Well

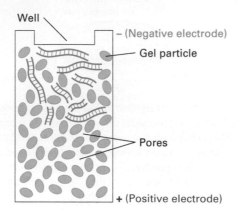

– (Negative electrode)

Gel particle

Pores

+ (Positive electrode)

↓ Molecules move through pores in gel at a rate inversely proportional to their chain length

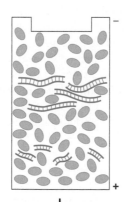

–

+

↓ Subject to autoradiography or incubate with fluorescent dye

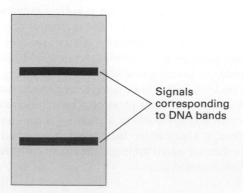

Signals corresponding to DNA bands

**(b)**

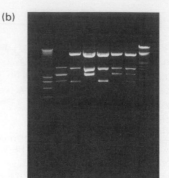

**EXPERIMENTAL FIGURE 5-19 Gel electrophoresis separates DNA molecules of different lengths.** (a) A gel is prepared by pouring a liquid containing either melted agarose or unpolymerized acrylamide between two glass plates a few millimeters apart. As the agarose solidifies or the acrylamide polymerizes into polyacrylamide, a gel matrix (orange ovals) forms consisting of long, tangled chains of polymers. The dimensions of the interconnecting channels, or pores, depend on the concentration of the agarose or acrylamide used to form the gel. The separated bands can be visualized by autoradiography (if the fragments are radiolabeled) or by addition of a fluorescent dye (e.g., ethidium bromide) that binds to DNA. (b) A photograph of a gel stained with ethidium bromide (EtBr). EtBr binds to DNA and fluoresces under UV light. The bands in the far left and far right lanes are known as DNA ladders—DNA fragments of known size that serve as a reference for determining the length of the DNA fragments in the experimental sample. [Part (b) Science Photo Library.]

## The Polymerase Chain Reaction Amplifies a Specific DNA Sequence from a Complex Mixture

If the nucleotide sequences at the ends of a particular DNA region are known, the intervening fragment can be amplified directly by the **polymerase chain reaction (PCR)**. Here we describe the basic PCR technique and three situations in which it is used.

The PCR depends on the ability to alternately denature (melt) double-stranded DNA molecules and hybridize complementary single strands in a controlled fashion. As outlined in Figure 5-20, a typical PCR procedure begins by heat-denaturation at 95 °C of a DNA sample into single strands. Next, two synthetic oligonucleotides complementary to the 3′ ends of the target DNA segment of interest are added in great excess to the denatured DNA, and the temperature is lowered to 50–60 °C. These specific oligonucleotides, which are at a very high concentration, will hybridize with their complementary sequences in the DNA sample, whereas the long strands of the sample DNA remain apart because of their low concentration. The hybridized oligonucleotides then serve as primers for DNA chain synthesis in the presence of deoxynucleotides (dNTPs) and a temperature-resistant DNA polymerase such as that from *Thermus aquaticus* (a bacterium that lives in hot springs). This enzyme, called Taq *polymerase*, can remain active even after being heated to 95 °C and can extend the primers at temperatures up to 72 °C. When synthesis is complete, the whole mixture

**FIGURE 5-20 The polymerase chain reaction (PCR) is widely used to amplify DNA regions of known sequences.** To amplify a specific region of DNA, an investigator will chemically synthesize two different oligonucleotide primers complementary to sequences of approximately 18 bases flanking the region of interest (designated as light blue and dark blue bars). The complete reaction is composed of a complex mixture of double-stranded DNA (usually genomic DNA containing the target sequence of interest), a stoichiometric excess of both primers, the four deoxynucleoside triphosphates, and a heat-stable DNA polymerase known as *Taq* polymerase. During each PCR cycle, the reaction mixture is first heated to separate the strands and then cooled to allow the primers to bind to complementary sequences flanking the region to be amplified. *Taq* polymerase then extends each primer from its 3′ end, generating newly synthesized strands that extend in the 3′ direction to the 5′ end of the template strand. During the third cycle, two double-stranded DNA molecules are generated equal in length to the sequence of the region to be amplified. In each successive cycle the target segment, which will anneal to the primers, is duplicated and will eventually vastly outnumber all other DNA segments in the reaction mixture. Successive PCR cycles can be automated by cycling the reaction for timed intervals at high temperature for DNA melting and at a defined lower temperature for the annealing and elongation parts of the cycle. A reaction that cycles 20 times will amplify the specific target sequence 1-million-fold.

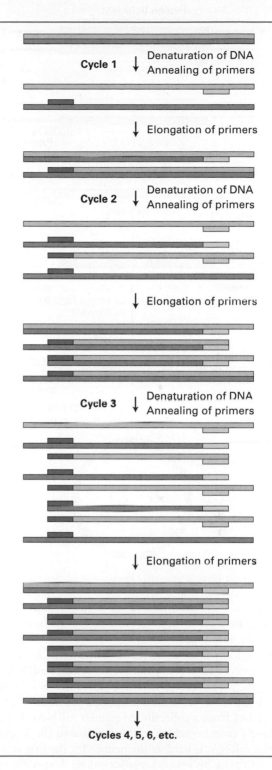

is then heated to 95 °C to denature the newly formed DNA duplexes. After the temperature is lowered again, another cycle of synthesis takes place because excess primer is still present. Repeated cycles of denaturation (heating) followed by hybridization and synthesis (cooling) quickly amplify the sequence of interest. At each cycle, the number of copies of the sequence between the primer sites is doubled; therefore, the desired sequence increases exponentially—about a million-fold after 20 cycles—whereas all other sequences in the original DNA sample remain unamplified.

**Direct Isolation of a Specific Segment of Genomic DNA** For organisms in which all or most of the genome has been sequenced, PCR amplification starting with the total genomic DNA often is the easiest way to obtain a specific DNA region of interest for cloning. In this application, the two oligonucleotide primers are designed to hybridize to sequences flanking the genomic region of interest and to include sequences that are recognized by specific restriction enzymes (Figure 5-21). After amplification of the desired target sequence for about 20 PCR cycles, cleavage with the appropriate restriction enzymes produces sticky ends that allow efficient ligation of the fragment into a plasmid vector cleaved by the same restriction enzymes in the polylinker. The resulting recombinant plasmids, all carrying the identical genomic DNA segment, can then be cloned in *E. coli* cells. With certain refinements of the PCR, even DNA segments greater than 10 kb in length can be amplified and cloned in this way.

Note that this method does not involve cloning of large numbers of restriction fragments derived from genomic DNA and their subsequent screening to identify the specific fragment of interest. In effect, the PCR method inverts this traditional approach and so avoids its most tedious aspects. The PCR method is useful for isolating gene sequences to be manipulated in a variety of useful ways described later. In

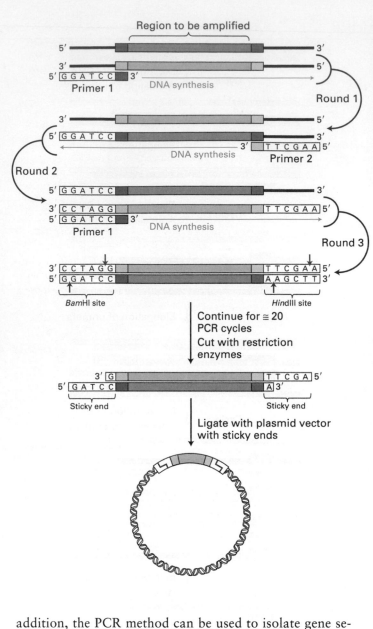

Region to be amplified

Primer 1

DNA synthesis

Round 1

Round 2

DNA synthesis

Primer 2

Primer 1

DNA synthesis

Round 3

BamHI site

HindIII site

Continue for ≅ 20 PCR cycles

Cut with restriction enzymes

Sticky end

Sticky end

Ligate with plasmid vector with sticky ends

To carry out a quantitative RT-PCR reaction, the amount of double-stranded DNA sequence produced by each amplification cycle is determined as the amplification of a particular mRNA sequence proceeds. By extrapolation from these amounts, an estimate of the amount of starting mRNA sequence can be obtained. Such quantitative RT-PCR carried out on tissues or whole organisms using primers targeted to genes of interest provide one of the most accurate means to follow changes in gene expression.

**Preparation of Probes** Earlier we mentioned that oligonucleotide probes for hybridization assays can be chemically synthesized. Preparation of such probes by PCR amplification requires chemical synthesis of only two relatively short primers corresponding to the two ends of the target sequence. The starting sample for PCR amplification of the target sequence can be a preparation of genomic DNA or a preparation of cDNA synthesized from the total cellular mRNA. To generate a radiolabeled product from PCR, $^{32}$P-labeled dNTPs are included during the last several amplification cycles or a fluorescently labeled product can be obtained by using fluorescently labeled dNTPs during the last amplification cycles. Because probes prepared by PCR are relatively long and have many radioactive or fluorescent nucleotides incorporated into them, these probes usually give a stronger and more specific signal than chemically synthesized probes.

**Tagging of Genes by Insertion Mutations** Another useful application of the PCR is to amplify a "tagged" gene from the genomic DNA of a mutant strain. This approach is a simpler method for identifying genes associated with a particular mutant phenotype than screening of a library by functional complementation (see Figure 5-18).

The key to this use of the PCR is the ability to produce mutations by insertion of a known DNA sequence into the genome of an experimental organism. Such insertion mutations can be generated by use of mobile DNA elements, which can move (or transpose) from one chromosomal site to another. As discussed in more detail in Chapter 6, these DNA sequences occur naturally in the genomes of most organisms and may give rise to loss-of-function mutations if they transpose into a protein-coding region.

addition, the PCR method can be used to isolate gene sequences from mutant organisms to determine how they differ from the wild type.

A variation on the PCR method allows PCR amplification of a specific cDNA sequence from cellular mRNAs. This method, known as *reverse transcriptase–PCR (RT-PCR)*, begins with the same procedure described previously for isolation of cDNA from a collection of cellular mRNAs. Typically, an oligo-dT primer, which will hybridize to the 3′ poly(A) tail of the mRNA, is used as the primer for the first strand of cDNA synthesis by reverse transcriptase. A specific cDNA can then be isolated from this complex mixture of cDNAs by PCR amplification using two oligonucleotide primers designed to match sequences at the 5′ and 3′ ends of the corresponding mRNA. As described previously, these primers could be designed to include restriction sites to facilitate the insertion of amplified cDNA into a suitable plasmid vector.

RT-PCR can be performed so that the starting amount of a particular cellular mRNA can be determined accurately.

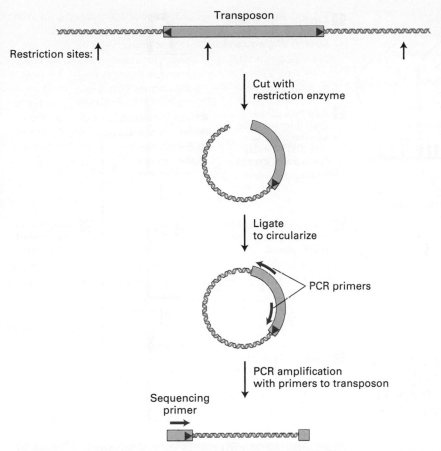

**Restriction sites:**

Transposon

Cut with restriction enzyme

Ligate to circularize

PCR primers

PCR amplification with primers to transposon

Sequencing primer

**FIGURE 5-22 The genomic sequence at the insertion site of a transposon is revealed by PCR amplification and sequencing.** To obtain the DNA sequence of the insertion site of a P-element transposon it is necessary to PCR-amplify the junction between known transposon sequences and unknown flanking chromosomal sequences. One method to achieve this is to cleave genomic DNA with a restriction enzyme that cleaves once within the transposon sequence. Ligation of the resulting restriction fragments will generate circular DNA molecules. By using appropriately designed DNA primers that match transposon sequences, it is possible to PCR-amplify the desired junction fragment. Finally, a DNA sequencing reaction (see Figures 5-23 and 5-24) is performed using the PCR-amplified fragment as a template and an oligonucleotide primer that matches sequences near the end of the transposon to obtain the sequence of the junction between the transposon and chromosome.

For example, researchers have modified a *Drosophila* mobile DNA element, known as the *P element,* to optimize its use in the experimental generation of insertion mutations. Once it has been demonstrated that insertion of a P element causes a mutation with an interesting phenotype, the genomic sequences adjacent to the insertion site can be amplified by a variation of the standard PCR protocol that uses synthetic primers complementary to the known P-element sequence but that allows unknown neighboring sequences to be amplified. One such method, depicted in Figure 5-22, begins by cleaving *Drosophila* genomic DNA containing a P-element insertion with a restriction enzyme that cleaves once within the P-element DNA. The collection of cleaved DNA fragments treated with DNA ligase yields circular molecules, some of which will contain P-element DNA. The chromosomal region flanking the P element can then be amplified by PCR using primers that match P-element sequences and are elongated in opposite directions. The sequence of the resulting amplified fragment can then be determined using a third DNA primer. The crucial sequence for identifying the site of P-element insertion is the junction between the end of the P-element and genomic sequences. Overall, this approach avoids the cloning of large numbers of DNA fragments and their screening to detect a cloned DNA corresponding to a mutated gene of interest.

Similar methods have been applied to other organisms for which insertion mutations can be generated using either mobile DNA elements or viruses with sequenced genomes that can insert randomly into the genome.

## Cloned DNA Molecules Are Sequenced Rapidly by Methods Based on PCR

The complete characterization of any cloned DNA fragment requires determination of its nucleotide sequence. The technology used to determine the sequence of a DNA segment represents one of the most rapidly developing fields in molecular biology. In the 1970s, F. Sanger and his colleagues developed the chain-termination procedure, which served as the basis for most DNA sequencing methods for the next 30 years. The idea behind this method is to synthesize from the DNA fragment to be sequenced a set of daughter strands that are labeled at one end and terminate at one of the four nucleotides. Separation of the truncated daughter strands by gel electrophoresis, which can resolve strands that differ in length by one nucleotide, can then reveal the length of all strands ending in G, A, T, or C. From these collections of strands of different lengths, the nucleotide sequence of the original DNA fragment can be established. The Sanger method has undergone many refinements and now can be fully automated, but because each new DNA sequence requires a separate individual sequencing reaction, the overall rate by which new DNA sequences can be produced by this method is limited by the total number of reactions that can be performed at one time.

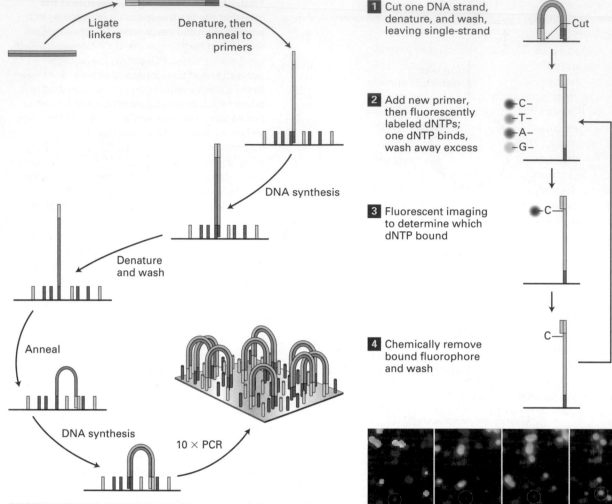

**1** Cut one DNA strand, denature, and wash, leaving single-strand — Cut

**2** Add new primer, then fluorescently labeled dNTPs; one dNTP binds, wash away excess

C–
T–
A–
G–

**3** Fluorescent imaging to determine which dNTP bound

C

**5** Repeat until DNA strand is replicated

**4** Chemically remove bound fluorophore and wash

C

Ligate linkers

Denature, then anneal to primers

DNA synthesis

Denature and wash

Anneal

DNA synthesis

10 × PCR

**EXPERIMENTAL FIGURE 5-23 Generation of clusters of identical DNA molecules attached to a solid support.** A large collection of DNA molecules to be sequenced is ligated to double-stranded linkers, which become attached to each end of the fragment. The DNA is then amplified by PCR using primers matching the sequences of the linkers that are covalently attached to a solid substrate. Ten cycles of amplification yield about 1000 identical copies of the DNA fragment localized in a small cluster, which is attached at both ends to the solid substrate. These reactions are optimized to produce as many as $3 \times 10^9$ discrete, non-overlapping clusters that are ready to be sequenced.

**EXPERIMENTAL FIGURE 5-24 Using fluorescent-tagged deoxyribonucleotide triphosphates for sequence determination.** The reaction begins by cleaving one strand of the clustered DNA. After melting, a single DNA strand remains attached to the flow cell. A synthetic oligodeoxynucleotide is used as the primer for the polymerization reaction that contains dNTPs, each fluorescently tagged with a different color. The fluorescent tag is designed to block the 3′ OH group on the dNTP so that once the fluorescent dNTP has been incorporated, further elongation is not possible. Because DNA polymerase will incorporate the same fluorescent dNTP into each of the ~1000 DNA copies in a cluster, the entire cluster will be uniformly labeled with the same fluorescent color, which can be imaged in a special microscope. Once all of the clusters have been imaged, the fluorescent tags are removed by a chemical reaction that leaves a new primer terminus available for the next cycle of fluorescently labeled dNTP addition. A typical sequencing reaction may carry out 100 polymerization cycles, allowing 100 bases of sequence for each cluster to be determined. Thus a total sequencing reaction of this type may generate as much as $3 \times 10^{11}$ bases of sequence information in about two days. [OpenWetWare user Andrea Loehr (http://openwetware.org/wiki/User:Andrea_Loehr).]

A breakthrough in sequencing technology occurred when methods were devised to allow a single sequencing instrument to carry out billons of sequencing reactions simultaneously by localizing them in tiny clusters on the surface of a solid substrate. Since 2007, when the so-called *next generation* sequencers became commercially available, the capacity for new sequence production increased enormously and since then has been doubling every year. In one popular sequencing method, billions of different DNA strands to be sequenced are prepared by ligating double-stranded linkers to their ends (Figure 5-23). Next the DNA fragments are amplified by PCR using primers that match the linker sequences. This PCR amplification reaction differs from the standard PCR amplification shown in Figure 5-20 in that the primers used are covalently attached to a solid substrate. Thus as the PCR amplification proceeds, one end of each daughter DNA strand is covalently linked to the substrate and at the end of the amplification ~1,000 identical PCR products are linked to the surface in in a tight cluster.

These clusters can then be sequenced by using a special microscope to image fluorescently labeled deoxyribonucleotides (dNTPs) as they are incorporated by DNA polymerase one at a time into a growing DNA chain (Figure 5-24). First, one strand is cut and washed out, leaving a single-stranded DNA template. Then sequencing is carried out on the ~1,000 identical templates in clusters, one nucleotide at a time. All four dNTPs are fluorescently labeled and added to the sequencing reaction. After they are allowed to anneal, the substrate is imaged and the color of each cluster is recorded. Next the fluorescent tag is chemically removed and a new dNTP is allowed to bind. This cycle is repeated about 100 times, resulting in billions of ~100 nucleotide sequences.

In order to sequence a long continuous region of genomic DNA or even the entire genome of an organism, researchers usually employ one of the strategies outlined in Figure 5-25. The first method requires the isolation of a collection of cloned DNA fragments whose sequences overlap. Once the sequence of one of these fragments is determined, oligonucleotides based on that sequence can be chemically synthesized for use as primers in sequencing the adjacent overlapping fragments. In this way, the sequence of a long stretch of DNA is determined incrementally by sequencing of the overlapping cloned DNA fragments that compose it. A second method, which is called *whole genome shotgun sequencing*, bypasses the time-consuming step of isolating an ordered collection of DNA segments that span the genome. This method involves simply sequencing random clones from a genomic library. A total number of clones are chosen for sequencing so that on average each segment of the genome is sequenced about 10 times. This degree of coverage ensures that each segment of the genome is sequenced more than once. The entire genomic sequence is then assembled using a computer algorithm that aligns all the sequences, using their regions of overlap. Whole genome shotgun sequencing is the fastest and most cost-effective method for sequencing long stretches of DNA, and most genomes, including the human genome, have been sequenced by this method.

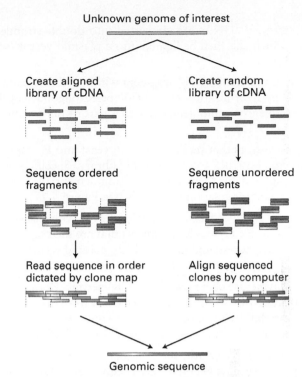

**EXPERIMENTAL FIGURE 5-25 Two strategies for assembling whole genome sequences.** One method (*left*) depends on isolating and assembling a set of cloned DNA segments that span the genome. This can be done by matching cloned segments by hybridization or by alignment of restriction-site maps. The DNA sequence of the ordered clones can then be assembled into a complete genomic sequence. The alternative method (*right*) depends on the relative ease of automated DNA sequencing and bypasses the laborious step of ordering the library. By sequencing enough random library clones so that each segment of the genome is represented from 3 to 10 times, it is possible to reconstruct the genomic sequence by computer alignment of the very large number of sequence fragments.

## KEY CONCEPTS of Section 5.2

### DNA Cloning and Characterization

• In DNA cloning, recombinant DNA molecules are formed in vitro by inserting DNA fragments into vector DNA molecules. The recombinant DNA molecules are then introduced into host cells, where they replicate, producing large numbers of recombinant DNA molecules.

• Restriction enzymes (endonucleases) typically cut DNA at specific 4- to 8-bp palindromic sequences, producing defined fragments that often have self-complementary single-stranded tails (sticky ends).

• Two restriction fragments with complementary ends can be joined with DNA ligase to form a recombinant DNA molecule (see Figure 5-12).

• *E. coli* cloning vectors are small circular DNA molecules (plasmids) that include three functional regions: an origin of replication, a drug-resistance gene, and a site where a DNA fragment can be inserted. Transformed cells carrying a vector grow into colonies on the selection medium (see Figure 5-13).

• A cDNA library is a set of cDNA clones prepared from the mRNAs isolated from a particular type of tissue. A genomic library is a set of clones carrying restriction fragments produced by cleavage of the entire genome.

• In cDNA cloning, expressed mRNAs are reverse-transcribed into complementary DNAs, or cDNAs. By a series of reactions, single-stranded cDNAs are converted into double-stranded DNAs, which can then be ligated into a plasmid vector (see Figure 5-15).

• A particular cloned DNA fragment within a library can be detected by hybridization to a radiolabeled oligonucleotide whose sequence is complementary to a part of the fragment (see Figure 5-16).

• Shuttle vectors that replicate in both yeast and *E. coli* can be used to construct a yeast genomic library. Specific genes can be isolated by their ability to complement the corresponding mutant genes in yeast cells (see Figure 5-17).

• Long cloned DNA fragments can be cleaved with restriction enzymes, producing smaller fragments that are then separated by gel electrophoresis and subcloned into plasmid vectors prior to sequencing or experimental manipulation.

• The polymerase chain reaction (PCR) permits exponential amplification of a specific segment of DNA from just a single initial template DNA molecule if the sequence flanking the DNA region to be amplified is known (see Figure 5-20).

• PCR is a highly versatile method that can be programmed to amplify a specific genomic DNA sequence, a cDNA, or a sequence at the junction between a transposable element and flanking chromosomal sequences.

• DNA fragments up to about 100 nucleotides long are sequenced by generating clusters of identical molecules by PCR and imaging fluorescently labeled nucleotide precursors incorporated by DNA polymerase (see Figures 5-23 and 5-24).

• Whole genome sequences can be assembled from the sequences of a large number of overlapping clones from a genomic library (see Figure 5-25).

## 5.3 Using Cloned DNA Fragments to Study Gene Expression

In the last section, we described the basic techniques for using recombinant DNA technology to isolate specific DNA clones, and ways in which the clones can be further characterized. Now we consider how an isolated DNA clone can be used to study gene expression. We discuss several widely used general techniques that rely on nucleic acid hybridization to elucidate when and where genes are expressed, as well as methods for generating large quantities of protein and otherwise manipulating amino acid sequences to determine their expression patterns, structure, and function. More specific applications of all these basic techniques are examined in the following sections.

### Hybridization Techniques Permit Detection of Specific DNA Fragments and mRNAs

Two very sensitive methods for detecting a particular DNA or RNA sequence within a complex mixture combine sepa-

ration by gel electrophoresis and hybridization with a complementary DNA probe that is either radioactively or fluorescently labeled. A third method involves hybridizing labeled probes directly onto a prepared tissue sample. We will encounter references to all three of these techniques, which have numerous applications, in other chapters.

**Southern Blotting** The first hybridization technique to detect DNA fragments of a specific sequence is known as **Southern blotting**, after its originator E. M. Southern. This technique is capable of detecting a single specific restriction fragment in the highly complex mixture of fragments produced by cleavage of the entire human genome with a restriction enzyme. When such a complex mixture is subjected to gel electrophoresis, so many different fragments of nearly the same length are present it is not possible to resolve any particular DNA fragments as a discrete band on the gel. Nevertheless, it is possible to identify a particular fragment migrating as a band on the gel by its ability to hybridize to a specific DNA probe. To accomplish this, the restriction fragments present in the gel are denatured with alkali and transferred onto a nitrocellulose filter or nylon membrane by blotting (Figure 5-26). This procedure preserves the distribution of the fragments in the gel, creating a replica of the gel on the filter. (The blot is used because probes do not readily diffuse into the original gel.) The filter then is incubated under hybridization conditions with a specific labeled DNA probe, which usually is generated from a cloned restriction fragment. The DNA restriction fragment that is complementary to the probe hybridizes, and its location on the filter can be revealed by autoradiography for a radiolabeled probe or by fluorescent imaging for a fluorescently labeled probe. Although PCR is most commonly used to detect the presence of a particular sequence in a complex mixture, Southern blotting is still useful for reconstructing the relationship between genomic sequences that are too far apart to be amplified by PCR in a single reaction.

**Northern Blotting** One of the most basic ways to characterize a cloned gene is to determine when and where in an organism the gene is expressed. Expression of a particular gene can be followed by assaying for the corresponding mRNA by **Northern blotting**, named, in a play on words, after the related method of Southern blotting. An RNA sample, often the total cellular RNA, is denatured by treatment with an agent such as formaldehyde that disrupts the hydrogen bonds between base pairs, ensuring that all the RNA molecules have an unfolded, linear conformation. The individual RNAs are separated according to size by gel electrophoresis and transferred to a nitrocellulose filter to which the extended denatured RNAs adhere. As in Southern blotting, the filter then is exposed to a labeled DNA probe that is complementary to the gene of interest; finally, the labeled filter is subjected to autoradiography. Because the amount of a specific RNA in a sample can be estimated from a Northern blot, the procedure is widely used to compare the amounts of a particular mRNA in cells under different conditions (Figure 5-27).

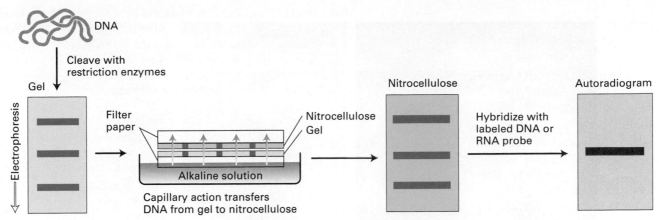

**EXPERIMENTAL FIGURE 5-26 Southern blot technique can detect a specific DNA fragment in a complex mixture of restriction fragments.** The diagram depicts three different restriction fragments in the gel, but the procedure can be applied to a mixture of millions of DNA fragments. Only fragments that hybridize to a labeled probe will give a signal on an autoradiogram. A similar technique called Northern blotting detects specific mRNAs within a mixture. [See E. M. Southern, 1975, *J. Mol. Biol.* **98**:508.]

**In Situ Hybridization** Northern blotting requires extracting the mRNA from a cell or mixture of cells, which means that the cells are removed from their normal location within an organism or tissue. As a result, the location of a cell and its relation to its neighbors is lost. To retain such positional information in precise studies of gene expression, a whole or sectioned tissue or even a whole permeabilized embryo may be subjected to **in situ hybridization** to detect the mRNA encoded by a particular gene. This technique allows gene transcription to be monitored in both time and space (Figure 5-28).

## DNA Microarrays Can Be Used to Evaluate the Expression of Many Genes at One Time

Monitoring the expression of thousands of genes simultaneously is possible with **DNA microarray** analysis, another technique based on the concept of nucleic acid hybridization. A DNA microarray consists of an organized array of thousands of individual, closely packed gene-specific sequences attached to the surface of a glass microscope slide. By coupling microarray analysis with the results from genome sequencing projects, researchers can analyze the global patterns of gene expression of an organism during specific physiological responses or developmental processes.

**Preparation of DNA Microarrays** In one method for preparing microarrays, an ≈1-kb part of the coding region of each gene analyzed is individually amplified by the PCR. A robotic device is used to apply each amplified DNA sample to the surface of a glass microscope slide, which then is chemically processed to permanently attach the DNA sequences to the glass surface and to denature them. A typical array might contain ≈6000 spots of DNA in a 2 × 2–cm grid.

In an alternative method, multiple DNA oligonucleotides, usually at least 20 nucleotides in length, are synthesized from an initial nucleotide that is covalently bound to the surface of a glass slide. The synthesis of an oligonucleotide of specific sequence can be programmed in a small region on the surface of the slide. Several oligonucleotide sequences from a single gene are thus synthesized in neighboring regions of the slide to analyze expression of that gene. With this method, oligonucleotides representing thousands of genes can be produced on a single glass slide. Because the methods for constructing these arrays of synthetic oligonucleotides were adapted from

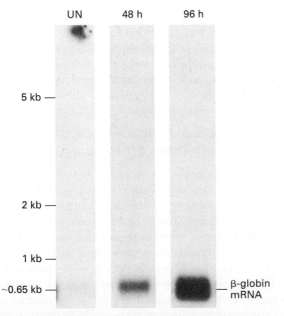

**EXPERIMENTAL FIGURE 5-27 Northern blot analysis reveals increased expression of β-globin mRNA in differentiated erythroleukemia cells.** The total mRNA in extracts of erythroleukemia cells that were growing but uninduced and in cells induced to stop growing and allowed to differentiate for 48 hours or 96 hours was analyzed by Northern blotting for β-globin mRNA. The density of a band is proportional to the amount of mRNA present. The β-globin mRNA is barely detectable in uninduced cells (UN lane) but increases more than 1000-fold by 96 hours after differentiation is induced. [Courtesy of L. Kole.]

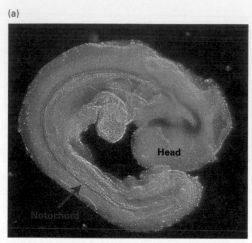

(a)

Head

Notochord

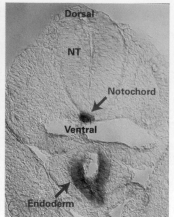

(b)

Dorsal

NT

Notochord

Ventral

Endoderm

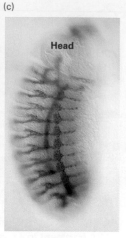

(c)

Head

**EXPERIMENTAL FIGURE 5-28 In situ hybridization can detect activity of specific genes in whole and sectioned embryos.** The specimen is permeabilized by treatment with detergent and a protease to expose the mRNA to the probe. A DNA or RNA probe, specific for the mRNA of interest, is made with nucleotide analogs containing chemical groups that can be recognized by antibodies. After the permeabilized specimen has been incubated with the probe under conditions that promote hybridization, the excess probe is removed with a series of washes. The specimen is then incubated in a solution containing an antibody that binds to the probe. This antibody is covalently joined to a reporter enzyme (e.g., horseradish peroxidase or alkaline phosphatase) that produces a colored reaction product. After excess antibody has been removed, substrate for the reporter enzyme is added. A colored precipitate forms where the probe has hybridized to the mRNA being detected. (a) A whole mouse embryo at about 10 days of development probed for *Sonic hedgehog* mRNA. The stain marks the notochord (red arrow), a rod of mesoderm running along the future spinal cord. (b) A section of a mouse embryo similar to that in part (a). The dorsal/ventral axis of the neural tube (NT) can be seen, with the *Sonic hedgehog*–expressing notochord (red arrow) below it and the endoderm (blue arrow) still farther ventral. (c) A whole *Drosophila* embryo probed for an mRNA produced during trachea development. The repeating pattern of body segments is visible. Anterior (head) is up; ventral is to the left. [Courtesy of L. Milenkovic and M. P. Scott.]

methods for manufacturing microscopic integrated circuits used in computers, these types of oligonucleotide microarrays are often called *DNA chips.*

### Using Microarrays to Compare Gene Expression Under Different Conditions

The initial step in a microarray expression study is to prepare fluorescently labeled cDNAs corresponding to the mRNAs expressed by the cells under study. When the cDNA preparation is applied to a microarray, spots representing genes that are expressed will hybridize under appropriate conditions to their complementary cDNAs in the labeled probe mix and can subsequently be detected in a scanning laser microscope.

Figure 5-29 depicts how this method can be applied to examine the changes in gene expression observed after starved human fibroblasts are transferred to a rich, serum-containing growth medium. In this type of experiment, the separate cDNA preparations from starved and serum-grown fibroblasts are labeled with differently colored fluorescent dyes. A DNA array comprising 8600 mammalian genes then is incubated with a mixture containing equal amounts of the two cDNA preparations under hybridization conditions. After unhybridized cDNA is washed away, the intensity of green and red fluorescence at each DNA spot is measured using a fluorescence microscope and stored in computer files under the name of each gene according to its known position on the slide. The relative intensities of red and green fluorescence signals at each spot are a measure of the relative level of expression of that gene in response to serum. Genes that are not transcribed under these growth conditions give no detectable signal. Genes that are transcribed at the same level under both conditions will hybridize equally to both red- and green-labeled cDNA preparations. Microarray analysis of gene expression in fibroblasts showed that transcription of about 500 of the 8600 genes examined changed substantially after addition of serum.

### Cluster Analysis of Multiple Expression Experiments Identifies Co-regulated Genes

Firm conclusions rarely can be drawn from a single microarray experiment about whether genes that exhibit similar changes in expression are co-regulated and hence likely to be closely related functionally. For example, many of the observed differences in gene expression just described in fibroblasts could be indirect consequences of the many different changes in cell physiology that occur when cells are transferred from one medium to another. In other words, genes that appear to be co-regulated in a single microarray expression experiment may undergo changes in expression for very different reasons and may actually have very different biological functions. A solution to this problem is to combine the information from a set of expression array experiments to find genes that are similarly regulated under a variety of conditions or over a period of time.

This more informative use of multiple expression array experiments is illustrated by examining the relative expression of the 8600 genes at different times after serum addition,

**EXPERIMENTAL FIGURE 5-29  DNA microarray analysis can reveal differences in gene expression in fibroblasts under different experimental conditions.** (a) In this example, cDNA prepared from mRNA isolated from fibroblasts either starved for serum or after serum addition is labeled with different fluorescent dyes. A microarray composed of DNA spots representing 8600 mammalian genes is exposed to an equal mixture of the two cDNA preparations under hybridization conditions. The ratio of the intensities of red and green fluorescence over each spot, detected with a scanning confocal laser microscope, indicates the relative expression of each gene in response to serum. (b) A micrograph of a small segment of an actual DNA microarray. Each spot in this 16 × 16 array contains DNA from a different gene hybridized to control and experimental cDNA samples labeled with red and green fluorescent dyes. (A yellow spot indicates equal hybridization of green and red fluorescence, indicating no change in gene expression.) [Part (b) Alfred Pasieka/Photo Researchers, Inc.]

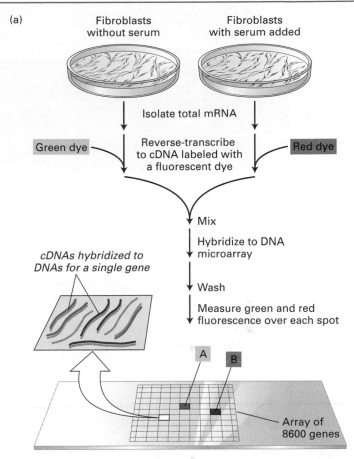

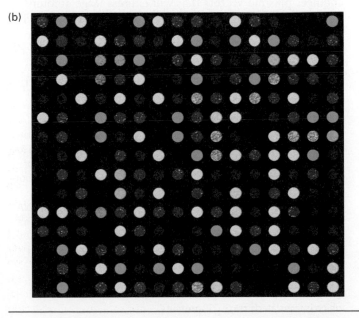

generating more than $10^4$ individual pieces of data. A computer program, related to the one used to determine the relatedness of different protein sequences, can organize these data and cluster genes that show similar expression over the time course after serum addition. Remarkably, such *cluster analysis* groups sets of genes whose encoded proteins participate in a common cellular process, such as cholesterol biosynthesis or the cell cycle (Figure 5-30).

In the future, microarray analysis will be a powerful diagnostic tool in medicine. For instance, particular sets of mRNAs have been found to distinguish tumors with a poor prognosis from those with a good prognosis. Previously indistinguishable disease variations are now detectable. Analysis of tumor biopsies for these distinguishing mRNAs will help physicians to select the most appropriate treatment. As more patterns of gene expression characteristic of various diseased tissues are recognized, the diagnostic use of DNA microarrays will be extended to other conditions. ■

## *E. coli* Expression Systems Can Produce Large Quantities of Proteins from Cloned Genes

Many protein hormones and other signaling or regulatory proteins are normally expressed at very low concentrations, precluding their isolation and purification in large quantities by standard biochemical techniques. Widespread therapeutic use of such proteins, as well as basic research on their structure and functions, depends on efficient procedures for producing them in large amounts at reasonable cost. Recombinant DNA techniques that turn *E. coli* cells into factories for synthesizing low-abundance proteins now are used to commercially produce granulocyte-colony-stimulating factor (G-CSF), insulin, growth hormone, and other human proteins with therapeutic uses. For example, G-CSF stimulates the production of granulocytes, the phagocytic white blood

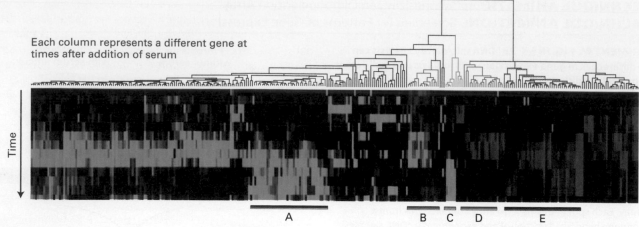

Time

A B C D E

**EXPERIMENTAL FIGURE 5-30 Cluster analysis of data from multiple microarray expression experiments can identify co-regulated genes.** The expression of 8600 mammalian genes was detected by microarray analysis at time intervals over a 24-hour period after serum-starved fibroblasts were provided with serum. The cluster diagram shown here is based on a computer algorithm that groups genes showing similar changes in expression compared with a serum-starved control sample over time. Each column of colored boxes represents a single gene, and each row represents a time point. A red box indicates an increase in expression relative to the control; a green box, a decrease in expression; and a black box, no significant change in expression. The "tree" diagram at the top shows how the expression patterns for individual genes can be organized in a hierarchical fashion to group together the genes with the greatest similarity in their patterns of expression over time. Five clusters of coordinately regulated genes were identified in this experiment, as indicated by the bars at the bottom. Each cluster contains multiple genes whose encoded proteins function in a particular cellular process: cholesterol biosynthesis (A), the cell cycle (B), the immediate-early response (C), signaling and angiogenesis (D), and wound healing and tissue remodeling (E). [Courtesy of Michael B. Eisen, Lawrence Berkeley National Laboratory.]

cells critical to defense against bacterial infections. Administration of G-CSF to cancer patients helps offset the reduction in granulocyte production caused by chemotherapeutic agents, thereby protecting patients against serious infection while they are receiving chemotherapy. ■

The first step in producing large amounts of a low-abundance protein is to obtain a cDNA clone encoding the full-length protein by methods discussed previously. The second step is to engineer plasmid vectors that will express large amounts of the encoded protein when it is inserted into *E. coli* cells. The key to designing such expression vectors is inclusion of a promoter, a DNA sequence from which transcription of the cDNA can begin. Consider, for example, the relatively simple system for expressing G-CSF shown in Figure 5-31. In this

**EXPERIMENTAL FIGURE 5-31 Some eukaryotic proteins can be produced in *E. coli* cells from plasmid vectors containing the *lac* promoter.** (a) The plasmid expression vector contains a fragment of the *E. coli* chromosome containing the *lac* promoter and the neighboring *lacZ* gene. In the presence of the lactose analog IPTG, RNA polymerase normally transcribes the *lacZ* gene, producing *lacZ* mRNA, which is translated into the encoded protein, β-galactosidase. (b) The *lacZ* gene can be cut out of the expression vector with restriction enzymes and replaced by a cloned cDNA, in this case one encoding granulocyte-colony-stimulating factor (G-CSF). When the resulting plasmid is transformed into *E. coli* cells, addition of IPTG and subsequent transcription from the *lac* promoter produce G-CSF mRNA, which is translated into G-CSF protein.

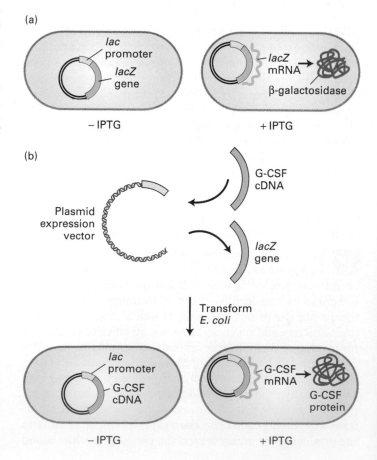

case, G-CSF is expressed in *E. coli* transformed with plasmid vectors that contain the *lac* promoter adjacent to the cloned cDNA encoding G-CSF. Transcription from the *lac* promoter occurs at high rates only when lactose, or a lactose analog such as isopropylthiogalactoside (IPTG), is added to the culture medium. Even larger quantities of a desired protein can be produced in more complicated *E. coli* expression systems.

To aid in purification of a eukaryotic protein produced in an *E. coli* expression system, researchers often modify the cDNA encoding the recombinant protein to facilitate its separation from endogenous *E. coli* proteins. A commonly used modification of this type is to add a short nucleotide sequence to the end of the cDNA, so that the expressed protein will have six histidine residues at the C-terminus. Proteins modified in this way bind tightly to an affinity matrix that contains chelated nickel atoms, whereas most *E. coli* proteins will not bind to such a matrix. The bound proteins can be released from the nickel atoms by decreasing the pH of the surrounding medium. In most cases, this procedure yields a pure recombinant protein that is functional, since addition of short amino acid sequences to either the C-terminus or the N-terminus of a protein usually does not interfere with the protein's biochemical activity.

## Plasmid Expression Vectors Can Be Designed for Use in Animal Cells

Although bacterial expression systems can be used successfully to create large quantities of some proteins, bacteria cannot be used in all cases. Many experiments to examine the function of a protein in an appropriate cellular context require expression of a genetically modified protein in cultured animal cells. Genes are cloned into specialized eukaryotic expression vectors and are introduced into cultured animal cells by a process called **transfection**. Two common methods for transfecting animal cells differ in whether the recombinant vector DNA is or is not integrated into the host-cell genomic DNA.

In both methods, cultured animal cells must be treated to facilitate their initial uptake of the recombinant plasmid vector. This can be done by exposing cells to a preparation of lipids that penetrate the plasma membrane, increasing its permeability to DNA. Alternatively, subjecting cells to a brief electric shock of several thousand volts, a technique known as *electroporation*, makes them transiently permeable to DNA. Usually the plasmid DNA is added in sufficient concentration to ensure that a large proportion of the cultured cells will receive at least one copy of the plasmid DNA. Researchers have also harnessed viruses for their use in the laboratory; viruses can be modified to contain DNA of interest, which is then introduced into host cells by simply infecting them with the recombinant virus.

**Transient Transfection** The simplest of the two expression methods, called *transient transfection*, employs a vector similar to the yeast shuttle vectors described previously. For use in mammalian cells, plasmid vectors are engineered also to carry an origin of replication derived from a virus that infects mammalian cells, a strong promoter recognized by mammalian RNA polymerase, and the cloned cDNA encoding the protein to be expressed adjacent to the promoter (Figure 5-32a). Once such a plasmid vector enters a mammalian cell, the viral origin of replication allows it to replicate efficiently, generating numerous plasmids from which the protein is expressed. However, during cell division such plasmids are not faithfully segregated into both daughter cells and in time a substantial fraction of the cells in a culture will not contain a plasmid, hence the name *transient transfection*.

**Stable Transfection (Transformation)** If an introduced vector integrates into the genome of the host cell, the genome is permanently altered and the cell is said to be *transformed*. Integration most likely is accomplished by mammalian enzymes that normally function in DNA repair and recombination. A commonly used selectable marker is the gene for neomycin phosphotransferase (designated *neo^r*), which confers resistance to a toxic compound chemically related to neomycin known as G-418. The basic procedure for expressing a cloned cDNA by *stable transfection* is outlined in Figure 5-32b. Only those cells that have integrated the expression vector into the host chromosome will survive and give rise to a clone in the presence of a high concentration of G-418. Because integration occurs at random sites in the genome, individual transformed clones resistant to G-418 will differ in their rates of transcribing the inserted cDNA. Therefore, the stable transfectants usually are screened to identify those that produce the protein of interest at the highest levels.

**Retroviral Expression Systems** Researchers have exploited the basic mechanisms used by viruses for introduction of genetic material into animal cells and subsequent insertion into chromosomal DNA to greatly increase the efficiency by which a modified gene can be stably expressed in animal cells. An example of one such viral expression is derived from a class of retroviruses known as *lentiviruses*. As shown in Figure 5-33, three different plasmids, introduced into cells by transient transfection, are used to produce recombinant lentivirus particles suitable for efficient introduction of a cloned gene into target animal cells. The first plasmid, known as the *vector plasmid*, contains a cloned gene of interest next to a selectable marker such as *neo^r* flanked by lentivirus LTR sequences. As described in Chapter 6, viral LTR sequences direct synthesis of a viral RNA molecule that on introduction into a virally infected target cell can be copied into DNA by reverse transcription and then integrated into chromosomal DNA. A second plasmid known as the packaging plasmid carries all the viral genes, except for the major viral envelope protein, necessary for packaging LTR-containing viral RNA into a functional lentivirus particle. The final plasmid allows expression of a viral envelope protein that when incorporated into a recombinant lentivirus will allow the resulting hybrid virus particles to infect a

**(a) Transient transfection**

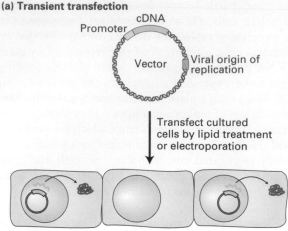

Protein is expressed from cDNA in plasmid DNA

**(b) Stable transfection (transformation)**

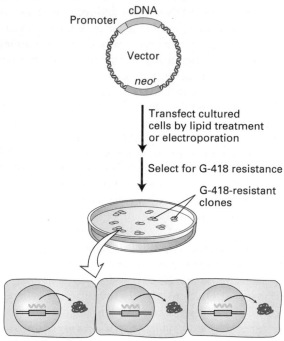

Protein is expressed from cDNA integrated into host chromosome

EXPERIMENTAL FIGURE 5-32 **Transient and stable transfection with specially designed plasmid vectors permits expression of cloned genes in cultured animal cells.** Both methods employ plasmid vectors that contain the usual elements—ORI, selectable marker (e.g., *amp*<sup>r</sup>), and polylinker—that permit propagation in *E. coli* and insertion of a cloned cDNA with an adjacent animal promoter. For simplicity, these elements are not depicted. (a) In transient transfection, the plasmid vector contains an origin of replication for a virus that can replicate in the cultured animal cells. Since the vector is not incorporated into the genome of the cultured cells, production of the cDNA-encoded protein continues only for a limited time. (b) In stable transfection, the vector carries a selectable marker such as *neo*<sup>r</sup>, which confers resistance to G-418. The relatively few transfected animal cells that integrate the exogenous DNA into their genomes are selected on medium containing G-418. Because the vector is integrated into the genome, these stably transfected, or transformed, cells will continue to produce the cDNA-encoded protein as long as the culture is maintained. See the text for discussion.

cells with a stably integrated cloned gene and *neo*<sup>r</sup> marker can be selected for by resistance to G-418. Many of the techniques for inactivating the function of specific genes (see Section 5.5) require that an entire population of cultured cells be genetically modified simultaneously. Engineered lentiviruses are particularly useful for such experiments because they infect cells with very high efficiency such that every cell in a population will receive at least one copy of the lentivirus-borne plasmid.

**Gene and Protein Tagging** Expression vectors can provide a way to study the expression and intracellular localization of eukaryotic proteins. This method often relies on using a reporter protein such as *green fluorescent protein* (*GFP*) that can conveniently be detected in cells. Here we describe two ways to create a hybrid gene that connects expression of the reporter protein to the protein of interest. When the hybrid gene is reintroduced into cells, either by transfection with a plasmid expression vector containing the modified gene or by creation of a transgenic animal as described in Section 5.5, the expression of the reporter protein can be used to determine where and when a gene is expressed. This method provides similar data to in situ hybridization experiments described previously but often with greater resolution and sensitivity.

Figure 5-34 illustrates the use of two different types of GFP-tagging experiments to study the expression of an odorant receptor protein in *C. elegans*. When the promoter for the odorant receptor is linked directly to the coding sequence of GFP in a configuration usually known as a *promoter-fusion*, GFP is expressed in specific neurons, filling the cytoplasm of those neurons. In contrast, when the hybrid gene is constructed by linking GFP to the coding sequence of the receptor, the resulting *protein-fusion* can be localized by GFP fluorescence at the distal cilia in sensory neurons, the site at which the receptor protein is normally located.

desired target cell type. A common envelope protein used in this context is the glycoprotein of the vesicular stomatitis virus (VSV-G protein), which can readily replace the normal lentivirus envelope protein on the surface of completed virus particles and will allow the resulting virus particles to infect a wide variety of mammalian cell types, including hematopoietic stem cells, neurons, and muscle and liver cells. After cell infection, the cloned gene flanked by viral LTR sequences is reverse-transcribed into DNA, which is transported into the nucleus and then integrated into the host genome. If necessary, as in the case for stable transfection,

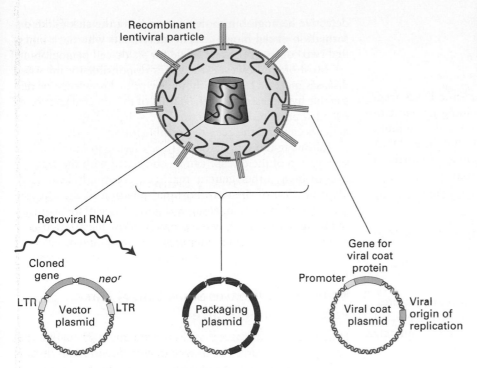

Recombinant
lentiviral particle

Retroviral RNA

Cloned
gene — neo$^r$

LTR — Vector
plasmid — LTR

Packaging
plasmid

Gene for
viral coat
protein

Promoter

Viral coat
plasmid

Viral
origin of
replication

**EXPERIMENTAL FIGURE 5-33** Retroviral vectors can be used for efficient integration of cloned genes into the mammalian genome. See the text for discussion.

An alternative to GFP tagging for detecting the intracellular location of a protein is to modify the gene of interest by fusing it with a short DNA sequence that encodes a short stretch of amino acids recognized by a known monoclonal antibody. Such a short peptide that can be bound by an antibody is called an epitope; hence this method is known as *epitope tagging*. After transfection with a plasmid expression vector containing the modified gene, the expressed epitope-tagged form of the protein can be detected by immunofluorescence labeling of the cells with the monoclonal antibody specific for the epitope. The choice of whether to use a short epitope or GFP to tag a given protein often depends on what types of modification a cloned gene can tolerate and still remain functional.

**(a) Promoter-fusion; ODR10 promoter fused to GFP**

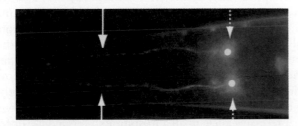

**(b) Protein-fusion; ODR10-GFP fusion protein**

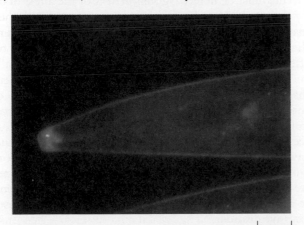

10 μm

**EXPERIMENTAL FIGURE 5-34** Gene and protein tagging facilitate cellular localization of proteins expressed from cloned genes. In this experiment, the gene encoding a chemical odorant receptor, Odr10, of *C. elegans* was fused to the gene sequence for green fluorescent protein (GFP). (a) A promoter-fusion was generated by linking GFP to the promoter and the first four amino acid codons of Odr10. This protein is expressed in the cytoplasm of specific sensory neurons in the head of *C. elegans*. Note that the cell body (dashed arrow) and sensory dendrites (solid arrow) are fluorescently labeled. (b) A protein-fusion was constructed by linking GFP to the end of the full-length Odr10 coding sequence. In this case, the Odr10-GFP fusion protein is targeted to the membrane at the tip of the sensory neurons and is only apparent at the distal end of the sensory cilia. The observed distribution can be inferred to reflect the normal location of Odr10 protein in specific neurons. Because the promoter-fusion shown in (a) lacks the Odr10 localization sequences, the expressed GFP fills the entire cell cytoplasm rather than being localized just to distal tip of the sensory cilia. [P. Sengupta et al., 1996, *Cell* **84**:899 (derived from Figures 4 and 5).]

## 5.4 Locating and Identifying Human Disease Genes

Inherited human diseases are the phenotypic consequence of defective human genes. Table 5-2 lists several of the most commonly occurring inherited diseases. Although a "disease" gene may result from a new mutation that arose in the preceding generation, most cases of inherited diseases are caused by preexisting mutant alleles that have been passed from one generation to the next for many generations. ∎

The typical first step in deciphering the underlying cause for any inherited human disease is to identify the affected gene and its encoded protein. Comparison of the sequences of a disease gene and its product with those of genes and proteins whose sequence and function are known can provide clues to the molecular and cellular cause of the disease. Historically, researchers have used whatever phenotypic clues might be relevant to make guesses about the molecular basis of inherited diseases. An early example of successful guesswork was the hypothesis that sickle-cell anemia, known to be a disease of blood cells, might be caused by defective hemoglobin. This idea led to identification of a specific amino acid substitution in hemoglobin that causes polymerization of the defective hemoglobin molecules, causing the sickle-like deformation of red blood cells in individuals who have inherited two copies of the $Hb^s$ allele for sickle-cell hemoglobin.

Most often, however, the genes responsible for inherited diseases must be found without any prior knowledge or reasonable hypotheses about the nature of the affected gene or its encoded protein. In this section, we will see how human geneticists can find the gene responsible for an inherited disease by following the segregation of the disease in families. The segregation of the disease can be correlated with the segregation of many other genetic markers, eventually leading to identification of the chromosomal position of the affected gene. This information, along with knowledge of the sequence of the human genome, can ultimately allow the affected gene and the disease-causing mutations to be pinpointed.

### Monogenic Diseases Show One of Three Patterns of Inheritance

Human genetic diseases that result from mutation in one specific gene are referred to as *monogenic diseases* and display different inheritance patterns depending on the nature and chromosomal location of the alleles that cause them. One characteristic pattern is that exhibited by a dominant allele in an autosome (that is, one of the 22 human chromosomes that is not a sex chromosome). Because an *autosomal dominant* allele is expressed in the heterozygote, usually at least one of the parents of an affected individual will also have the disease. It is often the case that the diseases caused by dominant alleles appear later in life after the reproductive age. If this were not the case, natural selection would have eliminated the allele during human evolution. An example of an autosomal dominant disease is Huntington's disease, a neural degenerative disease that generally strikes in mid- to late life. If either parent carries a mutant *HD* allele, each of his or her children (regardless of sex) has a 50 percent chance of inheriting the mutant allele and being affected (Figure 5-35a).

A recessive allele in an autosome exhibits a quite different segregation pattern. For an *autosomal recessive* allele, both parents must be heterozygous *carriers* of the allele in order for their children to be at risk of being affected with the disease. Each child of heterozygous parents has a 25 percent chance of receiving both recessive alleles and thus being affected, a 50 percent chance of receiving one normal and one mutant allele and thus being a carrier, and a 25 percent chance of receiving two normal alleles. A clear example of an autosomal recessive disease is cystic fibrosis, which results from a defective chloride-channel gene known as *CFTR* (Figure 5-35b). Related individuals (e.g., first or second cousins) have a relatively high probability of being carriers for the same recessive alleles. Thus children born to related parents are much more likely than those born to unrelated parents to be homozygous for, and therefore affected by, an autosomal recessive disorder.

The third common pattern of inheritance is that of an *X-linked recessive* allele. A recessive allele on the X chromo-

## TABLE 5-2 | Common Inherited Human Diseases

| Disease | Molecular and Cellular Defect | Incidence |
|---|---|---|
| **Autosomal Recessive** | | |
| Sickle-cell anemia | Abnormal hemoglobin causes deformation of red blood cells, which can become lodged in capillaries; also confers resistance to malaria. | 1/625 of sub-Saharan African origin |
| Cystic fibrosis | Defective chloride channel (CFTR) in epithelial cells leads to excessive mucus in lungs. | 1/2500 of European origin |
| Phenylketonuria (PKU) | Defective enzyme in phenylalanine metabolism (tyrosine hydroxylase) results in excess phenylalanine leading to mental retardation, unless restricted by diet. | 1/10,000 of European origin |
| Tay-Sachs disease | Defective hexosaminidase enzyme leads to accumulation of excess sphingolipids in the lysosomes of neurons, impairing neural development. | 1/1000 eastern European Jews |
| **Autosomal Dominant** | | |
| Huntington's disease | Defective neural protein (huntingtin) may assemble into aggregates, causing damage to neural tissue. | 1/10,000 of European origin |
| Hypercholesterolemia | Defective LDL receptor leads to excessive cholesterol in blood and early heart attacks. | 1/122 French Canadians |
| **X-Linked Recessive** | | |
| Duchenne muscular dystrophy (DMD) | Defective cytoskeletal protein dystrophin leads to impaired muscle function. | 1/3500 males |
| Hemophilia A | Defective blood clotting factor VIII leads to uncontrolled bleeding. | 1–2/10,000 males |

some will most often be expressed in males, who receive only one X chromosome from their mother, but not in females, who receive an X chromosome from both their mother and their father. This leads to a distinctive sex-linked segregation pattern where the disease is exhibited much more frequently in males than in females. For example, Duchenne muscular dystrophy (DMD), a muscle degenerative disease that specifically affects males, is caused by a recessive allele on the X chromosome. DMD exhibits the typical sex-linked segregation pattern in which mothers who are heterozygous and therefore phenotypically normal can act as carriers, transmitting the DMD allele and therefore the disease to 50 percent of their male progeny (Figure 5-35c).

## DNA Polymorphisms Are Used as Markers for Linkage-Mapping of Human Mutations

Once the mode of inheritance has been determined, the next step in determining the position of a disease allele is to genetically map its position with respect to known genetic markers using the basic principle of genetic linkage as described in Section 5.1. The presence of many different already mapped genetic traits, or markers, distributed along the length of a chromosome facilitates the mapping of a new mutation: one can assess its possible linkage to these marker genes in appropriate crosses. The more markers that are available, the more precisely a mutation can be mapped. The density of genetic markers needed for a high-resolution human genetic map is about one marker every 5 centimorgans (cM) (as discussed previously, one genetic map unit, or centimorgan, is defined as the distance between two positions along a chromosome that results in one recombinant individual in 100 progeny). Thus a high-resolution genetic map requires 25 or so genetic markers of known position spread along the length of each human chromosome.

In the experimental organisms commonly used in genetic studies, numerous markers with easily detectable phenotypes are available for genetic mapping of mutations. This is not the case for mapping genes whose mutant alleles are associated with inherited diseases in humans. However, recombinant

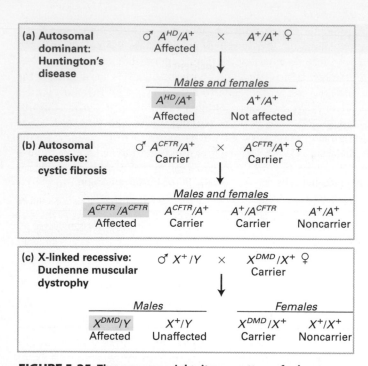

FIGURE 5-35 **Three common inheritance patterns for human genetic diseases.** Wild-type autosomal (A) and sex chromosomes (X and Y) are indicated by superscript plus signs. (a) In an autosomal dominant disorder such as Huntington's disease, only one mutant allele is needed to confer the disease. If either parent is heterozygous for the mutant *HD* allele, his or her children have a 50 percent chance of inheriting the mutant allele and getting the disease. (b) In an autosomal recessive disorder such as cystic fibrosis, two mutant alleles must be present to confer the disease. Both parents must be heterozygous carriers of the mutant *CFTR* gene for their children to be at risk of being affected or being carriers. (c) An X-linked recessive disease such as Duchenne muscular dystrophy is caused by a recessive mutation on the X chromosome and exhibits the typical sex-linked segregation pattern. Males born to mothers heterozygous for a mutant *DMD* allele have a 50 percent chance of inheriting the mutant allele and being affected. Females born to heterozygous mothers have a 50 percent chance of being carriers.

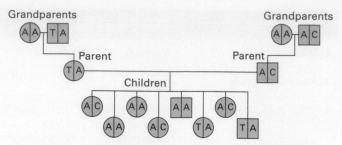

**EXPERIMENTAL FIGURE 5-36 Single-Nucleotide Polymorphisms (SNPs) can be followed like genetic markers.** Hypothetical pedigree based on SNP analysis of the DNA from a region of a chromosome. In this family, the SNP exists as an A, T, or C nucleotide. Each individual has two alleles: some contain an A on both chromosomes, and others are heterozygous at this site. Circles indicate females; squares indicate males. Blue indicates unaffected individuals; orange indicates individuals with the trait. Analysis reveals that the trait segregates with a C at the SNP.

DNA technology has made available a wealth of useful DNA-based molecular markers. Because most of the human genome does not code for protein, a large amount of sequence variation exists between individuals. Indeed, it has been estimated that nucleotide differences between unrelated individuals can be detected on an average of every $10^3$ nucleotides. If these variations in DNA sequence, referred to as *DNA polymorphisms*, can be followed from one generation to the next, they can serve as genetic markers for linkage studies. Currently, a panel of as many as $10^4$ different known polymorphisms whose locations have been mapped in the human genome is used for genetic linkage studies in humans.

*Single-nucleotide polymorphisms* (*SNPs*) constitute the most abundant type and are therefore useful for constructing genetic maps of maximum resolution (Figure 5-36). Another useful type of DNA polymorphism consists of a variable number of repetitions of a one-, two-, or three-base sequence. Such polymorphisms, known as simple-sequence repeats (SSRs) or *microsatellites*, presumably are formed by recombination or a slippage mechanism of either the template or newly synthesized strands during DNA replication. A useful property of SSRs is that different individuals will often have different numbers of repeats. The existence of multiple versions of an SSR makes it more likely to produce an informative segregation pattern in a given pedigree and therefore to be of more general use in mapping the positions of disease genes. These polymorphisms can be detected by PCR amplification and DNA sequencing.

## Linkage Studies Can Map Disease Genes with a Resolution of About 1 Centimorgan

Without going into all the technical considerations, let's see how the allele conferring a particular dominant trait (e.g., familial hypercholesterolemia) might be mapped. The first step is to obtain DNA samples from all the members of a family containing individuals that exhibit the disease. The DNA from each affected and unaffected individual then is analyzed to determine the identity of a large number of known DNA polymorphisms (either SSR or SNP markers can be used). The segregation pattern of each DNA polymorphism within the family is then compared with the segregation of the disease under study to find those polymorphisms that tend to segregate along with the disease. Finally, computer analysis of the segregation data is used to calculate the likelihood of linkage between each DNA polymorphism and the disease-causing allele.

In practice, segregation data are collected from different families exhibiting the same disease and pooled. The more families exhibiting a particular disease that can be examined, the greater the statistical significance of evidence for linkage that can be obtained and the greater the precision with which

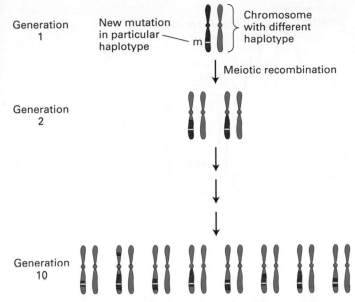

Generation 1 — New mutation in particular haplotype — m — Chromosome with different haplotype

Meiotic recombination

Generation 2

Generation 10

**FIGURE 5-37 Linkage-disequilibrium studies of human populations can be used to map genes at high resolution.** A new disease mutation will arise in the context of an ancestral chromosome among a set of polymorphisms known as the *haplotype* (indicated by red shading). After many generations, chromosomes that carry the disease mutation will also carry segments of the ancestral haplotype that have not been separated from the disease mutation by recombination. The blue segments of these chromosomes represent general haplotypes derived from the general population and not from the ancestral haplotype in which the mutation originally arose. This phenomenon is known as *linkage disequilibrium*. The position of the disease mutation can be located by scanning chromosomes containing the disease mutation for highly conserved polymorphisms corresponding to the ancestral haplotype.

the distance can be measured between a linked DNA polymorphism and a disease allele. Most family studies have a maximum of about 100 individuals in which linkage between a disease gene and a panel of DNA polymorphisms can be tested. This number of individuals sets the practical upper limit on the resolution of such a mapping study to about 1 centimorgan, or a physical distance of about $7.5 \times 10^5$ base pairs.

A phenomenon called *linkage disequilibrium* is the basis for an alternative strategy, which often can afford a higher degree of resolution in mapping studies. This approach depends on the particular circumstance in which a genetic disease commonly found in a particular population results from a single mutation that occurred many generations in the past. The DNA polymorphisms carried by this ancestral chromosome are collectively known as the *haplotype* of that chromosome. As the disease allele is passed from one generation to the next, only the polymorphisms that are closest to the disease gene will not be separated from it by recombination. After many generations, the region that contains the disease gene will be evident because this will be the only region of the chromosome that will carry the haplotype of the ancestral chromo-

some conserved through many generations (Figure 5-37). By assessing the distribution of specific markers in all the affected individuals in a population, geneticists can identify DNA markers tightly associated with the disease, thus localizing the disease-associated gene to a relatively small region. Under ideal circumstances, linkage disequilibrium studies can improve the resolution of mapping studies to less than 0.1 centimorgan. The resolving power of this method comes from the ability to determine whether a polymorphism and the disease allele were ever separated by a meiotic recombination event at any time since the disease allele first appeared on the ancestral chromosome—in some cases this can amount to finding markers that are so closely linked to the disease gene that even after hundreds of meioses, they have never been separated by recombination.

## Further Analysis Is Needed to Locate a Disease Gene in Cloned DNA

Although linkage mapping can usually locate a human disease gene to a region containing about $10^5$ base pairs, as many as 10 different genes may be located in a region of this size. The ultimate objective of a mapping study is to locate the gene within a cloned segment of DNA and then to determine the nucleotide sequence of this fragment. The relative scales of a chromosomal genetic map and physical maps corresponding to ordered sets of plasmid clones and the nucleotide sequence are shown in Figure 5-38.

One strategy for further localizing a disease gene within the genome is to identify mRNA encoded by DNA in the region of the gene under study. Comparison of gene expression in tissues from normal and affected individuals may suggest tissues in which a particular disease gene normally is expressed. For instance, a mutation that phenotypically affects muscle, but no other tissue, might be in a gene that is expressed only in muscle tissue. The expression of mRNA in both normal and affected individuals generally is determined by Northern blotting or in situ hybridization of labeled DNA or RNA to tissue sections. Northern blots, in situ hybridization, or microarray experiments permit comparison of both the level of expression and the size of mRNAs in mutant and wild-type tissues. Although the sensitivity of in situ hybridization is lower than that of Northern blot analysis, it can be very helpful in identifying an mRNA that is expressed at low levels in a given tissue but at very high levels in a subclass of cells within that tissue. An mRNA that is altered or missing in various individuals affected with a disease compared with wild-type individuals would be an excellent candidate for encoding the protein whose disrupted function causes that disease.

In many cases, point mutations that give rise to disease-causing alleles may result in no detectable change in the level of expression or electrophoretic mobility of mRNAs. Thus if comparison of the mRNAs expressed in normal and affected individuals reveals no detectable differences in the candidate mRNAs, a search for point mutations in the DNA regions

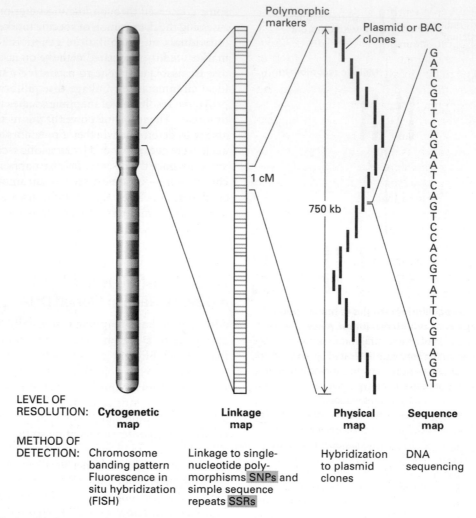

Polymorphic markers

Plasmid or BAC clones

1 cM

750 kb

```
G
A
T
C
G
T
T
C
A
G
A
A
T
C
A
G
T
C
C
A
C
G
T
A
T
T
C
G
T
A
G
G
T
```

| LEVEL OF RESOLUTION: | Cytogenetic map | Linkage map | Physical map | Sequence map |
|---|---|---|---|---|
| METHOD OF DETECTION: | Chromosome banding pattern Fluorescence in situ hybridization (FISH) | Linkage to single-nucleotide poly-morphisms SNPs and simple sequence repeats SSRs | Hybridization to plasmid clones | DNA sequencing |

**FIGURE 5-38 The relationship between the genetic and physical maps of a human chromosome.** The diagram depicts a human chromosome analyzed at different levels of detail. The chromosome as a whole can be viewed in the light microscope when it is in a condensed state that occurs at metaphase, and the approximate location of specific sequences can be determined by fluorescence in situ hybridization (FISH). At the next level of detail, genetic traits can be mapped relative to DNA-based genetic markers. Local segments of the chromosome can be analyzed at the level of DNA sequences identified by Southern hybridization or PCR. Finally, important genetic differences can be most precisely defined by differences in the nucleotide sequence of the chromosomal DNA. [Adapted from L. Hartwell et al., 2003, *Genetics: From Genes to Genomes*, 2d ed., McGraw-Hill.]

encoding the mRNAs is undertaken. Now that highly efficient methods for sequencing DNA are available, researchers frequently determine the sequence of candidate regions of DNA isolated from affected individuals to identify point mutations. The overall strategy is to search for a coding sequence that consistently shows possibly deleterious alterations in DNA from individuals that exhibit the disease. A limitation of this approach is that the region near the affected gene may carry naturally occurring polymorphisms unrelated to the gene of interest. Such polymorphisms, not functionally related to the disease, can lead to misidentification of the DNA fragment carrying the gene of interest. For this reason, the more mutant alleles available for analysis, the more likely that a gene will be correctly identified.

## Many Inherited Diseases Result from Multiple Genetic Defects

Most of the inherited human diseases that are now understood at the molecular level are monogenetic diseases; that is, a clearly discernible disease state is produced by a defect in a single gene. Monogenic diseases caused by mutation in one specific gene exhibit one of the characteristic inheritance patterns shown in Figure 5-35. The genes associated with most of the common monogenic diseases have already been mapped using DNA-based markers as described previously.

However, many other inherited diseases show more complicated patterns of inheritance, making the identification of the underlying genetic cause much more difficult. One type of added complexity that is frequently encountered is *genetic heterogeneity*. In such cases, mutations in any one of multiple different genes can cause the same disease. For example, retinitis pigmentosa, which is characterized by degeneration of the retina usually leading to blindness, can be caused by mutations in any one of more than 60 different genes. In human linkage studies, data from multiple families usually must be combined to determine whether a statistically significant linkage exists between a disease gene and known molecular markers. Genetic heterogeneity such as that exhibited by retinitis pigmentosa can confound such an approach because any statistical trend in the mapping data from one family tends to be canceled out by the data obtained from another family with an unrelated causative gene.

Human geneticists used two different approaches to identify the many genes associated with retinitis pigmentosa. The first approach relied on mapping studies in exceptionally large single families that contained a sufficient number of affected individuals to provide statistically significant evidence for linkage between known DNA polymorphisms and a single causative gene. The genes identified in such studies showed that several of the mutations that cause retinitis pigmentosa lie within genes that encode abundant proteins of the retina. Following up on this clue, geneticists concentrated their attention on those genes that are highly expressed in the retina when screening other individuals with retinitis pigmentosa. This approach of using additional information to focus screening efforts on a subset of candidate genes led to identification of additional rare causative mutations in many different genes encoding retinal proteins.

A further complication in the genetic dissection of human diseases is posed by diabetes, heart disease, obesity, predisposition to cancer, and a variety of mental disorders that have at least some heritable properties. These and many other diseases can be considered to be *polygenic diseases* in the sense that alleles of multiple genes, acting together within an individual, contribute to both the occurrence and the severity of disease. How to systematically map complex polygenic traits in humans is one of the most important and challenging problems in human genetics today.

One of the most promising methods to study diseases that exhibit genetic heterogeneity or are polygenic is to seek a statistical correlation between inheritance of a particular region of a chromosome and the propensity to have a disease using a procedure known as a genome-wide association study (GWAS). The identification of disease-causing genes by GWAS relies on the phenomenon of linkage disequilibrium described previously. If an allele that causes or even predisposes an individual to a disease has occurred relatively recently during human evolution, the disease-causing allele will tend to remain associated with the particular set of DNA-based markers in the neighborhood of its chromosomal location. By examining a large number of DNA markers in populations of individuals with a particular disease as well as in control populations of individuals without the disease, chromosomal regions that tend to be correlated with occurrence of the disease can be identified with

GWAS. Genomic sequencing and other methods can then be used to identify possible disease causing mutations in these regions. Although GWAS can be a powerful tool to identify candidate disease genes, much further work is needed to determine how an individual carrying a particular mutation might be predisposed to the disease.

Models of human disease in experimental organisms may also contribute to unraveling the genetics of complex traits such as obesity or diabetes. For instance, large-scale controlled breeding experiments in mice can identify mouse genes associated with diseases analogous to those in humans. The human orthologs of the mouse genes identified in such studies would be likely candidates for involvement in the corresponding human disease. DNA from human populations then could be examined to determine if particular alleles of the candidate genes show a tendency to be present in individuals affected with the disease but absent from unaffected individuals. This "candidate gene" approach is currently being used intensively to search for genes that may contribute to the major polygenic diseases in humans.

## KEY CONCEPTS of Section 5.4

### Locating and Identifying Human Disease Genes

- Inherited diseases and other traits in humans show three major patterns of inheritance: autosomal dominant, autosomal recessive, and X-linked recessive (see Figure 5-35).

- Genes for human diseases and other traits can be mapped by determining their co-segregation during meiosis with markers whose locations in the genome are known. The closer a gene is to a particular marker, the more likely they are to co-segregate.

- Mapping of human genes with great precision requires thousands of molecular markers distributed along the chromosomes. The most useful markers are differences in the DNA sequence (polymorphisms) between individuals in noncoding regions of the genome.

- DNA polymorphisms useful in mapping human genes include single-nucleotide polymorphisms (SNPs) and simple sequence-repeats (SSRs).

- Linkage mapping often can locate a human disease gene to a chromosomal region that includes as many as 10 genes. To identify the gene of interest within this candidate region typically requires expression analysis and comparison of DNA sequences between wild-type and disease-affected individuals.

- Some inherited diseases can result from mutations in different genes in different individuals (genetic heterogeneity). The occurrence and severity of other diseases depend on the presence of mutant alleles of multiple genes in the same individuals (polygenic traits). Mapping of the genes associated with such diseases can be achieved by finding a statistical correlation between the disease and a particular chromosomal location in a genome-wide association study.

## 5.5 Inactivating the Function of Specific Genes in Eukaryotes

The elucidation of DNA and protein sequences in recent years has led to identification of many genes, using sequence patterns in genomic DNA and the sequence similarity of the encoded proteins with proteins of known function. As discussed in Chapter 6, the general functions of proteins identified by sequence searches may be predicted by analogy with known proteins. However, the precise in vivo roles of such "new" proteins may be unclear in the absence of mutant forms of the corresponding genes. In this section, we describe several ways for disrupting the normal function of a specific gene in the genome of an organism. Analysis of the resulting mutant phenotype often helps reveal the in vivo function of the normal gene and its encoded protein.

Three basic approaches underlie these gene-inactivation techniques: (1) replacing a normal gene with other sequences, (2) introducing an allele whose encoded protein inhibits functioning of the expressed normal protein, and (3) promoting destruction of the mRNA expressed from a gene. The normal endogenous gene is modified in techniques based on the first approach but is not modified in the other approaches.

### Normal Yeast Genes Can Be Replaced with Mutant Alleles by Homologous Recombination

Modifying the genome of the yeast *S. cerevisiae* is particularly easy for two reasons: yeast cells readily take up exogenous DNA under certain conditions, and the introduced DNA is efficiently exchanged for the homologous chromosomal site in the recipient cell. This specific, targeted recombination of identical stretches of DNA allows any gene in yeast chromosomes to be replaced with a mutant allele. (As we discuss in Section 5.1, recombination between homologous chromosomes also occurs naturally during meiosis.)

In one popular method for disrupting yeast genes in this fashion, the PCR is used to generate a *disruption construct* containing a selectable marker that subsequently is transfected into yeast cells. As shown in Figure 5-39a, primers for PCR amplification of the selectable marker are designed to include about 20 nucleotides identical with sequences flanking the yeast gene to be replaced. The resulting amplified construct comprises the selectable marker (e.g., the *kanMX* gene, which like *neo^r* confers resistance to G-418) flanked by about 20 base pairs that match the ends of the target yeast gene. Transformed diploid yeast cells in which one of the two copies of the target endogenous gene has been replaced by the disruption construct are identified by their resistance to G-418 or other selectable phenotype. These heterozygous diploid yeast cells generally grow normally regardless of the function of the target gene, but half the haploid spores derived from these cells will carry only the disrupted allele (Figure 5-39b). If a gene is essential for viability, then spores carrying a disrupted allele will not survive.

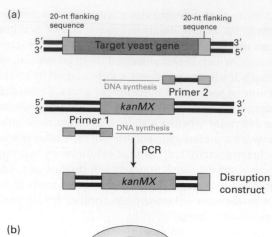

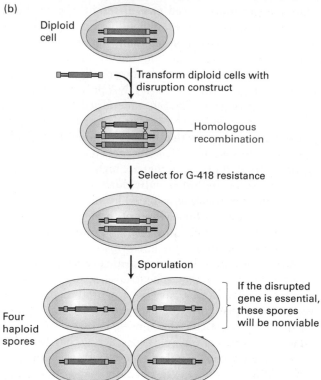

**EXPERIMENTAL FIGURE 5-39 Homologous recombination with transfected disruption constructs can inactivate specific target genes in yeast.** (a) A suitable construct for disrupting a target gene can be prepared by the PCR. The two primers designed for this purpose each contain a sequence of about 20 nucleotides (nt) that is homologous to one end of the target yeast gene as well as sequences needed to amplify a segment of DNA carrying a selectable marker gene such as *kanMX*, which confers resistance to G-418. (b) When recipient diploid *Saccharomyces* cells are transformed with the gene-disruption construct, homologous recombination between the ends of the construct and the corresponding chromosomal sequences will integrate the *kanMX* gene into the chromosome, replacing the target gene sequence. The recombinant diploid cells will grow on a medium containing G-418, whereas nontransformed cells will not. If the target gene is essential for viability, half the haploid spores that form after sporulation of recombinant diploid cells will be nonviable.

Disruption of yeast genes by this method is proving particularly useful in assessing the role of proteins identified by analysis of the entire genomic DNA sequence (see Chapter 6). A large consortium of scientists has replaced each of the approximately 6000 genes identified by this analysis with the *kanMX* disruption construct and determined which gene disruptions lead to nonviable haploid spores. These analyses have shown that about 4500 of the 6000 yeast genes are not required for viability, an unexpectedly large number of apparently nonessential genes. In some cases, disruption of a particular gene may give rise to subtle defects that do not compromise the viability of yeast cells growing under laboratory conditions. Alternatively, cells carrying a disrupted gene may be viable because of operation of backup or compensatory pathways. To investigate this possibility, yeast geneticists currently are searching for synthetic lethal mutations that might reveal nonessential genes with redundant functions (see Figure 5-9c).

## Transcription of Genes Ligated to a Regulated Promoter Can Be Controlled Experimentally

Although disruption of an essential gene required for cell growth will yield nonviable spores, this method provides little information about what the encoded protein actually does in cells. To learn more about how a specific gene contributes to cell growth and viability, investigators must be able to selectively inactivate the gene in a population of growing cells. One method for doing this employs a regulated promoter to selectively shut off transcription of an essential gene.

A useful promoter for this purpose is the yeast *GAL1* promoter, which is active in cells grown on galactose but completely inactive in cells grown on glucose. In this approach, the coding sequence of an essential gene (*X*) ligated to the *GAL1* promoter is inserted into a yeast shuttle vector (see Figure 5-17a). The recombinant vector then is introduced into haploid yeast cells in which gene *X* has been disrupted. Haploid cells that are transformed will grow on galactose medium since the normal copy of gene *X* on the vector is expressed in the presence of galactose. When the cells are transferred to a glucose-containing medium, gene *X* no longer is transcribed; as the cells divide, the amount of the encoded protein X will decline, eventually reaching a state of depletion that mimics a complete loss-of-function mutation. The observed changes in the phenotype of these cells after the shift to glucose medium may suggest which cell processes depend on the protein encoded by the essential gene *X*.

In an early application of this method, researchers explored the function of cytosolic *Hsc70* genes in yeast. Haploid cells with a disruption in all four redundant *Hsc70* genes were nonviable unless the cells carried a vector containing a copy of the *Hsc70* gene that could be expressed from the *GAL1* promoter on galactose medium. On transfer to glucose, the vector-carrying cells eventually stopped growing because of insufficient Hsc70 activity. Careful examination of these dying cells revealed that their secretory proteins could no longer enter the endoplasmic reticulum (ER). This study provided the first evidence for the unexpected role of Hsc70 protein in translocation of secretory proteins into the ER, a process examined in detail in Chapter 13.

## Specific Genes Can Be Permanently Inactivated in the Germ Line of Mice

Many of the methods for disrupting genes in yeast can be applied to genes of higher eukaryotes. These altered genes can be introduced into the germ line via homologous recombination to produce animals with a **gene knockout**, or simply "knockout." Knockout mice in which a specific gene is disrupted are a powerful experimental system for studying mammalian development, behavior, and physiology. They also are useful in studying the molecular basis of certain human genetic diseases.

Gene-targeted knockout mice are generated by a two-stage procedure. In the first stage, a DNA construct containing a disrupted allele of a particular target gene is introduced into embryonic stem (ES) cells. These cells, which are derived from the blastocyst, can be grown in culture through many generations (see Figure 21-7). In a small fraction of transfected cells, the introduced DNA undergoes homologous recombination with the target gene, although recombination at nonhomologous chromosomal sites occurs much more frequently. To select for cells in which homologous gene-targeted insertion occurs, the recombinant DNA construct introduced into ES cells needs to include two selectable marker genes (Figure 5-40). One of these genes (*neo^r*), which confers G-418 resistance, is inserted within the target gene (*X*), thereby disrupting it. The other selectable gene, the thymidine kinase gene from herpes simplex virus (*tk^{HSV}*), is inserted into the construct outside the target-gene sequence. ES cells that undergo recombination between the DNA construct and the homologous site on the chromosome will contain *neo^r* but will not incorporate *tk^{HSV}*. Because *tk^{HSV}* confers *sensitivity* to the cytotoxic nucleotide analog ganciclovir, the desired recombinant ES cells can be selected by their ability to survive in the presence of both G-418 and ganciclovir. In these cells, one allele of gene *X* will be disrupted.

In the second stage in production of knockout mice, ES cells heterozygous for a knockout mutation in gene *X* are injected into a recipient wild-type mouse blastocyst, which subsequently is transferred into a surrogate pseudopregnant female mouse (Figure 5-41). The resulting progeny will be chimeras, containing tissues derived from both the transplanted ES cells and the host cells. If the ES cells also are homozygous for a visible marker trait (e.g., coat color), then chimeric progeny in which the ES cells survived and proliferated can be identified easily. Chimeric mice are then mated with mice homozygous for another allele of the marker trait to determine if the knockout mutation is incorporated into the germ line. Finally, mating of mice, each heterozygous for the knockout allele, will produce progeny homozygous for the knockout mutation.

**Isolation of mouse ES cells with a gene-targeted disruption is the first stage in production of knockout mice.** (a) When exogenous DNA is introduced into embryonic stem (ES) cells, random insertion via nonhomologous recombination occurs much more frequently than gene-targeted insertion via homologous recombination. Recombinant cells in which one allele of gene *X* (orange and white) is disrupted can be obtained by using a recombinant vector that carries gene *X* disrupted with *neo*[r] (green), which confers resistance to G-418, and, outside the region of homology, *tk*[HSV] (yellow), the thymidine kinase gene from herpes simplex virus. The viral thymidine kinase, unlike the endogenous mouse enzyme, can convert the nucleotide analog ganciclovir into the monophosphate form; this is then modified to the triphosphate form, which inhibits cellular DNA replication in ES cells. Thus ganciclovir is cytotoxic for recombinant ES cells carrying the *tk*[HSV] gene. Nonhomologous insertion includes the *tk*[HSV] gene, whereas homologous insertion does not; therefore, only cells with nonhomologous insertion are sensitive to ganciclovir. (b) Recombinant cells are selected by treatment with G-418 since cells that fail to pick up DNA or integrate it into their genome are sensitive to this cytotoxic compound. The surviving recombinant cells are treated with ganciclovir. Only cells with a targeted disruption in gene *X*, and therefore lacking the *tk*[HSV] gene and its accompanying cytotoxicity, will survive. [See S. L. Mansour et al., 1988, *Nature* **336:**348.]

(a) **Formation of ES cells carrying a knockout mutation**

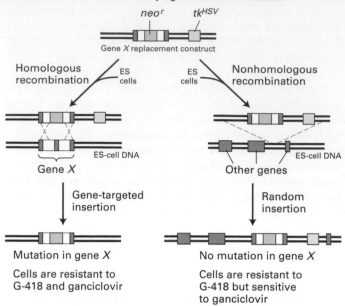

(b) **Positive and negative selection of recombinant ES cells**

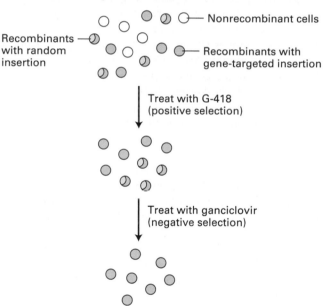

Development of knockout mice that mimic certain human diseases can be illustrated by cystic fibrosis. By methods discussed in Section 5.4, the recessive mutation that causes this disease eventually was shown to be located in a gene known as *CFTR*, which encodes a chloride channel. Using the cloned wild-type human *CFTR* gene, researchers isolated the homologous mouse gene and subsequently introduced mutations in it. The gene-knockout technique was then used to produce homozygous mutant mice, which showed symptoms (i.e., a phenotype), including disturbances to the functioning of epithelial cells, similar to those of humans with cystic fibrosis. These knockout mice are currently being used as a model system for studying this genetic disease and developing effective therapies. ■

## Somatic Cell Recombination Can Inactivate Genes in Specific Tissues

Investigators often are interested in examining the effects of knockout mutations in a particular tissue of the mouse, at a specific stage in development, or both. However, mice carrying a germ-line knockout may have defects in numerous tissues or die before the developmental stage of interest. To address this problem, mouse geneticists have devised a clever technique to inactivate target genes in specific types of somatic cells or at particular times during development.

This technique employs site-specific DNA recombination sites (called loxP *sites*) and the enzyme Cre that catalyzes recombination between them. The loxP-*Cre recombination system* is derived from bacteriophage P1, but this site-specific recombination system also functions when placed in mouse

cells. An essential feature of this technique is that expression of Cre is controlled by a cell-type-specific promoter. In *loxP*-Cre mice generated by the procedure depicted in Figure 5-42, inactivation of the gene of interest (*X*) occurs only in cells in which the promoter controlling the *cre* gene is active.

An early application of this technique provided strong evidence that a particular neurotransmitter receptor is important for learning and memory. Previous pharmacological and physiological studies had indicated that normal learning requires the NMDA class of glutamate receptors in the hippocampus, a region of the brain. But mice in which the gene

**EXPERIMENTAL FIGURE 5-41 ES cells heterozygous for a disrupted gene are used to produce gene-targeted knockout mice.** Step **1**: Embryonic stem (ES) cells heterozygous for a knockout mutation in a gene of interest ($X$) and homozygous for a dominant allele of a marker gene (here brown coat color, $A$) are transplanted into the blastocoel cavity of 4.5-day embryos that are homozygous for a recessive allele of the marker (here black coat color, $a$). Step **2**: The early embryos then are implanted into a pseudopregnant female. Those progeny containing ES-derived cells are chimeras, indicated by their mixed black and brown coats. Step **3**: Chimeric mice then are backcrossed to black mice; brown progeny from this mating have ES-derived cells in their germ line. Steps **4**–**6**: Analysis of DNA isolated from a small amount of tail tissue can identify brown mice heterozygous for the knockout allele. Intercrossing of these mice produces some individuals homozygous for the disrupted allele, that is, knockout mice. [Adapted from M. R. Capecchi, 1989, *Trends Genet.* **5**:70.]

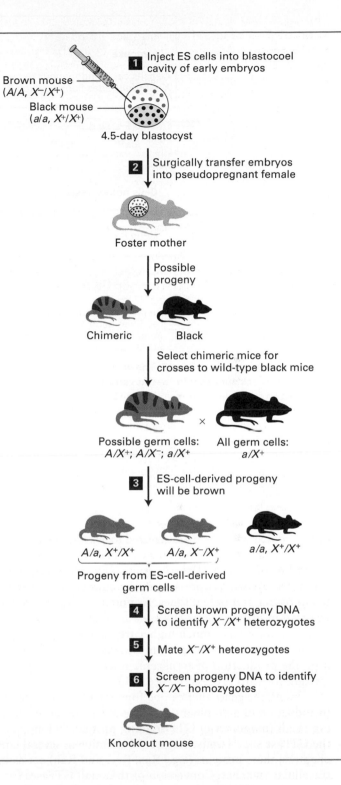

encoding an NMDA receptor subunit was knocked out died neonatally, precluding analysis of the receptor's role in learning. Following the protocol in Figure 5-42, researchers generated mice in which the receptor subunit gene was inactivated in the hippocampus but expressed in other tissues. These mice survived to adulthood and showed learning and memory defects, confirming a role for these receptors in the ability of mice to encode their experiences into memory.

## Dominant-Negative Alleles Can Functionally Inhibit Some Genes

In diploid organisms, as noted in Section 5.1, the phenotypic effect of a recessive allele is expressed only in homozygous individuals, whereas dominant alleles are expressed in heterozygotes. Thus an individual must carry two copies of a recessive allele but only one copy of a dominant allele to exhibit the corresponding phenotypes. We have seen how strains of mice that are homozygous for a given recessive knockout mutation can be produced by crossing individuals that are heterozygous for the same knockout mutation (see Figure 5-41). For experiments with cultured animal cells, however, it is usually difficult to disrupt both copies of a gene in order to produce a mutant phenotype. Moreover, the difficulty in producing strains with both copies of a gene mutated is often compounded by the presence of related genes of similar function that must also be inactivated in order to reveal an observable phenotype.

For certain genes, the difficulties in producing homozygous knockout mutants can be avoided by use of an allele carrying a dominant-negative mutation. These alleles are genetically dominant; that is, they produce a mutant phenotype even in cells carrying a wild-type copy of the gene. However, unlike other types of dominant alleles, dominant-negative alleles produce a phenotype equivalent to that of a loss-of-function mutation.

Useful dominant-negative alleles have been identified for a variety of genes and can be introduced into cultured cells by transfection or into the germ line of mice or other organisms. In both cases, the introduced gene is integrated into the genome by nonhomologous recombination. Such randomly inserted genes are called **transgenes**; the cells or

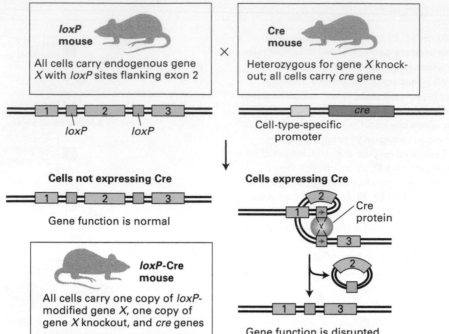

**EXPERIMENTAL FIGURE 5-42 The *loxP*-Cre recombination system can knock out genes in specific cell types.** A *loxP* site (purple) is inserted on each side of the essential exon 2 of the target gene *X* (blue) by homologous recombination, producing a *loxP* mouse. Since the *loxP* sites are in introns, they do not disrupt the function of *X*. The Cre mouse carries one gene *X* knockout allele and an introduced *cre* gene (orange) from bacteriophage P1 linked to a cell-type-specific promoter (yellow). The *cre* gene is incorporated into the mouse genome by nonhomologous recombination and does not affect the function of other genes. In the *loxP*-Cre mice that result from crossing, Cre protein is produced only in those cells in which the promoter is active. Thus these are the only cells in which recombination between the *loxP* sites catalyzed by Cre occurs, leading to deletion of exon 2. Since the other allele is a constitutive gene *X* knockout, deletion between the *loxP* sites results in complete loss of function of gene *X* in all cells expressing Cre. By using different promoters, researchers can study the effects of knocking out gene *X* in various types of cells.

organisms carrying them are referred to as transgenic. Transgenes carrying a dominant-negative allele usually are engineered so that the allele is controlled by a regulated promoter, allowing expression of the mutant protein in different tissues at different times. As noted above, the random integration of exogenous DNA via nonhomologous recombination occurs at a much higher frequency than insertion via homologous recombination. Because of this phenomenon, the production of transgenic mice is an efficient and straight forward process (Figure 5-43).

Among the genes that can be functionally inactivated by introduction of a dominant-negative allele are those encoding small (monomeric) GTP-binding proteins belonging to the GTPase superfamily. As we will examine in several later chapters, these proteins (e.g., Ras, Rac, and Rab) act as intracellular switches. Conversion of the small GTPases from an inactive GDP-bound state to an active GTP-bound state depends on their interacting with a corresponding guanine nucleotide exchange factor (GEF). A mutant small GTPase that permanently binds to the GEF protein will block conversion of endogenous wild-type small GTPases to the active GTP-bound state, thereby inhibiting them from performing their switching function (Figure 5-44).

## RNA Interference Causes Gene Inactivation by Destroying the Corresponding mRNA

The phenomenon known as **RNA interference (RNAi)** is perhaps the most straightforward method to inhibit the function of specific genes. This approach is technically simpler than the methods described above for disrupting genes. First observed in the roundworm *C. elegans*, RNAi refers to the ability of double-stranded RNA to block expression of its corresponding single-stranded mRNA but not that of mRNAs with a different sequence.

As described in Chapter 8, the phenomenon of RNAi rests on the general ability of eukaryotic cells to cleave double-stranded RNA into short (23-nt) double-stranded segments known as small inhibitory RNA (siRNA). The RNA endonuclease that catalyzes this reaction, known as Dicer, is found in all metazoans but not in simpler eukaryotes such as yeast. The siRNA molecules, in turn, can cause cleavage of mRNA molecules of matching sequence, in a reaction catalyzed by a protein complex known as *RISC*. RISC mediates recognition and hybridization between one strand of the siRNA and its complementary sequence on the target mRNA; subsequently, specific nucleases in the RISC complex

**EXPERIMENTAL FIGURE 5-43 Transgenic mice are produced by random integration of a foreign gene into the mouse germ line.** Foreign DNA injected into one of the two pronuclei (the male and female haploid nuclei contributed by the parents) has a good chance of being randomly integrated into the chromosomes of the diploid zygote. Because a transgene is integrated into the recipient genome by nonhomologous recombination, it does not disrupt endogenous genes. [See R. L. Brinster et al., 1981, *Cell* **27**:223.]

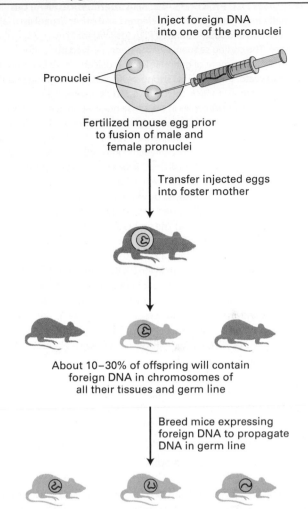

Inject foreign DNA
into one of the pronuclei

Pronuclei

Fertilized mouse egg prior
to fusion of male and
female pronuclei

Transfer injected eggs
into foster mother

About 10–30% of offspring will contain
foreign DNA in chromosomes of
all their tissues and germ line

Breed mice expressing
foreign DNA to propagate
DNA in germ line

cleave the mRNA-siRNA hybrid. This model accounts for the specificity of RNAi, since it depends on base pairing, and for its potency in silencing gene function, since the complementary mRNA is permanently destroyed by nucleolytic degradation. The normal function of both Dicer and RISC is to allow for gene regulation by small endogenous RNA molecules known as micro-RNAs, or miRNAs.

Researchers exploit the micro-RNA pathway for intentional silencing of a gene of interest by using either of two general methods for generating siRNAs of defined sequence. In the first method, a double-stranded RNA corresponding to the target-gene sequence is produced by in vitro transcription of both sense and antisense copies of this sequence (Figure 5-45a).

**(a) Cells expressing only wild-type alleles of a small GTPase**

Inactive          Active

GDP

GTP

GTP

GTP

GDP
Wild type

GEF

**(b) Cells expressing both wild-type alleles and a dominant-negative allele**

GDP    GDP

GDP    GDP

Dominant-negative
mutant

GEF

**FIGURE 5-44 Inactivation of the function of a wild-type GTPase by the action of a dominant-negative mutant allele.** (a) Small (monomeric) GTPases (purple) are activated by their interaction with a guanine nucleotide exchange factor (GEF), which catalyzes the exchange of GDP for GTP. (b) Introduction of a dominant-negative allele of a small GTPase gene into cultured cells or transgenic animals leads to expression of a mutant GTPase that binds to and inactivates the GEF. As a result, endogenous wild-type copies of the same small GTPase are trapped in the inactive GDP-bound state. A single dominant-negative allele thus causes a loss-of-function phenotype in heterozygotes similar to that seen in homozygotes carrying two recessive loss-of-function alleles.

This dsRNA is injected into the gonad of an adult worm, where it is converted to siRNA by Dicer in the developing embryos. In conjunction with the RISC complex, the siRNA molecules cause the corresponding mRNA molecules to be destroyed rapidly. The resulting worms display a phenotype similar to the one that would result from disruption of the corresponding gene itself. In some cases, entry of just a few molecules of a particular dsRNA into a cell is sufficient to inactivate many copies of the corresponding mRNA. Figure 5-45b illustrates the ability of an injected dsRNA to interfere with production of the corresponding endogenous mRNA in *C. elegans* embryos. In this experiment, the mRNA levels in embryos were determined by in situ hybridization, as described earlier, using a fluorescently labeled probe.

The second method is to produce a specific double-stranded RNA in vivo. An efficient way to do this is to express a synthetic gene that is designed to contain tandem segments of both sense and antisense sequences corresponding to the

**EXPERIMENTAL FIGURE 5-45  RNA interference (RNAi) can functionally inactivate genes in *C. elegans* and other organisms.** (a) In vitro production of double-stranded RNA (dsRNA) for RNAi of a specific target gene. The coding sequence of the gene, derived from either a cDNA clone or a segment of genomic DNA, is placed in two orientations in a plasmid vector adjacent to a strong promoter. Transcription of both constructs in vitro using RNA polymerase and ribonucleoside triphosphates yields many RNA copies in the sense orientation (identical with the mRNA sequence) or complementary antisense orientation. Under suitable conditions, these complementary RNA molecules will hybridize to form dsRNA. When the dsRNA is injected into cells, it is cleaved by Dicer into siRNAs. (b) Inhibition of *mex3* RNA expression in worm embryos by RNAi (see the text for the mechanism). (*Left*) Expression of *mex3* RNA in embryos was assayed by in situ hybridization with a probe specific for this mRNA, that is, linked to an enzyme that produces a colored (purple) product. (*Right*) The embryo derived from a worm injected with double-stranded *mex3* mRNA produces little or no endogenous *mex3* mRNA, as indicated by the absence of color. Each four-cell-stage embryo is ≈50 μm in length. (c) In vivo production of double-stranded RNA occurs via an engineered plasmid introduced directly into cells. The synthetic gene construct is a tandem arrangement of both sense and antisense sequences of the target gene. When it is transcribed, double-stranded small hairpin RNA forms (shRNA). The shRNA is cleaved by Dicer to form siRNA. [Part (b) from A. Fire et al., 1998, *Nature* **391:**806.]

**(a) In vitro production of double-stranded RNA**

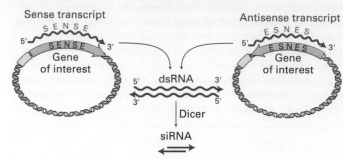

**(b)**

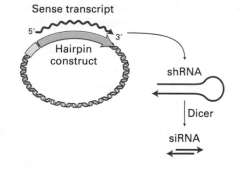

Noninjected | Injected

**(c) In vivo production of double-stranded RNA**

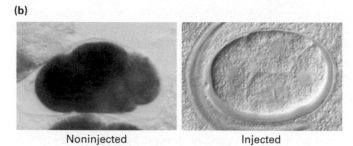

target gene (Figure 5-45c). When this gene is transcribed, a double-stranded RNA "hairpin" structure forms, known as *small hairpin RNA,* or shRNA. The shRNA will then be cleaved by Dicer to form siRNA molecules. The lentiviral expression vectors are particularly useful for introducing synthetic genes for the expression of shRNA constructs into animal cells.

Both RNAi methods lend themselves to systematic studies to inactivate each of the known genes in an organism and to observe what goes wrong. For example, in initial studies with *C. elegans,* RNA interference with 16,700 genes (about 86 percent of the genome) yielded 1722 visibly abnormal phenotypes. The genes whose functional inactivation causes particular abnormal phenotypes can be grouped into sets; each member of a set presumably controls the same signals or events. The regulatory relations between the genes in the set—for example, the genes that control muscle development—can then be worked out.

Other organisms in which RNAi-mediated gene inactivation has been successful include *Drosophila,* many kinds of plants, zebra fish, the frog *Xenopus,* and mice and are now the subjects of large-scale RNAi screens. For example, lentiviral vectors have been designed to inactivate by RNAi more than 10,000 different genes expressed in cultured mammalian cells. The function of the inactivated genes can be inferred from defects in growth or morphology of cell clones transfected with lentiviral vectors.

## KEY CONCEPTS of Section 5.5

### Inactivating the Function of Specific Genes in Eukaryotes

• Once a gene has been cloned, important clues about its normal function in vivo can be deduced from the observed phenotypic effects of mutating the gene.

• Genes can be disrupted in yeast by inserting a selectable marker gene into one allele of a wild-type gene via homologous recombination, producing a heterozygous mutant. When such a heterozygote is sporulated, disruption of an essential gene will produce two nonviable haploid spores (see Figure 5-39).

• A yeast gene can be inactivated in a controlled manner by using the *GAL1* promoter to shut off transcription of a gene when cells are transferred to glucose medium.

- In mice, modified genes can be incorporated into the germ line at their original genomic location by homologous recombination, producing knockouts (see Figures 5-40 and 5-41). Mouse knockouts can provide models for human genetic diseases such as cystic fibrosis.

- The *loxP*-Cre recombination system permits production of mice in which a gene is knocked out in a specific tissue.

- In the production of transgenic cells or organisms, exogenous DNA is integrated into the host genome by nonhomologous recombination (see Figure 5-43). Introduction of a dominant-negative allele in this way can functionally inactivate a gene without altering its sequence.

- In many organisms, including the roundworm *C. elegans*, double-stranded RNA triggers destruction of the all the mRNA molecules with the same sequence (see Figure 5-45). This phenomenon, known as *RNAi* (RNA interference), provides a specific and potent means of functionally inactivating genes without altering their structure.

## Perspectives for the Future

As the examples in this chapter and throughout the book illustrate, genetic analysis is the foundation of our understanding of many fundamental processes in cell biology. By examining the phenotypic consequences of mutations that inactivate a particular gene, geneticists are able to connect knowledge about the sequence, structure, and biochemical activity of the encoded protein to its function in the context of a living cell or multicellular organism. The classical approach to making these connections in both humans and simpler, experimentally accessible organisms has been to identify new mutations of interest based on their phenotypes and then to isolate the affected gene and its protein product.

Although scientists continue to use this classical genetic approach to dissect fundamental cellular processes and biochemical pathways, the availability of complete genomic sequence information for most of the common experimental organisms has fundamentally changed the way genetic experiments are conducted. Using various computational methods, scientists have identified the protein-coding gene sequences in most experimental organisms, including *E. coli*, yeast, *C. elegans*, *Drosophila*, *Arabidopsis*, mouse, and humans. The gene sequences, in turn, reveal the primary amino acid sequence of the encoded protein products, providing us with a nearly complete list of the proteins found in each of the major experimental organisms.

The approach taken by most researchers has thus shifted from discovering new genes and proteins to discovering the functions of genes and proteins whose sequences are already known. Once an interesting gene has been identified, genomic sequence information greatly speeds subsequent genetic manipulations of the gene, including its designed inactivation, so that more can be learned about its function.

Already sets of vectors for RNAi inactivation of most defined genes in the nematode *C. elegans* allow efficient genetic screens to be performed in this multicellular organism. These methods are now being applied to large collections of genes in cultured mammalian cells, and in the near future, either RNAi or knockout methods will have been used to inactivate every gene in the mouse.

In the past, a scientist might spend many years studying only a single gene, but nowadays scientists commonly study whole sets of genes at once. For example, with DNA microarrays the level of expression of all genes in an organism can be measured almost as easily as the expression of a single gene. One of the great challenges facing geneticists in the twenty-first century will be to exploit the vast amount of available data on the function and regulation of individual genes to understand how groups of genes are organized to form complex biochemical pathways and regulatory networks.

## Key Terms

| | |
|---|---|
| alleles 172 | phenotype 172 |
| clone 185 | plasmids 184 |
| complementary DNAs (cDNAs) 186 | point mutation 173 |
| DNA cloning 182 | polymerase chain reaction (PCR) 192 |
| DNA library 185 | probes 188 |
| DNA microarray 199 | recessive 173 |
| dominant 173 | recombinant DNA 182 |
| functional complementation 189 | recombination 180 |
| gene knockout 213 | restriction enzymes 183 |
| genomics 222 | RNA interference (RNAi) 216 |
| genotype 172 | segregation 175 |
| heterozygous 173 | Southern blotting 198 |
| homozygous 173 | temperature-sensitive mutations 176 |
| hybridization 188 | transfection 203 |
| in situ hybridization 199 | transformation 184 |
| linkage 182 | transgenes 215 |
| mutagen 172 | vector 182 |
| mutation 172 | wild type 172 |
| Northern blotting 198 | |

## Review the Concepts

1. Genetic mutations can provide insights into the mechanisms of complex cellular or developmental processes. How might your analysis of a genetic mutation be different depending on whether a particular mutation is recessive or dominant?

2. What is a temperature-sensitive mutation? Why are temperature-sensitive mutations useful for uncovering the function of a gene?

3. Describe how complementation analysis can be used to reveal whether two mutations are in the same or in different genes. Explain why complementation analysis will not work with dominant mutations.

4. Jane has isolated a mutant strain of yeast that forms red colonies instead of the normal white when grown on a plate. To determine the mutant gene, she decides to use functional complementation with a DNA library containing a lysine selection marker. In addition to the unknown gene mutation, the yeast are lacking the gene required to synthesize the amino acids leucine and lysine. What media will Jane grow her yeast on to ensure that they have acquired the library plasmids? How will she know when a library plasmid has complemented her yeast mutation?

5. Restriction enzymes and DNA ligase play essential roles in DNA cloning. How is it that a bacterium that produces a restriction enzyme does not cut its own DNA? Describe some general features of restriction-enzyme sites. What are the three types of DNA ends that can be generated after cutting DNA with restriction enzymes? What reaction is catalyzed by DNA ligase?

6. Bacterial plasmids often serve as cloning vectors. Describe the essential features of a plasmid vector. What are the advantages and applications of plasmids as cloning vectors?

7. A DNA library is a collection of clones, each containing a different fragment of DNA, inserted into a cloning vector. What is the difference between a cDNA and a genomic DNA library? You would like to clone gene X, a gene expressed only in neurons, into a vector using a library as the source of insert. If you have the following libraries at your disposal (genomic library from skin cells, cDNA library from skin cells, genomic library from neurons, cDNA library from neurons), which could you use and why?

8. In 1993, Kary Mullis won the Nobel Prize in Chemistry for his invention of the PCR process. Describe the three steps in each cycle of a PCR reaction. Why was the discovery of a thermostable DNA polymerase (e.g., *Taq* polymerase) so important for the development of PCR?

9. Southern and Northern blotting are powerful tools in molecular biology based on hybridization of nucleic acids. How are these techniques the same? How do they differ? Give some specific applications for each blotting technique.

10. A number of foreign proteins have been expressed in bacterial and mammalian cells. Describe the essential features of a recombinant plasmid that are required for expression of a foreign gene. How can you modify the foreign protein to facilitate its purification? What is the advantage of expressing a protein in mammalian cells versus bacteria?

11. Northern blotting, RT-PCR, and microarrays can be used to analyze gene expression. A lab studies yeast cells, comparing their growth in two different sugars, glucose and galactose. One student is comparing expression of the gene *HMG2* under the different conditions. Which technique(s) could he use and why? Another student wants to compare expression of all the genes on chromosome 4, of which there are approximately 800. What technique(s) could she use and why?

12. In determining the identity of the protein that corresponds to a newly discovered gene, it often helps to know the pattern of tissue expression for that gene. For example, researchers have found that a gene called *SERPINA6* is expressed in the liver, kidney, and pancreas but not in other tissues. What techniques might researchers use to find out which tissues express a particular gene?

13. DNA polymorphisms can be used as DNA markers. Describe the differences between SNP and SSR polymorphisms. How can these markers be used for DNA-mapping studies?

14. How can linkage-disequilibrium mapping sometimes provide a much higher resolution of gene location than classical linkage mapping?

15. Genetic linkage studies can usually only roughly locate the chromosomal position of a "disease" gene. How can expression analysis and DNA sequence analysis help locate a disease gene within the region identified by linkage mapping?

16. The ability to selectively modify the genome in the mouse has revolutionized mouse genetics. Outline the procedure for generating a knockout mouse at a specific genetic locus. How can the *loxP*-Cre system be used to conditionally knock out a gene? What is an important medical application of knockout mice?

17. Two methods for functionally inactivating a gene without altering the gene sequence are by dominant-negative mutations and RNA interference (RNAi). Describe how each method can inhibit expression of a gene.

## Analyze the Data

1. A culture of yeast that requires uracil for growth (*ura3⁻*) was mutagenized, and two mutant colonies, X and Y, have been isolated. Mating type **a** cells of mutant X are mated with mating type **α** cells of mutant Y to form diploid cells. The parental (*ura3⁻*), X, Y, and diploid cells are streaked onto agar plates containing uracil and incubated at 23 °C or 32 °C. Cell growth was monitored by the formation of colonies on the culture plates as shown in the figure below.

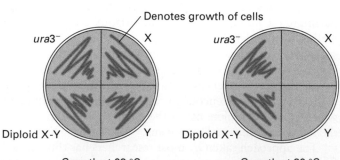

Growth at 23 °C            Growth at 32 °C

a. What can be deduced about mutants X and Y from the data provided?

b. A wild-type yeast cDNA library, prepared in a plasmid that contains the wild-type *URA3⁺* gene, is used to transform X cells, which are then cultured as indicated. Each black spot below represents a single clone growing on a petri plate. What are the molecular differences between the clones growing on the two plates? How can these results be used to identify the plasmid that contains a wild-type copy of gene X? Based on these results, how can the identity of gene X be uncovered?

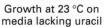

Growth at 23 °C on media lacking uracil

Growth at 32 °C on media lacking uracil

c. DNA is extracted from the parental cells, from X cells, and from Y cells. PCR primers are used to amplify the gene encoding X in both the parental and the X cells. The primers are complementary to regions of DNA just external to the gene encoding X. The PCR results are shown in the gel at the right. What can be deduced about the mutation in the X gene from these data?

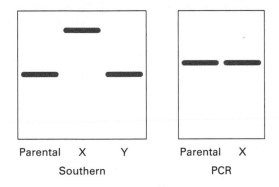

Parental    X    Y       Parental    X

Southern              PCR

d. A construct of the wild-type gene X is engineered to encode a fusion protein in which green fluorescent protein (GFP) is present at the N-terminus (GFP-X) or the C-terminus (X-GFP) of protein X. Both constructs, present on a *URA3+* plasmid, are used to transform X cells grown in the absence of uracil. The transformants are then monitored for growth at 32 °C, shown below at the left. At the right are typical fluorescent images of X-GFP and GFP-X cells grown at 23 °C in which green denotes the presence of green fluorescent protein. What is a reasonable explanation for growth of GFP-X but not X-GFP cells at 32 °C?

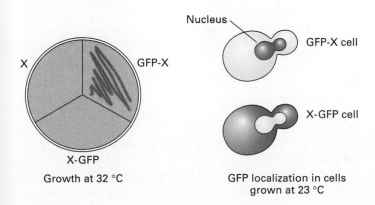

Growth at 32 °C

GFP localization in cells grown at 23 °C

e. Haploid offspring of the diploid cells from part (a) above are generated. XY double mutants constitute 1/4 of these offspring. Haploid X cells, Y cells, and XY cells in liquid culture are synchronized at a stage just prior to budding and then are shifted from 23 °C to 32 °C. Examination of the cells 24 hours later reveals that X cells are arrested with small buds, Y cells are arrested with large buds, and XY cells are arrested with small buds. What is the relationship between X and Y?

## References

### Genetic Analysis of Mutations to Identify and Study Genes

Adams, A. E. M., D. Botstein, and D. B. Drubin. 1989. A yeast actin-binding protein is encoded by *sac6*, a gene found by suppression of an actin mutation. *Science* **243**:231.

Griffiths, A. G. F., et al. 2000. *An Introduction to Genetic Analysis*, 7th ed. W. H. Freeman and Company.

Guarente, L. 1993. Synthetic enhancement in gene interaction: a genetic tool comes of age. *Trends Genet.* **9**:362–366.

Hartwell, L. H. 1967. Macromolecular synthesis of temperature-sensitive mutants of yeast. *J. Bacteriol.* **93**:1662.

Hartwell, L. H. 1974. Genetic control of the cell division cycle in yeast. *Science* **183**:46.

Nüsslein-Volhard, C., and E. Wieschaus. 1980. Mutations affecting segment number and polarity in *Drosophila*. *Nature* **287**:795–801.

Simon, M. A., et al. 1991. Ras1 and a putative guanine nucleotide exchange factor perform crucial steps in signaling by the sevenless protein tyrosine kinase. *Cell* **67**:701–716.

Tong, A. H., et al. 2001. Systematic genetic analysis with ordered arrays of yeast deletion mutants. *Science* **294**:2364–2368.

### DNA Cloning and Characterization

Ausubel, F. M., et al. 2002. *Current Protocols in Molecular Biology*. Wiley.

Gubler, U., and B. J. Hoffman. 1983. A simple and very efficient method for generating cDNA libraries. *Gene* **25**:263–289.

Han, J. H., C. Stratowa, and W. J. Rutter. 1987. Isolation of full-length putative rat lysophospholipase cDNA using improved methods for mRNA isolation and cDNA cloning. *Biochem.* **26**:1617–1632.

Itakura, K., J. J. Rossi, and R. B. Wallace. 1984. Synthesis and use of synthetic oligonucleotides. *Ann. Rev. Biochem.* **53**:323–356.

Maniatis, T., et al. 1978. The isolation of structural genes from libraries of eucaryotic DNA. *Cell* **15**:687–701.

Nasmyth, K. A., and S. I. Reed. 1980. Isolation of genes by complementation in yeast: molecular cloning of a cell-cycle gene. *Proc. Nat'l Acad. Sci. USA* **77**:2119–2123.

Nathans, D., and H. O. Smith. 1975. Restriction endonucleases in the analysis and restructuring of DNA molecules. *Ann. Rev. Biochem.* **44**:273–293.

Roberts, R. J., and D. Macelis. 1997. REBASE—restriction enzymes and methylases. *Nucl. Acids Res.* **25**:248–262. Information on accessing a continuously updated database on restriction and modification enzymes at http://www.neb.com/rebase.

### Using Cloned DNA Fragments to Study Gene Expression

Andrews, A. T. 1986. *Electrophoresis*, 2d ed. Oxford University Press.

Erlich, H., ed. 1992. *PCR Technology: Principles and Applications for DNA Amplification*. W. H. Freeman and Company.

Pellicer, A., M. Wigler, R. Axel, and S. Silverstein. 1978. The transfer and stable integration of the HSV thymidine kinase gene into mouse cells. *Cell* **41**:133–141.

Saiki, R. K., et al. 1988. Primer-directed enzymatic amplification of DNA with a thermostable DNA polymerase. *Science* **239**:487–491.

Sanger, F. 1981. Determination of nucleotide sequences in DNA. *Science* **214**:1205–1210.

Souza, L. M., et al. 1986. Recombinant human granulocyte-colony stimulating factor: effects on normal and leukemic myeloid cells. *Science* **232**:61–65.

Wahl, G. M., J. L. Meinkoth, and A. R. Kimmel. 1987. Northern and Southern blots. *Meth. Enzymol.* **152**:572–581.

Wallace, R. B., et al. 1981. The use of synthetic oligonucleotides as hybridization probes. II: Hybridization of oligonucleotides of mixed sequence to rabbit β-globin DNA. *Nucl. Acids Res.* **9**:879–887.

## Locating and Identifying Human Disease Genes

Botstein, D., et al. 1980. Construction of a genetic linkage map in man using restriction fragment length polymorphisms. *Am. J. Genet.* **32**:314–331.

Donis-Keller, H., et al. 1987. A genetic linkage map of the human genome. *Cell* **51**:319–337.

Hartwell, L., et al. 2006. *Genetics: From Genes to Genomes.* McGraw-Hill.

Hastbacka, T., et al. 1994. The diastrophic dysplasia gene encodes a novel sulfate transporter: positional cloning by fine-structure linkage disequilibrium mapping. *Cell* **78**:1073.

Orita, M., et al. 1989. Rapid and sensitive detection of point mutations and DNA polymorphisms using the polymerase chain reaction. *Genomics* **5**:874.

Tabor, H. K., N. J. Risch, and R. M. Myers. 2002. Opinion: candidate-gene approaches for studying complex genetic traits: practical considerations. *Nat. Rev. Genet.* **3**:391–397.

## Inactivating the Function of Specific Genes in Eukaryotes

Capecchi, M. R. 1989. Altering the genome by homologous recombination. *Science* **244**:1288–1292.

Deshaies, R. J., et al. 1988. A subfamily of stress proteins facilitates translocation of secretory and mitochondrial precursor polypeptides. *Nature* **332**:800–805.

Fire, A., et al. 1998. Potent and specific genetic interference by double-stranded RNA in *Caenorhabditis elegans*. *Nature* **391**:806–811.

Gu, H., et al. 1994. Deletion of a DNA polymerase beta gene segment in T cells using cell type-specific gene targeting. *Science* **265**:103–106.

Zamore, P. D., et al. 2000. RNAi: double-stranded RNA directs the ATP-dependent cleavage of mRNA at 21 to 23 nucleotide intervals. *Cell* **101**:25–33.

Zimmer, A. 1992. Manipulating the genome by homologous recombination in embryonic stem cells. *Ann. Rev. Neurosci.* **15**:115.

These brightly colored RxFISH-painted chromosomes are both beautiful and useful in revealing chromosome anomalies and in comparing karyotypes of different species. [© Department of Clinical Cytogenetics, Addenbrookes Hospital/Photo Researchers, Inc.]

# Genes, Genomics, and Chromosomes

In previous chapters we learned how the structure and composition of proteins allow them to perform a wide variety of cellular functions. We also examined another vital component of cells, the nucleic acids, and the process by which information encoded in the sequence of DNA is translated into protein. In this chapter, our focus again is on DNA and proteins as we consider the characteristics of eukaryotic nuclear and organellar genomes: the features of genes and the other DNA sequence that comprise the genome, and how this DNA is structured and organized by proteins within the cell.

By the beginning of the twenty-first century, molecular biologists had completed sequencing the entire genomes of hundreds of viruses, scores of bacteria, and one unicellular eukaryote, the budding yeast *S. cerevisiae*. By now, the vast majority of the genome sequence is also known for the fission yeast *S. pombe*, the simple plant *A. thaliana*, rice, and multiple multicellular animals (**metazoans**) including the roundworm *C. elegans*, the fruit fly *D. melanogaster*, mice, humans and at least one representative each of the ≈35 metazoan phyla. Detailed analysis of these sequencing data has revealed insights into evolution, genome organization, and gene function. It has allowed researchers to identify previously unknown genes and to estimate the total number of protein-coding genes in each genome. Comparisons between gene sequences often provide insight into possible functions of newly identified genes. Comparisons of genome sequence and organization between species also help us understand the evolution of organisms.

Surprisingly, DNA sequencing revealed that a large portion of the genomes of higher eukaryotes does not encode mRNAs or any other RNAs required by the organism. Remarkably, such noncoding DNA constitutes ≈98.5 percent of human chromosomal DNA! The noncoding DNA in multicellular organisms contains many regions that are similar but not identical. Variations within some stretches of this *repetitious DNA* between individuals are so great that every person can be distinguished by a DNA "fingerprint" based on these sequence variations. Moreover, some repetitious DNA sequences are not found in the same positions in the genomes of different individuals of the same species. At one time, all noncoding DNA was collectively termed "junk DNA" and was considered to serve no purpose. We now understand the evolutionary basis of all this extra DNA, and the variation in location of certain sequences between individuals. Cellular genomes harbor **transposable (mobile) DNA elements** that can copy themselves and move throughout the genome. Although transposable DNA elements seem to have little function in the life cycle of an individual organism, over evolutionary time they have shaped our genomes

## OUTLINE

and contributed to the rapid evolution of multicellular organisms.

In higher eukaryotes, DNA regions encoding proteins or functional RNAs—that is, **genes**—lie amidst this expanse of apparently nonfunctional DNA. In addition to the nonfunctional DNA *between* genes, noncoding **introns** are common *within* genes of multicellular plants and animals. Sequencing of the same protein-coding gene in a variety of eukaryotic species has shown that evolutionary pressure selects for maintenance of relatively similar sequences in the coding regions, or **exons.** In contrast, wide sequence variation, even including total loss, occurs among introns, suggesting that most intron sequences have little functional significance. However, as we shall see, although most of the DNA sequence of introns is not functional, the existence of introns has favored the evolution of multidomain proteins that are common in higher eukaryotes. It also allowed the rapid evolution of proteins with new combinations of functional domains. In addition, short noncoding RNAs called siRNAs and miRNAs that regulate translation and mRNA stability, and long noncoding RNAs that may regulate transcription through their influence on chromatin structure, can be processed from some introns (Chapters 7 and 8).

Mitochondria and chloroplasts also contain DNA that encodes proteins essential to the function of these vital organelles. We shall see that mitochondrial and chloroplast DNAs are evolutionary remnants of the origins of these organelles. Comparison of DNA sequences between different classes of bacteria and mitochondrial and chloroplast genomes has revealed that these organelles evolved from intracellular eubacteria that developed symbiotic relationships with ancient eukaryotic cells.

The sheer length of cellular DNA is a significant problem with which cells must contend. The DNA in a single human cell, which measures about 2 meters in total length, must be contained within cells with diameters of less than 10 μm, a compaction ratio of greater than $10^5$ to 1. In relative terms, if a cell were 1 centimeter in diameter (about the size of a pea), the length of DNA packed into its nucleus would be about 2 kilometers (1.2 miles)! Specialized eukaryotic proteins associated with nuclear DNA exquisitely fold and organize the DNA so that it fits into nuclei. And yet, at the same time, any given portion of this highly compacted DNA can be accessed readily for transcription, DNA replication, and repair of DNA damage without the long DNA molecules becoming tangled or broken. Furthermore, the integrity of DNA must be maintained during the process of cell division when it is partitioned into daughter cells. In eukaryotes, the complex of DNA and the proteins that organize it, called **chromatin**, can be visualized as individual **chromosomes** during mitosis. As we will see in this and the following chapter, the organization of DNA into chromatin allows a mechanism for regulation of gene expression that is not available in bacteria.

In the first five sections of this chapter, we provide an overview of the landscape of eukaryotic genes and genomes. First we discuss the structure of eukaryotic genes and the complexities that arise in higher organisms from the processing of

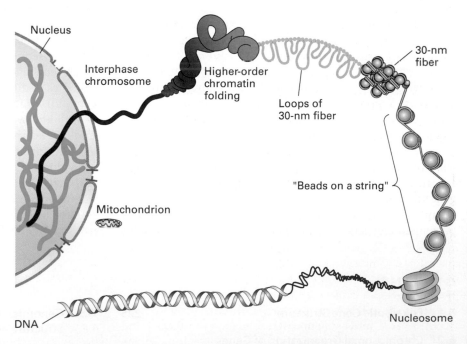

**FIGURE 6-1 Overview of the structure of genes and chromosomes.** DNA of higher eukaryotes consists of unique and repeated sequences. Only ≈1.5 percent of human DNA encodes proteins and functional RNAs and the regulatory sequences that control their expression; the remainder is merely spacer DNA between genes and introns within genes. Much of this DNA, ≈45 percent in humans, is derived from mobile DNA elements, genetic symbiotes that have contributed to the evolution of contemporary genomes. Each chromosome consists of a single, long molecule of DNA up to ≈280 Mb in humans, organized into increasing levels of condensation by the histone and nonhistone proteins with which it is intricately complexed. Much smaller DNA molecules are localized in mitochondria and chloroplasts.

| Major Types of DNA Sequence | |
| --- | --- |
| Single-copy genes | Simple-sequence DNA |
| Gene families | Transposable DNA elements |
| Tandemly repeated genes | Spacer DNA |
| Introns | |

mRNA precursors into alternatively spliced mRNAs. Next we discuss the main classes of eukaryotic DNA, including the special properties of transposable DNA elements and how they shaped contemporary genomes. We then consider organelle DNA and how it differs from nuclear DNA. This background prepares us to discuss **genomics**, computer-based methods for analyzing and interpreting vast amounts of sequence data. The final two sections of the chapter address how DNA is physically organized in eukaryotic cells. We consider the packaging of DNA and **histone** proteins into compact complexes called **nucleosomes** that are the fundamental building blocks of chromatin, the large-scale structure of chromosomes, and the functional elements required for chromosome duplication and segregation. Figure 6-1 provides an overview of these interrelated subjects. The understanding of genes, genomics, and chromosomes gained in this chapter will prepare us to explore current knowledge about how the synthesis and concentration of each protein and functional RNA in a cell is regulated in the following two chapters.

## 6.1 Eukaryotic Gene Structure

In molecular terms, a gene commonly is defined as *the entire nucleic acid sequence that is necessary for the synthesis of a functional gene product (polypeptide or RNA)*. According to this definition, a gene includes more than the nucleotides encoding an amino acid sequence or a functional RNA, referred to as the *coding region*. A gene also includes all the DNA sequences required for synthesis of a particular RNA transcript, no matter where those sequences are located in relation to the coding region. For example, in eukaryotic genes, transcription-control regions known as **enhancers** can lie 50 kb or more from the coding region. As we learned in Chapter 4, other critical noncoding regions in eukaryotic genes include the promoter, as well as sequences that specify 3′ cleavage and polyadenylation, known as *poly(A) sites,* and splicing of primary RNA transcripts, known as *splice sites* (see Figure 4-15). Mutations in these sequences, which control transcription initiation and RNA processing, affect the normal expression and function of RNAs, producing distinct phenotypes in mutant organisms. We examine these various control elements of genes in greater detail in Chapters 7 and 8.

Although most genes are transcribed into mRNAs, which encode proteins, some DNA sequences are transcribed into RNAs that do not encode proteins (e.g., tRNAs and rRNAs, described in Chapter 4, and miRNAs and siRNAs that regulate mRNA translation and stability, discussed in Chapter 8). Because the DNA that encodes tRNAs, rRNAs, miRNAs, and siRNAs can cause specific phenotypes when they are mutated, these DNA regions generally are referred to as tRNA, rRNA, miRNA, and siRNA *genes,* even though the final products of these genes are RNA molecules and not proteins.

In this section, we will examine the structure of genes in bacteria and eukaryotes and discuss how their respective gene structures influence gene expression and evolution.

## Most Eukaryotic Genes Contain Introns and Produce mRNAs Encoding Single Proteins

As discussed in Chapter 4, many bacterial mRNAs (e.g., the mRNA encoded by the *trp* operon) include the coding region for several proteins that function together in a biological process. Such mRNAs are said to be *polycistronic*. (A *cistron* is a genetic unit encoding a single polypeptide.) In contrast, most eukaryotic mRNAs are *monocistronic;* that is, each mRNA molecule encodes a single protein. This difference between polycistronic and monocistronic mRNAs correlates with a fundamental difference in their translation.

Within a bacterial polycistronic mRNA, a ribosome-binding site is located near the start site for each of the protein-coding regions, or cistrons, in the mRNA. Translation initiation can begin at any of these multiple internal sites, producing multiple proteins (see Figure 4-13a). In most eukaryotic mRNAs, however, the 5′-cap structure directs ribosome binding, and translation begins at the closest AUG start codon (see Figure 4-13b). As a result, translation begins only at this site. In many cases, the primary transcripts of eukaryotic protein-coding genes are processed into a single type of mRNA, which is translated to give a single type of polypeptide (see Figure 4-15).

Unlike bacterial and yeast genes, which generally lack introns, most genes in multicellular animals and plants contain introns, which are removed during RNA processing in the nucleus before the fully processed mRNA is exported to the cytosol for translation. In many cases, the introns in a gene are considerably longer than the exons. The median intron length in human genes is 3.3 kb. Some, however, are much longer: the longest known human intron is 17,106 bp and lies within the *titin* gene, encoding a structural protein in muscle cells. In comparison, most human exons contain only 50–200 base pairs. The typical human gene encoding an average-size protein is ≈50,000 bp long, but more than 95 percent of that sequence consists of introns and flanking noncoding 5′ and 3′ regions.

Many large proteins in higher organisms that have repeated domains are encoded by genes consisting of repeats of similar exons separated by introns of variable length. An example of this is fibronectin, a component of the extracellular matrix. The fibronectin gene contains multiple copies of five types of exons (see Figure 4-16). Such genes evolved by tandem duplication of the DNA encoding the repeated exon, probably by unequal crossing over during meiosis as shown in Figure 6-2a.

## Simple and Complex Transcription Units Are Found in Eukaryotic Genomes

The cluster of genes that forms a bacterial operon comprises a single **transcription unit** that is transcribed from a specific promoter in the DNA sequence to a termination site, producing a single primary transcript. In other words, genes and transcription units often are distinguishable in prokaryotes, since a single transcription unit contains several genes when they are part of an operon. In contrast, most eukaryotic genes

**(a) Exon duplication**

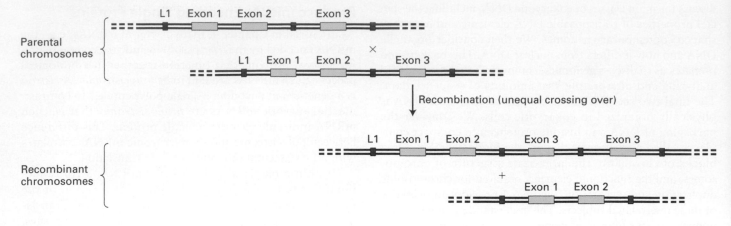

**(b) Gene duplication**

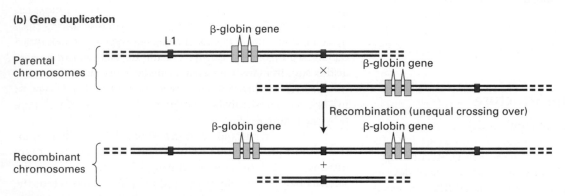

**FIGURE 6-2 Exon and Gene Duplication.** (a) Exon duplication results from unequal crossing over during meiosis. Each parental chromosome (*top*) contains one ancestral gene containing three exons and two introns. Homologous noncoding L1 repeated sequences lie 5′ and 3′ of the gene, and also in the intron between exons 2 and 3. As we shall see later in the chapter, L1 sequences have been repeatedly transposed to new sites in the genome over the course of human evolution, so that all chromosomes are peppered with them. The parental chromosomes are shown displaced relative to each other, so that the L1 sequences are aligned. Homologous recombination between L1 sequences as shown would generate one recombinant chromosome where the gene now has four exons (two copies of exon 3) and one chromosome where the gene is missing exon 3. (b) The same process can generate duplications of entire genes. Each parental chromosome (*top*) contains one ancestral β-globin gene. After the unequal recombination between L1 sequences, subsequent independent mutations in the resulting duplicated genes could lead to slight changes in sequence that might result in slightly different functional properties of the encoded proteins. Unequal crossing over also can result from rare recombinations between unrelated sequences. [Part (b) see D. H. A. Fitch et al., 1991, *Proc. Nat'l Acad. Sci. USA* **88**:7396.]

are expressed from separate transcription units, and each mRNA is translated into a single protein.

Eukaryotic transcription units, however, are classified into two types, depending on the fate of the primary transcript. The primary transcript produced from a *simple* transcription unit, such as the one encoding β-globin (Figure 4-15), is processed to yield a single type of mRNA, encoding a single protein. Mutations in exons, introns, and transcription-control regions all may influence expression of the protein encoded by a simple transcription unit (Figure 6-3a). In humans, simple transcription units such as β-globin are rare. Approximately 90 percent of human transcription units are *complex*. The primary RNA transcript can be processed in more than one way, leading to formation of mRNAs containing different exons. Each alternate mRNA, however, is monocistronic, being translated into a single polypeptide, with translation usually initiating at the first AUG in the mRNA. Multiple mRNAs can arise from a primary transcript in three ways, as shown in Figure 6-3b.

Examples of all three types of alternative RNA processing occur in the genes that regulate sexual differentiation in *Drosophila* (see Figure 8-16). Commonly, one mRNA is produced from a complex transcription unit in some cell types, and a different mRNA is made in other cell types. For example, **alternative splicing** of the primary fibronectin transcript in fibroblasts and hepatocytes determines whether or not the secreted protein includes domains that adhere to cell surfaces (see Figure 4-16). The phenomenon of alternative splicing greatly expands the number of proteins encoded in the genomes of higher organisms. It is estimated that ≈90 percent of human genes are contained within complex transcription units that give rise to alternatively spliced mRNAs encoding proteins with distinct functions, as for the fibroblast and hepatocyte forms of fibronectin.

## (a) Simple transcription unit

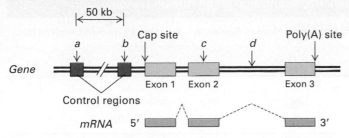

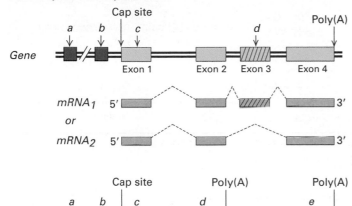

## (b) Complex transcription units

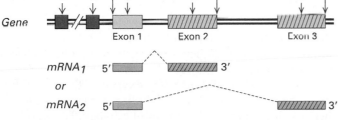

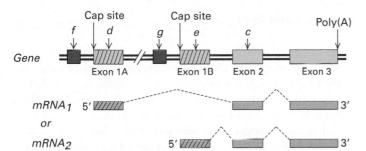

**FIGURE 6-3 Simple and complex eukaryotic transcription units.**
(a) A simple transcription unit includes a region that encodes one protein, extending from the 5′ cap site to the 3′ poly(A) site, and associated control regions. Introns lie between exons (light blue rectangles) and are removed during processing of the primary transcripts (dashed red lines); thus they do not occur in the functional monocistronic mRNA. Mutations in a transcription-control region (a, b) may reduce or prevent transcription, thus reducing or eliminating synthesis of the encoded protein. A mutation within an exon (c) may result in an abnormal protein with diminished activity. A mutation within an intron (d) that introduces a new splice site results in an abnormally spliced mRNA encoding a nonfunctional protein. (b) Complex transcription units produce primary transcripts that can be processed in alternative ways. (*Top*) If a primary transcript contains alternative splice sites, it can be processed into mRNAs with the same 5′ and 3′ exons but different internal exons. (*Middle*) If a primary transcript has two poly(A) sites, it can be processed into mRNAs with alternative 3′ exons. (*Bottom*) If alternative promoters (f or g) are active in different cell types, $mRNA_1$, produced in a cell type in which f is activated, has a different first exon (1A) than $mRNA_2$ has, which is produced in a cell type in which g is activated (and where exon 1B is used). Mutations in control regions (a and b) and those designated c within exons shared by the alternative mRNAs affect the proteins encoded by both alternatively processed mRNAs. In contrast, mutations (designated d and e) within exons unique to one of the alternatively processed mRNAs affect only the protein translated from that mRNA. For genes that are transcribed from different promoters in different cell types (bottom), mutations in different control regions (f and g) affect expression only in the cell type in which that control region is active.

The relationship between a mutation and a gene is not always straightforward when it comes to complex transcription units. A mutation in the control region or in an exon shared by alternatively spliced mRNAs will affect all the alternative proteins encoded by a given complex transcription unit. On the other hand, mutations in an exon present in only one of the alternative mRNAs will affect only the protein encoded by that mRNA. As explained in Chapter 5, **genetic complementation** tests commonly are used to determine if two mutations are in the same or different genes (see Figure 5-7). However, in the complex transcription unit shown in Figure 6-3b (*middle*), mutations d and e would comple-

ment each other in a genetic complementation test, even though they occur in the same gene. This is because a chromosome with mutation d can express a normal protein encoded by $mRNA_2$ and a chromosome with mutation e can express a normal protein encoded by $mRNA_1$. Both mRNAs produced from this gene would be present in a diploid cell carrying both mutations, generating both protein products and hence a wild-type phenotype. However, a chromosome with mutation c in an exon common to both mRNAs would not complement either mutation d or e. In other words, mutation c would be in the same complementation groups as mutations d and e, even though d and e themselves would not be in the same complementation group! Given these complications with the genetic definition of a gene, the genomic definition outlined at the beginning of this section is commonly used. In the case of protein-coding genes, a gene is the DNA sequence transcribed into a pre-mRNA precursor, equivalent to a transcription unit, plus any other regulatory elements required for synthesis of the primary transcript. The various proteins encoded by the alternatively spliced mRNAs expressed from one gene are called **isoforms**.

## Protein-Coding Genes May Be Solitary or Belong to a Gene Family

The nucleotide sequences within chromosomal DNA can be classified on the basis of structure and function, as shown in Table 6-1. We will examine the properties of each class, beginning with protein-coding genes, which comprise two groups.

| Class | Length | Copy Number in Human Genome | Fraction of Human Genome (%) |
|---|---|---|---|
| Protein-coding genes | 0.5–2200 kb | ≈25,000 | ≈55* (11.8)[†] |
| **Tandemly repeated genes** | | | |
| U2 snRNA | 6.1 kb[‡] | ≈20 | <0.001 |
| rRNAs | 43 kb[‡] | ≈300 | 0.4 |
| **Repetitious DNA** | | | |
| Simple-sequence DNA | 1–500 bp | Variable | ≈6 |
| Interspersed repeats (mobile DNA elements) | | | |
| DNA transposons | 2–3 kb | 300,000 | 3 |
| LTR retrotransposons | 6–11 kb | 440,000 | 8 |
| Non-LTR retrotransposons | | | |
| LINEs | 6–8 kb | 860,000 | 21 |
| SINEs | 100–400bp | 1,600,000 | 13 |
| Processed pseudogenes | Variable | 1–≈100 | ≈0.4 |
| Unclassified spacer DNA[§] | Variable | n.a. | ≈25 |

*Complete transcription units including introns.
[†]Transcription units not including introns. Protein-coding regions (exons) total 1.1% of the genome.
[‡]Length of each repeat in a tandemly repeated sequence.
[§]Sequences between transcription units that are not repeated in the genome; n.a. = not applicable.
SOURCE: International Human Genome Sequencing Consortium, 2001, *Nature* **409**:860 and 2004, *Nature* **431**:931.

In multicellular organisms, roughly 25–50 percent of the protein-coding genes are represented only once in the haploid genome and thus are termed *solitary* genes. A well-studied example of a solitary protein-coding gene is the chicken lysozyme gene. The 15-kb DNA sequence encoding chicken lysozyme constitutes a simple transcription unit containing four exons and three introns. The flanking regions, extending for about 20 kb upstream and downstream from the transcription unit, do not encode any detectable mRNAs. Lysozyme, an enzyme that cleaves the polysaccharides in bacterial cell walls, is an abundant component of chicken egg-white protein and also is found in human tears. Its activity helps to keep the surface of the eye and the chicken egg sterile.

*Duplicated* genes constitute the second group of protein-coding genes. These are genes with close but nonidentical sequences that often are located within 5–50 kb of one another. A set of duplicated genes that encodes proteins with similar but nonidentical amino acid sequences is called a **gene family;** the encoded, closely related, homologous proteins constitute a **protein family.** A few protein families, such as protein kinases, vertebrate immunoglobulins, and olfactory receptors, include hundreds of members. Most protein families, however, include from just a few to 30 or so members; common examples are cytoskeletal proteins, the myosin heavy chain, and the α- and β-globins in vertebrates.

The genes encoding the β-like globins are a good example of a gene family. As shown in Figure 6-4a, the β-like globin gene family contains five functional genes designated β, δ, $A_\gamma$, $G_\gamma$, and ε; the encoded polypeptides are similarly designated. Two identical β-like globin polypeptides combine with two identical α-globin polypeptides (encoded by another gene family) and four small heme groups to form a hemoglobin molecule (see Figure 3-13). All the hemoglobins formed from the different β-like globins carry oxygen in the blood, but they exhibit somewhat different properties that are suited to their specific functions in human physiology. For example, hemoglobins containing either the $A_\gamma$ or $G_\gamma$ polypeptides are expressed only during fetal life. Because these fetal hemoglobins have a higher affinity for oxygen than adult hemoglobins, they can effectively extract oxygen from the maternal circulation in the placenta. The lower oxygen affinity of adult hemoglobins, which are expressed after birth, permits better release of oxygen to the tissues, especially muscles, which have a high demand for oxygen during exercise.

The different β-globin genes arose by duplication of an ancestral gene, most likely as the result of an "unequal crossover" during meiotic recombination in a developing germ cell (egg or sperm, see Figure 6-2b). Over evolutionary time the two copies of the gene that resulted accumulated random mutations; beneficial mutations that conferred some refinement in the basic oxygen-carrying function of hemoglobin were retained by natural selection, resulting in *sequence drift*. Repeated gene duplications and subsequent sequence drift are thought to have generated the contemporary globin-like genes observed in humans and other mammals today.

Two regions in the human β-like globin gene cluster contain nonfunctional sequences, called **pseudogenes,** similar to those of the functional β-like globin genes (see Figure 6-4a). Sequence analysis shows that these pseudogenes have the same apparent exon–intron structure as the functional β-like globin

### (a) Human β-globin gene cluster (chromosome 11)

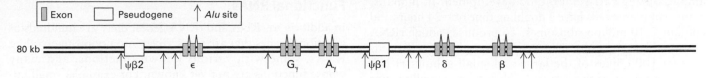

### (b) *S. cerevisiae* (chromosome III)

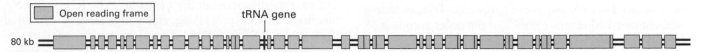

**FIGURE 6-4 Comparison of gene density in higher and lower eukaryotes.** (a) In the diagram of the β-globin gene cluster on human chromosome 11, the green boxes represent exons of β-globin–related genes. Exons spliced together to form one mRNA are connected by caret-like spikes. The human β-globin gene cluster contains two pseudogenes (white); these regions are related to the functional globin-type genes but are not transcribed. Each red arrow indicates the location of an *Alu* sequence, an ≈300-bp noncoding repeated sequence that is abundant in the human genome. (b) In the diagram of yeast DNA from chromosome III, the green boxes indicate open reading frames. Most of these potential protein-coding sequences are functional genes without introns. Note the much higher proportion of noncoding-to-coding sequences in the human DNA than in the yeast DNA. [Part (a), see F. S. Collins and S. M. Weissman, 1984, *Prog. Nucl. Acid Res. Mol. Biol.* **31**:315. Part (b), see S. G. Oliver et al., 1992, *Nature* **357**:28.]

genes, suggesting that they also arose by duplication of the same ancestral gene. However, there was little selective pressure to maintain the function of these genes. Consequently, sequence drift during evolution generated sequences that either terminate translation or block mRNA processing, rendering such regions nonfunctional. Because such pseudogenes are not deleterious, they remain in the genome and mark the location of a gene duplication that occurred in one of our ancestors.

Duplications of segments of a chromosome (called *segmental duplication*) occurred fairly often during the evolution of multicellular plants and animals. As a result, a large fraction of the genes in these organisms today has been duplicated, allowing the process of sequence drift to generate gene families and pseudogenes. The extent of sequence divergence between duplicated copies of the genome and characterization of the homologous genome sequences in related organisms allows an estimate of the time in evolutionary history when the duplication occurred. For example, the human fetal γ-globin genes ($G_\gamma$ and $A_\gamma$) evolved following the duplication of a 5.5-kb region in the β-globin locus that included the single γ-globin gene in the common ancestor of caterrhine primates (Old World monkeys, apes, and humans) and platyrrhine primates (New World monkeys) about 50 million years ago.

Although members of gene families that arose relatively recently in evolution, such as genes of the human β-globin locus, are often found near each other on the same chromosome, members of gene families may also be found on different chromosomes in the same organism. This is the case for the human α-globin genes, which were separated from the β-globin genes by an ancient chromosomal translocation. Both the α- and β-globin genes evolved from a single ancestral globin gene that was duplicated (see Figure 6-2b) to generate the predecessors of the contemporary α- and β-globin genes in mammals. Both the primordial α- and β-globin genes then underwent further duplications to generate the different genes of the α- and β-globin gene clusters found in mammals today.

Several different gene families encode the various proteins that make up the cytoskeleton. These proteins are present in varying amounts in almost all cells. In vertebrates, the major cytoskeletal proteins are the actins, tubulins, and intermediate filament proteins such as the keratins, discussed in Chapters 17, 18, and 20. We examine the origin of one such family, the tubulin family, in Section 6.5. Although the physiological rationale for the cytoskeletal protein families is not as obvious as it is for the globins, the different members of a family probably have similar but subtly different functions suited to the particular type of cell in which they are expressed.

## Heavily Used Gene Products Are Encoded by Multiple Copies of Genes

In vertebrates and invertebrates, the genes encoding ribosomal RNAs and some other nonprotein-coding RNAs such as those involved in RNA splicing occur as *tandemly repeated arrays*. These are distinguished from the duplicated genes of gene families in that the multiple tandemly repeated genes encode identical or nearly identical proteins or functional RNAs. Most often, copies of a sequence appear one after the other, in a head-to-tail fashion, over a long stretch of DNA. Within a tandem array of rRNA genes, each copy is nearly exactly like all the others. Although the transcribed portions of rRNA genes are the same in a given individual, the nontranscribed spacer regions between the transcribed regions can vary.

These tandemly repeated RNA genes are needed to meet the great cellular demand for their transcripts. To understand why, consider that a fixed maximal number of RNA copies can be produced from a single gene during one cell generation when the gene is fully loaded with RNA polymerase molecules. If more RNA is required than can be transcribed from one gene, multiple copies of the gene are necessary. For example, during early embryonic development in humans, many embryonic cells have a doubling time of ≈24 hours and contain 5–10 million ribosomes. To produce enough rRNA to form this many ribosomes, an embryonic human cell needs at least 100 copies of the large and small subunit rRNA genes, and most of these must be close to maximally active for the cell to divide every 24 hours. That is, multiple RNA polymerases must be transcribing each rRNA gene at the same time (see Figure 8-36). Indeed, all eukaryotes, including yeasts, contain 100 or more copies of the genes encoding 5S rRNA and the large and small subunit rRNAs.

Multiple copies of tRNA genes and the genes encoding the histone proteins also occur. As we will see later in this chapter, histones bind and organize nuclear DNA. Just as the cell requires multiple rRNA and tRNA genes to support efficient translation, multiple copies of the histone genes are required to produce sufficient histone protein to bind the large amount of nuclear DNA produced in each round of replication. While tRNA and histone genes often occur in clusters, they generally do not occur in tandem arrays in the human genome.

## Nonprotein-Coding Genes Encode Functional RNAs

In addition to rRNA and tRNA genes, there are hundreds of additional genes that are transcribed into nonprotein-coding RNAs, some with various known functions, and many whose functions are not yet known. For example, **small nuclear RNAs (snRNAs)** function in RNA splicing, and **small nucleolar RNAs (snoRNAs)** function in rRNA processing and base modification in the nucleolus. The RNase P RNA functions in tRNA processing, and a large family (≈1000) of short **micro RNAs (miRNAs)** regulates the translation and stability of specific mRNAs. The functions of these nonprotein-coding RNAs are discussed in Chapter 8. An RNA found in telomerase (see Figure 6-47) functions in maintaining the sequence at the ends of chromosomes, and the 7SL RNA functions in the import of secreted proteins and most membrane proteins into the endoplasmic reticulum (Chapter 13). These and other nonprotein-coding RNAs encoded in the human genome and their functions, when known, are listed in Table 6-2.

| TABLE 6-2 | Known Nonprotein-Coding RNAs and Their Functions | |
| --- | --- | --- |
| **RNA** | **Number of Genes in Human Genome** | **Function** |
| rRNAs | ≈300 | Protein synthesis |
| tRNAs | ≈500 | Protein synthesis |
| snRNAs | ≈80 | mRNA splicing |
| U7 snRNA | 1 | Histone mRNA 3′ processing |
| snoRNAs | ≈85 | Pre-rRNA processing and rRNA modification |
| miRNAs | ≈1000 | Regulation of gene expression |
| Xist | 1 | X-chromosome inactivation |
| 7SK | 1 | Transcription control |
| RNase P RNA | 1 | tRNA 5′ processing |
| 7SL RNA | 3 | Protein secretion (component of signal recognition particle, SRP) |
| RNase MRP RNA | 1 | rRNA processing |
| Telomerase RNA | 1 | Template for addition of telomeres |
| Vault RNAs | 3 | Components of Vault ribonucleoproteins (RNPs), function unknown |
| hY1, hY3, hY4, hY 5 | ≈30 | Components of ribonucleoproteins (RNPs), function unknown |
| H19 | 1 | Unknown |

SOURCE: International Human Genome Sequencing Consortium, 2001, *Nature* 409:860, and P.D. Zamore and B. Haley, 2005, Science 309:1519.

## KEY CONCEPTS of Section 6.1

### Eukaryotic Gene Structure

- In molecular terms, a gene is the entire DNA sequence required for synthesis of a functional protein or RNA molecule. In addition to the coding regions (exons), a gene includes control regions and sometimes introns.

- A simple eukaryotic transcription unit produces a single monocistronic mRNA, which is translated into a single protein.

- A complex eukaryotic transcription unit is transcribed into a primary transcript that can be processed into two or more different monocistronic mRNAs depending on the choice of splice sites or polyadenylation sites. A complex transcription unit with alternate promoters or alternate polyadenylation sites also generates two or more different mRNAs (see Figure 6-3b).

- Many complex transcription units (e.g., the fibronectin gene) express one mRNA in one cell type and an alternate mRNA in a different cell type.

- About half the protein-coding genes in vertebrate genomic DNA are solitary genes, each occurring only once in the haploid genome. The remainder are duplicated genes, which arose by duplication of an ancestral gene and subsequent independent mutations (see Figure 6-2b). The proteins encoded by a gene family have homologous but nonidentical amino acid sequences and exhibit similar but slightly different properties.

- In invertebrates and vertebrates, rRNAs are encoded by multiple copies of genes located in tandem arrays in genomic DNA. Multiple copies of tRNA and histone genes also occur, often in clusters, but not generally in tandem arrays.

- Many genes also encode functional RNAs that are not translated into protein but nonetheless perform significant functions, such as rRNA, tRNA, and snRNAs. Among these are micro RNAs, possibly up to 1000 in humans, whose biological significance in regulating gene expression has only recently been appreciated.

## 6.2 Chromosomal Organization of Genes and Noncoding DNA

Having reviewed the relationship between transcription units and genes, we now consider the organization of genes on chromosomes and the relationship of noncoding DNA sequences to coding sequences.

### Genomes of Many Organisms Contain Nonfunctional DNA

Comparisons of the total chromosomal DNA per cell in various species first suggested that much of the DNA in certain organisms does not encode RNA or have any apparent regulatory function. For example, yeasts, fruit flies, chickens, and humans have successively more DNA in their haploid chromosome sets (12; 180; 1300; and 3300 Mb, respectively), in keeping with what we perceive to be the increasing complexity of these organisms. Yet the vertebrates with the greatest amount of DNA per cell are amphibians, which are surely less complex than humans in their structure and behavior. Even more surprising, the unicellular protozoan species *Amoeba dubia* has 200 times more DNA per cell than humans. Many plant species also have considerably more DNA per cell than humans have. For example, tulips have 10 times as much DNA per cell as humans. The DNA content per cell also varies considerably between closely related species. All insects or all amphibians would appear to be similarly complex, but the amount of haploid DNA in species within each of these phylogenetic classes varies by a factor of 100.

Detailed sequencing and identification of exons in chromosomal DNA have provided direct evidence that the genomes of higher eukaryotes contain large amounts of noncoding DNA. For instance, only a small portion of the β-globin gene cluster of humans, about 80 kb long, encodes protein (see Figure 6-4a). In contrast, a typical 80-kb stretch of DNA from the yeast *S. cerevisiae*, a single-celled eukaryote, contains many closely spaced protein-coding sequences without introns and relatively much less noncoding DNA (see Figure 6-4b). Moreover, compared with other regions of vertebrate DNA, the β-globin gene cluster is unusually rich in protein-coding sequences, and the introns in globin genes are considerably shorter than those in most human genes. Globin proteins comprise >50 percent of the total protein in developing red blood cells (reticulocytes) and the globin genes are expressed at maximum rate, i.e., a new RNA polymerase initiates transcription as soon as the previous polymerase transcribes far enough from the promoter to allow a second RNA polymerase to bind and initiate. Consequently, there has been selective pressure to evolve small introns in the globin genes that are compatible with the required high rate of globin mRNA transcription and processing. However, the vast majority of human genes are expressed at much lower levels requiring production of one encoded mRNA on a time scale of only tens of minutes or hours. Consequently, there has been little selective pressure to reduce the sizes of introns in most human genes.

The density of genes varies greatly in different regions of human chromosomal DNA, from "gene-rich" regions, such as the β-globin cluster, to large gene-poor "gene deserts." Of the 96 percent of human genomic DNA that has been sequenced, only ≈1.5 percent corresponds to protein-coding sequences (exons). We learned in the previous section that the intron sequences of most human genes are significantly longer than the exon sequences. Approximately one-third of human genomic DNA is thought to be transcribed into pre-mRNA precursors or nonprotein-coding RNAs in one cell or another, but some 95 percent of this sequence is intronic, and thus removed by RNA splicing. This amounts to a large fraction of the total genome. The remaining two-thirds of human DNA is noncoding DNA between genes as well as regions of repeated DNA sequences that make up the centromeres and telomeres of the human chromosomes. Consequently, ≈98.5 percent of human DNA is noncoding.

Different selective pressures during evolution may account, at least in part, for the remarkable difference in the amount of nonfunctional DNA in different organisms. For example, many microorganisms must compete with other species of microorganisms in the same environment for limited amounts of available nutrients, and metabolic economy thus is a critical characteristic. Since synthesis of nonfunctional (i.e., noncoding) DNA requires time, nutrients, and energy, presumably there was selective pressure to lose nonfunctional DNA during the evolution of many rapidly growing microorganisms such as the yeast *S. cerevisiae*. On the other hand, natural selection in vertebrates depends largely on their behavior. The energy invested in DNA synthesis is trivial compared with the metabolic energy required for the movement of muscles; thus there may have been little selective pressure to eliminate nonfunctional DNA in vertebrates. Also, the replication time of cells in most multicellular organisms is much longer than in rapidly growing microorganisms, so there was little selective pressure to eliminate nonfunctional DNA in order to permit rapid cellular replication.

## Most Simple-Sequence DNAs Are Concentrated in Specific Chromosomal Locations

Besides duplicated protein-coding genes and tandemly repeated genes, eukaryotic cells contain multiple copies of other DNA sequences in the genome, generally referred to as repetitious DNA (see Table 6-1). Of the two main types of repetitious DNA, the less prevalent is **simple-sequence DNA** or *satellite DNA*, which constitutes about 6 percent of the human genome and is composed of perfect or nearly perfect repeats of relatively short sequences. The more common type of repetitious DNA, collectively called *interspersed repeats*, is composed of much longer sequences. These sequences, consisting of several types of transposable elements, are discussed in Section 6.3.

The length of each repeat in simple-sequence DNA can range from 1 to 500 base pairs. Simple-sequence DNAs in which the repeats contain 1–13 base pairs are often called *microsatellites*. Most microsatellite DNA has a repeat length of 1–4 base pairs and usually occurs in tandem repeats of 150 repeats or fewer. Microsatellites are thought to have originated by "backward slippage" of a daughter strand on its template strand during DNA replication so that the same short sequence is copied twice (Figure 6-5).

Microsatellites occasionally occur within transcription units. Some individuals are born with a larger number of repeats in specific genes than are observed in the general population, presumably because of daughter-strand slippage during DNA replication in a germ cell from which they developed. Such expanded microsatellites have been found to cause at least 14 different types of neuromuscular diseases, depending on the gene in which they occur. In some cases expanded microsatellites behave like a recessive mutation because they interfere

with just the function or expression of the encoded gene. But in the more common types of diseases associated with expanded microsatellite repeats, such as *Huntington disease* and *myotonic dystrophy type 1,* the expanded repeats behave like dominant mutations. In some microsatellite-repeat diseases, triplet repeats occur within a coding region, resulting in the formation of long polymers of a single amino acid that may aggregate over time in long-lived neuronal cells, eventually interfering with normal cellular function. For example, expansion of a CAG repeat in the first exon of the Huntington gene leads to synthesis of long stretches of polyglutamine that over several decades form toxic aggregates resulting in neuronal cell death in Huntington disease.

Pathogenic expanded repeats can also occur in the noncoding regions of some genes, where they are thought to

**(a) Normal replication**

**(b) Backward slippage**

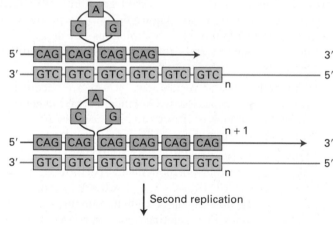

Second replication

**(c) Second replication**

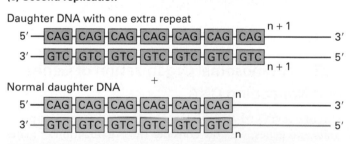

**FIGURE 6-5 Generation of microsatellite repeats by backward slippage of the nascent daughter strand during DNA replication.** If, during replication (a), the nascent daughter strand "slips" backward relative to the template strand by one repeat, one new copy of the repeat is added to the daughter strand when DNA replication continues (b). An extra copy of the repeat forms a single-stranded loop in the daughter strand of the daughter duplex DNA molecule. If this single-stranded loop is not removed by DNA repair proteins before the next round of DNA replication (c), the extra copy of the repeat is added to one of the double-stranded daughter DNA molecules.

function as dominant mutations because they interfere with RNA processing of a subset of mRNAs in the muscle cells and neurons where the affected genes are expressed. For example, in patients with myotonic dystrophy type 1, transcripts of the *DMPK* gene contain between 50 and 1500 repeats of the sequence CUG in the 3′ untranslated region, compared to 5 to 34 repeats in normal individuals. The extended stretch of CUG repeats in affected individuals is thought to form long RNA hairpins (see Figure 4-9), which bind and sequester nuclear RNA-binding proteins that normally regulate alternative RNA splicing of a subset of pre-mRNAs essential for muscle and nerve cell function. ■

Most simple-sequence satellite DNA is composed of repeats of 14–500 base pairs in tandem arrays of 20–100 kb. In situ hybridization studies with metaphase chromosomes have localized these simple-sequence DNAs to specific chromosomal regions. Much of this DNA lies near centromeres, the discrete chromosomal regions that attach to spindle microtubules during mitosis and meiosis (Figure 6-6). Experiments in the fission yeast *Schizosaccharomyces pombe* indicate that these sequences are required to form a specialized chromatin structure called *centromeric heterochromatin,* necessary for

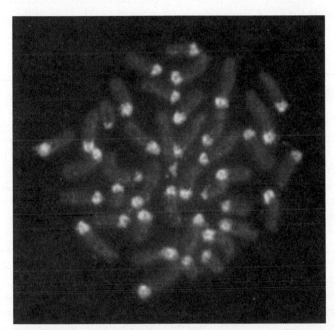

**EXPERIMENTAL FIGURE 6-6 Simple-sequence DNA is localized at the centromere in mouse chromosomes.** Purified simple-sequence DNA from mouse cells was copied in vitro using *E. coli* DNA polymerase I and fluorescently labeled dNTPs to generate a fluorescently labeled DNA "probe" for mouse simple-sequence DNA. Chromosomes from cultured mouse cells were fixed and denatured on a microscope slide, and then the chromosomal DNA was hybridized in situ to the labeled probe (light blue). The slide was also stained with DAPI, a DNA-binding dye, to visualize the full length of the chromosomes (dark blue). Fluorescence microscopy shows that the simple-sequence probe hybridizes primarily to one end of the telocentric mouse chromosomes (i.e., chromosomes in which the centromeres are located near one end). [Courtesy of Sabine Mai, Ph.D., Manitoba Institute of Cell Biology, Canada.]

the proper segregation of chromosomes to daughter cells during mitosis. Simple-sequence DNA is also found in long tandem repeats at the ends of chromosomes, the telomeres, where they function to maintain chromosome ends and prevent their joining to the ends of other DNA molecules, as discussed further in the last section of this chapter.

## DNA Fingerprinting Depends on Differences in Length of Simple-Sequence DNAs

Within a species, the nucleotide sequences of the repeat units composing simple-sequence DNA tandem arrays are highly conserved among individuals. In contrast, the *number* of repeats, and thus the length of simple-sequence tandem arrays containing the same repeat unit, is quite variable among individuals. These differences in length are thought to result from unequal crossing over within regions of simple-sequence DNA during meiosis. As a consequence of this unequal crossing over, the lengths of some tandem arrays are unique in each individual.

In humans and other mammals, some of the simple-sequence DNA exists in relatively short 1- to 5-kb regions made up of 20–50 repeat units, each containing ≈14–100 base pairs. These regions are called *minisatellites,* in contrast to microsatellites made up of tandem repeats of 1–13 base pairs. Even slight differences in the total lengths of various minisatellites from different individuals can be detected by Southern blotting (see Figure 5-26). This technique was exploited in the first application of *DNA fingerprinting,* which was developed to detect DNA *polymorphisms* (i.e., differences in sequence between individuals of the same species, Figure 6-7). Today, the far more sensitive technique of polymerase chain reaction (PCR, Figure 5-20) is generally used in forensic genetic testing. Microsatellite DNA sequences of short tandem repeats of four bases in ≈30–50 copies are usually analyzed today. The exact number of repeats generally varies on the two homologous chromosomes of an individual where they occur (one inherited from the mother and one from the father) and on the Y chromosome in males. A mixture of pairs of PCR primers that hybridize to unique sequences flanking 13 of these short tandem repeats and a Y-chromosome short tandem repeat are used to amplify DNA in a sample. The resulting mixture of PCR product lengths is unique in the human population, except for identical twins. The use of PCR methods allows analysis of minute amounts of DNA, and individuals can be distinguished more precisely and reliably than by conventional fingerprinting.

## Unclassified Spacer DNA Occupies a Significant Portion of the Genome

As Table 6.1 shows, ≈25 percent of human DNA lies between transcription units and is not repeated anywhere else in the genome. Much of this DNA probably arose from ancient transposable elements that have accumulated so many mutations over evolutionary time that they can no longer be recognized as having arisen from this source (see Section 6.3). Transcription-control regions on the order of 50–200 base

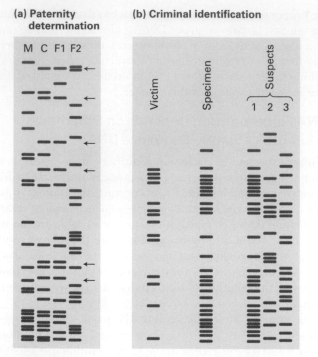

**(a) Paternity determination**

M  C  F1 F2

**(b) Criminal identification**

Victim  Specimen  Suspects  1 2 3

**FIGURE 6-7 Distinguishing individuals by DNA fingerprinting.**
(a) In this analysis of paternity, several minisatellite repeat lengths were determined by Southern blot analysis of restriction enzyme–digested genomic DNA and hybridization with a probe for a sequence shared by several minisatellite sequences. This generated hypervariable multiband patterns for each individual called "DNA fingerprints." Lane M shows the pattern of restriction fragment bands using the mother's DNA; C, using the child's DNA; and F1 and F2 using DNA from two potential fathers. The child has minisatellite repeat lengths inherited from either the mother or F1, demonstrating that F1 is the father. Arrows indicate restriction fragments from F1, but not F2, found in the child's DNA. (b) In these "DNA fingerprints" of a specimen isolated from a rape victim and three men suspected of the crime, it is clear that minisatellite repeat lengths in the specimen match those of suspect 1. The victim's DNA was included in the analysis to ensure that the specimen DNA was not contaminated with DNA from the victim. [From T. Strachan and A. P. Read, *Human Molecular Genetics 2*, 1999, John Wiley & Sons.]

pairs in length that help to regulate transcription from distant promoters also occur in these long stretches of unclassified spacer DNA. In some cases, sequences of this seemingly non-functional DNA are nonetheless conserved during evolution, indicating that they may perform a significant function that is not yet understood. For example, they may contribute to the structures of chromosomes discussed in Section 6.7.

## KEY CONCEPTS of Section 6.2

### Chromosomal Organization of Genes and Noncoding DNA

• In the genomes of prokaryotes and most lower eukaryotes, which contain few nonfunctional sequences, coding regions are densely arrayed along the genomic DNA.

• In contrast, vertebrate and higher plant genomes contain many sequences that do not code for RNAs or have any regulatory function. Much of this nonfunctional DNA is composed of repeated sequences. In humans, only about 1.5 percent of total DNA (the exons) actually encodes proteins or functional RNAs.

• Variation in the amount of nonfunctional DNA in the genomes of various species is largely responsible for the lack of a consistent relationship between the amount of DNA in the haploid chromosomes of an animal or plant and its phylogenetic complexity.

• Eukaryotic genomic DNA consists of three major classes of sequences: genes encoding proteins and functional RNAs, repetitious DNA, and spacer DNA (see Table 6-1).

• Simple-sequence DNA, short sequences repeated in long tandem arrays, is preferentially located in centromeres, telomeres, and specific locations within the arms of particular chromosomes.

• The length of a particular simple-sequence tandem array is quite variable between individuals in a species, probably because of unequal crossing over during meiosis. Differences in the lengths of some simple-sequence tandem arrays form the basis for DNA fingerprinting (see Figure 6.7).

## 6.3 Transposable (Mobile) DNA Elements

Interspersed repeats, the second type of repetitious DNA in eukaryotic genomes, is composed of a very large number of copies of relatively few sequence families (see Table 6-1). Also known as *moderately repeated DNA*, or *intermediate-repeat DNA*, these sequences are interspersed throughout mammalian genomes and make up ≈25–50 percent of mammalian DNA (≈45 percent of human DNA).

Because interspersed repeats have the unique ability to "move" in the genome, they are collectively referred to as transposable DNA elements or mobile DNA elements (we use these terms interchangeably). Although transposable DNA elements originally were discovered in eukaryotes, they also are found in prokaryotes. The process by which these sequences are copied and inserted into a new site in the genome is **transposition.** Transposable DNA elements are essentially molecular symbiotes that in most cases appear to have no specific function in the biology of their host organisms, but exist only to maintain themselves. For this reason, Francis Crick referred to them as "selfish DNA."

When transposition occurs in germ cells, the transposed sequences at their new sites are passed on to succeeding generations. In this way, mobile elements have multiplied and slowly accumulated in eukaryotic genomes over evolutionary time. Since mobile elements are eliminated from eukaryotic genomes very slowly, they now constitute a significant portion of the genomes of many eukaryotes.

Not only are mobile elements the source for much of the DNA in our genomes, they also provided a second mechanism, in addition to meiotic recombination, for bringing about chromosomal DNA rearrangements during evolution (see Figure 6-2). One reason for this is that during transposition of a particular mobile element, adjacent DNA sometimes also is mobilized (see Figure 6-19 later in the chapter). Transpositions occur rarely: in humans, about one new germ-line transposition for every eight individuals. Since 98.5 percent of our DNA is noncoding, most transpositions have no deleterious effects. But over time they played an essential part in the evolution of genes having multiple exons and of genes whose expression is restricted to specific cell types or developmental periods. In other words, although transposable elements probably evolved as cellular symbiotes, they have had an important function in the evolution of complex, multicellular organisms.

Transposition also may occur within a somatic cell; in this case the transposed sequence is transmitted only to the daughter cells derived from that cell. In rare cases, such somatic-cell transposition may lead to a somatic-cell mutation with detrimental phenotypic effects, for example, the inactivation of a tumor-suppressor gene (Chapter 24). In this section, we first describe the structure and transposition mechanisms of the major types of transposable DNA elements and then consider their likely role in evolution.

## Movement of Mobile Elements Involves a DNA or an RNA Intermediate

Barbara McClintock discovered the first mobile elements while doing classical genetic experiments in maize (corn) during the 1940s. She characterized genetic entities that could move into and back out of genes, changing the phenotype of corn kernels. Her theories were very controversial until similar mobile elements were discovered in bacteria, where they were characterized as specific DNA sequences, and the molecular basis of their transposition was deciphered.

As research on mobile elements progressed, they were found to fall into two categories: (1) those that transpose directly as DNA and (2) those that transpose via an RNA intermediate transcribed from the mobile element by an RNA polymerase and then converted back into double-stranded DNA by a **reverse transcriptase** (Figure 6-8). Mobile elements that transpose directly as DNA are generally referred to as **DNA transposons,** or simply **transposons.** Eukaryotic DNA transposons excise themselves from one place in the genome, leaving that site and moving to another. Mobile elements that transpose to new sites in the genome via an RNA intermediate are called **retrotransposons.** Retrotransposons make an RNA copy of themselves and introduce this new copy into another site in the genome, while also remaining at their original location. The movement of retrotransposons is analogous to the

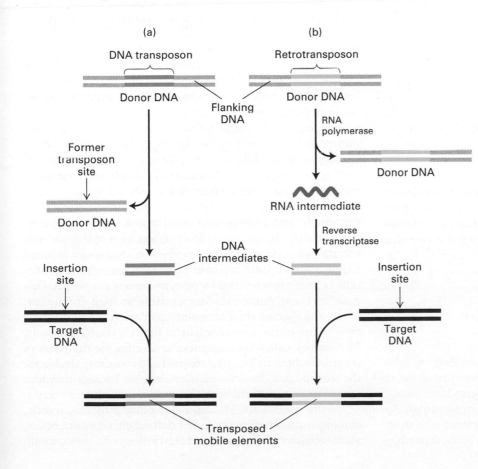

**FIGURE 6-8 Two major classes of mobile elements.** (a) Eukaryotic DNA transposons (orange) move via a DNA intermediate, which is excised from the donor site. (b) Retrotransposons (green) are first transcribed into an RNA molecule, which then is reverse-transcribed into double-stranded DNA. In both cases, the double-stranded DNA intermediate is integrated into the target-site DNA to complete movement. Thus DNA transposons move by a cut-and-paste mechanism, whereas retrotransposons move by a copy-and-paste mechanism.

infectious process of retroviruses (Figure 4-49). Indeed, retroviruses can be thought of as retrotransposons that evolved genes encoding viral coats, thus allowing them to transpose between cells. Retrotransposons can be further classified on the basis of their specific mechanism of transposition. To summarize, DNA transposons can be thought of as transposing by a "cut-and-paste" mechanism, while retrotransposons move by a "copy-and-paste" mechanism in which the copy is an RNA intermediate.

## DNA Transposons Are Present in Prokaryotes and Eukaryotes

Most mobile elements in bacteria transpose directly as DNA. In contrast, most mobile elements in eukaryotes are retrotransposons, but eukaryotic DNA transposons also occur. Indeed, the original mobile elements discovered by Barbara McClintock are DNA transposons.

**Bacterial Insertion Sequences** The first molecular understanding of mobile elements came from the study of certain *E. coli* mutations caused by the spontaneous insertion of a DNA sequence, ≈1–2 kb long, into the middle of a gene. These inserted stretches of DNA are called *insertion sequences,* or *IS elements.* So far, more than 20 different IS elements have been found in *E. coli* and other bacteria.

Transposition of an IS element is a very rare event, occurring in only one in $10^5$–$10^7$ cells per generation, depending on the IS element. Often, transpositions inactivate essential genes, killing the host cell and the IS elements it carries. Therefore, higher rates of transposition would probably result in too great a mutation rate for the host organism to survive. However, since IS elements transpose more or less randomly, some transposed sequences enter nonessential regions of the genome (e.g., regions between genes), allowing the cell to survive. At a very low rate of transposition, most host cells survive and therefore propagate the symbiotic IS element. IS elements also can insert into plasmids or lysogenic viruses, and thus be transferred to other cells. In this way, IS elements can transpose into the chromosomes of virgin cells.

The general structure of IS elements is diagrammed in Figure 6-9. An *inverted repeat* of ≈50 base pairs is invariably present at each end of an insertion sequence. In an inverted repeat, the 5′→3′ sequence on one strand is repeated on the other strand, such as

$$\overrightarrow{\phantom{aaaaaaaaaaaaaaaa}}$$
5′ **GAGC**————— GCTC 3′
3′ CTCG————— **CGAG** 5′
$$\overleftarrow{\phantom{aaaaaaaaaaaaaaaa}}$$

Between the inverted repeats is a region that encodes *transposase,* an enzyme required for transposition of the IS element to a new site. The transposase is expressed very rarely, accounting for the very low frequency of transposition. An important hallmark of IS elements is the presence of a short *direct-repeat sequence,* containing 5–11 base pairs, depending

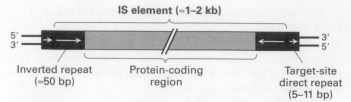

**FIGURE 6-9 General structure of bacterial IS elements.** The relatively large central region of an IS element, which encodes one or two enzymes required for transposition, is flanked by an inverted repeat at each end. The sequences of the inverted repeats are nearly identical, but they are oriented in opposite directions. The sequence is characteristic of a particular IS element. The 5′ and 3′ short *direct* (as opposed to *inverted*) repeats are not transposed with the insertion element; rather, they are insertion-site sequences that become duplicated, with one copy at each end, during insertion of a mobile element. The length of the direct repeats is constant for a given IS element, but their sequence depends on the site of insertion and therefore varies with each transposition of the IS element. Arrows indicate sequence orientation. The regions in this diagram are not to scale; the coding region makes up most of the length of an IS element.

on the IS element, immediately adjacent to both ends of the inserted element. The *length* of the direct repeat is characteristic of each type of IS element, but its *sequence* depends on the target site where a particular copy of the IS element inserted. When the sequence of a mutated gene containing an IS element is compared with the wild-type gene sequence, only one copy of the short direct-repeat sequence is found in the wild-type gene. Duplication of this target-site sequence to create the second direct repeat adjacent to an IS element occurs during the insertion process.

As depicted in Figure 6-10, transposition of an IS element occurs by a "cut-and-paste" mechanism. Transposase performs three functions in this process: it (1) precisely excises the IS element in the donor DNA, (2) makes staggered cuts in a short sequence in the target DNA, and (3) ligates the 3′ termini of the IS element to the 5′ ends of the cut donor DNA. Finally, a host-cell DNA polymerase fills in the single-stranded gaps, generating the short direct repeats that flank IS elements, and DNA ligase joins the free ends.

**Eukaryotic DNA Transposons** McClintock's original discovery of mobile elements came from observation of spontaneous mutations in maize that affect production of enzymes required to make anthocyanin, a purple pigment in maize kernels. Mutant kernels are white, and wild-type kernels are purple. One class of these mutations is revertible at high frequency, whereas a second class of mutations does not revert unless they occur in the presence of the first class of mutations. McClintock called the agent responsible for the first class of mutations the *activator (Ac) element* and those responsible for the second class *dissociation (Ds) elements* because they also tended to be associated with chromosome breaks.

Many years after McClintock's pioneering discoveries, cloning and sequencing revealed that Ac elements are equivalent to bacterial IS elements. Like IS elements, they contain

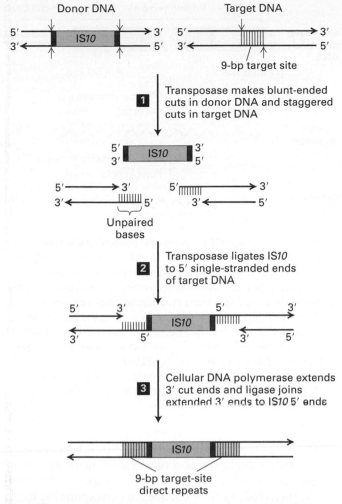

**FIGURE 6-10 Model for transposition of bacterial insertion sequences.** Step **1**: Transposase, which is encoded by the IS element (IS*10* in this example), cleaves both strands of the donor DNA next to the inverted repeats (dark red), excising the IS*10* element. At a largely random target site, transposase makes staggered cuts in the target DNA. In the case of IS*10*, the two cuts are 9 bp apart. Step **2**: Ligation of the 3′ ends of the excised IS element to the staggered sites in the target DNA also is catalyzed by transposase. Step **3**: The 9-bp gaps of single-stranded DNA left in the resulting intermediate are filled in by a cellular DNA polymerase; finally, cellular DNA ligase forms the 3′→5′ phosphodiester bonds between the 3′ ends of the extended target DNA strands and the 5′ ends of the IS*10* strands. This process results in duplication of the target-site sequence on each side of the inserted IS element. Note that the length of the target site and IS*10* are not to scale. [See H. W. Benjamin and N. Kleckner, 1989, *Cell* **59:**373; and 1992, *Proc. Nat'l Acad. Sci. USA* **89:**4648.]

inverted terminal repeat sequences that flank the coding region for a transposase, which recognizes the terminal repeats and catalyzes transposition to a new site in DNA. Ds elements are deleted forms of the Ac element in which a portion of the sequence encoding transposase is missing. Because it does not encode a functional transposase, a Ds element cannot move by itself. However, in plants that carry the Ac element and thus express a functional transposase, Ds elements can be

transposed because they retain the inverted terminal repeats recognized by the transposase.

Since McClintock's early work on mobile elements in corn, transposons have been identified in other eukaryotes. For instance, approximately half of all the spontaneous mutations observed in *Drosophila* are due to the insertion of mobile elements. Although most of the mobile elements in *Drosophila* function as retrotransposons, at least one—the *P element*—functions as a DNA transposon, moving by a mechanism similar to that used by bacterial insertion sequences. Current methods for constructing transgenic *Drosophila* depend on engineered, high-level expression of the P-element transposase and use of the P-element inverted terminal repeats as targets for transposition, as discussed in Chapter 5 (see Figure 5-22).

DNA transposition by the cut-and-paste mechanism can result in an increase in the copy number of a transposon if it occurs during the S phase of the cell cycle, when DNA synthesis occurs. An increase in the copy number happens when the donor DNA is from one of the two daughter DNA molecules in a region of a chromosome that has replicated and the target DNA is in the region that has not yet replicated. When DNA replication is complete at the end of the S phase, the target DNA in its new location is also replicated, resulting in a net increase in the total number of these transposons in the cell (Figure 6-11). When such a transposition occurs during the S phase preceding meiosis, one of the four germ cells produced contains the extra copy of the transposon. Repetition of this process over evolutionary time has resulted in the accumulation of large numbers of DNA transposons in the genomes of some organisms. Human DNA contains about 300,000 copies of full-length and deleted DNA transposons,

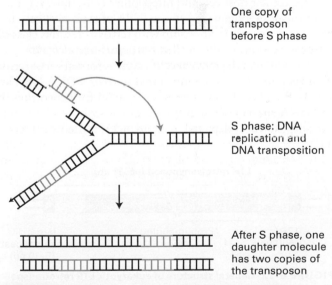

**FIGURE 6-11 Mechanism for increasing DNA-transposon copy number.** If a DNA transposon that transposes by a cut-and-paste mechanism (see Figure 6-10) transposes during S phase from a region of the chromosome that has replicated to a region that has not yet replicated, then, when chromosomal replication is completed, one of the two daughter chromosomes will have a net increase of one transposon insertion.

amounting to ≈3 percent of human DNA. As we will see shortly, this mechanism can lead to the transposition of genomic DNA as well as the transposon itself.

## LTR Retrotransposons Behave Like Intracellular Retroviruses

The genomes of all eukaryotes studied, from yeast to humans, contain retrotransposons, mobile DNA elements that transpose through an RNA intermediate utilizing a reverse transcriptase (see Figure 6-8b). These mobile elements are divided into two major categories, those containing and those lacking **long terminal repeats (LTRs)**. LTR retrotransposons, which we discuss here, are common in yeast (e.g., Ty elements) and in *Drosophila* (e.g., *copia* elements). Although less abundant in mammals than non-LTR retrotransposons, LTR retrotransposons nonetheless constitute ≈8 percent of human genomic DNA. In mammals, retrotransposons lacking LTRs are the most common type of mobile element; these are described in the next section.

The general structure of LTR retrotransposons found in eukaryotes is depicted in Figure 6-12. In addition to short 5′ and 3′ direct repeats typical of all transposons, these retrotransposons are marked by the presence of LTRs flanking the central protein-coding region. These *long direct terminal repeats*, containing ≈250–600 base pairs depending on the type of LTR retrotransposon, are characteristic of integrated retroviral DNA and are critical to the life cycle of retroviruses. In addition to sharing LTRs with retroviruses, LTR retrotransposons encode all the proteins of the most common type of retroviruses, except for the envelope proteins. Lacking these envelope proteins, LTR retrotransposons cannot bud from their host cell and infect other cells; however, they can transpose to new sites in the DNA of their host cell. Because of their clear relationship with retroviruses, this class of retrotransposons is often called *retrovirus-like elements*.

A key step in the retroviral life cycle is formation of retroviral genomic RNA from integrated retroviral DNA (see Figure 4-49). This process serves as a model for generation of the RNA intermediate during transposition of LTR retrotransposons. As depicted in Figure 6-13, the leftward retro-

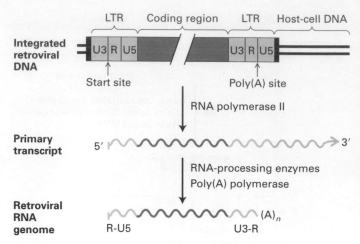

**FIGURE 6-13 Generation of retroviral genomic RNA from integrated retroviral DNA.** The left LTR directs cellular RNA polymerase to initiate transcription at the first nucleotide of the left R region. The resulting primary transcript extends beyond the right LTR. The right LTR, now present in the RNA primary transcript, directs cellular enzymes to cleave the primary transcript at the last nucleotide of the right R region and to add a poly(A) tail, yielding a retroviral RNA genome with the structure shown at the top of Figure 6-14. A similar mechanism is thought to generate the RNA intermediate during transposition of retrotransposons. The short direct-repeat sequences (black) of target-site DNA are generated during integration of the retroviral DNA into the host-cell genome.

viral LTR functions as a promoter that directs host-cell RNA polymerase to initiate transcription at the 5′ nucleotide of the R sequence. After the entire downstream retroviral DNA has been transcribed, the RNA sequence corresponding to the rightward LTR directs host-cell RNA-processing enzymes to cleave the primary transcript and add a poly(A) tail at the 3′ end of the R sequence. The resulting retroviral RNA genome, which lacks a complete LTR, exits the nucleus and is packaged into a virion that buds from the host cell.

After a retrovirus infects a cell, reverse transcription of its RNA genome by the retrovirus-encoded reverse transcriptase yields a double-stranded DNA containing complete LTRs (Figure 6-14). This DNA synthesis takes place in the

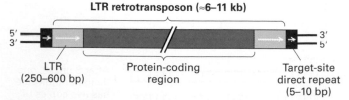

**FIGURE 6-12 General structure of eukaryotic LTR retrotransposons.** The central protein-coding region is flanked by two long terminal repeats (LTRs), which are element-specific direct repeats. Like other mobile elements, integrated retrotransposons have short target-site direct repeats at each end. Note that the different regions are not drawn to scale. The protein-coding region constitutes 80 percent or more of a retrotransposon and encodes reverse transcriptase, integrase, and other retroviral proteins.

**FIGURE 6-14 Model for reverse transcription of retroviral genomic RNA into DNA.** In this model, a complicated series of nine events generates a double-stranded DNA copy of the single-stranded RNA genome of a retrovirus. The genomic RNA is packaged in the virion with a retrovirus-specific cellular tRNA hybridized to a complementary sequence near its 5′ end called the *primer-binding site* (PBS). The retroviral RNA has a short direct-repeat terminal sequence (R) at each end. The overall reaction is carried out by reverse transcriptase, which catalyzes polymerization of deoxyribonucleotides. RNaseH digests the RNA strand in a DNA-RNA hybrid. The entire process yields a double-stranded DNA molecule that is longer than the template RNA and has a long terminal repeat (LTR) at each end. The different regions are not shown to scale. The PBS and R regions are actually much shorter than the U5 and U3 regions, and the central coding region is very much longer than the other regions. [See E. Gilboa et al., 1979, *Cell* **18**:93.]

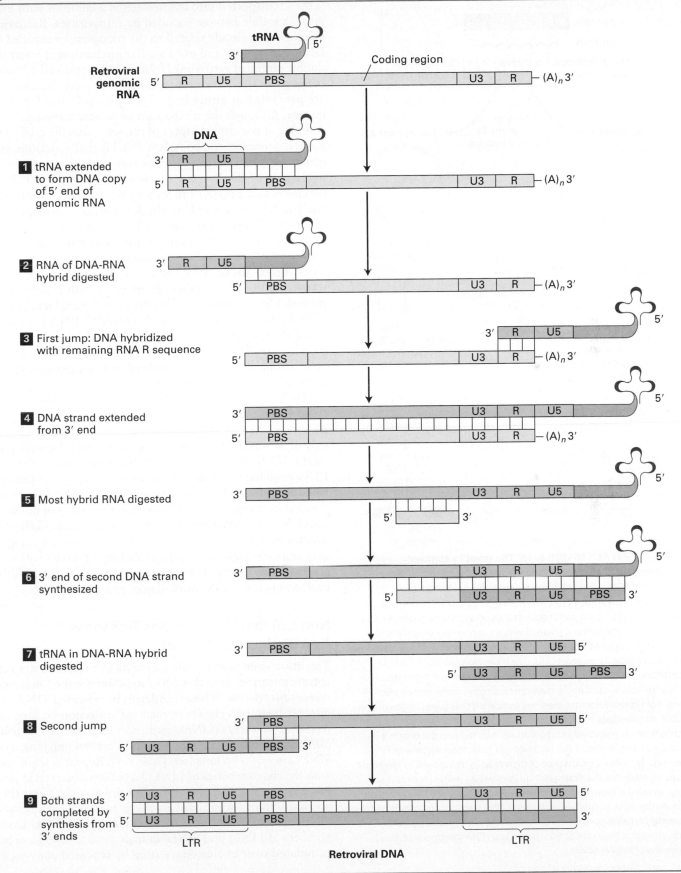

1  tRNA extended to form DNA copy of 5' end of genomic RNA

2  RNA of DNA-RNA hybrid digested

3  First jump: DNA hybridized with remaining RNA R sequence

4  DNA strand extended from 3' end

5  Most hybrid RNA digested

6  3' end of second DNA strand synthesized

7  tRNA in DNA-RNA hybrid digested

8  Second jump

9  Both strands completed by synthesis from 3' ends

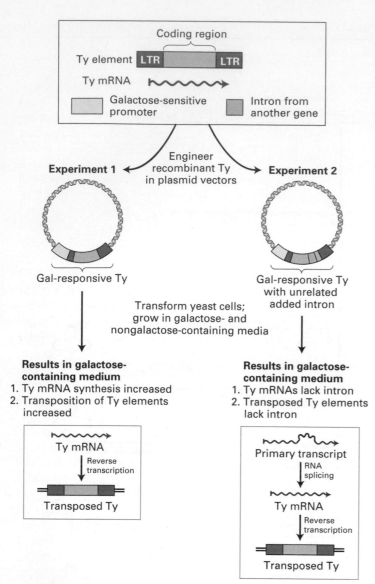

**EXPERIMENTAL FIGURE 6-15  The yeast Ty element transposes through an RNA intermediate.** When yeast cells are transformed with a Ty-containing plasmid, the Ty element can transpose to new sites, although normally this occurs at a low rate. Using the elements diagrammed at the top, researchers engineered two different recombinant plasmid vectors containing recombinant Ty elements adjacent to a galactose-sensitive promoter. These plasmids were transformed into yeast cells, which were grown in a galactose-containing and a nongalactose medium. In experiment 1, growth of cells in galactose-containing medium resulted in many more transpositions than in nongalactose medium, indicating that transcription into an mRNA intermediate is required for Ty transposition. In experiment 2, an intron from an unrelated yeast gene was inserted into the putative protein-coding region of the recombinant galactose-responsive Ty element. The observed absence of the intron in transposed Ty elements is strong evidence that transposition involves an mRNA intermediate from which the intron was removed by RNA splicing, as depicted in the box on the right. In contrast, eukaryotic DNA transposons, like the Ac element of maize, contain introns within the transposase gene, indicating that they do not transpose via an RNA intermediate. [See J. Boeke et al., 1985, *Cell* **40**:491.]

cytosol. The double-stranded DNA with an LTR at each end is then transported into the nucleus in a complex with integrase, another enzyme encoded by retroviruses. Retroviral integrases are closely related to the transposases encoded by DNA transposons and use a similar mechanism to insert the double-stranded retroviral DNA into the host-cell genome. In this process, short direct repeats of the target-site sequence are generated at either end of the inserted viral DNA sequence. Although the mechanism of reverse transcription is complex, it is a critical aspect of the retrovirus life cycle. The process generates the complete 5′ LTR that functions as a promoter for initiation of transcription precisely at the 5′ nucleotide of the R sequence, while the complete 3′ LTR functions as a poly(A) site leading to polyadenylation precisely at the 3′nucleotide of the R sequence. Consequently, no nucleotides are lost from an LTR retrotransposon as it undergoes successive rounds of insertion, transcription, reverse transcription, and re-insertion at a new site.

As noted above, LTR retrotransposons encode reverse transcriptase and integrase. By analogy with retroviruses, these mobile elements move by a "copy-and-paste" mechanism whereby reverse transcriptase converts an RNA copy of a donor element into DNA, which is inserted into a target site by integrase. The experiments depicted in Figure 6-15 provided strong evidence for the role of an RNA intermediate in transposition of Ty elements in yeast.

The most common LTR retrotransposons in humans are called *ERVs*, for *e*ndogenous *r*etro*v*iruses. Most of the 443,000 ERV-related DNA sequences in the human genome consist only of isolated LTRs. These are derived from full-length proviral DNA by homologous recombination between the two LTRs, resulting in deletion of the internal retroviral sequences. Isolated LTRs such as these cannot be transposed to a new position in the genome, but recombination between homologous LTRs at different positions in the genome have likely contributed to the chromosomal DNA rearrangements leading to gene and exon duplications, the evolution of proteins with new combinations of exons, and, as we will see in Chapter 7, the evolution of complex control of gene expression.

## Non-LTR Retrotransposons Transpose by a Distinct Mechanism

The most abundant mobile elements in mammals are retrotransposons that lack LTRs, sometimes called *nonviral retrotransposons*. These moderately repeated DNA sequences form two classes in mammalian genomes: *long interspersed elements* (*LINEs*) and *short interspersed elements* (*SINEs*). In humans, full-length LINEs are ≈6 kbp long, and SINEs are ≈300 bp long (see Table 6-1). Repeated sequences with the characteristics of LINEs have been observed in protozoans, insects, and plants, but for unknown reasons they are particularly abundant in the genomes of mammals. SINEs also are found primarily in mammalian DNA. Large numbers of LINEs and SINEs in higher eukaryotes have accumulated over evolutionary time by repeated copying of

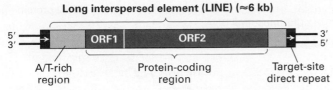

**Long interspersed element (LINE) (≈6 kb)**

A/T-rich region — Protein-coding region — Target-site direct repeat

**FIGURE 6-16 General structure of a LINE, a non-LTR retrotransposon.** Mammalian DNA carries two classes of non-LTR retrotransposons, LINEs and SINEs. LINE structure is depicted here. The length of the target-site direct repeats varies among copies of the element at different sites in the genome. Although the full-length L1 sequence is ≈6 kb long, variable amounts of the left end are absent at over 90 percent of the sites where this mobile element is found. The shorter open reading frame (ORF1), ≈1 kb in length, encodes an RNA-binding protein. The longer ORF2, ≈4 kb in length, encodes a bifunctional protein with reverse transcriptase and DNA endonuclease activity. Note that LINEs lack the long terminal repeats found in LTR retrotransposons.

sequences at a few positions in the genome and insertion of the copies into new positions.

**LINEs** Human DNA contains three major families of LINE sequences that are similar in their mechanism of transposition but differ in their sequences: L1, L2, and L3. Only members of the L1 family transpose in the contemporary human genome. Apparently there are no remaining functional copies of L2 or L3. LINE sequences are present at ≈900,000 sites in the human genome, accounting for a staggering 21 percent of total human DNA. The general structure of a complete LINE is diagrammed in Figure 6-16. LINEs usually are flanked by short direct repeats, the hallmark of mobile elements, and contain two long open reading frames (ORFs). ORF1, ≈1 kb long, encodes an RNA-binding protein. ORF2, ≈4 kb long, encodes a protein that has a long region of homology with the reverse transcriptases of retroviruses and LTR retrotransposons, but also exhibits DNA endonuclease activity.

Evidence for the mobility of L1 elements first came from analysis of DNA cloned from humans with certain genetic diseases such as hemophilia and myotonic dystrophy. DNA from these patients was found to carry mutations resulting from insertion of an L1 element into a gene, whereas no such element occurred within this gene in either parent. About 1 in 600 mutations that cause significant disease in humans are due to L1 transpositions or SINE transpositions that are catalyzed by L1-encoded proteins. Later experiments similar to those just described with yeast Ty elements (see Figure 6-15) confirmed that L1 elements transpose through an RNA intermediate. In these experiments, an intron was introduced into a cloned mouse L1 element, and the recombinant L1 element was stably transformed into cultured hamster cells. After several cell doublings, a PCR-amplified fragment corresponding to the L1 element but lacking the inserted intron was detected in the cells. This finding strongly suggests that, over time, the recombinant L1

element containing the inserted intron had transposed to new sites in the hamster genome through an RNA intermediate that underwent RNA splicing to remove the intron. ∎

Since LINEs do not contain LTRs, their mechanism of transposition through an RNA intermediate differs from that of LTR retrotransposons. ORF1 and ORF2 proteins are translated from a LINE RNA. In vitro studies indicate that transcription by RNA polymerase is directed by promoter sequences at the left end of integrated LINE DNA. LINE RNA is polyadenylated by the same post-transcriptional mechanism that polyadenylates other mRNAs. The LINE RNA then is exported into the cytosol, where it is translated into ORF1 and ORF2 proteins. Multiple copies of ORF1 protein then bind to the LINE RNA, and ORF2 protein binds to the poly(A) tail. The LINE RNA is then transported back into the nucleus as a complex with ORF1and ORF2 proteins, and is reverse transcribed into LINE DNA in the nucleus by ORF2. The mechanism involves staggered cleavage of cellular DNA at the insertion site, followed by priming of reverse transcription by the resulting cleaved cellular DNA as detailed in Figure 6-17. The complete process results in insertion of a copy of the original LINE retrotransposon into a new site in chromosomal DNA. A short direct repeat is generated at the insertion site because of the initial staggered cleavage of the two chromosomal DNA strands.

As noted already, the DNA form of an LTR retrotransposon is synthesized from its RNA form in the cytosol using a cellular tRNA as a primer for reverse transcription of the first strand of DNA (see Figure 6-14). The resulting double-stranded DNA with long terminal repeats is then transported into the nucleus, where it is integrated into chromosomal DNA by a retrotransposon-encoded integrase. In contrast, the DNA form of a non-LTR retrotransposon is synthesized in the nucleus. The synthesis of the first strand of the non-LTR retroviral DNA by ORF2, a reverse transcriptase, is primed by the 3′ end of cleaved chromosomal DNA, which base-pairs with the poly(A) tail of the non-LTR retroviral RNA (see Figure 6-17, step **2**). Since its synthesis is primed by the cut end of a cleaved chromosome, and synthesis of the other strand of the non-LTR retrotransposon DNA is primed by the 3′ end of chromosomal DNA on the other side of the initial cut in chromosomal DNA (step **6**), the mechanism of synthesis results in integration of the non-LTR retrotransposon DNA. There is no need for an integrase to insert the non-LTR retrotransposon DNA.

The vast majority of LINEs in the human genome are truncated at their 5′ end, suggesting that reverse transcription terminated before completion and the resulting fragments extending variable distances from the poly(A) tail were inserted. Because of this shortening, the average size of LINE elements is only about 900 base pairs, even though the full-length sequence is ≈6 kb long. Truncated LINE elements, once formed, probably are not further transposed because they lack a promoter for formation of the RNA intermediate in transposition. In addition to the fact that most L1 insertions

**FIGURE 6-17 Proposed mechanism of LINE reverse transcription and integration.** Only ORF2 protein is represented. Newly synthesized LINE DNA is shown in black. ORF1 and ORF2 proteins, produced by translation of LINE RNA in the cytoplasm, bind to LINE RNA and transport it into the nucleus. Step **1**: In the nucleus, ORF2 makes staggered cuts in AT-rich target-site DNA, generating the DNA 3′-OH ends indicated by blue arrowheads. Step **2**: The 3′ end of the T-rich DNA strand hybridizes to the poly(A) tail of the LINE RNA and primes DNA synthesis by ORF2. Step **3**: ORF2 extends the DNA strand using the LINE RNA as template. Steps **4** and **5**: When synthesis of the LINE DNA bottom strand reaches the 5′ end of the LINE RNA template, ORF2 extends the newly synthesized LINE DNA using as template the top-strand cellular DNA generated by the initial ORF2 staggered cleavage. Step **6**: A cellular DNA polymerase extends the 3′ end of the top strand generated by the initial ORF2 staggered cut, using the newly synthesized bottom-strand LINE DNA as template. The LINE RNA is digested as the DNA polymerase extends the upper-strand DNA, just as occurs during removal of lagging-strand primer RNA during cellular DNA synthesis (Figure 4-33). Step **7**: The 3′ end of the newly synthesized DNA strands are ligated to the 5′ ends of the cellular DNA strands as in lagging-strand cellular DNA synthesis. [Adapted from D. D. Luan et al., 1993, *Cell* **72**:595.]

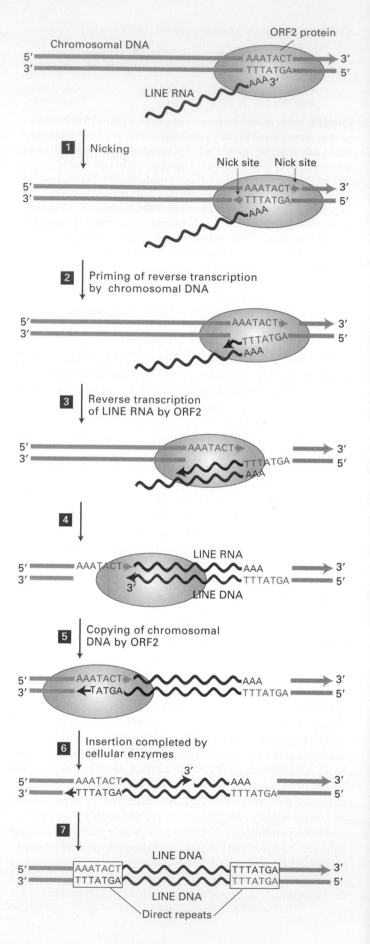

are truncated, nearly all the full-length elements contain stop codons and frameshift mutations in ORF1 and ORF2; these mutations probably have accumulated in most LINE sequences over evolutionary time. As a result, only ≈0.01 percent of the LINE sequences in the human genome are full-length, with intact open reading frames for ORF1 and ORF2, representing ≈60–100 in total number.

**SINEs** The second most abundant class of mobile elements in the human genome, SINEs constitute ≈13 percent of total human DNA. Varying in length from about 100 to 400 base pairs, these retrotransposons do not encode protein, but most contain a 3′ A/T-rich sequence similar to that in LINEs. SINEs are transcribed by the same nuclear RNA polymerase that transcribes genes encoding tRNAs, 5S rRNAs, and other small stable RNAs. Most likely, the ORF1 and ORF2 proteins expressed from full-length LINEs mediate reverse transcription and integration of SINEs by the mechanism depicted in Figure 6-17. Consequently, SINEs can be viewed as parasites of the LINE symbiotes, competing with LINE RNAs for binding and reverse transcription/integration by LINE encoded ORF1 and ORF2.

SINEs occur at about 1.6 million sites in the human genome. Of these, ≈1.1 million are *Alu elements,* so named because most of them contain a single recognition site for the restriction enzyme *Alu*I. *Alu* elements exhibit considerable sequence homology with and probably evolved from 7SL RNA, a cytosolic RNA in a ribonucleoprotein complex called the signal recognition particle. This abundant cytosolic ribonucleoprotein particle aids in targeting certain polypeptides to the membranes of the endoplasmic reticulum (Chapter 13). *Alu* elements are scattered throughout the human genome at sites where their insertion has not disrupted gene expression: between genes, within introns, and in the 3′ untranslated regions of some mRNAs. For instance, nine *Alu* elements are located within the human β-globin gene cluster (see Figure 6-4a). Of the one new germ-line non-LTR retrotransposition that is estimated to occur in about every eight individuals, ≈40 percent involve L1 elements and 60 percent involve SINEs, of which ≈90 percent are *Alu* elements.

Similar to other mobile elements, most SINEs have accumulated mutations from the time of their insertion in the germ line of an ancient ancestor of modern humans. Like LINEs, many SINEs also are truncated at their 5′ end.

## Other Retroposed RNAs Are Found in Genomic DNA

In addition to the mobile elements listed in Table 6-1, DNA copies of a wide variety of mRNAs appear to have integrated into chromosomal DNA. Since these sequences lack introns and do not have flanking sequences similar to those of the functional gene copies, they clearly are not simply duplicated genes that have drifted into nonfunctionality and become pseudogenes, as discussed earlier (Figure 6-4a). Instead, these DNA segments appear to be retrotransposed copies of spliced and polyadenylated mRNA. Compared with normal genes encoding mRNAs, these inserted segments generally contain multiple mutations, which are thought to have accumulated since their mRNAs were first reverse-transcribed and randomly integrated into the genome of a germ cell in an ancient ancestor. These nonfunctional genomic copies of mRNAs are referred to as *processed pseudogenes*. Most processed pseudogenes are flanked by short direct repeats, supporting the hypothesis that they were generated by rare retrotransposition events involving cellular mRNAs.

Other interspersed repeats representing partial or mutant copies of genes encoding small nuclear RNAs (snRNAs) and tRNAs are found in mammalian genomes. Like processed pseudogenes derived from mRNAs, these nonfunctional copies of small RNA genes are flanked by short direct repeats and most likely result from rare retrotransposition events that have accumulated through the course of evolution. Enzymes expressed from a LINE are thought to have carried out all these retrotransposition events involving mRNAs, snRNAs, and tRNAs.

## Mobile DNA Elements Have Significantly Influenced Evolution

Although mobile DNA elements appear to have no direct function other than to maintain their own existence, their presence has had a profound impact on the evolution of modern-day organisms. As mentioned earlier, about half the spontaneous mutations in *Drosophila* result from insertion of a mobile DNA element into or near a transcription unit. In mammals, mobile elements cause a much smaller proportion of spontaneous mutations: ≈10 percent in mice, and only 0.1–0.2 percent in humans. Still, mobile elements have been found in mutant alleles associated with several human genetic diseases. For example, insertions into the clotting factor IX gene cause hemophilia, and insertions into the gene encoding the muscle protein dystrophin lead to Duchenne muscular dystrophy. The genes encoding factor IX and dystrophin are both on the X chromosome. Because the male genome has only one copy of the X chromosome, transposition insertions into these genes predominantly affect males.

In lineages leading to higher eukaryotes, homologous recombination between mobile DNA elements dispersed throughout ancestral genomes may have generated gene duplications and other DNA rearrangements during evolution (see Figure 6-2b). For instance, cloning and sequencing of the β-globin gene cluster from various primate species has provided strong evidence that the human $G_\gamma$ and $A_\gamma$ genes arose from an unequal homologous crossover between two L1 sequences flanking an ancestral globin gene. Subsequent divergence of such duplicated genes could lead to acquisition of distinct, beneficial functions associated with each member of a gene family. Unequal crossing over between mobile elements located within introns of a particular gene could lead to the duplication of exons within that gene (see Figure 6-2a). This process most likely influenced the evolution of genes that contain multiple copies of similar exons encoding similar protein domains, such as the fibronectin gene (see Figure 4-16).

Some evidence suggests that during the evolution of higher eukaryotes, recombination between mobile DNA elements (e.g., *Alu* elements) in introns of *two separate* genes also occurred, generating new genes made from novel combinations of preexisting exons (Figure 6-18). This evolutionary process, termed **exon shuffling**, may have occurred during evolution of the genes encoding tissue plasminogen activator,

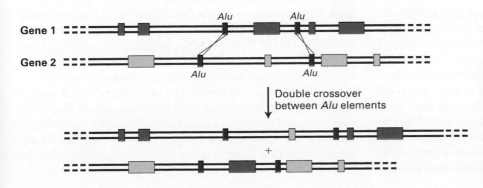

**FIGURE 6-18 Exon shuffling via recombination between homologous interspersed repeats.** Recombination between interspersed repeats in the introns of separate genes produces transcription units with a new combination of exons. In the example shown here, a double crossover between two sets of *Alu* repeats results in an exchange of exons between the two genes.

Gene 1

Gene 2

Double crossover between *Alu* elements

+

**FIGURE 6-19 Exon shuffling by transposition.** (a) Transposition of an exon flanked by homologous DNA transposons into an intron on a second gene. As we saw in Figure 6-10, step **1**, transposase can recognize and cleave the DNA at the ends of the transposon inverted repeats. In gene 1, if the transposase cleaves at the left end of the transposon on the left and at the right end of the transposon on the right, it can transpose all the intervening DNA, including the exon from gene 1, to a new site in an intron of gene 2. The net result is an insertion of the exon from gene 1 into gene 2. (b) Integration of an exon into another gene via LINE transposition. Some LINEs have weak poly(A) signals. If such a LINE is in the 3'-most intron of gene 1, during transposition its transcription may continue beyond its own poly(A) signals and extend into the 3' exon, transcribing the cleavage and polyadenylation signals of gene 1 itself. This RNA can then be reverse-transcribed and integrated by the LINE ORF2 protein (Figure 6-17) into an intron on gene 2, introducing a new 3' exon (from gene 1) into gene 2.

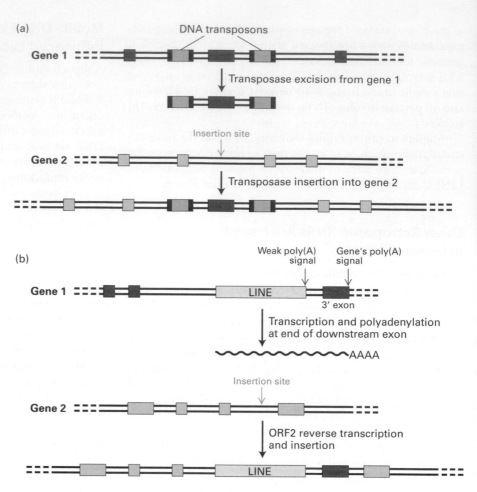

the Neu receptor, and epidermal growth factor, which all contain an EGF domain (see Figure 3-11). In this case, exon shuffling presumably resulted in insertion of an EGF domain–encoding exon into an intron of the ancestral form of each of these genes.

Both DNA transposons and LINE retrotransposons have been shown to occasionally carry unrelated flanking sequences when they transpose to new sites by the mechanisms diagrammed in Figure 6-19. These mechanisms likely also contributed to exon shuffling during the evolution of contemporary genes.

In addition to causing changes in coding sequences in the genome, recombination between mobile elements and transposition of DNA adjacent to DNA transposons and retrotransposons likely played a significant role in the evolution of regulatory sequences that control gene expression. As noted earlier, eukaryotic genes have transcription-control regions called enhancers that can operate over distances of tens of thousands of base pairs. Transcription of many genes is controlled through the combined effects of several enhancer elements. Insertion of mobile elements near such transcription-control regions probably contributed to the evolution of new combinations of enhancer sequences. These in turn control which specific genes are expressed in particular cell types and the amount of the encoded protein

produced in modern organisms, as we discuss in the next chapter.

These considerations suggest that the early view of mobile DNA elements as completely selfish molecular parasites misses the mark. Rather, they have contributed profoundly to the evolution of higher organisms by promoting (1) the generation of gene families via gene duplication, (2) the creation of new genes via shuffling of preexisting exons, and (3) formation of more complex regulatory regions that provide multifaceted control of gene expression. Today, researchers are attempting to harness transposition mechanisms for inserting therapeutic genes into patients as a form of gene therapy.

A process analogous to that shown in Figure 6-19a is largely responsible for the rapid spread of antibiotic resistance among pathogenic bacteria, a major problem in modern medicine. Bacterial genes encoding enzymes that inactivate antibiotics (drug resistance genes) have been flanked by insertion sequences generating drug resistance transposons. The widespread use of antibiotics in medicine, often unnecessarily in the treatment of viral infections where they have no effect, and to prevent infections of healthy agricultural animals, has led to the selection of such drug resistance transposons that have inserted into conjugating plasmids. Conjugating plasmids encode proteins that result in the replication and transfer of the plasmid to related bacteria through a complex macromolecular tube called a pilus. These plasmids, called R factors (for drug resistance), can contain multiple drug resistance genes introduced by transposition and selected in environments where antibiotics are used to sterilize surfaces, such as hospitals. These have led to the rapid spread of resistance to multiple antibiotics between pathogenic bacteria. Coping with the spread of R factors is a major challenge for modern medicine. ■

## KEY CONCEPTS of Section 6.3

### Transposable (Mobile) DNA Elements

• Transposable DNA elements are moderately repeated sequences interspersed at multiple sites throughout the genomes of higher eukaryotes. They are present less frequently in prokaryotic genomes.

• DNA transposons move to new sites directly as DNA; retrotransposons are first transcribed into an RNA copy of the element, which then is reverse-transcribed into DNA (see Figure 6-8).

• A common feature of all mobile elements is the presence of short direct repeats flanking the sequence.

• Enzymes encoded by transposons themselves catalyze insertion of these sequences at new sites in genomic DNA.

• Although DNA transposons, similar in structure to bacterial IS elements, occur in eukaryotes (e.g., the *Drosophila* P element), retrotransposons generally are much more abundant, especially in vertebrates.

• LTR retrotransposons are flanked by long terminal repeats (LTRs), similar to those in retroviral DNA; like retroviruses, they encode reverse transcriptase and integrase. They move in the genome by being transcribed into RNA, which then undergoes reverse transcription in the cytosol, nuclear import of the resulting DNA with LTRs, and integration into a host-cell chromosome (see Figure 6-14).

• Non-LTR retrotransposons, including long interspersed elements (LINEs) and short interspersed elements (SINEs), lack LTRs and have an A/T-rich stretch at one end. They are thought to move by a nonviral retrotransposition mechanism mediated by LINE-encoded proteins involving priming of reverse transcription by chromosomal DNA (see Figure 6-17).

• SINE sequences exhibit extensive homology with small cellular RNAs and are transcribed by the same RNA polymerase. *Alu* elements, the most common SINEs in humans, are ≈300-bp sequences found scattered throughout the human genome.

• Some interspersed repeats are derived from cellular RNAs that were reverse-transcribed and inserted into genomic DNA at some time in evolutionary history. Processed pseudogenes derived from mRNAs lack introns, a feature that distinguishes them from pseudogenes, which arose by sequence drift of duplicated genes.

• Mobile DNA elements most likely influenced evolution significantly by serving as recombination sites and by mobilizing adjacent DNA sequences.

## 6.4 Organelle DNAs

Although the vast majority of DNA in most eukaryotes is found in the nucleus, some DNA is present within the mitochondria of animals, plants, and fungi, and within the chloroplasts of plants. These organelles are the main cellular sites for ATP formation, during oxidative phosphorylation in mitochondria and photosynthesis in chloroplasts (Chapter 12). Many lines of evidence indicate that mitochondria and chloroplasts evolved from eubacteria that were engulfed into ancestral cells containing a eukaryotic nucleus, forming **endosymbiotes** (Figure 6-20). Over evolutionary time, most of the bacterial genes were lost from organellar DNAs. Some, such as genes encoding proteins involved in nucleotide, lipid, and amino acid biosynthesis, were lost because their functions were provided by genes in the nucleus of the host cell. Other genes encoding components of the present-day organelles were transferred to the nucleus. However, mitochondria and chloroplasts in today's eukaryotes retain DNAs encoding some proteins essential for organellar function, as well as the ribosomal and transfer RNAs required for synthesis of these proteins. Thus eukaryotic cells have multiple genetic systems: a predominant nuclear system and secondary systems with their own DNA, ribosomes, and tRNAs in mitochondria and chloroplasts.

### Mitochondria Contain Multiple mtDNA Molecules

Individual mitochondria are large enough to be seen under the light microscope, and even the mitochondrial DNA (mtDNA) can be detected by fluorescence microscopy. The mtDNA is located in the interior of the mitochondrion, the region known as the matrix (see Figure 12-6). As judged by the number of yellow fluorescent "dots" of mtDNA, a *Euglena gracilis* cell contains at least 30 mtDNA molecules (Figure 6-21).

Replication of mtDNA and division of the mitochondrial network can be followed in living cells using time-lapse microscopy. Such studies show that, in most organisms, mtDNA replicates throughout interphase. At mitosis, each daughter cell receives approximately the same number of mitochondria, but since there is no mechanism for apportioning exactly equal numbers of mitochondria to the daughter cells, some cells contain more mtDNA than others. By isolating mitochondria from cells and analyzing the DNA extracted from them, it can be seen that each mitochondrion contains

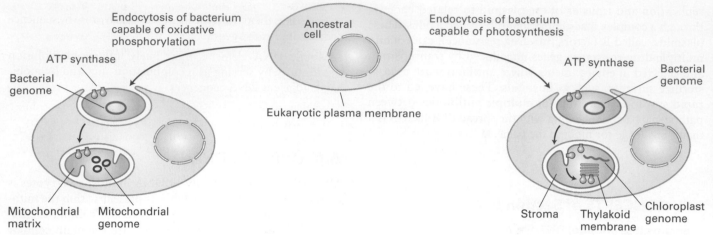

Endocytosis of bacterium capable of oxidative phosphorylation

Ancestral cell

Endocytosis of bacterium capable of photosynthesis

ATP synthase

Bacterial genome

Eukaryotic plasma membrane

ATP synthase

Bacterial genome

Mitochondrial matrix

Mitochondrial genome

Stroma

Thylakoid membrane

Chloroplast genome

**FIGURE 6-20 Endosymbiont hypothesis model of mitochondria and chloroplast evolution.** Endocytosis of a bacterium by an ancestral eukaryotic cell would generate an organelle with two membranes, the outer membrane derived from the eukaryotic plasma membrane and the inner one from the bacterial membrane. Proteins localized to the ancestral bacterial membrane retain their orientation, such that the portion of the protein once facing the extracellular space now faces the intermembrane space. Budding of vesicles from the inner chloroplast membrane, such as occurs during development of chloroplasts in contemporary plants, would generate the thylakoid membranes of chloroplasts. The organellar DNAs are indicated.

multiple mtDNA molecules. Thus the total amount of mtDNA in a cell depends on the number of mitochondria, the size of the mtDNA, and the number of mtDNA molecules per mitochondrion. Each of these parameters varies greatly between different cell types.

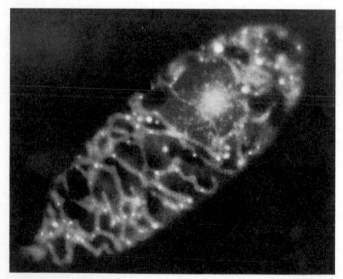

10 μm

**EXPERIMENTAL FIGURE 6-21 Dual staining reveals the multiple mitochondrial DNA molecules in a growing *Euglena gracilis* cell.** Cells were treated with a mixture of two dyes: ethidium bromide, which binds to DNA and emits a red fluorescence, and DiOC6, which is incorporated specifically into mitochondria and emits a green fluorescence. Thus the nucleus emits a red fluorescence, and areas rich in mitochondrial DNA fluoresce yellow—a combination of red DNA and green mitochondrial fluorescence. [From Y. Hayashi and K. Ueda, 1989, *J. Cell Sci.* **93**:565.]

## mtDNA Is Inherited Cytoplasmically

Studies of mutants in yeasts and other single-celled organisms first indicated that mitochondria exhibit *cytoplasmic inheritance* and thus must contain their own genetic system (Figure 6-22). For instance, *petite* yeast mutants exhibit structurally abnormal mitochondria and are incapable of oxidative phosphorylation. As a result, petite cells grow more slowly than wild-type yeasts and form smaller colonies. Genetic crosses between different (haploid) yeast strains showed that the *petite* mutation does not segregate with any known nuclear gene or chromosome. In later studies, most petite mutants were found to contain deletions of mtDNA.

In the mating by fusion of haploid yeast cells, both parents contribute equally to the cytoplasm of the resulting diploid; thus inheritance of mitochondria is biparental (see Figure 6-22a). In mammals and most other multicellular organisms, however, the sperm contributes little (if any) cytoplasm to the zygote, and virtually all the mitochondria in the embryo are derived from those in the egg, not the sperm. Studies in mice have shown that 99.99 percent of mtDNA is maternally inherited, but a small part (0.01 percent) is inherited from the male parent. In higher plants, mtDNA is inherited exclusively in a uniparental fashion through the female parent (egg), not the male (pollen).

## The Size, Structure, and Coding Capacity of mtDNA Vary Considerably Between Organisms

Surprisingly, the size of the mtDNA, the number and nature of the proteins it encodes, and even the mitochondrial genetic code itself vary greatly between different organisms. The mtDNAs of most multicellular animals are ≈16-kb circular molecules that encode intron-less genes compactly arranged on both DNA strands. Vertebrate mtDNAs encode

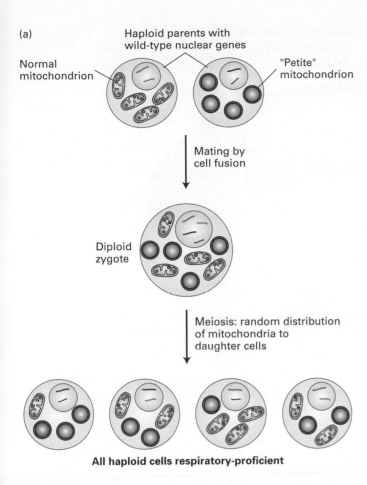

(a)

Normal mitochondrion — Haploid parents with wild-type nuclear genes — "Petite" mitochondrion

Mating by cell fusion

Diploid zygote

Meiosis: random distribution of mitochondria to daughter cells

**All haploid cells respiratory-proficient**

(b)

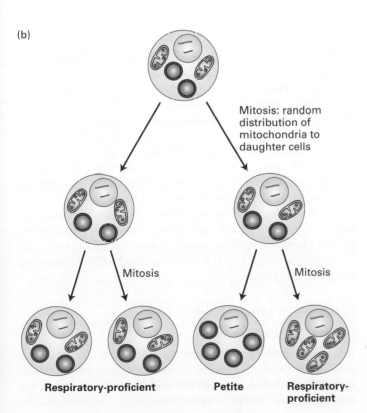

Mitosis: random distribution of mitochondria to daughter cells

Mitosis                    Mitosis

**Respiratory-proficient**    **Petite**    **Respiratory-proficient**

**FIGURE 6-22 Cytoplasmic inheritance of an mtDNA *petite* mutation in yeast.** Petite-strain mitochondria are defective in oxidative phosphorylation owing to a deletion in mtDNA. (a) Haploid cells fuse to produce a diploid cell that undergoes meiosis, during which random segregation of parental chromosomes and mitochondria containing mtDNA occurs. Note that alleles for genes in nuclear DNA (represented by large and small nuclear chromosomes colored red and blue) segregate 2:2 during meiosis (see Figure 5-5). In contrast, since yeast normally contain ≈50 mtDNA molecules per cell, all products of meiosis usually contain both normal and petite mtDNAs and are capable of respiration. (b) As these haploid cells grow and divide mitotically, the cytoplasm (including the mitochondria) is randomly distributed to the daughter cells. Occasionally, a cell is generated that contains only defective petite mtDNA and yields a petite colony. Thus formation of such petite cells is independent of any nuclear genetic marker.

the two rRNAs found in mitochondrial ribosomes, the 22 tRNAs used to translate mitochondrial mRNAs, and 13 proteins involved in electron transport and ATP synthesis (Chapter 12). The smallest mitochondrial genomes known are in *Plasmodium,* single-celled obligate intracellular parasites that cause malaria in humans. *Plasmodium* mtDNAs are only ≈6 kb, encoding five proteins and the mitochondrial rRNAs.

The mitochondrial genomes from a number of different metazoan organisms (i.e., multicellular animals) have now been cloned and sequenced, and mtDNAs from all these sources encode essential mitochondrial proteins (Figure 6-23). All proteins encoded by mtDNA are synthesized on mitochondrial ribosomes. Most mitochondrially synthesized polypeptides identified thus far are subunits of multimeric complexes used in electron transport, ATP synthesis, or insertion of proteins into the inner mitochondrial membrane or intermembrane space. However, most of the proteins localized in mitochondria, such as those involved in the processes listed at the top of Figure 6-23, are encoded by nuclear genes, synthesized on cytosolic ribosomes, and imported into the organelle by processes discussed in Chapter 13.

In contrast to metazoan mtDNAs, plant mtDNAs are many times larger, and most of the DNA does not encode protein. For instance, the mtDNA in the important model plant *Arabidopsis thaliana* is 366,924 base pairs, and the largest known mtDNA is ≈2 Mb, found in cucurbit plants (e.g., melon and cucumber). Most plant mtDNA consists of long introns, pseudogenes, mobile DNA elements restricted to the mitochondrial compartment, and pieces of foreign (chloroplast, nuclear, and viral) DNA that were probably inserted into plant mitochondrial genomes during their evolution. Duplicated sequences also contribute to the greater length of plant mtDNAs.

Differences in the number of genes encoded by the mtDNA from various organisms most likely reflect the movement of DNA between mitochondria and the nucleus during evolution. Direct evidence for this movement comes from the observation that several proteins encoded by mtDNA

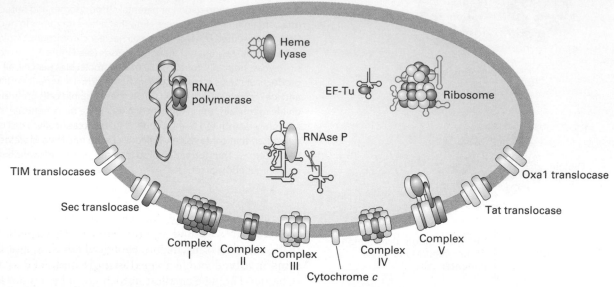

Lipid metabolism
Nucleotide metabolism
Amino acid metabolism

Carbohydrate metabolism
Heme synthesis
Fe-S synthesis

Ubiquinone synthesis
Co-factor synthesis
Proteases

Chaperones
Signaling pathways
DNA repair, replication, etc.

**FIGURE 6-23 Proteins encoded in mitochondrial DNA and their involvement in mitochondrial processes.** Only the mitochondrial matrix and inner membrane are depicted. Most mitochondrial components are encoded by the nucleus (blue); those highlighted in pink are encoded by mtDNA in some eukaryotes but by the nuclear genome in other eukaryotes, whereas a small portion are invariably specified by mtDNA (orange). Mitochondrial processes that have exclusively nucleus-encoded components are listed at the top. Complexes I–V are involved in electron transport and oxidative phosphorylation. TIM, Sec, Tat, and Oxa1 translocases are involved in protein import and export, and insertion of proteins into the inner membrane (see Chapter 13). RNase P is a ribozyme that processes the 5′ end of tRNAs (discussed in Chapter 8). It should be noted that the majority of eukaryotes have a multisubunit Complex I as depicted, with three subunits invariantly encoded by mtDNA. However, in a few organisms (*Saccharomyces, Schizosaccharomyces,* and *Plasmodium*), this complex is replaced by a nucleus-encoded, single-polypeptide enzyme. For more details on mitochondrial metabolism and transport, see Chapters 12 and 13. [Adapted from G. Burger et al., 2003, *Trends Genet.* **19:**709.]

in some species are encoded by nuclear DNA in other, closely related species. The most striking example of this phenomenon involves the *cox II* gene, which encodes subunit 2 of cytochrome *c* oxidase, which constitutes complex IV in the mitochondrial electron-transport chain (see Figure 12-16). This gene is found in mtDNA in all multicellular plants studied except for certain related species of legumes, including the mung bean and the soybean, in which the *cox II* gene is nuclear. The *cox II* gene is completely missing from mung bean mtDNA, but a defective *cox II* pseudogene that has accumulated many mutations can still be recognized in soybean mtDNA.

Many RNA transcripts of plant mitochondrial genes are edited, mainly by the enzyme-catalyzed conversion of selected C residues to U, and occasionally U to C. (RNA editing is discussed in Chapter 8.) The nuclear *cox II* gene of mung bean corresponds more closely to the edited *cox II* RNA transcripts than to the mitochondrial *cox II* genes found in other legumes. These observations are strong evidence that the *cox II* gene moved from the mitochondrion to the nucleus during mung bean evolution by a process that involved an RNA intermediate. Presumably this movement involved a reverse-transcription mechanism similar to that

by which processed pseudogenes are generated in the nuclear genome from nucleus-encoded mRNAs.

In addition to the large differences in the sizes of mtDNAs in different eukaryotes, the structure of the mtDNA also varies greatly. As mentioned above, mtDNA in most animals is a circular molecule ≈16 kb. However, the mtDNA of many organisms such as the protist *Tetrahymena* exists as linear head-to-tail concatemers of repeating sequence. In the most extreme examples, the mtDNA of the protist *Amoebidium parasiticum* is composed of several hundred distinct short linear molecules. And the mtDNA of *Trypanosoma* is comprised of multiple *maxicircles* concatenated (interlocked) to thousands of *minicircles* encoding *guide RNAs* involved in editing the sequence of the mitochondrial mRNAs encoded in the maxicircles.

## Products of Mitochondrial Genes Are Not Exported

As far as is known, all RNA transcripts of mtDNA and their translation products remain in the mitochondrion in which they are produced, and all mtDNA-encoded proteins are synthesized on mitochondrial ribosomes. Mitochondrial

DNA encodes the rRNAs that form mitochondrial ribosomes, although most of the ribosomal proteins are imported from the cytosol. In animals and fungi, all the tRNAs used for protein synthesis in mitochondria also are encoded by mtDNAs. However, in plants and many protozoans, most mitochondrial tRNAs are encoded by the nuclear DNA and imported into the mitochondrion.

Reflecting the bacterial ancestry of mitochondria, mitochondrial ribosomes resemble bacterial ribosomes and differ from eukaryotic cytosolic ribosomes in their RNA and protein compositions, their size, and their sensitivity to certain antibiotics (see Figure 4-22). For instance, chloramphenicol blocks protein synthesis by bacterial and mitochondrial ribosomes from most organisms, but cycloheximide, which inhibits protein synthesis on eukaryotic cytosolic ribosomes, does not affect mitochondrial ribosomes. This sensitivity of mitochondrial ribosomes to the important aminoglycoside class of antibiotics that includes chloramphenicol is the main cause of the toxicity that these antibiotics can cause. ■

## Mitochondria Evolved from a Single Endosymbiotic Event Involving a *Rickettsia*-like Bacterium

Analysis of the mtDNA sequences from various eukaryotes, including single-celled protists that diverged from other eukaryotes early in evolution, provides strong support for the idea that the mitochondrion had a single origin. Mitochondria most likely arose from a bacterial symbiote whose closest contemporary relatives are in the *Rickettsiaceae* group. Bacteria in this group are obligate intracellular parasites. Thus, the ancestor of the mitochondrion probably also had an intracellular lifestyle, putting it in a good location for

evolving into an intracellular symbiote. The mtDNA with the largest number of encoded genes so far found is in the protist species *Reclinomonas americana*. All other mtDNAs have a subset of the *R. americana* genes, strongly implying that they evolved from a common ancestor with *R. americana*, losing different groups of mitochondrial genes by deletion and/or transfer to the nucleus over time.

In organisms whose mtDNA includes only a limited number of genes, the same set of mitochondrial genes is retained, independent of the phyla that include these organisms (see Figure 6-23, orange proteins). One hypothesis for why these genes were never successfully transferred to the nuclear genome is that their encoded polypeptides are too hydrophobic to cross the outer mitochondrial membrane, and therefore would not be imported back into the mitochondria if they were synthesized in the cytosol. Similarly, the large size of rRNAs may interfere with their transport from the nucleus through the cytosol into mitochondria. Alternatively, these genes may not have been transferred to the nucleus during evolution because regulation of their expression in response to conditions within individual mitochondria may be advantageous. If these genes were located in the nucleus, conditions within each mitochondrion could not influence the expression of proteins found in that mitochondrion.

## Mitochondrial Genetic Codes Differ from the Standard Nuclear Code

The genetic code used in animal and fungal mitochondria is different from the standard code used in all prokaryotic and eukaryotic nuclear genes; remarkably, the code even differs in mitochondria from different species (Table 6-3). Why and how these differences arose during evolution is mysterious. UGA, for example, is normally a stop codon, but is read as tryptophan by human and fungal mitochondrial translation

| TABLE 6-3 | Alterations in the Standard Genetic Code in Mitochondria | | | | | |
|---|---|---|---|---|---|---|
| | | Mitochondria | | | | |
| Codon | Standard Code * | Mammals | *Drosophila* | *Neurospora* | Yeasts | Plants |
| UGA | Stop | Trp | Trp | Trp | Trp | Stop |
| AGA, AGG | Arg | Stop | Ser | Arg | Arg | Arg |
| AUA | Ile | Met | Met | Ile | Met | Ile |
| AUU | Ile | Met | Met | Met | Met | Ile |
| CUU, CUC, CUA, CUG | Leu | Leu | Leu | Leu | Thr | Leu |

*For nuclear-encoded proteins.

SOURCES: S. Anderson et al., 1981, *Nature* **290**:457; P. Borst, in *International Cell Biology 1980–1981*, H. G. Schweiger, ed., Springer-Verlag, p. 239; C. Breitenberger and U. L. Raj Bhandary, 1985, *Trends Biochem. Sci.* **10**:478–483; V. K. Eckenrode and C. S. Levings, 1986, *In Vitro Cell Dev. Biol.* **22**:169–176; J. M. Gualber et al., 1989, *Nature* **341**:660–662; and P. S. Covello and M. W. Gray, 1989, *Nature* **341**:662–666.

systems; however, in plant mitochondria, UGA is still recognized as a stop codon. AGA and AGG, the standard nuclear codons for arginine, also code for arginine in fungal and plant mtDNA, but they are stop codons in mammalian mtDNA and serine codons in *Drosophila* mtDNA.

As shown in Table 6-3, plant mitochondria appear to utilize the standard genetic code. However, comparisons of the amino acid sequences of plant mitochondrial proteins with the nucleotide sequences of plant mtDNAs suggested that CGG could code for *either* arginine (the "standard" amino acid) or tryptophan. This apparent nonspecificity of the plant mitochondrial code is explained by editing of mitochondrial RNA transcripts, which can convert cytosine residues to uracil residues. If a CGG sequence is edited to UGG, the codon specifies tryptophan, the standard amino acid for UGG, whereas unedited CGG codons encode the standard arginine. Thus the translation system in plant mitochondria does utilize the standard genetic code. ∎

## Mutations in Mitochondrial DNA Cause Several Genetic Diseases in Humans

The severity of disease caused by a mutation in mtDNA depends on the nature of the mutation and on the proportion of mutant and wild-type mtDNAs present in a particular cell type. Generally, when mutations in mtDNA are found, cells contain mixtures of wild-type and mutant mtDNAs—a condition known as *heteroplasmy.* Each time a mammalian somatic or germ-line cell divides, the mutant and wild-type mtDNAs segregate randomly into the daughter cells, as occurs in yeast cells (see Figure 6-22b). Thus, the mtDNA genotype, which fluctuates from one generation and from one cell division to the next, can drift toward predominantly wild-type or predominantly mutant mtDNAs. Since all enzymes required for the replication and growth of mammalian mitochondria, such as the mitochondrial DNA and RNA polymerases, are encoded in the nucleus and imported from the cytosol, a mutant mtDNA should not be at a "replication disadvantage"; mutants that involve large deletions of mtDNA might even be at a selective advantage in replication, because they can replicate faster.

Recent research suggests that the accumulation of mutations in mtDNA is an important component of aging in mammals. Mutations in mtDNA have been observed to accumulate with aging, probably because mammalian mtDNA is not repaired in response to DNA damage. To study this hypothesis, researchers used gene "knock-in" techniques to replace the nuclear gene encoding mitochondrial DNA polymerase with normal proofreading activity (see Figure 4-34) with a mutant gene encoding a polymerase defective in proofreading. Mutations in mtDNA accumulated much more rapidly in homozygous mutant mice than in wild-type mice, and the mutant mice aged at a highly accelerated rate (Figure 6-24).

With few exceptions, all human cells have mitochondria, yet mutations in mtDNA affect only some tissues. Those most commonly affected are tissues that have a high

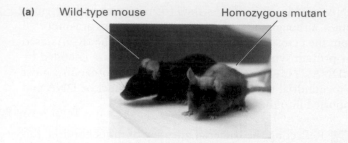

**(a)** Wild-type mouse    Homozygous mutant

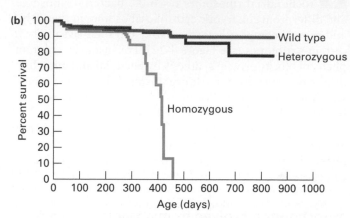

**(b)**

**EXPERIMENTAL FIGURE 6-24 Mice with a mitochondrial DNA polymerase defective for proofreading exhibit premature aging.** A line of "knock-in" mice were prepared by methods discussed in Chapter 5 with an aspartic acid-to-alanine mutation in the gene encoding mitochondrial DNA polymerase (D257A), inactivating the polymerase's proofreading function. (a) Wild-type and homozygous mutant mice at 390 days old (13 months). The mutant mouse displays many of the features of an aged mouse (>720 days, or 24 months of age). (b) Plot of survival versus time of wild-type (+/+), heterozygous (D257A/+) and homozygous (D257A/D257A) mice. [From G. C. Kujoth et al., 2005, *Science* **309:**481. Part (a) courtesy of Jeff Miller/University of Wisconsin-Madison and Gregory Kujoth, Ph.D.]

requirement for ATP produced by oxidative phosphorylation and tissues that require most or all of the mtDNA in the cell to synthesize sufficient amounts of functional mitochondrial proteins. For instance, *Leber's hereditary optic neuropathy* (degeneration of the optic nerve) is caused by a missense mutation in the mtDNA gene encoding subunit 4 of the NADH-CoQ reductase (complex I), a protein required for ATP production by mitochondria (see Figure 12-16). Any of several large deletions in mtDNA causes another set of diseases, including *chronic progressive external ophthalmoplegia,* characterized by eye defects, and *Kearns-Sayre syndrome,* characterized by eye defects, abnormal heartbeat, and central nervous system degeneration. A third condition, causing "ragged" muscle fibers (with improperly assembled mitochondria) and associated uncontrolled jerky movements, is due to a single mutation in the TΨCG loop of the mitochondrial lysine tRNA. As a result of this mutation, the translation of several mitochondrial proteins apparently is inhibited. ∎

## Chloroplasts Contain Large DNAs Often Encoding More Than a Hundred Proteins

Like mitochondria, chloroplasts are thought to have evolved from an ancestral endosymbiotic photosynthetic bacterium (see Figure 6-20). However, the endosymbiotic event giving rise to chloroplasts occurred more recently (1.2–1.5 billion years ago) than the event leading to the evolution of mitochondria (1.5–2.2 billion years ago). Consequently, contemporary chloroplast DNAs show less structural diversity than do mtDNAs. Also similar to mitochondria, chloroplasts contain multiple copies of the organellar DNA and ribosomes, which synthesize some chloroplast-encoded proteins using the standard genetic code. Like plant mtDNA, chloroplast DNA is inherited exclusively in a uniparental fashion through the female parent (egg). Other chloroplast proteins are encoded by nuclear genes, synthesized on cytosolic ribosomes, and then incorporated into the organelle (Chapter 13). ■

In higher plants, chloroplast DNAs are 120–160 kb long, depending on the species. They initially were thought to be circular DNA molecules because in genetically tractable organisms such as the model plant protozoan *Chlamydomonas reinhardtii*, the genetic map is circular. However, recent studies have revealed that plant chloroplast DNAs are actually long head-to-tail linear concatemers plus recombination intermediates between these long linear molecules. In these studies, researchers have used techniques that minimize mechanical breakage of long DNA molecules during isolation and gel electrophoresis, permitting analysis of megabase-size DNA.

The complete sequences of several chloroplast DNAs from higher plants have been determined. They contain 120–135 genes, 130 in the important model plant *Arabidopsis thaliana*. *A. thaliana* chloroplast DNA encodes 76 protein-coding genes and 54 genes with RNA products such as rRNAs and tRNAs. Chloroplast DNAs encode the subunits of a bacterial-like RNA polymerase and express many of their genes from polycistronic operons as in bacteria (see Figure 4-13a). Some chloroplast genes contain introns, but these are similar to the specialized introns found in some bacterial genes and in mitochondrial genes from fungi and protozoans, rather than the introns of nuclear genes. As in the evolution of mitochondrial genomes, many genes in the ancestral chloroplast endosymbiote that were redundant with nuclear genes have been lost from chloroplast DNA. Also, many genes essential for chloroplast function have been transferred to the nuclear genome of plants over evolutionary time. Recent estimates from sequence analysis of the *A. thaliana* and cyanobacterial genomes indicate that ≈4500 genes have been transferred from the original endosymbiote to the nuclear genome.

Methods similar to those used for the transformation of yeast cells (Chapter 5) have been developed for stably introducing foreign DNA into the chloroplasts of higher plants. The large number of chloroplast DNA molecules per cell permits the introduction of thousands of copies of an engineered gene into each cell, resulting in extraordinarily high levels of foreign protein production. Chloroplast transformation has recently led to the engineering of plants that are resistant to bacterial and fungal infections, drought, and herbicides. The level of production of foreign proteins is comparable with that achieved with engineered bacteria, making it likely that chloroplast transformation will be used for the production of human pharmaceuticals and possibly for the engineering of food crops containing high levels of all the amino acids essential to humans. ■

## KEY CONCEPTS of Section 6.4

### Organelle DNAs

• Mitochondria and chloroplasts most likely evolved from bacteria that formed a symbiotic relationship with ancestral cells containing a eukaryotic nucleus (see Figure 6-20).

• Most of the genes originally within mitochondria and chloroplasts were either lost because their functions were redundant with nuclear genes or moved to the nuclear genome over evolutionary time, leaving different gene sets in the organellar DNAs of different organisms (see Figure 6-23).

• Animal mtDNAs are circular molecules, reflecting their probable bacterial origin. Plant mtDNAs and chloroplast DNAs generally are longer than mtDNAs from other eukaryotes, largely because they contain more noncoding regions and repetitive sequences.

• All mtDNAs and chloroplast DNAs encode rRNAs and some of the proteins involved in mitochondrial or photosynthetic electron transport and ATP synthesis. Most animal mtDNAs and chloroplast DNAs also encode the tRNAs necessary to translate the organellar mRNAs.

• Because most mtDNA is inherited from egg cells rather than sperm, mutations in mtDNA exhibit a maternal cytoplasmic pattern of inheritance. Similarly, chloroplast DNA is exclusively inherited from the maternal parent.

• Mitochondrial ribosomes resemble bacterial ribosomes in their structure, sensitivity to chloramphenicol, and resistance to cycloheximide.

• The genetic code of animal and fungal mtDNAs differs slightly from that of bacteria and the nuclear genome and varies among different animals and fungi (see Table 6-3). In contrast, plant mtDNAs and chloroplast DNAs appear to conform to the standard genetic code.

• Several human neuromuscular disorders result from mutations in mtDNA. Patients generally have a mixture of wild-type and mutant mtDNA in their cells (heteroplasmy): the higher the fraction of mutant mtDNA, the more severe is the mutant phenotype.

## 6.5 Genomics: Genome-wide Analysis of Gene Structure and Expression

Using automated DNA sequencing techniques and computer algorithms to piece together the sequence data, researchers have determined vast amounts of DNA sequence including nearly the entire genomic sequence of humans and many key experimental organisms. This enormous volume of data, which is growing at a rapid pace, has been stored and organized in two primary data banks: the GenBank at the National Institutes of Health, Bethesda, Maryland, and the EMBL Sequence Data Base at the European Molecular Biology Laboratory in Heidelberg, Germany. These databases continuously exchange newly reported sequences and make them available to scientists throughout the world on the Internet. By now, the genome sequences have been completely, or nearly completely, determined for hundreds of viruses and bacteria, scores of archaea, yeasts (eukaryotes), plants including rice and maize, important model multicellular eukaryotes such as the roundworm *C. elegans*, the fruit fly *Drosophila melanogaster*, mice, humans, and representatives of the ≈35 metazoan phyla. The cost and speed of sequencing a megabase of DNA has fallen so low that the entire genome in cancer cells has been sequenced and compared to the genome in normal cells from the same patient in order to determine all the mutations that have accumulated in that patient's tumor cells. This approach may reveal genes that are commonly mutated in all cancers, as well as genes that are commonly mutated in tumor cells from different patients with the same type of cancer (e.g., breast versus colon cancer). This approach may eventually lead to highly individualized cancer treatments tailored to the specific mutations in the tumor cells of a particular patient. The latest automated DNA sequencing techniques are so powerful that a project known as the "1000 Genomes Project" is currently underway with the goal of sequencing most of the genomes of 1000–2000 randomly chosen individuals from all over the world in order to determine the extent of human genetic variation as a basis for investigating the relationship between genotype and phenotype in humans. Moreover, privately owned companies have been founded that will sequence much of an individual's genome for ≈$100 in order to search for sequence variations that may influence the probability of developing specific diseases.

In this section, we examine some of the ways researchers are mining this treasure trove of data to provide insights about gene function and evolutionary relationships, to identify new genes whose encoded proteins have never been isolated, and to determine when and where genes are expressed. This use of computers to analyze sequence data has led to the emergence of a new field of biology: *bioinformatics*.

### Stored Sequences Suggest Functions of Newly Identified Genes and Proteins

As discussed in Chapter 3, proteins with similar functions often contain similar amino acid sequences that correspond to important functional domains in the three-dimensional structure of the proteins. By comparing the amino acid sequence of the protein encoded by a newly cloned gene with the sequences of proteins of known function, an investigator can look for sequence similarities that provide clues to the function of the encoded protein. Because of the degeneracy in the genetic code, related proteins invariably exhibit more sequence similarity than the genes encoding them. For this reason, protein sequences rather than the corresponding DNA sequences are usually compared.

The most widely used computer program for this purpose is known as BLAST (*basic local alignment search tool*). The BLAST algorithm divides the "new" protein sequence (known as the *query sequence*) into shorter segments and then searches the database for significant matches to any of the stored sequences. The matching program assigns a high score to identically matched amino acids and a lower score to matches between amino acids that are related (e.g., hydrophobic, polar, positively charged, negatively charged) but not identical. When a significant match is found for a segment, the BLAST algorithm will search locally to extend the region of similarity. After searching is completed, the program ranks the matches between the query protein and various known proteins according to their *p-values*. This parameter is a measure of the probability of finding such a degree of similarity between two protein sequences by chance. The lower the *p*-value, the greater is the sequence similarity between two sequences. A *p*-value less than about $10^{-3}$ usually is considered as significant evidence that two proteins share a common ancestor. Many alternative computer programs have been developed in addition to BLAST that can detect relationships between proteins that are more distantly related to each other than can be detected by BLAST. The development of such methods is currently an active area of bioinformatics research.

To illustrate the power of this approach, we consider the human gene *NF1*. Mutations in *NF1* are associated with the inherited disease neurofibromatosis 1, in which multiple tumors develop in the peripheral nervous system, causing large protuberances in the skin. After a cDNA clone of *NF1* was isolated and sequenced, the deduced sequence of the NF1 protein was checked against all other protein sequences in GenBank. A region of NF1 protein was discovered to have considerable homology to a portion of the yeast protein called Ira (Figure 6-25). Previous studies had shown that Ira is a GTPase-activating protein (GAP) that modulates the GTPase activity of the monomeric G protein called Ras (see Figure 3-32). As we examine in detail in Chapter 16, GAP and Ras proteins normally function to control cell replication and differentiation in response to signals from neighboring cells. Functional studies on the normal NF1 protein, obtained by expression of the cloned wild-type gene, showed that it did, indeed, regulate Ras activity, as suggested by its homology with Ira. These findings suggest that patients with neurofibromatosis express a mutant NF1 protein in cells of the peripheral nervous system, leading to inappropriate cell division and formation of the tumors characteristic of the disease. ■

```
NF1   841  T R A T F M E V L T K I L Q Q G T E F D T L A E T V L A D R F E R L V E L V T M M G D Q G E L P I A  890
Ira  1500  I R I A F L R V F I D I V . . . T N Y P V N P E K H E M D K M L A I D D F L K Y I I K N P I L A F F  1546

      891  M A L A N V V P C S Q W D E L A R V L V T L F D S R H L L Y Q L L W N M F S K E V E L A D S M Q T L  940
     1547  G S L A . . C S P A D V D L Y A G G F L N A F D T R N A S H I L V T E L L K Q E I K R A A R S D D I  1594

      941  F R G N S L A S K I M T F C F K V Y G A T Y L Q K L L D P L L R I V I T S S D W Q H V S F E V D P T  990
     1595  L R R N S C A T R A L S L Y T R S R G N K Y L I K T L R P V L Q G I V D N K E . . . . S F E I D . .  1638

      991  R L E P S E S L E E N Q R N L L Q M T E K F . . . . F H A I I S S S S E F P P Q L R S V C H C L Y Q  1036
     1639  K M K P G . . . S E N S E K M L D L F E K Y M T R L I D A I T S S I D D F P I E L V D I C K T I Y N  1685

     1037  V V S Q R F P Q N S I G A V G S A M F L R F I N P A I V S P Y E A G I L D K K P P P R I E R G L K L  1086
     1686  A A S V N F P E Y A Y I A V G S F V F L R F I G P A L V S P D S E N I I . I V T H A H D R K P F I T  1734

     1087  M S K I L Q S I A N . . . . . . . H V L F T K E E H M R P F N D . . . . F V K S N F D A A R R F F  1124
     1735  L A K V I Q S L A N G R E N I F K K D I L V S K E E F L K T C S D K I F N F L S E L C K I P T N N F  1784

     1125  L D I A S D C P T S D A V N H S L . . . . . . . . . . . . . S F I S D G N V L A L H R L L W N N .  1159
     1785  T V N V R E D P T P I S F D Y S F L H K F F Y L N E F T I R K E I I N E S K L P G E F S F L K N T V  1834

     1160  . . Q E K I G Q Y L S S N R D H K A V G R R P F . . . . D K M A T L L A Y L G P P E H K P V A  1200
     1835  M L N D K I L G V L G Q P S M E I K N E I P P F V V E N R E K Y P S L Y E F M S R Y A F K K V D  1882
```

**FIGURE 6-25 Comparison of the regions of human NF1 protein and *S. cerevisiae* Ira protein that show significant sequence similarity.** The NF1 and the Ira sequences are shown on the top and bottom lines of each row, respectively, in the one-letter amino acid code (see Figure 2-14). Amino acids that are identical in the two proteins are highlighted in yellow. Amino acids with chemically similar but nonidentical side chains are connected by a blue dot. Amino acid numbers in the protein sequences are shown at the left and right ends of each row. Black dots indicate "gaps" in the protein sequence inserted in order to maximize the alignment of homologous amino acids. The BLAST *p*-value for these two sequences is $10^{-28}$, indicating a high degree of similarity. [From G. Xu et al., 1990, *Cell* **62**:599.]

Even when a protein shows no significant similarity to other proteins with the BLAST algorithm, it may nevertheless share a short sequence that is functionally important. Such short segments recurring in many different proteins, referred to as **structural motifs,** generally have similar functions. Several such motifs are described in Chapter 3 and illustrated in Figure 3-9. To search for these and other motifs in a new protein, researchers compare the query protein sequence with a database of known motif sequences.

## Comparison of Related Sequences from Different Species Can Give Clues to Evolutionary Relationships Among Proteins

BLAST searches for related protein sequences may reveal that proteins belong to a protein family. Earlier, we considered gene families in a single organism, using the β-globin genes in humans as an example (see Figure 6-4a). But in a database that includes the genome sequences of multiple organisms, protein families also can be recognized as being shared among related organisms. Consider, for example, the **tubulin** proteins; these are the basic subunits of microtubules, which are important components of the cytoskeleton (Chapter 18). According to the simplified scheme in Figure 6-26a, the earliest eukaryotic cells are thought to have contained a single tubulin gene that was duplicated early in evolution; subsequent divergence of the different copies of the original tubulin gene formed the ancestral versions of the α- and β-tubulin genes. As different species diverged from these early eukaryotic cells, each of these gene sequences further diverged, giving rise to the slightly different forms of α-tubulin and β-tubulin now found in each species.

All the different members of the tubulin family of genes (or proteins) are sufficiently similar in sequence to suggest a common ancestral sequence. Thus all these sequences are considered to be *homologous*. More specifically, sequences that presumably diverged as a result of gene duplication (e.g., the α- and β-tubulin sequences) are described as *paralogous*. Sequences that arose because of speciation (e.g., the α-tubulin genes in different species) are described as *orthologous*. From the degree of sequence relatedness of the tubulins present in different organisms today, evolutionary relationships can be deduced, as illustrated in Figure 6-26b. Of the three types of sequence relationships, orthologous sequences are the most likely to share the same function.

## Genes Can Be Identified Within Genomic DNA Sequences

The complete genomic sequence of an organism contains within it the information needed to deduce the sequence of every protein made by the cells of that organism. For organisms such as bacteria and yeast, whose genomes have few introns and short intergenic regions, most protein-coding sequences can be found simply by scanning the genomic sequence for **open reading frames (ORFs)** of significant length. An ORF usually is defined as a stretch of DNA containing at least 100 codons that begins with a start codon and ends with a stop codon. Because the probability that a random DNA sequence will contain no stop codons for 100 codons in a row is very small, most ORFs encode protein.

ORF analysis correctly identifies more than 90 percent of the genes in yeast and bacteria. Some of the very shortest genes,

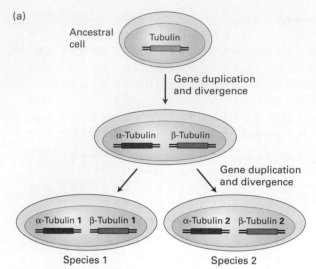

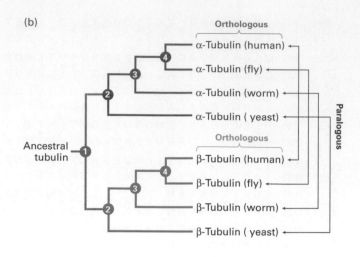

**FIGURE 6-26 Generation of diverse tubulin sequences during the evolution of eukaryotes.** (a) Probable mechanism giving rise to the tubulin genes found in existing species. It is possible to deduce that a gene duplication event occurred before speciation because the α-tubulin sequences from different species (e.g., humans and yeast) are more alike than are the α-tubulin and β-tubulin sequences within a species. (b) A phylogenetic tree representing the relationship between the tubulin sequences. The branch points (nodes), indicated by small numbers, represent common ancestral genes at the time that two sequences diverged. For example, node 1 represents the duplication event that gave rise to the α-tubulin and β-tubulin families, and node 2 represents the divergence of yeast from multicellular species. Braces and arrows indicate, respectively, the orthologous tubulin genes, which differ as a result of speciation, and the paralogous genes, which differ as a result of gene duplication. This diagram is simplified somewhat because flies, worms, and humans actually contain multiple α-tubulin and β-tubulin genes that arose from later gene duplication events.

however, are missed by this method, and occasionally long open reading frames that are not actually genes arise by chance. Both types of mis-assignments can be corrected by more sophisticated analysis of the sequence and by genetic tests for gene function. Of the *Saccharomyces* genes identified in this manner, about half were already known by some functional criterion such as mutant phenotype. The functions of some of the proteins encoded by the remaining putative (suspected) genes identified by ORF analysis have been assigned based on their sequence similarity to known proteins in other organisms.

Identification of genes in organisms with a more complex genome structure requires more sophisticated algorithms than searching for open reading frames. Because most genes in higher eukaryotes are composed of multiple, relatively short exons separated by often quite long noncoding introns, scanning for ORFs is a poor method for finding genes. The best gene-finding algorithms combine all the available data that might suggest the presence of a gene at a particular genomic site. Relevant data include alignment or hybridization of the query sequence to a full-length cDNA; alignment to a partial cDNA sequence, generally 200–400 bp in length, known as an *expressed sequence tag (EST);* fitting to models for exon, intron, and splice site sequences; and sequence similarity to other organisms. Using these computer-based bioinformatic methods, computational biologists have identified approximately ≈19,800 protein coding genes in the human genome.

A particularly powerful method for identifying human genes is to compare the human genomic sequence with that of the mouse. Humans and mice are sufficiently related to have most genes in common, although largely nonfunctional DNA sequences, such as intergenic regions and introns, will tend to be very different because these sequences are not under strong selective pressure. Thus corresponding segments of the human and mouse genome that exhibit high sequence similarity are likely to be functionally important: exons, transcription-control regions, or sequences with other functions that are not yet understood.

## The Number of Protein-Coding Genes in an Organism's Genome Is Not Directly Related to Its Biological Complexity

The combination of genomic sequencing and gene-finding computer algorithms has yielded the complete inventory of protein-coding genes for a variety of organisms. Figure 6-27 shows the total number of protein-coding genes in several eukaryotic genomes that have been completely sequenced. The functions of about half the proteins encoded in these genomes are known or have been predicted on the basis of sequence comparisons. One of the surprising features of this comparison is that the number of protein-coding genes within different organisms does not seem proportional to our intuitive sense of their biological complexity. For example, the roundworm *C. elegans* apparently has more genes than the fruit fly *Drosophila*, which has a much more complex body plan and more complex behavior. And humans have fewer than one and one-half the number of genes as *C. elegans.* When it first became apparent that humans have fewer than

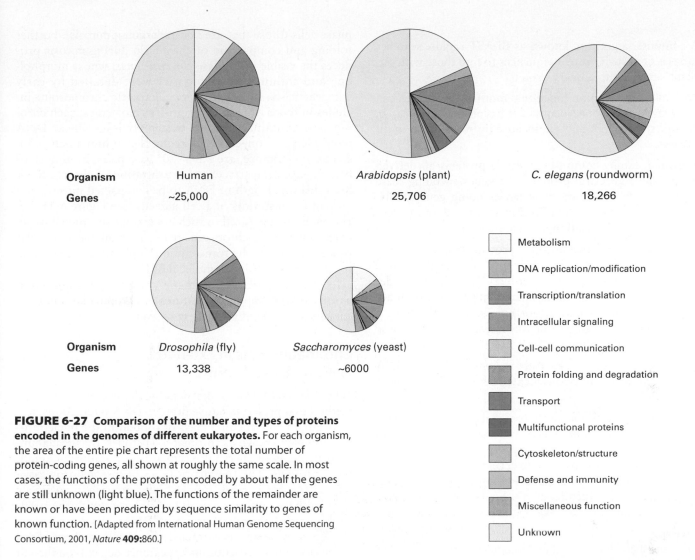

**Organism** | Human | *Arabidopsis* (plant) | *C. elegans* (roundworm)
**Genes** | ~25,000 | 25,706 | 18,266

**Organism** | *Drosophila* (fly) | *Saccharomyces* (yeast)
**Genes** | 13,338 | ~6000

Legend:
- Metabolism
- DNA replication/modification
- Transcription/translation
- Intracellular signaling
- Cell-cell communication
- Protein folding and degradation
- Transport
- Multifunctional proteins
- Cytoskeleton/structure
- Defense and immunity
- Miscellaneous function
- Unknown

**FIGURE 6-27 Comparison of the number and types of proteins encoded in the genomes of different eukaryotes.** For each organism, the area of the entire pie chart represents the total number of protein-coding genes, all shown at roughly the same scale. In most cases, the functions of the proteins encoded by about half the genes are still unknown (light blue). The functions of the remainder are known or have been predicted by sequence similarity to genes of known function. [Adapted from International Human Genome Sequencing Consortium, 2001, *Nature* **409**:860.]

twice the number of protein-coding genes as the simple roundworm, it was difficult to understand how such a small increase in the number of proteins could generate such a staggering difference in complexity.

Clearly, simple quantitative differences in the number of genes in the genomes of different organisms are inadequate for explaining differences in biological complexity. However, several phenomena can generate more complexity in the expressed proteins of higher eukaryotes than is predicted from their genomes. First, alternative splicing of a pre-mRNA can yield multiple functional mRNAs corresponding to a particular gene (Chapter 8). Second, variations in the post-translational modification of some proteins may produce functional differences. Finally, increased biological complexity results from increased numbers of cells built of the same kinds of proteins. Larger numbers of cells can interact in more complex combinations, as in comparing the cerebral cortex from mouse to man. Similar cells are present in both the mouse and human cerebral cortex, but in humans more of them make more complex connections. Evolution of the increasing biological complexity of multicellular organisms likely required increasingly complex regulation of cell replication and gene expression, leading to increasing complexity of embryological development.

The specific functions of many genes and proteins identified by analysis of genomic sequences still have not been determined. As researchers unravel the functions of individual proteins in different organisms and further detail their interactions with other proteins, the resulting advances will become immediately applicable to all homologous proteins in other organisms. When the function of every protein is known, no doubt, a more sophisticated understanding of the molecular basis of complex biological systems will emerge.

## KEY CONCEPTS of Section 6.5

### Genomics: Genome-wide Analysis of Gene Structure and Expression

- The function of a protein that has not been isolated (a query protein) often can be predicted on the basis of similarity of its amino acid sequence to the sequences of proteins of known function.

- A computer algorithm known as BLAST rapidly searches databases of known protein sequences to find those with significant similarity to a query protein.

- Proteins with common functional motifs, which often can be quite short, may not be identified in a typical BLAST search. Such short sequences may be located by searches of motif databases.

- A protein family comprises multiple proteins all derived from the same ancestral protein. The genes encoding these proteins, which constitute the corresponding gene family, arose by an initial gene duplication event and subsequent divergence during speciation (see Figure 6-26).

- Related genes and their encoded proteins that derive from a gene duplication event are paralogous, such as the $\alpha$- and $\beta$-globins that combine in hemoglobin ($\alpha_2\beta_2$); those that derive from mutations that accumulated during speciation are orthologous. Proteins that are orthologous usually have a similar function in different organisms, such as the mouse and human adult $\beta$-globins.

- Open reading frames (ORFs) are regions of genomic DNA containing at least 100 codons located between a start codon and stop codon.

- Computer search of the entire bacterial and yeast genomic sequences for open reading frames (ORFs) correctly identifies most protein-coding genes. Several types of additional data must be used to identify probable (putative) genes in the genomic sequences of humans and other higher eukaryotes because of their more complex gene structure, in which relatively short coding exons are separated by relatively long, noncoding introns.

- Analysis of the complete genome sequences for several different organisms indicates that biological complexity is not directly related to the number of protein-coding genes (see Figure 6-27).

# 6.6 Structural Organization of Eukaryotic Chromosomes

Now that we have examined the various types of DNA sequences found in eukaryotic genomes and how they are organized within it, we turn to the question of how DNA molecules as a whole are organized within eukaryotic cells. Because the total length of cellular DNA is up to a hundred thousand times a cell's diameter, the packing of DNA is crucial to cell architecture. It is also essential to prevent the long DNA molecules from getting knotted or tangled with each other during cell division, when they must be precisely segregated to daughter cells. The task of compacting and organizing chromosomal DNA is performed by abundant nuclear proteins called **histones**. The complex of histones and DNA is called **chromatin**.

Chromatin, which is about half DNA and half protein by mass, is dispersed throughout much of the nucleus in inter-

phase cells (those that are not undergoing mitosis). Further folding and compaction of chromatin during mitosis produces the visible *metaphase chromosomes*, whose morphology and staining characteristics were detailed by early cytogeneticists. Although every eukaryotic chromosome includes millions of individual protein molecules, each chromosome contains just one, extremely long, linear DNA molecule. The longest DNA molecules in human chromosomes, for instance, are $2.8 \times 10^8$ base pairs, or almost 10 cm, in length! The structural organization of chromatin allows this vast length of DNA to be compacted into the microscopic constraints of a cell nucleus (see Figure 6-1). Yet chromatin is organized in such a way that specific DNA sequences within the chromatin are readily available for cellular processes such as the transcription, replication, repair, and recombination of DNA molecules. In this section, we consider the properties of chromatin and its organization into chromosomes. Important features of chromosomes in their entirety are covered in the next section.

## Chromatin Exists in Extended and Condensed Forms

When the DNA from eukaryotic nuclei is isolated using a method that preserves native protein–DNA interactions, it is associated with an equal mass of protein in the nucleoprotein complex known as chromatin. Histones, the most abundant proteins in chromatin, constitute a family of small, basic proteins. The five major types of histone proteins—termed *H1, H2A, H2B, H3,* and *H4*—are rich in positively charged basic amino acids, which interact with the negatively charged phosphate groups in DNA.

When chromatin is extracted from nuclei and examined in the electron microscope, its appearance depends on the salt concentration to which it is exposed. At low salt concentration in the absence of divalent cations such as $Mg^{+2}$, isolated chromatin resembles "beads on a string" (Figure 6-28a). In this extended form, the string is composed of free DNA called "linker" DNA connecting beadlike structures termed **nucleosomes**. Composed of DNA and histones, nucleosomes are about 10 nm in diameter and are the primary structural units of chromatin. If chromatin is isolated at physiological salt concentration, it assumes a more condensed fiberlike form that is 30 nm in diameter (Figure 6-28b).

**Structure of Nucleosomes** The DNA component of nucleosomes is much less susceptible to nuclease digestion than is the linker DNA between them. If nuclease treatment is carefully controlled, all the linker DNA can be digested, releasing individual nucleosomes with their DNA component. A nucleosome consists of a protein core with DNA wound around its surface like thread around a spool. The core is an octamer containing two copies each of histones H2A, H2B, H3, and H4. X-ray crystallography has shown that the octameric histone core is a roughly disk-shaped structure made of interlocking histone subunits (Figure 6-29). Nucleosomes from all eukaryotes contain $\approx 147$ base pairs of DNA wrapped one

(a)

(b)

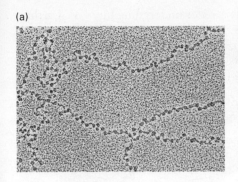 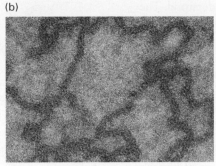

**EXPERIMENTAL FIGURE 6-28  The extended and condensed forms of extracted chromatin have very different appearances in electron micrographs.** (a) Chromatin isolated in low-ionic-strength buffer has an extended "beads-on-a-string" appearance. The "beads" are nucleosomes (10-nm diameter) and the "string" is connecting (linker) DNA. (b) Chromatin isolated in buffer with a physiological ionic strength (0.15 M KCl) appears as a condensed fiber 30 nm in diameter. [Part (a) courtesy of S. McKnight and O. Miller, Jr. Part (b) courtesy of B. Hamkalo and J. B. Rattner.]

and two-thirds turns around the protein core. The length of the linker DNA is more variable among species, and even between different cells of one organism, ranging from about 10 to 90 base pairs. During cell replication, DNA is assembled into nucleosomes shortly after the replication fork passes (see Figure 4-33). This process depends on specific histone **chaperones** that bind to histones and assemble them together with newly replicated DNA into nucleosomes.

**Structure of the 30-nm Fiber**  When extracted from cells in isotonic buffers (i.e., buffers with the same salt concentration found in cells, $\approx$0.15 M KCl, 0.004 M $MgCl_2$), most chromatin appears as fibers $\approx$30 nm in diameter (see Figure 6-28b). Current research, including x-ray crystallography of

nucleosomes assembled from recombinant histones, indicates that the 30-nm fiber has a "zig-zag ribbon" structure that is wound into a "two-start" helix made from two "strands" of nucleosomes stacked on top of each other like coins. The two "strands" of stacked nucleosomes are then wound into a double helix similarly to the two strands in a DNA double helix, except that the helix is left handed, rather than right handed as it is in DNA (Figure 6-30). The 30-nm fibers also include H1, the fifth major histone. H1 is bound to the DNA as it enters and exits the nucleosome core, but its structure in the 30-nm fiber is not known at atomic resolution.

The chromatin in chromosomal regions that are not being transcribed or replicated exists predominantly in the condensed, 30-nm fiber form and in higher-order folded structures

(a)

(b)

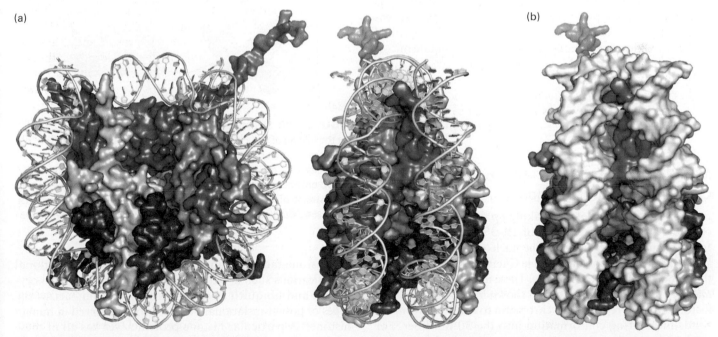

**FIGURE 6-29  Structure of the nucleosome based on x-ray crystallography.** (a) Nucleosome with space-filling model of the histones. The sugar-phosphate backbones of the DNA strands are represented as white tubes to allow better visualization of the histones. Nucleosome shown from the top (*left*) and from the side (*right*, rotated clockwise 90°). H2A subunits are yellow; H2Bs are red; H3s are blue; H4s are green. The N-terminal tails of the eight histones and the H2A and H2B C-terminal tails, involved in condensation of the chromatin, are not visible because they are disordered in the crystal. (b) Space-filling model of histones and DNA (white) viewed from the side of the nucleosome. [Parts (a) and (b) after K. Luger et al., 1997, *Nature* **389**:251.]

whose detailed conformation is not currently understood. The regions of chromatin actively being transcribed are thought to assume the extended beads-on-a-string form.

**Conservation of Chromatin Structure** The general structure of chromatin is remarkably similar in the cells of all eukaryotes, including fungi, plants, and animals, indicating that the structure of chromatin was optimized early in the evolution of eukaryotic cells. The amino acid sequences for four histones (H2A, H2B, H3, and H4) are highly conserved between distantly related species. For example, the sequences of histone H3 from sea urchin tissue and calf thymus differ by only a single amino acid, and H3 from the garden pea and calf thymus differ only in four amino acids. Apparently, significant deviations from the histone amino acid sequences were selected against strongly during evolution. The amino acid sequence of H1, however, varies more from organism to organism than do the sequences of the other major histones. The similarity in sequence among histones from all eukaryotes suggests that they fold into very similar three-dimensional conformations, which were optimized for histone function early in evolution in a common ancestor of all modern eukaryotes.

Minor histone variants encoded by genes that differ from the highly conserved major types also exist, particularly in vertebrates. For example, a special form of H2A, designated H2AX, is incorporated into nucleosomes in place of H2A in a small fraction of nucleosomes in all regions of chromatin. At sites of DNA double-stranded breaks in chromosomal DNA, H2AX becomes phosphorylated and participates in the chromosome-repair process, probably by functioning as a binding site for repair proteins. In the nucleosomes at centromeres, H3 is replaced by another variant histone called CENP-A, which participates in the binding of spindle microtubules during mitosis. Most minor histone variants differ only slightly in sequence from the major histones. These slight changes in histone sequence may influence the stability of the nucleosome as well as its tendency to fold into the 30-nm fiber and other higher-order structures.

## Modifications of Histone Tails Control Chromatin Condensation and Function

Each of the histone proteins making up the nucleosome core contains a flexible N-terminus of 19–39 residues extending from the globular structure of the nucleosome; the H2A and H2B proteins also contain a flexible C-terminus extending from the globular histone octamer core. These termini, called *histone tails,* are represented in the model shown in Figure 6-31a. The histone tails are required for chromatin to condense from the beads-on-a-string conformation into the 30-nm fiber. For example, recent experiments indicate that the N-terminal tails of histone H4, particularly lysine16, are critical for forming the 30-nm fiber. This positively charged lysine interacts with a negative patch at the H2A–H2B interface of the next nucleosome in the stacked nucleosomes of the 30-nm fiber (see Figure 6-30).

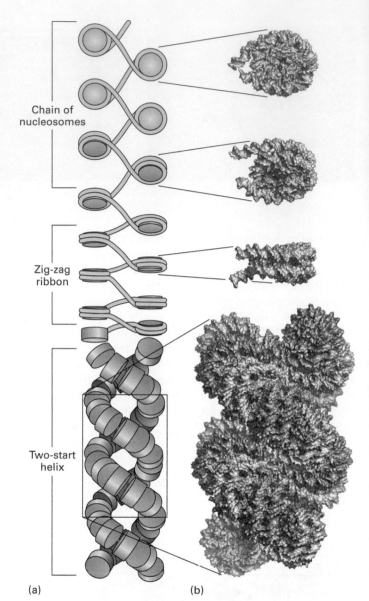

**FIGURE 6-30 Structure of the 30-nm chromatin fiber.** (a) Model for the folding of a nucleosomal chain at top into a "zig-zag ribbon" of nucleosomes that then folds into a two-start helix at bottom. For simplicity, DNA is not represented in the two-start helix. (b) Model of the 30-nm fiber based on x-ray crystallography of a tetranucleosome (a short stretch of four nucleosomes). [Part (a) adapted from C. L. F. Woodcock et al., 1984, *J. Cell Biol.* **99**:42. Part (b) from T. Schalch et al., 2005, *Nature* **436**:138.]

Histone tails are subject to multiple post-translational modifications such as acetylation, methylation, phosphorylation, and ubiquitination. Figure 6-31b summarizes the types of post-translational modifications observed in human histones. A particular histone protein never has all of these modifications simultaneously, but the histones in a single nucleosome usually contain several of these modifications simultaneously. The particular combinations of post-transcriptional modifications found in different regions of chromatin have been suggested to constitute a *histone code* that influences chromatin function by creating or removing binding sites for

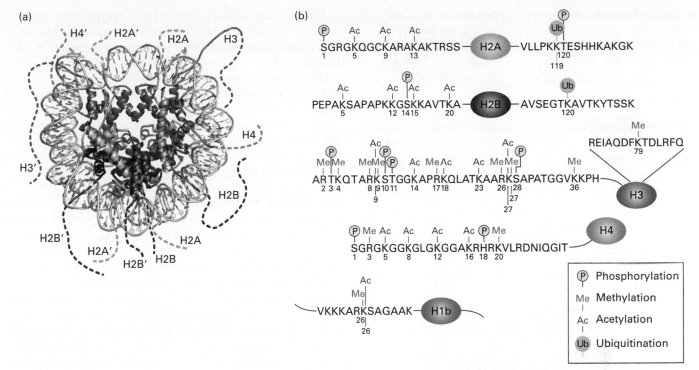

**FIGURE 6-31 Post-translational modifications observed on human histones.** (a) Model of a nucleosome viewed from the top with histones shown as ribbon diagrams. This model depicts the lengths of the histone tails (dotted lines), which are not visible in the crystal structure (see Figure 6-29). The H2A N-terminal tails are at the bottom, and the H2A C-terminal tails are at the top. The H2B N-terminal tails are on the right and left, and C-terminal tails are at the bottom center. Histones H3 and H4 have short C-terminal tails that are not modified.

(b) Summary of post-translational modifications observed in human histones. Histone-tail sequences are shown in the one-letter amino acid code (see Figure 2-14). The main portion of each histone is depicted as an oval. These modifications do not all occur simultaneously on a single histone molecule. Rather, specific combinations of a few these modifications are observed on any one histone. [Part (a) from K. Luger and T. J. Richmond, 1998, *Curr. Opin. Genet. Devel.* **8:**140. Part (b) adapted from R. Margueron et al., 2005, *Curr. Opin. Genet. Devel.* **15:**163.]

chromatin-associated proteins dependent on the specific combinations of these modifications present. Here we describe the most abundant kinds of modifications found in histone tails and how these modifications control chromatin condensation and function. We end with a discussion of a special case of chromatin condensation, the inactivation of X chromosomes in female mammals.

**Histone Acetylation** Histone-tail lysines undergo reversible acetylation and deacetylation by enzymes that act on specific lysines in the N-termini. In the acetylated form, the positive charge of the lysine ε-amino group is neutralized. As mentioned above, lysine 16 in histone H4 is particularly important for the folding of the 30-nm fiber because it interacts with a negatively charged patch on the surface of the neighboring nucleosome in the fiber. Consequently, when H4 lysine 16 is acetylated, the chromatin tends to form the less condensed "beads-on-a-string" conformation conducive for transcription and replication.

Histone acetylation at other sites in H4 and in other histones (see Figure 6-31a) is correlated with increased sensitivity of chromatin DNA to digestion by nucleases. This phenomenon can be demonstrated by digesting isolated nuclei with DNase I. Following digestion, the DNA is completely separated

from chromatin protein, digested to completion with a restriction enzyme, and analyzed by Southern blotting. An intact gene treated with a restriction enzyme yields fragments of characteristic sizes. If isolated nuclei are exposed first to DNase, the gene may be cleaved at random sites within the boundaries of the restriction enzyme cut sites. Consequently, any Southern blot bands normally seen with that gene will be lost. This method was first used to show that the transcriptionally inactive β-globin gene in nonerythroid cells, where it is associated with relatively unacetylated histones, is much more resistant to DNase I than is the active, transcribed β-globin gene in erythroid precursor cells, where it is associated with acetylated histones (Figure 6-32). These results indicate that the chromatin structure of nontranscribed DNA in *hypoacetylated* chromatin makes the DNA less accessible to the small DNase I enzyme ($\approx$10 kD) than it is in transcribed, *hyperacetylated* chromatin. This is thought to be because chromatin containing the repressed gene is folded into condensed structures that sterically inhibit access of the associated DNA to the nuclease. In contrast, the transcribed gene is associated with a more unfolded form of chromatin that allows better access of the nuclease to the associated DNA. Presumably, the condensed chromatin structure in nonerythroid cells also sterically inhibits access of the promoter

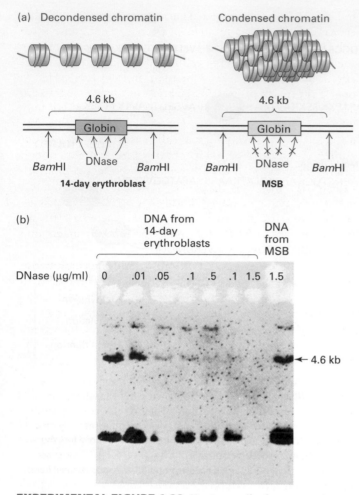

**(a)** Decondensed chromatin    Condensed chromatin

4.6 kb          4.6 kb

Globin              Globin

*Bam*HI  DNase  *Bam*HI    *Bam*HI  DNase  *Bam*HI

**14-day erythroblast**        **MSB**

**(b)**        DNA from 14-day erythroblasts    DNA from MSB

DNase (μg/ml)  0  .01  .05  .1  .5  .1  1.5    1.5

← 4.6 kb

**EXPERIMENTAL FIGURE 6-32 Nontranscribed genes are less susceptible to DNase I digestion than active genes.** Chick embryo erythroblasts at 14 days actively synthesize globin, whereas undifferentiated chicken lymphoblastic leukemia (MSB) cells do not. (a) Nuclei from each type of cell were isolated and exposed to Increasing concentrations of DNase I. The nuclear DNA was then extracted and treated with the restriction enzyme *Bam*HI, which cleaves the DNA around the globin sequence and normally releases a 4.6-kb globin fragment. (b) The DNase I- and *Bam*HI-digested DNA was subjected to Southern blot analysis with a probe of labeled cloned adult globin DNA, which hybridizes to the 4.6-kb *Bam*HI fragment. If the globin gene is susceptible to the initial DNase digestion, it would be cleaved repeatedly and would not be expected to show this fragment. As seen in the Southern blot, the transcriptionally active DNA from the 14-day globin-synthesizing cells was sensitive to DNase I digestion, indicated by the absence of the 4.6-kb band at higher nuclease concentrations. In contrast, the inactive DNA from MSB cells was resistant to digestion. These results suggest that the inactive DNA is in a more condensed form of chromatin in which the globin gene is shielded from DNase digestion. [See J. Stalder et al., 1980, *Cell* **19**:973; photograph courtesy of H. Weintraub.]

and other transcription control sequences in DNA to the proteins involved in transcription, contributing to transcriptional repression (Chapter 7).

Genetic studies in yeast indicated that *histone acetyl transferases (HATs)*, which acetylate specific lysine residues in histones, are required for the full activation of transcription of a number of genes. These enzymes are now known to have other substrates that influence gene expression in addition to histones. Consequently, they are more generally known as *nuclear lysine acetyl transferases, or KATs* because *K* represents lysine in the single-letter code for amino acids (Figure 2-14). Conversely, early genetic studies in yeast indicated that complete repression of many yeast genes requires the action of *histone deacetylases (HDACs)* that remove acetyl groups of acetylated lysines from histone tails, as discussed further in Chapter 7.

**Other Histone Modifications** As shown in Figure 6-31b, histone tails in chromatin can undergo a variety of other covalent modifications at specific amino acids. Lysine ε-amino groups can be methylated, a process that prevents acetylation, thus maintaining their positive charge. Moreover, the N of lysine ε-amino groups can be methylated once, twice, or three times. Arginine side chains can also be methylated. The O in hydroxyl groups (—OH) of serine and threonine side chains can be reversibly phosphorylated, introducing two negative charges. Each of these post-translational modifications contributes to the binding of chromatin-associated proteins that participate in the control of chromatin folding and the ability of DNA and RNA polymerases to replicate or transcribe the associated DNA. Finally, a single 76-amino-acid ubiquitin molecule can be reversibly added to a lysine in the C-terminal tails of H2A and H2B. Recall that addition of multiple linked ubiquitin molecules to a protein can mark it for degradation by the proteasome (see Figure 3-29b). In this case, however, the addition of a single ubiquitin molecule does not affect the stability of a histone, although it does influence chromatin structure.

As mentioned previously, it is the precise combination of modified amino acids in histone tails that helps control the condensation, or compaction, of chromatin and its ability to be transcribed, replicated, and repaired. This can be observed by electron microscopy and light microscopy using dyes that bind DNA. Condensed regions of chromatin known as **heterochromatin** stain much more darkly than less condensed chromatin, known as **euchromatin** (Figure 6-33a). Heterochromatin does not fully decondense following mitosis, remaining in a compacted state during interphase and usually associating with the nuclear envelope, nucleoli, and additional distinct foci. Heterochromatin includes centromeres and telomeres of chromosomes, as well as transcriptionally inactive genes. In contrast, areas of euchromatin, which are in a less compacted state during interphase, stain lightly with DNA dyes. Most transcribed regions of DNA are found in euchromatin. Heterochromatin usually contains histone H3 modified by methylation of lysine 9 or 27, while euchromatin generally contains histone H3 extensively acetylated on lysine 9 and 14, and to a lesser extent at other H3 lysines, methylation of lysine 4, and phosphorylation of serine 10 (Figure 6-33b). Other histone tails are also specifically modified in euchromatin versus heterochromatin. For example, H4 lysine 16 is generally unacetylated in

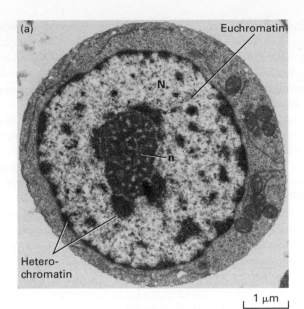

Euchromatin

N

n

Hetero-
chromatin

1 μm

(b)

**Heterochromatin (inactive/condensed)**

Me₃
|
H3   ARTKQTARKSTGGKAPRKQLATKAARKSAPAT
              9

                                    Me₃
                                    |
H3   ARTKQTARKSTGGKAPRKQLATKAARKSAPAT
                                    27

**Euchromatin (active/open)**

Me₃      Ac(P)  Ac   Ac              Ac
|        | |    |    |               |
H3   ARTKQTARKSTGGKAPRKQLATKAARKSAPAT
     4       9 10   14   18          27

**FIGURE 6-33 Heterochromatin versus euchromatin.** (a) In this
electron micrograph of a bone marrow stem cell, the dark-staining
areas in the nucleus (N) outside the nucleolus (n) are heterochromatin.
The light-staining, whitish areas are euchromatin. (b) The modifications
of histone N-terminal tails in heterochromatin and euchromatin differ,
as illustrated here for histone H3. Note in particular that histone tails
are generally much more extensively acetylated in euchromatin
compared with heterochromatin. Heterochromatin is much more
condensed (thus less accessible to proteins) and is much less transcriptionally active than is euchromatin. [Part (a) P. C. Cross and K. L. Mercer,
1993, *Cell and Tissue Ultrastructure,* W. H. Freeman and Company, p. 165. Part
(b) adapted from T. Jenuwein and C. D. Allis, 2001, *Science* **293:**1074.]

heterochromatin, allowing it to interact with neighboring
nucleosomes and stabilize chromatin folding into the 30-nm
fiber (Figure 6-30).

**Reading the Histone Code**  The histone code of modified
amino acids in the histone tails is "read" by proteins that
bind to the modified tails and in turn promote condensation
or decondensation of chromatin, forming "closed" or
"open" chromatin structures, as judged by their sensitivity

to DNase I digestion in isolated nuclei (see Figure 6-32). Higher
eukaryotes express a number of proteins containing a so-
called *chromodomain* that binds to histone tails when they
are methylated at specific lysines. One example is *heterochromatin protein 1 (HP1).* In addition to histones, HP1 is
one of the major proteins associated with heterochromatin.
The HP1 chromodomain binds the H3 N-terminal tail only
when it is tri-methylated at lysine 9 (see Figure 6-33b). HP1
also contains a second domain called a *chromoshadow domain* because it is frequently found in proteins that contain
a chromodomain. The chromoshadow domain binds to
other chromoshadow domains. Consequently, chromatin
containing H3 tri-methylated at lysine 9 (H3K9Me₃) is assembled into a condensed chromatin structure by HP1, although the structure of this chromatin is not well understood
(Figure 6-34a).

In addition to binding to itself, the chromoshadow domain also binds the enzyme that methylates H3 lysine 9, an
H3K9 *histone methyl transferase (HMT).* As a consequence,
nucleosomes adjacent to a region of HP1-containing heterochromatin also become methylated at lysine 9 (Figure 6-34b).
This creates a binding site for another HP1 that can bind the
H3K9 histone methyl transferase, resulting in "spreading"
of the heterochromatin structure along the chromosome
until a *boundary element* is encountered that blocks further
spreading. Boundary elements so far characterized are generally regions in chromatin where several nonhistone proteins
bind to DNA, possibly blocking histone methylation on the
other side of the boundary.

Significantly, the model of heterochromatin formation in
Figure 6-34b provides an explanation for how heterochromatic regions of a chromosome are reestablished following
DNA replication during the S phase of the cell cycle. When
DNA in heterochromatin is replicated, the histone octamers
that are tri-methylated at H3 lysine 9 become distributed to
both daughter chromosomes along with an equal number of
newly assembled histone octamers. The H3K9 histone
methyl transferase associated with the H3K9 tri-methylated
nucleosomes methylate lysine 9 of the newly assembled nucleosomes, regenerating the heterochromatin in both daughter chromosomes. Consequently, heterochromatin is marked
with an *epigenetic code,* so called because it does not depend
on the sequence of bases in DNA that maintains the repression of associated genes in replicated daughter cells.

Other protein domains associate with histone-tail modifications typical of euchromatin. For example, the *bromodo-
main* binds to acetylated histone tails and therefore is
associated with transcriptionally active chromatin. Several
proteins involved in stimulating gene transcription contain
bromodomains, such as the largest subunit of TFIID (see
Chapter 7). This *transcription factor* contains two closely
spaced bromodomains that probably help TFIID to associate
with transcriptionally active chromatin (i.e., euchromatin).
This protein and other bromodomain-containing proteins
also have histone acetylase activity, which helps to maintain
the chromatin in a hyperacetylated state conducive to transcription. Consequently, an epigenetic code associated with

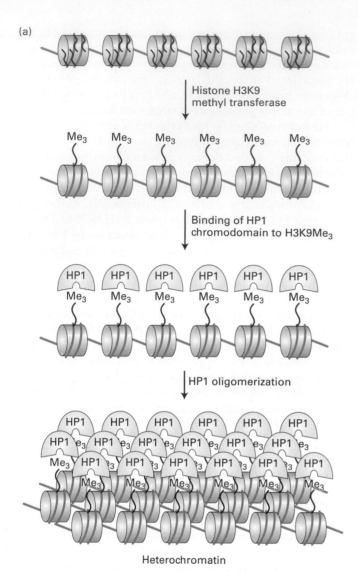

(a)

Histone H3K9
methyl transferase

Me₃  Me₃  Me₃  Me₃  Me₃  Me₃

Binding of HP1
chromodomain to H3K9Me₃

HP1  HP1  HP1  HP1  HP1  HP1
Me₃  Me₃  Me₃  Me₃  Me₃  Me₃

HP1 oligomerization

Heterochromatin

(b)

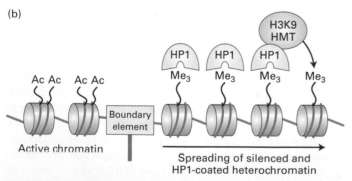

Ac Ac    Ac Ac

Boundary
element

Active chromatin

Spreading of silenced and
HP1-coated heterochromatin

**FIGURE 6-34 Model for the formation of heterochromatin by the binding of HP1 to histone H3 trimethylated at lysine 9.** (a) HP1 contributes to the condensation of heterochromatin by binding to histone H3 N-terminal tails tri-methylated at lysine 9, followed by association of the histone-bound HP1. (b) Heterochromatin condensation can spread along a chromosome because HP1 binds a histone methyltransferase (HMT) that methylates lysine 9 of histone H3. This creates a binding site for HP1 on the neighboring nucleosome. The spreading process continues until a "boundary element" is encountered. [Part (a) adapted from G. Thiel et al., 2004, *Eur. J. Biochem.* **271:**2855. Part (b) adapted from A. J. Bannister et al., 2001, *Nature* **410:**120.]

euchromatin helps to maintain the transcriptional activity of genes in euchromatin through successive cell divisions. These epigenetic codes for heterochromatin and euchromatin help to maintain the patterns of gene expression established in different cell types during early embryonic development as specific differentiated cell types increase in numbers by cell division. Importantly, abnormal alterations in these epigenetic codes have been found to contribute to the pathogenic replication and behavior of cancer cells (Chapter 24).

In summary, multiple types of covalent modifications of histone tails can influence chromatin structure by altering nucleosome–nucleosome interactions and interactions with additional proteins that participate in or regulate processes such as transcription and DNA replication. The mechanisms and molecular processes governing chromatin modifications that regulate transcription are discussed in greater detail in the next chapter.

**X-Chromosome Inactivation in Mammalian Females** One important example of epigenetic gene control through repression by heterochromatin is the random inactivation and condensation of one of the two X chromosomes in female mammals. Each female mammal has two X chromosomes, one contributed by the egg from which it developed ($X_m$) and one contributed by the sperm ($X_p$). Early during embryonic development, random inactivation of either the $X_m$ or the $X_p$ chromosome occurs in each somatic cell. In the female embryo, about half the cells have an inactive $X_m$, and the other half have an inactive $X_p$. All subsequent daughter cells maintain the same inactive X chromosomes as their parent cells. As a result, the adult female is a mosaic of clones, some expressing the genes from the $X_m$ and the rest expressing the genes from the $X_p$. This inactivation of one X chromosome in female mammals results in *dosage compensation,* the process that ensures that cells of females express the same level of proteins encoded on the X chromosome as the cells of males, which have only one X chromosome.

Histones associated with the inactive X chromosome have post-translational modifications characteristic of other regions of heterochromatin: hypoacetylation of lysines, di- and tri-methylation of histone H3 lysine 9, tri-methylation of H3 lysine 27, and a lack of methylation at histone H3 lysine 4 (see Figure 6-33b). X-chromosome inactivation at an early stage in embryonic development is controlled by the X-inactivation center, a complex locus on the X chromosome that determines which of the two X chromosomes will be inactivated and in which cells. The X-inactivation center also contains the *Xist* gene, which encodes a remarkable, long, non-protein-coding RNA that coats only the X chromosome it was transcribed from, thereby triggering silencing of the chromosome.

Although the mechanism of X-chromosome inactivation is not fully understood, it involves several processes including the action of *Polycomb* protein complexes that are discussed further in Chapter 7. One subunit of the Polycomb complex contains a chromodomain that binds to histone H3 tails when they are tri-methylated at lysine 27. The Polycomb complex also contains a histone methyl transferase specific for H3 lysine 27. This finding helps to explain how the X-inactivation process spreads along large regions of the X chromosome and how it is maintained through DNA replication, similar to heterochromatization by the binding of HP1 to histone H3 tails methylated at lysine 9 (see Figure 6-34b).

X-chromosome inactivation is another example of an epigenetic process, that is, a process that affects the expression of specific genes and is inherited by daughter cells but is not the result of a change in DNA sequence. Instead, the activity of genes on the X chromosome in female mammals is controlled by chromatin structure rather then the nucleotide sequence of the underlying DNA. And the inactivated X chromosome (either $X_m$ or $X_p$) is maintained as the inactive chromosome in the progeny of all future cell divisions because the histones are modified in a specific, repressing manner that is faithfully inherited through each cell division.

## Nonhistone Proteins Organize Long Chromatin Loops

Although histones are the predominant proteins in chromatin, less abundant, nonhistone chromatin-associated proteins, and the DNA molecule itself, are also crucial to chromosome structure. Recent results indicate that it is not protein alone that gives a metaphase chromosome its structure. Micromechanical studies of large metaphase chromosomes from newts in the presence of proteases or nucleases indicate that DNA, not protein, is responsible for the mechanical integrity of a metaphase chromosome when it is pulled from its ends. These results are inconsistent with a continuous protein scaffold at the chromosome axis. Rather, the integrity of chromosome structure requires the complete chromatin complex of DNA, histone octamers, and nonhistone chromatin-associated proteins.

In situ hybridization experiments with several different fluorescent-labeled probes to the DNA of one chromosome in human interphase cells support a model in which chromatin is arranged in large loops. In these experiments, some probe sequences separated by millions of base pairs in linear DNA appeared reproducibly very close to one another in interphase nuclei from different cells of the same type (Figure 6-35). These closely spaced probe sites are postulated to lie close to regions of chromatin, called *scaffold-associated regions (SARs)* or *matrix-attachment regions (MARs)*, located at the bases of the DNA loops. SARs/MARs have been mapped by digesting histone-depleted chromosomes with restriction enzymes and then recovering the fragments that remain associated with the digested histone-depleted preparation. The measured distances between probes are consistent with chromatin loops ranging in size from 1 million to 4 million base pairs in mammalian interphase cells. Loops of chromatin are also directly visualized by light microscopy in the active chromatin of growing amphibian oocytes ("lampbrush chromosomes," shown in the figure at the beginning of Chapter 8). These cells are enormous compared to most cells (≈1 mm in diameter) because they stockpile all

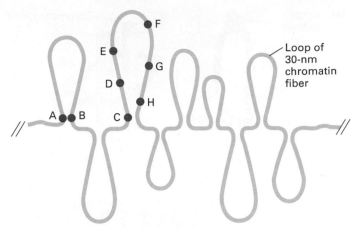

**EXPERIMENTAL FIGURE 6-35 Fluorescent-labeled probes hybridized to interphase chromosomes demonstrate chromatin loops and permit their measurement.** In situ hybridization of interphase cells was carried out with several different probes specific for sequences separated by known distances in linear, cloned DNA. Lettered circles represent probes. Measurement of the distances between different hybridized probes, which could be distinguished by their color, showed that some sequences (e.g., A and B), separated from one another by millions of base pairs, appear located near one another within nuclei. For some sets of sequences, the measured distances in nuclei between one probe (e.g., C) and sequences successively farther away initially appear to increase (e.g., D, E, and F) and then appear to decrease (e.g., G and H). [Adapted from H. Yokota et al., 1995, *J. Cell Biol.* **130**:1239.]

of the nuclear and cytoplasmic material required for division of the fertilized egg into the thousands of differentiated cells required to generate a feeding embryo that can ingest additional nutrients.

In general, SARs/MARs are found between transcription units, and genes are located primarily within the chromatin loops. As discussed below, the loops are tethered at their bases by a mechanism that does not break the duplex DNA molecule that extends the entire length of the chromosome. Evidence indicates that SARs/MARs may insulate neighboring genes. Some SARs/MARs function as **insulators,** that is, DNA sequences of tens to hundreds of base pairs that separate transcription units from each other. Proteins regulating transcription of one gene cannot influence the transcription of a neighboring gene that is separated from it by an insulator.

**Ringlike Structure of SMC Protein Complexes** The bases of chromatin loops (see Figure 6-35) in interphase chromosomes may be held in place by proteins called *structural maintenance of chromosome proteins,* or SMC proteins. These nonhistone proteins are critical for maintaining the morphological structure of condensed chromosomes during mitosis. In extracts prepared from the large nuclei of *Xenopus laevus* (African frog) eggs, chromosomes can be induced to condense as they do in intact cells as they enter the prophase period of mitosis. This condensation fails to occur when one type of SMC protein is depleted from the extract with specific

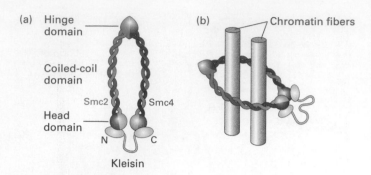

(a) Hinge domain
Coiled-coil domain
Smc2   Smc4
Head domain
N   C
Kleisin

(b) Chromatin fibers

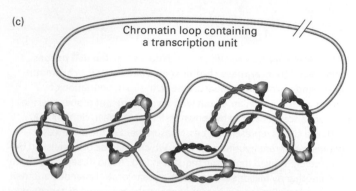

(c)
Chromatin loop containing a transcription unit

**FIGURE 6-36 Model of SMC complexes bound to chromatin.**
(a) Model of an SMC protein complex. (b) Model of SMC complex topologically linking two chromatin fibers represented by cylinders with the diameter of a nucleosome relative to the dimensions of the SMC complex. (c) Model for the binding of SMC complexes to the base of a loop of transcribed chromatin. [Adapted from K. Nasmyth and C. H. Haering, 2005, *Ann. Rev. Biochem.* **74:**595.]

antibodies. Yeast with mutations in certain SMC proteins fail to properly associate daughter chromatids following DNA replication in the S phase. As a result, chromosomes do not properly segregate to daughter cells during mitosis. Related SMC proteins are required for proper segregation of chromosomes in bacteria and archaea, indicating that this is an ancient class of proteins vital to chromosome structure and segregation in all kingdoms of life.

Each SMC monomer contains a hinge region where the polypeptide folds back on itself, forming a very long, coiled coil region and bringing the N- and C-termini together so they can interact to form a globular head domain (Figure 6-36a).

**EXPERIMENTAL FIGURE 6-37 During interphase, human chromosomes remain in nonoverlapping territories in the nucleus.**
Fixed interphase human fibroblasts were hybridized in situ to fluorescently labeled probes specific for sequences along the full length of human chromosomes 7 (cyan) and 8 (purple). DNA is stained blue with DAPI. In the diploid cell, each of the two chromosome 7s and two chromosome 8s is restricted to a territory or domain within the nucleus, rather than stretching throughout the entire nucleus. (b) Similar to (a) except that chromosome paint probes specific for each chromosome were hybridized to reveal the location of nearly all of the chromosomes in a fibroblast from a human male. Some of the chromosomes are not observed in this confocal slice through the nucleus. [Part (a) courtesy of Drs. I. Solovei and T. Cremer. Part (b) from A. Bolzer et al., 2005, *PLOS Biol* **3:**826.]

The hinge domain of one monomer (blue) binds to the hinge domain of a second monomer (red), forming a roughly U-shaped dimeric complex. The head domains of the monomers have ATPase activity and are linked by members of another small protein family called *kleisins*. The overall SMC complex is a ring with a diameter large enough to accommodate two 30-nm chromatin fibers (Figure 6-36b), and is capable of linking two circular DNA molecules in vitro. SMC proteins are proposed to form the base of chromatin loops by forming topologically constrained knots in 30-nm chromatin fibers, as diagrammed in Figure 6-36c. This can explain why cleavage of the DNA at a relatively small number of sites leads to rapid dissolution of condensed metaphase chromosome structure, whereas protease cleavage of proteins has only a minor effect on chromosome structure until most of the protein is digested: When the DNA is cut anywhere in a long region of chromatin containing several chromatin loops, the

(a)

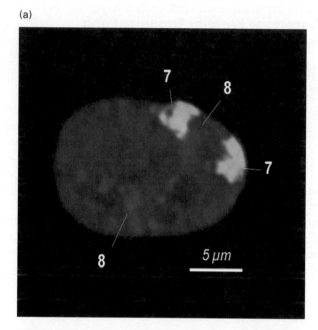

(b)

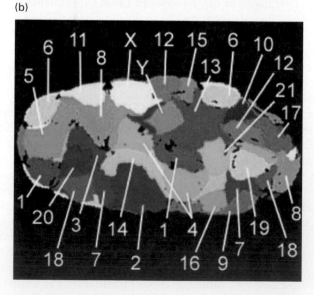

broken ends can slip through the SMC protein rings, "untying" the topological knots that constrain the loops of chromatin. In contrast, most of the individual rings of SMC proteins must be broken before the topological constraints holding the base of the loops together is released.

**Interphase Chromosome Territories** In the small nuclei of most cells, individual interphase chromosomes, which are less condensed than metaphase chromosomes, cannot be resolved by standard microscopy or electron microscopy. Nonetheless, the chromatin of one chromosome in interphase cells is not spread throughout the nucleus. Rather, interphase chromatin is organized into *chromosome territories*. As illustrated in Figure 6-37, in situ hybridization of interphase nuclei with chromosome-specific fluorescent-labeled probes shows that the probes are visualized within restricted regions of the nucleus rather than appearing throughout the nucleus. Use of probes specific for different chromosomes shows that there is little overlap between chromosomes in interphase nuclei. However, the precise positions of chromosomes are not reproducible between cells.

**Metaphase Chromosome Structure** Condensation of chromosomes during prophase may involve the formation of many more loops of chromatin, so that the length of each loop is greatly reduced compared to that in interphase cells. However, the folding of chromatin in metaphase chromosomes is not well understood. Microscopic analysis of mammalian chromosomes as they condense during prophase indicates that the 30-nm fiber folds into a 100- to 130-nm fiber called a

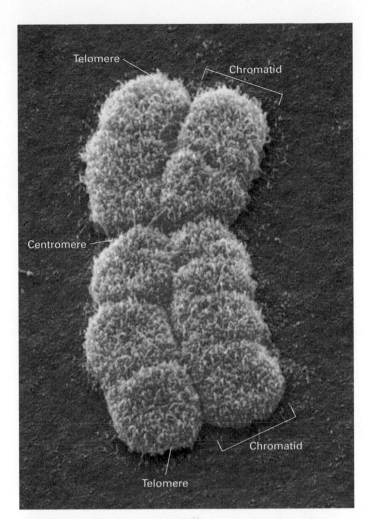

**FIGURE 6-39 Typical metaphase chromosome.** As seen in this scanning electron micrograph, the chromosome has replicated and comprises two chromatids, each containing one of two identical DNA molecules. The centromere, where chromatids are attached at a constriction, is required for their separation late in mitosis. Special telomere sequences at the ends function in preventing chromosome shortening. [Andrew Syred/Photo Researchers, Inc.]

*chromonema* fiber. As depicted in Figure 6-38, a chromonema fiber then folds into a structure with a diameter of 200–250 nm called a *middle prophase chromatid*, which then folds into the 500- to 750-nm-diameter chromatids observed during metaphase. Ultimately, the full lengths of two associated daughter chromosomes generated by DNA replication during the previous S phase of the cell cycle condense into bar-shaped structures that in most eukaryotes are linked at the central constriction called the centromere (Figure 6-39).

## Additional Nonhistone Proteins Regulate Transcription and Replication

The total mass of the histones associated with DNA in chromatin is about equal to that of the DNA. Interphase chromatin and metaphase chromosomes also contain small amounts of a complex set of other proteins. For instance, thousands

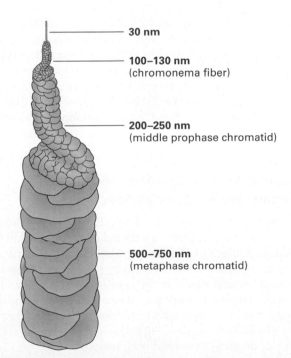

**FIGURE 6-38 Model for the folding of the 30-nm chromatin fiber in a metaphase chromosome.** A single chromatid of a metaphase chromosome is depicted. [Adapted from N. Kireeva et al., 2004, *J. Cell Biol.* **166**:775.]

30 nm

100–130 nm
(chromonema fiber)

200–250 nm
(middle prophase chromatid)

500–750 nm
(metaphase chromatid)

of different **transcription factors** are associated with interphase chromatin. The structure and function of these critical non-histone proteins, which regulate transcription, are examined in Chapter 7. Other low-abundance nonhistone proteins associated with chromatin regulate DNA replication during the eukaryotic cell cycle (Chapter 20).

A few other nonhistone DNA-binding proteins are present in much larger amounts than the transcription or replication factors. Some of these exhibit high mobility during electrophoretic separation and thus have been designated *HMG (high-mobility group) proteins*. When genes encoding the most abundant HMG proteins are deleted from yeast cells, normal transcription is disturbed in most genes examined. Some HMG proteins have been found to assist in the cooperative binding of several transcription factors to specific DNA sequences that are close to each other, stabilizing multiprotein complexes that regulate transcription of a neighboring gene, as discussed in Chapter 7.

## KEY CONCEPTS of Section 6.6

### Structural Organization of Eukaryotic Chromosomes

• In eukaryotic cells, DNA is associated with about an equal mass of histone proteins in a highly condensed nucleoprotein complex called chromatin. The building block of chromatin is the nucleosome, consisting of a histone octamer around which is wrapped 147 bp of DNA (see Figure 6-29).

• The chromatin in transcriptionally inactive regions of DNA within cells is thought to exist in a condensed, 30-nm fiber form and higher-order structures built from it (see Figure 6-30 and 6-38).

• The chromatin in transcriptionally active regions of DNA within cells is thought to exist in an open, extended form.

• The histone H4 tails, particularly H4 lysine 16, are required for beads-on-a-string chromatin (the 10-nm chromatin fiber) to fold into a 30-nm fiber.

• Histone tails can be modified by acetylation, methylation, phosphorylation, and monoubiquitination (see Figure 6-31). These modifications influence chromatin structure by regulating the binding of histone tails to other, less abundant chromatin-associated proteins.

• The reversible acetylation and deacetylation of lysine residues in the N-termini of the core histones regulates chromatin condensation. Proteins involved in transcription, replication, and repair, and enzymes such as DNaseI can more easily access chromatin with hyperacetylated histone tails (euchromatin) than chromatin with hypoacetylated histone tails (heterochromatin).

• When metaphase chromosomes decondense during interphase, areas of heterochromatin remain much more condensed than regions of euchromatin.

• Heterochromatin protein 1 (HP1) uses a chromodomain to bind to histone H3 tri-methylated on lysine 9. The chro-moshadow domain of HP1 also associates with itself and with the histone methyl transferase that methylates H3 lysine 9. These interactions cause condensation of the 30-nm chromatin fiber and spreading of the heterochromatic structure along the chromosome until a boundary element is encountered (see Figure 6-34).

• One X chromosome in nearly every cell of mammalian females is highly condensed heterochromatin, resulting in repression of expression of nearly all genes on the inactive chromosome. This inactivation results in dosage compensation so that genes on the X chromosome are expressed at the same level in both males and females.

• Each eukaryotic chromosome contains a single DNA molecule packaged into nucleosomes and folded into a 30-nm chromatin fiber, which is associated with a protein scaffold made up in part of structural maintenance of chromosome (SMC) proteins at sites between transcription units (see Figure 6-36c). Additional folding of the scaffold further compacts the structure into the highly condensed form of metaphase chromosomes (see Figure 6-38).

## 6.7 Morphology and Functional Elements of Eukaryotic Chromosomes

Having examined the detailed structural organization of chromosomes in the previous section, we now view them from a more global perspective. Early microscopic observations on the number and size of chromosomes and their staining patterns led to the discovery of many important general characteristics of chromosome structure. Researchers subsequently identified specific chromosomal regions critical to their replication and segregation to daughter cells during cell division. In this section we discuss these functional elements of chromosomes and consider how chromosomes evolved through rare rearrangements of ancestral chromosomes.

### Chromosome Number, Size, and Shape at Metaphase Are Species-Specific

As noted previously, in nondividing cells, individual chromosomes are not visible, even with the aid of histologic stains for DNA (e.g., Feulgen or Giemsa stains) or electron microscopy. During mitosis and meiosis, however, the chromosomes condense and become visible in the light microscope. Therefore, almost all cytogenetic work (i.e., studies of chromosome morphology) has been done with condensed metaphase chromosomes obtained from dividing cells—either somatic cells in mitosis or dividing gametes during meiosis.

The condensation of metaphase chromosomes probably results from several orders of folding of 30-nm chromatin fibers (see Figure 6-38). At the time of mitosis, cells have already progressed through the S phase of the cell cycle and have replicated their DNA. Consequently, the chromosomes that become visible during metaphase are *duplicated* structures. Each metaphase chromosome consists of two sister **chromatids,** which are linked at a constricted region, the centromere (see Figure 6-39). The number, sizes, and shapes of the metaphase chromosomes constitute the **karyotype,** which is distinctive for each species. In most organisms, all somatic cells have the same karyotype. However, species that appear quite similar can have very different karyotypes, indicating that similar genetic potential can be organized on chromosomes in very different ways. For example, two species of small deer—the Indian muntjac and Reeves muntjac—contain about the same total amount of genomic DNA. In one species, this DNA is organized into 22 pairs of homologous **autosomes** and two physically separate sex chromosomes. In contrast, the other species contains the smallest number of chromosomes of any mammal, only three pairs of autosomes; one sex chromosome is physically separate, but the other is joined to the end of one autosome.

## During Metaphase, Chromosomes Can Be Distinguished by Banding Patterns and Chromosome Painting

Certain dyes selectively stain some regions of metaphase chromosomes more intensely than other regions, producing characteristic banding patterns that are specific for individual chromosomes. The regularity of chromosomal bands serves as useful visible landmarks along the length of each chromosome and can help to distinguish chromosomes of similar size and shape.

Today, the method of *chromosome painting* greatly simplifies differentiating chromosomes of similar size and shape. This technique, a variation of **fluorescence in situ hybridization (FISH),** makes use of probes specific for sites scattered along the length of each chromosome. The probes are labeled with several different fluorescent dyes with distinct excitation and emission wavelengths. Probes specific for each chromosome are labeled with a predetermined fraction of each of the dyes. After the probes are hybridized to chromosomes and the excess removed, the sample is observed with a fluorescence microscope in which a detector determines the fraction of each dye present at each fluorescing position in the microscopic field. This information is conveyed to a computer, and a special program assigns a false-color image to each type of chromosome (Figure 6-40, *left*). Computer graphics allows the two homologs of each chromosome to be placed next to each other and named according to their decreasing size. Such a display clearly displays the cell's karyotype. Figure 6-40 shows a normal human male karyotype. Chromosomal painting is a powerful method for detecting an abnormal number of chromosomes, such as chromosome 21 trisomy in patients with Down syndrome, or chromosomal translocations that occur in rare individuals and in cancer cells (Figure 6-41). The use of probes with different ratios of fluorescent dyes that hybridize to distinct positions along each normal human chromosome allows finer structure analysis of the chromosomes that can more readily reveal deletions or duplications of chromosomal regions. The figure at the beginning of the chapter illustrates the use of such *multicolor FISH* in analysis of the karyotype of a normal human female.

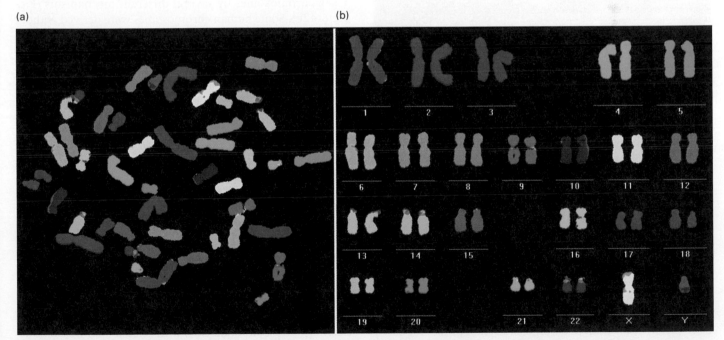

(a) (b)

**EXPERIMENTAL FIGURE 6-40 Human chromosomes are readily identified by chromosome painting.** (a) Fluorescence in situ hybridization (FISH) of human chromosomes from a male cell in mitosis using chromosome paint probes. (b) Alignment of these painted chromosomes by computer graphics to reveal the normal human male karyotype. [Courtesy of M. R. Speicher.]

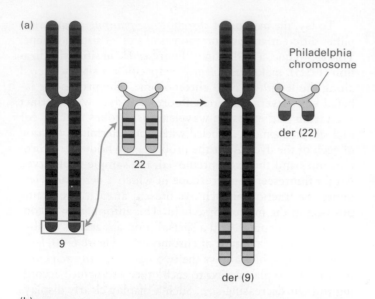

(a)

Philadelphia
chromosome

22

9

der (22)

der (9)

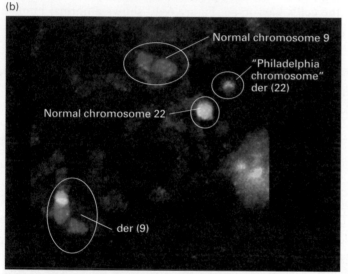

(b)

Normal chromosome 9

"Philadelphia
chromosome"
der (22)

Normal chromosome 22

der (9)

**EXPERIMENTAL FIGURE 6-41 Chromosomal translocations can be analyzed using chromosome paint probes for FISH.**
Characteristic chromosomal translocations are associated with certain genetic disorders and specific types of cancers. For example, in nearly all patients with chronic myelogenous leukemia, the leukemic cells contain the Philadelphia chromosome, a shortened chromosome 22 [der (22)], and an abnormally long chromosome 9 [der (9)] ("der" stands for derivative). These result from a translocation between normal chromosomes 9 and 22. This translocation can be detected by classical banding analysis diagrammed in (a) and FISH with chromosome paint probes (b). [Part (a) from J. Kuby, 1997, *Immunology*, 3d ed., W. H. Freeman and Company, p. 578. Part (b) courtesy of J. Rowley and R. Espinosa.]

## Chromosome Painting and DNA Sequencing Reveal the Evolution of Chromosomes

Analysis of chromosomes from different species has provided considerable insight about how chromosomes evolved. For example, hybridization of chromosome paint probes for chromosome 16 of the tree shrew *(Tupaia belangeri)* to tree shrew metaphase chromosomes revealed the two copies of chromosome 16, as expected (Figure 6-42a). However, when the same chromosome paint probes were hybridized to human metaphase chromosomes, most of the probes hybridized to the long arm of chromosome 10 (Figure 6-42b). Further, when multiple probes from the long arm of human chromosome 10 with different fluorescent dye labels were hybridized to human chromosome 10 and tree shrew metaphase chromosomes, tree shrew sequences homologous to each of these probes were found along tree shrew chromosome 16 in the same order that they occur on human chromosome 10.

These results indicate that during the evolution of humans and tree shrews from a common ancestor that lived ≈85 million years ago, a long, continuous DNA sequence on one of the ancestral chromosomes became chromosome 16 in tree shrews, but evolved into the long arm of chromosome 10 in humans. The phenomenon of genes occurring in the same order on a chromosome in two different species is referred to as conserved **synteny** (derived from Latin for "on the same ribbon"). The presence of two or more genes in a common chromosomal region in two or more species indicates a conserved syntenic segment.

The relationships between the chromosomes of many primates have been determined by cross-species hybridizations of chromosome paint probes as shown for human and tree shrew in Figure 6-42a, b. From these relationships and higher-resolution analyses of regions of synteny by DNA sequencing and other methods, it has been possible to propose the karyotype of the common ancestor of all primates based on the minimum number of chromosomal rearrangements necessary to generate the regions of synteny in chromosomes of contemporary primates.

Human chromosomes are thought to have derived from a common primate ancestor with 23 autosomes plus the X and Y sex chromosomes by several different mechanisms (Figure 6-42c). Some human chromosomes were derived without large-scale rearrangements of chromosome structure. Others are thought to have evolved by breakage of an ancestral chromosome into two chromosomes or, conversely, by fusion of two ancestral chromosomes. Still other human chromosomes appear to have been generated by exchanges of parts of the arms of distinct chromosomes, that is, by reciprocal translocation involving two ancestral chromosomes. Analysis of regions of conserved synteny between the chromosomes of many mammals indicates that chromosomal rearrangements such as breakage, fusion, and translocations occurred rarely in mammalian evolution, about once every five million years. When such chromosomal rearrangements did occur, they very likely contributed to the evolution of new species that could not interbreed with the species from which they evolved.

Chromosomal rearrangements similar to those inferred for the primate lineage have been inferred for other groups of related organisms, including invertebrate, plant, and fungi lineages. The excellent agreement between predictions of evolutionary relationships based on analysis of syntenic regions of chromosomes from organisms with related anatomical structure (i.e., among mammals, among insects with similar body organization, among similar plants, etc.) and the evolutionary relationships based on the fossil record and on the extent of

(a)

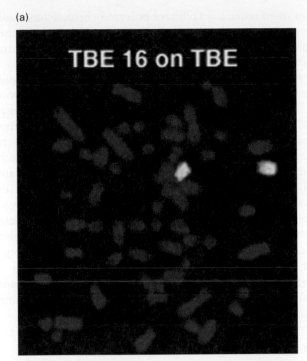

(b)

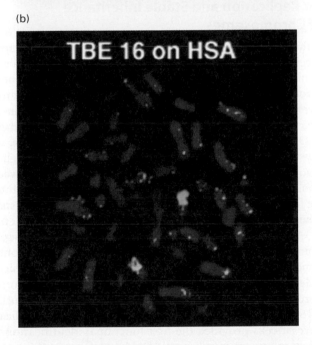

(c)

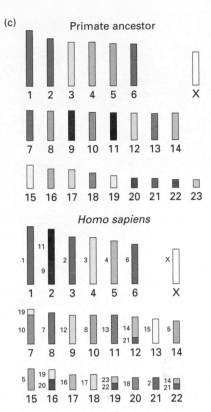

Primate ancestor

Homo sapiens

**FIGURE 6-42 Evolution of primate chromosomes.** (a) Chromosome paint probes for chromosome 16 of the tree shrew (*T. belangeri*, distantly related to humans) were hybridized (yellow) to tree shrew metaphase chromosomes (red). (b) The same tree shrew chromosome 16 paint probes were hybridized to human metaphase chromosomes. (c) Proposed evolution of human chromosomes (*bottom*) from the chromosomes of the common ancestor of all primates (*top*). The proposed common primate ancestor chromosomes are numbered according to their sizes, with each chromosome represented by a different color. The human chromosomes are also numbered according to their relative sizes with colors taken from the colors of the proposed common primate ancestor chromosomes from which they were derived. Small numbers to the left of the colored regions of the human chromosomes indicate the number of the ancestral chromosome from which the region was derived. Human chromosomes were derived from the proposed chromosomes of the common primate ancestor without significant rearrangements (e.g., human chromosome 1), by fusion (e.g., human chromosome 2 by fusion of ancestral chromosomes 9 and 11), breakage (e.g., human chromosomes 14 and 15 by breakage of ancestral chromosome 5), or chromosomal translocations (e.g., a reciprocal translocation between ancestral chromosomes 14 and 21 generated human chromosomes 12 and 22). [Parts (a) and (b) Muller et al. 1999. *Cromosoma* **108**:393. Part (c) derived from L. Froenicke, 2005, *Cytogenet. Genome Res.* **108**:122.]

divergence of DNA sequences for homologous genes is a strong argument for the validity of evolution as the process that generated the diversity of contemporary organisms.

## Interphase Polytene Chromosomes Arise by DNA Amplification

The larval salivary glands of *Drosophila* species and other dipteran insects contain enlarged interphase chromosomes that are visible in the light microscope. When fixed and stained with a dye that stains DNA, these **polytene chromo-** somes are characterized by a large number of reproducible, well-demarcated bands that have been assigned standardized numbers (Figure 6-43a). The densely staining bands represent regions where the chromatin is more condensed, and the light, interband regions, where chromatin is less condensed.

(a)

Chromocenter

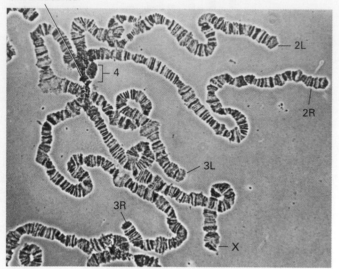

2L

4

2R

3L

3R

X

(b)

Centromere

Telomere

Telomere

**EXPERIMENTAL FIGURE 6-43 Banding on *Drosophila* polytene salivary gland chromosomes.** (a) In this light micrograph of *Drosophila melanogaster* larval salivary gland chromosomes, four chromosomes can be observed (X, 2, 3, and 4), with a total of approximately 5000 distinguishable bands. The banding pattern results from reproducible packing of DNA and protein within each amplified site along the chromosome. Dark bands are regions of more highly compacted chromatin. The centromeres of all four chromosomes often appear fused at the chromocenter. The tips of chromosomes 2 and 3 are labeled (L = left arm; R = right arm), as is the tip of the X chromosome. (b) The pattern of amplification of one chromosome during five replications. Double-stranded DNA is represented by a single line. Telomere and centromere DNA are not amplified. In salivary gland polytene chromosomes, each parental chromosome undergoes ≈10 replications ($2^{10}$ = 1024 strands). [Part (a) courtesy of J. Gall. Part (b) adapted from C. D. Laird et al., 1973, *Cold Spring Harbor Symp. Quant. Biol.* **38**:311.]

Although the molecular mechanisms that control the formation of bands in polytene chromosomes are not yet understood, the highly reproducible banding pattern seen in *Drosophila* salivary gland chromosomes provides an extremely powerful method for locating specific DNA sequences along the lengths of the chromosomes in this species. Chromosomal translocations and inversions are readily detectable in polytene chromosomes, and specific chromosomal proteins can be localized on interphase polytene chromosomes by immunostaining with specific antibodies raised against

them (see Figure 7-13). Insect polytene chromosomes offer one of the only experimental systems in all of nature where such immuno-localization studies on decondensed interphase chromosomes are possible.

A generalized amplification of DNA gives rise to the polytene chromosomes found in the salivary glands of *Drosophila*. This process, termed *polytenization*, occurs when the DNA repeatedly replicates everywhere except at the telomeres and centromere, but the daughter chromosomes do not separate. The result is an enlarged chromosome composed of many parallel copies of itself, 1024 resulting from ten such replications in *Drosophila melanogaster* salivary glands (Figure 6-43b). The amplification of chromosomal DNA greatly increases gene copy number, presumably to supply sufficient mRNA for protein synthesis in the massive salivary gland cells. The bands in *Drosophila* polytene chromosomes represent ≈50,000–100,000 base pairs, and the banding pattern reveals that the condensation of DNA varies greatly along these relatively short regions of an interphase chromosome.

## Three Functional Elements Are Required for Replication and Stable Inheritance of Chromosomes

Although chromosomes differ in length and number among species, cytogenetic studies have shown that they all behave similarly at the time of cell division. Moreover, any eukaryotic chromosome must contain three functional elements in order to replicate and segregate correctly: (1) **replication origins** at which DNA polymerases and other proteins initiate synthesis of DNA (see Figures 4-31 and 4-33); (2) the **centromere**, the constricted region required for proper segregation of daughter chromosomes; and (3) the two ends, or **telomeres.** The yeast transformation studies depicted in Figure 6-44 demonstrated the functions of these three chromosomal elements and established their importance for chromosome function.

As discussed in Chapter 4, replication of DNA begins from sites that are scattered throughout eukaryotic chromosomes. The yeast genome contains many ≈100-bp sequences, called *autonomously replicating sequences (ARSs)*, that act as replication origins. The observation that insertion of an ARS into a circular plasmid allows the plasmid to replicate in yeast cells provided the first functional identification of origin sequences in eukaryotic DNA (see Figure 6-44a).

Even though circular ARS-containing plasmids can replicate in yeast cells, only about 5–20 percent of progeny cells contain the plasmid because mitotic segregation of the plasmids is faulty. However, plasmids that also carry a CEN sequence, derived from the centromeres of yeast chromosomes, segregate equally or nearly so to both mother and daughter cells during mitosis (see Figure 6-44b).

If circular plasmids containing an ARS and CEN sequence are cut once with a restriction enzyme, the resulting linear plasmids do not produce *LEU*⁺ colonies unless they contain special telomeric (TEL) sequences ligated to their ends (see Figure 6-44c). The first successful experiments involving

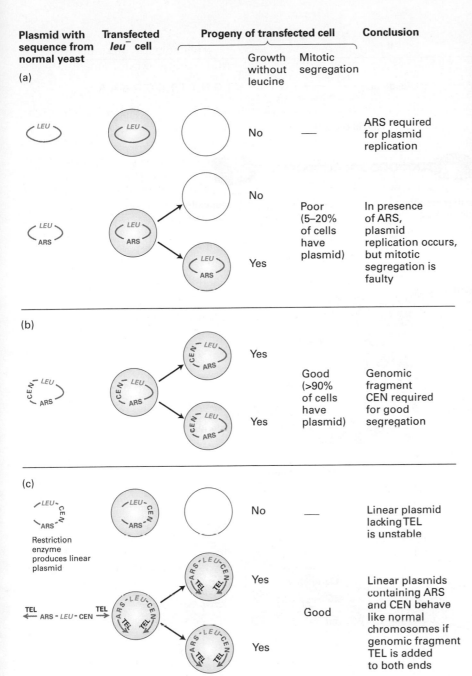

| Plasmid with sequence from normal yeast | Transfected leu⁻ cell | Progeny of transfected cell | | Conclusion |
|---|---|---|---|---|
| | | Growth without leucine | Mitotic segregation | |
| **(a)** | | | | |
| LEU | LEU | No | — | ARS required for plasmid replication |
| LEU / ARS | LEU / ARS | No / Yes | Poor (5–20% of cells have plasmid) | In presence of ARS, plasmid replication occurs, but mitotic segregation is faulty |
| **(b)** | | | | |
| CEN LEU / ARS | CEN LEU / ARS | Yes / Yes | Good (>90% of cells have plasmid) | Genomic fragment CEN required for good segregation |
| **(c)** | | | | |
| LEU–CEN / ARS  Restriction enzyme produces linear plasmid | LEU–CEN / ARS | No | — | Linear plasmid lacking TEL is unstable |
| TEL ← ARS – LEU – CEN → TEL | ARS·LEU·CEN TEL TEL | Yes / Yes | Good | Linear plasmids containing ARS and CEN behave like normal chromosomes if genomic fragment TEL is added to both ends |

**EXPERIMENTAL FIGURE 6-44 Yeast transfection experiments identify the functional chromosomal elements necessary for normal chromosome replication and segregation.** In these experiments, plasmids containing the *LEU* gene from normal yeast cells are constructed and introduced into *leu⁻* cells by transfection. If the plasmid is maintained in the *leu⁻* cells, they are transformed to *LEU⁺* by the *LEU* gene on the plasmid and can form colonies on medium lacking leucine. (a) Sequences that allow autonomous replication (ARS) of a plasmid were identified because their insertion into a plasmid vector containing a cloned *LEU* gene resulted in a high frequency of transformation to *LEU⁺*. However, even plasmids with ARS exhibit poor segregation during mitosis, and therefore do not appear in each of the daughter cells. (b) When randomly broken pieces of genomic yeast DNA are inserted into plasmids containing ARS and *LEU,* some of the subsequently transfected cells produce large colonies, indicating that a high rate of mitotic segregation among their plasmids is facilitating the continuous growth of daughter cells. The DNA recovered from plasmids in these large colonies contains yeast centromere (CEN) sequences. (c) When *leu⁻* yeast cells are transfected with linearized plasmids containing *LEU*, ARS, and CEN, no colonies grow. Addition of telomere (TEL) sequences to the ends of the linear DNA gives the linearized plasmids the ability to replicate as new chromosomes that behave very much like a normal chromosome in both mitosis and meiosis.
[See A. W. Murray and J. W. Szostak, 1983, *Nature* **305:**89, and L. Clarke and J. Carbon, 1985, *Ann. Rev. Genet.* **19:**29.]

transfection of yeast cells with linear plasmids were achieved by using the ends of a DNA molecule that was known to replicate as a linear molecule in the ciliated protozoan *Tetrahymena*. During part of the life cycle of *Tetrahymena*, much of the nuclear DNA is repeatedly copied in short pieces to form a so-called *macronucleus*. One of these repeated fragments was identified as a dimer of ribosomal DNA, the ends of which contained a repeated sequence $(G_4T_2)_n$. When a section of this repeated TEL sequence was ligated to the ends of linear yeast plasmids containing ARS and CEN, replication and good segregation of the linear plasmids occurred. This first cloning and characterization of telomeres garnered the Nobel Prize in Medicine and Physiology in 2009.

## Centromere Sequences Vary Greatly in Length and Complexity

Once the yeast centromere regions that confer mitotic segregation were cloned, their sequences could be determined and compared, revealing three regions (I, II, and III) that are conserved between the centromeres on different yeast chromosomes (Figure 6-45a). Short, fairly well conserved nucleotide sequences are present in regions I and III. Region II does not have a specific sequence, but is A-T rich with a fairly constant length, probably so that regions I and III lie on the same side of a specialized centromere-associated histone octamer. This specialized centromere-associated histone octamer contains

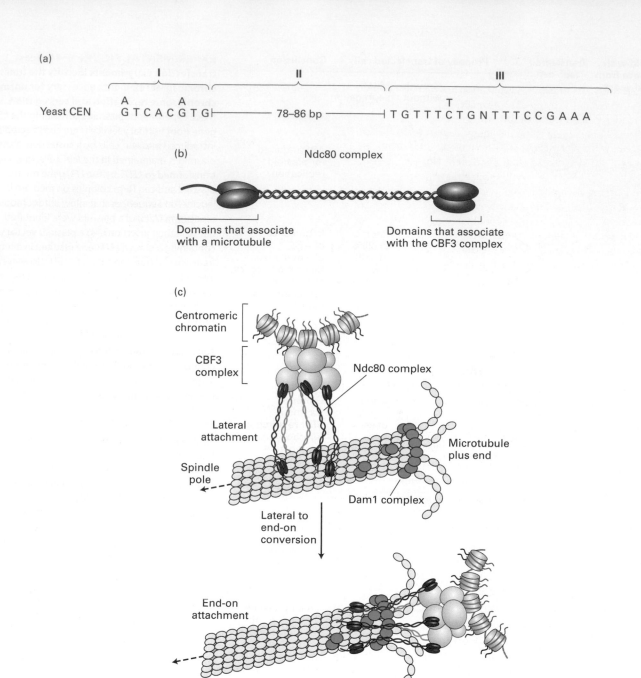

**FIGURE 6-45 Kinetochore-microtubule interaction in**
*S. cerevisiae.* (a) Sequence of the simple centromeres of *S. cerevisiae*.
(b) The Ndc80 complexes associate with both the microtubule and the
CBF3 complex. (c) Diagram of the centromere-associated CBF3
complex and its associated Ndc80 complexes that associate with a ring
of Dam1 proteins at the end of a spindle microtubule. The Ndc80
complexes initially make lateral interactions with the side of a spindle
microtubule (*top*) and then associate with the Dam1 ring, making an
end-on attachment (*bottom*) to the microtubule. [Part (a) from L. Clarke
and J. Carbon, 1985, *Ann. Rev. Genet.* **19**:29. Parts (b) and (c) adapted from
T. U. Tanaka, 2010, *EMBO J.* **29**:4070.]

the usual histones H2A, H2B, and H4, but a variant form of
histone H3. Centromeres from all eukaryotes similarly contain nucleosomes with a specialized, centromere-specific form
of histone H3, called CENP-A in humans. In *S. cerevisiae* the
CBF3 complex of proteins associates with this specialized nucleosome. The CBF3 complex in turn associates with multiprotein elongated Ndc80 complexes (Figure 6-45b) that
initially make lateral interactions with a spindle microtubule

and subsequently interact with a Dam1 complex that forms a
ring around the end of the microtubule (Figure 6-45c). This
results in an end-on interaction of the centromere with the
spindle microtubule. *S. cerevisiae* has by far the simplest centromere known in nature.

In the fission yeast *S. pombe*, centromeres are ≈40–100
kb in length and are composed of repeated copies of sequences
similar to those in *S. cerevisiae* centromeres. Multiple copies

**FIGURE 6-46 Standard DNA replication leads to loss of DNA at the 5′ end of each strand of a linear DNA molecule.** Replication of the right end of a linear DNA is shown; the same process occurs at the left end (shown by inverting the figure). As the replication fork approaches the end of the parental DNA molecule, the leading strand can be synthesized all the way to the end of the parental template strand without the loss of deoxyribonucleotides. However, since synthesis of the lagging strand requires RNA primers, the right end of the lagging daughter DNA strand would remain as ribonucleotides which are removed and therefore cannot serve as the template for a replicative DNA polymerase. Alternative mechanisms must be utilized to prevent successive shortening of the lagging strand with each round of replication. [Adapted from the Nobel Assembly at the Karolinska Institute.]

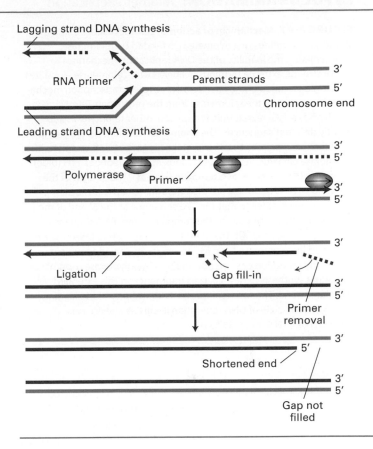

of proteins homologous to those that interact with *S. cerevisiae* centromeres bind to these complex *S. pombe* centromeres and in turn bind the much longer *S. pombe* chromosomes to several microtubules of the mitotic spindle apparatus. In plants and animals, centromeres are megabases in length and are composed of multiple repeats of simple-sequence DNA. In humans, centromeres contain 2- to 4-megabase arrays of a 171-bp simple-sequence DNA called *alphoid* DNA that is bound by nucleosomes containing the CENP-A histone H3 variant, as well as other repeated simple-sequence DNA.

In higher eukaryotes, a complex protein structure called the **kinetochore** assembles at centromeres and associates with multiple mitotic spindle fibers during mitosis (see Figure 18-39). Homologs of many of the centromere-associated proteins found in the yeasts occur in humans and other higher eukaryotes. For those yeast proteins where clear homologs are not evident in higher cells based on amino acid sequence comparisons (such as the Dam1 complex), alternative complexes with similar properties have been proposed to function at kinetochores that are bound to multiple spindle microtubules. The function of the centromere and kinetochore proteins that bind to it during the segregation of sister chromatids in mitosis and meiosis is described in Chapters 18 and 19.

## Addition of Telomeric Sequences by Telomerase Prevents Shortening of Chromosomes

Sequencing of telomeres from multiple organisms, including humans, has shown that most are repetitive oligomers with a high G content in the strand with its 3′ end at the end of the chromosome. The telomere repeat sequence in humans and other vertebrates is TTAGGG. These simple sequences are repeated at the very termini of chromosomes for a total of a few hundred base pairs in yeasts and protozoans, and a few thousand base pairs in vertebrates. The 3′ end of the G-rich strand extends 12–16 nucleotides beyond the 5′ end of the complementary C-rich strand. This region is bound by specific proteins that protect the ends of linear chromosomes from attack by exonucleases.

The need for a specialized region at the ends of eukaryotic chromosomes is apparent when we consider that all known DNA polymerases elongate DNA chains at the 3′ end, and all require an RNA or DNA primer. As the replication fork approaches the end of a linear chromosome, synthesis of the leading strand continues to the end of the DNA template strand, completing one daughter DNA double helix. However, because the lagging-strand template is copied in a discontinuous fashion, it cannot be replicated in its entirety (Figure 6-46). When the final RNA primer is removed, there is no upstream strand onto which DNA polymerase can build to fill the resulting gap. Without some special mechanism, the daughter DNA strand resulting from lagging-strand synthesis would be shortened at each cell division.

The problem of telomere shortening is solved by an enzyme that adds telomeric (TEL) repeat sequences to the ends of each chromosome. The enzyme is a protein–RNA complex called *telomere terminal transferase,* or *telomerase.* Because the sequence of the telomerase-associated RNA, as we will see, serves as the template for addition of deoxyribonucleotides to the ends of telomeres, the source of the enzyme and not the source of the telomeric DNA primer determines the sequence added. This was proven by transforming *Tetrahymena* with a mutated form of the gene encoding the telomerase-associated RNA. The resulting telomerase added a DNA sequence complementary to the mutated RNA sequence to the

**FIGURE 6-47 Mechanism of action of telomerase.** The single-stranded 3′ terminus of a telomere is extended by telomerase, counteracting the inability of the DNA replication mechanism to synthesize the extreme terminus of linear DNA. Telomerase elongates this single-stranded end by a reiterative reverse-transcription mechanism. The action of the telomerase from the protozoan *Tetrahymena*, which adds a $T_2G_4$ repeat unit, is depicted; other telomerases add slightly different sequences. The telomerase contains an RNA template (red) that base-pairs to the 3′ end of the lagging-strand template. The telomerase catalytic site then adds deoxyribonucleotides TTG (blue) using the RNA molecule as a template (step **1**). The strands of the resulting DNA-RNA duplex are then thought to slip (translocate) relative to each other so that the TTG sequence at the 3′ end of the replicating DNA base-pairs to the complementary RNA sequence in the telomerase RNA (step **2**). The 3′ end of the replicating DNA is again extended by telomerase (step **3**). Telomerases can add multiple repeats by repetition of steps **2** and **3**. DNA polymerase α-primase can prime synthesis of new Okazaki fragments on this extended template strand. The net result prevents shortening of the lagging strand at each cycle of DNA replication [From C.W. Greider and E. H. Blackburn, 1989, *Nature* **337:**331.]

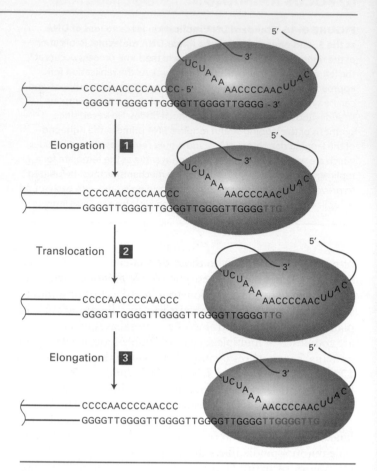

ends of telomeric primers. Thus telomerase is a specialized form of a reverse transcriptase that carries its own internal RNA template to direct DNA synthesis. These experiments also earned the Nobel Prize in Physiology and Medicine for the discovery and characterization of the mechanism of telomerase.

Figure 6-47 depicts how telomerase, by reverse transcription of its associated RNA, elongates the 3′ end of the single-stranded DNA at the end of the G-rich strand mentioned above. Cells from knockout mice that cannot produce the telomerase-associated RNA exhibit no telomerase activity, and their telomeres shorten successively with each cell generation. Such mice can breed and reproduce normally for three generations before the long telomere repeats become substantially eroded. Then, the absence of telomere DNA results in adverse effects, including fusion of chromosome termini and chromosomal loss. By the fourth generation, the reproductive potential of these knockout mice declines, and they cannot produce offspring after the sixth generation.

The human genes expressing the telomerase protein and the telomerase-associated RNA are active in germ cells and stem cells, but are turned off in most cells of adult tissues that replicate only a limited number of times, or will never replicate again (such cells are called *postmitotic*). However, these genes are activated in most human cancer cells, where telomerase is required for the multiple cell divisions necessary to form a tumor. This phenomenon has stimulated a search for inhibitors of human telomerase as potential therapeutic agents for treating cancer. ■

While telomerase prevents telomere shortening in most eukaryotes, some organisms use alternative strategies. *Drosophila*

species maintain telomere lengths by the regulated insertion of non-LTR retrotransposons into telomeres. This is one of the few instances in which a mobile element has a specific function in its host organism.

## KEY CONCEPTS of Section 6.7

### Morphology and Functional Elements of Eukaryotic Chromosomes

• During metaphase, eukaryotic chromosomes become sufficiently condensed that they can be visualized individually in the light microscope.

• The chromosomal karyotype is characteristic of each species. Closely related species can have dramatically different karyotypes, indicating that similar genetic information can be organized on chromosomes in different ways.

• Banding analysis and chromosome painting are used to identify the different human metaphase chromosomes and to detect translocations and deletions (see Figure 6-41).

• Analysis of chromosomal rearrangements and regions of conserved synteny between related species allows scientists to make predictions about the evolution of chromosomes (see Figure 6-42c). The evolutionary relationships between organisms indicated by these studies are consistent with proposed evolutionary relationships based on the fossil record and DNA sequence analysis.

- The highly reproducible banding patterns of polytene chromosomes make it possible to visualize chromosomal deletions and rearrangements as changes in the normal pattern of bands.

- Three types of DNA sequences are required for a long linear DNA molecule to function as a chromosome: a replication origin, called ARS in yeast; a centromere (CEN) sequence; and two telomere (TEL) sequences at the ends of the DNA (see Figure 6-44).

- Telomerase, a protein–RNA complex, has a special reverse transcriptase activity that completes replication of telomeres during DNA synthesis (see Figure 6-47). In the absence of telomerase, the daughter DNA strand resulting from lagging-strand synthesis would be shortened at each cell division in most eukaryotes (see Figure 6-46).

inserted into new sites in individuals throughout our history. Large numbers of these interspersed repeats are polymorphic within populations, occurring at a particular site in some individuals and not others. Individuals sharing an insertion at a particular site descended from a common ancestor that developed from an egg or sperm in which that insertion occurred. The time elapsed from the initial insertion can be estimated by the differences in sequences of the element that arose from the accumulation of random mutations. Further analysis of retrotransposon polymorphisms will undoubtedly add immensely to our understanding both of human migrations since *Homo sapiens* first evolved, as well as the history of contemporary populations.

As described in Chapter 5, the *Drosophila* P-element DNA transposon was exploited for the facile stable transformation of genes into the *Drosophila* germ line. This has been a powerful method for molecular cell biology experimentation in this organism. An active area of current research is to use mammalian transposons and retrotransposons for the transformation of human cells for gene therapy. This promises to be an exciting area of medicine in the future treatment of genetic diseases such as sickle-cell anemia and cystic fibrosis as well as more common diseases, especially when coupled with the recently developed techniques for generating pluripotent stem cells (iPS cells) from differentiated cells of a pediatric or adult patient (Chapter 21).

## Perspectives for the Future

The human genome sequence is a goldmine for new discoveries in molecular cell biology, in identifying new proteins that may be the basis of effective therapies of human diseases, and for understanding early human history and evolution. However, finding new genes is like finding a needle in a haystack, because only ≈1.5 percent of the sequence encodes proteins or functional RNA. Identification of genes in bacterial genome sequences is relatively simple because of the scarcity of introns; simply searching for long open reading frames free of stop codons identifies most genes. In contrast, the search for human genes is complicated by the structure of human genes, most of which are composed of multiple, relatively short exons separated by much longer, noncoding introns. Identification of complex transcription units by analysis of genomic DNA sequences alone is extremely challenging. Future improvements in bioinformatic methods for gene identification, and characterization of cDNA copies of mRNAs isolated from the hundreds of human cell types, will likely lead to the discovery of new proteins, to a better understanding of biological processes, and may lead to applications in medicine and agriculture.

We have seen that although most transposons do not function directly in cellular processes, they have helped to shape modern genomes by promoting gene duplications, exon shuffling, the generation of new combinations of transcription-control sequences, and other aspects of contemporary genomes. They are also teaching us about our own history and origins, because L1 and *Alu* retrotransposons have

## Key Terms

| | |
|---|---|
| bioinformatics 252 | matrix-associated regions (MARs) 263 |
| centromere 270 | monocistronic 225 |
| chromatid 267 | nucleosome 225 |
| chromatin 224 | open reading frame (ORF) 253 |
| cytoplasmic inheritance 246 | polytene chromosome 269 |
| DNA transposon 235 | protein family 228 |
| epigenetic code 261 | pseudogene 228 |
| euchromatin 260 | retrotransposon 235 |
| exon shuffling 243 | repetitious DNA 223 |
| fluorescence in situ hybridization (FISH) 267 | scaffold-associated regions (SARs) 263 |
| gene family 228 | simple-sequence (satellite) DNA 232 |
| genomics 225 | SINEs 240 |
| heterochromatin 260 | SMC proteins 263 |
| histones 225 | telomere 270 |
| histone code 258 | transcription unit 225 |
| karyotype 267 | transposable DNA element 223 |
| LINEs 240 | |
| long terminal repeats (LTRs) 238 | |

## Review the Concepts

**1.** Genes can be transcribed into mRNA for protein-coding genes or into RNA for genes such as ribosomal or transfer RNAs. Define a gene. For the following characteristics, state whether they apply to (a) continuous, (b) simple, or (c) complex transcription units.

  (i) Found in eukaryotes

  (ii) Contain introns

  (iii) Only capable of making a single protein from a given gene

**2.** Sequencing of the human genome has revealed much about the organization of genes. Describe the differences between solitary genes, gene families, pseudogenes, and tandemly repeated genes.

**3.** Much of the human genome consists of repetitious DNA. Describe the difference between microsatellite and minisatellite DNA. How is this repetitious DNA useful for identifying individuals by the technique of DNA fingerprinting?

**4.** Mobile DNA elements that can move or transpose to a new site directly as DNA are called DNA transposons. Describe the mechanism by which a bacterial DNA transposon, called an insertion sequence, can transpose.

**5.** Retrotransposons are a class of mobile elements that transpose via an RNA intermediate. Contrast the mechanism of transposition between retrotransposons that contain long terminal repeats (LTRs) and those that lack LTRs.

**6.** Discuss the role that transposons may have played in the evolution of modern organisms. What is exon shuffling? What role do transposons play in the process of exon shuffling?

**7.** Mitochondria and chloroplasts are thought to have evolved from symbiotic bacteria present in nucleated cells. What is the experimental evidence from this chapter that supports this hypothesis?

**8.** What are paralogous and orthologous genes? What are some of the explanations for the finding that humans are a much more complex organism than the roundworm *C. elegans*, yet have only fewer than one and a half as many genes (25,000 versus 18,000)?

**9.** The DNA in a cell associates with proteins to form chromatin. What is a nucleosome? What role do histones play in nucleosomes? How are nucleosomes arranged in condensed 30-nm fibers?

**10.** How do chromatin modifications regulate transcription? What modifications are observed in regions of the genome that are actively being transcribed? What about for regions that are not actively transcribed?

**11.** Describe the general organization of a eukaryotic chromosome. What structural role do scaffold-associated regions (SARs) or matrix attachment regions (MARs) play? Why does it make sense that protein-coding genes are not located in these regions?

**12.** What is FISH? Briefly describe how it works. How is FISH used to characterize chromosomal translocations associated with certain genetic disorders and specific types of cancers?

**13.** What is chromosome painting, and how is this technique useful? How can chromosome paint probes be used to analyze the evolution of mammalian chromosomes?

**14.** Certain organisms contain cells that possess polytene chromosomes. What are polytene chromosomes, where are they found, and what function do they serve?

**15.** Replication and segregation of eukaryotic chromosomes require three functional elements: replication origins, a centromere, and telomeres. How would a chromosome be affected if it lacked (a) replication origins or (b) a centromere?

**16.** Describe the problem that occurs during DNA replication at the ends of chromosomes. How are telomeres related to this problem?

## Analyze the Data

**1.** To determine whether gene transfer from an organelle genome to the nucleus can be observed in the laboratory, a chloroplast transformation vector was constructed that contained two selectable antibiotic-resistance markers, each with its own promoter: the spectinomycin-resistance gene and the kanamycin-resistance gene (see S. Stegemann et al., 2003, *Proc. Nat'l Acad. Sci. USA* **100**:8828–8833). The spectinomycin-resistance gene was controlled by a chloroplast promoter, yielding a chloroplast-specific selectable marker. Plants grown on spectinomycin are white unless they express the spectinomycin-resistance gene in the chloroplast. The kanamycin-resistance gene, inserted into the plasmid adjacent to the spectinomycin-resistance gene, was under the control of a strong nuclear promoter. Transgenic, spectinomycin-resistant tobacco plants were selected following transformation with this plasmid by identifying green plants grown on medium with spectinomycin. These plants contain the two antibiotic-resistance genes inserted into the chloroplast genome by a recombination event; however, kanamycin resistance is not expressed because it is under the control of a nuclear promoter. These spectinomycin-resistant plants were grown for multiple generations and used in the following studies.

  **a.** Leaves from the spectinomycin-resistant transgenic plants were placed in a plant-regeneration medium containing kanamycin. Some of the leaf cells were resistant to kanamycin, and grew into kanamycin-resistant plants. Pollen (paternal) from kanamycin-resistant plants was used to pollinate wild-type (nontransgenic) plants. In tobacco, no chloroplasts are inherited from pollen. The resulting seeds were germinated on media with and without kanamycin. Half of the resulting seedlings were kanamycin resistant. When these kanamycin-resistant plants were allowed to self-pollinate, the offspring exhibited a 3:1 ratio of kanamycin-resistant to sensitive phenotypes. What can be deduced from these data about the location of the kanamycin-resistance gene?

  **b.** To determine whether transfer of the kanamycin-resistance gene to the nucleus was mediated via DNA or an RNA intermediate, DNA was extracted from 10 seedling plants germinated from seeds produced by a wild-type plant pollinated with a kanamycin-resistant plant. The 10 seedling plants, numbered 1–10 in the corresponding gel lanes in the figure below, consist of 5 kanamycin-resistant (+) and 5 kanamycin-sensitive (−) plants. Each DNA sample was subjected to PCR analysis using primers to amplify the kanamycin-resistance gene (gel at left) or the spectinomycin-resistance gene (gel at right). The lane marked M shows molecular weight markers. What does the correspondence between the presence or absence of PCR products generated in the same plant with both sets of primers suggest about the mode of transfer of the kanamycin gene to the nucleus?

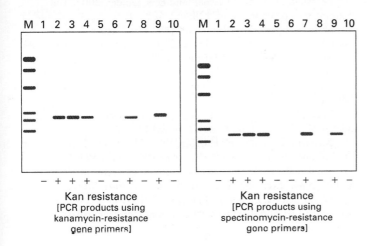

Kan resistance
[PCR products using
kanamycin-resistance
gene primers]

Kan resistance
[PCR products using
spectinomycin-resistance
gene primers]

c. When the original transgenic plants, which were selected on spectinomycin but not on kanamycin, were used to pollinate wild-type plants, none of the offspring were kanamycin resistant. What can be deduced from these observations?

2. Satellite DNA is a known component of our genome and can be found in both coding and noncoding DNA. When it is found in coding DNA, the number of repeats can result in altered proteins. But the effect of these repeats in noncoding DNA is not as well understood. To determine whether repeats in the promoter region can alter gene expression and chromatin compaction, Vinces et al. (Vinces et al., 2009, *Science* 324:1213–1216) searched the *Sacharomyces cerevisiae* genome for the presence of repetitive DNA in promoters and examined how altering the number of repeats affected gene expression and DNA packaging.

a. The group first searched for satellite DNA in the genome of various *S. cerevisiae* strains and found that 25 percent of promoters contained at least one repeat region. In addition, a single promoter in different strains often contained distinct numbers of repeats in each region of satellite DNA. What is the proposed mechanism by which the number of repeats in a given region of satellite DNA can increase?

b. To determine whether there is a correlation between the number of repeats and gene expression, transcription at the SDT1 gene was analyzed. The SDT1 promoter contains satellite DNA, and the number of repeats in this region was modified, ranging between 0 repeats to 60 repeats. It was discovered that SDT1 expression, analyzed by quantitative reverse transcriptase PCR (Q RT-PCR , see Chapter 5), increased when the repeat number increased from 0 to 13 repeats. SDT1

expression then progressively decreased as the repeat number increased from 13 to 60. What conclusion can be made based on these experimental results? Explain how Q RT-PCR can be used to analyze SDT1 gene expression.

c. The conclusion from part (b) then led the group to determine whether a cell can adapt to its surroundings by altering the number of repeats in promoter satellite DNA. The promoter from the SDT1 gene (containing 48 repeats) was attached to the URA3 open reading frame, a gene responsible for synthesis of the nucleotide uracil. Cells with this hybrid protein were then placed on media lacking uracil or media containing uracil. After growth on each media, the number of repeats in the promoter driving URA3 expression was analyzed. Those that were grown on the media lacking uracil showed a decrease in the number of repeats in the promoter, while those grown with uracil still had on average 48 repeats. What conclusion can be made from these results? Based on the data in part (b), how many repeats were likely found in the URA3 promoters for cells grown without uracil?

d. The location of the repeat DNA in the promoters was compared to the location of nucleosomes, and it was found that the nucleosome density in a promoter was inversely correlated with the number of repeats at that location of DNA. From this, what would you conclude about repetitive DNA and chromatin packaging? What effect would decreasing the repeat regions have on histone core binding at that DNA?

# References

### Eukaryotic Gene Structure

Black, D. L. 2003. Mechanisms of alternative pre-messenger RNA splicing. *Ann. Rev. Biochem.* 72:291–336.

Davuluri, R. V., et al. 2008. The functional consequences of alternative promoter use in mammalian genomes. *Trends Genet.* 24:167–177.

Wang, E. T., et al. 2008. Alternative isoform regulation in human tissue transcriptomes. *Nature* 456:470–476.

### Chromosomal Organization of Genes and Noncoding DNA

Celniker, S. E., and G. M. Rubin. 2003. The *Drosophila melanogaster* genome. *Ann. Rev. Genomics Hum. Genet.* 4:89–117.

Crook, Z. R., and D. Housman. 2011. Huntington's disease: can mice lead the way to treatment? *Neuron* 69:423–435.

Feuillet, C., et al. 2011. Crop genome sequencing: lessons and rationales. *Trends Plant Sci.* 16:77–88.

International Human Genome Sequencing Consortium. 2004. Finishing the euchromatic sequence of the human genome. *Nature* 431:931–945.

Giardina, E., A. Spinella, and G. Novelli. 2011. Past, present and future of forensic DNA typing. *Nanomedicine (Lond.)* 6:257–270.

Hannan, A. J. 2010. TRPing up the genome: tandem repeat polymorphisms as dynamic sources of genetic variability in health and disease. *Discov. Med.* 10:314–321.

Jobling, M. A., and P. Gill. 2004. Encoded evidence: DNA in forensic analysis. *Nat. Rev. Genet.* 5:739–751.

Lander, E. S., et al. 2001. Initial sequencing and analysis of the human genome. *Nature* 409:860–921.

Todd, P. K., and H. L. Paulson. 2010. RNA-mediated neurodegeneration in repeat expansion disorders. *Ann. Neurol.* 67:291–300.

Venter, J. C., et al. 2001. The sequence of the human genome. *Science* 291:1304–1351.

## Transposable (Mobile) DNA Elements

Curcio, M. J., and K. M. Derbyshire. 2003. The outs and ins of transposition: from mu to kangaroo. *Nat. Rev. Mol. Cell Biol.* **4**:865–877.

Goodier, J. L., and H. H. Kazazian, Jr. 2008. Retrotransposons revisited: the restraint and rehabilitation of parasites. *Cell* **135**:23–35.

Jones, R. N. 2005. McClintock's controlling elements: the full story. *Cytogenet. Genome Res.* **109**:90–103.

Lisch, D. 2009. Epigenetic regulation of transposable elements in plants. *Ann. Rev. Plant Biol.* **60**:43–66.

## Organelle DNAs

Bendich, A. J. 2004. Circular chloroplast chromosomes: the grand illusion. *Plant Cell* **16**:1661–1666.

Bonawitz, N. D., D. A. Clayton, and G. S. Shadel. 2006. Initiation and beyond: multiple functions of the human mitochondrial transcription machinery. *Mol. Cell* **24**:813–825.

Chan, D. C. 2006. Mitochondria: dynamic organelles in disease, aging, and development. *Cell* **125**:1241–1252.

## Genomics: Genome-wide Analysis of Gene Structure and Expression

BLAST Information can be found at: http://blast.ncbi.nlm.nih.gov/Blast.cgi

1000 Genomes Project Consortium, G.P. 2010. A map of human genome variation from population-scale sequencing. *Nature* **467**:1061–1073.

Alkan, C., B. P. Coe, and E. E. Eichler. 2011. Genome structural variation discovery and genotyping. *Nat. Rev. Genet.* **12**:363–376.

Chimpanzee Sequencing and Analysis Consortium. 2005. Initial sequence of the chimpanzee genome and comparison with the human genome. *Nature* **437**:69–87.

du Plessis, L., N. Skunca, and C. Dessimoz. 2011. The what, where, how and why of gene ontology—a primer for bioinformaticians. *Brief Bioinform.* 2011 Feb 17. [Epub ahead of print]

Ideker, T., J. Dutkowski, and L. Hood. 2011. Boosting signal-to-noise in complex biology: prior knowledge is power. *Cell* **144**:860–863.

International Human Genome Sequencing Consortium. 2004. Finishing the euchromatic sequence of the human genome. *Nature* **431**:931–945.

Lander, E. S. 2011. Initial impact of the sequencing of the human genome. *Nature* **470**:187–197.

Mills, R. E., et al. 2011. Mapping copy number variation by population-scale genome sequencing. *Nature* **470**:59–65.

Picardi, E., and G. Pesole. 2010. Computational methods for ab initio and comparative gene finding. *Meth. Mol. Biol.* **609**:269–284.

Ramskold, D., et al. 2009. An abundance of ubiquitously expressed genes revealed by tissue transcriptome sequence data. *PLoS Comput. Biol.* **5**:e1000598.

Raney, B. J., et al. 2011. ENCODE whole-genome data in the UCSC genome browser (2011 update). *Nucl. Acids Res.* **39**:D871–D875.

Sleator, R. D. 2010. An overview of the current status of eukaryote gene prediction strategies. *Gene* **461**:1–4.

Sonah, H., et al. 2011. Genomic resources in horticultural crops: status, utility and challenges. *Biotechnol. Adv.* **29**:199–209.

Stratton, M. R. 2011. Exploring the genomes of cancer cells: progress and promise. *Science* **331**:1553–1558.

Venter, J. C. 2011. Genome-sequencing anniversary. The human genome at 10: successes and challenges. *Science* **331**:546–547.

## Structural Organization of Eukaryotic Chromosomes

Bannister, A. J., and T. Kouzarides. 2011. Regulation of chromatin by histone modifications. *Cell Res.* **21**:381–395.

Bernstein, B. E., A. Meissner, and E. S. Lander. 2007. The mammalian epigenome. *Cell* **128**:669–681.

Horn, P. J., and C. L. Peterson. 2006. Heterochromatin assembly: a new twist on an old model. *Chromosome Res.* **14**:83–94.

Kurdistani, S. K. 2011. Histone modifications in cancer biology and prognosis. *Prog. Drug Res.* **67**:91–106.

Luger, K. 2006. Dynamic nucleosomes. *Chromosome Res.* **14**:5–16.

Luger, K., and T. J. Richmond. 1998. The histone tails of the nucleosome. *Curr. Opin. Genet. Dev.* **8**:140–146.

Nasmyth, K., and C. H. Haering. 2005. The structure and function of SMC and kleisin complexes. *Ann. Rev. Biochem.* **74**:595–648.

Schalch, T., et al. 2005. X-ray structure of a tetranucleosome and its implications for the chromatin fibre. *Nature* **436**:138–141.

Woodcock, C. L., and R. P. Ghosh. 2010. Chromatin higher-order structure and dynamics. *Cold Spring Harbor Perspect. Biol.* **2**:a000596.

## Morphology and Functional Elements of Eukaryotic Chromosomes

Armanios, M., and C. W. Greider. 2005. Telomerase and cancer stem cells. *Cold Spring Harbor Symp. Quant. Biol.* **70**:205–208.

Belmont, A. S. 2006. Mitotic chromosome structure and condensation. *Curr. Opin. Cell Biol.* **18**:632–638.

Blackburn, E. H. 2005. Telomeres and telomerase: their mechanisms of action and the effects of altering their functions. *FEBS Lett.* **579**:859–862.

Cvetic, C., and J. C. Walter. 2005. Eukaryotic origins of DNA replication: could you please be more specific? *Semin. Cell Dev. Biol.* **16**:343–353.

Froenicke, L. 2005. Origins of primate chromosomes as delineated by Zoo-FISH and alignments of human and mouse draft genome sequences. *Cytogenet. Genome Res.* **108**:122–138.

MacAlpine, D. M., and S. P. Bell. 2005. A genomic view of eukaryotic DNA replication. *Chromosome Res.* **13**:309–326.

Ohta, S., et al. 2011. Building mitotic chromosomes. *Curr. Opin. Cell Biol.* **23**:114–121.

Tanaka, T. U. 2010. Kinetochore-microtubule interactions: steps towards bi-orientation. *EMBO J.* **29**:4070–4082.

# Transcriptional Control of Gene Expression

*Drosophila* polytene chromosomes stained with antibodies against a chromatin-remodeling ATPase called Kismet (blue), RNA polymerase II with low CTD phosphorylation (red), and RNA polymerase II with high CTD phosphorylation (green). [Courtesy of John Tamkun; see S. Srinivasan et al., 2005, *Development* **132**:1623.]

I n previous chapters we have seen that the properties and functions of each cell type are determined by the proteins it contains. In this and the next chapter, we consider how the kinds and amounts of the various proteins produced by a particular cell type in a multicellular organism are regulated. This regulation of **gene expression** is the fundamental process that controls the development of a multicellular organism such as ourselves from a single fertilized egg cell into the thousands of cell types from which we are made. When gene expression goes awry, cellular properties are altered, a process that all too often leads to the development of cancer. As discussed further in Chapter 25, genes encoding proteins that restrain cell growth are abnormally repressed in cancer cells, whereas genes encoding proteins that promote cell growth and replication are inappropriately activated in cancer cells. Abnormalities in gene expression also result in developmental defects such as cleft palate, tetralogy of Fallot (a serious developmental defect of the heart that can be treated surgically), and many others. Regulation of gene expression also plays a vital role in bacteria and other single-celled microorganisms, where it allows cells to adjust their enzymatic machinery and structural components in response to their changing nutritional and physical environment. Consequently, to understand how microorganisms respond to their environment and how multicellular organisms normally develop, as well as how pathological abnormalities of gene expression occur, it is essential to understand the molecular interactions that control protein production.

The basic steps in gene expression, i.e., the entire process whereby the information encoded in a particular gene is decoded into a particular protein, are reviewed in Chapter 4. Synthesis of mRNA requires that an **RNA polymerase** initiate transcription (**initiation**), polymerize ribonucleoside triphosphates complementary to the DNA coding strand (**elongation**), and then terminate transcription (**termination**) (see Figure 4-11). In bacteria, ribosomes and translation-initiation factors have immediate access to newly formed RNA transcripts, which function as mRNA without further modification. In eukaryotes, however, the initial RNA transcript is subjected to processing that yields a functional mRNA (see Figure 4-15). The mRNA then is transported from its site of synthesis in the nucleus to the cytoplasm, where it is translated into protein with the aid of ribosomes, tRNAs, and translation factors (see Figures 4-24, 4-25, and 4-27).

Regulation may occur at several of the various steps in gene expression outlined above: transcription initiation, elongation, RNA processing, mRNA export from the nucleus, and translation into protein. This results in *differential* production of proteins in different cell types or developmental stages or in response to external conditions. Although examples of

## OUTLINE

regulation at each step in gene expression have been found, control of transcription initiation and elongation—the first two steps—are the most important mechanisms for determining whether most genes are expressed and how much of the encoded mRNAs and, consequently, proteins are produced. The molecular mechanisms that regulate transcription initiation and elongation are critical to numerous biological phenomena, including the development of a multicellular organism from a single fertilized egg cell as mentioned above, the immune responses that protect us from pathogenic microorganisms, and

neurological processes such as learning and memory. When these regulatory mechanisms controlling transcription function improperly, pathological processes may occur. For example, reduced activity of the *Pax6* gene causes *aniridia,* failure to develop an iris (Figure 7-1a). Pax6 is a **transcription factor** that normally regulates transcription of genes involved in eye development. In other organisms, mutations in transcription factors cause an extra pair of wings to develop in *Drosophila* (Figure 7-1b), alter the structures of flowers in plants (Figure 7-1c), and are responsible for multiple other developmental abnormalities.

**FIGURE 7-1 Phenotypes of mutations in genes encoding transcription factors.** (a) A mutation that inactivates one copy of the *Pax6* gene on either the maternal or paternal chromosome 9 results in failure to develop an iris, or *aniridia.* (b) Homozygous mutations that prevent expression of the *Ubx* gene in the third thoracic segment of *Drosophila* result in transformation of the third segment, which normally has a balancing organ called a haltere, into a second copy of the thoracic segment that develops wings. (c) Mutations in *Arabidopsis*

*thaliana* that inactivate both copies of three *floral organ-identity* genes transform the normal parts of the flower into leaflike structures. In each case, these mutations affect master regulatory transcription factors that regulate multiple genes, including many genes encoding other transcription factors. [Part (a), left, © Simon Fraser/Photo Researchers, Inc.; right, Visuals Unlimited. Part (b) from E. B. Lewis, 1978, *Nature* **276:**565. Part (c) from D. Wiegel and E. M. Meyerowitz, 1994, *Cell* **78:**203.]

Transcription is a complex process involving many layers of regulation. In this chapter, we focus on the molecular events that determine when transcription of a gene occurs. First, we consider the mechanisms of gene expression in bacteria, where DNA is not bound by histones and packaged into nucleosomes. **Repressor** and **activator** proteins recognize and bind to specific regions of DNA to control the transcription of a nearby gene. The remainder of the chapter focuses on eukaryotic transcription regulation and how the basic tenets of bacterial regulation are applied in more complex ways in higher organisms. These mechanisms also make use of the association of DNA with histone octamers, forming chromatin structures with varying degrees of condensation and post-translational modifications such as acetylation and methylation to regulate transcription. Figure 7-2 provides an overview of transcription regulation in metazoans (multicellular animals) and the processes outlined in this chapter. We discuss how specific DNA sequences function as **transcription-control regions** by serving

as the binding sites for transcription factors (repressors and activators) and how the RNA polymerases responsible for transcription bind to **promoter** sequences to initiate the synthesis of an RNA molecule complementary to template DNA. Next, we consider how activators and repressors influence transcription through interactions with large, multiprotein complexes. Some of these multiprotein complexes modify chromatin condensation, altering access of chromosomal DNA to transcription factors and RNA polymerases. Other complexes influence the rate at which RNA polymerase binds to DNA at the site of transcription initiation, as well as the frequency of initiation. Very recent research has revealed that, in multicellular animals, for many genes, the RNA polymerase pauses after transcribing a short RNA and transcription regulation involves a release of the paused polymerase, allowing it to transcribe through the rest of the gene. We discuss how transcription of specific genes can be specified by particular combinations of the ≈2000 transcription factors encoded in

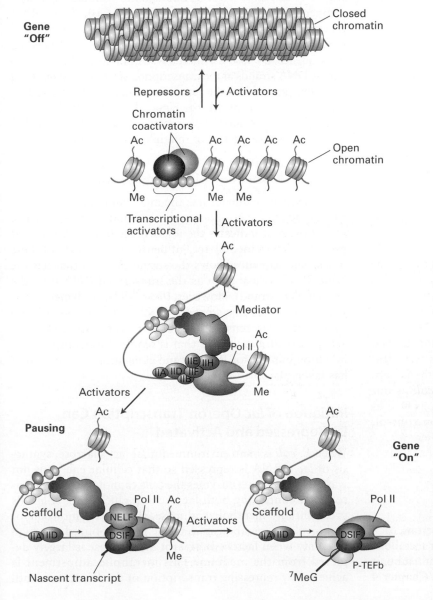

**FIGURE 7-2 Overview of eukaryotic transcription control.** Inactive genes are assembled into regions of condensed chromatin that inhibit RNA polymerases and their associated general transcription factors (GTFs) from interacting with promoters. Activator proteins bind to specific DNA sequence-control elements in chromatin and interact with multiprotein chromatin co-activator complexes to decondense chromatin and the multisubunit mediator to assemble RNA polymerase and general transcription factors on promoters. Alternatively, repressor proteins bind to other control elements to inhibit initiation by RNA polymerase and interact with multiprotein co-repressor complexes to condense chromatin. RNA polymerase initiates transcription but pauses after transcribing 20–50 nucleotides due to the action of elongation inhibitors. Activators promote the association of elongation factors that release the elongation inhibitors and allow productive elongation through the gene. DSIF is the DRB sensitivity-inducing factor, NELF is the negative elongation factor, and P-TEFb is a protein kinase comprised of CDK9 and cyclin T. [Adapted from S. Malik and R. G. Roeder, 2010, *Nat. Rev. Genet.* **11**:761.]

the human genome, giving rise to cell-type-specific gene expression. We will consider the various ways in which the activities of transcription factors themselves are controlled to ensure that genes are expressed only at the right time and in the right place. We will also discuss recent studies revealing that RNA-protein complexes in the nucleus can regulate transcription. New methods for sequencing DNA coupled with reverse transcription of RNA into DNA in vitro have revealed that much of the genome of eukaryotes is transcribed into low-abundance RNAs that do not encode protein, raising the possibility that transcription control by such noncoding RNAs may be a much more general process than is currently understood. RNA processing and various post-transcriptional mechanisms for controlling eukaryotic gene expression are covered in the next chapter. Subsequent chapters, particularly Chapters 15, 16, and 21, provide examples of how transcription is regulated by interactions between cells and how the resulting **gene control** contributes to the development and function of specific types of cells in multicellular organisms.

## 7.1 Control of Gene Expression in Bacteria

Since the structure and function of a cell are determined by the proteins it contains, the control of gene expression is a fundamental aspect of molecular cell biology. Most commonly, the "decision" to transcribe the gene encoding a particular protein is the major mechanism for controlling production of the encoded protein in a cell. By controlling transcription, a cell can regulate which proteins it produces and how rapidly. When transcription of a gene is *repressed,* the corresponding mRNA and encoded protein or proteins are synthesized at low rates. Conversely, when transcription of a gene is *activated,* both the mRNA and encoded protein or proteins are produced at much higher rates.

In most bacteria and other single-celled organisms, gene expression is highly regulated in order to adjust the cell's enzymatic machinery and structural components to changes in the nutritional and physical environment. Thus, at any given time, a bacterial cell normally synthesizes only those proteins of its entire proteome that are required for survival under particular conditions. Here we describe the basic features of transcription control in bacteria, using the *lac* operon and the glutamine synthetase gene in *E. coli* as our primary examples. Many of the same processes, as well as others, are involved in eukaryotic transcription control, which is the subject of the remainder of this chapter.

### Transcription Initiation by Bacterial RNA Polymerase Requires Association with a Sigma Factor

In *E. coli,* about half the genes are clustered into **operons,** each of which encodes enzymes involved in a particular metabolic pathway or proteins that interact to form one multisubunit protein. For instance, the *trp* operon discussed in Chapter 4

encodes five polypeptides needed in the biosynthesis of tryptophan (see Figure 4-13). Similarly, the *lac* operon encodes three proteins required for the metabolism of lactose, a sugar present in milk. Since a bacterial operon is transcribed from one start site into a single mRNA, all the genes within an operon are **coordinately regulated;** that is, they are all activated or repressed to the same extent.

Transcription of operons, as well as of isolated genes, is controlled by interplay between RNA polymerase and specific repressor and activator proteins. In order to initiate transcription, however, *E. coli* RNA polymerase must be associated with one of a small number of $\sigma$ *(sigma) factors.* The most common one in eubacterial cells is $\sigma^{70}$. $\sigma^{70}$ binds to RNA polymerase and to promoter DNA sequences, bringing the RNA polymerase enzyme to a promoter. $\sigma^{70}$ recognizes and binds to both a six-base-pair sequence centered at $\approx -10$ and a seven-base-pair sequence centered at $\approx -35$ from the $+1$ transcription start. Consequently, the $-10$ plus the $-35$ sequences constitute a promoter for *E. coli* RNA polymerase associated with $\sigma^{70}$ (see Figure 4-10b). Although the promoter sequences contacted by $\sigma^{70}$ are located at $-35$ and $-10$, *E. coli* RNA polymerase binds to the promoter region DNA from $\approx -50$ to $\approx +20$ through interactions with DNA that do not depend on the sequence. $\sigma^{70}$ also assists the RNA polymerase in separating the DNA strands at the transcription start site and inserting the coding strand into the active site of the polymerase so that transcription starts at $+1$ (see Figure 4-11, step **2**). The optimal $\sigma^{70}$-RNA polymerase promoter sequence, determined as the "consensus sequence" of multiple strong promoters, is

$$-35 \text{ region} \qquad -10 \text{ region}$$
$$\text{{\small T}T{\small G}AC{\small A}T}\text{——}15\text{–}17 \text{ bp}\text{——}\text{{\small TAT}AA{\small T}}$$

The consensus sequence has the most commonly occurring base at each of the positions in the $-35$ and $-10$ regions. The size of the font indicates the importance of the base at that position, determined by the influence of mutations of these bases. The sequence shows the strand of DNA that has the same $5' \rightarrow 3'$ orientation as the transcribed RNA (i.e., the nontemplate strand). However, the $\sigma^{70}$-RNA polymerase initially binds to double-stranded DNA. After the polymerase transcribes a few tens of base pairs, $\sigma^{70}$ is released. Thus $\sigma^{70}$ acts as an *initiation factor* that is required for transcription initiation but not for RNA strand elongation once initiation has taken place.

### Initiation of *lac* Operon Transcription Can Be Repressed and Activated

When *E. coli* is in an environment that lacks lactose, synthesis of *lac* mRNA is repressed so that cellular energy is not wasted synthesizing enzymes the cells cannot use. In an environment containing both lactose and glucose, *E. coli* cells preferentially metabolize glucose, the central molecule of carbohydrate metabolism. Lactose is metabolized at a high rate only when lactose is present and glucose is largely depleted from the medium. This metabolic adjustment is achieved by repressing transcription of the *lac* operon until

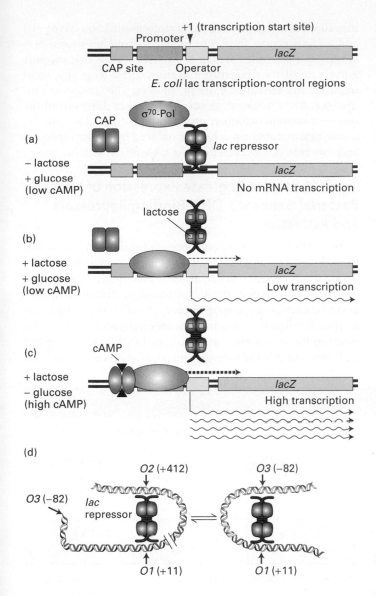

+1 (transcription start site)

Promoter ▼

CAP site    Operator    lacZ

*E. coli* lac transcription-control regions

(a)
− lactose
+ glucose
(low cAMP)

CAP    σ⁷⁰-Pol

*lac* repressor

lacZ

No mRNA transcription

(b)
+ lactose
+ glucose
(low cAMP)

lactose

lacZ

Low transcription

(c)
+ lactose
− glucose
(high cAMP)

cAMP

lacZ

High transcription

(d)

O2 (+412)          O3 (−82)

O3 (−82)
*lac*
repressor

O1 (+11)          O1 (+11)

**FIGURE 7-3 Regulation of transcription from the *lac* operon of *E. coli*.** (*Top*) The transcription-control region, composed of ≈100 base pairs, includes three protein-binding regions: the CAP site, which binds catabolite activator protein; the *lac* promoter, which binds the σ⁷⁰–RNA polymerase complex; and the *lac* operator, which binds *lac* repressor. The *lacZ* gene encoding the enzyme β-galactosidase, the first of three genes in the operon, is shown to the right. (a) In the absence of lactose, very little *lac* mRNA is produced because the *lac* repressor binds to the operator, inhibiting transcription initiation by σ⁷⁰–RNA polymerase. (b) In the presence of glucose and lactose, *lac* repressor binds lactose and dissociates from the operator, allowing σ⁷⁰–RNA polymerase to initiate transcription at a low rate. (c) Maximal transcription of the *lac* operon occurs in the presence of lactose and absence of glucose. In this situation, cAMP increases in response to the low glucose concentration and forms the CAP-cAMP complex, which binds to the CAP site, where it interacts with RNA polymerase to stimulate the rate of transcription initiation. (d) The tetrameric *lac* repressor binds to the primary *lac* operator (*O1*) and one of two secondary operators (*O2* or *O3*) simultaneously. The two structures are in equilibrium. [Part (d) adapted from B. Muller-Hill, 1998, *Curr. Opin. Microbiol.* **1**:145.]

operator. As a result, the polymerase can bind to the promoter and initiate transcription of the *lac* operon. However, when glucose also is present, the rate of transcription initiation (i.e., the number of times per minute different RNA polymerase molecules initiate transcription) is very low, resulting in synthesis of only low levels of *lac* mRNA and the proteins encoded in the *lac* operon (Figure 7-3b). The frequency of transcription initiation is low because the −35 and −10 sequences in the *lac* promoter differ from the ideal σ⁷⁰-binding sequences shown previously.

Once glucose is depleted from the medium and the intracellular glucose concentration falls, *E. coli* cells respond by synthesizing cyclic AMP, or cAMP. As the concentration of cAMP increases, it binds to a site in each subunit of the dimeric CAP protein, causing a conformational change that allows the protein to bind to the CAP site in the *lac* transcription-control region. The bound CAP-cAMP complex interacts with the polymerase bound to the promoter, greatly stimulating the rate of transcription initiation. This activation leads to synthesis of high levels of *lac* mRNA and subsequently of the enzymes encoded by the *lac* operon (Figure 7-3c).

In fact, the *lac* operon is more complex than depicted in the simplified model of Figure 7-3, parts (a)–(c). The tetrameric *lac* repressor actually binds to two sites simultaneously, one at the primary operator (*lacO1*) that overlaps the region of DNA bound by RNA polymerase at the promoter and at one of two secondary operators centered at +412 (*lacO2*) and −82 (*lacO3*) (Figure 7-3d). The *lac* repressor tetramer is a dimer of dimers. Each dimer binds to one operator. Simultaneous binding of the tetrameric *lac* repressor to the primary *lac* operator O1 and one of the two secondary operators is possible because DNA is quite flexible, as we saw in the wrapping of DNA around the surface of a histone octamer in the nucleosomes of eukaryotes (Figure 6-29).

lactose is present and allowing synthesis of only low levels of *lac* mRNA until the cytosolic concentration of glucose falls to low levels. Transcription of the *lac* operon under different conditions is controlled by *lac* repressor and *catabolite activator protein* (CAP) (also called CRP, for *catabolite receptor protein*), each of which binds to a specific DNA sequence in the *lac* transcription-control region called the **operator** and the **CAP site,** respectively (Figure 7-3, *top*).

For transcription of the *lac* operon to begin, the σ⁷⁰ subunit of the RNA polymerase must bind to the *lac* promoter at the −35 and −10 promoter sequences. When no lactose is present, the *lac* repressor binds to the *lac* operator, which overlaps the transcription start site. Therefore, *lac* repressor bound to the operator site blocks σ⁷⁰ binding and hence transcription initiation by RNA polymerase (Figure 7-3a). When lactose is present, it binds to specific binding sites in each subunit of the tetrameric *lac* repressor, causing a conformational change in the protein that makes it dissociate from the *lac*

These secondary operators function to increase the local concentration of *lac* repressor in the micro-vicinity of the primary operator where repressor binding blocks RNA polymerase binding. Since the equilibrium of binding reactions depends on the concentrations of the binding partners, the resulting increased local concentration of *lac* repressor in the vicinity of *O1* increases repressor binding to *O1*. There are approximately 10 *lac* repressor tetramers per *E. coli* cell. Because of binding to *O2* and *O3*, there is nearly always a *lac* repressor tetramer much closer to *O1* than would otherwise be the case if the 10 repressors were diffusing randomly through the cell. If both *O2* and *O3* are mutated so that the *lac* repressor no longer binds to them with high affinity, repression at the *lac* promoter is reduced by a factor of 70. Mutation of only *O2* or only *O3* reduces repression twofold, indicating that either one of these secondary operators provides most of the stimulation of repression.

Although the promoters for different *E. coli* genes exhibit considerable homology, their exact sequences differ. The promoter sequence determines the intrinsic rate at which an RNA polymerase–σ complex initiates transcription of a gene in the absence of a repressor or activator protein. Promoters that support a high rate of transcription initiation have −10 and −35 sequences similar to the ideal promoter shown previously and are called *strong promoters*. Those that support a low rate of transcription initiation differ from this ideal sequence and are called *weak promoters*. The *lac* operon, for instance, has a weak promoter. Its sequence differs from the consensus strong promoter at several positions. This low intrinsic rate of initiation is further reduced by the *lac* repressor and substantially increased by the cAMP-CAP activator.

## Small Molecules Regulate Expression of Many Bacterial Genes via DNA-Binding Repressors and Activators

Transcription of most *E. coli* genes is regulated by processes similar to those described for the *lac* operon, although the detailed interactions differ at each promoter. The general mechanism involves a specific repressor that binds to the operator region of a gene or operon, thereby blocking transcription initiation. A small-molecule ligand (or ligands) binds to the repressor, controlling its DNA-binding activity and consequently the rate of transcription as appropriate for

| TABLE 7-1 | Sigma Factors of *E. coli* | | |
|---|---|---|---|
| | | **Promoter Consensus** | |
| **Sigma Factor** | **Promoters Recognized** | **−35 Region** | **−10 Region** |
| $\sigma^{70}$ ($\sigma^D$) | Housekeeping genes, most genes in exponentially replicating cells | TTGACA | TATAAT |
| $\sigma^S$ ($\sigma^{38}$) | Stationary-phase genes and general stress response | TTGACA | TATAAT |
| $\sigma^{32}$ ($\sigma^H$) | Induced by unfolded proteins in the cytoplasm; genes encoding chaperones that refold unfolded proteins and protease systems leading to the degradation of unfolded proteins in the cytoplasm | TCTCNCCCTTGAA | CCCCATNTA |
| $\sigma^E$ ($\sigma^{24}$) | Activated by unfolded proteins in the periplasmic space and cell membrane; genes encoding proteins that restore integrity to the cellular envelope | GAACTT | TCTGA |
| $\sigma^F$ ($\sigma^{28}$) | Genes involved in flagellum assembly | CTAAA | CCGATAT |
| FecI ($\sigma^{18}$) | Genes required for iron uptake | TTGGAAA | GTAATG |
| | | **−24 REGION** | **−12 REGION** |
| $\sigma^{54}$ ($\sigma^N$) | Genes for nitrogen metabolism and other functions | CTGGNA | TTGCA |

SOURCES: T. M. Gruber and C. A. Gross, 2003, *Ann. Rev. Microbiol.* **57**:441, S. L. McKnight and K. R. Yamamoto, eds., Cold Spring Harbor Laboratory Press; R. L. Gourse, W. Ross, and S. T. Rutherford, 2006, *J. Bacteriol.* **188**:4627; U. K. Sharma and D. Chatterji, 2010, *FEMS Microbiol. Rev.* **34**:646.

the needs of the cell. As for the *lac* operon, many eubacterial transcription-control regions contain one or more secondary operators that contribute to the level of repression.

Specific activator proteins, such as CAP in the *lac* operon, also control transcription of a subset of bacterial genes that have binding sites for the activator. Like CAP, other activators bind to DNA together with RNA polymerase, stimulating transcription from a specific promoter. The DNA-binding activity of an activator can be modulated in response to cellular needs by binding specific small-molecule ligands (e.g., cAMP) or by post-translational modifications, such as phosphorylation, that alter the conformation of the activator.

## Transcription Initiation from Some Promoters Requires Alternative Sigma Factors

Most *E. coli* promoters interact with $\sigma^{70}$-RNA polymerase, the major initiating form of the bacterial enzyme. Transcription of certain groups of genes, however, is initiated by *E. coli* RNA polymerases containing one of several alternative sigma factors that recognize different consensus promoter sequences than $\sigma^{70}$ does (Table 7-1). These alternative $\sigma$-factors are required for the transcription of sets of genes with related functions such as those involved in the response to heat shock or nutrient deprivation, motility, or sporulation in gram-positive eubacteria. In *E. coli* there are six alternative $\sigma$-factors in addition to the major "housekeeping" $\sigma$-factor, $\sigma^{70}$. The genome of the gram-positive, sporulating bacterium *Streptomyces coelicolor* encodes 63 $\sigma$-factors, the current record, based on sequence analysis of hundreds of eubacterial genomes. Most are structurally and functionally related to $\sigma^{70}$. But one class is unrelated, represented in *E. coli* by $\sigma^{54}$. Transcription initiation by RNA polymerases containing $\sigma^{70}$-like factors is regulated by repressors and activators that bind to DNA near the region where the polymerase binds, similar to initiation by $\sigma^{70}$-RNA polymerase itself.

## Transcription by $\sigma^{54}$-RNA Polymerase Is Controlled by Activators That Bind Far from the Promoter

The sequence of one *E. coli* sigma factor, $\sigma^{54}$, is distinctly different from that of all the $\sigma^{70}$-like factors. Transcription of genes by RNA polymerases containing $\sigma^{54}$ is regulated solely by activators whose binding sites in DNA, referred to as **enhancers,** generally are located 80–160 base pairs upstream from the start site. Even when enhancers are moved more than a kilobase away from a start site, $\sigma^{54}$-activators can activate transcription.

The best-characterized $\sigma^{54}$-activator—the NtrC protein (nitrogen regulatory protein C)—stimulates transcription of the *glnA* gene. *glnA* encodes the enzyme glutamine synthetase, which synthesizes the amino acid glutamine from glutamic acid and ammonia. The $\sigma^{54}$-RNA polymerase binds to the *glnA* promoter but does not melt the DNA strands and initiate transcription until it is activated by NtrC, a dimeric protein. NtrC, in turn, is regulated by a protein kinase called

NtrB. In response to low levels of glutamine, NtrB phosphorylates dimeric NtrC, which then binds to an enhancer upstream of the *glnA* promoter. Enhancer-bound phosphorylated NtrC then stimulates the $\sigma^{54}$-polymerase bound at the promoter to separate the DNA strands and initiate transcription.

Electron microscopy studies have shown that phosphorylated NtrC bound at enhancers and $\sigma^{54}$-polymerase bound at the promoter directly interact, forming a loop in the DNA between the binding sites (Figure 7-4). As discussed later in this chapter, this activation mechanism resembles the predominant mechanism of transcriptional activation in eukaryotes.

NtrC has ATPase activity, and ATP hydrolysis is required for activation of bound $\sigma^{54}$-polymerase by phosphorylated NtrC. Evidence for this is that mutants with an NtrC defective in ATP hydrolysis are invariably defective in stimulating the $\sigma^{54}$-polymerase to melt the DNA strands at the transcription start site. It is postulated that ATP hydrolysis supplies the energy required for melting the DNA strands. In contrast, the $\sigma^{70}$-polymerase does not require ATP hydrolysis to separate the strands at a start site.

## Many Bacterial Responses Are Controlled by Two-Component Regulatory Systems

As we have just seen, control of the *E. coli glnA* gene depends on two proteins, NtrC and NtrB. Such two-component regulatory systems control many responses of bacteria to changes in their environment. At high concentrations of glutamine, glutamine binds to a sensor domain of NtrB, causing a conformational change in the protein that inhibits its histidine kinase activity (Figure 7-5a). At the same time, the regulatory domain of NtrC blocks the DNA-binding domain from binding the *glnA* enhancers. Under conditions of low glutamine, glutamine dissociates from the sensor domain in the NtrB protein, leading to activation of a histidine kinase *transmitter* domain in NtrB that transfers the $\gamma$-phosphate of ATP to a histidine residue (H) in the *transmitter* domain. This phosphohistidine then transfers the phosphate to an aspartic acid residue (D) in the NtrC protein. This causes a conformational change in NtrC that unmasks the NtrC DNA-binding domain so that it can bind to the *glnA* enhancers.

Many other bacterial responses are regulated by two proteins with homology to NtrB and NtrC (Figure 7-5b). In each of these regulatory systems, one protein, called a *histidine kinase sensor,* contains a latent histidine kinase transmitter domain that is regulated in response to environmental changes detected by a sensor domain. When activated, the transmitter domain transfers the $\gamma$-phosphate of ATP to a histidine residue in the transmitter domain. The second protein, called a *response regulator,* contains a *receiver* domain homologous to the region of NtrC containing the aspartic acid residue that is phosphorylated by activated NtrB. The response regulator contains a second functional domain that is regulated by phosphorylation of the receiver domain. In many cases this domain of the response regulator is a sequence-specific DNA-binding domain that binds to related DNA sequences and functions either as a repressor, like the *lac*

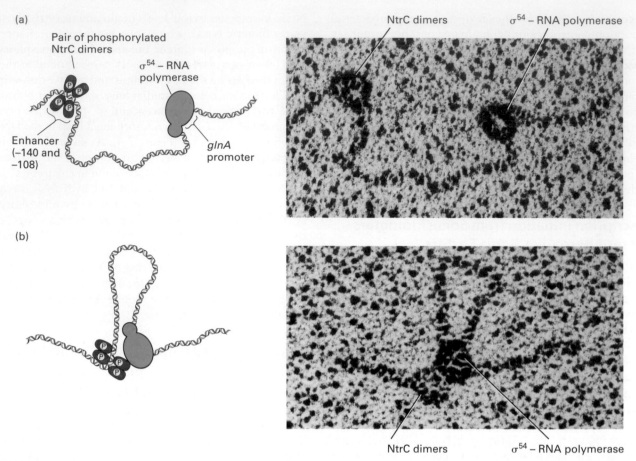

(a)

Pair of phosphorylated
NtrC dimers

σ⁵⁴ – RNA
polymerase

Enhancer
(–140 and
–108)

*glnA*
promoter

(b)

NtrC dimers

σ⁵⁴ – RNA polymerase

NtrC dimers

σ⁵⁴ – RNA polymerase

**EXPERIMENTAL FIGURE 7-4  DNA looping permits interaction of bound NtrC and σ⁵⁴–RNA polymerase.** (a) Drawing (*left*) and electron micrograph (*right*) of DNA restriction fragment with phosphorylated NtrC dimers binding to the enhancer region near one end and σ⁵⁴–RNA polymerase bound to the *glnA* promoter near the other end.

(b) Drawing (*left*) and electron micrograph (*right*) of the same fragment preparation showing NtrC dimers and σ⁵⁴–RNA polymerase binding to each other with the intervening DNA forming a loop between them. [Micrographs from W. Su et al., 1990, *Proc. Nat'l Acad. Sci. USA* **87**:5505; courtesy of S. Kustu.]

repressor, or as an activator, like CAP or NtrC, regulating the transcription of specific genes. However, the effector domain can have other functions as well, such as controlling the direction in which the bacterium swims in response to a concentration gradient of nutrients. Although all transmitter domains are homologous (as are receiver domains), the transmitter domain of a specific sensor protein will phosphorylate only the receiver domains of specific response regulators, allowing specific responses to different environmental changes. Similar two-component histidyl-aspartyl phospho-relay regulatory systems are also found in plants.

## Control of Transcription Elongation

In addition to regulation of transcription initiation by activators and repressors, expression of many bacterial operons is controlled by regulation of transcriptional elongation in the promoter-proximal region. This was first discovered in studies of *Trp* operon transcription in *E. coli* (Figure 4-13). *Trp* operon transcription is repressed by the Trp repressor when the concentration of tryptophan in the cytoplasm is high. But the low level of transcription initiation that still

occurs is further controlled by a process called *attenuation* when the concentration of charged tRNA^Trp is sufficient to support a high rate of protein synthesis. The first 140 nt of the Trp operon does not encode proteins required for tryptophan biosynthesis, but rather consists of a leader sequence as diagrammed in Figure 7-6a. Region 1 of this leader sequence contains two successive Trp codons. Region 3 can base-pair with both regions 2 and 4. A ribosome follows closely behind the RNA polymerase, initiating translation of the leader peptide shortly after the 5′ end of the Trp leader sequence emerges from the RNA polymerase. When the concentration of tRNA^Trp is sufficient to support a high rate of protein synthesis, the ribosome translates through region 1 into region 2, blocking the ability of region 2 to base-pair with region 3 as it emerges from the surface of the transcribing RNA polymerase (Figure 7-6b, *left*). Instead, region 3 base-pairs with region 4 as soon as it emerges from the surface of the polymerase, forming an RNA hairpin (see Figure 4-9a) followed by several uracils, which is a signal for bacterial RNA polymerase to pause transcription and terminate. As a consequence, the remainder of the long *Trp* operon is not transcribed, and the cell does not waste energy required for

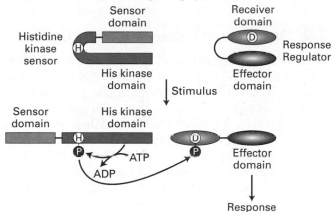

**(a) Two-component system regulating response to low Gln**

NtrB

High [Gln]

Sensor domain

Gln

His kinase transmitter domain

NtrC

Regulatory domain

D

DNA-binding domain

Low [Gln]

Sensor domain    His kinase transmitter domain

H

P

ATP

ADP

DNA-binding domain

D

P

Gln enhancer

**(b) General two-component signaling system**

Histidine kinase sensor

Sensor domain

H

His kinase domain

Receiver domain

D

Response Regulator

Effector domain

Stimulus

Sensor domain

His kinase domain

H

P

ATP

ADP

D

P

Effector domain

Response

**FIGURE 7-5 Two-component regulatory systems.** At low cytoplasmic concentrations of glutamine, glutamine dissociates from NtrB, resulting in a conformational change that activates a protein kinase transmitter domain that transfers an ATP γ-phosphate to a conserved histidine (H) in the transmitter domain. This phosphate is then transferred to an aspartic acid (D) in the regulatory domain of the response regulator NtrC. This converts Ntrc into its activated form, which binds the enhancer sites upstream of the *glnA* promoter (Figure 7-4). (b) General organization of two-component histidyl-aspartyl phospho-relay regulatory systems in bacteria and plants. [Adapted from A. H. West and A. M. Stock, 2001, *Trends Biochem. Sci.* **26:**369.]

its synthesis or for the translation of the encoded proteins when the concentration of tryptophan is high.

However, when the concentration of tRNA$^{Trp}$ is not sufficient to support a high rate of protein synthesis, the ribosome stalls at the two successive Trp codons in region 1 (Figure 7-6b, *right*). As a consequence, region 2 base-pairs with region 3 as soon as it emerges from the transcribing RNA polymerase. This prevents region 3 from base-pairing with region 4, so the 3-4 hairpin does not form and does not cause pausing by RNA polymerase or transcription termination. As a result, the proteins required for tryptophan synthesis are translated by ribosomes that initiate translation at the start codons for each of these proteins in the long polycistronic Trp mRNA.

Attenuation of transcription elongation also occurs at some operons and single genes encoding enzymes involved in the biosynthesis of other amino acids and metabolites through the function of *riboswitches*. Riboswitches form RNA tertiary structures that can bind small molecules when they are present at sufficiently high concentration. In some

**(a) *trp* leader RNA**

Translation start codon

1

50

100

140

5'

1

2

3

4

UUUUU 3'

**(b) Translation of *trp* leader**

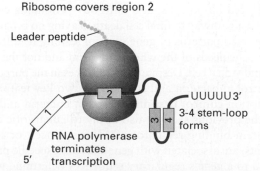

**High tryptophan**
Ribosome covers region 2

Leader peptide

2

1

UUUUU 3'

3  4

3-4 stem-loop forms

RNA polymerase terminates transcription

5'

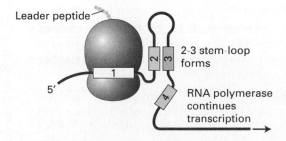

**Low tryptophan**
Ribosome is stalled at trp codons in region 1

Leader peptide

2  3

2-3 stem-loop forms

1

5'

4

RNA polymerase continues transcription

**FIGURE 7-6 Transcription control by regulation of RNA polymerase elongation and termination in the *E. coli Trp* operon.** (a) Diagram of the 140-nucleotide *trp* leader RNA. Colored regions are critical to the control of attenuation. (b) Translation of the *trp* leader sequence begins from the 5' end soon after it is synthesized, while synthesis of the rest of the polycistronic *trp* mRNA molecule continues.

At high concentrations of amino-acylated tRNA$^{Trp}$, formation of the 3-4 stem-loop followed by a series of Us causes termination of transcription. At low amino-acylated tRNA$^{Trp}$, region 3 is sequestered in the 2-3 stem-loop and cannot base-pair with region 4. In the absence of the stem-loop structure required for termination, transcription of the *trp* operon continues. [See C. Yanofsky, 1981, *Nature* **289:**751.]

cases this results in the formation of hairpin structures that lead to early termination of transcription as in the *Trp* operon. When the concentration of these small-molecule ligands is lower, the metabolites are not bound by the RNA and alternative RNA structures form that do not induce transcription termination. As discussed below, although the mechanism of transcriptional pausing and termination in eukaryotes is different, regulation of promoter-proximal transcriptional pausing and termination has recently been discovered to occur frequently in the regulation of gene expression in multicellular organisms as well.

## KEY CONCEPTS of Section 7.1

### Control of Gene Expression in Bacteria

- Gene expression in both prokaryotes and eukaryotes is regulated primarily by mechanisms that control the initiation of transcription.

- The first step in the initiation of transcription in *E. coli* is binding of the σ subunit complexed with an RNA polymerase to a promoter.

- The nucleotide sequence of a promoter determines its strength, that is, how frequently different RNA polymerase molecules can bind and initiate transcription per minute.

- Repressors are proteins that bind to operator sequences that overlap or lie adjacent to promoters. Binding of a repressor to an operator inhibits transcription initiation.

- The DNA-binding activity of most bacterial repressors is modulated by small-molecule ligands. This allows bacterial cells to regulate transcription of specific genes in response to changes in the concentration of various nutrients in the environment and metabolites in the cytoplasm.

- The *lac* operon and some other bacterial genes also are regulated by activator proteins that bind next to promoters and increase the rate of transcription initiation by interacting directly with RNA polymerase bound to an adjacent promoter.

- The major sigma factor in *E. coli* is $\sigma^{70}$, but several other, less abundant sigma factors are also found, each recognizing different consensus promoter sequences or interacting with different activators.

- Transcription initiation by all *E. coli* RNA polymerases, except those containing $\sigma^{54}$, can be regulated by repressors and activators that bind near the transcription start site (see Figure 7-3).

- Genes transcribed by $\sigma^{54}$-RNA polymerase are regulated by activators that bind to enhancers located ≈100 base pairs upstream from the start site. When the activator and $\sigma^{54}$-RNA polymerase interact, the DNA between their binding sites forms a loop (see Figure 7-4).

- In two-component regulatory systems, one protein acts as a sensor, monitoring the level of nutrients or other components in the environment. Under appropriate conditions, the γ-phosphate of an ATP is transferred first to a histidine in the sensor protein and then to an aspartic acid in a second protein, the response regulator. The phosphorylated response regulator then performs a specific function in response to the stimulus, such as binding to DNA regulatory sequences, thereby stimulating or repressing transcription of specific genes (see Figure 7-5).

- Transcription in bacteria can also be regulated by controlling transcriptional elongation in the promoter-proximal region. This can be regulated by ribosome binding to the nascent mRNA as in the case of the *Trp* operon (Figure 7-6), or by riboswitches, RNA tertiary structures that bind small molecules, to determine whether a stem-loop followed by a string of uracils forms, causing the bacterial RNA polymerase to pause and terminate transcription.

## 7.2 Overview of Eukaryotic Gene Control

In bacteria, gene control serves mainly to allow a single cell to adjust to changes in its environment so that its growth and division can be optimized. In multicellular organisms, environmental changes also induce changes in gene expression. An example is the response to low oxygen (hypoxia) in which a specific set of genes is rapidly induced that helps the cell survive under the hypoxic conditions. These include secreted angiogenic proteins that stimulate the growth and penetration of new capillaries into the surrounding tissue. However, the most characteristic and biologically far-reaching purpose of gene control in multicellular organisms is execution of the genetic program that underlies embryological development. Generation of the many different cell types that collectively form a multicellular organism depends on the right genes being activated in the right cells at the right time during the developmental period.

In most cases, once a developmental step has been taken by a cell, it is not reversed. Thus these decisions are fundamentally different from the reversible activation and repression of bacterial genes in response to environmental conditions. In executing their genetic programs, many differentiated cells (e.g., skin cells, red blood cells, and antibody-producing cells) march down a pathway to final cell death, leaving no progeny behind. The fixed patterns of gene control leading to differentiation serve the needs of the whole organism and not the survival of an individual cell. Despite the differences in the purposes of gene control in bacteria and eukaryotes, two key features of transcription control first discovered in bacteria and described in the previous section also apply to eukaryotic cells. First, protein-binding regulatory DNA sequences, or control elements, are associated with genes. Second, specific proteins that bind to a gene's regulatory sequences determine where transcription will start and either activate or repress its transcription. A fundamental difference between transcription control in bacteria and eukaryotes is a consequence of the association of eukaryotic chromosomal DNA with histone octamers, forming nucleosomes that associate into chromatin fibers that further associate into chromatin of varying degrees of condensation (Figures 6-29, 6-30, 6-32, and 6-33). Eukaryotic cells exploit chromatin structure to regulate transcription, a mechanism of transcription control that is not available to bacteria. As represented in Figure 7-2, in multicellular eukaryotes, many inactive genes are assembled into condensed chromatin, which inhibits the binding of RNA polymerases and general transcription factors required for transcription initiation. Activator proteins bind to control elements near the transcription start site of a gene as well as kilobases away and promote chromatin decondensation, binding of RNA polymerase to the promoter, and transcriptional elongation through chromatin. Repressor proteins bind to alternative control elements, causing condensation of chromatin and inhibition of polymerase binding or elongation. In this section, we discuss general principles of eukaryotic gene control and point out some similarities and differences between bacterial and eukaryotic systems. Subsequent sections of this chapter will address specific aspects of eukaryotic transcription in greater detail.

## Regulatory Elements in Eukaryotic DNA Are Found Both Close to and Many Kilobases Away from Transcription Start Sites

Direct measurements of the transcription rates of multiple genes in different cell types have shown that regulation of transcription, either at the initiation step or during elongation away from the transcription start site, is the most widespread form of gene control in eukaryotes, as it is in bacteria. In eukaryotes, as in bacteria, a DNA sequence that specifies where RNA polymerase binds and initiates transcription of a gene is called a promoter. Transcription from a particular promoter is controlled by DNA-binding proteins that are functionally equivalent to bacterial repressors and activators. Recent results suggest that the intrinsic ability of the DNA sequence of a promoter region to associate with histone octamers also influences transcription. Since transcriptional regulatory proteins can often function either to activate or to repress transcription, depending on their association with other proteins, they are more generally called *transcription factors*. The DNA control elements in eukaryotic genomes that bind transcription factors often are located much farther from the promoter they regulate than is the case in prokaryotic genomes. In some cases, transcription factors that regulate expression of protein-coding genes in higher eukaryotes bind at regulatory sites tens of thousands of base pairs either **upstream** (opposite to the direction of transcription) or **downstream** (in the same direction as transcription) from the promoter. As a result of this arrangement, transcription of a single gene may be regulated by the binding of multiple different transcription factors to alternative control elements, directing expression of the same gene in different types of cells and at different times during development.

For example, several separate transcription-control DNA sequences regulate expression of the mammalian gene encoding the transcription factor *Pax6*. As mentioned earlier, Pax6 protein is required for development of the eye. Pax6 is also required for the development of certain regions of the brain and spinal cord, and the cells in the pancreas that secrete hormones such as insulin. As also mentioned earlier, heterozygous humans with only one functional *Pax6* gene are born with *aniridia*, a lack of irises in the eyes (Figure 7-1a). The *Pax6* gene is expressed from at least three alternative promoters that function in different cell types and at different times during embryogenesis (Figure 7-7a).

Researchers often analyze gene control regions by preparing recombinant DNA molecules that can contain a fragment of DNA to be tested with the coding region for a **reporter gene** that is easy to assay. Typical reporter genes are luciferase, which generates light that can be assayed with great sensitivity and over many orders of magnitude of intensity using a luminometer. Other frequently used reporter genes encode a green fluorescent protein, which can be visualized by fluorescence microscopy (see Figures 9-8d and 9-15) and *E. coli* β-*galactosidase*, which generates an intensely blue insoluble precipitate when incubated with the colorless soluble lactose analog X-gal. When transgenic mice are prepared (see Figure 5-43) containing a β-galactosidase reporter gene fused to 8 kb of DNA upstream from *Pax6* exon 0, β-galactosidase is observed in the developing lens, cornea, and pancreas of the embryo halfway through gestation (Figure 7-7b). Analysis of transgenic mice with smaller fragments of DNA from this region allowed the mapping of separate transcription-control regions regulating transcription in the pancreas and in the lens and cornea. Transgenic mice with other reporter gene constructs revealed additional transcription-control regions (Figure 7-7a). These control transcription in the developing retina and different regions of the developing brain (encephalon). Some of these transcription-control regions are in introns between exons 4 and 5 and between exons 7 and 8. For example, a reporter gene under control of the region labeled *retina* in Figure 7-7a between exons 4 and 5 led to reporter gene expression specifically in the retina (Figure 7-7c).

Control regions for many genes are found several hundreds of kilobases away from the coding exons of the gene. One method for identifying such distant control regions is to compare the sequences of distantly related organisms. Transcription-control regions for a conserved gene are also often conserved and can be recognized in the background of a nonfunctional sequence that diverges during evolution. For example, there is a human DNA sequence ≈500 kilobases downstream of the *SALL1* gene that is highly conserved in mice, frogs, chickens, and fish (Figure 7-8a). *SALL1* encodes a transcription repressor required for normal development of the lower intestine, kidneys, limbs, and ears. When transgenic mice were generated containing this conserved DNA sequence linked to a β-galactosidase reporter gene (Figure 7-8b), the transgenic embryos expressed a very high level of the β-galactosidase reporter gene specifically in the developing limb buds (Figure 7-8c). Human patients with deletions in this region of the genome develop with limb abnormalities. These results indicate that this conserved region directs transcription of the *SALL1* gene in the developing limb. Presumably, other enhancers control expression of this gene in other

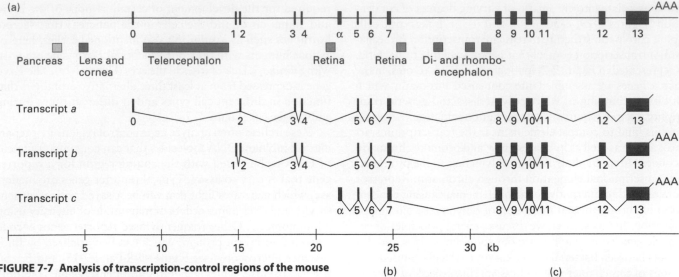

**FIGURE 7-7 Analysis of transcription-control regions of the mouse *Pax6* gene in transgenic mice.** (a) Three alternative *Pax6* promoters are utilized at distinct times during embryogenesis in different specific tissues of the developing embryo. Transcription-control regions regulating expression of *Pax6* in different tissues are indicated by colored rectangles. The telencephalon-specific control region in intron 1 between exons 0 and 1 has not been mapped to high resolution. The other control regions shown are ≈200–500 base pairs in length. (b) β-galactosidase expressed in tissues of a mouse embryo with a β-galactosidase reporter transgene 10.5 days after fertilization. The genome of the mouse embryo contained a transgene with 8 kb of DNA upstream from exon 0 fused to the β-galactosidase coding region. Lens pit (LP) is the tissue that will develop into the lens of the eye. Expression was also observed in tissue that will develop into the pancreas (P). (c) β-galactosidase expression in a 13.5-day embryo with a β-galactosidase reporter gene under control of the sequence in part (a) between exons 4 and 5 marked Retina. Arrow points to nasal and temporal regions of the developing retina. *Pax6* transcription-control regions have also been found ≈17 kb downstream from the 3′ exon in an intron of the neighboring gene. [Part (a) adapted from B. Kammendal et al., 1999, *Dev. Biol.* **205:**79. Parts (b) and (c) courtesy of Peter Gruss.]

types of cells, where it functions in the normal development of ears, the lower intestine, and kidneys. After discussing the proteins that carry out transcription in eukaryotic cells and eukaryotic promoters, we will return to a discussion of how such distant transcription-control regions, called **enhancers,** are thought to function.

## Three Eukaryotic RNA Polymerases Catalyze Formation of Different RNAs

The nuclei of all eukaryotic cells examined so far (e.g., vertebrate, *Drosophila,* yeast, and plant cells) contain three different RNA polymerases, designated I, II, and III. These enzymes are eluted at different salt concentrations during ion-exchange chromatography, reflecting the polymerases' various net charges. The three polymerases also differ in their sensitivity to α-amanitin, a poisonous cyclic octapeptide produced by

some mushrooms (Figure 7-9). RNA polymerase I is insensitive to α-amanitin, but RNA polymerase II is very sensitive—the drug binds near the active site of the enzyme and inhibits translocation of the enzyme along the DNA template. RNA polymerase III has intermediate sensitivity.

Each eukaryotic RNA polymerase catalyzes transcription of genes encoding different classes of RNA (Table 7-2). *RNA polymerase I* (Pol I), located in the nucleolus, transcribes genes encoding precursor rRNA (**pre-rRNA**), which is processed into 28S, 5.8S, and 18S rRNAs. *RNA polymerase III* (Pol III) transcribes genes encoding tRNAs, 5S rRNA, and an array of small, stable RNAs, including one involved in RNA splicing (U6) and the RNA component of the signal-recognition particle (SRP) involved in directing nascent proteins to the endoplasmic reticulum (Chapter 13). *RNA polymerase II* (Pol II) transcribes all protein-coding genes: that is, it functions in production of mRNAs. RNA polymerase II also produces four of the five small nuclear RNAs that take part in RNA splicing and micro-RNAs (miRNAs) involved in translation control as well as the closely related endogenous small interfering RNAs (siRNAs) (see Chapter 8).

Each of the three eukaryotic RNA polymerases is more complex than *E. coli* RNA polymerase, although their structures are similar (Figure 7-10a, b). All three contain two large subunits and 10–14 smaller subunits, some of which are

(a) Comparative analysis

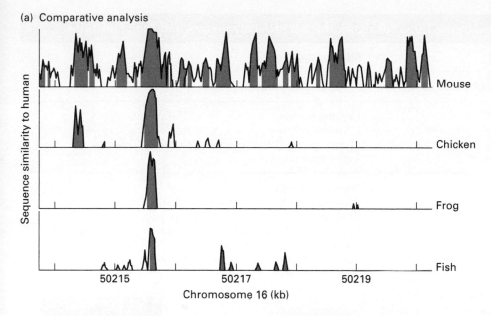

Mouse

Chicken

Frog

Fish

50215        50217        50219

Chromosome 16 (kb)

(b) Mouse egg microinjection

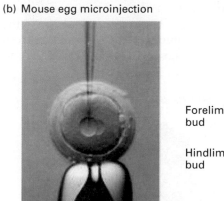

(c) E11.5 reporter staining

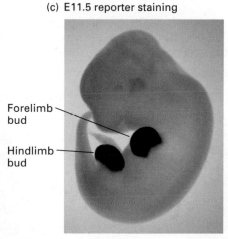

Forelimb bud

Hindlimb bud

EXPERIMENTAL FIGURE 7-8 The human *SALL1* gene enhancer activates expression of a reporter gene in limb buds of the developing mouse embryo.
(a) Graphic representation of the conservation of DNA sequence in a region of the human genome (from 50214–50220.5 kb of the chromosome 16 sequence) ≈500 kb downstream from the *SALL1* gene encoding a zinc-finger transcription repressor. A region of ≈500 bp of noncoding sequence is conserved from fish to human. Nine hundred base pairs including this conserved region were inserted into a plasmid next to the coding region for *E. coli* β-galactosidase. (b) The plasmid was microinjected into a pronucleus of a fertilized mouse egg and implanted in the uterus of a pseudo-pregnant mouse to generate a transgenic mouse embryo with the "reporter gene" on the injected plasmid incorporated into its genome (see Figure 5-43). (c) After 11.5 days of development, when limb buds develop, the fixed and permeabilized embryo was incubated in X-gal, which is converted by β-galactosidase into an insoluble, intensely blue compound. The ≈900-bp region of human DNA contained an enhancer that stimulated strong transcription of the β-galactosidase reporter gene in limb buds specifically. [From the VISTA Enhancer Browser, http://enhancer. lbl.gov. Parts (b) and (c) courtesy of Len A. Pennacchio, Joint Genome Institute, Lawrence Berkeley National Laboratory.]

common between two or all three of the polymerases. The best-characterized eukaryotic RNA polymerases are from the yeast *Saccharomyces cerevisiae*. Each of the yeast genes encoding the polymerase subunits has been subjected to gene-knockout mutations and the resulting phenotypes char- acterized. In addition, the three-dimensional structure of yeast RNA polymerase II has been determined (Figure 7-10b, c). The three nuclear RNA polymerases from all eukaryotes so far examined are very similar to those of yeast. Plants contain two additional nuclear RNA polymerases (RNA polymerases IV and V), which are closely related to their RNA polymerase II but have a unique large subunit and some additional unique subunits. These function in transcriptional

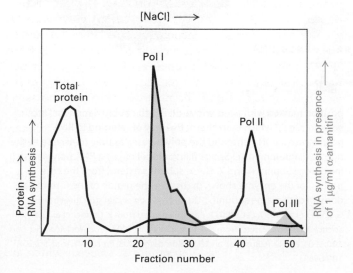

EXPERIMENTAL FIGURE 7-9 Column chromatography separates and identifies the three eukaryotic RNA polymerases, each with its own sensitivity to α-amanitin. A protein extract from the nuclei of cultured eukaryotic cells was passed through a DEAE Sephadex column and adsorbed protein eluted (black curve) with a solution of constantly increasing NaCl concentration. Fractions from the eluate were assayed for RNA polymerase activity (red curve). At a concentration of 1 μg/ml, α-amanitin inhibits polymerase II activity but has no effect on polymerases I and III (green shading). Polymerase III is inhibited by 10 μg/ml of α-amanitin, whereas polymerase I is unaffected even at this higher concentration. [See R. G. Roeder, 1974, *J. Biol. Chem.* **249**:241.]

| TABLE 7-2 | Classes of RNA Transcribed by the Three Eukaryotic Nuclear RNA Polymerases and Their Functions | |
|---|---|---|
| **Polymerase** | **RNA Transcribed** | **RNA Function** |
| RNA polymerase I | Pre r-RNA (28S, 18S, 5.8S rRNAs) | Ribosome components, protein synthesis |
| RNA polymerase II | mRNA | Encodes protein |
| | snRNAs | RNA splicing |
| | siRNAs | Chromatin-mediated repression, translation control |
| | miRNAs | Translation control |
| RNA polymerase III | tRNAs | Protein synthesis |
| | 5S rRNA | Ribosome component, protein synthesis |
| | snRNA U6 | RNA splicing |
| | 7S RNA | Signal-recognition particle for insertion of polypeptides into the endoplasmic reticulum |
| | Other stable short RNAs | Various functions, unknown for many |

repression directed by nuclear siRNAs in plants, discussed toward the end of this chapter.

The two large subunits of all three eukaryotic RNA polymerases (and RNA polymerases IV and V of plants) are related to each other and are similar to the *E. coli* β′ and β subunits, respectively (Figure 7-10). Each of the eukaryotic polymerases also contains an ω-like and two nonidentical α-like subunits (Figure 7-11). The extensive similarity in the structures of these core subunits in RNA polymerases from various sources indicates that this enzyme arose early in

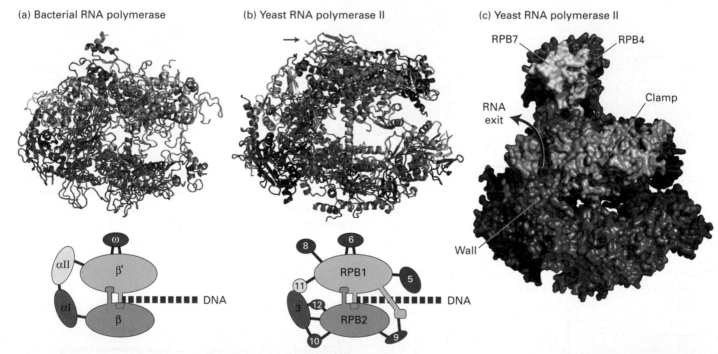

**FIGURE 7-10 Comparison of three-dimensional structures of bacterial and eukaryotic RNA polymerases.** (a, b) These $C_\alpha$ trace models are based on x-ray crystallographic analysis of RNA polymerase from the bacterium *T. aquaticus* and core RNA polymerase II from *S. cerevisiae*. (a) The five subunits of the bacterial enzyme are distinguished by color. Only the N-terminal domains of the α subunits are included in this model. (b) Ten of the 12 subunits constituting yeast RNA polymerase II are shown in this model. Subunits that are similar in conformation to those in the bacterial enzyme are shown in the same colors. The C-terminal domain of the large subunit RPB1 was not observed in the crystal structure, but it is known to extend from the

position marked with a red arrow. (RPB is the abbreviation for "*RNA polymerase B*," which is an alternative way of referring to RNA polymerase II.) DNA entering the polymerases as they transcribe to the right is diagrammed. (c) Space-filling model of yeast RNA polymerase II including subunits 4 and 7. These subunits extend from the core portion of the enzyme shown in (b) near the region of the C-terminal domain of the large subunit. [Part (a) based on crystal structures from G. Zhang et al., 1999, *Cell* **98**:811. Part (b) adapted from P. Cramer et al., 2001, *Science* **292**:1863. Part (c) from K. J. Armache et al., 2003, *Proc. Nat'l Acad. Sci. USA* **100**:6964, and D. A. Bushnell and R. D. Kornberg, 2003, *Proc. Nat'l Acad. Sci. USA* **100**:6969.]

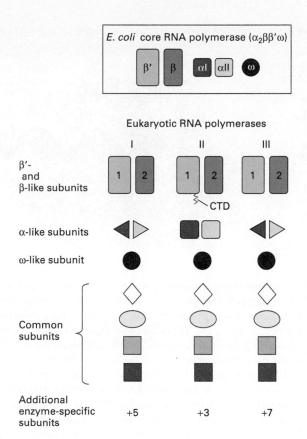

E. coli core RNA polymerase ($\alpha_2\beta\beta'\omega$)

$\beta'$

$\beta$

$\alpha$I

$\alpha$II

$\omega$

Eukaryotic RNA polymerases

I

II

III

$\beta'$-
and
$\beta$-like subunits

1  2

1  2

CTD

1  2

$\alpha$-like subunits

$\omega$-like subunit

Common
subunits

Additional
enzyme-specific
subunits

+5

+3

+7

**FIGURE 7-11 Schematic representation of the subunit structure of the *E. coli* RNA core polymerase and yeast nuclear RNA polymerases.** All three yeast polymerases have five core subunits homologous to the β, β′, two α, and ω subunits of *E. coli* RNA polymerase. The largest subunit (RPB1) of RNA polymerase II also contains an essential C-terminal domain (CTD). RNA polymerases I and III contain the same two nonidentical α-like subunits, whereas RNA polymerase II contains two other nonidentical α-like subunits. All three polymerases share the same ω-like subunit and four other common subunits. In addition, each yeast polymerase contains three to seven unique smaller subunits.

evolution and was largely conserved. This seems logical for an enzyme catalyzing a process so fundamental as copying RNA from DNA. In addition to their core subunits related to the *E. coli* RNA polymerase subunits, all three yeast RNA polymerases contain four additional small subunits, common to them but not to the bacterial RNA polymerase. Finally, each eukaryotic nuclear RNA polymerase has several enzyme-specific subunits that are not present in the other two nuclear RNA polymerases (Figure 7-11). Three of these additional subunits of Pol I and Pol III are homologous to the three additional Pol II-specific subunits. The other two Pol I-specific subunits are homologous to the Pol II general transcription factor TFIIF, discussed later, and the four additional subunits of Pol III are homologous to the Pol II general transcription factors TFIIF and TFIIE.

The clamp domain of RPBI is so designated because it has been observed in two different positions in crystals of the free enzyme (Figure 7-12a) and a complex that mimics the elongating form of the enzyme (Figure 7-12b, c). This domain rotates on a hinge that is probably open when downstream DNA (dark blue template strand, cyan nontemplate strand) is inserted into this region of the polymerase, and then swings shut when the enzyme is in its elongation mode. RNA base-paired to the template strand is red in Figure 7-12b and c. It is postulated that when the 8–9 base-pair RNA-DNA hybrid region near the active site (Figure 7-12c) is bound between RBP1 and RBP2 and nascent RNA fills the exit channel, the clamp is locked in its closed position, anchoring the polymerase to the downstream double-stranded DNA. Also, a

transcription elongation factor called DSIF, discussed later, associates with the elongating polymerase, holding the clamp in its closed conformation. As a consequence, the polymerase is extraordinarily processive, which is to say that it continues to polymerize ribonucleotides until it terminates transcription. After termination and RNA is released from the exit channel, the clamp can swing open, releasing the enzyme from the template DNA. This can explain how human RNA polymerase II can transcribe the longest human gene encoding dystrophin (*DMD*), which is ≈2 million base pairs in length, without dissociating and terminating transcription. Since transcription elongation proceeds at 1–2 kb per minute, transcription of the *DMD* gene requires approximately one day!

Gene-knockout experiments in yeast indicate that most of the subunits of the three nuclear RNA polymerases are essential for cell viability. Disruption of the few polymerase subunit genes that are not absolutely essential for viability (e.g., subunits 4 and 7 of RNA polymerase II) nevertheless results in very poorly growing cells. Thus, all the subunits are necessary for eukaryotic RNA polymerases to function normally. Archaea, like eubacteria, have a single type of RNA polymerase involved in gene transcription. But the archaeal RNA polymerases, like the eukaryotic nuclear RNA polymerases, have on the order of a dozen subunits. Archaea also have related general transcription factors, discussed later, consistent with their closer evolutionary relationship to eukaryotes than to eubacteria (Figure 1-1a).

## The Largest Subunit in RNA Polymerase II Has an Essential Carboxyl-Terminal Repeat

The carboxyl end of the largest subunit of RNA polymerase II (RPB1) contains a stretch of seven amino acids that is nearly precisely repeated multiple times. Neither RNA polymerase I nor III contains these repeating units. This heptapeptide repeat, with a consensus sequence of Tyr-Ser-Pro-Thr-Ser-Pro-Ser, is known as the *carboxyl-terminal domain (CTD)* (Figure 7-10b, extending from the red arrow). Yeast RNA polymerase II contains 26 or more repeats, vertebrate enzymes have 52 repeats, and an intermediate number of repeats occur in RNA polymerase II from nearly all other eukaryotes. The CTD is critical for viability, and at least 10 copies of the repeat must be present for yeast to survive.

In vitro experiments with model promoters first showed that RNA polymerase II molecules that initiate transcription

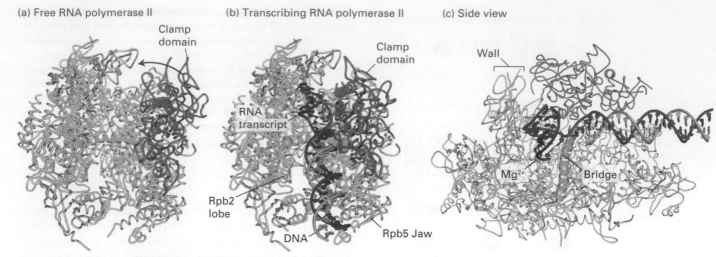

**FIGURE 7-12  The clamp domain of RPBI.** The structures of the free (a) and transcribing (b) RNA polymerase II differ mainly in the position of a clamp domain in RPB1 (orange), which swings over the cleft between the jaws of the polymerase during formation of the transcribing complex, trapping the template DNA strand and transcript. Binding of the clamp domain to the 8–9-base-pair RNA-DNA hybrid may help couple clamp closure to the presence of RNA, stabilizing the closed, elongating complex. RNA is shown in red, the template DNA strand in dark blue, and the downstream nontemplate DNA strand in cyan in this model of an elongating complex. (c) The clamp closes over the incoming downstream DNA. This model is shown with portions of RBP2 that form one side of the cleft removed so that the nucleic acids can be better visualized. The $Mg^{2+}$ ion that participates in catalysis of

phosphodiester bond formation is shown in green. Wall is the domain of RPB2 that forces the template DNA entering the jaws of the polymerase to bend before it exits the polymerase. The bridge α helix shown in green extends across the cleft in the polymerase (see Figure 7-10b) and is postulated to bend and straighten as the polymerase translocates one base down the template strand. The nontemplate strand is thought to form a flexible single-stranded region above the cleft (not shown) extending from three bases downstream of the template base-paired to the 3′ base of the growing RNA and extending to the template strand as it exits the polymerase, where it hybridizes with the template strand to generate the transcription bubble. [Adapted from A. L. Gnatt et al., 2001, *Science* **292**:1876.]

have an unphosphorylated CTD. Once the polymerase initiates transcription and begins to move away from the promoter, many of the serine and some tyrosine residues in the CTD are phosphorylated. Analysis of polytene chromosomes from *Drosophila* salivary glands prepared just before molting of the larva, a time of active transcription, indicate that the CTD also is phosphorylated during in vivo tran-

scription. The large chromosomal "puffs" induced at this time in development are regions where the genome is very actively transcribed. Staining with antibodies specific for the phosphorylated or unphosphorylated CTD demonstrated that RNA polymerase II associated with the highly transcribed puffed regions contains a phosphorylated CTD (Figure 7-13).

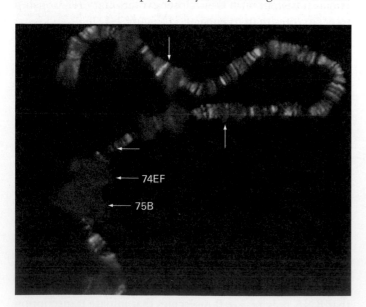

**EXPERIMENTAL FIGURE 7-13  Antibody staining demonstrates that the carboxyl-terminal domain (CTD) of RNA polymerase II is phosphorylated during in vivo transcription.** Salivary-gland polytene chromosomes were prepared from *Drosophila* larvae just before molting. The preparation was treated with a rabbit antibody specific for phosphorylated CTD and with a goat antibody specific for unphosphorylated CTD. The preparation then was stained with fluorescein-labeled anti-goat antibody (green) and rhodamine-labeled anti-rabbit antibody (red). Thus polymerase molecules with an unphosphorylated CTD stain green, and those with a phosphorylated CTD stain red. The molting hormone ecdysone induces very high rates of transcription in the puffed regions labeled 74EF and 75B; note that only phosphorylated CTD is present in these regions. Smaller puffed regions transcribed at high rates also are visible. Nonpuffed sites that stain red (up arrow) or green (horizontal arrow) also are indicated, as is a site staining both red and green, producing a yellow color (down arrow). [From J. R. Weeks et al., 1993, *Genes Dev.* **7**:2329; courtesy of J. R. Weeks and A. L. Greenleaf.]

## KEY CONCEPTS of Section 7.2

### Overview of Eukaryotic Gene Control

- The primary purpose of gene control in multicellular organisms is the execution of precise developmental decisions so that the proper genes are expressed in the proper cells during embryologic development and cellular differentiation.

- Transcriptional control is the primary means of regulating gene expression in eukaryotes, as it is in bacteria.

- In eukaryotic genomes, DNA transcription-control elements may be located many kilobases away from the promoter they regulate. Different control regions can control transcription of the same gene in different cell types.

- Eukaryotes contain three types of nuclear RNA polymerases. All three contain two large and three smaller core subunits with homology to the $\beta'$, $\beta$, $\alpha$, and $\omega$ subunits of *E. coli* RNA polymerase, as well as several additional small subunits (see Figure 7-11).

- RNA polymerase I synthesizes only pre-rRNA. RNA polymerase II synthesizes mRNAs, some of the small nuclear RNAs that participate in mRNA splicing, micro-RNAs (miRNAs) that regulate translation of complementary mRNAs, and small interfering RNAs (siRNAs) that regulate the stability of complementary mRNAs. RNA polymerase III synthesizes tRNAs, 5S rRNA, and several other relatively short, stable RNAs (see Table 7-2).

- The carboxyl-terminal domain (CTD) in the largest subunit of RNA polymerase II becomes phosphorylated during transcription initiation and remains phosphorylated as the enzyme transcribes the template.

# 7.3 RNA Polymerase II Promoters and General Transcription Factors

The mechanisms that regulate transcription initiation and elongation by RNA polymerase II have been studied extensively, because this is the polymerase that transcribes mRNAs. Transcription initiation and elongation by RNA polymerase II are the initial biochemical processes required for the expression of protein-coding genes and are the steps in gene expression that are most frequently regulated to determine when and in which cells specific proteins are synthesized. As noted in the previous section, expression of eukaryotic protein-coding genes is regulated by multiple protein-binding DNA sequences, generically referred to as transcription-control regions. These include promoters, which determine where transcription of the DNA template begins, and other types of control elements located near transcription start sites as well as sequences located far from the genes they regulate, which control the type of cell in which the gene is transcribed and how frequently it is transcribed. In this section, we take a closer look at the properties of various control elements found in eukaryotic protein-coding genes and some techniques used to identify them.

## RNA Polymerase II Initiates Transcription at DNA Sequences Corresponding to the 5′ Cap of mRNAs

In vitro transcription experiments using purified RNA polymerase II, a protein extract prepared from the nuclei of cultured cells, and DNA templates containing sequences encoding the 5′ ends of mRNAs for a number of abundantly expressed genes revealed that the transcripts produced always contained a cap structure at their 5′ ends identical with that present at the 5′ end of the spliced mRNA expressed from the gene (see Figure 4-14). In these experiments, the 5′ cap was added to the 5′ end of the nascent RNA by enzymes in the nuclear extract, which can only add a cap to an RNA that has a 5′ tri- or diphosphate. Because a 5′ end generated by cleavage of a longer RNA would have a 5′ monophosphate, it would not be capped. Consequently, researchers concluded that the capped nucleotides generated in the in vitro transcription reactions must have been the nucleotides with which transcription was initiated. Sequence analysis revealed that, for a given gene, the sequence at the 5′ end of the RNA transcripts produced in vitro is the same as that at the 5′ end of the mRNAs isolated from cells, confirming that the capped nucleotide of eukaryotic mRNAs coincides with the transcription start site. Today, the transcription start site for a newly characterized mRNA generally is determined simply by identifying the DNA sequence encoding the 5′-capped nucleotide of the encoded mRNA.

## The TATA Box, Initiators, and CpG Islands Function as Promoters in Eukaryotic DNA

Several different DNA sequences can function as promoters for RNA polymerase II, directing the polymerase where to initiate transcription of an RNA complementary to the template strand of a double-stranded DNA. These include **TATA boxes, initiators,** and **CpG islands.**

**TATA Boxes** The first genes to be sequenced and studied through in vitro transcription systems were viral genes and cellular protein-coding genes that are very actively transcribed either at particular times of the cell cycle or in specific differentiated cell types. In all these highly transcribed genes, a conserved sequence called the **TATA box** was found ≈26–31 base pairs upstream of the transcription start site (Figure 7-14). Mutagenesis studies have shown that a single-base change in this nucleotide sequence drastically decreases in vitro transcription by RNA polymerase II of genes adjacent to a TATA box. If the base pairs between the TATA box and the normal transcription start site are deleted, transcription of the altered, shortened template begins at a new site ≈25 base pairs downstream from the TATA box. Consequently, the TATA box acts similarly to an *E. coli* promoter

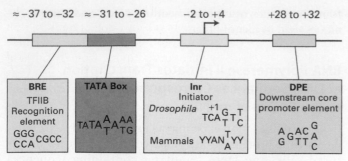

**FIGURE 7-14 Core promoter elements of non-CpG island promoters in metazoans.** The sequence of the strand with the 5′ end at the left and the 3′ end at the right is shown. The most frequently observed bases in TATA-box promoters are shown in larger font. $A^{+1}$ is the base at which transcription starts, Y is a pyrimidine (C or T), N is any of the four bases. [Adapted from S. T. Smale and J. T. Kadonaga, 2003, *Ann. Rev. Biochem.* **72**:449.]

to position RNA polymerase II for transcription initiation (see Figure 4-12).

**Initiator Sequences** Instead of a TATA box, some eukaryotic genes contain an alternative promoter element called an *initiator*. Most naturally occurring initiator elements have a cytosine (C) at the −1 position and an adenine (A) residue at the transcription start site (+1). Directed mutagenesis of mammalian genes with an initiator-containing promoter revealed that the nucleotide sequence immediately surrounding the start site determines the strength of such promoters. Unlike the conserved TATA box sequence, however, only an extremely degenerate initiator consensus sequence has been defined:

$$(5') \text{ Y-Y-A}^{+1}\text{-N-T/A-Y-Y-Y } (3')$$

where $A^{+1}$ is the base at which transcription starts, Y is a pyrimidine (C or T), N is any of the four bases, and T/A is T or A at position +3. As we shall see after discussing general transcription factors required for RNA polymerase II initiation, other specific DNA sequences designated BRE and DPE can be bound by these proteins and influence promoter strength (Figure 7-14).

**CpG Islands** Transcription of genes with promoters containing a TATA box or initiator element begins at a well-defined initiation site. However, transcription of most protein-coding genes in mammals ($\approx$60–70 percent) occurs at a lower rate than TATA box and Initiator-containing promoters, and initiates at several alternative start sites within regions of $\approx$100–1000 base pairs that have an unusually high frequency of CG sequences. Such genes often encode proteins that are not required in large numbers (e.g., enzymes involved in basic metabolic processes required in all cells, often called "housekeeping genes"). These promoter regions are called **CpG islands** (where "p" represents the phosphate between the C and G nucleotides) because they occur relatively rarely in the genome sequence of mammals.

In mammals, most Cs followed by a G that are not associated with CpG island promoters are methylated at position 5 of the pyrimidine ring (5-methyl C, represented $C^{Me}$; see Figure 2-17). CG sequences are thought to be underrepresented in mammalian genomes because spontaneous deamination of 5-methyl C generates thymidine. Over the time scale of mammalian evolution, this is thought to have led to the conversion of most CGs to TG by DNA-repair mechanisms. As a consequence, the frequency of CG in the human genome is only 21 percent of the expected frequency if Cs were randomly followed by a G. However, the Cs in active CpG island promoters are unmethylated. Consequently, when they deaminate spontaneously, they are converted to U, a base that is recognized by DNA repair enzymes and converted back to C. As a result, the frequency of CG sequences in CpG island promoters is close to that expected if C were followed by any of the other three nucleotides randomly.

CG-rich sequences are bound by histone octamers more weakly than CG-poor sequences because more energy is required to bend them into the small-diameter loops required to wrap around the histone octamer forming a nucleosome (Figure 6-29). As a consequence, CpG islands coincide with nucleosome-free regions of DNA. Much remains to be learned about the molecular mechanisms that control transcription from CpG island promoters, but a current hypothesis is that the general transcription factors discussed in the next section can bind to them because CpG islands exclude nucleosomes.

**Divergent Transcription from CpG Island Promoters** Another remarkable feature of CpG islands is that transcription is initiated in both directions, even though only transcription of the sense strand yields an mRNA. By a mechanism(s) that remains to be elucidated, most RNA polymerase II molecules transcribing in the "wrong" direction, i.e., transcribing the non-sense strand, pause or terminate by $\approx$1 kb from the transcription start site. This was discovered by taking advantage of the stability of the elongation complex, presumably conferred by the RNA polymerase II clamp domain when an RNA-DNA hybrid is bound near the active site (Figure 7-12b, c).

Nuclei were isolated from cultured human cells and incubated in a buffered solution containing a concentration of salt and mild detergent that removes RNA polymerases except for those in the process of elongation because of their stable association with template DNA. Nucleotide triphosphates were then added with UTP substituted by bromo-UTP containing uracil with a Br atom at the 5 position on the pyrimidine ring (Figure 2-17). The nuclei were then incubated at 37 °C long enough for $\approx$100 nucleotides to be polymerized by the RNA polymerase II (Pol II) molecules that were in the process of transcription elongation at the time the nuclei were isolated. RNA was then isolated and RNA containing bromo-U was immunoprecipitated with antibody specific for RNA labeled with bromo-U. Thirty-three nucleotides at the 5′ ends of these RNAs were then sequenced by massively parallel DNA sequencing of reverse transcripts, and the sequences were mapped on the human genome.

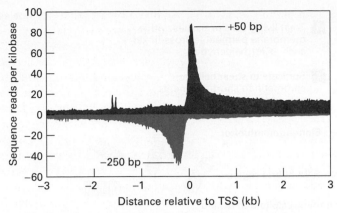

**EXPERIMENTAL FIGURE 7-15 Analysis of elongating RNA polymerase II molecules in human fibroblasts.** Nuclei from cultured fibroblasts were isolated and incubated in a buffer with a nonionic detergent that prevents RNA polymerase II from initiating transcription. Treated nuclei were incubated with ATP, CTP, GTP, and Br-UTP for 5 minutes at 30 °C, a time sufficient to incorporate ≈100 nucleotides. RNA was then isolated and fragmented to ≈100 nucleotides by controlled incubation at high pH. Specific RNA oligonucleotides were ligated to the 5′ and 3′ ends of the RNA fragments, which were then subjected to reverse transcription. The resulting DNA was amplified by polymerase chain reaction and subjected to massively parallel DNA sequencing. The sequences determined were aligned to the transcription start sites (TSS) of all known human genes and the number of sequence reads per kilobase of total sequenced DNA was plotted for 10 base-pair intervals of sense transcripts (red) and antisense transcripts (blue). See text for discussion. [From L. J. Core, J. J. Waterfall, and J. T. Lis, 2008, *Science* **322**:1845.]

Figure 7-15 shows a plot of the number of sequence reads per kilobase of total BrU-labeled RNA relative to the major transcription start sites (TSS) of all currently known human protein-coding genes. The results show that approximately equal numbers of RNA polymerase molecules transcribed most promoters (mostly CpG island promoters) in both the sense direction (red, plotted upward to indicate transcription in the sense direction) and the antisense direction (blue, plotted downward to represent transcription in the opposite, antisense direction). A peak of sense transcripts was observed at ≈ +50 relative to the major transcription start site (TSS), indicating that Pol II pauses in the +50 to +250 region before elongating further. A peak at −250 to −500 relative to the major sense transcription start site of Pol II transcribing in the antisense direction also was observed, revealing paused RNA polymerase II molecules at the other end of the nucleosomes-free regions in CpG island promoters. Note that the number of sequence reads, and therefore the number of elongating polymerases, is lower for polymerases transcribing in the antisense direction more than 1 kb from the transcription start compared to polymerases transcribing more than 1 kb from the transcription start site in the sense direction. The molecular mechanism(s) accounting for this difference is currently an intense area of investigation. Note that a low number of sequence reads was also observed transcribing in the "wrong" direction upstream of the major transcription start sites (red

sequences reads to the left of 0 and blue sequence reads to the right of 0), indicating that there is a low level of transcription from seemingly random sites throughout the genome. These recent discoveries of divergent transcription from CpG island promoters and low-level transcription of most of the genomes of eukaryotes have been a great surprise to most researchers.

**Chromatin Immunoprecipitation** The technique of chromatin immunoprecipitation outlined in Figure 7-16a provided additional data supporting the occurrence of divergent transcription from most CpG island promoters in mammals. The data from this analysis are reported as the number of times a specific sequence from this region of the genome was identified per million total sequences analyzed (Figure 7-16b). At divergently transcribed genes, such as the *Hsd17b12* gene encoding an enzyme involved in intermediary metabolism, two peaks of immunoprecipitated DNA were detected, corresponding to Pol II transcribing in the sense and antisense directions. However, Pol II was only detected >1 kb from the start site in the sense direction. The number of counts per million from this region of the genome was very low because the gene is transcribed at low frequency. However, the number of counts per million at the start-site regions for both sense and antisense transcription was much higher, reflecting the fact that Pol II molecules had initiated transcription in both directions at this promoter, but paused before transcribing >500 base pairs from the start sites in each direction. In contrast, the *Rpl6* gene encoding a large ribosomal subunit protein that was abundantly transcribed in these proliferating cells was transcribed almost exclusively in the sense direction. The number of sequence counts per million >1 kb downstream from the transcription start site was much higher, reflecting the high rate of transcription of this gene.

Transcription start-site-associated RNAs (TSSa RNAs, red and blue arrowheads at the bottom of Figure 7-16b) represent sequences of short RNAs isolated from these cells, thought to result from degradation of nascent RNAs released from paused Pol II molecules that terminate. Note that they include transcripts of both the sense (blue arrows) and antisense (red arrows) from the divergently transcribed gene, whereas only sense TSSa RNAs were found for the unidirectionally transcribed gene. The observation of these TSSa RNAs from CpG island promoters further support the conclusion that they are transcribed in both directions.

## General Transcription Factors Position RNA Polymerase II at Start Sites and Assist in Initiation

Initiation by RNA polymerase II requires several initiation factors. These initiation factors position Pol II molecules at transcription start sites and help to separate the DNA strands so that the template strand can enter the active site of the enzyme. They are called *general transcription factors* because they are required at most, if not all, promoters transcribed by RNA polymerase II. These proteins are designated

(a)

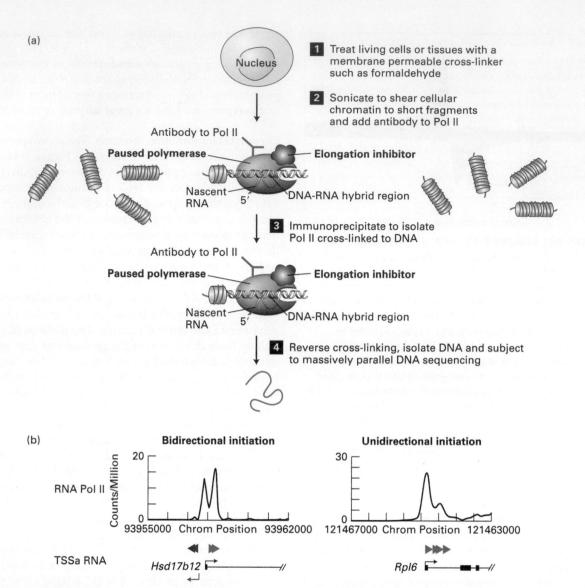

1 Treat living cells or tissues with a membrane permeable cross-linker such as formaldehyde

2 Sonicate to shear cellular chromatin to short fragments and add antibody to Pol II

Antibody to Pol II

**Paused polymerase**

**Elongation inhibitor**

Nascent RNA

5′

DNA-RNA hybrid region

3 Immunoprecipitate to isolate Pol II cross-linked to DNA

Antibody to Pol II

**Paused polymerase**

**Elongation inhibitor**

Nascent RNA

5′

DNA-RNA hybrid region

4 Reverse cross-linking, isolate DNA and subject to massively parallel DNA sequencing

(b)

**Bidirectional initiation**

**Unidirectional initiation**

RNA Pol II

Counts/Million

93955000  Chrom Position  93962000

121467000  Chrom Position  121463000

TSSa RNA

*Hsd17b12*

*Rpl6*

**EXPERIMENTAL FIGURE 7-16 Chromatin immunoprecipitation technique.** (a) Step **1**: Live cultured cells or tissues are incubated in 1% formaldehyde to covalently cross-link protein to DNA and proteins to proteins. Step **2**: The preparation is then subjected to sonication to solubilize and shear chromatin to fragments of 200 to 500 base pairs of DNA. Step **3**: An antibody to a protein of interest, here RNA polymerase II, is added, and DNA covalently linked to the protein of interest is immunoprecipitated. Step **4**: The covalent cross-linking is then reversed and DNA is isolated. The isolated DNA can be analyzed by polymerase chain reaction with primers for a sequence of interest. Alternatively, total recovered DNA can be amplified, labeled by incorporation of a fluorescently labeled nucleotide, and hybridized to a microarray (Figure 5-29) or subjected to massively parallel DNA

sequencing. (b) Results from DNA sequencing of chromatin from mouse embryonic stem cells immunoprecipitated with antibody to RNA polymerase II are shown for a gene that is divergently transcribed (*left*) and a gene that is transcribed only in the sense direction (*right*). Data are plotted as the number of times a DNA sequence in a 50-base-pair interval was observed per million base pairs sequenced. The region encoding the 5′ end of the gene is shown below, with exons shown as rectangles and introns as lines. TSSa RNAs (red and blue arrowheads) represent RNAs of ≈20–50 nucleotides that were isolated from the same cells. Blue indicates RNAs transcribed in the sense direction, and red indicates RNAs transcribed in the antisense direction. [Part (a), see A. Hecht and M. Grunstein, 1999, *Methods Enzymol.* **304**:399. Part (b) adapted from P. B. Rahl et al., 2010, *Cell* **141**:432.]

*TFIIA, TFIIB,* etc., and most are multimeric proteins. The largest is TFIID, which consists of a single 38-kDa *TATA box binding protein (TBP)* and 13 TBP-associated factors (TAFs). General transcription factors with similar activities and homologous sequences are found in all eukaryotes. The complex of Pol II and its general transcription factors bound

to a promoter and ready to initiate transcription is called a *preinitiation complex.* Figure 7-17 summarizes the stepwise assembly of the Pol II transcription preinitiation complex in vitro on a promoter containing a TATA box. The TBP subunit of TFIID rather than the intact TFIID complex was used in the studies that revealed the order of general transcription

**FIGURE 7-17 In vitro assembly of RNA polymerase II preinitiation complex.** The indicated general transcription factors and purified RNA polymerase II (Pol II) bind sequentially to TATA-box DNA to form a preinitiation complex. ATP hydrolysis then provides the energy for unwinding of DNA at the start site by a TFIIH subunit. As Pol II initiates transcription in the resulting open complex, the polymerase moves away from the promoter and its CTD becomes phosphorylated. In vitro, the general transcription factors (except for TBP) dissociate from the TBP-promoter complex, but it is not yet known which factors remain associated with promoter regions following each round of transcription initiation in vivo.

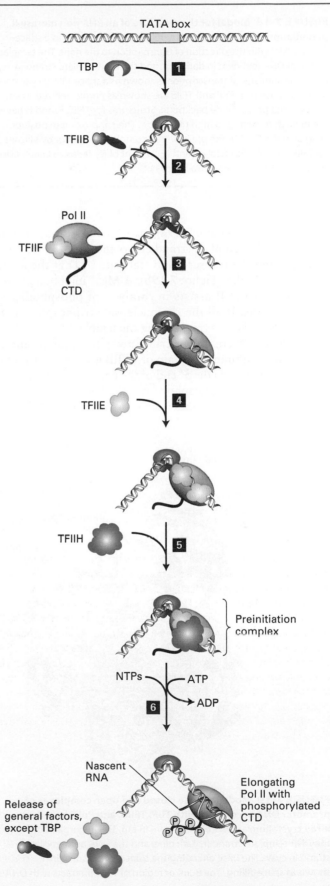

factor and RNA polymerase II assembly because it can be expressed at a high level in *E. coli* and readily purified, while intact TFIID is difficult to purify from eukaryotic cells.

TBP is the first protein to bind to a TATA box promoter. All eukaryotic TBPs analyzed to date have very similar C-terminal domains of 180 residues. This domain of TBP folds into a saddle-shaped structure; the two halves of the molecule exhibit an overall dyad symmetry but are not identical. TBP interacts with the minor groove in DNA, bending the helix considerably (see Figure 4-5). The DNA-binding surface of TBP is conserved in all eukaryotes, explaining the high conservation of the TATA box promoter element (see Figure 7-14).

Once TBP has bound to the TATA box, TFIIB can bind. TFIIB is a monomeric protein, slightly smaller than TBP. The C-terminal domain of TFIIB makes contact with both TBP and DNA on either side of the TATA box. During transcription initiation, its N-terminal domain is inserted into the RNA exit channel of RNA polymerase II (see Figure 7-10). The TFIIB N-terminal domain assists Pol II in melting the DNA strands at the transcription start site and interacts with the template strand near the Pol II active site. Following TFIIB binding, a preformed complex of TFIIF (a heterodimer of two different subunits in mammals) and Pol II binds, positioning the polymerase over the start site. Two more general transcription factors must bind before the DNA duplex can be separated to expose the template strand. First to bind is TFIIE, a heterodimer of two different subunits. TFIIE creates a docking site for TFIIH, another multimeric factor containing 10 different subunits. Binding of TFIIH completes assembly of the transcription preinitiation complex in vitro (Figure 7-17). Figure 7-18 shows a current model for the structure of a preinitiation complex.

The **helicase** activity of one of the TFIIH subunits uses energy from ATP hydrolysis to help unwind the DNA duplex at the start site, allowing Pol II to form an *open* complex in which the DNA duplex surrounding the start site is melted and the template strand is bound at the polymerase active site. Figure 7-19 shows molecular models based on x-ray crystallography of the complex of TBP (purple), TFIIB (red), and Pol II (gold) associated with promoter DNA before the strands near the transcription start site are separated

**FIGURE 7-18 Model for the structure of an RNA polymerase II preinitiation complex.** Yeast RNA polymerase II is shown as a space-filling model with the direction of transcription to the right. The template strand of DNA is shown in dark blue and the nontemplate strand in cyan. The start site of transcription is shown as a space-filling cyan and dark blue base pair. TBP and TFIIB are shown as purple and red worm traces of the polypeptide backbone. Structures for TFIIE, F, and H have not been determined to high resolution. Their approximate positions lying over the DNA in the preinitiation complex are shown by ellipses for TFIIE (green), TFIIF (violet), and TFIIH (light blue). [Adapted from G. Miller and S. Hahn, 2006, *Nat. Struct. Biol.* **13**:603.]

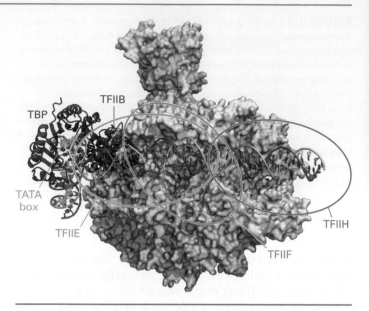

(closed complex, Figure 7-19a) and after the strands are separated and the template strand enters the Pol II-TFIIB complex, placing the transcription start site (+1) at the active site (open complex, Figure 7-19b). A Mg$^{2+}$ ion bound at the active site of Pol II assists in catalysis of phosphodiester bond synthesis. If all the ribonucleoside triphosphates are present, Pol II begins transcribing the template strand.

As the polymerase transcribes away from the promoter region, the N-terminal domain of TFIIB is released from the RNA exit channel as the 5′ end of the nascent RNA enters it. A subunit of TFIIH phosphorylates the Pol II CTD multiple times on the serine 5 (underlined) of the Tyr-Ser-Pro-Thr-

Ser-Pro-Ser repeat that comprises the CTD. As we shall discuss further in Chapter 8, the CTD that is multiply phosphorylated on serine 5 is a docking site for the enzymes that form the cap structure (Figure 4-14) on the 5′ end of RNAs transcribed by RNA polymerase II. In the minimal in vitro transcription

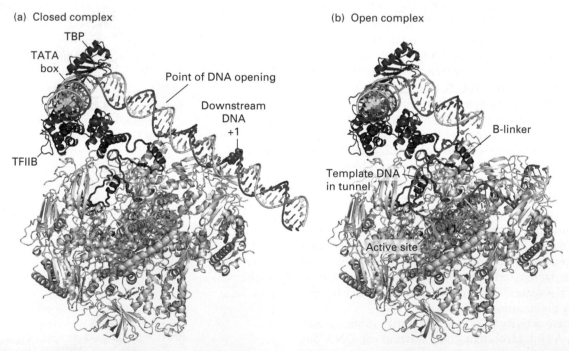

**FIGURE 7-19 Models for the closed and open complexes of promoter DNA in complex with TBP, TFIIB, and Pol II based on x-ray crystallography.** Pol II is shown in tan, TBP in purple, TFIIB in red, the DNA template strand in dark blue, and the DNA nontemplate strand in cyan. The base encoding the transcription start site (+1) is shown as space-filling. The B-linker region of TFIIB interacts with DNA in the closed complex (a), where the strands are initially separated (point of DNA opening). The Mg$^{2+}$ ion at the active site is shown as a green sphere. The nontemplate strand of the transcription bubble in the open complex (b) is not visualized in crystal structures of models of the open complex because it has alternative conformations in different complexes. [Adapted from D. Kostrewa et al., 2009, *Nature* **462**:323.]

assay containing only these general transcription factors and purified RNA polymerase II, TBP remains bound to the TATA box as the polymerase transcribes away from the promoter region, but the other general transcription factors dissociate.

Remarkably, the first subunits of TFIIH to be cloned from humans were identified because mutations in them cause defects in the repair of damaged DNA. In normal individuals, when a transcribing RNA polymerase becomes stalled at a region of damaged template DNA, a subcomplex composed of several subunits of TFIIH, including the helicase subunit mentioned above, recognizes the stalled polymerase and then associates with other proteins that function with TFIIH in repairing the damaged DNA region. In patients with mutant forms of these TFIIH subunits, such repair of damaged DNA in transcriptionally active genes is impaired. As a result, affected individuals have extreme skin sensitivity to sunlight (a common cause of DNA damage is ultraviolet light) and exhibit a high incidence of cancer. Consequently, these subunits of TFIIH serve two functions in the cell, one in the process of transcription initiation and a second function in the repair of DNA. Depending on the severity of the defect in TFIIH function, these individuals may suffer from diseases such as xeroderma pigmentosum and Cockayne's syndrome (Chapter 24). ■

## In Vivo Transcription Initiation by RNA Polymerase II Requires Additional Proteins

Although the general transcription factors discussed above allow RNA polymerase to initiate transcription in vitro, another general transcription factor, TFIIA, is required for initiation by Pol II in vivo. Purified TFIIA forms a complex with TBP and TATA box DNA. X-ray crystallography of this complex shows that TFIIA interacts with the side of TBP that is upstream from the direction of transcription on promoters containing a TATA box. In metazoans (multicellular animals), TFIIA and TFIID, with its multiple TAF subunits, bind first to TATA box DNA, and then the other general transcription factors subsequently bind as indicated in Figure 7-17.

The TAF subunits of TFIID function in initiating transcription from promoters that lack a TATA box. For instance, some TAF subunits contact the initiator element in promoters where it occurs, probably explaining how such sequences can replace a TATA box. Additional TFIID TAF subunits can bind to a consensus sequence A/G-G-A/T-C/T-G/A/C centered ≈30 base pairs downstream from the transcription start site in many genes that lack a TATA box promoter. Because of its position, this regulatory sequence is called the *downstream promoter element (DPE)* (Figure 7-14). The DPE facilitates transcription of TATA-less genes that contain it by increasing TFIID binding. Also, an α helix of TFIIB binds to the major groove of DNA upstream of the TATA-box (see Figure 7-19), and the strongest promoters contain the optimal sequence for this interaction, the BRE shown in Figure 7-14.

Chromatin immunoprecipitation assays (Figure 7-16) using antibodies to TBP show that it binds in the region between the sense and antisense transcription start sites in CpG island promoters. Consequently, the same general transcription factors probably are required for initiation from the weaker CpG island promoters as from promoters containing a TATA box. The absence of the promoter elements summarized in Figure 7-14 may account for the divergent transcription from multiple transcription start sites observed from these promoters, since cues from the DNA sequence are not present to orient the preinitiation complex. TFIID and the other general transcription factors may choose among alternative, nearly equivalent weak binding sites in this class of promoters, potentially explaining the low frequency of transcription initiation as well as the alternative transcription start sites in divergent directions generally observed from CpG island promoters.

## Elongation Factors Regulate the Initial Stages of Transcription in the Promoter-Proximal Region

In metazoans, at most promoters, Pol II pauses after transcribing ≈20–50 nucleotides, due to the binding of a five-subunit protein called NELF (*negative elongation factor*). This is followed by the binding of a two-subunit elongation factor called DSIF (*DRB sensitivity-inducing factor*), so named because an ATP analog called DRB inhibits further transcription elongation in its presence. The inhibition of Pol II elongation that results from NELF binding is relieved when DSIF, NELF, and serine 2 of the Pol II CTD repeat (Tyr-Ser-Pro-Thr-Ser-Pro-Ser) are phosphorylated by a protein kinase with two subunits, CDK9-cyclin T, also called P-TEFb, which associates with the Pol II, NELF, DSIF complex. The same elongation factors regulate transcription from CpG island promoters. These factors that regulate elongation in the promoter-proximal region provide a mechanism for controlling gene transcription in addition to the regulation of transcription initiation. This overall strategy for regulating transcription at both the steps of initiation and elongation in the promoter-proximal region is similar to the regulation of the *Trp* operon in *E. coli* (Figure 7-6), although the molecular mechanisms involved are distinct.

Transcription of HIV (human immunodeficiency virus), the cause of AIDS, is dependent on the activation of CDK9-cyclin T by a small viral protein called **Tat**. Cells infected with *tat*⁻ mutants produce short viral transcripts ≈50 nucleotides long. In contrast, cells infected with wild-type HIV synthesize long viral transcripts that extend throughout the integrated proviral genome (see Figure 4-49 and Figure 6-13). Thus Tat protein functions as an *antitermination factor*, permitting RNA polymerase II to read through a transcriptional block. (Tat is initially made by rare transcripts that fail to terminate when the HIV promoter is transcribed at high rate in "activated" T-lymphocytes, one type of white blood cell; see Chapter 23). Tat is a sequence-specific RNA-binding protein. It binds to the RNA copy of a sequence called TAR, which forms a stem-loop structure near the 5' end

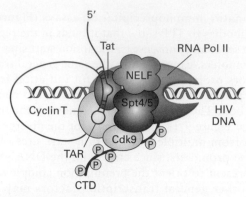

**FIGURE 7-20 Model of antitermination complex composed of HIV Tat protein and several cellular proteins.** The TAR element in the HIV transcript contains sequences recognized by Tat and the cellular protein cyclin T. Cyclin T activates and helps position the protein kinase CDK9 near its substrate, the CTD of RNA polymerase II. CTD phosphorylation at serine 2 of the Pol II CTD heptad repeat is required for transcription elongation. Cellular proteins DSIF (also called Spt4/5) and the NELF complex are also involved in regulating Pol II elongation, as discussed in the text. [See T. Wada et al., 1998, *Genes Dev.* **12**:343; Y. Yamaguchi et al., 1999, *Cell* **97**:451; T. Yamada et al., 2006, *Mol Cell* **21**:227]

of the HIV transcript (Figure 7-20). TAR also binds cyclin T, holding the CDK9-cyclin T complex close to the polymerase, where it efficiently phosphorylates its substrates, resulting in transcription elongation. Chromatin immunoprecipitation assays done after treating cells with specific inhibitors of CDK9 indicate that the transcription of ≈30 percent of mammalian genes is regulated by controlling the activity of CDK9-cyclin T (P-TEFb), although this is probably done most frequently by sequence-specific DNA-binding transcription factors rather an RNA-binding protein, as in the case of HIV Tat. ∎

## KEY CONCEPTS of Section 7.3

### RNA Polymerase II Promoters and General Transcription Factors

• RNA polymerase II initiates transcription of genes at the nucleotide in the DNA template that corresponds to the 5' nucleotide that is capped in the encoded mRNA.

• Transcription of protein-coding genes by Pol II can be initiated in vitro by sequential binding of the following in the indicated order: TBP, which binds to TATA box DNA; TFIIB; a complex of Pol II and TFIIF; TFIIE; and finally, TFIIH (see Figure 7-17).

• The helicase activity of a TFIIH subunit helps to separate the template strands at the start site in most promoters, a process that requires hydrolysis of ATP. As Pol II begins transcribing away from the start site, its CTD is phosphorylated on serine 5 of the heptapeptide CTD by another TFIIH subunit.

• In vivo transcription initiation by Pol II also requires TFIIA and, in metazoans, a complete TFIID protein complex, including its multiple TAF subunits as well as the TBP subunit.

• In metazoans, NELF associates with Pol II after initiation, inhibiting elongation ≈50–200 base pairs from the transcription start site. Inhibition of elongation is relieved when the heterodimeric elongation factors DSIF and CDK9-cyclin T (P-TEFb) associate with the elongation complex and CDK9 phosphorylates subunits of NELF, DSIF, and serine 2 of the Pol II CTD heptapeptide repeat.

## 7.4 Regulatory Sequences in Protein-Coding Genes and the Proteins Through Which They Function

As noted in the previous section, expression of eukaryotic protein-coding genes is regulated by multiple protein-binding DNA sequences, generically referred to as transcription-control regions. These include promoters and other types of control elements located near transcription start sites, as well as sequences located far from the genes they regulate. In this section, we take a closer look at the properties of various control elements found in eukaryotic protein-coding genes and the proteins that bind to them.

### Promoter-Proximal Elements Help Regulate Eukaryotic Genes

Recombinant DNA techniques have been used to systematically mutate the nucleotide sequences of various eukaryotic genes in order to identify transcription-control regions. For example, *linker scanning mutations* can pinpoint the sequences within a regulatory region that function to control transcription. In this approach, a set of constructs with contiguous overlapping mutations are assayed for their effect on expression of a reporter gene or production of a specific mRNA (Figure 7-21a). This type of analysis identified **promoter-proximal elements** of the thymidine kinase (*tk*) gene from herpes simplex type I virus (HSV-I). The results demonstrated that the DNA region upstream of the HSV *tk* gene contains three separate transcription-control sequences: a TATA box in the interval from −32 to −16 and two other control elements farther upstream (Figure 7-21b). Experiments using mutants containing single-base-pair changes in promoter-proximal control elements revealed that they are generally ≈6–10 base pairs long. Recent results indicate that they are found both upstream and downstream of the transcription start site for human genes at equal frequency. While, strictly speaking, the term *promoter* refers to the DNA sequence that determines where a polymerase initiates transcription, the term is often used to refer to both a promoter and its associated promoter-proximal control elements.

To test the spacing constraints on control elements in the HSV *tk* promoter region identified by analysis of linker scanning mutations, researchers prepared and assayed constructs containing small deletions and insertions between

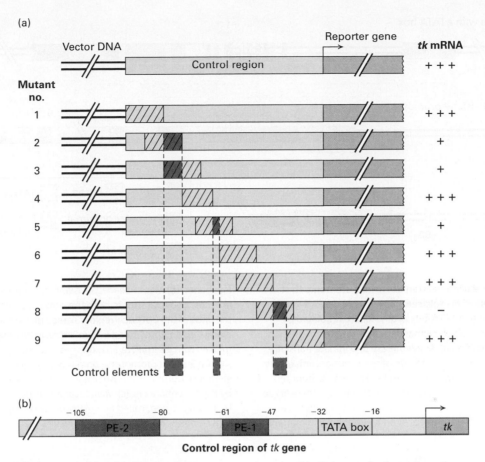

(a)

(b)

**Control region of *tk* gene**

**EXPERIMENTAL FIGURE 7-21 Linker scanning mutations identify transcription-control elements.** (a) A region of eukaryotic DNA (tan) that supports high-level expression of a reporter gene (light purple) is cloned in a plasmid vector as diagrammed at the top. Overlapping linker scanning (LS) mutations (crosshatch) are introduced from one end of the region being analyzed to the other. These mutations result from scrambling the nucleotide sequence in a short stretch of the DNA. After the mutant plasmids are transfected separately into cultured cells, the activity of the reporter-gene product is assayed. In the example shown here, the sequence from −120 to +1 of the herpes simplex virus thymidine kinase gene, LS mutations 1, 4, 6, 7, and 9 have little or no effect on expression of the reporter gene, indicating that the regions altered in these mutants contain no control elements. Reporter-gene expression is significantly reduced in mutants 2, 3, 5, and 8, indicating that control elements (brown) lie in the intervals shown at the bottom. (b) Analysis of these LS mutations identified a TATA box and two promoter-proximal elements (PE-1 and PE-2). [Part (b), see S. L. McKnight and R. Kingsbury, 1982, *Science* **217**:316.]

the elements. Changes in spacing between the promoter and promoter-proximal control elements of 20 nucleotides or fewer had little effect. However, insertions of 30 to 50 base pairs between the HSV-I *tk* promoter-proximal elements and the TATA box was equivalent to deleting the element. Similar analyses of other eukaryotic promoters have also indicated that considerable flexibility in the spacing between promoter-proximal elements is generally tolerated, but separations of several tens of base pairs may decrease transcription.

## Distant Enhancers Often Stimulate Transcription by RNA Polymerase II

As noted earlier, transcription from many eukaryotic promoters can be stimulated by control elements located thousands of base pairs away from the start site. Such long-distance transcription-control elements, referred to as enhancers, are common in eukaryotic genomes but fairly rare in bacterial genomes. Procedures such as linker scanning mutagenesis have indicated that enhancers, usually on the order of ≈200 base pairs, like promoter-proximal elements, are composed of several functional sequence elements of ≈6–10 base pairs. As discussed later, each of these regulatory elements is a binding site for a sequence-specific DNA-binding transcription factor.

Analyses of many different eukaryotic cellular enhancers have shown that in metazoans, they can occur with equal probability upstream from a promoter or downstream from a promoter within an intron, or even downstream from the final exon of a gene, as in the case of the *Sall1* gene (see Figure 7-8a). Many enhancers are cell-type specific. For example, an enhancer controlling *Pax6* expression in the retina was characterized in the intron between exons 4 and 5 (see Figure 7-7a), whereas an enhancer controlling *Pax6* expression in the hormone-secreting cells of the pancreas is located

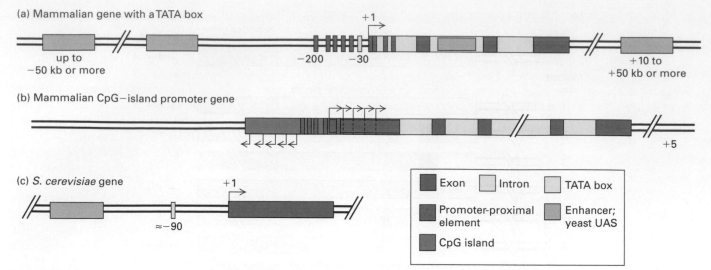

**(a) Mammalian gene with a TATA box**

up to
−50 kb or more

+1

−200    −30

+10 to
+50 kb or more

**(b) Mammalian CpG−island promoter gene**

+5

**(c) *S. cerevisiae* gene**

+1

≈−90

| Exon | Intron | TATA box |
| Promoter-proximal element | Enhancer; yeast UAS | |
| CpG island | | |

**FIGURE 7-22 General organization of control elements that regulate gene expression in multicellular eukaryotes and yeast.** (a) Mammalian genes with a TATA-box promoter are regulated by promoter-proximal elements and enhancers. Promoter elements shown in Figure 7-14 position RNA polymerase II to initiate transcription at the start site and influence the rate of transcription. Enhancers may be either upstream or downstream and as far away as hundreds of kilobases from the transcription start site. In some cases, enhancers lie within introns. Promoter-proximal elements are found upstream and downstream of transcription start sites at equal frequency in mamma-

lian genes. (b) Mammalian CpG-island promoters. Transcription initiates at several sites in both the sense and antisense directions from the ends of the CpG-rich region. Transcripts in the sense direction are elongated and processed into mRNAs by RNA splicing. They express mRNAs with alternative 5′ exons determined by the transcription start site. CpG-island promoters contain promoter-proximal control elements. Currently, it is not clear whether they are also regulated by distant enhancers. (c) Most *S. cerevisiae* genes contain only one regulatory region, called an *upstream activating sequence (UAS)*, and a TATA box, which is ≈90 base pairs upstream from the start site.

in an ≈200-base-pair region upstream of exon 0 (so named because it was discovered after the exon called "exon 1"). In the important model organism *Saccharomyces cerevisiae* (budding yeast), genes are closely spaced (Figure 6-4b) and few genes contain introns. In this organism, enhancers usually lie within ≈200 base pairs upstream of the promoters of the genes they regulate and are referred to by the term *upstream activating sequence (UAS)*.

## Most Eukaryotic Genes Are Regulated by Multiple Transcription-Control Elements

Initially, enhancers and promoter-proximal elements were thought to be distinct types of transcription-control elements. However, as more enhancers and promoter-proximal elements were analyzed, the distinctions between them became less clear. For example, both types of element generally can stimulate transcription even when inverted, and both types often are cell-type specific. The general consensus now is that a spectrum of control elements regulates transcription by RNA polymerase II. At one extreme are enhancers, which can stimulate transcription from a promoter tens of thousands of base pairs away. At the other extreme are promoter-proximal elements, such as the upstream elements controlling the HSV *tk* gene, which lose their influence when moved an additional 30–50 base pairs farther from the promoter. Researchers have identified a large number of transcription-

control elements that can stimulate transcription from distances between these two extremes.

Figure 7-22a summarizes the locations of transcription-control sequences for a hypothetical mammalian gene with a promoter containing a TATA box. The start site at which transcription initiates encodes the first (5′) nucleotide of the first exon of an mRNA, the nucleotide that is capped. In addition to the TATA box at ≈−31 to −26, promoter-proximal elements, which are relatively short (≈6–10 base pairs), are located within the first ≈200 base pairs either upstream or downstream of the start site. Enhancers, in contrast, usually are about 50–200 base pairs long and are composed of multiple elements of ≈6–10 base pairs. Enhancers may be located up to 50 kilobases or more upstream or downstream from the start site or within an intron. As for the *Pax6* gene, many mammalian genes are controlled by more than one enhancer region that function in different types of cells.

Figure 7-22b summarizes the promoter region of a mammalian gene with a CpG island promoter. About 60–70 percent of mammalian genes are expressed from CpG island promoters, usually at much lower levels than genes with TATA box promoters. Multiple alternative transcription start sites are used, generating mRNAs with alternative 5′ ends for the first exon derived from each start site. Transcription occurs in both directions, but Pol II molecules transcribing in the sense direction are elongated to >1 kb much more efficiently than transcripts in the antisense direction.

The *S. cerevisiae* genome contains regulatory elements called **upstream activating sequences (UASs),** which function similarly to enhancers and promoter-proximal elements in higher eukaryotes. Most yeast genes contain only one UAS, which generally lies within a few hundred base pairs of the start site. In addition, *S. cerevisiae* genes contain a TATA box ≈90 base pairs upstream from the transcription start site (Figure 7-22c).

## Footprinting and Gel-Shift Assays Detect Protein-DNA Interactions

The various transcription-control elements found in eukaryotic DNA are binding sites for regulatory proteins generally called **transcription factors.** The simplest eukaryotic cells encode hundreds of transcription factors, and the human genome encodes over 2000. The transcription of each gene in the genome is independently regulated by combinations of *specific transcription factors* that bind to its transcription-control regions. The number of possible combinations of this many transcription factors is astronomical, sufficient to generate unique controls for every gene encoded in the genome.

In yeast, *Drosophila,* and other genetically tractable eukaryotes, numerous genes encoding transcriptional activators and repressors have been identified by classical genetic analyses like those described in Chapter 5. However, in mammals and other vertebrates, which are less amenable to such genetic analysis, most transcription factors have been detected initially and subsequently purified by biochemical techniques. In this approach, a DNA regulatory element that has been identified by the kinds of mutational analyses described above is used to identify *cognate* proteins that bind specifically to it. Two common techniques for detecting such cognate proteins are DNase I footprinting and the electrophoretic mobility shift assay.

*DNase I footprinting* takes advantage of the fact that when a protein is bound to a region of DNA, it protects that DNA sequence from digestion by nucleases. As illustrated in Figure 7-23a, samples of a DNA fragment that is labeled at one end are digested under carefully controlled conditions in the presence and absence of a DNA-binding protein, and then denatured, electrophoresed, and the resulting gel subjected to autoradiography. The region protected by the bound protein appears as a gap, or "footprint," in the array of bands resulting from digestion in the absence of protein. When footprinting is performed with a DNA fragment containing a known DNA control element, the appearance of a footprint indicates the presence of a transcription factor that binds that control element in the protein sample being assayed. Footprinting also identifies the specific DNA sequence to which the transcription factor binds.

For example, DNase I footprinting of the strong adenovirus late promoter shows a protected region over the TATA box when TBP is added to the labeled DNA before DNase I digestion (Figure 7-23b). DNase I does not digest all phosphodiester bonds in a duplex DNA at equal rate. Consequently, in the absence of added protein (lanes 1, 6, and 9), a particular

pattern of bands is observed that depends on the DNA sequence and results from cleavage at some phosphodiester bonds and not others. However, when increasing amounts of TBP are incubated with the end-labeled DNA before digestion with DNase I, TBP binds to the TATA box and protects the region from ≈−35 to −20 from digestion when sufficient TBP is added to bind all the labeled DNA molecules. In contrast, increasing amounts of TFIID (lanes 7 and 8) protect both the TATA box region from DNase I digestion, as well as regions near −7, +1 to +5, +10 to +15, and +20, producing a different "footprint" from TBP. Results such as this tell us that other subunits of TFIID (the TBP-associated factors or TAFs) also bind to the DNA in the region downstream from the TATA box.

The *electrophoretic mobility shift assay (EMSA),* also called the *gel-shift* or *band-shift assay,* is more useful than the footprinting assay for quantitative analysis of DNA-binding proteins. In general, the electrophoretic mobility of a DNA fragment is reduced when it is complexed to protein, causing a shift in the location of the fragment band. This assay can be used to detect a transcription factor in protein fractions incubated with a radiolabeled DNA fragment containing a known control element (Figure 7-24). The more of the transcription factor that is added to the binding reaction, the more labeled probe is shifted to the position of the DNA-protein complex.

In the biochemical isolation of a transcription factor, an extract of cell nuclei commonly is subjected sequentially to several types of column chromatography (Chapter 3). Fractions eluted from the columns are assayed by DNase I footprinting or EMSA using DNA fragments containing an identified regulatory element (see Figure 7-21). Fractions containing a protein that binds to the regulatory element in these assays probably contain a putative transcription factor. A powerful technique that is commonly used for the final step in purifying transcription factors is *sequence-specific DNA affinity chromatography,* a particular type of affinity chromatography in which long DNA strands containing multiple copies of the transcription factor–binding site are coupled to a column matrix.

Once a transcription factor is isolated and purified, its partial amino acid sequence can be determined and used to clone the gene or cDNA encoding it, as outlined in Chapter 5. The isolated gene can then be used to test the ability of the encoded protein to activate or repress transcription in an in vivo transfection assay (Figure 7-25).

## Activators Promote Transcription and Are Composed of Distinct Functional Domains

Studies with a yeast transcription activator called GAL4 provided early insight into the domain structure of transcription factors. The gene encoding the GAL4 protein, which promotes expression of enzymes needed to metabolize galactose, was identified by complementation analysis of *gal4* mutants that cannot form colonies on an agar medium in which galactose is the only source of carbon and energy (Chapter 5). Directed mutagenesis studies like those described previously

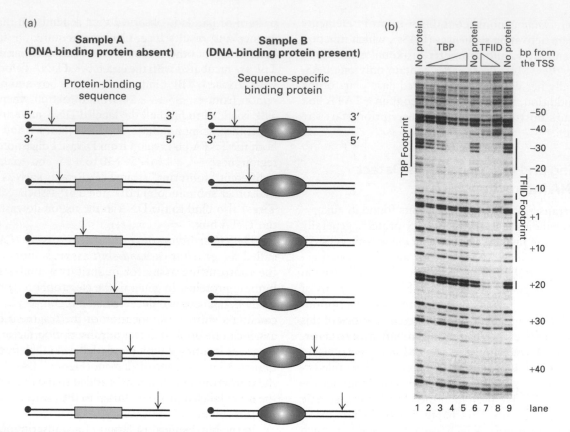

## EXPERIMENTAL FIGURE 7-23 DNase I footprinting reveals the region of a DNA sequence where a transcription factor binds.

(a) A DNA fragment known to contain a control element is labeled at one end with $^{32}$P (red dot). Portions of the labeled DNA sample then are digested with DNase I in the presence and absence of protein samples containing a sequence-specific DNA-binding protein. DNase I hydrolyzes the phosphodiester bonds of DNA between the 3′ oxygen on the deoxyribose of one nucleotide and the 5′ phosphate of the next nucleotide. A low concentration of DNase I is used so that, on average, each DNA molecule is cleaved just once (vertical arrows). If the protein sample does not contain a cognate DNA-binding protein, the DNA fragment is cleaved at multiple positions between the labeled and unlabeled ends of the original fragment, as in sample A (*left*). If the protein sample contains a protein that binds to a specific sequence in the labeled DNA, as in sample B (*right*), the protein binds to the DNA, thereby protecting a portion of the fragment from digestion. Following DNase treatment, the DNA is separated from protein, denatured to separate the strands, and electrophoresed. Autoradiography of the resulting gel detects only labeled strands and reveals fragments extending from the labeled end to the site of cleavage by DNase I. Cleavage fragments containing the control sequence show up on the gel for sample A but are missing in sample B because the bound cognate protein blocked cleavages within that sequence and thus production of the corresponding fragments. The missing bands on the gel constitute the footprint. (b) Footprints produced by increasing amounts of TBP (indicated by the triangle) and of TFIID on the strong adenovirus major late promoter. [Part (b) from Q. Zhou et al., 1992, *Genes Dev.* **6**:1964.]

## EXPERIMENTAL FIGURE 7-24 Electrophoretic mobility shift assay can be used to detect transcription factors during purification.

In this example, protein fractions separated by column chromatography were assayed for their ability to bind to a radiolabeled DNA-fragment probe containing a known regulatory element. After an aliquot of the protein sample was loaded onto the column (ON) and successive column fractions (numbers) were incubated with the labeled probe, the samples were electrophoresed under conditions that do not disrupt protein-DNA interactions. The free probe not bound to protein migrated to the bottom of the gel. A protein in the preparation applied to the column and in fractions 7 and 8 bound to the probe, forming a DNA-protein complex that migrated more slowly than the free probe. These fractions therefore likely contain the regulatory protein being sought. [From S. Yoshinaga et al., 1989, *J. Biol. Chem.* **264**:10529.]

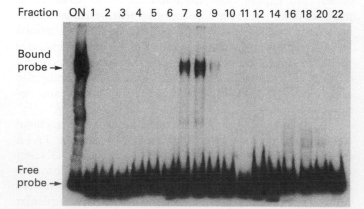

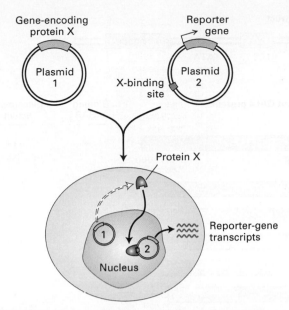

**EXPERIMENTAL FIGURE 7-25 In vivo transfection assay measures transcription activity to evaluate proteins believed to be transcription factors.** The assay system requires two plasmids. One plasmid contains the gene encoding the putative transcription factor (protein X). The second plasmid contains a reporter gene (e.g., luciferase) and one or more binding sites for protein X. Both plasmids are simultaneously introduced into cells that lack the gene encoding protein X. The production of reporter-gene RNA transcripts is measured; alternatively, the activity of the encoded protein can be assayed. If reporter-gene transcription is greater in the presence of the X-encoding plasmid than in its absence, then the protein is an activator; if transcription is less, then it is a repressor. By use of plasmids encoding a mutated or rearranged transcription factor, important domains of the protein can be identified.

identified UASs for the genes activated by GAL4. Each of these UASs was found to contain one or more copies of a related 17-bp sequence called UAS$_{GAL}$. DNase I footprinting assays with recombinant GAL4 protein produced in *E. coli* from the yeast *GAL4* gene showed that GAL4 protein binds to UAS$_{GAL}$ sequences. When a copy of UAS$_{GAL}$ was cloned upstream of a TATA box followed by a β-galactosidase reporter gene, expression of β-galactosidase was activated in galactose media in wild-type cells but not in *gal4* mutants. These results showed that UAS$_{GAL}$ is a transcription-control element activated by the GAL4 protein in galactose media.

A remarkable set of experiments with *gal4* deletion mutants demonstrated that the GAL4 transcription factor is composed of separable functional domains: an N-terminal **DNA-binding domain,** which binds to specific DNA sequences, and a C-terminal **activation domain,** which interacts with other proteins to stimulate transcription from a nearby promoter (Figure 7-26). When the N-terminal DNA-binding domain of GAL4 was fused directly to various portions of its own C-terminal region, the resulting truncated proteins retained the ability to stimulate expression of a reporter gene in

an in vivo assay like that depicted in Figure 7-25. Thus the internal portion of the protein is not required for functioning of GAL4 as a transcription factor. Similar experiments with another yeast transcription factor, GCN4, which regulates genes required for synthesis of many amino acids, indicated that it contains an ≈50–amino acid DNA-binding domain at its C-terminus and an ≈20–amino acid activation domain near the middle of its sequence.

Further evidence for the existence of distinct activation domains in GAL4 and GCN4 came from experiments in which their activation domains were fused to a DNA-binding domain from an entirely unrelated *E. coli* DNA-binding protein. When these fusion proteins were assayed in vivo, they activated transcription of a reporter gene containing the cognate site for the *E. coli* protein. Thus functional transcription factors can be constructed from entirely novel combinations of prokaryotic and eukaryotic elements.

Studies such as these have now been carried out with many eukaryotic activators. The structural model of eukaryotic activators that has emerged from these studies is a modular one in which one or more activation domains are connected to a sequence-specific DNA-binding domain through flexible protein domains (Figure 7-27). In some cases, amino acids included in the DNA-binding domain also contribute to transcriptional activation. As discussed in a later section, activation domains are thought to function by binding other proteins involved in transcription. The presence of flexible domains connecting the DNA-binding domains to activation domains may explain why alterations in the spacing between control elements are so well tolerated in eukaryotic control regions. Thus even when the positions of transcription factors bound to DNA are shifted relative to each other, their activation domains may still be able to interact because they are attached to their DNA-binding domains through flexible protein regions.

## Repressors Inhibit Transcription and Are the Functional Converse of Activators

Eukaryotic transcription is regulated by repressors as well as activators. For example, geneticists have identified mutations in yeast that result in continuously high expression of certain genes. This type of unregulated, abnormally high expression is called **constitutive** expression and results from the inactivation of a repressor that normally inhibits the transcription of these genes. Similarly, mutants of *Drosophila* and *Caenorhabditis elegans* have been isolated that are defective in embryonic development because they express genes in embryonic cells where those genes are normally repressed. The mutations in these mutants inactivate repressors, leading to abnormal development.

Repressor-binding sites in DNA have been identified by systematic linker scanning mutation analysis similar to that depicted in Figure 7-21. In this type of analysis, mutation of an activator-binding site leads to decreased expression of the linked reporter gene, whereas mutation of a repressor-binding site leads to increased expression of a reporter gene. Repressor proteins that bind such sites can be purified and

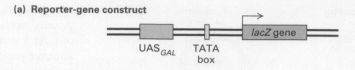

**(a) Reporter-gene construct**

UAS$_{GAL}$    TATA box    *lacZ* gene

**EXPERIMENTAL FIGURE 7-26 Deletion mutants of the *GAL4* gene in yeast with a UAS$_{GAL}$ reporter-gene construct demonstrate the separate functional domains in an activator.** (a) Diagram of DNA construct containing a *lacZ* reporter gene (encoding β-galactosidase) and TATA box ligated to UAS$_{GAL}$, a regulatory element that contains several GAL4-binding sites. The reporter-gene construct and DNA encoding wild-type or mutant (deleted) GAL4 were simultaneously introduced into mutant (*gal4*) yeast cells, and the activity of β-galactosidase expressed from *lacZ* was assayed. Activity will be high if the introduced *GAL4* DNA encodes a functional protein. (b) Schematic diagrams of wild-type GAL4 and various mutant forms. Small numbers refer to positions in the wild-type sequence. Deletion of 50 amino acids from the N-terminal end destroyed the ability of GAL4 to bind to UAS$_{GAL}$ and to stimulate expression of β-galactosidase from the reporter gene. Proteins with extensive deletions from the C-terminal end still bound to UAS$_{GAL}$. These results localize the DNA-binding domain to the N-terminal end of GAL4. The ability to activate β-galactosidase expression was not entirely eliminated unless somewhere between 126 and 189 or more amino acids were deleted from the C-terminal end. Thus the activation domain lies in the C-terminal region of GAL4. Proteins with internal deletions (*bottom*) also were able to stimulate expression of β-galactosidase, indicating that the central region of GAL4 is not crucial for its function in this assay. [See J. Ma and M. Ptashne, 1987, *Cell* **48**:847; I. A. Hope and K. Struhl, 1986, *Cell* **46**:885; and R. Brent and M. Ptashne, 1985, *Cell* **43**:729.]

**(b) Wild-type and mutant GAL4 proteins**

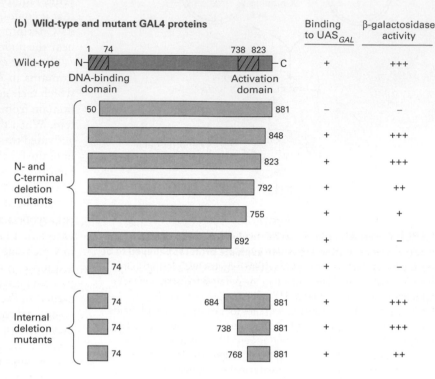

| | Binding to UAS$_{GAL}$ | β-galactosidase activity |
|---|---|---|
| Wild-type | + | +++ |
| 50–881 | – | – |
| –848 | + | +++ |
| –823 | + | +++ |
| –792 | + | ++ |
| –755 | + | + |
| –692 | + | – |
| 74 | + | – |
| 74 / 684–881 | + | +++ |
| 74 / 738–881 | + | +++ |
| 74 / 768–881 | + | ++ |

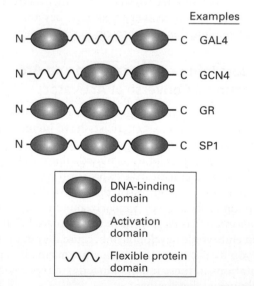

Examples: GAL4, GCN4, GR, SP1

- DNA-binding domain
- Activation domain
- ∿ Flexible protein domain

**FIGURE 7-27 Schematic diagrams illustrating the modular structure of eukaryotic transcription activators.** Transcription factors may contain more than one activation domain but rarely contain more than one DNA-binding domain. GAL4 and GCN4 are yeast transcription activators. The glucocorticoid receptor (GR) promotes transcription of target genes when certain hormones are bound to the C-terminal activation domain. SP1 binds to GC-rich promoter elements in a large number of mammalian genes.

assayed using the same biochemical techniques described earlier for activator proteins.

Eukaryotic transcription repressors are the functional converse of activators. They can inhibit transcription from a gene they do not normally regulate when their cognate binding sites are placed within tens of base pairs to many kilobases of the gene's start site. Like activators, most eukaryotic repressors are modular proteins that have two functional domains: a DNA-binding domain and a **repression domain.** Similar to activation domains, repression domains continue to function when fused to another type of DNA-binding domain. If binding sites for this second DNA-binding domain are inserted within a few hundred base pairs of a promoter, expression of the fusion protein inhibits transcription from the promoter. Also like activation domains, repression domains function by interacting with other proteins, as discussed later in this chapter.

## DNA-Binding Domains Can Be Classified into Numerous Structural Types

The DNA-binding domains of eukaryotic activators and repressors contain a variety of structural motifs that bind specific DNA sequences. The ability of DNA-binding proteins to bind to specific DNA sequences commonly results from noncovalent interactions between atoms in an α helix in the

(a)        (b)

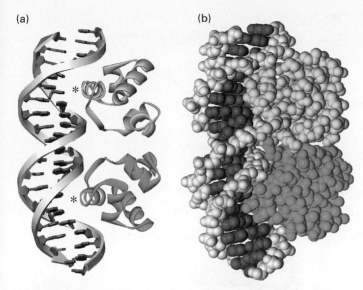

**FIGURE 7-28 Interaction of bacteriophage 434 repressor with DNA.** (a) Ribbon diagram of 434 repressor bound to its specific operator DNA. Repressor monomers are in yellow and green. The recognition helices are indicated by asterisks. A space-filling model of the repressor-operator complex (b) shows how the protein interacts intimately with one side of the DNA molecule over a length of 1.5 turns. [Adapted from A. K. Aggarwal et al., 1988, *Science* **242**:899.]

DNA-binding domain and atoms on the edges of the bases within a major groove in the DNA. Ionic interactions between positively charged residues arginine and lysine and negatively charged phosphates in the sugar phosphate backbone and, in some cases, interactions with atoms in a DNA minor groove also contribute to binding.

The principles of specific protein-DNA interactions were first discovered during the study of bacterial repressors. Many bacterial repressors are dimeric proteins in which an α helix from each monomer inserts into a major groove in the DNA helix (Figure 7-28). This α helix is referred to as the *recognition helix* or *sequence-reading helix,* because most of the amino acid side chains that contact DNA extend from this helix. The recognition helix that protrudes from the surface of bacterial repressors to enter the DNA major groove and make multiple, specific interactions with atoms in the DNA is usually supported in the protein structure in part by hydrophobic interactions with a second α helix just N-terminal to it. This structural element, which is present in many bacterial repressors, is called a *helix-turn-helix* motif.

Many additional motifs that can present an α helix to the major groove of DNA are found in eukaryotic transcription factors, which often are classified according to the type of DNA-binding domain they contain. Because most of these motifs have characteristic consensus amino acid sequences, potential transcription factors can be recognized among the cDNA sequences from various tissues that have been characterized in humans and other species. The human genome, for instance, encodes ≈2000 transcription factors.

Here we introduce several common classes of DNA-binding proteins whose three-dimensional structures have been determined. In all these examples and many other transcription factors, at least one α helix is inserted into a major groove of DNA. However, some transcription factors contain alternative structural motifs (e.g., β strands and loops, see NFAT in Figure 7-32 as an example) that interact with DNA.

**Homeodomain Proteins** Many eukaryotic transcription factors that function during development contain a conserved 60-residue DNA-binding motif, called a **homeodomain,** that is similar to the helix-turn-helix motif of bacterial repressors. These transcription factors were first identified in *Drosophila* mutants in which one body part was transformed into another during development (see Figure 7-1b). The conserved homeodomain sequence has also been found in vertebrate transcription factors, including those that have similar master-control functions in human development.

**Zinc-Finger Proteins** A number of different eukaryotic proteins have regions that fold around a central $Zn^{2+}$ ion, producing a compact domain from a relatively short length of the polypeptide chain. Termed a **zinc finger,** this structural motif was first recognized in DNA-binding domains but now is known to occur also in proteins that do not bind to DNA. Here we describe two of the several classes of zinc-finger motifs that have been identified in eukaryotic transcription factors.

The *$C_2H_2$ zinc finger* is the most common DNA-binding motif encoded in the human genome and the genomes of most other multicellular animals. It is also common in multicellular plants but is not the dominant type of DNA-binding domain in plants as it is in animals. This motif has a 23- to 26-residue consensus sequence containing two conserved cysteine (C) and two conserved histidine (H) residues, whose side chains bind one $Zn^{2+}$ ion (Figure 3-9c). The name "zinc finger" was coined because a two-dimensional diagram of the structure resembles a finger. When the three-dimensional structure was solved, it became clear that the binding of the $Zn^{2+}$ ion by the two cysteine and two histidine residues folds the relatively short polypeptide sequence into a compact domain, which can insert its α helix into the major groove of DNA. Many transcription factors contain multiple $C_2H_2$ zinc fingers, which interact with successive groups of base pairs, within the major groove, as the protein wraps around the DNA double helix (Figure 7-29a).

A second type of zinc-finger structure, designated the $C_4$ *zinc finger* (because it has four conserved cysteines in contact with the $Zn^{2+}$), is found in ≈50 human transcription factors. The first members of this class were identified as specific intracellular high-affinity binding proteins, or "receptors," for steroid hormones, leading to the name *steroid receptor superfamily.* Because similar intracellular receptors for nonsteroid hormones subsequently were found, these transcription factors are now commonly called **nuclear receptors.** The characteristic feature of $C_4$ zinc fingers is the presence of two groups of four critical cysteines, one toward each end of the 55- or 56-residue domain. Although the $C_4$ zinc finger initially was named by analogy with the $C_2H_2$ zinc finger, the

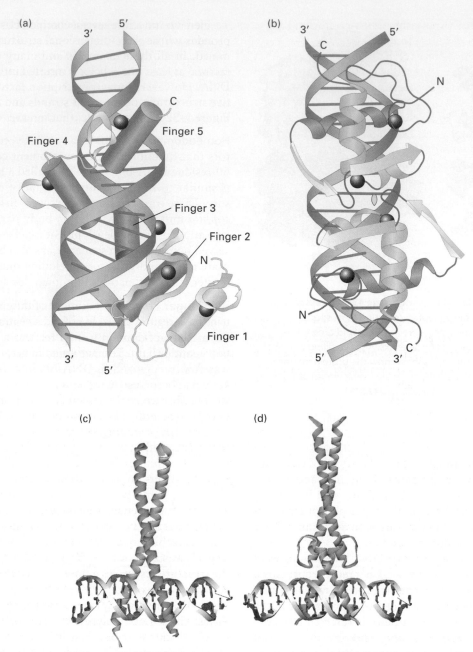

**FIGURE 7-29 Eukaryotic DNA-binding domains that use an α helix to interact with the major groove of specific DNA sequences.** (a) The GL1 DNA-binding domain is monomeric and contains five $C_2H_2$ zinc fingers. The α helices are shown as cylinders, the $Zn^{2+}$ ions as spheres. Finger 1 does not interact with DNA, whereas the other four fingers do. (b) The glucocorticoid receptor is a homodimeric $C_4$ zinc-finger protein. The α helices are shown as purple ribbons, the β-strands as green arrows, the $Zn^{2+}$ ions as spheres. Two α helices (darker shade), one in each monomer, interact with the DNA. Like all $C_4$ zinc-finger homodimers, this transcription factor has twofold rotational symmetry; the center of symmetry is shown by the yellow ellipse. (c) In leucine-zipper proteins, basic residues in the extended α-helical regions of the monomers interact with the DNA backbone at adjacent major grooves. The coiled-coil dimerization domain is stabilized by hydrophobic interactions between the monomers. (d) In bHLH proteins, the DNA-binding helices at the bottom (N-termini of the monomers) are separated by nonhelical loops from a leucine-zipper-like region containing a coiled-coil dimerization domain. [Part (a), see N. P. Pavletich and C. O. Pabo, 1993, *Science* **261:**1701. Part (b), see B. F. Luisi et al., 1991, *Nature* **352:**497. Part (c), see T. E. Ellenberger et al., 1992, *Cell* **71:**1223. Part (d), see A. R. Ferre-D'Amare et al., 1993, *Nature* **363:**38.]

three-dimensional structures of proteins containing these DNA-binding motifs later were found to be quite distinct. A particularly important difference between the two is that $C_2H_2$ zinc-finger proteins generally contain three or more repeating finger units and bind as monomers, whereas $C_4$ zinc-finger proteins generally contain only two finger units and generally bind to DNA as homodimers or heterodimers. Homodimers of $C_4$ zinc-finger DNA-binding domains have twofold rotational symmetry (Figure 7-29b). Consequently, homodimeric nuclear receptors bind to consensus DNA sequences that are inverted repeats.

**Leucine-Zipper Proteins** Another structural motif present in the DNA-binding domains of a large class of transcription factors contains the hydrophobic amino acid leucine at every seventh position in the sequence. These proteins bind to DNA as dimers, and mutagenesis of the leucines showed that they were required for dimerization. Consequently, the name **leucine zipper** was coined to denote this structural motif.

The DNA-binding domain of the yeast GCN4 transcription factor mentioned earlier is a leucine-zipper domain. X-ray crystallographic analysis of complexes between DNA and the GCN4 DNA-binding domain has shown that the dimeric protein contains two extended α helices that "grip" the DNA molecule, much like a pair of scissors, at two adjacent major grooves separated by about half a turn of the double helix (Figure 7-29c). The portions of the α helices contacting the DNA include positively charged (basic) residues that interact with phosphates in the DNA backbone and additional residues that interact with specific bases in the major groove.

GCN4 forms dimers via hydrophobic interactions between the C-terminal regions of the α helices, forming a **coiled-coil** structure. This structure is common in proteins containing amphipathic α helices in which hydrophobic amino acid residues are regularly spaced alternately three or four positions apart in the sequence, forming a stripe down one side of the α helix. These hydrophobic stripes make up the interacting surfaces between the α-helical monomers in a coiled-coil dimer (see Figure 3-9a).

Although the first leucine-zipper transcription factors to be analyzed contained leucine residues at every seventh position in the dimerization region, additional DNA-binding proteins containing other hydrophobic amino acids in these positions subsequently were identified. Like leucine-zipper proteins, they form dimers containing a C-terminal coiled-coil dimerization region and an N-terminal DNA-binding domain. The term *basic zipper (bZIP)* now is frequently used to refer to all proteins with these common structural features. Many basic-zipper transcription factors are heterodimers of two different polypeptide chains, each containing one basic-zipper domain.

**Basic Helix-Loop-Helix (bHLH) Proteins** The DNA-binding domain of another class of dimeric transcription factors contains a structural motif very similar to the basic-zipper motif except that a nonhelical loop of the polypeptide chain separates two α-helical regions in each monomer (Figure 7-29d).

Termed a **basic helix-loop-helix (bHLH)**, this motif was predicted from the amino acid sequences of these proteins, which contain an N-terminal α helix with basic residues that interact with DNA, a middle loop region, and a C-terminal region with hydrophobic amino acids spaced at intervals characteristic of an amphipathic α helix. As with basic-zipper proteins, different bHLH proteins can form heterodimers.

## Structurally Diverse Activation and Repression Domains Regulate Transcription

Experiments with fusion proteins composed of the GAL4 DNA-binding domain and random segments of *E. coli* proteins demonstrated that a diverse group of amino acid sequences can function as activation domains, ≈1 percent of all *E. coli* sequences, even though they evolved to perform other functions. Many transcription factors contain activation domains marked by an unusually high percentage of particular amino acids. GAL4, GCN4, and most other yeast transcription factors, for instance, have activation domains that are rich in acidic amino acids (aspartic and glutamic acids). These so-called *acidic activation domains* generally are capable of stimulating transcription in nearly all types of eukaryotic cells—fungal, animal, and plant cells. Activation domains from some *Drosophila* and mammalian transcription factors are glutamine-rich, and some are proline-rich; still others are rich in the closely related amino acids serine and threonine, both of which have hydroxyl groups. However, some strong activation domains are not particularly rich in any specific amino acid.

Biophysical studies indicate that acidic activation domains have an unstructured, random-coil conformation. These domains stimulate transcription when they are bound to a protein *co-activator*. The interaction with a co-activator causes the activation domain to assume a more structured α-helical conformation in the activation domain–co-activator complex. A well-studied example of a transcription factor with an acidic activation domain is the mammalian CREB protein, which is phosphorylated in response to increased levels of cAMP. This regulated phosphorylation is required for CREB to bind to its co-activator CBP (CREB *b*inding *p*rotein), resulting in the transcription of genes whose control regions contain a CREB-binding site (see Figure 15-32). When the phosphorylated random coil activation domain of CREB interacts with CBP, it undergoes a conformational change to form two α helices linked by a short loop, which wrap around the interacting domain of CBP (Figure 7-30a).

Some activation domains are larger and more highly structured than acidic activation domains. For example, the ligand-binding domains of nuclear receptors function as activation domains when they bind their specific ligand (Figure 7-30b, c). Binding of ligand induces a large conformational change that allows the ligand-binding domain with bound hormone to interact with a short α helix in nuclear-receptor co-activators; the resulting complex then can activate transcription of genes whose control regions bind the nuclear receptor.

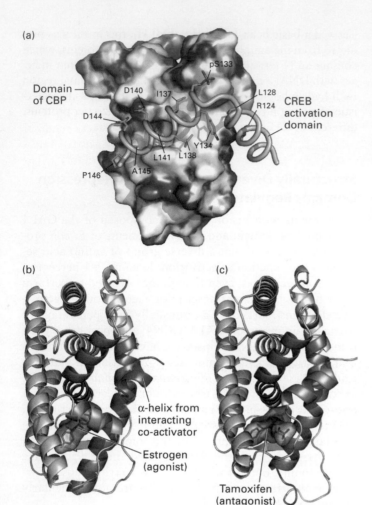

**(a)**

Domain of CBP

pS133

D140

I137

D144

L128

R124

CREB activation domain

Y134

L141    L138

P146    A145

**(b)**

α-helix from interacting co-activator

Estrogen (agonist)

**(c)**

Tamoxifen (antagonist)

**FIGURE 7-30 Activation domains may be random coils until they interact with co-activator proteins or folded protein domains.**
(a) The activation domain of CREB (*cyclic AMP response element-binding* protein) is activated by phosphorylation at serine 123. It is a random coil until it interacts with a domain of the CBP co-activator (shown as a space-filling surface model with negatively charged regions in red and positively charged regions in blue). When the CREB activation domain binds to CBP, it folds into two amphipathic α helices. Side chains in the activation domain that interact with the surface of the CBP domain are shown. (b) The ligand-binding activation domain of the estrogen receptor is a folded-protein domain. When estrogen is bound to the domain, the green α helix interacts with the ligand, generating a hydrophobic groove in the ligand-binding domain (dark brown helices), which binds an amphipathic α helix in a co-activator subunit (blue). (c) The conformation of the estrogen receptor in the absence of hormone is stabilized by binding of the estrogen antagonist tamoxifen. In this conformation, the green helix of the receptor folds into a conformation that interacts with the co-activator–binding groove of the active receptor, sterically blocking binding of co-activators. [Part (a) from I. Radhakrishnan et al. (1997) *Cell* **91**:741, courtesy of Peter Wright. Parts (b) and (c) from A. K. Shiau et al., 1998, *Cell* **95**:927.]

Thus the acidic activation domain in CREB and the ligand-binding activation domains in nuclear receptors represent two structural extremes. The CREB acidic activation domain is a random coil that folds into two α helices when it binds to the surface of a globular domain in a co-activator. In contrast, the

nuclear-receptor ligand-binding activation domain is a structured globular domain that interacts with a short α helix in a co-activator, which probably is a random coil before it is bound. In both cases, however, specific protein-protein interactions between co-activators and the activation domains permit the transcription factors to stimulate gene expression.

Currently, less is known about the structure of repression domains. The globular ligand-binding domains of some nuclear receptors function as repression domains in the absence of their specific hormone ligand. Like activation domains, repression domains may be relatively short, comprising 15 or fewer amino acids. Biochemical and genetic studies indicate that repression domains also mediate protein-protein interactions and bind to *co-repressor* proteins, forming a complex that inhibits transcription initiation by mechanisms that are discussed later in the chapter.

## Transcription Factor Interactions Increase Gene-Control Options

Two types of DNA-binding proteins discussed previously—basic-zipper proteins and bHLH proteins—often exist in alternative heterodimeric combinations of monomers. Other classes of transcription factors not discussed here also form heterodimeric proteins. In some heterodimeric transcription factors, each monomer recognizes the same sequence. In these proteins, the formation of alternative heterodimers does not increase the number of different sites on which the monomers can act, but rather allows the activation domains associated with each monomer to be brought together in alternative combinations that bind to the same site (Figure 7-31a). As we will see later, and in subsequent chapters, the activities of individual transcription factors can be regulated by multiple mechanisms. Consequently, a single bZIP or bHLH DNA regulatory element in the control region of a gene may elicit different transcriptional responses depending on which bZIP or bHLH monomers that bind to that site are expressed in a particular cell at a particular time and how their activities are regulated.

In some heterodimeric transcription factors, however, each monomer has a different DNA-binding specificity. The resulting combinatorial possibilities increase the number of potential DNA sequences that a family of transcription factors can bind. Three different factor monomers theoretically could combine to form six homo- and heterodimeric factors, as illustrated in Figure 7-31b. Four different factor monomers could form a total of 10 dimeric factors, five monomers, 16 dimeric factors, and so forth. In addition, inhibitory factors are known that bind to some basic-zipper and bHLH monomers, thereby blocking their binding to DNA. When these inhibitory factors are expressed, they repress transcriptional activation by the factors with which they interact (Figure 7-31c). The rules governing the interactions of members of a heterodimeric transcription factor class are complex. This combinatorial complexity expands both the number of DNA sites from which these factors can activate transcription and the ways in which they can be regulated.

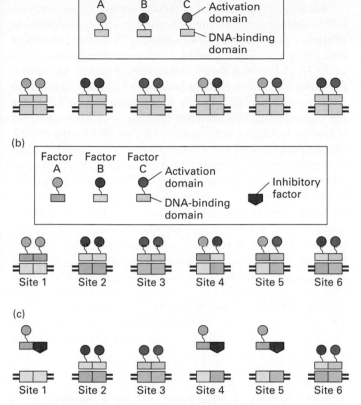

(a)

Factor A | Factor B | Factor C — Activation domain / DNA-binding domain

(b)

Factor A | Factor B | Factor C — Activation domain / DNA-binding domain | Inhibitory factor

Site 1 | Site 2 | Site 3 | Site 4 | Site 5 | Site 6

(c)

Site 1 | Site 2 | Site 3 | Site 4 | Site 5 | Site 6

**FIGURE 7-31 Combinatorial possibilities due to formation of heterodimeric transcription factors.** (a) In some heterodimeric transcription factors, each monomer recognizes the same DNA sequence. In the hypothetical example shown, transcription factors A, B, and C can all interact with one another, creating six different alternative combinations of activation domains that can all bind at the same site. Each composite binding site is divided into two half-sites, and each heterodimeric factor contains the activation domains of its two constituent monomers. (b) When transcription-factor monomers recognize different DNA sequences, alternative combinations of the three factors bind to six different DNA sequences (sites 1–6), each with a unique combination of activation domains. (c) Expression of an inhibitory factor (red) that interacts only with factor A inhibits binding; hence transcriptional activation at sites 1, 4, and 5 is inhibited, but activation at sites 2, 3, and 6 is unaffected.

Similar combinatorial transcriptional regulation is achieved through the interaction of structurally unrelated transcription factors bound to closely spaced binding sites in DNA. An example is the interaction of two transcription factors, NFAT and AP1, which bind to neighboring sites in a composite promoter-proximal element regulating the gene encoding interleukin-2 (IL-2). Expression of the *IL-2* gene is critical to the immune response, but abnormal expression of IL-2 can lead to autoimmune diseases such as rheumatoid arthritis. Neither NFAT nor AP1 binds to its site in the *IL-2* control region in the absence of the other. The affinities of the factors for these particular DNA sequences are too low for the individual factors to form a stable complex with

DNA. However, when both NFAT and AP1 are present, protein-protein interactions between them stabilize the DNA ternary complex composed of NFAT, AP1, and DNA (Figure 7-32a). Such *cooperative DNA binding* of various transcription factors results in considerable combinatorial complexity of transcription control. As a result, the ≈2000 transcription factors encoded in the human genome can bind to DNA through a much larger number of cooperative interactions, resulting in unique transcriptional control for each of

(a)

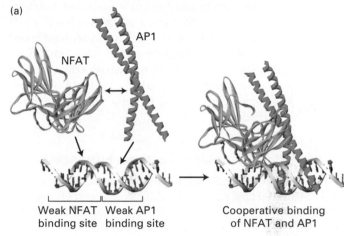

AP1 / NFAT

Weak NFAT binding site | Weak AP1 binding site → Cooperative binding of NFAT and AP1

(b)

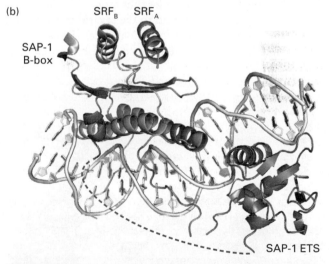

SRF$_B$  SRF$_A$ / SAP-1 B-box / SAP-1 ETS

**FIGURE 7-32 Cooperative binding of two unrelated transcription factors to neighboring sites in a composite control element.** (a) By themselves, both monomeric NFAT and heterodimeric AP1 transcription factors have low affinity for their respective binding sites in the *IL-2* promoter-proximal region. Protein-protein interactions between NFAT and AP1 add to the overall stability of the NFAT-AP1-DNA complex, so that the two proteins bind to the composite site cooperatively. (b) Cooperative DNA binding by dimeric SRF and monomeric SAP-1 can occur when their binding sites are separated by 5 to ≈30 bp and when the SAP-1 binding site is inverted because the B-box domain of SAP-1 that interacts with SRF is connected to the ETS DNA-binding domain of SAP-1 by a flexible linker region of the SAP-1 polypeptide chain (dotted line). [(a) See L. Chen et al., 1998, *Nature* **392**:42; (b) see M. Hassler and T. J. Richmond, 2001, *EMBO J.* **20**:3018.]

the ≈25,000 human genes. In the case of *IL-2*, transcription occurs only when both NFAT is activated, resulting in its transport from the cytoplasm to the nucleus, and the two subunits of AP1 are synthesized. These events are controlled by distinct signal transduction pathways (Chapters 15 and 16), allowing stringent control of IL-2 expression.

Cooperative binding by NFAT and AP1 occurs only when their weak binding sites are positioned quite close to each other in DNA. The sites must be located at a precise distance from each other for effective binding. The requirements for cooperative binding are not so stringent in the case of some other transcription factors and control regions. For example, the *EGR-1* control region contains a composite binding site to which the SRF and SAP1 transcription factors bind cooperatively (see Figure 7-32b). Because a SAP1 has a long, flexible domain that interacts with SRF, the two proteins can bind cooperatively when their individual sites in DNA are separated by any distance up to ≈30 base pairs or are inverted relative to each other.

## Multiprotein Complexes Form on Enhancers

As noted previously, enhancers generally range in length from about 50 to 200 base pairs and include binding sites for several transcription factors. Analysis of the ≈50-bp enhancer that regulates expression of β-interferon, an important protein in defense against viral infections in vertebrates, provides a good example of one of the few examples thus far of the structure of the DNA-binding domains bound to the several transcription factor–binding sites that comprise an enhancer (Figure 7-33). The term **enhanceosome** has been coined to describe such large

DNA-protein complexes that assemble from transcription factors as they bind to their multiple binding sites in an enhancer.

Because of the presence of flexible regions connecting the DNA-binding domains and activation or repression domains in transcription factors (see Figure 7-27), and the ability of interacting proteins bound to distant sites to produce loops in the DNA between their binding sites (Figure 7-4), considerable leeway in the spacing between regulatory elements in transcription-control regions is permissible. This tolerance for variable spacing between binding sites for regulatory transcription factors and promoter-binding sites for the general transcription factors and Pol II probably contributed to rapid evolution of gene control in eukaryotes. Transposition of DNA sequences and recombination between repeated sequences over evolutionary time likely created new combinations of control elements that were subjected to natural selection and retained if they proved beneficial. The latitude in spacing between regulatory elements probably allowed many more functional combinations to be subjected to this evolutionary experimentation than would be the case if constraints on the spacing between regulatory elements were strict, as for most genes in bacteria.

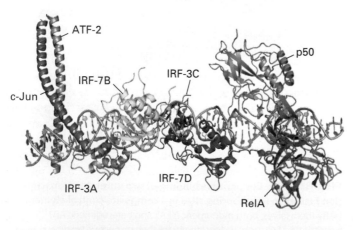

**FIGURE 7-33 Model of the enhanceosome that forms on the β-interferon enhancer.** Two heterodimeric factors, Jun/ATF-2 and p50/ RelA (NF-κB), and two copies each of the monomeric transcription factors IRF-3 and IRF-7, bind to the six overlapping binding sites in this enhancer. [Adapted from D. Penne, T. Manniatis, and S. Harrison, 2007, *Cell* **129**:1111.]

- Transcription activators and repressors are generally modular proteins containing a single DNA-binding domain and one or a few activation domains (for activators) or repression domains (for repressors). The different domains frequently are linked through flexible polypeptide regions (see Figure 7-27).

- Among the most common structural motifs found in the DNA-binding domains of eukaryotic transcription factors are the $C_2H_2$ zinc finger, homeodomain, basic helix-loop-helix (bHLH), and basic zipper (leucine zipper). All these and many other DNA-binding motifs contain one or more α helices that interact with major grooves in their cognate site in DNA.

- Activation and repression domains in transcription factors exhibit a variety of amino acid sequences and three-dimensional structures. In general, these functional domains interact with co-activators or co-repressors, which are critical to the ability of transcription factors to modulate gene expression.

- The transcription-control regions of most genes contain binding sites for multiple transcription factors. Transcription of such genes varies depending on the particular repertoire of transcription factors that are expressed and activated in a particular cell at a particular time.

- Combinatorial complexity in transcription control results from alternative combinations of monomers that form heterodimeric transcription factors (see Figure 7-31) and from cooperative binding of transcription factors to composite control sites (see Figure 7-32).

- Binding of multiple activators to nearby sites in an enhancer forms a multiprotein complex called an enhanceosome (see Figure 7-33).

# 7.5 Molecular Mechanisms of Transcription Repression and Activation

The repressors and activators that bind to specific sites in DNA and regulate expression of the associated protein-coding genes do so by three general mechanisms. First, these regulatory proteins act in concert with other proteins to modulate chromatin structure, inhibiting or stimulating the ability of general transcription factors to bind to promoters. Recall from Chapter 6 that the DNA in eukaryotic cells is not free but is associated with a roughly equal mass of protein in the form of **chromatin.** The basic structural unit of chromatin is the **nucleosome,** which is composed of ≈147 base pairs of DNA wrapped tightly around a disk-shaped core of **histone** proteins. Residues within the N-terminal region of each histone, and the C-terminal regions of histones H2A and H2B, called *histone tails,* extend from the surface of the nucleosome and can be reversibly modified (see Figure 6-31b). Such modifications influence the relative condensation of chromatin and thus its accessibility to proteins required for transcription initiation. In addition to their role in such chromatin-mediated transcriptional control, activators and repressors interact with a large multiprotein complex called the *mediator of transcription complex,* or simply **mediator.** This complex in turn binds to Pol II and directly regulates assembly of transcription preinitiation complexes. In addition, some activation domains interact with TFIID-TAF subunits or other components of the preinitiation complex, interactions that contribute to preinitiation complex assembly. Finally, activation domains may also interact with the elongation factor P-TEFb (CDK9-cyclin T) and other as yet unknown factors to stimulate Pol II elongation away from the promoter region.

In this section, we review the current understanding of how repressors and activators control chromatin structure and preinitiation complex assembly. In the next section of the chapter, we discuss how the concentrations and activities of activators and repressors themselves are controlled, so that gene expression is precisely attuned to the needs of the cell and organism.

## Formation of Heterochromatin Silences Gene Expression at Telomeres, Near Centromeres, and in Other Regions

For many years it has been clear that inactive genes in eukaryotic cells are often associated with **heterochromatin,** regions of chromatin that are more highly condensed and stain more darkly with DNA dyes than **euchromatin,** where most transcribed genes are located (see Figure 6-33a). Regions of chromosomes near the centromeres and telomeres and additional specific regions that vary in different cell types are organized into heterochromatin. The DNA in heterochromatin is less accessible to externally added proteins than DNA in euchromatin and consequently is often referred to as "closed" chromatin. For instance, in an experiment described in Chapter 6, the DNA of inactive genes was found to be far more resistant to digestion by DNase I than the DNA of transcribed genes (see Figure 6-32).

Study of DNA regions in *S. cerevisiae* that behave like the heterochromatin of higher eukaryotes provided early insight into the *chromatin-mediated repression* of transcription. This yeast can grow either as haploid or diploid cells. Haploid cells exhibit one of two possible mating types, called **a** and α. Cells of different mating type can "mate," or fuse, to generate a diploid cell. When a haploid cell divides by budding, the larger "mother" cell switches its mating type. Genetic and molecular analyses have revealed that three genetic loci on yeast chromosome III control the mating type of yeast cells (Figure 7-34). Only the central mating-type locus, termed *MAT,* is actively transcribed and expresses transcription factors (a1, or α1 and α2) that regulate genes controlling the mating type. In any one cell, either an **a** or α DNA sequence is located at the *MAT.* The two additional loci, termed *HML* and *HMR,* near the left and right telomere, respectively, contain "silent" (nontranscribed) copies of the **a** or α genes. These sequences are transferred alternately from *HMLα* or *HMRa* into the *MAT locus* by a type of nonreciprocal recombination between sister chromatids during cell division. When the

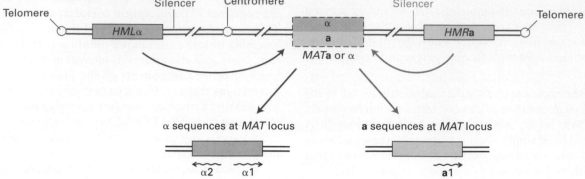

**FIGURE 7-34 Arrangement of mating-type loci on chromosome III in the yeast *S. cerevisiae*.** Silent (unexpressed) mating-type genes (either **a** or α, depending on the strain) are located at the *HML* locus. The opposite mating-type gene is present at the silent *HMR* locus. When the α or **a** sequences are present at the *MAT* locus, they can be transcribed into mRNAs whose encoded proteins specify the mating-type phenotype of the cell. The silencer sequences near *HML* and *HMR* bind proteins that are critical for repression of these silent loci. Haploid cells can switch mating types in a process that transfers the DNA sequence from *HML* or *HMR* to the transcriptionally active *MAT* locus.

*MAT* locus contains the DNA sequence from *HMLα*, the cells behave as α cells. When the *MAT* locus contains the DNA sequence from *HMRa*, the cells behave like **a** cells.

Our interest here is how transcription of the silent mating-type loci at *HML* and *HMR* is repressed. If the genes at these loci are expressed, as they are in yeast mutants with defects in the repressing mechanism, both **a** and α proteins are expressed, causing the cells to behave like diploid cells, which cannot mate. The promoters and UASs controlling transcription of the **a** and α genes lie near the center of the DNA sequence that is transferred and are identical whether the sequences are at the *MAT* locus or at one of the silent loci. This indicates that the function of the transcription factors that interact with these sequences must somehow be blocked at *HML* and *HMR* but not at the *MAT* locus. This repression of the silent loci depends on **silencer sequences** located next to the region of transferred DNA at *HML* and *HMR* (Figure 7-34). If the silencer is deleted, the adjacent locus is transcribed. Remarkably, any gene placed near the yeast mating-type silencer sequence by recombinant DNA techniques is repressed, or "silenced," even a tRNA gene transcribed by RNA polymerase III, which uses a different set of general transcription factors than RNA polymerase II uses, as discussed later.

Several lines of evidence indicate that repression of the *HML* and *HMR* loci results from a condensed chromatin structure that sterically blocks transcription factors from interacting with the DNA. In one telling experiment, the gene encoding an *E. coli* enzyme that methylates adenine residues in GATC sequences was introduced into yeast cells under the control of a yeast promoter so that the enzyme was expressed. Researchers found that GATC sequences within the *MAT* locus and most other regions of the genome in these cells were methylated, but not those within the *HML* and *HMR* loci. These results indicate that the DNA of the silent loci is inaccessible to the *E. coli* methylase and presumably to proteins in general, including transcription factors and RNA

polymerase. Similar experiments conducted with various yeast histone mutants indicated that specific interactions involving the histone tails of H3 and H4 are required for formation of a fully repressed chromatin structure. Other studies have shown that the telomeres of every yeast chromosome also behave like silencer sequences. For instance, when a gene is placed within a few kilobases of any yeast telomere, its expression is repressed. In addition, this repression is relieved by the same mutations in the H3 and H4 histone tails that interfere with repression at the silent mating-type loci.

Genetic studies led to identification of several proteins, RAP1 and three SIR proteins, that are required for repression of the silent mating-type loci and the telomeres in yeast. RAP1 was found to bind within the DNA silencer sequences associated with *HML* and *HMR* and to a sequence that is repeated multiple times at each yeast chromosome telomere. Further biochemical studies showed that the SIR2 protein is a *histone deacetylase;* it removes acetyl groups on lysines of the histone tails. Also, the RAP1, and SIR2, 3, and 4 proteins bind to one another, and SIR3 and SIR4 bind to the N-terminal tails of histones H3 and H4 that are maintained in a largely unacetylated state by the deacetylase activity of SIR2. Several experiments using fluorescence confocal microscopy of yeast cells either stained with fluorescent-labeled antibody to any one of the SIR proteins or RAP1 or hybridized to a labeled telomere-specific DNA probe revealed that these proteins form large, condensed telomeric nucleoprotein structures resembling the heterochromatin found in higher eukaryotes (Figure 7-35a, b, c).

Figure 7-35d depicts a model for the chromatin-mediated silencing at yeast telomeres based on these and other studies. Formation of heterochromatin at telomeres is nucleated by multiple RAP1 proteins bound to repeated sequences in a nucleosome-free region at the extreme end of a telomere. A network of protein-protein interactions involving telomere-bound RAP1, three SIR proteins (2, 3, and 4), and hypoacetylated histones H3 and H4 creates a higher-order nucleoprotein

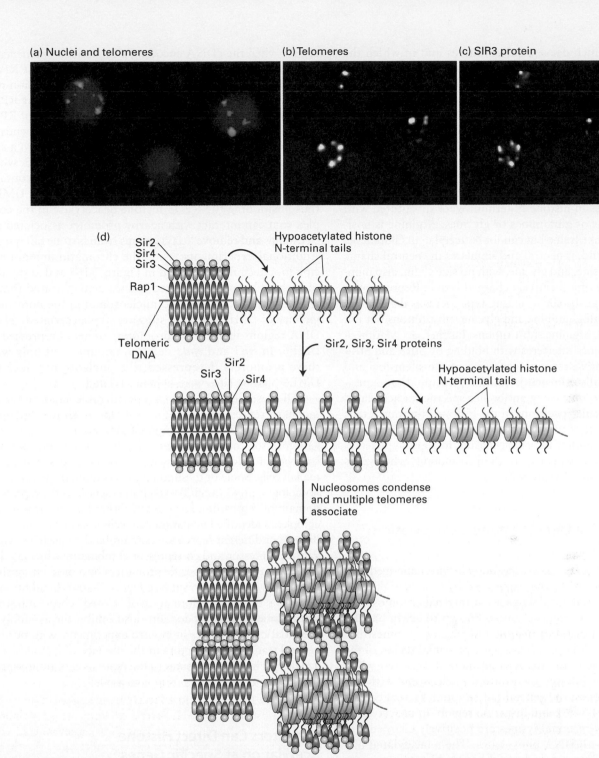

(a) Nuclei and telomeres

(b) Telomeres

(c) SIR3 protein

(d)

Sir2
Sir4
Sir3

Rap1

Telomeric
DNA

Hypoacetylated histone
N-terminal tails

Sir2, Sir3, Sir4 proteins

Sir2
Sir3    Sir4

Hypoacetylated histone
N-terminal tails

Nucleosomes condense
and multiple telomeres
associate

**EXPERIMENTAL FIGURE 7-35  Antibody and DNA probes colocalize SIR3 protein with telomeric heterochromatin in yeast nuclei.** (a) Confocal micrograph 0.3 mm thick through three diploid yeast cells, each containing 68 telomeres. Telomeres were labeled by hybridization to a fluorescent telomere-specific probe (yellow). DNA was stained red to reveal the nuclei. The 68 telomeres coalesce into a much smaller number of regions near the nuclear periphery. (b, c) Confocal micrographs of yeast cells labeled with a telomere-specific hybridization probe (b) and a fluorescent-labeled antibody specific for SIR3 (c). Note that SIR3 is localized in the repressed telomeric hetero-chromatin. Similar experiments with RAP1, SIR2, and SIR4 have shown that these proteins also colocalize with the repressed telomeric heterochromatin. (d) Schematic model of silencing mechanism at yeast telomeres. (*Top left*) Multiple copies of RAP1 bind to a simple repeated

sequence at each telomere region that lacks nucleosomes. SIR3 and SIR4 bind to RAP1, and SIR2 binds to SIR4. SIR2 is a histone deacetylase that deacetylates the tails on the histones neighboring the repeated RAP1-binding site. (*Middle*) The hypoacetylated histone tails are also binding sites for SIR3 and SIR4, which in turn bind additional SIR2, deacetylating neighboring histones. Repetition of this process results in spreading of the region of hypoacetylated histones with associated SIR2, SIR3, and SIR4. (*Bottom*) Interactions between complexes of SIR2, SIR3, and SIR4 cause the chromatin to condense and several telomeres to associate, as shown in a–c. The higher-order chromatin structure generated sterically blocks other proteins from interacting with the underlying DNA. [Parts (a)–(c) from M. Gotta et al., 1996, *J. Cell Biol.* **134:**1349; courtesy of M. Gotta, T. Laroche, and S. M. Gasser. Part (d) adapted from M. Grunstein, 1997, *Curr. Opin. Cell Biol.* **9:**383.]

complex that includes several telomeres and in which the DNA is largely inaccessible to external proteins. One additional protein, SIR1, is also required for silencing of the mating-type loci. It binds to the silencer regions associated with *HML* and *HMR* together with RAP1 and other proteins to initiate assembly of a similar multiprotein silencing complex that encompasses *HML* and *HMR*.

An important feature of this model is the dependence of repression on *hypoacetylation* of the histone tails. This was shown in experiments with yeast mutants expressing histones in which lysines in histone N-termini were substituted with either arginines or glutamines or glycines. Arginine is positively charged like lysine but cannot be acetylated. Glutamine, on the other hand, is neutral and simulates the neutral charge of acetylated lysine, and glycine, with no side chain, also mimics the absence of a positively charged lysine. Repression at telomeres and at the silent mating-type loci was defective in the mutants with glutamine and glycine substitutions but not in mutants with arginine substitutions. Further, acetylation of H3 and H4 lysines interferes with binding by SIR3 and SIR4 and consequently prevents repression at the silent loci and telomeres. Finally, chromatin immunoprecipitation experiments (Figure 7-16a) using antibodies specific for acetylated lysines at particular positions in the histone N-terminal tails (Figure 6-31a) confirmed that histones in repressed regions near telomeres and at the silent mating loci are hypoacetylated, but become hyperacetylated in *sir* mutants when genes in these regions are derepressed.

## Repressors Can Direct Histone Deacetylation at Specific Genes

The importance of *histone deacetylation* in chromatin-mediated gene repression was further supported by studies of eukaryotic repressors that regulate genes at internal chromosomal positions. These proteins are now known to act in part by causing deacetylation of histone tails in nucleosomes that bind to the TATA box and promoter-proximal region of the genes they repress. In vitro studies have shown that when promoter DNA is assembled onto a nucleosome with unacetylated histones, the general transcription factors cannot bind to the TATA box and initiation region. In unacetylated histones, the N-terminal lysines are positively charged and may interact with DNA phosphates. The unacetylated histone tails also interact with neighboring histone octamers and other chromatin-associated proteins, favoring the folding of chromatin into condensed, higher-order structures whose precise conformation is not well understood. The net effect is that general transcription factors cannot assemble into a preinitiation complex on a promoter associated with hypoacetylated histones. In contrast, binding of general transcription factors is repressed much less by histones with hyperacetylated tails in which the positively charged lysines are neutralized and electrostatic interactions are eliminated.

The connection between histone deacetylation and repression of transcription at specific yeast promoters became

clearer when the cDNA encoding a human *histone deacetylase* was found to have high homology to the yeast *RPD3* gene, known to be required for the normal repression of a number of yeast genes. Further work showed that RPD3 protein has histone deacetylase activity. The ability of RPD3 to deacetylate histones at a number of promoters depends on two other proteins: UME6, a repressor that binds to a specific upstream regulatory sequence (URS1), and SIN3, which is part of a large, multiprotein complex that also contains RPD3. SIN3 also binds to the repression domain of UME6, thus positioning the RPD3 histone deacetylase in the complex so it can interact with nearby promoter-associated nucleosomes and remove acetyl groups from histone tail lysines. Additional experiments, using the chromatin immunoprecipitation technique outlined in Figure 7-16a and antibodies to specific histone acetylated lysines demonstrated that in wild-type yeast, one or two nucleosomes in the immediate vicinity of UME6-binding sites are hypoacetylated. These DNA regions include the promoters of genes repressed by UME6. In *sin3* and *rpd3* deletion mutants, not only were these promoters derepressed, the nucleosomes near the UME6-binding sites were hyperacetylated.

All these findings provide considerable support for the model of repressor-directed deacetylation shown in Figure 7-36a. The SIN3-RPD3 complex functions as a *co-repressor*. Co-repressor complexes containing histone deacetylases also have been found associated with many repressors from mammalian cells. Some of these complexes contain the mammalian homolog of SIN3 (mSin3), which interacts with the repression domain of repressors, as in yeast. Other histone deacetylase complexes identified in mammalian cells appear to contain additional or different repression domain-binding proteins. These various repressor and co-repressor combinations mediate histone deacetylation at specific promoters by a mechanism similar to the yeast mechanism (see Figure 7-36a). In addition to repression through the formation of "closed" chromatin structures, some repression domains also inhibit the assembly of preinitiation complexes in in vitro experiments with purified general transcription factors in the absence of histones. This activity probably contributes to the repression of transcription by these repression domains in vivo as well.

## Activators Can Direct Histone Acetylation at Specific Genes

Just as repressors function through co-repressors that bind to their repression domains, the activation domains of DNA-binding activators function by binding multisubunit *co-activator* complexes. One of the first co-activator complexes to be characterized was the yeast *SAGA* complex, which functions with the GCN4 activator protein described in Section 7.4. Early genetic studies indicated that full activity of the GCN4 activator required a protein called GCN5. The clue to GCN5's function came from biochemical studies of a *histone acetylase* purified from the protozoan *Tetrahymena*, the first histone acetylase to be purified. Sequence analysis revealed homology

## (a) Repressor-directed histone deacetylation

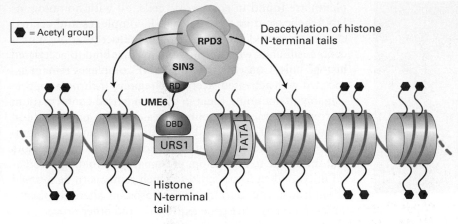

## (b) Activator-directed histone hyperacetylation

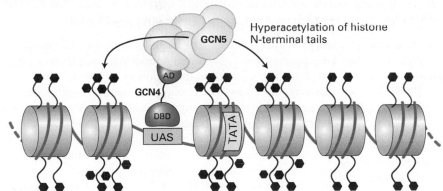

**FIGURE 7-36 Proposed mechanism of histone deacetylation and hyperacetylation in yeast transcription control.** (a) Repressor-directed deacetylation of histone N-terminal tails. The DNA-binding domain (DBD) of the repressor UME6 interacts with a specific upstream control element (URS1) of the genes it regulates. The UME6 repression domain (RD) binds SIN3, a subunit of a multiprotein complex that includes RPD3, a histone deacetylase. Deacetylation of histone N-terminal tails on nucleosomes in the region of the UME6-binding site inhibits binding of general transcription factors at the TATA box, thereby repressing gene expression. (b) Activator-directed hyperacetylation of histone N-terminal tails. The DNA-binding domain of the activator GCN4 interacts with specific upstream activating sequences (UAS) of the genes it regulates. The GCN4 activation domain (AD) then interacts with a multiprotein histone acetylase complex that includes the GCN5 catalytic subunit. Subsequent hyperacetylation of histone N-terminal tails on nucleosomes in the vicinity of the GCN4-binding site facilitates access of the general transcription factors required for initiation. Repression and activation of many genes in higher eukaryotes occurs by similar mechanisms.

between the *Tetrahymena* protein and yeast GCN5, which was soon shown to have histone acetylase activity as well. Further genetic and biochemical studies revealed that GCN5 is one subunit of a multiprotein co-activator complex, named the SAGA complex after genes encoding some of the subunits. Another subunit of this histone acetylase complex binds to activation domains in multiple yeast activator proteins, including GCN4. The model shown in Figure 7-36b is consistent with the observation that nucleosomes near the promoter region of a gene regulated by the GCN4 activator are specifically hyperacetylated compared to most histones in the cell. This activator-directed hyperacetylation of nucleosomes near a promoter region opens the chromatin structure so as to facilitate the binding of other proteins required for transcription initiation. The chromatin structure is less condensed compared to most chromatin, as indicated by its sensitivity to digestion with nucleases in isolated nuclei.

In addition to leading to the decondensation of chromatin, the acetylation of specific histone lysines generates binding sites for proteins with *bromodomains* that bind them. For example, a subunit of the general transcription factor TFIID contains two bromodomains that bind to acetylated nucleosomes with high affinity. Recall that TFIID binding to a promoter initiates assembly of an RNA polymerase II pre-

initiation complex (see Figure 7-17). Nucleosomes at promoter regions of virtually all active genes are hyperacetylated.

A similar activation mechanism operates in higher eukaryotes. Mammalian cells contain multisubunit histone acetylase co-activator complexes homologous to the yeast SAGA complex. They also express two related ≈300-kDa, multidomain proteins called *CBP* and *P300*, which function similarly. As noted earlier, one domain of CBP binds the phosphorylated acidic activation domain in the CREB transcription factor. Other domains of CBP interact with different activation domains in other activators. Yet another domain of CBP has histone acetylase activity, and another CBP domain associates with additional multisubunit histone acetylase complexes. CREB and many other mammalian activators function in part by directing CBP and the associated histone acetylase complex to specific nucleosomes, where they acetylate histone tails, facilitating the interaction of general transcription factors with promoter DNA.

## Chromatin-Remodeling Factors Help Activate or Repress Transcription

In addition to histone acetylase complexes, multiprotein chromatin-remodeling complexes also are required for activation

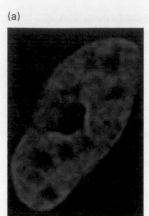
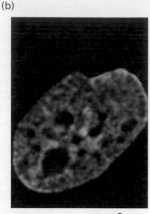

(a)                    (b)

2 µm

**FIGURE 7-37 Expression of fusion proteins demonstrates chromatin decondensation in response to an activation domain.** A cultured hamster cell line was engineered to contain multiple copies of a tandem array of *E. coli* lac operator sequences integrated into a chromosome in a region of heterochromatin. (a) When an expression vector for the lac repressor was transfected into these cells, lac repressors bound to the lac operator sites could be visualized in a region of condensed chromatin using an antibody against the lac repressor (red). DNA was visualized by staining with DAPI (blue), revealing the nucleus. (b) When an expression vector for the lac repressor fused to an activation domain was transfected into these cells, staining as in (a) revealed that the activation domain causes this region of chromatin to decondense into a thinner chromatin fiber that fills a much larger volume of the nucleus. [Courtesy of Andrew S. Belmont, 1999, *J. Cell Biol.* **145**:1341.]

at many promoters. The first of these characterized was the yeast SWI/SNF chromatin-remodeling complex. One of the SWI/SNF subunits has homology to DNA helicases, enzymes that use energy from ATP hydrolysis to disrupt interactions between base-paired nucleic acids or between nucleic acids and proteins. In vitro, the SWI/SNF complex is thought to pump or push DNA into the nucleosome so that DNA bound to the surface of the histone octamer transiently dissociates from the surface and translocates, causing the nucleosomes to "slide" along the DNA. The net result of such chromatin remodeling is to facilitate the binding of transcription factors to specific DNA sequences in chromatin. Many activation domains bind to chromatin-remodeling complexes, and this binding stimulates in vitro transcription from chromatin templates (DNA bound to nucleosomes). Thus the SWI/SNF complex represents another type of co-activator complex. The experiment shown in Figure 7-37 demonstrates dramatically how an activation domain can cause decondensation of a region of chromatin. This results from association of the activation domain with chromatin-remodeling and histone acetylase complexes.

Chromatin-remodeling complexes are required for many processes involving DNA in eukaryotic cells, including transcription control, DNA replication, recombination, and

DNA repair. Several types of chromatin-remodeling complexes are found in eukaryotic cells, all with homologous DNA helicase domains. SWI/SNF complexes and related chromatin-remodeling complexes in multicellular organisms contain subunits with bromodomains that bind to acetylated histone tails. Consequently, SWI/SNF complexes remain associated with activated, acetylated regions of chromatin, presumably maintaining them in a decondensed conformation. Chromatin-remodeling complexes can also participate in transcriptional repression. These chromatin-remodeling complexes bind to transcription repression domains of repressors and contribute to repression, presumably by folding chromatin into condensed structures. Much remains to be learned about how this important class of proteins alters chromatin structure to influence gene expression and other processes.

## The Mediator Complex Forms a Molecular Bridge Between Activation Domains and Pol II

Once the interaction of activation domains with histone acetylase complexes and chromatin remodeling complexes converts the chromatin of a promoter region to an "open" chromatin structure that allows the binding of general transcription factors, activation domains interact with another multisubunit co-activator complex, the **mediator** (Figure 7-38). Activation domain–mediator interactions stimulate assembly of the pre-initiation complex on the promoter. The head and middle domains of the mediator complex are proposed to interact directly with subunits RBP3, 4, 7, and 11 of Pol II. Several mediator subunits bind to activation domains in various activator proteins. Thus mediator can form a molecular bridge between an activator bound to its cognate site in DNA and Pol II at a promoter.

Experiments with temperature-sensitive yeast mutants indicate that some mediator subunits are required for transcription of virtually all yeast genes. These subunits most likely help maintain the overall structure of the mediator complex or bind to Pol II and therefore are required for activation by all activators. In contrast, other mediator subunits are required for normal activation or repression of specific subsets of genes. DNA microarray analysis of yeast gene expression in mutants with defects in these mediator subunits indicates that each such subunit influences transcription of ≈3–10 percent of all genes to the extent that its deletion either increases or decreases mRNA expression by a factor of twofold or more (see Figure 5-29 for DNA microarray technique). These mediator subunits are thought to interact with specific activation domains; thus when one subunit is defective, transcription of genes regulated by activators that bind to that subunit is severely depressed, but transcription of other genes is unaffected. Recent studies suggest that most activation domains may interact with more than one mediator subunit.

The various experimental results indicating that individual mediator subunits bind to specific activation domains suggest that multiple activators influence transcription from a single promoter by interacting with a mediator complex simultaneously

(a) Yeast mediator-Pol II complex

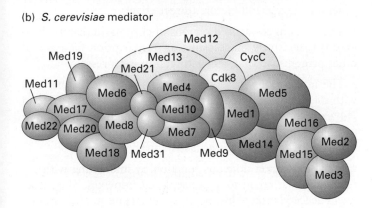

(b) *S. cerevisiae* mediator

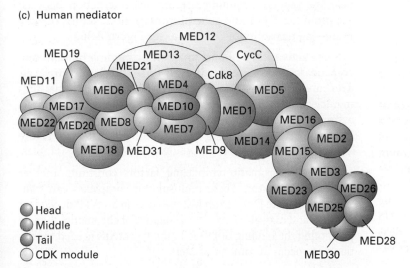

(c) Human mediator

Head
Middle
Tail
CDK module

**FIGURE 7-38 Structure of yeast and human mediator complexes.** (a) Reconstructed image of mediator from *S. cerevisiae* bound to Pol II. Multiple electron microscopy images were aligned and computer-processed to produce this average image in which the three-dimensional Pol II structure (light orange) is shown associated with the yeast mediator complex (dark blue). (b) Diagrammatic representation of mediator subunits from *S. cerevisiae*. Subunits shown in the same color are thought to form a module. Mutations in one subunit of a module may inhibit association of other subunits in the same module with the rest of the complex. (c) Diagrammatic representation of human mediator subunits. [Part (a), from S. Hahn, 2004, *Nat. Struct. Mol. Biol.* **11:**394, based on J. Davis et al., 2002, *Mol. Cell* **10:**409. Part (b), from B. Guglielmi et al., 2004, *Nucl. Acids Res.* **32:**5379. Part (c), adapted from S. Malik and R. G. Roeder, 2010, *Nat. Rev. Genet.* **11:**761. See H. M. Bourbon, 2008, *Nucl. Acids Res.* **36:**3993.]

(Figure 7-39). Activators bound at enhancers or promoter-proximal elements can interact with mediator associated with a promoter because chromatin, like DNA, is flexible and can form a loop bringing the regulatory regions and the promoter close together, as observed for the *E. coli* NtrC activator and $\sigma^{54}$-RNA polymerase (see Figure 7-4). The multiprotein nucleoprotein complexes that form on eukaryotic promoters may comprise as many as 100 polypeptides with a total mass of ≈3 megadaltons (MDa), as large as a ribosome.

## The Yeast Two-Hybrid System

A powerful molecular genetic method called the *yeast two-hybrid system* exploits the flexibility in activator structures to identify genes whose products bind to a specific protein of interest. Because of the importance of protein-protein interactions in virtually every biological process, the yeast two-hybrid system is used widely in biological research.

This method employs a yeast vector for expressing a DNA-binding domain and flexible linker region without the

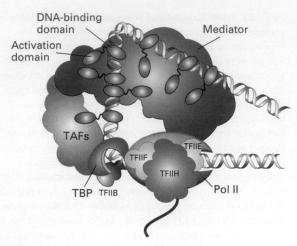

**FIGURE 7-39 Model of several DNA-bound activators interacting with a single mediator complex.** The ability of different mediator subunits to interact with specific activation domains may contribute to the integration of signals from several activators at a single promoter. See the text for discussion.

associated activation domain, such as the deleted GAL4-containing amino acids 1–692 (see Figure 7-26b). A cDNA sequence encoding a protein or protein domain of interest, called the *bait domain,* is fused in frame to the flexible linker region so that the vector will express a hybrid protein composed of the DNA-binding domain, linker region, and bait domain (Figure 7-40a, *left*). A cDNA library is cloned into multiple copies of a second yeast vector that encodes a strong activation domain and flexible linker to produce a vector library expressing multiple hybrid proteins, each containing a different *fish domain* (Figure 7-40a, *right*).

The bait vector and library of fish vectors are then transfected into engineered yeast cells in which the only copy of a gene required for histidine synthesis (*HIS*) is under control of a UAS with binding sites for the DNA-binding domain of the hybrid bait protein. Transcription of the *HIS* gene requires activation by proteins bound to the UAS. Transformed cells that express the bait hybrid and an *interacting* fish hybrid will be able to activate transcription of the *HIS* gene (Figure 7-40b). This system works because of the flexibility in the spacing between the DNA-binding and activation domains of eukaryotic activators.

A two-step selection process is used (Figure 7-40c). The bait vector also expresses a wild-type *TRP* gene, and the hybrid vector expresses a wild-type *LEU* gene. Transfected cells are first grown in a medium that lacks tryptophan and leucine but contains histidine. Only cells that have taken up the bait vector and one of the fish plasmids will survive in this medium. The cells that survive then are plated on a medium that lacks histidine. Those cells expressing a fish hybrid that does not bind to the bait hybrid cannot transcribe the *HIS* gene and consequently will not form a colony on medium lacking histidine. The few cells that express a bait-binding fish hybrid will grow and form colonies in the absence of histidine. Recovery of the fish vectors from these colonies

yields cDNAs encoding protein domains that interact with the bait domain.

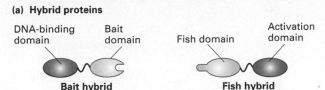

**(a) Hybrid proteins**

DNA-binding domain
Bait domain

Fish domain
Activation domain

**Bait hybrid**

**Fish hybrid**

**(b) Transcriptional activation by hybrid proteins in yeast**

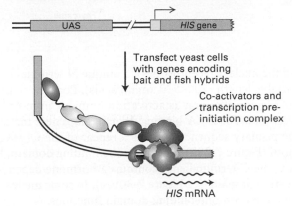

UAS

*HIS* gene

Transfect yeast cells with genes encoding bait and fish hybrids

Co-activators and transcription pre-initiation complex

*HIS* mRNA

**(c) Fishing for proteins that interact with bait domain**

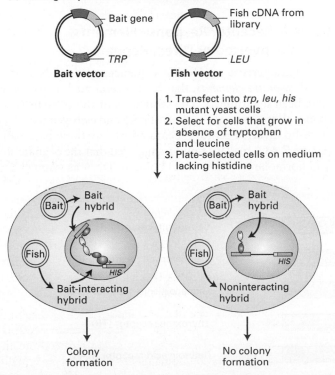

Bait gene

Fish cDNA from library

*TRP*

*LEU*

**Bait vector**

**Fish vector**

1. Transfect into *trp*, *leu*, *his* mutant yeast cells
2. Select for cells that grow in absence of tryptophan and leucine
3. Plate-selected cells on medium lacking histidine

Bait

Bait hybrid

Fish

Bait-interacting hybrid

*HIS*

Bait

Bait hybrid

Fish

Noninteracting hybrid

*HIS*

Colony formation

No colony formation

**EXPERIMENTAL FIGURE 7-40 The yeast two-hybrid system provides a way of screening a cDNA library for clones encoding proteins that interact with a specific protein of interest.** (a) Two vectors are constructed containing genes that encode hybrid (chimeric) proteins. In one vector (*left*), the coding sequence for the DNA-binding domain of a transcription factor is fused to the sequences for a known protein, referred to as the "bait" domain (light blue). The second vector (*right*) expresses an activation domain fused to a "fish" domain (green) that interacts with the bait domain. (b) If yeast cells are transformed with vectors expressing both hybrids, the bait and fish portions of the chimeric proteins interact to produce a functional transcriptional activator. In this example, the activator promotes transcription of a *HIS* gene. One end of this protein complex binds to the upstream activating sequence (UAS) of the *HIS3* gene; the other end, consisting of the activation domain, stimulates assembly of the transcription preinitiation complex at the promoter (yellow). (c) To screen a cDNA library for clones encoding proteins that interact with a particular bait protein of interest, the library is cloned into the vector encoding the activation domain so that hybrid proteins are expressed. The bait vector and fish vectors contain wild-type selectable genes (e.g., a *TRP* or *LEU* gene). The only transformed cells that survive the indicated selection scheme are those that express the bait hybrid and a fish hybrid that interacts with it. See the text for discussion. [See S. Fields and O. Song, 1989, *Nature* **340:**245.]

# 7.6 Regulation of Transcription-Factor Activity

We have seen in the preceding discussion how combinations of activators and repressors that bind to specific DNA regulatory sequences control transcription of eukaryotic genes. Whether or not a specific gene in a multicellular organism is expressed in a particular cell at a particular time is largely a consequence of the nuclear concentrations and activities of the transcription factors that interact with the regulatory sequences of that gene. (Exceptions are due to "transcriptional memory" of the functions of activators and repressors expressed in embryonic cells from which the cell has descended as the result of *epigenetic* mechanisms discussed in the next section.) Which transcription factors are expressed in a particular cell type, and the amounts produced, are determined by multiple regulatory interactions between transcription-factor genes that occur during the development and differentiation of a particular cell type.

In addition to controlling the expression of thousands of specific transcription factors, cells also regulate the activities of many of the transcription factors expressed in a particular cell type. For example, transcription factors are often regulated in response to extracellular signals. Interactions between the extracellular domains of transmembrane receptor proteins on the surface of the cell and specific protein ligands for these receptors activate protein domains associated with the intracellular domains of these transmembrane proteins,

the specific set of activators required for transcription of a particular gene in order to express that gene.

• The yeast two-hybrid system is widely used to detect cDNAs encoding protein domains that bind to a specific protein of interest (see Figure 7-40).

Cortisol

Retinoic acid

Thyroxine

transducing the signal received on the outside of the cell to a signal on the inside of the cell that eventually reaches transcription factors in the nucleus. In Chapter 16, we describe the major types of cell-surface receptors and intracellular signaling pathways that regulate transcription-factor activity.

In this section, we discuss the second major group of extracellular signals, the small, lipid-soluble hormones—including many different steroid hormones, retinoids, and thyroid hormones—that can diffuse through plasma and nuclear membranes and interact directly with the transcription factors they control (Figure 7-41). As noted earlier, the intracellular receptors for most of these lipid-soluble hormones, which constitute the *nuclear-receptor superfamily,* function as transcription activators when bound to their ligands.

## All Nuclear Receptors Share a Common Domain Structure

Sequencing of cDNAs derived from mRNAs encoding various nuclear receptors revealed a remarkable conservation in their amino acid sequences and three functional regions (Figure 7-42).

All the nuclear receptors have a unique N-terminal region of variable length (100–500 amino acids). Portions of this variable region function as activation domains in most nuclear receptors. The DNA-binding domain maps near the center of the primary sequence and has a repeat of the $C_4$ zinc-finger motif (Figure 7-29b). The hormone-binding domain, located near the C-terminal end, contains a hormone-dependent activation domain (see Figure 7-30b, c). In some nuclear receptors, the hormone-binding domain functions as a repression domain in the absence of ligand.

## Nuclear-Receptor Response Elements Contain Inverted or Direct Repeats

The characteristic nucleotide sequences of the DNA sites, called *response elements,* that bind several nuclear receptors have been determined. The sequences of the consensus response elements for the glucocorticoid and estrogen receptors are 6-bp inverted repeats separated by any three base pairs (Figure 7-43a, b). This finding suggested that the cognate steroid hormone receptors would bind to DNA as symmetrical

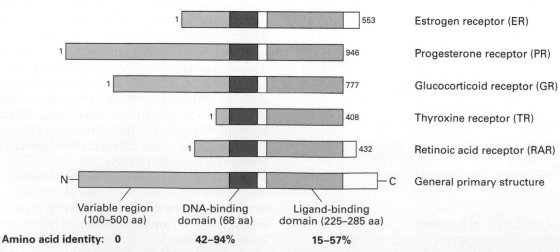

**FIGURE 7-42 General design of transcription factors in the nuclear-receptor superfamily.** The centrally located DNA-binding domain exhibits considerable sequence homology among different receptors and contains two copies of the $C_4$ zinc-finger motif (see Figure 7-29b). The C-terminal hormone-binding domain exhibits somewhat less homology. The N-terminal regions in various receptors vary in length, have unique sequences, and may contain one or more activation domains. [See R. M. Evans, 1988, *Science* **240:**889.]

(a) **GRE**
5′ AGAACA(N)$_3$TGTTCT 3′
3′ TCTTGT(N)$_3$ACAAGA 5′

(b) **ERE**
5′ AGGTCA(N)$_3$TGACCT 3′
3′ TCCAGT(N)$_3$ACTGGA 5′

(c) **VDRE**
5′ AGGTCA(N)$_3$AGGTCA 3′
3′ TCCAGT(N)$_3$TCCAGT 5′

(d) **TRE**
5′ AGGTCA(N)$_4$AGGTCA 3′
3′ TCCAGT(N)$_4$TCCAGT 5′

(e) **RARE**
5′ AGGTCA(N)$_5$AGGTCA 3′
3′ TCCAGT(N)$_5$TCCAGT 5′

**FIGURE 7-43 Consensus sequences of DNA response elements that bind three nuclear receptors.** The response elements for the glucocorticoid receptor (GRE) and estrogen receptor (ERE) contain inverted repeats that bind these homodimeric proteins. The response elements for heterodimeric receptors contain a common direct repeat separated by three to five base pairs for the vitamin D$_3$ receptor (VDRE), thyroid hormone receptor (TRE), and retinoic acid receptor (RARE). The repeat sequences are indicated by red arrows. [See K. Umesono et al., 1991, *Cell* **65:**1255, and A. M. Naar et al., 1991, *Cell* **65:**1267.]

dimers, as was later shown from the x-ray crystallographic analysis of the homodimeric glucocorticoid receptor's C$_4$ zinc-finger DNA-binding domain (see Figure 7-29b).

Some nuclear-receptor response elements, such as those for the receptors that bind vitamin D$_3$, thyroid hormone, and retinoic acid, are direct repeats of the same sequence recognized by the estrogen receptor, separated by three to five base pairs (Figure 7-43c–e). The specificity for responding to these different hormones by binding distinct receptors is determined by the spacing between the repeats. The receptors that bind to such direct-repeat response elements do so as heterodimers with a common nuclear-receptor monomer called RXR. The vitamin D$_3$ response element, for example, is bound by the RXR-VDR heterodimer, and the retinoic acid response element is bound by RXR-RAR. The monomers composing these heterodimers interact with each other in such a way that the two DNA-binding domains lie in the same rather than inverted orientation, allowing the RXR heterodimers to bind to direct repeats of the binding site for each monomer. In contrast, the monomers in homodimeric nuclear receptors (e.g., GRE and ERE) have an inverted orientation.

## Hormone Binding to a Nuclear Receptor Regulates Its Activity as a Transcription Factor

The mechanism whereby hormone binding controls the activity of nuclear receptors differs for heterodimeric and homodimeric

receptors. Heterodimeric nuclear receptors (e.g., RXR-VDR, RXR-TR, and RXR-RAR) are located exclusively in the nucleus. In the absence of their hormone ligand, they repress transcription when bound to their cognate sites in DNA. They do so by directing histone deacetylation at nearby nucleosomes by the mechanism described earlier (see Figure 7-36a). In the ligand-bound conformation, heterodimeric nuclear receptors containing RXR can direct hyperacetylation of histones in nearby nucleosomes, thereby reversing the repressing effects of the free ligand-binding domain. In the presence of ligand, ligand-binding domains of nuclear receptors also bind mediator, stimulating preinitiation complex assembly.

In contrast to heterodimeric nuclear receptors, homomeric receptors are found in the cytoplasm in the absence of their ligands. Hormone binding to these receptors leads to their translocation to the nucleus. The hormone-dependent translocation of the homodimeric glucocorticoid receptor (GR) was demonstrated in the transfection experiments shown in Figure 7-44. The GR hormone-binding domain alone mediates this transport. Subsequent studies showed that, in the absence of hormone, GR is anchored in the cytoplasm in a large protein complex with inhibitor proteins, including Hsp90, a protein related to Hsp70, the major heat-shock chaperone in eukaryotic cells. As long as the receptor is confined to the cytoplasm, it cannot interact with target genes and hence cannot activate transcription. Hormone binding to a homodimeric nuclear receptor releases the inhibitor proteins, allowing the receptor to enter the nucleus, where it can bind to response elements associated with target genes (Figure 7-44d). Once the receptor with bound hormone binds to a response element, it activates transcription by interacting with chromatin-remodeling and histone acetylase complexes and mediator.

## Metazoans Regulate the Pol II Transition from Initiation to Elongation

A recent unexpected discovery that resulted from application of the chromatin immunoprecipitation technique is that a large fraction of genes in metazoans have a paused elongating Pol II within ≈200 base pairs of the transcription start site (Figure 7-16). Thus expression of the encoded protein is controlled not only by transcription initiation, but also by transcription elongation early in the transcription unit. The first genes discovered to be regulated by controlling transcription elongation were *heat-shock genes* (e.g., *hsp70*) encoding protein chaperonins that help to refold denatured proteins and other proteins that help the cell to deal with denatured proteins. When heat shock occurs, the heat-shock transcription factor (HSTF) is activated. Binding of activated HSTF to specific sites in the promoter-proximal region of heat-shock genes stimulates the paused polymerase to continue chain elongation and promotes rapid reinitiation by additional RNA polymerase II molecules, leading to many transcription initiations per minute. This mechanism of transcriptional control permits a rapid response: these genes are always paused in a state of suspended transcription and therefore, when an emergency arises, require no time to

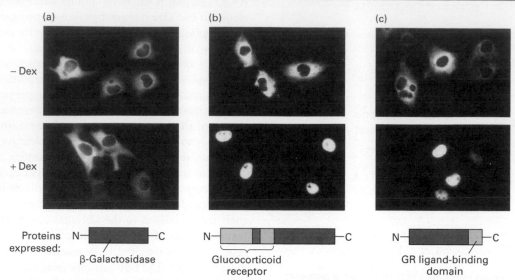

**EXPERIMENTAL FIGURE 7-44 Fusion proteins from expression vectors demonstrate that the hormone-binding domain of the glucocorticoid receptor (GR) mediates translocation to the nucleus in the presence of hormone.** Cultured animal cells were transfected with expression vectors encoding the proteins diagrammed at the bottom. Immunofluorescence with a labeled antibody specific for β-galactosidase was used to detect the expressed proteins in transfected cells. (a) In cells that expressed β-galactosidase alone, the enzyme was localized to the cytoplasm in the presence and absence of the glucocorticoid hormone dexamethasone (Dex). (b) In cells that expressed a fusion protein consisting of β-galactosidase and the entire glucocorticoid receptor (GR), the fusion protein was present in the cytoplasm in the absence of hormone but was transported to the nucleus in the presence of hormone. (c) Cells that expressed a fusion protein composed of β-galactosidase and just the GR ligand-binding domain (light purple) also exhibited hormone-dependent transport of the fusion protein to the nucleus. (d) Model of hormone-dependent gene activation by a homodimeric nuclear receptor. In the absence of hormone, the receptor is kept in the cytoplasm by interaction between its ligand-binding domain (LBD) and inhibitor proteins. When hormone is present, it diffuses through the plasma membrane and binds to the ligand-binding domain, causing a conformational change that releases the receptor from the inhibitor proteins. The receptor with bound ligand is then translocated into the nucleus, where its DNA-binding domain (DBD) binds to response elements, allowing the ligand-binding domain and an additional activation domain (AD) at the N-terminus to stimulate transcription of target genes. [Parts (a)–(c) from D. Picard and K. R. Yamamoto, 1987, *EMBO J.* **6**:3333; courtesy of the authors.]

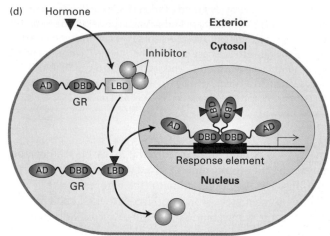

capable of differentiation into any cell type. The ability to induce differentiated cells to convert to pluripotent stem cells has elicited enormous research interest because of its potential for the development of therapeutic treatments for traumatic injuries to the nervous system and degenerative diseases (Chapter 21).

## Pol II Termination Is Also Regulated

Once Pol II has transcribed ≈200 nucleotides from the transcription start site, elongation through most genes is highly processive, although the chromatin immunoprecipitation with antibody to Pol II indicates that the amount of Pol II at various positions in a transcription unit in a population of cells varies greatly (Figure 7-16b, *right*). This indicates that the enzyme can elongate through some regions much more rapidly than others. In most cases, Pol II does not terminate until after a sequence is transcribed that directs cleavage and

remodel and acetylate chromatin over the promoter and assemble a transcription preinitiation complex.

Another transcription factor shown to regulate transcription by controlling elongation of Pol II paused near the transcription start site is MYC, which functions in the regulation of cell growth and division. MYC is often expressed at high level in cancer cells and is a key transcription factor in the reprogramming of somatic cells into pluripotent stem cells

polyadenylation of the RNA at the sequence that forms the 3′ end of the encoded mRNA. RNA polymerase II then can terminate at multiple sites located over a distance of 0.5–2 kb beyond this poly(A) addition site. Experiments with mutant genes show that termination is coupled to the process that cleaves and polyadenylates the 3′ end of a transcript, which is discussed in the next chapter.

## KEY CONCEPTS of Section 7.6

### Regulation of Transcription-Factor Activity

• The activities of many transcription factors are indirectly regulated by binding of extracellular proteins and peptides to cell-surface receptors. These receptors activate intracellular signal-transduction pathways that regulate specific transcription factors through a variety of mechanisms discussed in Chapter 16.

• Nuclear receptors constitute a superfamily of dimeric $C_4$ zinc-finger transcription factors that bind lipid-soluble hormones and interact with specific response elements in DNA (see Figures 7-41–43).

• Hormone binding to nuclear receptors induces conformational changes that modify their interactions with other proteins (Figure 7-30b, c).

• Heterodimeric nuclear receptors (e.g., those for retinoids, vitamin D, and thyroid hormone) are found only in the nucleus. In the absence of hormone, they repress transcription of target genes with the corresponding response element. When bound to their ligands, they activate transcription.

• Steroid hormone receptors are homodimeric nuclear receptors. In the absence of hormone, they are trapped in the cytoplasm by inhibitor proteins. When bound to their ligands, they can translocate to the nucleus and activate transcription of target genes (see Figure 7-44).

## 7.7 Epigenetic Regulation of Transcription

The term **epigenetic** refers to inherited changes in the phenotype of a cell that do not result from changes in DNA sequence. For example, during the differentiation of bone marrow stem cells into the several different types of blood cells, a hematopoietic stem cell (HSC) divides into two daughter cells, one of which continues to have the properties of an HSC with the potential to differentiate into all of the different types of blood cells. But the other daughter cell becomes either a lymphoid stem cell or a myeloid stem cell (see Figure 21-18). Lymphoid stem cells generate daughter cells that differentiate into lymphocytes, which perform many of the functions involved in immune responses to pathogens (Chapter 23). Myeloid stem cells divide into daughter cells that are committed to differentiating into red blood cells, different kinds of phagocytic white

blood cells, or the cells that generate platelets involved in blood clotting. Lymphoid and myeloid stem cells both have the identical DNA sequence as the zygote generated by fertilization of the egg cell by a sperm cell from which all cells develop, but they have restricted developmental potential because of epigenetic differences between them. Such epigenetic changes are initially the consequence of the expression of specific transcription factors that are master regulators of cellular differentiation, controlling the expression of other genes encoding transcription factors and proteins involved in cell-cell communication in complex networks of gene control that are currently the subject of intense investigation. Changes in gene expression initiated by transcription factors are often reinforced and maintained over multiple cell divisions by post-translational modifications of histones and methylation of DNA at position 5 of the cytosine pyrimidine ring (Figure 2-17) that are maintained and propagated to daughter cells when cells divide. Consequently, the term *epigenetic* is used to refer to such post-translational modifications of histones and 5-methyl C modification of DNA.

### Epigenetic Repression by DNA Methylation

As mentioned earlier, most promoters in mammals fall into the CpG island class. Active CpG island promoters have Cs in CG sequences that are unmethylated. Unmethylated CpG island promoters are generally depleted of histone octamers, but nucleosomes immediately neighboring the unmethylated CpG island promoters are modified by histone H3 lysine 4 di- or trimethylation and have associated Pol II molecules that are paused during transcription of both the sense and non-sense template DNA strands, as discussed earlier (Figures 7-16 and 17). Recent research indicates that methylation of histone H3 lysine 4 occurs in mouse cells because a protein named Cfp1 (CXXC finger protein 1) binds unmethylated CpG-rich DNA through a zinc-finger domain (CXXC) and associates with a histone methylase specific for histone H3 lysine 4 (Setd1). Chromatin-remodeling complexes and the general transcription factor TFIID, which initiates Pol II preinitiation complex assembly (Figure 7-17), associate with nucleosomes bearing the H3 lysine 4 trimethyl mark, promoting Pol II transcription initiation.

However, in differentiated cells, a few percent of specific CpG island promoters, depending on the cell type, have CpGs marked by 5-methyl C. This modification of CpG island DNA triggers chromatin condensation. A family of proteins that bind to DNA rich in 5 methyl-C modified CpGs (methyl CpG-binding proteins, MBDs) associate with histone deacetylases and repressing chromatin-remodeling complexes that condense chromatin, resulting in transcriptional repression. These methyl groups are added by de novo DNA methyl transferases named DNMT3a and DNMT3b. Much remains to be learned about how these enzymes are directed to specific CpG islands, but once they have methylated a DNA sequence, methylation is passed on through DNA replication through the action of the ubiquitous *maintenance* methyl transferase DNMT1:

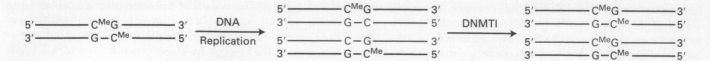

(Red indicates daughter strands.) DNMT1 also maintains methylation of the Cs in CpG sequences that are statistically underrepresented throughout most of the genome. As discussed above, the CG sequence is underrepresented in most of the sequence of mammalian genomes, probably because spontaneous deamination of 5-methyl C generates thymidine, leading to the substitution of CpGs with TpGs over the period of mammalian evolution, unless there is selection against the resulting mutation, as probably occurs when CpG island promoters are mutated. This mechanism of epigenetic repression is intensely investigated because tumor-suppressor genes encoding proteins that function to suppress the development of cancer are often inactivated in cancer cells by abnormal CpG methylation of their promoter regions, as discussed further in Chapter 24.

## Histone Methylation at Other Specific Lysines Are Linked to Epigenetic Mechanisms of Gene Repression

Figure 6-31b summarizes the different types of post-translational modifications that are found on histones, including acetylation of lysines and methylation of lysines on the nitrogen atom of the terminal ε-amino group of the lysine

side chain (see Figure 2-14). Lysines can be modified by the addition of one, two, or three methyl groups to this terminal nitrogen atom, generating mono-, di-, and trimethylated lysine, all of which carry a single positive charge. Pulse-chase radiolabeling experiments have shown that acetyl groups on histone lysines turn over rapidly, whereas methyl groups are much more stable. The acetylation state at a specific histone lysine on a particular nucleosome results from a dynamic equilibrium between acetylation and deacetylation by histone acetylases and histone deacetylases, respectively. Acetylation of histones in a localized region of chromatin predominates when local DNA-bound activators transiently bind histone acetylase complexes. De-acetylation predominates when repressors transiently bind histone deacetylase complexes.

In contrast to acetyl groups, methyl groups on histone lysines are much more stable and turn over much less rapidly than acetyl groups. Histone lysine methyl groups can be removed by *histone lysine demethylases*. But the resulting turnover of histone lysine methyl groups is much slower than the turnover of histone lysine acetyl groups, making them appropriate post-translational modifications for propagating epigenetic information. Several other post-translational modifications have been characterized on histones (Figure 6-31b). These all have the potential to positively or

| TABLE 7-3 | Histone Post-Translational Modifications Associated with Active and Repressed Genes | |
|---|---|---|
| **Modification** | **Sites of Modification** | **Effect on Transcription** |
| Acetylated lysine | H3 (K9, K14, K18, K27, K56)<br>H4 (K5, K8, K13, K16)<br>H2A (K5, K9, K13)<br>H2B (K5, K12, K15, K20) | Activation |
| Hypoacetylated lysine | | Repression |
| Phosphorylated serine/threonine | H3 (T3, S10, S28)<br>H2A (S1, T120)<br>H2B (S14) | Activation |
| Methylated arginine | H3 (R17, R23)<br>H4 (R3) | Activation |
| Methylated lysine | H3 (K4) Me3 in promoter region<br>H3 (K4) Me1 in enhancers<br>H3 (K36, K79) in transcribed region<br>H3 (K9, K27)<br>H4 (K20) | Activation<br><br>Elongation<br>Repression |
| Ubiquitinated lysine | H2B (K120 in mammals, K123 in *S. cerevisiae*)<br>H2A (K119 in mammals) | Activation<br>Repression |

negatively regulate the binding of proteins that interact with the chromatin fiber to regulate transcription and other processes such as chromosome folding into the highly condensed structures that form during mitosis (Figures 6-39 and 6-40). A picture of chromatin has emerged in which histone tails extending as random coils from the chromatin fiber are post-translationally modified to generate one of many possible combinations of modifications that regulate transcription and other processes by regulating the binding of a large number of different protein complexes. This control of protein interactions with specific regions of chromatin resulting from the combined influences of various post-translational modifications of histones has been called a *histone code*. Some of these modifications, such as histone lysine acetylation, are rapidly reversible, whereas others, such as histone lysine methylation, can be templated through chromatin replication, generating epigenetic inheritance in addition to inheritance of DNA sequence. Table 7-3 summarizes the influence that post-translational modifications of specific histone amino acid residues usually have in transcription.

**Histone H3 Lysine 9 Methylation in Heterochromatin** In most eukaryotes, some co-repressor complexes contain histone

methyl transferase subunits that methylate histone H3 at lysine 9, generating di- and trimethyl lysines. These methylated lysines are binding sites for isoforms of HP1 protein that function in the condensation of heterochromatin, as discussed in Chapter 6 (see Figure 6-34). For example, the KAP1 co-repressor complex functions with a class of more than 200 zinc-finger transcription factors encoded in the human genome. This co-repressor complex includes an H3 lysine 9 methyl transferase that methylates nucleosomes over the promoter region of repressed genes, leading to HP1 binding and repression of transcription. An integrated transgene in cultured mouse fibroblasts that was repressed through the action of the KAP1 co-repressor associated with heterochromatin in most cells, whereas the active form of the same transgene associated with euchromatin (Figure 7-45). Chromatin immunoprecipitation assays (see Figure 7-16) showed that the repressed gene was associated with histone H3 methylated at lysine 9 and HP1, whereas the active gene was not.

Importantly, H3 lysine 9 methylation is maintained following chromosome replication in S phase by the mechanism diagrammed in Figure 7-46. When chromosomes replicate in S phase, the nucleosomes associated with the parental DNA are randomly distributed to the daughter DNA molecules. New

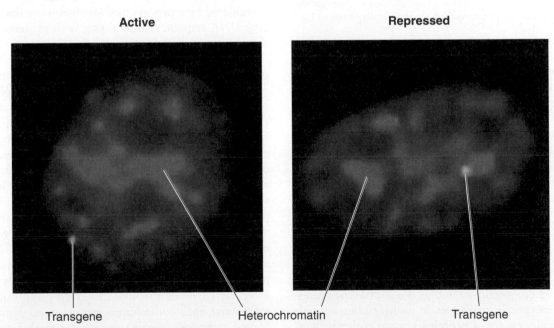

**Active**  **Repressed**

Transgene  Heterochromatin  Transgene

**FIGURE 7-45 Association of a repressed transgene with heterochromatin.** Mouse fibroblasts were stably transformed with a transgene with binding sites for an engineered repressor. The repressor was a fusion between a DNA-binding domain, a repression domain that interacts with the KAP1 co-repressor complex, and the ligand-binding domain of a nuclear receptor that allows the nuclear import of the fusion protein to be controlled experimentally (see Figure 7-44). DNA was stained blue with the dye DAPI. Brighter-staining regions are regions of heterochromatin, where the DNA concentration is higher than in euchromatin. The transgene was detected by hybridization of a fluorescently labeled complementary probe (green). When the recombinant repressor was retained in the cytoplasm, the transgene was transcribed (*left*) and was associated with euchromatin in most cells. When hormone was added so that the recombinant repressor entered the nucleus, the transgene was repressed (*right*) and associated with heterochromatin. Chromatin immunoprecipitation assays (see Figure 7-16) showed that the repressed gene was associated with histone H3 methylated at lysine 9 and HP1, whereas the active gene was not. [Courtesy of Frank Rauscher, from Ayyanathan et al., 2003, *Genes Dev.* **17**:1855.]

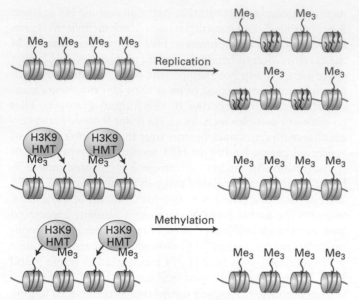

**FIGURE 7-46 Maintenance of histone H3 lysine 9 methylation during chromosome replication.** When chromosomal DNA is replicated, the parental histones randomly associate with the two daughter DNA molecules while unmethylated histones synthesized during S phase comprise other nucleosomes on the sister chromosomes. Association of histone H3 lysine 9 methyl transferases (H3K9 HMT) with parental nucleosomes bearing the histone 3 lysine 9 di- or trimethylation mark methylate the newly added unmodified nucleosomes. Consequently, histone H3 lysine 9 methylations are maintained during repeated cell divisions unless they are specifically removed by a histone demethylase.

histone octamers that are not methylated on lysine 9 associate with the new daughter chromosomes, but since the parental nucleosomes are associated with both daughter chromosomes, approximately half of them are methylated on lysine 9. Association of histone H3 lysine methyl transferases (directly or indirectly) with the parental methylated nucleosomes leads to methylation of the newly assembled histone octamers. Repetition of this process with each cell division results in maintenance of H3 lysine 9 methylation of this region of the chromosome.

## Epigenetic Control by Polycomb and Trithorax Complexes

Another kind of epigenetic mark that is essential for repression of genes in specific cell types in multicellular animals and plants involves a set of proteins known collectively as Polycomb proteins and a counteracting set of proteins known as Trithorax proteins, after phenotypes of mutations in the genes encoding them in *Drosophila,* where they were first discovered. The Polycomb repression mechanism is essential for maintaining the repression of genes in specific types of cells and in all of the subsequent cells that develop from them throughout the life of an organism. Important genes regulated by Polycomb proteins include the *Hox* genes, encoding master regulatory transcription factors. Different

combinations of Hox transcription factors help to direct the development of specific tissues and organs in a developing embryo. Early in embryogenesis, expression of *Hox* genes is controlled by typical activator and repressor proteins. However, the expression of these activators and repressors stops at an early point in embryogenesis. Correct expression of the *Hox* genes in the descendants of the early embryonic cells is then maintained throughout the remainder of embryogenesis and on into adult life by the Polycomb proteins which maintain the *repression* of specific *Hox* genes. Trithorax proteins perform the opposing function to Polycomb proteins, maintaining the *expression* of the *Hox* genes that were expressed in a specific cell early in embryogenesis and in all the subsequent descendants of that cell. Polycomb and Trithorax proteins control thousands of genes, including genes that regulate cell growth and division (i.e., the cell cycle, as discussed in Chapter 19). Polycomb and Trithorax genes are often mutated in cancer cells, contributing importantly to the abnormal properties of these cells (Chapter 24).

Remarkably, virtually all cells in the developing embryo and adult express a similar set of Polycomb and Trithorax proteins, and all cells contain the same set of *Hox* genes. Yet only the *Hox* genes in cells where they were initially repressed in early embryogenesis remain repressed, even though the same *Hox* genes in other cells remain active in the presence of the same Polycomb proteins. Consequently, as in the case of the yeast silent mating-type loci, the expression of *Hox* genes is regulated by a process that involves more than simply specific DNA sequences interacting with proteins that diffuse through the nucleoplasm.

A current model for repression by Polycomb proteins is depicted in Figure 7-47. Most Polycomb proteins are subunits of one of two classes of multiprotein complexes, PRC1-type complexes and PRC2 complexes. PRC2 complexes are thought to act initially by associating with specific repressors bound to their cognate DNA sequences early in embryogenesis, or ribonucleoprotein complexes containing long noncoding RNAs as discussed in a later section. The PRC2 complexes contain histone deacetylases that inhibit transcription as discussed above. They also contain a subunit [E(z) in *Drosophila,* EZH2 in mammals] with a *SET domain,* the enzymatically active domain of several histone methyl transferases. This SET domain in PRC2 complexes methylates histone H3 on lysine 27, generating di- and trimethyl lysines. The PRC1 complex then binds the methylated nucleosomes through dimeric Pc subunits (CBXs in mammals), each containing a methyl-lysine binding domain (called a *chromodomain*) specific for methylated H3 lysine 27. Binding of the dimeric Pc to neighboring nucleosomes is proposed to condense the chromatin into a structure that inhibits transcription. This is supported by electron microscopy studies showing that PRC1 complexes cause nucleosomes to associate in vitro (Figure 7-47d, e).

PRC1 complexes also contain a ubiquitin ligase that monoubiquitinates histone H2A at lysine 119 in the H2A C-terminal tail. This modification of H2A inhibits Pol II elongation through chromatin by inhibiting the association

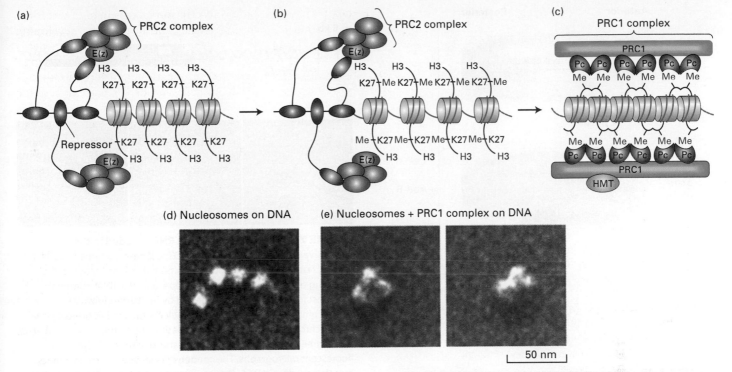

**FIGURE 7-47 Model for repression by Polycomb complexes.**
(a) During early embryogenesis, repressors associate with the PRC2 complex. (b) This results in methylation (Me) of neighboring nucleosomes on histone H3 lysine 27 (K27) by the SET-domain-containing subunit E(z). (c) The PRC1 complexes bind nucleosomes methylated at H3 lysine 27 through a dimeric, chromodomain-containing subunit Pc. The PRC1 complex condenses the chromatin into a repressed chromatin structure. PRC2 complexes associate with PRC1 complexes to maintain H3 lysine 27 methylation of neighboring histones. As a consequence, PRC1 and PRC2 association with the region is maintained when expression of the repressor proteins in (a) ceases. (d, e) Electron micrograph of an ≈1-kb fragment of DNA bound by four nucleosomes in the absence (d) and presence (e) of one PRC1 complex per five nucleosomes. [Parts (a)–(c), adapted from A. H. Lund and M. van Lohuizen, 2004, *Curr. Opin. Cell Biol.* **16**:239. Parts (d, e), from N. J. Francis, R. E. Kingston, and C. L. Woodcock, 2004, *Science* **306**:1574.]

of a histone chaperone required to remove histone octamers from DNA as Pol II transcribes through a nucleosome and then replaces them as the polymerase passes. PRC2 complexes are postulated to associate with nucleosomes bearing the histone H3 lysine 27 trimethylation mark, maintaining methylation of H3 lysine 27 in nucleosomes in the region. This results in association of the chromatin with PRC1 and PRC2 complexes even after expression of the initial repressor proteins in Figure 7-47a, b has ceased. This would also maintain histone H3 lysine 27 methylation and histone H2A monoubiquitination following DNA replication, by a mechanism analogous to that diagrammed in Figure 7-46. This is a key feature of Polycomb repression, which is maintained through successive cell divisions for the life of an organism (≈100 years for some vertebrates, 2000 years for a sugar cone pine!).

Trithorax proteins counteract the repressive mechanism of Polycomb proteins, as shown in studies of expression of the Hox transcription factor Abd-B in the *Drosophila* embryo (Figure 7-48). When the Polycomb system is defective, Abd-B is derepressed in all cells of the embryo. When the Trithorax system is defective and cannot counteract repression by the Polycomb system, Abd-B is repressed in most cells, except those in the very posterior of the embryo. Trithorax complexes include a histone methyl transferase that trimethylates histone H3 lysine 4, a histone methylation associated with the promoters of actively transcribed genes. This histone modification creates a binding site for histone acetylase and chromatin-remodeling complexes that promote transcription, as well as TFIID, the general transcription factor that initiates preinitiation-complex assembly (Figure 7-17). Nucleosomes with the H3 lysine 4 methyl modifications are also binding sites for specific histone demethylases that prevent methylation of histone H3 at lysine 9, preventing the binding of HP1, and at lysine 27, preventing the binding of the PRC-repressing complexes. Likewise, a histone demethylase specific for histone H3 lysine 4 associates with PRC2 complexes. Nucleosomes marked with histone H3 lysine 4 methylation also are thought to be distributed to both daughter DNA molecules during DNA replication, resulting in maintenance of this epigenetic mark by a strategy similar to that diagrammed in Figure 7-46.

## Noncoding RNAs Direct Epigenetic Repression in Metazoans

Repressing complexes also have been discovered that are complexes of proteins bound to RNA molecules. In some cases, this results in repression of genes on the same chromosome from

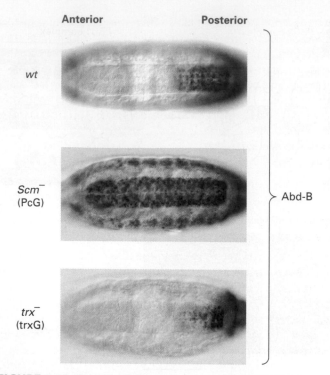

**FIGURE 7-48 Opposing influence of Polycomb and Trithorax complexes on expression of the Hox transcription factor Abd-B in Drosophila embryos.** At the stage of *Drosophila* embryogenesis shown, Abd-B is normally expressed only in posterior segments of the developing embryo, as shown at the top by immunostaining with a specific anti–Abd-B antibody. In embryos with homozygous mutations of *Scm*, a Polycomb gene (PcG) encoding a protein associated with the PRC1 complex, Abd-B expression is de-repressed in all embryo segments. In contrast, in homozygous mutants of *trx*, a Trithorax gene (trxG), Abd-B repression is increased so that it is only expressed to high level in the most posterior segment. [Courtesy of Juerg Mueller, European Molecular Biology Laboratory.]

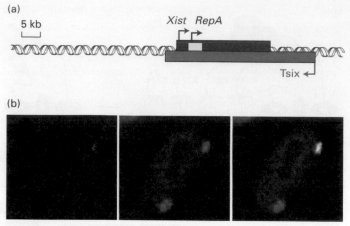

**FIGURE 7-49 The Xist noncoding RNA encoded in the X-inactivation center coats the inactive X chromosome in cells of human females.** (a) The region of the human X-inactivation center encoding the noncoding RNAs *Xist*, *RepA*, and *Tsix*. (b) A cultured fibroblast from a female was analyzed by in situ hybridization with a probe complementary to *Xist* RNA labeled with a red fluorescent dye (*left*), a chromosome paint set of probes for the X chromosome labeled with a green fluorescent dye (*center*), and an overlay of the two fluorescent micrographs. The condensed inactive X chromosome is associated with *Xist* RNA. [Part (a) adapted from J. T. Lee, 2010, *Cold Spring Harbor Perspect. Biol.* **2**:a003749. Part (b) from C. M. Clemson et al., 1996, *J. Cell Biol.* **132**:259.]

which the RNA is transcribed, as in the case of X-chromosome inactivation in female mammals. In other cases, these repressing RNA-protein complexes can be targeted to genes transcribed from other chromosomes by base-pairing with nascent RNAs as they are being transcribed.

**X-Chromosome Inactivation in Mammals** The phenomenon of X-chromosome inactivation in female mammals is one of the most intensely studied examples of epigenetic repression mediated by a long, non-protein-coding RNA. X inactivation is controlled by an ≈100-kb domain on the X chromosome called the X-inactivation center. Remarkably, the X-inactivation center does not express proteins, but rather several noncoding RNAs (ncRNAs) that participate in the random inactivation of one entire X chromosome early in the development of female mammals. The ncRNAs whose functions are partially understood are transcribed from the complementary DNA strands near the middle of the X-inactivation center: the 40-kb Tsix RNA, the Xist RNA which is spliced into an RNA of ≈17 kb, and the shorter 1.6-kb RepA RNA from the 5′ region of the Xist RNA (Figure 7-49a).

In differentiated female cells, the inactive X chromosome is associated with Xist RNA-protein complexes along its entire length. Targeted deletion of the *Xist* gene (Figure 5-42) in cultured embryonic stem cells showed that it is required for X inactivation. As opposed to most protein-coding genes on the inactive X chromosome, Xist is transcribed from the X-inactivation center of the otherwise mostly inactive X chromosome. The Xist RNA-protein complexes do not diffuse to interact with the active X chromosome, but remain associated with the inactive X chromosome. Since the full length of the inactive X becomes coated by Xist RNA-protein complexes (Figure 7-49b), these complexes must spread along the chromosome from the X-inactivation center where Xist is transcribed. The inactive X chromosome is also associated with Polycomb PRC2 complexes that catalyze the trimethylation of histone H3 lysine 27. This results in association of the PRC1 complex and transcriptional repression as discussed above.

In the early female embryo comprised of embryonic stem cells capable of differentiating into all cell types (see Chapter 21), genes on both X chromosomes are transcribed and the 40-kb Tsix ncRNA is transcribed from the X-inactivation center of both copies of the X chromosome. Experiments employing engineered deletions in the X-inactivation center have shown that Tsix transcription prevents significant transcription of the 17-kb Xist RNA from the complementary DNA strand. Later in development of the early embryo, as cells begin to differentiate, Tsix becomes transcribed only from the active X chromosome. The mechanism(s) controlling this asymmetric transcription of Tsix are not yet understood. However, the process occurs randomly on the two X chromosomes.

In a current model of X inactivation, inhibition of Tsix transcription allows transcription of RepA RNA from the complementary DNA strand (Figure 7-49a). RepA RNA has a repeating sequence that forms stem-loop secondary structures that are bound directly by subunits of the Polycomb PRC2 complex. This interaction occurs on nascent RepA transcripts that are tethered to the X chromosome during transcription and leads to methylation of histone H3 at lysine 27 in the surrounding chromatin. By mechanisms that are not yet understood, this activates transcription from the nearby Xist promoter. The transcribed Xist RNA contains RNA sequences that by unknown mechanisms cause it to spread along the X chromosome. The RepA repeated sequence near the 5' end of the Xist RNA binds the PRC2 polycomb complex leading to H3K27 di- and trimethylation along the entire length of the X chromosome. This in turn results in binding of the PRC1 polycomb complex and transcriptional repression as discussed earlier. At the same time, continued transcription of Tsix from the other, active X chromosome continues, represses Xist transcription from that X chromosome, and consequently prevents Xist-mediated repression of the active X. A short time later in development, the DNA of the inactive X also becomes methylated at most of its associated CpG island promoters, probably contributing to its stable inactivation through the multiple cell divisions that occur later during embryogenesis and throughout adult life.

*Trans* Repression by Long Noncoding RNAs   Another example of transcriptional repression by a long noncoding RNA was discovered recently by researchers studying the function of noncoding RNAs transcribed from a region encoding a cluster of *HOX* genes, the *HOXC* locus, in cultured human fibroblasts. Depletion of a 2.2-kb noncoding RNA expressed from the *HOXC* locus by siRNA (Figure 5-45) unexpectedly led to de-repression of the *HOXD* locus in these cells, an ≈40-kb region on another chromosome encoding several HOX proteins and multiple other noncoding RNAs. Assays similar to chromatin immunoprecipitation showed that this noncoding RNA, named HOTAIR for HOX Antisense Intergenic RNA, associates with the *HOXD* loci and with Polycomb PRC2 complexes. This results in histone H3 lysine 27 di- and trimethylation, PRC1 association, histone H3 lysine 4 demethylation, and transcriptional repression. This is similar to the recruitment of Polycomb complexes by Xist RNA except that Xist RNA functions *in cis,* remaining in association with the chromosome from which it is transcribed, whereas HOTAIR leads to Polycomb repression *in trans* on both copies of another chromosome.

Recently, characterization of DNA associated with the histone H3 lysine 4 trimethylation mark associated with promoter regions and H3 lysine 36 methylation associated with Pol II transcriptional elongation led to the discovery of ≈1600 long noncoding RNAs transcribed from intergenic regions between protein-coding genes that are evolutionarily conserved between mammals. This conservation of sequence strongly suggests that these noncoding RNAs have important functions. The examples of Xist, HOTAIR, and two other recently discovered ncRNAs that target Polycomb repression mechanisms to specific genes raise the possibility that many of these may also target Polycomb repression. Consequently, the study of these conserved noncoding RNAs is another area of intense current investigation.

## Plants and Fission Yeast Use Short RNA-Directed Methylation of Histones and DNA

Centromeres (Figure 6-45c) of the fission yeast *Schizosaccharomyces pombe* are composed of multiple sequence repeats as they are in multicellular organisms. Proper functioning of these centromeres during chromosome segregation in mitosis and meiosis (Figures 18-36, 5-10a, 19-38) requires centromeres to form heterochromatin. Heterochromatin formation at *S. pombe* centromeres is directed by short interfering RNAs (siRNAs), initially discovered in *C. elegans* for their function in the cytoplasm, where they direct degradation of mRNAs to which they hybridize (Figure 5-45 and discussed further in Chapter 8). RNA polymerase II transcribes low levels of noncoding transcripts from the centromeric repeats (cenRNA, Figure 7-50). This is converted into double stranded RNA by an RNA-dependent RNA polymerase found in plants and many fungi (but not in the budding yeast *S. cerevisiae,* where the siRNA system does not occur, and not in mammals, where this mechanism of transcriptional repression may not occur). The resulting long double-stranded RNAs are cleaved by a double-strand RNA specific ribonuclease called *Dicer* into 22-nucleotide fragments with two-nucleotide 3' overhangs. One strand of these Dicer fragments is bound by a member of a protein family called *Argonaut* proteins that associate with siRNAs in both translational and transcriptional repression mechanisms. The *S. pombe* Argonaut protein, Ago1, associates with two other proteins to form the RITS complex (for RNA-induced transcriptional silencing).

The RITS complex associates with centromeric regions by base-pairing between the siRNA associated with its Ago1 subunit and nascent transcripts from the region and interactions of its Chp1 (chromodomain protein 1) subunit which contains a methyl lysine-binding chromodomain specific for binding histone H3 di- and trimethyl lysine 9 associated with heterochromatin. The RITS complex also associates with an RNA-dependent RNA polymerase-containing complex, RDRC. Since multiple siRNAs are generated from the double-stranded RNA, this results in a positive feedback loop that increases the association of RITS complexes with centromeric heterochromatin. The RITS complex also associates with a histone H3 lysine 9 methyl transferase. The resulting histone H3 lysine 9 methyl marks on the centromeric chromatin are binding sites for *S. pombe* HP1 proteins and a histone deacetylase (HDAC), leading to the condensation of the centromere region into heterochromatin.

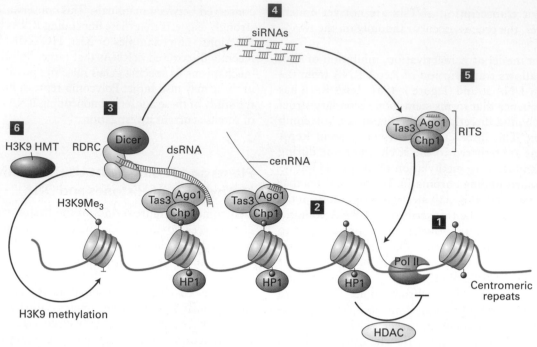

**FIGURE 7-50 Model for the generation of heterochromatin at
*S. pombe* centromeres by noncoding RNAs.** Step (**1**): Pol II transcripts of the repeated nonprotein-coding sequences of the centromere are transcribed at a low level. Step (**2**): The nascent RNA is bound
by the RITS complex by base-pairing of the complementary short
interfering RNA (siRNA) associated with the Ago1 subunit of the RITS
complex and the interaction of the Chp1 subunit with histone H3
methylated on lysine 9. Step (**3**): The RITS complex associates with the
RDRC complex, which includes an RNA-dependent RNA polymerase
that converts the nascent Pol II transcript into double-stranded RNA.
Step (**4**): The double-stranded RNA is cleaved by the Dicer double-
strand-specific ribonuclease into double-stranded fragments of ≈22
nucleotides with two base overhangs at the 3′ end of each strand.
Step (**5**): One of the two ≈22 nucleotide strands generated is bound
by the Ago1 subunit of a RITS complex. Since multiple siRNAs
associated with RITS complexes are generated from each Pol II
transcript, this results in a positive-feedback loop that concentrates
RITS complexes at the centromere region. Step (**6**): The RITS complex
also associates with a histone H3 lysine 9 methyl transferase (H3K9
HMT), which methylates histone H3 in the centromeric region. This
generates a binding site for *S. pombe* HP1 proteins, as well as the Chp1
subunit of the RITS complex. Binding of HP1 condenses the region into
heterochromatin as diagrammed in Figure 6-35a. [Adapted from
D. Moazed, 2009, *Nature* **457**:413.]

**5-Methyl C Induction by ncRNAs in Plants** The model
plant *Arabidopsis thaliana* uses DNA methylation extensively to repress transcription of transposons and retrotransposons (discussed in Chapter 6) and certain specific
genes. In addition to methylating C at the 5 position in the
sequence CG, plants also methylate genes at CHG (where H
is any of the other nucleotides) and CHH. There is a degree of
redundancy, but the DNA methyl transferase MET1 largely
carries out CpG methylation and is functionally similar to
DNMT1 in multicellular animals. CMT3 (chromomethylase
3) methylates CHG, and DRM2 is the primary methyl transferase of CHH. Methylation of CpG and CHG sequences are
maintained following DNA replication by MET1 and CMT3,
respectively, by recognition of the methyl C in the parental
strand of newly replicated DNA and methylation of the
daughter strand C, as discussed above for human DNMT1.
However, one of the daughter chromosomes of a CHH methylation site has an unmodified G at the position complementary to the methylated C, and hence carries no DNA
modification that can be recognized by the DRM2 methyl
transferase. Consequently, CHH methylation sites must be
maintained through cell division by an alternative mechanism.

The *FWA* gene encodes a homeodomain transcription
factor involved in regulation of the flowering time in response
to temperature, so that plants do not flower until the warm
days of spring. In wild-type *A. thaliana*, *FWA* is repressed by
CHH methylation of its promoter region. Failure to methylate the *FWA* promoter results in an easily recognized late-
flowering phenotype, allowing the isolation of *A. thaliana*
mutants in multiple genes that fail to methylate CHH sequences. These genes have been cloned by methods described
in Chapter 5, revealing a complex mechanism of RNA-di-
rected DNA methylation that involves the plant-specific RNA
polymerases IV and V mentioned earlier (Figure 7-51) and
plant-specific nuclear siRNAs that are 24 nucleotides long.

The *FWA* gene has a direct duplication in its promoter
region, and multiple copies of transposons are present in
plant genomes. By a mechanism yet to be elucidated, Pol IV
is directed to transcribe repeated DNA no matter what its
sequence. An RNA-dependent RNA polymerase (RDR2)
converts the single-stranded Pol IV transcript into double-
stranded RNA, which is cleaved by Dicer ribonucleases, especially DCL3, into 24 nucleotide double-stranded fragments
with two base overhangs. One strand of these RNA fragments

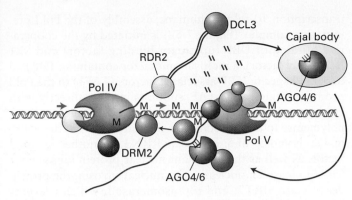

**FIGURE 7-51 Model of the mechanism of DNA methylation at Cs in CHH sequences in *A. thaliana*.** The plant-specific RNA polymerase IV transcribes repeated sequences such as transposons and the promoter-proximal region of the *FWA* gene (blue DNA, with the duplicated region indicated by blue arrows). The RNA-dependent RNA polymerase RDR2 converts this to double-stranded RNA, which is cleaved by the Dicer enzyme DCL3 into 24-nucleotide double-stranded RNA fragments with two base overhangs. One strand is bound by the Argonaut protein AGO4 or AGO6 and base-pairs with transcripts of repeated DNA transcribed by the plant-specific RNA polymerase V. This leads to methylation of Cs (M) by the DNA methyl transferase DRM2. Several other proteins that participate in this elaborate process are represented by colored circles. They were identified because mutations in them produce a late-flowering phenotype and they fail to methylate Cs in the *FWA* promoter region. [Adapted from M. V. C. Greenberg et al., 2011, *Epigenetics* **6**:344.]

is bound by an Argonaut protein (AGO4 or 6) in dense bodies in the nucleus called Cajal bodies, after the Spanish biologist who first described them early in the twentieth century. The 24-nucleotide single-stranded RNA in these Argonaut complexes then base-pairs with a nascent transcript of repetitive DNA synthesized by Pol V. This directs the DRM2 DNA-methyl transferase to methylate Cs in the repeated DNA. As in metazoans, a histone deacetylase interacts with the methylated Cs, leading to hypoacetylation of nucleosomes associated with repeated DNA and repression of transcription by Pol II. ∎

## KEY CONCEPTS of Section 7.7

### Epigenetic Regulation of Transcription

• The term *epigenetic* control of transcription refers to repression or activation that is maintained after cells replicate as the result of DNA methylation and/or post-translation modification of histones, especially histone methylation.

• Methylation of CpG sequences in CpG island promoters in mammals generates binding sites for a family of methyl-binding proteins (MBTs) that associate with histone deacetylases, inducing hypoacetylation of the promoter regions and transcriptional repression.

• Histone H3 lysine 9 di- and trimethylation creates binding sites for the heterochromatin-associated protein HP1, which results in the condensation of chromatin and transcriptional repression. These post-translational modifications are perpetuated following chromosome replication because the methylated histones are randomly associated with the daughter DNA molecules and associate with histone H3 lysine 9 methyl transferases that methylate histone 3 lysine 9 on newly synthesized histone H3 assembled on the daughter DNA.

• Polycomb complexes maintain repression of genes initially repressed by sequence-specific binding transcription factor repressors expressed early during embryogenesis. One class of Polycomb repression complexes, PRC2 complexes, is thought to associate with these repressors in early embryonic cells, resulting in methylation of histone H3 lysine 27. This creates binding sites for subunits in the PRC2 complex and PRC1 type complexes that inhibit the assembly of Pol II initiation complexes or inhibit transcription elongation. Since parental histone octamers with H3 methylated at lysine 27 are distributed to both daughter DNA molecules following DNA replication, PRC2 complexes that associate with these nucleosomes maintain histone H3 lysine 27 methylation through cell division.

• Trithorax complexes oppose repression by Polycomb complexes by methylating H3 at lysine 4 and maintaining this activating mark through chromosome replication.

• X-chromosome inactivation in female mammals requires a long noncoding RNA (ncRNA) called Xist that is transcribed from the X-inactivation center and then spreads by a poorly understood mechanism along the length of the same chromosome. Xist is bound by PRC2 complexes at an early stage of embryogenesis, initiating X inactivation that is maintained throughout the remainder of embryogenesis and adult life.

• Long ncRNAs also have been discovered that lead to repression of genes in *trans*, as opposed to the *cis* inactivation imposed by Xist. Repression is initiated by their interaction with PRC2 complexes. Much remains to be learned about how they are targeted to specific chromosomal regions, but the discovery of ≈1600 long ncRNAs conserved between mammals raises the possibility that this is a widely utilized mechanism of repression.

• In many fungi and plants, RNA-dependent RNA polymerases generate double-stranded RNAs from nascent transcripts of repeated sequences. These double-stranded RNAs are processed by Dicer ribonucleases into 22- or 24-nucleotide siRNAs bound by Argonaut proteins. The siRNAs base-pair with nascent transcripts from the repeated DNA sequences, inducing histone H3 lysine 9 methylation at centromeric repeats in the fission yeast *S. pombe*, and DNA methylation in plants, resulting in the formation of transcriptionally repressed heterochromatin.

## 7.8 Other Eukaryotic Transcription Systems

We conclude this chapter with a brief discussion of transcription initiation by the other two eukaryotic nuclear RNA polymerases, Pol I and Pol III, and by the distinct polymerases that transcribe mitochondrial and chloroplast DNA. Although these systems, particularly their regulation, are less thoroughly understood than transcription by RNA polymerase II, they are equally as fundamental to the life of eukaryotic cells.

### Transcription Initiation by Pol I and Pol III Is Analogous to That by Pol II

The formation of transcription-initiation complexes involving Pol I and Pol III is similar in some respects to assembly of Pol II initiation complexes (see Figure 7-17). However, each of the three eukaryotic nuclear RNA polymerases requires its own polymerase-specific general transcription factors and recognizes different DNA control elements. Moreover, neither Pol I nor Pol III requires ATP hydrolysis by a DNA helicase to help melt the DNA template strands to initiate transcription, whereas Pol II does. Transcription initiation by Pol I, which synthesizes pre-rRNA, and by Pol III, which synthesizes tRNAs, 5S rRNA, and other short, stable RNAs (see Table 7-2), is tightly coupled to the rate of cell growth and proliferation.

**Initiation by Pol I** The regulatory elements directing Pol I initiation are similarly located relative to the transcription start site in both yeast and mammals. A *core element* spanning the transcription start site from $-40$ to $+5$ is essential for Pol I transcription. An additional *upstream control element* extending from roughly $-155$ to $-60$ stimulates in vitro Pol I transcription 10-fold. In humans, assembly of the Pol I pre-initiation complex (Figure 7-52) is initiated by the cooperative binding of UBF (upstream binding factor) and SL1 (selectivity factor), a multisubunit factor containing TBP and four Pol I-specific TBP-associated factors (TAF$_I$s) to the Pol I promoter region. The TAF$_I$ subunits interact directly with Pol I-specific subunits, directing this specific nuclear RNA polymerase to the transcription start site. TIF-1A, the mammalian homolog of *S. cerevisiae* RRN3, is another required factor, as well as the abundant nuclear protein kinase CK2 (casein kinase 2), nuclear actin, nuclear myosin, the protein deacetylase SIRT7, and topoisomerase I, which prevents DNA supercoils (Figure 4-8) from forming during rapid Pol I transcription of the $\approx 14$-kb transcription unit.

Transcription of the $\approx 14$-kb precursor of 18S, 5.8S, and 28S rRNAs (see Chapter 8) is highly regulated to coordinate ribosome synthesis with cell growth and division. This is achieved through regulation of the activities of the Pol I initiation factors by post-translational modifications including phosphorylation and acetylation at specific sites, control of the rate of Pol I elongation, and control of the number of the $\approx 300$ human rRNA genes that are transcriptionally active by epigenetic mechanisms that assemble inactive copies into heterochromatin. Switching between the active and heterochromatic silent copies of rRNA genes is accomplished by a multisubunit chromatin-remodeling complex called NoRC ("No" for nucleolus, the site of rRNA transcription within nuclei). NoRC localizes a nucleosome over the Pol I transcription start site, blocking preinitiation complex assembly. It also interacts with a DNA methyl transferase that methylates a critical CpG in the upstream control element, inhibiting binding by UBF, as well as histone methyltransferases

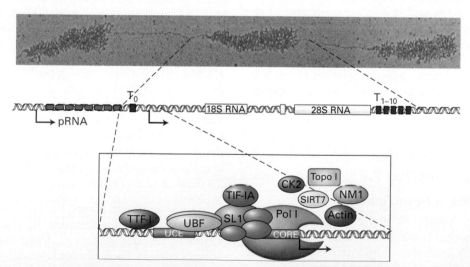

**FIGURE 7-52 Transcription of the rRNA precursor RNA by RNA polymerase I.** *Top:* electron micrograph of RNA protein complexes transcribed from repeated rRNA genes. One Pol I transcription unit is diagrammed in the middle. Enhancers that stimulate Pol I transcription from a single transcription start site are represented by blue boxes. Pol I transcription termination sites ($T_0$, $T_1$–$T_{10}$) bound by the Pol I–specific termination factor TTF-1 are shown as red rectangles. pRNA indicates transcription of the noncoding pRNA required for transcriptional silencing. Regions of DNA shown in blue are contained in the primary transcript, but are removed and degraded during rRNA processing. The core promoter element and upstream control element are diagrammed below with the location of Pol I and its general transcription factors UBF, SL1, and TIF-1A represented, as well as other proteins required for Pol I elongation and control. [Adapted from I. Grummt, 2010, *FEBS J.* **277:**4626.]

that di- and trimethylate histone H3 lysine 9, creating binding sites for heterochromatic HP1, and histone deacetylases. Moreover, an ≈250-nucleotide noncoding RNA called pRNA (for promoter associated) transcribed by Pol I from ≈2 kb upstream of the rRNA transcription unit (red arrow in Figure 7-52) is bound by a subunit of NoRC and is required for transcriptional silencing. pRNA is believed to target NoRC to Pol I promoter regions by forming an RNA:DNA triplex with the $T_0$ terminator sequence. This creates a binding site for the DNA methyl transferase DNMT3b that methylates the critical CpG in the upstream promoter element.

**Initiation by Pol III** Unlike protein-coding genes and pre-rRNA genes, the promoter regions of tRNA and 5S-rRNA genes lie entirely within the transcribed sequence (Figure 7-53a, b). Two such *internal* promoter elements, termed the *A box* and the *B box,* are present in all tRNA genes. These highly conserved sequences not only function as promoters but also encode two invariant portions of eukaryotic tRNAs that are required for protein synthesis. In 5S-rRNA genes, a single internal control region, the *C box,* acts as a promoter.

Three general transcription factors are required for Pol III to initiate transcription of tRNA and 5S-rRNA genes in vitro. Two multimeric factors, TFIIIC and TFIIIB, participate in initiation at both tRNA and 5S-rRNA promoters; a third factor, TFIIIA, is required for initiation at 5S-rRNA promoters. As with assembly of Pol I and Pol II initiation complexes, the Pol III general transcription factors bind to promoter DNA in a defined sequence.

The N-terminal half of one TFIIIB subunit, called *BRF* (for TFII*B-r*elated *f*actor), is similar in sequence to TFIIB (a Pol II factor). This similarity suggests that BRF and TFIIB perform a similar function in initiation, namely, to assist in separating the template DNA strands at the transcription start site (Figure 7-19). Once TFIIIB has bound to either a tRNA or 5S-rRNA gene, Pol III can bind and initiate transcription in the presence of ribonucleoside triphosphates. The BRF subunit of TFIIIB interacts specifically with one of the polymerase subunits unique to Pol III, accounting for initiation by this specific nuclear RNA polymerase.

Another of the three subunits composing TFIIIB is TBP, which we can now see is a component of a general transcription factor for all three eukaryotic nuclear RNA polymerases. The finding that TBP participates in transcription initiation by Pol I and Pol III was surprising, since the promoters recognized by these enzymes often do not contain TATA boxes. Nonetheless, in the case of Pol III transcription, the TBP subunit of TFIIIB interacts with DNA similarly to the way it interacts with TATA boxes.

Pol III also transcribes genes for small, stable RNAs with upstream promoters containing a TATA box. One example is the U6 snRNA involved in pre-mRNA splicing, as discussed in Chapter 8. In mammals, this gene contains an upstream promoter element called the PSE in addition to the TATA box (Figure 7-53c), which is bound by a multisubunit complex called SNAP$_C$, while the TATA box is bound by the TBP subunit of a specialized form of TFIIIB containing an alternative BRF subunit.

MAF1 is a specific inhibitor of Pol III transcription that functions by interacting with the BRF subunit of TFIIIB and Pol III. Its function is regulated by controlling its import from the cytoplasm into nuclei by phosphorylations at specific sites in response to signal transduction protein kinase cascades that respond to cell stress and nutrient deprivation (see Chapters 16 and 24). In mammals, Pol III transcription is also repressed by the critical tumor suppressors p53 and the retinblastoma (RB) family. In humans there are two genes encoding subunit RPC32. One of these is expressed specifically in replicating cells, and its forced expression can contribute to oncogenic transformation of cultured human fibroblasts.

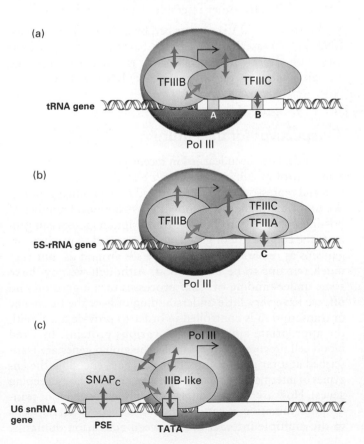

**FIGURE 7-53 Transcription-control elements in genes transcribed by RNA polymerase III.** Both tRNA (a) and 5S-rRNA (b) genes contain internal promoter elements (yellow) located downstream from the start site and named A, B, and C boxes, as indicated. Assembly of transcription initiation complexes on these genes begins with the binding of Pol III–specific general transcription factors TFIIIA, TFIIIB, and TFIIIC to these control elements. Green arrows indicate strong, sequence-specific protein-DNA interactions. Blue arrows indicate interactions between general transcription factors. Purple arrows indicate interactions between general transcription factors and Pol III. (c) Transcription of the U6 snRNA gene in mammals is controlled by an upstream promoter with a TATA box bound by the TBP subunit of a specialized form of TFIIIB with an alternative BRF subunit and an upstream regulatory element called the PSE bound by a multisubunit factor called SNAP$_C$. [From L. Schramm and N. Hernandez, 2002, *Genes Dev.* **16**:2593.]

## Mitochondrial and Chloroplast DNAs Are Transcribed by Organelle-Specific RNA Polymerases

As discussed in Chapter 6, mitochondria and chloroplasts probably evolved from eubacteria that were endocytosed into ancestral cells containing a eukaryotic nucleus. In modern-day eukaryotes, both organelles contain distinct DNAs that encode some of the proteins essential to their specific functions. Interestingly, the RNA polymerases that transcribe mitochondrial (mt) DNA and chloroplast DNA are similar to polymerases from eubacteria and bacteriophages, reflecting their evolutionary origins.

**Mitochondrial Transcription**  The RNA polymerase that transcribes mtDNA is encoded in nuclear DNA. After synthesis of the enzyme in the cytosol, it is imported into the mitochondrial matrix by mechanisms described in Chapter 13. The mitochondrial RNA polymerases from *S. cerevisiae* and the frog *Xenopus laevis* both consist of a large subunit with ribonucleotide-polymerizing activity and a small B subunit (TFBM). In mammals, another matrix protein, mitochondrial transcription factor A (TFAM), binds to mtDNA promoters and is essential for initiating transcription at the start sites used in the cell. The large subunit of yeast mitochondrial RNA polymerase clearly is related to the monomeric RNA polymerases of bacteriophage T7 and similar bacteriophages. However, the mitochondrial enzyme is functionally distinct from the bacteriophage enzyme in its dependence on two other polypeptides for transcription from the proper start sites.

The promoter sequences recognized by mitochondrial RNA polymerases include the transcription start site. These promoter sequences, which are rich in A residues, have been characterized in the mtDNA from yeast, plants, and animals. The circular human mitochondrial genome contains two related 15-bp promoter sequences, one for the transcription of each strand. Each strand is transcribed in its entirety; the long primary transcripts are then processed by cleavage at tRNA genes that separate each of the mitochondrial mRNAs and rRNAs. A second promoter appears to be responsible for transcribing additional copies of the rRNAs. Currently, there is relatively little understanding of how transcription of the mitochondrial genome is regulated to coordinate the production of the few mitochondrial proteins it encodes with synthesis and import of the thousands of nuclear DNA-encoded proteins that comprise the mitochondria.

**Chloroplast Transcription**  Chloroplast DNA is transcribed by two types of RNA polymerases, one multisubunit protein similar to bacterial RNA polymerases and one similar to the single subunit enzymes of bacteriophages and mitochondria. The core subunits of the bacterial-type enzyme, α, β, β′, and ω subunits, are encoded in the chloroplast DNAs of higher plants, whereas six $\sigma^{70}$-like σ factors are encoded in the nuclear DNA of higher plants. This is another example of the transfer of genes from organellar genomes to nuclear genomes during evolution. In this case, genes encoding the

regulatory transcription initiation factors have been transferred to the nucleus, where the control of their transcription by nuclear RNA polymerase II likely indirectly controls the expression of sets of chloroplast genes. The bacterial-like chloroplast RNA polymerase is called the plastid polymerase because its catalytic core is encoded by the chloroplast genome. Most chloroplast genes are transcribed by these enzymes and have −35 and −10 control regions similar to promoters in cyanobacteria, from which they evolved. The chloroplast T7-like RNA polymerase is also encoded in the nuclear genome of higher plants. It transcribes a different set of chloroplast genes. Curiously, this includes genes encoding subunits of the bacterial-like multisubunit plastid polymerase. Recent results indicate that transcription by the multisubunit polymerase is regulated by sigma factors whose activities are regulated by light and metabolic stress.

## KEY CONCEPTS of Section 7.8

### Other Eukaryotic Transcription Systems

- The process of transcription initiation by Pol I and Pol III is similar to that by Pol II but requires different general transcription factors, is directed by different promoter elements, and does not require hydrolysis of ATP β-γ phosphodiester bonds to separate the DNA strands at the start site.

- Mitochondrial DNA is transcribed by a nuclear-encoded RNA polymerase composed of two subunits. One subunit is homologous to the monomeric RNA polymerase from bacteriophage T7; the other resembles bacterial σ factors.

- Chloroplast DNA is transcribed by a chloroplast-encoded RNA polymerase homologous to bacterial RNA polymerases, with several alternative nuclear encoded σ-factors, and a single subunit bacteriophage T7-like RNA polymerase.

## Perspectives for the Future

A great deal has been learned in recent years about transcription control in eukaryotes. Genes encoding ≈2000 activators and repressors can be recognized in the human genome. We now have a glimpse of how the astronomical number of possible combinations of these transcription factors can generate the complexity of gene control required to produce organisms as remarkable as those we see around us. But very much remains to be understood. Although we now have some understanding of what processes turn a gene on and off, we have very little understanding of how the frequency of transcription is controlled in order to provide a cell with the appropriate amounts of its various proteins. In a red blood cell precursor, for example, the globin genes are transcribed at a far greater rate than the genes encoding the enzymes of intermediary metabolism (the so-called housekeeping genes). How are the vast differences in the frequency of transcription initiation at various genes achieved? What happens to the multiple interactions between activation domains, co-activator complexes, general transcription factors, and RNA polymerase II when the polymerase initiates transcription and transcribes away from the promoter region? Do these completely dissociate at promoters that are transcribed infrequently, so that the combination of multiple factors required for transcription must be reassembled anew for each round of transcription? Do complexes of activators with their multiple interacting co-activators remain assembled at promoters from which reinitiation takes place at a high rate, so that the entire assembly does not have to be reconstructed each time a polymerase initiates?

Much remains to be learned about the structure of chromatin and how that structure influences transcription. What additional components besides HP1 and methylated histone H3 lysine 9 are required to direct certain regions of chromatin to form heterochromatin, where transcription is repressed? Precisely how is the structure of chromatin changed by activators and repressors, and how does this promote or inhibit transcription? Once chromatin-remodeling complexes and histone acetylase complexes become associated with a promoter region, how do they remain associated? Current models suggest that certain subunits of these complexes associate with modified histone tails so that the combination of binding to a specific histone tail modification plus modification of neighboring histone tails in the same way results in retention of the modifying complex at an activated promoter region. In some cases, this type of assembly mechanism causes the complexes to spread along the length of a chromatin fiber. What controls when such complexes spread and how far they will spread?

Single activation domains have been discovered to interact with several co-activator complexes. Are these interactions transient, so that the same activation domain can interact with several co-activators sequentially? Is a specific order of co-activator interaction required? How does the interaction of activation domains with mediator stimulate transcription? Do these interactions simply stimulate the assembly of a preinitiation complex, or do they also influence the rate at which RNA polymerase II initiates transcription from an assembled preinitiation complex?

Transcriptional activation is a highly cooperative process so that genes expressed in a specific type of cell are expressed only when the complete set of activators that control that gene are expressed and activated. As mentioned earlier, some of the transcription factors that control expression of the *TTR* gene in the liver are also expressed in intestinal and kidney cells. Yet the *TTR* gene is not expressed in these other tissues, since its transcription requires two additional transcription factors expressed only in the liver. What mechanisms account for this highly cooperative action of transcription factors that is critical to cell-type-specific gene expression?

The discovery that long noncoding RNAs can repress transcription of specific target genes has stimulated tremendous interest. Do these always repress transcription by targeting Polycomb complexes? Can long noncoding RNAs also activate transcription of specific target genes? How are they targeted to specific genes? Do the ≈1600 long noncoding RNAs that are conserved between mammals all function to regulate transcription of specific target genes, adding to the complexity of transcription control by sequence-specific DNA-binding proteins? Research to address these questions will be an exciting area of investigation in the coming years.

A thorough understanding of normal development and of abnormal processes associated with disease will require answers to these and many related questions. As further understanding of the principles of transcription control are discovered, applications of the knowledge will likely be made. This understanding may allow fine control of the expression of therapeutic genes introduced by gene therapy vectors as they are developed. Detailed understanding of the molecular interactions that regulate transcription may provide new targets for the development of therapeutic drugs that inhibit or stimulate the expression of specific genes. A more complete understanding of the mechanisms of transcriptional control may allow improved engineering of crops with desirable characteristics. Certainly, further advances in the area of transcription control will help to satisfy our desire to understand how complex organisms such as ourselves develop and function.

## Key Terms

| | |
|---|---|
| activation domain 307 | *MAT* locus (in yeast) 315 |
| activators 281 | mediator 315 |
| antitermination factor 301 | nuclear receptors 309 |
| bromodomain 319 | promoter 281 |
| carboxyl-terminal domain (CTD) 293 | promoter-proximal elements 302 |
| chromatin-mediated repression 315 | repression domain 308 |
| chromodomain 330 | repressors 281 |
| co-activator 311 | RNA polymerase II 290 |
| co-repressor 312 | silencer sequences 316 |
| DNase I footprinting 305 | specific transcription factors 305 |
| enhanceosome 314 | TATA box 295 |
| enhancers 285 | TATA box-binding protein (TBP) 298 |
| general transcription factors 297 | upstream activating sequences (UASs) 305 |
| heat-shock genes 325 | yeast two-hybrid system 321 |
| histone deacetylation 318 | zinc finger 309 |
| leucine zipper 311 | |

## Review the Concepts

1. Describe the molecular events that occur at the *lac* operon when *E. coli* cells are shifted from a glucose-containing medium to a lactose-containing medium.

2. The concentration of free glutamine affects transcription of the enzyme glutamine synthetase in *E. coli*. Describe the mechanism for this.

3. What types of genes are transcribed by RNA polymerases I, II, and III? Design an experiment to determine whether a specific gene is transcribed by RNA polymerase II.

4. The CTD of the largest subunit of RNA polymerase II can be phosphorylated at multiple serine residues. What are the conditions that lead to the phosphorylated versus un-phosphorylated RNA polymerase II CTD?

5. What do TATA boxes, initiators, and CpG islands have in common? Which was the first of these to be identified? Why?

6. Describe the methods used to identify the location of DNA-control elements in promoter-proximal regions of genes.

7. What is the difference between a promoter-proximal element and a distal enhancer? What are the similarities?

8. Describe the methods used to identify the location of DNA-binding proteins in the regulatory regions of genes.

9. Describe the structural features of transcriptional activator and repressor proteins.

10. Give two examples of how gene expression may be repressed without altering the gene-coding sequence.

11. Using CREB and nuclear receptors as examples, compare and contrast the structural changes that take place when these transcription factors bind to their co-activators.

12. What general transcription factors associate with an RNA polymerase II promoter in addition to the polymerase? In what order do they bind in vitro? What structural change occurs in the DNA when an "open" transcription-initiation complex is formed?

13. Expression of recombinant proteins in yeast is an important tool for biotechnology companies that produce new drugs for human use. In an attempt to get a new gene X expressed in yeast, a researcher has integrated gene X into the yeast genome near a telomere. Will this strategy result in good expression of gene X? Why or why not? Would the outcome of this experiment differ if the experiment had been performed in a yeast line containing mutations in the H3 or H4 histone tails?

14. You have isolated a new protein called STICKY. You can predict from comparisons with other known proteins that STICKY contains a bHLH domain and a Sin3-interacting domain. Predict the function of STICKY and rationale for the importance of these domains in STICKY function.

15. The yeast two-hybrid method is a powerful molecular genetic method to identify a protein(s) that interacts with a known protein or protein domain. You have isolated the glucocorticoid receptor (GR) and have evidence that it is a modular protein containing an activation domain, a DNA-binding domain, and a second ligand-binding activation domain. Further analysis reveals that in pituitary cells, the protein is anchored in the cytoplasm in the absence of its hormone ligand, a result leading you to speculate that it binds to other inhibitory proteins. Describe how a two-hybrid analysis could be used to identify the protein(s) with which GR interacts. How would you specifically identify the domain in the GR that binds the inhibitor(s)?

16. Prokaryotes and lower eukaryotes such as yeast have DNA-regulatory elements called upstream activating sequences. What are the comparable sequences found in higher eukaryotic species?

17. Recall that the Trp repressor binds to a site in the operator region of tryptophan-producing genes when tryptophan is abundant, thereby preventing transcription. What would happen to the expression of the tryptophan biosynthetic enzyme genes in the following scenarios? Fill in the blanks with one of the following phrases:

**never be expressed/always (constitutively) be expressed**

   a. The cell produces a mutant Trp repressor that cannot bind to the operator. The enzyme genes will _____.

   b. The cell produces a mutant Trp repressor that binds to its operator site even if no Tryptophan is present. The enzyme genes will _____.

   c. The cell produces a mutant sigma factor that cannot bind the promoter region. The enzyme genes will _____.

   d. Elongation of the leader sequence is always stalled after transcription of region 1. The enzyme genes will _____.

18. Compare/contrast bacterial and eukaryotic gene expression mechanisms.

19. You are curious to identify the region of gene X sequence that serves as an enhancer for gene expression. Design an experiment to investigate this issue.

20. Some organisms have mechanisms in place that will override transcription termination. One such mechanism using the Tat protein is employed by the HIV retrovirus. Explain why Tat is therefore a good target for HIV vaccination.

21. Upon identification of the DNA regulatory sequence responsible for translating a given gene, you note that it is enriched with CG sequences. Is the corresponding gene likely to be a highly expressed transcript?

22. Name four major classes of DNA-binding proteins that are responsible for controlling transcription, and describe their structural features.

## Analyze the Data

In eukaryotes, the three RNA polymerases, Pol I, II, and III, each transcribes unique genes required for the synthesis of ribosomes: 25S and 18S rRNAs (Pol I), 5S rRNA (Pol III), and mRNAs for ribosomal proteins (Pol II). Researchers have long speculated that the activities of the three RNA polymerases are coordinately regulated according to the demand for ribosome synthesis: high in replicating cells in rich nutrient conditions and low when nutrients are scarce. To determine whether the activities of the three polymerases are coordinated, Laferte and colleagues engineered a strain of yeast to be partially resistant to the inhibition of cell growth by the drug rapamycin (2006, *Genes Dev.* 20:2030–2040). As discussed in Chapter 8, rapamycin inhibits a protein kinase (called TOR, for *t*arget *o*f *r*apamycin) that regulates the overall rate of protein synthesis and ribosome synthesis. When TOR is inhibited by rapamycin, the transcription of rRNAs by Pol I and Pol III and ribosomal protein mRNAs by RNA polymerase II are all rapidly repressed. Part of the inhibition of Pol I rRNA synthesis results from the dissociation of the Pol I transcription factor Rrn3 from Pol I. In the strain constructed by Laferte and colleagues, the wild-type *Rrn3* gene and the wild-type *A43* gene, encoding the Pol I subunit to which Rrn3 binds, were replaced with a gene encoding a fusion protein of the A43 Pol I subunit with Rrn3. The idea was that the covalent fusion of the two proteins would prevent the Rrn3 dissociation from Pol I otherwise caused by rapamycin treatment. The resulting CARA (*constitutive association of Rrn3 and A43*) strain was found to be partially resistant to rapamycin. In the absence of rapamycin, the CARA strain grew at the same rate and had equal numbers of ribosomes as wild-type cells.

a. To analyze rRNA transcription by Pol I, total RNA was isolated from rapidly growing wild-type (WT) and CARA cells at various times following the addition of rapamycin. The concentration of the 35S rRNA precursor transcribed by Pol I (see Figure 8-38) was assayed by the primer-extension method. Since the 5′ end of the 35S rRNA precursor is degraded during the processing of 25S and 18S rRNA, this method measures the relatively short-lived pre-rRNA precursor. This is an indirect measure of the rate of rRNA transcription by Pol I. The results of this primer extension assay are shown below. How does the CARA Pol I–Rrn3 fusion affect the response of Pol I transcription to rapamycin?

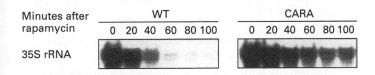

b. The concentrations of four mRNAs encoding ribosomal proteins, RPL30, RPS6a, RPL7a, and RPL5, and the mRNA for actin (ACT1), a protein present in the cytoskeleton, were assessed in wild-type and CARA cells by Northern blotting at various times after addition of rapamycin to rapidly growing cells (upper autoradiograms). 5S rRNA transcription was assayed by pulse labeling rapidly growing WT

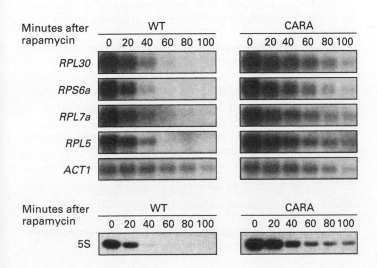

and CARA cells with [3]H uracil (for 20 minutes) at various times after addition of rapamycin to the media. Total cellular RNA was isolated and subjected to gel electrophoresis and autoradiography. The lower autoradiogram shows the region of the gel containing 5S rRNA. Based on these data, what can be concluded about the influence of Pol I transcription on the transcription of ribosomal protein genes by Pol II and 5S rRNA by Pol III?

c. To determine whether the difference in behavior of wild-type and CARA cells can be observed under normal physiological conditions (i.e., without drug treatment), cells were subjected to a shift in their food source, from nutrient-rich media to nutrient-poor media. Under these conditions, in wild-type cells, the TOR protein kinase becomes inactive. Consequently, shifting cells from nutrient-rich media to nutrient-poor media should result in a normal physiological response that is equivalent to treating cells with rapamycin, which inhibits TOR. To determine how the CARA fusion protein affected the response to this media shift, RNA was extracted from wild-type and CARA cells and used to probe microarrays containing all yeast open reading frames. The extent of RNA hybridization with the arrays was quantified and is expressed in the graphs as $\log_2$ of the ratio of CARA-cell RNA concentration to wild-type-cell RNA concentration for each open reading frame. A value of zero indicates that the two strains of yeast exhibit the same level of expression for those specific RNAs. A value of 1 indicates that the CARA cells contain twice as much of that particular RNA as do wild-type cells. The graphs below show the number of open reading frames (y axis) that have values for $\log_2$ of this ratio, indicated by the x axis. The results of hybridization to open reading frames encoding mRNAs for ribosomal proteins are shown by black bars, those for all other mRNAs by white bars. The graph on the left gives results for cells grown in nutrient-rich medium, the graph on the right for cells shifted to nutrient-poor medium for 90 minutes. What do these data suggest about the regulation of ribosomal protein gene transcription by Pol II?

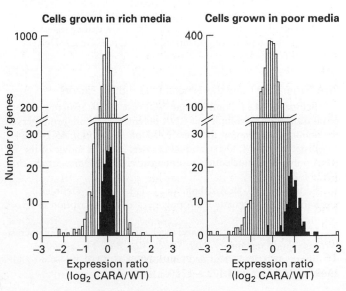

# References

## Control of Gene Expression in Bacteria

Campbell, E. A., L. F. Westblade, and S. A. Darst. 2008. Regulation of bacterial RNA polymerase sigma factor activity: a structural perspective. *Curr. Opin. Microbiol.* **11**:121–127.

Casino, P., V. Rubio, and A. Marina. 2010. The mechanism of signal transduction by two-component systems. *Curr. Opin. Struct. Biol.* **20**:763–771.

Halford, S. E., and J. F. Marko. 2004. How do site-specific DNA-binding proteins find their targets? *Nucl. Acids Res.* **32**:3040–3052.

Hsieh, Y. J., and B. L. Wanner. 2010. Global regulation by the seven-component Pi signaling system. *Curr. Opin. Microbiol.* **13**:198–203.

Lawson, C. L., et al. 2004. Catabolite activator protein: DNA binding and transcription activation. *Curr. Opin. Struct. Biol.* **14**:10–20.

Muller-Hill, B. 1998. Some repressors of bacterial transcription. *Curr. Opin. Microbiol.* **1**:145–151.

Murakami, K. S., and S. A. Darst. 2003. Bacterial RNA polymerases: the whole story. *Curr. Opin. Struct. Biol.* **13**:31–39.

Sharma, U. K., and D. Chatterji. 2010. Transcriptional switching in *Escherichia coli* during stress and starvation by modulation of sigma activity. *FEMS Microbiol. Rev.* **34**:646–657.

Wigneshweraraj, S. R., et al. 2008. Modus operandi of the bacterial RNA polymerase containing the sigma54 promoter-specificity factor. *Mol. Microbiol.* **68**:538–546.

## Overview of Eukaryotic Gene Control and RNA Polymerases

Brenner, S., et al. 2002. Conserved regulation of the lymphocyte-specific expression of lck in the Fugu and mammals. *Proc. Natl Acad. Sci. USA* **99**:2936–2941.

Cramer P., et al. 2008. Structure of eukaryotic RNA polymerases. *Ann. Rev. Biophys.* **37**:337–352.

Maston, G. A., S. K. Evans, and M. R. Green. 2006. Transcriptional regulatory elements in the human genome. *Ann. Rev. Genomics Hum. Genet.* **7**:29–59.

Ptashne, M., and A. Gann. 2001. Transcription initiation: imposing specificity by localization. *Essays Biochem.* **37**:1–15.

Struhl, K. 1999. Fundamentally different logic of gene regulation in eukaryotes and prokaryotes. *Cell* **98**:1–4.

Visel, A., E. M. Rubin, and L. A. Pennacchio. 2009. Genomic views of distant-acting enhancers. *Nature* **461**:199–205.

Wallace, J. A., and G. Felsenfeld. 2007. We gather together: insulators and genome organization. *Curr. Opin. Genet. Dev.* **17**:400–407.

## RNA Polymerase II and the General Transcription Factors

Baumann, M., J. Pontiller, and W. Ernst. 2010. Structure and basal transcription complex of RNA polymerase II core promoters in the mammalian genome: an overview. *Mol. Biotechnol.* **45**:241–247.

Brueckner, F., J. Ortiz, and P. Cramer. 2009. A movie of the RNA polymerase nucleotide addition cycle. *Curr. Opin. Struct. Biol.* **19**:294–299.

Fuda, N. J., M. B. Ardehali, and J. T. Lis. 2009. Defining mechanisms that regulate RNA polymerase II transcription in vivo. *Nature* **461**:186–192.

Hahn, S. 2004. Structure and mechanism of the RNA polymerase II transcription machinery. *Nat. Struct. Mol. Biol.* **11**:394–403.

Illingworth, R. S., and A. P. Bird. 2009. CpG islands—'a rough guide'. *FEBS Lett.* **583**:1713–1720.

Jun, S. H., et al. 2011. Archaeal RNA polymerase and transcription regulation. *Crit. Rev. Biochem. Mol. Biol.* **46**:27–40.

Kornberg, R. D. 2007. The molecular basis of eukaryotic transcription. *Proc. Natl Acad. Sci. USA.* **104**:12955–12961.

Müller, F., A. Zaucker, and L. Tora. 2010. Developmental regulation of transcription initiation: more than just changing the actors. *Curr. Opin. Genet. Dev.* **20**:533–540.

Papai, G., P. A. Weil, and P. Schultz. 2011. New insights into the function of transcription factor TFIID from recent structural studies. *Curr. Opin. Genet. Dev.* **21**:219–224.

Price, D. H. 2008. Poised polymerases: on your mark . . . get set . . . go! *Mol. Cell* **30**:7–10.

Roeder, R. G. 1996. The role of general initiation factors in transcription by RNA polymerase II. *Trends Biochem. Sci.* **21**:327–335.

Sandelin, A., et al. 2007. Mammalian RNA polymerase II core promoters: insights from genome-wide studies. *Nat. Rev. Genet.* **8**:424–436.

Seila, A. C., et al. 2009. Divergent transcription: a new feature of active promoters. *Cell Cycle* **8**:2557–2564.

Selth, L. A., S. Sigurdsson, and J. Q. Svejstrup. 2010. Transcript elongation by RNA polymerase II. *Ann. Rev. Biochem.* **79**:271–293.

Sikorski, T. W., and S. Buratowski. 2009. The basal initiation machinery: beyond the general transcription factors. *Curr. Opin. Cell Biol.* **21**:344–351.

Thomas, M. C., and C. M. Chiang. 2006. The general transcription machinery and general cofactors. *Crit. Rev. Biochem. Mol. Biol.* **41**:105–178.

Wade, J. T., and K. Struhl. 2008. The transition from transcriptional initiation to elongation. *Curr. Opin. Genet. Dev.* **18**:130–136.

Yamada, T., et al. 2006. P-TEFb-mediated phosphorylation of hSpt5 C-terminal repeats is critical for processive transcription elongation. *Mol. Cell* **21**:227–237.

## Regulatory Sequences in Protein-Coding Genes and the Proteins Through Which They Function

Fuxreiter, M., et al. 2008. Malleable machines take shape in eukaryotic transcriptional regulation. *Nat. Chem. Biol.* **4**:728–737.

Garvie, C. W., and C. Wolberger. 2001. Recognition of specific DNA sequences. *Mol. Cell.* **8**:937–946.

Kadonaga, J. T. 2004. Regulation of RNA polymerase II transcription by sequence-specific DNA binding factors. *Cell* **116**:247–257.

Kaufmann, K., A. Pajoro, and G. C. Angenent. 2010. Regulation of transcription in plants: mechanisms controlling developmental switches. *Nat. Rev. Genet.* **11**:830–842.

Riechmann, J. L., et al. 2000. *Arabidopsis* transcription factors: genome-wide comparative analysis among eukaryotes. *Science* **290**:2105–2110.

Tupler, R., G. Perini, and M. R. Green. 2001. Expressing the human genome. *Nature* **409**:832–833.

## Molecular Mechanisms of Transcription Repression and Activation

Bannister, A. J., and T. Kouzarides. 2011. Regulation of chromatin by histone modifications. *Cell Res.* **21**:381–395.

Bulger, M., and M. Groudine. 2011. Functional and mechanistic diversity of distal transcription enhancers. *Cell* **144**:327–339.

Cairns, B. R. 2009. The logic of chromatin architecture and remodelling at promoters. *Nature* **461**:193–198.

Conaway, R. C., and J. W. Conaway. 2011. Function and regulation of the Mediator complex. *Curr. Opin. Genet. Dev.* **21**:225–230.

Courey, A. J., and S. Jia. 2001. Transcriptional repression: the long and the short of it. *Genes Dev.* **15**:2786–2796.

Deaton, A. M., and A. Bird. 2011. CpG islands and the regulation of transcription. *Genes Dev.* **25**:1010–1022.

Hargreaves, D. C., and G. R. Crabtree. 2011. ATP-dependent chromatin remodeling: genetics, genomics and mechanisms. *Cell Res.* **21**:396–420.

Kornberg, R. D. 2005. Mediator and the mechanism of transcriptional activation. *Trends Biochem. Sci.* **30**:235–239.

Li, B., M. Carey, and J. L. Workman. 2007. The role of chromatin during transcription. *Cell* **128**:707–719.

Malik, S., and R. G. Roeder. 2010. The metazoan Mediator co-activator complex as an integrative hub for transcriptional regulation. *Nat. Rev. Genet.* **11**:761–772.

Mohrmann, L., and C. P. Verrijzer. 2005. Composition and functional specificity of SWI2/SNF2 class chromatin remodeling complexes. *Biochim. Biophys. Acta* **1681**:59–73.

Perissi, V., et al. 2010. Deconstructing repression: evolving models of co-repressor action. *Nat. Rev. Genet.* **11**:109–123.

Smith, C. L., and C. L. Peterson. 2005. ATP-dependent chromatin remodeling. *Curr. Top. Dev. Biol.* **65**:115–148.

Taatjes, D. J. 2010. The human Mediator complex: a versatile, genome-wide regulator of transcription. *Trends Biochem. Sci.* **35**:315–322.

Venters, B. J., and B. F. Pugh. 2009. How eukaryotic genes are transcribed. *Crit. Rev. Biochem. Mol. Biol.* **44**:117–141.

Yun, M., et al. 2011. Readers of histone modifications. *Cell Res.* **21**:564–578.

### Regulation of Transcription-Factor Activity

Altarejos, J. Y., and M. Montminy. 2011. CREB and the CRTC co-activators: sensors for hormonal and metabolic signals. *Nat. Rev. Mol. Cell Biol.* **12**:141–151.

Brivanlou, A. H., and J. E. Darnell, Jr. 2002. Signal transduction and the control of gene expression. *Science* **295**:813–818.

Chen, H., M. Tini, and R. M. Evans. 2001. HATs on and beyond chromatin. *Curr. Opin. Cell Biol.* **13**:218–224.

Echeverria, P. C., and D. Picard. 2010. Molecular chaperones, essential partners of steroid hormone receptors for activity and mobility. *Biochim. Biophys. Acta* **1803**:641–649.

Lefstin, J. A., and K. R. Yamamoto. 1998. Allosteric effects of DNA on transcriptional regulators. *Nature* **392**:885–888.

Perissi, V., and M. G. Rosenfeld. 2005. Controlling nuclear receptors: the circular logic of cofactor cycles. *Nat. Rev. Mol. Cell Biol.* **6**:542–554.

Wu, S. C., and Y. Zhang. 2009. Minireview: role of protein methylation and demethylation in nuclear hormone signaling. *Mol. Endocrinol.* **23**:1323–1334.

York, B., and B. W. O'Malley. 2010. Steroid receptor coactivator (SRC) family: masters of systems biology. *J. Biol. Chem.* **285**:38743–3850.

### Epigenetic Regulation of Transcription

Beisel, C., and R. Paro. 2011. Silencing chromatin: comparing modes and mechanisms. *Nat .Rev. Genet.* **12**:123–135.

Black, J. C., and J. R. Whetstine. 2011. Chromatin landscape: methylation beyond transcription. *Epigenetics* **6**:9–15.

Clouaire, T., and I. Stancheva. 2008. Methyl-CpG binding proteins: specialized transcriptional repressors or structural components of chromatin? *Cell Mol. Life Sci.* **65**:1509–1522.

Grewal, S. I. 2010. RNAi-dependent formation of heterochromatin and its diverse functions. *Curr. Opin. Genet. Dev.* **20**:134–141.

Lee, J. T. 2010. The X as model for RNA's niche in epigenomic regulation. *Cold Spring Harbor Perspect. Biol.* **2**(9):a003749 (1–12).

Minks, J., and C. J. Brown. 2009. Getting to the center of X-chromosome inactivation: the role of transgenes. *Biochem. Cell. Biol.* **87**:759–766.

Moazed, D. 2009. Small RNAs in transcriptional gene silencing and genome defence. *Nature* **457**:413–420.

Simon, J. A., and R. E. Kingston. 2009. Mechanisms of polycomb gene silencing: knowns and unknowns. *Nat. Rev. Mol. Cell Biol.* **10**:697–708

Vermaak, D., and H. S. Malik. 2009. Multiple roles for heterochromatin protein 1 genes in *Drosophila*. *Ann. Rev. Genet.* **43**:467–492.

Wutz, A., and J. Gribnau. 2007. X inactivation Xplained. *Curr. Opin. Genet. Dev.* **17**:387–393.

### Other Eukaryotic Transcription Systems

Bonawitz, N. D., D. A. Clayton, and G. S. Shadel. 2006. Initiation and beyond: multiple functions of the human mitochondrial transcription machinery. *Mol. Cell* **24**:813–825.

Dumay-Odelot, H., et al. 2010. Cell growth- and differentiation-dependent regulation of RNA polymerase III transcription. *Cell Cycle* **9**:3687–3699.

Grummt, I. 2010. Wisely chosen paths—regulation of rRNA synthesis. *FEBS J.* **277**:4626–4639.

Leigh-Brown, S., J. A. Enriquez, and D. T. Odom. 2010. Nuclear transcription factors in mammalian mitochondria. *Genome Biol.* **11**:215:1–9.

Schweer, J., et al. 2010. Role and regulation of plastid sigma factors and their functional interactors during chloroplast transcription—recent lessons from *Arabidopsis thaliana*. *Eur. J. Cell Biol.* **89**:940–946.

Willis, I. M., and R. D. Moir. 2007. Integration of nutritional and stress signaling pathways by Maf1. *Trends Biochem. Sci.* **32**:51–53.

# Post-transcriptional Gene Control

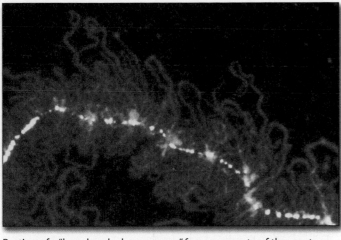

Portion of a "lampbrush chromosome" from an oocyte of the newt *Nophthalmus viridescens;* hnRNP protein associated with nascent RNA transcripts fluoresces red after staining with a monoclonal antibody. [Courtesy of M. Roth and J. Gall.]

I n the previous chapter, we saw that most genes are regulated at the first step in gene expression, transcription, by regulating the assembly of the transcription pre-initiation complex on a promoter DNA sequence and regulating transcription elongation in the promoter proximal region. Once transcription has been initiated, synthesis of the encoded RNA requires that RNA polymerase transcribe the entire gene and not terminate prematurely. Moreover, the initial **primary transcripts** produced from eukaryotic genes must undergo various processing reactions to yield the corresponding functional RNAs. For mRNAs, the 5' cap structure necessary for translation must be added (see Figure 4-14), introns must be spliced out of pre-mRNAs (Table 8-1), and the 3' end must be polyadenylated (see Figure 4-15). Once formed in the nucleus, mature, functional RNAs are exported to the cytoplasm as components of ribonucleoproteins. Both processing of RNAs and their export from the nucleus offer opportunities for further regulating gene expression after the initiation of transcription.

Recently, the vast amount of sequence data on human mRNAs expressed in different tissues and at various times during embryogenesis and cellular differentiation has revealed that ~95 percent of human genes give rise to alternatively spliced mRNAs. These alternatively spliced mRNAs encode related proteins with differences in sequences limited to specific functional domains. In many cases, alternative RNA splicing is regulated to meet the need for a specific protein isoform in a specific cell type. Given the complexity of pre-mRNA splicing, it is not surprising that mistakes are occasionally made, giving rise to mRNA precursors with improperly spliced exons. However, eukaryotic cells have evolved RNA surveillance mechanisms that prevent the transport of incorrectly processed RNAs to the cytoplasm or lead to their degradation if they are transported.

Additional control of gene expression can occur in the cytoplasm. In the case of protein-coding genes, for instance, the amount of protein produced depends on the stability of the corresponding mRNAs in the cytoplasm and the rate of their translation. For example, during an immune response, lymphocytes communicate by secreting polypeptide hormones called cytokines that signal neighboring lymphocytes through cytokine receptors that span their plasma membranes (Chapter 23). It is important for lymphocytes to synthesize and secrete cytokines in short bursts. This is possible

## OUTLINE

| TABLE 8-1 | RNAs Discussed in Chapter 8 |
|---|---|
| mRNA | Fully processed messenger RNA with 5′ cap, introns removed by RNA splicing, and a poly(A) tail |
| pre-mRNA | An mRNA precursor containing introns and not cleaved at the poly(A) site |
| hnRNA | Heterogeneous nuclear RNAs. These include pre-mRNAs and RNA processing intermediates containing one or more introns. |
| snRNA | Five small nuclear RNAs that function in the removal of introns from pre-mRNAs by RNA splicing, plus two small nuclear RNAs that substitute for the first two at rare introns |
| pre-tRNA | A tRNA precursor containing additional transcribed bases at the 5′ and 3′ ends compared to the mature tRNA. Some pre-tRNAs also contain an intron in the anti-codon loop. |
| pre-rRNA | The precursor to mature 18S, 5.8S, and 28S ribosomal RNAs. The mature rRNAs are processed from this long precursor RNA molecule by cleavage, removal of bases from the ends of the cleaved products, and modification of specific bases. |
| snoRNA | Small nucleolar RNAs. These base-pair with complementary regions of the pre-RNA molecule, directing cleavage of the RNA chain and modification of bases during maturation of the rRNAs. |
| siRNA | Short interfering RNAs, ~22 bases long, that are each perfectly complementary to a sequence in an mRNA. Together with associated proteins, siRNAs cause cleavage of the "target" RNA, leading to its rapid degradation. |
| miRNA | Micro RNAs, ~22 bases long, that base-pair extensively, but not completely, with mRNAs, especially over the six base pairs at the 5′ end of the miRNA. This inhibits translation of the "target" mRNA. |

because cytokine mRNAs are extremely unstable. Consequently, the concentration of the mRNA in the cytoplasm falls rapidly once its synthesis is stopped. In contrast, mRNAs encoding proteins required in large amounts that function over long periods, such as ribosomal proteins, are extremely stable so that multiple polypeptides are transcribed from each mRNA.

In addition to regulation of pre-mRNA processing, nuclear export, and translation, the cellular locations of many, if not most, mRNAs are regulated so that newly synthesized protein is concentrated where it is needed. Particularly striking examples of this occur in the nervous systems of multicellular animals. Some neurons in the human brain generate more than 1000 separate synapses with other neurons. During the process of learning, synapses that fire more frequently than others increase in size many times, while other synapses made by the same neuron do not. This can occur because mRNAs encoding proteins critical for synapse enlargement are stored at all synapses, but translation of these localized, stored mRNAs is regulated at each synapse independently by the frequency at which it initiates firing. In this way, synthesis of synapse-associated proteins can be regulated independently at each of the many synapses made by the same neuron.

Another type of gene regulation that has recently come to light involves micro RNAs (miRNAs), which regulate the stability and translation of specific target mRNAs in multicellular animals and plants. Analyses of these short miRNAs in various human tissues indicate that there are ~500 miRNAs expressed in the multiple types of human cells. Although some have recently been discovered to function through inhibition of target gene expression in the appropriate tissue and at the appropriate time in development, the functions of the vast majority of human miRNAs are unknown and are the subject of a growing new area of research. If most miRNAs do indeed have significant functions, miRNA genes constitute an important subset of the ~25,000 human genes. A closely related process called RNA interference (RNAi) leads to the degradation of viral RNAs in infected cells and the degradation of transposon-encoded RNAs in many eukaryotes. This is of tremendous significance to biological researchers because it is possible to design short interfering RNAs (siRNA) to inhibit the translation of specific mRNAs experimentally by a process called *RNA knockdown*. This makes it possible to inhibit the function of any desired gene, even in organisms that are not amenable to classic genetic methods for isolating mutants.

We refer to all the mechanisms that regulate gene expression following transcription as *post-transcriptional gene control* (Figure 8-1). Since the stability and translation rate of an mRNA contribute to the amount of protein expressed from a gene, these post-transcriptional processes are important components of **gene control**. Indeed, the protein output of a gene is regulated at every step in the life of an mRNA from the initiation of its synthesis to its degradation. Thus genetic regulatory processes act on RNA as well as DNA. In this chapter, we consider the events that occur in the processing of mRNA following transcription initiation and promoter proximal

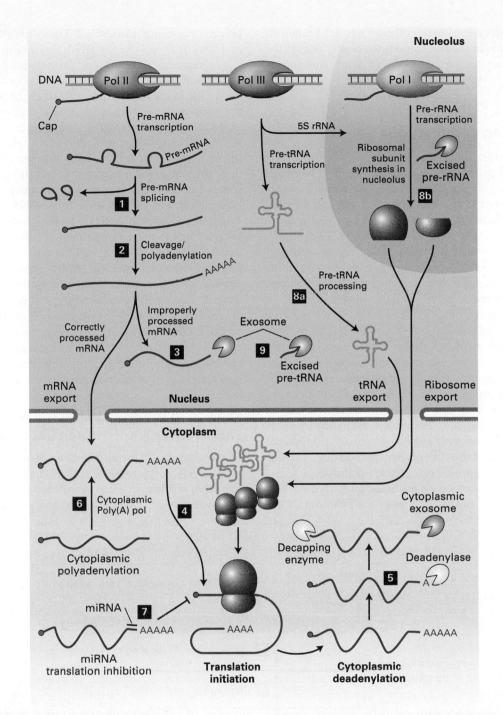

**FIGURE 8-1 Overview of RNA processing and post-transcriptional gene control.** Nearly all cytoplasmic RNAs are processed from primary transcripts in the nucleus before they are exported to the cytoplasm. For protein-coding genes transcribed by RNA polymerase II, gene control can be exerted through **1** the choice of alternative exons during pre-mRNA splicing and **2** the choice of alternative poly(A) sites. Improperly processed mRNAs are blocked from export to the cytoplasm and degraded **3** by a large complex called the exosome that contains multiple ribonucleases. Once exported to the cytoplasm, **4** translation initiation factors bind to the mRNA 5′-cap cooperatively with poly(A)-binding protein I bound to the poly(A) tail and initiate translation (see Figure 4-28). **5** mRNA is degraded in the cytoplasm by de-adenylation and decapping followed by degradation by cytoplasmic exosomes. The degradation rate of each mRNA is controlled, thereby regulating the mRNA concentration and, consequently, the amount of protein translated. Some mRNAs are synthesized without long poly(A) tails. Their translation is regulated by **6** controlling the synthesis of a long poly(A) tail by a cytoplasmic poly(A) polymerase. **7** Translation is also regulated by other mechanisms including miRNAs. When expressed, these ~22-nucleotide RNAs inhibit translation of mRNAs to which they hybridize, usually in the 3′-untranslated region. tRNAs and rRNAs are also synthesized as precursor RNAs that must be **8** processed before they are functional. Regions of precursors cleaved from the mature RNAs are degraded by nuclear exosomes **9**. [Adapted from Houseley, et. al., 2006, *Nat. Rev. Mol. Cell Biol.* **7:**529.]

elongation and the various mechanisms that are known to regulate these events. In the last section, we briefly discuss the processing of primary transcripts produced from genes encoding rRNAs and tRNAs.

## 8.1 Processing of Eukaryotic Pre-mRNA

In this section, we take a closer look at how eukaryotic cells convert the initial primary transcript synthesized by RNA polymerase II into a functional mRNA. Three major events occur during the process: *5′ capping, 3′ cleavage/polyadenylation,* and *RNA splicing* (Figure 8-2). Adding these specific modifications to the 5′ and 3′ ends of the pre-mRNA is important to protect it from enzymes that quickly digest uncapped RNAs generated by RNA processing, such as spliced-out introns and RNA transcribed downstream from a polyadenylation site. The 5′ cap and 3′ poly(A) tail distinguish pre-mRNA molecules from the many other kinds of RNAs in the nucleus. Pre-mRNA molecules (see Table 8-1) are bound by nuclear proteins that function in mRNA export to the cytoplasm. After mRNAs are exported to the cytoplasm, they are bound by a set of cytoplasmic proteins that stimulate translation and are critical for mRNA stability in the cytoplasm. Prior to nuclear export, introns must be removed to generate the correct coding region of the mRNA. In higher eukaryotes, including humans, alternative splicing is intricately regulated in order to substitute different functional domains into proteins, producing a considerable expansion of the proteome of these organisms.

The pre-mRNA processing events of capping, splicing, and polyadenylation occur in the nucleus as the nascent mRNA precursor is being transcribed. Thus pre-mRNA processing is *co-transcriptional.* As the RNA emerges from the surface of RNA polymerase II, its 5′ end is immediately modified by the addition of the 5′ cap structure found on all mRNAs (see Figure 4-14). As the nascent pre-mRNA continues to emerge from the surface of the polymerase, it is immediately bound by members of a complex group of RNA-binding proteins that assist in RNA splicing and export of the fully processed mRNA through nuclear pore complexes into the cytoplasm. Some of these proteins remain associated with the mRNA in the cytoplasm, but most either remain in the nucleus or shuttle back into the nucleus shortly after the mRNA is exported to the cytoplasm. Cytoplasmic RNA-binding proteins are exchanged for the nuclear ones. Consequently, mRNAs never occur as free RNA molecules in the cell but are always associated with protein as **ribonucleoprotein (RNP) complexes,** first as nascent *pre-mRNPs* that are capped and spliced as they are transcribed. Then, following cleavage and polyadenylation, they are referred to as *nuclear mRNPs.* Following the exchange of proteins that accompany export to the cytoplasm, they are called *cytoplasmic mRNPs.* Although we frequently refer to pre-mRNAs and mRNAs, it is important to remember that they are always associated with proteins as RNP complexes.

### The 5′ Cap Is Added to Nascent RNAs Shortly After Transcription Initiation

As the nascent RNA transcript emerges from the RNA channel of RNA polymerase II and reaches a length of ~25 nucleotides, a protective cap composed of 7-methylguanosine and methylated riboses is added to the 5′ end of eukaryotic

🎬 **ANIMATION:** Life Cycle of an mRNA

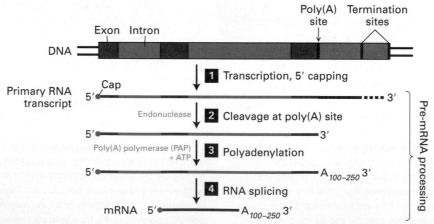

**FIGURE 8-2 Overview of mRNA processing in eukaryotes.** Shortly after RNA polymerase II initiates transcription at the first nucleotide of the first exon of a gene, the 5′ end of the nascent RNA is capped with 7-methylguanylate (step **1**). Transcription by RNA polymerase II terminates at any one of multiple termination sites downstream from the poly(A) site, which is located at the 3′ end of the final exon. After the primary transcript is cleaved at the poly(A) site (step **2**), a string of adenosine (A) residues is added (step **3**). The poly(A) tail contains ~250 A residues in mammals, ~150 in insects, and ~100 in yeasts. For short primary transcripts with few introns, splicing (step **4**) usually follows cleavage and polyadenylation, as shown. For large genes with multiple introns, introns often are spliced out of the nascent RNA during its transcription, i.e., before transcription of the gene is complete. Note that the 5′ cap and sequence adjacent to the poly(A) tail are retained in mature mRNAs.

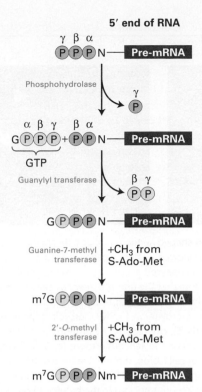

**5' end of RNA**

$\gamma \; \beta \; \alpha$

(P)(P)(P)N — **Pre-mRNA**

Phosphohydrolase

$\gamma$
(P)

$\alpha \; \beta \; \gamma \quad\quad \beta \; \alpha$
G(P)(P)(P) + (P)(P)N — **Pre-mRNA**

GTP

Guanylyl transferase

$\beta \; \gamma$
(P)(P)

G(P)(P)(P)N — **Pre-mRNA**

Guanine-7-methyl transferase | +CH$_3$ from S-Ado-Met

m$^7$G(P)(P)(P)N — **Pre-mRNA**

2'-O-methyl transferase | +CH$_3$ from S-Ado-Met

m$^7$G(P)(P)(P)Nm — **Pre-mRNA**

**FIGURE 8-3 Synthesis of 5'-cap on eukaryotic mRNAs.** The 5' end of a nascent RNA contains a 5'-triphosphate from the initiating NTP. The $\gamma$-phosphate is removed in the first step of capping, while the remaining $\alpha$- and $\beta$-phosphates (orange) remain associated with the cap. The third phosphate of the 5',5'-triphosphate bond is derived from the $\alpha$-phosphate of the GTP that donates the guanine. The methyl donor for methylation of the cap guanine and the first one or two riboses of the mRNA is *S*-adenosylmethionine (S-Ado-Met). [From S. Venkatesan and B. Moss, 1982, *Proc. Natl. Acad. Sci. USA* **79**:304.]

mRNAs (see Figure 4-14). The *5' cap* marks RNA molecules as mRNA precursors and protects them from RNA-digesting enzymes (5'-exoribonucleases) in the nucleus and cytoplasm. This initial step in RNA processing is catalyzed by a dimeric capping enzyme, which associates with the phosphorylated carboxyl-terminal domain (CTD) of RNA polymerase II. Recall that the CTD becomes phosphorylated by the TFIIII general transcription factor at multiple serines at the 5 position in the CTD heptapeptide repeat during transcription initiation (see Figure 7-17). Binding to the phosphorylated CTD stimulates the activity of the capping enzymes so that they are focused on RNAs containing a 5'-triphosphate that emerge from RNA polymerase II and not on RNAs transcribed by RNA polymerases I or III, which do not have a CTD. This is important because pre-mRNA synthesis accounts for only ~80 percent of the total RNA synthesized in replicating cells. About 20 percent is preribosomal RNA, which is transcribed by RNA polymerase I, and 5S rRNA, tRNAs, and other stable small RNAs, which are transcribed by RNA polymerase III. The two mechanisms of (1) binding of the capping enzyme to initiated RNA polymerase II specifically through its unique, phosphorylated CTD and (2)

activation of the capping enzyme by the phosphorylated CTD result in specific capping of RNAs transcribed by RNA polymerase II.

One subunit of the capping enzyme removes the $\gamma$-phosphate from the 5' end of the nascent RNA (Figure 8-3). Another domain of this subunit transfers the GMP moiety from GTP to the 5'-diphosphate of the nascent transcript, creating the unusual guanosine 5'-5'-triphosphate structure. In the final steps, separate enzymes transfer methyl groups from *S*-adenosylmethionine to the N$_7$ position of the guanine and one or two 2' oxygens of riboses at the 5' end of the nascent RNA.

Considerable evidence indicates that capping of the nascent transcript is coupled to elongation by RNA polymerase II so that all of its transcripts are capped during the earliest phase of elongation. As discussed in Chapter 7, in metazoans, during the initial phase of transcription the polymerase elongates the nascent transcript very slowly due to association of NELF (*negative elongation factor*) with RNA polymerase II in the promoter proximal region (see Figure 7-20). Once the 5' end of the nascent RNA is capped, phosphorylation of the RNA polymerase CTD at position 2 in the heptapeptide repeat and of NELF and DSIF by the CDK9-cyclin T protein kinase causes the release of NELF. This allows RNA polymerase II to enter into a faster mode of elongation that rapidly transcribes away from the promoter. The net effect of this mechanism is that the polymerase waits for the nascent RNA to be capped before elongating at a rapid rate.

## A Diverse Set of Proteins with Conserved RNA-Binding Domains Associate with Pre-mRNAs

As noted earlier, neither nascent RNA transcripts from protein-coding genes nor the intermediates of mRNA processing, collectively referred to as **pre-mRNA**, exist as free RNA molecules in the nuclei of eukaryotic cells. From the time nascent transcripts first emerge from RNA polymerase II until mature mRNAs are transported into the cytoplasm, the RNA molecules are associated with an abundant set of nuclear proteins. These are the major protein components of *heterogeneous ribonucleoprotein particles (hnRNPs)*, which contain *heterogeneous nuclear RNA (hnRNA)*, a collective term referring to pre-mRNA and other nuclear RNAs of various sizes. These hnRNP proteins contribute to further steps in RNA processing, including splicing, polyadenylation, and export through nuclear pore complexes to the cytoplasm.

Researchers identified hnRNP proteins by first exposing cultured cells to high-dose UV irradiation, which causes covalent cross-links to form between RNA bases and closely associated proteins. Chromatography of nuclear extracts from treated cells on an oligo-dT cellulose column, which binds RNAs with a poly(A) tail, was used to recover the proteins that had become cross-linked to nuclear polyadenylated RNA. Subsequent treatment of cell extracts from un-irradiated cells with monoclonal antibodies specific for the major proteins identified by this

(a)

(b)

(c)

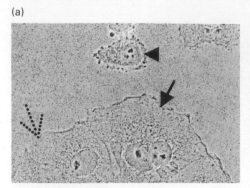

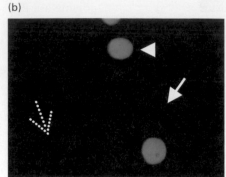

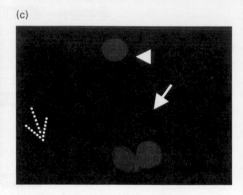

**FIGURE 8-4  Human hnRNP A1 protein can cycle in and out of the cytoplasm, but human hnRNP C protein cannot.** Cultured HeLa cells and *Xenopus* cells were fused by treatment with polyethylene glycol, producing heterokaryons containing nuclei from each cell type. The hybrid cells were treated with cycloheximide immediately after fusion to prevent protein synthesis. After 2 hours, the cells were fixed and stained with fluorescent-labeled antibodies specific for human hnRNP C and A1 proteins. These antibodies do not bind to the homologous *Xenopus* proteins. (a) A fixed preparation viewed by phase-contrast microscopy includes unfused HeLa cells (arrowhead) and *Xenopus* cells (dotted arrow), as well as fused heterokaryons (solid arrow). In the heterokaryon in this micrograph, the round HeLa-cell nucleus is to the right of the oval-shaped *Xenopus* nucleus. (b, c) When the same preparation was viewed by fluorescence microscopy, the stained hnRNP C protein appeared green and the stained hnRNP A1 protein appeared red. Note that the unfused *Xenopus* cell on the left is unstained, confirming that the antibodies are specific for the human proteins. In the heterokaryon, hnRNP C protein appears only in the HeLa-cell nucleus (b), whereas the A1 protein appears in both the HeLa-cell nucleus and the *Xenopus* nucleus (c). Since protein synthesis was blocked after cell fusion, some of the human hnRNP A1 protein must have left the HeLa-cell nucleus, moved through the cytoplasm, and entered the *Xenopus* nucleus in the heterokaryon. [See S. Pinol-Roma and G. Dreyfuss, 1992, *Nature* **355**:730; courtesy of G. Dreyfuss.]

cross-linking technique revealed a complex set of abundant hnRNP proteins ranging in size from ~30 to ~120 kDa.

Like transcription factors, most hnRNP proteins have a modular structure. They contain one or more *RNA-binding domains* and at least one other domain that interacts with other proteins. Several different RNA-binding motifs have been identified by creating hnRNP proteins with missing amino acid sequences and testing their ability to bind RNA.

**Functions of hnRNP Proteins**  The association of pre-mRNAs with hnRNP proteins prevents the pre-mRNAs from forming short secondary structures dependent on base pairing of complementary regions, thereby making the pre-mRNAs accessible for interaction with other RNA molecules or proteins. Pre-mRNAs associated with hnRNP proteins present a more uniform substrate for subsequent processing steps than would free, unbound pre-mRNAs, in which each mRNA forms a unique secondary structure due to its specific sequence.

Binding studies with purified hnRNP proteins indicate that different hnRNP proteins associate with different regions of a newly made pre-mRNA molecule. For example, the hnRNP proteins A1, C, and D bind preferentially to the pyrimidine-rich sequences at the 3′ ends of introns (see Figure 8-7). Some hnRNP proteins interact with the RNA sequences that specify RNA splicing or cleavage/polyadenylation and contribute to the structure recognized by RNA-processing factors. Finally, cell-fusion experiments have shown that some hnRNP proteins remain localized in the nucleus, whereas others cycle in and

out of the cytoplasm, suggesting that they function in the transport of mRNA (Figure 8-4).

**Conserved RNA-Binding Motifs**  The *RNA recognition motif (RRM)*, also called the RNP motif and the RNA-binding domain (RBD), is the most common RNA-binding domain in hnRNP proteins. This ~80-residue domain, which occurs in many other RNA-binding proteins, contains two highly conserved sequences (RNP1 and RNP2) that are found across organisms ranging from yeast to human—indicating that like many DNA-binding domains, they evolved early in eukaryotic evolution.

Structural analyses have shown that the RRM domain consists of a four-stranded β sheet flanked on one side by two α helices. To interact with the negatively charged RNA phosphates, the β sheet forms a positively charged surface. The conserved RNP1 and RNP2 sequences lie side by side on the two central β strands, and their side chains make multiple contacts with a single-stranded region of RNA that lies across the surface of the β sheet (Figure 8-5).

The 45-residue *KH motif* is found in the hnRNP K protein and several other RNA-binding proteins. The three-dimensional structure of representative KH domains is similar to that of the RRM domain but smaller, consisting of a three-stranded β sheet supported from one side by two α helices. Nonetheless, the KH domain interacts with RNA much differently than does the RRM domain. RNA binds to the KH domain by interacting with a hydrophobic surface

(a) RNA recognition motif (RRM)          (b) Sex-lethal (Sxl) RRM domains          (c) Polypyrimidine tract binding protein (PTB)

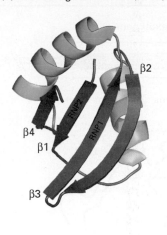
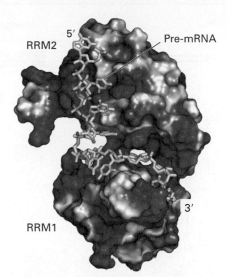
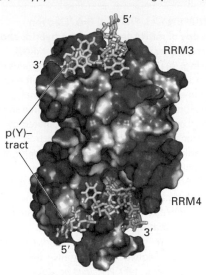

**FIGURE 8-5 Structure of the RRM domain and its interaction with RNA.** (a) Diagram of the RRM domain showing the two α helices (green) and four β strands (red) that characterize this motif. The conserved RNP1 and RNP2 regions are located in the two central β strands. (b) Surface representation of the two RRM domains in *Drosophila* Sex-lethal (Sxl) protein, which bind a nine-base sequence in transformer pre-mRNA (yellow). The two RRMs are oriented like the two parts of an open pair of castanets, with the β sheet of RRM1 facing upward and the β sheet of RRM2 facing downward. Positively charged regions in Sxl protein are shown in shades of blue; negatively charged regions, in shades of red. The pre-mRNA is bound to the surfaces of the positively charged β sheets, making most of its contacts with the RNP1 and RNP2 regions of each RRM. (c) Strikingly different orientation of RRM domains in a different hnRNP, the polypyrimidine tract binding (PTB) protein, illustrating that RRMs are oriented in different relative positions in different hnRNPs; colors are as in (b). Polypyrimidine (p(Y)) single-stranded RNA is bound to the upward (RRM3) and downward (RRM4) facing β-sheets. RNA is shown in yellow. [Part (a) adapted from K. Nagai et al., 1995, *Trends Biochem. Sci.* **20**:235. Part (b) after N. Harada et al., 1999, *Nature* **398**:579. Part (c) after F. C. Oberstrass et al., 2006, *Science* **309**:2054.]

formed by the α helices and one β strand. The *RGG box*, another RNA-binding motif found in hnRNP proteins, contains five Arg-Gly-Gly (RGG) repeats with several interspersed aromatic amino acids. Although the structure of this motif has not yet been determined, its arginine-rich nature is similar to the RNA-binding domains of the HIV Tat protein. KH domains and RGG repeats are often interspersed in two or more sets in a single RNA-binding protein.

## Splicing Occurs at Short, Conserved Sequences in Pre-mRNAs via Two Transesterification Reactions

During formation of a mature, functional mRNA, the **introns** are removed and **exons** are spliced together. For short transcription units, **RNA splicing** usually follows cleavage and polyadenylation of the 3′ end of the primary transcript, as depicted in Figure 8-2. However, for long transcription units containing multiple exons, splicing of exons in the nascent RNA begins before transcription of the gene is complete.

Early pioneering research on the nuclear processing of mRNAs revealed that mRNAs are initially transcribed as much longer RNA molecules than the mature mRNAs in the cytoplasm. It was also shown that RNA sequences near the 5′ cap added shortly after transcription initiation are retained in the mature mRNA, and that RNA sequences near the polyadenylated end of mRNA processing intermediates in hnRNA are retained in the mature mRNA in the cytoplasm. The solution to this apparent conundrum came from the discovery of introns by electron microscopy of RNA-DNA hybrids of adenovirus DNA and the mRNA encoding hexon, a major virion capsid protein (Figure 8-6). Other studies revealed nuclear viral RNAs that were colinear with the viral DNA (primary transcripts), and RNAs with one or two of the introns removed (processing intermediates). These results, together with the earlier findings that the 5′ cap and 3′ poly(A) tail at each end of long mRNA precursors are retained in shorter mature cytoplasmic mRNAs, led to the realization that introns are removed from primary transcripts as exons are spliced together.

The location of *splice sites*—that is, exon-intron junctions—in a pre-mRNA can be determined by comparing the sequence of genomic DNA with that of the cDNA prepared from the corresponding mRNA (see Figure 5-15). Sequences that are present in the genomic DNA but absent from the cDNA represent introns and indicate the positions of splice sites. Such analysis of a large number of different mRNAs revealed moderately conserved, short consensus sequences at the splice sites flanking introns in eukaryotic pre-mRNAs; a pyrimidine-rich region just upstream of the 3′ splice site also is common (Figure 8-7). Studies of mutant genes with deletions introduced into introns have shown that much of the

**EXPERIMENTAL FIGURE 8-6 Electron microscopy of mRNA-template DNA hybrids shows that introns are spliced out during pre-mRNA processing.** (a) Diagram of the *Eco*RI A fragment of adenovirus DNA, which extends from the left end of the genome to just before the end of the final exon of the hexon gene. The gene consists of three short exons and one long (~3.5 kb) exon separated by three introns of ~1, 2.5, and 9 kb. (b) Electron micrograph (*left*) and schematic drawing (*right*) of a hybrid between an *Eco*RI A DNA fragment and a hexon mRNA. The loops marked A, B, and C correspond to the introns indicated in (a). Since these intron sequences in the viral genomic DNA are not present in the mature hexon mRNA, they loop out between the exon sequences that hybridize to their complementary sequences in the mRNA. [Micrograph from S. M. Berget et al., 1977, *Proc. Nat'l. Acad. Sci. USA* **74**:3171; courtesy of P. A. Sharp.]

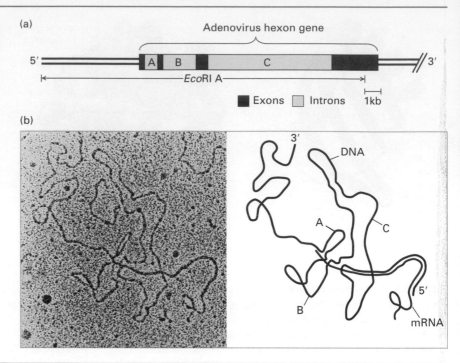

(a)

Adenovirus hexon gene

5' 　　　A　B　　　　C　　　　3'

*Eco*RI A

■ Exons ☐ Introns 　1kb

(b)

3'
DNA
A
C
B
5'
mRNA

center portion of introns can be removed without affecting splicing; generally only 30–40 nucleotides at each end of an intron are necessary for splicing to occur at normal rates.

Analysis of the intermediates formed during splicing of pre-mRNAs in vitro led to the discovery that splicing of exons proceeds via two sequential *transesterification reactions* (Figure 8-8). Introns are removed as a lariat structure in which the 5' G of the intron is joined in an unusual 2',5'-phosphodiester bond to an adenosine near the 3' end of the intron. This A residue is called the *branch point A* because it forms an RNA branch in the lariat structure. In each transesterification reaction, one phosphoester bond is exchanged for another. Since the number of phosphoester bonds in the molecule is not changed in either reaction, no energy is consumed. The net result of these two reactions is that two exons are ligated and the intervening intron is released as a branched lariat structure.

## During Splicing, snRNAs Base-Pair with Pre-mRNA

Splicing requires the presence of **small nuclear RNAs (snRNAs)**, important for base pairing with the pre-mRNA, and ~170 associated proteins. Five U-rich snRNAs, designated U1, U2, U4, U5, and U6, participate in pre-mRNA splicing. Ranging in length from 107–210 nucleotides, these snRNAs are associated with 6–10 proteins each in the many

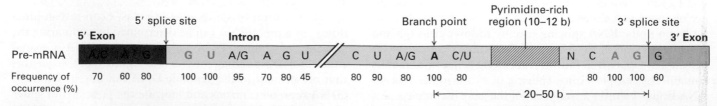

| | 5' splice site | | | | Intron | | | | | | | Branch point | | | Pyrimidine-rich region (10–12 b) | | 3' splice site | | | |
|---|---|---|---|---|---|---|---|---|---|---|---|---|---|---|---|---|---|---|---|---|
| | **5' Exon** | | | | | | | | | | | | | | | | | | | **3' Exon** |
| Pre-mRNA | A/C | A | A/G | G | | G | U | A/G | A | G | U | // C | U | A/G | **A** | C/U | | | N C | A G | G |
| Frequency of occurrence (%) | 70 | 60 | 80 | | | 100 | 100 | 95 | 70 | 80 | 45 | 80 | 90 | 80 | 100 | 80 | | | 80 100 | 100 | 60 |

20–50 b

**FIGURE 8-7 Consensus sequences around splice sites in vertebrate pre-mRNAs.** The only nearly invariant bases are the 5' GU and the 3' AG of the intron (blue), although the flanking bases indicated are found at frequencies higher than expected based on a random distribution. A pyrimidine-rich region (hatch marked) near the 3' end of the intron is found in most cases. The branch-point

adenosine, also invariant, usually is 20–50 bases from the 3' splice site. The central region of the intron, which may range from 40 bases to 50 kilobases in length, generally is unnecessary for splicing to occur. [See R. A. Padgett et al., 1986, *Ann. Rev. Biochem.* **55**:1119, and E. B. Keller and W. A. Noon, 1984, *Proc. Nat'l. Acad. Sci. USA* **81**:7417.]

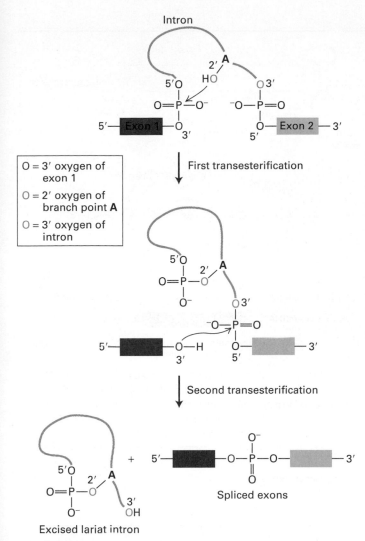

**FIGURE 8-8 Two transesterification reactions that result in splicing of exons in pre-mRNA.** In the first reaction, the ester bond between the 5′ phosphorus of the intron and the 3′ oxygen (dark red) of exon 1 is exchanged for an ester bond with the 2′ oxygen (blue) of the branch-site **A** residue. In the second reaction, the ester bond between the 5′ phosphorus of exon 2 and the 3′ oxygen (orange) of the intron is exchanged for an ester bond with the 3′ oxygen of exon 1, releasing the intron as a lariat structure and joining the two exons. Arrows show where activated hydroxyl oxygens react with phosphorus atoms.

small nuclear ribonucleoprotein particles (snRNPs) in the nuclei of eukaryotic cells.

Definitive evidence for the role of U1 snRNA in splicing came from experiments indicating that base pairing between the 5′ splice site of a pre-mRNA and the 5′ region of U1 snRNA is required for RNA splicing (Figure 8-9a). In vitro experiments showed that a synthetic oligonucleotide that hybridizes with the 5′-end region of U1 snRNA blocks RNA splicing. In vivo experiments showed that base-pairing-disrupting mutations in the pre-mRNA 5′ splice site also block RNA splicing; in this case, however, splicing can be restored by expression of a mutant U1 snRNA with a compensating mutation that restores base pairing to the mutant pre-mRNA 5′ splice site

(Figure 8-9b). Involvement of U2 snRNA in splicing initially was suspected when it was found to have an internal sequence that is largely complementary to the consensus sequence flanking the branch point in pre-mRNAs (see Figure 8-7). Compensating mutation experiments, similar to those conducted with U1 snRNA and 5′ splice sites, demonstrated that base pairing between U2 snRNA and the branch-point sequence in pre-mRNA also is critical to splicing.

Figure 8-9a illustrates the general structures of the U1 and U2 snRNAs and how they base-pair with pre-mRNA during splicing. Significantly, the branch-point A itself, which is not base-paired to U2 snRNA, "bulges out" (Figure 8-10a), allowing its 2′ hydroxyl to participate in the first transesterification reaction of RNA splicing (see Figure 8-8).

Similar studies with other snRNAs demonstrated that base pairing between them also occurs during splicing. Moreover, rearrangements in these RNA-RNA interactions are critical in the splicing pathway, as we describe next.

## Spliceosomes, Assembled from snRNPs and a Pre-mRNA, Carry Out Splicing

The five splicing snRNPs and other proteins involved in splicing assemble on a pre-mRNA, forming a large ribonucleoprotein complex called a **spliceosome** (Figure 8-11). The spliceosome has a mass similar to that of a ribosome. Assembly of a spliceosome begins with the base pairing of the U1 snRNA to the 5′ splice site and the cooperative binding of protein SF1 (splicing factor 1) to the branch point A and the heterodimeric protein U2AF (U2 associated factor) to the pyrimidine tract and the 3′ AG of the intron via its large and small subunits, respectively. The U2 snRNP then base-pairs with the branch point region (Figure 8-9a) as SF1 is released. Extensive base pairing between the snRNAs in the U4 and U6 snRNPs forms a complex that associates with U5 snRNP. The U4/U6/U5 "tri-snRNP" then associates with the previously formed U1/U2/pre-mRNA complex to generate a spliceosome.

After formation of the spliceosome, extensive rearrangements in the pairing of snRNAs and the pre-mRNA lead to release of the U1 snRNP. Figure 8-10b shows a cryoelectron-microscopy structure of this intermediate in the splicing process. A further rearrangement of spliceosomal components occurs with the loss of U4 snRNP. This generates a complex that catalyzes the first transesterification reaction that forms the 2′,5′-phosphodiester bond between the 2′ hydroxyl on the branch point A and the phosphate at the 5′ end of the intron (Figure 8-9). Following another rearrangement of the snRNPs, the second transesterification reaction ligates the two exons in a standard 3′,5′-phosphodiester bond, releasing the intron as a lariat structure associated with the snRNPs. This final intron-snRNP complex rapidly dissociates, and the individual snRNPs released can participate in a new cycle of splicing. The excised intron is then rapidly degraded by a debranching enzyme and other nuclear RNases discussed later.

As mentioned above, a spliceosome is roughly the size of a ribosome and is composed of ~170 proteins, including ~100 "splicing factors" in addition to the proteins associated with the

(a)

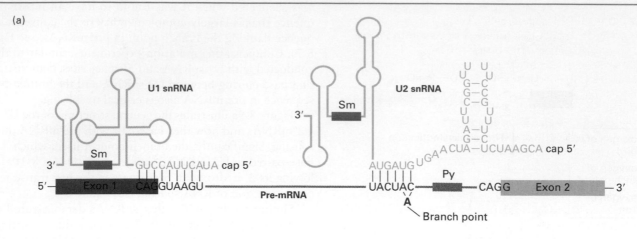

(b)

Mutation in pre-mRNA 5′ splice site
blocks splicing

Compensatory mutation in U1
restores splicing

**FIGURE 8-9 Base pairing between pre-mRNA, U1 snRNA, and U2 snRNA early in the splicing process.** (a) In this diagram, secondary structures in the snRNAs that are not altered during splicing are depicted schematically. The yeast branch-point sequence is shown here. Note that U2 snRNA base-pairs with a sequence that includes the branch-point A, although this residue is not base-paired. The purple rectangles represent sequences that bind snRNP proteins recognized by anti-Sm antibodies. For unknown reasons, antisera from patients with the autoimmune disease systemic lupus erythematosus (SLE) contain antibodies to snRNP proteins, which have been useful in

characterizing components of the splicing reaction. (b) Only the 5′ ends of U1 snRNAs and 5′ splice sites in pre-mRNAs are shown. *(Left)* A mutation (A) in a pre-mRNA splice site that interferes with base pairing to the 5′ end of U1 snRNA blocks splicing. *(Right)* Expression of a U1 snRNA with a compensating mutation (U) that restores base pairing also restores splicing of the mutant pre-mRNA. [Part (a) adapted from M. J. Moore et al., 1993, in R. Gesteland and J. Atkins, eds., *The RNA World,* Cold Spring Harbor Press, pp. 303–357. Part (b) see Y. Zhuang and A. M. Weiner, 1986, *Cell* **46:**827.]

---

**FIGURE 8-10 Structures of a bulged A in an RNA-RNA helix and an intermediate in the splicing process.** (a) The structure of an RNA duplex with the sequence shown, containing bulged A residues (red) at position 5 in the RNA helix, was determined by x-ray crystallography. (b) The bulged A residues extend from the side of an A-form RNA-RNA helix. The phosphate backbone of one strand is shown in green; the other strand in blue. The structure on the right is turned 90 degrees for a view down the axis of the helix. (c) 40 Å resolution structure of a spliceosomal splicing intermediate containing U2, U4, U5, and U6 snRNPs, determined by cryoelectron microscopy and image reconstruction. The U4/U6/U5 tri-snRNP complex has a similar structure to the triangular body of this complex, suggesting that these snRNPs are at the bottom of the structure shown here and that the head is composed largely of U2 snRNP. [Parts (a) and (b) from J. A. Berglund et al., 2001, *RNA* **7:**682. Part (c) from D. Boehringer et al., 2004, *Nat. Struct. Mol. Biol.* **11:**463. See also H. Stark and R. Luhrmann, 2006, *Annu. Rev. Biophys. Biomol. Struct.* **35:**435.]

(a) Self-complementary sequence with bulging A

$$5'UACU \overset{A}{\phantom{|}} ACGU \; AGUA$$
$$AUGA \; UGCA \underset{A}{\phantom{|}} UCAU5'$$

(b) X-ray crystallography structure

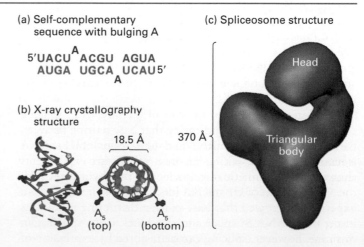

(c) Spliceosome structure

**FIGURE 8-11 Model of spliceosome-mediated splicing of pre-mRNA.** Step **1**: After U1 base pairs with the consensus 5′splice site, SF1 (splicing factor 1) binds the branch point A; U2AF (U2 snRNP associated factor) associates with the polypyrimidine tract and 3′splice site; and the U2 snRNP associates with the branch point A via base-pairing interactions shown in Figure 8-9, displacing SF1. Step **2**: A trimeric snRNP complex of U4, U5, and U6 joins the initial complex to form the spliceosome. Step **3**: Rearrangements of base-pairing interactions between snRNAs converts the spliceosome into a catalytically active conformation and destabilizes the U1 and U4 snRNPs, which are released. Step **4**: The catalytic core, thought to be formed by U6 and U2, then catalyzes the first transesterification reaction, forming the intermediate containing a 2′,5′-phosphodiester bond as shown in Figure 8-8. Step **5**: Following further rearrangements between the snRNPs, the second transesterification reaction joins the two exons by a standard 3′,5′-phosphodiester bond and releases the intron as a lariat structure and the remaining snRNPs. Step **6**: The excised lariat intron is converted into a linear RNA by a debranching enzyme. [Adapted from T. Villa et al., 2002, *Cell* **109**:149.]

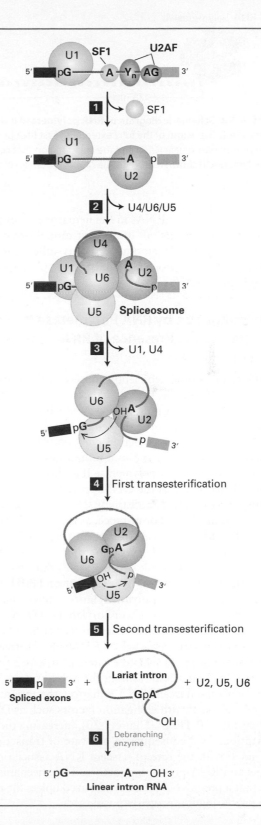

five snRNPs. This makes RNA splicing comparable in complexity to initiation of transcription and protein synthesis. Some of the splicing factors are associated with snRNPs, but others are not. For instance, the 65-kD subunit of the U2-associated factor (U2AF) binds to the pyrimidine-rich region near the 3′ end of introns and to the U2 snRNP. The 35-kD subunit of U2AF binds to the AG dinucleotide at the 3′ end of the intron and also interacts with the larger U2AF subunit bound nearby. These two U2AF subunits act together with SF1 to help specify the 3′ splice site by promoting interaction of U2 snRNP with the branch point (see Figure 8-11, step **1**). Some splicing factors also exhibit sequence homologies to known RNA helicases; these are probably necessary for the base-pairing rearrangements that occur in snRNAs during the spliceosomal splicing cycle.

Following RNA splicing, a specific set of hnRNP proteins remain bound to the spliced RNA approximately 20 nucleotides 5′ to each exon-exon junction, thus forming an *exon-junction complex*. One of the hnRNP proteins associated with the exon-junction complex is the *RNA export factor* (REF), which functions in the export of fully processed mRNPs from the nucleus to the cytoplasm, as discussed in Section 8.3. Other proteins associated with the exon-junction complex function in a quality-control mechanism that leads to the degradation of improperly spliced mRNAs, known as nonsense-mediated decay (Section 8.4).

A small fraction of pre-mRNAs (<1% in humans) contain introns whose splice sites do not conform to the standard consensus sequence. This class of introns begins with AU and ends with AC rather than following the usual "GU-AG rule" (see Figure 8-7). Splicing of this special class of introns occurs via a splicing cycle analogous to that shown in Figure 8-11, except that four novel, low-abundance snRNPs, together with the standard U5 snRNP, are involved.

Nearly all functional mRNAs in vertebrate, insect, and plant cells are derived from a single molecule of the corresponding pre-mRNA by removal of internal introns and splicing of exons. However, in two types of protozoans—trypanosomes and euglenoids—mRNAs are constructed by splicing together separate RNA molecules. This process, referred to as *trans-splicing*, is also used in the synthesis of

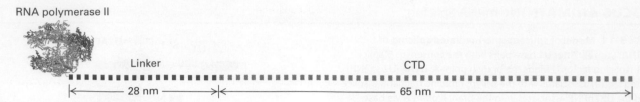

RNA polymerase II

Linker

CTD

|← 28 nm →|← 65 nm →|

**FIGURE 8-12 Schematic diagram of RNA polymerase II with the CTD extended.** The length of the fully extended yeast RNA polymerase II carboxyl-terminal domain (CTD) and the linker region that connects it to the polymerase is shown relative to the globular domain of the polymerase. The CTD of mammalian RNA polymerase II is twice as long. In its extended form, the CTD can associate with multiple RNA-processing factors simultaneously. [From P. Cramer, D. A. Bushnell, and R. D. Kornberg, 2001, *Science* **292**:1863.]

10–15 percent of the mRNAs in the nematode (roundworm) *Caenorhabditis elegans,* an important model organism for studying embryonic development. Trans-splicing is carried out by snRNPs by a process similar to the splicing of exons in a single pre-mRNA.

## Chain Elongation by RNA Polymerase II Is Coupled to the Presence of RNA-Processing Factors

How is RNA processing efficiently coupled with the transcription of a pre-mRNA? The key lies in the long carboxyl-terminal domain (CTD) of RNA polymerase II, which, as discussed in Chapter 7, is composed of multiple repeats of a seven-residue (heptapeptide) sequence. When fully extended, the CTD domain in the yeast enzyme is about 65 nm long (Figure 8-12); the CTD in human RNA polymerase II is about twice as long. The remarkable length of the CTD apparently allows multiple proteins to associate simultaneously with a single RNA polymerase II molecule. For instance, as mentioned earlier, the enzymes that add the 5′ cap to nascent transcripts associate with the CTD phosphorylated on multiple serines at the fifth position in the heptapeptide repeat (Ser-5) during or shortly after transcription initiation by a subunit of TFIIH. In addition, RNA splicing and polyadenylation factors have been found to associate with the phosphorylated CTD. As a consequence, these processing factors are present at high local concentrations when splice sites and poly(A) signals are transcribed by the polymerase, enhancing the rate and specificity of RNA processing. In a reciprocal fashion, the association of hnRNP proteins with the nascent RNA enhances the interaction of RNA polymerase II with elongation factors such as DSIF and CDK9-cyclin T (P-TEFb) (Figure 7-20), increasing the rate of transcription. As a consequence, the rate of transcription is coordinated with the rate of nascent RNA association with hnRNPs and RNA-processing factors. This mechanism may ensure that a pre-mRNA is not synthesized unless the machinery for processing it is properly positioned.

## SR Proteins Contribute to Exon Definition in Long Pre-mRNAs

The average length of an exon in the human genome is ~150 bases, whereas the average length of an intron is much longer (~3500 bases). The longest introns contain upward of 500 kb! Because the sequences of 5′ and 3′ splice sites and branch points are so degenerate, multiple copies are likely to occur randomly in long introns. Consequently, additional sequence information is required to define the exons that should be spliced together in higher organisms with long introns.

The information for defining the splice sites that demarcate exons is encoded within the sequences of the exons. A family of RNA-binding proteins, the *SR proteins,* interact with sequences within exons called *exonic splicing enhancers.* SR proteins are a subset of the hnRNP proteins discussed earlier and so contain one or more RRM RNA-binding domains. They also contain several protein-protein interaction domains rich in arginine (R) and serine (S) residues called RS domains. When bound to exonic splicing enhancers, SR proteins mediate the cooperative binding of U1 snRNP to a true 5′ splice site and U2 snRNP to a branch point through a network of protein-protein interactions that span an exon (Figure 8-13). The complex of SR proteins, snRNPs, and other splicing factors (e.g., U2AF) that assemble across an exon, which has been called a **cross-exon recognition complex,** permits precise specification of exons in long pre-mRNAs.

In the transcription units of higher organisms with long introns, exons not only encode the amino acid sequences of different portions of a protein but also contain binding sites for SR proteins. Mutations that interfere with the binding of an SR protein to an exonic splicing enhancer, even if they do not change the encoded amino acid sequence, prevent formation of the cross-exon recognition complex. As a result, the affected exon is "skipped" during splicing and is not included in the final processed mRNA. The truncated mRNA produced in this case is either degraded or translated into a mutant, abnormally functioning protein. This type of mutation occurs in some human genetic diseases. For example, *spinal muscular atrophy* is one of the most common genetic causes of childhood mortality. This disease results from mutations in a region of the genome containing two closely related genes, *SMN1* and *SMN2,* that arose by gene duplication. *SMN2* encodes a protein identical with *SMN1.* SMN2 is expressed at a much lower level because a silent mutation in one exon interferes with the binding of an SR protein. This leads to exon skipping in most of the *SMN2* mRNAs. The homologous *SMN* gene in the mouse, where there is

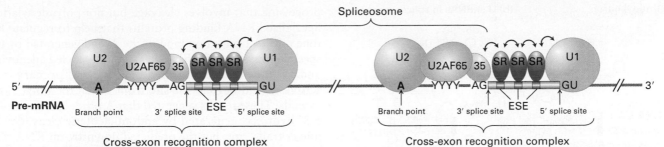

**FIGURE 8-13 Exon recognition through cooperative binding of SR proteins and splicing factors to pre-mRNA.** The correct 5′ GU and 3′ AG splice sites are recognized by splicing factors on the basis of their proximity to exons. The exons contain exonic splicing enhancers (ESEs) that are binding sites for SR proteins. When bound to ESEs, the SR proteins interact with one another and promote the cooperative binding of the U1 snRNP to the 5′ splice site of the downstream intron, SF1 and then the U2 snRNP to the branch point of the upstream intron, the 65- and 35-kD subunits of U2AF to the pyrimidine-rich region and AG 3′ splice site of the upstream intron, and other splicing factors (not shown). The resulting RNA-protein cross-exon recognition complex spans an exon and activates the correct splice sites for RNA splicing. Note that the U1 and U2 snRNPs in this unit do not become part of the same spliceosome. The U2 snRNP on the right forms a spliceosome with the U1 snRNP bound to the 5′ end of the same intron. The U1 snRNP shown on the right forms a spliceosome with the U2 snRNP bound to the branch point of the downstream intron (not shown), and the U2 snRNP on the left forms a spliceosome with a U1 snRNP bound to the 5′ splice site of the upstream intron (not shown). Double-headed arrows indicate protein-protein interactions. [Adapted from T. Maniatis, 2002, *Nature* **418**:236; see also S. M. Berget, 1995, *J. Biol. Chem.* **270**:2411.]

only a single copy, is essential for cell viability. Spinal muscular atrophy in humans results from homozygous mutations that inactivate *SMN1*. The low level of protein translated from the small fraction of *SMN2* mRNAs that are correctly spliced is sufficient to maintain cell viability during embryogenesis and fetal development, but it is not sufficient to maintain viability of spinal cord motor neurons in childhood, resulting in their death and the associated disease.

Approximately 15 percent of the single-base-pair mutations that cause human genetic diseases interfere with proper *exon definition*. Some of these mutations occur in 5′ or 3′ splice sites, often resulting in the use of nearby alternative "cryptic" splice sites present in the normal gene sequence. In the absence of the normal splice site, the cross-exon recognition complex recognizes these alternative sites. Other mutations that cause abnormal splicing result in a new consensus splice-site sequence that becomes recognized in place of the normal splice site. Finally, some mutations can interfere with the binding of specific SR proteins to pre-mRNAs. These mutations inhibit splicing at normal splice sites, as in the case of the *SMN2* gene, and thus lead to exon skipping. ■

## Self-Splicing Group II Introns Provide Clues to the Evolution of snRNAs

Under certain nonphysiological in vitro conditions, pure preparations of some RNA transcripts slowly splice out introns in the absence of any protein. This observation led to the recognition that some introns are *self-splicing*. Two types of self-splicing introns have been discovered: *group I introns*, present in nuclear rRNA genes of protozoans, and *group II introns*, present in protein-coding genes and some rRNA and tRNA genes in mitochondria and chloroplasts of plants and fungi. Discovery of the catalytic activity of self-splicing introns revolutionized concepts about the functions of RNA. As discussed in Chapter 4, RNA is now known to catalyze peptide-bond formation during protein synthesis in ribosomes. Here we discuss the probable role of group II introns, now found only in mitochondrial and chloroplast DNA, in the evolution of snRNAs; the functioning of group I introns is considered in the later section on rRNA processing.

Even though their precise sequences are not highly conserved, all group II introns fold into a conserved, complex secondary structure containing numerous stem loops (Figure 8-14a). Self-splicing by a group II intron occurs via two transesterification reactions, involving intermediates and products analogous to those found in nuclear pre-mRNA splicing. The mechanistic similarities between group II intron self-splicing and spliceosomal splicing led to the hypothesis that snRNAs function analogously to the stem loops in the secondary structure of group II introns. According to this hypothesis, snRNAs interact with 5′ and 3′ splice sites of pre-mRNAs and with each other to produce a three-dimensional RNA structure functionally analogous to that of group II self-splicing introns (Figure 8-14b).

An extension of this hypothesis is that introns in ancient pre-mRNAs evolved from group II self-splicing introns through the progressive loss of internal RNA structures, which concurrently evolved into trans-acting snRNAs that perform the same functions. Support for this type of evolutionary model comes from experiments with group II intron mutants in which domain V and part of domain I are deleted. RNA transcripts containing such mutant introns are defective in self-splicing, but when RNA molecules equivalent to the deleted regions are added to the in vitro reaction, self-splicing occurs. This finding demonstrates that these domains in group II introns can be trans-acting, like snRNAs.

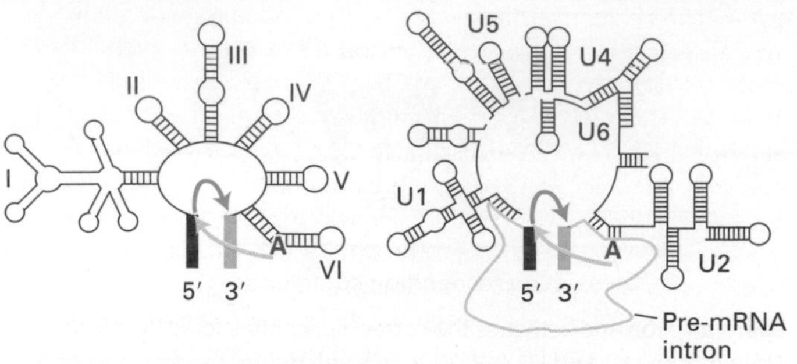

(a) Group II intron

(b) U snRNAs in spliceosome

Pre-mRNA
intron

**FIGURE 8-14 Comparison of group II self-splicing introns and the spliceosome.** The schematic diagrams compare the secondary structures of (a) group II self-splicing introns and (b) U snRNAs present in the spliceosome. The first transesterification reaction is indicated by light green arrows; the second reaction, by blue arrows. The branch-point A is boldfaced. The similarity in these structures suggests that the spliceosomal snRNAs evolved from group II introns, with the trans-acting snRNAs being functionally analogous to the corresponding domains in group II introns. The colored bars flanking the introns in (a) and (b) represent exons. [Adapted from P. A. Sharp, 1991, *Science* **254**:663.]

The similarity in the mechanisms of group II intron self-splicing and spliceosomal splicing of pre-mRNAs also suggests that the splicing reaction is catalyzed by the snRNA, not the protein, components of spliceosomes. Although group II introns can self-splice in vitro at elevated temperatures and $Mg^{2+}$ concentrations, under in vivo conditions proteins called *maturases,* which bind to group II intron RNA, are required for rapid splicing. Maturases are thought to stabilize the precise three-dimensional interactions of the intron RNA required to catalyze the two splicing transesterification reactions. By analogy, snRNP proteins in spliceosomes are thought to stabilize the precise geometry of snRNAs and intron nucleotides required to catalyze pre-mRNA splicing.

The evolution of snRNAs may have been an important step in the rapid evolution of higher eukaryotes. As internal intron sequences were lost and their functions in RNA splicing supplanted by trans-acting snRNAs, the remaining intron sequences would be free to diverge. This in turn likely facilitated the evolution of new genes through **exon shuffling** since there are few constraints on the sequence of new introns generated in the process (see Figures 6-18 and 6-19). It also permitted the increase in protein diversity that results from alternative RNA splicing and an additional level of gene control resulting from regulated RNA splicing.

## 3′ Cleavage and Polyadenylation of Pre-mRNAs Are Tightly Coupled

In eukaryotic cells, all mRNAs, except histone mRNAs, have a 3′ *poly(A) tail.* (The major histone mRNAs are transcribed from repeated genes at prodigious levels in replicating cells during the S phase. They undergo a special form of 3′-end processing that involves cleavage but not polyadenylation. Specialized RNA-binding proteins that help to regulate histone mRNA translation bind to the 3′ end generated by this specialized system.) Early studies of pulse-labeled adenovirus and SV40 RNA demonstrated that the viral primary transcripts extend beyond the site from which the poly(A) tail extends. These results suggested that A residues are added to a 3′ hydroxyl generated by endonucleolytic cleavage of a longer transcript, but the predicted downstream RNA fragments never were detected in vivo, presumably because of their rapid degradation. However, detection of both predicted cleavage products was observed in in vitro processing reactions performed with nuclear extracts of cultured human cells. The cleavage/polyadenylation process and degradation of the RNA downstream of the cleavage site occurs much more slowly in these in vitro reactions, simplifying detection of the downstream cleavage product.

Early sequencing of cDNA clones from animal cells showed that nearly all mRNAs contain the sequence AAUAAA 10–35 nucleotides *upstream* from the poly(A) tail (Figure 8-15). Polyadenylation of RNA transcripts is virtually eliminated when the corresponding sequence in the template DNA is mutated to any other sequence except one encoding a closely related sequence (AUUAAA). The unprocessed RNA transcripts produced from such mutant templates do not accumulate in nuclei but are rapidly degraded. Further mutagenesis studies revealed that a second signal downstream from the cleavage site is required for efficient cleavage and polyadenylation of most pre-mRNAs in animal cells. This downstream signal is not a specific sequence but rather a GU-rich or simply a U-rich region within ~50 nucleotides of the cleavage site.

Identification and purification of the proteins required for cleavage and polyadenylation of pre-mRNA have led to the model shown in Figure 8-15. A 360-kDa cleavage and polyadenylation specificity factor (CPSF), composed of five different polypeptides, first forms an unstable complex with the upstream AAUAAA poly(A) signal. Then at least three additional proteins bind to the CPSF-RNA complex: a 200-kDa heterotrimer called *cleavage stimulatory factor (CStF),* which interacts with the G/U-rich sequence; a 150-kDa heterotetramer called *cleavage factor I (CFI);* and a second heterodimeric cleavage factor (CFII). An ~150-kDa protein called *symplekin* is thought to form a scaffold on which these cleavage/polyadenylation factors assemble. Finally, *poly(A) polymerase (PAP)* binds to the complex *before* cleavage can occur. This requirement for PAP binding links cleavage and polyadenylation, so that the free 3′ end generated is rapidly polyadenylated and no essential information is lost to exonuclease degradation of an unprotected 3′ end.

Assembly of the large, multiprotein **cleavage/polyadenylation complex** around the AU-rich poly(A) signal in a pre-mRNA is analogous in many ways to formation of the transcription-preinitiation complex at the AT-rich TATA box of a template DNA molecule (see Figure 7-16). In both cases, multiprotein complexes assemble cooperatively through a network of specific protein–nucleic acid and protein-protein interactions.

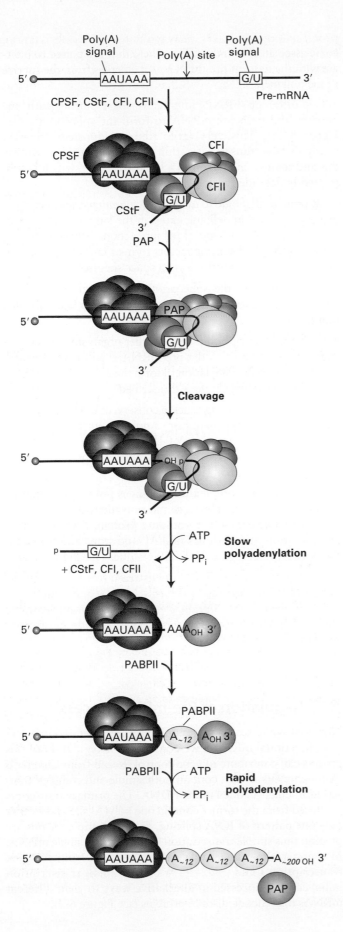

Cleavage and polyadenylation specificity factor (CPSF) binds to the upstream AAUAAA poly(A) signal. CStF interacts with a downstream GU- or U-rich sequence and with bound CPSF, forming a loop in the RNA; binding of CFI and CFII helps stabilize the complex. Binding of poly(A) polymerase (PAP) then stimulates cleavage at a poly(A) site, which usually is 10–35 nucleotides 3′ of the upstream poly(A) signal. The cleavage factors are released, as is the downstream RNA cleavage product, which is rapidly degraded. Bound PAP then adds ~12 A residues at a slow rate to the 3′-hydroxyl group generated by the cleavage reaction. Binding of poly(A)-binding protein II (PABPII) to the initial short poly(A) tail accelerates the rate of addition by PAP. After 200–250 A residues have been added, PABPII signals PAP to stop polymerization.

Following cleavage at the poly(A) site, polyadenylation proceeds in two phases. Addition of the first 12 or so A residues occurs slowly, followed by rapid addition of up to 200–250 more A residues. The rapid phase requires the binding of multiple copies of a *poly(A)-binding protein* containing the RRM motif. This protein is designated *PABPII* to distinguish it from the poly(A)-binding protein present in the cytoplasm. PABPII binds to the short A tail initially added by PAP, stimulating the rate of polymerization of additional A residues by PAP, resulting in the fast phase of polyadenylation (Figure 8-15). PABPII is also responsible for signaling poly(A) polymerase to terminate polymerization when the poly(A) tail reaches a length of 200–250 residues, although the mechanism for controlling the length of the tail is not yet understood. Binding of PABPII to the poly(A) tail is essential for mRNA export into the cytoplasm.

## Nuclear Exonucleases Degrade RNA That Is Processed Out of Pre-mRNAs

Because the human genome contains long introns, only ~5 percent of the nucleotides that are polymerized by RNA polymerase II during transcription are retained in mature, processed mRNAs. Although this process appears inefficient, it probably evolved in multicellular organisms because the process of exon shuffling facilitated the evolution of new genes in organisms with long introns (Chapter 6). The introns that are spliced out and the region downstream from the cleavage and polyadenylation site are degraded by nuclear exoribonucleases that hydrolyze one base at a time from either the 5′ or 3′ end of an RNA strand.

As mentioned earlier, the 2′,5′-phosphodiester bond in excised introns is hydrolyzed by a debranching enzyme, yielding a linear molecule with unprotected ends that can be attacked by exonucleases (see Figure 8-11). The predominant nuclear decay pathway is 3′→5′ hydrolysis by 11 exonucleases that associate with one another in a large protein complex called the **exosome**. Other proteins in the complex include RNA helicases that disrupt base pairing and RNA-protein interactions that would otherwise impede the exonucleases. Exosomes also function in the cytoplasm, as discussed later. In addition to introns, the exosome appears to degrade pre-mRNAs that have not been properly spliced or polyadenylated. It is not yet clear how the exosome recognizes improperly processed pre-mRNAs. But in yeast cells with temperature-sensitive mutant poly(A) polymerase (Figure 8-15), pre-mRNAs are retained at their sites of transcription in the nucleus at the nonpermissive temperature. These abnormally processed pre-mRNAs are released in cells with a second mutation in a subunit of the exosome found only in nuclear and not in cytoplasmic exosomes (PM-Scl; 100 kD in humans). Also, exosomes are found concentrated at sites of transcription in *Drosophila* polytene chromosomes, where they are associated with RNA polymerase II elongation factors. These results suggest that the exosome participates in an as yet poorly understood quality-control mechanism that recognizes aberrantly processed pre-mRNAs, preventing their export to the cytoplasm and ultimately leading to their degradation.

To avoid being degraded by nuclear exonucleases, nascent transcripts, pre-mRNA processing intermediates, and mature mRNAs in the nucleus must have their ends protected. As discussed above, the 5' end of a nascent transcript is protected by addition of the 5' cap structure as soon as the 5' end emerges from the polymerase. The 5' cap is protected because it is bound by a *nuclear cap-binding complex*, which protects it from 5' exonucleases and also functions in export of mRNA to the cytoplasm. The 3' end of a nascent transcript lies within the RNA polymerase and thus is inaccessible to exonucleases (see Figure 4-12). As discussed previously, the free 3' end generated by cleavage of a pre-mRNA downstream from the poly(A) signal is rapidly polyadenylated by the poly(A) polymerase associated with the other 3' processing factors, and the resulting poly(A) tail is bound by PABPII (Figure 8-15). This tight coupling of cleavage and polyadenylation protects the 3' end from exonuclease attack.

## KEY CONCEPTS of Section 8.1

### Processing of Eukaryotic Pre-mRNA

• In the nucleus of eukaryotic cells, pre-mRNAs are associated with hnRNP proteins and processed by 5' capping, 3' cleavage and polyadenylation, and splicing before being transported to the cytoplasm (see Figure 8-2).

• Shortly after transcription initiation, capping enzymes associated with the carboxyl-terminal domain (CTD) of RNA polymerase II, which is phosphorylated multiple times at serine 5 of the heptapeptide repeat by TFIIH during transcription initiation, add the 5' cap to the nascent transcript. Other RNA-processing factors involved in RNA splicing, 3' cleavage, and polyadenylation associate with the CTD when it is phosphorylated at serine 2 of the heptapeptide repeat, increasing the rate of transcription elongation. Consequently, transcription does not proceed at a high rate until RNA-processing factors become associated with the CTD, where they are poised to interact with the nascent pre-mRNA as it emerges from the surface of the polymerase.

• Five different snRNPs interact via base pairing with one another and with pre-mRNA to form the spliceosome (see Figure 8-11). This very large ribonucleoprotein complex catalyzes two transesterification reactions that join two exons and remove the intron as a lariat structure, which is subsequently degraded (see Figure 8-8).

• SR proteins that bind to *exonic splicing enhancer* sequences in exons are critical in defining exons in the large pre-mRNAs of higher organisms. A network of interactions between SR proteins, snRNPs, and splicing factors forms a cross-exon recognition complex that specifies correct splice sites (see Figure 8-13).

• The snRNAs in the spliceosome are thought to have an overall tertiary structure similar to that of group II self-splicing introns.

• For long transcription units in higher organisms, splicing of exons usually begins as the pre-mRNA is still being formed. Cleavage and polyadenylation to form the 3' end of the mRNA occur after the poly(A) site is transcribed.

• In most protein-coding genes, a conserved AAUAAA poly(A) signal lies slightly upstream from a poly(A) site where cleavage and polyadenylation occur. A GU- or U-rich sequence downstream from the poly(A) site contributes to the efficiency of cleavage and polyadenylation.

• A multiprotein complex that includes poly(A) polymerase (PAP) carries out the cleavage and polyadenylation of a pre-mRNA. A nuclear poly(A)-binding protein, PABPII, stimulates addition of A residues by PAP and stops addition once the poly(A) tail reaches 200–250 residues (see Figure 8-15).

• Excised introns and RNA downstream from the cleavage/poly(A) site are degraded primarily by exosomes, multiprotein complexes that contain eleven 3'→5' exonucleases as well as RNA helicases. Exosomes also degrade improperly processed pre-mRNAs.

## 8.2 Regulation of Pre-mRNA Processing

Now that we've seen how pre-mRNAs are processed into mature, functional mRNAs, we consider how regulation of this process can contribute to gene control. Recall from Chapter 6 that higher eukaryotes contain both simple and complex transcription units encoded in their DNA. The primary transcripts produced from the former contain one poly(A) site and exhibit only one pattern of RNA splicing, even if multiple introns are present; thus simple transcription units encode a single mRNA. In contrast, the primary transcripts produced from complex transcription units (~92–94% of all human transcription units) can be processed in alternative ways to yield different mRNAs that encode distinct proteins (see Figure 6-3).

## Alternative Splicing Generates Transcripts with Different Combinations of Exons

The discovery that a large fraction of transcription units in higher organisms encode alternatively spliced mRNAs and that differently spliced mRNAs are expressed in different cell types revealed that regulation of RNA splicing is an important gene-control mechanism in higher eukaryotes. Although many examples of cleavage at alternative poly(A) sites in pre-mRNAs are known, **alternative splicing** of different exons is the more common mechanism for expressing different proteins from one complex transcription unit. In Chapter 4, for example, we mentioned that fibroblasts produce one type of the extracellular protein fibronectin, whereas hepatocytes produce another type. Both fibronectin **isoforms** are encoded by the same transcription unit, which is spliced differently in the two cell types to yield two different mRNAs (see Figure 4-16). In other cases, alternative processing may occur simultaneously in the same cell type in response to different developmental or environmental signals. First we discuss one of the best-understood examples of regulated RNA processing and then we briefly consider the consequences of RNA splicing in the development of the nervous system.

## A Cascade of Regulated RNA Splicing Controls *Drosophila* Sexual Differentiation

One of the earliest examples of regulated alternative splicing of pre-mRNA came from studies of sexual differentiation in *Drosophila*. Genes required for normal *Drosophila* sexual differentiation were first characterized by isolating *Drosophila* mutants defective in the process. When the proteins encoded by the wild-type genes were characterized biochemically, two of them were found to regulate a cascade of alternative RNA splicing in *Drosophila* embryos. More recent research has provided insight into how these proteins regulate RNA processing and ultimately lead to the creation of two different sex-specific transcriptional repressors that suppress the development of characteristics of the opposite sex.

The Sxl protein, encoded by the *sex-lethal* gene, is the first protein to act in the cascade (Figure 8-16). The Sxl protein is present only in female embryos. Early in development, the gene is transcribed from a promoter that functions only in female embryos. Later in development, this female-specific promoter is shut off and another promoter for *sex-lethal* becomes active in both male and female embryos. However, in the absence of early Sxl protein, the *sex-lethal* pre-mRNA in

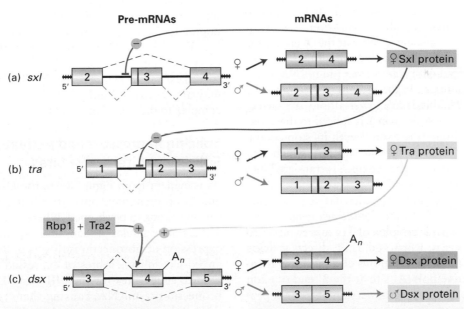

**FIGURE 8-16 Cascade of regulated splicing that controls sex determination in *Drosophila* embryos.** For clarity, only the exons (boxes) and introns (black lines) where regulated splicing occurs are shown. Splicing is indicated by red dashed lines above (female) and blue dashed lines below (male) the pre-mRNAs. Vertical red lines in exons indicate in-frame stop codons, which prevent synthesis of functional protein. Only female embryos produce functional Sxl protein, which *represses* splicing between exons 2 and 3 in *sxl* pre-mRNA (a) and between exons 1 and 2 in *tra* pre-mRNA (b). (c) In contrast, the cooperative binding of Tra protein and two SR proteins, Rbp1 and Tra2, *activates* splicing between exons 3 and 4 and cleavage/polyadenylation($A_n$) at the 3′ end of exon 4 in *dsx* pre-mRNA in female embryos. In male embryos, which lack functional Tra, the SR proteins do not bind to exon 4, and consequently exon 3 is spliced to exon 5. The distinct Dsx proteins produced in female and male embryos as the result of this cascade of regulated splicing repress transcription of genes required for sexual differentiation of the opposite sex. [Adapted from M. J. Moore et al., 1993, in R. Gesteland and J. Atkins, eds., *The RNA World*, Cold Spring Harbor Press, pp. 303–357.]

male embryos is spliced to produce an mRNA that contains a stop codon early in the sequence. The net result is that male embryos produce no functional Sxl protein either early or later in development.

In contrast, the Sxl protein expressed in early female embryos directs splicing of the *sex-lethal* pre-mRNA so that a functional *sex-lethal* mRNA is produced (Figure 8-16a). Sxl accomplishes this by binding to a sequence in the pre-mRNA near the 3′ end of the intron between exon 2 and exon 3, thereby blocking the proper association of U2AF and U2 snRNP (Figure 8-11). As a consequence, the U1 snRNP bound to the 3′ end of exon 2 assembles into a spliceosome with U2 snRNP bound to the branch point at the 3′ end of the intron between exons 3 and 4, leading to splicing of exon 2 to exon 4 and skipping of exon 3. The binding site for Sxl in the Sxl pre-mRNA is called an *intronic splicing silencer* because of its location in an intron and its function in blocking, or "silencing," use of a splice site. The resulting female-specific *sex-lethal* mRNA is translated into functional Sxl protein, which reinforces its own expression in female embryos by continuing to cause skipping of exon 3. The absence of Sxl protein in male embryos allows the inclusion of exon 3 and, consequently, of the stop codon that prevents translation of functional Sxl protein.

Sxl protein also regulates alternative RNA splicing of the *transformer* gene pre-mRNA (Figure 8-16b). In male embryos, where no Sxl is expressed, exon 1 is spliced to exon 2, which contains a stop codon that prevents synthesis of a functional *transformer* protein. In female embryos, however, binding of Sxl protein to an intronic splicing silencer at the 3′ end of the intron between exons 1 and 2 blocks binding of U2AF at this site. The interaction of Sxl with *transformer* pre-mRNA is mediated by two RRM domains in the protein (see Figure 8-5). When Sxl is bound, U2AF binds to a lower-affinity site farther 3′ in the pre-mRNA; as a result exon 1 is spliced to this alternative 3′ splice site, eliminating exon 2 with its stop codon. The resulting female-specific *transformer* mRNA, which contains additional constitutively spliced exons, is translated into functional Transformer (Tra) protein.

Finally, Tra protein regulates the alternative processing of pre-mRNA transcribed from the *double-sex* gene (Figure 8-16c). In female embryos, a complex of Tra and two constitutively expressed proteins, Rbp1 and Tra2, directs splicing of exon 3 to exon 4 and also promotes cleavage/polyadenylation at the alternative poly(A) site at the 3′ end of exon 4—leading to a short, female-specific version of the Dsx protein. In male embryos, which produce no Tra protein, exon 4 is skipped, so that exon 3 is spliced to exon 5. Exon 5 is constitutively spliced to exon 6, which is polyadenylated at its 3′ end—leading to a longer, male-specific version of the Dsx protein. The RNA sequence to which Tra binds in exon 4 is called an *exonic splicing enhancer* since it enhances splicing at a nearby splice site.

As a result of the cascade of regulated RNA processing depicted in Figure 8-16, different Dsx proteins are expressed in male and female embryos. The male Dsx protein is a transcriptional repressor that inhibits the expression of genes required

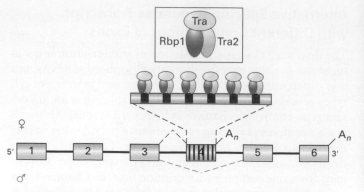

**FIGURE 8-17 Model of splicing activation by Tra protein and the SR proteins Rbp1 and Tra2.** In female *Drosophila* embryos, splicing of exons 3 and 4 in *dsx* pre-mRNA is activated by binding of Tra/Tra2/Rbp1 complexes to six sites in exon 4. Because Rbp1 and Tra2 cannot bind to the pre-mRNA in the absence of Tra, exon 4 is skipped in male embryos. See the text for discussion. $A_n$ = polyadenylation. [Adapted from T. Maniatis and B. Tasic, 2002, *Nature* **418**:236.]

for female development. Conversely, the female Dsx protein represses transcription of genes required for male development.

Figure 8-17 illustrates how the Tra/Tra2/Rbp1 complex is thought to interact with *double-sex (dsx)* pre-mRNA. Rbp1 and Tra2 are SR proteins, but they do not interact with exon 4 in the absence of the Tra protein. Tra protein interacts with Rbp1 and Tra2, resulting in the cooperative binding of all three proteins to six exonic splicing enhancers in exon 4. The bound Tra2 and Rbp1 proteins then promote the binding of U2AF and U2 snRNP to the 3′ end of the intron between exons 3 and 4, just as other SR proteins do for constitutively spliced exons (see Figure 8-13). The Tra/Tra2/Rbp1 complexes may also enhance binding of the cleavage/polyadenylation complex to the 3′ end of exon 4.

## Splicing Repressors and Activators Control Splicing at Alternative Sites

As is evident from Figure 8-16, the *Drosophila* Sxl protein and Tra protein have opposite effects: Sxl prevents splicing, causing exons to be skipped, whereas Tra promotes splicing. The action of similar proteins may explain the cell-type-specific expression of fibronectin isoforms in humans. For instance, an Sxl-like splicing repressor expressed in hepatocytes might bind to splice sites for the EIIIA and EIIIB exons in the fibronectin pre-mRNA, causing them to be skipped during RNA splicing (see Figure 4-16). Alternatively, a Tra-like splicing activator expressed in fibroblasts might activate the splice sites associated with the fibronectin EIIIA and EIIIB exons, leading to inclusion of these exons in the mature mRNA. Experimental examination in some systems has revealed that inclusion of an exon in some cell types versus skipping of the same exon in other cell types results from the combined influence of several splicing repressors and enhancers. RNA binding sites for repressors, usually hnRNP proteins, can also occur in exons, where they are called *exonic splicing silencers*. And binding sites for splicing activators, usually SR proteins, can also occur in introns, where they are called *intronic splicing enhancers*.

Alternative splicing of exons is especially common in the nervous system, generating multiple isoforms of many proteins required for neuronal development and function in both vertebrates and invertebrates. The primary transcripts from these genes often show complex splicing patterns that can generate several different mRNAs, with different spliced forms expressed in different anatomical locations within the central nervous system. We consider two remarkable examples that illustrate the critical role of this process in neural function.

### Expression of K⁺-Channel Proteins in Vertebrate Hair Cells

**Expression of $K^+$-Channel Proteins in Vertebrate Hair Cells** In the inner ear of vertebrates, individual "hair cells," which are ciliated neurons, respond most strongly to a specific frequency of sound. Cells tuned to low frequency ($\sim 50$ Hz) are found at one end of the tubular cochlea that makes up the inner ear; cells responding to high frequency ($\sim 5000$ Hz) are found at the other end (Figure 8-18a). Cells in between the ends respond to a gradient of frequencies between these extremes. One component in the tuning of hair cells in reptiles and birds is the opening of $K^+$ ion channels in response to increased intracellular $Ca^{2+}$ concentrations. The $Ca^{2+}$ concentration at which the channel opens determines the frequency with which the membrane potential oscillates and hence the frequency to which the cell is tuned.

The gene encoding this $Ca^{2+}$-activated $K^+$ channel is expressed as multiple, alternatively spliced mRNAs. The various proteins encoded by these alternative mRNAs open at different $Ca^{2+}$ concentrations. Hair cells with different response frequencies express different isoforms of the channel protein depending on their position along the length of the cochlea. The sequence variation in the protein is very complex: there are at least eight regions in the mRNA where alternative exons are utilized, permitting the expression of 576 possible isoforms (Figure 8-18b). PCR analysis of mRNAs from individual hair cells has shown that each hair cell expresses a mixture of different alternative $Ca^{2+}$-activated $K^+$-channel mRNAs, with different forms predominating in different cells according to their position along the cochlea. This remarkable arrangement suggests that splicing of the $Ca^{2+}$-activated $K^+$-channel pre-mRNA is regulated in response to extracellular signals that inform the cell of its position along the cochlea.

Other studies demonstrated that splicing at one of the alternative splice sites in the $Ca^{2+}$-activated $K^+$-channel pre-mRNA in the rat is suppressed when a specific protein kinase is activated by neuron depolarization in response to synaptic activity from interacting neurons. This observation raises the possibility that a splicing repressor specific for this site may be activated when it is phosphorylated by this protein kinase, whose activity in turn is regulated by synaptic activity. Since hnRNP and SR proteins are extensively modified by phosphorylation and other post-translational modifications, it seems likely that complex regulation of alternative RNA splicing through post-translational modifications of splicing factors plays a significant role in modulating neuron function.

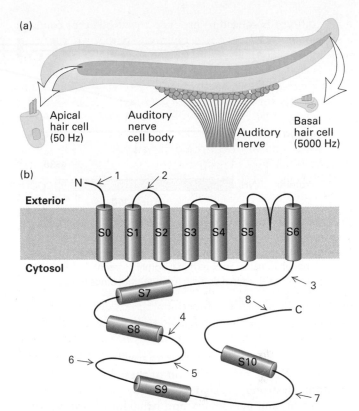

**FIGURE 8-18 Role of alternative splicing in the perception of sounds of different frequency.** (a) The chicken cochlea, a 5-mm-long tube, contains an epithelium of auditory hair cells that are tuned to a gradient of vibrational frequencies from 50 Hz at the apical end (*left*) to 5000 Hz at the basal end (*right*). (b) The $Ca^{2+}$-activated $K^+$ channel contains seven transmembrane α helices (S0–S6), which associate to form the channel. The cytosolic domain, which includes four hydrophobic regions (S7–S10), regulates opening of the channel in response to $Ca^{2+}$. Isoforms of the channel, encoded by alternatively spliced mRNAs produced from the same primary transcript, open at different $Ca^{2+}$ concentrations and thus respond to different frequencies. Red numbers refer to regions where alternative splicing produces different amino acid sequences in the various isoforms. [Adapted from K. P. Rosenblatt et al., 1997, *Neuron* **19**:1061.]

Many examples of genes similar to those that encode the cochlear $K^+$-channel have been observed in vertebrate neurons; alternatively spliced mRNAs co-expressed from a specific gene in one type of neuron are expressed at different relative concentrations in different regions of the central nervous system. Expansions in the number of microsatellite repeats within the transcribed regions of genes expressed in neurons can cause an alteration in the relative concentrations of alternatively spliced mRNAs transcribed from multiple genes. In Chapter 6, we discussed how backward slippage during DNA replication can lead to expansion of a microsatellite repeat (see Figure 6-5). At least 14 different types of neurological disease result from expansion of microsatellite regions within transcription units expressed in neurons. The resulting long regions of repeated simple sequences in nuclear RNAs of these neurons result in abnormalities in

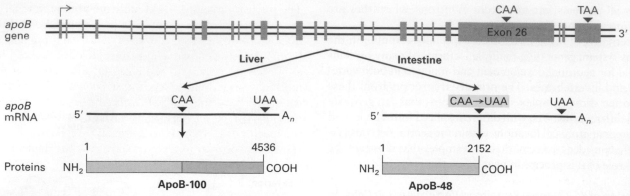

**FIGURE 8-19 RNA editing of *apo-B* pre-mRNA.** The *apoB* mRNA produced in the liver has the same sequence as the exons in the primary transcript. This mRNA is translated into apoB-100, which has two functional domains: an N-terminal domain (green) that associates with lipids and a C-terminal domain (orange) that binds to LDL receptors on cell membranes. In the *apo-B* mRNA produced in the intestine, the CAA codon in exon 26 is edited to a UAA stop codon. As a result, intestinal cells produce apoB-48, which corresponds to the N-terminal domain of apoB-100. [Adapted from P. Hodges and J. Scott, 1992, *Trends Biochem. Sci.* **17**:77.]

the relative concentrations of alternatively spliced mRNAs. For example, the most common of these types of diseases, myotonic dystrophy, is characterized by paralysis, cognitive impairment, and personality and behavior disorders. Myotonic dystrophy results from increased copies of either CUG repeats in one transcript, in some patients, or CCUG repeats in another transcript, in other patients. When the number of these repeats increases to 10 or more times the normal number of repeats in these genes, abnormalities are observed in the level of two hnRNP proteins that bind to these repeated sequences. This probably results because these hnRNPs are bound by the abnormally high concentration of this RNA sequence in the nuclei of neurons in such patients. The abnormal concentrations of these hnRNP proteins are thought to lead to alterations in the rate of splicing of different alternative splice sites in multiple pre-mRNAs normally regulated by these hnRNP proteins. ■

**Expression of *Dscam* Isoforms in *Drosophila* Retinal Neurons**
The most extreme example of regulated alternative RNA processing yet uncovered occurs in expression of the *Dscam* gene in *Drosophila*. Mutations in this gene interfere with the normal synaptic connections made between axons and dendrites during fly development. Analysis of the *Dscam* gene showed that it contains 95 alternatively spliced exons that could be spliced to generate 38,016 possible isoforms! Recent results have shown that *Drosophila* mutants with a version of the gene that can be spliced in only about 22,000 different ways have specific defects in connectivity between neurons. These results indicate that expression of most of the possible *Dscam* isoforms through regulated RNA splicing helps to specify the tens of millions of different specific synaptic connections between neurons in the *Drosophila* brain. In other words, the correct wiring of neurons in the brain requires regulated RNA splicing.

## RNA Editing Alters the Sequences of Some Pre-mRNAs

In the mid-1980s, sequencing of numerous cDNA clones and corresponding genomic DNAs from multiple organisms led to the unexpected discovery of another type of pre-mRNA processing. In this type of processing, called **RNA editing**, the sequence of a pre-mRNA is altered; as a result, the sequence of the corresponding mature mRNA differs from the exons encoding it in genomic DNA.

RNA editing is widespread in the mitochondria of protozoans and plants and also in chloroplasts. In the mitochondria of certain pathogenic trypanosomes, more than half the sequence of some mRNAs is altered from the sequence of the corresponding primary transcripts. Additions and deletions of specific numbers of U's follows templates provided by base-paired short "guide" RNAs. These RNAs are encoded by thousands of mini-mitochondrial DNA circles catenated to many fewer large mitochondrial DNA molecules. The reason for this baroque mechanism for encoding mitochondrial proteins in such protozoans is not clear. But this system does represent a potential target for drugs to inhibit the complex processing enzymes essential to the microbe that do not exist in the cells of their human or other vertebrate hosts.

In higher eukaryotes, RNA editing is much rarer, and thus far, only single-base changes have been observed. Such minor editing, however, turns out to have significant functional consequences in some cases. An important example of RNA editing in mammals involves the *apoB* gene. This gene encodes two alternative forms of a serum protein central to the uptake and transport of cholesterol. Consequently, it is important in the pathogenic processes that lead to *atherosclerosis*, the arterial disease that is the major cause of death in the developed world. The *apoB* gene expresses both the serum protein apolipoprotein B-100 (apoB-100) in hepatocytes, the major cell type in the liver, and apoB-48, expressed in intestinal epithelial cells. The ~240-kDa apoB-48 corresponds to the N-terminal region of the ~500-kDa apoB-100. As we detail in Chapter 10, both apoB proteins are components of large lipoprotein complexes that transport lipids in the serum. However, only low-density lipoprotein (LDL) complexes, which contain apoB-100 on their surface, deliver cholesterol to body tissues by binding to the LDL receptor present on all cells.

The cell-type-specific expression of the two forms of apoB results from editing of *apoB* pre-mRNA so as to change the nucleotide at position 6666 in the sequence from a C to a U. This alteration, which occurs only in intestinal cells, converts a CAA codon for glutamine to a UAA stop codon, leading to synthesis of the shorter apoB-48 (Figure 8-19). Studies with the partially purified enzyme that performs the post-transcriptional deamination of $C_{6666}$ to U shows that it can recognize and edit an RNA as short as 26 nucleotides with the sequence surrounding $C_{6666}$ in the *apoB* primary transcript.

## KEY CONCEPTS of Section 8.2

### Regulation of Pre-mRNA Processing

- Because of alternative splicing of primary transcripts, the use of alternative promoters, and cleavage at different poly(A) sites, different mRNAs may be expressed from the same gene in different cell types or at different developmental stages (see Figure 6-3 and Figure 8-16).

- Alternative splicing can be regulated by RNA-binding proteins that bind to specific sequences near regulated splice sites. Splicing repressors may sterically block the binding of splicing factors to specific sites in pre-mRNAs or inhibit their function. Splicing activators enhance splicing by interacting with splicing factors, thus promoting their association with a regulated splice site. The RNA sequences bound by splicing repressors are called intronic or exonic splicing silencers, depending on their location in an intron or exon. RNA sequences bound by splicing activators are called intronic or exonic splicing enhancers.

- In RNA editing the nucleotide sequence of a pre-mRNA is altered in the nucleus. In vertebrates, this process is fairly rare and entails deamination of a single base in the mRNA sequence, resulting in a change in the amino acid specified by the corresponding codon and production of a functionally different protein (see Figure 8-19).

## 8.3 Transport of mRNA Across the Nuclear Envelope

Fully processed mRNAs in the nucleus remain bound by hnRNP proteins in complexes referred to as *nuclear mRNPs*. Before an mRNA can be translated into its encoded protein, it must be exported out of the nucleus into the cytoplasm.

The **nuclear envelope** is a double membrane that separates the nucleus from the cytoplasm (see Figure 9-32). Like the plasma membrane surrounding cells, each nuclear membrane consists of a water-impermeable phospholipid bilayer and multiple associated proteins. mRNPs and other macromolecules including tRNAs and ribosomal subunits traverse the nuclear envelope through *nuclear pores*. This section will focus on the export of mRNPs through the nuclear pore and the mechanisms that allow some level of regulation of this step. Transport of other cargoes across the nuclear pore is discussed in Chapter 13.

### Macromolecules Exit and Enter the Nucleus Through Nuclear Pore Complexes

**Nuclear pore complexes (NPCs)** are large, symmetrical structures composed of multiple copies of approximately 30 different proteins called **nucleoporins**. Embedded in the nuclear envelope, NPCs are cylindrical in shape with a diameter of ~30 nm (Figure 8-20a). A special class of nucleoporins called **FG-nucleoporins** line the central channel through the NPC. FG-nucleoporins form a semi-permeable barrier that allows small molecules to diffuse freely, but restricts the passage of larger molecules. The FG-nucleoporin globular domain anchors the protein in the NPC scaffold. From these anchors, long random-coil stretches extend into the channel. These extensions are composed of hydrophilic amino acid sequences punctuated by hydrophobic *FG-repeats,* short sequences rich in hydrophobic phenylalanine (F) and glycine (G). These FG-repeat domains form a cloud of polypeptide chains in the central channel and extending into the nucleoplasm and cytoplasm, effectively limiting the free diffusion of macromolecules across the channel. Water, ions, metabolites, and small globular proteins up to ~40–60 kDa can diffuse through the cloud of FG-repeat domains. However, the FG-domains in the central channel form a barrier restricting the diffusion of larger macromolecules between the cytoplasm and nucleus.

Proteins and RNPs larger than ~40–60 kDa must be selectively transported across the nuclear envelope with the assistance of soluble transporter proteins that bind them and also interact reversibly with the FG-repeats of FG-nucleoporins. As a consequence of these reversible interactions with FG-domains, the transporter and its stably bound cargo can be passed from FG-domain to FG-domain, allowing them both to diffuse down a concentration gradient from the nucleus to the cytoplasm.

mRNPs are transported through the NPC by the *mRNP exporter*, a heterodimer consisting of a large subunit, called *n*uclear *e*xport *f*actor 1 (NXF1), and a small subunit, *n*uclear *e*xport *t*ransporter 1 (NXT1) (Figure 8-20b). NXF1 binds nuclear mRNPs through associations with both RNA and other proteins in the mRNP complex. One of the most important of these is REF (*R*NA *e*xport *f*actor), a component of the exon-junction complexes discussed earlier, which is bound approximately 20 nucleotides 5′ to each exon-exon junction (see Figure 8-21). The NXF1/NXT1 mRNP exporter also associates with SR proteins bound to exonic splicing

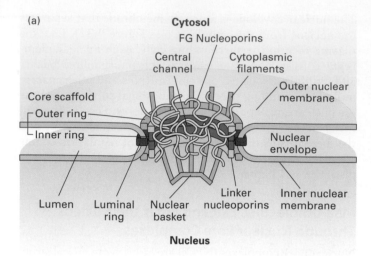

(a)

Cytosol

FG Nucleoporins

Central channel

Cytoplasmic filaments

Outer nuclear membrane

Core scaffold
Outer ring
Inner ring

Nuclear envelope

Lumen  Luminal ring  Nuclear basket  Linker nucleoporins  Inner nuclear membrane

Nucleus

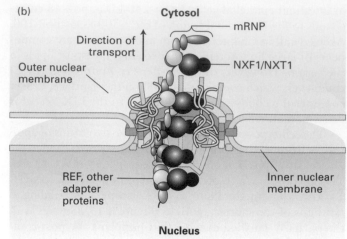

(b)

Cytosol

Direction of transport

mRNP

NXF1/NXT1

Outer nuclear membrane

REF, other adapter proteins

Inner nuclear membrane

Nucleus

**FIGURE 8-20 Model of transporter passage through an NPC.**
(a) Diagram of NPC structure. The nuclear envelope lipid bilayers are represented in green. Transmembrane proteins represented in red form a luminal ring around which the nuclear envelope membrane folds to form the inner and outer nuclear membranes. Transmembrane domains of these proteins are connected to globular domains on the inside of the pore to which the other nucleoporins bind, forming the core scaffold. The globular domains of the FG-nucleoporins are shown in purple. The FG-domains of FG-nucleoporins (tan worms) have an extended, random-coil conformation that forms a molecular cloud of continuously moving random-coil polypeptide. (b) Nuclear transporters (NXF1/NXT1) have hydrophobic regions on their surface that bind reversibly to the FG-domains in the FG-nucleoporins. As a consequence, they can penetrate the molecular cloud in the NPC central channel and diffuse in and out of the nucleus. [Adapted from D. Grünwald, R. H. Singer, and M. Rout, 2011, *Nature* **475**:333.]

enhancers. Thus SR proteins associated with exons function to direct both the splicing of pre-mRNAs and the export of fully processed mRNAs through NPCs to the cytoplasm. mRNPs are probably bound along their length by multiple NXF1/NXT1 mRNP exporters, which interact with the FG-domains of FG-nucleoporins to facilitate export of mRNPs through the NPC central channel (Figure 8-20b).

Protein filaments extend from the core scaffold into the nucleoplasm forming a "nuclear basket" (see Figure 8-20a). Protein filaments also extend into the cytoplasm. These filaments assist in mRNP export. Gle2, an adapter protein that reversibly binds both NXF1 and a nucleoporin in the nuclear basket, brings nuclear mRNPs to the pore in preparation for export. A nucleoporin in the cytoplasmic filaments of the NPC

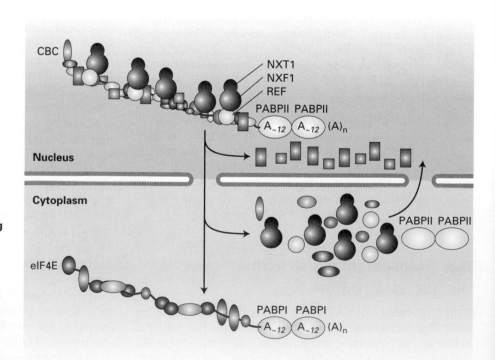

CBC

NXT1
NXF1
REF

PABPII  PABPII
A~12  A~12  (A)n

Nucleus

Cytoplasm

eIF4E

PABPII  PABPII

PABPI  PABPI
A~12  A~12  (A)n

**FIGURE 8-21 Remodeling of mRNPs during nuclear export.** Some mRNP proteins (rectangles) dissociate from nuclear mRNP complexes before export through an NPC. Some (ovals) are exported through the NPC associated with the mRNP but dissociate in the cytoplasm and are shuttled back into the nucleus through an NPC. In the cytoplasm, translation initiation factor eIF4E replaces CBC bound to the 5′ cap and PABPI replaces PABPII.

binds an RNA helicase (Dbp5) that functions in the dissociation of NXF1/NXT1 and other hnRNP proteins from the mRNP as it reaches the cytoplasm.

In a process called *mRNP remodeling,* the proteins associated with an mRNA in the nuclear mRNP complex are exchanged for a different set of proteins as the mRNP is transported through the NPC (Figure 8-21). Some nuclear mRNP proteins dissociate early in transport, remaining in the nucleus to bind to newly synthesized nascent pre-mRNA. Other nuclear mRNP proteins remain with the mRNP complex as it traverses the pore and do not dissociate from the mRNP until the complex reaches the cytoplasm. Proteins in this category include the NXF1/NXT1 mRNP exporter, cap-binding complex (CBC) bound to the 5′ cap, and PABPII bound to the poly(A) tail. They dissociate from the mRNP on the cytoplasmic side of the NPC through the action of the Dbp5 RNA helicase that associates with cytoplasmic NPC filaments, as discussed above. These proteins are then imported back into the nucleus as discussed for other nuclear proteins in Chapter 13, where they can function in the export of another mRNP. In the cytoplasm, the cap-binding translation initiation factor eIF4E replaces CBC bound to the 5′ cap of nuclear mRNPs (Figure 4-24). In vertebrates, the nuclear poly(A)-binding protein PABPII is replaced with the cytoplasmic poly(A)-binding protein PABPI (so named because it was discovered before PABPII). Only a single PABP is found in budding yeast, in both the nucleus and the cytoplasm.

**Yeast SR Protein** Studies of *S. cerevisiae* indicate that the direction of mRNP export from the nucleus into the cytoplasm is controlled by phosphorylation and dephosphorylation of mRNP adapter proteins such as REF that assist in the binding of the NXF1/NXT1 exporter to mRNPs. In one case, a yeast SR protein (Npl3) functions as an adapter protein that promotes the binding of the yeast mRNP exporter (Figure 8-22). The SR protein initially binds to nascent pre-mRNA in its phosphorylated form. When 3′ cleavage and polyadenylation are completed, the adapter protein is dephosphorylated by a specific nuclear protein phosphatase essential for mRNP export. Only the dephosphorylated adapter protein can bind the mRNP exporter, thereby coupling mRNP export to correct polyadenylation. This is one form of mRNA "quality control." If the nascent mRNP is not correctly processed, it is not recognized by the phosphatase that dephosphorylates Npl3. Consequently, it is not bound by the mRNA exporter and not exported from the nucleus. Instead it is degraded by exosomes, the multiprotein complexes that degrade unprotected RNAs in the nucleus and cytoplasm (see Figure 8-1).

Following export to the cytoplasm, the Npl3 SR protein is phosphorylated by a specific cytoplasmic protein kinase. This causes it to dissociate from the mRNP, along with the mRNP exporter. In this way, dephosphorylation of adapter mRNP proteins in the nucleus once RNA processing is complete and their phosphorylation and resulting dissociation in the cytoplasm results in a higher concentration of mRNP exporter–mRNP complexes in the nucleus, where they form, and a lower concentration of these complexes in the cyto-

plasm, where they dissociate. As a result, the direction of mRNP export may be driven by simple diffusion down a concentration gradient of the transport-competent mRNP exporter–mRNP complex across the NPC, from high in the nucleus to low in the cytoplasm.

**Nuclear Export of Balbiani Ring mRNPs** The larval salivary glands of the insect *Chironomous tentans* provide a good model system for electron microscopy studies of the formation of hnRNPs and their export through NPCs. In these larvae, genes in large chromosomal puffs called Balbiani rings are abundantly transcribed into nascent pre-mRNAs that associate with hnRNP proteins and are processed into coiled mRNPs with a final mRNA length of ~75 kb (Figure 8-23a, b). These giant mRNAs encode large glue proteins that adhere the developing larvae to a leaf. After processing of the pre-mRNA in Balbiani ring hnRNPs, the resulting mRNPs move through nuclear pores to the cytoplasm. Electron micrographs of sections of these cells show mRNPs that appear to uncoil during their passage through nuclear pores and then bind to ribosomes as they enter the cytoplasm. This uncoiling is probably a consequence of the remodeling of mRNPs as the result of phosphorylation of mRNP proteins by cytoplasmic kinases and the action of the RNA helicase associated with NPC cytoplasmic filaments, as discussed in the previous section. The observation that mRNPs become associated with ribosomes during transport indicates that the 5′ end leads the way through the nuclear pore complex. Detailed electron microscopic studies of the transport of Balbiani ring mRNPs through nuclear pore complexes led to the model depicted in Figure 8-23c.

## Pre-mRNAs in Spliceosomes Are Not Exported from the Nucleus

It is critical that only fully processed mature mRNAs be exported from the nucleus because translation of incompletely processed pre-mRNAs containing introns would produce defective proteins that might interfere with the functioning of the cell. To prevent this, pre-mRNAs associated with snRNPs in spliceosomes usually are prevented from being transported to the cytoplasm.

In one type of experiment demonstrating this restriction, a gene encoding a pre-mRNA with a single intron that normally is spliced out was mutated to introduce deviations from the consensus splice-site sequences. Mutation of either the 5′ or the 3′ invariant splice-site bases at the ends of the intron resulted in pre-mRNAs that were bound by snRNPs to form spliceosomes; however, RNA splicing was blocked, and the pre-mRNA was retained in the nucleus. In contrast, mutation of *both* the 5′ and 3′ splice sites in the same pre-mRNA resulted in export of the unspliced pre-mRNA, although less efficiently than for the spliced mRNA. When both splice sites were mutated, the pre-mRNAs were not efficiently bound by snRNPs, and, consequently, their export was not blocked.

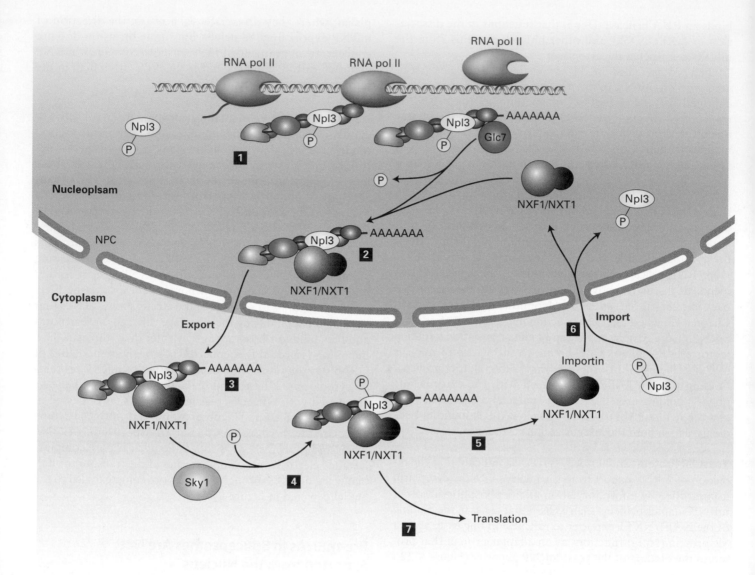

**FIGURE 8-22 Reversible phosphorylation and direction of mRNP nuclear export.** Step **1**: The yeast SR protein Npl3 binds nascent pre-mRNAs in its phosphorylated form. Step **2**: When polyadenylation has occurred successfully, the Glc7 nuclear phosphatase essential for mRNP export dephosphorylates Npl3, promoting the binding of the yeast mRNP exporter, NXF1/NXT1. Step **3**: The mRNP exporter allows diffusion of the mRNP complex through the central channel of the nuclear pore complex (NPC). Step **4**: The cytoplasmic protein kinase Sky1 phosphorylates Npl3 in the cytoplasm, causing **5** dissociation of the mRNP exporter and phosphorylated Npl3, probably through the action of an RNA helicase associated with NPC cytoplasmic filaments. **6** The mRNA transporter and phosphorylated Npl3 are transported back into the nucleus through NPCs. **7** Transported mRNA is available for translation in the cytoplasm. [From E. Izaurralde, 2004, *Nat. Struct. Mol. Biol.* **11**:210–212. See W. Gilbert and C. Guthrie, 2004, *Mol. Cell* **13**:201–212.]

Recent studies in yeast have shown that a nuclear protein that associates with a nucleoporin in the NPC nuclear basket is required to retain pre-mRNAs associated with snRNPs in the nucleus. If either this protein or the nucleoporin to which it binds is deleted, unspliced pre-mRNAs are exported.

Many cases of thalassemia, an inherited disease that results in abnormally low levels of globin proteins, are due to mutations in globin-gene splice sites that decrease the efficiency of splicing but do not prevent association of the pre-mRNA with snRNPs. The resulting unspliced globin pre-mRNAs are retained in reticulocyte nuclei and are rapidly degraded. ■

## HIV Rev Protein Regulates the Transport of Unspliced Viral mRNAs

As discussed earlier, transport of mRNPs containing mature, functional mRNAs from the nucleus to the cytoplasm entails a complex mechanism that is crucial to gene expression (see Figures 8-21, 8-22, and 8-23). Regulation of this transport theoretically could provide another means of gene control, although it appears to be relatively rare. Indeed, the only known examples of regulated mRNA export occur during the cellular response to conditions (e.g., heat shock) that cause protein denaturation or during viral infection when

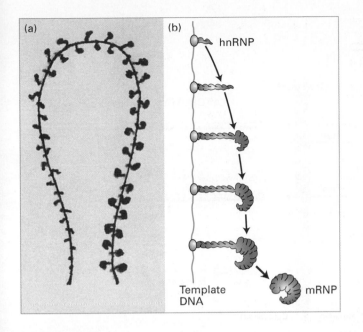

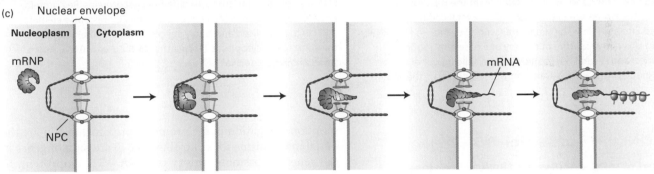

**FIGURE 8-23 Formation of heterogeneous ribonucleoprotein particles (hnRNPs) and export of mRNPs from the nucleus.**
(a) Model of a single chromatin transcription loop and assembly of Balbiani ring (BR) mRNP in *Chironomous tentans*. Nascent RNA transcripts produced from the template DNA rapidly associate with proteins, forming hnRNPs. The gradual increase in size of the hnRNPs reflects the increasing length of RNA transcripts at greater distances from the transcription start site. The model was reconstructed from electron micrographs of serial thin sections of salivary gland cells. (b) Schematic diagram of the biogenesis of hnRNPs. Following processing of the pre-mRNA, the resulting ribonucleoprotein particle is referred to as an mRNP. (c) Model for the transport of BR mRNPs through the nuclear pore complex (NPC) based on electron microscopic studies. Note that the curved mRNPs appear to uncoil as they pass through nuclear pores. As the mRNA enters the cytoplasm, it rapidly associates with ribosomes, indicating that the 5′ end passes through the NPC first. [Part (a) from C. Erricson et al., 1989, *Cell* **56:**631; courtesy of B. Daneholt. Parts (b) and (c) adapted from B. Daneholt, 1997, *Cell* **88:**585. See also B. Daneholt, 2001, *Proc. Nat'l. Acad. Sci. USA* **98:**7012.]

virus-induced alterations in nuclear transport maximize viral replication. Here we describe the regulation of mRNP export mediated by a protein encoded by human immunodeficiency virus (HIV).

A retrovirus, HIV integrates a DNA copy of its RNA genome into the host-cell DNA (see Figure 4-49). The integrated viral DNA, or provirus, contains a single transcription unit, which is transcribed into a single primary transcript by cellular RNA polymerase II. The HIV transcript can be spliced in alternative ways to yield three classes of mRNAs: a 9-kb unspliced mRNA; ~4-kb mRNAs formed by removal of one intron; and ~2-kb mRNAs formed by removal of two or more introns (Figure 8-24). After their synthesis in the host-cell nucleus, all three classes of HIV mRNAs are transported to the cytoplasm and translated into viral proteins; some of the 9-kb unspliced RNA is used as the viral genome in progeny virions that bud from the cell surface.

Since the 9-kb and 4-kb HIV mRNAs contain splice sites, they can be viewed as incompletely spliced mRNAs. As discussed earlier, association of such incompletely spliced mRNAs with snRNPs in spliceosomes normally blocks their export from the nucleus. Thus HIV, as well as other retroviruses, must

have some mechanism for overcoming this block, permitting export of the longer viral mRNAs. Some retroviruses have evolved a sequence called the *constitutive transport element (CTE),* which binds to the NXF1/NXT1 mRNP exporter with high affinity, thereby permitting export of unspliced retroviral RNA into the cytoplasm. HIV solved the problem differently.

Studies with HIV mutants showed that transport of unspliced 9-kb and singly spliced 4-kb viral mRNAs from the nucleus to the cytoplasm requires the virus-encoded Rev protein. Subsequent biochemical experiments demonstrated that Rev binds to a specific Rev-response element (RRE) present in HIV RNA. In cells infected with HIV mutants lacking the RRE, unspliced and singly spliced viral mRNAs remain in the nucleus, demonstrating that the RRE is required for Rev-mediated stimulation of nuclear export. Early in an infection, before any Rev protein is synthesized, only the multiply spliced 2-kb mRNAs can be exported. One of these 2-kb mRNAs encodes Rev, which contains a leucine-rich nuclear export signal that interacts with transporter Exportin1. As discussed in Chapter 13, translation and nuclear import of Rev results in export of the larger unspliced and singly spliced HIV mRNAs through the nuclear pore complex.

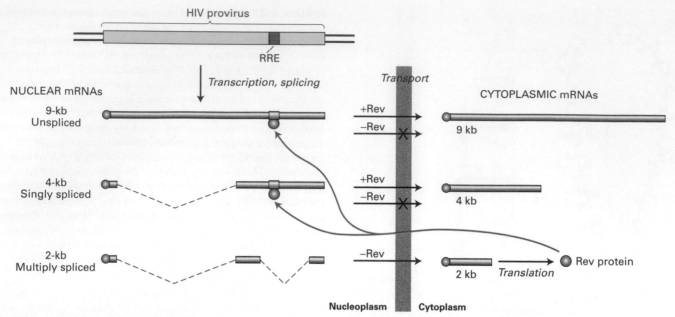

**FIGURE 8-24 Transport of HIV mRNAs from the nucleus to the cytoplasm.** The HIV genome, which contains several coding regions, is transcribed into a single 9-kb primary transcript. Several ~4-kb mRNAs result from alternative splicing out of any one of several introns (dashed lines) and several ~2-kb mRNAs from splicing out of two or more alternative introns. After transport to the cytoplasm, the various RNA species are translated into different viral proteins. Rev protein, encoded by a 2-kb mRNA, interacts with the Rev-response element (RRE) in the unspliced and singly spliced mRNAs, stimulating their transport to the cytoplasm. [Adapted from B. R. Cullen and M. H. Malim, 1991, *Trends Biochem. Sci.* **16:**346.]

## KEY CONCEPTS of Section 8.3

### Transport of mRNA Across the Nuclear Envelope

- Most mRNPs are exported from the nucleus by a heterodimeric mRNP exporter that interacts with FG-repeats of FG-nucleoporins (see Figure 8-20). The direction of transport (nucleus→cytoplasm) may result from dissociation of the exporter-mRNP complex in the cytoplasm by phosphorylation of mRNP proteins by cytoplasmic kinases and the action of an RNA helicase associated with cytoplasmic filaments of the nuclear pore complexes.

- The mRNP exporter binds to most mRNAs cooperatively with SR proteins bound to exons and with REF associated with the exon-junction complexes that bind to mRNAs following RNA splicing, and also binds to additional mRNP proteins.

- Pre-mRNAs bound by a spliceosome normally are not exported from the nucleus, ensuring that only fully processed, functional mRNAs reach the cytoplasm for translation.

## 8.4 Cytoplasmic Mechanisms of Post-transcriptional Control

Before proceeding, let's quickly review the steps in gene expression at which control is exerted. We saw in the previous chapter that regulation of transcription initiation and transcription elongation in the promoter proximal region are the initial mechanisms for controlling the expression of genes in the gene expression pathway of DNA→RNA→protein. In preceding sections of this chapter, we learned that the expression of protein isoforms is controlled by regulating alternative RNA splicing and cleavage and polyadenylation at alternative poly(A) sites. Although nuclear export of fully and correctly processed mRNPs to the cytoplasm is rarely regulated, the export of improperly processed or aberrantly remodeled pre-mRNPs is prevented, and such abnormal transcripts are degraded by the exosome. However, retroviruses, including HIV, have evolved mechanisms that permit pre-mRNAs that retain splice sites to be exported and translated.

In this section we consider other mechanisms of post-transcriptional control that contribute to regulating the expression of some genes. Most of these mechanisms operate in the cytoplasm, controlling the stability or localization of mRNA or its translation into protein. We begin by discussing two recently discovered and related mechanisms of gene control that provide powerful new techniques for manipulating the expression of specific genes for experimental and therapeutic purposes. These mechanisms are controlled by short, ~22-nucleotide, single-stranded RNAs called **micro RNAs (miRNAs)** and **short interfering (siRNAs)**. Both base-pair with specific target mRNAs, either inhibiting their translation (miRNAs) or causing their degradation (siRNAs). Humans express ~500 miRNAs. Most of these are expressed in specific cell types at particular times during embryogenesis and after birth. Many miRNAs can target more than one

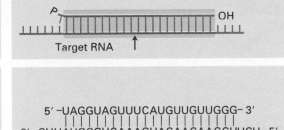

(a) miRNA → translation inhibition

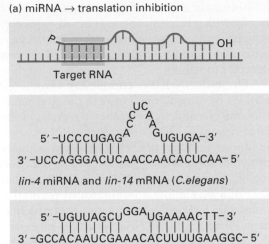

5′ –UCCCUGAG A C C UC A A A G GUGUGA– 3′
     | | | | | | | |
3′ –UCCAGGGACUCAACCAACACUCAA– 5′

*lin-4* miRNA and *lin-14* mRNA (*C.elegans*)

5′ –UGUUAGCU GGA UGAAAACTT– 3′
     | | | | | | | |    | | | | | | | | |
3′ –GCCACAAUCGAAACACUUUUGAAGGC– 5′

CXCR4 miRNA and target mRNA (*H. sapiens*)

(b) siRNA → RNA cleavage

5′ –UAGGUAGUUUCAUGUUGUUGGG– 3′
     | | | | | | | | | | | | | | | | | | | |
3′ –CUUAUCCGUCAAAGUACAACAACCUUCU– 5′

miR-196a and *HOXB8* mRNA (*H. sapiens*)

5′ –UCGGACCAGGCUUCAUUCC CC – 3′
     | | | | | | | | | | | | | | | | | | | |
3′ –UUAGGCCUGGUCCGAAGUAGGGUUAGU– 5′

miR-166 and *PHAVOLUTA* mRNA (*A. thaliana*)

**FIGURE 8-25 Base pairing with target RNAs distinguishes miRNA and siRNA.** (a) miRNAs hybridize imperfectly with their target mRNAs, repressing translation of the mRNA. Nucleotides 2 to 7 of an miRNA (highlighted blue) are the most critical for targeting it to a specific mRNA. (b) siRNA hybridizes perfectly with its target mRNA, causing cleavage of the mRNA at the position indicated by the red arrow, triggering its rapid degradation. [Adapted from P. D. Zamore and B. Haley, 2005, *Science* **309**:1519.]

mRNA. Consequently, these newly discovered mechanisms contribute significantly to the regulation of gene expression. siRNAs, involved in the process called RNA interference, are also an important cellular defense against viral infection and excessive transposition by transposons.

## Micro RNAs Repress Translation of Specific mRNAs

Micro RNAs (miRNAs) were first discovered during analysis of mutations in the *lin-4* and *let-7* genes of the nematode *C. elegans*, which influence development of the organism. Cloning and analysis of wild-type *lin-4* and *let-7* revealed that they encode not protein products but rather RNAs only 21 and 22 nucleotides long, respectively. The RNAs hybridize to the 3′ untranslated regions of specific target mRNAs. For example, the *lin-4* miRNA, which is expressed early in embryogenesis, hybridizes to the 3′ untranslated regions of both the *lin-14* and *lin-28* mRNAs in the cytoplasm, thereby repressing their translation by a mechanism discussed below. Expression of *lin-4* miRNA ceases later in development, allowing translation of newly synthesized *lin-14* and *lin-28* mRNAs at that time. Expression of *let-7* miRNA occurs at comparable times during embryogenesis of all bilaterally symmetric animals.

miRNA regulation of translation appears to be widespread in all multicellular plants and animals. In the past few years, small RNAs of 20–26 nucleotides have been isolated, cloned, and sequenced from various tissues of multiple model organisms. Recent estimates suggest the expression of one-third of all human genes is regulated by ~500 human miRNAs isolated from various tissues. The potential for regulation of multiple mRNAs by one miRNA is great because base pairing between the miRNA and the sequence in the 3′ ends of mRNAs

that they regulate need not be perfect (Figure 8-25). In fact, considerable experimentation with synthetic miRNAs has shown that complementarity between the six or seven 5′ nucleotides of an miRNA and its target mRNA 3′ untranslated region are most critical for target mRNA selection.

Most miRNAs are processed from RNA polymerase II transcripts of several hundred to thousands of nucleotides in length called pri (for *primary* transcript)-miRNAs (Figure 8-26). Pri-miRNAs can contain the sequence of one or more miRNAs. miRNAs are also processed out of some excised introns and from 3′ untranslated regions of some pre-mRNAs. Within these long transcripts are sequences that fold into hairpin structures of ~70 nucleotides in length with imperfect base pairing in the stem. A nuclear RNase specific for double-stranded RNA called *Drosha* acts with a nuclear double-stranded RNA-binding protein called DGCR8 in humans (Pasha in *Drosophila*) and cleaves the hairpin region out of the long precursor RNA, generating a pre-miRNA. Pre-miRNAs are recognized and bound by a specific nuclear transporter, *Exportin5*, which interacts with the FG-domains of nucleoporins, allowing the complex to diffuse through the inner channel of the nuclear pore complex, as discussed above (see Figure 8-20, and Chapter 13). Once in the cytoplasm, a cytoplasmic double-stranded RNA-specific RNase called *Dicer* acts with a cytoplasmic double-stranded RNA-binding protein called TRBP in humans (for Tar binding protein; called *Loquacious* in *Drosophila*) to further process the pre-miRNA into a double-stranded miRNA. The double-stranded miRNA is approximately two turns of an A-form RNA helix in length, with strands 21–23 nucleotides long and two unpaired 3′-nucleotides at each end. Finally, one of the two strands is selected for assembly into a mature **RNA-induced silencing complex (RISC)** containing a single-stranded mature miRNA bound by a

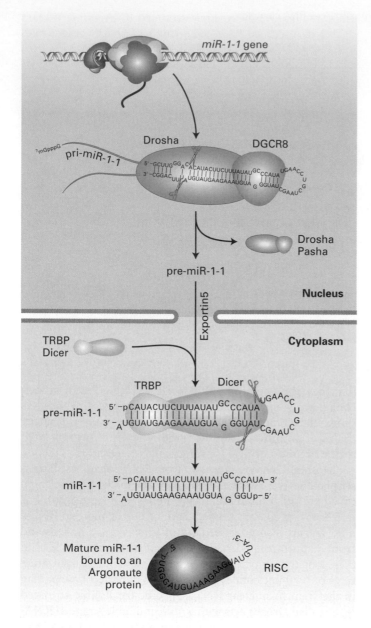

**FIGURE 8-26 miRNA processing.** This diagram shows transcription and processing of the miR-1-1 miRNA. The primary miRNA transcript (pri-miRNA) is transcribed by RNA polymerase II. The nuclear double-stranded RNA-specific endoribonuclease Drosha, with its partner double-stranded RNA-binding protein DGCR8 (Pasha in *Drosophila*), make the initial cleavages in the pri-miRNA, generating a ~70-nucleotide pre-miRNA that is exported to the cytoplasm by nuclear transporter Exportin5. The pre-miRNA is further processed in the cytoplasm to a double-stranded miRNA with a two-base single-stranded 3' end by Dicer in conjunction with the double-stranded RNA-binding protein TRBP (Loquatious in *Drosophila*). Finally, one of the two strands is incorporated into an RISC complex, where it is bound by an Argonaute protein. [Adapted from P. D. Zamore and B. Haley, 2005, *Science* **309:**1519.]

multidomain *Argonaute* protein, a member of a protein family with a recognizable conserved sequence. Several Argonaute proteins are expressed in some organisms, especially plants, and are found in distinct RISC complexes with different functions.

The miRNA-RISC complexes associate with target mRNPs by base pairing between the Argonaute-bound mature miRNA and complementary regions in the 3' untranslated regions (3' UTRs) of target mRNAs (see Figure 8-25). Inhibition of target mRNA translation requires the binding of two or more RISC complexes to distinct complementary regions in the target mRNA 3' UTR. It has been suggested that this may allow combinatorial regulation of mRNA translation by separately regulating the transcription of two or more different pri-miRNAs, which are processed to miRNAs that are required in combination to suppress the translation of a specific target mRNA.

The binding of several RISC complexes to an mRNA inhibits translation initiation by a mechanism currently being analyzed. Binding of RISC complexes causes the bound mRNPs to associate with dense cytoplasmic domains many times the size of a ribosome called cytoplasmic RNA-processing bodies, or simply P bodies. **P bodies**, which will be described in greater detail below, are sites of RNA degradation that contain no ribosomes or translation factors, potentially explaining the inhibition of translation. The association with P bodies may also explain why expression of an miRNA often decreases the stability of a targeted mRNA.

As mentioned earlier, approximately 500 different human miRNAs have been observed, many of them expressed only in specific cell types. Determining the function of these miRNAs is currently a highly active area of research. In one example, a specific miRNA called miR-133 is induced when myoblasts differentiate into muscle cells. miR-133 suppresses the translation of PTB, a regulatory splicing factor that functions similarly to Sxl in *Drosophila* (see Figure 8-16). PTB binds to the 3' splice-site region in the pre-mRNAs of many genes, leading to exon skipping or use of alternative 3' splice sites. When miR-133 is expressed in differentiating myoblasts, the PTB concentration falls without a significant decrease in the concentration of PTB mRNA. As a result, alternative isoforms of multiple proteins important for muscle-cell function are expressed in the differentiated cells.

Other examples of miRNA regulation in various organisms are being discovered at a rapid pace. Knocking out the *dicer* gene eliminates the generation of miRNA in mammals. This causes embryonic death early in mouse development. However, when *dicer* is knocked out only in limb primordia, the influence of miRNA on the development of the nonessential limbs can be observed (Figure 8-27). Although all major cell types differentiate and the fundamental aspects of limb patterning are maintained, development is abnormal—demonstrating the importance of miRNAs in regulating the proper level of translation of multiple mRNAs. Of the ~500 human miRNAs, 53 appear to be unique to primates. It seems likely that new miRNAs arose readily during evolution by the duplication of a pri-miRNA gene followed by mutation

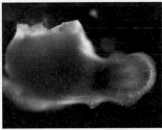

**EXPERIMENTAL FIGURE 8-27  miRNA function in limb development.** Micrographs comparing normal (*left*) and Dicer knockout (*right*) limbs of embryonic development day-13 mouse embryos immunostained for the Gd5 protein, a marker of joint formation. Dicer is knocked out in developing mouse embryos by conditional expression of Cre to induce deletion of the Dicer gene only in these cells (see Figure 5-42). [From B. D. Harfe et al., 2005, *Proc. Nat'l. Acad. Sci. USA* **102**:10898.]

of bases encoding the mature miRNA. miRNAs are particularly abundant in plants—more than 1.5 million distinct miRNAs have been characterized in *Arabidopsis thaliana!*

## RNA Interference Induces Degradation of Precisely Complementary mRNAs

**RNA interference (RNAi)** was discovered unexpectedly during attempts to experimentally manipulate the expression of specific genes. Researchers tried to inhibit the expression of a gene in *C. elegans* by microinjecting a single-stranded, complementary RNA that would hybridize to the encoded mRNA and prevent its translation, a method called antisense inhibition. But in control experiments, perfectly base-paired double-stranded RNA a few hundred base pairs long was much more effective at inhibiting expression of the gene than the antisense strand alone (see Figure 5-45). Similar inhibition of gene expression by an introduced double-stranded RNA soon was observed in plants. In each case, the double-stranded RNA induced degradation of all cellular RNAs containing a sequence that was exactly the same as one strand of the double-stranded RNA. Because of the specificity of RNA interference in targeting mRNAs for destruction, it has become a powerful experimental tool for studying gene function.

Subsequent biochemical studies with extracts of *Drosophila* embryos showed that a long double-stranded RNA that mediates interference is initially processed into a double-stranded short interfering RNA (siRNA). The strands in siRNA contain 21–23 nucleotides hybridized to each other so that the two bases at the 3' end of each strand are single-stranded. Further studies revealed that the cytoplasmic double-stranded RNA-specific ribonuclease that cleaves long double-stranded RNA into siRNAs is the same Dicer enzyme involved in processing pre-miRNAs after their nuclear export to the cytoplasm (see Figure 8-26). This discovery led to the realization that RNA interference and miRNA-mediated translational repression are related processes. In both cases, the mature short single-stranded RNA, either mature siRNA or mature miRNA, is assembled into RISC complexes in which the short RNAs are bound by an Argonaute protein. What distinguishes a RISC complex containing an siRNA from one containing an miRNA is that the siRNA base-pairs extensively with its target RNA and induces its cleavage, whereas a RISC complex associated with an miRNA recognizes its target through imperfect base-pairing and results in inhibition of translation.

The Argonaute protein is responsible for cleavage of target RNA; one domain of the Argonaute protein is homologous to RNase H enzymes that degrade the RNA of an RNA-DNA hybrid (see Figure 6-14). When the 5' end of the short RNA of a RISC complex base-pairs precisely with a target mRNA over a distance of one turn of an RNA helix (10–12 base pairs), this domain of Argonaute cleaves the phosphodiester bond of the target RNA across from nucleotides 10 and 11 of the siRNA (see Figure 8-25). The cleaved RNAs are released and subsequently degraded by cytoplasmic exosomes and 5' exoribonucleases. If base pairing is not perfect, the Argonaute domain does not cleave or release the target mRNA. Instead, if several miRNA-RISC complexes associate with a target mRNA, its translation is inhibited and the mRNA becomes associated with P bodies, where, as mentioned earlier, it is probably degraded by a different and slower mechanism than the degradation pathway initiated by RISC cleavage of a perfectly complementary target RNA.

When double-stranded RNA is introduced into the cytoplasm of eukaryotic cells, it enters the pathway for assembly of siRNAs into a RISC complex because it is recognized by the cytoplasmic Dicer enzyme and TRBP double-stranded RNA-binding protein that process pre-miRNAs (see Figure 8-26). This process of RNA interference is believed to be an ancient cellular defense against certain viruses and mobile genetic elements in both plants and animals. Plants with mutations in the genes encoding Dicer and RISC proteins exhibit increased sensitivity to infection by RNA viruses and increased movement of **transposons** within their genomes. The double-stranded RNA intermediates generated during replication of RNA viruses are thought to be recognized by the Dicer ribonuclease, inducing an RNAi response that ultimately degrades viral mRNAs. During transposition, transposons are inserted into cellular genes in a random orientation, and their transcription from different promoters produces complementary RNAs that can hybridize with each other, initiating the RNAi system that then interferes with the expression of transposon proteins required for additional transpositions.

In plants and *C. elegans* the RNAi response can be induced in all cells of the organism by introduction of double-stranded RNA into just a few cells. Such organism-wide induction requires production of a protein that is homologous to the RNA replicases of RNA viruses. It has been revealed that double-stranded siRNAs are replicated and then transferred to other cells in these organisms. In plants, transfer of siRNAs might occur through **plasmodesmata**, the cytoplasmic connections between plant cells that traverse the cell walls between them (see Figure 20-38). Organism-wide induction of RNA interference does not occur in *Drosophila* or mammals, presumably because their genomes do not encode RNA replicase homologs.

In mammalian cells, the introduction of long RNA-RNA duplex molecules into the cytoplasm results in the generalized inhibition of protein synthesis via the PKR pathway, discussed further below. This greatly limits the use of long double-stranded RNAs to experimentally induce an RNAi response against a specific targeted mRNA. Fortunately, researchers discovered that one strand of double-stranded siRNAs 21–23 nucleotides in length with two-base 3′ single-stranded regions leads to the generation of mature siRNA RISC complexes without inducing the generalized inhibition of protein synthesis. This has allowed researchers to use synthetic double-stranded siRNAs to "knock down" the expression of specific genes in human cells as well as in other mammals. This method of **siRNA knockdown** is now widely used in studies of diverse processes, including the RNAi pathway itself.

## Cytoplasmic Polyadenylation Promotes Translation of Some mRNAs

In addition to repression of translation by miRNAs, other protein-mediated translational controls help regulate expression of some genes. Regulatory sequences, or elements, in mRNAs that interact with specific proteins to control translation generally are present in the untranslated region (UTR) at the 3′ or 5′ end of an mRNA. Here we discuss a type of protein-mediated translational control involving 3′ regulatory elements. A different mechanism involving RNA-binding proteins that interact with 5′ regulatory elements is discussed later.

Translation of many eukaryotic mRNAs is regulated by sequence-specific RNA-binding proteins that bind cooperatively to neighboring sites in 3′ UTRs. This allows them to function in a combinatorial manner, similar to the cooperative binding of transcription factors to regulatory sites in an enhancer or promoter region. In most cases studied, translation is repressed by protein binding to 3′ regulatory elements and regulation results from derepression at the appropriate time or place in a cell or developing embryo. The mechanism of such repression is best understood for mRNAs that must undergo *cytoplasmic polyadenylation* before they can be translated.

Cytoplasmic polyadenylation is a critical aspect of gene expression in the early embryo of animals. The egg cells (oocytes) of multicellular animals contain many mRNAs, encoding numerous different proteins that are not translated until after the egg is fertilized by a sperm cell. Some of these "stored" mRNAs have a short poly(A) tail, consisting of only ~20–40 A residues, to which just a few molecules of cytoplasmic poly(A)-binding protein (PABPI) can bind. As discussed in Chapter 4, multiple PABPI molecules bound to the long poly(A) tail of an mRNA interact with the eIF4G initiation factor, thereby stabilizing the interaction of the mRNA 5′ cap with eIF4E, which is required for translation initiation (see Figure 4-24). Because this stabilization cannot occur with mRNAs that have short poly(A) tails, such

mRNAs stored in oocytes are not translated efficiently. At the appropriate time during oocyte maturation or after fertilization of an egg cell, usually in response to an external signal, approximately 150 A residues are added to the short poly(A) tails on these mRNAs in the cytoplasm, stimulating their translation.

Studies with mRNAs stored in *Xenopus* oocytes have helped elucidate the mechanism of this type of translational control. Experiments in which short-tailed mRNAs are injected into oocytes have shown that two sequences in their 3′ UTR are required for their polyadenylation in the cytoplasm: the AAUAAA poly(A) signal that is also required for the nuclear polyadenylation of pre-mRNAs, and one or more copies of an upstream U-rich *cytoplasmic polyadenylation element (CPE)*. This regulatory element is bound by a highly conserved *CPE-binding protein (CPEB)* that contains an RRM domain and a zinc-finger domain.

According to the current model, in the absence of a stimulatory signal, CPEB bound to the U-rich CPE interacts with the protein Maskin, which in turn binds to eIF4E associated with the mRNA 5′ cap (Figure 8-28, *left*). As a result, eIF4E cannot interact with other initiation factors and the 40S ribosomal subunit, so translation initiation is blocked. During oocyte maturation, a specific CPEB serine is phosphorylated, causing Maskin to dissociate from the complex. This allows cytoplasmic forms of the cleavage and polyadenylation specificity factor (CPSF) and poly(A) polymerase to bind to the mRNA cooperatively with CPEB. Once the poly(A) polymerase catalyzes the addition of A residues, PABPI can bind to the lengthened poly(A) tail, leading to the stabilized interaction of all the factors needed to initiate translation (Figure 8-28, *right*; see also Figure 4-24). In the case of *Xenopus* oocyte maturation, the protein kinase that phosphorylates CPEB is activated in response to the hormone progesterone. Thus timing of the translation of stored mRNAs encoding proteins needed for oocyte maturation is regulated by this external signal.

Considerable evidence indicates that a similar mechanism of translational control plays a role in learning and memory. In the central nervous system, the axons from a thousand or so neurons can make connections (synapses) with the dendrites of a single postsynaptic neuron (Figure 22-23). When one of these axons is stimulated, the postsynaptic neuron "remembers" which one of these thousands of synapses was stimulated. The next time that synapse is stimulated, the strength of the response triggered in the postsynaptic cell differs from the first time. This change in response has been shown to result largely from the translational activation of mRNAs stored in the region of the synapse, leading to the local synthesis of new proteins that increase the size and alter the neurophysiological characteristics of the synapse. The finding that CPEB is present in neuronal dendrites has led to the proposal that cytoplasmic polyadenylation stimulates translation of specific mRNAs in dendrites, much as it does in oocytes. In this case, presumably, synaptic activity (rather than a hormone) is the signal that induces phosphorylation of CPEB and subsequent activation of translation.

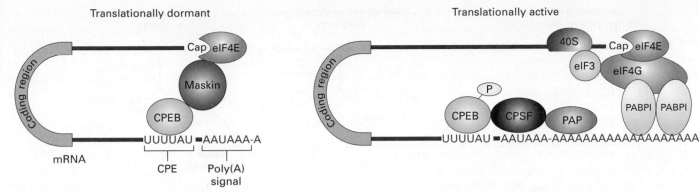

**Translationally dormant**

Coding region
mRNA

Cap eIF4E
Maskin
CPEB
UUUUAU ▬ AAUAAA-A

CPE    Poly(A)
signal

**Translationally active**

Coding region

40S
eIF3
Cap eIF4E
eIF4G
P
CPEB CPSF PAP
PABPI PABPI
UUUUAU ▬ AAUAAA-AAAAAAAAAAAAAAAAAAA

**FIGURE 8-28  Model for control of cytoplasmic polyadenylation and translation initiation.** *(Left)* In immature oocytes, mRNAs containing the U-rich cytoplasmic polyadenylation element (CPE) have short poly(A) tails. CPE-binding protein (CPEB) mediates repression of translation through the interactions depicted, which prevent assembly of an initiation complex at the 5′ end of the mRNA. *(Right)* Hormone stimulation of oocytes activates a protein kinase that phosphorylates CPEB, causing it to release Maskin. The cleavage and polyadenylation specificity factor (CPSF) then binds to the poly(A) site, interacting with both bound CPEB and the cytoplasmic form of poly(A) polymerase (PAP). After the poly(A) tail is lengthened, multiple copies of the cytoplasmic poly(A)-binding protein I (PABPI) can bind to it and interact with eIF4G, which functions with other initiation factors to bind the 40S ribosome subunit and initiate translation. [Adapted from R. Mendez and J. D. Richter, 2001, *Nature Rev. Mol. Cell Biol.* **2**:521.]

## Degradation of mRNAs in the Cytoplasm Occurs by Several Mechanisms

The concentration of an mRNA is a function of both its rate of synthesis and its rate of degradation. For this reason, if two genes are transcribed at the same rate, the steady-state concentration of the corresponding mRNA that is more stable will be higher than the concentration of the other. The stability of an mRNA also determines how rapidly synthesis of the encoded protein can be shut down. For a stable mRNA, synthesis of the encoded protein persists long after transcription of the gene is repressed. Most bacterial mRNAs are unstable, decaying exponentially with a typical half-life of a few minutes. For this reason, a bacterial cell can rapidly adjust the synthesis of proteins to accommodate changes in the cellular environment. Most cells in multicellular organisms, on the other hand, exist in a fairly constant environment and carry out a specific set of functions over periods of days to months or even the lifetime of the organism (nerve cells, for example). Accordingly, most mRNAs of higher eukaryotes have half-lives of many hours.

However, some proteins in eukaryotic cells are required only for short periods and must be expressed in bursts. For example, as discussed in the chapter introduction, certain signaling molecules called **cytokines**, which are involved in the immune response of mammals, are synthesized and secreted in short bursts (see Chapter 23). Similarly, many of the transcription factors that regulate the onset of the S phase of the cell cycle, such as c-Fos and c-Jun, are synthesized for brief periods only (Chapter 19). Expression of such proteins occurs in short bursts because transcription of their genes can be rapidly turned on and off, and their mRNAs have unusually short half-lives, on the order of 30 minutes or less.

Cytoplasmic mRNAs are degraded by one of the three pathways shown in Figure 8-29. For most mRNAs, the *deadenylation-dependent pathway* is followed: the length of the poly(A) tail gradually decreases with time through the action of a deadenylating nuclease. When it is shortened sufficiently, PABPI molecules can no longer bind and stabilize the interaction of the 5′ cap and translation initiation factors (see Figure 4-24). The exposed cap then is removed by a decapping enzyme (Dcp1/Dcp2 in *S. cerevisiae*), and the unprotected mRNA is degraded by a 5′→3′ exonuclease (Xrn1 in *S. cerevisiae*). Removal of the poly(A) tail also makes mRNAs susceptible to degradation by cytoplasmic exosomes containing 3′→5′ exonucleases. The 5′→3′ exonucleases predominate in yeast, and the 3′→5′ exosome predominates in mammalian cells. The decapping enzymes and 5′→3′ exonuclease are concentrated in the P bodies, regions of the cytoplasm of unusually high density.

Some mRNAs are degraded primarily by a *deadenylation-independent decapping pathway* (see Figure 8-29). This is because certain sequences at the 5′ end of an mRNA seem to make the cap sensitive to the decapping enzyme. For these mRNAs, the rate at which they are decapped controls the rate at which they are degraded because once the 5′ cap is removed, the RNA is rapidly hydrolyzed by the 5′→3′ exonuclease.

The rate of mRNA deadenylation varies inversely with the frequency of translation initiation for an mRNA: the higher the frequency of initiation, the slower the rate of deadenylation. This relation probably is due to the reciprocal interactions between translation initiation factors bound at the 5′ cap and PABPI bound to the poly(A) tail. For an mRNA that is translated at a high rate, initiation factors are bound to the cap much of the time, stabilizing the binding of PABPI and thereby protecting the poly(A) tail from the deadenylation exonuclease.

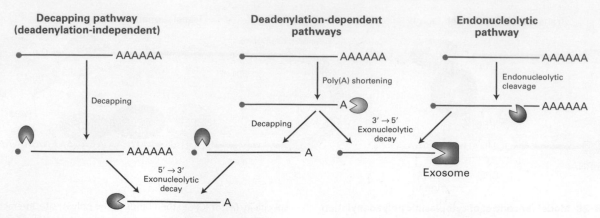

**Decapping pathway (deadenylation-independent)** | **Deadenylation-dependent pathways** | **Endonucleolytic pathway**

**FIGURE 8-29 Pathways for degradation of eukaryotic mRNAs.** In the deadenylation-dependent (*middle*) pathways, the poly(A) tail is progressively shortened by a deadenylase (orange) until it reaches a length of 20 or fewer A residues, at which point the interaction with PABPI is destabilized, leading to weakened interactions between the 5′ cap and translation-initiation factors. The deadenylated mRNA then may either (1) be decapped and degraded by a 5′→3′ exonuclease or (2) be degraded by a 3′→5′ exonuclease in cytoplasmic exosomes. Some mRNAs (*right*) are cleaved internally by an endonuclease and the fragments degraded by an exosome. Other mRNAs (*left*) are decapped before they are deadenylated and then degraded by a 5′→3′ exonuclease. [Adapted from M. Tucker and R. Parker, 2000, *Ann. Rev. Biochem.* **69:**571.]

Many short-lived mRNAs in mammalian cells contain multiple, sometimes overlapping copies of the sequence AUUUA in their 3′ untranslated region. Specific RNA-binding proteins have been found that both bind to these 3′ AU-rich sequences and also interact with a deadenylating enzyme and with the exosome. This causes rapid deadenylation and subsequent 3′→5′ degradation of these mRNAs. In this mechanism, the rate of mRNA degradation is uncoupled from the frequency of translation. Thus mRNAs containing the AUUUA sequence can be translated at high frequency yet also be degraded rapidly, allowing the encoded proteins to be expressed in short bursts.

As shown in Figure 8-29, some mRNAs are degraded by an *endonucleolytic pathway* that does not involve decapping or significant deadenylation. One example of this type of pathway is the RNAi pathway discussed above (see Figure 8-25). Each siRNA-RISC complex can degrade thousands of targeted RNA molecules. The fragments generated by internal cleavage then are degraded by exonucleases.

**P Bodies** As mentioned above, P bodies are sites of translational repression of mRNAs bound by miRNA-RISC complexes. They are also the major sites of mRNA degradation in the cytoplasm. These dense regions of cytoplasm contain the decapping enzyme (Dcp1/Dcp2 in yeast), activators of decapping (Dhh, Pat1, Lsm1-7 in yeast), the major 5′→3′ exonuclease (Xrn1), as well as densely associated mRNAs. P bodies are dynamic structures that grow and shrink in size depending on the rate at which mRNPs associate with them, the rate at which mRNAs are degraded, and the rate at which mRNPs exit P bodies and reenter the pool of translated mRNPs. mRNAs whose translation is inhibited by imperfect base-pairing of miRNAs (Figure 8-25) are major components of P-bodies.

## Protein Synthesis Can Be Globally Regulated

Like proteins involved in other processes, translation initiation factors and ribosomal proteins can be regulated by post-translational modifications such as phosphorylation. Such mechanisms affect the translation rate of most mRNAs and hence the overall rate of cellular protein synthesis.

**TOR Pathway** The TOR pathway was discovered through research into the mechanism of action of rapamycin, an antibiotic produced by a strain of *Streptomyces* bacteria, which is useful for suppressing the immune response in organ transplant patients. The *target of rapamycin (TOR)* was identified by isolating yeast mutants resistant to rapamycin inhibition of cell growth. TOR is a large (~2400 amino acid residue) protein kinase that regulates several cellular processes in yeast cells in response to nutritional status. In multicellular eukaryotes, *metazoan TOR (mTOR)* also responds to multiple signals from cell-surface-signaling proteins to coordinate cell growth with developmental programs as well as nutritional status.

Current understanding of the mTOR pathway is summarized in Figure 8-30. Active mTOR stimulates the overall rate of protein synthesis by phosphorylating two critical proteins that regulate translation directly. mTOR also activates transcription factors that control expression of ribosomal components, tRNAs, and translation factors, further activating protein synthesis and cell growth.

Recall that the first step in translation of a eukaryotic mRNA is binding of the eIF4 initiation complex to the 5′ cap via its eIF4E cap-binding subunit (see Figure 4-24). The concentration of active eIF4E is regulated by a small family of homologous *eIF4E-binding proteins (4E-BPs)* that inhibit the interaction of eIF4E with mRNA 5′ caps. 4E-BPs are direct targets of mTOR. When phosphorylated by mTOR, 4E-BPs release eIF4E, stimulating translation initiation. mTOR also phosphorylates and activates another protein kinase (S6K) that phosphorylates the small ribosomal subunit protein S6 and probably additional substrates, leading to a further increase in the rate of protein synthesis.

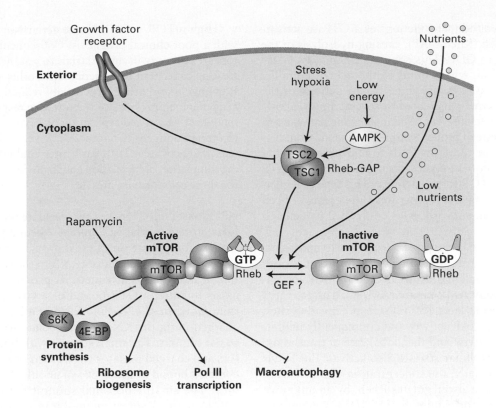

**FIGURE 8-30 mTOR pathway.** mTOR is an active protein kinase when bound by a complex of Rheb and an associated GTP (*lower left*). In contrast, mTOR is inactive when bound by a complex of Rheb associated with GDP (*lower right*). When active, the TSC1/TSC2 Rheb-GTPase activating protein (Rheb-GAP) causes hydrolysis of Rheb-bound GTP to GDP, thereby inactivating mTOR. The TSC1/TSC2 Rheb-GAP is activated (arrows) by phosphorylation by AMP kinase (AMPK) when cellular energy charge is low and by other cellular stress responses. Signal-transduction pathways activated by cell-surface growth factor receptors lead to phosphorylation of inactivating sites on TSC1/TSC2, inhibiting its GAP activity. Consequently, they leave a higher fraction of cellular Rheb in the GTP conformation that activates mTOR protein kinase activity. Low nutrient concentration also regulates Rheb GTPase activity, by a mechanism that does not require TSC1/TSC2. Active mTOR phosphorylates 4E-BP, causing it to release eIF4E, stimulating translation initiation. It also phosphorylates and activates S6 kinase (S6K), which in turn phosphorylates ribosomal proteins, stimulating translation. Activated mTOR also activates transcription factors for RNA polymerases I, II, and III, leading to synthesis and assembly of ribosomes, tRNAs, and translation factors. In the absence of mTOR activity, all of these processes are inhibited. In contrast, activated mTOR inhibits macroautophagy, which is stimulated in cells with inactive mTOR. [Adapted from S. Wullschleger et al., 2006, *Cell* **124**:471.]

Translation of a specific subset of mRNAs that have a string of pyrimidines in their 5′ untranslated regions (called TOP mRNAs for *t*ract of *o*ligo*p*yrimidine) is stimulated particularly strongly by mTOR. The TOP mRNAs encode ribosomal proteins and translation elongation factors. mTOR also activates the RNA polymerase I transcription factor TIF-1A, stimulating transcription of the large rRNA precursor (see Figure 7-52).

mTOR activates transcription by RNA polymerase III as well, by phosphorylating and thereby activating protein kinases that phosphorylate MAF1, a protein inhibitor of RNA polymerase III transcription. MAF1 phosphorylation causes it to be exported from the nucleus, relieving repression of RNA polymerase III transcription. When mTOR activity falls, MAF1 in the cytoplasm is rapidly dephosphorylated and imported into the nucleus where it represses transcription by RNA polymerase III.

In addition, mTOR activates two RNA polymerase II activators that stimulate transcription of ribosomal protein and translation factor genes. Finally, mTOR stimulates processing of the rRNA precursor (Section 8.5). As a consequence of phosphorylation of these several mTOR substrates, the synthesis and assembly of ribosomes as well as the synthesis of translation factors and tRNAs are greatly increased. Alternatively, when mTOR kinase activity is inhibited, these substrates become dephosphorylated, greatly decreasing the rate of protein synthesis and the production of ribosomes, translation factors, and tRNAs, thus halting cell growth.

mTOR activity is regulated by a **monomeric small G protein** in the Ras protein family called Rheb. Like other small G proteins, Rheb is in its active conformation when it is bound to GTP. Rheb·GTP binds the mTOR complex, stimulating mTOR kinase activity, probably by inducing a conformation change in its kinase domain. Rheb is in turn regulated by a heterodimer composed of subunits TSC1 and TSC2, named for their involvement in the medical syndrome *t*uberous *s*clerosis *c*omplex, as discussed below. In the active conformation, the TSC1/TSC2 heterodimer functions as a GTPase activating protein for Rheb (Rheb-GAP), causing hydrolysis of the Rheb-bound GTP to GDP. This converts Rheb to its GDP-bound conformation, which binds to the mTOR complex and inhibits its kinase activity. Finally, the activity of the TSC1/TSC2 Rheb-GAP is regulated by several inputs, allowing the cell to integrate different cellular signaling pathways to control the overall rate of protein synthesis. Signaling from cell-surface growth factor receptors leads to phosphorylation of TSC1/TSC2 at inhibitory sites, causing an increase in Rheb·GTP and activation of mTOR kinase activity. This type of regulation through cell-surface receptors links the control of cell growth to developmental processes controlled by cell-cell interactions.

mTOR activity also is regulated in response to nutritional status. When energy from nutrients is not sufficient for cell growth, the resulting fall in the ratio of ATP to AMP concentrations is detected by AMP kinase (AMPK). The activated AMP kinase phosphorylates TSC1/TSC2 at activating sites, stimulating its Rheb-GAP activity and consequently inhibiting mTOR kinase activity and the global rate of translation. Hypoxia and other cellular stresses also activate the TSC1/TSC2 Rheb-GAP. Finally, the concentration of nutrients in the extracellular space also regulates Rheb, by an unknown mechanism that does not require the TSC1/TSC2 complex.

In addition to regulating the global rate of cellular protein synthesis and the production of ribosomes, tRNAs, and translation factors, mTOR regulates at least one other process involved in the response to low levels of nutrients: macroautophagy (or simply **autophagy**). Starved cells degrade cytoplasmic constituents, including whole organelles, to supply energy and precursors for essential cellular processes. During this process a large, double-membrane structure engulfs a region of cytoplasm to form an autophagosome, which then fuses with a lysosome where the entrapped proteins, lipids, and other macromolecules are degraded, completing the process of macroautophagy. Activated mTOR inhibits macroautophagy in growing cells when nutrients are plentiful. Macroautophagy is stimulated when mTOR activity falls in nutrient-deprived cells.

Genes encoding components of the mTOR pathway are mutated in many human cancers, resulting in cell growth in the absence of normal growth signals. TSC1 and TSC2 (see Figure 8-30) were initially identified because one or the other of the proteins is mutant in a rare human genetic syndrome: *tuberous sclerosis complex*. Patients with this disorder develop benign tumors in multiple tissues. The disease results because inactivation of either TSC1 or TSC2 eliminates the Rheb-GAP activity of the TSC1/TSC2 heterodimer, resulting in an abnormally high and unregulated level of Rheb·GTP and the resulting high, unregulated activity of mTOR. Mutations in components of cell-surface receptor signal-transduction pathways that lead to inhibition of TSC1/TSC2 Rheb-GAP activity are also common in human tumors and contribute to cell growth and replication in the absence of normal signals for growth and proliferation.

High mTOR protein kinase activity in tumors correlates with a poor clinical prognosis. Consequently, mTOR inhibitors are currently in clinical trials to test their effectiveness for treating cancers in conjunction with other modes of therapy. Rapamycin and other structurally related mTOR inhibitors are potent suppressors of the immune response because they inhibit activation and replication of T lymphocytes in response to foreign antigens (Chapter 23). Several viruses encode proteins that activate mTOR early after viral infection. The resulting stimulation of translation has an obvious selective advantage for these cellular parasites. ∎

**eIF2 Kinases** eIF2 kinases also regulate the global rate of cellular protein synthesis. Figure 4-24 summarizes the steps in translation initiation. Translation initiation factor eIF2 brings the charged initiator tRNA to the small ribosome subunit P site. eIF2 is a **trimeric G protein** and consequently exists in either a GTP-bound or a GDP-bound conformation. Only the GTP-bound form of eIF2 is able to bind the charged initiator tRNA and associate with the small ribosomal subunit. The small subunit with bound initiation factors and charged initiator tRNA then interacts with the eIF4 complex bound to the 5′ cap of an mRNA via its eIF4E subunit. The small ribosomal subunit then scans down the mRNA in the 3′ direction until it reaches an AUG initiation codon that can base-pair with the initiator tRNA in its P site. When this occurs, the GTP bound by eIF2 is hydrolyzed to GDP and the resulting eIF2·GDP complex is released. GTP hydrolysis results in an irreversible "proofreading" step that prepares the small ribosomal subunit to associate with the large subunit only when an initiator tRNA is properly bound in the P site and is properly base-paired with the AUG start codon. Before eIF2 can participate in another round of initiation, its bound GDP must be replaced with a GTP. This process is catalyzed by the translation initiation factor eIF2B, a guanine nucleotide exchange factor (GEF) specific for eIF2.

A mechanism for inhibiting general protein synthesis in stressed cells involves phosphorylation of the eIF2 α subunit at a specific serine. Phosphorylation at this site does not interfere with eIF2 function in protein synthesis directly. Rather, phosphorylated eIF2 has very high affinity for the eIF2 guanine nucleotide exchange factor, eIF2B, which cannot release the phosphorylated eIF2 and consequently is blocked from catalyzing GTP exchange of additional eIF2 factors. Since there is an excess of eIF2 over eIF2B, phosphorylation of a fraction of eIF2 results in inhibition of all the cellular eIF2B. The remaining eIF2 accumulates in its GDP-bound form, which cannot participate in protein synthesis, thereby inhibiting nearly all cellular protein synthesis. However, some mRNAs have 5′ regions that allow translation initiation at the low eIF2-GTP concentration that results from eIF2 phosphorylation. These mRNAs include those for chaperone proteins that function to refold cellular proteins denatured as the result of cellular stress, additional proteins that help the cell to cope with stress, and transcription factors that activate transcription of the genes encoding these stress-induced proteins.

Human cells contain four eIF2 kinases that phosphorylate the same inhibitory eIF2α serine. Each of these is regulated by a different type of cellular stress, inhibiting protein synthesis and allowing cells to divert the large fraction of cellular resources usually devoted to protein synthesis in growing cells for use in responding to the stress.

The GCN2 (general control non-derepressible 2) eIF2-kinase is activated by binding uncharged tRNAs. The concentration of uncharged tRNAs increases when cells are starved for amino acids, activating GCN2 eIF2-kinase activity and greatly inhibiting protein synthesis.

PEK (pancreatic eIF2 kinase) is activated when proteins translocated into the endoplasmic reticulum (ER) do not fold properly because of abnormalities in the ER lumen environment. Inducers include abnormal carbohydrate concentration, because this inhibits the glycosylation of many ER proteins, and inactivating mutations in an ER chaperone required for proper folding of many ER proteins (Chapters 13 and 14).

Heme-regulated inhibitor (HRI) is activated in developing red blood cells when the supply of the heme prosthetic group is too low to accommodate the rate of globin protein synthesis. This negative feedback loop lowers the rate of globin protein synthesis until it matches the rate of heme synthesis. HRI is also activated in other types of cells in response to oxidative stress or heat shock.

Finally, protein kinase RNA activated (PKR) is activated by double-stranded RNAs longer than ~30 base pairs. Under normal circumstances in mammalian cells, such double-stranded RNAs are produced only during a viral infection. Long regions of double-stranded RNA are generated in replication intermediates of RNA viruses or from hybridization of complementary regions of RNA transcribed from both strands of DNA virus genomes. Inhibition of protein synthesis prevents the production of progeny virions, protecting neighboring cells from infection. Interestingly, adenoviruses evolved a defense against PKR: they express prodigious amounts of an ~160-nucleotide virus-associated (VA) RNA with long double-stranded hairpin regions. VA RNA is transcribed by RNA polymerase III and exported from the nucleus by Exportin5, the exportin for pre-miRNAs (see Figure 8-27). VA RNA binds to PKR with high affinity, inhibiting its protein kinase activity and preventing the inhibition of protein synthesis observed in cells infected with a mutant adenovirus from which the VA gene was deleted.

## Sequence-Specific RNA-Binding Proteins Control Specific mRNA Translation

In contrast to global mRNA regulation, mechanisms have also evolved for controlling the translation of certain specific mRNAs. This is usually done by sequence-specific RNA-binding proteins that bind to a particular sequence or RNA structure in the mRNA. When binding is in the 5′ untranslated region (5′ UTR) of an mRNA, the ribosome's ability to scan to the first initiation codon is blocked, inhibiting translation initiation. Binding in other regions can either promote or inhibit mRNA degradation.

Control of intracellular iron concentration by the *iron response element–binding protein (IRE-BP)* is an elegant example of a single protein that regulates the translation of one mRNA and the degradation of another. Precise regulation of cellular iron ion concentration is critical to the cell. Multiple enzymes and proteins contain $Fe^{2+}$ as a cofactor, such as enzymes of the Krebs cycle (see Figure 12-10) and electron-carrying proteins involved in the generation of ATP by mitochondria and chloroplasts (Chapter 12). On the other hand, excess $Fe^{2+}$ generates free radicals that react with and damage cellular macromolecules. When intracellular iron stores are low, a dual-control system operates to increase the level of cellular iron; when iron is in excess, the system operates to prevent accumulation of toxic levels of free ions.

One component in this system is the regulation of the production of ferritin, an intracellular iron-binding protein that binds and stores excess cellular iron. The 5′ untranslated region of ferritin mRNA contains iron-response elements (IREs) that have a stem-loop structure. The IRE-binding protein (IRE-BP) recognizes five specific bases in the IRE loop and the duplex nature of the stem. At low iron concentrations, IRE-BP is in an active conformation that binds to the IREs (Figure 8-31a). The bound IRE-BP blocks the small ribosomal subunit from scanning for the AUG start codon (see Figure 4-24), thereby inhibiting translation initiation. The resulting decrease in ferritin means less iron is complexed with the ferritin and therefore more is available to iron-requiring enzymes. At high iron concentrations, IRE-BP is in an inactive conformation that does not bind to the 5′ IREs, so translation initiation can proceed. The newly synthesized ferritin then binds free iron ions, preventing their accumulation to harmful levels.

The other part of this regulatory system controls the import of iron into cells. In vertebrates, ingested iron is carried through the circulatory system bound to a protein called transferrin. After binding to the transferrin receptor (TfR) in the plasma membrane, the transferrin-iron complex is brought into cells by receptor-mediated endocytosis (Chapter 14). The 3′ untranslated region of TfR mRNA contains IREs whose stems have AU-rich destabilizing sequences (Figure 8-31b). At high iron concentrations, when the IRE-BP is in the inactive, nonbinding conformation, these AU-rich sequences promote degradation of TfR mRNA by the same mechanism that leads to rapid degradation of other short-lived mRNAs, as described previously. The resulting decrease in production of the transferrin receptor quickly reduces iron import, thus protecting the cell from excess iron. At low iron concentrations, however, IRE-BP can bind to the 3′ IREs in TfR mRNA. The bound IRE-BP blocks recognition of the destabilizing AU-rich sequences by the proteins that would otherwise rapidly degrade the mRNAs. As a result, production of the transferrin receptor increases and more iron is brought into the cell.

Other regulated RNA-binding proteins may also function to control mRNA translation or degradation, much like the dual-acting IRE-BP. For example, a heme-sensitive RNA-binding protein controls translation of the mRNA encoding

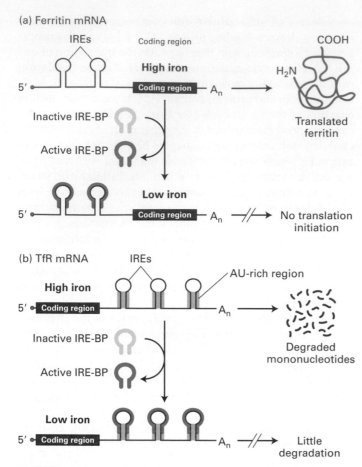

(a) Ferritin mRNA

(b) TfR mRNA

**FIGURE 8-31 Iron-dependent regulation of mRNA translation and degradation.** The iron response element–binding protein (IRE-BP) controls (a) translation of ferritin mRNA and (b) degradation of transferrin-receptor (TfR) mRNA. At low intracellular iron concentrations IRE-BP binds to iron-response elements (IREs) in the 5' or 3' untranslated region of these mRNAs. At high iron concentrations, IRE-BP undergoes a conformational change and cannot bind either mRNA. The dual control by IRE-BP precisely regulates the level of free iron ions within cells. See the text for discussion.

aminolevulinate (ALA) synthase, a key enzyme in the synthesis of heme. Similarly, in vitro studies have shown that the mRNA encoding the milk protein casein is stabilized by the hormone prolactin and rapidly degraded in its absence.

## Surveillance Mechanisms Prevent Translation of Improperly Processed mRNAs

Translation of an improperly processed mRNA could lead to production of an abnormal protein that interferes with the gene's normal function. This effect is equivalent to that resulting from dominant-negative mutations, discussed in Chapter 5 (Figure 5-44). Several mechanisms collectively termed **mRNA surveillance** help cells avoid the translation of improperly processed mRNA molecules. We have previously mentioned two such surveillance mechanisms: the recognition of improperly processed pre-mRNAs in the nucleus and

their degradation by the exosome, and the general restriction against nuclear export of incompletely spliced pre-mRNAs that remain associated with a spliceosome.

Another mechanism called nonsense-mediated decay (NMD) causes degradation of mRNAs in which one or more exons have been incorrectly spliced. Such incorrect splicing often will alter the open reading frame of the mRNA 3' to the improper exon junction, resulting in introduction of an out-of-frame missense mutation and an incorrect stop codon. For nearly all properly spliced mRNAs, the stop codon is in the last exon. The process of nonsense-mediated decay results in the rapid degradation of mRNAs with stop codons that occur before the last splice junction in the mRNA since in most cases, such mRNAs arise from errors in RNA splicing. However, NMD can also result from a mutation creating a stop codon within a gene or a frame-shifting deletion or insertion. NMD was initially discovered during the study of patients with $\beta^0$-thalasemia, who produce a low level of $\beta$-globin protein associated with a low level of $\beta$-globin mRNA (Figure 8-32a, b).

A search for possible molecular signals that might indicate the positions of splice junctions in a processed mRNA led to the discovery of exon-junction complexes. As noted already, these complexes of several proteins (including Y14, Magoh, eIF4IIIA, UPF2, UPF3, and REF), bind ~20 nucleotides 5' to an exon-exon junction following RNA splicing (Figure 8-32c), stimulate export of mRNPs from the nucleus by interacting with the mRNA exporter (see Figure 8-21). Analysis of yeast mutants indicated that one of the proteins in exon-junction complexes (UPF3) functions in nonsense-mediated decay. In the cytoplasm, this component of exon-junction complexes interacts with a protein (UPF1) and a protein kinase (SMG1) that phosphorylates UPF1, causing the mRNA to associate with P bodies, repressing translation of the mRNA. An additional protein (UPF2) associated with the mRNP complex binds a P-body associated deadenylase that rapidly removes the poly(A) tail from an associated mRNA, leading to its rapid decapping and degradation by the P-body associated 5'→3' exonuclease (see Figure 8-29). In the case of properly spliced mRNAs, the exon-junction complexes associate with the nuclear cap-binding complex (CBP80, CBP20) as the mRNP is transported through a nuclear pore complex, thereby protecting the mRNA from degradation. The exon-junction complexes are thought to be dislodged from the mRNA by passage of the first "pioneer" ribosome to translate the mRNA. However, for mRNAs with a stop codon before the final exon junction, one or more exon-junction complexes remain associated with the mRNA, resulting in nonsense-mediated decay (Figure 8-32).

## Localization of mRNAs Permits Production of Proteins at Specific Regions Within the Cytoplasm

Many cellular processes depend on localization of particular proteins to specific structures or regions of the cell. In later chapters we examine how some proteins are transported *after* their synthesis to their proper cellular location. Alternatively,

(a)

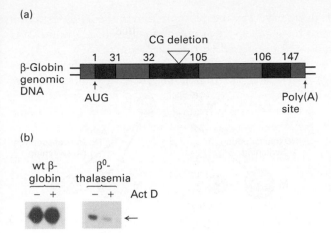

(b)

(c)

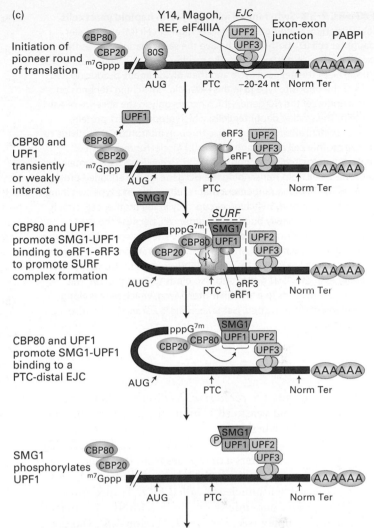

Translational repression and mRNA decay

**FIGURE 8-32 Discovery of nonsense-mediated mRNA decay (NMD).** (a) Patients with β⁰-thalasemia express very low levels of β-globin mRNA. A common cause of this syndrome is a single-base-pair deletion in exon 1 or exon 2 of the β-globin gene. Ribosomes translating the mutant mRNA read out of frame following the deletion and encounter a stop codon in the wrong reading frame before they translate across the last exon junction in the mRNA. Consequently, they do not displace an exon-junction complex (EJC) from the mRNA. Cytoplasmic proteins associate with the EJC and induce degradation of the mRNA. (b) Bone marrow was obtained from a patient with a wild-type β-globin gene and from a patient with β⁰-thalasemia. RNA was isolated from the bone marrow cells shortly after collection, or 30 min after incubation in media with Actinomycin D, a drug that inhibits transcription. The amount of β-globin RNA was measured using the S1-nuclease protection method (arrow). The patient with

β⁰-thalasemia had much less β-globin mRNA than the patient with a wild-type β-globin gene (− Act D). The mutant β-globin mRNA decayed rapidly when transcription was inhibited (+ Act D), whereas the wild-type β-globin mRNA remained stable. (c) Current model of NMD. PTC, premature termination codon. Norm Term, the normal termination codon. SURF, complex of protein kinase SMG1, UPF1, and translation termination factors eRF1 and eRF3. Formation of the SURF complex leads to phosphorylation of UPF1 and association of phospho-UPF1 with a UPF2-UPF3 complex bound to any exon-exon junction complexes that were not displaced from the mRNA by the first, pioneer ribosome to translate the message. This leads to the association of the PTC-containing mRNA with P-bodies, removal of the poly(A) tail, and degradation of the mRNA. [Part (b) from L. E. Maquat et al., 1981, *Cell* **27**:543. Part (c) adapted from J. Hwang et al., 2010, *Mol. Cell* **39**:396.]

protein localization can be achieved by localization of mRNAs to specific regions of the cytoplasm in which their encoded proteins function. In most cases examined thus far, such mRNA localization is specified by sequences in the 3′ untranslated region of the mRNA. A recent genomic-level study of mRNA localization in *Drosophila* embryos revealed that ~70 percent

of 3000 mRNAs analyzed were localized to specific subcellular regions, raising the possibility that this is a much more general phenomenon than previously appreciated.

**Localization of mRNAs to the bud in *S. cerevisiae*** The most thoroughly understood example of mRNA localization occurs

**FIGURE 8-33 Switching of mating type in haploid yeast cells.**
(a) Division by budding forms a larger mother cell (M) and smaller daughter cell (D), both of which have the same mating type as the original cell (α in this example). The mother cell can switch mating type during $G_1$ of the next cell cycle and then divide again, producing two cells of the opposite type (**a** in this example). Switching depends on transcription of the *HO* gene, which occurs only in the absence of Ash1 protein. The smaller daughter cells, which produce Ash1 protein, cannot switch; after growing in size through interphase, they divide to form a mother cell and daughter cell. (b) Model for restriction of mating-type switching to mother cells in *S. cerevisiae*. Ash1 protein prevents a cell from transcribing the *HO* gene whose encoded protein initiates the DNA rearrangement that results in mating-type switching from **a** to α or α to **a**. Switching occurs only in the mother cell, after it separates from a newly budded daughter cell, because the Ash1 protein is present only in the daughter cell. The molecular basis for this differential localization of Ash1 is the one-way transport of *ASH1* mRNA into the bud. A linking protein, She2, binds to specific 3′untranslated sequences in the *ASH1* mRNA and also binds to She3 protein. This protein in turn binds to a myosin motor, Myo4, which moves along actin filaments into the bud. [See S. Koon and B. J. Schnapp, 2001, *Curr. Biology* **11**:R166.]

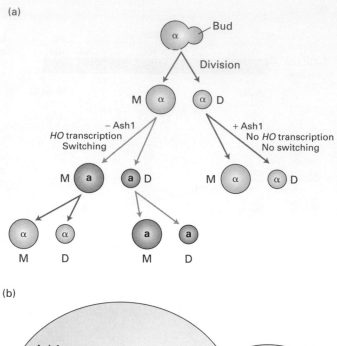

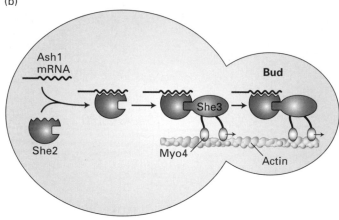

in the budding yeast *S. cerevisiae*. As discussed in Chapter 7, whether a haploid yeast cell exhibits the **a** or α mating type is determined by whether **a** or α genes are present at the expressed *MAT* locus on chromosome III (see Figure 7-33). The process that transfers **a** or α genes from the silent mating-type locus to the expressed *MAT* locus is initiated by a sequence-specific endonuclease called HO. Transcription of the *HO* gene is dependent on the SWI/SNF chromatin-remodeling complex (see Chapter 7, Section 7.5). Daughter yeast cells arising by budding from mother cells contain a transcriptional repressor called Ash1 (for *A*symmetric *s*ynthesis of *HO*) that prevents recruitment of the SWI/SNF complex to the *HO* gene, thereby preventing its transcription. The absence of Ash1 from mother cells allows them to transcribe the *HO* gene. As a consequence mother cells switch their mating type, while daughter cells generated by budding do not (Figure 8-33a).

Ash1 protein accumulates only in daughter cells because the mRNA encoding it is localized to daughter cells. The localization process requires three proteins: She2 (for *SWI-dependent HO expression*), an RNA-binding protein that binds specifically to a localization signal with a specific RNA structure in the ASH1 mRNA; Myo4, a myosin motor protein that moves cargos on actin filaments (see Chapter 17); and She3, which links She2 and therefore ASH1 mRNA to Myo4 (Figure 8-33b). ASH1 mRNA is transcribed in the nucleus of the mother cell before mitosis. Movement of Myo4 with its bound ASH1 mRNA along actin filaments that extend from the mother cell into the bud carries the ASH1 mRNA into the growing bud before cell division.

At least 23 other mRNAs were found to be transported by the She2, She3, Myo4 system. All have an RNA localiza-

tion signal to which She2 binds, usually in their 3′ UTR. The process can be visualized in live cells by the experiment shown in Figure 8-34. RNAs can be fluorescently labeled by including in their sequence high-affinity binding sites for RNA-binding proteins, such as bacteriophage proteins MS2 coat protein and bacteriophage λN protein, that bind to different stem loops of specific sequence (Figure 8-34a). When such engineered mRNAs are expressed in budding yeast cells along with the bacteriophage proteins fused to proteins that fluoresce different colors, the fusion proteins bind to these specific RNA sequences, thereby labeling the RNAs that contain them with different colors. In the experiment shown in Figure 8-34b, ASH1 mRNA was labeled by the binding of green fluorescent protein fused to λN. Another mRNA localized to the bud by this system, the IST2 mRNA encoding a component of the growing bud membrane, was labeled by the binding of red fluorescent protein fused to MS2 coat protein. Video of a budding cell showed that the differently labeled ASH1 and IST2 mRNAs accumulated in the same large cytoplasmic RNP particle containing multiple mRNAs in the mother cell cytoplasm, as can be seen from the merge

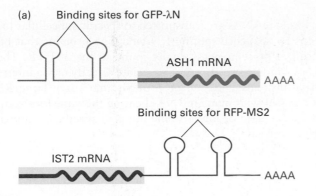

(a) Binding sites for GFP-λN

ASH1 mRNA

AAAA

Binding sites for RFP-MS2

IST2 mRNA

AAAA

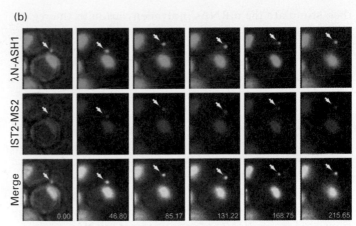

(b)

λN-ASH1

IST2-MS2

Merge

0.00   46.80   85.17   131.22   168.75   215.65

**EXPERIMENTAL FIGURE 8-34 Transport of mRNP particles from a yeast mother cell into the bud.** (a) Yeast cells were engineered to express an ASH1 mRNA with binding sites for the bacteriophage λN protein in its 5′ untranslated region and an IST2 mRNA with binding sites for bacteriophage MS2 coat protein in its 3′ untranslated region. A fusion of green fluorescent protein to λN protein (GFP-λN) and a fusion of red fluorescent protein to MS2 coat protein (RFP-MS2) also were expressed in the same cells. In other experiments, these fluorescently tagged sequence-specific RNA-binding proteins were shown to bind to their own specific binding sites engineered into the ASH1 and IST2 mRNAs, and not to each others' binding sites. Both fluorescently tagged proteins also contained a nuclear localization signal so that the fluorescent proteins that were not bound to their high-affinity binding sites in these mRNAs were transported into nuclei through nuclear pore complexes (see Chapter 13). This was necessary to prevent high fluorescence from excess GFP-λN and RFP-MS2 in the cytoplasm. At

the right, GFP-λN and RFP-MS2 were independently visualized by using millisecond alternating laser excitation of GFP and RFP. (b) Frames from a video of fluorescing cells are shown. The nucleus next to the large vacuole in the mother cell near the center of the micrographs, as well as nuclei in neighboring cells, was observed by green and red fluorescence as shown in the top and middle rows. A merge of the two images is shown in the bottom row, which also indicates the time elapsed between images. An RNP particle containing both the ASH1 mRNA with λN-binding sites and the IST2 mRNA with MS2-binding sites was observed in the mother cell cytoplasm in the left column of images (arrow). The particle increased in intensity between 0.00 and 46.80 seconds, indicating that more of these mRNAs joined the RNP particle. The RNP particle was transported into the bud between 46.80 and 85.17 seconds and then became localized to the bud tip.
[From S. Lange et al., 2008, *Traffic* **9:**1256. See this paper to view the video.]

of the green and red fluorescent signals. The RNP particle was then transported into the bud within about one minute.

**Localization of mRNAs to synapses in the mammalian nervous system** As mentioned earlier, in neurons, localization of specific mRNAs at synapses far from the nucleus in the cell body plays an essential function in learning and memory (Figure 8-35). Like the localized mRNAs in yeast, these mRNAs contain RNA localization signals in their 3′ untranslated region. Some of these mRNAs are initially synthesized with

short poly(A) tails that do not allow translation initiation. Once again, large RNP particles containing multiple mRNAs bearing localization signals form in the cytoplasm near the cell nucleus. In this case, the RNP particles are transported down the axon to synapses by kinesin motor proteins that travel down microtubules extending the length of the axon (see Chapter 18). Electrical activity at a given synapse may

**EXPERIMENTAL FIGURE 8-35 A specific neuronal mRNA localizes to synapses.** Sensory neurons from the sea slug *Aplysia californica* were cultured with target motor neurons so that processes from the sensory neurons formed synapses with processes from the motor neurons. The micrograph at the left shows motor neuron processes visualized with a blue fluorescent dye. GFP-VAMP (green) was expressed in sensory neurons and marks the location of synapses formed between sensory and motor neuron processes (arrows). The micrograph on the right shows red fluorescence from in situ hybridization of an antisensorin mRNA probe. Sensorin is a neurotransmitter expressed by the sensory neuron only; sensory neuron processes are not otherwise visualized in this preparation, but they lie adjacent to the motor neuron processes. The in situ hybridization results indicate that sensorin mRNA is localized to synapses. [From V. Lyles, Y. Zhao, and K. C. Martin, 2006, *Neuron* **49:**323.]

then stimulate the mRNAs' polyadenylation in the region of the synapse, activating the translation of encoded proteins that increase the size and alter the neurophysiological properties of the one synapse while leaving unaffected the hundreds to thousands of other synapses made by the neuron.

## KEY CONCEPTS of Section 8.4

### Cytoplasmic Mechanisms of Post-transcriptional Control

• Translation can be repressed by micro RNAs (miRNAs), which form imperfect hybrids with sequences in the 3′ untranslated region (UTR) of specific target mRNAs.

• The related phenomenon of RNA interference, which probably evolved as an early defense system against viruses and transposons, leads to degradation of mRNAs that form perfect hybrids with short interfering RNAs (siRNAs).

• Both miRNAs and siRNAs contain 21–23 nucleotides, are generated from longer precursor molecules, and are bound by an Argonaute protein and assembled into a multiprotein RNA-induced silencing complex (RISC) that either represses translation of target mRNAs or cleaves them (see Figures 8-25 and 8-26).

• Cytoplasmic polyadenylation is required for translation of mRNAs with a short poly(A) tail. Binding of a specific protein to regulatory elements in their 3′ UTRs represses translation of these mRNAs. Phosphorylation of this RNA-binding protein, induced by an external signal, leads to lengthening of the 3′ poly(A) tail and thus translation (see Figure 8-28).

• Most mRNAs are degraded as the result of the gradual shortening of their poly(A) tail (deadenylation) followed by exosome-mediated 3′→5′ digestion, or removal of the 5′ cap and digestion by a 5′→3′ exonuclease (see Figure 8-29).

• Eukaryotic mRNAs encoding proteins that are expressed in short bursts generally have repeated copies of an AU-rich sequence in their 3′ UTR. Specific proteins that bind to these elements also interact with the deadenylating enzyme and cytoplasmic exosomes, promoting rapid RNA degradation.

• Binding of various proteins to regulatory elements in the 3′ or 5′ UTRs of mRNAs regulates the translation or degradation of many mRNAs in the cytoplasm.

• Translation of ferritin mRNA and degradation of transferrin receptor (TfR) mRNA are both regulated by the same iron-sensitive RNA-binding protein. At low iron concentrations, this protein is in a conformation that binds to specific elements in the mRNAs, inhibiting ferritin mRNA translation or degradation of TfR mRNA (see Figure 8-31). This dual control precisely regulates the iron level within cells.

• Nonsense-mediated decay and other mRNA surveillance mechanisms prevent the translation of improperly processed mRNAs encoding abnormal proteins that might interfere with functioning of the corresponding normal proteins.

• Many mRNAs are transported to specific subcellular locations by sequence-specific RNA-binding proteins that bind localization sequences usually found in the 3′ UTR. These RNA-binding proteins then associate directly or via intermediary proteins with motor proteins that carry large RNP complexes with many mRNAs bearing the same localization signal on actin or microtubule fibers to specific locations in the cytoplasm.

## 8.5 Processing of rRNA and tRNA

Approximately 80 percent of the total RNA in rapidly growing mammalian cells (e.g., cultured HeLa cells) is rRNA, and 15 percent is tRNA; protein-coding mRNA thus constitutes only a small portion of the total RNA. The primary transcripts produced from most rRNA genes and from tRNA genes, like pre-mRNAs, are extensively processed to yield the mature, functional forms of these RNAs.

The ribosome is a highly evolved, complex structure (see Figure 4-23), optimized for its function in protein synthesis. Ribosome synthesis requires the function and coordination of all three nuclear RNA polymerases. The 28S and 5.8S rRNAs associated with the large ribosomal subunit and the single 18S rRNA of the small subunit are transcribed by RNA polymerase I. The 5S rRNA of the large subunit is transcribed by RNA polymerase III, and the mRNAs encoding the ribosomal proteins are transcribed by RNA polymerase II. In addition to the four rRNAs and ~70 ribosomal proteins, at least 150 other RNAs and proteins interact transiently with the two ribosomal subunits during their assembly through a series of coordinated steps. Furthermore, multiple specific bases and riboses of the mature rRNAs are modified to optimize their function in protein synthesis. Although most of the steps in ribosomal subunit synthesis and assembly occur in the **nucleolus** (a subcompartment of the nucleus not bounded by a membrane), some occur in the nucleoplasm during passage from the nucleolus to nuclear pore complexes. A quality-control step occurs before nuclear export so that only fully functional subunits are exported to the cytoplasm, where the final steps of ribosome subunit maturation occur. tRNAs also are processed from precursor primary transcripts in the nucleus and modified extensively before they are exported to the cytoplasm and used in protein synthesis. First we'll discuss the processing and modification of rRNA and the assembly and nuclear export of ribosomes. Then we'll consider the processing and modification of tRNAs.

### Pre-rRNA Genes Function as Nucleolar Organizers and Are Similar in All Eukaryotes

The 28S and 5.8S rRNAs associated with the large (60S) ribosomal subunit and the 18S rRNA associated with the small (40S) ribosomal subunit in higher eukaryotes (and the

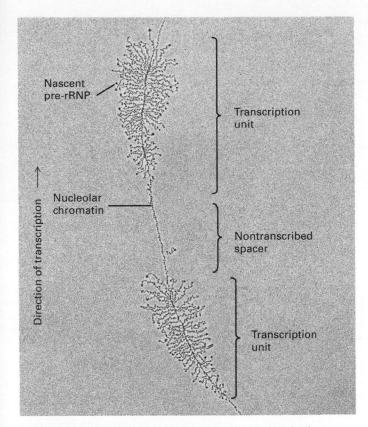

**EXPERIMENTAL FIGURE 8-36 Electron micrograph of pre-rRNA transcription units from the nucleolus of a frog oocyte.** Each "feather" represents multiple pre-rRNA molecules associated with protein in a pre-ribonucleoprotein complex (pre-rRNP) emerging from a transcription unit. Note the dense "knob" at the 5' end of each nascent pre-RNP thought to be a processome. Pre-rRNA transcription units are arranged in tandem, separated by nontranscribed spacer regions of nucleolar chromatin. [Courtesy of Y. Osheim and O. J. Miller, Jr.]

The synthesis and most of the processing of pre-rRNA occurs in the *nucleolus*. When pre-rRNA genes initially were identified in the nucleolus by in situ hybridization, it was not known whether any other DNA was required to form the nucleolus. Subsequent experiments with transgenic *Drosophila* strains demonstrated that a single complete pre-rRNA transcription unit induces formation of a small nucleolus. Thus a single pre-rRNA gene is sufficient to be a *nucleolar organizer,* and all the other components of the ribosome diffuse to the newly formed pre-rRNA. The structure of the nucleolus observed by light and electron microscopy results from the processing of pre-RNA and the assembly of ribosomal subunits.

## Small Nucleolar RNAs Assist in Processing Pre-rRNAs

Ribosomal subunit assembly, maturation, and export to the cytoplasm are best understood in the yeast *S. cerevisiae.* However, nearly all the proteins and RNAs involved are highly conserved in multicellular eukaryotes, where the fundamental aspects of ribosome biosynthesis are likely to be the same. As for pre-mRNAs, nascent pre-rRNA transcripts are immediately bound by proteins, forming preribosomal ribonucleoprotein particles (pre-rRNPs). For reasons not yet known, cleavage of the pre-rRNA does not begin until transcription of the pre-rRNA is nearly complete. In yeast, it takes approximately six minutes for a pre-rRNA to be transcribed. Once transcription is complete, the rRNA is cleaved, and bases and riboses are modified in about 10 seconds. In a rapidly growing yeast cell, ~40 pairs of ribosomal subunits are synthesized, processed, and transported to the cytoplasm every second. This extremely high rate of ribosome synthesis despite the seemingly long period required to transcribe a pre-rRNA is possible because pre-rRNA genes are packed

functionally equivalent rRNAs in all other eukaryotes) are encoded by a single type of **pre-rRNA** transcription unit. In human cells, transcription by RNA polymerase I yields a 45S (~13.7 kb) primary transcript (pre-rRNA), which is processed into the mature 28S, 18S, and 5.8S rRNAs found in cytoplasmic ribosomes. The fourth rRNA, 5S, is encoded separately and transcribed outside the nucleolus. Sequencing of the DNA encoding the 45S pre-rRNA from many species showed that this DNA shares several properties in all eukaryotes. First, the pre-rRNA genes are arranged in long tandem arrays separated by nontranscribed spacer regions ranging in length from ~2 kb in frogs to ~30 kb in humans (Figure 8-36). Second, the genomic regions corresponding to the three mature rRNAs are always arranged in the same 5'→3' order: 18S, 5.8S, and 28S. Third, in all eukaryotic cells (and even in bacteria), the pre-rRNA gene codes for regions that are removed during processing and rapidly degraded. These regions probably contribute to proper folding of the rRNAs but are not required once the folding has occurred. The general structure of pre-rRNAs is diagrammed in Figure 8-37.

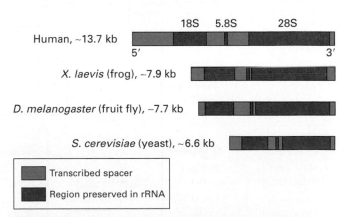

**FIGURE 8-37 General structure of eukaryotic pre-rRNA transcription units.** The three coding regions (red) encode the 18S, 5.8S, and 28S rRNAs found in ribosomes of higher eukaryotes or their equivalents in other species. The order of these coding regions in the genome is always 5'→3'. Variations in the lengths of the transcribed spacer regions (blue) account for the major difference in the lengths of pre-rRNA transcription units in different organisms.

with RNA polymerase I molecules transcribing the same gene simultaneously (see Figure 8-36) and because there are 100–200 such genes on chromosome XII, the yeast nucleolar organizer.

The primary transcript of ~7 kb is cut in a series of cleavage and exonucleolytic steps that ultimately yield the mature rRNAs found in ribosomes (Figure 8-38). During processing, pre-rRNA also is extensively modified, mostly by methylation of the 2'-hydroxyl group of specific riboses and conversion of specific uridine residues to pseudouridine. These post-transcriptional modifications of rRNA are probably important for protein synthesis, because they are highly conserved. Virtually all of these modifications occur in the most conserved core structure of the ribosome, which is directly involved in protein synthesis. The positions of the specific sites of 2'-O-methylation and pseudouridine formation are determined by approximately 150 different small nucleolus-restricted RNA species, called **small nucleolar RNAs (snoRNAs)**, which hybridize transiently to pre-rRNA molecules. Like the snRNAs that function in pre-mRNA processing, snoRNAs associate with proteins, forming ribonucleoprotein particles called snoRNPs. One class of more than 40 snoRNPs (containing box C+D snoRNAs) positions a methyltransferase enzyme near methylation sites in the pre-mRNA. The multiple different box C+D snoRNAs direct methylation at multiple sites through a similar mechanism. They share common sequence and structural features and are bound by a common set of proteins. One or two regions

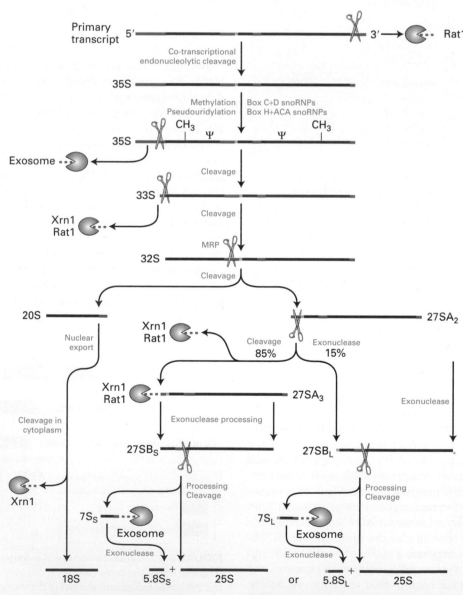

**FIGURE 8-38 rRNA processing.** Endoribonucleases that make internal cleavages are represented as scissors. Exoribonucleases that digest from one end, either 5' or 3', are shown as Pac-Men. Most 2'-O-ribose methylation ($CH_3$) and generation of pseudouridines in the rRNAs occurs following the initial cleavage at the 3' end, before the initial cleavage at the 5' end. Proteins and snoRNPs known to participate in these steps are indicated. [From J. Venema and D. Tollervey, 1999, *Ann. Rev. Genetics* **33**:261.]

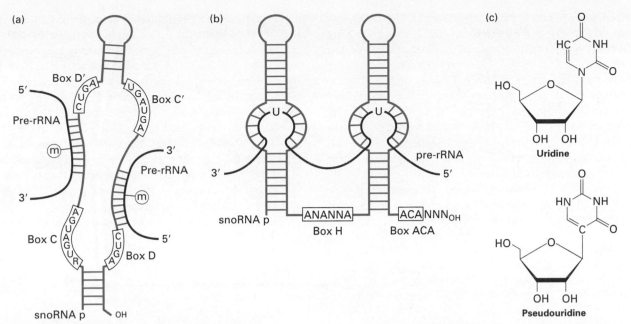

**FIGURE 8-39 snoRNP-directed modification of pre-rRNA.** (a) A snoRNA called box C+D snoRNA is involved in ribose 2'-*O*-methylation. Sequences in this snoRNA hybridize to two different regions in the pre-rRNA, directing methylation at the indicated sites. (b) Box H+ACA snoRNAs fold into two stem loops with internal single-stranded bulges in the stems. Pre-rRNA hybridizes to the single-stranded bulges, demarcating a site of pseudouridylation. (c) Conversion from uridine to pseudouridine directed by the box H+ACA snoRNAs of part (b). [Part (a) from T. Kiss, 2001, *EMBO J.* **20**:3617. Part (b) from U. T. Meier, 2005, *Chromosoma* **114**:1.]

of each of these snoRNAs are precisely complementary to sites on the pre-rRNA and direct the methyltransferase to specific riboses in the hybrid region (Figure 8-39a). A second major class of snoRNPs (containing box H+ACA snoRNAs) positions the enzyme that converts uridine to pseudouridine (Figure 8-39b). This conversion involves rotation of the pyrimidine ring (Figure 8-39c). Bases on either side of the modified uridine in the pre-rRNA base-pair with bases in the bulge of a stem in the H+ACA snoRNAs, leaving the modified uridine bulged out of the helical double-stranded region, like the branch point A bulges out in pre-mRNA spliceosomal splicing (see Figure 8-10). Other modifications of pre-rRNA nucleotides, such as adenine dimethylation, are carried out by specific proteins without the assistance of guiding snoRNAs.

The U3 snoRNA is assembled into a large snoRNP containing ~72 proteins called the small subunit (SSU) processome, which specifies cleavage at site $A_0$, the initial cut near the 5' end of the pre-rRNA (see Figure 8-38). U3 snoRNA base-pairs with an upstream region of the pre-rRNA to specify the location of the cleavage. The processome is thought to form the "5' knob" visible in electron micrographs of pre-rRNPs (see Figure 8-36). Base pairing of other snoRNPs specify additional cleavage reactions that remove transcribed spacer regions. The first cleavage to initiate processing of the 5.8S and 25S rRNAs of the large subunit is performed by RNase MRP, a complex of nine proteins with an RNA. Once cleaved from pre-rRNAs, these sequences are degraded by the same exosome-associated 3'→5' nuclear exonucleases that degrade introns spliced from pre-mRNAs. Nuclear 5'→3' exoribonucleases (Rat1; Xrn1) also remove some regions of 5' spacer.

Some snoRNAs are expressed from their own promoters by RNA polymerase II or III. Remarkably, however, the large majority of snoRNAs are processed from spliced-out introns of genes encoding functional mRNAs for proteins involved in ribosome synthesis or translation. Some snoRNAs are processed from introns spliced from apparently nonfunctional mRNAs. The genes encoding these mRNAs seem to exist only to express snoRNAs from excised introns.

Unlike 18S, 5.8S, and 28S genes, 5S rRNA genes are transcribed by RNA polymerase III in the nucleoplasm outside the nucleolus. With only minor additional processing to remove nucleotides at the 3' end, 5S rRNA diffuses to the nucleolus, where it assembles with the pre-rRNA precursor and remains associated with the region that is cleaved into the precursor of the large ribosomal subunit.

Most of the ribosomal proteins of the small 40S ribosomal subunit associate with the nascent pre-rRNA during transcription (Figure 8-40). Cleavage of the full-length pre-rRNA in the 90S RNP precursor releases a pre-40S particle that requires only a few more remodeling steps before it is transported to the cytoplasm. Once the pre-40S particle leaves the nucleolus, it traverses the nucleoplasm quickly and is exported through nuclear pore complexes (NPCs), as discussed below. Final maturation of the small ribosomal subunit occurs in the cytoplasm: exonucleolytic processing of the 20S rRNA into mature small subunit 18S rRNA by the cytoplasmic 5'→3' exoribonuclease Xrn1 and the dimethylation of two adjacent adenines near the 3' end of 18S rRNA by the cytoplasmic enzyme Dim1.

In contrast to the pre-40S particle, the precursor of the large subunit requires considerably more remodeling through

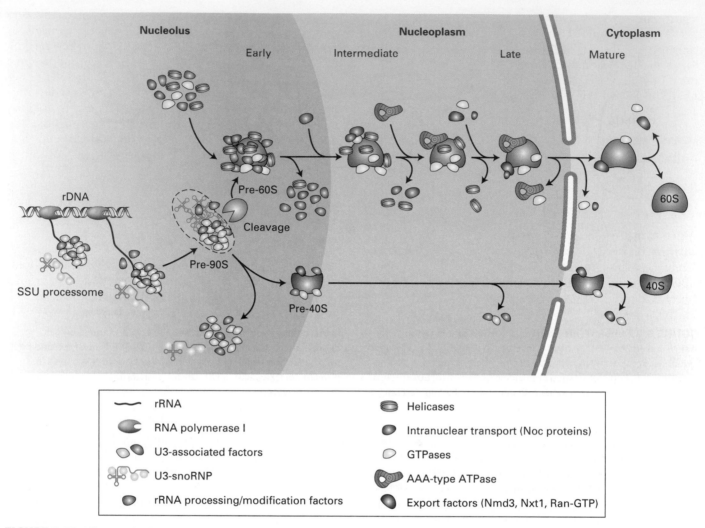

**Nucleolus**        **Nucleoplasm**        **Cytoplasm**

Early     Intermediate     Late     Mature

rDNA

Pre-60S

Cleavage

Pre-90S

SSU processome

Pre-40S

60S

40S

| | | |
|---|---|---|
| ⎯⎯⎯ | rRNA | |
| | RNA polymerase I | |
| | U3-associated factors | |
| | U3-snoRNP | |
| | rRNA processing/modification factors | |

| | | |
|---|---|---|
| | Helicases | |
| | Intranuclear transport (Noc proteins) | |
| | GTPases | |
| | AAA-type ATPase | |
| | Export factors (Nmd3, Nxt1, Ran-GTP) | |

**FIGURE 8-40 Ribosomal subunit assembly.** Ribosomal proteins and RNAs in the maturing small and large ribosomal subunits are depicted in blue, with a shape similar to the icons for the mature subunits in the cytoplasm. Other factors that associate transiently with the maturing subunits are depicted in different colors, as shown in the key. [From H. Tschochner and E. Hurt, 2003, *Trends Cell Biol.* **13**:255.]

many more transient interactions with nonribosomal proteins before it is sufficiently mature for export to the cytoplasm. Consequently, it takes a considerably longer period for the maturing 60S subunit to exit the nucleus (30 minutes compared to 5 minutes for export of the 40S subunit in cultured human cells). Multiple presumptive RNA helicases and small G proteins are associated with the maturing pre-60S subunits. Some RNA helicases are necessary to dislodge the snoRNPs that base-pair perfectly with pre-rRNA over up to 30 base pairs. Other RNA helicases may function in the disruption of protein-RNA interactions. The requirement for so many GTPases suggests that there are many quality-control checkpoints in the assembly and remodeling of the large subunit RNP, where one step must be completed before a GTPase is activated to allow the next step to proceed. Members of the **AAA ATPase** family are also bound transiently. This class of proteins is often involved in large molecular movements and may be required to fold the complex, large rRNA into the proper conformation. Some steps in 60S subunit maturation

occur in the nucleoplasm, during passage from the nucleolus to nuclear pore complexes (see Figure 8-40). Much remains to be learned about the complex, fascinating, and essential remodeling processes that occur during formation of the ribosomal subunits.

The large ribosomal subunit is one of the largest structures to pass through nuclear pore complexes. Maturation of the large subunit in the nucleoplasm leads to the generation of binding sites for a nuclear export adapter called Nmd3. Nmd3 is bound by the nuclear transporter Exportin1 (also called Crm1). This is another quality-control step, because only correctly assembled subunits can bind Nmd3 and be exported. The small subunit of the mRNP exporter (NXT1) also becomes associated with the nearly mature large ribosomal subunit. These nuclear transporters interact with FG-domains of FG-nucleoporins. This mechanism allows penetration of the molecular "cloud" that fills most of the central channel of the NPC (see Figure 8-20). Several specific nucleoporins without FG-domains are also required for ribosomal subunit export and may have additional functions specific for this task. The dimensions of ribosomal subunits (~25–30 nm in diameter) and the central channel of the NPC are comparable, so passage may not require distortion of either the ribosomal subunit or the channel. Final maturation of the large subunit in the cytoplasm includes removal of these export factors. As for the export of most macromolecules from the nucleus, including tRNAs and pre-miRNAs (but not most mRNPs), ribosome subunit export requires the function of a small G protein called Ran, as discussed in Chapter 13.

## Self-Splicing Group I Introns Were the First Examples of Catalytic RNA

During the 1970s, the pre-rRNA genes of the protozoan *Tetrahymena thermophila* were discovered to contain an intron. Careful searches failed to uncover even one pre-rRNA gene without the extra sequence, indicating that splicing is required to produce mature rRNA in these organisms. In 1982, in vitro studies showing that the pre-rRNA was spliced at the correct sites in the absence of any protein provided the first indication that RNA can function as a catalyst, like enzymes.

A whole raft of self-splicing sequences subsequently were found in pre-rRNAs from other single-celled organisms, in mitochondrial and chloroplast pre-rRNAs, in several pre-mRNAs from certain *E. coli* bacteriophages, and in some bacterial tRNA primary transcripts. The self-splicing sequences in all these precursors, referred to as *group I introns*, use guanosine as a cofactor and can fold by internal base pairing to juxtapose closely the two exons that must be joined. As discussed earlier, certain mitochondrial and chloroplast pre-mRNAs and tRNAs contain a second type of self-splicing intron, designated group II.

The splicing mechanisms used by group I introns, group II introns, and spliceosomes are generally similar, involving two transesterification reactions, which require no input of energy (Figure 8-41). Structural studies of the group I intron from *Tetrahymena* pre-rRNA combined with mutational and biochemical experiments have revealed that the RNA folds into a precise three-dimensional structure that, like protein enzymes, contains deep grooves for binding substrates and solvent-inaccessible regions that function in catalysis. The group I intron functions like a metalloenzyme to precisely orient the atoms that participate in the two transesterification

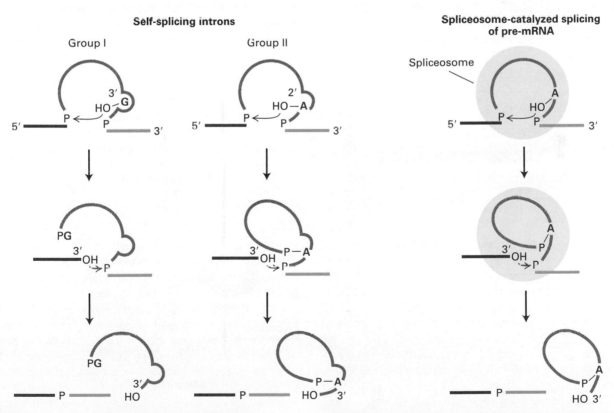

**FIGURE 8-41 Splicing mechanisms in group I and group II self-splicing introns and spliceosome-catalyzed splicing of pre-mRNA.** The intron is shown in gray, the exons to be joined in red. In group I introns, a guanosine cofactor (G) that is not part of the RNA chain associates with the active site. The 3'-hydroxyl group of this guanosine participates in a transesterification reaction with the phosphate at the 5' end of the intron; this reaction is analogous to that involving the 2'-hydroxyl groups of branch-site As in group II introns and pre-mRNA introns spliced in spliceosomes (see Figure 8-8). The subsequent transesterification that links the 5' and 3' exons is similar in all three splicing mechanisms. Note that spliced-out group I introns are linear structures, unlike the branched intron products in the other two cases. [Adapted from P. A. Sharp, 1987, *Science* **235**:769.]

reactions adjacent to catalytic $Mg^{2+}$ ions. Considerable evidence now indicates that splicing by group II introns and by snRNAs in the spliceosome also involves bound catalytic $Mg^{2+}$ ions. In both the group I and II self-splicing introns and probably in the spliceosome, RNA functions as a **ribozyme**, an RNA sequence with catalytic ability.

## Pre-tRNAs Undergo Extensive Modification in the Nucleus

Mature cytosolic tRNAs, which average 75–80 nucleotides in length, are produced from larger precursors (pre-tRNAs) synthesized by RNA polymerase III in the nucleoplasm. Mature tRNAs also contain numerous modified bases that are not present in tRNA primary transcripts. Cleavage and base modification occur during processing of all pre-tRNAs; some pre-tRNAs also are spliced during processing. All of these processing and modification events occur in the nucleus.

A 5' sequence of variable length that is absent from mature tRNAs is present in all pre-tRNAs (Figure 8-42). This occurs because the 5' end of mature tRNAs is generated by an endonucleolytic cleavage specified by the tRNA three-dimensional structure rather than the start site of transcription. These extra 5' nucleotides are removed by ribonuclease P (RNase P), a ribonucleoprotein endonuclease. Studies with *E. coli* RNase P indicate that at high $Mg^{2+}$ concentrations, the RNA component alone can recognize and cleave *E. coli* pre-tRNAs. The

RNase P polypeptide increases the rate of cleavage by the RNA, allowing it to proceed at physiological $Mg^{2+}$ concentrations. A comparable RNase P functions in eukaryotes.

About 10 percent of the bases in pre-tRNAs are modified enzymatically during processing. Three classes of base modifications occur (Figure 8-42): (1) U residues at the 3' end of pre-tRNA are replaced with a CCA sequence. The CCA sequence is found at the 3' end of all tRNAs and is required for their charging by aminoacyl-tRNA synthetases during protein synthesis. This step in tRNA synthesis likely functions as a quality-control point, since only properly folded tRNAs are recognized by the CCA addition enzyme. (2) Methyl and isopentenyl groups are added to the heterocyclic ring of purine bases, and the 2'-OH groups in the ribose of specific residues are methylated. (3) Specific uridines are converted to dihydrouridine, pseudouridine, or ribothymidine residues. The functions of these base and ribose modifications are not well understood, but since they are highly conserved, they probably have a positive influence on protein synthesis.

As shown in Figure 8-42, the pre-tRNA expressed from the yeast tyrosine tRNA (tRNA$^{Tyr}$) gene contains a 14-base intron that is not present in mature tRNA$^{Tyr}$. Some other eukaryotic tRNA genes and some archaeal tRNA genes also contain introns. The introns in nuclear pre-tRNAs are shorter than those in pre-mRNAs and lack the consensus splice-site sequences found in pre-mRNAs (see Figure 8-7). Pre-tRNA introns also are clearly distinct from the much longer self-splicing group I

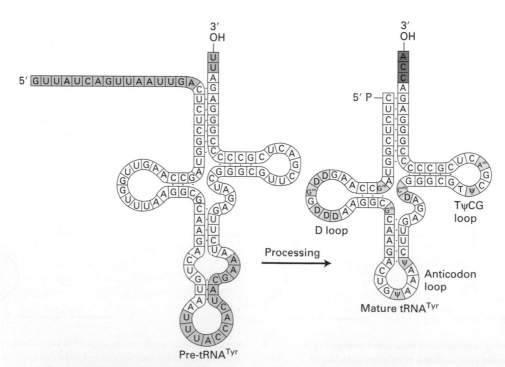

**FIGURE 8-42 Changes that occur during the processing of tyrosine pre-tRNA.** A 14-nucleotide intron (blue) in the anticodon loop is removed by splicing. A 16-nucleotide sequence (green) at the 5' end is cleaved by RNase P. U residues at the 3' end are replaced by the CCA sequence (red) found in all mature tRNAs. Numerous bases in the stem loops are converted to characteristic modified bases (yellow). Not all pre-tRNAs contain introns that are spliced out during processing, but they all undergo the other types of changes shown here. D = dihydrouridine; ψ = pseudouridine.

and group II introns found in chloroplast and mitochondrial pre-rRNAs. The mechanism of pre-tRNA splicing differs in three fundamental ways from the mechanisms utilized by self-splicing introns and spliceosomes (see Figure 8-41). First, splicing of pre-tRNAs is catalyzed by proteins, not by RNAs. Second, a pre-tRNA intron is excised in one step that entails simultaneous cleavage at both ends of the intron. Finally, hydrolysis of GTP and ATP is required to join the two tRNA halves generated by cleavage on either side of the intron.

After pre-tRNAs are processed in the nucleoplasm, the mature tRNAs are transported to the cytoplasm through nuclear pore complexes by Exportin-t, as discussed previously. In the cytoplasm, tRNAs are passed between aminoacyl-tRNA synthetases, elongation factors, and ribosomes during protein synthesis (Chapter 4). Thus tRNAs generally are associated with proteins and spend little time free in the cell, as is also the case for mRNAs and rRNAs.

## Nuclear Bodies Are Functionally Specialized Nuclear Domains

High-resolution visualization of plant- and animal-cell nuclei by electron microscopy and subsequent staining with fluorescently labeled antibodies has revealed domains in nuclei in addition to chromosome territories and nucleoli. These specialized nuclear domains, called **nuclear bodies**, are not surrounded by membranes but are nonetheless regions of high concentrations of specific proteins and RNAs that form distinct, roughly spherical structures within the nucleus. The most prominent nuclear bodies are nucleoli, the sites of ribosomal subunit synthesis and assembly discussed earlier. Several other types of nuclear bodies also have been described in structural studies.

Experiments with fluorescently labeled nuclear proteins have shown that the nucleus is a highly dynamic environment, with rapid diffusion of proteins through the nucleoplasm. Proteins associated with nuclear bodies are often also observed at lower concentrations in the nucleoplasm outside the nuclear bodies, and fluorescence studies indicate that they diffuse in and out of the nuclear bodies. Based on these measurements of molecular mobility in living cells, nuclear bodies can be mathematically modeled as the expected steady state for diffusing proteins that interact with sufficient affinity to form self-organized regions of high concentrations of specific proteins but with low enough affinity for each other to be able to diffuse in and out of the structure. In electron micrographs these structures appear to be a heterogeneous, spongelike network of interacting components. We discuss a few examples of nuclear bodies here.

**Cajal Bodies** Cajal bodies are ~0.2–1 μm spherical structures that have been observed in large nuclei for more than a century (Figure 8-43). Current research indicates that like nucleoli, Cajal bodies are centers of RNP-complex assembly for spliceosomal snRNPs and other RNPs. Like rRNAs, snRNAs undergo specific modifications, such as the conversion of specific uridine residues to pseudouridine and addition of

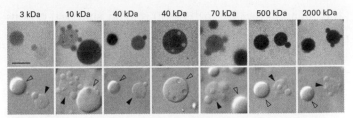

**FIGURE 8-43 Nuclear bodies are differentially permeable to molecules in the bulk nucleoplasm.** Each pair of panels shows a single area through a living *Xenopus* oocyte nucleus that was previously injected with fluorescent dextran of the indicated molecular mass (3–2000 kDa). Each section of the upper panel is a confocal image in which the intensity of fluorescence is a measure of dextran concentration (i.e., darker areas show regions where dextran has been excluded). Each section of the lower panel is a differential interference contrast image of the same field. Open arrowheads indicate nucleoli, closed arrowheads Cajal bodies (CBs) with attached nuclear speckles which are much larger in *Xenopus* oocytes than in most somatic cells. Dextrans of low molecular mass (e.g., 3 kDa) almost completely penetrated CBs but were excluded more from nuclear speckles and nucleoli. Exclusion of dextran increased with molecular mass. Bar = 10 μm. [From K. E. Handwerger et al., 2005, *Mol. Biol. Cell* **16**:202.]

methyl groups to the 2′-hydroxyl groups of specific riboses. These post-transcriptional modifications are important for the proper assembly and function of snRNPs in pre-mRNA splicing. These modifications occur in Cajal bodies, where they are directed by a class of snoRNA-like guide RNA molecules called *scaRNAs* (small Cajal body–associated RNAs). There is also evidence that the Cajal body is the site of reassembly of the U4/U6/U5 tri-snRNP complexes required for pre-mRNA splicing from the free U4, U5, and U6 snRNPs released during the removal of each intron (see Figure 8-11). Since Cajal bodies also contain a high concentration of the U7 snRNP involved in the specialized 3′-end processing of the major histone mRNAs, it is likely that this process also occurs in Cajal bodies, as may the assembly of the telomerase RNP.

**Nuclear Speckles** Nuclear speckles were observed, using fluorescently labeled antibodies to snRNP proteins and other proteins involved in pre-mRNA splicing, as approximately 25–50 irregular, amorphous structures 0.5–2 μm in diameter that are distributed through the nucleoplasm of vertebrate cells. Since speckles are not located at sites of co-transcriptional pre-mRNA splicing, which are associated closely with chromatin, they are thought to be storage regions for snRNPs and proteins involved in pre-mRNA splicing that are released into the nucleoplasm when required.

**Promyelocytic Leukemia (PML) Nuclear Bodies** The *PML* gene was originally discovered when chromosomal translocations within the gene were observed in the leukemic cells of patients with the rare disease promyelocytic leukemia (PML). When antibodies specific for the PML protein were used in immunofluorescence microscopy studies, the protein was found to localize to ~10–30 roughly spherical regions 0.3–1 μm in diameter in the nuclei of mammalian cells. Multiple functions have been proposed for these PML nuclear bodies, but a consensus is emerging that they function as sites for the assembly and modification of protein complexes involved in DNA repair and the induction of apoptosis. For example, the important p53 tumor suppressor protein appears to be post-translationally modified by phosphorylation and acetylation in PML nuclear bodies in response to DNA damage, increasing its ability to activate the expression of DNA-damage response genes. PML nuclear bodies are also required for cellular defenses against DNA viruses that are induced by interferons, proteins secreted by virus-infected cells and T-lymphocytes involved in the immune response (see Chapter 23).

PML nuclear bodies are also sites of protein post-translational modification through the addition of a small, ubiquitin-like protein called *SUMO1* (*s*mall *u*biquitin-like *m*oiety-1), which can control the activity and subcellular localization of the modified protein. Many transcriptional activators are inhibited when they are sumoylated, and mutation of their site of sumoylation increases their activity in stimulating transcription. These observations indicate that PML nuclear bodies are involved in a mechanism of transcriptional repression that remains to be studied and thoroughly understood.

### Nucleolar Functions in Addition to Ribosomal Subunit Synthesis
The first nuclear bodies to be observed, the nucleoli, may have specialized regions of substructure that are dedicated to functions other than ribosome biogenesis. There is evidence that immature SRP ribonucleoprotein complexes involved in protein secretion and ER membrane insertion (Chapter 13) are assembled in nucleoli and then exported to the cytoplasm, where their final maturation takes place. The Cdc14 protein phosphatase that regulates processes in the final stages of mitosis is sequestered in nucleoli in yeast cells until chromosomes have been properly segregated into the bud (Chapter 19). Also, a tumor suppressor protein called ARF, which is involved in the regulation of the protein encoded by the most frequently mutated gene in human cancers, p53, is sequestered in nucleoli and released in response to DNA damage (Chapter 24). In addition, heterochromatin often forms on the surface of nucleoli (Figure 6-33), suggesting that proteins associated with nucleoli also participate in the formation of this repressing chromatin structure.

### KEY CONCEPTS of Section 8.5

#### Processing of rRNA and tRNA

- A large precursor pre-rRNA (13.7 kb in humans) synthesized by RNA polymerase I undergoes cleavage, exonucleolytic digestion, and base modifications to yield mature 28S, 18S, and 5.8S rRNAs, which associate with ribosomal proteins into ribosomal subunits.

- Synthesis and processing of pre-rRNA occur in the nucleolus. The 5S rRNA component of the large ribosomal subunit is synthesized in the nucleoplasm by RNA polymerase III.

- Approximately 150 snoRNAs, associated with proteins in snoRNPs, base-pair with specific sites in pre-rRNA where they direct ribose methylation, modification of uridine to pseudouridine, and cleavage at specific sites during rRNA processing in the nucleolus.

- Group I and group II self-splicing introns and probably snRNAs in spliceosomes all function as ribozymes, or catalytically active RNA sequences, that carry out splicing by analogous transesterification reactions requiring bound $Mg^{2+}$ ions (see Figure 8-41).

- Pre-tRNAs synthesized by RNA polymerase III in the nucleoplasm are processed by removal of the 5′-end sequence, addition of CCA to the 3′ end, and modification of multiple internal bases (see Figure 8-42).

- Some pre-tRNAs contain a short intron that is removed by a protein-catalyzed mechanism distinct from the splicing of pre-mRNA and self-splicing introns.

- All species of RNA molecules are associated with proteins in various types of ribonucleoprotein particles, both in the nucleus and after export to the cytoplasm.

- Nuclear bodies are functionally specialized regions in the nucleus where interacting proteins form self-organized structures. Many of these, like the nucleolus, are regions of assembly of RNP complexes.

## Perspectives for the Future

In this and the previous chapter, we have seen that in eukaryotic cells, mRNAs are synthesized and processed in the nucleus, transported through nuclear pore complexes to the cytoplasm, and then, in some cases, transported to specific areas of the cytoplasm before being translated by ribosomes. Each of these fundamental processes is carried out by complex macromolecular machines composed of scores of proteins and in many cases RNAs as well. The complexity of these macromolecular machines ensures accuracy in finding promoters and splice sites in the long length of DNA and RNA sequences and provides various avenues for regulating synthesis of a polypeptide chain. Much remains to be learned about the structure, operation, and regulation of such complex machines as spliceosomes and the cleavage/polyadenylation apparatus.

Recent examples of the regulation of pre-mRNA splicing raise the question of how extracellular signals might control such events, especially in the nervous system of vertebrates. A case in point is the remarkable situation in the chick inner ear, where multiple isoforms of the $Ca^{2+}$-activated $K^+$ channel called *Slo* are produced by alternative RNA splicing. Cell-cell interactions appear to inform cells of their position in the cochlea, leading to alternative splicing of Slo pre-mRNA. The challenging task facing researchers is to discover how such cell-cell interactions regulate the activity of RNA-processing factors.

The mechanism of mRNP transport through nuclear pore complexes poses many intriguing questions. Future research will likely reveal additional activities of hnRNP and nuclear mRNP proteins and clarify their mechanisms of action. For instance, there is a small gene family encoding proteins homologous to the large subunit of the mRNA exporter. What are the functions of these related proteins? Do they participate in the transport of overlapping sets of mRNPs? Some hnRNP proteins contain nuclear-retention signals that prevent nuclear export when fused to hnRNP proteins with nuclear-export signals (NESs). How are these hnRNP proteins selectively removed from processed mRNAs in the nucleus, allowing the mRNAs to be transported to the cytoplasm?

The localization of certain mRNAs to specific subcellular locations is fundamental to the development of multicellular organisms. As we will discuss in Chapter 21, during development an individual cell frequently divides into daughter cells that function differently from each other. In the language of developmental biology, the two daughter cells are said to have different developmental fates. In many cases, this difference in developmental fate results from the localization of an mRNA to one region of the cell before mitosis so that after cell division, it is present in one daughter cell and not the other. Much exciting work remains to be done to fully understand the molecular mechanisms controlling mRNA localization that are critical for the normal development of multicellular organisms.

Some of the most exciting and unanticipated discoveries in molecular cell biology in recent years have concerned the existence and function of miRNAs and the process of RNA interference. RNA interference (RNAi) provides molecular cell biologists with a powerful method for studying gene function. The discovery of ~500 miRNAs in humans and other organisms suggests that multiple significant examples of translational control by this mechanism await characterization. Recent studies in *S. pombe* and plants link similar short nuclear RNAs to the control of DNA methylation and the formation of heterochromatin. Will similar processes control gene expression through the assembly of heterochromatin in humans and other animals? What other regulatory processes might be directed by other kinds of small RNAs? Since control by these mechanisms depends on base pairing between miRNAs and target mRNAs or genes, genomic and bioinformatic methods will probably suggest genes that may be controlled by these mechanisms. What other processes in addition to translation control, mRNA degradation, and heterochromatin assembly might be controlled by miRNAs?

These are just a few of the fascinating questions concerning RNA processing, post-transcriptional control, and nuclear transport that will challenge molecular cell biologists in the coming decades. The astounding discoveries of entirely unanticipated mechanisms of gene control by miRNAs remind us that many more surprises are likely in the future.

## Key Terms

| | |
|---|---|
| 5′ cap 349 | poly(A) tail 358 |
| alternative splicing 361 | pre-mRNA 349 |
| cleavage/polyadenylation complex 358 | pre-rRNA 385 |
| cross-exon recognition complex 356 | ribozyme 390 |
| Dicer 371 | RNA editing 364 |
| Drosha 371 | RNA-induced silencing complex (RISC) 371 |
| exosome 359 | RNA interference (RNAi) 373 |
| FG-nucleoporins 365 | RNA splicing 348 |
| group I introns 357 | short interfering RNAs (siRNA) 346 |
| group II introns 357 | siRNA knockdown 374 |
| iron-response element–binding protein (IRE-BP) 379 | small nuclear RNAs (snRNAs) 352 |
| micro RNAs (miRNAs) 370 | small nucleolar RNAs (snoRNAs) 386 |
| mRNA surveillance 380 | spliceosome 353 |
| mRNP exporter 365 | SR proteins 356 |
| nuclear pore complex (NPC) 365 | |

## Review the Concepts

1. Describe three types of post-transcriptional regulation of protein-coding genes.

2. True or False? The CTD is responsible for mRNA-processing steps that are specific for mRNA, and not other forms of RNA. Explain why you chose true or false.

3. There are a number of conserved sequences found in an mRNA which dictate where splicing occurs. Where are these sequences found relative to the exon/intron junctions? What is the significance of these sequences in the splicing process? One of these important regions is the branch point A found in the intron. What is the role of the branch point A in the splicing process, and can this be accomplished with the OH group on either the 2′ or the 3′ carbon?

4. What is the difference between hnRNAs, snRNAs, miRNAs, siRNAs, and snoRNAs?

5. What are the mechanistic similarities between group II intron self-splicing and spliceosomal splicing? What is the evidence that there may be an evolutionary relationship between the two?

6. You obtain the sequence of a gene containing 10 exons, 9 introns, and a 3′ UTR containing a polyadenylation consensus sequence. The fifth intron also contains a polyadenylation site. To test whether both polyadenylation sites are used, you isolate mRNA and find a longer transcript from muscle tissue and a shorter transcript from all other tissues. Speculate about the mechanism involved in the production of these different transcripts.

7. RNA editing is a common process occurring in the mitochondria of trypanosomes and plants, in chloroplasts, and in rare cases in higher eukaryotes. What is RNA editing, and what benefit does it demonstrate in the documented example of apoB in humans?

8. As DNA is found in the nucleus, transcription is a nuclear-localized process. Ribosomes responsible for protein synthesis are found in the cytoplasm. Why is hnRNP trafficking to the cytoplasm restricted to the nuclear pore complexes? How do the FG-repeats of the nuclear pore complexes act as a specificity barrier in nuclear transport?

9. A protein complex in the nucleus is responsible for transporting mRNA molecules into the cytoplasm. Describe the proteins that form this exporter. What two protein groups are likely behind the mechanism involved in the directional movement of the mRNP and exporter into the cytosol.

10. RNA knockdown has become a powerful tool in the arsenal of methods to deregulate gene expression. Briefly describe how gene expression can be knocked down. What effect would introducing siRNAs to TSC1 have on human cells?

11. Speculate about why plants deficient in Dicer activity show increased sensitivity to infection by RNA viruses.

12. mRNA stability is a key regulator of protein levels in a cell. Briefly describe the three mRNA degradation pathways. A yeast cell has a mutation in the DCP1 gene, resulting in decreased uncapping activity. Would you expect to see a change in the P bodies found in this mutant cell?

13. mRNA localization now appears to be a common phenomenon. What benefit does mRNA localization have for a cell? What is the evidence that some mRNAs are directed to accumulate in specific subcellular locations?

## Analyze the Data

Most humans are infected with herpes simplex virus-1 (HSV-1), the causative agent of cold sores. The HSV-1 genome comprises about 100 genes, most of which are expressed in infected host cells at the site of oral sores. The infectious process involves replication of viral DNA, transcription and translation of viral genes, assembly of new viral particles, and death of the host cell as the viral progeny are released. Unlike most other types of viruses, herpesvirus also has a latent phase, in which the virus remains hidden in neurons. These latently infected neurons are the source of active infections, causing cold sores when latency is overcome.

Interestingly, only a single viral transcript is expressed during latency. This transcript, *LAT* (*latency-associated transcript*), does not encode a protein, and neurons infected with mutant HSV-1 lacking the *LAT* gene undergo cell death by apoptosis at a rate twice that of cells infected with wild-type HSV-1. To determine if *LAT* functions to block apoptosis by encoding a miRNA, the following studies were done (see Gupta et al., 2006, *Nature* 442:82–85).

a. A cell line was transfected (a process in which foreign DNA is inserted into a cell) with an expression vector that expresses a Pst-Mlu fragment of the *LAT* gene (see diagram in part b). The percentage of these transfected cells that then underwent drug-induced cell death was compared to that of control cells. The experiment was repeated in cells in which Dicer expression was knocked down using Dicer siRNA. The data obtained are shown in the graph below. What conclusions can be drawn from these data? Why did the scientists who conducted this study examine the effects of silencing Dicer?

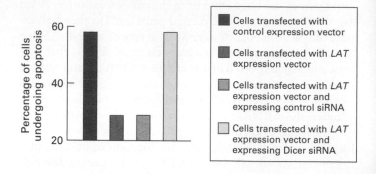

b. Cells were transfected with an expression vector expressing the Pst-Mlu fragment of the *LAT* gene from which the region between the two Sty restriction sites was deleted (DSty; diagram below). When these cells were induced to undergo apoptosis, they died at the same rate as did non-transfected cells. In additional studies, cells were transfected with an expression vector expressing the Sty-Sty region of the *LAT* gene. These cells exhibited the same resistance to apoptosis as did cells transfected with the Pst-Mlu fragment. What can be deduced from these findings about the region of the *LAT* gene required to protect cells from apoptosis?

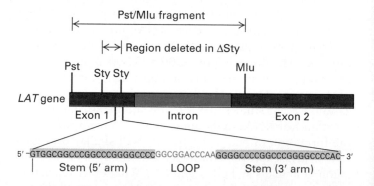

c. RNA encoded within the Sty-Sty region is predicted to form a stem loop (see diagram in part b). Northern blot analysis was performed on total-cell RNA isolated from control cells (mock), cells infected with wild-type HSV-1, cells infected with an HSV-1 deletion mutant from which the sequence between the two Sty sites in the *LAT* gene was deleted (DSty), and cells infected with a rescued DSty virus into which the deleted region was re-inserted into the viral genome (StyR). The probe used for the Northern blot was the labeled 3′ stem region of the *LAT* RNA in the Sty-Sty region, as diagrammed in part (b). The RNAs recognized by this probe were either ~55 nucleotides or 20 nucleotides, as shown in the Northern blot below. Why were two different-sized RNAs detected? When a second probe was used that was the labeled 5′ stem region of the RNA sequence shown in part (b), only the ~55-nucleotide RNA was detected. What can you deduce about the processing of RNA expressed from the *LAT* gene? What enzyme likely produced the ~55-nucleotide RNA? In what part of the cell? What enzyme likely produced the 20-nucleotide RNA? In what part of the cell?

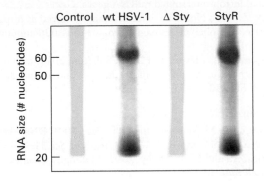

**d.** TGF-β mRNA encodes a protein, transforming growth factor β, that inhibits cell growth and induces apoptosis. The 3′ untranslated region (3′ UTR) of TGF-β mRNA can form an imperfect duplex with miRNA encoded by the 5′ stem region of the *LAT* Sty-Sty domain (miR-LAT), as shown below. In what way might the expression levels of TGF-β differ in cells infected with wild-type HSV-1 compared to uninfected cells? What can you infer about latent HSV-1 infections from these studies?

miR-LAT

C

C–C–G–G–G–G–C–C–C–G–G–C–C–C–G–G G–G–U 5′

5′ C–G–C–C–C–C–G–G G–C–C–C–G–G–C–C–C–C–A

CAG    TGF-β-3′UTR

# References

### Processing of Eukaryotic Pre-mRNA

Bergkessel, M., G. M. Wilmes, and C. Guthrie. 2009. SnapShot: formation of mRNPs. *Cell* **136**:794.

Blanc, V., and N. O. Davidson. 2010. APOBEC-1-mediated RNA editing. *Wiley Interdiscip. Rev. Syst. Biol. Med.* **2**:594–602.

de Almeida, S. F., and M. Carmo-Fonseca. 2008. The CTD role in cotranscriptional RNA processing and surveillance. *FEBS Lett.* **582**:1971–1976.

Gu, M., and C. D. Lima. 2005. Processing the message: structural insights into capping and decapping mRNA. *Curr. Opin. Struct. Biol.* **15**:99–106.

Hocine, S., R. H. Singer, and D. Grünwald. 2010. RNA processing and export. *Cold Spring Harb. Perspect. Biol.* **2**(12):a000752.

Houseley, J., and D. Tollervey. 2009. The many pathways of RNA degradation. *Cell* **136**:763–776.

Lambowitz, A. M., and S. Zimmerly. 2004. Mobile group II introns. *Annu. Rev. Genet.* **38**:1–35.

Moore, M. J., and N. J. Proudfoot. 2009. Pre-mRNA processing reaches back to transcription and ahead to translation. *Cell* **136**:688–700.

Pawlicki, J. M., and J. A. Steitz. 2010. Nuclear networking fashions pre-messenger RNA and primary microRNA transcripts for function. *Trends Cell Biol.* **20**:52–61.

Perales, R., and D. Bentley. 2009. "Cotranscriptionality": the transcription elongation complex as a nexus for nuclear transactions. *Mol. Cell.* **36**:178–191.

Pyle, A. M. 2010. The tertiary structure of group II introns: implications for biological function and evolution. *Crit. Rev. Biochem. Mol. Biol.* **45**:215–232.

Sharp, P. A. 2005. The discovery of split genes and RNA splicing. *Trends Biochem. Sci.* **30**:279–281.

Valadkhan, S. 2010. Role of the snRNAs in spliceosomal active site. *RNA Biol.* **7**:345–353.

Wahl, M. C., C. L. Will, and R. Lührmann. 2009. The spliceosome: design principles of a dynamic RNP machine. *Cell* **136**:701–718.

### Regulation of Pre-mRNA Processing

Black, D. L. 2003. Mechanisms of alternative pre-mRNA splicing. *Ann. Rev. Biochem.* **72**:291–336.

Chen, M., and J. L. Manley. 2009. Mechanisms of alternative splicing regulation: insights from molecular and genomics approaches. *Nat. Rev. Mol. Cell Biol.* **10**:741–754.

Licatalosi, D. D., and R. B. Darnell. 2010. RNA processing and its regulation: global insights into biological networks. *Nat. Rev. Genet.* **11**:75–87.

Maniatis, T., and B. Tasic. 2002. Alternative pre-mRNA splicing and proteome expansion in metazoans. *Nature* **418**:236–243.

Raponi, M., and D. Baralle. 2010. Alternative splicing: good and bad effects of translationally silent substitutions. *FEBS J.* **277**:836–840.

Wang, E. T., et al. 2008. Alternative isoform regulation in human tissue transcriptomes. *Nature* **456**:470–476.

Zhong, X. Y., et al. 2009. SR proteins in vertical integration of gene expression from transcription to RNA processing to translation. *Curr. Opin. Genet. Dev.* **19**:424–436.

### Transport of mRNA Across the Nuclear Envelope

Cole, C. N., and J. J. Scarcelli. 2006. Transport of messenger RNA from the nucleus to the cytoplasm. *Curr. Opin. Cell Biol.* **18**:299–306.

Grünwald, D., R. H. Singer, and M. Rout. 2011. Nuclear export dynamics of RNA-protein complexes. *Nature* **475**:333–341.

Iglesias, N., and F. Stutz. 2008. Regulation of mRNP dynamics along the export pathway. *FEBS Lett.* **582**:1987–1996.

Katahira, J., and Y. Yoneda. 2009. Roles of the TREX complex in nuclear export of mRNA. *RNA Biol.* **6**:149–152.

Rodríguez-Navarro, S., and E. Hurt. 2011. Linking gene regulation to mRNA production and export. *Curr. Opin. Cell Biol.* **23**:302–309.

Stewart, M. 2010. Nuclear export of mRNA. *Trends Biochem. Sci.* **35**:609–617.

Wente, S. R., and M. P. Rout. 2010. The nuclear pore complex and nuclear transport. *Cold Spring Harb. Perspect. Biol.* **2**(10):a000562.

### Cytoplasmic Mechanisms of Post-transcriptional Control

Ambros, V. 2004. The functions of animal microRNAs. *Nature* **431**:350–355.

Buchan, J. R., and R. Parker. 2009. Eukaryotic stress granules: the ins and outs of translation. *Mol. Cell* **36**:932–941.

Carthew, R. W., and E. J. Sontheimer. 2009. Origins and mechanisms of miRNAs and siRNAs. *Cell* 136:642–655.

Doma, M. K., and R. Parker. 2007. RNA quality control in eukaryotes. *Cell* 131:660–668.

Eulalio, A., I. Behm-Ansmant, and E. Izaurralde. 2007. P bodies: at the crossroads of post-transcriptional pathways. *Nat. Rev. Mol. Cell Biol.* 8:9–22.

Fabian, M. R., N. Sonenberg, and W. Filipowicz. 2010. Regulation of mRNA translation and stability by microRNAs. *Annu. Rev. Biochem.* 79:351–379.

Ghildiyal, M., and P. D. Zamore. 2009. Small silencing RNAs: an expanding universe. *Nat. Rev. Genet.* 10:94–108.

Groppo, R., and J. D. Richter. 2009. Translational control from head to tail. *Curr. Opin. Cell Biol.* 21:444–451.

Hirokawa, N. 2006. mRNA transport in dendrites: RNA granules, motors, and tracks. *J. Neurosci.* 26:7139–7142.

Huntzinger, E., and E. Izaurralde. 2011. Gene silencing by microRNAs: contributions of translational repression and mRNA decay. *Nat. Rev. Genet.* 12:99–110.

Jobson, R. W., and Y. L. Qiu. 2008. Did RNA editing in plant organellar genomes originate under natural selection or through genetic drift? *Biol. Direct.* 3:43.

Kidner, C. A., and R. A. Martienssen. 2005. The developmental role of microRNA in plants. *Curr. Opin. Plant Biol.* 8:38–44.

Leung, A. K., and P. A. Sharp. 2010. MicroRNA functions in stress responses. *Mol. Cell.* 40:205–215.

Lodish, H. F., et al. 2008. Micromanagement of the immune system by microRNAs. *Nat. Rev. Immunol.* 8:120–130.

Maquat, L. E., W. Y. Tarn, and O. Isken. 2010. The pioneer round of translation: features and functions. *Cell* 142:368–374.

Martin, K. C., and A. Ephrussi. 2009. mRNA localization: gene expression in the spatial dimension. *Cell* 136:719–730.

Mello, C. C., and D. Conte Jr. 2004. Revealing the world of RNA interference. *Nature* 431:338–342. C. C. Mello's Nobel Prize lecture can be viewed at http://nobelprize.org/nobel_prizes/medicine/laureates/2006/announcement.html

Michels, A. A. 2011. MAF1: a new target of mTORC1. *Biochem. Soc. Trans.* 39:487–491.

Mihaylova, M. M., and R. J. Shaw. 2011. The AMPK signalling pathway coordinates cell growth, autophagy and metabolism. *Nature Cell Biol.* 13:1016–1023.

Parker, R., and H. Song. 2004. The enzymes and control of eukaryotic mRNA turnover. *Nat. Struct. Mol. Biol.* 11:121–127.

Richter, J. D., and N. Sonenberg. 2005. Regulation of cap-dependent translation by eIF4E inhibitory proteins. *Nature* 433:477–480.

Ruvkun, G. B. 2004. The tiny RNA world. *Harvey Lect.* 99:1–21.

Shaw, R. J. 2008. mTOR signaling: RAG GTPases transmit the amino acid signal. *Trends Biochem. Sci.* 33:565–568.

Simpson, L., et al. 2004. Mitochondrial proteins and complexes in Leishmania and Trypanosoma involved in U-insertion/deletion RNA editing. *RNA* 10:159–170.

Siomi, M. C., et al. 2011. PIWI-interacting small RNAs: the vanguard of genome defence. *Nat. Rev. Mol. Cell Biol.* 12:246–258.

Willis, I. M., and R. D. Moir. 2007. Integration of nutritional and stress signaling pathways by Maf1. *Trends Biochem. Sci.* 32:51–53.

Wullschleger, S., R. Loewith, and M. N. Hall. 2006. TOR signaling in growth and metabolism. *Cell* 124:471–484.

Zhang, H., J. M. Maniar, and A. Z. Fire. 2011. 'lnc-miRs': functional intron-interrupted miRNA genes. *Genes Dev.* 25:1589–1594. A. Z. Fire's Nobel Prize lecture can be viewed at http://nobelprize.org/nobel_prizes/medicine/laureates/2006/announcement.html

Zoncu, R., A. Efeyan, and D. M. Sabatini. 2011. mTOR: from growth signal integration to cancer, diabetes and ageing. *Nat. Rev. Mol. Cell Biol.* 12:21–35.

### Processing of rRNA and tRNA

Evans, D., S. M. Marquez, and N. R. Pace. 2006. RNase P: interface of the RNA and protein worlds. *Trends Biochem. Sci.* 31:333–341.

Fatica, A., and D. Tollervey. 2002. Making ribosomes. *Curr. Opin. Cell Biol.* 14:313–318.

Hage, A. E., and D. Tollervey. 2004. A surfeit of factors: why is ribosome assembly so much more complicated in eukaryotes than bacteria? *RNA Biol.* 1:10–15.

Hamma, T., and A. R. Ferré-D'Amaré. 2010. The box H/ACA ribonucleoprotein complex: interplay of RNA and protein structures in post-transcriptional RNA modification. *J. Biol. Chem.* 285:805–809.

Handwerger, K. E., and J. G. Gall. 2006. Subnuclear organelles: new insights into form and function. *Trends Cell Biol.* 16:19–26.

Kressler, D., E. Hurt, and J. Bassler. 2010. Driving ribosome assembly. *Biochim. Biophys. Acta.* 1803:673–683.

Liang, B., and H. Li. 2011. Structures of ribonucleoprotein particle modification enzymes. *Q. Rev. Biophys.* 44:95–122.

Marvin, M. C., and D. R. Engelke. 2009. RNase P: increased versatility through protein complexity? *RNA Biol.* 6:40–42.

Nizami, Z., S. Deryusheva, and J. G. Gall. 2010. The Cajal body and histone locus body. *Cold Spring Harb. Perspect. Biol.* 2(7):a000653.

Phizicky, E. M., and A. K. Hopper. 2010. tRNA biology charges to the front. *Genes Dev.* 24:1832–1860.

Schmid, M., and T. H. Jensen. 2008. The exosome: a multipurpose RNA-decay machine. *Trends Biochem. Sci.* 33:501–510.

Stahley, M. R., and S. A. Strobel. 2006. RNA splicing: group I intron crystal structures reveal the basis of splice site selection and metal ion catalysis. *Curr. Opin. Struct. Biol.* 16:319–326.

Tschochner, H., and E. Hurt. 2003. Pre-ribosomes on the road from the nucleolus to the cytoplasm. *Trends Cell Biol.* 13:255–263.

| Nuclei | Microtubules | Golgi | Actin fibers | Mitochondria |

20 μm

Fluorescence microscopy shows the location of DNA and multiple proteins within the same cell. Here fluorescent tagging and staining techniques using different fluorescent molecules reveal the cytoskeletal proteins α-tubulin (green) and actin (red), DNA (blue), the Golgi complex (yellow), and mitochondria (purple). The images along the top are false-colored images of each structure stained individually. The larger image merges these separate images to depict the full cell.
[From B. N. G. Giepmans, et al., 2006, *Science* **312**:217.]

# CHAPTER
# 9

# Culturing, Visualizing, and Perturbing Cells

I t is difficult to believe that 400 years ago, it was not yet known that all living things are made of cells. In 1655, Robert Hooke used a primitive microscope to examine a piece of cork and saw an orderly arrangement of rectangles—the walls of the dead plant cells—that reminded him of monk cells in a monastery, so he coined the term *cells*. Shortly after this, Antonie van Leeuwenhook described the microorganisms that he saw in his simple microscope, the first description of living cells. Two hundred years later, Matthias Schleiden and Theodore Schwann observed that individual cells constitute the fundamental unit of life in a variety of plants, animals, and single-celled organisms. Collectively, these were some of the greatest discoveries in biology and posed the question of how cells are organized and function. However, many technical constraints hamper studies of cells in intact animals and plants. One alternative is the use of intact organs that are removed from animals and treated to maintain their physiologic integrity and function. However, the organization of

organs, even isolated ones, is sufficiently complex to pose numerous problems for research. Thus molecular cell biologists often conduct experimental studies on cells isolated from an organism. In Section 9.1, we learn how to maintain and grow diverse cell types and how to isolate specific types of cells from complex mixtures. However, cells in culture are not in their native setting, so we discuss how researchers are now growing and examining cells in three-dimensional environments to more closely mimic their situation in an animal.

In many cases, isolated cells can be maintained in the laboratory under conditions that permit their survival and growth, a procedure known as *culturing*. Cultured cells have several advantages over intact organisms for cell biology research. Cells of a single specific type can be grown in culture, experimental conditions can be better controlled, and in many cases a single cell can be readily grown into a colony of many identical cells. The resulting strain of cells, which is genetically homogeneous, is called a **clone**.

## OUTLINE

Discoveries about cellular organization have been intimately tied with developments in both light and electron microscopy. This is as true today as it was 400 years ago. Light microscopy initially revealed the beautiful internal organization of cells, and today highly sophisticated microscopes are continually being improved to probe deeper and deeper to reveal the molecular mechanism by which cells function. In Section 9.2, we discuss light microscopy and the different technologies available, long-standing but still valuable methods, and then trace through several clever methods that have been developed since, culminating with the newest, cutting-edge technologies. A major advance came in the 1960s and 1970s with the development of *immunofluorescence microscopy* to allow the localization of specific proteins within fixed cells, thus providing a static image of their location, as illustrated in the opening figure. Such studies led to the important concept that the membranes and interior spaces of each type of organelle contain a distinctive group of proteins that are essential for the organelle to carry out its unique functions. A major advance came in the mid-1990s with the simple idea of expressing *chimeric proteins*—consisting of a protein of interest covalently linked to a naturally fluorescent protein—to enable biologists to visualize the movements of individual proteins in live cells. Suddenly, the dynamic nature of cells could be appreciated, which changed the view of cells from the previously available static images. In addition, it presented a technological challenge—the more sensitive a microscope could be made to detect the fluorescent protein, the more information the investigator could glean from the data. It also opened up the development of fluorescent techniques to monitor protein-protein interactions in living cells, as well as a myriad of other sophisticated molecular technologies, some of which we also discuss in this section.

Despite the amazing developments in light microscopy, visible light provides too low a resolution to examine cells in ultrastructural detail. The electron microscope gives a much higher resolution, but the technology generally requires that the cell be fixed and sectioned and so all cell movements are frozen in time. Electron microscopy also allows investigators to examine the structure of macromolecular complexes or single macromolecules. In Section 9.3, we outline the various approaches for preparing specimens for observation in the electron microscope and describe the type of information that can be derived from them.

Light and electron microscopy revealed that all eukaryotic cells—whether of fungal, plant, or animal origin—contain a similar repertoire of membrane-limited compartments termed **organelles**. In Section 9.4, we provide a simple introduction to the basic structure and function of the major organelles in animal and plant cells, as a prelude to their detailed description in subsequent chapters. In parallel with the developments in microscopy, subcellular fractionation methods were developed that have enabled cell biologists to isolate individual organelles to a high degree of purity. These techniques, also detailed in Section 9.4, continue to provide important information about the protein composition and biochemical function of organelles.

Microscopy and organelle characterization are inherently descriptive technologies. How can one investigate the molecular mechanisms underlying cell biological processes? If we want to understand how a car works, we can explore the effects of removing or interfering with individual components to see what happens. This, of course, is in principle the concept of genetic analysis we described in Chapter 5: how interfering with a specific component can be used to explore its function. In Section 9.5, we describe how small molecules that interfere with the function of specific proteins can also be used to dissect cellular processes. Finally, we describe how the discovery of small interfering RNAs that target specific mRNAs for destruction has been exploited to suppress expression of specific proteins in both cultured cells and whole animals to expand and complement the use of classical genetics in the analysis of biological processes.

# 9.1 Growing Cells in Culture

The study of cells is greatly facilitated by growing them in culture, where they can be examined by microscopy and subjected to specific treatments under controlled conditions. It is generally quite easy to grow unicellular bacterial, fungal, or protist cells; for example, by placing them in a rich medium that supports their growth. However, animal cells come from multicellular organisms, making it more difficult to culture single or small groups of cells. In this section, we discuss how animal cells are grown in culture and how different cell types can be purified for study.

## Culture of Animal Cells Requires Nutrient-Rich Media and Special Solid Surfaces

To permit the survival and normal function of cultured tissues or cells, the temperature, pH, ionic strength, and access to essential nutrients must simulate as closely as possible the conditions within an intact organism. Isolated animal cells are typically placed in a nutrient-rich liquid, called the culture medium, within specially coated plastic dishes or flasks. The cultures are kept in incubators in which the temperature, atmosphere, and humidity can be controlled. To reduce the chances of bacterial or fungal contamination, antibiotics are often added to the culture medium. To further guard against contamination, investigators usually transfer cells between dishes, add reagents to the culture medium, and otherwise manipulate the specimens within special sterile cabinets containing circulating air that is filtered to remove microorganisms and other airborne contaminants.

Media for culturing animal cells must supply the nine amino acids (phenylalanine, valine, threonine, tryptophan, isoleucine, methionine, leucine, lysine, and histidine) that cannot be synthesized by adult vertebrate animal cells. In addition, most cultured cells require three other amino acids (cysteine, tyrosine, and arginine) that are synthesized only by specialized cells in intact animals, as well as glutamine, which serves as a nitrogen source. The other necessary components of a medium for culturing animal cells are vitamins, various salts, fatty acids, glucose, and serum—the fluid remaining after the noncellular part of blood (plasma) has been allowed to clot. Serum contains various protein factors that are needed for the proliferation of mammalian cells in culture, including the polypeptide hormone insulin; transferrin, which supplies iron in a bioaccessible form; and numerous growth factors. In addition, certain cell types require specialized protein growth factors not present in serum. For instance, progenitors of red blood cells require erythropoietin, and T lymphocytes require interleukin 2 (see Chapter 16). A few mammalian cell types can be grown in a chemically defined, serum-free medium containing amino acids, glucose, vitamins, and salts plus certain trace minerals, specific protein growth factors, and other components.

Unlike bacterial and yeast cells, which can be grown in suspension, most animal cell types will grow only attached to a solid surface. This requirement highlights the importance of cell-surface proteins, called **cell-adhesion molecules (CAMs)**, that cells use to bind to adjacent cells and to components of the extracellular matrix (ECM) such as collagen or fibronectin (see Chapter 20). These ECM proteins coat the solid surface (usually glass or plastic) and either come from the serum or are secreted by the cells in culture. A single cell cultured on a glass or a plastic dish proliferates to form a visible mass, or *colony*, containing thousands of genetically identical cells in 4 to 14 days, depending on the growth rate. Some specialized blood cells and tumor cells can be maintained or grown in suspension as single cells.

## Primary Cell Cultures and Cell Strains Have a Finite Life Span

Normal animal tissues (e.g., skin, kidney, liver) or whole embryos are commonly used to establish *primary cell cultures*. To prepare individual tissue cells for a primary culture, the cell-cell and cell-matrix interactions must be broken. To do so, tissue fragments are treated with a combination of a protease (e.g., trypsin, the collagen-hydrolyzing enzyme collagenase, or both) and a divalent cation chelator (e.g., EDTA) that depletes the medium of free $Ca^{2+}$. Many cell-adhesion molecules require calcium and are thus inactivated when calcium is removed; other cell-adhesion molecules that are not calcium dependent need to be proteolyzed for the cells to separate. The released cells are then placed in dishes in a nutrient-rich, serum-supplemented medium, where they can adhere to the surface and to one another. The same protease-chelator solution is used to remove adherent cells from a culture dish for biochemical studies or subculturing (transfer to another dish).

**Fibroblasts** are the predominant cells in connective tissue and normally produce ECM components such as collagen that bind to cell-adhesion molecules, thereby anchoring cells to a surface. In culture, fibroblasts usually divide more rapidly than other cells from a tissue, eventually becoming the predominant cell type in a primary culture unless special precautions are taken to remove them when isolating other types of cells.

When cells removed from an embryo or an adult animal are cultured, most of the adherent cells will divide a finite number of times and then cease growing (cell senescence). For instance, human fetal fibroblasts divide about 50 times before they cease growth (Figure 9-1a). Starting with $10^6$ cells, 50 doublings has the potential to produce $10^6 \times 2^{50}$, or more than $10^{20}$ cells, which is equivalent to the weight of about 1000 people. Normally, only a very small fraction of these cells are used in any one experiment. Thus, even though its lifetime is limited, a single culture, if carefully maintained, can be studied through many cell generations. Such a lineage of cells originating from one initial primary culture is called a **cell strain**.

One important exception to the finite life of normal cells is the *embryonic stem cell*, which, as its name implies, is

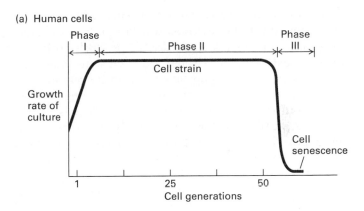

(a) Human cells

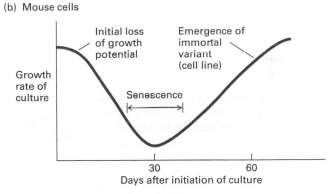

(b) Mouse cells

**FIGURE 9-1 Stages in the establishment of a cell culture.**
(a) When cells isolated from human tissue are initially cultured, some cells die and others (mainly fibroblasts) start to grow; overall, the growth rate increases (phase I). If the remaining cells are harvested, diluted, and replated into dishes again and again, the cell strain continues to divide at a constant rate for about 50 cell generations (phase II), after which the growth rate falls rapidly. In the ensuing period (phase III), all the cells in the culture stop growing (senescence). (b) In a culture prepared from mouse or other rodent cells, initial cell death (not shown) is coupled with the emergence of healthy growing cells. As these dividing cells are diluted and allowed to continue growth, they soon begin to lose growth potential, and most stop growing (i.e., the culture goes into senescence). Very rare cells undergo oncogenic mutations that allow them to survive and continue dividing until their progeny overgrow the culture. These cells constitute a cell line, which will grow indefinitely if it is appropriately diluted and fed with nutrients. Such cells are said to be immortal.

derived from an embryo and will divide and give rise to all tissues during development. As we discuss in Chapter 21, embryonic stems cells can be cultured indefinitely under the appropriate conditions.

Research with cell strains is simplified by the ability to freeze and successfully thaw them at a later time for experimental analysis. Cell strains can be frozen in a state of suspended animation and stored for extended periods at liquid nitrogen temperature, provided that a preservative that prevents the formation of damaging ice crystals is used. Although not all cells survive thawing, many do survive and resume growth.

## Transformed Cells Can Grow Indefinitely in Culture

To be able to clone individual cells, modify cell behavior, or select mutants, biologists often want to maintain cell cultures for many more than 50 doublings. Such prolonged growth is exhibited by cells derived from some tumors. In addition, rare cells in a population of primary cells may undergo spontaneous oncogenic mutations, leading to oncogenic **transformation** (see Chapter 24). Such cells, said to be oncogenically transformed or simply *transformed*, are able to grow indefinitely. A culture of cells with an indefinite life span is considered immortal and is called a **cell line**.

Primary cell cultures of normal rodent cells commonly undergo spontaneous transformation into a cell line. After rodent cells are grown in culture for several generations, the culture goes into senescence (Figure 9-1b). During this pe-

riod, most of the cells stop growing, but often a rapidly dividing transformed cell arises spontaneously and takes over, or overgrows, the culture. A cell line derived from such a transformed variant will grow indefinitely if provided with the necessary nutrients. In contrast to rodent cells, normal human cells rarely undergo spontaneous transformation into a cell line. The HeLa cell line, the first human cell line established, was originally obtained in 1952 from a malignant tumor (carcinoma) of the uterine cervix. Other human cell lines are often derived from cancers, and others have been rendered immortal by transforming them to express oncogenes.

Regardless of the source, cells in immortalized lines often have chromosomes with abnormal DNA sequences. In addition, the number of chromosomes in such cells is usually greater than that in the normal cell from which they arose, and the chromosome number changes as the cells continue to divide in culture. A noteworthy exception is the Chinese hamster ovary (CHO) line and its derivatives, which have fewer chromosomes than their hamster progenitors. Cells with an abnormal number of chromosomes are said to be *aneuploid*.

## Flow Cytometry Separates Different Cell Types

Some cell types differ sufficiently in density that they can be separated on the basis of this physical property. White blood cells (leukocytes) and red blood cells (erythrocytes), for instance, have very different densities because erythrocytes have no nucleus; thus these cells can be separated by equilibrium

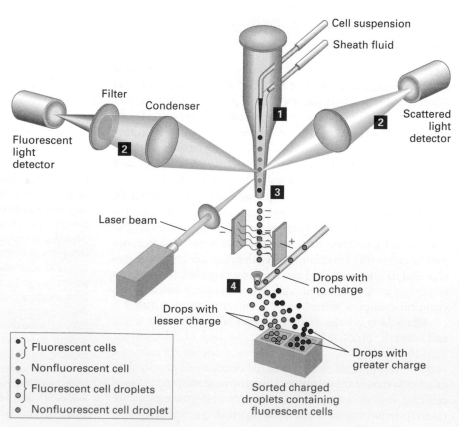

**FIGURE 9-2 Fluorescence-activated cell sorter (FACS) separates cells that are labeled differentially with a fluorescent reagent.** Step **1**: A concentrated suspension of labeled cells is mixed with a buffer (the sheath fluid) so that the cells pass single-file through a laser light beam. Step **2**: Both the fluorescent light emitted and the light scattered by each cell are measured; from measurements of the scattered light, the size and shape of the cell can be determined. Step **3**: The suspension is then forced through a nozzle, which forms tiny droplets containing at most a single cell. At the time of formation at the nozzle tip, each droplet containing a cell is given a negative electric charge proportional to the fluorescence of that cell determined from the earlier measurement. Step **4**: Droplets now pass through an electric field, so that those with no charge are discarded, whereas those with different electric charges are separated and collected. Because it takes only milliseconds to sort each droplet, as many as 10 million cells per hour can pass through the machine. [Adapted from D. R. Parks and L. A. Herzenberg, 1982, *Meth. Cell Biol.* **26**:283.]

density centrifugation (described in Section 9.4). Because most cell types cannot be differentiated so easily, other techniques such as flow cytometry must be used to separate them.

To identify one type of cell from a complex mixture, it is necessary to have some way to mark and then sort out the desired cells. Different cell types often express different molecules on their cell surface. If a particular surface molecule is only expressed on the desired cell type, this can be used to mark those cells. The cell mixture can be incubated with a fluorescent dye linked to an antibody to the specific cell-surface molecule, thus rendering just the desired cells fluorescent. The cells can be analyzed in a *flow cytometer*. This machine flows cells past a laser beam that measures the light that they scatter and the fluorescence that they emit; thus it can quantify the numbers of cells of the desired type from a mixture. A *fluorescence-activated cell sorter (FACS)*, which is based on flow cytometry, can both analyze the cells and select one or a few cells from thousands of others and sort them into a separate culture dish (Figure 9-2). To sort the cells, their concentration has to be adjusted so that the tiny droplets that the FACS machine makes and analyzes contain only one cell each. A stream of droplets is analyzed for fluorescence, and those that have the desired signal are sorted away from those that do not. Having been sorted from other cells, the selected cells can be grown in culture.

The FACS procedure is commonly used to purify the different types of white blood cells, each of which bears on its surface one or more distinctive proteins and so will bind monoclonal antibodies specific for that protein. Only the T cells of the immune system, for instance, have both CD3 and Thy1.2 proteins on their surfaces. The presence of these surface proteins allows T cells to be separated easily from other types of blood cells or spleen cells (Figure 9-3).

Other uses of flow cytometry include the measurement of a cell's DNA and RNA content and the determination of its general shape and size. The FACS can make simultaneous measurements of the size of a cell (from the amount of scattered light) and the amount of DNA that it contains (from the amount of fluorescence emitted from a DNA-binding dye). Measurements of the DNA content of individual cells are used to follow replication of DNA as the cells progress through the cell cycle (see Chapter 19).

An alternative method for separating specific types of cells uses small magnetic beads coupled to antibodies for the specific surface molecule. For example, to isolate T cells, the beads are coated with a monoclonal antibody specific for a surface protein such as CD3 or Thy1.2. Only cells with these proteins will stick to the beads and can be recovered from the preparation by adhesion to a small magnet on the side of the test tube.

## Growth of Cells in Two-Dimensional and Three-Dimensional Culture Mimics the In Vivo Environment

While much has been learned using cells grown on a plastic or glass surface, these surfaces are far removed from cells' normal tissue environment. As detailed in Chapter 20, many

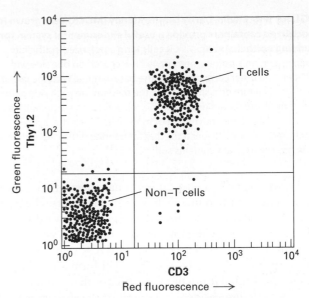

**EXPERIMENTAL FIGURE 9-3  T cells bound to fluorescence-tagged antibodies to two cell-surface proteins are separated from other white blood cells by FACS.** Spleen cells from a mouse were treated with a red fluorescent monoclonal antibody specific for the CD3 cell-surface protein and with a green fluorescent monoclonal antibody specific for a second cell-surface protein, Thy1.2. As the cells were passed through a FACS machine, the intensity of the green and red fluorescence emitted by each cell was recorded. Each dot represents a single cell. This plot of the green fluorescence (vertical axis) versus red fluorescence (horizontal axis) for thousands of spleen cells shows that about half of them—the T cells—express both CD3 and Thy1.2 proteins on their surfaces (upper-right quadrant). The remaining cells, which exhibit low fluorescence (lower-left quadrant), express only background levels of these proteins and are other types of white blood cells. Note the logarithmic scale on both axes. [Courtesy of Chengcheng Zhang, Whitehead Institute.]

cell types function only when closely linked to other cells. Key examples are the sheet-like layers of epithelial tissue, called **epithelia** (singular, **epithelium**), which cover the external and internal surfaces of organs. Typically, the distinct surfaces of a polarized epithelial cell are called the **apical** (top), **basal** (base or bottom), and **lateral** (side) surfaces (see Figure 20-10). The basal surface usually contacts an underlying extracellular matrix called the **basal lamina**, whose composition and function are discussed in Section 20.3. Epithelial cells often function to transport specific classes of molecules across the epithelial sheet; for example, the epithelial lining of the intestine transports nutrients into the cell through the apical surface and out toward the bloodstream across the basolateral surface. When grown on plastic or glass, epithelial cells cannot easily perform this function. Therefore, special containers have been designed with a porous surface that acts as the basal lamina to which epithelial cells attach and form a uniform two-dimensional sheet (Figure 9-4). A commonly used cultured cell line derived from dog kidney epithelium is called *Madin-Darby canine kidney (MDCK) cells* and is often used to study the formation and function of epithelial sheets.

**FIGURE 9-4 Madin-Darby canine kidney (MDCK) cells grown in specialized containers provide a useful experimental system for studying epithelial cells.** MDCK cells form a polarized epithelium when grown on a porous membrane filter coated on one side with collagen and other components of the basal lamina. With the use of the special culture dish shown here, the medium on each side of the filter (apical and basal sides of the monolayer) can be experimentally manipulated and the movement of molecules across the layer monitored. Several cell junctions that interconnect the cells form only if the growth medium contains sufficient Ca$^{2+}$.

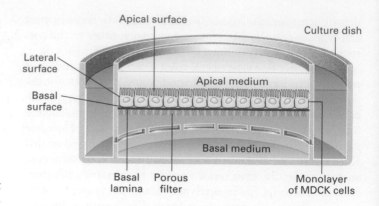

However, even a two-dimensional sheet often does not allow cells to fully mimic behavior in their normal environment. Methods have now been developed to grow cells in three dimensions by providing a support infiltrated with components of the extracellular matrix. If MDCK cells are cultured under appropriate conditions, they will form a tubular sheet mimicking a tubular organ or the duct of the secretory gland. In these three-dimensional structures, the apical aspect of the epithelial sheet lines the lumen, whereas the basal side of each cell is in contact with the extracellular matrix (Figure 9-5).

## Hybrid Cells Called Hybridomas Produce Abundant Monoclonal Antibodies

In addition to serving as research models for studies on cell function, cultured cells can be converted into "factories" for producing specific proteins. In Chapter 5, we described how introducing genes encoding insulin, growth factors, and other therapeutically useful proteins into bacterial or eukaryotic cells can be used to express and recover these proteins (see Figures 5-31 and 5-32). Here we consider the use of special cultured cells to generate monoclonal antibodies, which are experimental tools widely used in many aspects of cell biological research. Increasingly, they are being used for diagnostic and therapeutic purposes in medicine, as we discuss in later chapters.

To understand the challenge of generating monoclonal antibodies, we need to briefly review how mammals produce antibodies; more detail is provided in Chapter 23. Recall that antibodies are proteins secreted by white blood cells that bind with high affinity to their antigen (see Figure 3-19). Each normal antibody-producing B lymphocyte in a mammal is capable

(a)

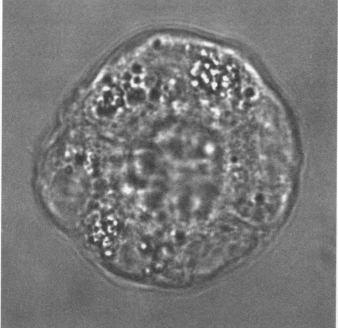

(b)

**EXPERIMENTAL FIGURE 9-5 MDCK cells can form cysts in culture.** (a) MDCK cells grown in a supported extracellular matrix will form groups of cells that polarize to form a spherical single layer of cells with a lumen in the middle, called a cyst. (b) By examining the localization of proteins found in the apical (red) and basolateral membranes (green), these cells can be seen to be fully polarized with the apical side facing the lumen, which recapitulates their organization in the kidney tubules from which they are derived. The nuclear DNA is stained blue. [Parts (a) and (b) from D. M. Bryant et al., 2010, *Nat. Cell Biol.* **12**:1035.]

of producing a single type of antibody that can bind to a particular **determinant** or **epitope** on an antigen molecule. An epitope is generally a small region on the antigen, for example, consisting of just a few amino acids. If an animal is injected with an antigen, the B lymphocytes that make antibodies recognizing that antigen are stimulated to grow and secrete the antibodies. Each antigen-activated B lymphocyte forms a clone of cells in the spleen or lymph nodes, with each cell of the clone producing the identical antibody—that is, a *monoclonal antibody*. Because most natural antigens contain multiple epitopes, exposure of an animal to an antigen usually stimulates the formation of multiple different B-lymphocyte clones, each producing a different specific antibody. The resulting mixture of antibodies from the many B-lymphocyte clones that recognize different epitopes on the same antigen is said to be *polyclonal*. Such polyclonal antibodies circulate in the blood and can be isolated as a group.

Although polyclonal antibodies are very useful, monoclonal antibodies are suitable for many types of experiments and medical applications when you need a reagent that binds to just one site on a protein; for example, one that competes with a ligand on a cell-surface receptor. Unfortunately, the biochemical purification of any one type of monoclonal antibody from blood is not feasible for two main reasons: the concentration of any given antibody is quite low, and all antibodies have the same basic molecular architecture (see Figure 3-19).

To produce and then purify monoclonal antibodies, one first needs to be able to grow the appropriate B-lymphocyte clone. However, primary cultures of normal B lymphocytes are of limited usefulness for the production of monoclonal antibodies because they have a limited life span. Thus the first step in producing a monoclonal antibody is to generate immortal, antibody-producing cells (Figure 9-6). This immortality is achieved by fusing normal B lymphocytes from an immunized animal with transformed, immortal lymphocytes called *myeloma cells* that themselves synthesize neither the heavy nor the light polypeptides that constitute all antibodies (see Figure 3-19). Treatment with certain viral glycoproteins or the chemical polyethylene glycol promotes the plasma membranes of two cells to fuse, allowing their cytosols and organelles to intermingle. Some of the fused cells undergo division, and their nuclei eventually coalesce, producing viable *hybrid cells* with a single nucleus that contains chromosomes from both "parents." The fusion of two cells that are genetically different can yield a hybrid cell with novel characteristics. For instance, the fusion of a myeloma cell with a normal antibody-producing cell from a rat or mouse spleen yields a hybrid that proliferates into a clone called a **hybridoma**. Like myeloma cells, hybridoma cells grow rapidly and are immortal. Each hybridoma produces the monoclonal antibody encoded by its B-lymphocyte parent.

The second step in this procedure for producing monoclonal antibody is to separate, or select, the hybridoma cells from the unfused parental cells and the self-fused cells generated by the fusion reaction. This selection is usually performed by incubating the mixture of cells in a special culture medium, called *selection medium*, that permits the growth of only the

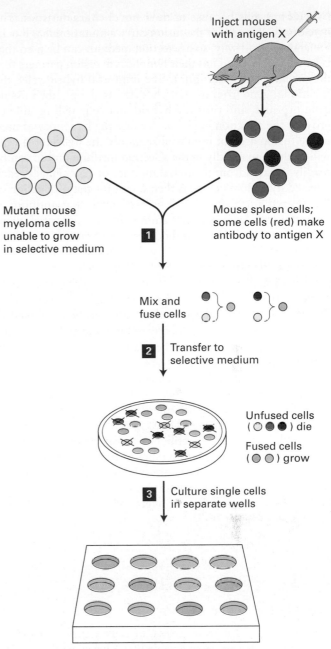

**FIGURE 9-6 Use of cell fusion and selection to obtain hybridomas producing monoclonal antibody to a specific protein.** Step **1**: Immortal myeloma cells that cannot synthesize purines under special conditions because they lack thymidine kinase are fused with normal antibody-producing spleen cells from an animal that was immunized with antigen X. Step **2**: When cultured in/on a special selective medium, unfused and self-fused cells do not grow: the mutant myeloma cells do not grow because the selective medium does not contain purines and the spleen cells because they have a limited life span in culture. Thus only fused cells formed from a myeloma cell *and* a spleen cell survive in the special medium, proliferating into clones called hybridomas. Each hybridoma produces a single antibody. Step **3**: Testing of individual clones identifies those that recognize antigen X. After a hybridoma that produces a desired antibody has been identified, the clone can be cultured to yield large amounts of that antibody.

hybridoma cells because of their novel characteristics. The myeloma cells used for the fusion carry a mutation that blocks a metabolic pathway, so a selection medium can be used that is lethal to them and not their lymphocyte fusion partners that do not have the mutation. In the immortal hybrid cells, the functional gene from the lymphocyte can supply the missing gene product, and thus the hybridoma cells will be able to grow in the selection medium. Because the lymphocytes used in the fusion are not immortalized, only the hybridoma cells will proliferate rapidly in the selection medium and so can be readily isolated from the initial mixture of cells. Finally, each selected hybridoma clone is then tested for the production of the desired antibody; any clone producing that antibody is then grown in large cultures, from which a substantial quantity of pure monoclonal antibody can be obtained.

Monoclonal antibodies have become very valuable reagents as specific research tools. They are commonly employed in affinity chromatography to isolate and purify proteins from complex mixtures (see Figure 3-38c). As we discuss later in this chapter, they can also be employed in immunofluorescence microscopy to bind and so locate a particular protein within cells. They can also be used to identify specific proteins in cell fractions with the use of immunoblotting (see Figure 3-39). Monoclonal antibodies have become important diagnostic and therapeutic tools in medicine; for example, monoclonal antibodies that bind to and inactivate toxins secreted by bacterial pathogens are used to treat diseases. Other monoclonal antibodies are specific for cell-surface proteins expressed by certain types of tumor cells. Several of these anti-tumor antibodies are widely used in cancer therapy, including monoclonal antibody against a mutant form of the Her2 receptor that is overexpressed in some breast cancers (see Figure 16-7).

## KEY CONCEPTS of Section 9.1

### Growing Cells in Culture

• Animal cells have to be grown in culture under conditions that mimic their natural environment, which generally requires them to be supplied with necessary amino acids and growth factor supplements.

• Most animal cells need to adhere to a solid surface to grow.

• Primary cells—those isolated directly from tissue—have a finite life span.

• Transformed cells, like cells derived from tumors, can grow indefinitely in culture.

• Cells that can be grown indefinitely are called a cell line.

• Many cells lines are aneuploid, having a different number of chromosomes than the parent animal from which they were derived.

• Different cells express different marker proteins on their cell surface, which can be used to distinguish them.

• Using fluorescent antibodies to cell-surface molecules, a machine called a fluorescent-activated cell sorter can sort out cells with different surface markers.

• To mimic growth in tissues, epithelial cells are often grown is special containers to mimic their functional polarity. Cells can also be grown in three-dimensional matrices to more accurately reflect their normal environment.

• Monoclonal antibodies, reagents that bind one epitope on an antigen, can be secreted by cultured cells called hybridomas. These hybrid cells are made by fusing an antibody-producing B-cell with an immortalized myeloma cell and then identifying those clones that produce the antibody. Monoclonal antibodies are important for basic research and as therapeutic agents.

## 9.2 Light Microscopy: Exploring Cell Structure and Visualizing Proteins Within Cells

The existence of the cellular basis of life was first appreciated using primitive light microscopes. Since then, progress in cell biology has paralleled and often been driven by technological advances in light microscopy (Figure 9-7). Here we discuss each of these major developments and how they advanced the study of cellular processes. First we describe basic uses of a light microscope to observe unstained cells and structures. Next we describe the development of fluorescence microscopy and its use to localize specific proteins in fixed cells. By using molecular genetic approaches to express a fusion between a protein of interest and a naturally fluorescent protein, it is possible to follow the localization of specific proteins in living cells—an ability that revealed how dynamic the organization of living cells is. In parallel with these advances in specimen preparation, optical advances were being made to enhance and sharpen the images provided by fluorescence microscopy to reveal cellular structure in unprecedented clarity. Many specialized technologies have emerged from these advances, and we describe some of the more important ones.

### The Resolution of the Light Microscope Is About 0.2 μm

All microscopes produce a magnified image of a small object, but the nature of the image depends on the type of microscope employed and on the way the specimen is prepared. The compound microscope, used in conventional *bright-field light microscopy*, contains several lenses that magnify the image of a specimen under study (Figure 9-8a). The total magnification is a product of the magnification of the individual lenses: if the *objective lens*, the lens closest to the specimen, magnifies 100-fold (a 100× lens, the maximum usually employed) and the *projection lens*, sometimes called

(a)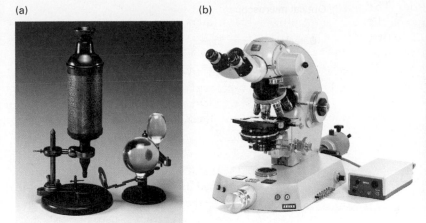

(b)

(c)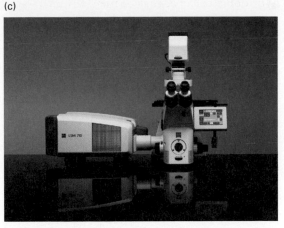

**FIGURE 9-7 Development of the light microscope.** (a) Early microscopes, like ones used by Robert Hooke in the 1660s, used lenses or a mirror to illuminate the specimen. (b) Optics in general and light microscopes in particular developed enormously during the nineteenth century, and by the middle of the twentieth century highly sophisticated microscopes limited only by the resolution of light were common. (c) In the second half of the twentieth century, fluorescence microscopy and digital imaging together with confocal techniques were developed to yield the versatile microscopes of today. [Part (a) SSPL via Getty Images; part (b) courtesy of Carl Zeiss Archive; part (c) Zeiss.com.]

the ocular or eyepiece, magnifies 10-fold, the final magnification recorded by the human eye or on a camera will be 1000-fold.

However, the most important property of any microscope is not its magnification but its resolving power, or **resolution**—the ability to distinguish between two very closely positioned objects. Merely enlarging the image of a specimen accomplishes nothing if the image is blurred. The resolution of a microscope lens is numerically equivalent to $D$, the minimum distance between two distinguishable objects. The smaller the value of $D$, the better the resolution. The value of $D$ is given by the equation

$$D = \frac{0.61\lambda}{N \sin\alpha} \qquad (9\text{-}1)$$

where $\alpha$ is the angular aperture, or half-angle, of the cone of light entering the objective lens from the specimen (see Figure 9-8a), $N$ is the refractive index of the medium between the specimen and the objective lens (i.e., the relative velocity of light in the medium compared with the velocity in air), and $\lambda$ is the wavelength of the incident light. Resolution is improved by using shorter wavelengths of light (decreasing the value of $\lambda$) or gathering more light (increasing either $N$ or $\alpha$). Lenses for high-resolution microscopy are designed to work with oil between the lens and the specimen since oil has a higher refractive index (1.56, compared with 1.0 for air and 1.3 for water). To maximize the angle $\alpha$, and hence $\sin\alpha$, the lenses are also designed to focus very close to the thin coverslip covering the specimen. The term $N \sin\alpha$ is known as the *numerical aperture (NA)* and is usually marked on the objective lens. A good high-magnification lens has an NA of about 1.4 and the very best lenses a value approaching 1.7, and costing as much as a medium-size car! Notice that the magnification is not part of this equation.

Owing to limitations in the values of $\alpha$, $\lambda$, and $N$ based on the physical properties of light, the *limit of resolution* of a light microscope using visible light is about 0.2 μm (200 nm). No matter how many times the image is magnified, a conventional light microscope can never resolve objects that are less than ~0.2 μm apart or reveal details smaller than ~0.2 μm in size. However, some new sophisticated technologies have been devised to 'beat' this resolution barrier and can resolve objects just a few nanometers apart; we discuss such a super-resolution microscope in a later section.

Despite this lack of resolution, a conventional microscope can track a single object to within a few nanometers. If we know the precise size and shape of an object—say, a 5-nm sphere of gold that is attached to an antibody in turn bound to a cell-surface protein on a living cell—and if we use a camera to rapidly take multiple digital images, then a computer can calculate the average position to reveal the center of the object to within a few nanometers. In this way, computer algorithms can be used to locate single objects at a more precise level—in this case the location and movement with time of a cell-surface protein labeled with the gold-tagged antibody—than would be possible based on the light microscope's resolution alone. This technique has been used to measure nanometer-size steps as molecules and vesicles move along cytoskeletal filaments (see Figures 17-29 and 17-30).

## Phase-Contrast and Differential-Interference-Contrast Microscopy Visualize Unstained Living Cells

Cells are about 70 percent water, 15 percent protein, 6 percent RNA, and smaller amounts of lipids, DNA, and small molecules. Since none of these major classes of molecules are colored, other methods have to be used to see cells in a microscope. The simplest microscope views cells under *bright-field* optics

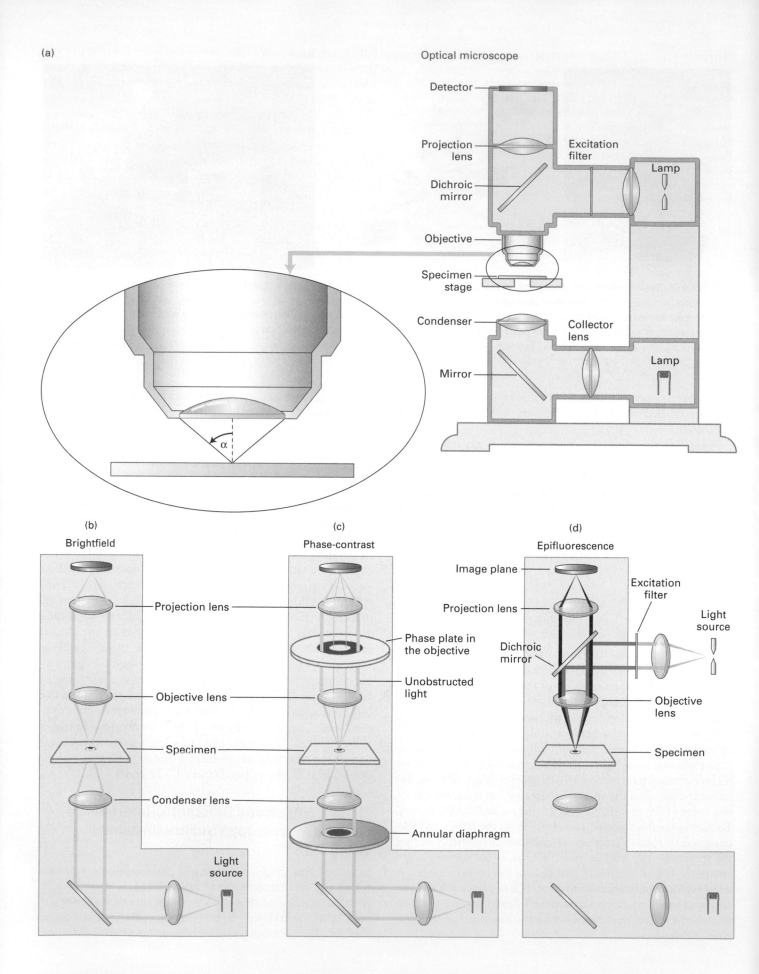

**(a)**

Optical microscope

Detector

Projection lens

Excitation filter

Dichroic mirror

Lamp

Objective

Specimen stage

Condenser

Collector lens

Mirror

Lamp

α

**(b)**

Brightfield

Projection lens

Objective lens

Specimen

Condenser lens

Light source

**(c)**

Phase-contrast

Phase plate in the objective

Unobstructed light

Annular diaphragm

**(d)**

Epifluorescence

Image plane

Excitation filter

Projection lens

Light source

Dichroic mirror

Objective lens

Specimen

**FIGURE 9-8 Optical microscopes are commonly configured for bright-field (transmitted), phase-contrast, and epifluorescence microscopy.** (a) In a typical light microscope, the specimen is usually mounted on a transparent glass slide and positioned on the movable specimen stage. (b) In bright-field light microscopy, light from a tungsten lamp is focused on the specimen by a condenser lens below the stage; the light travels the pathway shown in yellow. (c) In phase-contrast microscopy, incident light passes through an annular diaphragm, which focuses a circular annulus (ring) of light on the sample. Light that passes unobstructed through the specimen is focused by the objective lens onto the thicker gray ring of the phase plate, which absorbs some of the direct light and alters its phase by one-quarter of a wavelength. If a specimen refracts (bends) or diffracts the light, the phase of some light waves is altered (green lines) and the light waves pass through the clear region of the phase plate. The refracted and unrefracted light are recombined at the image plane to form the image. (d) In epifluorescence microscopy, a beam of light from a mercury lamp (gray lines) is directed to the excitation filter that allows just the correct wavelength light to pass (green lines). The light is then reflected off a dichroic filter and through the objective that focuses it on the sample. The fluorescent light emitted by the sample (red lines) passes up through the objective, then through the dichroic mirror and is focused and recorded on the detector at the image plane.

(Figure 9-8b), and little detail can be seen (Figure 9-9). Two common methods for imaging live cells and unstained tissues to generate contrast takes advantage of differences in the refractive index and thickness of cellular materials. These methods, called *phase-contrast microscopy* and *differential-interference-contrast (DIC) microscopy* (or Nomarski interference microscopy), produce images that differ in appearance and reveal different features of cell architecture. Figure 9-9 compares images of live, cultured cells obtained with these two methods and standard bright-field microscopy. Since optical microscopes are expensive, they are often set up to perform many different types of microscopy on the same microscope stand (see Figure 9-8a–d).

Phase-contrast microscopy generates an image in which the degree of darkness or brightness of a region of the sample depends on the *refractive index* of that region. Light moves more slowly in a medium of higher refractive index. Thus a beam of light is refracted (bent) once as it passes from the medium into a transparent object and again when it departs. In a phase-contrast microscope, a cone of light generated by an annular diaphragm in the condenser illuminates the specimen (see Figure 9-8c). The light passes through the specimen into the objective, and the unobstructed direct light passes through a region of the phase plate that both transmits only a small percentage of the light and changes its phase slightly. The part of a light wave that passes through a specimen will be refracted and will be out of phase (out of synchrony) with the part of the wave that does not pass through the specimen. How much their phases differ depends on the difference in refractive index along the two paths and on the thickness of the specimen. The refracted and unrefracted light are recombined at the image plane to form the image. If the two parts of the light wave are recombined, the resultant light will be brighter if they are in phase and less bright if they are out of phase. Phase-contrast microscopy is suitable for observing single cells or thin cell layers but not thick tissues. It is particularly useful for examining the location and movement of larger organelles in live cells.

DIC microscopy is based on interference between polarized light and is the method of choice for visualizing extremely small details and thick objects. Contrast is generated by differences in the refractive index of the object and its surrounding medium. In DIC images, objects appear to cast a shadow to one side. The "shadow" primarily represents a difference in the refractive index of a specimen rather than its topography. DIC microscopy easily defines the outlines of large organelles, such as the nucleus and vacuole. In addition to having a "relief"-like appearance, a DIC image is a thin *optical section*, or slice, through the object (Figure 9-9, *right*). Thus details of the nucleus in thick specimens (e.g., an intact *Caenorhabditis elegans* roundworm; see Figure 21-31) can be observed in a series of such optical sections, and the three-dimensional structure of the object can be reconstructed by combining the individual DIC images.

Both phase-contrast and DIC microscopy can be used in *time-lapse microscopy*, in which the same cell is photographed at regular intervals over time to generate a movie.

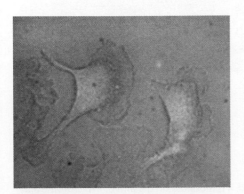 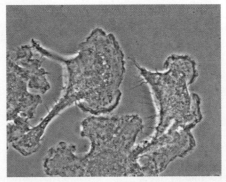 

**FIGURE 9-9 Live cells can be visualized by microscopy techniques that generate contrast by interference.** These micrographs show live, cultured macrophage cells viewed by bright-field microscopy (*left*), phase-contrast microscopy (*middle*), and differential-interference-contrast (DIC) microscopy (*right*). In a phase-contrast image, cells are surrounded by alternating dark and light bands; in-focus and out-of-focus details are simultaneously imaged in a phase-contrast microscope. In a DIC image, cells appear in pseudo-relief. Because only a narrow in-focus region is imaged, a DIC image is an optical slice through the object. [Courtesy of N. Watson and J. Evans.]

This procedure allows the observer to study cell movement, provided the microscope's stage can control the temperature of the specimen and the appropriate environment.

## Imaging Subcellular Details Often Requires That the Samples Be Fixed, Sectioned, and Stained

Live cells and tissues generally lack compounds that absorb light and so are nearly invisible in a light microscope. Although such specimens can be visualized by the special techniques we just discussed, these methods do not reveal the fine details of structure.

Specimens for light and electron microscopy are commonly fixed with a solution containing chemicals that cross-link most proteins and nucleic acids. Formaldehyde, a common fixative, cross-links amino groups on adjacent molecules; these covalent bonds stabilize protein-protein and protein–nucleic acid interactions and render the molecules insoluble and stable for subsequent procedures. After fixation, a tissue sample for examination by light microscopy is usually embedded in paraffin and cut into sections about 50 μm thick (Figure 9-10a). Cultured cells growing on glass coverslips, as described above, are thin enough so they can be fixed in situ and visualized by light microscopy without the need for sectioning.

A final step in preparing a specimen for light microscopy is to stain it so as to visualize the main structural features of the cell or tissue. Many chemical stains bind to molecules that have specific features. For example, histological samples are often stained with *hematoxylin* and *eosin* ("H&E stain").

Hematoxylin binds to basic amino acids (lysine and arginine) on many different kinds of proteins, whereas eosin binds to acidic molecules (such as DNA and side chains of aspartate and glutamate). Because of their different binding properties, these dyes stain various cell types sufficiently differently that they are distinguishable visually (Figure 9-10b). If an enzyme catalyzes a reaction that produces a colored or otherwise visible precipitate from a colorless precursor, the enzyme can be detected in cell sections by their colored reaction products. Such staining techniques, although once quite common, have been largely replaced by other techniques for visualizing particular proteins, as we discuss next.

## Fluorescence Microscopy Can Localize and Quantify Specific Molecules in Live Cells

Perhaps the most versatile and powerful technique for localizing molecules within a cell by light microscopy is **fluorescent staining** of cells and observation by *fluorescence microscopy*. A chemical is said to be fluorescent if it absorbs light at one wavelength (the excitation wavelength) and emits light (fluoresces) at a specific and longer wavelength. Modern microscopes for observing fluorescent samples are configured to pass the excitation light through the objective into the sample and then selectively observe the emitted fluorescent light coming back through the objective from the sample. This is achieved by reflecting the excitation light on a special type of filter called a dichroic mirror into the sample and allowing the light emitted at the longer wavelength to pass through to the observer (see Figure 9-8d).

(a)

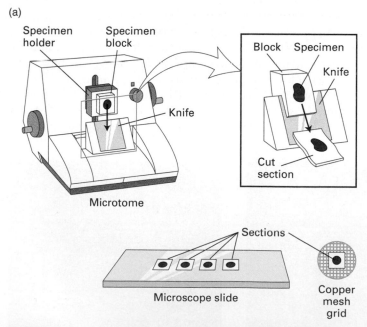

(b)

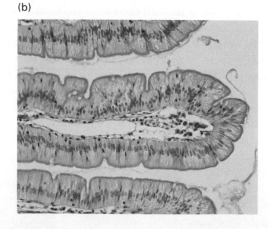

**FIGURE 9-10 Tissues for light microscopy are commonly fixed, embedded in a solid medium, and cut into thin sections.** (a) A fixed tissue is dehydrated by soaking in a series of alcohol-water solutions, ending with an organic solvent compatible with the embedding medium. To embed the tissue for sectioning, the tissue is placed in liquid paraffin for light microscopy. After the block containing the specimen has hardened, it is mounted on the arm of a microtome and slices are cut with a knife. Typical sections cut for light microscopy are 0.5 to 50 μm thick. The sections are collected on microscope slides and stained with an appropriate agent. (b) A section of mouse intestine stained with H&E. [Part (b) © Dr. Gladden Willis/Visuals Unlimited/Corbis.]

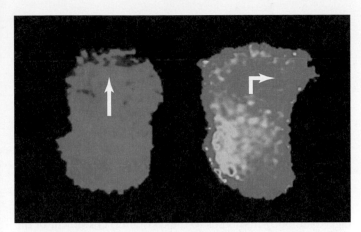

**EXPERIMENTAL FIGURE 9-11 Fura-2, a Ca²⁺-sensitive fluorochrome, can be used to monitor the relative concentrations of cytosolic Ca²⁺ in different regions of live cells.** (*Left*) In a moving leukocyte, a Ca²⁺ gradient is established. The highest levels (green) are at the rear of the cell, where cortical contractions take place, and the lowest levels (blue) are at the cell front, where actin undergoes polymerization. (*Right*) When a pipette filled with chemotactic molecules placed to the side of the cell induces the cell to turn, the Ca²⁺ concentration momentarily increases throughout the cytoplasm and a new gradient is established. The gradient is oriented such that the region of lowest Ca²⁺ (blue) lies in the direction that the cell will turn, whereas a region of high Ca²⁺ (yellow) always forms at the site that will become the rear of the cell. [From R. A. Brundage et al., 1991, *Science* **254:**703; courtesy of F. Fay.]

## Determination of Intracellular Ca²⁺ and H⁺ Levels with Ion-Sensitive Fluorescent Dyes

The concentration of $Ca^{2+}$ or $H^+$ within live cells can be measured with the aid of fluorescent dyes, or *fluorochromes*, whose fluorescence depends on the concentration of these ions. As discussed in later chapters, intracellular $Ca^{2+}$ and $H^+$ concentrations have pronounced effects on many cellular processes. For instance, many hormones and other stimuli cause a rise in cytosolic $Ca^{2+}$ from the resting level of about $10^{-7}$ M to $10^{-6}$ M, which induces various cellular responses such as the contraction of muscle.

The fluorescent dye *fura-2*, which is sensitive to $Ca^{2+}$, contains five carboxylate groups that form ester linkages with ethanol. The resulting fura-2 ester is lipophilic and can diffuse from the medium across the plasma membrane into cells. Within the cytosol, esterases hydrolyze fura-2 ester, yielding fura-2, whose free carboxylate groups render the molecule nonlipophilic and thus unable to cross cellular membranes, so it remains in the cytosol. Inside cells, each fura-2 molecule can bind a single $Ca^{2+}$ ion but no other cellular cation. This binding, which is proportional to the cytosolic $Ca^{2+}$ concentration over a certain range, increases the fluorescence of fura-2 at one particular wavelength. At a second wavelength, the fluorescence of fura-2 is the same whether or not $Ca^{2+}$ is bound and provides a measure of the total amount of fura-2 in a region of the cell. By examining cells continuously in the fluorescence microscope and measuring rapid changes in the ratio of fura-2 fluorescence at these two wavelengths, one can quantify rapid changes in the

fraction of fura-2 that has a bound $Ca^{2+}$ ion and thus in the concentration of cytosolic $Ca^{2+}$ (Figure 9-11).

Fluorescent dyes (e.g., SNARF-1) that are sensitive to the $H^+$ concentration can similarly be used to monitor the cytosolic pH of living cells. Other useful probes consist of a fluorochrome linked to a weak base that is only partially protonated at neutral pH and that can freely permeate cell membranes. In acidic organelles, however, these probes become protonated; because the protonated probes cannot recross the organelle membrane, they accumulate in the lumen in concentrations manyfold greater than in the cytosol. Thus this type of fluorescent dye can be used to specifically stain mitochondria and lysosomes in living cells (Figure 9-12).

## Immunofluorescence Microscopy Can Detect Specific Proteins in Fixed Cells

The common chemical dyes mentioned above stain nucleic acids or broad classes of proteins, but it is much more informative to detect the presence and location of specific proteins. *Immunofluorescence microscopy* is the most widely used method to detect specific proteins with an antibody to which a fluorescent dye has been covalently attached. To do this, you first need to generate antibodies to your specific protein. As discussed briefly in Section 9.1 and in detail in Chapter 23, as part of the response to infection the vertebrate immune

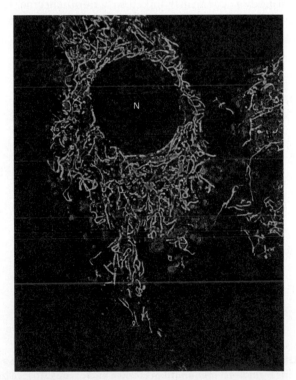

**EXPERIMENTAL FIGURE 9-12 Location of lysosomes and mitochondria in a cultured living bovine pulmonary artery endothelial cell.** The cell was stained with a green-fluorescing dye that is specifically bound to mitochondria and a red-fluorescing dye that is specifically incorporated into lysosomes. The image was sharpened using a deconvolution computer program discussed later in the chapter. N, nucleus. [Courtesy Invitrogen/Molecular Probes Inc.]

system generates proteins called antibodies that bind specifically to the infectious agent. Cell biologists have made use of this immunological response to generate antibodies to specific proteins. Consider you have purified protein X and then inject it into an experimental animal so that it responds to the protein as a foreign molecule. Over a period of weeks, the animal will mount an immune response and make antibodies to protein X (the "antigen"). If you collect the blood from the animal, it will have antibodies to protein X mixed in with antibodies to many other different antigens, together with all the other blood proteins. You can now covalently bind protein X to a resin and, using affinity chromatography, bind and selectively retain just those antibodies specific to protein X. The antibodies can be eluted from the resin, and now you have a reagent that binds specifically to protein X. This approach generates *polyclonal antibodies* since many different cells in the animal have contributed the antibodies. Alternatively, as we described earlier in this chapter, it is possible to generate a clonal cell line that secretes antibodies to a specific epitope on protein X; these are called *monoclonal antibodies*.

To use either type of antibody to localize the protein, the cells or tissue must first be fixed to ensure that all components remain in place and the cell permeabilized to allow entry of the antibody, commonly done by incubating the cells with a non-ionic detergent or extracting the lipids with an organic solvent. In one version of immunofluorescence microscopy, the antibody is covalently linked to a fluorochrome. Commonly used fluorochromes include rhodamine and Texas red,

which emit red light; Cy3, which emits orange light; and fluorescein, which emits green light. When a fluorochrome-antibody complex is added to a permeabilized cell or tissue section, the complex will bind to the corresponding antigen, then light up when illuminated by the exciting wavelength. Staining a specimen with different dyes that fluoresce at different wavelengths allows multiple proteins as well as DNA to be localized within the same cell (see chapter opening figure).

The most commonly used variation of this technique is called *indirect immunofluorescence microscopy* since the specific antibody is detected indirectly. In this technique, an unlabeled monoclonal or polyclonal antibody is applied to the cells or fixed tissue section, followed by a second fluorochrome-tagged antibody that binds to the constant (Fc) segment of the first antibody. For example, a "second" antibody can be generated by immunizing a goat with the Fc segment that is common to all rabbit IgG antibodies; when coupled to a fluorochrome, this second antibody preparation (called "goat anti-rabbit") will detect any rabbit antibody used to stain a tissue or cell (Figure 9-13). Because several goat anti-rabbit antibody molecules can bind to a single rabbit antibody molecule in a section, the fluorescence is generally much brighter than if a directly coupled single fluorochrome antibody is used. This approach is often extended to do *double-label fluorescence microscopy*, in which two proteins can be visualized simultaneously. For example, both proteins can be visualized by indirect immunofluorescence microscopy using first antibodies made in different animals (e.g., rabbit and chicken) and the second antibodies

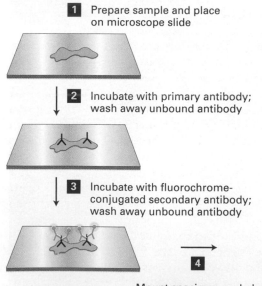

1 Prepare sample and place on microscope slide

2 Incubate with primary antibody; wash away unbound antibody

3 Incubate with fluorochrome-conjugated secondary antibody; wash away unbound antibody

4

Mount specimen and observe in fluorescence microscope

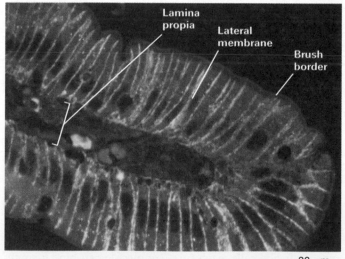

**FIGURE 9-13 A specific protein can be localized in fixed tissue sections by indirect immunofluorescence microscopy.** To localize a protein by immunofluorescence microscopy, a tissue section, or sample of cells, has to be chemically fixed and made permeable to antibodies (step **1**). The sample is then incubated with a primary antibody that binds specifically to the antigen of interest and then unbound antibody removed by washing (step **2**). The sample is next incubated with a fluorochrome-labeled secondary antibody that specifically binds to the primary antibody, and again excess secondary antibody is removed by washing (step **3**). The sample is then mounted in specialized mounting

medium and examined in a fluorescence microscope (step **4**). In this example, a section of the rat intestinal wall was stained with Evans blue, which generates a nonspecific red fluorescence, and GLUT2, a glucose transport protein, was localized by indirect immunofluorescence microscopy. GLUT2 is seen to be present in the basal and lateral sides of the intestinal cells but is absent from the brush border, composed of closely packed microvilli on the apical surface facing the intestinal lumen. Capillaries run through the lamina propria, a loose connective tissue beneath the epithelial layer. [B. Thorens et al., 1990, *Am. J. Physiol.* **259**:C279; courtesy of B. Thorens.]

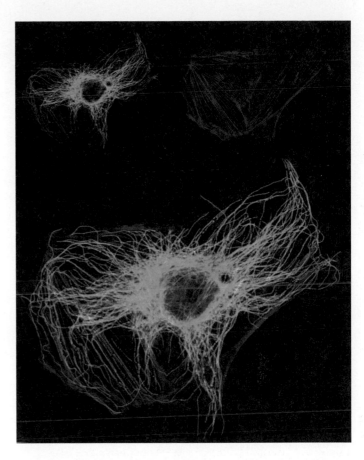

Rhodamine-labeled phalloidin
(fluorochrome-conjugated
drug that binds actin filaments, red)

Microtubule

Actin filament

Primary antibodies (rabbit, black)
that recognize microtubules and
fluorochrome-conjugated secondary
antibodies (goat–anti-rabbit, green)

**EXPERIMENTAL FIGURE 9-14 Double-label fluorescence microscopy can visualize the relative distributions of two proteins.** In double-label fluorescence microscopy, each protein has to be labeled specifically with different fluorochromes. The diagram at left shows how this can be done: a cultured cell was fixed and permeabilized and then incubated with Rhodamine-labeled phalloidin, a reagent that specifically binds to filamentous actin. It was also incubated with rabbit antibodies to tubulin, the major component of microtubules, followed by a fluorescein-labeled second goat–anti-rabbit antibody. The upper panels on the right show the fluorescein-stained tubulin (*left*) and Rhodamine-stained actin (*right*) and the lower panel the electronically merged images. [Part (right) Courtesy of A. Bretscher.]

(e.g., goat–anti-rabbit and sheep–anti-chicken) labeled with different fluorochromes. In another variation, one protein can be visualized by indirect immunofluorescence microscopy and the second protein by a dye that specifically binds to it. Once the individual images are taken on the fluorescence microscope, their images can be merged electronically (Figure 9-14).

In another widely used version of this technology, molecular biology techniques are used to make a cDNA encoding a recombinant protein to which is fused a short sequence of amino acids called an *epitope tag*. When expressed in cells, this cDNA will generate the protein linked to the specific tag. Two commonly used epitope tags are called FLAG, encoding the amino acid sequence DYKDDDDK (single-letter code), and myc, encoding the sequence EQKLISEEDL. Commercial fluorochrome-coupled monoclonal antibodies to the FLAG or myc epitopes can then be used to detect the recombinant protein in the cell. In an extension of this technology to allow the simultaneous visualization of two proteins, one protein can be tagged with FLAG and a different protein with myc. Each tagged protein is then visualized with a different color, for example, with a rhodamine-labeled antibody to the myc epitope and a fluorescein-labeled antibody to the FLAG epitope.

## Tagging with Fluorescent Proteins Allows the Visualization of Specific Proteins in Living Cells

The jellyfish *Aequorea victoria* expresses a naturally fluorescent protein, called *green fluorescent protein (GFP, ~27 kD)*. GFP contains a serine, tyrosine, and glycine sequence whose side chains spontaneously cyclize to form a green-fluorescing

chromophore when illuminated with blue light. Using recombinant DNA technologies, it is possible to make a DNA construct in which the coding sequence of GFP is fused to the coding sequence of a protein of interest. When introduced and expressed in cells, a GFP "tagged" protein is made in which the protein of interest is covalently linked to GFP as part of the same polypeptide. Although GFP is a moderate-size protein, the function of the protein of interest is often not changed by fusing it to GFP. This now allows one to visualize GFP—and hence the protein of interest. Not only can one immediately see the localization of the GFP-tagged protein, but one can view its distribution in a living cell over time and thereby assess its dynamics or track its localization following various cell treatments. The simple idea of tagging specific proteins with GFP has revolutionized cell biology and led to the development of many different fluorescent proteins (Figure 9-15). One use of this colorful variety of fluorescent proteins allows one to visualize two or more proteins simultaneously if they are each tagged with a different-colored fluorescent protein. We describe additional techniques that exploit fluorescent proteins in later sections.

## Deconvolution and Confocal Microscopy Enhance Visualization of Three-Dimensional Fluorescent Objects

Conventional fluorescence microscopy has two major limitations. First, the fluorescent light emitted by a sample comes not only from the plane of focus but also from molecules above and below it; thus the observer sees a blurred image

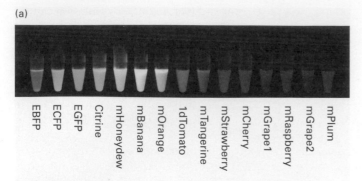

(a)

EBFP
ECFP
EGFP
Citrine
mHoneydew
mBanana
mOrange
1dTomato
mTangerine
mStrawberry
mCherry
mGrape1
mRaspberry
mGrape2
mPlum

(b)

**FIGURE 9-15 Many different colors of fluorescent proteins are now available.** (a) Tubes show the emission colors and names of many different fluorescent proteins, and (b) an agar dish is illuminated to show growing bacteria expressing several different-colored fluorescent proteins. [R. Tsien.]

caused by the superposition of fluorescent images from molecules at many depths in the cell. The blurring effect makes it difficult to determine the actual molecular arrangements. Second, to visualize thick specimens, consecutive (serial) images at various depths through the sample must be collected and then aligned to reconstruct structures in the original thick tissue. Two general approaches have been developed to obtain high-resolution three-dimensional information. Both these methods require the image to be collected electronically so that it can then be computationally manipulated as necessary.

The first approach is called *deconvolution microscopy*, which uses computational methods to remove fluorescence contributed from out-of-focus parts of the sample. Consider a three-dimensional sample in which images from three different focal planes are recorded. Since the whole sample is illuminated, the image from plane 2 will contain out-of-focus

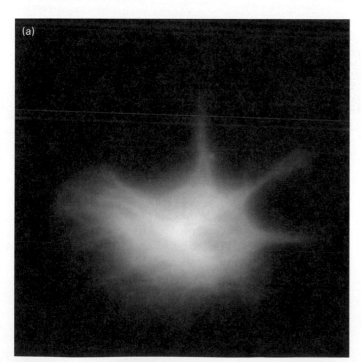

(a)

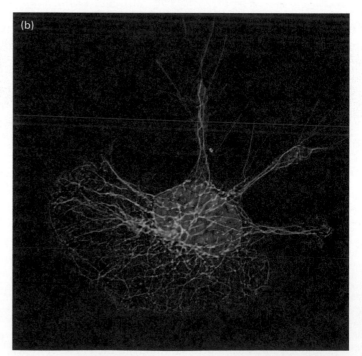

(b)

**EXPERIMENTAL FIGURE 9-16 Deconvolution fluorescence microscopy yields high-resolution optical sections that can be reconstructed into one three-dimensional image.** A macrophage cell was stained with fluorochrome-labeled reagents specific for DNA (blue), microtubules (green), and actin microfilaments (red). The series of fluorescent images obtained at consecutive focal planes (optical sections) through the cell were recombined in three dimensions. (a) In this three-dimensional reconstruction of the raw images, the DNA, microtubules, and actin appear as diffuse zones in the cell. (b) After application of the deconvolution algorithm to the images, the fibrillar organization of microtubules and the localization of actin to adhesions become readily visible in the reconstruction. [Courtesy of J. Evans, Whitehead Institute.]

fluorescence from planes 1 and 3. If you knew exactly how out-of-focus fluorescence from planes 1 and 3 contributed to light collected in plane 2, you could computationally remove it. To gain this information for a particular microscope, a series of images of focal planes are made from a test slide containing tiny fluorescent beads. Each bead represents a pinpoint of light that becomes a blurred object outside its focal plane; from these images a *point spread function* is determined that enables the investigator to calculate the distribution of fluorescent point sources that contributed to the "blur" when out of focus. Having calibrated the microscope in this manner, the experimental series of images can be computationally deconvolved. Images restored by deconvolution display impressive detail without any blurring, as illustrated in Figure 9-16.

The second approach to obtaining better three-dimensional information is called *confocal microscopy* because it uses optical methods to obtain images from a specific focal plane and exclude light from other planes. By collecting a series of images focused through the vertical depth of the sample, an accurate three-dimensional representation can be computationally generated. Two types of confocal microscopes are in common use today, a *point-scanning* confocal microscope (also known as a laser-scanning confocal microscope, or LSCM) and a *spinning disk* confocal microscope. The idea behind each microscope is to both illuminate and collect emitted fluorescent light in just one small area of a focal plane at a time in such a way that out-of-focus light is excluded. This can be achieved by collecting the emitted light through a pinhole before reaching the detector—light from the focal plane passes through, whereas light from other focal planes is largely excluded. The illuminated area is then moved across the whole focal plane to build up the image electronically. The two types of microscopes differ in how they cover the image. The point-scanning microscope uses a point laser light source at the excitation wavelength to rapidly scan the focal plane in a raster pattern, with the emitted fluorescence collected by a photomultiplier tube, and thereby build up an image (Figure 9-17a). It can then take a

(a) Laser-Scanning Confocal Microscope

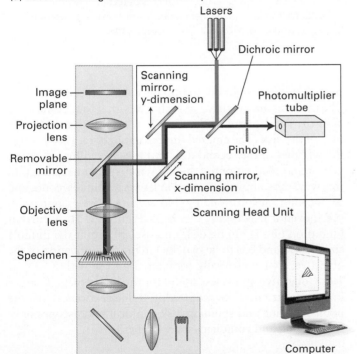

(b) Spinning Disk Confocal Microscope

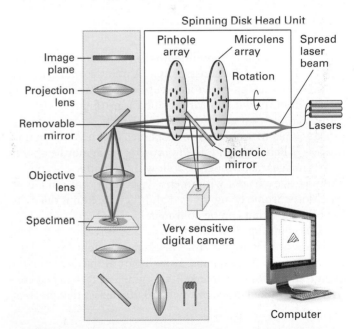

**FIGURE 9-17 Light paths for two types of confocal microscopy.** Both types of microscopy are assembled around a conventional fluorescence microscope (yellow shading). The diagram in (a) depicts the light path in a laser scanning confocal microscope. A single-wavelength point of light from an appropriate laser is reflected off a dichroic mirror and bounces off two scanning mirrors and from there through the objective to illuminate a spot in the specimen. The scanning mirrors rock back and forth in such a way that the light scans the sample in a raster fashion (see green lines in the sample). The fluorescence emitted by the sample passes back through the objective and is bounced off the scanning mirrors onto the dichroic mirror. This allows the light to pass through toward the pinhole. This pinhole excludes light from out-of-focus focal planes, so the light reaching the photomultiplier

tube comes almost exclusively from the illuminated spot in the focal plane. A computer then takes these signals and reconstructs the image. The diagram in (b) depicts the light path in a spinning disk confocal microscope. Here instead of using two scanning mirrors, the beam from the laser is spread to focus on pinholes in a spinning disk. The excitation light passes through the objective to provide point illumination of a number of spots in the sample. The fluorescence emitted passes back through the objective, through the holes in the spinning disk, and is then bounced off a dichroic mirror into a sensitive digital camera. The pinholes in the disk are arranged so that as it spins, it rapidly illuminates all parts of the sample several times. As the disk spins fast, for example at 3,000 rpm, very dynamic events in living cells can be recorded.

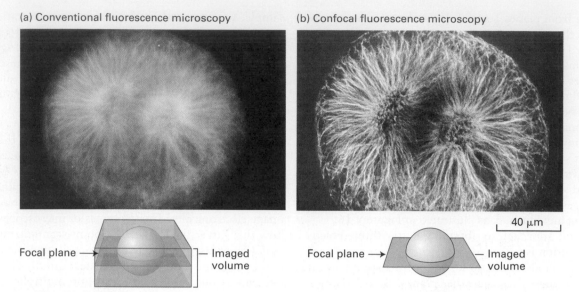

(a) Conventional fluorescence microscopy    (b) Confocal fluorescence microscopy

40 μm

Focal plane → | Imaged volume    Focal plane → | Imaged volume

**FIGURE 9-18 Confocal microscopy produces an in-focus optical section through thick cells.** A mitotic fertilized egg from a sea urchin (*Psammechinus*) was lysed with a detergent, exposed to an anti-tubulin antibody, and then exposed to a fluorescein-tagged antibody that binds to the anti-tubulin antibody. (a) When viewed by conventional fluorescence microscopy, the mitotic spindle is blurred. This blurring occurs because background fluorescence is detected from tubulin above and below the focal plane as depicted in the sketch. (b) The confocal microscopic image is sharp, particularly in the center of the mitotic spindle. In this case, fluorescence is detected only from molecules in the focal plane, generating a very thin optical section. [Micrographs from J. G. White et al., 1987, *J. Cell Biol.* **104**:41.]

series of images at different depths in the sample to generate a three-dimensional reconstruction. A point-scanning confocal microscope can provide exceptionally high-resolution images in both two and three dimensions (Figure 9-18), although it has two minor limitations. First, it can take significant time to scan each focal plane, so if a very dynamic process is being imaged, the microscope may not be able to collect images fast enough to follow the dynamics. Second, it illuminates each spot with intense laser light that can bleach the fluorochrome being imaged and therefore limit the number of images that can be collected.

The spinning disk microscope circumvents these two problems (see Figure 9-17b). The excitation light from a laser is spread out and illuminates a small part of the disk spinning at high speed, for example at 3000 rpm. The disk in fact consists of two linked disks: one with 20,000 lenses that precisely focuses the laser light on 20,000 pinholes of the second disk. The pinholes are arranged in such a way that they completely scan the focal plane of the sample several times with each turn of the disk. The emitted fluorescent light returns through the pinholes of the second disk and is reflected by a dichroic mirror and focused onto a highly sensitive digital camera. In this way, the sample is scanned in less than a millisecond, and so the real-time location of a fluorescent reporter can be captured even if it is highly dynamic (Figure 9-19). A current limitation of a spinning disk microscope is that the pinhole size is fixed and has to be matched to the magnification of the objective, so it is generally configured for use with a 63× or 100× objective and is less useful for the lower-magnification imaging that might be required in tissue sections. Thus the point-scanning and spinning disk confocal microscopes have overlapping and complementary strengths.

**VIDEO:** Microtubule Dynamics in Fission Yeast

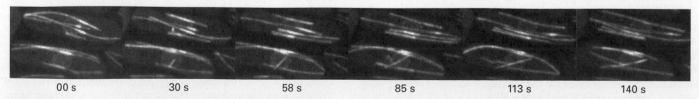

00 s          30 s          58 s          85 s          113 s          140 s

**EXPERIMENTAL FIGURE 9-19 The dynamics of microtubules can be imaged on the spinning disk confocal microscope.** Six frames from a movie of GFP-tubulin in two of the rod-shaped cells of fission yeast are shown. [Courtesy of Fred Chang.]

(a)

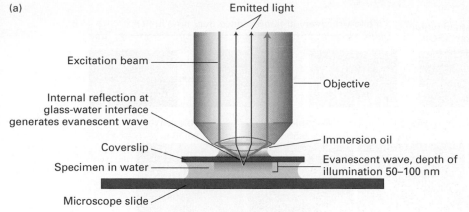

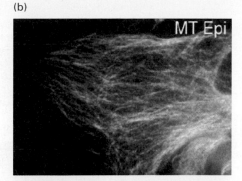

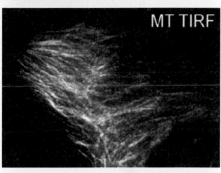

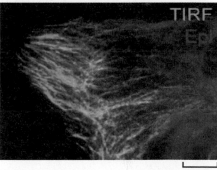

(b)

5 μm

**EXPERIMENTAL FIGURE 9-20 Fluorescent samples in a restricted focal plane can be imaged by total internal reflection (TIRF) microscopy.** (a) In TIRF microscopy only about 50 to 100 nm adjacent to the coverslip is illuminated so that fluorescent molecules in the rest of the sample are not excited. This limited illumination is achieved by directing the illuminating light at an angle where it is reflected from the glass-water interface of the coverslip rather than passing through it. Whereas most of the light is reflected, it also generates a very small region of illumination called the evanescent wave (depicted in light blue). (b) Immunofluorescence microscopy with tubulin antibody has been used to visualize microtubules viewed by conventional fluorescence microscopy (top panel), TIRF (middle panel), and a merged image. The two images were collected and false colored red and green so that the merge could highlight those microtubules that are close to the coverslip (green). [Part (b) from J. B. Manneville et al., 2010, *J. Cell Biol.* **191**:585.]

## TIRF Microscopy Provides Exceptional Imaging in One Focal Plane

The confocal microscopes we have just described provide amazing and informative images. However, they are not perfect, and some experimental situations call for fluorescence imaging in a thin focal plane adjacent to a surface. For example, confocal imaging is not ideal for exploring the details of proteins at adhesion sites between a cell and a coverslip or to follow the kinetics of assembly of microtubules attached to a coverslip. Both these situations can be imaged at high sensitivity using *total internal reflection fluorescence (TIRF)* microscopy. In the most common configuration of TIRF microscopy, the excitation beam of light comes through the objective lens (Figure 9-20a). However, the angle at which the light arrives at the coverslip is adjusted to the critical angle so that the light is reflected off the coverslip and returns up through the objective. This generates a narrow band, called an *evanescent wave*, that illuminates only about 50 to 100 nm of the sample adjacent to the coverslip, with no illumination to the rest of the sample. Thus if you have a complex mixture of fluorescent structures in a specimen, in the TIRF microscope you will see only those that are within 50 to100 nm (2 to 4 times the thickness of a microtubule) of the coverslip. When cells are grown on a coverslip,

TIRF has been exceptionally useful in identifying structures on the bottom of cells and therefore close to the coverslip (Figure 9-20b) and for measuring the kinetics of assembly and disassembly of structures such as microtubules and actin filaments (see Chapters 17 and 18).

## FRAP Reveals the Dynamics of Cellular Components

Live cell fluorescence imaging reveals the location and bulk dynamics of populations of fluorescent molecules, but it doesn't tell you how dynamic individual molecules are. For example, if we see that a GFP-labeled protein forms a patch at the surface of a cell, does this represent a stable collection of fluorescent protein molecules or a dynamic equilibrium with fluorescent proteins coming in and out of the patch? We can investigate this question by observing the dynamics of the molecules in the patch (Figure 9-21). If we use a high-intensity light to permanently bleach the fluorochrome (e.g., GFP) just in the patch, there will initially be no fluorescence coming

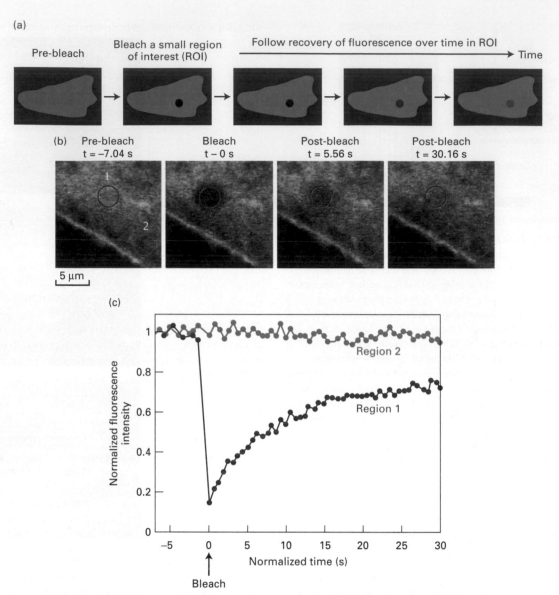

(a) Pre-bleach → Bleach a small region of interest (ROI) → Follow recovery of fluorescence over time in ROI → Time

(b) Pre-bleach t = −7.04 s | Bleach t − 0 s | Post-bleach t = 5.56 s | Post-bleach t = 30.16 s

5 μm

(c)

Region 2

Region 1

Bleach

Normalized fluorescence intensity

Normalized time (s)

**EXPERIMENTAL FIGURE 9-21 Fluorescence recovery after photobleaching (FRAP) reveals the dynamics of molecules.** In a living cell, following the distribution of a GFP-labeled protein provides a view of the overall distribution of the protein, but it doesn't tell you how dynamic populations of individual molecules might be. This can be determined by FRAP. (a) In this technique, the GFP signal is bleached by a short burst of strong laser light focused on the region of interest (ROI). This rapidly bleaches the molecules irreversibly, so they are not detected again. Restoration of fluorescence into the region represents unbleached molecules that have moved into the ROI. (b) The mobility of GFP-serotonin receptors on the cell surface observed using FRAP. The fluorescence in two regions is followed—one that is bleached (region 1) and a control region that is not (region 2). (c) By quantitating the recovery, the dynamic properties of the serotonin receptor can be established. [Part (b) from S. Kalipatnapu, 2007, Membrane Organization and Dynamics of the Serotonin 1A Receptor Monitored using Fluorescence Microscopic Approaches, in *Serotonin Receptors in Neurobiology*, A. Chattopadhyay, ed. CRC press.]

from it and it will look dark in the fluorescence microscope. However, if the components in the patch are in dynamic equilibrium with unbleached molecules elsewhere in the cell, the bleached molecules will be replaced by unbleached ones, and the fluorescence will begin to come back. The rate of fluorescence recovery is a measure of the dynamics of the molecules. This technique, known as *fluorescence recovery after photobleaching (FRAP)*, has revealed how very dynamic many components in cells are. For example, it has been used to determine the diffusion coefficient of membrane proteins (see Figure 10-10), the dynamics of specific components of the secretory pathway.

## FRET Measures Distance Between Chromophores

Fluorescence microscopy can also be used to determine if two proteins interact in vivo using a technique called *Förster resonance energy transfer (FRET)*. This technique utilizes two fluorescent proteins in which the emission wavelength of the first is the same as the excitation wavelength of the second (Figure

(a)

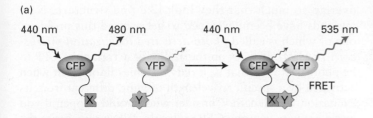

(b)

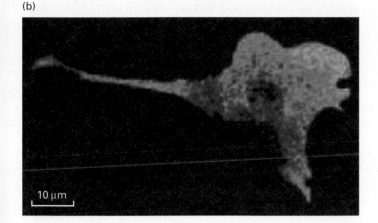

FIGURE 9-22 Protein-protein interactions can be visualized by FRET. The idea behind FRET is to use two different fluorescent proteins so that when one is excited, its emission will excite the second fluorescent protein provided they are sufficiently close (upper panel). In this example, cyan fluorescent protein (CFP) is fused to protein X, yellow fluorescent protein (YFP) is fused to protein Y, and both proteins are expressed in a living cell. If the cell is now illuminated with 440-nm light, the CFP will emit a fluorescent signal at 480 nm. If YFP is not close by, it will not absorb the 480-nm light and no 535-nm light will be emitted. However, if protein X interacts with protein Y (as shown), it will bring CFP close to YFP, the emitted 480-nm light will be captured by YFP, and it will emit light at 535 nm. (b) In this mouse fibroblast, FRET has been used to reveal that the interaction between an active regulatory protein (Rac) and its binding partner is localized to the front of a migrating cell. [Part (b) from R. B. Sekar and Periasamy, 2003, *J. Cell Biol.* **160**:629.]

sensitive to small changes in distance and in practice is not detectable at >10 nm. Thus by illuminating an appropriate sample with 440-nm light and observing at 535 nm, one can tell if proteins separately tagged with CFP and YFP are in very close proximity. For example, FRET sensors have been developed to determine where in a cell signaling between a small GTP-binding protein and its effector occurs (Figure 9-22b).

A modified version of this technique can be used to measure conformational changes in a protein. For example, a sensor called cameleon has been designed to measure intracellular levels of $Ca^{2+}$. Cameleon consists of a single polypeptide containing both CFP and YFP joined by a piece of polypeptide capable of binding $Ca^{2+}$ (Figure 9-23a). In the absence of $Ca^{2+}$, the CFP

9-22). For example, when cyan fluorescent protein (CFP) is excited with 440-nm wavelength light, it fluoresces and emits light at 480 nm. If yellow fluorescent protein (YFP) is close by, it will absorb the 480-nm light and emit light at 535 nm. The efficiency of energy transfer is proportional to $R^6$, where $R$ is the distance between the fluorophores—it is therefore very

EXPERIMENTAL FIGURE 9-23 FRET biosensors can detect local biochemical environments. (a) A FRET biosensor is a fusion protein containing two fluorescent proteins linked by a region sensitive to the environment under study. In this example, a protein construct called cameleon consists of CFP linked to YFP by a sequence based on the protein calmodulin that undergoes a large conformational change when it binds $Ca^{2+}$. In the absence of $Ca^{2+}$, the two fluorescent proteins are too far apart to undergo FRET, whereas when the local $Ca^{2+}$ rises, they are brought sufficiently close to undergo FRET. (b) An example of the use of cameleon reveals the oscillation in local $Ca^{2+}$ levels in an *Arabidopsis* growing root tip.

[Part (b) from G. B. Monshausen et al., 2008, *Plant Physiol.* **147**:1690–1698.]

(a)

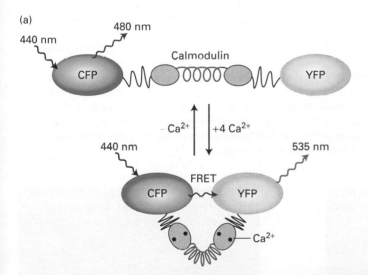

(b)

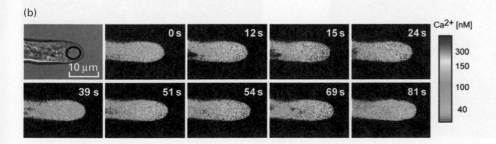

and YFP are not close enough for FRET to occur. However, in the presence of an appropriate local concentration of $Ca^{2+}$, cameleon binds $Ca^{2+}$ and undergoes a conformational change that brings CFP and YFP in close proximity, and they can now undergo FRET. Sensors such as cameleon can be used to measure the level of $Ca^{2+}$ in living cells, for example in a growing root tip (Figure 9-23b). Creative researchers are developing FRET sensors to illuminate many different types of local environments; for example, it is possible to make a probe that undergoes FRET only when it becomes phosphorylated by a specific kinase and thereby reveal where in the cell the active kinase is localized.

## Super-Resolution Microscopy Can Localize Proteins to Nanometer Accuracy

As we discussed earlier, the theoretical resolution limit of the fluorescence microscope is about 0.2 μm (200 nm). To understand why this is, consider two fluorescent structures separated by 100 nm. When you try to image them, they each generate a Gaussian distribution of fluorescence, which

overlap so much that they look like one structure. New methods have been developed to get around this problem, one of which is called *photo-activated localization microscopy (PALM)*. It relies on the ability of a variant of GFP to be photoactivated; that is, it only becomes fluorescent when activated by a specific wavelength of light, different from its excitation wavelength. Consider what would happen if you could activate just one GFP molecule. When you excite the sample, the one activated GFP would emit many hundreds of photons, giving rise to a Gaussian distribution (Figure 9-24a). Although analysis of each photon does not tell you precisely where the GFP is, the center of the peak can tell you where the GFP is located with nanometer accuracy. If you now activate another GFP, you can localize it individually with the same precision. In PALM, a small percentage of GFPs are activated and each localized with high precision, and then another set is activated and localized, and as additional cycles of activation and localization are recorded, a high-resolution image emerges. For example, the three-dimensional distribution of microtubules can be seen with much greater clarity than with any other light-microscopic method (Figure 9-24b) and a clathrin-coated pit—about 100 nm in diameter—

(a)

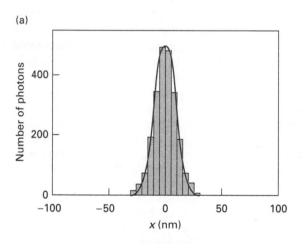

**EXPERIMENTAL FIGURE 9-24 Super-resolution microscopy can generate light-microscope images with nanometer resolution.** The theoretical resolution of the light microscope can be circumvented by super-resolution microscopy, where single molecules are imaged individually/separately to generate a composite image. One version of this technology images proteins fused to photo-activatable GFP in fixed samples. (a) When a GFP is activated and then excited, it will emit thousands of photons that can be collected. This generates a Gaussian curve centered around the location of the emitting GFP; the center provides the location of the GFP to nanometer accuracy. This process is reiterated hundreds of times to excite other GFP molecules, and a high-resolution image emerges. (b) A confocal image of microtubules (*left*) is compared with a corresponding super-resolution image (*right*) in which the three-dimensional arrangement of the microtubules is color coded. (c) The circular nature of a clathrin-coated pit (discussed in Chapter 14) is shown—a confocal image of this structure would look like two bright spots without any detail visible. [Parts (b) and (c) from B. Huang et al., 2008, *Science* **319**:810–813.]

(b)

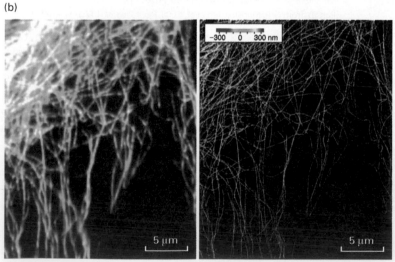

(c)

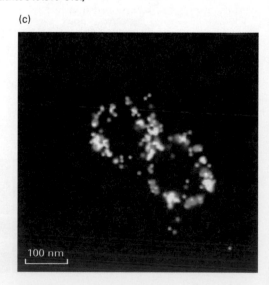

can be seen in remarkable detail (Figure 9-24c). These types of images can take up to an hour to generate, so they are restricted to fixed images and currently cannot be used for imaging proteins in live cells.

## KEY CONCEPTS of Section 9.2

### Light Microscopy: Exploring Cell Structure and Visualizing Proteins Within Cells

- The resolution of the light microscope, about 0.2 μm, is limited by the wavelength of light.

- Because most cellular components are not colored, differences in refractive index can be used to observe parts of single cells employing phase-contrast and interference-contrast microscopy.

- Tissues generally have to be fixed, sectioned, and stained for cells and subcellular structures to be observed.

- Fluorescence microscopy makes use of compounds that absorb light at one wavelength and emit it at a longer wavelength.

- Ion-sensitive fluorescent dyes can measure intracellular concentrations of ions, such as $Ca^{2+}$.

- Immunofluorescence microscopy makes use of antibodies to localize specific components in fixed and permeablized cells.

- Indirect immunofluorescence microscopy uses an unlabeled primary antibody, followed by a fluorescently labeled secondary antibody, that recognizes the primary one and allows it to be localized.

- Short sequences encoding epitope tags can be appended to protein-encoding sequences to allow localization of the expressed protein using an antibody to the epitope tag.

- Green fluorescence protein (GFP) and its derivatives are naturally occurring fluorescent proteins.

- Fusing GFP to a protein of interest allows its localization and dynamics to be explored in a living cell.

- Deconvolution and confocal microscopy provide greatly improved clarity in fluorescent images by removing out-of-focus fluorescent light.

- Total internal reflection (TIRF) microscopy allows fluorescent samples adjacent to a coverslip to be seen with great clarity.

- Fluorescence recovery after photobleaching (FRAP) of a GFP fusion protein allows the dynamics of the population of molecules to be analyzed.

- Förster resonance energy transfer (FRET) is a technique in which light energy is transferred from one fluorescent protein to another when the proteins are very close, thereby revealing when two molecules are close in the cell.

- Super-resolution microscopy allows for detailed fluorescent images at nanometer resolution.

## 9.3 Electron Microscopy: High-Resolution Imaging

Electron microscopy of biological samples, such as single proteins, organelles, cells, and tissues, offers a much higher resolution of ultrastructure than can be obtained by light microscopy. The short wavelength of electrons means that the limit of resolution for a transmission electron microscope is theoretically 0.005 nm (less than the diameter of a single atom), or 40,000 times better than the resolution of a light microscope and 2 million times better than that of the unaided human eye. However, the effective resolution of the transmission electron microscope in the study of biological systems is considerably less than this ideal. Under optimal conditions, a resolution of 0.10 nm can be obtained with transmission electron microscopes, about 2000 times better than with conventional high-resolution light microscopes.

The fundamental principles of electron microscopy are similar to those of light microscopy; the major difference is that electromagnetic lenses focus a high-velocity electron beam instead of the visible light used by optical lenses. In the *transmission electron microscope (TEM)*, electrons are emitted from a filament and accelerated in an electric field. A condenser lens focuses the electron beam onto the sample; objective and projector lenses focus the electrons that pass through the specimen and project them onto a viewing screen or other detector (Figure 9-25, *left*). Because atoms in air absorb electrons, the entire tube between the electron source and the detector is maintained under an ultrahigh vacuum. Thus living material cannot be imaged by electron microscopy.

In this section, we describe various different approaches to viewing biological material by electron microscopy. The most widely used instrument is the transmission electron microscope, but also in common usage is the *scanning electron microscope (SEM)*, which provides complementary information as we discuss at the end of this section.

### Single Molecules or Structures Can Be Imaged After a Negative Stain or Metal Shadowing

It is common in biology to explore the detailed shape of single macromolecules, such as proteins or nucleic acids, or of structures, such as viruses and the filaments that make up the cytoskeleton. It is relatively easy to view these in the transmission electron microscope provided they are stained with a heavy metal that scatters the incident electrons. To prepare a sample, it is first absorbed to a 3-mm *electron microscope grid* (Figure 9-26a) coated with a thin film of plastic and carbon. The sample is then bathed in a solution of a heavy metal, such as uranyl acetate, and excess solution removed (Figure 9-26b). As a result of this procedure, the uranyl acetate coats the grid but is excluded from the regions where the sample has adhered. When viewed in the TEM, you observe where the stain has been excluded, and so the sample is said to be *negatively stained*. Because the stain can precisely reveal the

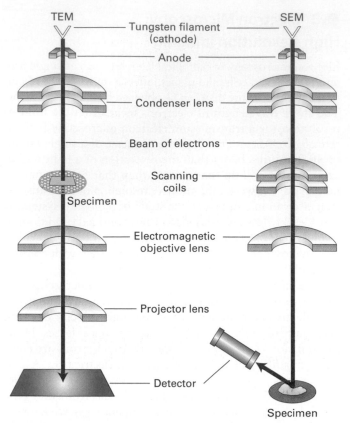

**FIGURE 9-25 In electron microscopy, images are formed from electrons that pass through a specimen or are scattered from a metal-coated specimen.** In a transmission electron microscope (TEM), electrons are extracted from a heated filament, accelerated by an electric field, and focused on the specimen by a magnetic condenser lens. Electrons that pass through the specimen are focused by a series of magnetic objective and projector lenses to form a magnified image of the specimen on a detector, which may be a fluorescent viewing screen, a photographic film, or a charged-couple-device (CCD) camera. In a scanning electron microscope (SEM), electrons are focused by condenser and objective lenses on a metal-coated specimen. Scanning coils move the beam across the specimen, and electrons scattered from the metal are collected by a photomultiplier tube detector. In both types of microscopes, because electrons are easily scattered by air molecules, the entire column is maintained at a very high vacuum.

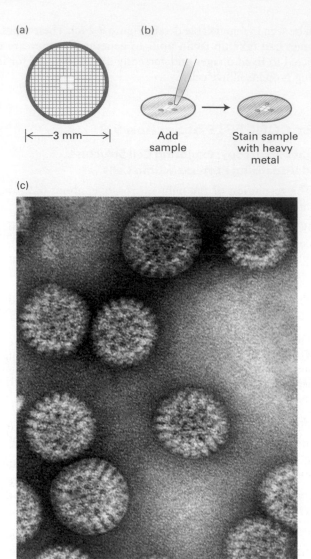

**FIGURE 9-26 Transmission electron microscopy of negatively stained samples reveals fine features.** (a) Samples for transmission electron microscopy (TEM) are usually mounted on a small copper or gold grid. The grid is usually covered with a very thin film of plastic and carbon to which a sample can adhere. (b) The specimen is then incubated in a heavy metal, such as uranyl acetate, and excess stain removed. (c) When observed in the TEM, the sample excludes the stain, so it is seen in negative outline. The example in (c) is a negative stain of rotaviruses. [Part (c) ISM/Phototake.]

topology of the sample, a high-resolution image can be obtained (Figure 9-26c).

Samples can also be prepared by *metal shadowing*. In this technique, the sample is absorbed to a small piece of mica, then coated with a thin film of platinum by evaporation of the metal, followed by dissolving the sample with acid or bleach. The platinum coating can be generated from a fixed angle or at a low angle as the sample is rotated, in which case it is called *low-angle rotary shadowing*. When the sample is transferred to a grid and examined in the TEM, these techniques provide information about the three-dimensional topology of the sample (Figure 9-27).

## Cells and Tissues Are Cut into Thin Sections for Viewing by Electron Microscopy

Single cells and pieces of tissue are too thick to be viewed directly in the standard transmission electron microscope. To overcome this, methods were developed to prepare and cut *thin sections* of cells and tissues. When these were examined

(a)

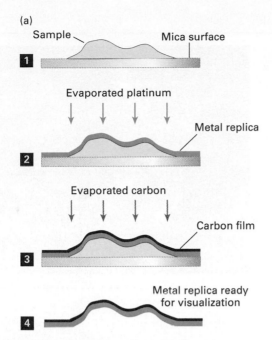

Sample     Mica surface

**1**

Evaporated platinum

Metal replica

**2**

Evaporated carbon

Carbon film

**3**

Metal replica ready
for visualization

**4**

(b)

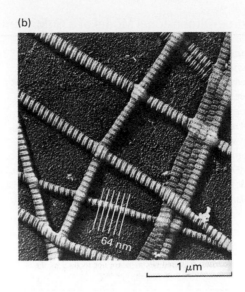

64 nm

1 μm

**FIGURE 9-27  Metal shadowing makes surface details on very small objects visible by transmission electron microscopy.** (a) The sample is spread on a mica surface and then dried in a vacuum evaporator (step **1**). The sample grid is coated with a thin film of a heavy metal, such as platinum or gold, evaporated from an electrically heated metal filament (step **2**). To stabilize the replica, the specimen is then coated with a carbon film evaporated from an overhead electrode (step **3**). The biological material is then dissolved by acid and bleach (step **4**), and the remaining metal replica is viewed in a TEM. In electron micrographs of such preparations, the carbon-coated areas appear light—the reverse of micrographs of simple metal-stained preparations, in which the areas of heaviest metal staining appear the darkest. (b) A platinum-shadowed replica of the substructural fibers of calfskin collagen, the major structural protein of tendons, bone, and similar tissues. The fibers are about 200 nm thick; a characteristic 64-nm repeated pattern (white parallel lines) is visible along the length of each fiber. [Courtesy R. Kessel and R. Kardon.]

in the electron microscope, the organization, beauty, and complexity of the cell interior was revealed and led to a revolution in cell biology—for the first time, new organelles and the first glimpses of the cytoskeleton were seen.

To prepare thin sections, it is necessary to chemically fix the sample, dehydrate it, impregnate it with a liquid plastic that hardens (similar to Plexiglas), and then cut sections of about 5 to 100 nm in thickness. For structures to be seen, the sample has to be stained with heavy metals such as uranium and lead salts, which can be done either before embedding in the plastic or after sections are cut. Examples of cells and tissues viewed by thin-section electron microscopy appear throughout this book (see, for example, Figure 9-33). It is important to realize that the images obtained represent just a thin slice through a cell, so to get a three-dimensional view, it is necessary to cut *serial sections* through the sample and reconstruct the sample from a series of sequential images (Figure 9-28).

## Immunoelectron Microscopy Localizes Proteins at the Ultrastructural Level

Just like immunofluorescence microscopy for localizing proteins at the light-microscope level, methods have been developed to use antibodies to localize proteins in thin sections at the electron microscope level. However, the harsh procedures used to prepare traditional thin sections—chemical fixation and embedding in plastic—can denature or modify the antigens so that they are no longer recognized by the specific antibodies. Gentler methods, such as a light fixation, sectioning material after freezing at the temperature of liquid nitrogen, followed by antibody incubations at room temperature, have been developed. To make the antibody visible in the electron microscope, it has to be attached to an electron-dense marker. One way to do this is to use electron-dense gold particles coated with protein A, a bacterial protein that binds the Fc segment of all antibody molecules (Figure 9-29). Because the gold particles diffract incident electrons, they appear as dark spots.

## Cryoelectron Microscopy Allows Visualization of Specimens Without Fixation or Staining

Standard transmission electron microscopy cannot be used to study live cells, and the absence of water causes macromolecules to become denatured and nonfunctional. However,

**FIGURE 9-28 Model of the Golgi complex based on three-dimensional reconstruction of electron microscopy images.** Transport vesicles (white spheres) that have budded off the rough ER fuse with the *cis* membranes (light blue) of the Golgi complex. By mechanisms described in Chapter 14, proteins move from the *cis* region to the *medial* region and finally to the *trans* region of the Golgi complex. Eventually, vesicles bud off the *trans*-Golgi membranes (orange and red); some move to the cell surface and others move to lysosomes. The Golgi complex, like the rough endoplasmic reticulum, is especially prominent in secretory cells. [Brad J. Marsh & Katheryn E. Howell, *Nature Reviews Molecular Cell Biology* 3, 789-785 (2002).]

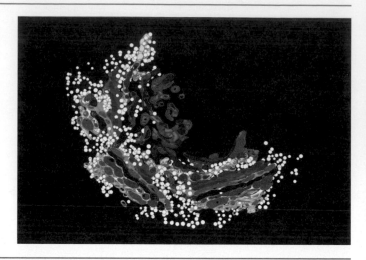

hydrated, unfixed, and unstained biological specimens can be viewed directly in a transmission electron microscope if the sample is frozen. In this technique of *cryoelectron microscopy*, an aqueous suspension of a sample is applied to a grid in an extremely thin film, frozen in liquid nitrogen, and maintained in this state by means of a special mount. The frozen sample then is placed in the electron microscope. The very low temperature (−196 °C) keeps water from evaporating, even in a vacuum. Thus the sample can be observed in detail in its native, hydrated state without fixing or heavy metal staining. By computer-based averaging of hundreds of images, a three-dimensional model can be generated almost to atomic resolution. For example, this method has been used to generate models of ribosomes, the muscle calcium pump discussed in Chapter 11, and other large proteins that are difficult to crystallize.

Many viruses have coats, or capsids, that contain multiple copies of one or a few proteins arranged in a symmetric array. In a cryoelectron microscope, images of these particles can be viewed from a number of angles. A computer analysis of multiple images can make use of the symmetry of the particle to calculate the three-dimensional structure of the capsid to about 5-nm resolution. Examples of such images are shown in Figure 4-44.

An extension of this technique, *cryoelectron tomography*, allows researchers to determine the three-dimensional

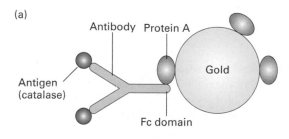

(a)

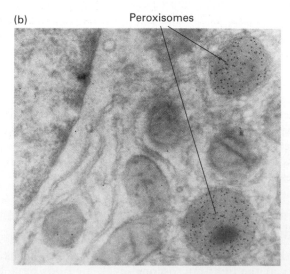

**FIGURE 9-29 Gold particles coated with protein A are used to detect an antibody-bound protein by transmission electron microscopy.** (a) First, antibodies are allowed to interact with their specific antigen (e.g., catalase) in a section of fixed tissue. Then the section is treated with electron-dense gold particles coated with protein A from the bacterium *S. aureus*. Binding of the bound protein A to the Fc domains of the antibody molecules makes the location of the target protein, catalase in this case, visible in the electron microscope. (b) A slice of liver tissue was fixed with glutaraldehyde, sectioned, and then treated as described in part (a) to localize catalase. The gold particles (black dots) indicating the presence of catalase are located exclusively in peroxisomes. [From H. J. Geuze et al., 1981, *J. Cell Biol.* **89**:653. Reproduced from the *Journal of Cell Biology* by copyright permission of The Rockefeller University Press.]

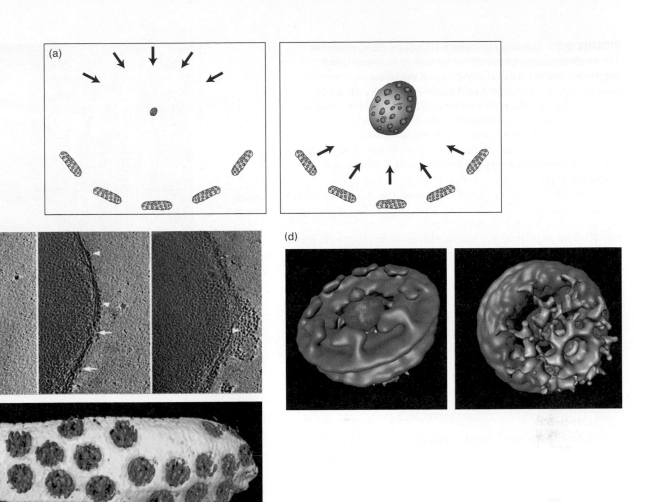

**FIGURE 9-30 Structure of the nuclear pore complex (NPC) by cryoelectron tomography.** (a) In electron tomography, a semicircular series of two-dimensional projection images is recorded from the three-dimensional specimen that is located at the center; the specimen is tilted while the electron optics and detector remain stationary. The three-dimensional structure is computed from the individual two-dimensional images that are obtained when the object is imaged by electrons coming from different directions (arrows in left panel). These individual images are used to generate a three-dimensional image of the object (arrows, right panel). (b) Isolated nuclei from the cellular slime mold *Dictyostelium discoideum* were quick-frozen in liquid nitrogen and maintained in this state as the sample was observed in the electron microscope. The panel shows three sequential tilted images. Different orientations of NPCs (arrows) are shown in top view (*left and center*) and side view (*right*). Ribosomes connected to the outer nuclear membrane are visible, as is a patch of rough ER (arrowheads). (c) Computer-generated surface-rendered representation of a segment of the nuclear envelope membrane (yellow) studded with NPCs (blue). (d) By averaging the images of multiple nuclear pores, much more detail can be discerned. [Part (a) after S. Nickell et al., 2006, *Nature Rev. Mol. Cell. Biol.* **7**:225. Parts (b), (c), and (d) from M. Beck et al., 2004, *Science* **306**:1387.]

architecture of organelles or even whole cells embedded in ice, that is, in a state close to life. In this technique, the specimen holder is tilted in small increments around the axis perpendicular to the electron beam; thus images of the object viewed from different directions are obtained (Figure 9-30a, b). The images are then merged computationally into a three-dimensional reconstruction termed a *tomogram* (Figure 9-30c, d). A disadvantage of cryoelectron tomography is that the samples must be relatively thin, about 200 nm; this is much thinner than the samples (200 μm thick) that can be studied by confocal light microscopy.

## Scanning Electron Microscopy of Metal-Coated Specimens Reveals Surface Features

*Scanning electron microscopy (SEM)* allows investigators to view the surfaces of unsectioned metal-coated specimens. An intense electron beam inside the microscope scans rapidly over the sample. Molecules in the coating are excited and release secondary electrons that are focused onto a scintillation detector; the resulting signal is displayed on a cathode-ray tube much like a conventional television (see Figure 9-25,

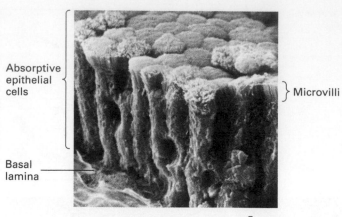

**FIGURE 9-31 Scanning electron microscopy (SEM) produces a three-dimensional image of the surface of an unsectioned specimen.** Shown here is an SEM image of the epithelium lining the lumen of the intestine. Abundant fingerlike microvilli extend from the lumen-facing surface of each cell. The basal lamina beneath the epithelium helps support and anchor it to the underlying connective tissue. Compare this image of intestinal cells with that in Figure 9-13, a fluorescence micrograph. [From R. Kessel and R. Kardon, 1979, *Tissues and Organs: A Text-Atlas of Scanning Electron Microscopy*, W. H. Freeman and Company, p. 176.]

*right*). The resulting scanning electron micrograph has a three-dimensional appearance because the number of secondary electrons produced by any one point on the sample depends on the angle of the electron beam in relation to the surface (Figure 9-31). The resolving power of scanning electron microscopes, which is limited by the thickness of the metal coating, is only about 10 nm, much less than that of transmission instruments.

## KEY CONCEPTS of Section 9.3

### Electron Microscopy: High-Resolution Imaging

- Electron microscopy provides very high-resolution images because of the short wavelength of the high-energy electrons used to image the sample.

- Simple specimens, such as proteins or viruses, can be negatively stained or shadowed with heavy metals for examination in a transmission electron microscope (TEM).

- Thicker sections generally must be fixed, dehydrated, embedded in plastic, sectioned, and then stained with electron-dense heavy metals before viewing by TEM.

- Specific proteins can be localized by TEM by employing specific antibodies associated with a heavy metal marker, such as small gold particles.

- Cryoelectron microscopy allows examination of hydrated, unfixed, and unstained biological specimens in the TEM by maintaining them at a few degrees above absolute zero.

- Scanning electron microscopy (SEM) of metal-shadowed material reveals the surface features of specimens.

## 9.4 Isolation and Characterization of Cell Organelles

The examination of cells by light and electron microscopy led to the appreciation that eukaryotic cells contain a common set of organelles. Most organelles are enclosed in a lipid bilayer and perform a specific function. To perform this function, each type of organelle has a recognizable structure and contains a

specific set of proteins to perform the function. Cell biologists use this fact to identify specific organelles. For example, as discussed in Chapter 12, most of the ATP in a cell is made by ATP synthase, which converts ADP to ATP and is localized in mitochondria, and ATP synthase is a good marker for mitochondria. As we will discuss below, the availability of specific markers for organelles has helped in the development of organelle purification.

In this section, we first give a brief overview of the organelles of eukaryotic cells as a prelude to their discussion in much more detail in later chapters. We then discuss methods that are used to open up cells for the purification of organelles. We end with recent advances in proteomics aimed at defining the complete protein inventory of organelles.

### Organelles of the Eukaryotic Cell

The major organelles in animal and plant cells are depicted in Figure 9-32, and some are shown in more detail in Figure 9-33.

The *plasma membrane* encloses the cell and is a vitally important barrier since it is the interface a cell has with its environment, separating the outside world from the internal **cytoplasm**. It is made up of about an equal mass of lipids and proteins. Although the physical properties of the plasma membrane are largely determined by its lipid composition, a membrane's complement of proteins is primarily responsible for the membrane's functional properties. As we discuss in Chapter 13, the plasma membrane is a permeability barrier that contains specific **membrane transport proteins** necessary to bring ions and metabolites into the cell. It is also the site of reception of chemical signals from other cells, so it contains **receptors** that sense these signals and transmits the information across the plasma membrane into the cytosol, a topic we discuss in Chapter 15. The plasma membrane defines the shape of a cell, and so it is intimately associated with the cytoskeleton and with other cells and the cellular matrix, interactions we cover in Chapters 17 through 19.

Larger molecules than ions and metabolites can be taken up by *endocytosis*, a process involving the invagination of the plasma membrane, which then pinches off to form an

Animal cell

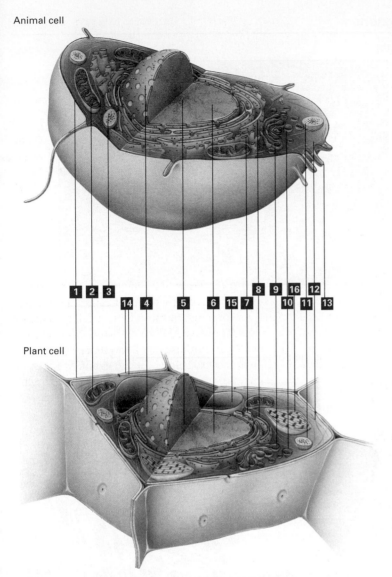

**1** Plasma membrane controls movement of molecules in and out of the cell and functions in cell-cell signaling and cell adhesion.

**2** Mitochondria, which are surrounded by a double membrane, generate ATP by oxidation of glucose and fatty acids.

**3** Lysosomes, which have an acidic lumen, degrade material internalized by the cell and worn-out cellular membranes and organelles.

**4** Nuclear envelope, a double membrane, encloses the contents of the nucleus; the outer nuclear membrane is continuous with the rough ER.

**5** Nucleolus is a nuclear subcompartment where most of the cell's rRNA is synthesized.

**6** Nucleus is filled with chromatin composed of DNA and proteins; site of mRNA and tRNA synthesis.

**7** Smooth endoplasmic reticulum (ER) synthesizes lipids and detoxifies certain hydrophobic compounds.

**8** Rough endoplasmic reticulum (ER) functions in the synthesis, processing, and sorting of secreted proteins, lysosomal proteins, and certain membrane proteins.

**9** Golgi complex processes and sorts secreted proteins, lysosomal proteins, and membrane proteins synthesized on the rough ER.

**10** Secretory vesicles store secreted proteins and fuse with the plasma membrane to release their contents.

**11** Peroxisomes detoxify various molecules and also break down fatty acids to produce acetyl groups for biosynthesis.

**12** Cytoskeletal fibers form networks and bundles that support cellular membranes, help organize organelles, and participate in cell movement.

**13** Microvilli increase surface area for absorption of nutrients from surrounding medium.

**14** Cell wall, composed largely of cellulose, helps maintain the cell's shape and provides protection against mechanical stress.

**15** Vacuole stores water, ions, and nutrients, degrades macromolecules, and functions in cell elongation during growth.

**16** Chloroplasts, which carry out photosynthesis, are surrounded by a double membrane and contain a network of internal membrane-bounded sacs.

Plant cell

**FIGURE 9-32 Schematic overview of a "typical" animal cell (top) and plant cell (bottom) and their major substructures.** Not every cell will contain all the organelles, granules, and fibrous structures shown here, and other substructures can be present in some. Cells also differ considerably in shape and in the prominence of various organelles and substructures.

*endosome* in the cytoplasm. During the best-studied form of endocytosis, special regions of the plasma membrane called *coated pits* are formed in which receptors collect and bring specific molecules or particles into the cell (Figure 9-33a). This process is known as **receptor-mediated endocytosis**. Once materials are internalized, they are sorted and can either be returned to the plasma membrane or delivered to *lysosomes* for degradation. Lysosomes contain a battery of degradative enzymes that can break down essentially any biological molecule into smaller components. The lumen of lysosomes has an acidic pH of 4.5; this helps to denature proteins and the degradative enzymes—collectively known as *acid hydrolases*—can withstand this environment and in fact work optimally at this pH.

The largest internal membrane system is an organelle known as the **endoplasmic reticulum (ER)**, consisting of an extensive interconnected network of flattened membrane-bound sacs and tubules. The ER can be divided into the *smooth endoplasmic reticulum*, so called because the membrane has a smooth surface, and the *rough endoplasmic reticulum*, which is studded with ribosomes (Figure 9-33b). The smooth endoplasmic reticulum is the site of synthesis of fatty acids and phospholipids. In contrast, the rough endoplasmic reticulum with its associated ribosomes is the site of synthesis of membrane proteins and proteins that will be secreted out of a cell, accounting for about one-third of all the different types of proteins synthesized by a cell. After synthesis on the ER, proteins destined for the plasma membrane or for secretion are first transported to the **Golgi complex**, a stack of flattened membranes called *cisternae* (Figure 9-33b), in which the proteins are modified and sorted before being transported to their destination at the plasma membrane

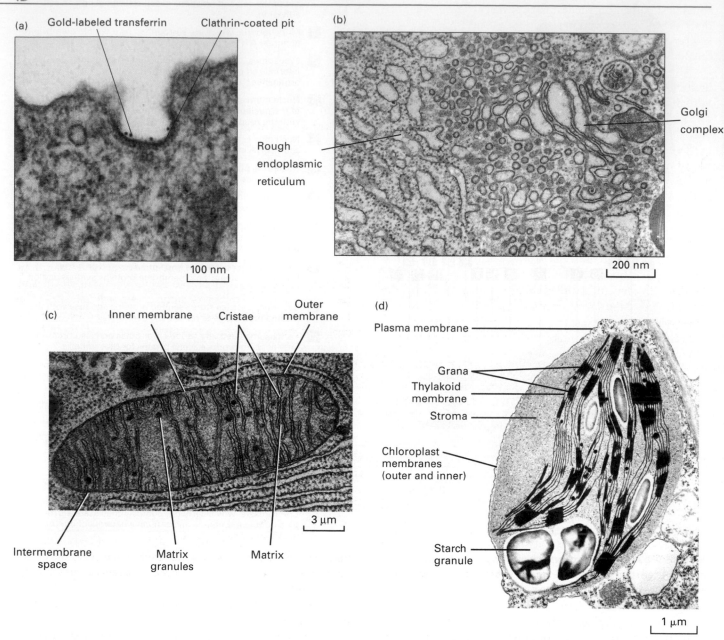

**FIGURE 9-33 Examples of organelles viewed by transmission electron microscopy of thin sections.** (a) The plasma membrane contains a clathrin-coated pit. Before fixation, the cells were incubated with colloidal gold-labeled transferrin, a protein involved in iron uptake through a receptor localized in coated pits (see Chapter 14). (b) A section through a secretory cell shows the ribosome-studded endoplasmic reticulum and the Golgi complex. (c) The two membranes of a mitochondrion and the membrane infoldings called cristae are shown.

(d) Plants contain chloroplasts, another double-membrane organelle. The thylakoid membranes contain the enzymes of the photosynthetic pathway that involves the conversion of light energy into ATP. [Part (a) from C. Lamaze et al., 1997, *Journal of Biological Chemistry* **272**:20332; part (b) from G. Palade collection; part (c) from D. W. Fawcett, 1981, *The Cell,* 2d ed., Saunders, p. 415; part (d) courtesy of Biophoto Associates/M. C. Ledbetter/ Brookhaven National Laboratory.]

or, in some cases, delivered to endosomes. Because proteins destined for secretion are made on the endoplasmic reticulum, transported through the Golgi complex, and released from the cell, this whole process is collectively known as the *secretory pathway,* although it also includes the synthesis and transport of

membrane proteins that will remain in the endoplasmic reticulum, the Golgi complex, and the plasma membrane and are therefore not secreted. These so-called *membrane-trafficking pathways,* encompassing both the endocytic and the secretory pathways, are discussed in detail in Chapter 14.

Another common organelle is the **peroxisome**, a class of roughly spherical organelles that contain *oxidases*—enzymes that use molecular oxygen to oxidize toxins to produce harmless products and for the oxidation of fatty acids for the production of acetyl groups, a topic we come to in Chapter 12.

All the organelles discussed so far are enclosed by a single lipid bilayer membrane. Some organelles, namely the **nucleus**, **mitochondria**, and, in plant cells, **chloroplasts**, have an additional membrane that serves various functions, as we describe below.

The nucleus contains the DNA of the genome and is the site of transcription of the DNA into messenger RNA. The nucleus has an inner membrane that defines the nucleus itself. It also contains an outer membrane that is continuous with the membrane of the endoplasmic reticulum in such a way that the space between the inner and outer nuclear membranes is continuous with the endoplasmic reticulum (see Figure 9-32). Access in and out of the nucleus is provided by tubular connections between the inner and outer membrane stabilized by structures called **nuclear pores**. Nuclear pores not only define the site of transport across the nuclear membranes but act as gatekeepers, only allowing transport of specific macromolecules in and out of the nucleus—an important and fascinating topic we mentioned in Chapter 8 and will discuss in more detail in Chapter 13.

Mitochondria and chloroplasts are believed to have evolved from an event a long time ago when a eukaryotic cell engulfed one type of bacterium that gave rise to mitochondria and a different kind that gave rise to chloroplasts. That mitochondria and chloroplasts have two membranes is evidence supporting this hypothesis. The inner membrane is most likely derived from the original bacterial membrane, whereas the outer membrane is a vestige of the plasma membrane from the engulfment event. There is a lot of evidence for the bacterial original of these organelles, including the fact that mitochondria and chloroplasts both have their own DNA genome and the biosynthesis of proteins in these organelles is more similar to bacterial protein synthesis than eukaryotic protein synthesis.

Mitochondria can occupy as much as 25 percent of the volume of the cytoplasm. They are thread-like organelles in which the outer membrane contains porin proteins that render the membrane permeable to molecules up to a molecular weight of 10,000. The inner mitochondrial membrane is highly convoluted with infoldings called *cristae* that protrude into the central space, called the *matrix* (Figure 9-33c). A major function of mitochondria is to complete the terminal stages of degradation of glucose by oxidation to generate most of the ATP supply of the cell. Thus mitochondria can be regarded as the "power plants" of the cell.

All plants and green algae are characterized by the presence of chloroplasts (Figure 9-33d), organelles that use *photosynthesis* to capture the energy of light with colored pigments, including the green pigment *chlorophyll*, and ultimately store the captured energy in the form of ATP. The processes by which ATP is made in mitochondria and chloroplasts is described in Chapter 12.

## Disruption of Cells Releases Their Organelles and Other Contents

The initial step in purifying subcellular structures is to release the cell's contents by rupturing the plasma membrane and the cell wall, if present. First, the cells are suspended in a solution of appropriate pH and salt content, usually isotonic sucrose (0.25 M) or a combination of salts similar in composition to those in the cell's interior. Many cells can then be broken by stirring the cell suspension in a high-speed blender or by exposing it to ultrahigh-frequency sound (*sonication*). Alternatively, plasma membranes can be sheared by special pressurized tissue homogenizers in which the cells are forced through a very narrow space between a plunger and the vessel wall; the pressure of being forced between the wall of the vessel and the plunger ruptures the cell.

Recall that water flows into cells when they are placed in a hypotonic solution, that is, one with a lower concentration of ions and small molecules than found inside the cell. This osmotic flow can be used to cause cells to swell, weakening the plasma membrane and facilitating its rupture. Generally, the cell solution is kept at 0 °C to best preserve enzymes and other constituents after their release from the stabilizing forces of the cell.

Disrupting the cell produces a mix of suspended cellular components, the homogenate, from which the desired organelles can be retrieved. Because rat liver contains an abundance of a single cell type, this tissue has been used in many classic studies of cell organelles. However, the same isolation principles apply to virtually all cells and tissues, and modifications of these cell-fractionation techniques can be used to separate and purify any desired components.

## Centrifugation Can Separate Many Types of Organelles

In Chapter 3, we considered the principles of centrifugation and the uses of centrifugation techniques for separating proteins and nucleic acids. Similar approaches are used for separating and purifying the various organelles, which differ in both size and density and thus undergo sedimentation at different rates.

Most cell-fractionation procedures begin with *differential centrifugation* of a filtered cell homogenate at increasingly higher speeds (Figure 9-34). After centrifugation at each speed for an appropriate time, the liquid that remains at the top of the vessel, called the supernatant, is poured off and centrifuged at higher speed. The pelleted fractions obtained by differential centrifugation generally contain a mixture of organelles, although nuclei and viral particles can sometimes be purified completely by this procedure.

An impure organelle fraction obtained by differential centrifugation can be further purified by *equilibrium density-gradient centrifugation*, which separates cellular components according to their density. After the fraction is resuspended, it is layered on top of a solution that contains a gradient of a dense non-ionic substance (e.g., sucrose or glycerol). The tube is centrifuged at a high speed (about 40,000 rpm) for

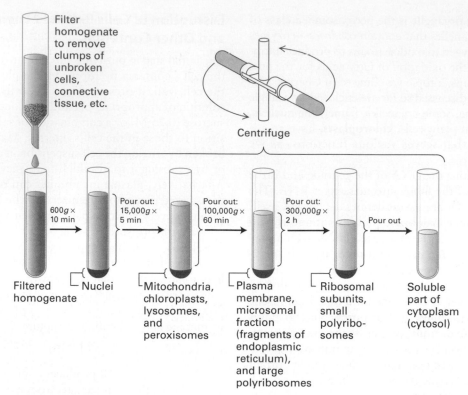

Filter homogenate to remove clumps of unbroken cells, connective tissue, etc.

Centrifuge

600g × 10 min

Pour out: 15,000g × 5 min

Pour out: 100,000g × 60 min

Pour out: 300,000g × 2 h

Pour out

Filtered homogenate

Nuclei

Mitochondria, chloroplasts, lysosomes, and peroxisomes

Plasma membrane, microsomal fraction (fragments of endoplasmic reticulum), and large polyribosomes

Ribosomal subunits, small polyribosomes

Soluble part of cytoplasm (cytosol)

**FIGURE 9-34 Differential centrifugation is a common first step in fractionating a cell homogenate.** The homogenate resulting from disrupting cells is usually filtered to remove unbroken cells and then centrifuged at a fairly low speed to selectively pellet the nucleus—the largest organelle. The undeposited material (the supernatant) is next centrifuged at a higher speed to sediment the mitochondria, chloroplasts, lysosomes, and peroxisomes.

Subsequent centrifugation in the ultracentrifuge at 100,000g for 60 minutes results in deposition of the plasma membrane, fragments of the endoplasmic reticulum, and large polyribosomes. The recovery of ribosomal subunits, small polyribosomes, and particles such as complexes of enzymes requires additional centrifugation at still higher speeds. Only the cytosol—the soluble aqueous part of the cytoplasm— remains in the supernatant after centrifugation at 300,000g for 2 hours.

several hours, allowing each particle to migrate to an equilibrium position where the density of the surrounding liquid is equal to the density of the particle (Figure 9-35). The different layers of liquid are then recovered by pumping out the contents of the centrifuge tube through a narrow piece of tubing and collecting fractions.

Because each organelle has unique morphological features, the purity of organelle preparations can be assessed by examination in an electron microscope. Alternatively, organ-

elle-specific marker molecules can be quantified. For example, the protein cytochrome $c$ is present only in mitochondria, so the presence of this protein in a fraction of lysosomes would indicate its contamination by mitochondria. Similarly, catalase is present only in peroxisomes; acid phosphatase, only in

**FIGURE 9-35 A mixed-organelle fraction can be further separated by equilibrium density-gradient centrifugation.** In this example, utilizing rat liver, material in the pellet from centrifugation at 15,000g (see Figure 9-34) is resuspended and layered on a gradient of increasingly dense sucrose solutions in a centrifuge tube. During centrifugation for several hours, each organelle migrates to its appropriate equilibrium density and remains there. To obtain a good separation of lysosomes from mitochondria, the liver is perfused with a solution containing a small amount of detergent before the tissue is disrupted. During this perfusion period, detergent is taken into the cells by endocytosis and transferred to the lysosomes, making them less dense than they would normally be and permitting a "clean" separation of lysosomes from mitochondria.

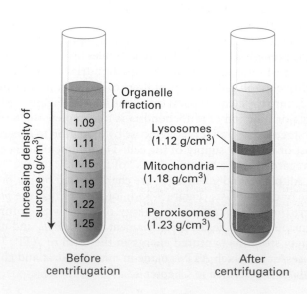

Increasing density of sucrose (g/cm³)

1.09
1.11
1.15
1.19
1.22
1.25

Organelle fraction

Lysosomes (1.12 g/cm³)

Mitochondria (1.18 g/cm³)

Peroxisomes (1.23 g/cm³)

Before centrifugation

After centrifugation

lysosomes; and ribosomes, only in the rough endoplasmic reticulum or the cytosol.

## Organelle-Specific Antibodies Are Useful in Preparing Highly Purified Organelles

Cell fractions remaining after differential and equilibrium density-gradient centrifugation usually contain more than one type of organelle. Monoclonal antibodies for various organelle-specific membrane proteins are a powerful tool for further purifying such fractions. One example is the purification of vesicles whose outer surface is covered with the protein **clathrin**; these coated vesicles are derived from coated pits at the plasma membrane during receptor-mediated endocytosis (see Figure 9-33a), a topic discussed in detail in Chapter 14. An antibody to clathrin, bound to a bacterial carrier, can selectively bind these vesicles in a crude preparation of membranes, and the whole antibody complex can then be isolated by low-speed centrifugation (Figure 9-36). A related technique uses tiny metallic beads coated with specific antibodies. Organelles that bind to the antibodies, and are thus linked to the metallic beads, are recovered from the preparation by adhesion to a small magnet on the side of the test tube.

All cells contain a dozen or more different types of small membrane-limited vesicles of about the same size (50 to 100 nm in diameter) and density, which makes them difficult to separate from one another by centrifugation techniques. Immunological techniques are particularly useful for purifying specific classes of such vesicles. Fat and muscle cells, for instance, contain a particular glucose transporter (GLUT4) that is localized to the membrane of one of these types of vesicle. When insulin is added to the cells, these vesicles fuse with the plasma membrane and increase the number of glucose transporters able to take up glucose from the blood. As we will see in Chapter 15, this process is critical to maintaining the appropriate concentration of sugar in the blood. The GLUT4-containing vesicles can be purified by using an antibody that binds to a segment of the GLUT4 protein that faces the cytosol. Likewise, the various transport vesicles discussed in Chapter 14 are characterized by unique surface proteins that permit their separation with the aid of specific antibodies.

A variation of this technique is employed when no antibody specific for the organelle under study is available. A gene encoding an organelle-specific membrane protein is modified by the addition of a segment encoding an epitope tag; the tag is placed on a segment of the protein that faces the cytosol. Following stable expression of the recombinant protein in the cell under study, an anti-epitope monoclonal antibody (described above) can be used to purify the organelle.

(a)

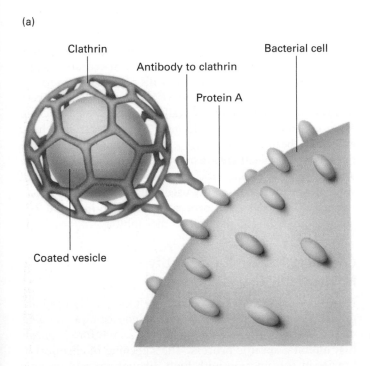

(b)

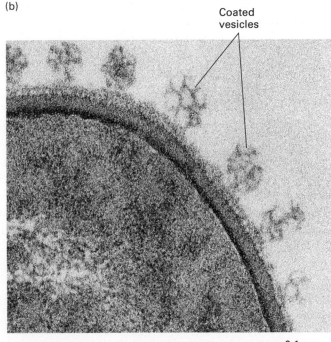

**FIGURE 9-36 Small coated vesicles can be purified by binding of antibody specific for a vesicle surface protein and linkage to bacterial cells.** In this example, a suspension of membranes from rat liver is incubated with an antibody specific for clathrin, a protein that coats the outer surface of certain cytosolic vesicles. To this mixture is added a suspension of killed *Staphylococcus aureus* bacteria, whose surface membrane contains protein A, which binds to the constant (Fc) region of antibodies. (a) Interaction of protein A with antibodies bound to clathrin-coated vesicles links the vesicles to the bacterial cells. The vesicle-bacteria complexes can then be recovered by low-speed centrifugation. (b) A thin-section electron micrograph reveals clathrin-coated vesicles bound to an *S. aureus* cell. [See E. Merisko et al., 1982, *J. Cell Biol.* **93**:846. Micrograph courtesy of G. Palade.]

## Proteomics Reveals the Protein Composition of Organelles

To identify all the proteins in an organelle requires three steps. First, one has to be able to obtain the organelle in high purity. Second, one has to have a way to identify all the sequences of the proteins in the organelle. This identification is generally done by digesting all the proteins with a protease such as trypsin, which cleaves all polypeptides at lysine and arginine residues, and then determining the mass and sequence of all these peptides by mass spectrometry. Third, one has to have a genomic sequence to identify the proteins from which all the peptides came. In this way, the "proteome" of many organelles has been determined. As one example, a recent proteomic study on mitochondria purified from mouse brain, heart, kidney, and liver revealed 591 mitochondrial proteins, including 163 proteins not previously known to be associated with this organelle. Several proteins were found in mitochondria only in specific cell types. Determining the functions associated with these newly identified mitochondrial proteins is a major objective of current research on this organelle.

## KEY CONCEPTS of Section 9.4

### Isolation and Characterization of Cell Organelles

- Microscopy has revealed a common set of organelles present in eukaryotic cells (see Figure 9-32).

- Disruption of cells by vigorous homogenization, sonication, or other techniques releases their organelles. Swelling of cells in a hypotonic solution weakens the plasma membrane, making it easier to rupture.

- Sequential differential centrifugation of a cell homogenate yields fractions of partly purified organelles that differ in mass and density.

- Equilibrium density-gradient centrifugation, which separates cellular components according to their densities, can further purify cell fractions obtained by differential centrifugation.

- Immunological techniques using antibodies against organelle-specific membrane proteins are particularly useful in purifying organelles and vesicles of similar sizes and densities.

- Proteomic analysis can identify all the protein components in a preparation of a purified organelle.

## 9.5 Perturbing Specific Cell Functions

What general approaches have scientists used to understand the function of specific proteins in cell biological processes? We have already discussed in Chapter 3 how proteins can be purified and their properties characterized in detail. In many cases, this has led to the in vitro biochemical reconstitution of complicated biochemical processes, such as DNA replication or protein synthesis. These biochemical approaches have been complemented by genetic approaches, and as we have seen in Chapter 5, mutations can be used to identify genes whose products play specific functions. As we will see in Chapters 14 and 19, classic genetic screens in yeast were used to identify proteins that participate in the secretory pathway and the cell cycle, respectively. Genetic approaches in other organisms, such as the nematode worm, the fruit fly, and the mouse, have contributed immensely to uncovering basic aspects of cell biology and development (see Chapter 1).

Over the last few years, additional new and very powerful approaches have been developed to perturb specific components in living cells and thereby shed light on their functions. In this section, we discuss two of these approaches: the use of specific chemicals to perturb cell function and the use of interfering RNA to suppress the expression of specific genes.

## Drugs Are Commonly Used in Cell Biology

Naturally occurring drugs have been used for centuries, but how they worked was often not known. For example, extracts of the meadow saffron were used to treat gout, a painful disease resulting from inflammation of joints. Today we know that the extract contains colchicine, a drug that depolymerizes microtubules and interferes with the ability of white blood cells to move to the sites of inflammation (see Chapter 18). Alexander Fleming discovered that certain fungi secrete compounds that kill bacteria (antibiotics), resulting in the discovery of penicillin. Only later was it was discovered that penicillin inhibits bacterial cell division by blocking the assembly of the cell walls of certain bacteria.

Many examples like these have resulted in the discovery of a very wide range of drugs available to inhibit specific and essential processes of cells. In most cases, researchers have eventually been able to identify the molecular target of the drug. For example, there are many other antibiotic drugs that affect aspects of prokaryotic protein synthesis. A selection of some of the more commonly used drugs that affect a broad variety of cell biological processes are listed in Table 9-1, grouped according to the process they inhibit.

## Chemical Screens Can Identify New Specific Drugs

How does one discover a new drug? One widely used approach makes use of *chemical libraries* consisting of 10,000s to 100,000s of different compounds to search for chemicals that inhibit a specific process. The screening of chemical libraries in conjunction with high-throughput microscopic techniques has now become one of the major routes for new leads in drug discovery. Here we give just one case to illustrate how this type of approach works.

In our example (Figure 9-37a), researchers wanted to identify compounds that inhibit mitosis, the process where duplicated chromosomes are accurately segregated by a

## TABLE 9-1 | Selected Set of Small Molecules Used in Cell Biological Research

Some of the following molecules have broad specificity, whereas others are highly specific. More information about many of these compounds can be found in the relevant chapters in this text.

| | |
|---|---|
| **DNA replication inhibitors** | Aphidicolin (eukaryotic DNA polymerase inhibitor); camptothecin, etoposide (eukaryotic topoisomerase inhibitors) |
| **Transcription inhibitors** | $\alpha$-Amanitin (eukaryotic RNA polymerase II inhibitor); actinomycin D (eukaryotic transcription elongation inhibitor); rifampicin (bacterial RNA polymerase inhibitor); thiolutin (bacterial and yeast RNA polymerase inhibitor) |
| **Protein synthesis inhibitors—block general protein production; toxic after extended exposure** | Cycloheximide (translational inhibitor in eukaryotes); geneticin/G418, hygromycin, puromycin (translation inhibitors in bacteria and eukaryotes); chloramphenicol (translation inhibitor in bacteria and mitochondria); tetracycline (translation inhibitor in bacteria) |
| **Protease inhibitors—block protein degradation** | MG-132, lactacystin (proteasome inhibitors); E-64, leupeptin (serine and/or cysteine protease inhibitors); phenylmethanesulfonylfluoride (PMSF) (serine proteases inhibitor); tosyl-ʟ-lysine chloromethyl ketone (TLCK) (trypsin-like serine protease inhibitor) |
| **Compounds affecting the cytoskeleton** | Phalloidin, jasplakinolide (F-actin stabilizer); latrunculin, cytochalasin (F-actin polymerization inhibitors); taxol (microtubule stabilizer); colchicine, nocodazole, vinblastine, podophyllotoxin (microtubule polymerization inhibitors); monastrol (kinesin-5 inhibitor) |
| **Compounds affecting membrane traffic, intracellular movement and the secretory pathway, protein glycosylation** | Brefeldin A (secretion inhibitor); leptomycin B (nuclear protein export inhibitor); dynasore (dynamin inhibitor); tunicamycin (N-linked glycosylation inhibitor) |
| **Kinase inhibitors** | Genistein, rapamycin, gleevec (tyrosine kinase inhibitors with various specificities); wortmannin, LY294002 (PI3 kinase inhibitors); staurosporine (protein kinase inhibitor); roscovitine (cell cycle CDK1 and CDK2 inhibitors) |
| **Phosphatase inhibitors** | Cyclosporine A, FK506, calyculin (protein phosphatase inhibitors with various specificities); okadaic acid (general inhibitor of serine/threonine phosphatases); phenylarsine oxide, sodium orthovanadate (tyrosine phosphatase inhibitors) |
| **Compounds affecting intracellular cAMP levels** | Forskolin (adenylate cyclase activator) |
| **Compounds affecting ions (e.g., K⁺ Ca²⁺)** | A23187 ($Ca^{2+}$ ionophore); valinomycin ($K^+$ ionophore); BAPTA (divalent cation (e.g., $Ca^{2+}$) binding/sequestering agent); thapsigargin (endoplasmic reticulum $Ca^{2+}$ ATPase inhibitor); ouabain ($Na^+/K^+$ ATPase inhibitor) |
| **Some drugs used in medicine** | Propranolol ($\beta$–adrenergic receptor antagonist), statins (HMG-CoA reductase inhibitors, block cholesterol synthesis) |

microtubule-based machine called the mitotic spindle (discussed in Chapter 18). It was known that if spindle assembly is compromised, cells arrest in mitosis. Therefore, the screen first used an automated robotic method to look for compounds that arrest cells in mitosis. The basis for the inhibition of the candidate compounds was then explored to see if they affected assembly of the microtubules. Since inhibition of microtubule assembly was not of interest, the effect of the remaining candidates on the structure of the spindle was determined by immunofluorescence microscopy with antibodies to tubulin, the major protein of microtubules. Over 16,000 compounds were screened, and a compound was identified that resulted in cells with abnormal spindles—instead of having two asters, they had a single aster, what is called a mono-astral array (Figure 9-37b). This drug, now called *monastrol*, was found to interfere with the assembly of the spindle by inhibiting a microtubule-based motor called kinesin-5 (see Chapter 18 for more details about the mitotic spindle). Derivatives of monastrol are now being tested as anti-tumor agents for the treatment of certain cancers.

**FIGURE 9-37 Screening for drugs that affect specific biological processes.** In this example, a chemical library of 16,320 different chemicals was subject to a series of screens for inhibitors of mitosis. Since such an inhibitor is expected to arrest cells at the mitotic stage of the cell cycle, the first screen **1** was to see if any of the chemicals enhanced the level of a marker for mitotic cells, and this yielded 139 candidates. Microtubules make up the structure of the mitotic spindle, and the researchers were not interested in new drugs that target microtubules, so in the second screen **2** they tested the 139 compounds for their ability to affect microtubule assembly, and this eliminated 53 candidates. Immunofluorescence microscopy with antibodies to tubulin (the major subunit of microtubules) together with a stain for DNA was then used in the third screen **3** to identify compounds that disrupt the structure of the spindle. (b) Localization of tubulin (green) and DNA (blue) are shown for an untreated mitotic spindle and one treated with one of the recovered compounds, now called monastrol. Monastrol inhibits a microtubule-based motor called kinesin-5, discussed in Chapter 18, necessary to separate the poles of the mitotic spindle. When kinesin-5 is inhibited, the two poles remain associated to give a monopolar spindle. [Part (b) T. U. Mayer et al., *Science* **286:**971–974.]

(a)

16,320 Chemical Compounds

**1** Screen for those that arrest cells in mitosis

139

**2** Screen for those that do not affect microtubule assembly in vitro

86

**3** Screen for those that specifically affect spindle morphology

5

(b)

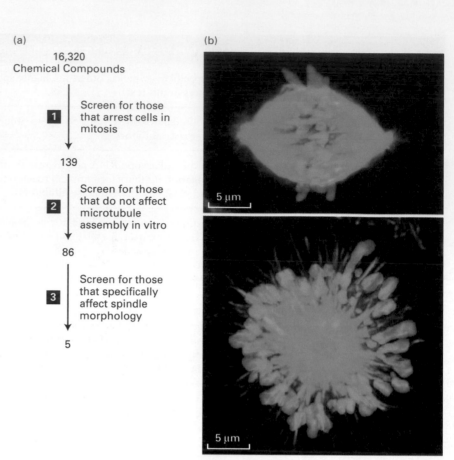

5 μm

5 μm

## Small Interfering RNAs (siRNAs) Can Knock Down Expression of Specific Proteins

RNA interference (RNAi) is a mechanism that cells use to suppress the expression of genes by either blocking translation of specific mRNAs through micro-RNAs (miRNAs) or degradation of specific mRNAs targeted by small interfering RNA (siRNA). The extensive use of miRNAs to regulate gene expression, especially during development, has been discussed in Chapter 8. Here we focus on the experimental use of siRNA technology to suppress the expression of genes in animal cells.

The discovery of the siRNA pathway arose from many different observations. For example, it was discovered in plants that recombinant expression of a gene could lead to the down-regulation of the target gene rather than the expected result of enhanced expression. A similar type of result was seen in the nematode *C. elegans*. Investigating this phenomenon, Andrew Fire and Craig Mello reported in 1998 that suppression could not be achieved by expressing either sense or antisense mRNA, but that it required expression of double-stranded RNA. Fire and Mello were awarded the Nobel Prize in Physiology or Medicine in 2006 for this discovery. Subsequent work in a number of systems showed that the double-stranded RNA has to be cleaved by a protein called Dicer to produce double-stranded fragments of 21 to 23 base pairs with a two-nucleotide overhang at each of the 3' ends. This double-stranded RNA is recognized by RISC (RNA-induced silencing complex), and one of the strands is degraded by the associated argonaute protein. If the single-stranded siRNA sequence can base pair exactly with a target mRNA sequence, the argonaute protein–RNA complex cleaves the target mRNA, which is then degraded (Figure 9-38a). Although this system probably evolved as a defense mechanism against invading viruses, it has provided researchers with a very powerful tool to experimentally suppress the expression of particular genes and explore the resulting consequences. It has been used very effectively in many different systems, as we summarize below.

**siRNA Knockdown in Cultured Cells** Since the discovery in 2001 that treatment of cultured cells with siRNAs suppresses gene expression by degrading the target mRNA, siRNAs have been used in thousands of studies to suppress—or "knock down"—the levels of target proteins. To do this, researchers use computer programs to identify a ~21-base sequence in the mRNA that is unique to the target gene and has the characteristics optimal for siRNA. Double-stranded RNA is then synthesized and applied to cells in culture (Figure 9-38a). If effective, this will result in degradation of the specific mRNA, and no new target protein will be synthesized. However, the target protein is present at the beginning of the experiment, so the cells have to be able to grow to allow the endogenous protein to undergo its normal turnover as well as get diluted by cell division—this usually takes 24 to 72 hours. The level of the target protein is generally determined by immunoblotting (see Figure 3-39), and if significantly reduced, the phenotype of the cells is examined.

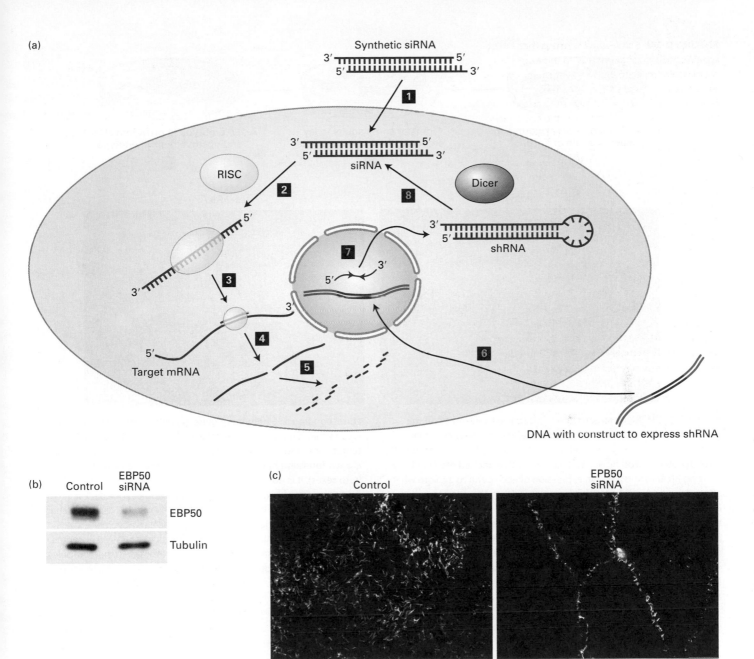

(a) Synthetic siRNA

**FIGURE 9-38 siRNA and DNA expressing shRNA can target the degradation of specific mRNAs in cultured cells.** In the first step (**1**), a double-stranded siRNA that has homology to the target mRNA is introduced into cells by transfection. This double-stranded RNA is recognized by the RISC complex (**2**), which degrades one strand of the RNA and targets the mRNA with the homologous sequence (**3**). The target mRNA is cleaved (**4**) and degraded (**5**). In an alternative strategy, a DNA construct containing a sequence that when transcribed will form a hairpin is introduced into the cell (**6**). This DNA can either be introduced by transfection or carried in a virus particle. In either case, it is engineered to carry with it a drug-selectable marker (not shown) so that cells in which this DNA is integrated into the genome can be selected. When transcribed (**7**) and transported out of the

nucleus, the RNA hairpin becomes a substrate of the nuclease Dicer to generate the appropriate siRNA. (b) As an example of this technology, researchers wanted to examine the effects of knocking down a protein, called EBP50, that is a component of cell-surface microvilli (see Figure 17-21d). siRNAs were designed and their ability to knock down EBP50 in cultured cells assessed by doing an immunoblot with EBP50 antibodies and tubulin antibodies as a control. (c) They then examined the cells for microvilli by staining for the microvillar-specific protein ezrin. In these confocal sections at the top of the cell, untreated cells have abundant microvilli, whereas cells in which EBP50 has been knocked down only have a few microvilli around the cell periphery. [Parts (b) and (c) from Hanono et al., J. Cell Biol. **175**:803.]

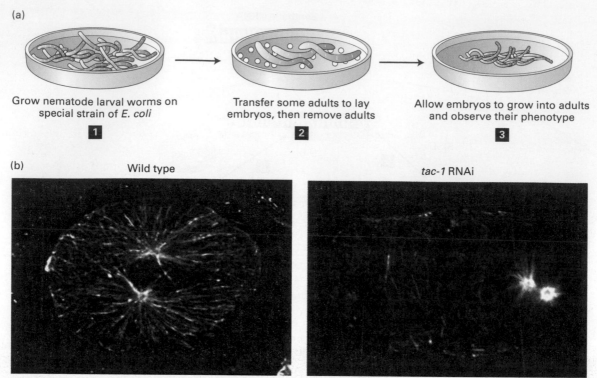

(a)

Grow nematode larval worms on special strain of *E. coli*
**1**

Transfer some adults to lay embryos, then remove adults
**2**

Allow embryos to grow into adults and observe their phenotype
**3**

(b)
Wild type            *tac-1* RNAi

**FIGURE 9-39 RNAi screens can explore the function of all the genes in the nematode *Caenorhabditis elegans*.** The determination of the *C. elegans* genome sequence in 1998 revealed it contains about 20,000 protein-encoding genes. This information opened the possibility of using RNAi to knock down expression of each gene to explore what effect it would have. Today this is routinely done. The nematode can live by eating the bacterium *Escherichia coli*, and it is possible to express in the bacterium a long stretch of double-stranded RNA corresponding to a single nematode gene. Remarkably, when the nematode eats the bacteria, the double-stranded RNA enters the cells of the gut and is recognized by Dicer and processed into siRNAs that spread to almost all the cells in the animal. Therefore, researchers have made a library of *E. coli* strains each expressing double-stranded RNA corresponding to a nematode gene. (a) In this approach, a specific *E. coli* strain is fed to larval nematodes **1** and the expression of the target gene is suppressed in the germ line. Because these nematodes are self-fertilizing hermaphrodites (having the reproductive organs of both sexes), it is not necessary to mate them but merely to let the adult worms lay eggs **2**. When these grow into adults, the effect of RNAi on the target gene can be assessed. (b) In this example, the researchers were screening for genes necessary for nuclear movement. They identified a gene called TAC-1, whose product is located at centrosomes and is necessary for the normal distribution of microtubules, as revealed by immunofluorescence microscopy with tubulin antibodies. [Part (b) from N. Le Bot et al., *Current Biology* **13**:1499 (2003).]

Suppression of gene expression by siRNAs has become a standard technique; an example is shown in Figure 9-38b and 9-38c, and many other examples can be found throughout this book.

An alternative strategy to knockdown protein expression is to introduce appropriate DNA constructs into cells that will generate siRNAs when they are transcribed (see Figure 9-38a). To achieve this, the target sequence is present as an inverted repeat in the DNA sequence. When transcribed, the mRNA will form a double-stranded short hairpin RNA (shRNA), which is recognized and cleaved by Dicer to generate siRNAs. This approach has the advantage that once the shRNA is expressed, the siRNAs are always made, resulting in permanent knockdown of the target protein. This will not work if the protein is essential, in which case treating cells with siRNAs is the method of choice. The DNA construct to express shRNAs can be introduced into cells by simple addition of the DNA under the appropriate conditions that allow the cells to take it up or by use of viral vectors that more efficiently introduce the DNA.

Massive efforts are currently under way to explore the effects of knocking down expression of each gene in cultured cell lines and then examining the effects on specific pathways. This effort is in its infancy, so future refinements and analysis will provide a "systems biology" view of cell organization and function.

## Genomic Screens Using siRNA in the Nematode *C. elegans*

When the annotated genome sequence of the nematode worm *Caenorhabditis elegans* was determined in 1998, this provided the first catalog of all the genes present in an animal. It also

(a)

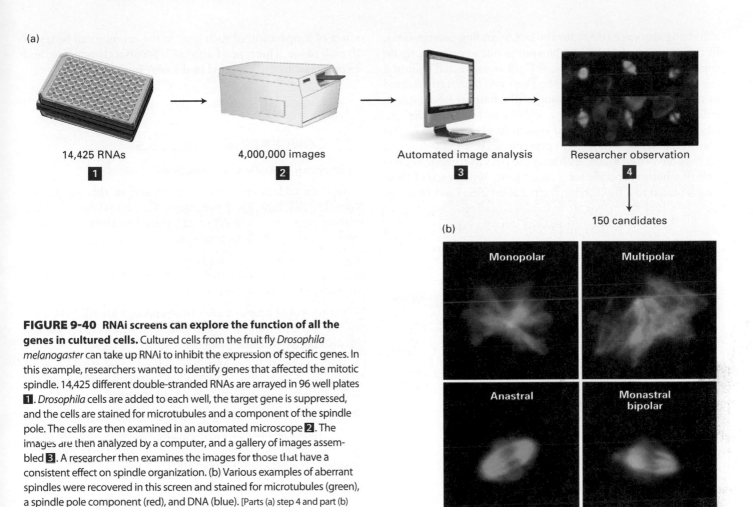

14,425 RNAs
**1**

4,000,000 images
**2**

Automated image analysis
**3**

Researcher observation
**4**

150 candidates

(b)

Monopolar

Multipolar

Anastral

Monastral
bipolar

**FIGURE 9-40 RNAi screens can explore the function of all the genes in cultured cells.** Cultured cells from the fruit fly *Drosophila melanogaster* can take up RNAi to inhibit the expression of specific genes. In this example, researchers wanted to identify genes that affected the mitotic spindle. 14,425 different double-stranded RNAs are arrayed in 96 well plates **1**. *Drosophila* cells are added to each well, the target gene is suppressed, and the cells are stained for microtubules and a component of the spindle pole. The cells are then examined in an automated microscope **2**. The images are then analyzed by a computer, and a gallery of images assembled **3**. A researcher then examines the images for those that have a consistent effect on spindle organization. (b) Various examples of aberrant spindles were recovered in this screen and stained for microtubules (green), a spindle pole component (red), and DNA (blue). [Parts (a) step 4 and part (b) from G. Goshima et al., 2007, *Science* **316**:417.]

provided the possibility to explore the function of each gene using RNAi to suppress expression of each individual gene. In fact, this nematode was the first animal in which a genomic RNAi screen was attempted. *C. elegans* can live by eating the bacterium *E. coli*. Remarkably, if the bacterium expresses a double-stranded RNA homologous to a nematode gene, when the bacteria are eaten, they are broken open and the RNA is absorbed through the intestine, then processed by Dicer to suppress expression of the target gene. Since there are about 20,000 different genes in the nematode, this many different *E. coli* strains were generated, each expressing double-stranded RNA targeted to a specific nematode gene. In a typical experiment to suppress expression of a single gene (Figure 9-39a), nematodes are grown on the bacteria expressing the specific double-stranded RNA, and this suppresses expression of the target gene in their embryos. After the adult nematodes have laid eggs, they are removed and the effect on the growing embryos is examined (Figure 9-39b).

**Genomic Screens Using siRNA in Fruit Flies** The fruit fly has about 14,000 protein-encoding genes, and techniques have

been developed to explore the consequence of using siRNA to suppress each of these in cultured fruit fly cells. With such a large number to be tested, automatic high-throughput approaches were developed (Figure 9-40a). For example, about 150 96 well plates are made, with each well containing one double-stranded RNA for a specific gene. Cells are added, and the double-stranded RNA is taken up and processed by Dicer into siRNAs, which then suppress expression of the target gene. The cells can then be examined for a specific phenotype. In the example shown in Figure 9-40b, the investigators explored the effect of gene suppression on cells arrested in mitosis. Since this is a morphological screen, they stained cells with appropriate markers and used a robotic microscope to take pictures and a computer program to analyze them. In this way they identified about 150 new genes whose products contribute to mitosis and are therefore excellent subjects for further in-depth studies.

Unlike in the nematode described above, it is not possible to suppress expression of genes by feeding fly larvae double-stranded RNA. However, it is possible to use RNAi for tissue-specific suppression. This is achieved by making a fly in which a specific hairpin RNA is expressed behind an upstream-activating sequence (UAS) for the DNA-binding protein Gal4. In the absence of Gal4, the hairpin is not expressed, so no suppression occurs. However, if Gal4 is expressed behind a tissue-specific promoter, it will be expressed in that tissue, bind to the UAS, and drive expression of the hairpin RNA, which will be processed by Dicer into siRNAs and suppress the specific gene (Figure 9-41a). The system has been set up in this way so that it can be done on a genomic scale: 14,000 different fly lines have been made, each with the UAS to regulate expression of a hairpin specific for a target gene. When each of these flies is mated to a fly carrying the tissue-specific Gal4 gene, the effect of suppression of each gene in the genome can be tested in each tissue. This type of approach is currently ongoing, and exciting results are expected in the near future.

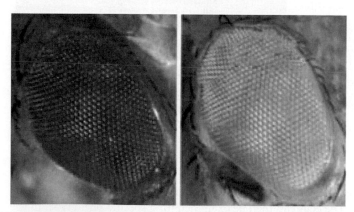

FIGURE 9-41 **RNAi can be used to suppress genes in a tissue-specific manner in the fruit fly.** Genetic approaches have been used to develop methods to suppress target gene expression in specific tissues. This involves making use of two large sets of flies. In the first set of about 14,000 different flies each designed to target one gene (set A), flies are generated in which a specific hairpin RNA for the target gene is under the control of an upstream activating sequence (UAS) for the transcriptional activator Gal4. In each of the second set of flies (set B), expression of GAL4 is controlled by a tissue-specific promoter. When a fly from set A is mated with a fly from set B, Gal4 will be expressed in one tissue and therefore the RNAi hairpin will be expressed and the gene silenced in this one tissue. (b) A wild-type fly eye (*left*), and one in which the *white* gene has been specifically suppressed in the eye (*right*). [Part (b) from N. Perrimon et al., 2010, *Cold Spring Harbor Perspect. Biol.* **2**:a003640.]

## KEY CONCEPTS of Section 9.5

### Perturbing Specific Cell Functions

• Genetic techniques have been critical in the analysis of complex cell biological pathways. Genetic approaches are now being extended and complemented by chemical screens and the use of RNAi technology.

• Large chemical libraries can be screened for compounds that target specific processes to study those processes and to identify new components.

• Treatment of cultured cells with appropriate siRNAs leads to the destruction of target mRNAs and hence "knockdown" of the encoded protein.

• With the availability of annotated genomes, RNAi screens can be used to explore the effect of suppressing expression of each individual gene in an organism. This has been achieved in the nematode worm and in cultured cells from the fruit fly.

• RNAi can be used to suppress expression of specific genes in a tissue-specific manner; this technique is currently being applied on a genomic scale in the fruit fly.

## Perspectives for the Future

This chapter has introduced many aspects of technology currently used by cell biologists. Science is driven by the technology available, and with each development we can peer more deeply into the mysteries of life.

The ability to grow cells in culture was a tremendous advance in technology—it allowed researchers to examine and explore the inner workings of cells. Techniques in cell culture are still developing; for example, they are currently contributing to the exciting developments in stem-cell research (see Chapter 21). Although most studies have used flat dishes to grow these cells, in the body they form a three-dimensional structure. Major areas of research are now examining the functions of cells in three-dimensional environments and generating three-dimensional cell organizations, such as epithelial tubes, in supported culture systems.

The discovery and use of GFP and other fluorescent proteins has revolutionized cell biology. By tagging proteins with GFP and following their localization in live cells, it has become apparent that the cytoplasm of cells is far more dynamic than previously envisaged. Every year brings new technologies associated with fluorescent proteins; approaches such as FRAP, FRET, and TIRF have become widespread tools to explore the dynamics and molecular mechanisms of proteins, either in vivo or in vitro. At the time of this writing, we are seeing a revolution in super-resolution microscopy, opening up the ability to localize molecules by light microscopy several times more accurately than was believed to be possible. Super-resolution microscopy can currently be done only on fixed samples; optimists believe it will soon be possible to achieve this level of resolution in living cells and thereby open up the possibility of watching dynamic processes at the molecular level. As these techniques develop, fewer people need to use electron microscopy, and so the expertise in this important area is waning.

RNAi has provided an awesome and unanticipated new technology to the arsenal of techniques available to cell and developmental biologists. The ability to perform genome-wide screens in both the nematode worm and fruit fly has made traditionally excellent genetic systems even more powerful. Coupling these technologies with visual screens opens up yet another dimension. Consider the following problem: which genes in the nematode affect the organization of a small subset of neurons? A few years ago, this would have been a technically challenging problem. Now it is possible to make a nematode in which just those neurons are marked with GFP and then subject them to a visual genomics RNAi screen to see which gene products are necessary for the normal morphology of those neurons. More and more imaginative approaches are being developed combining RNAi with visual and functional screens in both nematode worms and the fruit fly, permitting an ever-deepening understanding of life processes. In addition, efforts are currently under way to explore the effects of knocking down expression of each gene in cultured cell lines and then examining the effects on specific pathways. This effort is in its infancy, so future refinements and analysis will provide a systems biology view of cell organization and function.

Can RNAi technology be used in medicine? Could it be used to suppress expression of oncogenes in the treatment of cancer? The technological delivery problems are significant since the siRNA needs to be delivered to the right cells, remain stable in the patient, and be effective at knocking down the appropriate protein. Currently, at least a dozen clinical trials are testing the feasibility of this approach. If the technical hurdles can be overcome, RNAi might become a major class of therapeutic agent.

What new technologies will the next decade bring? In the last decade, RNAi and GFP have revolutionized cell biology. No doubt the next decade will bring about unexpected new developments, so we should be excited about what is to come.

## Key Terms

bright-field light
    microscopy 404
cell line 400
cell strain 399
chimeric proteins 398
chloroplast 427

clone 398
confocal microscopy 413
cryoelectron microscopy
    422
culturing 397
cytoplasm 424

deconvolution microscopy
    412
differential centrifugation
    427
differential-interference-
    contrast (DIC) micros-
    copy 427
endoplasmic reticulum
    (ER) 425
endosome 425
equilibrium density-gradient
    centrifugation 427
fluorescence-activated cell
    sorter (FACS) 401
fluorescence recovery after
    photobleaching
    (FRAP) 416
fluorescent staining 408
Förster resonance energy
    transfer (FRET) 416
Golgi complex 425
hybridoma 403
immunofluorescence
    microscopy 398

indirect immunofluores-
    cence microscopy 410
lysosome 425
membrane transport
    protein 424
metal shadowing 420
mitochondria 427
monoclonal antibody 403
organelles 398
peroxisome 427
phase-contrast
    microscopy 407
photo-activated localization
    microscopy (PALM) 418
polyclonal antibody 403
resolution 405
scanning electron
    microscope (SEM) 419
total internal reflection
    fluorescence (TIRF)
    microscopy 415
transmission electron
    microscope (TEM) 419

## Review the Concepts

1. Both light and electron microscopy are commonly used to visualize cells, cell structures, and the location of specific molecules. Explain why a scientist may choose one or the other microscopy technique for use in research.

2. The magnification possible with any type of microscope is an important property, but its resolution, the ability to distinguish between two very closely apposed objects, is even more critical. Describe why the resolving power of a microscope is more important for seeing finer details than its magnification. What is the formula to describe the resolution of a microscope lens and what are the limitations placed on the values in the formula?

3. Why are chemical stains required for visualizing cells and tissues with the basic light microscope? What advantage do fluorescent dyes and fluorescence microscopy provide in comparison to the chemical dyes used to stain specimens for light microscopy? What advantages do confocal scanning microscopy and deconvolution microscopy provide in comparison to conventional fluorescence microscopy?

4. In certain electron microscopy methods, the specimen is not directly imaged. How do these methods provide information about cellular structure, and what types of structures do they visualize? What limitation applies to most forms of electron microscopy?

5. What is the difference between a cell strain, a cell line, and a clone?

6. Explain why the process of cell fusion is necessary to produce monoclonal antibodies used for research.

7. Much of what we know about cellular function depends on experiments utilizing specific cells and specific parts (e.g., organelles) of cells. What techniques do scientists commonly use to isolate cells and organelles from complex mixtures, and how do these techniques work?

8. Hoechst 33258 is a chemical dye that binds specifically to DNA in live cells, and when excited by UV light, it fluoresces in the visible spectrum. Name and describe one specific method, employing Hoechst 33258, an investigator would use to isolate fibroblasts in the $G_2$ phase of the cell cycle from those fibroblasts in interphase.

9. shRNAs and siRNAs can be used to successfully knock down the expression of any specific protein in a given cell line or organism. The utility of one over the other is debatable, but there are merits to using one for therapeutic applications in a living organism. Which of the two methods is likely to be more advantageous in the long-term and what is one of its limitations?

## Analyze the Data

1. Mouse liver cells were homogenized and the homogenate subjected to equilibrium density-gradient centrifugation with sucrose gradients. Fractions obtained from these gradients were assayed for *marker molecules* (i.e., molecules that are limited to specific organelles). The results of these assays are shown in the figure. The marker molecules have the following functions: cytochrome oxidase is an enzyme involved in the process by which ATP is formed in the complete aerobic degradation of glucose or fatty acids; ribosomal RNA forms part of the protein-synthesizing ribosomes; catalase catalyzes decomposition of hydrogen peroxide; acid phosphatase hydrolyzes monophosphoric esters at acid pH; cytidylyl transferase is involved in phospholipid biosynthesis; and amino acid permease aids in transport of amino acids across membranes.

**a.** Name the marker molecule and give the number of the fraction that is *most* enriched for each of the following cell components: lysosomes; peroxisomes; mitochondria; plasma membrane; rough endoplasmic reticulum; smooth endoplasmic reticulum.

**b.** Is the rough endoplasmic reticulum more or less dense than the smooth endoplasmic reticulum? Why?

**c.** Describe an alternative approach by which you could identify which fraction was enriched for which organelle.

**d.** How would addition of a detergent to the homogenate, which disrupts membranes by solubilizing their lipid and protein components, affect the equilibrium density-gradient results?

2. The nematode worm *C. elegans* is amenable as a model for siRNA studies. In this experiment, adult nematodes are fed bacteria expressing a double-stranded RNA to suppress the expression of the unc18 gene, whose mammalian homolog encodes a protein that participates in the integration of GLUT4 storage vesicles into the plasma membrane. Before evaluating the effects on the organism itself, it is necessary to use RT-PCR to analyze if the siRNA experiment worked. The samples are from embryos from five adults fed the bacteria, and the results are those following the RT-PCR using primers to amplify mRNA from the two different genes, *unc18* and *GLUT4*.

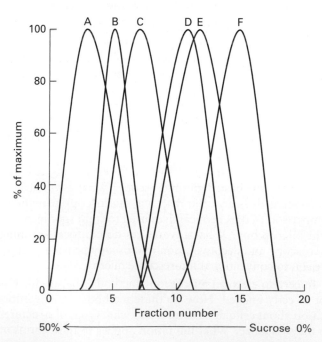

Curve A = cytochrome oxidase
Curve B = ribosomal RNA
Curve C = catalase
Curve D = acid phosphatase
Curve E = cytidylyl transferase
Curve F = amino acid permease

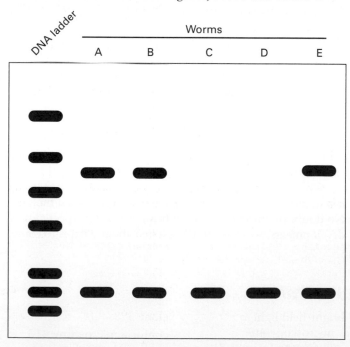

**a.** Of the embryos examined, which samples show that the adults successfully took up bacteria containing the double-stranded unc18 RNA? What made you come to those conclusions?

**b.** Label which set of bands is the result of the amplification by the unc18 primers and which by the GLUT4 primers. Why was RT-PCR with GLUT4 primers used as the positive control? What does this tell you about the relationship between the knock down of the unc18 and GLUT4 mRNAs?

**c.** With antibodies against unc18, draw a representative Western blot showing the expected results from protein samples from each of these five samples.

**d.** To further investigate the relationship between the unc18 and GLUT4 proteins, the siRNA experiment is repeated, but in embryonic cells expressing a GFP-tagged GLUT4 protein. Draw a cell including its mitochondria, nucleus, and rough endoplasmic reticulum and show where, if using a laser scanning confocal microscope, the GFP fluorescence would localize in the cells expressing the unc18 siRNAs. Do not forget to draw the cell representing the control.

**e.** The ability of a protein to colocalize with another protein suggests but does not prove that the two physically interact with each other. Having each protein labeled, in this case unc18 to green fluorescent protein and GLUT4 to red fluorescent protein, provides the experimenter with reagents to tease apart the question of colocalization versus interaction. Using these reagents, describe one technique that would simply demonstrate unc18 colocalizes with GLUT4 and another technique that proves the two proteins interact physically.

# References

### Growing Cells in Culture

Bissell, M. J., A. Rizki, and I. S. Mian. 2003. Tissue architecture: the ultimate regulator of breast epithelial function. *Curr. Opin. Cell Biol.* **15**:753–762.

Battye, F. L., and K. Shortman. 1991. Flow cytometry and cell-separation procedures. *Curr. Opin. Immunol.* **3**:238–241.

Davis, J. M., ed. 1994. *Basic Cell Culture: A Practical Approach*. IRL Press.

Edwards, B., et al. 2004. Flow cytometry for high-throughput, high-content screening. *Curr. Opin. Chem. Biol.* **8**:392–398.

Goding, J. W. 1996. *Monoclonal Antibodies: Principles and Practice. Production and Application of Monoclonal Antibodies in Cell Biology, Biochemistry, and Immunology*, 3d ed. Academic Press.

Griffith, L. G., and M. A. Swartz. 2006. Capturing complex 3D tissue physiology in vitro. *Nature Rev. Mol. Cell. Biol.* **7**:211–224.

Krutzik, P., et al. 2004. Analysis of protein phosphorylation and cellular signaling events by flow cytometry: techniques and clinical applications. *Clin. Immunol.* **110**:206–221.

Paszek, M. J., and V. M. Weaver. 2004. The tension mounts: mechanics meets morphogenesis and malignancy. *J. Mammary Gland Biol. Neoplasia* **9**:325–342.

Shaw, A. J., ed. 1996. *Epithelial Cell Culture*. IRL Press.

Tyson, C. A., and J. A. Frazier, eds. 1993. *Methods in Toxicology. Vol. I (Part A): In Vitro Biological Systems*. Academic Press. Describes methods for growing many types of primary cells in culture.

### Light Microscopy: Exploring Cell Structure and Visualizing Proteins within Cells

Chen, X., M. Velliste, and R. F. Murphy. 2006. Automated interpretation of subcellular patterns in fluorescence microscope images for location proteomics. *Cytometry* (Part A) **69A**:631–640.

Egner, A., and S. Hell. 2005. Fluorescence microscopy with super-resolved optical sections. *Trends Cell Biol.* **15**:207–215.

Gaietta, G., et al. 2002. Multicolor and electron microscopic imaging of connexin trafficking. *Science* **296**:503–507.

Giepmans, B. N. G., et al. 2006. The fluorescent toolbox for assessing protein location and function. *Science* **312**:217–224.

Gilroy, S. 1997. Fluorescence microscopy of living plant cells. *Ann. Rev. Plant Physiol. Plant Mol. Biol.* **48**:165–190.

Huang, B., H. Babcock, and X. Zhuang. 2010. Breaking the diffraction barrier: super-resolution imaging of cells. *Cell* **143**:1047–1058.

Inoué, S., and K. Spring. 1997. *Video Microscopy*, 2d ed. Plenum Press.

Lippincott-Schwartz, J. 2010. Imaging: visualizing the possibilities. *J. Cell Science* **123**:3619–3620.

Lippincott-Schwartz, J. 2011. Emerging in vivo analyses of cell function using fluorescence imaging. *Ann. Rev. Biochem.* **80**:327–332.

Matsumoto, B., ed. 2002. *Methods in Cell Biology*. Vol. 70: *Cell Biological Applications of Confocal Microscopy*. Academic Press.

Mayor, S., and S. Bilgrami. 2007. Fretting about FRET in cell and structural biology. In *Evaluating Techniques in Biochemical Research*, D. Zuk, ed. Cell Press.

Misteli, T., and D. L. Spector. 1997. Applications of the green fluorescent protein in cell biology and biotechnology. *Nature Biotech.* **15**:961–964.

Pepperkok, R., and Ellenberg, J. 2006. High-throughput fluorescence microscopy for systems biology. *Nature Rev. Mol. Cell Biol.* AOP, published online July 19, 2006 (doi:10.1038/nrm1979).

Roukos, V., T. Misteli, and C. K. Schmidt. 2010. Descriptive no more: the dawn of high-throughput microscopy. *Trends in Cell Biology* **20**:503–506.

Sako, Y., S. Minoguchi, and T. Yanagida. 2000. Single-molecule imaging of EGFR signalling on the surface of living cells. *Nature Cell Biol.* **2**:168–172.

Simon, S., and J. Jaiswal. 2004. Potentials and pitfalls of fluorescent quantum dots for biological imaging. *Trends Cell Biol.* **14**:497–504.

Sluder, G., and D. Wolf, eds. 1998. *Methods in Cell Biology*. Vol. 56: *Video Microscopy*. Academic Press.

So, P. T. C., et al. 2000. Two-photon excitation fluorescence microscopy. *Ann. Rev. Biomed. Eng.* **2**:399–429.

Tsien, R. Y. 2009. Indicators based on fluorescence resonance energy transfer (FRET). *Cold Spring Harbor Protoc.*, doi:10.1101/pdb.top57.

Willig, K. I., et al. 2006. STED microscopy reveals that synaptotagmin remains clustered after synaptic vesicle exocytosis. *Nature* **440**:935–939.

### Electron Microscopy: High-Resolution Imaging

Beck, M., et al. 2004. Nuclear pore complex structure and dynamics revealed by cryoelectron tomography. *Science* **306**:1387–1390.

Frey, T. G., G. A. Perkins, and M. H. Ellisman. 2006. Electron tomography of membrane-bound cellular organelles. *Ann. Rev. Biophy. Biomol. Struc.* **35**:199–224.

Hyatt, M. A. *Principles and Techniques of Electron Microscopy*, 4th ed. 2000. Cambridge University Press.

Koster, A., and J. Klumperman. 2003. Electron microscopy in cell biology: integrating structure and function. *Nature Rev. Mol. Cell Biol.* **4**:SS6–SS10.

Lučić, V., et al. 2005. Structural studies by electron tomography: from cells to molecules. *Ann. Rev. Biochem.* **74**:833–865.

Medalia, O., et al. 2002. Macromolecular architecture in eukaryotic cells visualized by cryoelectron tomography. *Science* **298**:1209–1213.

Nickell, S., et al. 2006. A visual approach to proteomics. *Nature Rev. Mol. Cell Biol.* **7**:225–230.

## Isolation and Characterization of Cell Organelles

Bainton, D. 1981. The discovery of lysosomes. *J. Cell Biol.* **91**:66s–76s.

Cuervo, A. M., and J. F. Dice. 1998. Lysosomes: a meeting point of proteins, chaperones, and proteases. *J. Mol. Med.* **76**:6–12.

de Duve, C. 1996. The peroxisome in retrospect. *Ann. NY Acad. Sci.* **804**:1–10.

de Duve, C. 1975. Exploring cells with a centrifuge. *Science* **189**:186–194. The Nobel Prize lecture of a pioneer in the study of cellular organelles.

de Duve, C., and H. Beaufay. 1981. A short history of tissue fractionation. *J. Cell Biol.* **91**:293s–299s.

Foster, L. J., et al. 2006. A mammalian organelle map by protein correlation profiling. *Cell* **125**:187–199.

Holtzman, E. 1989. *Lysosomes.* Plenum Press.

Howell, K. E., E. Devaney, and J. Gruenberg. 1989. Subcellular fractionation of tissue culture cells. *Trends Biochem. Sci.* **14**:44–48.

Lamond, A., and W. Earnshaw. 1998. Structure and function in the nucleus. *Science* **280**:547–553.

Mootha, V. K., et al. 2003. Integrated analysis of protein composition, tissue diversity, and gene regulation in mouse mitochondria. *Cell* **115**:629–640.

Palade, G. 1975. Intracellular aspects of the process of protein synthesis. *Science* **189**:347–358. The Nobel Prize lecture of a pioneer in the study of cellular organelles.

Ormerod, M. G., ed. 1990. *Flow Cytometry: A Practical Approach.* IRL Press.

Rickwood, D. 1992. *Preparative Centrifugation: A Practical Approach.* IRL Press.

Wanders, R., and H. R. Waterham. 2006. Biochemistry of mammalian peroxisomes revisited. *Ann. Rev. Biochem.* **75**:295–332.

## Perturbing Specific Cell Functions

Eggert, U. S., and T. J. Mitchison. 2006. Small molecule screening by imaging. *Curr. Opin. Chem. Biol.* **10**:232–237.

Elbashir, S. M., et al. 2001. Duplexes of 21-nucleotide RNAs mediate RNA interference in cultured mammalian cells. *Nature* **411**:494–498.

Fire, A., et al. 1998. Potent and specific genetic interference by double-stranded RNA in *Caenorhabditis elegans*. *Nature* **391**:806–811.

Goshima, G., et al. 2007. Genes required for mitotic spindle assembly in *Drosophila* S2 cells. *Science* **316**:417–421.

Kamath, R. S., et al. 2003. Systematic functional analysis of the *Caenorhabditis elegans* genome using RNAi. *Nature* **421**:231–237.

Mayer, T. U., et al. 1999. Small molecule inhibitor of mitotic spindle bipolarity identified in a phenotype-based screen. *Science* **286**:971–974.

Meister, G., and T. Tuschl 2004. Mechanisms of gene silencing by double-stranded RNA. *Nature* **431**:343–349.

Mohr, S., et al. 2010. Genomic screening with RNAi: result and challenges. *Ann. Rev. Biochem.* **79**:37–64.

Perrimon, N., et al. 2010. In vivo RNAi: today and tomorrow. *Cold Spring Harbor Perspect. Biol.* **2**:a003640.

# Separating Organelles

H. Beaufay et al., 1964, *Biochemical Journal* **92**:191

In the 1950s and 1960s, scientists used two techniques to study cell organelles: microscopy and fractionation. Christian de Duve was at the forefront of cell fractionation. In the early 1950s, he used centrifugation to distinguish a new organelle, the lysosome, from previously characterized fractions: the nucleus, the mitochondrial-rich fraction, and the microsomes. Soon thereafter, he used equilibrium-density centrifugation to uncover yet another organelle.

## Background

Eukaryotic cells are highly organized and composed of cell structures known as organelles that perform specific functions. Although microscopy has allowed biologists to describe the location and appearance of various organelles, it is of limited use in uncovering an organelle's function. To do this, cell biologists have relied on a technique known as cell fractionation. Here cells are broken open and the cellular components are separated on the basis of size, mass, and density using a variety of centrifugation techniques. Scientists could then isolate and analyze cell components of different densities, called *fractions*. Using this method, biologists had divided the cell into four fractions: nuclei, mitochondrial-rich fraction, microsomes, and cell sap.

De Duve was a biochemist interested in the subcellular locations of metabolic enzymes. He had already completed a large body of work on the fractionation of liver cells, in which he had determined the subcellular location of numerous enzymes. By locating these enzymes in specific cell fractions, he could begin to elucidate the function of the organelle. He noted that his work was guided by two hypotheses: the "postulate of biochemical homogeneity" and "the postulate of single location." In short, these hypotheses propose that the entire composition of a subcellular population will contain the same enzymes and that each enzyme is located at a discrete site within the cell. Armed with these hypotheses and the powerful tool of centrifugation, de Duve further subdivided the mitochondrial-rich fraction. First, he identified the light mitochondrial fraction, which is made up of hydrolytic enzymes that are now known to compose the lysosome. Then, in a series of experiments described here, he identified another discrete subcellular fraction, which he called the peroxisome, within the mitochondrial-rich fraction.

## The Experiment

De Duve studied the distribution of enzymes in rat liver cells. Highly active in energy metabolism, the liver contains a number of useful enzymes to study. To look for the presence of various enzymes during the fractionation, de Duve relied on known tests, called enzyme assays, for enzyme activity. To retain maximum enzyme activity, he had to take precautions, which included performing all fractionation steps at 0 °C to reduce protease activity.

De Duve used rate-zonal centrifugation to separate cellular components by successive centrifugation steps. He removed the rat's liver and broke it apart by homogenization. The crude preparation of homogenized cells was then subjected to relatively low-speed centrifugation. This initial step separated the cell nucleus, which collects as sediment at the bottom of the tube, from the cytoplasmic extract, which remains in the supernatant. Next, de Duve further subdivided the cytoplasmic extract into heavy mitochondrial fraction, light mitochondrial fraction, and microsomal fraction. He accomplished separating the cytoplasm by employing successive centrifugation steps of increasing force. At each step he collected and stored the fractions for subsequent enzyme analysis. Once the fractionation was complete, de Duve performed enzyme assays to determine the subcellular distribution of each enzyme. He then graphically plotted the distribution of the enzyme throughout the cell. As had been shown previously, the activity of cytochrome oxidase, an important enzyme in the electron-transfer system, was found primarily in the heavy mitochondrial fractions. The microsomal fraction was shown to contain another previously characterized enzyme, glucose-6-phosphatase. The light mitochondrial fraction, which is made up of the lysosome, showed the characteristic acid phosphatase activity. Unexpectedly, de Duve observed a fourth pattern when he assayed uricase activity. Rather than following the pattern of the reference enzymes, uricase activity was sharply concentrated within the light mitochondrial fraction. This sharp concentration, in contrast to the broad distribution, suggested to de Duve that the uricase might be secluded in another subcellular population separate from the lysosomal enzymes.

To test this theory, de Duve employed a technique known as equilibrium density-gradient centrifugation, which separates macromolecules on the basis of density. Equilibrium density-gradient centrifugation can be performed using a number of different gradients, including sucrose and glycogen. In addition, the gradient can be made up in either water or "heavy water," which contains the hydrogen isotope deuterium in place of hydrogen. In his experiment, de Duve separated the mitochondrial-rich fraction prepared by rate-zonal centrifugation in each of these different gradients (see Figure 9-35). If uricase were part of a separate subcellular compartment, it would separate from the lysosomal enzymes in each gradient tested. De Duve performed the fractionations in this series of gradients, then performed

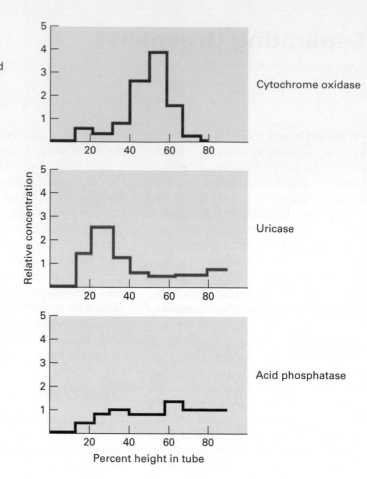

**FIGURE 1 Graphical representation of the enzyme analysis of products from a sucrose gradient.** The mitochondrial-rich fraction was separated as depicted in Figure 9-35, and then enzyme assays were performed. The relative concentration of active enzyme is plotted on the y axis; the height in the tube is plotted on the x axis. The peak activities of cytochrome oxidase (*top*) and acid phosphatase (*bottom*) are observed near the top of the tube. The peak activity of uricase (*middle*) migrates to the bottom of the tube.

enzyme assays as before. In each case, he found uricase in a separate population than the lysosomal enzyme acid phosphatase and the mitochondrial enzyme cytochrome oxidase (Figure 1). By repeatedly observing uricase activity in a distinct fraction from the activity of the lysosomal and mitochondrial enzymes, de Duve concluded that uricase was part of a separate organelle. The experiment also showed that two other enzymes, catalase and D–amino acid oxidase, segregated into the same fractions as uricase. Because each of these enzymes either produced or used

hydrogen peroxide, de Duve proposed that this fraction represented an organelle responsible for the peroxide metabolism and dubbed it the peroxisome.

## Discussion

De Duve's work on cellular fractionation provided an insight into the function of cell structures as he sought to map the location of known enzymes. Examining the inventory of enzymes in a given cell fraction gave him *clues* to its function. His careful work resulted in the uncovering of two organelles:

the lysosome and the **peroxisome**. His work also provided important clues to the organelles' function. The lysosome, where de Duve found so many potentially destructive enzymes, is now known to be an important site for degradation of biomolecules. The **peroxisome** has been shown to be the site of fatty acid and amino acid oxidation, reactions that produce a large amount of hydrogen peroxide. In 1974, de Duve received the Nobel Prize for Physiology or Medicine in recognition of his pioneering work.

# Biomembrane Structure

Molecular model of a lipid bilayer with embedded membrane proteins. Integral membrane proteins have distinct exoplasmic, cytosolic, and membrane-spanning domains. Shown here are portions of the insulin receptor, which regulates cell metabolism. [Ramon Andrade 3Dciencia/ Science Photo Library.]

Membranes participate in many aspects of cell structure and function. The **plasma membrane** defines the cell and separates the inside from the outside. In eukaryotes, membranes also define the intracellular organelles such as the nucleus, mitochondrion, and lysosome. These biomembranes all have the same basic architecture—a phospholipid bilayer in which proteins are embedded (Figure 10-1). By preventing the unassisted movement of most water-soluble substances from one side of the membrane to the other, the phospholipid bilayer serves as a permeability barrier, helping to maintain the characteristic differences between the inside and outside of the cell or organelle; in turn, the embedded proteins endow the membrane with specific functions, such as regulated transport of substances from one side to the other. Each cellular membrane has its own set of proteins that allow it to carry out a multitude of different functions.

Prokaryotes, the simplest and smallest cells, are about 1–2 μm in length and are surrounded by a single plasma membrane; in most cases they contain no internal membrane-limited subcompartments (see Figure 1-11). However, this single plasma membrane contains hundreds of different types of proteins that are integral to the function of the cell. Some of these proteins catalyze ATP synthesis and initiation of DNA replication, for instance. Others include the many types of **membrane transport proteins** that enable specific ions, sugars, amino acids, and vitamins to cross the otherwise impermeable phospholipid bilayer to enter the cell and that allow specific metabolic products to exit. **Receptors** in the plasma membrane are proteins that allow the cell to recognize chemical signals present in its environment and adjust its metabolism or pattern of gene expression in response.

Eukaryotes also have a plasma membrane studded with a multitude of proteins that perform a variety of functions, including membrane transport, cell signaling, and connecting cells into tissues. In addition, eukaryotic cells—which are generally much larger than prokaryotes—also have a variety of internal membrane-bound organelles (see Figure 9-32). Each organelle membrane has a unique complement of proteins that enable it to carry out its characteristic cellular functions, such as ATP generation (in mitochondria) and DNA

## OUTLINE

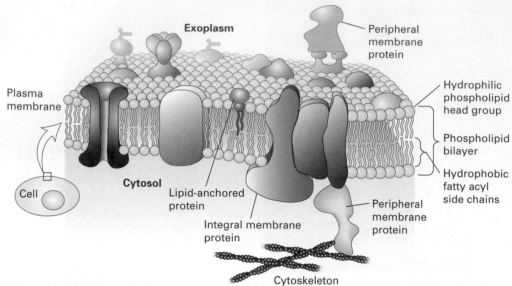

**Exoplasm**

Peripheral membrane protein

Plasma membrane

Hydrophilic phospholipid head group

Phospholipid bilayer

Hydrophobic fatty acyl side chains

**Cytosol**

Cell

Lipid-anchored protein

Integral membrane protein

Peripheral membrane protein

Cytoskeleton

**FIGURE 10-1 Fluid mosaic model of biomembranes.** A bilayer of phospholipids ~3 nm thick provides the basic architecture of all cellular membranes; membrane proteins give each cellular membrane its unique set of functions. Individual phospholipids can move laterally and spin within the plane of the membrane, giving the membrane a fluidlike consistency similar to that of olive oil. Noncovalent interactions between phospholipids, and between phospholipids and proteins, lend strength and resilience to the membrane, while the hydrophobic core of the bilayer prevents the unassisted movement of water-soluble substances from one side to the other. Integral (transmembrane) proteins span the bilayer and often form dimers and higher-order oligomers. Lipid-anchored proteins are tethered to one leaflet by a covalently attached hydrocarbon chain. Peripheral proteins associate with the membrane primarily by specific noncovalent interactions with integral proteins or membrane lipids. Proteins in the plasma membrane also make extensive contact with the cytoskeleton. [After D. Engelman, 2005, *Nature* **438**:578–580.]

synthesis (in the nucleus). Many plasma membrane proteins also bind components of the **cytoskeleton,** a dense network of protein filaments that crisscrosses the cytosol to provide mechanical support for cellular membranes, interactions that are essential for the cell to assume its specific shape and for many types of cell movements.

Despite playing a structural role in cells, membranes are not rigid structures. They can bend and flex in three dimensions while still maintaining their integrity, due in part to abundant noncovalent interactions that hold lipids and proteins together. Moreover, within the plane of the membrane, there is considerable mobility of individual lipids and proteins. According to the *fluid mosaic model* of biomembranes, first proposed by researchers in the 1970s, the lipid bilayer behaves in some respects like a two-dimensional fluid, with individual lipid molecules able to move past one another as well as spin in place. Such fluidity and flexibility not only allows organelles to assume their typical shapes, but also enables the dynamic

(a)

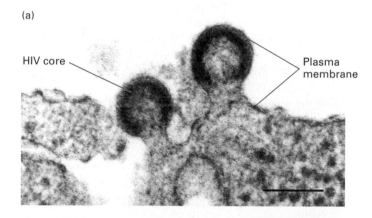

HIV core

Plasma membrane

(b)

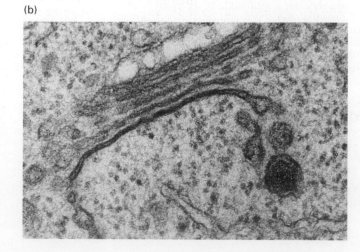

**FIGURE 10-2 Eukaryotic cell membranes are dynamic structures.** (a) An electron micrograph of the plasma membrane of an HIV-infected cell, showing HIV particles budding into the culture medium. As the virus core buds from the cell, it becomes enveloped by a membrane derived from the cell's plasma membrane that contains specific viral proteins. (b) Stacked membranes of the Golgi complex with budding vesicles. Note the irregular shape and curvature of these membranes. [Part (a) from W. Sundquist and U. von Schwedler, University of Utah; part (b) from Biology Pics/Photo Researchers, Inc.]

property of membrane budding and fusion, such as occurs when viruses are released from an infected cell (Figure 10-2a) and when the internal cellular membranes of the Golgi complex bud into vesicles in the cytosol (Figure 10-2b) and then fuse with other membranes to transport their contents from one organelle to another (Chapter 14).

We begin our examination of biomembranes by considering their lipid components. These not only affect membrane shape and function but also help anchor proteins to the membrane, modify membrane protein activities, and transduce signals to the cytoplasm. We then consider the structure of membrane proteins. Many of these proteins have large segments that are embedded in the hydrocarbon core of the phospholipid bilayer, and we will focus on the principal classes of such membrane proteins. Finally, we consider how lipids such as phospholipids and cholesterol are synthesized in cells and distributed to the many membranes and organelles. Cholesterol is an essential component of the plasma membrane of all animal cells but is toxic to the organism if present in excess.

## 10.1 The Lipid Bilayer: Composition and Structural Organization

In Chapter 2 we learned that phospholipids are the principal building blocks of biomembranes. The most common phospholipids in membranes are the phosphoglycerides (see Figure 2-20), but as we will see in this chapter, there are multiple types of phospholipid. All phospholipids are **amphipathic** molecules—they consist of two segments with very different chemical properties: a fatty acid-based (fatty acyl) hydrocarbon "tail" that is **hydrophobic** and partitions away from water, and a polar "head group" that is strongly **hydrophilic**, or water loving, and tends to interact with water molecules. The interactions of phospholipids with each other and with water largely determine the structure of biomembranes.

Besides phospholipids, biomembranes contain smaller amounts of other amphipathic lipids, such as glycolipids and cholesterol, which contribute to membrane function in important ways. We first consider the structure and properties of pure phospholipid bilayers and then discuss the composition and behavior of natural cell membranes. We will see how the precise lipid composition of a given membrane influences its physical properties.

### Phospholipids Spontaneously Form Bilayers

The amphipathic nature of phospholipids, which governs their interactions, is critical to the structure of biomembranes. When a suspension of phospholipids is mechanically dispersed in aqueous solution, the phospholipids aggregate into one of three forms: spherical **micelles** and **liposomes**, or sheetlike **phospholipid bilayers,** which are two molecules thick (Figure 10-3). The type of structure formed by pure phospholipids or

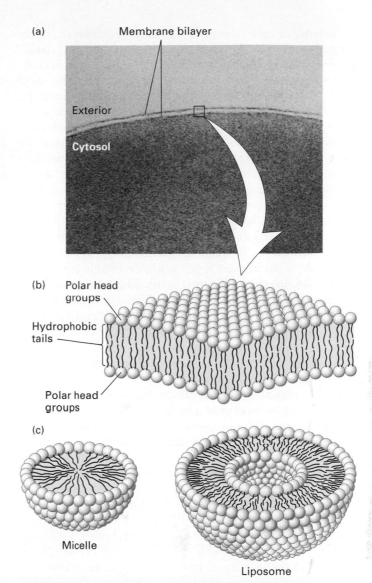

**FIGURE 10-3  The bilayer structure of biomembranes.** (a) Electron micrograph of a thin section through an erythrocyte membrane stained with osmium tetroxide. The characteristic "railroad track" appearance of the membrane indicates the presence of two polar layers, consistent with the bilayer structure of phospholipid membranes. (b) Schematic interpretation of the phospholipid bilayer in which polar groups face outward to shield the hydrophobic fatty acyl tails from water. The hydrophobic effect and van der Waals interactions between the fatty acyl tails drive the assembly of the bilayer (Chapter 2). (c) Cross-sectional views of two other structures formed by dispersal of phospholipids in water. A spherical micelle has a hydrophobic interior composed entirely of fatty acyl chains; a spherical liposome consists of a phospholipid bilayer surrounding an aqueous center.
[Part (a) courtesy of J. D. Robertson.]

a mixture of phospholipids depends on several factors, including the length of the fatty acyl chains in the hydrophobic tail, their degree of saturation (i.e., the number of C—C and C=C bonds), and temperature. In all three structures, the hydrophobic effect causes the fatty acyl chains to aggregate and exclude water molecules from the "core." Micelles are rarely formed from natural phospholipids, whose fatty acyl chains generally are too bulky to fit into the interior of a micelle. However, micelles are formed if one of the two fatty acyl chains that make up the tail of a phospholipid is removed by hydrolysis, forming a lysophospholipid, as occurs upon treatment with the enzyme phospholipase. In aqueous solution, common detergents and soaps form micelles that behave like the balls in tiny ball bearings, thus giving soap solutions their slippery feel and lubricating properties.

Phospholipids of the composition present in cells spontaneously form symmetric phospholipid bilayers. Each phospholipid layer in this lamellar structure is called a *leaflet*. The hydrophobic fatty acyl chains in each leaflet minimize their contact with water by aligning themselves tightly together in the center of the bilayer, forming a hydrophobic core that is about 3–4 nm thick (Figure 10-3b). The close packing of these nonpolar tails is stabilized by van der Waals interactions between the hydrocarbon chains. Ionic and hydrogen bonds stabilize the interactions of the phospholipid polar head groups with one another and with water. Electron microscopy of thin membrane sections of cells stained with osmium tetroxide, which binds strongly to the polar head groups of phospholipids, reveals the bilayer structure (Figure 10-3a). A cross section of a single membrane stained with osmium tetroxide looks like a railroad track: two thin dark lines (the stained head group complexes) with a uniform light space of about 2 nm between them (the hydrophobic tails).

A phospholipid bilayer can be of almost unlimited size—from micrometers (μm) to millimeters (mm) in length or width—and can contain tens of millions of phospholipid molecules. The phospholipid bilayer is the basic structural unit of nearly all biological membranes. Its hydrophobic core prevents most water-soluble substances from crossing from one side of the membrane to the other. Although biomembranes contain other molecules (e.g., cholesterol, glycolipids, proteins), it is the phospholipid bilayer that separates two aqueous solutions and acts as a permeability barrier. The lipid bilayer thus defines cellular compartments and allows a separation of the cell's interior from the outside world.

## Phospholipid Bilayers Form a Sealed Compartment Surrounding an Internal Aqueous Space

Phospholipid bilayers can be generated in the laboratory by simple means, using either chemically pure phospholipids or lipid mixtures of the composition found in cell membranes (Figure 10-4). Such synthetic bilayers possess three important properties. First, they are virtually impermeable to water-soluble (hydrophilic) solutes, which do not readily diffuse across the bilayer. This includes salts, sugars, and most other small hydrophilic molecules—including water itself. The second property of the bilayer is its stability. Hydrophobic and van der Waals interactions between the fatty acyl chains maintain the integrity of the interior of the bilayer structure.

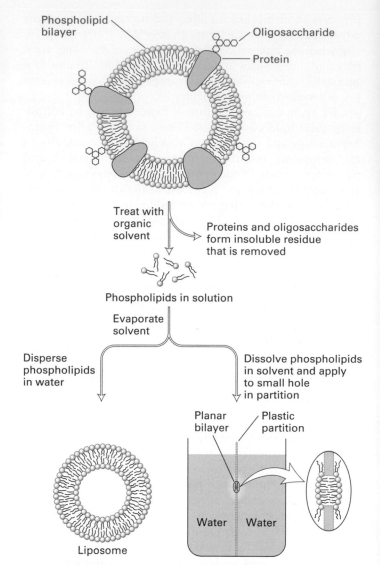

**EXPERIMENTAL FIGURE 10-4 Formation and study of pure phospholipid bilayers.** (*Top*) A preparation of biological membranes is treated with an organic solvent, such as a mixture of chloroform and methanol (3:1), which selectively solubilizes the phospholipids and cholesterol. Proteins and carbohydrates remain in an insoluble residue. The solvent is removed by evaporation. (*Bottom left*) If the lipids are mechanically dispersed in water, they spontaneously form a liposome, shown in cross section, with an internal aqueous compartment. (*Bottom right*) A planar bilayer, also shown in cross section, can form over a small hole in a partition separating two aqueous phases; such a system can be used to study the physical properties of bilayers, such as their permeability to solutes.

Even though the exterior aqueous environment can vary widely in ionic strength and pH, the bilayer has the strength to retain its characteristic architecture. Third, all phospholipid bilayers can spontaneously form sealed closed compartments where the aqueous space on the inside is separated from that on the outside. An "edge" of a phospholipid bilayer, as depicted in Figure 10-3b, with the hydrocarbon core of the bilayer exposed to an aqueous solution, is unstable; the

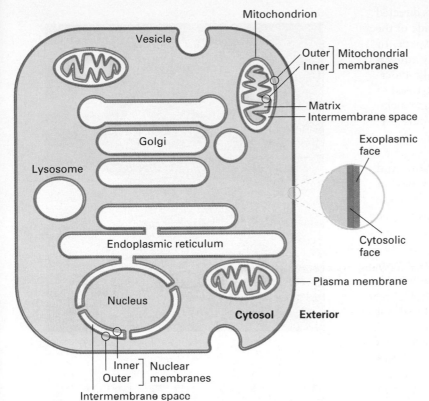

Vesicle

Mitochondrion

Outer ] Mitochondrial
Inner ] membranes

Matrix
Intermembrane space

Golgi

Lysosome

Exoplasmic
face

Endoplasmic reticulum

Cytosolic
face

Plasma membrane

Nucleus

Cytosol      Exterior

Inner ] Nuclear
Outer ] membranes

Intermembrane space

**FIGURE 10-5  The faces of cellular membranes.** The plasma membrane, a single bilayer membrane, encloses the cell. In this highly schematic representation, internal cytosol (tan) and external environment (white) define the cytosolic (red) and exoplasmic (gray) faces of the bilayer. Vesicles and some organelles have a single membrane and their internal aqueous space (white) is topologically equivalent to the outside of the cell. Three organelles—the nucleus, mitochondrion, and chloroplast (which is not shown)—are enclosed by two membranes separated by a small intermembrane space. The exoplasmic faces of the inner and outer membranes around these organelles border the intermembrane space between them. For simplicity, the hydrophobic membrane interior is not indicated in this diagram.

exposed fatty acyl side chains would be in an energetically much more stable state if they were not adjacent to water molecules but surrounded by other fatty acyl chains (hydrophobic effect; Chapter 2). Thus in aqueous solution, sheets of phospholipid bilayers spontaneously seal their edges, forming a spherical bilayer that encloses an aqueous central compartment. The liposome depicted in Figure 10-3c is an example of such a structure viewed in cross section.

This physical chemical property of a phospholipid bilayer has important implications for cellular membranes: no membrane in a cell can have an "edge" with exposed hydrocarbon fatty acyl chains. All membranes form closed compartments, similar in basic architecture to liposomes. Because all cellular membranes enclose an entire cell or an internal compartment, they have an *internal face* (the surface oriented toward the interior of the compartment) and an *external face* (the surface presented to the environment). More commonly, we designate the two surfaces of a cellular membrane as the **cytosolic face** and the **exoplasmic face.** This nomenclature is useful in highlighting the topological equivalence of the faces in different membranes, as diagrammed in Figures 10-5 and 10-6. For example, the exoplasmic face of the plasma membrane is directed away from the cytosol, toward the extracellular space or external environment, and defines the outer limit of the cell. The cytosolic face of the plasma membrane faces the cytosol. Similarly for organelles and vesicles surrounded by a single membrane, the cytosolic

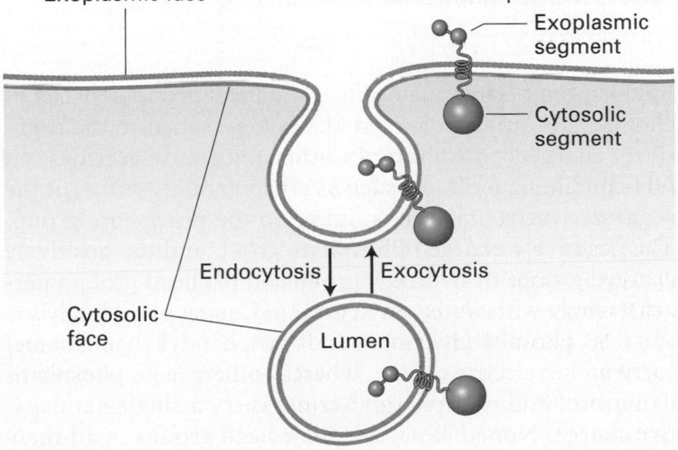

Exoplasmic face

Membrane protein

Exoplasmic
segment

Cytosolic
segment

Endocytosis    Exocytosis

Cytosolic
face

Lumen

**FIGURE 10-6  Faces of cellular membranes are conserved during membrane budding and fusion.** Red membrane surfaces are cytosolic faces; gray are exoplasmic faces. During endocytosis a segment of the plasma membrane buds inward toward the cytosol and eventually pinches off a separate vesicle. During this process the cytosolic face of the plasma membrane remains facing the cytosol and the exoplasmic face of the new vesicle membrane faces the vesicle lumen. During exocytosis an intracellular vesicle fuses with the plasma membrane, and the lumen of the vesicle (exoplasmic face) connects with the extracellular medium. Proteins that span the membrane retain their asymmetric orientation during vesicle budding and fusion; in particular the same segment always faces the cytosol.

face faces the cytosol. The exoplasmic face is always directed away from the cytosol and in this case is on the inside of the organelle in contact with the internal aqueous space, or **lumen**. The lumen of these vesicles is topologically equivalent to the extracellular space, a concept most easily understood for vesicles that arise by invagination (endocytosis) of the plasma membrane. The external face of the plasma membrane becomes the internal face of the vesicle membrane, while in the vesicle the cytosolic face of the plasma membrane still faces the cytosol (Figure 10-6).

Three organelles—the nucleus, mitochondrion, and chloroplast—are surrounded not by a single membrane, but by two. The exoplasmic surface of each membrane faces the space between the two membranes. This can perhaps best be understood by reference to the *endosymbiont hypothesis,* discussed in Chapter 6, which posits that mitochondria and chloroplasts arose early in the evolution of eukaryotic cells by the engulfment of bacteria capable of oxidative phosphorylation or photosynthesis, respectively (see Figure 6-20).

Natural membranes from different cell types exhibit a variety of shapes, which complement a cell's function. The smooth, flexible surface of the erythrocyte plasma membrane allows the cell to squeeze through narrow blood capillaries (Figure 10-7a). Some cells have a long, slender extension of the plasma membrane, called a **cilium** or **flagellum,** which beats in a whiplike manner (Figure 10-7b). This motion causes fluid to flow across the surface of a sheet of cells, or a sperm cell to swim toward an egg. The differing shapes and properties of biomembranes raise a key question in cell biology, namely how the composition of biological membranes is regulated to establish and maintain the identity of the different membrane structures and membrane-delimited compartments. We return to this question in Section 10.3 and in Chapter 14.

## Biomembranes Contain Three Principal Classes of Lipids

The term *phospholipid* is a somewhat generic term, encompassing multiple distinct molecules from multiple classes. It refers to any amphipathic lipid with a phosphate-based head group and a two-pronged hydrophobic tail. A typical biomembrane actually contains three classes of amphipathic lipids: phosphoglycerides, sphingolipids, and sterols, which differ in their chemical structures, abundance, and functions in the membrane (Figure 10-8). While all phosphoglycerides are phospholipids, only certain sphingolipids are, and no sterols are.

**Phosphoglycerides,** the most abundant class of phospholipids in most membranes, are derivatives of glycerol 3-phosphate (see Figure 10-8a). A typical phosphoglyceride molecule consists of a hydrophobic tail composed of two fatty acid-based (acyl) chains esterified to the two hydroxyl groups in glycerol phosphate and a polar head group attached to the phosphate group. The two fatty acyl chains may differ in the number of carbons that they contain (commonly 16 or 18) and their degree of saturation (0, 1, or 2 double bonds). A phosphoglyceride is classified according to the nature of its head group. In phosphatidylcholines, the most abundant phospho-

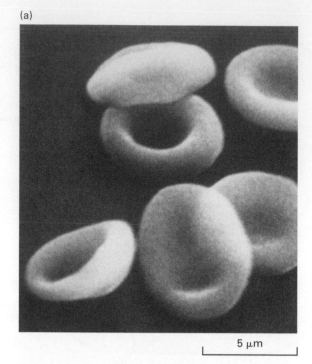

(a)

5 μm

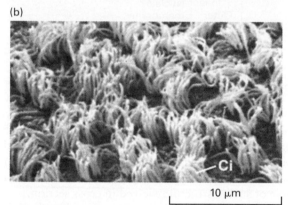

(b)

Ci

10 μm

**FIGURE 10-7 Variation in biomembranes in different cell types.** (a) A smooth, flexible membrane covers the surface of the discoid erythrocyte cell as seen in this scanning electron micrograph. (b) Tufts of cilia (Ci) project from the ependymal cells that line the brain ventricles. [Part (a) Copyright © Omi Kron/Photo Researchers, Inc. Part (b) from R. G. Kessel and R. H. Kardon, 1979, Tissues and Organs: A Text-Atlas of Scanning Electron Microscopy, W. H. Freeman and Company.]

lipids in the plasma membrane, the head group consists of choline, a positively charged alcohol, esterified to the negatively charged phosphate. In other phosphoglycerides, an OH-containing molecule such as ethanolamine, serine, or the sugar derivative inositol is linked to the phosphate group. The negatively charged phosphate group and the positively charged groups or hydroxyl groups on the head group interact strongly with water. At neutral pH, some phosphoglycerides (e.g., phosphatidylcholine and phosphatidylethanolamine) carry no net electric charge, whereas others (e.g., phosphatidylinositol and phosphatidylserine) carry a single net negative charge. Nonetheless, the polar head groups in all these

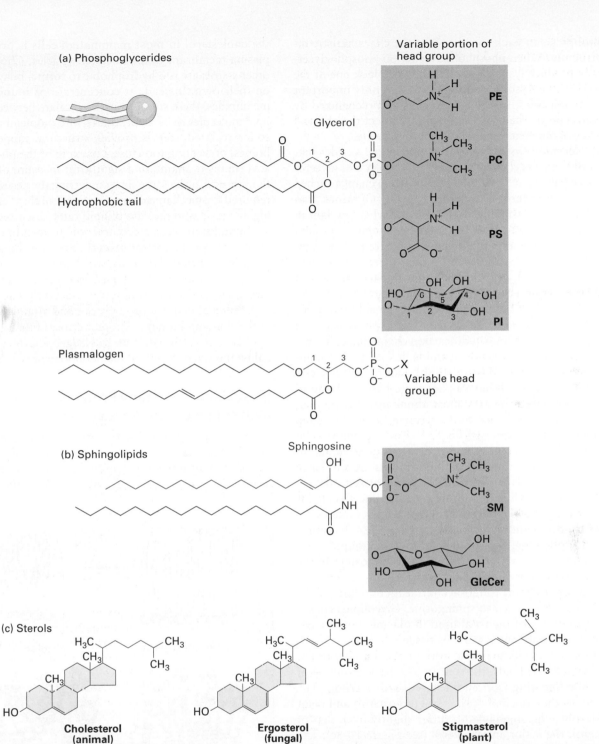

**FIGURE 10-8 Three classes of membrane lipids.** (a) Most phosphoglycerides are derivatives of glycerol 3-phosphate (red), which contains two esterified fatty acyl chains that constitute the hydrophobic "tail" and a polar "head group" esterified to the phosphate. The fatty acids can vary in length and be saturated (no double bonds) or unsaturated (one, two, or three double bonds). In phosphatidylcholine (PC), the head group is choline. Also shown are the molecules attached to the phosphate group in three other common phosphoglycerides: phosphatidylethanolamine (PE), phosphatidylserine (PS), and phosphatidylinositol (PI). Plasmalogens contain one fatty acyl chain attached to glycerol by an ester linkage and one attached by an ether linkage; these contain similar head groups as other phosphoglycerides. (b) Sphingolipids are derivatives of sphingosine (red), an amino alcohol with a long hydrocarbon chain. Various fatty acyl chains are connected to sphingosine by an amide bond. The sphingomyelins (SM), which contain a phosphocholine head group, are phospholipids. Other sphingolipids are glycolipids in which a single sugar residue or branched oligosaccharide is attached to the sphingosine backbone. For instance, the simple glycolipid glucosylcerebroside (GlcCer) has a glucose head group. (c) The major sterols in animals (cholesterol), fungi (ergosterol), and plants (stigmasterol) differ slightly in structure, but all serve as key components of cellular membranes. The basic structure of sterols is a four-ring hydrocarbon (yellow). Like other membrane lipids, sterols are amphipathic. The single hydroxyl group is equivalent to the polar head group in other lipids; the conjugated ring and short hydrocarbon chain form the hydrophobic tail. [See H. Sprong et al., 2001, *Nature Rev. Mol. Cell Biol.* **2**:504.]

phospholipids can pack together into the characteristic bilayer structure. When phospholipases act on phosphoglycerides, they produce lysophospholipids, which lack one of the two acyl chains. Lysophospholipids are not only important signaling molecules, released from cells and recognized by specific receptors; their presence can also affect the physical properties of the membranes in which they reside.

The *plasmalogens* are a group of phosphoglycerides that contain one fatty acyl chain attached to carbon 2 of glycerol by an ester linkage and one long hydrocarbon chain attached to carbon 1 of glycerol by an ether (C—O—C) rather than an ester linkage. Plasmalogens are particularly abundant in human brain and heart tissue. The greater chemical stability of the ether linkage in plasmalogens, compared to the ester linkage, or the subtle differences in their three-dimensional structure compared with that of other phosphoglycerides may have as yet unrecognized physiologic significance.

A second class of membrane lipid is the **sphingolipids.** All of these compounds are derived from sphingosine, an amino alcohol with a long hydrocarbon chain, and contain a long-chain fatty acid attached in amide linkage to the sphingosine amino group (see Figure 10-8b). Like phosphoglycerides, some sphingolipids have a phosphate-based polar head group. In sphingomyelin, the most abundant sphingolipid, phosphocholine is attached to the terminal hydroxyl group of sphingosine (see Figure 10-8b, SM). Thus sphingomyelin is a phospholipid, and its overall structure is quite similar to that of phosphatidylcholine. Sphingomyelins are similar in shape to phosphoglycerides and can form mixed bilayers with them. Other sphingolipids are amphipathic **glycolipids** whose polar head groups are sugars that are not linked via a phosphate group (and so technically are not phospholipids). Glucosylcerebroside, the simplest glycosphingolipid, contains a single glucose unit attached to sphingosine. In the complex glycosphingolipids called *gangliosides,* one or two branched sugar chains (oligosaccharides) containing sialic acid groups are attached to sphingosine. Glycolipids constitute 2–10 percent of the total lipid in plasma membranes; they are most abundant in nervous tissue.

**Cholesterol** and its analogs constitute the third important class of membrane lipids, the **sterols.** The basic structure of sterols is a four-ring isoprenoid-based hydrocarbon. The structures of the principal yeast sterol (ergosterol) and plant phytosterols (e.g., stigmasterol) differ slightly from that of cholesterol, the major animal sterol (see Figure 10-8c). The small differences in the biosynthetic pathways and structures of fungal and animal sterols are the basis of most antifungal drugs currently in use. Cholesterol, like the two other sterols, has a hydroxyl substituent on one ring. Although cholesterol is almost entirely hydrocarbon in composition, it is amphipathic because its hydroxyl group can interact with water. Because it lacks a phosphate-based head group, it is not a phospholipid. Cholesterol is especially abundant in the plasma membranes of mammalian cells but is absent from most prokaryotic and all plant cells. As much as 30–50 percent of the lipids in plant plasma membranes consists of certain steroids unique to plants. Between 50 and 90 percent of

the cholesterol in most mammalian cells is present in the plasma membrane and associated vesicles. Cholesterol and other sterols are too hydrophobic to form a bilayer structure on their own. Instead, at concentrations found in natural membranes, these sterols must intercalate between phospholipid molecules to be incorporated into biomembranes. When so intercalated, sterols provide structural support to membranes, preventing too close a packing of the phospholipids' acyl chains to maintain a significant measure of membrane fluidity, and at the same time conferring the necessary rigidity required for mechanical support. Some of these effects can be highly local, as in the case of lipid rafts, discussed below.

In addition to its structural role in membranes, cholesterol is the precursor for several important bioactive molecules. They include *bile acids,* which are made in the liver and help emulsify dietary fats for digestion and absorption in the intestines; steroid hormones produced by endocrine cells (e.g., adrenal gland, ovary, testes); and vitamin D produced in the skin and kidneys. Another critical function of cholesterol is its covalent addition to Hedgehog protein, a key signaling molecule in embryonic development (Chapter 16).

## Most Lipids and Many Proteins Are Laterally Mobile in Biomembranes

In the two-dimensional plane of a bilayer, thermal motion permits lipid molecules to rotate freely around their long axes and to diffuse laterally within each leaflet. Because such movements are lateral or rotational, the fatty acyl chains

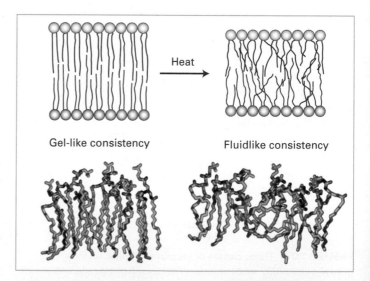

**FIGURE 10-9 Gel and fluid forms of the phospholipid bilayer.** (*Top*) Depiction of gel-to-fluid transition. Phospholipids with long saturated fatty acyl chains tend to assemble into a highly ordered, gel-like bilayer in which there is little overlap of the nonpolar tails in the two leaflets. Heat disorders the nonpolar tails and induces a transition from a gel to a fluid within a temperature range of only a few degrees. As the chains become disordered, the bilayer also decreases in thickness. (*Bottom*) Molecular models of phospholipid monolayers in gel and fluid states, as determined by molecular dynamics calculations. [Bottom based on H. Heller et al., 1993, *J. Phys. Chem.* **97**:8343.]

remain in the hydrophobic interior of the bilayer. In both natural and artificial membranes, a typical lipid molecule exchanges places with its neighbors in a leaflet about $10^7$ times per second and diffuses several micrometers per second at 37 °C. These diffusion rates indicate that the bilayer is 100 times more viscous than water—about the same as the viscosity of olive oil. Even though lipids diffuse more slowly in the bilayer than in an aqueous solvent, a membrane lipid could diffuse the length of a typical bacterial cell (1 μm) in only 1 second and the length of an animal cell in about 20 seconds. When artificial pure phospholipid membranes are cooled below 37 °C, the lipids can undergo a *phase transition* from a liquidlike (fluid) state to a gel-like (semisolid) state, analogous to the liquid-solid transition when liquid water freezes (Figure 10-9). Below the phase-transition temperature, the rate of diffusion of the lipids drops precipitously. At usual physiologic temperatures, the hydrophobic interior of natural membranes generally has a low viscosity and a fluidlike consistency, in contrast to the gel-like consistency observed at lower temperatures.

In pure membrane bilayers (i.e., in the absence of protein), phospholipids and sphingolipids rotate and move laterally, but they do not spontaneously migrate, or flip-flop, from one leaflet to the other. The energetic barrier is too high; migration would require moving the polar head group from its aqueous environment through the hydrocarbon core of the bilayer to the aqueous solution on the other side. Special membrane proteins discussed in Chapter 11 are required to flip membrane lipids and other polar molecules from one leaflet to the other.

The lateral movements of specific plasma-membrane proteins and lipids can be quantified by a technique called *fluorescence recovery after photobleaching (FRAP)*. Phospholipids containing a fluorescent substituent are used to monitor lipid movement. For proteins, a fragment of a monoclonal antibody that is specific for the exoplasmic domain of the desired protein and that has only a single antigen-binding site is tagged with a fluorescent dye. With this method, described in Figure 10-10, the rate at which membrane molecules move—the diffusion coefficient—can be

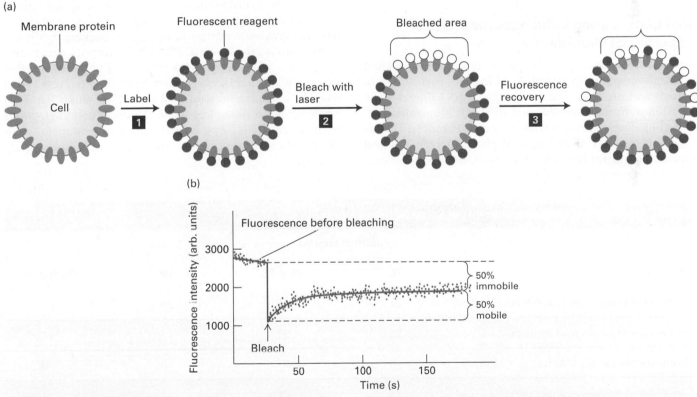

**EXPERIMENTAL FIGURE 10-10 Fluorescence recovery after photobleaching (FRAP) experiments can quantify the lateral movement of proteins and lipids within the plasma membrane.** (a) Experimental protocol. Step **1** Cells are first labeled with a fluorescent reagent that binds uniformly to a specific membrane lipid or protein. Step **2** A laser light is then focused on a small area of the surface, irreversibly bleaching the bound reagent and thus reducing the fluorescence in the illuminated area. Step **3** In time, the fluorescence of the bleached patch increases as unbleached fluorescent surface molecules diffuse into it and bleached ones diffuse outward. The extent of recovery of fluorescence in the bleached patch is proportional to the fraction of labeled molecules that are mobile in the membrane. (b) Results of a FRAP experiment with human hepatoma cells treated with a fluorescent antibody specific for the asialoglycoprotein receptor protein. The finding that 50 percent of the fluorescence returned to the bleached area indicates that 50 percent of the receptor molecules in the illuminated membrane patch were mobile and 50 percent were immobile. Because the rate of fluorescence recovery is proportional to the rate at which labeled molecules move into the bleached region, the diffusion coefficient of a protein or lipid in the membrane can be calculated from such data. [See Y. I. Henis et al., 1990, *J. Cell Biol.* **111:**1409.]

determined, as well as the proportion of the molecules that are laterally mobile.

The results of FRAP studies with fluorescence-labeled phospholipids have shown that in fibroblast plasma membranes, all the phospholipids are freely mobile over distances of about 0.5 μm, but most cannot diffuse over much longer distances. These findings suggest that protein-rich regions of the plasma membrane about 1 μm in diameter separate lipid-rich regions containing the bulk of the membrane phospholipid. Phospholipids are free to diffuse within such regions but not from one lipid-rich region to an adjacent one. Furthermore, the rate of lateral diffusion of lipids in the plasma membrane is nearly an order of magnitude slower than in pure phospholipid bilayers: diffusion constants of $10^{-8}$ cm²/s and $10^{-7}$ cm²/s are characteristic of the plasma membrane and a lipid bilayer, respectively. This difference suggests that lipids may be tightly but not irreversibly bound to certain integral proteins in some membranes, as indeed has recently been demonstrated (see discussion of annular phospholipids, below).

## Lipid Composition Influences the Physical Properties of Membranes

A typical cell contains many different types of membranes, each with unique properties derived from its particular mix of lipids and proteins. The data in Table 10-1 illustrate the variation in lipid composition in different biomembranes. Several phenomena contribute to these differences. For instance, the relative abundances of phosphoglycerides and sphingolipids differ between membranes in the endoplasmic

reticulum (ER), where phospholipids are synthesized, and the Golgi, where sphingolipids are synthesized. The proportion of sphingomyelin as a percentage of total membrane lipid phosphorus is about six times as high in Golgi membranes as it is in ER membranes. In other cases, the movement of membranes from one cellular compartment to another can selectively enrich certain membranes in lipids such as cholesterol. In responding to differing environments throughout an organism, different types of cells generate membranes with differing lipid compositions. In the cells that line the intestinal tract, for example, the membranes that face the harsh environment in which dietary nutrients are digested have a sphingolipid-to-phosphoglyceride-to-cholesterol ratio of 1:1:1 rather than the 0.5:1.5:1 ratio found in cells subject to less stress. The relatively high concentration of sphingolipid in this intestinal membrane may increase its stability because of extensive hydrogen bonding by the free —OH group in the sphingosine moiety (see Figure 10-8).

The degree of bilayer fluidity depends on the lipid composition, the structure of the phospholipid hydrophobic tails, and temperature. As already noted, van der Waals interactions and the hydrophobic effect cause the nonpolar tails of phospholipids to aggregate. Long, saturated fatty acyl chains have the greatest tendency to aggregate, packing tightly together into a gel-like state. Phospholipids with short fatty acyl chains, which have less surface area and therefore fewer van der Waals interactions, form more fluid bilayers. Likewise, the kinks in cis-unsaturated fatty acyl chains (Chapter 2) result in their forming less stable van der Waals interactions with other lipids, and hence more fluid bilayers, than do straight saturated chains, which can pack more tightly together.

| TABLE 10-1 | Major Lipid Components of Selected Biomembranes | | | |
|---|---|---|---|---|
| | Composition (mol %) | | | |
| Source/Location | PC | PE + PS | SM | Cholesterol |
| Plasma membrane (human erythrocytes) | 21 | 29 | 21 | 26 |
| Myelin membrane (human neurons) | 16 | 37 | 13 | 34 |
| Plasma membrane (*E. coli*) | 0 | 85 | 0 | 0 |
| Endoplasmic reticulum membrane (rat) | 54 | 26 | 5 | 7 |
| Golgi membrane (rat) | 45 | 20 | 13 | 13 |
| Inner mitochondrial membrane (rat) | 45 | 45 | 2 | 7 |
| Outer mitochondrial membrane (rat) | 34 | 46 | 2 | 11 |
| Primary leaflet location | Exoplasmic | Cytosolic | Exoplasmic | Both |

PC = phosphatidylcholine; PE = phosphatidylethanolamine; PS = phosphatidylserine; SM = sphingomyelin.
SOURCE: W. Dowhan and M. Bogdanov, 2002, in D. E. Vance and J. E. Vance, eds., *Biochemistry of Lipids, Lipoproteins, and Membranes,* Elsevier.

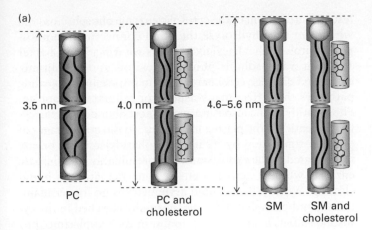

(a)

3.5 nm | 4.0 nm | 4.6–5.6 nm

PC | PC and cholesterol | SM | SM and cholesterol

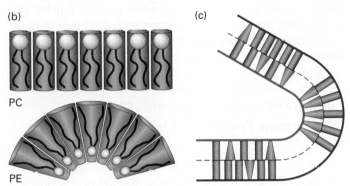

(b)

PC

PE

(c)

**FIGURE 10-11 Effect of lipid composition on bilayer thickness and curvature.** (a) A pure sphingomyelin (SM) bilayer is thicker than one formed from a phosphoglyceride such as phosphatidylcholine (PC). Cholesterol has a lipid-ordering effect on phosphoglyceride bilayers that increases their thickness, but it does not affect the thickness of the more-ordered SM bilayer. (b) Phospholipids such as PC have a cylindrical shape and form essentially flat monolayers, whereas those with smaller head groups such as phosphatidylethanolamine (PE) have a conical shape. (c) A bilayer enriched with PC in the exoplasmic leaflet and with PE in the cytosolic face, as in many plasma membranes, would have a natural curvature. [Adapted from H. Sprong et al., 2001, *Nature Rev. Mol. Cell Biol.* **2**:504.]

Cholesterol is important in maintaining the appropriate fluidity of natural membranes, a property that appears to be essential for normal cell growth and reproduction. Cholesterol restricts the random movement of phospholipid head groups at the outer surfaces of the leaflets, but its effect on the movement of long phospholipid tails depends on concentration. At cholesterol concentrations present in the plasma membrane, the interaction of the steroid ring with the long hydrophobic tails of phospholipids tends to immobilize these lipids and thus decrease biomembrane fluidity. It is this property that can help organize the plasma membrane into discrete subdomains of unique lipid and protein composition. At lower cholesterol concentrations, however, the steroid ring separates and disperses phospholipid tails, causing the inner regions of the membrane to become slightly more fluid.

The lipid composition of a bilayer also influences its thickness, which in turn may influence the distribution of other membrane components, such as proteins, in a particular membrane. It has been argued that relatively short transmembrane segments of certain Golgi-resident enzymes (glycosyltransferases) are an adaptation to the lipid composition of the Golgi membrane and contribute to the retention of these enzymes in the Golgi apparatus. The results of biophysical studies on artificial membranes demonstrate that sphingomyelin associates into a more gel-like and thicker bilayer than phosphoglycerides do (Figure 10-11a). Cholesterol and other molecules that decrease membrane fluidity also increase membrane thickness. Because sphingomyelin tails are already optimally stabilized, the addition of cholesterol has no effect on the thickness of a sphingomyelin bilayer.

Another property dependent on the lipid composition of a bilayer is its curvature, which depends on the relative sizes of the polar head groups and nonpolar tails of its constituent phospholipids. Lipids with long tails and large head groups are cylindrical in shape; those with small head groups are cone shaped (Figure 10-11b). As a result, bilayers composed of cylindrical lipids are relatively flat, whereas those containing large numbers of cone-shaped lipids form curved bilayers (Figure 10-11c). This effect of lipid composition on bilayer curvature may play a role in the formation of highly curved membranes, such as sites of viral budding (see Figure 10-2) and formation of internal vesicles from the plasma membrane (see Figure 10-6), and in specialized stable membrane structures such as microvilli. Several proteins bind to the surface of phospholipid bilayers and cause the membrane to curve; such proteins are important in formation of transport vesicles that bud from a donor membrane (Chapter 14).

## Lipid Composition Is Different in the Exoplasmic and Cytosolic Leaflets

A characteristic of all biomembranes is an asymmetry in lipid composition across the bilayer. Although most phospholipids are present in both membrane leaflets, some are commonly more abundant in one or the other leaflet. For instance, in plasma membranes from human erythrocytes and Madin Darby canine kidney (MDCK) cells grown in culture, almost all the sphingomyelin and phosphatidylcholine, both of which form less fluid bilayers, are found in the exoplasmic leaflet. In contrast, phosphatidylethanolamine, phosphatidylserine, and phosphatidylinositol, which form more fluid bilayers, are preferentially located in the cytosolic leaflet. Because phosphatidylserine and phosphatidylinositol carry a net negative charge, the stretch of amino acids on the cytoplasmic face of a single-pass membrane protein, in close proximity to the transmembrane segment, is often enriched in positively charged (Lys, Arg) residues, the "inside positive" rule. This segregation of lipids across the bilayer may influence membrane curvature (see Figure 10-11c). Unlike particular phospholipids, cholesterol is relatively evenly distributed in both leaflets of cellular membranes. The relative abundance of a particular phospholipid in the two leaflets of a plasma membrane can be determined experimentally on the basis of the susceptibility of phospholipids to hydrolysis

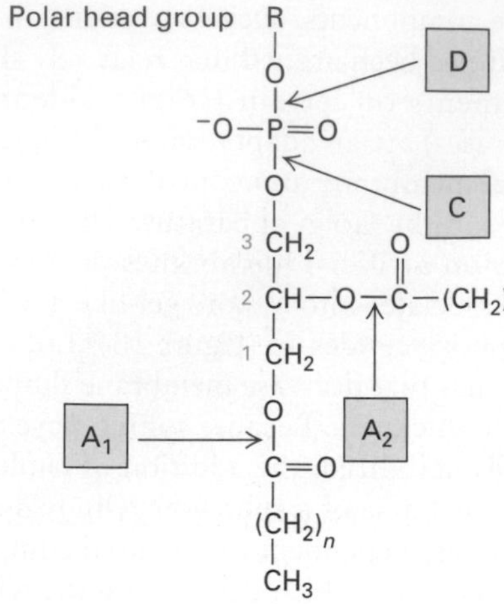

Polar head group

**FIGURE 10-12 Specificity of phospholipases.** Each type of phospholipase cleaves one of the susceptible bonds shown in red. The glycerol carbon atoms are indicated by small numbers. In intact cells, only phospholipids in the exoplasmic leaflet of the plasma membrane are cleaved by phospholipases in the surrounding medium. Phospholipase C, a cytosolic enzyme, cleaves certain phospholipids in the cytosolic leaflet of the plasma membrane.

by **phospholipases,** enzymes that cleave the ester bonds via which acyl chains and head groups are connected to the lipid molecule (Figure 10-12). When added to the external medium, phospholipases cannot cross the membrane, and thus they cleave off the head groups of only those lipids present in the exoplasmic face; phospholipids in the cytosolic leaflet are resistant to hydrolysis because the enzymes cannot penetrate to the cytosolic face of the plasma membrane.

How the asymmetric distribution of phospholipids in membrane leaflets arises is still unclear. As noted, in pure bilayers phospholipids do not spontaneously migrate, or flip-flop, from one leaflet to the other. In part, the asymmetry in phospholipid distribution may reflect where these lipids are synthesized in the endoplasmic reticulum and Golgi. Sphingomyelin is synthesized on the luminal (exoplasmic) face of the Golgi, which becomes the exoplasmic face of the plasma membrane. In contrast, phosphoglycerides are synthesized on the cytosolic face of the ER membrane, which is topologically equivalent to the cytosolic face of the plasma membrane (see Figure 10-5). Clearly, however, this explanation does not account for the preferential location of phosphatidylcholine (a phosphoglyceride) in the exoplasmic leaflet. Movement of this phosphoglyceride, and perhaps others, from one leaflet to the other in some natural membranes is most likely catalyzed by ATP-powered transport proteins called **flippases,** which are discussed in Chapter 11.

The preferential location of lipids on one face of the bilayer is necessary for a variety of membrane-based functions. For example, the head groups of all phosphorylated forms of phosphatidylinositol (see Figure 10-8; PI), an important source of second messengers, face the cytosol. Stimulation of many cell-surface receptors by their corresponding ligand results in activation of the cytosolic enzyme phospholipase C, which can then hydrolyze the bond connecting the phosphoinositols to the diacylglycerol. As we will see in Chapter 15, both water-soluble phosphoinositols and membrane-embedded diacylglycerol participate in intracellular signaling pathways that affect many aspects of cellular metabolism. Phosphatidylserine also is normally most abundant in the cytosolic leaflet of the plasma membrane. In the initial stages of platelet stimulation by serum, phosphatidylserine is briefly translocated to the exoplasmic face, presumably by a flippase enzyme, where it activates enzymes participating in blood clotting. When cells die, lipid asymmetry is no longer maintained, and phosphatidylserine, normally enriched in the cytosolic leaflet, is increasingly found in the exoplasmic one. This increased exposure is detected by use of a labeled version of Annexin V, a protein that specifically binds to phosphatidylserine, to measure the onset of programmed cell death (apoptosis).

## Cholesterol and Sphingolipids Cluster with Specific Proteins in Membrane Microdomains

Membrane lipids are not randomly distributed (evenly mixed) in each leaflet of a bilayer. One hint that lipids may be organized within the leaflets was the discovery that the lipids remaining after the extraction (solubilization) of plasma membranes with nonionic detergents such as Triton-X100 predominantly contain two species: cholesterol and sphingomyelin. Because these two lipids are found in more ordered, less fluid bilayers, researchers hypothesized that they form microdomains, termed **lipid rafts,** surrounded by other, more fluid phospholipids that are more readily extracted by nonionic detergents. (We discuss more fully the role of ionic and nonionic detergents in extracting membrane proteins in Section 10.2.)

Some biochemical and microscopic evidence supports the existence of lipid rafts, which in natural membranes are typically 50 nm in diameter. Rafts can be disrupted by methyl-β-cyclodextrin, which specifically extracts cholesterol out of membranes, or by antibiotics, such as filipin, that sequester cholesterol into aggregates within the membrane. Such findings indicate the importance of cholesterol in maintaining the integrity of these rafts. These raft fractions, defined by their insolubility in nonionic detergents, contain a subset of plasma membrane proteins, many of which are implicated in sensing extracellular signals and transmitting them into the cytosol. Because raft fractions are enriched in glycolipids, an important tool for microscopic visualization of raft-type structures in intact cells is the use of fluorescently labeled cholera toxin, a protein that specifically binds to certain gangliosides. By bringing many key proteins into close proximity and stabilizing their interactions, lipid rafts may facilitate signaling by cell-surface receptors and the subsequent activation of cytosolic events. However, much remains to be learned about the structure and biological function of lipid rafts.

## Cells Store Excess Lipids in Lipid Droplets

**Lipid droplets** are vesicular structures, composed of triglycerides and cholesterol esters, that originate from the ER and serve a lipid-storage function. When a cell's supply of lipids exceeds the immediate need for membrane construction, excess lipids are relegated to these lipid droplets, readily visualized in living cells by staining with a lipophilic dye such as Congo red. Feeding cells with oleic acid, a type of fatty acid, enhances lipid droplet formation. Lipid droplets are not only storage compartments for triglycerides and cholesterol esters, but may also serve as platforms for storage of proteins targeted for degradation. The biogenesis of lipid droplets starts with delamination of the lipid bilayer of the ER, through insertion of triglycerides and cholesterol esters (Figure 10-13). The lipid "lens" continues to grow by insertion of more lipid, until finally a lipid droplet is hatched by scission from the ER. The resulting cytoplasmic droplet is thereby enwrapped by a phospholipid monolayer. The details of lipid droplet biogenesis as well as their functions remain to be defined more clearly.

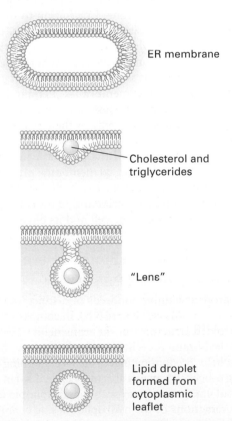

ER membrane

Cholesterol and triglycerides

"Lens"

Lipid droplet formed from cytoplasmic leaflet

**FIGURE 10-13 Lipid droplets form by budding and scission from the ER membrane.** Lipid droplet formation begins with the accumulation of cholesterol esters and triglycerides (yellow) within the hydrophobic core of the lipid bilayer. The resulting delamination of the two lipid monolayers causes a "lens" to form, the further growth of which creates a spherical droplet that is then released by scission at the neck. The newly formed droplet is surrounded by a lipid monolayer, derived from the cytosolic leaflet of the ER membrane.

## KEY CONCEPTS of Section 10.1

### The Lipid Bilayer: Composition and Structural Organization

• Membranes are crucial to cell structure and function. The eukaryotic cell is demarcated from the external environment by the plasma membrane and organized into membrane-limited internal compartments (organelles and vesicles).

• The phospholipid bilayer, the basic structural unit of all biomembranes, is a two-dimensional lipid sheet with hydrophilic faces and a hydrophobic core, which is impermeable to water-soluble molecules and ions; proteins embedded in the bilayer endow the membrane with specific functions (see Figure 10-1).

• The primary lipid components of biomembranes are phosphoglycerides, sphingolipids, and sterols such as cholesterol (see Figure 10-8). The term "phospholipid" applies to any amphipathic lipid molecule with a fatty acyl hydrocarbon tail and a phosphate-based polar head group.

• Phospholipids spontaneously form bilayers and sealed compartments surrounding an aqueous space (see Figure 10-3).

• As bilayers, all membranes have an internal (cytosolic) face and an external (exoplasmic) face (see Figure 10-5). Some organelles are surrounded by two, rather than one, membrane bilayer.

• Most lipids and many proteins are laterally mobile in biomembranes (see Figure 10-10). Membranes can undergo phase transitions from fluid- to gel-like states depending on the temperature and composition of the membrane (see Figure 10-9).

• Different cellular membranes vary in lipid composition (see Table 10-1). Phospholipids and sphingolipids are asymmetrically distributed in the two leaflets of the bilayer, whereas cholesterol is fairly evenly distributed in both leaflets.

• Natural biomembranes generally have a viscous consistency with fluidlike properties. In general, membrane fluidity is decreased by sphingolipids and cholesterol and increased by phosphoglycerides. The lipid composition of a membrane also influences its thickness and curvature (see Figure 10-11).

• Lipid rafts are microdomains containing cholesterol, sphingolipids, and certain membrane proteins that form in the plane of the bilayer. These lipid–protein aggregates might facilitate signaling by certain plasma membrane receptors.

• Lipid droplets are storage vesicles for lipids, originating in the ER (see Figure 10-13).

## 10.2 Membrane Proteins: Structure and Basic Functions

Membrane proteins are defined by their location within or at the surface of a phospholipid bilayer. Although every biological membrane has the same basic bilayer structure, the proteins associated with a particular membrane are responsible for its distinctive activities. The kinds and amounts of proteins associated with biomembranes vary depending on cell type and subcellular location. For example, the inner mitochondrial membrane is 76 percent protein; the myelin membrane that surrounds nerve axons, only 18 percent. The high phospholipid content of myelin allows it to electrically insulate the nerve from its environment, as we discuss in Chapter 22. The importance of membrane proteins is evident from the finding that approximately a third of all yeast genes encode a membrane protein. The relative abundance of genes for membrane proteins is greater in multicellular organisms, in which membrane proteins have additional functions in cell adhesion.

The lipid bilayer presents a distinctive two-dimensional hydrophobic environment for membrane proteins. Some proteins contain segments that are embedded within the hydrophobic core of the phospholipid bilayer; other proteins are associated with the exoplasmic or cytosolic leaflet of the bilayer. Protein domains on the extracellular surface of the plasma membrane generally bind to extracellular molecules, including external signaling proteins, ions, and small metabolites (e.g., glucose, fatty acids), as well as proteins on other cells or in the external environment. Segments of proteins within the plasma membrane perform multiple functions, such as forming the channels and pores through which molecules and ions move into and out of cells. Intramembrane segments also serve to organize multiple membrane proteins into larger assemblies within the plane of the membrane. Domains lying along the cytosolic face of the plasma membrane have a wide range of functions, from anchoring cytoskeletal proteins to the membrane to triggering intracellular signaling pathways.

In many cases, the function of a membrane protein and the topology of its polypeptide chain in the membrane can be predicted on the basis of its similarity with other well-characterized proteins. In this section, we examine the characteristic structural features of membrane proteins and some of their basic functions. We will describe the structures of several proteins to help you get a feel for the way membrane proteins interact with membranes. More complete characterization of the properties of various types of membrane proteins is presented in later chapters that focus on their structures and activities in the context of their cellular functions.

## Proteins Interact with Membranes in Three Different Ways

Membrane proteins can be classified into three categories—integral, lipid-anchored, and peripheral—on the basis of their position with respect to the membrane (see Figure 10-1). **Integral membrane proteins,** also called *transmembrane proteins,* span a phospholipid bilayer and comprise three segments. The cytosolic and exoplasmic domains have hydrophilic exterior surfaces that interact with the aqueous environment on the cytosolic and exoplasmic faces of the membrane. These domains resemble segments of other water-soluble proteins in their amino acid composition and structure. In contrast, the membrane-spanning segments usually contain many hydrophobic amino acids whose side chains protrude outward and interact with the hydrophobic hydrocarbon core of the phospholipid bilayer. In all transmembrane proteins examined to date, the membrane-spanning domains consist of one or more α helices or of multiple β strands. We discussed the ribosomal synthesis and post-translational processing of soluble cytosolic proteins in Chapters 4 and 8; the process by which integral membrane proteins are inserted into membranes as part of their synthesis is discussed in Chapter 13.

**Lipid-anchored membrane proteins** are bound covalently to one or more lipid molecules. The hydrophobic segment of the attached lipid is embedded in one leaflet of the membrane and anchors the protein to the membrane. The polypeptide chain itself does not enter the phospholipid bilayer.

**Peripheral membrane proteins** do not directly contact the hydrophobic core of the phospholipid bilayer. Instead they are bound to the membrane either indirectly by interactions with integral or lipid-anchored membrane proteins or directly by interactions with lipid head groups. Peripheral proteins can be bound to either the cytosolic or the exoplasmic face of the plasma membrane. In addition to these proteins, which are closely associated with the bilayer, cytoskeletal filaments can be more loosely associated with the cytosolic face, usually through one or more peripheral (adapter) proteins. Such associations with the cytoskeleton provide support for various cellular membranes, helping to determine cell shape and mechanical properties, and play a role in the two-way communication between the cell interior and the exterior, as we learn in Chapter 17. Finally, peripheral proteins on the outer surface of the plasma membrane and the exoplasmic domains of integral membrane proteins are often attached to components of the extracellular matrix or to the cell wall surrounding bacterial and plant cells, providing a crucial interface between the cell and its environment.

## Most Transmembrane Proteins Have Membrane-Spanning α Helices

Soluble proteins exhibit hundreds of distinct localized folded structures, or motifs (see Figure 3-9). In comparison, the repertoire of folded structures in the transmembrane domains of integral membrane proteins is quite limited, with the hydrophobic α helix predominating. Proteins containing membrane-spanning α-helical domains are stably embedded in membranes because of energetically favorable hydrophobic and van der Waals interactions of the hydrophobic side chains in the domain with specific lipids and probably also by ionic interactions with the polar head groups of the phospholipids.

A single α-helical domain is sufficient to incorporate an integral membrane protein into a membrane. However, many proteins have more than one transmembrane α helix. Typically, a membrane-embedded α helix is composed of a continuous segment of 20–25 hydrophobic (uncharged)

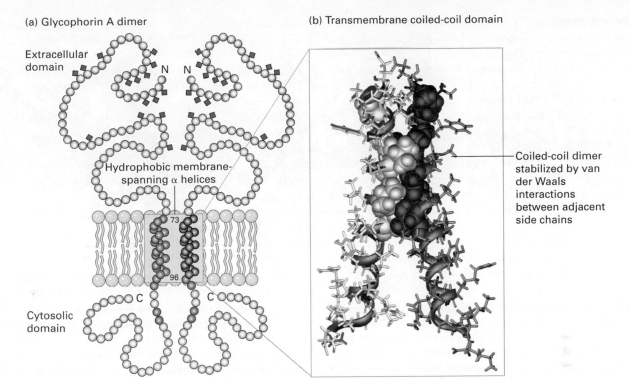

**(a) Glycophorin A dimer**

Extracellular domain

N   N

Hydrophobic membrane-spanning α helices

73

96

C   C

Cytosolic domain

**(b) Transmembrane coiled-coil domain**

Coiled-coil dimer stabilized by van der Waals interactions between adjacent side chains

**FIGURE 10-14 Structure of glycophorin A, a typical single-pass transmembrane protein.** (a) Diagram of dimeric glycophorin showing major sequence features and its relation to the membrane. The single 23-residue membrane-spanning α helix in each monomer is composed of amino acids with hydrophobic (uncharged) side chains (red and green spheres). By binding negatively charged phospholipid head groups, the positively charged arginine and lysine residues (blue spheres) near the cytosolic side of the helix help anchor glycophorin in the membrane. Both the extracellular and the cytosolic domains are rich in charged residues and polar uncharged residues; the extracellular domain is heavily glycosylated, with the carbohydrate side chains (green diamonds) attached to specific serine, threonine, and asparagine residues. (b) Molecular model of the transmembrane domain of dimeric glycophorin corresponding to residues 73–96. The hydrophobic side chains of the α helix in one monomer are shown in pink; those in the other monomer, in green. Residues depicted as space-filling structures participate in van der Waals interactions that stabilize the coiled-coil dimer. Note how the hydrophobic side chains project outward from the helix, toward what would be the surrounding fatty acyl chains. [Part (b) adapted from K. R. MacKenzie et al., 1997, *Science* **276**:131.]

amino acids (see Figure 2-14). The predicted length of such an α helix (3.75 nm) is just sufficient to span the hydrocarbon core of a phospholipid bilayer. In many membrane proteins, these helices are perpendicular to the plane of the membrane, whereas in others, the helices traverse the membrane at an oblique angle. The hydrophobic side chains protrude outward from the helix and form van der Waals interactions with the fatty acyl chains in the bilayer. In contrast, the hydrophilic amide peptide bonds are in the interior of the α helix (see Figure 3-4); each carbonyl (C=O) group forms a hydrogen bond with the amide hydrogen atom of the amino acid four residues toward the C-terminus of the helix. These polar groups are shielded from the hydrophobic interior of the membrane.

To help you get a better sense of the structures of proteins with α-helical domains, we will briefly discuss four different kinds of such proteins: glycophorin A, G protein–coupled receptors, aquaporins (water/glycerol channels), and T-cell receptor for antigen.

Glycophorin A, the major protein in the erythrocyte plasma membrane, is a representative *single-pass* transmembrane protein, which contains only one membrane-spanning α helix (Figure 10-14). The 23-residue membrane-spanning α helix is composed of amino acids with hydrophobic (uncharged) side chains, which interact with fatty acyl chains in the surrounding bilayer. In cells, glycophorin A typically forms dimers: the transmembrane helix of one glycophorin A polypeptide associates with the corresponding transmembrane helix in a second glycophorin A to form a coiled-coil structure (Figure 10-14b). Such interactions of membrane-spanning α helices are a common mechanism for creating dimeric membrane proteins, and many membrane proteins form oligomers (two or more polypeptides bound together noncovalently) by interactions between their membrane-spanning helices.

A large and important group of integral proteins is defined by the presence of seven membrane-spanning α helices; this includes the large family of G protein–coupled cell-surface receptors discussed in Chapter 15, several of which have been crystallized. One such *multipass* transmembrane protein of known structure is bacteriorhodopsin, a protein found in the membrane of certain photosynthetic bacteria; it

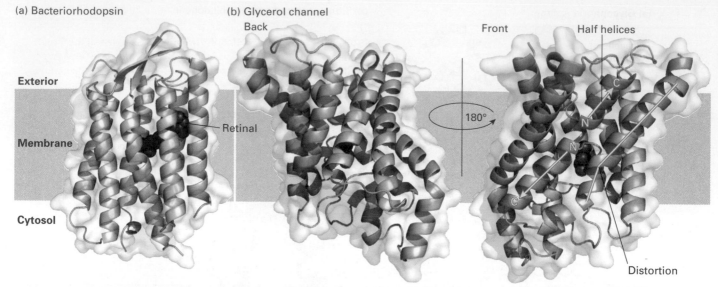

(a) Bacteriorhodopsin

(b) Glycerol channel

Exterior

Membrane

Cytosol

Back

Retinal

Front

Half helices

180°

C

N

N

C

Distortion

**FIGURE 10-15 Structural models of two multipass membrane proteins.** (a) Bacteriorhodopsin, a photoreceptor in certain bacteria. The seven hydrophobic α helices in bacteriorhodopsin traverse the lipid bilayer roughly perpendicular to the plane of the membrane. A retinal molecule (black) covalently attached to one helix absorbs light. The large class of G protein–coupled receptors in eukaryotic cells also has seven membrane-spanning α helices; their three-dimensional structure is thought to be similar to that of bacteriorhodopsin. (b) Two views of the glycerol channel Glpf, rotated 180° with respect to each other along an axis perpendicular to the plane of the membrane. Note several membrane-spanning α helices that are at oblique angles, the two helices that penetrate only halfway through the membrane (purple with yellow arrows), and one long membrane-spanning helix with a "break" or distortion in the middle (purple with yellow line). The glycerol molecule in the hydrophilic "core" is colored red. The structure was approximately positioned in the hydrocarbon core of the membrane by finding the most hydrophobic 3-μm slab of the protein perpendicular to the membrane plane. [Part (a) after H. Luecke et al., 1999, *J. Mol. Biol.* **291**:899. Part (b) after J. Bowie, 2005, *Nature* **438**:581–589, and D. Fu et al., 2000, *Science* **290**:481–486.]

illustrates the general structure of all these proteins (Figure 10-15a). Absorption of light by the retinal group covalently attached to this protein causes a conformational change in the protein that results in the pumping of protons from the cytosol across the bacterial membrane to the extracellular space. The proton concentration gradient thus generated across the membrane is used to synthesize ATP during photosynthesis (Chapter 12). In the high-resolution structure of bacteriorhodopsin the positions of all the individual amino acids, retinal, and the surrounding lipids are clearly defined. As might be expected, virtually all of the amino acids on the exterior of the membrane-spanning segments of bacteriorhodopsin are hydrophobic, permitting energetically favorable interactions with the hydrocarbon core of the surrounding lipid bilayer.

The **aquaporins** are a large family of highly conserved proteins that transport water, glycerol, and other hydrophilic molecules across biomembranes. They illustrate several aspects of the structure of multipass transmembrane proteins. Aquaporins are tetramers of four identical subunits. Each of the four subunits has six membrane-spanning α helixes, some of which traverse the membrane at oblique angles rather than perpendicularly. Because the aquaporins have similar structures, we will focus on one, the glycerol channel Glpf, that has an especially well-defined structure determined by x-ray diffraction studies (Figure 10-15b). This aquaporin has one long transmembrane helix with a bend in the middle, and more strikingly, there are two α helices that penetrate only *halfway* through the membrane. The N-termini of these helices face each other (yellow N's in the figure), and together they span the membrane at an oblique angle. Thus some membrane-embedded helices—and other, nonhelical, structures we will encounter later—do not traverse the entire bilayer. As we will see in Chapter 11, these short helices in aquaporins form part of the glycerol/water-selective pore in the middle of each subunit. This highlights the considerable diversity in the ways membrane-spanning α helices interact with the lipid bilayer and with other segments of the protein.

The specificity of phospholipid-protein interactions is evident from the structure of a different aquaporin, aquaporin 0 (Figure 10-16). Aquaporin 0 is the most abundant protein in the plasma membrane of the fiber cells that make up the bulk of the lens of the mammalian eye. Like other aquaporins, it is a tetramer of identical subunits. The protein's surface is not covered by a set of uniform binding sites for phospholipid molecules. Instead, fatty acyl side chains pack tightly against the irregular hydrophobic outer surface of the protein; these lipids are referred to as annular phospholipids, because they from a tight ring (annulus) of lipids that exchange less easily with bulk phospholipids in the bilayer. Some of the fatty acyl chains are straight, in the *all-trans* conformation (Chapter 2),

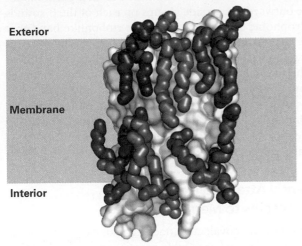

**FIGURE 10-16 Annular phospholipids.** Side view of the three-dimensional structure of one subunit of the lens-specific aquaporin 0 homotetramer, crystallized in the presence of the phospholipid dimyristoylphosphatidylcholine, a phospholipid with 14 carbon-saturated fatty acyl chains. Note the lipid molecules forming a bilayer shell around the protein. The protein is shown as a surface plot (the lighter background molecule). The lipid molecules are shown in space-fill format; the polar lipid head groups (grey and red) and the lipid fatty acyl chains (black and grey) form a bilayer with almost uniform thickness around the protein. Presumably, in the membrane, lipid fatty acyl chains will cover the whole of the hydrophobic surface of the protein; only the most ordered of the lipid molecules will be resolved in the crystallographic structure. [After A. Lee, 2005, *Nature* **438**:569–570, and T. Gonen et al., 2005, *Nature* **438**:633–688.]

whereas others are kinked in order to interact with bulky hydrophilic side chains on the surface of the protein. Some of the lipid head groups are parallel to the surface of the membrane, as is the case in purified phospholipid bilayers. Others, however, are oriented almost at right angles to the plane of the membrane. Thus there can be specific interactions between phospholipids and membrane-spanning proteins, and the function of many membrane proteins can be affected by the specific types of phospholipid present in the bilayer.

In addition to the predominantly hydrophobic (uncharged) residues that serve to embed integral membrane proteins in the bilayer, many such α-helical transmembrane segments do contain polar and/or charged residues. Their amino acid side chains can be used to guide the assembly and stabilization of multimeric membrane proteins. The T-cell receptor for antigen is a case in point: it is composed of four separate dimers, the interactions of which are driven by charge-charge interactions between α helices at the appropriate "depth" in the hydrocarbon core of the lipid bilayer (Figure 10-17). The electrostatic attraction of positive and negative charges on each dimer helps the dimers to "find each other." Thus charged residues in otherwise hydrophobic transmembrane segments can help guide assembly of multimeric membrane proteins.

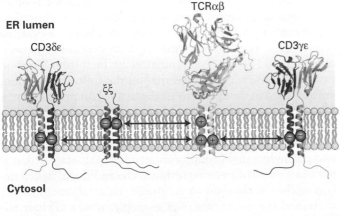

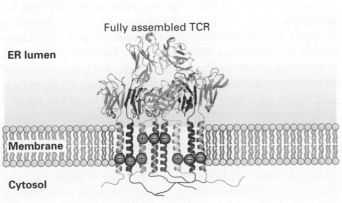

**FIGURE 10-17 Charged residues can orchestrate assembly of multimeric membrane proteins.** The T-cell receptor (TCR) for antigen is composed of four separate dimers: an αβ pair directly responsible for antigen recognition, and accessory subunits collectively referred to as the CD3 complex. These accessories include the γ, δ, ε, and ζ subunits. The ζ subunits form a disulfide-linked homodimer. The γ and δ subunits occur in complex with an ε subunit, to generate a γε and a δε pair. The transmembrane segments of the TCR α and β chains each contain positively charged residues (blue). These allow recruitment of corresponding δε and γε heterodimers, which carry negative charges (red) at the appropriate depth in the hydrophobic core of the bilayer. The ζ homodimer docks onto the charges in the TCR α chain (dark green), while the γε and δε subunit pairs find their corresponding partners deeper down in the hydrophobic core on both the TCR α and TCR β chain (light green). Charged residues in otherwise nonpolar transmembrane segments can thus guide assembly of higher order structures. [After K.W. Wucherpfennig, et al., 2010, *Cold Spring Harb Perspect Biol*, **2**.]

## Multiple β Strands in Porins Form Membrane-Spanning "Barrels"

The **porins** are a class of transmembrane proteins whose structure differs radically from that of other integral proteins based on α-helical transmembrane domains. Several types of porins are found in the outer membrane of gram-negative bacteria such as *E. coli* and in the outer membranes of mitochondria and chloroplasts. The outer membrane protects an intestinal bacterium from harmful agents (e.g., antibiotics, bile salts, and proteases) but permits the uptake and disposal of small hydrophilic molecules, including nutrients and waste products. Different types of porins in the outer membrane of an *E. coli* cell provide channels for the passage of specific types of disaccharides or other small molecules as well as of ions such as phosphate. The amino acid sequences of porins contain none of the long, continuous hydrophobic segments typical of integral proteins with α-helical membrane-spanning domains. Rather, it is the entire outer surface of the fully folded porin that displays its hydrophobic character to the hydrocarbon core of the lipid bilayer. X-ray crystallography shows that porins are trimers of identical subunits. In each subunit, 16 β strands form a sheet that twists into a barrel-shaped structure with a pore in the center (Figure 10-18). Unlike a typical water-soluble globular

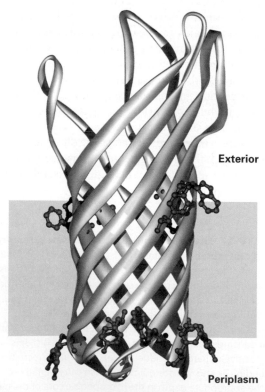

**FIGURE 10-18 Structural model of one subunit of OmpX, a porin found in the outer membrane of *E. coli*.** All porins are trimeric transmembrane proteins. Each subunit is barrel shaped, with β strands forming the wall and a transmembrane pore in the center. A band of aliphatic (hydrophobic and noncyclic) side chains (yellow) and a border of aromatic (ring-containing) side chains (red) position the protein in the bilayer. [After G. E. Schulz, 2000, *Curr. Opin. Struc. Biol.* **10**:443.]

protein, a porin has a hydrophilic interior and a hydrophobic exterior; in this sense, porins are inside out. In a porin monomer, the outward-facing side groups on each of the β strands are hydrophobic and form a nonpolar ribbonlike band that encircles the outside of the barrel. This hydrophobic band interacts with the fatty acyl groups of the membrane lipids or with other porin monomers. The side groups facing the inside of a porin monomer are predominantly hydrophilic; they line the pore through which small water-soluble molecules cross the membrane. (Note that the aquaporins discussed above, despite their name, are not porins and contain multiple transmembrane α helices.)

## Covalently Attached Lipids Anchor Some Proteins to Membranes

In eukaryotic cells, covalently attached lipids can anchor some otherwise typically water-soluble proteins to one or the other leaflet of the membrane. In such lipid-anchored proteins, the lipid hydrocarbon chains are embedded in the bilayer, but the protein itself does not enter the bilayer. The lipid anchors used to anchor proteins to the cytosolic face are not used for the exoplasmic face and vice versa.

One group of cytosolic proteins are anchored to the cytosolic face of a membrane by a fatty acyl group (e.g., myristate or palmitate) covalently attached to an N-terminal glycine residue, a process called *acylation* (Figure 10-19a). Retention of such proteins at the membrane by the N-terminal acyl anchor may play an important role in a membrane-associated function. For example, v-Src, a mutant form of a cellular tyrosine kinase, induces abnormal cellular growth that can lead to cancer but does so only when it has a myristylated N-terminus.

A second group of cytosolic proteins are anchored to membranes by a hydrocarbon chain attached to a cysteine residue at or near the C-terminus, a process called *prenylation* (Figure 10-19b). Prenyl anchors are built from 5-carbon isoprene units, which, as detailed in the following section, are also used in the synthesis of cholesterol. In prenylation, a 15-carbon farnesyl or 20-carbon geranylgeranyl group is bound through a thioether bond to the —SH group of a C-terminal cysteine residue of the protein. In some cases, a second geranylgeranyl group or a fatty acyl palmitate group is linked to a nearby cysteine residue. The additional hydrocarbon anchor is thought to reinforce the attachment of the protein to the membrane. For example, Ras, a GTPase superfamily protein that functions in intracellular signaling (Chapter 15), is recruited to the cytosolic face of the plasma membrane by such a double anchor. Rab proteins, which also belong to the GTPase superfamily, are similarly bound to the cytosolic surface of intracellular vesicles by prenyl anchors; these proteins are required for the fusion of vesicles with their target membranes (Chapter 14).

Some cell-surface proteins and specialized proteins with distinctive covalently attached polysaccharides called proteoglycans (Chapter 20) are bound to the exoplasmic face of the plasma membrane by a third type of anchor group, glycosylphosphatidylinositol (GPI). The exact structures of

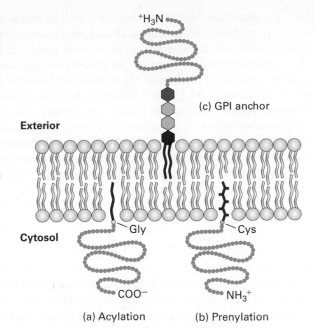

**FIGURE 10-19 Anchoring of plasma-membrane proteins to the bilayer by covalently linked hydrocarbon groups.** (a) Cytosolic proteins such as v-Src are associated with the plasma membrane through a single fatty acyl chain attached to the N-terminal glycine (Gly) residue of the polypeptide. Myristate (C14) and palmitate (C16) are common acyl anchors. (b) Other cytosolic proteins (e.g., Ras and Rab proteins) are anchored to the membrane by prenylation of one or two cysteine (Cys) residues, at or near the C-terminus. The anchors are farnesyl (C15) and geranylgeranyl (C20) groups, both of which are unsaturated. (c) The lipid anchor on the exoplasmic surface of the plasma membrane is glycosylphosphatidylinositol (GPI). The phosphatidylinositol part (red) of this anchor contains two fatty acyl chains that extend into the bilayer. The phosphoethanolamine unit (purple) in the anchor links it to the protein. The two green hexagons represent sugar units, which vary in number, nature, and arrangement in different GPI anchors. The complete structure of a yeast GPI anchor is shown in Figure 13-15. [Adapted from H. Sprong et al., 2001, *Nature Rev. Mol. Cell Biol.* **2**:504.]

*GPI anchors* vary greatly in different cell types, but they always contain phosphatidylinositol (PI), whose two fatty acyl chains extend into the lipid bilayer just like those of typical membrane phospholipids; phosphoethanolamine, which covalently links the anchor to the C-terminus of a protein; and several sugar residues (Figure 10-19c). Therefore GPI anchors are glycolipids. The GPI anchor is both necessary and sufficient for binding proteins to the membrane. For instance, treatment of cells with phospholipase C, which cleaves the phosphate-glycerol bond in phospholipids and in GPI anchors (see Figure 10-12), releases GPI-anchored proteins such as Thy-1 and placental alkaline phosphatase (PLAP) from the cell surface.

## All Transmembrane Proteins and Glycolipids Are Asymmetrically Oriented in the Bilayer

Every type of transmembrane protein has a specific orientation, known as its *topology*, with respect to the membrane faces. Its cytosolic segments are always facing the cytoplasm,

and exoplasmic segments always face the opposite side of the membrane. This asymmetry in protein orientation confers different properties on the two membrane faces. The orientation of different types of transmembrane proteins is established during their synthesis, as we describe in Chapter 13. Membrane proteins have never been observed to flip-flop across a membrane; such movement, requiring a transient movement of hydrophilic amino acid residues through the hydrophobic interior of the membrane, would be energetically unfavorable. Accordingly, the asymmetric topology of a transmembrane protein, which is established during its biosynthetic insertion into a membrane, is maintained throughout the protein's lifetime. As Figure 10-6 shows, membrane proteins retain their asymmetric orientation in the membrane during membrane budding and fusion events; the same segment always faces the cytosol and the same segment is always exposed to the exoplasmic face. In proteins with multiple transmembrane segments (multipass or polytopic membrane proteins), orientation of individual transmembrane segments can be affected by changes in phospholipid composition.

Many transmembrane proteins contain carbohydrate chains covalently linked to serine, threonine, or asparagine side chains of the polypeptide. Such transmembrane **glycoproteins** are always oriented so that all carbohydrate chains are in the exoplasmic domain (see Figure 10-14 for the example of glycophorin A). Likewise, glycolipids, in which a carbohydrate chain is attached to the glycerol or sphingosine backbone of a membrane lipid, are always located in the exoplasmic leaflet with the carbohydrate chain protruding from the membrane surface. The biosynthetic basis for the asymmetric glycosylation of proteins is described in Chapter 14. Both glycoproteins and glycolipids are especially abundant in the plasma membranes of eukaryotic cells and in the membranes of the intracellular compartments that establish the secretory and endocytic pathways; they are absent from the inner mitochondrial membrane, chloroplast lamellae, and several other intracellular membranes. Because the carbohydrate chains of glycoproteins and glycolipids in the plasma membrane extend into the extracellular space, they are available to interact with components of the extracellular matrix as well as **lectins** (proteins that bind specific sugars), growth factors, and antibodies.

One important consequence of such interactions is illustrated by the A, B, and O blood group antigens. These three structurally related oligosaccharide components of certain glycoproteins and glycolipids are expressed on the surfaces of human red blood cells and many other cell types (Figure 10-20). All humans have the enzymes for synthesizing O antigen. Persons with type A blood also have a glycosyltransferase enzyme that adds an extra modified monosaccharide called N-acetylgalactosamine to O antigen to form A antigen. Those with type B blood have a different transferase that adds an extra galactose to O antigen to form B antigen. People with both transferases produce both A and B antigen (AB blood type); those who lack these transferases produce O antigen only (O blood type).

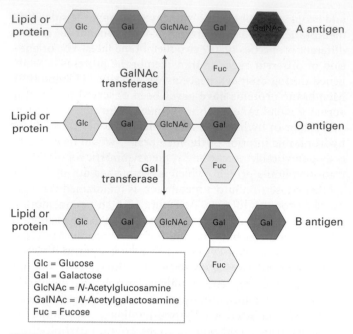

**FIGURE 10-20 Human ABO blood group antigens.** These antigens are oligosaccharide chains covalently attached to glycolipids or glycoproteins in the plasma membrane. The terminal oligosaccharide sugars distinguish the three antigens. The presence or absence of the glycosyltransferases that add galactose (Gal) or *N*-acetylgalactosamine (GalNAc) to O antigen determine a person's blood type.

People whose erythrocytes lack the A antigen, the B antigen, or both on their surface normally have antibodies against the missing antigen(s) in their serum. Thus if a type A or O person receives a transfusion of type B blood, antibodies against the B antigen will bind to the introduced red cells and trigger their destruction. To prevent such harmful reactions, blood group typing and appropriate matching of blood donors and recipients are required in all transfusions (Table 10-2). ■

## Lipid-Binding Motifs Help Target Peripheral Proteins to the Membrane

Many water-soluble enzymes use phospholipids as their substrates and thus must bind to membrane surfaces. Phospho-lipases, for example, hydrolyze various bonds in the head groups of phospholipids (see Figure 10-12), and thereby play a variety of roles in cells—helping to degrade damaged or aged cell membranes, generating precursors to signaling molecules, and even serving as the active components in many snake venoms. Many such enzymes, including phospholipases, initially bind to the polar head groups of membrane phospholipids to carry out their catalytic functions. The mechanism of action of phospholipase $A_2$ illustrates how such water-soluble enzymes can reversibly interact with membranes and catalyze reactions at the interface of an aqueous solution and lipid surface. When this enzyme is in aqueous solution, its $Ca^{2+}$-containing active site is buried in a channel lined with hydrophobic amino acids. The enzyme binds with greatest affinity to bilayers composed of negatively charged phospholipids (e.g., phosphotidylserine). This finding suggests that the rim of positively charged lysine and arginine residues around the entrance to the catalytic channel is particularly important in binding (Figure 10-21a). Binding induces a conformational change in phospholipase $A_2$ that strengthens its binding to the phospholipid heads and opens the hydrophobic channel. As a phospholipid molecule moves from the bilayer into the channel, the enzyme-bound $Ca^{2+}$ binds to the phosphate in the head group, thereby positioning the ester bond to be cleaved in the catalytic site (Figure 10-21b) and so releasing the acyl chain.

## Proteins Can Be Removed from Membranes by Detergents or High-Salt Solutions

Membrane proteins are often difficult to purify and study, mostly because of their tight association with membrane lipids and other membrane proteins. *Detergents* are amphipathic molecules that disrupt membranes by intercalating into phospholipid bilayers and can thus be used to solubilize lipids and many membrane proteins. The hydrophobic part of a detergent molecule is attracted to the phospholipid hydrocarbons and mingles with them readily; the hydrophilic part is strongly attracted to water. Some detergents such as the bile salts are natural products, but most are synthetic molecules developed for cleaning and for dispersing mixtures of oil and water in the food industry (e.g., creamy peanut butter)

| TABLE 10-2 | ABO Blood Groups | | |
|---|---|---|---|
| **Blood Group** | **Antigens on RBCs*** | **Serum Antibodies** | **Can Receive Blood Types** |
| A | A | Anti-B | A and O |
| B | B | Anti-A | B and O |
| AB | A and B | None | All |
| O | O | Anti-A and anti-B | O |

*See Figure 10-20 for antigen structures.

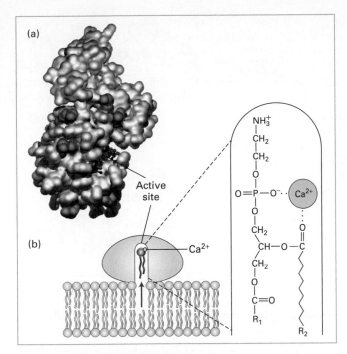

**FIGURE 10-21 Lipid-binding surface and mechanism of action of phospholipase A₂.** (a) A structural model of the enzyme showing the surface that interacts with a membrane. This lipid-binding surface contains a rim of positively charged arginine and lysine residues, shown in blue surrounding the cavity of the catalytic active site, in which a substrate lipid (red stick structure) is bound. (b) Diagram of catalysis by phospholipase A₂. When docked on a model lipid membrane, positively charged residues of the binding site bind to negatively charged polar groups at the membrane surface. This binding triggers a small conformational change, opening a channel lined with hydrophobic amino acids that leads from the bilayer to the catalytic site. As a phospholipid moves into the channel, an enzyme-bound Ca²⁺ ion (green) binds to the head group, positioning the ester bond to be cleaved (red) next to the catalytic site. [Part (a) adapted from M. H. Gelb et al., 1999, *Curr. Opin. Struc. Biol.* **9**:428. Part (b), see D. Blow, 1991, *Nature* **351**:444.]

(Figure 10-22). Ionic detergents, such as sodium deoxycholate and sodium dodecylsulfate (SDS), contain a charged group; nonionic detergents, such as Triton X-100 and octylglucoside, lack a charged group. At very low concentrations, detergents dissolve in pure water as isolated molecules. As the concentration increases, the molecules begin to form micelles—small, spherical aggregates in which the hydrophilic parts of the molecules face outward and the hydrophobic parts cluster in the center (see Figure 10-3c). The *critical micelle concentration* (*CMC*) at which micelles form is characteristic of each detergent and is a function of the structures of its hydrophobic and hydrophilic parts.

Ionic and nonionic detergents interact differently with proteins and have different uses in the lab. Ionic detergents bind to the exposed hydrophobic regions of membrane proteins as well as to the hydrophobic cores of water-soluble proteins. Because of their charge, these detergents can also disrupt ionic and hydrogen bonds. At high concentrations, for example, sodium dodecylsulfate completely denatures proteins by binding to every side chain, a property that is exploited in SDS gel electrophoresis (see Figure 3-36). Nonionic detergents generally do not denature proteins and are thus useful in extracting proteins in their folded and active form from membranes before the proteins are purified. Protein-protein interactions, especially the weaker ones, can be sensitive to both ionic and nonionic detergents.

At high concentrations (above the CMC), nonionic detergents solubilize biological membranes by forming mixed micelles of detergent, phospholipid, and integral membrane proteins, bulky hydrophobic structures that do not dissolve in aqueous solution (Figure 10-23, top). At low concentrations (below the CMC), these detergents bind to the hydrophobic regions of most integral membrane proteins, but without forming micelles, allowing them to remain soluble in aqueous solution (Figure 10-23, bottom). Creating such an aqueous solution of integral membrane proteins is a necessary first step in protein purification.

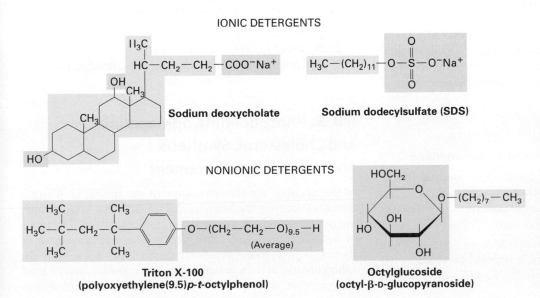

**FIGURE 10-22 Structures of four common detergents.** The hydrophobic part of each molecule is shown in yellow; the hydrophilic part, in blue. The bile salt sodium deoxycholate is a natural product; the others are synthetic. Although ionic detergents commonly cause denaturation of proteins, nonionic detergents do not and are thus useful in solubilizing integral membrane proteins.

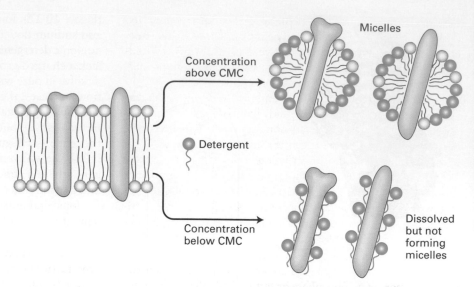

**FIGURE 10-23 Solubilization of integral membrane proteins by nonionic detergents.** At a concentration higher than its critical micelle concentration (CMC), a detergent solubilizes lipids and integral membrane proteins, forming mixed micelles containing detergent, protein, and lipid molecules. At concentrations below the CMC, nonionic detergents (e.g., octylglucoside, Triton X-100) can dissolve membrane proteins without forming micelles by coating the membrane-spanning regions.

Treatment of cultured cells with a buffered salt solution containing a nonionic detergent such as Triton X-100 extracts water-soluble proteins as well as integral membrane proteins. As noted earlier, the exoplasmic and cytosolic domains of integral membrane proteins are generally hydrophilic and soluble in water. The membrane-spanning domains, however, are rich in hydrophobic and uncharged residues (see Figure 10-14). When separated from membranes, these exposed hydrophobic segments tend to interact with one another, causing the protein molecules to aggregate and precipitate from aqueous solutions. The hydrophobic parts of nonionic detergent molecules preferentially bind to the hydrophobic segments of transmembrane proteins, preventing protein aggregation and allowing the proteins to remain in the aqueous solution. Detergent-solubilized transmembrane proteins can then be purified by affinity chromatography and other techniques used in purifying water-soluble proteins (see Chapter 3).

As discussed previously, most peripheral membrane proteins are bound to specific transmembrane proteins or membrane phospholipids by ionic or other weak noncovalent interactions. Generally, peripheral proteins can be removed from the membrane by solutions of high ionic strength (high salt concentrations), which disrupt ionic bonds, or by chemicals that bind divalent cations such as $Mg^{2+}$. Unlike integral proteins, most peripheral proteins are soluble in aqueous solution and need not be solubilized by nonionic detergents.

surrounding the cytosolic and exoplasmic faces of the membrane (see Figures 10-14, 10-15, and 10-17).

- Fatty acyl side chains as well as the polar head groups of membrane lipids pack tightly and irregularly around the hydrophobic segments of integral membrane proteins (see Figure 10-16).

- The porins, unlike other integral proteins, contain membrane-spanning β sheets that form a barrel-like channel through the bilayer (see Figure 10-18).

- Long-chain lipids attached to certain amino acids anchor some proteins to one or the other membrane leaflet (see Figure 10-19).

- All transmembrane proteins and glycolipids are asymmetrically oriented in the bilayer. Invariably, carbohydrate chains are present only on the exoplasmic surface of a glycoprotein or glycolipid.

- Many water-soluble enzymes (e.g., phospholipases) use phospholipids as their substrates and must bind to the membrane surface to carry out their function. Such binding is often due to the attraction between positive charges on basic residues in the protein and negative charges on phospholipid head groups in the bilayer.

- Transmembrane proteins are selectively extracted (solubilized) and purified with the use of nonionic detergents.

## KEY CONCEPTS of Section 10.2

### Membrane Proteins: Structure and Basic Functions

- Biological membranes usually contain both integral (transmembrane) proteins and peripheral membrane proteins, which do not enter the hydrophobic core of the bilayer (see Figure 10-1).

- Most integral membrane proteins contain one or more membrane-spanning hydrophobic α helices bracketed by hydrophilic domains that extend into the aqueous environment

# 10.3 Phospholipids, Sphingolipids, and Cholesterol: Synthesis and Intracellular Movement

In this section, we consider some of the special challenges that a cell faces in synthesizing and transporting lipids, which are poorly soluble in the aqueous interior of cells. The focus of our discussion will be the biosynthesis and movement of the major lipids found in cellular membranes—phospholipids, sphingolipids, and cholesterol—and their precursors. In lipid biosynthesis, water-soluble precursors are assembled into membrane-associated intermediates that are then converted into membrane lipid products. The movement of lipids, especially membrane components, between different organelles is critical for maintaining the proper composition and properties of membranes and overall cell structure.

A fundamental principle of membrane biosynthesis is that cells synthesize new membranes only by the expansion of existing membranes. (The one exception may be autophagy, where new membrane is formed first through the formation of an autophagic crescent, the construction of which involves modification of phosphatidylethanolamine with the ubiquitin-like modifier Atg8 [see Figure 14-35].) Although some early steps in the synthesis of membrane lipids take place in the cytoplasm, the final steps are catalyzed by enzymes bound to preexisting cellular membranes, and the products are incorporated into the membranes as they are generated. Evidence for this phenomenon is seen when cells are briefly exposed to radioactive precursors (e.g., phosphate or fatty acids): all the phospholipids and sphingolipids incorporating these precursor substances are associated with intracellular membranes; as expected from the hydrophobicity of the fatty acyl chains, none are found free in the cytosol. After they are formed, membrane lipids must be distributed appropriately both in leaflets of a given membrane and among the independent membranes of different organelles in eukaryotic cells, as well as in the plasma membrane. Here, we consider how this precise lipid distribution is accomplished; in Chapters 13 and 14 we discuss how membrane proteins are inserted into cell membranes and trafficked to their appropriate location within the cell.

## Fatty Acids Are Assembled from Two-Carbon Building Blocks by Several Important Enzymes

Fatty acids (Chapter 2) play a number of important roles in cells. In addition to being a cellular fuel source (see discussion of aerobic oxidation in Chapter 12), fatty acids are key components of both the phospholipids and sphingolipids making up cell membranes; they also anchor some proteins to cellular membranes (see Figure 10-19). Thus the regulation of fatty acid synthesis plays a key role in the regulation of membrane synthesis as a whole. The major fatty acids in phospholipids contain 14, 16, 18, or 20 carbon atoms and include both saturated and unsaturated chains. The fatty acyl chains found on sphingolipids can be longer than those in the phosphoglycerides, containing up to 26 carbon atoms, and may bear other chemical modifications (e.g., hydroxylation) as well.

Fatty acids are synthesized from the two-carbon building block acetate, $CH_3COO^-$. In cells, both acetate and the intermediates in fatty acid biosynthesis are esterified to the large water-soluble molecule coenzyme A (CoA), as exemplified by the structure of **acetyl CoA**:

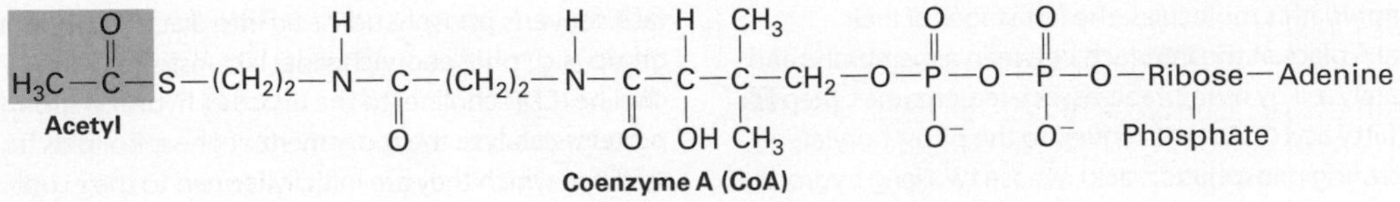

Acetyl CoA is an important intermediate in the metabolism of glucose, fatty acids, and many amino acids, as detailed in Chapter 12. It also contributes acetyl groups in many biosynthetic pathways. **Saturated** fatty acids (no carbon-carbon double bonds) containing 14 or 16 carbon atoms are made from acetyl CoA by two enzymes, *acetyl-CoA carboxylase* and *fatty acid synthase*. In animal cells, these enzymes are found in the cytosol; in plants, they are found in chloroplasts. Palmitoyl CoA (16-carbon fatty acyl group linked to CoA) can be elongated to 18–24 carbons by the sequential addition of two-carbon units in the endoplasmic reticulum (ER) or sometimes in the mitochondrion. Desaturase enzymes, also located in the ER, introduce double bonds at specific positions in some fatty acids, yielding **unsaturated** fatty acids. Oleyl CoA (oleate linked to CoA, see Table 2-4), for example, is formed by removal of two H atoms from stearyl CoA. In contrast to free fatty acids, fatty acyl CoA derivatives are soluble in aqueous solutions because of the hydrophilicity of the CoA segment.

## Small Cytosolic Proteins Facilitate Movement of Fatty Acids

In order to be transported through the cell cytoplasm, free, or unesterified, fatty acids (those unlinked to a CoA), commonly are bound by *fatty-acid-binding proteins* (FABPs), which belong to a group of small cytosolic proteins that facilitate the intracellular movement of many lipids. These proteins contain a hydrophobic pocket lined by β sheets (Figure 10-24). A long-chain fatty acid can fit into this pocket and interact noncovalently with the surrounding protein.

The expression of cellular FABPs is regulated coordinately with cellular requirements for the uptake and release of fatty acids. Thus FABP levels are high in active muscles that are using fatty acids for generation of ATP, and in adipocytes (fat-storing cells) when they are either taking up fatty acids to be stored as triglycerides or releasing fatty acids for use by other cells. The importance of FABPs in fatty acid metabolism is highlighted by the observations that they can compose as much as 5 percent of all cytosolic proteins in the liver and that genetic inactivation of cardiac muscle FABP converts the heart from a muscle that primarily burns fatty acids for energy into one that primarily burns glucose.

## Fatty Acids Are Incorporated into Phospholipids Primarily on the ER Membrane

Fatty acids are not directly incorporated into phospholipids; rather, in eukaryotic cells they are first converted into CoA esters. The subsequent synthesis of phospholipids such as the phosphoglycerides is carried out by enzymes associated with

Coenzyme A (CoA)

**FIGURE 10-24 Binding of a fatty acid to the hydrophobic pocket of a fatty-acid-binding protein (FABP).** The crystal structure of adipocyte FABP (ribbon diagram) reveals that the hydrophobic binding pocket is generated from two β sheets that are nearly at right angles to each other, forming a clam-shell-like structure. A fatty acid (carbons yellow; oxygens red) interacts noncovalently with hydrophobic amino acid residues within this pocket. [See A. Reese-Wagoner et al., 1999, *Biochim. Biophys. Acta* **23**:1441(2–3):106–116.]

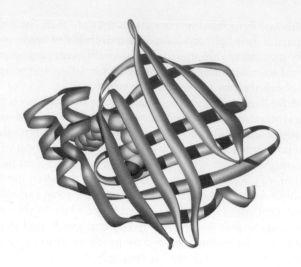

the cytosolic face of the ER membrane, usually the smooth ER, in animal cells; through a series of steps, fatty acyl CoAs, glycerol 3-phosphate, and polar head-group precursors are linked together and then inserted into the ER membrane (Figure 10-25). The fact that these enzymes are located on the cytosolic side of the membrane means that there is an inherent asymmetry in membrane biogenesis: new membranes are initially synthesized only on one leaflet—a fact with important consequences for the asymmetric distribution of lipids in membrane leaflets. Once synthesized on the ER, phospholipids are transported to other organelles and to the plasma membrane. Mitochondria synthesize some of their own membrane lipids and import others.

Sphingolipids are also synthesized indirectly from multiple precursors. Sphingosine, the building block of these lipids, is made in the ER, beginning with the coupling of a palmitoyl group from palmitoyl CoA to serine; the subsequent addition of a second fatty acyl group to form N-acyl sphingosine

(ceramide) also takes place in the ER. Later, in the Golgi, a polar head group is added to ceramide yielding *sphingomyelin*, whose head group is phosphorylcholine, and various *glycosphingolipids*, in which the head group may be a monosaccharide or a more complex oligosaccharide (see Figure 10-8b). Some sphingolipid synthesis can also take place in mitochondria. In addition to serving as the backbone for sphingolipids, ceramide and its metabolic products are important signaling molecules that can influence cell growth, proliferation, endocytosis, resistance to stress, and programmed cell death (apoptosis).

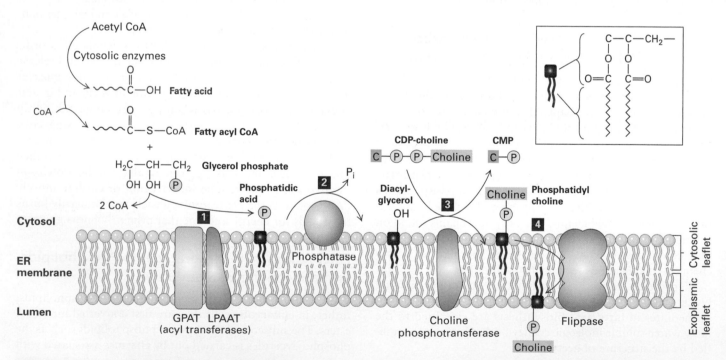

**FIGURE 10-25 Phospholipid synthesis in ER membrane.** Because phospholipids are amphipathic molecules, the last stages of their multistep synthesis take place at the interface between a membrane and the cytosol and are catalyzed by membrane-associated enzymes. Step **1**: Two fatty acids from fatty acyl CoA are esterified to the phosphorylated glycerol backbone, forming phosphatidic acid, whose two long hydrocarbon chains anchor the molecule to the membrane. Step **2**: A phosphatase converts phosphatidic acid into diacylglycerol. Step **3**: A polar head group (e.g., phosphorylcholine) is transferred from cytosine diphosphocholine (CDP-choline) to the exposed hydroxyl group. Step **4**: Flippase proteins catalyze the movement of phospholipids from the cytosolic leaflet in which they are initially formed to the exoplasmic leaflet.

After their synthesis is completed in the Golgi, sphingo-lipids are transported to other cellular compartments through vesicle-mediated mechanisms similar to those for the transport of proteins, discussed in Chapter 14. Any type of vesicular transport results in movement not only of the protein payload, but also of the lipids that compose the ve-sicular membrane. In addition, phospholipids such as phos-phoglycerides, as well as cholesterol, can move between organelles by different mechanisms, described below.

## Flippases Move Phospholipids from One Membrane Leaflet to the Opposite Leaflet

Even though phospholipids are initially incorporated into the cytosolic leaflet of the ER membrane, various phospho-lipids are asymmetrically distributed in the two leaflets of the ER membrane and of other cellular membranes. As noted above, phospholipids do not readily flip-flop from one leaflet to the other. For the ER membrane to expand by growth of both leaflets and have asymmetrically distributed phospholipids, its phospholipid components must be able to move from one membrane leaflet to the other. Although the mechanisms employed to generate and maintain membrane

phospholipid asymmetry are not well understood, it is clear that flippases play a key role. As described in Chapter 11, these integral membrane proteins use the energy of ATP hydrolysis to facilitate the movement of phospholipid mole-cules from one leaflet to the other (see Figure 11-15).

## Cholesterol Is Synthesized by Enzymes in the Cytosol and ER Membrane

Next we focus on cholesterol, the principal sterol in animal cells. Cholesterol is synthesized mainly in the liver. The first steps of cholesterol synthesis (Figure 10-26)—conversion of three acetyl groups linked to CoA (acetyl CoA) forming the 6-carbon molecule β-hydroxy-β-methylglutaryl linked to CoA (HMG-CoA)—take place in the cytosol. The conversion of HMG-CoA into mevalonate, the key rate-controlling step in cholesterol biosynthesis, is catalyzed by *HMG-CoA reductase,* an ER integral membrane protein, even though both its sub-strate and its product are water soluble. The water-soluble catalytic domain of HMG-CoA reductase extends into the cy-tosol, but its eight transmembrane α helices firmly embed the enzyme in the ER membrane. Five of the transmembrane α he-lices compose the so-called *sterol-sensing domain* and regulate

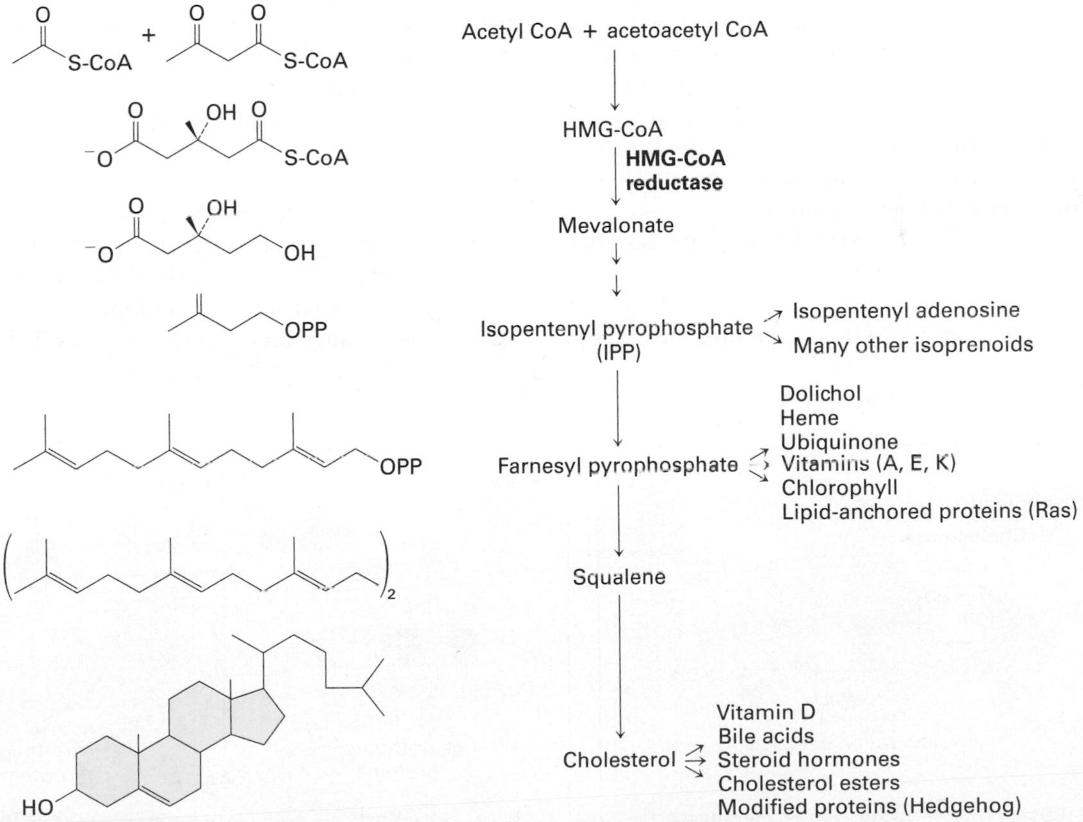

**FIGURE 10-26 Cholesterol biosynthetic pathway.** The regulated rate-controlling step in cholesterol biosynthesis is the conversion of β-hydroxy-β-methylglutaryl CoA (HMG-CoA) into mevalonic acid by HMG-CoA reductase, an ER-membrane protein. Mevalonate is then converted into isopentenyl pyrophosphate (IPP), which has the basic five-carbon isoprenoid structure. IPP can be converted into cholesterol and into many other lipids, often through the polyisoprenoid interme-diates shown here. Some of the numerous compounds derived from isoprenoid intermediates and cholesterol itself are indicated.

enzyme stability. When levels of cholesterol in the ER membrane are high, binding of cholesterol to this domain causes the protein to bind to two other integral ER membrane proteins, Insig-1 and Insig-2. This in turn induces ubiquitination (see Figure 3-29) of HMG-CoA reductase and its degradation by the proteasome pathway, reducing the production of mevalonate, the key intermediate in cholesterol biosynthesis.

**Atherosclerosis,** frequently called cholesterol-dependent clogging of the arteries, is characterized by the progressive deposition of cholesterol and other lipids, cells, and extracellular matrix material in the inner layer of the wall of an artery. The resulting distortion of the artery's wall can lead, either alone or in combination with a blood clot, to major blockage of blood flow. Atherosclerosis accounts for 75 percent of deaths due to cardiovascular disease in the United States.

Perhaps the most successful anti-atherosclerosis medications are the *statins*. These drugs bind to HMG-CoA reductase and directly inhibit its activity, thereby lowering cholesterol biosynthesis. As a consequence, the amount of low-density lipoproteins (see Figure 14-27)—the small, membrane-enveloped particles containing cholesterol esterified to fatty acids that often and rightly are called "bad cholesterol"—drops in the blood, reducing the formation of atherosclerotic plaques. ■

Mevalonate, the six-carbon product formed by HMG-CoA reductase, is converted in several steps into the five-carbon isoprenoid compound isopentenyl pyrophosphate (IPP) and its stereoisomer, dimethylallyl pyrophosphate (DMPP) (see Figure 10-26). These reactions are catalyzed by cytosolic enzymes, as are the subsequent reactions in the cholesterol synthesis pathway, in which six IPP units condense to yield squalene, a branched-chain 30-carbon intermediate. Enzymes bound to the ER membrane catalyze the multiple reactions that convert squalene into cholesterol in mammals or into related sterols in other species. One of the intermediates in this pathway, farnesyl pyrophosphate, is the precursor of the prenyl lipid that anchors Ras and related proteins to the cytosolic surface of the plasma membrane (see Figure 10-19) as well as other important biomolecules (see Figure 10-26).

## Cholesterol and Phospholipids Are Transported Between Organelles by Several Mechanisms

As already noted, the final steps in the synthesis of cholesterol and phospholipids take place primarily in the ER. Thus the plasma membrane and the membranes bounding other organelles must obtain these lipids by means of one or more intracellular transport processes. Membrane lipids can and do accompany both soluble and membrane proteins during the secretory pathway described in Chapter 14; membrane vesicles bud from the ER and fuse with membranes in the Golgi complex, and other membrane vesicles bud from the Golgi complex and fuse with the plasma membrane (Figure 10-27a). However, several lines of evidence suggest that there is substantial inter-organelle movement of cholesterol and phospholipids through other mechanisms. For example, chemical inhibitors of the classic secretory pathway and mutations that impede vesicular traffic in this pathway do not prevent cholesterol or phospholipid transport between membranes.

A second mechanism entails direct protein-mediated contact of ER or ER-derived membranes with membranes of other organelles (Figure 10-27b). In the third mechanism, small lipid-transfer proteins facilitate the exchange of phospholipids or cholesterol between different membranes (Figure 10-27c). Although such transfer proteins have been identified in assays in vitro, their role in intracellular movements of most phospholipids is not well defined. For instance, mice with a knockout mutation in the gene encoding the phosphatidylcholine-transfer protein appear to be normal in most respects, indicating that this protein is not essential for cellular phospholipid metabolism.

As noted earlier, the lipid compositions of different organelle membranes vary considerably (see Table 10-1). Some

(a)

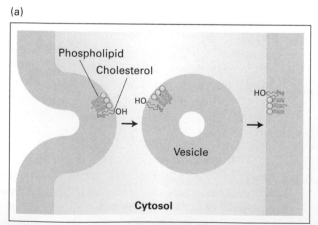

(b)

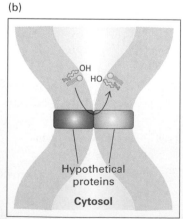

(c)

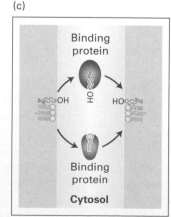

**FIGURE 10-27 Proposed mechanisms of transport of cholesterol and phospholipids between membranes.** In mechanism (a), vesicles transfer lipids between membranes. In mechanism (b), lipid transfer is a consequence of direct contact between membranes that is mediated by membrane-embedded proteins. In mechanism (c), transfer is mediated by small, soluble lipid-transfer proteins. [Adapted from F. R. Maxfield and D. Wustner, 2002, *J. Clin. Invest.* **110:**891.]

of these differences are due to different sites of synthesis. For example, a phospholipid called cardiolipin, which is localized to the mitochondrial membrane, is made only in mitochondria and little is transferred to other organelles. Differential transport of lipids also plays a role in determining the lipid compositions of different cellular membranes. For instance, even though cholesterol is made in the ER, the cholesterol concentration (cholesterol-to-phospholipid molar ratio) is ~1.5–13-fold higher in the plasma membrane than in other organelles (ER, Golgi, mitochondrion, lysosome). Although the mechanisms responsible for establishing and maintaining these differences are not well understood, we have seen that the distinctive lipid composition of each membrane has a major influence on its physical and biological properties.

## KEY CONCEPTS of Section 10.3

### Phospholipids, Sphingolipids, and Cholesterol: Synthesis and Intracellular Movement

- Saturated and unsaturated fatty acids of various chain lengths are components of phospholipids and sphingolipids.

- Fatty acids are synthesized from acetyl CoA by water-soluble enzymes and modified by elongation and desaturation in the endoplasmic reticulum (ER).

- Free fatty acids are transported within cells by fatty-acid-binding proteins (FABPs).

- Fatty acids are incorporated into phospholipids through a multi-step process. The final steps in the synthesis of phosphoglycerides and sphingolipids are catalyzed by membrane-associated enzymes primarily on the cytosolic face of the ER (see Figure 10-25).

- Each type of newly synthesized lipid is incorporated into the preexisting membranes on which it is made; thus, membranes are themselves the platform for the synthesis of new membrane material.

- Most membrane phospholipids are preferentially distributed in either the exoplasmic or the cytosolic leaflet. This asymmetry results in part from the action of phospholipid flippases, which flip lipids from one leaflet to the other.

- The initial steps in cholesterol biosynthesis take place in the cytosol, whereas the last steps are catalyzed by enzymes associated with the ER membrane.

- The rate-controlling step in cholesterol biosynthesis is catalyzed by HMG-CoA reductase, whose transmembrane segments are embedded in the ER membrane and contain a sterol-sensing domain.

- Considerable evidence indicates that Golgi-independent vesicular transport, direct protein-mediated contacts between different membranes, soluble protein carriers, or all three may account for some inter-organelle transport of cholesterol and phospholipids (see Figure 10-27).

## Perspectives for the Future

One fundamental question in lipid biology concerns the generation, maintenance, and function of the asymmetric distribution of lipids within the leaflets of one membrane and the variation in lipid composition among the membranes of different organelles. What are the mechanisms underlying this complexity, and why is such complexity needed? We already know that certain lipids can specifically interact with and influence the activity of some proteins. For example, the large multimeric proteins that participate in oxidative phosphorylation in the inner mitochondrial membrane appear to assemble into supercomplexes whose stability may depend on the physical properties and binding of specialized phospholipids such as cardiolipin (see Chapter 12).

The existence of lipid rafts in biological membranes and their function in cell signaling remains a topic of heated debate. Many biochemical studies using model membranes show that stable lateral assemblies of sphingolipids and cholesterol—lipid rafts—can facilitate selective protein-protein interactions by excluding or including specific proteins. But whether or not lipid rafts exist in natural biological membranes, as well as their dimensions and dynamics, is under intense investigation. New biophysical and microscopic tools are beginning to provide a more solid basis for their existence, size, and behavior.

Despite considerable progress in our understanding of the cellular metabolism and movement of lipids, the mechanisms for transporting cholesterol and phospholipids between organelle membranes remain poorly characterized. In particular, we lack a detailed understanding of how various transport proteins move lipids from one membrane leaflet to another (flippase activity) and into and out of cells. Such understanding will require a determination of high-resolution structures of these molecules, their capture in various stages of the transport process, and careful kinetic and other biophysical analyses of their function, similar to the approaches discussed in Chapter 11 for elucidating the operation of ion channels and ATP-powered pumps.

Recent advances in solubilizing and crystallizing integral membrane proteins have led to the delineation of the molecular structures of many important types of proteins, such as ion channels, G protein–coupled receptors, ATP-powered ion pumps, and aquaporins, as we will see in Chapter 11. However, many important classes of membrane proteins have proven recalcitrant to even these new approaches. For example, we lack the structure of any protein that transports glucose into a eukaryotic cell. As we will learn in Chapters 15 and 16, many classes of receptors span the plasma membrane with one or more $\alpha$ helixes. Perhaps surprisingly, we lack the molecular structure of the transmembrane segment of any single-pass eukaryotic cell-surface receptor, and so many aspects of the function of these proteins are still mysterious. The transmittal of information across the membrane, as it occurs when a single-pass receptor binds an appropriate ligand, remains to be described at adequate molecular resolution. Elucidating the molecular structures of these and many other types of membrane proteins will clarify many aspects of molecular cell biology.

## Key Terms

| | |
|---|---|
| amphipathic 445 | lipid raft 454 |
| aquaporin 458 | liposome 445 |
| atherosclerosis 468 | lumen 448 |
| cholesterol 450 | membrane transport |
| cilium 448 |   protein 443 |
| cytoskeleton 443 | micelle 445 |
| cytosolic face 447 | peripheral membrane |
| exoplasmic face 447 |   protein 456 |
| flagellum 448 | phosphoglyceride 448 |
| flippase 454 | phospholipase 454 |
| glycolipid 450 | phospholipid bilayer 445 |
| glycoprotein 461 | plasma membrane 443 |
| hydrophilic 445 | porin 458 |
| hydrophobic 445 | receptor protein 443 |
| integral membrane | saturated 465 |
|   protein 456 | sphingolipid 450 |
| lectin 461 | statin 468 |
| lipid-anchored membrane | sterol 450 |
|   protein 456 | unsaturated 465 |
| lipid droplet 455 | |

## Review the Concepts

1. When viewed by electron microscopy, the lipid bilayer is often described as looking like a railroad track. Explain how the structure of the bilayer creates this image.

2. Explain the following statement: The structure of all biomembranes depends on the chemical properties of phospholipids, whereas the function of each specific biomembrane depends on the specific proteins associated with that membrane.

3. Biomembranes contain many different types of lipid molecules. What are the three main types of lipid molecules found in biomembranes? How are the three types similar, and how are they different?

4. Lipid bilayers are considered to be two-dimensional fluids. What does this mean? What drives the movement of lipid molecules and proteins within the bilayer? How can such movement be measured? What factors affect the degree of membrane fluidity?

5. Why are water-soluble substances unable to freely cross the lipid bilayer of the cell membrane? How does the cell overcome this permeability barrier?

6. Name the three groups into which membrane-associated proteins may be classified. Explain the mechanism by which each group associates with a biomembrane.

7. Identify the following membrane-associated proteins based on their structure: (a) tetramers of identical subunits, each with six membrane-spanning α helices; (b) trimers of identical subunits, each with 16 β sheets forming a barrel-like structure.

8. Proteins may be bound to the exoplasmic or cytosolic face of the plasma membrane by way of covalently attached lipids. What are the three types of lipid anchors responsible for tethering proteins to the plasma membrane bilayer, and which type is used by cell-surface proteins that face the external medium and by glycosylated proteoglycans?

9. Although both faces of a biomembrane are composed of the same general types of macromolecules, principally lipids and proteins, the two faces of the bilayer are not identical. What accounts for the asymmetry between the two faces?

10. What are detergents? How do ionic and nonionic detergents differ in their ability to disrupt cell membrane structure?

11. What is the likely identity of these membrane-associated proteins: (a) released from the membrane with a high-salt solution causing disruption of ionic linkages; (b) not released from the membrane upon exposure to a high-salt solution alone, but released when incubated with an enzyme that cleaves phosphate-glycerol bonds and covalent linkages are disrupted; (c) not released from the membrane upon exposure to a high-salt solution, but released after addition of the detergent sodium dodecyl sulfate (SDS). Will the activity of the protein released in part (c) be preserved following release?

12. Following the production of membrane extracts using the nonionic detergent Triton X-100, you analyze the membrane lysates via mass spectrometry and note a high content of cholesterol and sphingolipids. Furthermore, biochemical analysis of the lysates reveals potential kinase activity. What have you likely isolated?

13. Phospholipid biosynthesis at the interface between the endoplasmic reticulum (ER) and the cytosol presents a number of challenges that must be solved by the cell. Explain how each of the following is handled.

    a. The substrates for phospholipid biosynthesis are all water soluble, yet the end products are not.

    b. The immediate site of incorporation of all newly synthesized phospholipids is the cytosolic leaflet of the ER membrane, yet phospholipids must be incorporated into both leaflets.

    c. Many membrane systems in the cell, for example the plasma membrane, are unable to synthesize their own phospholipids, yet these membranes must also expand if the cell is to grow and divide.

14. What are the common fatty acid chains in phosphoglycerides, and why do these fatty acid chains differ in their number of carbon atoms by multiples of 2?

15. Fatty acids must associate with lipid chaperones in order to move within the cell. Why are these chaperones needed, and what is the name given to a group of proteins that are responsible for this intracellular trafficking of fatty acids? What is the key distinguishing feature of these proteins that allows fatty acids to move within the cell?

**16.** The biosynthesis of cholesterol is a highly regulated process. What is the key regulated enzyme in cholesterol biosynthesis? This enzyme is subject to feedback inhibition. What is feedback inhibition? How does this enzyme sense cholesterol levels in a cell?

**17.** Phospholipids and cholesterol must be transported from their site of synthesis to various membrane systems within cells. One way of doing this is through vesicular transport, as is the case for many proteins in the secretory pathway (Chapter 14). However, most phospholipid and cholesterol membrane-to-membrane transport in cells is not by vesicular transport. What is the evidence for this statement? What appear to be the major mechanisms for phospholipid and cholesterol transport?

**18.** Explain the mechanism by which statins lower "bad" cholesterol.

## Analyze the Data

**1.** The behavior of receptor X (XR), a transmembrane protein present in the plasma membrane of mammalian cells, is being investigated. The protein has been engineered as a fusion protein containing the green fluorescent protein (GFP) at its N-terminus. GFP-XR is a functional protein and can replace XR in cells.

**a.** Cells expressing GFP-XR or artificial lipid vesicles (liposomes) containing GFP-XR are subjected to fluorescence recovery after photobleaching (FRAP). The intensity of the fluorescence of a small spot on the surface of the cells (solid line) or on the surface of the liposomes (dashed line) is measured prior to and following laser bleaching (arrow). The data are shown below.

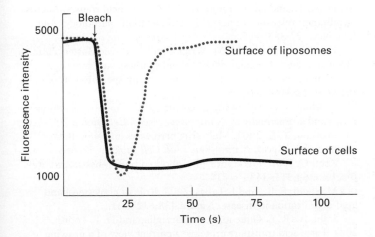

What explanation could account for the differing behavior of GFP-XR in liposomes versus in the plasma membrane of a cell?

**b.** Tiny gold particles can be attached to individual molecules and their movement then followed in a light microscope by single-particle tracking. This method allows one to observe the behavior of individual proteins in a membrane.

The tracks generated during a 5-second observational period by a gold particle attached to XR present in a cell (left) or in a liposome (middle) or to XR adhered to a microscope slide (right) are shown below.

XR present in a cell

XR present in a liposome

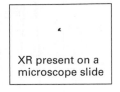

XR present on a microscope slide

What additional information do these data provide beyond what can be determined from the FRAP data?

**c.** Fluorescence resonance energy transfer (FRET) is a technique by which a fluorescent molecule, following its excitation with the appropriate wavelength of light, can transfer its emission energy to and excite a nearby different fluorescent molecule (sec Figure 15-18). Cyan fluorescent protein (CFP) and yellow fluorescent protein (YFP) are related to GFP but fluoresce at cyan and yellow wavelengths rather than at green. If CFP is excited with the appropriate wavelength of light and a YFP molecule is very near, then energy can be transferred from CFP emission and used to excite YFP, as indicated by a loss of emission of cyan fluorescence and an increase in emission of yellow fluorescence. CFP-XR and YFP-XR are expressed together in a cell line or are both incorporated into liposomes. The number of molecules of YFP-XR and CFP-XR per cm² of membrane is equivalent in the cells and the liposomes. The cells and liposomes are then irradiated with a wavelength of light that causes CFP but not YFP to fluoresce. The amount of cyan (CFP) and yellow (YFP) fluorescence emitted by the cells (solid line) or liposomes (dashed line) is then monitored, as shown below.

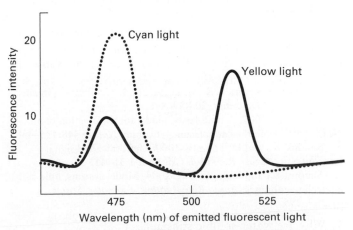

What can be deduced about XR from these data?

**2.** After performing differential scanning calorimetry (a procedure used to determine the transition temperature of a given membrane by recording the amount of heat absorbed prior to phase transition [solid to fluid state]) on membranes from three different organisms, you obtain the following results:

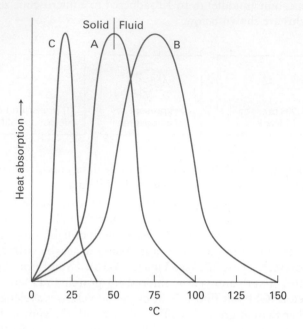

Which of the following statements is likely true about the lipid composition of membrane C?

i. It has high levels of saturated hydrocarbons and long hydrocarbon tails compared to A and B.

ii. It has high levels of saturated hydrocarbons and short hydrocarbon tails compared to A and B.

iii. It has high levels of unsaturated hydrocarbons and long hydrocarbon tails compared to A and B.

iv. It has high levels of unsaturated hydrocarbons and short hydrocarbon tails compared to A and B.

# References

The Lipid Bilayer: Composition and Structural Organization

McMahon, H., and J. L. Gallop. 2005. Membrane curvature and mechanisms of dynamic cell membrane remodeling. *Nature* 438:590–596.

Mukherjee, S., and F. R. Maxfield. 2004. Membrane domains. *Annu. Rev. Cell Dev. Biol.* 20:839–866.

Ploegh, H. 2007. A lipid-based model for the creation of an escape hatch from the endoplasmic reticulum. *Nature* 448:435–438.

Simons, K., and D. Toomre. 2000. Lipid rafts and signal transduction. *Nature Rev. Mol. Cell Biol.* 1:31–41.

Simons, K., and W. L. C. Vaz. 2004. Model systems, lipid rafts, and cell membranes. *Annu. Rev. Biophys. Biomolec. Struct.* 33:269–295.

Tamm, L. K., V. K. Kiessling, and M. L. Wagner. 2001. Membrane dynamics. *Encyclopedia of Life Sciences.* Nature Publishing Group.

Vance, D. E., and J. E. Vance. 2002. *Biochemistry of Lipids, Lipoproteins, and Membranes,* 4th ed. Elsevier.

Van Meer, G. 2006. Cellular lipidomics. *EMBO J.* 24:3159–3165.

Yeager, P. L. 2001. Lipids. *Encyclopedia of Life Sciences.* Nature Publishing Group.

Zimmerberg, J., and M. M. Kozlov. 2006. How proteins produce cellular membrane curvature. *Nature Rev. Mol. Cell Biol.* 7:9–19.

Membrane Proteins: Structure and Basic Functions

Bowie, J. 2005. Solving the membrane protein folding problem. *Nature* 438:581–589.

Cullen, P. J., G. E. Cozier, G. Banting, and H. Mellor. 2001. Modular phosphoinositide-binding domains: their role in signalling and membrane trafficking. *Curr. Biol.* 11:R882–R893.

Engelman, D. Membranes are more mosaic than fluid. 2005. *Nature* 438:578–580.

Lanyi, J. K., and H. Luecke. 2001. Bacteriorhodopsin. *Curr. Opin. Struc. Biol.* 11:415–519.

Lee, A. G. 2005. A greasy grip. *Nature* 438:569–570.

MacKenzie, K. R., J. H. Prestegard, and D. M. Engelman. 1997. A transmembrane helix dimer: structure and implications. *Science* 276:131–133.

McIntosh, T. J., and S. A. Simon. 2006. Roles of bilayer material properties in function and distribution of membrane proteins *Ann. Rev. Biophys. Biomolec. Struct.* 35:177–198.

Wucherpfennig, K.W., E. Gagnon, M .J. Call, E. S. Huseby, and M. E. Call. 2010. *Cold Spring Harb Perspect Biol*, 2.:a005140.

Schulz, G. E. 2000. β-Barrel membrane proteins. *Curr. Opin. Struc. Biol.* 10:443–447.

Phospholipids, Sphingolipids, and Cholesterol: Synthesis and Intracellular Movement

Bloch, K. 1965. The biological synthesis of cholesterol. *Science* 150:19–28.

Daleke, D. L., and J. V. Lyles. 2000. Identification and purification of aminophospholipid flippases. *Biochim. Biophys. Acta* 1486:108–127.

Futerman, A., and H. Riezman. 2005. The ins and outs of sphingolipid synthesis. *Trends Cell Biol.* 15:312–318.

Hajri, T., and N. A. Abumrad. 2002. Fatty acid transport across membranes: relevance to nutrition and metabolic pathology. *Ann. Rev. Nutr.* 22:383–415.

Henneberry, A. L., M. M. Wright, and C. R. McMaster. 2002. The major sites of cellular phospholipid synthesis and molecular determinants of fatty acid and lipid head group specificity. *Mol. Biol. Cell* 13:3148–3161.

Holthuis, J. C. M., and T. P. Levine. 2005. Lipid traffic: floppy drives and a superhighway *Nature Rev. Molec. Cell Biol.* 6:209–220.

Ioannou, Y. A. 2001. Multidrug permeases and subcellular cholesterol transport. *Nature Rev. Mol. Cell Biol.* 2:657–668.

Kent, C. 1995. Eukaryotic phospholipid biosynthesis. *Ann. Rev. Biochem.* 64:315–343.

Maxfield, F. R., and I. Tabas. 2005. Role of cholesterol and lipid organization in disease. *Nature* 438:612–621.

Stahl, A., R. E. Gimeno, L. A. Tartaglia, and H. F. Lodish. 2001. Fatty acid transport proteins: a current view of a growing family. *Trends Endocrinol. Metab.* 12(6):266–273.

van Meer, G., and H. Sprong. 2004. Membrane lipids and vesicular traffic. *Curr. Opin. Cell Biol.* 16:373–378.

# Transmembrane Transport of Ions and Small Molecules

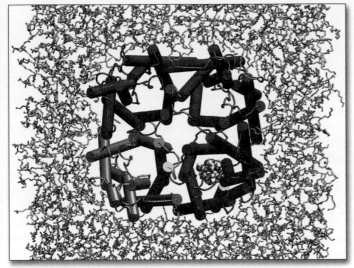

Outside–in view of a bacterial aquaporin protein, which transports water and glycerol into and out of the cell, embedded in a phospholipid membrane. The four identical monomers are colored individually; each has a channel in its center. [After M. Ø. Jensen et al., 2002, *Proc. Nat'l Acad. Sci. USA* **99**:6731–6736.]

In all cells, the plasma membrane forms the permeability barrier that separates the cytoplasm from the exterior environment, thus defining a cell's physical and chemical boundaries. By preventing the unimpeded movement of molecules and ions into and out of cells, the plasma membrane maintains essential differences between the composition of the extracellular fluid and that of the cytosol; for example, the concentration of NaCl in the blood and extracellular fluids of animals is generally above 150 mM, similar to that of the seawater in which all cells are thought to have evolved, whereas the $Na^+$ concentration in the cytosol is tenfold lower. In contrast, the potassium ion ($K^+$) concentration is higher in the cytosol than outside.

Organelle membranes, which separate the cytosol from the interior of the organelle, also form permeability barriers. For example, the proton concentration in the lysosome interior, pH 5, is about 100-fold greater than that of the cytosol, and many specific metabolites accumulate at higher concentrations in the interior of other organelles, such as the endoplasmic reticulum or the Golgi complex, than in the cytosol.

All cellular membranes, both plasma and organelle, consist of a bilayer of phospholipids into which other lipids and specific types of proteins are embedded. It is this combination of lipids and proteins that gives cellular membranes their distinctive permeability qualities. If cellular membranes were pure phospholipid bilayers (see Figure 10-4), they would be excellent chemical barriers, impermeable to virtually all ions, amino acids, sugars, and other water-soluble molecules. In fact, only a few gases and uncharged, small, water-soluble molecules can readily diffuse across a pure phospholipid bilayer (Figure 11-1). But cellular membranes must serve not only as barriers; they must also act as conduits, selectively transporting molecules and ions from one side of the membrane to the other. Energy-rich glucose, for example, must be imported into the cell, and wastes must be shipped out.

## OUTLINE

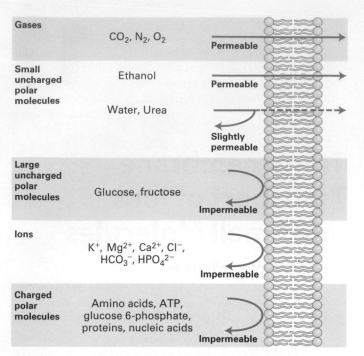

**FIGURE 11-1 Relative permeability of a pure phospholipid bilayer to various molecules and ions.** A bilayer is permeable to many gases and to small, uncharged, water-soluble (polar) molecules. It is slightly permeable to water, and essentially impermeable to ions and to large polar molecules.

Movement of virtually all small molecules and ions across cell membranes is mediated by **membrane transport proteins**—integral membrane proteins embedded by multiple transmembrane domains in cellular membranes. These membrane-spanning proteins act variously as shuttles, channels, or pumps for transporting molecules and ions through a membrane's hydrophobic interior. In some cases, molecules or ions are transported from a higher to a lower concentration, a thermodynamically favored process powered by an increase in entropy. Examples include the transport of water or glucose from the blood into most body cells. In other cases, molecules or ions must be pumped from a lower to a higher concentration, a thermodynamically unfavorable process that can only occur when an external source of energy is available to push the molecules "uphill" against a concentration gradient. An example is the cell's ability to concentrate protons within lysosomes to generate a low pH in the lumen. Often the required energy is provided by mechanistically coupling the energy-releasing hydrolysis of the terminal phosphoanhydride bond in ATP with the movement of a molecule or ion across the membrane. Other proteins couple the movement of one molecule or ion against its concentration gradient with the movement of another down its gradient, using the energy released by the downhill movement of one molecule or ion to thermodynamically drive the uphill movement of another. Proper functioning of any cell relies on a precise balance between such import and export of various molecules and ions.

We begin our discussion of membrane transport proteins by reviewing some of the general principles of transport across membranes and distinguishing between three major classes of such proteins. In subsequent sections, we describe the structure and operation of specific examples of each class and show how members of families of homologous transport proteins have different properties that enable different cell types to function appropriately. We also explain how both the plasma membrane and organellar membranes contain specific combinations of transport proteins that enable cells to carry out essential physiological processes, including the maintenance of cytosolic pH, the accumulation of sucrose and salts in plant cell vacuoles, and the directed flow of water in both plants and animals. The cell's resting membrane potential is an important consequence of selective ion transport across membranes, and we consider how this potential arises. Epithelial cells, such as those lining the small intestine, use a combination of membrane proteins to transport ions, sugars, and other small molecules and water from one side of the cell to the other. We will see how this understanding has led to the development of sports drinks as well as therapies for cholera.

Note that in this chapter we cover only transport of small molecules and ions; transport of larger molecules, such as proteins and oligosaccharides, is covered in Chapters 13 and 14.

## 11.1 Overview of Transmembrane Transport

In this section, we first describe the factors that influence the permeability of lipid membranes, and then briefly describe the three major classes of membrane transport proteins that allow molecules and ions to cross them. Different kinds of membrane-embedded proteins accomplish the task of moving molecules and ions in different ways.

### Only Gases and Small Uncharged Molecules Cross Membranes by Simple Diffusion

With its dense hydrophobic core, a phospholipid bilayer is largely impermeable to water-soluble molecules and ions. Only gases, such as $O_2$ and $CO_2$, and small uncharged polar molecules, such as urea and ethanol, can readily move by **simple diffusion** across an artificial membrane composed of pure phospholipid or of phospholipid and cholesterol (see Figure 11-1). Such molecules also can diffuse across cellular membranes without the aid of transport proteins. No metabolic energy is expended because movement is from a high to a low concentration of the molecule, down its chemical concentration gradient. As noted in Chapter 2, such movements are spontaneous because they have a positive $\Delta S$ value (increase in entropy) and thus a negative $\Delta G$ (decrease in free energy).

The relative diffusion rate of any substance across a pure phospholipid bilayer is proportional to its concentration gradient across the bilayer and to its hydrophobicity and size;

the movement of charged molecules is also affected by any electric potential across the membrane. When a pure phospholipid bilayer separates two aqueous spaces, or "compartments," membrane permeability can be easily determined by adding a small amount of radioactive material to one compartment and measuring its rate of appearance in the other compartment. The greater the concentration gradient of the substance, the faster its rate of movement across a bilayer.

The hydrophobicity of a substance is determined by measuring its partition coefficient $K$, the equilibrium constant for its partition between oil and water. The higher a substance's partition coefficient (the greater the fraction found in oil relative to water), the more lipid soluble it is and, therefore, the faster its rate of movement across a bilayer. The first and rate-limiting step in transport by simple diffusion is movement of a molecule from the aqueous solution into the hydrophobic interior of the phospholipid bilayer, which resembles olive oil in its chemical properties. This is the reason that the more hydrophobic a molecule is, the faster it diffuses across a pure phospholipid bilayer. For example, diethylurea, with an ethyl group attached to each nitrogen atom:

$$CH_3-CH_2-NH-\overset{\displaystyle O}{\overset{\|}{C}}-NH-CH_2-CH_3$$

has a $K$ of 0.01, whereas urea

$$NH_2-\overset{\displaystyle O}{\overset{\|}{C}}-NH_2$$

has a $K$ of 0.0002. Diethylurea, which is 50 times (0.01/0.0002) more hydrophobic than urea, will therefore diffuse through phospholipid bilayer membranes about 50 times faster than urea. Similarly, fatty acids with longer hydrocarbon chains are more hydrophobic than those with shorter chains and at all concentrations will diffuse more rapidly across a pure phospholipid bilayer.

If a substance carries a net charge, its movement across a membrane is influenced by both its concentration gradient and the **membrane potential,** the electric potential (voltage) across the membrane. The combination of these two forces, called the **electrochemical gradient,** determines the energetically favorable direction of movement of a charged molecule across a membrane. The electric potential that exists across most cellular membranes results from a small imbalance in the concentration of positively and negatively charged ions on the two sides of the membrane. We discuss how this ionic imbalance, and resulting potential, arise and are maintained in Sections 11.4 and 11.5.

## Three Main Classes of Membrane Proteins Transport Molecules and Ions Across Biomembranes

As is evident from Figure 11-1, very few molecules and no ions can cross a pure phospholipid bilayer at appreciable rates by simple diffusion. Thus transport of most molecules into and out of cells requires the assistance of specialized membrane proteins. Even in the case of molecules with relatively large partition coefficients (e.g., urea, fatty acids), and certain gases such as $CO_2$ (carbon dioxide) and $NH_3$ (ammonia), transport is frequently accelerated by specific proteins because simple diffusion usually does not occur rapidly enough to meet cellular needs.

All transport proteins are transmembrane proteins containing multiple membrane-spanning segments that generally are α helices. By forming a protein-lined pathway across the membrane, transport proteins are thought to allow movement of hydrophilic substances without their coming into contact with the hydrophobic interior of the membrane. Here we introduce the three main types of membrane transport proteins covered in this chapter (Figure 11-2).

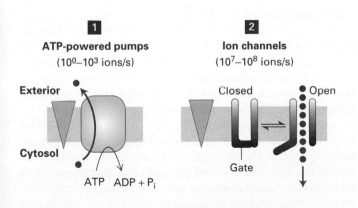

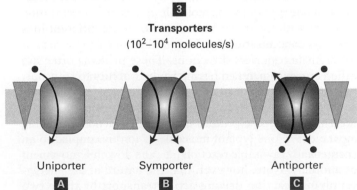

**FIGURE 11-2 Overview of membrane transport proteins.** Gradients are indicated by triangles with the tip pointing toward lower concentration, electric potential, or both. **1** Pumps utilize the energy released by ATP hydrolysis to power movement of specific ions or small molecules (red circles) against their electrochemical gradient. **2** Channels permit movement of specific ions (or water) down their electrochemical gradient. **3** Transporters, which fall into three groups, facilitate movement of specific small molecules or ions. Uniporters transport a single type of molecule down its concentration gradient **3A**. Cotransport proteins (symporters, **3B**, and antiporters, **3C**) catalyze the movement of one molecule *against* its concentration gradient (black circles), driven by movement of one or more ions down an electrochemical gradient (red circles). Differences in the mechanisms of transport by these three major classes of proteins account for their varying rates of solute movement.

ATP-powered pumps (or simply pumps) are ATPases that use the energy of ATP hydrolysis to move ions or small molecules across a membrane *against* a chemical concentration gradient, an electric potential, or both. This process, referred to as **active transport,** is an example of a coupled chemical reaction (Chapter 2). In this case, transport of ions or small molecules "uphill" against an electrochemical gradient, which requires energy, is coupled to the hydrolysis of ATP, which releases energy. The overall reaction—ATP hydrolysis and the "uphill" movement of ions or small molecules—is energetically favorable.

**Channels** transport water, specific ions, or hydrophilic small molecules across membranes *down* their concentration or electric potential gradients. Because this process requires transport proteins but not energy, it is sometimes referred to as "passive transport" or "facilitated diffusion," but it is more properly called **facilitated transport.** Channels form a hydrophilic "tube" or passageway across the membrane through which multiple water molecules or ions move simultaneously, single file, at a very rapid rate. Some channels are open much of the time; these are referred to as *nongated* channels. Most ion channels, however, open only in response to specific chemical or electric signals. These are referred to as *gated* channels because a protein "gate" alternatively blocks the channel or moves out of the way to open the channel (see Figure 11-2). Channels, like all transport proteins, are very selective for the type of molecule they transport.

**Transporters** (also called carriers) move a wide variety of ions and molecules across cell membranes, but at a much slower rate than channels. Three types of transporters have been identified. *Uniporters* transport a single type of molecule *down* its concentration gradient. Glucose and amino acids cross the plasma membrane into most mammalian cells with the aid of uniporters. Collectively, channels and uniporters are sometimes called *facilitated transporters,* indicating movement down a concentration or electrochemical gradient.

In contrast, *antiporters* and *symporters* couple the movement of one type of ion or molecule *against* its concentration gradient with the movement of one or more different ions *down* its concentration gradient, in the same (symporter) or different (antiporter) directions. These proteins often are called *cotransporters,* referring to their ability to transport two or more different solutes simultaneously.

Like ATP pumps, cotransporters mediate coupled reactions in which an energetically unfavorable reaction (i.e., uphill movement of one type of molecule or ion) is coupled to an energetically favorable reaction (i.e., the downhill movement of another). Note, however, that the nature of the energy-supplying reaction driving active transport by these two classes of proteins differs. ATP pumps use energy from hydrolysis of ATP, whereas cotransporters use the energy stored in an electrochemical gradient. This latter process sometimes is referred to as *secondary active transport.*

Conformational changes are essential to the function of all transport proteins. ATP-powered pumps and transporters undergo a cycle of conformational change exposing a binding site (or sites) to one side of the membrane in one conformation and

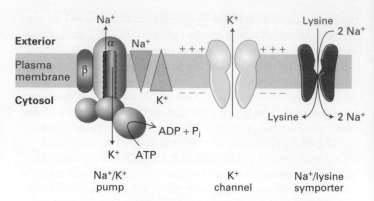

**FIGURE 11-3 Multiple membrane transport proteins function together in the plasma membrane of metazoan cells.** Gradients are indicated by triangles with the tip pointing toward lower concentration. The $Na^+/K^+$ ATPase in the plasma membrane uses energy released by ATP hydrolysis to pump $Na^+$ out of the cell and $K^+$ inward; this creates a concentration gradient of $Na^+$ that is greater outside the cell than inside and one of $K^+$ that is greater inside than outside. Movement of positively charged $K^+$ ions out of the cell through membrane $K^+$ channel proteins creates an electric potential across the plasma membrane—the cytosolic face is negative with respect to the extracellular face. A $Na^+$/lysine transporter, a typical sodium/amino acid cotransporter, moves 2 $Na^+$ ions together with one lysine from the extracellular medium into the cell. "Uphill" movement of the amino acid is powered by "downhill" movement of $Na^+$ ions, powered both by the outside-greater-than-inside $Na^+$ concentration gradient and by the negative potential on the inside of the cell membrane, which attracts the positively charged $Na^+$ ions. The ultimate source of the energy to power amino acid uptake comes from the ATP hydrolyzed by the $Na^+/K^+$ ATPase, since this pump creates both the $Na^+$ ion concentration gradient and, via the $K^+$ channels, the membrane potential, which together power influx of $Na^+$ ions.

to the other side in a second conformation. Because each such cycle results in movement of only one (or a few) substrate molecules, these proteins are characterized by relatively slow rates of transport ranging from $10^0$ to $10^4$ ions or molecules per second (see Figure 11-2). Most ion channels shuttle between a closed state and an open state, but many ions can pass through an open channel without any further conformational change. For this reason, channels are characterized by very fast rates of transport, up to $10^8$ ions per second.

Frequently, several different types of transport proteins work in concert to achieve a physiological function. An example is seen in Figure 11-3, where an ATPase pumps $Na^+$ out of the cell and $K^+$ ions inward; this pump, which is found in virtually all metazoan cells, establishes the oppositely directed concentration gradients of $Na^+$ and $K^+$ ions across the plasma membrane (relatively high concentrations of $K^+$ inside and $Na^+$ outside of cells) that are used to power the import of amino acids. The human genome encodes hundreds of different types of transport proteins that use the energy stored across the plasma membrane in the $Na^+$ concentration gradient and its associated electric potential to transport a wide variety of molecules into cells against their concentration gradients.

| TABLE 11-1 | Mechanisms for Transporting Ions and Small Molecules Across Cell Membranes | | | |
|---|---|---|---|---|
| Property | Simple Diffusion | Facilitated Transport | Active Transport | Cotransport* |
| Requires specific protein | − | + | + | + |
| Solute transported against its gradient | − | − | + | + |
| Coupled to ATP hydrolysis | − | − | + | − |
| Driven by movement of a cotransported ion down its gradient | − | − | − | + |
| Examples of molecules transported | $O_2$, $CO_2$, steroid hormones, many drugs | Glucose and amino acids (uniporters); ions and water (channels) | Ions, small hydrophilic molecules, lipids (ATP-powered pumps) | Glucose and amino acids (symporters); various ions and sucrose (antiporters) |

*Also called *secondary active transport*.

Table 11-1 summarizes the four mechanisms by which small molecules and ions are transported across cellular membranes. In the next section we consider some of the simplest membrane transport proteins, those responsible for the transport of glucose and water.

## KEY CONCEPTS of Section 11.1

### Overview of Transmembrane Transport

• Cellular membranes regulate the traffic of molecules and ions into and out of cells and their organelles. The rate of simple diffusion of a substance across a membrane is proportional to its concentration gradient and hydrophobicity.

• With the exception of gases (e.g., $O_2$ and $CO_2$) and small, uncharged, water-soluble molecules, most molecules cannot diffuse across a pure phospholipid bilayer at rates sufficient to meet cellular needs.

• Membrane transport proteins provide a hydrophilic passageway for molecules and ions to travel through the hydrophobic interior of a membrane.

• Three classes of transmembrane proteins mediate transport of ions, sugars, amino acids, and other metabolites across cell membranes: ATP-powered pumps, channels, and transporters (see Figure 11-2).

• ATP-powered pumps couple the movement of a substrate *against* its concentration gradient to ATP hydrolysis, a process known as active transport.

• Channels form a hydrophilic "tube" through which water or ions move *down* a concentration gradient, a process known as facilitated transport or facilitated diffusion.

• Transporters fall into three groups: Uniporters transport a molecule down its concentration gradient (facilitated transport); symporters and antiporters couple movement of a substrate against its concentration gradient to the movement of a second substrate down its concentration gradient, a process known as secondary active transport or cotransport (see Table 11-1).

• Conformational changes are essential to the function of all membrane transport proteins; speed of transport depends on the number of substrates that can pass through a protein at once.

## 11.2 Facilitated Transport of Glucose and Water

Most animal cells utilize glucose as a substrate for ATP production; they usually employ a glucose uniporter to take up glucose from the blood or other extracellular fluid. Many cells utilize channel-like membrane transport proteins called aquaporins to increase the rate of water movement across their surface membranes. Here, we discuss the structure and function of these and other facilitated transporters.

### Uniport Transport Is Faster and More Specific than Simple Diffusion

The protein-mediated transport of a single type of molecule, such as glucose or other small hydrophilic molecules, down a concentration gradient across a cellular membrane is known as **uniport**. Several features distinguish uniport from simple diffusion:

1. The rate of substrate movement by uniporters is far higher than simple diffusion through a pure phospholipid bilayer.

2. Because the transported molecule never enters the hydrophobic core of the phospholipid bilayer, its partition coefficient $K$ is irrelevant.

3. Transport occurs via a limited number of uniporter molecules. Consequently, there is a maximum transport rate $V_{max}$ that depends on the number of uniporters in the membrane. $V_{max}$ is achieved when the concentration gradient across the membrane is very large and each uniporter is working at its maximal rate.

4. Transport is reversible, and the direction of transport will change if the direction of the concentration gradient changes.

5. Transport is specific. Each uniporter transports only a single type of molecule or a single group of closely related molecules. A measure of the affinity of a transporter for its substrate is $K_m$, which is the concentration of substrate at which transport is half maximal.

These properties also apply to transport mediated by the other classes of proteins depicted in Figure 11-2.

One of the best-understood uniporters is the glucose transporter called *GLUT1* found in the plasma membrane of most mammalian cells. GLUT1 is especially abundant in the erythrocyte plasma membrane. Because these cells have a single membrane and no nucleus or other internal organelles (see Figure 10-7a), the properties of GLUT1 and many other transport proteins from mature erythrocytes have been extensively studied. The simplified structure of these cells makes isolating and purifying a transport protein a fairly straightforward procedure.

Figure 11-4 shows that glucose uptake by erythrocytes and liver cells exhibits kinetics similar to that of a simple enzyme-catalyzed reaction involving a single substrate. The kinetics of transport reactions mediated by other types of proteins are more complicated than those for uniporters. Nonetheless, all protein-assisted (facilitated) transport reactions occur faster than simple diffusion across the bilayer, are substrate-specific, and exhibit a maximal rate ($V_{max}$).

## The Low $K_m$ of the GLUT1 Uniporter Enables It to Transport Glucose into Most Mammalian Cells

Like other uniporters, GLUT1 alternates between two conformational states: in one, a glucose-binding site faces the outside of the cell; in the other, a glucose-binding site faces the cytosol. Since the glucose concentration usually is higher in the extracellular medium (blood, in the case of erythrocytes) than in the cell, the GLUT1 uniporter generally catalyzes the net import of glucose from the extracellular medium

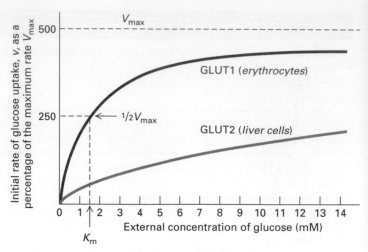

**EXPERIMENTAL FIGURE 11-4 Cellular uptake of glucose mediated by GLUT proteins exhibits simple enzyme kinetics.** The initial rate of glucose uptake, $\upsilon$ (measured as micromoles per milliliter of cells per hour), in the first few seconds is plotted as a percentage of the maximum rate, $V_{max}$, against increasing glucose concentration in the extracellular medium. In this experiment, the initial concentration of glucose in the cells is always zero. Both GLUT1, expressed by erythrocytes, and GLUT2, expressed by liver cells, catalyze glucose uptake (burgundy and tan curves). Like enzyme-catalyzed reactions, GLUT-facilitated uptake of glucose exhibits a maximum rate ($V_{max}$). The $K_m$ is the concentration at which the rate of glucose uptake is half maximal. GLUT2, with a $K_m$ of about 20 mM (not shown), has a much lower affinity for glucose than GLUT1, with a $K_m$ of about 1.5 mM.

into the cell. Figure 11-5 depicts the sequence of events occurring during the unidirectional transport of glucose from the cell exterior inward to the cytosol. GLUT1 also can catalyze the net export of glucose from the cytosol to the extracellular medium, when the glucose concentration is higher inside the cell than outside.

The kinetics of the unidirectional transport of glucose from the outside of a cell inward via GLUT1 can be described by the same type of equation used to describe a simple enzyme-catalyzed chemical reaction. For simplicity, let's assume that the substrate glucose, S, is present initially only on the outside of the cell; this can be achieved by first incubating the cells in a medium lacking glucose so their internal stores are depleted. In this case, we can write

$$S_{out} + GLUT1 \overset{K_m}{\rightleftharpoons} S_{out} - GLUT1 \overset{V_{max}}{\rightleftharpoons} S_{in} + GLUT1$$

where $S_{out} - GLUT1$ represents GLUT1 in the outward-facing conformation with a bound glucose. This equation is similar to the one describing the path of a simple enzyme-catalyzed reaction where the protein binds a single substrate and then transforms it into a different molecule. Here, however, no chemical modification occurs to the GLUT1-bound sugar; rather, it is moved across a cellular membrane. Nonetheless, the kinetics of this transport reaction are similar to those of simple enzyme-catalyzed reactions, and we can use the same

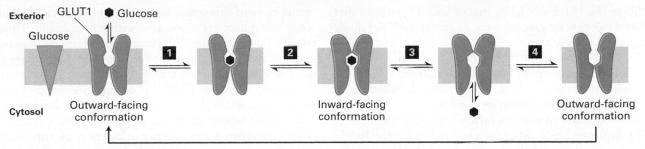

**Exterior**   GLUT1   ● Glucose

**Glucose**

**Cytosol**

Outward-facing conformation

**1**

**2**

Inward-facing conformation

**3**

**4**

Outward-facing conformation

**FIGURE 11-5 Model of uniport transport by GLUT1.** In one conformation, the glucose-binding site faces outward; in the other, the binding site faces inward. Binding of glucose to the outward-facing site (step **1**) triggers a conformational change in the transporter such that the binding site now faces inward toward the cytosol (step **2**). Glucose then is released to the inside of the cell (step **3**). Finally, the

transporter undergoes the reverse conformational change, regenerating the outward-facing binding site (step **4**). If the concentration of glucose is higher inside the cell than outside, the cycle will work in reverse (step **4** → step **1**), resulting in net movement of glucose out of the cell. The actual conformational changes are probably smaller than those depicted here.

derivation as that of the Michaelis-Menten equation in Chapter 3 to derive the following expression for $v_0$, the initial transport rate for S into the cell catalyzed by GLUT1:

$$v_0 = \frac{V_{max}}{1 + \dfrac{K_m}{C}} \qquad (11\text{-}1)$$

where $C$ is the concentration of $S_{out}$ (initially, the concentration of $S_{in} = 0$). $V_{max}$, the rate of transport when all molecules of GLUT1 contain a bound S, occurs at an infinitely high $S_{out}$ concentration. The lower the value of $K_m$, the more tightly the substrate binds to the transporter. Equation 11-1 describes the curve for glucose uptake by erythrocytes shown in Figure 11-4 as well as similar curves for other uniporters.

For GLUT1 in the human erythrocyte membrane, the $K_m$ for glucose transport is 1.5 mM. Thus when the extracellular glucose concentration is 1.5 mM, roughly half the GLUT1 transporters with outward-facing binding sites will have a bound glucose and transport will occur at 50 percent of the maximal rate. Since blood glucose is normally 5 mM, the erythrocyte glucose transporter usually is functioning at 77 percent of its maximal rate, as can be seen from Equation 11-1. The GLUT1 transporter (or the very similar GLUT3 glucose transporter) is expressed by all body cells that need to take up glucose from the blood continuously at high rates; the rate of glucose uptake by such cells will remain high regardless of small changes in the concentration of blood glucose, because the blood concentration remains much higher than the $K_m$ and the intracellular glucose concentration is kept low by metabolism.

In addition to glucose, the isomeric sugars D-mannose and D-galactose, which differ from D-glucose in their configuration at only one carbon atom, are transported by GLUT1 at measurable rates. However, the $K_m$ for glucose (1.5 mM) is much lower than it is for D-mannose (20 mM) or D-galactose (30 mM). Thus GLUT1 is quite specific, having a much higher affinity (indicated by a lower $K_m$) for the normal substrate D-glucose than for other substrates.

GLUT1 accounts for 2 percent of the protein in the plasma membrane of erythrocytes. After glucose is transported into the erythrocyte, it is rapidly phosphorylated, forming glucose 6-phosphate, which cannot leave the cell. Because this reaction, the first step in the metabolism of glucose (see Figure 12-3), is rapid and occurs at a constant rate, the intracellular concentration of glucose is kept low even when glucose is imported from the extracellular environment. Consequently the concentration gradient of glucose (outside greater than inside the cell) is maintained sufficiently high to support continuous, rapid import of additional glucose molecules and provide sufficient glucose for cellular metabolism.

## The Human Genome Encodes a Family of Sugar-Transporting GLUT Proteins

The human genome encodes at least 14 highly homologous **GLUT proteins**, GLUT1–GLUT14, that are all thought to contain 12 membrane-spanning α helices, suggesting that they evolved from a single ancestral transport protein. Although no three-dimensional structure of any GLUT protein is available, detailed biochemical studies on GLUT1 have shown that the amino acid residues in the transmembrane α helices are predominantly hydrophobic; several helices, however, bear amino acid residues (e.g., serine, threonine, asparagine, and glutamine) whose side chains can form hydrogen bonds with the hydroxyl groups on glucose. These residues are thought to form the inward-facing and outward-facing glucose-binding sites in the interior of the protein (see Figure 11-5).

The structures of all GLUT isoforms are thought to be quite similar, and all transport sugars. Nonetheless, their differential expression in various cell types, the regulation of the number of GLUT transporters on cell surfaces, and isoform-specific functional properties enable different body cells to regulate glucose metabolism differently and at the same time allow a constant concentration of glucose in the blood to be maintained. For instance, GLUT3 is found in neuronal cells of the brain. Neurons depend on a constant influx of glucose for metabolism, and the low $K_m$ of GLUT3 for glucose ($K_m = 1.5$ mM), like that of GLUT1, ensures that these cells incorporate glucose from brain extracellular fluids at a high and constant rate.

GLUT2, expressed in liver cells and the insulin-secreting islet β cells of the pancreas, has a $K_m$ of ~20 mM, about 13 times higher than the $K_m$ of GLUT1. As a result, when blood glucose rises after a meal from its basal level of 5 mM to 10 mM or so, the rate of glucose influx will almost double in GLUT2-expressing cells, whereas it will increase only slightly in GLUT1-expressing cells (see Figure 11-4). In liver, the "excess" glucose brought into the cell is stored as the polymer glycogen. In islet β cells, the rise in glucose triggers secretion of the hormone insulin (see Figure 16-38), which in turn lowers blood glucose by increasing glucose uptake and metabolism in muscle and by inhibiting glucose production in liver (see Figure 15-38). Indeed, cell-specific inactivation of GLUT2 in pancreatic β-cells prevents glucose-stimulated insulin secretion and in liver cells (hepatocytes) disrupts the regulated expression of glucose-sensitive genes.

Another GLUT isoform, GLUT4, is expressed only in fat and muscle cells, the cells that respond to insulin by increasing their uptake of glucose, thereby removing glucose from the blood. In the absence of insulin, GLUT4 resides in intracellular membranes, not the plasma membrane, and is unable to facilitate glucose uptake from the extracellular fluid. By a process detailed in Figure 16-39, insulin causes these GLUT4-rich internal membranes to fuse with the plasma membrane, increasing the number of GLUT4 molecules present on the cell surface and thus the rate of glucose uptake. This is one principal mechanism by which insulin lowers blood glucose; defects in the movement of GLUT4 to the plasma membrane is one of the causes of adult onset, or type II, diabetes, a disease marked by continuously high blood glucose.

GLUT5 is the only GLUT protein with a high specificity (preference) for fructose; its principal site of expression is the apical membrane of intestinal epithelial cells, where it transports dietary fructose from the intestinal lumen to inside the cells.

## Transport Proteins Can Be Studied Using Artificial Membranes and Recombinant Cells

There are a variety of approaches for studying the intrinsic properties of transport proteins, such as defining the $V_{max}$ and $K_m$ parameters and identifying key residues responsible for binding. Most cellular membranes contain many different types of transport proteins but a relatively low concentration of any particular one, making functional studies of a single protein difficult. To facilitate such studies, researchers use two approaches for enriching a transport protein of interest so that it predominates in the membrane: purification and reconstitution into artificial membranes, and overexpression in recombinant cells.

In the first approach, a specific transport protein is extracted from its membrane with detergent and purified. Although transport proteins can be isolated from membranes and purified, their functional properties (i.e., their role in the movement of substrates across membranes) can be studied only when they are associated with a membrane. Thus, the purified protein is usually reincorporated into pure phospholipid bilayer membranes, such as liposomes (see Figure 10-3), across which substrate transport can be readily measured. One good source of GLUT1 is erythrocyte membranes. Another is recombinant cultured mammalian cells expressing a GLUT1 transgene, often one expressing a modified GLUT1 that contains an epitope tag (a portion of a molecule to which a monoclonal antibody [see Chapter 9] can bind) fused to its N- or C-terminus. All of the integral proteins in either of these two types of cells can be solubilized by a nonionic detergent, such as octylglucoside. The glucose uniporter GLUT1 can be purified from the solubilized mixture by antibody affinity chromatography (Chapter 3) on a column containing either a GLUT1-specific monoclonal antibody or an antibody specific for the epitope tag, and then incorporated into liposomes made of pure phospholipids.

Alternatively, the gene encoding a specific transport protein can be expressed at high levels in a cell type that normally does not express it. The difference in transport of a substance by the transfected and by control nontransfected cells will be due to the expressed transport protein. In these systems, the functional properties of the various membrane proteins can be examined without ambiguity caused, for instance, by partial protein denaturation during isolation and purification procedures. As an example, overexpressing GLUT1 in lines of cultured fibroblasts increases severalfold their rate of uptake of glucose, and expression of mutant GLUT1 proteins with specific amino acid alterations can identify residues important for substrate binding.

## Osmotic Pressure Causes Water to Move Across Membranes

Movement of water into and out of cells is an important feature of the life of microorganisms, plants, and animals. **Aquaporins** are a family of membrane proteins that allow water and a few other small uncharged molecules, such as glycerol, to cross biomembranes efficiently. But before discussing these transport proteins, we need to review osmosis, the force that powers the movement of water across membranes.

Water spontaneously moves "downhill" across a semipermeable membrane from a solution of low solute concentration (relatively higher water concentration) to one of high solute concentration (relatively lower water concentration), a process termed **osmosis,** or osmotic flow. In effect, osmosis is equivalent to "diffusion" of water across the semipermeable membrane. Osmotic pressure is defined as the hydrostatic pressure required to stop the net flow of water across a membrane separating solutions of different water concentrations (Figure 11-6). In other words, the osmotic pressure balances the entropy-driven thermodynamic force of the water concentration gradient. In this context, the "membrane" may be a layer of cells or a plasma membrane that is permeable to water but not to the solutes. The osmotic pressure is directly proportional to the difference in the concentration

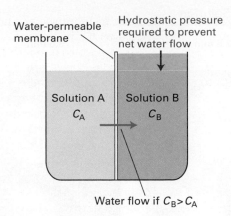

Water-permeable membrane

Hydrostatic pressure required to prevent net water flow

Solution A
$C_A$

Solution B
$C_B$

Water flow if $C_B > C_A$

**FIGURE 11-6 Osmotic pressure.** Solutions A and B are separated by a membrane that is permeable to water but impermeable to all solutes. If $C_B$ (the total concentration of solutes in solution B) is greater than $C_A$, water will tend to flow across the membrane from solution A to solution B. The osmotic pressure π between the solutions is the hydrostatic pressure that would have to be applied to solution B to prevent this water flow. From the van't Hoff equation, osmotic pressure is given by $\pi = RT(C_B - C_A)$, where $R$ is the gas constant and $T$ is the absolute temperature.

of the total number of solute molecules on each side of the membrane. For example, a 0.5 M NaCl solution is actually 0.5 M $Na^+$ ions and 0.5 M $Cl^-$ ions and has the same osmotic pressure as a 1 M solution of glucose or sucrose.

The movement of water across the plasma membrane determines the volume of individual cells, which must be regulated to avoid damage to the cell. Small changes in extracellular osmotic conditions cause most animal cells to swell or shrink rapidly. When placed in a **hypotonic** solution (i.e., one in which the concentration of solutes is *lower* than in the cytosol), animal cells swell owing to the osmotic flow of water inward. Conversely, when placed in a **hypertonic** solution (i.e., one in which the concentration of solutes is *higher* than in the cytosol), animal cells shrink as cytosolic water leaves the cell by osmotic flow. Consequently, cultured animal cells must be maintained in an **isotonic** medium, which has a solute concentration and thus osmotic strength similar to that of the cell cytosol.

In vascular plants, water and minerals are absorbed from the soil by the roots and move up the plant through conducting tubes (the xylem); water loss from the plant, mainly by evaporation from the leaves, drives this movement of water. Unlike animal cells, plant, algal, fungal, and bacterial cells are surrounded by a rigid cell wall, which resists the expansion of the volume of the cell when the intracellular osmotic pressure increases. Without such a wall, animal cells expand when internal osmotic pressure increases—if that pressure rises too much, the cells will burst like overinflated balloons. Because of the cell wall in plants, the osmotic influx of water that occurs when such cells are placed in a hypotonic solution (even pure water) leads to an increase in intracellular pressure but not in cell volume. In plant cells, the concentration of solutes (e.g.,

sugars and salts) usually is higher in the vacuole (see Figure 9-32) than in the cytosol, which in turn has a higher solute concentration than the extracellular space. The osmotic pressure, called *turgor pressure,* generated from the entry of water into the cytosol and then into the vacuole pushes the cytosol and the plasma membrane against the resistant cell wall. Plant cells can harness this pressure to help them stand upright, and also grow. Cell elongation during growth occurs by a hormone-induced localized loosening of a defined region of the cell wall, followed by influx of water into the vacuole, increasing its size and thus the size of the cell. ■

Although most protozoans (like animal cells) do not have a rigid cell wall, many contain a **contractile vacuole** that permits them to avoid osmotic lysis. A contractile vacuole takes up water from the cytosol and, unlike a plant vacuole, periodically discharges its contents through fusion with the plasma membrane. Thus even though water continuously enters the protozoan cell by osmotic flow, the contractile vacuole prevents too much water from accumulating in the cell and swelling it to the bursting point.

## Aquaporins Increase the Water Permeability of Cell Membranes

The natural tendency of water to flow across cell membranes as a result of osmotic pressure raises an obvious question: why don't the cells of fresh-water animals burst in water? For example, frogs lay their eggs in pond water (a hypotonic solution), but frog oocytes and eggs do not swell with water even though their internal salt (mainly KCl) concentration is comparable to that of other cells (~150 mM KCl). These observations were what first led investigators to suspect that the plasma membranes of most cell types, but not frog oocytes, contain water-channel proteins that accelerate the osmotic flow of water. The experimental results shown in Figure 11-7 demonstrate that an aquaporin from the erythrocyte plasma membrane functions as a water channel.

In its functional form, aquaporin is a tetramer of identical 28-kDa subunits (Figure 11-8a). Each subunit contains six membrane-spanning α helices that form a central pore through which water can move in either direction, depending on the osmotic gradient (Figure 11-8b, c). At the center of each monomer, the ~2-nm-long water-selective channel, or pore, is only 0.28 nm in diameter—only slightly larger than the diameter of a water molecule. The molecular sieving properties of the constriction are determined by several conserved hydrophilic amino acid residues whose side-chain and carbonyl groups extend into the middle of the channel and by a relatively hydrophobic wall that lines one side of the channel. Several water molecules move simultaneously through the channel, each molecule sequentially forming specific hydrogen bonds with channel-lining amino acids and displacing another water molecule downstream. Since aquaporins do not undergo conformational changes during water transport, they transport water orders of magnitude faster than GLUT1

0.5 min        1.5 min        2.5 min        3.5 min

**EXPERIMENTAL FIGURE 11-7 Expression of aquaporin by frog oocytes increases their permeability to water.** Frog oocytes, which normally are impermeable to water and do not express an aquaporin protein, were microinjected with mRNA encoding aquaporin. These photographs show control oocytes (bottom cell in each panel) and microinjected oocytes (top cell in each panel) at the indicated times after transfer from an isotonic salt solution (0.1 M) to a hypotonic salt solution (0.035 M). The volume of the control oocytes remained unchanged because they are poorly permeable to water. In contrast, the microinjected oocytes expressing aquaporin swelled and then burst because of an osmotic influx of water, indicating that aquaporin is a water-channel protein. [Courtesy of Gregory M. Preston and Peter Agre, Johns Hopkins University School of Medicine. See L. S. King, D. Kozono, and P. Agre, 2004, *Nat. Rev. Mol. Cell Biol.* **5**:687–698.]

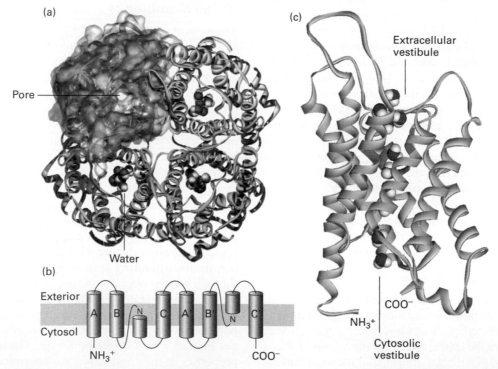

**FIGURE 11-8 Structure of the water-channel protein aquaporin.** (a) Structural model of the tetrameric protein comprising four identical subunits. Each subunit forms a water channel, as seen in this view looking down on the protein from the exoplasmic side. One of the monomers is shown with a molecular surface in which the pore entrance can be seen. (b) Schematic diagram of the topology of a single aquaporin subunit in relation to the membrane. Three pairs of homologous transmembrane α helices (A and A′, B and B′, and C and C′) are oriented in the opposite direction with respect to the membrane and are connected by two hydrophilic loops containing short non-membrane-spanning helices and conserved asparagine (N) residues. The loops bend into the cavity formed by the six transmembrane helices, meeting in the middle to form part of the water-selective gate. (c) Side view of the pore in a single aquaporin subunit in which several water molecules (red oxygens and white hydrogens) are seen within the 2-nm-long water-selective gate that separates the water-filled cytosolic and extracellular vestibules. The gate contains highly conserved arginine and histidine residues, as well as the two asparagine residues whose side chains form hydrogen bonds with transported water molecules. (Key gate residues, including the two asparagines, are highlighted in blue.) Transported water molecules also form hydrogen bonds to the main-chain carbonyl group of a cysteine residue. The arrangement of these hydrogen bonds and the narrow pore diameter of 0.28 nm prevent passage of protons (i.e., $H_3O^+$) or other ions. [After H. Sui et al., 2001, *Nature* **414**:872. See also T. Zeuthen, 2001, *Trends Biochem. Sci.* **26**:77, and K. Murata et al., 2000, *Nature* **407**:599.]

transports glucose. The formation of hydrogen bonds between the oxygen atom of water and the amino groups of two amino acid side chains ensures that only uncharged water (i.e., $H_2O$, but not $H_3O^+$) passes through the channel; the orientations of the water molecules in the channel prevent protons from jumping from one to the next and thus prevent the net movement of protons through the channel. As a consequence ionic gradients are maintained across membranes even when water is flowing across them through aquaporins.

Mammals express a family of aquaporins; 11 such genes are known in humans. Aquaporin 1 is expressed in abundance in erythrocytes, and the homologous aquaporin 2 is found in the kidney epithelial cells that resorb water from the urine, thus controlling the amount of water in the body. The activity of aquaporin 2 is regulated by vasopressin, also called antidiuretic hormone. The regulation of the activity of aquaporin 2 in resting kidney cells resembles that of GLUT4 in fat and muscle in that when its activity is not required, when the cells are in their resting state and water is excreted to form urine, aquaporin 2 is sequestered in intracellular vesicle membranes and so is unable to mediate water import into the cell. When the polypeptide hormone vasopressin binds to the cell-surface vasopressin receptor, it activates a signaling pathway using cAMP as the intracellular signal (detailed in Chapter 15) that causes these aquaporin 2–containing vesicles to fuse with the plasma membrane, increasing the rate of water uptake and its return into the circulation instead of the urine. Inactivating mutations in either the vasopressin receptor or the aquaporin 2 gene cause *diabetes insipidus,* a disease marked by excretion of large volumes of dilute urine. This finding demonstrates that the level of aquaporin 2 is rate limiting for water resorption from urine being formed by the kidney. ∎

Other members of the aquaporin family transport hydroxyl-containing molecules such as glycerol rather than water. Human aquaporin 3, for instance, transports glycerol and is similar in amino acid sequence and structure to the *Escherichia coli* glycerol transport protein GlpF.

## KEY CONCEPTS of Section 11.2

### Facilitated Transport of Glucose and Water

• Protein-catalyzed transport of biological solutes across a membrane occurs much faster than simple diffusion, exhibits a $V_{max}$ when the limited number of transporter molecules are saturated with substrate, and is highly specific for substrate (see Figure 11-4).

• Uniport proteins, such as the glucose transporters (GLUTs), are thought to shuttle between two conformational states, one in which the substrate-binding site faces outward and one in which the binding site faces inward (see Figure 11-5).

• All members of the GLUT protein family transport sugars and have similar structures. Differences in their $K_m$ values,

expression in different cell types, and substrate specificities are important for proper sugar metabolism in the body.

• Two common experimental systems for studying the functions of transport proteins are liposomes containing a purified transport protein and cells transfected with the gene encoding a particular transport protein.

• Most biological membranes are semipermeable, more permeable to water than to ions or most other solutes. Water moves by osmosis across membranes from a solution of lower solute concentration to one of higher solute concentration.

• The rigid cell wall surrounding plant cells prevents their swelling and leads to generation of turgor pressure in response to the osmotic influx of water.

• Aquaporins are water-channel proteins that specifically increase the permeability of biomembranes to water (see Figure 11-8).

• Aquaporin 2 in the plasma membrane of certain kidney cells is essential for resorption of water from urine being formed; the absence of aquaporin 2 leads to the medical condition diabetes insipidus.

## 11.3 ATP-Powered Pumps and the Intracellular Ionic Environment

In previous sections, we focused on transport proteins that move molecules down their concentration gradients (facilitated transport). Here we focus our attention on a major class of proteins—the ATP-powered pumps—that use the energy released by hydrolysis of the terminal phosphoanhydride bond of ATP to transport ions and various small molecules across membranes *against* their concentration gradients. All ATP-powered pumps are transmembrane proteins with one or more binding sites for ATP located on subunits or segments of the protein that always face the cytosol. These proteins commonly are called *ATPases* and they normally do not hydrolyze ATP into ADP and $P_i$ unless ions or other molecules are simultaneously transported. Because of this tight *coupling* between ATP hydrolysis and transport, the energy stored in the phosphoanhydride bond is not dissipated as heat but rather is used to move ions or other molecules uphill against an electrochemical gradient.

### There are Four Main Classes of ATP-Powered Pumps

The general structures of the four classes of ATP-powered pumps are depicted in Figure 11-9, with specific examples in each class listed below the figure. Note that the members of three of the classes (P, F, and V) only transport ions, as do some members of the fourth class, the ABC superfamily. Most members of the ABC superfamily transport small molecules such as amino acids, sugars, peptides, lipids, and other small molecules including many types of drugs.

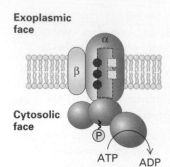

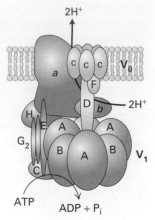

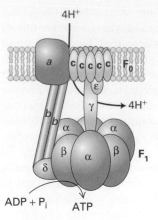

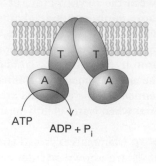

**P-class pumps**

Plasma membrane of plants and fungi (H$^+$ pump)

Plasma membrane of higher eukaryotes (Na$^+$/K$^+$ pump)

Apical plasma membrane of mammalian stomach (H$^+$/K$^+$ pump)

Plasma membrane of all eukaryotic cells (Ca$^{2+}$ pump)

Sarcoplasmic reticulum membrane in muscle cells (Ca$^{2+}$ pump)

**V-class proton pumps**

Vacuolar membranes in plants, yeast, other fungi

Endosomal and lysosmal membranes in animal cells

Plasma membrane of osteoclasts and some kidney tubule cells

**F-class proton pumps**

Bacterial plasma membrane

Inner mitochondrial membrane

Thylakoid membrane of chloroplast

**ABC superfamily**

Bacterial plasma membranes (amino acid, sugar, and peptide transporters)

Mammalian plasma membranes (transporters of phospholipids, small lipophilic drugs, cholesterol, other small molecules)

**FIGURE 11-9 The four classes of ATP-powered transport proteins.** The locations of specific pumps are indicated below each class. P-class pumps are composed of two catalytic α subunits, which become phosphorylated as part of the transport cycle. Two β subunits, present in some of these pumps, may regulate transport. Only one α and β subunit are depicted. V-class and F-class pumps do not form phosphoprotein intermediates and almost all transport only protons. Their structures are similar and contain similar proteins, but none of their subunits are related to those of P-class pumps. V-class pumps couple ATP hydrolysis to transport of protons against a concentration gradient, whereas F-class pumps normally operate in the reverse direction to utilize energy in a proton concentration or voltage gradient to synthesize ATP. All members of the large ABC superfamily of proteins contain two transmembrane (T) domains and two cytosolic ATP-binding (A) domains, which couple ATP hydrolysis to solute movement. These core domains are present as separate subunits in some ABC proteins (depicted here) but are fused into a single polypeptide in other ABC proteins. [See T. Nishi and M. Forgac, 2002, *Nature Rev. Mol. Cell Biol.* **3:**94; C. Toyoshima et al., 2000, *Nature* **405:**647; D. McIntosh, 2000, *Nature Struc. Biol.* **7:**532; and T. Elston, H. Wang, and G. Oster, 1998, *Nature* **391:**510.]

All *P-class ion pumps* possess two identical catalytic α subunits, each of which contains an ATP-binding site. Most also have two smaller β subunits that usually have regulatory functions. During transport, at least one of the α subunits becomes phosphorylated (hence the name "P" class), and the transported ions move through the phosphorylated subunit. The amino acid sequences around the phosphorylated residues are homologous in different pumps. This class includes the Na$^+$/K$^+$ ATPase in the plasma membrane, which generates the low cytosolic Na$^+$ and high cytosolic K$^+$ concentrations typical of animal cells (see Figure 11-3). Certain Ca$^{2+}$ ATPases pump Ca$^{2+}$ ions out of the cytosol into the external medium; others pump Ca$^{2+}$ from the cytosol into the endoplasmic reticulum or into the specialized ER called the **sarcoplasmic reticulum** that is found in muscle cells. Another member of the P class, found in acid-secreting cells of the mammalian stomach, transports protons (H$^+$ ions) out of and K$^+$ ions into the cell.

The structures of *V-class* and *F-class ion pumps* are similar to one another but unrelated to, and more complicated than, P-class pumps. V- and F-class pumps contain several different transmembrane and cytosolic subunits. Virtually all known V and F pumps transport only protons and do so in a process that does not involve a phosphoprotein intermediate. V-class pumps generally function to generate the low pH of plant vacuoles and of lysosomes and other acidic vesicles in animal cells by pumping protons from the cytosolic to the exoplasmic face of the membrane against a proton electrochemical gradient. In contrast, the H$^+$ pumps that generate and maintain the plasma membrane electric potential in plant, fungal, and many bacterial cells belong to the P-class of proton pumps.

F-class pumps are found in bacterial plasma membranes and in mitochondria and chloroplasts. In contrast to V-class pumps, they generally function as reverse proton pumps, in which the energy released by the energetically favored movement of protons from the exoplasmic to the cytosolic face of the membrane *down* the proton electrochemical gradient is used to power the energetically unfavorable synthesis of ATP from ADP and P$_i$. Because of their importance in ATP synthesis in chloroplasts and mitochondria, F-class proton pumps, commonly called ATP synthases, are treated separately in Chapter 12 (Cellular Energetics).

The final class of ATP-powered pumps is a large family of multiple members that are more diverse in function than those of the other classes. Referred to as the **ABC** (*ATP-binding cassette*) **superfamily,** this class includes several hundred different transport proteins found in organisms ranging from bacteria to humans. As detailed below, some of these transport proteins were first identified as multidrug-resistance proteins that, when overexpressed in cancer cells, export anticancer drugs and render tumors resistant to their action. Each ABC protein is specific for a single substrate or group of related substrates, which may be ions, sugars, amino acids, phospholipids, cholesterol, peptides, polysaccharides, or even proteins. All ABC transport proteins share a structural organization consisting of four "core" domains: two transmembrane (T) domains, forming the passageway through which transported molecules cross the membrane, and two cytosolic ATP-binding (A) domains. In some ABC proteins, mostly those in bacteria, the core domains are present in four separate polypeptides; in others, the core domains are fused into one or two multidomain polypeptides.

## ATP-Powered Ion Pumps Generate and Maintain Ionic Gradients Across Cellular Membranes

The specific ionic composition of the cytosol usually differs greatly from that of the surrounding extracellular fluid. In virtually all cells—including microbial, plant, and animal cells—the cytosolic pH is kept near 7.2 *regardless* of the extracellular pH. In the most extreme case, there is a million-fold difference in $H^+$ concentration between the pH of the cytosol of the epithelial cells lining the stomach and the pH of the stomach contents after a meal. Also, the cytosolic concentration of $K^+$ is much higher than that of $Na^+$. In both invertebrates and vertebrates the concentration of $K^+$ is 20–40 times higher in the cytosol than in the blood, while the concentration of $Na^+$ is 8–12 times lower in the cytosol than in the blood (Table 11-2). Some $Ca^{2+}$ in the cytosol is bound to the negatively charged groups in ATP and in proteins and other molecules, but it is the concentration of unbound (or "free") $Ca^{2+}$ that is critical to its functions in signaling pathways and muscle contraction. The concentration of free $Ca^{2+}$ in the cytosol is generally less than 0.2 micromolar ($2 \times 10^{-7}$ M), a thousand or more times lower than that in the blood. Plant cells and many microorganisms maintain similarly high cytosolic concentrations of $K^+$ and low concentrations of $Ca^{2+}$ and $Na^+$ even if the cells are cultured in very dilute salt solutions.

The ion pumps discussed in this section are largely responsible for establishing and maintaining the usual ionic gradients across the plasma and intracellular membranes. In carrying out this task, cells expend considerable energy. For example, up to 25 percent of the ATP produced by nerve and kidney cells is used for ion transport, and human erythrocytes consume up to 50 percent of their available ATP for this purpose; in both cases, most of this ATP is used to power the $Na^+/K^+$ pump (see Figure 11-3). The resultant $Na^+$ and $K^+$ gradients in nerve cells are essential for their

| Ion | Cell (mM) | Blood (mM) |
|---|---|---|
| **Squid Axon (invertebrate)*** | | |
| $K^+$ | 400 | 20 |
| $Na^+$ | 50 | 440 |
| $Cl^-$ | 40–150 | 560 |
| $Ca^{2+}$ | 0.0003 | 10 |
| $X^{-\dagger}$ | 300–400 | 5–10 |
| **Mammalian Cell (vertebrate)** | | |
| $K^+$ | 139 | 4 |
| $Na^+$ | 12 | 145 |
| $Cl^-$ | 4 | 116 |
| $HCO_3^-$ | 12 | 29 |
| $X^-$ | 138 | 9 |
| $Mg^{2+}$ | 0.8 | 1.5 |
| $Ca^{2+}$ | <0.0002 | 1.8 |

**TABLE 11-2** Typical Intracellular and Extracellular Ion Concentrations

*The large nerve axon of the squid has been widely used in studies of the mechanism of conduction of electric impulses.
†$X^-$ represents proteins, which have a net negative charge at the neutral pH of blood and cells.

ability to conduct electric signals rapidly and efficiently, as we detail in Chapter 22. Certain enzymes required for protein synthesis in all cells require a high $K^+$ concentration and are inhibited by high concentrations of $Na^+$; these would cease to function without the operation of the $Na^+/K^+$ pump. In cells treated with poisons that inhibit the production of ATP (e.g., 2,4-dinitrophenol in aerobic cells), the pumping stops and the ion concentrations inside the cell gradually approach those of the exterior environment as ions spontaneously move through channels in the plasma membrane down their electrochemical gradients. Eventually poison-treated cells die: partly because protein synthesis requires a high concentration of $K^+$ ions and partly because in the absence of a $Na^+$ gradient across the cell membrane, a cell cannot import certain nutrients such as amino acids (see Figure 11-3). Studies on the effects of such poisons provided early evidence for the existence and significance of ion pumps.

## Muscle Relaxation Depends on Ca²⁺ ATPases That Pump Ca²⁺ from the Cytosol into the Sarcoplasmic Reticulum

In skeletal muscle cells, $Ca^{2+}$ ions are concentrated and stored in the **sarcoplasmic reticulum** (SR), a specialized type of endoplasmic reticulum (ER). The release via ion channels of stored $Ca^{2+}$ ions from the SR lumen into the cytosol causes muscle contraction, as discussed in Chapter 17. A P-class $Ca^{2+}$ ATPase located in the SR membrane of skeletal muscle pumps $Ca^{2+}$ from the cytosol back into the lumen of the SR, thereby inducing muscle relaxation.

In the cytosol of muscle cells, the free $Ca^{2+}$ concentration ranges from $10^{-7}$ M (resting cells) to more than $10^{-6}$ M (contracting cells), whereas the *total* $Ca^{2+}$ concentration in the SR lumen can be as high as $10^{-2}$ M. The lumen of the SR contains two abundant proteins, calsequestrin and the so-called high-affinity $Ca^{2+}$ binding protein, each of which binds multiple $Ca^{2+}$ ions at high affinity. By binding much of the $Ca^{2+}$ in the SR lumen these proteins reduce the concentration of "free" $Ca^{2+}$ ions in the SR vesicles. This reduces the $Ca^{2+}$ concentration gradient between the cytosol and the SR lumen and consequently reduces the energy needed to pump $Ca^{2+}$ ions into the SR from the cytosol. The activity of the muscle $Ca^{2+}$ ATPase increases as the free $Ca^{2+}$ concentration in the cytosol rises. In skeletal muscle cells the calcium pump in the SR membrane works in concert with a similar $Ca^{2+}$ pump located in the plasma membrane to ensure that the cytosolic concentration of free $Ca^{2+}$ in resting muscle remains below 0.1 μM.

## The Mechanism of Action of the Ca²⁺ Pump Is Known in Detail

Because the calcium pump constitutes more than 80 percent of the integral protein in muscle SR membranes, it is easily purified from other membrane proteins and has been studied extensively. Determination of the three-dimensional structure of this protein in several conformational states representing different steps in the pumping process has revealed much about its mechanism of action, and serves as a paradigm for understanding many P-class ATPase pumps.

The current model for the mechanism of action of the $Ca^{2+}$ ATPase in the SR membrane involves multiple conformational states. For simplicity, we group these into E1 states, in which the two binding sites for $Ca^{2+}$, located in the center of the membrane-spanning domain, face the cytosol, and E2 states, in which these binding sites face the exoplasmic face of the membrane, pointing into the lumen of the SR. Coupling of ATP hydrolysis with ion pumping involves several conformational changes in the protein that must occur in a defined order, as shown in Figure 11-10. When the protein is in the E1 conformation, two $Ca^{2+}$ ions bind to two high-affinity binding sites accessible from the cytosolic side; even though the cytosolic $Ca^{2+}$ concentration is low (see Table 11-2), calcium ions still fill these sites.

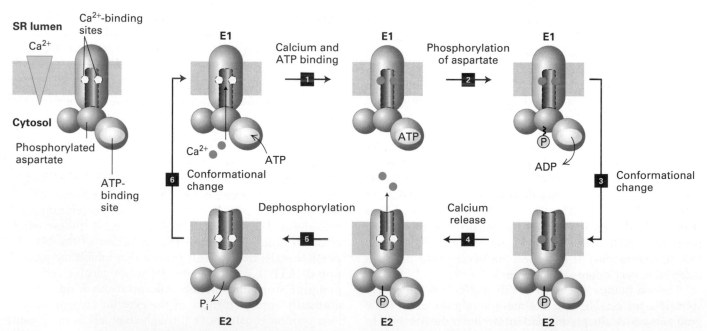

**FIGURE 11-10 Operational model of the Ca²⁺ ATPase in the SR membrane of skeletal muscle cells.** Only one of the two catalytic α subunits of this P-class pump is depicted. E1 and E2 are alternative conformations of the protein in which the Ca²⁺-binding sites are accessible to the cytosolic and exoplasmic (SR lumen) faces, respectively. An ordered sequence of steps, as diagrammed here, is essential for coupling ATP hydrolysis and the transport of Ca²⁺ ions across the membrane. In the figure, ~P indicates a high-energy aspartyl phosphate bond; –P indicates a low-energy bond. Because the affinity of Ca²⁺ for the cytosolic-facing binding sites in E1 is a thousandfold greater than the affinity of Ca²⁺ for the exoplasmic-facing sites in E2, this pump transports Ca²⁺ unidirectionally from the cytosol to the SR lumen. See the text and Figure 11-11 for more details. [See C. Toyoshima and G. Inesi, 2004, *Ann. Rev. Biochem.* **73**:269–292.]

Next, an ATP binds to a site on the cytosolic surface (step **1**). The bound ATP is hydrolyzed to ADP in a reaction that requires $Mg^{2+}$, and the liberated phosphate is transferred to a specific aspartate residue in the protein, forming the high-energy acyl phosphate bond denoted by E1 $\sim$ P (step **2**). The protein then undergoes a conformational change that generates E2, in which the affinity of the two $Ca^{2+}$-binding sites is reduced (shown in detail in the next figure) and in which these sites are now accessible to the SR lumen (step **3**). The free energy of hydrolysis of the aspartyl-phosphate bond in E1 $\sim$P is greater than that in E2–P, and this reduction in free energy of the aspartyl-phosphate bond can be said to power the E1 $\rightarrow$ E2 conformational change.

The $Ca^{2+}$ ions spontaneously dissociate from the low-affinity sites to enter the SR lumen, because even though the $Ca^{2+}$ concentration there is higher than in the cytosol, it is lower than the $K_d$ for $Ca^{2+}$ binding in the low-affinity state (step **4**). Finally, the aspartyl-phosphate bond is hydrolyzed (step **5**). This dephosphorylation, coupled with subsequent binding of cytosolic $Ca^{2+}$ to the high-affinity E1 $Ca^{2+}$ binding sites, stabilizes the E1 conformational state relative to E2, and can be said to power the E2 $\rightarrow$ E1 conformational change (step **6**). Now E1 is ready to transport two more $Ca^{2+}$ ions. Thus the cycle is complete and hydrolysis of a phosphoanhydride bond in ATP has powered the pumping of 2 $Ca^{2+}$ ions against its concentration gradient into the SR lumen.

Much structural and biophysical evidence supports the model depicted in Figure 11-10. For instance, the muscle calcium pump has been isolated with phosphate linked to the key aspartate residue, and spectroscopic studies have detected slight alterations in protein conformation during the E1 $\rightarrow$ E2 conversion. The two phosphorylated states can also be distinguished biochemically; addition of ADP to phosphorylated E1 results in synthesis of ATP, the reverse of step **2** in Figure 11-10, whereas addition of ADP to phosphorylated E2 does not. Each principal conformational state of the reaction cycle can also be characterized by a different susceptibility to various proteolytic enzymes such as trypsin.

Figure 11-11 shows the three-dimensional structure of the $Ca^{2+}$ pump in the E1 state. As can be seen in the right two panels of part c in the bottom half of the figure, the 10 membrane-spanning $\alpha$ helices in the catalytic subunit form the passageway through which $Ca^{2+}$ ions move. Amino acids in four of these helices form the two high-affinity E1 $Ca^{2+}$-binding sites (Figure 11-11a, *left*). One site is formed out of negatively charged oxygen atoms from the carboxyl groups ($COO^-$) of glutamate and aspartate side chains, as well as from water molecules. The other site is formed from side- and main-chain oxygen atoms. Thus, as $Ca^{2+}$-ions bind to the $Ca^{2+}$ pump they lose the water molecules that normally surround a $Ca^{2+}$ ion in aqueous solution (see Figure 2-7), but these waters are replaced by oxygen atoms with a similar geometry that are part of the transport protein. In contrast, in the E2 state (Figure 11-11a, *right*), several of these binding side chains have moved fractions of a nanometer and are unable to interact with bound $Ca^{2+}$ ions, accounting for the low affinity of the E2 state for $Ca^{2+}$ ions.

Binding of $Ca^{2+}$ ions to the $Ca^{2+}$ pump illustrates a general principle of ion binding to channel and transport proteins that we will encounter repeatedly in this chapter: as ions bind they lose most of their waters of hydration but interact with oxygen atoms in the protein that are in a similar geometry to the water oxygens that are bound to the ion in aqueous solution. This reduces the thermodynamic barrier for ion binding to the protein and allows tight binding of the ion even from solutions of relatively low concentrations.

The cytoplasmic region of the $Ca^{2+}$ pump consists of three domains that are well separated from each other in the E1 state (Figure 11-11b). Each of these domains is connected to the membrane-spanning helices by short segments of amino acids. Movements of these cytosolic domains during the pumping cycle cause movements of the connecting segments that are transmitted into movements of the attached membrane-spanning $\alpha$ helices. For example, the phosphorylated residue, Asp 351, is located in the P domain. The adenosine moiety of ATP binds to the N domain, but the $\gamma$-phosphate of ATP binds to specific residues on the P domain, requiring movements of both the N and P domains. Thus, following ATP and $Ca^{2+}$ binding, the $\gamma$ phosphate of the bound ATP sits adjacent to the aspartate on the P domain that is to receive the phosphate. Although the precise details of these and other protein conformational changes are not yet clear, the movements of the N and P domains are transmitted by lever-like motions of the connecting segments into rearrangements of several membrane-spanning $\alpha$ helices. These changes are especially apparent in the four helices that contain the two $Ca^{2+}$-binding sites: the changes prevent the bound $Ca^{2+}$ ions from moving back into the cytosol when released but enable them to dissociate into the exoplasmic space (lumen).

All P-class ion pumps, regardless of which ion they transport, are phosphorylated on a highly conserved aspartate residue during the transport process. As deduced from cDNA sequences, the catalytic $\alpha$ subunits of all the P pumps examined to date have similar amino acid sequences and thus are presumed to have similar arrangements of transmembrane $\alpha$ helices and cytosol-facing A, P, and N domains (see Figure 11-10). These findings strongly suggest that all such proteins evolved from a common precursor although they now transport different ions. This suggestion is borne out by the similarities of the three-dimensional structures of the membrane-spanning segments of the $Na^+/K^+$ ATPase with that of the $Ca^{2+}$ pump (Figure 11-12); the molecular structures of the three cytoplasmic domains are also very similar. Thus, the operational model in Figure 11-11 is generally applicable to all of the P-class ATP-powered ion pumps.

## Calmodulin Regulates the Plasma Membrane Pumps That Control Cytosolic $Ca^{2+}$ Concentrations

As we explain in Chapter 15, in many types of cells in addition to muscle cells, small increases in the concentration of free $Ca^{2+}$ ions in the cytosol trigger a variety of cellular responses.

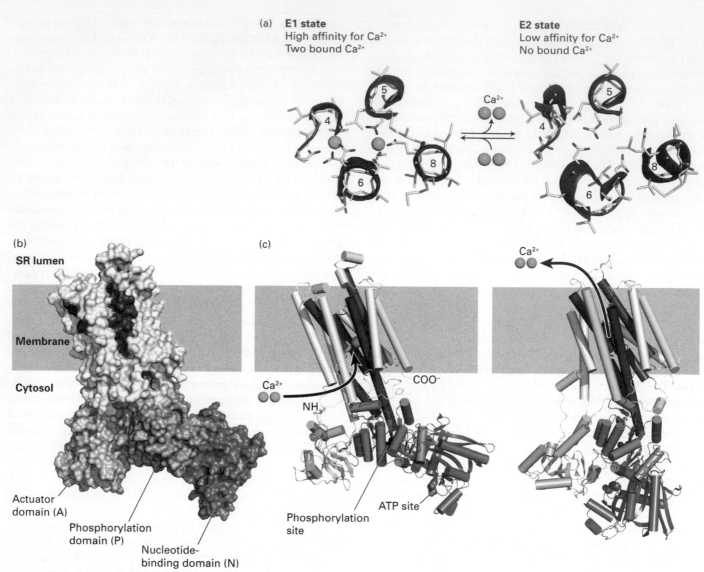

**(a)** **E1 state**
High affinity for Ca²⁺
Two bound Ca²⁺

**E2 state**
Low affinity for Ca²⁺
No bound Ca²⁺

**(b)**
SR lumen
Membrane
Cytosol
Actuator domain (A)
Phosphorylation domain (P)
Nucleotide-binding domain (N)

**(c)**
Ca²⁺
COO⁻
NH₃⁺
Phosphorylation site
ATP site
Ca²⁺

**FIGURE 11-11 Structure of the catalytic α subunit of the muscle Ca²⁺ ATPase.** (a) Ca²⁺-binding sites in the E1 state (*left*), with two bound calcium ions, and the low-affinity E2 state (*right*), without bound ions. Side chains of key amino acids are white, and the oxygen atoms on the glutamate and aspartate side chains are red. In the high-affinity E1 conformation, Ca²⁺ ions bind at two sites between helices 4, 5, 6, and 8 inside the membrane. One site is formed out of negatively charged oxygen atoms from glutamate and aspartate side chains and of water molecules (not shown), and the other is formed out of side- and main-chain oxygen atoms. Seven oxygen atoms surround the Ca²⁺ ion in both sites. (b) Three-dimensional model of the protein in the E1 state based on the structure determined by x-ray crystallography. There are 10 transmembrane α helices, four of which (purple) contain residues that participate in Ca²⁺ binding. The cytosolic segment forms three domains: the nucleotide-binding domain N (blue), the phosphorylation domain P (green), and the actuator domain A (beige) that connects two of the membrane-spanning helices. (c) Models of the pump in the E1 state (left) and E2 state (right). Note the differences between the E1 and E2 states in the conformations of the nucleotide-binding and actuator domains; these movements power the conformational changes of the membrane-spanning α helices (purple) that constitute the Ca²⁺-binding sites, converting them from one in which the Ca²⁺-binding sites are accessible to the cytosolic face (E1 state) to one in which the now loosely bound Ca²⁺ ions are accessible to the exoplasmic face (E2 state). [Adapted from C. Toyoshima and H. Nomura, 2002, *Nature* **418**:605–611; C. Toyoshima and G. Inesi, 2004, *Ann. Rev. Biochem.* **73**:269–292; and E. Gouaux and R. MacKinnon, 2005, *Science* **310**:1461.]

In order for Ca²⁺ to function in intracellular signaling, the concentration of Ca²⁺ ions free in the cytosol usually must be kept below 0.1–0.2 μM. Animal, yeast, and probably plant cells express plasma membrane Ca²⁺ ATPases that transport Ca²⁺ out of the cell against its electrochemical gradient. The catalytic α subunit of these P-class pumps is similar in structure and sequence to the α subunit of the muscle SR Ca²⁺ pump.

The activity of plasma membrane Ca²⁺ ATPases is regulated by **calmodulin,** a cytosolic Ca²⁺-binding protein (see Figure 3-31). A rise in cytosolic Ca²⁺ induces the binding of Ca²⁺ ions to calmodulin, which triggers activation of the Ca²⁺ ATPase. As a result, the export of Ca²⁺ ions from the cell accelerates, quickly restoring the low concentration of free cytosolic Ca²⁺ characteristic of the resting cell.

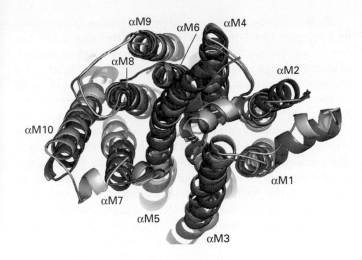

αM9  αM6  αM4
αM8
αM2
αM10
αM7
αM5
αM3

**FIGURE 11-12  Structural comparison of Na⁺/K⁺ ATPase and muscle Ca²⁺ ATPase.** Three-dimensional structure of the Na⁺/K⁺ ATPase (gold) compared to that of the muscle Ca²⁺ ATPase (purple), as seen from the cytoplasmic surface. αM1–10 denote the 10 membrane-spanning α-helices of the Na⁺/K⁺ ATPase. [After J. P. North et. al., 2007, *Nature* **450**:1043 and H. Ogawa et. al., 2009, *Proc. Nat'l Acad. Sci. USA* **106**:13742]

## Na⁺/K⁺ ATPase Maintains the Intracellular Na⁺ and K⁺ Concentrations in Animal Cells

An important P-class ion pump present in the plasma membrane of all animal cells is the **Na⁺/K⁺ ATPase.** This ion pump is a tetramer of subunit composition $\alpha_2\beta_2$, and shares structural homology with the Ca²⁺ pump (see Figure 11-12). The small, glycosylated β transmembrane polypeptide apparently is not involved directly in ion pumping. During the catalytic cycle of the Na⁺/K⁺ ATPase it moves three Na⁺ ions *out* of and two K⁺ ions *into* the cell per ATP molecule hydrolyzed. The mechanism of action of the Na⁺/K⁺ ATPase, outlined in Figure 11-13, is similar to that of the muscle SR calcium pump, except that ions are pumped in *both* directions across the membrane, with each ion moving *against* its concentration gradient. In its E1 conformation, the Na⁺/K⁺ ATPase has three high-affinity Na⁺-binding sites and two low-affinity K⁺-binding sites accessible to the cytosolic surface of the protein. The $K_m$ for binding of Na⁺ to these cytosolic sites is 0.6 mM, a value considerably lower than the intracellular Na⁺ concentration of ~12 mM; as a result, Na⁺ ions normally will fully occupy these sites. Conversely, the affinity of the cytosolic K⁺-binding sites is low enough that K⁺ ions, transported inward through the protein, dissociate from E1 into the cytosol despite the high intracellular K⁺ concentration. During the E1 → E2 transition, the three bound Na⁺ ions

🔊 **OVERVIEW ANIMATION:** Biological Energy Interconversions

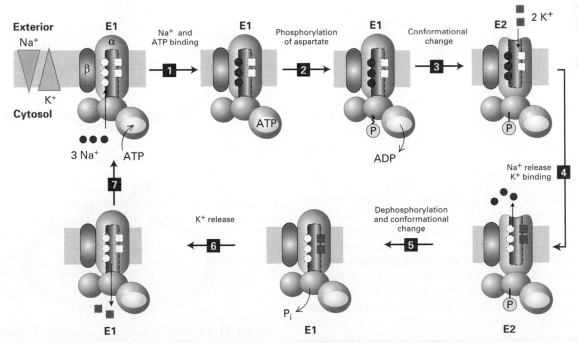

**FIGURE 11-13  Operational model of the plasma membrane Na⁺/K⁺ ATPase.** Only one of the two catalytic α subunits of this P-class pump is depicted. It is not known whether just one or both subunits in a single ATPase molecule transport ions. Ion pumping by the Na⁺/K⁺ ATPase involves phosphorylation, dephosphorylation, and conformational changes similar to those in the muscle Ca²⁺ ATPase (see Figure 11-11). In this case, hydrolysis of the E2–P intermediate powers the E2 → E1 conformational change and concomitant transport of two K⁺ ions inward. Na⁺ ions are indicated by red circles; K⁺ ions, by purple squares; high-energy acyl phosphate bond, by ~P; low-energy phosphoester bond, by –P.

become accessible to the exoplasmic face, and simultaneously the affinity of the three $Na^+$-binding sites drops. The three $Na^+$ ions now bound to low-affinity $Na^+$ sites dissociate one at a time into the extracellular medium despite the high extracellular $Na^+$ concentration. Transition to the E2 conformation also generates two high-affinity $K^+$ sites accessible to the exoplasmic face. Because the $K_m$ for $K^+$ binding to these sites (0.2 mM) is lower than the extracellular $K^+$ concentration (4 mM), these sites will fill with $K^+$ ions as the $Na^+$ ions dissociate. Similarly, during the subsequent E2 → E1 transition, the two bound $K^+$ ions are transported inward and then released into the cytosol.

Certain drugs (e.g., ouabain and digoxin) bind to the exoplasmic domain of the plasma membrane $Na^+/K^+$ ATPase and specifically inhibit its ATPase activity. The resulting disruption in the $Na^+/K^+$ balance of cells is strong evidence for the critical role of this ion pump in maintaining the normal $K^+$ and $Na^+$ ion concentration gradients. Classic Experiment 11.1, which directly follows this chapter, describes the discovery of this important enzyme, which is required for life.

## V-Class $H^+$ ATPases Maintain the Acidity of Lysosomes and Vacuoles

All V-class ATPases transport only $H^+$ ions. These proton pumps, present in the membranes of lysosomes, endosomes, and plant vacuoles, function to acidify the lumen of these organelles. The pH of the lysosomal lumen can be measured precisely in living cells by use of particles labeled with a pH-sensitive fluorescent dye. When these particles are added to the extracellular fluid, the cells engulf and internalize them (phagocytosis; see Chapter 17), ultimately transporting them into lysosomes. The lysosomal pH can be calculated from the spectrum of the fluorescence emitted. The DNA encoding a naturally fluorescent protein whose fluorescence depends on the pH can be modified (by adding DNA segments encoding "signal sequences," detailed in Chapters 13 and 14) such that the protein is targeted to the lysosome lumen; fluorescence measurements can then be used to determine the pH in the organelle lumen. Maintenance of the hundredfold or more proton gradient between the lysosomal lumen (pH ~4.5–5.0) and the cytosol (pH ~7.0) depends on a V-class ATPase and thus ATP production by the cell. The low lysosomal pH is necessary for optimal function of the many proteases, nucleases, and other hydrolytic enzymes in the lumen; on the other hand, a cytosolic pH of 5 would disrupt the functions of many proteins optimized to act at pH 7 and lead to death of the cell.

Pumping of relatively few protons is required to acidify an intracellular vesicle. To understand why, recall that a solution of pH 4 has an $H^+$ ion concentration of $10^{-4}$ moles per liter, or $10^{-7}$ moles of $H^+$ ions per milliliter. Since there are $6.02 \times 10^{23}$ atoms of H per mole (Avogadro's number), then a milliliter of a pH 4 solution contains $6.02 \times 10^{16}$ $H^+$ ions. Thus at pH 4, a primary spherical lysosome with a volume of $4.18 \times 10^{-15}$ ml (diameter of 0.2 μm) will contain just 252 protons. At pH 7, the same organelle would have an average of only 0.2 protons in its lumen, and thus pumping of only approximately 250 protons is necessary for lysosome acidification.

By themselves, ATP-powered proton pumps cannot acidify the lumen of an organelle (or the extracellular space) because these pumps are *electrogenic;* that is, a net movement of electric charge occurs during transport. Pumping of just a few protons causes a buildup of positively charged $H^+$ ions on the exoplasmic (inside) face of the organelle membrane. For each $H^+$ pumped across, a negative ion (e.g., $OH^-$ or $Cl^-$) will be "left behind" on the cytosolic face, causing a buildup of negatively charged ions there. These oppositely charged ions attract each other on opposite faces of the membrane, generating a charge separation, or electric potential, across the membrane. The lysosome membrane thus functions as a capacitor in an electric circuit, storing opposing charges (anions and cations) on opposite sides of a barrier impermeable to movement of charged particles.

As more and more protons are pumped and build up excess positive charge on the exoplasmic face, the energy required to move additional protons against this rising electric potential gradient increases dramatically and prevents pumping of additional protons long before a significant transmembrane $H^+$ concentration gradient is established (Figure 11-14a). In fact, this is the way that P-class $H^+$ pumps generate a cytosol-negative potential across plant and yeast plasma membranes.

In order for an organelle lumen or an extracellular space (e.g., the lumen of the stomach) to become acidic, movement

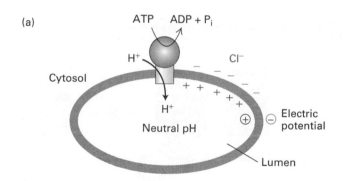

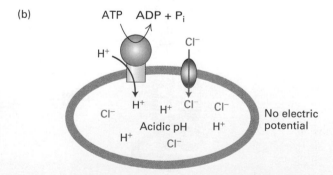

**FIGURE 11-14 Effect of V-class $H^+$ pumps on $H^+$ concentration gradients and electric potential gradients across cellular membranes.** (a) If an intracellular organelle contains only V-class pumps, proton pumping generates an electric potential across the membrane (cytosol-facing side negative and luminal-side positive) but no significant change in the intraluminal pH. (b) If the organelle membrane also contains $Cl^-$ channels, anions passively follow the pumped protons, resulting in an accumulation of $H^+$ and $Cl^-$ ions in the lumen (low luminal pH) but no electric potential across the membrane.

of protons must be accompanied either by (1) movement of an equal number of anions (e.g., $Cl^-$) in the same direction or by (2) movement of equal numbers of a different cation in the opposite direction. The first process occurs in lysosomes and plant vacuoles, whose membranes contain V-class $H^+$ ATPases and anion channels through which accompanying $Cl^-$ ions move (Figure 11-14b). The second process occurs in the lining of the stomach, which contains a P-class $H^+/K^+$ ATPase that is not electrogenic and pumps one $H^+$ outward and one $K^+$ inward. Operation of this pump is discussed later in the chapter.

The ATP-powered proton pumps in lysosomal and vacuolar membranes have been solubilized, purified, and incorporated into liposomes. As shown in Figure 11-9, these V-class proton pumps contain two discrete domains: a cytosolic hydrophilic domain ($V_1$) and a transmembrane domain ($V_0$) with multiple subunits forming each domain. Binding and hydrolysis of ATP by the B subunits in $V_1$ provide the energy for pumping of $H^+$ ions through the proton-conducting channel formed by the 'c' and 'a' subunits in $V_0$. Unlike P-class ion pumps, V-class proton pumps are not phosphorylated and dephosphorylated during proton transport. The structurally similar F-class proton pumps, which we describe in Chapter 12, normally operate in the "reverse" direction to generate ATP rather than pump protons; their structure and mechanism of action is understood in great detail.

## ABC Proteins Export a Wide Variety of Drugs and Toxins from the Cell

As noted earlier, all members of the very large and diverse ABC superfamily of transport proteins contain two transmembrane (T) domains and two cytosolic ATP-binding (A) domains (see Figure 11-9). The T domains, each built of 10 membrane-spanning α helices, form the pathway through which the transported substance (substrate) crosses the membrane (Figure 11-15a) and determine the substrate specificity of each ABC protein. The sequences of the A domains are approximately 30–40 percent homologous in all members of this superfamily, indicating a common evolutionary origin.

Discovery of the first eukaryotic ABC protein to be recognized came from studies on tumor cells and cultured cells that exhibited resistance to several drugs with unrelated chemical structures. Such cells eventually were shown to express elevated levels of a *multidrug-resistance (MDR) transport protein* originally called *MDR1* and now known as ABCB1. This protein uses the energy derived from ATP hydrolysis to *export* a large variety of drugs from the cytosol to the extracellular medium. The *Mdr1* gene is frequently amplified in multidrug-resistant cells, resulting in a large overproduction of the MDR1 protein. In contrast to bacterial ABC proteins, which are built of four discrete subunits, all four domains of mammalian ABCB1 are fused into a single 170,000-MW protein.

The substrates of mammalian ABCB1 are primarily planar, lipid-soluble molecules with one or more positive charges; they all compete with one another for transport, suggesting that they bind to the same or overlapping sites on the protein. Many drugs transported by ABCB1 diffuse from the extracellular medium across the plasma membrane,

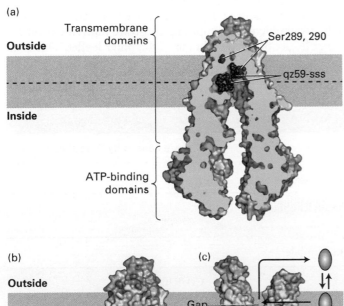

(a)

Transmembrane domains
Outside
Ser289, 290
qz59-sss
Inside
ATP-binding domains

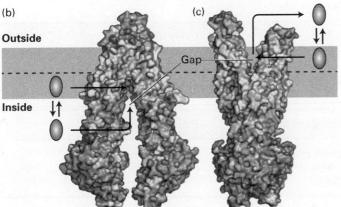

(b)
Outside
Inside
Gap

(c)

**FIGURE 11-15 The multidrug transporter ABCB1 (MDR1): structure and model of ligand export.** (a) Cross-sectional view through the center of an ABCB1 protein bound to two molecules of a drug analog qz59-sss (black) reveals the central location of the ligand-binding site in relation to the phospholipid bilayer: the central ligand-binding cavity is close to the leaflet–leaflet interface of the membrane. During transport, this binding cavity is alternately exposed to the exoplasmic and the cytosolic surface of the membrane. Serines 289 and 290 affect the ligand specificity of the transporter, and are shown as red spheres to highlight their juxtaposition to the bound ligand. Surface residues are colored yellow to denote hydrophobic and blue to denote hydrophilic amino acids. (b) Three-dimensional structure of ABCB1 with its ligand-binding site facing inward toward the cytosol. In this conformation a hydrophilic ligand can bind directly from the cytosol. A more hydrophobic ligand can partition into the inner leaflet of the plasma membrane and then enter the ligand-binding site through a gap in the protein that is accessible directly to the hydrophobic core of the inner leaflet. (c) Model for the structure of ABCB1 with its ligand-binding site facing outward, based on the structures of homologous bacterial ABC proteins. When the transporter assumes this conformation the ligand can either diffuse into the exoplasmic leaflet or directly into the aqueous extracellular medium. [After D. Gutman et. al., 2009, *Trends Biochem. Sci.* **35**:36–42. Structures from S. G. Aller et al., 2009, *Science* **323**:1718–1722.]

unaided by transport proteins, into the cell cytosol, where they block various cellular functions. Two such drugs are colchicine and vinblastine, which block assembly of microtubules (Chapter 18). ATP-powered export of such drugs by MDR1 reduces their concentration in the cytosol. As a result, a much higher extracellular drug concentration is required to kill cells that express ABCB1 than those that do not. That ABCB1 is an ATP-powered small-molecule pump has been demonstrated with liposomes containing the purified protein. Different drugs enhance the ATPase activity of these liposomes in a dose-dependent manner corresponding to their ability to be transported by ABCB1.

The three-dimensional structures of ABCB1, together with those of homologous bacterial ABC transporters, revealed its mechanism of transport as well as its ability to bind and transport a wide array of hydrophilic and hydrophobic substrates (Figure 11-15). The two T domains form a binding site in the center of the membrane that alternates between an inward (Figure 11-15b) and an outward (Figure 11-15c) facing orientation. The alternation between these two conformational states of the protein is powered by ATP binding to the two A subunits and subsequent hydrolysis to ADP and $P_i$, but precisely how this happens is not known.

The substrate-binding cavity formed by ABCB1 is large. Some of the amino acids that line the cavity have aromatic side chains, mainly tyrosine and phenylalanine, allowing ABCB1 to bind multiple types of hydrophobic ligands. Other segments of the cavity are lined with hydrophilic residues, allowing hydrophilic or amphipathic molecules to bind. In the inward-facing conformation the binding site is open directly to the surrounding aqueous solutions, allowing hydrophilic molecules to enter the binding site directly from the cytosol. A gap in the protein is accessible directly from the hydrophobic core of the inner leaflet of the membrane bilayer; this allows hydrophobic molecules to enter the binding site directly from the inner leaflet of the phospholipid bilayer (Figure 11-15b). After the ATP-powered change to the outward-facing conformation, molecules can exit the binding site into the exoplasmic membrane leaflet or directly into the extracellular medium (Figure 11-15c).

About 50 different mammalian ABC transport proteins are now recognized (Table 11-3). Several are expressed in abundance in the liver, intestines, and kidney—sites where natural toxic and waste products are removed from the body. Substrates for these ABC proteins include sugars, amino acids, cholesterol, bile acids, phospholipids, peptides, proteins, toxins, and foreign substances. The normal function of ABCB1 most likely is to transport various natural and metabolic toxins into the bile or intestinal lumen for excretion or into the urine being formed in the kidney. During the course of its evolution, ABCB1 appears to have acquired the ability to transport drugs whose structures are similar to those of these endogenous toxins. Tumors derived from MDR-expressing cell types, such as hepatomas (liver cancers), frequently are resistant to virtually all chemotherapeutic agents and are thus difficult to treat, presumably because the tumors exhibit increased expression of ABCB1 or a related ABC protein.

## Certain ABC Proteins "Flip" Phospholipids and Other Lipid-Soluble Substrates from One Membrane Leaflet to the Other

As shown in Figure 11-15b and c, ABCB1 can move, or "flip," a hydrophobic or amphipathic substrate molecule from the inner leaflet of the membrane to the outer leaflet. This is an otherwise energetically unfavorable reaction powered by the

| TABLE 11-3 | Selected Human ABC Proteins | | |
|---|---|---|---|
| Protein | Tissue Expression | Function | Disease Caused by Defective Protein |
| ABCB1 (MDR1) | Adrenal, kidney, brain | Exports lipophilic drugs | |
| ABCB4 (MDR2) | Liver | Exports phosphatidylcholine into bile | |
| ABCB11 | Liver | Exports bile salts into bile | |
| CFTR | Exocrine tissue | Transports Cl ions | Cystic fibrosis |
| ABCDI | Ubiquitous in peroxisomal membrane | Influences activity of peroxisomal enzyme that oxidizes very long chain fatty acids | Adrenoleukodystrophy (ADL) |
| ABCG5/8 | Liver, intestine | Exports cholesterol and other sterols | β-Sitosterolemia |
| ABCA1 | Ubiquitous | Exports cholesterol and phospholipid for uptake into high-density lipoprotein (HDL) | Tangier's disease |

coupled ATPase activity of the protein. Support for this so-called *flippase* model of transport by ABCB1 comes from experiments on ABCB4 (originally called MDR2), a protein homologous to ABCB1 that is present in the region of the liver-cell plasma membrane that faces the bile canaliculi. ABCB4 moves phosphatidylcholine from the cytosolic to the exoplasmic leaflet of the plasma membrane for subsequent release into the bile in combination with cholesterol and bile acids, which themselves are transported by other ABC family members. Several other ABC superfamily members participate in the cellular export of various lipids, presumably by mechanisms similar to that of ABCB1 (see Table 11-3).

ABCB4 was first suspected of having phospholipid flippase activity because mice with homozygous loss-of-function mutations in the *ABCB4* gene exhibited defects in the secretion of phosphatidylcholine into bile. To determine directly if ABCB4 was in fact a flippase, researchers performed experiments on a homogeneous population of purified vesicles isolated from special mutant yeast cells with ABCB4 in the membrane and with the cytosolic face directed outward (Figure 11-16). After

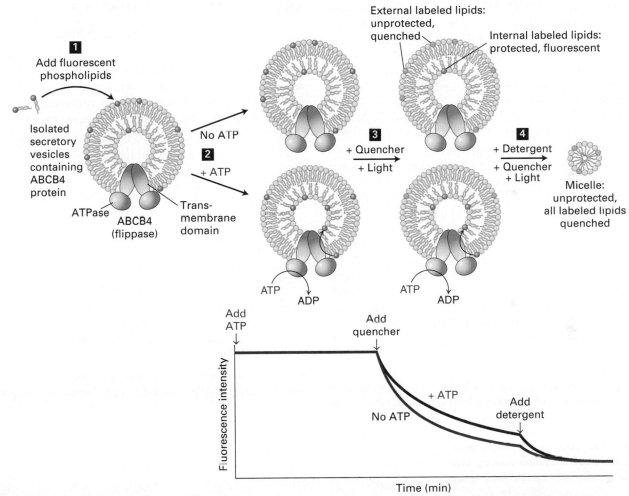

**EXPERIMENTAL FIGURE 11-16 In vitro fluorescence-quenching assay can detect phospholipid flippase activity of ABCB4.** A homogeneous population of secretory vesicles containing ABCB4 protein was obtained by introducing the cDNA encoding mammalian ABCB4 into a temperature-sensitive yeast *sec* mutant such that ABCB4 was localized to intracellular endoplasmic reticulum vesicles in its normal orientation and with the cytosolic face of the vesicles facing outward (see Figure 14-4). Step **1**: Synthetic phospholipids containing a fluorescently modified head group (blue) were incorporated primarily into the outer, cytosolic leaflets of the purified vesicles. Step **2**: If ABCB4 acted as a flippase, then on addition of ATP to the outside of the vesicles, a small fraction of the outward-facing labeled phospholipids would be flipped to the inside leaflet. Step **3**: Flipping was detected by adding a membrane-impermeable quenching compound called dithionite to the medium surrounding the vesicles. Dithionite reacts with the fluorescent head group, destroying its ability to fluoresce (gray). In the presence of the quencher, only labeled phospholipid in the protected environment on the inner leaflet will fluoresce. Subsequent to the addition of the quenching agent, the total fluorescence decreases with time until it plateaus at the point at which all external fluorescence is quenched and only the internal phospholipid fluorescence can be detected. The observation of greater fluorescence (less quenching) in the presence of ATP than in its absence indicates that ABCB4 has flipped some of the labeled phospholipid to the inside. Not shown here are "control" vesicles isolated from cells that did not express ABCB4 and that exhibited no flippase activity. Step **4**: Addition of detergent to the vesicles generates micelles and makes all fluorescent lipids accessible to the quenching agent, lowering the fluorescence to baseline values. [Adapted from S. Ruetz and P. Gros, 1994, *Cell* **77**:1071.]

purifying these vesicles, investigators labeled them in vitro with a fluorescent phosphatidylcholine derivative. The fluorescence-quenching assay outlined in Figure 11-16 was used to demonstrate that the vesicles containing ABCB4 exhibited an ATP-dependent flippase activity.

## The ABC Cystic Fibrosis Transmembrane Regulator (CFTR) Is a Chloride Channel, Not a Pump

Several human genetic diseases are associated with defective ABC proteins (see Table 11-3). The best studied and most widespread is cystic fibrosis (CF), caused by a mutation in the gene encoding the *cystic fibrosis transmembrane regulator (CFTR, also called ABCC7)*. Like other ABC proteins, CFTR has two transmembrane T domains and two cytosolic A, or ATP-binding, domains. CFTR contains an additional R (regulatory) domain on the cytosolic face; R links the two homologous halves of the protein, creating an overall domain organization of T1–A1–R–T2–A2. But CFTR is a $Cl^-$ channel protein, not an ion pump. It is expressed in the apical plasma membranes of epithelial cells in the lung, sweat glands, pancreas, and other tissues. For instance, CFTR protein is important for reuptake, into the cells of sweat glands, of $Cl^-$ lost by sweating; babies with cystic fibrosis, if licked, often taste "salty" because this reuptake is inhibited.

The $Cl^-$ channel of CFTR is normally shut. Channel opening is activated by phosphorylation of the R domain by a protein kinase (PKA, discussed in Chapter 15) that in turn is activated by an increase in cyclic AMP (cAMP), a small intracellular signaling molecule. Opening of the channel also requires sequential binding of two ATP molecules to the two A domains (Figure 11-17).

About two-thirds of all CF disease cases can be attributed to a single mutation in CFTR: deletion of Phe 508 in the ATP-binding A1 domain. At body temperature, the mutant protein fails to fold properly and to move to the cell surface where it normally functions. Interestingly, if cells expressing the mutant protein are incubated at room temperature, the protein accumulates normally on the plasma membrane, where it functions nearly as well as the wild-type CFTR channel. Much effort is now being directed toward identifying small molecules that might allow the mutant protein in CF patients to traffic normally to the cell surface and thus reverse the effects of the disease. ∎

**FIGURE 11-17 Structure and function of the cystic fibrosis transmembrane regulator (CFTR).** Structural interpretation of the ATP-dependent gating cycle of phosphorylated CFTR channels. The regulatory (R) domain (not depicted) must be phosphorylated before ATP is able to support channel opening. One ATP (yellow circle) becomes tightly bound to the A1 domain (green). Binding of the second ATP to the A2 domain (blue) is followed by formation of a tight intramolecular A1–A2 heterodimer and slow channel opening. The relatively stable open state becomes destabilized by hydrolysis of the ATP bound at A2 to ADP (red crescent) and $P_i$. The ensuing disruption of the tight A1–A2 dimer interface leads to channel closure. T=transmembrane domain; A=cytosolic ATP-binding domain. [After D. C. Gadsby et. al., 2006, *Nature* **440**:477 and S. G. Aller, 2009, *Science* **323**:1718.]

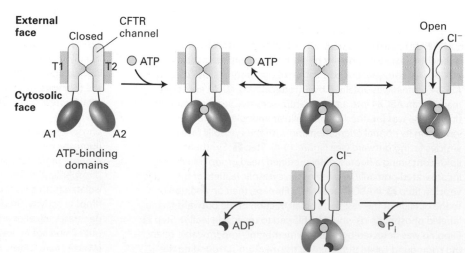

for maintaining a lower pH inside the organelles than in the surrounding cytosol (see Figure 11-14).

- All members of the large and diverse ABC superfamily of transport proteins contain four core domains: two transmembrane domains, which form a pathway for solute movement and determine substrate specificity, and two cytosolic ATP-binding domains (see Figure 11-15).

- The ABC superfamily includes bacterial amino acid and sugar permeases and about 50 mammalian proteins (e.g., ABCB1, ABCA1) that transport a wide array of substrates, including toxins, drugs, phospholipids, peptides, and proteins, into or out of the cell.

- The two T domains of the multidrug transporter ABCB1 form a ligand-binding site in the middle of the plane of the membrane; ligands can bind directly from the cytosol or from the inner membrane leaflet through a gap in the protein.

- Biochemical experiments directly demonstrate that ABCB4 (MDR2) possesses phospholipid flippase activity (see Figure 11-16).

- CFTR, an ABC protein, is a $Cl^-$ channel protein, not an ion pump. Channel opening is triggered by protein phosphorylation and by binding of ATP to the two A domains (Figure 11-17).

## 11.4 Nongated Ion Channels and the Resting Membrane Potential

In addition to ATP-powered ion pumps, which transport ions *against* their concentration gradients, the plasma membrane contains channel proteins that allow the principal cellular ions ($Na^+$, $K^+$, $Ca^{2+}$, and $Cl^-$) to move through them at different rates *down* their concentration gradients. Ion concentration gradients generated by pumps and selective movements of ions through channels constitute the principal mechanism by which a difference in voltage, or electric potential, is generated across the plasma membrane. In other words, ATP-powered ion pumps generate differences in ion concentrations across the plasma membrane, and ion channels utilize these concentration gradients to generate a tightly controlled electric potential across the membrane (see Figure 11-3).

In all cells the magnitude of this electric potential generally is ~70 millivolts (mV), with the *inside* cytosolic face of the cell membrane always *negative* with respect to the exoplasmic face. This value does not seem like much until we consider that the thickness of the plasma membrane is only ~3.5 nm. Thus the voltage gradient across the plasma membrane is 0.07 V per $3.5 \times 10^{-7}$ cm, or 200,000 volts per centimeter! (To appreciate what this means, consider that high-voltage transmission lines for electricity utilize gradients of about 200,000 volts per kilometer, $10^5$-fold less!)

The ionic gradients and electric potential across the plasma membrane play crucial roles in many biological processes. As noted previously, a rise in the cytosolic $Ca^{2+}$ concentration is an important regulatory signal, initiating contraction in muscle cells and triggering in many cells secretion of proteins, such as digestive enzymes from pancreatic cells. In many animal cells, the combined force of the $Na^+$ concentration gradient and membrane electric potential drives the uptake of amino acids and other molecules against their concentration gradients by symport and antiport proteins (see Figure 11-3 and Section 11.5). Furthermore, the electrical signaling by nerve cells depends on the opening and closing of ion channels in response to changes in the membrane electric potential (Chapter 22).

Here, we discuss the origin of the membrane electric potential in resting non-neuronal cells, often called the cell's "resting potential"; how ion channels mediate the selective movement of ions across a membrane; and useful experimental techniques for characterizing the functional properties of channel proteins.

### Selective Movement of Ions Creates a Transmembrane Electric Gradient

To help explain how an electric potential across the plasma membrane can arise, we first consider a set of simplified experimental systems in which a membrane separates a 150 mM NaCl/15 mM KCl solution (similar to the extracellular medium surrounding metazoan cells) on the right from a 15 mM NaCl/150 mM KCl solution (similar to that of the cytosol) on the left (Figure 11-18a). A potentiometer (voltmeter) is connected to both solutions to measure any difference in electric potential across the membrane. If the membrane is impermeable to all ions, no ions will flow across it. Initially both solutions contain an equal number of positive and negative ions. Furthermore, there will be no difference in voltage, or electric potential gradient, across the membrane, as shown in Figure 11-18a.

Now suppose that the membrane contains $Na^+$-channel proteins that accommodate $Na^+$ ions but exclude $K^+$ and $Cl^-$ ions (Figure 11-18b). $Na^+$ ions then tend to move down their concentration gradient from the right side to the left, leaving an excess of negative $Cl^-$ ions compared with $Na^+$ ions on the right side and generating an excess of positive $Na^+$ ions compared with $Cl^-$ ions on the left side. The excess $Na^+$ on the left and $Cl^-$ on the right remain near the respective surfaces of the membrane because the excess positive charges on one side of the membrane are attracted to the excess negative charges on the other side. The resulting separation of charge across the membrane constitutes an electric potential, or voltage, with the left (cytosolic) side of the membrane having excess positive charge with respect to the right.

As more and more $Na^+$ ions move through channels across the membrane, the magnitude of this charge difference (i.e., voltage) increases. However, continued right-to-left movement of the $Na^+$ ions eventually is inhibited by the mutual repulsion between the excess positive ($Na^+$) charges accumulated on the left side of the membrane and by the attraction of $Na^+$ ions to the excess negative charges built up

(a) Membrane impermeable to Na⁺, K⁺, and Cl⁻

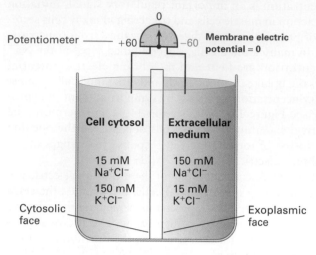

(b) Membrane permeable only to Na⁺

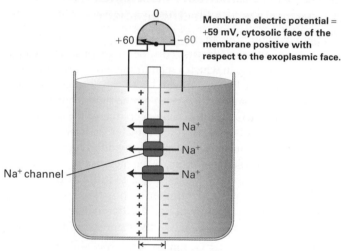

Charge separation across membrane

(c) Membrane permeable only to K⁺

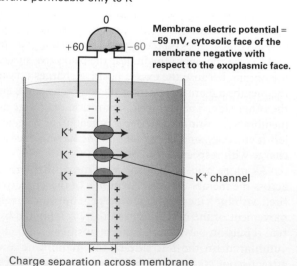

Charge separation across membrane

**EXPERIMENTAL FIGURE 11-18 Generation of a transmembrane electric potential (voltage) depends on the selective movement of ions across a semipermeable membrane.** In this experimental system, a membrane separates a 15 mM NaCl/150 mM KCl solution (*left*) from a 150 mM NaCl/15 mM KCl solution (*right*); these ion concentrations are similar to those in cytosol and blood, respectively. If the membrane separating the two solutions is impermeable to all ions (a), no ions can move across the membrane and no difference in electric potential is registered on the potentiometer connecting the two solutions. If the membrane is selectively permeable only to Na⁺ (b) or to K⁺ (c), then diffusion of ions through their respective channels leads to a separation of charge across the membrane. At equilibrium, the membrane potential caused by the charge separation becomes equal to the Nernst potential $E_{Na}$ or $E_K$ registered on the potentiometer. See the text for further explanation.

on the right side. The system soon reaches an equilibrium point at which the two opposing factors that determine the movement of Na⁺ ions—the membrane electric potential and the ion concentration gradient—balance each other out. At equilibrium, no net movement of Na⁺ ions occurs across the membrane. Thus this membrane, like all biological membranes, acts as a *capacitor*—a device consisting of a thin sheet of nonconducting material (the hydrophobic interior) surrounded on both sides by electrically conducting material (the polar phospholipid head groups and the ions in the surrounding aqueous solution) that can store positive charges on one side and negative charges on the other.

If a membrane is permeable only to Na⁺ ions, then at equilibrium the measured electric potential across the membrane equals the sodium equilibrium potential in volts, $E_{Na}$. The magnitude of $E_{Na}$ is given by the *Nernst equation*, which is derived from basic principles of physical chemistry:

$$E_{Na} = \frac{RT}{ZF} \ln \frac{[Na_{right}]}{[Na_{left}]} \qquad (11\text{-}2)$$

where $R$ (the gas constant) = 1.987 cal/(degree · mol), or 8.28 joules/(degree · mol); $T$ (the absolute temperature in degrees Kelvin) = 293 °K at 20 °C; $Z$ (the charge, also called the valency) here equal to +1; $F$ (the Faraday constant) = 23,062 cal/(mol · V), or 96,000 coulombs/(mol · V); and $[Na_{left}]$ and $[Na_{right}]$ are the Na⁺ concentrations on the left and right sides, respectively, at equilibrium. By convention the potential is expressed as the *cytosolic* face of the membrane relative to the *exoplasmic* face, and the equation is written with the concentration of ion in the extracellular solution (here the right side of the membrane) placed in the numerator and that of the cytosol in the denominator.

At 20 °C, Equation 11-2 reduces to

$$E_{Na} = 0.059 \log_{10} \frac{[Na_{right}]}{[Na_{left}]} \qquad (11\text{-}3)$$

If $[Na_{right}]/[Na_{left}] = 10$, a tenfold ratio of concentrations as in Figure 11-18b, then $E_{Na} = +0.059$ V (or +59 mV), with the left, cytosolic side positive with respect to the exoplasmic right side.

If the membrane is permeable only to $K^+$ ions and not to $Na^+$ or $Cl^-$ ions, then a similar equation describes the potassium equilibrium potential $E_K$:

$$E_K = 0.059 \log_{10} \frac{[K_{right}]}{[K_{left}]} \qquad (11\text{-}4)$$

The *magnitude* of the membrane electric potential is the same (59 mV, for a tenfold difference in ion concentrations), except that the left, cytosolic side is now *negative* with respect to the right (Figure 11-18c), opposite to the polarity obtained across a membrane selectively permeable to $Na^+$ ions.

## The Resting Membrane Potential in Animal Cells Depends Largely on the Outward Flow of $K^+$ Ions Through Open $K^+$ Channels

The plasma membranes of animal cells contain many open $K^+$ channels but few open $Na^+$, $Cl^-$, or $Ca^{2+}$ channels. As a result, the major ionic movement across the plasma membrane is that of $K^+$ from the *inside outward*, powered by the $K^+$ concentration gradient, leaving an excess of *negative* charge on the cytosolic face of the plasma membrane and creating an excess of *positive* charge on the exoplasmic face, similar to the experimental system shown in Figure 11-18c. This outward flow of $K^+$ ions through these channels, called **resting $K^+$ channels,** is the major determinant of the inside-negative membrane potential. Like all channels, these alternate between an open and a closed state (Figure 11-2), but since their opening and closing are not affected by the membrane potential or by small signaling molecules, these channels are called *nongated*. The various gated channels in nerve cells (Chapter 22) open only in response to specific ligands or to changes in membrane potential.

Quantitatively, the usual resting membrane potential of −70 mV is close to the potassium equilibrium potential, calculated from the Nernst equation and the $K^+$ concentrations in cells and surrounding media depicted in Table 11-2. Usually the potential is lower (less negative) than that calculated from the Nernst equation because of the presence of a few open $Na^+$ channels. These open $Na^+$ channels allow the net *inward* flow of $Na^+$ ions, making the cytosolic face of the plasma membrane more positive, that is, less negative, than predicted by the Nernst equation for $K^+$. The $K^+$ concentration gradient that drives the flow of ions through resting $K^+$ channels is generated by the $Na^+/K^+$ ATPase described previously (see Figures 11-3 and 11-13). In the absence of this pump, or when it is inhibited, the $K^+$ concentration gradient cannot be maintained, the magnitude of the membrane potential falls to zero, and the cell eventually dies.

Although resting $K^+$ channels play the dominant role in generating the electric potential across the plasma membrane of animal cells, this is not the case in bacterial, plant, and fungal cells. The inside-negative membrane potential in plant and fungal cells is generated by transport of positively charged protons ($H^+$) out of the cell by ATP-powered pro-

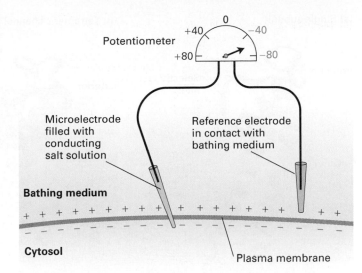

**EXPERIMENTAL FIGURE 11-19** **The electric potential across the plasma membrane of living cells can be measured.** A microelectrode, constructed by filling a glass tube of extremely small diameter with a conducting fluid such as a KCl solution, is inserted into a cell in such a way that the surface membrane seals itself around the tip of the electrode. A reference electrode is placed in the bathing medium. A potentiometer connecting the two electrodes registers the potential, in this case −60 mV with the cytosolic face *negative* with respect to the exoplasmic face of the membrane. A potential difference is registered only when the microelectrode is inserted into the cell; no potential is registered if the microelectrode is in the bathing fluid.

ton pumps, a process similar to what occurs in lysosomal membranes lacking $Cl^-$ channels (see Figure 11-14a): each $H^+$ pumped out of the cell leaves behind a $Cl^-$ ion, generating an electric potential gradient (cytosolic face negative) across the membrane. In aerobic bacterial cells the inside negative potential is generated by outward pumping of protons during electron transport, a process similar to proton pumping in mitochondrial inner membranes that will be discussed in detail in Chapter 12 (see Figure 12-16).

The potential across the plasma membrane of large cells can be measured with a microelectrode inserted inside the cell and a reference electrode placed in the extracellular fluid. The two are connected to a potentiometer capable of measuring small potential differences (Figure 11-19). The potential across the surface membrane of most animal cells generally does not vary with time. In contrast, neurons and muscle cells—the principal types of electrically active cells—undergo controlled changes in their membrane potential, as we discuss in Chapter 22.

## Ion Channels Are Selective for Certain Ions by Virtue of a Molecular "Selectivity Filter"

All ion channels exhibit specificity for particular ions: $K^+$ channels allow $K^+$ but not closely related $Na^+$ ions to enter, whereas $Na^+$ channels admit $Na^+$ but not $K^+$. Determination of the three-dimensional structure of a bacterial $K^+$ channel first revealed how this exquisite ion selectivity is achieved. Comparisons of the sequences and structures of other $K^+$ channels from organisms as diverse as bacteria, fungi, and

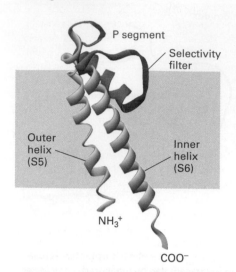

(a) Single subunit

P segment

Selectivity filter

Outer helix (S5)

Inner helix (S6)

$NH_3^+$

$COO^-$

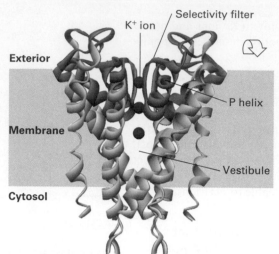

(b) Tetrameric channel

Selectivity filter

$K^+$ ion

Exterior

P helix

Membrane

Vestibule

Cytosol

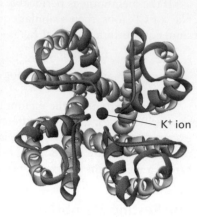

$K^+$ ion

**FIGURE 11-20 Structure of a resting $K^+$ channel from the bacterium *Streptomyces lividans*.** All $K^+$ channel proteins are tetramers comprising four identical subunits, each containing two conserved membrane-spanning α helices, called by convention S5 and S6, and a shorter P, or pore segment. (a) One of the subunits, viewed from the side, with key structural features indicated. (b) The complete tetrameric channel viewed from the side (*left*) and the top, or extracellular end (*right*). The P segments (pink) are located near the exoplasmic surface and connect the S5 and S6 α helices (yellow and silver); they consist of a nonhelical "turret," which lines the upper part of the pore; a short α helix; and an extended loop that protrudes into the narrowest part of the pore and forms the ion-selectivity filter. This filter allows $K^+$ (purple spheres) but not other ions to pass. Below the filter is the central cavity or vestibule lined by the inner, or S6, α helixes. The subunits in gated $K^+$ channels, which open and close in response to specific stimuli, contain additional transmembrane helices not shown here; these are discussed in Chapter 22. [See Y. Zhou et al., 2001, *Nature* **414**:43.]

humans established that all share a common structure and probably evolved from a single type of channel protein.

Like all other $K^+$ channels, bacterial $K^+$ channels are built of four identical transmembrane subunits symmetrically arranged around a central pore (Figure 11-20). Each subunit contains two membrane-spanning α helices (S5 and S6) and a short P (pore) segment that partly penetrates the membrane bilayer from the exoplasmic surface. In the tetrameric $K^+$ channel, the eight transmembrane α helices (two from each subunit) form an inverted cone, generating a water-filled cavity called the *vestibule* in the central portion of the channel that extends halfway through the membrane toward the cytosolic side. Four extended loops that are part of the four P segments form the actual *ion-selectivity filter* in the narrow part of the pore near the exoplasmic surface, above the vestibule.

Several related pieces of evidence support the role of P segments in ion selection. First, the amino acid sequence of the P segment is highly homologous in all known $K^+$ channels and is different from that in other ion channels. Second, mutation of certain amino acids in this segment alters the ability of a $K^+$ channel to distinguish $Na^+$ from $K^+$. Finally, replacing the P segment of a bacterial $K^+$ channel with the homologous segment from a mammalian $K^+$ channel yields a chimeric protein that exhibits normal selectivity for $K^+$ over other ions. Thus all $K^+$ channels are thought to use the same mechanism to distinguish $K^+$ from other ions.

$Na^+$ ions are smaller than $K^+$ ions. How, then, can a channel protein exclude smaller $Na^+$ ions, yet allow passage of larger $K^+$? The ability of the ion-selectivity filter in $K^+$ channels

to select $K^+$ over $Na^+$ is due mainly to backbone carbonyl oxygens on residues located in a Gly-Tyr-Gly sequence that is found in an analogous position in the P segment in every known $K^+$ channel. As a $K^+$ ion enters the narrow selectivity filter—the space between the P segment filter sequences contributed by the four adjacent subunits—it loses its eight waters of hydration but becomes bound in the same geometry to eight backbone carbonyl oxygens, two from the extended loop in each of the four P segments lining the channel (Figure 11-21a, bottom left). Thus little energy is required to strip off the eight waters of hydration of a $K^+$ ion, and as a result, a relatively low activation energy is required for passage of $K^+$ ions into the channel from an aqueous solution. A dehydrated $Na^+$ ion is too small to bind to all eight carbonyl oxygens that line the selectivity filter with the same geometry as a $Na^+$ ion surrounded by its normal eight water molecules in aqueous solution. As a result, $Na^+$ ions would "prefer" to remain in water rather than enter the selectivity filter, and thus the change in free energy for entry of $Na^+$ ions into the channel is relatively high (Figure 11-21a, right). This difference in free energies favors passage of $K^+$ ions through the channel over $Na^+$ by a factor of 1000. Like $Na^+$, the dehydrated $Ca^{2+}$ ion is smaller than the dehydrated $K^+$ ion and cannot interact properly with the oxygen atoms in the selectivity filter. Also, because a $Ca^{2+}$ ion has two positive charges and binds water oxygens more tightly than does a single positive $Na^+$ or $K^+$ ion, more energy is required to strip the waters of hydration from $Ca^{2+}$ than from $K^+$ or $Na^+$.

Recent x-ray crystallographic studies reveal that both when open and when closed, the channel contains $K^+$ ions

### (a) K⁺ and Na⁺ ions in the pore of a K⁺ channel (top view)

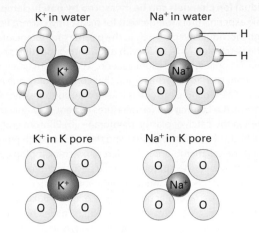

### (b) K⁺ ions in the pore of a K⁺ channel (side view)

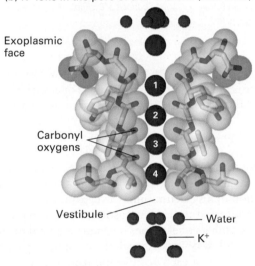

### (c) Ion movement through selectivity filter

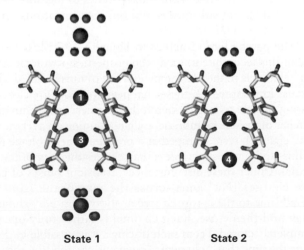

State 1                    State 2

**FIGURE 11-21 Mechanism of ion selectivity and transport in resting K⁺ channels.** (a) Schematic diagrams of K⁺ and Na⁺ ions hydrated in solution and in the pore of a K⁺ channel. As K⁺ ions pass through the selectivity filter, they lose their bound water molecules and become coordinated instead to eight backbone carbonyl oxygens, four of which are shown, that are part of the conserved amino acids in the channel-lining selectivity filter loop of each P segment. The smaller Na⁺ ions with their tighter shell of water molecules cannot perfectly coordinate with the channel oxygen atoms and therefore pass through the channel only rarely. (b) High-resolution electron density map obtained from x-ray crystallography showing K⁺ ions (purple spheres) passing through the selectivity filter. Only two of the diagonally opposed channel subunits are shown. Within the selectivity filter each unhydrated K⁺ ion interacts with eight carbonyl oxygen atoms (red sticks) lining the channel, two from each of the four subunits, as if to mimic the eight waters of hydration. (c) Interpretation of the electron density map showing the two alternating states by which K⁺ ions move through the channel. In state 1, numbered top-to-bottom from the exoplasmic side of the channel inward, one sees a hydrated K⁺ ion with its eight bound water molecules, K⁺ ions at positions 1 and 3 within the selectivity filter, and a fully hydrated K⁺ ion within the vestibule. During K⁺ movement each ion in state 1 moves one step inward, forming state 2. Thus in state 2 the K⁺ ion on the exoplasmic side of the channel has lost four of its eight waters, the ion at position 1 in state 1 has moved to position 2, and the ion at position 3 in state 1 has moved to position 4. In going from state 2 to state 1, the K⁺ at position 4 moves into the vestibule and picks up eight water molecules, while another hydrated K⁺ ion moves into the channel opening and the other K⁺ ions move down one step. Note that K⁺ ions are shown here moving from the exoplasmic side of the channel to the cytosolic side because that is the normal direction of movement in bacteria. In animal cells the direction of K⁺ movement is typically the reverse—from inside to outside. [Part (a) adapted from C. Armstrong, 1998, *Science* **280**:56. Parts (b) and (c) adapted from Y. Zhou et al., 2001, *Nature* **414**:43.]

within the selectivity filter; without these ions the channel probably would collapse. The K⁺ ions are thought to be present either at positions 1 and 3 or at 2 and 4, each surrounded by eight carbonyl oxygen atoms (Figure 11-21b and c). Several K⁺ ions move simultaneously through the channel such that when the ion on the exoplasmic face that has been partially stripped of its water of hydration moves into position 1, the ion at position 2 jumps to position 3 and the one at position 4 exits the channel (Figure 11-21c).

Although the amino acid sequences of the P segments in Na⁺ and K⁺ channels differ somewhat, they are similar enough to suggest that the general structure of the ion-selectivity filters are comparable in both types of channels. Presumably the diameter of the filter in Na⁺ channels is small enough that it permits dehydrated Na⁺ ions to bind to the backbone carbonyl oxygens but excludes the larger K⁺ ions from entering, but the first three-dimensional structure of a sodium channel was determined only in late 2011 and the mechanism of ion selectivity is only now being determined.

## Patch Clamps Permit Measurement of Ion Movements Through Single Channels

Once it was realized that in most cells there are only one or a few ion channels per square micrometer of plasma membrane,

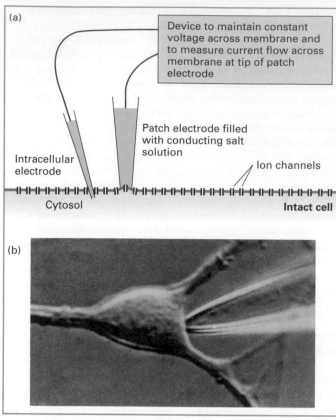

(a)

Device to maintain constant voltage across membrane and to measure current flow across membrane at tip of patch electrode

Patch electrode filled with conducting salt solution

Intracellular electrode

Ion channels

Cytosol

**Intact cell**

(b)

(c)

Tip of micropipette

Ion channel

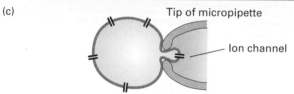

**On-cell patch** measures indirect effect of extracellular solutes on channels within membrane patch on intact cell

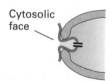

Cytosolic face

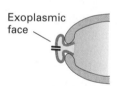

Exoplasmic face

**Inside-out detached patch** measures effects of intracellular solutes on channels within isolated patch

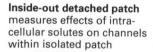

**Outside-out detached patch** measures effects of extracellular solutes on channels within isolated patch

**EXPERIMENTAL FIGURE 11-22 Current flow through individual ion channels can be measured by patch-clamping.**
(a) Basic experimental arrangement for measuring current flow through individual ion channels in the plasma membrane of a living cell. The patch electrode, filled with a current-conducting saline solution, is applied, with a slight suction, to the plasma membrane. The 0.5-μm-diameter tip covers a region that contains only one or a few ion channels. The second electrode is inserted through the membrane into the cytosol. A recording device measures current flow only through the channels in the patch of plasma membrane. (b) Photomicrograph of the cell body of a cultured neuron and the tip of a patch pipette touching the cell membrane. (c) Different patch-clamping configurations. Isolated, detached patches are the best configurations for studying the effects on channels of different ion concentrations and solutes such as extracellular hormones and intracellular second messengers (e.g., cAMP). An inside-out patch, in which an on-cell patch is formed and then the cell pulled away, is used in the experiment in Figure 11-23. [Part (b) from B. Sakmann, 1992, *Neuron* **8**:613 (Nobel lecture); also published in E. Neher and B. Sakmann, 1992, *Sci. Am.* **266**(3):44. Part (c) adapted from B. Hille, 1992, *Ion Channels of Excitable Membranes*, 2d ed., Sinauer Associates, p. 89.]

The inward or outward movement of ions across a patch of membrane is quantified from the amount of electric current needed to maintain the membrane potential at a particular "clamped" value (Figure 11-22a and b). To preserve electroneutrality and to keep the membrane potential constant even though ions are moving through channels in the membrane patch, the entry of each positive ion (e.g., a $Na^+$ ion) into the cell through a channel in the patch of membrane is balanced by the addition of an electron into the cytosol through a microelectrode inserted into the cytosol; an electronic device measures the numbers of electrons (current) required to counterbalance the inflow of ions through the membrane channels. Conversely, the exit of each positive ion from the cell (e.g., a $K^+$ ion) is balanced by the withdrawal of an electron from the cytosol. The patch-clamping technique can be employed on whole cells or isolated membrane patches to measure the effects of different substances and ion concentrations on ion flow (Figure 11-22c).

The patch-clamp tracings in Figure 11-23 illustrate the use of this technique to study the properties of voltage-gated $Na^+$ channels in the plasma membrane of muscle cells. As we discuss in Chapter 22, these channels normally are closed in resting muscle cells and open following nervous stimulation. Patches of muscle membrane, each containing on average one $Na^+$ channel, were clamped at a predetermined voltage that, in this study, was slightly less than the resting membrane potential. Under these circumstances, transient pulses of positive charges ($Na^+$ ions) cross the membrane from the exoplasmic to the cytosolic face as individual $Na^+$ channels open and then close. Each channel is either fully open or completely closed. From such tracings, it is possible to determine the time that a channel is open and the ion flux through it. For the channels measured in Figure 11-23, the flux is about 10 million $Na^+$ ions per channel per second, a typical value for ion channels. Replacement of the NaCl within the

it became possible to record ion movements through single ion channels, and to measure the rates at which these channels open and close and conduct specific ions, using a technique known as **patch clamping** or voltage clamping. As illustrated in Figure 11-22, a tiny pipette is tightly applied to the surface of a cell; the segment of the plasma membrane within the tip will contain only one or a few ion channels. An electrical recording device detects ion flow, measured as electric current, through the channels; this usually occurs in small bursts when the channel is open. An electrical device "clamps" or locks the electric potential across the membrane at a predetermined value (hence the term patch clamping).

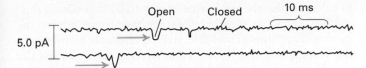

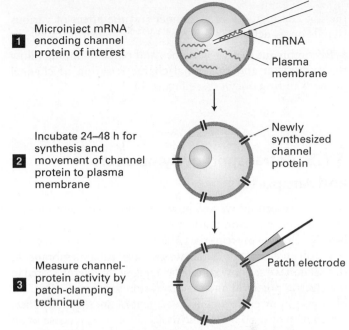

**EXPERIMENTAL FIGURE 11-23 Ion flux through individual Na⁺ channels can be calculated from patch clamp tracings.** Two inside-out patches of muscle plasma membrane were clamped at a potential of slightly less than that of the resting membrane potential. The patch electrode contained NaCl. The transient pulses of electric current in picoamperes (pA), recorded as large downward deviations (blue arrows), indicate the opening of a Na⁺ channel and movement of positive charges (Na⁺ ions) inward across the membrane. The smaller deviations in current represent background noise. The average current through an open channel is 1.6 pA, or $1.6 \times 10^{-12}$ amperes. Since 1 ampere = 1 coulomb (C) of charge per second, this current is equivalent to the movement of about 9900 Na⁺ ions per channel per millisecond: $(1.6 \times 10^{-12}$ C/s$)(10^{-3}$ s/ms$)(6 \times 10^{23}$ molecules/mol$) \div$ 96,500 C/mol. [See F. J. Sigworth and E. Neher, 1980, *Nature* **287**:447.]

**EXPERIMENTAL FIGURE 11-24 Oocyte expression assay is useful in comparing the function of normal and mutant forms of a channel protein.** A follicular frog oocyte is first treated with collagenase to remove the surrounding follicle cells, leaving a denuded oocyte, which is microinjected with mRNA encoding the channel protein under study. [Adapted from T. P. Smith, 1988, *Trends Neurosci.* **11**:250.]

patch pipette (corresponding to the outside of the cell) with KCl or choline chloride abolishes current through the channels, confirming that they conduct only Na⁺ ions, not K⁺ or other ions.

## Novel Ion Channels Can Be Characterized by a Combination of Oocyte Expression and Patch Clamping

Cloning of human-disease-causing genes and sequencing of the human genome have identified many genes encoding putative channel proteins, including 67 putative K⁺ channel proteins. One way of characterizing the function of these proteins is to transcribe a cloned cDNA in a cell-free system to produce the corresponding mRNA. Injecting this mRNA into frog oocytes and taking patch clamp measurements of the newly synthesized channel protein can often reveal its function (Figure 11-24). This experimental approach is especially useful because frog oocytes normally do not express any channel proteins on their surface membrane, so only the channel under study is present in the membrane. In addition, because of the large size of frog oocytes, patch-clamping studies are technically easier to perform on them than on smaller cells.

## KEY CONCEPTS of Section 11.4

### Nongated Ion Channels and the Resting Membrane Potential

- An inside-negative electric potential (voltage) of about −70 mV exists across the plasma membrane of all cells.

- The resting membrane potential in animal cells is the result of the combined action of the ATP-powered Na⁺/K⁺ pump, which establishes Na⁺ and K⁺ concentration gradients across the membrane, and resting K⁺ channels which permit selective movement only of K⁺ ions back down their concentration gradient to the external medium (see Figure 11-3).

- Unlike the more common gated ion channels, which open only in response to various signals, these nongated K⁺ channels are usually open.

- The electric potential generated by the selective flow of ions across a membrane can be calculated using the Nernst equation (see Equation 11-2).

- In plants and fungi, the membrane potential is maintained by the ATP-driven pumping of protons from the cytosol to the exterior of the cell.

- K⁺ channels are assembled from four identical subunits, each of which has at least two conserved membrane-spanning α helices and a nonhelical P segment that lines the ion pore and forms the selectivity filter (see Figure 11-20).

- The ion specificity of K⁺ channel proteins is due mainly to coordination of the selected ion with eight carbonyl oxygen atoms of specific amino acids in the P segments, which lowers the activation energy for passage of the selected K⁺ compared with Na⁺ or other ions (see Figure 11-21).

- Patch-clamping techniques, which permit measurement of ion movements through single channels, are used to determine the ion conductivity of a channel and the effect of various signals on its activity (see Figure 11-22).

• Recombinant DNA techniques and patch clamping allow the expression and functional characterization of channel proteins in frog oocytes (see Figure 11-24).

## 11.5 Cotransport by Symporters and Antiporters

In previous sections we saw how ATP-powered pumps generate ion concentration gradients across cell membranes and how $K^+$ channel proteins use the $K^+$ concentration gradient to establish an electric potential across the plasma membrane. In this section we see how cotransporters use the energy stored in the electric potential and concentration gradients of $Na^+$ or $H^+$ ions to power the uphill movement of another substance, which may be a small organic molecule such as glucose or an amino acid or a different ion. An important feature of such **cotransport** is that neither substance can move alone; movement of both substances together is obligatory, or *coupled.*

Cotransporters share common features with uniporters such as the GLUT proteins. The two types of transporters exhibit certain structural similarities, operate at equivalent rates, and undergo cyclical conformational changes during transport of their substrates. They differ in that uniporters can only accelerate thermodynamically favorable transport down a concentration gradient, whereas cotransporters can harness the energy released when one substance moves down its concentration gradient to drive the movement of another substance against its concentration gradient.

When the transported molecule and cotransported ion move in the same direction, the process is called **symport;** when they move in opposite directions, the process is called **antiport** (see Figure 11-2). Some cotransporters transport only positive ions (cations), while others transport only negative ions (anions). Yet other cotransporters mediate movement of both cations and anions together. Cotransporters are present in all organisms, including bacteria, plants, and animals, and in this section we describe the operation and function of several physiologically important symporters and antiporters.

### Na$^+$ Entry into Mammalian Cells Is Thermodynamically Favored

Mammalian cells express many types of $Na^+$-coupled symporters. The human genome encodes literally hundreds of different types of transport proteins that use the energy stored across the plasma membrane in the $Na^+$ concentration gradient and in the inside-negative electric potential across the membrane to transport a wide variety of molecules into cells against their concentration gradients. To see why such transporters allow cells to accumulate substrates against a considerable concentration gradient we first need

to calculate the change in free energy ($\Delta G$) that occurs during $Na^+$ entry. As mentioned earlier, two forces govern the movement of ions across selectively permeable membranes: the voltage and the ion concentration gradient across the membrane. The sum of these forces constitutes the electrochemical gradient. To calculate the free-energy change, $\Delta G$, corresponding to the transport of any ion across a membrane, we need to consider the independent contributions from each of the forces to the electrochemical gradient.

For example, when $Na^+$ moves from outside to inside the cell, the free-energy change generated from the $Na^+$ concentration gradient is given by

$$\Delta G_c = RT \ln \frac{[Na_{in}]}{[Na_{out}]} \qquad (11\text{-}5)$$

At the concentrations of $Na_{in}$ and $Na_{out}$ shown in Figure 11-25, which are typical for many mammalian cells, $\Delta G_c$, the change in free energy due to the concentration gradient, is $-1.45$ kcal for transport of 1 mol of $Na^+$ ions from outside to inside a cell, assuming there is no membrane electric potential. Note the free energy is negative, indicating spontaneous movement of $Na^+$ into the cell down its concentration gradient.

The free-energy change generated from the membrane electric potential is given by

$$\Delta G_m = FE \qquad (11\text{-}6)$$

where $F$ is the Faraday constant [$= 23{,}062$ cal/(mol · V)]; and $E$ is the membrane electric potential. If $E = -70$ mV, then $\Delta G_m$, the free-energy change due to the membrane potential, is $-1.61$ kcal for transport of 1 mol of $Na^+$ ions from outside to inside a cell, assuming there is no $Na^+$ concentration gradient. Since both forces do in fact act on $Na^+$ ions, the total $\Delta G$ is the sum of the two partial values:

$$\Delta G = \Delta G_c + \Delta G_m = (-1.45) + (-1.61) = -3.06 \text{ kcal/mole}$$

In this example, the $Na^+$ concentration gradient and the membrane electric potential contribute almost equally to the total $\Delta G$ for transport of $Na^+$ ions. Since $\Delta G$ is $<0$, the inward movement of $Na^+$ ions is thermodynamically favored. As discussed in the next section, the inward movement of $Na^+$ is used to power the uphill movement of other ions and several types of small molecules into or out of animal cells. The rapid, energetically favorable movement of $Na^+$ ions through gated $Na^+$ channels also is critical in generating action potentials in nerve and muscle cells, as we discuss in Chapter 22.

### Na$^+$-Linked Symporters Enable Animal Cells to Import Glucose and Amino Acids Against High Concentration Gradients

Most body cells import glucose from the blood *down* a concentration gradient of glucose, utilizing GLUT proteins to facilitate this transport. However, certain cells, such as those

**FIGURE 11-25 Transmembrane forces acting on Na⁺ ions.** As with all ions, the movement of Na⁺ ions across the plasma membrane is governed by the sum of two separate forces—the ion concentration gradient and the membrane electric potential. At the internal and external Na⁺ concentrations typical of mammalian cells, these forces usually act in the same direction, making the inward movement of Na⁺ ions energetically favorable.

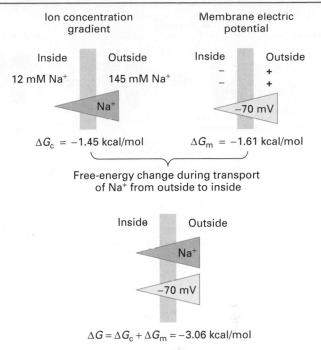

lining the small intestine and the kidney tubules, need to import glucose from extracellular fluids (digestive products or urine) against a very large concentration gradient (glucose concentration higher inside the cell). Such cells utilize a *two-Na⁺/one-glucose symporter,* a protein that couples import of one glucose molecule to the import of two Na⁺ ions:

$$2\ Na^+_{out} + glucose_{out} \rightleftharpoons 2\ Na^+_{in} + glucose_{in}$$

Quantitatively, the free-energy change for the symport transport of two Na⁺ ions and one glucose molecule can be written

$$\Delta G = RT \ln \frac{[glucose_{in}]}{[glucose_{out}]} + 2RT \ln \frac{[Na^+_{in}]}{[Na^+_{out}]} + 2FE \quad (11\text{-}7)$$

Thus the $\Delta G$ for the overall reaction is the sum of the free-energy changes generated by the glucose concentration gradient (1 molecule transported), the Na⁺ concentration gradient (2 Na⁺ ions transported), and the membrane potential (2 Na⁺ ions transported). As illustrated in Figure 11-25, the free energy released by movement of 1 mole of Na⁺ ions into mammalian cells down its electrochemical gradient has a free-energy change, $\Delta G$, of about $-3$ kcal per mole of Na⁺ transported. Thus the $\Delta G$ for transport of two moles of Na⁺ inward would be twice this amount, or about $-6$ kcal. This negative free-energy change of sodium import is coupled to the uphill transport of glucose, a process with a positive $\Delta G$. We can calculate the glucose concentration gradient, inside greater than outside, that can be established by the action of this Na⁺-powered symporter by realizing that at equilibrium for sodium-coupled glucose import, $\Delta G = 0$. By substituting the values for sodium import into Equation 11-7 and setting $\Delta G = 0$, we see that

$$0 = RT \ln \frac{[glucose_{in}]}{[glucose_{out}]} - 6\ kcal$$

and we can calculate that at equilibrium, the ratio of glucose_in/glucose_out = ~30,000. Thus the inward flow of two moles of Na⁺ can generate an intracellular glucose concentration that is ~30,000 times greater than the exterior concentration. If only one Na⁺ ion were imported ($\Delta G$ of approximately $-3$ kcal/mol) per glucose molecule, then the available energy could generate a glucose concentration gradient (inside/outside) of only about 170-fold. Thus by coupling the transport of two Na⁺ ions to the transport of one glucose, the two-Na⁺/

one-glucose symporter permits cells to accumulate a very high concentration of glucose relative to the external concentration. This means that glucose present even at very low concentrations in the lumen of the intestine or in the kidney tubules can be efficiently transported into the lining cells and not lost from the body.

The two-Na⁺/one-glucose symporter is thought to contain 14 transmembrane α helices with both its N- and C-termini extending into the cytosol. A truncated recombinant protein consisting of only the five C-terminal transmembrane α helices can transport glucose independently of Na⁺ across the plasma membrane, *down* its concentration gradient. This portion of the molecule thus functions as a glucose uniporter. The N-terminal portion of the protein, including helices 1–9, is required to couple Na⁺ binding and influx to the transport of glucose against a concentration gradient.

Figure 11-26 depicts the current model of transport by Na⁺/glucose symporters. This model entails conformational changes in the protein analogous to those that occur in uniport transporters, such as GLUT1, which do not require a cotransported ion (compare to Figure 11-5). Binding of all substrates to their sites on the extracellular domain is required before the protein undergoes the conformational change that converts the substrate-binding sites from outward to inward facing; this ensures that inward transport of glucose and Na⁺ ions are coupled.

Note that cells use comparable Na⁺-powered symporters to transport substances other than glucose into the cell against high concentration gradients. For example, several types of Na⁺/amino acid symporters allow cells to import many amino acids into the cell.

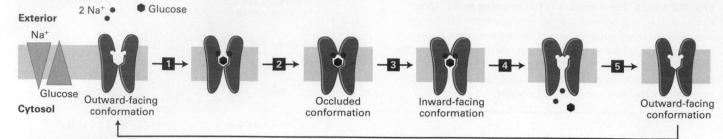

**FIGURE 11-26 Operational model for the two-Na$^+$/one-glucose symporter.** Simultaneous binding of Na$^+$ and glucose to the conformation with outward-facing binding sites (step **1**) causes a conformational change in the protein such that the bound substrates are transiently occluded, unable to dissociate into either medium (step **2**). In step **3** the protein assumes a third conformation with inward-facing sites. Dissociation of the bound Na$^+$ and glucose into the cytosol (step **4**) allows the protein to revert to its original outward-facing conformation (step **5**), ready to transport additional substrate. [See H. Krishnamurthy et al., 2009, *Nature* **459:**347–355 for details on the structure and function of this and related Na$^+$-coupled transporters.]

## A Bacterial Na$^+$/Amino Acid Symporter Reveals How Symport Works

No three-dimensional structure has yet been determined for any mammalian sodium symporter, but the structures of several homologous bacterial sodium/substrate symporters have provided considerable information about symport function. The bacterial two-Na$^+$/one-leucine symporter shown in Figure 11-27a consists of 12 membrane-spanning α helices. Two of the helices (numbers 1 and 6) have nonhelical segments in the middle of the membrane that form part of the leucine-binding site.

Amino acid residues involved in binding the leucine and the two Na$^+$ ions are located in the middle of the membrane-spanning segment (as depicted for the two-Na$^+$/one-glucose symporter in Figure 11-26) and are close together in three-dimensional space. This demonstrates that the coupling of amino acid and ion transport in these transporters is the consequence of direct or nearly direct physical interactions of the substrates. Indeed, one of the Na$^+$ ions is bound to the carboxyl group of the transported leucine (Figure 11-27b). Thus neither substance can bind to the transporter without the other, indicating how transport of sodium and leucine are coupled. Each of the two Na$^+$ ions is bound to six oxygen atoms. Sodium 1, for example, is bound to carbonyl oxygens of several transporter amino acids as well as to carbonyl oxygens and the hydroxyl oxygen of one threonine. Equally importantly, there are no water molecules surrounding either of the bound Na$^+$ atoms, as is the case for K$^+$ ions in potassium channels (see Figure 11-21). Thus as the Na$^+$ ions lose their water of hydration in binding to the transporter, they bind to six oxygen atoms with a similar geometry. This reduces the energy change required for binding of Na$^+$ ions and prevents other ions, such as K$^+$, from binding in place of Na$^+$.

One striking feature of the structure depicted in Figure 11-27 is that the bound Na$^+$ ions and leucine are *occluded*—that is, they cannot diffuse out of the protein to either the surrounding extracellular or cytoplasmic media. This structure represents an intermediate in the transport process (see Figure 11-26) in which the protein appears to be changing from a conformation with an exoplasmic- to one with a cytosolic-facing binding site.

## A Na$^+$-Linked Ca$^{2+}$ Antiporter Regulates the Strength of Cardiac Muscle Contraction

In all muscle cells, a rise in the cytosolic Ca$^{2+}$ concentration triggers contraction. In cardiac muscle cells a *three-Na$^+$/one-Ca$^{2+}$ antiporter*, rather than the plasma membrane Ca$^{2+}$ ATPase discussed earlier, plays the principal role in maintaining a low concentration of Ca$^{2+}$ in the cytosol. The transport reaction mediated by this *cation antiporter* can be written

$$3 \text{ Na}^+{}_{out} + \text{Ca}^{2+}{}_{in} \rightleftharpoons 3 \text{ Na}^+{}_{in} + \text{Ca}^{2+}{}_{out}$$

Note that the inward movement of three Na$^+$ ions is required to power the export of one Ca$^{2+}$ ion from the cytosol, with a [Ca$^{2+}$] of ~$2 \times 10^{-7}$ M, to the extracellular medium, with a [Ca$^{2+}$] of ~$2 \times 10^{-3}$ M, a gradient of some 10,000-fold (higher on the outside). By lowering cytosolic Ca$^{2+}$, operation of the Na$^+$/Ca$^{2+}$ antiporter reduces the strength of heart muscle contraction.

The Na$^+$/K$^+$ ATPase in the plasma membrane of cardiac muscle cells, as in other body cells, creates the Na$^+$ concentration gradient necessary for export of Ca$^{2+}$ by the Na$^+$-linked Ca$^{2+}$ antiporter. As mentioned earlier, inhibition of the Na$^+$/K$^+$ ATPase by the drugs ouabain and digoxin lowers the cytosolic K$^+$ concentration and, more relevant here, simultaneously increases cytosolic Na$^+$. The resulting reduced Na$^+$ electrochemical gradient across the membrane causes the Na$^+$-linked Ca$^{2+}$ antiporter to function less efficiently. As a result, fewer Ca$^{2+}$ ions are exported and the

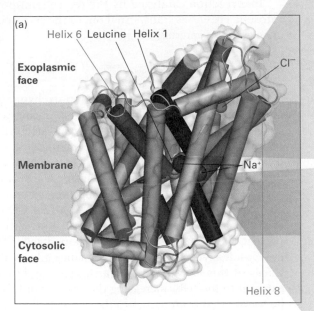

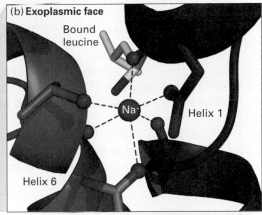

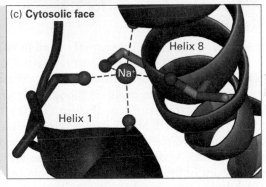

**FIGURE 11-27 Three-dimensional structure of the two-Na⁺/one-leucine symporter from the bacterium *Aquifex aeolicus*.** (a) The bound L-leucine, two Na⁺ ions, and a Cl⁻ ion are shown in yellow, purple and green, respectively. The three membrane-spanning α helices that bind the Na⁺ or leucine are colored brown, blue, and orange. (b,c) Binding of the two Na⁺ ions to carbonyl main-chain oxygen atoms or carboxyl side-chain oxygens (red) that are part of helices 1 (brown), 6 (blue), or 8 (orange). It is important that one of the sodium ions is also bound to the carboxyl group of the transported leucine (part b). [From A. Yamashita et al., 2005, *Nature* **437**:215; see also H. Krishnamurthy et al., 2009, *Nature* **459**:347–355 for details on the structure and function of this and related Na⁺-coupled transporters.]

cytosolic $Ca^{2+}$ concentration increases, causing the muscle to contract more strongly. Because of their ability to increase the force of heart muscle contractions, drugs such as ouabain and digoxin that inhibit the $Na^+/K^+$ ATPase are widely used in the treatment of congestive heart failure. ∎

## Several Cotransporters Regulate Cytosolic pH

The anaerobic metabolism of glucose yields lactic acid, and aerobic metabolism yields $CO_2$, which combines with water to form carbonic acid ($H_2CO_3$). These weak acids dissociate, yielding $H^+$ ions (protons); if these excess protons were not removed from cells, the cytosolic pH would drop precipitously, endangering cellular functions. Two types of cotransport proteins help remove some of the "excess" protons generated during metabolism in animal cells. One is a $Na^+HCO_3^-/Cl^-$ antiporter, which imports one $Na^+$ ion together with one $HCO_3^-$, in exchange for export of one $Cl^-$ ion. The cytosolic enzyme *carbonic anhydrase* catalyzes dissociation of the imported $HCO_3^-$ ions into $CO_2$ and an $OH^-$ (hydroxyl) ion:

$$HCO^-_3 \underset{\text{Carbonic anhydrase}}{\rightleftharpoons} CO_2 + OH^-$$

The $OH^-$ ions combine with intracellular protons, forming water, and the $CO_2$ diffuses out of the cell. Thus the overall action of this transporter is to *consume* cytosolic $H^+$ ions, thereby *raising* the cytosolic pH. Also important in raising cytosolic pH is a *$Na^+/H^+$ antiporter*, which couples entry of one $Na^+$ ion into the cell down its concentration gradient to the export of one $H^+$ ion.

Under certain circumstances, the cytosolic pH can rise beyond the normal range of 7.2–7.5. To cope with the excess $OH^-$ ions associated with elevated pH, many animal cells utilize an *anion antiporter* that catalyzes the one-for-one exchange of $HCO_3^-$ and $Cl^-$ across the plasma membrane. At high pH, this *$Cl^-/HCO_3^-$ antiporter* exports one molecule of

$HCO_3^-$ (which can be viewed as a "complex" of $OH^-$ and $CO_2$) in exchange for import of one molecule of $Cl^-$, thus lowering the cytosolic pH. The import of $Cl^-$ down its concentration gradient ($Cl^-_{medium} > Cl^-_{cytosol}$; see Table 11-2) powers the export of $HCO_3^-$.

The activity of all three of these antiport proteins is regulated by the cytosolic pH, providing cells with a finely tuned mechanism for controlling cytosolic pH. The two antiporters that operate to increase cytosolic pH are activated when the pH of the cytosol falls. Similarly, a rise in pH above 7.2 stimulates the $Cl^-/HCO_3^-$ antiporter, leading to a more rapid export of $HCO_3^-$ and a drop in the cytosolic pH. In this manner, the cytosolic pH of growing cells is maintained very close to pH 7.4.

## An Anion Antiporter Is Essential for Transport of $CO_2$ by Red Blood Cells

Transmembrane anion exchange is essential for an important function of erythrocytes—the transport of waste $CO_2$ from peripheral tissues to the lungs for exhalation. Waste $CO_2$ released from cells into the capillary blood freely diffuses across the erythrocyte membrane (Figure 11-28a). In its gaseous form, $CO_2$ dissolves poorly in aqueous solutions, such as the cytosol or blood plasma, as is apparent to anyone who has opened a bottle of a carbonated beverage. However, the large amount of the potent enzyme carbonic anhydrase inside the erythrocyte combines $CO_2$ with hydroxyl ions ($OH^-$) to form water-soluble bicarbonate ($HCO_3^-$) anions. This process occurs while red cells are in systemic (tissue) capillaries and releasing oxygen into the blood plasma. Release of oxygen from hemoglobin induces a change in its

conformation that enables a histidine side chain of a globin polypeptide to bind a proton. Thus when red cells are in systemic capillaries water is split into a proton that binds hemoglobin and an $OH^-$ that reacts with $CO_2$ to form an $HCO_3^-$ anion.

In a reaction catalyzed by the red cell antiporter AE1, cytosolic $HCO_3^-$ is transported out of the erythrocyte in exchange for an entering $Cl^-$ anion:

$$HCO_3^-{}_{in} + Cl^-{}_{out} \rightleftharpoons HCO_3^-{}_{out} + Cl^-{}_{in}$$

(see Figure 11-28a). The entire anion-exchange process is completed within 50 milliseconds (ms), during which time $5 \times 10^9$ $HCO_3^-$ ions are exported from each cell down its concentration gradient. If anion exchange did not occur, during periods such as exercise, when much $CO_2$ is generated, $HCO_3^-$ would accumulate inside the erythrocyte to toxic levels as the cytosol would become alkaline. The exchange of $HCO_3^-$ (equal to $OH^- + CO_2$) for $Cl^-$ causes the cytosolic pH to remain nearly neutral. Normally about 80 percent of the $CO_2$ in blood is transported as $HCO_3^-$ generated inside erythrocytes; anion exchange allows about two-thirds of this $HCO_3^-$ to be transported by blood plasma external to the cells, increasing the amount of $CO_2$ that can be transported from tissues to the lungs. In the lungs, where carbon dioxide leaves the body, the overall direction of this anion-exchange process is reversed (Figure 11-28b).

AE1 catalyzes the precise one-for-one sequential exchange of anions on opposite sides of the membrane required to preserve electroneutrality in the cell; only once every 10,000 or so transport cycles does an anion move unidirectionally from one side of the membrane to the other. AE1 is

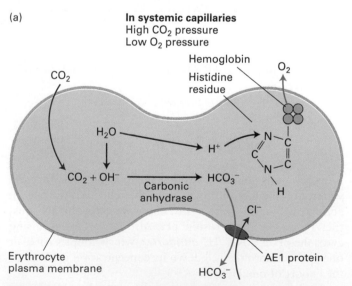

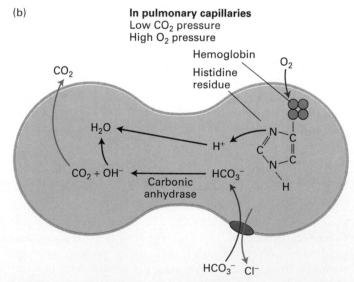

**FIGURE 11-28 Carbon dioxide transport in blood requires a $Cl^-/HCO_3^-$ antiporter.** (a) In systemic capillaries, carbon dioxide gas diffuses across the erythrocyte plasma membrane and is converted into soluble $HCO_3^-$ by the enzyme carbonic anhydrase; at the same time, oxygen leaves the cell and hemoglobin binds a proton. The anion antiporter AE1 (purple) catalyzes the reversible exchange of $Cl^-$ and

$HCO_3^-$ ions across the membrane. The overall reaction causes $HCO_3^-$ to be released from the cell, which is essential for maximal $CO_2$ transport from the tissues to the lungs, and for maintaining pH neutrality in the blood cell. (b) In the lungs, where carbon dioxide is excreted, the overall reaction is reversed. See text for additional discussion.

composed of a membrane-embedded domain, folded into at least 12 transmembrane α helices, that catalyzes anion transport, and a cytosolic-facing domain that anchors certain cytoskeletal proteins to the membrane (see Figure 17-21).

## Numerous Transport Proteins Enable Plant Vacuoles to Accumulate Metabolites and Ions

The lumen of plant vacuoles is much more acidic (pH 3–6) than is the cytosol (pH 7.5). The acidity of vacuoles is maintained by a V-class ATP-powered proton pump (see Figure 11-9) and by a pyrophosphate-powered proton pump that is unique to plants. Both of these pumps, located in the vacuolar membrane, import $H^+$ ions into the vacuolar lumen against a concentration gradient. The vacuolar membrane also contains $Cl^-$ and $NO_3^-$ channels that transport these anions from the cytosol into the vacuole. Entry of these anions against their concentration gradients is driven by the inside-positive potential generated by the $H^+$ pumps. The combined operation of these proton pumps and anion channels produces an inside-positive electric potential of about 20 mV across the vacuolar membrane and also a substantial pH gradient (Figure 11-29).

The proton electrochemical gradient across the plant vacuole membrane is used in much the same way as the $Na^+$ electrochemical gradient across the animal-cell plasma membrane: to power the selective uptake or extrusion of ions and small molecules by various antiporters. In the leaf, for example, excess sucrose generated during photosynthesis in the day is stored in the vacuole; during the night, the stored sucrose moves into the cytoplasm and is metabolized to $CO_2$ and $H_2O$ with concomitant generation of ATP from ADP and $P_i$. A *proton/sucrose antiporter* in the vacuolar membrane operates to accumulate sucrose in plant vacuoles. The inward movement of sucrose is powered by the outward movement of $H^+$, which is favored by its concentration gradient (lumen > cytosol) and by the cytosolic-negative potential across the vacuolar membrane (see Figure 11-29). Uptake of $Ca^{2+}$ and $Na^+$ into the vacuole from the cytosol against their concentration gradients is similarly mediated by proton antiporters.

Understanding of the transport proteins in plant vacuolar membranes has the potential for increasing agricultural production in high-salt (NaCl) soils, which are found throughout the world. Because most agriculturally useful crops cannot grow in such saline soils, agricultural scientists have long sought to develop salt-tolerant plants by traditional breeding methods. With the availability of the cloned gene encoding the vacuolar $Na^+/H^+$ antiporter, researchers can now produce transgenic plants that overexpress this transport protein, leading to increased sequestration of $Na^+$ in the vacuole. For instance, transgenic tomato plants that overexpress the vacuolar $Na^+/H^+$ antiporter can grow, flower, and produce fruit in the presence of soil NaCl concentrations that kill wild-type plants. Interestingly, although the leaves of these transgenic tomato plants accumulate large amounts of salt, the fruit has a very low salt content. ■

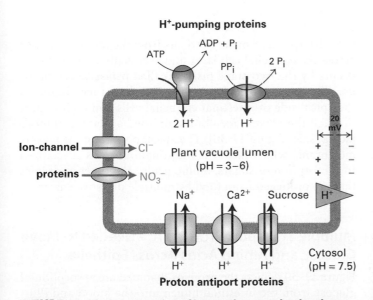

**FIGURE 11-29 Concentration of ions and sucrose by the plant vacuole.** The vacuolar membrane contains two types of proton pumps (orange): a V-class $H^+$ ATPase (*left*) and a pyrophosphate-hydrolyzing proton pump (*right*) that differs from all other ion transport proteins and probably is unique to plants. These pumps generate a low luminal pH as well as an inside-positive electric potential across the vacuolar membrane owing to the inward pumping of $H^+$ ions. The inside-positive potential powers the movement of $Cl^-$ and $NO_3^-$ from the cytosol through separate channel proteins (purple). Proton antiporters (green), powered by the $H^+$ gradient, accumulate $Na^+$, $Ca^{2+}$, and sucrose inside the vacuole. [After B. J. Barkla and O. Pantoja, 1996, *Rev. Plant Physiol. Plant Mol. Biol.* **47**:159–184 and P. A. Rea et al., 1992, *Trends Biochem. Sci.* **17**:348.]

## KEY CONCEPTS of Section 11.5

### Cotransport by Symporters and Antiporters

• The electrochemical gradient across a semipermeable membrane determines the direction of ion movement through transmembrane proteins. The two forces constituting the electrochemical gradient—the membrane electric potential and the ion concentration gradient—may act in the same or opposite directions (see Figure 11-25).

• Cotransporters use the energy released by movement of an ion (usually $H^+$ or $Na^+$) down its electrochemical gradient to power the import or export of a small molecule or different ion against its concentration gradient.

• The cells lining the small intestine and kidney tubules contain symport proteins that couple the energetically favorable entry of $Na^+$ to the import of glucose against its concentration gradient (see Figure 11-26). Amino acids also enter cells by $Na^+$-coupled symporters.

- The molecular structure of a bacterial $Na^+$/amino acid symporter reveals how binding of $Na^+$ and leucine are coupled and provides a snapshot of an occluded transport intermediate in which the bound substrates cannot diffuse out of the protein (see Figure 11-27).

- In cardiac muscle cells, the export of $Ca^{2+}$ is coupled to and powered by the import of $Na^+$ by a cation antiporter, which transports three $Na^+$ ions inward for each $Ca^{2+}$ ion exported.

- Two cotransporters that are activated at low pH help maintain the cytosolic pH in animal cells very close to 7.4 despite metabolic production of carbonic and lactic acids. One, a $Na^+/H^+$ antiporter, exports excess protons. The other, a $Na^+HCO_3^-/Cl^-$ cotransporter, imports $HCO_3^-$, which dissociates in the cytosol to yield pH-raising $OH^-$ ions.

- A $Cl^-/HCO_3^-$ antiporter that is activated at high pH functions to export $HCO_3^-$ when the cytosolic pH rises above normal and causes a decrease in pH.

- AE1, a $Cl^-/HCO_3^-$ antiporter in the erythrocyte membrane, increases the ability of blood to transport $CO_2$ from tissues to the lungs (see Figure 11-28).

- Uptake of sucrose, $Na^+$, $Ca^{2+}$, and other substances into plant vacuoles is carried out by proton antiporters in the vacuolar membrane. Ion channels and proton pumps in the membrane are critical in generating a large enough proton concentration gradient to power accumulation of ions and metabolites in vacuoles by these proton antiporters (see Figure 11-29).

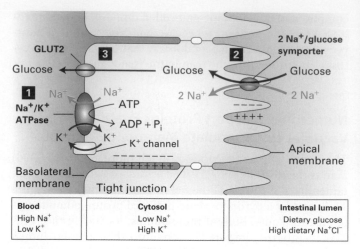

**FIGURE 11-30 Transcellular transport of glucose from the intestinal lumen into the blood.** The $Na^+/K^+$ ATPase in the basolateral surface membrane generates $Na^+$ and $K^+$ concentration gradients (step **1**). The outward movement of $K^+$ ions through nongated $K^+$ channels generates an inside-negative membrane potential across the entire plasma membrane. Both the $Na^+$ concentration gradient and the membrane potential are used to drive the uptake of glucose from the intestinal lumen by the two-$Na^+$/one-glucose symporter located in the apical surface membrane (step **2**). Glucose leaves the cell via facilitated diffusion catalyzed by GLUT2, a glucose uniporter located in the basolateral membrane (step **3**).

# 11.6 Transcellular Transport

Previous sections illustrated how several types of transporters function together to carry out important cell functions. Here, we extend this concept by focusing on the transport of several types of molecules and ions across polarized cells, which are cells that are asymmetric (have different "sides") and thus have biochemically distinct regions of the plasma membrane. A particularly well-studied class of polarized cells are the epithelial cells that form sheetlike layers (epithelia) covering most external and internal surfaces of body organs. Epithelial cells are discussed in greater detail in Chapter 20. Like many epithelial cells, an intestinal epithelial cell involved in absorbing nutrients from the gastrointestinal tract has a plasma membrane organized into two major discrete regions: the surface that faces the outside of the organism, called the **apical**, or top, surface, and the surface that faces the inside (or bloodstream-facing side) of the organism, called the **basolateral** surface (see Figure 20-10).

Specialized regions of the epithelial-cell plasma membrane, called **tight junctions**, separate the apical and basolateral membranes and prevent many, but not all, water-soluble substances on one side from moving across to the other side through the extracellular space between cells. For this reason, absorption of many nutrients from the intestinal lumen across the epithelial cell layer and eventually into the blood occurs by the two-stage process called *transcellular transport*: import of molecules through the plasma membrane on the apical side of intestinal epithelial cells and their export through the plasma membrane on the basolateral (blood-facing) side (Figure 11-30). The apical portion of the plasma membrane, which faces the intestinal lumen, is specialized for absorption of sugars, amino acids, and other molecules that are produced from food by multiple digestive enzymes.

## Multiple Transport Proteins Are Needed to Move Glucose and Amino Acids Across Epithelia

Figure 11-30 depicts the proteins that mediate absorption of glucose from the intestinal lumen into the blood and illustrates the important concept that different types of proteins are localized to the apical and basolateral membranes of epithelial cells. In the first stage of this process, a two-$Na^+$/one-glucose symporter located in the apical membrane imports glucose, against its concentration gradient, from the intestinal lumen across the apical surface of the epithelial cells. As noted above, this symporter couples the energetically unfavorable inward movement of one glucose molecule to the energetically favorable inward transport of two $Na^+$ ions (see Figure 11-26). In the steady state, all the $Na^+$ ions transported from the intestinal lumen into the cell during

Na$^+$/glucose symport, or the similar process of Na$^+$/amino acid symport that also takes place on the apical membrane, are pumped out across the basolateral membrane, which faces the blood. Thus the low intracellular Na$^+$ concentration is maintained. The Na$^+$/K$^+$ ATPase that accomplishes this is found exclusively in the basolateral membrane of intestinal epithelial cells. The coordinated operation of these two transport proteins allows uphill movement of glucose and amino acids from the intestine into the cell. This first stage in transcellular transport ultimately is powered by ATP hydrolysis by the Na$^+$/K$^+$ ATPase.

In the second stage, glucose and amino acids concentrated inside intestinal cells by apical symporters are exported down their concentration gradients into the blood via uniport proteins in the basolateral membrane. In the case of glucose, this movement is mediated by GLUT2 (see Figure 11-30). As noted earlier, this GLUT isoform has a relatively low affinity for glucose but increases its rate of transport substantially when the glucose gradient across the membrane rises (see Figure 11-4).

The net result of this two-stage process is movement of Na$^+$ ions, glucose, and amino acids from the intestinal lumen across the intestinal epithelium into the extracellular medium that surrounds the basolateral surface of intestinal epithelial cells, and eventually into the blood. Tight junctions between the epithelial cells prevent these molecules from diffusing back into the intestinal lumen. The increased osmotic pressure created by transcellular transport of salt, glucose, and amino acids across the intestinal epithelium draws water from the intestinal lumen, mainly through the tight junctions, into the extracellular medium that surrounds the basolateral surface; aquaporins do not appear to play a major role. In a sense, salts, glucose, and amino acids "carry" the water along with them.

## Simple Rehydration Therapy Depends on the Osmotic Gradient Created by Absorption of Glucose and Na$^+$

An understanding of osmosis and the intestinal absorption of salt and glucose forms the basis for a simple therapy that saves millions of lives each year, particularly in less-developed countries. In these countries, cholera and other intestinal pathogens are major causes of death of young children. A toxin released by the bacteria activates chloride secretion from the apical surface of the intestinal epithelial cells into the lumen; water follows osmotically, and the resultant massive loss of water causes diarrhea, dehydration, and ultimately death. A cure demands not only killing the bacteria with antibiotics but also *rehydration*—replacement of the water that is lost from the blood and other tissues.

Simply drinking water does not help, because it is excreted from the gastrointestinal tract almost as soon as it enters. However, as we have just learned, the coordinated transport of glucose and Na$^+$ across the intestinal epithelium creates a transepithelial osmotic gradient, forcing movement of water from the intestinal lumen across the epithelial cell layer and ultimately into the blood. Thus giving affected children a solution of sugar and salt to drink (but not sugar or salt alone) causes increased sodium and sugar transepithelial transport and consequently increased osmotic flow of water into the blood from the intestinal lumen, leading to rehydration. Similar sugar-salt solutions are the basis of popular drinks used by athletes to get sugar as well as water into the body quickly and efficiently. ■

## Parietal Cells Acidify the Stomach Contents While Maintaining a Neutral Cytosolic pH

The mammalian stomach contains a 0.1 M solution of hydrochloric acid (HCl). This strongly acidic medium kills many ingested pathogens and denatures many ingested proteins so that they can be degraded by proteolytic enzymes (e.g., pepsin) that function at acidic pH. Hydrochloric acid is secreted into the stomach by specialized epithelial cells called *parietal cells* (also known as *oxyntic cells*) in the stomach lining. These cells contain an *H$^+$/K$^+$ ATPase* in their apical membrane, which faces the stomach lumen and generates a millionfold H$^+$ concentration gradient: pH ~1.0 in the stomach lumen versus pH ~7.2 in the cell cytosol. This transport protein is a P-class ATP-powered ion pump similar in structure and function to the plasma membrane Na$^+$/K$^+$ ATPase discussed earlier. The numerous mitochondria in parietal cells produce abundant ATP for use by the H$^+$/K$^+$ ATPase.

If parietal cells simply exported H$^+$ ions in exchange for K$^+$ ions, the loss of protons would lead to a rise in the concentration of OH$^-$ ions in the cytosol and thus a marked increase in cytosolic pH. (Recall that [H$^+$] × [OH$^-$] always is a constant, $10^{-14}$ M$^2$.) Parietal cells avoid this rise in cytosolic pH in conjunction with acidification of the stomach lumen by using Cl$^-$/HCO$_3$$^-$ antiporters in the basolateral membrane to export the "excess" OH$^-$ ions from the cytosol to the blood. As noted earlier, this anion antiporter is activated at high cytosolic pH.

The overall process by which parietal cells acidify the stomach lumen is illustrated in Figure 11-31. In a reaction catalyzed by carbonic anhydrase, the "excess" cytosolic OH$^-$ combines with CO$_2$ that diffuses in from the blood, forming HCO$_3$$^-$. Catalyzed by the basolateral anion Cl$^-$/HCO$_3$$^-$ antiporter, this bicarbonate ion is exported across the basolateral membrane (and ultimately into the blood) in exchange for a Cl$^-$ ion. The Cl$^-$ ions then exit through Cl$^-$ channels in the apical membrane, entering the stomach lumen. To preserve electroneutrality, each Cl$^-$ ion that moves into the stomach lumen across the apical membrane is accompanied by a K$^+$ ion that moves outward through a separate K$^+$ channel. In this way, the excess K$^+$ ions pumped inward by the H$^+$/K$^+$ ATPase are returned to the stomach lumen, thus maintaining the normal intracellular K$^+$ concentration. The net result is secretion of equal amounts of H$^+$ and Cl$^-$ ions (i.e., HCl) into the stomach lumen, while the pH of the cytosol remains neutral and the excess OH$^-$ ions, as HCO$_3$$^-$, are transported into the blood.

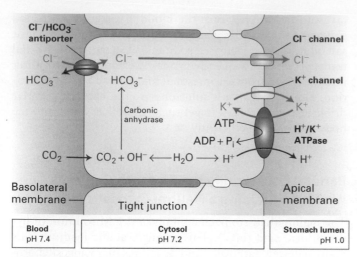

**FIGURE 11-31 Acidification of the stomach lumen by parietal cells in the gastric lining.** The apical membrane of parietal cells contains an $H^+/K^+$ ATPase (a P-class pump) as well as $Cl^-$ and $K^+$ channel proteins. Note the cyclic $K^+$ transport across the apical membrane: $K^+$ ions are pumped inward by the $H^+/K^+$ ATPase and exit via a $K^+$ channel. The basolateral membrane contains an anion antiporter that exchanges $HCO_3^-$ and $Cl^-$ ions. The combined operation of these four different transport proteins and carbonic anhydrase acidifies the stomach lumen while maintaining the neutral pH of the cytosol.

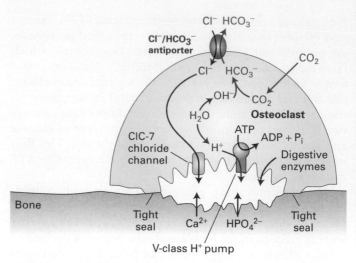

**FIGURE 11-32 Dissolution of bone by polarized osteoclast cells requires a V-class proton pump and the ClC-7 chloride channel protein.** The osteoclast plasma membrane is divided into two domains separated by the tight seal between a ring of membrane and the bone surface. The membrane domain facing the bone contains V-class proton pumps and ClC-7 $Cl^-$ channels. The opposing membrane domain contains anion antiporters that exchange $HCO_3^-$ and $Cl^-$ ions. The combined operation of these three transport proteins and carbonic anhydrase acidifies the enclosed space and allows bone resorption while maintaining the neutral pH of the cytosol.
[See R. Planells-Cases and T. Jentsch, 2009, *Biochim. Biophys. Acta* **1792**:173 for discussion of ClC-7.]

## Bone Resorption Requires Coordinated Function of a V-Class Proton Pump and a Specific Chloride Channel Protein

Net bone growth in mammals subsides just after puberty, but a finely balanced, highly dynamic process of disassembly (resorption) and reassembly (bone formation) goes on throughout adulthood. Such continual bone *remodeling* permits the repair of damaged bones and can release calcium, phosphate, and other ions from mineralized bone into the blood for use elsewhere in the body.

*Osteoclasts,* the bone-dissolving cells, are a type of macrophage best known for their role in protecting the body from infections. Osteoclasts are polarized cells that form specialized, very tight seals between themselves and bone, creating an enclosed extracellular space (Figure 11-32). An adhered osteoclast then secretes into this space a corrosive mixture of HCl and proteases that dissolves the inorganic components of the bone into $Ca^{2+}$ and phosphate and digests its protein components. The mechanism of HCl secretion is similar to that used by the stomach to generate digestive juice (see Figure 11-31). As in gastric HCl secretion, carbonic anhydrase and an anion antiport protein are important for osteoclast function. Osteoclasts employ a V-type proton pump to export $H^+$ ions into the bone-facing space rather than the P-class ATP-powered $H^+/K^+$ pump used by gastric epithelial cells

 The rare hereditary disease *osteopetrosis,* marked by increased bone density, is due to abnormally low bone

resorption. Many patients have an inactivating mutation in the gene encoding ClC-7, the chloride channel protein localized to the domain of the osteoclast plasma membrane that faces the space near the bone. As with lysosomes (see Figure 11-14), in the absence of a chloride channel the proton pump cannot acidify the enclosed extracellular space and thus bone resorption is defective. ■

## KEY CONCEPTS of Section 11.6

### Transcellular Transport

• The apical and basolateral plasma membrane domains of epithelial cells contain different transport proteins and carry out quite different transport processes.

• In the intestinal epithelial cell, the coordinated operation of $Na^+$-linked symporters in the apical membrane and $Na^+/K^+$ ATPases and uniporters in the basolateral membrane mediates transcellular transport of amino acids and glucose from the intestinal lumen to the blood (see Figure 11-30).

• The increased osmotic pressure created by transcellular transport of salt, glucose, and amino acids across the intestinal epithelium draws water from the intestinal lumen into the body, a phenomenon that serves as the basis for rehydration therapy using sugar-salt solutions.

• The combined action of carbonic anhydrase and four different transport proteins permits parietal cells in the stomach lining to secrete HCl into the lumen while maintaining their cytosolic pH near neutrality (see Figure 11-31).

- Bone resorption requires coordinated function in osteoclasts of a V-class proton pump and the ClC-7 chloride channel protein (Figure 11-32).

## Perspectives for the Future

In this chapter, we have explained the action of specific membrane transport proteins and their impact on certain aspects of human physiology; such a molecular physiology approach has many medical applications. Even today, specific inhibitors or activators of channels, pumps, and transporters constitute the largest single class of drugs. For instance, an inhibitor of the gastric $H^+/K^+$ ATPase that acidifies the stomach is the most widely used drug for treating stomach ulcers and gastric reflux syndrome. Inhibitors of channel proteins in the kidney are widely used to control hypertension (high blood pressure); by blocking resorption of water into the blood from urine forming in the kidneys, these drugs reduce blood volume and thus blood pressure. Calcium-channel blockers are widely employed to control the intensity of contraction of the heart. Drugs that inhibit a particular potassium channel in β islet cells enhance secretion of insulin (see Figure 16-38), and are widely used to treat adult-onset (type II) diabetes.

With the completion of the human genome project, we have in hand the sequences of all human membrane-transport proteins. Already we know that mutations in many of them cause disease—cystic fibrosis, due to mutations in CFTR, is one example, and osteopetrosis, caused by mutations in the ClC-7 chloride channel, is another. More recently it was shown that loss-of-function mutations in either subunit of a different chloride channel, ClC-K, cause both salt loss by the kidney and deafness. This explosion of basic knowledge, associating specific genetic diseases with specific transport proteins, will enable researchers to identify new types of compounds that inhibit or activate just one of these membrane transport proteins and not its homologs. An important challenge, however, is to understand the role of an individual transport protein in each of the several tissues in which it is expressed.

Another major challenge is to understand how each channel, transporter, and pump is regulated to meet the needs of the cell. Like other cellular proteins, many of these proteins undergo reversible phosphorylation, ubiquitination, and other covalent modifications that affect their activity, but in the vast majority of cases, we do not understand how this regulation affects cellular function. Many channels, transporters, and pumps normally reside on intracellular membranes, not on the plasma membrane, and move to the plasma membrane only when a particular hormone is present. The addition of insulin to muscle, for instance, causes the GLUT4 glucose transporter to move from intracellular membranes to the plasma membrane, increasing the rate of glucose uptake. We noted earlier that the addition of vaso-

pressin to certain kidney cells similarly causes an aquaporin to traffic to the plasma membrane, increasing the rate of water transport. But despite much research, the underlying cellular mechanisms by which hormones stimulate the movement of transport proteins to and from the plasma membrane, and the regulation of these processes, remain obscure.

## Key Terms

| | |
|---|---|
| ABC superfamily 485 | membrane potential 475 |
| active transport 476 | $Na^+/K^+$ ATPase 489 |
| antiport 502 | patch clamping 500 |
| aquaporins 480 | P-class pump 484 |
| ATP-powered pump 476 | resting $K^+$ channel 497 |
| cotransport 502 | resting potential 495 |
| electrochemical gradient 475 | sarcoplasmic reticulum 486 |
| facilitated transport 476 | simple diffusion 474 |
| F-class pump 484 | symport 502 |
| flippase 493 | tight junction 508 |
| gated channel 476 | transporter 476 |
| GLUT proteins 479 | transcellular transport 508 |
| hypertonic 481 | uniport 478 |
| hypotonic 481 | V-class pump 484 |
| isotonic 481 | |

## Review the Concepts

**1.** Nitric oxide (NO) is a gaseous molecule with lipid solubility similar to that of $O_2$ and $CO_2$. Endothelial cells lining arteries use NO to signal surrounding smooth muscle cells to relax, thereby increasing blood flow. What mechanism or mechanisms would transport NO from where it is produced in the cytoplasm of an endothelial cell into the cytoplasm of a smooth muscle cell where it acts?

**2.** Acetic acid (a weak acid with a $pK_a$ of 4.75) and ethanol (an alcohol) are each composed of two carbons, hydrogen, and oxygen, and both enter cells by passive diffusion. At pH 7, one is much more membrane permeable than the other. Which is more permeable and why? Predict how the permeability of each is altered when the pH is reduced to 1.0, a value typical of the stomach.

**3.** Uniporters and ion channels support facilitated diffusion across biomembranes. Although both are examples of facilitated diffusion, the rates of ion movement via an ion channel are roughly $10^4$- to $10^5$-fold faster than that of molecule movement via a uniporter. What key mechanistic difference results in this large difference in transport rate? What contribution to free energy ($\Delta G$) determines the direction of transport?

**4.** Name the three classes of transporters. Explain which one or more of these classes is able to move glucose and which move bicarbonate ($HCO_3^-$) against an electrochemical gradient. In the case of bicarbonate, but not glucose, the $\Delta G$ of the transport process has two terms. What are these two terms, and why does the second not apply to glucose? Why are cotransporters often referred to as examples of secondary active transport?

5. An $H^+$ ion is smaller than an $H_2O$ molecule, and a glycerol molecule, a three-carbon alcohol, is much larger. Both readily dissolve in $H_2O$. Why do aquaporins fail to transport $H^+$ whereas some can transport glycerol?

6. GLUT1, found in the plasma membrane of erythrocytes, is a classic example of a uniporter.

   a. Design a set of experiments to prove that GLUT1 is indeed a glucose-specific uniporter rather than a galactose- or mannose-specific uniporter.

   b. Glucose is a 6-carbon sugar while ribose is a 5-carbon sugar. Despite this smaller size, ribose is not efficiently transported by GLUT1. How can this be explained?

   c. A drop in blood sugar from 5 mM to 2.8 mM or below can cause confusion and fainting. Calculate the effect of this drop on glucose transport into cells expressing GLUT1.

   d. How do liver and muscle cells maximize glucose uptake without changing $V_{max}$?

   e. Tumor cells expressing GLUT1 often have a higher $V_{max}$ for glucose transport than do normal cells of the same type. How could these cells increase the $V_{max}$?

   f. Fat and muscle cells modulate the $V_{max}$ for glucose uptake in response to insulin signaling. How?

7. Name the four classes of ATP-powered pumps that produce active transport of ions and molecules. Indicate which of these classes transport ions only and which transport primarily small organic molecules. The initial discovery of one class of these ATP-powered pumps came from studying the transport of not a natural substrate but rather artificial substrates used as cancer chemotherapy drugs. What do investigators now think are common examples of the natural substrates of this particular class of ATP-powered pumps?

8. Explain why the coupled reaction $ATP \rightarrow ADP + P_i$ in the P-class ion pump mechanism does not involve direct hydrolysis of the phosphoanhydride bond.

9. Describe a negative feedback mechanism for controlling rising cytoplasmic $Ca^{2+}$ concentration in cells that require rapid changes in $Ca^{2+}$ concentration for normal functioning. How would a drug that inhibits calmodulin activity affect cytoplasmic $Ca^{2+}$ concentration regulated by this mechanism? What would be the effect on the function of, for example, a skeletal muscle cell?

10. Certain proton pump inhibitors inhibit secretion of stomach acid and are among the most widely sold drugs in the world today. What pump does this type of drug inhibit, and where is this pump located?

11. The membrane potential in animal cells, but not in plants, depends largely on resting $K^+$ channels. How do these channels contribute to the resting potential? Why are these channels considered to be nongated channels? How do these channels achieve selectivity for $K^+$ versus $Na^+$, which is smaller than $K^+$?

12. Patch clamping can be used to measure the conductance properties of individual ion channels. Describe how patch clamping can be used to determine whether or not the gene coding for a putative $K^+$ channel actually codes for a $K^+$ or $Na^+$ channel.

13. Plants use the proton electrochemical gradient across the vacuole membrane to power the accumulation of salts and sugars in the organelle. This creates a hypertonic situation. Why does this not result in the plant cell swelling and bursting? Even under isotonic conditions, there is a slow leakage of ions into animal cells. How does the plasma membrane $Na^+/K^+$ ATPase enable animal cells to avoid osmotic lysis under isotonic conditions?

14. In the case of the bacterial sodium/leucine transporter, what is the key distinguishing feature about the bound sodium ions that ensures that other ions, particularly $K^+$, do not bind?

15. Describe the symport process by which cells lining the small intestine import glucose. What ion is responsible for the transport, and what two particular features facilitate the energetically favored movement of this ion across the plasma membrane?

16. Movement of glucose from one side to the other side of the intestinal epithelium is a major example of transcellular transport. How does the $Na^+/K^+$ ATPase power the process? Why are tight junctions essential for the process? Why is localization of the transporters specifically in the apical and basolateral membrane crucial for transcellular transport? Rehydration supplements such as sport drinks include a sugar and a salt. Why are both important to rehydration?

## Analyze the Data

Imagine that you are investigating the transepithelial transport of radioactive glucose. Intestinal epithelial cells are grown in culture to form a complete sheet so that the fluid bathing the apical domain of the cells (the apical medium) is completely separated from the fluid bathing the basolateral domain of the cells (the basolateral medium). Radioactive ($^{14}C$-labeled) glucose is added to the apical medium, and the appearance of radioactivity in the basolateral medium is monitored in terms of counts per minute per milliliter (cpm/ml), a measure of radioactivity per unit volume.

Treatment 1: The apical and basolateral media each contain 150 mM $Na^+$ (curve 1).

Treatment 2: The apical medium contains 1 mM $Na^+$, and the basolateral medium contains 150 mM $Na^+$ (curve 2).

Treatment 3: The apical medium contains 150 mM $Na^+$, and the basolateral medium contains 1 mM $Na^+$ (curve 3).

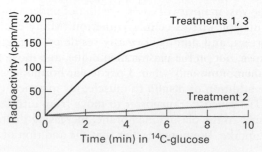

**Radioactivity in Basolateral Medium**

**a.** What is a likely explanation for the different results obtained in treatments 1 and 3 versus treatment 2?

In additional studies, the drug ouabain, which inhibits $Na^+/K^+$ ATPases, is included as noted.

Treatment 4: The apical and basolateral media contain 150 mM $Na^+$ and the apical media contains ouabain (curve 4).

Treatment 5: The apical and basolateral media contain 150 mM $Na^+$ and the basolateral media contains ouabain (curve 5).

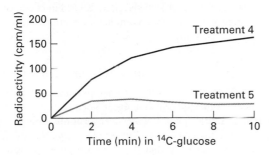

**Radioactivity in Basolateral Medium**

**b.** What is a likely explanation for the different results obtained in treatment 4 versus treatment 5?

**c.** Certain natural compounds and drugs being tested as treatments for diabetes lower glucose transport in intestinal or kidney epithelial cells, and thereby lower blood sugar levels. Addition of one such drug to the apical medium yields transport similar to that in Treatment 5, whereas its addition to the basolateral medium yields transport similar to that in Treatment 4. What is the most likely target of this drug, and what is the effect on this target?

# References

## Facilitated Transport of Glucose and Water

Engel, A., Y. Fujiyoshi, and P. Agre. 2000. The importance of aquaporin water channel protein structures. *EMBO J.* 19:800–806.

Hedfalk, K., et al. 2006. Aquaporin gating. *Curr. Opinion Structural Biology* 16:1–10.

Hruz, P. W., and M. M. Mueckler. 2001. Structural analysis of the GLUT1 facilitative glucose transporter (review). *Mol. Memb. Biol.* 18:183–193.

King, L. S., D. Kozono, and P. Agre. 2004. From structure to disease: the evolving tale of aquaporin biology. *Nat. Rev. Mol. Cell Biol.* 5:687–698.

Thorens, B., and M. Mueckler, Glucose transporters in the 21st century. *Am. J. Physiol.-Endoc. M.* 298:E141–E145, 2010.

Verkman, A. S. 2009. Knock-out models reveal new aquaporin functions. *Handb. Exp. Pharmacol.* 190:359–381.

Wang, Y., K. Schulten, and E. Tajkhorshid. 2005. What makes an aquaporin a glycerol channel? A comparative study of AqpZ and GlpF structure. *Structure* 13:1107–1118.

## ATP-Powered Pumps and the Intracellular Ionic Environment

Aller, S., et al. 2009. Structure of P-glycoprotein reveals a molecular basis for poly-specific drug binding. *Science* 323: 1718–1722.

Gottesman, M. M., and V. Ling. 2006. The molecular basis of multidrug resistance in cancer: the early years of P-glycoprotein research. *FEBS Lett.* 580:998–1009.

Guerini, D., L. Coletto, and E. Carafoli. 2005. Exporting calcium from cells. *Cell Calcium* 38:281–289.

Guttmann, D., et al. 2009. Understanding polyspecificity of multidrug ABC transporters: closing in on the gaps in ABCB1. *Trends Biochem. Sci.* 35:36–42.

Hall, M., et al. 2009. Is resistance useless? Multidrug resistance and collateral sensitivity. *Trends Pharmacol. Sci.* 30:546–556.

Jencks, W. P. 1995. The mechanism of coupling chemical and physical reactions by the calcium ATPase of sarcoplasmic reticulum and other coupled vectorial systems. *Biosci. Rept.* 15:283–287.

Locher, K. P., A. Lee, and D. C. Rees. 2002. The *E. coli* BtuCD structure: a framework for ABC transporter architecture and mechanism. *Science* 296:1091.

Ogawa, H., et. al. 2009. Crystal structure of the sodium-potassium pump ($Na^+,K^+$-ATPase) with bound potassium and ouabain. *Proc. Nat'l Acad. Sci. USA* 106:13742–13747.

Raggers, R. J., et al. 2000. Lipid traffic: the ABC of transbilayer movement. *Traffic* 1:226–234.

Riordan, J. 2005. Assembly of functional CFTR chloride channels. *Ann. Rev. Physiol.* 67:701–718.

Shinoda, T., et. al. 2009. Crystal structure of the sodium-potassium pump at 2.4 Å resolution. *Nature* 459:446–450.

Toel, M., R. Saum, and M. Forgac. 2010. Regulation and isoform function of the V-ATPases. *Biochemistry* 49:4715–4723.

Toyoshima, C. 2009. How $Ca^{2+}$-ATPase pumps ions across the sarcoplasmic reticulum membrane. *Biochim. Biophys. Acta* 1793: 941–946.

Verkman, A. S., G. L. Lukacs, and L. J. Galietta. 2006. CFTR chloride channel drug discovery—inhibitors as antidiarrheals and activators for therapy of cystic fibrosis. *Curr. Pharm. Des.* 12:2235–2247.

## Nongated Ion Channels and the Resting Membrane Potential

Dutzler, R., et al. 2002. X-ray structure of a ClC chloride channel at 3.0 Å reveals the molecular basis of anion selectivity. *Nature* 415:287–294.

Hibino, H., et al. 2010. Inwardly rectifying potassium channels: their structure, function, and physiological roles. *Physiol. Rev.* 90:291–366.

Hille, B. 2001. *Ion Channels of Excitable Membranes*, 3d ed. Sinauer Associates.

Jentsch, T.J. 2008. ClC chloride channels and transporters: from genes to protein structure, pathology and physiology. *Crit. Rev. Biochem. Mol.* 43:3–36.

Jouhaux, F., and R. Mackinnon. 2005. Principles of selective ion transport in channels and pumps. *Science* 310:1461–1465.

MacKinnon, R. 2004. Potassium channels and the atomic basis of selective ion conduction. Nobel Lecture reprinted in *Biosci. Rep.* 24:75–100.

Montello, C., L. Birnbaumer, and V. Flickers. 2002. The TRP channels, a remarkably functional family. *Cell* 108:595–598.

Neher, E. 1992. Ion channels for communication between and within cells. Nobel Lecture reprinted in *Neuron* 8:605–612 and *Science* 256:498–502.

Neher, E., and B. Sakmann. 1992. The patch clamp technique. *Sci. Am.* 266(3):28–35.

Planells-Cases, R., and T. J. Jentsch. 2009. Chloride channelopathies. *Biochim. Biophys. Acta* 1792:173–189.

Roux, B. 2005. Ion conduction and selectivity in $K^+$ channels. 2005. *Ann. Rev. Biophys. Biomol. Struct.* 34:153–171.

Zhou, Y., et al. 2001. Chemistry of ion coordination and hydration revealed by a $K^+$ channel–Fab complex at 2 Å resolution. *Nature* 414:43–48.

## Cotransport by Symporters and Antiporters

Alper, S. L. 2009. Molecular physiology and genetics of $Na^+$-independent SLC4 anion exchangers. *J. Exp. Biol.* 212:1672–1683.

Barkla, B. J., R. Vera-Estrella, and O. Pantoja. 1999. Towards the production of salt-tolerant crops. *Adv. Exp. Med. Biol.* 464:77–89.

Diallinas, G. 2008. An almost- complete movie: structural snapshots of transporter proteins reveal how they transport species across membranes. *Science* 322:1644–1645.

Gao, X., et al. 2009. Structure and mechanism of an amino acid antiporter. *Science* 324:1565–1568.

Gouaux, E. 2009. Review: The molecular logic of sodium-coupled neurotransmitter transporters. *Phil. Trans. R. Soc. Lond. B Biol. Sci.* 364:149–154.

Krishnamurthy, H., C. L. Piscitelli, and E. Gouaux. 2009. Unlocking the molecular secrets of sodium-coupled transporters. *Nature* 459:347–355.

Orlowski, J., and S. Grinstein. 2007. Emerging roles of alkali cation/proton exchangers in organellar homeostasis. *Curr. Opin. Cell Biol.* 19:483–492.

Shabala, S., and T. A. Cuin. 2008. Potassium transport and plant salt tolerance. *Physiol. Plant* 133:651–669.

Wakabayashi, S., M. Shigekawa, and J. Pouyssegur. 1997. Molecular physiology of vertebrate $Na^+/H^+$ exchangers. *Physiol. Rev.* 77:51–74.

Wright, E. M. 2004. The sodium/glucose cotransport family SLC5. *Pflugers Arch.* 447:510–518.

Wright, E. M., and D. D. Loo. 2000. Coupling between $Na^+$, sugar, and water transport across the intestine. *Ann. NY Acad. Sci.* 915:54–66.

## Transcellular Transport

Anderson, J. M., and C. M. Van Itallie. 2009. Physiology and function of the tight junction. *Cold Spring Harbor Perspect. Biol.* 1:a002584.

Elkouby-Naor, L., and T. Ben-Yosef. 2010. Functions of claudin tight junction proteins and their complex interactions in various physiological systems. *Int. Rev. Cell Mol. Biol.* 279:1–32.

Hubner, C. A., and T. J. Jentsch. 2008. Channelopathies of transepithelial transport and vesicular function. *Adv. Genet.* 63:113–152.

Rao, M. 2004. Oral rehydration therapy: new explanations for an old remedy. *Ann. Rev. Physiol.* 66:385–417.

Schafer, J. A. 2004. Renal water reabsorption: a physiologic retrospective in a molecular era. *Kidney Int. Suppl.* 91:S20–27.

Schultz, S. G. 2001. Epithelial water absorption: osmosis or cotransport? *Proc. Nat'l. Acad. Sci. USA* 98:3628–3630.

# Stumbling upon Active Transport

J. Skou, 1957, *Biochem. Biophys. Acta* **23**:394

In the mid-1950s Jens Skou was a young physician researching the effects of local anesthetics on isolated lipid bilayers. He needed an easily assayed membrane-associated enzyme to use as a marker in his studies. What he discovered was an enzyme critical to the maintenance of membrane potential, the $Na^+/K^+$ ATPase, a molecular pump that catalyzes active transport.

## Background

During the 1950s many researchers around the world were actively investigating the physiology of the cell membrane, which plays a role in a number of biological processes. It was well known that the concentration of many ions differs inside and outside the cell. For example, the cell maintains a lower intracellular sodium ($Na^+$) concentration and higher intracellular potassium ($K^+$) concentration than is found outside the cell. Somehow the membrane can regulate intracellular salt concentrations. Additionally, movement of ions across cell membranes had been observed, suggesting that some sort of transport system is present. To maintain normal intracellular $Na^+$ and $K^+$ concentrations, the transport system could not rely on passive diffusion because both ions must move across the membrane against their concentration gradients. This energy-requiring process was termed active transport.

At the time of Skou's experiments, the mechanism of active transport was still unclear. Surprisingly, Skou had no intention of helping to clarify the field. He found the $Na^+/K^+$ ATPase completely by accident in his search for an abundant, easily measured enzyme activity associated with lipid membranes. A recent study had shown that membranes derived from squid axons contained a membrane-associated enzyme that could hydrolyze ATP. Thinking

that this would be an ideal enzyme for his purposes, Skou set out to isolate such an ATPase from a more readily available source, crab leg neurons. It was during his characterization of this enzyme that he discovered the protein's function.

## The Experiment

Since the original goal of his study was to characterize the ATPase for use in subsequent studies, Skou wanted to know under what experimental condition its activity was both robust and reproducible. As often is the case with the characterization of a new enzyme, this requires careful titration of the various components of the reaction. Before this can be done, one must be sure the system is free from outside sources of contamination.

In order to study the influence of various cations, including three that are critical for the reaction—$Na^+$, $K^+$, and $Mg^{2+}$—Skou had to make sure that no contaminating ions were brought into the reaction from another source. Therefore all buffers used in the purification of the enzyme were prepared from salts that did not contain these cations. An additional source of contaminating cations was the ATP substrate, which contains three phosphate groups, giving it an overall negative charge. Because stock solutions of ATP often included a cation to balance the charge, Skou converted the ATP used in his reactions to the acid form so that balancing cations would not affect the experiments. Once he had a well-controlled environment, he could characterize the enzyme activity. These precautions were fundamental to his discovery.

Skou first showed that his enzyme could indeed catalyze the cleavage of ATP into ADP and inorganic phosphate. He then moved on to look for the optimal conditions for this activity by varying the pH of the reaction, and

the concentrations of salts and other cofactors, which bring cations into the reaction. He could easily determine a pH optimum as well as an optimal concentration of $Mg^{2+}$, but optimizing $Na^+$ and $K^+$ proved to be more difficult. Regardless of the amount of $K^+$ added to the reaction, the enzyme was inactive without $Na^+$. Similarly, without $K^+$, Skou observed only a low-level ATPase activity that did not increase with increasing amounts of $Na^+$.

These results suggested that the enzyme required both $Na^+$ and $K^+$ for optimal activity. To demonstrate that this was the case, Skou performed a series of experiments in which he measured the enzyme activity as he varied both the $Na^+$ and $K^+$ concentrations in the reaction (Figure 1). Although both cations clearly were required for significant activity, something interesting occurred at high concentrations of each cation. At the optimal concentration of $Na^+$ and $K^+$, the ATPase activity reached a peak. Once at that peak, further increasing the concentration did not affect the ATPase activity. $Na^+$ thus behaved like a classic enzyme substrate, with increasing input leading to increased activity until a saturating concentration was achieved, at which the activity plateaued. $K^+$, on the other hand, behaved differently. When the $K^+$ concentration was increased beyond the optimum, ATPase activity declined. Thus while $K^+$ was required for optimal activity, at high concentrations it inhibited the enzyme. Skou reasoned that the enzyme must have separate binding sites for $Na^+$ and $K^+$. For optimal ATPase activity, both must be filled. However, at high concentrations $K^+$ could compete for the $Na^+$-binding site, leading to enzyme inhibition. He hypothesized that this enzyme was involved in active transport, that is, the pumping of $Na^+$ out of the cell, coupled to the import of $K^+$ into the cell. Later studies would

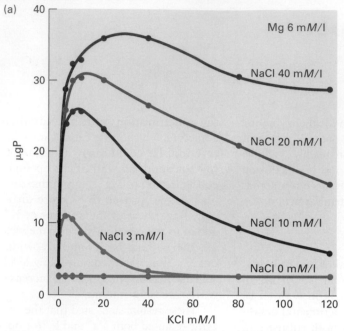

(a)

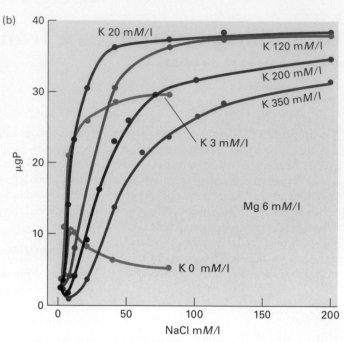

(b)

**FIGURE 1** **Demonstration of the dependence of Na⁺/K⁺ ATPase activity on the concentration of each ion.** The graph on the left shows that increasing $K^+$ leads to an inhibition of the ATPase activity. The graph on the right shows that with increasing $Na^+$, the enzyme activity increases up to a peak and then levels out. This graph also demonstrates the dependence of the activity on low levels of $K^+$. [Adapted from J. Skou, 1957, *Biochem. Biophys. Acta* **23:**394.]

prove that this enzyme was indeed the pump that catalyzed active transport. This finding was so exciting that Skou devoted his subsequent research to studying the enzyme, never using it as a marker, as he initially intended.

## Discussion

Skou's finding that a membrane ATPase used both $Na^+$ and $K^+$ as substrates was the first step in understanding active transport on a molecular level. How did Skou know to test both $Na^+$ and $K^+$? In his Nobel lecture in 1997, he explained that in his first attempts at characterizing the ATPase, he took no precautions to avoid the use of buffers and ATP stock solutions that contained $Na^+$ and $K^+$. Pondering the puzzling and unreproducible results that he obtained led to the realization that contaminating salts might be influencing the reaction. When he repeated the experiments, this time avoiding contamination by $Na^+$ and $K^+$ at all stages, he obtained clear-cut, reproducible results.

The discovery of the $Na^+/K^+$ ATPase had an enormous impact on membrane biology, leading to a better understanding of the membrane potential. The generation and disruption of membrane potential forms the basis of many biological processes, including neurotransmission and the coupling of chemical and electrical energy. For this fundamental discovery, Skou was awarded the Nobel Prize for Chemistry in 1997.

# Cellular Energetics

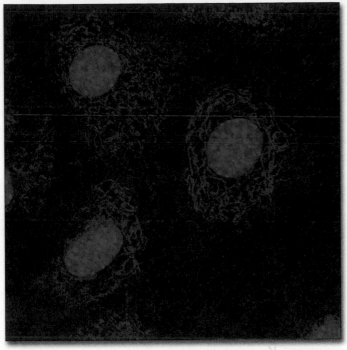

Immunofluorescence micrograph showing the intertwined network of mitochondria (red) in cultured human HeLa cells. The nuclei of the cells are stained purple. [Dr. Gopal Murti/Photo Researchers.]

From the growth and division of a cell to the beating of a heart to the electrical activity of a neuron that underlies thinking, life requires energy. Energy is defined as the capacity to do work, and on a cellular level that work includes conducting and regulating a multitude of chemical reactions and transport processes, growing and dividing, generating and maintaining a highly organized structure, and interacting with other cells. This chapter describes the molecular mechanisms by which cells use sunlight or chemical nutrients as sources of energy, with a special focus on how cells convert these external sources of energy into a biologically universal, intracellular, chemical energy carrier, **adenosine triphosphate,** or **ATP** (Figure 12-1). ATP, found in all types of organisms and presumably present in the earliest life-forms, is generated from the chemical addition of inorganic phosphate ($HPO_4^{2-}$, often abbreviated as $P_i$) to adenosine diphosphate, or ADP, a process called phosphorylation. Cells use the energy released during hydrolysis of the terminal phosphoanhydride bond in ATP (see Figure 2-31) to

power many otherwise energetically unfavorable processes. Examples include the synthesis of proteins from amino acids and of nucleic acids from nucleotides (Chapter 4), transport of molecules against a concentration gradient by ATP-powered pumps (Chapter 11), contraction of muscle (Chapter 17), and beating of cilia (Chapter 18). A key theme of cellular energetics is the use of proteins to use, or "couple," energy released from one process (e.g., ATP hydrolysis) to drive another process (e.g., movement of molecules across membranes) that otherwise would be thermodynamically unfavorable.

The energy to drive ATP synthesis from ADP ($\Delta G^{\circ\prime} = 7.3$ kcal/mol) derives primarily from two sources: the energy in the chemical bonds of nutrients and the energy in sunlight (Figure 12-1). The two processes primarily responsible for converting these energy sources into ATP are **aerobic oxidation** (also known as **aerobic respiration**), which occurs in mitochondria in nearly all eukaryotic cells (Figure 12-1, *top*), and **photosynthesis,** which occurs in chloroplasts only in leaf cells of plants (Figure 12-1, *bottom*) and certain single-cell

## OUTLINE

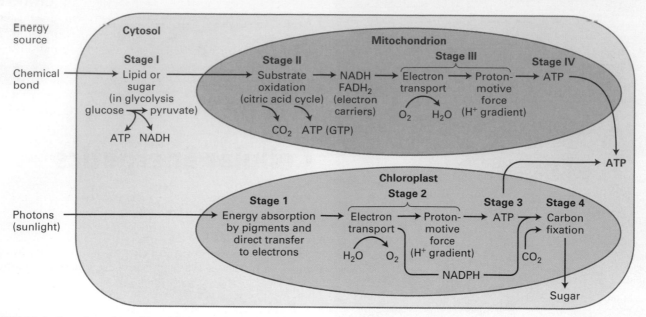

**FIGURE 12-1 Overview of aerobic oxidation and photosynthesis.** Eukaryotic cells use two fundamental mechanisms to convert external sources of energy into ATP. *(Top)* In aerobic oxidation, "fuel" molecules (primarily sugars and fatty acids) undergo preliminary processing in the cytosol, e.g., breakdown of glucose to pyruvate (**stage I**), and are then transferred into mitochondria, where they are converted by oxidation with $O_2$ to carbon dioxide and water (**stages II** and **III**) and ATP is generated (**stage IV**). *(Bottom)* In photosynthesis, which occurs in chloroplasts, the radiant energy of light is absorbed by specialized pigments (**stage 1**); the absorbed energy is used to both oxidize water to $O_2$ and establish conditions (**stage 2**) necessary for the generation of ATP (**stage 3**) and carbohydrates from $CO_2$ (carbon fixation, **stage 4**). Both mechanisms involve the production of reduced high-energy electron carriers (NADH, NADPH, $FADH_2$) and movement of electrons down an electric potential gradient in an electron transport chain through specialized membranes. Energy from these electrons is released and captured as a proton electrochemical gradient (proton-motive force) that is then used to drive ATP synthesis. Bacteria utilize comparable processes.

organisms, such as algae and cyanobacteria. Two additional processes, glycolysis and the citric acid cycle (Figure 12-1, *top*), are also important direct or indirect sources of ATP in both animal and plant cells.

In aerobic oxidation, breakdown products of sugars (carbohydrates) and fatty acids (hydrocarbons)—both derived in animals from the digestion of food—are converted by oxidation with $O_2$ to carbon dioxide and water. The energy released from this overall reaction is transformed into the chemical energy of phosphoanhydride bonds in ATP. This is analogous to burning wood (carbohydrates) or oil (hydrocarbons) to generate heat in furnaces or motion in automobile engines: both consume $O_2$ and generate carbon dioxide and water. The key difference is that cells break the overall reaction down into many intermediate steps, with the amount of energy released in any given step closely matched to the amount of energy that can be stored—for example as ATP—or that is required for the next intermediate step. If there were not such a close match, excess released energy would be lost as heat (which would be very inefficient) or not enough energy would be released to generate energy storage molecules such as ATP or to drive the next step in the process (which would be ineffective).

In photosynthesis, the radiant energy of light is absorbed by pigments such as chlorophyll and used to make ATP and

carbohydrates—primarily sucrose and starch. Unlike aerobic oxidation, which uses carbohydrates and $O_2$ to generate $CO_2$, photosynthesis uses $CO_2$ as a substrate and generates $O_2$ and carbohydrates as products.

This reciprocal relationship between aerobic oxidation occurring in mitochondria and photosynthesis in chloroplasts underlies a profound symbiotic relationship between photosynthetic and nonphotosynthetic organisms. The oxygen generated during photosynthesis is the source of virtually all the oxygen in the air, and the carbohydrates produced are the ultimate source of energy for virtually all nonphotosynthetic organisms on earth. (An exception is bacteria living in deep ocean vents—and the organisms that feed on them—that obtain energy for converting $CO_2$ into carbohydrates by oxidation of geologically generated reduced inorganic compounds released by the vents.)

At first glance, it might seem that the molecular mechanisms of photosynthesis and aerobic oxidation have little in common, besides the fact that they both produce ATP. However, a revolutionary discovery in cell biology established that bacteria, mitochondria, and chloroplasts all use the same mechanism, known as **chemiosmosis,** to generate ATP from ADP and $P_i$. In chemiosmosis (also known as chemiosmotic coupling), a proton electrochemical gradient is first generated across a membrane, driven by energy released as

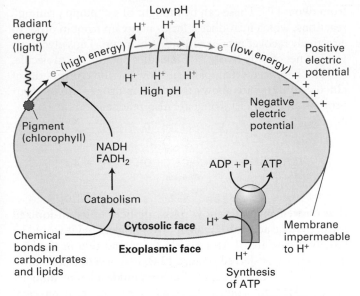

**FIGURE 12-2 The proton-motive force powers ATP synthesis.**
Transmembrane proton concentration and electrical (voltage) gradients, collectively called the *proton-motive force,* are generated during aerobic oxidation and photosynthesis in eukaryotes and prokaryotes (bacteria). High-energy electrons generated by light absorption by pigments (e.g., chlorophyll), or held in the reduced form of electron carriers (e.g., NADH, $FADH_2$) made during the catabolism of sugars and lipids, pass down an electron transport chain (blue arrows), releasing energy throughout the process. The released energy is used to pump protons across the membrane (red arrows), generating the proton-motive force. In chemiosmotic coupling, the energy released when protons flow down the gradient through ATP synthase drives the synthesis of ATP. The proton-motive force can also power other processes, such as the transport of metabolites across the membrane against their concentration gradient and rotation of bacterial flagella.

electrons travel down their electric potential gradient through an **electron transport chain.** The energy stored in this proton electrochemical gradient, called the **proton-motive force,** is then used to power the synthesis of ATP (Figure 12-2) or other energy-requiring processes. As protons move down their electrochemical gradient through the ATP synthesis enzyme, ATP is synthesized from ADP and $P_i$, a process that is the reverse of the ATP-powered ion pumps discussed in the previous chapter. In this chapter, we explore the molecular mechanisms of the two processes that share this central mechanism, focusing first on aerobic oxidation and then on photosynthesis.

# 12.1 First Step of Harvesting Energy from Glucose: Glycolysis

In an automobile engine, hydrocarbon fuel is oxidatively and explosively converted in an essentially one-step process to mechanical work (i.e., driving a piston) plus the products $CO_2$

and water. The process is relatively inefficient in that substantial amounts of the chemical energy stored in the fuel are wasted as they are converted to unused heat, and substantial amounts of fuel are only partially oxidized and are released as carbonaceous, sometimes toxic, exhaust. In the competition to survive, organisms cannot afford to squander their sometimes limited energy sources on an equivalently inefficient process, and have therefore evolved a more efficient mechanism for converting fuel into work. That mechanism, known as aerobic oxidation, provides the following advantages:

- By dividing the energy conversion process into multiple steps that generate several energy-carrying intermediates, chemical bond energy is efficiently channeled into the synthesis of ATP, with less energy lost as heat.

- Different fuels (sugars and fatty acids) are reduced to common intermediates that can then share subsequent pathways for combustion and ATP synthesis.

- Because the total energy stored in the bonds of the initial fuel molecules is substantially greater than that required to drive the synthesis of a single ATP molecule ($\sim$7.3 kcal/mol), many ATP molecules are produced.

An important feature of ATP production from the breakdown of nutrient fuels into $CO_2$ and water (see Figure 12-1, *top*) is a set of reactions, called **respiration,** involving a series of oxidation and reduction reactions called an *electron transport chain.* The combination of these reactions with phosphorylation of ADP to form ATP is called **oxidative phosphorylation** and occurs in mitochondria in nearly all eukaryotic cells. When oxygen is available and used as the final recipient of the electrons transported via the electron transport chain, the respiratory process that converts nutrient energy into ATP is called *aerobic respiration* or *aerobic oxidation.* Aerobic respiration is an especially efficient way to maximize the conversion of nutrient energy into ATP because oxygen is a relatively strong oxidant. If some molecule other than oxygen, for example the weaker oxidant sulfate ($SO_4^{2-}$) or nitrate ($NO^{3-}$), is the final recipient of the electrons in the electron transport chain, the process is called **anaerobic respiration.** Anaerobic respiration is typical of some prokaryotic microorganisms. Although there are exceptions, most known multicellular (metazoan) eukaryotic organisms use aerobic respiration to generate most of their ATP.

In our discussion of aerobic oxidation, we will be tracing the fate of the two main cellular fuels: sugars (principally glucose) and fatty acids. Under certain conditions amino acids also feed into these metabolic pathways. We first consider glucose oxidation, and then turn to fatty acids.

The complete aerobic oxidation of one molecule of glucose yields 6 molecules of $CO_2$ and the energy released is coupled to the synthesis of as many as 30 molecules of ATP. The overall reaction is

$$C_6H_{12}O_6 + 6\ O_2 + 30\ P_i^{2-} + 30\ ADP^{3-} + 30\ H^+ \rightarrow$$
$$6\ CO_2 + 30\ ATP^{4-} + 36\ H_2O$$

Glucose oxidation in eukaryotes takes place in four stages (see Figure 12-1, *top*):

**Stage I: Glycolysis** In the cytosol, one 6-carbon glucose molecule is converted by a series of reactions to two 3-carbon pyruvate molecules; a net of 2 ATPs are produced for each glucose molecule.

**Stage II: Citric Acid Cycle** In the mitochondrion, pyruvate oxidation to $CO_2$ is coupled to the generation of the high-energy electron carriers NADH and $FADH_2$, which store the energy for later use.

**Stage III: Electron Transport Chain** High-energy electrons flow down their electric potential gradient from NADH and $FADH_2$ to $O_2$ via membrane proteins that convert the energy released into a proton-motive force ($H^+$ gradient).

**Stage IV: ATP Synthesis** The proton-motive force powers the synthesis of ATP as protons flow down their concentration and voltage gradients through the ATP synthesis enzyme. For each original glucose molecule, an estimated 28 additional ATPs are produced by this mechanism of oxidative phosphorylation.

In this section, we discuss stage I: the biochemical pathways that break down glucose into pyruvate in the cytosol. We also discuss how these pathways are regulated, and contrast the metabolism of glucose under anaerobic and aerobic conditions. The ultimate fate of pyruvate, once it enters mitochondria, is discussed in Section 12.2.

## During Glycolysis (Stage I), Cytosolic Enzymes Convert Glucose to Pyruvate

**Glycolysis,** the first stage of glucose oxidation, occurs in the cytosol in both eukaryotes and prokaryotes; it does not require molecular oxygen ($O_2$), and is thus an anaerobic process. Glycolysis is an example of **catabolism,** the biological breakdown of complex substances into simpler ones. A set of 10 water-soluble cytosolic enzymes catalyze the reactions constituting the *glycolytic pathway* (*glyco,* "sweet"; *lysis,* "split"), in which one molecule of glucose is converted to two molecules of pyruvate (Figure 12-3). All the reaction intermediates produced by these enzymes are water-soluble, phosphorylated compounds called *metabolic intermediates.* In addition to chemically converting one glucose molecule into two pyruvates, the glycolytic pathway generates four ATP molecules by phosphorylation of four ADPs (reactions 7 and 10). ATP is formed directly through the enzyme-catalyzed joining of ADP and $P_i$ that is derived from phosphorylated metabolic intermediates; this process is called **substrate-level phosphorylation** (to distinguish it from the *oxidative phosphorylation* that generates ATP in stages III and IV). Substrate-level phosphorylation in glycolysis, which does not involve the use of a proton-motive force, requires the prior addition (in reactions 1 and 3) of two phosphates

from two ATPs. These can be thought of as "pump priming" reactions, which introduce a little energy up front in order to effectively recover more energy downstream. Thus glycolysis yields a net of only two ATP molecules per glucose molecule.

The balanced chemical equation for the conversion of glucose to pyruvate shows that four hydrogen atoms (four protons and four electrons) are also released:

$$C_6H_{12}O_6 \longrightarrow 2\ CH_3-\overset{\overset{O}{\|}}{C}-\overset{\overset{O}{\|}}{C}-OH + 4\ H^+ + 4\ e^-$$

Glucose            Pyruvate

(For convenience, we show pyruvate here in its un-ionized form, pyruvic acid, although at physiological pH it would be largely dissociated.) All four electrons and two of the four protons are transferred (Figure 12-3, reaction 6) to two molecules of the oxidized form of **nicotinamide adenine dinucleotide (NAD$^+$)** to produce the reduced form of the coenzyme, NADH (see Figure 2-33):

$$2H^+ + 4\ e^- + 2\ NAD^+ \rightarrow 2\ NADH$$

Later we will see that the energy carried by the electrons in NADH and the analogous electron carrier $FADH_2$, the reduced form of the coenzyme **flavin adenine dinucleotide (FAD),** can be used to make additional ATPs via the electron transport chain. The overall chemical equation for this first stage of glucose metabolism is

$$C_6H_{12}O_6 + 2\ NAD^+ + 2\ ADP^{3-} + 2\ P_i^{2-} \rightarrow$$
$$2\ C_3H_4O_3 + 2\ NADH + 2\ ATP^{4-}$$

After glycolysis, only a fraction of the energy available in glucose has been extracted and converted to ATP and NADH. The rest remains trapped in the covalent bonds of the two pyruvate molecules. The ability to efficiently convert the energy remaining in pyruvate to ATP depends on the presence of molecular oxygen. As we will see, energy conversion is substantially more efficient under aerobic conditions than under anaerobic conditions.

## The Rate of Glycolysis Is Adjusted to Meet the Cell's Need for ATP

To maintain appropriate levels of ATP, cells must control the rate of glucose catabolism. The operation of the glycolytic pathway (stage I), as well as the citric acid cycle (stage II), is continuously regulated, primarily by allosteric mechanisms (see Chapter 3 for general principles of allosteric control). Three allosteric enzymes involved in glycolysis play a key role in regulating the entire glycolytic pathway. *Hexokinase* (Figure 12-3, step **1**) is inhibited by its reaction product, glucose

**FIGURE 12-3 The glycolytic pathway.** A series of ten reactions degrades glucose to pyruvate. Two reactions consume ATP, forming ADP and phosphorylated sugars (red), two generate ATP from ADP by substrate-level phosphorylation (green), and one yields NADH by reduction of $NAD^+$ (yellow). Note that all the intermediates between glucose and pyruvate are phosphorylated compounds. Reactions 1, 3, and 10, with single arrows, are essentially irreversible (large negative $\Delta G$ values) under ordinary conditions in cells.

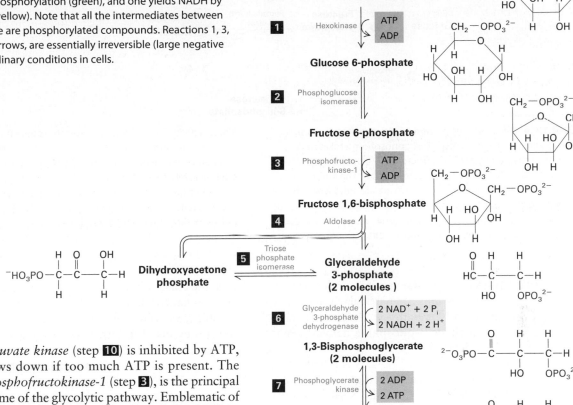

6-phosphate. *Pyruvate kinase* (step **10**) is inhibited by ATP, so glycolysis slows down if too much ATP is present. The third enzyme, *phosphofructokinase-1* (step **3**), is the principal rate-limiting enzyme of the glycolytic pathway. Emblematic of its critical role in regulating the rate of glycolysis, this enzyme is allosterically controlled by several molecules (Figure 12-4).

For example, phosphofructokinase-1 is allosterically *inhibited* by ATP and allosterically *activated* by adenosine monophosphate (AMP). As a result, the rate of glycolysis is very sensitive to the cell's **energy charge**, a measure of the fraction of total adenosine phosphates that have "high-energy" phosphoanhydride bonds, which is equal to ([ATP] + 0.5 [ADP])/([ATP] + [ADP] + [AMP]). The allosteric inhibition of phosphofructokinase-1 by ATP may seem unusual, because ATP is also a substrate of this enzyme. But the affinity of the substrate-binding site for ATP is much higher (has a lower $K_m$) than that of the allosteric site. Thus at low concentrations, ATP binds to the catalytic but not to the inhibitory allosteric site, and enzymatic catalysis proceeds at near maximal rates. At high concentrations, ATP also binds to the allosteric site, inducing a conformational change that reduces the affinity of the enzyme for the other substrate, fructose 6-phosphate, and thus reduces the rate of this reaction and the overall rate of glycolysis.

Another important allosteric activator of phosphofructokinase-1 is *fructose 2,6-bisphosphate*. This metabolite is formed from fructose 6-phosphate by an enzyme called *phosphofructokinase-2*. Fructose 6-phosphate accelerates the formation of fructose 2,6-bisphosphate, which in turn activates phosphofructokinase-1. This type of control is known as *feed-forward activation*, in which the abundance of a metabolite (here, fructose 6-phosphate) accelerates its subsequent metabolism. Fructose 2,6-bisphosphate allosterically activates phosphofructokinase-1 in liver cells by decreasing the inhibitory effect of high ATP and by increasing the affinity of phosphofructokinase-1 for one of its substrates, fructose 6-phosphate.

The three glycolytic enzymes that are regulated by allostery catalyze reactions with large negative $\Delta G^{\circ\prime}$ values—

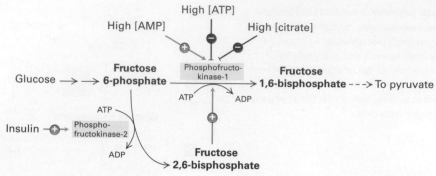

**FIGURE 12-4  Allosteric regulation of glucose metabolism.**
The key regulatory enzyme in glycolysis, phosphofructokinase-1, is allosterically activated by AMP and fructose 2,6-bisphosphate, which are elevated when the cell's energy stores are low. The enzyme is inhibited by ATP and citrate, both of which are elevated when the cell is actively oxidizing glucose to $CO_2$ (i.e., when energy stores are high). Later we will see that citrate is generated during stage II of glucose oxidation. Phosphofructokinase-2 (PFK2) is a bifunctional enzyme: its kinase activity forms fructose 2,6-bisphosphate from fructose 6-phosphate, and its phosphatase activity catalyzes the reverse reaction. Insulin, which is released by the pancreas when blood glucose levels are high, promotes PFK2 kinase activity and thus stimulates glycolysis. At low blood glucose, glucagon is released by the pancreas and promotes PFK2 phosphatase activity in the liver, indirectly slowing down glycolysis.

reactions that are essentially irreversible under ordinary conditions. These enzymes thus are particularly suitable for regulating the entire glycolytic pathway. Additional control is exerted by glyceraldehyde 3-phosphate dehydrogenase, which catalyzes the reduction of $NAD^+$ to NADH (see Figure 12-3, step **6**). As we shall see, NADH is a high-energy electron carrier used subsequently during oxidative phosphorylation in mitochondria. If cytosolic NADH builds up owing to a slowdown in mitochondrial oxidation, step **6** becomes thermodynamically less favorable.

Glucose metabolism is controlled differently in various mammalian tissues to meet the metabolic needs of the organism as a whole. During periods of carbohydrate starvation, for instance, it is necessary for the liver to release glucose into the bloodstream. To do this, the liver converts the polymer glycogen, a storage form of glucose (Chapter 2), directly to glucose 6-phosphate (without involvement of hexokinase, step **1**). Under these conditions, there is a reduction in fructose 2,6-bisphosphate levels and decreased phosphofructokinase-1 activity (Figure 12-4). As a result, glucose 6-phosphate derived from glycogen is not metabolized to pyruvate; rather, it is converted to glucose by a phosphatase and released into the blood to nourish the brain and red blood cells, which depend primarily on glucose for their energy. In all cases, the activity of these regulated enzymes is controlled by the level of small-molecule metabolites, generally by allosteric interactions, or by hormone-mediated phosphorylation and dephosphorylation reactions. (Chapter 15 gives a more detailed discussion of hormonal control of glucose metabolism in liver and muscle.)

## Glucose Is Fermented When Oxygen Is Scarce

Many eukaryotes, including humans, are *obligate aerobes:* they grow only in the presence of molecular oxygen and can metabolize glucose (or related sugars) completely to $CO_2$, with the concomitant production of a large amount of ATP. Most eukaryotes, however, can generate some ATP by anaerobic metabolism. A few eukaryotes are *facultative anaerobes:* they grow in either the presence or the absence of oxygen. For example, annelids, mollusks, and some yeasts can survive without oxygen, relying on the ATP produced by fermentation.

In the absence of oxygen, yeasts convert the pyruvate produced by glycolysis to one molecule each of ethanol and $CO_2$; in these reactions two NADH molecules are oxidized to $NAD^+$ for each two pyruvates converted to ethanol, thereby regenerating the supply of $NAD^+$, which is necessary for glycolysis to continue (Figure 12-5a, *left*). This anaerobic catabolism of glucose, called **fermentation**, is the basis of beer and wine production.

Fermentation also occurs in animal cells, although lactic acid rather than alcohol is the product. During prolonged contraction of mammalian skeletal muscle cells—for example, during exercise—oxygen within the muscle tissue can become scarce. As a consequence, glucose catabolism is limited to glycolysis and muscle cells convert pyruvate to two molecules of lactic acid by a reduction reaction that also oxidizes two NADHs to two $NAD^+$s (Figure 12-5a, *right*). Although the lactic acid is released from the muscle into the blood, if the contractions are sufficiently rapid and strong, the lactic acid can transiently accumulate in the tissue and contribute to muscle and joint pain during exercise. Once it is secreted into the blood, some of the lactic acid passes into the liver, where it is reoxidized to pyruvate and either further metabolized to $CO_2$ aerobically or converted back to glucose. Much lactate is metabolized to $CO_2$ by the heart, which is highly perfused by blood and can continue aerobic metabolism at times when exercising, oxygen-poor skeletal muscles secrete lactate. Lactic acid bacteria (the organisms that spoil milk) and other prokaryotes also generate ATP by the fermentation of glucose to lactic acid.

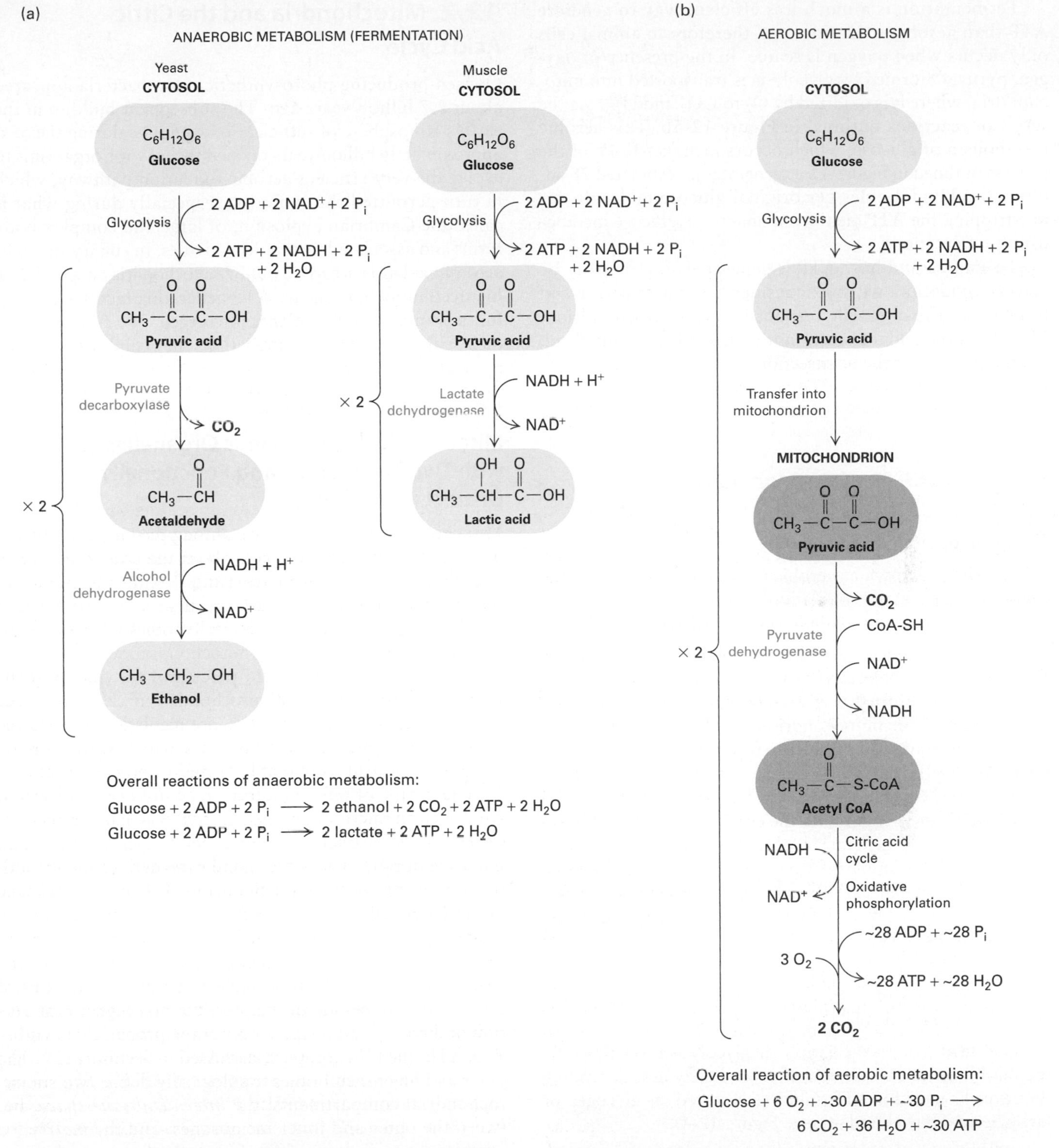

**FIGURE 12-5 Anaerobic versus aerobic metabolism of glucose.**
The ultimate fate of pyruvate formed during glycolysis depends on the presence or absence of oxygen. (a) In the absence of oxygen, pyruvate is only partially degraded and no further ATP is made. However, two electrons are transferred from each NADH molecule produced during glycolysis to an acceptor molecule to regenerate $NAD^+$, which is required for continued glycolysis. In yeast (left), acetaldehyde is the electron acceptor and ethanol is the product. This process is called *alcoholic fermentation*. When oxygen is scarce in muscle cells (right),

NADH reduces pyruvate to form lactic acid, regenerating $NAD^+$, a process called *lactic acid fermentation*. (b) In the presence of oxygen, pyruvate is transported into mitochondria, where first it is converted by pyruvate dehydrogenase into one molecule of $CO_2$ and one of acetic acid, the latter linked to coenzyme A (CoA-SH) to form acetyl CoA, concomitant with reduction of one molecule of $NAD^+$ to NADH. Further metabolism of acetyl CoA and NADH generates approximately an additional 28 molecules of ATP per glucose molecule oxidized.

Fermentation is a much less efficient way to generate ATP than aerobic oxidation, and therefore in animal cells only occurs when oxygen is scarce. In the presence of oxygen, pyruvate formed by glycolysis is transported into mitochondria, where it is oxidized by $O_2$ to $CO_2$ and $H_2O$ via the series of reactions outlined in Figure 12-5b. This aerobic metabolism of glucose, which occurs in stages II–IV of the process outlined in Figure 12-1, generates an estimated 28 additional ATP molecules per original glucose molecule, far outstripping the ATP yield from anaerobic glucose metabolism (fermentation).

To understand how ATP is generated so efficiently by aerobic oxidation, we must consider first the structure and function of the organelle responsible, the mitochondrion. Mitochondria, and the reactions that take place within them, are the subjects of the next section.

## KEY CONCEPTS of Section 12.1

### First Step of Harvesting Energy from Glucose: Glycolysis

- In a process known as aerobic oxidation, cells convert the energy released by the oxidation ("burning") of glucose or fatty acids into the terminal phosphoanhydride bond of ATP.

- The complete aerobic oxidation of each molecule of glucose produces six molecules of $CO_2$ and approximately 30 ATP molecules. The entire process, which starts in the cytosol and is completed in the mitochondrion, can be divided into four stages: (I) degradation of glucose to pyruvate in the cytosol (glycolysis); (II) pyruvate oxidation to $CO_2$ in the mitochondrion coupled to generation of the high-energy electron carriers NADH and $FADH_2$ (via the citric acid cycle); (III) electron transport to generate a proton-motive force together with conversion of molecular oxygen to water; and (IV) ATP synthesis (see Figure 12-1). From each glucose molecule two ATPs are generated by glycolysis (stage I) and approximately 28 from stages II–IV.

- In glycolysis (stage I), cytosolic enzymes convert glucose to two molecules of pyruvate and generate two molecules each of NADH and ATP (see Figure 12-3).

- The rate of glucose oxidation via glycolysis is regulated by the inhibition or stimulation of several enzymes, depending on the cell's need for ATP. Glucose is stored (as glycogen or fat) when ATP is abundant (see Figure 12-4).

- In the absence of oxygen (anaerobic conditions), cells can metabolize pyruvate to lactic acid or (in the case of yeast) to ethanol and $CO_2$, in the process converting NADH back to $NAD^+$, which is necessary for continued glycolysis. In the presence of oxygen (aerobic conditions), pyruvate is transported into the mitochondrion, where it is metabolized to $CO_2$, in the process generating abundant ATP (see Figure 12-5).

## 12.2 Mitochondria and the Citric Acid Cycle

Oxygen-producing photosynthetic cyanobacteria appeared about 2.7 billion years ago. The subsequent buildup in the earth's atmosphere of sufficient oxygen ($O_2$) during the next approximately billion years opened the way for organisms to evolve the very efficient aerobic oxidation pathway, which in turn permitted the evolution, especially during what is called the Cambrian explosion, of large and complex body forms and associated metabolic activities. In eukaryotic cells, aerobic oxidation is carried out by mitochondria (stages II–IV). In effect, mitochondria are ATP-generating factories, taking full advantage of this plentiful oxygen. We first describe their structure and then the reactions they employ to degrade pyruvate and make ATP.

### Mitochondria Are Dynamic Organelles with Two Structurally and Functionally Distinct Membranes

Mitochondria (Figure 12-6) are among the larger organelles in the cell. A mitochondrion is about the size of an *E. coli* bacterium, which is not surprising, because bacteria are thought to be the evolutionary precursors of mitochondria (see discussion of the endosymbiont hypothesis, below). Most eukaryotic cells contain many mitochondria, which may collectively occupy as much as 25 percent of the volume of the cytoplasm. The numbers of mitochondria in a cell, hundreds to thousands in mammalian cells, are regulated to match the cell's requirements for ATP (e.g., stomach cells, which use a lot of ATP for acid secretion, have many mitochondria).

The details of mitochondrial structure can be observed with electron microscopy (see Figure 9-33). Each mitochondrion has two distinct, concentric membranes: the inner and outer membrane. The outer membrane defines the smooth outer perimeter of the mitochondrion. The inner membrane lies immediately underneath the outer membrane and has numerous invaginations, called *cristae*, which extend from the perimeter of the inner membrane into the center of the mitochondrion. The high membrane curvature at the tips of the cristae may be due to the presence of a high concentration of dimers of an integral membrane protein that synthesizes ATP (the $F_0F_1$ complex, discussed in Section 12.3). The outer and inner membranes topologically define two submitochondrial compartments: the *intermembrane space,* between the outer and inner membranes, and the *matrix,* or central compartment, which forms the lumen within the inner membrane. The invaginating cristae greatly expand the surface area of the inner mitochondrial membrane, thus increasing the ability to synthesize ATP. In typical liver mitochondria, for example, the area of the inner membrane, including cristae, is about five times that of the outer membrane. In fact, the total area of all inner mitochondrial membranes in liver cells is about 17 times that of the plasma

(a)

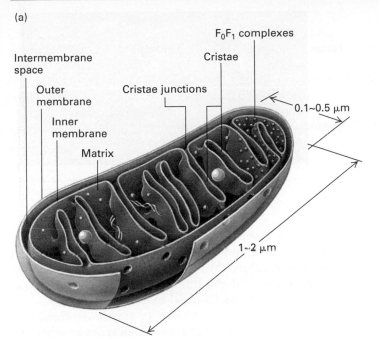

Intermembrane space

Outer membrane

Inner membrane

Matrix

$F_0F_1$ complexes

Cristae

Cristae junctions

0.1~0.5 μm

1~2 μm

(b)

**FIGURE 12-6 Internal structure of a mitochondrion.** (a) Schematic diagram showing the principal membranes and compartments. The smooth outer membrane forms the outside boundary of the mitochondrion. The inner membrane is distinct from the outer membrane and is highly invaginated to form sheets and tubes called cristae. The relatively small uniform tubular structures that connect the cristae to the portions of the inner membrane that are juxtaposed to the outer membrane are called *crista junctions.* The intermembrane space is continuous with the lumen of each crista. The $F_0F_1$ complexes (small red spheres), which synthesize ATP, are intramembrane particles that protrude from the cristae and inner membrane into the matrix. The matrix contains the mitochondrial DNA (blue strand), ribosomes (small blue spheres), and granules (large yellow spheres). (b) Computer-generated model of a section of a mitochondrion from chicken brain. This model is based on a three-dimensional electron microscopic image calculated from a series of two-dimensional electron micrographs recorded at regular intervals. This technique is analogous to a three-dimensional x-ray tomogram or CAT scan used in medical imaging. Note the tightly packed cristae (yellow-green), the inner membrane (light blue), and the outer membrane (dark blue). [Part (b) courtesy of T. Frey, from T. Frey and C. Mannella, 2000, *Trends Biochem. Sci.* **25:**319.]

membrane. The mitochondria in heart and skeletal muscles contain three times as many cristae as are found in typical liver mitochondria—presumably reflecting the greater demand for ATP by muscle cells.

Analysis of fluorescently labeled mitochondria in living cells has shown that mitochondria are highly dynamic. They undergo frequent fusions and fissions that generate tubular, sometimes branched networks (Figure 12-7), which may account for the wide variety of mitochondrial morphologies seen in different types of cells. When individual mitochondria fuse, each of the two membranes fuse (inner with inner, and outer with outer) and each of their distinct compartments intermix (matrix with matrix, intermembrane space with intermembrane space). Fusions and fissions apparently play a functional role as well, because genetic disruptions in several GTPase superfamily genes that are required for these dynamic processes can disrupt mitochondrial function, such as maintenance of proper inner membrane electric potential,

and cause human disease. An example is the inherited neuromuscular disease Charcot-Marie-Tooth subtype 2A, in which defects in peripheral nerve function lead to progressive muscle weakness, mainly in the feet and hands. The ongoing fusion and fission process appears to protect mitochondrial DNA from accumulating mutations, and may permit the isolation of dysfunctional or damaged segments of mitochondria that can be specifically targeted for destruction in the cell by a process called autophagy (see Chapter 14).

Fractionation and purification of mitochondrial membranes and compartments have made it possible to determine their protein, DNA, and phospholipid compositions and to localize each enzyme-catalyzed reaction to a specific membrane or compartment. Over 1000 different types of polypeptides are required to make and maintain mitochondria and permit them to function. Detailed biochemical analysis has established that there are at least 1098 proteins in mammalian mitochondria and perhaps as many at 1500. Only a small number of these—

**EXPERIMENTAL FIGURE 12-7 Mitochondria undergo rapid fusion and fission in living cells.** Mitochondria labeled with a fluorescent protein in a living normal murine embryonic fibroblast were observed using time-lapse fluorescence microscopy. Several mitochondria undergoing fusion (*top*) or fission (*bottom*) are artificially highlighted in blue and with arrows. [Modified from D. C. Chan, 2006, *Cell* **125**(7):1241–1252.]

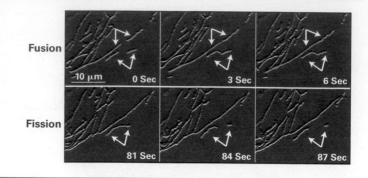

13 in humans—are encoded by mitochondrial DNA genes and synthesized inside the mitochondrial matrix space. The remaining proteins are encoded by nuclear genes (Chapter 6), synthesized in the cytosol, and then imported into mitochondria (Chapter 13). Defective functioning of the mitochondrial-associated proteins, due for example to inherited genetic mutations, leads to over 150 human diseases. The most common of these are electron transport chain diseases, which result from mutations in any one of 92 protein-encoding genes and exhibit a very wide variety of clinical abnormalities affecting muscles, the heart, the nervous system, and the liver, among other physiologic systems. Other mitochondrial-associated diseases include Miller syndrome, which results in multiple anatomical malformations, and connective tissue defects.

The most abundant protein in the outer membrane is mitochondrial **porin**, a transmembrane channel protein similar in structure to bacterial porins (see Figure 10-18). Ions and most small molecules (up to about 5000 Da) can readily pass through these channel proteins when they are open. Although there may be metabolic regulation of the opening of mitochondrial porins and thus the flow of metabolites across the outer membrane, the inner membrane and its cristae are the major permeability barriers between the cytosol and the mitochondrial matrix, limiting the rate of mitochondrial oxidation and ATP generation.

Protein constitutes 76 percent of the total mass of the inner mitochondrial membrane—a higher fraction than in any other cellular membrane. Many of these proteins are key participants in oxidative phosphorylation. They include ATP synthase, proteins responsible for electron transport, and a wide variety of transport proteins that permit the movement of metabolites between the cytosol and the mitochondrial matrix. The human genome encodes 48 members of a family of mitochondrial transport proteins. One of these is called the ADP/ATP carrier, an antiporter that moves newly synthesized ATP out of the matrix and into the inner membrane space (and subsequently the cytosol) in exchange for ADP originating from the cytosol. Without this essential antiporter, the energy trapped in the chemical bonds in mitochondrial ATP would not be available to the rest of the cell.

Note that plants have mitochondria and perform aerobic oxidation as well. In plants, stored carbohydrates, mostly in the form of starch, are hydrolyzed to glucose. Glycolysis then produces pyruvate that is transported into mitochondria, as in animal cells. Mitochondrial oxidation of pyruvate and concomitant formation of ATP occur in photosynthetic cells during dark periods when photosynthesis is not possible, and in roots and other nonphotosynthetic tissues at all times.

The mitochondrial inner membrane, cristae, and matrix are the sites of most reactions involving the oxidation of pyruvate and fatty acids to $CO_2$ and $H_2O$ and the coupled synthesis of ATP from ADP and $P_i$, with each reaction occurring in a discrete membrane or space in the mitochondrion (Figure 12-8).

We now continue our detailed discussion of glucose oxidation and ATP generation, exploring what happens to the pyruvate generated during glycolysis (stage I) after it is transported into the mitochondrial matrix. The last three of the four stages of glucose oxidation are:

- **Stage II.** Stage II can be subdivided into two distinct parts: (1) the conversion of pyruvate to acetyl CoA, followed by (2) oxidation of acetyl CoA to $CO_2$ in the citric acid cycle. These oxidations are coupled to reduction of $NAD^+$ to NADH and of FAD to $FADH_2$. (Fatty acid oxidation follows a similar route, with conversion of fatty acyl CoA to acetyl CoA.) Most of the reactions occur in or on the membrane facing the matrix.

- **Stage III.** Electron transfer from NADH and $FADH_2$ to $O_2$ via an electron transport chain within the inner membrane, which generates a proton-motive force across that membrane.

- **Stage IV.** Harnessing the energy of the proton-motive force for ATP synthesis in the mitochondrial inner membrane. Stages III and IV are together called oxidative phosphorylation.

## In the First Part of Stage II, Pyruvate Is Converted to Acetyl CoA and High-Energy Electrons

Within the mitochondrial matrix, pyruvate reacts with coenzyme A, forming $CO_2$, acetyl CoA, and NADH (Figure 12-8, Stage II *left*). This reaction, catalyzed by *pyruvate*

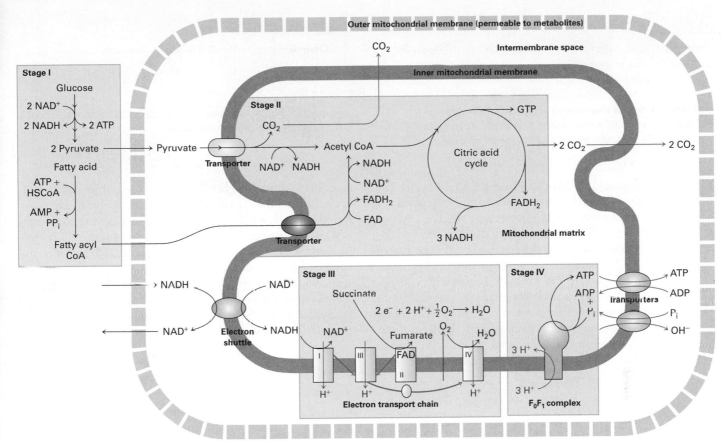

**FIGURE 12-8 Summary of aerobic oxidation of glucose and fatty acids. Stage I:** In the cytosol, glucose is converted to pyruvate (glycolysis) and fatty acid to fatty acyl CoA. Pyruvate and fatty acyl CoA then move into the mitochondrion. Mitochondrial porins make the outer membrane permeable to these metabolites, but specific transport proteins (colored ovals) in the inner membrane are required to import pyruvate (yellow) and fatty acids (blue) into the matrix. Fatty acyl groups are transferred from fatty acyl CoA to an intermediate carrier, transported across the inner membrane (blue oval), and then reattached to CoA on the matrix side. **Stage II:** In the mitochondrial matrix, pyruvate and fatty acyl CoA are converted to acetyl CoA and then oxidized, releasing $CO_2$. Pyruvate is converted to acetyl CoA with the formation of NADH and $CO_2$; two carbons from fatty acyl CoA are converted to acetyl CoA with the formation of $FADH_2$ and NADH. Oxidation of acetyl CoA in the citric acid cycle generates NADH and $FADH_2$, GTP, and $CO_2$. **Stage III:** Electron transport reduces oxygen to water and generates a proton-motive force. Electrons (blue) from reduced coenzymes are transferred via electron-transport complexes (blue boxes) to $O_2$ concomitant with transport of $H^+$ ions (red) from the matrix to the intermembrane space, generating the proton-motive force. Electrons from NADH flow directly from complex I to complex III, bypassing complex II. Electrons from $FADH_2$ flow directly from complex II to complex III, bypassing complex I. **Stage IV:** ATP synthase, the $F_0F_1$ complex (orange), harnesses the proton-motive force to synthesize ATP in the matrix. Antiporter proteins (purple and green ovals) transport ADP and $P_i$ into the matrix and export hydroxyl groups and ATP. NADH generated in the cytosol is not transported directly to the matrix because the inner membrane is impermeable to $NAD^+$ and NADH; instead, a shuttle system (red) transports electrons from cytosolic NADH to $NAD^+$ in the matrix. $O_2$ diffuses into the matrix, and $CO_2$ diffuses out.

*dehydrogenase*, is highly exergonic ($\Delta G^{\circ\prime} = -8.0$ kcal/mol) and essentially irreversible.

**Acetyl CoA** is a molecule consisting of a two-carbon acetyl group covalently linked to a longer molecule known as coenzyme A (CoA) (Figure 12-9). It plays a central role in the oxidation of pyruvate, fatty acids, and amino acids. In addition, it is an intermediate in numerous biosynthetic reactions, including transfer of an acetyl group to histone proteins and many mammalian proteins, and synthesis of lipids such as cholesterol. In respiring mitochondria, however, the two-carbon acetyl group of acetyl CoA is almost always oxidized to $CO_2$ via the citric acid cycle. Note that the two carbons in the acetyl group come from pyruvate; the third carbon of pyruvate is released as carbon dioxide.

## In the Second Part of Stage II, the Citric Acid Cycle Oxidizes the Acetyl Group in Acetyl CoA to $CO_2$ and Generates High-Energy Electrons

Nine sequential reactions operate in a cycle to oxidize the acetyl group of acetyl CoA to $CO_2$ (Figure 12-8, Stage II *right*). The cycle is referred to by several names: the **citric acid cycle,**

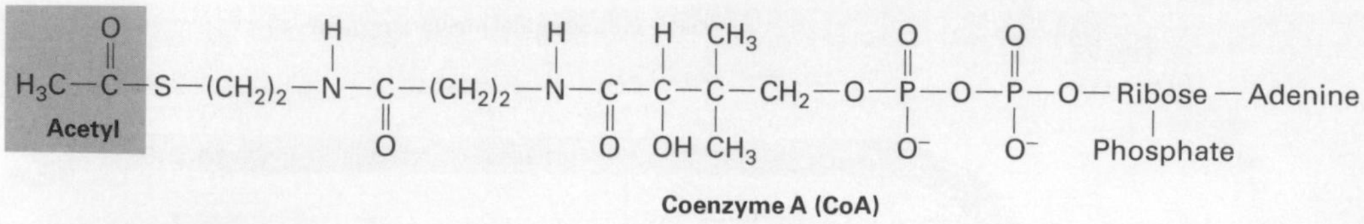

**FIGURE 12-9 The structure of acetyl CoA.** This compound, consisting of an acetyl group covalently linked to a coenzyme A (CoA) molecule, is an important intermediate in the aerobic oxidation of pyruvate, fatty acids, and many amino acids. It also contributes acetyl groups in many biosynthetic pathways.

the tricarboxylic acid (TCA) cycle, and the Krebs cycle. The net result is that for each acetyl group entering the cycle as acetyl CoA, two molecules of $CO_2$, three of NADH, and one each of $FADH_2$ and GTP are produced. NADH and $FADH_2$ are high-energy electron carriers that will play a major role in stage III of mitochondrial oxidation: electron transport.

As shown in Figure 12-10, the cycle begins with condensation of the two-carbon acetyl group from acetyl CoA with the four-carbon molecule *oxaloacetate* to yield the six-carbon *citric acid* (hence the name citric acid cycle). In both reactions 4 and 5, a $CO_2$ molecule is released and $NAD^+$ is reduced to NADH. Reduction of $NAD^+$ to NADH also occurs during

reaction 9; thus three NADHs are generated per turn of the cycle. In reaction 7, two electrons and two protons are transferred to FAD, yielding the reduced form of this coenzyme, $FADH_2$. Reaction 7 is distinctive because not only is it an intrinsic part of the citric acid cycle (stage II), but it is also catalyzed by a membrane-attached enzyme that, as we shall see, also plays an important role in stage III. In reaction 6, hydrolysis of the high-energy thioester bond in succinyl CoA is coupled to synthesis of one GTP by substrate-level phosphorylation. Because GTP and ATP are interconvertible,

$$GTP + ADP \rightleftharpoons GDP + ATP$$

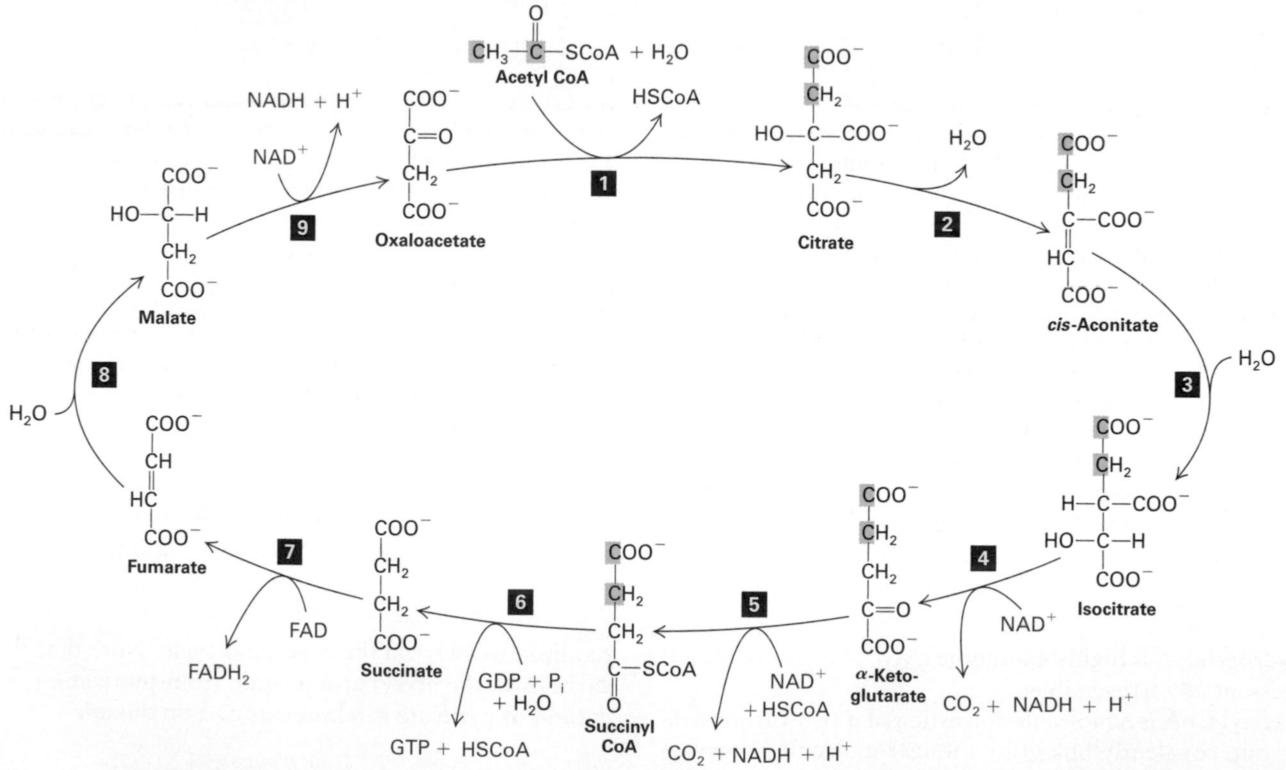

**FIGURE 12-10 The citric acid cycle.** Acetyl CoA is metabolized to $CO_2$ and the high-energy electron carriers NADH and $FADH_2$. In reaction 1, a two-carbon acetyl residue from acetyl CoA condenses with the four-carbon molecule oxaloacetate to form the six-carbon citrate. In the remaining reactions (2–9) each molecule of citrate is eventually converted back to oxaloacetate, losing two $CO_2$ molecules in the process. In each turn of the cycle, four pairs of electrons are removed from carbon atoms, forming three molecules of NADH, one molecule of $FADH_2$, and one molecule of GTP. The two carbon atoms that enter the cycle with acetyl CoA are highlighted in blue through succinyl CoA. In succinate and fumarate, which are symmetric molecules, they can no longer be specifically denoted. Isotope-labeling studies have shown that these carbon atoms are *not* lost in the turn of the cycle in which they enter; on average, one will be lost as $CO_2$ during the next turn of the cycle and the other in subsequent turns.

| TABLE 12-1 | Net Result of the Glycolytic Pathway and the Citric Acid Cycle | | | |
|---|---|---|---|---|
| Reaction | $CO_2$ Molecules Produced | $NAD^+$ Molecules Reduced to NADH | FAD Molecules Reduced to $FADH_2$ | ATP (OR GTP) |
| 1 glucose molecule to 2 pyruvate molecules | 0 | 2 | 0 | 2 |
| 2 pyruvates to 2 acetyl CoA molecules | 2 | 2 | 0 | 0 |
| 2 acetyl CoA to 4 $CO_2$ molecules | 4 | 6 | 2 | 2 |
| Total | 6 | 10 | 2 | 4 |

this can be considered an ATP-generating step. Reaction 9 regenerates oxaloacetate, so the cycle can begin again. Note that molecular $O_2$ does not participate in the citric acid cycle.

Most enzymes and small molecules involved in the citric acid cycle are soluble in the aqueous mitochondrial matrix. These include CoA, acetyl CoA, succinyl CoA, $NAD^+$, and NADH, as well as most of the eight cycle enzymes. *Succinate dehydrogenase* (reaction 7), however, is a component of an integral membrane protein in the inner membrane, with its active site facing the matrix. When mitochondria are disrupted by gentle ultrasonic vibration or by osmotic lysis, non-membrane-bound enzymes in the citric acid cycle are released as very large multiprotein complexes. It is believed that, within such complexes, the reaction product of one enzyme passes directly to the next enzyme without diffusing through the solution. Much work is needed to determine the structures of these large enzyme complexes as they exist in the cell.

Because glycolysis of one glucose molecule generates two acetyl CoA molecules, the reactions in the glycolytic pathway and citric acid cycle produce six $CO_2$ molecules, 10 NADH molecules, and two $FADH_2$ molecules per glucose molecule (Table 12-1). Although these reactions also generate four high-energy phosphoanhydride bonds in the form of two ATP and two GTP molecules, this represents only a small fraction of the available energy released in the complete aerobic oxidation of glucose. The remaining energy is stored as high-energy electrons in the reduced coenzymes NADH and $FADH_2$, which can be thought of as "electron carriers." The goal of stages III and IV is to recover this energy in the form of ATP.

## Transporters in the Inner Mitochondrial Membrane Help Maintain Appropriate Cytosolic and Matrix Concentrations of $NAD^+$ and NADH

In the cytosol $NAD^+$ is required for step **6** of glycolysis (see Figure 12-3), and in the mitochondrial matrix $NAD^+$ is required for conversion of pyruvate to acetyl CoA and for three steps in the citric acid cycle (**4**, **5**, and **9** in Figure 12-10). In each case, NADH is a product of the reaction. If glycolysis and oxidation of pyruvate are to continue, $NAD^+$ must be regenerated by oxidation of NADH. (Similarly, the $FADH_2$ generated in stage II reactions must be reoxidized to FAD if

FAD-dependent reactions are to continue.) As we will see in the next section, the electron transport chain *within* the mitochondrial inner membrane converts NADH to $NAD^+$ and $FADH_2$ to FAD as it reduces $O_2$ to water and converts the energy stored in the high-energy electrons in the reduced forms of these molecules into a proton-motive force (stage III). Even though $O_2$ is not involved in any reaction of the citric acid cycle, in the absence of $O_2$ this cycle soon stops operating as the intramitochondrial supplies of $NAD^+$ and FAD dwindle due to the inability of the electron transport chain to oxidize NADH and $FADH_2$. These observations raise the question of how a supply of $NAD^+$ in the cytosol is regenerated.

If the NADH from the cytosol could move into the mitochondrial matrix and be oxidized by the electron transport chain and if the $NAD^+$ product could be transported back into the cytosol, regeneration of cytosolic $NAD^+$ would be simple. However, the inner mitochondrial membrane is impermeable to NADH. To bypass this problem and permit the electrons from cytosolic NADH to be transferred *indirectly* to $O_2$ via the mitochondrial electron transport chain, cells use several *electron shuttles* to transfer electrons from cytosolic NADH to $NAD^+$ in the matrix. Operation of the most widespread shuttle—the *malate-aspartate shuttle*—is depicted in Figure 12-11.

For every complete "turn" of the cycle, there is no overall change in the numbers of NADH and $NAD^+$ molecules or the intermediates aspartate or malate used by the shuttle. However, in the cytosol, NADH is oxidized to $NAD^+$, which can be used for glycolysis, and in the matrix, $NAD^+$ is reduced to NADH, which can be used for electron transport:

$$NADH_{cytosol} + NAD^+_{matrix} \rightarrow NAD^+_{cytosol} + NADH_{matrix}$$

## Mitochondrial Oxidation of Fatty Acids Generates ATP

Up to now, we have focused mainly on the oxidation of carbohydrates, namely glucose, for ATP generation. Fatty acids are another important source of cellular energy. Cells can take up either glucose or fatty acids from the extracellular space with the help of specific transporter proteins (Chapter 11). Should a cell not need to immediately burn these molecules, it

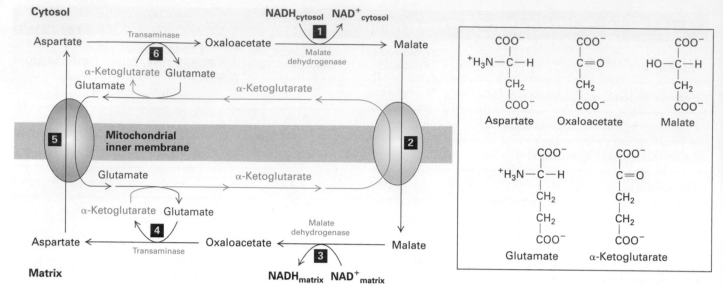

**FIGURE 12-11 The malate-aspartate shuttle.** This cyclical series of reactions transfers electrons from NADH in the cytosol (intermembrane space) across the inner mitochondrial membrane, which is impermeable to NADH itself, to NAD$^+$ in the matrix. The net result is the replacement of cytosolic NADH with NAD$^+$ and matrix NAD$^+$ with NADH. Step **1**: Cytosolic malate dehydrogenase transfers electrons from cytosolic NADH to oxaloacetate, forming malate. Step **2**: An antiporter (blue oval) in the inner mitochondrial membrane transports malate into the matrix in exchange for α-ketoglutarate. Step **3**: Mitochondrial malate dehydrogenase converts malate back to oxaloacetate, reducing NAD$^+$ in the matrix to NADH in the process. Step **4**: Oxaloacetate, which cannot directly cross the inner membrane, is converted to aspartate by addition of an amino group from glutamate. In this transaminase-catalyzed reaction in the matrix, glutamate is converted to α-ketoglutarate. Step **5**: A second antiporter (red oval) exports aspartate to the cytosol in exchange for glutamate. Step **6**: A cytosolic transaminase converts aspartate to oxaloacetate and α-ketoglutarate to glutamate, completing the cycle. The blue arrows reflect the movement of the α-ketoglutarate, the red arrows the movement of glutamate, and the black arrows that of aspartate/malate. It is noteworthy that, as aspartate and malate cycle clockwise, glutamate and α-ketoglutarate cycle in the opposite direction.

can store them as a polymer of glucose called glycogen (especially in muscle or liver) or as a trimer of fatty acids covalently linked to glycerol, called a **triacylglycerol** or **triglyceride.** In some cells, excess glucose is converted into fatty acids and then triacylglycerols for storage. However, unlike microorganisms, animals are unable to convert fatty acids to glucose. When the cells need to burn these energy stores to make ATP (e.g., when a resting muscle begins to do work and needs to burn glucose or fatty acids as fuel), enzymes break down glycogen to glucose or hydrolyze triacylglycerols to fatty acids, which are then oxidized to generate ATP:

CH$_3$—(CH$_2$)$_n$—C—O—CH$_2$
CH$_3$—(CH$_2$)$_n$—C—O—CH + 3 H$_2$O ⟶
CH$_3$—(CH$_2$)$_n$—C—O—CH$_2$
**Triacylglycerol**

3 CH$_3$—(CH$_2$)$_n$—C—OH + HO—CH$_2$ HO—CH HO—CH$_2$
**Fatty acid**  **Glycerol**

Fatty acids are the major energy source for some tissues, particularly adult heart muscle. In humans, in fact, more ATP is generated by the oxidation of fats than the oxidation

of glucose. The oxidation of 1 g of triacylglyceride to CO$_2$ generates about six times as much ATP as does the oxidation of 1 g of hydrated glycogen. Thus triglycerides are more efficient than carbohydrates for storage of energy, in part because they are stored in anhydrous form and can yield more energy when oxidized and in part because they are intrinsically more reduced (have more hydrogens) than carbohydrates. In mammals, the primary site of storage of triacylglycerides is fat (adipose) tissue, whereas the primary sites for glycogen storage are muscle and the liver.

Just as there are four stages in the oxidation of glucose, there are four stages in the oxidation of fatty acids. To optimize the efficiency of ATP generation, part of stage II (citric acid cycle oxidation of acetyl CoA) and all of stages III and IV of fatty acid oxidation are identical to those of glucose oxidation. The differences lie in the cytosolic stage I and the first part of the mitochondrial stage II. In stage I, fatty acids are converted to a fatty acyl CoA in the cytosol in a reaction coupled to the hydrolysis of ATP to AMP and PP$_i$ (inorganic pyrophosphate) (see Figure 12-8):

R—C—O$^-$ + HSCoA + ATP ⟶
**Fatty acid**

R—C—SCoA + AMP + PP$_i$
**Fatty acyl CoA**

Subsequent hydrolysis of $PP_i$ to two molecules of $P_i$ releases energy that drives this reaction to completion. To transfer the fatty acyl group into the mitochondrial matrix, it is covalently transferred to a molecule called carnitine and moved across the inner mitochondrial membrane by an acylcarnitine transporter protein (see Figure 12-8, blue oval); then, on the matrix side, the fatty acyl group is released from carnitine and reattached to another CoA molecule. The activity of the acylcarnitine transporter is regulated to prevent oxidation of fatty acids when cells have adequate energy (ATP) supplies.

In the first part of stage II, each molecule of a fatty acyl CoA in the mitochondrion is oxidized in a cyclical sequence of four reactions in which all the carbon atoms are converted two at a time to acetyl CoA with generation of $FADH_2$ and NADH (Figure 12-12a). For example, mitochondrial oxidation of each molecule of the 18-carbon stearic acid, $CH_3(CH_2)_{16}COOH$, yields nine molecules of acetyl CoA and eight molecules each of NADH and $FADH_2$. In the second part of stage II, as with acetyl CoA generated from pyruvate, these acetyl groups

enter the citric acid cycle and are oxidized to $CO_2$. As will be described in detail in the next section, the reduced NADH and $FADH_2$ with their high-energy electrons will be used in stage III to generate a proton-motive force that in turn is used in stage IV to power ATP synthesis.

## Peroxisomal Oxidation of Fatty Acids Generates No ATP

Mitochondrial oxidation of fatty acids is the major source of ATP in mammalian liver cells, and biochemists at one time believed this was true in all cell types. However, rats treated with clofibrate, a drug that affects many features of lipid metabolism, were found to exhibit an increased rate of fatty acid oxidation and a large increase in the number of peroxisomes in their liver cells. This finding suggested that peroxisomes, as well as mitochondria, can oxidize fatty acids. These small organelles, $\sim 0.2$–$1$ $\mu m$ in diameter, are lined by a single membrane (see Figure 9-32). They are present in all

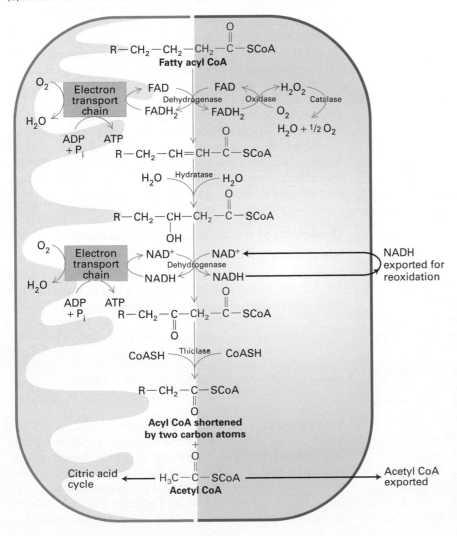

(a) MITOCHONDRIAL OXIDATION    (b) PEROXISOMAL OXIDATION

**FIGURE 12-12 Oxidation of fatty acids in mitochondria and peroxisomes.** In both mitochondrial oxidation (a) and peroxisomal oxidation (b), fatty acids are converted to acetyl CoA by a series of four enzyme-catalyzed reactions (shown down the center of the figure). A fatty acyl CoA molecule is converted to acetyl CoA and a fatty acyl CoA shortened by two carbon atoms. Concomitantly, one FAD molecule is reduced to $FADH_2$ and one $NAD^+$ molecule is reduced to NADH. The cycle is repeated on the shortened acyl CoA until fatty acids with an even number of carbon atoms are completely converted to acetyl CoA. In mitochondria, electrons from $FADH_2$ and NADH enter the electron transport chain and ultimately are used to generate ATP; the acetyl CoA generated is oxidized in the citric acid cycle, resulting in release of $CO_2$ and ultimately the synthesis of additional ATP. Because peroxisomes lack the electron transport complexes composing the electron transport chain and the enzymes of the citric acid cycle, oxidation of fatty acids in these organelles yields no ATP. [Adapted from D. L. Nelson and M. M. Cox, *Lehninger Principles of Biochemistry*, 3d ed., 2000, Worth Publishers.]

mammalian cells except erythrocytes and are also found in plant cells, yeasts, and probably most other eukaryotic cells.

Mitochondria preferentially oxidize short-chain (fewer than 8 carbons in the fatty acyl chain, or $<C_8$), medium-chain ($C_8$–$C_{12}$), and long-chain ($C_{14}$–$C_{20}$) fatty acids, whereas peroxisomes preferentially oxidize very long-chain fatty acids (VLCFAs, $>C_{20}$), which cannot be oxidized by mitochondria. Most dietary fatty acids have long chains, which means they are oxidized mostly in mitochondria. In contrast to mitochondrial oxidation of fatty acids, which is coupled to generation of ATP, peroxisomal oxidation of fatty acids is not linked to ATP formation, and energy is released as heat.

The reaction pathway by which fatty acids are degraded to acetyl CoA in peroxisomes is similar to that used in mitochondria (Figure 12-12b). However, peroxisomes lack an electron transport chain, and electrons from the $FADH_2$ produced during the oxidation of fatty acids are immediately transferred to $O_2$ by *oxidases*, regenerating FAD and forming hydrogen peroxide ($H_2O_2$). In addition to oxidases, peroxisomes contain abundant *catalase*, which quickly decomposes the $H_2O_2$, a highly cytotoxic metabolite. NADH produced during oxidation of fatty acids is exported and reoxidized in the cytosol; there is no need for a malate-aspartate shuttle here. Peroxisomes also lack the citric acid cycle, so acetyl CoA generated during peroxisomal degradation of fatty acids cannot be oxidized further; instead it is transported into the cytosol for use in the synthesis of cholesterol (Chapter 10) and other metabolites.

## KEY CONCEPTS of Section 12.2

### Mitochondria and the Citric Acid Cycle

• The mitochondrion has two distinct membranes (outer and inner) and two distinct subcompartments (intermembrane space between the two membranes, and the matrix surrounded by the inner membrane). Aerobic oxidation occurs in the mitochondrial matrix and on the inner mitochondrial membrane (see Figure 12-6).

• In stage II of glucose oxidation, the three-carbon pyruvate molecule is first oxidized to generate one molecule each of $CO_2$, NADH, and acetyl CoA. The acetyl group of acetyl CoA is then oxidized to $CO_2$ by the citric acid cycle (see Figure 12-8).

• Each turn of the citric acid cycle releases two molecules of $CO_2$ and generates three NADH molecules, one $FADH_2$ molecule, and one GTP (see Figure 12-10).

• Most of the energy released in stages I and II of glucose oxidation is temporarily stored in the reduced coenzymes NADH or $FADH_2$, which carry high-energy electrons that subsequently drive the electron transport chain (stage III).

• Neither glycolysis nor the citric acid cycle directly use molecular oxygen ($O_2$).

• The malate-aspartate shuttle regenerates the supply of cytosolic $NAD^+$ necessary for continued glycolysis (see Figure 12-11).

• Like glucose oxidation, the oxidation of fatty acids takes place in four stages. In stage I, fatty acids are converted to fatty acyl CoA in the cytosol. In stage II, the fatty acyl CoA is first converted into multiple acetyl CoA molecules with generation of NADH and $FADH_2$. Then, as in glucose oxidation, the acetyl CoA enters the citric acid cycle. Stages III and IV are identical for fatty acid and glucose oxidation (see Figure 12-8).

• In most eukaryotic cells, oxidation of short- to long-chain fatty acids occurs in mitochondria with production of ATP, whereas oxidation of very long-chain fatty acids occurs primarily in peroxisomes and is not linked to ATP production (see Figure 12-12); the energy released during peroxisomal oxidation of fatty acids is converted to heat.

## 12.3 The Electron Transport Chain and Generation of the Proton-Motive Force

Most of the energy released during the oxidation of glucose and fatty acids to $CO_2$ (stages I and II) is converted into high-energy electrons in the reduced coenzymes NADH and $FADH_2$. We now turn to stage III, in which the energy transiently stored in these reduced coenzymes is converted by an electron transport chain, also known as the **respiratory chain**, into the proton-motive force. We first describe the logic and components of the electron transport chain. Next we follow the path of electrons as they flow through the chain, and the mechanism of proton pumping across the mitochondrial inner membrane. We conclude this section with a discussion of the magnitude of the proton-motive force produced by electron transport and proton pumping. In Section 12.4, we will see how the proton-motive force is used to synthesize ATP.

### Oxidation of NADH and $FADH_2$ Releases a Significant Amount of Energy

During electron transport, electrons are released from NADH and $FADH_2$ and eventually transferred to $O_2$, forming $H_2O$ according to the following overall reactions:

$$NADH + H^+ + \tfrac{1}{2} O_2 \longrightarrow NAD^+ + H_2O,$$
$$\Delta G = -52.6 \text{ kcal/mol}$$

$$FADH_2 + \tfrac{1}{2} O_2 \longrightarrow FAD + H_2O,$$
$$\Delta G = -43.4 \text{ kcal/mol}$$

Recall that the conversion of 1 glucose molecule to $CO_2$ via the glycolytic pathway and citric acid cycle yields 10 NADH and 2 $FADH_2$ molecules (see Table 12-1). Oxidation of these reduced coenzymes has a total $\Delta G^{\circ\prime}$ of $-613$ kcal/mol [$10(-52.6) + 2(-43.4)$]. Thus of the potential free energy present in the chemical bonds of glucose ($-686$ kcal/mol), about 90 percent is conserved in the reduced coenzymes.

Why should there be two different coenzymes, NADH and $FADH_2$? Although many of the reactions involved in glucose and fatty acid oxidation are sufficiently energetic to reduce $NAD^+$, several are not, so those reactions are coupled to reduction of FAD, which requires less energy.

The energy carried in the reduced coenzymes can be released by oxidizing them. The biochemical challenge faced by the mitochondrion is to transfer, as efficiently as possible, the energy released by this oxidation into the energy in the terminal phosphoanhydride bond in ATP.

$$P_i^{2-} + H^+ + ADP^{3-} \rightarrow ATP^{4-} + H_2O,$$
$$\Delta G = +7.3 \text{ kcal/mol}$$

A relatively simple one-to-one reaction involving reduction of one coenzyme molecule and synthesis of one ATP would be terribly inefficient, because the $\Delta G^{\circ\prime}$ for ATP generation from ADP and $P_i$ is substantially less than that for the coenzyme oxidation and much energy would be lost as heat. To efficiently recover the energy, the mitochondrion converts the energy of coenzyme oxidation into a proton-motive force using a series of electron carriers, all but one of which are integral components of the inner membrane (see Figure 12-8). The proton-motive force can then be used to very efficiently generate ATP.

## Electron Transport in Mitochondria Is Coupled to Proton Pumping

During electron transport from NADH and $FADH_2$ to $O_2$, protons from the mitochondrial matrix are pumped across the inner membrane. This pumping raises the pH of the mitochondrial matrix relative to the intermembrane space and cytosol and also makes the matrix more negative with respect

to the intermembrane space. In other words, the free energy released during the oxidation of NADH or $FADH_2$ is stored both as a proton concentration gradient and an electrical gradient across the membrane—collectively, the proton-motive force (see Figure 12-2). As we will see in Section 12.4, the movement of protons back across the inner membrane, driven by this force, is coupled to the synthesis of ATP from ADP and $P_i$ by ATP synthase (stage IV).

The synthesis of ATP from ADP and $P_i$, driven by the energy released by transfer of electrons from NADH or $FADH_2$ to $O_2$, is the major source of ATP in aerobic nonphotosynthetic cells. Much evidence shows that in mitochondria and bacteria this process of *oxidative phosphorylation* depends on generation of a proton-motive force across the inner membrane (mitochondria) or bacterial plasma membrane, with electron transport, proton pumping, and ATP formation occurring simultaneously. In the laboratory, for instance, addition of $O_2$ and an oxidizable substrate such as pyruvate or succinate to isolated intact mitochondria results in a net synthesis of ATP if the inner mitochondrial membrane is intact. In the presence of minute amounts of detergents that make the membrane leaky, electron transport and the oxidation of these metabolites by $O_2$ still occurs. However, under these conditions no ATP is made, because the proton leak prevents the maintenance of the proton-motive force.

The coupling between electron transport from NADH (or $FADH_2$) to $O_2$ and proton transport across the inner mitochondrial membrane can be demonstrated experimentally with isolated, intact mitochondria (Figure 12-13). As soon as $O_2$ is added to a suspension of mitochondria in an otherwise $O_2$-free solution that contains NADH, the medium outside the mitochondria transiently becomes more acidic (increased proton concentration), because the mitochondrial outer membrane is freely permeable to protons. (Remember

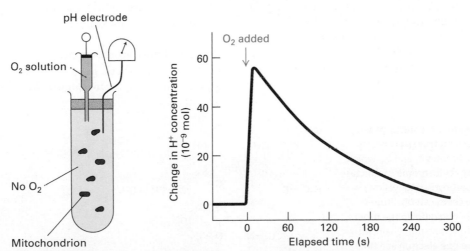

**EXPERIMENTAL FIGURE 12-13 Electron transfer from NADH to $O_2$ is coupled to proton transport across the mitochondrial membrane.** If NADH is added to a suspension of mitochondria depleted of $O_2$, no NADH is oxidized. When a small amount of $O_2$ is added to the system (arrow), there is a sharp rise in the concentration of pro-tons in the surrounding medium outside the mitochondria (decrease in pH). Thus the oxidation of NADH by $O_2$ is coupled to the movement of protons out of the matrix. Once the $O_2$ is depleted, the excess protons slowly move back into the mitochondria (powering the synthesis of ATP) and the pH of the extracellular medium returns to its initial value.

that malate-aspartate and other shuttles can convert the NADH in the solution into NADH in the matrix.) Once the $O_2$ is depleted by its reduction, the excess protons in the medium slowly leak back into the matrix. From analysis of the measured pH change in such experiments, one can calculate that about 10 protons are transported out of the matrix for every electron pair transferred from NADH to $O_2$.

To obtain numbers for $FADH_2$, the above experiment can be repeated using succinate instead of NADH as the substrate. (Recall that oxidation of succinate to fumarate in the citric acid cycle generates $FADH_2$; see Figure 12-10). The amount of succinate added can be adjusted so that the amount of $FADH_2$ generated is equivalent to the amount of NADH in the first experiment. As in the first experiment, addition of oxygen causes the medium outside the mitochondria to become acidic, but less so than with NADH. This is not surprising because electrons in $FADH_2$ have less potential energy (43.4 kcal/mol) than electrons in NADH (52.6 kcal/mol), and thus it drives the translocation of fewer protons from the matrix and a smaller change in pH.

## Electrons Flow "Downhill" Through a Series of Electron Carriers

We now examine more closely the energetically favored movement of electrons from NADH and $FADH_2$ to the final electron acceptor, $O_2$. For simplicity, we will focus our discussion on NADH. In respiring mitochondria, each NADH molecule releases two electrons to the electron transport chain; these electrons ultimately reduce one oxygen atom (half of an $O_2$ molecule), forming one molecule of water:

$$NADH \rightarrow NAD^+ + H^+ + 2\ e^-$$

$$2\ e^- + 2\ H^+ + \tfrac{1}{2}\ O_2 \rightarrow H_2O$$

As electrons move from NADH to $O_2$, their electric potential declines by 1.14 V, which corresponds to 26.2 kcal/mol of electrons transferred, or ~53 kcal/mol for a pair of electrons. As noted earlier, much of this energy is conserved in the proton-motive force generated across the inner mitochondrial membrane.

| TABLE 12-2 | Electron-Carrying Prosthetic Groups in the Respiratory Chain |
|---|---|
| **Protein Component** | **Prosthetic Groups*** |
| NADH-CoQ reductase (complex I) | FMN<br>Fe-S |
| Succinate-CoQ reductase (complex II) | FAD<br>Fe-S |
| $CoQH_2$–cytochrome $c$ reductase (complex III) | Heme $b_L$<br>Heme $b_H$<br>Fe-S<br>Heme $c_1$ |
| Cytochrome $c$ | Heme $c$ |
| Cytochrome $c$ oxidase (complex IV) | $Cu_a^{2+}$<br>Heme $a$<br>$Cu_b^{2+}$<br>Heme $a_3$ |

*Not included is coenzyme Q, an electron carrier that is not permanently bound to a protein complex.
SOURCE: J. W. De Pierre and L Ernster, 1977, *Ann. Rev. Biochem.* 46:201.

Four large multiprotein complexes (complexes I–IV) compose an electron transport chain in the inner mitochondrial membrane that is responsible for the generation of the proton-motive force (see Figure 12-8, stage III). Each complex contains several *prosthetic groups* that participate in the process of moving electrons from donor molecules to acceptor molecules in coupled oxidation-reduction reactions (see Chapter 2). These small nonpeptide organic molecules or metal ions are tightly and specifically associated with the multiprotein complexes (Table 12-2).

**Heme and the Cytochromes** Several types of *heme*, an iron-containing prosthetic group similar to that found in hemoglobin and myoglobin (Figure 12-14a), are tightly bound

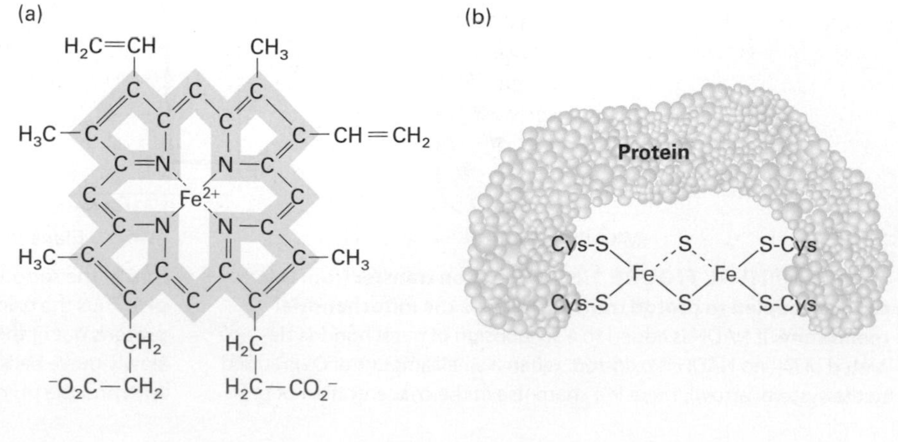

**FIGURE 12-14 Heme and iron-sulfur prosthetic groups in the electron transport chain.**
(a) Heme portion of cytochromes $b_L$ and $b_H$, which are components of $CoQH_2$–cytochrome $c$ reductase (complex III). The same porphyrin ring (yellow) is present in all hemes. The chemical substituents attached to the porphyrin ring differ in the other cytochromes in the electron transport chain. All hemes accept and release one electron at a time. (b) Dimeric iron-sulfur cluster (Fe-S). Each Fe atom is bonded to four S atoms: two are inorganic sulfur and two are in cysteine side chains of the associated protein. All Fe-S clusters accept and release one electron at a time.

(a)

(b)

(covalently or noncovalently) to a set of mitochondrial proteins called **cytochromes**. Each cytochrome is designated by a letter, such as $a$, $b$, $c$, or $c_1$. Electron flow through the cytochromes occurs by oxidation and reduction of the Fe atom in the center of the heme molecule:

$$Fe^{3+} + e^- \rightleftharpoons Fe^{2+}$$

Because the heme ring in cytochromes consists of alternating double- and single-bonded atoms, a large number of resonance hybrid forms exist. These allow the extra electron delivered to the cytochrome to be delocalized throughout the heme carbon and nitrogen atoms as well as the Fe ion.

The various cytochromes have slightly different heme groups and surrounding atoms (called axial ligands), which generate different environments for the Fe ion. Therefore, each cytochrome has a different reduction potential, or tendency to accept an electron—an important property dictating the unidirectional "downhill" electron flow along the chain. Just as water spontaneously flows downhill from a higher to lower potential energy state—but not uphill—so too do electrons flow in only one direction from one heme (or other prosthetic group) to another due to their differing reduction potentials. (For more on the concept of reduction potential, $E$, see Chapter 2.) All the cytochromes, except cytochrome $c$, are components of integral membrane multiprotein complexes in the inner mitochondrial membrane.

**Iron-Sulfur Clusters** *Iron-sulfur clusters* are nonheme, iron-containing prosthetic groups consisting of Fe atoms bonded both to inorganic S atoms and to S atoms on cysteine residues in a protein (Figure 12-14b). Some Fe atoms in the cluster bear a +2 charge; others have a +3 charge. However, the net charge of each Fe atom is actually between +2 and +3, because electrons in their outermost orbitals together with the extra electron delivered via the transport chain are dispersed among the Fe atoms and move rapidly from one atom to another. Iron-sulfur clusters accept and release electrons one at a time.

**Coenzyme Q (CoQ)** *Coenzyme Q* (CoQ), also called *ubiquinone*, is the only small-molecule electron carrier in the chain that is not an essentially irreversibly protein-bound prosthetic group (Figure 12-15). It is a carrier of both protons and electrons. The oxidized quinone form of CoQ can accept a single electron to form a semiquinone, a charged free radical denoted by $CoQ^{\bullet-}$. Addition of a second electron and two protons (thus a total of two hydrogen atoms) to $CoQ^{\bullet-}$ forms dihydroubiquinone ($CoQH_2$), the fully reduced form. Both CoQ and $CoQH_2$ are soluble in phospholipids and diffuse freely in the hydrophobic center of the inner mitochondrial membrane. This is how it participates in the electron transport chain—carrying electrons and protons between the protein complexes of the chain.

We now consider in detail the multiprotein complexes which use these prosthetic groups, and the paths taken by the electrons and protons as they pass through the complexes.

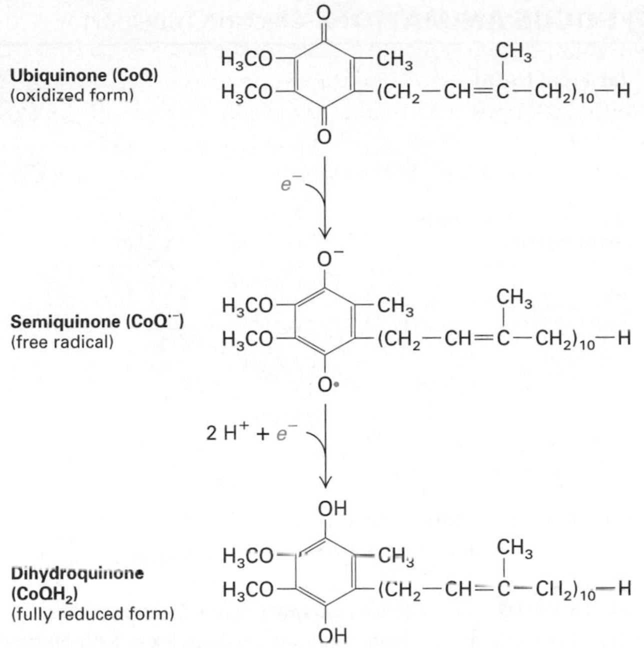

**FIGURE 12-15 Oxidized and reduced forms of coenzyme Q (CoQ), which can carry two protons and two electrons.** Because of its long hydrocarbon "tail" of isoprene units, CoQ, also called ubiquinone, is soluble in the hydrophobic core of phospholipid bilayers and is very mobile. Reduction of CoQ to the fully reduced form, $QH_2$ (dihydroquinone), occurs in two steps with a half-reduced free-radical intermediate, called semiquinone.

## Four Large Multiprotein Complexes Couple Electron Transport to Proton Pumping Across the Mitochondrial Inner Membrane

As electrons flow downhill from one electron carrier to the next in the electron transport chain, the energy released is used to power the pumping of protons against their electrochemical gradient across the inner mitochondrial membrane. Four large multiprotein complexes couple the movement of electrons to proton pumping: *NADH-CoQ reductase* (complex I, >40 subunits), *succinate-CoQ reductase* (complex II, 4 subunits), *CoQH$_2$-cytochrome c reductase* (complex III, 11 subunits), and *cytochrome c oxidase* (complex IV, 13 subunits) (Figure 12-16). Electrons from NADH flow from complex I via CoQ/CoQH$_2$ to complex III and then via the soluble protein cytochrome $c$ (cyt $c$) to complex IV to reduce molecular oxygen (complex II is bypassed) (see Figure 12-16a); alternatively, electrons from FADH$_2$ flow from complex II via CoQ/CoQH$_2$ to complex III and then via cytochrome $c$ to complex IV to reduce molecular oxygen (complex I is bypassed) (see Figure 12-16b).

As shown in Figure 12-16, CoQ accepts electrons released from NADH-CoQ reductase (complex I) or succinate-CoQ reductase (complex II) and donates them to CoQH$_2$-cytochrome c reductase (complex III). Protons are simultaneously transported from the matrix (also called the cytosolic) side of the membrane to the intermembrane space (also called the exoplasmic side). Whenever CoQ accepts electrons, it does so at a

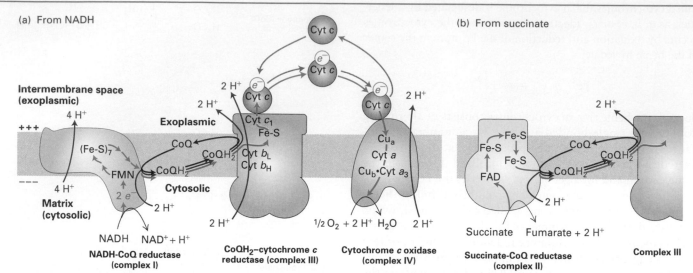

**FIGURE 12-16 The electron transport chain.** Electrons (blue arrows) flow through four major multiprotein complexes (I–IV). Electron movement between complexes is mediated either by the lipid-soluble molecule coenzyme Q (CoQ, oxidized form; CoQH$_2$, reduced form) or the water-soluble protein cytochrome c (Cyt c). The multiple multiprotein complexes use the energy released from passing electrons to pump protons from the matrix to the intermembrane space (red arrows). **(a)** Pathway from NADH. Electrons from NADH flow through complex I, initially via a flavin mononucleotide (FMN) and then via seven iron-sulfur clusters (Fe-S), to CoQ, to which two protons bind, forming CoQH$_2$. Conformational changes in complex I that accompany the electron flow drive proton pumping from the matrix to the intramembrane space (red arrows). Electrons then flow via the released (and recycled) CoQH$_2$ to complex III, and then via Cyt c to complex IV.

A total of 10 protons are translocated per pair of electrons that flow from NADH to O$_2$. The protons released into the matrix space during oxidation of NADH by complex I are consumed in the formation of water from O$_2$ by complex IV, resulting in no net proton translocation from these reactions. **(b)** Pathway from succinate. Electrons flow from succinate to complex II via FAD/FADH$_2$ and iron-sulfur clusters (Fe-S), from complex II to complex III via CoQ/CoQH$_2$, and then to complex IV via Cyt c. Electrons released during oxidation of succinate to fumarate in complex II are used to reduce CoQ to CoQH$_2$ without translocating additional protons. The remainder of electron transport from CoQH$_2$ proceeds by the same pathway as for the NADH pathway in (a). Thus for every pair of electrons transported from succinate to O$_2$, six protons are translocated by complexes III and IV.

binding site on the matrix side of a protein complex, always picking up protons from the medium there. Whenever CoQH$_2$ releases its electrons, it does so at a site on the intermembrane space side of a protein complex, releasing protons into the fluid of the intermembrane space. Thus transport of each pair of electrons by CoQ is obligatorily coupled to movement of two protons from the matrix to the intermembrane space.

**NADH-CoQ Reductase (Complex I)** Electrons are transferred from NADH to CoQ by NADH-CoQ reductase (see Figure 12-16a). Electron microscopy and x-ray crystallography of complex I from both bacteria (mass ~500 kDa, with 14 subunits) and eukaryotes (~1 MDa, with 14 highly conserved core subunits shared with bacteria plus about 26–32 accessory subunits) established that it is L-shaped (Figure 12-17a). The membrane-embedded arm of the L is slightly curved, ~180 Å long, and comprises proteins with more than 60 transmembrane alpha helices. This arm has four subdomains, three of which have proteins that are members of a family of cation antiporters. The hydrophilic peripheral arm extends over 130 Å away from the membrane into the cytosolic space.

NAD$^+$ is exclusively a two-electron carrier: it accepts or releases a pair of electrons simultaneously. In NADH-CoQ reductase (complex I), the NADH-binding site is at the tip of the peripheral arm (see Figure 12-17a); electrons released from NADH first flow to FMN (flavin mononucleotide), a cofactor related to FAD, then are shuttled ~95 Å down that arm through seven iron-sulfur clusters and finally to CoQ, which is bound at a site at least partially in the plane of the membrane. FMN, like FAD, can accept two electrons but does so one electron at a time.

Each transported electron undergoes a drop in potential of ~360 mV, equivalent to a $\Delta G°'$ of −16.6 kcal/mol for the two electrons transported. Much of this released energy is used to transport four protons across the inner membrane per molecule of NADH oxidized by complex I. Those four protons are distinct from the two protons that are transferred to the CoQ as illustrated in Figures 12-15, 12-16a, and 12-17a. The structure of complex I suggests that the energy released by the electron transport in the peripheral arm is used to change the conformation of subunits in the membrane arm and thus mediate movement of four protons across the membrane. Three protons are likely to pass through the three cation antiporter

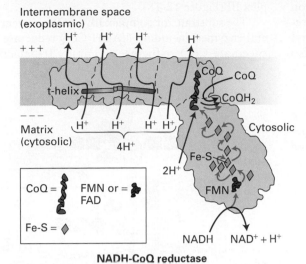

**(a) Complex I**

Intermembrane space
(exoplasmic)

$H^+$ $H^+$ $H^+$ $H^+$

t-helix

CoQ
CoQ
CoQH₂

$H^+$ $H^+$ $H^+$ $H^+$

Matrix
(cytosolic)

$4H^+$

Cytosolic

Fe-S

$2H^+$

FMN

| CoQ = | FMN or = FAD |
|---|---|
| Fe-S = ◆ | |

NADH    NAD⁺ + H⁺

**NADH-CoQ reductase
(complex I)**

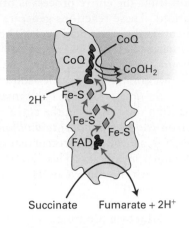

**(b) Complex II**

CoQ

CoQ

CoQH₂

Fe-S

$2H^+$

Fe-S

Fe-S

FAD

Succinate    Fumarate + 2H⁺

**Succinate-CoQ reductase
(complex II)**

**FIGURE 12-17 Electron and proton transport through complexes I and II.** (a) Model of complex I based on its three-dimensional structure. The outline of the shape of complex I, as determined by x-ray crystallography, is shown in light blue, and distinct structural subunits are indicated by thin dashed black lines. From NADH electrons flow first to a flavin mononucleotide (FMN) and then, via seven of the nine iron-sulfur clusters (Fe-S, blue diamonds), to CoQ, to which two protons from the matrix bind (red arrow) to form CoQH₂. Conformational changes due to the electron flow, which probably include a pistonlike horizontal movement of the t-helix, drive proton

pumping through the transmembrane subunits from the matrix to the intramembrane space (red arrows). (b) Model of complex II based on its three-dimensional structure. Electrons flow from succinate to complex II via FAD/FADH₂ and iron-sulfur clusters (Fe-S, blue diamonds), and from complex II to complex III via CoQ/CoQH₂. Electrons released during oxidation of succinate to fumarate in complex II are used to reduce CoQ to CoQH₂ without translocating additional protons. [Structure of complex I adapted from C. Hunte, V. Zickermann, and U. Brandt, 2010, *Science* **329**:448–451 and R. G. Efremov, R. Baradaran, and L. A. Sazanov, 2010, *Nature* **465**:441–445.]

domains while the route of the forth is through a different type of domain. An ~110 Å long, kinked, transverse alpha helix (t-helix) in the membrane arm runs parallel to the plane of the membrane, potentially mechanically linking the antiporter domains to the peripheral arm (Figure 12-17a) and thereby transmitting electron-transport-induced conformational changes in the peripheral arm to the distant antiporter domains to drive proton transport.

The overall reaction catalyzed by this complex is

$$NADH + CoQ + 6\ H^+_{in} \rightarrow$$
(Reduced) (Oxidized)

$$NAD^+ + H^+_{in} + CoQH_2 + 4\ H^+_{out}$$
(Oxidized)          (Reduced)

**Succinate-CoQ Reductase (Complex II)** Succinate dehydrogenase, the enzyme that oxidizes a molecule of succinate to fumarate in the citric acid cycle (and in the process generates the reduced coenzyme FADH₂), is one of the four subunits of complex II. Thus the citric acid cycle is physically as well as functionally linked to the electron transport chain. The two electrons released in conversion of succinate to fumarate are transferred first to FAD in succinate dehydrogenase, then to iron-sulfur clusters—regenerating FAD—and finally to CoQ, which binds to a cleft on the matrix side of the transmembrane portions of complex II (Figures 12-16b and 12-17b).

The pathway is somewhat reminiscent of that in complex I (Figure 12-17a).

The overall reaction catalyzed by this complex is

$$Succinate + CoQ \rightarrow fumarate + CoQH_2$$
(Reduced)   (Oxidized)      (Oxidized)      (Reduced)

Although the $\Delta G^{\circ\prime}$ for this reaction is negative, the released energy is insufficient for proton pumping in addition to reduction of CoQ to form CoQH₂. Thus no protons are translocated directly across the membrane by the succinate-CoQ reductase complex, and no proton-motive force is generated in this part of the respiratory chain. Shortly we will see how the protons and electrons in the CoQH₂ molecules generated by complexes I and II contribute to the generation of the proton-motive force.

Complex II generates CoQH₂ from succinate via FAD/FADH₂-mediated redox reactions. Another set of proteins in the matrix and inner mitochondrial membrane performs a comparable set of FAD/FADH₂-mediated redox reactions to generate CoQH₂ from fatty acyl CoA. *Fatty acyl–CoA dehydrogenase,* which is a water-soluble enzyme, catalyzes the first step of the oxidation of fatty acyl CoA in the mitochondrial matrix (see Figure 12-12). There are several fatty acyl–CoA dehydrogenase enzymes with specificities for fatty acyl chains of different lengths. These enzymes mediate the initial step in a four-step process that removes two carbons from the fatty acyl group by oxidizing the carbon in the β position of the fatty acyl chain (thus the entire process is often referred to as β-oxidation). These reactions generate acetyl CoA, which in turn enters the citric acid cycle. They also generate an $FADH_2$ intermediate and NADH. The $FADH_2$ generated remains bound to the enzyme during the redox reaction, as is the case for complex II. A water-soluble protein called *electron transfer flavoprotein (ETF)* transfers the high-energy electrons from the $FADH_2$ in the acyl-CoA dehydrogenase to *electron transfer flavoprotein:ubiquinone oxidoreductase (ETF:QO)*, a membrane protein that reduces CoQ to $CoQH_2$ in the inner membrane. This $CoQH_2$ intermixes in the membrane with the other $CoQH_2$ molecules generated by complexes I and II.

## $CoQH_2$–Cytochrome *c* Reductase (Complex III)

A $CoQH_2$ generated either by complex I, complex II, or ETF:QO donates two electrons to $CoQH_2$–cytochrome *c* reductase (complex III), regenerating oxidized CoQ. Concomitantly it releases into the intermembrane space two protons previously picked up by CoQ on the matrix face, generating part of the proton-motive force (see Figure 12-16). Within complex III, the released electrons first are transferred to an iron-sulfur cluster within the complex and then to cytochrome $c_1$ or to two *b*-type cytochromes ($b_L$ and $b_H$, see Q cycle below). Finally, the two electrons are transferred sequentially to two molecules of the oxidized form of cytochrome *c*, a water-soluble peripheral protein that diffuses in the intermembrane space. For each pair of electrons transferred, the overall reaction catalyzed by the $CoQH_2$–cytochrome *c* reductase complex is

$$CoQH_2 + 2\ \text{Cyt}\ c^{3+} + 2\ H^+_{in} \rightarrow CoQ + 4\ H^+_{out} + 2\ \text{Cyt}\ c^{2+}$$

(Reduced)   (Oxidized)          (Oxidized)          (Reduced)

The $\Delta G^{\circ\prime}$ for this reaction is sufficiently negative that two protons in addition to those from $CoQH_2$ are translocated from the mitochondrial matrix across the inner membrane for each pair of electrons transferred; this involves the proton-motive Q cycle, discussed below. The heme protein cytochrome *c* and the small lipid-soluble molecule CoQ play similar roles in the electron transport chain in that they both serve as mobile electron shuttles, transferring electrons (and thus energy) between the complexes of the electron transport chain.

## The Q Cycle

Experiments have shown that four protons are translocated across the membrane per electron pair transported from $CoQH_2$ through $CoQH_2$–cytochrome c reductase (complex III). These four protons are those carried on two $CoQH_2$ molecules, which are converted to two CoQ molecules during the cycle. However, another CoQ molecule receives two other protons from the matrix space and is converted to one $CoQH_2$ molecule. Thus the net overall reaction involves the conversion of only one $CoQH_2$ molecule to CoQ as two electrons are transferred one at a time to two molecules of the acceptor cytochrome c. An evolutionarily conserved mechanism, called the *Q cycle*, is responsible for

the two-for-one transport of protons and electrons by complex III (Figure 12-18).

The substrate for complex III, $CoQH_2$, is generated by several enzymes, including NADH-CoQ reductase (complex I), succinate-CoQ reductase (complex II), *electron transfer*

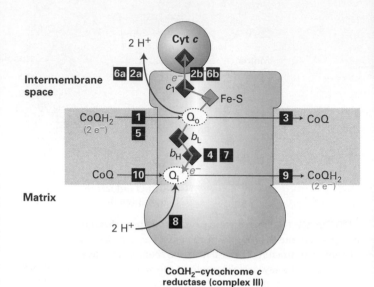

**$CoQH_2$–cytochrome *c* reductase (complex III)**

**At $Q_o$ site:** $2\ CoQH_2 + 2\ \text{Cyt}\ c^{3+} \longrightarrow$
(4 $H^+$, 4 $e^-$)
$2\ CoQ + 2\ \text{Cyt}\ c^{2+} + 2\ e^- + 4\ H^+_{(outside)}$
(2 $e^-$)

**At $Q_i$ site:** $CoQ + 2\ e^- + 2\ H^+_{(matrix\ side)} \longrightarrow CoQH_2$
(2 $H^+$, 2 $e^-$)

**Net Q cycle (sum of reactions at $Q_o$ and $Q_i$):**
$CoQH_2 + 2\ \text{Cyt}\ c^{3+} + 2\ H^+_{(matrix\ side)} \longrightarrow$
(2 $H^+$, 2 $e^-$)
$CoQ + 2\ \text{Cyt}\ c^{2+} + 4\ H^+_{(outside)}$
(2 $e^-$)

Per 2 $e^-$ transferred through complex III to cytochrome *c*, 4 $H^+$ released to the intermembrane space

**FIGURE 12-18 The Q cycle.** The Q cycle of complex III uses the net oxidation of one $CoQH_2$ molecule to transfer four protons into the intermembrane space and two electrons to two cytochrome c molecules. The cycle begins when a molecule from the combined pool of reduced $CoQH_2$ in the membrane binds to the $Q_o$ site on the *intermembrane space (outer) side* of the transmembrane portion of complex III (step **1**). There, $CoQH_2$ releases two protons into the intermembrane space (step **2a**), and two electrons and the resulting CoQ dissociate (step **3**). One of the electrons is transported, via an iron-sulfur protein and cytochrome $c_1$, directly to cytochrome c (step **2b**). (Recall that each cytochrome c shuttles one electron from complex III to complex IV.) The other electron moves through cytochromes $b_L$ and $b_H$ and partially reduces an oxidized CoQ molecule bound to the second, $Q_i$, site on the *matrix (inner) side* of the complex, forming a CoQ semiquinone anion, $Q^{\cdot-}$ (step **4**). The process is repeated with the binding of a second $CoQH_2$ at the $Q_o$ site (step **5**), proton release (step **6a**), reduction of another cytochrome c (step **6b**), and addition of the other electron to the $Q^{\cdot-}$ bound at the $Q_i$ site (step **7**). There, the addition of two protons from the matrix yields a fully reduced $CoQH_2$ molecule at the $Q_i$ site, which then dissociates (steps **8** and **9**), freeing the $Q_i$ to bind a new molecule of CoQ (step **10**) and begin the Q cycle over again. [Adapted from B. Trumpower, 1990, *J. Biol. Chem.* **265**:11409, and E. Darrouzet et al., 2001, *Trends Biochem. Sci.* **26**:445.]

*flavoprotein:ubiquinone oxidoreductase* (ETF:QO, during β-oxidation), and, as we shall see, by complex III itself.

As shown in Figure 12-18, in one turn of the Q cycle, two molecules of $CoQH_2$ are oxidized to CoQ at the $Q_o$ site and release a total of four protons into the intermembrane space, but at the $Q_i$ site one molecule of $CoQH_2$ is regenerated from CoQ and two additional proteins from the matrix space. The translocated protons are all derived from $CoQH_2$, which obtained its protons from the matrix as described above. Although seemingly cumbersome, the Q cycle optimizes the number of protons pumped per pair of electrons moving through complex III. The Q cycle is found in all plants and animals, as well as in bacteria. Its formation at a very early stage of cellular evolution was likely essential for the success of all life-forms, as a way of converting the potential energy in reduced coenzyme Q into the maximum proton-motive force across a membrane. In turn this maximizes the number of ATP molecules synthesized from each electron that moves down the electron transport chain from NADH or $FADH_2$ to oxygen.

How are the two electrons released from $CoQH_2$ at the $Q_o$ site directed to different acceptors, either to Fe-S, cytochrome $c_1$, and then cytochrome $c$ (upward pathway in Figure 12-18), or alternatively to cytochrome $b_L$, cytochrome $b_H$, and then CoQ at the $Q_i$ site (downward pathway in Figure 12-18)? The mechanism involves a flexible hinge in the Fe-S-containing protein subunit of complex III. Initially the Fe-S cluster is close enough to the $Q_o$ site to pick up an electron from $CoQH_2$ bound there. Once this happens, a segment of the protein containing this Fe-S cluster swings the cluster away from the $Q_o$ site to a position near enough to the heme on cytochrome $c_1$ for electron transfer to occur. With the Fe-S subunit in this alternate conformation, the second electron released from $CoQH_2$ bound to the $Q_o$ site cannot move to the Fe-S cluster—it is too far away, so it takes an alternative path open to it via a somewhat less thermodynamically favored route to cytochrome $b_L$ and through cytochrome $b_H$ to the CoQ at the $Q_i$ site.

**Cytochrome c Oxidase (Complex IV)** Cytochrome $c$, after being reduced by one electron from $CoQH_2$–cytochrome $c$ reductase (complex III), is reoxidized as it transports its electron to cytochrome $c$ oxidase (complex IV) (see Figure 12-16). Mitochondrial cytochrome $c$ oxidases contain 13 different subunits, but the catalytic core of the enzyme consists of only three subunits. The functions of the remaining subunits are not well understood. Bacterial cytochrome $c$ oxidases contain only the three catalytic subunits. In both mitochondria and bacteria, four molecules of reduced cytochrome $c$ bind, one at a time, to the oxidase. An electron is transferred from the heme of each cytochrome $c$, first to the pair of copper ions called $Cu_a^{2+}$, then to the heme in cytochrome $a$, and next to the $Cu_b^{2+}$ and the heme in cytochrome $a_3$ that together make up the oxygen reduction center. The four electrons are finally passed to $O_2$, the ultimate electron acceptor, yielding four $H_2O$, which together with $CO_2$ is one of the end products of the overall oxidation pathway. (Note that, for

simplicity, Figure 12-16 shows only two electrons moving and ½ $O_2$ being reduced.) Proposed intermediates in oxygen reduction include the peroxide anion ($O_2^{2-}$) and probably the hydroxyl radical (OH·), as well as unusual complexes of iron and oxygen atoms. These intermediates would be harmful to the cell if they escaped from complex IV, but they do so only rarely (see the discussion of reactive oxygen species below). During transport of four electrons through the cytochrome $c$ oxidase complex, four protons from the matrix space are translocated across the membrane. Thus, complex IV transports only one proton per electron transferred, whereas complex II, using the Q cycle, transports two protons per electron transferred. However, the mechanism by which complex IV translocates these protons is not known.

For each four electrons transferred, the overall reaction catalyzed by cytochrome $c$ oxidase is

$$4 \text{ Cyt } c^{2+} + 8 \text{ H}^+_{\text{in}} + O_2 \rightarrow 4 \text{ Cyt } c^{3+} + 2 \text{ H}_2O + 4 \text{ H}^+_{\text{out}}$$
   (Reduced)                       (Oxidized)

The poison cyanide, which has been used as a chemical warfare agent, by spies to commit suicide when captured, in gas chambers to execute prisoners, and by the Nazis (Zyklon B gas) for the mass murder of Jews and others, is toxic because it binds to the heme $a_3$ in mitochondrial cytochrome $c$ oxidase (complex IV), inhibiting electron transport and thus oxidative phosphorylation and production of ATP. Cyanide is one of many toxic small molecules that interfere with energy production in mitochondria. ∎

## Reduction Potentials of Electron Carriers in the Electron Transport Chain Favor Electron Flow from NADH to $O_2$

As we saw in Chapter 2, the **reduction potential $E$** for a partial reduction reaction

$$\text{Oxidized molecule} + e^- \rightleftharpoons \text{reduced molecule}$$

is a measure of the equilibrium constant of that partial reaction. With the exception of the $b$ cytochromes in the $CoQH_2$–cytochrome $c$ reductase complex, the standard reduction potential $E°'$ of the electron carriers in the mitochondrial respiratory chain increases steadily from NADH to $O_2$. For instance, for the partial reaction

$$\text{NAD}^+ + \text{H}^+ + 2 e^- \rightleftharpoons \text{NADH}$$

the value of the standard reduction potential is $-320$ mV, which is equivalent to a $\Delta G°'$ of $+14.8$ kcal/mol for transfer of two electrons. Thus this partial reaction tends to proceed toward the left, that is, toward the oxidation of NADH to $NAD^+$.

By contrast, the standard reduction potential for the partial reaction

$$\text{Cytochrome } c_{\text{ox}} \text{ (Fe}^{3+}) + e^- \rightleftharpoons \text{cytochrome } c_{\text{red}} \text{ (Fe}^{2+})$$

**FIGURE 12-19 Changes in redox potential and free energy during stepwise flow of electrons through the respiratory chain.** Blue arrows indicate electron flow; red arrows, translocation of protons across the inner mitochondrial membrane. Electrons pass through the multiprotein complexes from those at a lower reduction potential to those with a higher (more positive) reduction potential (left scale), with a corresponding reduction in free energy (right scale). The energy released as electrons flow through three of the complexes is sufficient to power the pumping of $H^+$ ions across the membrane, establishing a proton-motive force.

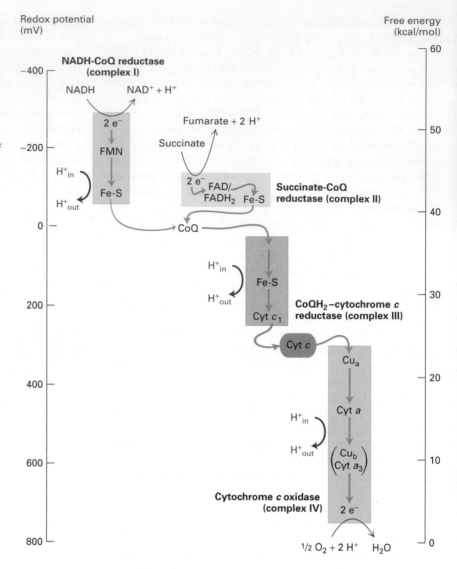

is $+220$ mV ($\Delta G^{\circ\prime} = -5.1$ kcal/mol) for transfer of one electron. Thus this partial reaction tends to proceed toward the right, that is, toward the reduction of cytochrome $c$ ($Fe^{3+}$) to cytochrome $c$ ($Fe^{2+}$).

The final reaction in the respiratory chain, the reduction of $O_2$ to $H_2O$

$$2\ H^+ + \tfrac{1}{2}\ O_2 + 2\ e^- \rightarrow H_2O$$

has a standard reduction potential of $+816$ mV ($\Delta G^{\circ\prime} = -37.8$ kcal/mol for transfer of two electrons), the most positive in the whole series; thus this reaction also tends to proceed toward the right.

As illustrated in Figure 12-19, the steady increase in $E^{\circ\prime}$ values, and the corresponding decrease in $\Delta G^{\circ\prime}$ values, of the carriers in the electron transport chain favors the flow of electrons from NADH and $FADH_2$ (generated from succinate) to oxygen. The energy released as electrons flow "downhill" energetically through the electron transport chain complexes drives the pumping of protons against their concentration gradient across the mitochondrial inner membrane.

## The Multiprotein Complexes of the Electron Transport Chain Assemble into Supercomplexes

Over 50 years ago Britton Chance proposed that electron transport complexes might assemble into large supercomplexes. Doing so would bring the complexes into close and highly organized proximity, which might improve the speed and efficiency of the overall process. Indeed, genetic, biochemical, and biophysical studies have provided very strong evidence for the existence of electron transport chain supercomplexes. These studies involved gel electrophoretic methods called blue native (BN)-PAGE and colorless native (CN)-PAGE, which permit separation of very large macromolecular protein complexes, and electron microscopic analysis of their three-dimensional structures. One such supercomplex contains one copy of complex I, a dimer of complex III ($III_2$), and one or more copies of complex IV (Figure 12-20). One supercomplex that contains all of the components thought to play a role in respiration—complexes I–IV, ubiquinone (CoQ), and cytochrome $c$—was isolated from BN-PAGE gels and shown to transfer electrons from NADH

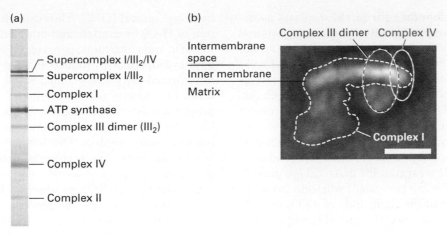

**EXPERIMENTAL FIGURE 12-20 Electrophoresis and electron microscopic imaging identifies an electron transport chain supercomplex containing complexes I, III, and IV.** (a) Membrane proteins In Isolated bovine heart mitochondria were solubilized with a detergent, and the complexes and supercomplexes were separated by gel electrophoresis using the blue native (BN)-PAGE method. Each blue-stained band within the gel represents the indicated protein complex or supercomplex, with $III_2$ representing a dimer of complex III. Intensity of the blue stain is approximately proportional to the amount of complex or supercomplex present. (b) Supercomplex $I/III_2/IV$ was extracted from the gel, and the particles were negatively stained with 1% uranyl acetate and visualized by transmission electron microscopy. Images of 228 particles were combined at a resolution of ~3.4 nm to generate an averaged image of the complex viewed from the side in the plane of the membrane. Approximate locations of the complex III dimer and complex IV are indicated by dashed ovals; the outline of complex I is also indicated by a dashed line (white). Scale bar is 10 nm. [Adapted from E. Schafer et al., 2006, *J. Biol. Chem.* **281**(22):15370–15375.]

to $O_2$; in other words, this supercomplex can respire—it is a respirasome.

The unique phospholipid *cardiolipin* (diphosphatidyl glycerol) appears to play an important role in the assembly and function of these supercomplexes. Generally not observed

Cardiolipin

in other membranes of eukaryotic cells, cardiolipin has been observed to bind to integral membrane proteins of the inner membrane (e.g., complex II). Genetic and biochemical studies in yeast mutants in which cardiolipin synthesis is blocked have established that cardiolipin contributes to the formation and activity of mitochondrial supercomplexes, and thus it has been called the glue that holds together the electron transport chain, though the precise mechanism remains to be defined. In addition, there is evidence that cardiolipin may influence the inner membrane's binding and permeability to protons and consequently the proton-motive force.

## Reactive Oxygen Species (ROS) Are Toxic By-products of Electron Transport That Can Damage Cells

About 1–2 percent of the oxygen metabolized by aerobic organisms, rather than being converted to water, is partially reduced to the superoxide anion radical ($O_2^{\bullet-}$, where the "dot" represents an unpaired electron). Radicals are atoms that have one or more unpaired electrons in an outer (valence) shell, or molecules that contain such an atom. Many, though not all, radicals are generally highly chemically reactive, altering the structures and properties of those molecules with which they react. The products of such reactions often are themselves radicals and thus can propagate a chain reaction that alters many additional molecules. Superoxide and other highly reactive oxygen-containing molecules, both radicals (e.g., $O_2^{\bullet-}$) and non-radicals (hydrogen peroxide, $H_2O_2$) are called *reactive oxygen species* (ROS). ROS are of great interest because they can react with and thus damage many key biological molecules, including lipids (particularly unsaturated fatty acids and their derivatives), proteins, and DNA, and thus severely interfere with their normal functions. At moderate to high levels, ROS contribute to what is often called *cellular oxidative stress* and can be highly toxic. Indeed, ROS are purposefully generated by body defense cells (e.g., macrophages, neutrophils) to kill pathogens. In humans, excessive or inappropriate generation of ROS has been implicated in many diverse diseases, including heart failure, neurodegenerative diseases, alcohol-induced liver disease, diabetes, and aging.

Although there are several mechanisms for generating ROS in cells, the major source in eukaryotic cells is electron transport in the mitochondria (or in chloroplasts as described below). Electrons passing through the mitochondrial electron transport chain can have sufficient energy to reduce molecular oxygen ($O_2$) to form superoxide anions (Figure 12-21, *top*). This can only occur, however, when molecular oxygen comes in close contact with the reduced electron carriers (iron, FMN, $CoQH_2$) in the chain. Usually such contact is prevented by sequestration of the carriers within the proteins involved. However, there are some sites (particularly in complex I and $CoQ^{\bullet-}$, see Figure 12-15) and some conditions (e.g., high $NADH/NAD+$ ratio in the matrix, high proton-motive force when ATP is not generated) when electrons can more readily "leak" out of the chain and reduce $O_2$ to $O_2^{\bullet-}$.

Superoxide anion is an especially unstable and reactive ROS. Mitochondria have evolved several defense mechanisms that help protect against $O_2^{\bullet-}$ toxicity, including the use of enzymes that inactivate superoxide, first by converting it to $H_2O_2$ (Mn-containing superoxide dismutase, SOD) and then to $H_2O$ (catalase) (Figure 12-21). Because $O_2^{\bullet-}$ is so highly reactive and toxic, SOD and catalase are some of the fastest enzymes known. SOD is found within mitochondria and other cellular compartments. Hydrogen peroxide itself is a ROS that can diffuse readily across membranes and react with molecules throughout the cell. It can also be converted by certain metals such as $Fe^{2+}$ into the even more dangerous

hydroxyl radical ($OH^{\bullet}$). Thus cells depend on the inactivation of $H_2O_2$ by catalase and other enzymes, such as peroxiredoxin and glutathione peroxidase, which also detoxify the lipid hydroperoxide products formed when ROS react with unsaturated fatty acyl groups. Small molecule antioxidant radical scavengers such as vitamin E and α-lipoic acid also protect against oxidative stress. Although in many cells catalase is located only in peroxisomes, in heart muscle cells it is found in mitochondria. This is not surprising, because the heart is the most oxygen-consuming organ per gram weight in mammals.

As the rate of ROS production by mitochondria and chloroplasts reflects the metabolic state of these organelles (e.g., size of proton-motive force, $NADH/NAD^+$ ratio), cells have developed ROS-sensing systems, such as ROS/redox-sensitive transcription factors, to monitor the metabolic state of these organelles and respond accordingly, for example by changing the rate of transcription of nuclear genes that encode organelle-specific proteins. ■

## Experiments Using Purified Electron Transport Chain Complexes Established the Stoichiometry of Proton Pumping

The multiprotein complexes of the electron transport chain that are responsible for proton pumping have been identified by selectively extracting mitochondrial membranes with detergents, isolating each of the complexes in nearly pure form, and then preparing artificial phospholipid vesicles (liposomes) containing each complex. When an appropriate electron donor and electron acceptor are added to such liposomes, a change in pH of the medium will occur if the embedded complex transports protons (Figure 12-22). Studies of this type indicate that NADH-CoQ reductase (complex I) translocates four protons per pair of electrons transported, whereas cytochrome *c* oxidase (complex IV) translocates two protons per electron pair transported.

Current evidence suggests that a total of 10 protons are transported from the matrix space across the inner mitochondrial membrane for every electron pair that is transferred from NADH to $O_2$ (see Figure 12-16). Because succinate-CoQ reductase (complex II) does not transport protons and complex I is bypassed when the electrons come from succinate-derived $FADH_2$, only six protons are transported across the membrane for every electron pair that is transferred from this $FADH_2$ to $O_2$.

## The Proton-Motive Force in Mitochondria Is Due Largely to a Voltage Gradient Across the Inner Membrane

The main result of the electron transport chain is the generation of the proton-motive force (pmf), which is the sum of a transmembrane proton concentration (pH) gradient and electric potential, or voltage gradient, across the mitochondrial inner membrane. The relative contribution of the two

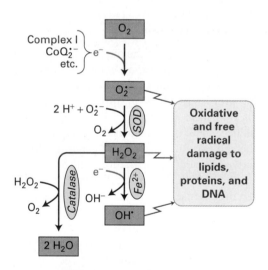

**FIGURE 12-21 Generation and inactivation of toxic reactive oxygen species.** Electrons from the electron transport chains of mitochondria and chloroplasts as well as some generated through other enzymatic reactions reduce molecular oxygen ($O_2$), forming the highly reactive radical anion superoxide ($O_2^{\bullet-}$). Superoxide is rapidly converted by superoxide dismutase (SOD) to hydrogen peroxide ($H_2O_2$), which in turn can be converted by metal ions such as $Fe^{2+}$ to hydroxyl radicals ($OH^{\bullet}$) or inactivated to water by enzymes such as catalase. Because of their high chemical reactivity $O_2^{\bullet-}$, $H_2O_2$, $OH^{\bullet}$, and similar molecules are called reactive oxygen species (ROS). They cause oxidative and free radical damage to many biomolecules, including lipids, proteins, and DNA. This damage leads to cellular oxidative stress that can cause disease and, if sufficiently severe, can kill cells.

(a)

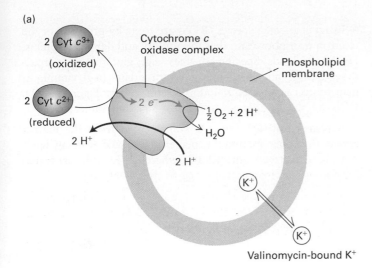

(b)

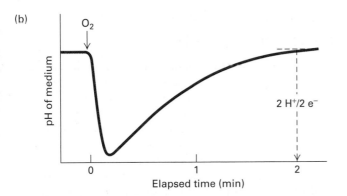

**EXPERIMENTAL FIGURE 12-22 Electron transfer from reduced cytochrome c to O₂ via cytochrome c oxidase (complex IV) is coupled to proton transport.** The oxidase complex is incorporated into liposomes with the binding site for cytochrome c positioned on the outer surface. (a) When $O_2$ and reduced cytochrome c are added, electrons are transferred to $O_2$ to form $H_2O$ and protons are transported from the inside to the medium outside of the vesicles. A drug called valinomycin was added to the medium to dissipate the voltage gradient generated by the translocation of $H^+$, which would otherwise reduce the number of protons moved across the membrane. (b) Monitoring of the medium's pH reveals a sharp drop in pH following addition of $O_2$. As the reduced cytochrome c becomes fully oxidized, protons leak back into the vesicles, and the pH of the medium returns to its initial value. Measurements show that two protons are transported per O atom reduced. Two electrons are needed to reduce one O atom, but cytochrome c transfers only one electron; thus two molecules of Cyt $c^{2+}$ are oxidized for each O reduced. [Adapted from B. Reynafarje et al., 1986, *J. Biol. Chem.* **261**:8254.]

components to the total pmf has been determined experimentally, and depends on the permeability of the membrane to ions other than $H^+$. A significant voltage gradient can develop only if the membrane is poorly permeable to other cations and to anions. Otherwise, anions would leak across from the matrix to the intermembrane space along with the protons and prevent a voltage gradient from forming. Similarly, cations leaking across from the intermembrane space

to the matrix (exchange of like charge) would also short-circuit voltage gradient formation. Indeed, the inner mitochondrial membrane is poorly permeable to other ions. Thus proton pumping generates a voltage gradient that makes it energetically difficult for additional protons to move across because of charge repulsion. As a consequence, proton pumping by the electron transport chain establishes a robust voltage gradient in the context of a rather small pH gradient.

Because mitochondria are much too small to be impaled with electrodes, the electric potential and pH gradient across the inner mitochondrial membrane cannot be directly measured. However, the electric potential can be measured indirectly by adding radioactive $^{42}K^+$ ions and a trace amount of valinomycin to a suspension of respiring mitochondria and measuring the amount of radioactivity that accumulates in the matrix. Although the inner membrane is normally impermeable to $K^+$, valinomycin is an *ionophore*, a small lipid-soluble molecule that selectively binds a specific ion (in this case, $K^+$) and carries it across otherwise impermeable membranes. In the presence of valinomycin, $^{42}K^+$ equilibrates across the inner membrane of isolated mitochondria in accordance with the electric potential: the more negative the matrix side of the membrane, the more $^{42}K^+$ will be attracted to and accumulate in the matrix.

At equilibrium, the measured concentration of radioactive $K^+$ ions in the matrix, $[K_{in}]$, is about 500 times greater than that in the surrounding medium, $[K_{out}]$. Substitution of this value into the Nernst equation (Chapter 11) shows that the electric potential $E$ (in mV) across the inner membrane in respiring mitochondria is $-160$ mV, with the matrix (inside) negative:

$$E = -59 \log \frac{[K_{in}]}{[K_{out}]} = -59 \log 500 = -160 \text{ mV}$$

Researchers can measure the matrix (inside) pH by trapping pH-sensitive fluorescent dyes inside vesicles formed from the inner mitochondrial membrane, with the matrix side of the membrane facing inward. They also can measure the pH outside of the vesicles (equivalent to the intermembrane space) and thus determine the pH gradient ($\Delta$pH), which turns out to be ~1 pH unit. Because a difference of one pH unit represents a tenfold difference in $H^+$ concentration, according to the Nernst equation a pH gradient of one unit across a membrane is equivalent to an electric potential of 59 mV at 20 °C. Thus, knowing the voltage and pH gradients, we can calculate the proton-motive force, pmf, as

$$\text{pmf} = \Psi - \left(\frac{RT}{F} \times \Delta\text{pH}\right) = \Psi - 59 \, \Delta\text{pH}$$

where $R$ is the gas constant of 1.987 cal/(degree · mol), $T$ is the temperature (in degrees Kelvin), $F$ is the Faraday constant [23,062 cal/(V · mol)], and $\Psi$ is the transmembrane electric potential; $\Psi$ and pmf are measured in millivolts. The electric potential $\Psi$ across the inner membrane is $-160$ mV (negative inside matrix) and $\Delta$pH is equivalent to ~60 mV. Thus the total pmf is $-220$ mV, with the transmembrane electric potential responsible for about 73 percent of the total.

## KEY CONCEPTS of Section 12.3

### The Electron Transport Chain and Generation of the Proton-Motive Force

- By the end of the citric acid cycle (stage II), much of the energy originally present in the covalent bonds of glucose and fatty acids is converted into high-energy electrons in the reduced coenzymes NADH and $FADH_2$. The energy from these electrons is used to generate the proton-motive force.

- In the mitochondrion, the proton-motive force is generated by coupling electron flow (from NADH and $FADH_2$ to $O_2$) to the uphill transport of protons from the matrix across the inner membrane to the intermembrane space. This process together with the synthesis of ATP from ADP and $P_i$ driven by the proton-motive force is called oxidative phosphorylation.

- As electrons flow from $FADH_2$ and NADH to $O_2$ they pass through multiprotein complexes. The four major complexes are NADH-CoQ reductase (complex I), succinate-CoQ reductase (complex II), $CoQH_2$–cytochrome *c* reductase (complex III), and cytochrome *c* oxidase (complex IV) (see Figure 12-16).

- Each complex contains one or more electron-carrying prosthetic groups: iron-sulfur clusters, flavins, heme groups, and copper ions (see Table 12-2). Cytochrome *c*, which contains heme, and coenzyme Q (CoQ), a lipid-soluble small molecule, are mobile carriers that shuttle electrons between the complexes.

- Complexes I, III, and IV pump protons from the matrix into the intermembrane space. Complexes I and II reduce CoQ to $CoQH_2$, which carries protons and high-energy electrons to complex III. The heme protein cytochrome *c* carries electrons from complex III to complex IV, which uses them to pump protons and reduce molecular oxygen to water.

- The high-energy electrons from NADH enter the electron transport chain through complex I, whereas the high-energy electrons from $FADH_2$ (derived from succinate in the citric acid cycle) enter the electron transport chain through complex II. Additional electrons derived from $FADH_2$ by the initial step of fatty acyl–CoA β-oxidation increase the supply of $CoQH_2$ available for electron transport.

- The Q cycle allows four protons to be translocated per pair of electrons moving through complex III (see Figure 12-18).

- Each electron carrier in the chain accepts an electron or electron pair from a carrier with a less positive reduction potential and transfers the electron to a carrier with a more positive reduction potential. Thus the reduction potentials of electron carriers favor unidirectional, "downhill," electron flow from NADH and $FADH_2$ to $O_2$ (see Figure 12-19).

- Within the inner membrane, electron transport complexes assemble into supercomplexes held together by cardiolipin, a specialized phospholipid. Supercomplex formation may enhance the speed and efficiency of generation of the proton-motive force.

- Reactive oxygen species (ROS) are toxic by-products of the electron transport chain that can modify and damage proteins, DNA, and lipids. Specific enzymes (e.g., glutathinone peroxidase, catalase) and small molecule antioxidants (e.g., vitamin E) help protect against ROS-induced damage (see Figure 12-21). ROS can also be used as intracellular signaling molecules.

- A total of 10 $H^+$ ions are translocated from the matrix across the inner membrane per electron pair flowing from NADH to $O_2$ (see Figure 12-16), whereas 6 $H^+$ ions are translocated per electron pair flowing from $FADH_2$ to $O_2$.

- The proton-motive force is due largely to a voltage gradient across the inner membrane produced by proton pumping; the pH gradient plays a quantitatively less important role.

## 12.4 Harnessing the Proton-Motive Force to Synthesize ATP

The hypothesis that a proton-motive force across the inner mitochondrial membrane is the immediate source of energy for ATP synthesis was proposed in 1961 by Peter Mitchell. Virtually all researchers studying oxidative phosphorylation and photosynthesis initially rejected his *chemiosmotic hypothesis*. They favored a mechanism similar to the then well-elucidated substrate-level phosphorylation in glycolysis, in which chemical transformation of a substrate molecule (in the case of glycolysis, phosphoenolpyruvate) is directly coupled to ATP synthesis. Despite intense efforts by a large number of investigators, however, compelling evidence for such a substrate-level phosphorylation-mediated mechanism was never observed.

Definitive evidence supporting Mitchell's hypothesis depended on developing techniques to purify and reconstitute organelle membranes and membrane proteins. The experiment with vesicles made from chloroplast thylakoid membranes (equivalent to the inner membrane of mitochondria) containing **ATP synthase,** outlined in Figure 12-23, was one of several demonstrating that this protein is an ATP-generating enzyme and that ATP generation is dependent on proton movement down an electrochemical gradient. It turns out that the protons actually move *through* the ATP synthase as they traverse the membrane.

As we shall see, the ATP synthase is a multiprotein complex that can be subdivided into two subcomplexes called $F_0$ (containing the transmembrane portions of the complex) and $F_1$ (containing the globular portions of the complex that sit above the membrane and point toward the matrix space in mitochondria). Thus the ATP synthase is often also called the **$F_0F_1$ complex**; we will use the terms interchangeably.

### The Mechanism of ATP Synthesis Is Shared Among Bacteria, Mitochondria, and Chloroplasts

Although bacteria lack internal membranes, aerobic bacteria nonetheless carry out oxidative phosphorylation by the same processes that occur in eukaryotic mitochondria and

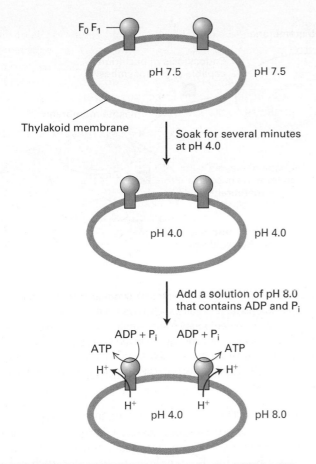

**EXPERIMENTAL FIGURE 12-23  Synthesis of ATP by ATP synthase depends on a pH gradient across the membrane.** Isolated chloroplast thylakoid vesicles containing ATP synthase ($F_0F_1$ particles) were equilibrated in the dark with a buffered solution at pH 4.0. When the pH in the thylakoid lumen became 4.0, the vesicles were rapidly mixed with a solution at pH 8.0 containing ADP and $P_i$. A burst of ATP synthesis accompanied the transmembrane movement of protons driven by the 10,000-fold $H^+$ concentration gradient ($10^{-4}$ M versus $10^{-8}$ M). In similar experiments using "inside-out" preparations of mitochondrial membrane vesicles, an artificially generated membrane electric potential also resulted in ATP synthesis.

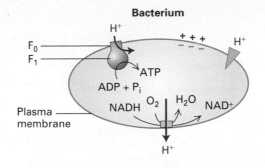

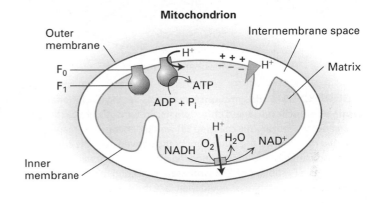

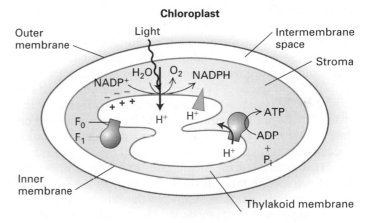

**FIGURE 12-24  ATP synthesis by chemiosmosis is similar in bacteria, mitochondria, and chloroplasts.** In chemiosmosis, a proton-motive force generated by proton pumping across a membrane is used to power ATP synthesis. The mechanism and membrane orientation of the process are similar in bacteria, mitochondria, and chloroplasts. In each illustration, the membrane surface facing a shaded area is a cytosolic face; the surface facing an unshaded, white area is an exoplasmic face. Note that the cytosolic face of the bacterial plasma membrane, the matrix face of the inner mitochondrial membrane, and the stromal face of the thylakoid membrane are all equivalent. During electron transport, protons are always pumped from the cytosolic face to the exoplasmic face, creating a proton concentration gradient (exoplasmic face > cytosolic face) and an electric potential (negative cytosolic face and positive exoplasmic face) across the membrane. During the synthesis of ATP, protons flow in the reverse direction (down their electrochemical gradient) through ATP synthase ($F_0F_1$ complex), which protrudes in a knob at the cytosolic face in all cases.

chloroplasts (Figure 12-24). Enzymes that catalyze the reactions of both the glycolytic pathway and the citric acid cycle are present in the cytosol of bacteria; enzymes that oxidize NADH to NAD$^+$ and transfer the electrons to the ultimate acceptor $O_2$ reside in the bacterial plasma membrane. The movement of electrons through these membrane carriers is coupled to the pumping of protons out of the cell. The movement of protons back into the cell, down their concentration gradient through ATP synthase, drives the synthesis of ATP. The bacterial ATP synthase ($F_0F_1$ complex) is essentially identical in structure and function to the mitochondrial and chloroplast ATP synthases but is simpler to purify and study.

Why is the mechanism of ATP synthesis shared among both prokaryotic organisms and eukaryotic organelles? Primitive aerobic bacteria were probably the progenitors of both mitochondria and chloroplasts in eukaryotic cells (Figure 12-25). According to this *endosymbiont hypothesis*, the

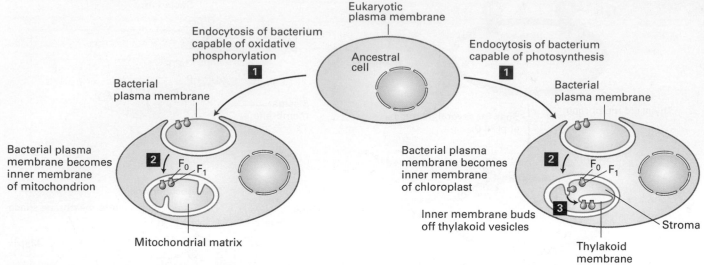

**FIGURE 12-25 Endosymbiont hypothesis for the evolutionary origin of mitochondria and chloroplasts.** Endocytosis of a bacterium by an ancestral eukaryotic cell (step **1**) would generate an organelle with two membranes, the outer membrane derived from the eukaryotic plasma membrane and the inner one from the bacterial membrane (step **2**). The $F_1$ subunit of ATP synthase, localized to the cytosolic face of the bacterial membrane, would then face the matrix of the evolving mitochondrion (*left*) or chloroplast (*right*). Budding of vesicles from the inner chloroplast membrane, such as occurs during development of chloroplasts in contemporary plants, would generate the thylakoid membranes with the $F_1$ subunit remaining on the cytosolic face, facing the chloroplast stroma (step **3**). Membrane surfaces facing a shaded area are cytosolic faces; surfaces facing an unshaded area are exoplasmic faces.

inner mitochondrial membrane would be derived from the bacterial plasma membrane with its cytosolic face pointing toward what became the matrix space of the mitochondrion. Similarly, in plants the progenitor's plasma membrane became the chloroplast's thylakoid membrane and its cytosolic face pointed toward what became the stromal space of the chloroplast. In all cases, ATP synthase is positioned with the globular $F_1$ domain, which catalyzes ATP synthesis, on the cytosolic face of the membrane, so ATP is always formed on the cytosolic face of the membrane (see Figure 12-24). Protons always flow through ATP synthase from the exoplasmic to the cytosolic face of the membrane. This flow is driven by the proton motive force. Invariably, the cytosolic face has a negative electric potential relative to the exoplasmic face.

In addition to ATP synthesis, the proton-motive force across the bacterial plasma membrane is used to power other processes, including the uptake of nutrients such as sugars (using proton/sugar symporters) and the rotation of bacterial flagella. Chemiosmotic coupling thus illustrates an important principle introduced in our discussion of active transport in Chapter 11: *the membrane potential, the concentration gradients of protons (and other ions) across a membrane, and the phosphoanhydride bonds in ATP are equivalent and interconvertible forms of chemical potential energy.* Indeed, ATP synthesis through ATP synthase can be thought of as active transport in reverse.

## ATP Synthase Comprises $F_0$ and $F_1$ Multiprotein Complexes

With general acceptance of Mitchell's chemiosmotic mechanism, researchers turned their attention to the structure and operation of ATP synthase. The complex has two principal components, $F_0$ and $F_1$, both of which are multimeric proteins (Figure 12-26a). The $F_0$ component contains three types of integral membrane proteins, designated **a**, **b**, and **c**. In bacteria and in yeast mitochondria the most common subunit stoichiometry is $a_1b_2c_{10}$, but $F_0$ complexes in animal mitochondria have 12 **c** subunits and those in chloroplasts have 14. In all cases the **c** subunits form a doughnut-shaped ring ("**c** ring") in the plane of the membrane. The **a** and two **b** subunits are rigidly linked to one another but not to the **c** ring, a critical feature of the protein to which we will return shortly.

The $F_1$ portion is a water-soluble complex of five distinct polypeptides with the composition $\alpha_3\beta_3\gamma\delta\epsilon$ that is normally firmly bound to the $F_0$ subcomplex at the surface of the membrane. The lower end of the rodlike $\gamma$ subunit of the $F_1$ subcomplex is a coiled coil that fits into the center of the **c**-subunit ring of $F_0$ and appears rigidly attached to it. Thus when the **c**-subunit ring rotates, the rodlike $\gamma$ subunit moves with it. The $F_1$ $\epsilon$ subunit is rigidly attached to $\gamma$ and also forms tight contacts with several of the **c** subunits of $F_0$. The $\alpha$ and $\beta$ subunits are responsible for the overall globular shape of the $F_1$ subcomplex and associate in alternating order to form a hexamer, $\alpha\beta\alpha\beta\alpha\beta$, or $(\alpha\beta)_3$, which rests atop the single long $\gamma$ subunit. The $F_1$ $\delta$ subunit is permanently linked to one of the $F_1$ $\alpha$ subunits and also binds to the **b** subunit of $F_0$. Thus the $F_0$ **a** and **b** subunits and the $\delta$ subunit and $(\alpha\beta)_3$ hexamer of the $F_1$ complex form a rigid structure anchored in the membrane. The rodlike **b** subunits form a "stator" that prevents the $(\alpha\beta)_3$ hexamer from moving while it rests on the $\gamma$ subunit, whose rotation together with the **c** subunits of $F_0$ plays an essential role in the ATP synthesis mechanism described below.

When ATP synthase is embedded in a membrane, the $F_1$ component forms a knob that protrudes from the cytosolic (in the mitochondrion this is the matrix) face. Because $F_1$ separated from membranes is capable of catalyzing ATP

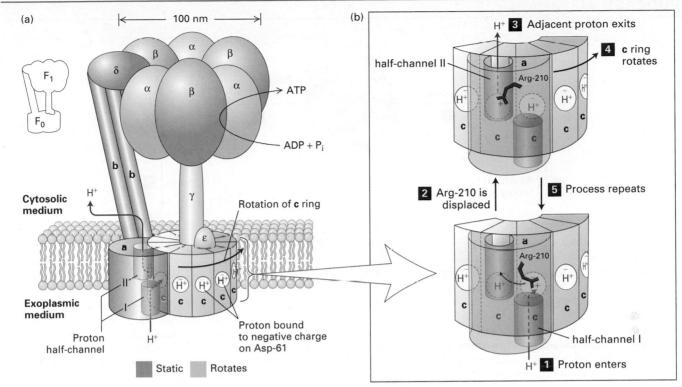

**FIGURE 12-26 Structure of ATP synthase (the $F_0F_1$ complex) in the bacterial plasma membrane and mechanism of proton translocation across the membrane.** (a) The $F_0$ membrane-embedded portion of ATP synthase is built of three integral membrane proteins: one copy of **a**, two copies of **b**, and on average 10 copies of **c** arranged in a ring in the plane of the membrane. Two proton half-channels in subunit **a** mediate proton movement across the membrane (proton path indicated by red arrows). Half-channel I allows protons to move one at a time from the exoplasmic medium to the negatively charged side chain of Asp-61 in the center of a **c** subunit near the middle of the membrane. The proton-binding site in each **c** subunit is represented as a white circle with a blue "−" representing the negative charge on the side chain of Asp-61. Half-channel II permits protons to move from the Asp-61 of an adjacent **c** subunit into the cytosolic medium. The $F_1$ portion of ATP synthase contains three copies each of subunits α and β that form a hexamer resting atop the single rod-shaped γ subunit, which is inserted into the **c** ring of $F_0$. The ε subunit is rigidly attached to the γ subunit and also to several of the **c** subunits. The δ subunit permanently links one of the α subunits in the $F_1$ complex to the **b** subunit of $F_0$. Thus the $F_0$ **a** and **b**

subunits and the $F_1$ δ subunit and $(\alpha\beta)_3$ hexamer form a rigid structure anchored in the membrane (orange). During proton flow, the **c** ring and the attached $F_1$ ε and γ subunits rotate as a unit (green), causing conformation changes in the $F_1$ β subunits, leading to ATP synthesis. (b) Potential mechanism of proton translocation. Step **1**: A proton from the exoplasmic space enters half-channel I and moves toward the "empty" (unprotonated) Asp-61 proton-binding site. The negative charge (blue "−") on the unprotonated side chain Asp-61 is balanced, in part, by a positive charge on the side chain of Arg-210 (red "+"). Step **2**: The proton fills the empty proton-binding site and simultaneously displaces the Arg-210 side chain, which swings over to the filled proton-binding site on the adjacent **c** subunit. As a consequence the proton bound at that adjacent site is displaced. Step **3**: The displaced adjacent proton moves through half-channel II and is released into the cytosolic space, leaving an empty proton-binding site on Asp-61. Step **4**: Counterclockwise rotation of the entire **c** ring moves the "empty" **c** subunit over half-channel I. Step **5**: the process is repeated. [Adapted from M. J. Schnitzer, 2001, *Nature* **410**:878; P. D. Boyer, 1999, *Nature* **402**:247; and C. von Ballmoos, A. Wiedenmann, and P. Dimroth, 2009, *Ann. Rev. Biochem.* **78**:649–672.]

---

hydrolysis (ATP conversion to ADP plus $P_i$) in the absence of the $F_0$ component, it has been called the $F_1$ ATPase; however, its function in cells is the reverse, to synthesize ATP. ATP hydrolysis is a spontaneous process ($\Delta G < 0$); thus energy is required to drive the ATPase "in reverse" and generate ATP.

## Rotation of the $F_1$ γ Subunit, Driven by Proton Movement Through $F_0$, Powers ATP Synthesis

Each of the three β subunits in the globular $F_1$ portion of the complete $F_0F_1$ complex can bind ADP and $P_i$ and catalyze the endergonic synthesis of ATP when coupled to the flow of

protons from the exoplasmic medium (intermembrane space in the mitochondrion) to the cytosolic (matrix) medium. However, the coupling between proton flow and ATP synthesis must not occur in the same portions of the protein, because the nucleotide-binding sites on the β subunits of $F_1$, where ATP synthesis occurs, are 9–10 nm from the surface of the mitochondrial membrane. The most widely accepted model for ATP synthesis by the $F_0F_1$ complex—the *binding-change mechanism*—posits just such an indirect coupling (Figure 12-27).

According to this mechanism, energy released by the "downhill" movement of protons through $F_0$ directly powers rotation of the c-subunit ring together with its attached γ and

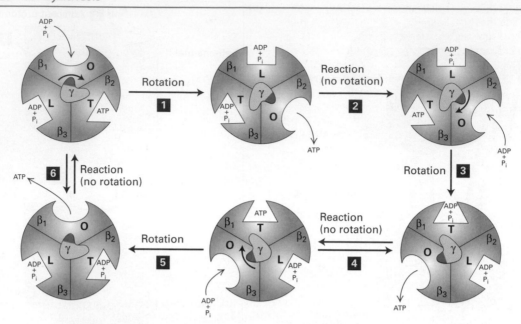

**FIGURE 12-27 The binding-change mechanism of ATP synthesis from ADP and P$_i$.** This view is looking up at F$_1$ from the membrane surface (see Figure 12-26). As the γ subunit rotates by 120° in the center, each of the otherwise identical F$_1$ β subunits alternates between three conformational states (O, open with oval representation of the binding site; L, loose with a rectangular binding site; T, tight with a triangular site) that differ in their binding affinities for ATP, ADP, and P$_i$. The cycle begins (upper left) when ADP and P$_i$ bind loosely to one of the three β subunits (here, arbitrarily designated β$_1$) whose nucleotide-binding site is in the O (open) conformation. Proton flux through the F$_0$ portion of the protein powers a 120° rotation of the γ subunit (relative to the fixed β subunits) (step **1**). This causes the rotating γ subunit, which is asymmetric, to push differentially against the β subunits, resulting in a conformational change and an increase in the binding affinity of the β$_1$ subunit for ADP and P$_i$ (from O → L), an increase in the binding affinity of the β$_3$ subunit for ADP and P$_i$ that were previously bound (from L → T),

and a decrease in the binding affinity of the β$_2$ subunit for a previously bound ATP (from T → O), causing release of the bound ATP. Step **2**: Without additional rotation the ADP and P$_i$ in the T site (here the β$_3$ subunit) form ATP, a reaction that does not require an input of additional energy due to the special environment in the active site of the T state. At the same time a new ADP and P$_i$ bind loosely to the unoccupied O site on β$_2$. Step **3**: Proton flux powers another 120° rotation of the γ subunit, consequent conformational changes in the binding sites (L → T, O → L, T → O), and release of ATP from β$_3$. Step **4**: Without additional rotation the ADP and P$_i$ in the T site of β$_1$ form ATP, and additional ADP and P$_i$ bind to the unoccupied O site on β$_3$. The process continues with rotation (step **5**) and ATP formation (step **6**) until the cycle is complete, with three ATPs having been produced for every 360° rotation of γ. [Adapted from P. Boyer, 1989, *FASEB J.* **3**:2164; Y. Zhou et al., 1997, *Proc. N at'l. Acad. Sci. USA* **94**:10583; and M. Yoshida, E. Muneyuki, and T. Hisabori, 2001, *Nat. Rev. Mol. Cell Biol.* **2**:669–677.]

ε subunits (see Figure 12-26a). The γ subunit acts as a cam, or nonsymmetrical rotating shaft, whose rotation within the center of the static (αβ)$_3$ hexamer of F$_1$ causes it to push sequentially against each of the β subunits and thus cause cyclical changes in their conformations between three different states. As schematically depicted in a view of the bottom of the (αβ)$_3$ hexamer's globular structure in Figure 12-27, rotation of the γ subunit relative to the fixed (αβ)$_3$ hexamer causes the nucleotide-binding site of each β subunit to cycle through three conformational states in the following order:

1. An O (open) state that binds ATP very poorly and ADP and P$_i$ weakly

2. An L (loose) state that binds ADP and P$_i$ more strongly but cannot bind ATP

3. A T (tight) state that binds ADP and P$_i$ so tightly that they spontaneously react and form ATP

In the T state the ATP produced is bound so tightly that it cannot readily dissociate from the site—it is trapped until another rotation of the γ subunit returns that β subunit to the O state, thereby releasing ATP and beginning the cycle again. ATP or ADP also binds to regulatory or allosteric sites on the three α subunits; this binding modifies the rate of ATP synthesis according to the level of ATP and ADP in the matrix, but is not directly involved in synthesis of ATP from ADP and P$_i$.

Several types of evidence support the binding-change mechanism. First, biochemical studies showed that one of the three β subunits on isolated F$_1$ particles can tightly bind ADP and P$_i$ and then form ATP, which remains tightly bound. The measured ΔG for this reaction is near zero, indicating that once

**EXPERIMENTAL FIGURE 12-28 The γ subunit of the F₁ complex rotates relative to the (αβ)₃ hexamer.** F₁ complexes were engineered that contained β subunits with an additional His-6 sequence, which causes them to adhere to a glass plate coated with a metal reagent that binds polyhistidine. The γ subunit in the engineered F₁ complexes was linked covalently to a fluorescently labeled actin filament. When viewed in a fluorescence microscope, the actin filaments were seen to rotate counterclockwise in discrete 120° steps in the presence of ATP, powered by ATP hydrolysis by the β subunits. [Adapted from H. Noji et al., 1997, *Nature* **386**:299, and R. Yasuda et al., 1998, *Cell* **93**:1117.]

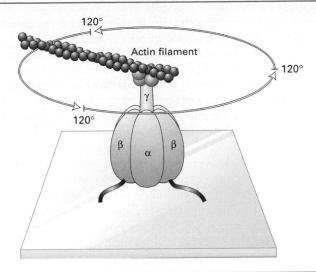

ADP and P$_i$ are bound to the T state of a β subunit, they spontaneously form ATP. Importantly, dissociation of the bound ATP from the β subunit on isolated F₁ particles occurs extremely slowly. This finding suggested that dissociation of ATP would have to be powered by a conformational change in the β subunit, which in turn would be caused by proton movement.

X-ray crystallographic analysis of the (αβ)₃ hexamer yielded a striking conclusion: although the three β subunits are identical in sequence and overall structure, the ADP/ATP-binding sites have different conformations in each subunit. The most reasonable conclusion was that the three β subunits cycle in an energy-dependent reaction between three conformational states (O, L, T), in which the nucleotide-binding site has substantially different structures.

In other studies, intact F₀F₁ complexes were treated with chemical cross-linking agents that covalently linked the γ and ε subunits and the c-subunit ring. The observation that such treated complexes could synthesize ATP or use ATP to power proton pumping indicates that the cross-linked proteins normally rotate together.

Finally, rotation of the γ subunit relative to the fixed (αβ)₃ hexamer, as proposed in the binding-change mechanism, was observed directly in the clever experiment depicted in Figure 12-28. In one modification of this experiment in which tiny gold particles, rather than an actin filament, were attached to the γ subunit, rotation rates of 134 revolutions per second were observed. Hydrolysis of three ATPs, which you recall is the reverse reaction catalyzed by the same enzyme, is thought to power one revolution; this result is close to the experimentally determined rate of ATP hydrolysis by F₀F₁ complexes: about 400 ATPs per second. In a related experiment, a γ subunit linked to an ε subunit and a ring of c subunits was seen to rotate relative to the fixed (αβ)₃ hexamer. Rotation of the γ subunit in these experiments was powered by ATP hydrolysis, the reverse of the normal process in which proton movement through the F₀ complex drives rotation of the γ subunit.

These observations established that the γ subunit, along with the attached c ring and ε subunit, does indeed rotate, thereby driving the conformational changes in the β subunits that are required for binding of ADP and P$_i$, followed by synthesis and subsequent release of ATP.

## Multiple Protons Must Pass Through ATP Synthase to Synthesize One ATP

A simple calculation indicates that the passage of more than one proton is required to synthesize one molecule of ATP from ADP and P$_i$. Although the ΔG for this reaction under standard conditions is +7.3 kcal/mol, at the concentrations of reactants in the mitochondrion, ΔG is probably higher (+10 to +12 kcal/mol). We can calculate the amount of free energy released by the passage of 1 mol of protons down an electrochemical gradient of 220 mV (0.22 V) from the Nernst equation, setting $n = 1$ and measuring ΔE in volts:

$$\Delta G(\text{cal/mol}) = -nF\Delta E = -(23{,}062 \text{ cal} \cdot \text{V}^{-1} \cdot \text{mol}^{-1})\Delta E$$
$$= (23{,}062 \text{ cal} \cdot \text{V}^{-1} \cdot \text{mol}^{-1})(0.22 \text{ V})$$
$$= -5074 \text{ cal/mol, or } -5.1 \text{ kcal/mol}$$

Because the downhill movement of 1 mol of protons releases just over 5 kcal of free energy, the passage of at least two protons is required for synthesis of each molecule of ATP from ADP and P$_i$.

## F₀ c Ring Rotation Is Driven by Protons Flowing Through Transmembrane Channels

Each copy of subunit c contains two membrane-spanning α helices that form a hairpin-like structure. An aspartate residue, Asp-61 (*E. coli* ATPase numbering), in the center of one of these helices in each subunit is thought to play a key role in proton movement by binding and releasing protons as they traverse the membrane. Chemical modification of this aspartate by the poison dicyclohexylcarbodiimide or its mutation to alanine specifically blocks proton movement through $F_0$. According to one current model, the protons traverse the membrane via two staggered, proton half-channels, I and II (see Figure 12-26a and b). They are called half-channels because each only extends halfway across the membrane; the intramembrane termini of the channels are at the level of Asp-61 in the middle of the membrane. Half-channel I is open only to the exoplasmic surface and II is open only to the cytosolic face. Prior to rotation, each of the Asp-61 carboxylate side chains in the c subunits are bound to a proton, except that on the c subunit in contact with half-channel I. The negative charge on that unprotonated carboxylate (the "empty" proton-binding site; see Figure 12-26b, *bottom*) is neutralized by interaction with the positively charged side chain of Arg-210 from the a subunit. Proton translocation across the membrane begins when a proton from the exoplasmic medium moves upwards through half-channel I (Figure 12-26b, step **1**). As that proton moves into the empty proton-binding site, it displaces the Arg-210 side chain, which swings toward the filled proton-binding site of the adjacent c subunit in contact with half-channel II (step **2**). As a consequence, the positive side chain of Arg-210 displaces the proton bound to Asp-61 of the adjacent c subunit. This displaced proton is now free to travel up half-channel II and out into the cytosolic medium (step **3**). Thus, when one proton entering from half-channel I binds to the c ring, a different proton is released to the opposite side of the membrane via half-channel II. Rotation of the entire c ring due to thermal/Brownian motion (step **4**) then allows the newly unprotonated c subunit to move into alignment above half-channel I as an adjacent, protonated c subunit rotates in to take its place under half-channel II. The entire cycle is then repeated (step **5**), as additional protons move down their electrochemical gradient from the exoplasmic medium to the cytosolic medium. During each partial rotation (360° divided by the number of c subunits in the ring), the c ring rotation is ratcheted in that net movement of the ring only occurs in one direction. The energy driving the protons across the membrane, and thus rotation of the c ring, comes from the electric potential and pH gradient across the membrane. If the direction of proton flow is reversed, which can be done by experimentally reversing the direction of the proton gradient and proton-motive force, the direction of c ring rotation is reversed.

Because the γ subunit of $F_1$ is tightly attached to the c ring of $F_0$, rotation of the c ring associated with proton movement causes rotation of the γ subunit. According to the binding-change mechanism, a 120° rotation of γ powers synthesis of one ATP (see Figure 12-27). Thus complete rotation of the c ring by 360° would generate three ATPs. In *E. coli*, where the $F_0$ composition is $a_1b_2c_{10}$, movement of 10 protons drives one complete rotation and thus synthesis of three ATPs. This value is consistent with experimental data on proton flux during ATP synthesis, providing indirect support for the model coupling proton movement to c-ring rotation depicted in Figure 12-26. The $F_0$ from chloroplasts contains 14 c subunits per ring, and movement of 14 protons would be needed for synthesis of three ATPs. Why these otherwise similar $F_0F_1$ complexes have evolved to have different $H^+$:ATP ratios is not clear.

## ATP-ADP Exchange Across the Inner Mitochondrial Membrane Is Powered by the Proton-Motive Force

The proton-motive force is used to power multiple energy-requiring processes in cells. In addition to powering ATP synthesis, the proton-motive force across the inner mitochondrial membrane powers the exchange of ATP formed by oxidative phosphorylation inside the mitochondrion for ADP and $P_i$ in the cytosol. This exchange, which is required to supply ADP and $P_i$ substrate for oxidative phosphorylation to continue, is mediated by two proteins in the inner membrane: a *phosphate transporter* ($HPO_4^{2-}/OH^-$ antiporter) that mediates the import of one $HPO_4^{2-}$ coupled to the export of one $OH^-$, and an *ATP/ADP antiporter* (Figure 12-29).

The ATP/ADP antiporter allows one molecule of ADP to enter only if one molecule of ATP exits simultaneously. The ATP/ADP antiporter, a dimer of two 30,000-Da subunits, makes up 10–15 percent of the protein in the inner membrane,

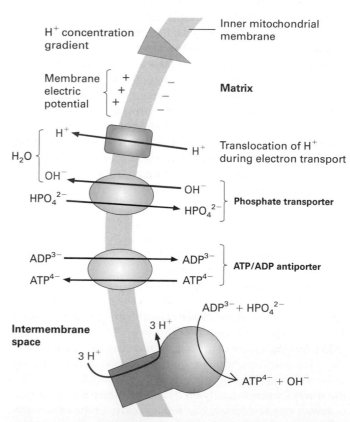

**FIGURE 12-29 The phosphate and ATP/ADP transport system in the inner mitochondrial membrane.** The coordinated action of two antiporters (purple and green) results in the uptake of one $ADP^{3-}$ and one $HPO_4^{2-}$ in exchange for one $ATP^{4-}$ and one hydroxyl, powered by the outward translocation of one proton (mediated by the proteins of the electron transport chain, blue) during electron transport. The outer membrane is not shown here because it is permeable to molecules smaller than 5000 Da.

so it is one of the more abundant mitochondrial proteins. Functioning of the two antiporters together produces an influx of one $ADP^{3-}$ and one $P_i^{2-}$ and efflux of one $ATP^{4-}$ together with one $OH^-$. Each $OH^-$ transported outward combines with a proton, translocated during electron transport to the intermembrane space, to form $H_2O$. This drives the overall reaction in the direction of ATP export and ADP and $P_i$ import.

Because some of the protons translocated out of the mitochondrion during electron transport provide the power (by combining with the exported $OH^-$) for the ATP-ADP exchange, fewer protons are available for ATP synthesis. It is estimated that for every four protons translocated out, three are used to synthesize one ATP molecule and one is used to power the export of ATP from the mitochondrion in exchange for ADP and $P_i$. This expenditure of energy from the proton concentration gradient to export ATP from the mitochondrion in exchange for ADP and $P_i$ ensures a high ratio of ATP to ADP in the cytosol, where hydrolysis of the high-energy phosphoanhydride bond of ATP is utilized to power many energy-requiring reactions.

Studies of what turned out to be ATP/ADP antiporter activity were first recorded about 2000 years ago, when Dioscorides (~AD 40–90) described a poisonous herb from the thistle *Atractylis gummifera,* found commonly in the Mediterranean region. The same agent is found in the traditional Zulu multipurpose herbal remedy *impila (Callilepis laureola).* In Zulu *impila* means "health," although it has been associated with numerous poisonings. In 1962 the active agent in the herb, atractyloside, which inhibits the ATP/ADP antiporter, was shown to inhibit oxidative phosphorylation of extramitochondrial ADP but not intramitochondrial ADP. This demonstrated the importance of the ATP/ADP antiporter and has provided a powerful tool to study the mechanism by which this transporter functions.

Dioscorides lived near Tarsus, at the time a province of Rome in southeastern Asia Minor in what is now Turkey. His five-volume *De Materia Medica (The Materials of Medicine)* "on the preparation, properties, and testing of drugs" described the medicinal properties of about 1000 natural products and 4740 medicinal usages of them. For approximately 1600 years it was the basic reference in medicine from northern Europe to the Indian Ocean, comparable to today's *Physicians' Desk Reference* as a guide for using drugs. ■

## Rate of Mitochondrial Oxidation Normally Depends on ADP Levels

If intact isolated mitochondria are provided with NADH (or a source of $FADH_2$ such as succinate) plus $O_2$ and $P_i$, but not ADP, the oxidation of NADH and the reduction of $O_2$ rapidly cease, because the amount of endogenous ADP is depleted by ATP formation. If ADP is then added, the oxidation of NADH is rapidly restored. Thus mitochondria can oxidize $FADH_2$ and NADH only as long as there is a source of ADP and $P_i$ to generate ATP. This phenomenon, termed **respiratory control**, occurs because oxidation of NADH and succinate ($FADH_2$) is obligatorily coupled to proton transport across the inner

mitochondrial membrane. If the resulting proton-motive force is not dissipated during the synthesis of ATP from ADP and $P_i$ (or during other energy-requiring processes), both the transmembrane proton concentration gradient and the membrane electric potential will increase to very high levels. At this point, pumping of additional protons across the inner membrane requires so much energy that it eventually ceases, blocking the coupled oxidation of NADH and other substrates.

## Brown-Fat Mitochondria Use the Proton-Motive Force to Generate Heat

*Brown-fat tissue,* whose color is due to the presence of abundant mitochondria, is specialized for the generation of heat. In contrast, *white-fat tissue* is specialized for the storage of fat and contains relatively few mitochondria.

The inner membrane of brown-fat mitochondria contains *thermogenin,* a protein that functions as a natural **uncoupler** of oxidative phosphorylation and generation of a proton-motive force. Thermogenin, or UCP1, is one of several uncoupling proteins (UCPs) found in most eukaryotes (but not in fermentative yeasts). Thermogenin dissipates the proton-motive force by rendering the inner mitochondrial membrane permeable to protons. As a consequence the energy released by NADH oxidation in the electron transport chain and used to create a proton gradient is not then used to synthesize ATP via ATP synthase. Instead, when protons move back into the matrix down their concentration gradient via thermogenin, the energy is released as heat. Thermogenin is a proton transporter, not a proton channel, and shuttles protons across the membrane at a rate that is a millionfold slower than that of typical ion channels (see Figure 11-2). Thermogenin is similar in sequence to the mitochondrial ATP/ADP transporter, as are many other mitochondrial transporter proteins that compose the ATP/ADP transporter family. Certain small-molecule poisons also function as uncouplers by rendering the inner mitochondrial membrane permeable to protons. One example is the lipid-soluble chemical 2,4-dinitrophenol (DNP), which can reversibly bind to and release protons and shuttle them across the inner membrane from the intermembrane space into the matrix.

Environmental conditions regulate the amount of thermogenin in brown-fat mitochondria. For instance, during the adaptation of rats to cold, the ability of their tissues to generate heat is increased by the induction of thermogenin synthesis. In cold-adapted animals, thermogenin may constitute up to 15 percent of the total protein in the inner mitochondrial membrane.

For many years it was known that small animals and human infants expressed significant amounts of brown fat, but there was scant evidence for it playing a significant role in adult humans. In the newborn human, thermogenesis by brown-fat mitochondria is vital to survival, as it is in hibernating mammals. In fur seals and other animals naturally acclimated to the cold, muscle-cell mitochondria contain thermogenin; as a result, much of the proton-motive force is used for generating heat, thereby maintaining body temperature. Recently investigators have used sophisticated functional imaging methods (for example, positron-emission tomography) to definitively establish the presence of brown fat in adult humans in the neck, clavicle, and other sites, the levels of which are significantly increased on cold exposure.

## KEY CONCEPTS of Section 12.4

### Harnessing the Proton-Motive Force to Synthesize ATP

• Peter Mitchell proposed the chemiosmotic hypothesis that a proton-motive force across the inner mitochondrial membrane is the immediate source of energy for ATP synthesis.

• Bacteria, mitochondria, and chloroplasts all use the same chemiosmotic mechanism and a similar ATP synthase to generate ATP (see Figure 12-24).

• ATP synthase (the $F_0F_1$ complex) catalyzes ATP synthesis as protons flow through the inner mitochondrial membrane (plasma membrane in bacteria) down their electrochemical proton gradient.

• $F_0$ contains a ring of 10–14 **c** subunits that is rigidly linked to the rod-shaped $\gamma$ subunit and the $\varepsilon$ subunit of $F_1$. Together they rotate during ATP synthesis. Resting atop the $\gamma$ subunit is the hexameric knob of $F_1$ $[(\alpha\beta)_3]$, which protrudes into the mitochondrial matrix (cytosol in bacteria). The three $\beta$ subunits are the sites of ATP synthesis (see Figure 12-26).

• Movement of protons across the membrane via two half-channels at the interface of the $F_0$ **a** subunit and the **c** ring powers rotation of the **c** ring with its attached $F_1$ $\varepsilon$ and $\gamma$ subunits.

• Rotation of the $F_1$ $\gamma$ subunit, which is inserted in the center of the nonrotating $(\alpha\beta)_3$ hexamer and operates like a camshaft, leads to changes in the conformation of the nucleotide-binding sites in the three $F_1$ $\beta$ subunits (see Figure 12-27). By means of this binding-change mechanism, the $\beta$ subunits bind ADP and $P_i$, condense them to form ATP, and then release the ATP. Three ATPs are made for each revolution made by the assembly of **c**, $\gamma$, and $\varepsilon$ subunits.

• The proton-motive force also powers the uptake of $P_i$ and ADP from the cytosol in exchange for mitochondrial ATP and $OH^-$, thus reducing some of the energy available for ATP synthesis. The ATP/ADP antiporter that participates in this exchange is one of the most abundant proteins in the inner mitochondrial membrane (see Figure 12-29).

• Continued mitochondrial oxidation of NADH and reduction of $O_2$ are dependent on sufficient ADP being present in the matrix. This phenomenon, termed respiratory control, is an important mechanism for coordinating oxidation and ATP synthesis in mitochondria.

• In brown fat, the inner mitochondrial membrane contains the uncoupler protein thermogenin, a proton transporter that dissipates the proton-motive force into heat. Certain chemicals also function as uncouplers (e.g., DNP) and have the same effect, uncoupling oxidative phosphorylation from electron transport.

## 12.5 Photosynthesis and Light-Absorbing Pigments

We now shift our attention to photosynthesis, the second main process for synthesizing ATP. In plants, photosynthesis occurs in chloroplasts, large organelles found mainly in leaf cells. During photosynthesis, chloroplasts capture the energy of sunlight, convert it into chemical energy in the form of ATP and NADPH, and then use this energy to make complex carboydrates out of carbon dioxide and water. The principal carbohydrates produced are polymers of hexose (six-carbon) sugars: sucrose, a glucose-fructose disaccharide (see Figure 2-19), and leaf **starch,** a mixture of two types of a large insoluble glucose polymer called amylose and amylopectin. Starch is the primary storage carbohydrate in plants (Figure 12-30). Leaf starch is synthesized and stored in the chloroplast. Sucrose is synthesized in the leaf cytosol from three-carbon precursors generated in the chloroplast; it is transported to nonphotosynthetic (nongreen) plant tissues (e.g., roots and seeds), which metabolize sucrose for energy by the pathways described in the previous sections. Photosynthesis in plants, as well as in eukaryotic single-celled algae and in several photosynthetic bacteria (e.g., the cyanobacteria and prochlorophytes), also generates oxygen. The overall reaction of oxygen-generating photosynthesis,

$$6\ CO_2 + 6\ H_2O \rightarrow 6\ O_2 + C_6H_{12}O_6$$

is the reverse of the overall reaction by which carbohydrates are oxidized to $CO_2$ and $H_2O$. In effect, photosynthesis in chloroplasts produces energy-rich sugars that are broken down and harvested for energy by mitochondria using oxidative phosphorylation.

Although green and purple bacteria also carry out photosynthesis, they use a process that does not generate oxygen. As discussed in Section 12.6, detailed analysis of the photosynthetic system in these bacteria has helped elucidate the first stages in the more common process of oxygen-generating photosynthesis. In this section, we provide an overview of the stages in oxygen-generating photosynthesis and introduce the main molecular components, including the **chlorophylls,** the principal light-absorbing pigments. ■

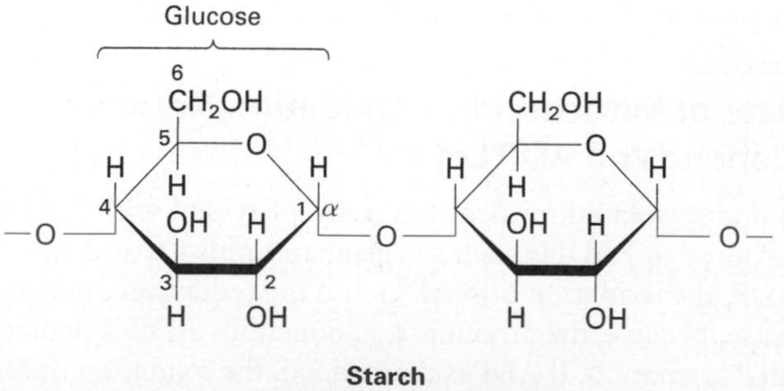

**FIGURE 12-30 Structure of starch.** This large glucose polymer and the disaccharide sucrose (see Figure 2-19) are the principal end products of photosynthesis. Both are built of six-carbon sugars (hexoses).

## Thylakoid Membranes in Chloroplasts Are the Sites of Photosynthesis in Plants

Chloroplasts are lens shaped with a diameter of approximately 5 μm and a width of approximately 2.5 μm. They contain about 3000 different proteins, 95 percent of which are encoded in the nucleus, made in the cytosol, imported into the organelle, and then transported to their appropriate membrane or space (Chapter 13). They are bounded by two membranes, which do not contain chlorophyll and do not participate directly in the generation of ATP and NADPH driven by light (Figure 12-31). As in mitochondria, the outer membrane of chloroplasts contains porins and thus is permeable to metabolites of small molecular weight. The inner membrane forms a permeability barrier that contains transport proteins for regulating the movement of metabolites into and out of the organelle.

Unlike mitochondria, chloroplasts contain a third membrane—the *thylakoid membrane*—on which the light-driven generation of ATP and NADPH occurs. The chloroplast thylakoid membrane is believed to constitute a single sheet that forms numerous small, interconnected flattened structures, the **thylakoids**, which commonly are arranged in stacks termed *grana* (Figure 12-31). The spaces within all the thylakoids constitute a single continuous compartment, the *thylakoid lumen*. The thylakoid membrane contains a number of integral membrane proteins to which are bound several important prosthetic groups and light-absorbing pigments, most notably chlorophyll. Starch synthesis and storage occurs in the *stroma*, the soluble phase between the thylakoid membrane and the inner membrane. In photosynthetic bacteria extensive invaginations of the plasma membrane form a set of internal membranes, also termed thylakoid membranes, where photosynthesis occurs.

## Three of the Four Stages in Photosynthesis Occur Only During Illumination

The photosynthetic process in plants can be divided into four stages (Figure 12-32), each localized to a defined area of the chloroplast: (1) absorption of light, generation of a high-energy electrons, and formation of $O_2$ from $H_2O$; (2) electron transport leading to reduction of $NADP^+$ to NADPH, and generation of a proton-motive force; (3) synthesis of ATP; and (4) conversion of $CO_2$ into carbohydrates, commonly referred to as **carbon fixation.** All four stages of photosynthesis are tightly coupled and controlled so as to produce the amount of carbohydrate required by the plant. All the reactions in stages 1–3 are catalyzed by multiprotein complexes in the thylakoid membrane. The generation of a pmf and the use of the pmf to synthesize ATP resemble stages III and IV of mitochondrial oxidative phosphorylation. The enzymes that incorporate $CO_2$ into chemical intermediates and then convert them to starch are soluble constituents of the chloroplast stroma; the enzymes that form sucrose from three-carbon intermediates are in the cytosol.

**Stage 1: Absorption of Light Energy, Generation of High-Energy Electrons, and $O_2$ Formation** The initial step in photosynthesis is the absorption of light by chlorophylls attached

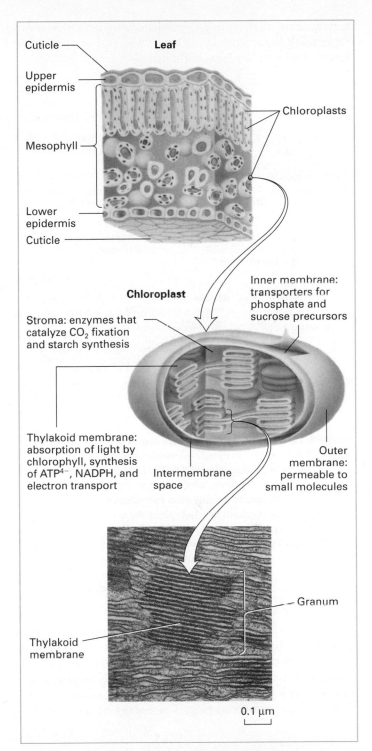

**FIGURE 12-31 Cellular structure of a leaf and chloroplast.** Like mitochondria, plant chloroplasts are bounded by a double membrane separated by an intermembrane space. Photosynthesis occurs on a third membrane, the thylakoid membrane, which is surrounded by the inner membrane and forms a series of flattened vesicles (thylakoids) that enclose a single interconnected *luminal* space. The green color of plants is due to the green color of chlorophyll, all of which is localized to the thylakoid membrane. A granum is a stack of adjacent thylakoids. The stroma is the space enclosed by the inner membrane and surrounding the thylakoids. [Photomicrograph courtesy of Katherine Esau, University of California, Davis.]

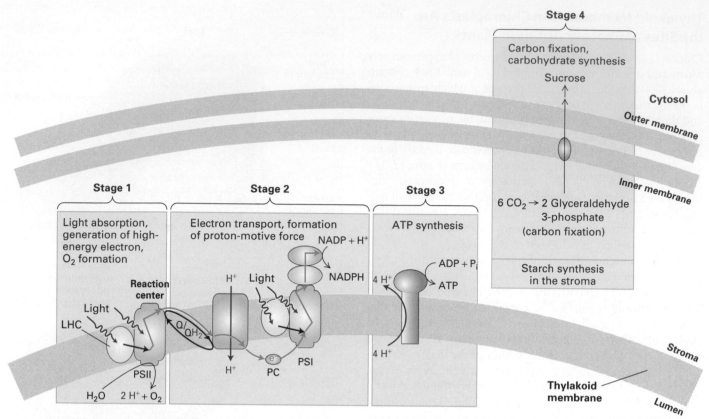

**FIGURE 12-32 Overview of the four stages of photosynthesis.** In **stage 1,** light is absorbed by light-harvesting complexes (LHC) and the reaction center of photosystem II (PSII). The LHCs transfer the absorbed energy to the reaction centers, which use it, or the energy absorbed directly from a photon, to oxidize water to molecular oxygen and generate high-energy electrons (electron paths shown by blue arrows). In **stage 2,** these electrons move down an electron transport chain, which uses either lipid-soluble (Q/QH$_2$) or water-soluble (plastocyanin, PC) electron carriers to shuttle electrons between multiple protein complexes. As electrons move down the chain, they release energy that the complexes use to generate a proton-motive force and, after additional energy is introduced by absorption of light in photosystem I (PSI), to synthesize the high-energy electron carrier NADPH. In **stage 3,** flow of protons down their concentration and voltage gradient through the F$_0$F$_1$ ATP synthase drives ATP synthesis. **Stages 1–3** in plants take place in the thylakoid membrane of the chloroplast. In **stage 4,** in the chloroplast stroma, the energy stored in NADPH and ATP is used to convert CO$_2$ initially into three-carbon molecules (glyceraldehyde 3-phosphate), a process known as carbon fixation. These molecules are then transported to the cytosol of the cell for conversion to hexose sugars in the form of sucrose. Glyceraldehyde 3-phosphate is also used to make starch within the chloroplast.

to proteins in the thylakoid membranes. Like the heme component of cytochromes, chlorophylls consist of a porphyrin ring attached to a long hydrocarbon side chain (Figure 12-33). In contrast to the hemes (see Figure 12-14), chlorophylls contain a central Mg$^{2+}$ ion (rather than Fe$^{2+}$) and have an additional five-membered ring. The energy of the absorbed

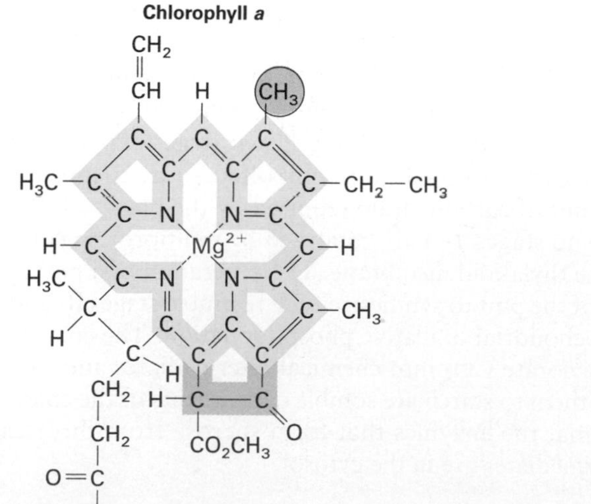

**FIGURE 12-33 Structure of chlorophyll *a*, the principal pigment that traps light energy.** Electrons are delocalized among three of chlorophyll *a*'s four central rings (yellow) and the atoms that interconnect them. In chlorophyll, a Mg$^{2+}$ ion, rather than the Fe$^{2+}$ ion found in heme, sits at the center of the porphyrin ring and an additional five-membered ring (blue) is present; otherwise, the structure of chlorophyll is similar to that of heme, found in molecules such as hemoglobin and cytochromes (see Figure 12-14a). The hydro-carbon phytol "tail" facilitates binding of chlorophyll to hydrophobic regions of chlorophyll-binding proteins. The CH$_3$ group (green) is replaced by a formaldehyde (CHO) group in chlorophyll *b*.

light ultimately is used to remove electrons from a donor (water in the case of green plants), forming oxygen:

$$2 H_2O \xrightarrow{light} O_2 + 4 H^+ + 4 e^-$$

The electrons are transferred to a *primary electron acceptor,* a quinone designated Q, which is similar to CoQ in mitochondria. In plants the oxidation of water takes place in a multiprotein complex called *photosystem II (PSII).*

**Stage 2: Electron Transport and Generation of a Proton-Motive Force** Electrons move from the quinone primary electron acceptor through a series of electron carriers until they reach the ultimate electron acceptor, usually the oxidized form of **nicotinamide adenine dinucleotide phosphate (NADP⁺)**, reducing it to NADPH. The structure of $NADP^+$ is identical to that of $NAD^+$ except for the presence of an additional phosphate group. Both molecules gain and lose electrons in the same way (see Figure 2-33). In plants the reduction of $NADP^+$ takes place in a complex called *photosystem I (PSI).* The transport of electrons in the thylakoid membrane is coupled to the movement of protons from the stroma to the thylakoid lumen, forming a pH gradient across the membrane ($pH_{lumen} < pH_{stroma}$). This process is analogous to generation of a proton-motive force across the inner mitochondrial membrane and in bacterial membranes during electron transport (see Figure 12-23).

Thus the overall reaction of stages 1 and 2 can be summarized as

$$2 H_2O + 2 NADP^+ \xrightarrow{light} 2 H^+ + 2 NADPH + O_2$$

**Stage 3: Synthesis of ATP** Protons move down their concentration gradient from the thylakoid lumen to the stroma through the $F_0F_1$ complex (ATP synthase), which couples proton movement to the synthesis of ATP from ADP and $P_i$. The chloroplast ATP synthase works similarly to the synthases of mitochondria and bacteria (see Figures 12-26 and 12-27).

**Stage 4: Carbon Fixation** The NADPH and ATP generated by the second and third stages of photosynthesis provide the energy and the electrons to drive the synthesis of polymers of six-carbon sugars from $CO_2$ and $H_2O$. The overall chemical equation is written as

$$6 CO_2 + 18 ATP^{4-} + 12 NADPH + 12 H_2O \rightarrow$$
$$C_6H_{12}O_6 + 18 ADP^{3-} + 18 P_i^{2-} + 12 NADP^+ + 6 H^+$$

The reactions that generate the ATP and NADPH used in carbon fixation are directly dependent on light energy; thus stages 1–3 are called the *light reactions* of photosynthesis. The reactions in stage 4 are indirectly dependent on light energy; they are sometimes called the *dark reactions* of photosynthesis because they can occur in the dark, utilizing the supplies of ATP and NADPH generated by light energy. However, the reactions in stage 4 are not confined to the dark; in fact, they occur primarily during illumination.

## Each Photon of Light Has a Defined Amount of Energy

Quantum mechanics established that light, a form of electromagnetic radiation, has properties of both waves and particles. When light interacts with matter, it behaves as discrete packets of energy (quanta) called *photons.* The energy of a photon, $\varepsilon$, is proportional to the frequency of the light wave: $\varepsilon = h\gamma$, where $h$ is Planck's constant ($1.58 \times 10^{-34}$ cal $\cdot$ s, or $6.63 \times 10^{-34}$ J $\cdot$ s) and $\gamma$ is the frequency of the light wave. It is customary in biology to refer to the wavelength of the light wave, $\lambda$, rather than to its frequency, $\gamma$. The two are related by the simple equation $\gamma = c \div \lambda$, where $c$ is the velocity of light ($3 \times 10^{10}$ cm/s in a vacuum). Note that photons of *shorter* wavelength have *higher* energies. Also, the energy in 1 mol of photons can be denoted by $E = N\varepsilon$, where $N$ is Avogadro's number ($6.02 \times 10^{23}$ molecules or photons/mol). Thus

$$E = Nh\gamma = \frac{Nhc}{\lambda}$$

The energy of light is considerable, as we can calculate for light with a wavelength of 550 nm ($550 \times 10^{-7}$ cm), typical of sunlight:

$$E = \frac{(6.02 \times 10^{23} \text{photons/mol})(1.58 \times 10^{-34} \text{cal} \cdot \text{s})(3 \times 10^{10} \text{cm/s})}{550 \times 10^{-7} \text{cm}}$$
$$= 51,881 \text{ cal/mol}$$

or about 52 kcal/mol. This is enough energy to synthesize several moles of ATP from ADP and $P_i$ if all the energy were used for this purpose.

## Photosystems Comprise a Reaction Center and Associated Light-Harvesting Complexes

The absorption of light energy and its conversion into chemical energy occurs in multiprotein complexes called **photosystems.** Found in all photosynthetic organisms, both eukaryotic and prokaryotic, photosystems consist of two closely linked components: a *reaction center,* where the primary events of photosynthesis—generation of high-energy electrons—occur; and an antenna complex consisting of numerous protein complexes, including internal antenna proteins within the photosystem proper and external antenna complexes termed *light-harvesting complexes (LHCs),* made up of specialized proteins which capture light energy and transmit it to the reaction center (see Figure 12-32).

Both reaction centers and antennas contain tightly bound light-absorbing pigment molecules. Chlorophyll *a* is the principal pigment involved in photosynthesis, being present in both reaction centers and antennas. In addition to chlorophyll *a*, antennas contain other light-absorbing pigments: *chlorophyll b* in vascular plants and *carotenoids* in both plants and photosynthetic bacteria. Carotenoids consist of long branched hydrocarbon chains with alternating single

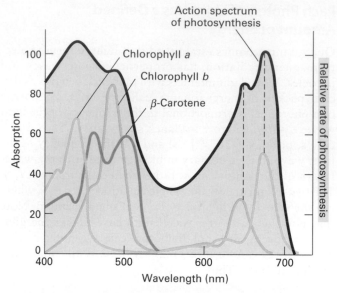

**EXPERIMENTAL FIGURE 12-34 The rate of photosynthesis is greatest at wavelengths of light absorbed by three pigments.** The action spectrum of photosynthesis in plants (the ability of light of different wavelengths to support photosynthesis) is shown in black. The energy from light can be converted into ATP only if it can be absorbed by pigments in the chloroplast. Absorption spectra (showing how well light of different wavelengths is absorbed) for three photosynthetic pigments present in the antennas of plant photosystems are shown in color. Comparison of the action spectrum with the individual absorption spectra suggests that photosynthesis at 680 nm is primarily due to light absorbed by chlorophyll *a*; at 650 nm, to light absorbed by chlorophyll *b*; and at shorter wavelengths, to light absorbed by chlorophylls *a* and *b* and by carotenoid pigments, including β-carotene.

and double bonds; they are similar in structure to the visual pigment retinal, which absorbs light in the eye. The presence of various antenna pigments, which absorb light at different wavelengths, greatly extends the range of light that can be absorbed and used for photosynthesis.

One of the strongest pieces of evidence for the involvement of chlorophylls and carotenoids in photosynthesis is that the absorption spectrum of these pigments is similar to the action spectrum of photosynthesis (Figure 12-34). The latter is a measure of the relative ability of light of different wavelengths to support photosynthesis.

When chlorophyll *a* (or any other molecule) absorbs visible light, the absorbed light energy raises electrons in the chlorophyll *a* to a higher-energy (excited) state. This state differs from the ground (unexcited) state largely in the distribution of the electrons around the C and N atoms of the porphyrin ring. Excited states are unstable, and the electrons return to the ground state by one of several competing processes. For chlorophyll *a* molecules dissolved in organic solvents such as ethanol, the principal reactions that dissipate the excited-state energy are the emission of light (fluorescence and phosphorescence) and thermal emission (heat). When the same chloro-

phyll *a* is bound in the unique protein environment of the reaction center, dissipation of excited-state energy occurs by a quite different process that is the key to photosynthesis.

## Photoelectron Transport from Energized Reaction-Center Chlorophyll *a* Produces a Charge Separation

The absorption of a photon of light of wavelength ~680 nm by one of the two "special-pair" chlorophyll *a* molecules in the reaction center increases the molecules' energy by 42 kcal/mol (the first excited state). Such an energized chlorophyll *a* molecule in a plant reaction center rapidly donates an electron to an intermediate acceptor, and the electron is rapidly passed on to the primary electron acceptor, quinone Q, near the stromal surface of the thylakoid membrane (Figure 12-35). This light-driven electron transfer, called **photoelectron transport,** depends on the unique environment of both the chlorophylls and the acceptor within the reaction center. Photoelectron transport, which occurs nearly every time a photon is absorbed, leaves a positive charge on the chlorophyll *a* close to the luminal surface of the thylakoid membrane (opposite side from the stroma) and generates a reduced, negatively charged acceptor ($Q^-$) near the stromal surface.

The $Q^-$ produced by photoelectron transport is a powerful reducing agent with a strong tendency to transfer an electron to another molecule, ultimately to $NADP^+$. The positively charged chlorophyll $a^+$, a strong oxidizing agent, attracts an electron from an electron donor on the luminal surface to regenerate the original chlorophyll *a*. In plants, the oxidizing power of four chlorophyll $a^+$ molecules is used, by way of intermediates, to remove four electrons from 2 $H_2O$ molecules bound to a site on the luminal surface to form $O_2$:

$$2\ H_2O + 4\ \text{chlorophyll}\ a^+ \rightarrow 4\ H^+ + O_2 + 4\ \text{chlorophyll}\ a$$

These potent biological reductants and oxidants provide all the energy needed to drive all subsequent reactions of photosynthesis: electron transport (stage 2), ATP synthesis (stage 3), and $CO_2$ fixation (stage 4).

Chlorophyll *a* also absorbs light at discrete wavelengths shorter than 680 nm (see Figure 12-34). Such absorption raises the molecule into one of several excited states, whose energies are higher than that of the first excited state described above and which decay by releasing energy within $2 \times 10^{-12}$ seconds (2 picoseconds, ps) to the lower-energy first excited state with loss of the extra energy as heat. Because photoelectron transport and the resulting charge separation occur only from the first excited state of the reaction-center chlorophyll *a*, the quantum yield—the amount of photosynthesis per absorbed photon—is the same for all wavelengths of visible light shorter (and therefore of higher energy) than 680 nm. How closely the wavelength of light matches the absorption spectra of the pigments determines how likely it is that the photon will be absorbed. Once absorbed, the photon's exact wavelength is

**FIGURE 12-35 Photoelectron transport, the primary event in photosynthesis.** After absorption of a photon of light, one of the excited special pair of chlorophyll *a* molecules in the reaction center (*left*) donates via several intermediates (not shown) an electron to a loosely bound acceptor molecule, the quinone Q, on the stromal surface of the thylakoid membrane, creating an essentially irreversible charge separation across the membrane (*right*). Subsequent transfers of this electron release energy that is used to generate ATP and NADPH (Figures 12-38 and 12-39). The positively charged chlorophyll $a^+$ generated when the light-excited electron moves to Q is eventually neutralized by the transfer to the chlorophyll $a^+$ of another electron. In plants the oxidation of water to molecular oxygen provides this neutralizing electron and takes place in a multiprotein complex called photosystem II (Figure 12-39). The complex photosystem I uses a similar photoelectron transport pathway, but instead of oxidizing water, it receives an electron from a protein carrier called plastocyanin to neutralize the positive charge on chlorophyll $a^+$ (Figure 12-39).

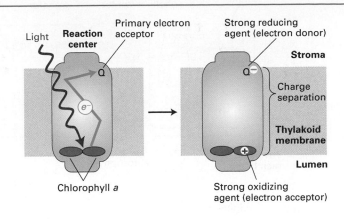

not critical, provided it is at least energetic enough to push the chlorophyll into the first excited state.

## Internal Antenna and Light-Harvesting Complexes Increase the Efficiency of Photosynthesis

Although chlorophyll *a* molecules within a reaction center that are involved directly with charge separation and electron transfer are capable of directly absorbing light and initiating photosynthesis, they most commonly are energized indirectly by energy transferred to them from other light-absorbing and energy-transferring pigments. These other pigments, which include many other chlorophyll molecules, are involved with absorption of photons and passing the energy to the chlorophyll *a* molecules in the reaction center. Some are bound to protein subunits that are considered to be intrinsic components of the photosystem and thus are called internal antennas; others are bound to protein complexes that bind to, but are distinct from, the photosystem core proteins and are called light-harvesting complexes (LHCs). Even at the maximum light intensity encountered by photosynthetic organisms (tropical noontime sunlight), each reaction-center chlorophyll *a* molecule absorbs only about one photon per second, which is not enough to support photosynthesis sufficient for the needs of the plant. The involvement of internal antennas and LHCs greatly increases the efficiency of photosynthesis, especially at more typical light intensities, by increasing absorption of 680-nm light and by extending the range of wavelengths of light that can be absorbed by other antenna pigments.

Photons can be absorbed by any of the pigment molecules in internal antennas or LHCs. The absorbed energy is then rapidly transferred (in $<10^{-9}$ seconds) to one of the two "special-pair" chlorophyll *a* molecules in the associated reaction center, where it promotes the primary photosynthetic charge separation (Figure 12-35). Photosystem core proteins and LHC proteins maintain the pigment molecules in the precise orientation and position optimal for light absorption and energy transfer, thereby maximizing the very rapid and efficient *resonance transfer* of energy from antenna pigments to reaction-center chlorophylls. Resonance energy transfer does not involve the transfer of an electron. Studies on one of the two photosystems in cyanobacteria, which are similar to those in multicellular, seed-bearing plants, suggest that energy from absorbed light is funneled first to a "bridging" chlorophyll in each LHC and then to the special pair of reaction-center chlorophylls (Figure 12-36a). Surprisingly, however, the molecular structures of LHCs from plants and cyanobacteria are completely different from those in green and purple bacteria, even though both types contain carotenoids and chlorophylls in a clustered arrangement within the membrane. Figure 12-36b shows the distribution of the chlorophyll pigments in photosystem I from *Pisum sativum* (garden pea) together with those from peripheral LHC antennas. The large number of internal and LHC antenna chlorophylls surround the core reaction center to permit efficient transfer of absorbed light energy to the special chlorophylls in the reaction center.

Although LHC antenna chlorophylls can transfer light energy absorbed from a photon, they cannot release an electron. As we've seen already, this function resides in the two reaction-center chlorophylls. To understand their electron-releasing ability, we examine the structure and function of the reaction center in bacterial and plant photosystems in the next section.

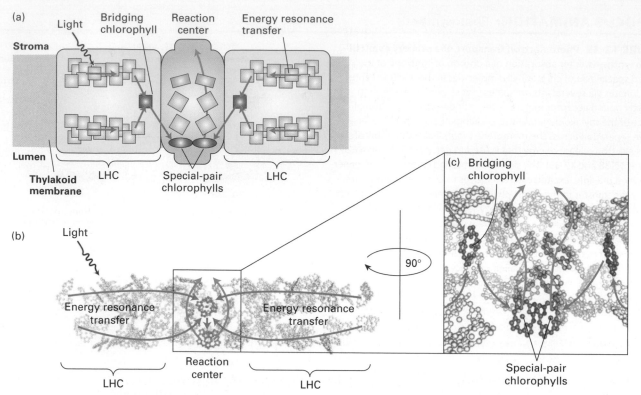

**FIGURE 12-36 Light-harvesting complexes and photosystems in cyanobacteria and plants.** (a) Diagram of the membrane of a cyanobacterium, in which the multiprotein light-harvesting complex (LHC) contains 90 chlorophyll molecules (green) and 31 other small molecules, all held in a specific geometric arrangement for optimal light absorption and energy transfer. Of the six chlorophyll molecules in the reaction center, two constitute the special-pair chlorophylls (ovals, dark green) that can initiate photoelectron transport when excited (blue arrow). Resonance transfer of energy (red arrows) rapidly funnels energy from absorbed light to one of two "bridging" chlorophylls (squares, dark green) and thence to chlorophylls in the reaction center. (b) Three-dimensional organization of the photosystem I (PSI) and associated LHCs of *Pisum sativum* (garden pea), as determined by x-ray crystallography and seen from the plane of the membrane. Only the chlorophylls together with the reaction-center electron carriers are shown. (c) Expanded view of the reaction center from (b), rotated 90° about a vertical axis. [Part (a) adapted from W. Kühlbrandt, 2001, *Nature* **411**:896, and P. Jordan et al., 2001, *Nature* **411**:909. Parts (b) and (c) based on the structural determination by A. Ben-Sham et al., 2003, *Nature* **426**:630.]

## KEY CONCEPTS of Section 12.5

### Photosynthesis and Light-Absorbing Pigments

• The principal end products of photosynthesis in plants are molecular oxygen and polymers of six-carbon sugars (starch and sucrose).

• The light-capturing and ATP-generating reactions of photosynthesis occur in the thylakoid membrane located within chloroplasts. The permeable outer membrane and inner membrane surrounding chloroplasts do not participate directly in photosynthesis (see Figure 12-31).

• There are four stages in photosynthesis: (1) absorption of light, generation of a high-energy electrons, and formation of $O_2$ from $H_2O$; (2) electron transport leading to reduction of $NADP^+$ to NADPH, and to generation of a proton-motive force; (3) synthesis of ATP; and (4) conversion of $CO_2$ into carbohydrates (carbon fixation).

• In stage 1 of photosynthesis, light energy is absorbed by one of two "special-pair" chlorophyll *a* molecules bound to reaction-center proteins in the thylakoid membrane. The energized chlorophylls donate, via intermediates, an electron to a quinone on the opposite side of the membrane, creating a charge separation (see Figure 12-35). In green plants, the positively charged chlorophylls then remove electrons from water, forming molecular oxygen ($O_2$).

• In stage 2, electrons are transported from the reduced quinone via carriers in the thylakoid membrane until they reach the ultimate electron acceptor, usually $NADP^+$, reducing it to NADPH. Electron transport is coupled to movement of protons across the membrane from the stroma to the thylakoid lumen, forming a pH gradient (proton-motive force) across the thylakoid membrane.

• In stage 3, movement of protons down their electrochemical gradient through $F_0F_1$ complexes (ATP synthase) powers the synthesis of ATP from ADP and $P_i$.

• In stage 4, the NADPH and ATP generated in stages 2 and 3 provide the energy and the electrons to drive the fixation of $CO_2$, which results in the synthesis of carbohydrates. These reactions occur in the thylakoid stroma and cytosol.

• Associated with each reaction center are multiple internal antenna and light-harvesting complexes (LHCs), which contain chlorophylls *a* and *b*, carotenoids, and other pigments that absorb light at multiple wavelengths. Energy, but not an electron, is transferred from the internal antenna and LHC chlorophyll molecules to reaction-center chlorophylls by resonance energy transfer (see Figure 12-36).

## 12.6 Molecular Analysis of Photosystems

As noted in the previous section, photosynthesis in the green and purple bacteria does not generate oxygen, whereas photosynthesis in cyanobacteria, algae, and plants does.* This difference is attributable to the presence of two types of photosystem (PS) in the latter organisms: PSI reduces $NADP^+$ to NADPH, and PSII forms $O_2$ from $H_2O$. In contrast, the green and purple bacteria have only one type of photosystem, which cannot form $O_2$. We first discuss the simpler photosystem of purple bacteria and then consider the more complicated photosynthetic machinery in chloroplasts.

### The Single Photosystem of Purple Bacteria Generates a Proton-Motive Force but No $O_2$

The three-dimensional structures of photosynthetic reaction centers have been determined, permitting scientists to trace in detail the paths of electrons during and after the absorption of light. The reaction center of purple bacteria contains three protein subunits (L, M, and H) located in the plasma membrane (Figure 12-37). Bound to these proteins are the prosthetic groups that absorb light and transport electrons during photosynthesis. The prosthetic groups include a "special pair" of bacteriochlorophyll *a* molecules equivalent to the reaction-center chlorophyll *a* molecules in plants, as well as several other pigments and two quinones, termed $Q_A$ and $Q_B$, that are structurally similar to mitochondrial ubiquinone.

**Initial Charge Separation** The mechanism of charge separation in the photosystem of purple bacteria is identical to that in plants outlined earlier; that is, energy from absorbed light is used to strip an electron from a reaction-center bacteriochlorophyll *a* molecule and transfer it, via several different pigments, to the primary electron acceptor $Q_B$, which is loosely bound to a site on the cytosolic membrane face. The

---

*A very different type of mechanism used to harvest the energy of light, which occurs only in certain archaebacteria, is not discussed here because it is very different from reaction center mechanisms described here. In this other mechanism, the plasma-membrane protein that absorbs a photon of light, called bacteriorhodopsin, also pumps one proton from the cytosol to the extracellular space for every photon of light absorbed.

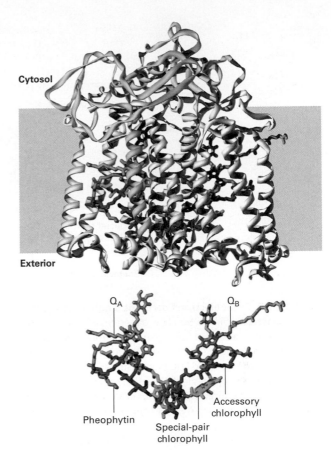

**FIGURE 12-37 Three-dimensional structure of the photosynthetic reaction center from the purple bacterium *Rhodobacter spheroides*.** *(Top)* The L subunit (yellow) and M subunit (gray) each form five transmembrane α helices and have a very similar structure overall; the H subunit (light blue) is anchored to the membrane by a single transmembrane α helix. A fourth subunit (not shown) is a peripheral protein that binds to the exoplasmic segments of the other subunits. *(Bottom)* Within each reaction center, but not easily distinguished in the top image, is a special pair of bacteriochlorophyll *a* molecules (green), capable of initiating photoelectron transport; two accessory chlorophylls (purple); two pheophytins (dark blue), and two quinones, $Q_A$ and $Q_B$ (orange). $Q_B$ is the primary electron acceptor during photosynthesis. [After M. H. Stowell et al., 1997, *Science* **276**:812.]

chlorophyll thereby acquires a positive charge, and $Q_B$ acquires a negative charge. To determine the pathway traversed by electrons through the bacterial reaction center, researchers exploited the fact that each pigment absorbs light of only certain wavelengths, and its absorption spectrum changes when it possesses an extra electron. Because these electron movements are completed in less than 1 millisecond (ms), a special technique called *picosecond absorption spectroscopy* is required to monitor the changes in the absorption spectra of the various pigments as a function of time shortly after the absorption of a light photon.

When a preparation of bacterial membrane vesicles is exposed to an intense pulse of laser light lasting less than 1 ps, each reaction center absorbs one photon (Figure 12-38). Light absorbed by the chlorophyll *a* molecules in each reaction center

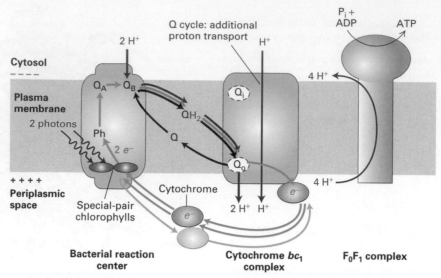

**FIGURE 12-38 Cyclic electron flow in the single photosystem of purple bacteria.** Cyclic electron flow generates a proton-motive force but no $O_2$. Blue arrows indicate flow of electrons; red arrows indicate proton movement. *(Left)* Energy absorbed directly from light or funneled from an associated LHC (not illustrated here) energizes one of the special-pair chlorophylls in the reaction center. Photoelectron transport from the energized chlorophyll, via an accessory chlorophyll, pheophytin (Ph), and quinone A ($Q_A$), to quinone B ($Q_B$) forms the semiquinone $Q^{\bullet-}$ and leaves a positive charge on the chlorophyll. Following absorption of a second photon and transfer of a second electron to the semiquinone, the quinone rapidly picks up two protons from the

cytosol to form $QH_2$. *(Center)* After diffusing through the membrane and binding to the $Q_o$ site on the exoplasmic face of the cytochrome $bc_1$ complex, $QH_2$ donates two electrons and simultaneously gives up two protons to the external medium in the periplasmic space, generating a proton electrochemical gradient (proton-motive force). Electrons are transported back to the reaction-center chlorophyll via a soluble cytochrome, which diffuses in the periplasmic space. Note the cyclic path (blue) of electrons. Operation of a Q cycle in the cytochrome $bc_1$ complex pumps additional protons across the membrane to the external medium, as in mitochondria. [Adapted from J. Deisenhofer and H. Michael, 1991, *Ann. Rev. Cell Biol.* **7**:1.]

converts them to the excited state, and the subsequent electron transfer processes are synchronized in all reaction centers in the experimental sample. Within $4 \times 10^{-12}$ seconds (4 ps), an electron moves, via the accessory bacterial chlorophyll (see Figure 12-37, *bottom*) as an intermediate, to the pheophytin molecules (Ph), leaving a positive charge on the chlorophyll *a*. This state exists for about 200 ps before the electron moves to $Q_A$, and then, in the slowest step, it takes ~200 μs for it to move to $Q_B$. This pathway of electron flow is traced in the left part of Figure 12-38. The later steps are slower than inherently rapid electron movements because they involve relatively slow protein conformational changes.

**Subsequent Electron Flow and Coupled Proton Movement** After the primary electron acceptor, $Q_B$, in the bacterial reaction center accepts one electron, forming $Q_B^{\bullet-}$, it accepts a second electron from the same reaction-center chlorophyll following its re-excitation (e.g., by absorption of a second photon or transfer of energy from antenna molecules). The quinone then binds two protons from the cytosol, forming the reduced quinone ($QH_2$), which is released from the reaction center (Figure 12-38). $QH_2$ diffuses within the bacterial membrane to the $Q_o$ site on the exoplasmic face of a cytochrome $bc_1$ electron transport complex similar in structure to complex III in mitochondria. There it releases its

two protons into the periplasmic space (the space between the plasma membrane and the bacterial cell wall). This process moves protons from the cytosol to the outside of the cell, generating a proton-motive force across the plasma membrane. Simultaneously, $QH_2$ releases its two electrons, which move through the cytochrome $bc_1$ complex exactly as depicted for the mitochondrial complex III ($CoQH_2$–cytochrome *c* reductase) in Figure 12-18. The Q cycle in the bacterial reaction center, like the Q cycle in mitochondria, pumps additional protons from the cytosol to the intermembrane space, thereby increasing the proton-motive force.

The acceptor for electrons transferred through the cytochrome $bc_1$ complex is a soluble cytochrome, a one-electron carrier, in the periplasmic space, which is reduced from the $Fe^{3+}$ to the $Fe^{2+}$ state. The reduced cytochrome (analogous to cytochrome *c* in mitochondria) then diffuses to a reaction center, where it releases its electron to a positively charged chlorophyll $a^+$, returning that chlorophyll to the uncharged ground state and the cytochrome to the $Fe^{3+}$ state. This *cyclic* electron flow generates no oxygen and no reduced coenzymes, but it has generated a proton-motive force.

As in other systems, this proton-motive force is used by the $F_0F_1$ complex located in the bacterial plasma membrane to synthesize ATP and also to transport molecules across the membrane against a concentration gradient.

## Chloroplasts Contain Two Functionally and Spatially Distinct Photosystems

In the 1940s, biophysicist R. Emerson discovered that the rate of plant photosynthesis generated by light of wavelength 700 nm can be greatly enhanced by adding light of shorter wavelength (higher energy). He found that a combination of light at, say, 600 and 700 nm supports a greater rate of photosynthesis than the sum of the rates for the two separate wavelengths. This so-called *Emerson effect* led researchers to conclude that photosynthesis in plants involves the interaction of two separate photosystems, referred to as *PSI* and *PSII*. PSI is driven by light of wavelength 700 nm or less; PSII, only by shorter-wavelength light (<680 nm).

In chloroplasts, the special-pair reaction-center chlorophylls that initiate photoelectron transport in PSI and PSII differ in their light-absorption maxima because of differences in their protein environments. For this reason, these chlorophylls are often denoted $P_{680}$ (PSII) and $P_{700}$ (PSI). Like a bacterial reaction center, each chloroplast reaction center is associated with multiple internal antenna and light-harvesting complexes (LHCs); the LHCs associated with PSII (e.g., LHCII) and PSI (e.g., LHCI) contain different proteins.

The two photosystems also are distributed differently in thylakoid membranes: PSII primarily in stacked regions (grana, see Figure 12-31) and PSI primarily in unstacked regions. The stacking of the thylakoid membranes may be due to the binding properties of the proteins associated with PSII, especially LCHII. Evidence for this distribution came from studies in which thylakoid membranes were gently fragmented into vesicles by ultrasound. Stacked and unstacked thylakoid vesicles were then fractionated by density-gradient centrifugation. The stacked fractions contained primarily PSII protein and the unstacked fractions PSI.

Finally, and most importantly, the two chloroplast photosystems differ significantly in their functions (Figure 12-39): only PSII oxidizes water to form molecular oxygen, whereas only PSI transfers electrons to the final electron acceptor, $NADP^+$. Photosynthesis in chloroplasts can follow a linear or cyclic pathway. The linear pathway, which we discuss first, can support carbon fixation as well as ATP synthesis. In contrast, the cyclic pathway supports only ATP synthesis and generates no reduced NADPH for use in carbon fixation. Photosynthetic algae and cyanobacteria contain two photosystems analogous to those in chloroplasts. Similar proteins and pigments compose photosystems I and II of plants and photosynthetic bacteria.

## Linear Electron Flow Through Both Plant Photosystems, PSII and PSI, Generates a Proton-Motive Force, $O_2$, and NADPH

Linear electron flow in chloroplasts involves PSII and PSI in an obligate series in which electrons are transferred from $H_2O$ to $NADP^+$. The process begins with absorption of a

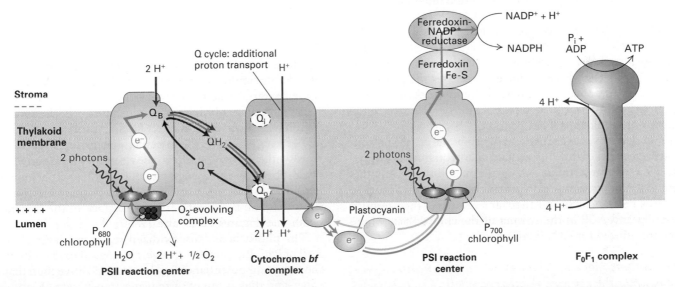

**FIGURE 12-39 Linear electron flow in plants, which requires both chloroplast photosystems PSI and PSII.** Blue arrows indicate flow of electrons; red arrows indicate proton movement. LHCs are not shown. *(Left)* In the PSII reaction center, two sequential light-induced excitations of the same $P_{680}$ chlorophylls result in reduction of the primary electron acceptor $Q_B$ to $QH_2$. On the luminal side of PSII, electrons removed from $H_2O$ in the thylakoid lumen are transferred to $P_{680}^+$, restoring the reaction-center chlorophylls to the ground state and generating $O_2$. *(Center)* The cytochrome *bf* complex then accepts electrons from $QH_2$, coupled to the release of two protons into the lumen. Operation of a Q cycle in the cytochrome *bf* complex translocates

additional protons across the membrane to the thylakoid lumen, increasing the proton motive force. *(Right)* In the PSI reaction center, each electron released from light-excited $P_{700}$ chlorophylls moves via a series of carriers in the reaction center to the stromal surface, where soluble ferredoxin (an Fe-S protein) transfers the electron to ferredoxin-$NADP^+$ reductase (FNR). This enzyme uses the prosthetic group flavin adenine dinucleotide (FAD) and a proton to reduce $NADP^+$, forming NADPH. $P_{700}^+$ is restored to its ground state by addition of an electron carried from PSII via the cytochrome *bf* complex and plastocyanin, a soluble electron carrier.

photon by PSII, causing an electron to move from a $P_{680}$ chlorophyll $a$ to an acceptor plastoquinone ($Q_B$) on the stromal surface (Figure 12-39). The resulting oxidized $P_{680}^+$ strips one electron from the relatively unwilling donor $H_2O$, forming an intermediate in $O_2$ formation and a proton, which remains in the thylakoid lumen and contributes to the proton-motive force. After $P_{680}$ absorbs a second photon, the semiquinone $Q^{\cdot -}$ accepts a second electron and picks up two protons from the stromal space, generating $QH_2$. After diffusing in the membrane, $QH_2$ binds to the $Q_o$ site on a cytochrome $bf$ complex that is analogous to the bacterial cytochrome $bc_1$ complex and to the mitochondrial complex III. As in these systems, a Q cycle operates, thereby increasing the proton-motive force generated by electron transport. After the cytochrome $bf$ complex accepts electrons from $QH_2$, it transfers them, one at a time, to the $Cu^{2+}$ form of the soluble electron carrier plastocyanin (analogous to cytochrome $c$), reducing it to the $Cu^{1+}$ form. Reduced plastocyanin then diffuses in the thylakoid lumen, carrying the electron to PSI.

Absorption of a photon by PSI leads to removal of an electron from the reaction-center chlorophyll $a$, $P_{700}$ (Figure 12-39). The resulting oxidized $P_{700}^+$ is reduced by an electron passed from the PSII reaction center via the cytochrome $bf$ complex and plastocyanin. Again, this is analogous to the situation in mitochondria, where cytochrome $c$ acts as a single-electron shuttle from complex III to complex IV (see Figure 12-16). The electron taken up at the luminal surface by the $P_{700}$ energized by photon absorption moves within PSI via several carriers to the stromal surface of the thylakoid membrane, where it is accepted by ferredoxin, an iron-sulfur (Fe-S) protein. In linear electron flow electrons excited in PSI are transferred from ferredoxin via the enzyme ferredoxin-$NADP^+$ reductase (FNR). This enzyme uses the prosthetic group FAD as an electron carrier to reduce $NADP^+$, forming, together with one proton picked up from the stroma, the reduced molecule NADPH.

$F_0F_1$ complexes in the thylakoid membrane use the proton-motive force generated during linear electron flow to synthesize ATP on the stromal side of the membrane. Thus this pathway exploits the energy from multiple photons absorbed by both PSII and PSI and their antennas to generate both NADPH and ATP in the stroma of the chloroplast, where they are utilized for $CO_2$ fixation.

## An Oxygen-Evolving Complex Is Located on the Luminal Surface of the PSII Reaction Center

Somewhat surprisingly, the structure of the PSII reaction center, which removes electrons from $H_2O$ to form $O_2$, resembles that of the reaction center of photosynthetic purple bacteria, which does not form $O_2$. Like the bacterial reaction center, the PSII reaction center contains two molecules of chlorophyll $a$ ($P_{680}$), as well as two other accessory chlorophylls, two pheophytins, two quinones ($Q_A$ and $Q_B$), and one nonheme iron atom. These small molecules are bound to

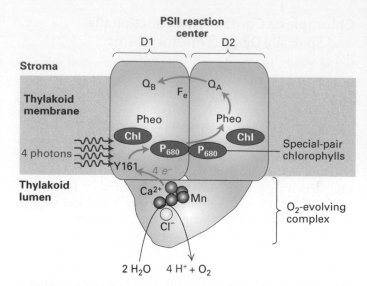

**FIGURE 12-40 Electron flow and $O_2$ evolution in chloroplast PSII.** The PSII reaction center, comprising the two integral proteins D1 and D2, special-pair chlorophylls ($P_{680}$), and other electron carriers, is associated with an oxygen-evolving complex on the luminal surface. Bound to the three extrinsic proteins (33, 23, and 17 kDa) of the oxygen-evolving complex are four manganese ions (Mn, red), a $Ca^{2+}$ ion (blue), and a $Cl^-$ ion (yellow). These bound ions function in the splitting of $H_2O$ and maintain the environment essential for high rates of $O_2$ evolution. Tyrosine-161 (Y161) of the D1 polypeptide conducts electrons from the Mn ions to the oxidized reaction-center chlorophyll ($P_{680}^+$), reducing it to the ground state $P_{680}$. [Adapted from C. Hoganson and G. Babcock, 1997, *Science* **277**:1953.]

two proteins in PSII, called D1 and D2, whose sequences are remarkably similar to the sequences of the L and M subunits of the bacterial reaction center (Figure 12-37), attesting to their common evolutionary origins. When PSII absorbs a photon with a wavelength of <680 nm, it triggers the loss of an electron from a $P_{680}$ molecule, generating $P_{680}^+$. As in photosynthetic purple bacteria, the electron is transported rapidly, probably via an accessory chlorophyll, to a pheophytin, then to a quinone ($Q_A$), and then to the primary electron acceptor, $Q_B$, on the outer (stromal) surface of the thylakoid membrane (Figures 12-39 and 12-40).

The photochemically oxidized reaction-center chlorophyll of PSII, $P_{680}^+$, is the *strongest* biological oxidant known. The reduction potential of $P_{680}^+$ is more positive than that of water, and thus it can oxidize water to generate $O_2$ and $H^+$ ions. Photosynthetic bacteria cannot oxidize water because the excited chlorophyll $a^+$ in the bacterial reaction center is not a sufficiently strong oxidant. Thus they use other sources of electrons, such as $H_2S$ and $H_2$.

The oxidation of $H_2O$, which provides the electrons for reduction of $P_{680}^+$ in PSII, is catalyzed by a three-protein complex, the *oxygen-evolving complex*, located on the luminal surface of PSII in the thylakoid membrane. The oxygen-evolving complex contains four manganese (Mn) ions connected by bridging oxygen atoms, as well as bound $Cl^-$ and

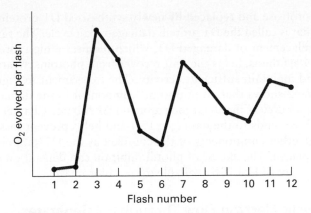

**EXPERIMENTAL FIGURE 12-41 A single PSII absorbs a photon and transfers an electron four times to generate one O₂.** Dark-adapted chloroplasts were exposed to a series of closely spaced, short (5 μs) pulses of light that activated virtually all the PSIIs in the preparation. The peaks in O₂ evolution occurred after every fourth pulse, indicating that absorption of four photons by one PSII is required to generate each O₂ molecule. Because the dark-adapted chloroplasts were initially in a partially reduced state, the peaks in O₂ evolution occurred after flashes 3, 7, and 11. [From J. Berg et al., 2002, *Biochemistry*, 5th ed., W. H. Freeman and Company.]

Herbicides that inhibit photosynthesis not only are very important in agriculture but also have proved useful in dissecting the pathway of photoelectron transport in plants. One such class of herbicides, the *s*-triazines (e.g., atrazine), binds specifically to the D1 subunit in the PSII reaction center, thus inhibiting binding of oxidized $Q_B$ to its site on the stromal surface of the thylakoid membrane. When added to illuminated chloroplasts, *s*-triazines cause all downstream electron carriers to accumulate in the oxidized form, because no electrons can be released from PSII. In atrazine-resistant mutants, a single amino acid change in D1 renders it unable to bind the herbicide, so photosynthesis proceeds at normal rates. Such resistant weeds are prevalent and present a major agricultural problem. ■

$Ca^{2+}$ ions (Figure 12-40); this is one of the very few cases in which manganese plays a role in a biological system. These manganese ions together with the three extrinsic proteins can be removed from the reaction center by treatment with solutions of concentrated salts; this abolishes O₂ formation but does not affect light absorption or the initial stages of electron transport.

The oxidation of two molecules of $H_2O$ to form O₂ requires the removal of four electrons, but absorption of each photon by PSII results in the transfer of just one electron. A simple experiment, described in Figure 12-41, resolved whether the formation of O₂ depends on a single PSII or multiple ones acting in concert. The results indicated that a single PSII must lose an electron and then oxidize the oxygen-evolving complex four times in a row for an O₂ molecule to be formed.

Manganese is known to exist in multiple oxidation states with from two to five positive charges. Indeed, spectroscopic studies showed that the bound Mn ions in the oxygen-evolving complex cycle through five different oxidation states, $S_0$–$S_4$. In this S cycle, a total of two $H_2O$ molecules are split to generate four protons, four electrons, and one O₂ molecule. Channels in the structure of the oxygen-evolving complex have been proposed to serve as conduits for the delivery of $H_2O$ to and the removal of O₂ from the active site through the surrounding protein of the oxygen-evolving complex. The electrons released from $H_2O$ are transferred, one at a time, via the Mn ions and a nearby tyrosine side chain on the D1 subunit to the reaction-center $P_{680}^+$, where they regenerate the reduced chlorophyll, $P_{680}$, ground state by replacing the electron that was removed by light absorption. The protons released from $H_2O$ remain in the thylakoid lumen.

## Multiple Mechanisms Protect Cells Against Damage from Reactive Oxygen Species During Photoelectron Transport

As we saw earlier in the case of mitochondria, ROS generated during electron transport through the electron transport chain (see Figure 12-21) can both serve as signals to regulate organelle function and cause damage to a variety of biomolecules. The same is true for chloroplasts. Even though the PSI and PSII photosystems with their associated light-harvesting complexes are remarkably efficient at converting radiant energy to useful chemical energy in the form of ATP and NADPH, they are not perfect. Depending on the intensity of the light and the physiologic conditions of the cells, a relatively small—but significant—amount of energy absorbed by chlorophylls in the light-harvesting antennas and reaction centers results in the chlorophyll being converted to an activated state called *"triplet"* chlorophyll. In this state, the chlorophyll can transfer some of its energy to molecular oxygen (O₂), converting it from its normal, *relatively* unreactive ground state, called triplet oxygen ($^3O_2$), to a very highly reactive (ROS) singlet state form, $^1O_2$. Some of this $^1O_2$ can be used for signaling to the nucleus to communicate the metabolic state of the chloroplast to the rest of the cell. However, if the majority of the $^1O_2$ is not quickly quenched by reacting with specialized $^1O_2$ "scavenger molecules," it will react with and usually damage nearby molecules. This damage can suppress the efficiency of thylakoid activity and is called *photoinhibition*. Carotenoids (polymers of unsaturated isoprene groups, including beta-carotene, which gives carrots their orange color) and α-tocopherol (a form of vitamin E) are hydrophobic small molecules that play important roles as $^1O_2$ quenchers to protect plants. For example, inhibition of tocopherol synthesis in the unicellular green alga *Chlamydomonas reinhardtii* by the herbicide pyrazolynate can result in greater light-induced photoinhibition. The carotenoids, which very efficiently siphon off energy from the dangerous triplet chlorophyll when they are in close proximity, are the quantitatively most important molecules for preventing $^1O_2$ formation. There are about 11 carotenoid molecules and 35 chlorophylls in the PSII monomer from the cyanobacterium *Thermosynechococcus elongatus*.

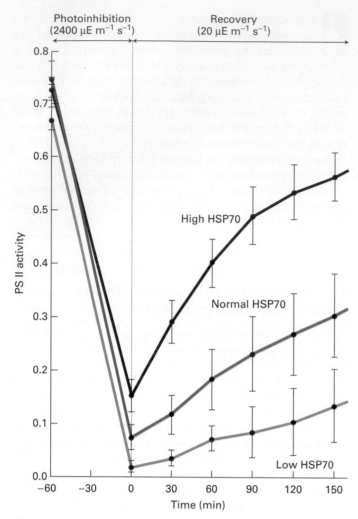

Photoinhibition
(2400 µE m⁻¹ s⁻¹)
Recovery
(20 µE m⁻¹ s⁻¹)

**EXPERIMENTAL FIGURE 12-42** **The chaperone HSP70B helps PSII recover from photoinhibition after exposure to intense light.** The unicellular green alga *Chlamydomonas reinhardtii* was genetically manipulated so that it had abnormally high or low levels of the chaperone protein HSP70B. The high, low, and normal strains were then exposed to high-intensity light (2400 µE m⁻² s⁻¹) for 60 minutes to induce photoinhibition followed by exposure to low light (20 µE m⁻² s⁻¹) for up to 150 minutes. The effects of photoinhibition by the high-intensity light and the ability of PSII to recover from the photoinhibition were measured using fluorescence spectroscopy to determine PSII activity. The ability of the cells to recover PSII activity depends on the levels of HSP70B—the more HSP70B available, the more rapid the recovery—due to HSP70B protection of the PSII reaction centers that had withstood $^1O_2$-induced D1 subunit damage. [From Schroda et al., 1999, *Plant Cell* **11**:1165.]

Under intense illumination, photosystem PSII is especially prone to generating $^1O_2$, whereas PSI will produce other ROS, including superoxide, hydrogen peroxide, and hydroxyl radicals. The D1 subunit in the PSII reaction center (see Figure 12-40) is, even under low light conditions, subjected to almost constant $^1O_2$-mediated damage. A damaged reaction center moves from the grana to the unstacked regions of the thylakoid, where the D1 subunit is degraded by

a protease and replaced by newly synthesized D1 protein in what is called the D1 protein damage-repair cycle. The rapid replacement of damaged D1, which requires a high rate of D1 synthesis, helps the PSII recover from photoinactivation and maintain sufficient activity. The experiment in Figure 12-42 shows that an important component in the damage-repair cycle is the chaperone protein HSP70B (see Chapter 3), which binds to the damaged PSII and helps prevent loss of the other components of the complex as the D1 subunit is replaced. The extent of photoinhibition can depend on the amount of HSP70B available to the chloroplasts.

## Cyclic Electron Flow Through PSI Generates a Proton-Motive Force but No NADPH or $O_2$

As we've seen, electrons from reduced ferredoxin in PSI are transferred to $NADP^+$ during linear electron flow, resulting in production of NADPH (see Figure 12-39). In some circumstances, such as drought, high light intensity, or low carbon dioxide levels, cells must generate relatively greater amounts of ATP relative to NADPH than that produced by linear electron flow. To do this, they photosynthetically produce ATP from PSI without concomitant NADPH production. This is accomplished using a PSII-independent process called *cyclic photophosphorylation*, or *cyclic electron flow*. In this process electrons cycle between PSI, ferredoxin, plastoquinone (Q), and the cytochrome *bf* complex (Figure 12-43); thus no *net* NADPH is generated, and there is no need to oxidize water and produce $O_2$. There are two distinct cyclic electron flow pathways: the *NAD(P)H dehydrogenase (Ndh)-dependent* (shown in Figure 12-43) and *Ndh-independent* pathways. Ndh is an enzyme complex, very similar to the mitochondrial complex I (see Figure 12-16), that oxidizes NADPH or NADH while reducing Q to $QH_2$ and thus contributes to the proton motive force by transporting protons. During cyclic electron flow, the substrate for the Ndh is the NADPH generated by light absorption by PSI, ferredoxin, and ferredoxin-NADP reductase (FNR). The $QH_2$ formed by Ndh then diffuses through the thylakoid membrane to the $Q_o$ binding site on the luminal surface of the cytochrome *bf* complex. There it releases two electrons to the cytochrome *bf* complex and two protons to the thylakoid lumen, generating a proton-motive force. As in linear electron flow, these electrons return to PSI via plastocyanin. This cyclic electron flow is similar to the cyclic process that occurs in the single photosystem of purple bacteria (see Figure 12-38). A Q cycle operates in the cytochrome *bf* complex during cyclic electron flow, leading to transport of two additional protons into the lumen for each pair of electrons transported and a greater proton-motive force.

In Ndh-independent cyclic electron flow, the mechanism of which has not yet been completely defined, electrons from the ferredoxin are used to reduce Q, either via a hypothetical membrane-associated ferredoxin:plastoquinone oxidoreductase (FQR) or via the $Q_i$ site, which is part of the Q cycle in the cytochrome *bf* complex. Genetic analysis of *Arabidopsis thaliana* has identified several genes involved in

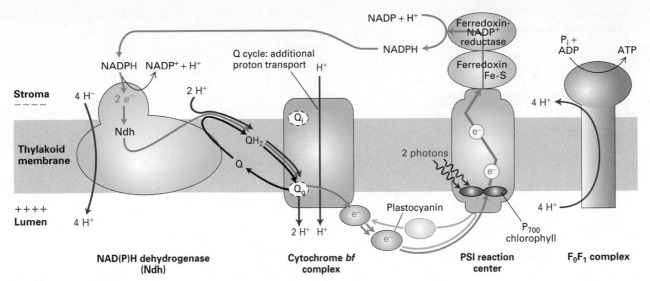

**FIGURE 12-43 Cyclic electron flow in plants, which generates a proton-motive force and ATP but no oxygen or net NADPH.** In the NAD(P)H-dehydrogenase (Ndh)-dependent pathway for cyclic electron flow, light energy is used by PSI to transport electrons in a cycle to generate a proton-motive force and ATP without oxidizing water. The NADPH formed via the PSI/ferredoxin/FNR—instead of being used to fix carbon—is oxidized by Ndh. The released electrons are transferred to plastoquinone (Q) within the membrane to generate QH₂, which then transfers the electrons to the cytochrome *bf* complex, then to plastocyanin, and finally back to PSI, as is the case for the linear electron flow pathway (see Figure 12-39).

Ndh-independent cyclic electron flow, including the integral membrane protein PGRL1.

## Relative Activities of Photosystems I and II Are Regulated

In order for PSII, which is preferentially located in the stacked grana, and PSI, which is preferentially located in the unstacked thylakoid membranes, to act in sequence during linear electron flow, the amount of light energy delivered to the two reaction centers must be controlled so that each center activates the same number of electrons. This balanced condition is called state 1 (Figure 12-44a). If the two photosystems are not equally excited, then cyclic electron flow occurs in PSI and PSII becomes less active (state 2). Variations in the wavelengths and intensities of ambient light (as a consequence of the time of day, cloudiness, etc.) can change the relative activation of the two photosystems, potentially upsetting the appropriate relative amounts of linear and cyclic electron flow necessary for production of optimal ratios of ATP and NADPH.

One mechanism for regulating the relative contributions of PSI and PSII, in response to varying lighting conditions and thus the relative amounts of linear and cyclic electron flow, entails redistributing the light-harvesting complex LHCII between the two photosystems. The more LHCII associated with a particular photosystem, the more efficiently that system will be activated by light and the greater its contribution to electron flow. The distribution of LHCII between PSI and PSII is mediated by reversible phosphorylation and dephosphorylation of LHCII by a regulated, membrane-associated kinase and an apparently constitutively active phosphatase. LHCII's unphosphorylated form is preferentially associated with PSII, and the phosphorylated form diffuses in the thylakoid membrane from the grana to the unstacked region and associates with PSI more than the unphosphorylated form. Light conditions in which there is preferential absorption of light by PSII result in the production of high levels of QH₂ that bind to the cytochrome *bf* complex (see Figure 12-39). Consequent conformational changes in this complex are apparently responsible for activation of the LHCII kinase, increased LHCII phosphorylation, compensatory increased activation of PSI relative to PSII, and thus an increase in cyclic electron flow in state 2 (Figure 12-44a). When the green alga *Chlamydomonas reinhardtii* was forced into state 2, it was possible to isolate a "super-supercomplex" containing PSI, LHCI, LHCII, Cyt *bf*, ferredoxin (Fd), NADPH oxidoreductase (FNR), and the integral membrane protein PGRL1 that participates in *Ndh-independent cyclic electron flow* (Figure 12-44b). Thus it appears that the efficient operation of electron transport chains has involved the evolution of functional complexes of increasing size and complexity, from individual proteins to complexes to supercomplexes to super-supercomplexes.

Regulating the supramolecular organization of the photosystems in plants has the effect of directing them toward ATP production (state 2) or toward generation of reducing equivalents (NADPH) and ATP (state 1), depending on ambient light conditions and the metabolic needs of the plant. Both NADPH and ATP are required to convert $CO_2$ to sucrose or starch, the fourth stage in photosynthesis, which we cover in the last section of this chapter.

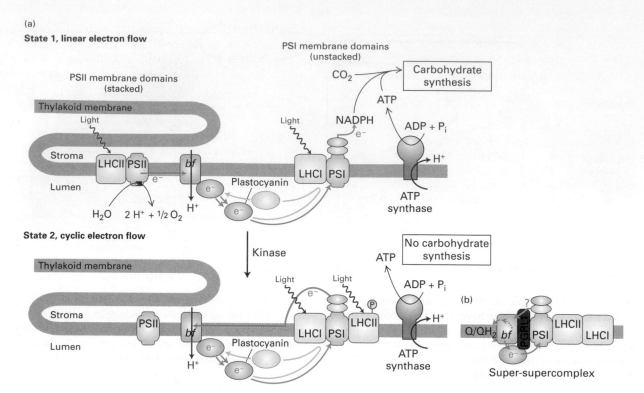

**(a)**

**State 1, linear electron flow**

PSII membrane domains (stacked)

PSI membrane domains (unstacked)

**State 2, cyclic electron flow**

Kinase

**(b)**

Super-supercomplex

**FIGURE 12-44 Phosphorylation of LHCII and the regulation of linear versus cyclic electron flow. (a,** *top)* In normal sunlight, PSI and PSII are equally activated, and the photosystems are organized in state 1. In this arrangement, light-harvesting complex II (LHCII) is not phosphorylated and six copies of LHCII trimers together with several other light-harvesting proteins encircle a dimeric PSII reaction center in a tightly associated supercomplex in the grana (for clarity, molecular details of the supercomplexes not shown). As a result, PSII and PSI can function in parallel in linear electron flow. **(a,** *bottom)* When light excitation of the two photosystems is unbalanced (e.g., too much via PSII), LHCII becomes phosphorylated, dissociates from PSII, and diffuses into the unstacked membranes, where it associates with PSI and its permanently associated LHCI. In this alternative supramolecular organization (state 2), most of the absorbed light energy is transferred to PSI, supporting cyclic electron flow and ATP production but no formation of NADPH and thus no $CO_2$ fixation. **(b)** Model of a PSI "super-supercomplex" involved with *Ndh-independent cyclic electron flow* that was isolated from green algae in stage 2. The super-supercomplex contains multiple complexes, including the integral membrane protein PGRL1 that was identified by genetic analysis. [Adapted from F. A. Wollman, 2001, *EMBO J.* **20:**3623; and M. Iwai, et al., 2010, *Nature* **464:**1210–1213.]

## KEY CONCEPTS of Section 12.6

### Molecular Analysis of Photosystems

• In the single photosystem of purple bacteria, cyclic electron flow from light-excited, special-pair chlorophyll *a* molecules in the reaction center generates a proton-motive force, which is used mainly to power ATP synthesis by the $F_0F_1$ complex in the plasma membrane (see Figure 12-38).

• Plants contain two photosystems, PSI and PSII, which have different functions and are physically separated in the thylakoid membrane. PSII converts $H_2O$ into $O_2$, and PSI reduces $NADP^+$ to NADPH. Cyanobacteria have two analogous photosystems.

• In chloroplasts, light energy absorbed by light-harvesting complexes (LHCs) is transferred to chlorophyll *a* molecules in the reaction centers ($P_{680}$ in PSII and $P_{700}$ in PSI).

• Electrons flow through PSII via the same carriers that are present in the bacterial photosystem. In contrast to the bacterial system, photochemically oxidized $P_{680}^+$ in PSII is regenerated to $P_{680}$ by electrons derived from the evolution of $O_2$ from $H_2O$ (see Figure 12-39, *left*).

• In linear electron flow, photochemically oxidized $P_{700}^+$ in PSI is reduced, regenerating $P_{700}$, by electrons transferred from PSII via the cytochrome *bf* complex and soluble plastocyanin. Electrons released from $P_{700}$ following excitation of PSI are transported via several carriers ultimately to $NADP^+$, generating NADPH (see Figure 12-39, *right*).

• The absorption of light by pigments in the chloroplast can generate reactive oxygen species (ROS), including singlet oxygen, $^1O_2$, and hydrogen peroxide, $H_2O_2$. In small amounts they are not toxic and are used as intracellular signaling molecules to control cellular metabolism. In larger amounts they can be toxic. Small molecule scavengers and antioxidant enzymes help to protect against ROS-induced damage; however, singlet oxygen damage to the D1 subunit of PSII still occurs, causing photoinhibition. An HSP70 chaperone helps PSII recover from the damage.

- In contrast to linear electron flow, which requires both PSII and PSI, cyclic electron flow in plants involves only PSI. In this pathway, neither net NADPH nor $O_2$ is formed although a proton-motive force is generated. Very large super-supercomplexes can be involved in cyclic electron flow.

- Reversible phosphorylation and dephosphorylation of the light-harvesting complex II (LHCII) control the functional organization of the photosynthetic apparatus in thylakoid membranes. State 1 favors linear electron flow, whereas state 2 favors cyclic electron flow (see Figure 12-44).

## 12.7 $CO_2$ Metabolism During Photosynthesis

Chloroplasts perform many metabolic reactions in green leaves. In addition to $CO_2$ fixation—incorporation of gaseous $CO_2$ into small organic molecules and then sugars—the synthesis of almost all amino acids, all fatty acids and carotenes, all pyrimidines, and probably all purines occurs in chloroplasts. However, the synthesis of sugars from $CO_2$ is the most extensively studied biosynthetic pathway in plant cells. We first consider the unique pathway, known as the **Calvin cycle** (after discoverer Melvin Calvin), that fixes $CO_2$ into three-carbon compounds, powered by energy released during ATP hydrolysis and oxidation of NADPH.

### Rubisco Fixes $CO_2$ in the Chloroplast Stroma

The enzyme **ribulose 1,5-bisphosphate carboxylase,** or **rubisco,** fixes $CO_2$ into precursor molecules that are subsequently converted into carbohydrates. Rubisco is located in the stromal space of the chloroplast. This enzyme adds $CO_2$ to the five-carbon sugar ribulose 1,5-bisphosphate to form two molecules of the three-carbon-containing 3-phosphoglycerate (Figure 12-45). Rubisco is a large enzyme ($\sim$500 kDa), with the most common form composed of eight identical large and eight identical small subunits. One subunit is encoded in chloroplast DNA; the other, in nuclear DNA. Because the catalytic rate of rubisco is quite low, many copies of the enzyme are needed to fix sufficient $CO_2$. Indeed, this enzyme makes up almost 50 percent of the chloroplast soluble protein and is believed to be the most abundant protein on earth. It is estimated that rubisco fixes more than $10^{11}$ tons of atmospheric $CO_2$ each year.

When photosynthetic algae are exposed to a brief pulse of $^{14}$C-labeled $CO_2$ and the cells are then quickly disrupted, 3-phosphoglycerate is radiolabeled most rapidly, and all the radioactivity is found in the carboxyl group. Because $CO_2$ is initially incorporated into a three-carbon compound, the Calvin cycle is also called the $C_3$ *pathway* of carbon fixation (Figure 12-46).

The fate of 3-phosphoglycerate formed by rubisco is complex: some is converted to hexoses incorporated into starch or sucrose, but some is used to regenerate ribulose 1,5-bisphosphate. At least nine enzymes are required to regenerate ribulose 1,5-bisphosphate from 3-phosphoglycerate. Quantitatively, for every 12 molecules of 3-phosphoglycerate generated by rubisco (a total of 36 C atoms), 2 of them (6 C atoms) are converted to 2 molecules of glyceraldehyde 3-phosphate (and later to 1 hexose), whereas 10 molecules (30 C atoms) are converted to 6 molecules of ribulose 1,5-bisphosphate (Figure 12-46, *top*). The fixation of 6 $CO_2$ molecules and the net formation of 2 glyceraldehyde 3-phosphate molecules require the consumption of 18 ATPs and 12 NADPHs, generated by the light-requiring processes of photosynthesis.

### Synthesis of Sucrose Using Fixed $CO_2$ Is Completed in the Cytosol

After its formation in the chloroplast stroma, glyceraldehyde 3-phosphate is transported to the cytosol in exchange for phosphate. The final steps of sucrose synthesis (Figure 12-46, *bottom*) occur in the cytosol of leaf cells.

An antiporter transport protein in the chloroplast membrane brings fixed $CO_2$ (as glyceraldehyde 3-phosphate) into the cytosol when the cell is exporting sucrose vigorously. No

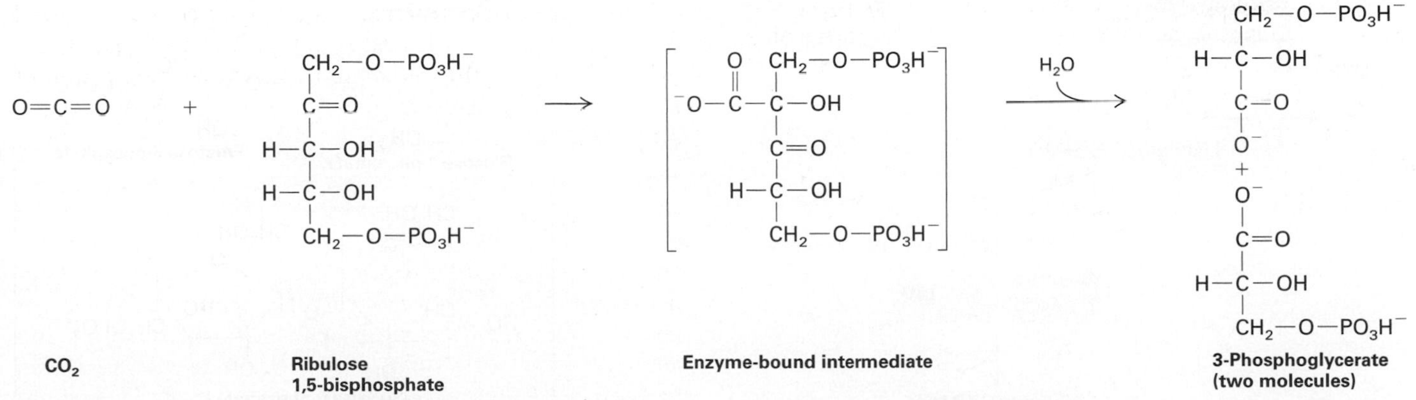

**FIGURE 12-45  The initial reaction of rubisco that fixes $CO_2$ into organic compounds.** In this reaction, catalyzed by ribulose 1,5-bisphosphate carboxylase (rubisco), $CO_2$ condenses with the five-carbon sugar ribulose 1,5-bisphosphate. The products are two molecules of 3-phosphoglycerate.

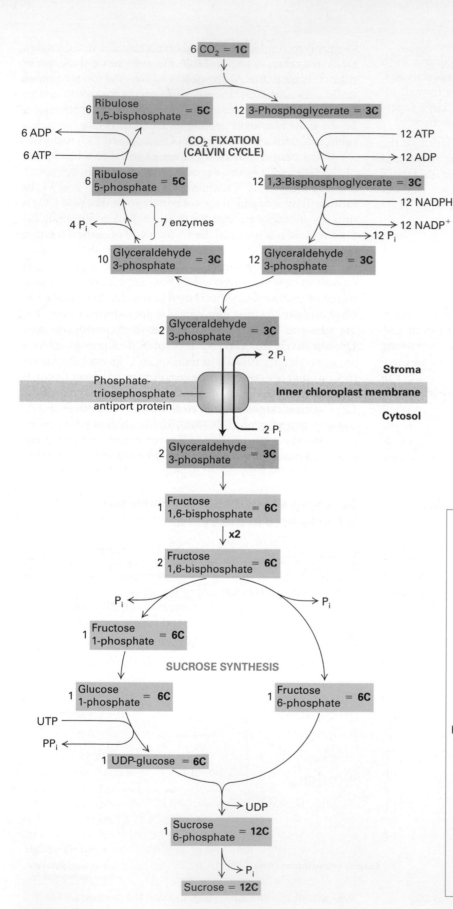

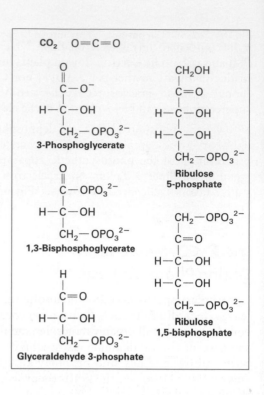

**FIGURE 12-46 The pathway of carbon during photosynthesis.** *(Top)* Six molecules of $CO_2$ are converted into two molecules of glyceraldehyde 3-phosphate. These reactions, which constitute the Calvin cycle, occur in the stroma of the chloroplast. Via the phosphate/triosephosphate antiporter, some glyceraldehyde 3-phosphate is transported to the cytosol in exchange for phosphate. *(Bottom)* In the cytosol, an exergonic series of reactions converts glyceraldehyde 3-phosphate to fructose 1,6-bisphosphate. Two molecules of fructose 1,6-bisphosphate are used to synthesize one of the disaccharide sucrose. Some glyceraldehyde 3-phosphate (not shown here) is also converted to amino acids and fats, compounds essential for plant growth.

fixed $CO_2$ leaves the chloroplast unless phosphate is fed into it to replace the phosphate carried out of the stroma in the form of glyceraldehyde 3-phosphate. During the synthesis of sucrose from glyceraldehyde 3-phosphate, inorganic phosphate groups are released (Figure 12-46, *bottom left*). Thus the synthesis of sucrose facilitates the transport of additional glyceraldehyde 3-phosphate from the chloroplast to the cytosol by providing phosphate for the antiporter. It is worth noting that glyceraldehyde 3-phosphate is a glycolytic intermediate and that the mechanism of the conversion of glyceraldehyde 3-phosphate to hexoses is almost the reverse of that in glycolysis.

The synthesis of starch is more complex. The key monomer substrate used to build large starch polymers is ADP-glucose. This polymerization takes place in the stroma and starch polymers are stored in densely packed crystalline aggregates called granules. The enzymes that generate ADP-glucose from glucose-1 phosphate and ATP are found in both the stroma and the cytosol, indicating that hexoses of various structures in the cytosol are imported into to stroma for starch synthesis.

## Light and Rubisco Activase Stimulate $CO_2$ Fixation

The Calvin cycle enzymes that catalyze $CO_2$ fixation are rapidly inactivated in the dark, thereby conserving ATP that is generated in the dark (for example by the breakdown of starch) for other synthetic reactions, such as lipid and amino acid biosynthesis. One mechanism that contributes to this control is the pH dependence of several Calvin cycle enzymes. Because protons are transported from the stroma into the thylakoid lumen during photoelectron transport (see Figure 12-39), the pH of the stroma increases from ~7 in the dark to ~8 in the light. The increased activity of several Calvin cycle enzymes at the higher pH promotes $CO_2$ fixation in the light.

A stromal protein called *thioredoxin (Tx)* also plays a role in controlling some Calvin cycle enzymes. In the dark, thioredoxin contains a disulfide bond; in the light, electrons are transferred from PSI, via ferredoxin, to thioredoxin, reducing its disulfide bond:

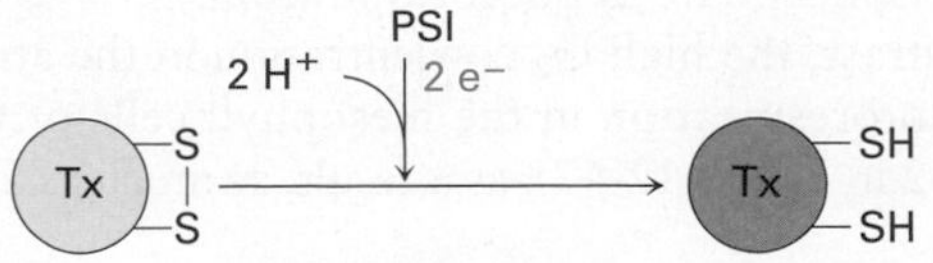

Reduced thioredoxin then activates several Calvin cycle enzymes by reducing their disulfide bonds. In the dark, when thioredoxin becomes reoxidized, these enzymes are reoxidized and so inactivated. Thus these enzymes are sensitive to the redox state of the stroma, which in turn is light sensitive—an elegant mechanism for regulating enzymatic activity by light.

Rubisco is one such light/redox-sensitive enzyme, although its regulation is very complex and not yet fully understood. Rubisco is spontaneously activated in the presence of high $CO_2$ and $Mg^{2+}$ concentrations. The activating reaction entails covalent addition of $CO_2$ to the side-chain amino group of a lysine in the active site, forming a carbamate group that then binds a $Mg^{2+}$ ion required for enzymatic activity. Under normal conditions, however, with ambient levels of $CO_2$, the reaction is slow and usually requires catalysis by *rubisco activase*, a member of the AAA+ family of ATPases. Rubisco activase hydrolyzes ATP and uses the energy released to clear the active site of rubisco so that $CO_2$ can be added to its active site lysine. Rubisco activase also accelerates an activating conformational change in rubisco (inactive-closed to active-opened state). The regulation of rubisco activase by thioredoxin is, at least in part in some species, responsible for rubisco's light/redox sensitivity. Furthermore, rubisco activase's activity is sensitive to the ratio of ATP:ADP. If that ratio is low (relatively high ADP), then the activase will not activate rubisco (and so the cell will expend less of its scarce ATP to fix carbon). Photosynthesis is sensitive to a variety of typical plant stresses—moderate heat, cool temperatures, drought (limited water), high salt, high light intensity, and UV radiation. At least some of these influence $CO_2$ fixation by reducing the activity of rubisco activase and thus rubisco. Inhibition of $CO_2$ fixation reduces consumption of NADPH. Under strong light conditions the excess $NADPH/NADP^+$ ratio can reduce electron flow to $NADP^+$ and increase leakage to $O_2$, resulting in increased ROS formation, which can interfere with a variety of cellular processes. Given the key role of rubisco in controlling energy utilization and carbon flux—both in an individual chloroplast and, in a sense, throughout the entire biosphere—it is not surprising that its activity is tightly regulated.

## Photorespiration Competes with Carbon Fixation and Is Reduced in $C_4$ Plants

As noted above, rubisco catalyzes the incorporation of $CO_2$ into ribulose 1,5-bisphosphate as part of photosynthesis. It can catalyze a second, distinct, and *competing* reaction with the same substrate—ribulose 1,5-bisphosphate—but with $O_2$ in place of $CO_2$ as a second substrate, in a process known as **photorespiration** (Figure 12-47). The products of the second reaction are one molecule of 3-phosphoglycerate and one molecule of the two-carbon compound phosphoglycolate. The carbon-fixing reaction is favored when the ambient $CO_2$ concentration is relatively high, whereas photorespiration is favored when $CO_2$ is low and $O_2$ relatively high. Photorespiration takes place in light, consumes $O_2$, and converts ribulose

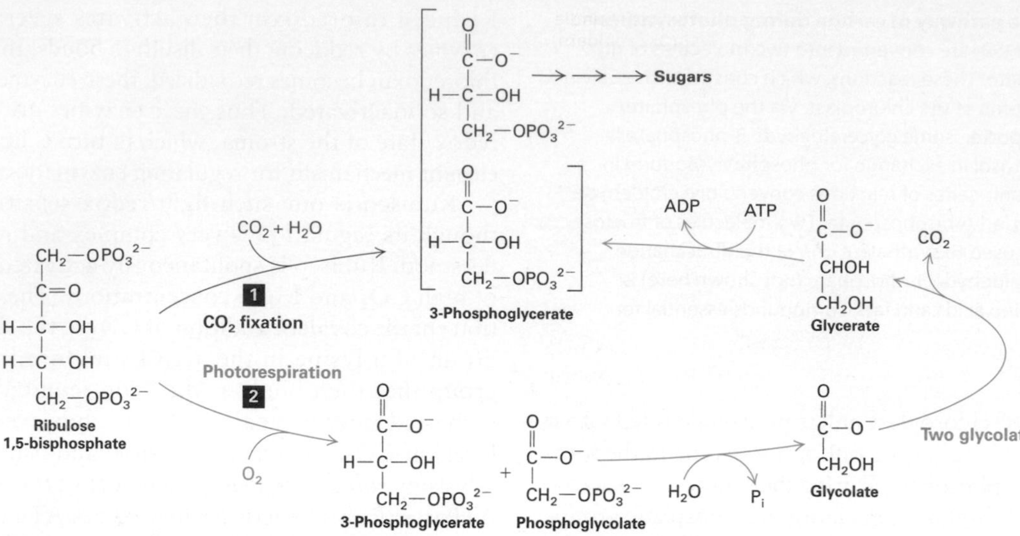

**FIGURE 12-47 $CO_2$ fixation and photorespiration.** These competing pathways are both initiated by ribulose 1,5-bisphosphate carboxylase (rubisco), and both utilize ribulose 1,5-bisphosphate. $CO_2$ fixation, pathway 1, is favored by high $CO_2$ and low $O_2$ pressures; photorespiration, pathway 2, occurs at low $CO_2$ and high $O_2$ pressures (that is, under normal atmospheric conditions). Phosphoglycolate is recycled via a complex set of reactions that take place in peroxisomes and mitochondria, as well as chloroplasts. The net result: for every two molecules of phosphoglycolate formed by photorespiration (four C atoms), one molecule of 3-phosphoglycerate is ultimately formed and recycled and one molecule of $CO_2$ is lost.

1,5-bisphosphate in part to $CO_2$. As Figure 12-47 shows, photorespiration is wasteful to the energy economy of the plant: it consumes ATP and $O_2$, and it generates $CO_2$ without fixing carbon. Indeed, when $CO_2$ is low and $O_2$ is high, much of the $CO_2$ fixed by the Calvin cycle is lost as the result of photorespiration. Recent studies have suggested that this surprising, wasteful alternative reaction catalyzed by rubisco may be a consequence of the inherent difficulty the enzyme has in specifically binding the relatively featureless $CO_2$ molecule and of the ability of both $CO_2$ and $O_2$ to react and form distinct products with the same initial enzyme/ribulose 1,5-bisphosphate intermediate.

Excessive photorespiration could become a problem for plants in a hot, dry environment, because they must keep the gas-exchange pores (stomata) in their leaves closed much of the time to prevent excessive loss of moisture. As a consequence, the $CO_2$ level inside the leaf can fall below the $K_m$ of rubisco for $CO_2$. Under these conditions, the rate of photosynthesis is slowed, photorespiration is greatly favored, and the plant might be in danger of fixing inadequate amounts of $CO_2$. Corn, sugarcane, crabgrass, and other plants that can grow in hot, dry environments have evolved a way to avoid this problem by utilizing a two-step pathway of $CO_2$ fixation in which a $CO_2$-hoarding step precedes the Calvin cycle. The pathway has been named the $C_4$ *pathway* because [$^{14}$C]$CO_2$ labeling showed that the first radioactive molecules formed during photosynthesis in this pathway are four-carbon compounds, such as oxaloacetate and malate, rather than the three-carbon molecules that initiate the Calvin cycle ($C_3$ pathway).

The $C_4$ pathway involves two types of cells: *mesophyll cells*, which are adjacent to the air spaces in the leaf interior, and *bundle sheath cells*, which surround the vascular tissue and are sequestered away from the high oxygen levels to which mesophyll cells are exposed (Figure 12-48a). In the mesophyll cells of $C_4$ plants, phosphoenolpyruvate, a three-carbon molecule derived from pyruvate, reacts with $CO_2$ to generate oxaloacetate, a four-carbon compound (Figure 12-48b). The enzyme that catalyzes this reaction, *phosphoenolpyruvate carboxylase*, is found almost exclusively in $C_4$ plants and unlike rubisco is insensitive to $O_2$. The overall reaction from pyruvate to oxaloacetate involves the hydrolysis of one ATP and has a negative $\Delta G$. Therefore, $CO_2$ fixation will proceed even when the $CO_2$ concentration is low. The oxaloacetate formed in mesophyll cells is reduced to malate, which is transferred by a special transporter to the bundle sheath cells, where the $CO_2$ released by decarboxylation enters the Calvin cycle (Figure 12-48b).

Because of the transport of $CO_2$ from mesophyll cells, the $CO_2$ concentration in the bundle sheath cells of $C_4$ plants is much higher than it is in the normal atmosphere. Bundle sheath cells are also unusual in that they lack PSII and carry out only cyclic electron flow catalyzed by PSI, so no $O_2$ is evolved. The high $CO_2$ and reduced $O_2$ concentrations in the bundle sheath cells favor the fixation of $CO_2$ by rubisco to form 3-phosphoglycerate and inhibit the utilization of ribulose 1,5-bisphosphate in photorespiration.

In contrast, the high $O_2$ concentration in the atmosphere favors photorespiration in the mesophyll cells of $C_3$ plants (pathway 2 in Figure 12-47); as a result, as much as 50 percent

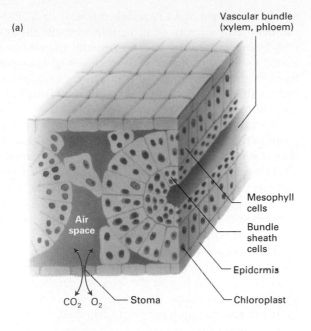

(a)

Vascular bundle
(xylem, phloem)

Mesophyll
cells

Bundle
sheath
cells

Epidermis

Chloroplast

Air
space

$CO_2$  $O_2$    Stoma

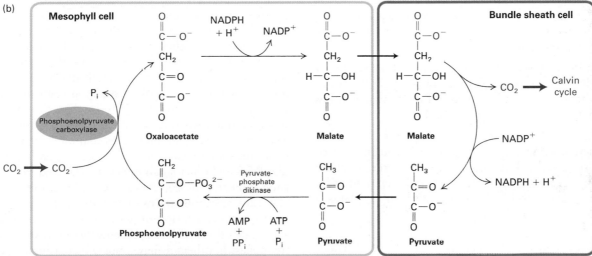

(b)

**FIGURE 12-48  Leaf anatomy of $C_4$ plants and the $C_4$ pathway.**
(a) In $C_4$ plants, bundle sheath cells line the vascular bundles containing the xylem and phloem. Mesophyll cells, which are adjacent to the substomal air spaces, can assimilate $CO_2$ into four-carbon molecules at low ambient $CO_2$ and deliver it to the interior bundle sheath cells. Bundle sheath cells contain abundant chloroplasts and are the sites of photosynthesis and sucrose synthesis. Sucrose is carried to the rest of the plant via the phloem. In $C_3$ plants, which lack bundle sheath cells, the Calvin cycle operates in the mesophyll cells to fix $CO_2$. (b) The key enzyme in the $C_4$ pathway is phosphoenolpyruvate carboxylase, which assimilates $CO_2$ to form oxaloacetate in mesophyll cells. Decarboxylation of malate or other $C_4$ intermediates in bundle sheath cells releases $CO_2$, which enters the standard Calvin cycle (see Figure 12-46, *top*).

of the carbon fixed by rubisco may be reoxidized to $CO_2$ in $C_3$ plants. $C_4$ plants are superior to $C_3$ plants in utilizing the available $CO_2$, because the $C_4$ enzyme phosphoenolpyruvate carboxylase has a higher affinity for $CO_2$ than does rubisco in the Calvin cycle. However, one ATP is converted to one AMP in the cyclic $C_4$ process (to generate phosphoenolpyruvate from pyruvate); thus the overall efficiency of the photosynthetic production of sugars from NADPH and ATP is lower than it is in $C_3$ plants, which use only the Calvin cycle for $CO_2$ fixation. Nonetheless, the net rates of photosynthesis for $C_4$ grasses, such as corn or sugarcane, can be two to three times the rates for otherwise similar $C_3$ grasses, such as wheat, rice, or oats, owing to the elimination of losses from photorespiration.

Of the two carbohydrate products of photosynthesis, starch remains in the mesophyll cells of $C_3$ plants and the bundle sheaf cells in $C_4$ plants. In these cells, starch is subjected to glycolysis, mainly in the dark, forming ATP, NADH, and small molecules that are used as building blocks for the synthesis of amino acids, lipids, and other cellular constituents. Sucrose, in contrast, is exported from the photosynthetic cells and transported throughout the plant.

## KEY CONCEPTS of Section 12.7

### $CO_2$ Metabolism During Photosynthesis

• In the Calvin cycle, $CO_2$ is fixed into organic molecules in a series of reactions that occur in the chloroplast stroma. The initial reaction, catalyzed by rubisco, forms a three-carbon intermediate. Some of the glyceraldehyde 3-phosphate generated in the cycle is transported to the cytosol and converted to sucrose (see Figure 12-46).

• The light-dependent activation of several Calvin cycle enzymes and other mechanisms increases fixation of $CO_2$ in the light. The redox state of the stroma plays a key role in this regulation as does the regulation of the activity of rubisco by rubisco activase.

• In $C_3$ plants, a substantial fraction of the $CO_2$ fixed by the Calvin cycle can be lost as the result of photorespiration, a wasteful reaction catalyzed by rubisco that is favored at low $CO_2$ and high $O_2$ levels (see Figure 12-47).

• In $C_4$ plants, $CO_2$ is fixed initially in the outer mesophyll cells by reaction with phosphoenolpyruvate. The four-carbon molecules so generated are shuttled to the interior bundle sheath cells, where the $CO_2$ is released and then used in the Calvin cycle. The rate of photorespiration in $C_4$ plants is much lower than it is in $C_3$ plants.

## Perspectives for the Future

Although the overall processes of photosynthesis and mitochondrial oxidation are well understood, many important details remain to be uncovered. For example, while increasingly high-resolution structures of complexes and supercomplexes are being determined, many of the mechanistic details underlying the function and regulation of electron transport chains and their associated reactions (proton translocation, oxygen generation, etc.) remain to be established. Moving beyond this static picture of these remarkably complex structures requires additional biophysical analysis of the dynamics underlying their activities. For example, we do not know with certainty the pathway taken by protons during proton pumping in some of the electron transport complexes.

Although the binding-change mechanism for ATP synthesis by the $F_0F_1$ complex is now generally accepted, we do not understand precisely how conformational changes in each β subunit are coupled to the cyclical binding of ADP

and $P_i$, formation of ATP, and then release of ATP. Nor has the detailed pathway of proton movement though the c ring been defined. In addition, many questions remain about the precise mechanism of action of transport proteins in the inner mitochondrial and chloroplast membranes that play key roles in oxidative phosphorylation and photosynthesis.

We now know that release of cytochrome c and other proteins from the intermembrane space of mitochondria into the cytosol plays a major role in triggering apoptosis (Chapter 21). Certain members of the Bcl-2 family of apoptotic proteins and ion channels localized in part to the outer mitochondrial membrane participate in this process. The connections between energy metabolism and mechanisms underlying apoptosis remain to be clearly defined.

The recognition over the past decade of the importance of mitochondrial dynamics (e.g., fusion and fission) to mitochondrial function has set the stage for detailed genetic molecular analysis of these processes. Several of the key players in fusion and fission have been identified, but many additional components have yet to be discovered, and the mechanisms of these complex processes, such as the coordinated fusion of inner membranes with each other and outer membranes with each other, are waiting to be elucidated.

The role of reactive oxygen species (ROS) in cell biology is an active area of research. ROS-mediated cellular stress is now thought to play a role in many diseases and will likely continue to be a major area of research in the coming years. In addition to their role in cellular oxidative stress, ROS can also serve as signaling molecules that alter nuclear gene expression, sometimes called retrograde signaling. It appears that ROS and other small molecules released from the mitochondrion and chloroplast can be used to inform the nucleus about the metabolic status of each organelle and thus permit appropriate regulation of gene expression in response. In some cases this involves compensatory activation of protective genes. In others it may involve increasing or decreasing the production of nuclear-encoded proteins to insure proper organelle functioning. The mechanisms of these signaling pathways, which in some cases involve redox reactions with thiols on signaling molecules, remain to be determined.

As we better understand the mechanisms underlying photosynthesis, particularly the action of rubisco—both its regulation and its influence on photosynthesis and overall chloroplast metabolism—it is possible that we will be able to exploit these insights to improve crop yields to provide abundant and inexpensive food to all who need it.

## Key Terms

| | |
|---|---|
| aerobic oxidation 517 | Calvin cycle 567 |
| ATP synthase 544 | carbon fixation 553 |
| binding-change mechanism 547 | catabolism 520 |
| | $C_4$ pathway 570 |

## Review the Concepts

**1.** The proton-motive force (pmf) is essential for both mitochondrial and chloroplast function. What produces the pmf, and what is its relationship to ATP? The compound 2,4-dinitrophenol (DNP), which was used in diet pills in the 1930s but later shown to have dangerous side effects, allows protons to diffuse across membranes. Why is it dangerous to consume DNP?

**2.** The mitochondrial inner membrane exhibits all of the fundamental characteristics of a typical cell membrane, but it also has several unique characteristics that are closely associated with its role in oxidative phosphorylation. What are these unique characteristics? How does each contribute to the function of the inner membrane?

**3.** Maximal production of ATP from glucose involves the reactions of glycolysis, the citric acid cycle, and the electron transport chain. Which of these reactions requires $O_2$, and why? Which, in certain organisms or physiological conditions, can proceed in the absence of $O_2$?

**4.** Fermentation permits the continued extraction of energy from glucose in the absence of oxygen. If glucose catabolism is anaerobic, why is fermentation necessary for glycolysis to continue?

**5.** Describe the step-by-step process by which electrons from glucose catabolism in the cytoplasm are transferred to the electron transport chain in the mitochondrial inner membrane. In your answer, note whether the electron transfer at each step is direct or indirect.

**6.** Mitochondrial oxidation of fatty acids is a major source of ATP, yet fatty acids can be oxidized elsewhere. What organelle, besides the mitochondrion, can oxidize fatty acids? What is the fundamental difference between oxidation occurring in this organelle and mitochondrial oxidation?

**7.** Each of the cytochromes in the mitochondrion contains prosthetic groups. What is a prosthetic group? Which type of prosthetic group is associated with the cytochromes? What property of the various cytochromes ensures unidirectional electron flow along the electron transport chain?

**8.** The electron transport chain consists of a number of multiprotein complexes, which work in conjunction to pass electrons from an electron carrier, like NADH, to $O_2$. What is the role of these complexes in ATP synthesis? It has been demonstrated that respiration supercomplexes contain all the protein components necessary for respiration. Why is this beneficial for ATP synthesis, and what is one way that the existence of supercomplexes has been demonstrated experimentally? Coenzyme Q (CoQ) is not a protein, but a small, hydrophobic molecule. Why is it important for the functioning of the electron transport chain that CoQ is a hydrophobic molecule?

**9.** It is estimated that each electron pair donated by NADH leads to the synthesis of approximately three ATP molecules, whereas each electron pair donated by $FADH_2$ leads to the synthesis of approximately two ATP molecules. What is the underlying reason for the difference in yield for electrons donated by $FADH_2$ versus NADH?

**10.** Describe the main functions of the different components of ATP synthase enzyme in the mitochondrion. A structurally similar enzyme is responsible for the acidification of lysosomes and endosomes. Given what you know about the mechanism of ATP synthesis, explain how this acidification might occur.

**11.** Much of our understanding of ATP synthase is derived from research on aerobic bacteria. What makes these organisms useful for this research? Where do the reactions of glycolysis, the citric acid cycle, and the electron transport chain occur in these organisms? Where is the pmf generated in aerobic bacteria? What other cellular processes depend on the pmf in these organisms?

**12.** An important function of the mitochondrial inner membrane is to provide a selectively permeable barrier to the movement of water-soluble molecules and thus generate different chemical environments on either side of the membrane. However, many of the substrates and products of oxidative phosphorylation are water soluble and must cross the inner membrane. How does this transport occur?

**13.** The Q cycle plays a major role in the electron transport chain of mitochondria, chloroplasts, and bacteria. What is the function of the Q cycle, and how does it carry out this function? What electron transport components participate in the Q cycle in mitochondria, in purple bacteria, and in chloroplasts?

**14.** True or False: Since ATP is generated in chloroplasts, cells capable of undergoing photosynthesis do not require mitochondria. Explain. Name and describe the idea that explains how mitochondria and chloroplasts are thought to have originated in eukaryotic cells.

15. Write the overall reaction of oxygen-generating photosynthesis. Explain the following statement: the $O_2$ generated by photosynthesis is simply a by-product of the pathway's generation of carbohydrates and ATP.

16. Photosynthesis can be divided into multiple stages. What are the stages of photosynthesis, and where does each occur within the chloroplast? Where is the sucrose produced by photosynthesis generated?

17. The photosystems responsible for absorption of light energy are each composed of two linked components, the reaction center and an antenna complex. What is the pigment composition and role of each component in the process of light absorption? What evidence exists that the pigments found in these components are involved in photosynthesis?

18. Photosynthesis in green and purple bacteria does not produce $O_2$. Why? How can these organisms still use photosynthesis to produce ATP? What molecules serve as electron donors in these organisms?

19. Chloroplasts contain two photosystems. What is the function of each? For linear electron flow, diagram the flow of electrons from photon absorption to NADPH formation. What does the energy stored in the form of NADPH synthesize?

20. The Calvin cycle reactions that fix $CO_2$ do not function in the dark. What are the likely reasons for this? How are these reactions regulated by light?

21. Rubisco, which may be the most abundant protein on earth, plays a key role in the synthesis of carbohydrates in organisms that use photosynthesis. What is rubisco, where is it located, and what function does it serve?

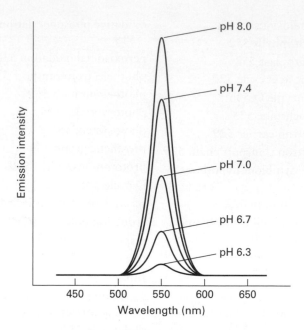

c. After the vesicles were incubated in buffer containing ADP, $P_i$, and $O_2$ for a period of time, addition of dinitrophenol caused an increase in BCECF fluorescence. In contrast, addition of valinomycin produced only a small transient effect. Explain these findings.

d. What result would you expect to see if the source of the mitochondrial membrane were brown-fat mitochondria? Explain.

e. Chloroplasts could also be used as a source of membranes in a similar experiment (as in part a) involving BCECF. In this case, the BCECF would be surrounded by what membrane? How would the fluorescence change upon addition of light, ADP, and $P_i$?

## Analyze the Data

A proton gradient can be analyzed with fluorescent dyes whose emission-intensity profiles depend on pH. One of the most useful dyes for measuring the pH gradient across mitochondrial membranes is the membrane-impermeant, water-soluble fluorophore 2′,7′-bis-(2-carboxyethyl)-5(6)-carboxyfluorescein (BCECF). The effect of pH on the emission intensity of BCECF, excited at 505 nm, is shown in the accompanying figure. In one study, sealed vesicles containing this compound were prepared by mixing unsealed, isolated inner mitochondrial membranes with BCECF; after resealing of the membranes, the vesicles were collected by centrifugation and then resuspended in nonfluorescent medium.

a. When these vesicles were incubated in a physiological buffer containing NADH, ADP, $P_i$, and $O_2$, the fluorescence of BCECF trapped inside gradually decreased in intensity. What does this decrease in fluorescent intensity suggest about this vesicular preparation?

b. How would you expect the concentrations of ADP, $P_i$, and $O_2$ to change during the course of the experiment described in part a? Why?

## References

First Step of Harvesting Energy from Glucose: Glycolysis

Berg, J., J. Tymoczko, and L. Stryer. 2002. *Biochemistry*, 5th ed. W. H. Freeman and Company, chaps. 16 and 17.

Depre, C., M. Rider, and L. Hue. 1998. Mechanisms of control of heart glycolysis. *Eur. J. Biochem.* 258:277–290.

Fersht, A. 1999. *Structure and Mechanism in Protein Science: A Guide to Enzyme Catalysis and Protein Folding*. W. H. Freeman and Company.

Fothergill-Gilmore, L. A., and P. A. Michels. 1993. Evolution of glycolysis. *Prog. Biophys. Mol. Biol.* 59:105–135.

Nelson, D. L., and M. M. Cox. 2000. *Lehninger Principles of Biochemistry*. Worth, chaps. 14–17, 19.

Pilkis, S. J., T. H. Claus, I. J. Kurland, and A. J. Lange. 1995. 6-Phosphofructo-2-kinase/fructose-2,6-bisphosphatase: a metabolic signaling enzyme. *Ann. Rev. Biochem.* 64:799–835.

Mitochondria and the Citric Acid Cycle

Canfield, D. E. 2005. The early history of atmospheric oxygen: homage to Robert M. Garrels. *Annu. Rev. Earth Planet. Sci.* 33:1–36.

Chan, D. C. 2006. Mitochondria: dynamic organelles in disease, aging, and development. *Cell* 125(7):1241–1252.

Eaton, S., K. Bartlett, and M. Pourfarzam. 1996. Mammalian mitochondrial beta-oxidation. *Biochem. J.* 320 (Part 2):345–557.

Guest, J. R., and G. C. Russell. 1992. Complexes and complexities of the citric acid cycle in *Escherichia coli*. *Curr. Top. Cell Reg.* 33:231–247.

Krebs, H. A. 1970. The history of the tricarboxylic acid cycle. *Perspect. Biol. Med.* 14:154–170.

Rasmussen, B., and R. Wolfe. 1999. Regulation of fatty acid oxidation in skeletal muscle. *Ann. Rev. Nutrition* 19:463–484.

Velot, C., M. Mixon, M. Teige, and P. Srere. 1997. Model of a quinary structure between Krebs TCA cycle enzymes: a model for the metabolon. *Biochemistry* 36:14271–14276.

Wanders, R. J., and H. R. Waterham. 2006. Biochemistry of mammalian peroxisomes revisited. *Annu. Rev. Biochem.* 75:295–332.

## The Electron Transport Chain and Generation of the Proton-Motive Force

Acin-Pérez, R., P. Fernandez-Silva, M. L. Peleato, A. Pérez-Martos, and J. A. Enriquez. 2008. Respiratory active mitochondrial supercomplexes. *Mol. Cell* 32:529–539.

Babcock, G. 1999. How oxygen is activated and reduced in respiration. *Proc. Nat'l. Acad. Sci. USA* 96:12971–12973.

Beinert, H., R. Holm, and E. Münck. 1997. Iron-sulfur clusters: nature's modular, multipurpose structures. *Science* 277:653–659.

Brandt, U. 2006. Energy Converting NADH:quinone oxidoreductase (complex I). *Annu. Rev. Biochem.* 75:165–187.

Brandt, U., and B. Trumpower. 1994. The protonmotive Q cycle in mitochondria and bacteria. *Crit. Rev. Biochem. Mol. Biol.* 29:165–197.

Daiber, A. 2010. Redox signaling (cross-talk) from and to mitochondria involves mitochondrial pores and reactive oxygen species. *Biochim. Biophys. Acta* 6-7:897–906.

Darrouzet, E., C. Moser, P. L. Dutton, and F. Daldal. 2001. Large scale domain movement in cytochrome bc1: a new device for electron transfer in proteins. *Trends Biochem. Sci.* 26:445–451.

Dickinson, B. C., D. Srikun, and C. J. Chang. 2010. Mitochondrial-targeted fluorescent probes for reactive oxygen species. *Curr. Opin. Chem. Biol.* 14:50–56.

Efremov, R. G., R. Baradaran, and L. A. Sazanov. 2010. The architecture of respiratory complex I. *Nature* 465:441–445.

Finkel, T. 2011. Signal transduction by reactive oxygen species. *Journal Cell Biology* 194:7–15.

Grigorieff, N. 1999. Structure of the respiratory NADH:ubiquinone oxidoreductase (complex I). *Curr. Opin. Struc. Biol.* 9:476–483.

Hosler, J. P., S. Ferguson Miller, and D. A. Mills. 2006. Energy transduction: proton transfer through the respiratory complexes. *Annu. Rev. Biochem.* 75:165–187.

Hunte, C., V. Zickermann, and U. Brandt. 2010. Functional modules and structural basis of conformational coupling in mitochondrial complex I. *Science* 329:448–451.

Hyde, B. B., G. Twig, and O. S. Shirihai. 2010. Organellar vs cellular control of mitochondrial dynamics. *Semin. Cell Dev. Biol.* 21:575–581.

Koopman, W. J., et al. 2010. Mammalian mitochondrial complex I: biogenesis, regulation, and reactive oxygen species generation. *Antioxid. Redox Signal.* 12:1431–1470.

Michel, H., J. Behr, A. Harrenga, and A. Kannt. 1998. Cytochrome c oxidase. *Ann. Rev. Biophys. Biomol. Struc.* 27:329–356.

Mitchell, P. 1979. Keilin's respiratory chain concept and its chemiosmotic consequences. *Science* 206:1148–1159. (Nobel Prize Lecture.)

Murphy, M. P. 2009. How mitochondria produce reactive oxygen species. *Biochem. J.* 417:1–13.

Ramirez, B. E., B. Malmström, J. R. Winkler, and H. B. Gray. 1995. The currents of life: the terminal electron-transfer complex of respiration. *Proc. Nat'l. Acad. Sci. USA* 92:11949–11951.

Ruitenberg, M., et al. 2002. Reduction of cytochrome c oxidase by a second electron leads to proton translocation. *Nature* 417:99–102.

Saraste, M. 1999. Oxidative phosphorylation at the fin de siècle. *Science* 283:1488–1492.

Schafer, E., et al. 2006. Architecture of active mammalian respiratory chain supercomplexes. *J. Biol. Chem.* 281(22):15370–15375.

Schultz, B., and S. Chan. 2001. Structures and proton-pumping strategies of mitochondrial respiratory enzymes. *Ann. Rev. Biophys. Biomol. Struc.* 30:23–65.

Sheeran, F. L., and S. Pepe. 2006. Energy deficiency in the failing heart: linking increased reactive oxygen species and disruption of oxidative phosphorylation rate. *Biochim. Biophys. Acta* 1757(5–6):543–552.

Tsukihara, T., et al. 1996. The whole structure of the 13-subunit oxidized cytochrome c oxidase at 2.8 Å. *Science* 272:1136–1144.

Walker, J. E. 1995. Determination of the structures of respiratory enzyme complexes from mammalian mitochondria. *Biochim. Biophys. Acta* 1271:221–227.

Wallace, D. C. 2005. A mitochondrial paradigm of metabolic and degenerative diseases, aging, and cancer: a dawn for evolutionary medicine. *Annu. Rev. Genet.* 39:359–407.

Xia, D., et al. 1997. Crystal structure of the cytochrome bc1 complex from bovine heart mitochondria. *Science* 277:60–66.

Zaslavsky, D., and R. Gennis. 2000. Proton pumping by cytochrome oxidase: progress and postulates. *Biochim. Biophys. Acta* 1458:164–179.

Zhang, M., E. Mileykovskaya, and W. Dowhan. 2005. Cardiolipin is essential for organization of complexes III and IV into a supercomplex in intact yeast mitochondria. *J. Biol. Chem.* 280(33):29403–29408.

Zhang, Z., et al. 1998. Electron transfer by domain movement in cytochrome bc1. *Nature* 392:677–684.

## Harnessing the Proton-Motive Force to Synthesize ATP

Aksimentiev, A., I. A. Balabin, R. H. Fillingame, and K. Schulten. 2004. Insights into the molecular mechanism of rotation in the $F_0$ sector of ATP synthase. *Biophys. J.* 86(3):1332–1344.

Bianchet, M. A., J. Hullihen, P. Pedersen, and M. Amzel. 1998. The 2.8 Å structure of rat liver F1-ATPase: configuration of a critical intermediate in ATP synthesis/hydrolysis. *Proc. Nat'l. Acad. Sci. USA* 95:11065–11070.

Boyer, P. D. 1997. The ATP synthase—a splendid molecular machine. *Ann. Rev. Biochem.* 66:717–749.

Capaldi, R., and R. Aggeler. 2002. Mechanism of the $F_0F_1$-type ATP synthase—a biological rotary motor. *Trends Biochem. Sci.* 27:154–160.

Elston, T., H. Wang, and G. Oster. 1998. Energy transduction in ATP synthase. *Nature* 391:510–512.

Hinkle, P. C. 2005. P/O ratios of mitochondrial oxidative phosphorylation. *Biochim. Biophys. Acta* 1706(1–2):1–11.

Junge, W., S. Hendrik, and S. Engelbrecht. 2009. Torque generation and elastic power transmission in the rotary $F_0F_1$-ATPase. *Nature* 459:364–370.

Kinosita, K., et al. 1998. F1-ATPase: a rotary motor made of a single molecule. *Cell* 93:21–24.

Klingenberg, M., and S. Huang. 1999. Structure and function of the uncoupling protein from brown adipose tissue. *Biochim. Biophys. Acta* 1415:271–296.

Nury, H., et al. 2006. Relations between structure and function of the mitochondrial ADP/ATP carrier. *Annu. Rev. Biochem.* 75:713–741.

Tsunoda, S., R. Aggeler, M. Yoshida, and R. Capaldi. 2001. Rotation of the c subunit oligomer in fully functional $F_0F_1$ ATP synthase. *Proc. Nat'l. Acad. Sci. USA* **98**:898–902.

Vercesi, A. E., et al. 2006. Plant uncoupling mitochondrial proteins. *Annu. Rev. Plant Biol.* **57**:383–404.

von Ballmoos, C., A. Wiedenmann, and P. Dimroth. 2009. Essentials for ATP synthesis by $F_1F_0$ ATP synthases. *Annu. Rev. Biochem.* **78**:649–672.

Yasuda, R., et al. 2001. Resolution of distinct rotational substeps by submillisecond kinetic analysis of F1-ATPase. *Nature* **410**:898–904.

## Photosynthesis and Light-Absorbing Pigments

Ben-Shem, A., F. Frolow, and N. Nelson. 2003. Crystal structure of plant photosystem I. *Nature* **426**(6967):630–635.

Blankenship, R. E. 2002. *Molecular Mechanisms of Photosynthesis*. Blackwell.

Deisenhofer, J., and J. R. Norris, eds. 1993. *The Photosynthetic Reaction Center*, vols. 1 and 2. Academic Press.

McDermott, G., et al. 1995. Crystal structure of an integral membrane light-harvesting complex from photosynthetic bacteria. *Nature* **364**:517.

Nelson, N., and C. F. Yocum. 2006. Structure and function of photosystems I and II. *Annu. Rev. Plant Biol.* **57**:521–565.

Prince, R. 1996. Photosynthesis: the Z-scheme revisited. *Trends Biochem. Sci.* **21**:121–122.

Wollman, F. A. 2001. State transitions reveal the dynamics and flexibility of the photosynthetic apparatus. *EMBO J.* **20**:3623–3630.

## Molecular Analysis of Photosystems

Allen, J. F. 2002. Photosynthesis of ATP—electrons, proton pumps, rotors, and poise. *Cell* **110**:273–276.

Amunts, A., H. Toporik, A. Borovikova, and N. Nelson. 2010. Structure determination and improved model of plant photosystem I. *J. Biol. Chem.* **285**:3478–3486.

Aro, E. M., I. Virgin, and B. Andersson. 1993. Photoinhibition of photosystem II: Inactivation, protein damage, and turnover. *Biochim. Biophys. Acta* **1143**:113–134.

Deisenhofer, J., and H. Michel. 1989. The photosynthetic reaction center from the purple bacterium *Rhodopseudomonas viridis*. *Science* **245**:1463–1473. (Nobel Prize Lecture.)

Deisenhofer, J., and H. Michel. 1991. Structures of bacterial photosynthetic reaction centers. *Ann. Rev. Cell Biol.* **7**:1–23.

Dekker, J. P., and E. J. Boekema. 2005. Supramolecular organization of thylakoid membrane proteins in green plants. *Biochim. Biophys. Acta* **1706**(1–2):12–39.

Finazzi, G. 2005. The central role of the green alga *Chlamydomonas reinhardtii* in revealing the mechanism of state transitions *J. Exp. Bot.* **56**(411):383–388.

Guskov, A., et al. 2010. Recent progress in the crystallographic studies of photosystem II. *Chemphyschem.* **11**(6):1160–1171.

Haldrup, A., P. Jensen, C. Lunde, and H. Scheller. 2001. Balance of power: a view of the mechanism of photosynthetic state transitions. *Trends Plant Sci.* **6**:301–305.

Hankamer, B., J. Barber, and E. Boekema. 1997. Structure and membrane organization of photosystem II from green plants. *Ann. Rev. Plant Physiol. Plant Mol. Biol.* **48**:641–672.

Heathcote, P., P. Fyfe, and M. Jones. 2002. Reaction centres: the structure and evolution of biological solar power. *Trends Biochem. Sci.* **27**:79–87.

Horton, P., A. Ruban, and R. Walters. 1996. Regulation of light harvesting in green plants. *Ann. Rev. Plant Physiol. Plant Mol. Biol.* **47**:655–684.

Iwai, M., et al. 2010. Isolation of the elusive supercomplex that drives cyclic electron flow in photosynthesis. *Nature* **464**:1210–1213.

Joliot, P., and A. Joliot. 2005. Quantification of cyclic and linear flows in plants. *Proc. Natl. Acad. Sci. USA* **102**(13):4913–4918.

Jordan, P., et al. 2001. Three-dimensional structure of cyanobacterial photosystem I at 2.5 Å resolution. *Nature* **411**:909–917.

Kühlbrandt, W. 2001. Chlorophylls galore. *Nature* **411**:896–898.

Martin, J. L., and M. H. Vos. 1992. Femtosecond biology. *Ann. Rev. Biophys. Biomol. Struc.* **21**:199–222.

Penner-Hahn, J. 1998. Structural characterization of the Mn site in the photosynthetic oxygen-evolving complex. *Struc. Bonding* **90**:1–36.

Tommos, C., and G. Babcock. 1998. Oxygen production in nature: a light-driven metalloradical enzyme process. *Acc. Chem. Res.* **31**:18–25.

## $CO_2$ Metabolism During Photosynthesis

Buchanan, B. B. 1991. Regulation of $CO_2$ assimilation in oxygenic photosynthesis: the ferredoxin/thioredoxin system. Perspective on its discovery, present status, and future development. *Arch. Biochem. Biophys.* **288**:1–9.

Gutteridge, S., and J. Pierce. 2006. A unified theory for the basis of the limitations of the primary reaction of photosynthetic $CO_2$ fixation: was Dr. Pangloss right? *Proc. Natl. Acad. Sci. USA* **103**:7203–7204.

Portis, A. 1992. Regulation of ribulose 1,5-bisphosphate carboxylase/oxygenase activity. *Ann. Rev. Plant Physiol. Plant Mol. Biol.* **43**:415–437.

Rawsthorne, S. 1992. Towards an understanding of $C_3$-$C_4$ photosynthesis. *Essays Biochem.* **27**:135–146.

Rokka, A., I. Zhang, and E.-M. Aro. 2001. Rubisco activase: an enzyme with a temperature-dependent dual function? *Plant J.* **25**:463–472.

Sage, R., and J. Colemana. 2001. Effects of low atmospheric $CO_2$ on plants: more than a thing of the past. *Trends Plant Sci.* **6**:18–24.

Schneider, G., Y. Lindqvist, and C. I. Branden. 1992. Rubisco: structure and mechanism. *Ann. Rev. Biophys. Biomol. Struc.* **21**:119–153.

Tcherkez, G. G., G. D. Farquhar, and T. J. Andrews. 2006. Despite slow catalysis and confused substrate specificity, all ribulose bisphosphate carboxylases may be nearly perfectly optimized. *Proc. Natl. Acad. Sci. USA* **103**(19):7246–7251.

Wolosiuk, R. A., M. A. Ballicora, and K. Hagelin. 1993. The reductive pentose phosphate cycle for photosynthetic $CO_2$ assimilation: enzyme modulation. *FASEB J.* **7**:622–637.

# Moving Proteins into Membranes and Organelles

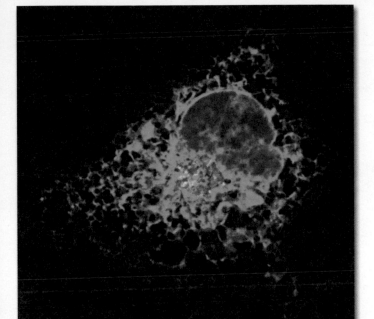

Fluorescence micrograph of a cultured mammalian (COS-7) cell showing the distribution of endoplasmic reticulum (green), Golgi apparatus (red), and nucleus (blue). Newly synthesized secretory proteins are first targeted to the ER, where they are folded and modified before being exported to the Golgi for sorting to downstream destinations. [Courtesy of Jennifer Lippincott-Schwartz and Prasanna Satpute]

A typical mammalian cell contains up to 10,000 different kinds of proteins; a yeast cell, about 5000. The vast majority of these proteins are synthesized by cytosolic ribosomes, and many remain within the cytosol (Chapter 4). However, as many as half of the different kinds of proteins produced in a typical cell are delivered to one or another of the various membrane-bounded organelles within the cell or to the cell surface. For example, many receptor proteins and transporter proteins must be delivered to the plasma membrane, some water-soluble enzymes such as RNA and DNA polymerases must be targeted to the nucleus, and components of the extracellular matrix as well as digestive enzymes and polypeptide signaling molecules must be directed to the cell surface for secretion from the cell. These and all the other proteins produced by a cell must reach their correct locations for the cell to function properly.

The delivery of newly synthesized proteins to their proper cellular destinations, usually referred to as *protein targeting* or *protein sorting*, encompasses two very different kinds of processes: signal-based targeting and vesicle-based trafficking. The first general process involves targeting of a newly synthesized protein from the cytoplasm to an intracellular organelle. Targeting can occur during translation or soon after synthesis of the protein is complete. For membrane proteins, targeting leads to insertion of the protein into the lipid bilayer of the membrane, whereas for water-soluble proteins, targeting leads to translocation of the entire protein across the membrane into the aqueous interior of the organelle. Proteins are sorted to the endoplasmic reticulum (ER), mitochondria, chloroplasts, peroxisomes, and nucleus by this general process (Figure 13-1).

The second general sorting process is known as the **secretory pathway,** and involves transport of proteins from the ER to their final destination within membrane-enclosed vesicles. For many proteins, including those that make up the extracellular matrix, the final destination is the outside of the cell (hence the name); integral membrane proteins are also

## OUTLINE

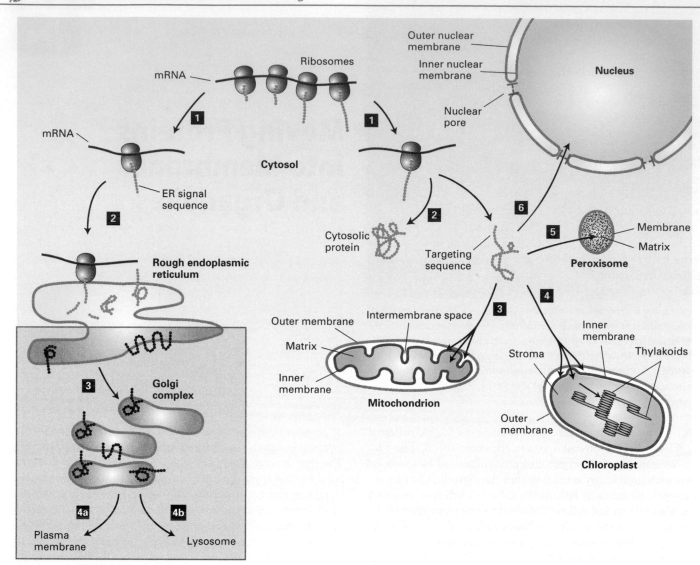

**SECRETORY PATHWAY**

**FIGURE 13-1 Overview of major protein-sorting pathways in eukaryotes.** All nuclear-encoded mRNAs are translated on cytosolic ribosomes. *Right (nonsecretory pathways):* Synthesis of proteins lacking an ER signal sequence is completed on free ribosomes (step **1**). Those proteins that contain no targeting sequence are released into the cytosol and remain there (step **2**). Proteins with an organelle-specific targeting sequence (pink) first are released into the cytosol (step **2**) but then are imported into mitochondria, chloroplasts, peroxisomes, or the nucleus (steps **3**–**6**). Mitochondrial and chloroplast proteins typically pass through the outer and inner membranes to enter the matrix or stromal space, respectively. Other proteins are sorted to other subcompartments of these organelles by additional sorting steps. Nuclear proteins enter and exit through visible pores in the nuclear envelope. *Left (secretory pathway):* Ribosomes synthesizing nascent proteins in the secretory pathway are directed to the rough endoplasmic reticulum (ER) by an ER signal sequence (pink; steps **1**, **2**). After translation is completed on the ER, these proteins can move via transport vesicles to the Golgi complex (step **3**). Further sorting delivers proteins either to the plasma membrane or to lysosomes (steps **4a**, **4b**). The vesicle-based processes underlying the secretory pathway (steps **3**, **4**, *shaded box*) are discussed in Chapter 14.

transported to the Golgi, lysosome, and plasma membrane by this process. The secretory pathway begins in the ER; thus all proteins slated to enter the secretory pathway are initially targeted to this organelle.

Targeting to the ER generally involves *nascent* proteins still in the process of being synthesized on a ribosome. Newly made proteins are thus extruded from the ribosome directly into the ER membrane. Once translocated across the ER membrane, proteins are assembled into their native conformation by protein-folding catalysts present in the lumen of the ER. Indeed, the ER is the location where about one-third of the proteins in a typical cell fold into their native conformations, and most of the resident ER proteins either directly or indirectly contribute to the folding process. As part of the folding process, proteins also undergo specific post-translational modifications in the ER. These processes are monitored carefully, and only after their folding and assembly is complete are proteins permitted to be transported out of the ER to other destinations. Proteins whose final destination is the Golgi, lysosome, plasma membrane, or cell exterior are transported along the secretory pathway by the action of small vesicles that bud from the membrane of one organelle and then fuse with the membrane of another (see Figure 13-1, *shaded box*). We discuss vesicle-based protein trafficking in the next chapter because mechanistically it differs significantly from non-vesicle-based protein targeting to intracellular organelles.

In this chapter, we examine how proteins are targeted to five intracellular organelles: ER, mitochondria, chloroplast, peroxisome, and nucleus. Two features of this protein-targeting process initially were quite baffling: how a given protein could be directed to only one specific membrane, and how relatively large hydrophilic protein molecules could be translocated across a hydrophobic membrane without disrupting the bilayer as a barrier to ions and small molecules. Using a combination of biochemical purification methods and genetic screens for identifying mutants unable to execute particular translocation steps, cell biologists have identified many of the cellular components required for translocation across each of the different intracellular membranes. In addition, many of the major translocation processes in the cell have been reconstituted using the purified protein components incorporated into artificial lipid bilayers. Such in vitro systems can be freely manipulated experimentally.

These studies have shown that, despite some variations, the same basic mechanisms govern protein sorting to all the various intracellular organelles. We now know, for instance, that the information to target a protein to a particular organelle destination is encoded within the amino acid sequence of the protein itself, usually within sequences of about 20 amino acids, known generically as **signal sequences** (see Figure 13-1); these are also called *uptake-targeting sequences* or *signal peptides*. Such targeting sequences usually occur at the N-terminus of a protein and are thus the first part of a protein to be synthesized. More rarely, targeting sequences can occur at either the C-terminus or within the interior of a protein sequence. Each organelle carries a set of receptor proteins that bind only to specific kinds of signal sequences, thus ensuring that the information encoded in a signal sequence governs the specificity of targeting. Once a protein containing a signal sequence has interacted with the corresponding receptor, the protein chain is transferred to some kind of *translocation channel* that allows the protein to pass into or through the membrane bilayer. The unidirectional transfer of a protein into an organelle, without sliding back out into the cytoplasm, is usually achieved by coupling translocation to an energetically favorable process such as hydrolysis of GTP or ATP. Some proteins are subsequently sorted further to reach a subcompartment within the target organelle; such sorting depends on yet other signal sequences and other receptor proteins. Finally, signal sequences often are removed from the mature protein by specific proteases once translocation across the membrane is completed.

For each of the protein-targeting events discussed in this chapter, we will seek to answer four fundamental questions:

1. What is the nature of the *signal sequence,* and what distinguishes it from other types of signal sequences?

2. What is the *receptor* for the signal sequence?

3. What is the structure of the *translocation channel* that allows transfer of proteins across the membrane bilayer? In particular, is the channel so narrow that proteins can pass through only in an unfolded state, or will it accommodate folded protein domains?

4. What is the source of *energy* that drives unidirectional transfer across the membrane?

In the first part of the chapter, we cover targeting of proteins to the ER, including the post-translational modifications that occur to proteins as they enter the secretory pathway. Targeting of proteins to the ER is the best-understood example of protein targeting, and will serve as an exemplar of the process in general. We then describe targeting of proteins to mitochondria, chloroplasts, and peroxisomes. Finally, we cover the transport of proteins into and out of the nucleus through nuclear pores.

## 13.1 Targeting Proteins to and Across the ER Membrane

All eukaryotic cells have an endoplasmic reticulum (ER). The ER is a large, convoluted organelle made up of tubules and flattened sacs, whose membrane is continuous with the membrane of the nucleus. The ER membrane is where cellular lipids are synthesized (Chapter 10), and the ER is where most membrane proteins are assembled, including those of the plasma membrane and the membrane of the lysosomes, ER, and Golgi. In addition, all soluble proteins that will eventually be secreted from the cell—as well as those destined for the lumen of the ER, Golgi, or lysosomes—are initially delivered to the ER lumen (see Figure 13-1). Since the ER plays such an important role in protein secretion, we refer to the pathway of protein trafficking that flows through the ER as the "secretory pathway." For simplicity, we will refer to all proteins initially targeted to the ER as "secretory proteins," *but* keep in mind that not all proteins that are targeted to the ER are actually secreted from the cell.

In this first section, we discuss how proteins are initially identified as secretory proteins, and how such proteins are translocated across the ER membrane. We deal first with soluble proteins—those that pass all the way through the ER membrane, into the lumen. In the next section, we discuss integral membrane proteins, which are inserted into the ER membrane.

### Pulse-Labeling Experiments with Purified ER Membranes Demonstrated That Secreted Proteins Cross the ER Membrane

Although all cells secrete a variety of proteins (e.g., extracellular matrix proteins), certain types of cells are specialized for secretion of large amounts of specific proteins. Pancreatic

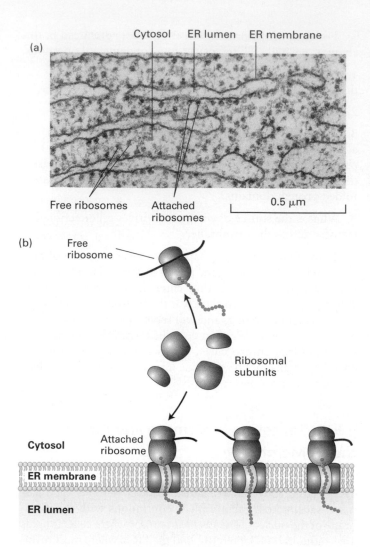

(a)

Cytosol  ER lumen  ER membrane

Free ribosomes   Attached ribosomes    0.5 μm

(b)

Free ribosome

Ribosomal subunits

Cytosol

Attached ribosome

ER membrane

ER lumen

**FIGURE 13-2 Structure of the rough ER.** (a) Electron micrograph of ribosomes attached to the rough ER in a pancreatic acinar cell. Most of the proteins synthesized by this type of cell are to be secreted and are formed on membrane-attached ribosomes. A few membrane-unattached (free) ribosomes are evident; presumably, these are synthesizing cytosolic or other nonsecretory proteins. (b) Schematic representation of protein synthesis on the ER. Note that membrane-bound and free cytosolic ribosomes are identical. Membrane-bound ribosomes get recruited to the endoplasmic reticulum during protein synthesis of a polypeptide containing an ER signal sequence. [Part (a) courtesy of G. Palade.]

acinar cells, for instance, synthesize large quantities of several digestive enzymes that are secreted into ductules that lead to the intestine. Because such secretory cells contain the organelles of the secretory pathway (e.g., ER and Golgi) in great abundance, they have been widely used in studying this pathway, including the initial steps that occur at the ER membrane.

The sequence of events that occurs immediately after the synthesis of a secretory protein were first elucidated by pulse-labeling experiments with pancreatic acinar cells. In such cells, radioactively labeled amino acids are incorporated into secretory proteins as they are synthesized on ribosomes that are bound to the surface of the ER. The portion of the ER

that receives proteins entering the secretory pathway is known as the *rough ER* because it is so densely studded with ribosomes that its surface appears morphologically distinct from other ER membranes (Figure 13-2). From these experiments, it became clear that during or immediately after their synthesis on the ribosome, secretory proteins translocate across the ER membrane into the lumen of the ER.

To delineate the steps in the translocation process, it was necessary to isolate the ER from the rest of the cell. Isolation of intact ER with its delicate lacelike structure and interconnectedness with other organelles is not feasible. However, scientists discovered that after cells are homogenized, the rough ER breaks up into small closed vesicles with ribosomes on the outside, termed *rough microsomes*, which retain most of the biochemical properties of the ER, including the capability of protein translocation. The experiments depicted in Figure 13-3, in which microsomes isolated from pulse-labeled cells are treated with a protease, demonstrate that although secretory proteins are synthesized on ribosomes bound to the cytosolic face of the ER membrane, the polypeptides produced by these ribosomes end up within the lumen of ER vesicles. Experiments such as this raised the question of how polypeptides are recognized as secretory proteins shortly after their synthesis begins and how a nascent secretory protein is threaded across the ER membrane.

## A Hydrophobic N-Terminal Signal Sequence Targets Nascent Secretory Proteins to the ER

After synthesis of a secretory protein begins on free ribosomes in the cytosol, a 16- to 30-residue ER signal sequence in the nascent protein directs the ribosome to the ER membrane and initiates translocation of the growing polypeptide across the ER membrane (see Figure 13-1, *left*). An ER signal sequence typically is located at the N-terminus of the protein, the first part of the protein to be synthesized. The signal sequences of different secretory proteins all contain one or more positively charged amino acids adjacent to a continuous stretch of 6–12 hydrophobic residues (known as the hydrophobic core), but otherwise they have little in common. For most secretory proteins, the signal sequence is cleaved from the protein while it is still elongating on the ribosome; thus signal sequences are usually not present in the "mature" proteins found in cells.

The hydrophobic core of ER signal sequences is essential for their function. For instance, the specific deletion of several of the hydrophobic amino acids from a signal sequence or the introduction of charged amino acids into the hydrophobic core by mutation can abolish the ability of the N-terminus of a protein to function as a signal sequence. As a consequence, the modified protein remains in the cytosol, unable to cross the ER membrane into the lumen. Conversely, signal sequences can be added to normally cytosolic proteins using recombinant DNA techniques. Provided the added sequence is sufficiently long and hydrophobic, such a modified cytosolic protein acquires the ability to be translocated to the ER lumen. Thus the hydrophobic residues in the core of ER signal sequences form a binding site that is critical

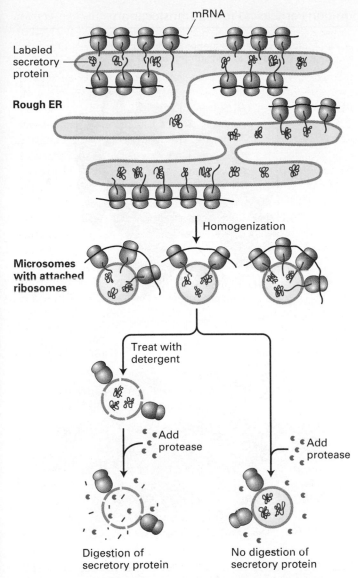

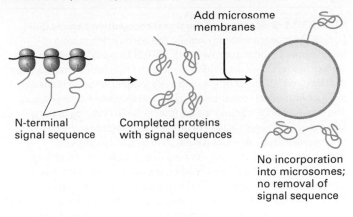

Add microsome membranes

N-terminal signal sequence

Completed proteins with signal sequences

No incorporation into microsomes; no removal of signal sequence

(b) Cell-free protein synthesis; microsomes present

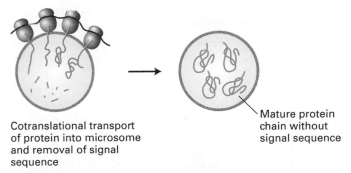

Cotranslational transport of protein into microsome and removal of signal sequence

Mature protein chain without signal sequence

**EXPERIMENTAL FIGURE 13-4 Translation and translocation occur simultaneously.** Cell-free experiments demonstrate that translocation of secretory proteins into microsomes is coupled to translation. Treatment of microsomes with EDTA, which chelates $Mg^{2+}$ ions, strips them of associated ribosomes, allowing isolation of ribosome-free microsomes, which are equivalent to ER membranes (see Figure 13-3). Protein synthesis is carried out in a cell-free system containing functional ribosomes, tRNAs, ATP, GTP, and cytosolic enzymes to which mRNA encoding a secretory protein is added. The secretory protein is synthesized in the absence of microsomes (a) but is translocated across the vesicle membrane and loses its signal sequence (resulting in a decrease in molecular weight) only if microsomes are present during protein synthesis (b).

**EXPERIMENTAL FIGURE 13-3 Secretory proteins enter the ER.** Labeling experiments demonstrate that secretory proteins are localized to the ER lumen shortly after synthesis. Cells are incubated for a brief time with radiolabeled amino acids so that only newly synthesized proteins become labeled. The cells then are homogenized, fracturing the plasma membrane and shearing the rough ER into small vesicles called *microsomes*. Because they have bound ribosomes, microsomes have a much greater buoyant density than other membranous organelles and can be separated from them by a combination of differential and sucrose density-gradient centrifugation (Chapter 9). The purified microsomes are treated with a protease in the presence or absence of a detergent. The labeled secretory proteins associated with the microsomes are digested by the protease only if the permeability barrier of the microsomal membrane is first destroyed by treatment with detergent. This finding indicates that the newly made proteins are inside the microsomes, equivalent to the lumen of the rough ER.

for the interaction of signal sequences with the machinery responsible for targeting the protein to the ER membrane.

Biochemical studies utilizing a cell-free protein-synthesizing system, mRNA encoding a secretory protein, and microsomes stripped of their own bound ribosomes have elucidated the

function and fate of ER signal sequences. Initial experiments with this system demonstrated that a typical secretory protein is incorporated into microsomes and has its signal sequence removed only if the microsomes are present during protein synthesis. If microsomes are added to the system after protein synthesis is completed, no protein transport into the microsomes occurs (Figure 13-4). Subsequent experiments were designed to determine the precise stage of protein synthesis at which microsomes must be present in order for translocation to occur. In these experiments, microsomes were added to the reaction mixtures at different times after protein synthesis had begun. These experiments showed that microsomes must be added before the first 70 or so amino acids are translated in order for the completed secretory protein to be localized in the microsomal lumen. At this point, the first 40 amino acids or so protrude from

**FIGURE 13-5 Structure of the signal-recognition particle (SRP).**
(a) Signal-sequence binding domain: The bacterial Ffh protein is homologous to the portion of P54 that binds ER signal sequences. This surface model shows the binding domain in Ffh, which contains a large cleft lined with hydrophobic amino acids (purple) whose side chains interact with signal sequences. (b) GTP- and receptor-binding domain: The structure of GTP bound to FtsY (the bacterial homolog of the α subunit of SRP receptor) and Ffh proteins illustrates how the interaction between these proteins is controlled by GTP binding and hydrolysis. Ffh and FtsY each can bind to one molecule of GTP, and when Ffh and FtsY bind to each other, the two bound molecules of GTP fit in the interface between the protein subunits and stabilize the dimer. Assembly of the semisymmetrical dimer allows formation of two active sites for the hydrolysis of both bound GTP molecules. Hydrolysis to GDP destabilizes the interface, causing disassembly of the dimer.
[Part (a) adapted from R. J. Keenan et al., 1998, *Cell* **94:**181. Part (b) adapted from P. J. Focia et al., 2004, *Science* **303:**373.]

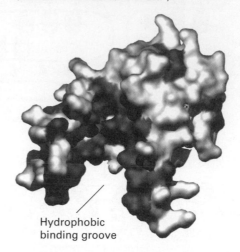

(a) Ffh signal-sequence-binding domain (related to P54 subunit of SRP)

Hydrophobic binding groove

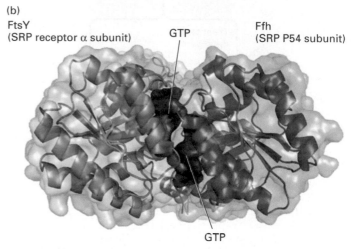

(b)

FtsY (SRP receptor α subunit)   GTP   Ffh (SRP P54 subunit)

GTP

the ribosome, including the signal sequence that later will be cleaved off, and the next 30 or so amino acids are still buried within a channel in the ribosome (see Figure 4-26). Thus the transport of most secretory proteins into the ER lumen begins while the incompletely synthesized (nascent) protein is still bound to the ribosome, a process referred to as **cotranslational translocation.**

## Cotranslational Translocation Is Initiated by Two GTP-Hydrolyzing Proteins

Since secretory proteins are synthesized in association with the ER membrane but not with any other cellular membrane, a signal-sequence recognition mechanism must target them there. The two key components in this targeting are the **signal-recognition particle (SRP)** and its receptor, located in the ER membrane. The SRP is a cytosolic ribonucleoprotein particle that transiently binds to both the ER signal sequence in a nascent protein as well as the large ribosomal subunit, forming a large complex; SRP then targets the nascent protein-ribosome complex to the ER membrane by binding to the SRP receptor on the membrane.

The SRP is made up of six proteins bound to a 300-nucleotide RNA, which acts as a scaffold for the hexamer. One of the SRP proteins (P54) can be chemically cross-linked to ER signal sequences, showing that this is the subunit that binds to the signal sequence in a nascent secretory protein. A region of P54 known as the M domain, containing many methionine and other amino acid residues with hydrophobic side chains, contains a cleft whose inner surface is lined by hydrophobic side chains (Figure 13-5a). The hydrophobic core of the signal peptide binds to this cleft via hydrophobic interactions. Other polypeptides in the SRP interact with the ribosome or are required for protein translocation into the ER lumen.

The SRP brings the nascent chain-ribosome complex to the ER membrane by docking with the SRP receptor, an integral protein of the ER membrane made up of two subunits: an α subunit and a smaller β subunit. Interaction of the SRP/nascent chain/ribosome complex with the SRP receptor is strengthened when both the P54 subunit of SRP and the α subunit of the SRP receptor are bound to GTP. The structure of the P54 subunit of SRP and the SRP receptor α subunit (FtsY), from the archaebacteria *Thermus aquaticus,* provides insight into how a cycle of GTP binding and hydrolysis can drive the binding and dissociation of these proteins. Figure 13-5b shows that the P54 and FtsY each bound to a single molecule of GTP come together to form a pseudo-symmetrical heterodimer. Neither subunit alone contains a complete active site for the hydrolysis of GTP, but when the two proteins come together, they form two complete active sites that are capable of hydrolyzing both bound GTP molecules.

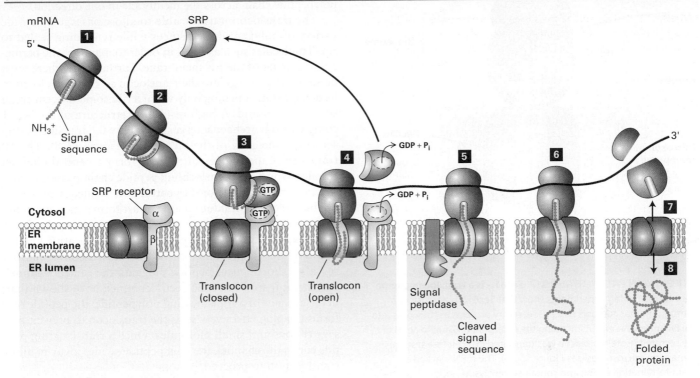

**FIGURE 13-6 Cotranslational translocation.** Steps **1**, **2**: Once the ER signal sequence emerges from the ribosome, it is bound by a signal-recognition particle (SRP). Step **3**: The SRP delivers the ribosome/nascent polypeptide complex to the SRP receptor in the ER membrane. This interaction is strengthened by binding of GTP to both the SRP and its receptor. Step **4**: Transfer of the ribosome/nascent polypeptide to the translocon leads to opening of this translocation channel and insertion of the signal sequence and adjacent segment of the growing polypeptide into the central pore. Both the SRP and SRP receptor, once dissociated from the translocon, hydrolyze their bound GTP and then are ready to initiate the insertion of another polypeptide chain. Step **5**: As the polypeptide chain elongates, it passes through the translocon channel into the ER lumen, where the signal sequence is cleaved by signal peptidase and is rapidly degraded. Step **6**: The peptide chain continues to elongate as the mRNA is translated toward the 3' end. Because the ribosome is attached to the translocon, the growing chain is extruded through the translocon into the ER lumen. Steps **7**, **8**: Once translation is complete, the ribosome is released, the remainder of the protein is drawn into the ER lumen, the translocon closes, and the protein assumes its native folded conformation.

Figure 13-6 summarizes our current understanding of secretory protein synthesis and the role of the SRP and its receptor in this process. Hydrolysis of the bound GTP accompanies disassembly of the SRP and SRP receptor and, in a manner that is not understood, initiates transfer of the nascent chain and ribosome to a site on the ER membrane, where translocation can take place. After dissociating from each other, SRP and its receptor each release their bound GDP, SRP recycles back to the cytosol, and both are ready to initiate another round of interaction between ribosomes synthesizing nascent secretory proteins and the ER membrane.

## Passage of Growing Polypeptides Through the Translocon Is Driven by Translation

Once the SRP and its receptor have targeted a ribosome synthesizing a secretory protein to the ER membrane, the ribosome and nascent chain are rapidly transferred to the **translocon**, a complex of proteins that forms a channel embedded within the ER membrane. As translation continues, the elongating chain passes directly from the large ribosomal subunit into the central pore of the translocon. The 60S ribosomal subunit is aligned with the pore of the translocon in such a way that the growing chain is never exposed to the cytoplasm and is prevented from folding until it reaches the ER lumen (see Figure 13-6).

The translocon was first identified by mutations in the yeast gene encoding Sec61α, which caused a block in the translocation of secretory proteins into the lumen of the ER. Subsequently, three proteins called the *Sec61 complex* were found to form the mammalian translocon: Sec61α, an integral membrane protein with 10 membrane-spanning α helices, and two smaller proteins, termed Sec61β and Sec61γ. Chemical cross-linking experiments—in which amino acid side chains from a nascent secretory protein can become covalently attached to the Sec61α subunit—demonstrated that

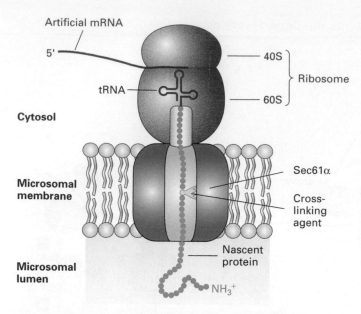

**Artificial mRNA**

5'

40S
Ribosome
tRNA
60S

**Cytosol**

**Microsomal membrane**

Sec61α

Cross-linking agent

Nascent protein

**Microsomal lumen**

NH₃⁺

**EXPERIMENTAL FIGURE 13-7 Sec61α is a translocon component.** Cross-linking experiments show that Sec61α is a translocon component that contacts nascent secretory proteins as they pass into the ER lumen. An mRNA encoding the N-terminal 70 amino acids of the secreted protein prolactin was translated in a cell-free system containing microsomes (see Figure 13-4b). The mRNA lacked a chain-termination codon and contained one lysine codon, near the middle of the sequence. The reactions contained a chemically modified lysyl-tRNA in which a light-activated cross-linking reagent was attached to the lysine side chain. Although the entire mRNA was translated, the completed polypeptide could not be released from the ribosome without a chain-termination codon and thus became "stuck" crossing the ER membrane. The reaction mixtures then were exposed to an intense light, causing the nascent chain to become covalently bound to whatever proteins were near it in the translocon. When the experiment was performed using microsomes from mammalian cells, the nascent chain became covalently linked to Sec61α. Different versions of the prolactin mRNA were created so that the modified lysine residue would be placed at different distances from the ribosome; cross-linking to Sec61α was observed only when the modified lysine was positioned within the translocation channel. [Adapted from T. A. Rapoport, 1992, *Science* **258**:931, and D. Görlich and T. A. Rapoport, 1993, *Cell* **75**:615.]

the translocating polypeptide chain comes into contact with the Sec61α protein, confirming its identity as the translocon pore (Figure 13-7).

When microsomes in the cell-free translocation system were replaced with reconstituted phospholipid vesicles containing only the SRP receptor and Sec61 complex, nascent secretory protein was translocated from its SRP/ribosome complex into the vesicles. This finding indicates that the SRP receptor and the Sec61 complex are the only ER-membrane proteins that are absolutely required for translocation. Because neither of these can hydrolyze ATP or otherwise provide energy to drive the translocation, the energy derived from chain

elongation at the ribosome appears to be sufficient to push the polypeptide chain across the membrane in one direction.

The translocon must be able to allow passage of a wide variety of polypeptide sequences while remaining sealed to small molecules such as ATP, in order to maintain the permeability barrier of the ER membrane. Furthermore, there must be some way to regulate the translocon so that it is closed in its default state, opening only when a ribosome-nascent chain complex is bound. A high-resolution structure of the Sec61 complex from archaebacteria shows how the translocon preserves the integrity of the membrane (Figure 13-8). The 10 transmembrane helices of Sec61α form a central channel through which the translocating peptide chain passes. A constriction in the middle of the central pore is lined with hydrophobic isoleucine residues that in effect form a gasket around the translocating peptide. The structural model of the Sec61 complex, which was isolated without a translocating peptide and therefore is presumed to be in a closed conformation, reveals a short helical peptide plugging the central channel. Biochemical studies of the Sec61 complex have shown that, in the absence of a translocating polypeptide, the peptide that forms the plug effectively seals the translocon to prevent passage of ions and small molecules. Once a translocating peptide enters the channel, the plug peptide swings away to allow translocation to proceed.

As the growing polypeptide chain enters the lumen of the ER, the signal sequence is cleaved by *signal peptidase*, which is a transmembrane ER protein associated with the translocon (see Figure 13-6, step **5**). Signal peptidase recognizes a sequence on the C-terminal side of the hydrophobic core of the signal peptide and cleaves the chain specifically at this sequence once it has emerged into the luminal space of the ER. After the signal sequence has been cleaved, the growing polypeptide moves through the translocon into the ER lumen. The translocon remains open until translation is completed and the entire polypeptide chain has moved into the ER lumen. After translocation is complete, the plug helix returns to the pore to reseal the translocon channel.

## ATP Hydrolysis Powers Post-translational Translocation of Some Secretory Proteins in Yeast

In most eukaryotes, secretory proteins enter the ER by cotranslational translocation. In yeast, however, some secretory proteins enter the ER lumen after translation has been completed. In such *post-translational translocation*, the translocating protein passes through the same Sec61 translocon that is used in cotranslational translocation. However, the SRP and SRP receptor are not involved in post-translational translocation, and in such cases a direct interaction between the translocon and the signal sequence of the completed protein appears to be sufficient for targeting to the ER membrane. In addition, the driving force for unidirectional translocation across the ER membrane is provided by an additional protein complex known as the *Sec63 complex*

**EXPERIMENTAL FIGURE 13-8 Structure of a bacterial Sec61 complex.** The structure of the detergent-solubilized Sec61 complex from the archaebacterium *M. jannaschii* (also known as the SecY complex) was determined by x-ray crystallography. (a) A side view shows the hourglass-shaped channel through the center of the pore. A ring of isoleucine residues at the constricted waist of the pore forms a gasket that keeps the channel sealed to small molecules even as a translocating polypeptide passes through the channel. When no translocating peptide is present, the channel is closed by a short helical plug (red). This plug moves out of the channel during translocation. In this view the front half of protein has been removed to better show the pore. (b) A view looking through the center of the channel shows a region (on the left side) where helices may separate, allowing lateral passage of a hydrophobic transmembrane domain into the lipid bilayer. [Adapted from A. R. Osborne et al., 2005, *Ann. Rev. Cell Dev. Biol.* **21**:529.]

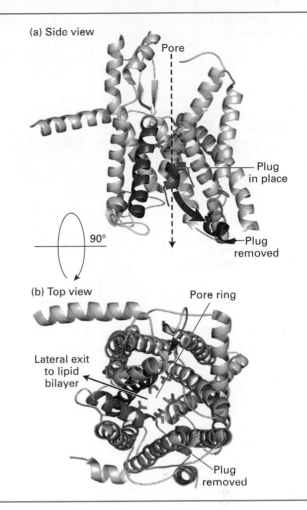

(a) Side view

Pore

Plug in place

Plug removed

90°

(b) Top view

Pore ring

Lateral exit to lipid bilayer

Plug removed

and a member of the Hsc70 family of **molecular chaperones** known as *BiP* (see Chapter 3 for further discussion of molecular chaperones). The tetrameric Sec63 complex is embedded in the ER membrane in the vicinity of the translocon, whereas BiP is within the ER lumen. Like other members of the Hsc70 family, BiP has a peptide-binding domain and an ATPase domain. These chaperones bind and stabilize unfolded or partially folded proteins (see Figure 3-16).

The current model for post-translational translocation of a protein into the ER is outlined in Figure 13-9. Once the N-terminal segment of the protein enters the ER lumen, signal peptidase cleaves the signal sequence just as in cotranslational translocation (step **1**). Interaction of BiP·ATP with the luminal portion of the Sec63 complex causes hydrolysis of the bound ATP, producing a conformational change in BiP that promotes its binding to an exposed polypeptide chain (step **2**). Since the Sec63 complex is located near the translocon, BiP is thus activated at sites where nascent polypeptides can enter the ER. Certain experiments suggest that, in the absence of binding to BiP, an unfolded polypeptide slides back and forth within the translocon channel. Such random sliding motions rarely result in the entire polypeptide's crossing the ER membrane. Binding of a molecule of BiP·ADP to the luminal portion of the polypeptide prevents backsliding of the polypeptide out of the ER. As further inward random sliding exposes more of the polypeptide on the luminal side of the ER membrane, successive binding of BiP·ADP molecules to the polypeptide chain acts as a ratchet, ultimately drawing the entire polypeptide into the ER within a few seconds (steps **3** and **4**). On a slower time scale, the BiP molecules spontaneously exchange their bound ADP for ATP, leading to release of the polypeptide, which can then fold into its native conformation (steps **5** and **6**). The recycled BiP·ATP then is ready for another interaction with Sec63. BiP and the Sec63 complex are also required for cotranslational translocation. The details of

their role in this process are not well understood, but they are thought to act at an early stage of the process, such as threading the signal peptide into the pore of the translocon.

The overall reaction carried out by BiP is an important example of how the chemical energy released by the hydrolysis of ATP can power the mechanical movement of a protein across a membrane. Bacterial cells also use an ATP-driven process for translocating completed proteins across the plasma membrane—in this case to be released from the cell. In bacteria the driving force for translocation comes from a cytosolic ATPase known as the SecA protein. SecA binds to the cytoplasmic side of the translocon and hydrolyzes cytosolic ATP. By a mechanism that resembles the needle on a sewing machine, the SecA protein pushes segments of the polypeptide through the membrane in a mechanical cycle coupled to the hydrolysis of ATP.

As we will see, translocation of proteins across other eukaryotic organelle membranes, such as those of mitochondria and chloroplasts, also typically occurs by post-translational translocation. This explains why ribosomes are not found bound to these other organelles, as they are to the rough ER.

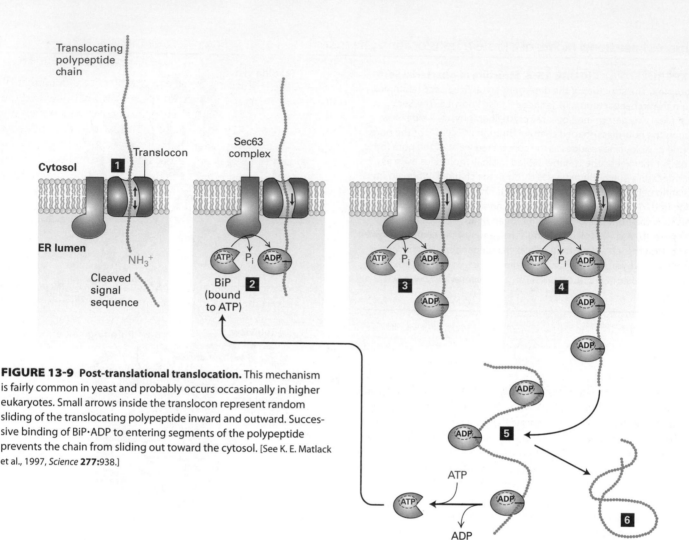

**FIGURE 13-9 Post-translational translocation.** This mechanism is fairly common in yeast and probably occurs occasionally in higher eukaryotes. Small arrows inside the translocon represent random sliding of the translocating polypeptide inward and outward. Successive binding of BiP·ADP to entering segments of the polypeptide prevents the chain from sliding out toward the cytosol. [See K. E. Matlack et al., 1997, *Science* **277:**938.]

## KEY CONCEPTS of Section 13.1

### Targeting Proteins to and Across the ER Membrane

• Synthesis of secreted proteins, integral plasma-membrane proteins, and proteins destined for the ER, Golgi complex, or lysosome begins on cytosolic ribosomes, which become attached to the membrane of the ER, forming the rough ER (see Figure 13-1, *left*).

• The ER signal sequence on a nascent secretory protein consists of a segment of hydrophobic amino acids located at the N-terminus.

• In cotranslational translocation, the signal-recognition particle (SRP) first recognizes and binds the ER signal sequence on a nascent secretory protein and in turn is bound by an SRP receptor on the ER membrane, thereby targeting the ribosome/nascent chain complex to the ER.

• The SRP and SRP receptor then mediate insertion of the nascent secretory protein into the translocon (Sec61 complex). Hydrolysis of two molecules of GTP by the SRP and its receptor cause the dissociation of SRP (see Figures 13-5 and 13-6). As the ribosome attached to the translocon continues

translation, the unfolded protein chain is extruded into the ER lumen. No additional energy is required for translocation.

• The translocon contains a central channel lined with hydrophobic residues that allows transit of an unfolded protein chain while remaining sealed to ions and small hydrophilic molecules. In addition, the channel is gated so that it is open only when a polypeptide is being translocated.

• In post-translational translocation, a completed secretory protein is targeted to the ER membrane by interaction of the signal sequence with the translocon. The polypeptide chain is then pulled into the ER by a ratcheting mechanism that requires ATP hydrolysis by the chaperone BiP, which stabilizes the entering polypeptide (see Figure 13-9). In bacteria, the driving force for post-translational translocation comes from SecA, a cytosolic ATPase that pushes polypeptides through the translocon channel.

• In both cotranslational and post-translational translocation, a signal peptidase in the ER membrane cleaves the ER signal sequence from a secretory protein soon after the N-terminus enters the lumen.

## 13.2 Insertion of Membrane Proteins into the ER

In previous chapters we have encountered many of the vast array of integral (transmembrane) proteins that are present throughout the cell. Each such protein has a unique orientation with respect to the membrane's phospholipid bilayer. Integral membrane proteins located in the ER, Golgi, and lysosomes, and also proteins in the plasma membrane, which are all synthesized on the rough ER, remain embedded in the membrane in their unique orientation as they move to their final destinations along the same pathway followed by soluble secretory proteins (see Figure 13-1, left). During this transport, the orientation of a membrane protein is preserved; that is, the same segments of the protein always face the cytosol, whereas other segments always face in the opposite direction. Thus the final orientation of these membrane proteins is established during their biosynthesis on the ER membrane. In this section, we first see how integral proteins can interact with membranes and then examine how several types of sequences, known collectively as **topogenic sequences,** direct the membrane insertion and orientation of various classes of integral proteins. These processes occur via modifications of the basic mechanism used to translocate soluble secretory proteins across the ER membrane.

### Several Topological Classes of Integral Membrane Proteins Are Synthesized on the ER

The *topology* of a membrane protein refers to the number of times that its polypeptide chain spans the membrane and the orientation of these membrane-spanning segments within the membrane. The key elements of a protein that determine its topology are membrane-spanning segments themselves, which usually are α helices containing 20–25 hydrophobic amino acids that contribute to energetically favorable interactions within the hydrophobic interior of the phospholipid bilayer.

Most integral membrane proteins fall into one of the five topological classes illustrated in Figure 13-10. Topological classes I, II, III, and tail-anchored proteins comprise *single-pass* proteins, which have only one membrane-spanning α-helical segment. Type I proteins have a cleaved N-terminal ER signal sequence and are anchored in the membrane with their hydrophilic N-terminal region on the luminal face (also known as the exoplasmic face) and their hydrophilic C-terminal region on the cytosolic face. Type II proteins do not contain a cleavable ER signal sequence and are oriented with their hydrophilic N-terminal region on the cytosolic face and their hydrophilic C-terminal region on the exoplasmic face (i.e., opposite to type I proteins). Type III proteins have a hydrophobic membrane-spanning segment at their N-terminus and thus have the same orientation as type I proteins but do not contain

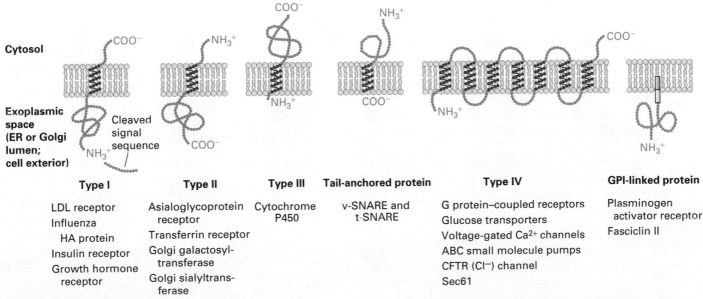

**FIGURE 13-10 ER membrane proteins.** Five topological classes of integral membrane proteins are synthesized on the rough ER as well as a sixth type tethered to the membrane by a phospholipid anchor. Membrane proteins are classified by their orientation in the membrane and the types of signals they contain to direct them there. For the integral membrane proteins, hydrophobic segments of the protein chain form α helices embedded in the membrane bilayer; the regions outside the membrane are hydrophilic and fold into various conformations. All type IV proteins have multiple transmembrane α helices. The type IV topology depicted here corresponds to that of G protein–coupled receptors: seven α helices, the N-terminus on the exoplasmic side of the membrane, and the C-terminus on the cytosolic side. Other type IV proteins may have a different number of helices and various orientations of the N-terminus and C-terminus. [See E. Hartmann et al., 1989, *Proc. Nat'l Acad. Sci. USA* **86**:5786, and C. A. Brown and S. D. Black, 1989, *J. Biol. Chem.* **264**:4442.]

a cleavable signal sequence. Finally, tail-anchored proteins have a hydrophobic segment at their C-terminus that spans the membrane. These different topologies reflect distinct mechanisms used by the cell to establish the membrane orientation of transmembrane segments, as discussed in the next section.

The proteins forming topological class IV contain two or more membrane-spanning segments and are sometimes called *multipass* proteins. For example, many of the membrane transport proteins discussed in Chapter 11 and the numerous G protein–coupled receptors covered in Chapter 15 belong to this class. A final type of membrane protein lacks a hydrophobic membrane-spanning segment altogether; instead, these proteins are linked to an amphipathic phospholipid anchor that is embedded in the membrane (Figure 13-10, *right*).

## Internal Stop-Transfer and Signal-Anchor Sequences Determine Topology of Single-Pass Proteins

We begin our discussion of how membrane protein topology is determined with the membrane insertion of integral proteins that contain a single, hydrophobic membrane-spanning segment. Two sequences are involved in targeting and orienting type I proteins in the ER membrane, whereas type II and type III proteins contain a single, internal topogenic sequence. As we will see, there are three main types of topogenic

sequences that are used to direct proteins to the ER membrane and to orient them within it. We have already introduced one, the N-terminal signal sequence. The other two, introduced here, are internal sequences known as *stop-transfer anchor sequences* and *signal-anchor sequences*. Unlike signal sequences, the two types of internal topogenic sequences end up in the mature protein as membrane-spanning segments. However, the two types of internal topogenic sequences differ in their final orientation in the membrane.

**Type I Proteins** All type I transmembrane proteins possess an N-terminal signal sequence that targets them to the ER as well as an internal hydrophobic sequence that becomes the membrane-spanning α helix. The N-terminal signal sequence on a nascent type I protein, like that of a soluble secretory protein, initiates cotranslational translocation of the protein through the combined action of the SRP and SRP receptor. Once the N-terminus of the growing polypeptide enters the lumen of the ER, the signal sequence is cleaved, and the growing chain continues to be extruded across the ER membrane. However, unlike the case with soluble secretory proteins, when the sequence of approximately 22 hydrophobic amino acids that will become a transmembrane domain of the nascent chain enters the translocon, it stops transfer of the protein through the channel (Figure 13-11). The Sec61 complex is then able to open like a clamshell, allowing the hydrophobic

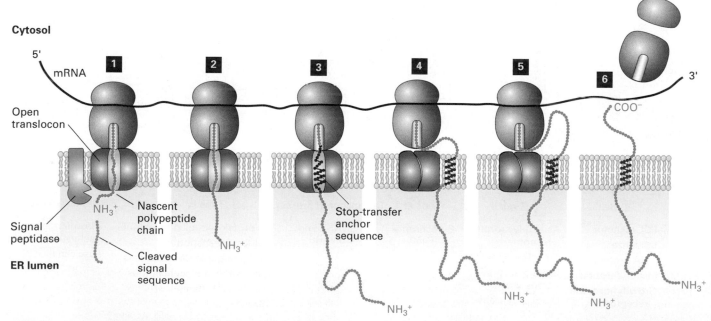

**FIGURE 13-11 Positioning type I single-pass proteins.** Step **1**: After the ribosome/nascent chain complex becomes associated with a translocon in the ER membrane, the N-terminal signal sequence is cleaved. This process occurs by the same mechanism as the one for soluble secretory proteins (see Figure 13-6). Steps **2**, **3**: The chain is elongated until the hydrophobic stop-transfer anchor sequence is synthesized and enters the translocon, where it prevents the nascent chain from extruding farther into the ER lumen. Step **4**: The stop-transfer anchor sequence moves laterally between the translocon subunits and becomes anchored in the phospholipid bilayer. At this time, the translocon probably closes. Step **5**: As synthesis continues, the elongating chain may loop out into the cytosol through the small space between the ribosome and translocon. Step **6**: When synthesis is complete, the ribosomal subunits are released into the cytosol, leaving the protein free to diffuse in the membrane. [See H. Do et al., 1996, *Cell* **85:**369, and W. Mothes et al., 1997, *Cell* **89:**523.]

transmembrane segment of the translocating peptide to move laterally between the protein domains constituting the translocon wall (see Figure 13-8). When the peptide exits the translocon in this manner, it becomes anchored in the phospholipid bilayer of the membrane. Because of the dual function of such a sequence to both stop passage of the polypeptide chain through the translocon and to become a hydrophobic transmembrane segment in the membrane bilayer, it is called a *stop-transfer anchor sequence*.

Once translocation is interrupted, translation continues at the ribosome, which is still anchored to the now unoccupied and closed translocon. As the C-terminus of the protein chain is synthesized, it loops out on the cytosolic side of the membrane. When translation is completed, the ribosome is released from the translocon and the C-terminus of the newly synthesized type I protein remains in the cytosol.

Support for this mechanism has come from studies in which cDNAs encoding various mutant receptors for human growth hormone (HGH) are expressed in cultured mammalian cells. The wild-type HGH receptor, a typical type I protein, is transported normally to the plasma membrane. However, a mutant receptor that has charged residues inserted into the

single α-helical membrane-spanning segment or that is missing most of this segment is translocated entirely into the ER lumen and is eventually secreted from the cell as a soluble protein. These kinds of experiments establish that the hydrophobic membrane-spanning α helix of the HGH receptor and of other type I proteins functions both as a stop-transfer sequence and as a membrane anchor that prevents the C-terminus of the protein from crossing the ER membrane.

**Type II and Type III Proteins** Unlike type I proteins, type II and type III proteins lack a cleavable N-terminal ER signal sequence. Instead, both possess a single internal hydrophobic *signal-anchor sequence* that functions as both an ER signal sequence and a membrane-anchor sequence. Recall that type II and type III proteins have opposite orientations in the membrane (see Figure 13-10); this difference depends on the orientation that their respective signal-anchor sequences assume within the translocon. The internal signal-anchor sequence in type II proteins directs insertion of the nascent chain into the ER membrane so that the N-terminus of the chain faces the cytosol, using the same SRP-dependent mechanism described for signal sequences (Figure 13-12a). However, the internal

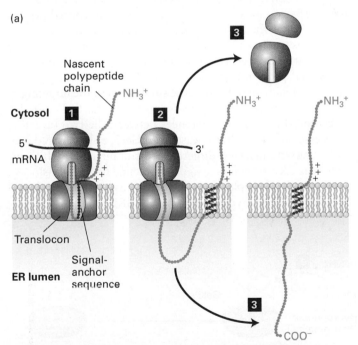

(a)

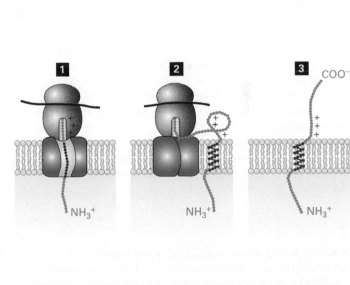

(b)

**FIGURE 13-12 Positioning type II and type III single-pass proteins.** (a) Type II proteins. Step **1**: After the internal signal-anchor sequence is synthesized on a cytosolic ribosome, it is bound by an SRP (not shown), which directs the ribosome/nascent chain complex to the ER membrane. This is similar to targeting of soluble secretory proteins except that the hydrophobic signal sequence is not located at the N-terminus and is not subsequently cleaved. The nascent chain becomes oriented in the translocon with its N-terminal portion toward the cytosol. This orientation is believed to be mediated by the positively charged residues shown N-terminal to the signal-anchor sequence. Step **2**: As the chain is elongated and extruded into the lumen, the internal signal-anchor moves laterally out of the translocon

and anchors the chain in the phospholipid bilayer. Step **3**: Once protein synthesis is completed, the C-terminus of the polypeptide is released into the lumen, and the ribosomal subunits are released into the cytosol. (b) Type III proteins. Step **1**: Assembly is by a similar pathway to that of type II proteins except that positively charged residues on the C-terminal side of the signal-anchor sequence cause the transmembrane segment to be oriented within the translocon with its C-terminal portion oriented to the cytosol and the N-terminal side of the protein in the ER lumen. Steps **2**, **3**: Chain elongation of the C-terminal portion of the protein is completed in the cytosol, and ribosomal subunits are released. [See M. Spiess and H. F. Lodish, 1986, *Cell* **44**:177, and H. Do et al., 1996, *Cell* **85**:369.]

signal-anchor sequence is *not* cleaved and moves laterally between the protein domains of the translocon wall into the phospholipid bilayer, where it functions as a membrane anchor. As elongation continues, the C-terminal region of the growing chain is extruded through the translocon into the ER lumen by cotranslational translocation.

In the case of type III proteins, the signal-anchor sequence, which is located near the N-terminus, inserts the nascent chain into the ER membrane with its N-terminus facing the lumen, in the opposite orientation of the signal anchor in type II proteins. The signal-anchor sequence of type III proteins also functions like a stop-transfer sequence and prevents further extrusion of the nascent chain into the ER lumen (Figure 13-12b). Continued elongation of the chain C-terminal to the signal-anchor/stop-transfer sequence proceeds as it does for type I proteins, with the hydrophobic sequence moving laterally between the translocon subunits to anchor the polypeptide in the ER membrane (see Figure 13-11).

One of the features of signal-anchor sequences that appears to determine their insertion orientation is a high density of positively charged amino acids adjacent to one end of the hydrophobic segment. These positively charged residues tend to remain on the cytosolic side of the membrane, not traversing the membrane into the ER lumen. Thus the position of the charged residues dictates the orientation of the signal-anchor sequence within the translocon as well as whether the rest of the polypeptide chain continues to pass into the ER lumen: type II proteins tend to have positively charged residues on the N-terminal side of their signal-anchor sequence, orienting the N-terminus in the cytosol and allowing passage of the C-terminal side into the ER (Figure 13-12a), whereas type III proteins tend to have positively charged residues on the C-terminal side of their signal-anchor sequence, inserting the N-terminus into the translocon and restricting the C-terminus to the cytosol (Figure 13-12b).

A striking experimental demonstration of the importance of the flanking charge in determining membrane orientation is provided by neuraminidase, a type II protein in the surface coat of influenza virus. Three arginine residues are located just N-terminal to the internal signal-anchor sequence in neuraminidase. Mutation of these three positively charged residues to negatively charged glutamate residues causes neuraminidase to acquire the reverse orientation. Similar experiments have shown that other proteins, with either type II or type III orientation, can be made to "flip" their orientation in the ER membrane by mutating charged residues that flank the internal signal-anchor segment.

**Tail-Anchored Proteins** For all topological classes of proteins we have considered so far, membrane insertion begins when SRP recognizes a hydrophobic topogenic peptide as it emerges from the ribosome. Recognition of tail-anchored proteins, which have a single hydrophobic topogenic sequence at the C-terminus, present a unique challenge since the hydrophobic C-terminus only becomes available for recognition after completion of translation and the protein has been released from the ribosome. Insertion of tail-anchored proteins into the ER membrane does not employ SRP, SRP receptor, or the translocon, but instead depends on a pathway dedicated for this purpose as depicted in Figure 13-13. Targeting of tail-anchored proteins involves an ATPase known as Get3, which binds to the C-terminal hydrophobic segment of tail-anchored proteins. The complex of Get3 bound to a tail-anchored protein is recruited to the ER by a dimeric integral membrane receptor known as Get1/Get2, and the tail-anchored protein is released from Get3 for insertion into the membrane. Insertion of tail-anchored protein into the ER membranes by Get proteins shares fundamental mechanistic similarities to the targeting of signal sequence–bearing proteins to the ER by SRP and SRP receptor. Two major differences between the two targeting processes are that, after release from Get3, tail-anchored proteins may be inserted directly into the membrane bilayer, whereas SRP transfers a signal sequence to the translocon, and that Get3 couples targeting and transfer of tail-anchored proteins to ATP hydrolysis, whereas SRP couples secretory protein targeting to GTP hydrolysis.

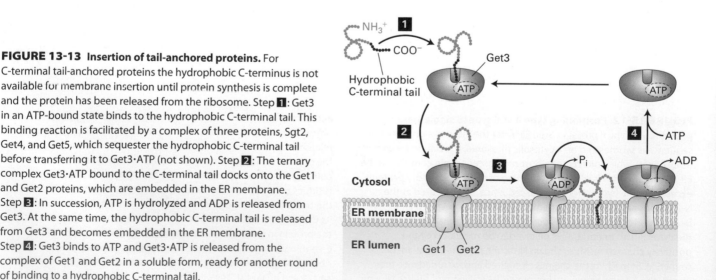

**FIGURE 13-13 Insertion of tail-anchored proteins.** For C-terminal tail-anchored proteins the hydrophobic C-terminus is not available for membrane insertion until protein synthesis is complete and the protein has been released from the ribosome. Step **1**: Get3 in an ATP-bound state binds to the hydrophobic C-terminal tail. This binding reaction is facilitated by a complex of three proteins, Sgt2, Get4, and Get5, which sequester the hydrophobic C-terminal tail before transferring it to Get3·ATP (not shown). Step **2**: The ternary complex Get3·ATP bound to the C-terminal tail docks onto the Get1 and Get2 proteins, which are embedded in the ER membrane. Step **3**: In succession, ATP is hydrolyzed and ADP is released from Get3. At the same time, the hydrophobic C-terminal tail is released from Get3 and becomes embedded in the ER membrane. Step **4**: Get3 binds to ATP and Get3·ATP is released from the complex of Get1 and Get2 in a soluble form, ready for another round of binding to a hydrophobic C-terminal tail.

## Multipass Proteins Have Multiple Internal Topogenic Sequences

Figure 13-14 summarizes the arrangements of topogenic sequences in single-pass and multipass transmembrane proteins. In multipass (type IV) proteins, each of the membrane-spanning α helices acts as a topogenic sequence in the ways that we have already discussed: they can act to direct the protein to the ER, to anchor the protein in the ER membrane, or to stop transfer of the protein through the membrane. Multipass proteins fall into one of two types, depending on whether the N-terminus extends into the cytosol or the exoplasmic space (e.g., the ER lumen, cell exterior). This N-terminal topology usually is determined by the hydrophobic segment closest to the N-terminus and the charge of the sequences flanking it. If a type IV protein has an *even* number of transmembrane α helices, both its N-terminus and C-terminus will be oriented toward the same side of the membrane (Figure 13-14d). Conversely, if a type IV protein has an *odd* number of α helices, its two ends will have opposite orientations (Figure 13-14e).

**Type IV Proteins with N-Terminus in Cytosol**  Among the multipass proteins whose N-terminus extends into the cytosol are the various glucose transporters (GLUTs) and most ion-channel proteins, discussed in Chapter 11. In these proteins, the hydrophobic segment closest to the N-terminus

initiates insertion of the nascent chain into the ER membrane with the N-terminus oriented toward the cytosol; thus this α-helical segment functions like the internal signal-anchor sequence of a type II protein (see Figure 13-12a). As the nascent chain following the first α helix elongates, it moves through the translocon until the second hydrophobic α helix is formed. This helix prevents further extrusion of the nascent chain through the translocon; thus its function is similar to that of the stop-transfer anchor sequence in a type I protein (see Figure 13-11).

After synthesis of the first two transmembrane α helices, both ends of the nascent chain face the cytosol and the loop between them extends into the ER lumen. The C-terminus of the nascent chain then continues to grow into the cytosol, as it does in synthesis of type I and type III proteins. According to this mechanism, the third α helix acts as another type II signal-anchor sequence and the fourth as another stop-transfer anchor sequence (Figure 13-14d). Apparently, once the first topogenic sequence of a multipass polypeptide initiates association with the translocon, the ribosome remains attached to the translocon, and topogenic sequences that subsequently emerge from the ribosome are threaded into the translocon without the need for the SRP and the SRP receptor.

Experiments that use recombinant DNA techniques to exchange hydrophobic α helices have provided insight into the functioning of the topogenic sequences in type IV-A multipass proteins. These experiments indicate that the order of

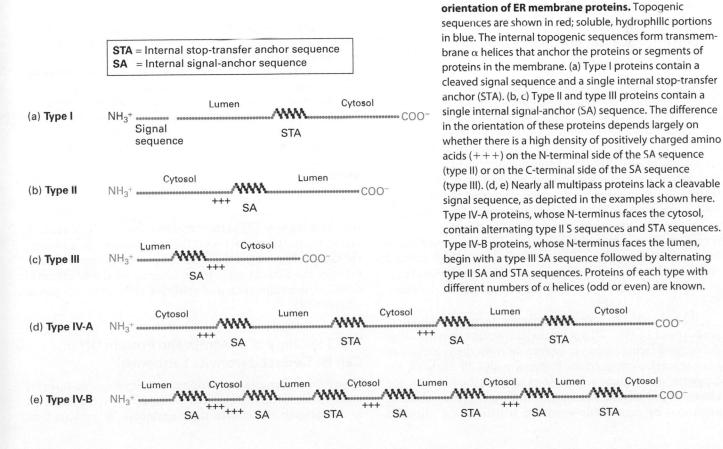

**FIGURE 13-14 Topogenic sequences determine orientation of ER membrane proteins.** Topogenic sequences are shown in red; soluble, hydrophilic portions in blue. The internal topogenic sequences form transmembrane α helices that anchor the proteins or segments of proteins in the membrane. (a) Type I proteins contain a cleaved signal sequence and a single internal stop-transfer anchor (STA). (b, c) Type II and type III proteins contain a single internal signal-anchor (SA) sequence. The difference in the orientation of these proteins depends largely on whether there is a high density of positively charged amino acids (+++) on the N-terminal side of the SA sequence (type II) or on the C-terminal side of the SA sequence (type III). (d, e) Nearly all multipass proteins lack a cleavable signal sequence, as depicted in the examples shown here. Type IV-A proteins, whose N-terminus faces the cytosol, contain alternating type II S sequences and STA sequences. Type IV-B proteins, whose N-terminus faces the lumen, begin with a type III SA sequence followed by alternating type II SA and STA sequences. Proteins of each type with different numbers of α helices (odd or even) are known.

the hydrophobic α helices relative to each other in the growing chain largely determines whether a given helix functions as a signal-anchor sequence or a stop-transfer anchor sequence. Other than its hydrophobicity, the specific amino acid sequence of a particular helix has little bearing on its function. Thus the first N-terminal α helix and the subsequent odd-numbered ones function as signal-anchor sequences, whereas the intervening even-numbered helices function as stop-transfer anchor sequences. This odd-even relationship among signal-anchor and stop-transfer anchor sequences is dictated by the fact that the transmembrane α helices assume alternating orientations as a multipass protein is woven back and forth across the membrane; signal anchor sequences are oriented with their N-termini toward the cytoplasmic side of the bilayer, whereas stop-transfer sequences have their N-termini oriented toward the expolasmic side of the bilayer.

**Type IV Proteins with N-Terminus in the Exoplasmic Space** The large family of G protein–coupled receptors, all of which contain seven transmembrane α helices, constitute the most numerous type IV-B proteins, whose N-terminus extends into the exoplasmic space. In these proteins, the hydrophobic α helix closest to the N-terminus often is followed by a cluster of positively charged amino acids, similar to a type III signal-anchor sequence (see Figure 13-12b). As a result, the first α helix inserts the nascent chain into the translocon with the N-terminus extending into the lumen (see Figure 13-14e). As the chain is elongated, it is inserted into the ER membrane by alternating type II signal-anchor sequences and stop-transfer sequences, as just described for type IV-A proteins.

## A Phospholipid Anchor Tethers Some Cell-Surface Proteins to the Membrane

Some cell-surface proteins are anchored to the phospholipid bilayer not by a sequence of hydrophobic amino acids but by a covalently attached amphipathic molecule, *glycosylphosphatidylinositol (GPI)* (Figure 13-15a and Chapter 10). These proteins are synthesized and initially anchored to the ER membrane exactly like type I transmembrane proteins, with a cleaved N-terminal signal sequence and an internal stop-transfer anchor sequence directing the process (see Figure 13-11). However, a short sequence of amino acids in the luminal domain, adjacent to the membrane-spanning domain, is recognized by a transamidase located within the ER membrane. This enzyme simultaneously cleaves off the original stop-transfer anchor sequence and transfers the luminal portion of the protein to a preformed GPI anchor in the membrane (Figure 13-15b).

Why change one type of membrane anchor for another? Attachment of the GPI anchor, which results in removal of the cytosol-facing hydrophilic domain from the protein, can have several consequences. Proteins with GPI anchors, for example, can diffuse relatively rapidly in the plane of the phospholipid bilayer membrane. In contrast, many proteins anchored by membrane-spanning α helices are impeded

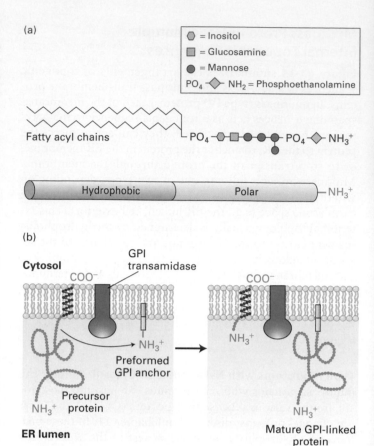

**FIGURE 13-15 GPI-anchored proteins.** (a) Structure of a glycosylphosphatidylinositol (GPI) from yeast. The hydrophobic portion of the molecule is composed of fatty acyl chains, whereas the polar (hydrophilic) portion of the molecule is composed of carbohydrate residues and phosphate groups. In other organisms, both the length of the acyl chains and the carbohydrate moieties may vary somewhat from the structure shown. (b) Formation of GPI-anchored proteins in the ER membrane. The protein is synthesized and initially inserted into the ER membrane as shown in Figure 13-11. A specific transamidase simultaneously cleaves the precursor protein within the exoplasmic-facing domain, near the stop-transfer anchor sequence (red), and transfers the carboxyl group of the new C-terminus to the terminal amino group of a preformed GPI anchor. [See C. Abeijon and C. B. Hirschberg, 1992, *Trends Biochem. Sci.* **17**:32, and K. Kodukula et al., 1992, *Proc. Nat'l Acad. Sci. USA* **89**:4982.]

from moving laterally in the membrane because their cytosol-facing segments interact with the cytoskeleton. In addition, the GPI anchor targets the attached protein to the apical domain of the plasma membrane (instead of the basolateral domain) in certain polarized epithelial cells, as we discuss in Chapter 14.

## The Topology of a Membrane Protein Often Can Be Deduced from Its Sequence

As we have seen, various topogenic sequences in integral membrane proteins synthesized on the ER govern interaction of the nascent chain with the translocon. When scientists begin to study a protein of unknown function, the identification of potential topogenic sequences within the corresponding gene sequence can provide important clues about the protein's topological class and function. Suppose, for example, that the gene for a protein known to be required for a cell-to-cell signaling pathway contains nucleotide sequences that encode an apparent N-terminal signal sequence and an internal hydrophobic sequence. These findings suggest that the protein is a type I integral membrane protein and therefore may be a cell-surface receptor for an extracellular ligand. Furthermore, the implied type I topology suggests that the N-terminal segment that lies between the signal sequence and the internal hydrophobic sequence constitutes the extracellular domain which would likely have a part in ligand binding, whereas the C-terminal segment that lies after the internal hydrophobic sequence would likely be cytosolic and may have a part in intracellular signalling.

Identification of topogenic sequences requires a way to scan sequence databases for segments that are sufficiently hydrophobic to be either a signal sequence or a transmembrane anchor sequence. Topogenic sequences can often be identified with the aid of computer programs that generate a *hydropathy profile* for the protein of interest. The first step is to assign a value known as the *hydropathic index* to each amino acid in the protein. By convention, hydrophobic amino acids are assigned positive values and hydrophilic amino acids negative values. Although different scales for the hydropathic index exist, all assign the most positive values to amino acids with side chains made up of mostly hydrocarbon residues (e.g., phenylalanine and methionine) and the most negative values to charged amino acids (e.g., arginine and aspartate). The second

step is to identify longer segments of sufficient overall hydrophobicity to be N-terminal signal sequences or internal stop-transfer sequences and signal-anchor sequences. To accomplish this, the total hydropathic index for each successive segment of 20 consecutive amino acids is calculated along the entire length of the protein. Plots of these calculated values against position in the amino acid sequence yield a hydropathy profile.

Figure 13-16 shows the hydropathy profiles for three different membrane proteins. The prominent peaks in such plots identify probable topogenic sequences as well as their position and approximate length. For example, the hydropathy profile of the human growth hormone receptor reveals the presence of both a hydrophobic signal sequence at the extreme N-terminus of the protein and an internal hydrophobic stop-transfer sequence (Figure 13-16a). On the basis of this profile, we can deduce, correctly, that the human growth hormone receptor is a type I integral membrane protein. The hydropathy profile of the asialoglycoprotein receptor, a cell-surface protein that mediates removal of abnormal extracellular glycoproteins, reveals a prominent internal hydrophobic signal-anchor sequence but gives no indication of a hydrophobic N-terminal signal sequence (Figure 13-16b). Thus we can predict that the asialoglycoprotein receptor is a type II or type III membrane protein. The distribution of charged residues on either side of the signal-anchor sequence often can differentiate between these possibilities, since positively charged amino acids flanking a membrane-spanning segment usually are oriented toward the cytosolic face of the membrane. For instance, in the case of the asialoglycoprotein receptor, examination of the residues flanking the signal-anchor sequence reveals that the residues on the N-terminal

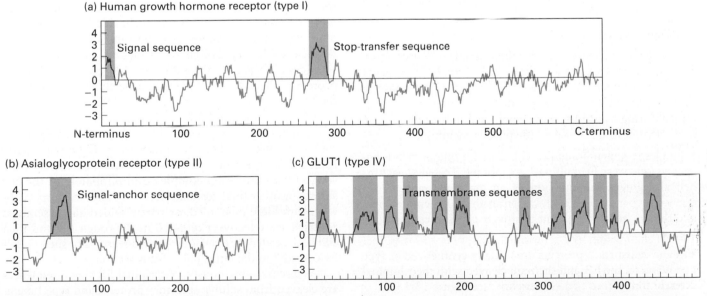

**EXPERIMENTAL FIGURE 13-16 Hydropathy profiles.**
Hydropathy profiles can identify likely topogenic sequences in integral membrane proteins. They are generated by plotting the total hydrophobicity of each segment of 20 contiguous amino acids along the length of a protein. Positive values indicate relatively hydrophobic portions; negative values, relatively polar portions of the protein. Probable topogenic sequences are marked. The complex profiles for multipass (type IV) proteins, such as GLUT1 in part (c), often must be supplemented with other analyses to determine the topology of these proteins.

side carry a net positive charge, thus correctly predicting that this is a type II protein.

The hydropathy profile of the GLUT1 glucose transporter, a multipass membrane protein, shows the presence of many segments that are sufficiently hydrophobic to be membrane-spanning helices (Figure 13-16c). The complexity of this profile illustrates the difficulty both in unambiguously identifying all the membrane-spanning segments in a multipass protein and in predicting the topology of individual signal-anchor and stop-transfer sequences. More sophisticated computer algorithms have been developed that take into account the presence of positively charged amino acids adjacent to hydrophobic segments as well as the length of and spacing between segments. Using all this information, the best algorithms can predict the complex topology of multipass proteins with an accuracy of greater than 75 percent.

Finally, sequence homology to a known protein may permit accurate prediction of the topology of a newly discovered multipass protein. For example, the genomes of multicellular organisms encode a very large number of multipass proteins with seven transmembrane α helices. The similarities between the sequences of these proteins strongly suggest that all have the same topology as the well-studied G protein–coupled receptors, which have the N-terminus oriented to the exoplasmic side and the C-terminus oriented to the cytosolic side of the membrane.

## KEY CONCEPTS of Section 13.2

### Insertion of Membrane Proteins into the ER

• Integral membrane proteins synthesized on the rough ER fall into five topological classes as well as a lipid-linked type (see Figure 13-10).

• Topogenic sequences—N-terminal signal sequences, internal stop-transfer anchor sequences, and internal signal-anchor sequences—direct the insertion and orientation of nascent proteins within the ER membrane. This orientation is retained during transport of the completed membrane protein to its final destination—e.g., the plasma membrane.

• Single-pass membrane proteins contain one or two topogenic sequences. In multipass membrane proteins, each α-helical segment can function as an internal topogenic sequence, depending on its location in the polypeptide chain and the presence of adjacent positively charged residues (see Figure 13-14).

• Some cell-surface proteins are initially synthesized as type I proteins on the ER and then are cleaved with their luminal domain transferred to a GPI anchor (see Figure 13-15).

• The topology of membrane proteins can often be correctly predicted by computer programs that identify hydrophobic topogenic segments within the amino acid sequence and generate hydropathy profiles (see Figure 13-16).

## 13.3 Protein Modifications, Folding, and Quality Control in the ER

Membrane and soluble secretory proteins synthesized on the rough ER undergo four principal modifications before they reach their final destinations: (1) covalent addition and processing of carbohydrates (*glycosylation*) in the ER and Golgi, (2) formation of disulfide bonds in the ER, (3) proper folding of polypeptide chains and assembly of multisubunit proteins in the ER, and (4) specific proteolytic cleavages in the ER, Golgi, and secretory vesicles. Generally speaking, these modifications promote folding of secretory proteins into their native structures and add structural stability to proteins exposed to the extracellular environment. Modifications such as glycosylation also allow the cell to produce a vast array of chemically distinct molecules at the cell surface that are the basis of specific molecular interactions used in cell-to-cell adhesion and communication.

One or more carbohydrate chains are added to the majority of proteins that are synthesized on the rough ER; indeed, glycosylation is the principal chemical modification to most of these proteins. Proteins with attached carbohydrates are known as **glycoproteins.** Carbohydrate chains in glycoproteins attached to the hydroxyl (—OH) group in serine and threonine residues are referred to as **O-linked oligosaccharides,** and carbohydrate chains attached to the amide nitrogen of asparagine are referred to as **N-linked oligosaccharides.** The various types of O-linked oligosaccharides include the mucin-type O-linked chains (named after abundant glycoproteins found in mucus) and the carbohydrate modifications on proteoglycans described in Chapter 20. O-linked chains typically contain only one to four sugar residues, which are added to proteins by enzymes known as glycosyltransferases, located in the lumen of the Golgi complex. The more common N-linked oligosaccharides are larger and more complex, containing several branches. In this section we focus on N-linked oligosaccharides, whose initial synthesis occurs in the ER. After the initial N-glycosylation of a protein in the ER, the oligosaccharide chain is modified in the ER and commonly in the Golgi as well.

Disulfide bond formation, protein folding, and assembly of multimeric proteins, which take place exclusively in the rough ER, also are discussed in this section. Only properly folded and assembled proteins are transported from the rough ER to the Golgi complex and ultimately to the cell surface or other final destination through the secretory pathway. Unfolded, misfolded, or partly folded and assembled proteins are selectively retained in the rough ER and then can be degraded. We consider several features of such "quality control" in the latter part of this section.

As discussed previously, N-terminal ER signal sequences are cleaved from soluble secretory proteins and type I membrane proteins in the ER. Some proteins also undergo other specific proteolytic cleavages in the Golgi complex or secretory vesicles. We cover these cleavages, as well as carbohydrate modifications that occur primarily or exclusively in the Golgi complex, in the next chapter.

## A Preformed N-Linked Oligosaccharide Is Added to Many Proteins in the Rough ER

Biosynthesis of all N-linked oligosaccharides begins in the rough ER with addition of a preformed oligosaccharide precursor containing 14 residues (Figure 13-17). The structure of this precursor is the same in plants, animals, and single-celled eukaryotes: a branched oligosaccharide, containing three glucose (Glc), nine mannose (Man), and two N-acetylglucosamine (GlcNAc) molecules, which can be written as $Glc_3Man_9(GlcNAc)_2$. Once added to a protein, this branched carbohydrate structure is modified by addition or removal of monosaccharides in the ER and Golgi compartments. The modifications to N-linked chains differ from one glycoprotein to another and differ among different organisms, but a core of 5 of the 14 residues is conserved in the structures of all N-linked oligosaccharides on secretory and membrane proteins.

Prior to transfer to a nascent chain in the lumen of the ER, the precursor oligosaccharide is assembled on a membrane-attached anchor called *dolichol phosphate*, a long-chain polyisoprenoid lipid (Chapter 10). After the first sugar, GlcNAc, is attached to the dolichol phosphate by a pyrophosphate bond, the other sugars are added by glycosidic bonds in a complex set of reactions catalyzed by enzymes attached to the cytosolic or luminal faces of the rough ER membrane (Figure 13-17). The final dolichol pyrophosphoryl oligosaccharide is oriented so that the oligosaccharide portion faces the ER lumen.

The entire 14-residue precursor is transferred from the dolichol carrier to an asparagine residue on a nascent polypeptide as it emerges into the ER lumen (Figure 13-18, step **1**). Only asparagine residues in the tripeptide sequences Asn-X-Ser and Asn-X-Thr (where X is any amino acid except proline) are substrates for *oligosaccharyl transferase,* the enzyme that catalyzes this reaction. Two of the three subunits of this enzyme are ER membrane proteins whose cytosol-facing domains bind to the ribosome, localizing a third subunit of the transferase, the catalytic subunit, near the growing polypeptide chain in the ER lumen. Not all Asn-X-Ser/Thr sequences become glycosylated, and it is not possible to predict from the amino acid sequence alone which potential N-linked glycosylation sites will be modified; for instance, rapid folding of a segment of a protein containing an Asn-X-Ser/Thr sequence may prevent transfer of the oligosaccharide precursor to it.

Immediately after the entire precursor, $Glc_3Man_9(GlcNAc)_2$, is transferred to a nascent polypeptide, three different enzymes, called glycosidases, remove all three glucose residues and one particular mannose residue (Figure 13-18, steps **2**–**4**). The three glucose residues, which are the last residues added during synthesis of the precursor on the dolichol carrier, appear to act as a signal that the oligosaccharide is complete and ready to be transferred to a protein.

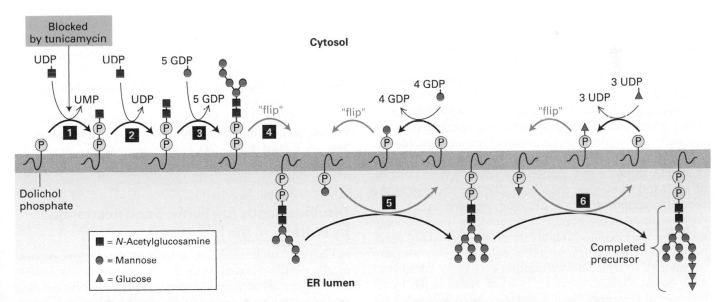

**FIGURE 13-17 Biosynthesis of the oligosaccharide precursor.** Dolichol phosphate is a strongly hydrophobic lipid, containing 75–95 carbon atoms, that is embedded in the ER membrane. Two N-acetylglucosamine (GlcNAc) and five mannose residues are added one at a time to a dolichol phosphate on the cytosolic face of the ER membrane (steps **1**–**3**). The nucleotide-sugar donors in these and later reactions are synthesized in the cytosol. Note that the first sugar residue is attached to dolichol by a high-energy pyrophosphate linkage. Tunicamycin, which blocks the first enzyme in this pathway, inhibits the synthesis of all N-linked oligosaccharides in cells. After the seven-residue dolichol pyrophosphoryl intermediate is flipped to the luminal face (step **4**), the remaining four mannose and all three glucose residues are added one at a time (steps **5**, **6**). In the later reactions, the sugar to be added is first transferred from a nucleotide-sugar to a carrier dolichol phosphate on the cytosolic face of the ER; the carrier is then flipped to the luminal face, where the sugar is transferred to the growing oligosaccharide, after which the "empty" carrier is flipped back to the cytosolic face. [After C. Abeijon and C. B. Hirschberg, 1992, *Trends Biochem. Sci.* **17**:32.]

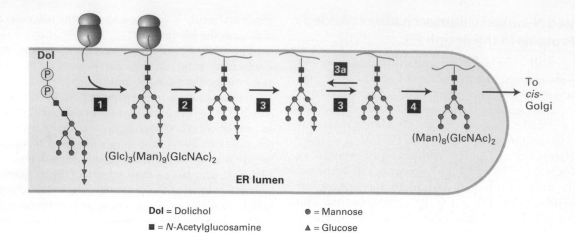

Dol = Dolichol
■ = *N*-Acetylglucosamine
● = Mannose
▲ = Glucose

**FIGURE 13-18 Addition and initial processing of *N*-linked oligosaccharides.** In the rough ER of vertebrate cells, the $Glc_3Man_9(GlcNAc)_2$ precursor is transferred from the dolichol carrier to a susceptible asparagine residue on a nascent protein as soon as the asparagine crosses to the luminal side of the ER (step **1**). In three separate reactions, first one glucose residue (step **2**), then two glucose residues (step **3**), and finally one mannose residue (step **4**) are removed. Re-addition of one glucose residue (step **3a**) plays a role in the correct folding of many proteins in the ER, as discussed later. The process of N-linked glycosylation of a soluble secretory protein is shown here, but the luminal portions of an integral membrane protein can be modified on asparagine residues by the same mechanism. [See R. Kornfeld and S. Kornfeld, 1985, *Ann. Rev. Biochem.* **45**:631, and M. Sousa and A. J. Parodi, 1995, *EMBO J.* **14**:4196.]

## Oligosaccharide Side Chains May Promote Folding and Stability of Glycoproteins

The oligosaccharides attached to glycoproteins serve various functions. For example, some proteins require *N*-linked oligosaccharides in order to fold properly in the ER. This function has been demonstrated in studies with the antibiotic tunicamycin, which blocks the first step in the formation of the dolichol-linked oligosaccharide precursor and therefore inhibits synthesis of all *N*-linked oligosaccharides in cells (see Figure 13-17, *top left*). For example, in the presence of tunicamycin, the flu virus hemagglutinin precursor polypeptide ($HA_0$) is synthesized, but it cannot fold properly and form a normal trimer; in this case, the protein remains, misfolded, in the rough ER. Moreover, mutation of a particular asparagine in the HA sequence to a glutamine residue prevents addition of an *N*-linked oligosaccharide to that site and causes the protein to accumulate in the ER in an unfolded state.

In addition to promoting proper folding, *N*-linked oligosaccharides also confer stability on many secreted glycoproteins. Many secretory proteins fold properly and are transported to their final destination even if the addition of all *N*-linked oligosaccharides is blocked, for example, by tunicamycin. However, such nonglycosylated proteins have been shown to be less stable than their glycosylated forms. For instance, glycosylated fibronectin, a normal component of the extracellular matrix, is degraded much more slowly by tissue proteases than is nonglycosylated fibronectin.

Oligosaccharides on certain cell-surface glycoproteins also play a role in cell-cell adhesion. For example, the plasma membrane of white blood cells (leukocytes) contains cell-adhesion molecules (CAMs) that are extensively glycosylated. The oligosaccharides in these molecules interact with a sugar-binding domain in certain CAMs found on endothelial cells lining blood vessels. This interaction tethers the leukocytes to the endothelium and assists in their movement into tissues during an inflammatory response to infection (see Figure 20-39). Other cell-surface glycoproteins possess oligosaccharide side chains that can induce an immune response. A common example is the A, B, O blood-group antigens, which are O-linked oligosaccharides attached to glycoproteins and glycolipids on the surface of erythrocytes and other cell types (Figure 10-20). In both cases, oligosaccharides are added to the luminal face of these membrane proteins, in a manner similar to what is shown in Figure 13-18 for soluble proteins. The luminal face of these membrane proteins is topologically equivalent to the exterior face of the plasma membrane, where these proteins eventually end up.

## Disulfide Bonds Are Formed and Rearranged by Proteins in the ER Lumen

In Chapter 3 we learned that both intramolecular and intermolecular **disulfide bonds** (—S—S—) help stabilize the tertiary and quaternary structure of many proteins. These covalent bonds form by the oxidative linkage of **sulfhydryl groups** (—**SH**), also known as *thiol* groups, on two cysteine residues in the same or different polypeptide chains. This reaction can proceed spontaneously only when a suitable oxidant is present. In eukaryotic cells, disulfide bonds are formed only in the lumen of the rough ER. Thus disulfide bonds are found only in soluble secretory proteins and in the exoplasmic domains of membrane proteins. Cytosolic proteins and organelle proteins synthesized on free ribosomes (i.e., those destined for mitochondria, chloroplasts, peroxisomes, etc.) usually lack disulfide bonds.

The efficient formation of disulfide bonds in the lumen of the ER depends on the enzyme *protein disulfide isomerase* (PDI), which is present in all eukaryotic cells. This enzyme is especially abundant in the ER of secretory cells in such organs as the liver and pancreas, where large quantities of proteins that contain disulfide bonds are produced. As shown in Figure 13-19a, the disulfide bond in the active site of PDI can be readily transferred to a protein by two sequential thiol-disulfide transfer reactions. The reduced PDI generated by this reaction is returned to an oxidized form by the action of an ER-resident protein, called *Ero1*, which carries a disulfide bond that can be transferred to PDI. Ero1 itself becomes oxidized by reaction with molecular oxygen that has diffused into the ER.

In proteins that contain more than one disulfide bond, the proper pairing of cysteine residues is essential for normal structure and activity. Disulfide bonds commonly are formed between cysteines that occur sequentially in the amino acid sequence while a polypeptide is still growing on the ribosome. Such sequential formation, however, sometimes yields disulfide bonds between the wrong cysteines. For example, proinsulin, a precursor to the peptide hormone insulin, has three disulfide bonds that link cysteines 1 and 4, 2 and 6, and 3 and 5. In this case, a disulfide bond that initially formed sequentially (e.g., between cysteines 1 and 2) would have to be rearranged for the protein to achieve its proper folded conformation. In cells, the rearrangement of disulfide bonds also is accelerated by PDI, which acts on a broad range of protein substrates, allowing them to reach their thermodynamically most stable conformations (Figure 13-19b). Disulfide bonds generally form in a specific order, first stabilizing small domains of a polypeptide, then stabilizing the interactions of more distant segments; this phenomenon is illustrated by the folding of influenza hemagglutinin (HA) protein, discussed in the next section.

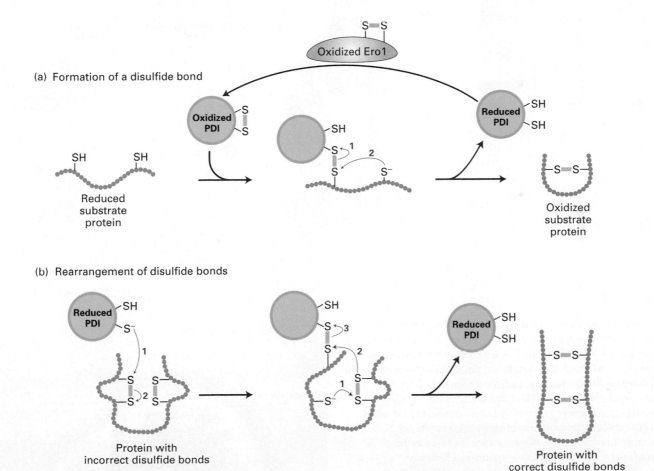

**FIGURE 13-19 Action of protein disulfide isomerase (PDI).** PDI forms and rearranges disulfide bonds via an active site with two closely spaced cysteine residues that are easily interconverted between the reduced dithiol form and the oxidized disulfide form. Numbered red arrows indicate the sequence of electron transfers. Yellow bars represent disulfide bonds. (a) In the formation of disulfide bonds, the ionized (–S⁻) form of a cysteine thiol in the substrate protein reacts with the disulfide (S—S) bond in oxidized PDI to form a disulfide-bonded PDI–substrate protein intermediate. A second ionized thiol in the substrate then reacts with this intermediate, forming a disulfide bond within the substrate protein and releasing reduced PDI. PDI, in turn, transfers electrons to a disulfide bond in the luminal protein Ero1, thereby regenerating the oxidized form of PDI. (b) Reduced PDI can catalyze rearrangement of improperly formed disulfide bonds by similar thiol-disulfide transfer reactions. In this case, reduced PDI both initiates and is regenerated in the reaction pathway. These reactions are repeated until the most stable conformation of the protein is achieved. [See M. M. Lyles and H. F. Gilbert, 1991, *Biochemistry* **30**:619.]

## Chaperones and Other ER Proteins Facilitate Folding and Assembly of Proteins

Although many denatured proteins can spontaneously refold into their native state in vitro, such refolding usually requires hours to reach completion. Yet new soluble and membrane proteins produced in the ER generally fold into their proper conformation within minutes after their synthesis. The rapid folding of these newly synthesized proteins in cells depends on the sequential action of several proteins present within the ER lumen. We have already seen how the molecular chaperone BiP can drive post-translational translocation in yeast by binding fully synthesized polypeptides as they enter the ER (see Figure 13-9). BiP can also bind transiently to nascent chains as they enter the ER during cotranslational translocation. Bound BiP is thought to prevent segments of a nascent chain from misfolding or forming aggregates, thereby promoting folding of the entire polypeptide into the proper conformation. Protein disulfide isomerase (PDI) also contributes to proper folding, because correct 3-D conformation is stabilized by disulfide bonds in many proteins.

As illustrated in Figure 13-20, two other ER proteins, the homologous **lectins** (carbohydrate-binding proteins) *calnexin* and *calreticulin*, bind selectively to certain *N*-linked oligosaccharides on growing nascent chains. The ligand for these two lectins, which resembles the *N*-linked oligosaccharide precursor but that has only a single glucose residue

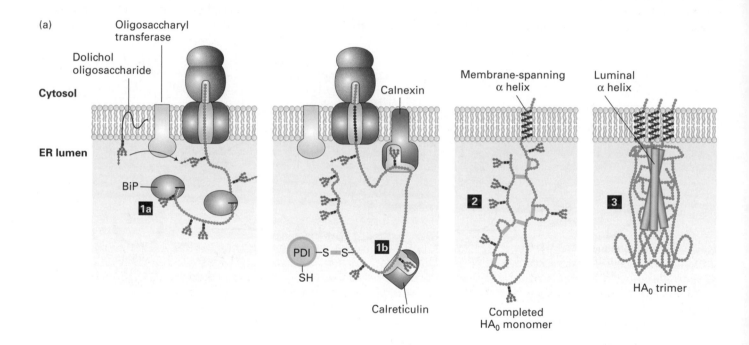

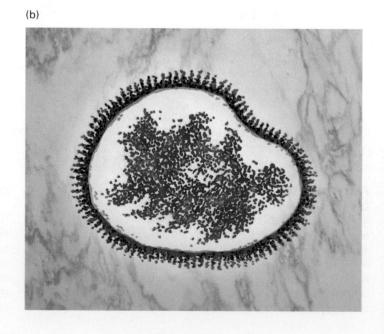

**FIGURE 13-20 Hemagglutinin folding and assembly.**
(a) Mechanism of (HA$_0$) trimer assembly. Transient binding of the chaperone BiP (step **1a**) to the nascent chain and of two lectins, calnexin and calreticulin, to certain oligosaccharide chains (step **1b**) promotes proper folding of adjacent segments. A total of seven *N*-linked oligosaccharide chains are added to the luminal portion of the nascent chain during cotranslational translocation, and PDI catalyzes the formation of six disulfide bonds per monomer. Completed HA$_0$ monomers are anchored in the membrane by a single membrane-spanning α helix with the N-terminus in the lumen (step **2**). Interaction of three HA$_0$ chains with one another, initially via their transmembrane α helices, apparently triggers formation of a long stem containing one α helix from the luminal part of each HA$_0$ polypeptide. Finally, interactions occur among the three globular heads, generating a stable HA$_0$ trimer (step **3**). (b) Electron micrograph of a complete influenza virion showing trimers of HA protein protruding as spikes from the surface of the viral membrane. [Part (a), see U. Tatu et al., 1995, *EMBO J.* **14**:1340, and D. Hebert et al., 1997, *J. Cell Biol.* **139**:613. Part (b), Chris Bjornberg/Photo Researchers, Inc.]

$[Glc_1Man_9(GlcNAc)_2]$, is generated by a specific glucosyltransferase in the ER lumen (see Figure 13-18, step **3a**). This enzyme acts only on polypeptide chains that are unfolded or misfolded, and in this respect the glucosyltransferase acts as one of the primary surveillance mechanisms to ensure quality control of protein folding in the ER, but the mechanism by which the glucosyltransferase distinguishes folded and unfolded proteins is not yet understood. Binding of calnexin and calreticulin to unfolded nascent chains marked with glucosylated N-linked oligosaccharides prevents aggregation of adjacent segments of a protein as it is being made on the ER. Thus calnexin and calreticulin, like BiP, help prevent premature, incorrect folding of segments of a newly made protein.

Other important protein-folding catalysts in the ER lumen are *peptidyl-prolyl isomerases,* a family of enzymes that accelerate the rotation about peptidyl-prolyl bonds at proline residues in unfolded segments of a polypeptide:

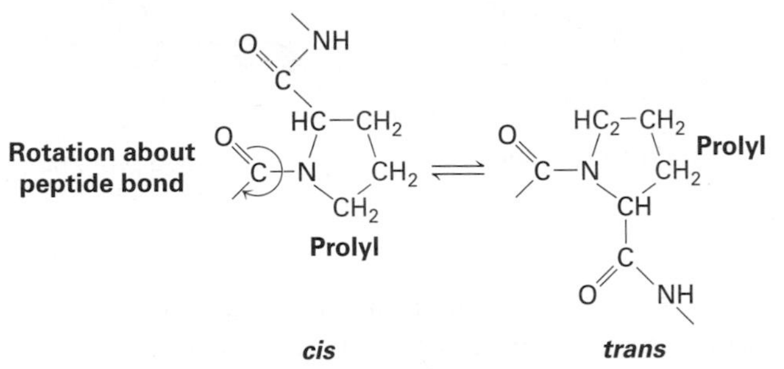

Such isomerizations sometimes are the rate-limiting step in the folding of protein domains. Many peptidyl-prolyl isomerases can catalyze the rotation of exposed peptidyl-prolyl bonds indiscriminately in numerous proteins, but some have very specific protein substrates.

Many important soluble secretory and membrane proteins synthesized on the ER are built of two or more polypeptide subunits. In all cases, the assembly of subunits constituting these multisubunit (multimeric) proteins occurs in the ER. An important class of multimeric secreted proteins is the immunoglobulins, which contain two heavy (H) and two light (L) chains, all linked by intrachain disulfide bonds. Hemagglutinin (HA) is another multimeric protein that provides a good illustration of folding and subunit assembly (see Figure 13-20). This trimeric protein forms the spikes that protrude from the surface of an influenza virus particle. The HA trimer is formed within the ER of an infected host cell from three copies of a precursor protein termed $HA_0$, which has a single membrane-spanning α helix. In the Golgi complex, each of the three $HA_0$ proteins is cleaved to form two polypeptides, $HA_1$ and $HA_2$; thus each HA molecule that eventually resides on the viral surface contains three copies of $HA_1$ and three of $HA_2$ (see Figure 3-10). The trimer is stabilized by interactions between the large exoplasmic domains of the constituent polypeptides, which extend into the ER lumen; after HA is transported to the cell surface, these domains extend into the extracellular space. Interactions between the smaller cytosolic and membrane-spanning portions of the HA subunits also help stabilize the trimeric protein. Studies have shown that it takes just 10 minutes for the $HA_0$ polypeptides to fold and assemble into their proper trimeric conformation.

## Improperly Folded Proteins in the ER Induce Expression of Protein-Folding Catalysts

Wild-type proteins that are synthesized on the rough ER cannot exit this compartment until they achieve their completely folded conformation. Likewise, almost any mutation that prevents proper folding of a protein in the ER also blocks movement of the polypeptide from the ER lumen or membrane to the Golgi complex. The mechanisms for retaining unfolded or incompletely folded proteins within the ER probably increase the overall efficiency of folding by keeping intermediate forms in proximity to folding catalysts, which are most abundant in the ER. Improperly folded proteins retained within the ER generally are seen bound to the ER chaperones BiP and calnexin. Thus these luminal folding catalysts perform two related functions: assisting in the folding of normal proteins by preventing their aggregation and binding to misfolded proteins to retain them in the ER.

Both mammalian cells and yeasts respond to the presence of unfolded proteins in the rough ER by increasing transcription of several genes encoding ER chaperones and other folding catalysts. A key participant in this *unfolded-protein response* is Ire1, an ER membrane protein that exists as both a monomer and a dimer. The dimeric form, but not the monomeric form, promotes formation of Hac1, a transcription factor in yeast that activates expression of the genes induced in the unfolded-protein response. As depicted in Figure 13-21, binding of BiP to the luminal domain of monomeric Ire1 prevents formation of the Ire1 dimer. Thus the quantity of free BiP in the ER lumen determines the relative proportion of monomeric and dimeric Ire1. Accumulation of unfolded proteins within the ER lumen sequesters BiP molecules, making them unavailable for binding to Ire1. As a result, the level of dimeric Ire1 increases, leading to an increase in the level of Hac1 and production of proteins that assist in protein folding.

Mammalian cells contain an additional regulatory pathway that operates in response to unfolded proteins in the ER. In this pathway, accumulation of unfolded proteins in the ER triggers proteolysis of ATF6, a transmembrane protein in the ER membrane, at a site within the membrane-spanning segment. The cytosolic domain of ATF6 released by proteolysis then moves to the nucleus, where it stimulates transcription of the genes encoding ER chaperones. Activation of a transcription factor by such *regulated intramembrane proteolysis* also occurs in the Notch signaling pathway and during activation of the cholesterol-responsive transcription factor SREBP (see Figures 16-35 and 16-37).

A hereditary form of emphysema illustrates the detrimental effects that can result from misfolding of proteins in the ER. This disease is caused by a point mutation in $\alpha_1$-antitrypsin, which normally is secreted by hepatocytes and

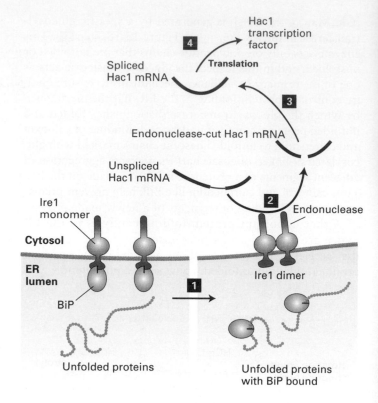

**FIGURE 13-21 The unfolded-protein response.** Ire1, a transmembrane protein in the ER membrane, has a binding site for BiP on its luminal domain; the cytosolic domain contains a specific RNA endonuclease. Step **1**: Accumulating unfolded proteins in the ER lumen bind BiP molecules, releasing them from monomeric Ire1. Dimerization of Ire1 then activates its endonuclease activity. Steps **2**, **3**: The unspliced mRNA precursor encoding the transcription factor Hac1 is cleaved by dimeric Ire1, and the two exons are joined to form functional Hac1 mRNA. Current evidence indicates that this processing occurs in the cytosol, although pre-mRNA processing generally occurs in the nucleus. Step **4**: Hac1 is translated into Hac1 protein, which then moves back into the nucleus and activates transcription of genes encoding several protein-folding catalysts. [See U. Ruegsegger et al., 2001, *Cell* **107**:103; A. Bertolotti et al., 2000, *Nat. Cell Biol.* **2**:326; and C. Sidrauski and P. Walter, 1997, *Cell* **90**:1031.]

macrophages. The wild-type protein binds to and inhibits trypsin and also the blood protease elastase. In the absence of $\alpha_1$-antitrypsin, elastase degrades the fine tissue in the lung that participates in the absorption of oxygen, eventually producing the symptoms of emphysema. Although the mutant $\alpha_1$-antitrypsin is synthesized in the rough ER, it does not fold properly, forming an almost crystalline aggregate that is not exported from the ER. In hepatocytes, the secretion of other proteins also becomes impaired as the rough ER is filled with aggregated $\alpha_1$-antitrypsin. ■

## Unassembled or Misfolded Proteins in the ER Are Often Transported to the Cytosol for Degradation

Misfolded soluble secretory and membrane proteins, as well as the unassembled subunits of multimeric proteins, often are degraded within an hour or two after their synthesis in the rough ER. For many years, researchers thought that proteolytic enzymes within the ER lumen catalyzed degradation of misfolded or unassembled polypeptides, but such proteases were never found. More recent studies have shown that misfolded secretory proteins are recognized by specific ER membrane proteins and are targeted for transport from the ER lumen into the cytosol, by a process known as *dislocation*.

The dislocation of misfolded proteins out of the ER depends on a set of proteins located in the ER membrane and in the cytosol that perform three basic functions. The first function is recognition of misfolded proteins that will be substrates for the dislocation reaction. One mechanism for recognition involves trimming of N-linked carbohydrate chains by the enzyme α-*mannosidase I* (Figure 13-22). Trimmed glycans of the structure $Man_8(GlcNAc)_2$ are recognized by the lectinlike protein known as EDEM, and glycans that are further trimmed to $Man_7(GlcNAc)_2$ are recognized by the lectinlike protein OS-9. Both EDEM and OS-9 target the trimmed glycoprotein to the dislocation complex for degradation. It is not known precisely how α-mannosidase I distinguishes proteins that cannot fold properly, and are thus legitimate substrates for the dislocation process, from normal

proteins that have transient partially folded states as they acquire their fully folded conformation. One possibility is that trimming of N-linked carbohydrate chains by the α-mannosidase I may occur slowly, such that only those glycoproteins that remain misfolded in the ER lumen for a sufficiently long time are trimmed and therefore targeted for degradation. Luminal proteins that lack carbohydrate chains altogether can also be targeted for degradation, indicating that other processes for the recognition of unfolded proteins must also exist. Other mechanisms to recognize unfolded proteins that do not involve trimming of N-linked carbohydrate chains must exist because misfolded membrane proteins that lack N-linked carbohydrate chains altogether can nevertheless be targeted for degradation.

Once an unfolded protein has been identified, it is targeted for dislocation across the ER membrane. Some kind of channel must exist for dislocation of misfolded proteins across the ER membrane, and a complex of at least four integral membrane proteins, known as the ERAD (*ER-a*ssociated *d*egradation) complex, appears to serve this function. The structure of the dislocation channel and the mechanism by which misfolded proteins traverse the ER membrane are not yet known.

As segments of the dislocated polypeptide are exposed to the cytosol, they encounter cytosolic enzymes that drive dislocation. One of these enzymes is an ATPase called p97, a member of a protein family known as the **AAA ATPase family,** which couple the energy of ATP hydrolysis to disassembly of protein complexes. In retrotranslocation, hydrolysis of ATP by p97 may provide the driving force to pull misfolded proteins from the ER membrane into the cytosol. As misfolded

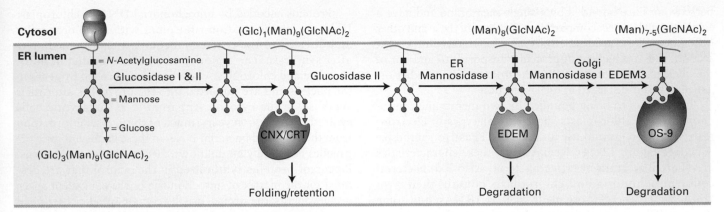

**FIGURE 13-22 Modifications of *N*-linked oligosaccharides are used to monitor folding and quality control.** After removal of three glucose residues from the *N*-linked oligosaccharides in the ER, a single glucose can be re-added by a glucosyl transferase to form $Glc_1Man_9(GlcNAc)_2$ (see Figure 13-18, step **3a**). This modified *N*-linked carbohydrate binds the lectins calnexin (CNX) and calreticulin (CRT) for retention in the ER and engagement of folding chaperones. Proteins that cannot fold and are therefore retained in the ER for longer times undergo mannose trimming by mannosidase I to form $Man_8(GlcNAc)_2$, which is recognized by the lectin EDEM, or further trimming to $Man_{7-5}(GlcNAc)_2$, which is recognized by OS-9. Recognition by either EDEM or OS-9 leads to dislocation of the misfolded protin out of the ER, ubiquitination, and degradation by the proteasome.

proteins reenter the cytosol, specific ubiquitin ligase enzymes in the ER membrane add ubiquitin residues to the dislocated peptide. Like the action of p97, the ubiquitination reaction is coupled to ATP hydrolysis; this release of energy possibly also contributes to trapping proteins in the cytosol. The resulting polyubiquitinated polypeptides, now fully in the cytosol, are removed from the cell altogether by degradation in the proteasome. The role of polyubiquitination in targeting proteins to the proteasome is discussed more fully in Chapter 3 (see Figure 3-29 and Figure 3-34).

## KEY CONCEPTS of Section 13.3

### Protein Modifications, Folding, and Quality Control in the ER

• All *N*-linked oligosaccharides, which are bound to asparagine residues, contain a core of two *N*-acetylglucosamine and at least three mannose residues and usually have several branches. *O*-linked oligosaccharides, which are bound to serine or threonine residues, are generally short, often containing only one to four sugar residues.

• Formation of *N*-linked oligosaccharides begins with assembly of a conserved 14-residue high-mannose precursor on dolichol, a lipid in the membrane of the rough ER (see Figure 13-17). After this preformed oligosaccharide is transferred to specific asparagine residues of nascent polypeptide chains in the ER lumen, three glucose residues and one mannose residue are removed (see Figure 13-18).

• Oligosaccharide side chains may assist in the proper folding of glycoproteins, help protect the mature proteins from proteolysis, participate in cell-cell adhesion, and function as antigens.

• Disulfide bonds are added to many soluble secretory proteins and the exoplasmic domain of membrane proteins in the ER. Protein disulfide isomerase (PDI), present in the ER lumen, catalyzes both the formation and the rearrangement of disulfide bonds (see Figure 13-19).

• The chaperone BiP, the lectins calnexin and calreticulin, and peptidyl-prolyl isomerases work together to ensure proper folding of newly made secretory and membrane proteins in the ER. The subunits of multimeric proteins also assemble in the ER (see Figure 13-20).

• Only properly folded proteins and assembled subunits are transported from the rough ER to the Golgi complex in vesicles.

• The accumulation of abnormally folded proteins and unassembled subunits in the ER can induce increased expression of ER protein-folding catalysts via the unfolded-protein response (see Figure 13-21).

• Unassembled or misfolded proteins in the ER often are transported back to the cytosol, where they are degraded in the ubiquitin/proteasome pathway (see Figure 13-22).

## 13.4 Targeting of Proteins to Mitochondria and Chloroplasts

In the remainder of this chapter, we examine how proteins synthesized on cytosolic ribosomes are sorted to mitochondria, chloroplasts, peroxisomes, and the nucleus (see Figure 13-1). In both mitochondria and chloroplasts, an internal lumen called the *matrix* is surrounded by a double membrane, and internal subcompartments exist within the matrix. In contrast, peroxisomes are bounded by a single membrane and have a single luminal matrix compartment. Because of these and other differences, we consider peroxisomes separately in the next section. The mechanism of protein transport into and out of the nucleus differs in many respects from sorting to other organelles; this is discussed in the last section.

In addition to being bounded by two membranes, mitochondria and chloroplasts share similar types of electron-transport proteins and use an F-class ATPase to synthesize ATP (see Figure 12-24). Remarkably, these characteristics are shared by gram-negative bacteria. Also like bacterial cells, mitochondria and chloroplasts contain their own DNA, which encodes organelle rRNAs, tRNAs, and some proteins (Chapter 6). Moreover, growth and division of mitochondria and chloroplasts are not coupled to nuclear division. Rather, these organelles grow by the incorporation of cellular proteins and lipids, and new organelles form by division of preexisting organelles. The numerous similarities of free-living bacterial cells with mitochondria and chloroplasts have led to the understanding that these organelles arose by the incorporation of bacteria into ancestral eukaryotic cells, forming endosymbiotic organelles (see Figure 6-20). The sequence similarity of many membrane translocation proteins shared by mitochondria, chloroplasts, and bacteria provides the most striking evidence for this ancient evolutionary relationship. In this section we will examine these membrane translocation proteins in detail.

Proteins encoded by mitochondrial DNA or chloroplast DNA are synthesized on ribosomes within the organelles and directed to the correct subcompartment immediately after synthesis. The majority of proteins located in mitochondria and chloroplasts, however, are encoded by genes in the nucleus and are imported into the organelles after their synthesis in the cytosol. Apparently, as eukaryotic cells evolved over a billion years, much of the genetic information from the ancestral bacterial DNA in these endosymbiotic organelles moved, by an unknown mechanism, to the nucleus. Precursor proteins synthesized in the cytosol that are destined for the matrix of mitochondria or the equivalent space in chloroplasts, the stroma, usually contain specific N-terminal uptake-targeting sequences that specify binding to receptor proteins on the organelle surface. Generally, this sequence is cleaved once it reaches the matrix or stroma. Clearly, these uptake-targeting sequences are similar in their location and general function to the signal sequences that direct nascent proteins to the ER lumen. Although the three types of signals share some common sequence features, their specific sequences differ considerably, as summarized in Table 13-1.

In both mitochondria and chloroplasts, protein import requires energy and occurs at points where the outer and inner organelle membranes are in close contact. Because mitochondria and chloroplasts contain multiple membranes and membrane-limited spaces, sorting of many proteins to their correct location often requires the sequential action of

| TABLE 13-1 | Uptake-Targeting Sequences That Direct Proteins from the Cytosol to Organelles* | | |
|---|---|---|---|
| **Target Organelle** | **Location of Sequence Within Protein** | **Removal of Sequence** | **Nature of Sequence** |
| Endoplasmic reticulum (lumen) | N-terminus | Yes | Core of 6–12 hydrophobic amino acids, often preceded by one or more basic amino acids (Arg, Lys) |
| Mitochondrion (matrix) | N-terminus | Yes | Amphipathic helix, 20–50 residues in length, with Arg and Lys residues on one side and hydrophobic residues on the other |
| Chloroplast (stroma) | N-terminus | Yes | No common motifs; generally rich in Ser, Thr, and small hydrophobic residues and poor in Glu and Asp |
| Peroxisome (matrix) | C-terminus (most proteins); N-terminus (few proteins) | No | PTS1 signal (Ser-Lys-Leu) at extreme C-terminus; PTS2 signal at N-terminus |
| Nucleus (nucleoplasm) | Varies | No | Multiple different kinds; a common motif includes a short segment rich in Lys and Arg residues |

*Different or additional sequences target proteins to organelle membranes and subcompartments.

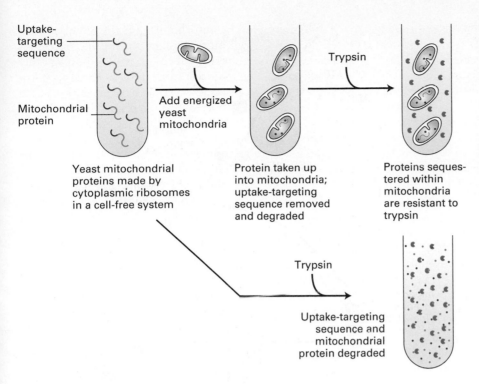

**EXPERIMENTAL FIGURE 13-23 Import of mitochondrial precursor proteins is assayed in a cell-free system.** Mitochondrial precursor proteins with attached uptake-targeting signals can be synthesized on ribosomes in a cell-free reaction. When respiring mitochondria are added to the synthesized mitochondrial precursor proteins (*top*), the proteins are taken up by mitochondria. Inside mitochondria, proteins are protected from the action of proteases such as trypsin. When no mitochondria are present (*bottom*), mitochondrial proteins are degraded by added protease. Protein uptake occurs only with energized (respiring) mitochondria, which have a proton electrochemical gradient (proton-motive force) across the inner membrane. The imported protein must contain an appropriate uptake-targeting sequence. Uptake also requires ATP and a cytosolic extract containing chaperone proteins that maintain the precursor proteins in an unfolded conformation. This assay has been used to study targeting sequences and other features of the translocation process.

two targeting sequences and two membrane-bound translocation systems: one to direct the protein into the organelle and the other to direct it into the correct organellar compartment or membrane. As we will see, the mechanisms for sorting various proteins to mitochondria and chloroplasts are related to some of the mechanisms discussed previously.

## Amphipathic N-Terminal Signal Sequences Direct Proteins to the Mitochondrial Matrix

All proteins that travel from the cytosol to the same mitochondrial destination have targeting signals that share common motifs, although the signal sequences are generally not identical. Thus the receptors that recognize such signals are able to bind to a number of different but related sequences. The most extensively studied sequences for localizing proteins to mitochondria are the *matrix-targeting sequences*. These sequences, located at the N-terminus, are usually 20–50 amino acids in length. They are rich in hydrophobic amino acids, positively charged basic amino acids (arginine and lysine), and hydroxylated ones (serine and threonine) but tend to lack negatively charged acidic residues (aspartate and glutamate).

Mitochondrial matrix-targeting sequences are thought to assume an α-helical conformation in which positively charged amino acids predominate on one side of the helix and hydrophobic amino acids predominate on the other side. Sequences such as these that contain both hydrophobic and hydrophilic regions are said to be **amphipathic**. Mutations that disrupt this amphipathic character usually disrupt targeting to the matrix, although many other amino acid substitutions do not. These findings indicate that the amphipathicity of matrix-targeting sequences is critical to their function.

The cell-free assay outlined in Figure 13-23 has been widely used in studies to define the biochemical steps in the import of mitochondrial precursor proteins. In this system, respiring (energized) mitochondria extracted from cells can incorporate mitochondrial precursor proteins carrying appropriate uptake-targeting sequences that have been synthesized in the absence of mitochondria. Successful incorporation of the precursor into the organelle can be assayed either by resistance to digestion by an added protease such as trypsin. In other assays, successful import of a precursor protein can be shown by the proper cleavage of the N-terminal targeting sequences by specific mitochondrial proteases. The uptake of completely presynthesized mitochondrial precursor proteins by the organelle in this system contrasts with the cell-free cotranslational translocation of secretory proteins into the ER, which generally occurs only when microsomal (ER-derived) membranes are present during synthesis (see Figure 13-4).

## Mitochondrial Protein Import Requires Outer-Membrane Receptors and Translocons in Both Membranes

Figure 13-24 presents an overview of protein import from the cytosol into the mitochondrial matrix, the route into the mitochondrion followed by most imported proteins. We will discuss in detail each step of protein transport into the matrix and then consider how some proteins subsequently are targeted to other compartments of the mitochondrion.

After synthesis in the cytosol, the soluble precursors of mitochondrial proteins (including hydrophobic integral membrane proteins) interact directly with the mitochondrial membrane. In general, only unfolded proteins can be imported into

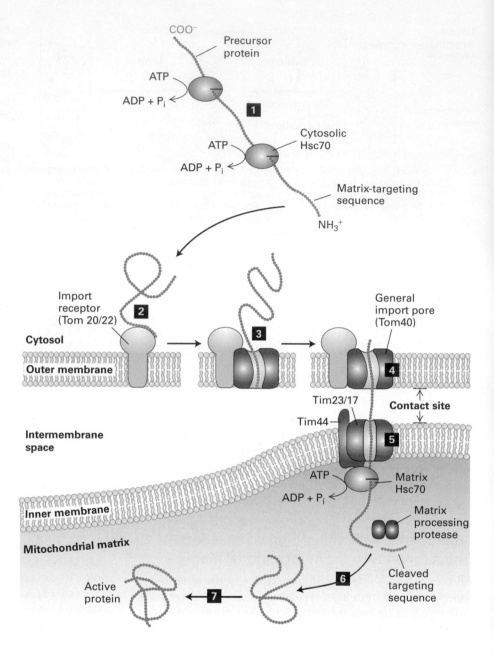

**FIGURE 13-24 Protein import into the mitochondrial matrix.** Precursor proteins synthesized on cytosolic ribosomes are maintained in an unfolded or partially folded state by bound chaperones, such as Hsc70 (step **1**). After a precursor protein binds to an import receptor near a site of contact with the inner membrane (step **2**), it is transferred into the general import pore (step **3**). The translocating protein then moves through this channel and an adjacent channel in the inner membrane (steps **4**, **5**). Note that translocation occurs at rare "contact sites" at which the inner and outer membranes appear to touch. Binding of the translocating protein by the matrix chaperone Hsc70 and subsequent ATP hydrolysis by Hsc70 helps drive import into the matrix. Once the uptake-targeting sequence is removed by a matrix protease and Hsc70 is released from the newly imported protein (step **6**), it folds into the mature, active conformation within the matrix (step **7**). Folding of some proteins depends on matrix chaperonins. [See G. Schatz, 1996, *J. Biol. Chem.* **271**:31763, and N. Pfanner et al., 1997, *Ann. Rev. Cell Devel. Biol.* **13**:25.]

the mitochondrion. Chaperone proteins such as cytosolic Hsc70 keep nascent and newly made proteins in an unfolded state so that they can be taken up by mitochondria. This process requires ATP hydrolysis. Import of an unfolded mitochondrial precursor is initiated by the binding of a mitochondrial targeting sequence to an *import receptor* in the outer mitochondrial membrane. These receptors were first identified by experiments in which antibodies to specific proteins of the outer mitochondrial membrane were shown to inhibit protein import into isolated mitochondria. Subsequent genetic experiments, in which the genes for specific mitochondrial outer-membrane proteins were mutated, showed that specific receptor proteins were responsible for the import of different classes of mitochondrial proteins. For example, N-terminal matrix-targeting sequences are recognized by Tom20 and Tom22. (Proteins in the outer mitochondrial membrane

involved in targeting and import are designated *Tom* proteins, for *t*ranslocon of the *o*uter *m*embrane.)

The import receptors subsequently transfer the precursor proteins to an import channel in the outer membrane. This channel, composed mainly of the Tom40 protein, is known as the *general import pore* because all known mitochondrial precursor proteins gain access to the interior compartments of the mitochondrion through this channel. When Tom40 is purified and incorporated into liposomes, it forms a transmembrane channel with a pore wide enough to accommodate an unfolded polypeptide chain. The general import pore forms a largely passive channel through the outer mitochondrial membrane, and the driving force for unidirectional transport into mitochondria comes from within the mitochondrion. In the case of precursors destined for the mitochondrial matrix, transfer through the outer membrane occurs simultaneously with transfer through an inner-membrane channel composed of the Tim23 and Tim17 proteins. (*Tim* stands for *t*ranslocon of the *i*nner *m*embrane.) Translocation into the matrix thus occurs at "contact sites" where the outer and inner membranes are in close proximity.

Soon after the N-terminal matrix-targeting sequence of a protein enters the mitochondrial matrix, it is removed by a protease that resides within the matrix. The emerging protein also is bound by matrix Hsc70, a chaperone that is localized to the translocation channels in the inner mitochondrial membrane by interacting with transmembrane protein Tim44. This interaction stimulates ATP hydrolysis by matrix Hsc70, and together these two proteins are thought to power translocation of proteins into the matrix.

Some imported proteins can fold into their final, active conformation without further assistance. Final folding of many matrix proteins, however, requires a **chapcronin**. As discussed in Chapter 3, chaperonin proteins actively facilitate protein folding in a process that depends on ATP. For instance, yeast mutants defective in Hsc60, a chaperonin in the mitochondrial matrix, can import matrix proteins and

cleave their uptake-targeting sequence normally, but the imported polypeptides fail to fold and assemble into the native tertiary and quaternary structures.

## Studies with Chimeric Proteins Demonstrate Important Features of Mitochondrial Import

Dramatic evidence for the ability of mitochondrial matrix-targeting sequences to direct import was obtained with chimeric proteins produced by recombinant DNA techniques. For example, the matrix-targeting sequence of alcohol dehydrogenase can be fused to the N-terminus of dihydrofolate reductase (DHFR), which normally resides in the cytosol. In the presence of chaperones, which prevent the C-terminal DHFR segment from folding in the cytosol, cell-free translocation assays show that the chimeric protein is transported into the matrix (Figure 13-25a). The inhibitor methotrexate, which binds tightly to the active site of DHFR and greatly stabilizes its folded conformation, renders the chimeric protein resistant to unfolding by cytosolic chaperones. When translocation assays are performed in the presence of methotrexate, the chimeric protein

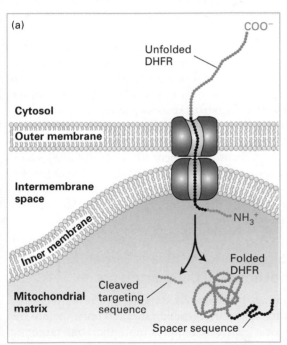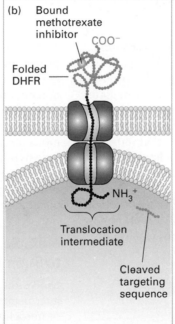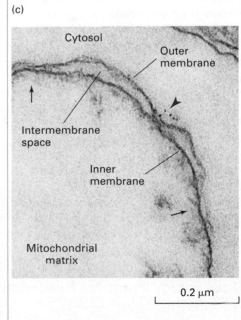

**EXPERIMENTAL FIGURE 13-25 Experiments with chimeric proteins elucidate mitochondrial protein import.** These experiments show that a matrix-targeting sequence alone directs proteins to the mitochondrial matrix and that only unfolded proteins are translocated across both membranes. The chimeric protein in these experiments contained a matrix-targeting signal at its N-terminus (red), followed by a spacer sequence of no particular function (black) and then by dihydrofolate reductase (DHFR), an enzyme normally present only in the cytosol. (a) When the DHFR segment is unfolded, the chimeric protein moves across both membranes to the matrix of energized mitochondria and the matrix-targeting signal then is removed. (b) When the C-terminus of the chimeric protein is locked in the folded state by binding of methotrexate, translocation is blocked. If

the spacer sequence is long enough to extend across both transport channels, a stable translocation intermediate, with the targeting sequence cleaved off, is generated in the presence of methotrexate, as shown here. (c) The C-terminus of the translocation intermediate in (b) can be detected by incubating the mitochondria with antibodies that bind to the DHFR segment, followed by gold particles coated with bacterial protein A, which binds nonspecifically to antibody molecules (see Figure 9-29). An electron micrograph of a sectioned sample reveals gold particles (red arrowhead) bound to the translocation intermediate at a contact site between the inner and outer membranes. Other contact sites (black arrows) also are evident. [Parts (a) and (b) adapted from J. Rassow et al., 1990, *FEBS Lett.* **275**:190. Part (c) from M. Schweiger et al., 1987, *J. Cell Biol.* **105**:235, courtesy of W. Neupert.]

does not completely enter the matrix. This finding demonstrates that a precursor must be unfolded in order to traverse the import pores in the mitochondrial membranes.

Additional studies revealed that if a sufficiently long spacer sequence separates the N-terminal matrix-targeting sequence and DHFR portion of the chimeric protein, then, in the presence of methotrexate, a translocation intermediate that spans both membranes can be trapped if enough of the polypeptide protrudes into the matrix to prevent the polypeptide chain from sliding back into the cytosol, possibly by stably associating with matrix Hsc70 (Figure 13-25b). In order for such a stable translocation intermediate to form, the spacer sequence must be long enough to span both membranes; a spacer of 50 amino acids extended to its maximum possible length is adequate to do so. If the chimera contains a shorter spacer—say, 35 amino acids—no stable translocation intermediate is obtained because the spacer cannot span both membranes. These observations provide further evidence that translocated proteins can span both inner and outer mitochondrial membranes and traverse these membranes in an unfolded state.

Microscopy studies of stable translocation intermediates show that they accumulate at sites where the inner and outer mitochondrial membranes are close together, evidence that precursor proteins enter only at such sites (Figure 13-25c). The distance from the cytosolic face of the outer membrane to the matrix face of the inner membrane at these *contact sites* is consistent with the length of an unfolded spacer sequence required for formation of a stable translocation intermediate. Moreover, stable translocation intermediates can be chemically cross-linked to the protein subunits that comprise the translocation channels of both the outer and inner membranes. This finding demonstrates that imported proteins can simultaneously engage channels in both the outer and inner mitochondrial membrane, as depicted in Figure 13-24. Since roughly 1000 stuck chimeric proteins can be observed in a typical yeast mitochondrion, it is thought that mitochondria have approximately 1000 general import pores for the uptake of mitochondrial proteins.

## Three Energy Inputs Are Needed to Import Proteins into Mitochondria

As noted previously and indicated in Figure 13-24, ATP hydrolysis by Hsc70 chaperone proteins in both the cytosol and the mitochondrial matrix is required for import of mitochondrial proteins. Cytosolic Hsc70 expends energy to maintain bound precursor proteins in an unfolded state that is competent for translocation into the matrix. The importance of ATP to this function was demonstrated in studies in which a mitochondrial precursor protein was purified and then denatured (unfolded) by urea. When tested in the cell-free mitochondrial translocation system, the denatured protein was incorporated into the matrix in the absence of ATP. In contrast, import of the native, undenatured precursor required ATP for the normal unfolding function of cytosolic chaperones.

The sequential binding and ATP-driven release of multiple matrix Hsc70 molecules to a translocating protein may simply trap the unfolded protein in the matrix. Alternatively, the matrix Hsc70, anchored to the membrane by the Tim44 protein, may act as a molecular motor to pull the protein into the matrix (see Figure 13-24). In this case, the functions of matrix Hsc70 and Tim44 would be analogous to those of the chaperone BiP and Sec63 complex, respectively, in post-translational translocation into the ER lumen (see Figure 13-9).

The third energy input required for mitochondrial protein import is a $H^+$ electrochemical gradient, or *proton-motive force,* across the inner membrane. Recall from Chapter 12 that protons are pumped from the matrix into the intermembrane space during electron transport, creating a transmembrane potential across the inner membrane. In general, only mitochondria that are actively undergoing respiration, and therefore have generated a proton-motive force across the inner membrane, are able to translocate precursor proteins from the cytosol into the mitochondrial matrix. Treatment of mitochondria with inhibitors or uncouplers of oxidative phosphorylation, such as cyanide or dinitrophenol, dissipates this proton-motive force. Although precursor proteins still can bind tightly to receptors on such poisoned mitochondria, the proteins cannot be imported, either in intact cells or in cell-free systems, even in the presence of ATP and chaperone proteins. Scientists do not fully understand how the proton-motive force is used to facilitate entry of a precursor protein into the matrix. Once a protein is partially inserted into the inner membrane, it is subjected to a transmembrane potential of 200 mV (matrix space negative). This seemingly small potential difference is established across the very narrow hydrophobic core of the lipid bilayer, which gives an enormous electric gradient, equivalent to about 400,000 V/cm. One hypothesis is that the positive charges in the amphipathic matrix-targeting sequence could simply be "electrophoresed," or pulled, into the matrix space by the inside-negative membrane electric potential.

## Multiple Signals and Pathways Target Proteins to Submitochondrial Compartments

Unlike targeting to the matrix, targeting of proteins to the intermembrane space, inner membrane, and outer membrane of mitochondria generally requires more than one targeting sequence and occurs via one of several pathways. Figure 13-26 summarizes the organization of targeting sequences in proteins sorted to different mitochondrial locations.

**Inner-Membrane Proteins** Three separate pathways are known to target proteins to the inner mitochondrial membrane. One pathway makes use of the same machinery that is used for targeting of matrix proteins (Figure 13-27, path A). A cytochrome oxidase subunit called CoxVa is a protein transported by this pathway. The precursor form of CoxVa, which contains an N-terminal matrix-targeting sequence recognized by the Tom20/22 import receptor, is transferred through the Tom40 general import pore of the outer membrane and the

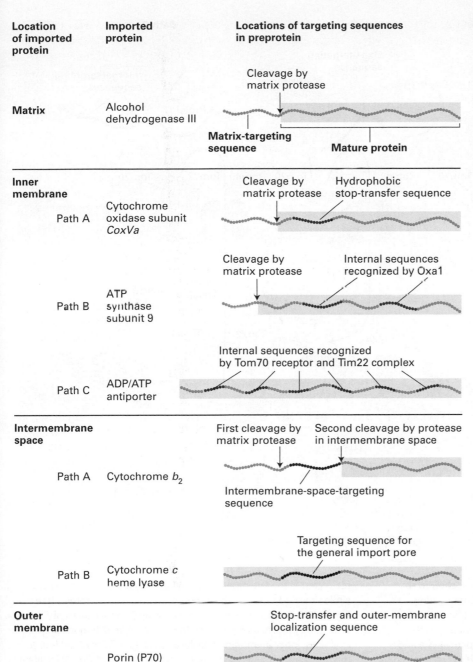

| Location of imported protein | Imported protein | Locations of targeting sequences in preprotein |
|---|---|---|
| **Matrix** | Alcohol dehydrogenase III | Cleavage by matrix protease / Matrix-targeting sequence / Mature protein |
| **Inner membrane** Path A | Cytochrome oxidase subunit *CoxVa* | Cleavage by matrix protease / Hydrophobic stop-transfer sequence |
| Path B | ATP synthase subunit 9 | Cleavage by matrix protease / Internal sequences recognized by Oxa1 |
| Path C | ADP/ATP antiporter | Internal sequences recognized by Tom70 receptor and Tim22 complex |
| **Intermembrane space** Path A | Cytochrome $b_2$ | First cleavage by matrix protease / Second cleavage by protease in intermembrane space / Intermembrane-space-targeting sequence |
| Path B | Cytochrome *c* heme lyase | Targeting sequence for the general import pore |
| **Outer membrane** | Porin (P70) | Stop-transfer and outer-membrane localization sequence |

**FIGURE 13-26 Targeting sequences in imported mitochondrial proteins.** Most mitochondrial proteins have an N-terminal matrix-targeting sequence (pink) that is similar but not identical in different proteins. Proteins destined for the inner membrane, the intermembrane space, or the outer membrane have one or more additional targeting sequences that function to direct the proteins to these locations by several different pathways. The lettered pathways correspond to those illustrated in Figures 13-26 and 13-27. [See W. Neupert, 1997, *Ann. Rev. Biochem.* **66**:863.]

inner-membrane Tim23/17 translocation complex. In addition to the matrix-targeting sequence, which is cleaved during import, CoxVa contains a hydrophobic stop-transfer sequence. As the protein passes through the Tim23/17 channel, the stop-transfer sequence blocks translocation of the C-terminus across the inner membrane. The membrane-anchored intermediate is then transferred laterally into the bilayer of the inner membrane much as type I integral membrane proteins are incorporated into the ER membrane (see Figure 13-11).

A second pathway to the inner membrane is followed by proteins (e.g., ATP synthase subunit 9) whose precursors contain both a matrix-targeting sequence and internal hydrophobic

domains recognized by an inner-membrane protein termed *Oxa1*. This pathway is thought to involve translocation of at least a portion of the precursor into the matrix via the Tom40 and Tim23/17 channels. After cleavage of the matrix-targeting sequence, the protein is inserted into the inner membrane by a process that requires interaction with Oxa1 and perhaps other inner-membrane proteins (Figure 13-27, path B). Oxa1 is related to a bacterial protein involved in inserting some cytoplasmic membrane proteins in bacteria. This relatedness suggests that Oxa1 may have descended from the translocation machinery in the endosymbiotic bacterium that eventually became the mitochondrion. However, the proteins forming the inner-membrane channels in mitochondria are not related to the

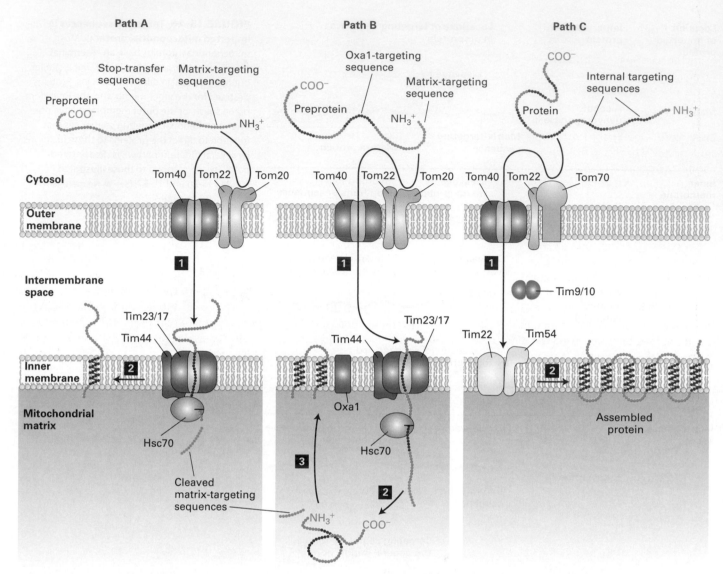

**Path A**

Preprotein
COO⁻
Stop-transfer sequence
Matrix-targeting sequence
NH₃⁺

**Path B**

Oxa1-targeting sequence
COO⁻
Preprotein
Matrix-targeting sequence
NH₃⁺

**Path C**

COO⁻
Protein
Internal targeting sequences
NH₃⁺

**Cytosol**

Tom40  Tom22  Tom20

Tom40  Tom22  Tom20

Tom40  Tom22  Tom70

**Outer membrane**

**Intermembrane space**

Tim9/10

Tim23/17
Tim44

Tim23/17
Tim44
Oxa1

Tim22  Tim54

**Inner membrane**

**Mitochondrial matrix**

Hsc70
Cleaved matrix-targeting sequences

Hsc70
NH₃⁺
COO⁻

Assembled protein

**FIGURE 13-27 Three pathways to the inner mitochondrial membrane from the cytosol.** Proteins with different targeting sequences are directed to the inner membrane via different pathways. In all three pathways, proteins cross the outer membrane via the Tom40 general import pore. Proteins delivered by pathways A and B contain an N-terminal matrix-targeting sequence that is recognized by the Tom20/22 import receptor in the outer membrane. Although both these pathways use the Tim23/17 inner-membrane channel, they differ in that the entire precursor protein enters the matrix and then is redirected to the inner membrane in pathway B. Matrix Hsc70 plays a role similar to its role in the import of soluble matrix proteins (see Figure 13-23). Proteins delivered by pathway C contain internal sequences that are recognized by the Tom70/Tom22 import receptor; a different inner-membrane translocation channel (Tim22/54) is used in this pathway. Two intermembrane proteins (Tim9 and Tim10) facilitate transfer between the outer and inner channels. See the text for discussion. [See R. E. Dalbey and A. Kuhn, 2000, *Ann. Rev. Cell Dev. Biol.* **16**:51, and N. Pfanner and A. Geissler, 2001, *Nature. Rev. Mol. Cell Biol.* **2**:339.]

proteins in bacterial translocons. Oxa1 also participates in the inner-membrane insertion of certain proteins (e.g., subunit II of cytochrome oxidase) that are encoded by mitochondrial DNA and synthesized in the matrix by mitochondrial ribosomes.

The final pathway for insertion in the inner mitochondrial membrane is followed by multipass proteins that contain six membrane-spanning domains, such as the ADP/ATP antiporter. These proteins, which lack the usual N-terminal matrix-targeting sequence, contain multiple internal mitochondrial targeting sequences. After the internal sequences are recognized by a second import receptor composed of outer-membrane proteins Tom70 and Tom22, the imported protein passes through the outer membrane via the general import pore (Figure 13-27, path C). The protein then is transferred to a second translocation complex in the inner membrane composed of the Tim22, Tim18, and Tim54 proteins. Transfer to the Tim22/18/54 complex depends on a multimeric complex of two small proteins, Tim9 and Tim10, which reside in the intermembrane space. The small Tim proteins are thought to act as chaperones, guiding imported protein precursors from the general import pore to the Tim22/18/54 complex in the inner membrane by binding to their hydrophobic regions, preventing them from forming insoluble aggregates in the aqueous environment of the intermembrane space. Ultimately the Tim22/18/54 complex is responsible for incorporating the multiple hydrophobic segments of the imported protein into the inner membrane.

**Intermembrane-Space Proteins** Two pathways deliver cytosolic proteins to the space between the inner and outer mitochondrial membranes. The major pathway is followed by proteins, such as cytochrome $b_2$, whose precursors carry two different N-terminal targeting sequences, both of which ultimately are cleaved. The most N-terminal of the two sequences is a matrix-targeting sequence, which is removed by the matrix protease. The second targeting sequence is a hydrophobic segment that blocks complete translocation of the protein across the inner membrane (Figure 13-28, path A). After the resulting membrane-embedded intermediate diffuses laterally away from the Tim23/17 translocation channel, a protease in the membrane cleaves the protein near the hydrophobic transmembrane segment, releasing the mature protein in a soluble form into the intermembrane space. Except for the second

proteolytic cleavage, this pathway is similar to that of inner-membrane proteins such as CoxVa (see Figure 13-27, path A).

The small Tim9 and Tim10 proteins, which reside in the intermembrane space, illustrate a second pathway for targeting to the intermembrane space. In this pathway, the imported proteins do not contain an N-terminal matrix-targeting sequence and are delivered directly to the intermembrane space via the general import pore without involvement of any inner-membrane translocation factors (Figure 13-28, path B). Translocation through the Tom40 general import pore does not seem to be coupled to any energetically favorable process; however, once located in the intermembrane space, Tim9 and Tim10 proteins acquire two disulfide bonds each and acquire compact stable folded structures. Apparently, the mechanism that drives unidirectional translocation through the outer membrane involves passive diffusion through the outer membrane, followed by folding and disulfide bond formation which irreversibly traps the protein in the intermembrane space. In many respects the process of disulfide bond formation in the intermembrane space resembles that of the ER lumen and involves a disulfide bond-generating protein Erv1 and a disulfide transfer protein Mia40.

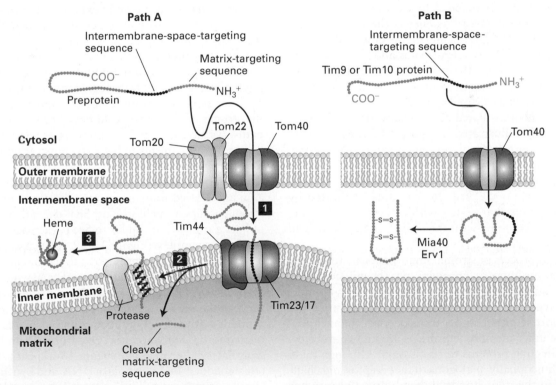

**FIGURE 13-28 Two pathways to the mitochondrial intermembrane space.** Pathway A, the major one for delivery of proteins |from the cytosol to the intermembrane space, is similar to pathway A for delivery to the inner membrane (see Figure 13-26). The major difference is that the internal targeting sequence in proteins such as cytochrome $b_2$ destined for the intermembrane space is recognized by an inner-membrane protease, which cleaves the protein on the intermembrane-space side of the membrane. The released protein then folds and binds to its heme cofactor within the intermembrane

space. Pathway B is a specialized pathway for delivery to the intermembrane space of the proteins Tim9 and Tim10. These proteins readily pass through the Tom40 general import pore and, once they are in the intermembrane space, fold and form disulfide bonds that prevent reverse translocation through Tom40. The disulfide bonds are generated by Erv1 and are transferred to Tim9 and Tim10 by Mia40. [See R. E. Dalbey and A. Kuhn, 2000, *Ann. Rev. Cell Dev. Biol.* **16**:51; N. Pfanner and A. Geissler, 2001, *Nat. Rev. Mol. Cell Biol.* **2**:339, and K. Tokatlidis, 2005, A disulfide relay system in mitochondria. *Cell* **121**:965–967.]

**Outer-Membrane Proteins** Many of the proteins that reside in the mitochondrial outer membrane, including the Tom40 pore itself and mitochondrial porin, have a β-barrel structure in which antiparallel strands form hydrophobic transmembrane segments surrounding a central channel. Such proteins are incorporated into the outer membrane by first interacting with the general import pore, Tom40, and then they are transferred to a complex known as the SAM (*sorting and assembly machinery*) complex, which is composed of at least three outer-membrane proteins. Presumably it is the very stable hydrophobic nature of β-barrel proteins that ultimately causes them to be stably incorporated into the outer membrane, but precisely how the SAM complex facilitates this process is not known.

## Targeting of Chloroplast Stromal Proteins Is Similar to Import of Mitochondrial Matrix Proteins

Among the proteins found in the chloroplast stroma are the enzymes of the Calvin cycle, which functions in fixing carbon dioxide into carbohydrates during photosynthesis (Chapter 12). The large (L) subunit of ribulose 1,5-bisphosphate carboxylase (rubisco) is encoded by chloroplast DNA and synthesized on chloroplast ribosomes in the stromal space. The small (S) subunit of rubisco and all the other Calvin-cycle enzymes are encoded by nuclear genes and transported to chloroplasts after their synthesis in the cytosol. The precursor forms of these stromal proteins contain an N-terminal *stromal-import* sequence (see Table 13-1).

Experiments with isolated chloroplasts, similar to those with mitochondria illustrated in Figure 13-23, have shown that they can import the S-subunit precursor after its synthesis. After the unfolded precursor enters the stromal space, it binds transiently to a stromal Hsc70 chaperone and the N-terminal sequence is cleaved. In reactions facilitated by Hsc60 chaperonins that reside within the stromal space, eight S subunits combine with the eight L subunits to yield the active rubisco enzyme.

The general process of stromal import appears to be very similar to that for importing proteins into the mitochondrial matrix (see Figure 13-24). At least three chloroplast outer-membrane proteins, including a receptor that binds the stromal-import sequence and a translocation channel protein, and five inner-membrane proteins are known to be essential for directing proteins to the stroma. Although these proteins are functionally analogous to the receptor and channel proteins in the mitochondrial membrane, they are not structurally homologous. The lack of sequence similarity between these chloroplast and mitochondrial proteins suggests that they may have arisen independently during evolution.

The available evidence suggests that chloroplast stromal proteins, like mitochondrial matrix proteins, are imported in the unfolded state. Import into the stroma depends on ATP hydrolysis catalyzed by a stromal Hsc70 chaperone whose function is similar to that of Hsc70 in the mitochondrial

matrix and BiP in the ER lumen. Unlike mitochondria, chloroplasts do not generate an electrochemical gradient (proton-motive force) across their inner membrane. Thus protein import into the chloroplast stroma appears to be powered solely by ATP hydrolysis.

## Proteins Are Targeted to Thylakoids by Mechanisms Related to Translocation Across the Bacterial Cytoplasmic Membrane

In addition to the double membrane that surrounds them, chloroplasts contain a series of internal interconnected membranous sacs, the **thylakoids** (see Figure 12-31). Proteins localized to the thylakoid membrane or lumen carry out photosynthesis. Many of these proteins are synthesized in the cytosol as precursors containing multiple targeting sequences. For example, plastocyanin and other proteins destined for the thylakoid lumen require the successive action of two uptake-targeting sequences. The first is an N-terminal stromal-import sequence that directs the protein to the stroma by the same pathway that imports the rubisco S subunit. The second sequence targets the protein from the stroma to the thylakoid lumen. The role of these targeting sequences has been shown in experiments measuring the uptake of mutant proteins generated by recombinant DNA techniques into isolated chloroplasts. For instance, mutant plastocyanin that lacks the thylakoid-targeting sequence but contains an intact stromal-import sequence accumulates in the stroma and is not transported into the thylakoid lumen.

Four separate pathways for transporting proteins from the stroma into the thylakoid have been identified. All four pathways have been found to be closely related to analogous transport mechanisms in bacteria, illustrating the close evolutionary relationship between the stromal membrane and the bacterial cytoplasmic membrane. Transport of plastocyanin and related proteins into the thylakoid lumen from the stroma occurs by a chloroplast SRP-dependent pathway that utilizes a translocon similar to SecY, the bacterial version of the Sec61 complex (Figure 13-29, *left*). A second pathway for transporting proteins into the thylakoid lumen involves a protein related to bacterial protein SecA, which uses the energy from ATP hydrolysis to drive protein translocation through the SecY translocon. A third pathway, which targets proteins to the thylakoid membrane, depends on a protein related to the mitochondrial Oxa1 protein and the homologous bacterial protein (see Figure 13-27, path B). Some proteins encoded by chloroplast DNA and synthesized in the stroma or transported into the stroma from the cytosol are inserted into the thylakoid membrane via this pathway.

Finally, thylakoid proteins that bind metal-containing cofactors follow another pathway into the thylakoid lumen (Figure 13-29, *right*). The unfolded precursors of these proteins are first targeted to the stroma, where the N-terminal stromal-import sequence is cleaved off and the protein then folds and binds its cofactor. A set of thylakoid-membrane proteins assists in translocating the folded protein and bound

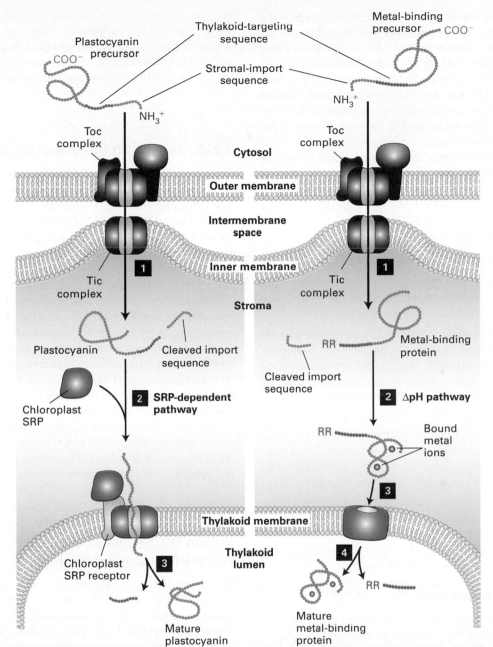

**FIGURE 13-29 Transporting proteins to chloroplast thylakoids.**
Two of the four pathways for transporting proteins from the cytosol to the thylakoid lumen are shown here. In these pathways, unfolded precursors are delivered to the stroma via the same outer-membrane proteins that import stromal-localized proteins. Cleavage of the N-terminal stromal-import sequence by a stromal protease then reveals the thylakoid-targeting sequence (step **1**). At this point the two pathways diverge. In the SRP-dependent pathway (*left*), plastocyanin and similar proteins are kept unfolded in the stromal space by a set of chaperones (not shown) and, directed by the thylakoid-targeting sequence, bind to proteins that are closely related to the bacterial SRP, SRP receptor, and SecY translocon, which mediate movement into the lumen (step **2**). After the thylakoid-targeting sequence is removed in

the thylakoid lumen by a separate endoprotease, the protein folds into its mature conformation (step **3**). In the pH-dependent pathway (*right*), metal-binding proteins fold in the stroma, and complex redox cofactors are added (step **2**). Two arginine residues (RR) at the N-terminus of the thylakoid-targeting sequence and a pH gradient across the inner membrane are required for transport of the folded protein into the thylakoid lumen (step **3**). The translocon in the thylakoid membrane is composed of at least four proteins related to proteins in the bacterial cytoplasmic membrane. The thylakoid targeting sequence containing the two arginine residues is cleaved in the thylakoid lumen (step **4**). [See R. Dalbey and C. Robinson, 1999, *Trends Biochem. Sci.* **24**:17; R. E. Dalbey and A. Kuhn, 2000, *Ann. Rev. Cell Dev. Biol.* **16**:51; and C. Robinson and A. Bolhuis, 2001, *Nat. Rev. Mol. Cell Biol.* **2**:350.]

cofactor into the thylakoid lumen, a process powered by the $H^+$ electrochemical gradient normally maintained across the thylakoid membrane. The thylakoid-targeting sequence that directs a protein to this pathway includes two closely spaced arginine residues that are crucial for recognition. Bacterial cells also have a mechanism for translocating folded proteins with a similar arginine-containing sequence across the cytoplasmic membrane, known as the Tat (*t*win-*a*rginine *t*ranslocation) pathway. The molecular mechanism whereby these large folded globular proteins can be translocated across the thylakoid membrane is currently under intense study.

## KEY CONCEPTS of Section 13.4

### Targeting of Proteins to Mitochondria and Chloroplasts

- Most mitochondrial and chloroplast proteins are encoded by nuclear genes, synthesized on cytosolic ribosomes, and imported post-translationally into the organelles.

- All the information required to target a precursor protein from the cytosol to the mitochondrial matrix or chloroplast stroma is contained within its N-terminal uptake-targeting sequence. After protein import, the uptake-targeting sequence is removed by proteases within the matrix or stroma.

- Cytosolic chaperones maintain the precursors of mitochondrial and chloroplast proteins in an unfolded state. Only unfolded proteins can be imported into the organelles. Translocation in mitochondria occurs at sites where the outer and inner membranes of the organelles are close together.

- Proteins destined for the mitochondrial matrix bind to receptors on the outer mitochondrial membrane and then are transferred to the general import pore (Tom40) in the outer membrane. Translocation occurs concurrently through the outer and inner membranes, driven by the proton-motive force across the inner membrane and ATP hydrolysis by the Hsc70 ATPase in the matrix (see Figure 13-24).

- Proteins sorted to mitochondrial destinations other than the matrix usually contain two or more targeting sequences, one of which may be an N-terminal matrix-targeting sequence (see Figure 13-26).

- Some mitochondrial proteins destined for the intermembrane space or inner membrane are first imported into the matrix and then redirected; others never enter the matrix but go directly to their final location.

- Protein import into the chloroplast stroma occurs through inner-membrane and outer-membrane translocation channels that are analogous in function to mitochondrial channels, but composed of proteins unrelated in sequence to the corresponding mitochondrial proteins.

- Proteins destined for the thylakoid have secondary targeting sequences. After entry of these proteins into the stroma, cleavage of the stromal-targeting sequences reveals the thylakoid-targeting sequences.

- The four known pathways for moving proteins from the chloroplast stroma to the thylakoid closely resemble translocation across the bacterial cytoplasmic membrane (see Figure 13-29). One of these systems can translocate folded proteins.

## 13.5 Targeting of Peroxisomal Proteins

Peroxisomes are small organelles bounded by a single membrane. Unlike mitochondria and chloroplasts, peroxisomes lack DNA and ribosomes. Thus all peroxisomal proteins are encoded by nuclear genes, synthesized on ribosomes free in the cytosol, and then incorporated into preexisting or newly generated peroxisomes. As peroxisomes are enlarged by addition of protein (and lipid), they eventually divide, forming new ones, as is the case with mitochondria and chloroplasts.

The size and enzyme composition of peroxisomes vary considerably in different kinds of cells. However, all peroxisomes contain enzymes that use molecular oxygen to oxidize various substrates such as amino acids and fatty acids, breaking them down into smaller components for use in biosynthetic pathways. The hydrogen peroxide ($H_2O_2$) generated by these oxidation reactions is extremely reactive and potentially harmful to cellular components; however, the peroxisome also contains enzymes, such catalase, that efficiently convert $H_2O_2$ into $H_2O$. In mammals, peroxisomes are most abundant in liver cells, where they constitute about 1–2 percent of the cell volume.

### Cytosolic Receptor Targets Proteins with an SKL Sequence at the C-Terminus into the Peroxisomal Matrix

Peroxisomal targeting signals were first identified by testing deletions of peroxisomal proteins for a specific defect in peroxisomal targeting. In one early study, the gene for firefly luciferase was expressed in cultured insect cells and the resulting protein was shown to be properly targeted to the peroxisome. However, expression of a truncated gene missing a small portion of the C-terminus of the protein led to luciferase that failed to be targeted to the peroxisome and remained in the cytoplasm. By testing various mutant luciferase proteins in this system, researchers discovered that the sequence Ser-Lys-Leu (SKL in one-letter code) or a related sequence at the C-terminus was necessary for peroxisomal targeting. Further, addition of the SKL sequence to the C-terminus of a normally cytosolic protein leads to uptake of the altered protein by peroxisomes in cultured cells. All but a few of the many different peroxisomal matrix proteins bear a sequence of this type, known as *peroxisomal-targeting sequence 1*, or simply *PTS1*.

The pathway for import of catalase and other PTS1-bearing proteins into the peroxisomal matrix is depicted in Figure 13-30. The PTS1 binds to a soluble carrier protein in the cytosol (Pex5), which in turn binds to a receptor in the peroxisome

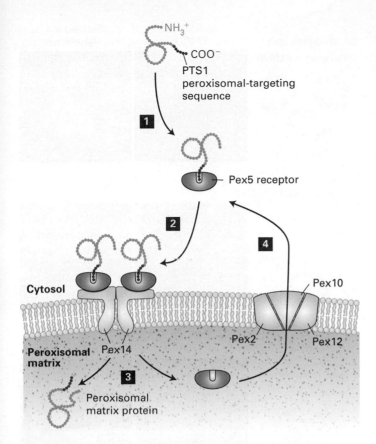

**FIGURE 13-30 PTS1-directed import of peroxisomal matrix proteins.** Step **1**: Most peroxisomal matrix proteins contain a C-terminal PTS1 uptake-targeting sequence (red) that binds to the cytosolic receptor Pex5. Step **2**: Pex5 with the bound matrix protein forms a multimeric complex with the Pex14 receptor located on the peroxisome membrane. Step **3**: By a process that is not well understood, the matrix protein-Pex5 complex is then transferred into the peroxisomal matrix, where Pex5 dissociates from the matrix protein. Step **4**: Pex5 is then returned to the cytosol by a process that involves the peroxisomal membrane proteins Pex2, Pex10, and Pex12, as well as additional membrane and cytosolic proteins not shown. Note that folded proteins can be imported into peroxisomes and that the targeting sequence is not removed in the matrix. [See P. E. Purdue and P. B. Lazarow, 2001, *Ann. Rev. Cell Dev. Biol.* **17**:701; S. Subramani et al., 2000, *Ann. Rev. Biochem.* **69**:399; and V. Dammai and S. Subramani, 2001, *Cell* **105**:187.]

membrane (Pex14). The protein to be imported then moves across the peroxisomal membrane while still bound to Pex5. The peroxisome import machinery, unlike most systems that mediate protein import into the ER, mitochondria, and chloroplasts, can translocate folded proteins across the membrane. For example, catalase assumes a folded conformation and binds to heme in the cytoplasm before traversing the peroxisomal membrane. Cell-free studies have shown that the peroxisome import machinery can transport large macromolecular objects, including gold particles of about 9 nm in diameter, as long as they have a PTS1 tag attached to them. However, peroxisomal membranes do not appear to contain large stable pore structures, such as the nuclear pore described in the next section. The fundamental mechanism of peroxisomal matrix

protein translocation is not well understood but probably involves the formation of oligomers of Pex5 bound to PTS1-bearing cargo molecules and the Pex14 receptor. There is evidence that the size of the oligomer adjusts according to the size of the PTS1-bearing cargo molecules and that the oligomers dissociate once the complex of Pex5 bound to PTS1-bearing cargo molecule enters the peroxisomal matrix. The dynamic formation of oligomers apparently is the key mechanism by which PTS1-bearing cargo molecules can be accommodated without the formation of large stable pores that would disrupt the integrity of the peroxisomal membrane.

Once the complex of a PTS1-bearing cargo molecule bound to Pex5 enters the matrix, Pex5 dissociates from the peroxisomal matrix protein for recycling back to the cytoplasm. The peroxisomal membrane proteins Pex10, Pex12, and Pex2 form a complex that is crucial for recycling of Pex5. Pex5 is modified by ubiquitination and then is deubiquitinated as part of the recycling process. Since ubiquitin modification of proteins ultimately requires ATP hydrolysis, the energy-dependent recycling of Pex5 may be the step in the import process that uses energy to power unidirectional translocation of cargo molecules across the peroxisomal membrane.

A few peroxisomal matrix proteins such as thiolase are synthesized as precursors with an N-terminal uptake-targeting sequence known as *PTS2*. These proteins bind to a different cytosolic receptor protein, but otherwise import is thought to occur by the same mechanism as for PTS1-containing proteins.

## Peroxisomal Membrane and Matrix Proteins Are Incorporated by Different Pathways

Autosomal recessive mutations that cause defective peroxisome assembly occur naturally in the human population. Such defects can lead to severe developmental defects often associated with craniofacial abnormalities. In *Zellweger syndrome* and related disorders, for example, the transport of many or all proteins into the peroxisomal matrix is impaired: newly synthesized peroxisomal enzymes remain in the cytosol and are eventually degraded. Genetic analyses of cultured cells from different Zellweger patients and of yeast cells carrying similar mutations have identified more than 20 genes that are required for peroxisome biogenesis. ∎

Studies with peroxisome-assembly mutants have shown that different pathways are used for importing peroxisomal matrix proteins versus inserting proteins into the peroxisomal membrane. For example, analysis of cells from some Zellweger patients led to identification of genes encoding the Pex5-recycling proteins Pex10, Pex12, and Pex2. Mutant cells defective in any one of these proteins cannot incorporate matrix proteins into peroxisomes; nonetheless, the cells contain empty peroxisomes that have a normal complement of peroxisomal membrane proteins (Figure 13-31b). Mutations in any one of three other genes were found to block insertion of peroxisomal membrane proteins as well as import of matrix proteins (Figure 13-31c). These findings demonstrate that one set of proteins translocates soluble proteins

**EXPERIMENTAL FIGURE 13-31 Studies reveal different pathways for incorporation of peroxisomal membrane and matrix proteins.** Cells were stained with fluorescent antibodies to PMP70, a peroxisomal membrane protein, or with fluorescent antibodies to catalase, a peroxisomal matrix protein, then viewed in a fluorescent microscope. (a) In wild-type cells, both peroxisomal membrane and matrix proteins are visible as bright foci in numerous peroxisomal bodies. (b) In cells from a Pex12-deficient patient, catalase is distributed uniformly throughout the cytosol, whereas PMP70 is localized normally to peroxisomal bodies. (c) In cells from a Pex3-deficient patient, peroxisomal membranes cannot assemble, and as a consequence, peroxisomal bodies do not form. Thus both catalase and PMP70 are mis-localized to the cytosol. [Courtesy of Stephen Gould, Johns Hopkins University.]

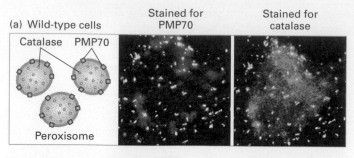

(a) Wild-type cells · Stained for PMP70 · Stained for catalase

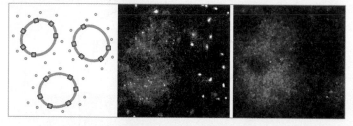

(b) Pex12 mutants (deficient in matrix protein import)

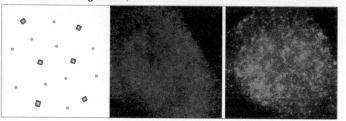

(c) Pex3 mutants (deficient in membrane biogenesis)

into the peroxisomal matrix, but a different set is required for insertion of proteins into the peroxisomal membrane. This situation differs markedly from that of the ER, mitochondrion, and chloroplast, for which, as we have seen, membrane proteins and soluble proteins share many of the same components for their insertion into these organelles.

Although most peroxisomes are generated by division of preexisting organelles, these organelles can arise de novo by the three-stage process depicted in Figure 13-32. In this case, peroxisome assembly begins in the ER. At least two peroxisomal membrane proteins, Pex3 and Pex16, are inserted into the ER membrane by the mechanisms described in Section 13.2. Pex3 and Pex16 then recruit Pex19 to form a specialized region of the ER membrane that can bud off of the ER to form a peroxisomal precursor membrane. Current evidence indicates that peroxisomal membrane protein assembly into mature peroxisomes may also follow the same Pex19-dependent pathway for the de novo formation of new peroxisomes from the ER. The insertion of peroxisomal membrane proteins generates membranes that have all the components necessary for import of matrix proteins, leading to the formation of mature, functional peroxisomes.

Division of mature peroxisomes, which largely determines the number of peroxisomes within a cell, depends on still another protein, Pex11. Overexpression of the Pex11 protein causes a large increase in the number of peroxisomes, suggesting that this protein controls the extent of peroxisome division. The small peroxisomes generated by division can be

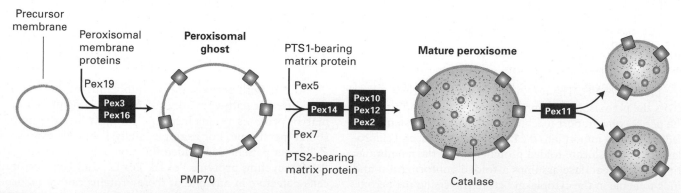

**FIGURE 13-32 Model of peroxisomal biogenesis and division.** The first stage in the de novo formation of peroxisomes is the incorporation of peroxisomal membrane proteins into precursor membranes derived from the ER. Pex19 acts as the receptor for membrane-targeting sequences. A complex of Pex3 and Pex16 is required for proper insertion of proteins (e.g., PMP70) into the forming peroxisomal membrane. Insertion of all peroxisomal membrane proteins produces a peroxisomal ghost, which is capable of importing proteins targeted to the matrix. The pathways for importing PTS1- and PTS2-bearing matrix proteins differ only in the identity of the cytosolic receptor (Pex5 and Pex7, respectively) that binds the targeting sequence (see Figure 13-30). Complete incorporation of matrix proteins yields a mature peroxisome. Although peroxisomes can form de novo as just described, under most conditions the proliferation of peroxisomes involves division of mature peroxisomes, a process that depends on the Pex11 protein.

enlarged by incorporation of additional matrix and membrane proteins via the same pathways described previously.

## KEY CONCEPTS of Section 13.5

### Targeting of Peroxisomal Proteins

- All peroxisomal proteins are synthesized on cytosolic ribosomes and incorporated into the organelle post-translationally.

- Most peroxisomal matrix proteins contain a C-terminal PTS1 targeting sequence; a few have an N-terminal PTS2 targeting sequence. Neither targeting sequence is cleaved after import.

- All proteins destined for the peroxisomal matrix bind to a cytosolic carrier protein, which differs for PTS1- and PTS2-bearing proteins, and then are directed to common import receptor and translocation machinery on the peroxisomal membrane (see Figure 13-30).

- Translocation of matrix proteins across the peroxisomal membrane depends on ATP hydrolysis. Unlike protein import to the ER, mitochondrion, and chloroplast, many peroxisomal matrix proteins fold in the cytosol and traverse the membrane in a folded conformation.

- Proteins destined for the peroxisomal membrane contain different targeting sequences than peroxisomal matrix proteins and are imported by a different pathway.

- Unlike mitochondria and chloroplasts, peroxisomes can arise de novo from precursor membranes probably derived from the ER as well as by division of preexisting organelles (see Figure 13-32).

## 13.6 Transport into and out of the Nucleus

The nucleus is separated from the cytoplasm by two membranes, which form the **nuclear envelope** (see Figure 9-32). The nuclear envelope is continuous with the ER and forms a part of it. Transport of proteins from the cytoplasm into the nucleus and movement of macromolecules, including mRNAs, tRNAs, and ribosomal subunits, out of the nucleus occur through *nuclear pores,* which span both membranes of the nuclear envelope. Import of proteins into the nucleus shares some fundamental features with protein import into other organelles. For example, imported nuclear proteins carry specific targeting sequences known as nuclear localization sequences, or NLSs. However, proteins are imported into the nucleus in a folded state, and thus nuclear import differs fundamentally from protein translocation across the membranes of the ER, mitochondrion, and chloroplast, where proteins are unfolded during translocation. In this section we discuss the main mechanism by which proteins and some ribonuclear proteins such as ribosomes enter and exit the nucleus. We also discuss how mRNAs and other ribonuclear protein complexes are exported from the nucleus by a process that differs mechanistically from nuclear protein import.

## Large and Small Molecules Enter and Leave the Nucleus via Nuclear Pore Complexes

Numerous pores perforate the nuclear envelope in all eukaryotic cells. Each nuclear pore is formed from an elaborate structure termed the **nuclear pore complex (NPC),** which is one of the largest protein assemblages in the cell. The total mass of the pore structure is 60–80 million Da in vertebrates, which is about 16 times larger than a ribosome. An NPC is made up of multiple copies of some 30 different proteins called **nucleoporins.** Electron micrographs of nuclear pore complexes reveal an octagonal, membrane-embedded ring structure that surrounds a largely aqueous pore (Figure 13-33). Eight approximately 100-nm-long filaments extend into the nucleoplasm with the distal ends of these filaments joined by a terminal ring, forming a structure called the *nuclear basket.* Cytoplasmic filaments extend from the cytoplasmic side of the NPC into the cytosol.

Ions, small metabolites, and globular proteins up to about 40 kDa can diffuse passively through the central aqueous region of the nuclear pore complex. However, large proteins and ribonucleoprotein complexes cannot diffuse in and out of the nucleus. Rather, these macromolecules are actively transported through the NPC with the assistance of soluble transporter proteins that bind macromolecules and also interact with nucleoporins. The capacity and efficiency of the NPC for such active transport is remarkable. In one minute, each NPC is estimated to transport 60,000 protein molecules into the nucleus, 50–250 mRNA molecules, 10–20 ribosomal subunits, and 1000 tRNAs out of the nucleus.

In general terms, the nuclearporins are of three types: *structural nucleoporins, membrane nucleoporins,* and *FG-nucleoporins.* The structural nucleoporins form the scaffold of the nuclear pore, which is a ring of eightfold symmetry that traverses both membranes of the nuclear envelope, creating an annulus. The membranes of the inner and outer leaflets of the nuclear envelope connect at the NPC by a highly curved region of membrane that contains the embedded membrane nucleoporins (see Figure 13-33b). A set of seven structural nucleoporins forms a Y-shaped structure about the size of the ribosome, known as the *Y-complex.* Sixteen copies of the Y-complex form the basic structural scaffold of the pore, which has bilateral symmetry across the nuclear envelope and eightfold rotational symmetry in the plane of the envelope (see Figure 13-33c). A structural motif repeated several times within the Y-complex is closely related to a structure found in the COPII proteins that drive formation of coated vesicles within cells (see Chapter 14). This primordial relationship between nuclear pore structural proteins and vesicle coat proteins suggests that the two types of membrane coat complexes share a common origin. The basic function of this element may be to form a protein lattice that, in a complex with membrane nucleoporins, deforms the membrane into a highly curved structure.

The FG-nucleoporins, which line the channel of the nuclear pore complex and also are found associated with the nuclear basket and the cytoplasmic filaments, contain multiple repeats of short hydrophobic sequences that are rich in

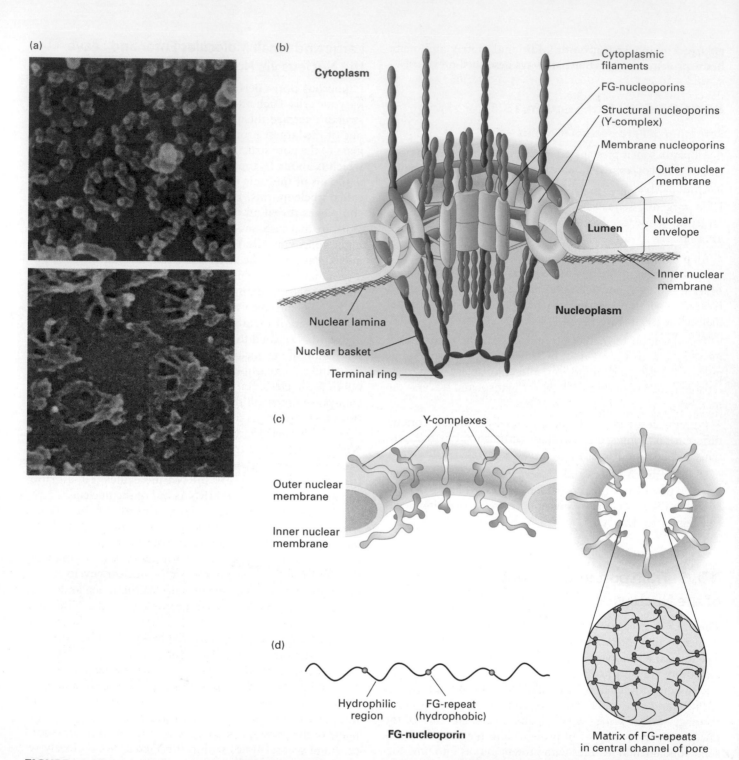

**FIGURE 13-33 Nuclear pore complex at different levels of resolution.** (a) Visualized by scanning electron microscopy, nuclear envelopes from the large nuclei of *Xenopus* oocytes. *Top:* View of the cytoplasmic face reveals octagonal shape of membrane-embedded portion of nuclear pore complexes. *Bottom:* View of the nucleoplasmic face shows the nuclear basket that extends from the membrane portion. (b) Cutaway model of the pore complex showing the major structural features formed by membrane nucleoporins, structural nucleoporins, and FG-nucleoporins. (c) Sixteen copies of the Y-complex forms a major part of the structural scaffold of the nuclear pore complex. The three-dimensional structure of the Y-complex is modeled into the pore structure. Note the twofold symmetry across the double membrane of the nucleus (*left*) and the eightfold rotational symmetry around the axis of the pore (*right*). (d) The FG-nucleoporins have extended disordered structures that are composed of repeats of the sequence Phe–Gly interspersed with hydrophilic regions (*left*). The FG-nucleoporins are most abundant in the central part of the pore, and the FG-repeat sequences are thought to fill the central channel with a gel-like matrix (*right*). [Part (a) from V. Doye and E. Hurt, 1997, *Curr. Opin. Cell Biol.* **9:**401, courtesy of M. W. Goldberg and T. D. Allen. Part (b) adapted from M. P. Rout and J. D. Atchison, 2001, *J. Biol. Chem.* **276:**16593. Part (c) courtesy of Thomas Schwartz. Part (d) adapted from K. Ribbeck and D. Görlich, 2001, *EMBO J.* **20:**1320–1330.]

phenylalanine (F) and glycine (G) residues (FG-repeats). The hydrophobic FG-repeats are thought to occur in regions of extended, otherwise hydrophilic polypeptide chains that fill the central transporter channel. The FG-nucleoporins are essential for the function of the NPC; however, the NPC remains functional even if up to half of the FG-repeats have been deleted. The FG-nucleoporins are thought to form a flexible gel-like matrix with bulk properties that allow the diffusion of small molecules while excluding unchaperoned hydrophilic proteins larger than 40 kDa (see Figure 13-33d).

## Nuclear Transport Receptors Escort Proteins Containing Nuclear-Localization Signals into the Nucleus

All proteins found in the nucleus—such as histones, transcription factors, and DNA and RNA polymerases—are synthesized in the cytoplasm and imported into the nucleus through nuclear pore complexes. Such proteins contain a *nuclear-localization signal (NLS)* that directs their selective transport into the nucleus. NLSs were first discovered through the analysis of mutants of the gene for large T-antigen encoded by simian virus 40 (SV40). The wild-type form of large T-antigen is localized to the nucleus in virus-infected cells, whereas some mutated forms of large T-antigen accumulate in the cytoplasm (Figure 13-34). The mutations responsible for this altered cellular localization all occur within a specific seven-residue sequence rich in basic

amino acids near the C-terminus of the protein: Pro-Lys-Lys-Lys-Arg-Lys-Val. Experiments with engineered hybrid proteins in which this sequence was fused to a cytosolic protein demonstrated that it directs transport into the nucleus and consequently functions as an NLS. NLS sequences subsequently were identified in numerous other proteins imported into the nucleus. Many of these are similar to the basic NLS in SV40 large T-antigen, whereas other NLSs are chemically quite different. For instance, an NLS in the RNA-binding protein hnRNP A1 is hydrophobic. Thus there must be multiple mechanisms for the recognition of these very different sequences.

Early work on the mechanism of nuclear import showed that proteins containing a basic NLS, similar to the one in SV40 large T-antigen, will be efficiently transported into isolated nuclei if they are provided with a cytosolic extract (Figure 13-35). Using this assay system, researchers purified two

(a) Effect of digitonin

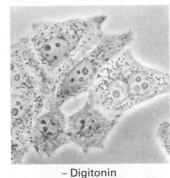 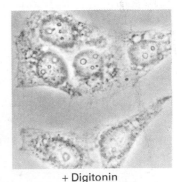

– Digitonin         + Digitonin

(b) Nuclear import by permeabilized cells

 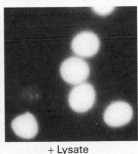

– Lysate         + Lysate

**EXPERIMENTAL FIGURE 13-35 Cytosolic proteins are required for nuclear transport.** The failure of nuclear transport to occur in permeabilized cultured cells in the absence of lysate demonstrates the involvement of soluble cytosolic components in the process. (a) Phase-contrast micrographs of untreated and digitonin-permeabilized HeLa cells. Treatment of a monolayer of cultured cells with the mild, nonionic detergent digitonin permeabilizes the plasma membrane so that cytosolic constituents leak out but leaves the nuclear envelope and NPCs intact. (b) Fluorescence micrographs of digitonin-permeabilized HeLa cells incubated with a fluorescent protein chemically coupled to a synthetic SV40 T-antigen NLS peptide in the presence and absence of cytosol (lysate). Accumulation of this transport substrate in the nucleus occurred only when cytosol was included in the incubation (*right*). [From S. Adam et al., 1990, *J. Cell Biol.* **111**:807, courtesy of Dr. Larry Gerace.]

(a)                   (b)

**EXPERIMENTAL FIGURE 13-34 Nuclear-localization signal (NLS) directs proteins to the cell nucleus.** Cytoplasmic proteins can be localized to the nucleus when they are fused to a nuclear localization signal. (a) Normal pyruvate kinase, visualized by immunofluorescence after cultured cells were treated with a specific antibody (yellow), is localized to the cytoplasm. This very large cytosolic protein functions in carbohydrate metabolism. (b) When a chimeric pyruvate kinase protein containing the SV40 NLS at its N-terminus was expressed in cells, it was localized to the nucleus. The chimeric protein was expressed from a transfected engineered gene produced by fusing a viral gene fragment encoding the SV40 NLS to the pyruvate kinase gene. [From D. Kalderon et al., 1984, *Cell* **39**:499, courtesy of Dr. Alan Smith.]

**FIGURE 13-36 Nuclear import.** Mechanism for nuclear import of "cargo" proteins. In the cytoplasm (*top*), a free nuclear transport receptor (importin) binds to the NLS of a cargo protein, forming a bimolecular cargo complex. The cargo complex diffuses through the NPC by transiently interacting with FG-nucleoporins. In the nucleoplasm, Ran·GTP binds to the importin, causing a conformational change that decreases its affinity for the NLS and releasing the cargo. To support another cycle of import, the exportin-Ran·GTP complex is transported back to the cytoplasm. A GTPase-accelerating protein (GAP) associated with the cytoplasmic filaments of the NPC stimulates Ran to hydrolyze the bound GTP. This generates a conformational change that causes dissociation from the nuclear transport receptor, which can then initiate another round of import. Ran·GDP is returned to the nucleoplasm, where a guanine nucleotide-exchange factor (GEF) causes release of GDP and rebinding of GTP.

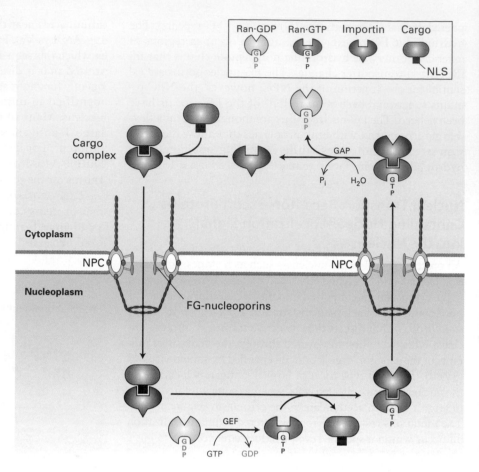

required cytosolic components: Ran and a nuclear transport receptor. *Ran* is a small monomeric G protein that exists in either GTP- or GDP-bound conformations (see Figure 3-32). The *nuclear transport receptor* binds to both the NLS on a cargo protein to be transported into the nucleus and to FG-repeats on nucleoporins. By a physical process that is not well understood, by binding transiently to FG-repeats, nuclear transport receptors have the ability to rapidly traverse the FG-repeat–containing matrix in the central channel of the nuclear pore, whereas proteins of similar size that lack this property are excluded from the central channel. Nuclear transport receptors can be monomeric, with a single polypeptide that can bind to both an NLS and FG-repeats, or they can be dimeric, with one subunit binding to the NLS and the other binding to FG-repeats.

The mechanism for import of cytoplasmic cargo proteins mediated by a nuclear import receptor is shown in Figure 13-36. Free nuclear transport receptor in the cytoplasm binds to its cognate NLS in a cargo protein, forming an importin-cargo complex. The cargo complex then translocates through the NPC channel as the nuclear transport receptor interacts with FG-repeats. The cargo complex rapidly reaches the nucleoplasm, and there the nuclear transport receptor interacts with Ran·GTP, causing a conformational change in the nuclear transport receptor that displaces the NLS, releasing the cargo protein into the nucleoplasm. The nuclear transport receptor-Ran·GTP complex then diffuses back through the NPC. Once the nuclear transport receptor-Ran·GTP complex

reaches the cytoplasmic side of the NPC, Ran interacts with a specific *GTPase-activating protein (Ran-GAP)* that is a component of the NPC cytoplasmic filaments. This stimulates Ran to hydrolyze its bound GTP to GDP, causing it to convert to a conformation that has low affinity for the nuclear transport receptor, so that the free nuclear transport receptor is released into the cytoplasm, where it can participate in another cycle of import. Ran·GDP travels back through the pore to the nucleoplasm, where it encounters a specific *guanine nucleotide-exchange factor (Ran-GEF)* that causes Ran to release its bound GDP in favor of GTP. The net result of this series of reactions is the coupling of the hydrolysis of GTP to the transfer of an NLS-bearing protein from the cytoplasm to the nuclear interior, thus providing a driving force for nuclear transport.

Although the nuclear transport receptor-cargo complex travels through the pore by random diffusion, the overall process of transport of cargo into the nucleus is unidirectional. Because of the rapid dissociation of the import complex when it reaches the nucleoplasm, there is a concentration gradient of the nuclear transport receptor-cargo complex across the NPC: high in the cytoplasm, where the complex assembles, and low in the nucleoplasm, where it dissociates. This concentration gradient is responsible for the unidirectional nature of nuclear import. A similar concentration gradient is responsible for driving the nuclear transport receptor in the nucleus back into the cytoplasm. The concentration of the nuclear transport receptor-Ran·GTP complex is higher in the nucleoplasm, where it assembles, than on the cytoplasmic side of the NPC, where it dissociates. Ultimately, the direction of the transport processes depends on the asymmetric distribution of the Ran-GEF and the Ran-GAP. Ran-GEF in the nucleoplasm maintains Ran in the Ran·GTP state, where it promotes dissociation of the cargo complex. Ran-GAP on the cytoplasmic side of the NPC converts Ran·GTP to Ran·GDP, dissociating the nuclear transport receptor-Ran·GTP complex and releasing free nuclear transport receptor into the cytosol.

## A Second Type of Nuclear Transport Receptors Escort Proteins Containing Nuclear-Export Signals out of the Nucleus

A very similar mechanism is used to export proteins, tRNAs, and ribosomal subunits from the nucleus to the cytoplasm. This mechanism initially was elucidated from studies of certain ribonuclear protein complexes that "shuttle" between the nucleus and the cytoplasm. Such "shuttling" proteins contain a *nuclear-export signal (NES)* that stimulates their export from the nucleus to the cytoplasm through nuclear pores, in addition to an NLS that results in their uptake into the nucleus. Experiments with engineered hybrid genes encoding a nucleus-restricted protein fused to various segments of a protein that shuttles in and out of the nucleus have identified at least three different classes of NESs: a leucine-rich sequence found in PKI (an inhibitor of protein kinase A) and in the Rev protein of human immunodeficiency virus (HIV), as well as two sequences identified in two different heterogeneous ribonucleoprotein particles (hnRNPs). The functionally significant structural features that specify nuclear export remain poorly understood.

The mechanism whereby shuttling proteins are exported from the nucleus is best understood for those containing a leucine-rich NES. According to the current model, shown in Figure 13-37a, a specific nuclear transport receptor, in the nucleus, called exportin 1, first forms a complex with Ran·GTP and then binds the NES in a cargo protein. Binding of exportin 1 to Ran·GTP causes a conformational change in exportin 1 that increases its affinity for the NES so that a *trimolecular cargo complex* is formed. Like other nuclear transport receptors, exportin 1 interacts transiently with FG-repeats in FG-nucleoporins and diffuses through the NPC. The cargo complex dissociates when it encounters the Ran-GAP in the NPC cytoplasmic filaments, which stimulates Ran to hydrolyze the bound GTP, shifting it into a conformation that has low affinity for exportin 1. The released exportin 1 changes conformation to a structure that has low affinity for the NES, releasing the cargo into the cytosol. The direction of the export process is driven by this dissociation of the cargo from exportin 1 in the cytoplasm, which causes a concentration gradient of the cargo complex across the NPC that is high in the nucleoplasm and low in the cytoplasm. Exportin 1 and the Ran·GDP are then transported back into the nucleus through an NPC.

By comparing this model for nuclear export with that in Figure 13-36 for nuclear import, we can see one obvious difference: Ran·GTP is part of the cargo complex during export but not during import. Apart from this difference, the two transport processes are remarkably similar. In both processes, association of a nuclear transport receptor with Ran·GTP in the nucleoplasm causes a conformational change that affects its affinity for the transport signal. During import, the interaction causes release of the cargo, whereas during export, the interaction promotes association with the cargo. In both export and import, stimulation of Ran·GTP hydrolysis in the cytoplasm by Ran-GAP produces a conformational change in Ran that releases the transport signal receptor. During nuclear export, the cargo is also released. Localization of the Ran-GAP and -GEF to the cytoplasm and nucleus, respectively, is the basis for the unidirectional transport of cargo proteins across the NPC.

In keeping with their similarity in function, the two types of nuclear transport receptors are highly homologous in sequence and structure. The family of nuclear transport receptors has 14 members in yeast and more than 20 in mammalian cells. The NESs or NLSs to which they bind have been determined for only a fraction of them. Some individual nuclear transport receptors function in both import and export.

A similar shuttling mechanism has been shown to export other cargoes from the nucleus. For example, exportin-t functions to export tRNAs. Exportin-t binds fully processed tRNAs in a complex with Ran·GTP that diffuses through NPCs and dissociates when it interacts with Ran-GAP in the NPC cytoplasmic filaments, releasing the tRNA into the cytosol. A Ran-dependent process is also required for the nuclear export of ribosomal subunits through NPCs once the protein and RNA components have been properly assembled in the nucleolus. Likewise, certain specific mRNAs that associate with particular hnRNP proteins can be exported by a Ran-dependent mechanism.

## Most mRNAs Are Exported from the Nucleus by a Ran-Independent Mechanism

Once the processing of an mRNA is completed in the nucleus, it remains associated with specific hnRNP proteins in a *messenger ribonuclear protein complex,* or mRNP. The principal transporter of mRNPs out of the nucleus is the **mRNP exporter**, a heterodimeric protein composed of a large subunit called *nuclear export factor 1 (NXF1)* and a small subunit, *nuclear export transporter 1 (NXT1)*. Multiple NXF1/NXT1 dimers bind to nuclear mRNPs through cooperative interactions with the RNA and other mRNP adapter proteins that associate with nascent pre-mRNAs during transcription elongation and pre-mRNA processing. In many respects, NXF1/NXT1 acts like a nuclear transport receptor that binds to an NLS or NES in the sense that both subunits interact with the FG-domains of FG-nucleoporins, allowing them to diffuse through the central channel of the NPC.

The process of mRNP export does not require Ran, and thus the unidirectional transport of mRNA out of the nucleus requires a source of energy other than GTP hydrolysis by Ran. Once the mRNP-NXF1/NXT1 complex reaches the

**FIGURE 13-37 Ran-dependent and Ran-independent nuclear export.** (a) Ran-dependent mechanism for nuclear export of cargo proteins containing a leucine-rich nuclear-export signal (NES). In the nucleoplasm (*bottom*), the protein exportin 1 binds cooperatively to the NES of the cargo protein to be transported and to Ran·GTP. After the resulting cargo complex diffuses through an NPC via transient interactions with FG repeats in FG-nucleoporins, the GAP associated with the NPC cytoplasmic filaments stimulates GTP hydrolysis, converting Ran·GTP to Ran·GDP. The accompanying conformational change in Ran leads to dissociation of the complex. The NES-containing cargo protein is released into the cytosol, whereas exportin 1 and Ran·GDP are transported back into the nucleus through NPCs. Ran-GEF in the nucleoplasm then stimulates conversion of Ran·GDP to Ran·GTP. (b) Ran-independent nuclear export of mRNAs. The heterodimeric NXF1/NXT1 complex binds to mRNA-protein complexes (mRNPs) in the nucleus. NXF1/NXT1 act as a nuclear export factor and directs the associated mRNP to the central channel of the NPC by transiently interacting with FG-nucleoporins. An RNA helicase (Dbp5) located on the cytoplasmic side of the NPC removes NXF1 and NXT1 from the mRNA in a reaction that is powered by ATP hydrolysis. Free NXF1 and NXT1 proteins are recycled back into the nucleus by the Ran-dependent import process depicted in Figure 13-36.

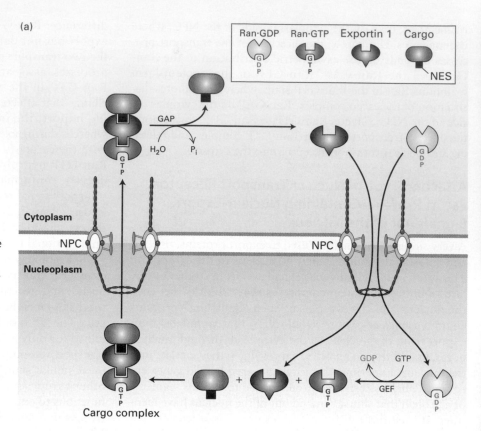

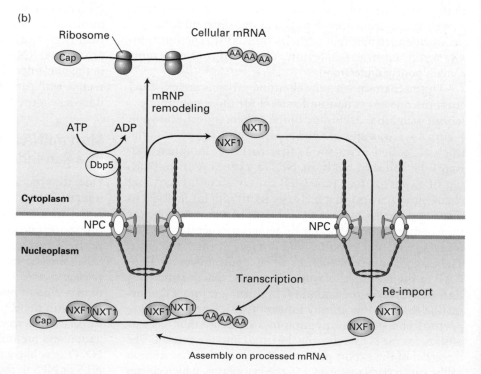

cytoplasmic side of the NPC, NXF1 and NXT1 dissociate from the mRNP with the help of the RNA helicase, Dbp5, which associates with cytoplasmic NPC filaments. Recall that RNA helicases use the energy derived from hydrolysis of ATP to move along RNA molecules, separating double-stranded RNA chains and dissociating RNA-protein complexes (Chapter 4). This leads to the simple idea that Dpb5, which associates with the cytoplasmic side of the nuclear pore complex, acts as an ATP-driven motor to remove NXF1/NXT1 from the mRNP complexes as they emerge on the cytoplasmic side of the NPC. The assembly of NXF1/NXT1 onto mRNPs on the nucleoplasmic side of the NPC and the subsequent ATP-dependent disassembly of NXF1/NXT1 from mRNPs on the cytoplasmic side of the NPC creates a concentration gradient of mRNP-NXF1/NXT1 which drives unidirectional export. After being removed from the mRNP, the free NXF1 and NXT1 proteins that have been stripped from the mRNA by Dbp5 helicase are imported back into the nucleus by a process that depends on Ran and a nuclear transport receptor (Figure 13-37b).

In Ran-dependent nuclear export (discussed in the previous subsection), hydrolysis of GTP by Ran on the cytoplasmic side of the NPC causes dissociation of the nuclear transport receptor from its cargo. In basic outline, the Ran-*independent* nuclear export discussed here operates by a similar mechanism except that Dbp5p on the cytosolic side of the NPC uses hydrolysis of ATP to dissociate the mRNP exporter from mRNA.

GTPase-activating protein (GAP) in the cytoplasm creates a gradient with high Ran·GTP in the nucleoplasm and Ran·GDP in the cytoplasm. The interaction of import cargo complexes with the Ran·GTP in the nucleoplasm causes dissociation of the complex, releasing the cargo into the nucleoplasm (see Figure 13-36), whereas the assembly of export cargo complexes is stimulated by interaction with Ran·GTP in the nucleoplasm (see Figure 13-37).

• Most mRNPs are exported from the nucleus by binding to a heterodimeric mRNP exporter in the nucleoplasm that interacts with FG-repeats. The direction of transport (nucleus to cytoplasm) results from the action of an RNA helicase associated with the cytoplasmic filaments of the nuclear pore complexes that removes the heterodimeric mRNP exporter once the transport complex has reached the cytoplasm.

## KEY CONCEPTS of Section 13.6

### Transport into and out of the Nucleus

• The nuclear envelope contains numerous nuclear pore complexes (NPCs), which are large, complicated structures composed of multiple copies of 30 proteins called *nucleoporins* (see Figure 13-33). FG-nucleoporins, which contain multiple repeats of a short hydrophobic sequence (FG-repeats), line the central transporter channel and play a role in transport of all macromolecules through nuclear pores.

• Transport of macromolecules larger than 20–40 kDa through nuclear pores requires the assistance of nuclear transport receptors that interact with both the transported molecule and FG-repeats of FG-nucleoporins.

• Proteins imported to or exported from the nucleus contain a specific amino acid sequence that functions as a nuclear-localization signal (NLS) or a nuclear-export signal (NES). Nucleus-restricted proteins contain an NLS but not an NES, whereas proteins that shuttle between the nucleus and cytoplasm contain both signals.

• Several different types of NES and NLS have been identified. Each type of nuclear-transport signal is thought to interact with a specific nuclear transport protein belonging to a family of homologous proteins.

• A cargo protein bearing an NES or NLS translocates through nuclear pores bound to its cognate nuclear transport protein. The transient interactions between nuclear transport receptors and FG-repeats allow very rapid diffusion of nuclear transport protein-cargo complex through the central channel of the NPC, which is filled with a hydrophobic matrix of FG-repeats.

• The unidirectional nature of protein export and import through nuclear pores results from participation of Ran, a monomeric G protein that exists in different conformations when bound to GTP or GDP. Localization of the Ran guanine nucleotide-exchange factor (GEF) in the nucleus and of Ran

## Perspectives for the Future

As we have seen in this chapter, we now understand many aspects of the basic processes responsible for selectively transporting proteins into the endoplasmic reticulum (ER), mitochondrion, chloroplast, peroxisome, and nucleus. Biochemical and genetic studies, for instance, have identified signal sequences responsible for targeting proteins to the correct organelle membrane and the membrane receptors that recognize these signal sequences. We also have learned much about the underlying mechanisms that translocate proteins across organelle membranes and have determined whether energy is used to push or pull proteins across the membrane in one direction, the type of channel through which proteins pass, and whether proteins are translocated in a folded or an unfolded state. Nonetheless, fundamental questions remain unanswered; probably the most puzzling is how fully folded proteins move across a membrane.

The peroxisomal import machinery provides one example of the translocation of folded proteins. It not only is capable of translocating fully folded proteins with bound cofactors into the peroxisomal matrix, it can even direct the import of a large gold particle decorated with a (PTS1) peroxisomal targeting peptide. Some researchers have speculated that the mechanism of peroxisomal import may be related to that of nuclear import, the best-understood example of post-translational translocation of folded proteins. Both the peroxisomal and nuclear import machinery can transport folded molecules of very divergent sizes, and both appear to involve a component that cycles between the cytosol and the organelle interior—the Pex5 PTS1 receptor in the case of peroxisomal import and the Ran-importin complex in the case of nuclear import. However, there also appear to be crucial differences between the two translocation processes. For example, nuclear pores represent large, stable macromolecular assemblies that are readily observed by electron microscopy, whereas analogous porelike structures have not been observed in the peroxisomal membrane. Moreover, small molecules can readily pass through nuclear pores, whereas peroxisomal membranes maintain a permanent barrier to the diffusion of small hydrophilic molecules. Taken together, these observations suggest that peroxisomal import may require an entirely new type of translocation mechanism.

The evolutionarily conserved mechanisms for translocating folded proteins across the cytoplasmic membrane of bacterial cells and across the thylakoid membrane of chloroplasts also are poorly understood. A better understanding of all of these processes for translocating folded proteins across a membrane will likely hinge on future development of in vitro translocation systems that allow investigators to define the biochemical mechanisms driving translocation and to identify the structures of trapped translocation intermediates.

Compared with our understanding of how soluble proteins are translocated into the ER lumen and mitochondrial matrix, our understanding of how targeting sequences specify the topology of multipass membrane proteins is quite elementary. For instance, we do not know how the translocon channel accommodates polypeptides that are oriented differently with respect to the membrane, nor do we understand how local polypeptide sequences interact with the translocon channel both to set the orientation of transmembrane spans and to signal for lateral passage into the membrane bilayer. A better understanding of how the amino acid sequences of membrane proteins can specify membrane topology will be crucial for decoding the vast amount of structural information for membrane proteins contained within databases of genomic sequences.

A more detailed understanding of all translocation processes should continue to emerge from genetic and biochemical studies, both in yeasts and in mammals. These studies will undoubtedly reveal additional key proteins involved in the recognition of targeting sequences and in the translocation of proteins across lipid bilayers. Finally, the structural studies of translocon channels will likely be extended in the future to reveal, at resolutions on the atomic scale, the conformational states that are associated with each step of the translocation cycle.

## Key Terms

cotranslational translocation 582
dislocation 600
dolichol phosphate 595
FG-nucleoporins 615
general import pore 604
hydropathy profile 593
microsomes 580
molecular chaperones 585
multipass membrane proteins 588

N-linked oligosaccharides 594
nuclear pore complex (NPC) 615
nuclear transport receptor 618
O-linked oligosaccharides 594
post-translational translocation 584
protein disulfide isomerase 597

Ran protein 618
rough ER 580
signal-anchor sequence 588
signal-recognition particle (SRP) 582
signal (uptake-targeting) sequences 579
single-pass membrane proteins 587

stop-transfer anchor sequence 588
topogenic sequences 587
topology (membrane protein) 587
translocon 583
trimolecular cargo complex 619
unfolded-protein response 599

## Review the Concepts

1. The following results were obtained in early studies on the translation of secretory proteins. Based on what we now know of this process, explain the reason why each result was observed. (a) An in vitro translation system consisting only of mRNA and ribosomes resulted in secretory proteins that were larger than the identical protein when translated in a cell. (b) A similar system that also included microsomes produced secretory proteins that were identical in size to those found in a cell. (c) When the microsomes were added after in vitro translation, the synthesized proteins were again larger than those made in a cell.

2. Describe the source or sources of energy needed for unidirectional translocation across the membrane in (a) cotranslational translocation into the endoplasmic reticulum (ER); (b) post-translational translocation into the ER; (c) translocation into the mitochondrial matrix.

3. Translocation into most organelles usually requires the activity of one or more cytosolic proteins. Describe the basic function of three different cytosolic factors required for translocation into the ER, mitochondria, and peroxisomes, respectively.

4. Describe the typical principles used to identify topogenic sequences within proteins and how these can be used to develop computer algorithms. How does the identification of topogenic sequences lead to prediction of the membrane arrangement of a multipass protein? What is the importance of the arrangement of positive charges relative to the membrane orientation of a signal-anchor sequence?

5. An abundance of misfolded proteins in the ER can result in the activation of the unfolded protein response (UPR) and ER-associated degradation (ERAD) pathways. UPR decreases the amount of unfolded proteins by altering gene expression of what type of genes? What is one manner in which ERAD may identify misfolded proteins? Why is dislocation of these misfolded proteins to the cytoplasm necessary?

6. Temperature-sensitive yeast mutants have been isolated that block each of the enzymatic steps in the synthesis of the dolichol-oligosaccharide precursor for N-linked glycosylation. Propose an explanation for why mutations that block synthesis of the intermediate with the structure dolichol-PP-$(GlcNAc)_2Man_5$ completely prevent addition of N-linked oligosaccharide chains to secretory proteins, whereas mutations

that block conversion of this intermediate into the completed precursor—dolichol-PP-(GlcNAc)$_2$Man$_9$Glc$_3$—allow the addition of N-linked oligosaccharide chains to secretory glycoproteins.

7. Name four different proteins that facilitate the modification and/or folding of secretory proteins within the lumen of the ER. Indicate which of these proteins covalently modifies substrate proteins and which brings about only conformational changes in substrate proteins.

8. Describe what would happen to the precursor of a mitochondrial matrix protein in the following types of mitochondrial mutants: (a) a mutation in the Tom22 signal receptor; (b) a mutation in the Tom70 signal receptor; (c) a mutation in the matrix Hsc70; and (d) a mutation in the matrix signal peptidase.

9. Describe the similarities and differences between the mechanism of import into the mitochondrial matrix and the chloroplast stroma.

10. Design a set of experiments using chimeric proteins, composed of a mitochondrial precursor protein fused to dihydrofolate reductase (DHFR), that could be used to determine how much of the precursor protein must protrude into the mitochondrial matrix in order for the matrix-targeting sequence to be cleaved by the matrix-processing protease.

11. Peroxisomes contain enzymes that use molecular oxygen to oxidize various substrates, but in the process hydrogen peroxide—a compound that can damage DNA and proteins—is formed. What is the name of the enzyme responsible for the breakdown of hydrogen peroxide to water? What is the mechanism of the import of this protein into the peroxisome, and what proteins are involved?

12. Suppose that you have identified a new mutant cell line that lacks functional peroxisomes. Describe how you could determine experimentally whether the mutant is primarily defective for insertion/assembly of peroxisomal membrane proteins or matrix proteins.

13. The nuclear import of proteins larger than 40 kDa requires the presence of what amino acid sequence? Describe the mechanism of nuclear import. How are nuclear transport receptors able to get through the nuclear pore complex?

14. Why is localization of Ran-GAP in the nucleus and Ran-GEF in the cytoplasm necessary for unidirectional transport of cargo proteins containing an NES?

## Analyze the Data

1. Imagine that you are evaluating the early steps in translocation and processing of the secretory protein prolactin. By using an experimental approach similar to that shown in Figure 13-7, you can use truncated prolactin mRNAs to control the length of nascent prolactin polypeptides that are synthesized. When prolactin mRNA that lacks a chain-termination (stop) codon is translated in vitro, the newly synthesized polypeptide ending with the last codon included on the mRNA will remain attached to the ribosome, thus allowing a polypeptide of defined length to extend from the ribosome. You have generated a set of mRNAs that encode segments of the N-terminus of prolactin of increasing length, and each mRNA can be translated in vitro by a cytosolic translation extract containing ribosomes, tRNAs, aminoacyl-tRNA synthetases, GTP, and translation initiation and elongation factors. When radiolabeled amino acids are included in the translation mixture, only the polypeptide encoded by the added mRNA will be labeled. After completion of translation, each reaction mixture was resolved by SDS poly-acrylamide gel electrophoresis, and the labeled polypeptides were identified by autoradiography.

a. The autoradiogram depicted below shows the results of an experiment in which each translation reaction was carried out either in the presence (+) or the absence (−) of microsomal membranes. Based on the gel mobility of peptides synthesized in the presence or absence of microsomes, deduce how long the prolactin nascent chain must be in order for the prolactin signal peptide to enter the ER lumen and to be cleaved by signal peptidase. (Note that microsomes carry significant quantities of SRP weakly bound to the membranes.)

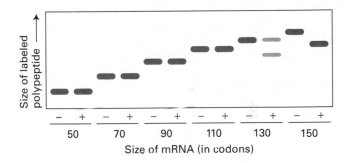

b. Given this length, what can you conclude about the conformational state(s) of the nascent prolactin polypeptide when it is cleaved by signal peptidase? The following lengths will be useful for your calculation: the prolactin signal sequence is cleaved after amino acid 31; the channel within the ribosome occupied by a nascent polypeptide is about 150 Å long; a membrane bilayer is about 50 Å thick; in polypeptides with an α-helical conformation, one residue extends 1.5 Å, whereas in fully extended polypeptides, one residue extends about 3.5 Å.

c. The experiment described in part (a) is carried out in an identical manner except that microsomal membranes are not present during translation but are added after translation is complete. In this case none of the samples shows a difference in mobility in the presence or absence of microsomes. What can you conclude about whether prolactin can be translocated into isolated microsomes post-translationally?

d. In another experiment, each translation reaction was carried out in the presence of microsomes, and then the microsomal membranes and bound ribosomes were separated from free ribosomes and soluble proteins by centrifugation. For each translation reaction, both the total reaction (T) and the membrane fraction (M) were resolved in neighboring gel lanes. Based on the amounts of labeled polypeptide in the membrane fractions in the autoradiogram depicted below, deduce how long the prolactin nascent chain must be in order for ribosomes engaged in translation to engage the SRP and thereby become bound to microsomal membranes.

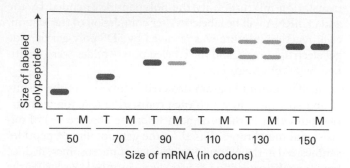

2. Recently, researchers discovered that treating mammalian cells with juniferdin, a plant-derived compound, affects protein secretion, and have reported that the target of this drug is *protein disulfide isomerase* (PDI). In the following experiment, cultured pancreatic β-cells were treated with juniferdin and protein lysates were isolated and compared to lysates from untreated cells using immunoblot analysis. Probing blots with antibodies against PDI (57 kDa), actin (43 kDa) and pro-insulin (9.8 kDa), show the following:

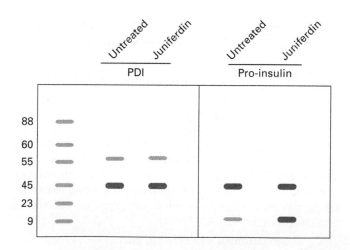

a. Given that approximately the same amount of protein was loaded in each lane, as evidenced by the actin signals, how do you explain the fact that the PDI levels also appear about the same, while most of the pro-insulin remains accumulated in the juniferdin-treated cells?

b. To confirm your results, protein lysates of juniferdin-treated and untreated cells were separated by SDS-PAGE and blotted to membranes and then probed with antibodies against Ire1 and Hac1.

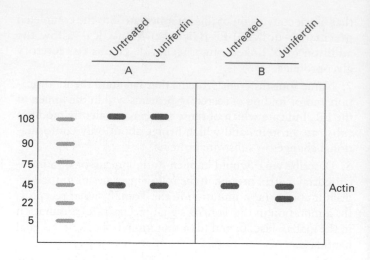

Label each blot with the antibody that was used for the analysis. How do you explain the increase in the intensity of the signal seen in the juniferdin-treated cells in blot B?

c. An immunocytochemistry and fluorescence microscopy analysis was undertaken with the antibody used in blot B and a secondary antibody labeled with rhodamine (red). Since juniferdinspecifically affects PDI, a resident rough ER (RER) protein, how do you explain the localization of the signal in the nucleus?

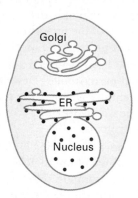

3. Antibody labeling of proteins like that used in immuno-fluorescence analysis can be applied to electron microscopy, but instead of using fluorescent labels attached to antibodies, investigators use gold particles that are electron dense and appear as uniform dots in an electron micrograph. Furthermore, by varying the size of the gold particles (e.g., 5 nm vs 10 nm), one can identify the localization of more than one protein in the cell.

a. Using this approach, investigators have determined the subcellular localization of Tim and Tom proteins used for protein import into mitochondria. In the drawing of the results below, label the gold particles showing the localization of Tim44 and Tom40. What made you come to these conclusions?

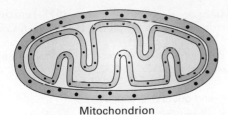

Mitochondrion

**b.** Genetically engineering the N-terminus of alcohol dehydrogenase (ADH) onto a cytosolic protein alters that protein's localization. The following blot was seen when cells were transfected with an ADH-actin chimeric construct, and proteins were isolated from the cytosol and mitochondria. Antibodies against actin (43 kDa), the cytosolic protein GAPDH (37 kDa), and the mitochondrial inner membrane protein succinate dehydrogenase A (72 kDa), were used in the analysis.

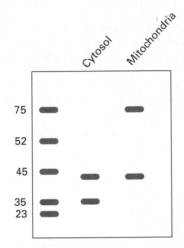

How do you explain the presence of actin in two distinct subcellular pools of protein? How do you explain it being the same molecular mass in both pools? If the blot were stripped and reprobed with an antibody against the N-terminus of alcohol dehydrogenase, where would you expect to find the signal?

**c.** Cyanide is toxic to cells because it inhibits a specific mitochondrial protein complex that is responsible for producing ATP. An experiment like that described above was repeated and the results compared to those for cells exposed to hydrogen cyanide. The following are the results from the immunoblot analysis:

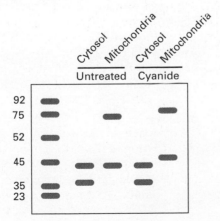

How do you explain the apparent shift in the molecular mass of actin in the mitochondrial fraction of cyanide-treated cells? Note that there is a similar shift in the mass of the succinate dehydrogenase A control. What would you expect to see on blots if cyanide-treated cells were supplemented with ATP?

# References

**Targeting Proteins to and Across the ER Membrane**

Egea, P. F., R. M. Stroud, and P. Walter. 2005. Targeting proteins to membranes: structure of the signal recognition particle. *Curr. Opin. Struc. Biol.* **15**:213–220.

Osborne, A. R., T. A. Rapoport, and B. van den Berg. 2005. Protein translocation by the Sec61/SecY channel. *Ann. Rev. Cell Dev. Biol.* **21**:529–550.

Wickner, W., and R. Schekman. 2005. Protein translocation across biological membranes. *Science* **310**:1452–1456.

**Insertion of Membrane Proteins into the ER**

Englund, P. T. 1993. The structure and biosynthesis of glycosylphosphatidylinositol protein anchors. *Ann. Rev. Biochem.* **62**:121–138.

Mothes, W., et al. 1997. Molecular mechanism of membrane protein integration into the endoplasmic reticulum. *Cell* **89**:523–533.

Shao, S., and R. S. Hegde. 2011. Membrane protein insertion at the endoplasmic reticulum. *Ann. Rev. Cell Dev. Biol.* **27**:25–56.

Wang, F., et al. 2011. The mechanism of tail-anchored protein insertion into the ER membrane. *Mol. Cell* **43**:738–750.

**Protein Modifications, Folding, and Quality Control in the ER**

Braakman, I., and N. J. Bulleid. 2011. Protein folding and modification in the mammalian endoplasmic reticulum. *Ann. Rev. Biochem.* **80**:71–99.

Hegde, R. S., and H. L. Ploegh. 2010. Quality and quantity control at the endoplasmic reticulum. *Curr. Opin. Cell Biol.* **22**:437–446.

Helenius, A., and M. Aebi. 2004. Roles of N-linked glycans in the endoplasmic reticulum. *Ann. Rev. Biochem.* **73**:1019–1049.

Kornfeld, R., and S. Kornfeld. 1985. Assembly of asparagine-linked oligosaccharides. *Ann. Rev. Biochem.* **45**:631–664.

Patil, C., and P. Walter. 2001. Intracellular signaling from the endoplasmic reticulum to the nucleus: the unfolded protein response in yeast and mammals. *Curr. Opin. Cell Biol.* **13**:349–355.

Meusser, B., et al. 2005. ERAD: the long road to destruction. *Nat. Cell Biol.* **7**:766–772.

Sevier, C. S., and C. A. Kaiser. 2002. Formation and transfer of disulphide bonds in living cells. *Nat. Rev. Mol. Cell Biol.* **3**:836–847.

Tsai, B., Y. Ye, and T. A. Rapoport. 2002. Retro-translocation of proteins from the endoplasmic reticulum into the cytosol. *Nat. Rev. Mol. Cell Biol.* **3**:246–255.

**Targeting of Proteins to Mitochondria and Chloroplasts**

Dalbey, R. E., and A. Kuhn. 2000. Evolutionarily related insertion pathways of bacterial, mitochondrial, and thylakoid membrane proteins. *Ann. Rev. Cell Dev. Biol.* **16**:51–87.

Dolezal, P., et al. 2006. Evolution of the molecular machines for protein import into mitochondria. *Science* **313**:314–318.

Koehler, C. M. 2004. New developments in mitochondrial assembly. *Ann. Rev. Cell Dev. Biol.* **20**:309–335.

Li, H.-M., and C.-C. Chiu. 2010. Protein transport into chloroplasts. *Ann. Rev. Plant Biol.* **61**:157–180.

Matouschek, A., N. Pfanner, and W. Voos. 2000. Protein unfolding by mitochondria: the Hsp70 import motor. *EMBO Rep.* 1:404–410.

Neupert, W., and M. Brunner. 2002. The protein import motor of mitochondria. *Nat. Rev. Mol. Cell Biol.* 3:555–565.

Rapaport, D. 2005. How does the TOM complex mediate insertion of precursor proteins into the mitochondrial outer membrane? *J. Cell Biol.* 171:419–423.

Robinson, C., and A. Bolhuis. 2001. Protein targeting by the twin-arginine translocation pathway. *Nat. Rev. Mol. Cell Biol.* 2:350–356.

Truscott, K. N., K. Brandner, and N. Pfanner. 2003. Mechanisms of protein import into mitochondria. *Curr. Biol.* 13:R326–R337.

## Targeting of Peroxisomal Proteins

Dammai, V., and S. Subramani. 2001. The human peroxisomal targeting signal receptor, Pex5p, is translocated into the peroxisomal matrix and recycled to the cytosol. *Cell* 105:187–196.

Gould, S. J., and C. S. Collins. 2002. Opinion: peroxisomal-protein import: is it really that complex? *Nat. Rev. Mol. Cell Biol.* 3:382–389.

Gould, S. J., and D. Valle. 2000. Peroxisome biogenesis disorders: genetics and cell biology. *Trends Genet.* 16:340–345.

Hoepfner, D., et al. 2005. Contribution of the endoplasmic reticulum to peroxisome formation. *Cell* 122:85–95.

Ma, C., G. Agrawal, and S. Subramani. 2011. Peroxisome assembly: matrix and membrane protein biogenesis. *J. Cell Biol.* 193:7–16.

Purdue, P. E., and P. B. Lazarow. 2001. Peroxisome biogenesis. *Ann. Rev. Cell Dev. Biol.* 17:701–752.

## Transport into and out of the Nucleus

Chook, Y. M., and G. Blobel. 2001. Karyopherins and nuclear import. *Curr. Opin. Struc. Biol.* 11:703–715.

Cole, C. N., and J. J. Scarcelli. 2006. Transport of messenger RNA from the nucleus to the cytoplasm. *Curr. Opin. Cell Biol.* 18:299–306.

Johnson, A. W., E. Lund, and J. Dahlberg. 2002. Nuclear export of ribosomal subunits. *Trends Biochem. Sci.* 27:580–585.

Ribbeck, K., and D. Gorlich. 2001. Kinetic analysis of translocation through nuclear pore complexes. *EMBO J.* 20:1320–1330.

Rout, M. P., and J. D. Aitchison. 2001. The nuclear pore complex as a transport machine. *J. Biol. Chem.* 276:16593–16596.

Schwartz, T. U. 2005. Modularity within the architecture of the nuclear pore complex. *Curr. Opin. Struc. Biol.* 15:221–226.

Stewart, M. 2010. Nuclear export of mRNA. *Trends Biochem. Sci.* 35:609–617.

Terry, L. J., and S. R. Wente. 2009. Flexible gates: dynamic topologies and functions for FG nucleoporins in nucleocytoplasmic transport. *Eukaryot. Cell* 8:1814–1827.

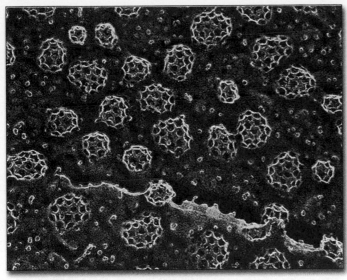

# Vesicular Traffic, Secretion, and Endocytosis

Scanning electron micrograph showing the formation of clathrin-coated vesicles on the cytosolic face of the plasma membrane. [John Heuser, Washington University School of Medicine.]

I n the previous chapter we explored how proteins are targeted to and translocated across the membranes of several different intracellular organelles, including the endoplasmic reticulum, mitochondria and chloroplasts, peroxisomes, and the nucleus. In this chapter we turn our attention to the **secretory pathway** and the mechanisms of vesicular traffic that allow proteins to be secreted from the cell or delivered to the plasma membrane and the lysosome. We will also discuss the related processes of endocytosis and autophagy, which deliver proteins and small molecules from either outside the cell or from the cytoplasm to the interior of the lysosome for degradation.

The secretory pathway carries both soluble and membrane proteins from the ER to their final destination at the cell surface or in the lysosome. Proteins delivered to the plasma membrane include cell-surface receptors, transporters for nutrient uptake, and ion channels that maintain the proper ionic and electrochemical balance across the plasma membrane. Such membrane proteins, once they reach the plasma membrane, become embedded within it. Soluble secreted proteins also follow the secretory pathway to the cell surface, but instead of remaining embedded in the membrane they are released into the aqueous extracellular environment. Examples of secreted proteins are digestive enzymes, peptide hormones, serum proteins, and collagen. As described in Chapter 9, the lysosome is an organelle with

an acidic interior that is generally used for degradation of unneeded proteins and the storage of small molecules such as amino acids. Accordingly, the types of proteins delivered to the lysosomal membrane include subunits of the V-class proton pump that pumps $H^+$ from the cytosol into the acidic lumen of the lysosome, as well as transporters that release small molecules stored in the lysosome into the cytoplasm. Soluble proteins delivered by this pathway include lysosomal digestive enzymes such as proteases, glycosidases, phosphatases, and lipases.

In contrast to the secretory pathway, which allows proteins to be targeted to the cell surface, the **endocytic pathway** is used to take up substances from the cell surface and move them into the interior of the cell. The endocytic pathway is used to ingest certain nutrients that are too large to be transported across the plasma membrane by one of the transport mechanisms discussed in Chapter 11. For example, the endocytic pathway is utilized in the uptake of cholesterol carried in LDL particles, and iron atoms carried by the iron-binding protein transferrin. In addition, the endocytic pathway can be used to remove receptor proteins from the cell surface as a way to down-regulate their activity.

A single unifying principle governs all protein trafficking in the secretory and endocytic pathways: transport of membrane and soluble proteins from one membrane-bounded compartment to another is mediated by **transport vesicles**

## OUTLINE

**FIGURE 14-1 Overview of the secretory and endocytic pathways of protein sorting.** *Secretory pathway:* Synthesis of proteins bearing an ER signal sequence is completed on the rough ER **1**, and the newly made polypeptide chains are inserted into the ER membrane or cross it into the lumen (Chapter 13). Some proteins (e.g., ER enzymes or structural proteins) remain within the ER. The remainder are packaged into transport vesicles **2** that bud from the ER and fuse to form new *cis*-Golgi cisternae. Missorted ER-resident proteins and vesicle membrane proteins that need to be reused are retrieved to the ER by vesicles **3** that bud from the *cis*-Golgi and fuse with the ER. Each *cis*-Golgi cisterna, with its protein content, physically moves from the *cis* to the *trans* face of the Golgi complex **4** by a nonvesicular process called cisternal maturation. Retrograde transport vesicles **5** move Golgi-resident proteins to the proper Golgi compartment. In all cells, certain soluble proteins move to the cell surface in transport vesicles **6** and are secreted continuously (constitutive secretion). In certain cell types, some soluble proteins are stored in secretory vesicles **7** and are released only after the cell receives an appropriate neural or hormonal signal (regulated secretion). Lysosome-destined membrane and soluble proteins, which are transported in vesicles that bud from the *trans*-Golgi **8**, first move to the late endosome and then to the lysosome. *Endocytic pathway:* Membrane and soluble extracellular proteins taken up in vesicles that bud from the plasma membrane **9** also can move to the lysosome via the endosome.

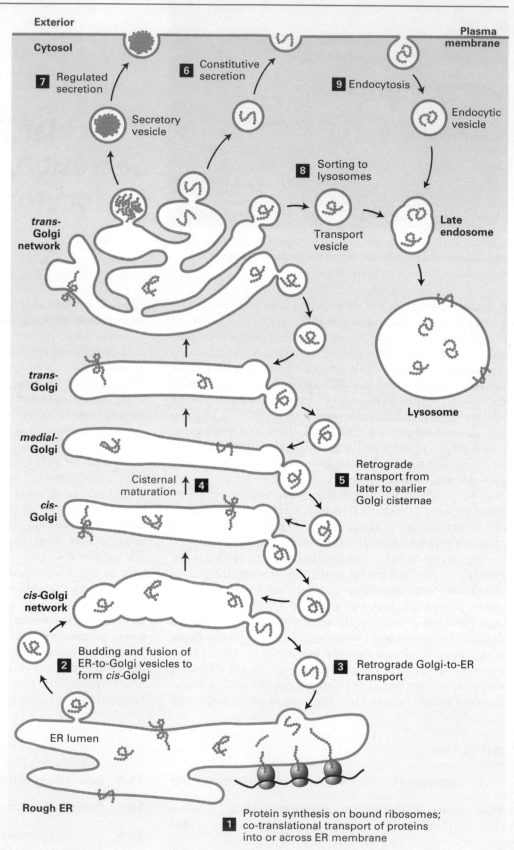

*Exterior*

*Cytosol*

*Plasma membrane*

**7** Regulated secretion

**6** Constitutive secretion

**9** Endocytosis

Secretory vesicle

Endocytic vesicle

**8** Sorting to lysosomes

*trans*-Golgi network

Transport vesicle

Late endosome

*trans*-Golgi

Lysosome

*medial*-Golgi

Cisternal maturation **4**

**5** Retrograde transport from later to earlier Golgi cisternae

*cis*-Golgi

*cis*-Golgi network

**2** Budding and fusion of ER-to-Golgi vesicles to form *cis*-Golgi

**3** Retrograde Golgi-to-ER transport

ER lumen

Rough ER

**1** Protein synthesis on bound ribosomes; co-translational transport of proteins into or across ER membrane

that collect "*cargo*" *proteins* in buds arising from the membrane of one compartment and then deliver these cargo proteins to the next compartment by fusing with the membrane of that compartment. Importantly, as transport vesicles bud from one membrane and fuse with the next, the same face of the membrane remains oriented toward the cytosol. Therefore once a protein has been inserted into the membrane or the lumen of the ER, the protein can be carried along the secretory pathway, moving from one organelle to the next without being translocated across another membrane or altering its orientation within the membrane. Similarly, the endocytic pathway uses vesicle traffic to transport proteins from the plasma membrane to the endosome and lysosome and thus preserves their orientation in the membrane of these organelles. Figure 14-1 outlines the main secretory and endocytic pathways in the cell.

Reduced to its simplest elements, the secretory pathway operates in two stages. The first stage takes place in the rough endoplasmic reticulum (ER) (Figure 14-1, step **1**). As described in Chapter 13, newly synthesized soluble and membrane proteins are translocated into the ER, where they fold into their proper conformation and receive covalent modifications such as N-linked and O-linked carbohydrates and disulfide bonds. Once newly synthesized proteins are properly folded and have received their correct modifications in the ER lumen, they progress to the second stage of the secretory pathway, transport to and through the Golgi.

The second stage of the secretory pathway can be summarized as follows. In the ER, cargo proteins are packaged into anterograde (forward-moving) transport vesicles (Figure 14-1, step **2**). These vesicles fuse with each other to form a flattened membrane-bounded compartment known as the *cis*-Golgi network or *cis*-Golgi **cisterna** (a "cistern" is a container for holding water or other liquid). Certain proteins, mainly proteins that function in the ER, can be retrieved from the *cis*-Golgi cisterna to the ER via a different set of retrograde (backward-moving) transport vesicles (step **3**). In a manner reminiscent of an assembly line, the new *cis*-Golgi cisterna with its cargo of proteins physically moves from the *cis* position (nearest the ER) to the *trans* position (farthest from the ER), successively becoming first a *medial*-Golgi cisterna and then a *trans*-Golgi cisterna (step **4**). This process, known as *cisternal maturation*, primarily involves retrograde transport vesicles (step **5**), which retrieve enzymes and other Golgi-resident proteins from later to earlier Golgi cisternae, thereby "maturing" the *cis*-Golgi to the *medial*-Golgi, and the *medial*-Golgi to the *trans*-Golgi. As secretory proteins move through the Golgi, they can receive further modifications to linked carbohydrates by specific glycosyl transferases that are housed in the different Golgi compartments.

Proteins in the secretory pathway are eventually delivered to a complex network of membranes and vesicles termed the ***trans*-Golgi network** (**TGN**). The TGN is a major branch point in the secretory pathway. It is at this point that proteins are loaded into different kinds of vesicles and thereby trafficked to different destinations. Depending on which kind of vesicle the protein is loaded into, it will be either transported to the plasma membrane and secreted immediately, stored for later release, or shipped to the lysosome (steps **6**–**8**). The process by which a vesicle moves to and fuses with the plasma membrane and releases its contents is known as **exocytosis**. In all cell types, at least some proteins are secreted continuously, while others are stored inside the cell until a signal for exocytosis causes them to be released. Secretory proteins destined for lysosomes are first transported by vesicles from the *trans*-Golgi network to a compartment usually called the **late endosome**; proteins then are transferred to the lysosome by direct fusion of the endosome with the lysosomal membrane.

**Endocytosis** is related mechanistically to the secretory pathway. In the endocytic pathway, vesicles bud inward from the plasma membrane, bringing membrane proteins and their bound ligands into the cell (see Figure 14-1, *right*). After being internalized by endocytosis, some proteins are transported to lysosomes via the late endosome, whereas others are recycled back to the cell surface.

In this chapter we first discuss the experimental techniques that have contributed to our knowledge of the secretory pathway and endocytosis. Then we focus on the general mechanisms of membrane budding and fusion. We will see that although different kinds of transport vesicles utilize distinct sets of proteins for their formation and fusion, all vesicles use the same general mechanism for budding, selection of particular sets of cargo molecules, and fusion with the appropriate target membrane. In the remaining sections of the chapter, we discuss both the early and late stages of the secretory pathway, including how specificity of targeting to different destinations is achieved, and conclude with a discussion of how proteins are transported to the lysosome by the endocytic pathway.

## 14.1 Techniques for Studying the Secretory Pathway

The key to understanding how proteins are transported through the organelles of the secretory pathway has been to develop a basic description of the function of transport vesicles. Many components required for the formation and fusion of transport vesicles have been identified in the past decade by a remarkable convergence of the genetic and biochemical approaches described in this section. All studies of intracellular protein trafficking employ some method for assaying the transport of a given protein from one compartment to another. We begin by describing how intracellular protein transport can be followed in living cells and then consider genetic and in vitro systems that have proved useful in elucidating the secretory pathway.

### Transport of a Protein Through the Secretory Pathway Can Be Assayed in Living Cells

The classic studies of G. Palade and his colleagues in the 1960s first established the order in which proteins move from one organelle to the next in the secretory pathway.

These early studies also showed that secretory proteins are never released into the cytosol, the first indication that transported proteins are always associated with some type of membrane-bounded intermediate. In these experiments, which combined pulse-chase labeling (see Figure 3-40) and autoradiography, radioactively labeled amino acids were injected into the pancreas of hamsters. At different times after injection, the animals were sacrificed and the pancreatic cells were immediately fixed with glutaraldehyde, sectioned, and subjected to autoradiography to visualize the location of the radiolabeled proteins. Because the radioactive amino acids were administered in a short pulse, only those proteins synthesized immediately after injection were labeled, forming a distinct cohort of labeled proteins whose transport could be followed. In addition, because pancreatic acinar cells are dedicated secretory cells, almost all of the labeled amino acids in these cells are incorporated into secretory proteins, facilitating the observation of transported proteins.

Although autoradiography is rarely used today to localize proteins within cells, these early experiments illustrate the two basic requirements for any assay of intercompartmental transport. First, it is necessary to label a cohort of proteins in an early compartment so that their subsequent transfer to later compartments can be followed with time. Second, it is necessary to have a way to identify the compartment in which a labeled protein resides. Here we describe two modern experimental procedures for observing the intracellular trafficking of a secretory protein in almost any type of cell.

In both procedures, a gene encoding an abundant membrane glycoprotein (G protein) from vesicular stomatitis virus (VSV) is introduced into cultured mammalian cells either by transfection or simply by infecting the cells with the virus. The treated cells, even those that are not specialized for secretion, rapidly synthesize the VSV G protein on the ER like normal cellular secretory proteins. Use of a mutant encoding a temperature-sensitive VSV G protein allows researchers to turn subsequent transport of this protein on and off. At the restrictive temperature of 40 °C, newly made VSV G protein is misfolded and therefore retained within the ER by quality-control mechanisms discussed in Chapter 13, whereas at the permissive temperature of 32 °C, the protein is correctly folded and is transported through the secretory pathway to the cell surface. Importantly, the misfolding of the temperature-sensitive VSV G protein is reversible; thus when cells synthesizing mutant VSV G protein are grown at 40 °C and then shifted to 32 °C, the misfolded mutant VSV G protein that had accumulated in the ER will refold and be transported normally. This clever use of a temperature-sensitive mutation in effect defines a protein cohort whose subsequent transport can be followed.

In two variations of this basic procedure, transport of VSV G protein is monitored by different techniques. Studies using both of these modern trafficking assays and Palade's early experiments all came to the same conclusion: in mammalian cells vesicle-mediated transport of a protein molecule from its site of synthesis on the rough ER to its arrival at the plasma membrane takes from 30 to 60 minutes.

**Microscopy of GFP-Labeled VSV G Protein** One approach for observing transport of VSV G protein employs a hybrid gene

---

🅥 **VIDEO:** Transport of VSVG-GFP Through the Secretory Pathway

---

(a)

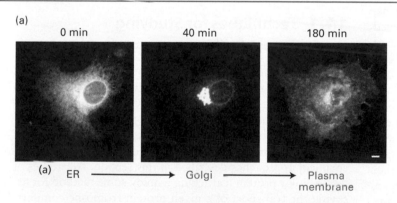

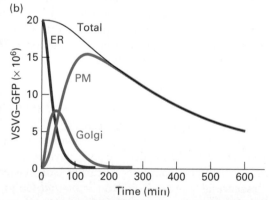

**EXPERIMENTAL FIGURE 14-2 Protein transport through the secretory pathway can be visualized by fluorescence microscopy of cells producing a GFP-tagged membrane protein.** Cultured cells were transfected with a hybrid gene encoding the viral membrane glycoprotein VSV G protein linked to the gene for green fluorescent protein (GFP). A mutant version of the viral gene was used so that newly made hybrid protein (VSVG-GFP) was retained in the ER at 40 °C but was released for transport at 32 °C. (a) Fluorescence micrographs of cells just before and two times after they were shifted to the lower temperature. Movement of VSVG-GFP from the ER to the Golgi and finally to the cell surface occurred within 180 minutes. The scale bar is 5 μm. (b) Plot of the levels of VSVG-GFP in the endoplasmic reticulum (ER), Golgi, and plasma membrane (PM) at different times after shift to lower temperature. The kinetics of transport from one organelle to another can be reconstructed from computer analysis of these data. The decrease in total fluorescence that occurs at later times probably results from slow inactivation of GFP fluorescence. [From Jennifer Lippincott-Schwartz and Koret Hirschberg, Metabolism Branch, National Institute of Child Health and Human Development.]

in which the viral gene is fused to the gene encoding *green fluorescent protein (GFP)*, a naturally fluorescent protein (Chapter 9). The hybrid gene is transfected into cultured cells by techniques described in Chapter 5. When cells expressing the temperature-sensitive form of the hybrid protein (VSVG-GFP) are grown at the restrictive temperature, VSVG-GFP accumulates in the ER, which appears as a lacy network of membranes when cells are observed in a fluorescent microscope. When the cells are subsequently shifted to a permissive temperature, the VSVG-GFP can be seen to move first to the membranes of the Golgi apparatus, which are densely concentrated at the edge of the nucleus, and then to the cell surface (Figure 14-2a). By observing the distribution of VSVG-GFP at different times after shifting cells to the permissive temperature, researchers have determined how long VSVG-GFP resides in each organelle of the secretory pathway (Figure 14-2b).

**Detection of Compartment-Specific Oligosaccharide Modifications** A second way to follow the transport of secretory proteins takes advantage of modifications to their carbohy-drate side chains that occur at different stages of the secretory pathway. To understand this approach, recall that many secretory proteins leaving the ER contain one or more copies of the N-linked oligosaccharide $Man_8(GlcNAc)_2$, which are synthesized and attached to secretory proteins in the ER (see Figure 13-18). As a protein moves through the Golgi complex, different enzymes localized to the *cis-*, *medial-*, and *trans-*Golgi cisternae catalyze an ordered series of reactions to these core $Man_8(GlcNAc)_2$ chains, as discussed in a later section of this chapter. For instance, glycosidases that reside specifically in the *cis-*Golgi compartment sequentially trim mannose residues off the core oligosaccharide to yield a "trimmed" form, $Man_5(GlcNAc)_2$. Scientists can use a specialized carbohydrate-cleaving enzyme known as endoglycosidase D to distinguish glycosylated proteins that remain in the ER from those that have entered the *cis-*Golgi: trimmed *cis-*Golgi-specific oligosaccharides are cleaved from proteins by endoglycosidase D, whereas the core (untrimmed) oligosaccharide chains on secretory proteins within the ER are resistant to cleavage by this enzyme (Figure 14-3a). Because

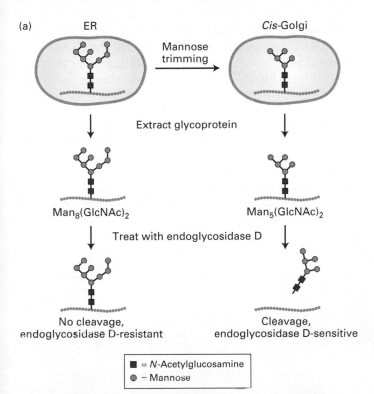

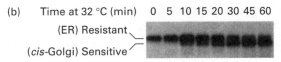

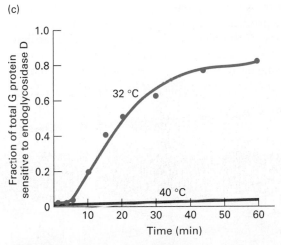

**EXPERIMENTAL FIGURE 14-3 Transport of a membrane glycoprotein from the ER to the Golgi can be assayed based on sensitivity to cleavage by endoglycosidase D.** Cells expressing a temperature-sensitive VSV G protein (VSVG) were labeled with a pulse of radioactive amino acids at the nonpermissive temperature so that labeled protein was retained in the ER. At periodic times after a return to the permissive temperature of 32 °C, VSVG was extracted from cells and digested with endoglycosidase D. (a) As proteins move to the *cis-*Golgi from the ER, the core oligosaccharide $Man_8(GlcNAc)_2$ is trimmed to $Man_5(GlcNAc)_2$ by enzymes that reside in the *cis-*Golgi compartment. Endoglycosidase D cleaves the oligosaccharide chains from proteins processed in the *cis-*Golgi but not from proteins in the

ER. (b) SDS gel electrophoresis of the digestion mixtures resolves the resistant, uncleaved (slower-migrating) and sensitive, cleaved (faster-migrating) forms of labeled VSVG. As this electrophoretogram shows, initially all of the VSVG was resistant to digestion, but with time an increasing fraction was sensitive to digestion, reflecting protein transported from the ER to the Golgi and processed there. In control cells kept at 40 °C, only slow-moving, digestion-resistant VSVG was detected after 60 minutes (not shown). (c) Plot of the proportion of VSVG that is sensitive to digestion, derived from electrophoretic data, reveals the time course of ER → Golgi transport. [From C. J. Beckers et al., 1987, *Cell* **50**:523.]

a deglycosylated protein produced by endoglycosidase D digestion moves faster on an SDS gel than the corresponding glycosylated protein, these proteins can be readily distinguished (Figure 14-3b).

This type of assay can be used to track movement of VSV G protein in virus-infected cells pulse-labeled with radioactive amino acids. Immediately after labeling, all the labeled VSV G protein is still in the ER and, upon extraction, is resistant to digestion by endoglycosidase D, but with time the fraction of the extracted glycoprotein that is sensitive to digestion increases. This conversion of VSV G protein from an endoglycosidase D–resistant form to an endoglycosidase D–sensitive form corresponds to vesicular transport of the protein from the ER to the *cis*-Golgi. Note that transport of VSV G protein from the ER to the Golgi takes about 30 minutes as measured by either the assay based on oligosaccharide processing or fluorescence microscopy of VSVG-GFP (Figure 14-3c). A variety of assays based on specific carbohydrate modifications that occur in later Golgi compartments have been developed to measure progression of VSV G protein through each stage of the Golgi apparatus.

## Yeast Mutants Define Major Stages and Many Components in Vesicular Transport

The general organization of the secretory pathway and many of the molecular components required for vesicle trafficking are similar in all eukaryotic cells. Because of this conservation, genetic studies with yeast have been useful in confirming the sequence of steps in the secretory pathway and in identifying many of the proteins that participate in vesicular traffic. Although yeasts secrete few proteins into the growth medium, they continuously secrete a number of enzymes that remain localized in the narrow space between the plasma membrane

and the cell wall. The best studied of these, invertase, hydrolyzes the disaccharide sucrose to glucose and fructose.

A large number of yeast mutants initially were identified based on their ability to secrete proteins at one temperature and inability to do so at a higher, nonpermissive temperature. When these temperature-sensitive *secretion (sec) mutants* are transferred from the lower to the higher temperature, they accumulate secretory proteins at the point in the pathway blocked by the mutation. Analysis of such mutants identified five classes (A–E) characterized by protein accumulation in the cytosol, rough ER, small vesicles taking proteins from the ER to the Golgi complex, Golgi cisternae, or constitutive secretory vesicles (Figure 14-4). Subsequent characterization of *sec* mutants in the various classes has helped elucidate the fundamental components and molecular mechanisms of vesicle trafficking that we discuss in later sections.

To determine the order of the steps in the pathway, researchers analyzed double *sec* mutants. For instance, when yeast cells contain mutations in both class B and class D functions, proteins accumulate in the rough ER, not in the Golgi cisternae. Since proteins accumulate at the earliest blocked step, this finding shows that class B mutations must act at an earlier point in the secretory pathway than class D mutations do. These studies confirmed that as a secreted protein is synthesized and processed, it moves sequentially from the cytosol → rough ER → ER-to-Golgi transport vesicles → Golgi cisternae → secretory vesicles and finally is exocytosed.

The three methods outlined in this section have delineated the major steps of the secretory pathway and have contributed to the identification of many of the proteins responsible for vesicle budding and fusion. Currently each of the individual steps in the secretory pathway is being studied in mechanistic detail, and increasingly, biochemical assays and molecular genetic studies are used to study each of these steps in terms of the function of individual protein molecules.

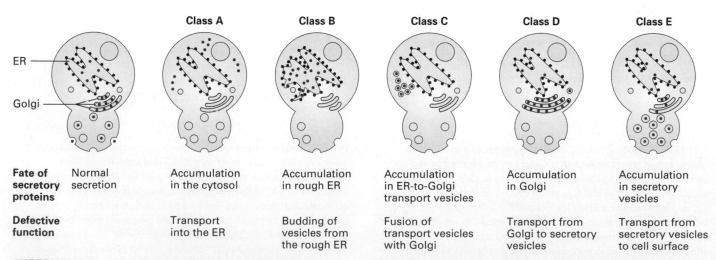

| | Class A | Class B | Class C | Class D | Class E |
|---|---|---|---|---|---|
| **Fate of secretory proteins** | Normal secretion | Accumulation in the cytosol | Accumulation in rough ER | Accumulation in ER-to-Golgi transport vesicles | Accumulation in Golgi | Accumulation in secretory vesicles |
| **Defective function** | | Transport into the ER | Budding of vesicles from the rough ER | Fusion of transport vesicles with Golgi | Transport from Golgi to secretory vesicles | Transport from secretory vesicles to cell surface |

**EXPERIMENTAL FIGURE 14-4 Phenotypes of yeast *sec* mutants identified five stages in the secretory pathway.** These temperature-sensitive mutants can be grouped into five classes based on the site where newly made secretory proteins (red dots) accumulate when cells are shifted from the permissive temperature to the higher, nonpermissive one. Analysis of double mutants permitted the sequential order of the steps to be determined. [See P. Novicket al., 1981, *Cell* **25**:461, and C. A. Kaiser and R. Schekman, 1990, *Cell* **61**:723.]

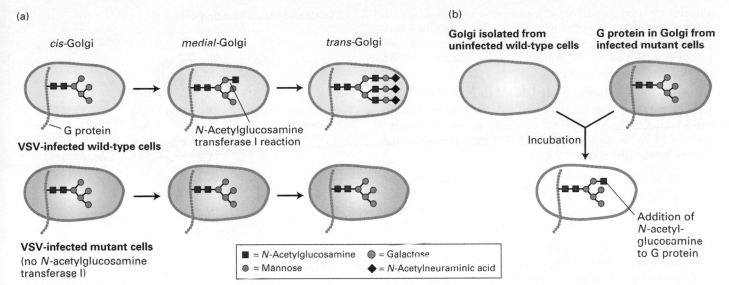

(a)

*cis*-Golgi     *medial*-Golgi     *trans*-Golgi

G protein

**VSV-infected wild-type cells**

*N*-Acetylglucosamine transferase I reaction

**VSV-infected mutant cells**
(no *N*-acetylglucosamine transferase I)

■ = *N*-Acetylglucosamine  ● = Galactose
● = Mannose  ◆ = *N*-Acetylneuraminic acid

(b)

**Golgi isolated from uninfected wild-type cells**

**G protein in Golgi from infected mutant cells**

Incubation

Addition of *N*-acetyl-glucosamine to G protein

**EXPERIMENTAL FIGURE 14-5 A cell-free assay demonstrates protein transport from one Golgi cisterna to another.** (a) A mutant line of cultured fibroblasts is essential in this type of assay. In this example, the cells lack the enzyme *N*-acetylglucosamine transferase I (step **2** in Figure 14-14). In wild-type cells, this enzyme is localized to the *medial*-Golgi and modifies N-linked oligosaccharides by the addition of one *N*-acetylglucosamine. In VSV-infected wild-type cells, the oligosaccharide on the viral G protein is modified to a typical complex oligosaccharide, as shown in the *trans*-Golgi panel. In infected mutant cells, however, the G protein reaches the cell surface with a simpler high-mannose oligosaccharide containing only two *N*-acetylglucosamine and five mannose residues. (b) When Golgi cisternae isolated from infected mutant cells are incubated with Golgi cisternae from normal, uninfected cells, the VSV G protein produced in vitro contains the additional *N*-acetylglucosamine. This modification is carried out by transferase enzyme that is moved by transport vesicles from the wild-type *medial*-Golgi cisternae to the mutant *cis*-Golgi cisternae in the reaction mixture. [See W. E. Balch et al., 1984, *Cell* **39:**405 and 525; W. A. Braell et al., 1984, *Cell* **39:**511; and J. E. Rothman and T. Söllner, 1997, *Science* **276:**1212.]

## Cell-Free Transport Assays Allow Dissection of Individual Steps in Vesicular Transport

In vitro assays for intercompartmental transport are powerful complementary approaches to studies with yeast *sec* mutants for identifying and analyzing the cellular components responsible for vesicular trafficking. In one application of this approach, cultured mutant cells lacking one of the enzymes that modify N-linked oligosaccharide chains in the Golgi are infected with vesicular stomatitis virus (VSV), and the fate of the VSV G protein is followed. For example, if infected cells lack *N*-acetylglucosamine transferase I, they produce abundant amounts of VSV G protein but cannot add *N*-acetylglucosamine residues to the oligosaccharide chains in the *medial*-Golgi as wild-type cells do (Figure 14-5a). When Golgi membranes isolated from such mutant cells are mixed with Golgi membranes from wild-type, uninfected cells, the addition of *N*-acetylglucosamine to VSV G protein is restored (Figure 14-5b). This modification is the consequence of vesicular transport of *N*-acetylglucosamine transferase I from the wild-type *medial*-Golgi to the *cis*-Golgi isolated from virally infected mutant cells. Successful intercompartmental transport in this cell-free system depends on requirements that are typical of a normal physiological process, including a cytosolic extract, a source of chemical energy in the form of ATP and GTP, and incubation at physiological temperatures.

In addition, under appropriate conditions a uniform population of the transport vesicles that move *N*-acetylglucosamine transferase I from the *medial*- to *cis*-Golgi can be purified away from the donor wild-type Golgi membranes by centrifugation. By examining the proteins that are enriched in these vesicles, scientists have been able to identify many of the integral membrane proteins and peripheral vesicle coat proteins that are the structural components of this type of vesicle. Moreover, fractionation of the cytosolic extract required for transport in cell-free reaction mixtures has permitted isolation of the various proteins required for formation of transport vesicles and of proteins required for the targeting and fusion of vesicles with appropriate acceptor membranes. In vitro assays similar in general design to the one shown in Figure 14-5 have been used to study various transport steps in the secretory pathway.

a temperature-sensitive mutant protein that is retained in the ER due to misfolding at the nonpermissive temperature will be released as a cohort for transport when cells are shifted to the permissive temperature.

- Transport of a fluorescently labeled protein along the secretory pathway can be observed by microscopy (see Figure 14-2). Transport of a radiolabeled protein commonly is tracked by following compartment-specific covalent modifications to the protein.

- Many of the components required for intracellular protein trafficking have been identified in yeast by analysis of temperature-sensitive *sec* mutants defective for the secretion of proteins at the nonpermissive temperature (see Figure 14-4).

- Cell-free assays for intercompartmental protein transport have allowed the biochemical dissection of individual steps of the secretory pathway. Such in vitro reactions can be used to produce pure transport vesicles and to test the biochemical function of individual transport proteins.

## 14.2 Molecular Mechanisms of Vesicle Budding and Fusion

Small membrane-bounded vesicles that transport proteins from one organelle to another are common elements in the secretory and endocytic pathways (see Figure 14-1). These vesicles bud from the membrane of a particular *"parent" (donor) organelle* and fuse with the membrane of a particular *"target" (destination) organelle*. Although each step in the secretory and endocytic pathways employs a different type of vesicle, studies employing genetic and biochemical techniques have revealed that each of the different vesicular transport steps is simply a variation on a common theme. In this section we explore the basic mechanisms underlying vesicle budding and fusion that all vesicle types have in common, before discussing the details unique to each pathway.

### Assembly of a Protein Coat Drives Vesicle Formation and Selection of Cargo Molecules

The budding of vesicles from their parent membrane is driven by the polymerization of soluble protein complexes onto the membrane to form a proteinaceous vesicle coat (Figure 14-6a). Interactions between the cytosolic portions of integral membrane proteins and the vesicle coat gather the appropriate cargo proteins into the forming vesicle. Thus the coat gives curvature to the membrane to form a vesicle and acts as the filter to determine which proteins are admitted into the vesicle.

The coat is also responsible for including in the vesicle fusion proteins known as **v-SNAREs**. After formation of a vesicle is completed, the coat is shed, exposing the vesicle's v-SNARE proteins. The specific joining of v-SNAREs in the vesicle membrane with cognate **t-SNAREs** in the target membrane to which the vesicle is docked brings the membranes into close apposition, allowing the two bilayers to

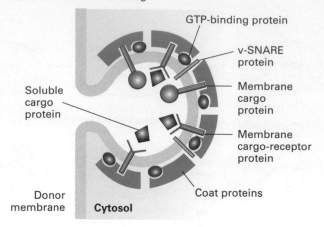

(a) Coated vesicle budding

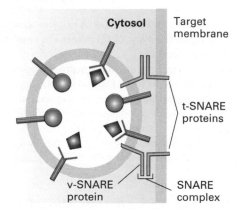

(b) Uncoated vesicle fusion

**FIGURE 14-6 Overview of vesicle budding and fusion with a target membrane.** (a) Budding is initiated by recruitment of a small GTP-binding protein to a patch of donor membrane. Complexes of coat proteins in the cytosol then bind to the cytosolic domain of membrane cargo proteins, some of which also act as receptors that bind soluble proteins in the lumen, thereby recruiting luminal cargo proteins into the budding vesicle. (b) After being released and shedding its coat, a vesicle fuses with its target membrane in a process that involves interaction of cognate SNARE proteins.

fuse (Figure 14-6b). Regardless of target organelle, all transport vesicles use v-SNARES and t-SNARES to bud and fuse.

Three major types of coated vesicles have been characterized, each with a different type of protein coat and each formed by reversible polymerization of a distinct set of protein subunits (Table 14-1). Each type of vesicle, named for its primary coat proteins, transports cargo proteins from particular parent organelles to particular destination organelles:

- **COPII** vesicles transport proteins from the ER to the Golgi.

- **COPI** vesicles mainly transport proteins in the retrograde direction between Golgi cisternae and from the *cis*-Golgi back to the ER.

- **Clathrin** vesicles transport proteins from the plasma membrane (cell surface) and the *trans*-Golgi network to late endosomes.

| TABLE 14-1 | Coated Vesicles Involved in Protein Trafficking | | |
|---|---|---|---|
| **Vesicle Type** | **Transport Step Mediated** | **Coat Proteins** | **Associated GTPase** |
| COPII | ER to *cis*-Golgi | Sec23/Sec24 and Sec13/Sec31 complexes, Sec16 | Sar1 |
| COPI | *cis*-Golgi to ER<br>Later to earlier Golgi cisternae | Coatomers containing seven different COP subunits | ARF |
| Clathrin and adapter proteins* | *trans*-Golgi to endosome | Clathrin + AP1 complexes | ARF |
| | *trans*-Golgi to endosome | Clathrin + GGA | ARF |
| | Plasma membrane to endosome | Clathrin + AP2 complexes | ARF |
| | Golgi to lysosome, melanosome, or platelet vesicles | AP3 complexes | ARF |

*Each type of AP complex consists of four different subunits. It is not known whether the coat of AP3 vesicles contains clathrin.

Every vesicle-mediated trafficking step is thought to utilize some kind of vesicle coat; however, a specific coat protein complex has not been identified for every type of vesicle. For example, vesicles that move proteins from the *trans*-Golgi to the plasma membrane during either constitutive or regulated secretion exhibit a uniform size and morphology suggesting that their formation is driven by assembly of a regular coat structure, yet researchers have not identified specific coat proteins surrounding these vesicles.

The general scheme of vesicle budding shown in Figure 14-6a applies to all three known types of coated vesicles. Experiments with isolated or artificial membranes and purified coat proteins have shown that polymerization of the coat proteins onto the cytosolic face of the parent membrane is necessary to produce the high curvature of the membrane that is typical of a transport vesicle about 50 nm in diameter. Electron micrographs of in vitro budding reactions often reveal structures that exhibit discrete regions of the parent membrane bearing a dense coat accompanied by the curvature characteristic of a completed vesicle (Figure 14-7). Such structures, usually called *vesicle buds,* appear to be intermediates that are visible after the coat has begun to polymerize but before the completed vesicle pinches off from the parent membrane. The polymerized coat proteins are thought to form a curved lattice that drives the formation of a vesicle bud by adhering to the cytosolic face of the membrane.

## A Conserved Set of GTPase Switch Proteins Controls Assembly of Different Vesicle Coats

Based on in vitro vesicle-budding reactions with isolated membranes and purified coat proteins, scientists have determined the minimum set of coat components required to form each of the three major types of vesicles. Although most of the coat proteins differ considerably from one type of vesicle to another, the coats of all three vesicles contain a small GTP-binding protein that acts as a regulatory subunit to control coat assembly (see Figure 14-6a). For both COPI and clathrin vesicles, this GTP-binding protein is known as *ARF protein.* A different but related GTP-binding protein known as *Sar1 protein* is present in the coat of COPII vesicles. Both ARF and Sar1 are monomeric proteins with an overall structure similar to that of Ras, a key intracellular signal-transducing protein (see Figure 16-19). ARF and Sar1 proteins, like Ras, belong to the **GTPase superfamily** of switch proteins that cycle between GDP-bound and GTP-bound forms (see Figure 3-32 to review the mechanism of GTPase "switch" proteins).

The cycle of GTP binding and hydrolysis by ARF and Sar1 are thought to control the initiation of coat assembly,

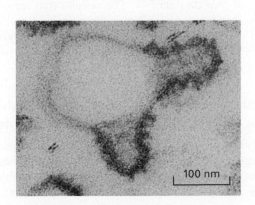

100 nm

**EXPERIMENTAL FIGURE 14-7 Vesicle buds can be visualized during in vitro budding reactions.** When purified COPII coat components are incubated with isolated ER vesicles or artificial phospholipid vesicles (liposomes), polymerization of the coat proteins on the vesicle surface induces emergence of highly curved buds. In this electron micrograph of an in vitro budding reaction, note the distinct membrane coat, visible as a dark protein layer, present on the vesicle buds. [From K. Matsuoka et al., 1988, *Cell* **93**(2):263.]

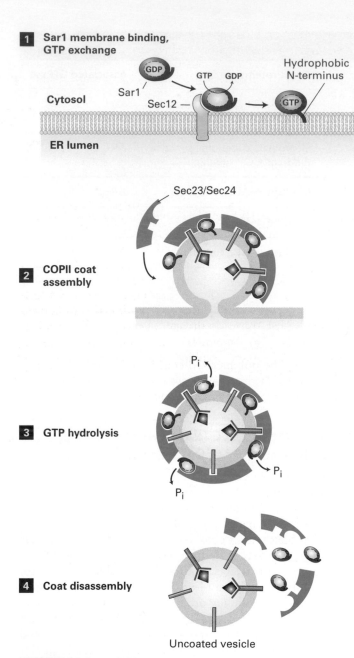

**1** **Sar1 membrane binding, GTP exchange**

Hydrophobic N-terminus

Sar1

Cytosol

Sec12

ER lumen

Sec23/Sec24

**2** **COPII coat assembly**

**3** **GTP hydrolysis**

Pi
Pi
Pi

**4** **Coat disassembly**

Uncoated vesicle

**FIGURE 14-8 Model for the role of Sar1 in the assembly and disassembly of COPII coats.** Step **1**: Interaction of soluble GDP-bound Sar1 with the exchange factor Sec12, an ER integral membrane protein, catalyzes exchange of GTP for GDP on Sar1. In the GTP-bound form of Sar1, its hydrophobic N-terminus extends outward from the protein's surface and anchors Sar1 to the ER membrane. Step **2**: Sar1 attached to the membrane serves as a binding site for the Sec23/Sec24 coat protein complex. Membrane cargo proteins are recruited to the forming vesicle bud by binding of specific short sequences (sorting signals) in their cytosolic regions to sites on the Sec23/Sec24 complex. Some membrane cargo proteins also act as receptors that bind soluble proteins in the lumen. The coat is completed by assembly of a second type of coat complex composed of Sec13 and Sec31 (not shown). Step **3**: After the vesicle coat is complete, the Sec23 coat subunit promotes GTP hydrolysis by Sar1. Step **4**: Release of Sar1·GDP from the vesicle membrane causes disassembly of the coat. [See S. Springer et al., 1999, *Cell* **97**:145.]

as schematically depicted for the assembly of COPII vesicles in Figure 14-8. First, an ER membrane protein known as Sec12 catalyzes release of GDP from cytosolic Sar1·GDP and binding of GTP. This *guanine nucleotide–exchange factor* apparently receives and integrates multiple as-yet-unknown signals, probably including the presence in the ER membrane of cargo proteins that are ready to be transported. Binding of GTP causes a conformational change in Sar1 that exposes its hydrophobic N-terminus, which then becomes embedded in the phospholipid bilayer and tethers Sar1·GTP to the ER membrane (Figure 14-8, step **1**). The membrane-attached Sar1·GTP drives polymerization of cytosolic complexes of COPII subunits on the membrane, eventually leading to formation of vesicle buds (step **2**). Once COPII vesicles are released from the donor membrane, the Sar1 GTPase activity hydrolyzes Sar1·GTP in the vesicle membrane to Sar1·GDP with the assistance of one of the coat subunits (step **3**). This hydrolysis triggers disassembly of the COPII coat (step **4**). Thus Sar1 couples a cycle of GTP binding and hydrolysis to the formation and then dissociation of the COPII coat.

ARF protein undergoes a similar cycle of nucleotide exchange and hydrolysis coupled to the assembly of vesicle coats composed either of COPI or of clathrin and other coat proteins (AP complexes), discussed later. A covalent protein modification known as a myristate anchor on the N-terminus of ARF protein weakly tethers ARF·GDP to the Golgi membrane. When GTP is exchanged for the bound GDP by a nucleotide-exchange factor attached to the Golgi membrane, the resulting conformational change in ARF allows hydrophobic residues in its N-terminal segment to insert into the membrane bilayer. The resulting tight association of ARF·GTP with the membrane serves as the foundation for further coat assembly.

Drawing on the structural similarities of Sar1 and ARF to other small GTPase switch proteins, researchers have constructed genes encoding mutant versions of the two proteins that have predictable effects on vesicular traffic when transfected into cultured cells. For example, in cells expressing mutant versions of Sar1 or ARF that cannot hydrolyze GTP, vesicle coats form and vesicle buds pinch off. However, because the mutant proteins cannot trigger disassembly of the coat, all available coat subunits eventually become permanently assembled into coated vesicles that are unable to fuse with target membranes. Addition of a nonhydrolyzable GTP analog to in vitro vesicle-budding reactions causes a similar blocking of coat disassembly. The vesicles that form in such reactions have coats that never dissociate, allowing their composition and structure to be more readily analyzed. The purified COPI vesicles shown in Figure 14-9 were produced in such a budding reaction.

## Targeting Sequences on Cargo Proteins Make Specific Molecular Contacts with Coat Proteins

In order for transport vesicles to move specific proteins from one compartment to the next, vesicle buds must be able to discriminate among potential membrane and soluble cargo proteins, accepting only those cargo proteins that should

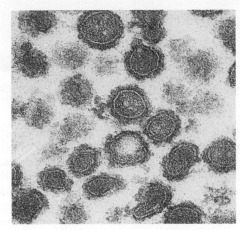

**EXPERIMENTAL FIGURE 14-9 Coated vesicles accumulate during in vitro budding reactions in the presence of a nonhydrolyzable analog of GTP.** When isolated Golgi membranes are incubated with a cytosolic extract containing COPI coat proteins, vesicles form and bud off from the membranes. Inclusion of a nonhydrolyzable analog of GTP in the budding reaction prevents disassembly of the coat after vesicle release. This micrograph shows COPI vesicles generated in such a reaction and separated from membranes by centrifugation. Coated vesicles prepared in this way can be analyzed to determine their components and properties. [Courtesy of L. Orci.]

60 nm

advance to the next compartment and excluding those that should remain as residents in the donor compartment. In addition to sculpting the curvature of a donor membrane, the vesicle coat functions in selecting specific proteins as cargo. The primary mechanism by which the vesicle coat selects cargo molecules is by directly binding to specific sequences, or **sorting signals,** in the cytosolic portion of membrane cargo

proteins (see Figure 14-6a). The polymerized coat thus acts as an affinity matrix to cluster selected membrane cargo proteins into forming vesicle buds. Since soluble proteins within the lumen of parent organelles cannot contact the coat directly, they require a different kind of sorting signal. Soluble luminal proteins often contain what can be thought of as *luminal sorting signals,* which bind to the luminal domains of certain membrane cargo proteins that act as receptors for luminal cargo proteins. The properties of several known sorting signals in membrane and soluble proteins are summarized in Table 14-2. We describe the role of these signals in more detail in later sections.

**TABLE 14-2** | **Known Sorting Signals That Direct Proteins to Specific Transport Vesicles**

| Signal Sequence* | Proteins with Signal | Signal Receptor | Vesicles That Incorporate Signal-Bearing Protein |
|---|---|---|---|
| **LUMINAL SORTING SIGNALS** | | | |
| Lys-Asp-Glu-Leu (KDEL) | ER-resident soluble proteins | KDEL receptor in *cis*-Golgi membrane | COPI |
| Mannose 6-phosphate (M6P) | Soluble lysosomal enzymes after processing in *cis*-Golgi | M6P receptor in *trans*-Golgi membrane | Clathrin/AP1 |
| | Secreted lysosomal enzymes | M6P receptor in plasma membrane | Clathrin/AP2 |
| **CYTOPLASMIC SORTING SIGNALS** | | | |
| Lys-Lys-X-X (KKXX) | ER-resident membrane proteins | COPI α and β subunits | COPI |
| Di-arginine (X-Arg-Arg-X) | ER-resident membrane proteins | COPI α and β subunits | COPI |
| Di-acidic (e.g., Asp-X-Glu) | Cargo membrane proteins in ER | COPII Sec24 subunit | COPII |
| Asn-Pro-X-Tyr (NPXY) | LDL receptor in plasma membrane | AP2 complex | Clathrin/AP2 |
| Tyr-X-X-Φ (YXXΦ) | Membrane proteins in *trans*-Golgi | AP1 (μ1 subunit) | Clathrin/AP1 |
| | Plasma membrane proteins | AP2 (μ2 subunit) | Clathrin/AP2 |
| Leu-Leu (LL) | Plasma membrane proteins | AP2 complexes | Clathrin/AP2 |

*X = any amino acid; Φ = hydrophobic amino acid. Single-letter amino acid abbreviations are in parentheses.

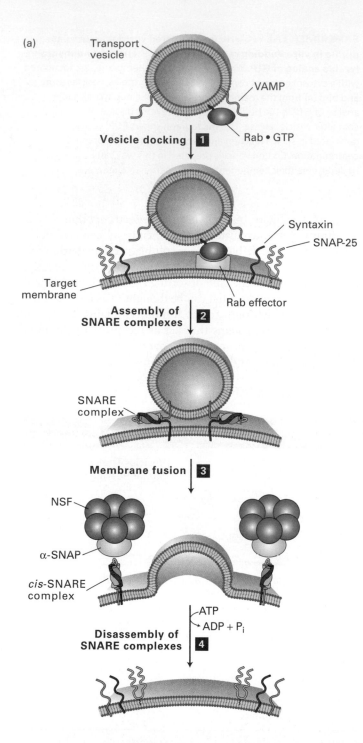

**(a)**

Transport vesicle

VAMP

Rab • GTP

**Vesicle docking** 1

Syntaxin

SNAP-25

Target membrane

Rab effector

**Assembly of SNARE complexes** 2

SNARE complex

**Membrane fusion** 3

NSF

α-SNAP

*cis*-SNARE complex

ATP

ADP + P$_i$

**Disassembly of SNARE complexes** 4

**(b) SNARE complex**

VAMP

Syntaxin

SNAP-25

**FIGURE 14-10 Model for docking and fusion of transport vesicles with their target membranes.** (a) The proteins shown in this example participate in fusion of secretory vesicles with the plasma membrane, but similar proteins mediate all vesicle-fusion events. Step 1: A Rab protein tethered via a lipid anchor to a secretory vesicle binds to an effector protein complex on the plasma membrane, thereby docking the transport vesicle on the appropriate target membrane. Step 2: A v-SNARE protein (in this case, VAMP) interacts with the cytosolic domains of the cognate t-SNAREs (in this case, syntaxin and SNAP-25). The very stable coiled-coil SNARE complexes that are formed hold the vesicle close to the target membrane. Step 3: Fusion of the two membranes immediately follows formation of SNARE complexes, but precisely how this occurs is not known. Step 4: Following membrane fusion, NSF in conjunction with α-SNAP protein binds to the SNARE complexes. The NSF-catalyzed hydrolysis of ATP then drives dissociation of the SNARE complexes, freeing the SNARE proteins for another round of vesicle fusion. Also at this time, Rab·GTP is hydrolyzed to Rab·GDP and dissociates from the Rab effector (not shown). (b) The SNARE complex. Numerous noncovalent interactions between four long α helices, two from SNAP-25 and one each from syntaxin and VAMP, stabilize the coiled-coil structure. [See J. E. Rothman and T. Söllner, 1997, *Science* **276**:1212, and W. Weis and R. Scheller, 1998, *Nature* **395**:328. Part (b) from Y. A. Chen and R. H. Scheller, 2001, *Nat. Rev. Mol. Cell Biol.* **2**(2):98.]

change in Rab that enables it to interact with a surface protein on a particular transport vesicle and insert its isoprenoid anchor into the vesicle membrane. Once Rab·GTP is tethered to the vesicle surface, it is thought to interact with one of a number of different large proteins, known as Rab effectors, attached to the target membrane. Binding of Rab·GTP to a Rab effector docks the vesicle on an appropriate target membrane (Figure 14-10, step 1). After vesicle fusion occurs, the GTP bound to the Rab protein is hydrolyzed to GDP, triggering the release of Rab·GDP, which then can undergo another cycle of GDP-GTP exchange, binding, and hydrolysis.

Several lines of evidence support the involvement of specific Rab proteins in vesicle-fusion events. For instance, the yeast *SEC4* gene encodes a Rab protein, and yeast cells expressing mutant Sec4 proteins accumulate secretory vesicles that are unable to fuse with the plasma membrane (class E mutants in Figure 14-4). In mammalian cells, Rab5 protein is localized to endocytic vesicles, also known as early endosomes. These uncoated vesicles form from clathrin-coated vesicles just after they bud from the plasma membrane during endocytosis (see Figure 14-1, step 9). The fusion of early endosomes with each other in cell-free systems requires the presence of Rab5, and addition of Rab5 and GTP to cell-free extracts accelerates the rate at which these vesicles fuse with

## Rab GTPases Control Docking of Vesicles on Target Membranes

A second set of small GTP-binding proteins, known as *Rab proteins,* participate in the targeting of vesicles to the appropriate target membrane. Like Sar1 and ARF, Rab proteins belong to the GTPase superfamily of switch proteins. Rab proteins also contain an isoprenoid anchor that allows them to become tethered to the vesicle membrane. Conversion of cytosolic Rab·GDP to Rab·GTP, catalyzed by a specific guanine nucleotide–exchange factor, induces a conformational

each other. A long coiled protein known as EEA1 (early endosome antigen 1), which resides on the membrane of the early endosome, functions as the effector for Rab5. In this case, Rab5·GTP on one endocytic vesicle is thought to specifically bind to EEA1 on the membrane of another endocytic vesicle, setting the stage for fusion of the two vesicles.

A different type of Rab effector appears to function for each vesicle type and at each step of the secretory pathway. Many questions remain about how Rab proteins are targeted to the correct membrane and how specific complexes form between the different Rab proteins and their corresponding effector proteins.

## Paired Sets of SNARE Proteins Mediate Fusion of Vesicles with Target Membranes

As noted previously, shortly after a vesicle buds off from the donor membrane, the vesicle coat disassembles to uncover a vesicle-specific membrane protein, a v-SNARE (see Figure 14-6b). Likewise, each type of target membrane in a cell contains t-SNARE membrane proteins, which specifically interact with v-SNAREs. After Rab-mediated docking of a vesicle on its target (destination) membrane, the interaction of cognate SNAREs brings the two membranes close enough together that they can fuse.

One of the best-understood examples of SNARE-mediated fusion occurs during exocytosis of secreted proteins (see Figure 14-10, steps **2** and **3**). In this case, the v-SNARE, known as *VAMP* (*v*esicle-*a*ssociated *m*embrane *p*rotein), is incorporated into secretory vesicles as they bud from the *trans*-Golgi network. The t-SNAREs are *syntaxin,* an integral membrane protein in the plasma membrane, and *SNAP-25,* which is attached to the plasma membrane by a hydrophobic lipid anchor in the middle of the protein. The cytosolic region in each of these three SNARE proteins contains a repeating heptad sequence that allows four α helices—one from VAMP, one from syntaxin, and two from SNAP-25—to coil around one another to form a four-helix bundle (Figure 14-10b). The unusual stability of this bundled SNARE complex is conferred by the arrangement of hydrophobic and charged amino residues in the heptad repeats. The hydrophobic amino acids are buried in the central core of the bundle, and amino acids of opposite charge are aligned to form favorable electrostatic interactions between helices. As the four-helix bundles form, the vesicle and target membranes are drawn into close apposition by the embedded transmembrane domains of VAMP and syntaxin.

In vitro experiments have shown that when liposomes containing purified VAMP are incubated with other liposomes containing syntaxin and SNAP-25, the two classes of membranes fuse, albeit slowly. This finding is strong evidence that the close apposition of membranes resulting from formation of SNARE complexes is sufficient to bring about membrane fusion. Fusion of a vesicle and target membrane occurs more rapidly and efficiently in the cell than it does in liposome experiments in which fusion is catalyzed only by SNARE

proteins. The likely explanation for this difference is that in the cell, other proteins such as Rab proteins and their effectors are involved in targeting vesicles to the correct membrane.

Yeast cells, like all eukaryotic cells, express more than 20 different related v-SNARE and t-SNARE proteins. Analyses of yeast mutants defective in each of the SNARE genes have identified specific membrane-fusion events in which each SNARE protein participates. For all fusion events that have been examined, the SNAREs form four-helix bundled complexes, similar to the VAMP/syntaxin/SNAP-25 complexes that mediate fusion of secretory vesicles with the plasma membrane. However, in other fusion events (e.g., fusion of COPII vesicles with the *cis*-Golgi network), each participating SNARE protein contributes only one α helix to the bundle (unlike SNAP-25, which contributes two helices); in these cases the SNARE complexes comprise one v-SNARE and three t-SNARE molecules.

Using the in vitro liposome fusion assay, researchers have tested the ability of various combinations of individual v-SNARE and t-SNARE proteins to mediate fusion of donor and target membranes. Of the very large number of different combinations tested, only a small number could efficiently mediate membrane fusion. To a remarkable degree, the functional combinations of v-SNAREs and t-SNAREs revealed in these in vitro experiments correspond to the actual SNARE protein interactions that mediate known membrane-fusion events in the yeast cell. Thus, together with the specificity of interaction between Rab and Rab effector proteins, the specificity of the interaction between SNARE proteins can account for most, if not all, of the specificity of fusion between a particular vesicle type and its target membrane.

## Dissociation of SNARE Complexes After Membrane Fusion Is Driven by ATP Hydrolysis

After a vesicle and its target membrane have fused, the SNARE complexes must dissociate to make the individual SNARE proteins available for additional fusion events. Because of the stability of SNARE complexes, which are held together by numerous noncovalent intermolecular interactions, their dissociation depends on additional proteins and the input of energy.

The first clue that dissociation of SNARE complexes required the assistance of other proteins came from in vitro transport reactions depleted of certain cytosolic proteins. The observed accumulation of vesicles in these reactions indicated that vesicles could form but were unable to fuse with a target membrane. Eventually two proteins, designated *NSF* and α-*SNAP*, were found to be required for ongoing vesicle fusion in the in vitro transport reaction. The function of NSF in vivo can be blocked selectively by *N*-ethylmaleimide (NEM), a chemical that reacts with an essential –SH group on NSF (hence the name, NEM-sensitive factor).

Yeast mutants have also contributed to our understanding of SNARE function. Among the class C yeast *sec* mutants are strains that lack functional Sec18 or Sec17, the yeast counterparts of mammalian NSF and α-SNAP, respectively. When these class C mutants are placed at the nonpermissive temperature, they accumulate ER-to-Golgi transport vesicles; when the cells are shifted to the lower, permissive temperature, the accumulated vesicles are able to fuse with the *cis*-Golgi.

Subsequent to the initial biochemical and genetic studies identifying NSF and α-SNAP, more sophisticated in vitro transport assays were developed. Using these newer assays, researchers have shown that NSF and α-SNAP proteins are not necessary for actual membrane fusion but rather are required for regeneration of free SNARE proteins. NSF, a hexamer of identical subunits, associates with a SNARE complex with the aid of α-SNAP (*soluble NSF attachment protein*). The bound NSF then hydrolyzes ATP, releasing sufficient energy to dissociate the SNARE complex (Figure 14-10, step ◼). Evidently, the defects in vesicle fusion observed in the earlier in vitro fusion assays and in the yeast mutants after a loss of Sec17 or Sec18 were a consequence of free SNARE proteins rapidly becoming sequestered in undissociated SNARE complexes and thus being unavailable to mediate membrane fusion.

## KEY CONCEPTS of Section 14.2

### Molecular Mechanisms of Vesicle Budding and Fusion

• The three well-characterized transport vesicles—COPI, COPII, and clathrin vesicles—are distinguished by the proteins that form their coats and the transport routes they mediate (see Table 14-1).

• All types of coated vesicles are formed by polymerization of cytosolic coat proteins onto a donor (parent) membrane to form vesicle buds that eventually pinch off from the membrane to release a complete vesicle. Shortly after vesicle release, the coat is shed, exposing proteins required for fusion with the target membrane (see Figure 14-6).

• Small GTP-binding proteins (ARF or Sar1) belonging to the GTPase superfamily control polymerization of coat proteins, the initial step in vesicle budding (see Figure 14-8). After vesicles are released from the donor membrane, hydrolysis of GTP bound to ARF or Sar1 triggers disassembly of the vesicle coats.

• Specific sorting signals in membrane and luminal proteins of donor organelles interact with coat proteins during vesicle budding, thereby recruiting cargo proteins to vesicles (see Table 14-2).

• A second set of GTP-binding proteins, the Rab proteins, regulate docking of vesicles with the correct target membrane. Each Rab appears to bind to a specific Rab effector associated with the target membrane.

• Each v-SNARE in a vesicular membrane specifically binds to a complex of cognate t-SNARE proteins in the target

membrane, inducing fusion of the two membranes. After fusion is completed, the SNARE complex is disassembled in an ATP-dependent reaction mediated by other cytosolic proteins (see Figure 14-10).

## 14.3 Early Stages of the Secretory Pathway

In this section we take a closer look at vesicular traffic between the ER and the Golgi and some of the evidence supporting the general mechanisms discussed in the previous section. Recall that *anterograde transport* from the ER to Golgi, the first vesicle trafficking step in the secretory pathway, is mediated by COPII vesicles. These vesicles contain newly synthesized proteins destined for the Golgi, cell surface, or lysosomes as well as vesicle components such as v-SNAREs that are required to target vesicles to the *cis*-Golgi membrane. Proper sorting of proteins between the ER and Golgi also requires *retrograde (reverse) transport* from the *cis*-Golgi to the ER, which is mediated by COPI vesicles (Figure 14-11). This retrograde vesicle transport serves to retrieve v-SNARE proteins and the membrane itself back to the ER to provide the necessary material for additional rounds of vesicle budding from the ER. COPI-mediated retrograde transport also retrieves missorted ER-resident proteins from the *cis*-Golgi to correct sorting mistakes.

We also discuss in this section the process by which proteins that have been correctly delivered to the Golgi advance through successive compartments of the Golgi, from *cis*- to *trans*-network. This process, known as cisternal maturation, involves budding and fusion of retrograde rather than anterograde transport vesicles.

## COPII Vesicles Mediate Transport from the ER to the Golgi

COPII vesicles were first recognized when cell-free extracts of yeast rough ER membranes were incubated with cytosol and a nonhydrolyzable analog of GTP. The vesicles that formed from the ER membranes had a distinct coat similar to that on COPI vesicles but composed of different proteins, designated COPII proteins. Yeast cells with mutations in the genes for COPII proteins are class B *sec* mutants and accumulate proteins in the rough ER (see Figure 14-4). Analysis of such mutants has revealed a set of proteins required for formation of COPII vesicles, including the proteins which comprise the COPII vesicle coat.

As described previously, formation of COPII vesicles is triggered when Sec12, a guanine nucleotide–exchange factor in the ER membrane, catalyzes the exchange of bound GDP for GTP on cytosolic Sar1. This exchange induces binding of Sar1 to the ER membrane followed by binding of a complex of Sec23 and Sec24 proteins (see Figure 14-8). The resulting

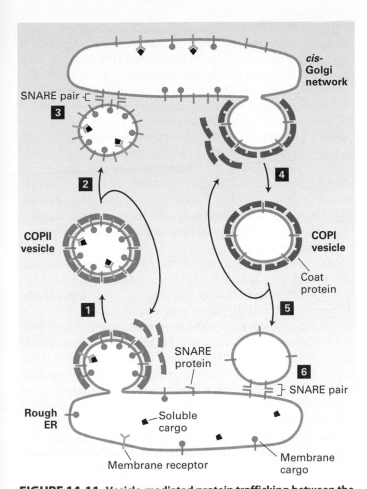

FIGURE 14-11 **Vesicle-mediated protein trafficking between the ER and *cis*-Golgi.** Steps **1**–**3**: Forward (anterograde) transport is mediated by COPII vesicles, which are formed by polymerization of soluble COPII coat protein complexes (green) on the ER membrane. v-SNAREs (orange) and other cargo proteins (blue) in the ER membrane are incorporated into the vesicle by interacting with coat proteins. Soluble cargo proteins (magenta) are recruited by binding to appropriate receptors in the membrane of budding vesicles. Dissociation of the coat recycles free coat complexes and exposes v-SNARE proteins on the vesicle surface. After the uncoated vesicle becomes tethered to the *cis*-Golgi membrane in a Rab-mediated process, pairing between the exposed v-SNAREs and cognate t-SNAREs in the Golgi membrane allows vesicle fusion, releasing the contents into the *cis*-Golgi compartment (see Figure 14-10). Steps **4**–**6**: Reverse (retrograde) transport, mediated by vesicles coated with COPI proteins (purple), recycles the membrane bilayer and certain proteins, such as v-SNAREs and missorted ER-resident proteins (not shown), from the *cis*-Golgi to the ER. All SNARE proteins are shown in orange although v-SNAREs and t-SNAREs are distinct proteins.

ternary complex formed between Sar1·GTP, Sec23, and Sec24 is shown in Figure 14-12. After this complex forms on the ER membrane, a second complex comprising Sec13 and Sec31 proteins binds to complete the coat structure. Pure Sec13 and Sec31 proteins can spontaneously assemble into cagelike lattices. It is thought that Sec13 and Sec31 form the structural scaffold for COPII vesicles. Finally, a large fibrous

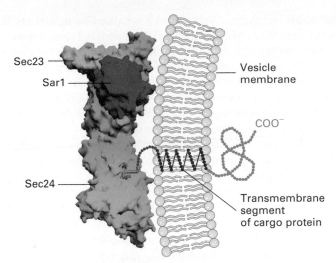

FIGURE 14-12 **Three-dimensional structure of the ternary complex comprising the COPII coat proteins Sec23 and Sec24 and Sar1·GTP.** Early in the formation of the COPII coat, Sec23 (orange)/Sec24 (green) complexes are recruited to the ER membrane by Sar1 (red) in its GTP-bound state. In order to form a stable ternary complex in solution for structural studies, the nonhydrolyzable GTP analog GppNHp is used. A cargo protein in the ER membrane can be recruited to COPII vesicles by interaction of a tripeptide di-acidic signal (purple) in the cargo's cytosolic domain with Sec24. The likely position of the COPII vesicle membrane and the transmembrane segment of the cargo protein are indicated. The N-terminal segment of Sar1 that tethers it to the membrane is not shown. [See X. Bi et al., 2002, *Nature* **419**:271; interaction with peptide courtesy of J. Goldberg.]

protein called Sec16, which is bound to the cytosolic surface of the ER, interacts with Sar1·GTP and the Sec13/31 and Sec23/24 complexes and acts to organize the other coat proteins, increasing the efficiency of coat polymerization. Similar to Sec 13/31, clathrin also has this ability to self-assemble into a coatlike structure, as will be discussed in Section 14.4.

Certain integral ER membrane proteins are specifically recruited into COPII vesicles for transport to the Golgi. The cytosolic segments of many of these proteins contain a *di-acidic sorting signal* (the key residues in this sequence are Asp-X-Glu, or DXE in the one-letter code) (see Table 14-2). This sorting signal binds to the Sec24 subunit of the COPII coat and is essential for the selective export of certain membrane proteins from the ER (see Figure 14-12). Biochemical and genetic studies currently are under way to identify additional signals that help direct membrane cargo proteins into COPII vesicles. Other ongoing studies seek to determine how soluble cargo proteins are selectively loaded into COPII vesicles. Although purified COPII vesicles from yeast cells have been found to contain a membrane protein that binds the soluble α mating factor, the receptors for other soluble cargo proteins such as invertase are not yet known.

The inherited disease cystic fibrosis is characterized by an imbalance in chloride and sodium ion transport in the epithelial cells of the lung, leading to fluid build-up and difficulty breathing. Cystic fibrosis is caused by mutations in a protein known as CFTR, which is synthesized as an integral membrane protein in the ER and is transported to the Golgi before being transported to the plasma membranes of epithelial cells, where it functions as a chloride channel. Researchers have recently shown that the CFTR protein contains a diacidic sorting signal that binds to the Sec24 subunit of the COPII coat and is necessary for transport of the CFTR protein out of the ER. The most common CFTR mutation is a deletion of a phenylalanine at position 508 in the protein sequence (known as ΔF508). This mutation prevents normal transport of CFTR to the plasma membrane by blocking its packaging into COPII vesicles budding from the ER. Although the ΔF508 mutation is not in the vicinity of the diacidic sorting signal, this mutation may change the conformation of the cytosolic portion of CFTR so that the di-acidic signal is unable to bind to Sec24. Interestingly, a folded CFTR with this mutation would still likely function properly as a normal chloride channel. However, it never reaches the membrane; the disease state is therefore caused by the absence of the channel rather than by a defective one. ■

The experiments described previously in which the transit of VSVG-GFP in cultured mammalian cells is followed by fluorescence microscopy (see Figure 14-2) provided insight into the intermediates in ER-to-Golgi transport. In some cells, small fluorescent vesicles containing VSVG-GFP could be seen to form from the ER, move less than 1 μm, and then fuse directly with the cis-Golgi. In other cells, in which the ER was located several micrometers from the Golgi complex, several ER-derived vesicles were seen to fuse with each other shortly after their formation, forming what is termed the ER-to-Golgi intermediate compartment or the cis-Golgi network. These larger structures then were transported along microtubules to the cis-Golgi, much in the way vesicles in nerve cells are transported from the cell body, where they are formed, down the long axon to the axon terminus (Chapter 18). Microtubules function much as "railroad tracks," enabling these large aggregates of transport vesicles to move long distances to their cis-Golgi destination. At the time the ER-to-Golgi intermediate compartment is formed, some COPI vesicles bud off from it, recycling some proteins back to the ER.

## COPI Vesicles Mediate Retrograde Transport Within the Golgi and from the Golgi to the ER

COPI vesicles were first discovered when isolated Golgi fractions were incubated in a solution containing cytosol and a nonhydrolyzable analog of GTP (see Figure 14-9). Subsequent analysis of these vesicles showed that the coat is formed from large cytosolic complexes, called coatomers, composed of seven polypeptide subunits. Yeast cells containing temperature-sensitive mutations in COPI proteins accumulate proteins in the rough ER at the nonpermissive temperature and thus are categorized as class B sec mutants (see Figure 14-4). Although discovery of these mutants initially

suggested that COPI vesicles mediate ER-to-Golgi transport, subsequent experiments showed that their main function is retrograde transport, both between Golgi cisternae and from the cis-Golgi to the rough ER (see Figure 14-11, right). Because COPI mutants cannot recycle key membrane proteins back to the rough ER, the ER gradually becomes depleted of ER proteins such as v-SNAREs necessary for COPII vesicle function. Eventually, vesicle formation from the rough ER grinds to a halt; secretory proteins continue to be synthesized but accumulate in the ER, the defining characteristic of class B sec mutants. The general ability of sec mutants involved in either COPI or COPII vesicle function to eventually block both anterograde and retrograde transport illustrates the fundamental interdependence of these two transport processes.

As discussed in Chapter 13, the ER contains several soluble proteins dedicated to the folding and modification of newly synthesized secretory proteins. These include the chaperone BiP and the enzyme protein disulfide isomerase, which are necessary for the ER to carry out its functions. Although such ER-resident luminal proteins are not specifically selected by COPII vesicles, their sheer abundance causes them to be continuously loaded passively into vesicles destined for the cis-Golgi. The transport of these soluble proteins back to the ER, mediated by COPI vesicles, prevents their eventual depletion.

Most soluble ER-resident proteins carry a Lys-Asp-Glu-Leu (KDEL in the one-letter code) sequence at their C-terminus (see Table 14-2). Several experiments demonstrated that this KDEL sorting signal is both necessary and sufficient to cause a protein bearing this sequence to be located in the ER. For instance, when a mutant protein disulfide isomerase lacking these four residues is synthesized in cultured fibroblasts, the protein is secreted. Moreover, if a protein that normally is secreted is altered so that it contains the KDEL signal at its C-terminus, the protein is located in the ER. The KDEL sorting signal is recognized and bound by the KDEL receptor, a transmembrane protein found primarily on small transport vesicles shuttling between the ER and the cis-Golgi and on the cis-Golgi reticulum. In addition, soluble ER-resident proteins that carry the KDEL signal have oligosaccharide chains with modifications that are catalyzed by enzymes found only in the cis-Golgi or cis-Golgi network; thus at some time these proteins must have left the ER and been transported at least as far as the cis-Golgi network. These findings indicate that the KDEL receptor acts mainly to retrieve soluble proteins containing the KDEL sorting signal that have escaped to the cis-Golgi network and return them to the ER (Figure 14-13). The KDEL receptor binds more tightly to its ligand at low pH, and it is thought that the receptor is able to bind KDEL peptides in the cis-Golgi but to release these peptides in the ER because the pH of the Golgi is slightly lower than that of the ER.

The KDEL receptor and other membrane proteins that are transported back to the ER from the Golgi contain a Lys-Lys-X-X sequence at the very end of their C-terminal segment, which faces the cytosol (see Table 14-2). This KKXX sorting signal, which binds to a complex of the COPI α and

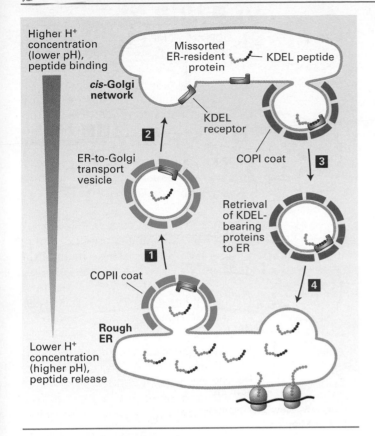

Higher H⁺ concentration (lower pH), peptide binding

Missorted ER-resident protein — KDEL peptide

**cis-Golgi network**

KDEL receptor

2

ER-to-Golgi transport vesicle

COPI coat

3

Retrieval of KDEL-bearing proteins to ER

1

COPII coat

4

**Rough ER**

Lower H⁺ concentration (higher pH), peptide release

**FIGURE 14-13 Role of the KDEL receptor in retrieval of ER-resident luminal proteins from the Golgi.** ER luminal proteins, especially those present at high levels, can be passively incorporated into COPII vesicles and transported to the Golgi (steps 1 and 2). Many such proteins bear a C-terminal KDEL (Lys-Asp-Glu-Leu) sequence (red) that allows them to be retrieved. The KDEL receptor, located mainly in the cis-Golgi network and in both COPII and COPI vesicles, binds proteins bearing the KDEL sorting signal and returns them to the ER (steps 3 and 4). This retrieval system prevents depletion of ER luminal proteins such as those needed for proper folding of newly made secretory proteins. The binding affinity of the KDEL receptor is very sensitive to pH. The small difference between the pH of the ER and that of the Golgi favors binding of KDEL-bearing proteins to the receptor in Golgi-derived vesicles and their release in the ER. [Adapted from J. Semenza et al., 1990, *Cell* **61**:1349.]

This basic question concerning correct membrane partitioning has recently been answered for COPII vesicles. After these vesicles form, the COPII coat proteins remain assembled long enough for the Sec23/Sec24 complex to interact with a specific tethering factor attached to the cis-Golgi membrane. Vesicle uncoating to expose the v-SNAREs is completed only after the COPII vesicle is already closely associated with the cis-Golgi membrane and the COPII v-SNAREs are in position to form complexes with their cognate t-SNAREs. Although COPII vesicles also carry COPI-specific v-SNARE proteins, which are being recycled back to the cis-Golgi, these COPI v-SNARE proteins included in COPII vesicles never have the opportunity to form SNARE complexes with cognate ER-localized t-SNARE proteins.

## Anterograde Transport Through the Golgi Occurs by Cisternal Maturation

The Golgi complex is organized into three or four subcompartments, often arranged in a stacked set of flattened sacs, called cisternae. The subcompartments of the Golgi differ from one another according to the enzymes they contain. Many of the enzymes are glycosidases and glycosyltransferases that are involved in modifying N-linked or O-linked carbohydrates attached to secretory proteins as they transit the Golgi stack. On the whole, the Golgi complex operates much like an assembly line, with proteins moving in sequence through the Golgi stack, the modified carbohydrate chains in one compartment serving as the substrates for the modifying enzymes of the next compartment (see Figure 14-14 for a representative sequence of modification steps).

For many years it was thought that the Golgi complex was an essentially static set of compartments with small transport vesicles carrying secretory proteins forward, from the cis- to the medial-Golgi and from the medial- to the trans-Golgi. Indeed, electron microscopy reveals many small vesicles associated with the Golgi complex that appear to move proteins

β subunits (two of the seven polypeptide subunits in the COPI coatomer), is both necessary and sufficient to incorporate membrane proteins into COPI vesicles for retrograde transport to the ER. Temperature-sensitive yeast mutants lacking COPIα or COPIβ not only are unable to bind the KKXX signal but also are unable to retrieve proteins bearing this signal back to the ER, indicating that COPI vesicles mediate retrograde Golgi-to-ER transport.

A second sorting signal that will target proteins to COPI vesicles and thus will enable recycling from the Golgi to the ER is a di-arginine sequence. Unlike the KKXX sorting signal, which must be located at the cytoplasmically oriented C-terminus of a protein, the di-arginine sorting signal can reside in any segment of a membrane protein that is on the cytoplasmic face of the membrane.

The partitioning of proteins between the ER and Golgi complex is a highly dynamic process depending on both COPII (anterograde) and COPI (retrograde) vesicles, with each type of vesicle responsible for recycling the components necessary for the function of the other type of vesicle. The organization of this partitioning process raises an interesting puzzle: how do vesicles preferentially use the v-SNAREs that will specify fusion with the correct target membrane instead of the v-SNAREs that are being recycled and would have specificity for fusion with the donor membrane?

**FIGURE 14-14 Processing of N-linked oligosaccharide chains on glycoproteins within *cis*-, *medial*-, and *trans*-Golgi cisternae in vertebrate cells.** The enzymes catalyzing each step are localized to the indicated compartments. After removal of three mannose residues in the *cis*-Golgi (step **1**), the protein moves by cisternal maturation to the *medial*-Golgi. Here, three *N*-acetylglucosamine (GlcNAc) residues are added (steps **2** and **4**), two more mannose residues are removed (step **3**), and a single fucose is added (step **5**). Processing is completed in the *trans*-Golgi by addition of three galactose residues (step **6**) and finally by linkage of an *N*-acetylneuraminic acid residue to each of the galactose residues (step **7**). Specific transferase enzymes add sugars to the oligosaccharide, one at a time, from sugar nucleotide precursors imported from the cytosol. This pathway represents the Golgi processing events for a typical mammalian glycoprotein. Variations in the structure of N-linked oligosaccharides can result from differences in processing steps in the Golgi. [See R. Kornfeld and S. Kornfeld, 1985, *Ann. Rev. Biochem.* **45**:631.]

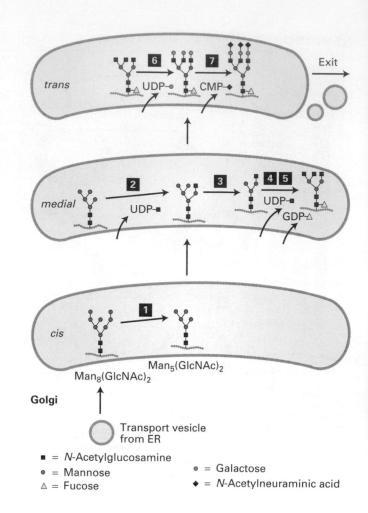

from one Golgi compartment to another (Figure 14-15). However, these vesicles are now known to mediate retrograde transport, retrieving ER or Golgi enzymes from a later compartment and transporting them to an earlier compartment in the secretory pathway. Thus the Golgi appears to have a highly dynamic organization, continually forming transport vesicles, though only in the retrograde direction. To see the effect this retrograde transport has on the organization of the Golgi, consider the net effect on the *medial*-Golgi compartment as enzymes from the *trans*-Golgi move to the *medial*-Golgi while enzymes from the *medial*-Golgi

**VIDEO:** 3-D Model of a Golgi Complex

**EXPERIMENTAL FIGURE 14-15 Electron micrograph of the Golgi complex in an exocrine pancreatic cell reveals secretory and retrograde transport vesicles.** A large secretory vesicle can be seen forming from the *trans*-Golgi network. Elements of the rough ER are on the bottom and left in this micrograph. Adjacent to the rough ER are transitional elements from which smooth protrusions appear to be budding. These buds form the small vesicles that transport secretory proteins from the rough ER to the Golgi complex. Interspersed among the Golgi cisternae are other small vesicles now known to function in retrograde, not anterograde, transport. [Courtesy G. Palade.]

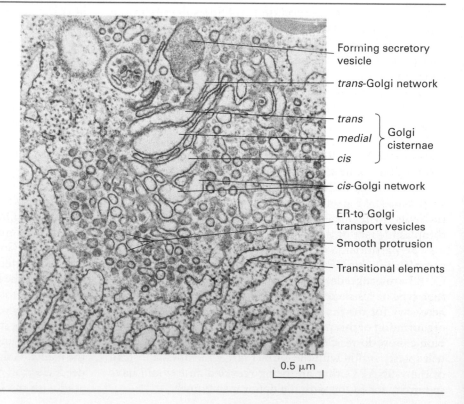

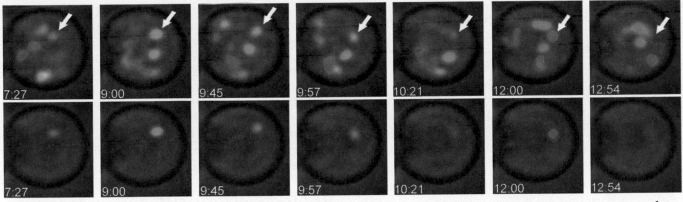

1 μm

**EXPERIMENTAL FIGURE 14-16 Fluorescence-tagged fusion proteins demonstrate Golgi cisternal maturation in a living yeast cell.** Yeast cells expressing the early Golgi protein Vrg4 fused to GFP (green fluorescence) and the late Golgi protein Sec7 fused to DsRed (red fluorescence) are imaged by time-lapse microscopy. The top series of images, taken approximately 1 minute apart, shows a collection of Golgi cisternae, which at any one time are labeled either with Vrg4 or Sec7. The bottom series of images show just one Golgi cisterna, isolated by digital processing of the image. First only Vrg4-GFP is located in the isolated cisterna and then Sec7-DsRed alone is located in the isolated cisterna, following a brief period in which both proteins are co-localized in this compartment. This experiment is a direct demonstration of the cisternal maturation hypothesis, showing that the composition of individual cisternae follow a process of maturation characterized by loss of early Golgi proteins and gain of late Golgi proteins. [From Losev et al., 2006, *Nature* **441**:1002.]

are transported to the *cis*-Golgi. As this process continues, the *medial*-Golgi acquires enzymes from the *trans*-Golgi while losing *medial*-Golgi enzymes to the *cis*-Golgi and thus progressively becomes a new *trans*-Golgi compartment. In this way, secretory cargo proteins acquire carbohydrate modification in the proper sequential order without being moved from one cisterna to another via anterograde vesicle transport.

The first evidence that the forward transport of cargo proteins from the *cis*- to the *trans*-Golgi occurs by such a progressive mechanism, called *cisternal maturation,* came from careful microscopic analysis of the synthesis of algal scales. These cell-wall glycoproteins are assembled in the *cis*-Golgi into large complexes visible in the electron microscope. Like other secretory proteins, newly made scales move from the *cis*- to the *trans*-Golgi, but they can be 20 times larger than the usual transport vesicles that bud from Golgi cisternae. Similarly, in the synthesis of collagen by fibroblasts, large aggregates of the procollagen precursor often form in the lumen of the *cis*-Golgi (see Figure 20-24). The procollagen aggregates are too large to be incorporated into small transport vesicles, and investigators could never find such aggregates in transport vesicles. These observations show that the forward movement of these and perhaps all secretory proteins from one Golgi compartment to another does *not* occur via small vesicles.

A particularly elegant demonstration of cisternal maturation in yeast takes advantage of different-colored fluorescent labels to image two different Golgi proteins simultaneously. Figure 14-16 shows how a *cis*-Golgi resident protein labeled with a green fluorescent protein and a *trans*-Golgi protein labeled with a red fluorescent protein behave in the same yeast cell. At any given moment individual Golgi cisternae appear to have a distinct compartmental identity, in the sense that they contain either the *cis*-Golgi protein or the *trans*-Golgi protein but only rarely contain both proteins. However, over time an individual cisterna labeled with the *cis*-Golgi protein can be seen to progressively lose this protein and acquire the *trans*-Golgi protein. This behavior is exactly that predicted for the cisternal maturation model, in which the composition of an individual cisterna changes as Golgi resident proteins move from later to earlier Golgi compartments.

Although most protein traffic appears to move through the Golgi complex by a cisternal maturation mechanism, there is evidence that at least some of the COPI transport vesicles that bud from Golgi membranes contain cargo proteins (rather than Golgi enzymes) and move in an anterograde (rather than retrograde) direction.

## KEY CONCEPTS of Section 14.3

### Early Stages of the Secretory Pathway

- COPII vesicles transport proteins from the rough ER to the *cis*-Golgi; COPI vesicles transport proteins in the reverse direction (see Figure 14-11).

- COPII coats comprise three components: the small GTP-binding protein Sar1, a Sec23/Sec24 complex, and a Sec13/Sec31 complex.

- Components of the COPII coat bind to membrane cargo proteins containing a di-acidic or other sorting signal in their cytosolic regions (see Figure 14-12). Soluble cargo proteins probably are targeted to COPII vesicles by binding to a membrane protein receptor.

- Many soluble ER-resident proteins contain a KDEL sorting signal. Binding of this retrieval sequence to a specific receptor protein in the *cis*-Golgi membrane recruits missorted ER proteins into retrograde COPI vesicles (see Figure 14-13).

- Membrane proteins needed to form COPII vesicles can be retrieved from the *cis*-Golgi by COPI vesicles. One of the sorting signals that directs membrane proteins into COPI vesicles is a KKXX sequence, which binds to subunits of the COPI coat. A distinct di-arginine sorting signal operates by a similar mechanism.

- COPI vesicles also carry Golgi-resident proteins from later to earlier compartments in the Golgi stack.

- Soluble and membrane proteins advance through the Golgi complex by cisternal maturation, a process of anterograde transport that depends on resident Golgi enzymes moving by COPI vesicular transport in a retrograde direction.

## 14.4 Later Stages of the Secretory Pathway

As cargo proteins move from the *cis*- to the *trans*-Golgi by cisternal maturation, modifications to their oligosaccharide chains are carried out by Golgi-resident enzymes. The retrograde trafficking of COPI vesicles from later to earlier Golgi compartments maintains sufficient levels of these carbohydrate-modifying enzymes in their functional compartments. Eventually, properly processed cargo proteins reach the *trans*-Golgi network, the most distal Golgi compartment. Here, they are sorted into one of a number of different kinds of vesicles for delivery to their final destination. In this section we discuss the different kinds of vesicles that bud from the *trans*-Golgi network, the mechanisms that segregate cargo proteins among them, and key processing events that occur late in the secretory pathway. The various types of vesicles that bud from the *trans*-Golgi are summarized in Figure 14-17.

## Vesicles Coated with Clathrin and/or Adapter Proteins Mediate Transport from the *trans*-Golgi

The best-characterized vesicles that bud from the *trans*-Golgi network (TGN) have a two-layered coat: an outer layer composed of the fibrous protein clathrin and an inner layer composed of *adapter protein (AP) complexes*. Purified clathrin molecules, which have a three-limbed shape, are called *triskelions,* from the Greek for "three-legged" (Figure 14-18a). Each limb contains one clathrin heavy chain (180,000 MW) and one clathrin light chain (~35,000–40,000 MW). Triskelions polymerize to form a polygonal lattice with an intrinsic curvature (Figure 14-18b). When clathrin polymerizes on a donor membrane, it does so in association with AP complexes, which fill the space between the clathrin lattice and the membrane. Each AP complex (340,000 MW) contains one copy each of four

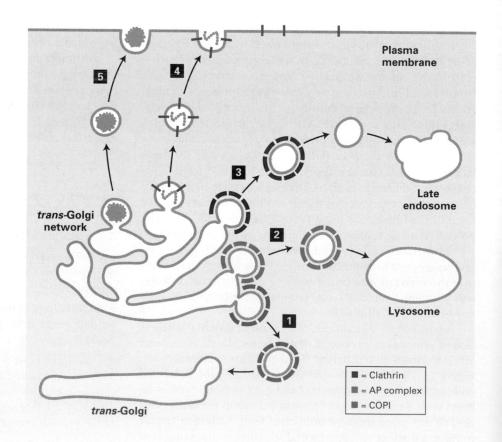

**FIGURE 14-17 Vesicle-mediated protein trafficking from the *trans*-Golgi network.** COPI (purple) vesicles mediate retrograde transport within the Golgi (**1**). Proteins that function in the lumen or in the membrane of the lysosome are first transported from the *trans*-Golgi network via clathrin-coated (red) vesicles (**3**); after uncoating, these vesicles fuse with late endosomes, which deliver their contents to the lysosome. The coat on most clathrin vesicles contains additional proteins (AP complexes) not indicated here. Some vesicles from the *trans*-Golgi carrying cargo destined for the lysosome fuse with the lysosome directly (**2**), bypassing the endosome. These vesicles are coated with a type of AP complex (blue); it is unknown whether these vesicles also contain clathrin. The coat proteins surrounding constitutive (**4**) and regulated (**5**) secretory vesicles are not yet characterized; these vesicles carry secreted proteins and plasma-membrane proteins from the *trans*-Golgi network to the cell surface.

■ = Clathrin
■ = AP complex
■ = COPI

**VIDEO:** Birth of a Clathrin Coat

**FIGURE 14-18 Structure of clathrin coats.** (a) A clathrin molecule, called a triskelion, is composed of three heavy and three light chains. It has an intrinsic curvature due to the bend in the heavy chains. (b) Clathrin coats were formed in vitro by mixing purified clathrin heavy and light chains with AP2 complexes in the absence of membranes. Cryoelectron micrographs of more than 1000 assembled hexagonal clathrin barrel particles were analyzed by digital image processing to generate an average structural representation. The processed image shows only the clathrin heavy chains in a structure composed of 36 triskelions. Three representative triskelions are highlighted in red, yellow, and green. Part of the AP2 complexes packed into the interior of the clathrin cage are also visible in this representation. [See B. Pishvaee and G. Payne, 1998, *Cell* **95**:443. Part (b) from Fotin et al., 2004, *Nature* **432**:573.]

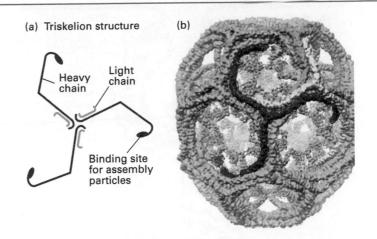

different adapter subunit proteins. A specific association between the globular domain at the end of each clathrin heavy chain in a triskelion and one subunit of the AP complex both promotes the co-assembly of clathrin triskelions with AP complexes and adds to the stability of the completed vesicle coat.

By binding to the cytosolic face of membrane proteins, adapter proteins determine which cargo proteins are specifically included in (or excluded from) a budding transport vesicle. Three different AP complexes are known (AP1, AP2, AP3), each with four subunits of different, though related, proteins. Recently, a second general type of adapter protein known as GGA has been shown to contain in a single 70,000 MW polypeptide both clathrin- and cargo-binding elements similar to those found in the much larger hetero-tetrameric AP complexes. Vesicles containing each type of adapter complex (AP or GGA) have been found to mediate specific transport steps (see Table 14-1). All vesicles whose coats contain one of these complexes utilize ARF to initiate coat assembly onto the donor membrane. As discussed previously, ARF also initiates assembly of COPI coats. The additional features of the membrane or protein factors that determine which type of coat will assemble after ARF attachment are not well understood at this time.

Vesicles that bud from the *trans*-Golgi network en route to the lysosome by way of the late endosome (see Figure 14-17, step **3**) have clathrin coats associated with either AP1 or GGA. Both AP1 and GGA bind to the cytosolic domain of cargo proteins in the donor membrane. Membrane proteins containing a Tyr-X-X-Φ sequence, where X is any amino acid and Φ is a bulky hydrophobic amino acid, are recruited into clathrin/AP1 vesicles budding from the *trans*-Golgi network. This *YXXΦ sorting signal* interacts with one of the AP1 subunits in the vesicle coat. As we discuss in the next section, vesicles with clathrin/AP2 coats, which bud from the plasma membrane during endocytosis, also can recognize the YXXΦ sorting signal. Vesicles coated with GGA proteins and clathrin bind cargo molecules with a different kind of sorting sequence. Cytosolic sorting signals that specifically bind to GGA adapter

proteins include Asp-X-Leu-Leu and Asp-Phe-Gly-X-Φ sequences (where X and Φ are defined as above).

Some vesicles that bud from the *trans*-Golgi network have coats composed of the AP3 complex. Although the AP3 complex does contain a binding site for clathrin similar to the AP1 and AP2 complexes, it is not clear whether clathrin is necessary for functioning of AP3-containing vesicles since mutant versions of AP3 that lack the clathrin binding site appear to be fully functional. AP3-coated vesicles mediate trafficking to the lysosome, but they appear to bypass the late endosome and fuse directly with the lysosomal membrane (see Figure 14-17, step **2**). In certain types of cells, such AP3 vesicles mediate protein transport to specialized storage compartments related to the lysosome. For example, AP3 is required for delivery of proteins to melanosomes, which contain the black pigment melanin in skin cells, and to platelet storage vesicles in megakaryocytes, large cells that fragment into dozens of platelets. Mice with mutations in either of two different subunits of AP3 not only have abnormal skin pigmentation but also exhibit bleeding disorders. The latter occur because tears in blood vessels cannot be repaired without platelets that contain normal storage vesicles.

## Dynamin Is Required for Pinching Off of Clathrin Vesicles

A fundamental step in the formation of a transport vesicle that we have not yet considered is how a vesicle bud is pinched off from the donor membrane. In the case of clathrin/AP-coated vesicles, a cytosolic protein called *dynamin* is essential for release of complete vesicles. At the later stages of bud formation, dynamin polymerizes around the neck portion and then hydrolyzes GTP. The energy derived from GTP hydrolysis is thought to drive a conformational change in dynamin that stretches the vesicle neck until the vesicle pinches off (Figure 14-19). Interestingly, COPI and COPII vesicles appear to pinch off from donor membranes without the aid of a GTPase such as

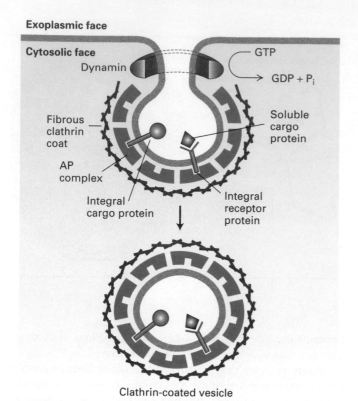

**FIGURE 14-19 Model for dynamin-mediated pinching off of clathrin/AP-coated vesicles.** After a vesicle bud forms, dynamin polymerizes over the neck. By a mechanism that is not well understood, dynamin-catalyzed hydrolysis of GTP leads to release of the vesicle from the donor membrane. Note that membrane proteins in the donor membrane are incorporated into vesicles by interacting with AP complexes in the coat. [Adapted from K. Takel et al., 1995, *Nature* **374**:186.]

Labels in figure: Exoplasmic face; Cytosolic face; Dynamin; GTP; GDP + Pᵢ; Fibrous clathrin coat; AP complex; Integral cargo protein; Integral receptor protein; Soluble cargo protein; Clathrin-coated vesicle

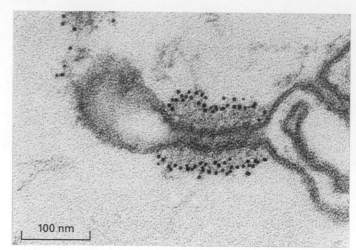

**EXPERIMENTAL FIGURE 14-20 GTP hydrolysis by dynamin is required for pinching off of clathrin-coated vesicles in cell-free extracts.** A preparation of nerve terminals, which undergo extensive endocytosis, was lysed by treatment with distilled water and incubated with GTP-γ-S, a nonhydrolyzable derivative of GTP. After sectioning, the preparation was treated with gold-tagged anti-dynamin antibody and viewed in the electron microscope. This image, which shows a long-necked clathrin/AP-coated bud with polymerized dynamin lining the neck, reveals that buds can form in the absence of GTP hydrolysis, but vesicles cannot pinch off. The extensive polymerization of dynamin that occurs in the presence of GTP-γ-S probably does not occur during the normal budding process. [From K. Takel et al., 1995, *Nature* **374**:186; courtesy of Pietro De Camilli.]

Scale bar: 100 nm

dynamin. In vitro budding experiments suggest that dimerization of ARF proteins drives the pinching off of COPI vesicles, but the mechanism is not understood.

Incubation of cell extracts with a nonhydrolyzable derivative of GTP provides dramatic evidence for the importance of dynamin in pinching off of clathrin/AP2 vesicles during endocytosis. Such treatment leads to accumulation of clathrin-coated vesicle buds with excessively long necks that are surrounded by polymeric dynamin but do not pinch off (Figure 14-20). Likewise, cells expressing mutant forms of dynamin that cannot bind GTP do not form clathrin-coated vesicles and instead accumulate similar long-necked vesicle buds encased with polymerized dynamin.

As with COPI and COPII vesicles, clathrin/AP vesicles normally lose their coat soon after their formation. Cytosolic Hsc70, a constitutive chaperone protein found in all eukaryotic cells, is thought to use energy derived from the hydrolysis of ATP to drive depolymerization of the clathrin coat into triskelions. Uncoating not only releases triskelions for reuse in the formation of additional vesicles but also exposes v-SNAREs for use in fusion with target membranes. Conformational changes that occur when ARF switches from the GTP-bound to GDP-bound state are thought to regulate the timing of clathrin coat depolymerization. How the action of Hsc70 might be coupled to ARF switching is not well understood.

## Mannose 6-Phosphate Residues Target Soluble Proteins to Lysosomes

As we have seen, many of the sorting signals that direct cargo protein trafficking in the secretory pathway are short amino acid sequences in the targeted protein. In contrast, the sorting signal that directs soluble lysosomal enzymes from the *trans*-Golgi network to the late endosome is a carbohydrate residue, *mannose 6-phosphate (M6P)*, which is formed in the *cis*-Golgi. The addition and initial processing of one or more preformed N-linked oligosaccharide precursors in the rough ER is the same for lysosomal enzymes as for membrane and secreted proteins, yielding core Man₈(GlcNAc)₂ chains (see Figure 13-18). In the *cis*-Golgi, the N-linked oligosaccharides present on most lysosomal enzymes undergo a two-step reaction sequence that generates M6P residues (Figure 14-21). The addition of M6P residues to the oligosaccharide chains of soluble lysosomal enzymes prevents these proteins from undergoing the further processing reactions characteristic of secreted and membrane proteins (see Figure 14-14).

As shown in Figure 14-22, the segregation of M6P-bearing lysosomal enzymes from secreted and membrane proteins occurs in the *trans*-Golgi network. Here, transmembrane

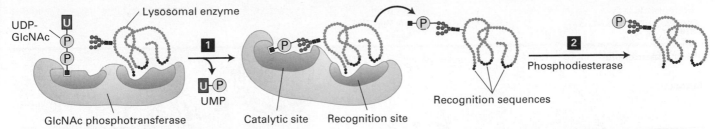

**FIGURE 14-21 Formation of mannose 6-phosphate (M6P) residues that target soluble enzymes to lysosomes.** The M6P residues that direct proteins to lysosomes are generated in the *cis*-Golgi by two Golgi-resident enzymes. Step **1**: An *N*-acetylglucosamine (GlcNAc) phosphotransferase transfers a phosphorylated GlcNAc group to carbon atom 6 of one or more mannose residues. Because only lysosomal enzymes contain sequences (red) that are recognized and bound by this enzyme, phosphorylated GlcNAc groups are added specifically to lysosomal enzymes. Step **2**: After release of a modified protein from the phosphotransferase, a phosphodiesterase removes the GlcNAc group, leaving a phosphorylated mannose residue on the lysosomal enzyme. [See A. B. Cantor et al., 1992, *J. Biol. Chem.* **267**:23349, and S. Kornfeld, 1987, *FASEB J.* **1**:462.]

*mannose 6-phosphate receptors* bind the M6P residues on lysosome-destined proteins very tightly and specifically. Clathrin/AP1 vesicles containing the M6P receptor and bound lysosomal enzymes then bud from the *trans*-Golgi network, lose their coats, and subsequently fuse with the late endosome by mechanisms described previously. Because M6P receptors can bind M6P at the slightly acidic pH (~6.5) of the *trans*-Golgi network but not at a pH less than 6, the bound lysosomal enzymes are released within late endosomes, which have an internal pH of 5.0–5.5. Furthermore, a phosphatase within late endosomes usually removes the phosphate from M6P residues on lysosomal enzymes, preventing any rebinding to the M6P receptor that might occur in spite of the low pH in endosomes. Vesicles budding from late endosomes recycle the M6P receptor back to the *trans*-Golgi network or, on occasion, to the cell surface. Eventually, mature late endosomes fuse with lysosomes, delivering the lysosomal enzymes to their final destination.

The sorting of soluble lysosomal enzymes in the *trans*-Golgi network (Figure 14-22, steps **1**–**4**) shares many of the features of trafficking between the ER and *cis*-Golgi compartments mediated by COPII and COPI vesicles. First, mannose 6-phosphate acts as a sorting signal by interacting with the luminal domain of a receptor protein in the donor membrane. Second, the membrane-embedded receptors with their bound ligands are incorporated into the appropriate vesicles—in this case, either GGA or AP1-containing clathrin vesicles—by interacting with the vesicle coat. Third, these transport vesicles fuse only with one specific organelle, here the late endosome, as the result of interactions between specific v-SNAREs and t-SNAREs. And finally, intracellular transport receptors are recycled after dissociating from their bound ligand.

## Study of Lysosomal Storage Diseases Revealed Key Components of the Lysosomal Sorting Pathway

 A group of genetic disorders termed *lysosomal storage diseases* are caused by the absence of one or more lysosomal enzymes. As a result, undigested glycolipids and extracellular components that would normally be degraded by lysosomal enzymes accumulate in lysosomes as large inclusions. Patients with lysosomal storage diseases can have a variety of developmental, physiological, and neurological abnormalities depending on the type and severity of the storage defect. *I-cell disease* is a particularly severe type of lysosomal storage disease in which multiple enzymes are missing from the lysosomes. Cells from affected individuals lack the *N*-acetylglucosamine phosphotransferase that is required for formation of M6P residues on lysosomal enzymes in the *cis*-Golgi (see Figure 14-21). Biochemical comparison of lysosomal enzymes from normal individuals with those from patients with I-cell disease led to the initial discovery of mannose 6-phosphate as the lysosomal sorting signal. Lacking the M6P sorting signal, the lysosomal enzymes in I-cell patients are secreted rather than being sorted to and sequestered in lysosomes.

When fibroblasts from patients with I-cell disease are grown in a medium containing lysosomal enzymes bearing M6P residues, the diseased cells acquire a nearly normal intracellular content of lysosomal enzymes. This finding indicates that the plasma membrane of these cells contains M6P receptors, which can internalize extracellular phosphorylated lysosomal enzymes by receptor-mediated endocytosis. This process, used by many cell-surface receptors to bring bound proteins or particles into the cell, is discussed in detail in the next section. It is now known that even in normal cells, some M6P receptors are transported to the plasma membrane and some phosphorylated lysosomal enzymes are secreted (see Figure 14-22). The secreted enzymes can be retrieved by receptor-mediated endocytosis and directed to lysosomes. This pathway thus scavenges any lysosomal enzymes that escape the usual M6P sorting pathway.

Hepatocytes from patients with I-cell disease contain a normal complement of lysosomal enzymes and no inclusions, even though these cells are defective in mannose phosphorylation. This finding implies that hepatocytes (the most abundant

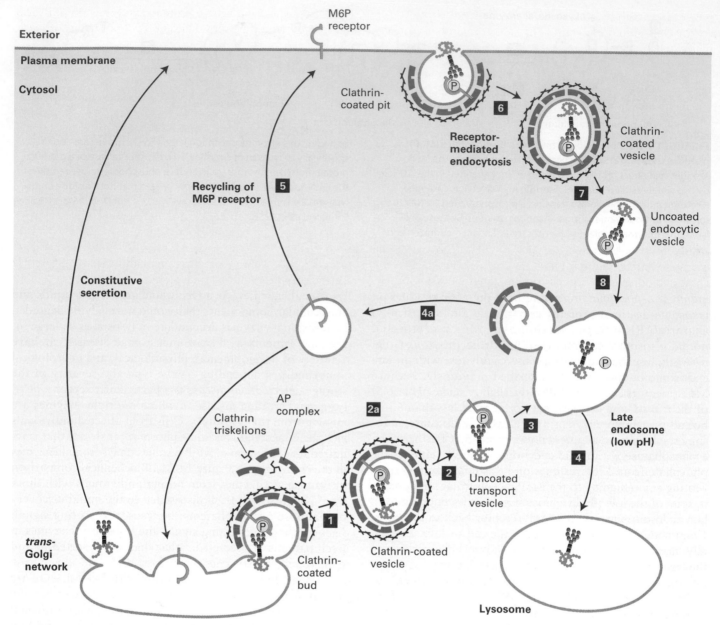

**FIGURE 14-22 Trafficking of soluble lysosomal enzymes from the *trans*-Golgi network and cell surface to lysosomes.** Newly synthesized lysosomal enzymes, produced in the ER, acquire mannose 6-phosphate (M6P) residues in the *cis*-Golgi (see Figure 14-21). For simplicity, only one phosphorylated oligosaccharide chain is depicted, although lysosomal enzymes typically have many such chains. In the *trans*-Golgi network, proteins that bear the M6P sorting signal interact with M6P receptors in the membrane and thereby are directed into clathrin/AP1 vesicles (step **1**). The coat surrounding released vesicles is rapidly depolymerized (step **2**), and the uncoated transport vesicles fuse with late endosomes (step **3**). After the phosphorylated enzymes dissociate from the M6P receptors and are dephosphorylated, late endosomes subsequently fuse with a lysosome (step **4**). Note that coat proteins and M6P receptors are recycled (steps **2a** and **4a**), and some receptors are delivered to the cell surface (step **5**). Phosphorylated lysosomal enzymes occasionally are sorted from the *trans*-Golgi to the cell surface and secreted. These secreted enzymes can be retrieved by receptor-mediated endocytosis (steps **6**–**8**), a process that closely parallels trafficking of lysosomal enzymes from the *trans*-Golgi network to lysosomes. [See G. Griffiths et al., 1988, *Cell* **52**:329; S. Kornfeld, 1992, *Ann. Rev. Biochem.* **61**:307; and G. Griffiths and J. Gruenberg, 1991, *Trends Cell Biol.* **1**:5.]

type of liver cell) employ an M6P-independent pathway for sorting lysosomal enzymes. The nature of this pathway, which also may operate in other cell types, is unknown. ■

do not associate with these proteins, and thus do not form aggregates, would be sorted into unregulated transport vesicles by default.

## Protein Aggregation in the *trans*-Golgi May Function in Sorting Proteins to Regulated Secretory Vesicles

As noted in the chapter introduction, all eukaryotic cells continuously secrete certain proteins, a process commonly called *constitutive secretion*. Specialized secretory cells also store other proteins in vesicles and secrete them only when triggered by a specific stimulus. One example of such *regulated secretion* occurs in pancreatic β cells, which store newly made insulin in special secretory vesicles and secrete insulin in response to an elevation in blood glucose (see Figure 16-38). These and other secretory cells simultaneously utilize two different types of vesicles to move proteins from the *trans*-Golgi network to the cell surface: regulated transport vesicles, often simply called secretory vesicles, and unregulated transport vesicles, also called constitutive secretory vesicles.

A common mechanism appears to sort regulated proteins as diverse as ACTH (adrenocorticotropic hormone), insulin, and trypsinogen into regulated secretory vesicles. Evidence for a common mechanism comes from experiments in which recombinant DNA techniques are used to induce the synthesis of insulin and trypsinogen in pituitary tumor cells already synthesizing ACTH. In these cells, which do not normally express insulin or trypsinogen, all three proteins segregate into the same regulated secretory vesicles and are secreted together when a hormone binds to a receptor on the pituitary cells and causes a rise in cytosolic Ca$^{2+}$. Although these three proteins share no identical amino acid sequences that might serve as a sorting sequence, they must have some common feature that signals their incorporation into regulated secretory vesicles.

Morphologic evidence suggests that sorting into the regulated pathway is controlled by selective protein aggregation. For instance, immature vesicles in this pathway—those that have just budded from the *trans*-Golgi network—contain diffuse aggregates of secreted protein that are visible in the electron microscope. These aggregates also are found in vesicles that are in the process of budding, indicating that proteins destined for regulated secretory vesicles selectively aggregate together before their incorporation into the vesicles.

Other studies have shown that regulated secretory vesicles from mammalian secretory cells contain three proteins, *chromogranin A, chromogranin B,* and *secretogranin II,* that together form aggregates when incubated at the ionic conditions (pH 6.5 and 1 mM Ca$^{2+}$) thought to occur in the *trans*-Golgi network; such aggregates do not form at the neutral pH of the ER. The selective aggregation of regulated secreted proteins together with chromogranin A, chromogranin B, or secretogranin II could be the basis for sorting of these proteins into regulated secretory vesicles. Secreted proteins that

## Some Proteins Undergo Proteolytic Processing After Leaving the *trans*-Golgi

For some secretory proteins (e.g., growth hormone) and certain viral membrane proteins (e.g., the VSV glycoprotein), removal of the N-terminal ER signal sequence from the nascent chain is the only known proteolytic cleavage required to convert the polypeptide to the mature, active species (see Figure 13-6). However, some membrane and many soluble secretory proteins initially are synthesized as relatively long-lived, inactive precursors, termed *proproteins,* that require further proteolytic processing to generate the mature, active proteins. Examples of proteins that undergo such processing are soluble lysosomal enzymes, many membrane proteins such as influenza hemagglutinin (HA), and secreted proteins such as serum albumin, insulin, glucagon, and the yeast α mating factor. In general, the proteolytic conversion of a proprotein to the corresponding mature protein occurs after the proprotein has been sorted in the *trans*-Golgi network to appropriate vesicles.

In the case of soluble lysosomal enzymes, the proproteins are called *proenzymes,* which are sorted by the M6P receptor as catalytically inactive enzymes. In the late endosome or lysosome a proenzyme undergoes a proteolytic cleavage that generates a smaller but enzymatically active polypeptide. Delaying the activation of lysosomal proenzymes until they reach the lysosome prevents them from digesting macromolecules in earlier compartments of the secretory pathway.

Normally, mature vesicles carrying secreted proteins to the cell surface are formed by fusion of several immature ones containing proprotein. Proteolytic cleavage of proproteins, such as proinsulin, occurs in vesicles after they move away from the *trans*-Golgi network (Figure 14-23). The proproteins of most constitutively secreted proteins (e.g., albumin) are cleaved only once at a site C-terminal to a dibasic recognition sequence such as Arg-Arg or Lys-Arg (Figure 14-24a). Proteolytic processing of proteins whose secretion is regulated generally entails additional cleavages. In the case of proinsulin, multiple cleavages of the single polypeptide chain yields the N-terminal B chain and the C-terminal A chain of mature insulin, which are linked by disulfide bonds, and the central C peptide, which is lost and subsequently degraded (Figure 14-24b).

The breakthrough in identifying the proteases responsible for such processing of secreted proteins came from analysis of yeast with a mutation in the *KEX2* gene. These mutant cells synthesized the precursor of the α mating factor but could not proteolytically process it to the functional form and thus were unable to mate with cells of the opposite mating type (see Figure 16-23). The wild-type *KEX2* gene encodes an endoprotease that cleaves the α-factor precursor at a site C-terminal to Arg-Arg and Lys-Arg residues. Mammals contain

**EXPERIMENTAL FIGURE 14-23 Proteolytic cleavage of proinsulin occurs in secretory vesicles after they have budded from the *trans*-Golgi network.** Serial sections of the Golgi region of an insulin-secreting cell were stained with (a) a monoclonal antibody that recognizes proinsulin but not insulin, or (b) a different antibody that recognizes insulin but not proinsulin. The antibodies, which were bound to electron-opaque gold particles, appear as dark dots in these electron micrographs (see Figure 9-29). Immature secretory vesicles (closed arrowheads) and vesicles budding from the *trans*-Golgi (arrows) stain with the proinsulin antibody but not with insulin antibody. These vesicles contain diffuse protein aggregates that include proinsulin and other regulated secreted proteins. Mature vesicles (open arrowheads) stain with insulin antibody but not with proinsulin antibody and have a dense core of almost crystalline insulin. Since budding and immature secretory vesicles contain proinsulin (not insulin), the proteolytic conversion of proinsulin to insulin must take place in these vesicles after they bud from the *trans*-Golgi network. The inset in (a) shows a proinsulin-rich secretory vesicle surrounded by a protein coat (dashed line). [From L. Orci et al., 1987, *Cell* **49**:865; courtesy of L. Orci.]

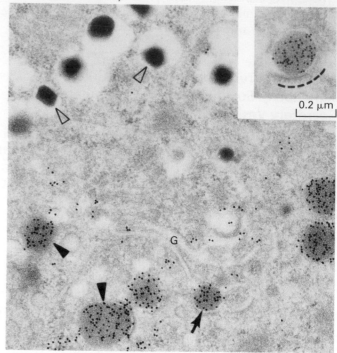

(a) Proinsulin antibody

0.2 μm

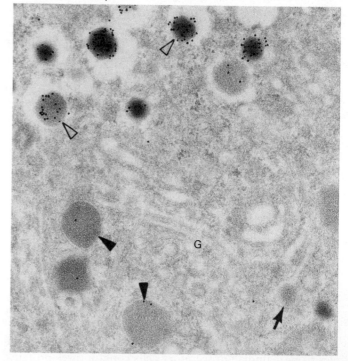

(b) Insulin antibody

0.5 μm

a family of endoproteases homologous to the yeast KEX2 protein, all of which cleave a protein chain on the C-terminal side of an Arg-Arg or Lys-Arg sequence. One, called *furin*, is found in all mammalian cells; it processes proteins such as albumin that are secreted by the continuous pathway. In contrast, the *PC2* and *PC3 endoproteases* are found only in cells that exhibit regulated secretion; these enzymes are localized to regulated secretory vesicles and proteolytically cleave the precursors of many hormones at specific sites.

## Several Pathways Sort Membrane Proteins to the Apical or Basolateral Region of Polarized Cells

The plasma membrane of polarized epithelial cells is divided into two domains, **apical** and **basolateral;** tight junctions located between the two domains prevent the movement of plasma-membrane proteins between the domains (see Figure 20-10). Several sorting mechanisms direct newly synthesized membrane proteins to either the apical or basolateral domain of epithelial cells, and any one protein may be sorted by more than one mechanism. As a result of this sorting and the restriction on protein movement within the plasma membrane due to tight junctions, distinct sets of proteins are found in the apical or basolateral domain. This preferential localization of certain transport proteins is critical to a variety of important physiological functions, such as absorption of nutrients from the intestinal lumen and acidification of the stomach lumen (see Figures 11-30 and 11-31).

Microscopic and cell-fractionation studies indicate that proteins destined for either the apical or the basolateral membranes are initially transported together to the membranes of the *trans*-Golgi network. In some cases, proteins destined for the apical membrane are sorted into their own transport vesicles that bud from the *trans*-Golgi network

and then move to the apical region, whereas proteins destined for the basolateral membrane are sorted into other vesicles that move to the basolateral region. The different vesicle types can be distinguished by their protein constituents, including distinct Rab and v-SNARE proteins, which

### (a) Constitutive secreted proteins

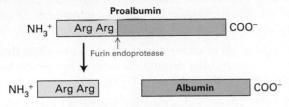

### (b) Regulated secreted proteins

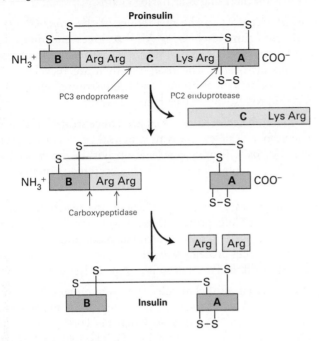

**FIGURE 14-24 Proteolytic processing of proproteins in the constitutive and regulated secretion pathways.** The processing of proalbumin and proinsulin is typical of the constitutive and regulated pathways, respectively. The endoproteases that function in such processing cleave at the C-terminal side of two consecutive amino acids. (a) The endoprotease furin acts on the precursors of constitutive secreted proteins. (b) Two endoproteases, PC2 and PC3, act on the precursors of regulated secreted proteins. The final processing of many such proteins is catalyzed by a carboxypeptidase that sequentially removes two basic amino acid residues at the C-terminus of a polypeptide. [See D. Steiner et al., 1992, *J. Biol. Chem.* **267**:23435.]

apparently target them to the appropriate plasma-membrane domain. In this mechanism, segregation of proteins destined for either the apical or basolateral membranes occurs as cargo proteins are incorporated into particular types of vesicles budding from the *trans*-Golgi network.

Such direct basolateral-apical sorting has been investigated in cultured Madin-Darby canine kidney (MDCK) cells, a line of cultured polarized epithelial cells (see Figure 9-4). In MDCK cells infected with the influenza virus, progeny viruses bud only from the apical membrane, whereas in cells infected with vesicular stomatitis virus (VSV), progeny viruses bud only from the basolateral membrane. This difference occurs because the HA glycoprotein of influenza virus is transported from the Golgi complex exclusively to the apical membrane and the VSV G protein is transported only to the basolateral membrane (Figure 14-25). Furthermore, when the gene encoding HA protein is introduced into uninfected cells by recombinant DNA techniques, all the expressed HA accumulates in the apical membrane, indicating that the sorting signal resides in the HA glycoprotein itself and not in other viral proteins produced during viral infection.

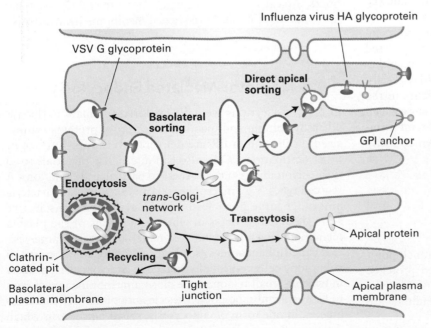

**FIGURE 14-25 Sorting of proteins destined for the apical and basolateral plasma membranes of polarized cells.** When cultured MDCK cells are infected simultaneously with VSV and influenza virus, the VSV G glycoprotein (purple) is found only on the basolateral membrane, whereas the influenza HA glycoprotein (green) is found only on the apical membrane. Some cellular proteins (orange circle), especially those with a GPI anchor, are likewise sorted directly to the apical membrane and others to the basolateral membrane (not shown) via specific transport vesicles that bud from the *trans*-Golgi network. In certain polarized cells, some apical and basolateral proteins are transported together to the basolateral surface; the apical proteins (yellow oval) then move selectively, by endocytosis and transcytosis, to the apical membrane. [After K. Simons and A. Wandinger-Ness, 1990, *Cell* **62**:207, and K. Mostov et al., 1992, *J. Cell Biol.* **116**:577.]

Among the cellular proteins that undergo similar apical-basolateral sorting in the Golgi are those with a *glycosylphosphatidylinositol (GPI) membrane anchor*. In MDCK cells and most other types of epithelial cells, GPI-anchored proteins are targeted to the apical membrane. In membranes GPI-anchored proteins are clustered into lipid rafts, which are rich in sphingolipids (see Chapter 10). This finding suggests that lipid rafts are localized to the apical membrane along with proteins that preferentially partition them in many cells. However, the GPI anchor is not an apical sorting signal in all polarized cells; in thyroid cells, for example, GPI-anchored proteins are targeted to the basolateral membrane. Other than GPI anchors, no unique sequences have been identified that are both necessary and sufficient to target proteins to either the apical or basolateral domain. Instead, each membrane protein may contain multiple sorting signals, any one of which can target it to the appropriate plasma-membrane domain. The identification of such complex signals and of the vesicle coat proteins that recognize them is currently being pursued for a number of different proteins that are sorted to specific plasma-membrane domains of polarized epithelial cells.

Another mechanism for sorting apical and basolateral proteins, also illustrated in Figure 14-25, operates in hepatocytes. The basolateral membranes of hepatocytes face the blood (as in intestinal epithelial cells), and the apical membranes line the small intercellular channels into which bile is secreted. In hepatocytes, newly made apical and basolateral proteins are first transported in vesicles from the *trans*-Golgi network to the basolateral region and incorporated into the plasma membrane by exocytosis (i.e., fusion of the vesicle membrane with the plasma membrane). From there, both basolateral and apical proteins are endocytosed in the same vesicles, but then their paths diverge. The endocytosed basolateral proteins are sorted into transport vesicles that recycle them to the basolateral membrane. In contrast, the apically destined endocytosed proteins are sorted into transport vesicles that move across the cell and fuse with the apical membrane, a process called **transcytosis**. This process also is used to move extracellular materials from one side of an epithelium to another. Even in epithelial cells, such as MDCK cells, in which apical-basolateral protein sorting occurs in the Golgi, transcytosis may provide an editing function by which an apical protein sorted incorrectly to the basolateral membrane would be subjected to endocytosis and then correctly delivered to the apical membrane.

## KEY CONCEPTS of Section 14.4

### Later Stages of the Secretory Pathway

• The *trans*-Golgi network (TGN) is a major branch point in the secretory pathway where soluble secreted proteins, lysosomal proteins, and in some cells membrane proteins destined for the basolateral or apical plasma membrane are segregated into different transport vesicles.

• Many vesicles that bud from the *trans*-Golgi network as well as endocytic vesicles bear a coat composed of AP (adapter protein) complexes and clathrin (see Figure 14-18).

• Pinching off of clathrin-coated vesicles requires dynamin, which forms a collar around the neck of the vesicle bud and hydrolyzes GTP (see Figure 14-19).

• Soluble enzymes destined for lysosomes are modified in the *cis*-Golgi, yielding multiple mannose 6-phosphate (M6P) residues on their oligosaccharide chains.

• M6P receptors in the membrane of the *trans*-Golgi network bind proteins bearing M6P residues and direct their transfer to late endosomes, where receptors and their ligand proteins dissociate. The receptors then are recycled to the Golgi or plasma membrane, and the lysosomal enzymes are delivered to lysosomes (see Figure 14-22).

• Regulated secreted proteins are concentrated and stored in secretory vesicles to await a neural or hormonal signal for exocytosis. Protein aggregation within the *trans*-Golgi network may play a role in sorting secreted proteins to the regulated pathway.

• Many proteins transported through the secretory pathway undergo post-Golgi proteolytic cleavages that yield the mature, active proteins. Generally, proteolytic maturation can occur in vesicles carrying proteins from the *trans*-Golgi network to the cell surface, in the late endosome, or in the lysosome.

• In polarized epithelial cells, membrane proteins destined for the apical or basolateral domains of the plasma membrane are sorted in the *trans*-Golgi network into different transport vesicles (see Figure 14-25). The GPI anchor is the only apical-basolateral sorting signal identified so far.

• In hepatocytes and some other polarized cells, all plasma-membrane proteins are directed first to the basolateral membrane. Apically destined proteins then are endocytosed and moved across the cell to the apical membrane (transcytosis).

## 14.5 Receptor-Mediated Endocytosis

In previous sections we have explored the main pathways whereby soluble and membrane secretory proteins synthesized on the rough ER are delivered to the cell surface or other destinations. Cells also can internalize materials from their surroundings and sort these to particular destinations. A few cell types (e.g., macrophages) can take up whole bacteria and other large particles by **phagocytosis**, a nonselective actin-mediated process in which extensions of the plasma membrane envelop the ingested material, forming large vesicles called phagosomes (see Figure 17-19). In contrast, all eukaryotic cells continually engage in endocytosis, a process in which a small region of the plasma membrane invaginates to form a membrane-limited vesicle about 0.05–0.1 μm in diameter. In one form of endocytosis, called *pinocytosis*, small droplets of extracellular fluid and any material dissolved in it are nonspecifically taken up. Our focus in this section, however, is on **receptor-mediated endocytosis,** in which a specific receptor on the cell surface binds tightly to an extracellular macromolecular ligand that it recognizes; the plasma membrane region containing the receptor-ligand complex then buds inward and pinches off, becoming a transport vesicle.

Among the common macromolecules that vertebrate cells internalize by receptor-mediated endocytosis are cholesterol-containing particles called low-density lipoprotein (LDL), the iron-carrying protein transferrin, many protein hormones (e.g., insulin), and certain glycoproteins. Receptor-mediated endocytosis of such ligands generally occurs via clathrin/AP2-coated pits and vesicles in a process similar to the packaging of lysosomal enzymes by binding of mannose 6-phosphate (M6P) in the *trans*-Golgi network (see Figure 14-22). As noted earlier, some M6P receptors are found on the cell surface, and these participate in the receptor-mediated endocytosis of lysosomal enzymes that are mistakenly secreted. In

general, the transmembrane receptor proteins that function in the uptake of extracellular ligands are internalized from the cell surface during endocytosis and are then sorted and recycled back to the cell surface, much like the recycling of M6P receptors to the plasma membrane and *trans*-Golgi. The rate at which a ligand is internalized is limited by the amount of its corresponding receptor on the cell surface.

Clathrin/AP2 pits make up about 2 percent of the surface of cells such as hepatocytes and fibroblasts. Many internalized ligands have been observed in these pits and vesicles, which are thought to function as intermediates in the endocytosis of most (though not all) ligands bound to cell-surface receptors (Figure 14-26). Some receptors are clustered over clathrin-coated pits even in the absence of ligand. Other receptors diffuse freely in the plane of the plasma membrane but undergo a conformational change when binding to ligand, so that when the receptor-ligand complex diffuses into a clathrin-coated pit, it is retained there. Two or more types

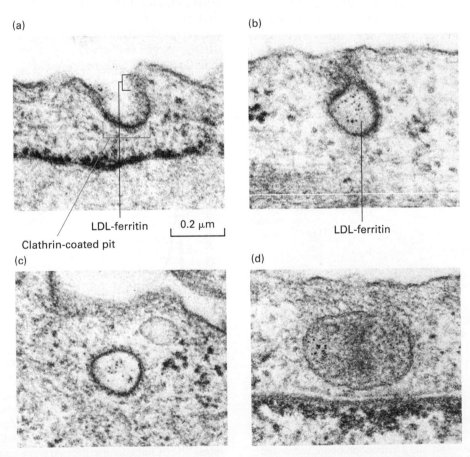

(a)

LDL-ferritin 0.2 μm

Clathrin-coated pit

(b)

LDL-ferritin

(c)

(d)

**EXPERIMENTAL FIGURE 14-26** **The initial stages of receptor-mediated endocytosis of low-density lipoprotein (LDL) particles are revealed by electron microscopy.** Cultured human fibroblasts were incubated in a medium containing LDL particles covalently linked to the electron-dense, iron-containing protein ferritin; each small iron particle in ferritin is visible as a small dot under the electron microscope. Cells initially were incubated at 4 °C; at this temperature LDL can bind to its receptor, but internalization does not occur. After excess LDL not bound to the cells was washed away, the cells were warmed to 37 °C and then prepared for microscopy at periodic intervals. (a) A coated pit, showing the clathrin coat on the inner (cytosolic) surface of the pit, soon after the temperature was raised. (b) A pit containing LDL apparently closing on itself to form a coated vesicle. (c) A coated vesicle containing ferritin-tagged LDL particles. (d) Ferritin-tagged LDL particles in a smooth-surfaced early endosome 6 minutes after internalization began. [Photographs courtesy of R. Anderson. Reprinted by permission from J. Goldstein et al., *Nature* **279**:679. Copyright 1979, Macmillan Journals Limited. See also M. S. Brown and J. Goldstein, 1986, *Science* **232**:34.]

of receptor-bound ligands, such as LDL and transferrin, can be seen in the same coated pit or vesicle.

## Cells Take Up Lipids from the Blood in the Form of Large, Well-Defined Lipoprotein Complexes

Lipids absorbed from the diet in the intestines or stored in adipose tissue can be distributed to cells throughout the body. To facilitate the mass transfer of lipids between cells, animals have evolved an efficient way to package from hundreds to thousands of lipid molecules into water-soluble, macromolecular carriers, called **lipoproteins,** that cells can take up from the circulation as an ensemble. A lipoprotein particle has a shell composed of proteins (*apolipoproteins*) and a cholesterol-containing phospholipid monolayer. The shell is amphipathic because its outer surface is hydrophilic, making these particles water soluble, and its inner surface is hydrophobic. Adjacent to the hydrophobic inner surface of the shell is a core of neutral lipids containing mostly cholesteryl esters, triglycerides, or both. Mammalian lipoproteins fall into different classes, defined by their differing buoyant densities. The class we will consider here is **low-density lipoprotein (LDL).** A typical LDL particle, depicted in Figure 14-27, is a sphere 20–25 nm in diameter. The amphipathic outer shell is composed of a phospholipid monolayer and a single molecule of a large protein known as *apoB-100;* the core of the particle is packed with cholesterol in the form of cholesteryl esters.

Two general experimental approaches have been used to study how LDL particles enter cells. The first method makes use of LDL that has been labeled by the covalent attachment of radioactive $^{125}$I to the side chains of tyrosine residues in apoB-100 on the surfaces of the LDL particles. After cultured cells are incubated for several hours with the labeled LDL, it is possible to determine how much LDL is bound to the surfaces of cells, how much is internalized, and how much of the apoB-100 component of the LDL is degraded by enzymatic hydrolysis to individual amino acids. The degradation of apoB-100 can be detected by the release of $^{125}$I-tyrosine into the culture medium. Figure 14-28 shows the time course of events in receptor-mediated cellular LDL processing, determined by pulse-chase experiments with a fixed concentration of $^{125}$I-labeled LDL. These experiments clearly demonstrate the order of events: surface binding of LDL → internalization → degradation. The second approach involves tagging LDL particles with an electron-dense label that can be detected by electron microscopy. Such studies can reveal the details of how LDL particles first bind to the surface of cells at clathrin-coated endocytic pits and then remain associated with the coated pits as they invaginate and bud off to form coated vesicles and finally are transported to endosomes (see Figure 14-26).

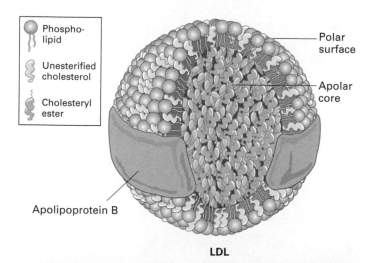

**LDL**

**FIGURE 14-27 Model of low-density lipoprotein (LDL).** This class and the other classes of lipoproteins have the same general structure: an amphipathic shell, composed of a phospholipid monolayer (not bilayer), cholesterol, and protein, and a hydrophobic core, composed mostly of cholesteryl esters or triglycerides or both but with minor amounts of other neutral lipids (e.g., some vitamins). This model of LDL is based on electron microscopy and other low-resolution biophysical methods. LDL is unique in that it contains only a single molecule of one type of apolipoprotein (apoB), which appears to wrap around the outside of the particle as a band of protein. The other lipoproteins contain multiple apolipoprotein molecules, often of different types. [Adapted from M. Krieger, 1995, in E. Haber, ed., *Molecular Cardiovascular Medicine,* Scientific American Medicine, pp. 31–47.]

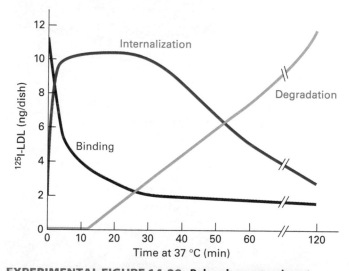

**EXPERIMENTAL FIGURE 14-28 Pulse-chase experiment demonstrates precursor-product relations in cellular uptake of LDL.** Cultured normal human skin fibroblasts were incubated in a medium containing $^{125}$I-LDL for 2 hours at 4 °C (the pulse). After excess $^{125}$I-LDL not bound to the cells was washed away, the cells were incubated at 37 °C for the indicated amounts of time in the absence of external LDL (the chase). The amounts of surface-bound, internalized, and degraded (hydrolyzed) $^{125}$I-LDL were measured. Binding but not internalization or hydrolysis of LDL apoB-100 occurs during the 4 °C pulse. The data show the very rapid disappearance of bound $^{125}$I-LDL from the surface as it is internalized after the cells have been warmed to allow membrane movements. After a lag period of 15–20 minutes, lysosomal degradation of the internalized $^{125}$I-LDL commences. [See M. S. Brown and J. L. Goldstein, 1976, *Cell* **9:**663.]

## Receptors for Low-Density Lipoprotein and Other Ligands Contain Sorting Signals That Target Them for Endocytosis

The key to understanding how LDL particles bind to the cell surface and are then taken up into endocytic vesicles came from discovery of the *LDL receptor (LDLR)*. The LDL receptor is an 839-residue glycoprotein with a single transmembrane segment; it has a short C-terminal cytosolic segment and a long N-terminal exoplasmic segment that contains the LDL-binding domain. Seven cysteine-rich repeats form the ligand-binding domain, which interacts with the apoB-100 molecule in an LDL particle. Figure 14-29 shows how LDL receptor proteins facilitate internalization of LDL particles by receptor-mediated endocytosis. After internalized LDL particles reach lysosomes, lysosomal proteases hydrolyze their surface apolipoproteins and lysosomal cholesteryl esterases hydrolyze their core cholesteryl esters. The unesterified cholesterol is then free to leave the lysosome and be used as necessary by the cell in the synthesis of membranes or various cholesterol derivatives.

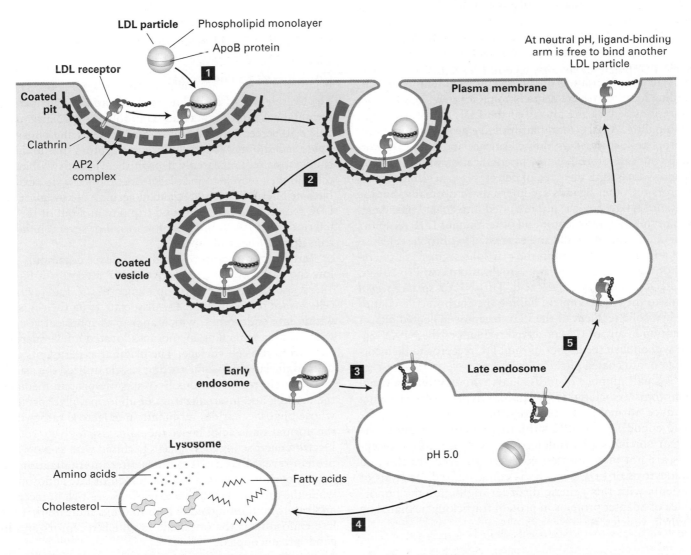

**FIGURE 14-29 Endocytic pathway for internalizing low-density lipoprotein (LDL).** Step **1**: Cell-surface LDL receptors bind to an apoB protein embedded in the phospholipid outer layer of LDL particles. Interaction between the NPXY sorting signal in the cytosolic tail of the LDL receptor and the AP2 complex incorporates the receptor-ligand complex into forming endocytic vesicles. Step **2**: Clathrin-coated pits (or buds) containing receptor-LDL complexes are pinched off by the same dynamin-mediated mechanism used to form clathrin/AP1 vesicles on the *trans*-Golgi network (see Figure 14-19). Step **3**: After the vesicle coat is shed, the uncoated endocytic vesicle (early endo-some) fuses with the late endosome. The acidic pH in this compartment causes a conformational change in the LDL receptor that leads to release of the bound LDL particle. Step **4**: The late endosome fuses with the lysosome, and the proteins and lipids of the free LDL particle are broken down to their constituent parts by enzymes in the lysosome. Step **5**: The LDL receptor recycles to the cell surface, where at the neutral pH of the exterior medium the receptor undergoes a conformational change so that it can bind another LDL particle. [See M. S. Brown and J. L. Goldstein, 1986, *Science* **232**:34, and G. Rudenko et al., 2002, *Science* **298**:2353.]

The discovery of the LDL receptor and an understanding of how it functions came from studying cells from patients with *familial hypercholesterolemia (FH)*, a hereditary disease that is marked by elevated plasma LDL cholesterol and is now known to be caused by mutations in the *LDLR* gene. In patients who have one normal and one defective copy of the *LDLR* gene (heterozygotes), LDL cholesterol in the blood is increased about twofold. Those with two defective *LDLR* genes (homozygotes) have LDL cholesterol levels that are from fourfold to sixfold as high as normal. FH heterozygotes commonly develop cardiovascular disease about 10 years earlier than normal people do, and FH homozygotes usually die of heart attacks before reaching their late 20s.

A variety of mutations in the gene encoding the LDL receptor can cause familial hypercholesterolemia. Some mutations prevent the synthesis of the LDLR protein; others prevent proper folding of the receptor protein in the ER, leading to its premature degradation (Chapter 13); still other mutations reduce the ability of the LDL receptor to bind LDL tightly. A particularly informative group of mutant receptors are expressed on the cell surface and bind LDL normally but cannot mediate the internalization of bound LDL. In individuals with this type of defect, plasma-membrane receptors for other ligands are internalized normally, but the mutant LDL receptor is not recruited into coated pits. Analysis of this mutant receptor and other mutant LDL receptors generated experimentally and expressed in fibroblasts identified a four-residue motif in the cytosolic segment of the receptor that is crucial for its internalization: Asn-Pro-X-Tyr, where X can be any amino acid. This *NPXY sorting signal* binds to the AP2 complex, linking the clathrin/AP2 coat to the cytosolic segment of the LDL receptor in coated pits. A mutation in any of the conserved residues of the NPXY signal will abolish the ability of the LDL receptor to be incorporated into coated pits.

A small number of individuals who exhibit the usual symptoms associated with familial hypercholesterolemia produce normal LDL receptors. In these individuals, the gene encoding the AP2 subunit protein that binds the NPXY sorting signal is defective. As a result, LDL receptors are not incorporated into clathrin/AP2 vesicles and endocytosis of LDL particles is compromised. Analysis of patients with this genetic disorder highlights the importance of adapter proteins in protein trafficking mediated by clathrin vesicles. ■

Mutational studies have shown that other cell-surface receptors can be directed into forming clathrin/AP2 pits by a YXXF sorting signal. Recall from our earlier discussion that this same sorting signal recruits membrane proteins into clathrin/AP1 vesicles that bud from the *trans*-Golgi network by binding to a subunit of AP1 (see Table 14-2). All these observations indicate that YXXF is a widely used signal for sorting membrane proteins to clathrin-coated vesicles.

In some cell-surface proteins, however, other sequences (e.g., Leu-Leu) or covalently linked ubiquitin molecules signal endocytosis. Among the proteins associated with clathrin/AP2 vesicles, several contain domains that specifically bind to ubiquitin, and it has been hypothesized that these vesicle-associated proteins mediate the selective incorporation of ubiquitinated membrane proteins into endocytic vesicles. As described later, the ubiquitin tag on endocytosed membrane proteins is also recognized at a later stage in the endocytic pathway and plays a role in delivering these proteins into the interior of the lysosome, where they are degraded.

## The Acidic pH of Late Endosomes Causes Most Receptor-Ligand Complexes to Dissociate

The overall rate of endocytic internalization of the plasma membrane is quite high: cultured fibroblasts regularly internalize 50 percent of their cell-surface proteins and phospholipids each hour. Most cell-surface receptors that undergo endocytosis will repeatedly deposit their ligands within the cell and then recycle to the plasma membrane, once again to mediate internalization of ligand molecules. For instance, the LDL receptor makes one round trip into and out of the cell interior every 10–20 minutes, for a total of several hundred trips in its 20-hour life span.

Internalized receptor-ligand complexes commonly follow the pathway depicted for the M6P receptor in Figure 14-22 and the LDL receptor in Figure 14-29. Endocytosed cell-surface receptors typically dissociate from their ligands within late endosomes, which appear as spherical vesicles with tubular branching membranes located a few micrometers from the cell surface. The original experiments that defined the late endosome sorting vesicle utilized the asialoglycoprotein receptor. This liver-specific protein mediates the binding and internalization of abnormal glycoproteins whose oligosaccharides terminate in galactose rather than the normal sialic acid; hence the name *asialo*glycoprotein. Electron microscopy of liver cells perfused with asialoglycoprotein reveal that 5–10 minutes after internalization, ligand molecules are found in the lumen of late endosomes, while the tubular membrane extensions are rich in receptor and rarely contain ligand. These findings indicate that the late endosome is the organelle in which receptors and ligands are uncoupled.

The dissociation of receptor-ligand complexes in late endosomes occurs not only in the endocytic pathway but also in the delivery of soluble lysosomal enzymes via the secretory pathway (see Figure 14-22). As discussed in Chapter 11, the membranes of late endosomes and lysosomes contain V-class proton pumps that act in concert with Cl⁻ channels to acidify the vesicle lumen (see Figure 11-14). Most receptors, including the M6P receptor and cell-surface receptors for LDL particles and asialoglycoprotein, bind their ligands tightly at

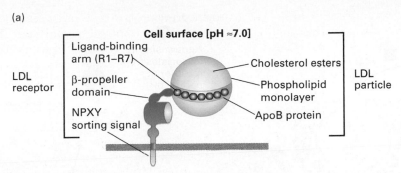

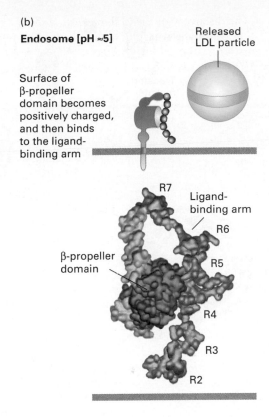

**FIGURE 14-30  Model for pH-dependent binding of LDL particles by the LDL receptor.** Schematic depiction of an LDL receptor at the neutral pH found at the cell surface (a) and at the acidic pH found in the interior of the late endosome (b). (a) At the cell surface, apoB-100 on the surface of an LDL particle binds tightly to the receptor. Of the seven repeats (R1–R7) in the ligand-binding arm, R4 and R5 appear to be most critical for LDL binding. (b, *top*) Within the endosome, histidine residues in the β-propeller domain of the LDL receptor become protonated. The positively charged propeller can bind with high affinity to the ligand-binding arm, which contains negatively charged residues, causing release of the LDL particle. (b, *bottom*) Experimental electron density and $C_\alpha$ trace model of the extracellular region of the LDL receptor at pH 5.3 based on x-ray crystallographic analysis. In this conformation, extensive hydrophobic and ionic interactions occur between the β propeller and the R4 and R5 repeats. [Part (b) from G. Rudenko et al., 2002, *Science* **298**:2353.]

neutral pH but release their ligands if the pH is lowered to 6.0 or below. The late endosome is the first vesicle encountered by receptor-ligand complexes whose luminal pH is sufficiently acidic to promote dissociation of most endocytosed receptors from their tightly bound ligands.

The mechanism by which the LDL receptor releases bound LDL particles is now understood in detail (Figure 14-30). At the endosomal pH of 5.0–5.5, histidine residues in a region known as the β-propeller domain of the receptor become protonated, forming a site that can bind with high affinity to the negatively charged repeats in the LDL-binding domain. This intramolecular interaction sequesters the repeats in a conformation that cannot simultaneously bind to apoB-100, thus causing release of the bound LDL particle.

## The Endocytic Pathway Delivers Iron to Cells Without Dissociation of the Receptor-Transferrin Complex in Endosomes

The endocytic pathway involving the transferrin receptor and its ligand differs from the LDL pathway in that the receptor-ligand complex does not dissociate in late endosomes. Nonetheless, changes in pH also mediate the sorting of receptors and ligands in the transferrin pathway, which functions to deliver iron to cells.

A major glycoprotein in the blood, transferrin transports iron to all tissue cells from the liver (the main site of iron storage in the body) and from the intestine (the site of iron absorption). The iron-free form, *apotransferrin*, binds two $Fe^{3+}$ ions very tightly to form *ferrotransferrin*. All mammalian cells contain cell-surface transferrin receptors that avidly bind ferrotransferrin at neutral pH, after which the receptor-bound ferrotransferrin is subjected to endocytosis. Like the components of an LDL particle, the two bound $Fe^{3+}$ atoms remain in the cell, but the apotransferrin part of the ligand does not dissociate from the receptor in the late endosome, and within minutes after being endocytosed, apotransferrin is returned to the cell surface and secreted from the cell.

As depicted in Figure 14-31, the explanation for the behavior of the transferrin receptor–ligand complex lies in the unique ability of apotransferrin to remain bound to the transferrin receptor at the low pH (5.0–5.5) of late endosomes. At a pH of less than 6.0, the two bound $Fe^{3+}$ atoms dissociate from ferrotransferrin, are reduced to $Fe^{2+}$ by an unknown mechanism, and then are exported into the cytosol by an endosomal transporter specific for divalent metal ions. The receptor-apotransferrin complex remaining after dissociation of the iron atoms is recycled back to the cell surface. Although apotransferrin binds tightly to its receptor at a pH of 5.0 or 6.0, it does not bind at neutral pH. Hence the bound apotransferrin dissociates from the transferrin receptor when the recycling vesicles fuse with the plasma membrane and the receptor-ligand complex encounters the neutral pH of the extracellular interstitial fluid or growth medium. The recycled receptor is then free to bind another

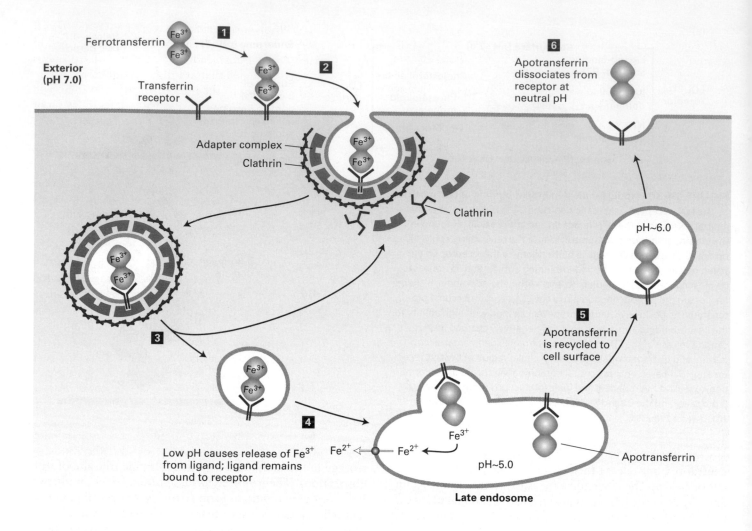

**FIGURE 14-31 The transferrin cycle, which operates in all growing mammalian cells.** Step **1**: The transferrin dimer carrying two bound atoms of $Fe^{3+}$, called ferrotransferrin, binds to the transferrin receptor at the cell surface. Step **2**: Interaction between the tail of the transferrin receptor and the AP2 adapter complex incorporates the receptor-ligand complex into endocytic clathrin-coated vesicles. Steps **3** and **4**: The vesicle coat is shed and the endocytic vesicles fuse with the membrane of the endosome. $Fe^{3+}$ is released from the receptor-ferrotransferrin complex in the acidic late endosome compartment. Step **5**: The apotransferrin protein remains bound to its receptor at this pH, and they recycle to the cell surface together. Step **6**: The neutral pH of the exterior medium causes release of the iron-free apotransferrin. [See A. Ciechanover et al., 1983, *J. Biol. Chem.* **258**:9681.]

molecule of ferrotransferrin, and the released apotransferrin is carried in the bloodstream to the liver or intestine to be reloaded with iron.

## KEY CONCEPTS of Section 14.5

### Receptor-Mediated Endocytosis

• Some extracellular ligands that bind to specific cell-surface receptors are internalized, along with their receptors, in clathrin-coated vesicles whose coats also contain AP2 complexes (see Figure 14-26).

• Sorting signals in the cytosolic domain of cell-surface receptors target them into clathrin/AP2-coated pits for internalization. Known signals include the Asn-Pro-X-Tyr, Tyr-X-X-Φ, and Leu-Leu sequences (see Table 14-2).

• The endocytic pathway delivers some ligands (e.g., LDL particles) to lysosomes, where they are degraded. Transport vesicles from the cell surface first fuse with late endosomes, which subsequently fuse with the lysosome.

• Most receptor-ligand complexes dissociate in the acidic milieu of the late endosome; the receptors are recycled to the plasma membrane, while the ligands are sorted to lysosomes (see Figure 14-29).

- Iron is imported into cells by an endocytic pathway in which $Fe^{3+}$ ions are released from ferrotransferrin in the late endosome. The receptor-apotransferrin complex is recycled to the cell surface, where the complex dissociates, releasing both the receptor and apotransferrin for reuse.

## 14.6 Directing Membrane Proteins and Cytosolic Materials to the Lysosome

The major function of lysosomes is to degrade extracellular materials taken up by the cell and intracellular components under certain conditions. Materials to be degraded must be delivered to the lumen of the lysosome, where the various degradative enzymes reside. As just discussed, endocytosed ligands (e.g., LDL particles) that dissociate from their receptors in the late endosome subsequently enter the lysosomal lumen when the membrane of the late endosome fuses with the membrane of the lysosome (see Figure 14-29). Likewise, phagosomes carrying bacteria or other particulate matter can fuse with lysosomes, releasing their contents into the lumen for degradation.

It is apparent how the general vesicular trafficking mechanism discussed in this chapter can be used to deliver the luminal contents of an endosomal organelle to the lumen of the lysosome for degradation. However, membrane proteins delivered to the lysosome by the typical vesicular trafficking process we have discussed in this chapter should ultimately be delivered to the membrane of the lysosome. How then are membrane proteins degraded by the lysosome? As we will see in this section, the cell has two different specialized pathways for delivery of materials to the lysosomal lumen for degradation, one for membrane proteins and one for cytosolic materials. The first pathway, used to degrade endocytosed membrane proteins, utilizes an unusual type of vesicle that buds into the lumen of the endosome to produce a multivesicular endosome. The second pathway, known as **autophagy**, involves the de novo formation of a double membrane organelle known as an autophagosome that envelops cytosolic material, such as soluble cytosolic proteins or sometimes organelles such as peroxisomes or mitochondria. Both pathways lead to fusion of either the multivesicular endosome or autophagosome with the lysosome, depositing the contents of these organelles into the lysosomal lumen for degradation.

### Multivesicular Endosomes Segregate Membrane Proteins Destined for the Lysosomal Membrane from Proteins Destined for Lysosomal Degradation

Resident lysosomal proteins, such as V-class proton pumps and amino acid transporters, can carry out their functions and remain in the lysosomal membrane, where they are protected from degradation by the soluble hydrolytic enzymes in the lumen. Such proteins are delivered to the lysosomal membrane by transport vesicles that bud from the *trans*-Golgi network by the same basic mechanisms described in earlier sections. In contrast, endocytosed membrane proteins such as receptor proteins that are to be degraded are transferred in their entirety to the interior of the lysosome by a specialized delivery mechanism. Lysosomal degradation of cell-surface receptors for extracellular signaling molecules is a common mechanism for controlling the sensitivity of cells to such signals (Chapter 15). Receptors that become damaged also are targeted for lysosomal degradation.

Early evidence that membranes can be delivered to the lumen of compartments came from electron micrographs showing membrane vesicles and fragments of membranes within endosomes and lysosomes. Parallel experiments in yeast revealed that endocytosed receptor proteins targeted to the vacuole (the yeast organelle equivalent to the lysosome) were primarily associated with membrane fragments and small vesicles within the interior of the vacuole rather than with the vacuole surface membrane.

These observations suggest that endocytosed membrane proteins can be incorporated into specialized vesicles that form at the endosomal membrane (Figure 14-32). Although these vesicles are similar in size and appearance to transport vesicles, they differ topologically. Transport vesicles bud *outward* from the surface of a donor organelle into the cytosol, whereas vesicles within the endosome bud *inward* from the surface into the lumen (away from the cytosol). Mature endosomes containing numerous vesicles in their interior are usually called *multivesicular endosomes* (or bodies). The surface membrane of a multivesicular endosome fuses with the membrane of a lysosome, thereby delivering its internal vesicles and the membrane proteins they contain into the lysosome interior for degradation. Thus the sorting of proteins in the endosomal membrane determines which ones will remain on the lysosome surface (e.g., pumps and transporters) and which ones will be incorporated into internal vesicles and ultimately degraded in lysosomes.

Many of the proteins required for inward budding of the endosomal membrane were first identified by mutations in yeast that blocked delivery of membrane proteins to the interior of the vacuole. More than 10 such "budding" proteins have been identified in yeast, most with significant similarities to mammalian proteins that evidently perform the same function in mammalian cells. The current model of endosomal budding to form multivesicular endosomes in mammalian cells is based primarily on studies in yeast (Figure 14-33). Most cargo proteins that enter the multivesicular endosome are tagged with ubiquitin. Cargo proteins destined to enter the multivesicular endosome usually receive their ubiquitin tag at the plasma membrane, the TGN, or the endosomal membrane. We have already seen how ubiquitin tagging can serve as a signal for degradation of cytosolic or misfolded ER proteins by the proteasome (see Chapters 3 and 13). When used as a signal for proteasomal degradation, the ubiquitin tag usually consists of a chain of covalently linked ubiquitin molecules (polyubiquitin), whereas ubiquitin used to tag proteins for entry into the multivesicular endosome usually takes the form of a single (monoubiquitin) molecule. In the membrane

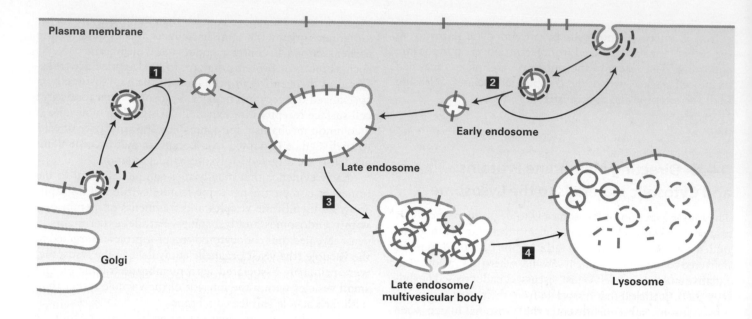

**Early endosome**

**Late endosome**

**Golgi**

**Late endosome/
multivesicular body**

**Lysosome**

**Plasma membrane**

**FIGURE 14-32 Delivery of plasma-membrane proteins to the lysosomal interior for degradation.** Early endosomes carrying endocytosed plasma-membrane proteins (blue) and vesicles carrying lysosomal membrane proteins (green) from the *trans*-Golgi network fuse with the late endosome, transferring their membrane proteins to the endosomal membrane (steps **1** and **2**). Proteins to be degraded, such as those from the early endosome, are incorporated into vesicles that bud *into* the interior of the late endosome, eventually forming a multivesicular endosome containing many such internal vesicles (step **3**). Fusion of a multivesicular endosome directly with a lysosome releases the internal vesicles into the lumen of the lysosome, where they can be degraded (step **4**). Because proton pumps and other lysosomal membrane proteins normally are not incorporated into internal endosomal vesicles, they are delivered to the lysosomal membrane and are protected from degradation. [See F. Reggiori and D. J. Klionsky, 2002, *Eukaryot. Cell* **1**:11, and D. J. Katzmann et al., 2002, *Nature Rev. Mol. Cell Biol.* **3**:893.]

of the endosome a ubiquitin-tagged peripheral membrane protein, known as Hrs, facilitates recruitment of a set of three different protein complexes to the membrane. These *ESCRT* (*endosomal sorting complexes required for transport*) *proteins* include the ubiquitin-binding protein Tsg101. The membrane-associated ESCRT proteins act to drive vesicle budding directed into the interior of the endosome as well as loading of specific monoubiquitinated membrane cargo proteins into the vesicle buds. Finally the ESCRT proteins pinch off the vesicle, releasing it and the specific membrane cargo proteins it carries into the interior of the endosome. An ATPase, known as Vps4, uses the energy from ATP hydrolysis to disassemble the ESCRT proteins, releasing them into the cytosol for another round of budding. In the fusion event that pinches off a completed endosomal vesicle, the ESCRT proteins and Vps4 may function like SNAREs and NSF, respectively, in

**FIGURE 14-33 Model of the mechanism for formation of multivesicular endosomes.** In endosomal budding, ubiquitinated Hrs on the endosomal membrane directs loading of specific membrane cargo proteins (blue) into vesicle buds and then recruits cytosolic ESCRT proteins to the membrane (step **1**). Note that both Hrs and the recruited cargo proteins are tagged with ubiquitin. After the set of bound ESCRT complexes mediate the completion and pinching off of the inwardly budding vesicles (step **2**), they are disassembled by the ATPase Vps4 and returned to the cytosol (step **3**). See text for discussion. [Adapted from O. Pornillos et al., 2002, *Trends Cell Biol.* **12**:569.]

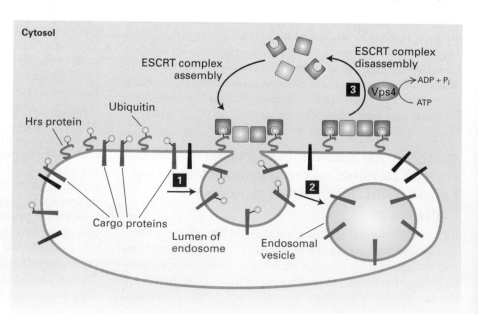

**Cytosol**

**ESCRT complex
assembly**

**ESCRT complex
disassembly**

**Hrs protein**

**Ubiquitin**

Vps4 → ADP + P$_i$ ← ATP

**Cargo proteins**

**Lumen of
endosome**

**Endosomal
vesicle**

the typical membrane-fusion process discussed previously (see Figure 14-10).

## Retroviruses Bud from the Plasma Membrane by a Process Similar to Formation of Multivesicular Endosomes

The vesicles that bud into the interior of endosomes have a topology similar to that of enveloped virus particles that bud from the plasma membrane of virus-infected cells. Moreover, recent experiments demonstrate that a common set of proteins is required for both types of membrane-budding events. In fact, the two processes so closely parallel each other in mechanistic detail as to suggest that enveloped viruses have evolved mechanisms to recruit the cellular proteins used in inward endosomal budding for their own purposes.

The human immunodeficiency virus (HIV) is an enveloped retrovirus that buds from the plasma membrane of infected cells in a process driven by viral Gag protein, the major structural component of completed virus particles. Gag protein binds to the plasma membrane of an infected cell and ~4000 Gag molecules polymerize into a spherical shell, producing a structure that looks like a vesicle bud protruding outward from the plasma membrane. Mutational studies with HIV have revealed that the N-terminal segment of Gag protein is required for association with the plasma membrane, whereas the C-terminal segment is required for pinching off of complete HIV particles. For instance, if the portion of the viral genome encoding the C-terminus of Gag is removed, HIV buds will form in infected cells, but pinching off does not occur, and thus no free virus particles are released.

The first indication that HIV budding employs the same molecular machinery as vesicle budding into endosomes came from the observation that Tsg101, an ESCRT protein, binds to the C-terminus of Gag protein. Subsequent findings have clearly established the mechanistic parallels between the two processes. For example, Gag is ubiquitinated as part of the process of virus budding, and in cells with mutations in Tsg101 or Vps4, HIV virus buds accumulate but cannot pinch off from the membrane (Figure 14-34). Moreover,

🎧 **PODCAST:** HIV Budding from the Plasma Membrane

**FIGURE 14-34 Mechanism for budding of HIV from the plasma membrane.** Proteins required for formation of multivesicular endosomes are exploited by HIV for virus budding from the plasma membrane. (a) Budding of HIV particles from HIV-infected cells occurs by a similar mechanism as in Figure 14-33, using the virally encoded Gag protein and cellular ESCRT and Vps4 (steps **1**–**3**). Ubiquitinated Gag near a budding particle functions like Hrs. See text for discussion. (b) In wild-type cells infected with HIV, virus particles bud from the plasma membrane and are rapidly released into the extracellular space. (c) In cells that lack the functional ESCRT protein Tsg101, the viral Gag protein forms dense viruslike structures, but budding of these structures from the plasma membrane cannot be completed and chains of incomplete viral buds still attached to the plasma membrane accumulate. [Wes Sundquist, University of Utah.]

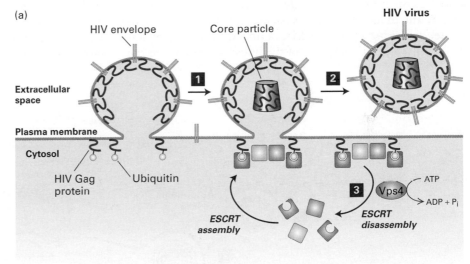

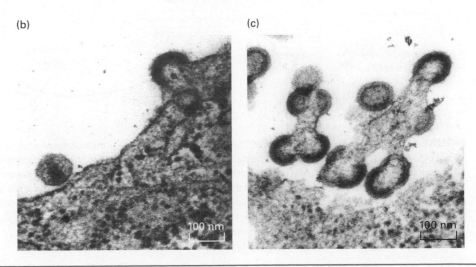

when a segment from the cellular Hrs protein is added to a truncated Gag protein by construction of the appropriate hybrid gene, proper budding and release of virus particles is restored. Taken together, these results indicate that Gag protein mimics the function of Hrs, redirecting ESCRT proteins to the plasma membrane, where they can function in the budding of virus particles.

Other enveloped retroviruses such as murine leukemia virus and Rous sarcoma virus also have been shown to require ESCRT complexes for their budding, although each virus appears to have evolved a somewhat different mechanism to recruit ESCRT complexes to the site of virus budding.

## The Autophagic Pathway Delivers Cytosolic Proteins or Entire Organelles to Lysosomes

When cells are placed under stress such as conditions of starvation, they have the capacity to recycle macromolecules for use as nutrients in a process of lysosomal degradation known as **autophagy** ("eating oneself"). The autophagic pathway involves the formation of a flattened double-membrane cup-shaped structure that envelops a region of the cytosol or an entire organelle (e.g., mitochondrion), forming an *autophagosome,* or *autophagic vesicle* (Figure 14-35). The outer membrane of an autophagic vesicle can fuse with the lysosome, delivering a large vesicle, bounded by a single membrane bilayer, to the interior of the lysosome. Similar to the

situation that occurs when the contents of multivesicular endosomes are delivered to the lysosome, lipases and proteases within the lysosome will degrade the autophagic vesicle and its contents into their molecular components. Amino acid permeases in the lysosomal membrane then allow for transport of free amino acids back into the cytosol for use in synthesis of new proteins.

By studying mutants defective in the autophagic pathway, scientists have identified processes other than recycling of cellular components during starvation that also depend on autophagy. Experiments carried out principally in *Drosophila* and mice have shown that autophagy participates in a type of quality control that removes organelles that have ceased to function properly. In particular, the autophagic pathway can target for destruction dysfunctional mitochondria that have lost their integrity and no longer have an electrochemical gradient across their inner membrane. In certain cell types, pathogenic bacteria and viruses that are multiplying in the cytosol of host cells can be targeted to the autophagic pathway for destruction in the lysosome as part of a host defense mechanism against infection.

For each of these processes and in all eukaryotic organisms the autophagic pathway takes place in three basic steps. Although the underlying mechanisms for each of these steps are relatively poorly understood, they are thought to be related to the basic mechanisms for vesicular trafficking discussed in this chapter.

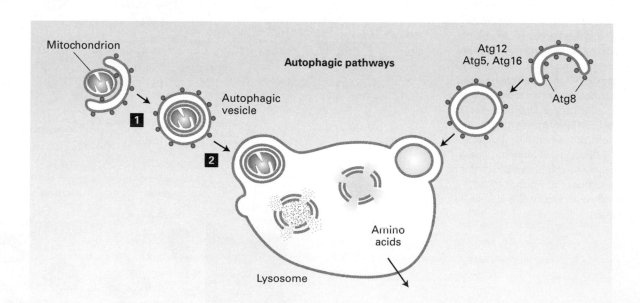

**FIGURE 14-35 The autophagic pathway.** The autophagic pathway allows cytosolic proteins and organelles to be delivered to the lysosomal interior for degradation. In the autophagic pathway, a cup-shaped structure forms around portions of the cytosol (*right*) or an organelle such as a mitochondrion as shown here (*left*). Continued addition of membrane eventually leads to the formation of an autophagosome vesicle that envelops its contents by two complete membranes (step **1**). Fusion of the outer membrane with the

membrane of a lysosome releases a single-layer vesicle and its contents into the lysosome interior (step **2**). After degradation of the protein and lipid components by hydrolases in the lysosome interior, the released amino acids are transported across the lysosomal membrane into the cytosol. Proteins known to participate in the autophagic pathway include Atg8, which forms a coat structure around the autophagosome.

**Autophagic Vesicle Nucleation** The autophagic vesicle is thought to originate from a fragment of a membrane-bounded organelle. The origin of this membrane has been difficult to trace because no known integral membrane proteins, which might serve to identify the source of this membrane, are known to be required for the formation of the autophagic vesicle. Studies in yeast have shown that some mutants defective in Golgi trafficking are also defective in autophagy, suggesting that the autophagic vesicle is initially derived from a fragment of the Golgi. Autophagy that is induced by starvation appears to be a nonspecific process in which a random portion of the cytoplasm, including organelles, becomes enveloped by an autophagosome. In these cases, the site of nucleation is probably random. In cases in which defective organelles are enveloped by the autophagosome, some type of signal or binding site must be present on the surface of the organelle to target nucleation of the autophagic vesicle.

**Autophagic Vesicle Growth and Completion** New membrane must be delivered to the autophagosome membrane in order for this cup-shaped organelle to grow. This growth is likely to occur by the fusion of transport vesicles with the membrane of the autophagosome. About 30 proteins that participate in the formation of autophagosomes have been identified in genetic screens for yeast mutants that are defective in autophagy. One of these proteins is Atg8, shown in Figure 14-35, which is covalently linked to the lipid phosphatidylethanolamine and thus becomes attached to the cytoplasmic leaflet of the autophagic vesicle. Association of Atg8 with a membrane vesicle appears to be the key step in enabling a vesicle to fuse with the growing autophagosome.

Fusion of Atg8-containing vesicles with the autophagosome involves the formation of a cytosolic assembly of Atg12, Atg5, and Atg16. Atg12 is similar in structure to ubiquitin, and a set of proteins related to ubiquitin-conjugating enzymes are responsible for covalently joining Atg12 to Atg5, by a process similar to that used for covalently joining ubiquitin to a target protein (see Figure 3-29). The covalently linked Atg12-Atg5 dimer then co-assembles with Atg16 to form a polymeric complex localized to the site of growing autophagosomes. By an unknown mechanism this cytosolic complex is thought to bring about the fusion of Atg8-containing vesicles into a cup-shaped autophagosome.

**Autophagic Vesicle Targeting and Fusion** The outer membrane of the completed autophagosome is thought to contain a set of proteins that target fusion with the membrane of the lysosome. Two vesicle-tethering proteins have been found to be required for autophagosome fusion with the lysosome, but the corresponding SNARE proteins have not been identified. Fusion of the autophagosome with the lysosome occurs after Atg8 has been released from the membrane by proteolytic cleavage, and this proteolysis step only occurs once the autophagic vesicle has completely formed a sealed double-membrane system. Thus Atg8 protein appears to mask fusion proteins and to prevent premature fusion of the autophagosome with the lysosome.

## KEY CONCEPTS of Section 14.6

### Directing Membrane Proteins and Cytosolic Materials to the Lysosome

• Endocytosed membrane proteins destined for degradation in the lysosome are incorporated into vesicles that bud into the interior of the endosome. Multivesicular endosomes, which contain many of these internal vesicles, can fuse with the lysosome to deliver the vesicles to the interior of the lysosome (see Figure 14-32).

• Some of the cellular components (e.g., ESCRT) that mediate inward budding of endosomal membranes are used in the budding and pinching off of enveloped viruses such as HIV from the plasma membrane of virus-infected cells (see Figures 14-33 and 14-34).

• A portion of the cytoplasm or an entire organelle (e.g., a mitochondrion) can be enveloped in a flattened membrane and eventually incorporated into a double-membrane autophagic vesicle. Fusion of the outer vesicle membrane with the lysosome delivers the enveloped contents to the interior of the lysosome for degradation (see Figure 14-35).

## Perspectives for the Future

The biochemical, genetic, and structural information presented in this chapter shows that we now have a basic understanding of how protein traffic flows from one membrane-bounded cellular compartment to another. Our understanding of these processes has come largely from experiments on the function of various types of transport vesicles. These studies have led to the identification of many vesicle components and the discovery of how these components work together to drive vesicle budding, to incorporate the correct set of cargo molecules from the donor organelle, and then to mediate fusion of a completed vesicle with the membrane of a target organelle.

Despite these advances, there is still much to learn about important stages of the secretory and endocytic pathways. For example, we do not yet know what types of proteins form the coats of either the regulated or the constitutive secretory vesicles that bud from the *trans*-Golgi network. In the same vein, we do not know what feature of the Golgi membrane determines whether a COPI-coated vesicle or a clathrin/AP-coated vesicle will bud from it. In both cases, binding of ARF protein to the Golgi membrane appears to initiate vesicle budding. Moreover, the types of signals on cargo proteins that might target them for packaging into secretory vesicles have not yet been defined.

Another baffling process is the formation of vesicles that bud away from the cytosol, such as the vesicles that enter multivesicular endosomes. Although some of the proteins that participate in formation of these "internal" endosome vesicles are known, we do not know what determines their shape or what type of process causes them to pinch off from the donor membrane. Similarly, the origin and growth of the membrane of the autophagic vesicle is also poorly understood. In the future, it should be possible for these and other poorly understood vesicle-trafficking steps to be dissected through the use of the same powerful combination of biochemical and genetic methods that have delineated the working parts of COPI, COPII, and clathrin/AP vesicles.

In addition to understanding the basic mechanisms that direct the trafficking of cargo proteins in the secretory and endocytic pathways, a major goal of research in this area is to define all of the signals that direct proteins to specific intracellular locations. Although a number of such sequences are known (see Table 14-2), we are only beginning to compile a catalog of these targeting signals and the context in which they are read. The ultimate goal will be to deduce, starting with just the primary coding sequence of any given gene, the trafficking pattern and intracellular location of the gene's protein product. Our ability to fully extract biological information from genomic sequences will be realized only when we have the capability of reading trafficking information from primary protein sequences.

## Key Terms

AP (adapter protein) complexes 646
anterograde transport 640
ARF protein 635
autophagy 661
cisternal maturation 629
clathrin 634
constitutive secretion 651
COPI 634
COPII 634
dynamin 647
endocytic pathway 627
ESCRT proteins 000
late endosome 629
low-density lipoprotein (LDL) 656
mannose 6-phosphate (M6P) 648

multivesicular endosomes 661
Rab proteins 638
receptor-mediated endocytosis 655
regulated secretion 651
retrograde transport 640
sec mutants 632
secretory pathway 627
sorting signals 637
transcytosis 654
trans-Golgi network (TGN) 629
transport vesicles 627
t-SNAREs 634
v-SNAREs 634

## Review the Concepts

1. The studies of Palade and colleagues used pulse-chase labeling with radioactively labeled amino acids and autoradiography to visualize the location of newly synthesized proteins in pancreatic acinar cells. These early experiments provided invaluable information on protein synthesis and intercompartmental transport. New methods have replaced these early approaches, but two basic requirements are still necessary for any assay to study this type of protein transport. What are they and how do recent experimental approaches meet these criteria?

2. Vesicle budding is associated with coat proteins. What is the role of coat proteins in vesicle budding? How are coat proteins recruited to membranes? What kinds of molecules are likely to be included or excluded from newly formed vesicles? What is the best-known example of a protein likely to be involved in vesicle pinching off?

3. Treatment of cells with the drug brefeldin A (BFA) has the effect of decoating Golgi apparatus membranes, resulting in a cell in which the vast majority of Golgi proteins are found in the ER. What inferences can be made from this observation regarding roles of coat proteins other than promoting vesicle formation? Predict what type of mutation in Arf1 might have the same effect as treating cells with BFA.

4. Microinjection of an antibody known as EAGE, which reacts with the "hinge" region of the β subunit of COPI, causes accumulation of Golgi enzymes in transport vesicles and inhibits anterograde transport of newly synthesized vesicles from the ER to the plasma membrane. What effect does the antibody have on COPI activity? Explain the results.

5. Specificity in fusion between vesicles involves two discrete and sequential processes. Describe the first of the two processes and its regulation by GTPase switch proteins. What effect on the size of early endosomes might result from overexpression of a mutant form of Rab5 that is stuck in the GTP-bound state?

6. *Sec18* is a yeast gene that encodes NSF. It is a class C mutant in the yeast secretory pathway. What is the mechanistic role of NSF in membrane trafficking? As indicated by its class C phenotype, why does an NSF mutation produce accumulation of vesicles at what appears to be only one stage of the secretory pathway?

7. What feature of procollagen synthesis provided early evidence for the Golgi cisternal maturation model?

8. Sorting signals that cause retrograde transport of a protein in the secretory pathway are sometimes known as retrieval sequences. List the two known examples of retrieval sequences for soluble and membrane proteins of the ER. How does the presence of a retrieval sequence on a soluble ER protein result in its retrieval from the *cis*-Golgi complex? Describe how the concept of a retrieval sequence is essential to the cisternal-maturation model.

9. Clathrin adapter protein (AP) complexes bind directly to the cytosolic face of membrane proteins and also interact with clathrin. What are the four known adapter protein complexes? What observation regarding AP3 suggests that clathrin is an accessory protein to a core coat composed of adapter proteins?

10. I-cell disease is a classic example of an inherited human defect in protein targeting that affects an entire class of proteins, soluble enzymes of the lysosome. What is the molecular defect in I-cell disease? Why does it affect the targeting of an entire class of proteins? What other types of mutations might produce the same phenotype?

11. The TGN, *trans*-Golgi network, is the site of multiple sorting processes as proteins and lipids exit the Golgi complex. Compare and contrast the sorting of proteins to lysosomes

versus the packaging of proteins into regulated secretory granules such as those containing insulin. Compare and contrast the sorting of proteins to the basolateral versus apical cell surfaces in MDCK cells versus hepatocytes.

12. What does the budding of influenza virus and vesicular stomatitis virus (VSV) from polarized MDCK cells reveal about the sorting of newly synthesized cell plasma membrane proteins to the apical or basolateral domains? Now consider the following result: A peptide with a sequence identical to that of the VSV G protein cytoplasmic domain inhibits targeting of the G protein to the basolateral surface and has no effect on HA targeting to the apical membrane, but a peptide in which the single tyrosine residue is mutated to an alanine has no effect on G protein basolateral targeting. What does this tell you about the sorting process?

13. Describe how pH plays a key role in regulating the interaction between mannose 6-phosphate and the mannose 6-phosphate receptor. Why does elevating endosomal pH lead to the secretion of newly synthesized lysosomal enzymes into the extracellular medium?

14. What mechanistic features are shared by (a) the formation of multivesicular endosomes by budding into the interior of the endosome and (b) the outward budding of HIV virus at the cell surface? You wish to design a peptide inhibitor/competitor of HIV budding and decide to mimic in a synthetic peptide a portion of the HIV Gag protein. Which portion of the HIV Gag protein would be a logical choice? What normal cellular process might this inhibitor block?

15. The phagocytic and autophagic pathways serve two fundamental roles, but both deliver their vesicles to the lysosome. What are the fundamental differences between the two pathways? Describe the three basic steps in the formation and fusion of autophagic vesicles.

16. Compare and contrast the location and pH sensitivity of receptor-ligand interaction in the LDL and transferrin receptor-mediated endocytosis (RME) pathways.

17. What do LDL receptor (LDLR) cytoplasmic domain mutations that cause familial hypercholesterolemia reveal about the receptor-mediated endocytosis (RME) pathway?

## Analyze the Data

1. In order to examine the specificity of membrane fusion conferred by specific v-SNAREs and t-SNAREs, researchers reconstituted liposomes (artificial lipid membranes) with specific t-SNARE complexes or with v-SNAREs (see McNew et al., 2000, *Nature* 407:153–159). To measure fusion, the v-SNARE liposomes also contained a fluorescent lipid at a relatively high concentration such that its fluorescence is quenched. (Quenching is reduced fluorescence relative to that expected. In this case, quenching occurs because the fluorescent lipids are too concentrated and interfere with each other's ability to become excited.) On fusion of these liposomes with those lacking the fluorescent lipid, the fluorescent lipids are diluted, and quenching is alleviated. Three sets of liposomes were pre-

pared using yeast t-SNARE complexes: those containing plasma membrane t-SNAREs, Golgi t-SNAREs, or vacuolar t-SNAREs. Each of these was mixed with fluorescent liposomes containing one of three different yeast v-SNAREs. The following data were obtained.

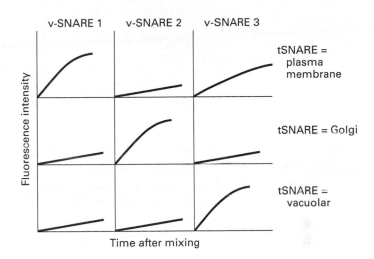

a. What can be deduced from these data about the specificity of membrane fusion events?

b. Where might you expect to find v-SNAREs 1, 2, and 3 in yeast?

c. What kind of experiment could be designed to determine where in the secretory pathway a given v-SNARE is required in vivo?

d. The cytoplasmic domain of v-SNARE 2 has been expressed and purified from *E. coli*. Various amounts of this domain are incubated either with the Golgi t-SNARE liposomes or with v-SNARE 2 liposomes. The liposomes are then washed free of unbound protein. The various liposomes are then mixed, as indicated below, and the fluorescence of each sample is measured 1 hour after mixing. How can the data be explained? What would you predict the outcome to be if yeast were to overexpress the cytoplasmic domain of v-SNARE 2?

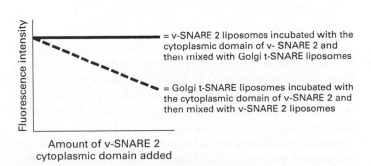

2. You have genetically engineered green fluorescent protein (GFP) containing a KDEL sequence. When the construct is transfected into normal human fibroblasts and examined using fluorescence microscopy, the fluorescence appears throughout the cytoplasm, as drawn below.

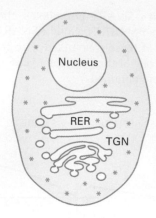

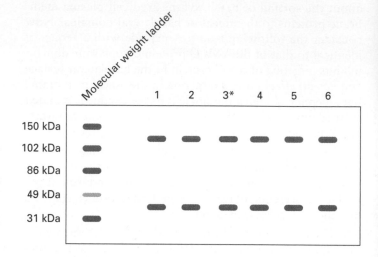

**a.** How would you explain this pattern given that KDEL is supposed to be an ER-specific sorting sequence?

**b.** To analyze the results further, fractions of different organelles and the cytoplasm were collected from cells expressing this KDEL-containing GFP construct and then examined on Western blots using antibodies against GFP (27 kDa) and protein disulfide isomerase (PDI), a resident-rough ER (RER) protein of approximately 55 kDa.

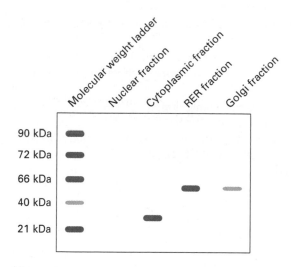

The blot confirms the presence of GFP exclusively in the cytoplasm, and as expected a PDI signal in the RER fraction. How would you explain the PDI band, albeit weak, in the Golgi fraction? Given the function of PDI proteins, what would you expect if both alleles of a PDI gene were knocked out in mice?

**c.** The antibodies used above do not detect any signals in the nuclear fraction, which indicates their specificity or the fact that no proteins were isolated and loaded in the nuclear fraction lane. What antibody could be used to show there were nuclear proteins present in this sample?

**3.** A child appears to be suffering from I-cell disease, but when a sample of his proteins (lane 3* below), isolated from skin fibroblasts, is compared to protein samples from fibroblasts of his healthy parents (lanes 1 and 2) and siblings (lanes 4–6) using Western blot analysis and antibodies against $N$-acetylglucosamine phosphotransferase (~145 kDa) and actin (loading control, ~43 kDa), the following is seen:

In a second set of experiments, $N$-acetylglucosamine phosphotransferase was isolated from cells from the afflicted child and from his healthy parents, and used in an assay with $^{32}P$ to measure enzyme activity and the production of mannose 6-phosphate. The assay yielded the following results:

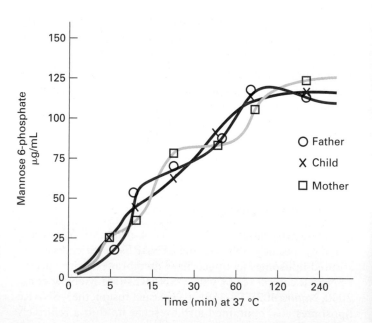

**a.** Using fibroblasts cultured from the child, design an experiment using the $N$-acetylglucosamine phosphotransferase antibody and fluorescence microscopy and draw the results that could explain why the child presents symptoms similar to I-cell disease.

**b.** Given the results from these three different experiments, how would you explain the I-cell symptoms seen in the child and what experiment would you propose to test your hypothesis?

**c.** A laser scanning confocal micrograph of MDCK cells labeled with an antibody against the mannose 6-phosphate receptor shows the following:

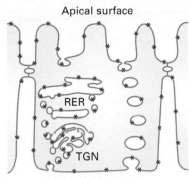

Apical surface

RER

TGN

Basal surface

How do you explain the labeling on the apical *and* basolateral surfaces, for a receptor whose function is to target enzymes from the *trans*-Golgi network (TGN) to the lysosome? Likewise, what explains the labeling seen at the RER?

## References

### Techniques for Studying the Secretory Pathway

Beckers, C. J., et al. 1987. Semi-intact cells permeable to macromolecules: use in reconstitution of protein transport from the endoplasmic reticulum to the Golgi complex. *Cell* 50:523–534.

Kaiser, C. A., and R. Schekman. 1990. Distinct sets of SEC genes govern transport vesicle formation and fusion early in the secretory pathway. *Cell* 61:723–733.

Novick, P., et al. 1981. Order of events in the yeast secretory pathway. *Cell* 25:461–469.

Lippincott-Schwartz, J., et al. 2001. Studying protein dynamics in living cells. *Nature Rev. Mol. Cell Biol.* 2:444–456.

Orci, L., et al. 1989. Dissection of a single round of vesicular transport: sequential intermediates for intercisternal movement in the Golgi stack. *Cell* 56:357–368.

Palade, G. 1975. Intracellular aspects of the process of protein synthesis. *Science* 189:347–358.

### Molecular Mechanisms of Vesicle Budding and Fusion

Bonifacino, J. S., and B. S. Glick. 2004. The mechanisms of vesicle budding and fusion. *Cell* 116:153–166.

Grosshans, B. L., D. Ortiz, and P. Novick. 2006. Rabs and their effectors: achieving specificity in membrane traffic. *Proc. Natl. Acad. Sci. USA* 103:11821–11827.

Jahn, R., et al. 2003. Membrane fusion. *Cell* 112:519–533.

Kirchhausen, T. 2000. Three ways to make a vesicle. *Nature Rev. Mol. Cell Biol.* 1:187–198.

McNew, J. A., et al. 2000. Compartmental specificity of cellular membrane fusion encoded in SNARE proteins. *Nature* 407:153–159.

Ostermann, J., et al. 1993. Stepwise assembly of functionally active transport vesicles. *Cell* 75:1015–1025.

Schimmöller, F., I. Simon, and S. Pfeffer. 1998. Rab GTPases, directors of vesicle docking. *J. Biol. Chem.* 273:22161–22164.

Weber, T., et al. 1998. SNAREpins: minimal machinery for membrane fusion. *Cell* 92:759–772.

Wickner, W., and A. Haas. 2000. Yeast homotypic vacuole fusion: a window on organelle trafficking mechanisms. *Ann. Rev. Biochem.* 69:247–275.

Zerial, M., and H. McBride. 2001. Rab proteins as membrane organizers. *Nature Rev. Mol. Cell Biol.* 2:107–117.

### Early Stages of the Secretory Pathway

Barlowe, C. 2003. Signals for COPII-dependent export from the ER: what's the ticket out? *Trends Cell Biol.* 13:295–300.

Behnia, R., and S. Munro. 2005. Organelle identity and the signposts for membrane traffic. *Nature* 438:597–604,

Bi, X., et al. 2002. Structure of the Sec23/24-Sar1 pre-budding complex of the COPII vesicle coat. *Nature* 419:271–277.

Gurkan, C., et al. 2006. The COPII cage: unifying principles of vesicle coat assembly. *Nat. Rev. Mol. Cell Biol.* 7:727–738.

Lee, M. C., et al. 2004. Bi-directional protein transport between the ER and Golgi. *Ann. Rev. Cell Dev. Biol.* 20:87–123.

Letourneur, F., et al. 1994. Coatomer is essential for retrieval of dilysine-tagged proteins to the endoplasmic reticulum. *Cell* 79:1199–1207.

Losev, E., et al. 2006. Golgi maturation visualized in living yeast. *Nature* 441:1002–1006.

Pelham, H. R. 1995. Sorting and retrieval between the endoplasmic reticulum and Golgi apparatus. *Curr. Opin. Cell Biol.* 7:530–535.

### Later Stages of the Secretory Pathway

Bonifacino, J. S. 2004. The GGA proteins: adaptors on the move. *Nat. Rev. Mol. Cell Biol.* 5:23–32.

Bonifacino, J. S., and E. C. Dell'Angelica. 1999. Molecular bases for the recognition of tyrosine-based sorting signals. *J. Cell Biol.* 145:923–926.

Edeling, M. A., C. Smith, and D. Owen. 2006. Life of a clathrin coat: insights from clathrin and AP structures. *Nat. Rev. Mol. Cell Biol.* 7:32–44.

Fotin, A., et al. 2004. Molecular model for a complete clathrin lattice from electron cryomicroscopy. *Nature* 432:573–579.

Ghosh, P., et al. 2003. Mannose 6-phosphate receptors: new twists in the tale. *Nature Rev. Mol. Cell Bio.* 4:202–213.

Mostov, K. E., M. Verges, and Y. Altschuler. 2000. Membrane traffic in polarized epithelial cells. *Curr. Opin. Cell Biol.* 12:483–490.

Schmid, S. 1997. Clathrin-coated vesicle formation and protein sorting: an integrated process. *Ann. Rev. Biochem.* 66:511–548.

Simons, K., and E. Ikonen. 1997. Functional rafts in cell membranes. *Nature* 387:569–572.

Song, B. D., and S. L. Schmid. 2003. A molecular motor or a regulator? Dynamin's in a class of its own. *Biochemistry* 42:1369–1376.

Steiner, D. F., et al. 1996. The role of prohormone convertases in insulin biosynthesis: evidence for inherited defects in their action in man and experimental animals. *Diabetes Metab.* 22:94–104.

Tooze, S. A., et al. 2001. Secretory granule biogenesis: rafting to the SNARE. *Trends Cell Biol.* 11:116–122.

### Receptor-Mediated Endocytosis

Brown, M. S., and J. L. Goldstein. 1986. Receptor-mediated pathway for cholesterol homeostasis. Nobel Prize Lecture. *Science* 232:34–47.

Kaksonen, M., C. P. Toret, and D. G. Drubin. 2006. Harnessing actin dynamics for clathrin-mediated endocytosis. *Nat. Rev. Mol. Cell Biol.* **7**:404–414.

Rudenko, G., et al. 2002. Structure of the LDL receptor extracellular domain at endosomal pH. *Science* **298**:2353–2358.

### Directing Membrane Proteins and Cytosolic Materials to the Lysosome

Geng, J., and D. J. Klionsky. 2008. The Atg8 and Atg12 ubiquitin-like conjugation systems in macroautophagy. *EMBO Rep.* **9**:859–864.

Henne, W. M., N. J. Buchkovich, and S. D. Emr. 2011. The ESCRT pathway. *Dev. Cell* **21**:77–91.

Katzmann, D. J., et al. 2002. Receptor downregulation and multivesicular-body sorting. *Nature Rev. Mol. Cell Biol.* **3**:893–905.

Lemmon, S. K., and L. M. Traub. 2000. Sorting in the endosomal system in yeast and animal cells. *Curr. Opin. Cell Biol.* **12**:457–466.

Pornillos, O., et al. 2002. Mechanisms of enveloped RNA virus budding. *Trends Cell Biol.* **12**:569–579.

Shintani, T., and D. J. Klionsky. 2004. Autophagy in health and disease: a double-edged sword. *Science* **306**:990–995.

# Following a Protein Out of the Cell

J. Jamieson and G. Palade, 1966, *Proc. Natl. Acad. Sci. USA* **55**(2):424–431

The advent of electron microscopy allowed researchers to see the cell and its structures at an unprecedented level of detail. George Palade utilized this tool not only to look at the fine details of the cell but also to analyze the process of secretion. By combining electron microscopy with pulse-chase experiments, Palade uncovered the path proteins follow to leave the cell.

## Background

In addition to synthesizing proteins to carry out cellular functions, many cells must also produce and secrete additional proteins that perform their duties outside the cell. Cell biologists, including Palade, wondered how secreted proteins make their passage from the inside to the outside of the cell. Early experiments suggesting that proteins destined for secretion are synthesized in a particular intracellular location and then follow a pathway to the cell surface employed methods to disrupt cells synthesizing a particular secreted protein and to separate their various organelles by centrifugation. These cell-fractionation studies showed that secreted proteins can be found in membrane-bounded vesicles derived from the endoplasmic reticulum (ER), where they are synthesized, and within zymogen granules, from which they are eventually released from the cell. Unfortunately, results from these studies were hard to interpret due to difficulties in obtaining clean separation of all of the different organelles that contain secretory proteins. To further clarify the pathway, Palade turned to a newly developed technique, high-resolution autoradiography, that allowed him to detect the position of radioactively labeled proteins in thin cell sections that had been prepared for electron microscopy of intracellular organelles. His work led to the seminal finding that secreted proteins travel within vesicles from the ER to the Golgi complex and then to the plasma membrane.

## The Experiment

Palade wanted to identify which cell structures and organelles participate in protein secretion. To study such a complex process, he carefully chose an appropriate model system for his studies, the pancreatic exocrine cell, which is responsible for producing and secreting large amounts of digestive enzymes. Because these cells have the unusual property of expressing only secretory proteins, a general label for newly synthesized protein, such as radioactively labeled leucine, will only be incorporated into protein molecules that are following the secretory pathway.

Palade first examined the protein secretion pathway in vivo by injecting live guinea pigs with [$^3$H]-leucine, which was incorporated into newly made proteins, thereby radioactively labeling them. At time points from 4 minutes to 15 hours, the animals were sacrificed, and the pancreatic tissue was fixed. By subjecting the specimens to autoradiography and viewing them in an electron microscope, Palade could trace where the labeled proteins were in cells at various times. As expected, the radioactivity localized in vesicles at the ER at time points immediately following the [$^3$H]-leucine injection and at the plasma membrane at the later time points. The surprise came in the middle time points. Rather than traveling straight from the ER to the plasma membrane, the radioactively labeled proteins appeared to stop off at the Golgi complex in the middle of their journey. In addition, there never was a time point where the radioactively labeled proteins were not confined to vesicles.

The observation that the Golgi complex was involved in protein secretion was both surprising and intriguing. To thoroughly address the role of this organelle in protein secretion, Palade turned to in vitro pulse-chase experiments, which permitted more precise monitoring of the fate of labeled proteins. In this labeling technique, cells are exposed to radiolabeled precursor, in this case [$^3$H]-leucine, for a short period known as the *pulse*. The radioactive precursor is then replaced with its nonlabeled form for a subsequent *chase* period. Proteins synthesized during the pulse period will be labeled and detected by autoradiography, whereas those synthesized during the chase period, which are nonlabeled, will not be detected. Palade began by cutting guinea pig pancreas into thick slices, which were then incubated for 3 minutes in media containing [$^3$H]-leucine. At the end of the pulse, he added excess unlabeled leucine. The tissue slices were then either fixed for autoradiography or used for cell fractionation. To ensure that his results were an accurate reflection of protein secretion in vivo, Palade meticulously characterized the system. Once convinced that his in vitro system accurately mimicked protein secretion in vivo, he proceeded to the critical experiment. He pulse-labeled tissue slices with [$^3$H]-leucine for 3 minutes, then chased the label for 7, 17, 37, 57, and 117 minutes with unlabeled leucine. Radioactivity, again confined in vesicles, began at the ER, then traveled in vesicles to the Golgi complex and remained in the vesicles as they passed through the Golgi and onto the plasma membrane (see Figure 1). As the vesicles traveled farther along the pathway, they became more densely packed with radioactive protein. From his remarkable series of autoradiograms at different chase times, Palade concluded that secreted proteins travel in vesicles from the ER to the Golgi and onto the plasma membrane and that throughout this process, they remain in vesicles and do not mix with the rest of the cell.

**FIGURE 1** **The synthesis and movement of guinea pig pancreatic secretory proteins as revealed by electron microscope autoradiography.** After a period of labeling with [³H]-leucine, the tissue is fixed, sectioned for electron microscopy, and subjected to autoradiography. The radioactive decay of [³H] in newly synthesized proteins produces autoradiographic grains in an emulsion placed over the cell section (which appear in the micrograph as dense, wormlike granules) that mark the position of newly synthesized proteins. (a) At the end of a 3-minute labeling period autoradiographic grains are over the rough ER. (b) Following a 7-minute chase period with unlabeled leucine, most of the labeled proteins have moved to the Golgi vesicles. (c) After a 37-minute chase, most of the proteins are over immature secretory vesicles. (d) After a 117-minute chase, the majority of the proteins are over mature zymogen granules. [Courtesy of J. Jamieson and G. Palade.]

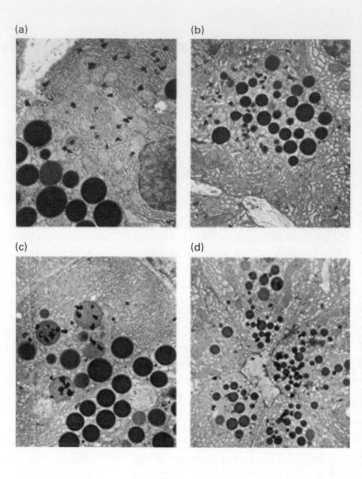

## Discussion

Palade's experiments gave biologists the first clear look at the stages of the secretory pathway. His studies on pancreatic exocrine cells yielded two fundamental observations. First, that secreted proteins pass through the Golgi complex on their way out of the cell. This was the first function assigned to the Golgi complex. Second, that secreted proteins never mix with cellular proteins in the cytosol; they are segregated into vesicles throughout the pathway. These findings were predicated from two important aspects of the experimental design. Palade's careful use of electron microscopy and autoradiography allowed him to look at the fine details of the pathway. Of equal importance was the choice of a cell type devoted to secretion, the pancreatic exocrine cell, as a model system. In a different cell type, significant amounts of nonsecreted proteins would have also been produced during the labeling, obscuring the fate of secretory proteins in particular.

Palade's work set the stage for more detailed studies. Once the secretory pathway was clearly described, entire fields of research were opened up to investigation of the synthesis and movement of both secreted and membrane proteins. For this groundbreaking work, Palade was awarded the Nobel Prize for Physiology and Medicine in 1974.

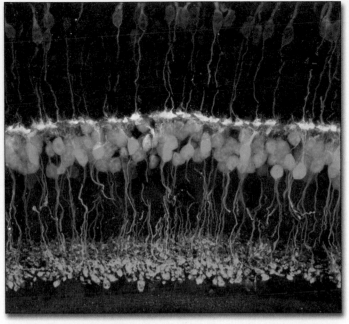

# SIGNAL TRANSDUCTION AND G PROTEIN–COUPLED RECEPTORS

The mouse retina contains photoreceptors (purple) that sense light using G protein–coupled receptors and four other types of neurons stained yellow, green, pink, and blue, which connect the photoreceptor cells to the brain. [Rachel Wong, University of Washington]

No cell lives in isolation. Cellular communication is a fundamental property of all cells and shapes the development and function of every living organism. Even single-celled eukaryotic microorganisms, such as yeasts, slime molds, and protozoans, communicate through extracellular signals: secreted molecules called **pheromones** coordinate the aggregation of free-living cells for sexual mating or differentiation under certain environmental conditions. More important in plants and animals are **hormones** and other extracellular signaling molecules that function *within* an organism to control a variety of processes, including the metabolism of sugars, fats, and amino acids; the growth and differentiation of tissues; the synthesis and secretion of proteins; and the composition of intracellular and extracellular fluids. Many types of cells also respond to signals from the external environment, including light, oxygen, odorants, and tastants in food.

In any system, for a signal to have an effect on a target, it has to be received. In cells, a signal produces a specific response only in target cells with **receptors** for that signal. For some receptors, this signal is a physical stimulus such as light, touch, or heat. For others, it is a chemical molecule. Many types of chemicals are used as signals: small molecules (e.g., amino acid or lipid derivatives, acetylcholine), gases (nitric oxide), peptides (e.g., ACTH and vasopressin), soluble proteins (e.g., insulin and growth hormone), and proteins that are tethered to the surface of a cell or bound to the extracellular matrix. Many of these extracellular signaling molecules are synthesized and released by specialized signaling cells within multicellular organisms. Most receptors bind a single molecule or a group of closely related molecules.

Some signaling molecules, especially hydrophobic molecules such as steroids, retinoids, and thyroxine, spontaneously diffuse through the plasma membrane and bind to intracellular receptors; signaling from such intracellular receptors is discussed in detail in Chapter 7.

Most extracellular signaling molecules, however, are too large and too hydrophilic to penetrate through the plasma membrane. How, then, can they affect intracellular processes?

## OUTLINE

**FIGURE 15-1 Overview of signaling by cell-surface receptors.** Communication by extracellular signals usually involves the following steps: synthesis of the signaling molecule by the signaling cell and its incorporation into small intracellular vesicles (step **1**), its release into the extracellular space by exocytosis (step **2**), and transport of the signal to the target cell (step **3**). Binding of the signaling molecule to a specific cell-surface receptor protein triggers a conformational change in the receptor, thus activating it (step **4**). The activated receptor then activates one or more downstream signal transduction proteins or small-molecule second messengers (step **5**), eventually leading to activation of one or more effector proteins (step **6**). The end result of a signaling cascade can be either a short-term change in cellular function, metabolism, or movement (step **7a**) or a long-term change in gene expression or development (step **7b**). Termination or down-modulation of the cellular response is caused by negative feedback from intracellular signaling molecules (step **8**) and by removal of the extracellular signal (step **9**).

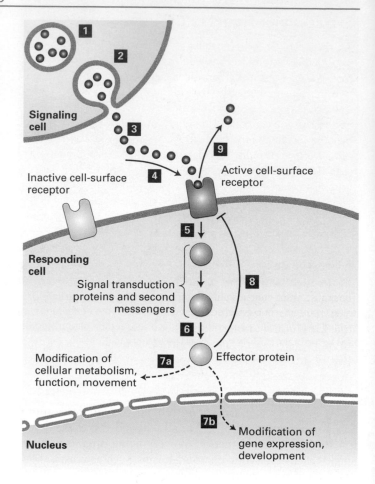

These signaling molecules bind to cell-surface receptors that are integral membrane proteins embedded in the plasma membrane. Cell-surface receptors generally consist of three discrete domains, or segments: an extracellular domain facing the extracellular fluid, a membrane-spanning (transmembrane) domain that spans the plasma membrane, and the intracellular domain segment facing the cytosol. The signaling molecule acts as a **ligand**, which binds to a structurally complementary site on the extracellular or the membrane-spanning domains of the receptor. Binding of the ligand to its site on the receptor induces a conformational change in the receptor that is transmitted through the membrane-spanning domain to the cytosolic domain, resulting in binding to and subsequent activation or inhibition of other proteins in the cytosol or attached to the plasma membrane. In many cases, these activated proteins catalyze the synthesis of certain small molecules or change the concentration of an intracellular ion such as $Ca^{2+}$. These intracellular proteins or small molecule **second messengers** then carry the signal to one or more effector proteins. The overall process of converting extracellular signals into intracellular responses, as well as the individual steps in this process, is termed **signal transduction** (Figure 15-1).

In eukaryotes, there are about a dozen classes of cell-surface receptors, which activate several types of intracellular signal transduction pathways. Our knowledge of signal transduction has advanced greatly in recent years, in large measure because these receptors and pathways are highly conserved and function in essentially the same way in organisms as diverse as worms, flies, mice, and humans. Genetic studies combined with biochemical analyses have enabled researchers to trace many entire signaling pathways from binding of ligand to final cellular response.

Perhaps the most numerous class of receptors—found in organisms from yeasts to humans—are **G protein–coupled receptors (GPCRs)**. As their name implies, G protein–coupled receptors consist of an integral membrane receptor protein coupled to an intracellular G protein that transmits signals to the interior of the cell. The human genome encodes about 900 G protein–coupled receptors, including receptors in the visual, olfactory (smell), and gustatory (taste) systems, many neurotransmitter receptors, and most of the receptors for hormones that control carbohydrate, amino acid, and fat metabolism and even behavior. Signal transduction through GPCRs usually induces short-term changes in cell function, such as a change in metabolism or movement. In contrast, activation of other cell-surface receptors primarily alters a cell's pattern of gene expression, leading to cell differentiation or division and other long-term consequences. These latter receptors and the intracellular signaling pathways they activate are explored in Chapter 16.

In this chapter, we first review some general principles of signal transduction, such as the molecular basis for ligand-receptor binding, and certain evolutionarily conserved components of signal transduction pathways. Next we describe how cell-surface receptors and signal transduction proteins are identified and characterized biochemically. We then turn to an in-depth discussion of G protein–coupled receptors, focusing first on their structure and mechanism of action and then on the signaling pathways activated by them. We show how these pathways affect many aspects of cell function, including glucose metabolism, muscle contraction, perception of light, and gene expression.

## 15.1 Signal Transduction: From Extracellular Signal to Cellular Response

As shown in Figure 15-1, signal transduction begins when extracellular signaling molecules bind to cell-surface receptors. Binding of signaling molecules to their receptors induces two major types of cellular responses: (1) changes in the activity or function of specific enzymes and other proteins that preexist in the cell and (2) changes in the amounts of specific proteins produced by a cell, most commonly by modification of **transcription factors** that stimulate or repress gene expression (see Figure 15-1, steps **7a** and **7b**). In general, the first type of response occurs more rapidly than the second. Transcription factors activated in the cytosol by these pathways move into the nucleus, where they stimulate (or occasionally repress) transcription of specific target genes.

The connection between an activated receptor and a cellular response is not direct and generally involves several intermediate proteins or small molecules. Collectively, this chain of intermediates is called a *signal transduction pathway* because it transduces, or converts, information from one form into another as a signal is relayed from a receptor to its targets. Some signal transduction pathways contain just two or three intermediates; others can involve over a dozen. Regardless, most pathways contain members of certain classes of signal transduction proteins that have been highly conserved throughout evolution.

In this section, we provide an overview of the major steps in signal transduction, starting with the signaling molecules themselves. We explore the molecular basis for ligand-receptor binding and the chain of events initiated in the target cell by binding of the signal to its receptor, focusing on a few components that are central to many signal transduction pathways.

### Signaling Molecules Can Act Locally or at a Distance

Cells respond to many different types of signals—some originating from outside the organism, some internally generated. Those that are generated internally can be described by how they reach their target. Some signaling molecules are transported long distances by the blood; others have more local effects. In animals, signaling by extracellular molecules can be classified into three types—endocrine, paracrine, or autocrine—based on the distance over which the signal acts (Figure 15-2a–c). In addition, certain membrane-bound proteins on one cell can directly signal an adjacent cell.

In **endocrine** signaling, the signaling molecules are synthesized and secreted by signaling cells (for example, those found in endocrine glands), transported through the circulatory system of the organism, and finally act on target cells distant from their site of synthesis. The term *hormone* generally refers to signaling molecules that mediate endocrine signaling. Insulin secreted by the pancreas and epinephrine secreted by the adrenal glands are examples of hormones that travel through the blood and thus mediate endocrine signaling.

In **paracrine** signaling, the signaling molecules released by a cell affect only those target cells in close proximity. A nerve cell releasing a neurotransmitter (e.g., acetylcholine) that acts on an adjacent nerve cell or on a muscle cell (inducing or inhibiting muscle contraction) is an example of paracrine signaling. In addition to neurotransmitters, many protein **growth factors** regulating development in multicellular organisms act at short range. Some of these growth-factor proteins bind tightly to components of the extracellular matrix

(a) Endocrine signaling

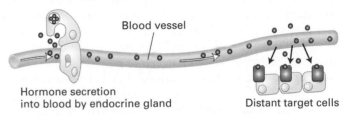

Blood vessel

Hormone secretion into blood by endocrine gland

Distant target cells

(b) Paracrine signaling

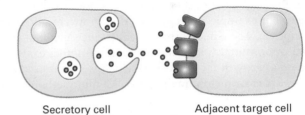

Secretory cell

Adjacent target cell

(c) Autocrine signaling

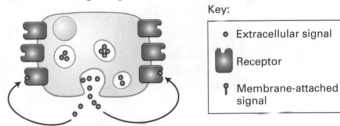

Target sites on same cell

Key:
- Extracellular signal
- Receptor
- Membrane-attached signal

(d) Signaling by plasma-membrane-attached proteins

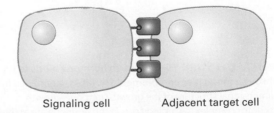

Signaling cell

Adjacent target cell

**FIGURE 15-2 Types of extracellular signaling.** (a–c) Cell-to-cell signaling by extracellular chemicals occurs over distances from a few micrometers in autocrine and paracrine signaling to several meters in endocrine signaling. (d) Proteins attached to the plasma membrane of one cell can interact directly with cell-surface receptors on adjacent cells.

and are unable to signal to adjacent cells; subsequent degradation of these matrix components, triggered by injury or infection, will release the active growth factor and enable it to signal. Many developmentally important signaling proteins diffuse away from the signaling cell, forming a concentration gradient and inducing different cellular responses depending on the concentration of the signaling protein.

In **autocrine** signaling, cells respond to substances that they themselves release. Some growth factors act in this fashion, and cultured cells often secrete growth factors that stimulate their own growth and proliferation. This type of signaling is particularly characteristic of tumor cells, many of which overproduce and release growth factors that stimulate inappropriate, unregulated self-proliferation, a process that may lead to formation of a tumor.

Integral membrane proteins located on the cell surface also play an important role in signaling (Figure 15-2d). In some cases, such membrane-bound signals on one cell bind receptors on the surface of an adjacent target cell, triggering its differentiation. In other cases, proteolytic cleavage of a membrane-bound signaling protein releases the extracellular segment, which functions as a soluble signaling molecule.

Some signaling molecules can act at both short and long ranges. For example, epinephrine (also known as adrenaline) functions as a systemic hormone (endocrine signaling) and as a neurotransmitter (paracrine signaling). Another example is epidermal growth factor (EGF), which is synthesized as an integral plasma membrane protein. Membrane-bound EGF can bind to receptors on an adjacent cell. In addition, cleavage by an extracellular protease releases a soluble form of EGF, which can signal in either an autocrine or a paracrine manner.

## Binding of Signaling Molecules Activates Receptors on Target Cells

Receptor proteins for all hydrophilic extracellular small molecules and protein signaling molecules are located on the surface of the target cell. The signaling molecule, or ligand, binds to a site on the extracellular domain of the receptor with high specificity and affinity. Each receptor generally binds only a single signaling molecule or a group of structurally very closely related molecules. The *binding specificity* of a receptor refers to its ability to bind or not bind closely related substances.

Ligand binding depends on weak, multiple noncovalent forces (i.e., ionic, van der Waals, and hydrophobic interactions) and **molecular complementarity** between the interacting surfaces of a receptor and ligand (see Figure 2-12). For example, the growth hormone receptor (Figure 15-3) binds to growth hormone but not to other hormones with very similar, though not identical, structures. Similarly, acetylcholine receptors bind only this small molecule and not others that differ only slightly in chemical structure, while the insulin receptor binds insulin and related hormones called insulin-like growth factors 1 and 2 (IGF-1 and IGF-2), but no other hormones.

Binding of ligand to receptor causes a conformational change in the receptor that initiates a sequence of reactions

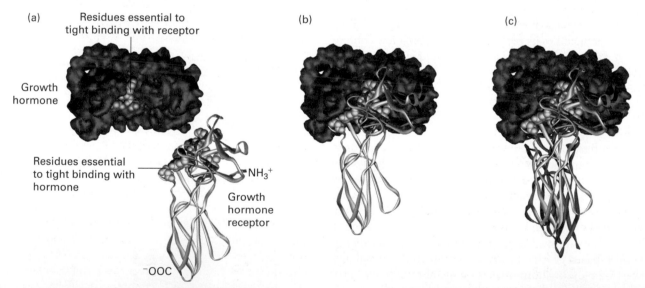

**EXPERIMENTAL FIGURE 15-3 Growth hormone binds to its receptor through molecular complementary.** (a) As determined from the three-dimensional structure of the growth hormone–growth hormone receptor complex, 28 amino acids in the hormone are at the binding interface with one receptor. To determine which amino acids are important in ligand-receptor binding, researchers mutated each of these amino acids one at a time, to alanine, and measured the effect on receptor binding. From this study, it was found that only eight amino acids on growth hormone (pink) contribute 85 percent of the energy that is responsible for tight receptor binding; these amino acids are distant from each other in the primary sequence but adjacent in the folded protein. Similar studies showed that two tryptophan residues (blue) in the receptor contribute most of the energy responsible for tight binding of growth hormone, although other amino acids at the interface with the hormone (yellow) are also important. (b) Binding of growth hormone to one receptor molecule is followed by (c) binding of a second receptor (purple) to the opposing side of the hormone; this involves the same set of yellow and blue amino acids on the receptor but different residues on the hormone. As we see in the next chapter, such hormone-induced receptor dimerization is a common mechanism for activation of receptors for protein hormones. [After B. Cunningham and J. Wells, 1993, *J. Mol. Biol.* **234**:554, and T. Clackson and J. Wells, 1995, *Science* **267**:383.]

leading to a specific response inside the cell. Organisms have evolved to be able to use a single ligand to stimulate different cells to respond in distinct ways. For example, different cell types may have different sets of receptors for the same ligand, each of which induces a different intracellular signal response pathway. Alternatively, the same receptor can be found on various cell types in an organism, but binding of a particular ligand to the receptor triggers a different response in each type of cell, given the unique complement of proteins expressed by the cell. In these ways, the same ligand can induce different cells to respond in a variety of ways. This is what's known as the *effector specificity* of the receptor-ligand complex.

For instance, the surfaces of skeletal muscle cells, heart muscle cells, and the pancreatic acinar cells that produce hydrolytic digestive enzymes each have different types of receptors for acetylcholine. In a skeletal muscle cell, release of acetylcholine from a motor neuron innervating the cell triggers muscle contraction by activating an acetylcholine-gated ion channel. In heart muscle, the release of acetylcholine by certain neurons activates a G protein–coupled receptor and slows the rate of contraction and thus the heart rate. Acetylcholine stimulation of pancreatic acinar cells triggers a rise in cytosolic $[Ca^{2+}]$ that induces exocytosis of the digestive enzymes stored in secretory granules to facilitate digestion of a meal. Thus formation of different acetylcholine–receptor complexes in different cell types leads to different cellular responses.

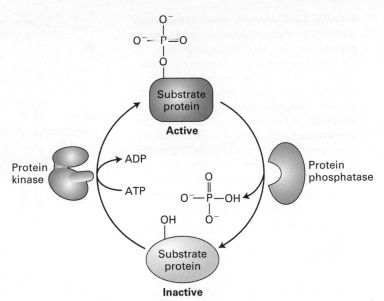

**FIGURE 15-4 Regulation of protein activity by a kinase/phosphatase switch.** The cyclic phosphorylation and dephosphorylation of a protein is a common cellular mechanism for regulating protein activity. In this example, the target, or substrate, protein is inactive (light green) when not phosphorylated and active (dark green) when phosphorylated; some proteins have the opposite pattern. Both the protein kinase and the phosphatase act only on specific target proteins, and their activities are usually highly regulated.

## Protein Kinases and Phosphatases Are Employed in Virtually All Signaling Pathways

Activation of virtually all cell-surface receptors leads directly or indirectly to changes in protein phosphorylation through the activation of protein **kinases**, which add phosphate groups to specific residues of specific target proteins. Some receptors activate protein **phosphatases**, which remove phosphate groups from specific residues on target proteins. Phosphatases act in concert with kinases to switch the function of various proteins on or off (Figure 15-4).

At last count, the human genome encodes about 600 protein kinases and 100 different phosphatases. In general, each protein kinase phosphorylates specific amino acid residues in a set of target, or substrate, proteins whose patterns of expression generally differ in different cell types. Animal cells contain two types of protein kinases: those that add phosphate to the hydroxyl group on tyrosine residues and those that add phosphate to the hydroxyl group on serine or threonine (or both) residues. All kinases also bind to specific amino acid sequences surrounding the phosphorylated residue, and thus one can look at amino acid sequences surrounding tyrosine, serine, and threonine residues in a protein and make a good guess as to which kinases might phosphorylate this residue.

In some signaling pathways, the receptor itself possesses intrinsic kinase activity or the receptor is tightly bound to a cytosolic kinase. Figure 15-5 illustrates a simple signal transduction pathway involving one kinase tightly bound to a receptor and one predominant target protein. In the absence of a bound ligand the kinase is held in the inactive state. Ligand binding triggers a conformational change in the receptor,

leading to activation of the appended kinase. The kinase then phosphorylates the monomeric, inactive form of a specific transcription factor, leading to its dimerization and movement from the cytosol into the nucleus, where it activates transcription of target genes. A phosphatase in the nucleus subsequently removes the phosphate group from the transcription factor, causing it to form two inactive monomers and then move back into the cytosol, where it can be reactivated by a receptor-associated kinase.

As this example illustrates, the activity of all protein kinases is opposed by the activity of protein phosphatases, some of which are themselves regulated by extracellular signals. Thus the activity of a protein in a cell can be a complex function of the activities of the usually multiple kinases and phosphatases that act on it, either directly or indirectly through phosphorylation of another protein. Several examples of this phenomenon occur in regulation of the cell cycle and are described in Chapter 19.

Many proteins are substrates for multiple kinases, each of which phosphorylates different amino acids. Each phosphorylation event can modify the activity of a particular target protein in different ways, some activating its function, others inhibiting it. An example we encounter later is glycogen phosphorylase kinase, a key regulatory enzyme in glucose metabolism. In many cases, addition of a phosphate group to an amino acid creates a binding surface that allows a second protein to bind; in the following chapter we will encounter many examples of such kinase-driven assembly of multiprotein complexes.

Commonly the catalytic activity of a protein kinase itself is modulated by phosphorylation by other kinases, by the

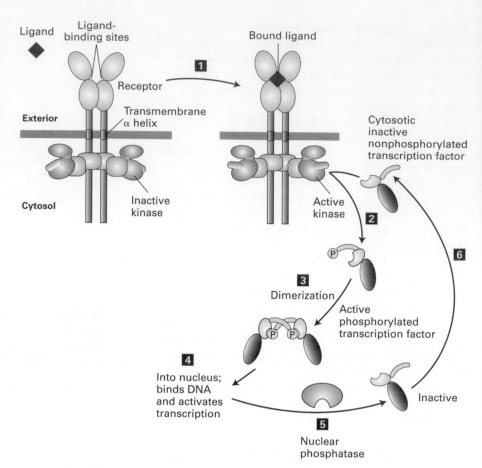

**FIGURE 15-5 A simple signal transduction pathway involving one kinase and one target protein.** The receptor is tightly bound to a protein kinase that, in the absence of a bound ligand, is held in the inactive state. Ligand binding triggers a conformational change in the receptor, leading to activation of the appended kinase (**1**). The kinase then phosphorylates the monomeric, inactive form of a specific transcription factor (**2**), leading to its dimerization (**3**) and movement from the cytosol into the nucleus (**4**), where it activates transcription of target genes. A phosphatase in the nucleus will remove the phosphate group from the transcription factor (**5**), causing it to form the inactive monomer and then move back into the cytosol (**6**).

binding of other proteins to it, and by changes in the levels of various small intracellular signaling molecules and metabolites. The resulting cascades of kinase activity are a common feature of many signaling pathways.

## GTP-Binding Proteins Are Frequently Used in Signal Transduction as On/Off Switches

Many signal transduction pathways utilize intracellular "switch" proteins that turn downstream proteins on or off. The most important group of intracellular switch proteins is the **GTPase superfamily**. All the GTPase switch proteins exist in two forms (Figure 15-6): (1) an active ("on") form with bound GTP (guanosine triphosphate) that modulates the activity of specific target proteins and (2) an inactive ("off") form with bound GDP (guanosine diphosphate).

Conversion of the inactive to active state is triggered by a signal (e.g., a hormone binding to a receptor) and is mediated by a *guanine nucleotide exchange factor* (*GEF*), which causes release of GDP from the switch protein. Subsequent binding of GTP, favored by its high intracellular concentration relative to its binding affinity, induces a conformational change to the active form. The principal conformational changes involve two highly conserved segments of the protein, termed switch I and switch II, that allow the protein to bind to and activate other downstream signaling proteins (Figure 15-7). Conversion of the active form back to the inactive state is

mediated by a GTPase, which slowly hydrolyzes the bound GTP to GDP and $P_i$, thus altering the conformation of the switch I and switch II segments so that they are unable to bind to the effector protein. The GTPase can be an intrinsic part of the G protein or a separate protein.

The rate of GTP hydrolysis regulates the length of time the switch protein remains in the active conformation and is

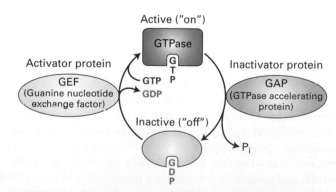

**FIGURE 15-6 GTPase switch proteins cycle between active and inactive forms.** The switch protein is active when it has bound GTP and inactive when it has bound GDP. Conversion of the active into the inactive form by hydrolysis of the bound GTP is accelerated by GAPs (GTPase-accelerating proteins) and other proteins. Reactivation is promoted by GEFs (guanine nucleotide exchange factors) that catalyze the dissociation of the bound GDP and its replacement by GTP.

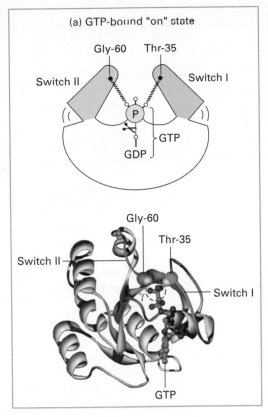

(a) GTP-bound "on" state

Switch II — Gly-60   Thr-35 — Switch I

GDP
GTP

Switch II — Gly-60
Thr-35
Switch I

GTP

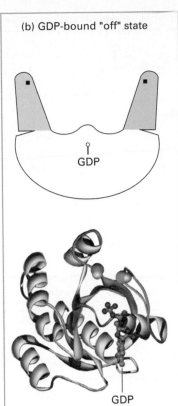

(b) GDP-bound "off" state

GDP

GDP

**FIGURE 15-7 Switching mechanism of G proteins.** The ability of a G protein to interact with other proteins and thus transduce a signal differs in the GTP-bound "on" state and GDP-bound "off" state. (a) In the active "on" state, two domains, termed switch I (green) and switch II (blue), are bound to the terminal gamma phosphate of GTP through interactions with the backbone amide groups of a conserved threonine and glycine residue. When bound to GTP in this way, the two switch domains are in a conformation such that they can bind to and thus activate specific downstream effector proteins. (b) Release of the gamma phosphate by GTPase-catalyzed hydrolysis causes switch I and switch II to relax into a different conformation, the inactive "off" state; in this state they are unable to bind to effector proteins. The ribbon models shown here represent both conformations of Ras, a monomeric G protein. A similar spring-loaded mechanism switches the alpha subunit in trimeric G proteins between the active and inactive conformations by movement of three switch segments. [Adapted from I. Vetter and A. Wittinghofer, 2001, *Science* **294**:1299.]

able to signal its downstream target proteins: the slower the rate of GTP hydrolysis, the longer the protein remains in the active state. The rate of GTP hydrolysis is often modulated by other proteins. For instance, both *GTPase-activating proteins (GAP)* and *regulator of G protein signaling (RGS) proteins* accelerate GTP hydrolysis. Many regulators of G protein activity are themselves controlled by extracellular signals.

Two large classes of GTPase switch proteins are used in signaling. **Trimeric (large) G proteins** directly bind to and are activated by certain cell-surface receptors. As we will see in Section 15.3, G protein–coupled receptors function as guanine nucleotide–exchange factors (GEFs)—triggering release of GDP and binding of GTP, thus activating the G protein. **Monomeric (small) G proteins**, such as Ras and various Ras-like proteins, are not bound to receptors but play crucial roles in many pathways that regulate cell division and cell motility, as is evidenced by the fact that mutations in genes encoding these G proteins frequently lead to cancer. Other members of both GTPase classes, by switching between GTP-bound "on" and GDP-bound "off" forms, function in protein synthesis, the transport of proteins between the nucleus and the cytoplasm, the formation of coated vesicles and their fusion with target membranes, and rearrangements of the actin cytoskeleton.

## Intracellular "Second Messengers" Transmit and Amplify Signals from Many Receptors

The binding of ligands ("first messengers") to many cell-surface receptors leads to a short-lived increase (or decrease) in the concentration of certain low-molecular-weight intracellular signaling molecules termed **second messengers**. These, in turn, bind to other proteins, modifying their activity.

One second messenger used in virtually all metazoan cells is $Ca^{2+}$ ions. We noted in Chapter 11 that the concentration of free $Ca^{2+}$ in the cytosol is kept very low ($<10^{-7}$ M) by ATP-powered pumps that continually transport $Ca^{2+}$ out of the cell or into the endoplasmic reticulum (ER). The cytosolic $Ca^{2+}$ level can increase from 10- to 100-fold by a signal-induced release of $Ca^{2+}$ from ER stores or by its import through calcium channels from the extracellular environment; this change can be detected by fluorescent dyes introduced into the cell (see Figure 9-11). In muscle, a signal-induced rise in cytosolic $Ca^{2+}$ triggers contraction (see Figure 17-35). In endocrine cells, a similar increase in $Ca^{2+}$ induces exocytosis of secretory vesicles containing hormones, which are thus released into the circulation. In nerve cells, an increase in cytosolic $Ca^{2+}$ leads to the exocytosis of neurotransmitter-containing vesicles (see Chapter 22). In all cells, this rise in cytosolic $Ca^{2+}$ is sensed by $Ca^{2+}$-binding proteins, particularly those of the *EF hand family*, such as *calmodulin*, all of which contain the helix-loop-helix motif (see Figure 3-9b). The binding of $Ca^{2+}$ to calmodulin and other EF hand proteins causes a conformational change that permits the protein to bind various target proteins, thereby switching their activities on or off (see Figure 3-31).

Another nearly universal second messenger is **cyclic AMP (cAMP)**. In many eukaryotic cells, a rise in cAMP triggers activation of a particular protein kinase, protein kinase A, that in turn phosphorylates specific target proteins to induce specific changes in cell metabolism. In some cells, cAMP

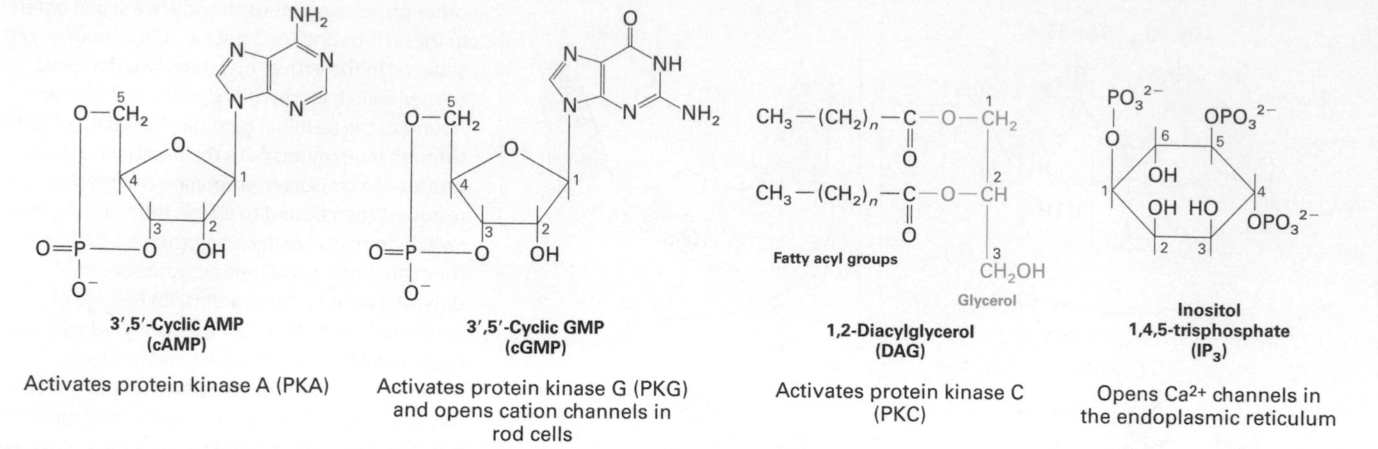

**FIGURE 15-8 Four common intracellular second messengers.** The major direct effect or effects of each compound are indicated below its structural formula. Calcium ions ($Ca^{2+}$) and several membrane-bound phosphatidylinositol derivatives also act as second messengers.

3′,5′-Cyclic AMP (cAMP) — Activates protein kinase A (PKA)

3′,5′-Cyclic GMP (cGMP) — Activates protein kinase G (PKG) and opens cation channels in rod cells

1,2-Diacylglycerol (DAG) — Activates protein kinase C (PKC)

Inositol 1,4,5-trisphosphate ($IP_3$) — Opens $Ca^{2+}$ channels in the endoplasmic reticulum

regulates the activity of certain ion channels. The structures of cAMP and three other common second messengers are shown in Figure 15-8. Later in this chapter, we examine the specific roles of second messengers in signaling pathways activated by various G protein–coupled receptors.

Because second messengers such as $Ca^{2+}$ and cAMP diffuse through the cytosol much faster than do proteins, they are employed in pathways where the downstream target is located in an intracellular organelle (such as a secretory vesicle or the nucleus) distant from the plasma membrane receptor where the messenger is generated.

Another advantage of second messengers is that they facilitate *amplification* of an extracellular signal. Activation of a *single* cell-surface receptor molecule can result in an increase in perhaps thousands of cAMP molecules or $Ca^{2+}$ ions in the cytosol. Each of these, in turn, by activating its target protein affects the activity of multiple downstream proteins. In many signal transduction pathways, amplification is necessary because cell surface receptors are typically low-abundance proteins, present in only a thousand or so copies per cell. Yet the cellular responses induced by the binding of a relatively small number of hormones to the available receptors often require production of tens or hundreds of thousands of activated effector molecules per cell. In the case of G protein–coupled hormone receptors, *signal amplification* is possible in part because a single receptor can activate multiple G proteins, each of which in turn activates an effector protein. For example, a single epinephrine-GPCR complex causes activation of up to 100 adenylyl cyclase molecules, each of which in turn catalyzes synthesis of many cAMP molecules during the time it remains in the active state. Two cAMP molecules activate one molecule of protein kinase A that in turn phosphorylates and activates multiple target product molecules (Figure 15-9). Later in this chapter, we see how this amplification cascade allows blood levels of epinephrine as low as $10^{-10}$ M to stimulate glycogenolysis (conversion of glycogen to glucose) by the liver and release of glucose into the blood.

**FIGURE 15-9 Amplification of an extracellular signal.** In this example, binding of a single epinephrine molecule to one G protein–coupled receptor molecule induces activation of several molecules of adenylyl cyclase, the enzyme that catalyzes the synthesis of cyclic AMP, and each of these enzyme synthesizes a large number of cAMP molecules, the first level of amplification. Two molecules of cAMP activate one molecule of protein kinase A (PKA), but each activated PKA phosphorylates and activates multiple target proteins. This second level of amplification may involve several sequential reactions in which the product of one reaction activates the enzyme catalyzing the next reaction. The more steps in such a cascade, the greater the signal amplification possible.

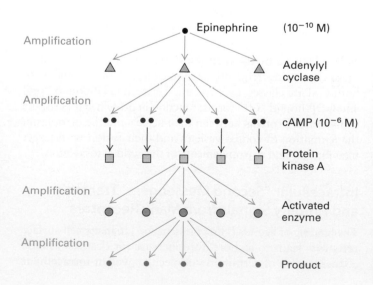

## KEY CONCEPTS of Section 15.1

### Signal Transduction: From Extracellular Signal to Cellular Response

- All cells communicate through extracellular signals. In unicellular organisms, extracellular signaling molecules regulate interactions between individuals, while in multicellular organisms, they regulate physiology and development.

- External signals include membrane-anchored and secreted proteins or peptides (e.g., vasopressin and insulin), small hydrophobic molecules (e.g., steroid hormones and thyroxine), small hydrophilic molecules (e.g., epinephrine), gases (e.g., $O_2$, nitric oxide), and physical stimuli (e.g., light).

- Binding of extracellular signaling molecules to cell-surface receptors triggers a conformational change in the receptor, which in turn leads to activation of intracellular signal transduction pathways that ultimately modulate cellular metabolism, function, or gene expression (see Figure 15-1).

- Signals from one cell act on distant cells in endocrine signaling, on nearby cells in paracrine signaling, or on the signaling cell itself in autocrine signaling (see Figure 15-2).

- Protein phosphorylation and de-phosphorylation, catalyzed by protein kinases and phosphatases, are employed in virtually all signaling pathways. The activities of kinases and phosphatases are highly regulated by many receptors and signal transduction proteins (see Figures 15-4 and 15-5).

- GTP-binding proteins of the GTPase superfamily act as switches regulating many signal transduction pathways (see Figures 15-6 and 15-7).

- $Ca^{2+}$, cAMP, and other nonprotein, low-molecular-weight intracellular molecules (see Figure 15-8) act as "second messengers," relaying and often amplifying the signal of the "first messenger," that is, the ligand. Binding of ligand to cell-surface receptors often results in a rapid increase (or, occasionally, decrease) in the intracellular concentration of these ions or molecules.

## 15.2 Studying Cell-Surface Receptors and Signal Transduction Proteins

The response of a cell to an external signal depends on the cell's complement of receptors that recognize the signal and the signal transduction pathways activated by those receptors. In this section, we explore the biochemical basis for the specificity of receptor-ligand binding, as well as the ability of different concentrations of ligand to activate a pathway. We also examine experimental techniques used to characterize receptor proteins. Many of these methods are also applicable to receptors that mediate endocytosis (see Chapter 14) or cell adhesion (see Chapter 20). We conclude the section with a discussion of techniques commonly used to measure the

activity of signal transduction components, such as kinases and GTP-binding "switch" proteins.

### The Dissociation Constant Is a Measure of the Affinity of a Receptor for Its Ligand

Ligand binding to a receptor usually can be viewed as a simple reversible reaction, where the receptor is represented as R, the ligand as L, and the receptor-ligand complex as RL:

$$R + L \underset{k_{on}}{\overset{k_{off}}{\rightleftharpoons}} RL \qquad (15\text{-}1)$$

$k_{off}$ is the rate constant for dissociation of a ligand from its receptor, and $k_{on}$ is the rate constant for formation of a receptor-ligand complex from free ligand and receptor.

At equilibrium, the rate of formation of the receptor-ligand complex is equal to the rate of its dissociation and can be described by the simple equilibrium-binding equation

$$K_d = \frac{[R][L]}{[RL]} \qquad (15\text{-}2)$$

where [R] and [L] are the concentrations of free receptor (that is, receptor without bound ligand) and ligand, respectively, at equilibrium, and [RL] is the concentration of the receptor-ligand complex. $K_d$, the **dissociation constant**, is a measure of the *affinity* (or tightness of binding) of the receptor for its ligand (see also Chapter 2). For a simple binding reaction, $K_d = k_{off}/k_{on}$. The *lower* $k_{off}$ is relative to $k_{on}$, the more *stable* the RL complex—the tighter the binding—and thus the *lower* the value of $K_d$. Another way of seeing this key point is that $K_d$ equals the concentration of ligand at which half of the receptors have a ligand bound when the system is at equilibrium; at this ligand concentration [R] = [RL] and thus, from Equation 15-2, $K_d = [L]$. The lower the $K_d$, the lower the ligand concentration required to bind 50 percent of the cell-surface receptors. The $K_d$ for a binding reaction here is essentially equivalent to the Michaelis constant $K_m$, which reflects the affinity of an enzyme for its substrate (see Chapter 3). Like all equilibrium constants, however, the value of $K_d$ does not depend on the *absolute* values of $k_{off}$ and $k_{on}$, only on their ratio. In the next section, we learn how $K_d$ values are experimentally determined.

Hormone receptors are characterized by their high affinity and specificity for their ligands. Because of their high affinity and great specificity for their target hormone, the extracellular, ligand-binding domains of cell surface receptors can be converted into powerful drugs. Consider the hormone tumor necrosis factor alpha (TNFα), which is secreted by a number of immune system cells. TNFα induces inflammation by recruiting various immune cells to a site of injury or infection; abnormal levels of TNFα cause the excessive inflammation seen in patients with autoimmune diseases such as the blistering skin disease psoriasis or the joint disease rheumatoid arthritis. These diseases are being treated with a chimeric "fusion" protein, generated by recombinant DNA, that contains the extracellular domain of a TNFα receptor fused to the constant (Fc) region of a human immunoglobin (see Figures 3-19 and 23-8). The drug binds tightly to free TNFα and prevents it from binding to its cell-surface receptors and causing inflammation; the fused Fc domain causes the protein to be stable when injected into the body. ■

## Binding Assays Are Used to Detect Receptors and Determine Their Affinity and Specificity for Ligands

Usually receptors are detected and measured by their ability to bind radioactive or fluorescent ligands to intact cells or to cell fragments. Figure 15-10 illustrates such a *binding assay* for interaction of the red-cell-forming hormone erythropoietin (Epo) with Epo receptors that are expressed by recombinant DNA in a line of cultured cells. The amounts of radioactive Epo bound to its receptor on growing cells (vertical axis) were measured as a function of increasing concentration of $^{125}$I-labeled Epo added to the extracellular fluid (horizontal axis). Both the number of ligand-binding sites per cell and the $K_d$ value are easily determined from the specific binding curve (curve C). Assuming each receptor binds just one ligand molecule, the total number of ligand-binding sites on a cell equals the number of active receptors per cell. In the example shown in Figure 15-10, the value of $K_d$ is about $1.1 \times 10^{-10}$ M, or 0.1 nM. In other words, an Epo concentration of $1.1 \times 10^{-10}$ M in the extracellular fluid is required for 50 percent of a cell's Epo receptors to have a bound Epo.

Direct binding assays like the one in Figure 15-10 are feasible with receptors that have a high affinity for their ligands, such as the erythropoietin receptor and the insulin receptor on liver cells ($K_d = 1.4 \times 10^{-10}$ M). However, many ligands, such as epinephrine and other catecholamines, bind to their receptors with much lower affinity. If the $K_d$ for binding is greater than $\sim 1 \times 10^{-7}$ M, a case when the rate constant $k_{off}$ is relatively large compared to $k_{on}$, then it is likely that during the seconds to minutes required to measure the amount of bound ligand, some of the receptor-bound ligand will dissociate and thus the observed binding values will be systematically too low.

One way to measure relatively weak binding of a ligand to its receptor is in a *competition assay* with another ligand that binds to the same receptor with high affinity (low $K_d$ value). In this type of assay, increasing amounts of an unlabeled, low-affinity ligand (the competitor) are added to a cell sample with a constant amount of the radiolabeled, high-affinity ligand (Figure 15-11). Binding of unlabeled competitor to the receptor blocks binding of the radioactive ligand to the receptor. The concentration dependence of this competition can be used together with the $K_d$ value of the radioactive ligand to calculate the inhibitory constant, $K_i$, which is very close to the $K_d$ value for binding of the competitor to the receptor. It is possible to accurately measure the amount

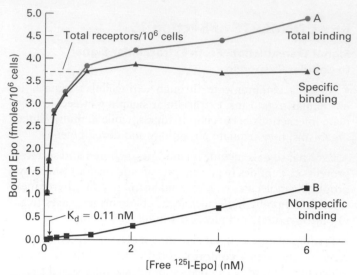

**EXPERIMENTAL FIGURE 15-10 Binding assays can determine the $K_d$ and the number of receptors per cell.** Shown here are data for erythropoietin-specific receptors on the surface of a cultured mouse cell line that expresses a recombinant human erythropoietin (Epo) receptor compared to control cells that do not normally express the receptor. A suspension of cells is incubated for 1 hour at 4 °C with increasing concentrations of $^{125}$I-labeled Epo; the low temperature is used to prevent endocytosis of the cell-surface receptors. The cells are separated from unbound $^{125}$I Epo, usually by centrifugation, and the amount of radioactivity bound to them is measured. The total binding curve A represents Epo specifically bound to high-affinity receptors as well as Epo nonspecifically bound with low affinity to other molecules on the cell surface. The contribution of nonspecific binding to total binding is determined by repeating the binding assay with the control cell line, where Epo binds only to nonspecific sites, yielding curve B. The specific binding curve C is calculated as the difference between curves A and B. As determined by the maximum of the specific binding curve C, the number of specific Epo-binding sites (surface receptors) per cell is about 2200 ($3.7 \times 10^{-15}$ moles $\times$ $6.02 \times 10^{23}$ molecules/mole/$10^6$ cells = 2227 molecules/cell). The $K_d$ is the concentration of Epo required to bind to 50 percent of the surface Epo receptors (in this case about 1050 receptors/cell). Thus the $K_d$ is about $1.1 \times 10^{-10}$ M, or 0.1 nM. [Courtesy Alec Gross; after A. Gross and H. Lodish, 2006, *J. Biol. Chem.* **281**:2024.]

of the high-affinity ligand bound in this assay because little dissociates during the experimental manipulations required for the measurement (relatively low $k_{off}$).

Competitive binding is often used to study synthetic analogs of natural hormones that activate or inhibit receptors. These analogs, which are widely used in research on cell-surface receptors and as drugs, fall into two classes: **agonists**, which mimic the function of a natural hormone by binding to its receptor and inducing the normal response, and **antagonists**, which bind to the receptor but induce no response. By occupying ligand-binding sites on a receptor, an antagonist can block binding of the natural hormone (or agonist) and thus reduce the usual physiological activity of the hormone. In other words, antagonists inhibit receptor signaling. ■

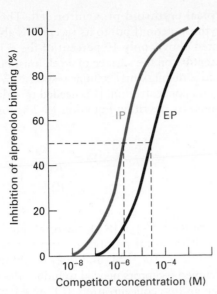

**EXPERIMENTAL FIGURE 15-11  For low-affinity ligands, binding can be detected in competition assays.** In this example, the synthetic ligand alprenolol, which binds with high affinity to the epinephrine receptor on liver cells ($K_d \sim 3 \times 10^{-9}$ M), is used to detect the binding of two low-affinity ligands, the natural hormone epinephrine (EP) and a synthetic ligand called isoproterenol (IP). Assays are performed as described in Figure 15-10 but in reactions containing a constant amount of [³H] alprenolol and increasing amounts of unlabeled epinephrine or isoproterenol. At each competitor concentration, the amount of bound labeled alprenolol is determined. In a plot of the inhibition of [³H] alprenolol binding versus epinephrine or isoproterenol concentration, such as shown here, the concentration of the competitor that inhibits alprenolol binding by 50 percent approximates the $K_d$ value for competitor binding. Note that the concentrations of competitors are plotted on a logarithmic scale. The $K_d$ for binding of epinephrine to its receptor on liver cells is only $\sim 5 \times 10^{-5}$ M and would not be measurable by a direct binding assay with [³H] epinephrine. The $K_d$ for binding of isoproterenol, which induces the normal cellular response, is more than tenfold lower.

Consider for instance the drug isoproterenol, used to treat asthma. Isoproterenol is made by the chemical addition of two methyl groups to epinephrine (see Figure 15-11, *right*). Isoproterenol, an agonist of the epinephrine-responsive G protein–coupled receptors on bronchial smooth muscle cells, binds about tenfold more strongly (tenfold lower $K_d$) than does epinephrine (see Figure 15-11, *left*). Because activation of these receptors promotes relaxation of bronchial smooth muscle and thus opening of the air passages in the lungs, isoproterenol is used in treating bronchial asthma, chronic bronchitis, and emphysema. In contrast, activation of a different type of epinephrine-responsive G protein–coupled receptors on cardiac muscle cells (called β-adrenergic receptors) increases the heart contraction rate. Antagonists of this receptor, such as alprenolol and related compounds, are referred to as *beta-blockers*; such antagonists are used to slow heart contractions in the treatment of cardiac arrhythmias and angina.

## Maximal Cellular Response to a Signaling Molecule Usually Does Not Require Activation of All Receptors

All signaling systems evolved such that a rise in the level of extracellular signaling molecules induces a proportional response in the responding cell. For this to happen, the binding affinity ($K_d$ value) of a cell-surface receptor for a signaling molecule must be greater than the normal (unstimulated) level of that molecule in the extracellular fluids or blood. We can see this principle in practice by comparing the levels of insulin present in the body and the $K_d$ for binding of insulin to its receptor on liver cells, $1.4 \times 10^{-10}$ M. Suppose, for instance, that the normal concentration of insulin in the blood is $5 \times 10^{-12}$ M. By substituting this value and the insulin $K_d$ into Equation 15-2, we can calculate the fraction of insulin receptors with bound insulin

$$[RL]/([RL] + [R])$$

at equilibrium as 0.0344; that is, about 3 percent of the total insulin receptors will be bound with insulin. If the insulin concentration rises fivefold to $2.5 \times 10^{-11}$ M, the number of receptor-hormone complexes will rise proportionately, almost fivefold, so that about 15 percent of the total receptors will have bound insulin. If the extent of the induced cellular response parallels the number of insulin-receptor complexes, [RL], as is often the case, then the cellular responses also will increase by about fivefold.

On the other hand, suppose that the normal concentration of insulin in the blood were the same as the $K_d$ value of $1.4 \times 10^{-10}$ M; in this case, 50 percent of the total receptors would have a bound insulin. A fivefold increase in the insulin concentration to $7 \times 10^{-10}$ M would result in 83 percent of all insulin receptors having insulin bound (a 66 percent increase). Thus, in order for a rise in hormone concentration

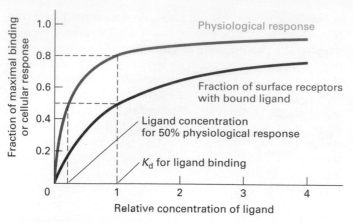

**EXPERIMENTAL FIGURE 15-12 The maximal physiological response to an external signal occurs when only a fraction of the receptors are occupied by ligand.** For signaling pathways that exhibit this behavior, plots of the extent of ligand binding to the receptor and of physiological response at different ligand concentrations differ. In the example shown here, 50 percent of the maximal physiological response is induced at a ligand concentration at which only 18 percent of the receptors are occupied. Likewise, 80 percent of the maximal response is induced when the ligand concentration equals the $K_d$ value, at which 50 percent of the receptors are occupied.

to cause a proportional increase in the fraction of receptors with bound ligand, the normal concentration of the hormone must be well below the $K_d$ value.

In general, the maximal cellular response to a particular ligand is induced when much less than 100 percent of its receptors are bound to the ligand. This phenomenon can be revealed by determining the extent of the response and of receptor-ligand binding at different concentrations of ligand (Figure 15-12). For example, a typical red blood (erythroid) progenitor cell has ~1000 surface receptors for erythropoietin, the protein hormone that induces these cells to proliferate and differentiate into red blood cells. Because only 100 of these receptors need to bind erythropoietin to induce division of a progenitor cell, the ligand concentration needed to induce 50 percent of the maximal cellular response is proportionally lower than the $K_d$ value for binding. In such cases, a plot of the percentage of maximal binding versus ligand concentration differs from a plot of the percentage of maximal cellular response versus ligand concentration.

## Sensitivity of a Cell to External Signals Is Determined by the Number of Surface Receptors and Their Affinity for Ligand

Because the cellular response to a particular signaling molecule depends on the *number* of receptor-ligand complexes, the fewer receptors present on the surface of a cell, the less *sensitive* the cell is to that ligand. As a consequence, a higher ligand concentration is necessary to induce the physiological response than would be the case if more receptors were present.

To illustrate the important relationship between receptor number and ligand sensitivity, let's extend our example of a

typical erythroid progenitor cell. The $K_d$ for binding of erythropoietin (Epo) to its receptor is about $10^{-10}$ M. As we noted above, only 10 percent of the ~1000 erythropoietin receptors on the surface of a cell must be bound to ligand to induce the maximal cellular response. We can determine the ligand concentration, [L], needed to induce the maximal response by rewriting Equation 15-2 as follows:

$$[L] = \frac{K_d}{\dfrac{R_T}{[RL]} - 1} \qquad (15-3)$$

where $R_T = [R] + [RL]$, the total number of receptors per cell. If the total number of Epo receptors per cell, $R_T$, is 1000, $K_d$ is $10^{-10}$ M, and [RL] is 100 (the number of Epo-occupied receptors needed to induce the maximal response), then an Epo concentration ([L]) of $1.1 \times 10^{-11}$ M will elicit the maximal response. If the total number of Epo receptors ($R_T$) is reduced to 200 per cell, then a ninefold-higher Epo concentration ($10^{-10}$ M) is required to occupy 100 receptors and induce the maximal response. Clearly, therefore, a cell's sensitivity to a signaling molecule is heavily influenced by the number of receptors for that ligand that are present as well as the $K_d$.

Epithelial growth factor (EGF), as its name implies, stimulates the proliferation of many types of epithelial cells, including those that line the ducts of the mammary gland. In about 25 percent of breast cancers, the tumor cells produce elevated levels of one particular EGF receptor called HER2. The overproduction of HER2 makes the cells hypersensitive to ambient levels of EGF that normally are too low to stimulate cell proliferation; as a consequence, growth of these tumor cells is inappropriately stimulated by EGF. We will see in Chapter 16 that an understanding of the role of HER2 in certain breast cancers led to development of monoclonal antibodies that bind HER2 and thereby block signaling by EGF; these antibodies have proved useful in treatment of these breast cancer patients. ∎

The HER2–breast cancer connection vividly demonstrates that regulation of the number of receptors for a given signaling molecule expressed by a cell plays a key role in directing physiological and developmental events. Such regulation can occur at the levels of transcription, translation, and post-translational processing or by controlling the rate of receptor degradation. Alternatively, endocytosis of receptors on the cell surface can sufficiently reduce the number present such that the cellular response is terminated. As we discuss in later sections, other mechanisms can reduce a receptor's affinity for ligand and so reduce the cell's response to a given concentration of ligand. Thus reduction in a cell's sensitivity to a particular ligand, called *desensitization*, can result from various mechanisms and is critical to the ability of cells to respond appropriately to external signals.

## Receptors Can Be Purified by Affinity Techniques

In order to fully understand how receptors function, it is necessary to purify them and analyze their biochemical properties. Determining their molecular structures with and without a bound ligand, for instance, can elucidate the conformational changes that occur on ligand binding that activate downstream signal transduction proteins. But this can be challenging. A "typical" mammalian cell has 1000 to 50,000 copies of a single type of cell-surface receptor. This may seem like a large number, but when you consider that this same cell contains $\sim 10^{10}$ total protein molecules and $\sim 10^6$ proteins in the plasma membrane alone, you realize that these receptors constitute only 0.1 to 5 percent of plasma-membrane proteins. This low abundance complicates the isolation and purification of cell-surface receptors. Purification of receptors is also difficult because these integral membrane proteins first must be solubilized from the membrane with a non-ionic detergent (see Figure 10-23) and then separated from other cellular proteins.

As we saw with the Epo receptor discussed earlier, recombinant DNA techniques can be used to generate cells that express large amounts of these proteins. But even when recombinant DNA techniques are used to generate cells that express receptors in large amounts, special techniques are necessary to isolate and purify them from other membrane proteins. One technique often used in purifying cell-surface receptors that retain their ligand-binding ability when solubilized by detergents is similar to *affinity chromatography* using antibodies (see Figure 3-38c). To purify a receptor by this technique, a ligand for the receptor of interest, rather than an antibody, is chemically linked to the beads used to form a column. A crude, detergent-solubilized preparation of membrane proteins is passed through the column; only the receptor binds, while other proteins are washed away. Passage of an excess of the soluble ligand through the column causes the bound receptor to be displaced from the beads and eluted from the column. In some cases, a receptor can be purified as much as 100,000-fold in a single affinity-chromatographic step.

## Immunoprecipitation Assays and Affinity Techniques Can Be Used to Study the Activity of Signal Transduction Proteins

Following ligand binding, receptors activate one or more signal transduction proteins that, in turn, can affect the activity of multiple effector proteins (see Figure 15-1); to understand a signaling cascade requires the researcher to be able to quantify the activity of these signal transduction proteins. Kinases and GTP-binding proteins are found in many signaling cascades, and in this section we describe several assays used for measuring their activities.

**Immunoprecipitation of Kinases** Kinases function in virtually all signaling pathways, and typical mammalian cells contain a hundred or more different kinases, each of which is highly regulated and can phosphorylate many target proteins. Immunoprecipitation assays are frequently used to measure the activity of a particular kinase in a cell extract. In one version of the method, an antibody specific for the desired kinase is first reacted with small beads coated with Protein A; this causes the antibody to bind to the beads via its Fc segment (see Figure 9-29). The beads are then mixed with a preparation of cell cytosol or nucleus, then recovered by centrifugation and washed extensively with a salt solution to remove weakly bound proteins that are unlikely to be binding specifically to the antibody. Thus only cell proteins that specifically bind to the antibody—the kinase itself and proteins tightly bound to the kinase—are present on the beads. The beads are then incubated in a buffered solution with a substrate protein and $\gamma$-[$^{32}$P] ATP, where only the $\gamma$ phosphate is labeled. The amount of [$^{32}$P] transferred to the substrate protein is a measure of kinase activity and can be quantified either by polyacrylamide gel electrophoresis followed by autoradiography (see Figure 3-36) or by immunoprecipitation with an antibody specific for the substrate followed by counting the radioactivity in the immunoprecipitate. By comparing extracts from cells before and after ligand addition, for example, one can readily determine whether or not a particular kinase is activated in the signal transduction pathway triggered by that ligand.

We noted that many proteins can be phosphorylated by several different kinases, usually on different serine, threonine, or tyrosine residues. Thus it is important to measure the extent of phosphorylation of a single amino acid side chain in a specific protein, say before and after hormone stimulation. Antibodies play a crucial role in detecting such phosphorylation events. To generate an antibody that can recognize a specific phosphorylated amino acid in a specific protein, one first chemically synthesizes an approximately 15 amino acid peptide that has the amino acid sequence surrounding the phosphorylated amino acid of the specific protein but where a phosphate group has been chemically linked to the desired serine, threonine, or tyrosine. After coupling this peptide to an adjuvant to increase its immunogenicity, it is used to generate a set of monoclonal antibodies (see Figure 9-6). One then selects a particular monoclonal antibody that reacts only with the phosphorylated, but not the nonphosphorylated peptide; such an antibody generally will bind to the parent protein only when this specific amino acid is phosphorylated. This specificity is possible because the antibody binds simultaneously to the phosphorylated amino acid and to side chains of adjacent amino acids. As an example of the use of such antibodies, Figure 15-13 shows that three signal transduction proteins in red-cell progenitors become phosphorylated on specific amino acid residues within 10 minutes of stimulation by varying concentrations of the hormone erythropoietin; phosphorylation increases with Epo concentration and is the first step in triggering the differentiation of these cells into red blood cells.

**Pulldown Assays of GTP-Binding Proteins** We've seen that the GTPase superfamily of intracellular-switch proteins cycle

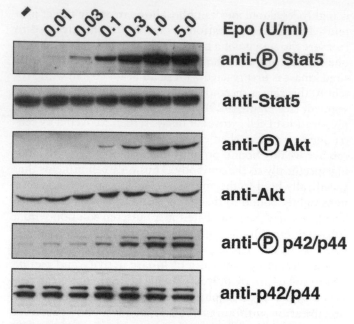

**EXPERIMENTAL FIGURE 15-13 Activation of three signal transduction proteins by phosphorylation.** Mouse erythrocyte progenitor cells were treated for 10 min with different concentrations of the hormone erythropoietin (Epo). Extracts of the cell were analyzed by Western blotting with three different antibodies specific for the phosphorylated forms of three signal transduction proteins and three that recognize a nonphosphorylated segment of amino acids in the same protein. The data show that with increasing concentration of Epo, the three proteins become phosphorylated. Treatment with 1 unit Epo per ml is sufficient to maximally phosphorylate and thus activate all three pathways. Stat 5 = transcription factor phosphorylated on tyrosine 694; Akt = kinase phosphorylated on serine 473; p42/p44 = p42/p44 MAP kinase phosphorylated on threonine 202 and tyrosine 204. [Courtesy Jing Zhang; Zhang et al., 2003, *Blood* **102**:3938.]

between an active ("on") form with bound GTP that modulates the activity of specific target proteins and an inactive ("off") form with bound GDP. The principal assay for measuring activation of this class of proteins takes advantage of the fact that each such protein has one or more targets to

**EXPERIMENTAL FIGURE 15-14 A pull-down assay shows that the small GTP-binding protein Rac1 is activated by platelet-derived growth factor (PDGF).** Like other small GTPases, Rac1 regulates molecular events by cycling between an inactive GDP-bound form and an active GTP-bound form. In its active (GTP-bound) state, Rac1 binds specifically to the p21-binding domain (PBD) of p21-activated protein kinase (PAK) to control downstream signaling cascades. (a) Assay principle: the Rac-binding PBD domain is generated by recombinant DNA techniques and attached to agarose beads, then mixed with cell extracts (step **1**). The beads are recovered by centrifugation (step **2**) and the amount of GTP-bound Rac1 is quantified by Western blotting using an anti-Rac1 antibody (step **3**). (b) Western blot showing activation of Rac1 after treatment of hematopoietic stem cells for 1 min with the hormone platelet-derived growth factor (PDGF). A Western blot for actin serves as a control that the same amount of total protein is loaded on each lane of the gel. [(a) After Cell Biolabs Inc.; (b) from G. Ghiaur et al., 2006, *Blood* **108**:2087–2094.]

which it binds only when it has a bound GTP; the target protein usually has a specific binding domain that binds to the switch segments of the GTP-binding protein. Pull-down assays used to quantify the activation of a specific GTP-binding protein are similar to immunoprecipitations except that the specific binding domain of the target protein is immobilized on small beads (Figure 15-14). The beads are mixed with a cell extract and then recovered by centrifugation; the amount of the GTP-binding protein on the beads is quantified by Western blotting. The example in Figure 15-14 shows that

(a) Assay Priciple

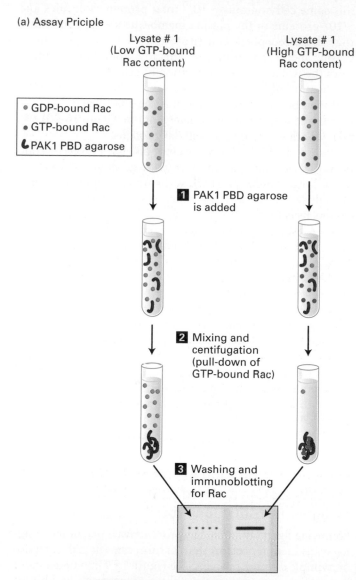

(b) Western blot of hematopoietic stem cells before and after treatment with PDGF

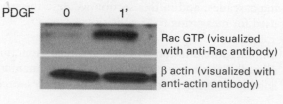

the fraction of the small GTPase Rac1 that has a bound GTP increases markedly after stimulation by the hormone platelet-derived growth factor (PDGF), indicating that Rac1 is a signal transduction protein activated by the PDGF receptor.

## 15.3 G Protein–Coupled Receptors: Structure and Mechanism

As noted above, perhaps the most numerous class of receptors are the G protein–coupled receptors (GPCRs). In humans, GPCRs are used to detect and respond to many different types of signals, including neurotransmitters, hormones involved in glycogen and fat metabolism, and even photons of light. All GPCR signal transduction pathways share the following common elements: (1) a receptor that contains seven membrane-spanning α helixes; (2) a coupled trimeric G protein, which functions as a switch by cycling between active and inactive forms; (3) a membrane-bound effector protein; and (4) proteins that participate in feedback regulation and desensitization of the signaling pathway. A second messenger also occurs in many GPCR pathways. GPCR pathways usually have short-term effects in the cell by quickly modifying existing proteins, either enzymes or ion channels. Thus these pathways allow cells to respond rapidly to a variety of signals, whether they are environmental stimuli such as light or hormonal stimuli such as epinephrine.

In this section, we discuss the basic structure and mechanism of GPCRs and their associated trimeric G proteins. In Sections 15.4 through 15.6, we describe GPCR pathways that activate several different effector proteins.

### All G Protein–Coupled Receptors Share the Same Basic Structure

All G protein–coupled receptors have the same orientation in the membrane and contain seven transmembrane α-helical regions (H1–H7), four extracellular segments, and four cytosolic segments (Figure 15-15). Invariably the N-terminus is on the exoplasmic face and the C-terminus is on the cytosolic face of the plasma membrane. The carboxyl-terminal segment (C4), the C3 loop, and, in some receptors, also the C2 loop are involved in interactions with a coupled trimeric G protein. Many subfamilies of G protein–coupled receptors have been conserved through evolution; members of these subfamilies are especially similar in amino acid sequence and structure.

G protein–coupled receptors are stably anchored in the hydrophobic core of the plasma membrane by many hydrophobic amino acids on the outer surfaces of the seven membrane-spanning segments. One group of G protein–coupled receptors whose structure is known in molecular detail is the **β-adrenergic receptors**, which bind hormones such as epinephrine and norepinephrine (Figure 15-16). In these and many other receptors, segments of several membrane-embedded α helices and extracellular loops form the ligand binding site that is open to the exoplasmic surface. The antagonist cyanopindolol, shown in Figure 15-16, binds with a much higher affinity to the receptor than most agonists, and the receptor-ligand complex has been crystallized and its structure determined. Side chains of 15 amino acids located in four transmembrane α helices and extracellular loop 2 make

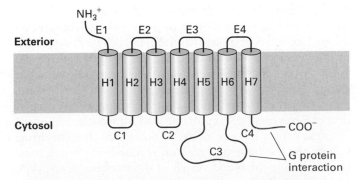

**FIGURE 15-15 General structure of G protein–coupled receptors.** All receptors of this type have the same orientation in the membrane and contain seven transmembrane α-helical regions (H1–H7), four extracellular segments (E1–E4), and four cytosolic segments (C1–C4). The carboxyl-terminal segment (C4), the C3 loop, and, in some receptors, also the C2 loop are involved in interactions with a coupled trimeric G protein.

**FIGURE 15-16 Structure of the turkey β₁-adrenergic receptor complexed with the antagonist cyanopindolol.** (a) Side view showing the approximate location of the membrane phospholipid bilayer. A ribbon representation of the receptor structure is in rainbow coloration (N-terminus, blue; C-terminus, red), with cyanopindolol as a gray space-filling model. The extracellular loop 2 (E2) and cytoplasmic loops 1 and 2 (C1, C2) are labeled. (b) View from external face showing a close-up of the ligand-binding pocket that is formed by amino acids in helices 3, 5, 6, and 7, as well as extracellular loop 2, located between helices 4 and 5. Cyanopindolol atoms are colored grey (carbon), blue (nitrogen), and red (oxygen). The ligand-binding pocket comprises 15 side chains from amino acid residues in four transmembrane α-helices and extracellular loop 2. As examples of specific binding interactions, the positively charged N atom in the amino group found both in cyanopindolol and in epinephrine forms an ionic bond with the carboxylate side chain of aspartate 121 (D¹²¹) in helix 3 and the carboxylate of asparagine 329 (N³²⁹) in helix 7. [From T. Wayne et al., 2008, *Nature* **454**:486.]

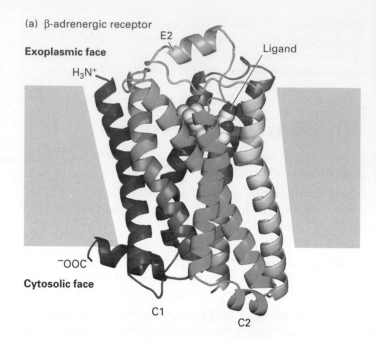

(a) β-adrenergic receptor

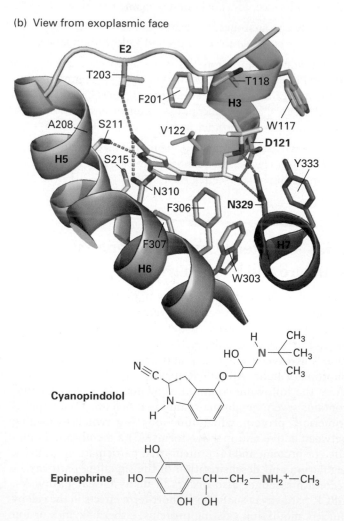

(b) View from exoplasmic face

Cyanopindolol

Epinephrine

noncovalent contacts with the ligand. The amino acids that form the interior of different G protein–coupled receptors are diverse, allowing different receptors to bind very different small molecules, whether they are hydrophilic such as epinephrine or hydrophobic such as many odorants.

While all G protein–coupled receptors share the same basic structure, different subtypes of GCPRs can bind the same hormone, with different cellular effects. To illustrate the versatility of these receptors, we will consider the set of G protein–coupled receptors for epinephrine found in different types of mammalian cells. The hormone **epinephrine** is particularly important in mediating the body's response to stress, also known as the fight-or-flight response. During moments of fear or heavy exercise, when tissues may have an increased need to catabolize glucose and fatty acids to produce ATP, epinephrine signals the rapid breakdown of glycogen to glucose in the liver and of triacylglycerols to fatty acids in adipose (fat) cells; within seconds these principal metabolic fuels are supplied to the blood. In mammals, the liberation of glucose and fatty acids is triggered by binding of epinephrine (or its derivative norepinephrine) to β-adrenergic receptors on the surface of hepatic (liver) and adipose cells.

Epinephrine has other bodily effects as well. Epinephrine bound to β adrenergic receptors on heart muscle cells, for example, increases the contraction rate, which increases the blood supply to the tissues. In contrast, epinephrine stimulation of β-adrenergic receptors on smooth muscle cells of the intestine causes them to relax. Another type of epinephrine GPCR, the *α-adrenergic receptor*, is found on smooth muscle cells lining the blood vessels in the intestinal tract, skin, and kidneys. Binding of epinephrine to these receptors causes the arteries to constrict, cutting off circulation to these organs. These diverse effects of epinephrine help orchestrate integrated responses throughout the body all directed to a common end: supplying energy to major locomotor muscles, while at the same time diverting it from other organs not as crucial in executing a response to bodily stress.

## Ligand-Activated G Protein–Coupled Receptors Catalyze Exchange of GTP for GDP on the α Subunit of a Trimeric G Protein

Trimeric G proteins contain three subunits designated α, β, and γ. Both the $G_\alpha$ and $G_\gamma$ subunits are linked to the membrane by covalently attached lipids. The β and γ subunits are always bound together and are usually referred to as the $G_{\beta\gamma}$ subunit. In the resting state, when no ligand is bound to the receptor, the $G_\alpha$ subunit has a bound GDP and is complexed with $G_{\beta\gamma}$. Binding of a ligand (e.g., epinephrine) or an agonist (e.g., isoproterenol) to a G protein–coupled receptor changes the conformation of its cytosol-facing loops and enables the receptor to bind to the $G_\alpha$ subunit (Figure 15-17, steps **1** and **2**). This binding releases the bound GDP; thus the activated ligand-bound receptor functions as a guanine nucleotide exchange factor (GEF) for the $G_\alpha$ subunit (step **3**). Next, GTP rapidly binds to the "empty" guanine nucleotide site in the $G_\alpha$ subunit, causing a change in the conformation of its switch segments (see Figure 15-7). These changes weaken the binding of $G_\alpha$ with both the receptor and the $G_{\beta\gamma}$ subunit (step **4**). In most cases, $G_\alpha \cdot GTP$, which remains anchored in the membrane, then interacts with and activates an effector protein, as depicted in Figure 15-17 (step **5**). In some cases, $G_\alpha \cdot GTP$ inhibits the effector. Moreover, depending on the type of cell and G protein, the $G_{\beta\gamma}$ subunit, freed

### OVERVIEW ANIMATION: Extracellular Signaling

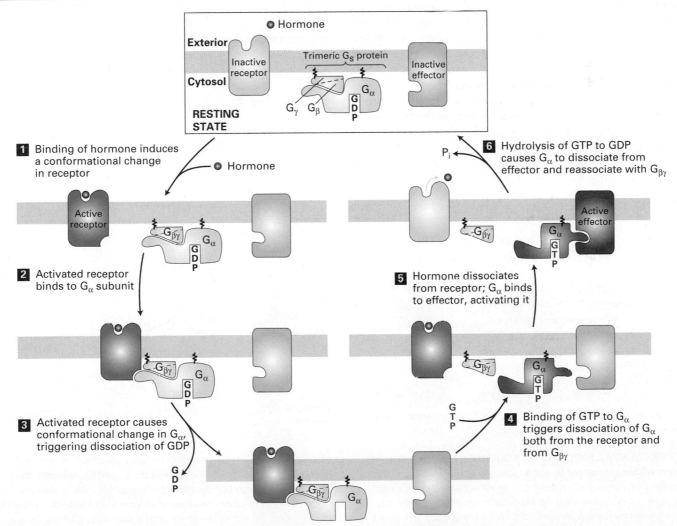

**FIGURE 15-17 General mechanism of the activation of effector proteins associated with G protein–coupled receptors.** The $G_\alpha$ and $G_{\beta\gamma}$ subunits of trimeric G proteins are tethered to the membrane by covalently attached lipid molecules (wiggly black lines). Following ligand binding, exchange of GDP with GTP, and dissociation of the G protein subunits (steps **1**–**4**), the free $G_\alpha \cdot GTP$ binds to and activates an effector protein (step **5**). Hydrolysis of GTP terminates signaling and leads to reassembly of the trimeric G protein, returning the system to the resting state (step **6**). Binding of another ligand molecule causes repetition of the cycle. In some pathways, the effector protein is activated by the free $G_{\beta\gamma}$ subunit. The *s* in *trimeric $G_s$ protein* stands for "stimulatory." [After W. Oldham and H. Hamm, 2006, *Quart. Rev. Biophys.* **39**:117.]

from its α subunit, will sometimes transduce a signal by interacting with an effector protein.

The active $G_\alpha \cdot$GTP state is short-lived because the bound GTP is hydrolyzed to GDP in minutes, catalyzed by the intrinsic GTPase activity of the $G_\alpha$ subunit (see Figure 15-17, step **6**). The conformation of the $G_\alpha$ thus switches back to the inactive $G_\alpha \cdot$GDP state, blocking any further activation of effector proteins. The rate of GTP hydrolysis is sometimes further enhanced by binding of the $G_\alpha \cdot$GTP complex to the effector; the effector thus functions as a GTPase-activating protein (GAP). This mechanism significantly reduces the duration of effector activation and avoids a cellular overreaction. In many cases, a second type of GAP protein called a regulator of G protein signaling (RGS) also accelerates GTP hydrolysis by the $G_\alpha$ subunit, further reducing the time during which the effector remains activated. The resulting $G_\alpha \cdot$GDP quickly reassociates with $G_{\beta\gamma}$ and the complex becomes ready to interact with an activated receptor and start the process all over again. Thus the GPCR signal transduction system contains a built-in feedback mechanism that ensures the effector protein becomes activated only for a few seconds or minutes following receptor activation; continual activation of receptors via ligand binding together with subsequent activation of the corresponding G protein is essential for prolonged activation of the effector.

Early evidence supporting the model shown in Figure 15-17 came from studies with compounds called GTP analogs that are structurally similar to GTP and so can bind to $G_\alpha$ subunits as well as GTP does but cannot be hydrolyzed by the intrinsic GTPase. In some of these compounds, the P–O–P phosphodiester linkage connecting the β and γ phosphates of GTP is replaced by a nonhydrolyzable $P-CH_2-P$ or P–NH–P linkage. Addition of such a GTP analog to a plasma membrane preparation in the presence of an agonist for a particular receptor results in a much longer-lived activation of the G protein and its associated effector protein than occurs with GTP. In this experiment, once the nonhydrolyzable GTP analog is exchanged for GDP bound to $G_\alpha$, it remains permanently bound to $G_\alpha$. Because the $G_\alpha \cdot$GTP-analog complex is as functional as the normal $G_\alpha \cdot$GTP complex in activating the effector protein, the effector remains permanently active.

GPCR-mediated dissociation of trimeric G proteins can be detected in living cells. These studies have exploited the phenomenon of *fluorescence energy transfer*, which changes the wavelength of emitted fluorescence when two fluorescent proteins interact (see Figure 9-22). Figure 15-18 shows how this experimental approach has demonstrated the dissociation of the $G_\alpha \cdot G_{\beta\gamma}$ complex within a few seconds of ligand addition, providing further evidence for the model of G protein cycling. This general experimental approach can be used to follow the formation and dissociation of other protein-protein complexes in living cells.

For many years, it was impossible to determine the structure of the same GPCR in the active and inactive states. This has now been accomplished with the β-adrenergic receptor (as

🎙 **PODCAST:** Activation of G Proteins Measured by Fluorescence Resonance Energy Transfer (FRET)

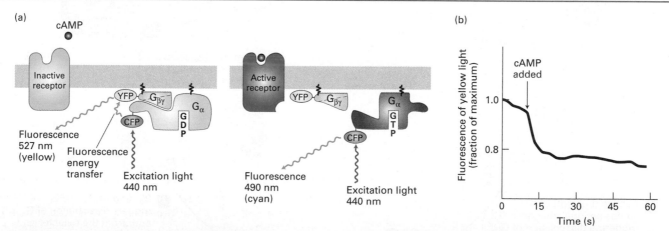

**EXPERIMENTAL FIGURE 15-18 Activation of G proteins occurs within seconds of ligand binding in amoeba cells.** In the amoeba *Dictyostelium discoideum* cell, cAMP acts as an extracellular signaling molecule and binds to a G protein–coupled receptor; it is not a second messenger. Amoeba cells were transfected with genes encoding two fusion proteins: a $G_\alpha$ fused to cyan fluorescent protein (CFP), a mutant form of green fluorescent protein (GFP), and a $G_\beta$ fused to another GFP variant, yellow fluorescent protein (YFP). CFP normally fluoresces 490-nm light; YFP, 527-nm light. (a) When CFP and YFP are nearby, as in the resting $G_\alpha \cdot G_{\beta\gamma}$ complex, fluorescence energy transfer can occur between CFP and YFP (*left*). As a result, irradiation of resting cells with 440-nm light (which directly excites CFP but not YFP) causes emission of 527-nm (yellow) light, characteristic of YFP. However, if ligand binding leads to dissociation of the $G_\alpha$ and $G_{\beta\gamma}$ subunits, then fluorescence energy transfer cannot occur. In this case, irradiation of cells at 440 nm causes emission of 490-nm light (cyan) characteristic of CFP (*right*). (b) Plot of the emission of yellow light (527 nm) from a single transfected amoeba cell before and after addition of extracellular cyclic AMP (arrow), the ligand for the G protein–coupled receptor in these cells. The drop in yellow fluorescence, which results from the dissociation of the $G_\alpha$-CFP fusion protein from the $G_\beta$-YFP fusion protein, occurs within seconds of cAMP addition. [Adapted from C. Janetopoulos et al., 2001, *Science* **291**:2408.]

## (a) Side view   β-adrenergic receptor

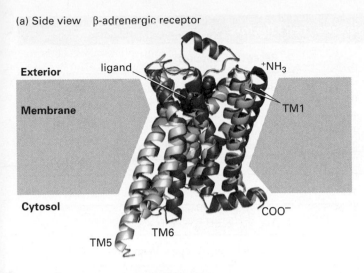

Exterior

Membrane

Cytosol

ligand

+NH₃

TM1

COO⁻

TM6

TM5

## (b) View from cytosolic surface

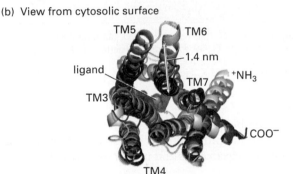

TM5    TM6

1.4 nm

ligand

TM7

+NH₃

TM3

COO⁻

TM4

## (c)

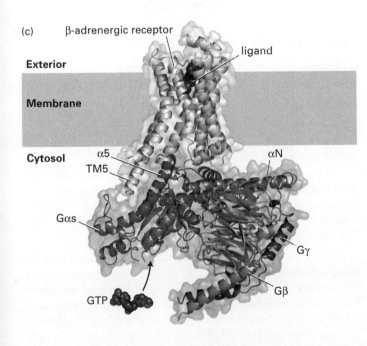

β-adrenergic receptor

ligand

Exterior

Membrane

Cytosol

α5

αN

TM5

Gαs

Gγ

GTP

Gβ

**FIGURE 15-19 Structure of the β-adrenergic receptor in the inactive and active states and with its associated trimeric G protein, G$_{\alpha s}$.** (a) Comparison of the three-dimensional structures of the activated β-adrenergic receptor (gold) bound to a strong agonist and the inactive receptor (purple) bound to an antagonist. (b) View from the cytosolic surface. Note the major changes seen in the conformations of the intracellular domains of transmembrane helices 5 (TM5) and 6 (TM6). In the active state, TM5 is extended by two helical turns, whereas TM6 is moved outward by 1.4 nm. (c) The overall structure of the active receptor complex shows the adrenergic receptor (gold) bound to an agonist (black and red spheres) and engaged in extensive interactions with a segment of G$_{\alpha s}$ (purple). G$_{\alpha s}$ together with G$_\beta$ (green) and G$_\gamma$ (red) constitute the heterotrimeric G protein G$_s$. [After S. Rasmussen et al., 2011, *Nature* **476**:387–390.]

change (Figure 15-19a) in which there are substantial movements of transmembrane helices 5 and 6 and changes in the structure of the C3 loop; together these create a surface that can now bind to a segment of the G$_{\alpha s}$ subunit (Figure 15-19b).

X-ray crystallographic studies of the complex of activated receptor and G$_s$ have also revealed how the subunits of a G protein interact with each other and provided clues about how binding of GTP leads to dissociation of the G$_\alpha$ from the G$_{\beta\gamma}$ subunit. As revealed in the structural model in Figure 15-19b, a large surface of G$_\alpha$·GDP interacts with the G$_\beta$ subunit; part of this surface is located in the αN alpha helix in the N-terminal segment of G$_\alpha$·GDP. Note that G$_\alpha$ directly contacts G$_\beta$ but not G$_\gamma$. Binding of the N-terminal alpha-helical segments αN and α5 of the G$_{\alpha s}$ protein to transmembrane helices 5 and 6 of the activated receptor (Figure 15-19b) will, as with other G proteins, be followed by opening of the G$_\alpha$ subunit, eviction of the bound GDP, and its replacement with GTP; this is immediately followed by conformational changes within switches I and II that disrupt the molecular interactions between G$_\alpha$ and G$_{\beta\gamma}$, leading to their dissociation.

## Different G Proteins Are Activated by Different GPCRs and In Turn Regulate Different Effector Proteins

All effector proteins in GPCR pathways are either membrane-bound ion channels or membrane-bound enzymes that catalyze formation of the second messengers shown in Figure 15-8. The variations on the theme of GPCR signaling that we examine in Sections 15.4 through 15.6 arise because multiple G proteins are encoded in eukaryotic genomes. At last count, humans have 21 different G$_\alpha$ subunits encoded by 16 genes, several of which undergo alternative splicing; six G$_\beta$ subunits; and 12 G$_\gamma$ subunits. So far as is known, the different G$_{\beta\gamma}$ subunits are essentially interchangeable in their functions, while the different G$_\alpha$ subunits afford the various G proteins their specificity. Thus we can refer to the entire three-subunit G protein by the name of its alpha subunit.

Table 15-1 summarizes the functions of the major classes of G proteins with different G$_\alpha$ subunits. For example, the different types of epinephrine receptors mentioned previously

well as with rhodopsin, discussed in Section 15.4). The seven membrane-embedded α helices of the β-adrenergic receptor completely surround a central segment to which an agonist or antagonist is noncovalently bound (Figure 15-19). Binding of an agonist to the receptor induces a major conformational

## TABLE 15-1 | Major Classes of Mammalian Trimeric G Proteins and Their Effectors[*]

| $G_\alpha$ Class | Associated Effector | 2nd Messenger | Receptor Examples |
|---|---|---|---|
| $G_{\alpha s}$ | Adenylyl cyclase | cAMP (increased) | β-Adrenergic (epinephrine) receptor; receptors for glucagon, serotonin, vasopressin |
| $G_{\alpha i}$ | Adenylyl cyclase K$^+$ channel ($G_{\beta\gamma}$ activates effector) | cAMP (decreased) Change in membrane potential | α$_2$-Adrenergic receptor Muscarinic acetylcholine receptor |
| $G_{\alpha olf}$ | Adenylyl cyclase | cAMP (increased) | Odorant receptors in nose |
| $G_{\alpha q}$ | Phospholipase C | IP$_3$, DAG (increased) | α$_1$-Adrenergic receptor |
| $G_{\alpha o}$ | Phospholipase C | IP$_3$, DAG (increased) | Acetylcholine receptor in endothelial cells |
| $G_{\alpha t}$ | cGMP phosphodiesterase | cGMP (decreased) | Rhodopsin (light receptor) in rod cells |

[*]A given $G_\alpha$ subclass may be associated with more than one effector protein. To date, only one major $G_{\alpha s}$ has been identified, but multiple $G_{\alpha q}$ and $G_{\alpha i}$ proteins have been described. Effector proteins commonly are regulated by $G_\alpha$ but in some cases by $G_{\beta\gamma}$ or the combined action of $G_\alpha$ and $G_{\beta\gamma}$. IP$_3$ = inositol 1,4,5-trisphosphate; DAG = 1,2-diacylglycerol.
SOURCES: See L. Birnbaumer, 1992, *Cell* **71**:1069; Z. Farfel et al., 1999, *New Eng. J. Med.* **340**:1012; and K. Pierce et al., 2002, *Nature Rev. Mol. Cell Biol.* **3**:639.

are coupled to different $G_\alpha$ subunits that influence effector proteins differently and so have distinct effects on cell behavior in a target cell. Both subtypes of β-adrenergic receptors, termed β$_1$ and β$_2$, are coupled to a *stimulatory* G protein (G$_s$) whose alpha subunit (G$_{\alpha s}$) activates a membrane-bound effector enzyme called **adenylyl cyclase**. Once activated, this enzyme catalyzes synthesis of the second messenger cAMP. In contrast, the α$_2$ subtype of β-adrenergic receptor is coupled to an *inhibitory* G protein (G$_i$) whose alpha subunit G$_{\alpha i}$ inhibits adenylyl cyclase, the same effector enzyme associated with β-adrenergic receptors. The G$_{\alpha q}$ subunit, which is coupled to the α$_1$-adrenergic receptor, activates a different effector enzyme, **phospholipase C**, which generates two other second messengers, DAG and IP$_3$ (see Figure 15-8). Examples of signaling pathways that use each of the $G_\alpha$ subunits listed in Table 15-1 are described in the following three sections.

Some bacterial toxins contain a subunit that penetrates the plasma membrane of target mammalian cells and in the cytosol catalyzes a chemical modification on $G_\alpha$ proteins that prevents hydrolysis of bound GTP to GDP. For example, toxins produced by the bacterium *Vibrio cholera*, which causes cholera, or certain strains of *E. coli*, modify the G$_{\alpha s}$ protein in intestinal epithelial cells. As a result, G$_{\alpha s}$ remains in the active state, continuously activating the effector adenylyl cyclase in the absence of hormonal stimulation. The resulting excessive rise in intracellular cAMP leads to the loss of electrolytes and water into the intestinal lumen, producing the watery diarrhea characteristic of infection by these bacteria. The toxin produced by *Bordetella pertussis*, a bacterium that commonly infects the respiratory tract and causes whooping cough, catalyzes a modification of G$_{\alpha i}$ that prevents release of bound GDP. As a result, G$_{\alpha i}$ is locked in the inactive state,

reducing the inhibition of adenylyl cyclase. The resulting increase in cAMP in epithelial cells of the airways promotes loss of fluids and electrolytes and mucus secretion. ∎

## KEY CONCEPTS of Section 15.3

### G Protein–Coupled Receptors: Structure and Mechanism

• G protein–coupled receptors (GPCRs) are a large and diverse family with a common structure of seven membrane-spanning α helices and an internal ligand-binding pocket that is specific for ligands (see Figures 15-15 and 15-16).

• GPCRs can have a range of cellular effects depending on the subtype of receptor that binds ligand. The hormone epinephrine, for example, which mediates the fight-or-flight response, binds to multiple subtypes of GPCRs in multiple cell types, with varying physiological effects.

• GPCRs are coupled to trimeric G proteins, which contain three subunits designated α, β, and γ. The $G_\alpha$ subunit is a GTPase switch protein that alternates between an active ("on") state with bound GTP and inactive ("off") state with GDP. The "on" form separates from the β and γ subunits and activates a membrane-bound effector. The β and γ subunits remain bound together and only occasionally transduce signals (see Figure 15-17).

• Ligand binding causes a conformational change in certain membrane-spanning helices and intracellular loops of the GPCR, allowing it to bind to and function as a guanine nucleotide exchange factor (GEF) for its coupled $G_\alpha$ subunit, catalyzing dissociation of GDP and allowing GTP to bind. The resulting change in conformation of switch regions in $G_\alpha$ causes it to dissociate from the $G_{\beta\gamma}$ subunit and the receptor and interact with an effector protein (see Figure 15-17).

- Fluorescence energy-transfer experiments demonstrate receptor-mediated dissociation of coupled $G_\alpha$ and $G_{\beta\gamma}$ subunits in living cells (see Figure 15-18).

- The effector proteins activated (or inactivated) by trimeric G proteins are either enzymes that form second messengers (e.g., adenylyl cyclase, phospholipase C) or ion channels (see Table 15-1). In each case, it is the $G_\alpha$ subunit that determines the function of the G protein and affords its specificity.

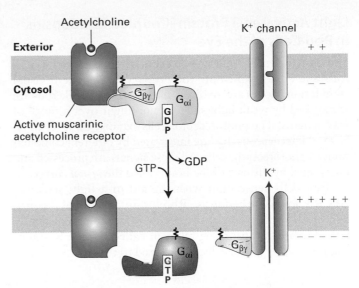

**FIGURE 15-20 Activation of the muscarinic acetylcholine receptor and its effector K$^+$ channel in heart muscle.** Binding of acetylcholine triggers activation of the $G_{\alpha i}$ subunit and its dissociation from the $G_{\beta\gamma}$ subunit in the usual way (see Figure 15-17). In this case, the released $G_{\beta\gamma}$ subunit (rather than $G_{\alpha i}$·GTP) binds to and opens the associated effector protein, a K$^+$ channel. The increase in K$^+$ permeability hyperpolarizes the membrane, which reduces the frequency of heart muscle contraction. Though not shown here, activation is terminated when the GTP bound to $G_{\alpha i}$ is hydrolyzed (by a GAP enzyme that is an intrinsic part of the $G_{\alpha i}$ subunit) to GDP and $G_{\alpha i}$·GDP recombines with $G_{\beta\gamma}$. [See K. Ho et al., 1993, *Nature* **362**:31, and Y. Kubo et al., 1993, *Nature* **362**:127.]

## 15.4 G Protein–Coupled Receptors That Regulate Ion Channels

One of the simplest cellular responses to a signal is the opening of ion channels essential for transmission of nerve impulses. Nerve impulses are essential to the sensory perception of environmental stimuli such as light and odors, to transmission of information to and from the brain, and to the stimulation of muscle movement. During transmission of nerve impulses, the opening and closing of ion channels causes changes in the membrane potential. Many neurotransmitter receptors are ligand-gated ion channels, which open in response to binding of a ligand. Such receptors include some types of glutamate, serotonin, and acetylcholine receptors, including the acetylcholine receptor found at nerve-muscle synapses. Ligand-gated ion channels that function as neurotransmitter receptors are covered in Chapter 22.

Many neurotransmitter receptors, however, are G protein–coupled receptors whose effector proteins are Na$^+$ or K$^+$ channels. Neurotransmitter binding to these receptors causes the associated ion channel to open or close, leading to changes in the membrane potential. Still other neurotransmitter receptors, as well as odorant receptors in the nose and photoreceptors in the eye, are G protein–coupled receptors that indirectly modulate the activity of ion channels via the action of second messengers. In this section, we consider two G protein–coupled receptors that illustrate the direct and indirect mechanisms for regulating ion channels: the muscarinic acetylcholine receptor of the heart and the light-activated rhodopsin protein in the eye.

### Acetylcholine Receptors in the Heart Muscle Activate a G Protein That Opens K$^+$ Channels

*Muscarinic acetylcholine receptors* are a type of GPCR found in cardiac muscle. When activated, these receptors *slow* the rate of heart muscle contraction. Because muscarine, an acetylcholine analog, also activates these receptors, they are termed "muscarinic." This type of acetylcholine receptor is coupled to a $G_{\alpha i}$ subunit, and ligand binding leads to opening of associated K$^+$ channels (the effector protein) in the plasma membrane (Figure 15-20). The subsequent efflux of K$^+$ ions from the cytosol causes an increase in the magnitude of the usual inside-negative potential across the plasma membrane that lasts for several seconds. This state of the membrane, called **hyperpolarization**, reduces the frequency of muscle contraction. This effect can be shown experimentally by adding acetylcholine to isolated heart muscle cells and measuring the membrane potential using a microelectrode inserted into the cell (see Figure 11-19).

As shown in Figure 15-20, the signal from activated muscarinic acetylcholine receptors is transduced to the effector channel protein by the released $G_{\beta\gamma}$ subunit rather than by $G_{\alpha i}$·GTP. That $G_{\beta\gamma}$ directly activates the K$^+$ channel was demonstrated by patch-clamping experiments, which can measure ion flow through a single ion channel in a small patch of membrane (see Figure 11-22). When purified $G_{\beta\gamma}$ protein was added to the cytosolic face of a patch of heart muscle plasma membrane, K$^+$ channels opened immediately, even in the absence of acetylcholine or other neurotransmitters—clearly indicating that it is the $G_{\beta\gamma}$ protein that is responsible for opening the effector K$^+$ channels and not $G_\alpha$·GTP.

## Light Activates G Protein–Coupled Rhodopsins in Rod Cells of the Eye

The human retina contains two types of photoreceptor cells, *rods* and *cones*, which are the primary recipients of visual stimulation. Cones are involved in color vision, while rods are stimulated by weak light such as moonlight over a range of wavelengths. The photoreceptor cells synapse on layer upon layer of interneurons that are innervated by different combinations of photoreceptor cells. All these signals are processed and interpreted by the part of the brain called the *visual cortex*.

Rod cells sense light with the aid of a light-sensitive GPCR known as *rhodopsin*. Rhodopsin consists of the protein opsin, which has the usual GPCR structure, covalently linked to a light-absorbing pigment called retinal. Rhodopsin, found only in rod cells, is localized to the thousand or so flattened membrane disks that make up the outer segment of these rod-shaped cells (Figure 15-21). A human rod cell contains about $4 \times 10^7$ molecules of rhodopsin. The trimeric G protein coupled to rhodopsin, called *transducin* ($G_t$), contains a $G_\alpha$ unit referred to as $G_{\alpha t}$; like rhodopsin, $G_{\alpha t}$ is found only in rod cells.

Rhodopsin differs from other GPCRs in that binding of a ligand is not what activates the receptor. Rather, absorption of a photon of light by the bound retinal is the activating signal. On absorption of a photon, the retinal moiety of rhodopsin is immediately converted from the cis form (known as 11-*cis*-retinal) to the all-trans isomer, causing a conformational change in the opsin protein (Figure 15-22). This is equivalent to the activating conformational change that occurs on ligand binding by other G protein–coupled receptors; this conformational

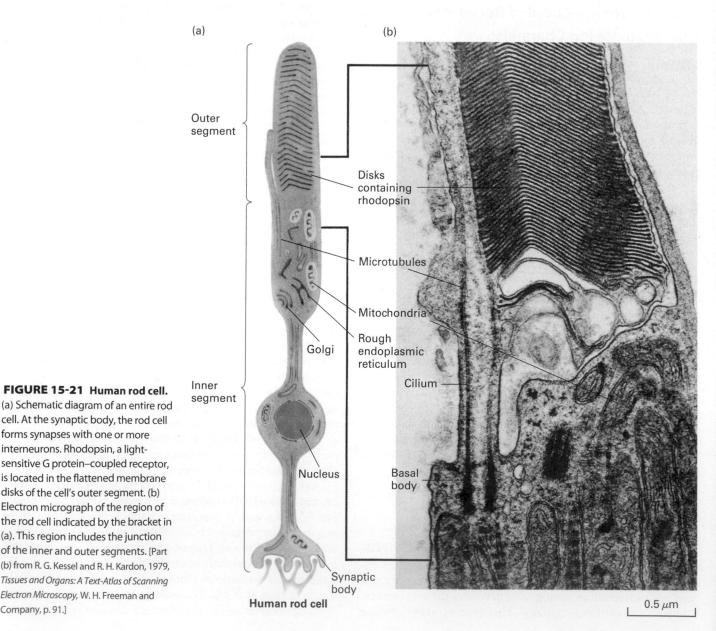

**FIGURE 15-21 Human rod cell.** (a) Schematic diagram of an entire rod cell. At the synaptic body, the rod cell forms synapses with one or more interneurons. Rhodopsin, a light-sensitive G protein–coupled receptor, is located in the flattened membrane disks of the cell's outer segment. (b) Electron micrograph of the region of the rod cell indicated by the bracket in (a). This region includes the junction of the inner and outer segments. [Part (b) from R. G. Kessel and R. H. Kardon, 1979, *Tissues and Organs: A Text-Atlas of Scanning Electron Microscopy,* W. H. Freeman and Company, p. 91.]

(a) / (b) Outer segment / Inner segment / Disks containing rhodopsin / Microtubules / Mitochondria / Rough endoplasmic reticulum / Golgi / Cilium / Nucleus / Basal body / Synaptic body / Human rod cell / 0.5 μm

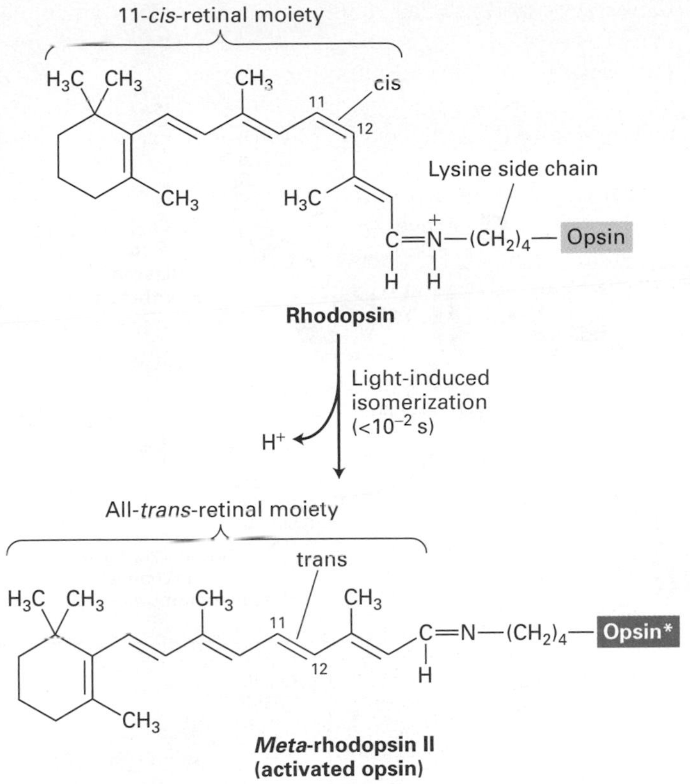

**11-*cis*-retinal moiety**

Rhodopsin

Light-induced
isomerization
($<10^{-2}$ s)

$H^+$

**All-*trans*-retinal moiety**

***Meta*-rhodopsin II**
**(activated opsin)**

**FIGURE 15-22 The light-triggered step in vision.** The light-absorbing pigment 11-*cis*-retinal is covalently bound to the amino group of a lysine residue in opsin, the protein part of rhodopsin. Absorption of light causes rapid photoisomerization of the bound *cis*-retinal to the all-*trans* isomer. This triggers a conformational change in the opsin protein, forming the unstable intermediate *meta*-rhodopsin II, or activated opsin (see Figure 15-23), which activates $G_t$ proteins. Within seconds, all-*trans*-retinal dissociates from opsin and is converted by an enzyme back to the *cis* isomer, which then rebinds to another opsin molecule. [See J. Nathans, 1992, *Biochemistry* **31**:4923.]

change allows rhodopsin to bind an adjacent $G_{\alpha t}$ subunit of the coupled G protein, triggering exchange of GTP for GDP. Activated rhodopsin, R*, is unstable and spontaneously dissociates into its component parts, releasing the covalently attached opsin, which can no longer bind to a $G_{\alpha t}$ subunit and all-*trans*-retinal, thereby terminating visual signaling. In the dark, free all-*trans*-retinal is converted back to 11-*cis*-retinal, which can then rebind to opsin, re-forming rhodopsin.

## Activation of Rhodopsin by Light Leads to Closing of cGMP-Gated Cation Channels

In the dark, the membrane potential of a rod cell is about $-30$ mV, considerably less than the resting potential ($-60$ to $-90$ mV) typical of neurons and other electrically active cells. This state of the membrane, called **depolarization**, causes rod cells in the dark to constantly secrete neurotransmitters, and thus the neurons with which they synapse are continually being stimulated. The depolarized state of the plasma

membrane of resting rod cells is due to the presence of a large number of open *nonselective* ion channels that admit $Na^+$ and $Ca^{2+}$; recall from Chapter 11 that movement of positively charged ions such as $Na^+$ and $Ca^{2+}$ from the outside of the cell to the inside will reduce the magnitude of the inside-negative membrane potential. Absorption of light by rhodopsin leads to closing of these channels, causing the membrane potential to become *more* inside negative.

The more photons absorbed by rhodopsin, the more channels are closed, the fewer $Na^+$ and $Ca^{2+}$ ions cross the membrane from the outside, the more negative the membrane potential becomes, and the less neurotransmitter is released. The reduction in neurotransmitter release is transmitted to the brain by a series of neurons, where it is perceived as light.

Unlike the muscarinic acetylcholine receptor discussed earlier, G proteins activated by rhodopsin do not act directly on ion channels. The closing of cation channels in the rod-cell plasma membrane requires changes in the concentration of the second messenger **cyclic GMP**, or **cGMP** (see Figure 15-8). Rod cell outer segments contain an unusually high concentration ($\sim0.07$ mM) of cGMP, which is continuously formed from GTP in a reaction catalyzed by guanylyl cyclase. However, light absorption by rhodopsin induces activation of a *cGMP phosphodiesterase* (PDE), which hydrolyzes cGMP to 5'-GMP. As a result, the cGMP concentration decreases on illumination. The high level of cGMP present in the dark acts to keep *cGMP-gated cation channels* open; the light-induced decrease in cGMP leads to channel closing, membrane hyperpolarization, and reduced neurotransmitter release.

As depicted in Figure 15-23, cGMP phosphodiesterase is the effector protein for $G_{\alpha t}$, both of which are localized to the disk membrane of the rod cell. The free $G_{\alpha t} \cdot$GTP complex that is generated after light absorption by rhodopsin binds to the two inhibitory $\gamma$ subunits of cGMP phosphodiesterase, releasing the active catalytic $\alpha$ and $\beta$ subunits, which then convert cGMP to GMP. This is a clear example of how signal-induced removal of an inhibitor can quickly activate an enzyme, a common mechanism in signaling pathways. In turn, the lowered concentration of cGMP leads to closing of the cGMP-gated ion channel in the rod cell plasma membrane, thereby reducing neurotransmitter release.

Direct support for the role of cGMP in rod-cell activity has been obtained in patch-clamping studies using isolated patches of rod outer-segment plasma membrane, which contains abundant cGMP-gated cation channels. When cGMP is added to the cytosolic surface of these patches, there is a rapid increase in the number of open ion channels; cGMP binds directly to a site on the channel protein to keep them open. Like the $K^+$ channels discussed in Chapter 11, the cGMP-gated channel protein contains four subunits (see Figure 11-20). In this case, each of the subunits is able to bind a cGMP molecule. Three or four cGMP molecules must bind per channel in order to open it; this allosteric interaction makes channel opening very sensitive to small changes in cGMP levels.

Conversion of active $G_{\alpha t} \cdot$GTP back to inactive $G_{\alpha t} \cdot$GDP is accelerated by a specific GTPase-activating protein (GAP).

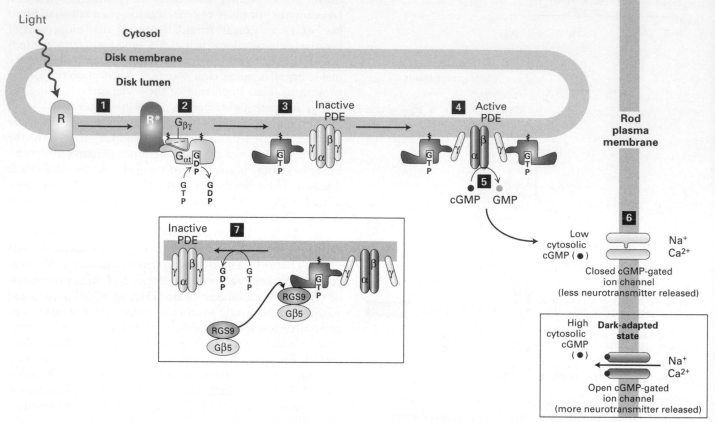

**FIGURE 15-23 Light-activated rhodopsin pathway and the closing of cation channels in rod cells.** In dark-adapted rod cells, a high level of cGMP keeps nucleotide-gated nonselective cation channels open, leading to depolarization of the plasma membrane and neurotransmitter release. Light absorption generates activated rhodopsin, R* (step **1**), which binds inactive GDP-bound $G_{\alpha t}$ protein and mediates replacement of GDP with GTP (step **2**). The free $G_{\alpha t}\cdot GTP$ generated then activates cGMP phosphodiesterase (PDE) by binding to its inhibitory $\gamma$ subunits (step **3**) and dissociating them from the catalytic $\alpha$ and $\beta$ subunits (step **4**). Relieved of their inhibition, the $\alpha$ and $\beta$ subunits of PDE hydrolyze cGMP to GMP (step **5**). The resulting decrease in cytosolic cGMP leads to dissociation of cGMP from the nucleotide-gated channels in the plasma membrane and closing of the channels (step **6**). The membrane then becomes transiently hyperpolarized, and neurotransmitter release is reduced. The complex of $G_{\alpha t}\cdot GTP$ and the PDE $\gamma$ subunits binds a GTPase activating complex termed RGS9-G$\beta$5 (step **7**); by hydrolyzing the bound GTP, this triggers the physiologically rapid inactivation of the phosphodiesterase. [Adapted from V. Arshavsky and E. Pugh, 1998, *Neuron* **20**:11, and V. Arshavsky, 2002, *Trends Neurosci.* **25**:124.]

In mammals, $G_{\alpha t}$ normally remains in the active GTP-bound state for only a fraction of a second. Thus cGMP phosphodiesterase rapidly becomes inactivated, and the cGMP level gradually rises to its original level when the light stimulus is removed. This allows rapid responses of the eye toward moving or changing objects.

## Signal Amplification Makes the Rhodopsin Signal Transduction Pathway Exquisitely Sensitive

Remarkably, a single photon absorbed by a resting rod cell produces a measurable response, a more inside-negative change in the membrane potential of about 1 mV, which in amphibians lasts a second or two. Humans are able to detect a flash of as few as five photons. The light-detecting system is so sensitive because the signal is greatly amplified during the signal transduction pathway. Each activated opsin in the disk membrane of the rod cell can activate 500 $G_{\alpha t}$ molecules, each of which in turn activates a cGMP phosphodiesterase. Each molecule of phosphodiesterase hydrolyzes hundreds of cGMP molecules during the fraction of a second it remains active. Thus absorbance of a single photon—yielding a single activated opsin molecule—can trigger closing of thousands of ion channels in the plasma membrane and a measurable change in the membrane potential of the cell.

## Rapid Termination of the Rhodopsin Signal Transduction Pathway Is Essential for Acute Vision

As in all G protein–coupled signaling pathways, timely termination of the rhodopsin signaling pathway requires that all the activated intermediates be inactivated rapidly, restoring the system to its basal state, ready for signaling again. Thus the three protein intermediates, activated rhodopsin (R*), $G_{\alpha t} \cdot GTP$, and activated cGMP phosphodiesterase (PDE), must all be inactivated, and the concentration of cytoplasmic messenger cGMP must be restored to its dark level by guanylyl cyclase. During a single photon response of a mammalian rod cell, the entire process of rhodopsin activation and inactivation is completed within ~50 milliseconds, enabling the eye to detect rapid movements or other changes in objects in our surroundings. Several mechanisms act together to make possible this very rapid response.

**GAP Proteins That Inactivate $G_{\alpha t} \cdot GTP$** The complex of the inhibitory γ phosphodiesterase subunit and $G_{\alpha t} \cdot GTP$ recruits a complex of two proteins, RGS9 and Gβ5, that together act as a GAP protein and hydrolyze the bound GTP to GDP. This releases the inhibitory γ subunit and terminates phosphodiesterase activation. Experiments with mice that have the RGS9 gene knocked out showed that this protein is essential for normal inactivation of the cascade in vivo. In single mouse rod cells, the time for recovery from a single flash increased from the normal 0.2 s to about 9 s in the mutant, a 45-fold increase, attesting to the importance of this GAP protein.

**$Ca^{2+}$-Sensing Proteins That Activate Guanylate Cyclase** Light-triggered closing of the cGMP-gated $Na^+$ and $Ca^{2+}$ channel will cause a drop in the cytosolic $Ca^{2+}$ concentration since $Ca^{2+}$ is continuously pumped out of the cell independently of the state of these channels. The fall in intracellular $Ca^{2+}$ is sensed by $Ca^{2+}$-binding proteins called guanylate-cyclase-activating proteins, or GCAPs. This results in a rapid stimulation of cGMP synthesis by guanylate cyclase, causing the ion channels to reopen.

**Rhodopsin Phosphorylation and Binding of Arrestin** A major process that down-modulates and terminates the visual response involves phosphorylation of rhodopsin in its active conformation (R*) but not in its inactive, or dark form (R) by *rhodopsin kinase* (Figure 15-24), a member of a class of GPCR kinases. Each opsin molecule has three principal serine phosphorylation sites on its cytosol-facing C-terminal C4 segment; the more sites that are phosphorylated, the less able R* is to activate $G_{\alpha t}$ and so induce closing of cGMP-gated cation channels. The protein *arrestin* binds to three phosphorylated serine residues on the C-terminal opsin segment. Bound arrestin completely prevents interaction of $G_{\alpha t}$ with phosphorylated R*, totally blocking formation of the active $G_{\alpha t} \cdot GTP$ complex and stopping additional activation of cGMP phosphodiesterase. During a single photon response of a mammalian rod cell, the entire process of multiple rhodopsin phosphorylation and arrestin binding is completed within ~50 milliseconds. The phosphates linked to rhodopsin are being continuously removed by phosphodiesterase enzymes, causing dissociation of arrestin and rapid restoration of rhodopsin to its original, native state.

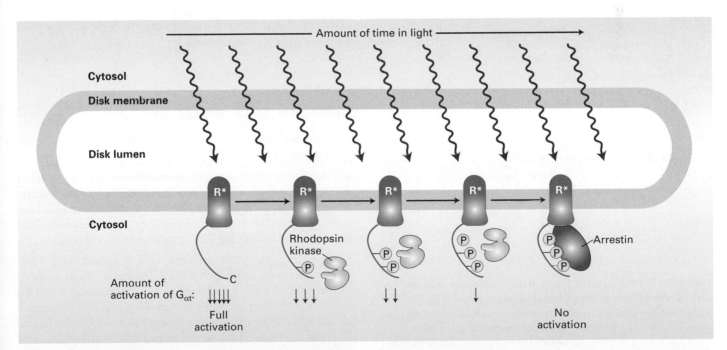

**FIGURE 15-24 Inhibition of rhodopsin signaling by rhodopsin kinase.** Light-activated rhodopsin (R*), but not dark-adapted rhodopsin, is a substrate for rhodopsin kinase. The extent of rhodopsin phosphorylation is proportional to the amount of time each rhodopsin molecule spends in the light-activated form and reduces the ability of R* to activate transducin. Arrestin binds to the completely phosphorylated opsin, forming a complex that cannot activate transducin at all. [See A. Mendez et al., 2000, *Neuron* **28**:153, and V. Arshavsky, 2002, *Trends Neurosci.* **25**:124.]

## Rod Cells Adapt to Varying Levels of Ambient Light by Intracellular Trafficking of Arrestin and Transducin

Cone cells are insensitive to low levels of illumination, and the activity of rod cells is inhibited at high light levels. Thus when we move from bright daylight into a dimly lighted room, we are initially blinded. Slowly, however, the rod cells become sensitive to the dim light, and we gradually are able to see and distinguish objects. During this interval, the rod cell has "turned up" its sensitivity to flashes of light, or contrast. Through this process of *visual adaptation*, a rod cell can perceive contrast over a 100,000-fold range of ambient

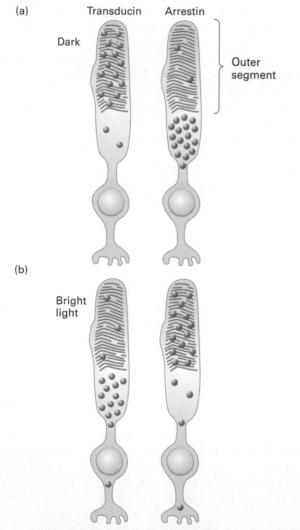

**FIGURE 15-25 Schematic illustration of transducin and arrestin distribution in dark-adapted and light-adapted rod cells.** (a) In the dark, most transducin is localized to the outer segment, while most arrestin is found in other parts of the cell; in this condition vision is most sensitive to very low light levels. (b) In bright light, little transducin is found in the outer segment and abundant arrestin is found there; in this condition vision is relatively insensitive to small changes in light. Coordinated movement of these proteins contributes to our ability to perceive images over a 100,000-fold range of ambient light levels. [After P. Calvert et al., 2006, *Trends Cell Biol.* **16**:560.]

light levels, from very dim to bright sunlight. This wide range of sensitivity is possible because differences in light levels in the visual field, rather than the absolute amount of absorbed light, are ultimately sensed by the brain and used to form visual images. A mechanism for light-dependent regulation of the rhodopsin-signaling pathway that involves subcellular trafficking of two key signal transduction proteins (Figure 15-25) is responsible for this extraordinarily wide sensitivity range.

In dark-adapted rod cells, ~80 to 90 percent of the $G_{\alpha t}$ and $G_{\beta\gamma}$ transducin subunits are in the outer segments, while less than 10 percent of arrestin is localized there (Figure 15-25). This allows maximal activation of the downstream effector cGMP phosphodiesterase and thus maximal sensitivity to small changes in light. But exposure for 10 minutes to moderate daytime intensities of light causes a complete redistribution of these proteins: over 80 percent of the $G_{\alpha t}$ and $G_{\beta\gamma}$ subunits move out of the outer segment into other parts of the cell while over 80 percent of the inhibitor arrestin moves into the outer segment. The mechanism by which these proteins move is not yet known but probably involves microtubule-attached motors that move attached proteins and particles outward and inward (see Chapter 18). The reduction in $G_{\alpha t}$ and $G_{\beta\gamma}$ in the outer segment means that $G_{\alpha t}$ proteins are physically unable to bind activated rhodopsin and so activate cGMP phosphodiesterase. At the same time, the increase in arrestin in the outer segment means that any activated rhodopsin will become inactivated more rapidly. Together, the drop in transducin and the increase in arrestin greatly reduce the ability of small increases in light levels to activate the downstream effector cGMP phosphodiesterase; thus only large changes in light levels will be sensed by the rod cells. These protein movements are reversed when the ambient light level is lowered.

Na$^+$/Ca$^{2+}$ channels, hyperpolarization of the membrane, and decreased release of neurotransmitter (see Figure 15-23).

- Several mechanisms act to terminate visual signaling: GAP proteins inactivate G$_{\alpha t}$·GTP, Ca$^{2+}$-sensing proteins activate guanylate cyclase, and rhodopsin phosphorylation and binding of arrestin inhibits activation of transducin.

- Adaptation to a wide range of ambient light levels is mediated by movements of transducin and arrestin into and out of the rod-cell outer segment, which together modulate the ability of small increases in light levels to activate the downstream effector cGMP phosphodiesterase and thus the sensitivity of the rod cell in different ambient levels of light.

## 15.5 G Protein–Coupled Receptors That Activate or Inhibit Adenylyl Cyclase

GPCR pathways that utilize **adenylyl cyclase** as an effector protein and cAMP as the second messenger are found in most mammalian cells, where they regulate cellular functions as diverse as metabolism of fats and sugars, synthesis and secretion of hormones, and muscle contraction. These pathways follow the general GPCR mechanism outlined in Figure 15-17: ligand binding to the receptor activates a coupled trimeric G protein that activates an effector protein—in this case, adenylyl cyclase, which synthesizes the diffusible second messenger cAMP from ATP (Figure 15-26). cAMP, in turn, activates a cAMP-dependent protein kinase that phosphorylates specific target proteins.

To explore this GPCR/cAMP pathway, we focus on the first such pathway discovered: the hormone-stimulated generation of glucose-1-phosphate from **glycogen**, a storage polymer of glucose. The breakdown of glycogen (*glycogenolysis*) occurs in muscle and liver cells in response to hormones such as epinephrine and glucagon and is a principal way that glucose is made available to cells in need of energy. This example shows how activation of a GPCR can stimulate the activity of a host of intracellular enzymes, all involved in a physiologically important task: glycogen metabolism.

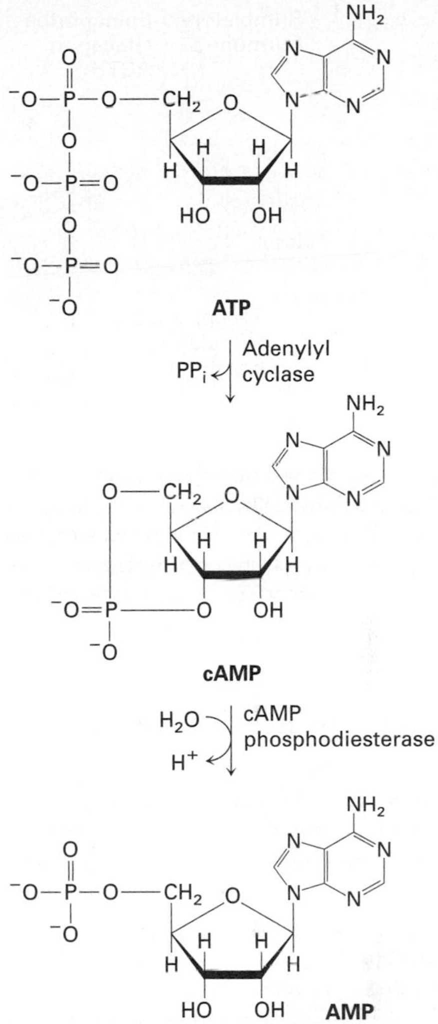

### Adenylyl Cyclase Is Stimulated and Inhibited by Different Receptor-Ligand Complexes

Under conditions where demand for glucose is high because of low blood sugar, *glucagon* is released by the alpha cells in the pancreatic islets; in case of sudden stress, epinephrine is released by the adrenal glands. Both glucagon and epinephrine signal liver and muscle cells to depolymerize glycogen, releasing individual glucose molecules. In the liver, glucagon and epinephrine bind to different G protein–coupled receptors, but both receptors interact with and activate the same stimulatory G$_s$ protein that activates adenylyl cyclase. Hence both hormones induce the same metabolic responses. Activation of adenylyl cyclase, and thus the cAMP level, is proportional to

**FIGURE 15-26 Synthesis and hydrolysis of cAMP by adenylyl cyclase and cAMP phosphodiesterase.** Similar reactions occur for production of cGMP from GTP and hydrolysis of cGMP.

the total concentration of G$_{\alpha s}$·GTP resulting from binding of both hormones to their respective receptors.

Positive (activation) and negative (inhibition) regulation of adenylyl cyclase activity occurs in many cell types, providing fine-tuned control of the cAMP level and so of the downstream cellular response (Figure 15-27). For example, in adipose cells the breakdown of triacylglycerols to fatty acids and glycerol (*lipolysis*) is stimulated by binding of epinephrine, glucagon, or adrenocorticotropic hormone (ACTH) to separate receptors that activate adenylyl cyclase. Conversely, binding of two other hormones, prostaglandin E$_1$ (PGE$_1$) or adenosine, to their respective G protein–coupled receptors inhibits adenylyl cyclase. The prostaglandin and adenosine receptors activate an inhibitory G$_i$ protein that contains the same β and γ subunits as the stimulatory G$_s$ protein but a different α subunit (G$_{\alpha i}$). After the active G$_{\alpha i}$·GTP complex dissociates from G$_{\beta \gamma}$, it binds to but inhibits (rather than stimulates) adenylyl cyclase, resulting in lower cAMP levels.

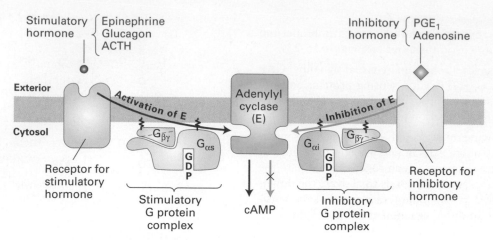

**FIGURE 15-27 Hormone-induced activation and inhibition of adenylyl cyclase in adipose cells.** Ligand binding to $G_{\alpha s}$-coupled receptors causes activation of adenylyl cyclase, whereas ligand binding to $G_{\alpha i}$-coupled receptors causes inhibition of the enzyme. The $G_{\beta\gamma}$ subunit in both stimulatory and inhibitory G proteins is identical; the $G_\alpha$ subunits and their corresponding receptors differ. Ligand-stimulated formation of active $G_\alpha \cdot$GTP complexes occurs by the same mechanism in both $G_{\alpha s}$ and $G_{\alpha i}$ proteins (see Figure 15-17). However, $G_{\alpha s} \cdot$GTP and $G_{\alpha i} \cdot$GTP interact differently with adenylyl cyclase, so that one stimulates and the other inhibits its catalytic activity. [See A. G. Gilman, 1984, *Cell* **36**:577.]

## Structural Studies Established How $G_{\alpha s} \cdot$GTP Binds to and Activates Adenylyl Cyclase

X-ray crystallographic analysis has pinpointed the regions in $G_{\alpha s} \cdot$GTP that interact with adenylyl cyclase. This enzyme is a multispanning transmembrane protein with two large cytosolic segments containing the catalytic domains that convert ATP to cAMP (Figure 15-28a). Because such transmembrane proteins are notoriously difficult to crystallize, scientists prepared two protein fragments encompassing the two catalytic domains of adenylyl cyclase that tightly associate with each other in a heterodimer. When these catalytic fragments are allowed to associate in the presence of $G_{\alpha s} \cdot$GTP and forskolin (a plant chemical that binds to and activates adenylyl cyclase), they are stabilized in their active conformations.

The resulting water-soluble complex (two adenylyl cyclase domain fragments/$G_{\alpha s} \cdot$GTP/forskolin) had an enzymatic activity synthesizing cAMP similar to that of intact full-length adenylyl cyclase. In this complex, two regions of

$G_{\alpha s} \cdot$GTP—the switch II helix and the $\alpha 3$–$\beta 5$ loop—contact the adenylyl cyclase fragments (Figure 15-28b). These contacts are thought to be responsible for the activation of the enzyme by $G_{\alpha s} \cdot$GTP. Recall that switch II is one of the segments of a $G_\alpha$ subunit whose conformation is different in the GTP-bound and GDP-bound states (see Figure 15-7). The GTP-induced conformation of $G_{\alpha s}$ that favors its dissociation from $G_{\beta\gamma}$ is precisely the conformation essential for binding of $G_{\alpha s}$ to adenylyl cyclase.

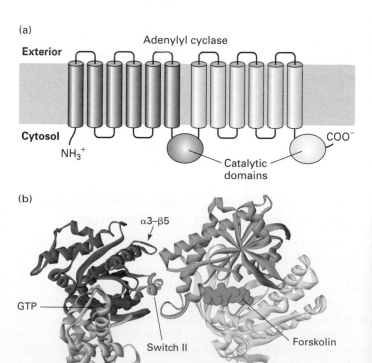

**FIGURE 15-28 Structure of mammalian adenylyl cyclase and its interaction with $G_{\alpha s} \cdot$GTP.** (a) Schematic diagram of mammalian adenylyl cyclase. The membrane-bound enzyme contains two similar catalytic domains, which convert ATP to cAMP, on the cytosolic face of the membrane, and two integral membrane domains, each of which is thought to contain six transmembrane $\alpha$ helices. (b) Model of the three-dimensional structure of $G_{\alpha s} \cdot$GTP complexed with two fragments encompassing one catalytic domain of adenylyl cyclase determined by x-ray crystallography. The $\alpha 3$-$\beta 5$ loop (gray) and the helix in the switch II region (blue) of $G_{\alpha s} \cdot$GTP interact simultaneously with a specific region of adenylyl cyclase. The darker-colored part of $G_{\alpha s}$ is the GTP-binding domain, which is similar in structure to Ras (see Figure 15-7); the lighter part is a helical domain. The two adenylyl cyclase fragments are shown in orange and yellow. Forskolin (green) locks the cyclase fragments in their active conformations. [Part (a), see W.-J. Tang and A. G. Gilman, 1992, *Cell* **70**:869; part (b) adapted from J. J. G. Tesmer et al., 1997, *Science* **278**:1907.]

## cAMP Activates Protein Kinase A by Releasing Inhibitory Subunits

The second messenger cAMP, synthesized by adenylyl cyclase, transduces a wide variety of physiological signals in different cell types in multicellular animals. Virtually all of the diverse effects of cAMP are mediated through activation of **protein kinase A (PKA)**, also called *cAMP-dependent protein kinase*, which phosphorylates different intracellular target proteins expressed in different cell types. Inactive PKA is a tetramer consisting of two regulatory (R) subunits and two catalytic (C) subunits (Figure 15-29a). Each R subunit binds to the active site in a catalytic domain and inhibits the activity of the catalytic subunits. Inactive PKA is turned on by binding of cAMP. Each R subunit has two distinct cAMP-binding sites, called CNB-A and CNB-B (Figure 15-29b). Binding of cAMP to both sites on an R subunit causes a conformational change in the R subunit that leads to release of the associated C subunit, unmasking its catalytic site and activating its kinase activity (Figure 15-29c).

Binding of cAMP by an R subunit of protein kinase A occurs in a cooperative fashion; that is, binding of the first cAMP molecule to CNB-B lowers the $K_d$ for binding of the second cAMP to CNB-A. Thus small changes in the level of cytosolic cAMP can cause proportionately large changes in the number of dissociated C subunits and, hence, in cellular kinase activity. Rapid activation of enzymes by hormone-triggered dissociation of an inhibitor is a common feature of many signaling pathways.

## Glycogen Metabolism Is Regulated by Hormone-Induced Activation of Protein Kinase A

Glycogen, a large glucose polymer, is the major storage form of glucose in animals. Like all biopolymers, glycogen is synthesized by one set of enzymes and degraded by another (Figure 15-30). Degradation of glycogen, or glycogenolysis, involves the stepwise removal of glucose residues from one end of the polymer by a phosphorolysis reaction, catalyzed by *glycogen phosphorylase*, yielding glucose 1-phosphate.

In both muscle and liver cells, glucose-1-phosphate produced from glycogen is converted to glucose-6-phosphate. In muscle cells, this metabolite enters the glycolytic pathway and is metabolized to generate ATP for use in powering muscle contraction (see Chapter 12). Unlike muscle cells, liver cells contain a phosphatase that hydrolyzes glucose-6-phosphate to glucose, which is exported from these cells mainly by a glucose transporter (GLUT2) in the plasma membrane (see Chapter 11). Thus glycogen stores in the liver are primarily broken down to glucose, which is immediately released into the blood and transported to other tissues, particularly the muscles and brain, to nourish them.

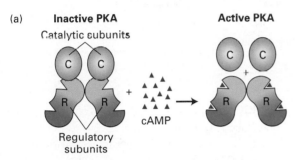

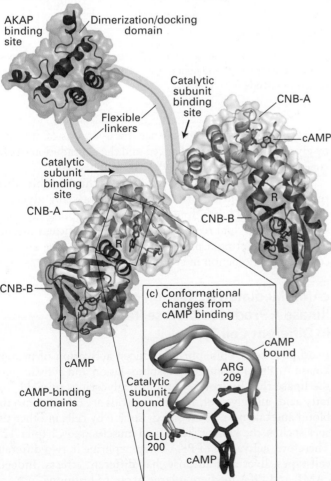

**FIGURE 15-29  Structure of protein kinase A (PKA) and its activation by cAMP.** (a) Protein kinase A (PKA) consists of two regulatory (R) subunits (green) and two catalytic (C) subunits. When cAMP (red triangle) binds to the regulatory subunit, the catalytic subunit is released, thus activating PKA. (b) The two regulatory subunits form a dimer, joined by a dimerization/docking domain and a flexible linker to which A-kinase-activating protein (AKAP; see Figure 15-33) can bind. Each R subunit has two cAMP-binding domains, CNB-A and CNB-B, and a binding site for a catalytic subunit (arrow). (c) Binding of cAMP to the CNB-A domain causes a subtle conformational change that displaces the catalytic subunit from R, leading to its activation. Without bound cAMP, one loop of the CNB-A domain (purple) is in a conformation that can bind the catalytic (C) subunit. A glutamate (E200) and arginine (R209) residue participate in binding of cAMP (red), which causes a conformational change (green) in the loop that prevents binding of the loop to the C subunit. [Part (b) after S. S. Taylor et al., 2005, *Biochim. Biophys. Acta* **1754**:25; part (c) after C. Kim, N. H. Xuong, and S. S. Taylor, 2005, *Science* **307**:690.]

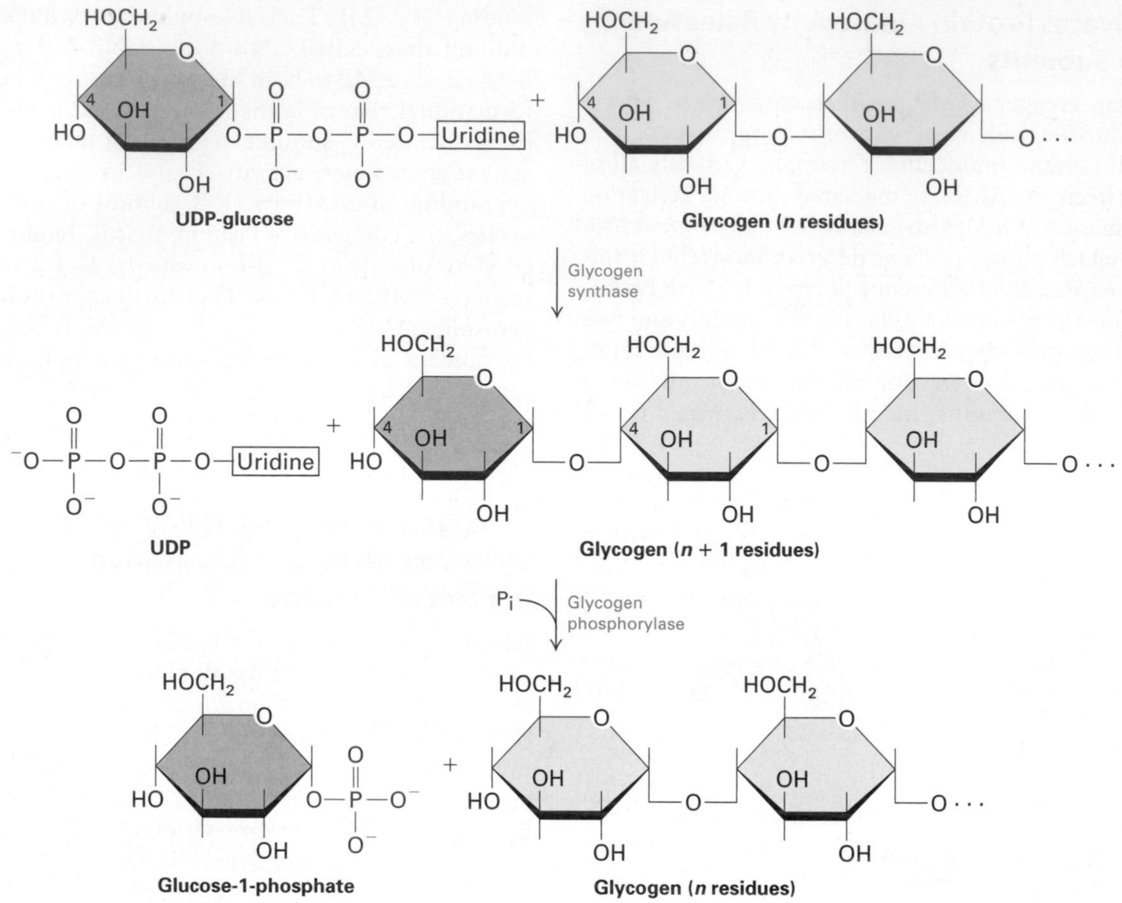

**FIGURE 15-30 Synthesis and degradation of glycogen.** Incorporation of glucose from UDP-glucose into glycogen is catalyzed by glycogen synthase. Removal of glucose units from glycogen is catalyzed by glycogen phosphorylase. Because two different enzymes catalyze the formation and degradation of glycogen, the two reactions can be independently regulated.

The epinephrine-stimulated activation of adenylyl cyclase, resulting increase in cAMP, and subsequent activation of protein kinase A (PKA) enhances the conversion of glycogen to glucose-1-phosphate in two ways: by *inhibiting* glycogen synthesis and by *stimulating* glycogen degradation (Figure 15-31a). PKA phosphorylates and in so doing inactivates glycogen synthase (GS), the enzyme that synthesizes glycogen. PKA promotes glycogen degradation indirectly by phosphorylating and thus activating an intermediate kinase, glycogen phosphorylase kinase (GPK), that in turn phosphorylates and activates glycogen phosphorylase (GP), the enzyme that degrades glycogen. These kinases are counteracted by a phosphatase called phosphoprotein phosphatase (PP). At high cAMP levels, PKA phosphorylates an inhibitor of phosphoprotein phosphatase (IP), which keeps this phosphatase in its inactive state (see Figure 15-31a, *right*).

The entire process is reversed when epinephrine is removed and the level of cAMP drops, inactivating protein kinase A (PKA). When PKA is inactive, it can no longer phosphorylate the inhibitor of phosphoprotein phosphatase (IP), so this phosphatase becomes active (Figure 15-31b). Phosphoprotein phosphatase (PP) removes the phosphate residues previously added by PKA to glycogen synthase (GS), glycogen phosphorylase kinase (GPK), and glycogen phos-

phorylase (GP). As a consequence, the synthesis of glycogen by glycogen synthase is enhanced and the degradation of glycogen by glycogen phosphorylase is inhibited.

Epinephrine-induced glycogenolysis thus exhibits dual regulation: activation of the enzymes catalyzing glycogen degradation and inhibition of enzymes promoting glycogen synthesis. Such dual regulation provides an efficient mechanism for regulating a particular cellular response and is a common phenomenon in cell biology.

## cAMP-Mediated Activation of Protein Kinase A Produces Diverse Responses in Different Cell Types

In adipose cells, epinephrine-induced activation of protein kinase A (PKA) promotes phosphorylation and activation of the lipase that hydrolyzes stored triglycerides to yield free fatty acids and glycerol. These fatty acids are released into the blood and taken up as an energy source by cells in other tissues such as the kidney, heart, and muscles (see Chapter 12). Therefore, activation of PKA by epinephrine in two different cell types, liver and adipose, has different effects. Indeed, cAMP and PKA mediate a large array of hormone-induced cellular responses in numerous tissues (Table 15-2).

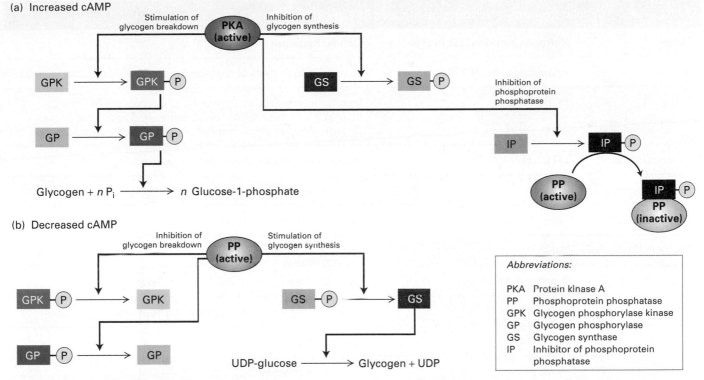

(a) Increased cAMP

Stimulation of glycogen breakdown — Inhibition of glycogen synthesis

Glycogen + $n$ P$_i$ ⟶ $n$ Glucose-1-phosphate

Inhibition of phosphoprotein phosphatase

(b) Decreased cAMP

Inhibition of glycogen breakdown — Stimulation of glycogen synthesis

UDP-glucose ⟶ Glycogen + UDP

Abbreviations:

PKA   Protein kinase A
PP    Phosphoprotein phosphatase
GPK   Glycogen phosphorylase kinase
GP    Glycogen phosphorylase
GS    Glycogen synthase
IP    Inhibitor of phosphoprotein phosphatase

**FIGURE 15-31 Regulation of glycogen metabolism by cAMP and PKA.** Active enzymes are highlighted in darker shades; inactive forms, in lighter shades. (a) An increase in cytosolic cAMP activates protein kinase A (PKA), which inhibits glycogen synthesis directly and promotes glycogen degradation via a protein kinase cascade. At high cAMP, PKA also phosphorylates an inhibitor of phosphoprotein phosphatase (PP). Binding of the phosphorylated inhibitor to PP prevents this phosphatase from dephosphorylating the activated enzymes in the kinase cascade or the inactive glycogen synthase. (b) A decrease in cAMP inactivates PKA, leading to release of the active form of PP. The action of this enzyme promotes glycogen synthesis and inhibits glycogen degradation.

Although protein kinase A acts on different substrates in different types of cells, it always phosphorylates a serine or threonine residue that occurs within the same sequence motif: X-Arg-(Arg/Lys)-X-(Ser/Thr)-Φ, where X denotes any amino acid and Φ denotes a hydrophobic amino acid. Other serine/threonine kinases phosphorylate target residues within other sequence motifs.

## Signal Amplification Occurs in the cAMP–Protein Kinase A Pathway

We've seen that receptors such as the β-adrenergic receptor are low-abundance proteins, typically present in only a few thousand copies per cell. Yet the cellular responses induced by a hormone such as epinephrine can require production of large numbers of cAMP and activated enzyme molecules per cell. As an example, following activation of G$_{\alpha s}$-coupled receptors, the intracellular concentration of cAMP will rise to about $10^{-6}$ M; in a typical cell that is a cube ~15 μm on a side, this comes to ~2 million molecules of cAMP produced per cell. Thus substantial amplification of the signal is necessary in order for the hormone to induce a significant cellular response. We have already seen how signal amplification occurs following photon absorbance in rod cells. In the case of G protein–coupled hormone receptors, signal amplification is possible in part because both receptors and G proteins can diffuse rapidly in the plasma membrane. A single epinephrine-GPCR complex causes con-

version of up to 100 inactive G$_{\alpha s}$ molecules to the active form before epinephrine dissociates from the receptor. Each active G$_{\alpha s}$·GTP, in turn, activates a single adenylyl cyclase molecule, which then catalyzes synthesis of many cAMP molecules during the time G$_{\alpha s}$·GTP is bound to it.

The amplification that occurs in such a signal transduction cascade depends on the number of steps in it and the relative concentrations of the various components. In the epinephrine-induced cascade shown in Figure 15-9, for example, blood levels of epinephrine as low as $10^{-10}$ M can stimulate liver glycogenolysis and release of glucose. An epinephrine stimulus of this magnitude generates an intracellular cAMP concentration of $10^{-6}$ M, an amplification of $10^4$-fold. Because three more catalytic steps precede the release of glucose, another $10^4$ amplification can occur, resulting in a $10^8$ amplification of the epinephrine signal. In striated muscle, the amplification is less dramatic because the concentrations of the three successive enzymes in the glycogenolytic cascade—protein kinase A, glycogen phosphorylase kinase, and glycogen phosphorylase—are in a 1:10:240 ratio (a potential 240-fold maximal amplification).

## CREB Links cAMP and Protein Kinase A to Activation of Gene Transcription

Activation of protein kinase A also stimulates the expression of many genes, leading to long-term effects on the cells that often enhance the short-term effects of activated protein kinase A.

## TABLE 15-2 Cellular Responses to Hormone-Induced Rise in cAMP in Various Tissues*

| Tissue | Hormone Inducing Rise in cAMP | Cellular Response |
|---|---|---|
| Adipose | Epinephrine; ACTH; glucagon | Increase in hydrolysis of triglyceride; decrease in amino acid uptake |
| Liver | Epinephrine; norepinephrine; glucagon | Increase in conversion of glycogen to glucose; inhibition of glycogen synthesis; increase in amino acid uptake; increase in gluconeogenesis (synthesis of glucose from amino acids) |
| Ovarian follicle | FSH; LH | Increase in synthesis of estrogen, progesterone |
| Adrenal cortex | ACTH | Increase in synthesis of aldosterone, cortisol |
| Cardiac muscle | Epinephrine | Increase in contraction rate |
| Thyroid gland | TSH | Secretion of thyroxine |
| Bone | Parathyroid hormone | Increase in resorption of calcium from bone |
| Skeletal muscle | Epinephrine | Conversion of glycogen to glucose-1-phosphate |
| Intestine | Epinephrine | Fluid secretion |
| Kidney | Vasopressin | Resorption of water |
| Blood platelets | Prostaglandin I | Inhibition of aggregation and secretion |

*Nearly all the effects of cAMP are mediated through protein kinase A (PKA), which is activated by binding of cAMP.
SOURCE: E. W. Sutherland, 1972, *Science* 177:401.

For instance, in liver cells, protein kinase A induces expression of several enzymes involved in gluconeogenesis—the conversion of three-carbon compounds such as pyruvate (see Figure 12-3) to glucose—thus increasing the level of glucose in the blood.

All genes regulated by protein kinase A contain a cis-acting DNA sequence, the *cAMP-response element* (*CRE*), that binds the phosphorylated form of a transcription factor called *CRE-binding (CREB) protein*, which is found only in the nucleus. Following the elevation of cAMP levels and the release of the active protein kinase A catalytic subunit, some of the catalytic subunits then translocate to the nucleus. There they phosphorylate serine-133 on the CREB protein. Phosphorylated CREB protein binds to CRE-containing target genes and also binds to a *co-activator* termed *CBP/300*. CBP/300 links CREB to RNA polymerase 2 and other gene regulatory proteins, thereby stimulating gene transcription (Figure 15-32).

Thus protein kinase A phosphorylates multiple types of proteins: some have relatively short-term effects on cellular metabolism, lasting seconds to minutes; other substrates such as CREB, by activating expression of specific genes, affect cellular metabolism over hours and days.

## Anchoring Proteins Localize Effects of cAMP to Specific Regions of the Cell

In many cell types, a rise in the cAMP level may produce a response that is required in one part of the cell but is unneeded, perhaps deleterious, in another. A family of anchoring proteins localizes isoforms of protein kinase A (PKA) to specific subcellular locations, thereby restricting cAMP-dependent responses to these locations. These proteins, referred to as *A kinase–associated proteins* (*AKAPs*), have a two-domain structure with one domain conferring a specific subcellular location and another that binds to the regulatory (R) subunit of protein kinase A (see Figure 15-29b).

One such anchoring protein (AKAP15) is tethered to the cytosolic face of the plasma membrane near a particular type of gated $Ca^{2+}$ channel in certain heart muscle cells. In the heart, activation of β-adrenergic receptors by epinephrine (as part of the fight-or-flight response) leads to PKA-catalyzed phosphorylation of these $Ca^{2+}$ channels, causing them to open; the resulting influx of $Ca^{2+}$ increases the rate of heart muscle contraction. The binding of AKAP15 to protein kinase A localizes the kinase next to these channels, thereby reducing the time that otherwise would be required for diffusion of PKA catalytic subunits from their sites of generation to their $Ca^{2+}$-channel substrates.

A different AKAP in heart muscle anchors both protein kinase A and cAMP phosphodiesterase (PDE)—the enzyme that hydrolyzes cAMP to AMP (see Figure 15-26)—to the outer nuclear membrane. Because of the close proximity of PDE to protein kinase A, negative feedback provides tight local control of the cAMP concentration and hence local PKA activity (Figure 15-33). As cAMP levels rise in response to hormone stimulation, PKA is activated. Activated PKA phosphorylates PDE, which in turn becomes more active and

**FIGURE 15-32 Activation of CREB transcription factor following ligand binding to G$_s$ protein–coupled receptors.** Receptor stimulation (**1**) leads to activation of protein kinase A (PKA) (**2**). Catalytic subunits of PKA translocate to the nucleus (**3**) and there phosphorylate and activate the CREB transcription factor (**4**). Phosphorylated CREB associates with the co-activator CBP/P300 (**5**) and other proteins to stimulate transcription of the various target genes controlled by the CRE regulatory element. [See K. A. Lee and N. Masson, 1993, *Biochim. Biophys. Acta* **1174**:221, and D. Parker et al., 1996, *Mol. Cell Biol.* **16**(2):694.]

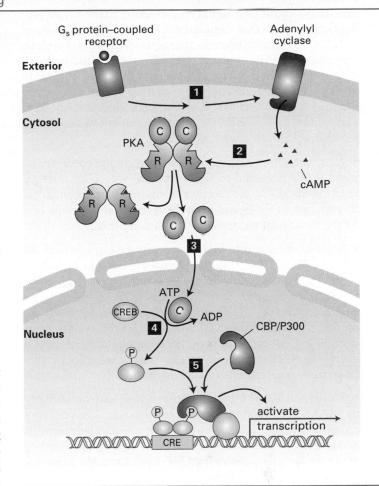

hydrolyzes cAMP, thus returning PKA to its inactive state. The localization of protein kinase A near the nuclear membrane also facilitates entry of its catalytic subunits into the nucleus, where they phosphorylate and activate the CREB transcription factor (see Figure 15-32).

## Multiple Mechanisms Down-Regulate Signaling from the GPCR/cAMP/PKA Pathway

For cells to respond effectively to changes in their environment, they must not only activate a signaling pathway but also down-modulate or terminate the response once it is no longer needed; otherwise signal transduction pathways would remain "on" too long or at too high a level and the cell would become overstimulated. Abnormal regulation of signaling pathways is very common in cancer cells, where mutant proteins that stimulate cell proliferation or prevent programmed cell death remain active even when the signals that normally activate them are absent.

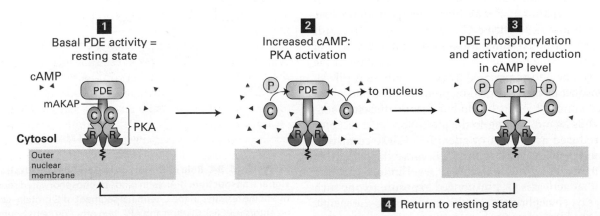

**4** Return to resting state

**FIGURE 15-33 Localization of protein kinase A (PKA) to the nuclear membrane in heart muscle by an A-kinase-associated protein (AKAP).** This member of the AKAP family, designated mAKAP, anchors both cAMP phosphodiesterase (PDE) and the regulatory subunit (R, see Figure 15-29b) of PKA to the nuclear membrane, maintaining them in a negative feedback loop that provides close local control of the cAMP level and PKA activity. Step **1**: The basal level of PDE activity in the absence of hormone (resting state) keeps cAMP levels below those necessary for PKA activation. Steps **2** and **3**: Activation of β-adrenergic receptors causes an increase in cAMP level in excess of that which can be degraded by PDE. The resulting binding of cAMP to the regulatory (R) subunits of PKA releases the active catalytic (C) subunits into the cytosol. Some C subunits enter into the nucleus, where they phosphorylate and thus activate certain transcription factors (see Figure 15-32). Other C subunits phosphorylate PDE, stimulating its catalytic activity. Active PDE hydrolyzes cAMP, thereby driving cAMP levels back to basal levels and causing re-formation of the inactive PKA-R complex. Step **4**: Subsequent dephosphorylation of PDE returns the complex to the resting state. [Adapted from K. L. Dodge et al., 2001, *EMBO J.* **20**:1921.]

Earlier we saw the multiple mechanisms that rapidly terminate the rhodopsin signal transduction pathway, including GAP proteins that stimulate the hydrolysis of GTP bound to $G_{\alpha t}$, $Ca^{2+}$-sensing proteins that activate guanylate cyclase, and phosphorylation of active rhodopsin by rhodopsin kinase followed by binding of arrestin (see Figure 15-24). In fact, most G protein–coupled receptors are modulated by multiple mechanisms that down-regulate their activity, as is exemplified by β-adrenergic receptors and others coupled to $G_{\alpha s}$ that activate adenylyl cyclase.

• First, the affinity of the receptor for its ligand decreases when the GDP bound to $G_{\alpha s}$ is replaced with GTP. This increase in the $K_d$ of the receptor-hormone complex enhances dissociation of the ligand from the receptor and thereby limits the number of $G_{\alpha s}$ proteins that are activated.

• Second, the intrinsic GTPase activity of $G_{\alpha s}$ converts the bound GTP to GDP, resulting in inactivation of $G_{\alpha s}$ and decreased activation of its downstream target adenylyl cyclase. Importantly, the rate of hydrolysis of GTP bound to $G_{\alpha s}$ is enhanced when $G_{\alpha s}$ binds to adenylyl cyclase, lessening the duration of cAMP production; thus adenylyl cyclase functions as a GAP for $G_{\alpha s}$. More generally, binding of most if not all $G_\alpha \cdot$GTP complexes to their respective effector proteins accelerates the rate of GTP hydrolysis.

• Finally, *cAMP phosphodiesterase* acts to hydrolyze cAMP to 5′-AMP, terminating the cellular response. Thus the continuous presence of hormone at a high enough concentration is required for continuous activation of adenylyl cyclase and maintenance of an elevated cAMP level. Once the hormone concentration falls sufficiently, the cellular response quickly terminates.

Most GPCRs are also down-regulated by *feedback repression*, a term describing the situation in which the end product of a signaling pathway blocks an early step in the pathway. For instance, when a $G_{\alpha s}$ protein–coupled receptor is exposed to hormonal stimulation for several hours, several serine and threonine residues in the cytosolic domain of the receptor become phosphorylated by protein kinase A (PKA), the end product of the $G_{\alpha s}$ signaling pathway. The phosphorylated receptor can bind its ligand but cannot efficiently activate $G_{\alpha s}$; thus ligand binding to the phosphorylated receptor leads to reduced activation of adenylyl cyclase compared with a nonphosphorylated receptor. Because the activity of PKA is enhanced by the high cAMP level induced by any hormone that activates $G_{\alpha s}$, prolonged exposure to one such hormone, say, epinephrine, desensitizes not only β-adrenergic receptors but also other $G_{\alpha s}$ protein–coupled receptors that bind different ligands (e.g., glucagon receptor in liver). This cross-regulation is called *heterologous desensitization*.

Similar to phosphorylation of activated rhodopsin by rhodopsin kinase, particular residues in the cytosolic domain of the β-adrenergic receptor, not those phosphorylated by PKA, can be phosphorylated by the related enzyme *β-adrenergic receptor kinase (BARK)*, but *only* when epinephrine or an agonist is bound to the receptor and thus the receptor is in its active

conformation. This process is called *homologous desensitization*, because only those receptors that are in their active conformations are subject to deactivation by phosphorylation.

We noted that binding of arrestin to extensively phosphorylated opsin completely inhibits activation of coupled G proteins by activated opsin (see Figure 15-24). In fact, a related protein termed *β-arrestin* plays a similar role in desensitizing other G protein–coupled receptors, including β-adrenergic receptors. An additional function of β-arrestin in regulating cell-surface receptors initially was suggested by the observation that disappearance of β-adrenergic receptors from the cell surface in response to ligand binding is stimulated by overexpression of BARK and β-arrestin. Subsequent studies revealed that β-arrestin binds not only to phosphorylated receptors but also to clathrin and an associated protein termed AP2, two key components of the coated vesicles that are involved in one type of endocytosis from the plasma membrane (Figure 15-34). These interactions promote the formation of coated pits and endocytosis of the associated receptors, thereby decreasing the number of receptors exposed on the cell surface. Eventually some of the internalized receptors are degraded intracellularly, and some are dephosphorylated in endosomes. Following dissociation of β-arrestin, the resensitized (dephosphorylated) receptors recycle to the cell surface, similar to recycling of the LDL receptor (see Chapter 14).

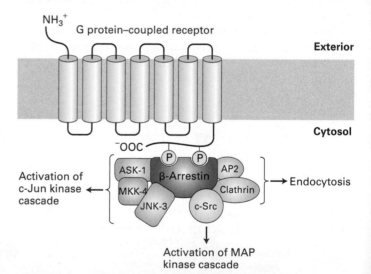

**FIGURE 15-34 Role of β-arrestin in GPCR desensitization and signal transduction.** β-Arrestin binds to phosphorylated serine and threonine residues in the C-terminal segment of G protein–coupled receptors (GPCRs). Clathrin and AP2, two other proteins bound by β-arrestin, promote endocytosis of the receptor. β-arrestin also functions in transducing signals from activated receptors by binding to and activating several cytosolic protein kinases. c-Src activates the MAP kinase pathway, leading to phosphorylation of key transcription factors (see Chapter 16). Interaction of β-arrestin with three other proteins, including JNK-3 (a Jun N-terminal kinase), results in phosphorylation and activation of another transcription factor, c-Jun. [Adapted from W. Miller and R. J. Lefkowitz, 2001, *Curr. Opin. Cell Biol.* **13**:139, and K. Pierce et al., 2002, *Nature Rev. Mol. Cell Biol.* **3**:639.]

In addition to its role in regulating receptor activity, β-arrestin functions as an adapter protein in transducing signals from G protein–coupled receptors to the nucleus (see Chapter 16). The GPCR-arrestin complex acts as a scaffold for binding and activating several cytosolic kinases (see Figure 15-34), which we discuss in detail in subsequent chapters. These include c-Src, a cytosolic protein tyrosine kinase that activates the MAP kinase pathway and other pathways leading to transcription of genes needed for cell division (see Chapter 19). A complex of three arrestin-bound proteins, including a Jun N-terminal kinase (JNK-3), initiates a kinase cascade that ultimately activates the c-Jun transcription factor, which promotes expression of certain growth-promoting enzymes and other proteins that help cells respond to stresses. Thus the BARK-β–arrestin pathway, originally just thought to suppress signaling by GPCRs, actually functions as a switch, turning off signaling by G proteins and turning on other signaling pathways. The multiple functions of β-arrestin illustrate the importance of adapter proteins in both regulating signaling and transducing signals from cell-surface receptors.

## KEY CONCEPTS of Section 15.5

### G Protein–Coupled Receptors That Activate or Inhibit Adenylyl Cyclase

- Ligand binding of G protein–coupled receptors that activate $G_{\alpha s}$ results in the activation of the membrane-bound enzyme adenylyl cyclase, which converts ATP to the second messenger cyclic AMP (cAMP; see Figure 15-26). Ligand binding of G protein–coupled receptors that activate $G_{\alpha i}$ results in the inhibition of adenylyl cyclase and lower levels of cAMP (see Figure 15-27).

- $G_{\alpha s}$·GTP and $G_{\alpha i}$·GTP bind to the heterodimeric active site domains in adenylyl cyclase to activate or inhibit the enzyme, respectively (see Figure 15-28).

- cAMP binds cooperatively to a regulatory subunit of protein kinase A (PKA), releasing the active kinase catalytic subunit (see Figure 15-29).

- PKA mediates the diverse effects of cAMP in most cells (see Table 15-2). The substrates for PKA, and thus the cellular response to hormone-induced activation of PKA, vary among cell types.

- In liver and muscle cells, activation of PKA induced by epinephrine and other hormones exerts a dual effect, inhibiting glycogen synthesis and stimulating glycogen breakdown via a kinase cascade (see Figure 15-31).

- The signal that activates the GPCR/adenylyl cyclase/cAMP/PKA signaling pathway is amplified tremendously by second messengers and kinase cascades (see Figure 15-9).

- Activation of PKA often leads to phosphorylation of nuclear CREB protein, which together with the CBP/300 co-activator stimulates transcription of genes, thus initiating a long-term change in the cell's protein composition (see Figure 15-32).

- Localization of PKA to specific regions of the cell by anchoring proteins restricts the effects of cAMP to particular subcellular locations (see Figure 15-33).

- Signaling from $G_s$-coupled receptors is down-regulated by multiple mechanisms: (1) the affinity of the receptor for its ligand decreases when the GDP bound to $G_{\alpha s}$ is replaced with GTP; (2) the intrinsic GTPase activity of $G_{\alpha s}$ that converts the bound GTP to GDP is enhanced when $G_{\alpha s}$ binds to adenylyl cyclase (this occurs when many $G_{\alpha}$·GTP complexes bind to their respective effector proteins); and (3) cAMP phosphodiesterase acts to hydrolyze cAMP to 5'-AMP, terminating the cellular response.

- Most GPCRs are also down-regulated by feedback repression, in which the end product of a pathway (e.g., PKA) blocks an early step in the pathway. As with opsin, binding of β-arrestin to phosphorylated β-adrenergic receptors completely inhibits activation of coupled G proteins (see Figure 15-24).

- β-adrenergic receptors are deactivated by β-adrenergic kinase (BARK), which phosphorylates cytosolic residues of the receptor in its active conformation. BARK phosphorylation of ligand-bound β-adrenergic receptors also leads to the binding of β-arrestin and endocytosis of the receptors. The consequent reduction in the number of cell-surface receptors renders the cell less sensitive to additional hormone.

- The GPCR–arrestin complex functions as a scaffold that activates several cytosolic kinases, initiating cascades that lead to transcriptional activation of many genes controlling cell growth (see Figure 15-34).

## 15.6 G Protein–Coupled Receptors That Trigger Elevations in Cytosolic $Ca^{2+}$

Calcium ions play an essential role in regulating cellular responses to many signals, and many GCPRs and other types of receptors exert their effects on cells by influencing the cytosolic concentration of $Ca^{2+}$. As we saw in Chapter 11, the level of $Ca^{2+}$ in the cytosol is maintained at a submicromolar level (~0.2 μM) by the continuous action of ATP-powered $Ca^{2+}$ pumps, which transport $Ca^{2+}$ ions across the plasma membrane to the cell exterior or into the lumen of the endoplasmic reticulum and other vesicles. Much intracellular $Ca^{2+}$ is also sequestered in mitochondria.

A small rise in cytosolic $Ca^{2+}$ induces a variety of cellular responses, including hormone secretion by endocrine cells, secretion of digestive enzymes by pancreatic exocrine cells, and contraction of muscle (Table 15-3). For example, acetylcholine stimulation of GPCRs in secretory cells of the pancreas and parotid (salivary) gland induces a rise in cytosolic $Ca^{2+}$ that triggers the fusion of secretory vesicles with the plasma membrane and release of their protein contents into the extracellular space. Thrombin, an enzyme in the blood-clotting cascade, binds to a GPCR on blood platelets and

## TABLE 15-3 Cellular Responses to Hormone-Induced Rise in Cytosolic $Ca^{2+}$ in Various Tissues*

| Tissue | Hormone Inducing Rise in $Ca^{2+}$ | Cellular Response |
|---|---|---|
| Pancreas (acinar cells) | Acetylcholine | Secretion of digestive enzymes, such as amylase and trypsinogen |
| Parotid (salivary) gland | Acetylcholine | Secretion of amylase |
| Vascular or stomach smooth muscle | Acetylcholine | Contraction |
| Liver | Vasopressin | Conversion of glycogen to glucose |
| Blood platelets | Thrombin | Aggregation, shape change, secretion of hormones |
| Mast cells | Antigen | Histamine secretion |
| Fibroblasts | Peptide growth factors | DNA synthesis, cell division (e.g., bombesin and PDGF) |

*Hormone stimulation leads to production of inositol 1,4,5-trisphosphate ($IP_3$), a second messenger that promotes release of $Ca^{2+}$ stored in the endoplasmic reticulum.
SOURCE: M. J. Berridge, 1987, *Ann. Rev. Biochem.* 56:159, and M. J. Berridge and R. F. Irvine, 1984, *Nature* 312:315.

triggers a rise in cytosolic $Ca^{2+}$ that, in turn, causes a conformational change in the platelets that leads to their aggregation, an important step in blood clotting to prevent leakage of blood out of damaged blood vessels.

In this section, we first discuss an important signal transduction pathway that results in an elevation of cytosolic $Ca^{2+}$ ions: the GPCR-stimulated activation of a **phospholipase C (PLC)**. Phospholipases C (PLCs) are a family of enzymes that hydrolyzes a phosphoester bond in certain phospholipids, yielding two second messengers that function in elevating the cytosolic $Ca^{2+}$ level and activating a family of kinases known as protein kinases C (PKCs); PKCs in turn affect many important cellular processes such as growth and differentiation. Some PLCs are activated by GPCRs, as we describe here; others, covered in the following chapter, are activated by other types of receptors. Phospholipase Cs also produce second messengers that are important for remodeling the actin cytoskeleton (see Chapter 17) and for binding of proteins important for endocytosis and vesicle fusions (see Chapter 14). Later in this section, we see how one PLC pathway leads to the synthesis of a gas, nitric oxide (NO), that in turn signals adjacent cells. In the final part of the section, we will see how second messengers such as $Ca^{2+}$ are used to help cells integrate their responses to more than one extracellular signal.

## Activated Phospholipase C Generates Two Key Second Messengers Derived from the Membrane Lipid Phosphatidylinositol

A number of important second messengers, used in several signal transduction pathways, are derived from the membrane lipid *phosphatidylinositol* (PI). The inositol group in this phospholipid, which always faces the cytosol, can be reversibly phosphorylated at one or more positions by the combined actions of various kinases and phosphatases discussed in Chapter 16. One derivative of PI, the lipid phosphatidyl inositol 4,5-bisphosphate ($PIP_2$), is cleaved by activated phospholipase C into two important second messengers: **1,2-diacylglycerol (DAG)**, a lipophilic molecule that remains associated with the membrane, and **inositol 1,4,5-trisphosphate ($IP_3$)**, which can freely diffuse in the cytosol (Figure 15-35). We refer to downstream events involving these two second messengers collectively as the *$IP_3$/DAG pathway*.

Phospholipase C is activated by G proteins containing either $G_{\alpha o}$ or $G_{\alpha q}$ subunits. In response to hormone activation of the GPCR, the $G_{\alpha o}$ or $G_{\alpha q}$ subunits bound to GTP separate from $G_{\beta\gamma}$ and bind to and activate phospholipase C in the membrane (Figure 15-36a, step **1**). In turn, activated phospholipase C cleaves $PIP_2$ into DAG, which remains associated with the membrane, and $IP_3$, which freely diffuses in the cytosol (Figure 15-36a, step **2**). The two second messengers trigger separate downstream effects.

**$Ca^{2+}$ Release from the ER Triggered by $IP_3$** G protein–coupled receptors that activate phospholipase C induce an elevation in cytosolic $Ca^{2+}$ even when $Ca^{2+}$ ions are absent from the surrounding extracellular fluid. In this case, $Ca^{2+}$ is released into the cytosol from the ER lumen through operation of the *$IP_3$-gated $Ca^{2+}$ channel* in the ER membrane, as depicted in Figure 15-36a (steps **3** and **4**). This large-channel protein is composed of four identical subunits, each of which contains an $IP_3$-binding site in the N-terminal cytosolic domain. $IP_3$ binding induces opening of the channel, allowing $Ca^{2+}$ to flow down its concentration gradient from the ER into the cytosol. When various phosphorylated inositols found in cells are added to preparations of ER vesicles, only $IP_3$ causes release of $Ca^{2+}$ ions from the vesicles. This simple experiment demonstrates the specificity of the $IP_3$ effect.

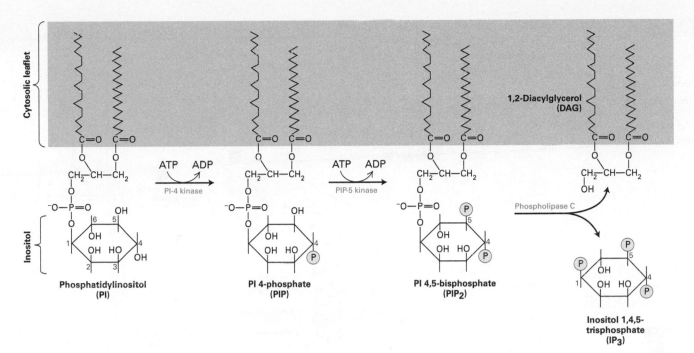

**FIGURE 15-35 Synthesis of second messengers DAG and IP₃ from phosphatidylinositol (PI).** Each membrane-bound PI kinase places a phosphate (yellow circles) on a specific hydroxyl group on the inositol ring, producing the phosphorylated derivatives PIP and PIP₂. Cleavage of PIP₂ by phospholipase C yields the two important second messengers DAG and IP₃. [See A. Toker and L. C. Cantley, 1997, *Nature* **387**:673, and C. L. Carpenter and L. C. Cantley, 1996, *Curr. Opin. Cell Biol.* **8**:153.]

The IP₃-mediated rise in the cytosolic Ca²⁺ level is transient because Ca²⁺ pumps located in the plasma membrane and ER membrane actively transport Ca²⁺ from the cytosol to the cell exterior and ER lumen, respectively. Furthermore, within a second of its generation, the phosphate linked to the carbon-5 of IP₃ (see Figure 15-35) is hydrolyzed, yielding inositol 1,4-bisphosphate. This compound cannot bind to the IP₃-gated Ca²⁺ channel protein and thus does not stimulate Ca²⁺ release from the ER.

Without some means for replenishing depleted stores of intracellular Ca²⁺, a cell would soon be unable to increase the cytosolic Ca²⁺ level in response to hormone-induced IP₃. Patch-clamping studies (see Figure 11-22) have revealed that a plasma membrane Ca²⁺ channel, called the *store-operated channel*, opens in response to depletion of ER Ca²⁺ stores. Studies in which each potential channel protein was knocked down one at a time with shRNAs established the identity of this channel protein as Orai1. The Ca²⁺-sensing protein is STIM, a transmembrane protein in the endoplasmic reticulum membrane (Figure 15-36b). An EF hand, similar to that in calmodulin (see Figure 3-31), on the luminal side of the ER membrane binds Ca²⁺ when its level in the lumen is high. As endoplasmic reticulum Ca²⁺ stores are depleted, the STIM proteins lose their bound Ca²⁺, oligomerize, and in an unknown manner relocalize to areas of the ER membrane near the plasma membrane (Figure 15-36b, *right*). There the STIM CAD domains bind to and trigger opening of Orai1, allowing influx of extracellular Ca²⁺. Combined overexpression of Orai and STIM in cultured cells leads to a marked increase in Ca²⁺ influx, establishing that these two proteins are the key components of the store-operated Ca²⁺ pathway.

Continuous activation of certain G protein–coupled receptors induces rapid, repeated spikes in the level of cytosolic Ca²⁺. These bursts in cytosolic Ca²⁺ levels are caused by a complex interaction between the cytosolic Ca²⁺ concentration and the IP₃-gated Ca²⁺-channel protein. The submicromolar level of cytosolic Ca²⁺ in the resting state potentiates opening of these channels by IP₃, thus facilitating the rapid rise in cytosolic Ca²⁺ following hormone stimulation of the cell-surface G protein–coupled receptor. However, the higher cytosolic Ca²⁺ levels reached at the peak of the spike inhibit IP₃-induced release of Ca²⁺ from intracellular stores by decreasing the affinity of the Ca²⁺ channels for IP₃. As a result, the channels close, and the cytosolic Ca²⁺ level drops rapidly. Thus cytosolic Ca²⁺ is a feedback inhibitor of the protein, the IP₃-gated Ca²⁺ channel, that when open triggers elevation in cytosolic Ca²⁺. Calcium ion spikes occur in the pituitary gland cells that secrete luteinizing hormone (LH), which plays an important role in controlling ovulation and thus female fertility. LH secretion is induced by binding of luteinizing hormone–releasing hormone (LHRH) to its G protein–coupled receptors on these cells; LHRH binding induces repeated Ca²⁺ spikes. Each Ca²⁺ spike induces exocytosis of a few LH-containing secretory vesicles, presumably those close to the plasma membrane.

**DAG Activation of Protein Kinase C** After its formation by phospholipase C–catalyzed hydrolysis of PIP₂, DAG remains

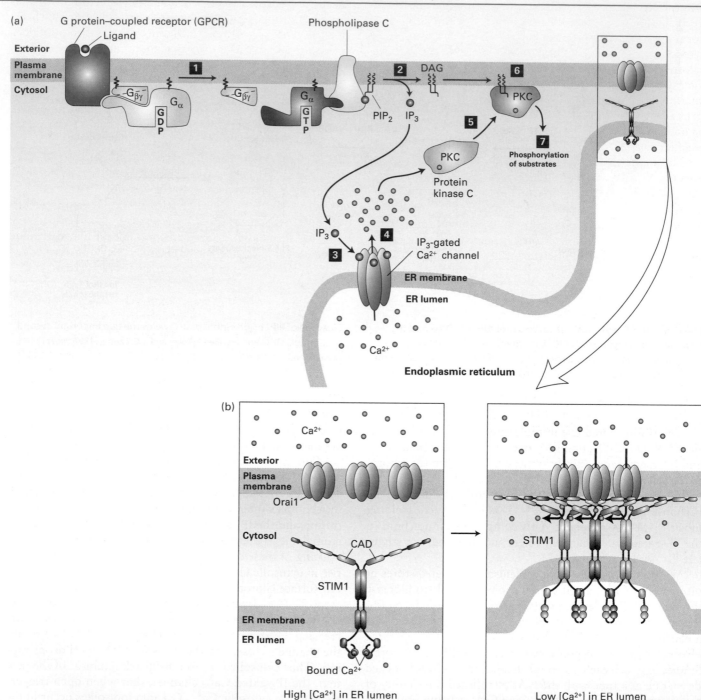

**FIGURE 15-36 IP3/DAG pathway and the elevation of cytosolic Ca²⁺.** (a) Opening of endoplasmic reticulum $Ca^{2+}$ channels. This pathway can be triggered by ligand binding to GPCRs that activate either the $G_{\alpha o}$ or $G_{\alpha q}$ alpha subunit leading to activation of phospholipase C (step **1**). Cleavage of $PIP_2$ by phospholipase C yields $IP_3$ and DAG (step **2**). After diffusing through the cytosol, $IP_3$ interacts with and opens $Ca^{2+}$ channels in the membrane of the endoplasmic reticulum (step **3**), causing release of stored $Ca^{2+}$ ions into the cytosol (step **4**). One of several cellular responses induced by a rise in cytosolic $Ca^{2+}$ is recruitment of protein kinase C (PKC) to the plasma membrane (step **5**), where it is activated by DAG (step **6**). The activated membrane-associated kinase can phosphorylate various cellular enzymes and receptors, thereby altering their activity (step **7**). (b) Opening of plasma membrane $Ca^{2+}$ channels. *Left:* In the resting cell, $Ca^{2+}$ levels in the endoplasmic reticulum lumen are high, and $Ca^{2+}$ ions (blue circles) bind to the EF hand domains of the transmembrane STIM proteins. *Right:* As endoplasmic reticulum $Ca^{2+}$ stores are depleted and $Ca^{2+}$ ions dissociate from the EF hands, STIMs undergo oligomerization and relocalization to areas of the ER membrane near the plasma membrane. There the STIM CRAC-activating domains (CAD, green) bind to and trigger opening of Orai1, the store-operated $Ca^{2+}$ channels in the plasma membrane, allowing influx of extracellular $Ca^{2+}$. [Adapted from J. W. Putney, 1999, *Proc. Nat'l. Acad. Sci. USA* **96:**14669; Y. Zhou, 2010, *Proc. Nat'l. Acad. Sci. USA* **107:**4896; and M. Cahalan, 2010, *Science* **130:**43.]

associated with the plasma membrane. The principal function of DAG is to activate a family of protein kinases collectively termed **protein kinase C (PKC)**. In the absence of hormone stimulation, protein kinase C is present as a soluble cytosolic protein that is catalytically inactive. A rise in the cytosolic $Ca^{2+}$ level causes protein kinase C to translocate to the cytosolic leaflet of the plasma membrane, where it can interact with membrane-associated DAG (see Figure 15-36a, steps **5** and **6**). Activation of protein kinase C thus depends on an increase of both $Ca^{2+}$ ions and DAG, suggesting an interaction between the two branches of the $IP_3$/DAG pathway.

The activation of protein kinase C in different cells results in a varied array of cellular responses, indicating that it plays a key role in many aspects of cellular growth and metabolism. In liver cells, for instance, protein kinase C helps regulate glycogen metabolism by phosphorylating and so inhibiting glycogen synthase. Protein kinase C also phosphorylates various transcription factors that in some cells activate genes necessary for cell division.

## The $Ca^{2+}$-Calmodulin Complex Mediates Many Cellular Responses to External Signals

The ubiquitous small cytosolic protein calmodulin functions as a multipurpose switch protein that mediates many cellular effects of $Ca^{2+}$ ions. Binding of $Ca^{2+}$ to four sites on calmodulin yields a complex that interacts with and modulates the activity of many enzymes and other proteins (see Figure 3-31). Because four $Ca^{2+}$ bind to calmodulin in a cooperative fashion, a small change in the level of cytosolic $Ca^{2+}$ leads to a large change in the level of active calmodulin. One well-studied enzyme activated by the $Ca^{2+}$-calmodulin complex is myosin light-chain kinase, which regulates the activity of myosin and thus contraction in muscle cells (see Chapter 17). Another is cAMP phosphodiesterase, the enzyme that degrades cAMP to $5'$-AMP and terminates its effects. This reaction thus links $Ca^{2+}$ and cAMP, one of many examples in which two second-messenger-mediated pathways interact to fine-tune a cellular response.

In many cells, the rise in cytosolic $Ca^{2+}$ following receptor signaling via phospholipase C–generated $IP_3$ leads to activation of specific transcription factors. In some cases, $Ca^{2+}$-calmodulin activates protein kinases that, in turn, phosphorylate transcription factors, thereby modifying their activity and regulating gene expression. In other cases, $Ca^{2+}$-calmodulin activates a phosphatase that removes phosphate groups from a transcription factor, thus activating it. An important example of this mechanism involves T cells of the immune system (see Chapter 23).

## Signal-Induced Relaxation of Vascular Smooth Muscle Is Mediated by a $Ca^{2+}$-Nitric Oxide-cGMP-Activated Protein Kinase G Pathway

Nitroglycerin has been used for over a century as a treatment for the intense chest pain of angina. It was known to slowly decompose in the body to *nitric oxide* (NO), which causes relaxation of the smooth muscle cells surrounding the blood vessels that "feed" the heart muscle itself, thereby increasing the diameter of the blood vessels and increasing the flow of oxygen-bearing blood to the heart muscle. One of the most intriguing discoveries in modern medicine is that NO, a toxic gas found in car exhaust, is in fact a natural signaling molecule. ■

Definitive evidence for the role of NO in inducing relaxation of smooth muscle came from a set of experiments in which acetylcholine was added to experimental preparations of the smooth muscle cells that surround blood vessels. Direct application of acetylcholine to these cells caused them to contract, the expected effect of acetylcholine on these muscle cells. But addition of acetylcholine to the lumen of small isolated blood vessels caused the underlying smooth muscles to relax, not contract. Subsequent studies showed that in response to acetylcholine, the endothelial cells that line the lumen of blood vessels were releasing some substance that in turn triggered muscle-cell relaxation. That substance turned out to be NO.

We now know that endothelial cells contain a $G_o$ protein–coupled receptor that binds acetylcholine and activates phospholipase C, leading to an increase in the level of cytosolic $Ca^{2+}$. After $Ca^{2+}$ binds to calmodulin, the resulting complex stimulates the activity of NO synthase, an enzyme that catalyzes formation of NO from $O_2$ and the amino acid arginine. Because NO has a short half-life (2–30 seconds), it can diffuse only locally in tissues from its site of synthesis. In particular, NO diffuses from the endothelial cell into neighboring smooth muscle cells, where it triggers muscle relaxation (Figure 15-37).

The effect of NO on smooth muscle is mediated by the second messenger cGMP, which is formed by an intracellular NO receptor expressed by smooth muscle cells. Binding of NO to the heme group in this receptor leads to a conformational change that increases its intrinsic guanylyl cyclase activity, leading to a rise in the cytosolic cGMP level. Most of the effects of cGMP are mediated by a cGMP-dependent protein kinase, also known as **protein kinase G (PKG)**. In vascular smooth muscle, protein kinase G activates a signaling pathway that results in inhibition of the actin-myosin complex, relaxation of the cell, and so dilation of the blood vessel. In this case, cGMP acts indirectly via protein kinase G, whereas in rod cells cGMP acts directly by binding to and opening cation channels in the plasma membrane (see Figure 15-23).

## Integration of $Ca^{2+}$ and cAMP Second Messengers Regulates Glycogenolysis

Just as no cell lives in isolation, no intracellular signaling pathway functions alone. All cells constantly receive multiple signals from their environment, including changes in hormone levels, metabolites, and gases such as NO and oxygen; these signals must be integrated. The breakdown of glycogen to glucose (glycogenolysis) provides an excellent example of how cells can integrate their responses to more than one signal. As discussed in Section 15.5, epinephrine stimulation of muscle and liver cells leads to a rise in the second messenger cAMP,

**FIGURE 15-37 The Ca²⁺/nitric oxide (NO)/ cGMP pathway and the relaxation of arterial smooth muscle.** Nitric oxide is synthesized in endothelial cells in response to activation of acetylcholine GPCRs, phospholipase C, and the subsequent elevation in cytosolic Ca²⁺ (steps **1**–**4**). NO diffuses locally through tissues and activates an intracellular NO receptor with guanylyl cyclase activity in nearby smooth muscle cells (**5**). The resulting rise in cGMP (**6**) activates protein kinase G (**7**), leading to relaxation of the muscle and thus vasodilation (**8**). PP$_i$ = pyrophosphate. [See C. S. Lowenstein et al., 1994, *Ann. Intern. Med.* **120**:227.]

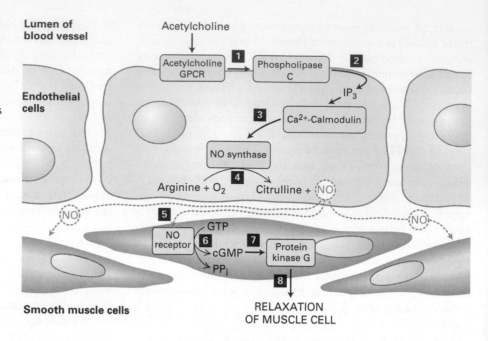

which promotes glycogen breakdown (see Figure 15-31a). In both muscle and liver cells, other second messengers also produce the same cellular response.

In muscle cells, stimulation by nerve impulses causes the release of Ca²⁺ ions from the sarcoplasmic reticulum and an increase in the cytosolic Ca²⁺ concentration, which triggers muscle contraction. The rise in cytosolic Ca²⁺ also activates glycogen phosphorylase kinase (GPK), thereby stimulating the degradation of glycogen to glucose-1-phosphate, which fuels prolonged contraction. Recall that phosphorylation by cAMP-dependent protein kinase A also activates glycogen phosphorylase kinase (see Figure 15-31). Thus this key regulatory enzyme in glycogenolysis is subject to both neural and hormonal regulation in muscle (Figure 15-38a).

In liver cells, hormone-induced activation of the effector protein phospholipase C also regulates glycogen breakdown by generating the second messengers DAG and IP₃. As we just learned, IP₃ induces an increase in cytosolic Ca²⁺, which

**FIGURE 15-38 Integrated regulation of glycogenolysis by Ca²⁺ and cAMP/PKA pathways.** (a) Neuronal stimulation of striated muscle cells or epinephrine binding to β-adrenergic receptors on their surfaces leads to increased cytosolic concentration of the second messengers Ca²⁺ or cAMP, respectively. The key regulatory enzyme glycogen phosphorylase kinase (GPK) is activated by binding Ca²⁺ ions and by phosphorylation by cAMP-dependent protein kinase A (PKA). (b) In liver cells, hormonal stimulation of β-adrenergic receptors leads to increased cytosolic concentrations of cAMP and two other second messengers, diacylglycerol (DAG) and inositol 1,4,5-trisphosphate (IP₃). Enzymes are marked by white boxes. (+) = activation of enzyme activity; (−) = inhibition.

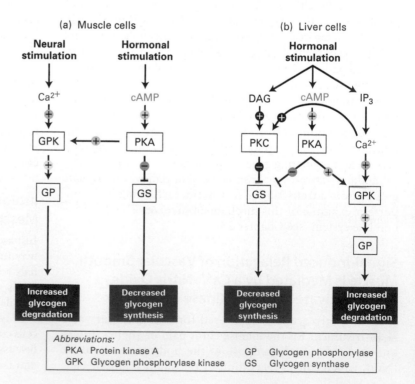

activates glycogen phosphorylase kinase as in muscle cells, leading to glycogen degradation. Moreover, the combined effect of DAG and increased $Ca^{2+}$ activates protein kinase C (see Figure 15-36). This kinase can phosphorylate glycogen synthase, thereby inhibiting the enzyme and reducing the rate of glycogen synthesis. In this case, multiple intracellular signal transduction pathways are activated by the same signal (Figure 15-38b).

The dual regulation of glycogen phosphorylase kinase by $Ca^{2+}$ and protein kinase A in both muscle and liver results from its multimeric subunit structure $(\alpha\beta\gamma\delta)_4$. The $\gamma$ subunit is the catalytic enzyme; the regulatory $\alpha$ and $\beta$ subunits, which are similar in structure, are phosphorylated by protein kinase A; and the $\delta$ subunit is the calcium sensor calmodulin. Glycogen phosphorylase kinase is maximally active when $Ca^{2+}$ ions are bound to the calmodulin subunit and the $\alpha$ subunit has been phosphorylated by protein kinase A. In fact, binding of $Ca^{2+}$ to the calmodulin subunit may be essential to the enzymatic activity of glycogen phosphorylase kinase. Phosphorylation of the $\alpha$ and also the $\beta$ subunits by protein kinase A increases the affinity of the calmodulin subunit for $Ca^{2+}$, allowing $Ca^{2+}$ ions to bind to the enzyme at the submicromolar $Ca^{2+}$ concentrations found in noncontracting cells. Thus increases in the cytosolic concentration of $Ca^{2+}$ or of cAMP (or both) induce incremental increases in the activity of glycogen phosphorylase kinase. As a result of the elevated level of cytosolic $Ca^{2+}$ after neuronal stimulation of muscle cells, glycogen phosphorylase kinase will be active even if it is unphosphorylated; therefore glycogen can be hydrolyzed to fuel continued muscle contraction even in the absence of hormone stimulation.

## KEY CONCEPTS of Section 15.6

### G Protein–Coupled Receptors That Trigger Elevations in Cytosolic $Ca^{2+}$

- A small rise in cytosolic $Ca^{2+}$ induces a variety of responses in different cells, including hormone secretion, contraction of muscle, and platelet aggregation (see Table 15-3).

- Many hormones bind GPCRs coupled to G proteins containing a $G_{\alpha o}$ or $G_{\alpha q}$ subunit. The effector protein activated by GTP-bound $G_{\alpha o}$ or $G_{\alpha q}$ is a phospholipase C enzyme.

- Phospholipase C cleaves a phospholipid known as $PIP_2$, generating two second messengers: diffusible $IP_3$ and membrane-bound DAG (see Figure 15-35).

- $IP_3$ triggers opening of $IP_3$-gated $Ca^{2+}$ channels in the endoplasmic reticulum and elevation of cytosolic free $Ca^{2+}$. In response to elevated cytosolic $Ca^{2+}$, protein kinase C is recruited to the plasma membrane, where it is activated by DAG (see Figure 15-36a).

- Depletion of ER $Ca^{2+}$ stores leads to opening of plasma membrane store-operated $Ca^{2+}$ channels and an influx of $Ca^{2+}$ from the extracellular medium (see Figure 15-36b).

- The $Ca^{2+}$-calmodulin complex regulates the activity of many different proteins, including cAMP phosphodiesterase and protein kinases and phosphatases that control the activity of various transcription factors.

- Stimulation of acetylcholine G protein–coupled receptors on endothelial cells induces an increase in cytosolic $Ca^{2+}$ and subsequent synthesis of NO. After diffusing into surrounding smooth muscle cells, NO activates an intracellular guanylate cyclase to synthesize cGMP. The resulting increase in cGMP leads to activation of protein kinase G, which triggers a pathway resulting in muscle relaxation and vasodilation (see Figure 15-37).

- Glycogen breakdown and synthesis is coordinately regulated by the second messengers $Ca^{2+}$ and cAMP, whose levels are regulated by neural and hormonal stimulation (see Figure 15-38).

## Perspectives for the Future

In this chapter, we focused primarily on signal transduction pathways activated by individual G protein–coupled receptors. However, even these relatively simple pathways presage the more complex situation within living cells. Many G protein–coupled receptors form homodimers or heterodimers with other G protein–coupled receptors that bind ligands with different specificities and affinities. Much current research is focused on determining the functions of these dimeric receptors in the body.

With ~900 members in total, the G protein–coupled receptors represent the largest protein family in the human genome. Approximately half of these genes are thought to encode sensory receptors; of these the majority are in the olfactory system and bind odorants. Of the remaining G protein receptors, the natural ligand has not been identified for many so-called *orphan GPCRs*—that is, putative GPCRs without known cognate ligands. Many of these orphan receptors are likely to bind heretofore unidentified signaling molecules, including new peptide hormones. G protein–coupled receptors already comprise the targets of more than 30 percent of all approved therapeutic drugs, and therefore orphan GPCRs represent a fruitful resource for drug discovery by the pharmaceutical industry.

One approach that has proved fruitful in identifying ligands of orphan GPCRs involves expressing the receptor genes in transfected cells and using them as a reporter system to detect substances in tissue extracts that activate signal transduction pathways in these cells. This approach has already led to stunning insights into human behavior. One example is two novel peptides termed orexin-A and orexin-B (from the Greek *orexis*, meaning "appetite") that were identified as the ligands for two orphan GPCRs. Further research showed that the *orexin* gene is expressed only in the hypothalamus, the part of the brain that regulates feeding. Injection of orexin into the brain ventricles caused animals to eat more, and expression of the *orexin* gene increased markedly during fasting. Both of these findings are consistent with orexin's role in increasing appetite. Strikingly, mice deficient for orexins suffer from narcolepsy, a disorder characterized in humans by excessive daytime sleepiness (for mice, nighttime sleepiness). Moreover, very recent reports suggest that the orexin system is dysfunctional in a majority of human narcolepsy patients: orexin peptides cannot be detected in their cerebrospinal fluid (although there is no evidence of mutation in their *orexin* genes). These findings firmly link orexin neuropeptides and their receptors to both feeding behavior and sleep in both animals and humans.

More recently a new neuropeptide, neuropeptide S, was identified as the ligand for another previously orphaned GPCR. Researchers then showed that this neuropeptide modulates a number of biological functions, including anxiety, arousal, locomotion, and memory. One can only wonder about what other peptides and small-molecule hormones remain to be discovered and the insights that study of these will provide for our understanding of human metabolism, growth, and behavior.

## Key Terms

adenylyl cyclase 692

agonist 682

antagonist 682

arrestin 697

autocrine 676

β-adrenergic receptors 687

calmodulin 679

competition assay 682

cyclic AMP (cAMP) 679

desensitization 684

endocrine 675

epinephrine 688

glucagon 699

glycogenolysis 699

G protein–coupled receptors (GPCRs) 674

GTPase superfamily 678

hormone 673

IP$_3$/DAG pathway 708

kinase 677

muscarinic acetylcholine receptors 693

nitric oxide 711

paracrine 675

phosphatase 677

phospholipase C (PLC) 708

protein kinase A (PKA) 701

protein kinase C (PKC) 711

protein kinase G (PKG) 711

rhodopsin 694

second messengers 674

signal amplification 680

signal transduction 674

transducin 694

trimeric G proteins 679

## Review the Concepts

**1.** What common features are shared by most cell signaling systems?

**2.** Signaling by soluble extracellular molecules can be classified as endocrine, paracrine, or autocrine. Describe how these three types of cellular signaling differ. Growth hormone is secreted from the pituitary, which is located at the base of the brain and acts through growth hormone receptors located on the liver. Is this an example of endocrine, paracrine, or autocrine signaling? Why?

**3.** A ligand binds two different receptors with a $K_d$ value of $10^{-7}$ M for receptor 1 and a $K_d$ value of $10^{-9}$ M for receptor 2. For which receptor does the ligand show the greater affinity? Calculate the fraction of receptors that have a bound ligand ($[RL]/R_T$) in the case of receptor 1 and receptor 2 if the concentration of free ligand is $10^{-8}$ M.

**4.** To understand how a signaling pathway works, it often is useful to isolate the cell-surface receptor and to measure the activity of downstream effector proteins under different conditions. How could you use affinity chromatography to isolate a cell-surface receptor? With what technique could you measure the amount of activated G protein (the GTP-bound form) in ligand-stimulated cells? Describe the approach you would take.

**5.** How do seven transmembrane domain G protein–coupled receptors transmit a signal across the plasma membrane? In your answer, include the conformational changes that occur in the receptor in response to ligand binding.

**6.** Signal-transducing trimeric G proteins consist of three subunits designated α, β, and γ. The $G_\alpha$ subunit is a GTPase switch protein that cycles between active and inactive states depending on whether it is bound to GTP or to GDP. Review the steps for ligand-induced activation of effector proteins mediated by the trimeric G proteins. Suppose that you have isolated a mutant $G_\alpha$ subunit that has an increased GTPase activity. What effect would this mutation have on the G protein and the effector protein?

**7.** Explain how FRET could be used to monitor the association of $G_{\alpha s}$ and adenylyl cyclase following activation of the epinephrine receptor.

**8.** Which of the following steps amplify the epinephrine signal response in cells: receptor activation of G protein, G protein activation of adenylyl cyclase (AC), cAMP activation of PKA, or PKA phosphorylation of glycogen phosphorylase kinase (GPK)? Which change will have a greater effect on signal amplification: an increase in the number of epinephrine receptors or an increase in the number of $G_{\alpha s}$ proteins?

**9.** The cholera toxin, produced by the bacterium *Vibrio cholera*, causes a watery diarrhea in infected individuals. What is the molecular basis for this effect of cholera toxin?

**10.** Both rhodopsin in vision and the muscarinic acetylcholine receptor system in cardiac muscle are coupled to ion channels via G proteins. Describe the similarities and differences between these two systems.

**11.** Epinephrine binds to both β-adrenergic and α-adrenergic receptors. Describe the opposite actions on the effector protein, adenylyl cyclase, elicited by the binding of epinephrine to these two types of receptors. Describe the effect of adding an agonist or antagonist to a β-adrenergic receptor on the activity of adenylyl cyclase.

**12.** In liver and muscle, epinephrine stimulation of the cAMP pathway activates glycogen breakdown and inhibits glycogen synthesis, whereas in adipose tissue, epinephrine activates hydrolysis of triglycerides and in other cells causes a diversity of other responses. What step in the cAMP signaling pathways in these cells specifies the cell response?

13. Continuous exposure of a $G_{\alpha s}$ protein–coupled receptor to its ligand leads to a phenomenon known as desensitization. Describe several molecular mechanisms for receptor desensitization. How can a receptor be reset to its original sensitized state? What effect would a mutant receptor lacking serine or threonine phosphorylation sites have on a cell?

14. What is the purpose of A kinase–associated proteins (AKAPs)? Describe how AKAPs work in heart muscle cells.

15. Inositol 1,4,5-trisphosphate ($IP_3$) and diacylglycerol (DAG) are second messenger molecules derived from the cleavage of the phosphatidylinositol 4,5-bisphosphate ($PIP_2$) by activated phospholipase C. Describe the role of $IP_3$ in causing a rise in cytosolic $Ca^{2+}$ concentration. How do cells restore resting levels of cytosolic $Ca^{2+}$? What is the principal function of DAG?

16. In Chapter 3, the $K_d$ of calmodulin's EF hands for binding $Ca^{2+}$ is given as $\sim 10^{-6}$ M. Many proteins have much higher affinities for their respective ligands. Why is the specific affinity of calmodulin important for $Ca^{2+}$ signaling processes such as that initiated by production of $IP_3$?

17. Most of the short-term physiological responses of cells to cAMP are mediated by activation of PKA. cGMP is another common second messenger. What are the targets of cGMP in rod and smooth muscle cells?

## Analyze the Data

1. Mutations in trimeric G proteins can cause many diseases in humans. Patients with acromegaly often have pituitary tumors that oversecrete growth hormone (GH). GH-releasing hormone (GHRH) stimulates GH release from the pituitary by binding to GHRH receptors and stimulating adenylyl cyclase. Researchers wanted to know whether mutations in $G_{\alpha s}$ played a roll in this condition. Cloning and sequencing of the wild-type and mutant $G_{\alpha s}$ gene from normal individuals and patients with the pituitary tumors revealed a missense mutation in the $G_{\alpha s}$ gene sequence.

a. To investigate the effect of the mutation on $G_{\alpha s}$ activity, wild-type and mutant $G_{\alpha s}$ cDNAs were transfected into cells that lack the $G_{\alpha s}$ gene. These cells express a $\beta_2$-adrenergic receptor, which can be activated by isoproterenol, a $\beta_2$-adrenergic receptor agonist. Membranes were isolated from transfected cells and assayed for adenylyl cyclase activity in the presence of GTP or the hydrolysis-resistant GTP analog, GTP-$\gamma$S. From the figure below, what do you conclude about the effect of the mutation on $G_{\alpha s}$ activity in the presence of GTP alone compared with GTP-$\gamma$S alone or GTP plus isoproterenol (iso)?

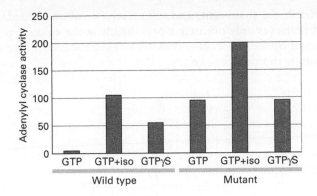

b. In the transfected cells described in part (a), what would you predict would be the cAMP levels in cells transfected with the wild-type $G_{\alpha s}$ and the mutant $G_{\alpha s}$? What effect might this have on the cells?

c. To further characterize the molecular defect caused by this mutation, the intrinsic GTPase activity present in both wild-type and mutant $G_{\alpha s}$ was assayed. Assays for GTPase activity showed that the mutation reduced the $k_{cat\text{-}GTP}$ (catalysis rate constant for GTP hydrolysis) from a wild-type value of 4.1 $min^{-1}$ to the mutant value of 0.1 $min^{-1}$. What do you conclude about the effect of the mutation on the GTPase activity present in the mutant $G_{\alpha s}$ subunit? How do these GTPase results explain the adenylyl cyclase results shown in part (a)?

2. The phosphorylation of a protein can influence its ability to interact with other proteins. These protein-protein interactions play a fundamental role in signal transduction pathways, and these interactions can be identified using numerous techniques, including fluorescence energy transfer (see Figure 15-18). Protein kinase A (PKA) has many substrates, one of which is glycogen phosphorylase kinase, which has a multimeric $(\alpha\beta\gamma\delta)_4$ structure containing two regulatory subunits ($\alpha$ and $\beta$), the catalytic $\gamma$ subunit, and the calcium sensor $\delta$ subunit.

a. You are using fluorescence energy transfer to investigate PKA interactions with glycogen phosphorylase kinase and have cloned cDNA fusion constructs for three of its four different subunits ($\gamma$, $\beta$, and $\delta$), all containing a fluorescent tag that when expressed excites at 480 nm and emits fluorescence at 535 nm. You also have cDNA encoding the catalytic domain of PKA fused to a tag that when expressed excites at 440 nm and emits at 480 nm. In the assay, if the PKA fusion protein interacts with one or more of the tagged glycogen phosphorylase kinase substrates, the transfer of energy from the PKA tag excites the tag on the substrate, causing it to emit fluorescence at 535 nm, and this can be detected.

Liver cells are transfected with the PKA fusion construct alone (control) or with the PKA fusion construct plus one of the three tagged glycogen phosphorylase kinase constructs and then later treated with epinephrine. The fluorescence emissions at 535 nm, resulting from the four different transfection experiments, repeated three different times, are shown in the graph below. Label the four bars on the graph, showing the emission of PKA by itself, PKA + the $\gamma$ subunit,

PKA + the β subunit, and PKA + the δ subunit. Explain why there is only one major peak and why the values represented by the other three bars are not significantly different from each other.

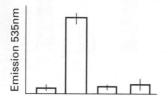

**b.** Which combination above would produce emission at 535 nm if the experiment was repeated but instead of epinephrine you used dibutyryl-cAMP, which freely crosses the plasma membrane?

**c.** As described above, there are two regulatory subunits of glycogen phosphorylase kinase, both of which are subjected to post-translational modifications. If the gene encoding the α subunit contained missense mutations whereby during translation all the serine, threonine and tyrosine residues were converted to some other amino acid, how would this affect the calcium sensor subunit of glycogen phosphorylase kinase? The activity of glycogen phosphorylase kinase in cells treated with epinephrine is shown in the following graph.

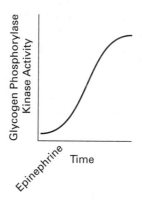

Draw what the activity would look like in comparison in epinephrine-treated cells expressing the α subunit containing the missense mutations described above.

**3.** cAMP is a second messenger that regulates many diverse cellular functions. In the intestinal lumen, cAMP is responsible for maintaining electrolyte and water balance. Certain bacterial toxins, including one produced by *Vibrio cholera*, can upset the levels of cAMP, leading to fatal dehydration.

**a.** Given what you know about the mechanism of *Vibrio cholera* toxin, label the graph below showing cAMP concentration in (1) normal intestinal epithelial cells treated with a GPCR agonist to activate $G_{\alpha s}$ and (2) cholera toxin–treated cells stimulated with the same GPCR agonist. Explain how you came to these conclusions.

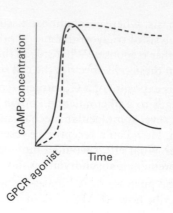

**b.** Would you expect lower or higher levels of PKA in cells treated with cholera toxin? Explain how you came to these conclusions.

## References

### Signal-Transduction: From Extracellular Signal to Cellular Response

Cabrera-Vera, T. M., et al. 2003. Insights into G protein structure, function, and regulation. *Endocr. Rev.* **24**:765–781.

Grecco, H., M. Schmick, and P. Bastiaens. 2011. Signaling from the living plasma membrane. *Cell* **144**:897–909.

Kornev, A., and S. S. Taylor. 2010. Defining the conserved internal architecture of a protein kinase. *Biochim. Biophys. Acta* **1804**:440–444.

Manning, G., et al. 2002. Evolution of protein kinase signaling from yeast to man. *Trends Biochem. Sci.* **27**:514–520.

Manning, G., et al. 2002. The protein kinase complement of the human genome. *Science* **298**:1912–1934.

Taylor, S. S., and A. Kornev. 2011. Protein kinases: evolution of dynamic regulatory proteins. *Trends Biochem. Sci.* **36**:65–77.

Vetter, I. R., and A. Wittinghofer. 2001. The guanine nucleotide-binding switch in three dimensions. *Science* **294**:1299–1304.

### Studying Cell-Surface Receptors and Signal Transduction Proteins

Gross, A., and H. F. Lodish. 2006. Cellular trafficking and degradation of erythropoietin and NESP. *J. Biol. Chem.* **281**: 2024–2032.

Lauffenburger, D., and J. Linderman. 1993. Receptors: models for binding, trafficking, and signaling. New York: Oxford University Press.

Selinger, Z. 2008. Discovery of G protein signaling. *Ann. Rev. Biochem.* **77**:1–13.

Tarrant, M., and P. Cole. 2009. The chemical biology of protein phosphorylation. *Ann. Rev. Biochem.* **78**:797–825.

### G Protein–Coupled Receptors: Structure and Mechanism

Birnbaumer, L. 2007. The discovery of signal transduction by G proteins: a personal account and an overview of the initial findings and contributions that led to our present understanding. *Biochim. Biophys. Acta* **1768**:756–771.

Oldham, W. M., and H. E. Hamm. 2008. Heterotrimeric G protein activation by G-protein-coupled receptors. *Nat. Rev. Mol. Cell Biol.* **9**:60–71.

Rosenbaum, D., S. Rasmussen, and B. Kobilka. 2008. The structure and function of G-protein-coupled receptors *Nature* **459**:356–363.

Schwartz, T., and W. Hubbell. 2008. Structural biology: a moving story of receptors. *Nature* **454**:473.

Sprang, S. 2011. Cell signaling: binding the receptor at both ends. *Nature* **469**:172–173.

Tesmer, J. 2010. The quest to understand heterotrimeric G protein signalling. *Nat. Struct. Mol. Biol.* **17**:650–652.

Warne, T., et. al. 2008 Structure of a β1-adrenergic G-protein-coupled receptor. *Nature* **454**: 486–491.

## G Protein–Coupled Receptors That Regulate Ion Channels

Burns, M., and V. Arshavsky. 2005. Beyond counting photons: trials and trends in vertebrate visual transduction. *Neuron* **48**:387–401.

Calvert, P., et al. 2006. Light-driven translocation of signaling proteins in vertebrate photoreceptors. *Trends Cell Biol.* **16**:560–568.

Hofmann, K. P., et al. 2009. A G protein-coupled receptor at work: the rhodopsin model. *Trends Biochem. Sci.* **34**:540–552.

Smith, S. O. 2010. Structure and activation of the visual pigment rhodopsin. *Ann. Rev. Biophys.* **39**:309–328.

## G Protein–Coupled Receptors That Activate or Inhibit Adenylyl Cyclase

Agius, L. 2010. Physiological control of liver glycogen metabolism: lessons from novel glycogen phosphorylase inhibitors. *Mini-Rev. Med. Chem.* **10**:1175–1187.

Carnegie, G., C. Means, and J. Scott. 2009. A-kinase anchoring proteins: from protein complexes to physiology and disease. *IUBMB Life* **61**(4):394–406.

Dessauer, C. 2009. Adenylyl cyclase–A-kinase anchoring protein complexes: the next dimension in cAMP signaling. *Mol. Pharmacol.* **76**:935–941.

DeWire, S., et al. 2007. β-Arrestins and cell signaling. *Ann. Rev. Physiol.* **69**:483–510.

Johnson, L. N. 1992. Glycogen phosphorylase: control by phosphorylation and allosteric effectors. *FASEB J.* **6**:2274–2282.

Lefkowitz, R. J., and S. K. Shenoy. 2005. Transduction of receptor signals by β-arrestins. *Science* **308**:512–517.

Rajagopal, S., K. Rajagopal, and R. J. Lefkowitz. 2010. Teaching old receptors new tricks: biasing seven-transmembrane receptors. *Nat. Rev. Drug Discov.* **9**:373–386.

Somsak, L., et al. 2008. New inhibitors of glycogen phosphorylase as potential antidiabetic agents. *Curr. Med. Chem.* **15**:2933–2983.

Taylor, S. S., et al. 2005. Dynamics of signaling by PKA. *Biochim. Biophys. Acta* **1754**:25–37.

Taylor, S. S., et al. 2008. Signaling through cAMP and cAMP-dependent protein kinase: diverse strategies for drug design. *Biochim. Biophys. Acta* **1784**:16–26.

## G Protein–Coupled Receptors That Trigger Elevations in Cytosolic Ca$^{2+}$

Cahalan, M. 2010. How to STIMulate calcium channels. *Science* **130**:43.

Chin, D., and A. R. Means. 2000. Calmodulin: a prototypical calcium sensor. *Trends Cell Biol.* **10**:322–328.

Duda, T. 2009. Atrial natriuretic factor-receptor guanylate cyclase signal transduction mechanism. *Mol. Cell Biochem.* **334**:37–51.

Hoeflich, K. P., and M. Ikura. 2002. Calmodulin in action: diversity in target recognition and activation mechanisms. *Cell* **108**:739–742.

Hogan, P. G., R. S. Lewis, and A. Rao. 2010. Molecular basis of calcium signaling in lymphocytes: STIM and ORAI. *Ann. Rev. Immunol.* **28**:491–533.

Parekh, A. 2011. Decoding cytosolic Ca$^{2+}$ oscillations. *Trends Biochem. Sci.* **36**:78–87.

Zhou, Y., et al. 2010. Pore architecture of the ORAI1 store-operated calcium channel. *Proc. Nat'l Acad. Sci. USA* **107**:4896–4901.

# The Infancy of Signal Transduction—
# GTP Stimulation of cAMP Synthesis

M. Rodbell et al., 1971, *J. Biol. Chem.* **246**:1877

In the late 1960s, the study of hormone action blossomed following the discovery that cyclic adenosine monophosphate (cAMP) functioned as a second messenger, coupling the hormone-mediated activation of a receptor to a cellular response. In setting up an experimental system to investigate the hormone-induced synthesis of cAMP, Martin Rodbell discovered an important new player in intracellular signaling—guanosine triphosphate (GTP).

## Background

The discovery of GTP's role in regulating signal transduction began with studies on how glucagon and other hormones send a signal across the plasma membrane that eventually evokes a cellular response. At the outset of Rodbell's studies, it was known that binding of glucagon to specific receptor proteins embedded in the membrane stimulates production of cAMP. The formation of cAMP from ATP is catalyzed by a membrane-bound enzyme called adenyl cyclase. It had been proposed that the action of glucagon, and other cAMP-stimulating hormones, relied on additional molecular components that couple receptor activation to the production of cAMP. However, in studies with isolated fat-cell membranes known as "ghosts," Rodbell and his coworkers were unable to provide any further insight into how glucagon binding leads to an increase in production of cAMP. Rodbell then began a series of studies with a newly developed cell-free system, purified rat liver membranes, which retained both membrane-bound and membrane-associated proteins. These experiments eventually led to the finding that GTP is required for the glucagon-induced stimulation of adenyl cyclase.

## The Experiment

One of Rodbell's first goals was to characterize the binding of glucagon to the glucagon receptor in the cell-free rat liver membrane system. First, purified rat liver membranes were incubated with glucagon labeled with the radioactive isotope of iodine ($^{125}$I). Membranes were then separated from the unbound [$^{125}$I] glucagon by centrifugation. Once it was established that labeled glucagon would indeed bind to the purified rat liver cell membranes, the study went on to determine if this binding led directly to activation of adenyl cyclase and production of cAMP in the purified rat liver cell membranes.

The production of cAMP in the cell-free system required the addition of ATP; the substrate for adenyl cyclase, $Mg^{2+}$; and an ATP-regenerating system consisting of creatine kinase and phosphocreatine. Surprisingly, when the glucagon-binding experiment was repeated in the presence of these additional factors, Rodbell observed a 50 percent decrease in glucagon binding. Full binding could be restored only when ATP was omitted from the reaction. This observation inspired an investigation of the effect of nucleoside triphosphates on the binding of glucagon to its receptor. It was shown that relatively high (i.e., millimolar) concentrations of not only ATP but also uridine triphosphate (UTP) and cytidine triphosphate (CTP) reduced the binding of labeled glucagon. In contrast, the reduction of glucagon binding in the presence of GTP occurred at far lower (micromolar) concentrations. Moreover, low concentrations of GTP were found to stimulate the dissociation of bound glucagon from the receptor. Taken together, these studies suggested that GTP alters the glucagon receptor in a manner that lowers its affinity for

glucagon. This decreased affinity both affects the ability of glucagon to bind to the receptor and encourages the dissociation of bound glucagon.

The observation that GTP was involved in the action of glucagon led to a second key question: Can GTP also exert an effect on adenyl cyclase? Addressing this question experimentally required the addition of both ATP, as a substrate for adenyl cyclase, and GTP, as the factor being examined, to the purified rat liver membranes. However, the previous study had shown that the concentration of ATP required as a substrate for adenyl cyclase could affect glucagon binding. Might it also stimulate adenyl cyclase? The concentration of ATP used in the experiment could not be reduced because ATP was readily hydrolyzed by ATPases present in the rat liver membrane. To get around this dilemma, Rodbell replaced ATP with an AMP analog, 59-adenyl-imidodiphosphate (AMP-PNP), which can be converted to cAMP by adenyl cyclase yet is resistant to hydrolysis by membrane ATPases. The critical experiment now could be performed. Purified rat liver membranes were treated with glucagon both in the presence and absence of GTP, and the production of cAMP from AMP-PNP was measured. The addition of GTP clearly stimulated the production of cAMP when compared to the addition of glucagon alone (Figure 1), indicating that GTP affects not only the binding of glucagon to its receptor but also stimulates the activation of adenylyl cyclase.

## Discussion

Two key factors led Rodbell and his colleagues to detect the role of GTP in signal transduction, whereas previous studies had failed to do so. First, by switching from fat-cell ghosts to the rat

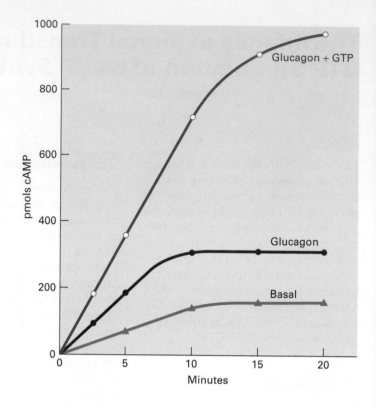

**FIGURE 1** **Effect of GTP on glucagon-stimulated cAMP production from AMP-PNP by purified rat liver membranes.** In the absence of GTP, glucagon stimulates cAMP formation about twofold over the basal level in the absence of added hormone. When GTP also is added, cAMP production increases another fivefold. [Adapted from M. Rodbell et al., 1971, *J. Biol. Chem.* **246**:1877.]

liver membrane system, the Rodbell researchers avoided contamination of their cell-free system with GTP, a problem associated with the procedure for isolating ghosts. Such contamination would mask the effects of GTP on glucagon binding and activation of adenyl cyclase. Second, when ATP was first shown to influence glucagon binding, Rodbell did not simply accept the plausible explanation that ATP, the substrate for adenyl cyclase, also affects binding of glucagon. Instead, he chose to test the effects on binding of the other common nucleoside triphosphates. Rodbell later noted that he knew commercial preparations of ATP often are contaminated with low concentra-

tions of other nucleoside triphosphates. The possibility of contamination suggested to him that small concentrations of GTP might exert large effects on glucagon binding and the stimulation of adenyl cyclase.

This critical series of experiments stimulated a large number of studies on the role of GTP in hormone action, eventually leading to the discovery of G proteins, the GTP-binding proteins that couple certain receptors to the adenyl cyclase. Subsequently, an enormous family of receptors that require G proteins to transduce their signals were identified in eukaryotes from yeast to humans. These G protein–coupled re-

ceptors are involved in the action of many hormones as well as in a number of other biological activities, including neurotransmission and the immune response. It is now known that binding of ligands to their cognate G protein–coupled receptors stimulates the associated G proteins to bind GTP. This binding causes transduction of a signal that stimulates adenyl cyclase to produce cAMP and also desensitization of the receptor, which then releases its ligand. Both of these effects were observed in Rodbell's experiments on glucagon action. For these seminal observations, Rodbell was awarded the Nobel Prize in Physiology or Medicine in 1994.

# Signaling Pathways That Control Gene Expression

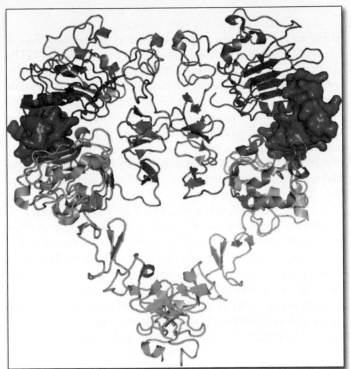

A molecular valentine—dimerized extracellular domain of the epidermal growth factor receptor (red, yellow, and green) bound to two molecules of epidermal growth factor (magenta). [Courtesy Jiahai Shi.]

xtracellular signals can have both short- and long-term effects on cells. Short-term effects are usually triggered by modification of existing proteins or enzymes, as we saw in Chapter 15. Many extracellular signals also affect gene expression and thus induce long-term changes in cell function. Long-term changes include alterations in cell division and differentiation, such as occur during development and cell fate determination. The body's production of red blood cells, white blood cells, and platelets in response to *cytokines* is a good example of signal-induced changes in gene expression that influence cell proliferation and differentiation. Changes in gene expression also enable differentiated cells to respond to their environment by changing their shape, metabolism, or movement. In immune system cells, for example, several hormones activate one type of transcription factor (NF-κB) that ultimately impacts expression of more than 150 genes involved in the immune response to infection. Given the extensive role of gene transcription in

mediating critical aspects of development, metabolism, and movement, it is not surprising that mutations in such signaling pathways cause many human diseases, including cancer, diabetes, and immune disorders.

Transcription of genes is influenced by chromatin structure, epigenetic modifications to histones and other nuclear proteins, and the cell's complement of transcription factors and other proteins (see Chapter 7). These properties determine which genes the cell can potentially transcribe at any given time; we think of these properties as the cell's "memory," determined by its history and response to previous signals. Importantly, many key regulatory transcription factors are held in an inactive state in the cytosol or nucleus and become activated only in response to external signals, thus inducing expression of a set of genes that are specific to this cell type.

In this chapter, we explore the main signaling pathways that cells use to influence gene expression. In eukaryotes, there are about a dozen classes of highly conserved cell-surface

## OUTLINE

receptors, and these activate several types of highly conserved intracellular signal transduction pathways. Many of these pathways consist of multiple proteins, small intracellular molecules, and ions such as $Ca^{2+}$, which together form a complex cascade. Given this complexity, cell signaling can seem a daunting subject to learn for the first time; the many names and abbreviations of molecules found in each pathway can indeed be challenging. The subject repays careful study, however: when one becomes familiar with these pathways, one understands in a profound way the regulatory mechanisms that control a vast array of biological processes.

For simplicity, signal transduction pathways can be grouped into several basic types, based on the sequence of intracellular events. In one very common type of signal transduction pathway (Figure 16-1a), ligand binding to a receptor triggers activation of a receptor-associated kinase. This kinase may be an intrinsic part of the receptor protein or be tightly bound to the receptor. These kinases often directly phosphorylate and activate a variety of signal transduction proteins, including transcription factors located in the cytosol (Figure 16-1a, **1**). Some receptor kinases also activate small GTP-binding "switch" proteins such as Ras (Figure 16-1a, **2**). Other receptors, mainly the seven spanning receptors introduced in Chapter 15, activate the larger GTP binding $G_\alpha$ proteins (Figure 16-1b). Both types of GTP-binding proteins can activate protein kinases that in turn phosphorylate multiple target proteins, including transcription factors. Many signal transduction pathways, such as those activated by Ras, involve several kinases in which one kinase phosphorylates and thus activates (or occasionally inhibits) the activity of another kinase.

In yet other signaling pathways, binding of a ligand to a receptor triggers disassembly of a multiprotein complex in the cytosol, releasing a transcription factor that then translocates into the nucleus (Figure 16-1c). Finally, in the last common type, proteolytic cleavage of an inhibitor or the receptor itself releases an active transcription factor, which then travels into the nucleus (Figure 16-1d). While every signaling pathway has its own subtleties and distinctions, nearly every one can be grouped into one of these basic types.

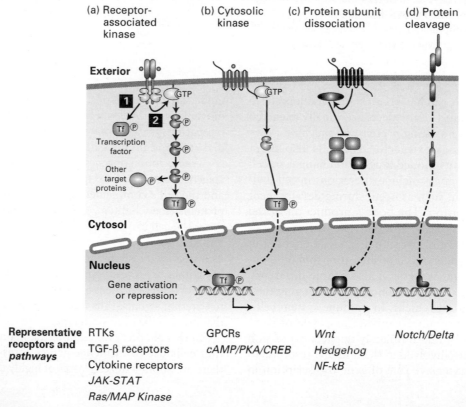

**FIGURE 16-1 Several common cell-surface receptors and signal transduction pathways.** (a) The cytosolic domains of many receptors contain protein kinase domains or are tightly associated with a cytosolic kinase; commonly the kinases are activated by ligand binding followed by receptor dimerization. Some of these kinases directly phosphorylate and activate transcription factors (**1**) or other signaling proteins. Many of these receptors also activate small GTP-binding "switch" proteins such as Ras (**2**). Many signal transduction pathways, such as those activated by Ras, involve several kinases in which one kinase phosphorylates and thus activates (or occasionally inhibits) the activity of another kinase. Many of the kinases in these pathways phosphorylate multiple protein targets that can be different in different cells, including transcription factors. (b) Other receptors, mainly the seven-spanning receptors, activate the larger GTP-binding $G_\alpha$ proteins, which in turn activate specific kinases or other signaling proteins. (c) Several signaling pathways involve disassembly of a multiprotein complex in the cytosol, releasing a transcription factor that then translocates into the nucleus. (d) Some signaling pathways are irreversible; in many cases proteolytic cleavage of a receptor releases an active transcription factor.

The pathways we discuss in this chapter have been conserved throughout evolution and operate in much the same manner in flies, worms, and humans. The substantial homology exhibited among proteins in these pathways has enabled researchers to study them in a variety of experimental systems. For instance, the secreted signaling protein Hedgehog (Hh) and its receptor were first identified in *Drosophila* mutants. Subsequently, the human and mouse homologs of these proteins were cloned and shown to participate in a number of important signaling events during differentiation, resulting in the discovery that abnormal activation of the Hh pathway occurs in several human tumors. Such discoveries illustrate the importance of studying signaling pathways both genetically—in flies, mice, worms, yeasts, and other organisms—and biochemically.

No signaling pathway acts in isolation. Many cells respond to multiple types of hormones and other signaling molecules; some mammalian cells express ~100 different types of cell-surface receptors, each of which binds a different ligand. Since many genes are regulated by multiple transcription factors that in turn are activated or repressed by different intracellular signaling pathways, expression of any one gene can be regulated by multiple extracellular signals. Especially during early development, such "cross talk" between signaling pathways and the resultant sequential alterations in the pattern of gene expression eventually can become so extensive that the cell assumes a different developmental fate. In this chapter, we will see how multiple signaling pathways interact to regulate crucial aspects of metabolism, such as the level of glucose in the blood and the formation of adipose cells.

## 16.1 Receptors That Activate Protein Tyrosine Kinases

We begin with a discussion of two large classes of receptors that activate protein tyrosine kinases. Protein tyrosine kinases, of which there are about 90 in the human genome, phosphorylate specific tyrosine residues on target proteins, usually in the context of a specific linear sequence of amino acids in which the tyrosine is embedded. The phosphorylated target proteins can then activate one or more signaling pathways. These pathways are noteworthy because they regulate most aspects of cell proliferation, differentiation, survival, and metabolism.

There are two broad categories of receptors that activate tyrosine kinases: (1) those in which the tyrosine kinase enzyme is an intrinsic part of the receptor's polypeptide chain (encoded by the same gene), called the **receptor tyrosine kinases (RTKs)**, and (2) those, such as **cytokine receptors**, in which the receptor and kinase are encoded by different genes yet bound tightly together. For cytokine receptors, the tightly bound kinase is known as a *JAK kinase*. Both classes of receptors activate similar intracellular signal transduction pathways, and we therefore consider them together in this section (Figure 16-2).

We will explore each of these important pathways in subsequent sections of the chapter. In this section, we focus on the receptors themselves, showing how ligand binding leads to kinase activation. We start with RTKs and then turn to cytokine receptors. After discussing how both types of receptors are activated, we explore some of the downstream molecules that are enlisted on activation. In the last part of the section, we discuss how signaling from RTKs and cytokine receptors is down-regulated.

### Numerous Factors Regulating Cell Division and Metabolism Are Ligands for Receptor Tyrosine Kinases

The signaling molecules that activate RTKs are soluble or membrane-bound peptide or protein hormones, including many that were initially identified as growth factors for specific types of cells. These RTK ligands include many, such as nerve growth factor (NGF), platelet-derived growth factor (PDGF), fibroblast growth factor (FGF), and epidermal

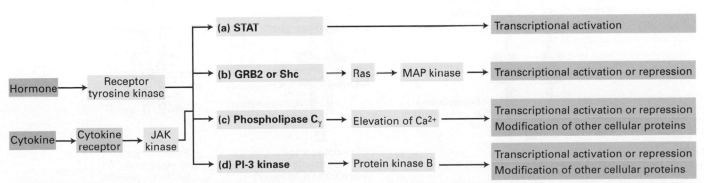

**FIGURE 16-2 Overview of signal transduction pathways triggered by receptors that activate protein tyrosine kinases.** Both RTKs and cytokine receptors activate multiple signal transduction pathways that ultimately regulate transcription of genes. (a) In the most direct pathway, mainly employed by cytokine receptors, a STAT transcription factor binds to the activated receptor, becomes phosphorylated, moves to the nucleus, and directly activates transcription (see Section 16.1).

(b) Binding of one type of adapter protein (GRB2 or Shc) to an activated receptor leads to activation of the Ras/MAP kinase pathway (see Section 16.2). (c, d) Two phosphoinositide pathways are triggered by recruitment of phospholipase C$_\gamma$ and PI-3 kinase to the membrane (see Section 16.3). Elevated levels of Ca$^{2+}$ and activated protein kinase B modulate the activity of transcription factors as well as of cytosolic proteins that are involved in metabolic pathways or cell movement or shape.

growth factor (EGF), that stimulate proliferation and differentiation of specific cell types. Others, such as insulin, regulate expression of multiple genes that control sugar and lipid metabolism in liver, muscle, and adipose (fat) cells. Many RTKs and their ligands were identified in studies of human cancers associated with mutant forms of growth-factor receptors that stimulate proliferation even in the absence of growth factor. The mutation "tricks" the receptor into behaving as though the ligand is present at all times and so the receptor is constantly in an active state (*constitutively* active). Other RTKs have been uncovered during analysis of developmental mutations that lead to blocks in differentiation of certain cell types in *C. elegans*, *Drosophila*, and the mouse.

## Binding of Ligand Promotes Dimerization of an RTK and Leads to Activation of Its Intrinsic Kinase

All RTKs have three essential components: an extracellular domain containing a ligand-binding site, a single hydrophobic transmembrane α helix, and a cytosolic segment that includes a domain with protein tyrosine kinase activity (Figure 16-3). Most RTKs are monomeric, and ligand binding to the extracellular domain induces formation of receptor dimers. The formation of functional dimers is a necessary step in activation of all RTKs. We term this process of two (or more) receptors joining together "activation by receptor oligomerization." Such oligomerization of cell-surface receptors is, we will see, a common mechanism for activating multiple types of receptors.

RTK activation can be summarized as follows: in the resting, unstimulated (no ligand bound) state, the intrinsic kinase activity of an RTK is very low (see Figure 16-3, step **1**). Like most other kinases, RTKs contain a flexible domain termed the *activation lip*. In the resting state, the activation lip is unphosphorylated and assumes a conformation that blocks kinase activity. In some receptors (e.g., the insulin receptor), it prevents binding of ATP. In others, (e.g., the FGF receptor), it prevents binding of substrate. Binding of ligand causes a conformational change that promotes dimerization of the extracellular domains of RTKs, which brings their transmembrane segments—and therefore their cytosolic domains—close together. The kinase in one subunit then phosphorylates a particular tyrosine residue in the activation lip in the other subunit (Figure 16-3, step **2**). This phosphorylation leads to a conformational change in the activation lip that unblocks and thus activates kinase activity by reducing the $K_m$ for ATP or the substrate to be phosphorylated. The resulting enhanced kinase activity can then phosphorylate additional tyrosine residues in the cytosolic domain of the receptor (Figure 16-3, step **3**) as well as phosphorylate other target proteins, leading to intracellular signaling.

Although dimerization is a necessary step in the activation of all RTKs, functional dimers can be formed in multiple ways. Binding of EGF, for example, to its RTK triggers a conformational change in the receptor extracellular domain so that it "clamps" down on the ligand. This action pushes out a loop located between the two EGF-binding domains, and interactions between the two extended ("activated") loop segments allow formation of the functional receptor dimer (Figure 16-4). In other cases, such as the fibroblast growth factor (FGF) receptor, each of the two ligands binds simultaneously to the extracellular domains of two receptor subunits. FGF also binds tightly to heparan sulfate, a negatively charged polysaccharide component of some cell-surface

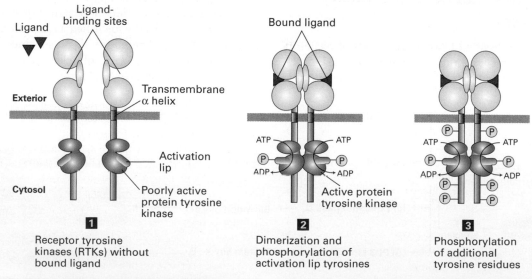

**1**
Receptor tyrosine kinases (RTKs) without bound ligand

**2**
Dimerization and phosphorylation of activation lip tyrosines

**3**
Phosphorylation of additional tyrosine residues

**FIGURE 16-3 General structure and activation of receptor tyrosine kinases (RTKs).** The cytosolic domain of RTKs contains an intrinsic protein tyrosine kinase catalytic site. In the absence of ligand (**1**), RTKs generally exist as monomers with poorly active kinases. Ligand binding causes a conformational change that promotes formation of a functional dimeric receptor, bringing together two poorly active kinases that then phosphorylate each other on a tyrosine residue in the activation lip (**2**). Phosphorylation causes the lip to move out of the kinase catalytic site, thus increasing the ability of ATP and the protein substrate to bind. The activated kinase then phosphorylates several tyrosine residues in the receptor's cytosolic domain (**3**). The resulting phosphotyrosines function as docking sites for various signal transduction proteins.

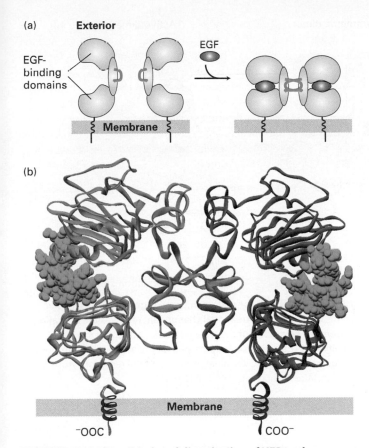

(a)

Exterior

EGF-binding domains

Membrane

EGF

(b)

−OOC

Membrane

COO−

**FIGURE 16-4 Ligand-induced dimerization of HER1, a human receptor for epidermal growth factor (EGF).** (a) Schematic depiction of the extracellular and transmembrane domains of HER1, which is a receptor tyrosine kinase. Binding of one EGF molecule to a monomeric receptor causes an alteration in the structure of a loop between the two EGF-binding domains. Dimerization of two identical ligand-bound receptor monomers in the plane of the membrane occurs primarily through interactions between the two "activated" loop segments. (b) Structure of the dimeric HER1 protein bound to transforming growth factor α (TGF-α), a member of the EGF family. The receptor's extracellular domains are shown in blue; the transmembrane domain is shown in red as an alpha helix, but its structure is not known in detail. The two smaller TGFα molecules are colored green. Note the interaction between the "activated" loop segments in the two receptor monomers. [Part (a) adapted from J. Schlessinger, 2002, *Cell* **110**:669; part (b) from T. Garrett et al., 2002, *Cell* **110**:763.]

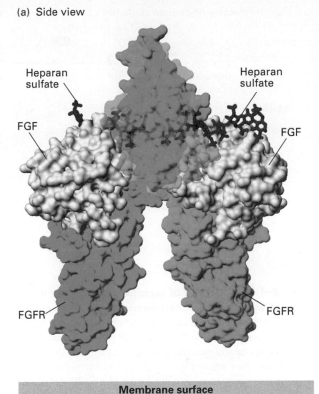

(a) Side view

Heparan sulfate

FGF

Heparan sulfate

FGF

FGFR

FGFR

Membrane surface

(b) Top-down view

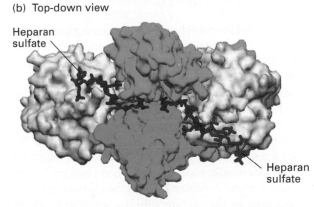

Heparan sulfate

Heparan sulfate

**FIGURE 16-5 Structure of the fibroblast growth factor (FGF) receptor, stabilized by heparan sulfate.** Shown here are side and top-down views of the complex comprising the extracellular domains of two FGF receptor (FGFR) monomers (green and blue), two bound FGF molecules (white), and two short heparan sulfate chains (purple), which bind tightly to FGF. (a) In the side view, the upper domain of one receptor monomer (blue) is seen situated behind that of the other (green); the plane of the plasma membrane is at the bottom. A small segment of the extracellular domain whose structure is not known connects to the membrane-spanning α-helical segment of each of the two receptor monomers (not shown) that protrude downward into the membrane. (b) In the top view, the heparan sulfate chains are seen threading between and making numerous contacts with the upper domains of both receptor monomers. These interactions promote binding of the ligand to the receptor and receptor dimerization. [Adapted from J. Schlessinger et al., 2000, *Mol. Cell* **6**:743.]

proteins and of the extracellular matrix (see Chapter 20); this association enhances ligand binding and formation of a dimeric receptor–ligand complex (Figure 16-5). The participation of the heparan sulfate is essential for efficient receptor activation. The ligands for some RTKs are dimeric, and their binding brings two receptor monomers together directly. Yet other RTKs, such as the insulin receptor, form disulfide-linked dimers even in the absence of hormone; binding of ligand to this type of RTK alters its conformation in such a way that the receptor kinase becomes activated. This last example highlights that simply having two receptor monomers in close contact is not sufficient for receptor activation—the

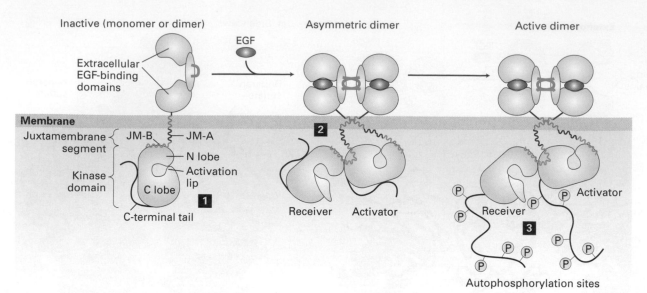

Inactive (monomer or dimer)                Asymmetric dimer                Active dimer

**FIGURE 16-6 Activation of EGF receptor by EGF results in the formation of an asymmetric kinase domain dimer.** In the inactive, monomeric state (**1**) the unstructured segment of the juxtamembrane domain (JM-B; green) binds to the upper, or N lobe of the kinase domain, causing a conformational change that positions the activation lip in the kinase active site and thus inhibits kinase activation. Receptor dimerization generates an asymmetric kinase dimer (**2**) such that the *activator kinase* binds the juxtamembrane segment of the *receiver kinase*, causing a conformational change that removes the activation lip from the kinase site of the receiver kinase, activating its kinase activity. (**3**) The active kinase then phosphorylates tyrosine residues (yellow circles) in the C-terminal segments of the receptor cytosolic domain. [After N. Jura et al., 2009, *Cell* **137**:1293.]

proper conformational changes must accompany receptor dimerization to lead to tyrosine kinase activation. Once an RTK is locked into a functional dimeric state, its associated tyrosine kinase becomes activated.

Exactly how dimerization leads to kinase activation is understood only for members of the EGF receptor family and was uncovered through structural studies of the receptor cytosolic domains in both active and inactive states. The kinase domains are separated from the transmembrane segment by a so-called juxtamembrane segment, whose two parts are colored red and green in Figure 16-6. In the inactive, monomeric state, one part of the juxtamembrane segment binds to the upper, or N, lobe of the adjacent kinase domain in the same molecule. This causes a conformational change such that the activation lip is localized in the active site of the kinase, blocking its activity. In this way the kinase is maintained in the "off" state (Figure 16-6, step **1**). Receptor dimerization generates an asymmetric kinase dimer (Figure 16-6, step **2**) such that one kinase domain—termed the activator—binds the juxtamembrane segment of the second kinase domain—the receiver. This changes the conformation of the N lobe of the receiver, causing the activation lip to move out of the kinase active site and allowing the kinase to function (step **3**). In a sense, an RTK can be thought of as an allosteric enzyme whose active site is inside the cell and whose allosteric effector—the ligand—binds to an extracellular regulatory site on the enzyme. Evolution has produced many variations on the theme of this simple ligand-RTK mechanism, as is exemplified by the families of EFG ligands and receptors discussed below.

## Homo- and Hetero-oligomers of Epidermal Growth Factor Receptors Bind Members of the Epidermal Growth Factor Superfamily

Four receptor tyrosine kinases (RTKs) participate in signaling by the many members of the **epidermal growth factor** (**EGF**) family of signaling molecules. In humans, the four members of the **HER** (*h*uman *e*pidermal growth factor *r*eceptor) **family** are denoted HER1, 2, 3, and 4. HER1 directly binds three EGF family members: EGF, heparin-binding EGF (HB-EGF), and tumor-derived growth factor alpha (TGF-α). Binding of any of these ligands to the extracellular domain of a HER1 monomer leads to homodimerization of the HER1 extracellular domain (Figure 16-7).

Two other members of the EGF family, neuregulins 1 and 2 (NRG1 and NRG2), bind to both HER3 and HER4; HB-EGF also binds to HER4. Importantly, HER2 does not directly bind a ligand but exists on the membrane in a preactivated conformation with the loop segment protruding outward and the ligand-binding domains in close proximity (Figure 16-7a). HER2, however, cannot form homodimers. It can signal only by forming heterocomplexes with ligand-bound HER1, HER3, or HER4. Thus it facilitates signaling by all EGF family members (Figure 16-7b); an increase in HER2 on the cell surface will make the cell more sensitive to signaling by many EGF family members because the rate at which the signaling heterodimers are formed after ligand binding will be enhanced. Even though HER3 lacks a functional kinase domain, it can still participate in signaling;

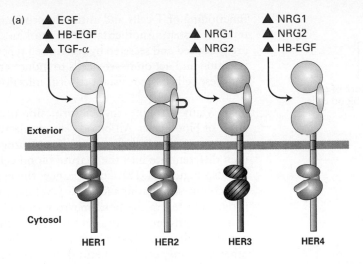

(a)

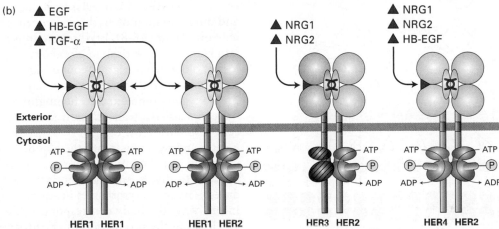

(b)

**FIGURE 16-7  The HER family of receptors and their ligands.**
Humans express four receptor tyrosine kinases—denoted HER1, 2, 3, and 4—that bind epidermal growth factor (EGF) and other EGF family members. (a) As shown, the HER proteins differentially bind EGF, heparin-binding EGF (HB-EGF), tumor-derived growth factor alpha (TGF-α), and neuregulins 1 and 2 (NRG1 and NRG2). Note that HER2, which does not directly bind a ligand, exists in the plasma surface membrane in a preactivated state indicated by a red hook.

(b) Ligand-bound HER1 can form activated homodimers bound together by loop segments (red hooks), as detailed in Figure 16-4. HER2 forms heterodimers with ligand-bound HER1, HER3, and HER4 and facilitates signaling by all EGF family members. HER3 has a very poorly active kinase domain and can signal only when complexed with HER2. [After N. E. Hynes and H. A. Lane, 2005, *Nature Rev. Cancer* **5**:341(erratum in *Nature Rev. Cancer* **5**:580), and A. B. Singh and R. C. Harris, 2005, *Cell Signal* **17**(Oct.):1183.]

after binding a ligand, it dimerizes with HER2 and becomes phosphorylated by the HER2 kinase. This activates downstream signal transduction pathways as indicated below.

Understanding the HERs has helped explain why a particular form of breast cancer is so dangerous and has led to an important drug therapy. Breast cancer can involve the abnormal growth of breast epithelial cells. Normal epithelial cells express a small amount of HER2 protein on their plasma membranes in a tissue-specific pattern, and they do not grow inappropriately. In tumor cells, errors in DNA replication often result in formation of multiple copies of a given gene on a single chromosome, an alteration known as gene amplification (see Chapter 24). Amplification of the *HER2*

gene occurs in approximately 25 percent of breast cancers, resulting in overexpression of HER2 protein in the tumor cells. Breast cancer patients with HER2 overexpression have a worse prognosis, including shortened survival, than do patients without this abnormality. As Figure 16-7 emphasizes, overexpression of HER2 makes the tumor cells sensitive to growth stimulation by low levels of any member of the EGF family of growth factors, levels that would not stimulate proliferation of cells with normal HER2 levels. Discovery of the role of HER2 overexpression in certain breast cancers led researchers to develop monoclonal antibodies specific for the HER2 protein. These have proved to be effective therapies for those breast cancer patients in which HER2 is overexpressed, reducing recurrence by about 50 percent in these patients. ∎

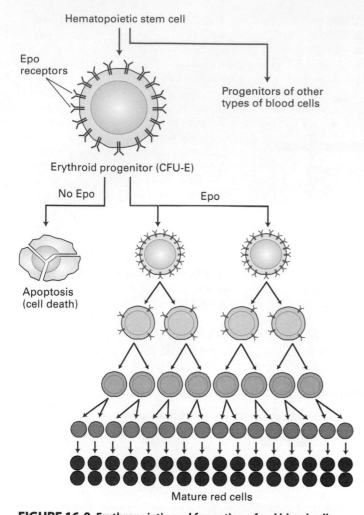

**FIGURE 16-8 Erythropoietin and formation of red blood cells (erythrocytes).** Erythroid progenitor cells, called colony-forming units erythroid (CFU-E), are derived from hematopoietic stem cells, which also give rise to progenitors of other blood cell types (see Figure 21-18). In the absence of erythropoietin (Epo), CFU-E cells undergo apoptosis. Binding of Epo to its receptors on a CFU-E induces transcription of several genes whose encoded proteins prevent programmed cell death (apoptosis), allowing the cell to survive. Other Epo-induced proteins trigger the developmental program of three to five terminal cell divisions. If CFU-E cells are cultured with Epo in a semisolid medium (e.g., containing methylcellulose), daughter cells cannot move away, and thus each CFU-E produces a colony of 30–100 erythroid cells; hence its name. [See M. Socolovsky et al., 2001, *Blood* **98**:3261.]

## Cytokines Influence Development of Many Cell Types

The cytokines form a family of relatively small, secreted signaling molecules (generally containing about 160 amino acids) that control growth and differentiation of specific types of cells. During pregnancy, for example, the cytokine *prolactin* induces epithelial cells lining the immature ductules of the mammary gland to differentiate into the acinar cells that produce milk proteins and secrete them into the ducts. Other cytokines, the **interleukins**, are essential for proliferation and

functioning of T cells and antibody-producing B cells of the immune system. Another family of cytokines, the **interferons**, are produced and secreted by certain cell types following virus infection and act on nearby cells to induce enzymes that render these cells more resistant to virus infection.

Many cytokines induce formation of important types of blood cells. All blood cells are derived from a common stem cell, which forms a series of progenitor cells that then differentiate into the mature blood cells (Figure 16-8; see also Figure 21-18). For instance, the cytokine granulocyte colony stimulating factor (G-CSF) induces a granulocyte progenitor cell in the bone marrow to divide several times and then differentiate into granulocytes, the type of white blood cell that inactivates bacteria and other pathogens. Another cytokine, **erythropoietin (Epo)**, triggers production of erythrocytes (red blood cells) by inducing the proliferation and differentiation of erythroid progenitor cells in the bone marrow (Figure 16-8). Erythropoietin is synthesized by certain kidney cells. A drop in blood oxygen, such as caused by loss of blood from a large wound, signifies a lower than optimal level of erythrocytes, whose major function is to transport oxygen complexed to hemoglobin. By means of the oxygen-sensitive transcription factor HIF-1$\alpha$, the kidney cells respond to low oxygen by synthesizing more erythropoietin and secreting it into the blood. As the level of erythropoietin rises, more and more erythroid progenitors are induced to divide and differentiate; each progenitor produces ~50 or so erythrocytes in a period of only a few days. In this way, the body can respond to the loss of blood by accelerating the production of erythrocytes. Both Epo and GCSF are produced commercially by recombinant expression in cultured mammalian cells. Patients with kidney disease, especially those undergoing dialysis, frequently are anemic (have a low red blood cell count) and therefore are treated with recombinant Epo to boost red cell levels. Epo and GCSF are used as adjuncts to certain cancer therapies since many cancer treatments affect the bone marrow and reduce production of red cells and granulocytes. ∎

## Binding of a Cytokine to Its Receptor Activates a Tightly Bound JAK Protein Tyrosine Kinase

All cytokines evolved from a common ancestral protein and have a similar tertiary structure consisting of four long conserved $\alpha$ helices folded together. Likewise, the various cytokine receptors undoubtedly evolved from a single common ancestor since all cytokine receptors have similar structures. Their extracellular domains are constructed of two subdomains, each of which contains seven conserved $\beta$ strands folded together in a characteristic fashion. The interaction of one erythropoietin molecule with two identical erythropoietin receptor (EpoR) proteins, depicted in Figure 16-9, exemplifies the binding of a cytokine to its receptor.

Cytokine receptors do not possess intrinsic enzymatic activity. Rather, a **JAK kinase** is tightly bound to the cytosolic

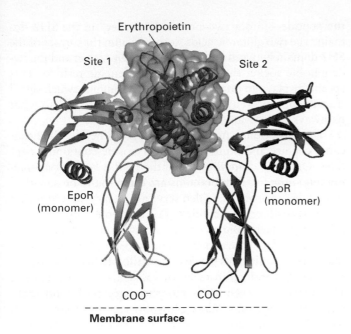

Erythropoietin

Site 1    Site 2

EpoR                    EpoR
(monomer)               (monomer)

COO⁻        COO⁻

**Membrane surface**

**FIGURE 16-9 Structure of erythropoietin bound to an erythropoietin receptor.** Erythropoietin (Epo) contains four conserved long α helices that are folded in a particular arrangement. The activated erythropoietin receptor (EpoR) is a dimer of identical subunits; the extracellular domain of each monomer is constructed of two subdomains each containing seven conserved β strands folded in a characteristic fashion. Side chains of residues on two of the α helices in Epo, termed site 1, contact loops on one EpoR monomer, while residues on the two other Epo α helices, site 2, bind to the same loop segments in a second receptor monomer, thereby stabilizing the dimeric receptor in a specific conformation. The structures of other cytokines and their receptors are similar to Epo and EpoR. [Courtesy Lucy Zhang; adapted from R. S. Syed et al., 1998, *Nature* **395:**511, and L. Zhang et al., 2009, *Mol. Cell* **33:**266–274.]

domain of all cytokine receptors (Figure 16-10). The four members of the JAK family of kinases contain an N-terminal receptor-binding domain, a C-terminal kinase domain that is normally poorly active catalytically, and a middle domain that regulates kinase activity by an unknown mechanism. (JAKs are so named because when they were cloned and characterized, their function was unknown; they were termed *just another kinase.*) As in RTKs, this kinase becomes activated after ligand binding and receptor dimerization (Figure 16-10, step **1**).

As a result of receptor dimerization, the associated JAKs are brought close enough together so that one can phosphory-

late the other on a critical tyrosine in the activation lip (Figure 16-10, step **2**). As with many other kinases, phosphorylation of the activation lip leads to a conformational change that enhances the affinity for ATP or the substrate to be phosphorylated, thereby increasing kinase activity (Figure 16-10, step **3**). One piece of evidence for this activation mechanism comes from study of a mutant JAK2 in which the critical tyrosine is mutated to phenylalanine. The mutant JAK2 binds normally to the EpoR but cannot be phosphorylated and is catalytically inactive. In erythroid cells, expression of this mutant JAK2 in greater than normal amounts totally blocks EpoR signaling because the mutant JAK2 binds to the majority of cytokine receptors, preventing binding and functioning of the wild-type JAK2 protein. This type of mutation, referred to as **dominant-negative,** causes loss of function even in cells that carry copies of the wild-type gene because the mutant protein prevents the normal protein from functioning (see Chapter 5).

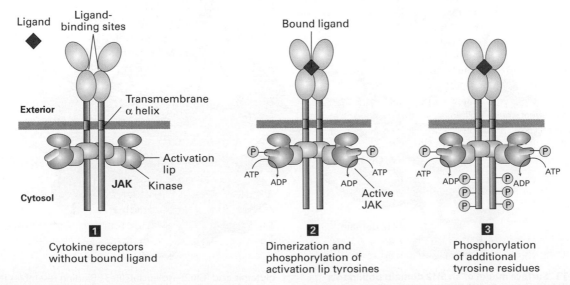

**1** Cytokine receptors without bound ligand

**2** Dimerization and phosphorylation of activation lip tyrosines

**3** Phosphorylation of additional tyrosine residues

**FIGURE 16-10 General structure and activation of cytokine receptors.** The cytosolic domain of cytokine receptors binds tightly and irreversibly to a JAK protein tyrosine kinase. In the absence of ligand (**1**), the receptors form a homodimer but the JAK kinases are poorly active. Ligand binding causes a conformational change that brings together the associated JAK kinase domains, which then phosphorylate each other on a tyrosine residue in the activation lip (**2**). Downstream signaling (**3**) then proceeds in a manner similar to that from receptor tyrosine kinases.

## Phosphotyrosine Residues Are Binding Surfaces for Multiple Proteins with Conserved Domains

Once the RTK kinases or JAK kinases become activated, they first phosphorylate several tyrosine residues on the cytosolic domain of the receptor (see Figures 16-3 and 16-10). Several of these phosphotyrosine residues then serve as binding sites for proteins that have conserved phosphotyrosine-binding domains. One such phosphotyrosine-binding domain is called the *SH2 domain*. The SH2 domain derived its full name, the Src homology 2 domain, from its homology with a region in the prototypical Src cytosolic tyrosine kinase encoded by the *src* gene. (*Src* is an acronym for *sarcoma*, and a mutant form of the cellular *src* gene was found in chickens with sarcomas, as Chapter 24 details.) The three-dimensional structures of SH2 domains in different proteins are very similar, but each binds to a distinct sequence of amino acids surrounding a phosphotyrosine residue. The unique amino acid sequence of each SH2 domain determines the specific phosphotyrosine residues it binds (Figure 16-11). Variations in the hydrophobic socket in the SH2 domains of different signal transduction proteins allow them to bind to phosphotyrosines adjacent to different sequences, accounting for differences in their binding partners. The SH2 domain of the Src tyrosine kinase, for example, binds strongly to any peptide containing a critical four-residue core sequence: phosphotyrosine–glutamic acid–glutamic acid–isoleucine (Figure 16-11). These four amino acids make intimate contact with the peptide-binding site in the Src SH2 domain. Binding resembles the insertion of a two-pronged "plug"—the phosphotyrosine and isoleucine side chains of

the peptide—into a two-pronged "socket" in the SH2 domain. The two glutamic acids fit snugly onto the surface of the SH2 domain between the phosphotyrosine socket and the hydrophobic socket that accepts the isoleucine residue. This specificity plays an important role in determining which signal transduction proteins bind to which receptors and so what pathways are activated.

There are other small protein domains besides SH2 that can recognize and bind to phosphotyrosine-containing peptides. One such domain is called the *PTB (phosphotyrosine-binding) domain*. PTB domains are often found on so-called *multidocking* proteins, which serve as docking sites for other signal transduction proteins. For example, when several RTKs (e.g., the insulin receptor) and cytokine receptors (e.g., the IL-4 receptor) are activated and tyrosine phosphorylated, they bind a multidocking protein called IRS-1 (discovered because it is an insulin receptor substrate) (Figure 16-12). The activated receptor then phosphorylates the bound docking protein, forming many phosphotyrosines that in turn serve as docking sites for SH2-containing signaling proteins. Some of these proteins in turn may also be phosphorylated by the activated receptor, and thus these multidocking proteins expand the number of intracellular signaling pathways that can be activated by the receptor.

## SH2 Domains in Action: JAK Kinases Activate STAT Transcription Factors

To illustrate how binding of SH2 domains to specific phosphotyrosine residues induces specific signaling pathways, here

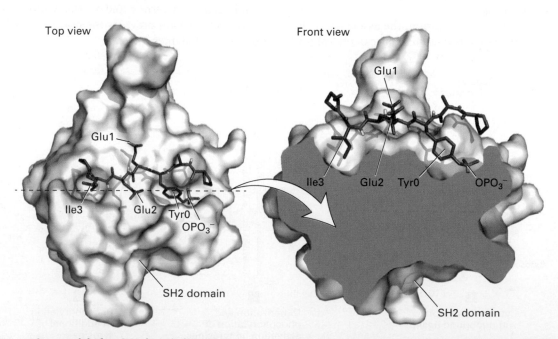

**FIGURE 16-11 Surface model of an SH2 domain bound to a phosphotyrosine-containing peptide.** The peptide bound by this SH2 domain from Src tyrosine kinase (blue backbone with red oxygen atoms) is shown in stick form. The SH2 domain binds strongly to shorttarget peptides containing a critical four-residue core sequence: phosphotyrosine (Tyr0 and $OPO_3^-$)–glutamic acid (Glu1)–

glutamic acid (Glu2)–isoleucine (Ile3). Binding resembles the insertion of a two-pronged "plug"—the phosphotyrosine and isoleucine side chains of the peptide—into a two-pronged "socket" in the SH2 domain. The two glutamate residues are bound to sites on the surface of the SH2 domain between the two sockets. [See G. Waksman et al., 1993, *Cell* **72**:779.]

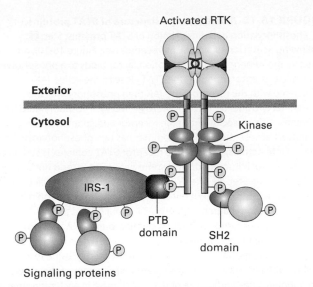

**Activated RTK**

Exterior

Cytosol

Kinase

IRS-1

PTB domain

SH2 domain

Signaling proteins

**FIGURE 16-12 Recruitment of intracellular signal transduction proteins to the cell membrane by binding to phosphotyrosine residues in receptors or receptor-associated proteins.** Cytosolic proteins with SH2 (purple) or PTB (maroon) domains can bind to specific phosphotyrosine residues in activated RTKs (shown here) or cytokine receptors. In some cases, these signal transduction proteins then are phosphorylated by the receptor's intrinsic or associated protein tyrosine kinase, enhancing their activity. Certain RTKs and cytokine receptors utilize multidocking proteins such as IRS-1 to increase the number of signaling proteins that are recruited and activated. Subsequent phosphorylation of a receptor-bound IRS-1 by the receptor kinase creates additional docking sites for SH2-containing signaling proteins.

we'll discuss the straightforward mechanism by which all JAK kinases and some RTKs directly activate members of the STAT family of transcription factors. All STAT proteins contain an N-terminal DNA-binding domain, an SH2 domain that binds to one or more specific phosphotyrosines in a cytokine receptor's cytosolic domain, and a C-terminal domain with a critical tyrosine residue. Once a monomeric STAT is bound to the receptor via its SH2 domain, the C-terminal tyrosine is phosphorylated by an associated JAK kinase (Figure 16-13a). This arrangement ensures that in a particular cell, only those STAT proteins with an SH2 domain that can bind to a particular receptor protein will be activated and only when that receptor is activated. The erythropoietin receptor, for example, activates STAT5 but not STATs 1, 2, 3, or 4; these are activated by other receptors. A phosphorylated STAT dissociates spontaneously from the receptor, and two phosphorylated STAT proteins form a dimer in which the SH2 domain on each binds to the phosphotyrosine in the other. Because dimerization involves conformational changes that expose the nuclear-localization signal (NLS), STAT dimers move into the nucleus, where they bind to specific **enhancers** (DNA regulatory sequences) controlling target genes (Figure 16-13b) and thus alter gene expression.

A given STAT can activate different genes in different cells depending on the "cell memory" discussed in the chap-ter introduction. Because different cell types have unique complements of transcription factors and unique epigenetic modifications on their chromatin, the genes that are available to be activated by any STAT are also different. For example, in mammary gland cells STAT5, the same STAT activated by the Epo receptor in erythroid cells, becomes activated following prolactin binding to the prolactin receptor and induces transcription of genes encoding certain milk proteins. In contrast, when STAT5 becomes activated in erythroid progenitor cells following binding of Epo to the Epo receptor, it induces transcription of the Bcl-x$_L$ gene. Bcl-x$_L$ prevents the programmed cell death, or **apoptosis**, of these progenitors, allowing them to proliferate and differentiate into red blood cells. Here we have a case of different cytokine receptors in different cells activating the same intermediate signaling molecule, STAT5, yet leading to the activation of different genes. Combinatorial diversity allows a relatively limited set of signaling pathways to control a vast array of cellular activities.

## Multiple Mechanisms Down-Regulate Signaling from RTKs and Cytokine Receptors

In the last chapter, we saw several ways in which signaling from G protein–coupled receptors is terminated. For instance, phosphorylation of receptors and downstream signaling proteins suppress signaling and this suppression can be reversed by the controlled action of phosphatases. Here we discuss several mechanisms by which RTK and cytokine receptor signaling is regulated.

**Receptor-Mediated Endocytosis** Prolonged treatment of cells with ligand often reduces the number of available cell-surface receptors such that the cells will have a less robust response to exposure to a given concentration of ligand than they did before the treatment. This desensitization response helps prevent inappropriately prolonged receptor activity. In the absence of epidermal growth factor (EGF), for instance, cell-surface HER1 receptors for this ligand are relatively long-lived, with an average half-life of 10 to 15 hours. Unbound receptors are internalized via clathrin-coated pits into endosomes at a relatively slow rate, on average once every 30 minutes, and often are returned rapidly to the plasma membrane so that there is little reduction in total surface receptor numbers. Following binding of an EGF ligand, the rate of endocytosis of HER1 is increased ~10-fold, and only a fraction of the internalized receptors return to the plasma membrane; the rest are degraded in lysosomes. Each time a HER1–EGF complex is internalized, via the process termed **receptor-mediated endocytosis** (see Figure 14-29), the receptor has about a 20 to 80 percent chance of being degraded, depending on the cell type. Exposure of a fibroblast cell to high levels of EGF for several hours induces several rounds of endocytosis, resulting in degradation of most cell-surface receptor molecules and thus a reduction in the cell's sensitivity to EGF. In this way, prolonged treatment with a given concentration of EGF desensitizes the cell to that level of hormone, though the cell may respond if the level of EGF is increased.

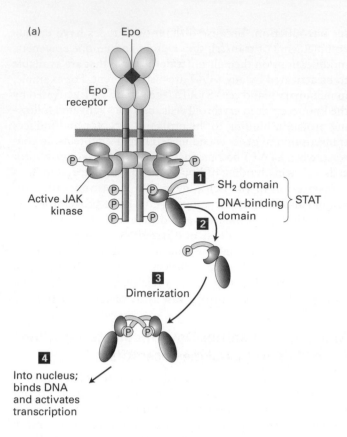

(a)

Epo

Epo receptor

Active JAK kinase

**1** SH₂ domain

DNA-binding domain } STAT

**2**

**3** Dimerization

**4**

Into nucleus; binds DNA and activates transcription

**FIGURE 16-13 Activation and structure of STAT proteins.**
(a) Phosphorylation and dimerization of STAT proteins. Step **1**: Following activation of a cytokine receptor (see Figure 16-10), an inactive monomeric STAT transcription factor binds to a phosphotyrosine in the receptor, bringing the STAT close to the active JAK associated with the receptor. The JAK then phosphorylates the C-terminal tyrosine in the STAT. Steps **2** and **3**: Phosphorylated STATs spontaneously dissociate from the receptor and spontaneously dimerize. Because the STAT homodimer has two phosphotyrosine-SH2 domain interactions, whereas the receptor-STAT complex is stabilized by only one such interaction, phosphorylated STATs tend not to rebind to the receptor. Step **4**: The STAT dimer moves into the nucleus, where it can bind to promoter sequences and activate transcription of target genes. (b) Ribbon diagram of the STAT1 dimer bound to DNA (black). The STAT1 dimer forms a C-shaped clamp around DNA that is stabilized by reciprocal and highly specific interactions between the SH2 domain (purple) of one monomer and the phosphorylated tyrosine residue (yellow with red oxygens) on the C-terminal segment of the other. The phosphotyrosine-binding site of the SH2 domain in each monomer is coupled structurally to the DNA-binding domain (magenta), suggesting a potential role for the SH2-phosphotyrosine interaction in the stabilization of DNA interacting elements. [Part (b) after X. Chen et al., 1998, *Cell* **93**:827.]

(b)

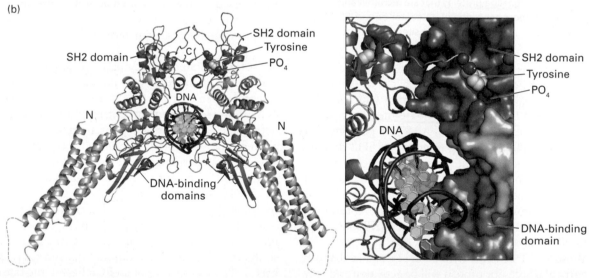

SH2 domain · C · SH2 domain · Tyrosine · PO₄ · DNA · N · N · DNA-binding domains · SH2 domain · Tyrosine · PO₄ · DNA · DNA-binding domain

HER1 mutants that lack kinase activity do not undergo accelerated endocytosis in the presence of ligand. It is likely that ligand-induced activation of the kinase activity in normal HER1 induces a conformational change in the cytosolic tail, exposing a sorting motif that facilitates receptor recruitment into clathrin-coated pits and subsequent internalization of the receptor-ligand complex. Despite extensive study of mutant HER1 cytosolic domains, the identity of these "sorting motifs" is controversial, and most likely multiple motifs function to enhance endocytosis. Interestingly, inter-

nalized receptors can continue to signal from endosomes or other intracellular compartments before their degradation, as evidenced by their binding to signaling proteins such as Grb-2 and Sos, which are discussed in the next section.

**Lysosomal Degradation** After internalization, some cell-surface receptors (e.g., the LDL receptor) are efficiently recycled to the surface (see Figure 14-29). As noted above, the fraction of activated HER1 receptors that are sorted to lysosomes can vary from 20 to 80 percent in different cell types.

There are several potential processes that can influence the recycling versus lysosomal degradation fates of surface receptors. One is covalent modification by the small protein ubiquitin (see Chapter 3). There is a strong correlation between monoubiquitination (addition of a single ubiquitin to a given lysine of a protein) of the HER1 cytosolic domain and HER1 degradation. The monubiquitination is mediated by the enzyme c-Cbl. An E3 ubiquitin ligase (see Figure 3-29), c-Cbl contains an EGFR-binding domain, which binds directly to phosphorylated EGF receptors, and a RING finger domain, which recruits ubiquitin-conjugating enzymes and mediates transfer of ubiquitin to the receptor. The ubiquitin functions as a "tag" on the receptor that stimulates its incorporation from endosomes into multivesicular bodies (see Figure 14-33) that ultimately are degraded inside lysosomes. A role for c-Cbl in EGF receptor trafficking emerged from genetic studies in *C. elegans*, which established that c-Cbl negatively regulates the function of the nematode EGF receptor (Let-23), probably by inducing its degradation. Similarly, knockout mice lacking c-Cbl show hyperproliferation of mammary gland epithelia, consistent with a role of c-Cbl as a negative regulator of EGF signaling.

Experiments with mutant cell lines demonstrate that internalization of RTKs plays an important role in regulating cellular responses to EGF and other growth factors. For instance, a mutation in the EGF receptor (HER1) that prevents it from being incorporated into coated pits makes it resistant to receptor-mediated (ligand-induced) endocytosis. As a result, this mutation leads to substantially above-normal numbers of EGF receptors on cells and thus increased sensitivity of cells to EGF as a mitogenic signal. Such mutant cells are prone to EGF-induced **transformation** into tumor cells (see Chapter 24). Interestingly, the other EGF family receptors—HER2, HER3, and HER4—do not undergo ligand-induced internalization, an observation that emphasizes how each receptor evolved to be regulated in its own appropriate manner.

**Phosphotyrosine Phosphatases** These dephosphorylating enzymes specifically hydrolyze phosphotyrosine linkages on specific target proteins. An excellent example of how phosphotyrosine phosphatase enzymes function to suppress the activity of protein tyrosine kinases is provided by SHP1, the phosphatase that negatively regulates signaling from several types of cytokine receptors. Its role was first identified from analysis of mice lacking this protein; they died because of excess production of erythrocytes and several other types of blood cells.

SHP1 dampens cytokine signaling by binding to a cytokine receptor and inactivating the associated JAK protein, as is depicted in Figure 16-14a. In addition to a phosphatase catalytic domain, SHP1 has two SH2 domains. When cells are in the resting state, unstimulated by a cytokine, one of the SH2 domains in SHP1 physically binds to and inactivates the catalytic site in the phosphatase domain. In the stimulated state, however, this blocking SH2 domain binds to a specific phosphotyrosine residue in the activated receptor. The conformational change that accompanies this binding

(a) Short-term regulation: JAK2 deactivation by SHP1 phosphatase

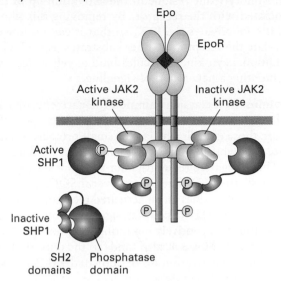

(b) Long-term regulation: signal blocking and protein degradation by SOCS proteins

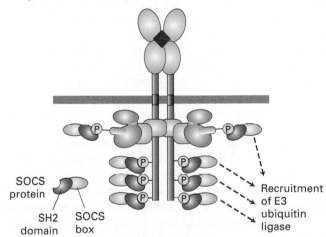

**FIGURE 16-14 Two mechanisms for terminating signal transduction from the erythropoietin receptor (EpoR).** (a) Short-term regulation: SHP1, a phosphotyrosine phosphatase, is present in an inactive form in unstimulated cells. Binding of an SH2 domain in SHP1 to a particular phosphotyrosine in the activated receptor unmasks its phosphatase catalytic site and positions it near the phosphorylated tyrosine in the lip region of JAK2. Removal of the phosphate from this tyrosine inactivates the JAK kinase. (b) Long-term regulation: SOCS proteins, whose expression is induced by STAT proteins in erythropoietin-stimulated erythroid cells, inhibit or permanently terminate signaling over longer time periods. Binding of SOCS to phosphotyrosine residues on EpoR or JAK2 blocks binding of other signaling proteins (*left*). The SOCS box can also target proteins such as JAK2 for degradation by the ubiquitin-proteasome pathway (*right*). Similar mechanisms regulate signaling from other cytokine receptors. [Part (a) adapted from S. Constantinescu et al., 1999, *Trends Endocrin. Metabol.* **10**:18; part (b) adapted from B. T. Kile and W. S. Alexander, 2001, *Cell. Mol. Life Sci.* **58**:1.]

unmasks the SHP1 catalytic site and also brings it adjacent to the phosphotyrosine residue in the activation lip of the JAK associated with the receptor. By removing this phosphate, SHP1 inactivates the JAK, so that it can no longer phosphorylate the receptor or other substrates (e.g. STATs) unless additional cytokine molecules bind to cell-surface receptors, initiating a new round of signaling.

**SOCS Proteins** In a classic example of negative feedback, among the genes whose transcription is induced by STAT proteins are those encoding a class of small proteins termed *SOCS proteins*, which terminate signaling from cytokine receptors. These negative regulators act in two ways (Figure 16-14b). First, the SH2 domain in several SOCS proteins binds to phosphotyrosines on an activated receptor, preventing binding of other SH2-containing signaling proteins (e.g., STATs) and thus competitively inhibiting receptor signaling. One SOCS protein, SOCS-1, also binds to the critical phosphotyrosine in the activation lip of activated JAK2 kinase, thereby inhibiting its catalytic activity. Second, all SOCS proteins contain a domain, called the SOCS box, that recruits components of E3 ubiquitin ligases (see Figure 3-29). As a result of binding SOCS-1, for instance, JAK2 becomes polyubiquitinated (a polymer of ubiquitins covalently attached to the side chain of a lysine) and is then degraded in **proteasomes**, thereby permanently turning off all JAK2-mediated signaling pathways until new JAK2 proteins can be made. The observation that proteasome inhibitors prolong JAK2 signal transduction supports this mechanism.

Studies with cultured mammalian cells have shown that the receptor for growth hormone, which belongs to the cytokine receptor superfamily, is down-regulated by another SOCS protein, SOCS-2. Strikingly, mice deficient in SOCS-2 grow significantly larger than their wild-type counterparts and have long bone lengths and proportionate enlargement of most organs. Thus SOCS proteins play an essential negative role in regulating intracellular signaling from the receptors for erythropoietin, growth hormone, and other cytokines.

## KEY CONCEPTS of Section 16.1

### Receptors That Activate Protein Tyrosine Kinases

• Two broad classes of receptors activate tyrosine kinases: (1) receptor tyrosine kinases (RTKs), in which the kinase is an intrinsic part of the receptor, and (2) cytokine receptors, in which the kinase is bound tightly to the cytosolic domain of the receptor. Signaling from receptor tyrosine kinases and cytokine receptors activate similar downstream signaling pathways (see Figure 16-2).

• Receptor tyrosine kinases, which bind to peptides and signaling proteins such as growth factors and insulin, may exist as preformed dimers or dimerize during binding to ligands. Ligand binding triggers formation of functional dimeric receptors, a necessary step in activation of the receptor-associated kinase.

• Activation of an RTK leads to phosphorylation of the activation lip in the protein tyrosine kinases that are an intrinsic part of their cytoplasmic domains, enhancing their catalytic activity (see Figure 16-3). The activated kinase then phosphorylates tyrosine residues in the receptor cytosolic domain and in other protein substrates.

• Humans express many RTKS, four of which (HER1-4) define the epidermal growth factor receptor family that mediates signaling from different members of the epidermal growth factor family of signaling molecules (see Figure 16-7). One of these receptors, HER2, does not bind ligand; it forms active heterodimers with ligand-bound monomers of the other three HER proteins. Overexpression of HER2 is implicated in about 25 percent of breast cancers.

• Cytokines play numerous roles in development. Erythropoietin, a cytokine secreted by kidney cells, promotes proliferation and differentiation of erythroid progenitor cells in the bone marrow (see Figure 16-8) to increase the number of mature red cells in the blood.

• All cytokine receptors have similar structures, and their cytosolic domains are tightly bound to a JAK protein tyrosine kinase, which becomes activated after cytokine binding and receptor dimerization (see Figure 16-10).

• In both RTKs and cytokine receptors, short amino acid sequences containing a phosphotyrosine residue are bound by proteins with conserved SH2 or PTB domains, which are found in many signal-transducing proteins. The sequence of amino acids surrounding the phosphorylated tyrosine determines which domain will bind to it. Such protein-protein interactions are important in many signaling pathways (see Figures 16-11 and 16-12).

• The *JAK/STAT pathway* operates downstream from all cytokine receptors and some RTKs. STAT monomers bound to phosphotyrosines on receptors are phosphorylated by receptor-associated JAKs, then dimerize and move to the nucleus, where they activate transcription (see Figure 16-13).

• Endocytosis of receptor-hormone complexes and their degradation in lysosomes is a principal way of reducing the number of receptor tyrosine kinases and cytokine receptors on the cell surface, thus decreasing the sensitivity of cells to many peptide hormones.

• Signaling from cytokine receptors is terminated by the phosphotyrosine phosphatase SHP1 and several SOCS proteins (see Figure 16-14).

## 16.2 The Ras/MAP Kinase Pathway

Almost all receptor tyrosine kinases and cytokine receptors activate the *Ras/MAP kinase pathway* (see Figure 16-2b). The **Ras protein**, a monomeric (small) G protein, belongs to the **GTPase superfamily** of intracellular switch proteins (see Figure 15-7). Activated Ras promotes formation, at the membrane, of signal transduction complexes containing three sequentially acting protein kinases. This *kinase cascade* culminates in activation of certain members of the **MAP kinase** family, which can translocate into the nucleus and phosphorylate many different proteins. Among the target proteins for MAP kinase are transcription factors that regulate expression of proteins with important roles in the cell cycle and in differentiation. Importantly, different types of extracellular signals often activate different signaling pathways that result in activation of different members of the MAP kinase family.

Because an activating mutation in a RTK, Ras, or a protein in the MAP kinase cascade is found in almost all types of human tumors, the RTK/Ras/MAP kinase pathway has been subjected to extensive study and a great deal is known about the components of this pathway. We begin our discussion by reviewing how Ras cycles between the active and inactive state. We then describe how Ras is activated and passes a signal to the MAP kinase pathway. Finally we examine recent studies indicating that both yeasts and cells of higher eukaryotes contain multiple MAP kinase pathways and consider the ways cells keep different MAP kinase pathways separate from one another through the use of scaffold proteins.

## Ras, a GTPase Switch Protein, Operates Downstream of Most RTKs and Cytokine Receptors

Like the $G_\alpha$ subunits in trimeric G proteins discussed in Chapter 15, the monomeric G protein known as Ras alternates between an active "on" state with a bound GTP and an inactive "off" state with a bound GDP (see Figure 15-6 to review this concept). Unlike trimeric G proteins, Ras is not directly linked to cell-surface receptors. Ras (~170 amino acids) is smaller than $G_\alpha$ proteins (~300 amino acids), but the GTP-binding domains of the two proteins have a similar structure (see Figure 15-7 to review the structure of Ras). Structural and biochemical studies show that $G_\alpha$ also contains a GTPase-activating protein (GAP) domain that increases the intrinsic rate of GTP hydrolysis by $G_\alpha$. Because this domain is not present in Ras, it has an intrinsically slower rate of GTP hydrolysis. Thus the average lifetime of a GTP bound to Ras is about 1 minute, which is much longer than the average lifetime of a $G_\alpha \cdot$GTP complex.

The activity of the Ras protein is regulated by several factors. Ras activation is accelerated by a *guanine nucleotide exchange factor* (GEF), which binds to the Ras·GDP complex, causing dissociation of the bound GDP (see Figure 15-6). Because GTP is present in cells at a higher concentration than GDP, GTP binds spontaneously to "empty" Ras molecules, with release of GEF and formation of the active Ras·GTP. Subsequent hydrolysis of the bound GTP to GDP deactivates Ras. Because the intrinsic GTPase activity of Ras·GTP is low compared to that of $G_\alpha \cdot$GTP, Ras·GTP requires the assistance of another protein, a GTPase-activating protein (GAP), to deactivate it. Binding of GAP to Ras·GTP accelerates the intrinsic GTPase activity of Ras by more than

a hundredfold; the actual hydrolysis of GTP is catalyzed by amino acids from both Ras and GAP. In particular, insertion of one of GAP's arginine side chains into the Ras active site stabilizes an intermediate in the hydrolysis reaction.

Mammalian Ras proteins have been studied in great detail because mutant Ras proteins are associated with many types of human cancer. These mutant proteins, which bind but cannot hydrolyze GTP, are permanently in the "on" state and contribute to oncogenic transformation (see Chapter 24). Determination of the three-dimensional structure of the Ras-GAP complex and tests of mutant forms of Ras explained the puzzling observation that most oncogenic, constitutively active Ras proteins ($Ras^D$) contain a mutation at position 12. Replacement of the normal glycine-12 with any other amino acid (except proline) blocks the functional binding of GAP and in essence "locks" Ras in the active GTP-bound state. ∎

The first indication that Ras functions downstream from RTKs in a common signaling pathway came from experiments in which cultured fibroblast cells were induced to proliferate by treatment with a mixture of two protein hormones: platelet-derived growth factor (PDGF) and epidermal growth factor (EGF). Microinjection of anti-Ras antibodies into these cells blocked cell proliferation. Conversely, injection of $Ras^D$, a constitutively active mutant Ras protein that hydrolyzes GTP very inefficiently and thus persists in the active state, caused the cells to proliferate in the absence of the growth factors. These findings are consistent with studies, using the pull-down assay method detailed in Figure 15-14, showing that addition of FGF to fibroblasts leads to a rapid increase in the proportion of Ras present in the GTP-bound active form. However, as we will see, an activated RTK (or cytokine receptor) cannot directly activate Ras. Instead, other proteins must first be recruited to the activated receptor and serve as adapters.

## Genetic Studies in *Drosophila* Identified Key Signal-Transducing Proteins in the Ras/MAP Kinase Pathway

Our knowledge of the proteins involved in the Ras/MAP kinase pathway came principally from genetic analyses of mutant fruit flies (*Drosophila*) and worms (*C. elegans*) that were blocked at particular stages of differentiation. To illustrate the power of this experimental approach, we consider development of a particular type of cell in the compound eye of *Drosophila*.

The compound eye of the fly is composed of some 800 individual eyes called *ommatidia* (Figure 16-15a). Each ommatidium consists of 22 cells, eight of which are photosensitive neurons called *retinula*, or R cells, designated R1–R8 (Figure 16-15b). An RTK called *Sevenless* (Sev) specifically regulates development of the R7 cell and is not essential for any other known function. In flies with a mutant *sevenless* (sev) gene, the R7 cell in each ommatidium does not form (Figure 16-15c,

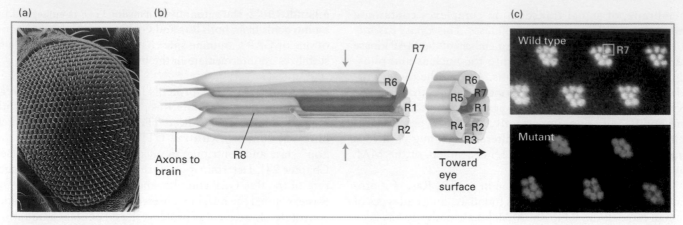

FIGURE 16-15 The compound eye of *Drosophila melanogaster.*
(a) Scanning electron micrograph showing individual ommatidia that compose the fruit fly eye. (b) Longitudinal and cutaway views of a single ommatidium. Each of these tubular structures contains eight photoreceptors, designated R1–R8, which are long, cylindrically shaped light-sensitive cells. R1–R6 (yellow) extend throughout the depth of the retina, whereas R7 (brown) is located toward the surface of the eye and R8 (blue) toward the back side, where the axons exit. (c) Comparison of eyes from wild-type and *sevenless* mutant flies viewed by a special technique that can distinguish the photoreceptors in an ommatidium. The plane of sectioning is indicated by the blue arrows in (b), and the R8 cell is out of the plane of these images. The seven photoreceptors in this plane are easily seen in the wild-type ommatidia (*top*), whereas only six are visible in the mutant ommatidia (*bottom*). Flies with the *sevenless* mutation lack the R7 cell in their eyes. [Part (a) from E. Hafen and K. Basler, 1991, *Development* **1**(suppl.):123; part (b) adapted from R. Reinke and S. L. Zipursky, 1988, *Cell* **55**:321; part (c) courtesy of U. Banerjee.]

*bottom*). Since the R7 photoreceptor is necessary only for flies to see in ultraviolet light, mutants that lack functional R7 cells but are otherwise normal are easily isolated. Therefore, fly R7 cells are an ideal genetic system for studying cell development.

During development of each ommatidium, a protein called *Boss* (*Bride of Sevenless*) is expressed on the surface of the R8 cell. This membrane-tethered protein is the ligand for the Sev RTK on the surface of the neighboring R7 precursor cell, signaling it to develop into a photosensitive neuron (Figure 16-16a). In mutant flies that do not express a functional Boss protein or Sev RTK, interaction between the Boss and Sev proteins cannot occur, and no R7 cells develop (Figure 16-16b); this is the origin of the name "Sevenless" for the RTK in the R7 cells.

To identify intracellular signal-transducing proteins in the Sev RTK pathway, investigators produced mutant flies expressing a temperature-sensitive Sev protein. When these flies were maintained at a permissive temperature, all their ommatidia contained R7 cells; when they were maintained at a nonpermissive temperature, no R7 cells developed. At a particular intermediate temperature, however, just enough of the Sev RTK was functional to mediate normal R7 development. The investigators reasoned that at this intermediate

EXPERIMENTAL FIGURE 16-16 Genetic studies reveal that activation of Ras induces development of R7 photoreceptors in the *Drosophila* eye. (a) During larval development of wild-type flies, the R8 cell in each developing ommatidium expresses a cell-surface protein, called Boss, which binds to the Sev RTK on the surface of its neighboring R7 precursor cell. This interaction induces changes in gene expression that result in differentiation of the precursor cell into a functional R7 neuron. (b) In fly embryos with a mutation in the *sevenless* (*sev*) gene, R7 precursor cells cannot bind Boss and therefore do not differentiate normally into R7 cells. Rather, the precursor cell enters an alternative developmental pathway and eventually becomes a cone cell. (c) Double-mutant larvae (*sev⁻*; *Ras^D*) express a constitutively active Ras (Ras^D) in the R7 precursor cell, which induces differentiation of R7 precursor cells in the absence of the Boss-mediated signal. This finding shows that activated Ras is sufficient to mediate induction of an R7 cell. [See M. A. Simon et al., 1991, *Cell* **67**:701, and M. E. Fortini et al., 1992, *Nature* **355**:559.]

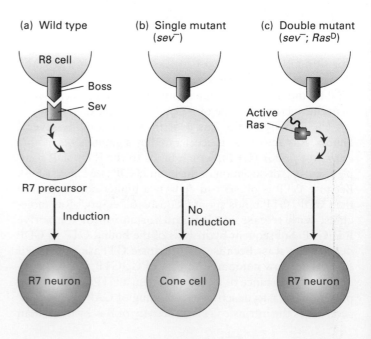

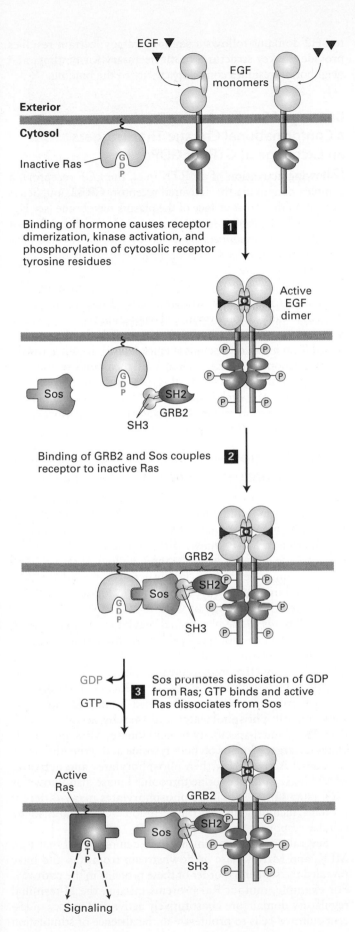

**Exterior**

**Cytosol**

Inactive Ras

EGF ▼

FGF monomers

**1** Binding of hormone causes receptor dimerization, kinase activation, and phosphorylation of cytosolic receptor tyrosine residues

Active EGF dimer

Sos

GRB2

SH2

SH3

**2** Binding of GRB2 and Sos couples receptor to inactive Ras

GRB2

Sos

SH2

SH3

GDP ←
GTP →

**3** Sos promotes dissociation of GDP from Ras; GTP binds and active Ras dissociates from Sos

Active Ras

GRB2

Sos

SH2

SH3

Signaling

**FIGURE 16-17 Activation of Ras following ligand binding to receptor tyrosine kinases (RTKs) or cytokine receptors.** The receptors for epidermal growth factor (EGF) and many other growth factors are RTKs. The cytosolic adapter protein GRB2 binds to a specific phosphotyrosine on an activated, ligand-bound receptor and to the cytosolic Sos protein, bringing it near the plasma membrane and to its substrate, the inactive Ras·GDP. The guanine nucleotide exchange factor (GEF) activity of Sos then promotes formation of active Ras·GTP. Note that Ras is tethered to the cytosolic surface of the plasma membrane by a hydrophobic farnesyl anchor (see Figure 10-19). [See J. Schlessinger, 2000, *Cell* **103**:211, and M. A. Simon, 2000, *Cell* **103**:13.]

temperature, the signaling pathway would become defective (and thus no R7 cells would develop) if the level of another protein involved in the pathway was reduced, thereby reducing the activity of the overall pathway below the level required to form an R7 cell. A recessive mutation affecting such a protein would have this effect because, in diploid organisms such as *Drosophila*, a heterozygote containing one wild-type and one mutant allele of a gene will produce half the normal amount of the gene product; hence, even if such a recessive mutation is in an essential gene, the organism will usually be viable. However, a fly carrying a temperature-sensitive mutation in the *sev* gene and a second mutation affecting another protein in the signaling pathway would be expected to lack R7 cells at the intermediate temperature.

By use of this screen, researchers identified three genes encoding important proteins in the Sev pathway: an SH2-containing adapter protein exhibiting 64 percent amino acid sequence identity to human GRB2 (growth factor *receptor-bound* protein 2), a guanine nucleotide exchange factor called Sos (Son of Sevenless) exhibiting 45 percent identity with its mouse counterpart, and a Ras protein exhibiting 80 percent identity with its mammalian counterparts. These three proteins later were found to function in other signaling pathways initiated by ligand binding to different RTKs and used at different times and places in the developing fly.

In subsequent studies, researchers introduced a mutant *ras*^D gene into fly embryos carrying the sevenless mutation. As noted earlier, the *ras*^D gene encodes a constitutively active Ras protein that is present in the active GTP-bound form even in the absence of a hormone signal. Although no functional Sev RTK was expressed in these double mutants (*sev*^−; *ras*^D), R7 cells formed normally, indicating that presence of an activated Ras protein is sufficient for induction of R7-cell development (Figure 16-16c). This finding, which is consistent with the results with cultured fibroblasts described earlier, supports the conclusion that activation of Ras is a principal step in intracellular signaling by most if not all RTKs and cytokine receptors.

## Receptor Tyrosine Kinases and JAK Kinases Are Linked to Ras by Adapter Proteins

In order for activated RTKs and cytokine receptors to activate Ras, two cytosolic proteins—GRB2 and Sos—must first be recruited to provide a link between the receptor and Ras (Figure 16-17). GRB2 is an *adapter protein*, meaning that it has no enzymatic activity and serves as a link, or scaffold, between two other proteins—in this case between the activated receptor and Sos. Sos is a guanine nucleotide exchange protein (GEF), which catalyzes conversion of inactive GDP-bound Ras to the active GTP-bound form.

GRB2 is able to serve as an adapter protein because of its SH2 domain, which binds to a specific phosphotyrosine residue in the activated RTK (or cytokine receptor). In addition to its SH2 domain, the GRB2 adapter protein contains two *SH3 domains*, which bind to Sos, the Ras guanine nucleotide exchange factor (Figure 16-17). Like phosphotyrosine-binding SH2 and PTB domains, SH3 domains are present in a large number of proteins involved in intracellular signaling. Although the three-dimensional structures of various SH3 domains are similar, their specific amino acid sequences differ. The SH3 domains in GRB2 selectively bind to proline-rich sequences in Sos; different SH3 domains in other proteins bind to proline-rich sequences distinct from those in Sos.

Proline residues play two roles in the interaction between an SH3 domain in an adapter protein (e.g., GRB2) and a proline-rich sequence in another protein (e.g., Sos). First, the proline-rich sequence assumes an extended conformation that permits extensive contacts with the SH3 domain, thereby facilitating interaction. Second, a subset of these prolines fit into binding "pockets" on the surface of the SH3 domain (Figure 16-18). Several nonproline residues also interact with the SH3 domain and are responsible for determining the binding specificity. Hence the binding of proteins to SH3 and to SH2 domains follows a similar strategy: certain residues provide the key structural motif necessary for binding, and neighboring residues confer specificity to the binding.

## Binding of Sos to Inactive Ras Causes a Conformational Change That Triggers an Exchange of GTP for GDP

Following activation of an RTK (e.g., the EGF receptor), a complex containing the activated receptor, GRB2, and Sos is formed on the cytosolic face of the plasma membrane (see Figure 16-17). Complex formation depends on the ability of GRB2 to bind *simultaneously* to the receptor and to Sos. Thus receptor activation leads to relocalization of Sos from the cytosol to the membrane, bringing Sos near to its substrate, namely, Ras·GDP that is already bound to the plasma membrane by means of a covalently attached lipid. Binding of Sos to Ras·GDP leads to conformational changes in the Switch I and Switch II segments of Ras, thereby opening the binding pocket for GDP so it can diffuse out (Figure 16-19). In other words, Sos is functions as a GEF for Ras. GTP then binds to and activates Ras. Binding of GTP to Ras, in turn, induces a specific conformation of Switch I and Switch II that allows Ras·GTP to activate the next protein in the Ras/MAP kinase pathway.

## Signals Pass from Activated Ras to a Cascade of Protein Kinases, Ending with MAP Kinase

Biochemical and genetic studies in yeast, *C. elegans*, *Drosophila*, and mammals have revealed that downstream of Ras is a highly conserved cascade of three protein kinases, culminating in MAP kinase. Although activation of the kinase cascade does not yield the same biological results in all cells, a common set of sequentially acting kinases defines the Ras/MAP kinase pathway, as outlined in Figure 16-20. Ras is activated by exchange of GDP for GTP (step **1**). Active Ras·GTP binds to the N-terminal regulatory domain of *Raf*, a serine/threonine (not tyrosine) kinase, thereby activating it (step **2**). In unstimulated cells, Raf is phosphorylated and bound in an inactive state to the phosphoserine-binding protein 14-3-3. Hydrolysis of Ras·GTP to Ras·GDP releases active Raf from its complex with 14-3-3 (step **3**), and Raf subsequently phosphorylates and thereby activates *MEK* (step **4**). (A dual-specificity protein kinase, MEK phosphorylates its target proteins on both tyrosine and serine/threonine residues.) Active MEK then phosphorylates and activates MAP kinase, another serine/threonine kinase also known as ERK (step **5**). MAP kinase phosphorylates many different proteins, including nuclear transcription factors that mediate cellular responses (step **6**).

Several types of experiments have demonstrated that Raf, MEK, and MAP kinase lie downstream from Ras and have revealed the sequential order of these proteins in the pathway. For example, mutant Raf proteins missing the N-terminal regulatory domain are constitutively active and induce quiescent cultured cells to proliferate in the absence of stimulation

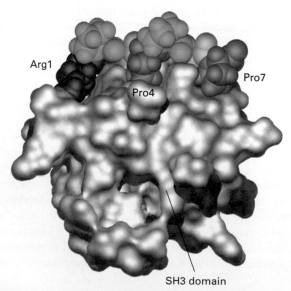

**FIGURE 16-18 Surface model of an SH3 domain bound to a target peptide.** The short, proline-rich target peptide is shown as a space-filling model. In this target peptide, two prolines (Pro4 and Pro7, dark blue) fit into binding pockets on the surface of the SH3 domain. Interactions involving an arginine (Arg1, red), two other prolines (light blue), and other residues in the target peptide (green) determine the specificity of binding. [After H. Yu et al., 1994, *Cell* **76**:933.]

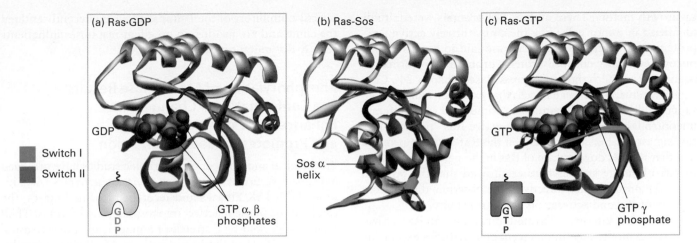

**FIGURE 16-19 Structures of Ras bound to GDP, Sos protein, and GTP.** (a) In Ras·GDP, the Switch I (green) and Switch II (blue) segments do not directly interact with GDP. (b) One α helix (brown) in Sos binds to both switch regions of Ras·GDP, leading to a massive conformational change in Ras. In effect, Sos pries Ras open by displacing the Switch I region, thereby allowing GDP to diffuse out. (c) GTP is thought to bind to the Ras-Sos complex first through its base (guanine); subsequent binding of the GTP phosphates completes the interaction. The resulting conformational change in Switch I and Switch II segments of Ras, allowing both to bind to the GTP γ phosphate, displaces Sos and promotes interaction of Ras·GTP with its effectors (discussed later). See Figure 15-8 for another depiction of Ras·GDP and Ras·GTP. [Adapted from P. A. Boriack-Sjodin and J. Kuriyan, 1998, *Nature* **394:**341.]

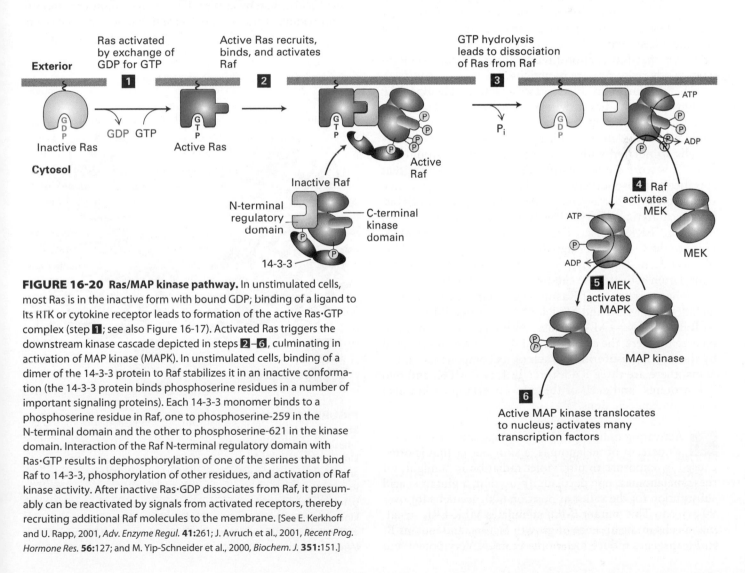

**FIGURE 16-20 Ras/MAP kinase pathway.** In unstimulated cells, most Ras is in the inactive form with bound GDP; binding of a ligand to its RTK or cytokine receptor leads to formation of the active Ras·GTP complex (step **1**; see also Figure 16-17). Activated Ras triggers the downstream kinase cascade depicted in steps **2**–**6**, culminating in activation of MAP kinase (MAPK). In unstimulated cells, binding of a dimer of the 14-3-3 protein to Raf stabilizes it in an inactive conformation (the 14-3-3 protein binds phosphoserine residues in a number of important signaling proteins). Each 14-3-3 monomer binds to a phosphoserine residue in Raf, one to phosphoserine-259 in the N-terminal domain and the other to phosphoserine-621 in the kinase domain. Interaction of the Raf N-terminal regulatory domain with Ras·GTP results in dephosphorylation of one of the serines that bind Raf to 14-3-3, phosphorylation of other residues, and activation of Raf kinase activity. After inactive Ras·GDP dissociates from Raf, it presumably can be reactivated by signals from activated receptors, thereby recruiting additional Raf molecules to the membrane. [See E. Kerkhoff and U. Rapp, 2001, *Adv. Enzyme Regul.* **41:**261; J. Avruch et al., 2001, *Recent Prog. Hormone Res.* **56:**127; and M. Yip-Schneider et al., 2000, *Biochem. J.* **351:**151.]

by growth factors. These mutant Raf proteins were initially identified in tumor cells; like the constitutively active Ras$^D$ protein, such mutant Raf proteins are said to be encoded by **oncogenes**, whose encoded proteins promote transformation of the cells in which they are expressed (see Chapter 24). Conversely, cultured mammalian cells that express a mutant, nonfunctional Raf protein cannot be stimulated to proliferate uncontrollably by a constitutively active Ras$^D$ protein. This finding established a link between the Raf and Ras proteins and that Raf lies downstream of Ras in the signaling pathway. In vitro binding studies further showed that the purified Ras·GTP protein binds directly to the N-terminal regulatory domain of Raf and activates its catalytic activity.

That MAP kinase is activated in response to Ras activation was demonstrated in quiescent cultured cells expressing a constitutively active Ras$^D$ protein. In these cells, activated MAP kinase is generated in the absence of stimulation by growth-promoting hormones. More importantly, R7 photoreceptors develop normally in the developing eye of *Drosophila* mutants that lack a functional Ras or Raf protein but express a constitutively active MAP kinase. This finding indicates that activation of MAP kinase is sufficient to transmit a proliferation or differentiation signal normally initiated by ligand binding to a receptor tyrosine kinase such as Sevenless (see Figure 16-16). Biochemical studies showed, however, that Raf cannot directly phosphorylate MAP kinase or otherwise activate its activity.

The final link in the kinase cascade activated by Ras·GTP emerged from studies in which scientists fractionated extracts of cultured cells searching for a kinase activity that could phosphorylate MAP kinase and that was present only in cells stimulated with growth factors, not unstimulated cells. This work led to identification of MEK, a kinase that specifically phosphorylates one threonine and one tyrosine residue on the activation lip of MAP kinase, thereby activating its catalytic activity. (The acronym *MEK* comes from *MAP* and *ERK* kinase.) Later studies showed that MEK binds to the C-terminal catalytic domain of Raf and is phosphorylated by the Raf serine/threonine kinase; this phosphorylation activates the catalytic activity of MEK.

Hence activation of Ras induces a kinase cascade that includes Raf, MEK, and MAP kinase: activated RTK → Ras → Raf → MEK → MAP kinase. Although we will not emphasize this here, the complexity of this pathway is increased by the multiple isoforms of each of its components. In humans, there are three RAS, three Raf, two MEK, and two Erk proteins, and each of these has overlapping but also nonredundant functions.

Activating mutations in the *B-Raf* gene occur in over 40 percent of melanomas, a skin cancer that is often caused by exposure to ultraviolet radiation in sunlight. Of these melanomas, one particular mutation, a glutamic acid substitution for the valine at position 600, accounts for over 90 percent. This mutant *B-Raf* stimulates MEK-ERK signaling in cells in the absence of growth factors, and mutant *B-Raf* transgenes induce melanoma in mice. Very potent and selective inhibitors of the *B-Raf* kinase have recently entered the clinic and are producing excellent responses in patients with *B-Raf* mutant melanoma. ■

## Phosphorylation of MAP Kinase Results in a Conformational Change That Enhances Its Catalytic Activity and Promotes Kinase Dimerization

Biochemical and x-ray crystallographic studies have provided a detailed picture of how phosphorylation activates MAP kinase. As in JAK kinases and receptor tyrosine kinases, the catalytic site in the inactive, unphosphorylated form of MAP kinase is blocked by a stretch of amino acids, the activation lip (Figure 16-21a). Binding of MEK to MAP kinase destabilizes the lip structure, resulting in exposure of tyrosine-185, which is buried in the inactive conformation. Following phosphorylation of this critical tyrosine, MEK phosphorylates the neighboring threonine-183 (Figure 16-21b).

Both the phosphorylated tyrosine and the phosphorylated threonine residue in MAP kinase interact with additional amino acids, thereby conferring an altered conformation to the lip region, which in turn permits binding of ATP to the

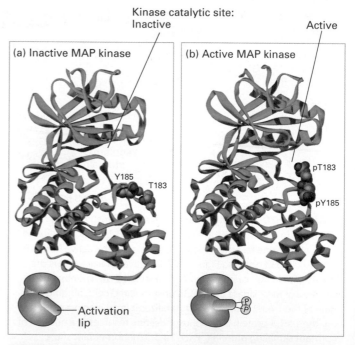

**FIGURE 16-21 Structures of inactive, unphosphorylated MAP kinase and the active, phosphorylated form.** (a) In inactive MAP kinase, the activation lip is in a conformation such that it blocks the kinase active site. (b) Phosphorylation by MEK at tyrosine-185 (Y185) and threonine-183 (T183) leads to a marked conformational change in the activation lip. This activating change both promotes binding of its substrates—ATP and its target proteins—to the kinase and MAP kinase dimerization. A similar phosphorylation-dependent mechanism activates JAK kinases and the intrinsic kinase activity of RTKs.
[After B. J. Canagarajah et al., 1997, *Cell* **90**:859.]

catalytic site, which, as in all kinases, is in the groove between the upper and lower kinase domains. The phosphotyrosine residue (pY185) also plays a key role in binding specific substrate proteins to the surface of MAP kinase. Phosphorylation promotes not only the catalytic activity of MAP kinase but also its dimerization. The dimeric form of MAP kinase is translocated to the nucleus, where it regulates the activity of many nuclear transcription factors.

## MAP Kinase Regulates the Activity of Many Transcription Factors Controlling Early Response Genes

Addition of a growth factor (e.g., EGF or PDGF) to quiescent (non-growing) cultured mammalian cells causes a rapid increase in the expression of as many as 100 different genes. These are called *early response genes* because they are induced well before cells enter the S phase and replicate their DNA (see Chapter 20). One important early response gene encodes the transcription factor c-Fos. Together with other transcription factors, such as c-Jun, c-Fos induces expression

of many genes encoding proteins necessary for cells to progress through the cell cycle. Most RTKs that bind growth factors utilize the MAP kinase pathway to activate genes encoding proteins such as c-Fos, which in turn propel the cell through the cell cycle.

The enhancer that regulates the c-*fos* gene contains a *serum response element* (*SRE*), so named because it is activated by many growth factors in serum. This complex enhancer contains DNA sequences that bind multiple transcription factors. As depicted in Figure 16-22, activated (phosphorylated) dimeric MAP kinase induces transcription of the c-*fos* gene by direct activation of one transcription factor, *ternary complex factor* (*TCF*), and indirect activation of another, *serum response factor* (*SRF*). In the cytosol, MAP kinase phosphorylates and activates a kinase called p90$^{RSK}$, which translocates to the nucleus, where it phosphorylates a specific serine in SRF. After translocating to the nucleus, MAP kinase directly phosphorylates specific serines in TCF. Association of phosphorylated TCF with two molecules of phosphorylated SRF forms an active trimeric factor that activates gene transcription.

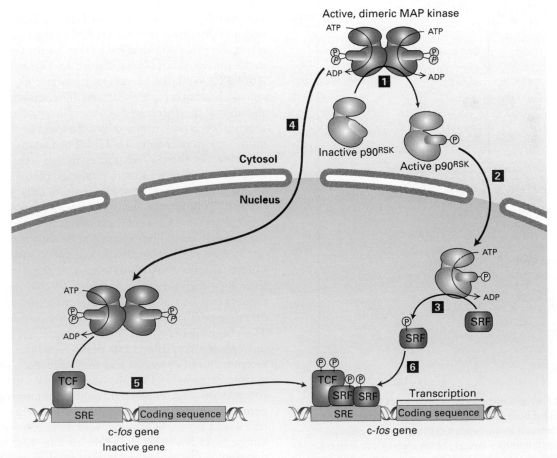

**FIGURE 16-22 Induction of gene transcription by MAP kinase.** Steps **1**–**3**: In the cytosol, MAP kinase phosphorylates and activates the kinase p90$^{RSK}$, which then moves into the nucleus and phosphorylates the SRF transcription factor. Steps **4** and **5**: After translocating into the nucleus, MAP kinase directly phosphorylates the transcription factor TCF that is already bound to the promoter of the c-*fos* gene. Step **6**: Phosphorylated TCF and SRF act together to stimulate transcription of genes (e.g., c-*fos*) that contain an SRE sequence in their promoter. See the text for details. [See R. Marais et al., 1993, *Cell* **73**:381, and V. M. Rivera et al., 1993, *Mol. Cell Biol.* **13**:6260.]

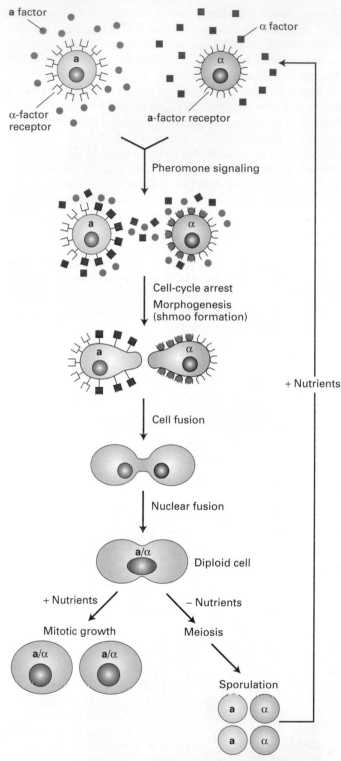

**FIGURE 16-23 Pheromone-induced mating of haploid yeast cells.** The α cells produce α mating factor and **a**-factor receptor; the **a** cells produce **a** factor and α-factor receptor. Both receptors are G protein–coupled receptors. Binding of the mating factors to their cognate receptors on cells of the opposite type leads to gene activation, resulting in mating and production of diploid cells. In the presence of sufficient nutrients, these cells will grow as diploids. Without sufficient nutrients, the cells will undergo meiosis and form four haploid spores.

Labels within figure:
- a factor
- α factor
- α-factor receptor
- **a**-factor receptor
- Pheromone signaling
- Cell-cycle arrest
- Morphogenesis (shmoo formation)
- Cell fusion
- Nuclear fusion
- a/α Diploid cell
- + Nutrients
- − Nutrients
- Mitotic growth
- Meiosis
- a/α   a/α
- Sporulation
- + Nutrients

## G Protein–Coupled Receptors Transmit Signals to MAP Kinase in Yeast Mating Pathways

Although in multicellular animals MAP kinase is often activated by RTKs or cytokine receptors, signaling from other receptors can activate MAP kinase in different eukaryotic cells (see Figure 15-34). To illustrate, we consider the mating pathway in *S. cerevisiae*, a well-studied example of a MAP kinase cascade linked to G protein–coupled receptors (GPCRs), in this case for two secreted peptide pheromones, the **a** and α factors.

Haploid yeast cells are either of the **a** or α mating type and secrete protein signals known as pheromones, which induce mating between haploid yeast cells of the opposite mating type, **a** or α. An **a** haploid cell secretes the **a** mating factor and has cell-surface receptors for the α factor; an α cell secretes the α factor and has cell-surface receptors for the **a** factor (see Figure 16-23). Thus each type of cell recognizes the mating factor produced by the opposite type. Activation of the MAP kinase pathway by either the **a** or α receptors induces transcription of genes that inhibit progression of the cell cycle and others that enable cells of opposite mating type to fuse together and ultimately form a diploid cell.

Ligand binding to either of the two yeast pheromone GPCRs triggers the exchange of GTP for GDP on the $G_\alpha$ subunit and dissociation of $G_\alpha \cdot$GTP from the $G_{\beta\gamma}$ complex. This activation process is identical to that for the GPCRs discussed in the previous chapter (see Figure 15-17). In many mammalian GPCR-initiated pathways, the active $G_\alpha$ transduces the signal. In contrast, mutant and biochemical studies have shown that the dissociated $G_{\beta\gamma}$ complex mediates all the physiological responses induced by activation of the yeast pheromone receptors (Figure 16-24a). For instance, in yeast cells that lack $G_\alpha$, the $G_{\beta\gamma}$ subunit is always free. Such cells can mate in the absence of mating factors; that is, the mating response is constitutively on. However, in cells defective for the $G_\beta$ or $G_\gamma$ subunit, the mating pathway cannot be induced at all. If dissociated $G_\alpha$ were the transducer, in these mutant cells the pathway would be expected to be constitutively active.

In yeast mating pathways, $G_{\beta\gamma}$ functions by triggering a kinase cascade that is analogous to the one downstream from Ras; each protein has a yeast-specific name but shares sequences with and is analogous in structure and function to the corresponding mammalian protein shown in Figure 16-20. The components of this cascade were uncovered mainly through analyses of mutants that possess functional **a** and α receptors and G proteins but are sterile (*Ste*), or defective in mating responses. The physical interactions between the components were assessed through immunoprecipitation experiments with extracts of yeast cells and other types of studies. Based on these studies, scientists have proposed the kinase cascade shown in Figure 16-24a. Free $G_{\beta\gamma}$, which is tethered to the membrane via the lipid bound to the γ subunit, binds the Ste5 protein, thus recruiting it and its bound kinases to the plasma membrane. Ste5 has no obvious catalytic function and acts as a scaffold for assembling other components in the cascade (Ste11, Ste7, and Fus3). $G_{\beta\gamma}$ also

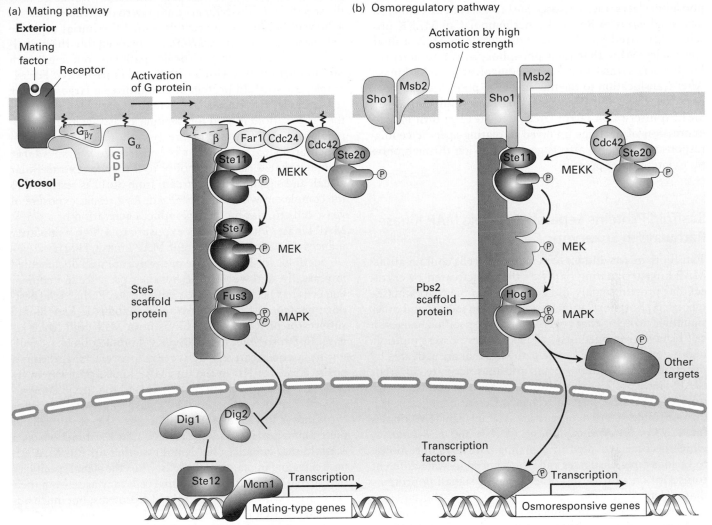

**FIGURE 16-24 Yeast MAP kinase cascades in the mating and osmoregulatory pathways.** In yeast, different receptors activate different MAP kinase pathways, two of which are outlined here. The two MEKs depicted, like all MEKs, are dual specificity threonine/tyrosine kinases; all of the others are serine/threonine kinases. (a) *Mating pathway:* The receptors for yeast α and **a** mating factors are coupled to the same trimeric G protein. Following ligand binding and dissociation of the G protein subunits, the membrane-tethered $G_{\beta\gamma}$ subunit binds the Ste5 scaffold to the plasma membrane. $G_{\beta\gamma}$ also activates Cdc24, a GEF for the Ras-like protein Cdc42; the active GTP-bound Cdc42 in turn binds to and activates the resident Ste20 kinase. Ste 20 then phosphorylates and activates Ste11, which is analogous to Raf and other mammalian MEK kinase (MEKK) proteins. Ste20 thus serves as a MAPKKK kinase. Ste11 initiates a kinase cascade in which the final component, Fus3, is functionally equivalent to MAP kinase (MAPK) in higher eukaryotes. Like other MAP kinases, activated Fus3 then translocates into the nucleus. There it phosphorylates two proteins, Dig1 and Dig2, relieving their inhibition of the Ste12 transcription

factor, allowing it to bind to DNA and initiate transcription of genes that inhibit progression of the cell cycle and others that enable cells of opposite mating type to fuse together and ultimately form a diploid cell. (b) *Osmoregulatory pathway:* Two plasma membrane proteins, Sho1 and Msb1, are activated in an unknown manner by exposure of yeast cells to media of high osmotic strength. Activated Sho1 recruits the Pbs2 scaffold protein, which contains a MEK domain, to the plasma membrane. Similar to the mating pathway, at the plasma membrane the Sho1 Msb1 complex also activates Cdc42, which in turn activates the resident Ste20 kinase. Ste20 in turn phosphorylates and activates Ste11, initiating a kinase cascade that activates Hog1, a MAP kinase. In the cytosol, Hog1 phosphorylates specific protein targets, including ion channels; after translocating to the nucleus, Hog1 phosphorylates several transcription factors and chromatin-modifying enzymes. Hog1 appears also to promote transcriptional elongation. Together, the newly synthesized and modified proteins support survival in high-osmotic-strength media. [After N. Dard and M. Peter, 2006, *BioEssays* **28:**146, and R. Chen and J. Thorner, 2007, *Biochim. Biophys. Acta* **1773:**1311.]

activates cdc24, a GEF for the Ras-like protein cdc 42; GTP·cdc42 in turn activates the Ste20 protein kinase. Ste20 phosphorylates and activates Ste11, a serine/threonine kinase analogous to Raf and other mammalian MEKK proteins. Activated Ste11 then phosphorylates Ste7, a dual-specificity MEK that then phosphorylates and activates Fus3, a serine/threonine kinase equivalent to MAP kinase. After translocation to the nucleus, Fus3 phosphorylates two proteins, Dig1 and Dig2, relieving their inhibition of the Ste12 transcription factor. Activated Ste12 in turn induces expression of proteins involved in mating-specific cellular responses. Fus3 also affects gene expression through phosphorylating other proteins.

## Scaffold Proteins Separate Multiple MAP Kinase Pathways in Eukaryotic Cells

Thus both yeasts and higher eukaryotic cells contain a Ras/MAP kinase signaling pathway that is activated by extracellular protein signals and culminates in the MAP kinase–mediated phosphorylation of transcription factors and other signaling proteins that together trigger specific changes in cell behavior. Importantly, all eukaryotes possess multiple highly conserved MAP kinase pathways that are activated by different extracellular signals and that activate different MAP kinase proteins that phosphorylate different transcription factors; these in turn trigger different changes in cell division, differentiation, or function. Mammalian MAP kinases include *Jun N-terminal kinases* (*JNKs*) and *p38 kinases*, which become activated by signaling pathways in response to various types of stresses and which phosphorylate different transcription factors and other types of signaling proteins that affect cell division.

Current genetic and biochemical studies in the mouse and *Drosophila* are aimed at determining which MAP kinases mediate which responses to which signals in higher eukaryotes. This has already been accomplished in large part for the simpler organism *S. cerevisiae*. Each of the six MAP kinases encoded in the *S. cerevisiae* genome has been assigned by genetic analyses to specific signaling pathways triggered by various extracellular signals, such as pheromones, high osmolarity, starvation, hypotonic shock, and carbon/nitrogen deprivation. A second yeast MAP kinase cascade, known as the osmoregulatory pathway, is shown in Figure 16-24b. Each yeast MAP kinase mediates very specific cellular responses, as exemplified by Fus3 in the mating pathway and Hog1 in the osmoregulatory pathway.

A complication arises because in both yeasts and higher eukaryotic cells, different MAP kinase cascades share some common components. For instance, the MEKK Ste11 functions in three yeast signaling pathways: the mating pathway, the osmoregulatory pathway, and the filamentous growth pathway, which is induced by starvation. Nevertheless, each pathway activates a distinct MAP kinase. Similarly, in mammalian cells, common upstream signal-transducing proteins participate in activating multiple JNK kinases.

Once the sharing of components among different MAP kinase pathways was recognized, researchers wondered how the specificity of the cellular responses to particular signals is achieved. Studies with yeast provided the initial evidence that pathway-specific *scaffold proteins* enable the signal-transducing kinases in a particular pathway to interact with one another but not with kinases in other pathways. For example, the scaffold protein Ste5 stabilizes a large complex that includes the kinases in the mating pathway; similarly, the Pbs2 scaffold is used for the kinase cascade in the osmoregulatory pathway (see Figure 16-24). In each pathway in which Ste11 participates, it is constrained within a large complex that forms in response to a specific extracellular signal, and signaling downstream from Ste11 is restricted to the complex in which it is localized. As a result, exposure of yeast cells to mating factors induces activation of a single MAP kinase, Fus3, whereas exposure to a high osmolarity induces activation of a different MAP kinase, Hog1.

Scaffolds for MAP kinase pathways are well documented in yeast, fly, and worm cells, but their presence in mammalian cells has been difficult to demonstrate. Perhaps the best-documented scaffold protein in metazoans is *Ksr* (*kinase suppressor of Ras*), which binds both MEK and MAP kinase. In *Drosophila*, loss of the Ksr homolog blocks signaling by a constitutively active Ras protein, suggesting a positive role for Ksr in the Ras/MAP kinase pathway in fly cells. Although knockout mice that lack Ksr are phenotypically normal, activation of MAP kinase by growth factors or cytokines is lower than normal in several types of cells in these animals. This finding suggests that Ksr functions as a scaffold that enhances but is not essential for Ras/MAP kinase signaling in mammalian cells. Thus the signal specificity of different MAP kinases in animal cells may arise from their association with various scaffold-like proteins, but much additional research is needed to test this possibility.

## KEY CONCEPTS of Section 16.2

### The Ras/MAP Kinase Pathway

• Ras is an intracellular GTPase switch protein that acts downstream from most RTKs and cytokine receptors. Like $G_\alpha$, Ras cycles between an inactive GDP-bound form and an active GTP-bound form. Ras cycling requires the assistance of two proteins: a guanine nucleotide exchange factor (GEF) and a GTPase-activating protein (GAP).

• RTKs are linked indirectly to Ras via two proteins: GRB2, an adapter protein, and Sos, which has GEF activity (see Figure 16-17).

• The SH2 domain in GRB2 binds to a phosphotyrosine in activated RTKs, while its two SH3 domains bind Sos, thereby bringing Sos close to membrane-bound Ras·GDP and activating its nucleotide-exchange activity.

- Binding of Sos to inactive Ras causes a large conformational change that permits release of GDP and binding of GTP, forming active Ras (see Figure 16-19).

- Activated Ras triggers a kinase cascade in which Raf, MEK, and MAP kinase are sequentially phosphorylated and thus activated. Activated MAP kinase then translocates to the nucleus (see Figure 16-20).

- Activation of MAP kinase following stimulation of a growth-factor receptor leads to phosphorylation and activation of two transcription factors, which associate into a trimeric complex that promotes transcription of various early response genes (see Figure 16-22).

- Different extracellular signals induce activation of different MAP kinase pathways, which regulate diverse cellular processes by phosphorylating different sets of transcription factors.

- The kinase components of each MAP kinase cascade assemble into a large pathway-specific complex stabilized by a scaffold protein (see Figure 16-24). This ensures that activation of one MAP kinase pathway by a particular extracellular signal does not lead to activation of other pathways containing shared components.

## 16.3 Phosphoinositide Signaling Pathways

In previous sections, we have seen how signal transduction from receptor tyrosine kinases (RTKs) and cytokine receptors begins with formation of multiprotein complexes associated with the plasma membrane (see Figures 16-12 and 16-13) and how these complexes initiate the Ras/MAP kinase pathway. Here we discuss how these same receptors initiate signaling pathways that involve as intermediates special phosphorylated phospholipids derived from phosphatidyl inositol. As discussed in Chapter 15, these membrane-bound lipids are collectively referred to as **phosphoinositides**. These phosphoinositide signaling pathways include several enzymes that synthesize different phosphoinositides and proteins with domains that can bind to these molecules and are thus recruited to the cytosolic surface of the plasma membrane. In addition to the short-term effects on cell metabolism we encountered in Chapter 15, these phosphoinositide pathways have long-term effects on the pattern of gene expression. We will see that phosphoinositide pathways end with a variety of kinases, including protein kinase C (PKC) and protein kinase B (PKB), that play key roles in cell growth and metabolism. As an example, later in the chapter we see how insulin activation of PKB plays a key role in stimulating glucose import into muscle.

### Phospholipase C$_\gamma$ Is Activated by Some RTKs and Cytokine Receptors

As discussed in Chapter 15, hormonal stimulation of some G protein–coupled receptors leads to activation of phospholipase C (PLC). This membrane-associated enzyme then cleaves phosphatidylinositol 4,5-bisphosphate (PIP$_2$) to generate two important second messengers: 1,2-diacylglycerol (DAG) and inositol 1,4,5-trisphosphate (IP$_3$). Signaling via the *IP$_3$/DAG pathway* leads to an increase in cytosolic Ca$^{2+}$ and to activation of protein kinase C (see Figure 15-36).

Although we did not mention it during our discussion of phospholipase C in Chapter 15, it is specifically the β isoform of this enzyme (PLC$_\beta$) that is activated by GPCRs. Many RTKs and cytokine receptors also can initiate the IP$_3$/DAG pathway by activating another isoform of phospholipase C, the γ isoform (PLC$_\gamma$), an isoform that contains SH2 domains. The SH2 domains of PLC$_\gamma$ bind to specific phosphotyrosines on the activated receptors, thus positioning the enzyme close to its membrane-bound substrate, phosphatidyl inositol 4,5-bisphosphate (PIP$_2$). In addition, the kinase activity associated with receptor activation phosphorylates tyrosine residues on the bound PLC$_\gamma$, enhancing its hydrolase activity. Thus activated RTKs and cytokine receptors promote PLC$_\gamma$ activity in two ways: by localizing the enzyme to the membrane and by phosphorylating it. As seen in Chapter 15, the IP3/DAG pathway initiated by PLC has multiple physiological effects.

### Recruitment of PI-3 Kinase to Activated Receptors Leads to Synthesis of Three Phosphorylated Phosphatidylinositols

Besides the IP$_3$/DAG pathway, many activated RTKs and cytokine receptors initiate another phosphoinositide pathway by recruiting the enzyme *phosphatidylinositol-3 (PI-3) kinase* to the membrane. PI-3 kinase is recruited to the plasma membrane by binding of its SH2 domain to phosphotyrosines on the cytosolic domain of many activated RTKs and cytokine receptors. This recruitment positions the catalytic domain of PI-3 kinase near its phosphoinositide substrates on the cytosolic face of the plasma membrane. Unlike kinases we have encountered earlier that phosphorylate proteins, PI-3 kinase adds a phosphate to the 3′ carbon in the lipid phosphatidylinositol, leading to formation of two separate phosphatidyl inositol 3-phosphates: PI 3,4-bisphosphate or PI 3,4,5-trisphosphate (Figure 16-25). By acting as docking sites for various signal-transducing proteins, these membrane-bound PI 3-phosphate products of the PI-3 kinase reactions in turn transduce signals downstream in several important pathways.

In some cells, this *PI-3 kinase pathway* can trigger cell division and prevent programmed cell death (apoptosis), thus ensuring cell survival. In other cells, this pathway induces specific changes in cell metabolism.

PI-3 kinase was first identified in studies of the polyoma virus, a DNA virus that transforms certain mammalian cells to uncontrolled growth. Transformation requires several viral-encoded oncoproteins, including one termed "middle T." In an attempt to discover how middle T functions, investigators uncovered PI-3 kinase protein in partially purified preparations of middle T, suggesting a specific interaction between the two. Then they set out to determine how PI-3 kinase might affect cell behavior.

**FIGURE 16-25 Generation of phosphatidylinositol 3-phosphates.**
The enzyme phosphatidylinositol-3 kinase (PI-3 kinase) is recruited to the membrane by many activated receptor tyrosine kinases (RTKs) and cytokine receptors. The 3-phosphate added by this enzyme, to yield PI 3,4-bisphosphate or PI 3,4,5-trisphosphate, is a binding site for various signal-transduction proteins, such as the PH domain of protein kinase B. PI 4,5-bisphosphate also is the substrate of phospholipase C (see Figure 15-35). [See L. Rameh and L. C. Cantley, 1999, *J. Biol. Chem.* **274**:8347.]

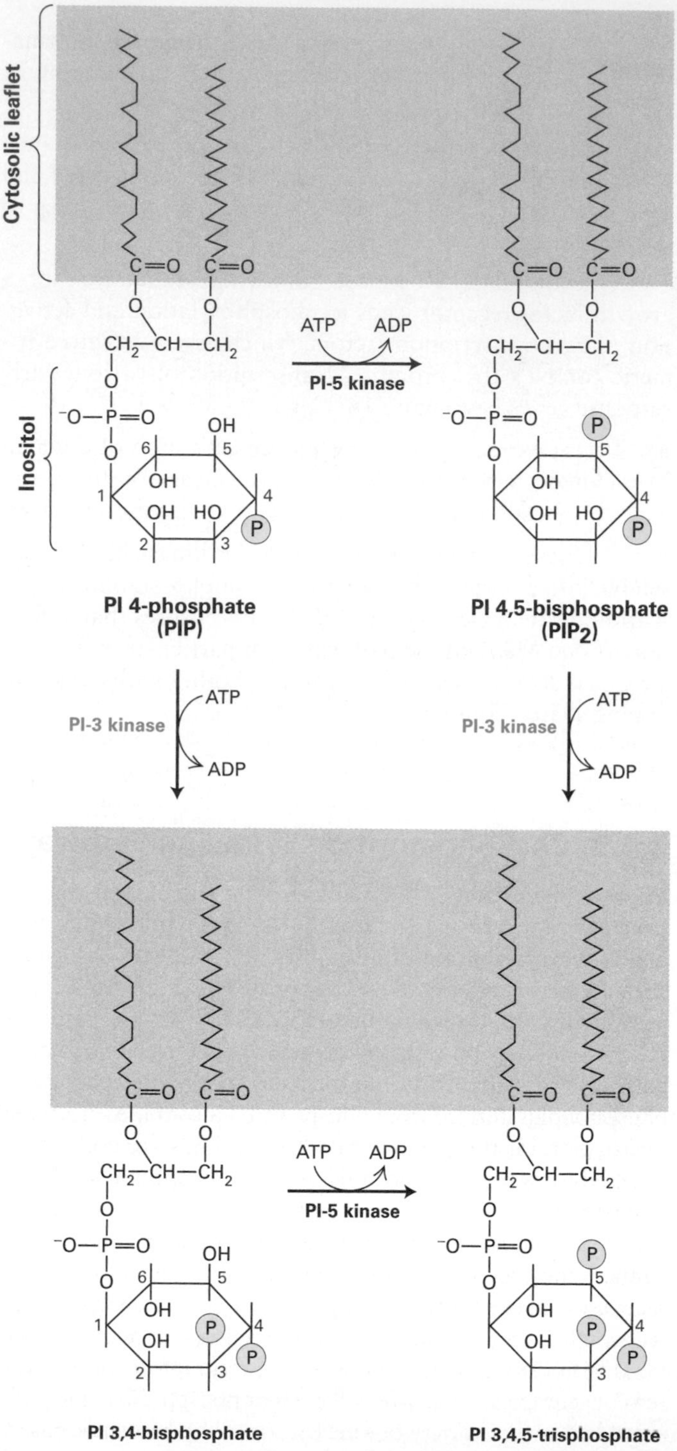

When an inactive, dominant-negative version of PI-3 kinase was expressed in polyoma virus–transformed cells, it inhibited the uncontrolled cell proliferation characteristic of virus-transformed cells. This finding suggested that the normal kinase is important in certain signaling pathways essential for cell proliferation or for the prevention of apoptosis. Subsequent work showed that PI-3 kinases participate in many signaling pathways related to cell growth and apoptosis. Of the nine PI-3 kinase homologs encoded by the human genome, the best characterized contains a p110 subunit with catalytic activity and a p85 subunit with an SH2 phosphotyrosine-binding domain.

## Accumulation of PI 3-Phosphates in the Plasma Membrane Leads to Activation of Several Kinases

Many protein kinases become activated by binding to phosphatidyl inositol 3-phosphates in the plasma membrane. In turn, these kinases affect the activity of many cellular proteins. One important kinase that binds to PI 3-phosphates is **protein kinase B (PKB)**, a serine/threonine kinase that is also called **Akt**. Besides its kinase domain, protein kinase B also contains a *PH domain*, a conserved protein domain present in a wide variety of signaling proteins that binds with high affinity to the 3-phosphates in both PI 3,4-bisphosphate and PI 3,4,5-trisphosphate. Since these inositol phosphates are present on the cytosolic face of the plasma membrane, binding recruits the entire protein to the cell membrane. In unstimulated, resting cells, the level of these phosphoinositides (collectively called PI 3-phosphates) is low, and protein kinase B is present in the cytosol in an inactive form (Figure 16-26). Following hormone stimulation and the resulting rise in PI 3-phosphates, protein kinase B binds to these membrane-bound molecules via its PH domain and becomes localized at the plasma membrane. Binding of protein kinase B to PI 3-phosphates not only recruits the enzyme to the plasma membrane but also releases inhibition of the catalytic site by the PH domain. However, maximal activation of protein kinase B depends on recruitment of two other kinases, named PDK1 and PDK2.

PDK1 is recruited to the plasma membrane via binding of its own PH domain to PI 3-phosphates. Both membrane-associated protein kinase B and PDK1 diffuse randomly in the plane of the membrane, eventually bringing them close enough together so that PDK1 can phosphorylate protein

kinase B on a critical threonine residue in its activation lip—yet another example of kinase activation by phosphorylation. Phosphorylation of a second serine, not in the lip segment, by PDK2 is necessary for maximal protein kinase B activity (Figure 16-26). Similar to the regulation of Raf activity (see Figure 16-20), release of an inhibitory domain and phosphorylation by other kinases regulate the activity of protein kinase B.

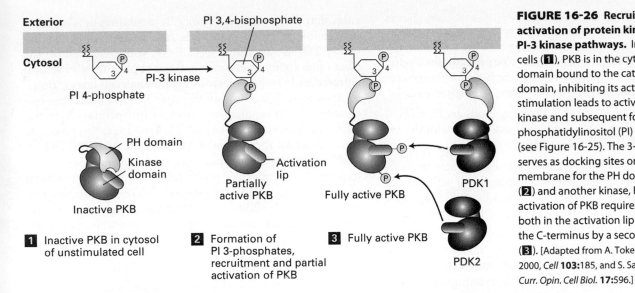

**FIGURE 16-26 Recruitment and activation of protein kinase B (PKB) in PI-3 kinase pathways.** In unstimulated cells (**1**), PKB is in the cytosol with its PH domain bound to the catalytic kinase domain, inhibiting its activity. Hormone stimulation leads to activation of PI-3 kinase and subsequent formation of phosphatidylinositol (PI) 3-phosphates (see Figure 16-25). The 3-phosphate group serves as docking sites on the plasma membrane for the PH domain of PKB (**2**) and another kinase, PDK1. Full activation of PKB requires phosphorylation both in the activation lip by PDK1 and at the C-terminus by a second kinase, PDK2 (**3**). [Adapted from A. Toker and A. Newton, 2000, *Cell* **103**:185, and S. Sarbassov et al., 2005, *Curr. Opin. Cell Biol.* **17**:596.]

## Activated Protein Kinase B Induces Many Cellular Responses

Once fully activated, protein kinase B can dissociate from the plasma membrane and phosphorylate its many target proteins throughout the cell, which have a wide range of effects on cell behavior. Activation of PKB takes only 5 to 10 minutes, yet its effects can last as long as several hours.

In many cells, activated protein kinase B directly phosphorylates and inactivates pro-apoptotic proteins such as Bad, a short-term effect that prevents activation of an apoptotic pathway leading to cell death (see Figure 21-38). Activated protein kinase B also promotes survival of many cultured cells by phosphorylating the Forkhead transcription factor FOXO3a on multiple serine/threonine residues, thereby reducing its ability to induce expression of several pro-apoptotic genes.

In the absence of growth factors, FOXO3a is unphosphorylated and mainly localizes to the nucleus, where it activates transcription of several genes encoding pro-apoptotic proteins. When growth factors are added to the cells, protein kinase B becomes active and phosphorylates FOXO3a. This allows the cytosolic phosphoserine-binding protein 14-3-3 to bind FOXO3a and thus sequester it in the cytosol. (Recall that 14-3-3 also retains phosphorylated Raf protein in an inactive state in the cytosol; see Figure 16-20.) A FOXO3a mutant in which the three serine target residues for protein kinase B are mutated to alanines is "constitutively active" and initiates apoptosis even in the presence of activated protein kinase B. This finding demonstrates the importance of FOXO3a and protein kinase B in controlling apoptosis of cultured cells. Deregulation of protein kinase B is implicated in the pathogenesis both of cancer and diabetes, and in Section 16.7 we will see how protein kinase B, activated downstream of the insulin RTK, promotes glucose uptake and storage in muscle and liver. This is another example of one signaling pathway controlling different cellular functions in different cells.

## The PI-3 Kinase Pathway Is Negatively Regulated by PTEN Phosphatase

Like virtually all intracellular signaling events, phosphorylation by PI-3 kinase is reversible. The relevant phosphatase, termed *PTEN phosphatase*, has an unusually broad specificity. Although PTEN can remove phosphate groups attached to serine, threonine, and tyrosine residues in proteins, its ability to remove the 3-phosphate from PI 3,4,5-trisphosphate is thought to be its major function in cells. Overexpression of PTEN in cultured mammalian cells promotes apoptosis by reducing the level of PI 3,4,5-trisphosphate and hence the activation and anti-apoptotic effect of protein kinase B.

The *PTEN* gene is deleted in multiple types of advanced human cancers. The resulting loss of PTEN protein contributes to the uncontrolled growth of cells. Indeed, cells lacking PTEN have elevated levels of PI 3,4,5-trisphosphate and PKB activity. Since protein kinase B exerts an anti-apoptotic effect, loss of PTEN indirectly reduces the programmed cell death that is the normal fate of many cells. In certain cells, such as neuronal stem cells, absence of PTEN not only prevents apoptosis but also leads to stimulation of cell cycle progression and an enhanced rate of proliferation. Knockout mice lacking PTEN have big brains with an excess numbers of neurons, attesting to PTEN's importance in control of normal development. ■

## KEY CONCEPTS of Section 16.3

### Phosphoinositide Signaling Pathways

• Many RTKs and cytokine receptors can initiate the $IP_3$/DAG signaling pathway by activating phospholipase $C_\gamma$ ($PLC_\gamma$), a different PLC isoform than the one activated by G protein–coupled receptors.

- Activated RTKs and cytokine receptors also can initiate another phosphoinositide pathway by binding a PI-3 kinase, thereby allowing the enzyme access to its membrane-bound phosphoinositide substrates, which then become phosphorylated at the 3 position (PI 3-phosphates; see Figure 16-26).

- The PH domain in various proteins binds to PI 3-phosphates, forming signaling complexes associated with the cytosolic face of the plasma membrane.

- Protein kinase B (PKB) becomes partially activated by binding to PI 3-phosphates with its PH domain. Full activation of PKB requires phosphorylation by another kinase, PDK1, which also is recruited to the membrane by binding to PI 3-phosphates and by a second kinase, PDK2 (see Figure 16-26).

- Activated protein kinase B promotes survival of many cells by directly phosphorylating and inactivating several pro-apoptotic proteins and by phosphorylating and inactivating the FOXO3a transcription factor, which otherwise induces synthesis of pro-apoptotic proteins.

- Signaling via the PI-3 kinase pathway is terminated by the PTEN phosphatase, which hydrolyzes the 3-phosphate in PI 3-phosphates. Loss of PTEN, a common occurrence in human tumors, promotes cell survival and proliferation.

## 16.4 Receptor Serine Kinases That Activate Smads

We have seen how many receptors activate kinases that phosphorylate target proteins on tyrosine residues and activate a conserved set of signal transduction pathways. We have also seen how several cytosolic kinases that phosphorylate target proteins on serine or threonine become activated and function in signaling pathways; these include PKA, PKB, and PKC family members as well as MAP kinases. In this section, we discuss an evolutionarily conserved family of receptor serine kinases (the TGF-β receptor superfamily) and the conserved large family of signaling molecules (the TGF-β superfamily) that binds to them. These receptors phosphorylate and thus trigger the activation of one conserved class of transcription factors (the Smads) that regulate several growth and differentiation pathways. In unstimulated cells, Smads are in the cytosol, but when activated, they move into the nucleus to regulate transcription. The TGF-β pathway has widely diverse effects in different types of cells because different members of the TGF-β superfamily activate different members of the TGF-β receptor family, which activate different members of the Smad class of transcription factors. In addition, as we've seen with other receptor-activated transcription factors such as the STATs, the same activated Smad protein will partner with different transcription factors in different cell types and thus activate different sets of genes in these cells.

The **transforming growth factor β (TGF-β)** superfamily includes a number of related extracellular signaling molecules that play widespread roles in regulating development in both invertebrates and vertebrates. The founding member of the TGF-β superfamily, TGF-β1, was identified on the basis of its ability to induce a malignant phenotype in several cultured early stage cancerous mammalian cell lines ("transforming growth factor"); in this case TGF-β1 promoted metastases, the spreading and invasion of primary tumors, which is discussed in Chapter 24. However, the principal function of all three human TGF-β isoforms, TGF-β1, 2, and 3, on most normal (non-cancerous) mammalian cells is to potently prevent their proliferation by inducing synthesis of proteins that inhibit the cell cycle. TGF-β is produced by many cells in the body and inhibits growth both of the secreting cell (autocrine signaling) and neighboring cells (paracrine signaling). Loss of TGF-β receptors or any of several intracellular signal transduction proteins in the TGF-β pathway releases cells from this growth inhibition and occurs frequently in early development of human tumors. TGF-β proteins also promote expression of cell-adhesion molecules and extracellular-matrix molecules, which play important roles in tissue organization (see Chapter 20). A *Drosophila* homolog of TGF-β, called Dpp protein, participates in dorsal-ventral patterning in fly embryos. Other mammalian members of the TGF-β superfamily, the activins and inhibins, affect early development of the genital tract.

Another member of this superfamily, *bone morphogenetic protein (BMP)*, initially was identified by its ability to induce bone formation in cultured cells. Now called BMP7, it is used clinically to strengthen bone after severe fractures. Of the numerous BMP proteins subsequently recognized, many induce key steps in development, including formation of mesoderm and the earliest blood-forming cells. Most have nothing to do with bones.

Most animal cell types produce and secrete members of the TGF-β superfamily in an inactive form that is stored nearby attached to specialized cell-surface molecules or in the extracellular matrix. Release of the active form from the matrix by protease digestion or inactivation of an inhibitor leads to quick activation of the signaling molecules already in place—an important feature of many signaling pathways. The monomeric form of TGF-β growth factors contains three conserved intramolecular disulfide linkages. An additional cysteine in the center of each monomer links TGF-β monomers into functional homodimers and heterodimers (Figure 16-27).

## Three Separate TGF-β Receptor Proteins Participate in Binding TGF-β and Activating Signal Transduction

Researchers soon identified TGF-β1 as a key growth inhibitory factor, but to understand the way it worked, they had to find the receptors to which it bound. The logic of how they went about their search is representative of typical biochemical approaches to identifying receptors (see Section 15.2). Investigators first reacted the purified growth factor with the radioisotope iodine-125 ($^{125}$I) under conditions such that the

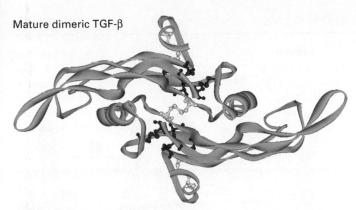

Mature dimeric TGF-β

**FIGURE 16-27 Structure of TGF-β superfamily of signaling molecules.** In this ribbon diagram of a mature TGF-β dimer, the two subunits are shown in green and blue. Disulfide-linked cysteine residues (yellow and red) are shown in ball-and-stick form. The three intrachain disulfide linkages (red) in each monomer form a cystine-knot domain, which is resistant to degradation. [From S. Daopin et al., 1992, *Science* **257**:369.]

iodine covalently bonds to exposed tyrosine residues—effectively tagging them with a radioactive label. The $^{125}$I-labeled TGF-β protein was incubated with cultured cells, and the incubation mixture then was treated with a chemical agent that covalently cross-linked the labeled TGF-β to its receptors on the cell surface. Purification of the $^{125}$I-labeled TGF-β-receptor complexes revealed three different polypeptides with molecular weights of 55, 85, and 280 kDa, referred to as types RI, RII, and RIII TGF-β receptors, respectively.

Figure 16-28 (steps **1** and **2**) depicts the relationship and function of the three TGF-β receptor proteins. The most abundant, RIII, also called β-glycan, is a cell-surface **proteoglycan**. A proteoglycan consists of a protein bound to **glycosaminoglycan (GAG)** chains such as heparin sulfate and chondroitin sulfate (see Figure 20-31). RIII, a transmembrane protein, binds and concentrates mature TGF-β molecules near the cell surface, facilitating their binding to RII receptors. The type I and type II receptors are dimeric transmembrane proteins with serine/threonine kinases as part of their cytosolic domains. RII exhibits *constitutive* kinase activity; that is, it is active even when not bound to TGF-β. Binding of TGF-β induces the formation of complexes containing two copies each of RI and RII, yet another example of ligand-induced oligomerization of cell-surface receptors. An RII subunit then phosphorylates serine and threonine residues in a highly conserved sequence of the RI subunit adjacent to the cytosolic face of the plasma membrane, thereby activating the RI kinase activity.

## Activated TGF-β Receptors Phosphorylate Smad Transcription Factors

Researchers identified the transcription factors downstream from TGF-β receptors from studies of *Drosophila* mutants. These transcription factors in *Drosophila* and the related vertebrate proteins are now called **Smads**. Three types of Smad proteins function in the TGF-β signaling pathway: *R-Smads* (receptor-regulated Smads; Smads 2 and 3), *co-Smads* (Smad4), and *I-Smads* (inhibitory Smads).

As illustrated in Figure 16-28, an R-Smad (Smad2 or Smad3) contains two domains, termed MH1 and MH2, separated by a flexible linker region. The N-terminal MH1 domain contains the specific DNA-binding segment and also a domain called the *nuclear-localization signal* (*NLS*). NLSs are present in virtually all transcription factors found in the cytosol and are required for their transport into the nucleus (see Chapter 13). However, when R-Smads are in their inactive, nonphosphorylated state, the NLS is masked and the MH1 and MH2 domains associate in such a way that they cannot bind to DNA or to a co-Smad. Phosphorylation of three serine residues near the C-terminus of an R-Smad by activated type I TGF-β receptors separates the domains, permitting binding of an importin (see Figure 13-36) to the NLS, which allows entrance of the Smad into the nucleus.

Simultaneously two of the phosphorylated serines in each Smad3 that were added by the RI receptor kinase bind to phosphoserine-binding sites in the MH2 domains in a Smad3 and a Smad4, forming a stable complex containing two molecules of Smad3 (or Smad2) and one molecule of a co-Smad (Smad4). The bound importin then mediates translocation of the heteromeric R-Smad/co-Smad complex into the nucleus. After importin dissociates inside the nucleus, the Smad3/Smad4 (or Smad2/Smad4) complex binds to other transcription factors to activate transcription of specific target genes.

Within the nucleus, R-Smads are continuously being dephosphorylated by a nuclear phosphatase, which results in the dissociation of the R-Smad/co-Smad complex and export of these Smads from the nucleus. Because of this continuous nucleocytoplasmic shuttling of the Smads, the concentration of active Smads within the nucleus closely reflects the levels of activated TGF-β receptors on the cell surface.

Virtually all mammalian cells secrete at least one TGF-β isoform, and most have TGF-β receptors on their surface. However, because different types of cells contain different sets of transcription factors to which the activated Smads can bind, the cellular responses induced by TGF-β vary among cell types. In epithelial cells and fibroblasts, for example, TGF-β induces expression of extracellular-matrix proteins (e.g., fibronectins and collagens; see Chapter 20) and proteins that inhibit serum proteases, which otherwise would degrade these extracellular-matrix proteins. This inhibition stabilizes the matrix, allowing cells to form stable tissues. The inhibitory proteins include plasminogen activator inhibitor 1 (PAI-1). Transcription of the *PAI-1* gene requires formation of a complex of the transcription factor TFE3 with the R-Smad/co-Smad (Smad3/Smad4) complex and binding of all these proteins to specific sequences within the regulatory region of the *PAI-1* gene (Figure 16-28, *bottom*). By partnering with other transcription factors, a R-Smad/co-Smad complex promotes expression of genes encoding other proteins such as p15, which arrests the cell cycle at the $G_1$ stage and thus blocks cell proliferation (see Chapter 19).

**FIGURE 16-28 TGF-β/Smad signaling pathway.** Step **1a**: In some cells, TGF-β binds to the type III TGF-β receptor (RIII), which increases the concentration of TGF-β near the cell surface and also presents TGF-β to the type II receptor (RII). Step **1b**: In other cells, TGF-β binds directly to RII, a constitutively phosphorylated and active kinase. Step **2**: Ligand-bound RII recruits and phosphory-lates the juxtamembrane segment of the type I receptor (RI), which does not directly bind TGF-β. This releases the inhibition of RI kinase activity that otherwise is imposed by the segment of RI between the membrane and its kinase domain. Step **3**: Activated RI then phosphorylates Smad2 or Smad 3 (shown here as Smad 2/3), causing a conforma-tional change that unmasks its nuclear-localization signal (NLS). Step **4**: Two phosphorylated molecules of Smad 2/3 bind to a co-Smad (Smad4) molecule, which is not phosphorylated, and with an importin, forming a large cytosolic complex. Steps **5** and **6**: After the entire complex translocates into the nucleus, Ran·GTP causes dissociation of the importin as discussed in Chapter 13. Step **7**: A nuclear transcription factor (e.g., TFE3) then associates with the Smad2/3/Smad4 complex, forming an activation complex that cooperatively binds in a precise geometry to regulatory se-quences of a target gene. Step **8**: This complex then recruits transcriptional co-activators and induces gene transcription (see Chapter 7). Smad 2/3 is dephosphorylated by a nuclear phosphatase (step **9**) and recycles through a nuclear pore to the cytosol (step **10**), where it can be reactivated by another TGF-β receptor complex. Shown at the bottom is the activation complex for the gene encoding plasminogen activator inhibitor (PAI-1), and similar transcription complexes activate expression of genes encoding other extracellular matrix proteins such as fibronectin. [See A. Moustakas and C.-H. Heldin, 2009, *Development* **136**:3699, and D. Clarke and X. Liu, 2008, *Trends Cell Biol.* **18**:430.]

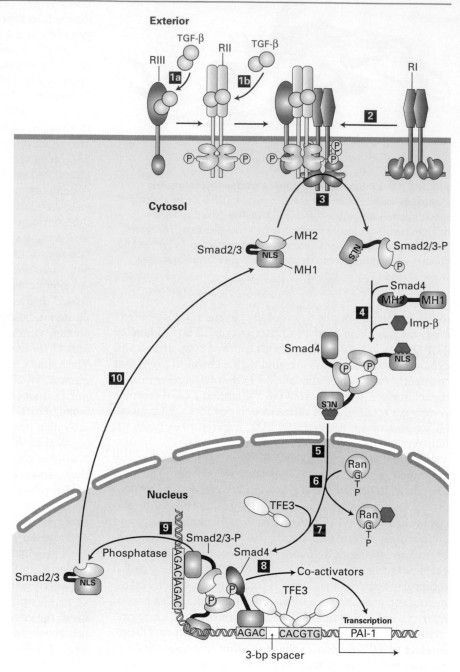

BMP proteins, which also belong to the TGF-β super-family, bind to and activate a different set of receptors that are similar to the TGF-β RI and RII proteins but phosphory-late other R-Smads. Two of these phosphorylated Smads then form a trimeric complex with Smad4, and this Smad complex activates different transcriptional responses than those induced by the TGF-β receptor.

Loss of TGF-β signaling plays a key role in early devel-opment of many cancers. Many human tumors contain inactivating mutations in either TGF-β receptors or Smad pro-teins and thus are resistant to growth inhibition by TGF-β (see Figure 24-24). Most human pancreatic cancers, for instance, contain a deletion in the gene encoding Smad4 and thus cannot induce cell cycle inhibitors in response to TGF-β. In fact,

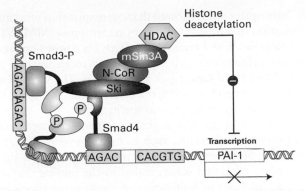

**FIGURE 16-29 Model of Ski-mediated down-regulation of Smad transcription-activating function.** Ski represses Smad function by binding directly to Smad4. Since the Ski-binding domain on Smad4 significantly overlaps with the Smad4 MH2 domain required for binding the phosphorylated tail of Smad3, binding of Ski disrupts the normal interactions between Smad3 and Smad4 necessary for transcriptional activation. In addition, Ski recruits the protein N-CoR, which binds directly to mSin3A; in turn, mSin3A interacts with histone deacetylase (HDAC), an enzyme that promotes histone deacetylation on nearby promoters, repressing gene expression (see Chapter 7). As a result of both processes, transcription activation induced by TGF-β and mediated by Smad complexes is shut down. The related protein SnoN functions similarly to Ski in repressing TGF-β signaling. [See J. Deheuninck and K. Luo, 2009, *Cell Res.* **19**:47.]

Smad4 was originally called *DPC* (*d*eleted in *p*ancreatic cancer). Retinoblastoma, colon and gastric cancer, hepatoma, and some T- and B-cell malignancies are also unresponsive to TGF-β growth inhibition. This loss of responsiveness correlates with loss of type I or type II TGF-β receptors; responsiveness to TGF-β can be restored by recombinant expression of the "missing" protein. Mutations in Smad2 also commonly occur in several types of human tumors. ∎

## Negative Feedback Loops Regulate TGF-β/Smad Signaling

In most signaling pathways, the response to a growth factor or other signaling molecule decreases with time (desensitization). This response is adaptive, preventing overreaction and making possible the fine-tuned control of cellular responses. Several intracellular proteins down-regulate TGF-β/Smad pathways, including two cytosolic proteins called *SnoN* and *Ski* (*Ski* stands for "*S*loan-*K*ettering Cancer *I*nstitute"). These proteins were originally identified as cancer-causing **oncoproteins** because expression of Ski or SnoN is elevated in many cancers, including melanomas and certain breast cancers. When overexpressed in cultured primary fibroblast cells, Ski or SnoN cause abnormal cell proliferation, and down-regulation of Ski in pancreatic cancers reduces tumor growth. How SnoN and Ski trigger abnormal cell proliferation was not understood until years later, when they were found to bind to both the co-Smad (Smad4) and phosphory-

lated R-Smads (Smad3) after TGF-β stimulation. SnoN and Ski do not prevent formation of an R-Smad/co-Smad complex or affect the ability of a Smad complex to bind to DNA-control regions. Rather, they block transcription activation by a bound Smad complex, in part by inducing deacetylation of histones in adjacent chromatin segments. This renders the cells resistant to the growth-inhibitory effects of TGF-β (Figure 16-29). Interestingly, stimulation by TGF-β causes the rapid degradation of Ski and SnoN, but after a few hours, expression of SnoN becomes strongly induced via binding of the Smad2/Smad4 complex to the promoter of the SnoN gene. The increased levels of these proteins are thought to dampen long-term signaling effects due to continued exposure to TGF-β. This is another example of negative feedback, where a gene induced by TGF signaling, in this case SnoN, inhibits further signaling by TGF-β.

Among the other proteins induced after TGF-β stimulation are the I-Smads, especially Smad7. Smad7 blocks the ability of activated type I receptors (RI) to phosphorylate R-Smad proteins, and it may also target TGF-β receptors for degradation. In these ways Smad7, like Ski and SnoN, participates in a negative feedback loop: its induction inhibits intracellular signaling by long-term exposure to the stimulating hormone.

## KEY CONCEPTS of Section 16.4

### Receptor Serine Kinases That Activate Smads

• The transforming growth factor β (TGF-β) superfamily includes a number of related extracellular signaling molecules that play widespread roles in regulating development.

• TGF-β monomers are stored in an inactive form on the cell surface or in the extracellular matrix; release of active monomers (e.g., by protease digestion) leads to formation of functional homodimers and heterodimers.

• TGF-β receptors consist of three types (RI, RII, RIII). Binding of members of the TGF-β superfamily to the RII receptor kinase allows RII to phosphorylate the cytosolic domain of the RI receptor and activate its intrinsic serine/threonine kinase activity. RI then phosphorylates an R-Smad, exposing a nuclear-localization signal (see Figure 16-28).

• After phosphorylated R-Smads bind a co-Smad, the resulting complex translocates into the nucleus, where it interacts with various transcription factors to induce expression of target genes (see Figure 16-28).

• Oncoproteins (e.g., Ski and SnoN) and I-Smads (e.g., Smad7) act as negative regulators of TGF-β signaling (Figure 16-29) by inhibiting transcription mediated by the Smad2/3/Smad4 complex.

• TGF-β signaling generally inhibits cell proliferation. Loss of various components of the signaling pathway contributes to abnormal cell proliferation and malignancy.

## 16.5 Signaling Pathways Controlled by Ubiquitination: Wnt, Hedgehog, and NF-κB

All the signaling pathways discussed so far are reversible and so can be turned off relatively quickly if the extracellular signal is removed. In this section, we discuss several irreversible or only slowly reversible pathways in which a critical component—either a transcription factor or an inhibitor of a transcription factor—is ubiquitinated and then proteolytically cleaved. First we discuss signaling by **Wnts** and **Hedgehogs**, two evolutionarily conserved families of signaling proteins that play key roles in many developmental pathways and often induce expression of genes required for a cell to acquire a new identify or fate. Although Wnt and Hedgehog signaling pathways use different sets of receptors and signaling proteins, they do share similarities, which is why we group them together:

• Wnt and Hedgehog bind to receptors that are similar in structure to seven-spanning G protein–coupled receptors but do not activate G proteins.

• In the resting state, key transcription factors in both pathways are ubiquitinated and targeted for proteolytic cleavage, rendering them inactive.

• Activation of each pathway involves disassembly of large cytosolic protein complexes, inhibition of ubiquitination, and release of the active transcription factor.

• Kinases including glycogen synthase kinase 3 (GSK3) play key roles in both signaling pathways.

Next we examine the *NF-κB pathway*, a third signaling pathway controlled by ubiquitination. In this case, an inhibitor of a transcription factor, rather than a transcription factor itself, is deactivated by ubiquitination. In the resting state, a key transcription factor termed *NF-κB* is sequestered in the cytosol bound to an inhibitor. Several stress-inducing conditions cause ubiquitination and immediate degradation of the inhibitor, allowing cells to respond immediately and vigorously by activating gene transcription. In learning how the NF-κB pathway is activated by one class of surface receptors, we also see a very different function of polyubiquitination: the formation of a scaffold to assemble a key signal transduction complex.

### Wnt Signaling Triggers Release of a Transcription Factor from a Cytosolic Protein Complex

The components of the Wnt as well as the Hedgehog signaling pathways were elucidated mainly through genetic analysis of developmental mutants in *Drosophila* but are operative in humans as well. Mutations in these pathways are thought to trigger several types of human cancers. In fact, the first vertebrate *Wnt* gene to be discovered, the mouse *Wnt-1* gene, attracted notice because it was overexpressed in certain mammary can-

cers. Subsequent work showed that overexpression was caused by insertion of a mouse mammary tumor virus (MMTV) genome near the *Wnt-1* gene. Hence *Wnt-1* is a **proto-oncogene**, a normal cellular gene whose inappropriate expression promotes the onset of cancer (see Chapter 24). The word *Wnt* is an amalgamation of *wingless*, the corresponding fly gene, with *int* for the retrovirus integration site in mice.

Activation of the *Wnt pathway* controls numerous critical developmental events, such as brain development, limb patterning, and organogenesis. A major role for Wnt signaling in bone formation was revealed by the finding that inactivating mutations in Wnt pathway components affects bone density in humans. Wnt signaling is now known to control formation of osteoblasts (bone-forming cells). Additionally, Wnt signals are important in controlling stem cells (see Chapter 21) and in many other aspects of development.

Because of the conservation of the Wnt signaling pathway in metazoan evolution, genetic studies in *Drosophila* and *C. elegans*, studies of mouse proto-oncogenes and tumor-suppressor genes, and studies of cell-junction components have all contributed to identifying various pathway components. Wnt proteins are secreted extracellular signaling molecules, which are modified by addition of a palmitate group near their N-termini. This hydrophobic group is thought to tether Wnt proteins to the plasma membrane of Wnt-secreting cells, thus limiting their range of action to adjacent cells. Wnt proteins act through two cell-surface receptor proteins: *Frizzled (Fz)*, which contains seven transmembrane α helices and directly binds Wnt, and a co-receptor designated *LRP*, which appears to associate with Frizzled in a Wnt signal–dependent manner (see Figure 20-30). Mutations in the genes encoding Wnt proteins, Frizzled, or LRP (called Arrow in *Drosophila*) all have similar effects on the development of embryos.

According to a current model of the *Wnt pathway*, the central player in intracellular Wnt signal transduction is called *β-catenin* in vertebrates and Armadillo in *Drosophila*. This multi-talented protein functions both as a transcriptional activator and as a membrane-cytoskeleton linker protein (see Figure 20-13). In the absence of a Wnt signal, the β-catenin that is not attached to cell-adhesion molecules in the membrane or cytoskeleton is bound to a complex based on the scaffold protein Axin and containing the adenomatous polyposis coli (APC) protein, so named because its loss may result in colorectal cancer. In the resting state, two kinases in the complex, casein kinase 1 (CK1) and GSK3, sequentially phosphorylate β-catenin on multiple serine and threonine residues. Some of these phosphorylated residues serve as binding sites for a ubiquitin-ligase protein named TrCP. β-catenin is then ubiquitinated and rapidly degraded by the 26S proteasome (Figure 16-30a; for more on ubiquitination, see Figures 3-29 and 3-34).

The complete pathway by which Wnt signaling blocks the degradation of β-catenin has not yet been identified. We do know that Wnt binding to both Fz and LRP leads to the phosphorylation of the LRP cytosolic domain, probably by free GSK3 or CK1. This enables Axin to bind to the cytosolic do-

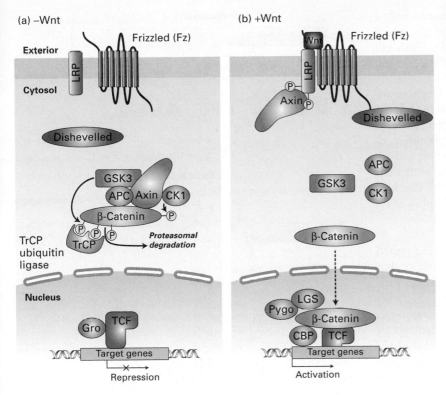

**FIGURE 16-30 Wnt signaling pathway.** (a) In the absence of Wnt, the transcription factor TCF is bound to promoters or enhancers of target genes, but its association with transcriptional repressors such as Groucho (Gro) inhibits gene activation. β-catenin is found in a complex with Axin (a scaffold protein), APC, and the kinases CK1 and GSK3, which sequentially phosphorylate β-catenin. Axin-mediated formation of this complex facilitates phosphorylation of β-catenin by GSK3 by an estimated factor of 20,000. The E3 TrCP ubiquitin ligase then binds to two of these phosphory-lated β-catenin residues, leading to β-catenin ubiquitination and degradation in proteasomes. (b) Binding of Wnt to its receptor Frizzled (Fz) and to the LRP co-receptor triggers phosphorylation of LRP by GSK3 and another kinase, allowing subsequent binding of Axin. This disrupts the Axin–APC–CK1–GSK3–β-catenin complex, preventing phosphorylation of β-catenin by CK1 and GSK3 and leading to accumula-tion of β-catenin in the cell. After translocation to the nucleus, β-catenin binds to TCF to displace the Gro repressor and recruits Pygo, LGS, and other proteins to activate gene expression. [After R. van Amerongen and R. Nusse, 2009, *Development* **136:**3205; F. Staal and J. Sen, 2008, *Eur. J. Immunol.* **38:**1788, and E. Verheyen and C. Gottardi, 2010, *Dev. Dyn.* **239:**34. See also the Wnt Homepage, www.stanford.edu/group/nussselab/cgi-bin/wnt/.]

main of the LRP co-receptor. This shift in Axin localization disrupts the interactions that stabilize the cytosolic complex containing Axin, GSK3, CK1, and β-catenin and thus pre-vents phosphorylation of β-catenin by CK1 and GSK3. This in turn prevents ubiquitination and subsequent degradation of β-catenin and stabilizes it in the cytosol (Figure 16-30b). This process requires the Dishevelled (Dsh) protein, which becomes bound to the cytosolic domain of the Frizzled receptor. The freed β-catenin translocates to the nucleus, where it associates with a transcription factor (TCF) and functions as a co-activator to induce expression of particular target genes, often including those that promote cell proliferation. (Recall that TCF also functions in the MAP kinase pathway; see Figure 16-22.)

Inappropriate activation of the Wnt pathway is charac-teristic of many human cancers. In many tumors, the level of free β-catenin is abnormally high, and this observation pro-vided one of the earliest clues that β-catenin can activate many growth-promoting genes. Inactivating mutations in genes encoding APC and Axin are found in multiple types of human cancers, as are mutations in β-catenin phosphoryla-tion sites for GSK3 or Ck1; these mutations reduce forma-tion of the cytosolic complex (Figure 16-30a), reduce β-catenin degradation, and allow β-catenin to activate gene expression in the absence of the normal Wnt signal.

Among the Wnt target genes are many that also control Wnt signaling, indicating a high degree of feedback regula-tion. The importance of β-catenin stability and location means that Wnt signals affect a critical balance between the three pools of β-catenin in the cell: the membrane-cytoskeleton interface, cytosol, and nucleus.

In order to signal, Wnt must also bind to cell-surface proteoglycans. Evidence for the participation of proteogly-cans in Wnt signaling comes from *Drosophila sugarless* (*sgl*) mutants, which lack a key enzyme needed to synthesize the GAGs heparin and chondroitin sulfate. These mutants have greatly depressed levels of Wingless (the fly Wnt protein) and exhibit other phenotypes associated with defects in Wnt signaling. How proteoglycans facilitate Wnt signaling is un-known, but perhaps binding of Wnt to specific glycosamino-glycan chains is required for it to bind to its receptor Fz or co-receptor LRP. This mechanism would be analogous to the binding of fibroblast growth factor (FGF) to heparan sul-fate, which enhances binding of FGF to its receptor tyrosine kinase (see Figure 16-5).

## Hedgehog Signaling Relieves Repression of Target Genes

The *Hedgehog (Hh) pathway* is similar to the Wnt pathway in that two membrane proteins, one with seven membrane-spanning segments, are required to receive and transduce a signal. The Hh pathway also involves disassembly of an intra-cellular complex containing a transcription factor, like the Wnt pathway. Unlike Wnt, the Hh protein undergoes distinc-tive post-translational processing, described below. Hh signal-ing also differs from Wnt signaling because its two membrane receptors move between the plasma membrane and intracel-lular vesicles, and in mammals Hh signaling is restricted to the primary cilium that protrudes from the cell surface.

Although Hedgehog is a secreted protein, it moves only a short distance from a signaling cell, on the order of 1 to 20 cells, and is bound by receptors on receiving cells. Thus Hh signals, like Wnt signals, have quite localized effects. As Hh diffuses farther and farther away from secreting cells, its concentration decreases, and different Hh concentrations induce different fates in target cells: cells that receive a large amount of Hh turn on certain genes and form certain structures; cells that receive a smaller amount turn on different genes and so form different structures. Signals that induce different cell fates depending on their concentration at their target cells are referred to as **morphogens**. During development, the production of Hedgehog and other morphogens is tightly regulated in time and space.

**Processing of Hh Precursor Protein** Hedgehog is formed from a precursor protein with autoproteolytic activity that enables the protein to cut itself in half. The cleavage produces an N-terminal fragment, which is subsequently secreted to signal to other cells, and a C-terminal fragment, which is degraded. As shown in Figure 16-31, cleavage of the precursor is accompanied by covalent addition of the lipid cholesterol to the new carboxyl terminus of the N-terminal fragment. The C-terminal domain of the precursor, which catalyzes these reactions, is found in other proteins and may promote similar autoproteolytic mechanisms.

A second modification to Hedgehog, the addition of a palmitoyl group to the N-terminus, makes the protein even more hydrophobic. Together, the two attached hydrophobic groups may cause secreted Hedgehog to bind nonspecifically and reversibly to cell plasma membranes, thereby limiting its diffusion and so its range of action in tissues. Spatial restriction plays a crucial role in constraining the effects of powerful signals such as Hh. Recall that the palmitoyl group also is added to Wnt proteins and likely also causes Wnt to bind reversibly to cells, thereby restricting Wnt signaling to cells adjacent to the signaling cell.

**Hh Pathway in *Drosophila*** Genetic studies in *Drosophila* indicate that two membrane proteins, *Smoothened* (*Smo*) and *Patched* (*Ptc*), are required to receive and transduce a Hedgehog signal to the cell interior. Smoothened has seven membrane-spanning α helices and is related in sequence to the Wnt receptor Fz. Patched is predicted to contain 12 transmembrane α helices and is most similar structurally to the Niemann-Pick C1 (NPC1) protein, a member of the **ABC superfamily** of membrane proteins (see Table 11-3).

Figure 16-32 depicts a current model of the *Hedgehog* (*Hh*) *pathway* in *Drosophila*. Evidence supporting this model initially came from study of fly embryos with loss-of-function mutations in the *hedgehog* (*hh*) or *smoothened* (*smo*) genes. Both types of mutant embryos have very similar developmental phenotypes. Moreover, both the *hh* and *smo* genes are required to activate transcription of the same target genes (e.g., *patched* and *wingless*) during embryonic development. In contrast, loss-of-function mutations in the *patched* (*ptc*) gene produce a quite different phenotype, one similar to the

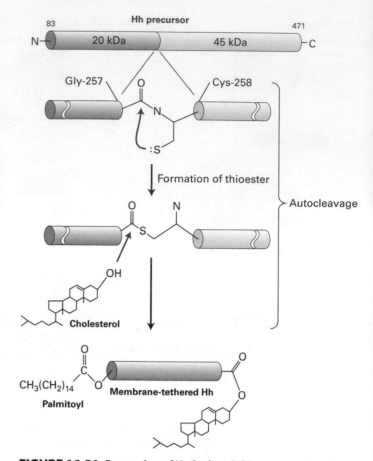

**FIGURE 16-31 Processing of Hedgehog (Hh) precursor protein.** Cells synthesize a 45-kDa Hh precursor, which undergoes a nucleophilic attack by the thiol side chain of cysteine 258 (Cys-258) on the carbonyl carbon of the adjacent residue glycine 257 (Gly-257), forming a high-energy thioester intermediate. An enzymatic activity in the C-terminal domain then catalyzes the formation of an ester bond between the β-3 hydroxyl group of cholesterol and glycine 257, cleaving the precursor into two fragments. The N-terminal signaling fragment (blue) retains the cholesterol moiety and is also modified by the addition of a palmitoyl group to the N-terminus. This processing is thought to occur mostly intracellularly. The two hydrophobic anchors may tether the secreted, processed Hh protein to the plasma membrane. [Adapted from J. A. Porter et al., 1996, *Science* **274**:255.]

effect of flooding the embryo with Hedgehog protein. Thus Patched appears to antagonize the actions of Hedgehog and vice versa. These findings suggest that, in the absence of Hedgehog, Patched represses target genes by inhibiting a signaling pathway needed for gene activation. The additional observation that Smoothened is required for the transcription of target genes in mutants lacking *patched* function places Smoothened downstream of Patched in the Hh pathway. The evidence indicates that Hedgehog binds directly to Patched and prevents Patched from blocking Smoothened action, thus activating the transcription of target genes.

In the absence of Hedgehog, Patched is enriched in the plasma membrane, but Smoothened is in membranes lining internal vesicles. The cytosolic protein complex in the Hh

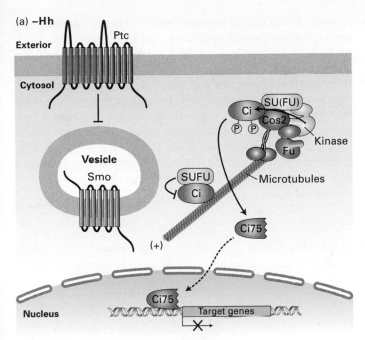

**(a) −Hh**

Exterior

Ptc

Cytosol

Vesicle

Smo

SU(FU)

Ci

Cos2

P P

Fu

Kinase

SUFU

Ci

Microtubules

(+)

Ci75

Nucleus

Ci75

Target genes

**(b) +Hh**

Hh

Ptc

Smo

Smo moves to
plasma membrane

P P

Cos2

SU(FU)

Fu

P

Ci

P

Ci*

CBP

Ci*

Target genes

**FIGURE 16-32 Hedgehog signaling in flies.** (a) In the absence of hedgehog (Hh), patched (Ptc) protein inhibits smoothened (Smo), which is present largely in the membrane of internal vesicles. A complex containing Fused (Fu), a kinase; other kinases; Costal-2 (Cos2), a kinesin-related motor protein; and Cubitis interruptus (Ci), a zinc-finger transcription factor, binds to microtubules. Ci is phosphorylated in a series of steps involving protein kinase A (PKA), glycogen synthase kinase 3 (GSK3), and casein kinase 1 (CK1). The phosphorylated Ci is then proteolytically cleaved by the ubiquitin-proteasome pathway, generating the fragment Ci75, which functions as a transcriptional repressor of Hh-responsive genes. Su(Fu) may also associate with full-length Ci to prevent its translocation to the nucleus. (b) Hh binds to

Ptc, causing some Ptc to move to internal compartments (not shown) and relieving the inhibition of Smo. Smo then moves to the plasma membrane, is phosphorylated, binds Cos2, and is stabilized from degradation. Both Fu and Cos2 become extensively phosphorylated, and most importantly the Fu-Cos2-Ci complex becomes dissociated. This leads to the stabilization of a full-length, alternately modified Ci, Ci*, which displaces the repressor Ci 75 from the promoter of target genes, recruits the CREB-binding activator protein (CBP), and induces expression of target genes. The exact membrane compartments in which Ptc and Smo respond to Hh and function are unknown.

[After S. Goetz and K. Anderson, 2010, *Nature Rev. Genet.* **11**:331.]

pathway consists of several proteins (Figure 16-32a), including Fused (Fu), a serine-threonine kinase; Costal-2 (Cos2), a microtubule-associated kinesin-like protein; and Cubitis interruptus (Ci), a transcription factor. This complex is bound to microtubules in the cytosol. Phosphorylation of Ci by at least three kinases causes binding of a component of a ubiquitin ligase complex that in turn directs ubiquitination of Ci and its targeting to proteasomes. There Ci undergoes proteolytic cleavage; the resulting Ci fragment, designated Ci75, translocates to the nucleus and represses expression of Hh target genes.

Following binding of Hedgehog to the receptor Patched, both proteins move from the cell surface into internal vesicles while Smoothened moves from internal vesicles to the plasma membrane; the binding of Hedgehog to Patched also inhibits its ability to inhibit Smoothened (Figure 16-32a). This triggers several cellular responses, including an increase in phosphorylation of Fu and Cos2. Importantly, the complex of Fu, Cos2, and Ci dissociates from microtubules, and Cos2 becomes associated with the C-terminal tail of Smoothened. The resulting disruption of the Fus/Cos2/Ci complex causes a reduction in both phosphorylation and cleavage of Ci. As a result, a modified form of full-length Ci is generated, termed Ci*, and translocates to the nucleus, where it binds to the

transcriptional co-activator CREB-binding protein (CBP), promoting the expression of target genes.

**Regulation of Hh Signaling** Feedback control of the Hh pathway is important because unrestrained Hh signaling can cause cancerous overgrowth or formation of the wrong cell types. In *Drosophila*, one of the genes induced by the Hh signal is *patched*. The subsequent increase in expression of Patched antagonizes the Hh signal in large measure by reducing the pool of active Smoothened protein. Thus the system is buffered: if during development too much Hh signal is made, a consequent increase in Patched will compensate; if too little Hh signal is made, the amount of Patched is decreased.

## Hedgehog Signaling in Vertebrates Involves Primary Cilia

Hh signaling pathway in vertebrates shares many features with the *Drosophila* pathway, but there are also some striking differences. First, mammalian genomes contain three *hh* genes and two *ptc* genes, which are expressed differentially among various tissues. Second, mammals express three Gli transcription factors that divide up the roles of the single Ci protein in *Drosophila*. All other components of the Hh pathway also are conserved.

The most fascinating aspect of the mammalian Hh pathway is the newly recognized importance of primary cilia. Cilia are long plasma membrane–enveloped structures that protrude from the cell surface. The roles of the abundant cilia in the trachea in moving materials along the tracheal surface and of flagella in sperm locomotion are well known (see Chapter 18). Most cells, however, have a single immotile cilium called the *primary cilium* (Figure 16-33). As we learn in Chapter 18, cilia are extended and maintained by the transport of proteins and particles along a bundle of microtubules in its center; different intraflagellar transport (IFT) proteins move proteins and particles from the base of the cilium to the tip and in the opposite direction. Some of the first evidence for a role of cilia in Hh signaling came from a screen for mutations that altered early mammalian development in a manner similar to that seen in embryos with altered Hh signaling: these phenotypes included loss of certain types of cells in the neural tube that require high levels of one Hh protein. Many of these mutations were in genes encoding IFT proteins, indicating a role for cilia (or flagella) in Hh signaling.

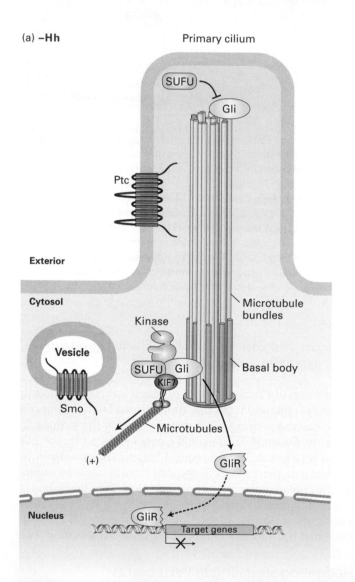

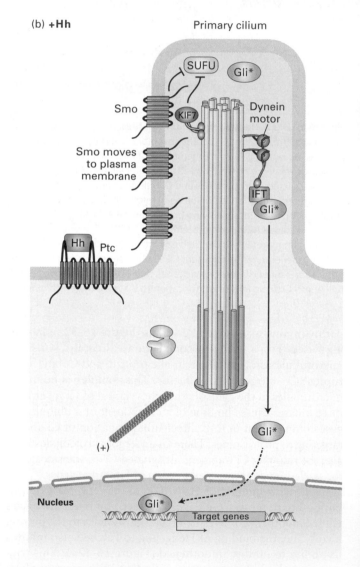

**FIGURE 16-33 Hedgehog signaling in vertebrates.** Hedgehog (Hh) signaling occurs in primary cilia, but otherwise the overall process is similar to that in flies. (a) In the absence of Hh, Patched is localized to the ciliary membrane and in an unknown manner blocks the entry of Smoothened into cilia; Smo is present mainly in the membrane of internal vesicles. The kinesin KIF7 (the Cos2 homolog) binds to microtubules at the cilium base, where it may form a complex with the transcription factor Gli (the vertebrate homolog of Ci), SuFu, and kinases. KIF7 prevents Gli enrichment within the cilium and promotes proteolytic processing of Gli to the Gli repressor GliR. (b) Hh binding triggers movement of Smo to the ciliary membrane and movement of the KIF7 motor protein up the microtubule to the ciliary tip, where Gli accumulates and gets activated by an as yet unknown mechanism. Activated Gli is then transported down the cilium and is released into the cytosol. [After S. Goetz and K. Anderson, 2010, *Nature Rev. Genet.* **11**:331.]

Subsequent analysis showed that, in the absence of Hh signaling, Ptc is localized to the membrane of the primary cilium and Smo is in internal vesicles near the base of the cilium (Figure 16-33a). After Hh addition, Smo becomes localized to the ciliary membrane while Ptc moves out of the ciliary membrane (Figure 16-33b). This movement of Smo involves phosphorylation of the receptor C-terminal cytosolic domain by *β-adrenergic receptor kinase (BARK)*, the same enzyme that modifies G protein–coupled receptors. β-arrestin then binds to Smo. In turn β-arrestin recruits the microtubule motor protein Kif3A, which binds to the microtubules in the core of the cilium and moves Smo up the ciliary membrane. At the same time degradation of Gli to a repressor fragment is blocked, and the motor protein Kif7 moves Gli to the tip of the cilium. There it becomes activated by Smo by a mechanism as yet unknown, and then another motor protein, a dynein, moves the activated Gli to the base of the cilium (Figure 16-31b) As in flies, this active transcription factor then moves into the nucleus, where it can activate expression of multiple target genes.

It is not clear why during vertebrate evolution primary cilia became necessary for Hh signaling since the same result—conversion of a transcription factor from a repressor to an activator of gene expression—occurs downstream of Hh signaling in both mammalian and invertebrate systems.

Inappropriate activation of Hh signaling is the cause of several types of human tumors, including medulloblastomas (cerebellum tumors) and rhabdomyosarcomas (muscle tumors). Primary cilia are essential for this abnormal Hh signaling, and drugs that inhibit the function of primary cilia are being tested on animal models of these cancers. For instance, expression of a mutant activated form of Smoothened in the postnatal mouse brain will cause medulloblastomas, but these tumors will not form if, simultaneously, a gene encoding an essential ciliary protein is inactivated. ■

## Degradation of an Inhibitor Protein Activates the NF-κB Transcription Factor

In the resting state of both the Wnt and Hedgehog pathways, a key transcription factor is ubiquitinated and subjected to proteolytic degradation; activation of the signaling pathway involves blockage of ubiquitination and release of the transcription factor in its active state. The *NF-κB pathway* works in the opposite manner: in the resting state, the *NF-κB* transcription factor is retained in the cytosol bound to an inhibitor; activation of the signaling pathway involves ubiquitination followed by degradation of the inhibitor, triggering release of the active transcription factor. This mechanism allows cells to respond to a variety of stress signals by immediately and vigorously activating gene transcription. The steps in the NF-κB pathway were revealed in studies with both mammalian cells and *Drosophila*.

NF-κB (an acronym for the somewhat unwieldy descriptor "nuclear-factor kappa-light-chain enhancer of activated B cells") is rapidly activated in mammalian immune-system cells in response to bacterial and viral infection, inflammation, and a number of other stressful situations, such as ionizing radiation. The NF-κB pathway is activated in some cells of the immune system when components of bacterial or fungal cell walls bind to certain *Toll-like receptors* on the cell surface (see Figure 23-23). This pathway is also activated by so-called inflammatory cytokines, such as *tumor necrosis factor alpha* (TNFα) and *interleukin 1* (IL-1), which are released by nearby cells in response to infection. In all cases, binding of ligand to its receptor induces assembly of a multiprotein complex in the cytosol near the plasma membrane that triggers a signaling pathway resulting in activation of the NF-κB transcription factor.

NF-κB was originally discovered on the basis of its transcriptional activation of the gene encoding the light chain of antibodies (immunoglobulins) in B cells. It is now thought to be the master transcriptional regulator of the immune system in mammals. Although flies do not make antibodies, NF-κB homologs in *Drosophila* induce synthesis of a large number of secreted antimicrobial peptides in response to bacterial and viral infection. This phenomenon indicates that the NF-κB regulatory system has been conserved during evolution and is more than half a billion years old.

Biochemical studies in mammalian cells and genetic studies in flies have provided important insights into the operation of the NF-κB pathway. The two subunits of heterodimeric NF-κB (p65 and p50) share a region of homology at their N-termini that is required for their dimerization and binding to DNA. In cells that are not undergoing a stress or responding to signs of an infection, direct binding to an inhibitor called I-κBα sequesters NF-κB in an inactive state in the cytosol. A single molecule of I-κBα binds to the paired N-terminal domains of the p50-p65 heterodimer, thereby masking their nuclear-localization signals (Figure 16-34a). A three-protein complex termed *I-κB kinase* operates immediately upstream of NF-κβ and is responsible for releasing it from sequestration. The β kinase subunit of I-κB kinase is the point of convergence of all of the extracellular signals that activate NF-κB noted above. Within minutes of stimulation of the cell by an infectious agent or inflammatory cytokine, the IKK kinase β subunit becomes activated by phosphorylation and then phosphorylates two N-terminal serine residues on I-κBα (Figure 16-34a, steps **1** and **2**). An E3 ubiquitin ligase then binds to these phosphoserines and polyubiquitinates I-κBα, triggering its immediate degradation by a proteasome (steps **3** and **4**). In cells expressing mutant forms of I-κBα in which these two serines have been changed to alanine and so cannot be phosphorylated, NF-κB is permanently inactive, demonstrating that phosphorylation of I-κBα is essential for pathway activation.

The degradation of I-κB exposes the nuclear-localization signals on NF-κB, which then translocates into the nucleus and activates transcription of a multitude of target genes (Figure 16-34a, steps **5** and **6**). Despite its activation by proteolysis, NF-κB signaling eventually is turned off by a negative feedback loop since one of the genes whose transcription is immediately induced by NF-κB encodes I-κBα. The resulting

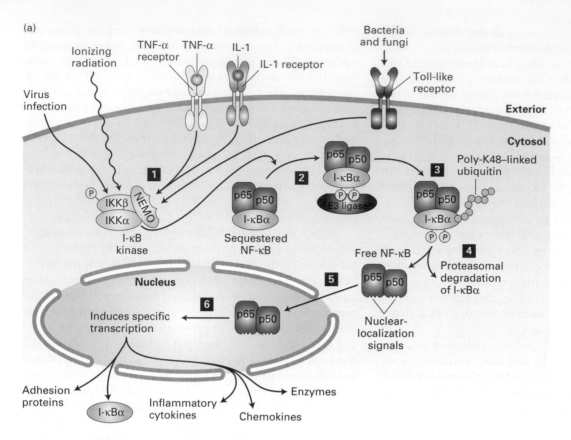

(a)

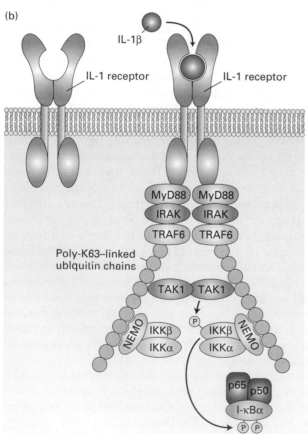

(b)

**FIGURE 16-34 Activation of the NF-κB signaling pathway.** (a) In resting cells, the dimeric transcription factor NF-κB, composed of p50 and p65 subunits, is sequestered in the cytosol, bound to the inhibitor I-κBα. Step **1**: Activation of the trimeric I-κB kinase is stimulated by many agents, including virus infection, ionizing radiation, binding of the pro-inflammatory cytokines TNFα or IL-1 to their respective receptors, or activation of any of several Toll-like receptors by components of invading bacteria or fungi. Step **2**: The β subunit of I-κB kinase then phosphorylates the inhibitor I-κBα, which then binds an E3 ubiquitin ligase. Steps **3** and **4**: Subsequent lysine 48–linked polyubiquitination of I-κBα targets it for degradation by proteasomes. Step **5**: The removal of I-κBα unmasks the nuclear-localization signals (NLS) in both subunits of NF-κB, allowing their translocation to the nucleus. Step **6**: In the nucleus, NF-κB activates transcription of numerous target genes, including the gene encoding I-κBα, which acts to terminate signaling, and genes encoding various inflammatory cytokines. (b) Binding of interleukin-1β (IL-1β) to IL-1 receptors (IL-1R) triggers receptor oligomerization and recruitment of several proteins to the receptor cytosolic domain, including TRAF6, an E3 ubiquitin ligase, which catalyzes synthesis of long lysine-63-linked polyubiquitin chains linked to TRAF6 and other proteins in the complex. The polyubiquitin chains function as a scaffold to recruit the kinase TAK1 and the NEMO subunit of the trimeric I-κB kinase complex. TAK1 then phosphorylates itself and the β subunit of I-κB kinase, activating its kinase activity and enabling it to phosphorylate I-κBα. [Part (a) after R. Khush et al., 2001, *Trends Immunol.* **22:**260, and J-L Luo et al., 2005, *J. Clin. Invest.* **115:**2625; part (b) after B. Skaug et al., 2009, *Ann. Rev. Biochem.* **78:**769.]

increased levels of the I-κBα protein bind active NF-κB in the nucleus and return it to the cytosol.

In many immune-system cells, NF-κB stimulates transcription of more than 150 genes, including those encoding cytokines and chemokines; the latter attract other immune-system cells and fibroblasts to sites of infection. NF-κB also promotes expression of receptor proteins that enable neutrophils (a type of white blood cell) to migrate from the blood into the underlying tissue (see Figure 20-39). In addition, NF-κB stimulates expression of iNOS, the inducible isoform of the enzyme that produces nitric oxide, which is toxic to bacterial cells, as well as expression of several anti-apoptotic proteins, which prevent cell death. Thus this single transcription factor coordinates and activates the body's defense either directly by responding to pathogens and stress or indirectly by responding to signaling molecules released from other infected or wounded tissues and cells.

## Polyubiquitin Chains Serve as Scaffolds Linking Receptors to Downstream Proteins in the NF-κB Pathway

Above we saw that the β kinase subunit of I-κB is the point of convergence for extracellular signals transmitted through multiple receptors, including Toll and IL-1 receptors. Since the cytosolic domains of the Toll and IL-1 receptors have no enzymatic activity, it was a mystery for many years how activation of these receptors led to phosphorylation and activation of the β kinase subunit of I-κB. Early work showed that the presence of IL-1 led to oligomerization of the IL-1 receptor and binding of several proteins to its cytosolic domain, including TRAF6, an E3 ubiquitin ligase that synthesizes polyubiquitin chains. Since all polyubiquitination was then thought to signal degradation by proteasomes, researchers looked for ubiquitinated target proteins that were quickly destroyed. Not finding these, scientists looked for other possible roles for polyubiquitin and soon found that depending on the specific E3 ubiquitin ligase, ubiquitin forms multiple types of polymers that have different structures and biological functions.

The E3 ubiquitin ligase that modifies I-κBα links the carboxyl terminus of one ubiquitin to lysine 48 (K48) on another; this poly-K48 ubiquitin targets the attached protein to the proteasome (Figure 16-36a). The E3 ligase TRAF6, in contrast, links the carboxyl terminus of one ubiquitin to lysine 63 (K63) on another (see Figure 3-34). The resultant poly-K63 ubiquitin chain does not target proteins for degradation; rather, these ubiquitin chains act as scaffolds that bind proteins with a *poly-K63 ubiquitin-binding domain*. One of these is the protein kinase TAK1, which becomes activated by binding to the polyubiquitin chain; another is the NEMO subunit of I-κB kinase. Binding to poly-K63 ubiquitin thus brings the kinase and its target, the β kinase subunit of I-κB kinase, into proximity so that TAK1 can phosphorylate and activate this downstream kinase (Figure 16-36b). As noted above, this kinase then phosphorylates I-κBα. Thus

different types of polyubiquitin chains participate in very different ways in transmitting the IL-1 signal to activation of the NF-κB transcription factors.

## KEY CONCEPTS of Section 16.5

### Signaling Pathways Controlled by Ubiquitination: Wnt, Hedgehog, and NF-κB

- Many signaling pathways involve ubiquitination and proteolysis of target protein and so are irreversible or only slowly reversible. Target proteins can be either a transcription factor or an inhibitor of a transcription factor.

- Wnt controls numerous critical developmental events, such as brain development, limb patterning, and organogenesis. Hedgehog functions as a morphogen during development. Activating mutations in both pathways can cause cancer.

- Both Hedgehog and Wnt are secreted proteins that contain lipid anchors that tether them to cell membranes, thereby reducing their signaling ranges.

- Wnt signals act through two cell-surface proteins, the receptor Frizzled and co-receptor LRP, and an intracellular complex containing β-catenin (see Figure 16-30). Binding of Wnt promotes the stability and nuclear localization of β-catenin, which either directly or indirectly promotes activation of the TCF transcription factor.

- The Hedgehog signal also acts through two cell-surface proteins, Smoothened and Patched, and an intracellular complex containing the Cubitis interruptus (Ci) transcription factor (see Figure 16-32). An activating form of Ci is generated in the presence of Hedgehog; a repressing Ci fragment is generated in the absence of Hedgehog. Both Patched and Smoothened change their subcellular location in response to Hedgehog binding to Patched.

- Hh signaling in vertebrates requires primary cilia and intraflagellar transport proteins. Patched localizes to the ciliary membrane in the absence of Hh and Smo moves to cilia when Hh is present (see Figure 16-33).

- The NF-κB transcription factor regulates many genes that permit cells to respond to infection and inflammation.

- In unstimulated cells, NF-κB is localized to the cytosol, bound to the inhibitor protein I-κBα. In response to many types of extracellular signals, phosphorylation-dependent ubiquitination and degradation of I-κBα in proteasomes releases active NF-κB, which translocates to the nucleus (see Figure 16-34a).

- Polyubiquitin chains linked to the activated IL-1 receptor form a scaffold that brings the TAK1 kinase near its substrate, a subunit of the I-κB kinase, and thus allows signals to be transmitted from the receptor to downstream components of the NF-κB pathway (see Figure 16-34b).

## 16.6 Signaling Pathways Controlled by Protein Cleavage: Notch/Delta, SREBP

In this section, we consider signaling pathways activated by protein cleavage in an extracellular space—often at the surface of the cell—generally by members of the *matrix metalloprotease (MMP) family*. In the *Notch/Delta pathway*, for instance, MMP cleavage of the extracellular part of the Notch receptor is followed by its cleavage within the plasma membrane by a different protease, releasing the cytosolic domain that functions as a transcription factor. This pathway determines the fates of many types of cells during development.

Earlier in the chapter, we saw that multiple growth factors signal through receptor tyrosine kinases. Many such growth factors, including members of the epidermal growth factor (EGF) family, are made as membrane-spanning precursors and can signal adjacent cells by binding to EGF receptors on their surfaces. But cleavage of these proteins by matrix metalloproteases releases the active growth factors into the extracellular medium, allowing them to signal cells much farther away and even the releasing cells themselves (autocrine signaling). Since this process involves a form of proteolytic cleavage similar to that which occurs in the Notch/Delta pathway, we consider it here as well. Activation and release of growth factors by protein cleavage goes awry in many cancers and may lead to an often-fatal enlargement of the heart. Inappropriate MMP cleavage of yet another membrane-spanning protein has been implicated in the pathology of Alzheimer's disease.

Regulated protein cleavage is also used in some *intra*cellular signaling pathways. Thus we conclude our discussion by describing one such pathway: the intramembrane cleavage of a transcription factor precursor within the Golgi membrane in response to low cholesterol levels. This pathway is essential for maintaining the proper balance of cholesterol and phospholipids for constructing cell membranes (see Chapter 10).

### On Binding Delta, the Notch Receptor Is Cleaved, Releasing a Component Transcription Factor

Both the receptor called Notch and its ligand Delta are single-spanning transmembrane proteins found on the cell surface. Notch also has other ligands, but the molecular mechanisms of activation are the same with each ligand. Delta on one cell binds to Notch on an adjacent cell (but not on the same cell), activating Notch so that it undergoes two cleavage events; these result in release of the Notch cytosolic domain, which functions as a transcription factor.

Notch protein is synthesized as a monomeric membrane protein in the endoplasmic reticulum. In the Golgi complex, it undergoes a proteolytic cleavage that generates an extracellular subunit and a transmembrane-cytosolic subunit; the two subunits remain noncovalently associated with each other. Following binding of Delta, the Notch protein on the responding cell undergoes two additional proteolytic cleavages in a proscribed order (Figure 16-35). The first is catalyzed by ADAM 10, a matrix metalloprotease. (The name ADAM stands for *a disintegrin and metalloprotease*; a disintegrin is a conserved

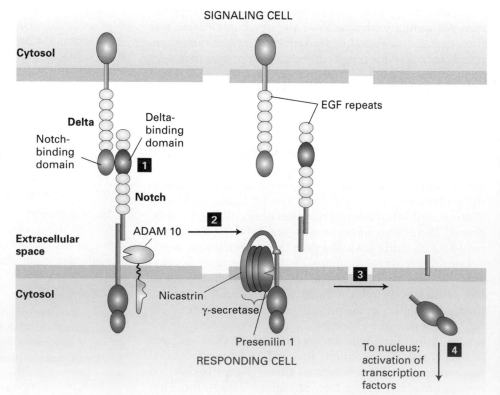

**FIGURE 16-35 Notch/Delta signaling pathway.** In the absence of Delta, the extracellular subunit of Notch on a responding cell is noncovalently associated with its transmembrane-cytosolic subunit. When Notch binds to its ligand Delta on an adjacent signaling cell (step **1**), Notch is first cleaved by the matrix metalloprotease ADAM 10, which is bound to the membrane, releasing the extracellular Notch segment (step **2**). Next the nicastrin subunit of the four protein γ-secretase complex binds to the stump generated by ADAM 10, and the presumed protease, presenilin 1, catalyzes an intramembrane cleavage that releases the cytosolic segment of Notch (step **3**). Following translocation to the nucleus, this Notch segment interacts with several transcription factors to affect expression of genes that in turn influence the determination of cell fate during development (step **4**). [See M. S. Brown et al., 2000, *Cell* **100**:391, and D. Seals and S. Courtneidge, 2003, *Genes Dev.* **17**:7.]

protein domain that binds integrins and disrupts cell-matrix interactions—see Chapter 20.) The second cleavage occurs within the hydrophobic membrane-spanning region of Notch and is catalyzed by a four-protein transmembrane complex termed γ-*secretase*. This cleavage releases the Notch cytosolic segment, which immediately translocates to the nucleus, where it affects transcription of various target genes. How peptide bond hydrolysis can occur within a hydrophobic intramembrane environment is not well understood. Such signal-induced *regulated intramembrane proteolysis (RIP)* is used in a variety of signaling systems, including the response of cells to low cholesterol (see below) and to the presence of unfolded proteins in the endoplasmic reticulum (see Chapter 13).

The γ-secretase complex contains a protein termed *presenilin 1* and three other essential subunits, aph-1, pen-2, and nicastrin. Presenilin 1 (PS1) was first identified as the product of a gene that commonly is mutated in patients with an early-onset autosomal dominant form of Alzheimer's disease. Studies on cells lacking nicastrin revealed why γ-secretase can only cleave proteins that have first been cleaved by an ADAM or other matrix metalloprotease. Nicastrin binds to the N-terminal extracellular stump of the membrane protein that is generated by the first protease (see Figure 16-35). Without this stump, nicastrin and thus the entire γ-secretase complex cannot interact with its target protein. We examine the role of ADAM proteins and γ-secretase in the development of Alzheimer's disease below.

The location of Notch and Delta in different, adjacent cells is essential because they participate in a highly conserved and important cell differentiation process in both invertebrates and vertebrates, called **lateral inhibition**. In this process, adjacent and initially developmentally equivalent cells assume completely different fates. In effect, one cell in a group of equivalent cells instructs the others around it to choose a different fate. As an example of how this occurs, in *Drosophila*, the released intracellular segment of Notch forms a complex with a DNA-binding protein called Suppressor of Hairless, or Su(H). This complex stimulates transcription of many genes whose net effect is to influence the determination of cell fate during development. One of the proteins increased in this manner is Notch itself, and Delta production is correspondingly reduced. Thus a cell that happens to have a bit more Notch than its neighbors will be stimulated to produce yet more Notch and less Delta and so assume a developmental fate different from the Delta-rich adjacent cells. Reciprocal regulation of the receptor and ligand in this fashion is an essential feature of the interaction between initially equivalent cells that causes them to assume different cell fates.

## Matrix Metalloproteases Catalyze Cleavage of Many Signaling Proteins from the Cell Surface

Many signaling molecules are synthesized as transmembrane proteins whose signal domain extends into the extracellular space. Such signaling proteins, like Delta described above, are often biologically active but can signal only by binding to receptors on adjacent cells. However, many growth factors and other protein signals are synthesized as transmembrane precursors whose cleavage releases the soluble, active signaling molecule into the extracellular space. This cleavage is often carried out by matrix metalloproteases (MMP), which are metal-containing enzymes that cleave the extracellular segments of target proteins near the outer surface of the plasma membrane. The human genome encodes 19 metalloproteases in the ADAM family, and many are involved in cleaving the precursors of signaling proteins just outside their transmembrane segment. This ADAM-mediated proteolysis of such precursors is similar to the cleavage of Notch by ADAM 10 (see Figure 16-35) except that the released extracellular segment has signaling activity. ADAM activity and hence the release of active signaling proteins must be tightly regulated by the cell, but how this happens is not yet

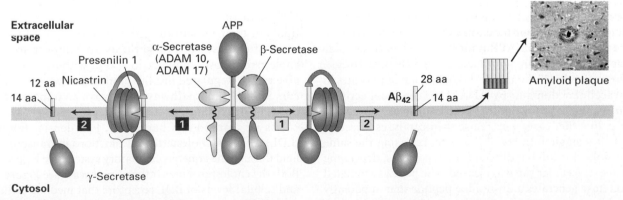

**FIGURE 16-36 Proteolytic cleavage of APP and Alzheimer's disease.** (*Left*) Sequential proteolytic cleavage by α-secretase (ADAM 10 or ADAM 17) (**1**) and γ-secretase (**2**) produces an innocuous membrane-embedded peptide of 26 amino acids. (*Right*) Cleavage in the extracellular domain by β-secretase (**1**) followed by cleavage within the membrane by γ-secretase (**2**) generates the 42-residue Aβ₄₂ peptide, which spontaneously forms oligomers, and then the larger amyloid plaques found in the brain of patients with Alzheimer's disease. In both pathways, the cytosolic segment of APP is released into the cytosol, but its function is not known. [See S. Lichtenthaler and C. Haass, 2004, *J. Clin. Invest.* **113**:1384, and V. Wilquet and B. De Strooper, 2004, *Curr. Opin. Neurobiol.* **14**:582. Inset © ISM/Phototake.]

clear. A breakdown in the mechanisms for regulating ADAM proteases can lead to abnormal cell proliferation.

Medically important examples of the regulated cleavage of signal protein precursors are members of the EGF family, including EGF, HB-EGF, TGF-α, NRG1, and NRG2 (see Figure 16-7). The increased activity of one or more ADAMs that is seen in many cancers can promote cancer development in two ways. First, heightened ADAM activity can lead to high levels of extracellular EGF family growth factors that stimulate secreting cells (autocrine signaling) or adjacent cells (paracrine signaling) to proliferate inappropriately. Second, by destroying components of the extracellular matrix, increased ADAM activity is thought to facilitate metastasis, the movement of tumor cells to other sites in the body.

ADAM proteases also are an important factor in heart disease. As we learned in the last chapter, epinephrine (adrenaline) stimulation of β-adrenergic receptors in heart muscle causes glycogenolysis and an increase in the rate of muscle contraction. Prolonged treatment of heart muscle cells with epinephrine, however, leads to activation of ADAM 9 by an unknown mechanism. This matrix metalloprotease cleaves the transmembrane precursor of HB-EGF. The released HB-EGF then binds to EGF receptors on the signaling heart muscle cells and stimulates their inappropriate growth. This excessive proliferation can lead to an enlarged but weakened heart—a condition known as cardiac hypertrophy, which may cause early death. ∎

## Inappropriate Cleavage of Amyloid Precursor Protein Can Lead to Alzheimer's Disease

Alzheimer's disease is another disorder marked by the inappropriate activity of matrix metalloproteases. A major pathologic change associated with Alzheimer's disease is accumulation in the brain of *amyloid plaques* containing aggregates of a small (42 residue-containing) peptide termed $A\beta_{42}$. This peptide is derived by proteolytic cleavage of *amyloid precursor protein (APP)*, a transmembrane cell-surface protein of still mysterious function expressed by neurons.

Like Notch protein, APP undergoes one extracellular cleavage and one intramembrane cleavage (Figure 16-36). First, the extracellular domain is cleaved at one of two sites in the extracellular domain: by ADAM10 (often collectively called α-*secretase*) or by another matrix protease termed β-*secretase*. In either case, γ-secretase then catalyzes a second cleavage at a single intramembrane site, releasing the same APP cytosolic domain but different small peptides, depending on which extracellular site was cleaved. The pathway initiated by α-secretase generates a 26-residue peptide that apparently does no harm. In contrast, the pathway initiated by β-secretase generates the pathologic $A\beta_{42}$ peptide, which spontaneously forms oligomers and then the larger amyloid plaques found in the brain of patients with Alzheimer's disease.

APP was recognized as a major player in Alzheimer's disease through a genetic analysis of the small percentage of patients with a family history of the disease. Many had mutations in the APP protein, and intriguingly these mutations are clustered around the cleavage sites of α-, β-, or γ-secretase depicted in Figure 16-36. Other cases of familial Alzheimer's disease involve missense mutations in presenilin 1, a subunit of γ-secretase that enhances the formation of the $A\beta_{42}$ peptide, leading to plaque formation and eventually to the death of neurons.

At one time chemical inhibitors of γ-secretase activity were proposed as ideal therapeutics for treatment of Alzheimer's disease, but as might be expected, they had many severe side effects due to the concomitant inhibition of cleavage of Notch and other transmembrane proteins. The recently discovered γ-*secretase-activating protein (GSAP)* drastically and selectively increases amyloid $A\beta_{42}$ production through a mechanism involving its interactions with both γ-secretase and its substrate, the APP carboxy-terminal fragment generated by β-secretase. Since GSAP does not affect cleavage of Notch by γ-secretase, chemicals that bind selectively to GRASP and inhibit its interactions with γ-secretase or its substrate, the APP carboxy-terminal fragment, offer promise as Alzheimer therapeutics. ∎

## Regulated Intramembrane Proteolysis of SREBP Releases a Transcription Factor That Acts to Maintain Phospholipid and Cholesterol Levels

Although this chapter is focused on signaling pathways initiated by extracellular molecules (e.g. growth factors), *intra*cellular signaling pathways that sense the levels of internal molecules and respond accordingly sometimes share principles of molecular regulation and even mechanisms with pathways initiated from outside the cell. One such case is the control of cellular membrane lipids. A cell would soon face a crisis if it did not have enough phospholipids to make adequate amounts of membranes or had so much cholesterol that large crystals formed and damaged cellular structures. Cells sense the relative amounts of cholesterol and phospholipids in their membranes; they respond by adjusting the rates of cholesterol biosynthesis and import so that the cholesterol:phospholipid ratio is kept within a narrow desirable range. Regulated intramembrane proteolysis, which occurs in the Notch pathway, also plays an important role in this cellular response to altered cholesterol levels.

As we learned in Chapter 14, low-density lipoprotein (LDL) is rich in cholesterol and functions in transporting this lipid through the aqueous circulatory system (see Figure 14-27). Both the cholesterol biosynthetic pathway (see Figure 10-26) and cellular levels of LDL receptors that mediate cellular uptake of LDL are down-regulated when cellular cholesterol levels are adequate. Since LDL is imported into cells via receptor-mediated endocytosis (see Figure 14-29), a decrease in the number of LDL receptors leads to reduced cellular import of cholesterol. Both cholesterol biosynthesis and import are regulated at the level of gene transcription. For example, when growing cultured cells that need new membrane for sustained division are incubated with an external source of cholesterol, for example, LDL added to the culture medium, the level and the activity of HMG-CoA reductase, the rate-controlling enzyme in cholesterol biosynthesis, is suppressed, whereas the activity of acyl:cholesterol acyl transferase (ACAT), which converts cholesterol into the esterified storage form, is increased. Thus energy is not wasted making unnecessary additional cholesterol and cholesterol homeostasis is achieved.

Genes whose expression is controlled by the level of sterols such as cholesterol often contain one or more 10-base-pair *sterol regulatory elements* (*SREs*), or SRE half-sites, in their promoters. (These SREs differ from the *serum* response elements that control many early response genes, discussed in Section 16.2.) The interaction of cholesterol-dependent transcription factors called **SRE-binding proteins** (**SREBPs**) with these response elements modulates the expression of the target genes. How do cells sense how much cholesterol they have and how

is this "signal" used to control the level of SREBPs in the nucleus and thus gene expression? The SREBP-mediated pathway begins in the membranes of the endoplasmic reticulum (ER) and includes at least two other proteins besides SREBP.

When cells have adequate concentrations of cholesterol, SREBP is found in the ER membrane complexed with SCAP (SREBP cleavage-activating protein), insig-1 (or its close homolog insig-2), and perhaps other proteins (Figure 16-37a). SREBP has three distinct domains: an N-terminal cytosolic domain, containing a basic helix-loop-helix (bHLH) DNA-binding motif (see Figure 7-29) that functions as a transcription factor when cleaved from the rest of SREBP; a central membrane-anchoring domain containing two transmembrane α helices; and a C-terminal cytosolic regulatory domain. SCAP has eight transmembrane α helices and a large C-terminal cytosolic domain that interacts with the regulatory domain of SREBP. Five of the transmembrane α helices in SCAP form a *sterol-sensing domain* similar to that in HMG-CoA reductase (Figure 16-37a; see Section 10.3). When

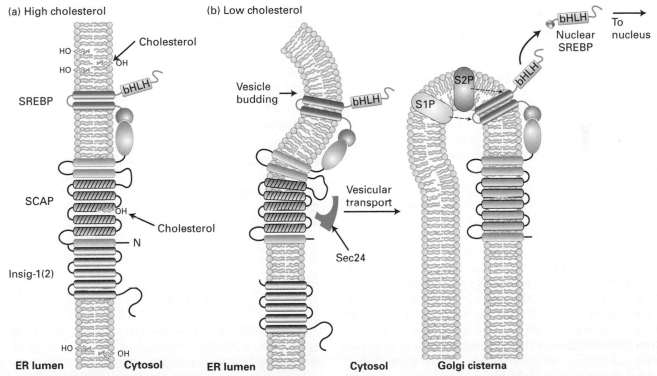

**FIGURE 16-37 Cholesterol-sensitive control of SREBP activation.** The cellular pool of cholesterol is monitored by the combined action of insig-1(2) and SCAP, both transmembrane proteins located in the ER membrane. Membrane-spanning helices 2-6 of SCAP (orange with black lines) form a sterol-binding domain, and a C-terminal segment binds to SREBP. (a) When cholesterol levels are high such that ER cholesterol exceeds 5 percent of total ER lipids, cholesterol binds to the sterol-sensing domain in SCAP, triggering a conformational change that enables the N-terminal SCAP domain to bind to insig-1(2), anchoring the SCAP–SREBP complex in the ER membrane. (b) At low cholesterol levels, cholesterol dissociates from the SCAP sterol-sensing domain, triggering a reverse conformational change that dissociates SCAP from insig-1(2) and enables SCAP to bind to Sec24, a subunit of the COPII complex (see Figure 14-8). This initiates movement of the SCAP-SREBP complex to the Golgi complex by vesicular transport. In the Golgi, the sequential cleavage of SREBP by the site 1 and site 2 proteases (S1P, S2P) releases the N-terminal bHLH domain of SREBP. After this released domain, called nuclear SREBP (nSREBP), translocates into the nucleus, it controls the transcription of genes containing sterol regulatory elements (SREs) in their promoters. [Adapted from A. Radhakrishnan, 2008, *Cell Metab.* **8**:451, and M. Brown and J. Goldstein, 2009, *J. Lipid. Res.* **50**:S15.]

the sterol-sensing domain in SCAP is bound to cholesterol, the protein also binds to insig-1(2). When insig-1(2) is tightly bound to the SCAP-cholesterol complex, it blocks the binding of SCAP to the Sec24 coat protein subunit of COPII vesicles, thereby preventing incorporation of the SCAP-SREBP complex into ER-to-Golgi transport vesicles (see Chapter 14). This occurs when cholesterol concentrations in the ER membrane exceed 5 percent of total ER membrane lipids. Thus the cholesterol-dependent binding of insig to the SCAP-cholesterol-SREBP complex traps that complex in the ER.

Cholesterol bound to SCAP is released when cellular cholesterol levels drop to less than 5 percent of ER lipids, a value that reflects total cellular cholesterol levels. Consequently, insig-1(2) no longer binds to the cholesterol-free SCAP, and the SCAP–SREBP complex moves from the ER to the Golgi apparatus via COPII vesicles (Figure 16-37b). In the Golgi, SREBP is cleaved sequentially at two sites by two membrane-bound proteases, S1P and S2P; the latter is an additional example of regulated intramembrane proteolysis. The second cleavage at site 2 releases the N-terminal bHLH-containing domain into the cytosol. This fragment, called *nSREBP* (*nuclear SREBP*), is rapidly translocated into the nucleus. There it activates transcription of genes containing *sterol regulatory elements* (SREs) in their promoters, such as those encoding the LDL receptor and HMG-CoA reductase. Thus a reduction in cellular cholesterol, by activating the *insig1(2)/SCAP/SREBP pathway*, triggers expression of genes encoding proteins that both import cholesterol into the cell (the LDL receptor) and synthesize cholesterol from small precursor molecules (HMG-CoA reductase).

After cleavage of SREBP in the Golgi, SCAP apparently recycles back to the ER, where it can interact with insig-1(2) and another intact SREBP molecule. High-level transcription of SRE-controlled genes requires the ongoing generation of new nSREBP because it is degraded fairly rapidly by the ubiquitin-mediated proteasomal pathway (see Chapter 3). The rapid generation and degradation of nSREBP help cells respond quickly to changes in levels of intracellular cholesterol.

Under some circumstances (e.g., during cell growth), cells need an increased supply of all the essential membrane lipids and their fatty acid precursors (coordinate regulation). But cells sometimes need greater amounts of some lipids, such as cholesterol, to make steroid hormones, than others, such as phospholipids (differential regulation). How is such differential production achieved? Mammals express three known isoforms of SREBP: SREBP-1a and SREBP-1c, which are generated from alternatively spliced RNAs produced from the same gene, and SREBP-2, which is encoded by a different gene. Together, these RIP-regulated transcription factors control expression of proteins that regulate availability not only of cholesterol but also of fatty acids and the triglycerides and phospholipids made from fatty acids. In mammalian cells, SREBP-1a and SREBP-1c exert a greater influence on fatty acid metabolism than on cholesterol metabolism, whereas the reverse is the case for SREBP-2.

Because the risk for atherosclerotic disease, the major cause of heart attacks, is directly proportional to the plasma levels of LDL cholesterol (the so-called bad cholesterol) and inversely proportional to those of HDL cholesterol, a major public health goal has been to lower LDL and raise HDL cholesterol levels. The most successful drugs for controlling the LDL:HDL ratio are the statins, which cause reductions in plasma LDL. As discussed in Chapter 10, these drugs bind to HMG-CoA reductase and directly inhibit its activity, thereby lowering cholesterol biosynthesis and the pool of cholesterol in the liver. Activation of SREBP in response to this cholesterol depletion promotes increased synthesis of HMG-CoA reductase and the LDL receptor. Of greatest importance here is the resulting increased numbers of hepatic LDL receptors, which mediate increased import of LDL cholesterol from the blood and so lower the level of LDL cholesterol in the circulation. Thus statins may inhibit atherosclerosis and heart disease by mechanisms independent of their inhibition of cholesterol biosynthesis, but these mechanisms are not well understood. ■

## KEY CONCEPTS of Section 16.6

### Signaling Pathways Controlled by Protein Cleavage: Notch/Delta, SREBP

• Many important growth factors and other signaling proteins such as EGFs are synthesized as transmembrane proteins; regulated cleavage of the precursor near the plasma membrane by members of the matrix metalloprotease (MMP) family releases the active molecule into the extracellular space to signal distant cells.

• On binding to its ligand Delta on the surface of an adjacent cell, the receptor Notch protein undergoes two proteolytic cleavages (see Figure 16-35). The released Notch cytosolic segment then translocates into the nucleus and modulates transcription of target genes critical in determining cell fate during development.

• Cleavage of membrane-bound precursors of members of the EGF family of signaling molecules is catalyzed by ADAM metalloproteases. Inappropriate cleavage of these precursors can result in abnormal cell proliferation, potentially leading to cancer, cardiac hypertrophy, and other diseases.

• γ-Secretase, which catalyzes the regulated intramembrane proteolysis of Notch, also participates in the cleavage of amyloid precursor protein (APP) into a peptide that forms plaques characteristic of Alzheimer's disease (see Figure 16-36).

• In the insig-1(2)/SCAP/SREBP pathway, the active nSREBP transcription factor is released from the Golgi membrane by intramembrane proteolysis when cellular cholesterol is low (see Figure 16-37). It then stimulates the expression of genes encoding proteins that function in cholesterol biosynthesis (e.g., HMG-CoA reductase) and cellular import of cholesterol (e.g., LDL receptor). When cholesterol is high, SREBP is retained in the ER membrane complexed with insig-1(2) and SCAP.

## 16.7 Integration of Cellular Responses to Multiple Signaling Pathways

In this section, we consider how multiple signal transduction pathways interact. We will focus on only one of the myriad of systems controlled by multiple signaling pathways—regulation of the body's needs for the metabolites glucose and fatty acids. We'll first consider several key cellular responses to variations in the demand for the key metabolite glucose. Then we'll turn to control of production of the one type of cell in the adult body that can increase in mass and number almost without limit—the adipose, or fat-storing cell. Cellular responses to changes in other nutrients and to oxygen, which are largely reflected in alterations in gene expression, are covered in Chapter 7.

### Insulin and Glucagon Work Together to Maintain a Stable Blood Glucose Level

During normal daily living, the maintenance of normal blood glucose concentrations depends on the balance between two peptide hormones, **insulin** and **glucagon**, which are made in distinct pancreatic islet cells and elicit different cellular responses. Insulin, which contains two polypeptide chains linked by disulfide bonds, is synthesized by the β cells in the islets (see Figures 14-23 and 14-24); glucagon, a monomeric peptide, is produced by the α islet cells. Insulin *reduces* the level of blood glucose, whereas glucagon *increases* blood glucose. The availability of blood glucose is regulated during periods of abundance (following a meal) or scarcity (following fasting) by the adjustment of insulin and glucagon concentrations in the blood.

After a meal, when blood glucose rises above its normal level of 5 mM, the pancreatic β cells respond to the rise in glucose (and amino acids) by releasing insulin into the blood (Figure 16-38). The released insulin circulates in the blood and binds to insulin receptors present on many different kinds of cells, including muscle and adipocyte cells. The insulin receptor, a receptor tyrosine kinase, activates several signal transduction pathways, including the one leading to the activation of protein kinase B (PKB; see Figure 16-26). In this case, the main actions of these signaling pathways are manifest within minutes. The active PKB phosphorylates a specific target protein that then triggers the rapid fusion of intracellular vesicles containing the glucose transporter GLUT4 with the plasma membrane (Figure 16-39). The resulting immediate tenfold increase in the number of GLUT4 molecules on the cell surface increases glucose influx proportionally, thus lowering blood glucose.

Insulin stimulation of muscle cells enhances within minutes the conversion of glucose into glycogen, and PKB, activated downstream of the insulin receptor, again plays a crucial role. Active PKB phosphorylates *glycogen synthase kinase 3* (GSK3, the same enzyme that functions in the Wnt and Hh pathways). Although GSK in non-insulin-stimulated cells can phosphorylate glycogen synthase and thus inhibit

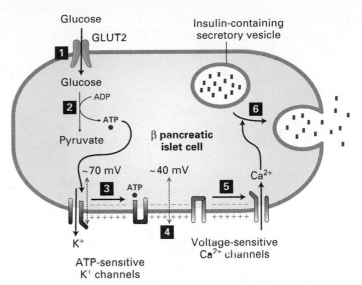

**FIGURE 16-38 Secretion of insulin in response to a rise in blood glucose.** The entry of glucose into pancreatic β cells is mediated by the GLUT2 glucose transporter (1). Because the $K_m$ for glucose of GLUT2 is ~20 mM, a rise in extracellular glucose from 5 mM, characteristic of the fasting state, causes a proportionate increase in the rate of glucose entry (see Figure 11-4). The conversion of glucose into pyruvate is thus accelerated, resulting in an increase in the concentration of ATP in the cytosol (2). The binding of ATP to ATP-sensitive $K^+$ channels closes these channels (3), thus reducing the efflux of $K^+$ ions from the cell. The resulting small depolarization of the plasma membrane (4) triggers the opening of voltage-sensitive $Ca^{2+}$ channels (5). The influx of $Ca^{2+}$ ions raises the cytosolic $Ca^{2+}$ concentration, triggering the fusion of insulin-containing secretory vesicles with the plasma membrane and the secretion of insulin (6). [Adapted from J. Q. Henquin, 2000, *Diabetes* **49**:1751.]

its activity, in insulin-treated muscle, GSK3 phosphorylated by PKB cannot phosphorylate glycogen synthase; thus insulin-stimulated activation of PKB results in net short-term activation of glycogen synthase and glycogen synthesis. Insulin also acts on hepatocytes (liver cells) to inhibit glucose synthesis from smaller molecules, such as lactate and acetate, and to enhance glycogen synthesis from glucose. Many of these effects are manifest at the level of gene transcription since insulin signaling reduces expression of genes whose encoded enzymes simulate synthesis of glucose from small metabolites such as pyruvic acid. The net effect of all these actions is to lower blood glucose back to the fasting concentration of about 5 mM while storing the excess glucose intracellularly as glycogen for future use.

As the blood glucose level drops, insulin secretion and blood levels drop, and insulin receptors are no longer being activated as strongly. In muscles, the response is that cell-surface GLUT4 becomes internalized by endocytosis, lowering the level of cell-surface GLUT4 and thus glucose import. If the blood glucose level falls below about 5 mM, for example due to sudden muscular activity, reduced insulin secretion

(a)

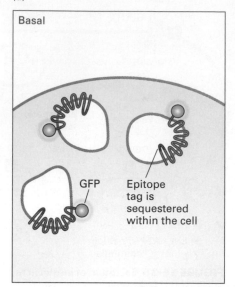

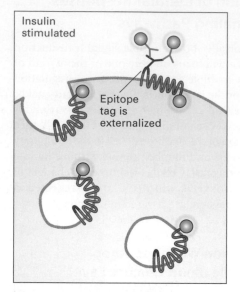

(b)

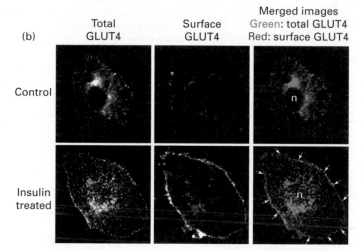

**EXPERIMENTAL FIGURE 16-39 Insulin stimulation of fat cells induces translocation of GLUT4 from intracellular vesicles to the plasma membrane.** (a) Outline of the experiment: Cultured adipose cells were engineered to express a chimeric protein whose N-terminal end corresponded to the GLUT4 sequence, followed by the entirety of the GFP sequence; inserted into the GLUT4 extracellular loop between helixes 1 and 2 is a "myc" epitope that is recognized by a red-fluorescing anti-epitope monoclonal antibody added to the outside of the cell. Thus green fluorescence monitors the total cellular GLUT4 while red fluorescence measures only cell surface GLUT4. (b) Cultured adipocytes expressing this recombinant GLUT4 protein were untreated (*top*) or treated with insulin (*bottom*), reacted with the red-fluorescing anti-epitope antibody, then fixed and viewed under a confocal fluorescence microscope. In the absence of insulin, virtually all of the GLUT4 is in intracellular membranes that are not connected to the plasma membrane; there is little surface staining. Insulin triggers fusion of the GLUT4-containing membranes with the plasma membrane, thereby moving GLUT4 to the cell surface and enabling it to transport glucose from the blood into the cell. Muscle cells also contain insulin-responsive GLUT4 transporters. Arrows highlight GLUT4 present at the plasma membrane; *N* indicates the position of the nuclei. [Courtesy of J. Bogan; see C. Yu et al., 2007, *J. Biol. Chem.* **282**:7710.]

from pancreatic β cells induces pancreatic α cells to increase their secretion of glucagon into the blood. Like the epinephrine receptor, the glucagon receptor, found primarily on liver cells, is coupled to the $G_{\alpha s}$ protein, whose effector protein is adenylyl cyclase. Glucagon stimulation of liver cells induces a rise in cAMP, leading to activation of protein kinase A, which inhibits glycogen synthesis and promotes

glycogenolysis, yielding glucose 1-phosphate (see Figures 15-31a and 15-38b). Liver cells convert glucose 1-phosphate into glucose, which is released into the blood, thus raising blood glucose back toward its normal fasting level.

 Unfortunately, these intricate and powerful control systems sometimes fail, causing serious, even life-threatening disease. *Diabetes mellitus* results from a deficiency in the amount of insulin released from the pancreas in response to rising blood glucose (type I) or from a decrease in the ability of muscle and fat cells to respond to insulin (type II). In both types, the regulation of blood glucose is impaired, leading to persistent elevated blood glucose concentrations (hyperglycemia) and other possible complications if left untreated. Type I diabetes is caused by an autoimmune process that destroys the insulin-producing β cells in the pancreas. Also called insulin-dependent diabetes, this form of the disease is generally responsive to insulin therapy. Most Americans with diabetes mellitus have type II, or insulin-independent diabetes. While the underlying cause of this form of the disease is not well understood, obesity is correlated with a huge increase in the incidence of diabetes. Further identification of the signaling pathways that control energy metabolism is expected to provide insight into the pathophysiology of diabetes, hopefully leading to new methods for its prevention and treatment. ■

## Multiple Signal Transduction Pathways Interact to Regulate Adipocyte Differentiation Through PPARγ, the Master Transcriptional Regulator

White adipocytes, commonly called "fat cells," are the major depots for storage of fats; mature adipocytes have a few triglyceride globules that occupy the bulk of the cell. Adipocytes are also endocrine cells and secrete several signaling proteins that affect the metabolic functions of muscle, liver, and other organs. Adipocytes are the one type of cell in the body that can increase both in number and size almost without limit. Readers in every country do not need to be reminded that obesity—an overabundance of adipocytes—is a growing public health problem, and obesity is a major risk factor not only for diabetes but also for cardiovascular diseases such as heart attacks and stroke and certain cancers. Thus much effort has gone into understanding the factors that regulate fat-cell formation, with the hope of developing drugs that can slow or reverse the process.

As we discuss in Chapter 21, several types of stem cells reside in vertebrates and are used to generate specific types of differentiated cells. The **mesenchymal stem cell** resides in the bone marrow and other organs and gives rise to progenitor cells that in turn can form either adipocytes, cartilage-producing cells, or bone-forming osteoblasts. The adipocyte progenitor, called the preadipocyte, has lost the potential to differentiate into other cell types. When treated with specific hormones, preadipocytes undergo terminal differentiation; they acquire the proteins that are necessary for lipid transport and synthesis, insulin responsiveness, and the secretion of adipocyte-specific proteins. Several lines of cultured preadipocytes can differentiate into adipocytes and express adipocyte-specific mRNAs and proteins, such as enzymes required for triglyceride synthesis.

The transcription factor *PPARγ*, a member of the nuclear-receptor superfamily, is the master transcriptional regulator of adipocyte differentiation. As evidence, recombinant expression of PPARγ in many fibroblast lines is sufficient to trigger their differentiation into adipocytes. Conversely, knocking down the gene for PPARγ in preadipocytes totally prevents their differentiation into adipocytes. Most hormones such as insulin that promote adipogenesis do so at least in part by activating expression of PPARγ. PPARγ, in turn, binds to the promoters of most adipocyte-specific genes and induces their expression, including genes encoding proteins needed in the insulin-signaling pathway such as the insulin receptor and GLUT4. Like other members of the nuclear-receptor superfamily, such as steroid hormone receptors (see Chapter 7), which become activated when they bind their ligand, PPARγ is also thought to bind a ligand, probably an oxidized derivative of a fatty acid.

Another transcription factor, C/EBPα, is induced during adipocyte differentiation and also directly induces many adipocyte genes. Importantly, C/EBPα induces expression of the PPARγ gene and PPARγ induces expression of C/EBPα, leading to a rapid increase in both proteins during the first two days of differentiation. PPARγ together with C/EBPα induces expression of all genes required for differentiation of preadipocytes into mature fat cells.

Many signaling proteins such as Wnt and TGF-β oppose the action of insulin and prevent preadipocyte differentiation into adipocytes. As Figure 16-40 shows, transcription factors activated by receptors for these hormones prevent expression of the PPARγ gene, in part by blocking the ability of C/EBPα to induce PPARγ gene expression. Thus multiple extracellular signals act in concert to regulate adipogenesis, and the signal transduction pathways activated by them intersect at the regulation of expression of one key "master" gene, encoding PPARγ.

## KEY CONCEPTS of Section 16.7

### Integration of Cellular Responses to Multiple Signaling Pathways

- A rise in blood glucose stimulates the release of insulin from pancreatic β cells (see Figure 16-38). Subsequent binding of insulin to its receptor on muscle cells and adipocytes leads to the activation of protein kinase B, which promotes glucose uptake and glycogen synthesis, resulting in a decrease in blood glucose (see Figure 16-39).

- A lowering of blood glucose stimulates glucagon release from pancreatic α cells. Binding of glucagon to its G protein–coupled receptor on liver cells promotes glycogenolysis by the cAMP-triggered kinase cascade (similar to epinephrine stimulation under stress conditions) and an increase in blood glucose (see Figures 15-31a and 15-38b).

- PPARγ, a member of the nuclear receptor superfamily, is the master transcriptional regulator of adipocyte differentiation.

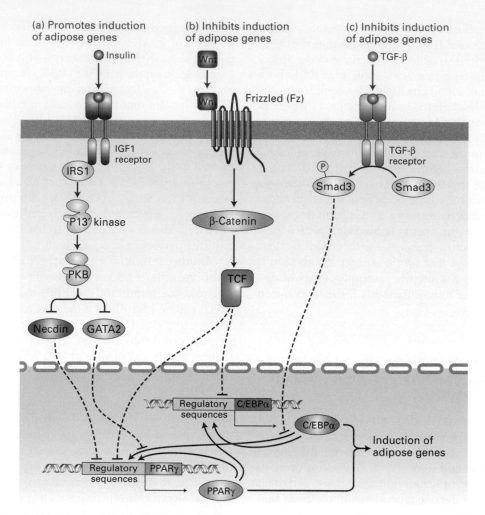

**FIGURE 16-40 Multiple signal transduction pathways interact to regulate adipocyte differentiation.** The transcription factor PPARγ (purple oval) is the master regulator of adipocyte differentiation; together with C/EBPα, it induces expression of all genes required for differentiation of preadipocytes into mature fat cells. Both PPARγ and C/EBPα are induced early in adipogenesis; each of them enhances the transcription of the other's gene (an arrow at the end of a line means enhancement of expression of target genes), leading to a rapid increase in expression of both proteins during the first two days of differentiation. Signals from hormones such as insulin and growth factors such as Wnt and TGF-β that activate or repress adipogenesis are integrated in the nucleus by transcription factors that regulate—directly or indirectly—expression of the PPARγ and C/EBPα genes. A *T* at the end of a line indicates inhibition of expression of the target gene. (a) Insulin activates adipogenesis by several pathways leading to activation of PPARγ expression, two of which are depicted here. Activation of protein kinase B (PKB) downstream of the IGF1 receptor tyrosine kinase and IRS1 leads to repression of Necdin expression; Necdin, by modulating other transcription factors, would otherwise repress expression of the PPARγ gene. PKB also phosphorylates and thus inactivates the transcription factor GATA2, which when unphosphorylated binds to the C/EBPα protein and prevents it from activating expression of the PPARγ gene. By inhibiting two repressors of the PPARγ gene, insulin thus stimulates PPARγ expression. (b) Wnt and TGF-β inhibit adipogenesis by reducing expression of the PPARγ gene. Wnt signaling triggers release of β-catenin from a cytoplasmic complex, and free β-catenin binds the transcription factor TCF (see Figure 16-30). Active TCF blocks expression of the PPARγ and C/EBPα genes, probably by binding to their regulatory sequences. (c) Smad3, activated by phosphorylation following TGF-β binding to the types I and II TGF-β receptors, binds to the C/EBPα protein and prevents it from activating expression of the PPARγ gene. [After E. Rosen and O. MacDougald, 2006, *Nature Rev. Mol. Cell. Biol.* **7**:885.]

• Extracellular hormones such as insulin that promote adipocyte differentiation induce signal transduction pathways that lead to enhanced production of PPARγ. Conversely, signaling proteins such as Wnt and TGF-β that prevent preadipocyte differentiation activate signaling pathways that prevent expression of the PPARγ gene (see Figure 16-40).

## Perspectives for the Future

The confluence of genetics, biochemistry, and structural biology has given us an increasingly detailed view of how signals are transmitted from the cell surface and transduced into changes in cellular behavior. The multitude of different extracellular signals, receptors for them, and intracellular signal transduction pathways fall into a relatively small number of classes, and one major goal is to understand how similar signaling pathways often regulate very different cellular processes. For instance, STAT5 activates very different sets of genes in erythroid precursor cells, following stimulation of the erythropoietin receptor, than in mammary epithelial cells, following stimulation of the prolactin receptor. Presumably STAT5 binds to different groups of transcription factors in these and other cell types, but the nature of these proteins and how they collaborate to induce cell-specific patterns of gene expression remain to be uncovered.

Conversely, activation of the same signal transduction component in the same cell through different receptors often elicits different cellular responses. One commonly held view is that the duration of activation of the MAP kinase and other signaling pathways affects the pattern of gene expression. But how this specificity is determined remains an outstanding question in signal transduction. Genetic and molecular studies in flies, worms, and mice will contribute to our understanding of the interplay between different pathway components and the underlying regulatory principles controlling specificity in multicellular organisms.

Researchers have determined the three-dimensional structures of various signaling proteins during the past several years, permitting more detailed analysis of several signal transduction pathways. The molecular structures of different kinases, for example, exhibit striking similarities and important variations that impart to them novel regulatory features. The activity of several kinases, such as Raf and protein kinase B (PKB), is controlled by inhibitory domains as well as by multiple phosphorylations catalyzed by several other kinases. Our understanding of how the activity of these and other kinases is precisely regulated to meet the cell's needs will require additional structural and cell biological studies.

Abnormalities in signal transduction underlie many different diseases, including the majority of cancers and many inflammatory conditions. Detailed knowledge of the signaling pathways involved and the structure of their protein components will continue to provide important molecular clues for the design of specific therapies. Despite the close structural relationship between different signaling molecules (e.g., kinases), recent studies suggest that inhibitors selective for specific subclasses can be designed. In many tumors of epithelial origin, the EGF receptor has undergone a specific mutation that increases its activity. Remarkably a small-molecule drug (Iressa) inhibits the kinase activity of the mutant EGF receptor but has no effect on the normal EGF receptor or other receptors. Thus the drug slows cancer growth only in patients with this particular mutation. Similarly, monoclonal antibodies or decoy receptors (soluble proteins that contain the ligand-binding domain of a receptor and so sequester the ligand) that prevent pro-inflammatory cytokines such as IL-1 and TNF-α from binding to their cognate receptors are now being used in treatment of several inflammatory diseases such as arthritis.

## Key Terms

| | |
|---|---|
| activation lip 724 | PI-3 kinase pathway 745 |
| adapter protein 737 | PPARγ 767 |
| constitutive 749 | presenilin 1 761 |
| cytokines 721 | primary cilium 756 |
| diabetes mellitus 767 | protein kinase B (PKB) 746 |
| erythropoietin (Epo) 728 | PTB (phosphotyrosine-binding) domain 730 |
| Hedgehog (Hh) pathway 753 | PTEN phosphatase 747 |
| HER family 726 | Ras protein 734 |
| insig-1(2)/SCAP/SREBP pathway 764 | receptor tyrosine kinases (RTKs) 723 |
| insulin 765 | regulated intramembrane proteolysis (RIP) 761 |
| JAK/STAT pathway 734 | scaffold proteins 744 |
| kinase cascade 735 | SH2 domains 730 |
| MAP kinase 735 | Smads 749 |
| matrix metalloprotease (MMP) family 760 | SRE-binding protein (SREBP) 763 |
| NF-κB pathway 752 | transforming growth factor β (TGF-β) 748 |
| Notch/Delta pathway 760 | Wnt pathway 752 |
| nuclear-localization signal (NLS) 749 | |
| phosphoinositides 745 | |

## Review the Concepts

1. Name three features common to the activation of cytokine receptors and receptor tyrosine kinases. Name one difference with respect to the enzymatic activity of these receptors.

2. Erythropoietin (Epo) is a hormone that is produced naturally in the body in response to low $O_2$ levels in the blood. The intracellular events that occur in response to Epo binding to its cell-surface receptor are well characterized. What molecule translocates from the cytosol to the nucleus after (a) JAK2 activates STAT5 and (b) GRB2 binds to the Epo receptor? Why did some endurance athletes use Epo to improve their performance ("blood doping") until it was banned by most sports?

3. Explain how expression of a dominant-negative mutant of JAK blocks the erythropoietin (Epo)-cytokine signaling pathway.

4. Even though GRB2 lacks intrinsic enzymatic activity, it is an essential component of the epidermal growth factor (EGF) signaling pathway that activates MAP kinase. What is the function of GRB2? What role do the SH2 and SH3 domains play in the function of GRB2? Many other signaling proteins possess SH2 domains. What determines the specificity of SH2 interactions with other molecules?

5. Once an activated signaling pathway has elicited the proper changes in target gene expression, the pathway must be inactivated. Otherwise, pathologic consequences may result, as exemplified by persistent growth factor–initiated signaling in many cancers. Many signaling pathways possess

intrinsic negative feedback by which a downstream event in a pathway turns off an upstream event. Describe the negative feedback that down-regulates signals induced by (a) erythropoietin and (b) TGF-β.

**6.** A mutation in the Ras protein renders Ras constitutively active (Ras^D). What is constitutive activation? How is constitutively active Ras cancer promoting? What type of mutation might render the following proteins constitutively active: (a) Smad3, (b) MAP kinase, and (c) NF-κB?

**7.** The enzyme Ste11 participates in several distinct MAP kinase signaling pathways in the budding yeast *S. cerevisiae*. What is the substrate for Ste11 in the mating factor signaling pathway? When a yeast cell is stimulated by mating factor, what prevents induction of osmolytes required for survival in high-osmotic-strength media since Ste11 also participates in the MAP kinase pathway initiated by high osmolarity?

**8.** Describe the events required for full activation of protein kinase B. Name two effects of insulin mediated by protein kinase B in muscle cells.

**9.** Describe the function of the PTEN phosphatase in the PI-3 kinase signaling pathway. Why is a loss-of-function mutation in PTEN cancer promoting? Predict the effect of constitutively active PTEN on cell growth and survival.

**10.** Binding of TGF-β to its receptors can elicit a variety of responses in different cell types. For example, TGF-β induces plasminogen activator inhibitor in epithelial cells and specific immunoglobulins in B cells. In both cell types, Smad3 is activated. Given the conservation of the signaling pathway, what accounts for the diversity of the response to TGF-β in various cell types?

**11.** How is the signal generated by binding of TGF-β to cell-surface receptors transmitted to the nucleus, where changes in target gene expression occur? What activity in the nucleus ensures that the concentration of active Smads closely reflects the level of activated TGF-β receptors on the cell surface?

**12.** The extracellular signaling protein Hedgehog can remain anchored to cell membranes. What modifications to Hedgehog enable it to be membrane bound? Why is this property useful?

**13.** Explain why loss-of-function *hedgehog* and *smoothened* mutations yield the same phenotype but a loss-of-function *patched* mutation yields the opposite phenotype in flies.

**14.** Most mammalian cells have a single immobile cilium called the primary cilium, in which intraflagellar transport (IFT) microtubule motor proteins (discussed in greater detail in Chapter 18) move elements of the Hedgehog (Hh) signaling pathway. What parts of the Hh signaling pathway would mutations in the IFT motor proteins Kif3A, Kif7, and dynein disrupt?

**15.** Why is the signaling pathway that activates NF-κB considered to be relatively irreversible compared with cytokine or RTK signaling pathways? Nonetheless, NF-κB signaling must be down-regulated eventually. How is the NF-κB signaling pathway turned off?

**16.** Describe two roles for polyubiquitination in the NK-κB signaling pathway.

**17.** What feature of Delta ensures that only neighboring cells are signaled?

**18.** What biochemical reaction is catalyzed by γ-secretase? Why was it proposed that a chemical inhibitor of this activity might be a useful drug for treating Alzheimer's disease? What possible side effects of such a drug would complicate this use?

## Analyze the Data

**1.** G. Johnson and colleagues have analyzed the MAP kinase cascade in which MEKK2 participates in mammalian cells. By a yeast two-hybrid screen (see Chapter 7), MEKK2 was found to bind MEK5, which can phosphorylate a MAP kinase. To elucidate the signaling pathway transduced by MEKK2 in vivo, the following studies were performed in human embryonic kidney (HEK293) cells in culture.

  **a.** HEK293 cells were transfected with a plasmid-encoding recombinant, tagged MEKK2, along with a plasmid-encoding MEK5 or a control vector that did not encode a protein (mock). Recombinant MEK5 was precipitated from the cell extract by absorption to a specific antibody. The immunoprecipitated material was then resolved by polyacrylamide gel electrophoresis, transferred to a membrane, and examined by Western blotting with an antibody that recognized tagged MEKK2. The results are shown in part (a) of the figure below. What information about this MAP kinase cascade do we learn from this experiment? Do the data in part (a) of the figure prove that MEKK2 activates MEK5 or vice versa?

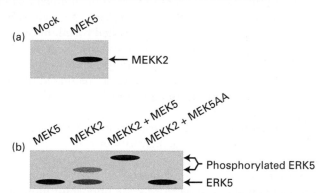

  **b.** ERK5 is a MAP kinase previously shown to be activated when phosphorylated by MEK5. When ERK5 is phosphorylated by MEK5, its migration on a polyacrylamide gel is retarded. In another experiment, HEK293 cells were transfected with a plasmid encoding ERK5 along with plasmids encoding MEK5, MEKK2, MEKK2, and MEK5 or MEKK2 and MEK5AA. MEK5AA is a mutant, inactive version of MEK5 that functions as a dominant-negative. Expression of MEK5AA in HEK293 cells prevents signaling through active, endogenous MEK5. Lysates of transfected cells were analyzed by Western blotting with an antibody against recombinant ERK5. From the data in part (b) of the figure, what can we conclude about the role of MEKK2 in the activation of ERK5? How do the data obtained when cells are cotransfected with ERK5, MEKK2, and MEK5AA help to elucidate the order of participants in this kinase cascade?

**2.** Scaffold proteins can segregate different MAPK signaling pathways that share common components. In the yeast mating pathway, the MEK (MAPKK) Ste7 phosphorylates and activates the Fus3 MAPK, whereas Ste7 phosphorylates and activates the Kss1 MAPK in the starvation pathway. The mating pathway is activated by mating factor receptor activation of a G-protein; $G_{\beta\gamma}$ recruits the Ste5 scaffold protein and the Ste11, Ste7, and Fus3 components of the kinase cascade. Mutation of the Ste5 binding sites for Ste11 and Ste7 disrupts the mating response, clearly demonstrating the importance of Ste5 for tethering the kinases together. Mutation of the Fus3 binding site on Ste5 gave a more complicated response, suggesting that Ste5-Fus3 interaction may involve more than just tethering. This possibility was investigated with yeast proteins expressed as recombinant proteins in bacteria or insect cells and then purified (see Good et al., 2009, *Cell* 136:1085–1097).

**a.** A fluorescence quenching assay was used to measure the activity of Fus3 and Kss1 MAP kinases using a substrate peptide that can be phosphorylated by both kinases. Phosphorylated peptide binds to gallium coupled to fluorescence beads and quenches the fluorescence. The rate of fluorescence quenching (loss of fluorescence) corresponds to the Fus3 or Kss1 kinase activity. Results of Ste7 phosphorylation and thereby activation of Kss1 and Fus3 in the presence and absence of Ste5 are shown below:

Quenching curves:
1. Fus3 or Kss1 alone (control)
2. Fus3 or Kss1 + Ste7
3. Fus3 or Kss1 + Ste7 + Ste5

Is Ste5 required for Ste7 activity? Does Ste7 activate Fus3 and Kss1 equivalently? What does the effect of the presence or absence of Ste5 in the assay tell you about Ste7 phosphorylation and activation of Fus3 and Kss1?

**b.** The protein sequences of Fus3 and Kss1 are 55 percent identical, but each has a unique so-called MAPK insertion loop near the activation domain that is phosphorylated by Ste7. Mutation of an isoleucine residue in the Fus3 insertion loop or replacement of the Fus3 insertion loop with the equivalent region of Kss1 both yield curves similar to Kss1 curve 2. What does this suggest about Fus3 and Kss1 as substrates for Ste7 and the role of Ste5 in stimulating Ste7 phosphorylation of Fus3?

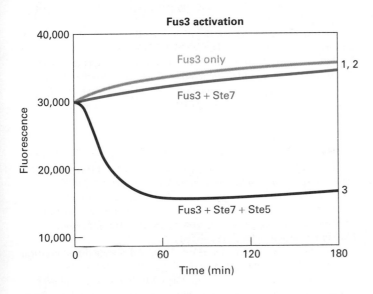

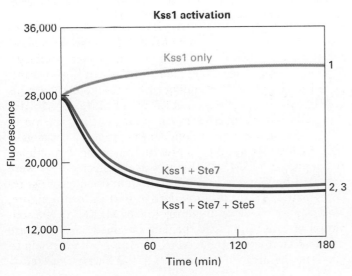

## References

### Receptors That Activate Protein Tyrosine Kinases

Brewer, M., 2009. The juxtamembrane region of the EGF receptor functions as an activation domain. *Mol. Cell* **34**:641–651.

Goh, K., et al. 2010. Multiple mechanisms collectively regulate clathrin-mediated endocytosis of the epidermal growth factor receptor. *J. Cell Biol.* **189**:871–883.

Jura, N., et. al. 2009. Mechanism for activation of the EGF receptor catalytic domain by the juxtamembrane segment. *Cell* **137**:1293–1307.

Lazzara, M. J., and D. A. Lauffenburger. 2009. Quantitative modeling perspectives on the ErbB system of cell regulatory processes. *Exp. Cell Res.* **315**:717–725.

Lemmon, M. A., and J. Schlessinger. 2010. Cell signaling by receptor tyrosine kinases. Cell **141**:1117–1134.

Lodish, H. F., et al. 2009. *Intracellular Signaling by the Erythropoietin Receptor in Erythropoiesis and Eythropoietins*, 2d ed. G. Molineux, M. A. Foote, and S. G. Elliott, eds. Birkhauser, pp. 155–174.

Pfeifer, A. C., J. Timmer, and U. Klingmuller. 2008. Systems biology of JAK/STAT signalling. *Essays Biochem.* **45**:109–120.

Schindler, C., D. E. Levy, and T. Decker. 2007. JAK-STAT signaling: from interferons to cytokines. *J. Biol. Chem.* **282**:20059–20063.

Wiley, H. S., S. Y. Shvartsman, and D. A. Lauffenburger. 2003. Computational modeling of the EGF-receptor system: a paradigm for systems biology. *Trends Cell Biol.* **13**:43–50.

### The Ras/MAP Kinase Pathway

Chen, R., and J. Thorner. 2007. Function and regulation in MAPK signaling pathways: lessons learned from the yeast *Saccharomyces cerevisiae. Biochim. Biophys. Acta* **1773**:1311–1340.

Chong, H., J. Lee, and K-L Guan. 2001. Positive and negative regulation of Raf kinase activity and function by phosphorylation. *EMBO J.* **20**:3716–3727.

Delpire, E. 2009. The mammalian family of sterile 20p-like protein kinases. *Pflugers Arch.—Eur. J. Physiol.* 458:953–967.

Gastel, M. 2006. MAPKAP kinases—MKs—two's company, three's a crowd. *Nature Rev. Mol. Cell Biol.* 7:211–224.

Nadal, E., and F. Posas. 2010. Multilayered control of gene expression by stress-activated protein kinases. *EMBO J.* 29:4–13.

Schwartz, M. A., and H. Madhani. 2004. Principles of MAP kinase signaling specificity in *Saccharomyces cerevisiae. Ann. Rev. Genet.* 38:725–748.

Wiley, H. S., S. Y. Shvartsman, and D. A. Lauffenburger. 2003. Computational modeling of the EGF-receptor system: a paradigm for systems biology. *Trends Cell Biol.* 13:43–50.

## Phosphoinositide Signaling Pathways

Engelman, J. A., J. Luo, and L. C. Cantley. 2006. The evolution of phosphatidylinositol 3-kinases as regulators of growth and metabolism. *Nat. Rev. Genet.* 7:606–619.

Fayard, E., et al. 2010. Protein kinase B (PKB/Akt), a key mediator of the PI3K signaling pathway. *Curr. Top. Microbiol. Immunol.* 346:31–56.

Manning, B. D., and L. C. Cantley. 2007. AKT/PKB signaling: navigating downstream. *Cell* 129:1261–1274.

Michell, R. H., et al. 2006. Phosphatidylinositol 3,5-bisphosphate: metabolism and cellular functions. *Trends Biochem. Sci.* 31:52–63.

Niggli, V. 2005. Regulation of protein activities by phosphoinositide phosphates. *Ann. Rev. Cell Devel. Biol.* 21:57–79.

Vogt, P. K., et al. 2010. Phosphatidylinositol 3-kinase: the oncoprotein. *Curr. Top. Microbiol. Immunol.* 347:79–104.

## Receptor Serine Kinases That Activate Smads

Clarke, D., and X. Liu. 2008. Decoding the quantitative nature of TGF-β/Smad signalling. *Trends Cell Biol.* 18:430–442.

Deheuninck, J., and K. Luo. 2009. Ski and SnoN, potent negative regulators of TGF-β signalling. *Cell Res.* 19:47–57.

Moustakas, A., and C.-H. Heldin. 2009. The regulation of TGFβ signal transduction. *Development* 136:3699–3714.

## Signaling Pathways Controlled by Ubiquitination: Wnt, Hedgehog, and NF-κB

Bianchi, K., and P. Meier. 2010. A tangled web of ubiquitin chains: breaking news in TNF-R1 signaling. *Mol. Cell* 36:736–742.

Goetz, S., and K. Anderson. 2010. The primary cilium: a signalling centre during vertebrate development. *Nature Rev. Genet.* 11:331–344.

Hayden, M., and S. Ghosh. 2008. Shared principles in NF-κB signaling *Cell* 132:344–362.

Iwai, K., and F. Tokunaga. 2009. Linear polyubiquitination: a new regulator of NF-κB activation *EMBO Reports* 10:706–713.

Skaug, B., X. Jiang, and Z. Chen. 2009. The role of ubiquitin in NF-κB regulatory pathways *Ann. Rev. Biochem.* 78:769–796.

Van Amerongen, R., and R. Nusse. 2009. Towards an integrated view of Wnt signaling in development. *Development* 136:3205–3214.

Verheyen, E., and C. Gottardi. 2010. Regulation of Wnt/β-catenin signaling by protein kinases. *Dev. Dyn.* 239:34–44.

Wan, F., and M. Lenardo. 2010. The nuclear signaling of NF-κB: current knowledge, new insights, and future perspectives *Cell Res.* 20:24–33.

Wu, D., and W. Pan. 2009. GSK3: a multifaceted kinase in Wnt signaling. *Trends Biochem. Sci.* 35:161–168.

## Signaling Pathways Controlled by Protein Cleavage: Notch/Delta, SREBP

Blobel, C., G. Carpenter, and M. Freeman. 2009. The role of protease activity in ErbB biology. *Exp. Cell Res.* 315:671–682.

Brown, M. S., and J. L. Goldstein. 2009. Cholesterol feedback: from Schoenheimer's bottle to Scap's MELADL. *J. Lipid Res.* 50:S15–S27.

De Strooper, B. 2005. Nicastrin: gatekeeper of the γ-secretase complex. *Cell* 122:318–320.

Goldstein, J., R. DeBose-Boyd, and M. Brown. 2006. Protein sensors for membrane sterols. *Cell* 124:35–46.

He, G., et al. 2010. Gamma-secretase activating protein is a therapeutic target for Alzheimer's disease. *Nature* 467:95–98.

Seals, D., and S. A. Courtneidge. 2003. The ADAMs family of metalloproteases: multidomain proteins with multiple functions. *Genes Dev.* 17:7–30.

## Integration of Cellular Responses to Multiple Signaling Pathways

Bogan, J., and K. Kandror. 2010. Biogenesis and regulation of insulin-responsive vesicles containing GLUT4. *Curr. Opin. Cell Biol.* 22:506–512.

Boura-Halfon, S., and Y. Zick. 2008. Phosphorylation of IRS proteins, insulin action, and insulin resistance. *Am. J. Physiol. Endocrinol. Metab.* 296:E581–E591.

Rosen, E., and O. MacDougald. 2006. Adipocyte differentiation from the inside out. *Nature Rev. Mol. Cell Biol.* 7:885–896.

Wang, Z., and D. Thurmond. 2009. Mechanisms of biphasic insulin-granule exocytosis—roles of the cytoskeleton, small GTPases and SNARE proteins. *J. Cell Sci.* 122:893–903.

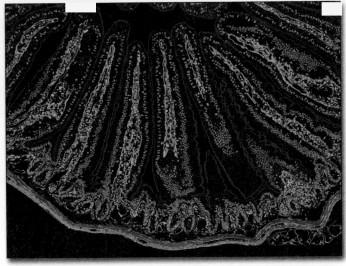

# Cell Organization and Movement I: Microfilaments

A section of mouse intestine stained for actin (red), the extracellular matrix protein laminin (green), and DNA (blue). Each blue dot of DNA indicates the presence of a cell. Actin in the microvilli on the apical end of the epithelial cells can be seen lining the surface facing the lumen (*top*). Actin can also be seen prominently in the smooth muscle that surrounds the intestine (*bottom*). [Micrograph courtesy of Thomas Deerinck and Mark Ellisman.]

When we look through a microscope at the wonderful diversity of cells in nature, the variety of cell shapes and movements we can discern is astonishing. At first we may notice that some cells, such as vertebrate sperm, ciliates such as *Tetrahymena,* or flagellates such as *Chlamydomonas,* swim rapidly, propelled by cilia and flagella. Other cells, such as amebas and human macrophages, move more sedately, propelled not by external appendages but by coordinated movement of the cell itself. We also might notice that some cells in tissues attach to one another, forming a pavementlike sheet, whereas other cells—neurons, for example—have long processes, up to 3 ft in length, and make selective contacts between cells. Looking more closely at the internal organization of cells, we see that organelles have characteristic locations; for example, the Golgi apparatus is generally near the central nucleus. How is this diversity of shape, cellular organization, and motility achieved? Why is it important for cells to have a distinct shape and clear internal organization? Let us first consider two examples of cells with very different functions and organizations.

The epithelial cells that line the intestine form a tight, pavementlike layer of brick-shaped cells, known as an epithelium

(Figure 17-1a, b). Their function is to import nutrients (such as glucose) from the intestinal lumen across the apical (top) plasma membrane and export them across the basolateral (bottom-side) plasma membrane toward the bloodstream. To perform this directional transport, the apical and basolateral plasma membranes of epithelial cells must have different protein compositions. Epithelial cells are attached and sealed together by cellular junctions (discussed in Chapter 20), which create a physical barrier between the apical and basolateral domains of the membrane. This separation allows the cell to place the correct transport proteins in the plasma membranes of the two surfaces. In addition, the apical membrane has a unique morphology, with numerous fingerlike projections called **microvilli** that increase the area of the plasma membrane available for nutrient absorption. To achieve this organization, epithelial cells must have an internal structure to give them shape and to deliver the appropriate proteins to the correct membrane surface.

Now consider macrophages, a type of white blood cell that seeks out infectious agents and destroys them by an engulfing process called phagocytosis. Bacteria release chemicals that attract the macrophage and guide it to the infection.

## OUTLINE

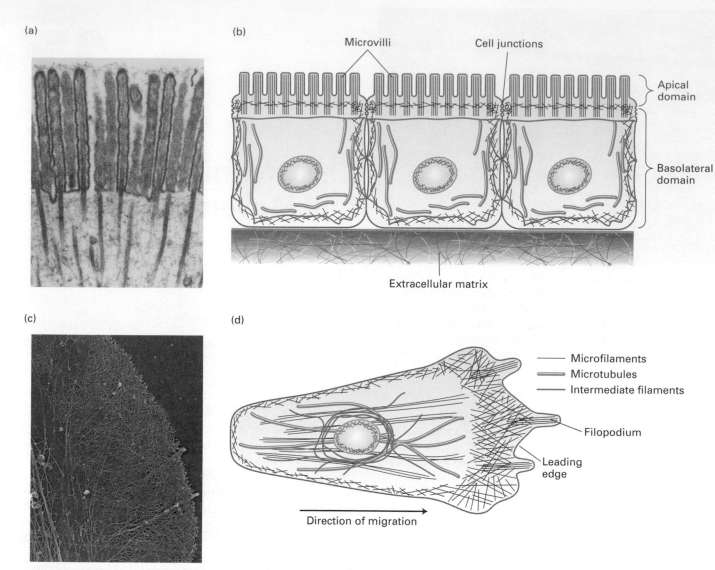

(a)

(b)

Microvilli   Cell junctions

Apical domain

Basolateral domain

Extracellular matrix

(c)

(d)

Microfilaments
Microtubules
Intermediate filaments

Filopodium

Leading edge

Direction of migration

**FIGURE 17-1 Overview of the cytoskeletons of an epithelial cell and a migrating cell.** (a) Transmission electron micrograph of a thin section of an epithelial cell from the small intestine, showing the cytoskeletal components of the microvilli. (b) Epithelial cells are highly polarized, with distinct apical and basolateral domains. An intestinal epithelial cell transports nutrients into the cell through the apical domain and out of the cell across the basolateral domain. (c) Transmission electron micrograph of part of the leading edge of a migrating cell. The cell was treated with a mild detergent to dissolve the membranes, which also allows solubilization of most cytoplasmic components. The remaining cytoskeleton was shadowed with platinum and visualized in the electron microscope. Note the meshwork of actin filaments visible in this micrograph. (d) A migrating cell, such as a fibroblast or a macrophage, has morphologically distinct domains, with a leading edge at the front. Microfilaments are indicated in red, microtubules in green, and intermediate filaments in dark blue. The position of the nucleus (light blue oval) is also shown. [Part (a) Courtesy of Mark Mooseker; Part (c) from T. M. Svitkina et al., 1999, *J. Cell Biol.* **145:**1009, courtesy of Tatyana Svitkina.]

As the macrophage follows the chemical gradient, twisting and turning to get to the bacteria and phagocytose them, it has to constantly reorganize its cell locomotion machinery. As we will see, the internal motile machinery of macrophages and other crawling cells is always oriented in the direction that they crawl (Figure 17-1c, d).

These are just two examples of **cell polarity**—the ability of cells to generate functionally distinct regions. In fact, as you think about all types of cells, you will realize that most of them have some form of cell polarity. An additional and fundamental example of cell polarity is the ability of cells to divide: they must first select an axis for cell division and

then set up the machinery to segregate their organelles along that axis.

A cell's shape, internal organization, and functional polarity are provided by a three-dimensional filamentous protein network called the **cytoskeleton**. The cytoskeleton can be isolated and visualized after treating cells with gentle detergents that solubilize the plasma membrane and internal organelles, releasing most of the cytoplasm (Figure 17-1c). The cytoskeleton extends throughout the cell and is attached to the plasma membrane and internal organelles, thus providing a framework for cellular organization. The term *cytoskeleton* may imply a fixed structure like a bone skeleton. In

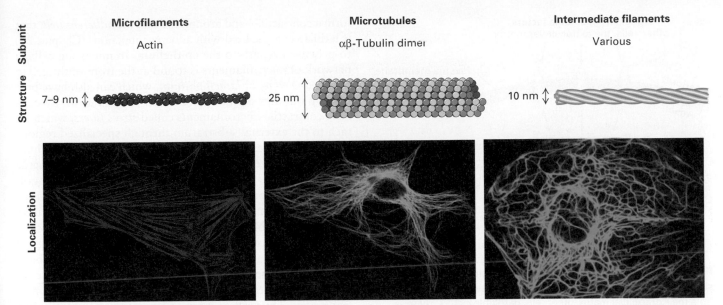

| | Microfilaments | Microtubules | Intermediate filaments |
|---|---|---|---|
| **Subunit** | Actin | αβ-Tubulin dimer | Various |
| **Structure** | 7–9 nm | 25 nm | 10 nm |
| **Localization** | | | |

**FIGURE 17-2 The components of the cytoskeleton.** Each filament type is assembled from specific subunits in a reversible process so that cells can assemble and disassemble filaments as needed. Bottom panels show localization of the three filament systems in cultured cells as seen by immunofluorescence microscopy of actin, tubulin, and an intermediate filament protein, respectively. [Actin and tubulin courtesy of D. Garbett and A. Bretscher; intermediate filaments Copyright Molecular Expressions, Nikon & FSU.]

fact, the cytoskeleton can be very dynamic, with components capable of reorganization in less than a minute, or it can be quite stable for hours at a time. As a result, the lengths and dynamics of filaments can vary greatly, filaments can be assembled into diverse types of structures, and they can be regulated locally in the cell.

The cytoskeleton is composed of three major filament systems, shown in Figure 17-1b, d as well as in Figure 17-2, all of which are organized and regulated in time and space. Each filament system is composed of a polymer of assembled subunits. The subunits that make up the filaments undergo regulated assembly and disassembly, giving the cell the flexibility to assemble or disassemble different types of structures as needed.

• **Microfilaments** are polymers of the protein *actin* organized into functional bundles and networks by actin-binding proteins. Microfilaments are especially important in the organization of the plasma membrane, giving shape to surface structures such as microvilli. Microfilaments can function on their own or serve as tracks for ATP-powered myosin **motor proteins,** which provide a contractile function (as in muscle) or ferry cargo along microfilaments.

• **Microtubules** are long tubes formed by the protein *tubulin* and organized by microtubule-associated proteins. They often extend throughout the cell, providing an organizational framework for associated organelles and structural support to cilia and flagella. They also make up the structure of the mitotic spindle, the machine for separating duplicated chromosomes at mitosis. Molecular motors called kinesins and dyneins transport cargo along microtubules and, like myosin, are also powered by ATP hydrolysis.

• **Intermediate filaments** are tissue-specific filamentous structures that serve a number of different functions, including lending structural support to the nuclear membrane, providing structural integrity to cells in tissues, and serving structural and barrier functions in skin, hair, and nails. Unlike the situation for microfilaments and microtubules, there are no motors that use intermediate filaments as tracks.

As we can see in Figure 17-1, cells can construct very different arrangements of their cytoskeletons. To establish these arrangements, cells must sense signals—either from soluble factors bathing the cell, from adjacent cells, or from the extracellular matrix—and interpret them (Figure 17-3). These signals are detected by cell-surface receptors that activate signal-transduction pathways that ultimately converge on factors that regulate cytoskeletal organization.

The importance of the cytoskeleton for normal cell function and motility is evident when a defect in a cytoskeletal component—or in cytoskeletal regulation—causes a disease. For example, about 1 in 500 people has a defect that affects the contractile apparatus of the heart, which results in cardiomyopathies varying in degree of severity. Many diseases of the red blood cell affect the cytoskeletal components that support these cells' plasma membranes. Metastatic cancer cells exhibit unregulated motility due to misregulation of the cytoskeleton, breaking away from their tissue of origin and migrating to new locations to form new colonies of uncontrolled growth.

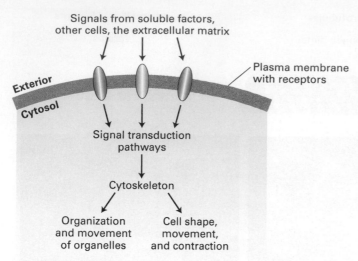

**FIGURE 17-3 Regulation of cytoskeleton function by cell signaling.** Cells use cell-surface receptors to sense external signals from the extracellular matrix, other cells, or soluble factors. These signals are transmitted across the plasma membrane and activate specific cytosolic signaling pathways. Signals—often integrated from more than one receptor—lead to the organization of the cytoskeleton to provide cells with their shape, as well as to determine organelle distribution and movement. In the absence of external signals, cells still organize their internal structure, but not in a polarized manner.

In this and the following chapter, we discuss the structure, function, and regulation of the cytoskeleton. We will see how a cell arranges its cytoskeleton to determine cell shape and polarity, to provide organization and motility to its organelles, and to be the structural framework for such processes as cell swimming and cell crawling. We will discuss how cells assemble the three different filament systems and how signal-transduction pathways regulate these structures both locally and globally. How the cytoskeleton is regulated during the cell cycle is discussed in Chapter 19, and how it participates in the functional organization of tissue is covered in Chapter 20. Our focus in this chapter is on microfilaments and actin-based structures. Although we initially examine the cytoskeletal systems separately, in the next chapter we will see that microfilaments cooperate with microtubules and intermediate filaments in the normal functioning of cells.

## 17.1 Microfilaments and Actin Structures

Microfilaments can assemble into a wide variety of different types of structures within a cell (Figure 17-4a). Each of these diverse structures underlies particular cellular functions. Microfilaments can exist as a tight bundle of filaments making up the core of the slender, fingerlike *microvilli,* but they can also be found in a less ordered network beneath the plasma membrane, known as the *cell cortex,* where they provide support and organization. In epithelial cells, microfilaments

form a contractile band around the cell, the *adherens belt,* that is intimately associated with adherens junctions (Chapter 20) to provide strength to the epithelium. In migrating cells, a network of microfilaments is found at the front of the cell in the *leading edge,* or *lamellipodium,* which can also have protruding bundles of filaments called *filopodia.* Many cells have contractile microfilaments called *stress fibers,* which attach to the external substratum through specialized regions called *focal adhesions* or *focal contacts* (discussed in Chapter 20). Specialized cells such as macrophages use contractile microfilaments in a process called *phagocytosis* to engulf and internalize pathogens (such as bacteria), which are then destroyed internally. Highly dynamic, short bursts of actin filament assembly can power the movement of *endocytic vesicles* away from the plasma membrane. At a late stage of cell division in animals, after all the organelles have been duplicated and segregated, a *contractile ring* forms and constricts to generate two daughter cells in a process known as *cytokinesis.* Thus cells use actin filaments in many ways: in a structural role, by harnessing the power of actin polymerization to do work, or as tracks for myosin motors. The electron micrograph in Figure 17-4b shows microfilaments in microvilli. Different arrangements of microfilaments often coexist within a single cell, as shown in Figure 17-4c, in this case a migrating fibroblast.

The basic building block of microfilaments is **actin,** a protein that has the remarkable property of reversibly assembling into a polarized filament with functionally distinct ends. These filaments are then molded into the various structures described in the previous paragraph by actin-binding proteins. The name *microfilament* refers to actin in its polymerized form, with its associated proteins. In this section, we look at the actin protein itself and the filaments into which it assembles.

## Actin Is Ancient, Abundant, and Highly Conserved

Actin is an abundant intracellular protein in eukaryotic cells. In muscle cells, for example, actin comprises 10 percent by weight of the total cell protein; even in nonmuscle cells, actin makes up 1–5 percent of the cellular protein. The cytosolic concentration of actin in nonmuscle cells ranges from 0.1 to 0.4 mM; in special structures such as microvilli, however, the local actin concentration can be as high as 5 mM. To grasp how much actin is present in cells, consider a typical liver cell, which has $2 \times 10^4$ insulin receptor molecules but approximately $5 \times 10^8$, or half a billion, actin molecules. Because they form structures that extend across large parts of the cell interior, cytoskeletal proteins are among the most abundant proteins in a cell.

Actin is encoded by a large family of genes that gives rise to some of the most conserved proteins within and across species. The protein sequences of actins from amebas and from animals are identical at 80 percent of the amino acid positions, despite about a billion years of evolution. The multiple actin genes found in modern eukaryotes are related

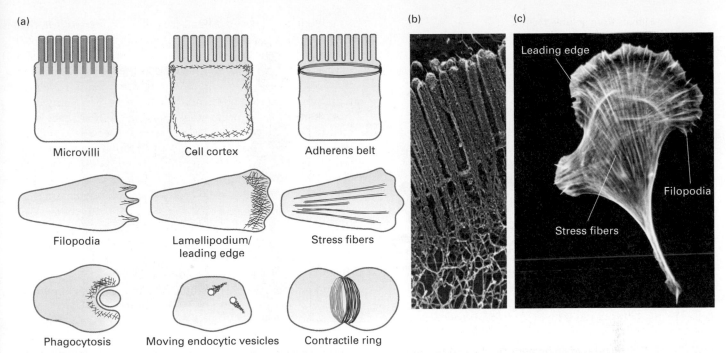

(a)

Microvilli

Cell cortex

Adherens belt

Filopodia

Lamellipodium/
leading edge

Stress fibers

Phagocytosis

Moving endocytic vesicles

Contractile ring

(b)

(c)

Leading edge

Filopodia

Stress fibers

**FIGURE 17-4 Examples of microfilament-based structures.** (a) In each panel, microfilaments are depicted in red. (b) Scanning electron micrograph of the apical region of a polarized epithelial cell, showing the bundles of actin filaments that make up the cores of the microvilli.

(c) A cell moving toward the top of the page, stained for actin with fluorescent phalloidin, a drug that specifically binds F-actin. Note how many different organizations can exist in one cell. [Part (b) courtesy of N. Hirokawa; Part (c) courtesy of J. V. Small.]

to a bacterial gene that has evolved to have a role in bacterial cell-wall synthesis. Some single-celled eukaryotes, such as yeasts and amebas, have one or two ancestral actin genes, whereas multicellular organisms often contain multiple actin genes. For instance, humans have six actin genes, and some plants have more than 60 actin genes (although most are pseudogenes, which do not encode functional actin proteins). Each functional actin gene encodes a different isoform of the protein. Actin isoforms can be classified into three groups: the $\alpha$-actins, $\beta$-actins, and $\gamma$-actins. In vertebrates, four actin isoforms are present in specific types of muscle cells, and two isoforms are found in nonmuscle cells. These six isoforms differ at only about 25 of the 375 residues in the complete protein, or show about 93 percent identity. Although these differences may seem minor, the three types of isoforms have different functions: $\alpha$-actin is associated with contractile structures, $\gamma$-actin accounts for filaments in stress fibers, and $\beta$-actin is enriched in the cell cortex and leading edge of motile cells.

## G-Actin Monomers Assemble into Long, Helical F-Actin Polymers

Actin exists as a globular monomer called **G-actin** and as a filamentous polymer called **F-actin,** which is a linear chain of G-actin subunits. Each actin molecule contains a $Mg^{2+}$ ion complexed with either ATP or ADP. The importance of the interconversion between the ATP and the ADP forms of actin is discussed later.

X-ray crystallographic analysis reveals that the G-actin monomer is separated into two lobes by a deep cleft (Figure 17-5a). At the base of the cleft is the *ATPase fold,* the site where ATP and $Mg^{2+}$ are bound. The floor of the cleft acts as a hinge that allows the lobes to flex relative to each other. When ATP or ADP is bound to G-actin, the nucleotide affects the conformation of the molecule; in fact, without a bound nucleotide, G-actin denatures very quickly. The addition of cations—$Mg^{2+}$, $K^+$, or $Na^+$—to a solution of G-actin will induce the polymerization of G-actin into F-actin filaments. The process is reversible: F-actin depolymerizes into G-actin when the ionic strength of the solution is lowered. The F-actin filaments that form in vitro are indistinguishable from microfilaments seen in cells, indicating that F-actin is the major component of microfilaments.

From the results of x-ray diffraction studies of actin filaments and the actin monomer structure shown in Figure 17-5a, scientists have determined that the subunits in an actin filament are arranged in a helical structure (Figure 17-5b). In this arrangement, the filament can be considered as two helical strands wound around each other. Each subunit in the structure contacts one subunit above, one below in one strand, and two in the other strand. The subunits in a single strand wind around the back of the other strand and repeat after 72 nm or 14 actin subunits. Since there are two strands, the actin filament appears to repeat every 36 nm (see Figure 17-5b). When F-actin is negatively stained with uranyl acetate for electron microscopy, it appears as a twisted string whose diameter varies between 7 and 9 nm (Figure 17-5c).

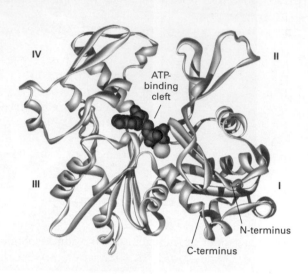

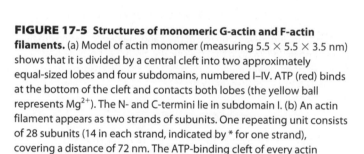

(a)

IV

II

ATP-binding cleft

III

I

N-terminus

C-terminus

(b)

(−) end

36 nm

36 nm

(+) end

(c)

**FIGURE 17-5 Structures of monomeric G-actin and F-actin filaments.** (a) Model of actin monomer (measuring 5.5 × 5.5 × 3.5 nm) shows that it is divided by a central cleft into two approximately equal-sized lobes and four subdomains, numbered I–IV. ATP (red) binds at the bottom of the cleft and contacts both lobes (the yellow ball represents $Mg^{2+}$). The N- and C-termini lie in subdomain I. (b) An actin filament appears as two strands of subunits. One repeating unit consists of 28 subunits (14 in each strand, indicated by * for one strand), covering a distance of 72 nm. The ATP-binding cleft of every actin subunit is oriented toward the same end of the filament. The end of a filament with an exposed binding cleft is the (−) end; the opposite end is the (+) end. (c) In the electron microscope, negatively stained actin filaments appear as long, flexible, and twisted strands of beaded subunits. Because of the twist, the filament appears alternately thinner (7-nm diameter) and thicker (9-nm diameter) (arrows). (The microfilaments visualized in a cell by electron microscopy are F-actin filaments plus any bound proteins.) [Part (a) adapted from C. E. Schutt et al., 1993, *Nature* **365**:810, courtesy of M. Rozycki. Part (c) courtesy of R. Craig.]

## F-Actin Has Structural and Functional Polarity

All subunits in an actin filament are oriented the same way. Consequently, a filament exhibits polarity; that is, one end differs from the other. As we will see, one end of the filament is favored for the addition of actin subunits and is designated the (+) end, whereas the other end is favored for subunit dissociation, designated the (−) end. At the (+) end, the ATP-binding cleft of the terminal actin subunit contacts the neighboring subunit, whereas on the (−) end, the cleft is exposed to the surrounding solution (see Figure 17-5b).

Without the atomic resolution afforded by x-ray crystallography, the cleft in an actin subunit and therefore the polarity of a filament is not detectable. However, the polarity of actin filaments can be demonstrated by electron microscopy in "decoration" experiments, which exploit the ability of the motor protein myosin to bind specifically to actin filaments. In this type of experiment, an excess of myosin S1, the actin-binding head domain of myosin, is mixed with actin filaments and binding is permitted to take place. Myosin attaches to the sides of a filament with a slight tilt. When all the actin subunits are bound by myosin, the filament appears "decorated" with arrowheads that all point toward one end of the filament (Figure 17-6).

The ability of the myosin S1 head to bind and coat F-actin is very useful experimentally—it has allowed researchers to identify the polarity of actin filaments, both in vitro

**EXPERIMENTAL FIGURE 17-6 Myosin S1 decoration demonstrates the polarity of an actin filament.** Myosin S1 head domains bind to actin subunits in a particular orientation. When bound to all the subunits in a filament, S1 appears to spiral around the filament. This coating of myosin heads produces a series of arrowheadlike decorations (arrows), most easily seen at the wide views of the filament. The polarity in decoration defines a pointed (−) end and a barbed (+) end. [Courtesy of R. Craig.]

(−) end

Pointed end

Barbed end

(+) end

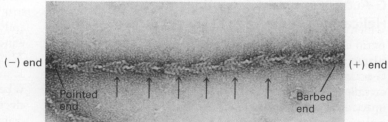

and in cells. The arrowhead points to the (−) end, and so the (−) end is often called the "pointed" end of an actin filament; the (+) end is known as the "barbed" end. Because myosin binds to actin filaments and not to microtubules or intermediate filaments, arrowhead decoration is one criterion by which actin filaments can be definitively identified among the other cytoskeletal fibers in electron micrographs of cells.

---

## KEY CONCEPTS of Section 17.1

### Microfilaments and Actin Structures

• Microfilaments can be assembled into diverse structures, many associated with the plasma membrane (see Figure 17-4a).

• Actin, the basic building block of microfilaments, is a major protein of eukaryotic cells and is highly conserved.

• Actin can reversibly assemble into filaments that consist of two helices of actin subunits.

• The actin subunits in a filament are all oriented in the same direction, with the nucleotide-binding site exposed on the (−) end (see Figure 17-5).

---

# 17.2 Dynamics of Actin Filaments

The actin cytoskeleton is not a static, unchanging structure consisting of bundles and networks of filaments. Although microfilaments may be relatively static in some structures, in others they are highly dynamic, growing or shrinking in length. These changes in the organization of actin filaments can generate forces that cause large changes in the shape of a cell or drive intracellular movements. In this section, we consider the mechanism and regulation of actin polymerization, which is largely responsible for the dynamic nature of cells. We will see that several actin-binding proteins make important contributions to these processes.

## Actin Polymerization in Vitro Proceeds in Three Steps

The in vitro polymerization of G-actin monomers to form F-actin filaments can be monitored by viscometry, sedimentation, fluorescence spectroscopy, or fluorescence microscopy (Chapter 9). When actin filaments grow long enough to become entangled, the viscosity of the solution increases, which is measured as a decrease in its flow rate in a viscometer. The basis of the sedimentation assay is the ability of ultracentrifugation (100,000g for 30 minutes) to sediment F-actin but not G-actin. The third assay makes use of G-actin covalently labeled with a fluorescent dye; the fluorescence spectrum of the labeled G-actin monomer changes when it is polymerized into F-actin. Finally, growth of the fluorescently

labeled filaments can be imaged with fluorescence video microscopy. These assays are useful for kinetic studies of actin polymerization and for characterization of actin-binding proteins to determine how they affect actin dynamics or how they cross-link actin filaments.

The mechanism of actin assembly has been studied extensively. Remarkably, one can purify G-actin at a high protein concentration without it forming filaments—provided it is maintained in a buffer with ATP and low levels of cations. However, as we saw earlier, if the cation level is increased (e.g., to 100 mM $K^+$ and 2 mM $Mg^{2+}$), G-actin will polymerize, with the kinetics of the reaction depending on the starting concentration of G-actin. The polymerization of pure G-actin in vitro proceeds in three sequential phases (Figure 17-7a):

1. The *nucleation phase* is marked by a lag period in which G-actin subunits combine into two or three subunits. When the oligomer reaches three subunits in length, it can act as a seed, or nucleus, for the next phase.

2. During the *elongation phase*, the short oligomer rapidly increases in length by the addition of actin monomers to both of its ends. As F-actin filaments grow, the concentration of G-actin monomers decreases until equilibrium is reached between the filament ends and monomers, and a steady state is reached.

3. In the *steady-state phase*, G-actin monomers exchange with subunits at the filament ends, but there is no net change in the total length of filaments.

The kinetic curves in Figure 17-7b, c show the state of filament mass during each phase of polymerization. In Figure 17-7c we see that the lag period is due to nucleation, because it can be eliminated by the addition of a small number of F-actin nuclei to the solution of G-actin.

How much G-actin is required for spontaneous filament assembly? Scientists have placed various concentrations of ATP–G-actin under polymerizing conditions and found that, below a certain concentration, filaments cannot assemble (Figure 17-8). Above this concentration, filaments begin to form, and when steady state is reached, the incorporation of more free subunits is balanced by the dissociation of subunits from filament ends to yield a mixture of filaments and monomers. The concentration at which filaments are formed is known as the overall *critical concentration*, $C_c$. Below $C_c$, filaments will not form; above $C_c$, filaments form. At steady state, the concentration of monomeric actin remains at the critical concentration (see Figure 17-8).

## Actin Filaments Grow Faster at (+) Ends Than at (−) Ends

We saw earlier that myosin S1 head decoration experiments reveal an inherent structural polarity of F-actin (see Figure 17-6). If free ATP–G-actin is added to a preexisting myosin-decorated filament, the two ends grow at very different rates

(a)

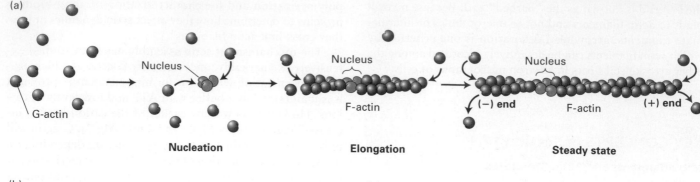

(b)

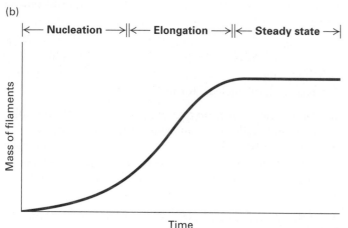

(c)

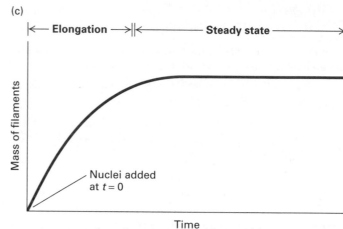

**FIGURE 17-7 The three phases of in vitro G-actin polymerization.**
(a) In the initial nucleation phase, ATP–G-actin monomers (red) slowly form stable complexes of actin (purple). These nuclei are rapidly elongated in the second phase by the addition of subunits to both ends of the filament. In the third phase, the ends of actin filaments are in a steady state with monomeric G-actin. (b) Time course of the in vitro polymerization reaction reveals the initial lag period associated with nucleation, the elongation phase, and steady state. (c) If some short, stable actin filament fragments are added at the start of the reaction to act as nuclei, elongation proceeds immediately, without any lag period.

(Figure 17-9). In fact, the rate of addition of ATP–G-actin is nearly 10 times faster at the (+) end than at the (−) end. The rate of addition is, of course, determined by the concentration of free ATP–G-actin. Kinetic experiments have shown that the rate of addition at the (+) end is about 12 $\mu M^{-1} s^{-1}$ and about 1.3 $\mu M^{-1} s^{-1}$ at the (−) end (Figure 17-10a). This means that if 1 $\mu M$ free ATP–G-actin is added to preformed filaments, on average 12 subunits will be added to the (+) end every second, whereas only 1.3 will be added at the (−) end every second. What about the rate of subunit loss from each end? By contrast, the rates of dissociation of ATP–G-actin subunits from the two ends are quite similar, about 1.4 $s^{-1}$

from the (+) end and 0.8 $s^{-1}$ from the (−) end. Since this dissociation is simply the rate at which subunits leave ends, it does not depend on the concentration of free ATP–G-actin.

What implications do these association and dissociation rates have for actin dynamics? First let's consider just one end, the (+) end. As we noted above, the rate of addition

**FIGURE 17-8 Determination of filament formation by actin concentration.** The critical concentration ($C_c$) is the concentration of G-actin monomers in equilibrium with actin filaments. At monomer concentrations below the $C_c$, no polymerization takes place. When polymerization is induced at monomer concentrations above the $C_c$, filaments assemble until steady state is reached and the monomer concentration falls to $C_c$.

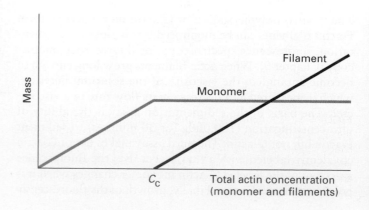

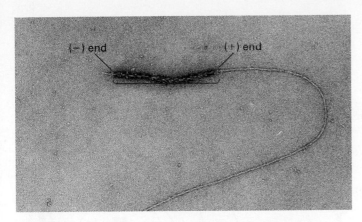

**EXPERIMENTAL FIGURE 17-9** **The two ends of a myosin-decorated actin filament grow unequally.** When short actin filaments are decorated with a myosin S1 head and then used to nucleate actin polymerization, the resulting actin subunits add much more efficiently to the (+) end than the (−) end of the nucleating filament. This result indicates that G-actin monomers are added much faster at the (+) end than at the (−) end. [Courtesy of T. Pollard.]

depends on the free ATP–G-actin concentration, whereas the rate of loss of subunits does not. Thus subunits will be added at high free ATP–G-actin concentrations, but as the concentration is lowered, a point will be reached at which the rate of addition is balanced by the rate of loss and no net growth occurs at that end. This is called the critical concentration $C^+_c$ for the (+) end, and we can calculate it by setting the rate of assembly equal to the rate of disassembly. Thus at the

critical concentration, the rate of assembly is $C^+_c$ times the measured rate of addition of 12 $\mu M^{-1} s^{-1}$ ($C^+_c$ 12 $s^{-1}$), whereas the rate of disassembly is independent of the free actin concentration, namely, 1.4 $s^{-1}$. Setting these equal to each other yields $C^+_c = 1.4 \ s^{-1}/12 \ \mu M^{-1} s^{-1}$ or 0.12 $\mu M$ for the (+) end. Above this free ATP-G-actin concentration, subunits add to the (+) end and net growth occurs, whereas below it, there is a net loss of subunits and shrinkage occurs.

Now let's consider just the (−) end. Because the rate of addition is much lower, 1.3 $\mu M^{-1} s^{-1}$, yet the rate of dissociation is about the same, 0.8 $s^{-1}$, we expect the critical concentration $C^-_c$ at the (−) end to be higher than $C^+_c$. Indeed, as we just did for the (+) end, we can calculate $C^-_c$ to be about 0.8 $s^{-1}/1.3 \ \mu M^{-1} s^{-1}$, or 0.6 $\mu M$. Thus at less than 0.6 $\mu M$ free ATP–G-actin, say, 0.3$\mu M$, the (−) end will lose subunits. But notice that at this concentration the (+) end will grow, since 0.3 $\mu M$ is above $C^+_c$. Because the critical concentrations are different, at steady state the free ATP–G-actin will be intermediate between $C^+_c$ and $C^-_c$, so the (+) end will grow and the (−) end will lose subunits. This phenomenon is known as *treadmilling*, because particular subunits, such those shown in blue in Figure 17-10b, appear to move through the filament.

The ability of actin filaments to treadmill is powered by hydrolysis of ATP. When ATP–G-actin binds to a (+) end, ATP is hydrolyzed to ADP and $P_i$. The $P_i$ is slowly released from the subunits in the filament, so that the filament becomes asymmetric, with ATP-actin subunits at the (+) end of the filament followed by a region with ADP-$P_i$–actin and then, after $P_i$ release, ADP-actin subunits toward the (−)

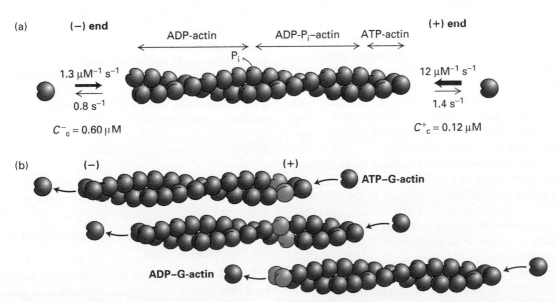

**FIGURE 17-10 Actin treadmilling.** ATP-actin subunits add faster at the (+) end than at the (−) end of an actin filament, resulting in a lower critical concentration and treadmilling at steady state. (a) The rate of addition of ATP–G-actin is much faster at the (+) end than at the (−) end, whereas the rate of dissociation of ADP–G-actin is similar at the two ends. This difference results in a lower critical concentration at the

(+) end. At steady state, ATP-actin is added preferentially at the (+) end, giving rise to a short region of the filament containing ATP-actin and regions containing ADP-$P_i$-actin and ADP-actin toward the (−) end. (b) At steady state, ATP–G-actin subunits add preferentially to the (+) end, while ADP–G-actin subunits disassemble from the (−) end, giving rise to treadmilling of subunits.

end (see Figure 17-10a). During hydrolysis of ATP and subsequent release of $P_i$ from subunits in a filament, actin undergoes a conformational change that is responsible for the different association and dissociation rates at the two ends. Here we have considered only the kinetics of ATP–G-actin, but in reality it is ADP–G-actin that dissociates from the (−) end. Our analysis also relies on a plentiful supply of ATP–G-actin, which, as we will see, turns out to be the case in vivo. Thus actin can use the power generated by hydrolysis of ATP to treadmill, and treadmilling filaments can do work in vivo, as we will see later.

## Actin Filament Treadmilling Is Accelerated by Profilin and Cofilin

Measurements of the rate of actin treadmilling in vivo show that it can be several times higher than can be achieved with pure actin in vitro under physiological conditions. Consistent with a treadmilling model, growth of actin filaments in vivo only ever occurs at the (+) end. How is enhanced treadmilling achieved, and how does the cell recharge the ADP-actin dissociating from the (−) end to ATP-actin for assembly at the (+) end? Two different actin-binding proteins make important contributions to these processes.

The first is *profilin*, a small protein that binds G-actin on the side opposite to the nucleotide-binding cleft. When profilin binds ADP-actin, it opens the cleft and greatly enhances the loss of ADP, which is replaced by the more abundant cellular ATP, yielding a profilin–ATP-actin complex. This complex cannot bind to the (−) end because profilin blocks the sites on G-actin for (−) end assembly. However, the profilin–ATP-actin complex can bind efficiently to the (+) end, and profilin dissociates after the new actin subunit is bound (Figure 17-11). This function of profilin on its own does not enhance treadmilling rate, but it does provide a supply of ATP-actin from released ADP-actin; as a consequence, essentially all the free G-actin in a cell has bound ATP.

Profilin has another important property: it can bind other proteins with sequences rich in proline residues at the same time as binding actin. We will see later how this property is important in actin filament assembly.

*Cofilin* is also a small protein involved in actin treadmilling, but it binds specifically to F-actin in which the subunits contain ADP, which are the older subunits in the filament toward the (−) end (see Figure 17-10a). Cofilin binds by bridging two actin monomers and inducing a small change in the twist of the filament. This small twist destabilizes the filament, breaking it into short pieces. By breaking the filament in this way, cofilin generates many more free (−) ends and therefore greatly enhances the disassembly of the (−) end of the filament (see Figure 17-11). The released ADP-actin subunits are then recharged by profilin and added to the (+) end as described above. In this way, profilin and cofilin can enhance treadmilling in vitro more than tenfold, up to levels seen in vivo. As might be anticipated, the cell

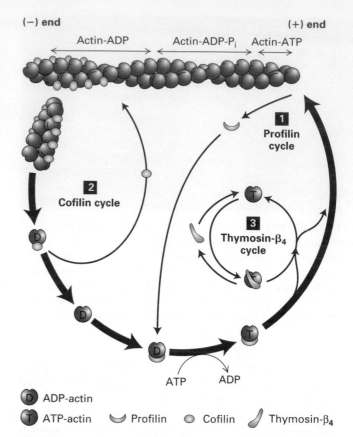

**FIGURE 17-11 Regulation of filament formation by actin-binding proteins.** Actin-binding proteins regulate the rate of assembly and disassembly as well as the availability of G-actin for polymerization. In the profilin cycle **1**, profilin binds ADP–G-actin and catalyzes the exchange of ADP for ATP. The ATP–G-actin–profilin complex can deliver actin to the (+) end of a filament with dissociation and recycling of profilin. In the cofilin cycle **2**, cofilin binds preferentially to filaments containing ADP-actin, inducing them to fragment and thus enhancing depolymerization by making more filament ends. In the thymosin-β4 cycle **3**, G-actin available from the actin-profilin equilibrium is bound by thymosin-β4, sequestering it from polymerization. As the free G-actin concentration is lowered by polymerization, G-actin–thymosin-β4 dissociates to make free G-actin available for association with profilin and further polymerization.

uses signal-transduction pathways to regulate both profilin and cofilin, and thereby the turnover of actin filaments.

## Thymosin-β4 Provides a Reservoir of Actin for Polymerization

It has long been known that cells often have a very large pool of unpolymerized actin, sometimes as much as half the actin in the cell. Since cellular actin levels can be as high as 100–400 μM, this means that there can be 50–200 μM unpolymerized actin in cells. Since the critical concentration in vitro is about 0.2 μM, why doesn't all this actin polymerize? The answer lies, at least in part, in the presence of actin monomer sequestering proteins. One of these is *thymosin-β4*, a small protein that binds to ATP–G-actin in such a way that it inhibits addition of the actin subunit to either end of the filament. Thymosin-β4 can be very plentiful, for example, in human blood platelets. These discoid-shaped cell fragments are very abundant in the blood, and when they are activated during blood clotting, they undergo a burst of actin assembly. Platelets are rich in actin: they are estimated to have a total concentration of about 550 μM actin, of which about 220 μM is in the unpolymerized form. They also contain about 500 μM thymosin-β4, which sequesters much of the free actin. However, as in any protein-protein interaction, free actin and free thymosin-β4 are in a dynamic equilibrium with the actin–thymosin-β4. If some of the free actin is used up for polymerization, more actin–thymosin-β4 will dissociate, providing more free actin for polymerization (see Figure 17-11). Thus thymosin-β4 behaves as a buffer of unpolymerized actin for when it is needed.

## Capping Proteins Block Assembly and Disassembly at Actin Filament Ends

The treadmilling and dynamics of actin filaments are further regulated in cells by *capping proteins* that specifically bind to the ends of the filaments. If this were not the case, actin filaments would continue to grow and disassemble in an uncontrolled manner. As one might expect, two classes of proteins have been discovered: ones that bind the (+) end and ones that bind the (−) end (Figure 17-12).

A protein known as *CapZ*, consisting of two closely related subunits, binds with a very high affinity (≈0.1 nM) to the (+) end of actin filaments, thereby inhibiting subunit addition or loss. The concentration of CapZ in cells is generally sufficient to rapidly cap any newly formed (+) ends. So how can filaments grow at their (+) ends? At least two mecha-

nisms regulate the activity of CapZ. First, the capping activity of CapZ is inhibited by the regulatory lipid PI(4,5)P$_2$, found in the plasma membrane (Chapter 16). Second, recent work has shown that certain regulatory proteins are able to bind the (+) end and simultaneously protect it from CapZ while still allowing assembly there. Thus cells have evolved an elaborate mechanism to block assembly of actin filaments at their (+) ends except when and where assembly is needed.

Another protein called *tropomodulin* binds to the (−) end of actin filaments, also inhibiting assembly and disassembly. This protein is found predominantly in cells in which actin filaments need to be highly stabilized. Two examples we will encounter later in this chapter are the short actin filaments in the cortex of the red blood cell and the actin filaments in muscle. As we will see, in both cases tropomodulin works with another protein, tropomyosin, which lies along the filament to stabilize it. Tropomodulin binds to both tropomyosin and actin at the (−) end to greatly stabilize the filament.

In addition to CapZ, another class of proteins can cap the (+) ends of actin filaments. These proteins also can sever actin filaments. One member of this family, *gelsolin*, is regulated by increased levels of Ca$^{2+}$ ions. On binding Ca$^{2+}$, gelsolin undergoes a conformational change that allows it to bind to the side of an actin filament and then insert itself between subunits of the helix, thereby breaking the filament. It then remains bound to and caps the (+) end, generating a new (−) end that can disassemble. As we discuss in a later section, actin cross-linking proteins can provide linkages between individual actin filaments to turn a solution of F-actin into a gel. If gelsolin is added to such a gel, and the level of Ca$^{2+}$ is elevated, gelsolin will sever the actin filaments and turn it back into a liquid solution. This ability to turn a gel into a sol is why the protein was named "gelsolin."

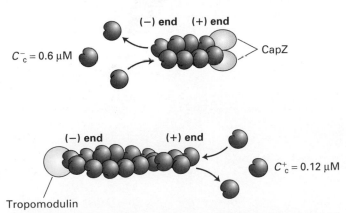

**FIGURE 17-12 Capping proteins.** Capping proteins block assembly and disassembly at filament ends. CapZ blocks the (+) end, which is where filaments normally grow, so its function is to limit actin dynamics to the (−) end. The capping protein tropomodulin blocks (−) ends, where filament disassembly normally occurs; thus the major function of tropomodulin is to stabilize filaments.

In the figure: (−) end, (+) end, $C_c^- = 0.6$ μM, CapZ; (−) end, (+) end, $C_c^+ = 0.12$ μM, Tropomodulin.

## KEY CONCEPTS of Section 17.2

### Dynamics of Actin Filaments

• The rate-limiting step in actin assembly is the formation of a short actin oligomer (nucleus) that can then be elongated into filaments.

• The critical concentration ($C_c$) is the concentration of free G-actin at which the assembly onto a filament end is balanced by loss from that end.

• When the concentration of G-actin is above the $C_c$, the filament end will grow; when it is less than the $C_c$, the filament will shrink (see Figure 17-8).

• ATP–G-actin adds much faster at the (+) end than at the (−) end, resulting in a lower critical concentration at the (+) end than at the (−) end.

• At steady state, actin subunits treadmill through a filament. ATP-actin is added at the (+) end, ATP is then hydrolyzed to ADP and P$_i$, P$_i$ is lost, and ADP-actin dissociates from the (−) end.

- The length and rate of turnover of actin filaments is regulated by specialized actin-binding proteins (see Figure 17-11). Profilin enhances the exchange of ADP for ATP on G-actin; cofilin enhances the rate of loss of ADP-actin from the filament (−) end, and thymosin-$\beta_4$ binds G-actin to provide reserve actin when it is needed. Capping proteins bind to filament ends, blocking assembly and disassembly.

## 17.3 Mechanisms of Actin Filament Assembly

The rate-limiting step of actin polymerization is the formation of an initial actin nucleus from which a filament can grow (see Figure 17-7a). In cells, this inherent property of actin is used as a control point to determine where actin filaments are assembled—this is how the different actin assemblies within a single cell are generated (see Figures 17-1 and 17-4). Two major classes of *actin nucleating proteins,* the *formin* protein family and the *Arp2/3 complex,* nucleate actin assembly under the control of signal-transduction pathways. Moreover, they nucleate the assembly of different actin organizations: formins lead to the assembly of long actin filaments, whereas the Arp2/3 complex leads to branched networks. We will discuss each separately and see how the power of actin polymerization can drive motile processes in a cell. We will then touch on the recent discoveries of new, specialized actin nucleating factors.

### Formins Assemble Unbranched Filaments

Formins are found in essentially all eukaryotic cells as quite a diverse family of proteins: seven different classes are present in vertebrates. Although they are diverse, all formin family members have two adjacent domains in common, the so-called FH1 and FH2 domains (formin-homology domains 1 and 2). Two FH2 domains from two individual monomers associate to form a doughnut-shaped complex (Figure 17-13a). This complex has the ability to nucleate actin assembly by binding two actin subunits, holding them so that the (+) end is toward the FH2 domains. The nascent filament can now grow at the (+) end while the FH2 domain dimer remains attached. How is this possible? As we saw earlier, an actin filament can be thought of as two intertwined strands of subunits. The FH2 dimer can bind to the two terminal subunits. It then probably rocks between the two end subunits, letting go of one to allow addition of a new subunit and then binding the newly added subunit and freeing up space for the addition of another subunit to the other strand. In this way, rocking between the two subunits on the end, it can remain attached while simultaneously allowing growth at the (+) end (see Figure 17-13a).

---

🎯 **FOCUS ANIMATION:** Elongation of Actin Filament by Formin FH2 Dimer

---

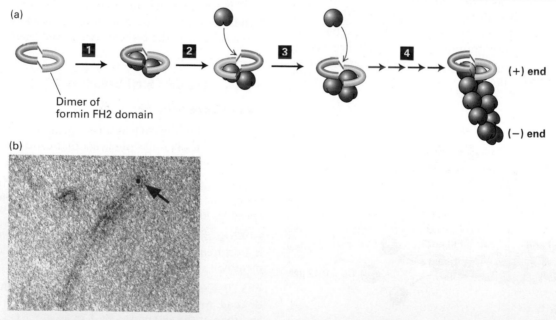

**FIGURE 17-13 Actin nucleation by the formin FH2 domain.** (a) Formins have a domain called FH2 that can form a dimer and nucleate filament assembly. The dimer binds two actin subunits (step **1**), and, by rocking back and forth (steps **2**–**4**), can allow insertion of additional subunits between the FH2 domain and the (+) end of the growing filament. The FH2 domain protects the (+) end from being capped by capping proteins. (b) The FH2 domain of a formin was labeled with colloidal gold (black dot) and used to nucleate assembly of an actin filament. The resulting filament was visualized by electron microscopy after staining with uranyl acetate. Formins assemble long unbranched filaments. [Part (b) from D. Pruyne et al., 2002, *Science* **297:**612.]

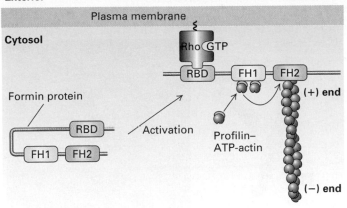

**Exterior**

Plasma membrane

Rho·GTP

**Cytosol**

RBD  FH1  FH2

(+) end

Profilin–
ATP-actin

(−) end

Formin protein

RBD

Activation

FH1  FH2

**FIGURE 17-14 Regulation of formins by an intramolecular interaction.** Some of the formin classes found in vertebrates are regulated by an intramolecular interaction. The inactive formin is activated by binding its Rho-binding domain (RBD) to membrane-bound active Rho-GTP, resulting in exposure of the formin's FH2 domain, which can then nucleate the assembly of a new filament. All formins have an FH1 domain adjacent to the FH2 domain; the proline-rich FH1 domain is a site for recruitment of profilin–ATP-G-actin complexes that can then be "fed" into the growing (+) end. For simplicity of representation, a single formin protein is shown, but as shown in Figure 17-13, the FH2 domain functions as a dimer to nucleate actin assembly. Regulation of the Rho family of small GTPases is detailed in Figure 17-42.

The FH1 domain adjacent to the FH2 domain also makes an important contribution to actin filament growth (Figure 17-14). This domain is rich in proline residues that are sites for the binding of several profilin molecules. We discussed earlier how profilin can exchange the ADP nucleotide on G-actin to generate profilin–ATP-actin. The FH1 domain behaves as a landing site to increase the local concentration of profilin–ATP-G-actin complexes. The actin from the localized profilin-actin complexes is fed into the FH2 domain to add actin to the (+) end of the filament with the concomitant release of profilin, thereby allowing rapid FH2-mediated filament assembly (see Figure 17-14). Since the formin allows addition of actin subunits to the (+) end, long filaments with a formin at their (+) end are generated (Figure 17-13b). In this manner, formins nucleate actin assembly and have the remarkable ability to remain bound to the (+) end while also allowing rapid assembly there. To ensure the continued growth of the filament, formins bind to the (+) end in such a way that precludes binding of a (+) end capping protein such as CapZ, which would normally terminate assembly.

To be useful to a cell, formin activity has to be regulated. Many formins exist in a folded inactive conformation as a result of an interaction between the first half of the protein and the C-terminal tail. These formins are activated by membrane-bound Rho-GTP, a Ras-related small GTPase (discussed in Section 17.7). When Rho is switched from the inactive Rho-GDP form into its activated Rho-GTP state, it can bind and activate the formin (see Figure 17-14).

Recent studies have shown that formins are responsible for the assembly of long actin filaments such as those found in stress fibers, filopodia, and in the contractile ring during cytokinesis (see Figure 17-4). The actin-nucleating role of formins was only discovered recently, so the roles performed by this diverse protein family are only now being uncovered. Since there are many different formin classes in animals, it is likely that formins will be found to assemble additional actin-based structures.

## The Arp2/3 Complex Nucleates Branched Filament Assembly

The Arp2/3 complex is a protein machine consisting of seven subunits, two of which are actin-related proteins ("Arp"), explaining its name (Figure 17-15a). It is found in essentially all eukaryotes, including plants, yeasts, and animal cells. The Arp2/3 complex alone is a very poor nucleator. To nucleate the assembly of branched actin, Arp2/3 needs to be activated by interacting with a *nucleation promoting factor (NPF)*, in addition to associating with the side of a preexisting actin filament. Although there are many different NPFs, the major family is characterized by the presence of a region called WCA (WH2, connector, acidic). Experiments have shown that if you add the WCA domain into an actin assembly assay together with preformed actin filaments, Arp2/3 becomes a potent nucleator of further actin assembly.

How do the Arp2/3 complex and NPF nucleate filaments? The NPF binds an actin subunit through its WH2 domain and activates the Arp2/3 complex through interaction with its acidic domain. In the inactive Arp2/3 complex, the two actin-related polypeptides—Arp2 and Arp3—are in the wrong configuration to nucleate filament assembly (see Figure 17-15a). When activated by the NPF, Arp2 and Arp3 move into the correct configuration, and the complex binds the side of preexisting actin filament. The actin subunit brought in by the WH2 domain of the NPF binds to the Arp2/3 template to nucleate filament assembly at the (+) end (Figure 17-15b). This new (+) end then grows as long as ATP–G-actin is available or until it is capped by a (+) end capping protein such as CapZ. The angle between the old filament and the new one is 70° (Figure 17-15c). This is also the angle observed experimentally in branched filaments at the leading edge of motile cells, which is believed to be formed by the action of the activated Arp2/3 complex (Figure 17-15d). As we discuss in subsequent sections, the Arp2/3 complex can be used to drive actin polymerization to power intracellular motility.

Actin nucleation by the Arp2/3 complex is exquisitely controlled, and the NPFs are part of that regulatory process. One NPF is called WASp, as it is defective in patients with Wiskott-Aldrich syndrome, an X-linked disease characterized by eczema, low platelet count, and immune deficiency. WASp exists in a folded inactive conformation, so that the WCA domain is not available (Figure 17-16). One mechanism to activate the protein involves the small Ras-related GTP-binding protein Cdc42 (discussed in Section 17.7), which in the GTP-bound state binds to and opens WASp, making the WH2 actin-binding and acidic activation domains accessible.

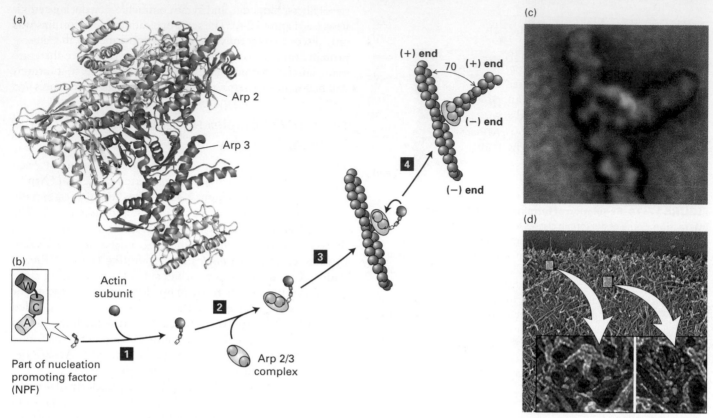

**FIGURE 17-15 Actin nucleation by the Arp2/3 complex.** (a) X-ray structure of the Arp2/3 complex, with five of the subunits in gray, and the Arp2 and Arp3 subunits in green and blue. (b) To nucleate actin assembly efficiently, the activating part of an NPF is shown with its W (WH2), C (connector), and A (acidic) domains. An actin subunit binds to the W domain (step **1**), and then the A domain binds the Arp2/3 complex (step **2**). This interaction induces a conformational change in the Arp2/3 complex, and after binding to the side of an actin filament, the actin subunit bound to the W domain binds to the Arp2/3 complex (step **3**), which then initiates the assembly of an actin filament at the available (+) end (step **4**). The Arp2/3 branch makes a characteristic 70° angle between the filaments. (c) Averaged image compiled from several electron micrographs of Arp2/3 at an actin branch. (d) Image of actin filaments in the leading edge, with a magnification and coloring of individual branched filaments. [Part (a) PDB ID 2P9L; part (c) from C. Egile et al., 2006, *PLoS Biol.* **3:**e383; part (d) from T. M. Svitkina and G. G. Borisy, 1999, *J. Cell Biol.* **145:**1009.]

Although formins and the Arp2/3 complex are found in fungi, plants, and animals, additional actin nucleators have recently been discovered in animal cells. One of these, called Spire, has four tandem WH2 domains, so it can bind four actin monomers. It does this in such a manner to allow the actins to assemble into a filament, although the exact mechanism remains to be understood. As actin filaments perform so many functions in cells, it is likely that additional nucleators will be discovered.

**FIGURE 17-16 Regulation of the Arp2/3 complex by WASp.** WASp is inactive due to an intramolecular interaction that masks the WCA domain. On binding the membrane-bound active small G protein Cdc42-GTP (a member of the Rho family) through its Rho-binding domain (RBD), the intramolecular interaction in WASp is relieved, exposing the W domain to bind actin and the acidic A domain for activation of the Arp2/3 complex. Regulation of the Rho family of small GTPases is detailed in Figure 17-42.

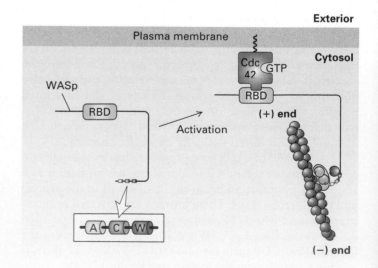

## Intracellular Movements Can Be Powered by Actin Polymerization

How can actin polymerization be harnessed to do work? As we have seen, actin polymerization involves the hydrolysis of actin-ATP to actin-ADP, which allows actin to grow preferentially at the (+) end and disassemble at the (−) end. If an actin filament were to become fixed in the meshwork of the cytoskeleton and you could bind and ride on the assembling (+) end, you would be transported across the cell. This is just what the intracellular bacterial parasite *Listeria monocytogenes* does to get around the cell. The study of *Listeria* motility

was, in fact, the way the nucleating activity of the Arp2/3 protein was discovered. As we shall see shortly, *Listeria* has hijacked a normal cell motility process for its own purposes; we discuss *Listeria* first, as it is currently much better understood than the normal processes that employ similar mechanisms.

*Listeria* is a food-borne pathogen that causes mild gastrointestinal symptoms in most adults but can be fatal to elderly or immunocompromised individuals. It enters animal cells and divides in the cytoplasm. To move from one host cell to another, it moves around the cell by polymerizing actin into a comet tail like the plume behind a rocket (Figure 17-17a, b), and when it runs into the plasma membrane, it pushes its way

**VIDEO:** In Vivo Assembly of Actin Tails in *Listeria*-Infected Bacteria

(a)

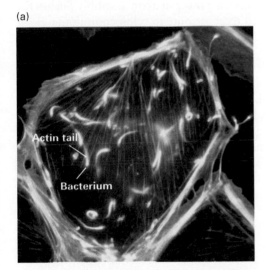

(b)

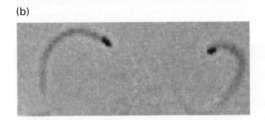

(c)

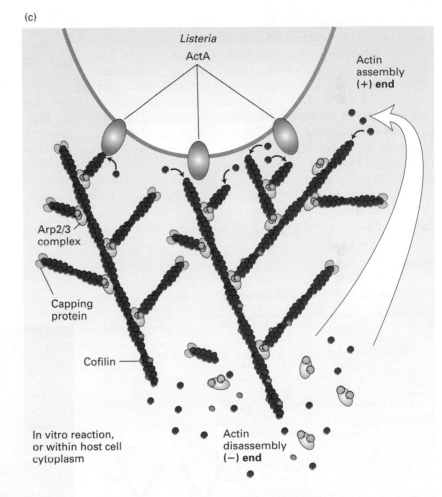

**EXPERIMENTAL FIGURE 17-17** *Listeria* **utilizes the power of actin polymerization for intracellular movement.** (a) Fluorescence microscopy of a cultured cell stained with an antibody to a bacterial surface protein (red) and fluorescent phalloidin to localize F-actin (green). Behind each *Listeria* bacterium is an actin "comet tail" that propels the bacterium forward by actin polymerization. When the bacterium runs into the plasma membrane, it pushes the membrane out into a structure like a filopodium, which protrudes into a neighboring cell. (b) *Listeria* motility can be reconstituted in vitro with bacteria and just four proteins: ATP–G-actin, Arp2/3 complex, CapZ, and cofilin.

This phase micrograph shows bacteria (black), behind which are the phase-dense actin tails. (c) A model of how *Listeria* moves using just four proteins. The ActA protein on the cell surface activates the Arp2/3 complex to nucleate new filament assembly from preexisting filaments. Filaments grow at their (+) end until capped by CapZ. Actin is recycled through the action of cofilin, which enhances depolymerization at the (−) end of the filaments. In this way, polymerization is confined to the back of the bacterium, which propels it forward. [Part (a) courtesy of J. Theriot and T. Michison; part (b) from T. P. Loisel et al., 1999, *Nature* **401**:613.]

into the adjacent cell to infect it. How does it recruit the host cell actin to propel itself? *Listeria* has on its surface a protein called ActA that mimics an NPF by having an actin-binding site and an acidic region to activate the Arp2/3 complex (Figure 17-17c). The ActA protein also binds a protein known as VASP, which has three important properties. First, VASP has a proline-rich region that can bind profilin–ATP-actin for enhancing ATP-actin assembly into the newly formed barbed ends generated by the Arp2/3 complex. Second, it can hold on to the end of the newly formed filament. Third, it can protect the (+) end of the growing filament from capping by CapZ. These properties allow VASP to enhance actin assembly and confine it to the rear of the bacterium. The assembling filaments then push on the bacterium. Since the filaments are embedded in the stationary cytoskeletal matrix of the cell, the *Listeria* cell is pushed forward, ahead of the polymerizing actin. Researchers have reconstituted *Listeria* motility in the test tube using purified proteins to see what the minimal requirements for *Listeria* motility are. Remarkably, the bacteria will move when just four proteins are added: ATP–G-actin, the Arp2/3 complex, CapZ, and cofilin (see Figure 17-17b, c). We have discussed the role of actin and Arp2/3, but why are CapZ and cofilin needed? As we have seen earlier, CapZ rapidly caps the free (+) end of actin filaments, so when a growing filament no longer contributes to bacterial movement, it is rapidly capped and inhibited from further elongation. In this way, assembly occurs only adjacent to the bacterium where ActA is stimulating the Arp2/3 complex. Cofilin is necessary to accelerate the disassembly of the (−) end of the actin filament, regenerating

free actin to keep the polymerization cycle going (see Figure 17-11). This minimal rate of motility can be increased by the presence of other proteins, such as VASP and profilin, as mentioned above.

To move inside cells, the *Listeria* bacterium, as well as other opportunistic pathogens such as the *Shigella* species that cause dysentery, hijacks a normal, regulated cellular process involved in cell locomotion. As we discuss in more detail later (Section 17.7), moving cells have a thin sheet of cytoplasm that protrudes from the front of the cell called the leading edge (see Figures 17-1c, 17-4, and 17-15d). This thin sheet of cytoplasm consists of a dense meshwork of actin filaments that are continually elongating at the front of the cell to push the membrane forward. Factors in the leading-edge membrane activate the Arp2/3 complex to nucleate these filaments. Thus the power of actin assembly pushes the membrane forward to contribute to cell locomotion.

## Microfilaments Function in Endocytosis

As we saw in Chapter 14, endocytosis describes the processes that cells use to take up particles, molecules, or fluid from the external medium by enclosing them in plasma membrane and then internalizing them. The uptake of molecules or liquid is called receptor-mediated or fluid-phase endocytosis, and the uptake of large particles is called phagocytosis ("cell eating"). Microfilaments participate in both of these processes.

Fluid-phase endocytosis is a very highly organized process, and recent studies have shown that the power of actin assembly contributes to this mechanism. Endocytosis assembly factors

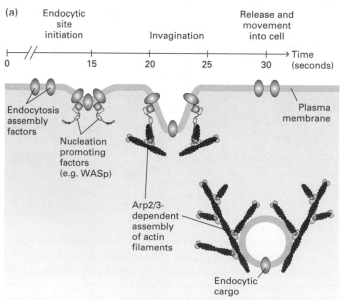

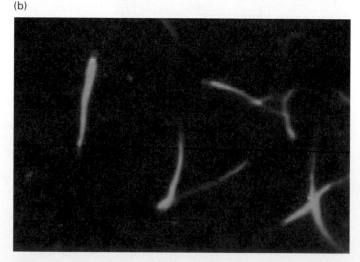

**FIGURE 17-18 Arp2/3-dependent actin assembly during endocytosis.** (a) Clathrin-mediated endocytosis is a rapid and ordered process. It has been best studied in yeast, where the temporal order of specific steps has been delineated. In vivo imaging has shown that endocytosis assembly factors recruit nucleation-promoting factors that activate the Arp2/3 complex. A burst of Arp2/3-dependent actin assembly drives internalized endocytic vesicles away from the

plasma membrane, just like the movement of *Listeria*. (b) Endosome movement can be reconstituted in vitro. Endosomes isolated from cells that had taken up fluorescently labeled transferrin (red) were added to a cell extract containing fluorescently labeled actin (green). The endosomes bind WASp, which then activates the Arp2/3 complex to assemble actin tails that propel them through the cytoplasm. [Part (b) from Taunton et al., 2000, *J. Cell Biol.* **148**:519].

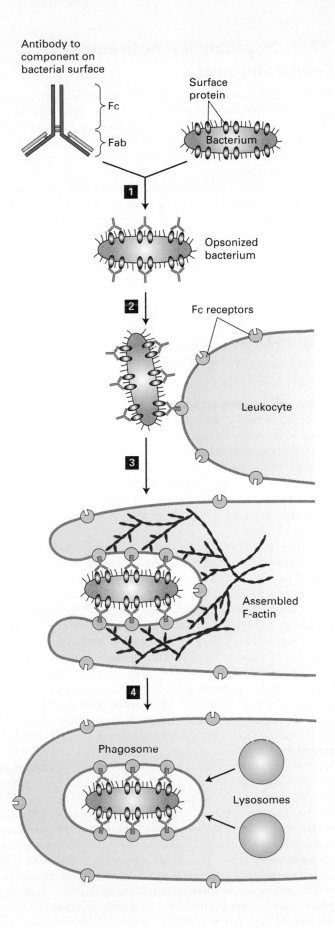

Antibody to
component on
bacterial surface

Surface
protein

Fc

Fab

Bacterium

**1**

Opsonized
bacterium

**2**

Fc receptors

Leukocyte

**3**

Assembled
F-actin

**4**

Phagosome

Lysosomes

**FIGURE 17-19 Phagocytosis and actin dynamics.** Actin assembly and contraction drives the internalization of phagocytic particles. Shown here is the phagocytosis and degradation of a bacterium by a leukocyte. An invading bacterium is coated by specific antibodies to a cell-surface protein in a process known as opsonization (step **1**). The Fc region of the bound antibodies is displayed on the bacterial surface and recognized by a specific receptor, the Fc receptor, on the leukocyte surface (step **2**). This interaction signals the cell to assemble a contractile actin structure that results in the internalization and engulfment of the bacterium (step **3**). Once it has been internalized into a phagosome, the bacterium is killed and degraded by enzymes delivered from lysosomes (step **4**).

recruit NPFs so that as the endocytic vesicles invaginate and pinch off from the membrane, they are then driven into the cytoplasm, powered by a rapid and very short-lived burst (a few seconds in duration) of actin polymerization driven by the Arp2/3 complex (Figure 17-18a). This actin-based movement of endocytic vesicles involving the Arp2/3 complex can be reconstituted in vitro (Figure 17-18b) and is mechanistically very similar to leading-edge formation and *Listeria* motility.

Phagocytosis is a vital process in the recognition and removal of pathogens, such as bacteria, by white blood cells. The immune system identifies the bacterium as foreign material and makes antibodies that recognize components on its surface. As we discussed in Chapter 3, each antibody has a region called the Fab domain that binds specifically to its antigen, in this case a component on the bacterial cell surface. As antibodies coat the bacterium through interaction between their Fab domains and the cell-surface antigen, a second antibody domain known as the Fc domain is exposed. This process is known as opsonization (Figure 17-19, step **1**). The white blood cells have a receptor on their cell surface, the Fc receptor, that recognizes the antibodies on the bacterium; this interaction signals the cells to bind and engulf the pathogen (steps **2** and **3**). The signal also tells the cell to assemble microfilaments at the interaction site with the bacterium, and the assembled microfilaments, together with myosin motor proteins, provide the force necessary to draw the bacterium into the cell, ultimately fully enclosing the pathogen in plasma membrane (step **4**). Once internalized, the newly formed phagosome fuses with lysosomes, where the pathogen is killed and degraded by lysosomal enzymes.

## Toxins That Perturb the Pool of Actin Monomers Are Useful for Studying Actin Dynamics

Certain fungi and sponges have developed toxins that target the polymerization cycle of actin and are therefore toxic to animal cells. Two types of toxins have been characterized. The first class is represented by two unrelated toxins, cytochalasin D and latrunculin, which promote the depolymerization of filaments, though by different mechanisms. Cytochalasin D, a fungal alkaloid, depolymerizes actin filaments by binding to the (+) end of F-actin, where it blocks further addition of subunits. Latrunculin, a toxin secreted by sponges, binds and sequesters G-actin, inhibiting it from adding to a filament end. Exposure to either toxin thus increases the monomer pool. When cytochalasin D or latrunculin is added to live cells, the actin cytoskeleton disassembles and cell movements such as locomotion and cytokinesis are inhibited. These observations were among the first to implicate actin filaments in cell motility. Latrunculin is especially useful because it binds actin monomers and prevents any new actin assembly. Thus if you add latrunculin to a cell, the rate at which actin-based structures disappear reflects their normal rate of turnover. This has revealed that some structures have half-lives of less than a minute, whereas others are much more stable. For example, experiments with latrunculin show that the leading edge of motile cells turns over every 30–180 seconds, and stress fibers turn over every 5–10 minutes.

In contrast, the monomer-polymer equilibrium is shifted in the direction of filaments by jasplakinolide, another sponge toxin, and by phalloidin, which is isolated from *Amanita phalloides* (the "angel of death" mushroom). Jasplakinolide enhances nucleation by binding and stabilizing actin dimers and thereby lowering the critical concentration. Phalloidin binds at the interface between subunits in F-actin, locking adjacent subunits together and preventing actin filaments from depolymerizing. Even when actin is diluted below its critical concentration, phalloidin-stabilized filaments will not depolymerize. Because many actin-based processes depend on actin filament turnover, the introduction of phalloidin into a cell paralyzes all these systems and the cell dies. However, phalloidin has been very useful to researchers, as fluorescent-labeled phalloidin, which binds only to F-actin, is commonly used to stain actin filaments for light microscopy (see Figure 17-4).

---

## KEY CONCEPTS of Section 17.3

### Mechanisms of Actin Filament Assembly

• Actin assembly is nucleated by two classes of proteins: formins nucleate the assembly of unbranched filaments (see Figure 17-13), whereas the Arp2/3 complex nucleates the assembly of branched actin networks (see Figure 17-15). The activities of formins and Arp2/3 are regulated by signal-transduction pathways.

• Functionally different actin-based structures are assembled by formins and Arp2/3 nucleators. Formins drive the assembly of stress fibers and the contractile ring, whereas Arp2/3 nucleates the assembly of branched actin filaments found in the leading edge of motile cells.

• The power of actin polymerization can be harnessed to do work, as is seen in the Arp2/3-dependent intracellular movement of pathogenic bacteria (see Figure 17-17) and inward movement of endocytic vesicles (see Figures 17-18 and 17-19).

• Several toxins affect the dynamics of actin polymerization; some, such as latrunculin, bind and sequester actin monomers, whereas others, such as phalloidin, stabilize filamentous actin. Fluorescently labeled phalloidin is useful for staining actin filaments.

## 17.4 Organization of Actin-Based Cellular Structures

We have seen that actin filaments are assembled into a wide variety of different arrangements and how many associated proteins nucleate actin assembly and regulate filament turnover. Dozens of proteins in a vertebrate cell organize these filaments into diverse functional structures. Here we discuss just a few of these proteins, giving examples of typical types of actin cross-linking proteins found in cells, and also discuss the proteins involved in making functional links between actin and membrane proteins. One fascinating problem, about which very little is known, is how cells assemble different actin-based structures within the same cytoplasm of a cell. Some of this organization must be due to local regulation, a topic we come to at the end of the chapter.

### Cross-Linking Proteins Organize Actin Filaments into Bundles or Networks

When one assembles actin filaments in a test tube, they form a tangled network. In cells, however, actin filaments are found in a variety of distinct structures, such as the highly ordered filament bundles in microvilli or the meshwork characteristic of the leading edge (see Figure 17-4a). These different organizations are determined by the presence of actin *cross-linking proteins*. To be able to organize actin, an actin cross-linking protein must have two F-actin–binding sites (Figure 17-20a).

Cross-linking of F-actin can be achieved by having two actin-binding sites within a single polypeptide, as with *fimbrin*, a protein found in microvilli that builds bundles of filaments all having the same polarity (Figure 17-20b). Other actin cross-linking proteins have a single actin-binding site in a polypeptide chain and then two chains associate to form dimers that bring together two actin-binding sites. These dimeric cross-linking proteins can assemble to generate a rigid rod connecting the two binding sites, as happens with α-*actinin*. Like fimbrin, α-actinin also bundles parallel actin filaments, but farther apart than fimbrin. Another protein, called *spectrin*, is a tetramer with two actin-binding sites; spectrin spans an even greater distance between actin filaments and makes networks under the plasma membrane (shown in Figure 17-21 and discussed in the next section). Other types of cross-linking proteins, such as *filamin*, have a highly flexible region between the two binding sites, functioning like a molecular leaf spring, so they can make stabilizing cross-links between filaments in a meshwork (Figure 17-20c), as is found in the leading edge of motile cells. The Arp2/3 complex, which we discussed in terms of its ability to nucleate actin filament assembly, is also an important cross-linking protein, attaching the (−) end of one filament to the side of another filament (see Figure 17-15).

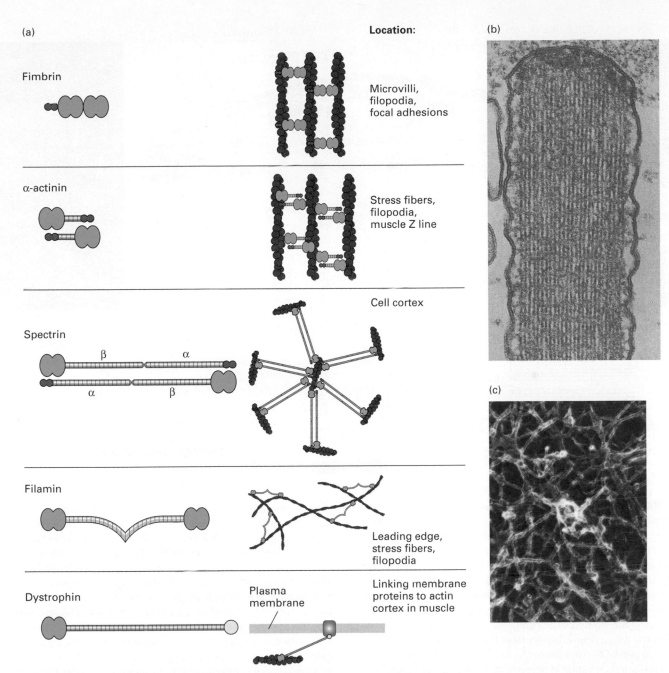

(a)

Fimbrin

Location:

Microvilli,
filopodia,
focal adhesions

α-actinin

Stress fibers,
filopodia,
muscle Z line

Cell cortex

Spectrin

β          α

α          β

Filamin

Leading edge,
stress fibers,
filopodia

Dystrophin

Plasma
membrane

Linking membrane
proteins to actin
cortex in muscle

(b)

(c)

**FIGURE 17-20 Actin cross-linking proteins.** Actin cross-linking proteins mold F-actin filaments into diverse structures. (a) Examples of four F-actin cross-linking proteins, all of which have two domains (blue) that bind F-actin. Some have a Ca²⁺-binding site (purple) that inhibits their activity at high levels of free Ca²⁺. Also shown is dystrophin, which has an actin-binding site on its N-terminal end and a C-terminal domain that binds the membrane protein dystroglycan. (b) Transmission electron micrograph of a thin section of a stereocilium (an unfortunate name, since it is really a giant microvillus) on a sensory hair cell in the inner ear. This structure contains a bundle of actin filaments cross-linked by fimbrin, a small cross-linking protein that allows for close and regular interaction of actin filaments. (c) Long cross-linking proteins such as filamin are flexible and can thus cross-link actin filaments into a loose network. [Part (b) from L. G. Tilney, 1983, *J. Cell Biol.* **96:**822; part (c) courtesy of J. Hartwig.]

## Adaptor Proteins Link Actin Filaments to Membranes

To contribute to the structure of cells and also to harness the power of actin polymerization, actin filaments are very often attached to membranes or are associated with intracellular structures. Actin filaments are especially abundant in the cell cortex underlying the plasma membrane, to which they give support. Actin filaments can interact with membranes either laterally or at their ends.

Our first example of actin filaments attached to membranes is the human erythrocyte—the red blood cell. The erythrocyte consists essentially of plasma membrane enclosing

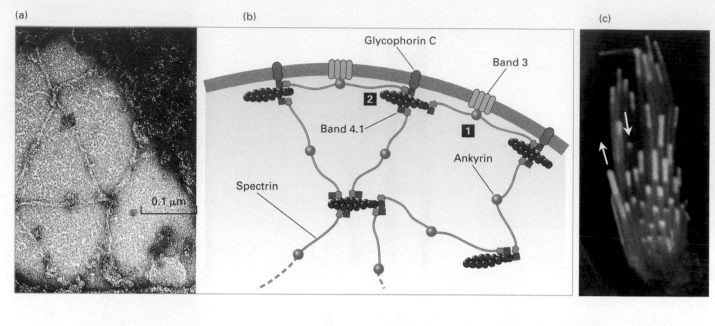

**(a)**

0.1 μm

**(b)**

Glycophorin C

Band 3

**2**

Band 4.1

**1**

Ankyrin

Spectrin

**(c)**

**(d)**

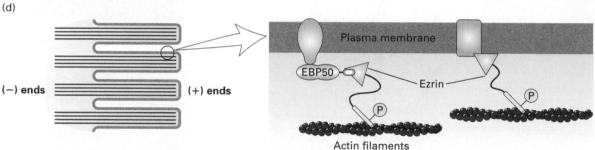

(−) ends                    (+) ends

Plasma membrane

EBP50

Ezrin

P

P

Actin filaments

**FIGURE 17-21 Lateral attachment of microfilaments to membranes.** (a) Electron micrograph of the erythrocyte membrane showing the spoke-and-hub organization of the cortical cytoskeleton supporting the plasma membrane in human erythrocytes. The long spokes are composed mainly of spectrin and can be seen to intersect at the hubs, or membrane-attachment sites. The darker spots along the spokes are ankyrin molecules, which cross-link spectrin to integral membrane proteins. (b) Diagram of the erythrocyte cytoskeleton, showing the two main types of membrane attachment: **1** ankyrin and **2** band 4.1. (c) Actin is incorporated into the tip of stereocilia (giant microvilli). Cells with stereocilia were transfected to express GFP-actin for a short period of time and then counterstained with rhodamine-phalloidin to stain all the F-actin. The experiment shows that new actin is incorporated at the tips of the stereocilia. (d) Ezrin, a member of the ezrin-radixin-moesin (ERM) family, links actin filaments laterally to the plasma membrane in surface structures such as microvilli; attachment can be direct or indirect. Ezrin, activated by phosphorylation (P), links directly to the cytoplasmic region of transmembrane proteins (*right*) or indirectly through a scaffolding protein such as EBP50 (*left*). [Part (a) from T. J. Byers and D. Branton, 1985, *Proc. Nat'l Acad. Sci. USA* **82:**6153, courtesy of D. Branton; part (b) adapted from S. E. Lux, 1979, *Nature* **281:**426, and E. J. Luna and A. L. Hitt, 1992, *Science* **258:**955; part (c) from A. K. Rzadzinska et al., 2004, *J. Cell Biol.* **164:**887; (d) adapted from R. G. Fehon et al., 2010, *Nature Rev. Mol. Cell Biol.* **11:**276.]

a high concentration of the protein hemoglobin to transport oxygen from the lungs to tissues and carbon dioxide from tissues back to the lung—all powered by the magnificent muscle known as the heart. Erythrocytes must be able to survive the raging torrents of blood flow in the heart, then flow down arteries and survive squeezing through narrow capillaries before being cycled through the lungs via the heart. To survive this grueling process for thousands of cycles, erythrocytes have a microfilament-based network underlying their plasma membrane that gives them both the tensile strength and the flexibility necessary for their journey. This network is based on short actin filaments of about 14 subunits in length, stabilized on

their sides by tropomyosin (discussed in more detail in Section 17.6) and by the capping protein tropomodulin on the (−) end. These short filaments serve as hubs for binding about six flexible spectrin molecules, generating a fishing-net type of structure (Figure 17-21a). This network gives the erythrocyte both its strength and flexibility. Spectrin is attached to membrane proteins through two mechanisms: through a protein called *ankyrin* to the bicarbonate transporter (a transmembrane protein also known as *band 3*) and through a spectrin and F-actin–binding protein called *band 4.1* to another transmembrane protein called *glycophorin* C (Figure 17-21b). Although this spectrin-based network is highly developed in the erythrocyte, similar types of linkages occur in many cell types. For example, a related type of ankyrin-spectrin attachment links the $Na^+/K^+$ ATPase to the actin cytoskeleton on the basolateral membrane of epithelial cells.

Genetic defects in proteins of the red blood cell cytoskeleton can result in cells that rupture easily, giving rise to diseases known as hereditary spherocytic anemias (*spherocytic* because the cells are rounder, *anemias* because there is a shortage of red blood cells) and hence a shorter life span. In human patients, mutations in spectrin, band 4.1, and ankyrin can cause this disease. ∎

In addition to the spectrin-based type of support in the cell cortex, microfilaments provide the support for cell-surface structures such as microvilli and membrane ruffles. If we look at a microvillus, it is clear that it must have an end-on attachment at the tip and lateral attachments down its length. What is the orientation of actin filaments in microvilli? Decoration of microvillar filaments by the S1 fragment of myosin show that it is the (+) end at the tip. Moreover, when fluorescent actin is added to a cell, it is incorporated at the tip of a microvillus, showing that not only is the (+) end there, actin filament assembly occurs there (Figure 17-21c). At present it is not known how actin filaments are attached at the microvillus tip, but a likely candidate is a formin protein. This (+) end orientation of actin filaments with respect to the plasma membrane is found almost universally—not just in microvilli but also, for example, in the leading edge of motile cells. The lateral attachments to the plasma membrane are provided, at least in part, by the *ERM (ezrin-radixin-moesin) family* of proteins. These are regulated proteins that exist in a folded, inactive form. When activated by phosphorylation in response to an external signal, F-actin and membrane-protein–binding sites of the ERM protein are exposed to provide a lateral linkage to actin filaments (Figure 17-21d). At the plasma membrane, ERM proteins can link the actin filaments directly or indirectly through scaffolding proteins to the cytoplasmic domain of membrane proteins.

The types of actin membrane linkages we have discussed so far do not involve areas of the plasma membrane attached directly to other cells in a tissue or to the extracellular matrix. Contact between epithelial cells is mediated by highly specialized regions of the plasma membrane called *adherens junctions* (see Figure 17-1b). Other specialized regions of association called *focal adhesions* mediate attachment of cells to the extracellular matrix. In turn, these specialized types of attachments connect to the cytoskeleton, as will be described in more detail when we discuss cell migration (Section 17.7) and cells in the context of tissues (Chapter 20).

Muscular dystrophies are genetic diseases that are often characterized by the progressive weakening of skeletal muscle. One of these genetic diseases, Duchenne muscular dystrophy, affects the protein dystrophin, whose gene is located on the X chromosome, and so the disease is much more prevalent in males. *Dystrophin* is a modular protein whose function is to link the cortical actin network of muscle cells to a complex of membrane proteins that link to the extracellular matrix. Thus dystrophin has an N-terminal actin-binding domain, followed by a series of spectrinlike repeats and terminating in a domain that binds the transmembrane dystroglycan complex to the extracellular matrix protein laminin (see Figure 17-20a). In the absence of dystrophin, the plasma membrane of muscle cells becomes weakened by cycles of muscle contraction and eventually ruptures, resulting in death of the muscle myofibril. ∎

## KEY CONCEPTS of Section 17.4

### Organization of Actin-Based Cellular Structures

• Actin filaments are organized by cross-linking proteins that have two F-actin–binding sites. Actin cross-linking proteins can be long or short, rigid or flexible, depending on the type of structure involved (see Figure 17-20).

• Actin filaments are attached laterally to the plasma membrane by specific classes of proteins, as seen in the red blood cell or in cell-surface structures such as microvilli (see Figure 17-21).

• The (+) end of actin filaments can also be attached to membranes, with assembly mediated between the filament end and the membrane.

• Several diseases have been traced to defects in the microfilament-based cortical cytoskeleton that underlies the plasma membrane.

## 17.5 Myosins: Actin-Based Motor Proteins

In Section 17.3 we discussed how actin polymerization nucleated by the Arp2/3 complex can be harnessed to do work, such as in the movement of vesicles during endocytosis, at the leading edge of motile cells, and the propulsion of *Listeria* bacterium across the eukaryotic cell. In addition to actin-polymerization–based motility, cells have a large family of motor proteins called **myosins** that can move along actin filaments. The first myosin discovered, myosin II, was isolated from skeletal muscle. For a long time, biologists thought that this was the only type of myosin found in nature. However, they then discovered other types of myosins and began to ask how many different functional classes might exist. Today we know that there are several different classes of myosins, in addition to the myosin II of skeletal muscle, that move along actin. Indeed, with the discovery and analysis of all these actin-based motors and the corresponding microtubule-based motors described in the next chapter, the former relatively static view of a cell has been replaced with the realization that the cytoplasm is incredibly dynamic—more like an organized but busy freeway system with motors busily ferrying components around.

Myosins have the amazing ability to convert the energy released by ATP hydrolysis into mechanical work (movement along actin). All myosins convert ATP hydrolysis into work, yet different myosins can perform very different types of functions. For example, many molecules of myosin II pull together on actin filaments to bring about muscle contraction, whereas myosin V binds to vesicular cargo to transport it along actin filaments. The other classes of myosin provide a myriad of functions, from moving organelles around cells to contributing to cell migration.

To begin to understand myosins, we first discuss their general domain organization. Armed with this information, we explore the diversity of myosins in different organisms and describe in more detail some of those that are common in eukaryotes. To understand how such diverse functions

can be accommodated by one type of motor mechanism, we will investigate the basic mechanism of how the energy released by ATP hydrolysis is converted into work and then see how this mechanism is modified to tailor the properties of specific myosin classes for their specific functions.

## Myosins Have Head, Neck, and Tail Domains with Distinct Functions

Much of what we know about myosins comes from studies of myosin II isolated from skeletal muscle. In skeletal muscle, hundreds of individual myosin II molecules are assembled into bundles called bipolar thick filaments (Figure 17-22a). In a later section we will discuss how these myosin filaments interdigitate with actin filaments to bring about

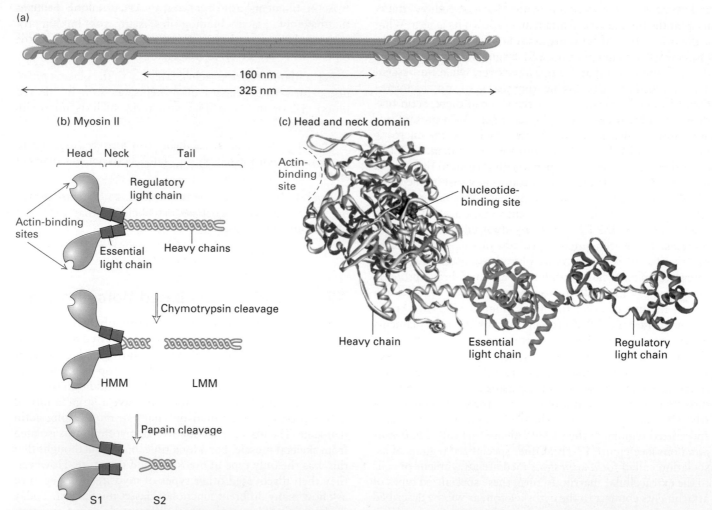

**FIGURE 17-22 Structure of myosin II.** (a) Organization of myosin II in filaments isolated from skeletal muscle. Myosin II assembles into bipolar filaments in which the tails form the shaft of the filament with heads exposed at the ends. Extraction of bipolar filaments with high salt and ATP disassembles the filament into individual myosin II molecules. (b) Myosin II molecules consist of two identical heavy chains (light blue) and four light chains (green and dark blue). The tail of the heavy chains forms a coiled coil to dimerize; the neck region of each heavy chain has two light chains associated with it. Limited proteolytic cleavage of myosin II generates tail fragments—LMM and S2—and the S1 motor domain. (c) Three-dimensional model of a single S1 head domain shows that it has a curved, elongated shape and is bisected by a cleft. The nucleotide-binding pocket lies on one side of this cleft, and the actin-binding site lies on the other side near the tip of the head. Wrapped around the shaft of the α-helical neck are two light chains. These chains stiffen the neck so that it can act as a lever arm for the head. Shown here is the ADP-bound conformation.

muscle contraction. Here, we first investigate the properties of the individual myosin molecule itself.

It is possible to dissolve the myosin thick filament in a solution of ATP and high salt, generating a pool of individual myosin II molecules. The soluble myosin II molecule is actually a protein complex consisting of six polypeptides. Two of the subunits are identical high-molecular-weight polypeptides known as myosin heavy chains. Each consists of a globular *head* domain and a long *tail* domain, connected by a flexible *neck* domain. The tails of the two myosin heavy chains intertwine, so that the head regions are in close proximity. The remaining four subunits of the myosin complex are smaller in size and are known as the light chains. There are two types of light chain, the *essential light chain* and the *regulatory light chain*. One light chain of each type associates with the neck region of each heavy chain (Figure 17-22b, *top*). The myosin heavy chain and the two types of light chains are encoded by three different genes.

The soluble myosin II molecule has ATPase activity, reflecting its ability to power movements by hydrolysis of ATP. But which part of the myosin complex is responsible for this activity? To identify functional domains in a protein, a standard approach is to cleave the protein into fragments with specific proteases and then ask which fragments have the activity. Soluble myosin II can be cleaved by gentle treatment with the protease chymotrypsin to yield two fragments, one called heavy meromyosin (HMM; *mero* means "part of") and the other, light meromyosin (LMM) (Figure 17-22b, *middle*). The heavy meromyosin can be further cleaved by the protease papain to yield subfragment 1 (S1) and subfragment 2 (S2) (Figure 17-22b, *bottom*). By analyzing the properties of the various fragments—S1, S2, and LMM—it was found that the intrinsic ATPase activity of myosin resides in the S1 fragment, as does its F-actin–binding site. Moreover, it was found that the ATPase activity of the S1 fragment was greatly enhanced by the presence of filamentous actin, so it is said to have an *actin-activated ATPase activity,* which is a hallmark of all myosins. The S1 fragment of myosin II consists of the head and neck domains with associated light chains, whereas the S2 and LMM regions make up the tail domain.

X-ray crystallographic analysis of the head and neck domains revealed its shape, the positions of the light chains, and the locations of the ATP-binding and actin-binding sites (Figure 17-22c). At the base of the myosin head is the α-helical neck, where two light-chain molecules wrap around the neck like C-clamps. In this position, the light chains stiffen the neck region. The actin-binding site is an exposed region at the tip of the head domain; the ATP-binding site is also in the head domain, within a cleft opposite the actin-binding site.

How much of myosin II is necessary and sufficient for "motor" activity? To answer this question, one needs a simple in vitro motility assay. In one such assay, the *sliding-filament assay,* myosin molecules are tethered to a coverslip to which is added stabilized, fluorescently labeled actin filaments. Because the myosin molecules are tethered, they cannot slide; thus any force generated by interaction of myosin heads with actin filaments forces the filaments to move relative to the myosin (Figure 17-23a). If ATP is present, added actin filaments can be seen to glide along the surface of the coverslip; if ATP is absent, no filament movement is observed. Using this assay, one can show that the S1 head of myosin II is sufficient to bring about movement of actin filaments. This movement is caused by the tethered myosin S1 fragments (bound to the coverslip) trying to "move" toward the (+) end of a filament; thus the filaments move with the (−) end leading. The rate at which myosin moves an actin filament can be determined from video recordings of sliding-filament assays (Figure 17-23b).

**⊙ TECHNIQUE ANIMATION:** In Vitro Motility Myosin Assay

**EXPERIMENTAL FIGURE 17-23 Sliding-filament assay is used to detect myosin-powered movement.** (a) After myosin molecules are adsorbed onto the surface of a glass coverslip, excess unbound myosin is removed; the coverslip then is placed myosin-side down on a glass slide to form a chamber through which solutions can flow. A solution of actin filaments, made visible and stable by staining with rhodamine-labeled phalloidin, is allowed to flow into the chamber. In the presence of ATP, the myosin heads walk toward the (+) end of filaments by the mechanism illustrated in Figure 17-26. Because myosin tails are immobilized, walking of the heads toward the (+) ends causes sliding of the filaments, which appear to be moving with their (−) ends leading the way. Movement of individual filaments can be observed in a fluorescence light microscope. (b) These photographs show the positions of three actin filaments (numbered 1, 2, 3) at 30-second intervals recorded by video microscopy. The rate of filament movement can be determined from such recordings. [Part (b) courtesy of M. Footer and S. Kron.]

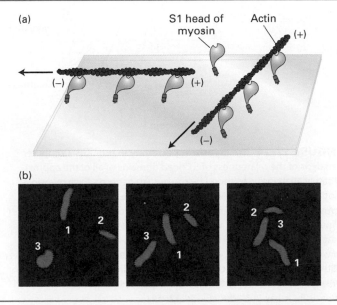

All myosins have a domain related to the S1 domain of myosin II, comprising the head and neck domains, which is responsible for their motor activity. However, as we will see in a later section, the length of the neck domain and the number and type of light chains associated with it varies in different myosin classes. The tail domain does not contribute to motility but rather defines what is moved by the S1-related domain. Thus, as might be expected, the tail domains can be very different and are tailored to bind specific cargoes.

## Myosins Make Up a Large Family of Mechanochemical Motor Proteins

Since all myosins have related S1-motor domains with considerable similarity in primary amino acid sequence, it is possible to determine how many myosin genes, and how many different classes of myosins, exist in a sequenced genome. There are about 40 myosin genes in the human genome (Figure 17-24), nine in *Drosophila,* and five in budding yeast. Computer analysis of the sequence relationships between the myosin head domains suggests that about 20 distinct classes

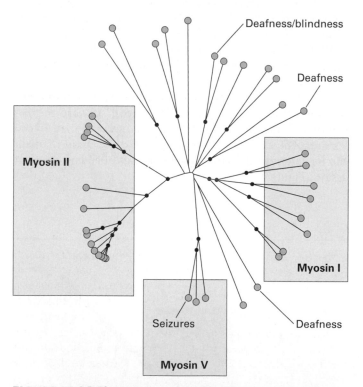

**FIGURE 17-24 The myosin superfamily in humans.** Computer analysis of the relatedness of S1 head domains of all of the approximately 40 myosins encoded by the human genome. Each myosin is indicated by a blue dot, with the length of the black lines indicating phylogenetic distance relationships. Thus myosins connected by short lines are closely related, whereas those separated by longer lines are more distantly related. Among these myosins are three classes—myosins I, II, and V—widely represented among eukaryotes, with others having more specialized functions. Indicated are examples in which loss of a specific myosin causes a disease. [Redrawn and modified from R. E. Cheney, 2001, *Mol. Biol. Cell* **12**:780.]

of myosins have evolved in eukaryotes, with greater sequence similarity within a class than between. As indicated in Figure 17-24, the genetic basis for some diseases has been traced to genes encoding myosins. All myosin head domains convert ATP hydrolysis into mechanical work using the same general mechanism. However, as we will see, subtle differences in this mechanism can have profound effects on the functional properties of different myosin classes. How do these different classes relate with respect to their tail domains? Amazingly, if one takes just the protein sequences of the tail domains of the myosins and uses this information to place them in classes, they fall into the same groupings as the motor domains. This implies that head domains with specific properties have co-evolved with specific classes of tail domains, which makes a lot of sense, suggesting that each class of myosin has evolved to carry out a specific function.

Among all these different classes of myosins are three especially well-studied ones, which are commonly found in animals and fungi: the so-called *myosin I, myosin II,* and *myosin V* families (Figure 17-25). In humans, eight genes encode heavy chains for the myosin I family, 14 for the myosin II family, and three for the myosin V family (see Figure 17-24).

The myosin II class assembles into bipolar filaments, with opposite orientations in each half of the bipolar filament so that there is a cluster of head domains at each end of the filament. This organization is important for its involvement in contraction; indeed, this is the only class of myosins involved in contractile functions. The large number of members in this class reflects the need for myosin II filaments with the slightly different contractile properties seen in different muscles (e.g., skeletal, cardiac, and various types of smooth muscle) as well as in nonmuscle cells.

The myosin II class is the only one that assembles into bipolar filaments. All myosin II members have a relatively short neck domain, with two light chains per heavy chain. The myosin I class is quite large, has a variable number of light chains associated with the neck region, and is the only one in which two heavy chains are not associated through their tail domains and so are single-headed. The large size and diversity of the myosin I class suggests that these myosins perform many functions, most of which remain to be determined, but some members of this family connect actin filaments to membranes, and others are implicated in endocytosis. Members of the myosin V class have two heavy chains, giving a motor with two heads, long neck regions with six light chains each, and tail regions that dimerize and terminate in domains that bind to specific organelles to be transported. As we will see shortly, the length of the neck region affects the rate of myosin movement.

In every case that has been tested so far, myosins move toward the (+) end of an actin filament—with one exception, the *myosin VI* found in animals. This remarkable myosin has an insert in its head domain to make it work in the opposite direction, and so motility is toward the (−) end of an actin filament. Myosin VI is believed to contribute to endocytosis by moving the endocytic vesicles along actin filaments away from the plasma membrane. Recall that membrane-associated

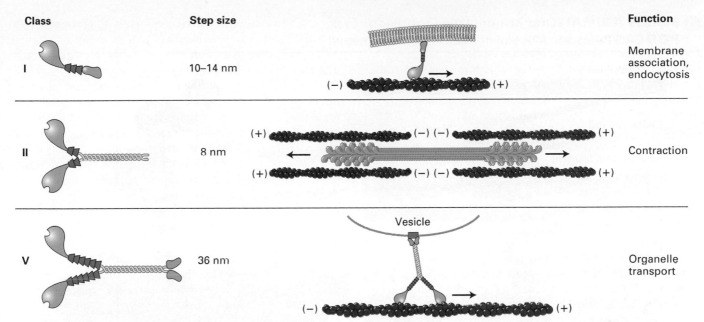

| Class | Step size | | Function |
|---|---|---|---|
| I | 10–14 nm | | Membrane association, endocytosis |
| II | 8 nm | | Contraction |
| V | 36 nm | | Organelle transport |

**FIGURE 17-25 Three common classes of myosin.** Myosin I consists of a head domain with a variable number of light chains associated with the neck domain. Members of the myosin I class are the only myosins to have a single head domain. Some of these myosins are believed to associate directly with membranes through lipid interactions. Myosin IIs have two head domains and two light chains per neck and are the only class that can assemble into bipolar filaments. Myosin Vs have two head domains and six light chains per neck. They bind specific receptors (brown box) on organelles, which they transport. All myosins in these three classes move toward the (+) end of actin filaments.

actin filaments have their (+) ends toward the membrane, so a motor directed toward the (−) end would take them away from the membrane toward the center of the cell.

## Conformational Changes in the Myosin Head Couple ATP Hydrolysis to Movement

Studies of muscle contraction provided the first evidence that myosin heads slide or walk along actin filaments. Unraveling the mechanism of muscle contraction was greatly aided by the development of in vitro motility assays and single-molecule force measurements. On the basis of information obtained with these techniques and the three-dimensional structure of the myosin head (see Figure 17-22c), researchers developed a general model for how myosin harnesses the energy released by ATP hydrolysis to move along an actin filament (Figure 17-26). Because all myosins are thought to use the same basic mechanism to generate movement, we will ignore whether the myosin tail is bound to a vesicle or is part of a thick filament, as it is in muscle. The most important aspect of this model is that the hydrolysis of a single ATP molecule is coupled to each step taken by a myosin molecule along an actin filament.

How can myosin convert the chemical energy released by ATP hydrolysis into mechanical work? This question has long intrigued biologists. It has been known for a long time that the S1 head of myosin is an ATPase, having the ability to hydrolyze ATP into ADP and $P_i$. Biochemical analysis revealed the mechanism of myosin movement (Figure 17-26a).

In the absence of ATP, the head of myosin binds very tightly to F-actin. When ATP binds, the affinity of the head for F-actin is greatly reduced and releases from actin. The myosin head then hydrolyzes the ATP, and the hydrolysis products, ADP and $P_i$, remain bound. The energy provided by the hydrolysis of ATP induces a conformational change in the head that results in the head domain rotating with respect to the neck. This is known as the "cocked" position of the head (Figure 17-26b, *top*). In the absence of F-actin, release of $P_i$ is exceptionally slow—the slowest part of the ATPase cycle. However, in the presence of actin, the head binds F-actin tightly, inducing both release of $P_i$ and rotation of the head back to its original position, thus moving the actin filament relative to the neck domain (Figure 17-26b, *bottom*). In this way, binding to F-actin induces the movement of the head and release of $P_i$, thereby coupling the two processes. This step is known as the *power stroke*. The head remains bound until the ADP leaves and a fresh ATP binds the head, releasing it from the filament. The cycle then repeats, and the myosin can move again against the filament.

How is hydrolysis of ATP in the nucleotide-binding pocket converted into force? The results of structural studies of myosin in the presence of nucleotides, and nucleotide analogs that mimic the various steps in the cycle, indicate that the binding and hydrolysis of a nucleotide cause a small conformational change in the head domain. This small movement is amplified by a "converter" region at the base of the head, acting like a fulcrum and causing the leverlike neck to rotate. This rotation is amplified by the rodlike lever arm,

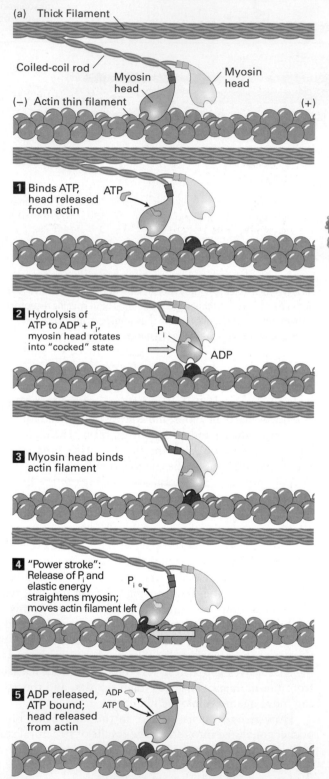

(a) Thick Filament

Coiled-coil rod

Myosin head

Myosin head

(−) Actin thin filament (+)

**1** Binds ATP, head released from actin

ATP

**2** Hydrolysis of ATP to ADP + P$_i$, myosin head rotates into "cocked" state

P$_i$

ADP

**3** Myosin head binds actin filament

**4** "Power stroke": Release of P$_i$ and elastic energy straightens myosin; moves actin filament left

P$_i$

**5** ADP released, ATP bound; head released from actin

ADP
ATP

(b)

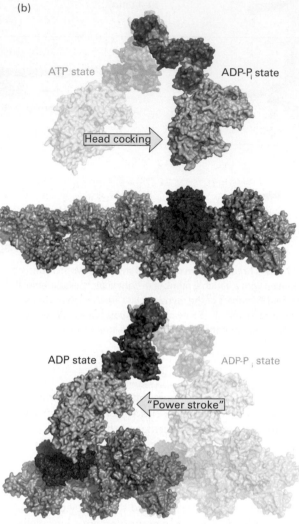

ATP state

ADP-P$_i$ state

Head cocking

ADP state

ADP-P$_i$ state

"Power stroke"

**FIGURE 17-26 ATP-driven myosin movement along actin filaments.** (a) In the absence of ATP, the myosin head is firmly attached to the actin filament. Although this state is very short-lived in living muscle, it is the state responsible for muscle stiffness in death (rigor mortis). Step **1**: On binding ATP, the myosin head releases from the actin filament. Step **2**: The head hydrolyzes the ATP to ADP and P$_i$, which induces a rotation in the head with respect to the neck. This "cocked state" stores the energy released by ATP hydrolysis as elastic energy, like a stretched spring. Step **3**: Myosin In the "cocked" state binds actin. Step **4**: When it is bound to actin, the myosin head couples release of P$_i$ with release of the elastic energy to move the actin filament. This is known as the "power stroke," as it involves moving the actin filament with respect to the end of the myosin neck domain. Step **5**: The head remains tightly bound to the filament as ADP is released and before fresh ATP is bound by the head. (b) Molecular models of the conformational changes in the myosin head involved in "cocking" the head (*upper panel*) and during the power stroke (*lower panel*). The myosin light chains are shown in dark blue and green; the rest of the myosin head and neck are colored in light blue, and actin is red [Part (a) adapted from R. D. Vale and R. A. Milligan, 2002, *Science* **288**:88; part (b) courtesy of Mike Geeves.]

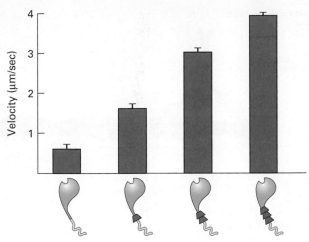

EXPERIMENTAL FIGURE 17-27 **The length of the myosin II neck domain determines the rate of movement.** To test the lever-arm model of myosin movement, investigators used recombinant DNA techniques to make myosin heads attached to different-length neck domains. The rate at which they moved on actin filaments was determined. The longer the lever arm, the faster the myosin moved, supporting the proposed mechanism. [Redrawn from K. A. Ruppel and J. A. Spudich, 1996, *Annu. Rev. Cell Mol. Biol.* **12**:543–573.]

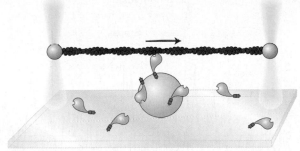

FIGURE 17-28 **Optical trapping of actin.** Optical trap techniques can be used to determine the step size and force generated by a single myosin molecule. In an optical trap, the beam of an infrared laser is focused by a light microscope on a latex bead (or any other object that does not absorb infrared light), which captures and holds the bead in the center of the beam. The strength of the force holding the bead is adjusted by increasing or decreasing the intensity of the laser beam. In this experiment, an actin filament is held between two optical traps. The actin filament is then lowered onto a third bead coated with a dilute concentration of myosin molecules. If the actin filament encounters a myosin molecule in the presence of ATP, the myosin will pull on the actin filament, which allows the investigators to measure both the force generated and the step size the myosin takes.

which constitutes the neck domain, so the actin filament moves by a few nanometers (see Figure 17-26b).

This model makes a strong prediction: the distance a myosin head moves along actin during hydrolysis of one ATP—the myosin *step size*—should be proportional to the length of the neck domain. To test this, mutant myosin molecules were constructed with different-length neck domains and the rate at which they moved down an actin filament was determined. Remarkably, there is an excellent correspondence between the length of the neck domain and the rate of movement (Figure 17-27).

## Myosin Heads Take Discrete Steps Along Actin Filaments

The most critical feature of myosin is its ability to generate a force that powers movements. Researchers have used *optical traps* to measure the forces generated by single myosin molecules (Figure 17-28). In this approach, myosin is immobilized on beads at a low density. An actin filament, held between two optical traps, is lowered toward the bead until it contacts a myosin molecule on the bead. When ATP is added, the myosin pulls on the actin filament. Using a mechanical feedback mechanism controlled by a computer, one can measure the distance pulled and the forces and duration of the movement. The results of optical trap studies show that myosin II does not interact with the actin filament continuously but rather binds, moves, and releases it. In fact, myosin II spends on average only about 10 percent of each ATPase cycle in contact with F-actin—it is said to have a *duty ratio* of 10 percent. This

will be important later when we consider that in contracting muscle, hundreds of myosin heads pull on actin filaments, so that at any one time, 10 percent of the heads are engaged to provide a smooth contraction.

When myosin II does contact F-actin, it takes discrete steps, which average out to about 8 nm (Figure 17-29, *top*), and generates 3–5 piconewtons (pN) of force, approximately the same force as that exerted by gravity on a single bacterium.

If we now look at a similar optical trap experiment with myosin V, the curves look completely different (Figure 17-29, *bottom*). Now we can easily discern clear steps of about 36 nm in length. This larger step size reflects the longer neck domain—the lever arm—of myosin V. Moreover, we see that the motor takes many sequential steps without releasing from the actin—it is said to move *processively*. This is because its ATPase cycle is modified to have a much higher duty ratio (>70 percent) by slowing the rate of ADP release; thus the head remains in contact with the actin filament for a much larger percentage of the cycle. Since a single myosin V molecule has two heads, a duty ratio of >50 percent ensures that one head is in contact at all times as it moves down an actin filament, so that it does not fall off.

## Myosin V Walks Hand over Hand down an Actin Filament

The next question is, how do the two heads of myosin V work together to move down a filament? One model proposes that the two heads walk down a filament hand over hand with alternately leading heads (Figure 17-30a). An

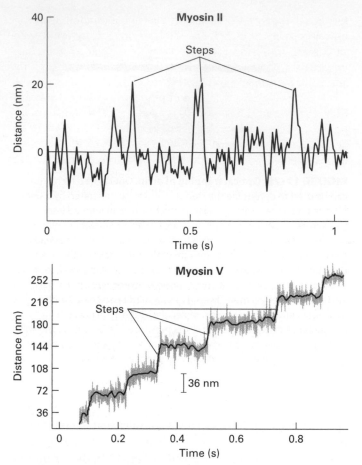

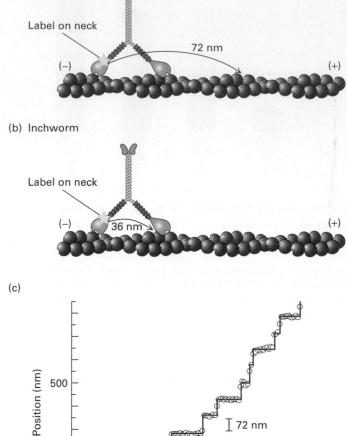

(a) Hand over hand

Label on neck

72 nm

(−)                                          (+)

(b) Inchworm

Label on neck

36 nm

(−)                                          (+)

(c)

72 nm

**EXPERIMENTAL FIGURE 17-29 Optical trap experiments measure the step size and processivity of myosins.** Using an optical trap setup similar to the one described in Figure 17-28, investigators have analyzed the behavior of myosin II (*top trace*) and myosin V (*bottom trace*). As shown by the peaks in the trace, myosin II takes erratic small steps (5–15 nm), which means it binds the actin filament, moves, and then lets go. It is therefore a nonprocessive motor. By contrast, single-headed myosin V takes clear 36-nm steps one after the other, so it has a step size of 36 nm and is highly processive—that is, it does not let go of the actin filament. [Part (a) from Finer et al., 1994, *Nature* **368**:113; part (b) from M. Rief et al., 2000, *Proc. Nat'l Acad. Sci. USA* **97**:9482.]

**EXPERIMENTAL FIGURE 17-30 Myosin V has a step size of 36 nm, yet each head moves in 72-nm steps, so it moves hand over hand.** Two models for myosin V movement down a filament have been suggested. (a) In the hand-over-hand model, one head binds an actin filament, and the other then swings around and binds a site 72 nm ahead. (b) In the inchworm model, the leading head moves 36 nm, then the lagging head moves up behind it, allowing the leading head to take another 36-nm step. (c) Two-headed myosin V labeled with a fluorescent tag on just one head appears to have a step size of 72 nm. Thus myosin V walks hand over hand. [Adapted from A. Yildiz et al., 2003, *Science* **300**:2061.]

alternative possibility is an inchworm model, in which the leading head takes a step, the second head is pulled up behind it, and then the leading head takes another step (Figure 17-30b). How can one distinguish between these models? In the inchworm model, each individual head takes 36-nm steps, whereas in the walking model, each takes 72-nm steps. Scientists have managed to attach a fluorescent probe to just one neck region of myosin V and watch it walk down an actin filament: it takes 72-nm steps (Figure 17-30c), and so it walks hand over hand down a filament. Why is the step size of myosin V so large? If we compare its step size of 36 nm to the structure of the actin filament, we see that it is the same as the length between helical repeats in the actin filament (see Figures 17-5b and 17-30a), so myosin V steps between equiva-

lent binding sites as it walks down one side of an actin filament. Myosin V has presumably evolved to take large steps the size of the helical repeat of actin and to do this very processively so that it rarely dissociates from an actin filament. These are exactly the properties one would expect for a motor designed to transport cargo along an actin filament.

## 17.6 Myosin-Powered Movements

We have already discussed how myosins have head and neck domains responsible for their motor properties. We now come to the tail regions, which define the cargoes that myosins move. The function of many of the newly discovered classes of myosins found in metazoans is not yet known. In this section, we give just two examples where we have a good idea of specific myosin functions. Our first example is skeletal muscle, which is where myosin II was discovered. In muscle, many myosin II heads bundled into bipolar filaments, each with a short duty cycle, work together to bring about contraction. Similarly organized contractile machineries function in the contraction of smooth muscle and in stress fibers, as well as in the contractile ring during cytokinesis. We then turn to the myosin V class, which has a long duty cycle that allows these myosins to transport cargoes over relatively long distances without dissociating from actin filaments.

### Myosin Thick Filaments and Actin Thin Filaments in Skeletal Muscle Slide Past One Another During Contraction

Muscle cells have evolved to carry out one highly specialized function: contraction. Muscle contractions must occur quickly and repetitively, and they must occur through long distances and with enough force to move large loads. A typical skeletal muscle cell is cylindrical, large (1–40 μm in length and 10–50 μm in width), and multinucleated (containing as many as 100 nuclei) (Figure 17-31a). Within each muscle cell are many **myofibrils** consisting of a regular repeating array of a specialized structure called a **sarcomere** (Figure 17-31b). A sarcomere, which is about 2 μm long in resting muscle, shortens by about 70 percent of its length during contraction. Electron microscopy and biochemical analysis have shown that each sarcomere contains two major types of filaments: *thick filaments,* composed of myosin II, and *thin filaments,* containing actin and associated proteins (Figure 17-31c).

The thick filaments are composed of myosin II bipolar filaments, in which the heads on each half of the filament have opposite orientations (see Figure 17-22a). The thin actin filaments are assembled with their (+) ends embedded in a densely staining structure known as the Z *disk,* so that the two sets of actin filaments in a sarcomere have opposite orientations (Figure 17-32). To understand how a muscle contracts, consider the interactions between one myosin head (among the hundreds in a thick filament) and a thin (actin) filament, as diagrammed in Figure 17-26. During these cyclical interactions, also called the *cross-bridge cycle,* the hydrolysis of ATP is coupled to the movement of a myosin head toward the Z disk, which corresponds to the (+) end of the actin thin filament. Because the thick filament is bipolar, the action of the myosin heads at opposite ends of the thick filament draws the thin filaments toward the center of the thick filament and therefore toward the center of the sarcomere (see Figure 17-32). This movement shortens the sarcomere until the ends of the thick filaments abut the Z disk. Contraction of an intact muscle results from the activity of hundreds of myosin heads on a single thick filament, amplified by the hundreds of thick and thin filaments in a sarcomere and thousands of sarcomeres in a muscle fiber. We can now see why myosin II is both nonprocessive and needs to have a short duty cycle: each head pulls a short distance on the actin filament and then lets go to allow other heads to pull, and so many heads working together allow the smooth contraction of the sarcomere.

The heart is an amazing contractile organ—it contracts without interruption about 3 million times a year, or a fifth of a billion times in a lifetime. The muscle cells of the heart contain contractile machinery very similar to that of skeletal muscle except that they are mono- and bi-nucleated cells. In each cell, the end sarcomeres insert into structures at the plasma membrane called intercalated disks, which link the cells into a contractile chain. Since heart muscle cells are only generated early in human life, they cannot be replaced in response to damage, such as occurs during a heart attack. Many different mutations in proteins of the heart contractile machinery give rise to *hypertrophic cardiomyopathies*—thickening of the heart wall muscle, which compromises its function. For example, many mutations have been documented in the cardiac myosin heavy-chain gene that compromise the protein's contractile function even in heterozygous individuals. In such individuals, the heart tries to compensate by hypertrophy (enlargement), often resulting in fatal heart arrhythmia (irregular beating). In addition to myosin heavy-chain defects, defects that result in cardiomyopathies have been traced to mutations in other components of the contractile machinery, including actin, myosin light chains, tropomyosin and troponin, and structural components such as titin (discussed below). ■

## FIGURE 17-31 Structure of the skeletal muscle sarcomere.

(a) Skeletal muscles consist of muscle fibers made of bundles of multinucleated cells. Each cell contains a bundle of myofibrils, which consist of thousands of repeating contractile structures called sarcomeres. (b) Electron micrograph of mouse striated muscle in longitudinal section, showing one sarcomere. On either side of the Z disks are the lightly stained I bands, composed entirely of actin thin filaments. These thin filaments extend from both sides of the Z disk to interdigitate with the dark-stained myosin thick filaments in the A band. (c) Diagram of the arrangement of myosin and actin filaments in a sarcomere. [Part (b) courtesy of S. P. Dadoune.]

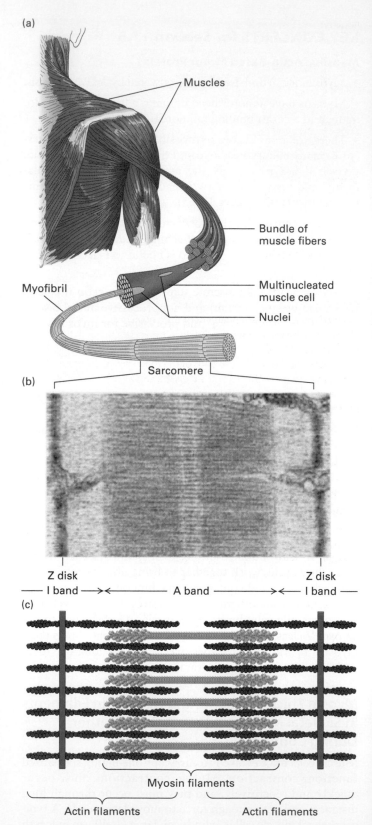

## Skeletal Muscle Is Structured by Stabilizing and Scaffolding Proteins

The structure of the sarcomere is maintained by a number of accessory proteins (Figure 17-33). The actin filaments are stabilized on their (+) ends by *CapZ* and on their (−) ends by *tropomodulin*. A giant protein known as *nebulin* extends along the thin actin filament all the way from the Z disk to tropomodulin, to which it binds. Nebulin consists of repeating domains that bind to the actin in the filament, and it is believed that the number of actin-binding repeats, and therefore the length of nebulin, determines the length of the thin filaments. Another giant protein, called *titin* (because it is so large), has its head associated with the Z disk and extends to the middle of the thick filament, where another titin molecule extends to the subsequent Z disk. Titin is believed to be an elastic molecule that holds the thick filaments in the middle of the sarcomere and also prevents overstretching to ensure that the thick filaments remain interdigitated between the thin filaments.

## Contraction of Skeletal Muscle Is Regulated by Ca$^{2+}$ and Actin-Binding Proteins

Like many cellular processes, skeletal muscle contraction is initiated by an increase in the cytosolic Ca$^{2+}$ concentration. As described in Chapter 11, the Ca$^{2+}$ concentration of the cytosol is normally kept low, below 0.1 μM. In skeletal muscle cells, a low cytosolic Ca$^{2+}$ level is maintained primarily by a unique Ca$^{2+}$ ATPase that continually pumps Ca$^{2+}$ ions from the cytosol containing the myofibrils into the **sarcoplasmic reticulum (SR)**, a specialized endoplasmic reticulum of the muscle cells (Figure 17-34). This activity establishes a reservoir of Ca$^{2+}$ in the SR.

The arrival of a nerve impulse (or *action potential*; see Chapter 22) at a neuromuscular junction triggers an action potential in the muscle-cell plasma membrane (also known as the *sarcolemma*). The action potential travels down invaginations of the plasma membrane known as *transverse tubules,* which penetrate the cell to lie around each myofibril. The arrival of the action potential in the transverse tubules stimulates the opening of voltage-gated Ca$^{2+}$ channels in the SR membrane, and the ensuing release of Ca$^{2+}$ from the SR raises the cytosolic Ca$^{2+}$ concentration in the myofi-

brils. This elevated Ca$^{2+}$ concentration induces a change in two accessory proteins, tropomyosin and troponin, which are bound to the actin thin filaments and normally block myosin binding. The change in position of these proteins on the actin thin filaments in turn permits the myosin-actin

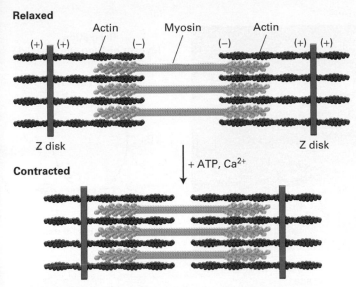

**FIGURE 17-32 The sliding-filament model of contraction in striated muscle.** The arrangement of thick myosin and thin actin filaments in the relaxed state is shown in the top diagram. In the presence of ATP and $Ca^{2+}$, the myosin heads extending from the thick filaments walk toward the (+) ends of the thin filaments. Because the thin filaments are anchored at the Z disks (purple), movement of myosin pulls the actin filaments toward the center of the sarcomere, shortening its length in the contracted state, as shown in the bottom diagram.

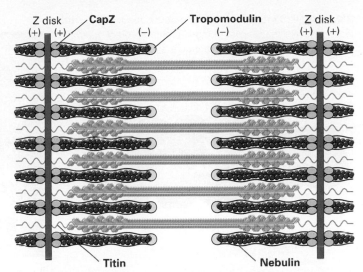

**FIGURE 17-33 Accessory proteins found in skeletal muscle.** To stabilize the actin filaments, CapZ caps the (+) end of the thin filaments at the Z disk, whereas tropomodulin caps the (−) end. The giant protein titin extends through the thick filaments and attaches to the Z disk. Nebulin binds actin subunits and determines the length of the thin filament.

interactions and hence contraction. This type of regulation is very rapid and is known as *thin-filament regulation*.

*Tropomyosin* (TM) is a ropelike molecule, about 40 nm in length, that binds to seven actin subunits in an actin filament. TM molecules are strung together head to tail, forming a continuous chain along each side of the actin thin filament (Figure 17-35a, b). Associated with each tropomyosin is *troponin* (TN), a complex of three subunits, TN-T, TN-I, and TN-C. Troponin-C is the calcium-binding subunit of troponin.

TN-C controls the position of TM on the surface of an actin filament through the TN-I and TN-T subunits.

Under the control of $Ca^{2+}$ and TN, TM can occupy two positions on a thin filament—switching from a state of muscle relaxation to contraction. In the absence of $Ca^{2+}$ (the relaxed state), TM blocks myosin's interaction with F-actin and the muscle is relaxed. Binding of $Ca^{2+}$ ions to TN-C triggers movement of TM to a new site on the filament, thereby exposing the myosin-binding sites on actin (see Figure 17-35b). Thus, at $Ca^{2+}$ concentrations greater than 1 μM, the inhibition exerted by the TM-TN complex is relieved and contraction occurs. The $Ca^{2+}$-dependent cycling between

(a)

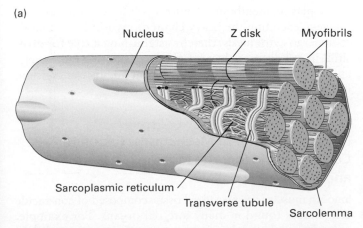

(b)

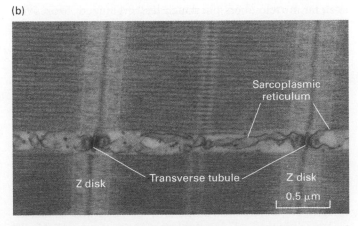

**FIGURE 17-34 The sarcoplasmic reticulum regulates the level of free $Ca^{2+}$ in myofibrils.** (a) When a nerve impulse stimulates a muscle cell, the action potential is transmitted down a transverse tubule (yellow), which is continuous with the plasma membrane (sarcolemma), leading to release of $Ca^{2+}$ from the adjacent sarcoplasmic reticulum (blue) into the myofibrils. (b) Thin-section electron micrograph of skeletal muscle, showing the intimate relationship of the sarcoplasmic reticulum to the muscle fibers. [Part (b) from K. R. Porter and C. Franzini-Armstrong, ASCB Image & Video Library, August 2006:FND-14. Available at: http://cellimages.ascb.org/u?/p4041coll1, 83.]

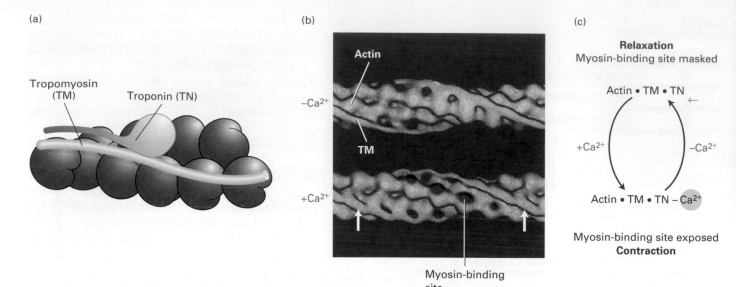

(a)

Tropomyosin (TM)

Troponin (TN)

(b)

Actin

−Ca²⁺

TM

+Ca²⁺

Myosin-binding site

(c)

**Relaxation**
Myosin-binding site masked

Actin • TM • TN

+Ca²⁺                    −Ca²⁺

Actin • TM • TN − Ca²⁺

Myosin-binding site exposed
**Contraction**

**FIGURE 17-35 Ca²⁺-dependent thin-filament regulation of skeletal muscle contraction.** (a) Model of the tropomyosin-troponin regulatory complex on a thin filament. Troponin is a protein complex that is bound to the long α-helical tropomyosin molecule. (b) Three-dimensional electron-microscopic reconstructions of the tropomyosin helix (yellow) on a muscle thin filament. Tropomyosin in the relaxed state (*top*) shifts to a new position (arrow) in the state inducing contraction (*bottom*) when the Ca²⁺ concentration increases. This movement exposes myosin-binding sites (red) on actin. (Troponin is not shown in this representation, but it remains bound to tropomyosin in both states.) (c) Summary of the regulation of skeletal muscle contraction by Ca²⁺ binding to troponin. [Part (b) adapted from W. Lehman, R. Craig, and P. Vibert, 1993, *Nature* **123**:313, courtesy of P. Vibert.]

relaxation and contraction states in skeletal muscle is summarized in Figure 17-35c.

## Actin and Myosin II Form Contractile Bundles in Nonmuscle Cells

In skeletal muscle, actin thin filaments and myosin II thick filaments assemble into contractile structures. Nonmuscle cells contain several types of related **contractile bundles** composed of actin and myosin II filaments, which are similar to skeletal muscle fibers but much less organized. Moreover, they lack the troponin regulatory system and are instead regulated by myosin phosphorylation, as we will discuss later.

In epithelial cells, contractile bundles are most commonly found as an *adherens belt,* also known as the circumferential belt, which encircles the inner surface of the cell at the level of the adherens junction (see Figure 17-4a) and are important in maintaining the integrity of the epithelium (discussed in Chapter 20). Stress fibers, which are seen along the lower surfaces of cells cultured on artificial (glass or plastic) surfaces or in extracellular matrices, are a second type of contractile bundle (see Figure 17-4a, c) important in cell adhesion, especially on deformable substrates. The ends of stress fibers terminate at integrin-containing focal adhesions, special structures that attach a cell to the underlying substratum (see Figure 17-41 and Chapter 20). Circumferential belts and stress fibers contain several proteins found in the contractile apparatus of smooth muscle and exhibit some organizational features resem-

bling those of muscle sarcomeres. A third type of contractile bundle, referred to as a contractile ring, is a transient structure that assembles at the equator of a dividing cell, encircling the cell midway between the poles of the mitotic spindle (Figure 17-36a). As the ring contracts, pulling the plasma membrane in, the cytoplasm is divided and eventually pinched into two parts in a process known as *cytokinesis,* giving rise to two daughter cells. Dividing cells stained with antibodies against myosin I and myosin II show that myosin II is localized to the contractile ring, whereas myosin I is at the distal regions, where it links the actin cortex to the plasma membrane (Figure 17-36b). Cells deleted for the gene encoding the heavy chain of myosin II are unable to undergo cytokinesis, thereby establishing a role for myosin II in cell division. Instead, these cells form a multinucleated syncytium because cytokinesis, but not nuclear division, is inhibited.

## Myosin-Dependent Mechanisms Regulate Contraction in Smooth Muscle and Nonmuscle Cells

Smooth muscle is a specialized tissue composed of contractile cells that is found in many internal organs. For example, smooth muscle surrounds blood vessels to regulate blood pressure, surrounds the intestine to move food through the gut, and restricts airway passages in the lung. Smooth muscle cells contain large, loosely aligned contractile bundles that

**EXPERIMENTAL FIGURE 17-36 Fluorescent antibodies reveal the localization of myosin I and myosin II during cytokinesis.**
(a) Diagram of a cell going through cytokinesis, showing the mitotic spindle (microtubules green, chromosomes blue) and the contractile ring with actin filaments (red). (b) Fluorescence micrograph of a *Dictyostelium* ameba during cytokinesis reveals that myosin II (orange) is concentrated in the contractile ring, also known as the cleavage furrow, whereas myosin I (green) is localized at the poles of the cell. The cell was stained with antibodies specific for myosin I and myosin II, with each antibody preparation linked to a different fluorescent dye. [Courtesy of Y. Fukui.]

resemble the contractile bundles in epithelial cells. The contractile apparatus of smooth muscle and its regulation constitute a valuable model for understanding how myosin activity is regulated in a nonmuscle cell. As we have just seen, skeletal muscle contraction is regulated by the tropomyosin-troponin complex bound to the actin thin filament switching between the contraction-inducing state in the presence of $Ca^{2+}$ and the relaxed state in its absence. In contrast, smooth muscle contraction is regulated by the cycling of myosin II between on and off states. Myosin II cycling, and thus contraction of smooth muscle and nonmuscle cells, is regulated in response to many extracellular signaling molecules.

Contraction of vertebrate smooth muscle is regulated primarily by a pathway in which the *myosin regulatory light chain (LC)* associated with the myosin II neck domain (see Figure 17-22b) undergoes phosphorylation and dephosphorylation. When the regulatory light chain is not phosphorylated, the smooth muscle myosin II adopts a folded conformation and its ATPase cycle is inactive. When the regulatory LC is phosphorylated by the enzyme *myosin LC kinase*, whose activity is regulated by the level of cytosolic free $Ca^{2+}$, the myosin II unfolds and assembles into active bipolar filaments and becomes active to induce contraction (Figure 17-37). The $Ca^{2+}$-dependent regulation of myosin LC kinase activity is mediated through the $Ca^{2+}$-binding protein calmodulin (see Figure 3-31). Calcium first binds to calmodulin, which induces a conformational change in the protein, and the $Ca^{2+}$/calmodulin complex then binds to myosin LC kinase and activates it. When the $Ca^{2+}$ returns to its resting level, myosin LC kinase becomes inactive and myosin light-chain phosphatase removes the phosphates to allow the system to return to its relaxed state. This mode of regulation relies on the diffusion of $Ca^{2+}$ over greater distances than in sarcomeres and on the action of protein kinases, so contraction is much slower in smooth muscle than in skeletal muscle. Because this regulation involves myosin, it is known as *thick-filament regulation*.

The role of activated myosin LC kinase can be demonstrated by microinjecting a kinase inhibitor into smooth muscle cells. Even though the inhibitor does not block the rise in the cytosolic $Ca^{2+}$ level that occurs following stimulation of the cell, injected cells cannot contract.

Unlike skeletal muscle, which is stimulated to contract solely by nerve impulses, smooth muscle cells and nonmuscle cells are regulated by many types of external signals. For example, norepinephrine, angiotensin, endothelin, histamine, and other signaling molecules can modulate or induce the contraction of smooth muscle, or elicit changes in the shape and adhesion of nonmuscle cells by triggering various signal-transduction pathways. Some of these pathways lead to an increase in the cytosolic $Ca^{2+}$ level; as previously described, this increase can stimulate myosin activity by activating myosin LC kinase (see Figure 17-37). As we will discuss below, other pathways activate *Rho kinase*, which is also able to activate myosin activity by phosphorylating the regulatory light chain, although in a $Ca^{2+}$-independent manner.

## Myosin-V-Bound Vesicles Are Carried Along Actin Filaments

In contrast to the contractile functions of myosin II filaments, the myosin V family of proteins are the most processive myosin motors known and transport cargo down actin filaments. In the next chapter we discuss how they can work together with microtubule motors to bring about transport of organelles. Although not a lot is known about their functions in mammalian cells, myosin V motors are not unimportant: defects in a

**FIGURE 17-37 Myosin phosphorylation mechanism for regulating smooth muscle contraction.** In vertebrate smooth muscle, phosphorylation of the myosin regulatory light chain (LC) activates contraction. At $Ca^{2+}$ concentrations $<10^{-6}$ M, the regulatory light chain is not phosphorylated, and the myosin adopts a folded conformation. When the $Ca^{2+}$ level rises, it binds calmodulin (CaM), which undergoes a conformational change (CaM*). The CaM*-$Ca^{2+}$ complex binds and activates myosin light-chain kinase (MLC kinase), which then phosphorylates the myosin LC. This phosphorylation event unfolds the myosin, which is now active and can assemble into bipolar filaments to participate in contraction. When the $Ca^{2+}$ levels drop, the myosin LC is dephosphorylated by myosin light-chain (MLC) phosphatase, which is not dependent on $Ca^{2+}$ for activity, causing muscle relaxation.

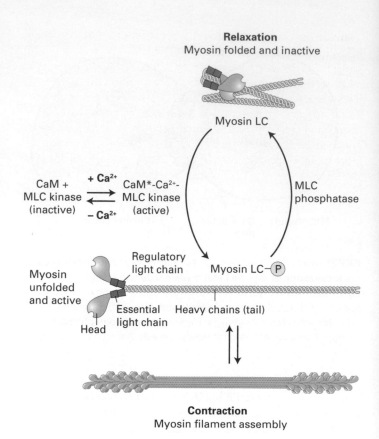

specific myosin V protein can cause severe diseases, such as seizures (see Figure 17-24).

Much more is known about myosin V motors in more experimentally accessible and simpler systems such as the budding yeast. This well-studied organism grows by budding, which requires its secretory machinery to target newly synthesized material to the growing bud (Figure 17-38a). Myosin V transports secretory vesicles along actin filaments at 3 μm/s into the bud. However, this is not the only function of myosin V proteins in yeast. At a later stage of the cell cycle, all the organelles have to be distributed between the mother and daughter cells. Remarkably, myosin Vs in yeast

---

🔁 **OVERVIEW ANIMATION:** Movement of Multiple Cargoes by Myosin V in Yeast

---

**FIGURE 17-38 Cargo movement by myosin Vs in budding yeast.** (a) The yeast *Saccharomyces cerevisiae* (used in making bread, beer, and wine) grows by budding. Secretory vesicles are transported into the bud, which swells to about the size of the mother cell. The cells then go through cytokinesis to form two daughter cells, and each divides again. (b) Diagram of a medium-sized bud showing how myosin Vs transport secretory vesicles (SV) down actin cables nucleated by formins (purple) located at the bud tip and bud neck. Myosins Vs are also used to segregate organelles, such as the vacuole (the yeast equivalent of a lysosome), peroxisomes, endoplasmic reticulum (ER), *trans*-Golgi network (TGN), and even selected mRNAs into the bud. Myosin V also binds the end of cytoplasmic microtubules (green) to orient the nucleus in preparation for mitosis. [Adapted from D. Pruyne et al., 2004, *Ann. Rev. Cell Biol.* **20**:559.]

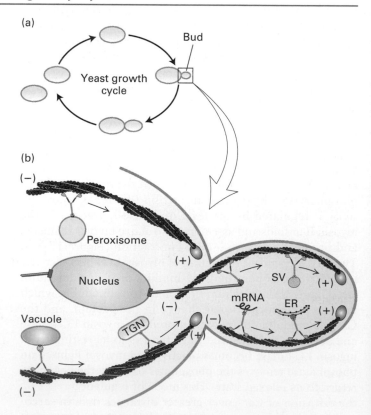

provide the transport system for segregation of many organelles, including peroxisomes, lysosomes (also known as vacuoles), endoplasmic reticulum, the *trans*-Golgi network, and even transport the ends of microtubules and some specific messenger RNAs into the bud (Figure 17-38b). Whereas budding yeast uses myosin V and polarized actin filaments in the transport of many organelles, animal cells, which are much larger, employ microtubules and their motors to transport many of these organelles over relatively long distances. We discuss these transport mechanisms in the next chapter.

Perhaps the most dramatic use of myosin Vs is seen in the giant green algae, such as *Nitella* and *Chara*. In these large cells, which can be 2 cm in length, cytosol flows rapidly, at a rate approaching 4.5 mm/min, in an endless loop around the inner circumference of the cell (Figure 17-39). This *cytoplasmic streaming* is a principal mechanism for distributing cellular metabolites, especially in large cells such as plant cells and amebas.

Close inspection of objects caught in the flowing cytosol, such as the endoplasmic reticulum (ER) and other membrane-bounded vesicles, shows that the velocity of streaming increases

(a)

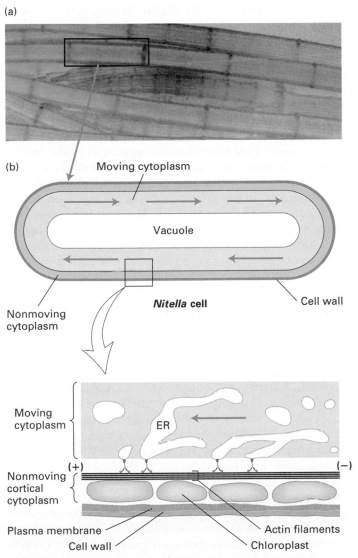

(c)

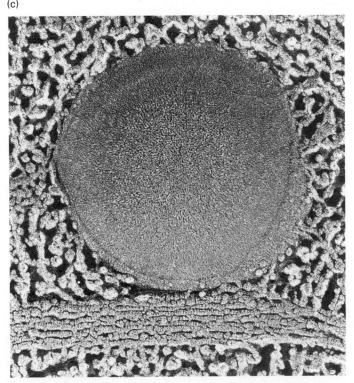

**FIGURE 17-39 Cytoplasmic streaming in cylindrical giant algae.**
(a) Cells of *Nitella*, a freshwater alga commonly found in ponds in the summer. The cytoplasmic movement, described below, is amazing and can readily be observed with a simple microscope, so go find some *Nitella* (or related alga) and watch this wonderful phenomenon! (b) The center of a *Nitella* cell is filled with a single large water-filled vacuole, which is surrounded by a layer of moving cytoplasm (blue arrows). A nonmoving layer of cortical cytoplasm filled with chloroplasts lies just under the plasma membrane (enlarged in bottom figure). On the inner side of this layer are bundles of stationary actin filaments (red), all oriented with the same polarity. A motor protein (blue), a plant myosin V, carries parts of the endoplasmic reticulum (ER) along the actin filaments. The movement of the ER network propels the entire viscous cytoplasm, including organelles that are enmeshed in the ER network. (c) Electron micrograph of the cortical cytoplasm showing a large vesicle connected to an underlying bundle of actin filaments. [Part (a) from James C. French; part (c) from B. Kachar.]

from the cell center (zero velocity) to the cell periphery. This gradient in the rate of flow is most easily explained if the motor generating the flow lies at the membrane. In electron micrographs, bundles of actin filaments can be seen aligned along the length of the cell, lying above stationary chloroplasts located adjacent to the membrane. The bulk cytosol is propelled by myosin V (also known as myosin XI in plants) attached to parts of the ER adjacent to the actin filaments. The flow rate of the cytosol in *Nitella* is about 15 times as fast as the movement produced by any other known myosin.

## KEY CONCEPTS of Section 17.6

### Myosin-Powered Movements

• In skeletal muscle, contractile myofibrils are composed of thousands of repeating units called sarcomeres. Each sarcomere consists of two interdigitating filament types: myosin thick filaments and actin thin filaments (see Figure 17-31).

• Skeletal muscle contraction involves the ATP-dependent sliding of myosin thick filaments along actin thin filaments to shorten the sarcomere and hence the myofibril (see Figure 17-32).

• The ends of the actin thin filaments in skeletal muscle are stabilized by CapZ at the (+) end and by tropomodulin at the (−) end. Two large proteins, nebulin associated with the thin filaments and titin with the thick filaments, also contribute to skeletal muscle organization.

• Skeletal muscle contraction is subject to thin-filament regulation. At low levels of free $Ca^{2+}$, the muscle is relaxed and tropomyosin blocks the interaction of myosin and F-actin. At elevated levels of free $Ca^{2+}$, the troponin complex associated with tropomyosin binds $Ca^{2+}$ and moves the tropomyosin to uncover the myosin-binding sites on actin, allowing contraction (see Figure 17-35).

• Smooth and nonmuscle cells have contractile bundles of actin and myosin filaments, with a similar organization as skeletal muscle but less well ordered.

• Contractile bundles are subject to thick-filament regulation. A myosin light chain is phosphorylated by myosin light-chain kinase, which activates myosin and hence induces contraction. The myosin light-chain kinase is activated by binding $Ca^{2+}$-calmodulin when the free $Ca^{2+}$ concentration rises (see Figure 17-37).

• Myosin V transports cargo by walking processively along actin filaments.

## 17.7 Cell Migration: Mechanism, Signaling, and Chemotaxis

We have now examined the different mechanisms used by cells to create movement—from the assembly of actin filaments and the formation of actin-filament bundles and networks to the contraction of bundles of actin and myosin and the transport of organelles by myosin molecules along actin filaments. Some of these same mechanisms constitute the major processes whereby cells generate the forces needed to migrate. *Cell migration* results from the coordination of motions generated in different parts of a cell, integrated with a directed endocytic cycle.

The study of cell migration is important to many fields of biology and medicine. For example, an essential feature of animal development is the migration of specific cells along predetermined paths. Epithelial cells in an adult animal migrate to heal a wound, and white blood cells migrate to sites of infection. Less obvious are the continual slow migration of intestinal epithelial cells along the villi in the intestine and the slow but constant migration of endothelial cells that line the blood vessels. The inappropriate migration of cancer cells after breaking away from their normal tissue results in metastasis.

Cell migration is initiated by the formation of a large, broad membrane protrusion at the leading edge of a cell. Video microscopy reveals that a major feature of this movement is the polymerization of actin at the membrane. Actin filaments at the leading edge are rapidly cross-linked into bundles and networks in a protruding region, called a *lamellipodium* in vertebrate cells. In some cases, slender, fingerlike membrane projections, called *filopodia*, also extend from the leading edge. These structures form stable contacts with the underlying surface (such as the extracellular matrix) that the cell moves across. In this section, we take a closer look at how cells coordinate various microfilament-based processes with endocytosis to move across a surface. We also consider the role of signaling pathways in coordinating and integrating the actions of the cytoskeleton, a major focus of current research.

### Cell Migration Coordinates Force Generation with Cell Adhesion and Membrane Recycling

A moving fibroblast (connective tissue cell) displays a characteristic sequence of events—initial extension of a membrane protrusion, attachment to the substratum, forward flow of cytosol, and retraction of the rear of the cell (Figure 17-40). These events occur in an ordered pattern in a slowly moving cell such as a fibroblast, but in rapidly moving cells, such as macrophages, all of them are occurring simultaneously in a coordinated manner. We first consider the role of the actin cytoskeleton, and then the involvement of the endocytic cycle.

**Membrane Extension** The network of actin filaments at the leading edge is a type of cellular engine that pushes the membrane forward in a manner very similar to the propulsion of *Listeria* by actin polymerization (Figure 17-41d; for *Listeria*, see Figure 17-17c). Thus at the membrane of the leading edge, actin is nucleated by the activated Arp2/3 complex and filaments are elongated by assembly onto (+) ends adjacent to the plasma membrane. As the actin meshwork is fixed with respect to the substratum, the front membrane is pushed out as the filaments elongate. This is very similar to *Listeria*, which "rides" on the polymerizing actin tail, which is also fixed within the cytoplasm. Actin turnover, and thus treadmilling, is mediated, as it is in the comet tails of *Listeria*, by the action of profilin and cofilin (see Figure 17-41d).

**FIGURE 17-40 Steps in cell movement.** Movement begins with the extension of one or more lamellipodia from the leading edge of the cell **1**; some lamellipodia adhere to the substratum by focal adhesions **2**. Then the bulk of the cytoplasm in the cell body flows forward due to contraction at the rear of the cell **3**. The trailing edge of the cell remains attached to the substratum until the tail eventually detaches and retracts into the cell body. During this cytoskeleton-based cycle, the endocytic cycle internalizes membrane and integrins at the rear of the cell and transports them to the front of the cell (arrows) for reuse in making new adhesions **4**.

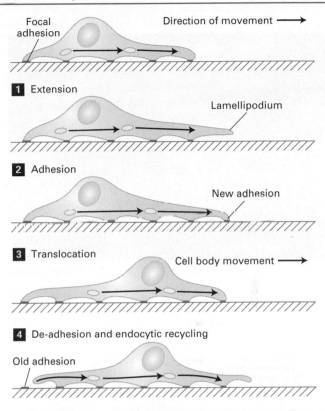

**Cell-Substrate Adhesions** When the membrane has been extended and the cytoskeleton has been assembled, the plasma membrane becomes firmly attached to the substratum. Time-lapse microscopy shows that actin bundles in the leading edge become anchored to structures known as *focal adhesions* (Figure 17-41c). The attachment serves two purposes: it prevents the leading lamella from retracting, and it attaches the cell to the substratum, allowing the cell to move forward. Given the importance of focal adhesions and their regulation during cell locomotion, it is not surprising that they have been found to be very rich in molecules involved in signal-transduction pathways. Focal adhesions are discussed in more detail in Chapter 20, when we discuss cell-matrix interactions.

The cell-adhesion molecules that mediate most cell-matrix interactions are membrane proteins called *integrins*. These proteins have an external domain that binds to specific components of the extracellular matrix, such as fibronectin and collagen, and a cytoplasmic domain that links them to the actin cytoskeleton (see Figure 17-41c and Chapter 20). The cell makes attachments at the front, and as the cell migrates forward, the adhesions eventually assume positions toward the back.

**Cell-Body Translocation** After the forward attachments have been made, the bulk contents of the cell body are translocated forward (see Figure 17-40, step **3**). It is believed that the nucleus and the other organelles embedded in the cytoskeleton are moved forward by myosin II–dependent cortical contraction in the rear part of the cell, like squeezing the lower half of a tube of toothpaste. Consistent with this model, myosin II is localized to the rear cell cortex.

**Breaking Cell Attachments** Finally, in the last step of movement (de-adhesion), the focal adhesions at the rear of the cell are broken, the integrins recycled, and the freed tail brought forward. In the light microscope, the tail is seen to "snap" loose from its connections—perhaps by the contraction of stress fibers in the tail or by elastic tension—and it sometimes leaves a little bit of its membrane behind, still firmly attached to the substratum.

The ability of a cell to move corresponds to a balance between the mechanical forces generated by the cytoskeleton and the resisting forces generated by cell adhesions. Cells cannot move if they are either too strongly attached or not attached to a surface. This relationship can be demonstrated by measuring the rate of movement in cells that express varying levels of integrins. Such measurements show that the fastest migration occurs at an intermediate level of adhesion, with the rate of movement falling off at high and low levels of adhesion. Cell locomotion thus results from traction forces exerted by the cell on the underlying substratum.

**Recycling Membrane and Integrins by Endocytosis** The dynamic changes in the actin cytoskeleton alone are not sufficient to drive cell migration; it is also dependent on endocytic recycling of membranes. The membrane needed during lamellipodium extension is provided from internal endosomes following their exocytosis. Adhesion molecules in focal adhesions at the rear of the cell are internalized from disassembling focal contacts and transported by an endocytic cycle to the front to make new substrate attachments (Figure 17-40, step **4**). This cycle of adhesion molecules in a migrating cell resembles the way a tank uses its treads to move forward. This movement of membrane internally through the cell also generates a rearwards membrane flow across the surface of the cell. Indeed, this type of flow may contribute to cell locomotion, as it has recently been found that white blood cells can move in a liquid ("swim") in the absence of attachment to a substrate.

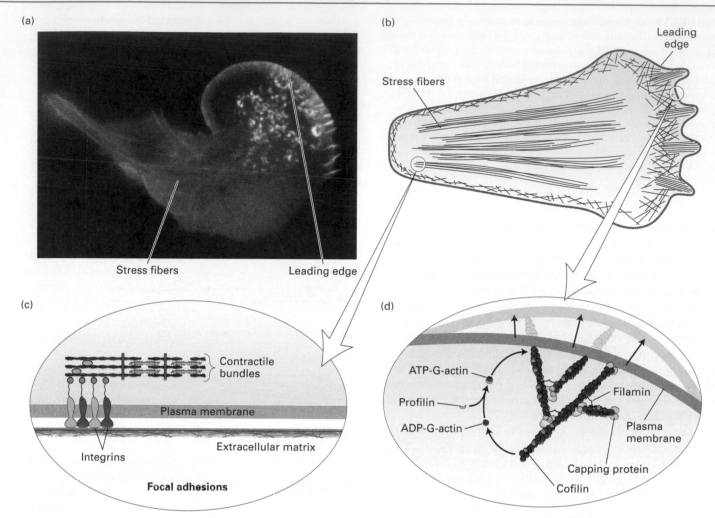

(a)

(b)

Leading edge

Stress fibers

Stress fibers

Leading edge

(c)

(d)

Contractile bundles

Plasma membrane

Integrins

Extracellular matrix

**Focal adhesions**

ATP-G-actin

Profilin

ADP-G-actin

Filamin

Plasma membrane

Capping protein

Cofilin

**FIGURE 17-41 Actin-based structures involved in cell locomotion.** (a) Localization of actin in a fibroblast expressing GFP-actin. (b) Diagram of the classes of microfilaments involved in cell migration. The meshwork of actin filaments in the leading edge advances the cell forward. Contractile fibers in the cell cortex squeeze the cell body forward, and stress fibers terminating in focal adhesions also pull the bulk of the cell body up as the rear adhesions are released.

(c) The structure of focal adhesions involves the attachment of the ends of stress fibers through integrins to the underlying extracellular matrix. Focal adhesions also contain many signaling molecules important for cell locomotion. (d) The dynamic actin meshwork in the leading edge is nucleated by the Arp2/3 complex and employs the same set of factors that control assembly and disassembly of actin filaments in the *Listeria* tail (see Figure 17-17). [Part (a) Courtesy of J. Vic Small.]

## The Small GTP-Binding Proteins Cdc42, Rac, and Rho Control Actin Organization

A striking feature of a moving cell is its polarity: a cell has a front and a back. When a cell makes a turn, a new leading edge forms in the new direction. If these extensions formed in all directions at once, the cell would be unable to pick a new direction of movement. To sustain movement in a particular direction, a cell requires signals to coordinate events at the front of the cell with events at the back and, indeed, signals to tell the cell where its front is. Understanding how this coordination occurs emerged from studies with growth factors.

Growth factors, such as epidermal growth factor (EGF) and platelet-derived growth factor (PDGF), bind to specific cell-surface receptors (Chapter 16) and stimulate cells to move and then to divide. For example, in a wound, blood platelets become activated by being exposed to collagen in the extracellular matrix at the wound edge, which helps the blood to clot. Activated platelets also secrete PDGF to attract fibroblasts and epithelial cells to enter the wound and repair it. It is possible to watch part of this process in vitro. If you grow cells in a culture dish and, after starving them of growth factors, you add some fresh growth factor, within a minute or two the cells respond by forming membrane ruffles. Membrane ruffles are very similar to the lamellipodia of migrating cells: they are a result of activation of the machinery that controls exocytosis of endosomes coupled with actin assembly.

Scientists knew that growth factors bind to very specific receptors on the cell surface and induce a signal-transduction pathway on the inner surface of the plasma membrane (Chapter 15), but how that linked up to the actin machinery was mysterious. Research then revealed that the signal-transduction pathway activates *Rac,* a member of the small GTPase superfamily of Ras-related proteins (Chapter 15). Rac is one member of a family of proteins that regulate microfilament organization; two others are *Cdc42* and *Rho.* Unfortunately, due to the history of their discovery, this family of proteins also has been collectively named "Rho proteins," of which Cdc42, Rac, and Rho are members. To understand how these proteins work, we first have to recall the way small GTP-binding proteins function.

Like all small GTPases of the Ras superfamily, Cdc42, Rac, and Rho act as molecular switches, inactive in the GDP-bound state and active in the GTP-bound state (Figure 17-42). In their GDP bound state, they exist free in the cytoplasm in an inactive form bound to a protein known as guanine nucleotide dissociation inhibitor (GDI). Growth factors can bind and activate their receptors to turn on specific membrane-bound regulatory proteins, guanine nucleotide exchange factors (GEFs), which activate Rho proteins at the membrane by releasing them from GDI and catalyzing the exchange of GDP for GTP. The GTP-bound active Rho protein associates with the plasma membrane, where it binds *effector proteins* to initiate the biological response. The small GTPase remains active until the GTP is hydrolyzed to GDP, which is stimulated by specific GTPase-activating proteins (GAPs). An important approach to unraveling the functions of Rho proteins has been to introduce into cells mutant proteins that are locked either in the active—Rho-GTP—state or in the inactive—Rho-GDP—state. A mutant small GTPase that is locked in the active state is said to be a *dominant-active* protein. Such a dominant-active protein binds the effector molecules constitutively, and one can then assess the biological outcomes. Alternatively, one can introduce a different mutant that is *dominant negative,* which binds and thereby inhibits the relevant GEF protein. Thus introduction of a dominant-negative protein interferes with the signal-transduction pathway, so one can now assess what processes are blocked.

Cdc42, Rac, and Rho were implicated in regulation of microfilament organizations because introduction of dominant-active mutants had dramatic effects on the actin cytoskeleton, even in the absence of growth factors. It was discovered that dominant-active Cdc42 resulted in the appearance of filopodia, dominant-active Rac resulted in the appearance of membrane ruffles, and dominant-active Rho resulted in the formation of stress fibers that then contracted (Figure 17-43). How can one tell if dominant-active Rac and growth-factor stimulation, both of which stimulate membrane ruffle formation, operate in the

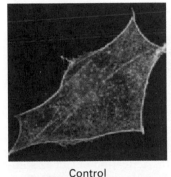

Control

Dominant active Rho

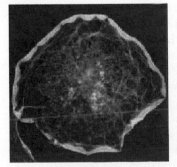
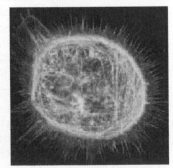

Dominant active Rac

Dominant active Cdc42

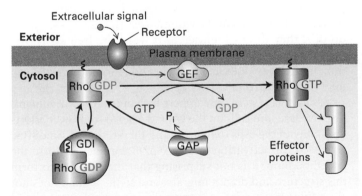

**FIGURE 17-42 Regulation of the Rho family of small GTPases.** The Rho family of small GTPases are molecular switches regulated by accessory proteins. Rho proteins exist in the Rho-GDP bound form complexed with a protein known as GDI (guanine nucleotide dissociation inhibitor), which retains them in an inactive state in the cytosol. Membrane-bound signaling pathways bring Rho proteins to the membrane and, through the action of a GEF (guanine nucleotide exchange factor), exchange the GDP for GTP, thus activating them. Membrane-bound activated Rho-GTP can then bind effector proteins that cause changes in the actin cytoskeleton. The Rho protein remains in the active Rho-GTP state until acted on by a GAP (GTPase-activating protein), which returns it to the cytoplasm. [Adapted from S. Etienne-Manneville and A. Hall, 2002, *Nature* **420:**629.]

**EXPERIMENTAL FIGURE 17-43 Dominant-active Rac, Rho, and Cdc42 induce different actin-containing structures.** To look at the effects of constitutively active Rac, Rho, and Cdc42, growth-factor-starved fibroblasts were microinjected with plasmids to express dominant-active versions of the three proteins. The cells were then treated with fluorescent phalloidin, which stains filamentous actin. Dominant-active Rac induces the formation of peripheral membrane ruffles, whereas dominant-active Rho induces abundant contractile stress fibers and dominant active Cdc42 induces filopodia. [From A. Hall, 1998, *Science* **279:**509.]

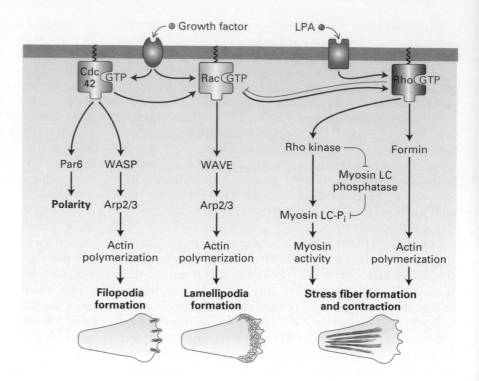

**FIGURE 17-44 Summary of signal-induced changes in the actin cytoskeleton.** Specific signals, such as growth factors and lysophosphatidic acid (LPA), are detected by cell-surface receptors. Detection leads to the activation of the small GTP-binding proteins, which then interact with effectors to bring about cytoskeletal changes as indicated.

same signal-transduction pathway? If growth-factor stimulation leads to Rac activation, introduction of a dominant-negative Rac protein into a cell should block the ability of a growth factor to induce membrane ruffling. This is precisely what is found. Using this and many other biochemical strategies, scientists have identified signaling pathways involving Cdc42, Rac, and Rho (Figure 17-44).

Some of the pathways that these proteins regulate contain proteins we are familiar with. Activation of Cdc42 stimulates actin assembly by Arp2/3 through activation of WASp, a nucleation-promoting-factor (NPF) protein (see Figure 17-16), resulting in the formation of filopodia. Activation of Rac also induces Arp2/3, mediated by the WAVE complex, leading to the assembly of branched actin filaments in the leading edge. Activation of Rho has at least two effects. First, it can activate a formin for unbranched actin filament assembly. Second, through activation of Rho kinase, it can phosphorylate the myosin light chain to activate nonmuscle myosin II and also inhibit light-chain dephosphorylation by phosphorylating myosin light-chain phosphatase to inhibit its activity. Both actions of Rho kinase lead to a higher level of myosin light-chain phosphorylation and therefore higher myosin activity and contraction. The three Rho proteins, Cdc42, Rac, and Rho, are also linked by activation and inhibition pathways, as shown in Figure 17-44.

## Cell Migration Involves the Coordinate Regulation of Cdc42, Rac, and Rho

How does each of these small GTP-binding proteins contribute to the regulation of cell migration? To answer this question, researchers developed an in vitro wound-healing assay (Figure 17-45a). Cells in culture are grown in a petri dish with growth factors, allowing them to grow until they are confluent and form a tight monolayer, at which point they stop dividing. The cell monolayer is then scratched with a needle to remove a swath of cells to generate a "wound" containing a free edge of cells. The cells on the edge sense the loss of their neighbors and, in response to components of the extracellular matrix now exposed on the dish surface, move to fill up the empty wound area (Figure 17-45b). To do this, they orient themselves toward the free area, first putting out a lamellipodium and then moving in that direction. In this way one can study the induction of directed cell migration in vitro.

Using this system, researchers have introduced dominant-negative Rac into cells on the wound edge to see how it affects the ability of the cells to migrate and fill the wound. Since Rac is needed for activation of the Arp2/3 complex to form the lamellipodium, it is not surprising that the cells fail to form this structure and do not migrate, and so the wound does not close (Figure 17-45c). A very interesting result is obtained when dominant-negative Cdc42 is introduced into the cells at the wound edge: they can form a leading edge but do not orient in the correct direction—in fact, they try to migrate in random directions. This suggests that Cdc42 is critical for regulating the overall polarity of the cell. Studies from yeast (where Cdc42 was first described), wounded-cell monolayers, epithelial cells, and neurons reveal that Cdc42 is a master regulator of polarity in many different systems. Part of this regulation in animals involves Cdc42 binding to its effector, Par6, a polarity protein that functions in nematodes (where it was first discovered), neurons, and epithelial cells. We explore these polarity pathways in more detail in Chapter 21.

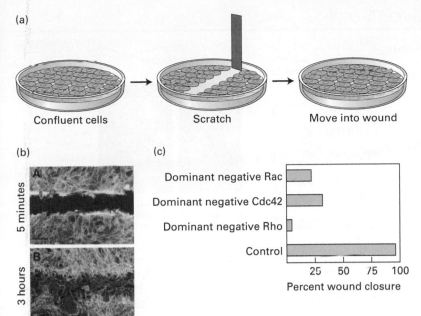

(a)

Confluent cells → Scratch → Move into wound

(b)

5 minutes — A

3 hours — B

(c)

Dominant negative Rac
Dominant negative Cdc42
Dominant negative Rho
Control

25    50    75    100
Percent wound closure

**EXPERIMENTAL FIGURE 17-45 The wounded-cell monolayer assay can be used to dissect signaling pathways in directed cell movement.** (a) A confluent layer of cells is scratched to remove a swath of about three cells wide to generate a free cell border. The cells detect the free space and newly exposed extracellular matrix and, over a period of hours, fill the area. (b) Localization of actin in a wounded monolayer 5 minutes and 3 hours after scratching; the cells have migrated into the wounded area. (c) Effect of introducing dominant-negative Cdc42, Rac, and Rho into cells at the wound edge; all affect wound closure.
[Parts (b) and (c) from C. D. Nobes and A. Hall, 1999, *J. Cell Biol.* **144:**1235.]

Studies such as these suggest a general model of how cell migration is controlled (Figure 17-46). Signals from the environment are transmitted to Cdc42, which orients the cell. The oriented cell has high Rac activity at the front, to induce the formation of the leading edge; Rho activity is high in the rear, to assemble contractile structures and activate the myosin-II-based contractile machinery. It is important to notice that different regions of the cell can have different levels of active Cdc42, Rac, or Rho, so these regulators are controlled locally within the cell. Part of this spatial regulation occurs because some small G proteins can work antagonistically. For example, active Rho can stimulate pathways that lead to the inactivation of Rac. This might help ensure that no leading-edge structures form at the rear of the cell.

## Migrating Cells Are Steered by Chemotactic Molecules

Under certain conditions, extracellular chemical cues guide the locomotion of a cell in a particular direction. In some cases, the movement is guided by insoluble molecules in the underlying substratum, as in the wound-healing assay described above. In other cases, the cell senses soluble molecules and follows them, along a concentration gradient, to their source—a process known as **chemotaxis**. For example, leukocytes (white blood cells) are guided toward an infection by a tripeptide secreted by many bacterial cells (Figure 17-47a). In another example, during the development of skeletal muscle, a secreted protein signal called *scatter factor* guides the migration of myoblasts to the proper locations in limb buds. One of the best-studied examples of chemotaxis is the migration of *Dictyostelium* amebas during their starvation response. When these soil amebas are stressed, they begin to secrete cAMP, which is an extracellular chemotactic agent in this organism. Other *Dictyostelium* cells move up the cAMP concentration gradient toward its source (see Figure 17-47a). Thus the amebas move toward each other, aggregate into a migratory slug, and then differentiate into a fruiting body in which starvation-resistant spores are formed.

Despite the variety of different chemotactic molecules—sugars, peptides, cell metabolites, cell-wall or membrane lipids—they all work through a common and familiar mechanism: binding to cell-surface receptors, activation of intracellular signaling pathways, and remodeling of the cytoskeleton through the activation or inhibition of various actin-binding

---

| Back: | | Front: | | Cdc42 |
| **Rho activation** Leading to myosin II activation | ← | **Rac activation** leading to Arp2/3 activation | ← | **activation at the front** |

Actin filament assembly and treadmilling in the leading edge

Contraction of myosin II filaments in both stress fibers and cell cortex

**FIGURE 17-46 Contribution of Cdc42, Rac, and Rho to cell movement.** The overall polarity of a migrating cell is controlled by Cdc42, which is activated at the front of a cell. Cdc42 activation leads to active Rac in the front of the cell, which generates the leading edge, and active Rho at the back of the cell, which causes myosin II activation and contraction. Active Rho inhibits Rac activation, ensuring the asymmetry of the two active G-proteins.

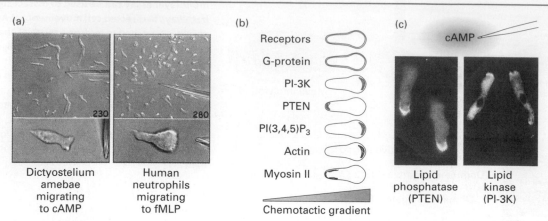

(a) Dictyostelium amebae migrating to cAMP | Human neutrophils migrating to fMLP

(b) Receptors / G-protein / PI-3K / PTEN / PI(3,4,5)P₃ / Actin / Myosin II — Chemotactic gradient

(c) cAMP — Lipid phosphatase (PTEN) | Lipid kinase (PI-3K)

**EXPERIMENTAL FIGURE 17-47 Chemotaxis involves elevated levels of signaling phosphoinositides, which signal to the actin cytoskeleton.** (a) *Dictyostelium* cells migrate toward a pipet of cAMP (*left*), and human neutrophils (a type of leukocyte) migrate toward a pipet of fMLP (formylated Met-Leu-Phe), a chemotactic peptide produced by bacteria (*right*). In the lower two panels are individual chemotaxing *Dictyostelium* and neutrophil cells that look remarkably similar, despite about 800 million years of evolution separating them. (b) Summary of results of studies exploring the localization of components of signaling pathways (green) in *Dictyostelium* cells undergoing chemotaxis toward cAMP. Also shown are the localization of actin and myosin (red). (c) The enzyme PI-3 kinase, which generates PI(3,4,5)P$_3$, is enriched at the front of chemotaxing cells, whereas PTEN, the phosphatase that hydrolyses PI(3,4,5)P$_3$, is enriched at the back. These distributions result in elevated PI(3,4,5)P$_3$ at the front of the cells, which signals the polarity for movement. [Part (a) from C. Parent, 2004, *Curr. Opin. Cell Biol.* **16**:4; part (c) from M. Iijima et al., 2002, *Dev. Cell* **3**:469.]

---

proteins. What is quite amazing is that just a 2 percent difference in the concentration of chemotactic molecules between the front and back of the cell is sufficient to induce directed cell migration. Equally amazing is the finding that the internal signal-transduction pathways used in chemotaxis have been conserved between *Dictyostelium* amebas and human leukocytes despite almost a billion years of evolution.

## Chemotactic Gradients Induce Altered Phosphoinositide Levels Between the Front and Back of a Cell

To investigate how *Dictyostelium* amebas sense a chemotactic gradient, investigators have studied the cell-surface receptors for extracellular cAMP and downstream signaling pathways in the expectation that these must somehow sense the concentration gradient. Before we discuss the details, let's consider how such a system might work. If a cell can sense a 2 percent difference in concentration across its length, it is unlikely that simply activating actin assembly 2 percent more at the front than at the back could lead to directed movement. Rather, there must be some mechanism that amplifies this small external signal difference into a large internal biochemical difference. One way to do this would be for the cell to subtract the average signal from the front and back and only respond to a *difference in signal*. It is believed that this is how the system works. To try to understand this mechanism, investigators have looked at the concentration of active components of the signaling pathway to see where the amplification occurs.

Micrographs of cAMP receptors tagged with green fluorescent protein (GFP) show that the receptors are distributed uniformly on the surface of an ameba cell (Figure 17-47b); therefore an internal gradient must be established by another component of the signaling pathway. Because cAMP receptors signal through trimeric G proteins (Chapter 16), a subunit of the trimeric G protein and other downstream signaling proteins were tagged with GFP to look at their distributions. Fluorescence micrographs show that the concentration of trimeric G proteins is also rather uniform. Downstream of the trimeric G proteins is PI-3 kinase, an enzyme that phosphorylates membrane-bound inositol phospholipids (phosphoinositides), such as PI4,5-biphosphate [PI(4,5)P$_2$], creating the signaling lipid PI3,4,5-triphosphate [PI(3,4,5)P$_3$] (see Figure 16-25). Remarkably, the enzyme PI-3 kinase is highly enriched at the front of a migrating cell, as are its products. PTEN, the phosphatase that dephosphorylates the signaling lipid PI(3,4,5)P$_3$ back to PI(4,5)P$_2$, is enriched in the tail of the migrating cell (Figure 17-47b, c). This asymmetry is believed to be established in the following way. Prior to the cell's exposure to a cAMP gradient, the phosphatase PTEN is associated uniformly with the plasma membrane. When a cell "sees" a gradient, PI 3-kinase is activated a bit more at the front than the back. This results in slightly higher levels of the signaling phospholipid at the front. The association of the phosphatase PTEN with the membrane is very sensitive to the level of PI(3,4,5)P$_3$, so it is preferentially depleted from the front. Since it is less effective at dephosphorylating the PI(3,4,5)P$_3$ at the front and more effective at dephosphorylating PI(3,4,5)P$_3$ at the rear, a strong asymmetry of PI(3,4,5)P$_3$ results. Thus the phosphatase PTEN contributes to the background subtraction necessary for a cell to sense a shallow gradient of chemoattractant.

The difference in local PI(3,4,5)P$_3$ concentration now signals to the actin cytoskeleton to assemble a leading edge at the front and contraction at the rear (Figure 17-47b), and the cell is on its way to the source of chemoattractant. A very similar mechanism has been implicated in the chemotaxis of leukocytes. This cell polarization is not stable in the absence of the chemotactic gradient, so if the gradient changes, as might happen with a leukocyte chasing a moving bacterium, the cell will also change its direction and follow the gradient to its source.

## KEY CONCEPTS of Section 17.7

### Cell Migration: Mechanism, Signaling, and Chemotaxis

- Cell migration involves the extension of an actin-rich leading edge at the front of the cell, the formation of adhesive contacts that move backward with respect to the cell, and their subsequent release, combined with rear contraction to push the cell forward (see Figure 17-40).

- Cell migration also involves a directed endocytic cycle, taking membrane and adhesion molecules from the rear of the cell and inserting them at the front.

- The assembly and function of actin filaments is controlled by signaling pathways through small GTP-binding proteins of the Rho family. Cdc42 regulates overall polarity and the formation of filopodia, Rac regulates actin meshwork formation through the Arp2/3 complex, and Rho regulates both actin filament formation by formins as well as contraction through regulation of myosin II (see Figure 17-44).

- Chemotaxis, the directed movement toward an attractant, involves signaling pathways that establish differences in phosphoinositides between the front and rear of the cell, which in turn regulate the actin cytoskeleton and direction of cell migration (see Figure 17-47).

## Perspectives for the Future

In this chapter, we have seen that cells have intricate mechanisms for the regulated spatial and temporal assembly and turnover of microfilaments to perform their many functions. Biochemical analyses of actin-binding proteins coupled with protein inventories provided by sequences of whole genomes have allowed the cataloging of many different classes of actin-binding proteins. To understand how this large group of proteins can assemble specific structures in a cell, it will be important to know the concentration of all the components, how they interact, and their regulation by signaling pathways. Although this might seem like a daunting task, new microscopic methods to detect the location of specific protein-protein interactions and the location of many of the key signaling pathways suggest that rapid progress will be made in this area.

The protein inventories provided by genomic sequences have also documented the large number of myosin families, yet the biochemical properties of many of these motors, or their biological functions, remain to be elucidated. Again, recent technical developments, including the ability to tag motors with fluorescent tracers such as GFP, or knocking down their expression with RNAi technologies are providing very powerful avenues to help reveal motor functions. However, some important aspects of motors remain largely unexplored. For example, a motor that transports an organelle down a filament first has to bind the organelle, then transport it, and then release it at the destination. However, little is known about how these different events are coordinated or how these types of myosin-based motors are returned to pick up new cargo.

Twenty years ago it was believed that all actin assembly was driven by activation of the Arp2/3 complex. Then the nucleating and capping activities of formins were discovered, and more recently, additional actin nucleators, with colorful names such as Spire, Cordon-bleu, WASH, and WHAMM have been discovered. It is likely that additional nucleators will continue to be discovered.

Finally, although we have generally discussed microfilaments without regard to tissue type—except for the specializations found in skeletal and smooth muscles—many actin-binding proteins show cell-type-specific expression, and so the array and relative levels of these proteins are tailored to specific functions of different cell types. This is clearly the case, as revealed by proteomic analysis of cell-specific protein expression and by the fact that many diseases are a consequence of tissue-specific expression of actin-binding proteins or myosins.

## Key Terms

| | |
|---|---|
| actin 775 | microtubules 775 |
| actin cross-linking proteins 790 | microvilli 773 |
| Arp2/3 complex 784 | motor protein 775 |
| CapZ protein 783 | myosin 793 |
| Cdc42 protein 811 | myosin light-chain kinase 805 |
| cell migration 808 | nucleation promoting factor (NPF) 785 |
| cell polarity 774 | |
| chemotaxis 813 | power stroke 797 |
| cofilin 782 | profilin 782 |
| contractile bundles 804 | Rac protein 811 |
| critical concentration, C$_c$ 779 | Rho protein 811 |
| | skeletal muscle 793 |
| cytoskeleton 774 | smooth muscle 804 |
| duty ratio 799 | step size 799 |
| F-actin 777 | stress fibers 776 |
| filopodia 776 | thick filaments 801 |
| formin 784 | thin filaments 801 |
| G-actin 777 | thymosin β$_4$ 783 |
| intermediate filaments 775 | treadmilling 781 |
| lamellipodium 776 | tropomodulin 783 |
| leading edge 776 | tropomyosin 803 |
| microfilaments 775 | WASp protein 785 |

## Review the Concepts

**1.** Three systems of cytoskeletal filaments exist in most eukaryotic cells. Compare them in terms of composition, function, and structure.

**2.** Actin filaments have a defined polarity. What is filament polarity? How is it generated at the subunit level? How is filament polarity detectable?

**3.** In cells, actin filaments form bundles and/or networks. How do cells form such structures, and what specifically determines whether actin filaments will form a bundle or a network?

**4.** Much of our understanding of actin assembly in the cell is derived from experiments using purified actin in vitro. What techniques can be used to study actin assembly in vitro? Explain how each of these techniques works. Which of these techniques would tell you whether the mass of actin filaments is comprised of many short actin filaments or fewer longer filaments?

**5.** The predominant forms of actin inside a cell are ATP–G-actin and ADP–F-actin. Explain how the interconversion of the nucleotide state is coupled to the assembly and disassembly of actin subunits. What would be the consequence for actin filament assembly/disassembly if a mutation prevented actin's ability to bind ATP? What would be the consequence if a mutation prevented actin's ability to hydrolyze ATP?

**6.** Actin filaments at the leading edge of a crawling cell are believed to undergo treadmilling. What is treadmilling, and what accounts for this assembly behavior?

**7.** Although purified actin can assemble reversibly in vitro, various actin-binding proteins regulate the assembly of actin filaments in the cell. Predict the effect on a cell's actin cytoskeleton if function-blocking antibodies against each of the following are independently microinjected into cells: profilin, thymosin-$\beta_4$, CapZ, and the Arp2/3 complex.

**8.** Predict how actin would polymerize on an arrowhead decorated seed (as shown in Figure 17-9) in the presence of CapZ, tropomodulin, or profilin-actin.

**9.** Compare and contrast the ways in which formin and WASp are activated, and how each stimulates actin filament formation.

**10.** There are at least 20 different types of myosin. What properties do all types share, and what makes them different? Why is myosin II the only myosin capable of producing contractile force?

**11.** The ability of myosin to walk along an actin filament may be observed with the aid of an appropriately equipped microscope. Describe how such assays are typically performed. Why is ATP required in these assays? How can such assays be used to determine the direction of myosin movement or the force produced by myosin?

**12.** Contractile bundles occur in nonmuscle cells; these structures are less organized than the sarcomeres of muscle cells. What is the purpose of nonmuscle contractile bundles? Which type of myosin is found in contractile bundles?

**13.** How does myosin convert the chemical energy released by ATP hydrolysis into mechanical work?

**14.** Myosin II has a duty ratio of 10 percent, and its step size is 8 nm. In contrast, myosin V has a much higher duty ratio (about 70 percent) and takes 36-nm steps as it walks down an actin filament. What differences between myosin II and myosin V account for their different properties? How do the different structures and properties of myosin II and myosin V reflect their different functions in cells?

**15.** Contraction of both skeletal and smooth muscle is triggered by an increase in cytosolic $Ca^{2+}$. Compare the mechanisms by which each type of muscle converts a rise in $Ca^{2+}$ into contraction.

**16.** Phosphorylation of myosin light-chain kinase (MLCK) by protein kinase A (PKA) inhibits MLCK activation by $Ca^{2+}$-calmodulin. Drugs such as albuterol bind to $\beta$-adrenergic receptor, which causes a rise in cAMP in cells and activation of PKA activity. Explain why albuterol is useful for treating the severe contraction of the smooth muscle cells surrounding airway passages involved in an asthma attack.

**17.** Several types of cells utilize the actin cytoskeleton to power locomotion across surfaces. How are different assemblies of actin filaments involved in locomotion?

**18.** To move in a specific direction, migrating cells must utilize extracellular cues to establish which portion of the cell will act as the front and which will act as the back. Describe how G proteins appear to be involved in the signaling pathways used by migrating cells to determine direction of movement.

**19.** Cell motility has been described as being like the motion of tank treads. At the leading edge, actin filaments form rapidly into bundles and networks that make protrusions and move the cell forward. At the rear, cell attachments are broken and the tail end of the cell is brought forward. What provides the traction for moving cells? How does cell-body translocation happen? How are cell attachments released as cells move forward?

## Analyze the Data

Myosin V is an abundant nonmuscle myosin that is responsible for the transport of cargo such as organelles in many cell types. Structurally, it consists of two identical polypeptide chains that dimerize to form a homodimer. The motor domains reside at the N-terminus of each chain and contain both ATP- and actin-binding sites. The motor domain is followed by a neck region containing six "IQ" motifs, each of which binds calmodulin, a $Ca^{2+}$-binding protein. The neck domain is followed by a region capable of forming coiled coils, via which the two chains dimerize. The final 400 amino acid residues form a globular tail domain (GTD), to which cargo binds. Myosin V would consume large amounts of ATP if its motor domain were always active, and a number of studies have been conducted to understand how this motor is regulated.

**a.** The rate of ATP hydrolysis (i.e., ATP molecules hydrolyzed per second per myosin V) was measured in the presence of increasing amounts of free $Ca^{2+}$. The concentration of cytosolic free $Ca^{2+}$ is normally less than $10^{-6}$ M but can be elevated in localized areas of the cell and is often elevated in response to a signaling event. What do these data suggest about myosin V regulation?

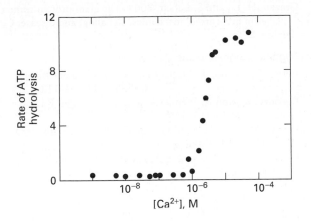

**b.** In additional studies, the ATPase activity of myosin V was measured in the presence of increasing amounts of F actin in the presence or absence of $10^{-6}$ M free $Ca^{2+}$. What additional information about myosin V regulation do these data provide?

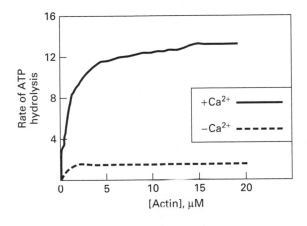

**c.** Next, the behavior of truncated myosin V, which lacks just its C-terminal globular tail, was examined and compared to the behavior of intact myosin V. From this experiment, what can you deduce about the mechanism by which myosin V is regulated?

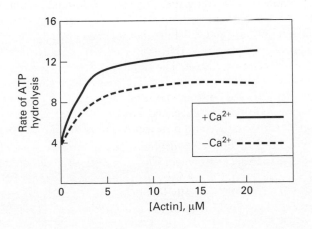

# References

## General References

Bray, D. 2001. *Cell Movements*. Garland.

Howard, J. 2001. *The Mechanics of Motor Proteins and the Cytoskeleton*. Sinauer.

Kreis, T., and R. Vale. 1999. *Guidebook to the Cytoskeletal and Motor Proteins*. Oxford University Press.

## Web Sites

The myosin home page:
http://www.mrc-lmb.cam.ac.uk/myosin/myosin.html

The kinesin home page:
http://www.cellbio.duke.edu/kinesin/

## Microfilaments and Actin Structures

Holmes, K. C., et al. 1990. Atomic model of the actin filament. *Nature* 347:44–49.

Kabsch, W., et al. 1990. Atomic structure of the actin:DNase I complex. *Nature* 347:37–44.

Pollard, T. D., and J. A. Cooper. 2009. Actin, a central player in cell shape and movement. *Science* 326:1208–1212.

Pollard, T. D., L. Blanchoin, and R. D. Mullins. 2000. Molecular mechanisms controlling actin filament dynamics in nonmuscle cells. *Ann. Rev. Biophys. Biomol. Struc.* 29:545–576.

## Dynamics of Actin Filaments

Paavilainen, V. O., et al. 2004. Regulation of cytoskeletal dynamics by actin-monomer-binding proteins. *Trends Cell Biol.* 14:386–394.

Theriot, J. A. 1997. Accelerating on a treadmill: ADF/cofilin promotes rapid actin filament turnover in the dynamic cytoskeleton. *J. Cell Biol.* 136:1165–1168.

## Mechanisms of Actin Filament Assembly

Campellone, K. G., and M. D. Welch. 2010. A nucleator arms race: cellular control of actin assembly. *Nat. Rev. Mol. Cell Biol.* 11:237–251.

Chesarone, M. A., et al. 2010. Unleashing formins to remodel the actin and microtubule cytoskeletons. *Nat. Rev. Mol. Cell Biol.* 11:62–74.

Goode, B. L., and M. J. Eck. 2007. Mechanism and function of formins in the control of actin assembly. *Ann. Rev. Biochem.* 76:593–627.

Gouin, E., M. D. Welch, and P. Cossart. 2005. Actin-based motility of intracellular pathogens. *Curr. Opin. Microbiol.* 8:35–45.

Higgs, H. N. 2005. Formin proteins: a domain-based approach. *Trends Biochem. Sci.* 30:342–353.

Pruyne, D., et al. 2002. Role of formins in actin assembly: nucleation and barbed end association. *Science* 297:612–615.

Rouiller, I., et al. 2008. The structural basis of actin filament branching by the Arp2/3 complex. *J. Cell Biol.* 180:887–895.

## Organization of Actin-Based Cellular Structures

Bennett, V., and A. J. Baines. 2001. Spectrin and ankyrin-based pathways: metazoan inventions for integrating cells into tissues. *Physiol. Rev.* 81:1353–1392.

Fehon, R.G., A. I. McClatchey, and A. Bretscher. 2010. Organizing the cell cortex: the role of ERM proteins. *Nat. Rev. Mol. Cell Biol.* 11:276–287.

McGough, A. 1998. F-actin-binding proteins. *Curr. Opin. Struc. Biol.* **8**:166–176.

Stossel, T. P., et al. 2001. Filamins as integrators of cell mechanics and signalling. *Nat. Rev. Mol. Cell Biol.* **2**:138–145.

### Myosins: Actin-Based Motor Proteins

Berg, J. S., B. C. Powell, and R. E. Cheney. 2001. A millennial myosin census. *Mol. Biol. Cell* **12**:780–794.

Mermall, V., P. L. Post, and M. S. Mooseker. 1998. Unconventional myosin in cell movement, membrane traffic, and signal transduction. *Science* **279**:527–533.

Rayment, I. 1996. The structural basis of the myosin ATPase activity. *J. Biol. Chem.* **271**:15850–15853.

Vale, R. D. 2003. The molecular motor toolbox for intracellular transport. *Cell* **112**:467–480.

Vale, R. D., and R. A. Milligan. 2000. The way things move: looking under the hood of molecular motor proteins. *Science* **288**:88–95.

### Myosin-Powered Movements

Bretscher A. 2003. Polarized growth and organelle segregation in yeast—the tracks, motors, and receptors. *J. Cell Biol.* **160**:811–816.

Clark, K. A., et al. 2002. Striated muscle cytoarchitecture: an intricate web of form and function. *Ann. Rev. Cell Dev. Biol.* **18**:637–706.

Grazier, H. L., and S. Labeit. 2004. The giant protein titin: a major player in myocardial mechanics, signaling, and disease. *Circ. Res.* **94**:284–295.

### Cell Migration: Signaling and Chemotaxis

Borisy, G. G., and T. M. Svitkina. 2000. Actin machinery: pushing the envelope. *Curr. Opin. Cell Biol.* **12**:104–112.

Burridge, K., and K. Wennerberg. 2004. Rho and Rac take center stage. *Cell* **116**:167–179.

Etienne-Manneville, S. 2004. Cdc42—the centre of polarity. *J. Cell Sci.* **117**:1291–1300.

Etienne-Manneville, S., and A. Hall. 2002. Rho GTPases in cell biology. *Nature* **420**:629–635.

Manahan, C. L., et al. 2004. Chemoattractant signaling in *Dictyostelium discoideum. Ann. Rev. Cell Dev. Biol.* **20**:223–253.

Pollard, T. D., and G. G. Borisy. 2003. Cellular motility driven by assembly and disassembly of actin filaments. *Cell* **112**:453–465.

Ridley, A. J., et al. 2003. Cell migration: integrating signals from the front to back. *Science* **302**:1704–1709.

Small, J. V., T. Strada, E. Vignal, and K. Rottner. 2002. The lamellipodium: where motility begins. *Trends Cell Biol.* **12**:112–120.

# Looking at Muscle Contraction

H. Huxley and J. Hanson, 1954, *Nature* **173:**973–976

The contraction and relaxation of striated muscles allow us to perform all of our daily tasks. How does this happen? Scientists have long looked to see how fused muscle cells, called myofibrils, differ from other cells that cannot perform powerful movement. In 1954, Jean Hanson and Hugh Huxley published their microscopy studies on muscle contraction, which demonstrated the mechanism by which it occurs.

## Background

The ability of muscles to perform work has long been a fascinating process. Voluntary muscle contraction is performed by striated muscles, which are named for their appearance when viewed under the microscope. By the 1950s, biologists studying myofibrils had named many of the structures they observed under the microscope. One contracting unit, called a sarcomere, is made up of two main regions called the A band and the I band. The A band contains two darkly colored thick striations and one thin striation. The I band is made up primarily of light-colored striations, which are divided by a darkly colored line known as the Z disk. Although these structures had been characterized, their role in muscle contraction remained unclear. At the same time, biochemists also tried to tackle this problem by looking for proteins that are more abundant in myofibrils than in nonmuscle cells. They found muscles to contain large amounts of the structural proteins actin and myosin in a complex with each other. Actin and myosin form polymers that can shorten when treated with adenosine triphosphate (ATP).

With these observations in mind, Hanson and Huxley began their study of cross striations in muscle. In a few short years, they united the biochemical

data with the microscopy observations and developed a model for muscle contraction that still holds true today.

## The Experiment

Hanson and Huxley primarily used phase-contrast microscopy in their studies of striated muscles that they isolated from rabbits. The technique allowed them to obtain clear pictures of the sarcomere and to take careful measurements of the A and the I bands. By treating the muscles with a variety of chemicals, then studying them under the phase-contrast microscope, they were able to successfully combine biochemistry with microscopy to describe muscle structure as well as the mechanism of contraction.

In their first set of studies, Hanson and Huxley employed chemicals that are known to specifically extract either myosin or actin from myofibrils. First, they treated myofibrils with a chemical that specifically removes myosin from muscle. They used phase-contrast microscopy to compare untreated myofibrils to myosin-extracted myofibrils. In the untreated muscle, they observed the previously identified sarcomeric structure, including the darkly colored A band. When they looked at the myosin-extracted cells, however, the darkly colored A band was not observed. Next, they extracted actin from the myosin-extracted muscle cells. When they extracted both myosin and actin from the myofibril, they could see no identifiable structure to the cell under phase-contrast microscopy. From these experiments, they concluded that myosin was located primarily in the A band, whereas actin is found throughout the myofibril.

With a better understanding of the biochemical nature of muscle structures, Huxley and Hanson went on to study the mechanism of muscle contraction. They isolated individual myo-

fibrils from muscle tissue and treated them with ATP, causing them to contract at a slow rate. Using this technique, they could take pictures of various stages of muscle contraction observed using phase-contrast microscopy. They could also mechanically induce stretching by manipulating the coverslip, which allowed them to also observe the relaxation process. With these techniques in hand, they examined how the structure of the myofibril changes during contraction and stretch.

First, Huxley and Hanson treated myofibrils with ATP, then photographed the images they observed under phase-contrast microscopy. These pictures allowed them to measure the lengths of both the A band and the I band at various stages of contraction. When they looked at myofibrils freely contracting, they noticed a consistent shortening of the lightly colored I band, whereas the length of the A band remained constant (Figure 1). Within the A band, they observed the formation of an increasingly dense area throughout the contraction.

Next, the two scientists examined how the myofibril structure changes during a simulated muscle stretch. They stretched isolated myofibrils mounted on glass slides by manipulating the coverslip. They again photographed phase-contrast microscopy images and measured the lengths of the A and the I bands. During stretch the length of the I band increased, rather than shortened, as it had in contraction. Once again, the length of the A band remained unchanged. The dense zone that formed in the A band during contraction became less dense during stretch.

From their observations, Hanson and Huxley developed a model for muscle contraction and stretch (Figure 1). In their model, the actin filaments in the I band are drawn up into the A band during contraction, and thus

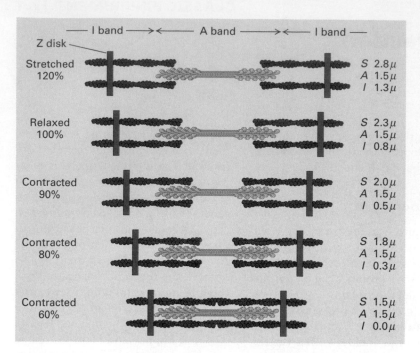

**FIGURE 1 Schematic diagram of muscle contraction and stretch observed by Hanson and Huxley.** The lengths of the sarcomere (S), the A band (A), and the I band (I) were measured in muscle samples contracted 60 percent in length relative to the relaxed muscle (*bottom*) or stretched to 120 percent (*top*). The lengths of the sarcomere, the I band, and the A band are noted on the right. Notice that from 120 percent stretch to 60 percent contraction the A band does not change in length. However, the length of the I band can stretch to 1.3 μm, and at 60 percent contraction, it disappears as the sarcomere shortens to the overall length of the A band. [Adapted from J. Hanson and H. E. Huxley, 1955, *Symp. Soc. Exp. Biol. Fibrous Proteins and Their Biological Significance* **9**:249.]

the I band becomes shorter. This allows for increased interaction between the myosin located in the A band and the actin filaments. As the muscle stretches, the actin filaments withdraw from the A band. From these data, Hanson and Huxley proposed that muscle contraction is driven by actin filaments moving in and out of a mass of stationary myosin filaments.

## Discussion

By combining microscopic observations with known biochemical treatments of muscle fibers, Hanson and Huxley were able to describe the biochemical nature of muscle structures and outline a mechanism for muscle contraction. A large body of research continues to focus on understanding the process of muscle contraction. Scientists now know that muscles contract by ATP hydrolysis, driving a conformational change in myosin that allows it to pull on actin. Researchers are continuing to uncover the molecular details of this process, whereas the mechanism of contraction proposed by Hanson and Huxley remains in place.

# Cell Organization and Movement II: Microtubules and Intermediate Filaments

Newt lung cell in mitosis stained for centrosomes (magenta), microtubules (green), chromosomes (blue), and keratin intermediate filaments (red). [Courtesy of A. Khodjakor, from *Nature* **408**:423–24 (2000).]

A s we learned in the previous chapter, three types of filaments make up the animal-cell cytoskeleton: microfilaments, microtubules, and intermediate filaments. Why have these three distinct types of filaments evolved? It seems likely that their physical properties are suited to different functions. In Chapter 17 we described how actin microfilaments are often cross-linked into networks of bundles to form flexible and dynamic structures and to serve as tracks for the many different classes of myosin motors. Likewise, **microtubules** are stiff tubes that can exist as a single structure extending up to 20 μm in cells, or in bundled arrangements such as those seen in specialized cell surface structures like cilia and flagella. A consequence of their tubular design is the ability of microtubules to generate pulling and pushing forces without buckling, a property that allows single tubules to extend large distances within a cell and bundles to slide past each other, as occurs in flagella and in the mitotic spindle. Microtubules' ability to extend long distances in the cell, together with their intrinsic polarity,

is exploited by microtubule-dependent motors, which use microtubules as tracks for long-range transport of organelles. Microtubules can be highly dynamic—being assembled and disassembled from their ends—providing the cell with the flexibility to alter microtubule organization as needed.

In contrast to microfilaments and microtubules, **intermediate filaments** have great tensile strength and have evolved to withstand much larger stresses and strains. With properties akin to strong molecular ropes, they are ideally suited to endow both cells and tissues with structural integrity and contribute to cellular organization. Intermediate filaments do not have an intrinsic polarity like microfilaments and microtubules, so it is not surprising that there are no known motor proteins that use intermediate filaments as tracks. Although we discuss microtubules and intermediate filaments together in this chapter—and their localization in the cytoplasm can look superficially quite similar—we will see their dynamics and functions are very different. A summary of the similarities and differences among the three cytoskeletal systems is presented in Figure 18-1.

## OUTLINE

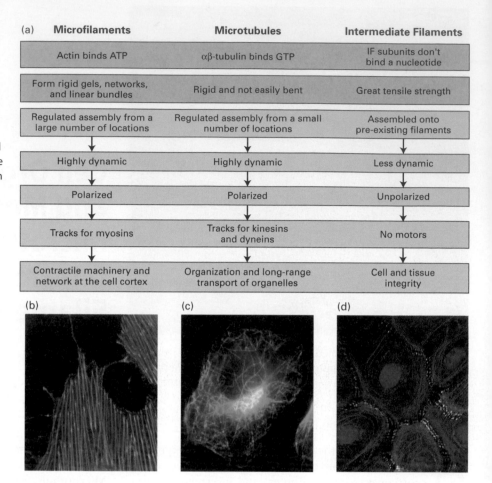

**FIGURE 18-1 Overview of the physical properties and functions of the three cytoskeletal systems in animal cells.**
(a) Biophysical and biochemical properties (orange) and biological properties (green) are shown for each filament type. The micrographs show examples of each filament type in a particular cellular context, but note that microtubules also make up other structures, and intermediate filaments also line the inner surface of the nucleus. (b) Cultured cells stained for actin (green) and sites of actin attachment to the substratum (orange). (c) Localization of microtubules (green) and the Golgi apparatus (yellow). Notice the central location of the Golgi apparatus, which is collected there by transport along microtubules. (d) Localization of cytokeratins (red), a type of intermediate filament, and a component of desmosomes (yellow) in epithelial cells. Cytokeratins from individual cells are attached to each other through the desmosomes. [Part (b) courtesy of K. Burridge. Part (c) courtesy of W. Brown. Part (d) courtesy of E. Fuchs.]

This chapter covers five main topics. First, we discuss the structure and dynamics of microtubules and their motor proteins. Second, we examine how microtubules and their motors contribute to the movement of cilia and flagella. Third, we discuss the role of microtubules in the mitotic spindle—a molecular machine that accurately segregates duplicated chromosomes. Fourth, we explore the roles of the different classes of intermediate filaments that provide structure to the nuclear envelope as well as strength and organization to cells and tissues. Although we consider microtubules, microfilaments, and intermediate filaments individually, the three cytoskeletal systems do not act independently of one another, and we consider some examples of this interdependence in the last section of the chapter.

## 18.1 Microtubule Structure and Organization

In the early days of electron microscopy, cell biologists noted long tubules in the cytoplasm that they called **microtubules.** Morphologically similar microtubules were seen making up the fibers of the mitotic spindle, as components of axons, and as the structural elements in cilia and flagella (Figure 18-2a, b).

A careful examination of single microtubules from various sources seen in transverse section indicated that they are all made up of 13 longitudinal repeating units (Figure 18-2c), now called *protofilaments,* suggesting the various microtubules all have a common structure. Microtubules purified from brain were then found to consist of a major protein, **tubulin,** and associated proteins, **microtubule-associated proteins (MAPs).** Purified tubulin alone can assemble into a microtubule under favorable conditions, proving that it is the structural component of the microtubule wall. MAPs, as we shall see, help mediate the assembly and dynamics of microtubules. In this section we consider the general structure and organization of microtubules, before turning to a more detailed discussion of their dynamics and regulation in Sections 18.2 and 18.3.

### Microtubule Walls Are Polarized Structures Built from αβ-Tubulin Dimers

Tubulin isolated in a pure and soluble form consists of two closely related subunits called α- and β-tubulin, each with a molecular weight of about 55,000 daltons. Genomic analyses reveal that genes encoding both α- and β-tubulins are present in all eukaryotes, with considerable expansion of the number of genes in multicellular organisms. For example, budding

(a)

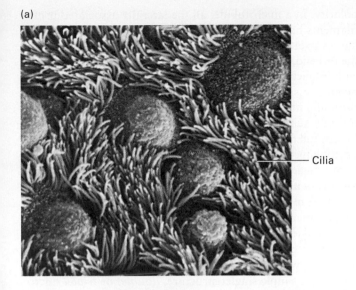

(b)

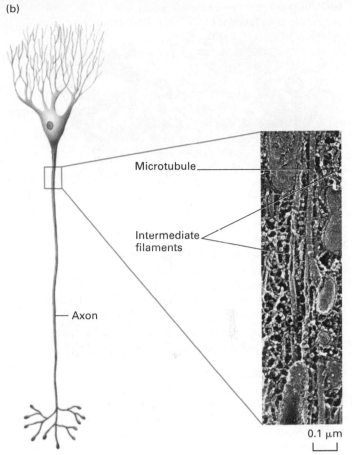

Cilia

Microtubule

Intermediate filaments

Axon

0.1 μm

(c)

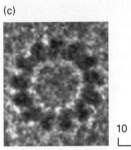

10 nm

**FIGURE 18-2 Microtubules are found in many different locations, and all have similar structures.** (a) Surface of the ciliated epithelium lining a rabbit oviduct viewed in a scanning electron microscope. Beating cilia, which have a core of microtubules, propel eggs down the oviduct. (b) Microtubules and intermediate filaments in a quick-frozen and deep-etched frog axon visualized in a transmission electron microscope. (c) High-magnification view of a single microtubule showing 13 repeating units known as protofilaments. [Part (a) from R. G. Kessels and R. H. Kardon, 1975, *Tissues and Organs,* W. H. Freeman and Company. Part (b) from N. Hirokawa, 1982, *J. Cell Biol.* **94:**129; courtesy of N. Hirokawa. Part (c) courtesy C. Bouchet-Marquis, 2007, *Biology of the Cell* **99:**45.]

yeast has two genes specifying α-tubulin and one for β-tubulin, whereas the soil nematode *Caenorhabditis elegans* has nine genes encoding α-tubulin and six for β-tubulin. In addition to α- and β-tubulin, all eukaryotes also have genes specifying a third tubulin, γ-tubulin, which is involved in microtubule assembly, as we discuss shortly. Additional isoforms of tubulin have also been discovered that are present only in organisms that possess cellular structures called centrioles and basal bodies, suggesting that these tubulin isoforms are important for those structures. As we'll learn in this chapter, centrioles and basal bodies are specialized structures that some organisms use to nucleate and organize microtubule assembly.

The α- and β-subunits of the tubulin dimer can each bind one molecule of GTP (Figure 18-3a). The GTP in the α-tubulin subunit is never hydrolyzed and is trapped by the interface between the α- and β-subunits. By contrast, the GTP-binding site on the β-subunit is at the surface of the dimer. GTP bound by the β-subunit can be hydrolyzed, and the resulting

GDP can be exchanged for free GTP. Under appropriate conditions, soluble tubulin dimers can assemble into microtubules (Figure 18-3b). As we saw in Chapter 17 for the polymerization of actin, ATP-G actin is preferentially added to one end of the filament, designated the (+) end because it is the end favored for assembly. Once incorporated into the filament, the bound ATP is hydrolyzed to ADP and $P_i$. In a similar manner, tubulin dimers in which the β-subunit has bound GTP add preferentially to one end of the microtubule, also designated the (+) end. As we will see, the GTP is hydrolyzed once tubulin is incorporated into the microtubule, but in contrast to the situation with ATP hydrolysis in an actin filament, this GTP hydrolysis has dramatic effects on the behavior of the microtubule (+) end.

Microtubules are composed of 13 laterally associated protofilaments that form a tubule whose external diameter is about 25 nm (see Figure 18-3b). Each of the 13 protofilaments is a string of αβ-tubulin dimers, longitudinally arranged

(a)

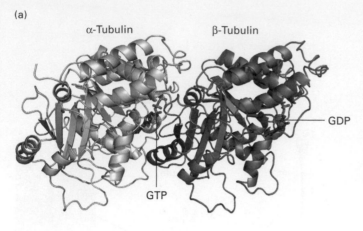

α-Tubulin    β-Tubulin

GDP

GTP

(b)

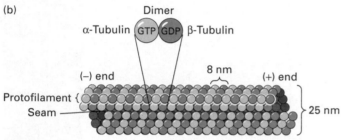

Dimer

α-Tubulin (GTP GDP) β-Tubulin

(−) end          8 nm          (+) end

Protofilament {
Seam

25 nm

**FIGURE 18-3 Structure of tubulin dimers and their organization into microtubules.** (a) Ribbon diagram of the tubulin dimer. The GTP bound to the α-tubulin monomer is nonexchangeable, whereas the GDP bound to the β-tubulin monomer is exchangeable with free GTP. (b) The organization of tubulin subunits in a microtubule. The dimers are aligned end to end into protofilaments, which pack side by side to form the wall of the microtubule. The protofilaments are slightly staggered so that α-tubulin in one protofilament is in contact with α-tubulin in the neighboring protofilaments, except at the seam, where an α-subunit contacts a β-subunit. The microtubule displays a structural polarity in that subunits are added preferentially at the end where β-tubulin monomers are exposed. This end of the microtubule is known as the (+) end. [Part (a) modified from E. Nogales et al., 1998, *Nature* **391**:199; courtesy of E. Nogales.]

**polarity.** In a microtubule, all the laterally associated protofilaments have the same polarity, thus the microtubule also has an overall polarity. The end with exposed β-subunits is the (+) end, while the end with exposed α-subunits is the (−) end. In microtubules, the heterodimers in adjacent protofilaments are staggered slightly, forming tilted rows of α- and β-tubulin monomers in the microtubule wall. If you follow a row of β-subunits, for example, spiraling around a microtubule for one full turn, you will end up precisely three subunits up the protofilament, abutting an α-subunit. Thus all microtubules have a single longitudinal *seam*, where an α-subunit in one protofilament meets a β-subunit in the adjacent protofilament.

Most microtubules in a cell consist of a simple tube, a *singlet* microtubule, built from 13 protofilaments. In rare cases, singlet microtubules contain more or fewer protofilaments; for example, certain microtubules in the neurons of nematode worms contain 11 or 15 protofilaments. In addition to the simple singlet structure, *doublet* or *triplet* microtubules are found in specialized structures such as cilia and flagella (doublet microtubules) and centrioles and basal bodies (triplet microtubules), structures we will explore later in the chapter. Each doublet or triplet contains one complete 13-protofilament microtubule (called the A tubule) and one or two additional tubules (B and C) consisting of 10 protofilaments each (Figure 18-4).

## Microtubules Are Assembled from MTOCs to Generate Diverse Organizations

With the identification of tubulin as the major structural component of microtubules, antibodies to tubulin were generated and used in immunofluorescence microscopy to localize microtubules in cells (Figure 18-5a, b). This approach, coupled with the description of microtubules seen by electron microscopy, showed that microtubules are assembled from specific sites to generate many different types of organization.

The nucleation phase of microtubule assembly is such an unfavorable reaction that spontaneous nucleation does not play a significant role in microtubule assembly in vivo. Rather, all microtubules are nucleated from structures known as **microtubule-organizing centers,** or **MTOCs.** In most cases the (−) end of the microtubule stays anchored in the MTOC while the (+) end extends away from it.

The **centrosome** is the main MTOC in animal cells. In nonmitotic cells, also known as *interphase* cells, the centrosome

so that the subunits alternate down a protofilament, with each subunit type repeating every 8 nm. Because the αβ-tubulin dimers in a protofilament are all oriented in the same way, each protofilament has an α-subunit at one end and a β-subunit at the other—so the protofilaments have an intrinsic

**FIGURE 18-4 Singlet, doublet, and triplet microtubules.** In cross section, a typical microtubule, a singlet, is a simple tube built from 13 protofilaments. In a doublet microtubule, an additional set of 10 protofilaments forms a second tubule (B) by fusing to the wall of a singlet (A) microtubule. Attachment of another 10 protofilaments to the (B) tubule of a doublet microtubule creates a (C) tubule and a triplet structure.

Singlet

Doublet
(cilia, flagella)

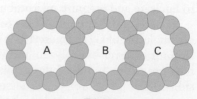

Triplet
(basal bodies, centrioles)

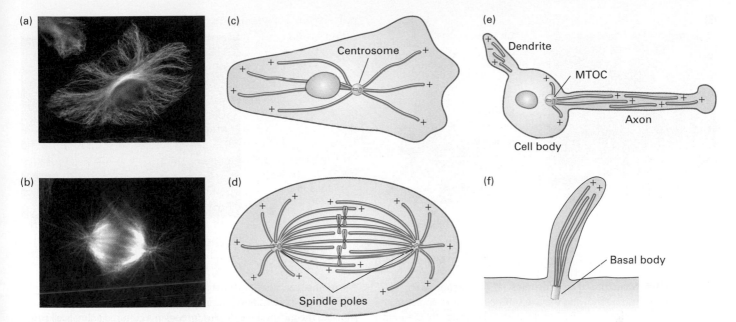

**FIGURE 18-5 Microtubules are assembled from microtubule organizing centers (MTOCs).** The distribution of microtubules in cultured cells as seen by immunofluorescence microscopy using antibodies to tubulin in an interphase cell (a) and a cell in mitosis (b). (c–f) Diagrams of the distribution of microtubules in cells and structures, all of which are assembled from distinct MTOCs. In an interphase cell (c), the MTOC is called a centrosome (the nucleus is indicated by a blue oval); in a mitotic cell (d), the two MTOCs are called spindle poles (the chromosomes are shown in blue); in a neuron (e), microtubules in both axons and dendrites are assembled from an MTOC in the cell body and then released from it; the microtubules that make up the shaft of a cilium or flagellum (f) are assembled from an MTOC known as a basal body. The polarity of microtubules is indicted by (+) and (−). [Part (a) courtesy of A. Bretscher. Part (b) courtesy of T. Wittmann.]

is generally located near the nucleus, producing an array of microtubules with their (+) ends radiating toward the cell periphery (Figure 18-5c). This radial display provides tracks for microtubule-based motor proteins to organize and transport membrane-bound compartments, such as those comprising the secretory and endocytic pathways. During mitosis, cells completely reorganize their microtubules to form a bipolar spindle, assembled from two centrosomes, also known as *spindle poles,* to accurately segregate copies of the duplicated chromosomes (Figure 18-5d). In another example, neurons have long processes called axons, in which organelles are transported in both directions along microtubules (Figure 18-5e). The microtubules in axons, which can be as long as 1 meter in length, are not continuous and have been released from the centrosome but nevertheless are all of the same polarity. In the same cells, the microtubules in the dendrites have mixed polarity, although the functional significance of this is not clear. In cilia and flagella (Figure 18-5f), microtubules are assembled from an MTOC called a *basal body.* As we mention later, plants do not have centrosomes and basal bodies but use other mechanisms to nucleate the assembly of microtubules.

Electron microscopy shows that centrosomes in animal cells consist of a pair of orthogonally arranged cylindrical **centrioles** surrounded by apparently amorphous material called **pericentriolar material** (Figure 18-6a, arrowheads). Centrioles, which are about 0.5 μm long and 0.2 μm in diameter,

are highly organized and stable structures that consist of nine sets of triplet microtubules and are closely related in structure to the basal bodies found at the base of cilia and flagella. It is not the centrioles themselves that nucleate the cytoplasmic microtubule array, but rather factors in the pericentriolar material. A critical component is the *γ-tubulin ring complex (γ-TuRC)* (Figures 18-6b and 18-7). γ-TuRC is located in the pericentriolar material and consists of many copies of γ-tubulin associated with several other proteins. It is believed that γ-TuRC acts like a split-washer template to bind αβ-tubulin dimers for the formation of a new microtubule, with the (−) end associated with γ-TuRC and the (+) end free for assembly. In addition to nucleating the assembly of microtubules, centrosomes anchor and regulate the dynamics of the (−) ends of the microtubules, which are located there.

Basal bodies have a structure similar to the centriole and are the MTOCs found at the base of cilia and flagella. The A and B tubules of their triplet microtubules provide a template for the assembly of the microtubules making up the core structure of cilia and flagella.

Recent work has uncovered an additional mechanism for the nucleation of microtubules in animal cells, also involving γ-TuRC. A protein complex called the *augmin complex*, consisting of eight polypeptides, can bind to the side of existing microtubules, then recruit γ-TuRC and nucleate the assembly of new ones. As we discuss in a later section, the augmin complex contributes to microtubule assembly in the mitotic spindle.

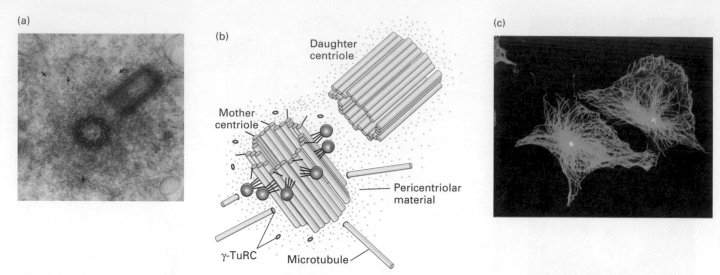

**FIGURE 18-6 Structure of centrosomes.** (a) Thin section of an animal-cell centrosome showing the two centrioles at right angles to each other surrounded by petricentriolar material (arrows). (b) Diagram of a centrosome showing the mother and daughter centrioles, each of which consists of nine linked outer triplet microtubules embedded in pericentriolar material that contains γ-TuRC nucleating structures.

The mother centriole is distinct from the daughter as it has distal appendages (blue spheres). (c) Immunofluorescence microscopy showing the microtubule display (green) in a cultured animal cell and the location of the MTOC, using an antibody to a centrosomal protein (yellow). [Parts (a) and (b) from G. Sluder, 2005, *Nature Rev. Mol. Cell Biol.* **6:**743. Part (c) courtesy of R. Kuriyama.]

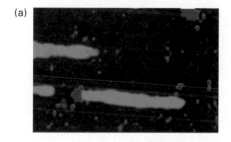

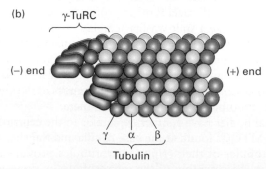

**FIGURE 18-7 The γ-tubulin ring complex (γ-TuRC) that nucleates microtubule assembly.** (a) An immunofluorescence micrograph of in vitro assembled microtubules is labeled green and a component of the γ-TuRC is labeled red, showing that it is located specifically at one end of the microtubule. (b) Model of how γ-TuRC may nucleate assembly of a microtubule by forming a template corresponding to the (−) end of a microtubule. [Part (a) Modified from T. J. Keating and G. G. Borisy, 2000, *Nature Cell Biol.* **2:**352; courtesy of T. J. Keating and G. G. Borisy.]

## KEY CONCEPTS of Section 18.1

### Microtubule Structure and Organization

- Tubulin is the major structural component of microtubules (see Figure 18-3) with which microtubule-associated proteins (MAPs) associate.

- Free tubulin exists as an αβ-dimer, with the α-subunit binding a trapped and nonhydrolyzable GTP and the β-subunit binding an exchangeable and hydrolyzable GTP.

- αβ-tubulin assembles into microtubules having 13 laterally associated protofilaments, with an α-subunit exposed at the (−) end and a β-subunit at the (+) end of each protofilament.

- In cilia and flagella, as well as in centrioles and basal bodies, doublet or triplet microtubules exist in which the additional microtubules have 10 protofilaments (see Figure 18-4).

- All microtubules are nucleated from microtubule-organizing centers (MTOCs), and many remain anchored with their (−) end there. Thus the end away from the MTOC is always the (+) end.

- The centrosome is the MTOC that nucleates the radial array of microtubules in non-mitotic animal cells; two centrosomes, or spindle poles, are the MTOCs that nucleate the microtubules of the mitotic spindle; and basal bodies are the MTOCs that assemble microtubules of cilia and flagella (see Figure 18-5).

- Centrosomes consist of two centrioles and the pericentriolar material that contains the γ–TuRC microtubule-nucleating complex (see Figures 18-6 and 18-7).

## 18.2 Microtubule Dynamics

Microtubules are dynamic structures due to assembly and disassembly at their ends. The degree of dynamics can vary enormously, with an average microtubule lifetime of less than 1 minute for cells in mitosis and about 5–10 minutes for the microtubules that make up the radial array seen in nonmitotic animal cells. Microtubule lifetime is longer in axons and much longer in cilia and flagella. To elucidate how these differences occur, we discuss the dynamic properties of microtubules and how this behavior contributes to their cellular organization.

### Individual Microtubules Exhibit Dynamic Instability

Early experiments revealed that most microtubules in animal cells will disassemble when the cells are cooled to 4 °C, and will repolymerize when the cells are warmed back to 37 °C. Researchers realized that this intrinsic property of microtubules could be exploited to purify the components of microtubules. Since brain tissue is rich in microtubules, soluble extracts of pig brains were prepared at 4 °C; these clarified extracts were then warmed to 37 °C to induce microtubule assembly. The assembled microtubules were collected into a pellet by centrifugation, separated from the supernatant, and then disassembled by adding buffer at 4 °C. After another cycle of assembly by warming, collection, and disassembly by cooling, researchers recovered *microtubular protein*, a collective term for αβ-tubulin and microtubule-associated proteins (MAPs). They were then able to fractionate the microtubular protein into pure αβ-tubulin and MAPs, to study their behaviors separately. Investigators found that polymerization of dimeric αβ-tubulin into microtubules is greatly catalyzed by the presence of the MAPs.

Although a tremendous amount of research effort was devoted to characterizing the bulk polymerization properties of microtubular protein in solution, its general relevance was superseded by subsequent studies examining the properties of individual microtubules. Nevertheless, some lessons learned from the earlier in vitro studies are important to microtubule biology. First, for assembly to occur, the αβ-tubulin concentration must be above the *critical concentration* ($C_c$), just as we saw for actin polymerization (see Figure 17-8). Second, at αβ-tubulin concentrations higher than the $C_c$ for polymerization, dimers add faster to one end of the microtubule than to the other (Figure 18-8). As with F-actin assembly, the preferred end for assembly is designated the (+) end, which is the end with β-tubulin exposed. The (−) end has α-tubulin exposed (see Figure 18-3b).

When studying the bulk properties of microtubule assembly, one assumes that all microtubules are behaving similarly. However, when researchers examined the behavior of individual microtubules within a population, they found this was not the case. Individual microtubule behavior was examined in a very simple experiment. Microtubules were assembled

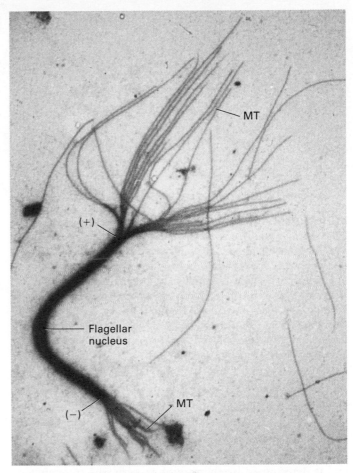

**EXPERIMENTAL FIGURE 18-8 Microtubules grow preferentially at the (+) end.** A fragment of a microtubule bundle from a flagellum was used as a nucleus for the in vitro addition of αβ-tubulin. The nucleating flagellar fragment is the thick bundle seen in this electron micrograph, with the newly formed microtubules (MT) radiating from its ends. The greater length of the microtubules at one end, the (+) end, indicates that tubulin subunits are added preferentially to this end. [Courtesy of G. Borisy.]

in vitro and then sheared to break them into shorter pieces whose individual lengths could be analyzed by microscopy. Under these conditions, one would expect all the short microtubules to either grow or shrink, depending on the free tubulin concentration. However, the investigators found that some of the microtubules grew in length, whereas others shortened very rapidly—thus indicating the existence of two distinct populations of microtubules. Further studies showed that individual microtubules could grow and then suddenly experience a *catastrophe* to a shrinking phase during which the microtubule undergoes rapid depolymerization. Moreover, sometimes a depolymerizing microtubule end could go through a *rescue* and begin growing again (Figure 18-9). Although this phenomenon was first seen in vitro, analysis of fluorescently labeled tubulin microinjected into live cells showed that microtubules in cells also undergo periods of

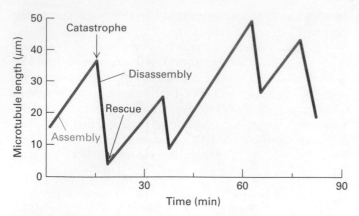

**FIGURE 18-9 Dynamic instability of microtubules in vitro.**
Individual microtubules can be observed in the light microscope and their lengths plotted at different times during assembly and disassembly. Assembly and disassembly each proceed at uniform rates, but there is a big difference between the rate of assembly and that of disassembly, as seen in the different slopes of the lines. Shortening of a microtubule is much more rapid (7 μm/min) than growth (1 μm/min). Notice the abrupt transitions to the shrinkage stage (catastrophe) and to the elongation stage (rescue). [Adapted from P. M. Bayley, K. K. Sharma, and S. R. Martin, 1994, in *Microtubules*, Wiley-Liss, p. 118.]

growth and shrinkage (Figure 18-10). This process of alternating between growing and shrinking states is known as *dynamic instability*. Thus the dynamic life of a microtubule end is determined by the rate of growth, the frequency of catastrophes, the rate of depolymerization, and the frequency of rescues. As we see later, these features of microtubule dynamics are controlled in vivo. Since the (−) ends of microtubules in animal cells are generally anchored on an MTOC, this dynamic nature is most relevant to the (+) end of the microtubule.

What is the molecular basis of dynamic instability? If you look carefully by electron microscopy at the ends of growing and shrinking microtubules, you can see they are quite different. A growing microtubule has a relatively blunt end, whereas a depolymerizing end has protofilaments peeling off like rams' horns (Figure 18-11). In fact, the growing microtubule end is not simply a blunt end but rather a short and flat sheet-like structure, made by the addition of tubulin dimers to the ends of protofilaments, that then rolls up along the seam to make the cylindrical microtubule.

Recent studies have provided a simple structural explanation for the two classes of microtubule ends. As we noted above, the β-subunit of the αβ-tubulin dimer is exposed on the (+) end of each protofilament. Using a GDP analog, researchers have found that artificially made *single* protofilaments—where there are no lateral interactions—made up of repeating αβ-tubulin dimers containing GDP-β-tubulin are curved, like a ram's horn. However, artificially made single protofilaments made up of αβ-tubulin dimers containing GTP-β-tubulin are straight. Thus growing microtubules with blunter ends terminate in GTP-β-tubulin, whereas shrinking ones with curled ends terminate in GDP-β-tubulin. Therefore, if the GTP molecules in the terminal β-tubulins become hydrolyzed on a microtubule that has stopped growing, a formerly blunt-end microtubule will curl and a catastrophe ensues. These relationships are summarized in Figure 18-11.

These results have an additional implication, and to understand this we have to consider the growing microtubule in more detail. The addition of a dimer to the (+) end of a protofilament on a growing microtubule involves an interaction between the preexisting terminal β-subunit and the new α-subunit. This interaction enhances the hydrolysis of the GTP to GDP in the formerly terminal β-subunit. However, the β-tubulin in the newly added dimer contains GTP. Thus each protofilament in a growing microtubule has mostly GDP-β-tubulin down its length and is capped by one or two terminal dimers containing GTP-β-tubulin. As we mentioned above, an *isolated* protofilament containing GDP-β-tubulin is curved along its length, so when it is present in a microtubule, why doesn't it break out and peel away? The lateral protofilament-protofilament interactions in the β-tubulin-GTP cap are sufficiently strong that they do

🎬 **VIDEO:** Assembly of Microtubules in Cultured Cells

**EXPERIMENTAL FIGURE 18-10 Fluorescence microscopy reveals growth and shrinkage of individual microtubules in vivo.** Fluorescently labeled tubulin was microinjected into cultured human fibroblasts. The cells were chilled to depolymerize preexisting microtubules into tubulin dimers and were then incubated at 37 °C to allow repolymerization, thus incorporating the fluorescent tubulin into all the cells' microtubules. A region of a cell periphery was viewed in the fluorescence microscope at 0 second, 27 seconds later, and 3 minutes 51 seconds later (left to right panels). In this period, several microtubules can be seen to have elongated and shortened. The dots labeled A, B, and C mark the position of the ends of three microtubules. [Modified from P. J. Sammak and G. Borisy, 1988, *Nature* **332**:724.]

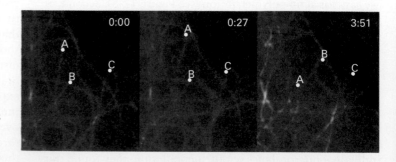

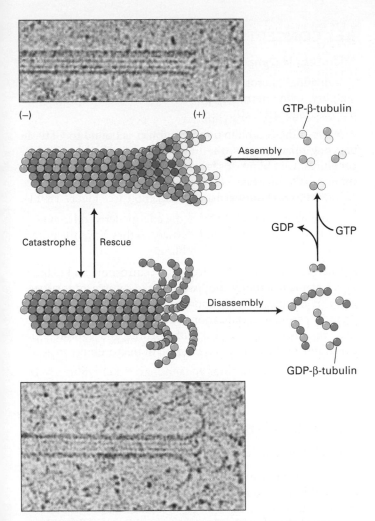

**FIGURE 18-11 Dynamic instability depends on the presence or absence of a GTP-β-tubulin cap.** Images taken in the electron microscope of frozen samples of a growing microtubule (upper) and a shrinking microtubule (lower). Notice that the end of the growing microtubule has a blunter end, whereas the shrinking one has curls like a ram's horns. The diagram shows that a microtubule with GTP-β-tubulin on the end of each protofilament is strongly favored to grow. However, a microtubule with GDP-β-tubulin at the ends of the protofilaments forms a curved structure and will undergo rapid disassembly. Switching between growing and shrinking phases, called rescues and catastrophes, can occur, and the rate of switching is regulated by associated proteins. [Images from E-M Mandelkow et al., 1991, *J. Cell Biol.* **114**:977. Diagram modified from A. Desai and T. J. Mitchison, 1997, *Annu. Rev. Cell Dev. Biol.* **13**:83–117.]

not allow the microtubule to unpeel at its end—and so the protofilaments behind the GTP-β-tubulin cap are constrained from unpeeling (see Figure 18-11). The energy released by GTP hydrolysis of the subunits behind the cap is stored within the lattice as structural strain waiting to be released when the GTP-β-tubulin cap is lost. If the GTP-β-tubulin cap is lost, the stored energy can do work if some structure, such as a chromosome, is attached to the disassembling microtubule end. As we will see, this stored energy

contributes to the movement of chromosomes during the anaphase stage of mitosis.

How can a disassembling microtubule suddenly be rescued to grow again? A possible answer to this perplexing problem has recently been suggested. Using an antibody that only recognizes GTP-β-tubulin and not GDP-β-tubulin, researchers have found that "islands" of GTP-β-tubulin can occur along the length of an assembled microtubule. It seems likely that when a disassembling microtubule encounters one of these GTP-β-tubulin islands, disassembly pauses and may provoke a rescue.

## Localized Assembly and "Search-and-Capture" Help Organize Microtubules

We have now presented two major concepts relating to microtubule organization and (+) end dynamics: microtubules are assembled from localized sites known as MTOCs, and individual microtubules can undergo dynamic instability. Together these two processes contribute to the distribution of microtubules in cells.

In an interphase cell growing in culture, microtubules are constantly being nucleated from the centrosome and spreading out, randomly "searching" the cytoplasmic space. The frequency of catastrophes and rescues, together with growth and shrinkage rates, determines the length of each microtubule—if the microtubule is subject to a high catastrophe frequency and low rescue, it will shrink back to the centrosome and disappear, whereas if it has few catastrophes and is readily rescued, it will continue to grow. If the searching microtubule encounters an appropriate target on a cell structure or organelle, the microtubule end may become attached to the structure. Organelle or cell structure "capture" by the microtubule stabilizes its (+) end and protects it from catastrophes, whereas unattached microtubules have a greater frequency of being disassembled. So the dynamics of the microtubule end is a very important determinant of microtubule life cycle and function. "Search and capture" is part of the mechanism determining the overall organization of microtubules in a cell. Moreover, by changing the rate of nucleation or local microtubule dynamics and capture sites, a cell can rapidly change its overall microtubule distribution. We will see later that this is what happens as cells enter mitosis.

## Drugs Affecting Tubulin Polymerization Are Useful Experimentally and in Treatment of Diseases

The conserved nature of tubulins and their essential involvement in critical processes such as mitosis make them prime targets for both naturally occurring and synthetic drugs that affect polymerization or depolymerization. Historically, the first known such drug was colchicine, present in extracts of the meadow saffron, which binds tubulin dimers so that they cannot polymerize into a microtubule. Since most microtubules are in a dynamic state between dimers and

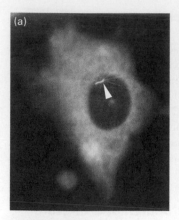
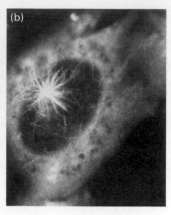

**EXPERIMENTAL FIGURE 18-12 Microtubules grow from the MTOC.** To investigate from where microtubules assemble in vivo, a cultured fibroblast was treated with colchicine until almost all the cytoplasmic microtubules were disassembled. The cell was then stained with antibodies to tubulin and viewed by immunofluorescence microscopy (a). The colchicine was then washed out to allow the reassembly of microtubules. Panel (b) shows the first stages of reassembly, revealing microtubules growing from the MTOC in the central region above the nucleus (dark areas). Note in panel (a) the remaining primary cilium (arrowhead; discussed in Section 18.5) associated with the centrosome; it is not depolymerized by colchicine treatment under these conditions. Note also the fluorescence from the cytoplasm, which is from unpolymerized αβ-tubulin dimers. [From M. Osborn and K. Weber, 1976, *Proc. Natl. Acad. Sci. USA* **73**:867–871.]

polymers, the addition of colchicine sequesters all free dimers in the cytoplasm, resulting in loss of microtubules due to their natural turnover. Treatment of cultured cells with colchicine for a short time results in the depolymerization of all the cytoplasmic microtubules, leaving the more stable tubulin-containing centrosome (Figure 18-12a). When the colchicine is washed out to allow regrowth of the microtubules, they can be seen to grow from the centrosome, revealing its ability to nucleate new microtubule assembly (Figure 18-12b).

Colchicine has been used for hundreds of years to relieve the joint pain of acute gout—a famous patient was King Henry VIII of England, who was treated with colchicine to relieve this ailment. A low level of colchicine relieves the inflammation caused in gout by reducing the microtubule dynamics of white blood cells, rendering them unable to migrate efficiently to the site of inflammation.

In addition to colchicine, a number of other drugs bind the tubulin dimer and restrain it from forming polymers. These include podophyllotoxin (from juniper) and nocodazole (a synthetic drug).

Taxol, a plant alkaloid from the Pacific yew tree, binds and stabilizes microtubules against depolymerization. Because taxol stops cells from dividing by inhibiting mitosis, it has been used to treat some cancers, such as those of the breast and ovary, where the cells are especially sensitive to the drug. ■

## KEY CONCEPTS of Section 18.2

### Microtubule Dynamics

• Individual microtubule (+) ends can undergo dynamic instability, with alternating periods of growth or rapid shrinkage (see Figure 18-10).

• Most of the β-tubulin in microtubules is bound to GDP. In growing microtubules, the (+) ends are capped by GTP-β-tubulin and are blunt or slightly splayed out. Shrinking microtubules have lost the GTP-β-tubulin cap, causing the protofilaments to peel outward and disassemble (see Figure 18-11).

• Growing microtubules store the energy derived from GTP hydrolysis in the microtubule lattice, so they have the potential to do work when disassembling.

• Microtubules assembled from the centrosome and exhibiting dynamic instability can "search" the cytoplasm for structures or organelles with appropriate targets and "capture" them, resulting in stabilization of the microtubule (+) end. In this way, assembly coupled with "search and capture" can contribute to the overall distribution of microtubules in a cell.

## 18.3 Regulation of Microtubule Structure and Dynamics

The wall of microtubules is built from αβ-tubulin dimers, and highly purified αβ-tubulin will assemble in vitro into microtubules. But assembly of microtubules in vitro can be greatly enhanced by the presence of stabilizing microtubule-associated proteins (MAPs). Stabilizing MAPs represent just one class of protein that interacts with tubulin in microtubules; other classes destabilize microtubules, or modify their growth properties. We discuss the various classes in this section. The regulation of microtubule structure and dynamics is critical for proper cell function. As we will see later, microtubules are the major organizers of organelles in animal cells, and their stability and dynamics are tailored for the specific function of the cell at any given time. For example, the dynamics of microtubules increase dramatically as cells enter mitosis to allow the cell to build a new microtubule organization, the mitotic spindle.

### Microtubules Are Stabilized by Side-Binding Proteins

Several different classes of proteins stabilize microtubules, many of them showing cell-type-specific expression. Among the best studied are the *tau* family of proteins, which includes tau itself, and proteins called MAP2 and MAP4. Tau and MAP2 are neuronal proteins, while MAP4 is expressed by other cell types and is generally not present in neurons. These proteins have a modular design with two key domains.

(a)

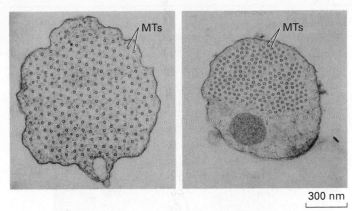

300 nm

(b)

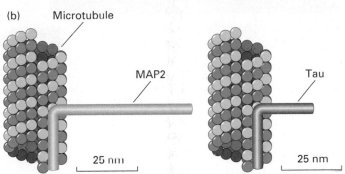

**EXPERIMENTAL FIGURE 18-13 Spacing of microtubules depends on the length of the projection domain of microtubule-associated proteins.** Insect cells transfected to express MAP2, which has a long arm, or to express tau protein, which has a short arm, grow long axonlike processes. (a) Electron micrographs of cross sections through the processes induced by the expression of MAP2 (*left*) or tau (*right*) in transfected cells. Note that the spacing between microtubules (MTs) in MAP2-containing cells is larger than in tau-containing cells. Both cell types contain approximately the same number of microtubules, but the effect of MAP2 is to enlarge the caliber of the axonlike process. (b) Diagrams of association between microtubules and MAPs. Note the difference in the lengths of the projection arms in MAP2 and tau. [Part (a) from J. Chen et al., 1992, *Nature* **360**:674.]

One domain consists of a positively charged 18-residue sequence, repeated three to four times, that binds to the negatively charged tubulin surface. The second domain projects out at a right angle from the microtubule (Figure 18-13). Tau proteins are believed to stabilize microtubules and also to act as spacers between them. MAP2 is found only in dendrites of neurons, where it forms fibrous cross-bridges between microtubules and links microtubules to intermediate filaments. Tau, which is much smaller than most other MAPs, is present in both axons and dendrites. The basis for this selectivity is still a mystery.

When stabilizing MAPs coat the outer wall of a microtubule, they can increase the growth rate of microtubules or suppress the catastrophe frequency. In many cases, the activity of the MAPs is regulated by the reversible phosphoryla-

tion of their projection domain. Phosphorylated MAPs are unable to bind to microtubules; thus phosphorylation promotes microtubule disassembly. For example, microtubule-affinity-regulating kinase (MARK/Par-1) is a key modulator of tau proteins. Some MAPs, like MAP4, are also phosphorylated by a *cyclin-dependent kinase (CDK)* that plays a major role in controlling the activities of proteins in the course of the cell cycle (Chapter 19).

## +TIPs Regulate the Properties and Functions of the Microtubule (+) End

In addition to the side-binding MAPs like the tau proteins, MAPs have been identified that associate with the (+) ends of microtubules. In many cases, they only associate with (+) ends that are growing, not shrinking (Figure 18-14a, b). The MAPs in this class are known as +TIPs, for plus-end tracking proteins. Although there are various mechanisms by which +TIPs recognize a growing (+) microtubule end, the association of a major +TIP called EB1 (end binding-1) is believed to be through interaction with a unique structure only present in growing microtubules (Figure 18-14c). The most obvious unique feature of a growing microtubule is the more blunt nature of its (+) end, so perhaps this is what EB1 recognizes. However, the localization of EB1 extends much farther back along the microtubule than just the very tip with the blunt end, suggesting that this simple model is not telling the whole story. Most other +TIPs associate with the (+) end by either binding EB1, or requiring EB1 for their association there, and are generally said to be "hitchhiking" on EB1 (see Figure 18-14d).

+TIPs are very important in the life of a microtubule, as they can modify its properties in several ways. First, proteins such as EB1 promote microtubule growth by enhancing polymerization at the (+) end. Second, other +TIPs can reduce the frequency of catastrophes, thereby also promoting microtubule growth. A third class links the microtubule (+) end to other cellular structures, such as the cell cortex, F-actin, and as we will see later during our discussion of mitosis, chromosomes; a key feature of this dynamic system is that when "searching" microtubules grow and a +TIP encounters an appropriate target, the microtubule can become "captured" and stabilized. Yet other +TIPs link microtubule (+) ends to membranes; for example, linkage to the endoplasmic reticulum transmembrane protein STIM promotes microtubule-dependent extension of the tubular endoplasmic reticulum (discussed in Section 18.4; see Figure 18-27).

## Other End-Binding Proteins Regulate Microtubule Disassembly

Mechanisms also exist for enhancing the disassembly of microtubules. Although most of the regulation of microtubule dynamics appears to happen at the (+) end, in some situations, such as in mitosis, it can occur at both ends.

(a)

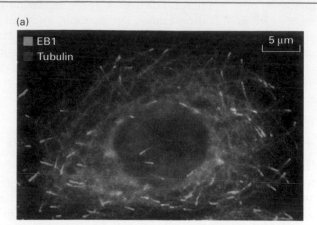

(b)

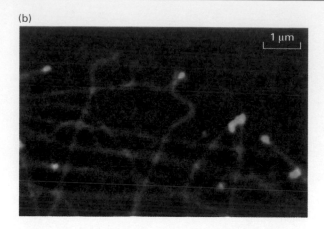

(c)

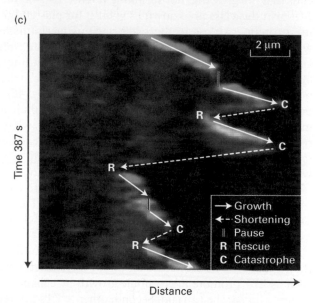

(d)

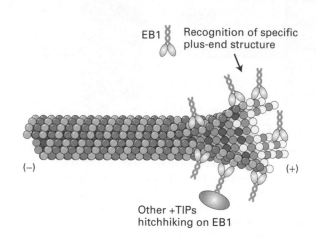

**EXPERIMENTAL FIGURE 18-14 The +TIP protein EB1 associates dynamically with the (+) ends of microtubules.** (a) A cultured cell stained with antibodies to tubulin (red) and the +TIP protein EB1 (green). EB1 is enriched in the region of the microtubule (+) end. (b) Edge of a living cell expressing EB3-GFP (green) and mCherry-α-tubulin (red). EB3, which is closely related to EB1, is found at the ends of some microtubules. (c) EB3-GFP selectively associates with growing microtubules as seen in this so-called "kymograph." In this figure, the dynamics of a single microtubule (red) and EB3 (green) in a live cell like that shown in (b) is followed by taking the same region of sequential frames from a movie and lining them up top to bottom. At the top, one sees the start of the movie with the microtubule capped by EB3. Moving down the figure, one can track the dynamics of the microtubule over time as it grows and shrinks. When the microtubule grows, it remains capped by EB3. When microtubule growth pauses or the microtubule shrinks, EB3 is no longer associated with the end, but it becomes reassociated when growth resumes. A diagrammatic summary of the microtubule dynamics is also shown. (d) A possible mechanism for how EB1 binds a growing microtubule and how other proteins can "hitchhike" on EB1. [Parts (a)-(c) Courtesy of Dr. A. Akhmanova, Cell Biology, Utrecht University, The Netherlands, and Dr. M. Steinmetz, Biomolecular Research, Paul Scherrer Institut, Villigen PSI, Switzerland.]

Various mechanisms for microtubule destabilization are known. One of these involves the kinesin-13 family of proteins. As we discuss in Section 18.4, most kinesins are molecular motors, but the kinesin-13 proteins are a distinct class that bind and curve the end of the tubulin protofilaments into the GDP-β-tubulin conformation. They then facilitate the removal of terminal tubulin dimers, thereby greatly enhancing the frequency of catastrophes (Figure 18-15a). They act catalytically in the sense that they need to hydrolyze ATP to sequentially remove terminal tubulin dimers.

Another protein, known as Op18/stathmin, also enhances the rate of catastrophes. It was originally identified as a protein highly overexpressed in certain cancers; hence part of its name (Oncoprotein 18). Op18/stathmin is a small protein that binds two tubulin dimers in a curved, GDP-β-tubulin-like

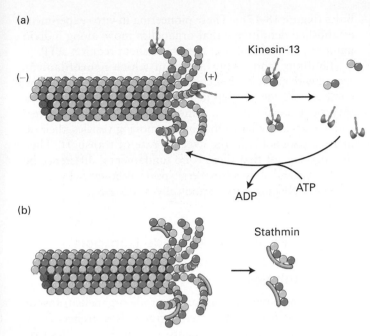

**(a)**

Kinesin-13

(−)  (+)

ADP  ATP

**(b)**

Stathmin

**FIGURE 18-15 Proteins that destabilize the ends of microtu-bules.** (a) A member of the kinesin-13 family enriched at microtubule ends can enhance the disassembly of that end. (Although depolymer-ization of the (+) end is shown, kinesin-13 can also depolymerize the (−) end.) These proteins are ATPases, and ATP enhances their activity by dissociating them from the αβ-tubulin dimer. (b) Stathmin binds selectively to curved protofilaments and enhances their dissociation from a microtubule end. Stathmin's activity is inhibited by phosphorylation.

conformation (Figure 18-15b). It may function by enhancing the hydrolysis of the GTP in the terminal tubulin dimer and aiding in its dissociation from the end of the microtubule. As might be expected for a regulator of microtubule ends, it is subject to negative regulation by phosphorylation by a large variety of kinases. In fact, it has been found that Op18/stath-min is inactivated by phosphorylation near the leading edge of motile cells, which contributes to preferential growth of mi-crotubules toward the front of the cell.

## KEY CONCEPTS of Section 18.3

### Regulation of Microtubule Structure and Dynamics

• Microtubules can be stabilized by side-binding microtubule-associated proteins (MAPs) (see Figure 18-13).

• Some MAPs, called +TIPs, bind selectively to growing (+) ends of microtubules and can alter the dynamic properties of the microtubule or localize components to the searching (+) end of the microtubule (see Figure 18-14).

• Microtubule ends can be destabilized by some proteins, such as the kinesin-13 family of proteins and Op18/stathmin, to enhance the frequency of catastrophes (see Figure 18-15).

## 18.4 Kinesins and Dyneins: Microtubule-Based Motor Proteins

Organelles in cells are frequently transported distances of many micrometers along well-defined routes in the cyto-plasm and delivered to particular intracellular locations. Dif-fusion alone cannot account for the rate, directionality, and destinations of such transport processes. Findings from early experiments with fish-scale pigment cells and nerve cells first demonstrated that microtubules function as tracks in the in-tracellular transport of various types of "cargo."

As already discussed, polymerization and depolymeriza-tion of microtubules can do work using the energy provided by GTP hydrolysis. In addition, **motor proteins** move along microtubules powered by ATP hydrolysis. Two main fami-lies of motor proteins—kinesins and dyneins—are known to mediate transport along microtubules. In this section we dis-cuss how these motor proteins work and the roles they per-form in interphase cells. In subsequent sections, we discuss their functions in cilia and flagella, and in mitosis.

## Organelles in Axons Are Transported Along Microtubules in Both Directions

A neuron must constantly supply new materials—proteins and membranes—to an axon terminal to replenish those lost in the exocytosis of neurotransmitters at the junction (syn-apse) with another cell (Chapter 22). Because proteins and membranes are primarily synthesized in the cell body, these materials must be transported down the axon, which can be as long as a meter in some neurons, to the synaptic region. This movement of materials is accomplished on microtu-bules, which are all oriented with their (+) ends toward the axon terminal (see Figure 18-5e).

The results of classic pulse-chase experiments, in which radioactive precursors were microinjected into the dorsal-root ganglia near the spinal cord and then tracked along their nerve axons, showed that **axonal transport** occurs from the cell body down the axon. Other experiments showed transport can also occur in the reverse direction, i.e., toward the cell body. *Anterograde* transport proceeds from the cell body to the synaptic terminals and is associated with axonal growth and the delivery of synaptic vesicles. In the opposite, *retrograde,* direction, "old" membranes from the synaptic terminals move along the axon rapidly toward the cell body, where they may be degraded in lysosomes. Findings from such experiments also revealed that different materials move at different speeds (Figure 18-16). The fastest-moving ma-terial, consisting of membrane-limited vesicles, has a velocity of about 3 μm/s, or 250 mm/day—requiring about four days to travel from a cell body in your back down an axon that terminates in your big toe. The slowest-moving material, comprising tubulin subunits and neurofilaments (the inter-mediate filaments found in neurons), moves only a fraction of a millimeter per day. Organelles such as mitochondria move down the axon at an intermediate rate.

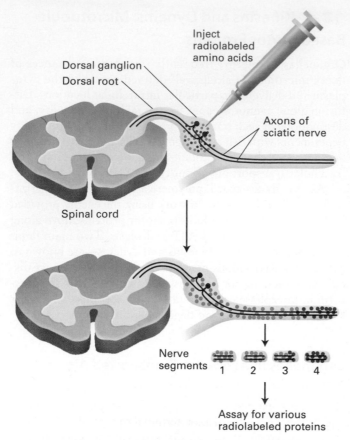

**EXPERIMENTAL FIGURE 18-16 The rate of axonal transport in vivo can be determined by radiolabeling and gel electrophoresis.** The cell bodies of neurons in the sciatic nerve are located in dorsal-root ganglia (near the spinal cord). Radioactive amino acids injected into these ganglia in experimental animals are incorporated into newly synthesized proteins, which are then transported down the axon to the synapse. Animals are sacrificed at various times after injection and the dissected sciatic nerve is cut into small segments to see how far radioactively labeled proteins have been transported; these proteins can be identified after gel electrophoresis and autoradiography. The red, blue, and purple dots represent groups of proteins that are transported down the axon at different rates, red most rapidly, purple least rapidly.

Neurobiologists have long made extensive use of the squid giant axon for studying organelle movement along microtubules. Involved in regulating the squid's water propulsion system, the aptly named giant axon can be up to 1 mm in diameter, which is about 100 times wider than the average mammalian axon. Moreover, squeezing the axon like a tube of toothpaste results in the extrusion of the cytoplasm (also known as axoplasm), which can then be observed by video microscopy. The movement of vesicles along microtubules in this cell-free system requires ATP, its rate is similar to that of fast axonal transport in intact cells, and it can proceed in both the anterograde and the retrograde directions (Figure 18-17a). Electron microscopy of the same region of the axon cytoplasm reveals organelles attached to individual microtu-

bules (Figure 18-17b). These pioneering in vitro experiments established definitively that organelles move along individual microtubules and that their movement requires ATP.

Findings from experiments in which neurofilaments tagged with green fluorescent protein (GFP) were injected into cultured cells suggest that neurofilaments pause frequently as they move down an axon. Although the peak velocity of neurofilaments is similar to that of fast-moving vesicles, their numerous pauses lower the average rate of transport. These findings suggest that there is no fundamental difference between fast and slow axonal transport, although why neurofilament transport stops periodically is unknown.

## Kinesin-1 Powers Anterograde Transport of Vesicles Down Axons Toward the (+) End of Microtubules

The protein responsible for anterograde organelle transport was first purified from axonal extracts. Researchers found that by mixing three components—purified organelles from squid axons, an organelle-free cytoplasmic axonal extract, and taxol-stabilized microtubules—organelles could be seen moving on the microtubules in an ATP-dependent manner. However, if they omitted the axonal extract, the organelles neither bound nor moved along the microtubules, suggesting that the extract contributes a protein that both attaches organelles to microtubules and transports them along it—i.e., a motor protein. The strategy for purifying the motor protein was based on additional observations of organelles moving on microtubules. It was known that if ATP was hydrolyzed to ADP, the organelles fell off the microtubules. However, if the nonhydrolyzable ATP analog AMPPNP was added, organelles remained associated with microtubules but did not move. This suggested that the motor linked the organelles to the microtubules very tightly in the presence of AMPPNP but then was released from the microtubule when the AMPPNP was replaced by ATP and its subsequent hydrolysis to ADP. Researchers used this clue to purify the motor.

Kinesin-1 isolated from squid giant axons is a dimer of two heavy chains, each associated with a light chain, with a total molecular weight of about 380,000. The molecule comprises a pair of globular *head domains* connected by a short flexible *linker domain* to a long *central stalk* and terminating in a pair of small globular *tail domains*, which associate with the light chains (Figure 18-18). Each domain carries out a particular function: the head domain binds microtubules and ATP and is responsible for the motor activity of kinesin; the linker domain is critical for forward motility; the stalk domain is involved in dimerization of the two heavy chains; and the tail domain is responsible for binding to receptors on the membrane of cargoes.

Kinesin-1-dependent movement of vesicles can be tracked by in vitro motility assays similar to those used to study myosin-dependent movements (see Figure 17-23). In one type of assay, a vesicle or a plastic bead coated with kinesin-1 is added to a glass slide along with a preparation of stabilized microtubules.

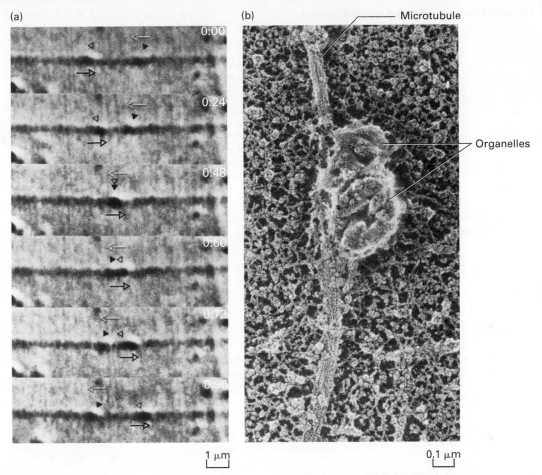

**EXPERIMENTAL FIGURE 18-17 DIC microscopy demonstrates microtubule-based vesicle transport in vitro.** (a) The cytoplasm was squeezed from a squid giant axon with a roller onto a glass coverslip. After buffer containing ATP was added to the preparation, it was viewed by differential interference contrast (DIC) microscopy, and the images were recorded on videotape. In the sequential images shown, the two organelles indicated by open and solid triangles move in opposite directions (indicated by colored arrows) along the same filament, pass each other, and continue in their original directions.

Elapsed time in seconds appears at the upper-right corner of each video frame. (b) A region of cytoplasm similar to that shown in part (a) was freeze-dried, rotary shadowed with platinum, and viewed in the electron microscope. Two large structures attached to one microtubule are visible; these structures presumably are small vesicles that were moving along the microtubule when the preparation was frozen.
[See B. J. Schnapp et al., 1985, *Cell* **40**:455; courtesy of B. J. Schnapp, R. D. Vale, M. P. Sheetz, and T. S. Reese.]

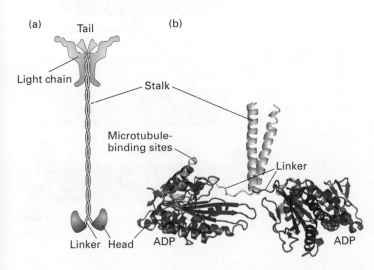

**FIGURE 18-18 Structure of kinesin-1.** (a) Representation of kinesin-1 showing its two intertwined heavy chains, each with a motor domain in the head region. Each head is attached to the coiled-coil stalk by a flexible linker domain. Two light chains associate with the tail of the heavy chain. (b) X-ray structure of the kinesin heads with the microtubule-binding and nucleotide-binding sites (containing ADP) indicated, including the linkers and the beginning of the stalk region.
[Part (a) modified from R. D. Vale, 2003, *Cell* **112**:467. Part (b) courtesy of E. Mandelkow and E. M. Mandelkow, adapted from M. Thormahlen et al., 1998, *J. Struc. Biol* **122**:30.]

**FIGURE 18-19 Model of kinesin-1-catalyzed vesicle transport.**
Kinesin-1 molecules, attached to receptors on the vesicle surface, transport the vesicles from the (−) end to the (+) end of a stationary microtubule. ATP is required for movement. [Adapted from R. D. Vale et al., 1985, *Cell* **40**:559, and T. Schroer et al., 1988, *J. Cell Biol.* **107**:1785.]

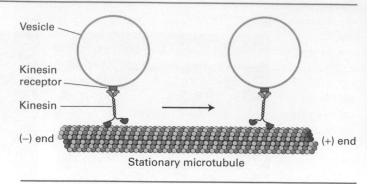

In the presence of ATP, the beads can be observed microscopically to move along a microtubule in one direction. Researchers found that the beads coated with kinesin-1 always moved from the (−) to the (+) end of a microtubule (Figure 18-19). Thus kinesin-1 is a (+) end–directed microtubule motor protein, and additional evidence shows that it mediates anterograde axonal transport.

## Kinesins Form a Large Protein Family with Diverse Functions

Following the discovery of kinesin-1, a number of proteins with similar motor domains were identified both in genetic screens and using molecular biology approaches. There are now 14 known classes of kinesins in animals, defined as sharing amino acid sequence homology with the motor domain of

kinesin-1. Proteins of the kinesin superfamily are encoded by about 45 genes in the human genome. Although the functions of all of these proteins have not yet been elucidated, some of the best-studied kinesins are involved in processes such as organelle, mRNA, and chromosome transport, microtubule sliding, and microtubule depolymerization.

As with the different classes of myosin motors, in the various kinesin families the conserved motor domain is fused to a variety of class-specific nonmotor domains (Figure 18-20). Whereas kinesin-1 has two identical heavy chains and two identical light chains, members of the kinesin-2 family (also

**FIGURE 18-20 Structure and function of selected members of the kinesin superfamily.** Kinesin-1, which includes the original kinesin isolated from squid axons, is a (+) end–directed microtubule motor involved in organelle transport. The kinesin-2 family has two different, but closely related, heavy chains, and a third cargo-binding subunit; this class also transports organelles in a (+) end-directed manner. The kinesin-5 family has four heavy chains assembled in a bipolar manner to interact with two antiparallel microtubules and also move toward the (+) end. Kinesin-13 family members have the motor domain in the middle of their heavy chains and do not have motor activity but they do destabilize microtubule ends (see also Figure 18-15a). Additional kinesin family members are mentioned in the text. Different kinesins have been given many different names; we use the unified nomenclature described in C. J. Lawrence et al., 2004, *J. Cell Biol.* **167**:19–22. [Diagrams modified from R. D. Vale, 2003, *Cell* **112**:467.]

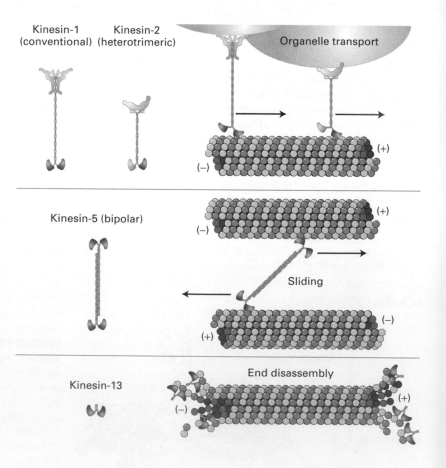

involved in organelle transport) have two different related heavy-chain motor domains and a third polypeptide that associates with the tail and binds cargo. Members of the bipolar kinesin-5 family have four heavy chains, forming bipolar motors that can cross-link antiparallel microtubules and, by walking toward the (+) end of each microtubule, slide them past each other. The kinesin-14 motors are the only known class to move toward the (−) end of a microtubule; this class functions in mitosis. Yet another type, the kinesin-13 family, has two subunits but with the conserved kinesin domain in the middle of the polypeptide. Kinesin-13 proteins do not have motor activity, but recall that these are special ATP-hydrolyzing proteins that can enhance the depolymerization of microtubule ends (see Figure 18-15).

## Kinesin-1 Is a Highly Processive Motor

How does kinesin-1 move down a microtubule? Optical trap and fluorescent-labeling techniques similar to those used to characterize myosin (see Figures 17-28, 17-29, and 17-30) have been used to study how kinesin-1 moves down a microtubule and how ATP hydrolysis is converted into mechanical work. Such experiments demonstrate that it is a very processive motor—taking hundreds of steps walking hand over hand down a microtubule without dissociating. During this process, the double-headed molecule takes 8-nm steps from one tubulin dimer to the next, tracking down the same protofilament in the microtubule. This entails each *individual* head taking 16-nm steps. The two heads work in a highly coordinated manner so that one is always attached to the microtubule.

The ATP cycle of kinesin-1 movement is most easily understood by first considering the cycle just after the motor has taken a step (Figure 8-21a). At this point the motor has a nucleotide-free leading head, under which conditions it is strongly bound to a tubulin dimer in a protofilament, and an ADP-bound trailing head that is weakly associated with the protofilament. ATP then binds to the leading head (Figure 18-21a, step **1**) and this binding induces a conformational change in the linker domain so that instead of pointing backwards, it swings forward and "docks" into its associated head. This swinging motion results in the linker domain rotating forward and, because it is attached to the trailing head, it swings the trailing head—like throwing a ballet dancer—into position to become the leading head (Figure 18-21a, step **2**). The new leading head finds the next binding site on the microtubule (Figure 18-21a, step **3** and Figure 18-21b). The binding of the leading head to the microtubule induces the leading head to release ADP while the trailing head hydrolyzes ATP to ADP and $P_i$, releasing $P_i$ (Figure 18-21a, step **4**). ATP can now bind to the leading head to repeat the cycle and allow the protein to take another step down the microtubule. Two features of this cycle ensure that one head is always firmly bound to the microtubule. First, the head domain binds tightly to the microtubule in the nucleotide-free, ATP and

ADP+$P_i$ states, but weakly in the ADP state. Second, the two heads communicate—when the leading head binds the microtubule and releases ADP, it is converted from a weak to a tight binding state. This message is communicated to the ATP-bound trailing head, tightly associated with the microtubule. The trailing head is stimulated to hydrolyze ATP, releasing $P_i$ and converting to a weak binding state. Because this cycle requires one head to always be firmly attached to a tubulin dimer in a protofilament, kinesin-1 can take thousands of steps along a microtubule without disassociating, and is therefore exceptionally processive as it moves down a microtubule.

When the x-ray structure of the kinesin head was determined, it revealed a major surprise—the catalytic core has the same overall structure as myosin's (Figure 18-22)! This occurs despite the fact that there is no amino acid sequence conservation between the two proteins, arguing strongly that convergent evolution twice generated a fold that can utilize the hydrolysis of ATP to generate work. Moreover, the same type of three-dimensional structure is seen in small GTP-binding proteins, such as Ras, that undergo a conformational change on GTP hydrolysis (see Figure 15-7).

## Dynein Motors Transport Organelles Toward the (−) End of Microtubules

In addition to kinesin motors, which primarily mediate anterograde (+) end–directed transport of organelles, cells use another motor, *cytoplasmic dynein*, to transport organelles in a retrograde fashion toward the (−) end of microtubules. This motor protein is very large, consisting of two large (>500 kDa), two intermediate, and two small subunits. It is responsible for the ATP-dependent retrograde transport of organelles toward the (−) ends of microtubules in axons, as well as many other functions we consider in the next sections. Compared to myosins and kinesins, the family of dynein-related proteins is not very diverse.

Like kinesin-1, cytoplasmic dynein is a two-headed molecule, built around two identical or nearly identical heavy chains. However, because of the enormous size of the motor domain, dynein has been less well characterized in terms of its mechanochemical activity. A single dynein heavy chain consists of a number of distinct domains (Figure 18-23). It consists of the *stem*, to which the other dynein subunits bind and which associates with its cargo through dynactin (see below). The next part of the heavy chain is a *linker* that, as we discuss later, plays a critical role during ATP-dependent motor activity. A large part of the heavy chain makes up the *head* containing the *AAA ATPase* domain, consisting of six repeats that assemble into a flowerlike structure, within which lies the ATPase activity. Embedded between the fourth and fifth AAA repeats is the *stalk*, which protrudes from the structure and contains the microtubule-binding region. Electron microscopy, combined with recently acquired x-ray images of the structure of a dynein heavy chain, provide a

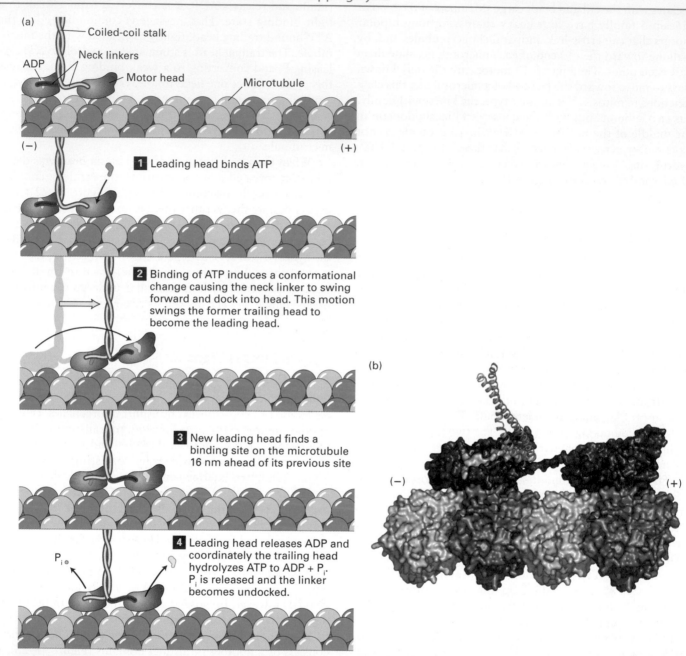

(a)

Coiled-coil stalk

ADP

Neck linkers

Motor head

Microtubule

(−)  (+)

**1** Leading head binds ATP

**2** Binding of ATP induces a conformational change causing the neck linker to swing forward and dock into head. This motion swings the former trailing head to become the leading head.

(b)

**3** New leading head finds a binding site on the microtubule 16 nm ahead of its previous site

(−)  (+)

P_i

**4** Leading head releases ADP and coordinately the trailing head hydrolyzes ATP to ADP + P_i. P_i is released and the linker becomes undocked.

**FIGURE 18-21 Kinesin-1 uses ATP to "walk" down a microtubule.**
(a) In this diagram, the two kinesin heads are shown with differently colored linker domains (yellow and red) to distinguish them. The cycle is shown starting after kinesin has taken a step, with the leading head tightly bound to the microtubule and not bound by any nucleotide, while the trailing head is weakly bound to the microtubule and has ADP bound. The leading head then binds ATP (step **1**), which induces a conformational change that causes the yellow linker region to swing forward and dock into its associated head domain, thereby thrusting the trailing head forward (step **2**). The new leading head now finds a binding site 16 nm down the microtubule, to which it binds weakly (step **3**). The leading head now releases ADP and binds tightly to the microtubule, which induces the trailing head to hydrolyze ATP to ADP and P_i (step **4**). P_i is released and the trailing head is converted into a weak binding state, and also releases the docked linker domain. The cycle now repeats itself for another step. (b) Structural model of two kinesin heads (purple) bound to a protofilament in a microtubule. The trailing head, at left, has bound ATP and has thrust the other head into the leading position. Notice how the linker domain (yellow) is docked into the trailing head, whereas the linker domain (red) of the leading head is still free. [Part (a) modified from R. D. Vale and R. A. Milligan, 2000, *Science* **288**:88. Part (b) based on E. P. Sablin and R. J. Fletterick, 2004, *J. Biol. Chem.* **279**:15707–15710.]

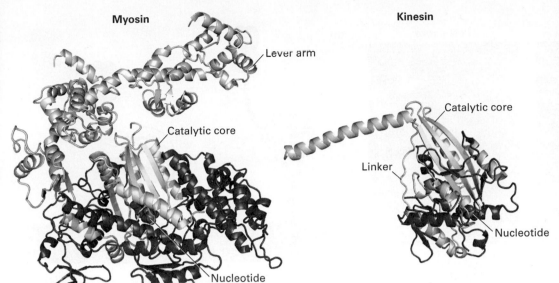

**Myosin**

Lever arm

Catalytic core

Nucleotide

**Kinesin**

Catalytic core

Linker

Nucleotide

**FIGURE 18-22 Convergent structural evolution of the ATP-binding cores of myosin and kinesin heads.** The common catalytic cores of myosin and kinesin are shown in yellow, the nucleotide in red, and the lever arm (for myosin-II) and linker domain (for kinesin-1) in light purple. [Modified from R. D. Vale and R. A. Milligan, 2000, *Science* **288**:88.]

glimpse of how dynein might work. Before a power stroke, the stem is attached to the linker that lies across the AAA domain and associates with the first and third AAA repeats (Figure 18-24a and b, left panels). Upon ATP binding and hydrolysis, the AAA ring changes conformation slightly and the linker domain becomes associated with the first and fifth

**(a)**

NH₂ | Stem | Linker | 1 | 2 | 3 | 4 | Stalk | 5 | 6 | COOH

Microtubule-binding domain

**(b)**

Stem

— Dynactin-binding region
— Dynein intermediate and light chains

Head

— Linker region

Stalk

— Microtubule-binding domain

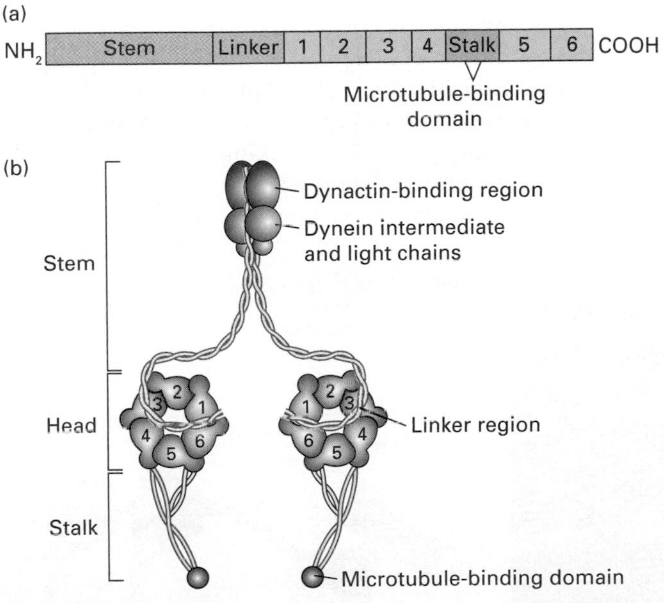

**FIGURE 18-23 The domain structure of cytoplasmic dynein.** (a) The dynein heavy chain, consisting of over 4000 amino acid residues, has several distinct domains. Following the stem and linker domains are six AAA repeats (peach, numbered 1-6), with the stalk and its microtubule-binding domain between repeats 4 and 5. The protein ends in an α-helical domain that supports the stalk (b). The six AAA repeats assume a structure like petals on a flower. Emerging from this is a coiled-coil stalk domain with a microtubule-binding site at the end. A number of additional subunits associate with the stem region and link dynein to cargo through dynactin. [Modified from R. D. Vale, 2003, *Cell* **112**:467.]

AAA repeat. This conformational change rotates the molecule to bring the stem and stalk closer, resulting in the transport of cargo toward the (−) end of the microtubule (Figure 18-24a and b, right panels).

Unlike kinesin-1, cytoplasmic dynein cannot mediate cargo transport by itself. Rather, dynein-related transport generally requires *dynactin,* a large protein complex that both links dynein to its cargo and regulates its activity (Figure 18-25). Dynactin consists of 11 different types of subunits, functionally organized into two domains. One domain is built around eight copies of the actin-related protein Arp1, which assembles into a short filament. The end corresponding to the (+) end of this filament is capped by CapZ, the same capping protein that binds the (+) end of an actin filament (see Figure 17-12); a number of subunits are associated with the (−) end. This Arp1-containing domain is responsible for binding cargo. The second domain of dynactin consists of a long protein called p150^Glued, which contains the dynein-binding site and also has a microtubule-binding site at one end. Holding the two dynactin domains together is a protein called *dynamitin*—so named because when it is overexpressed, it dissociates (or "blows apart") the two domains, making a nonfunctional complex. This feature has been very useful experimentally because it has allowed researchers to identify processes that are dependent on dynein-dynactin, which are disrupted in cells overexpressing dynamitin. The microtubule-binding site in dynactin allows it to hold loosely onto a microtubule as the dynein motor moves down the microtubule (Figure 18-25c), helping the dynactin-dynein complex remain associated with the microtubule. Thus, the major functions of dynactin are to bind cargo and to make dynein more processive.

How is dynein regulated? It has been found that the dynactin p150^Glued subunit binds +TIP EB1, allowing dynein to be associated with the growing (+) end of microtubules. Why would dynein associate with microtubule (+) ends if it is a (−) end–directed motor? Recent work suggests that when dynein is associated with the (+) end of microtubules

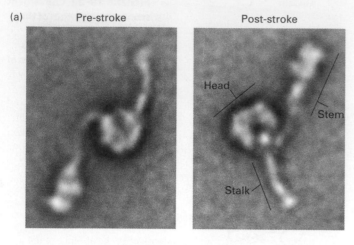

(a) Pre-stroke    Post-stroke

Head

Stem

Stalk

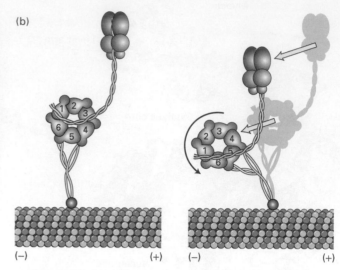

(b)

(−)              (+)      (−)              (+)

**FIGURE 18-24 The power stroke of dynein.** (a) Multiple images of purified single-headed dynein molecules in their pre-stroke and post-stroke states were recorded in an electron microscope and then averaged. The image at left shows dynein in the ADP+P state, which represents the pre-stroke state, and the image at right shows it in a nucleotide-free post-stroke state. (b) A comparison of the images combined with recently acquired structural data shows that the force-generation mechanism involves a change in orientation of the head relative to the stem, causing a movement of the microtubule-binding stalk. [Part (a) modified from S. A. Burgess et al., 2003, *Nature* **421:**715; courtesy of S. A. Burgess. Part (b) based on the dynein structure in A. P. Carter et al., 2011, *Science* **331:**1159–1165.]

via the dynactin-EB1 interaction, it is held in an inactive conformation. When the growing microtubule reaches the cell cortex, the inactive dynein and dynactin encounter an activator localized there. The dynein now becomes active, associating with the cortex and pulling on the microtubule that delivered it to the cortex! This mechanism has been shown to help orient the mitotic spindle in yeast, and likely applies to other situations as well.

 In addition to the regulatory p150^Glued subunit of dynactin, additional regulators of dynein activity exist. A group of proteins, two of which are LIS1 and NudE, are involved in regulating the activity of dynein. NudE links the dynein intermediate and light chains to LIS1 (Figure 18-26a). LIS1 then interacts with the ATPase domain of dynein to lengthen the power stroke, making the motor more processive under conditions of high load. Defects in LIS1 cause the fatal disease Miller-Dieker lissencephaly (from where the protein got its name); lissencephaly means "smooth brain," as cortical folds and grooves are lacking in the brains of patients with this condition (Figure 18-26b). Mutations in LIS1 result in defects in both neuronal mitoses and migration from

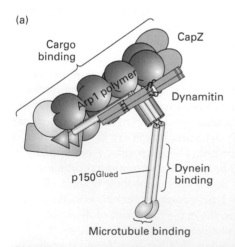

(a)

Cargo binding

CapZ

Arp1 polymer

Dynamitin

p150^Glued

Dynein binding

Microtubule binding

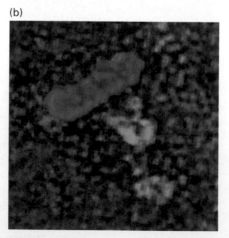

(b)

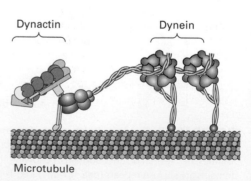

(c)

Dynactin          Dynein

Microtubule

**FIGURE 18-25 The dynactin complex linking dynein to cargo.** (a) One domain of the complex binds cargo and is built around a short filament made up of about eight subunits of the actin-related protein Arp1 capped by CapZ. Another domain consists of the protein p150^Glued, which has a microtubule-binding site on its distal end and is also involved in attaching cytoplasmic dynein to the complex. Dynamitin holds the two parts of the dynactin complex together. (b) Electron micrograph of a metal replica of the dynactin complex isolated from brain. The Arp1 minifilament (purple) and the dynamitin/p150^Glued side arm (teal) are highlighted. (c) Diagram of how the dynactin and dynein complex might interact with each other and with a microtubule. [Part (a) modified from T. A. Schroer, 2004, *Annu. Rev. Cell Dev. Biol.* **20:**759. Part (b) from D. M. Eckley et al, 1999, *J. Cell Biol.* **147:**307.]

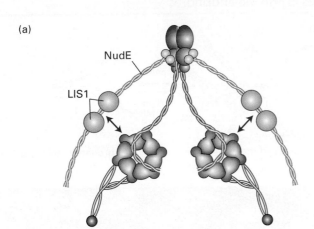

(a)

NudE

LIS1

**EXPERIMENTAL FIGURE 18-26** **The LIS1 protein regulates dynein and is required for brain development.** (a) Model of how NudE associates with dynein to allow LIS1 to interact with the ATPase domain of the dynein heavy chain. (b) Magnetic resonance images (MRIs) of a normal brain and a brain from a patient with Miller-Dieker

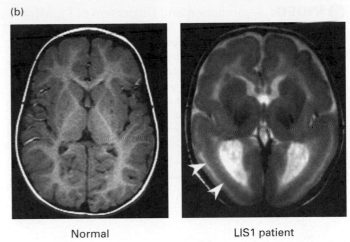

(b)

Normal          LIS1 patient

Lissencephaly, which lacks LIS1 function. Notice the absence of folding in the patient brain. [Part (a) modified from McKenny et al., 2010, *Cell* **141**:304–314. Part (b) from M. Kato and W. B. Dobyns, 2003, *Hum. Mol. Genet.* **12**:R89–R96.]

the ventricular zone to the cortical plate in early development, resulting in the smooth-brain phenotype and defects in mental development as well as many other abnormalities. ■

## Kinesins and Dyneins Cooperate in the Transport of Organelles Throughout the Cell

Both dynein and kinesin family members play important roles in the microtubule-dependent organization of organelles in

cells (Figure 18-27). Because the orientation of microtubules is fixed by the MTOC, the direction of transport—toward or away from the cell center—depends on the motor protein. For example, the Golgi apparatus collects in the vicinity of the centrosome, where the (−) ends of microtubules lie, and is driven there by dynein-dynactin. In addition, secretory cargo emerging from the endoplasmic reticulum is transported to the Golgi by dynein-dynactin. Conversely, the endoplasmic reticulum is spread throughout the cytoplasm and

**VIDEO:** Transport of Secretory Vesicles Along Microtubules
**VIDEO:** Transport of Vesicles Along Microtubules from the Endoplasmic Reticulum to the Golgi

**FIGURE 18-27** **Organelle transport by microtubule motors.** Cytoplasmic dyneins (red) mediate retrograde transport of organelles toward the (−) end of microtubules (cell center); kinesins (purple) mediate anterograde transport toward the (+) end (cell periphery). Most organelles have one or more microtubule-based motors associated with them. It should be noted that the association of motors with organelles varies by cell type, so some of these associations may not exist in all cells, whereas others not shown here also exist. ERGIC = ER-to-Golgi intermediate compartment.

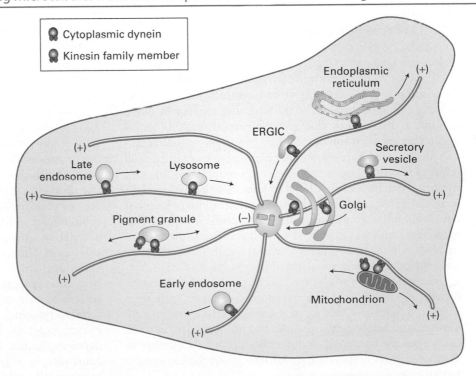

**FIGURE 18-28 Movement of pigment granules in frog melano-phores.** (a) Diagram of the microtubule-based reorganization of melanosomes according to the level of cAMP. Melanosomes are aggregated by cytoplasmic dynein and dispersed by kinesin-2. (b) Visualization of melanosomes in the dispersed state as seen by immunofluorescence microscopy for microtubules (green), the DNA in the nucleus (blue), and pigment granules (red). [Part (a) modified from V. Gelfand and S. Rogers, http://www.cellbio.duke.edu/kinesin/Pigment–aggregation.html. Part (b) from S. Rogers, www.itg.uiuc.edu/exhibits/gallery/ ]

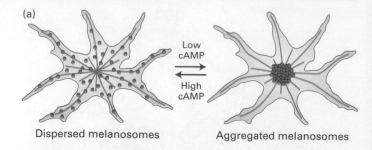

(a)

Dispersed melanosomes          Aggregated melanosomes

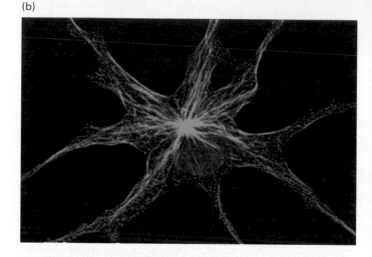

(b)

is transported there by kinesin-1, which moves toward the peripheral (+) ends of microtubules. Some organelles of the endocytic pathway are associated with dynein-dynactin, including late endosomes and lysosomes. Kinesins have been shown to transport mitochondria, as well as nonmembranous cargo such as specific mRNAs encoding proteins that need to be localized during development.

We have seen how kinesin-1 transports organelles in an anterograde fashion down axons. What happens to the motor when it gets to the end of the axon? The answer is that it is carried back in a retrograde fashion on organelles transported by cytoplasmic dynein. Thus kinesin-1 and dynein can associate with the same organelle and a mechanism must exist that turns one motor off while activating the other, although such mechanisms are not yet fully understood.

Much of what we know about the regulation of microtubule-based organelle transport comes from studies using fish (e.g., angelfish) or frog melanophores. Melanophores are cells of the vertebrate skin that contain hundreds of dark melanin-filled pigment granules called melanosomes. Melanophores either have their pigment granules dispersed, in which case they make the skin darker, or aggregated at the center, which makes the skin paler (Figure 18-28). These changes in skin color, mediated by neurotransmitters in the fish and regulated by hormones in the frog, serve to camouflage the fish and enhance social interactions in the frog. The movement of the granules is mediated by changes in intracellular cAMP and is dependent on microtubules. Studies investigating which motors are involved have shown that pigment granule dispersion requires kinesin-2, whereas aggregation requires cytoplasmic dynein-dynactin. The first hints of how these activities might be coordinated came from the finding that overexpression of dynamitin inhibited granule transport in both directions. This surprising result was explained when it was found that dynactin binds not only to cytoplasmic dynein but also to kinesin-2—and may coordinate the activity of the two motors.

The association of dynein and kinesin-2 with the same organelle is not limited to melanosomes; it has recently been suggested that these motors may cooperate to appropriately localize late endosomes/lysosomes and mitochondria in some cells. Thus the concept that organelles can have a number of distinct motors associated with them is not the exception, but an emerging theme.

## Tubulin Modifications Distinguish Different Microtubules and Their Accessibility to Motors

The stability and functions of different classes of microtubules are influenced by post-translational modifications. Although multiple types of modification have been detected, we restrict our discussion to those that are the best understood—lysine acetylation, detyrosylation, polyglutamylation, and polyglycylation (Figure 18-29a)—and their functional consequences.

Two of these modifications are only found on α-tubulin and are absent from β-tubulin. The first is the *acetylation* of the ε-amino group of a specific lysine residue of α-tubulin that lies on the inside of the microtubule; microtubules with this acetylated lysine are found in stable microtubule structures like centrioles, basal bodies, and primary cilia (primary cilia are discussed in Section 18.5). Indeed, cells unable to acetylate tubulin have defective primary cilia, whereas cells in which acetylation cannot be removed result in unusually stable primary cilia. The second modification of α-tubulin relates to its C-terminal tyrosine. This tyrosine can be specifically removed by a carboxypeptidase that only functions when bound to the surface of microtubules, where it sequentially removes the C-terminal tyrosines off α-tubulin subunits. These *detyrosylated* microtubules are more stable, as they are more resistant to depolymerization by the kinesin-13 family of depolymerizers. Moreover, in migrating cells, these more stable microtubules are generally oriented toward the front of the cell. When a

(a)

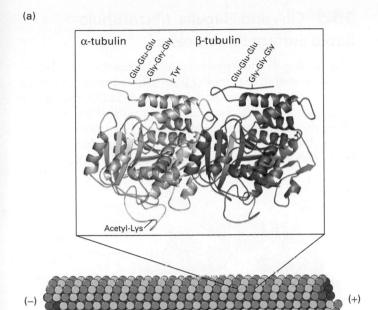

(b)

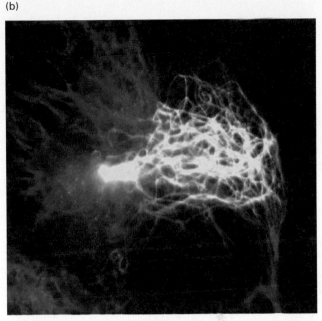

**EXPERIMENTAL FIGURE 18-29 Post-translational modifications of tubulin affect the stability and function of microtubules.** (a) Structure of α- and β-tubulin showing the sites of lysine acetylation on the inner surface of the microtubule, and polyglutamylation, polyglycylation, and detyrosylation on the outer surface. Note that polyglutamylation and polyglycylation are believed to be mutually exclusive, so they would not normally occur at the same time.

(b) Detyrosylated microtubules are preferentially oriented toward the leading edge of a moving cell. A cell migrating toward the right is stained for total microtubules (red) and detyrosylated microtubules (green). The resulting merge shows the detyrosylated microtubules enriched toward the front of the cell in yellow, which is a combination of red and green. [(a) Modified from J. W. Hamond, D. Cai, and K. J. Verhey, 2008, *Curr. Opin. Cell Biol.* **20**:71–76. (b) Courtesy of Greg Gundersen.]

stable microtubule depolymerizes, the α-tubulin subunit of the αβ-dimer has the C-terminal tyrosine added back by a tyrosine ligase that only acts on soluble tubulin, and the αβ-tubulin dimer can now be used for a new microtubule.

The C-terminal regions of both α- and β-tubulin are very rich in glutamic acid residues, and specific enzymes can modify these residues. Again, these modifications occur only after assembly of the microtubule. The tails can be modified by *polyglutamylation,* where a chain of glutamic acid residues is linked to a specific glutamate residue, or subject to *polyglycylation,* where a chain of glycine residues is added to a different glutamic residue. At present it is believed that these two modifications may be mutually exclusive, so that if a tubulin subunit is modified by polyglycylation, it is protected from polyglutamylation, and vice versa. Like the other modifications, polyglutamylation can enhance microtubule stability.

These post-translational modifications of tubulin not only affect microtubule stability, but they can also affect the ability of molecular motors to interact with microtubules (Figure 18-29b). Kinesin-1 associates preferentially with detyrosylated and acetylated microtubules, which may be important in recruiting this motor for axonal transport in neurons. As we mentioned in Figure 18-5e, neurons have different microtubule organizations in their dendrites and axons. The microtubules in the axon are stabilized by acetylation and detyrosylation and this allows kinesin-1 to preferentially associate with them for axonal transport. Polyglutamylation has a key role in the beating of cilia and flagella, which we discuss in the next section.

The research elucidating the effects of post-translational modifications of tubulin on microtubule function and microtubule-based motors is all quite recent; we can expect future studies to reveal multiple "codes" that distinguish different classes of microtubules and specialize them for specific functions.

## KEY CONCEPTS of Section 18.4

### Kinesins and Dyneins: Microtubule-Based Motor Proteins

- Kinesin-1 is a microtubule (+) end–directed ATP-dependent motor that transports membrane-bound organelles (see Figure 18-19).

- Kinesin-1 consists of two heavy chains, each with an N-terminal motor domain, and two light chains that associate with cargo (see Figure 18-18).

- The kinesin superfamily includes motors that function in interphase and mitotic cells, transporting organelles and sliding antiparallel microtubules past one another. The superfamily includes one class, kinesin-13, that is not motile but destabilizes microtubule ends (see Figure 18-20).

- Kinesin-1 is a highly processive motor because it coordinates ATP hydrolysis between its two heads so that one head is always firmly bound to a microtubule (see Figure 18-21).

- Cytoplasmic dynein is a microtubule (−) end–directed ATP-dependent motor that associates with the dynactin complex to transport cargo (see Figure 18-25).

- Kinesins and dyneins associate with many different organelles to organize their location in cells (see Figure 18-27).

- Post-translational modifications of tubulin can affect microtubule stability and regulate their ability to interact with microtubule-based motors.

## 18.5 Cilia and Flagella: Microtubule-Based Surface Structures

**Cilia** and **flagella** are related microtubule-based and membrane-bound extensions that project from many protozoa and most animal cells. Abundant motile cilia are found on the surface of specific epithelia, such as those that line the trachea, where they beat in an orchestrated wavelike fashion to move fluids. Animal-cell flagella, which are longer than cilia but have a very similar structure, can propel a cell, such as sperm,

(a)

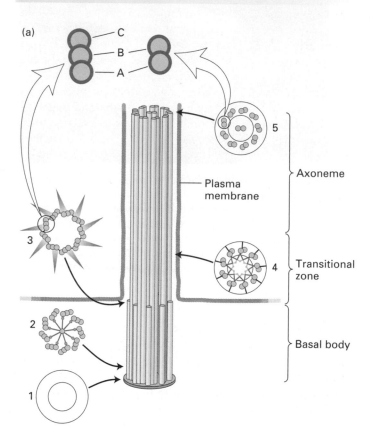

(b)

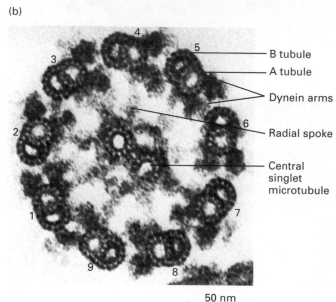

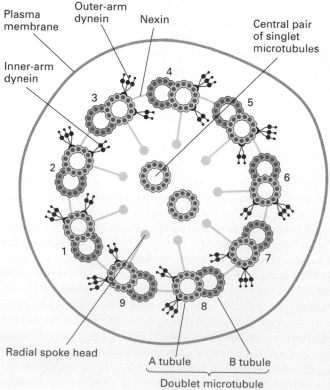

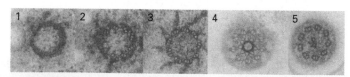

**FIGURE 18-30 Structural organization of cilia and flagella.**
(a) Cilia and flagella are assembled from a basal body, a structure built around nine linked triplet microtubules. Continuous with the A and B microtubules of the basal body are the A and B tubules of the axoneme—the membrane-bound core of the cilium or flagellum. Between the basal body and axoneme is the transitional zone. The diagram and accompanying transverse sections of the basal body, transitional zone, and axoneme show their intricate structures. (b) Thin section of a transverse section of a cilium (with plasma membrane removed) with a diagram to show the identity of the structures. [Part (a) modified from S. K. Dutcher, 2001, *Curr. Opin. Cell Biol.* **13**:49–54; courtesy of S. Dutcher. Part (b) courtesy of L. Tilney.]

through liquid. Cilia and flagella contain many different microtubule-based motors: axonemal dyneins are responsible for the beating of flagella and cilia, whereas kinesin-2 and cytoplasmic dynein are responsible for flagella and cilia assembly and turnover.

## Eukaryotic Cilia and Flagella Contain Long Doublet Microtubules Bridged by Dynein Motors

Cilia and flagella range in length from a few micrometers to more than 2 mm for some insect sperm flagella. They possess a central bundle of microtubules, called the **axoneme**, which consists of a so-called 9 + 2 arrangement of nine doublet microtubules surrounding a central pair of singlet, yet ultrastructurally distinct, microtubules (Figure 18-30a, b). Each of the nine outer doublets consists of an A microtubule with 13 protofilaments and a B microtubule with 10 protofilaments (see Figure 18-4). All the microtubules in cilia and flagella have the same polarity: the (+) ends are located at the distal tip. At its point of attachment in the cell, the axoneme connects with the **basal body**, a complicated structure containing nine triplet microtubules (see Figure 18-30a).

The structure of the axoneme is held together by three sets of protein cross-links (see Figure 18-30b). The two central singlet microtubules are connected to each other by periodic bridges, like rungs on a ladder. A second set of linkers, composed of the protein *nexin*, joins adjacent outer doublet microtubules to each other. *Radial spokes* project from each A tubule of the outer doublets toward the central pair.

The major motor protein present in cilia and flagella is *axonemal dynein*, a large, multisubunit protein related to cytoplasmic dynein. Two rows of dynein motors are attached periodically down the length of each A tubule of the outer doublet microtubules; these are called the *inner-arm* and *outer-arm* dyneins (see Figure 18-30b). It is these dynein motors interacting with the B tubule in the adjacent doublet that bring about cilia and flagella bending.

## Ciliary and Flagellar Beating Are Produced by Controlled Sliding of Outer Doublet Microtubules

Cilia and flagella are motile structures because activation of the axonemal dynein motors induces bending in them. A close examination of this motility using video microscopy reveals that a bend starts at the base of a cilium or flagellum and then propagates along the structure (Figure 18-31). A clue to how this occurs came from studies of isolated axonemes. In classic experiments, axonemes were gently treated with a protease that cleaves just the nexin links. When ATP was added to the treated axonemes, the doublet microtubules slid past one

**EXPERIMENTAL FIGURE 18-31 Video microscopy shows flagellar movements that propel sperm and *Chlamydomonas* forward.** In both cases, the cells are moving to the left. (a) In the typical sperm flagellum, successive waves of bending originate at the base and are propagated out toward the tip; these waves push against the water and propel the cell forward. Captured in this multiple-exposure sequence, a bend at the base of the sperm in the first (*top*) frame has moved distally halfway along the flagellum by the last frame. A pair of gold beads on the flagellum are seen to slide apart as the bend moves through their region. (b) Beating of the two flagella on *Chlamydomonas* occurs in two stages, called the *effective stroke* (top three frames) and the *recovery stroke* (remaining frames). The effective stroke pulls the organism through the water. During the recovery stroke, a different wave of bending moves outward from the bases of the flagella, pushing the flagella along the surface of the cell until they reach the position to initiate another effective stroke. Beating commonly occurs 5–10 times per second. [Part (a) from C. Brokaw, 1991, *J. Cell Biol.* **114**(6): cover photograph; courtesy of C. Brokaw. Part (b) courtesy of S. Goldstein.]

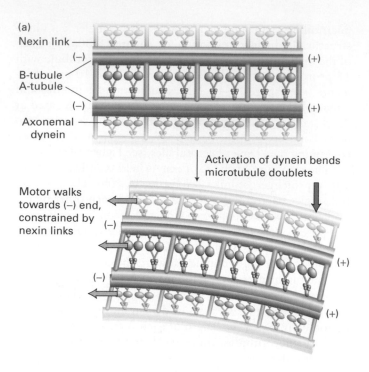

(a)

Nexin link

B-tubule
A-tubule

Axonemal
dynein

(+)
(+)

Activation of dynein bends
microtubule doublets

Motor walks
towards (−) end,
constrained by
nexin links

(b) Nexin links removed by protease

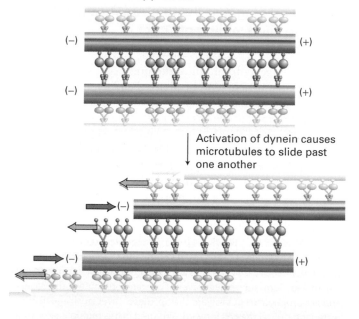

Activation of dynein causes
microtubules to slide past
one another

(c)

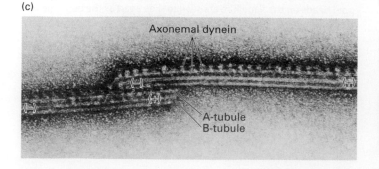

Axonemal dynein

A-tubule
B-tubule

**FIGURE 18-32 Cilia and flagella bending mediated by axonemal dynein.** (a) Axonemal dynein attached to an A tubule of an outer doublet pulls on the B tubule of the adjacent tubule trying to move to the (−) end. Because the adjacent tubules are tethered by nexin, the force generated by dynein bends the cilium or flagellum. (b) Experimental evidence for the model in (a). When the nexin links are cleaved with a protease and ATP added to induce dynein activity, the microtubule doublets slide past one another. (c) Electron micrograph of two doublet microtubules in a protease-treated axoneme incubated with ATP. In the absence of cross-linking proteins, doublet microtubules slide excessively. The dynein arms can be seen projecting from A tubules and interacting with B tubules of the left microtubule doublet. [Part (a) Courtesy of Wallace Marshall; part (c) courtesy of P. Satir.]

another as dynein, attached to the A tubule of one doublet, "walked" down the B tubule of the adjacent doublet (Figure 18-32b, c). In an axoneme with intact nexin links, the action of dynein induces flagellar bending as the microtubule doublets are connected to one another (Figure 18-32a).

How specific subsets of dynein are activated and how a wave of activation is propagated down the axoneme are not yet fully understood, but post-translational modifications of tubulin may play a role. Recall from Section 18.4 that post-translational modifications of tubulin subunits can affect the interactions between microtubules and motor proteins. The B tubules of the outer axoneme doublets are often polyglutamylated, and this modification strongly affects the interaction of the inner-arm dynein with the B tubule. Since inner-arm dynein motors mainly affect the waveform of the ciliary beat, it is this aspect of ciliary function that is compromised in mutants unable to undergo polyglutamylation.

## Intraflagellar Transport Moves Material up and down Cilia and Flagella

Although axonemal dynein is involved in bending flagella, another type of motility has been observed more recently. Careful examination of flagella on the biflagellate green alga *Chlamydomonas reinhardtii* revealed cytoplasmic particles moving at a constant speed of ∼2.5 μm/s toward the tip of the flagella (anterograde movement) and other particles moving at ∼4 μm/s from the tip to the base (retrograde movement). This transport is known as *intraflagellar transport (IFT)* and occurs in both cilia and flagella. Light and electron microscopy revealed that the particles move between the outer doublet microtubules and the plasma membrane (Figure 18-33). Analysis of algal mutants demonstrated that the anterograde movement is powered by kinesin-2 and the retrograde movement is powered by cytoplasmic dynein.

The anterograde and retrograde IFT particles transported in *Chlamydomonas* flagella have been isolated and their composition determined. They consist of two distinct protein complexes, called IFT complex A and IFT complex B. By analyzing the phenotypes of cells having mutations affecting complexes A or B, it has been found that complex B is necessary for anterograde IFT transport, whereas complex A is important for retrograde IFT transport. Despite this segregation of function, both complexes are transported in both directions. All the

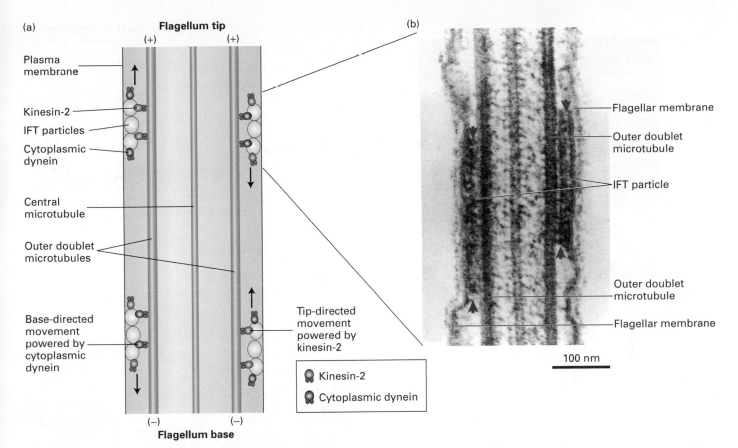

**FIGURE 18-33 Intraflagellar transport.** (a) Particles are transported between the plasma membrane and the outer doublet microtubules. Transport of the particles to the tip is dependent on kinesin-2, whereas transport toward the base is mediated by cytoplasmic dynein. (b) Thin-section electron micrograph shows IFT particles in a section of a *Chlamydomonas* flagellum. [Part (b) from J. L. Rosenbaum and G. B. Witman, 2003, *Nature Rev. Mol. Cell Biol.* **3**:813–825.]

components of IFT particles have homologs in organisms containing cilia, such as nematodes, fruit flies, mice, and humans, but these particles are absent from the genomes of yeasts and plants that lack cilia, suggesting they are specific to IFT.

What is the function of IFT? Because all the microtubules have their growing (+) ends at the flagellar tip, this is the site at which new tubulin subunits and flagellar structural proteins are added. Moreover, even in cells with flagella of uniform length, the microtubules are turning over with assembly and disassembly occurring at the flagellar tips. In cells defective for kinesin-2, flagella shrink, suggesting that IFT transports new material to the tip for growth. Since IFT is a continually occurring process, what happens to the kinesin-2 molecules when they get to the tip, and where do the dynein motors come from to transport the particles retrogradely? Remarkably, dynein is carried to the tip as cargo on the anterograde-moving particles, powered by kinesin-2, and then kinesin-2 becomes cargo as the particles are transported back to the base by dynein.

## Primary Cilia Are Sensory Organelles on Interphase Cells

Many vertebrate cells contain a solitary non-motile cilium, known as the *primary cilium*. The primary cilium is a stable structure that is much more resistant to drugs like colchicine that disassemble most microtubules; after colchicine treatment the only remaining microtubules are found in the centrioles and the primary cilium (see Figure 18-12). Moreover, the tubulin in the primary cilium is highly acetylated, so using antibodies that specifically recognize acetylated α-tubulin readily identifies the single primary cilium on each interphase cell (Figure 18-34a).

Terminally differentiated and dividing cells in interphase contain primary cilia, and in the latter case the presence of the cilium is tied to the duplication cycle of the centrioles (discussed in Section 18.6), with the "older" centriole functioning as the basal body for the assembly of the cilium (Figure 18-34b).

The primary cilium is non-motile because it lacks the central pair of microtubules and dynein side arms (Figure 18-34c). Recent work has led to the discovery that the primary cilium is instead a sensory organelle, acting as the cell's "antenna" by detecting extracellular signals. For example, the sense of smell is due to binding of odorants by receptors located in the primary cilium of olfactory sensory neurons in the nose (see Chapter 22). In another example, the rod and cone cells of the eye have a primary cilium with a greatly expanded tip to accommodate proteins involved in photoreception. The retinal protein opsin moves through the cilium at about 2000 molecules per minute, transported by kinesin-2 as part of the IFT system. Defects in this transport cause retinal degeneration.

(a)

(b)

Cilium

Chromosome
Microtubule

S phase

G₁ or G₀

Early G₁

Mitosis

☐ Mother centriole
☐ Daughter centriole
(previous cycle)
☐ Daughter centriole
(current cycle)

(c) Outer microtubule
doublet

B tubule   A tubule

Ciliary membrane

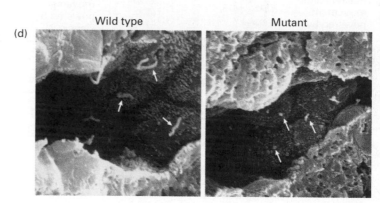

(d)            Wild type                    Mutant

**FIGURE 18-34 Many interphase cells contain a non-motile primary cilium.** (a) Fluorescence micrograph of mouse epithelial cells stained with antibodies to acetylated α-tubulin (green), which decorate the primary cilia; to pericentrin (magenta), which decorate the centrosome; and to ZO-1(red), which label the tight junctions that encircle each cell. (b) A diagram depicting how the presence of the primary cilium is tied to the centrioles, with one serving as a basal body. (c) A diagram depicting a section through a non-motile primary cilium, showing the lack of a central pair of microtubules and dynein arms typical of motile cilia and flagella. (d) Scanning electron micrographs of epithelial cells of a kidney collecting tubule from a wild-type mouse (*left*) and a mutant mouse defective in a component of the IFT particles. Arrows point to the primary cilia, which are short stubs in the mutant mouse. [(a) Courtesy of Wallace Marshall. (b and c) Modified from H. Ishikawa and W. F. Marshall, 2011, *Nature Rev. Mol Cell Biol.* **12**:222–234. (d) From G. Pazour et al., 2000, *J. Cell Biol.* **151**:709–718.]

## Defects in Primary Cilia Underlie Many Diseases

For many years the existence and function of the primary cilium was ignored. However, this situation has changed dramatically over the last decade as it has become appreciated that defects in intraflagellar transport result in the loss of primary cilia in mice (Figure 18-34d), and diseases have been traced to defects in primary cilia and IFT. One of the first clues came from the discovery that loss of a mammalian homolog of a *Chlamydomonas* IFT protein results in defects in the primary cilium and causes autosomal dominant polycystic kidney disease (ADPKD). It is believed that the primary cilia on the epithelial cells of the kidney collecting tubule act as mechanochemical sensors to measure the rate of fluid flow by the degree to which they are bent. In another example, patients with Bardet-Biedl syndrome have retinal degeneration, polydactyly (from the Greek for "many fingers"), and obesity; the syndrome can be caused by mutations in any one of 14 genes and has been traced to defects in the function of primary cilia. Many of these genes encode subunits of the BBsome, an octomeric complex that forms a coat with structural elements in common with COPI, COPII, and clathrin coats, and that traffics

membrane proteins to cilia. While defects in many of the BBsome components do not affect the structure of the primary cilium itself, they result in a lack of specific membrane receptors that are delivered to primary cilia through the interaction of the BBsome with the IFT apparatus. For example, the polydactyly seen in patients with Bardet-Biedl syndrome is due to a loss of localized Hedgehog signaling (see Chapter 16) in the primary cilium that is necessary for patterning during embryogenesis. ∎

## KEY CONCEPTS of Section 18.5

### Cilia and Flagella: Microtubule-Based Surface Structures

• Motile cilia and flagella are microtubule-based cell-surface structures with a characteristic central pair of singlet microtubules and nine sets of outer doublet microtubules (see Figure 18-30).

• All cilia and flagella grow from basal bodies, structures with nine sets of outer triplet microtubules and closely related to centrioles.

- Axonemal dyneins attached to the A tubule on one doublet microtubule interact with the B tubule of another to bend cilia and flagella.

- Cilia and flagella have a mechanism, intraflagellar transport (IFT), to transport material to their tips by kinesin-2 and from the tip back to the base by cytoplasmic dynein. This transport regulates the function and length of cilia and flagella.

- Many cells have on their surface a single non-motile primary cilium that lacks the normal central pair of microtubules and dynein side arms of motile cilia. The primary cilium functions as a sensory organelle, with receptors for extracellular signals localized to its plasma membrane. Due to its sensory function, many diseases result from defects in receptor localization or in the structure of the primary cilium itself.

## 18.6 Mitosis

Of all the events that permit the existence and perpetuation of life, perhaps the most critical is the ability of cells to accurately duplicate and then faithfully segregate their chromosomes at each cell division. During the **cell cycle,** a highly regulated process discussed in Chapter 19, cells duplicate their chromosomes precisely once during a period known as S phase (for DNA synthesis phase). Once the individual chromosomes are duplicated, they are held together by proteins called *cohesins.* The cells then pass through a period called $G_2$ (for gap 2) before entering **mitosis,** the process by which the duplicated chromosomes are segregated to the daughter cells. This process has to be very precise—loss or gain of a chromosome can either be lethal to the cell (in which case it is often not detected) or cause severe complications for the cell. It is estimated that yeast only mis-segregates one of its 16 chromosomes every 100,000 cell divisions, which makes mitosis one of the most accurate processes in biology. To achieve this accuracy, it is essential for the process to be highly regulated such that it proceeds in an orderly series of steps and that errors are not made. The timing and mechanisms that ensure the fidelity of the process are closely regulated by the cell-cycle circuitry that we discuss in detail in Chapter 19. Here we limit our discussion of this circuitry as it applies to microtubules and the mechanics of mitosis.

### Centrosomes Duplicate Early in the Cell Cycle in Preparation for Mitosis

In order to separate the chromosomes during mitosis, cells duplicate their MTOCs—their centrosomes—coordinately with the duplication of their chromosomes in S phase (Figure 18-35). The duplicated centrosomes separate and become the two spindle poles of the mitotic spindle. The number of centrosomes in animal cells has to be very carefully controlled. In fact, many tumor cells have more than two centrosomes, which contributes to genetic instability resulting from mis-segregation of chromosomes and hence **aneuploidy**

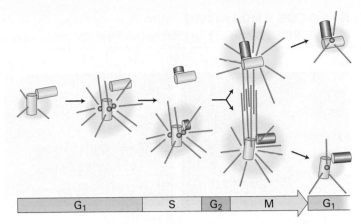

**FIGURE 18-35 Relation of centrosome duplication to the cell cycle.** After the pair of parent centrioles (green) separates slightly, a daughter centriole (blue) buds from each and elongates. By $G_2$, growth of the daughter centrioles is complete, but the two pairs remain within a single centrosomal complex. Early in mitosis, the centrosome splits, and each centriole pair migrates to opposite sides of the nucleus. The amount of pericentriolar material and the activity to nucleate microtubule assembly increases greatly in mitosis. In mitosis, these MTOCs are called spindle poles.

(unequal numbers of chromosomes). The reasons why aneuploidy results in cancer are discussed in detail in Chapter 24.

As cells enter mitosis, the activity of the two MTOCs—their ability to nucleate microtubules—increases greatly as they accumulate more pericentriolar material. Because the microtubules radiating from these two MTOCs now resemble stars, they are often called mitotic asters.

### Mitosis Can Be Divided into Six Phases

Mitosis has been divided up into several stages for ease of description (Figure 18-36a), but in reality it is a continuous process. Here, we review the major events of each stage.

The first stage of mitosis, called **prophase,** is signaled by a number of coordinated and dramatic events. First, the interphase array of microtubules is replaced as the duplicated centrosomes become more active in microtubule nucleation. This provides two sites of assembly for dynamic microtubules, forming the mitotic **asters.** Additionally, the dynamics of the growing microtubules themselves increase, due to changes in the activities of +TIPs at their (+) ends. The two asters are then moved to opposite sides of the nucleus by the action of bipolar kinesin-5 motors (see Figure 18-20) that push the intermingling astral microtubules apart. The separated centrosomes will become the two poles of the **mitotic spindle,** the microtubule-based structure that separates chromosomes. Second, protein synthesis is switched from being CAP-dependent to CAP-independent (see Chapter 4, Figure 4-24), and the internal order of membrane systems, normally dependent on the interphase array of microtubules, is disassembled. In addition, endocytosis and exocytosis are halted, and the microfilament organization is generally rearranged to give rise to a rounded cell. In the nucleus, the nucleolus breaks down and

(a)

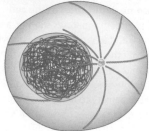 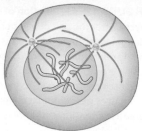 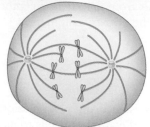 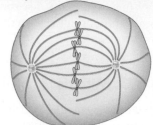

**Interphase**

**Prophase**

**Prometaphase**

**Metaphase**

Chromosome duplication and cohesion, Centrosome duplication

Breakdown of interphase microtubule display and its replacement by mitotic asters, Mitotic aster separation, Chromosome condensation, Kinetochore assembly

Nuclear envelope breakdown, Chromosomes captured, bi-oriented and brought to the spindle equator

Chromosomes aligned at the metaphase plate

(b)

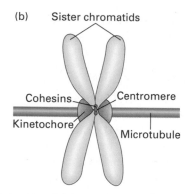

Sister chromatids

Cohesins  Centromere

Kinetochore  Microtubule

**FIGURE 18-36 The stages of mitosis.** (a) Upper panels show stages in cultured PtK2 cells stained blue for DNA and green for tubulin. Lower diagrams show the different stages and the events that occur in them. Mitosis is a continuous process, and it has simply been divided into stages for ease of description. (b) Parts of a condensed chromosome in mitosis. The duplicated chromosome has two sister chromatids (each is a single DNA duplex), held together by cohesins at a constricted region called the centromere. The centromere is also the site where the kinetochore forms, which makes attachments to the kinetochore microtubules. [Part (a) micrographs courtesy of T. Wittmann.]

chromosomes begin to condense. Cohesins holding together each pair of duplicated chromosomes, or **sister chromatids** as they are called at this stage, are degraded except at the centromeric region, where the two sister chromatids remain linked by intact cohesins (Figure 18-36b). Also during prophase, specialized structures called **kinetochores**, which will become sites of microtubule attachment, assemble at the centromeric region of each sister chromatid. As discussed in more detail in Chapter 19, all these events are coordinated by a rapid increase in the activity of the mitotic cyclin-CDK complex, which is a kinase that phosphorylates multiple proteins.

The next stage of mitosis, **prometaphase,** is initiated by the breakdown of the nuclear envelope and nuclear pores, and disassembly of the lamin-based nuclear lamina. Microtubules assembled from the spindle poles search and "capture"

chromosome pairs by associating with their kinetochores. Each chromatid has a kinetochore, so a sister chromatid pair has two kinetochores, each of which becomes attached to opposite spindle poles during prometaphase in a critical process we discuss in detail below. When sister chromatids are attached to both spindle poles, they are said to be *bi-oriented.* They then become aligned equidistant between the two spindle poles in a process known as *congression.* Prometaphase continues until all chromosomes have congressed, at which point the cell enters the next stage, **metaphase,** defined as the stage when all the chromosomes are aligned at the *metaphase plate.*

The next stage, **anaphase,** is induced by activation of the anaphase-promoting complex/cyclosome (APC/C) (discussed in Chapter 19). The activated APC/C (through several intermediary steps) ultimately leads to the destruction of the cohesins

**Anaphase**

**Telophase**

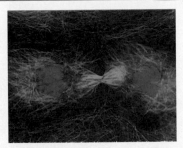

**Cytokinesis**

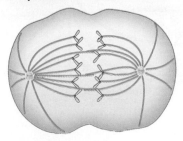

APC/C activated and
cohesins degraded
Anaphase A: Chromosome
movement to poles
Anaphase B:
Spindle pole separation

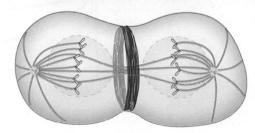

Nuclear envelope reassembly,
Assembly of contractile ring

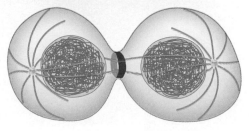

Reformation of interphase microtubule array,
Contractile ring forms cleavage furrow

that were holding the sister chromatids together, so that now each separated chromosome can be pulled to its respective pole by the microtubules attached to its kinetochore. This movement is known as *anaphase A*. A separate and distinct movement also occurs: the movement of the spindle poles farther apart in a process known as *anaphase B*. Now that the chromosomes have separated, the cell enters **telophase,** when the nuclear envelope reforms, the chromosomes decondense, and the cell is pinched into two daughter cells by the contractile ring during **cytokinesis.**

## The Mitotic Spindle Contains Three Classes of Microtubules

Before we discuss the mechanisms involved in the remarkable process of mitosis, it is important to understand the three distinct classes of microtubules that emanate from the spindle poles, which is where all their (−) ends are embedded. The first class is the *astral microtubules,* which extend from the spindle poles to the cell cortex (Figure 18-37). By interacting with the cortex, the astral microtubules perform the critical function of orienting the spindle with the axis of cell division. The second class, *kinetochore microtubules,* function by a search-and-capture mechanism to link the spindle poles to the kinetochores on sister chromatid pairs. During anaphase A, the kinetochore microtubules transport the newly separated chromosomes to their respective poles. The third set of microtubules extend from each spindle pole body

toward the opposite one and interact together in an antiparallel manner; these are called *polar microtubules.* These microtubules are responsible initially for pushing the duplicated centrosomes apart during prophase, then for maintaining the structure of the spindle, and then for pushing the spindle poles apart in anaphase B.

Note that all the microtubules in each half of the symmetrical spindle have the same orientation except for some polar microtubules, which extend beyond the midpoint and interdigitate with polar microtubules from the opposite pole.

## Microtubule Dynamics Increase Dramatically in Mitosis

Although we have drawn static images of the stages of mitosis, microtubules in all stages of mitosis are highly dynamic. As discussed previously, as cells enter mitosis, the ability of their centrosomes to nucleate assembly of microtubules increases significantly (see Figure 18-35). In addition, microtubules become much more dynamic. How was this determined? In principle, you could watch microtubules and follow their individual behaviors. However, in general, there are too many microtubules in a mitotic spindle to do this. To get an average value for how dynamic microtubules are, researchers have introduced fluorescent labeled tubulin into cells, which becomes incorporated randomly into all microtubules. They then bleach the fluorescent label in a small region of the mitotic spindle and measure the rate at which fluorescence

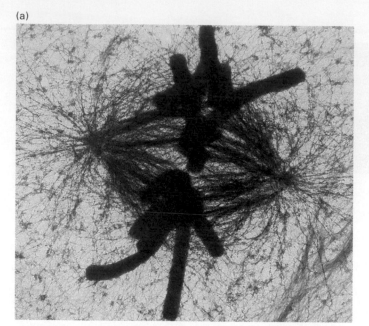

(a)

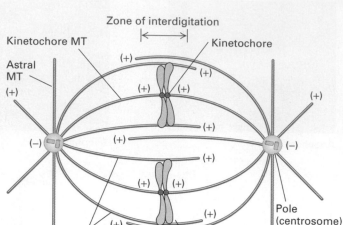

(b)

**FIGURE 18-37 Mitotic spindles have three distinct classes of microtubules.** (a) In this high-voltage electron micrograph, microtubules were stained with biotin-tagged anti-tubulin antibodies to increase their size. The large cylindrical objects are chromosomes. (b) Schematic diagram corresponding to the metaphase cell in (a). Three sets of microtubules (MTs) make up the mitotic apparatus. All the microtubules have their (−) ends at the poles. Astral microtubules project toward the cortex and are linked to it. Kinetochore microtubules are connected to chromosomes. Polar microtubules project toward the cell center with their distal (+) ends overlapping. The spindle pole and associated microtubules is also known as a mitotic aster. [Part (a) courtesy of J. R. McIntosh.]

comes back in a technique known as *fluorescence recovery after photobleaching (FRAP)* (see Figure 9-21). Since the recovery of fluorescence is due to assembly of new microtubules from soluble fluorescent tubulin dimers, this represents the average rate at which microtubules turn over. In a mitotic spindle, their half-life is about 15 seconds, whereas in an interphase cell, it is about 5 minutes. It should be noted that these are bulk measurements and individual microtubules can be more stable or dynamic, as we will see.

What makes microtubules more dynamic in mitosis? As we discussed earlier, dynamic instability is a measure of relative contributions of growth rates, shrinkage rates, catastrophes, and rescues (see Figure 18-9). Analysis of microtubule dynamics in vivo shows that the enhanced dynamics of individual microtubules in mitosis is mostly generated by increased catastrophes and fewer rescues, with little change in rates of growth (i.e., lengthening) or shrinkage (i.e., shortening). Studies with extracts from frog oocytes have suggested that the main factor enhancing catastrophes in both interphase and mitotic extracts is depolymerization by kinesin-13 proteins. This can be seen in an in vitro assay where microtubule assembly from pure tubulin is nucleated from purified centrosomes (Figure 18-38a). If kinesin-13 is added into the assay, many fewer microtubules are formed. However, if the stabilizing microtubule-associated protein called XMAP215 is added with the kinesin-13, many microtubules are formed due to a dramatic reduction in catastrophe frequency. It turns out that the activity of kinesin-13 does not change significantly during the cell cycle, whereas the activity of XMAP215

is inhibited by its phosphorylation during mitosis (Figure 18-38b). This results in much more unstable (more dynamic) microtubules as the cell enters mitosis (Figure 18-38c).

## Mitotic Asters Are Pushed Apart by Kinesin-5 and Oriented by Dynein

As the two mitotic asters form, they generate interdigitated microtubules of opposite polarity between them. During prophase, the bipolar kinesin-5 interacts with the antiparallel microtubules and, by moving toward the (+) end of each microtubule, slides them apart and thereby pushes the two asters apart. The (−) end–directed motor, cytoplasmic dynein, can also contribute to aster separation as well as orienting the spindle appropriately in the cell. Dynein does this by associating with the plasma membrane and pulling on microtubules nucleated from the mitotic asters. As we discuss shortly, this same mechanism is used to elongate the spindle during anaphase B (see Figure 18-42).

## Chromosomes Are Captured and Oriented During Prometaphase

Kinetochores, the structures that mediate attachment between chromosomes and microtubules, assemble on each sister chromatid at a region called the **centromere**. The centromere is a constricted region of the condensed chromosome defined by a centromeric DNA sequence. Centromeric DNA can vary enormously in size; in budding yeast it is

(a)

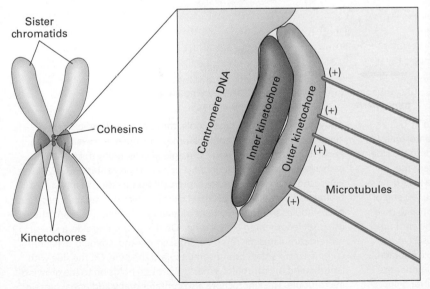

Tubulin alone

Tubulin + kinesin-13

Tubulin + kinesin-13 + XMAP215

10 µm

(b)

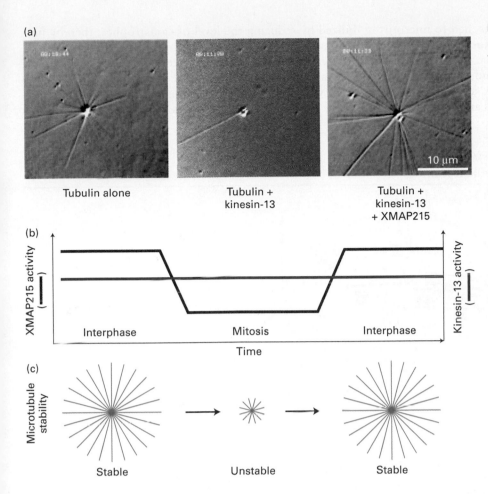

XMAP215 activity

Kinesin-13 activity

Interphase          Mitosis          Interphase

Time

(c)

Microtubule stability

Stable          Unstable          Stable

**EXPERIMENTAL FIGURE 18-38**

**Microtubule dynamics increase in mitosis due to loss of a stabilizing MAP.** (a) These three panels reveal the ability of centrosomes to assemble microtubules from pure tubulin (*left*); tubulin and the destabilizing protein kinesin-13 (*middle*); or tubulin, kinesin-13, and the stabilizing protein XMAP215 (*Xenopus* MAP of 215kD) (*right*). Further analysis shows that the major effect of XMAP215 is to suppress catastrophes induced by kinesin-13. (b) The increased dynamics of microtubules in mitosis is due to the inactivation of XMAP215 by phosphorylation. (c) Diagram relating the different stabilities of microtubules in interphase and mitosis. Note that in addition to differential *stability* between interphase and mitosis, the ability of MTOCs to *nucleate* microtubules also increases dramatically in mitosis. [Part (a) from Kinoshita et al., 2001, *Science* **294**:1340–1343. Part (b) from Kinoshita et al, 2002, *Trends Cell Biol.* **12**:267–273.]

about 125 base pairs, whereas in humans it is on the order of 1 Mbp (see Chapter 6). Kinetochores contain many protein complexes to facilitate the linkage between centromeric DNA and microtubules. In animal cells, the kinetochore consists of a centromeric DNA layer and inner and outer kinetochore layers, with the (+) ends of the kinetochore microtubules terminating in the outer layer (Figure 18-39). Yeast kinetochores are attached by a single microtubule to their spindle pole, human kinetochores are attached by about 30, and plant chromosomes by hundreds.

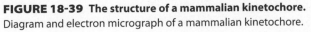

Sister chromatids

Cohesins

Kinetochores

Centromere DNA

Inner kinetochore

Outer kinetochore

(+)

(+)

(+)

(+)

Microtubules

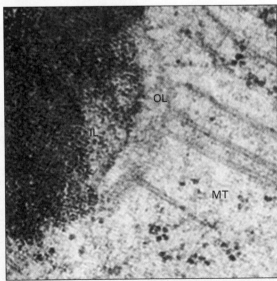

OL

IL

MT

**FIGURE 18-39  The structure of a mammalian kinetochore.** Diagram and electron micrograph of a mammalian kinetochore.

[Modified from B. McEwen et al., 1998, *Chromosoma* **107**:366; courtesy of B. McEwen.]

How does a kinetochore become attached to microtubules in prometaphase? Microtubules nucleated from the spindle poles are very dynamic, and when they contact the kinetochore, either laterally or at their end, this can lead to chromosomal attachment (Figure 18-40a, steps **1a** and **1b**). Microtubules "captured" by kinetochores are selectively stabilized by reducing the level of catastrophes, thereby promoting the chance that the attachment will persist.

Recent studies have uncovered a mechanism involving Ran, a small GTPase, that enhances the chance that microtubules will encounter kinetochores. Recall that during interphase the Ran GTPase cycle is involved in transport of

**VIDEO:** Microtubule Dynamics in Mitosis
**FOCUS ANIMATION:** Microtubule Dynamics

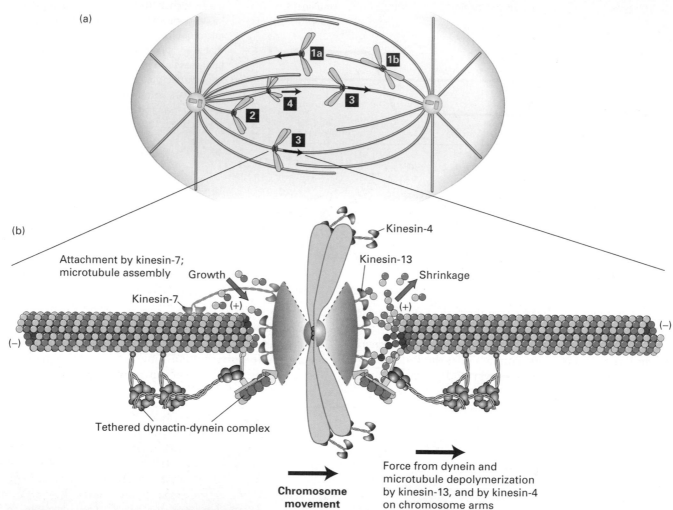

**FIGURE 18-40 Chromosome capture and congression in prometaphase.** (a) In the first stage of prometaphase, chromosomes become attached, either to the end of a microtubule (**1a**) or to the side of a microtubule (**1b**). The chromosome is then drawn toward the spindle pole by dynein-dynactin associated with one of the kinetochores of the chromosome, as this motor moves toward the (−) end of the microtubule (**2**). Eventually, a microtubule from the opposite pole finds and becomes attached to the free kinetochore, and the chromosome is now said to be bi-oriented (**3**). Once some chromosomes are bi-oriented, others, having established one kinetochore-pole interaction, use CENP-E/kinesin-7 on their free kinetochore to aid in orientation (**4**). The bi-oriented chromosomes then move to a central point between the spindle poles in a process known as chromosome congression. Note that during these steps, chromosome arms point

away from the closest spindle pole: this is due to chromokinesin/kinesin-4 motors on the chromosome arms moving toward the (+) ends of the polar microtubules. In animal cells, many microtubules associate with each kinetochore. For ease of presentation, only single kinetochore microtubules are shown. (b) Congression involves bidirectional oscillations of chromosomes, with one set of kinetochore microtubules shortening on one side of the chromosomes and the other set lengthening on the other. On the shortening side, a kinesin-13 protein stimulates microtubule disassembly and a dynein-dynactin complex moves the chromosome toward the pole. On the side with lengthening microtubules, kinesin-7 protein holds on to the growing microtubule. The kinetochore also contains many additional protein complexes not shown here. [Modified from Cleveland et al., 2003, *Cell* **112:**407–421.]

proteins into and out of the nucleus through nuclear pores (Chapter 13; see Figure 13-37). During mitosis, when the nuclear membrane and pores have disassembled, an exchange factor for the Ran GTPase is bound to chromosomes, thereby generating a higher local concentration of Ran-GTP in the vicinity of the chromosomes. Because the enzyme that stimulates GTP hydrolysis on Ran—the Ran GAP—is evenly distributed in the cytosol, this generates a gradient of Ran-GTP centered on the chromosomes. Ran-GTP induces the association of cytosolic microtubule-stabilizing factors with the microtubule, resulting in enhanced microtubule growth, and in this way biases growth of microtubules nucleated from spindle poles toward chromosomes.

Once a kinetochore is attached laterally or terminally to a microtubule, the motor protein dynein-dynactin associates with the kinetochore to move the duplicated chromosome down the microtubule toward the spindle pole. This eventually results in an end-on attachment of the microtubule to one kinetochore (Figure 18-40a, step **2**). This movement helps orient the sister chromatid so that the unoccupied kinetochore on the opposite side is pointing toward the distal spindle pole. Eventually a microtubule from the distal pole will capture the free kinetochore; at this point the sister chromatid pair is now said to be *bi-oriented* (Figure 18-40a, step **3**). With the two kinetochores attached to opposite poles, the duplicated chromosome is now under tension, being pulled in both directions by the two sets of kinetochore microtubules. When one or a few chromosomes are bi-oriented, other chromosomes use these existing kinetochore microtubules to contribute to their orientation and movement to the spindle center. This is mediated by kinesin-7 (also known as CENP-E) associated with the free kinetochore moving the chromosome to the (+) end of the kinetochore microtubule (Figure 18-40a, step **4**).

## Duplicated Chromosomes Are Aligned by Motors and Microtubule Dynamics

During prometaphase, the chromosomes come to lie at the midpoint between the two spindle poles, in a process known as chromosome *congression*. During this process, bi-oriented chromosome pairs often oscillate backward and forward before arriving at the midpoint. Chromosome congression involves the coordinated activity of several microtubule-based motors together with regulators of microtubule assembly and disassembly (Figure 18-40b). These regulators are localized at the kinetochores, but how they are maintained there is poorly understood—they are not part of the stable kinetochore complexes described in the next section. The oscillating behavior of chromosomes involves lengthening of microtubules attached to one kinetochore and shortening of microtubules attached to the other kinetochore, all without losing their attachments. In metazoans, several microtubule-based motors associated with the kinetochore contribute to this process. First, dynein-dynactin provides the strongest force pulling the chromosome pair toward the more *distant* pole. This movement requires simultaneous shortening of the microtubule, which is enhanced by kinetochore-localized kinesin-13. The microtubules associ-

ated with the other kinetochore have to grow as the chromosome moves. Anchored at this kinetochore is the kinesin-related motor kinesin-7, which holds on to the growing (+) end of the lengthening microtubule. Also contributing to congression is another kinesin, kinesin-4, which associates with the chromosome arms. Kinesin-4, a (+) end–directed motor, interacts with the polar microtubules to pull the chromosomes toward the center of the spindle. When the chromosomes have congressed to the metaphase plate, dynein-dynactin is released from the kinetochores and streams down the kinetochore microtubules to the poles. These different activities and opposing forces work together to bring all the chromosomes to the metaphase plate, at which point the cell is ready for anaphase.

## The Chromosomal Passenger Complex Regulates Microtubule Attachment at Kinetochores

We have mentioned that the segregation of chromosomes at mitosis must be very accurate, so it is crucial that all chromosomes are bi-oriented before anaphase begins. During the random kinetochore-to-microtubule attachment process, it is possible for mistakes to be made; for example, both kinetochores of a sister chromatid pair may attach to microtubules from the same spindle pole. If such attachments persisted during metaphase, it would result in one cell missing a chromosome and the other having an extra one, both lethal conditions. Cells have two important mechanisms to ensure that all chromosomes are correctly bi-oriented before anaphase begins.

The first mechanism ensures that the kinetochore-microtubule interactions are weak until bi-orientation occurs. When a chromosome is correctly bi-oriented, tension is produced across the chromosome and this tension leads to the kinetochore-microtubule attachments becoming stabilized. To understand how this works, we need to look a bit closer at the molecular components that link a kinetochore to a microtubule. As we discussed in Chapter 6, kinetochores assemble on regions of chromosomal DNA marked by a specific centromeric-specific H3 histone variant called CENP-A. This marks the site for kinetochore assembly, which is a very complicated process. About half a dozen distinct stable protein complexes consisting of more than 40 different proteins have been shown to associate with this centromeric region in yeast. Essentially all these protein complexes are conserved in humans, which is not surprising because of the fundamental importance of kinetochores. One of these, the so-called Ndc80 complex, is long and flexible, and many copies of it link the inner kinetochore with the (+) end of the microtubule in a sleeve type of arrangement (Figure 18-41a). The function of Ndc80 and many of the associated factors at the kinetochore is regulated by the *chromosomal passenger complex (CPC)*. This complex associates with the inner centromeric region of chromosomes early in mitosis, and among its components is a protein kinase called *Aurora B*. Recent work has shown that part of the initial mechanism for recruitment of the CPC is through its binding to phosphorylated CENP-A. Once associated with the centromeric region, Aurora B within the CPC can phosphorylate several components in the near vicinity, including the

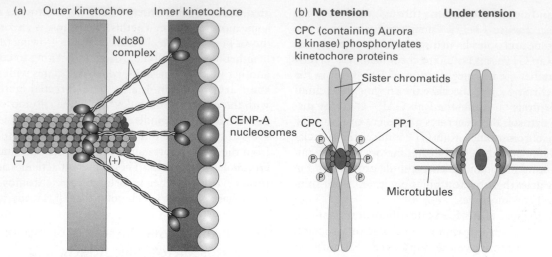

(a) Outer kinetochore    Inner kinetochore

Ndc80 complex

(−)    (+)

CENP-A nucleosomes

(b) **No tension**    **Under tension**

CPC (containing Aurora B kinase) phosphorylates kinetochore proteins

Sister chromatids

CPC    PP1

Microtubules

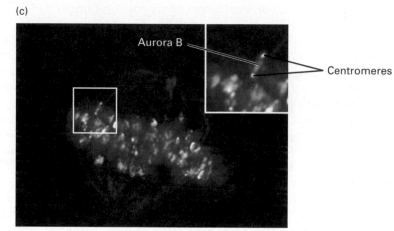

(c)

Aurora B

Centromeres

**FIGURE 18-41 CPC regulation of microtubule-kinetochore attachment.** The Ndc80 complex forms a critical and regulated attachment between the kinetochore and microtubule (+) ends, (a) Diagram showing the sleeve arrangement of Ndc80 complex linking the inner plate of the kinetochore to the (+) end of a microtubule embedded in the outer plate. (b) Diagram of the relationship between the chromosomal passenger complex (CPC), which contains the kinase Aurora B, and the outer kinetochore where the phosphatase PP1 binds. Notice that when both kinetochores are under tension the outer kinetochores move away from the CPC; as a result, Aurora B cannot phosphorylate components in the outer plate, which includes the microtubule-binding site of the Ndc80 complex. (c) Cell in the metaphase stage of mitosis stained for tubulin (red), DNA (blue), Aurora B kinase (green), and centromeres (magenta). Notice how the centromeres are pulled away from Aurora B (inset). [(a) Modified from S. Santaguida and A. Musacchio, 2009, *EMBO J.* **28:**2511–2531. (c) From S. Ruchaud, M. Carmena, and W. C. Earnshaw, 2007, *Nature Rev. Mol. Cell Biol.* **8:**798–812.]

Ndc80 complex, which loosens the attachment of Ndc80 to the microtubule. The phosphorylation of these components is not stable—another protein, the phosphatase PP1 associated with the outer kinetochore, can dephosphorylate them. Thus when the kinetochores on a pair of sister chromatids are not under tension, Ndc80 is continually phosphorylated by CPC and dephosphorylated by PP1. The result is a weak interaction between the kinetochore and microtubule. However, when bi-orientation occurs, the tension generated pulls on both kinetochores to move them away from the CPC (Figure 18-41b, c). Additionally, it is believed that the tension extends the flexible Ndc80 complex to increase the spacing between the inner and outer kinetochore plates. As a result of these movements, Ndc80 cannot be phosphorylated by Aurora B and the dephosphorylated state of Ndc80 renders it more firmly attached to the microtubule. In this way, microtubule attachments to bi-oriented chromosomes are selectively stabilized. While the CPC is important for bi-orientation of each individual chromosome, this does not ensure that *all* chromosomes are bi-oriented before anaphase begins.

The second mechanism to ensure correct chromosome segregation is the spindle assembly checkpoint, a signaling circuitry that stops progression of the cell cycle into anaphase until tension is present at *all* the kinetochores. This mechanism, discussed in detail in Chapter 19, guarantees that all the chromosomes are correctly bi-oriented before the cell proceeds into anaphase.

## Anaphase A Moves Chromosomes to Poles by Microtubule Shortening

The onset of anaphase A is one of the most dramatic movements that can be observed in the light microscope. When the spindle assembly checkpoint has been satisfied, APC/C activation induces proteolysis of the remaining cohesins holding the sister chromatids together. Suddenly, the two paired sister chromatids separate from each other and are drawn to their respective poles. The movement is sudden because the kinetochore microtubules are under tension, and as soon as the cohesin attachments between the chromatids are released, the separated chromosomes are free to move.

Experiments with isolated metaphase chromosomes have shown that anaphase A movement can be powered by microtubule shortening, utilizing the stored structural strain released by removing the GTP-bound tubulin subunits at the microtubule tip. This can be nicely demonstrated in vitro. When metaphase chromosomes arc added to purified microtubules, they bind preferentially to the (+) ends of the microtubules. Dilution of the mixture to reduce the concentration of free tubulin dimers results in the movement of the chromosomes toward the (−) ends by microtubule depolymerization at the chromosome-bound (+) end. In addition, recent experiments have shown that in *Drosophila* two kinesin-13 proteins, members of a class of microtubule-depolymerizing proteins (see Figure 18-15), also contribute to chromosome movement in anaphase A. One of the kinesin-13 proteins is localized at the kinetochore and enhances disassembly there (Figure 18-42, **A1**), and the other is localized at the spindle pole, enhancing depolymerization there (Figure 18-42, **A2**). Thus, at least in the fly, anaphase A is powered in part by kinesin-13 proteins specifically localized at the kinetochore

and spindle pole to shorten the kinetochore microtubules at both their (+) and (−) ends, drawing the chromosomes to the poles.

## Anaphase B Separates Poles by the Combined Action of Kinesins and Dynein

The second part of anaphase involves separation of the spindle poles in a process known as anaphase B. A major contributor to this movement is the involvement of the bipolar kinesin-5 proteins (Figure 18-42, **B1**). These motors associate with the overlapping polar microtubules, and since they are (+) end–directed motors, they push the poles apart. While this is happening, the polar microtubules have to grow to accommodate the increased distance between the spindle poles. Another motor—the microtubule (−) end–directed motor cytoplasmic dynein, localized and anchored on the

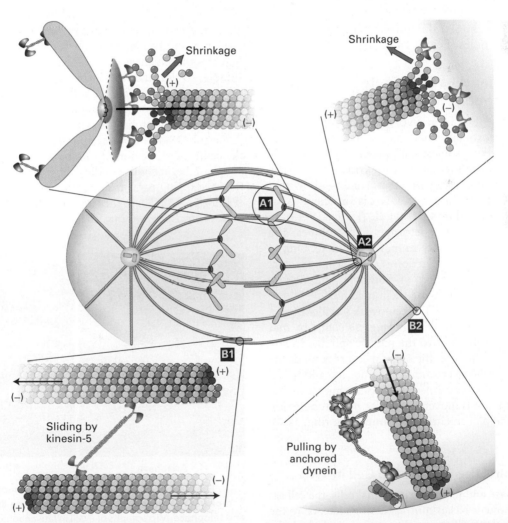

**FIGURE 18-42 Chromosome movement and spindle pole separation in anaphase.** Anaphase A movement is powered by microtubule-shortening kinesin-13 proteins at the kinetochore (**A1**) and at the spindle pole (**A2**). Note that the chromosome arms still point away from the spindle poles due to associated chromokinesin/ kinesin-4 members, so the depolymerization force has to be able to

overcome the force pulling the arms toward the center of the spindle. Anaphase B also has two components: sliding of antiparallel polar microtubules powered by a kinesin-5 (+) end–directed motor (**B1**), and pulling on astral microtubules by dynein-dynactin located at the cell cortex (**B2**). Arrows indicate the direction of movement generated by the respective forces. [Modified from Cleveland et al., 2003, *Cell* **112**:407–421.]

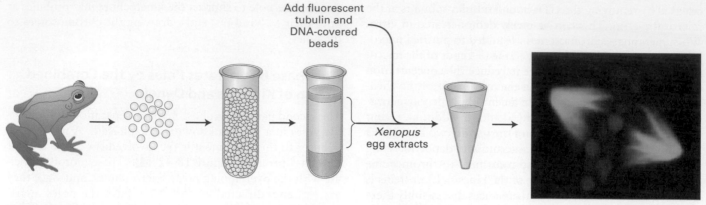

Add fluorescent tubulin and DNA-covered beads

*Xenopus* egg extracts

**EXPERIMENTAL FIGURE 18-43  Mitotic spindles can form in the absence of centrosomes.** Centrosome-free extracts can be isolated from frog oocytes arrested in mitosis by centrifuging eggs to separate a soluble material from the organelles and yolk. When fluorescently labeled tubulin (green) is added to extracts of the soluble material together with beads covered with DNA (red), mitotic spindles spontaneously form around the beads from randomly nucleated microtubules. [Modified from Kinoshita et al, 2002, *Trends Cell Biol.* **12:**267–273, and Antonio et al., 2000, *Cell* **102:**425.]

cell cortex—pulls on the astral microtubules and thus helps separate the spindle poles (Figure 18-42, **B2**).

## Additional Mechanisms Contribute to Spindle Formation

There are a number of cases in vivo where spindles form in the absence of centrosomes. This implies that nucleation of microtubules from centrosomes is not the only way a spindle can form. Studies exploiting mitotic extracts from frog eggs—extracts that do not contain centrosomes—show that the addition of beads covered with DNA is sufficient to assemble a relatively normal mitotic spindle (Figure 18-43). In this system, the beads recruit preformed microtubules, and factors in the extract cooperate to make a spindle. One of the factors necessary for this reaction is cytoplasmic dynein, which is proposed to bind to two microtubules and migrate to their (−) ends, thereby drawing them together.

As mentioned in Section 18.1, recent work has identified a new γ-TuRC-associated complex, the *augmin complex*, that also contributes microtubules to the mitotic spindle. In late prometaphase and metaphase, the augmin complex binds the sides of existing spindle microtubules to nucleate additional microtubules that contribute especially to polar microtubules, and therefore anaphase B movement. Polar microtubules are also involved in the placement of the contractile ring, as we discuss next.

## Cytokinesis Splits the Duplicated Cell in Two

During late anaphase and telophase in animal cells, the cell assembles a microfilament-based *contractile ring* attached to the plasma membrane that will eventually contract and pinch the cell into two, a process known as cytokinesis (see Figure 18-36). The contractile ring is a thin band of actin filaments of mixed polarity interspersed with myosin-II bipolar filaments (see Figure 17-36). On receiving a signal, the ring contracts first to generate a *cleavage furrow* and then to pinch the cell into two.

Two aspects of the contractile ring are essential to its function. First, it has to be appropriately placed in the cell. It is known that this placement is determined by signals provided by the spindle, so that the ring forms equidistant between the two spindle poles. The signal is provided, at least in part, by the chromosomal passenger complex (CPC) that regulates the attachment of microtubules to kinetochores during prometaphase (see Figure 18-41). Up to anaphase, the CPC is associated with the centromeric region of unseparated chromosomes. When anaphase begins, it leaves the centromeres and associates with the overlapping polar microtubules at the center of the spindle (Figure 18-44). There the CPC recruits another protein complex, *centralspindlin*, that includes a (+) end–directed kinesin

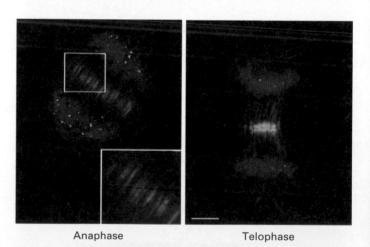

Anaphase                Telophase

**EXPERIMENTAL FIGURE 18-44  The Chromosomal Passenger Complex (CPC) remains at the spindle midzone during anaphase and telophase.** Micrographs of a cell in late anaphase (left) and telophase (right) showing microtubules (red), DNA (blue), Aurora B kinase (green), and chromosome centromeres (magenta). Notice how the Aurora B in the CPC complex concentrates in the region where the polar microtubules overlap and where the contractile ring will form. Scale bar = 5 μm. [From S. Ruchard, M. Carmena, and W. C. Earnshaw, 2007, *Nature Rev. Mol. Cell Biol.* **8:**798–812.]

motor protein, which concentrates at the middle of the spindle due to its motor activity. As anaphase B continues, centralspindlin recruits an exchange factor for RhoA. Recall from Chapter 17 that Rho proteins are small GTP-binding proteins that are activated by exchange factors to catalyze the exchange of GDP for GTP (see Figure 17-42). Once activated, the RhoA-GTP activates a formin protein to drive the nucleation and assembly of actin filaments that make up the contractile ring (see Figure 17-44). In this way, the position of the spindle directly defines the site of contractile ring formation, and hence cytokinesis.

The second important aspect of the contractile ring is the timing of its contraction—if it were to contract before all chromosomes have moved to their respective poles, disastrous genetic consequences would ensue. As we discuss in Chapter 19, a signaling pathway has been discovered in budding yeast called the *spindle position checkpoint* that pauses the cell cycle to ensure that cytokinesis does not occur until the spindle is appropriately oriented. The mechanism of this coordination in animal cells is still being unraveled.

## Plant Cells Reorganize Their Microtubules and Build a New Cell Wall in Mitosis

Interphase plant cells lack a central MTOC that organizes microtubules into the radiating interphase array typical of animal cells. Instead, numerous MTOCs containing γ-tubulin line the cortex of plant cells and nucleate the assembly of transverse bands of microtubules below the cell wall (Figure 18-45, *left*). The microtubules, which are of mixed polarity, are released from the cortical MTOCs by the action of katanin, a microtubule-severing protein; loss of katanin gives rise to very long microtubules and misshapen cells. The reason for this is that these cortical microtubules, which are crosslinked by plant-specific MAPs, aid in laying down extracellular cellulose microfibrils, the main component of the rigid cell wall (see Figure 20-40).

Although mitotic events in plant cells are generally similar to those in animal cells, formation of the spindle and cytokinesis have unique features in plants (Figure 18-45). Plant cells

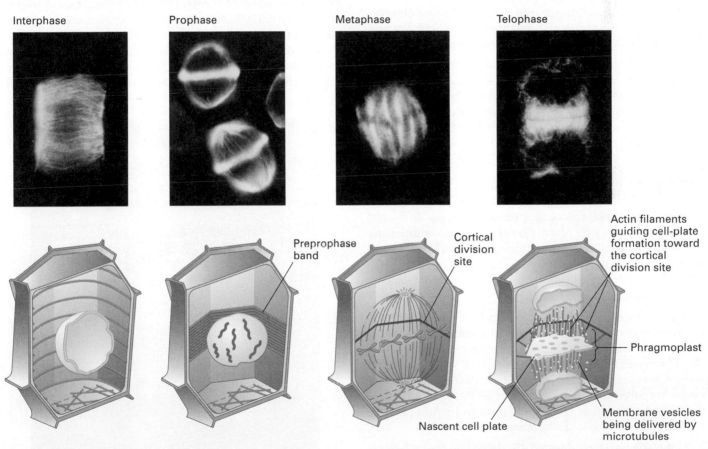

Interphase  Prophase  Metaphase  Telophase

Preprophase band

Cortical division site

Actin filaments guiding cell-plate formation toward the cortical division site

Phragmoplast

Membrane vesicles being delivered by microtubules

Nascent cell plate

**FIGURE 18-45 Mitosis in a higher plant cell.** Immunofluorescence micrographs (*top*) and corresponding diagrams (*bottom*) showing arrangement of microtubules in interphase and mitotic plant cells. A cortical array of microtubules girdles a cell during interphase. As the cell enters prophase, the microtubules (green), together with actin filaments (red) assemble under the cell cortex into the preprophase band that marks the future division site. As the cell enters prometaphase and metaphase, a spindle forms similar to that seen in animal cells. However, due to the cell wall, cytokinesis is very different in plants than in animal cells. Microtubules deliver membranes to assemble a nascent cell plate called a phragmoplast, whose organization is defined by actin filaments linked to the cortical division site. Eventually, the newly delivered membranes fuse at the cortical division site and become part of the plasma membranes of the two cells. Enzymes secreted with the vesicles then build a cell wall between the two daughter cells. [Adapted from G. Jurgens, 2005, *Ann. Rev. Plant Biol.* **56:**281–289; micrographs courtesy of Susan M. Wick.]

bundle their cortical microtubules and actin filaments into a *preprophase band* and reorganize them into a spindle at prophase without the aid of centrosomes. The site of the preprophase band defines the later division site. At metaphase, the mitotic apparatus appears much the same in plant and animal cells. However, the division of the cell into two is quite different from animal cells. Golgi-derived vesicles, which appear at telophase, are transported along microtubules to form the *nascent cell plate*. The cell plate expands and is guided toward the division site by actin filaments to form the **phragmoplast,** a membrane structure that replaces the animal-cell contractile ring. The membranes of the vesicles forming the phragmoplast become the plasma membranes of the daughter cells. The contents of these vesicles, such as polysaccharide precursors of cellulose and pectin, form the early cell plate, which develops into the new cell wall between the daughter cells.

## KEY CONCEPTS of Section 18.6

### Mitosis

• Mitosis—the accurate separation of duplicated chromosomes—involves a molecular machine comprising dynamic microtubules and microtubule-associated motors.

• The mitotic spindle has three classes of microtubules, all emanating from the spindle poles—kinetochore microtubules, which attach to chromosomes; polar microtubules from each spindle pole, which overlap in the middle of the spindle; and astral microtubules, which extend to the cell cortex (see Figure 18-37).

• In the first stage of mitosis, prophase, the nuclear chromosomes condense and the spindle poles move to either side of the nucleus (see Figure 18-36).

• At prometaphase, the nuclear envelope breaks down and microtubules emanating from the spindle poles capture sister chromatid pairs at their kinetochores. The two kinetochores (one on each chromatid) become attached to opposite spindle poles (bi-oriented), which allows the chromosome to congress to the middle of the spindle.

• The chromosomal passenger complex (CPC) associated with the inner kinetochore plate keeps microtubule attachments weak by the activity of its kinase component Aurora B, which phosphorylates critical kinetochore proteins. When a chromosome is bi-oriented, tension is generated and the Aurora B substrates are pulled away from the kinase (see Figure 18-41). Without phosphorylation of kinetochore proteins by Aurora B, the chromosome-kinetochore attachment becomes stable.

• At metaphase, chromosomes are aligned on the metaphase plate. The spindle assembly checkpoint system monitors unattached kinetochores and delays anaphase until all chromosomes are attached.

• At anaphase, duplicated chromosomes separate and move toward the spindle poles by shortening of the kinetochore microtubules at both the kinetochore and spindle pole (anaphase A). The spindle poles also move apart, pushed by bipolar kinesin-5

moving toward the (+) ends of the polar microtubules (anaphase B). Spindle separation is also facilitated by cortically located dynein pulling on astral microtubules (see Figure 18-42).

• Since the mitotic spindle has the ability to self-assemble in the absence of MTOCs, redundant mechanisms apparently contribute to the fidelity of mitosis.

• The actin-myosin–based contractile ring, the position of which is determined by the position of the spindle, contracts to pinch the cell in two during cytokinesis.

• In plants, cell division involves the delivery of membranes by microtubules to assemble the phragmoplast, which becomes the plasma membrane of the two daughter cells.

## 18.7 Intermediate Filaments

The third major filament system of eukaryotes is collectively called **intermediate filaments.** This name reflects their diameter of about 10 nm, which is intermediate between the 6–8 nm of microfilaments and the myosin thick filaments of skeletal muscle. Intermediate filaments extend throughout the cytoplasm as well as lining the inner nuclear envelope of interphase animal cells (Figure 18-46). Intermediate filaments have several

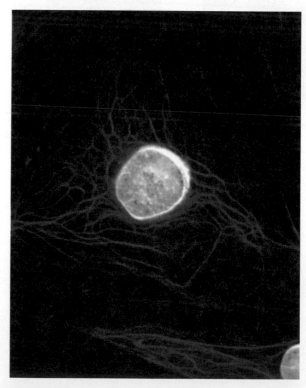

**EXPERIMENTAL FIGURE 18-46 Localization of two types of intermediate filaments in an epithelial cell.** Immunofluorescence micrograph of a PtK2 cell doubly stained with keratin (red) and lamin (blue) antibodies. A meshwork of lamin intermediate filaments can be seen underlying the nuclear membrane, whereas the keratin filaments extend from the nucleus to the plasma membrane. [Courtesy of R. D. Goldman.]

unique properties that distinguish them from microfilaments and microtubules. First, they are biochemically much more heterogeneous—that is, many different, but evolutionarily related, intermediate filament subunits exist—and are often expressed in a tissue-dependent manner. Second, they have great tensile strength, as clearly demonstrated by hair and nails, which consist primarily of the intermediate filaments of dead cells. Third, they do not have an intrinsic polarity like microfilaments and microtubules, and their constituent subunits do not bind a nucleotide. Fourth, because they have no intrinsic polarity, it is not surprising that there are no known motors that use them as tracks. Fifth, although they are dynamic in terms of subunit exchange, they are much more stable than microfilaments and microtubules because the exchange rate is much slower. Indeed, a standard way to purify intermediate filaments is to subject cells to harsh extraction conditions in a detergent so that all membranes, microfilaments, and microtubules are solubilized, leaving a residue that is almost exclusively intermediate filaments. Finally, intermediate filaments are not found in all eukaryotes. Fungi and plants do not have intermediate filaments, and insects only have one class, represented by two genes that express lamin A/C and B.

These properties make intermediate filaments unique and important structures of metazoans. The importance of intermediate filaments is underscored by the identification of more than 40 clinical disorders, some of which are discussed here, associated with defects in genes encoding intermediate filament proteins. To understand their contributions to cell and tissue structure, we first examine the structure of intermediate filament proteins and how they assemble into filaments. We then describe the different classes of intermediate filaments and the functions they perform.

## Intermediate Filaments Are Assembled from Subunit Dimers

Intermediate filaments (IFs) are encoded in the human genome by 70 different genes in at least five subfamilies. The defining feature of IF proteins is the presence of a conserved α-helical rod domain of about 310 residues that has the sequence features of a coiled-coil motif (see Figure 3-9a). Flanking the rod domain are nonhelical N- and C-terminal domains of different sizes, characteristic of each IF class.

The primary building block of intermediate filaments is a dimer held together through the rod domains that associate as a coiled-coil (Figure 18-47a). These dimers then associate in an offset fashion to make tetramers, where the two dimers are in opposite orientations (Figure 18-47b). Tetramers are assembled end-to-end and interlocked into long *protofilaments*. Four protofilaments associate into a *protofibril*, and four protofibrils associate side to side to generate the 10-nm filament. Thus an intermediate filament has 16 protofilaments in it (Figure 18-47c). Because the tetramer is symmetric, intermediate filaments have no polarity. This description of the filament is based on its structure rather than its mechanism of assembly: at present it is not yet clear how intermediate

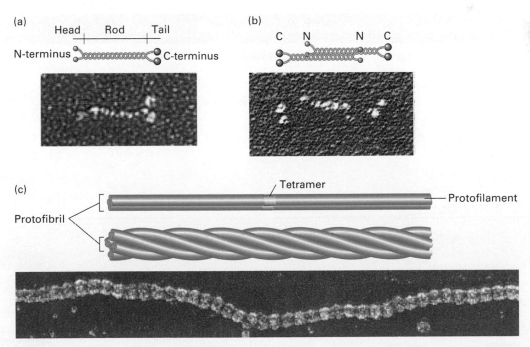

**FIGURE 18-47 Structure and assembly of intermediate filaments.** Electron micrographs and drawings of IF protein dimers, tetramers, and mature intermediate filaments from *Ascaris,* an intestinal parasitic worm. (a) IF proteins form parallel dimers through a highly conserved coiled-coil core domain. The globular heads and tails are quite variable in length and sequence between intermediate filament classes. (b) A tetramer is formed by antiparallel, staggered side-by-side aggregation of two identical dimers. (c) Tetramers aggregate end to end and laterally into a protofibril. In a mature filament, consisting of four protofibrils, the globular domains form beaded clusters on the surface. [Adapted from N. Geisler et al., 1998, *J. Mol. Biol.* **282:**601; courtesy of Ueli Aebi.]

filaments are assembled in vivo. Unlike microfilaments and microtubules, there are no known intermediate filament nucleating, sequestering, capping, or filament-severing proteins.

## Intermediate Filament Proteins Are Expressed in a Tissue-Specific Manner

Sequence analysis of IF proteins reveals that they fall into at least five distinct homology classes, with four classes showing a strong correspondence between the sequence class and the developmental origin of the cell type in which the IF protein is expressed (Table 18-1).

The **keratins** that make up classes I and II are found in epithelia; class III IF proteins are generally found in cells of mesodermal origin; and class IV IF proteins compose the **neurofilaments** found in neurons. The **lamins,** which make up class V, are found lining the nucleus of all animal tissues. We briefly summarize the five different homology classes and discuss their roles in specific tissues.

**Keratins** Keratins provide strength to epithelial cells. The first two homology classes are the so-called *acidic* and *basic keratins.* There are about 50 genes in the human genome encoding keratins, about evenly split between the acidic and basic classes. These keratin subunits assemble into an obligate dimer, so that each dimer consists of one basic chain and one acidic chain; these are then assembled into a filament as described in the previous section.

The keratins are by far the most diverse of the IF protein families, with keratin pairs showing different expression patterns between distinct epithelia and also showing differentiation-specific regulation. Among these are the so-called hard keratins that make up hair and nails. These keratins are rich in cysteine residues that become oxidized to form disulfide bridges, thereby strengthening them. This property is exploited by hair stylists: if you do not like the shape of your hair, the disulfide bonds in your hair keratin can be reduced, the hair reshaped, and the disulfide bonds reformed by oxidation—the result is "permanent" hair curling or hair straightening.

The so-called soft, or *cyto-keratins,* are found in epithelial cells. The epidermal-cell layers that make up the skin provide a good example of the function of keratins. The lowest layer of cells, the *basal layer,* which is in contact with the basal lamina, proliferates constantly, giving rise to cells called *keratinocytes.* After they leave the basal layer, the keratinocytes differentiate and express abundant cytokeratins. The cytokeratins associate with specialized attachment sites between cells, thereby providing sheets of cells that can withstand abrasion. The cells eventually die, leaving dead cells from which all cell organelles have disappeared. This dead cell layer provides an essential barrier to water evaporation, without which we could not survive. The life of a skin cell, from birth to its loss from the animal as a skin flake, is about one month.

In all epithelia, keratin filaments associate with desmosomes, which link adjacent cells together, and hemidesmosomes, which link cells to the extracellular matrix, thereby

| TABLE 18-1 | The Major Classes of Intermediate Filaments in Mammals | | |
|---|---|---|---|
| **Class** | **Protein** | **Distribution** | **Proposed Function** |
| I | Acidic keratins | Epithelial cells | Tissue strength and integrity |
| II | Basic keratins | Epithelial cells | |
| III | Desmin, GFAP, vimentin | Muscle, glial cells, mesenchymal cells | Sarcomere organization, integrity |
| IV | Neurofilaments (NFL, NFM, and NFH) | Neurons | Axon organization |
| V | Lamins | Nucleus | Nuclear structure and organization |

giving cells and tissues their strength. These structures are described in more detail in Chapter 20.

In addition to simply providing structural support, there is increasing evidence that keratin filaments provide some organization to organelles and participate in signal-transduction pathways. For example, in response to tissue injury, rapid cell growth is induced. In epithelial cells it has been shown that the growth signal requires an interaction between a cell-growth-signaling molecule and a specific keratin.

**Desmin** The class III intermediate filament proteins include vimentin, found in mesenchymal cells; GFAP (glial fibrillary acidic protein), found in glial cells; and *desmin,* found in muscle cells. Desmin provides strength and organization to muscle cells (see cartoons in Table 18-1).

In smooth muscle, desmin filaments link cytoplasmic *dense bodies,* to which the contractile myofibrils are also attached, to the plasma membrane to ensure that cells resist overstretching. In skeletal muscle, a lattice composed of a band of desmin filaments surrounds the sarcomere. The desmin filaments encircle the Z disk and are cross-linked to the plasma membrane. Longitudinal desmin filaments cross to neighboring Z disks within the myofibril, and connections between desmin filaments around Z disks in adjacent myofibrils serve to cross-link myofibrils into bundles within a muscle cell. The lattice is also attached to the sarcomere through interactions with myosin thick filaments. Because the desmin filaments lie outside the sarcomere, they do not actively participate in generating contractile forces. Rather, desmin plays an essential structural role in maintaining muscle integrity. In transgenic mice lacking desmin, for example, this supporting architecture is disrupted and Z disks are misaligned. The location and morphology of mitochondria in these mice are also abnormal, suggesting that these intermediate filaments may also contribute to organization of organelles.

**Neurofilaments** Type IV intermediate filaments consist of the three related subunits—NF-L, NF-M, and NF-H (for NF light, medium, and heavy)—that make up the neurofilaments found in the axons of nerve cells (see Figure 18-2). The three subunits differ mainly in the size of their C-terminal domains, and all form obligate heterodimers. Experiments with transgenic mice reveal that neurofilaments are necessary to establish the correct diameter of axons, which determines the rate at which nerve impulses are propagated down them.

**Lamins** The most widespread intermediate filaments are the class V lamins, which provide strength and support to the inner surface of the nuclear membrane (see Figure 18-46). Lamins are the progenitor of all IF proteins, with the cytoplasmic IFs arising by gene duplication and mutation. The lamins provide a two-dimensional network lying between the nuclear envelope and the chromatin in the nucleus. In humans, three genes encode lamins: one encodes A-type and C-type lamin and two encode B-type lamin. The B-type lamin appears to be the primordial gene and is expressed in all cells, whereas A- and C-type lamins are developmentally regulated. B-lamins are

post-translationally isoprenylated, which helps them associate with the inner nuclear envelope membrane. In addition, they bind inner nuclear membrane proteins such as emerin and lamin-associated polypeptides (LAP2). Lamins bind multiple proteins and have been proposed to play roles in large-scale chromatin organization and the spacing of nuclear pores. As cells enter mitosis, lamins become hyperphosphorylated and disassemble; in telophase they reassemble with the reassembling nuclear membrane (discussed in Chapter 19).

## Intermediate Filaments Are Dynamic

Although intermediate filaments are much more stable than microtubules and microfilaments, IF protein subunits have been shown to be in dynamic equilibrium with the existing IF cytoskeleton. In one experiment, a biotin-labeled type I keratin was injected into fibroblasts; within 2 hours, the labeled protein had been incorporated into the already existing keratin cytoskeleton (Figure 18-48). The results of this experiment and others demonstrate that IF subunits in a soluble pool are able to add themselves to preexisting filaments and that subunits are able to dissociate from intact filaments.

The relative stability of intermediate filaments presents special challenges in mitotic cells, which must reorganize all three cytoskeletal networks in the course of the cell cycle. In particular, breakdown of the nuclear envelope early in mitosis depends on the disassembly of the lamin filaments that form a meshwork supporting the membrane. As discussed in Chapter 19, the phosphorylation of nuclear lamins by a mitotic cyclin-dependent kinase that becomes active early in mitosis (prophase) induces the disassembly of intact filaments and prevents their reassembly. Later in mitosis (telophase), removal of these phosphates by specific phosphatases promotes lamin reassembly, which is critical to re-formation of a nuclear envelope around the daughter chromosomes. The opposing actions of kinases and phosphatases thus provide a rapid mechanism for controlling the assembly state of lamin intermediate filaments. Other intermediate filaments undergo similar disassembly and reassembly in the cell cycle.

## Defects in Lamins and Keratins Cause Many Diseases

There are about 50 known mutations in the human gene for type-A lamin that are known to cause diseases, collectively called laminopathies, many of which cause forms of Emery-Dreifuss muscular dystrophy (EDMD). Other mutations in the lamin-A gene cause dilated cardiomyopathy. It is not yet clear why these type-A lamin mutations cause EDMD, but perhaps in muscle tissues the fragile nuclei cannot stand the stress and strains of the tissue, so they are the first to show symptoms. Interestingly, other forms of EDMD have been traced to mutations in emerin, the lamin-binding membrane protein of the inner nuclear envelope. Yet other mutations in type-A lamin cause progeria—accelerated aging. The Hutchison-Gilford progeria ("prematurely old") syndrome is caused by a splicing error that results in a lamin A with a defective C-terminal domain.

**EXPERIMENTAL FIGURE 18-48 Keratin intermediate filaments are dynamic, as soluble keratin is incorporated into filaments.** Monomeric type I keratin was purified, chemically labeled with biotin, and microinjected into living epithelial cells. The cells were then fixed at different times after injection and stained with an antibody to biotin and with antibodies to keratin. (a) At 20 minutes after injection, the injected biotin-labeled keratin is concentrated in small foci scattered throughout the cytoplasm (*left*) and has not been integrated into the endogenous keratin cytoskeleton (*right*). (b) After 4 hours, the biotin-labeled keratin (*left*) and the keratin filaments (*right*) display identical patterns, indicating that the microinjected protein has become incorporated into the existing cytoskeleton. [From R. K. Miller, K. Vistrom, and R. D. Goldman, 1991, *J. Cell Biol.* **113**:843; courtesy of R. D. Goldman.]

(a) 20 minutes after injection

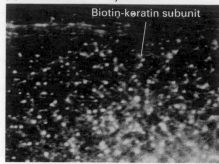

Biotin-keratin subunit

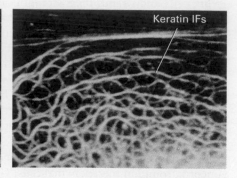

Keratin IFs

(b) 4 hours after injection

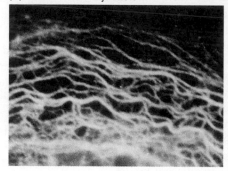

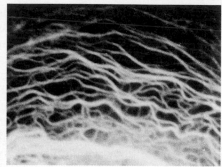

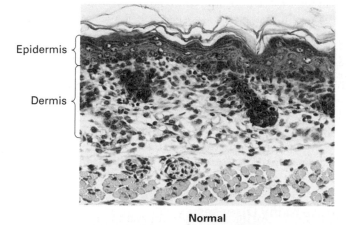

Epidermis

Dermis

**Normal**

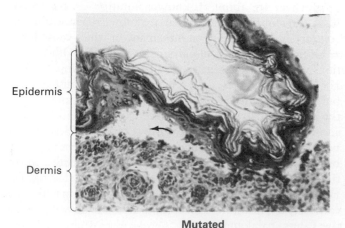

Epidermis

Dermis

**Mutated**

The structural integrity of the skin is essential in order to withstand abrasion. In humans and mice, the K4 and K14 keratin isoforms form heterodimers that assemble into protofilaments. A mutant K14 with deletions in either the N- or the C-terminal domain can form heterodimers in vitro but does not assemble into protofilaments. The expression of such mutant keratin proteins in cells causes IF networks to break down into aggregates. Transgenic mice that express a mutant K14 protein in the basal stem cells of the epidermis display gross skin abnormalities, primarily blistering of the epidermis, that resemble the human skin disease *epidermolysis bullosa simplex* (EBS). Histological examination of the blistered area reveals a high incidence of dead basal cells. Death of these cells appears to be caused by mechanical trauma from rubbing of the skin during movement of the limbs. Without their normal bundles of keratin filaments, the mutant basal cells become fragile and easily damaged, causing the overlying epidermal layers to delaminate and blister (Figure 18-49). Like the role of desmin filaments in supporting muscle tissue,

**EXPERIMENTAL FIGURE 18-49 Transgenic mice carrying a mutant keratin gene exhibit blistering similar to that in the human disease epidermolysis bullosa simplex.** Histological sections through the skin of a normal mouse and a transgenic mouse carrying a mutant K14 keratin gene are shown. In the normal mouse, the skin consists of a hard outer epidermal layer covering and in contact with the soft inner dermal layer. In the skin from the transgenic mouse, the two layers are separated (arrow) due to weakening of the cells at the base of the epidermis. [From P. Coulombe et al., 1991, *Cell* **66**:1301; courtesy of E. Fuchs.]

the general role of keratin filaments appears to be to maintain the structural integrity of epithelial tissues by mechanically reinforcing the connections between cells. ■

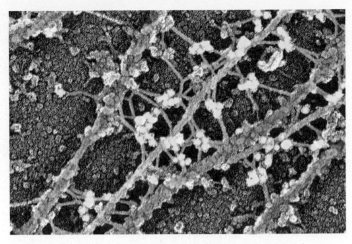

**EXPERIMENTAL FIGURE 18-50 Gold-labeled antibody identifies plectin cross-links between intermediate filaments and microtubules.** In this immunoelectron micrograph of a fibroblast cell, microtubules are highlighted in red, intermediate filaments in blue, and the short connecting fibers between them in green. Staining with gold-labeled antibodies to plectin (yellow) reveals that these connecting fibers contain plectin. [From T. M. Svitkina, A. B. Verkhovsky, and G. G. Borisy, 1996, *J. Cell Biol.* **135**:991; courtesy of T. M. Svitkina.]

# 18.8 Coordination and Cooperation Between Cytoskeletal Elements

So far, we have generally discussed the three cytoskeletal filament classes—microfilaments, microtubules, and intermediate filaments—as though they function independently of one another. However, the fact that the microtubule-based mitotic spindle determines the site of formation of the microfilament-based contractile ring is just one example of how these two cytoskeletal systems are coordinated. Here we mention some other examples of linkages, physical and regulatory, between cytoskeletal elements and their integration into other aspects of cellular organization.

## Intermediate Filament–Associated Proteins Contribute to Cellular Organization

A group of proteins collectively called *intermediate filament–associated proteins (IFAPs)* have been identified that co-purify with intermediate filaments. Among these are the family of

plakins, which are involved in attaching intermediate filaments to other structures. Some of these associate with keratin filaments to link them to desmosomes, which are junctions between epithelial cells that provide stability to a tissue, and hemidesmosomes, which are located at regions of the plasma membrane where intermediate filaments are linked to the extracellular matrix (these topics are covered in detail in Chapter 20). Other plakins are found along intermediate filaments and have binding sites for microfilaments and microtubules. One of these proteins, called plectin, can be seen by immunoelectron microscopy to provide connections between microtubules and intermediate filaments (Figure 18-50).

## Microfilaments and Microtubules Cooperate to Transport Melanosomes

Studies of mutant mice with light-colored coats have uncovered a pathway in which microtubules and microfilaments cooperate to transport pigment granules. The color pigment in the hair is produced in cells called melanocytes, cells very similar to the fish and frog melanophores discussed earlier (see Figure 18-28). Melanocytes are found in the hair follicle at the base of the hair shaft and contain pigment-laden granules called melanosomes. Melanosomes are transported to the dendritic extensions of melanocytes for subsequent exocytosis to the surrounding epithelial cells. Transport to the cell periphery is mediated, just as in frog melanophores, by a kinesin family member. At the periphery, they are then handed off to myosin V and delivered for exocytosis. If the myosin V system is defective, the melanosomes are not captured and stay in the cell body. Thus microtubules are responsible for the long-range transport of melanosomes, whereas microfilament-based myosin V is responsible for capture and delivery at the cell cortex. This type of division of labor—long-range transport by microtubules and short range by microfilaments—has been found in many different systems, from transport in filamentous fungi to transport along axons.

## Cdc42 Coordinates Microtubules and Microfilaments During Cell Migration

In Chapter 17, we discussed how the polarity of a migrating cell is regulated by Cdc42, which results in the formation of an actin-based leading edge at the front of the cell and contraction at the back (see Figure 17-46 and Figure 18-51, step **1**). It turns out that Cdc42 activation at the cell front also leads to polarization of the microtubule cytoskeleton. This was originally studied in wound-healing assays (see Figure 17-45), where it was noticed that when the cells at the edge of a scratch are induced to polarize and move to fill in the empty space, the Golgi complex is moved to the front of the nucleus toward the cell front. Golgi localization at the front of the cell indicates that the centrosome moves to lie in front of the nucleus (recall that Golgi localization is dependent on the location of the MTOC; see Figures 18-1c, 18-27). Recent studies have suggested how this happens. Cdc42 activation at the front of the cell binds the polarity factor Par6, which results in the recruitment of the dynein-dynactin complex (Figure 18-51, step **2**). Cortically localized dynein-dynactin then interacts with microtubules, pulling on them to orient the centrosome and hence the whole radial array of microtubules (Figure 18-51, step **3**). This reorientation of the microtubule system leads to the reorganization of the secretory pathway to deliver secretory products, especially integrins to bind the extracellular matrix, to the front of the cell for attachment to the substratum for cell migration (Figure 18-51, step **4**).

## Advancement of Neural Growth Cones Is Coordinated by Microfilaments and Microtubules

The nervous system depends on the integration and transmission of signals by neurons. Neurons have specialized structures called dendrites that receive signals, and a single axon that terminates in one or more synapses on a target cell or cells (for example, another neuron or a muscle cell) (see Figure 18-2). It is critical that neurons make the right connections, so how are axons guided to their correct destinations? As the axon extends, it has a terminal growth cone that senses signals from the extracellular matrix and other cells to guide it along the right path. Therefore, how the growth cone receives and interprets cues to direct axon growth is critical to the function of the nervous system. Growth cones are very rich in actin, and typically have a broad lamellipodium and multiple filopodia (Figure 18-52a). Also essential for the guidance of growth cones are microtubules. Recall that axons have microtubules of uniform polarity on which materials for growth of the growth cone are transported by axonal transport (see Figure 18-5e). These microtubules extend into the growth cone and, together with actin, are involved in guiding the direction of advancement of the growth cone. While actin is necessary for the advancement of the growth cone, the microtubules and actin together are necessary to steer growth in the correct direction. Although the mechanisms have not been fully elucidated, it has been found that a local growth signal alters local actin dynamics with the result that microtubules extend into that region. It has also been found that microtubules in the shaft of the axon have post-translational modifications, such as acetylation, that stabilizes them, whereas the more dynamic microtubules in the growth cone often do not (Figure 18-52b).

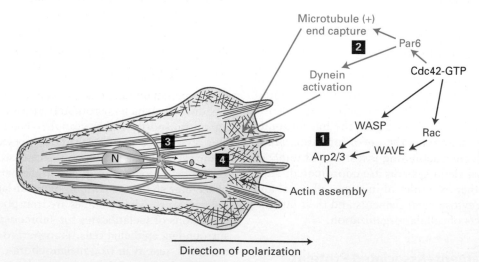

Direction of polarization

**FIGURE 18-51 Independent Cdc42 regulation of microfilaments and microtubules to polarize a migrating cell.** Active Cdc42-GTP at the front of the cell leads to Rac and WASP activation, which results in the assembly of a microfilament-based leading edge (step **1**). Independently, Cdc42-GTP also leads to the capture of microtubule (+) ends and the activation of dynein (step **2**). Together these pull on microtubules to orient the centrosome (step **3**) toward the front of the cell. This reorientation polarizes the secretory pathway for the delivery along microtubules of adhesion molecules carried in secretory vesicles (step **4**). [Based on studies in S. Etienne-Manneville et al., 2005, *J. Cell Biol.* **170**:895–901.]

(a)

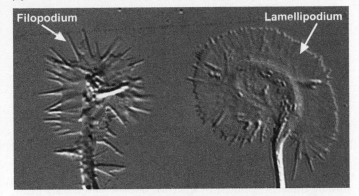

Filopodium

Lamellipodium

(b)

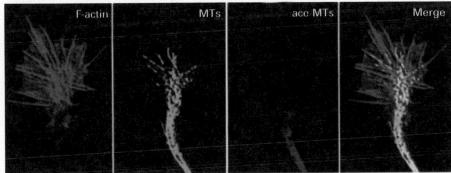

F-actin | MTs | acc MTs | Merge

**EXPERIMENTAL FIGURE 18-52** **The neuronal growth cone contains dynamic actin filaments and microtubules.** (a) Two growth cones viewed by DIC microscopy showing their regions of filopodia and lamellipodia. (b) Localization of actin (red), microtubules (green), and acetylated microtubules (blue) in a small growth cone. Notice how the stable acetylated microtubules are localized to the shaft of the axon and do not penetrate into the dynamic growth cone. [From E. W. Bent and F. B. Gertler, 2003, *Neuron* **40:**209–227].

## KEY CONCEPTS of Section 18.8

### Coordination and Cooperation Between Cytoskeletal Elements

• Intermediate filaments are linked both to specific attachment sites on the plasma membrane (called desmosomes and hemidesmosomes), and to microfilaments and microtubules (see Figure 18-50).

• In animal cells, microtubules are generally utilized for the long-range delivery of organelles, whereas microfilaments handle their local delivery.

• The signaling molecule Cdc42 coordinately regulates microfilaments and microtubules during cell migration.

• The advancement of growth cones in neurons requires the interplay of microfilaments and microtubules.

## Perspectives for the Future

In Chapters 17 and 18 we have seen how microfilaments, microtubules, and intermediate filaments provide structure and organization to cells. Without this elaborate system, cells would lack all order and hence all possibility of function or division. The name "cytoskeleton" suggests a relatively static structure on which the cell organization is hung. However, the cytoskeleton is actually a dynamic framework responding to signal-transduction pathways and operating both locally and globally to provide cells with order to undertake their functions.

In outline, we have elucidated many of the distinct and common functions of the three filament systems. We know most of the components and probably all the motors. However, in many ways this is just an exciting beginning. With the available sequenced genomes and, at least in principle, a complete inventory of the cytoskeletal components, we have a parts list. However, a parts list is just that; what we need is to understand how the parts come together in specific processes.

A very active area of research today is to use the parts list to systematically identify the localization (through GFP fusions), functions (through RNAi knockdown), and associated partners (through isolation of protein complexes) of all cytoskeletal components. Consider that there are about 45 genes in animals that encode members of the kinesin family, yet we only know what a small subset of them do or what cargo they carry and for what purpose. In each case it is reasonable to assume the motors are regulated, but very little is currently known about how. It will be important to understand how motors pick up the right cargo, and then dissociate from them when they arrive at their destination. As we begin to put all the pieces in place, it will be increasingly possible to reconstitute specific processes in vitro. Some aspects of the mitotic spindle have already been reconstituted, which is an encouraging beginning, but it will be some time before it is possible to reconstitute the whole process.

Structural biology is going to play a major role because it will allow us to see in detail how different components work. Consider the large number of proteins that associate with the microtubule (+) end, the so-called +TIPs. We know a bit about how they maintain their association at the tip, and recent work has suggested that associations can change in different parts of the cell—again, we are only just beginning to see how these processes are regulated.

Perhaps the biggest—and most exciting—challenge is to uncover how signal-transduction pathways coordinate functions between all the different cytoskeletal elements within single cells, and in different cellular contexts. How cells organize and regulate different processes in various regions within a single cell is only now beginning to be tackled. We are beginning to see glimpses of what is in store from the signal-transduction pathways that regulate cell polarity and allow cell migration.

Although all these studies are likely to be aimed at basic cell biology, as we can see from the studies of intraflagellar transport and intermediate filaments, such studies often open a window into the underlying basis of disease, from which strategies for treatments can be developed. The interplay between basic cell biology and medicine contributes immensely to the excitement and social value of working in this area.

## Key Terms

anaphase 850

anterograde 833

asters 849

axonal transport 833

axoneme 845

basal body 845

centromere 852

centrosome 824

cilia 844

cytokinesis 851

desmin 863

dynamic instability 828

dyneins 833

flagella 844

γ-tubulin ring complex (γ-TuRC) 825

intermediate filament 821

intermediate filament–associated proteins (IFAPs) 865

intraflagellar transport (IFT) 846

keratins 862

kinesins 833

kinetochores 850

lamins 862

metaphase 850

microtubule 821

microtubule-associated proteins (MAPs) 822

microtubule-organizing center (MTOC) 824

mitosis 849

mitotic spindle 849

neurofilaments 862

primary cilium 847

prophase 849

protofilament 822

retrograde 833

telophase 851

tubulin 822

## Review the Concepts

1. Microtubules are polar filaments; that is, one end is different from the other. What is the basis for this polarity, how is polarity related to microtubule organization within the cell, and how is polarity related to the intracellular movements powered by microtubule-dependent motors?

2. Microtubules both in vitro and in vivo undergo dynamic instability, and this type of assembly is thought to be intrinsic to the microtubule. What is the current model that accounts for dynamic instability?

3. In cells, microtubule assembly depends on other proteins as well as tubulin concentration and temperature. What types of proteins influence microtubule assembly in vivo, and how does each type affect assembly?

4. Microtubules within a cell appear to be arranged in specific arrays. What cellular structure is responsible for determining the arrangement of microtubules within a cell? How many of these structures are found in a typical cell? Describe how such structures serve to nucleate microtubule assembly.

5. Many drugs that inhibit mitosis bind specifically to tubulin, microtubules, or both. What diseases are such drugs used to treat? Functionally speaking, these drugs can be divided into two groups based on their effect on microtubule assembly. What are the two mechanisms by which such drugs alter microtubule structure?

6. Kinesin-1 was the first member of the kinesin motor family to be identified and therefore is perhaps the best-characterized family member. What fundamental property of kinesin was used to purify it?

7. Certain cellular components appear to move bidirectionally on microtubules. Describe how this is possible given that microtubule orientation is fixed by the MTOC.

8. The motile properties of kinesin motor proteins involve both the motor domain and the linker domain. Describe the role of each domain in kinesin movement, direction of movement, or both. Could kinesin-1 with one inactive head efficiently move a vesicle along a microtubule?

9. What features of the dynactin complex enable cytoplasmic dynein to transport cargo toward the microtubule (−) end? What effect could inhibition of dynactin interactions with the +TIP EB-1 have on spindle orientation in cells?

10. Cell swimming depends on appendages containing microtubules. What is the underlying structure of these appendages, and how do these structures generate the force required to produce swimming?

11. What effect would dynein inactivation have on kinesin-2-dependent IFT transport?

12. The mitotic spindle is often described as a microtubule-based cellular machine. The microtubules that constitute the mitotic spindle can be classified into three distinct types. What are the three types of spindle microtubules, and what is the function of each?

13. Mitotic spindle function relies heavily on microtubule motors. For each of the following motor proteins, predict the effect on spindle formation, function, or both of adding a drug that specifically inhibits only that motor: kinesin-5, kinesin-13, and kinesin-4.

14. The poleward movement of kinetochores, and hence chromatids, during anaphase A requires that kinetochores

maintain a hold on the shortening microtubules. How does a kinetochore hold on to shortening microtubules?

**15.** Anaphase B involves the separation of spindle poles. What forces have been proposed to drive this separation? What underlying molecular mechanisms are thought to provide these forces?

**16.** Cytokinesis, the process of cytoplasmic division, occurs shortly after the separated sister chromatids have neared the opposite spindle poles. How is the plane of cytokinesis determined? What are the respective roles of microtubules and actin filaments in cytokinesis?

**17.** The best strategy for treating a specific type of human tumor can depend on identifying the type of cell that became cancerous to give rise to the tumor. For some tumors that have metastasized (moved) to colonize a distant location, identifying the parental cell type can be difficult. Because the type of IF protein expressed is cell-type specific, using monoclonal antibodies that react with only one type of IF protein can help in this identification. Monoclonal antibodies against what IF proteins would you use to identify a) a sarcoma of muscle cell origin, b) an epithelial cell carcinoma, and c) an astrocytoma?

**18.** Explain why there are no known motors that use intermediate filaments as tracks.

**19.** Growth cones are highly mobile regions of developing neurons. What prevents the growth cone from moving or collapsing back into the main cell body like often occurs with lamellipodia?

## Analyze the Data

**1a.** Kinesin-1 contains two identical heavy chains and therefore has two identical motor domains. In contrast, kinesin-5 contains four identical heavy chains. Electron microscopic analysis of metal-shadowed kinesins results in the images shown in the top panel. Pretreatment of these kinesins with an antibody that binds specifically to the kinesin motor domain results in the images shown in the lower panel. All four images are at the same approximate magnification. What can you deduce about the structure of kinesin-5 from these data?

**b.** To determine if kinesin-5 is a (+) or (−) end microtubule motor, polarity-marked microtubules are generated by assembling short microtubules from brightly fluorescent tubulin and then elongating those short, bright microtubules using less fluorescent tubulin. As a result, the microtubules are very fluorescent at one end and less fluorescent along most of their length. A perfusion chamber is then coated with purified kinesin-5, which becomes immobilized on the glass surface. The chamber is then perfused with the polarity-labeled microtubules and ATP, and microtubule gliding with respect to the immobilized kinesin-5 is observed. The following sequence of images is obtained. Which end of these microtubules, the bright or the less-bright end, is the (+) end? Do these microtubules glide on kinesin-5 with their (+) or (−) end leading? Based on these data, is kinesin-5 a (+) or (−) end microtubule motor?

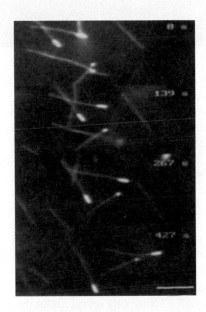

**c.** Kinesin-5 can cross-bridge adjacent microtubules. Polarity-marked microtubules are assembled in which tubulin attached to a red fluorescent dye is assembled to form short red microtubules, which are then elongated with tubulin attached to a green fluorescent dye. The microtubules are mixed with kinesin-5 and observed by fluorescence microscopy as ATP is added. The following images show a time sequence of two overlapping and cross-bridged microtubules as ATP is added. The arrowhead is in a fixed position. Can you explain what happens when ATP is added to microtubules cross-bridged by kinesin-5?

**d.** Eg5 is a kinesin-5 family member in *Xenopus*. To understand Eg5 function in vivo, cells are transfected with RNAi

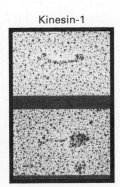

Kinesin-1

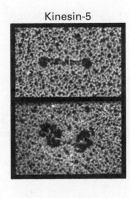

Kinesin-5

No antibody

Decorated with kinesin motor domain antibody

directed against this motor. The following images are obtained of mitotic cells. What function might Eg5 play in cells?

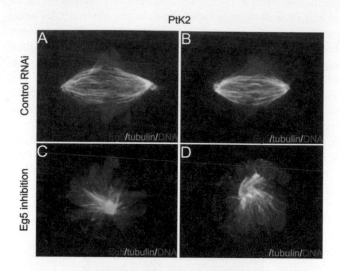

PtK2

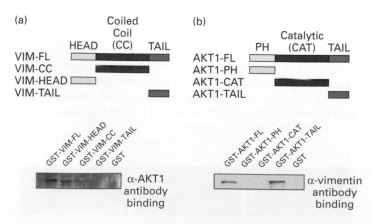

**2.** The PI3K/AKT signaling pathway is aberrant in a wide variety of cancers. In soft-tissue sarcoma (STS) cells, AKT1 activation induces cell motility and invasiveness, which leads to aggressive metastasis of the cells. AKT1 was shown to bind to vimentin (see Q-S Zhu et al., 2010, Vimentin is a novel AKT1 target mediating motility and invasion. *Oncogene* 30:457–470; doi:10.1038/onc.2010.421; published online 20 September 2010).

**a.** To map the vimentin and AKT interaction domains, the vimentin and AKT1 full-length and fragment constructs indicated below were expressed as GST-fusion proteins. Each of the GST-fusion constructs bound to glutathione beads was used to pull down associated proteins from crude STS cell lysates. What did the Western blot analysis of the pellet using an AKT1 antibody reveal about the AKT-binding domain in vimentin? What did analysis of the pellet using a vimentin antibody reveal about the vimentin-binding domain in AKT1?

**b.** AKT1 is a kinase and therefore likely to phosphorylate vimentin. Sequence analysis revealed that serines (S) at positions 39 and 325 in vimentin were likely sites for AKT1

phosphorylation. To test whether one or both of these sites is phosphorylated by AKT1, each site was mutated to alanine (A), which cannot be phosphorylated. Each alanine-mutated vimentin was mixed with AKT1 and tested for phosphorylation by Western blot analysis using an antibody that reacts with AKT1-phosphorylated serines (PAS antibody). Which site(s) does AKT1 phosphorylate?

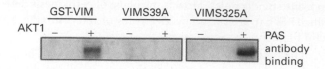

**c.** The propensity of cancer cells to metastasize can be measured with an invasion assay in which cells migrate through a filter coated with extracellular matrix (ECM) proteins. The more cells migrating through the ECM, the higher the propensity for metastasis. The invasion assay was used to monitor effects of expression of a permanently active AKT1 (AKT1DD) mutation and overexpression of wild-type vimentin, a vimentin mutation that cannot be phosphorylated by AKT1 (VIMS39A), and a phosphomimetic vimentin mutation (VIMS39D) in which mutation of S39 to an aspartate residue (D) mimics phosphorylation of the serine. What effect does vimentin overexpression and phosphorylation have on cell migration?

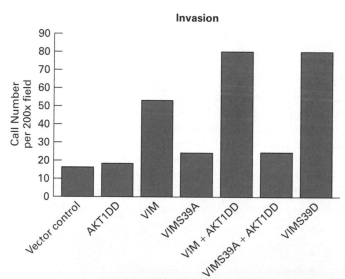

## References

### Microtubule Structure and Organization

Badano, J. L., T. M. Teslovich, and N. Katsanis. 2005. The centrosome in human genetic disease. *Nature Rev. Mol. Cell Biol.* 6:194–205.

Doxsey, S. 2001. Re-evaluating centrosome function. *Nature Rev. Mol. Cell Biol.* 2:688–698.

Dutcher, S. K. 2001. The tubulin fraternity: alpha to eta. *Curr. Opin. Cell Biol.* **13**:49–54.

Nogales, E., and H-W Wang. 2006. Structural intermediates in microtubule assembly and disassembly: how and why? *Curr. Opin. Cell Biol.* **18**:179–184.

## Microtubule Dynamics

Cassimeris, L. 2002. The oncoprotein 18/stathmin family of microtubule destabilizers. *Curr. Opin. Cell Biol.* **14**:18–24.

Desai, A., and T. J. Mitchison. 1997. Microtubule polymerization dynamics. *Annu. Rev. Cell Dev. Biol.* **13**:83–117.

Howard, J., and A. A. Hyman. 2003. Dynamics and mechanics of the microtubule plus end. *Nature* **422**:753–758.

## Regulation of Microtubule Structure and Dynamics

Akhmanova, A., and M. O. Steinmetz. 2010. Microtubule +TIPs at a glance. *J. Cell Science* **123**:3414–3418.

Galjart, N. 2005. CLIPs and CLASPs and cellular dynamics. *Nature Rev. Mol. Cell Biol.* **6**:487–498.

Hammond, J. W., D. Cai, and K. J. Verhey. 2008. Tubulin modifications and their cellular functions. *Curr. Opin. Cell Biol.* **20**:71–76.

Wloga, D., and J. Gaertig. 2010. Post-translational modifications of microtubules. *J. Cell Science* **123**:3447–3455.

## Kinesins and Dyneins: Microtubule-Based Motor Proteins

Web site: Kinesin Home Page, http://www.cellbio.duke.edu/kinesin/

Burgess, S. A., et al. 2003. Dynein structure and power stroke. *Nature* **421**:715–718.

Carter, A. P., C. Carol., L. Jin, and R. D. Vale. 2011. The crystal structure of dynein. *Science* **331**:1159–1165.

Dell, K. R. 2003. Dynactin polices two-way organelle traffic. *J. Cell Biol.* **160**:291–293.

Dujardin, D. L., and R. B. Vallee. 2002. Dynein at the cortex. *Curr. Opin. Cell Biol.* **14**:44–49.

Endow, S. A., F. J Kull, and H. Liu. 2010. Kinesins at a glance. *J. Cell Science* **123**:3420–3424.

Goldstein, L. S. 2001. Kinesin molecular motors: transport pathways, receptors, and human disease. *Proc. Nat'l. Acad. Sci. USA* **98**:6999–7003.

Hirokawa, N., N. Noda, Y. Tanaka, and S. Niwa. 2009. Kinesin superfamily motor proteins and intracellular transport. *Nature Rev. Mol. Cell Biol.* **10**:682–696.

Hirokawa, N., and R. Takemure. 2003. Biochemical and molecular characterization of diseases linked to motor proteins. *Trends Cell Biol.* **28**:558–565.

Kardon, J. R., and R. D. Vale. 2009. Regulators of the cytoplasmic dynein motor. *Nature Rev. Mol. Cell Biol.* **10**:854–865.

Lawrence, C. J., et al. 2004. A standardized kinesin nomenclature. *J. Cell Biol.* **167**:19–22.

McKenney, R. J., et al. 2010. LIS1 and NudE induce a persistent dynein force-producing state. *Cell* **141**:304–314.

Schroer, T. A. 2004. Dynactin. *Ann. Rev. Cell Dev. Biol.* **20**:759–779.

Vale, R. D. 2003. The molecular motor toolbox for intracellular transport. *Cell* **112**:467–480.

Vale, R. D., and R. A. Milligan. 2000. The way things move: looking under the hood of molecular motor proteins. *Science* **288**:88–95.

Verhey, K. J., and J. W. Hammond. 2009. Traffic control: regulation of kinesin motors. *Nature Rev. Mol. Cell Biol.* **10**:765–777.

Wordeman, L. 2005. Microtubule-depolymerizing kinesins. *Curr. Opin. Cell Biol.* **17**:82–88.

Yildız, A., M. Tomishige, R. D. Vale, and P. R. Selvin. 2004. Kinesin walks hand-over-hand. *Science* **303**:676–678.

## Cilia and Flagella: Microtubule-Based Surface Structures

Gerdes, J. M., E. E. Davis, and N. Katsanis. 2009. The vertebrate primary cilium in development, homeostasis, and disease. *Cell* **137**:32–45.

Ishkiawa, H., and W. F. Marshall. 2011. Ciliogenesis: building the cell's antenna. *Nature Mol. Cell Biol.* **12**:222–234.

Jin, H. et al. 2010. The conserved Bardet-Biedl syndrome proteins assemble a coat that traffics membrane proteins to cilia. *Cell* **141**:1208–1218.

Rosenbaum, J. L., and G. B. Witman. 2002. Intraflagellar transport. *Nature Rev. Mol. Cell Biol.* **3**:813–825.

Singla, V., and J. F. Reiter. 2006. The primary cilium as the cells' antenna: signaling at a sensory organelle. *Science* **313**:629–633.

## Mitosis

Web site: http://www.cellbio.duke.edu/kinesin/FxnSpindleMotility.html

Alushin, G.M., et al. 2010. The Ndc80 kinetochore complex forms oligomeric arrays along microtubules. *Nature* **467**:805–810.

Cheeseman, I. M., and A. Desai. 2008. Molecular architecture of the kinetochore-microtubule interface. *Nature Rev. Mol. Cell Biol.* **9**:33–46.

Cleveland, D. W., Y. Mao, and K. F. Sullivan. 2003. Centromeres and kinetochores: from epigenetics to mitotic checkpoint signaling. *Cell* **112**:407–421.

Gadde, S., and R. Heald. 2004. Mechanisms and molecules of the mitotic spindle. *Curr. Biol.* **14**:R797–R805.

Goshima, G., et al. 2009. Augmin: a protein complex required for centrosome-independent microtubule generation within the spindle. *J. Cell Biol.* **181**:421–429.

Heald, R., et al. 1997. Spindle assembly in *Xenopus* egg extracts: respective roles of centrosomes and microtubule self-organization. *J. Cell Biol.* **138**:615–628.

Kim, Y., A. J. Holland, W. Lan, and D. W. Cleveland. 2010. Aurora kinases and protein phosphatase 1 mediate chromosome congression through regulation of CENP-E. *Cell* **142**: 444–455.

Kinoshita, K., B. Habermann, and A. A. Hyman. 2002. XMAP215: a key component of the dynamic microtubule cytoskeleton. *Trends Cell Biol.* **12**:267–273.

Liu, D., et al. 2009. Sensing chromosome bi-orientation by spatial separation of Aurora B kinase from kinetochore substrates. *Science* **323**:1350–1353.

Mitchison, T. J., and E. D. Salmon. 2001. Mitosis: a history of division. *Nature Cell Biol.* **3**:E17–E21.

Rogers, G. C., et al. 2004. Two mitotic kinesins cooperate to drive sister chromatid separation during anaphase. *Nature* **427**:364–370.

Ruchaud, S., M. Carmena, and W. C. Earnshaw. 2007. Chromosomal passengers: conducting cell division. *Nature Rev. Mol. Cell Biol.* **8**:798–812.

Santaguida, S., and A. Musacchio. 2009. The life and miracle of kinetochores. *EMBO Journal* **28**:2511–2531.

Urges, G. 2005. Cytokinesis in higher plants. *Ann. Rev. Plant Biol.* **56**:281–299.

Wittmann, T., A. Hyman, and A. Desai. 2001. The spindle: a dynamic assembly of microtubules and motors. *Nature Cell Biol.* **3**:E28–E34.

## Intermediate Filaments

Intermediate Filaments Database: http://www.interfil.org/index.php

Colakoglu, G., and A. Brown. 2009. Intermediate filaments exchange subunits along their length and elongate by end-to-end annealing. *J. Cell Biol.* **185**:769–777.

Goldman, R. D., et al. 2002. Nuclear lamins: building blocks of nuclear architecture. *Genes Dev.* **16**:533–547.

Herrmann, H., and U. Aebi. 2000. Intermediate filaments and their associates: multi-talented structural elements specifying cytoarchitecture and cytodynamics. *Curr. Opin. Cell Biol.* **12**:79–90.

Mattout, A., et al. 2006. Nuclear lamins, disease and aging. *Curr. Opin. Cell Biol.* **18**:335–341.

## Coordination and Cooperation Between Cytoskeletal Elements

Web site: Melanophores, http://www.cellbio.duke.edu/kinesin/Melanophore.html

Chang, L., and R. D. Goldman. 2004. Intermediate filaments mediate cytoskeletal crosstalk. *Nature Rev. Mol. Cell Biol.* **5**:601–613.

Etienne-Manneville, S., et al. 2005. Cdc42 and Par6-PKCζ regulate the spatially localized association of Dlg1 and APC to control cell polarization. *J. Cell Biol.* **170**:895–901.

Kodama, A., T. Lechler, and E. Fuchs. 2004. Coordinating cytoskeletal tracks to polarize cellular movements. *J. Cell Biol.* **167**:203–207.

Schaefer, A. W., et al. 2008. Coordination of actin filament and microtubule dynamics during neurite outgrowth. *Dev. Cell* **15**:146–162.

Wu, X., X. Xiang, and J. A. Hammer III. 2006. Motor proteins at the microtubule plus-end. *Trends Cell Biol.* **16**:135–143.

# The Eukaryotic Cell Cycle

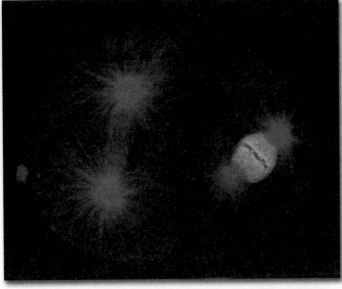

A two-cell C. *elegans* embryo stained with antibodies against tubulin (red) and CeBUB-1, a spindle checkpoint protein (green). DNA is stained with DAPI (blue). CeBUB-1 is localized on the chromosomes and kinetochore-attached spindle microtubules during metaphase in the smaller, posterior cell (*right*). It is presumed to monitor chromosome attachment and tension. The larger, anterior cell (*left*) has already entered anaphase, and CeBUB-1 is no longer detectable on the chromosomes or spindle microtubules. Thus asynchrony of this second cell cycle in the C. *elegans* embryo allows the observation of both the presence of a functional spindle checkpoint protein during metaphase and its absence after initiation of anaphase. [Encanada et al., 2005, *Mol. Biol. Cell* **16**:1056.]

P roper control of **cell division** is vital to all organisms. In unicellular organisms, cell division must be balanced with cell growth so that cell size is properly maintained. If several divisions occur before parental cells have reached the proper size, daughter cells eventually become too small to be viable. If cells grow too large before cell division, the cells function improperly and the number of cells increases slowly. In developing multicellular organisms, the replication of each cell must be precisely controlled and timed to faithfully and reproducibly complete the developmental program in every individual. Each type of cell in every tissue must control its replication precisely for normal development of complex organs such as the brain or the kidney. In a normal adult, cells divide only when and where they are needed. However, loss of normal controls on cell replication is the fundamental defect in cancer, an all-too-familiar disease that kills one in every six people in the developed world (see Chapter 24). The

molecular mechanisms regulating eukaryotic cell division discussed in this chapter have gone a long way in explaining how replication control goes awry in cancer cells. Appropriately, Leland Hartwell, Tim Hunt, and Paul Nurse were awarded the Nobel Prize in Physiology or Medicine in 2001 for the initial experiments that elucidated the master regulators of cell division in all eukaryotes.

The term **cell cycle** refers to the ordered series of events that lead to cell division and the production of two daughter cells, each containing chromosomes identical to those of the parental cell. Two main molecular processes take place during the cell cycle, with resting intervals in between: during the S phase of the cycle, each parental chromosome is duplicated to form two identical sister chromatids; in mitosis (M phase), the resulting sister chromatids are distributed to each daughter cell (Figure 19-1). Chromosome replication and segregation to daughter cells must occur in the proper order

## OUTLINE

**FIGURE 19-1 The fate of a single parental chromosome throughout the eukaryotic cell cycle.** Following mitosis (M), daughter cells contain 2*n* chromosomes in diploid organisms and 1*n* chromosomes in haploid organisms. In proliferating cells, $G_1$ is the period between the "birth" of a cell following mitosis and the initiation of DNA synthesis, which marks the beginning of the S phase. At the end of the S phase, cells enter $G_2$ containing twice the number of chromosomes as $G_1$ cells (4*n* in diploid organisms, 2*n* in haploid organisms). The end of $G_2$ is marked by the onset of mitosis, during which numerous events leading to cell division occur. The $G_1$, S, and $G_2$ phases are collectively referred to as *interphase*, the period between one mitosis and the next. Most nonproliferating cells in vertebrates leave the cell cycle in $G_1$, entering the $G_0$ state. Although chromosomes condense only during mitosis, here they are shown in condensed form throughout the cell cycle to emphasize the number of chromosomes at each stage. For simplicity, the nuclear envelope is not depicted.

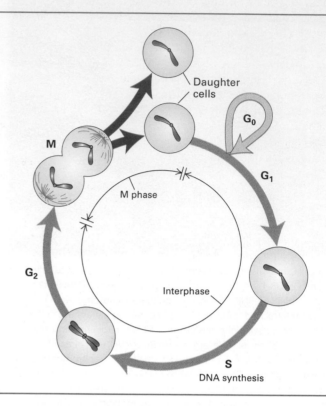

in every cell division. If a cell undergoes chromosome segregation before the replication of all chromosomes has been completed, at least one daughter cell will lose genetic information. Likewise, if a second round of replication occurs in one region of a chromosome before cell division occurs, the genes encoded in that region are increased in number out of proportion to other genes, a phenomenon that often leads to an imbalance of gene expression that is incompatible with viability.

High accuracy and fidelity are required to ensure that DNA replication is carried out correctly and that each daughter cell inherits the correct number of each chromosome. To achieve this, cell division is controlled by surveillance mechanisms known as **checkpoint pathways** that prevent initiation of each step in cell division until earlier steps on which it depends have been completed and mistakes that occurred during the process have been corrected. Mutations that inactivate or alter the normal operation of these checkpoint pathways contribute to the generation of cancer cells because they result in chromosomal rearrangements and abnormal numbers of chromosomes, which lead to further mutations and changes in gene expression level that cause uncontrolled cell growth (see Chapter 24).

In the late 1980s, it became clear that the molecular processes regulating the two key events in the cell cycle—chromosome replication and segregation—are fundamentally similar in all eukaryotic cells. Initially, it was surprising to many researchers that cells as diverse as budding yeast and developing human neurons use nearly identical proteins to regulate their division. However, like transcription

and protein synthesis, control of cell division appears to be a fundamental cellular process that evolved and was largely optimized early in eukaryotic evolution. Because of this similarity, research with diverse organisms, each with its own particular experimental advantages, has contributed to a growing understanding of how cell cycle events are coordinated and controlled. Biochemical, genetic, imaging, and micromanipulation techniques all have been employed in studying various aspects of the eukaryotic cell cycle. These studies have revealed that cell division is controlled primarily by regulating the timing of entry into the cell division cycle, nuclear DNA replication, and mitosis.

The master controllers of the cell cycle are a small number of *heterodimeric protein kinases* that contain a regulatory subunit (**cyclin**) and a catalytic subunit (**cyclin-dependent kinase**, or **CDK**). These heterodimeric kinases regulate the activities of multiple proteins involved in entry into the cell cycle, DNA replication, and mitosis by phosphorylating them at specific regulatory sites, activating some and inhibiting others to coordinate their activities. Regulated degradation of proteins also plays a prominent role in important cell cycle transitions. Since protein degradation is irreversible, this ensures that the processes move in only one direction through the cell cycle.

In this chapter, we first present an overview of the cell cycle and then describe the various experimental systems that contributed to our current understanding of it. We will then discuss cyclin-dependent kinases (CDKs) and the many different ways these key cell cycle controllers can be regulated. Next, we'll examine each cell cycle phase in greater detail with an emphasis on how control of CDK activity governs the events that take place in each phase. We will then discuss the system of checkpoint pathways that establish the order of the cell cycle and ensure that each cell cycle phase occurs with accuracy. In our discussion we will emphasize the general principles governing cell cycle progression and will use a species-spanning nomenclature when discussing the factors controlling each cell cycle phase. The chapter concludes with a discussion of meiosis, a special type of cell division that generates haploid germ cells (egg and sperm), and the molecular mechanisms that distinguish it from mitosis.

# 19.1 Overview of the Cell Cycle and Its Control

We begin our discussion by reviewing the stages of the eukaryotic cell cycle, presenting a summary of the current model of how the cycle is regulated. We will see how DNA replication leads to the creation of two identical DNA molecules during the DNA synthesis phase and how these DNA molecules are compacted and structured for their segregation into daughter cells. We will then introduce the master regulators of the cell cycle, the cyclin-dependent kinases, before concluding with an overview of the principles that govern the cell cycle to ensure that the process occurs in the correct temporal fashion and without mistakes.

## The Cell Cycle Is an Ordered Series of Events Leading to Cell Replication

As illustrated in Figure 19-1, the cell cycle is divided into four major phases. Cycling (replicating) mammalian somatic cells grow in size and synthesize RNAs and proteins required for DNA synthesis during the $G_1$ (**first gap**) **phase**. When cells have reached the appropriate size and have synthesized the required proteins, they enter the cell cycle by traversing a point in $G_1$ known as **START**. Once this point has been crossed, cells are committed to cell division. The first step toward successful cell division is entry into the **S** (**synthesis**) **phase**, the period in which cells actively replicate their chromosomes. After progressing through a second gap phase, the $G_2$ **phase**, cells begin the complicated process of mitosis, also called the **M** (**mitotic**) **phase**, which is divided into several stages (Figure 19-2).

In discussing mitosis, we commonly use the term **chromosome** for the *replicated* structures that condense and become visible in the light microscope during the early stages of mitosis. Thus each chromosome is composed of two identical DNA molecules resulting from DNA replication, plus the histones and other chromosome-associated proteins (see Figure 6-39). The two identical DNA molecules and associated chromosomal proteins that form one chromosome are called **sister chromatids**. Sister chromatids are attached to each other by protein cross-links along their lengths.

**FIGURE 19-2 The stages of mitosis.** During prophase, the nuclear envelope breaks down, microtubules form the mitotic spindle apparatus, and chromosomes condense. At metaphase, attachment of chromosomes to microtubules via their kinetochores is complete. During anaphase, microtubule motors and shortening of spindle microtubules pull the sister chromatids toward opposite spindle poles. After chromosome movement to the spindle poles, chromosomes decondense, and cells reassemble nuclear membranes around the daughter-cell nuclei and undergo cytokinesis.

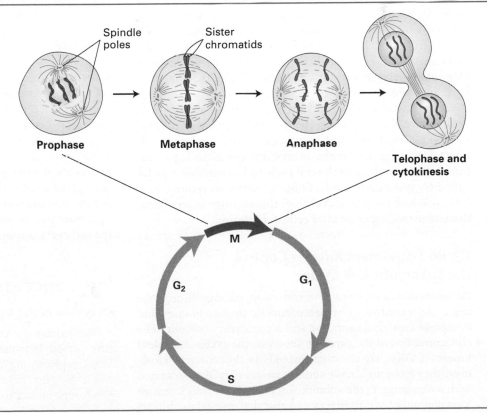

During **interphase**, the part of the cell cycle between the end of one M phase and the beginning of the next, the outer nuclear membrane is continuous with the endoplasmic reticulum. With the onset of mitosis in **prophase**, the nuclear envelope retracts into the endoplasmic reticulum in most cells from higher eukaryotes, and Golgi membranes break down into vesicles. This is necessary so that the microtubules, nucleated by the **centrosomes**, can interact with the chromosomes to form the **mitotic spindle**, consisting of a football-shaped bundle of microtubules with a star-shaped cluster of microtubules radiating from each end, or spindle pole. A multiprotein complex, the **kinetochore**, assembles at each **centromere**. After nuclear envelope breakdown, the kinetochores of sister chromatids associate with microtubules coming from opposite spindle poles (see Figure 18-37), and chromosomes align in a plane in the center of the cell at **metaphase**. During the **anaphase** period of mitosis, sister chromatids separate. They initially are pulled by microtubules toward the spindle poles and then are further separated as the spindle poles move away from each other (see Figure 19-2).

Once chromosome separation is complete, the mitotic spindle disassembles and chromosomes decondense during **telophase**. The nuclear envelope re-forms around the segregated chromosomes as they decondense. The physical division of the cytoplasm, called **cytokinesis**, yields two daughter cells. Following mitosis, cycling cells enter the $G_1$ phase, embarking on another turn of the cycle.

The progression of cell cycle stages is the same for all eukaryotes, though the time it takes to complete one turn of the cycle varies considerably between organisms. Rapidly replicating human cells progress through the full cell cycle in about 24 hours: $G_1$ takes 9 hours; the S phase, 10 hours; $G_2$, 4.5 hours; and mitosis, 30 minutes. In contrast, the full cycle takes only 90 minutes in rapidly growing yeast cells. The cell divisions that take place during early embryonic development of the fruit fly *Drosophila melanogaster* are completed in as little as 8 minutes!

In multicellular organisms, most differentiated cells "exit" the cell cycle and survive for days, weeks, or in some cases (e.g., nerve cells and cells of the eye lens) even the lifetime of the organism without dividing again. Such *postmitotic* cells generally exit the cell cycle in $G_1$, entering a phase called $G_0$ (see Figure 19-1). Some $G_0$ cells can return to the cell cycle and resume replicating; this re-entry is regulated, thereby providing control of cell proliferation.

## Cyclin-Dependent Kinases Control the Eukaryotic Cell Cycle

As mentioned in the chapter introduction, passage through the cell cycle is controlled by heterodimeric protein kinases that comprise a catalytic subunit and a regulatory subunit. The concentrations of the catalytic subunits, the **cyclin-dependent kinases (CDKs)**, are constant throughout the cell cycle. However, they have no kinase activity unless they are associated with a regulatory **cyclin** subunit. Each CDK can associate with a small number of different cyclins that determine the substrate specificity of the complex, that is, which proteins it phosphorylates. Each cyclin is only present and active during the cell cycle stage it promotes and hence restricts the kinase activity of the CDKs it binds to just that cell cycle stage. Cyclin-CDK complexes activate or inhibit hundreds of proteins involved in cell cycle progression by phosphorylating them at specific regulatory sites. Thus proper progression through the cell cycle is governed by activation of the appropriate cyclin-CDK complex at the appropriate time. As we will see, restricting cyclin expression to the appropriate cell cycle stage is one of the many mechanisms cells employ to regulate the activities of each cyclin-CDK heterodimer.

## Several Key Principles Govern the Cell Cycle

The goal of each cell division is to generate two daughter cells of identical genetic makeup. To achieve this, *cell cycle events must occur in the proper order*. DNA replication must always precede chromosome segregation. Today we know that the activity of the key proteins that promote cell cycle progression, the CDKs, *fluctuates during the cell cycle*. For example, CDKs that promote S phase are active during S phase but are inactive during mitosis. CDKs that promote mitosis are only active during mitosis. These *oscillations* in CDK activity are a fundamental aspect of eukaryotic cell cycle control, and we have gained some understanding over the last few years as to how these oscillations are generated. Oscillations are generated by **positive feedback mechanisms**, where specific CDKs promote their own activation. These positive feedback loops are coupled to subsequent **negative feedback mechanisms** where, indirectly or with a built-in delay, CDKs promote their own inactivation. Oscillators not only propel the cell cycle forward but also create abrupt transitions between different cell cycle states, which is essential to bring about distinct cell cycle states.

Laid on the cell cycle oscillator machinery is a system of surveillance mechanisms that further ensures that the next cell cycle event is not activated before the preceding one has been completed or before errors that occurred during the preceding step are corrected. These surveillance mechanisms are called **checkpoint pathways**, and their job is it to ensure accuracy of the chromosome replication and segregation processes. The system that ensures that chromosomes are segregated accurately is so efficient, a mis-segregation event occurs only once in $10^4$–$10^5$ divisions! These multiple layers of control put on the cell cycle control machinery ensure that the cell cycle is robust and error free.

## KEY CONCEPTS of Section 19.1

### Overview of the Cell Cycle and Its Control

• The eukaryotic cell cycle is divided into four phases: $G_1$ (the period between mitosis and the initiation of nuclear DNA replication), S (the period of nuclear DNA replication), $G_2$ (the period between the completion of nuclear DNA replication and mitosis), and M (mitosis).

- Cells commit to a new cell division at a specific point in $G_1$ known as START.
- Cyclin-CDK complexes, composed of a regulatory cyclin subunit and a catalytic cyclin-dependent kinase (CDK) subunit, drive progression of a cell through the cell cycle.
- Cyclins activate CDKs and are present only in the cell cycle stage that they promote.
- CDK activities oscillate during the cell cycle. Positive and negative feedback loops drive these oscillations.
- Surveillance mechanisms, called checkpoint pathways, guarantee that each cell cycle step is completed correctly before the next one is initiated.

# 19.2 Model Organisms and Methods to Study the Cell Cycle

The unraveling of the molecular mechanisms governing cell cycle progression in eukaryotes was remarkably speedy and fueled by the powerful combination of genetic and biochemical approaches. In this section we will discuss several model systems and their contribution to the discovery of the molecular mechanisms of cell division. The three most important systems employed to study the cell cycle are the single-celled yeasts *Saccharomyces cerevisiae* (budding yeast) and *Schizosaccharomyces pombe* (fission yeast) and oocytes and early embryos of the frog *Xenopus laevis*. We will also discuss the fruit fly *Drosophila melanogaster*, which proved extremely powerful in the study of the interplay between cell division and development as well as how the study of mammalian tissue culture cells led to the characterization of cell cycle control in mammals.

The studies of the cell division cycle in many different experimental systems also led to two remarkable discoveries about the general control of the cell cycle. First, complex molecular processes such as initiation of DNA replication and entry into mitosis are all regulated and coordinated by a small number of master cell cycle regulatory proteins. Second, these master regulators and the proteins that control them are highly conserved, so that cell cycle studies in fungi, sea urchins, insects, frogs, and other species are directly applicable to all eukaryotic cells, including human cells.

## Budding and Fission Yeast Are Powerful Systems for Genetic Analysis of the Cell Cycle

Budding and fission yeast have proved to be valuable systems for the study of the cell cycle. Although they both belong to the kingdom of fungi, they are only distantly related. Both organisms can exist in the haploid state, carrying only one copy of each chromosome. The fact that these yeasts can exist as haploid cells makes them powerful genetic systems. It is easy to generate mutations that inactivate genes in haploids because there is only one copy of each gene (a diploid

system would require an inactivating mutation in each of the two copies of the gene to render its activity nonfunctional). Haploid yeast can be easily employed to screen or select for mutants with specific defects, such as defects in cell proliferation. Additional advantages of the two systems are the relative ease with which one can manipulate the expression of individual genes, their fully sequenced genomes, and the ease with which they can be cultivated and manipulated so that cultures of yeast cells progress through the cell cycle in a synchronous manner.

Budding yeast cells are ovoid in shape and divide by budding (Figure 19-3a). The bud is the future daughter cell and begins to form concomitant with the initiation of DNA replication and continues to grow throughout the cell cycle (Figure 19-3b). Cell cycle stage can therefore be inferred from the size of the bud, which makes *S. cerevisiae* a useful system for identifying mutants that are blocked at specific steps in the cell cycle. Indeed, it was in this organism that Lee Hartwell and colleagues first identified mutants that were defective in progressing through specific cell cycle stages. Like mammalian cells, the budding yeast cell cycle has a long $G_1$ phase, and the study of the budding yeast cell cycle shaped our understanding of how the $G_1$–S phase transition is controlled.

Fission yeast cells are rod shaped and grow entirely by elongation at their ends (Figure 19-4a). After the completion of mitosis, cytokinesis occurs by the formation of a septum (Figure 19-4b). The molecular mechanisms governing $G_2$ and entry into mitosis are very similar between fission yeast and metazoan cells, and studies with this organism revealed the molecular events surrounding the $G_2$–M phase transition.

Budding and fission yeast are both useful for the isolation of mutants that are blocked at specific steps in the cell cycle or that exhibit altered regulation of the cycle. Since cell cycle progression is essential for viability, scientists isolated conditional mutants that encode proteins that are functional at one temperature but become inactive at a different, often elevated, temperature (e.g., due to protein misfolding at the non-permissive temperature). Mutants arrested at a particular cell cycle stage are easily distinguished from normal dividing cells by microscopic examination. Thus, in both of these yeasts, cells with **temperature-sensitive mutations** causing defects in specific proteins required to progress through the cell cycle were readily isolated (see Figure 5-6). Such cells are called *cdc* (*cell division cycle*) mutants.

How can one identify which gene is defective in a given *cdc* mutant? The wild-type alleles of recessive temperature-sensitive *cdc* mutant alleles can be isolated readily by transforming haploid mutant cells with a plasmid library prepared from wild-type cells and then plating the transformed cells at the non-permissive temperature (Figure 19-5). Haploid mutant cells cannot form colonies at the non-permissive temperature. However, a transformed mutant cell can grow into a colony if it also contains a plasmid that carries the wild-type allele that complements the recessive mutation; the plasmids bearing the wild-type allele can then be recovered from those cells, allowing the identification of the complementing gene. Because many of the proteins that regulate the cell

(a)

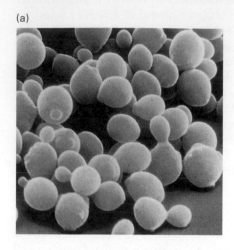

(b)

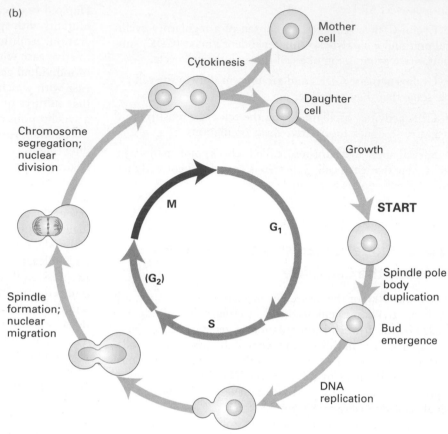

**FIGURE 19-3 The budding yeast *S. cerevisiae*.** (a) Scanning electron micrograph of *S. cerevisiae* cells at various stages of the cell cycle. The larger the bud, which emerges at the end of the $G_1$ phase, the farther along in the cycle the cell is. (b) Main events in the *S. cerevisiae* cell cycle. Daughter cells are born smaller than mother cells and must grow to a greater extent in $G_1$ before they are large enough to enter the S phase. START is the point in the cell cycle after which cells are irreversibly committed to undergoing a cell cycle. $G_2$ is not well defined in budding yeast and is therefore denoted in parentheses. Note that the nuclear envelope does not disassemble during mitosis in *S. cerevisiae* and other yeasts. The small *S. cerevisiae* chromosomes do not condense sufficiently to be visible by light microscopy. [Part (a) courtesy of E. Schachtbach and I. Herskowitz.]

cycle are highly conserved, human cDNAs cloned into yeast expression vectors often can complement yeast cell cycle mutants, leading to the rapid isolation of human genes encoding cell cycle control proteins. In fact, it was the ability of the human gene encoding CDK1 to complement the growth defects caused by inactivation of fission yeast CDK1 that led to the appreciation of the high degree of conservation that exists among eukaryotic cell cycle regulators.

## Frog Oocytes and Early Embryos Facilitate Biochemical Characterization of the Cell Cycle Engine

Biochemical studies require the preparation of cell extracts from many cells. For biochemical studies of the cell cycle, the eggs and early embryos of amphibians and marine invertebrates are particularly suitable. These organisms typi-

cally have large eggs, and fertilization is followed by multiple synchronous cell cycles. By isolating large numbers of eggs from females and fertilizing them simultaneously by addition of sperm (or treating them in ways that mimic fertilization), researchers can obtain extracts from cells at specific points in the cell cycle for analysis of proteins and enzymatic activities.

To understand how *X. laevis* oocytes and eggs can be used for the analysis of cell cycle progression, we must first lay out the events of oocyte maturation, which can be recapitulated in vitro. So far, we discussed mitotic division. Oocytes, however, undergo a meiotic division (see Figure 19-38 for an overview of meiosis). As oocytes develop in the frog ovary, they replicate their DNA and become arrested in $G_2$ for 8 months, during which time they grow in size to a diameter of 1 mm, stockpiling all the materials needed for the multiple cell divisions of the early embryo. When stimulated by a male, an adult female's ovarian cells secrete the steroid

(a)

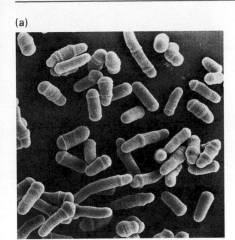

(b)

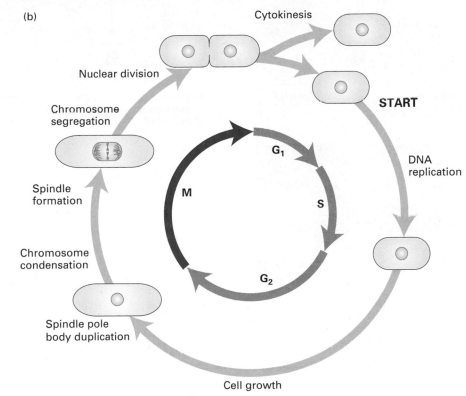

**FIGURE 19-4 The fission yeast *S. pombe*.** (a) Scanning electron micrograph of *S. pombe* cells at various stages of the cell cycle. Long cells are about to enter mitosis; short cells have just passed through cytokinesis. (b) Main events in the *S. pombe* cell cycle. START is the point in the cell cycle after which cells are irreversibly committed to undergoing a cell cycle. As in *S. cerevisiae*, the nuclear envelope does not break down during mitosis. [Part (a) courtesy of N. Hajibagheri.]

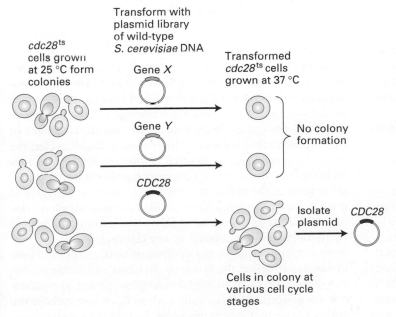

**EXPERIMENTAL FIGURE 19-5 Wild-type cell division cycle (*CDC*) genes can be isolated from a *S. cerevisiae* genomic library by functional complementation of *cdc* mutants.** Mutant cells with a temperature-sensitive mutation in a *CDC* gene are transformed with a genomic library prepared from wild-type cells and plated on nutrient agar at the non-permissive temperature (37 °C). Each transformed cell takes up a single plasmid containing one genomic DNA fragment. Most such fragments include genes (e.g., *X* and *Y*) that do not encode the defective Cdc protein; transformed cells that take up such fragments do not form colonies at the non-permissive temperature. The rare cell that takes up a plasmid containing the wild-type version of the mutant gene (in this case *CDC28*, a cyclin-dependent kinase) is complemented, allowing the cell to replicate and form a colony at the non-permissive temperature. Plasmid DNA isolated from this colony carries the wild-type *CDC* gene corresponding to the gene that is defective in the mutant cells. The same procedure is used to isolate wild-type *cdc*⁺ genes in *S. pombe*. See Figures 5-17 and 5-18 for more detailed illustrations of the construction and screening of a yeast genomic library.

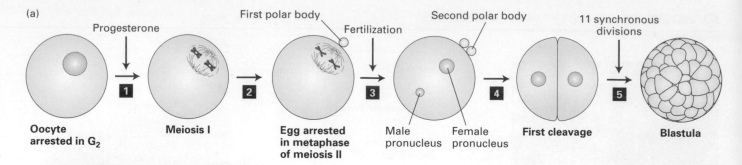

(a)

Progesterone

First polar body

Fertilization

Second polar body

11 synchronous divisions

**1**  **2**  **3**  **4**  **5**

**Oocyte arrested in G₂**   **Meiosis I**   **Egg arrested in metaphase of meiosis II**   Male pronucleus   Female pronucleus   **First cleavage**   **Blastula**

(b)

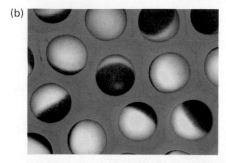

**FIGURE 19-6 Progesterone stimulates meiotic maturation of Xenopus oocytes.** (a) Step **1**: Progesterone treatment of G₂-arrested *Xenopus* oocytes surgically removed from the ovary of an adult female causes the oocytes to enter meiosis I. Two pairs of synapsed homologous chromosomes (blue) connected to meiotic spindle microtubules (green) are shown schematically to represent cells in metaphase of meiosis I. Step **2**: Segregation of homologous chromosomes and a highly asymmetrical cell division expels half the chromosomes into a small cell called the *first polar body*. The oocyte immediately commences meiosis II and arrests in metaphase II to yield an egg. Two chromosomes connected to spindle microtubules are shown

schematically to represent egg cells arrested in metaphase of meiosis II. Step **3**: Fertilization by sperm releases eggs from their metaphase arrest, allowing them to proceed through anaphase of meiosis II and undergo a second highly asymmetrical cell division that eliminates one chromatid of each chromosome in a second polar body. The resulting haploid female pronucleus fuses with the haploid sperm pronucleus to produce a diploid zygote. Step **4**: The zygote undergoes DNA replication and the first mitosis. Step **5**: The first mitosis is followed by 11 more synchronous divisions to form a blastula. (b) Micrograph of *Xenopus* eggs. [Part (b) copyright © ISM/Phototake.]

hormone progesterone, which induces the G₂-arrested oocytes to enter meiosis. As we will see in Section 19.8, meiosis consists of two consecutive chromosome segregation phases known as meiosis I and meiosis II. Progesterone triggers oocytes to undergo meiosis I and progress to the second meiotic metaphase, where they arrest and await fertilization (Figure 19-6). At this stage the cells are called eggs. When fertilized by sperm, the egg nucleus is released from its metaphase II arrest and completes meiosis. The resulting haploid egg nucleus then fuses with the haploid sperm nucleus, producing a diploid **zygote** nucleus. DNA replication follows, and the first mitotic division of embryogenesis begins. The resulting embryonic cells then proceed through 11 more rapid, synchronous cell cycles, generating a hollow sphere of cells called the blastula. Cell division then slows, and subsequent divisions are non-synchronous, with cells at different positions in the blastula dividing at different times.

The advantage of using *X. laevis* to study factors involved in mitosis is that large numbers of oocytes and eggs can be prepared that are all proceeding synchronously through the cell cycle events that follow progesterone treatment and fertilization. This makes it possible to prepare sufficient amounts of extract for biochemical experiments from cells that were all at the same point in the cell cycle. It was in this system that the cyclin-CDK complexes that trigger mitosis and the oscillatory nature of their activity was first dis-

covered. This activity was called **maturation-promoting factor (MPF)** because of its ability to induce entry into meiosis when injected into G₂-resting oocytes.

## Fruit Flies Reveal the Interplay Between Development and the Cell Cycle

The development of complex tissues often requires specific modifications to the cell cycle. Understanding the interplay between development and cell division is thus crucial if we want to understand how complex organisms are built. *Drosophila melanogaster* has established itself as the premier model system for studying the interplay between development and the cell cycle. Not only does the development of this organism involve several highly unusual cell cycles, the powerful genetic techniques that can be applied to fruit flies facilitated the discovery of genes involved in the developmental control of the cell cycle. The first 13 nuclear divisions of the fertilized *Drosophila* embryo all occur in a common cytoplasm and are rapid cycles of DNA replication and mitosis (with no gap phases), fueled by key cell cycle regulators that were stockpiled in the egg cytoplasm as it matured. These divisions are called the syncytial divisions and occur in unison (Figure 19-7). As maternal stockpiles run out, gap phases are introduced, first G₂, followed by G₁. Most cells in the embryo cease to divide at this point, form plasma membranes,

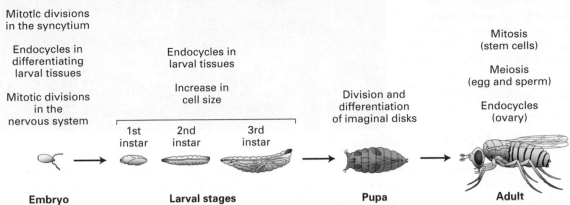

**FIGURE 19-7 Cell division patterns during the life cycle of *Drosophila melanogaster*.** After fertilization, nuclei in the embryo undergo 13 rapid S phase–M phase cycles. These are followed by three divisions that include a $G_2$ phase. All these nuclear divisions occur within a common cytoplasm and are therefore called the syncytial divisions. During late stages of embryogenesis and throughout larval development (with the exception of the nervous system), cells undergo endocycles. This leads to an increase in cellular ploidy and size and hence larval growth. In the pupa, during a process called metamorphosis, imaginal disks, the tissues that give rise to the adult organs, undergo mitotic divisions and then differentiate to form adult structures. Several types of divisions are seen in the adult fly. Stem cells undergo mitotic divisions, meiosis gives rise to sperm and egg, and endocycles create polyploid cells in the ovary. [Adapted from Lee and Orr-Weaver, 2003, *Ann. Rev. Genet.* **37:**545–578.]

and utilize a specialized cell cycle known as the endocycle. In the endocycle, cells replicate their DNA but do not undergo mitosis. This leads to an increase in gene dosage and fuels increased macromolecule biosynthesis, which allows individual cells to grow in size. Thus the embryo, which has now developed into a crawling larva, grows simply by an increase in cell size and not through cell multiplication. A select number of cells do not share this fate. These cells are in the imaginal disks, the organs that will give rise to the adult fly tissues during metamorphosis. Metamorphosis occurs during the pupa stage and transforms larvae into adult flies. The divisions that give rise to the adult fly are canonical cell cycles leading to the adult fly being a diploid organism.

## The Study of Tissue Culture Cells Uncovers Cell Cycle Regulation in Mammals

Cell cycle regulation in human cells is more complex than in other non-mammalian systems. To understand this increased level of complexity and to understand the cell cycle alterations that are the cause of cancer, it is important to study the cell cycle not only in model organisms but also in human cells. Researchers use normal or tumor cells grown in plastic dishes to study the properties of the human cell cycle, a method called tissue culture or cell culture. It is, however, important to note that many of the cell types used to study the human cell cycle themselves have altered cell cycle properties due to genetic alterations that occurred during their culturing or because they were isolated from human tumors. Furthermore, in vitro culture conditions do not resemble those found in the organism and could lead to altered behavior of cells. Although some aspects of mammalian cell division are not recapitulated in cell culture conditions—such as the importance of tissue

organization and developmental signals governing cell cycle control—cell culture systems nevertheless provide critical insights into the mammalian cell's intrinsic mechanisms governing cell division. Researchers also work toward establishing culture systems that more closely resemble the cell architecture in tissues. For example, polymers are currently being developed that allow scientists to grow cells in 3-D culture.

As we will see in Chapter 21, primary human cells and other mammalian cells have a finite life span when cultured in vitro. Normal human cells, for example, divide 25–50 times, but thereafter proliferation slows and eventually stops. This process is called *replicative senescence*. Cells can escape this process and become immortalized, allowing researchers to establish cell lines. Although these cell lines harbor genetic alterations that affect some aspects of cell proliferation, they are nevertheless a useful tool to study cell cycle progression in human cells. These cell lines provide an inexhaustible supply of cells that, as we will see next, can be manipulated to progress through the cell cycle in a synchronous manner, allowing for the analysis of protein levels and enzymatic activity at different stages of the cell cycle.

## Researchers Use Multiple Tools to Study the Cell Cycle

The experimental analysis of cell cycle properties requires that we are able to determine the cell cycle stage of individual cells. Light microscopy provides some estimate of cell cycle progression. For example, light microscopy allows a researcher to determine whether cultured mammalian cells are in interphase ($G_1$, S phase, and $G_2$) or in mitosis. Mammalian tissue culture cells are flat and adhere to the plastic dish during interphase but round up and form spherical structures as they undergo

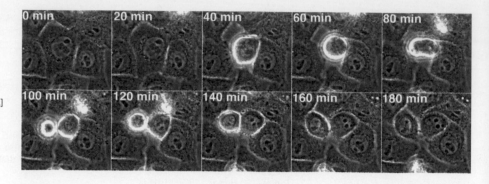

**FIGURE 19-8 Human cells undergoing mitosis.** HeLa Kyoto cells were filmed as they underwent mitosis. The images shown were taken every 20 minutes. Cells are flat during interphase, but as cells undergo mitosis, they round up and divide. Subsequently, they flatten out again. [Courtesy of Sejal Vyas and Paul Chang, MIT.]

mitosis (Figure 19-8). Fluorescence microscopy of cellular structures or the analysis of specific cell cycle markers, that is, proteins that are only present in certain cell cycle stages, allows for a more accurate determination of cell cycle stage.

In addition to microscopic tools, cell cycle researchers use flow cytometry to determine the DNA content of a cell population (Figure 19-9; see also Figure 9-2). Cells are treated with a DNA-binding fluorescent dye and the amount of dye that incorporates into the DNA of cells can then be quantitatively assessed using a flow cytometer. Cells are then sorted by their DNA content, and the percentage of cells in $G_1$, S phase, and $G_2$ or mitosis can be assessed in this manner. Cells in $G_1$ will have half as much DNA as cells in $G_2$ or mitosis. Cells undergoing DNA synthesis in S phase will have an intermediate amount of DNA.

To characterize different cell cycle events, it is essential to examine cell populations that progress through the cell cycle in unison. Researchers achieve this by *reversibly arresting* cells in a particular cell cycle stage. This cell cycle arrest is usually accomplished by restricting nutrients or adding anti-growth factors, which cause cells to arrest in $G_1$. In budding yeast, for example, cells treated with a mating pheromone arrest in $G_1$. When the pheromone is removed from cells (usually by washing them extensively), cells exit the $G_1$ arrest and progress through the cell cycle in a synchronous manner. In mammalian cells, removal of growth factors by removing serum from the culture medium (serum starvation) arrests cells in $G_0$. Re-addition of serum allows cells to re-enter the cell cycle. Other methods involve blocking a certain cell cycle step with chemicals. Hydroxyurea inhibits DNA replication, leading to an S phase arrest. On removal of the drug, cells will resume DNA synthesis in unison. Nocodazole disrupts the mitotic spindle and halts the cell cycle in mitosis. Once the drug is washed away, cells will resume progression through mitosis in a synchronous manner. In budding and fission yeast, the conditional cell division cycle (*cdc*) mutants introduced earlier have proved a powerful tool for creating synchronous cultures. Temperature-sensitive *cdc* mutants, when incubated at the non-permissive temperature, arrest in a particular cell cycle stage because they are defective in a certain key cell cycle protein. Returning cells to the permissive temperature allows cells to continue with the cell division cycle in a synchronous fashion.

**EXPERIMENTAL FIGURE 19-9 Analysis of DNA content by flow cytometry.** Haploid yeast cells were grown in culture and stained with propidium iodide, a fluorescent dye that incorporates into DNA. The x axis shows DNA content, the y axis the number of cells. The DNA content analysis shows two predominant populations of cells: cells with unreplicated DNA (1C) and with replicated DNA (2C). The cells between the two peaks represent cells that are in the process of undergoing DNA replication. [Courtesy of Heidi Blank.]

## KEY CONCEPTS of Section 19.2

### Model Organisms and Methods to Study the Cell Cycle

• The ability to isolate mutants and the powerful genetic tools of budding and fission yeast allowed for the isolation of key factors important for cell cycle regulation.

• Frog eggs and early embryos from synchronously fertilized eggs provide sources of extracts for biochemical studies of cell cycle events and identified the oscillatory nature of cyclin-CDK complexes.

• Fruit flies are a powerful system to investigate the interplay between cell division and the developmental programs responsible for building multicellular organisms.

• Human tissue culture cells are used to study the properties of the mammalian cell cycle.

- The generation of synchronized cell populations through reversibly arresting cells in a particular cell cycle stage allows the researcher to examine the behavior of proteins and cellular processes during the cell cycle.

## 19.3 Regulation of CDK Activity

In the following sections we describe the current model of eukaryotic cell cycle regulation summarized in Figure 19-10 and present some of the experiments that led to this understanding. As we will see, results obtained with different experimental systems and approaches have provided insights into each of the

transition points in the cell cycle. A key discovery in cell cycle studies was that cyclin-dependent kinases govern progression through the cell cycle. Three key features about these kinases are important to keep in mind throughout this chapter:

- Cyclin-dependent kinases (CDKs) are only active when bound to a regulatory cyclin subunit.

- Different types of cyclin-CDK complexes initiate different events. **$G_1$ CDKs** and **$G_1$/S phase CDKs** promote entry into the cell cycle, **S phase CDKs** trigger S phase, and **mitotic CDKs** initiate the events of mitosis (Figure 19-11).

- Multiple mechanisms are in place to ensure that the different CDKs are only active in the stages of the cell cycle they trigger.

**VIDEO:** Dynamic Behavior of Mitotic Cyclin in HeLa Cells

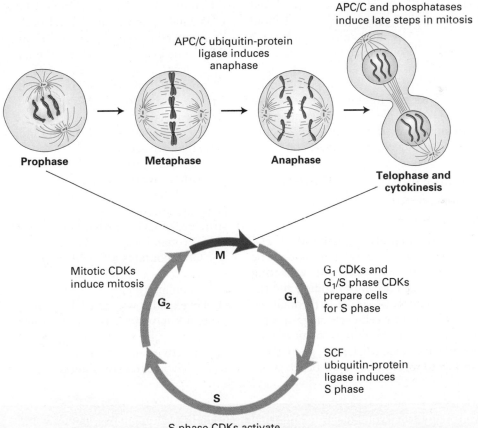

**FIGURE 19-10 Regulation of cell cycle transitions.** Cell cycle transitions are regulated by cyclin-CDK protein kinases, protein phosphatases, and ubiquitin-protein ligases. Here the cell cycle is diagrammed, with the major stages of mitosis shown at the top. In early $G_1$, no cyclin-CDKs are active. In mid-$G_1$, $G_1$/S phase CDKs activate transcription of genes required for DNA replication. S phase is initiated by the SCF ubiquitin-protein ligase that ubiquitinylates inhibitors of S phase CDKs, marking them for degradation by proteasomes. The S phase CDKs then activate DNA replication and DNA synthesis commences. Once DNA replication is complete, cells enter $G_2$. In late $G_2$, mitotic CDKs trigger entry into mitosis. During prophase, the nuclear envelope breaks down and chromosomes align on the mitotic spindle but they cannot separate until the anaphase-promoting complex (APC/C), a ubiquitin-protein ligase, ubiquitinylates the anaphase inhibitor protein securin, marking it for degradation by proteasomes. This results in degradation of protein complexes linking the sister chromatids and the onset of anaphase as sister chromatids separate. After chromosome movement to the spindle poles, the APC/C ubiquitinylates mitotic cyclins, causing their degradation by proteasomes. The resulting drop in mitotic CDK activity, along with the action of protein phosphatases, results in chromosome decondensation, reassembly of nuclear membranes around the daughter-cell nuclei, and cytokinesis.

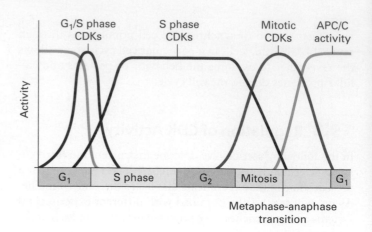

**FIGURE 19-11 An overview of how CDKs regulate cell cycle progression.** Cells harbor different types of CDKs that initiate different events of the cell cycle. Importantly, the CDKs are only active in the stages of the cell cycle they trigger. $G_1$/S phase CDKs are active at the $G_1$-S phase transition to trigger entry into the cell cycle. S phase CDKs are active during S phase and trigger S phase. Mitotic CDKs are active during mitosis and trigger mitosis. A ubiquitin ligase known as the anaphase-promoting complex or cyclosome (APC/C) catalyzes two key cell cycle transitions by ubiquitinylating proteins, hence targeting them for degradation. The APC/C initiates anaphase and exit from mitosis.

In this section we will first discuss the properties of CDKs and investigate the structural basis of their activation and regulation. We will then see how cyclins activate CDKs and investigate the multiple regulatory mechanisms that restrict the different cyclins to the appropriate cell cycle stage. We will see that protein degradation plays an essential part in this process. In addition, we will discuss how post-translational modifications to CDKs and inhibitory proteins that directly bind to cyclin-CDK complexes are essential additional control mechanisms in restricting different cyclin-CDK activities to the appropriate cell cycle stage.

## Cyclin-Dependent Kinases Are Small Protein Kinases That Require a Regulatory Cyclin Subunit for Their Activity

Cyclin-dependent kinases are a family of small (30–40 kD) serine/threonine kinases. They are not active in the monomeric form but, as mentioned previously, require an activating subunit to be active as a protein kinase. In budding and fission yeast, a single CDK controls progression through the cell cycle. Its activity is specified by cell-cycle-stage-specific cyclin subunits. Mammalian cells contain as many as nine CDKs, with four of them, CDK1, CDK2, CDK4, and CDK6, having clearly been shown to regulate cell cycle progression.

They bind to different types of cyclins and together promote different cell cycle transitions. CDK4 and CDK6 are $G_1$ CDKs and promote entry into the cell cycle, CDK2 functions as a $G_1$/S phase and S phase CDK, and CDK1 is the mitotic CDK. For historical reasons, the names of various cyclin-dependent kinases from yeasts and vertebrates differ. Whenever possible, we will use the general terms *$G_1$, $G_1$/S phase, S phase,* and *mitotic CDKs* to describe CDKs instead of the species-specific terminology. Table 19-1 lists the different names of the various CDKs and indicates when in the cell cycle they are active.

CDKs are not only regulated by cyclin binding but also by both activating and inhibitory phosphorylation. Together, these regulatory events ensure that CDKs are only active at the appropriate cell cycle stage. The three-dimensional structure of CDKs provides insight into how the activity of these protein kinases is regulated. Unphosphorylated, inactive CDK contains a flexible region, called the *T loop,* that blocks access of protein substrates to the active site where ATP is bound (Figure 19-12a). Steric blocking by the T loop largely explains why free CDK, unbound to cyclin, has little protein kinase activity. Unphosphorylated CDK bound to one of its cyclin partners has minimal but detectable protein kinase activity in vitro, although it may be essentially inactive in vivo. Extensive interactions between the cyclin and

| TABLE 19-1 | Cyclins and CDKs: Nomenclature and Their Roles in the Mammalian Cell Cycle | | |
|---|---|---|---|
| **CDK** | **Cyclin** | **Function** | **General Name** |
| CDK1 | Cyclin A, cyclin B | Mitosis | Mitotic CDKs |
| CDK2 | Cyclin E, cyclin A | Entry into the cell cycle<br>S phase | $G_1$/S phase CDKs<br>S phase CDKs |
| CDK4 | Cyclin D | $G_1$<br>Entry into the cell cycle | $G_1$ CDKs |
| CDK6 | Cyclin D | $G_1$<br>Entry into the cell cycle | $G_1$ CDKs |

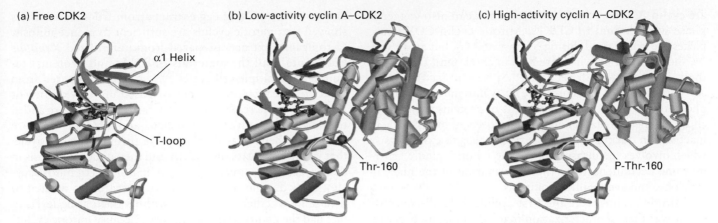

(a) Free CDK2  (b) Low-activity cyclin A–CDK2  (c) High-activity cyclin A–CDK2

α1 Helix

T-loop

Thr-160

P-Thr-160

**FIGURE 19-12 Structural models of human CDK2.** (a) Free, inactive CDK2 unbound to its cyclin subunit, cyclin A. In free CDK2, the T loop blocks access of protein substrates to the γ phosphate of the bound ATP, shown as a ball-and-stick model. The conformations of the regions highlighted in yellow are altered when CDK is bound to cyclin A. (b) Unphosphorylated, low-activity cyclin A–CDK2 complex. Conformational changes induced by binding of a domain of cyclin A (blue) cause the T loop to pull away from the active site of CDK2 so that substrate proteins can bind. The α1 helix in CDK2, which interacts extensively with cyclin A, moves several angstroms into the catalytic cleft, repositioning key catalytic side chains required for the phosphotransfer reaction. The red ball marks the position of the threonine (Thr-160) whose phosphorylation activates CDKs. (c) Phosphorylated, high-activity cyclin A–CDK2 complex. The conformational changes induced by phosphorylation of the activating threonine (red ball) alter the shape of the substrate-binding surface, greatly increasing the affinity for protein substrates. [Courtesy of P. D. Jeffrey. See A. A. Russo et al., 1996, *Nature Struct. Biol.* **3:**696.]

the T loop cause a dramatic shift in the position of the T loop, thereby exposing the CDK active site (Figure 19-12b). As we will see shortly, high activity of the cyclin-CDK complex requires phosphorylation of the activating threonine, in the T loop, causing additional conformational changes in the cyclin-CDK complex that greatly increase its affinity for protein substrates (Figure 19-12c). As a result, the kinase activity of the phosphorylated complex is a hundredfold greater than that of the unphosphorylated complex.

## Cyclins Determine the Activity of CDKs

Cyclins are so named because their levels change during the cell cycle. They form a family of proteins that is defined by three key features:

- Cyclins bind to and activate CDKs. The activity and substrate specificity of any given CDK is primarily defined by the particular cyclin to which it is bound.

- Cyclins are only present during the cell cycle stage that they trigger and are absent in other cell cycle stages.

- Cyclins not only regulate a particular cell cycle stage but also set in motion a series of events in preparation for the next cell cycle stage. In this way, they propel the cell cycle forward.

Cyclins are divided into four classes defined by their presence and activity during the cell cycle: $G_1$ cyclins, $G_1$/S cyclins, S phase cyclins, and mitotic cyclins (see Table 19-1). The different types of cyclins are quite distinct from each other in protein sequence, but all of them contain a conserved 100 amino acid region known as the cyclin box and possess similar three-dimensional structures.

The $G_1$ cyclins are the lynchpin in coordinating the cell cycle with extracellular events. Their activity is subject to regulation by signal transduction pathways that sense the presence of growth factors or cell proliferation inhibitory signals. In metazoans, $G_1$ cyclins are known as cyclin Ds, and they bind to CDK4 and CDK6. $G_1$ cyclins are unusual in that their levels do not fluctuate in a specific pattern during the cell cycle. Instead, in response to macromolecule biosynthesis and extracellular signals, their levels gradually increase throughout the cell cycle.

The $G_1$/S cyclins accumulate during late $G_1$, reach peak levels when cells enter S phase, and decline during S phase (see Figure 19-11). They are known as cyclin E in metazoans and bind to CDK2. The main function for cyclin E–CDK2 complexes, together with cyclin D–CDK4/6, is to trigger the $G_1$–S phase transition. This transition is known as START and is defined as the point at which cells are irreversibly committed to cell division and can no longer return to the $G_1$ state. In molecular terms, this means that cells initiate DNA replication as well as duplicate their centrosomes, which is the first step in the formation of the mitotic spindle that will be used during mitosis.

S phase cyclins are synthesized concomitantly with $G_1$ cyclins, but levels remain high throughout S phase and do not decline until early mitosis. Two types of S phase cyclins trigger S phase in metazoans: cyclin E, which can also promote entry into the cell cycle and is therefore also a $G_1$/S cyclin, and cyclin A. Both cyclins bind CDK2 (see Table 19-1) and are directly responsible for DNA synthesis. As we will see in Section 19.4, these protein kinases phosphorylate proteins that activate DNA helicases and load polymerases onto DNA.

Mitotic cyclins bind CDK1 to promote entry into and progression through mitosis. The metazoan mitotic cyclins are cyclin A and cyclin B (note cyclin A can also trigger S phase when bound to CDK2). Mitotic cyclin-CDK complexes are synthesized during S phase and G$_2$, but as we will see shortly, their activities are held in check until DNA synthesis is completed. In Section 19.5, we will see that once activated, mitotic CDKs promote entry into mitosis by phosphorylating and activating hundreds of proteins to promote chromosome segregation and other aspects of mitosis. Their inactivation during anaphase prompts cells to exit mitosis, which involves the disassembly of the mitotic spindle, chromosome decondensation, the re-formation of the nuclear envelope, and eventually cytokinesis.

Mitotic cyclins were the first cyclins to be discovered, and it was their characterization that led to the discovery of the oscillatory nature of the activities that govern cell cycle progression. The experiment that led to their discovery is described at the end of this chapter as a classic experiment in cell biology (see Classic Experiment 19.1, Figure 1).

The fact that cyclins are the rate-limiting proteins in triggering cell cycle transitions was discovered using the embryonic extract system. Using egg extracts from frogs, researchers showed that mitotic cyclins are sufficient to induce mitosis. Cytoplasmic extracts prepared from unfertilized *Xenopus* eggs contain all the materials (mRNAs and proteins) required for multiple cell cycles. When nuclei prepared from *Xenopus* sperm (sperm nuclei are used in this experiment because they are readily isolated in large numbers) are added to such an egg extract, the nucleus and DNA inside it are induced to behave as if progressing through the cell cycle; that is, they replicate their DNA and then undergo mitosis in the extract. Concomitant with the final stages of mitosis, cyclin levels decline (Figure 19-13a). Researchers wanted to determine whether this protein with its fluctuating levels in fact had the ability to induce mitosis. Because mitotic cyclins appeared to be unstable at the end of mitosis, the researchers reasoned that removing the mRNA from the egg extract would prevent the new synthesis of all unstable proteins in the next cell cycle, including that of the mitotic cyclins (all other stable proteins would still be present in the extract). To do this, they digested all mRNAs with a low concentration of

(a) Untreated extract

(b) RNase-treated extract

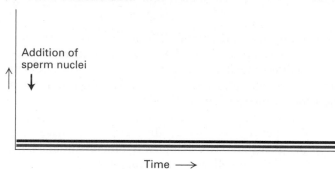

(c) RNase-treated extract + wild-type mitotic cyclin mRNA

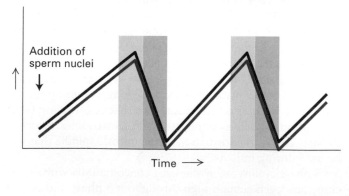

(d) RNase-treated extract + nondegradable mitotic cyclin mRNA

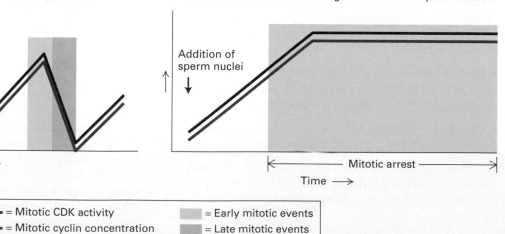

— = Mitotic CDK activity
— = Mitotic cyclin concentration
▓ = Early mitotic events
▓ = Late mitotic events

**EXPERIMENTAL FIGURE 19-13 Mitotic cyclins are rate limiting for mitosis.** In all cases, mitotic CDK activity and mitotic cyclin concentration were determined at various times after addition of sperm nuclei to a *Xenopus* egg extract treated as indicated in each panel. Microscopic observations determined the occurrence of early mitotic events (blue shading), including chromosome condensation and nuclear envelope disassembly, and of late events (orange shading), including chromosome decondensation and nuclear envelope reassembly. See text for discussion. [See A. W. Murray et al., 1989, *Nature* **339:**275; adapted from A. Murray and T. Hunt, 1993, *The Cell Cycle: An Introduction,* W. H. Freeman and Company.]

RNase, which then was inactivated by addition of a specific inhibitor. This treatment destroys mRNAs without affecting the tRNAs and rRNAs required for protein synthesis. When sperm nuclei were added to the RNase-treated extracts, the nuclei replicated their DNA, but they did not undergo mitosis. Furthermore, mitotic cyclin protein was not detected (Figure 19-13b). Addition of mitotic cyclin mRNA, produced in vitro from cloned mitotic cyclin cDNA, to the RNase-treated egg extract induced mitosis as observed with the untreated egg extract (Figure 19-13c). Since mitotic cyclin is the only newly synthesized protein under these conditions, these results demonstrate that the protein is the only one limiting for entering mitosis. Subsequent studies showed that $G_1$/S phase cyclins have similar properties. Their expression is sufficient to promote entry into the cell cycle, and therefore all the other proteins needed for cell cycle entry are present in unlimited amounts. It is thus clear that the regulation of cyclin levels is an essential aspect of the eukaryotic cell cycle. As we will see in the following section, cells utilize multiple mechanisms to restrict cyclins to the appropriate cell cycle stage and to keep them at the right concentration.

## Cyclin Levels Are Primarily Regulated by Protein Degradation

Multiple mechanisms ensure that CDKs are active in the right stage of the cell cycle. Table 19-2 lists these key regulators of CDKs. In this section we will discuss how the regulation of cyclin levels is brought about. The timely activation of CDKs depends, in part, on the presence of the appropriate cyclins in the cell cycle stage where they are needed. Transcriptional control of the cyclin subunits is one mechanism that ensures proper temporal expression of the cyclins. In somatic cells and yeast, waves of transcription factor activities help establish waves of cyclin activity. A general principle here is that an earlier wave of transcriptional activity helps produce the factors essential to generate a subsequent transcriptional wave. As we will see in Section 19.4, transcription of the $G_1$/S phase cyclins is promoted by the **E2F transcription factor complex**.

| TABLE 19-2 | Regulators of Cyclin-CDK Activity |
|---|---|
| **Type of Regulator** | **Function** |
| **Kinases and Phosphatases** | |
| CAK kinase | Activates CDKs |
| Wee1 kinase | Inhibits CDKs |
| Cdc25 phosphatase | Activates CDKs |
| Cdc14 phosphatase | Activates Cdh1 to degrade mitotic cyclins |
| Cdc25A phosphatase | Activates vertebrate S phase CDKs |
| Cdc25C phosphatase | Activates vertebrate mitotic CDKs |
| **Inhibitory Proteins** | |
| Sic1 | Binds and inhibits S phase CDKs |
| CKIs p27$^{KIP1}$, p57$^{KIP2}$, and p21$^{CIP}$ | Bind and inhibit CDKs |
| INK4 | Binds and inhibits $G_1$ CDKs |
| Rb | Binds E2Fs, preventing transcription of multiple cell cycle genes |
| **Ubiquitin-Protein Ligases** | |
| SCF | Degradation of phosphorylated Sic1 or p27$^{KIP1}$ to activate S phase CDKs |
| APC/C + Cdc20 | Degradation of securin, initiating anaphase. Induces degradation of B-type cyclins |
| APC/C + Cdh1 | Degradation of B-type cyclins in $G_1$ and geminin in metazoans to allow loading of replicative helicases on DNA replication origins |

Among the many other genes that E2F transcribes are the transcription factors that will promote the synthesis of mitotic cyclins.

The most important regulatory control that restricts cyclins to the appropriate cell cycle stage is ubiquitin-mediated, proteasome-dependent protein degradation. Because protein degradation is an irreversible process, in the sense that the protein can only be replenished through de novo protein synthesis, this regulatory mechanism is ideal to ensure that the cell cycle engine is driven forward and cells cannot "go backward" in the cell cycle. In other words, once a particular cyclin is degraded, the processes that it activated can no longer take place.

Recall that during ubiquitin-mediated protein degradation, ubiquitin-protein ligases polyubiquitinylate substrate proteins, marking them for degradation by the proteasome (see Figure 3-29). Cyclins are degraded through the action of two different ubiquitin-protein ligases, **SCF** (named after the first letters of its constituents, *Skp1, Cullin, F*-box proteins) and the **anaphase-promoting complex** or **cyclosome** (abbreviated as **APC/C** in this chapter). SCF controls the $G_1$–S phase transition by degrading $G_1$/S phase cyclins and, as we will see in detail shortly, CDK inhibitory proteins. The APC/C degrades S phase and mitotic cyclins, thereby promoting the exit from mitosis. APC/C also controls the onset of chromosome segregation at the metaphase-anaphase transition by degrading an anaphase inhibitory protein (discussed in Section 19.6).

The SCF and APC/C are multisubunit ubiquitin ligases that belong to the RING finger family of ubiquitin ligases. Despite the fact that SCF and APC/C belong to the same ubiquitin ligase family, their regulation is quite different. The SCF recognizes its substrates only when they are phosphorylated. The SCF is continuously active throughout the cell cycle, and cell-cycle-regulated phosphorylation of its substrates ensures that they are only degraded at certain stages of the cell cycle. In the case of APC/C-dependent protein degradation, the regulation is reversed. Substrates are recognizable throughout the cell cycle, but the activity of the APC/C is cell cycle regulated. APC/C is activated by phosphorylation at the metaphase-anaphase transition, through the action of mitotic CDKs and other protein kinases. The APC/C is then active throughout the rest of mitosis and during $G_1$ to promote the degradation of cyclins and other mitotic regulators (see Figure 19-11). Substrate specificity of active, phosphorylated APC/C is determined in part by its association with one of two related substrate-targeting factors called Cdc20 and Cdh1. During anaphase APC/C bound to Cdc20 ubiquitinylates proteins that lead to chromosome segregation, while during telophase and $G_1$, APC/C-Cdh1 targets different substrates for degradation. The substrates of the APC/C contain recognition motifs. The first to be identified is the **destruction box**. It is found in most S phase and mitotic cyclins. This destruction box is both necessary and sufficient to target proteins for degradation.

Today we know that cyclin degradation at the appropriate cell cycle transition is essential for cell cycle progression. The importance of cyclin degradation was again first demonstrated in frog egg extracts. Recall that the accumulation of mitotic cyclins not only coincided with entry into mitosis, but its disappearance occurred concomitantly with exit from mitosis. This finding raised the possibility that mitotic cyclin degradation was necessary for cells to exit from mitosis. To test this, researchers added an mRNA encoding a nondegradable mitotic cyclin (lacking the destruction box) to a mixture of RNase-treated *Xenopus* egg extract and sperm nuclei. As shown in Figure 19-13d, entry into mitosis occurred on schedule, but exit from mitosis did not. This experiment demonstrates that mitotic exit requires the degradation of mitotic cyclin. Later studies showed that inhibiting the degradation of other cyclins also severely affected cell cycle progression, indicating that ubiquitin-mediated degradation of cyclins is an essential aspect of the eukaryotic cell cycle.

## CDKs Are Regulated by Activating and Inhibitory Phosphorylation

Regulating the levels of cyclins is not the only mechanism that controls CDK activity. Activating and inhibitory phosphorylation events on the CDK subunit itself are essential to control cyclin-CDK activity. Phosphorylation of a threonine residue near the active site of the enzyme is required for CDK activity. This phosphorylation is mediated by the **CDK-activating kinase (CAK)**. In some organisms cyclin binding is a prerequisite for CAK phosphorylation, whereas in others this phosphorylation event occurs prior to cyclin binding. Although the sequence of assembling active CDKs differs between organisms, it is clear that CAK phosphorylation of CDK is not a rate-limiting step in CDK activation. CAK activity is constant throughout the cell cycle and phosphorylates the CDK as soon as a cyclin-CDK complex is formed.

Two inhibitory phosphorylations on CDK play a critical role in controlling CDK activity. In contrast to the CAK-induced activating phosphorylation, these inhibitory phosphorylations are regulated. A highly conserved tyrosine (Y15 in human CDKs) and an adjacent threonine (T14 in humans) are subject to regulated phosphorylation. Both residues are situated in the ATP binding pocket of the CDK, and their phosphorylation most likely interferes with positioning of ATP in the pocket. Changes in the phosphorylation of these sites are essential for the regulation of mitotic CDKs and have also been implicated in the control of $G_1$/S and S phase CDKs. As we will see in Section 19.5, a highly conserved kinase called Wee1 brings about this inhibitory phosphorylation, and a highly conserved phosphatase called Cdc25 mediates dephosphorylation.

## CDK Inhibitors Control Cyclin-CDK Activity

So far, we have discussed the importance of regulating cyclin protein levels and of CDK phosphorylation in controlling CDK activity. The final layer of control that is of critical importance in the regulation of CDKs is a family of proteins known as **CDK inhibitors**, or **CKIs**, that directly bind to the cyclin-CDK complex and inhibit its activity. As we will see in Section 19.4, these proteins play an especially important role in the regulation of the $G_1$–S phase transition and its integration with extracellular signals. It thus comes as no surprise that the genes encoding these CKIs are often found mutated in human cancers (discussed in Chapter 24).

All eukaryotes harbor CKIs involved in regulating S phase and mitotic CDKs. Although these inhibitors display little sequence similarity, they are all essential to prevent premature activation of S phase and M phase CDKs. Inhibitors of $G_1$ CDKs play an essential role in mediating a $G_1$ arrest in response to proliferation inhibitory signals. A class of CKIs called *INK4s* (*in*hibitors of *kinase 4*) includes several small, closely related proteins that interact only with the $G_1$ CDKs. Binding of INK4s to CDK4 and CDK6 blocks their interaction with cyclin D and hence their protein kinase activity. A second class of CKIs found in metazoan cells consists of three proteins—p21$^{CIP}$, p27$^{KIP1}$, and p57$^{KIP2}$. These CKIs inhibit $G_1$/S phase CDKs and S phase CDKs and must be degraded before DNA replication can begin. As we will discuss in Section 19.7, p21$^{CIP}$ plays an important role in the response of metazoan cells to DNA damage. CKIs regulating $G_1$ CDKs play a critical role in preventing tumor formation. For example, both copies of the INK4 gene, which encodes p16, are found inactivated in a large fraction of human cancers.

## Special CDK Alleles Led to the Discovery of CDK Functions

Different CDKs initiate different cell cycle phases by phosphorylating specific proteins. It is now clear that rather than phosphorylating a small number of proteins that in turn initiate a certain cell cycle stage, CDKs phosphorylate a myriad of substrates, thereby directly initiating all aspects of a given cell cycle phase. Analysis of a small number of substrates has provided examples that show how phosphorylation by mitotic CDKs mediates many of the early events of mitosis: chromosome condensation, formation of the mitotic spindle, and disassembly of the nuclear envelope. We will discuss these events in detail in the sections that follow.

In recent years systematic efforts to identify all CDK substrates have been initiated. The challenge in identifying substrates of a particular kinase is how to distinguish that kinase's phosphorylation events from those carried out by other kinases. A breakthrough in understanding which proteins are targets of CDKs was facilitated by the engineering of a CDK mutant in budding yeast that can utilize an analog of ATP that is not bound by other kinases (Figure 19-14). This ATP analog has a bulky benzyl group attached to $N_6$ of the adenine, which makes the analog too large to fit into the ATP-binding pocket of wild-type protein kinases. However, the ATP-binding pocket of the mutant CDK was modified to accommodate this large ATP analog. Consequently, only the mutant CDK can utilize this ATP analog as a substrate for transferring its $\gamma$ phosphate to a protein side chain. When the $N_6$-benzyl ATP analog with a labeled $\gamma$ phosphate was incubated with yeast cell extracts containing a recombinant mitotic CDK containing the altered ATP-binding pocket, multiple proteins were labeled. True yeast mitotic CDK substrates could be verified among these potential substrates by treatment of cells expressing the mutant CDK with a similar derivative of another ATP analog that binds the mutant CDK in the same way as the

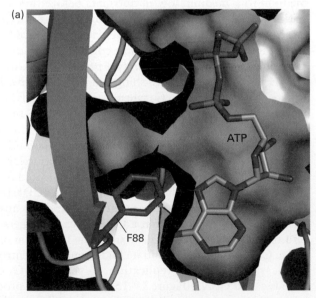 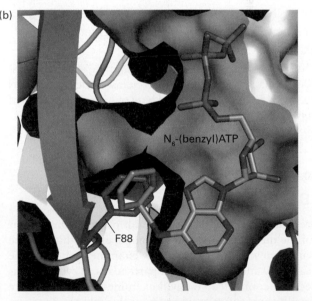

**FIGURE 19-14 ATP analog–dependent CDK mutant. (a)** Representation of the ATP-binding and catalytic sites of wild-type *S. cerevisiae* CDK1 (called Cdc28 in budding yeast). Bound ATP and a phenylalanine side chain (purple) in the vicinity of the binding pocket are shown in stick format. **(b)** Bulky ATP analogs such as those containing a benzyl group bound to the $N_6$ amino nitrogen are too large to fit into the ATP-binding pocket of wild-type protein kinases and thus cannot be utilized by them. In the *S. cerevisiae* CDK mutant, the phenylalanine at position 88 is changed to glycine, which lacks a large side chain. The mutant exhibits high protein kinase activity using $N_6$-(benzyl)ATP. These models of *S. cerevisiae* CDK are based on crystal structures of the PKA kinase domain, which shares extensive homology with the kinase domain of *S. cerevisiae* CDK. [See J. A. Ubersax et al., 2003, *Nature* **425**:859; K. Shah et al., 1997, *Proc. Nat'l. Acad. Sci. USA* **94**:3565.]

$N_6$-benzyl ATP analog but that cannot be used for substrate phosphorylation. Because of the bulky benzyl group, it is sterically blocked from binding to all other kinases, consequently inhibiting only the mutant CDK. When cells expressing the mutant CDK were treated with this specific CDK inhibitor, most of the putative mitotic CDK targets identified initially were found in a dephosphorylatyed state, indicating that these proteins are indeed phosphorylated by the CDK in vivo as well as in vitro. In yeast, this procedure identified most of the known CDK substrates plus more than 150 additional yeast proteins. Similar approaches have also been used in mammalian cells to identify CDK substrates. For example, a hunt for substrates of the S phase CDK cyclin A-CDK2 revealed 180 potential substrates. These are currently being analyzed for their functions in cell cycle processes.

## KEY CONCEPTS of Section 19.3

### Regulation of CDK Activity

- Cyclin-dependent kinases are activated by cyclin subunits. Their activity is controlled at multiple levels.

- Different cyclin subunits activate CDKs at different cell cycle stages. Cyclins are present only in the cell cycle stages that they promote.

- Protein degradation is the key mechanism responsible for restricting cyclins to the appropriate cell cycle stage. This degradation is mediated by the ubiquitin-proteasome system and the ubiquitin ligases APC/C and SCF.

- Activating and inhibitory phosphorylation on the CDK subunit contributes to the regulation of CDK activity.

- CDK inhibitors (CKIs) inhibit CDK activity by directly binding to the cyclin-CDK complex.

- CDKs initiate every aspect of each cell cycle stage by phosphorylating many different target proteins. Systematic efforts using protein kinases engineered to bind only modified forms of ATP have led to the identification of many of these substrates.

## 19.4 Commitment to the Cell Cycle and DNA Replication

The previous section described the multiple mechanisms that control the different cyclin-CDK complexes. In this and the following two sections we will now examine each cell cycle stage carefully and discuss how a particular cell cycle stage is induced and controlled. We will examine how cells initiate DNA replication and mitosis and how chromosomes are segregated. We will focus on how cyclin-CDK complexes and other key cell cycle regulators impact each cell cycle phase and examine the mechanisms that coordinate their activities.

This section investigates how cells decide whether or not to undergo cell division and how DNA replication is initiated. The process of cell cycle entry is well understood in budding yeast, and it was in this organism that the molecular mechanisms underlying this cell cycle transition were initially elucidated. We will therefore first examine the molecular events governing cell cycle entry in budding yeast. We will then investigate the striking similarities between the pathways that govern cell cycle entry in yeast and metazoan cells and discuss the realization that many genes involved in this decision are frequently found mutated in cancer. Next we will see that the decision to enter the cell cycle is influenced by extracellular events and learn about the signaling mechanisms that convey these environmental cues to the cell cycle machinery. Finally, we will discuss the molecular mechanisms that govern initiation of DNA replication. We will describe how degradation of an S phase CKI is essential for this process and discover how CDKs ensure that DNA replication occurs only during S phase and then only once.

### Cells Are Irreversibly Committed to Cell Division at a Cell Cycle Point Called START

In most eukaryotic cells, the key decision of whether or not a cell will divide is made at the point of whether or not to enter S phase. In most cases, once a cell has become committed to entering the cell cycle, it must complete it. The budding yeast *Saccharomyces cerevisiae* regulates its proliferation in this manner, and much of our current understanding of the molecular mechanisms controlling entry into the cell cycle originated with genetic studies of *S. cerevisiae*.

When *S. cerevisiae* cells in $G_1$ have grown sufficiently in size, they begin a program of gene expression that leads to entry into the cell cycle. If $G_1$ cells are shifted from a rich medium to a medium low in nutrients before they reach a critical size, they remain in $G_1$ and grow slowly until they are large enough to enter the cell cycle. However, once $G_1$ cells reach the critical size, they become committed to completing the cell cycle, entering S phase and proceeding through $G_2$ and mitosis, even if they are shifted to a medium low in nutrients. The point in late $G_1$ when *S. cerevisiae* cells become irrevocably committed to entering and traversing the entire cell cycle is called **START**.

CDK activity is essential for entry into S phase. This was first realized in budding yeast, where temperature-sensitive mutants in the gene encoding CDK1 arrest in $G_1$, failing to form a bud and to initiate DNA replication (CDK1 is the only CDK in the budding yeast genome and is known as *CDC28*). We now know that a CDK cascade triggers entry into the cell cycle. $G_1$ CDKs stimulate the formation of $G_1$/S phase CDKs, which then initiate bud formation, centrosome duplication, and DNA replication. In yeast, the $G_1$ cyclin gene is called *CLN3* (Figure 19-15a). Its mRNA is produced at a nearly constant level throughout the cell cycle, but its translation is regulated in response to nutrient levels and, as we will see shortly, *CLN3* is a lynchpin in coupling cell cycle entry to nutrient signals. Once sufficient Cln3 is synthesized from its mRNA, Cln3-CDK complexes phosphorylate and inactivate the transcriptional repressor Whi5. Phosphorylation of Whi5 promotes its export out of the nucleus, allowing

(a) *S. cerevisiae*

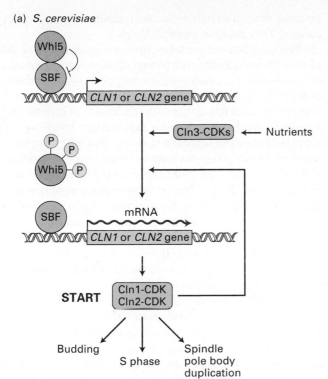

(b) Metazoans

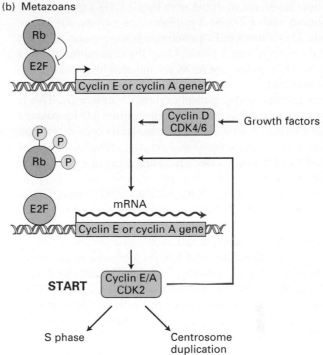

**FIGURE 19-15 Control of the G₁–S phase transition** (a) In budding yeast, Cln3-CDK activity rises during G₁ and is controlled by nutrient availability. Once sufficiently active, the kinase phosphorylates the transcriptional inhibitor Whi5, promoting its export from the nucleus. This causes the transcription factor complex SBF to induce the transcription of the G₁/S phase cyclins *CLN1* and *CLN2* and of other genes whose products are needed for DNA replication. G₁/S phase CDKs further phosphorylate Whi5, promoting further *CLN1* and *CLN2* transcription. Once sufficiently high levels of G₁/S phase CDKs have been produced, START is traversed. Cells enter the cell cycle: they initiate DNA replication,

bud formation, and spindle pole body duplication. (b) In vertebrates, G₁-CDK activity rises during G₁ and is stimulated by the presence of growth factors. When signaling from mitogens is sustained, the resulting cyclin D–CDK4/6 complexes begin phosphorylating Rb, releasing some E2F, which stimulates transcription of the genes encoding cyclin E, CDK2, and E2F itself. The cyclin E–CDK2 complexes further phosphorylate Rb, resulting in a positive feedback loop that leads to a rapid rise in the expression and activity of both E2F and cyclin E–CDK2. Once G₁/S phase CDK is sufficiently high, cells traverse START. They commence DNA replication and centrosome duplication.

the transcription factor complex SBF to induce transcription of the G₁/S phase cyclin genes *CLN1* and *CLN2* as well as other genes important for DNA replication. Once produced, Cln1/2-CDKs contribute to further Whi5 phosphorylation. This positive feedback loop ensures the rapid accumulation of G₁/S phase CDKs. Once a critical level of Cln1/2-CDKs is reached, these G₁/S phase CDKs promote bud formation, entry into S phase, and the duplication of the centrosome (also known as the spindle pole body, which later in the cell cycle will organize the mitotic spindle). This state of G₁/S phase CDK activity, sufficient to initiate S phase, bud formation, and centrosome duplication, is the molecular definition of START.

## The E2F Transcription Factor and Its Regulator Rb Control the G₁–S Phase Transition in Metazoans

The molecular events governing entry into S phase in mammalian—and in fact all metazoan—cells are remarkably similar to those of budding yeast (Figure 19-15b). G₁ cyclins are present throughout G₁ and are often found to be expressed at increased levels in response to growth factors. In turn, the G₁

CDKs activate members of a small family of related transcription factors, referred to collectively as *E2F transcription factors (E2Fs)*. During G₁, E2Fs are held inactive through their association with the retinoblastoma protein (Rb), and G₁ CDKs activate E2Fs by phosphorylating and inactivating Rb. E2Fs then activate genes encoding many of the proteins involved in DNA synthesis. They also stimulate transcription of genes encoding the G₁/S phase cyclins and the S phase cyclins. Thus the E2Fs function in late G₁ similarly to the *S. cerevisiae* transcription factor complex SBF.

Key to the regulation of E2F function is the **Rb protein**. When E2Fs are bound to Rb, they function as transcriptional repressors. This is because Rb recruits chromatin-modifying enzymes that promote deacetylation and methylation of specific histone lysines, causing chromatin to assume a condensed, transcriptionally inactive form. Rb was initially identified as the gene mutated in retinoblastoma, a childhood cancer of the retina. Subsequent studies found *RB* to be inactivated in almost all cancer cells either by mutations in both alleles of *RB* or by abnormal regulation of Rb phosphorylation.

Rb protein regulation by G₁ CDKs in mammalian cells is analogous to that of Cln3-CDK regulation of Whi5 in yeast.

Phosphorylation on multiple sites by $G_1$ CDKs prevents Rb association with E2Fs and promotes its export out of the nucleus. This allows E2Fs to activate transcription of genes required for entry into S phase. Once the expression of genes coding for $G_1$/S cyclins and CDK are induced by phosphorylation of some of the Rb molecules, the resulting $G_1$/S phase CDK complexes further phosphorylate Rb in late $G_1$. This is one of the principal biochemical events responsible for passage through START. Since E2F also stimulates its own expression and that of the $G_1$/S cyclin-CDKs, positive cross-regulation of E2F and $G_1$/S cyclin-CDKs produces a rapid rise of both activities in late $G_1$.

As they accumulate, S phase CDKs and mitotic CDKs maintain Rb protein in the phosphorylated state throughout the S, $G_2$, and early M phases. After cells complete anaphase and enter early $G_1$ or $G_0$, a fall in all cyclin-CDK activities leads to dephosphorylation of Rb. As a consequence, hypophosphorylated Rb is available to inhibit E2F activity during early $G_1$ of the next cycle and in $G_0$-arrested cells. Thus $G_1$/S phase CDK activity remains low until cells decide to enter a new cell cycle and $G_1$ CDKs break the inhibitory grasp of Rb over E2F.

## Extracellular Signals Govern Cell Cycle Entry

Whether or not cells enter the cell cycle is influenced by extracellular as well as intracellular signals. Unicellular organisms such as yeast, for example, enter the cell cycle only when they have reached an appropriate size, known as the **critical cell size**. This critical size, in turn, is controlled by nutrients available in the environment. This coordination between cell size and cell cycle entry will be discussed in Section 19.7. Here we restrict our discussion to the fact that $G_1$ cyclin synthesis is responsive to protein synthesis rate, which is in turn controlled by pathways that are regulated by nutrients in the environment. This link between the macromolecule biosynthesis and cell cycle machinery is best understood in budding yeast. In this organism, the $G_1$ cyclin transcript *CLN3* contains a short upstream open-reading frame that inhibits translation initiation at the Cln3 open reading frame when nutrients are limited. This inhibition is diminished when nutrients are in abundance. In the presence of sufficient nutrients, the TOR pathway, which senses nutrients and growth factor signals, is active and the pathway stimulates translational activity (see Figure 8-30). Since Cln3 is a highly unstable protein, its concentration fluctuates with the translation rate of its mRNA. Consequently, the amount and activity of Cln3-CDK complexes, which depend on the concentration of Cln3 protein, is largely regulated by nutrient levels.

In multicellular organisms, cells are surrounded by nutrients, and as such, usually nutrients are not rate limiting for proliferation. Rather, cell proliferation is controlled by the presence of growth-promoting factors (**mitogens**) and growth inhibitory factors (anti-mitogens) in the cell surroundings. Addition of mitogens to $G_0$-arrested mammalian cells induces— as we discussed in Chapter 16—receptor tyrosine-kinase-linked signal transduction pathways that initiate signal transduction cascades that ultimately influence transcription and cell cycle control. They do so in multiple ways.

Mitogens activate the transcription of multiple genes. Most of these genes fall into one of two classes—*early-response* or *delayed-response* genes—depending on how soon their encoded mRNAs appear. Transcription of early response genes is induced within a few minutes after addition of growth factors by signal transduction cascades that activate preexisting transcription factors in the cytosol or nucleus (see Chapter 16). Many of the early-response genes encode transcription factors, such as c-Fos and c-Jun, that stimulate transcription of the delayed response genes. The early response transcription factors AP-1 and Myc induce the transcription of genes encoding $G_1$ cyclins and CDKs. In addition to being controlled by transcription of the genes encoding the $G_1$ cyclins, $G_1$ CDKs are regulated by CKIs. The CKI p15INK4b is a potent CDK inhibitor. In some tissues mitogens inhibit the production of this CKI by inhibiting its transcription.

Cell proliferation in many tissues is not only regulated by proliferation promoting mitogens but also by anti-mitogens, which prevent entry into the cell cycle. Similarly, during differentiation, cells cease to divide and enter $G_0$. Some differentiated cells (e.g., fibroblasts and lymphocytes) can be stimulated to re-enter the cycle and replicate. Many postmitotic differentiated cells, however, never re-enter the cell cycle to replicate again. Anti-mitogens and differentiation pathways prevent the accumulation of $G_1$ CDKs. They antagonize the production of $G_1$ cyclins and induce the production of CKIs. Transforming growth factor β (TGF-β) is an important anti-mitogen. This hormone induces a signaling cascade that brings about $G_1$ arrest by inducing the expression of p15INK4b. As we will see in Chapter 24, the signaling pathways that regulate $G_1$ CDKs are found mutated in most human cancers.

## Degradation of an S Phase CDK Inhibitor Triggers DNA Replication

Entry into S phase is defined by the unwinding of origins of DNA replication. The molecular events leading to this event are best understood in *S. cerevisiae*. $G_1$/S phase CDKs play an essential role in this process. They turn off the degradation machinery that degrades S phase cyclins during exit from mitosis and $G_1$ and induce the degradation of a CKI that inhibits S phase CDKs.

One of the important substrates of the $G_1$/S phase cyclin-CDK complexes is Cdh1. During late anaphase, this substrate targeting factor directs the APC/C to ubiquitinylate substrate proteins including S phase and mitotic cyclins, marking them for proteolysis by proteasomes. The APC/C-Cdh1 remains active throughout $G_1$, preventing the premature accumulation of S phase and mitotic cyclins. Phosphorylation of Cdh1 by $G_1$/S cyclin-CDKs causes it to dissociate from the APC/C complex, inhibiting further ubiquitinylation of S phase and mitotic cyclins during late $G_1$ (Figure 19-16). This, combined with the induced transcription of S phase cyclins during late $G_1$, allows S phase cyclin proteins to accumulate as $G_1$/S cyclin-CDKs levels rise. Later in the cell cycle, S phase and mitotic CDKs

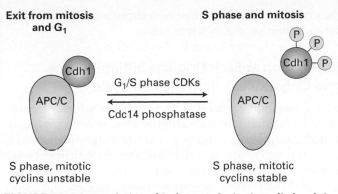

**Exit from mitosis and G₁**

**S phase and mitosis**

G₁/S phase CDKs →
← Cdc14 phosphatase

S phase, mitotic cyclins unstable

S phase, mitotic cyclins stable

**FIGURE 19-16 Regulation of S phase and mitotic cyclin levels in budding yeast.** In late anaphase, the anaphase-promoting complex (APC/C) ubiquitinylates S phase and mitotic cyclins. APC/C activity is directed toward mitotic cyclins by a specificity factor, called *Cdh1*. Cdh1 activity is regulated by phosphorylation. During exit from mitosis and G₁, the specificity factor is dephosphorylated and active; during S phase and mitosis, Cdh1 is phosphorylated and dissociates from the APC/C, and the ubiquitin ligase is inactive. The G₁/S-phase CDKs, which themselves are not APC/C-Cdh1 substrates, phosphorylate Cdh1 at the G₁–S phase transition. A specific phosphatase called *Cdc14* removes the regulatory phosphate from the specificity factor late in anaphase.

take over to maintain Cdh1 in the phosphorylated and hence inactive state. Only as mitotic CDKs decline and a protein phosphatase known as Cdc14 is activated are these inhibitory phosphates removed from Cdh1, leading to its reactivation. In mammalian cells similar mechanisms are responsible for stabilizing S phase and mitotic cyclins, but the phosphatase involved in dephosphorylation of Cdh1 has not been identified.

In *S. cerevisiae*, as the S phase cyclin-CDK heterodimers accumulate in late G₁ following the inactivation of APC/C-Cdh1, they are immediately inactivated by binding of a CKI called *Sic1* that is expressed late in mitosis and in early G₁ (Figure 19-17). Because Sic1 specifically inhibits S phase and M phase CDK complexes but has no effect on the G₁ CDK

and G₁/S phase CDK complexes, it functions as an S phase inhibitor. Initiation of DNA replication occurs when the Sic1 inhibitor is precipitously degraded following its ubiquitinylation by the SCF ubiquitin-protein ligase.

Degradation of Sic1 is induced by its phosphorylation by G₁/S phase CDKs (see Figure 19-17). It must be phosphorylated at at least six sites, which are relatively poor substrates for the G₁/S phase CDKs, before it is bound sufficiently well by SCF to be ubiquitinylated. Multiple, poor G₁/S phase CDK phosphorylation sites lead to an ultrasensitive, switch-like response in Sic1 degradation and hence precipitous activation of S phase CDKs (Figure 19-18). If Sic1 were inactivated following the phosphorylation of a single site, Sic1 molecules would start to get phosphorylated as soon as the levels of G₁/S phase CDK activity begin to rise, leading to a gradual decrease in Sic1 levels. In contrast, when several sites need to be phosphorylated, at low levels of G₁/S phase CDK activity only a few sites are phosphorylated and Sic1 is not destroyed. Only when G₁/S phase CDK levels are high is Sic1 sufficiently phosphorylated at multiple sites to target it for degradation. Thus Sic1 degradation occurs only when G₁/S phase CDK activity has reached its peak and virtually all other G₁/S phase CDK substrates have been phosphorylated.

Once Sic1 is degraded, the S phase cyclin-CDK complexes induce DNA replication by, as we will see shortly, phosphorylating several proteins involved in activating the replicative helicases. This mechanism for activating the S phase cyclin-CDK complexes—that is, inhibiting them as the cyclins are synthesized and then precipitously degrading the inhibitor—permits the sudden initiation of replication at large numbers of replication origins. An obvious advantage of proteolysis for controlling passage through this critical point in the cell cycle is that protein degradation is an *irreversible process*, ensuring that cells proceed in one direction through the cycle. The dependence of this event on phosphorylation of multiple poor phosphorylation sites makes the degradation of Sic1, and hence the activation of DNA replication, abrupt.

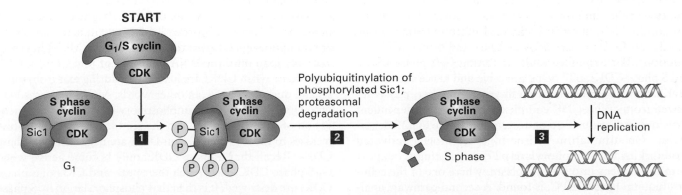

**START**

G₁/S cyclin — CDK

S phase cyclin — Sic1 — CDK

G₁

**1** → P P Sic1 CDK S phase cyclin (P P P)

Polyubiquitinylation of phosphorylated Sic1; proteasomal degradation **2** →

S phase cyclin CDK — S phase

**3** → DNA replication

**FIGURE 19-17 Control of S phase onset in *S. cerevisiae* by regulated proteolysis of the S phase inhibitor, Sic1.** The S phase cyclin-CDK complexes begin to accumulate in G₁ but are inhibited by Sic1. This inhibition prevents initiation of DNA replication until the cells have completed all G₁ events. G₁/S phase CDKs assembled in late G₁ phosphorylate Sic1 at multiple sites (step **1**), marking it for

ubiquitinylation by the SCF ubiquitin ligase and subsequent proteasomal degradation (step **2**). The active S phase CDKs then trigger initiation of DNA synthesis (step **3**) by phosphorylating and recruiting MCM helicase activators to DNA replication origins. [Adapted from R. W. King et al., 1996, *Science* **274**:1652.]

(a) One optimal G$_1$/S phase CDK site in Sic1

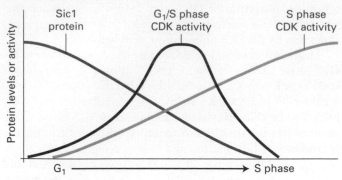

(b) Six sub-optimal G$_1$/S phase CDK site in Sic1

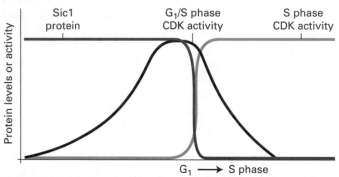

**FIGURE 19-18 Six suboptimal G$_1$/S phase CDK phosphorylation sites in Sic1 create a switch-like cell cycle entry.** (a) A single optimal G$_1$/S phase CDK site in Sic1 would result in a sluggish G$_1$–S phase transition. As G$_1$/S phase CDKs accumulate during G$_1$, Sic1 would get progressively degraded. As a result, S phase CDKs would slowly rise. Instead of an abrupt rise in S phase CDK activity, initiation of S phase would be a drawn-out event. (b) Six suboptimal phosphorylation sites ensure that Sic1 is only fully phosphorylated and hence recognized by the SCF when G$_1$/S phase CDKs have reached high levels. This also ensures that Sic1 degradation occurs rapidly and when G$_1$/S phase CDKs have accomplished all their other G$_1$ tasks. [Adapted from Nash et al., 2001, *Nature* **414:**514–521, and D. O. Morgan, 2006.]

Entry into S phase in metazoan cells is regulated in a similar manner as in budding yeast. Like Sic1, the CKI p27 prevents the premature activation of S phase CDKs during G$_1$. Unlike Sic1, however, this CKI inhibits both S phase CDKs and G$_1$/S phase CDKs and has additional cell cycle functions. For example, while inhibiting G$_1$/S phase CDKs and S phase CDKs, p27 helps assemble and hence activate G$_1$ CDKs. Similar to the situation in yeast, however, p27 is removed from cyclin-CDK complexes by ubiquitin-dependent protein degradation. Two pathways contribute to p27 degradation. On stimulation with mitogens, mitogen-activated protein kinases phosphorylate p27, promoting its export from the nucleus into the cytoplasm, where one of the cellular ubiquitin ligases, KPC, is found. A second pathway, analogous to the one operating on Sic1, targets p27 for degradation at the G$_1$-S phase transition. As G$_1$/S phase CDKs and S phase CDKs reach high levels during late G$_1$ and early S phase, they begin to phosphorylate p27, targeting the CKI for ubiquitinylation by SCF. Degradation of p27 causes activation of G$_1$/S phase and S phase CDKs. These kinases

then initiate S phase by phosphorylating proteins important for the initiation of DNA replication.

## Replication at Each Origin Is Initiated Once and Only Once During the Cell Cycle

As discussed in Chapter 4, eukaryotic chromosomes are replicated from multiple replication origins. Initiation of replication from these origins occurs throughout S phase. However, no eukaryotic origin initiates more than once per S phase. Moreover, S phase continues until replication from multiple origins along the length of each chromosome results in complete replication of the entire chromosome. These two factors ensure that the correct gene copy number is maintained as cells proliferate.

S phase CDKs play an essential role in the regulation of DNA replication. The kinases initiate DNA replication only at the G$_1$–S phase transition and prevent re-initiation from origins that have already fired. We will first discuss how initiation of DNA replication is controlled and the role of S phase CDKs in the process before turning to the mechanisms whereby these kinases prevent re-initiation.

The mechanisms underlying the initiation of DNA replication are best understood in budding yeast, so we will focus our discussion on this organism. However, it is important to note that the proteins and mechanisms controlling the initiation of DNA synthesis are essentially the same in all eukaryotic species. A protein complex known as the *origin-recognition complex (ORC)* is associated with all DNA replication origins. In budding yeast, replication origins contain an 11-base-pair conserved core sequence to which ORC binds. In multicellular organisms, origins of DNA replication lack a recognizable consensus sequence. Instead, chromatin-associated factors target ORCs to the DNA. ORC and two additional replication initiation factors, Cdc6 and Cdt1, associate with the ORC at origins during G$_1$ to load the replicative helicases known as the MCM helicase complex onto DNA (Figure 19-19, step **1**). The MCM helicases function to unwind the DNA during the initiation of DNA replication.

To ensure that origins fire only once at the beginning of S phase, MCM helicase complex loading and its activation occur at two opposing phosphorylation states. The MCM helicases can only load onto the DNA in a state of low CDK activity that occurs when CDKs are inactivated during exit from mitosis and during early G$_1$. In other words, MCM helicases load onto DNA when they are unphosphorylated. In contrast, activation of MCM helicases and the recruitment of DNA polymerases to the unwound origin DNA are triggered by S phase CDKs. Recall that S phase CDKs only become active when G$_1$/S phase CDK levels reach their peak and CKIs of S phase CDKs are destroyed. It is then that phosphorylation by S phase CDKs and a second heterodimeric protein kinase, DDK, activates the MCM helicase and recruits the polymerases to the sites of replication initiation (Figure 19-19, steps **2** and **3**).

So how do S phase CDKs and DDK collaborate to initiate DNA replication? ORC and the two other initiation factors, Cdc6 and Cdt1, recruit the MCM helicases to sites of

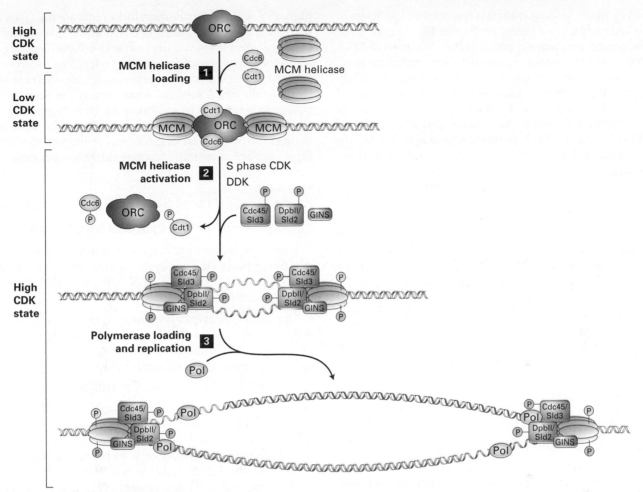

**FIGURE 19-19 The molecular mechanisms governing the initiation of DNA replication.** Step **1**: During exit from mitosis and early G₁, when CDK activity is low, the MCM loading factors ORC, Cdc6, and Cdt1 load the replicative helicase, the MCM complex, onto origin DNA. Step **2**: Activation of S phase CDKs and DDK mark the onset of S phase. They phosphorylate the MCM helicase, Sld2 and Sld3 (depicted as green phosphorylation events), to facilitate the loading of MCM helicase activators—the Cdc45-Sld3 and the GINS complexes— onto sites of replication initiation. This leads MCM helicases to unwind

DNA. S phase CDKs also prevent reloading of MCM helicases by phosphorylating the pre-RC components Cdc6 and Cdt1 (shown as yellow phosphorylation events), promoting their release from origins and degradation by the SCF. S phase CDKs also phosphorylate MCM helicases, which leads to their export from the nucleus when the helicases disengage from the DNA when replication is complete. Step **3**: DNA polymerases are recruited to origins, which leads to the initiation of DNA synthesis (see Figure 4-31).

replication initiation during G₁, when CDK activity is low (see Figure 19-19, step **1**). When DDK and S phase CDKs are activated in late G₁, DDK phosphorylates two subunits of the MCM helicase. The S phase CDKs phosphorylate two proteins called Sld2 and Sld3. These phosphorylation events have an activating effect, promoting the recruitment of MCM helicase activators to sites of replication initiation (the phosphorylation events shown in green in Figure 19-19, steps **2** and **3**). The helicase activators are called the Cdc45-Sld3 complex and the GINS complex. Exactly how they promote activation of the MCM helicases is not yet clear. In addition to activating the MCM helicase to unwind DNA, the Cdc45-Sld3 complex and the GINS complex recruit polymerases to the DNA, polymerase ε to synthesize the leading strand and polymerase δ to synthesize the lagging strand (see Figure 19-19, step **3**). The replication machinery then initiates DNA synthesis.

S phase CDKs are not only essential to initiate DNA replication, they are also responsible for ensuring that each origin fires only once during S phase. Preventing re-firing of origins during S phase is brought about by phosphorylation of several components of the MCM helicase loading machinery and the MCM helicase itself. To distinguish these phosphorylation events from the ones required for the initiation of DNA replication, they are depicted in yellow in Figure 19-19. Concomitant with activation of the MCM helicase, Cdc6 and Cdt1 dissociate from the sites of DNA replication initiation. Once they dissociate, their phosphorylation leads to their degradation by the SCF ubiquitin ligase. Phosphorylation of the MCM helicase leads to the export of these proteins from the nucleus after they dissociate from the DNA on completion of DNA replication. Thus only after CDK activity is lowered by APC/C-Cdh1 during exit from mitosis can the MCM helicases be reloaded onto

DNA. As a result, helicase loading is restricted to late stages of mitosis and early $G_1$ (see Figure 19-19, step **1**).

The general mechanisms governing the initiation of DNA replication in metazoan cells parallel those in *S. cerevisiae*, although small differences are found in vertebrates. Phosphorylation of MCM helicase activators by $G_1$/S phase CDKs and S phase CDKs likely promotes their recruitment to sites of DNA replication initiation. As in yeast, phosphorylation of MCM loading factors likely prevents reloading of MCM helicases until the cell passes through mitosis, thereby ensuring that replication from each origin occurs only once during each cell cycle. In metazoans, a second small protein, geminin, contributes to the inhibition of re-initiation at origins until cells complete a full cell cycle. Geminin is expressed in late $G_1$; it binds to and inhibits MCM helicase loading factors as they are released from origins once DNA replication is initiated during S phase (see Figure 19-19, step **2**), contributing to the inhibition of re-initiation at an origin. Geminin contains a destruction box at its N-terminus that is recognized by the APC/C-Cdh1, causing it to be ubiquitinylated in late anaphase and degraded by proteasomes. This frees the MCM helicase loading factors, which are also dephosphorylated as CDK activity declines, to bind to ORC on replication origins and load MCM helicases during the following $G_1$ phase.

## Duplicated DNA Strands Become Linked During Replication

During S phase, as chromosomes are duplicated to form sister chromatids, they become tethered to each other by protein links. The linkages between sister chromatids established during S phase will be essential for their accurate segregation during mitosis.

The protein complexes that establish cohesion between sister chromatids are called **cohesins**. The cohesin complex is composed of four subunits: Smc1 (sometimes also called *Rad21*), Smc3, Scc1, and Scc3. Smc1 and Smc3 are members of the SMC protein family, which is characterized by long coiled-coil domains that are flanked by a globular domain containing ATPase activity. The ATPase domains interact with Scc1 and Scc3 and, together, form a ring structure. The structural mechanism by which cohesins link sister chromatids together is not understood, but it is likely that rings of cohesin embrace one or both copies of the replicated DNA. However, it is clear that cohesins are essential to hold the replicated DNA molecules together. When *Xenopus* egg extracts were depleted of cohesin by treatment with antibodies specific for the cohesin SMC proteins, the depleted extracts were able to replicate the DNA in added sperm nuclei, but the resulting sister chromatids did not associate properly with each other.

Some details of how cohesins are loaded onto DNA to mediate sister chromatid cohesion are being unraveled. We know that establishment of cohesion between sister chromatids is tightly tied to DNA replication. Cohesins associate with chromosomes during $G_1$ (Figure 19-20, step **1**). During DNA replication, they are loaded onto chromosomes in such a way that they can hold sister chromatids together. Most

likely this occurs as replication forks replicate the DNA (Figure 19-20, step **2**). Converting DNA-bound $G_1$ cohesins into cohesive complexes requires several cohesin loading factors—including a protein complex related to proteins that load the sliding clamp at the replication fork—and the acetylation of the Smc3 subunit. As we will see in Section 19.6, cohesins are essential to accurately attach the replicated sister chromatids onto the mitotic spindle and for their segregation during mitosis. Cells lacking cohesins or the factors that load them onto chromosomes segregate chromosomes randomly.

## KEY CONCEPTS of Section 19.4

### Commitment to the Cell Cycle and DNA Replication

• START defines a stage in $G_1$ after which cells are irreversibly committed to the cell cycle. Molecularly, it is defined as the point when $G_1$/S phase CDK activity reaches levels sufficient to initiate S phase.

• The molecular events promoting entry into the cell cycle are conserved across species. $G_1$ CDKs inhibit a transcriptional inhibitor. This allows the transcription of $G_1$/S phase cyclins and other genes important for S phase.

• Extracellular signals such as nutritional state (in yeast) and the presence of mitogens and anti-mitogens (in vertebrates) regulate entry into the cell cycle.

• Various polypeptide growth factors called mitogens stimulate cultured mammalian cells to proliferate by inducing expression of early-response genes. Many of these encode transcription factors that stimulate expression of delayed response genes encoding the $G_1$/S phase cyclins and E2F transcription factors.

• The $G_1$/S phase CDKs phosphorylate and inhibit Cdh1, the specificity factor that directs the anaphase-promoting complex (APC/C) to ubiquitinylate S phase and M phase cyclins. This allows S phase cyclins to accumulate in late $G_1$.

• In yeast, S phase CDKs initially are inhibited by Sic1. Phosphorylation marks Sic1 for ubiquitinylation by the SCF ubiquitin-protein ligase and proteosomal degradation, releasing activated S phase CDKs that trigger onset of the S phase (see Figure 19-17).

• DNA replication is initiated from helicase loading sites known as origins of replication.

• Loading and activation of the MCM helicase occur in mutually exclusive cell cycle states: MCM helicase loading can only occur when CDK activity is low (during early $G_1$); MCM helicases are activated when CDK activity is high.

• S phase CDKs and DDK trigger the initiation of DNA replication by the recruitment of MCM helicase activators to origins (see Figure 19-19).

• Initiation of DNA replication occurs at each origin only once during the cell cycle. This is achieved because S phase CDKs activate the helicases and at the same time prevent additional helicases from loading onto DNA.

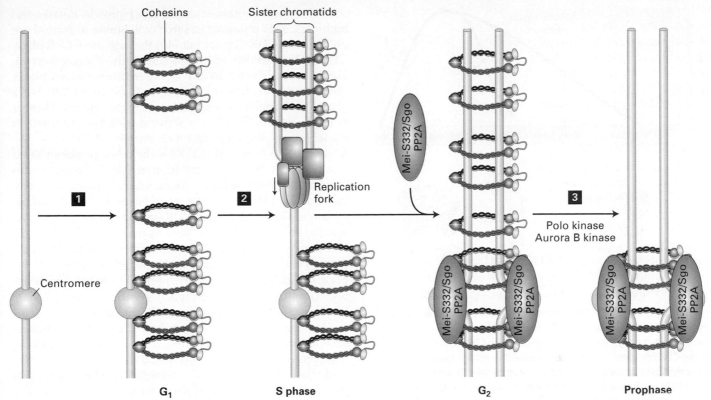

**Cohesins**

**Sister chromatids**

**1**

**Centromere**

**2**

Replication fork

Mei-S332/Sgo PP2A

**3**

Polo kinase
Aurora B kinase

Mei-S332/Sgo PP2A

Mei-S332/Sgo PP2A

Mei-S332/Sgo PP2A

Mei-S332/Sgo PP2A

**G₁**      **S phase**      **G₂**      **Prophase**

**FIGURE 19-20 Model for establishing cohesin linkage of sister chromatids.** There is strong evidence that the cohesin complex is circular, but it is not known whether a single cohesin ring links sister chromatids or whether two rings, one each around the separate sister chromatids, are linked to each other like links in a chain. For simplicity, only one ring is shown here. Step **1**: Cohesins are loaded onto chromosomes during G₁, but they do not possess cohesive properties (indicated as cohesins laterally associated with chromosomes). Step **2**: Concomitant with DNA replication, and most likely closely behind the replication fork, cohesins are converted into cohesive molecules, able to hold sister chromatids together (indicated as cohesin rings encircling the replicated sister chromatids). This conversion into cohesive cohesins requires cohesin loading factors. During G₂, sister chromatids are replicated and linked along their entire length by cohesins. During this time, the Mei-S332/Sgo proteins recruit the protein phosphatase 2A (PP2A) to centromeric regions. Step **3**: In vertebrate cells, cohesins are released from chromosome arms during prophase and early metaphase by the action of Polo kinase and Aurora B kinase. By the end of metaphase, cohesins are retained only in the region of the centromere, where PP2A prevents cohesin phosphorylation and hence dissociation through recruiting PP2A.

- Cohesins establish linkages between the replicated DNA molecules. This linking mechanism is coupled to DNA replication.

## 19.5 Entry into Mitosis

Once S phase has been completed and the entire genome has been duplicated, the pairs of duplicated DNA chromosomes—the sister chromatids—are segregated to the future daughter cells. This process requires not only the formation of the apparatus that facilitates this segregation—the mitotic spindle—but essentially a complete remodeling of the cell. Chromosomes condense and attach to the mitotic spindle, the nuclear envelope is disassembled, and almost all organelles are rebuilt or modified. All these events are triggered by mitotic CDKs. This section will first discuss how the mitotic CDKs are precipitously activated after the completion of DNA replication, during G₂. We will then see how these pro-

tein kinases bring about the dramatic changes in the cell necessary to facilitate sister chromatid segregation during anaphase, focusing on the events as they occur in metazoans.

### Precipitous Activation of Mitotic CDKs Initiates Mitosis

Mitotic CDKs initiate mitosis. Whereas levels of the catalytic CDK subunit are constant throughout the cell cycle, mitotic cyclins gradually accumulate during S phase. Most eukaryotes contain multiple mitotic cyclins, which for historical reasons are called cyclin A and cyclin B. As they assemble, mitotic CDK complexes are maintained in an inactive state through inhibitory phosphorylation on the CDK subunit. Recall from Section 19-3 that two highly conserved tyrosine and threonine residues in mammalian CDKs are subject to regulated phosphorylation. In CDK1, the mitotic CDK, phosphorylation of tyrosine 15 and threonine 14 maintains mitotic cyclin-CDK complexes in an inactivate state. The phosphorylation state of T14 and Y15 is controlled by a dual-specificity protein kinase

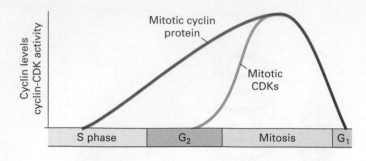

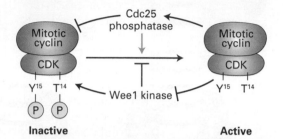

**FIGURE 19-21 Phosphorylation on the CDK subunit restrains mitotic CDK activity during S phase and G₂.** Mitotic cyclins are synthesized during S phase and G₂ and bind to CDK1. However, the cyclin-CDK complex is not active because threonine 14 and tyrosine 15 of the CDK1 subunit are phosphorylated by the protein kinase Wee1. Once DNA replication has been completed, the protein phosphatase Cdc25 is activated, and the phosphatase dephosphorylates CDK1. Active mitotic CDKs further stimulate Cdc25. At the same time, mitotic CDKs inhibit Wee1, the protein kinase that places the inhibitory phosphorylation on the CDK subunit. Ongoing DNA replication inhibits Cdc25 activity. How Cdc25 is initially activated upon completion of DNA replication to put in motion these feedback loops is not yet known.

known as **Wee1** and a dual-specificity phosphatase Cdc25 (Figure 19-21). The regulation of mitotic CDKs by these activities underlies the abrupt activation of their kinase activity at the G₂-M phase transition and explains the observation that although mitotic cyclins gradually accumulate during S phase and G₂, mitotic CDKs are not active until cells enter mitosis.

Studies in the fission yeast *Schizosaccharomyces pombe* unraveled the mechanisms that lead to the precipitous activation of mitotic CDKs during G₂. The dual-specificity protein kinase Wee1 phosphorylates CDKs on the inhibitory tyrosine 15. (Threonine 14 is not phosphorylated in *S. pombe* CDK1.) Yeast cells with a defective *wee1⁺* gene activate mitotic CDKs prematurely and hence experience premature entry into mitosis. Wee1 mutants not only enter mitosis prematurely but are also smaller. This is because unlike most other eukaryotes, which coordinate cell size and cell division during G₁, this coordination occurs during G₂ in fission yeast. Fission yeast cells carrying a mutant CDK1 gene in which the tyrosine 15 residue is replaced by phenylalanine (phenylalanine is structurally similar to tyrosine but cannot be phosphorylated) show the same premature mitotic CDK activation and entry into mitosis. The phosphatase that opposes Wee1 is Cdc25. Fission yeast cells carrying mutations in the *cdc25⁺* gene arrest in G₂, indicating that the phosphatase is essential for entry into mitosis.

Vertebrates contain multiple Wee1 protein kinases and multiple **Cdc25 phosphatases** that collaborate to control not only mitotic CDK activity but also the activity of G₁/S phase CDKs. One member of the Cdc25 family of phosphatases, Cdc25A, is activated in late G₁. It removes the inhibitory phosphorylation on tyrosine 15 of the G₁/S phase CDKs and S phase CDK catalytic subunit to activate the kinases. Another family member, Cdc25C, is active during G₂ and removes the inhibitory phosphorylation on mitotic CDKs.

Activation of mitotic CDKs is the consequence of rapid inactivation of Wee1 and activation of Cdc25. Central to this rapid transition are feedback loops, where mitotic CDKs activate Cdc25 and inactivate Wee1 (see Figure 19-21). Phosphorylation of Cdc25 by mitotic CDKs stimulates its phosphatase activity; phosphorylation of Wee1 by mitotic CDKs inhibits its kinase activity. Ongoing DNA replication inhibits Cdc25 activity. A critical question that we know little about is how this positive feedback loop is started once DNA replication has been completed. CDKs that function earlier in the cell cycle have been suggested to start the positive feedback loop.

Although it is not yet known how the precipitous activation of mitotic CDKs is initiated, it is clear that once active, these protein kinases set in motion all the events necessary to ready the cell for chromosome segregation. The activation of mitotic CDKs is associated with changes in the subcellular localization of these kinases. Mitotic CDKs initially associate with centrosomes, where they are thought to facilitate centrosome maturation. They then enter the nucleus, where they bring about chromosome condensation and nuclear envelope breakdown. In what follows we will discuss how the mitotic CDKs accomplish the coordinated execution of mitosis.

During DNA replication initiation, S phase CDKs work together with DDK to promote MCM helicase activation. As in the initiation of DNA replication, CDKs collaborate with other protein kinases to bring about the mitotic events. The **Polo kinase** family is critical for formation of the mitotic spindle as well as chromosome segregation. The family of **Aurora kinases** plays key roles in mitotic spindle formation and ensuring that chromosomes attach to the mitotic spindle in the correct way so that they are segregated accurately during mitosis. Their contribution to the various mitotic events will also be discussed.

## Mitotic CDKs Promote Nuclear Envelope Breakdown

During interphase, chromosomes are surrounded by the nuclear envelope. The centrosomes that form the mitotic spindle are located in the cytoplasm. For chromosomes to interact with the microtubules nucleated by the centrosomes, the nuclear envelope needs to be dismantled.

The nuclear envelope is a double-membrane extension of the endoplasmic reticulum containing many nuclear pore complexes (see Figures 9-32 and 13-33). The lipid bilayer of the inner nuclear membrane is associated with the **nuclear lamina**, a meshwork of lamin filaments adjacent to the inside face of the nuclear envelope (Figure 19-22a). The three nuclear **lamins** (A, B, and C) present in vertebrate cells be-

(a)

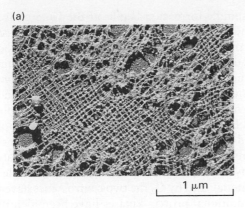

1 μm

(b)

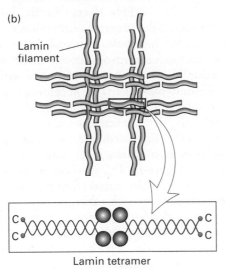

Lamin filament

Lamin tetramer

MPF

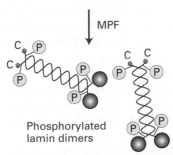

Phosphorylated lamin dimers

**FIGURE 19-22 The nuclear lamina and its regulation by phosphorylation.** (a) Electron micrograph of the nuclear lamina from a *Xenopus* oocyte. Note the regular mesh-like network of lamin intermediate filaments. This structure lies adjacent to the inner nuclear membrane (see Figure 18-46). (b) Schematic diagrams of the structure of the nuclear lamina. Two perpendicular sets of 10-nm-diameter filaments built of lamins A, B, and C form the nuclear lamina (*top*). Individual lamin filaments are formed by end-to-end polymerization of lamin tetramers, which consist of two lamin coiled-coil dimers (*middle*). The red and blue circles represent the globular N-terminal and C-terminal domains, respectively. Phosphorylation of specific serine residues near the ends of the rod-like central section of lamin dimers causes the tetramers to depolymerize (*bottom*). As a result, the nuclear lamina disintegrates. [Part (a) from U. Aebi et al., 1986, *Nature* **323**:560; courtesy of U. Aebi; part (b) adapted from A. Murray and T. Hunt, 1993, *The Cell Cycle: An Introduction*, W. H. Freeman and Company.]

long to a class of cytoskeletal proteins, the intermediate filaments, that are critical in supporting cellular membranes. Lamins A and C, which are encoded by the same transcription unit and produced by alternative splicing of a single pre-mRNA, are identical except for a 133-residue region at the C-terminus of lamin A, which is absent in lamin C. Lamin B, encoded by a different transcription unit, is modified post-transcriptionally by the addition of a hydrophobic isoprenyl group near its carboxyl terminus. This fatty acid becomes embedded in the inner nuclear membrane, thereby anchoring the nuclear lamina to the membrane (see Figure 10-19). All three nuclear lamins form dimers containing a rod-like α-helical coiled-coil central section and globular head and tail domains; polymerization of these dimers through head-to-head and tail-to-tail associations generates the intermediate filaments that compose the nuclear lamina (Figure 19-22b).

Once mitotic CDKs are activated at the end of $G_2$, they phosphorylate specific serine residues in all three nuclear lamins. This causes depolymerization of the lamin intermediate filaments (see Figure 19-22b). The phosphorylated lamin A and C dimers are released into solution, whereas the phosphorylated lamin B dimers remain associated with the nuclear membrane via their isoprenyl anchor. Depolymerization of the nuclear lamins leads to disintegration of the nuclear lamina meshwork and contributes to disassembly of the nuclear envelope.

Mitotic CDKs also affect other nuclear envelope components (Figure 19-23). The CDKs phosphorylate specific **nucleoporins**, which causes nuclear pore complexes to dissociate into subcomplexes during prophase. Phosphorylation of integral membrane proteins of the inner nuclear membrane is thought to decrease their affinity for chromatin and further contributes to the disassembly of the nuclear envelope. The weakening of the associations between the inner nuclear membrane proteins and the nuclear lamina and chromatin allows sheets of inner nuclear membrane to retract into the endoplasmic reticulum, which is continuous with the outer nuclear membrane.

## Mitotic CDKs Promote Mitotic Spindle Formation

A key function of mitotic CDKs is to induce the formation of the mitotic spindle, also known as the mitotic apparatus. As we saw in Chapter 18, the mitotic spindle is made of microtubules that attach to chromosomes via specialized protein structures associated with chromosomes known as **kinetochores**. In most organisms, the mitotic spindle is organized by **centrosomes**, sometimes called **spindle pole bodies**. They contain a specialized tubulin, γ tubulin, which together with associated proteins nucleates microtubules. Notable exceptions to centrosome-based spindle assembly mechanisms are higher plants and metazoan oocytes. In these cells, (−) ends of microtubules are cross-linked and the microtubules self-assemble into a spindle.

The function of the mitotic spindle is to segregate chromosomes so that the sister chromatids separate from each other and are moved to opposite poles of the mitotic spindle (see Figure 18-37). To achieve this, chromosomes have to attach to the

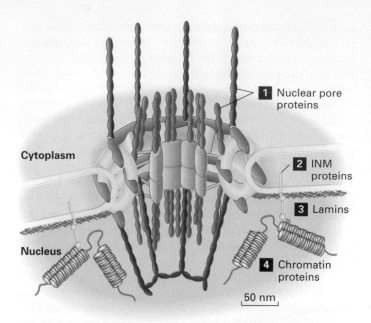

1 Nuclear pore proteins

Cytoplasm

2 INM proteins

3 Lamins

Nucleus

4 Chromatin proteins

50 nm

**FIGURE 19-23 Nuclear envelope proteins phosphorylated by mitotic CDKs.** Step 1: Components of the nuclear pore complex (NPC) are phosphorylated by mitotic CDKs in prophase, causing NPCs to dissociate into soluble and membrane-associated NPC subcomplexes. Step 2: Mitotic CDK phosphorylation of inner nuclear membrane (INM) proteins inhibits their interactions with the nuclear lamina and chromatin. Step 3: Mitotic CDK phosphorylation of nuclear lamins causes their depolymerization and dissolution of the nuclear lamina. Step 4: Mitotic CDK phosphorylation of chromatin proteins induces chromatin condensation and inhibits interactions between chromatin and the nuclear envelope. [Adapted from B. Burke and J. Ellenberg, 2002, *Nat. Rev. Mol. Cell Biol.* **3**:487.]

Mitotic spindle formation begins at the $G_1$–S phase transition with the duplication of centrosomes. The mechanism whereby this duplication occurs is poorly understood, but at the heart of this process is the duplication of the pair of **centrioles**, short microtubules arranged orthogonally to each other. As discussed in Chapter 18, $G_1$ cells contain a single pair of centrioles. Concomitant with entry into S phase and triggered by the $G_1$/S phase CDKs, the two centrioles split apart, and each centriole begins to grow a daughter centriole (see Figure 18-35). The new centrioles grow and mature during S phase, each centriole pair begins to assemble centrosomal material, and by $G_2$ the two centrosomes have formed. Several additional protein kinases have been identified that control centrosome duplication. Chief among them is a member of the conserved Polo kinase family, Plk4. How $G_1$/S phase CDKs and Plk4 promote centrosome duplication is not yet understood but is thought to involve the phosphorylation of multiple centrosome components, which facilitates their duplication and growth. As we will see, the Polo kinases not only play a key role in centrosome duplication but also participate in essentially all aspects of mitosis.

The key initiating step of mitotic spindle formation is the severing of the ties that link the duplicated centrosomes. This **centrosome disjunction** occurs in $G_2$ and is triggered by mitotic CDKs (see Figure 18-35). As soon as this separation occurs, microtubules are nucleated from both centrosomes and they move away from each other, pulled by the motor protein dynein. The specifics of microtubule array formation and mitotic spindle assembly were discussed in Chapter 18. Here we will briefly consider how chromosomes attach to the mitotic spindle and how mistakes in the process are corrected.

For chromosomes to be accurately segregated during mitosis, the sister chromatid pair has to be stably bi-oriented on the mitotic spindle (Figure 19-24). How is this accomplished? Once centrosomes have moved apart from each other, microtubules, in a search-and-capture mechanism, begin to interact with kinetochores of sister chromatid pairs. Initially, chromosomes glide along the length of microtubules propelled by motor proteins. When the chromosome reaches the end of a

mitotic spindle so that one kinetochore of each sister chromatid pair attaches to microtubules emanating from opposite poles. Sister chromatids are then said to be **bi-oriented**. In what follows, we will discuss how the mitotic spindle forms, how chromosomes attach to it, and how cells correct faulty attachments.

During $G_1$, cells contain a single centrosome that functions as the major microtubule nucleating center of the cell.

⏵ **VIDEO:** Normal Cell Division in *C. elegans* Embryogenesis

**FIGURE 19-24 Chromosome attachment to the mitotic spindle.** Chromosomes attach to the mitotic spindle and congress to the spindle center. They then attach via their kinetochores to the ends of microtubules (called end-on attachments), and these attachments are stabilized by additional microtubules. The final chromosome attachment where the chromosome has stably bi-oriented on the mitotic spindle is shown.

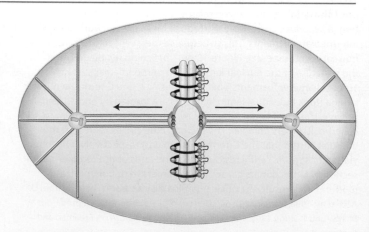

**FIGURE 19-25 Stable and unstable chromosome attachments.**
When sister kinetochores attach to microtubules emanating from opposite spindle poles, they are stably attached. This configuration is called amphitelic attachment (a). Microtubules (green) pull kinetochores (tan); cohesins resist this pulling force. This leads to kinetochores being pulled away from the protein kinase Aurora B (red zone), which localizes to the inter sister kinetochore space. As a result, Aurora B can no longer phosphorylate the microtubule binding factors at the kinetochore and kinetochore-microtubule attachments are stable. When one kinetochore attaches to microtubules emanating from two opposite spindle poles (merotelic attachment, b), or both sister kinetochores attach to microtubules emanating from the same spindle pole (syntelic attachment, c), or only one of the two sister kinetochores attaches to microtubules (monotelic attachment, d), kinetochores are not pulled out of the Aurora B zone. As a result, Aurora B phosphorylates the microtubule binding subunits of the kinetochore and microtubule attachments are destabilized.

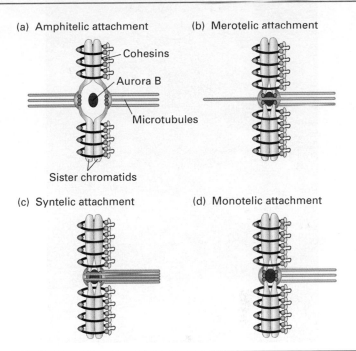

(a) Amphitelic attachment

Cohesins

Aurora B

Microtubules

Sister chromatids

(b) Merotelic attachment

(c) Syntelic attachment

(d) Monotelic attachment

microtubule, the kinetochores attach to microtubules in an end-on attachment, the final configuration in which chromosomes are linked to the mitotic spindle. Kinetochores of sister chromatids then bind microtubules emanating from the opposite spindle pole. In the end, the sister chromatid pair is said to be stably bi-oriented on the mitotic spindle.

The ultimate goal of chromosome attachment to the mitotic spindle is that each and every chromosome is attached to the mitotic spindle in a bi-oriented manner (also known as **amphitelic attachment**; Figure 19-25a). How does the cell "know" that this has occurred? Microscopic analysis of chromosome attachment has shown that initially many chromosomes attach to microtubules in faulty ways. A kinetochore can attach to microtubules emanating from both poles, a situation called **merotelic attachment** (Figure 19-25b). Kinetochores of a sister chromatid pair can attach to microtubules from the same pole (**syntelic attachment**; Figure 19-25c) or only one kinetochore can attach to microtubules (**monotelic attachment**; Figure 19-25d). Clearly, none of these attachments are productive in the sense that they would not result in accurate chromosome segregation. Thus mechanisms must be in place that detect and correct such faulty attachments.

The sensing mechanism used by cells to detect incorrect attachments is based on tension. When sister chromatids are correctly attached to microtubules, their kinetochores are under tension (see Figure 19-25a). Microtubules attached to the kinetochores pull at them and the cohesin molecules that hold the sister chromatids together withstand these forces, creating tension at kinetochores. Merotelic, syntelic, or monotelic attachment leads to insufficient tension at kinetochores, allowing the cell to distinguish these faulty forms of attachment from the correct amphitelic one.

How does the cell sense whether or not kinetochores are under tension? The protein kinase **Aurora B** and its associated regulatory factors, together known as the chromosomal passenger complex (CPC), sense kinetochores that are not under tension and sever these microtubule attachments, giving cells a

second chance to get the attachment right. The molecular basis for this sensing mechanism is partly understood. Aurora B phosphorylates a number of kinetochore components involved in microtubule binding. When phosphorylated, these proteins lose their microtubule-binding activity. Aurora B localizes to the region of sister chromatids *between* the two kinetochores. When kinetochores are not under tension, they are in close proximity to Aurora B and the protein kinase can phosphorylate the microtubule-binding subunits of the kinetochores, destabilizing any kinetochore-microtubule attachments (see Figure 19-25b–d). When microtubules are attached correctly to kinetochores, microtubule forces pull kinetochores away from Aurora B and the kinase can no longer reach its targets at kinetochores (see Figure 19-25a).

Microtubules continuously pull on chromosomes. Once all chromosomes have attached to microtubules in an amphitelic manner, the only things that prevent chromosomes from segregating to the poles and that hold them back in the middle of the spindle are the cohesin complexes (see Figure 19-25a). As we will see in Section 19.6, it is the severing of these cohesin molecules that initiates anaphase chromosome segregation.

## Chromosome Condensation Facilitates Chromosome Segregation

Chromosome segregation not only requires the building of the apparatus that segregates chromosomes, it also requires that the DNA is compacted into travel-friendly structures. An attempt to segregate the long and intertwined DNA-protein complexes present in interphase cells would lead to breakage of the DNA and hence loss of genetic material. To

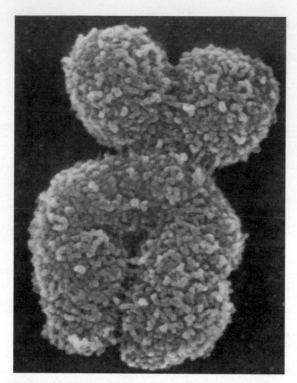

**FIGURE 19-26 Scanning electron micrograph of a metaphase chromosome.** During metaphase, the chromosomes are fully condensed and the two individual sister chromatids are visible. [Biophoto Associates/Photo Researchers.]

avoid this fate, cells compact their chromosomes during prophase into the dense structures we have become acquainted with through light or electron microscopy (Figure 19-26).

Chromosome condensation results in a dramatic reduction in chromosome length, up to 10,000-fold in vertebrates. The second key aspect of the compaction process is the untangling of the intertwined sister chromatids. This process is called **sister chromatid resolution** and is mediated in part by the decatenation activity of topoisomerase II and goes hand in hand with the condensation process.

Condensation of chromosomes is not very well understood, but central to the process is a protein complex known as the **condensin complex**. This protein complex is related to the cohesin complex that links sister chromatids together after DNA replication and was first identified based on its ability to promote chromosome condensation in frog extracts. Like the cohesin complex, condensins are composed of two large coiled-coil SMC protein subunits that associate through their ATPase domains with non-SMC subunits. When condensin function is lost in cells, chromosomes do not condense and sister chromatid tangles are not resolved. How condensins compact chromosomes is not understood, but in analogy to cohesin rings, condensins may create intrachromosomal linkages that package chromosomes into the characteristic loops seen in electron micrographs of compacted chromosomes.

Like all early mitotic events, chromosome condensation is ultimately triggered by mitotic CDKs. The exact mechanisms whereby these kinases induce chromosome condensation is not known, but it appears that mitotic CDKs induce the process at least in part through activating condensins. Two of the non-SMC condensin subunits are targets of mitotic CDKs. Their phosphorylation stimulates condensins to supercoil DNA in vitro.

Finally, processes acting in parallel to condensins also contribute to chromosome compaction. The dissociation of cohesins from chromosomes is such a parallel process. A large fraction of cohesins is removed from chromosomes during prophase (see Figure 19-20, step **3**). This process is mediated by phosphorylation of cohesins by Polo kinase and Aurora B kinase. In most organisms, cohesins are only maintained around centromeres. Cohesins around centromeres are protected from phosphorylation-dependent removal by protein phosphatase 2A (PP2A). This phosphatase is recruited to centromeric regions by a member of a family of PP2A targeting factors known as the Mei-S332/Shugoshin family of proteins. The protected pool of cohesins provides the resistance to the pulling force exerted by microtubules necessary to establish tension at bi-oriented kinetochores. As we will see in Section 19.8, this protection mechanism also plays an essential role in establishing the meiotic chromosome segregation pattern.

## KEY CONCEPTS of Section 19.5

### Entry into Mitosis

• Mitotic CDKs induce entry into mitosis in all eukaryotes.

• Mitotic CDKs are held inactive until the completion of DNA replication by inhibitory phosphorylation on the CDK subunit.

• Mitotic CDKs promote their own activation through positive feedback loops leading to the rapid inactivation of Wee1 kinase and activation of Cdc25 phosphatase.

• Mitotic CDKs induce nuclear envelope breakdown in most eukaryotes by phosphorylating lamins.

• Centrosome duplication occurs during S phase. Mitotic CDKs induce the separation of the duplicated centrosomes, which initiates mitotic spindle formation.

• Sister chromatids attach to the mitotic spindle via their kinetochores in a bi-oriented manner, with one sister kinetochore attaching to microtubules emanating from one spindle pole and the other one to microtubules nucleated by the other spindle pole.

• Cells sense bi-orientation of sister chromatids through a tension-based mechanism. When kinetochores are not under tension, the protein kinase Aurora B phosphorylates the microtubule binding subunits of the kinetochore, which decreases their microtubule binding affinity.

• Chromosomes must be compacted for segregation.

• Condensins, protein complexes that are related to cohesins, facilitate chromosome condensation and are activated by mitotic CDKs.

## 19.6 Completion of Mitosis: Chromosome Segregation and Exit from Mitosis

Once all chromosomes have condensed and correctly attached to the mitotic spindle, chromosome segregation commences. In this section, we will discuss how cleavage of cohesins by a protease known as separase triggers anaphase chromosome movement and how this cleavage is regulated. We will then explore how the machinery that initiates cohesin cleavage at the metaphase-anaphase transition, the APC/C, also initiates mitotic CDK inactivation. Next, we will see how phosphatases activated at the end of mitosis participate in mitotic CDK inactivation, bringing about the disassembly of mitotic structures and the resetting of the cell to the G₁ state. We will end this section with a discussion of cytokinesis, the process that produces two daughter cells.

### Separase-Mediated Cleavage of Cohesins Initiates Chromosome Segregation

As mentioned in the previous section, each sister chromatid of a metaphase chromosome is attached to microtubules via its kinetochore (see Figure 19-24). At metaphase, the spindle is in a state of tension, with forces pulling the two kinetochores toward the opposite spindle poles. Sister chromatids do not separate because they are held together at their centromeres by cohesin complexes. In all organisms analyzed to date, loss of cohesins from chromosomes triggers anaphase chromosome movement. The mechanism that brings about this loss of cohesins from chromosomes is conserved too. A protease known as **separase** cleaves a subunit of cohesin called Scc1 or Rad21, breaking the protein circles linking sister chromatids (Figure 19-27). Once this link is broken, anaphase begins as poleward force exerted on kinetochores moves the split sister chromatids toward spindle poles.

Cohesin cleavage was discovered in budding yeast. Analysis of the Scc1 subunit by Western blot analysis showed that the protein migrated in SDS PAGE gels according to its predicted molecular weight from G₁ until metaphase, but during anaphase the protein ran considerably faster in SDS PAGE, indicating the protein became somehow smaller. Subsequent studies showed that the faster-migrating form of Scc1 was indeed a cleavage product. Insight into the identity of the protein that was responsible for the cleavage of cohesin came from the analysis of previously identified yeast mutants that failed to segregate chromosomes during anaphase. A mutant in the gene encoding *ESP1*—what we now know to be separase—failed to produce the cleavage fragment. Subsequent analyses revealed not only that separase is a protease but that cleavage of cohesin is essential for chromosome segregation. Cells expressing a form of Scc1 with its cleavage sites mutated fail to segregate their chromosomes. Given the irreversible nature of Scc1 cleavage, it is absolutely essential that separase activity is tightly controlled. In what follows we will discuss this regulation.

### The APC/C Activates Separase Through Securin Ubiquitinylation

Prior to anaphase, a protein known as **securin** binds to and inhibits separase (see Figure 19-27). Once all kinetochores have attached to spindle microtubules in the correct bi-oriented manner, the APC/C ubiquitin ligase directed by specificity factor Cdc20 ubiquitinylates securin (note that this specificity factor is distinct from Cdh1, which targets APC/C substrates for degradation later during mitosis). Polyubiquitinylated securin is rapidly degraded by proteasomes, thereby releasing separase.

APC/C^{Cdc20} is activated in prophase by mitotic CDK phosphorylation of several APC/C subunits. However, this

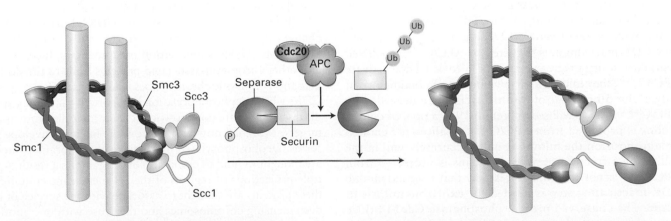

**FIGURE 19-27 Regulation of cohesin cleavage.** Separase, a protease that can cleave the Scc1 subunit of cohesin complexes, is inhibited before anaphase by the binding of securin. Mitotic CDKs also inhibit separase by phosphorylating it. When all the kinetochores have attached to spindle microtubules and the spindle apparatus is properly assembled and oriented, the Cdc20 specificity factor associated with the APC/C directs it to ubiquitinylate securin and mitotic cyclins. Following securin degradation and a decrease in mitotic CDK activity, the released and dephosphorylated separase cleaves the Scc1 subunit, breaking the cohesin circles and allowing sister chromatids to be pulled apart by the spindle apparatus that is pulling them toward opposite spindle poles.

phosphorylated APC/C$^{Cdc20}$ is not active until all chromosomes have bi-oriented on the mitotic spindle. As we will see in Section 19.7, APC/C$^{Cdc20}$ is inhibited by a checkpoint pathway that ensures that mitosis does not continue until all chromosomes have achieved proper attachment to the mitotic apparatus. Cdc20 is inhibited until every kinetochore has attached to microtubules and tension is applied to the kinetochores of all sister chromatids, pulling them toward opposite spindle poles. In vertebrate cells, separase is also regulated by phosphorylation. Mitotic CDK activity inhibits separase during prophase and metaphase. Only when mitotic CDK activity begins to decline at the metaphase-anaphase transition through APC/C$^{Cdc20}$-mediated protein degradation can separase become active and trigger chromosome segregation.

Once cohesins are cleaved, anaphase chromosome movement ensues. As discussed in Chapter 18, chromosome segregation is mediated by microtubule depolymerization and motor proteins as spindle poles move away from each other. Decline in mitotic CDK activity is important for these anaphase chromosome movements. When mitotic CDK inactivation is inhibited, anaphase does occur, but it is abnormal. Dephosphorylation of a number of microtubule-associated proteins that affect microtubule dynamics appears important for this process. In budding yeast, this dephosphorylation is brought about by the protein phosphatase Cdc14, which we will see, plays an essential role in the final cell cycle stage, exit from mitosis.

## Mitotic CDK Inactivation Triggers Exit from Mitosis

Anaphase spindle elongation and the events associated with exit from mitosis—mitotic spindle disassembly, chromosome decondensation, and nuclear envelope re-formation—are brought about by the dephosphorylation of CDK substrates. In other words, exit from mitosis can be viewed as a reversal of entry into mitosis. The phosphorylation events that triggered the different mitotic events need to be undone for the cell to reverse to the G$_1$ state.

Dephosphorylation of mitotic CDK substrates is caused by the inactivation of mitotic CDKs. In most organisms, mitotic CDK inactivation is triggered by APC/C$^{Cdc20}$-mediated degradation of mitotic cyclins. As mitotic CDKs activate APC/C$^{Cdc20}$, they initiate their own demise. In budding yeast only about 50 percent of mitotic cyclins are degraded by APC/C$^{Cdc20}$. As we will see in Section 19.7, a pool of mitotic cyclins is protected from APC/C$^{Cdc20}$ to allow for enough time to position the mitotic spindle accurately within the cell. How a fraction of mitotic cyclins is protected from APC/C$^{Cdc20}$ is not known, but it is clear that a second mitotic CDK-inactivating step is needed for exit from mitosis to occur. The conserved protein phosphatase **Cdc14** brings about this second step in mitotic CDK inhibition.

In budding yeast, complete inactivation of mitotic CDKs requires the destruction of mitotic cyclins by APC/C$^{Cdh1}$ and the accumulation of the CDK inhibitor Sic1, which—recall—holds S phase CDKs in check until cells enter the cell cycle. Both APC/C$^{Cdh1}$ and Sic1 are inhibited by mitotic

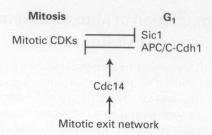

**FIGURE 19-28 The protein phosphatase Cdc14 triggers exit from mitosis in budding yeast.** During mitosis, mitotic CDK activity inhibits its inhibitors, APC/C$^{Cdh1}$ and Sic1. During G$_1$, APC/C$^{Cdh1}$ and Sic1 inhibit mitotic CDKs. During exit from mitosis, the protein phosphatase Cdc14 throws the switch between these two antagonistic states. The mitotic exit network activates the phosphatase during anaphase, allowing it to dephosphorylate APC/C$^{Cdh1}$, thereby activating it. The phosphatase also promotes the accumulation of Sic1. In addition, Cdc14 dephosphorylates the many mitotic CDK substrates, which leads to rapid exit from mitosis.

CDKs. Conversely, APC/C$^{Cdh1}$ and Sic1 inhibit mitotic CDKs (Figure 19-28). The protein phosphatase Cdc14 throws the switch between these two mutually antagonistic states during anaphase. Cdc14 is kept inactive during most of the cell cycle, but the phosphatase is activated during anaphase by a GTPase signaling pathway known as the mitotic exit network (MEN). This signaling cascade is, as we will see in Section 19.7, responsive to spindle position and only becomes active in anaphase, when the anaphase spindle is properly positioned within the cell. Once activated during anaphase, Cdc14 dephosphorylates APC/C$^{Cdh1}$ and Sic1 to promote mitotic cyclin degradation and mitotic CDK inactivation, respectively. This leads to exit from mitosis.

Phosphatase activity is also essential for exit from mitosis in vertebrates. Simple inactivation of mitotic CDKs is not sufficient to trigger the timely exit from mitosis. It is not yet clear which phosphatase dephosphorylates CDK substrates to reset the cell to the G$_1$ stage. Both protein phosphatase 1 and protein phosphatase 2A have been implicated in the process.

Ultimately, reversal of mitotic CDK phosphorylation changes the activities of many proteins back to their interphase state. Dephosphorylation of condensins, histone H1, and other chromatin-associated proteins leads to the decondensation of mitotic chromosomes in telophase. The targets of CDKs whose dephosphorylation is important for mitotic spindle disassembly are not known, but likely multiple proteins are targets. More is known about how the nuclear envelope reforms. Dephosphorylated inner nuclear membrane proteins are thought to bind to chromatin once again. As a result, multiple projections of regions of the ER membrane containing these proteins are thought to associate with the surface of the decondensing chromosomes and then fuse with one another directed by an unknown mechanism to form a continuous double membrane around each chromosome (Figure 19-29). Dephosphorylation of nuclear pore subcomplexes allows them to reassemble into complete NPCs traversing the inner and outer membranes soon after fusion of the ER projections. Ran-GTP, required for driving most nuclear import and

**FIGURE 19-29 Model for reassembly of the nuclear envelope during telophase.** Extensions of the endoplasmic reticulum (ER) associate with each decondensing chromosome and then fuse with one another, forming a double membrane around the chromosome. Dephosphorylated nuclear pore subcomplexes reassemble into nuclear pores, forming individual mini-nuclei called *karyomeres*. The enclosed chromosome further decondenses, and subsequent fusion of the nuclear envelopes of all the karyomeres at each spindle pole forms a single nucleus containing a full set of chromosomes. NPC, nuclear pore complex. [Adapted from B. Burke and J. Ellenberg, 2002, *Nature Rev. Mol. Cell Biol.* **3**:487.]

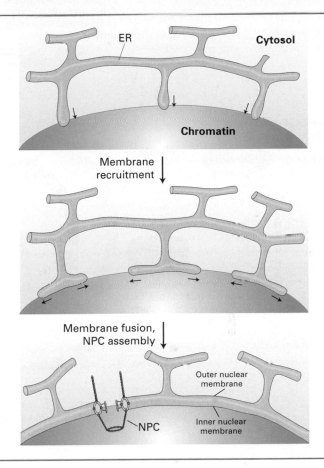

export (see Chapter 13), stimulates both fusion of the ER projections to form daughter nuclear envelopes and assembly of NPCs (see Figure 19-29). The Ran-GTP concentration is highest in the microvicinity of the decondensing chromosomes because the Ran-guanine nucleotide-exchange factor (Ran-GEF) is bound to chromatin. Consequently, membrane fusion is stimulated at the surfaces of decondensing chromosomes, forming sheets of nuclear membrane with inserted NPCs.

## Cytokinesis Creates Two Daughter Cells

When chromosome segregation is completed, the cytoplasm and organelles are distributed between the two future daughter cells. This process is called **cytokinesis**. With the exception of higher plants, the division of the cell is brought about by a **contractile ring** made of actin and the actin motor myosin (see Figure 17-36). During cytokinesis, the ring contracts in a manner similar to muscle contraction, pulling the membrane inward and eventually closing the connection between the two daughter cells.

Cytokinesis must be coordinated with other cell cycle events in space and time. For cell division to produce two daughter cells each containing the components necessary for survival, the division plane must be placed so that each cell receives approximately half of the mother cell's cytoplasmic content and *exactly* half of the genetic content. Cytokinesis must also be coordinated with the temporal sequence of cell cycle events, the completion of mitosis. In what follows, we will explore both these aspects of cytokinesis regulation.

In animal cells, the contractile ring forms during anaphase and is placed in the middle of the anaphase spindle. This ensures that each daughter cell receives half of the genetic material. Despite the importance of this coordination, surprisingly little is known about it. Some experiments support the idea that signals sent from the **spindle midzone** to the cell cortex are important to coordinate the site of cytokinesis with spindle position. Other research suggests that microtubules of the spindle interact with the cell cortex, positioning the **cleavage furrow** with respect to spindle pole position. Most likely, a combination of these pathways governs the formation of the cleavage furrow during cytokinesis.

In budding yeast and fission yeast, the site of cytokinesis is determined prior to mitosis. In budding yeast, this occurs during $G_1$, when the bud site is determined. In fission yeast, proteins important for contractile ring formation accumulate in the middle of the cell during $G_2$. Irrespective of the sequence of events—site of mitosis first, then site of cytokinesis or vice versa—it is clear that these events must be tightly coordinated. As we will see in Section 19.7, cells have developed surveillance mechanisms that ensure that the site of cytokinesis is coordinated with spindle position. This is especially important during **asymmetric cell divisions**, which give rise to cells of different size and/or fate. Such cell divisions are essential during development and in stem cell divisions (see Chapter 21). Cytokinesis must also be coordinated with other cell cycle events. The major signal for cytokinesis is the inactivation of mitotic CDKs. Cells expressing a stabilized version of mitotic cyclins progress through anaphase but do not undergo cytokinesis. The CDK targets in the cytokinesis machinery have not yet been discovered.

This concludes our discussion of the molecular events of cell division. As we saw, cyclin-dependent kinases and ubiquitin-mediated protein degradation are at the center of this control (Figure 19-30). In what follows, we will discuss the mechanisms that ensure a subsequent cell cycle stage is not initiated until the previous has been completed and that each cell cycle step occurs accurately.

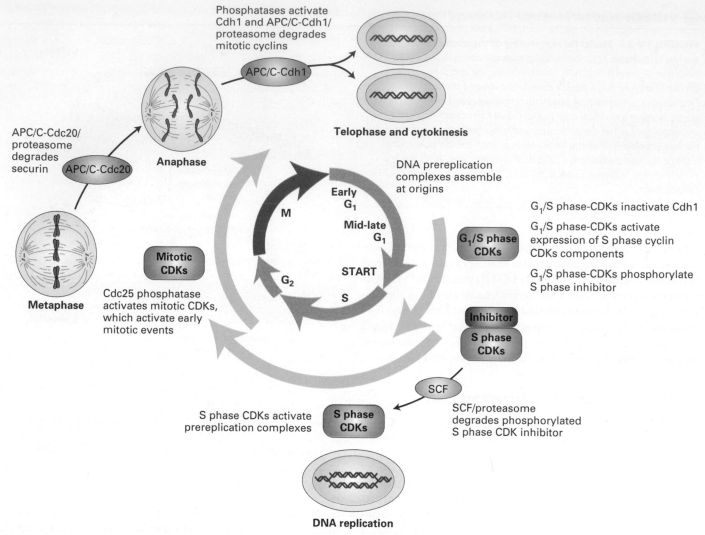

**FIGURE 19-30 Fundamental processes in the eukaryotic cell cycle.** See text for discussion.

The following labels appear in the figure:

Phosphatases activate Cdh1 and APC/C-Cdh1/proteasome degrades mitotic cyclins

APC/C-Cdh1

Telophase and cytokinesis

APC/C-Cdc20/proteasome degrades securin

APC/C-Cdc20

Anaphase

DNA prereplication complexes assemble at origins

Early G₁

Mid-late G₁

START

S

G₂

M

Mitotic CDKs

Metaphase

Cdc25 phosphatase activates mitotic CDKs, which activate early mitotic events

G₁/S phase CDKs

G₁/S phase-CDKs inactivate Cdh1

G₁/S phase-CDKs activate expression of S phase cyclin CDKs components

G₁/S phase-CDKs phosphorylate S phase inhibitor

Inhibitor

S phase CDKs

SCF

SCF/proteasome degrades phosphorylated S phase CDK inhibitor

S phase CDKs activate prereplication complexes

S phase CDKs

DNA replication

# 19.7 Surveillance Mechanisms in Cell Cycle Regulation

Surveillance mechanisms known as **checkpoint pathways** operate to ensure that the next cell cycle event is not initiated until the previous one has been completed. Checkpoint pathways consist of **sensors** that monitor a particular cellular event, a **signaling cascade** that initiates the response, and an **effector** that halts cell cycle progression and activates repair pathways when necessary. Cell cycle events monitored by checkpoint pathways include growth, DNA replication, DNA damage, kinetochore attachment to the mitotic spindle, and position of the spindle within the cell. They are responsible for the extraordinarily high fidelity of cell division, ensuring that each daughter cell receives the correct number of accurately replicated chromosomes. They function by controlling the protein kinase activities of the cyclin-CDKs through a variety of mechanisms: regulation of the synthesis and degradation of cyclins, phosphorylation of CDKs at inhibitory sites, regulation of the synthesis and stability of

CDK inhibitors (CKIs) that inactivate cyclin-CDK complexes, and regulation of the APC/C ubiquitin-protein ligase.

## Checkpoint Pathways Establish Dependencies and Prevent Errors in the Cell Cycle

The experiments that led to the idea of surveillance mechanisms or checkpoint pathways establishing dependencies in the cell cycle were simple and beautiful in their interpretation. Recall that Lee Hartwell and colleagues isolated temperature-sensitive *cdc* mutants in *S. cerevisiae*. Their characterization and the study of the genes affected in these mutants profoundly shaped our understanding of the eukaryotic cell cycle. It was the characterization of one of these *cdc* mutants, *cdc13*, that led Hartwell and colleagues to formulate the checkpoint pathway concept.

For the purpose of our discussion here, the only thing that is important to know about *CDC13* function is that the gene is required for telomere replication and, in its absence, large stretches of incompletely replicated telomeric DNA persist in cells. Cells carrying a temperature-sensitive allele of the *cdc13* gene as the sole source of *CDC13* arrest with a $G_2$ DNA content when shifted to the restrictive temperature (Figure 19-31a). This arrest is indicative of a defect in late S phase or entry into mitosis. When the cells are returned to

the permissive temperature, the temperature-sensitive protein is functional again, and cells continue to proliferate. Thus, although the *cdc13* mutant cells failed to divide at the restrictive temperature, they retained their viability and so were able to resume proliferation once they were returned to the permissive temperature where the temperature-sensitive cdc13 protein was functional again.

To characterize the *cdc13* mutant in more detail, Hartwell and colleagues examined the effects of introducing a second mutation in another gene, a deletion in the *RAD9* gene. The *RAD9* gene is not essential for viability, but when it is deleted, cells are extremely sensitive to DNA damaging agents such as x-rays. This mutation on its own does not affect growth of cells at any temperature but had a dramatic effect on *cdc13* mutants. When the researchers examined the *cdc13 rad9* double mutant at the restrictive temperature, they observed that the mutant no longer arrested in $G_2$ but instead the cells continued to divide for a few divisions (Figure 19-31b). When they returned these cells to the permissive temperature, the double mutant failed to resume proliferation. This indicates that while the cells continued to divide a few times at the restrictive temperature, they lost viability.

Hartwell and colleagues proposed the following explanation for this observation: *cdc13* mutants arrest at the restrictive temperature because they harbor incompletely replicated DNA. This damaged DNA signals the cell to arrest cell cycle progression and induce repair of the damage because mitosis of cells with damaged DNA would almost certainly lead to cell death. The *RAD9* gene is part of the machinery that conveys this cell cycle arrest. In cells lacking *RAD9*, the "halt cell cycle progression" signal does not work and cells undergo mitosis despite incompletely replicated DNA. This kills the cells. Hartwell and his colleagues called this surveillance mechanism a **checkpoint pathway**.

Today we know that cells harbor multiple checkpoint pathways to ensure that a cell cycle phase does not commence before the previous one has been completed. In addition to establishing dependencies, the checkpoint pathways ensure that each aspect of chromosome replication and division occurs with accuracy. For example, a single kinetochore that fails to attach to the mitotic spindle can halt cell cycle progression in metaphase by activating the spindle assembly checkpoint.

Each checkpoint is built in the same manner. A sensor detects a defect in a particular cellular process and in response to this defect activates a signal transduction pathway. Effectors activated by the signaling pathway initiate repair of the defect and halt cell cycle progression until the defect is corrected. In what follows we will discuss the major checkpoint pathways that govern cell cycle progression.

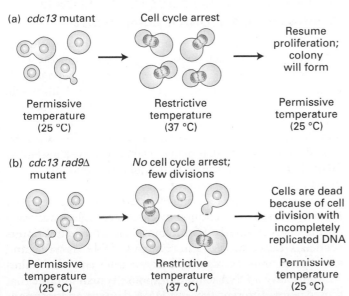

**EXPERIMENTAL FIGURE 19-31 An experiment that led to the checkpoint pathway concept.** (a) When shifted to the restrictive temperature, *cdc13* mutants arrest cell cycle progression because of incomplete DNA replication. When the cells are returned to the permissive temperature, they resume proliferation because they maintained viability during the cell cycle arrest. (b) *cdc13 rad9Δ* double mutants do not arrest when shifted to the restrictive temperature because these cells cannot sense that their DNA is incompletely replicated. Cells undergo mitosis, and this leads to cell death because genetic information is lost. Therefore, cells quickly lose viability at the restrictive temperature and can no longer resume proliferation when they are returned to the permissive temperature.

## The Growth Checkpoint Pathway Ensures That Cells Only Enter the Cell Cycle After Sufficient Macromolecule Biosynthesis

Cell proliferation requires that cells multiply through the process of cell division and that the individual cells grow through macromolecule biosynthesis. Cell growth and division are separate processes, but for cells to maintain a constant size as they multiply, cell growth and division must be tightly coordinated. For example, when nutrients are limited, cells reduce their growth rate, and cell division must be down-regulated accordingly. This type of coordination between cell growth and division is especially important in unicellular organisms that experience changes in nutrient availability as part of their natural life cycle. It is therefore not surprising that surveillance mechanisms exist that adjust cell division rate according to growth rate.

In budding yeast, cell growth and division are coordinated in $G_1$. In this stage of the cell cycle, the growth and division cycles are linked by making the activity of $G_1$ CDKs dependent on cell growth. Which aspect of growth is linked to the cell cycle? Classical experiments using protein synthesis inhibitors indicate that growth rate and hence cell cycle control by growth is determined by protein synthesis. How protein synthesis controls $G_1$ CDK activity is an area of active investigation. The $G_1$ cyclin Cln3 is subject to translational control, making levels of this cyclin especially sensitive to protein synthesis rate. However, it is clear that this is not the only mechanism. Although the molecular pathways that coordinate cell division with cell growth are not yet understood, it appears that this control is highly plastic. The length of $G_1$ and the *critical cell size*, that is, the size at which cells enter the cell cycle, change with nutrient availability.

*S. pombe* grows as a rod-shaped cell that increases in length as it grows and then divides in the middle during mitosis to produce two daughter cells of equal size (see Figure 19-4). Unlike budding yeast and most metazoan cells that grow primarily during $G_1$, this yeast does most of its growing during the $G_2$ phase of the cell cycle and it is entry into mitosis that is carefully regulated in response to cell size. Recall entry into mitosis is regulated by the protein kinase Wee1, which inhibits CDK1 by phosphorylating tyrosine 15. When nutrients are limited, Wee1 phosphorylates CDK1; hence cells remain in $G_2$ until they reach the critical size for mitotic entry. This size control is brought about by regulated protein localization. The protein kinase Pom1 forms a gradient from each pole toward the middle of the cell. The CDK inhibitor Wee1 and its inhibitor Cdr2 localize in patches to the cell cortex in the middle of the cell (Figure 19-32). Pom1 prevents Cdr2 from inhibiting Wee1. When cells are small, Pom1 inhibits Cdr2. Wee1 is active and prevents entry into mitosis. As cells grow, the local concentration of Pom1 in the middle of the cell declines and Cdr2 becomes active and inhibits Wee1. Cells can now enter mitosis. Hence in this organism, cell length is measured by a protein gradient.

Nutrients are usually not limiting in multicellular organisms. Instead, cell growth is controlled by growth factor signaling pathways such as the Ras, AMPK, and TOR pathways (see Chapters 8 and 16). These pathways also appear to be important for coordinating cell growth and division. Mutations in key components of growth factor signaling pathways, such as Myc, cause dramatic changes in cell size in *Drosophila*. Myc regulates the transcription of many genes important for macromolecule biosynthesis and also, more

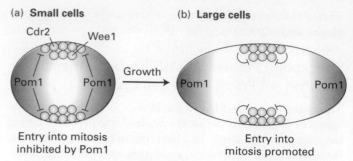

**FIGURE 19-32 A protein gradient measures cell length in *S. pombe*.** The protein kinase Pom1 forms a gradient from each pole toward the middle of the cell. Wee1 and its inhibitor Cdr2 are localized to patches of cell cortex in the middle of the cell. Pom1 inhibits Cdr2, and hence the pathway that inhibits Wee1. (a) In small cells, the concentration of Pom1 in the middle of the cell is higher. Cdr2 is inhibited, allowing Wee1 to remain active and prohibit entry into mitosis. (b) As cells grow in length, the concentration of Pom1 in the middle of the cell declines. Pom1 inhibition of Cdr2 decreases, and the Cdr2 signaling pathway is now able to inhibit Wee1, promoting entry into mitosis. [Adapted from Moseley et al., 2009; *Nature* **459**:857–860.]

indirectly, $G_1$ CDKs. Thus this transcription factor appears to integrate cell growth and division. However, the details of this coordination remain to be elucidated.

## The DNA Damage Response Halts Cell Cycle Progression When DNA Is Compromised

The complete and accurate duplication of the genetic material is essential for cell division. If cells enter mitosis when DNA is not completely replicated or otherwise damaged, genetic changes occur. In many instances, those changes will lead to cell death, but as we will see in Chapter 24, they can also lead to genetic alterations that result in loss of growth and division control and eventually cancer. This is underscored by the finding that many proteins involved in sensing DNA damage and its repair are frequently found mutated in human cancers.

The enzymes that replicate DNA are highly accurate, but this exactness is not enough to ensure complete accuracy during DNA synthesis. Furthermore, environmental insults such as x-rays and UV light can cause DNA damage, and this damage must be repaired before a cell's entry into mitosis. Cells have a **DNA damage response system** in place that senses many different types of DNA damage and responds by activating repair pathways and halting cell cycle progression until the damage has been repaired. Cell cycle arrest can occur either in $G_1$, S phase, or $G_2$, depending on whether DNA damage occurred before cell cycle entry or during DNA replication. In multicellular organisms, the strategy to deal with particularly severe DNA damage is different. Rather than attempting to repair the damage, cells undergo **programmed cell death** or **apoptosis**, a mechanism that we will discuss in detail in Chapter 21.

DNA damage exists in many different forms and ranges in severity. A break of the DNA helix, known as a **double-strand**

**break**, is perhaps the most severe form of damage since such a lesion would almost certainly lead to DNA loss if mitosis ensued in its presence. More subtle defects include single-strand breaks, structural changes in nucleotides, or DNA mismatches. For our discussion here, it is important to note that cells have sensors for all these different types of damage. These sensors scan the genome and, when they detect a lesion, assemble signaling and repair factors onto the site of the lesion.

Central to the detection of the different lesions is a pair of homologous protein kinases called **ATM** and **ATR**. These protein kinases get recruited to sites of DNA damage. They then initiate the sequential recruitment of adapter proteins and another set of protein kinases called Chk1 and Chk2. These kinases then activate repair mechanisms and cause cell cycle arrest or apoptosis in animals (Figure 19-33). ATR and ATM recognize different types of DNA damage. ATM is very specialized in that it only responds to double-strand breaks. ATR is able to recognize more diverse types of DNA damage, such as stalled replication forks, damaged nucleotides, and double-strand breaks. ATR recognizes these diverse types of damage because all of them contain some amount of *single-stranded DNA*, either as part of the damage itself or because repair enzymes create single-stranded DNA as part of the repair process. Stalled replication forks, for example, are recognized by ATR. The association of ATR with stalled forks is

thought to activate its protein kinase activity, leading to the recruitment of adapter proteins whose function is to recruit and help activate the Chk1 kinase. Active Chk1 then induces repair pathways and inhibits cell cycle progression.

Chk1 and Chk2 halt the cell cycle. The protein kinases phosphorylate Cdc25, thereby inactivating it (see Figure 19-33). When the DNA damage occurs during $G_1$, Cdc25A inhibition results in inhibition of $G_1$/S phase CDKs and S phase CDKs (Figure 19-34). As a result, these kinases cannot initiate DNA replication. When the DNA damage occurs during S phase or in $G_2$, Cdc25C inhibition by Chk1/2 results in the inhibition of mitotic CDKs and hence arrest in $G_2$. Active DNA replication also inhibits entry into mitosis. ATR continues to inhibit Cdc25C via Chk1 until all replication forks complete DNA replication and disassemble. This mechanism makes the initiation of mitosis *dependent* on the completion of chromosome replication. Finally, cells also sense DNA replication stress that results in stalled or slowing of replication forks. It triggers activation of the ATR-Chk1 checkpoint pathway and results in down-regulation of S phase CDK activity and prevents firing of late-replicating origins.

Chk1-mediated inhibition of the Cdc25 family of phosphatases is not the only mechanism whereby DNA damage or incomplete replication inhibits cell cycle progression. As we will see below, DNA damage leads to the activation of the transcription factor p53, which transcribes the CDK inhibitor p21. p21 binds to and inhibits all metazoan cyclin-CDK complexes. As a result, cells are arrested in $G_1$ and $G_2$ (see Figure 19-34).

ATM recognizes double-strand breaks (see Figure 19-33). The protein kinase gets directly recruited to the DNA ends by a complex known as the MRN complex, which binds to broken ends and holds them together. Activated ATM then phosphorylates and activates Chk2 and recruits repair proteins. These repair proteins initiate **homologous recombination**, as discussed in Chapter 4. This process involves the creation of single-stranded overhangs, which in turn recruit and activate ATR and its effectors, further enhancing the DNA damage response. ATM can also recruit an alternative repair pathway where two double-strand breaks are directly fused with each other in a repair process known as **nonhomologous end joining**. Like ATR, ATM activation also halts cell cycle progression by a Chk2-mediated inhibition of Cdc25, thus preventing activation of CDKs. This inhibition can occur in $G_1$ or $G_2$.

A key effector of the DNA damage response in metazoan cells is the transcription factor **p53** (see Figure 19-33). It is known as a tumor suppressor because its normal function is to limit cell proliferation in the face of DNA damage. The protein is extremely unstable and generally does not accumulate to high enough levels to stimulate transcription under normal conditions. The instability of p53 results from its ubiquitinylation by a ubiquitin-protein ligase called *Mdm2* and subsequent proteasomal degradation. The rapid degradation of p53 is inhibited by ATM and ATR, which phosphorylate p53 at a site that interferes with Mdm2 binding. This and other modifications of p53 in response to DNA damage greatly increase its ability to activate transcription of

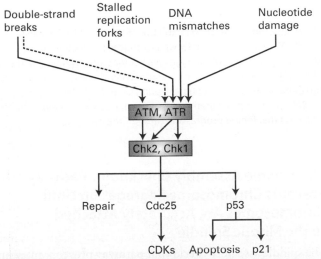

**FIGURE 19-33 The DNA damage response system.** The protein kinases ATM and ATR are activated by damaged DNA. ATR responds to a variety of DNA damage—most likely to the single-stranded DNA that exists either as a result of the damage itself or as a result of repair. ATM is specifically activated by double-strand breaks. Because double-strand breaks are converted into single-stranded DNA as a result of repair, they also, albeit indirectly and hence depicted as a dashed line, activate ATR. ATM and ATR, once activated by DNA damage, activate another pair of related protein kinases, Chk1 and Chk2. These kinases then induce the DNA repair machinery and cause cell cycle arrest by inhibiting Cdc25. In metazoan cells, when the DNA damage is severe, Chk1 and Chk2 also activate the transcription factor p53. p53 induces cell cycle arrest by inducing transcription of the CKI p21 and apoptosis.

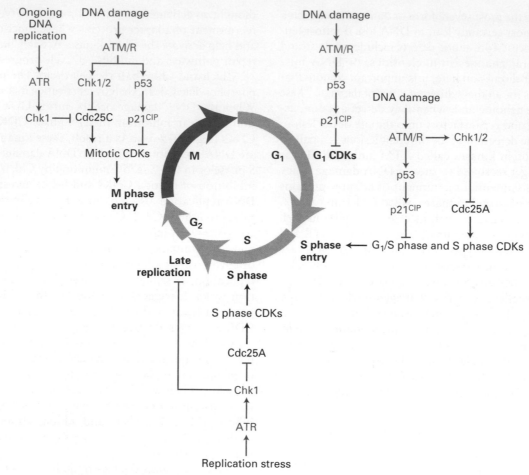

**FIGURE 19-34 Overview of DNA damage checkpoint controls in the cell cycle.** During G₁, the p53-p21^CIP pathway inhibits G₁ CDKs. During ongoing DNA replication and in response to replication stress (slow DNA replication fork movement or DNA replication fork collapse), the ATR-Chk1 protein kinase cascade phosphorylates and inactivates Cdc25C, thereby preventing the activation of mitotic CDKs and inhibiting entry into mitosis. In response to DNA damage, the ATM or ATR protein kinases (ATM/R) inhibit Cdc25 via the Chk1/2 protein kinases. They also activate p53, which induces production of the CKI p21. During G₁, the DNA damage checkpoint pathway inhibits Cdc25A inhibiting G₁/S phase CDKs and S phase CDKs, thereby blocking entry into or passage through S phase. During G₂, ATM/R-Chk1/2 inhibit Cdc25C. The p53-p21^CIP pathway is also activated. Red symbols indicate pathways that inhibit progression through the cell cycle.

specific genes that help the cell cope with DNA damage. One of these genes encodes the CKI p21 (see Figure 19-34).

Under some circumstances, such as when DNA damage is extensive, p53 also activates expression of genes that lead to apoptosis, the process of programmed cell death that normally occurs in specific cells during the development of multicellular animals. In metazoans, the p53 response evolved to induce apoptosis in the face of extensive DNA damage, presumably to prevent the accumulation of multiple mutations that might convert a normal cell into a cancer cell. The dual role of p53 in both cell cycle arrest and the induction of apoptosis may account for the observation that nearly all cancer cells have mutations in both alleles of the *p53* gene or in the pathways that stabilize p53 in response to DNA damage (see Chapter 24). The consequences of mutations in *p53*, *ATM*, and *Chk2* provide dramatic examples of the significance of cell cycle checkpoint pathways to the health of a multicellular organism.

## The Spindle Assembly Checkpoint Pathway Prevents Chromosome Segregation Until Chromosomes Are Accurately Attached to the Mitotic Spindle

The **spindle assembly checkpoint pathway** prevents entry into anaphase until every kinetochore of every chromatid is properly attached to spindle microtubules. If even a single kinetochore is unattached or not under tension, anaphase is inhibited. Clues about how this checkpoint operates initially came from the isolation of yeast mutants with a defective response to benomyl, a microtubule-depolymerizing drug. Low concentrations of benomyl increase the time required for yeast cells to assemble the mitotic spindle and attach kinetochores to microtubules. Wild-type cells exposed to benomyl do not begin anaphase until these processes are completed and then proceed on through mitosis, producing normal daughter cells. In contrast, mutants defective in the spindle assembly checkpoint pathway proceed through anaphase before assembly of the spindle and attachment of kinetochores is complete; consequently, they mis-segregate their chromosomes, producing abnormal daughter cells that die.

We now know that cells harbor a surveillance mechanism that prevents anaphase entry in the presence of unattached kinetochores. Components of the spindle assembly checkpoint recognize and bind to unoccupied microtubule binding sites at kinetochores and create an anaphase inhibitory signal (Figure 19-35a). A protein known as Mad2 (*mitotic arrest defective 2*) is central to the creation of this inhibitory signal. Mad2 regulates Cdc20, the specificity factor required to target the APC/C to securin. Recall that APC/C$^{Cdc20}$-mediated ubiquitinylation of securin and its subsequent degradation is required for activation of separase and entry into anaphase (see Figure 19-27). Mad2 is recruited to kinetochores that are not attached to microtubules. Mad1 appears critical for this process. Kinetochore-bound Mad2 rapidly exchanges with a soluble form of Mad2 that inhibits all the Cdc20 in the cell (Figure 19-35a). When microtubules attach to kinetochores, the kinetochores release the bound Mad2 and cease the process by which the inhibitory, soluble form of Mad2 is produced (Figure 19-35b). However, when even a single kinetochore is unattached to microtubules from the opposite spindle pole of its sister, sufficient soluble inhibitory Mad2 is produced at the unattached kinetochore to inhibit all the Cdc20 in the cell. This elegant model for the spindle assembly checkpoint pathway can account for the ability of a single unattached kinetochore to inhibit all the cellular Cdc20 until the kinetochore becomes properly associated with spindle microtubules.

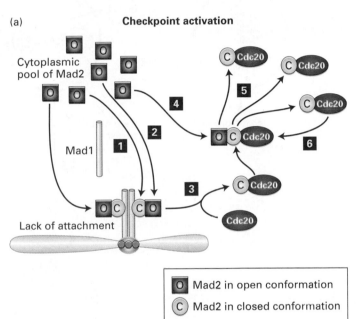

(a) **Checkpoint activation**

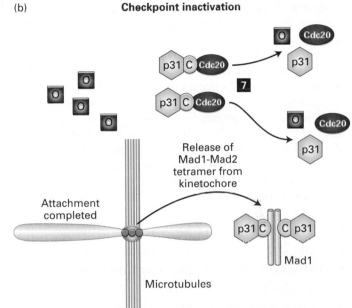

(b) **Checkpoint inactivation**

**FIGURE 19-35 The spindle assembly checkpoint pathway.** The spindle assembly checkpoint pathway is active until every single kinetochore has attached properly to spindle microtubules. (a) The Mad2 protein exists in two conformations, one "open" (red squares) and the other "closed" (orange circles). According to the current model, Mad1 and the closed Mad2 form a tetramer that binds to unattached kinetochores via the Mad1 subunit (step **1**). Open Mad2 can bind transiently to closed Mad2 bound to Mad1 at the kinetochore (step **2**). This interaction with closed Mad2 stimulates open Mad2 to bind a Cdc20. Open Mad2 can bind Cdc20 only while it is interacting with a closed Mad2. This converts the open Mad2 protein to the closed conformation, causing it to dissociate from the closed Mad2 at the kinetochore (step **3**). The stable interaction of closed Mad2 with Cdc20 prevents Cdc20 from binding to the APC/C. Further, the closed Mad2 bound to Cdc20 can interact transiently with another Mad2 in the open conformation (step **4**), causing it to bind another Cdc20 molecule. This converts the open Mad2 to the closed conformation bound to Cdc20. This newly formed closed Mad2-Cdc20 complex dissociates from the first Mad2-Cdc20 pair, generating two Mad2-Cdc20 complexes (step **5**). Thus free Mad2 in the open conformation is quickly converted to closed Mad2 bound to Cdc20

as this cycle repeats (step **6**). The source of closed Mad2 that initiates this chain reaction is the closed Mad2 bound to Mad1 associated with a kinetochore, explaining how a single unattached kinetochore can cause inactivation of all the Cdc20 in the cell through the formation of closed Mad2-Cdc20 complexes. How are unattached kinetochores that recruit Mad1-Mad2 complexes generated? Either microtubules fail to attach or Aurora B severs kinetochore-microtubule attachments that are not under tension (see Figure 19-25). (b) Silencing of the spindle assembly checkpoint pathway: Attachment of microtubules (green) to kinetochores causes the displacement of the Mad1-Mad2 tetramer. Mad2 in the displaced tetramer cannot interact with open Mad2 but rather binds to p31$^{comet}$. p31$^{comet}$ is always active and disassembles Mad2-Cdc20 complexes, releasing active Cdc20 (step **7**). However, a small number of Mad1-Mad2 tetramers bound to kinetochores can generate enough Mad2-Cdc20 complexes by the mechanism shown in (a) to overcome the activity of p31. Once all kinetochores have attached to microtubules, causing the release of all Mad1-Mad2 tetramers, p31 activity predominates, releasing active Cdc20, which binds to the APC/C, resulting in ubiquitinylation and proteasomal degradation of securin and the onset of anaphase. [Modified from A. De Antoni et al., 2005, *Curr. Biol.* **15**:214.]

Entry into anaphase is also inhibited when the attachments of microtubules to kinetochores are faulty. As we saw in Section 19.5, kinetochores of sister chromatids often attach to microtubules emanating from the same pole (syntelic attachment) or a single kinetochore attaches to microtubules that originate from two different poles (merotelic attachment). These faulty attachments result in insufficient or no tension at sister kinetochores, and such microtubule-kinetochore interactions are quickly destabilized by Aurora B phosphorylation of the microtubule-binding proteins of the kinetochore. This leads to the generation of unattached kinetochores, which are recognized by the spindle assembly checkpoint. In this manner Aurora B and the spindle assembly checkpoint collaborate during every cell cycle to accurately attach every single pair of sister chromatids on the mitotic spindle in the correct, bi-oriented manner.

The spindle assembly checkpoint pathway is essential for viability in mice, highlighting the importance of this quality-control pathway during every cell division. If anaphase is initiated before both kinetochores of a replicated chromosome become attached to microtubules from opposite spindle poles, daughter cells are produced that have missing or extra chromosomes, an outcome called *nondisjunction*. When nondisjunction occurs in mitotic cells, it can lead to the misregulation of genes and contribute to the development of cancer. When nondisjunction occurs during the meiotic division that generates a human egg or sperm, trisomy of any chromosome can occur. Down syndrome is caused by trisomy of chromosome 21, resulting in developmental abnormalities and mental retardation. Every other trisomy results in embryonic lethality or death shortly after birth.

## The Spindle Position Checkpoint Pathway Ensures That the Nucleus Is Accurately Partitioned Between Two Daughter Cells

The coordination of the site of nuclear division with that of cytokinesis is essential for the production of two identical daughter cells. If cytokinesis occurred in such a way that each daughter cell failed to receive a complete genetic complement, chromosome loss or gain would ensue. In many systems, surveillance mechanisms have been described that ensure cytokinesis does not occur when the mitotic spindle is not correctly positioned in the cell. This surveillance mechanism, known as the **spindle position checkpoint pathway**, is best understood in budding yeast. In budding yeast, the site of bud formation and therefore the site of cytokinesis is determined during $G_1$. Thus the axis of division is defined prior to mitosis and the mitotic spindle must be aligned along this mother-bud axis during every cell division (Figure 19-36, step **1**). When this process fails, the spindle position checkpoint prevents mitotic CDK inactivation, giving the cell an opportunity to reposition the spindle prior to spindle disassembly and cytokinesis (Figure 19-36, step **2**). If the spindle position checkpoint fails, cells that misposition their spindles give rise to mitotic products with too many or too few nuclei (Figure 19-36, step **3**).

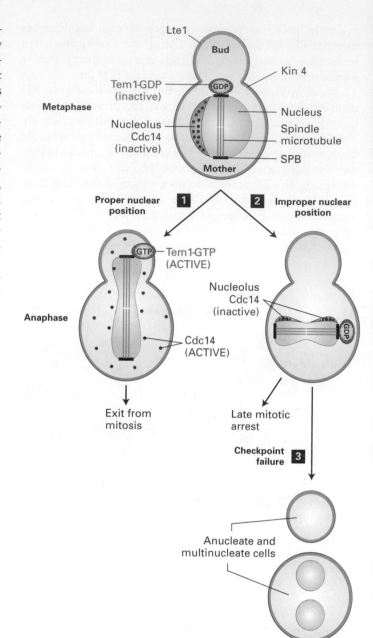

**FIGURE 19-36 The spindle position checkpoint in budding yeast.** Cdc14 phosphatase activity is required for exit from mitosis. *Top*: In *S. cerevisiae*, during interphase and early mitosis, Cdc14 (red dots) is sequestered and inactivated in the nucleolus. Inactive Tem1-GDP (purple) associates with the spindle pole body (SPB) nearest to the bud as soon as the mitotic spindle forms. If chromosome segregation occurs properly (step **1**), extension of the spindle microtubules inserts the daughter SPB into the bud, causing Tem1 to be activated by an unknown mechanism. Tem1-GTP activates a protein kinase cascade, which then promotes the release of active Cdc14 from the nucleolus and exit from mitosis. If the spindle apparatus fails to place the daughter SPB in the bud (step **2**), Kin4 (cyan), an inhibitor of Tem1, is recruited from the mother cell cortex to the mother-cell-located SPB and maintains Tem1 in the GDP bound form and mitotic exit does not occur. Lte1 (orange) is an inhibitor of Kin4 and localized to the bud. Lte1 prevents Kin4 that leaks into the bud from inhibiting Tem1. If the checkpoint fails (step **3**), cells with mispositioned spindles inappropriately exit from mitosis and produce anucleate and multi-nucleate cells.

Recall that in budding yeast, a pool of mitotic cyclins are spared from degradation by APC/C$^{Cdc20}$ to facilitate the sometimes difficult process of aligning the mitotic spindle in such a way that half the nucleus squeezes through the tiny bud neck during anaphase spindle elongation. Recall further that inactivation of this protected pool of mitotic cyclin-CDK complexes is triggered by the protein phosphatase Cdc14, which in turn is activated by a signal transduction pathway known as the *mitotic exit network* (see Figure 19-28). The mitotic exit network is controlled by a small (monomeric) GTPase called *Tem1*. This member of the **GTPase superfamily** of switch proteins controls the activity of a protein kinase cascade similarly to the way Ras controls MAP kinase pathways (see Chapter 16). Tem1 associates with spindle pole bodies (SPBs) as soon as they form. An inhibitor of the GTPase called Kin4 localizes to the mother cell but is absent from the bud (see Figure 19-36). An inhibitor of Kin4 called Lte1 localizes to the bud but is absent from the mother cell and inhibits any residual Kin4 that leaks into the bud. When spindle microtubule elongation at the end of anaphase has correctly positioned segregating daughter chromosomes into the bud, Tem1 inhibition by Kin4 is relieved. As a consequence, Tem1 is converted into its active GTP-bound state, activating the protein kinase signaling cascade. The terminal kinase in the cascade then phosphorylates the nucleolar anchor that binds and inhibits Cdc14, releasing the **Cdc14 phosphatase** into the cytoplasm and nucleoplasm in both the bud and mother cell (see Figure 19-36, step **1**). Once active Cdc14 is available, mitotic CDKs are inactivated and cells exit from mitosis. When the spindle fails to position correctly, the Tem1-bearing spindle pole body fails to enter the bud and the mitotic exit network cannot be activated. Cells arrest in anaphase. Thus spatial restriction of inhibitors and activators of a signal transduction pathway allows the cell to sense a spatial cue, spindle position, and translate it into regulation of a signal transduction pathway.

## KEY CONCEPTS of Section 19.7

### Surveillance Mechanisms in Cell Cycle Regulation

• Surveillance mechanisms known as checkpoint pathways establish dependencies among cell cycle events and ensure that cell cycle progression does not occur prior to the completion of a preceding event.

• Checkpoint pathways consist of sensors that monitor a particular cellular event or defect therein, a signaling pathway, and an effector that halts cell cycle progression and activates repair pathways when necessary.

• Growth and cell division are integrated during G$_1$ in most systems. Reduced macromolecule biosynthesis delays cell cycle entry.

• Cells are able to detect and respond to a wide variety of DNA damage, and the response differs depending on the cell cycle stage that cells are in.

• In response to DNA damage two related protein kinases ATM and ATR are recruited to the site of damage, where they activate signaling pathways that lead to cell cycle arrest, repair, and, under some circumstances, apoptosis.

• The spindle assembly checkpoint pathway, which prevents premature initiation of anaphase, utilizes Mad2 and other proteins to regulate the APC/C specificity factor Cdc20, which targets securin and mitotic cyclins for ubiquitinylation.

• The spindle position checkpoint pathway prevents mitotic CDK inactivation when the spindle is mispositioned. In this pathway localized activators and inhibitors and a sensor that shuttles between them allows the cells to sense spindle position.

# 19.8 Meiosis: A Special Type of Cell Division

In nearly all diploid eukaryotes, **meiosis** generates haploid germ cells (eggs and sperm), which can then fuse with a germ cell from another individual to generate a diploid zygote that develops into a new individual. Meiosis is a fundamental aspect of the biology and evolution of all eukaryotes because it results in the reassortment of the chromosome sets received from an individual's two parents. Both chromosome reassortment and **homologous recombination** between parental DNA molecules during meiosis guarantees that each haploid germ cell generated will receive a unique combination of gene alleles that is distinct from each parent as well as from every other haploid germ cell formed.

The mechanisms of meiosis are analogous to those of mitosis. However, several key differences in meiosis allow this process to generate haploid cells with genetic diversity (see Figure 5-3). In mitosis, each S phase is followed by chromosome segregation and cell division. In contrast, during meiosis, one round of DNA replication is followed by *two consecutive chromosome segregation phases*. This leads to the formation of haploid rather than diploid daughter cells. During the two divisions, maternal and paternal chromosomes are shuffled and divided so that the daughter cells are different in genetic makeup from the parent cell. In this section, we will discuss the similarities between mitosis and meiosis as well as the meiosis-specific mechanisms that transform the canonical mitotic cell cycle machinery so that it brings about the unusual cell division that leads to the formation of haploid daughter cells.

## Extracellular and Intracellular Cues Regulate Entry into Meiosis

The signals triggering entry into the meiotic divisions in metazoans are a very active area of research and much is still unknown. However, the same basic principles govern the decision to enter the meiotic program in all organisms where this has been studied. Extracellular signals induce a transcriptional program that produces meiosis-specific cell cycle factors that bring about the unusual meiotic cell divisions. This modification of the cell cycle goes hand in hand with a developmental program that induces the features characteristic of gametes, such as the development of a flagellum in sperm or the production of a stress-resistant cell wall during spore formation in fungi. At least one of the extracellular signals inducing entry into the meiotic divisions in mammals is retinoic acid, a signaling molecule that by binding to the transcription factor retionoic acid receptor (RAR) functions in many different developmental processes. The cellular targets of this hormone and how it functions to specify the meiotic fate remain to be discovered.

The molecular mechanisms underlying the decision to enter the meiotic divisions are well understood in *S. cerevisiae*. The decision to enter the meiotic divisions is made in $G_1$. Depletion of nitrogen and carbon sources induces diploid cells to undergo meiosis instead of mitosis, yielding haploid spores (see Figure 1-17). During the meiotic divisions, budding is repressed and pre-meiotic S phase and the two meiotic divisions occur within the confines of the mother cell. Spore walls are then produced around the four meiotic products. Recall that budding and the initiation of DNA replication are induced by $G_1$/S phase CDKs. Their expression needs to be inhibited to prevent budding. Nutritional starvation represses expression of $G_1$/S phase cyclins, inhibiting budding. However, DNA replication also relies on $G_1$/S phase CDKs. How can pre-meiotic DNA replication occur in the absence of $G_1$/S phase CDKs? The sporulation-specific protein kinase Ime2 takes over the role of $G_1$/S phase CDKs

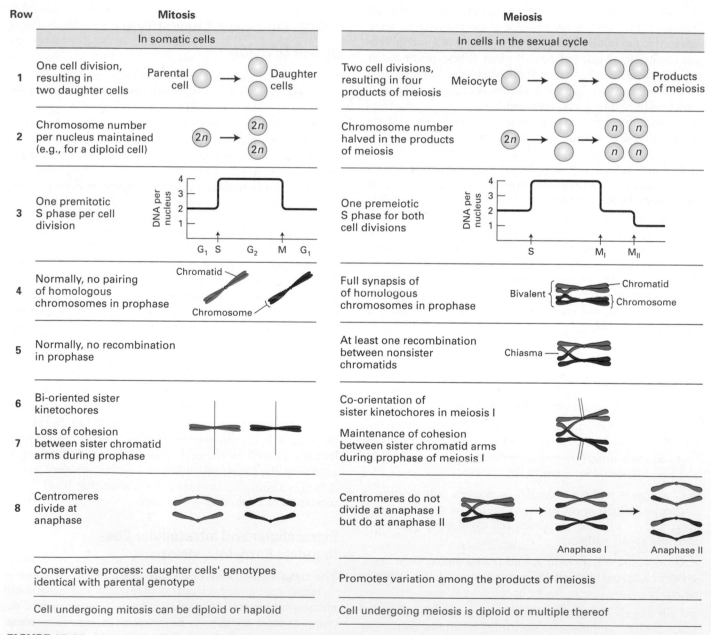

**FIGURE 19-37 Comparison of the main features of mitosis and meiosis.** [Adapted from A. J. F. Griffiths et al., 1999, *Modern Genetic Analysis*, W. H. Freeman and Company.]

in promoting DNA replication. Ime2 promotes (1) phosphorylation of the APC/C specificity factor Cdh1, inactivating it so that S phase and M phase cyclins can accumulate; (2) phosphorylation of transcription factors to induce genes required for S phase, including DNA polymerases and S phase cyclins; and (3) phosphorylation of the S phase CDK inhibitor Sic1, leading to release of active S phase CDKs and the onset of pre-meiotic DNA replication.

## Several Key Features Distinguish Meiosis from Mitosis

The meiotic divisions differ in several fundamental aspects from the mitotic divisions. These are summarized in Figure 19-37. During meiosis, a single round of DNA replication is followed by two cycles of cell division, termed *meiosis I* and *meiosis II* (Figure 19-38). Meiosis II resembles mitosis in that sister chromatids are segregated. However, meiosis I is very different. During this division, homologous chromosomes—the chromosome inherited from your mother and the same chromosome inherited from your father—are segregated. This unusual chromosome segregation requires three meiosis-specific modifications to the chromosome segregation machinery. In what follows we will discuss these and explain why they are needed.

The tension-based sensing mechanism responsible for accurately attaching chromosomes to the spindle during mitosis is also responsible for segregating chromosomes during meiosis I. Thus homologous chromosomes must be linked so that the tension-based mechanism of accurate attachment to the spindle can function. **Homologous recombination** between homologous chromosomes creates these linkages (see Figure 19-38). The molecular mechanisms of homologous recombination are discussed in detail in Chapter 4. Here, we will restrict our discussion to the importance of homologous recombination for the meiotic divisions to occur successfully.

In $G_2$ and prophase of meiosis I, the two replicated chromatids of each chromosome are associated with each other by cohesin complexes along the full length of the chromosome arms, just as they are following DNA replication in a mitotic cell cycle (see Figure 19-38). In prophase of meiosis I, homologous chromosomes (i.e., the maternal and paternal chromosome 1, the maternal and paternal chromosome 2, etc.) pair with each other and undergo homologous recombination. Significantly, at least one recombination event occurs between a maternal and a paternal chromosome. The **crossing over** of chromatids produced by recombination can be observed microscopically in the first meiotic prophase and metaphase as structures called *chiasmata* (singular, *chiasma*). In contrast, no pairing between homologous chromosomes occurs during mitosis, and recombination between nonsister chromatids is rare. Concomitant with homologous recombination, homologous chromosomes associate with each other in a process known as *synapsis*. In most organisms this synapsis is mediated by a proteinaceous complex known as the **synaptonemal complex (SC)**. Homologous chromosomes linked through chiasmata are called bivalents

(see Figure 19-38). The chiasmata now provide the resistance to the pulling force exerted by microtubules on the metaphase I spindle (Figure 19-39).

The recombination between nonsister chromatids that occurs in prophase of meiosis I has at least two functional consequences: First, it holds homologous chromosomes together during meiosis I metaphase. Second, it contributes to genetic diversity among individuals of a species by ensuring new combinations of gene alleles in different individuals. (Note: genetic diversity primarily arises from the independent reassortment of maternal and paternal homologs during the meiotic divisions). The homologs connected through at least one chiasma, which formed during prophase of meiosis I, must now align on the meiosis I spindle so that maternal and paternal chromosomes are segregated away from each other during anaphase of meiosis I. This requires that the kinetochores of sister chromatids attach to spindle fibers emanating from the *same* spindle pole rather than from opposite spindle poles as in mitosis (see Figure 19-39). Sister chromatids are said to be **co-oriented**. However, the kinetochores of the maternal and paternal chromosomes of each bivalent attach to spindle microtubules from opposite spindle poles; they are **bi-oriented**.

Finally, to facilitate two consecutive chromosome segregation phases, cohesins have to be lost from chromosomes in a stepwise manner. Recall that during mitosis, all cohesins are lost by the onset of anaphase (Figure 19-40a). In contrast, during meiosis, cohesins are lost from chromosome arms by the end of meiosis I, but a pool of cohesins around kinetochores is protected from removal (Figure 19-40b). This pool of cohesins persists throughout meiosis I but is removed at the onset of anaphase II. As we will see next, loss of cohesins from chromosome arms is required for homologous chromosomes to segregate away from each other during meiosis I.

The mechanisms that remove cohesins during meiosis are the same as during mitosis. Securin degradation releases separase, which then cleaves the cohesins holding the chromosome arms together. This allows the recombined maternal and paternal chromosomes to separate, but each pair of chromatids remains associated at the centromere. During metaphase II, sister chromatids align on the metaphase II spindle and separase is activated yet again, cleaving the residual cohesin around centromeres, facilitating anaphase II (see Figure 19-40b).

## Recombination and a Meiosis-Specific Cohesin Subunit Are Necessary for the Specialized Chromosome Segregation in Meiosis I

As discussed earlier, in metaphase of meiosis I, both sister chromatids in one (replicated) chromosome associate with microtubules emanating from the *same* spindle pole rather than from opposite poles as they do in mitosis (see Figure 19-39). Two physical links between homologous chromosomes resist the pulling force of the spindle until anaphase: (a) crossing over between chromatids, one from each pair of homologous

**FIGURE 19-38 Meiosis.** Pre-meiotic cells have two copies of each chromosome (2n), one derived from the paternal parent and one from the maternal parent. For simplicity, the paternal and maternal homologs of only one chromosome are diagrammed. Step **1**: All chromosomes are replicated during S phase before the first meiotic division, giving a 4n chromosomal complement. Cohesin complexes (not shown) link the sister chromatids composing each replicated chromosome along their full lengths. Step **2**: As chromosomes condense during the first meiotic prophase, replicated homologs pair and undergo homologous recombination, leading to at least one crossover event. At metaphase, shown here, both chromatids of one chromosome associate with microtubules emanating from one spindle pole, but each member of a homologous chromosome pair associates with microtubules emanating from opposite poles. Step **3**: During anaphase of meiosis I, the homologous chromosomes, each consisting of two chromatids, are pulled to opposite spindle poles. Step **4**: Cytokinesis yields the two daughter cells (now 2n), which enter meiosis II without undergoing DNA replication. At metaphase of meiosis II, shown here, the sister chromatids associate with spindle microtubules from opposite spindle poles, as they do in mitosis. Steps **5** and **6**: Segregation of sister chromatids to opposite spindle poles during the second meiotic anaphase followed by cytokinesis generates haploid gametes (1n) containing one copy of each chromosome. Micrographs on the left show meiotic metaphase I and metaphase II in developing gametes from *Lilium* (lily) ovules. Chromosomes are aligned at the metaphase plate. [Photos courtesy of Ed Reschke/Peter Arnold, Inc.]

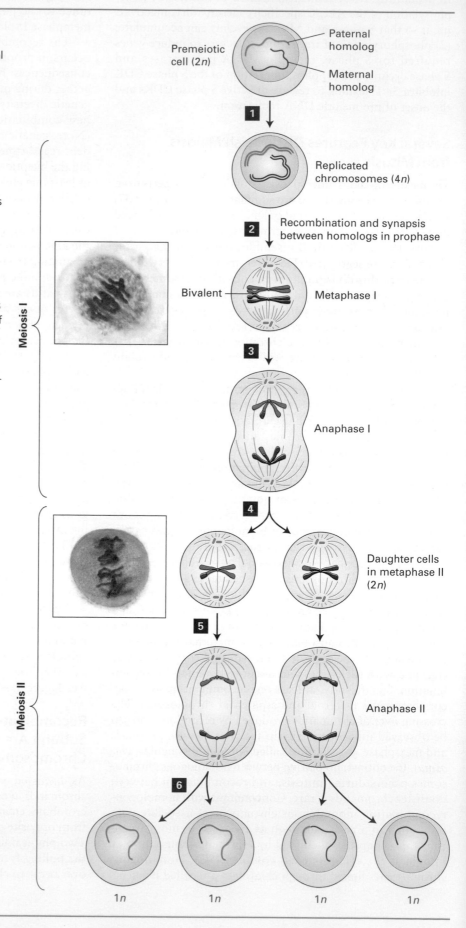

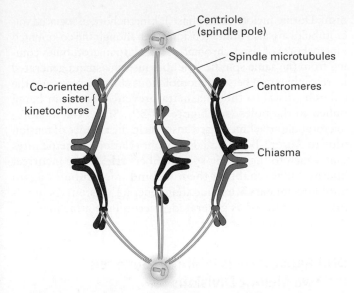

**FIGURE 19-39 Chiasmata and cohesins distal to them link homologous chromosomes in meiosis I metaphase.** Connections between chromosomes during meiosis I are most easily visualized in organisms with acrocentric centromeres, such as the grasshopper. The kinetochores at the centromeres of sister chromatids attach to spindle microtubules emanating from the same spindle pole, with the kinetochores of the maternal (red) and paternal (blue) chromosomes attaching to spindle microtubules from opposite spindle poles. The maternal and paternal chromosomes are attached at the chiasmata which are formed by recombination between them and the cohesion between sister chromatid arms that persists throughout meiosis I metaphase. Note that elimination of cohesion between sister chromatid arms is all that is required for the homologous chromosomes to separate at anaphase. [Adapted from L. V. Paliulis and R. B. Nicklas, 2000, *J. Cell Biol.* **150:**1223.]

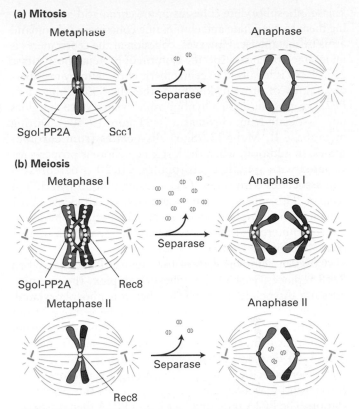

**FIGURE 19-40 Cohesin function during mitosis and meiosis.** (a) During mitosis, sister chromatids generated by DNA replication in S phase are initially linked by cohesin complexes along the length of the chromatids. During chromosome condensation, cohesin complexes (yellow) become restricted to the region of the centromere at metaphase. Mei-S332/Sgo1 (purple) recruits PP2A to centromeres, where they antagonize Polo kinase and Aurora B, preventing the dissociation of cohesins from centromeric regions. Dissociation of Mei-S332/Sgo1 from centromeres and activation of separase leads to removal of cohesins at the centromere. Sister chromatids now separate, marking the onset of anaphase. (b) In prophase of meiosis I, maternal and paternal chromatids establish linkages between each other by homologous recombination. By metaphase I, the chromatids of each replicated chromosome are cross-linked by cohesin complexes along their full length. Rec8, a meiosis-specific homolog of Scc1, is cleaved along chromosome arms but not around the centromere, allowing homologous chromosome pairs to segregate to daughter cells. Centromeric Rec8 is protected from cleavage by PP2A, recruited to centromeric regions by the PP2A regulator Mei-S332/Sgo1 (shown in purple). By metaphase II, the Mei-S332/Sgo1-PP2A complex dissociates from chromosomes. Cohesins can now be cleaved during meiosis II, allowing sister chromatids to segregate. [Modified from F. Uhlmann, 2001, *Curr. Opin. Cell Biol.* **13:**754.]

chromosomes, and (b) cohesins distal to the crossover point (see Figure 19-40b, *top*). Evidence for the function of recombination during meiosis comes from the observation that when recombination is blocked by mutations in proteins essential for the process, chromosomes segregate randomly during meiosis I; that is, homologous chromosomes do not necessarily segregate to opposite spindle poles.

At the onset of meiotic anaphase I, cohesins between chromosome arms are cleaved by separase. This cleavage is required for homologous chromosomes to segregate. If cohesins were not lost from chromosome arms, the recombined sister chromatid would rip during anaphase I. The maintenance of centromeric cohesion during meiosis I is necessary for the proper segregation of sister chromatids during meiosis II.

Studies in many organisms have shown that a specialized cohesin subunit, *Rec8*, is necessary for the stepwise loss of cohesins from chromosomes during meiosis. Expressed only during meiosis, Rec8 is homologous to Scc1, the cohesin subunit that closes the cohesin ring in the cohesin complex of mitotic cells. Immunolocalization experiments revealed that during early anaphase of meiosis I, Rec8 is lost from chromosome arms but is retained at centromeres. However, during early anaphase of meiosis II, centromeric Rec8 is cleaved by separase, so the sister chromatids can segregate, as they do in mitosis (see Figure 19-40b, *bottom*). Consequently, understanding the regulation of Rec8-cohesin complex cleavage is central to understanding chromosome segregation in meiosis I.

The mechanism that protects Rec8 from cleavage at centromeres during meiosis I is similar to the mechanism that protects Scc1 at centromeres during mitosis. Recall that during mitotic prophase, protein kinases, chief among them Polo kinase, phosphorylate cohesins in the chromatid arms, causing them to dissociate and eliminating cohesion on chromatid arms by metaphase. However, cohesion at the centromeres is maintained because a specific isoform of protein phosphatase 2A (PP2A) is localized to centromeric chromatin by members of a family of proteins known as the Mei-S332/Shugoshin. PP2A keeps cohesin in a hypophosphorylated state that does not dissociate from chromatin (see Figure 19-40a). During metaphase II, Mei-S332/Sgo1 dissociates from chromosomes. In addition, when the last kinetochore is properly associated with spindle microtubules, Cdc20 is derepressed and associates with the APC/C, causing ubiquitinylation of securin. This releases separase activity, which cleaves Scc1 whether it is phosphorylated or not, eliminating cohesion at the centromere and allowing chromatid separation in anaphase (see Figure 19-40a).

Cohesin removal differs for meiosis I because when Rec8 replaces Scc1 in the cohesin complex, the complex does not dissociate in prophase when it is phosphorylated. The meiotic cohesin complex can only be removed from chromatin via the action of separase. Rec8 also differs from Scc1 in that it must be phosphorylated by several protein kinases to be cleaved by separase. During meiosis I, the centromere-specific isoform of PP2A targeted to centromeric chromatin by Mei-S332/Shugoshin prevents this phosphorylation. The PP2A targeting factor and PP2A then dissociate from chromosomes by metaphase II, allowing separase cleavage of Rec8.

## Co-orienting Sister Kinetochores Is Critical for Meiosis I Chromosome Segregation

As discussed earlier, in mitosis and meiosis II, sister kinetochores attach to spindle microtubules emanating from *opposite* spindle poles; the kinetochores are said to be *bi-oriented*. This is essential for segregation of sister chromatids to different daughter cells. In contrast, at meiosis I metaphase, sister kinetochores attach to spindle microtubules emanating from the *same* spindle pole; the sister kinetochores are said to be *co-oriented* (see Figure 19-39). Obviously, attachment of sister kinetochores to the proper microtubules in meiosis I and II is critical for correct meiotic segregation of chromosomes.

Proteins required for meiosis I sister kinetochore co-orientation were first identified in *S. cerevisiae*. We now know that a protein complex known as the **monopolin complex** associates with kinetochores during meiosis I and links sister kinetochores to favor attachment to microtubules emanating from the same spindle pole. In organisms where kinetochores attach to multiple microtubules, Rec8-containing cohesins are essential for sister kinetochore co-orientation. These meiosis-specific cohesins impose a rigid kinetochore structure, restricting the movement of sister kinetochores and thereby favoring attachment to microtubules from the same spindle pole.

Like during mitosis and meiosis II, correct attachment of meiosis I chromosomes is mediated by a tension-based mechanism. During meiotic metaphase I, kinetochore-associated microtubules are also under tension (even though the co-oriented kinetochores of sister chromatids attach to microtubules coming from the same spindle pole) because chiasmata generated by recombination between homologous chromosomes and the cohesins distal to the chiasmata prevent them from being pulled to the poles (see Figure 19-39). Since kinetochore-microtubule attachments are unstable in the absence of tension (due to Aurora B–mediated phosphorylation), kinetochores that attach to the wrong spindle fibers release the incorrect microtubules, enabling them to bind microtubules again until attachments are made that generate tension. As in mitosis, once tension is generated, microtubule attachment to the kinetochores is stabilized.

## DNA Replication Is Inhibited Between the Two Meiotic Divisions

The mechanism by which DNA replication is suppressed between meiosis I and II is currently an active area of investigation, but it is thought that a change in the regulation of CDK activity is at least in part responsible for this suppression. The same S phase CDKs that promote DNA replication prior to mitosis are needed for pre-meiotic DNA replication. The same mitotic CDKs that promote mitosis also promote the meiotic divisions, except we now call mitotic CDKs meiotic CDKs since they promote the meiotic divisions and not mitosis.

So how is DNA replication prevented between the two meiotic divisions? Following anaphase of meiosis I, meiotic CDK activity does not fall as low as it does following mitotic anaphase. This partial drop in CDK activity is thought to be sufficient to promote the disassembly of the meiosis I spindle but insufficient to promote MCM helicase loading (recall a state of very low or no CDK activity is needed to load MCM helicases). During prophase of meiosis II, meiotic CDK activity rises again and the meiosis II spindle forms. After all sister kinetochores have attached to microtubules from opposite spindle poles, separase is activated and cells proceed through meiosis II anaphase, telophase, and cytokinesis to generate haploid germ cells.

## KEY CONCEPTS of Section 19.8

### Meiosis: A Special Type of Cell Division

• Meiosis is a specialized division in which meiosis-specific gene products modulate the mitotic cell division program (see Figure 19-38).

• The meiotic division comprises one cycle of chromosome replication followed by two cycles of cell division to produce haploid germ cells from a diploid pre-meiotic cell. During meiosis I, homologous chromosomes are segregated; during meiosis II sister chromatids separate.

• Specialized environmental conditions induce a developmental program that leads to the meiotic divisions.

- During prophase of meiosis I, homologous chromosomes undergo recombination. At least one recombination event occurs between chromatids of homologous chromosomes, linking the homologous chromosomes.

- Cohesins distal to the chiasma are responsible for holding the homologous chromosomes together during prophase and metaphase of meiosis I.

- At the onset of anaphase of meiosis I, cohesins on chromosome arms are phosphorylated and as a result cleaved by separase, but cohesins in the region of the centromere are protected from phosphorylation and cleavage. This protection is brought about by a meiosis-specific cohesin subunit and a protein phosphatase that associates with centromeres. As a result, the chromatids of homologous chromosomes remain associated during segregation in meiosis I.

- Cleavage of centromeric cohesins during anaphase of meiosis II allows individual chromatids to segregate into germ cells.

- A complex of meiosis-specific kinetochore proteins, known as the monopolin complex, promote the co-orientation of sister chromatids during meiosis I. Both sister chromatids attach to microtubules emanating from the same spindle pole.

- Incomplete CDK inactivation between the two meiotic divisions inhibits DNA replication.

the spindle assembly checkpoint. Many questions remain about how the plane of cytokinesis and the localization of daughter chromosomes are determined in cells that divide symmetrically and asymmetrically as is often seen as part of development of complex tissues and organ structures. How the cell cycle machinery is modulated by developmental cues to bring about specialized divisions is also an intense area of investigation.

Understanding these detailed aspects of cell cycle control will have significant consequences, particularly for the treatment of cancers. Cancer cells often have defects in cell cycle checkpoints that led to the accumulation of multiple mutations and DNA rearrangements that result in the cancer phenotype. However, the absence of these checkpoints can make specific types of cancers particularly vulnerable to extensive DNA damage induced with radiation therapy or chemotherapy. Normal cells activate cell cycle checkpoints that arrest the cell cycle until the DNA damage is repaired. But these cancer cells do not and as a consequence suffer sufficient genetic damage to induce apoptosis. If more were understood about cell cycle controls and checkpoint pathways, it might be possible to design ever more effective therapeutic strategies, especially against types of cancer that are largely resistant to today's conventional therapies. It seems very likely that better understanding of the molecular processes involved will allow the design of more effective treatments in the future.

## Perspectives for the Future

The remarkable pace of cell cycle research over the last 25 years has led to a detailed model of eukaryotic cell cycle control. A beautiful logic underlies these molecular controls. Each regulatory event has two important functions: to activate a step of the cell cycle and to prepare the cell for the next event of the cycle. This strategy ensures that the phases of the cycle occur in the proper order.

Although the general logic of cell cycle regulation now seems well established, many critical details remain to be discovered. For instance, how cell growth and division are coordinated and how the metabolic state of a cell feeds into its cell cycle machinery remain to be discovered. A number of key nutrient and growth-factor-sensing pathways such as the AMPK, Ras, and TOR pathways have been identified and their inner workings recently revealed. Understanding how they impact the cell cycle machinery will be a critical question to be addressed in the years to come. Substantial progress has been made recently in identifying substrates phosphorylated by different CDKs, but much work remains to be done to understand how the modification of these proteins leads to the multiple events that these CDKs trigger.

Much has been discovered recently about the operation of cell cycle checkpoint pathways, but the mechanisms that activate ATM and ATR in the DNA damage checkpoint are poorly understood. Likewise, much remains to be learned about the control and mechanism of regulation of Mad2 in

## Key Terms

anaphase-promoting complex/cyclosome (APC/C) 888

APC/C specificity factor 888

ATM/ATR 909

Aurora B kinase 901

Cdc14 phosphatase 913

Cdc25 phosphatase 898

CDK-activating kinases (CAK) 888

CDK inhibitor (CKIs) 888

checkpoint pathway 874

cohesin 896

condensin 902

critical cell size 892

crossing over (or cross over) 915

cyclin 874

cyclin-dependent kinase (CDK) 874

E2F transcription factor 887

$G_1$ CDKs 883

maturation-promoting factor (MPF) 880

meiosis 913

mitogen 892

mitosis 875

mitotic CDKs 883

monopolin complex 918

p53 protein 909

Polo kinases 898

Rb protein 891

SCF (Skp1, Cullin, F-box proteins) 888

S phase CDKs 883

START 875

securin 903

sensor 906

separase 903

sister chromatids 875

synaptonemal complex 915

Wee1 protein-tyrosine kinase 898

## Review the Concepts

**1.** What cellular mechanism(s) ensure that passage through the cell cycle is unidirectional and irreversible? What molecular machinery underlies these mechanism(s)?

**2.** What types of experimental strategies do researchers employ to study cell cycle progression? How can these strategies differ based on genetic versus biochemical approaches?

**3.** Tim Hunt shared the 2001 Nobel Prize in Physiology or Medicine for his work in the discovery and characterization of cyclin proteins in eggs and embryos. Describe the experimental steps that led him to his discovery of cyclins.

**4.** What experimental evidence indicates that cyclin B is required for a cell to enter mitosis? What evidence indicates that cyclin B must be destroyed for a cell to exit mitosis?

**5.** What physiological differences between *S. pombe* and *S. cerevisiae* make them useful yet complementary tools for studying the molecular mechanisms involved in cell cycle regulation and control?

**6.** In *Xenopus*, one of the substrates of mitotic CDKs is the Cdc25 phosphatase. When phosphorylated by mitotic CDKs, Cdc25 is activated. What is the substrate of Cdc25? How does this information help to explain the rapid rise in mitotic CDK activity as cells enter mitosis?

**7.** Explain how CDK activity is modulated by the following proteins: (a) cyclin, (b) CAK, (c) Wee1, (d) p21.

**8.** Explain the role of CDK inhibitors. If cyclin-CDK complexes are required to allow regulated progression through the eukaryotic cell cycle, what would be the physiological rationale for CDK inhibitors?

**9.** What is the functional definition of START? Cancer cells typically lose START control. Explain how the following mutations, which are found in some cancer cells, lead to a bypass of START controls: (a) overexpression of cyclin D, (b) loss of Rb function, (c) loss of p16 function, (d) hyperactive E2F.

**10.** The Rb protein has been called the "master brake" of the cell cycle. Describe how the Rb protein acts as a cell cycle brake. How is the brake released in mid- to late $G_1$ to allow the cell to proceed to the S phase?

**11.** A common feature of cell cycle regulation is that the events of one phase ensure progression into a subsequent phase. In *S. cerevisiae*, $G_1$ and $G_1$/S phase CDKs promote S-phase entry. Name two ways in which they promote the activation of S phase.

**12.** For S phase to be completed in a timely manner, DNA replication initiates from multiple origins in eukaryotes. In *S. cerevisiae*, what role do S-phase CDKs and DDK complexes play to ensure that the entire genome is replicated once and only once per cell cycle?

**13.** In 2001, the Nobel Prize in Physiology or Medicine was awarded to three cell-cycle scientists. Paul Nurse was recognized for his studies with the fission yeast *S. pombe*, in particular for the discovery and characterization of the *wee1*⁺ gene. What did the characterization of the *wee1*⁺ gene tell us about cell cycle control?

**14.** Describe how cells know whether sister kinetochores are properly attached to the mitotic spindle.

**15.** Describe the series of events by which the APC/C promotes the separation of sister chromatids at anaphase.

**16.** Meiosis and mitosis are overall analogous processes involving many of the same proteins. However, some proteins function uniquely in each of these cell-division events. Explain the meiosis-specific function of the following: (a) Ime2, (b) Rec8, (c) monopolin.

**17.** Leland Hartwell, the third recipient of the 2001 Nobel Prize in Physiology or Medicine, was acknowledged for his characterization of cell cycle checkpoint pathways in the budding yeast *S. cerevisiae*. What is a cell cycle checkpoint pathway? When during the cell cycle do checkpoint pathways function? How do cell cycle checkpoint pathways help to preserve the genome?

**18.** What role do tumor suppressors, including p53, play in mediating cell cycle arrest for cells with DNA damage?

**19.** Individuals with the hereditary disorder ataxia telangiectasia suffer from neurodegeneration, immunodeficiency, and increased incidence of cancer. The genetic basis for ataxia telangiectasia is a loss-of-function mutation in the ATM gene (*ATM 5*; *a*taxia *t*elangiectasia *m*utated). Besides p53, what other substrate is phosphorylated by ATM? How does the phosphorylation of this substrate lead to inactivation of CDKs to enforce cell cycle arrest?

## Analyze the Data

**1.** Many of the proteins that regulate transit through the cell cycle have been characterized. Xnf7, identified in extracts of *Xenopus* eggs, binds to the anaphase-promoting complex/cyclosome (APC/C). To elucidate the function of this protein, studies have been undertaken in which Xnf7 either has been depleted from extracts using an antibody raised against it or has been augmented in the extracts through addition of extra Xnf7. The consequences on transit through mitosis were then assessed (see J. B. Casaletto et al., 2005, *J. Cell Biol.* **169**:61–71).

    **a.** *Xenopus* egg extracts, arrested in metaphase, were either depleted of Xnf7 or were mock depleted (subjected to the same treatment as the first sample but without Xnf7 antibody), then released from metaphase arrest by addition of $Ca^{2+}$. Aliquots of the extract were then sampled at various times after $Ca^{2+}$ addition and the amounts of mitotic cyclin determined, as shown on the Western blot below. What information do these data provide about a possible function for Xnf7?

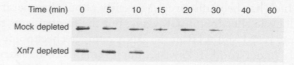

**b.** In additional studies, exogenous Xnf7 was added to *Xenopus* egg extracts, arrested at metaphase, so that the total amount of this protein in the extracts was higher than normal. The extracts, released from arrest by $Ca^{2+}$ addition, were then assessed at various times after release for mitotic cyclin ubiquitinylation (cyclin-Ub conjugates). What is the rationale for examining ubiquitinylation? Using the following figure, determine what information these studies add beyond that obtained from part (a).

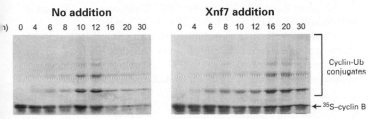

**No addition**　　**Xnf7 addition**

0　4　6　8　10　12　16　20　30　　　0　4　6　8　10　12　16　20　30

Cyclin-Ub conjugates

← $^{35}S$–cyclin B

**c.** The spindle checkpoint pathway prevents cells with unattached kinetochores from proceeding into anaphase. Thus cells in which this checkpoint has been activated do not enter anaphase and do not degrade mitotic cyclin. Nocodazole, a drug that prevents microtubule assembly, can be used to activate the spindle checkpoint. Cells in nocodazole become arrested in early mitosis because they cannot form a spindle, and so all kinetochores remain unattached. To determine if Xnf7 is required for a functional spindle checkpoint, *Xenopus* egg extracts, arrested in metaphase, were subjected to various protocols (see the following figure): untreated (no nocodazole) or treated with nocodazole and either mock depleted (preimmune) or immunodepleted of Xnf7 (α-Xnf7). The extracts were then treated with $Ca^{2+}$ to overcome arrest, and aliquots of the extracts were assessed at various times for mitotic cyclin, as shown on the Western blot below. What can you conclude about Xnf7 from these data?

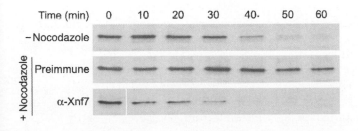

Time (min)　0　10　20　30　40　50　60

– Nocodazole

+ Nocodazole { Preimmune

α-Xnf7

**2.** In this chapter, we have discussed that cyclins are a required component of cyclin-CDK complexes for regulated progression through the eukaryotic cell cycle. Most cyclins are progressively synthesized and then degraded systematically in a temporal fashion at various points in the cell cycle. As we discussed in Chapter 7, cellular protein expression can be regulated at several different points starting with initiation of gene transcription.

**a.** What type of assay(s) could be employed to determine whether the expression of cyclin B is regulated at the transcriptional or translational level, if either?

**b.** Based on what you've learned in Chapter 19, is it possible that the activity of cyclin B could be regulated at the post-translational level? Describe a cellular mechanism through which this might happen.

**c.** How might it be possible for cyclin B expression and/or activity to be at least partially regulated by events in the cell's external environment?

## References

### Overview of the Cell Cycle and Its Control

Morgan, D. O. 2006. *The Cell Cycle: Principles of Control.* New Science Press.

### Regulation of CDK Activity

Doree, M., and T. Hunt. 2002. From Cdc2 to Cdk1: when did the cell cycle kinase join its cyclin partner? *J. Cell Sci.* **115**:2461–2464.

Masui, Y. 2001. From oocyte maturation to the in vitro cell cycle: the history of discoveries of Maturation-Promoting Factor (MPF) and Cytostatic Factor (CSF). *Differentiation* **69**:1–17.

Nurse, P. 2002. Cyclin dependent kinases and cell cycle control (Nobel lecture). *Chembiochem.* **3**:596–603.

### Commitment to the Cell Cycle and DNA Replication

Blow, J. J., and A. Dutta. 2005. Preventing re-replication of chromosomal DNA. *Nat. Rev. Mol. Cell Biol.* **6**(6):476–486.

Cardozo, T., and Pagano, M. 2004. The SCF ubiquitin ligase: insights into a molecular machine. *Nat. Rev. Mol. Cell Biol.* **9**:739–751.

Chen, H. Z., S. Y. Tsai, and G. Leone. 2009. Emerging roles of E2Fs in cancer: an exit from cell cycle control. *Nat. Rev. Cancer* **9**(11):785–797.

Hirano, T. 2006. At the heart of the chromosome: SMC proteins in action. *Nat. Rev. Mol. Cell Biol.* **7**(5):311–322.

Nasmyth, K., and C. H. Haering. 2009. Cohesin: its roles and mechanisms. *Ann. Rev. Genet.* **43**:525–558.

Remus, D., and J. F. Diffley. 2009. Eukaryotic DNA replication control: lock and load, then fire. *Curr. Opin. Cell Biol.* **21**(6):771–777.

Sears, R. C., and J. R. Nevins. 2002. Signaling networks that link cell proliferation and cell fate. *J. Biol. Chem.* **277**:11617–11620.

Sherr, C. J., and J. M. Roberts. 2004. Living with or without cyclins and cyclin-dependent kinases. *Genes & Dev.* **18**(22):2699–2711.

### Entry into Mitosis

Barr, F. A. 2004. Golgi inheritance: shaken but not stirred. *J. Cell Biol.* **164**:955–958.

Ferrell, J. E., Jr., et al. 2009. Simple, realistic models of complex biological processes: positive feedback and bistability in a cell fate switch and a cell cycle oscillator. *FEBS Lett.* **583**(24):3999–4005.

Nigg, E. A. 2001. Mitotic kinases as regulators of cell division and its checkpoints. *Nature Rev. Mol. Cell Biol.* **2**:21–32.

Roux, K. J., and B. Burke. 2006. From pore to kinetochore and back: regulating envelope assembly. *Dev. Cell* **11**:276–278.

Santaguida, S., and A. Musacchio. 2009. The life and miracles of kinetochores. *EMBO J.* **28**(17):2511–2531.

## Completion of Mitosis: Chromosome Segregation and Exit from Mitosis

Pesin, J. A., and T. L. Orr-Weaver. 2008. Regulation of APC/C activators in mitosis and meiosis. *Ann. Rev. Cell Dev. Biol.* **24**:475–499.

Stegmeier, F., and A. Amon. 2004. Closing mitosis: the functions of the Cdc14 phosphatase and its regulation. *Ann. Rev. Genet.* **38**:203–232.

Uhlmann, F. 2003. Separase regulation during mitosis. *Biochem. Soc. Symp.* **70**:243–251.

Wirth, K. G., et al. 2006. Separase: a universal trigger for sister chromatid disjunction but not chromosome cycle progression. *J. Cell Biol.* **172**:847–860.

## Surveillance Mechanisms in Cell Cycle Regulation

Jorgensen, P., and M. Tyers. 2004. How cells coordinate growth and division. *Curr. Biol.* **14**(23):R1014–1027.

Bartek, J., and J. Lukas. 2007. DNA damage checkpoints: from initiation to recovery or adaptation. *Curr. Opin. Cell Biol.* **19**(2):238–245.

Burke, D. J. 2009. Interpreting spatial information and regulating mitosis in response to spindle orientation. *Genes Dev.* **23**(14):1613–1618.

Harrison, J. C., and J. E. Haber. 2006. Surviving the breakup: the DNA damage checkpoint. *Ann. Rev. Genet.* **40**:209–235.

Kastan, M. B., and J. Bartek. 2004. Cell-cycle checkpoints and cancer. *Nature* **432**:316–323.

Musacchio, A., and E. D. Salmon. 2007. The spindle-assembly checkpoint in space and time. *Nat. Rev. Mol. Cell Biol.* **8**(5):379–393.

## Meiosis: A Special Type of Cell Division

Ishiguro, K., and Y. Watanabe. 2007. Chromosome cohesion in mitosis and meiosis. *J. Cell Sci.* **120**(Pt. 3):367–369.

Marston, A. L., and A. Amon. 2004. Meiosis: cell-cycle controls shuffle and deal. *Nature Rev. Mol. Cell Biol.* **5**:983–997.

Zickler, D., and N. Kleckner. 1999. Meiotic chromosomes: integrating structure and function. *Ann. Rev. Genet.* **33**:603–754.

# Cell Biology Emerging from the Sea: The Discovery of Cyclins

T. Evans et al., 1983, *Cell* **33**:391

From the first cell divisions after fertilization to aberrant divisions that occur in cancers, biologists have long been interested in how cells control when they divide. The processes of cell division have been separated into stages known collectively as the *cell cycle*. While studying early development in marine invertebrates in the early 1980s, Joan Ruderman and Tim Hunt discovered the cyclins, key regulators of the cell cycle.

## Background

The question of how an organism develops from a fertilized egg continues to drive a large body of scientific research. Whereas such research was classically the concern of embryologists, the developing understanding of gene expression in the 1980s brought new approaches to answer this question. One such approach was to examine the pattern of gene expression in both the oocyte and the newly fertilized egg. Ruderman and Hunt were among the biologists who took this approach to the study of early development.

Biologists had well characterized the early development of a number of marine invertebrate systems. Their eggs are fertilized externally, allowing researchers to study their development in a plastic dish. During the early stages of development, the embryonic cells divide synchronously, which allows an entire population of cells to be studied at the same stage of the cell cycle. Researchers had established that a large part of the mRNA in the unfertilized oocyte is not translated. On fertilization, these maternal mRNA are rapidly translated. Previous studies had shown that when fertilized eggs are treated with drugs that inhibit protein synthesis, cell division could not take place. This suggested that the initial burst of protein synthesis from the maternal mRNA is required at the earliest stages of development. Ruderman and Hunt, while teaching a physiology course at the Marine Biological Laboratory in Woods Hole, Massachusetts, began a set of experiments designed to uncover the genes that were expressed at this point as well as the mechanism by which this burst of protein synthesis was controlled.

## The Experiment

In a collaborative project, Ruderman and Hunt looked at regulation of gene expression in the fertilized egg of the surf clam *Spisula solidissima*. Whereas it was known that overall protein synthesis rapidly increased on fertilization, they wanted to find out whether the proteins expressed in the earliest stage of development, the two-cell embryo, were different from those expressed in the unfertilized egg. When either eggs or two-cell clam embryos are treated with radioactively labeled amino acids, the cell takes up the amino acids, which are subsequently incorporated into newly synthesized proteins. Using this technique, Ruderman and Hunt monitored the pattern of protein synthesis by breaking open the cells, separating the proteins using SDS-polyacrylamide gel electrophoresis (SDS-PAGE), and then visualizing the radioactively labeled proteins by autoradiography. When they compared the pattern of protein synthesis in the egg with that in the two-cell embryo, they saw that three different proteins that were either not expressed or expressed at an extremely low levels in the egg were highly expressed in the embryo. In a subsequent study, Ruderman examined the pattern of protein expression in the oocytes of the starfish *Asterias forbesi* as they mature. She again observed the increased expression of three proteins of similar size to those that she and Hunt had seen in surf clam embryos.

Soon afterward, in a third study, Hunt examined the changes in protein expression during the maturation and fertilization of sea urchin oocytes. This time he performed the experiment in a slightly different manner. Rather than treating the oocytes and embryo with radioactively labeled amino acids for a set time period, he labeled the cells continuously for more than 2 hours, removing samples for analysis at 10-minute intervals. Now he could monitor the changes in protein expression throughout the early stages of development. As had been shown in other organisms, the pattern of protein synthesis was altered when the sea urchin oocyte was fertilized. Three proteins—represented by three prominent bands on an autoradiograph—were expressed in the embryos, but not in the oocytes. Interestingly, the intensity of one of these bands changed over time: the band was intense at the early time points, then barely visible after 85 minutes. It increased in intensity again between 95 and 105 minutes. The intensity of the band, representing the amount of the protein in the cell, appeared to be oscillating over time (Figure 1a). This suggested that the protein had been quickly degraded and then synthesized again.

Because the time frame of the experiment coincided with early embryonic cell divisions, Hunt next asked whether the synthesis and destruction of the protein was correlated with progression of the cell cycle. He examined a portion of cells from each time point under a microscope, counting the number of cells dividing at each time point where samples had been taken for protein analysis. Hunt then correlated the amount of the protein present in the cell with the proportion of cells dividing at each time point. He noticed that the level of expression of one of the proteins was highest before the cell divided and lowest on cell division

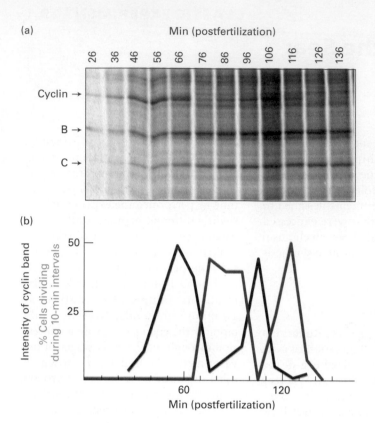

**(a)**

Min (postfertilization)

26 36 46 56 66 76 86 96 106 116 126 136

Cyclin →

B →

C →

**(b)**

**FIGURE 1** **Autoradiography permits the detection of cyclical synthesis and destruction of mitotic cyclin in sea urchin embryos.** A suspension of sea urchin eggs was synchronously fertilized by the addition of sea urchin sperm, and $^{35}$S-methionine was added. At 10-minute intervals beginning 26 minutes after fertilization, samples were taken for protein analysis on an SDS-polyacrylamide gel and for detection of cell cleavage by microscopy. (a) Autoradiogram of the SDS gel showing samples removed at each time point. Most proteins, such as B and C, continuously increased in intensity. In contrast, cyclin suddenly decreased in intensity at 76 minutes after fertilization and then began increasing again at 86 minutes. The cyclin band peaked again at 106 min and decreased again at 126 min. (b) Plot of the intensity of the cyclin band (red line) and the fraction of cells that had undergone cleavage during the previous 10-minute interval (cyan line). Note that the amount of cyclin fell precipitously just before cell cleavage. [From T. Evans et al., 1983, *Cell* **33**:389; courtesy of R. Timothy Hunt, Imperial Cancer Research Fund.]

(Figure 1b), suggesting a correlation with the stage of the cell cycle. When the same experiment was performed in the surf clam, Hunt saw that two of the proteins that he and Ruderman had described previously displayed the same pattern of synthesis and destruction. Hunt called these proteins *cyclins* to reflect their changing expression through the cell cycle.

## Discussion

The discovery of the cyclins heralded an explosion of investigation into the cell cycle. It is now known that these proteins regulate the cell cycle by associating with cyclin-dependent kinases, which in turn regulate the activities of a variety of transcription and replication factors, as well as other proteins involved in the complex alterations in cell architecture and chromosome structure that occur during mitosis. In brief, cyclin-CDK complexes direct and regulate progression through the cell cycle. As with so many key regulators of cellular functions, it was soon shown that the cyclins discovered in sea urchins and surf clams are conserved in eukaryotes from yeast to humans. Since the identification of the first cyclins, scientists have identified at least 15 other cyclins that regulate all phases of the cell cycle.

In addition to the basic research interest in these proteins, the cyclins' central role in cell division has made them a focal point in cancer research. Cyclins are involved in the regulation of several genes that are known to play prominent roles in tumor development. Scientists have shown that at least one cyclin, cyclin D1, is overexpressed in a number of tumors. The role of these proteins in both normal and aberrant cell division continues to be an active and exciting area of research today.

# Integrating Cells Into Tissues

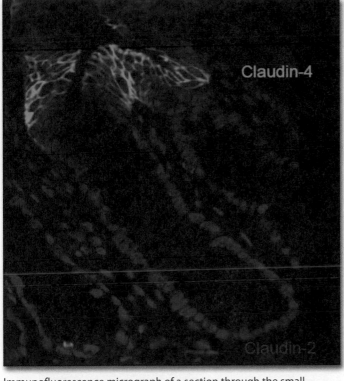

Immunofluorescence micrograph of a section through the small intestine of a mouse in which the adhesion proteins Claudin-2 and Claudin-4 are stained red and green, respectively, and the nuclei of the cells are stained blue. The claudins are a family of tight junctional adhesion proteins that also define the selective permeability of pores though which small molecules and ions may move between the cells of an epithelium. Claudin-2 is expressed in the crypts (lower parts) of the intestinal epithelium and is thought to be involved in the transport of cations such as calcium. Claudin-4 is expressed only in the upper villus/surface region and is thought to function as a barrier to cation transport. [Image from Christoph Rahner, Yale School of Medicine, and J. M. Anderson, University of North Carolina.]

In the development of complex multicellular organisms such as plants and animals, progenitor cells differentiate into distinct "types" that have characteristic compositions, structures, and functions. Cells of a given type often aggregate into a *tissue* to cooperatively perform a common function: muscle contracts; nervous tissue conducts electrical impulses; xylem tissue in plants transports water. Different tissues can be organized into an *organ*, again to perform one or more specific functions. For instance, the muscles, valves, and blood vessels of a heart work together to pump blood. The coordinated functioning of many types of cells and tissues permits the organism to move, metabolize, reproduce, and carry out other essential activities. Indeed, the complex and diverse morphologies of plants and animals are examples of the whole being greater than the sum of the individual parts, more technically described as the emergent properties of a complex system.

Vertebrates have hundreds of different cell types, including leukocytes (white blood cells) and erythrocytes (red blood

## OUTLINE

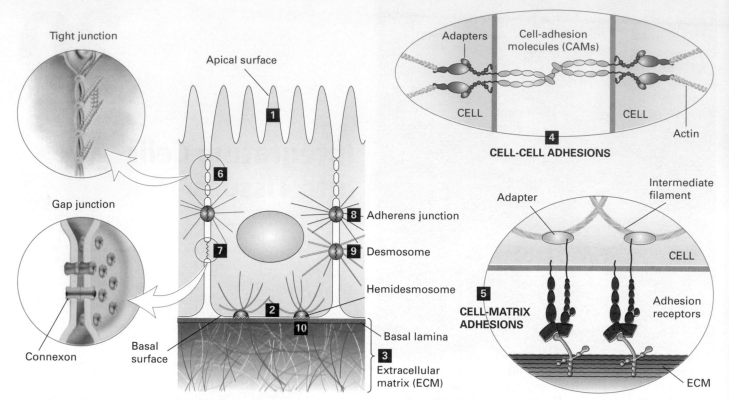

**FIGURE 20-1 Overview of major cell-cell and cell-matrix adhesive interactions.** Schematic cutaway drawing of a typical epithelial tissue, such as in the intestines. The apical (upper) surface of these cells is packed with fingerlike microvilli **1** that project into the intestinal lumen, and the basal (bottom) surface **2** rests on extracellular matrix (ECM). The ECM associated with epithelial cells is usually organized into various interconnected layers (e.g., the basal lamina, connecting fibers, connective tissue) in which large, interdigitating ECM macromolecules bind to one another and to the cells **3**. Cell-adhesion molecules (CAMs) bind to CAMs on other cells, mediating cell-cell adhesions **4**, and adhesion receptors bind to various components of the ECM, mediating cell-matrix adhesions **5**. Both types of cell-surface adhesion molecules are usually integral membrane proteins whose cytosolic domains often bind to multiple intracellular adapter proteins. These adapters, directly or indirectly, link the CAM to the cytoskeleton (actin or intermediate filaments) and to intracellular

signaling pathways. As a consequence, information can be transferred by CAMs and the macromolecules to which they bind from the cell exterior into the intracellular environment and vice versa. In some cases, a complex aggregate of CAMs, adapters, and associated proteins is assembled. Specific localized aggregates of CAMs or adhesion receptors form various types of cell junctions that play important roles in holding tissues together and facilitating communication between cells and their environment. Tight junctions **6**, lying just under the microvilli, prevent the diffusion of many substances through the extracellular spaces between the cells. Through connexon channels, gap junctions **7** allow the movement of small molecules and ions between the cytosols of adjacent cells. The remaining three types of junctions, adherens junctions **8**, spot desmosomes **9**, and hemidesmosomes **10**, link the cytoskeleton of a cell to other cells or to the ECM. [See V. Vasioukhin and E. Fuchs, 2001, *Curr. Opin. Cell Biol.* **13**:76.]

cells), photoreceptors in the retina, fat-storing adipocytes, fibroblasts in connective tissue, and hundreds of different subtypes of neurons in the human brain. Even simple animals exhibit complex tissue organization. The adult form of the roundworm *Caenorhabditis elegans* contains a mere 959 cells, yet these cells fall into 12 different general cell types and many distinct subtypes. Despite their diverse forms and functions, all animal cells can be classified as being components of just five main classes of tissue: *epithelial tissue, connective tissue, muscular tissue, nervous tissue,* and *blood.* Various cell types are arranged in precise patterns of staggering complexity to generate tissues and organs. The costs of such complexity include increased requirements for information, material, energy, and time during the development of an individual

organism. Although the physiological costs of complex tissues and organs are high, they confer the ability to thrive in varied and variable environments—a major evolutionary advantage.

One of the defining characteristics of animals with complex tissues and organs (metazoans) such as ourselves is that the external and internal surfaces of most of their tissues and organs, and indeed the exterior of the entire organism, are built from tightly packed sheet-like layers of cells known as **epithelia**. The formation of an epithelium and its subsequent remodeling into more complex collections of epithelial and nonepithelial tissues is a hallmark of the development of metazoans. Sheets of tightly attached epithelial cells act as regulatable, selective permeability barriers, which permit the generation of chemically and functionally distinct compartments in an organism, such as the stomach and bloodstream. As a result, distinct and sometimes opposite functions (e.g., digestion and synthesis) can efficiently proceed simultaneously within an organism. Such compartmentalization also permits more sophisticated regulation of diverse biological functions. In many ways, the roles of complex tissues and organs in an organism are analogous to those of organelles and membranes in individual cells.

The assembly of distinct tissues and their organization into organs are determined by molecular interactions at the cellular level (Figure 20-1) and would not be possible without the temporally, spatially, and functionally regulated expression of a wide array of adhesive molecules. Cells in tissues can adhere directly to one another (*cell-cell adhesion*) through specialized membrane proteins called **cell-adhesion molecules (CAMs)** that often cluster into specialized cell junctions. In the fruit fly *Drosophila melanogaster*, at least 500 genes (~4 percent of the total) are estimated to be involved in cell adhesion. Cells in animal tissues also adhere indirectly (*cell-matrix adhesion*) through the binding of **adhesion receptors** in the plasma membrane to components of the surrounding **extracellular matrix (ECM)**, a complex interdigitating meshwork of proteins and polysaccharides secreted by cells into the spaces between them. Some adhesion receptors can also function as CAMs, mediating direct interaction between cells.

Cell-cell and cell-matrix adhesions not only allow cells to aggregate into distinct tissues but also provide a means for the bidirectional transfer of information between the exterior and the interior of cells. As we will see, both types of adhesions are intrinsically associated with the cytoskeleton and cellular signaling pathways. As a result, a cell's surroundings influence its shape and functional properties ("outside-in" effects); likewise, cellular shape and function influence a cell's surroundings ("inside-out" effects). Thus *connectivity* and *communication* are intimately related properties of cells in tissues. Such information transfer is important to many biological processes, including cell survival, proliferation, differentiation, and migration. Therefore it is not surprising that defects that interfere with adhesive interactions and the associated flow of information can cause or contribute to diseases, including a wide variety of neuromuscular and skeletal disorders and cancer.

In this chapter, we examine various types of adhesive molecules found on the surfaces of cells and in the surrounding extracellular matrix. Interactions between these molecules allow organization of cells into tissues and have profound impacts on tissue development, function, and pathology. Many adhesion molecules are members of families or superfamilies of related proteins. While each individual adhesion molecule performs a distinct role, we will focus on the common features shared by members of some of these families to illustrate the general principles underlying their structures and functions. Because of the particularly well-understood nature of the adhesive molecules in tissues that form tight epithelia, as well as their very early evolutionary development, we will initially focus on epithelial tissues, such as the walls of the intestinal tract and those that form skin. Epithelial cells are normally nonmotile (sessile); however, during development, wound healing, and in certain pathologic states (e.g., cancer), epithe-

lial cells can transform into more motile cells. Changes in expression and function of adhesive molecules play a key role in this transformation, as they do in normal biological processes involving cell movement, such as the crawling of white blood cells into sites of infection. We therefore follow the discussion of epithelial tissues with a discussion of adhesion in nonepithelial, developing, and motile tissues.

The evolution of plants and animals diverged before multicellular organisms arose. Thus multicellularity and the molecular means for assembling tissues and organs must have arisen independently in animal and plant lineages. Not surprisingly, then, animals and plants exhibit many differences in the organization and development of tissues. For this reason, we first consider the organization of tissues in animals and then deal separately with plants.

## 20.1 Cell-Cell and Cell-Matrix Adhesion: An Overview

There are many different types of cells in the body that dynamically interact with each other in a myriad of ways. These interactions, achieved via adhesion molecules, must be precisely and carefully controlled in time and space to correctly determine the structures and functions of tissues in a complex organism. It is not surprising, therefore, that cell-cell and cell-ECM adhesion molecules exhibit diverse structures and their expression levels vary in different cells and tissues. As a consequence, they mediate both the very specific and distinctive cell-cell and cell-ECM interactions that hold tissues together, as well as essential communication between cells and their environment. We begin this overview with a brief orientation to the various types of adhesive molecules present on cells and within the extracellular matrix, their major functions in organisms, and their evolutionary origin. In subsequent sections, we examine in detail the unique structures and properties of the various participants in cell-cell and cell-matrix interactions.

### Cell-Adhesion Molecules Bind to One Another and to Intracellular Proteins

Cell-cell adhesion is mediated through membrane proteins called **cell-adhesion molecules (CAMs)**. Most CAMs fall into four major families: the cadherins, immunoglobulin (Ig) superfamily, integrins, and selectins. As the schematic structures in Figure 20-2 illustrate, CAMs are often mosaics of multiple distinct domains, many of which can be found in more than one kind of protein. Some of these domains confer the binding specificity that characterizes a particular protein. Other membrane proteins, whose structures do not belong to any of the major classes of CAMs, also participate in cell-cell adhesion in various tissues. As we will see later, integrins can function both as CAMs and, as depicted in Figure 20-2, adhesion receptors that bind to ECM components. Some Ig-superfamily CAMs can play this dual role as well.

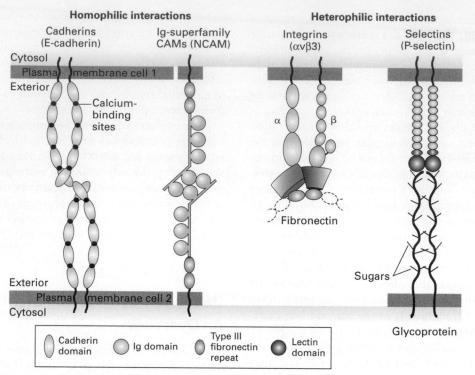

**Homophilic interactions**

Cadherins (E-cadherin)  Ig-superfamily CAMs (NCAM)

**Heterophilic interactions**

Integrins (αvβ3)  Selectins (P-selectin)

Cytosol
Plasma membrane cell 1
Exterior

Calcium-binding sites

α  β

Fibronectin

Sugars

Exterior
Plasma membrane cell 2
Cytosol

Glycoprotein

Cadherin domain  Ig domain  Type III fibronectin repeat  Lectin domain

**FIGURE 20-2 Major families of cell-adhesion molecules (CAMs) and adhesion receptors.** Dimeric E-cadherins most commonly form homophilic (self) cross-bridges with E-cadherins on adjacent cells. Members of the immunoglobulin (Ig) superfamily of CAMs can function as adhesion receptors or as CAMs that form both homophilic linkages (shown here) and heterophilic (nonself) linkages. Heterodimeric integrins (for example, αv and β3 chains) function as CAMs or as adhesion receptors (shown here) that bind to very large, multi-adhesive matrix proteins such as fibronectin, only a small part of which is shown here. Selectins, shown as dimers, contain a carbohydrate-binding lectin domain that recognizes specialized sugar structures on glycoproteins (shown here) or glycolipids on adjacent cells. Note that CAMs often form higher-order oligomers within the plane of the plasma membrane. Many adhesive molecules contain multiple distinct domains, some of which are found in more than one kind of CAM. The cytoplasmic domains of these proteins are often associated with adapter proteins that link them to the cytoskeleton or to signaling pathways. [See R. O. Hynes, 1999, *Trends Cell Biol.* **9**(12):M33, and R. O. Hynes, 2002, *Cell* **110**:673–687.]

CAMs mediate, through their extracellular domains, adhesive interactions between cells of the same type (*homotypic* adhesion) or between cells of different types (*heterotypic* adhesion). A CAM on one cell can directly bind to the same kind of CAM on an adjacent cell (*homophilic* binding) or to a different class of CAM (*heterophilic* binding). CAMs can be broadly distributed along the regions of plasma membranes that contact other cells or clustered in discrete patches or spots called **cell junctions**. Cell-cell adhesions can be tight and long lasting or relatively weak and transient. For example, the associations between nerve cells in the spinal cord or the metabolic cells in the liver exhibit tight adhesion. In contrast, immune-system cells in the blood often exhibit only weak, short-lasting interactions, allowing them to roll along and pass through a blood vessel wall on their way to fight an infection within a tissue.

The cytosol-facing domains of CAMs recruit sets of multifunctional **adapter proteins** (see Figure 20-1). These adapters act as linkers that directly or indirectly connect CAMs to elements of the cytoskeleton (see Chapters 17 and 18); they can also recruit intracellular molecules that function in signaling pathways to control gene expression and the activity of the CAMs themselves or other intracellular proteins (see

Chapters 15 and 16). In many cases, a complex aggregate of CAMs, adapter proteins, and other associated proteins is assembled at the inner surface of the plasma membrane. These complexes facilitate two-way "outside-in" and "inside-out" communication between cells and their surroundings.

The formation of many cell-cell adhesions often entails two types of molecular interactions (Figure 20-3). First, CAM monomers on one cell can bind to the same or different CAMs on an adjacent cell; these interactions are called *intercellular, adhesive, or trans* interactions. Second, monomeric CAMs on one cell may cluster together in the cell's plasma membrane, forming homodimers or higher-order oligomers through their extracellular domains, cytosolic domains, or both; these interactions are called *intracellular, lateral, or cis* interactions. The lateral monomer clustering in one cell may increase the probability of monomer-to-monomer or oligomer-to-oligomer trans interactions with clustered CAMs on an adjacent cell. In addition, formation of monomer-to-monomer trans interactions can induce lateral clustering that strengthens the adhesive interactions. In many cases, the type of initial interaction—trans or cis—that leads to adhesion has not been established. In the case of the CAM called

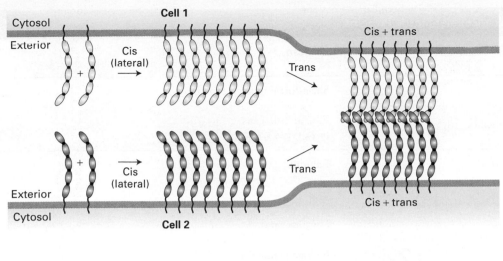

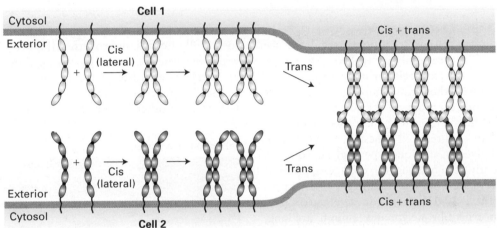

**FIGURE 20-3 Model for the generation of cell-cell adhesions.**
Lateral interactions between cell-adhesion molecules (CAMs) within the plasma membrane of a cell can form clusters of monomers (*top left*) or directly bound dimers and larger oligomers (*bottom left*). The parts of the molecules that participate in these cis interactions vary among the different CAMs. Subsequent trans interactions between distal domains of CAMs on adjacent cells—monomer-to-monomer trans interactions (*top right*) or oligomer-to-oligomer trans interactions (*bottom right*)—generate a strong, Velcro-like adhesion between the cells. [Adapted from M. S. Steinberg and P. M. McNutt, 1999, *Curr. Opin. Cell Biol.* **11**:554.]

E-cadherin, biophysical studies suggest that formation of relatively weak monomer-to-monomer trans interactions may precede lateral clustering that strengthens the adhesion.

Adhesive interactions between cells vary considerably, depending on the tissue and the particular CAMs participating. Just like Velcro, very tight adhesion can be generated when many weak interactions are combined, and this is especially the case when CAMs are concentrated in small, well-defined areas such as cellular junctions. Some CAMs require calcium ions to form effective adhesions; others do not. Furthermore, the association of intracellular molecules with the cytosolic domains of CAMs can dramatically influence the intermolecular interactions of CAMs by promoting their cis association (clustering) or by altering their conformation. Among the many variables that determine the nature of adhesion between two cells are the binding affinity of the interacting molecules (thermodynamic properties), the overall "on" and "off" rates of association and dissociation for each interacting molecule (kinetic properties), the spatial distribution or density of adhesion molecules (ensemble properties), the active versus inactive states of CAMs with respect to adhesion (biochemical properties), and external forces such as stretching and pulling in muscle or the laminar and turbulent flow of cells and surrounding fluids in the circulatory system (mechanical properties).

## The Extracellular Matrix Participates in Adhesion, Signaling, and Other Functions

The **extracellular matrix (ECM)** is a complex combination of secreted proteins that is involved in holding cells and tissues together. The composition and physical properties of the ECM, which can vary depending on the tissue type, its location, and its physiologic state, can be sensed by cell-adhesion receptors that then instruct cells to behave appropriately in response to their environments. Just as the expression of specific

| TABLE 20-1 | Extracellular Matrix Proteins | |
| --- | --- | --- |
| Proteoglycans | Perlecan | |
| Collagens | Sheet forming (e.g., type IV) | |
| | Fibular collagens (e.g., types I, II, and III) | |
| Multi-adhesive matrix proteins | Laminin | |
| | Fibronectin | |
| | Nidogen/entactin | |

adhesion molecules at the cell surface is tightly regulated, the composition of the ECM is carefully controlled.

The components of the extracellular matrix form a network by binding to each other and communicate with cells by binding to **adhesion receptors** on the cell surface. As a consequence of ECM interactions with receptors on adjacent cells, the ECM mediates indirect cell adhesion. ECM components include proteoglycans, a unique type of glycoprotein (a protein with covalently attached sugars); collagens, proteins that often form fibers; soluble multi-adhesive matrix proteins; and others (Table 20-1). Multi-adhesive matrix proteins, such as the proteins fibronectin and laminin, are long, flexible molecules that contain multiple domains. They are responsible for binding various types of collagen, other matrix proteins, polysaccharides, cell-surface adhesion receptors, and extracellular signaling molecules. These proteins are important organizers of the extracellular matrix. Through their interactions with adhesion receptors, they also regulate cell-matrix adhesion—and thus cell migration and cell shape.

The relative volumes of cells and their surrounding matrix vary greatly among different animal tissues. Some connective tissue, for instance, is mostly matrix, whereas many tissues such as epithelia are composed of very densely packed cells with relatively little matrix (Figure 20-4). The density of packing of the molecules within the ECM itself can also vary greatly.

H. V. Wilson's classic studies of adhesion in marine sponge cells showed conclusively that one primary function of ECM is to literally hold tissue together. Figures 20-5a and 20-5b, which re-create Wilson's classic work, show that when sponges are mechanically dissociated and individual cells from two species of sponge are mixed, the cells of one species will adhere to each other but not to those from the other species.

(a) Connective tissue

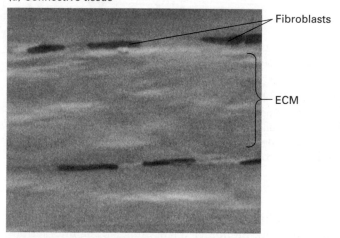

(b) Tightly packed epithelial cells

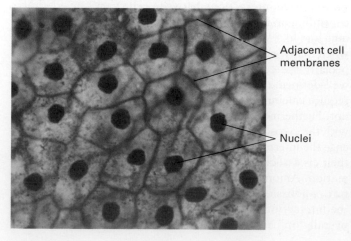

**FIGURE 20-4 Variation in the relative density of cells and ECM in different tissues.** (a) Dense connective tissue contains mostly matrix of tightly packed ECM fibers (pink) interspersed by rows of relatively sparse fibroblasts, the cells that synthesized this ECM (purple). (b) Squamous epithelium viewed from the top, showing epithelial cells tightly packed into a quilt-like pattern with the plasma membranes of adjacent cells close to one another and little ECM between the cells (see also Figure 20-9b). [(a) from Biophoto Associates (b) from Science Photo Library.]

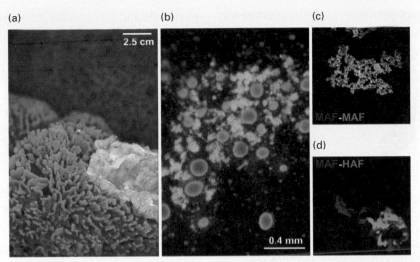

(a)  (b)  (c)

2.5 cm

MAF-MAF

(d)

MAF-HAF

0.4 mm

**EXPERIMENTAL FIGURE 20-5 Mechanically separated marine sponges reassemble through species-specific homotypic cell adhesion.** (a) Two sponges, *Microciona prolifera* (orange) and *Halichondria panicea* (yellow), growing in the wild. (b) After mechanical disruption and mixing of both types of intact sponges, the individual cells were allowed to reassociate for ∼30 minutes with gentle stirring. The cells aggregate with species-specific homotypic adhesion, forming micro-clumps of *Microciona prolifera* cells (orange) and *Halichondria panicea* cells (yellow). (c) and (d) Red or green fluorescently labeled beads were coated with the proteoglycan aggregation factor (AF) from the ECM of either *Microciona prolifera* (MAF) or *Halichondria panicea* (HAF). Panel (c) shows that when both colored beads were coated only with MAF, they all aggregated together, forming yellow aggregates (combination of red and green). Panel (d) shows that MAF (red) and HAF (green) coated beads do not readily form mixed aggregates but rather assemble into distinct clumps held together by homotypic adhesion. (magnification, 40x) [Adapted from X. Fernandez-Busquets and M. M. Burger, 2003, *Cell Mol. Life Sci.* **60:**88–112, and J. Jarchow and M. M. Burger, 1998, *Cell Adhes. Commun.* **6:**405–414.]

This specificity is due, in part, to different adhesive proteins in the ECM that bind to the cells via surface receptors. These adhesive proteins can be purified and used to coat colored beads, which, when mixed, aggregate with each other with a specificity similar to that of intact sponge cells (Figure 20-5c, d).

The ECM plays a multitude of other roles in addition to facilitating cell adhesion. Different combinations of ECM components tailor the extracellular matrix for specific purposes at different anatomical sites: strength in a tendon, tooth, or bone; cushioning in cartilage; and adhesion in most tissues. The composition of the matrix also provides positional information for cells, letting a cell know where it is and what it should do. Changes in ECM components, which are constantly being remodeled, degraded, and resynthesized locally, can modulate the interactions of a cell with its environment. Furthermore, the matrix serves as a reservoir for many extracellular signaling molecules that control cell growth and differentiation. In addition, the matrix provides a lattice through or on which cells can move, particularly in the early stages of tissue assembly. Morphogenesis—the stage of embryonic development in which tissues, organs, and body parts are formed by cell movements and rearrangements—is critically dependent on cell-matrix adhesion as well as cell-cell adhesion. For example, cell-matrix interactions are required for branching morphogenesis (formation of branching structures) to form blood vessels, the air sacs in the lung, mammary and salivary glands, and other structures (Figure 20-6).

Disruptions in cell-matrix and cell-cell interactions can have devastating consequences for the development of tissues. Figure 20-7 shows the dramatic changes in the skeletal system of embryonic mice when the genes for either of two key ECM molecules, collagen II or perlecan, are inactivated. Disruptions in adhesion also are characteristic of various diseases, such as metastatic cancer, in which cancerous cells leave their normal locations and spread throughout the body.

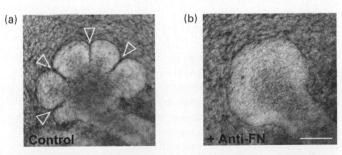

(a)  (b)

Control  + Anti-FN

**EXPERIMENTAL FIGURE 20-6 Antibodies to fibronectin block branching morphogenesis in developing mouse tissues.** Immature salivary glands were isolated from murine embryos and allowed to undergo branching morphogenesis in vitro for 10 hours in the absence (a) or presence (b) of an antibody that binds to and blocks the activity of the ECM molecule fibronectin. Anti-fibronectin antibody (Anti-FN) treatment blocked branch formation (arrowheads). Inhibition of fibronectin's adhesion receptor (an integrin) also blocks branch formation (not shown). [Takayoshi et al., 2003, *Nature* **423:**876–881.]

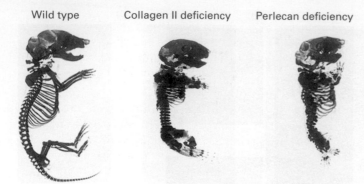

Wild type          Collagen II deficiency          Perlecan deficiency

**EXPERIMENTAL FIGURE 20-7 Inactivating the genes for some ECM proteins results in defective skeletal development in mice.** These photographs show skeletons of normal (*left*), collagen II–deficient (*center*), or perlecan-deficient (*right*) murine embryos that were isolated and stained to visualize the cartilage (blue) and bone (red). Absence of these key ECM components leads to dwarfism, with many skeletal elements shortened and disfigured. [From E. Gustafsson et al., 2003, *Ann. NY Acad. Sci.* **995:**140–150.]

Although many CAMs and adhesion receptors were initially identified and characterized because of their adhesive properties, they also play a major role in signaling, using many of the pathways discussed in Chapters 15 and 16. Figure 20-8 illustrates how one adhesion receptor, integrin,

physically and functionally interacts via adapters and signaling kinases with a broad array of intracellular signaling pathways to influence cell survival, gene transcription, cytoskeletal organization, cell motility, and cell proliferation. Conversely, changes in the activities of signaling pathways inside cells can influence the structures of CAMs and adhesion receptors and so modulate their ability to interact with other cells and with the ECM. Thus, outside-in and inside-out signaling involves numerous interconnected pathways.

## The Evolution of Multifaceted Adhesion Molecules Made Possible the Evolution of Diverse Animal Tissues

Cell-cell and cell-matrix adhesions are responsible for the formation, composition, architecture, and function of animal tissues. Not surprisingly, some adhesion molecules are evolutionarily ancient and are among the most highly conserved proteins in multicellular organisms. Sponges, the most primitive multicellular organisms, express certain CAMs and multi-adhesive ECM molecules whose structures are strikingly similar to those of the corresponding human proteins. The evolution of organisms with complex tissues and organs (metazoans) has depended on the evolution of diverse adhesion molecules with novel properties and functions, whose levels of

**FIGURE 20-8 Integrin adhesion receptor-mediated signaling pathways that control diverse cell functions.** Binding of integrins to their ligands (outside-in signaling) induces conformational changes in their cytoplasmic domains, directly or indirectly altering their interactions with cytoplasmic proteins. These include adapter proteins (e.g., talins, kindlins, paxillin, vinculin) and signaling kinases (src-family kinases, focal adhesion kinase [FAK], integrin-linked kinase [ILK]) that transmit signals via diverse signaling pathways, thereby influencing cell proliferation, cell survival, cytoskeletal organization, cell migration, and gene transcription. Components of several signaling pathways, some of which are associated directly with the plasma membrane, are shown in green boxes. Many of the components of the pathways shown here are shared with other cell-surface-activated signaling pathways (e.g., receptor tyrosine kinases shown on the right) and are discussed in Chapters 15 and 16. In turn, intracellular signaling pathways can, via adapter proteins, modify the ability of integrins to bind to their extracellular ligands (inside-out signaling). [Modified from W. Guo and F. G. Giancotti, 2004, *Nat. Rev. Mol. Cell Biol.* **5:**816–826, and R. O. Hynes, 2002, *Cell* **110:**673–687.]

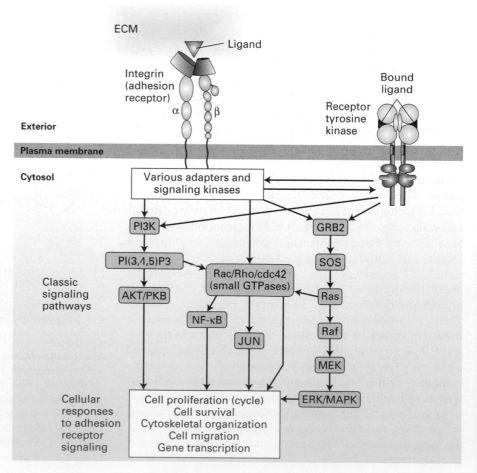

expression differ in different types of cells. Some CAMs and adhesion receptors (e.g., cadherins, integrins, and immunoglobulin-superfamily CAMs such as L1CAM) and ECM components (type IV collagen, laminin, nidogen/entactin, and perlecan-like proteoglycans) are highly conserved because they play crucial roles in many different organisms, whereas other adhesion molecules are less conserved. For example, fruit flies do not have certain types of collagen or the ECM protein fibronectin that play important roles in mammals. A common feature of adhesive proteins is repeating domains forming very large proteins. The overall length of these molecules, combined with their ability to bind numerous ligands via distinct functional domains, likely played a role in their evolution.

The diversity of adhesive molecules arises in large part from two phenomena that can generate numerous closely related proteins, called isoforms, that constitute a protein family. In some cases, the different members of a protein family are encoded by multiple genes that arose from a common ancestor by gene duplication and divergent evolution (see Chapter 6). In other cases, a single gene produces an RNA transcript that can undergo alternative splicing to yield multiple mRNAs, each encoding a distinct protein isoform (see Chapter 8). Both phenomena contribute to the diversity of some protein families such as the cadherins. Particular isoforms of an adhesive protein are often expressed in some cell types and tissues but not others.

## KEY CONCEPTS of Section 20.1

### Cell-Cell and Cell-Matrix Adhesion: An Overview

• Cell-cell and cell–extracellular matrix (ECM) interactions are critical for assembling cells into tissues, controlling cell shape and function, and determining the developmental fate of cells and tissues. Diseases may result from abnormalities in the structures or expression of adhesion molecules.

• Cell-adhesion molecules (CAMs) mediate direct cell-cell adhesions (homotypic and heterotypic), and cell-surface adhesion receptors mediate cell-matrix adhesions (see Figure 20-1). These interactions bind cells into tissues and facilitate communication between cells and their environments.

• The cytosolic domains of CAMs and adhesion receptors bind adapter proteins that mediate interaction with cytoskeletal fibers and intracellular signaling proteins.

• The major families of CAMs are the cadherins, selectins, Ig-superfamily CAMs, and integrins (see Figure 20-2). Members of the integrin and Ig-CAM superfamilies can also function as adhesion receptors.

• Tight cell-cell adhesions entail both cis (lateral or intracellular) oligomerization of CAMs and trans (intercellular) interaction of like (homophilic) or different (heterophilic) CAMs (see Figure 20-3). The combination of cis and trans interactions produces a Velcro-like adhesion between cells.

• The extracellular matrix (ECM) is a complex meshwork of proteins and polysaccharides that contributes to the structure and function of a tissue. The major classes of ECM molecules are proteoglycans, collagens, and multi-adhesive matrix proteins (fibronectin, laminin).

• The evolution of adhesion molecules with specialized structures and functions permits cells to assemble into diverse classes of tissues with varying functions.

## 20.2 Cell-Cell and Cell-ECM Junctions and Their Adhesion Molecules

Cells in epithelial and nonepithelial tissues use many, but not all, of the same cell-cell and cell-matrix adhesion molecules. Because of the relatively simple organization of epithelia, as well as their fundamental role in evolution and development, we begin our detailed discussion of adhesion with the epithelium. In this section we will focus on regions of the cell surface that contain clusters of adhesion molecules in discrete patches or spots called anchoring junctions, tight junctions, and gap junctions. Anchoring and tight junctions play critical roles in mediating cell-cell and cell-ECM adhesion, and all three types of junctions mediate intercellular and/or cell-ECM communication.

### Epithelial Cells Have Distinct Apical, Lateral, and Basal Surfaces

Cells that form epithelial tissues are said to be **polarized** because their plasma membranes are organized into discrete regions. Typically, the distinct surfaces of a polarized epithelial cell are called the **apical** (top), **lateral** (side), and **basal** (base or bottom) surfaces (see Figure 20-1 and Figure 20-9). The area of the apical surface is often greatly expanded by the formation of microvilli. Adhesion molecules play essential roles in generating and maintaining these distinct surfaces.

Epithelia in different body locations have characteristic morphologies and functions (see Figure 20-9). Stratified (multilayered) epithelia commonly serve as barriers and protective surfaces (e.g., the skin), whereas simple, single-layer epithelia often selectively move ions and small molecules from one side of the layer to the other. For instance, the simple columnar epithelium lining the stomach secretes hydrochloric acid into the lumen; a similar epithelium lining the small intestine transports products of digestion from the lumen of the intestine across the basolateral surface into the blood (see Figure 11-19).

In simple columnar epithelia, adhesive interactions between the lateral surfaces hold the cells together into a two-dimensional sheet, whereas those at the basal surface connect the cells to a specialized underlying extracellular matrix called the **basal lamina**. Often the basal and lateral surfaces are similar in composition and together are called the **basolateral** surface. The basolateral surfaces of most simple epithelia are

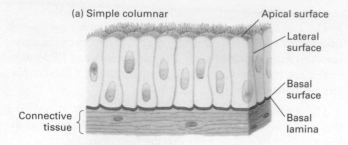

(a) Simple columnar

Apical surface

Lateral surface

Basal surface

Connective tissue

Basal lamina

(b) Simple squamous

(c) Transitional

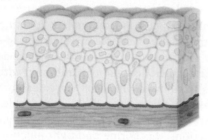

(d) Stratified squamous (nonkeratinized)

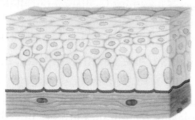

**FIGURE 20-9 Principal types of epithelia.** The apical, lateral, and basal surfaces of epithelial cells exhibit distinctive characteristics. Often the basal and lateral sides of cells are not distinguishable and are collectively known as the basolateral surface. (a) Simple columnar epithelia consist of elongated cells, including mucus-secreting cells (in the lining of the stomach and cervical tract) and absorptive cells (in the lining of the small intestine). (b) Simple squamous epithelia, composed of thin cells, line the blood vessels (endothelial cells/ endothelium) and many body cavities. (c) Transitional epithelia, composed of several layers of cells with different shapes, line certain cavities subject to expansion and contraction (e.g., the urinary bladder). (d) Stratified squamous (nonkeratinized) epithelia line surfaces such as the mouth and vagina; these linings resist abrasion and generally do not participate in the absorption or secretion of materials into or out of the cavity. The basal lamina, a thin fibrous network of collagen and other ECM components, supports all epithelia and connects them to the underlying connective tissue.

usually on the side of the cell closest to the blood vessels, whereas the apical surface is not in direct contact with other cells or the ECM. In animals with closed circulatory systems, blood flows through vessels whose inner lining is composed of flattened epithelial cells called endothelial cells. In general,

epithelial cells are sessile, immobile cells, in that adhesion molecules firmly and stably attach them to one another and their associated ECM. One especially important mechanism used to generate strong, stable adhesions is to concentrate subsets of these molecules into clusters called cell junctions.

## Three Types of Junctions Mediate Many Cell-Cell and Cell-ECM Interactions

All epithelial cells in a sheet are connected to one another and the extracellular matrix by specialized junctions. Although hundreds of individual dispersed adhesion-molecule-mediated interactions are sufficient to cause cells to adhere, the clustered groups of adhesion molecules at cell junctions play special roles in imparting strength and rigidity to a tissue, transmitting information between the extracellular and the intracellular space, controlling the passage of ions and molecules across cell layers, and serving as conduits for the movement of ions and molecules from the cytoplasm of one cell to that of its immediate neighbor. Particularly important to epithelial sheets is the formation of junctions that help form tight seals between the cells and thus allow the sheet to serve as a barrier to the flow of molecules from one side of the sheet to the other.

Three major classes of animal cell junctions are prominent features of simple columnar epithelia (Figure 20-10 and Table 20-2). **Anchoring junctions** and **tight junctions** perform the key task of holding the tissue together. Tight junctions also control the flow of solutes through the extracellular spaces between the cells forming an epithelial sheet. Tight junctions are found primarily in epithelial cells, whereas anchoring junctions can be seen in both epithelial and nonepithelial cells. These junctions are organized into three parts: (1) adhesive proteins in the plasma membrane that connect one cell to another cell on the lateral surfaces (CAMs) or to the extracellular matrix on the basal surfaces (adhesion receptors); (2) adapter proteins, which connect the CAMs or adhesion receptors to cytoskeletal filaments and signaling molecules; and (3) the cytoskeletal filaments themselves. The third class of junctions, **gap junctions**, permits the rapid diffusion of small, water-soluble molecules between the cytoplasms of adjacent cells. Along with anchoring and tight junctions, gap junctions share the role of helping a cell communicate with its environment. However, they are structurally very different from anchoring junctions and tight junctions and do not play a key role in strengthening cell-cell and cell-ECM adhesions. Found in both epithelial and nonepithelial cells, gap junctions resemble cell-cell junctions in plants called plasmodesmata, which we discuss in Section 20.6.

Three types of anchoring junctions are present in cells. Two participate in cell-cell adhesion, whereas the third participates in cell-matrix adhesion. *Adherens junctions* connect the lateral membranes of adjacent epithelial cells and are usually located near the apical surface, just below the tight junctions (see Figure 20-10). A circumferential belt of actin and myosin filaments in a complex with the adherens junctions functions as a tension cable that can internally brace the cell and thereby control its shape. Epithelial and some

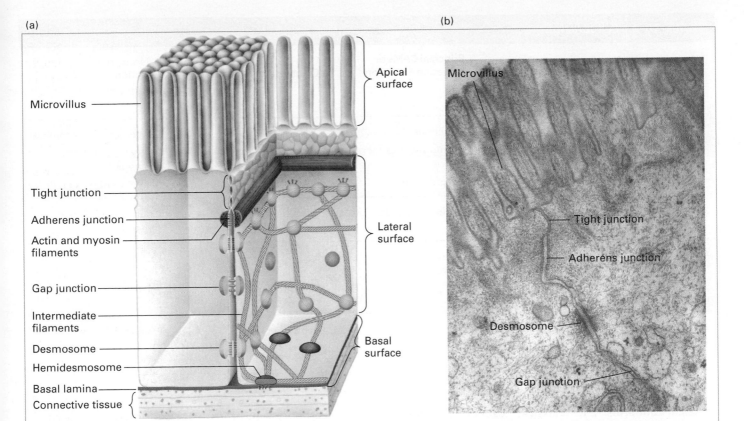

(a)

Microvillus

Tight junction

Adherens junction

Actin and myosin filaments

Gap junction

Intermediate filaments

Desmosome

Hemidesmosome

Basal lamina

Connective tissue

Apical surface

Lateral surface

Basal surface

(b)

Microvillus

Tight junction

Adherens junction

Desmosome

Gap junction

**FIGURE 20-10 Principal types of cell junctions connecting the columnar epithelial cells lining the small intestine.** (a) Schematic cutaway drawing of intestinal epithelial cells. The basal surface of the cells rests on a basal lamina, and the apical surface is packed with fingerlike microvilli that project into the intestinal lumen. Tight junctions, lying just under the microvilli, prevent the diffusion of many substances between the intestinal lumen and the blood through the extracellular space between cells. Gap junctions allow the movement of small molecules and ions between the cytosols of adjacent cells. The remaining three types of junctions—adherens junctions, spot desmosomes, and hemidesmosomes—are critical to cell-cell and cell-matrix adhesion and signaling. (b) Electron micrograph of a thin section of intestinal epithelial cells, showing relative locations of the different junctions. [Part (b) C. Jacobson et al., 2001, *J. Cell Biol.* **152**:435–450.]

other types of cells, such as smooth muscle and heart cells, are also bound tightly together by *desmosomes*, snaplike points of contact sometimes called spot desmosomes. *Hemidesmosomes*, found mainly on the basal surface of epithelial cells, anchor an epithelium to components of the underlying extracellular matrix, much like nails holding down a carpet. Adherens junctions and desmosomes are found in many different types of cells; hemidesmosomes appear to be restricted to epithelial cells.

Bundles of intermediate filaments running parallel to the cell surface or through the cell interconnect spot desmosomes and hemidesmosomes, imparting shape and rigidity to the cell. This close interaction between these junctions and the cytoskeleton helps transmit shear forces from one region of a cell layer to the epithelium as a whole, providing strength and rigidity to the entire epithelial cell layer. Desmosomes and hemidesmosomes are especially important in maintaining the integrity of skin epithelia. For instance, mutations that interfere with hemidesmosomal anchoring in the skin can lead to blistering in which the epithelium becomes detached from its matrix foundation and extracellular fluid accumulates at the basolateral surface, forcing the skin to balloon outward.

## Cadherins Mediate Cell-Cell Adhesions in Adherens Junctions and Desmosomes

The primary CAMs in adherens junctions and desmosomes belong to the **cadherin** family. In vertebrates, this protein family of more than 100 members can be grouped into at least six subfamilies, including *classical cadherins* and *desmosomal cadherins*, which we will describe below. The diversity of cadherins arises from the presence of multiple cadherin genes and alternative RNA splicing. It is not surprising that there are many different types of cadherins in vertebrates, because many different types of cells in widely diverse tissues use these CAMs to mediate adhesion and communication. The brain expresses the largest number of different cadherins, presumably owing to the necessity of forming many specific cell-cell contacts to help establish its complex wiring pattern. Invertebrates, however, are able to function with fewer than 20 cadherins.

**TABLE 20-2** | Cell Junctions

| Junction | Adhesion Type | Principal CAMs or Adhesion Receptors | Cytoskeletal Attachment | Function |
|---|---|---|---|---|
| **Anchoring junctions** | | | | |
| 1. Adherens junctions | Cell-cell | Cadherins | Actin filaments | Shape, tension, signaling |
| 2. Desmosomes | Cell-cell | Desmosomal cadherins | Intermediate filaments | Strength, durability, signaling |
| 3. Hemidesmosomes | Cell-matrix | Integrin (α6β4) | Intermediate filaments | Shape, rigidity, signaling |
| **Tight junctions** | Cell-cell | Occludin, claudin, JAMs | Actin filaments | Controlling solute flow, signaling |
| **Gap junctions** | Cell-cell | Connexins, innexins, pannexins | Possible indirect connections to cytoskeleton through adapters to other junctions | Communication; small-molecule transport between cells |
| **Plasmodesmata (plants only)** | Cell-cell | Undefined | Actin filaments | Communication; molecule transport between cells |

**Classical Cadherins** The "classical" cadherins include E-, N-, and P-cadherins, named for the type of tissues in which they were initially identified (epithelial, neural, and placental). E- and N-cadherins are the most widely expressed, particularly during early differentiation. Sheets of polarized epithelial cells, such as those that line the small intestine or kidney tubules, contain abundant E-cadherin along their lateral surfaces. Although E-cadherin is concentrated in adherens junctions, it is present throughout the lateral surfaces, where it is thought to link adjacent cell membranes. The results of experiments with L cells, a line of cultured mouse fibroblasts, demonstrated that E-cadherins preferentially mediate homophilic interactions. L cells express no cadherins and adhere poorly to themselves or to other cells. When the E-cadherin gene was introduced into L cells, the engineered cadherin-expressing L cells were found to adhere preferentially to other cells expressing E-cadherin (Figure 20-11). These L cells expressing E-cadherin formed epithelial-like aggregates with one another and with epithelial cells isolated from lungs. Although most E-cadherins exhibit primarily homophilic binding, some mediate heterophilic interactions.

The adhesiveness of cadherins depends on the presence of extracellular $Ca^{2+}$, the property that gave rise to their name (calcium *adhering*). For example, the adhesion of L cells expressing E-cadherin is prevented when the cells are bathed in a solution that is low in $Ca^{2+}$ (see Figure 20-11). Some adhesion molecules require some minimal amount of $Ca^{2+}$ in the extracellular fluid to function properly, whereas other molecules such as IgCAMs are $Ca^{2+}$ independent.

The role of E-cadherin in adhesion can also be demonstrated in experiments with cultured epithelial cells called *Madin-Darby canine kidney (MDCK)* cells (see Figure 9-4).

A green fluorescent-protein-labeled form of E-cadherin has been used in these cells to show that clusters of E-cadherin mediate the initial attachment and subsequent zippering up of the cells into sheets (Figure 20-12). In this experimental system, the addition of an antibody that binds to E-cadherin, preventing its homophilic interactions, blocks the $Ca^{2+}$-dependent attachment of MDCK cells to each other and the subsequent formation of intercellular adherens junctions.

Each classical cadherin contains a single transmembrane domain, a relatively short C-terminal cytosolic domain, and five extracellular "cadherin" domains (see Figure 20-2). The extracellular domains are necessary for $Ca^{2+}$ binding and cadherin-mediated cell-cell adhesion. Cadherin-mediated ad-

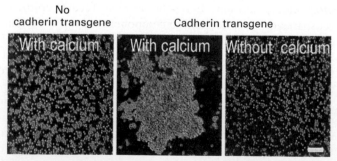

**EXPERIMENTAL FIGURE 20-11 E-cadherin mediates $Ca^{2+}$-dependent adhesion of L cells.** Under standard cell culture conditions in the presence of calcium in the extracellular fluid, L cells do not aggregate into sheets (*left*). Introduction of a gene that causes the expression of E-cadherin in these cells results in their aggregation into epithelial-like clumps in the presence of calcium (*center*) but not in its absence (*right*). Bar, 60 μm. [From Cynthia L. Adams et al., 1998, *J. Cell Biol.* **142**(4):1105–1119.]

**EXPERIMENTAL FIGURE 20-12 E-cadherin mediates adhesive connections in cultured MDCK epithelial cells.** An E-cadherin gene fused to green fluorescent protein (GFP) was introduced into cultured MDCK cells. The cells were then mixed together in a calcium-containing medium and the distribution of fluorescent E-cadherin was visualized over time (shown in hours). Clusters of E-cadherin mediate the initial attachment and subsequent zippering up of the epithelial cells and the formation of junctions (bicellular junctions are where two cells join and appear as lines; tricellular junctions are the sites of intersection of three cells). [From Cynthia L. Adams et al., 1998, *J. Cell Biol.* **142**:1105–1119.]

Time after mixing cells (h):

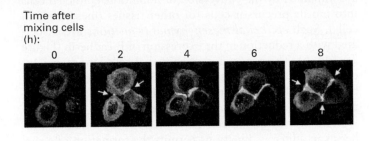

hesion entails both cis lateral clustering (intracellular) and trans adhesive (intercellular) molecular interactions (see Figure 20-3). The Ca$^{2+}$-binding sites, located between the cadherin repeats, may help stabilize cis or trans interactions. The cadherin clusters form intercellular complexes to generate cell-cell adhesion and then additional lateral contacts, resulting in a "zippering up" of cadherins into adhesive arrays. In this way, multiple low-affinity interactions sum to produce a very tight intercellular adhesion.

The results of domain swap experiments, in which an extracellular domain of one kind of cadherin is replaced with the corresponding domain of a different cadherin, have indicated that the specificity of binding resides, at least in part, in the most distal (farthest from the membrane) extracellular domain, the N-terminal domain. Cadherin-mediated trans adhesion is commonly thought to require only head-to-head interactions between the N-terminal domains of cadherin oligomers on adjacent cells, as depicted in Figure 20-13.

The C-terminal cytosolic domain of classical cadherins is linked to the actin cytoskeleton by adapter proteins (see Figure 20-13). These linkages are essential for strong adhesion, owing primarily to their contributing to increased lateral associations. For example, disruption of the interactions between classical cadherins and α- or β-catenin—two common adapter proteins that link classical cadherins to actin filaments—dramatically reduces cadherin-mediated cell-cell adhesion. This disruption occurs spontaneously in tumor cells, which sometimes fail to express α-catenin, and can be induced experimentally by depleting the cytosolic pool of accessible β-catenin. The cytosolic domains of cadherins also interact with intracellular signaling molecules such as p120-catenin. Interestingly, β-catenin plays a dual role: it not only mediates cytoskeletal attachment but also serves as a signaling molecule, translocating to the nucleus and altering gene transcription in the Wnt signaling pathway (see Figure 16-30).

Classical cadherins play a critical role during tissue differentiation. Each classical cadherin has a characteristic tissue distribution. In the course of differentiation, the amount or nature of the cell-surface cadherins changes, affecting many aspects of cell-cell adhesion and cell migration. For instance,

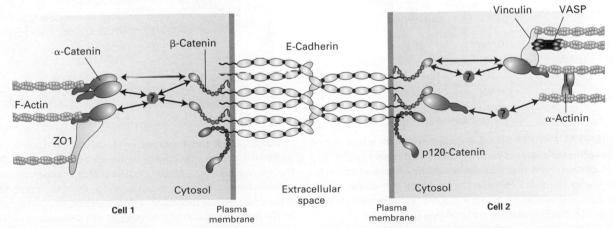

**FIGURE 20-13 Protein constituents of typical adherens junctions.** The exoplasmic domains of E-cadherin dimers clustered at adherens junctions on adjacent cells form Ca$^{2+}$-dependent homophilic interactions. The cytosolic domains of the E-cadherins bind directly or indirectly to multiple adapter proteins (e.g., β-catenin) that connect the junctions to actin filaments (F-actin) of the cytoskeleton and participate in intracellular signaling pathways. Somewhat different sets of adapter proteins are illustrated in the two cells to emphasize that a variety of adapters can interact with adherens junctions. Some of these adapters, such as ZO1, can interact with several different CAMs. There is some controversy about whether α-catenin mediates cadherin/β-catenin interaction with actin directly, via intermediate binding proteins (indicated in the figure by question marks) or via more complex mechanisms. [Adapted from V. Vasioukhin and E. Fuchs, 2001, *Curr. Opin. Cell Biol.* **13**:76.]

the reorganization of tissues during morphogenesis is often accompanied by the conversion of nonmotile epithelial cells into motile precursor cells for other tissues (mesenchymal cells). Such *epithelial-mesenchymal transitions* are associated with a reduction in the expression of E-cadherin (Figure 20-14a, b). The conversion of epithelial cells into malignant carcinoma cells, such as in certain ductal breast tumors or hereditary diffuse gastric cancer (Figure 20-14c), is also marked by a loss of E-cadherin activity.

The firm epithelial cell-cell adhesions mediated by cadherins in adherens junctions permits the formation of a second class of intercellular junctions in epithelia—tight junctions, which we will turn to shortly.

**Desmosomal Cadherins** Desmosomes (Figure 20-15) contain two specialized cadherin proteins, *desmoglein* and *desmocollin*, whose cytosolic domains are distinct from those in the classical cadherins. The cytosolic domains of desmosomal cadherins interact with adapter proteins such as plakoglobin (similar in structure to β-catenin), plakophilins, and a member of the plakin family of adapters called desmoplakin. These adapters, which form the thick cytoplasmic plaques characteristic of desmosomes, in turn interact with intermediate filaments.

(a) Adherent epithelial cells   (b) Motile mesenchymal cells

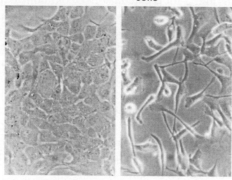

(c) Cancerous cells, no cadherin

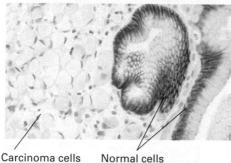

Carcinoma cells        Normal cells in epithelial lining of gastric glands express cadherin

**EXPERIMENTAL FIGURE 20-14 E-cadherin activity is lost during the epithelial-mesenchymal transition and cancer progression.** A protein called Snail that suppresses the expression of E-cadherin is associated with epithelial-mesenchymal transitions. (a) Normal epithelial MDCK cells grown in culture. (b) Expression of the *snail* gene in MDCK cells causes them to undergo an epithelial-mesenchymal transition. (c) Distribution of E-cadherin detected by immunohistochemical staining (dark brown) in thin sections of tissue from a patient with hereditary diffuse gastric cancer. E-cadherin is seen at the intercellular borders of normal stomach gastric gland epithelial cells (*right*); no E-cadherin is seen at the borders of underlying invasive carcinoma cells. [Panels (a) and (b) from Alfonso Martinez Arias, 2001, *Cell* **105**:425–431; images courtesy of M. A. Nieto; panel (c) from F. Carneiro et al., 2004, *J. Pathol.* **203**:681–687.]

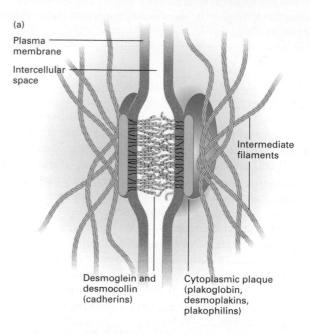

(a)

Plasma membrane

Intercellular space

Intermediate filaments

Desmoglein and desmocollin (cadherins)        Cytoplasmic plaque (plakoglobin, desmoplakins, plakophilins)

(b)   Intermediate filaments        Cytoplasmic plaques

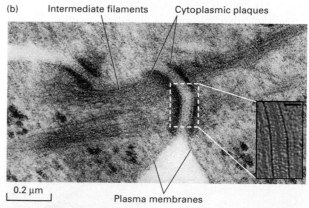

0.2 μm        Plasma membranes

**FIGURE 20-15 Desmosomes.** (a) Model of a desmosome between epithelial cells with attachments to the sides of intermediate filaments. The transmembrane CAMs desmoglein and desmocollin belong to the cadherin family. Adapter proteins bound to the cytoplasmic domains of the CAMs include plakoglobin, desmoplakins, and plakophilins. (b) Electron micrograph of a thin section of a desmosome connecting two cultured differentiated human keratinocytes. Bundles of intermediate filaments radiate from the two darkly staining cytoplasmic plaques that line the inner surface of the adjacent plasma membranes. Inset: Electron microscopic tomograph of a desmosome linking two human epidermal cells (plasma membranes, pink; CAMs, blue; bar, 35 nm). [Part (a), see B. M. Gumbiner, 1993, *Neuron* **11**:551, and D. R. Garrod, 1993, *Curr. Opin. Cell Biol.* **5**:30. Part (b) courtesy of R. van Buskirk. Inset from A. Al-Amoudi, D. C. Diez, M. J. Betts, and A. S. Frangakis, 2007, *Nature* **450**:832–837.]

The cadherin desmoglein was identified through studies of an unusual but revealing skin disease called *pemphigus vulgaris*, an autoimmune disease. Patients with autoimmune disorders synthesize self-attacking, or "auto," antibodies that bind to a normal body protein. In pemphigus vulgaris the auto-antibodies disrupt adhesion between epithelial cells, causing blisters of the skin and mucous membranes. The predominant auto-antibody was shown to be specific for desmoglein; indeed, the addition of such antibodies to normal skin induces the formation of blisters and disruption of cell adhesion. ∎

## Integrins Mediate Cell-ECM Adhesions, Including Those in Epithelial Cell Hemidesmosomes

To be stably anchored to solid tissue and organs, simple columnar epithelial sheets must be firmly attached via their basal surfaces to the underlying extracellular matrix (basal lamina). This attachment occurs via adhesion receptors called **integrins** (see Figure 20-2), which are located both within and outside of anchoring junctions called *hemidesmosomes* (see Figure 20-10a). Hemidesmosomes comprise several integral membrane proteins linked via cytoplasmic adapter proteins (e.g., plakins) to keratin-based intermediate filaments. The principal ECM adhesion receptor in epithelial hemidesmosomes is integrin α6β4.

Integrins function as adhesion receptors and CAMs in a wide variety of epithelial and nonepithelial cells, mediating many cell-matrix and cell-cell interactions (Table 20-3). In vertebrates, at least 24 integrin heterodimers, composed of 18 types of α subunits and eight types of β subunits in various combinations, are known. A single β chain can interact with any one of multiple α chains, forming integrins that bind different ligands. This phenomenon of *combinatorial diversity* allows a relatively small number of components to serve a large number of distinct functions. Although most cells express several distinct integrins that bind the same or different ligands, many integrins are expressed predominantly in certain types of cells. Not only do many integrins bind more than one ligand but several of their ligands bind to multiple integrins.

All integrins appear to have evolved from two ancient general subgroups: those that bind proteins containing the tripeptide sequence Arg-Gly-Asp, usually called the *RGD sequence* (fibronectin is one such protein), and those that bind laminin. Several integrin α subunits contain a distinctive inserted domain, the *I-domain*, which can mediate binding of certain integrins to various collagens in the ECM. Some integrins with I-domains are expressed exclusively on leukocytes (white blood cells) and red and white blood cell precursor (hematopoietic) cells. I-domains also recognize cell-adhesion molecules on other cells, including members of the Ig superfamily (e.g., ICAMs, VCAMs), and thus participate in cell-cell adhesion.

Integrins typically exhibit low affinities for their ligands, with dissociation constants $K_d$ between $10^{-6}$ and $10^{-7}$ mol/L. However, the multiple weak interactions generated by the

| TABLE 20-3 | Selected Vertebrate Integrins* | |
|---|---|---|
| **Subunit Composition** | **Primary Cellular Distribution** | **Ligands** |
| α1β1 | Many types | Mainly collagens |
| α2β1 | Many types | Mainly collagens; also laminins |
| α3β1 | Many types | Laminins |
| α4β1 | Hematopoietic cells | Fibronectin; VCAM-1 |
| α5β1 | Fibroblasts | Fibronectin |
| α6β1 | Many types | Laminins |
| αLβ2 | T lymphocytes | ICAM-1, ICAM-2 |
| αMβ2 | Monocytes | Serum proteins (e.g., C3b, fibrinogen, factor X); ICAM-1 |
| αIIbβ3 | Platelets | Serum proteins (e.g., fibrinogen, von Willebrand factor, vitronectin); fibronectin |
| α6β4 | Epithelial cells | Laminin |

*The integrins are grouped into subfamilies having a common β subunit. Ligands shown in red are CAMs; all others are ECM or serum proteins. Some subunits can have multiple spliced isoforms with different cytosolic domains.
SOURCE: R. O. Hynes, 1992, *Cell* **69**:11.

**FIGURE 20-16 Tight junctions.** (a) Freeze-fracture preparation of tight junction zone between two intestinal epithelial cells. The fracture plane passes through the plasma membrane of one of the two adjacent cells. A honeycomb-like network of ridges and grooves below the microvilli constitutes the tight junction zone. (b) Schematic drawing shows how a tight junction might be formed by the linkage of rows of protein particles in adjacent cells. In the inset micrograph of an ultrathin sectional view of a tight junction, the adjacent cells can be seen in close contact where the rows of proteins interact. [Part (a) courtesy of L. A. Staehelin. Drawing in part (b) adapted from L. A. Staehelin and B. E. Hull, 1978, *Sci. Am.* **238:**140, and D. Goodenough, 1999, *Proc. Nat'l. Acad. Sci. USA* **96:**319. Photograph in part (b) courtesy of S. Tsukita et al., 2001, *Nat. Rev. Mol. Cell Biol.* **2:**285.]

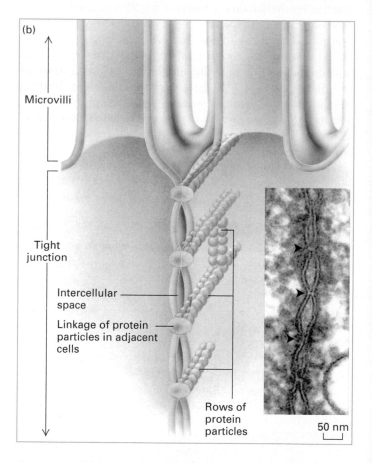

binding of hundreds or thousands of integrin molecules to their ligands on cells or in the extracellular matrix allow a cell to remain firmly anchored to its ligand-expressing target.

Parts of both the α and the β subunits of an integrin molecule contribute to the primary extracellular ligand-binding site (see Figure 20-2). Ligand binding to integrins also requires the simultaneous binding of divalent cations. Like other cell-surface adhesive molecules, the cytosolic region of integrins interacts with adapter proteins that in turn bind to the cytoskeleton and intracellular signaling molecules. Most integrins are linked to the actin cytoskeleton, including two of the integrins that connect the basal surface of epithelial cells to the basal lamina via the ECM molecule laminin. However, some integrin heterodimers interact with intermediate filaments. The cytosolic domain of the β4 chain in the α6β4 integrin in hemidesmosomes, which is much longer than the cytosolic domains of other β integrins, binds to specialized adapter proteins that in turn interact with keratin-based intermediate filaments (see Table 20-3).

As we will see, the diversity of integrins and their ECM ligands allows integrins to participate in a wide array of key biological processes, including the inflammatory response and the migration of cells to their correct locations in the formation of the body plan of an embryo (morphogenesis). The importance of integrins in diverse processes is highlighted by the defects exhibited by knockout mice engineered to have mutations in various integrin subunit genes. These defects include major abnormalities in development, blood vessel formation, leukocyte function, inflammation, bone remodeling, and hemostasis. Despite their differences, all these processes depend on integrin-mediated interactions between the cytoskeleton and either the ECM or CAMs on other cells.

In addition to their adhesion function, integrins can mediate outside-in and inside-out signaling (see Figure 20-8). The engagement of integrins by their extracellular ligands can, through adapter proteins bound to the integrin's cytosolic region, influence the cytoskeleton and intracellular signaling pathways (outside-in signaling). Conversely, intracellular signaling pathways can alter, from the cytoplasm, the structure of integrins and consequently their abilities to adhere to their extracellular ligands and mediate cell-cell and cell-matrix interactions (inside-out signaling). Integrin-mediated signaling pathways influence processes as diverse as cell survival, cell proliferation, and programmed cell death (see Chapter 21).

## Tight Junctions Seal Off Body Cavities and Restrict Diffusion of Membrane Components

For polarized epithelial cells to function as barriers and mediators of selective transport, extracellular fluids surrounding their apical and basolateral membranes must be kept separate. Tight junctions between adjacent epithelial cells are usually located in a band surrounding the cell just below the apical surface and help establish and maintain cell polarity (Figures 20-10 and 20-16). These specialized junctions form a barrier that seals off body cavities such as the intestinal lumen and separates the blood from the cerebral spinal fluid of the central nervous system (i.e., the blood-brain barrier).

Tight junctions prevent the diffusion of macromolecules and, to varying degrees, small water-soluble molecules and ions across an epithelial sheet via the spaces between cells. They also help maintain the polarity of epithelial cells by preventing the diffusion of membrane proteins and glycolipids between the apical and the basolateral regions of the plasma membrane, ensuring that these regions contain different membrane components. Indeed, the lipid compositions of the apical and basolateral regions of the exoplasmic leaflet are distinct. Essentially all glycolipids are restricted to the exoplasmic face of the apical membrane, as are all proteins linked to the membrane by a glycosylphosphatidylinositol (GPI) anchor (see Figure 10-19). In contrast, the apical and basolateral regions of the cytosolic leaflet have uniform membrane composition in epithelial cells; their lipids and proteins can apparently diffuse laterally from one region of the membrane to the other.

Tight junctions are composed of thin bands of plasma-membrane proteins that completely encircle the cell and are in contact with similar thin bands on adjacent cells. When thin sections of cells are viewed in an electron microscope, the lateral surfaces of adjacent cells appear to touch each other at intervals and even to fuse in the zone just below the apical surface (see Figure 20-10b). In freeze-fracture preparations, tight junctions appear as an interlocking network of ridges and grooves in the plasma membrane (Figure 20-16a). Very high magnification reveals that rows of protein particles 3–4 nm in diameter form the ridges seen in freeze-fracture micrographs of tight junctions. In the model shown in Figure 20-16b, the tight junction is formed by a double row of these particles, one row donated by each cell. Treatment of an epithelium with the protease trypsin destroys the tight junctions, supporting the proposal that proteins are essential structural components of these junctions.

The two principal integral membrane proteins found in tight junctions are *occludin* and *claudin* (from the Latin *claudere*, "to close"). When investigators engineered mice with mutations inactivating the occludin gene, which was thought to be essential for tight junction formation, the mice surprisingly still had morphologically distinct tight junctions. Further analysis led to the discovery of claudin. Each of these proteins has four membrane-spanning α helices (Figure 20-17). Another tight junction protein, *tricellulin*, also has this structure. Tricellulin is concentrated at junctions where three cells intersect (see Figure 20-12 and Figure 20-17). The claudin multi-gene family encodes at least 24 homologous proteins that exhibit distinct tissue-specific patterns of expression. A group of *junction adhesion molecules* (JAMs) have also been found to contribute to homophilic adhesion and other functions of tight junctions. JAMs and another junctional protein, the *coxsackievirus and adenovirus receptor* (CAR), contain a single transmembrane α helix and belong to the Ig superfamily of CAMs. The extracellular domains of rows of occludin, claudin, and JAM proteins in the plasma membrane of one cell apparently form extremely tight links with similar rows of the same proteins in an adjacent cell, creating a tight seal. $Ca^{2+}$-dependent

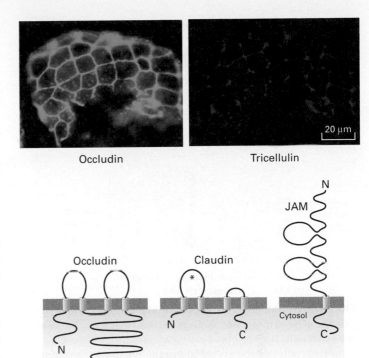

**FIGURE 20-17 Proteins mediating tight junctions.** As shown in these schematic drawings of the major proteins in tight junctions, both occludin and claudins contain four transmembrane helices, whereas the junction adhesion molecule (JAM) has a single transmembrane domain and a large extracellular region. The extracellular loop of the claudins that significantly contributes to paracellular ion selectivity is indicated by an asterisk. Inset: Immunofluorescence localization of occludin (green) and tricellulin (red) in mouse intestinal epithelium. Note how tricellulin is predominantly concentrated in tricellular junctions. Bar, 20 μm. [Bottom drawing adapted from S. Tsukita et al., 2001, *Nature Rev. Mol. Cell Biol.* **2**:285. Inset from J. Ikenouchi et al., 2005, *J. Cell Biol.* **171**:939–945. Tricellulin constitutes a novel barrier at tricellular contacts of epithelial cells.]

cadherin-mediated adhesion also plays an important role in tight junction formation, stability, and function.

The long C-terminal cytosolic segment of occludin binds to PDZ domains in some large cytosolic adapter proteins. *PDZ domains* are about 80 to 90 amino acids long and are found in various cytosolic proteins; they mediate binding to other cytosolic proteins or to the C-termini of particular plasma-membrane proteins. Cytosolic proteins containing a PDZ domain often have more than one of them. In the human genome, there are ~250 PDZ domains found in about 100 proteins. The multiple PDZ domain proteins can serve as scaffolds to assemble proteins into larger functional complexes. The PDZ-containing adapter proteins associated with occludin are bound, in turn, to other cytoskeletal and signaling proteins and to actin fibers. These interactions appear to stabilize the linkage between occludin and claudin molecules that is essential for maintaining the integrity of tight junctions. The C-termini of claudins also bind to the intracellular, multiple-PDZ-domain-containing adapter protein ZO-1, which is also found in adherens junctions (see Figure 20-13).

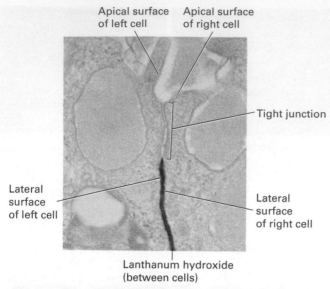

EXPERIMENTAL FIGURE 20-18 **Tight junctions prevent passage of large molecules through extracellular space between epithelial cells.** Tight junctions in the pancreas are impermeable to the large water-soluble colloid lanthanum hydroxide (dark stain) administered from the basolateral side of the epithelium. [Courtesy of D. Friend.]

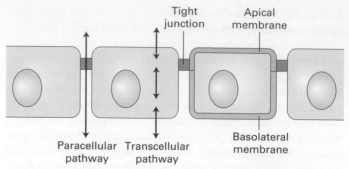

FIGURE 20-19 **Transcellular and paracellular pathways of transepithelial transport.** Transcellular transport requires the cellular uptake of molecules on one side and subsequent release on the opposite side by mechanisms discussed in Chapter 11. In paracellular transport, molecules move extracellularly through parts of tight junctions, whose permeability to small molecules and ions depends on the composition of the junctional components and the physiologic state of the epithelial cells. [Adapted from S. Tsukita et al., 2001, *Nat. Rev. Mol. Cell Biol.* **2**:285.]

Thus, as is the case for adherens junctions and desmosomes, cytosolic adapter proteins and their connections to the cytoskeleton are critical components of tight junctions.

A simple experiment demonstrates the impermeability of tight junctions to some water-soluble substances. In this experiment, lanthanum hydroxide (an electron-dense colloid of high molecular weight) is injected into the pancreatic blood vessel of an experimental animal; a few minutes later, the pancreatic epithelial acinar cells are fixed and prepared for microscopy. As shown in Figure 20-18, the lanthanum hydroxide diffuses from the blood into the space that separates the lateral surfaces of adjacent acinar cells but cannot penetrate past the tight junction.

As a consequence of tight junctions, movement of many nutrients across the intestinal epithelium cannot occur between cells and is in large part achieved through the *transcellular pathway* via specific membrane-bound transport proteins (see Figures 11-30 and 20-19). The barrier to diffusion provided by tight junctions, however, is not absolute, for they exhibit size- and ion-selective permeability. The importance of this selective permeability is highlighted by the evolutionary conservation of the molecules that establish it and diseases that arise when it is disrupted. For example, murine embryos cannot develop properly if this selective permeability is disrupted because fluid balance on distinct sides of epithelia cannot be maintained properly. Similarly, the kidneys depend on proper tight junctional permeability to establish the ion gradients necessary for normal regulation of body fluids and waste removal. Owing at least in part to the varying properties of the different types of claudin molecules located in different tight junctions, the permeability of the

tight junctions to ions, small molecules, and water varies enormously among different epithelial tissues. In epithelia with selectively permeable tight junctions, certain small molecules and ions can move from one side of the cell layer to the other through the *paracellular pathway* in addition to the transcellular pathway (Figure 20-19). One of the extracellular loops in the claudins (see Figure 20-17) is thought to play a major role in defining the selective permeability conferred on tight junctions by specific claudin isoforms.

The permeability of tight junctions can be altered by intracellular signaling pathways, especially G protein and cyclic AMP–coupled pathways (see Chapter 15). The regulation of tight junction permeability is often studied by measuring ion flux (electrical resistance) or the movement of radioactive or fluorescent molecules across monolayers of MDCK or other epithelial cells (transepithelial resistance).

The importance of paracellular transport is apparent in several human diseases. In hereditary hypomagnesemia, defects in the *claudin16* gene prevent the normal paracellular flow of magnesium in the kidney. This results in an abnormally low blood level of magnesium, which can lead to convulsions. Furthermore, a mutation in the *claudin14* gene causes hereditary deafness, apparently by altering transport around hair-cell epithelia in the cochlea of the inner ear.

Some pathogens have evolved to exploit the molecules in tight junctions. Some use junctional proteins as co-receptors to initially attach to cells prior to infecting them (e.g., hepatitis C virus uses *claudin1* and *occludin*, together with two other "co-receptors," to enter liver cells). Others break down the junctional barrier and cross epithelia via paracellular movement, and others produce toxins that alter barrier function (either from the extracellular or intracellular spaces). For example, toxins produced by *Vibrio cholerae*, the enteric bacteria that causes cholera, alter the permeability barrier of the intestinal epithelium by altering the composition or activity of tight junctions. *Vibrio cholerae* also releases a protease that disrupts tight junctions by degrading the extracellular part of occludin. Other bacterial toxins can affect the ion-pumping activity of membrane transport proteins in intestinal epithelial cells. Toxin-induced changes in tight junction permeability (increased paracellular transport) and in protein-mediated ion pumping (increased transcellular transport) can result in massive loss of internal body ions and water into the gastrointestinal tract, which in turn leads to diarrhea and potentially lethal dehydration (see Chapter 11). ■

## Gap Junctions Composed of Connexins Allow Small Molecules to Pass Directly Between Adjacent Cells

Early electron micrographs of cells in tissues revealed sites of cell-cell contact with a characteristic intercellular gap (Figure 20-20a). This feature was found in virtually all contacting animal cells and prompted early morphologists to call these regions gap junctions. In retrospect, the most important feature of these junctions is not the ~2–4-nm gap itself but a well-defined set of cylindrical particles that cross the gap and compose pores connecting the cytoplasms of adjacent cells.

In many tissues, anywhere from a few to thousands of gap junctional particles cluster together in patches (e.g., along the lateral surfaces of epithelial cells; see Figure 20-10). When the plasma membrane is purified and then sheared into small fragments, some pieces mainly containing patches of gap junctions are generated. Owing to their relatively high protein content, these fragments have a higher density than that of the bulk of the plasma membrane and can be purified by equilibrium density-gradient centrifugation (see Figure 9-26). When these preparations are viewed perpendicular to the membrane, the gap junctions appear as arrays of hexagonal particles that enclose water-filled channels (Figure 20-20b).

The effective pore size of gap junctions can be measured by injecting a cell with a fluorescent dye covalently linked to membrane bilayer impermeable molecules of various sizes and observing with a fluorescence microscope whether the dye passes into neighboring cells. Gap junctions between mammalian cells permit the passage of molecules as large as 1.2 nm in diameter. In insects, these junctions are permeable to molecules as large as 2 nm in diameter. Generally speaking, molecules smaller than 1200 Da pass freely and those larger than 2000 Da do not pass; the passage of intermediate-size molecules is variable and limited. Thus ions, many low-molecular-weight precursors of cellular macromolecules, products of intermediary metabolism, and small intracellular signaling molecules can pass from cell to cell through gap junctions.

In nervous tissue, some neurons are connected by gap junctions through which ions pass rapidly, thereby allowing very rapid transmission of electric signals. Impulse transmission through these connections, called electrical synapses, is almost a thousandfold as rapid as at chemical synapses (see Chapter 22). Gap junctions are also present in many nonneuronal tissues, where they help to integrate the electrical and metabolic activities of many cells. In the heart, for instance, gap junctions rapidly pass ionic signals among muscle cells, which are tightly interconnected via desmosomes, and thus contribute to the electrically stimulated coordinate contraction of cardiac muscle cells during a beat. As discussed in Chapter 15, some extracellular hormonal signals induce the production or release of small intracellular signaling molecules called **second messengers** (e.g., cyclic AMP, $IP_3$, and $Ca^{2+}$) that regulate cellular metabolism. Because second messengers can be transferred between cells through gap junctions, hormonal stimulation of one cell can trigger a coordinated response by that same cell as well as many of its neighbors. Such gap-junction-mediated signaling plays an important role, for example, in the secretion of digestive enzymes by the pancreas and in the coordinated muscular contractile waves (peristalsis) in the intestine. Another vivid example of gap-junction-mediated transport is the phenomenon of *metabolic coupling*, or *metabolic cooperation*, in which a cell transfers nutrients or intermediary metabolites to a neighboring cell that is itself unable to synthesize them. Gap junctions play critical roles in the development of egg cells in the ovary by mediating the movement of both metabolites and signaling molecules between an oocyte and its surrounding granulosa cells as well as between neighboring granulosa cells.

A current model of the structure of the gap junction is shown in Figure 20-20c, d, e. Vertebrate gap junctions are composed of **connexins**, a family of structurally related transmembrane proteins with molecular weights between 26,000 and 60,000. A completely different family of proteins, the innexins, forms the gap junctions in invertebrates. A third family of innexin-like proteins, called pannexins, has been found in both vertebrates and invertebrates. Each vertebrate hexagonal particle consists of 12 noncovalently associated connexin molecules: six form a cylindrical connexon hemichannel in one plasma membrane that is joined to a connexon hemichannel in the adjacent cell membrane, forming the continuous aqueous channel (diameter ~14 Å) between the cells. Each individual connexin molecule has four membrane-spanning α-helices with a topology similar to that of occludin (see Figure 20-17), resulting in 24 transmembrane α-helices in each hemichannel. Pannexins are capable of forming intercellular channels as well; however, pannexin hemichannels may also function to permit direct exchange between the intracellular and extracellular spaces.

There are 21 different connexin genes in humans, with different sets of connexins expressed in different cell types. This diversity, together with the generation of mutant mice with inactivating mutations in connexin genes, has highlighted the importance of connexins in a wide variety of cellular systems. Some cells express a single connexin that forms homotypic channels. Most cells, however, express at least two connexins; these different proteins assemble into heteromeric connexon, which in turn form heterotypic gap junction channels. Diversity in channel composition leads to differences in

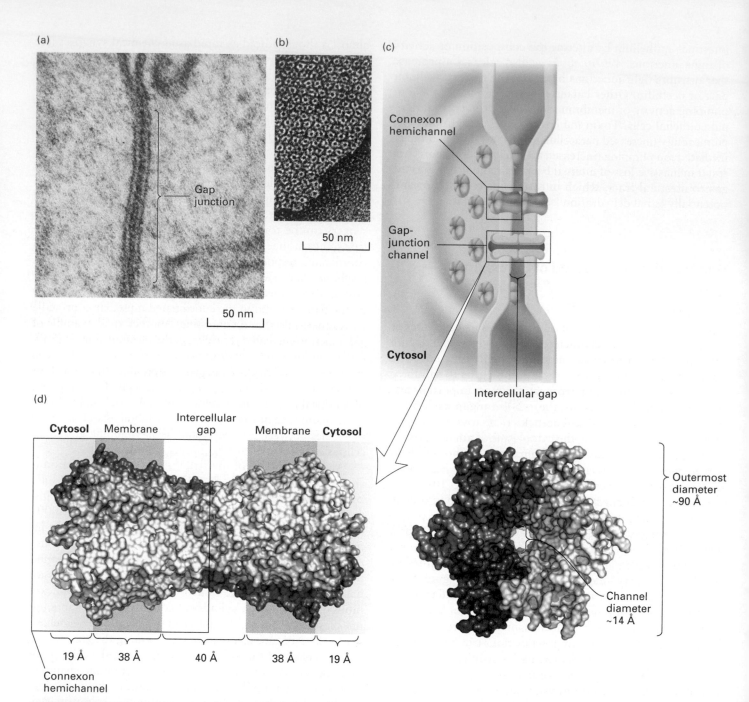

**FIGURE 20-20 Gap junctions.** (a) In this thin section through a gap junction connecting two mouse liver cells, the two plasma membranes are closely associated for a distance of several hundred nanometers, separated by a "gap" of 2–3 nm. (b) Numerous roughly hexagonal particles are visible in this perpendicular view of the cytosolic face of a region of plasma membrane enriched in gap junctions. Each particle aligns with a similar particle on an adjacent cell, forming a channel connecting the two cells. (c) Schematic model of a gap junction connecting two plasma membranes. Both membranes contain connexon hemichannels, cylinders of six dumbbell-shaped connexin molecules. Two connexons join in the gap between the cells to form a gap-junction channel, 1.5–2.0 nm in diameter, that connects the cytosols of the two cells. (d) Model of recombinant human Cx26 gap junction connexon channel determined by x-ray crystallography (3.5-Å resolution). (*Left*) Space filling model of a side view of the complete structure of two attached hemichannels oriented as in part (c). Each of the six connexins that comprise a connexon hemichannel has four transmembrane helices and a distinct color. The structures of the loops connecting the transmembrane helices are not well defined and not shown. (*Right*) View from the cytosol perpendicular to the membrane bilayers, looking down on the connexon with its central pore. The diameter of the pore's channel is ∼14 Å and it is lined by many polar/charged amino acids. [Part (a) courtesy of D. Goodenough; part (b) courtesy of N. Gilula; part (d) adapted from S. Nakagawa et al., 2010, *Curr. Opin. Struct. Biol.* **20**(4):423–430.]

channel permeability. For example, channels made from a 43-kDa connexin isoform, Cx43—the most ubiquitously expressed connexin—are more than 100-fold as permeable to ADP and ATP as those made from Cx32 (32 kDa).

The permeability of gap junctions is regulated by posttranslational modification of connexins (e.g., phosphorylation) and is sensitive to environmental changes such as intracellular pH and $Ca^{2+}$ concentration, membrane potential, and difference in the intercellular potential between adjacent interconnected cells ("voltage gating"). The N-termini of connexins appear to be especially important in the gating mechanism. One example of the physiological regulation of gap junctions occurs during mammalian childbirth. The muscle cells in the mammalian uterus must contract strongly and synchronously during labor to expel the fetus. To facilitate this coordinate activity, immediately prior to and during labor there is an approximately five- to tenfold increase in the amount of the major myometrial connexin, Cx43, and an increase in the number and size of gap junctions, which decrease rapidly postpartum.

Assembly of connexins, their trafficking within cells, and formation of functional gap junctions apparently depend on N-cadherin and its associated adapter proteins (e.g., α- and β-catenins, ZO-1, and ZO-2) as well as desmosomal proteins (plakoglobin, desmoplakin, and plakophilin-2). PDZ domains in ZO-1 and ZO-2 bind to the C-terminus of Cx43 and mediate its interaction with catenins and N-cadherin. The relevance of these relationships is particularly evident in the heart, which depends on gap junctions for rapid coordinated electrical coupling and on adjacent adherens junctions and desmosomes for mechanical coupling between cardiomyocytes to achieve the intercellular integration of electrical activity and movement required for normal cardiac function. It is noteworthy that ZO-1 serves as an adapter for adherens (see Figure 20-13), tight, and gap junctions, suggesting this and other adapters can help integrate the formation and functions of these diverse junctions.

Mutations in connexin genes cause at least eight human diseases, including neurosensory deafness (Cx26 and Cx31), cataract or heart malformations (Cx43, Cx46, and Cx50), and the X-linked form of Charcot-Marie-Tooth disease (Cx32), which is marked by progressive degeneration of peripheral nerves. ■

## KEY CONCEPTS of Section 20.2

### Cell-Cell and Cell-ECM Junctions and Their Adhesion Molecules

• Polarized epithelial cells have distinct apical, basal, and lateral surfaces. Microvilli projecting from the apical surfaces of many epithelial cells considerably expand the cells' surface areas.

• Three major classes of cell junctions—anchoring junctions, tight junctions, and gap junctions—assemble epithelial cells into sheets and mediate communication between them (see Figures 20-1 and 20-10). Anchoring junctions can be further subdivided into adherens junctions, desmosomes, and hemidesmosomes.

• Adherens junctions and desmosomes are cadherin-containing anchoring junctions that bind the membranes of adjacent cells, giving strength and rigidity to the entire tissue.

• Cadherins are cell-adhesion molecules (CAMs) responsible for $Ca^{2+}$-dependent interactions between cells in epithelial and other tissues. They promote strong cell-cell adhesion by mediating both lateral intracellular and intercellular interactions.

• Adapter proteins that bind to the cytosolic domain of cadherins and other CAMs and adhesion receptors mediate the association of cytoskeletal and signaling molecules with the plasma membrane (see Figures 20-8 and 20-13). Strong cell-cell adhesion depends on the linkage of the interacting CAMs to the cytoskeleton.

• Hemidesmosomes are integrin-containing anchoring junctions that attach cells to elements of the underlying extracellular matrix.

• Integrins are a large family of αβ heterodimeric cell-surface proteins that mediate both cell-cell and cell-matrix adhesions and inside-out and outside-in signaling in numerous tissues.

• Tight junctions block the diffusion of proteins and some lipids in the plane of the plasma membrane, contributing to the polarity of epithelial cells. They also limit and regulate the extracellular (paracellular) flow of water and solutes from one side of the epithelium to the other (see Figure 20-19). The two principal integral membrane proteins found in tight junctions are *occludin* and *claudin*.

• Gap junctions are constructed of multiple copies of connexin proteins, assembled into a transmembrane channel that interconnects the cytoplasms of two adjacent cells (see Figure 20-20). Small molecules and ions can pass through gap junctions, permitting metabolic and electrical coupling of adjacent cells.

## 20.3 The Extracellular Matrix I: The Basal Lamina

In animals, the extracellular matrix (ECM) helps organize cells into tissues and coordinates their cellular functions by activating intracellular signaling pathways that control cell growth, proliferation, and gene expression. ECM can directly influence cellular and tissue structure and function. In addition, ECM can serve as a repository for inactive or inaccessible signaling molecules (e.g., growth factors) that are released to function when the ECM is disassembled or remodeled by hydrolyases, such as proteases. Indeed, hydrolyzed fragments of ECM macromolecules can themselves have independent biological activity. Many functions of the matrix require transmembrane adhesion receptors, including the integrins that bind directly to ECM components and that also interact, through adapter proteins, with the cytoskeleton.

Adhesion receptors bind to three types of molecules abundant in the extracellular matrix of all tissues (see Table 20-1):

• **Proteoglycans,** a group of glycoproteins that cushion cells and bind a wide variety of extracellular molecules;

• **Collagen** fibers, which provide structural integrity, mechanical strength, and resilience;

• Soluble **multi-adhesive matrix proteins,** such as laminin and fibronectin, which bind to and cross-link cell-surface adhesion receptors and other ECM components.

We begin our description of the structures and functions of these major ECM components in the context of the basal lamina—the specialized extracellular matrix sheet that plays a particularly important role in determining the overall architecture and function of epithelial tissue. In the next section, we discuss the ECM molecules commonly found in nonepithelial tissues, including connective tissue.

## The Basal Lamina Provides a Foundation for Assembly of Cells into Tissues

In animals, most organized groups of cells in epithelial and nonepithelial tissues are underlain or surrounded by the basal lamina, a sheet-like meshwork of ECM components usually no more than 60–120 nm thick (Figure 20-21). The basal lamina is structured differently in different tissues. In columnar and other epithelia such as intestinal lining and skin, it is a foundation on which only one surface of the cells rests. In other tissues, such as muscle or fat, the basal lamina surrounds each cell. Basal laminae play important roles in regeneration after tissue damage and in embryonic development. For instance, the basal lamina helps four- and eight-celled embryos adhere together in a ball. In the development of the nervous system, neurons migrate along ECM pathways that contain basal lamina components. In higher animals, two distinct basal laminae are employed to form a tight barrier that limits diffusion of molecules between the blood and the brain (blood-brain barrier), and in the kidney a specialized basal lamina serves as a selective permeability blood filter. In muscle, the basal lamina helps protect the cell membranes from damage during contraction/relaxation and prevents muscular dystrophies. Thus the basal lamina is important for organizing cells into tissues and distinct compartments, tissue repair, forming permeability barriers, and

(a)

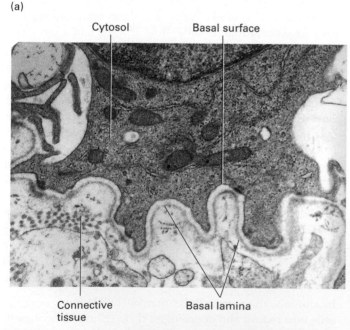

(b)

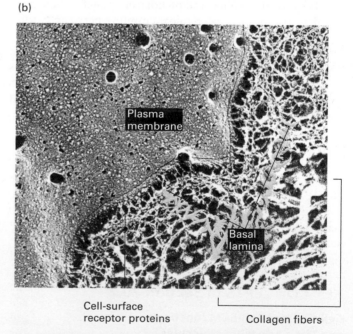

**FIGURE 20-21 The basal lamina separating epithelial cells and some other cells from connective tissue.** (a) Transmission electron micrograph of a thin section of cells (*top*) and underlying connective tissue (*bottom*). The electron-dense layer of the basal lamina can be seen to follow the undulation of the basal surface of the cells. (b) Electron micrograph of a quick-freeze deep-etch preparation of skeletal muscle showing the relation of the plasma membrane, basal lamina, and surrounding connective tissue. In this preparation, the basal lamina is revealed as a meshwork of filamentous proteins that associates with the plasma membrane and the thicker collagen fibers of the connective tissue. [Part (a) courtesy of P. FitzGerald. Part (b) from D. W. Fawcett, 1981, *The Cell,* 2d ed., Saunders/ Photo Researchers; courtesy of John Heuser.]

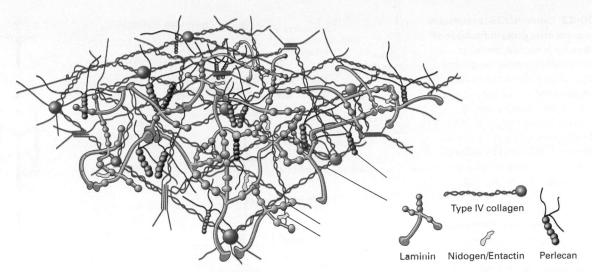

**FIGURE 20-22 Major protein components of the basal lamina.**
Type IV collagen and laminin each form two-dimensional networks, which are cross-linked by entactin and perlecan molecules. [Adapted from B. Alberts et al., 1994, *Molecular Biology of the Cell*, 3d ed., Garland, p. 991.]

guiding migrating cells during development. It is therefore not surprising that basal lamina components have been highly conserved throughout evolution.

Most of the ECM components in the basal lamina are synthesized by the cells that rest on it. Four ubiquitous protein components are found in basal laminae (Figure 20-22):

• *Type IV collagen*, trimeric molecules with both rodlike and globular domains that form a two-dimensional network;

• *Laminins*, a family of multi-adhesive, cross-shaped proteins that form a fibrous two-dimensional network with type IV collagen and that also bind to integrins and other adhesion receptors;

• *Perlecan*, a large multidomain proteoglycan that binds to and cross-links many ECM components and cell-surface molecules;

• *Nidogen* (also called *entactin*), a rodlike molecule that cross-links type IV collagen, perlecan, and laminin and helps incorporate other components into the ECM and also stabilize basal laminae.

Other ECM molecules, such as members of the evolutionarily ancient family of glycoproteins called fibulins, are incorporated into various basal laminae, depending on the tissue and particular functional requirements of the basal lamina.

As depicted in Figure 20-1, one side of the basal lamina is linked to cells by adhesion receptors, including integrins in hemidesmosomes, which bind to laminin in the basal lamina. The other side of the basal lamina is anchored to the adjacent connective tissue by a layer of collagen fibers embedded in a proteoglycan-rich matrix. In stratified squamous epithelia (e.g., skin), this linkage is mediated by anchoring fibrils of type VII collagen. Together, the basal lamina and collagen-anchoring fibrils form the structure called the *basement membrane*.

## Laminin, a Multi-adhesive Matrix Protein, Helps Cross-link Components of the Basal Lamina

**Laminin**, the principal multi-adhesive matrix protein in basal laminae, is a heterotrimeric protein comprising α, β, and γ chains. At least 16 laminin isoforms in vertebrates are assembled from five α, three β, and three γ chains and numbered to reflect the chain composition: laminin-αβγ (e.g., laminin-111 or laminin-511). Each laminin exhibits a distinctive pattern of tissue- and developmental-stage-specific expression. As shown in Figure 20-23 many laminins are large, cross-shaped proteins (molecular weight of about 820,000), although some are Y or rod shaped. Globular domains at the N-terminus of each subunit bind to one another and thus mediate the self-assembly of laminins into meshlike networks that, together with their binding to laminin receptors on cells, is essential for the assembly of basal laminae. Five globular *LG domains* at the C-terminus of the laminin α subunit mediate $Ca^{2+}$-dependent binding to cell-surface laminin receptors, including certain integrins (see Table 20-3) as well as sulfated glycolipids, syndecan, and dystroglycan, which will be described further in Section 20.4. Some of these interactions are via negatively charged carbohydrates on the receptors. LG domains are found in a wide variety of other proteins and can mediate binding to steroids and proteins as well as carbohydrates. Laminin is the principal basal laminal ligand of integrins.

## Sheet-Forming Type IV Collagen Is a Major Structural Component of the Basal Lamina

Type IV collagen is, together with laminin, a principal structural component of all basal laminae and can bind to adhesion receptors, including some integrins. Collagen IV is one of at least 28 types of collagen in humans that participate in

**FIGURE 20-23 Laminin, a heterotrimeric multi-adhesive matrix protein found in all basal laminae.** (a) Schematic model of cross-shaped laminin showing the general shape, location of globular domains, and coiled-coil region in which laminin's three chains are covalently linked by several disulfide bonds. Different regions of laminin bind to cell-surface receptors and various matrix components (indicated by arrows). Inset: laminins assemble into a lattice via interactions between their N-terminal globular domains. (b) Electron micrographs of intact laminin molecule, showing its characteristic cross appearance (*left*) and the carbohydrate-binding LG domains near the C-terminus (*right*). [Part (a) adapted from G. R. Martin and R. Timpl, 1987, *Ann. Rev. Cell Biol.* **3**:57; Durbeej M. Laminins, 2010, *Cell Tissue Res.* **339**(1):259–268; and S. Meinen, P. Barzaghi, S. Lin, H. Lochmüller, and M. A. Ruegg, 2007, *J. Cell Biol.* **176**(7):979–993. Part (b) from R. Timpl et al., 2000, *Matrix Biol.* **19**:309; photograph at right courtesy of Jürgen Engel.]

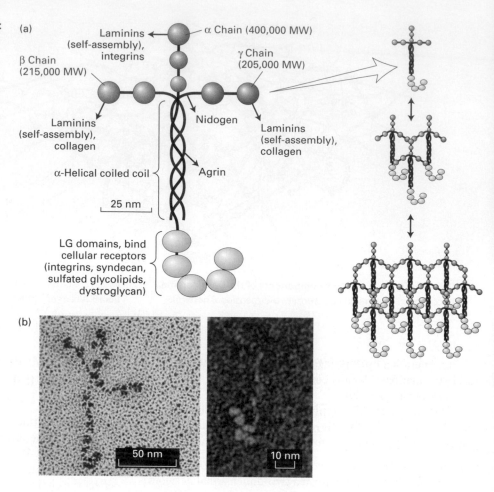

the formation of distinct extracellular matrices in various tissues (Table 20-4). There are also at least 20 additional collagen-like proteins (such as host defense collagens) in the human proteome. Although collagens differ in certain structural features and tissue distribution, all collagens are trimeric proteins made from three polypeptides, each encoded by one of at least 43 genes in humans, usually called collagen α chains. All three α chains can be identical (homotrimeric) or different (heterotrimeric). All or parts of the three-stranded collagen molecule can twist together into a special triple helix called a *collagenous* triple helix. When there is more than one triple helical segment, these segments are joined by nonhelical regions of the protein. Within a helical segment, each of the three α chains twists into a left-handed helix, and the three chains then wrap around each other to form a right-handed triple helix (Figure 20-24).

**FIGURE 20-24 The collagen triple helix.** (a) (*Left*) Side view of the crystal structure of a polypeptide fragment whose sequence is based on repeating sets of three amino acids, Gly-X-Y, characteristic of collagen α chains. (*Center*) Each chain is twisted into a left-handed helix, and three chains wrap around each other to form a right-handed triple helix. The schematic model (*right*) clearly illustrates the triple helical nature of the structure and shows the left-handed twist of the individual collagen α chains (red line). (b) View down the axis of the triple helix. The proton side chains of the glycine residues (orange) point into the very narrow space between the polypeptide chains in the center of the triple helix. In collagen mutations in which other amino acids replace glycine, the proton in glycine is replaced by larger groups that disrupt the packing of the chains and destabilize the triple-helical structure. [Adapted from R. Z. Kramer et al., 2001, *J. Mol. Biol.* **311**:131.]

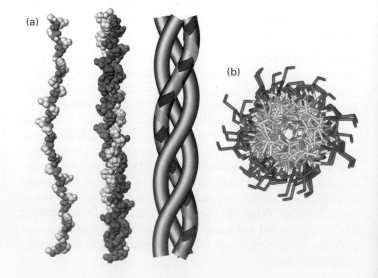

## TABLE 20-4 | Selected Collagens

| Type | Molecule Composition | Structural Features | Representative Tissues |
|---|---|---|---|
| **Fibrillar Collagens** | | | |
| I | $[\alpha 1(I)]_2[\alpha 2(I)]$ | 300-nm-long fibrils | Skin, tendon, bone, ligaments, dentin, interstitial tissues |
| II | $[\alpha 1(II)]_3$ | 300-nm-long fibrils | Cartilage, vitreous humor |
| III | $[\alpha 1(III)]_3$ | 300-nm-long fibrils; often with type I | Skin, muscle, blood vessels |
| V | $[\alpha 1(V)]_2[\alpha 2(V)]$, $[\alpha 1(V)]_3$ | 390-nm-long fibrils with globular N-terminal extension; often with type I | Cornea, teeth, bone, placenta, skin, smooth muscle |
| **Fibril-Associated Collagens** | | | |
| VI | $[\alpha 1(VI)][\alpha 2(VI)][\alpha 3(VI)]$ | Lateral association with type I; periodic globular domains | Most interstitial tissues |
| IX | $[\alpha 1(IX)][\alpha 2(IX)][\alpha 3(IX)]$ | Lateral association with type II; N-terminal globular domain; bound GAG | Cartilage, vitreous humor |
| **Sheet-Forming and Anchoring Collagens** | | | |
| IV | $[\alpha 1(IV)]_2[\alpha 2(IV)]$ | Two-dimensional network | All basal laminae |
| VII | $[\alpha 1(VII)]_3$ | Long fibrils | Below basal lamina of the skin |
| XV | $[\alpha 1(XV)]_3$ | Core protein of chondroitin sulfate proteoglycan | Widespread; near basal lamina in muscle |
| **Transmembrane Collagens** | | | |
| XIII | $[\alpha 1(XIII)]_3$ | Integral membrane protein | Hemidesmosomes in skin |
| XVII | $[\alpha 1(XVII)]_3$ | Integral membrane protein | Hemidesmosomes in skin |
| **Host Defense Collagens** | | | |
| Collectins | | Oligomers of triple helix; lectin domains | Blood, alveolar space |
| C1q | | Oligomers of triple helix | Blood (complement) |
| Class A scavenger receptors | | Homotrimeric membrane proteins | Macrophages |

SOURCES: K. Kuhn, 1987, in R. Mayne and R. Burgeson, eds., *Structure and Function of Collagen Types*, Academic Press, p. 2, and M. van der Rest and R. Garrone, 1991, *FASEB J.* 5:2814.

The collagen triple helix can form because of an unusual abundance of three amino acids: glycine, proline, and a modified form of proline called hydroxyproline (see Figure 2-15). They make up the characteristic repeating sequence motif Gly-X-Y, where X and Y can be any amino acid but are often proline and hydroxyproline and less often lysine and hydroxylysine. Glycine is essential because its small side chain, a hydrogen atom, is the only one that can fit into the crowded center of the three-stranded helix (see Figure 20-24). Hydrogen bonds help hold the three chains together. Although the rigid peptidyl-proline and peptidyl-hydroxyproline linkages are not compatible with formation of a classic single-stranded α helix, they stabilize the distinctive three-stranded collagen helix. The hydroxyl group in hydroxyproline helps hold its ring in a conformation that stabilizes the three-stranded helix.

There are several distinct cell-surface receptors for collagen IV and other types of collagen, discussed in the next section. These cell-surface receptors include certain integrins, discoidin domain receptors 1 and 2 (tyrosine kinase receptors), glycoprotein VI (on platelets), leukocyte-associated Ig-like receptor-1, members of the mannose receptor family, and a modified form of the protein CD44. They can play critical roles in helping to assemble ECM and in integrating cellular activity with the ECM.

The unique properties of each type of collagen are due mainly to differences in (1) the number and lengths of the

**FIGURE 20-25 Structure and assembly of type IV collagen.**
(a) Schematic representation of type IV collagen. This 400-nm-long molecule has a small noncollagenous globular domain at the N-terminus and a large globular domain at the C-terminus. The triple helix is interrupted by nonhelical segments that introduce flexible kinks in the molecule. Lateral interactions between triple helical segments, as well as head-to-head and tail-to-tail interactions between the globular domains, form dimers, tetramers, and higher-order complexes, yielding a sheet-like network. Multiple, unusual sulfilimine (—S═N—) or thioether bonds between hydroxylysine (or lysine) and methionine residues covalently cross-link some adjacent C-terminal domains and contribute to the stability of the network. (b) Electron micrograph of type IV collagen network formed in vitro. The lacy appearance results from the flexibility of the molecule, the side-to-side binding between triple-helical segments (white arrows), and the interactions between C-terminal globular domains (yellow arrows). [Part (a) adapted from A. Boutaud, 2000, *J. Biol. Chem.* **275**:30716. Part (b) courtesy of P. Yurchenco; see P. Yurchenco and G. C. Ruben, 1987, *J. Cell Biol.* **105**:2559.]

(a)

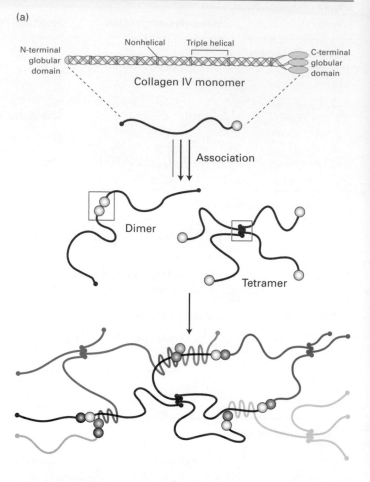

Collagen IV monomer

Association

Dimer

Tetramer

(b) Type IV network

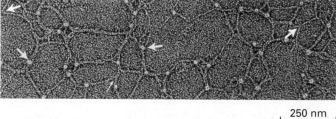

250 nm

collagenous, triple-helical segments; (2) the segments that flank or interrupt the triple-helical segments and that fold into other kinds of three-dimensional structures; and (3) the covalent modification of the α chains (e.g., hydroxylation, glycosylation, oxidation, cross-linking). For example, the chains in type IV collagen are designated IVα chains. Mammals express six homologous IVα chains, which assemble into three different heterotrimeric type IV collagens with distinct properties. All subtypes of type IV collagen, however, form a 400-nm-long triple helix (Figure 20-25) that is interrupted about 24 times with nonhelical segments and flanked by large globular domains at the C-termini of the chains and smaller globular domains at the N-termini. The nonhelical regions introduce flexibility into the molecule. Through both lateral associations and interactions entailing the globular N- and C-termini, type IV collagen molecules assemble into a branching, irregular two-dimensional fibrous network that forms a lattice on which, together with the laminin lattice, the basal lamina is built (see Figures 20-22 and 20-25).

In the kidney, a double basal lamina called the glomerular basement membrane separates the epithelium that lines the urinary space from the endothelium that lines the surrounding blood-filled capillaries. Defects in this structure, which is responsible for ultrafiltration of the blood and initial urine formation, can lead to renal failure. For instance, mutations that alter the C-terminal globular domain of certain IVα chains are associated with progressive renal failure as well as sensorineural hearing loss and ocular abnormalities, a condition known as *Alport's syndrome*. In *Goodpasture's syndrome*, a relatively rare autoimmune disease, antibodies bind to the α3 chains of type IV collagen found in the glomerular basement membrane and lungs. This binding sets off an immune response that causes cellular damage resulting in progressive renal failure and pulmonary hemorrhage. ■

## Perlecan, a Proteoglycan, Cross-links Components of the Basal Lamina and Cell-Surface Receptors

**Perlecan**, the major secreted proteoglycan in basal laminae, consists of a large multidomain core protein (~470 kDa) made up of repeats of five distinct domains, including laminin-like LG domains and Ig domains. The many globular repeats give it the appearance of a string of pearls when visualized by electron microscopy; hence the name perlecan. Perlecan is a glycoprotein that contains three types of covalent polysaccharide chains: N-linked (see Chapter 14), O-linked, and glycosaminoglycans (GAGs) (O-linked sugars and GAGs are discussed further in Section 20.4). GAGs are long polymers of repeating disaccharides. Glycoproteins containing covalently attached GAG chains are called **proteoglycans**. The protein constituent of a proteoglycan is usually called the *core* protein, onto which the GAG chains are attached. Both the protein and the GAG components of perlecan contribute to its ability to incorporate into and define the structure and function of basal laminae. Because of its multiple domains and polysaccharide chains that have distinctive binding properties, perlecan binds to dozens of other molecules, including other ECM components (e.g., laminin, nidogen), cell-surface receptors, and polypeptide growth factors. Simultaneous binding to these molecules results in perlecan-mediated cross-linking. Perlecan can be found in basal laminae and non-basal-laminae ECM. The adhesion receptor dystroglycan can bind perlecan directly, via its LG domains, and indirectly, via its binding to laminin. In humans, mutations in the perlecan gene can lead either to dwarfism or to muscle abnormalities, apparently due to dysfunction of the neuromuscular junction that controls muscle firing.

## KEY CONCEPTS of Section 20.3

### The Extracellular Matrix I: The Basal Lamina

- The basal lamina, a thin meshwork of extracellular matrix (ECM) molecules, separates most epithelia and other organized groups of cells from adjacent connective tissue. Together, the basal lamina and the immediately adjacent collagen network form a structure called the basement membrane.

- Four ECM proteins are found in all basal laminae (see Figure 20-22): laminin (a multi-adhesive matrix protein), type IV collagen, perlecan (a proteoglycan), and nidogen/enactin.

- Cell-surface adhesion receptors such as integrin anchor cells to the basal lamina, which in turn is connected to other ECM components (see Figure 20-1). Laminin in the basal lamina is the principal ligand of $\alpha6\beta4$ integrin (see Table 20-3).

- Laminin and other multi-adhesive matrix proteins are multidomain molecules that bind multiple adhesion receptors and ECM components.

- The large, flexible molecules of type IV collagen interact end to end and laterally to form a meshlike scaffold to which other ECM components and adhesion receptors can bind (see Figures 20-22 and 20-25).

- Type IV collagen is a member of the collagen family of proteins distinguished by the presence of repeating tripeptide sequences of Gly-X-Y that give rise to the collagen triple-helical structure (see Figure 20-24). Different collagens are distinguished by the length and chemical modifications of

their $\alpha$ chains and by the presence of segments that interrupt or flank their triple-helical regions.

- Perlecan, a large secreted proteoglycan present primarily in the basal lamina, binds many ECM components and adhesion receptors. Proteoglycans consist of membrane-associated or secreted core proteins covalently linked to one or more specialized polysaccharide chains called glycosaminoglycans (GAGs).

## 20.4 The Extracellular Matrix II: Connective Tissue

Connective tissue such as tendon and cartilage differs from other solid tissues in that most of its volume is made up of extracellular matrix rather than cells. This matrix is packed with insoluble protein fibers. Key ECM components in connective tissue, some of which are found in other types of tissues as well, are:

- *Collagens*, trimeric molecules that are often bundled together into fibers *(fibrillar collagens)*;

- *Glycosaminoglycans* *(GAGs)*, specialized linear polysaccharide chains of specific repeating disaccharides that can be highly hydrated and confer diverse binding and physical properties (e.g., resistance to compression);

- *Proteoglycans*, glycoproteins containing one or more covalently bound GAG chains

- *Multi-adhesive proteins*, large multidomain proteins often comprising many copies ("repeats") of a few distinctive domains that bind to and cross-link a variety of adhesive receptors and ECM components;

- *Elastin*, a protein that forms the amorphous core of elastic fibers.

Collagen is the most abundant fibrous protein in connective tissue. Rubber-like elastin fibers, which can be stretched and relaxed, are also present in deformable sites (e.g., skin, tendons, heart). The fibronectins, a family of multi-adhesive matrix proteins, form their own distinct fibrils in the matrix of most connective tissues. Although several types of cells are found in connective tissues, the various ECM components are produced largely by cells called **fibroblasts**. In this section we explore the structure and function of the various components in the ECM of connective tissue, as well as how the ECM is degraded and remodeled by a variety of specialized proteases.

### Fibrillar Collagens Are the Major Fibrous Proteins in the ECM of Connective Tissues

About 80–90 percent of the collagen in the body consists of *fibrillar collagens* (types I, II, and III), located primarily in connective tissues (see Table 20-4). Because of its abundance in tendon-rich tissue such as rat tail, type I collagen is easy to isolate and was the first collagen to be characterized. Its

**FIGURE 20-26 Biosynthesis of fibrillar collagens.**

Step **1**: Procollagen α chains are synthesized on ribosomes associated with the endoplasmic reticulum (rough ER), and in the ER asparagine-linked oligosaccharides are added to the C-terminal propeptide.

Step **2**: Propeptides associate to form trimers and are covalently linked by disulfide bonds, and selected residues in the Gly-X-Y triplet repeats are covalently modified (certain prolines and lysines are hydroxylated, galactose or galactose-glucose [hexagons] are attached to some hydroxylysines, prolines are cis → trans isomerized).

Step **3**: The modifications facilitate zipperlike formation, stabilization of triple helices, and binding by the chaperone protein Hsp47 (see Chapter 13), which may stabilize the helices or prevent premature aggregation of the trimers or both. Steps **4** and **5**: The folded procollagens are transported to and through the Golgi apparatus, where some lateral association into small bundles takes place. The chains are then secreted (step **6**), the N- and C-terminal propeptides are removed (step **7**), and the trimers assemble into fibrils and are covalently cross-linked (step **8**). The 67-nm staggering of the trimers gives the fibrils a striated appearance in electron micrographs (*inset*). [Adapted from A. V. Persikov and B. Brodsky, 2002, *Proc. Nat'l. Acad. Sci. USA* **99**:1101–1103.]

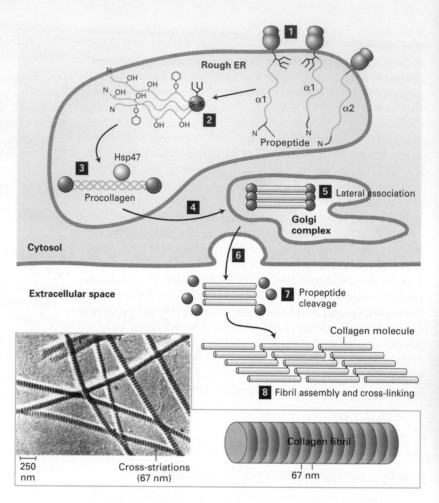

fundamental structural unit is a long (300-nm), thin (1.5-nm-diameter) triple helix (see Figure 20-24) consisting of two α1(I) chains and one α2(I) chain, each 1050 amino acids in length. The triple-stranded molecules pack tightly together and wrap around each other forming *microfibrils* that associate into higher-order polymers called collagen *fibrils*, which in turn often aggregate into larger bundles called collagen *fibers* (Figure 20-26).

Classes of collagen that are less abundant, but nevertheless important, include *fibril-associated collagens*, which link the fibrillar collagens to one another or to other ECM components; *sheet-forming and anchoring collagens*, which form two-dimensional networks in basal laminae (type IV) and connect the basal lamina in skin to the underlying connective tissue (type VII); *transmembrane collagens*, which function as adhesion receptors; and *host defense collagens*, which help the body recognize and eliminate pathogens. Interestingly, several collagens (e.g., types IX, XVIII, and XV) are also proteoglycans with covalently attached GAGs (see Table 20-4).

## Fibrillar Collagen Is Secreted and Assembled into Fibrils Outside the Cell

Fibrillar collagens are secreted proteins, produced primarily by fibroblasts in the ECM. Collagen biosynthesis and secretion follow the normal pathway for a secreted protein, described in

detail in Chapters 13 and 14. The collagen α chains are synthesized as longer precursors, called pro-α chains, by ribosomes attached to the endoplasmic reticulum (ER). The pro-α chains undergo a series of covalent modifications and fold into triple-helical *procollagen* molecules before their release from cells (see Figure 20-26).

After the secretion of procollagen from the cell, extracellular peptidases remove the N-terminal and C-terminal propeptides. In fibrillar collagens, the resulting molecules, which consist almost entirely of a triple-stranded helix, associate laterally to generate fibrils with a diameter of 50–200 nm. In fibrils, adjacent collagen molecules are displaced from one another by 67 nm, about one-quarter of their length. This staggered array produces a striated effect that can be seen in both light and electron microscopic images of collagen fibrils (see Figure 20-26, *inset*). The unique properties of the fibrillar collagens are mainly due to the formation of fibrils.

Short non-triple-helical segments at either end of the fibrillar collagen α chains are of particular importance in the formation of collagen fibrils. Lysine and hydroxylysine side chains in these segments are covalently modified by extracellular lysyl oxidases to form aldehydes in place of the amine group at the end of the side chain. These reactive aldehyde groups form covalent cross-links with lysine, hydroxylysine, and histidine residues in adjacent molecules. The cross-links stabilize the side-by-side packing of collagen molecules and generate a very

**FIGURE 20-27 Interactions of fibrous collagens with nonfibrous fibril-associated collagens.** (a) In tendons, type I fibrils are all oriented in the direction of the stress applied to the tendon. Proteoglycans and type VI collagen bind noncovalently to fibrils, coating the surface. The microfibrils of type VI collagen, which contain globular and triple-helical segments, bind to type I fibrils and link them together into thicker fibers. (b) In cartilage, type IX collagen molecules are covalently bound at regular intervals along type II fibrils. A chondroitin sulfate chain, covalently linked to the α2(IX) chain at the flexible kink, projects outward from the fibril, as does the globular N-terminal region. [Part (a), see R. R. Bruns et al., 1986, *J. Cell Biol.* **103**:393. Part (b), see L. M. Shaw and B. Olson, 1991, *Trends Biochem. Sci.* **18**:191.]

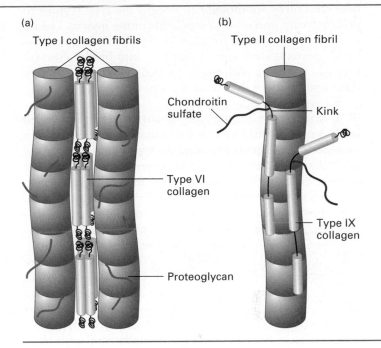

strong fibril. The removal of the propeptides and covalent cross-linking take place in the extracellular space to prevent the potentially catastrophic assembly of fibrils within the cell.

The post-translational modifications of pro-α chains are crucial for the formation of mature collagen molecules and their assembly into fibrils. Defects in these modifications have serious consequences, as ancient mariners frequently experienced. For example, ascorbic acid (vitamin C) is an essential cofactor for the hydroxylases responsible for adding hydroxyl groups to proline and lysine residues in pro-α chains. In cells deprived of ascorbate, as in the disease *scurvy*, the pro-α chains are not hydroxylated sufficiently to form stable triple-helical procollagen at normal body temperature, and the procollagen that forms cannot assemble into normal fibrils. Without the structural support of collagen, blood vessels, tendons, and skin become fragile. Fresh fruit in the diet can supply sufficient vitamin C to support the formation of normal collagen. Historically, British sailors were provided with limes to prevent scurvy, leading to their being called "limeys." Mutations in lysyl hydroxylase genes also can cause connective-tissue defects. ■

## Type I and II Collagens Associate with Nonfibrillar Collagens to Form Diverse Structures

Collagens differ in the structures of the fibers they form and how these fibers are organized into networks. Of the predominant types of collagen found in connective tissues, type I collagen forms long fibers, whereas networks of type II collagen are more meshlike. In tendons, for instance, the long type I collagen fibers connect muscles to bones and must withstand enormous forces. Because type I collagen fibers have great tensile strength, tendons usually can be stretched without being broken. Indeed, gram for gram, type I collagen is stronger than steel. Two quantitatively minor fibrillar collagens, type V and type XI, co-assemble into fibers with type I collagen, thereby regulating the structures and properties of the fibers. Incorporation of type V collagen, for example, results in smaller-diameter fibers.

Type I collagen fibrils are also used as the reinforcing rods in the construction of bone. Bones and teeth are hard and strong because they contain large amounts of dahllite, a crystalline calcium- and phosphate-containing mineral. Most bones are about 70 percent mineral and 30 percent protein, the vast majority of which is type I collagen. Bones form when certain cells (chondrocytes and osteoblasts) secrete collagen fibrils that are then mineralized by deposition of small dahllite crystals.

In many connective tissues, particularly skeletal muscle, proteoglycans and a fibril-associated collagen called type VI collagen are noncovalently bound to the sides of type I fibrils and may bind the fibrils together to form thicker collagen fibers (Figure 20-27a). Type VI collagen is unusual in that the molecule consists of a relatively short triple helix with globular domains at both ends. The lateral association of two type VI monomers generates an "antiparallel" dimer. The end-to-end association of these dimers through their globular domains forms type VI "microfibrils." These microfibrils have a beads-on-a-string appearance, with about 60-nm-long triple-helical regions separated by 40-nm-long globular domains.

The fibrils of type II collagen, the major collagen in cartilage, are smaller in diameter than type I fibrils and are oriented randomly in a viscous proteoglycan matrix. The rigid collagen fibrils impart strength to the matrix and allow it to resist large deformations in shape. Type II fibrils are cross-linked to matrix proteoglycans by type IX collagen, another fibril-associated collagen. Type IX collagen and several related types have two or three triple-helical segments connected by flexible kinks and an N-terminal globular segment (Figure 20-27b). The globular N-terminal segment of type IX collagen extends from the fibrils at the end of one of its helical segments, as does a GAG chain that is sometimes linked to one of the type IX chains. These protruding nonhelical structures are thought to anchor the type II fibril to proteoglycans and other components of the matrix. The interrupted triple-helical structure of type IX and related collagens prevents them from assembling into fibrils, although they can associate with fibrils formed from other collagen types and form covalent cross-links to them.

Mutations affecting type I collagen and its associated proteins cause a variety of human diseases. Certain mutations in the genes encoding the type I collagen α1(I) or α2(I) chains lead to *osteogenesis imperfecta*, or brittle-bone disease. Because every third position in a collagen α chain must be a glycine for the triple helix to form (see Figure 20-24), mutations of glycine to almost any other amino acid are deleterious, resulting in poorly formed and unstable helices. Only one defective α chain of the three in a collagen molecule can disrupt the whole molecule's triple-helical structure and function. A mutation in a single copy (allele) of either the α1(I) gene or the α2(I) gene, which are located on nonsex chromosomes (autosomes), can cause this disorder. Thus it normally shows autosomal dominant inheritance (see Chapter 5).

Absence or malfunctioning of collagen-fibril-associated microfibrils in muscle tissue due to mutations in the type VI collagen genes cause dominant or recessive congenital muscular dystrophies with generalized muscle weakness, respiratory insufficiency, muscle wasting, and muscle-related joint abnormalities. Skin abnormalities have also been reported with type VI collagen disease. ∎

## Proteoglycans and Their Constituent GAGs Play Diverse Roles in the ECM

As we saw with perlecan in the basal lamina, proteoglycans play an important role in cell-ECM adhesion. Proteoglycans are a subset of secreted or cell-surface-attached glycoproteins containing covalently linked specialized polysaccharide chains called **glycosaminoglycans (GAGs)**. GAGs are long linear polymers of specific repeating disaccharides. Usually one sugar is either a uronic acid (D-glucuronic acid or L-iduronic acid) or D-galactose; the other sugar is *N*-acetylglucosamine or *N*-acetylgalactosamine (Figure 20-28). One or both of the sugars contain at least one anionic group (carboxylate or sulfate). Thus each GAG chain bears many negative charges. GAGs are classified into several major types based on the nature of the repeating disaccharide unit: heparan sulfate, chondroitin sulfate, dermatan sulfate, keratan sulfate, and hyaluronan. A hypersulfated form of heparan sulfate called *heparin*, produced mostly by mast cells, plays a key role in allergic reactions. It is also used medically as an anticlotting drug because of its ability to activate a natural clotting inhibitor called antithrombin III.

With the exception of hyaluronan, which is discussed below, all the major GAGs occur naturally as components of proteoglycans. Like other secreted and transmembrane glycoproteins, proteoglycan core proteins are synthesized in the endoplasmic reticulum, and the GAG chains are assembled on and covalently attached to these cores in the Golgi complex.

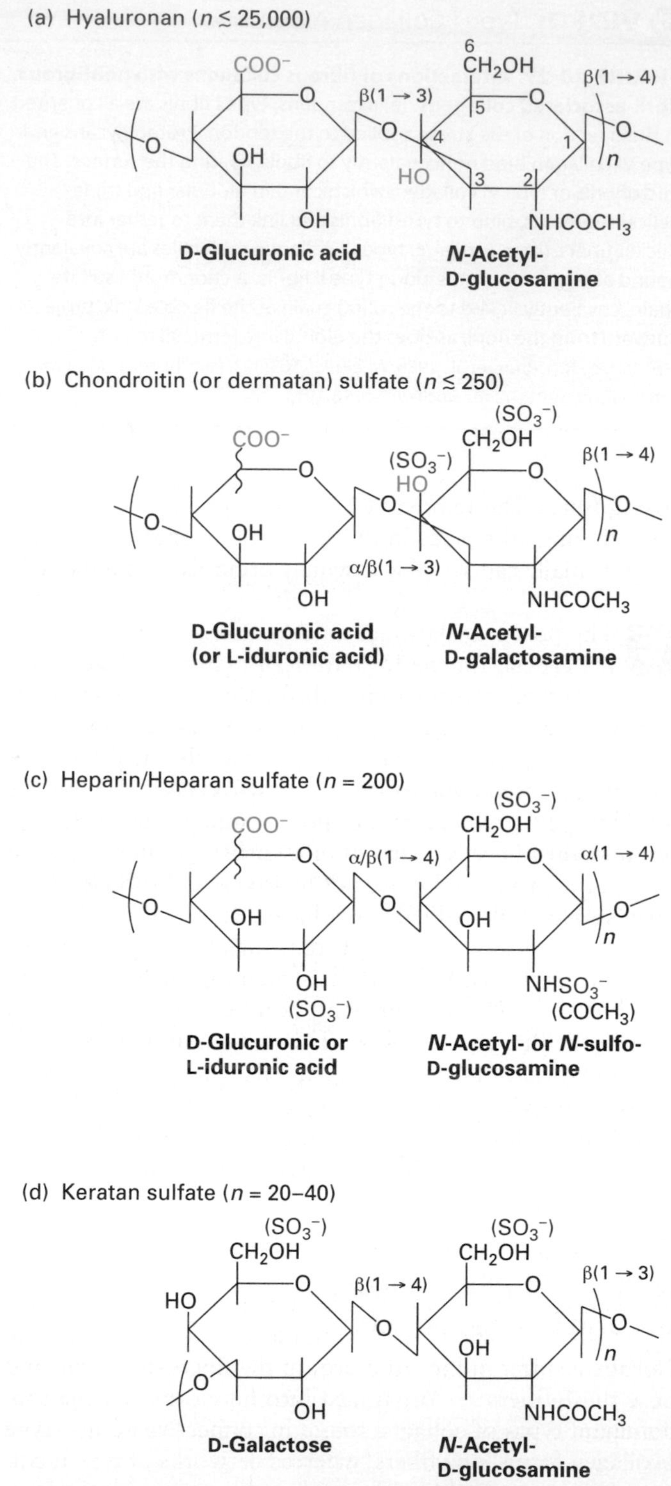

(a) Hyaluronan ($n \leq 25,000$)

(b) Chondroitin (or dermatan) sulfate ($n \leq 250$)

(c) Heparin/Heparan sulfate ($n = 200$)

(d) Keratan sulfate ($n = 20–40$)

**FIGURE 20-28 The repeating disaccharides of glycosaminoglycans (GAGs), the polysaccharide components of proteoglycans.** Each of the four classes of GAGs is formed by polymerization of monomer units into repeats of a particular disaccharide and subsequent modifications, including addition of sulfate groups and inversion (epimerization) of the carboxyl group on carbon 5 of D-glucuronic acid to yield L-iduronic acid. The squiggly lines represent covalent bonds that are oriented either above (D-glucuronic acid) or below (L-iduronic acid) the ring. Heparin is generated by hypersulfation of heparan sulfate, whereas hyaluronan is unsulfated.

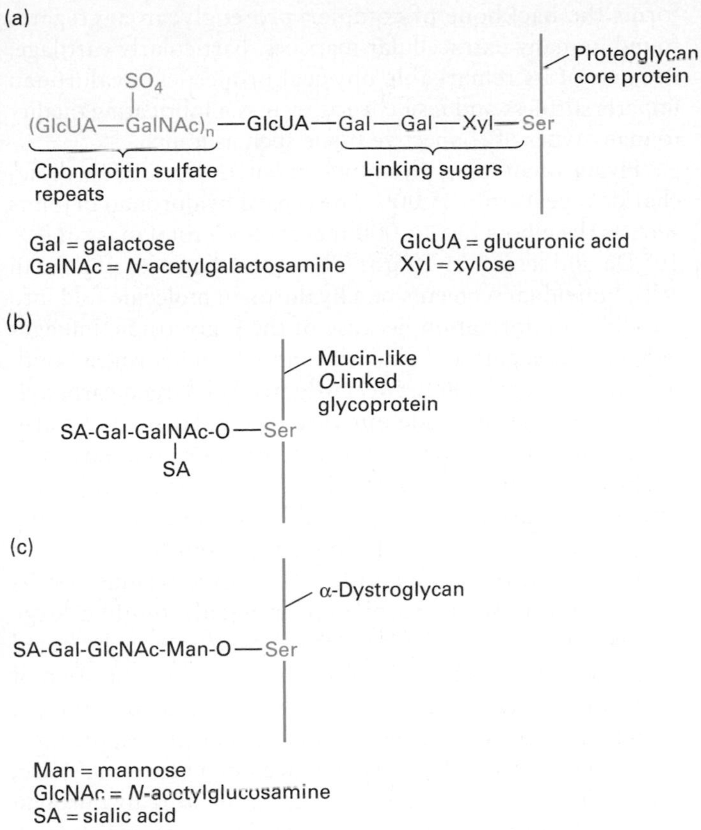

**FIGURE 20-29 Hydroxyl (O-) linked polysaccharides.** (a) Synthesis of a glycosaminoglycan (GAG), in this case chondroitin sulfate, is initiated by transfer of a xylose residue to a serine residue in the core protein, most likely in the Golgi complex, followed by sequential addition of two galactose residues. Glucuronic acid and N-acetylgalactosamine residues are then added sequentially to these linking sugars, forming the chondroitin sulfate chain. Heparan sulfate chains are connected to core proteins by the same three-sugar linker. (b) Mucin-type O-linked chains are covalently bound to glycoproteins via an N-acetylgalactosamine (GalNAc) monosaccharide to which are covalently attached a variety of other sugars. (c) Certain specialized O-linked oligosaccharides, such as those found in the protein dystroglycan, are bound to proteins via mannose (Man) monosaccharides.

To generate heparan or chondroitin sulfate chains, a three-sugar "linker" is first attached to the hydroxyl side chains of certain serine residues in a core protein; thus the linker is called an **O-linked oligosaccharide** (Figure 20-29). In contrast, the linkers for the addition of keratan sulfate chains are oligosaccharide chains attached to asparagine residues; such **N-linked oligosaccharides** are present in many glycoproteins (see Chapter 14), although only a subset carry GAG chains. All GAG chains are elongated by the alternating addition of sugar monomers to form the disaccharide repeats characteristic of a particular GAG; the chains are often modified subsequently by the covalent linkage of small molecules such as sulfate. The mechanisms responsible for determining which proteins are modified with GAGs, the sequence of disaccharides to be added, the sites to be sulfated, and the lengths of the GAG chains are unknown. The ratio of poly-

saccharide to protein in all proteoglycans is much higher than that in most other glycoproteins.

**Function of GAG Chain Modifications** As is the case with the sequence of amino acids in proteins, the arrangement of the sugar residues in GAG chains and the modification of specific sugars in the chains can determine their function and that of the proteoglycans containing them. For example, groupings of certain modified sugars in the GAG chains of heparin sulfate proteoglycans can control the binding of growth factors to certain receptors or the activities of proteins in the blood-clotting cascade.

In the past, the chemical and structural complexity of proteoglycans posed a daunting barrier to analyzing and understanding their structures and many diverse functions. In recent years, investigators employing classical and state-of-the-art biochemical techniques, mass spectrometry, and genetics have begun to elucidate the detailed structures and functions of these ubiquitous ECM molecules. The results of ongoing studies suggest that sets of sugar-residue sequences containing common modifications, rather than single unique sequences, are responsible for specifying distinct GAG functions. A case in point is a set of five-residue (pentasaccharide) sequences found in a subset of heparin GAGs that controls the activity of antithrombin III (ATIII), an inhibitor of the key blood-clotting protease thrombin. When these pentasaccharide sequences in heparin are sulfated at two specific positions (Figure 20-30), heparin can activate ATIII, thereby inhibiting clot formation. Several other sulfates can be present in the active pentasaccharide in various combinations, but they are not essential for the anticlotting activity of heparin. The rationale for generating sets of similar active sequences rather than a single unique sequence is not well understood.

**Diversity of Proteoglycans** The proteoglycans constitute a remarkably diverse group of molecules that are abundant in the extracellular matrix of all animal tissues and are also

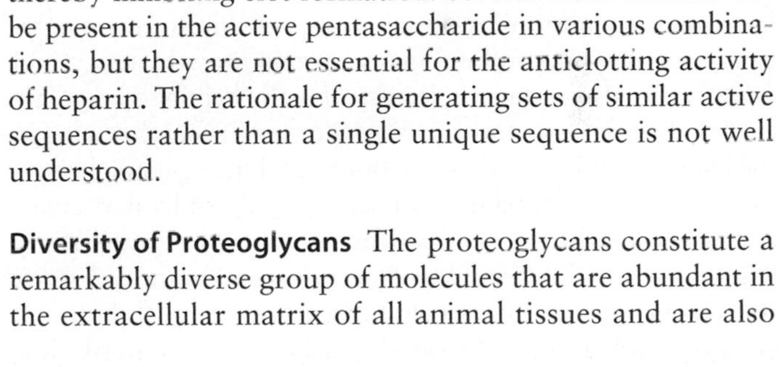

**FIGURE 20-30 Pentasaccharide GAG sequence that regulates the activity of antithrombin III (ATIII).** Sets of modified five-residue sequences in the much longer GAG called heparin with the composition shown here bind to ATIII and activate it, thereby inhibiting blood clotting. The sulfate groups in red type are essential for this heparin function; the modifications in blue type may be present but are not essential. Other sets of modified GAG sequences are thought to regulate the activity of other target proteins.

expressed on the cell surface. For example, of the five major classes of heparan sulfate proteoglycans, three are located in the extracellular matrix (perlecan, agrin, and type XVIII collagen) and two are cell-surface proteins. The latter include integral membrane proteins (syndecans) and GPI-anchored proteins (glypicans); the GAG chains in both types of cell-surface proteoglycans extend into the extracellular space. The sequences and lengths of proteoglycan core proteins vary considerably, and the number of attached GAG chains ranges from just a few to more than 100. Moreover, a core protein is often linked to two different types of GAG chains, generating a "hybrid" proteoglycan. The basal laminal proteoglycan perlecan is primarily a heparan sulfate proteoglycan (HSPG) with three to four GAG chains, although it sometimes can have a bound chondroitin sulfate chain. Additional diversity in proteoglycans occurs because the numbers of chains, compositions, and sequences of the GAGs attached to otherwise identical core proteins can differ considerably. Laboratory generation and analysis of mutants with defects in proteoglycan production in *Drosophila melanogaster* (fruit fly), *C. elegans* (roundworm), and mice have clearly shown that proteoglycans play critical roles in development, most likely as modulators of various signaling pathways.

**Syndecans** are cell-surface proteoglycans expressed by epithelial and nonepithelial cells that bind to collagens and multiadhesive matrix proteins such as fibronectin, anchoring cells to the extracellular matrix. Like that of many integral membrane proteins, the cytosolic domain of syndecan interacts with the actin cytoskeleton and in some cases with intracellular regulatory molecules. In addition, cell-surface proteoglycans such as syndecan bind many protein growth factors and other external signaling molecules, thereby helping to regulate cellular metabolism and function. For instance, syndecans in the hypothalamic region of the brain modulate feeding behavior in response to food deprivation. They do so by participating in the binding to cell-surface receptors of antisatiety peptides that help control feeding behavior. In the fed state, the syndecan extracellular domain decorated with heparan sulfate chains is released from the surface by proteolysis, thus suppressing the activity of the antisatiety peptides and feeding behavior. In mice engineered to overexpress the *syndecan-1* gene in the hypothalamic region of the brain and other tissues, normal control of feeding by antisatiety peptides is disrupted and the animals overeat and become obese.

## Hyaluronan Resists Compression, Facilitates Cell Migration, and Gives Cartilage Its Gel-like Properties

**Hyaluronan**, also called hyaluronic acid (HA) or hyaluronate, is a nonsulfated GAG (see Figure 20-28a) made by a plasma-membrane-bound enzyme called HA synthase and is directly secreted into the extracellular space as it is synthesized. (A similar approach is used by plant cells to make their ECM component cellulose.) HA is a major component of the extracellular matrix that surrounds migrating and proliferating cells, particularly in embryonic tissues. In addition, hyaluronan forms the backbone of complex proteoglycan aggregates found in many extracellular matrices, particularly cartilage. Because of its remarkable physical properties, hyaluronan imparts stiffness and resilience as well as a lubricating quality to many types of connective tissue such as joints.

Hyaluronan molecules range in length from a few disaccharide repeats to ~25,000. The typical hyaluronan in joints such as the elbow has 10,000 repeats for a total mass of $4 \times 10^6$ Da and length of 10 μm (about the diameter of a small cell). Individual segments of a hyaluronan molecule fold into a rodlike conformation because of the β glycosidic linkages between the sugars and extensive intrachain hydrogen bonding. Mutual repulsion between negatively charged carboxylate groups that protrude outward at regular intervals also contributes to these local rigid structures. Overall, however, hyaluronan is not a long, rigid rod as is fibrillar collagen; rather, in solution it is very flexible, bending and twisting into many conformations, forming a random coil.

Because of the large number of anionic residues on its surface, the typical hyaluronan molecule binds a large amount of water and behaves as if it were a large hydrated sphere with a diameter of ~500 nm. As the concentration of hyaluronan increases, the long chains begin to entangle, forming a viscous gel. Even at low concentrations, hyaluronan forms a hydrated gel; when placed in a confining space, such as in a matrix between two cells, the long hyaluronan molecules will tend to push outward. This outward pushing creates a swelling, or *turgor pressure*, within the extracellular space. In addition, the binding of cations by carboxylate ($COO^-$) groups on the surface of hyaluronan increases the concentration of ions and thus the osmotic pressure in the gel. As a result, large amounts of water are taken up into the matrix, contributing to the turgor pressure. These swelling forces give connective tissues their ability to resist compression forces, in contrast with collagen fibers, which are able to resist stretching forces.

Hyaluronan is bound to the surface of many migrating cells by a number of adhesion receptors, such as the receptor called CD44, containing HA-binding domains, each with a similar three-dimensional conformation. Because of its loose, hydrated, porous nature, the hyaluronan "coat" bound to cells appears to keep cells apart from one another, giving them the freedom to move about and proliferate. The cessation of cell movement and the initiation of cell-cell attachments are frequently correlated with a decrease in hyaluronan, a decrease in HA receptors, and an increase in the extracellular enzyme hyaluronidase, which degrades hyaluronan in the matrix. These alterations of hyaluronan are particularly important during the many cell migrations that facilitate differentiation and in the release of a mammalian egg cell (oocyte) from its surrounding cells after ovulation.

The predominant proteoglycan in cartilage, called *aggrecan*, assembles with hyaluronan into very large aggregates, illustrative of the complex structures that proteoglycans sometimes form. The backbone of the cartilage proteoglycan aggregate is a long molecule of hyaluronan to which multiple aggrecan molecules are bound tightly but noncovalently

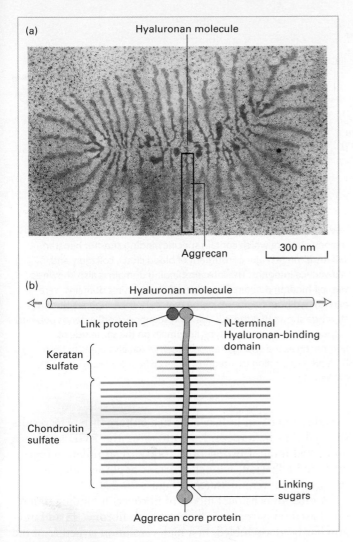

**(a)**

Hyaluronan molecule

Aggrecan

300 nm

**(b)**

Hyaluronan molecule

Link protein

N-terminal
Hyaluronan-binding
domain

Keratan
sulfate

Chondroitin
sulfate

Linking
sugars

Aggrecan core protein

**FIGURE 20-31 Structure of proteoglycan aggregate from cartilage.** (a) Electron micrograph of an aggrecan aggregate from fetal bovine epiphyseal cartilage. Aggrecan core proteins are bound at ~40-nm intervals to a molecule of hyaluronan. (b) Schematic representation of an aggrecan monomer bound to hyaluronan (yellow). In aggrecan, both keratan sulfate (green) and chondroitin sulfate (orange) chains are attached to the core protein. The N-terminal domain of the core protein binds noncovalently to a hyaluronan molecule. Binding is facilitated by a link protein, which binds to both the hyaluronan molecule and the aggrecan core protein. Each aggrecan core protein has 127 Ser-Gly sequences at which GAG chains can be added. The molecular weight of an aggrecan monomer averages $2 \times 10^6$. The entire aggregate, which may contain upward of 100 aggrecan monomers, has a molecular weight in excess of $2 \times 10^8$ and is about as large as the bacterium *E. coli*. [Part (a) from J. A. Buckwalter and L. Rosenberg, 1983, *Coll. Rel. Res.* **3**:489; courtesy of L. Rosenberg.]

(Figure 20-31). A single aggrecan aggregate, one of the largest macromolecular complexes known, can be more than 4 mm long and have a volume larger than that of a bacterial cell. These aggregates give cartilage its unique gel-like properties and its resistance to deformation, essential for distributing the load in weight-bearing joints.

The aggrecan core protein (~250,000 MW) has one N-terminal globular domain that binds with high affinity to a specific decasaccharide sequence within hyaluronan. This specific sequence is generated by covalent modification of some of the repeating disaccharides in the hyaluronan chain. The interaction between aggrecan and hyaluronan is facilitated by a link protein that binds to both the aggrecan core protein and hyaluronan (Figure 20-31b). Aggrecan and the link protein have in common a "link" domain, ~100 amino acids long, that is found in numerous matrix and cell-surface hyaluronan-binding proteins in both cartilaginous and noncartilaginous tissues. Almost certainly these proteins arose in the course of evolution from a single ancestral gene that encoded just this domain.

## Fibronectins Interconnect Cells and Matrix, Influencing Cell Shape, Differentiation, and Movement

Many different cell types synthesize **fibronectin**, an abundant multi-adhesive matrix protein found in all vertebrates. The discovery that fibronectin functions as an adhesive molecule stemmed from observations that it is present on the surfaces of normal fibroblastic cells, which adhere tightly to petri dishes in laboratory experiments, but is absent from the surfaces of tumorigenic (i.e., cancerous) cells, which adhere weakly. The 20 or so isoforms of fibronectin are generated by alternative splicing of the RNA transcript produced from a single gene (see Figure 4-16). Fibronectins are essential for the migration and differentiation of many cell types in embryogenesis. These proteins are also important for wound healing because they promote blood clotting and facilitate the migration of macrophages and other immune cells into the affected area.

Fibronectins help attach cells to the extracellular matrix by binding to other ECM components, particularly fibrous collagens and heparan sulfate proteoglycans, and to cell-surface adhesion receptors such as integrins (see Figure 20-2). Through their interactions with adhesion receptors, fibronectins influence the shape and movement of cells and the organization of the cytoskeleton. Conversely, by regulating their receptor-mediated attachments to fibronectin and other ECM components, cells can sculpt the immediate ECM environment to suit their needs.

Fibronectins are dimers of two similar polypeptides linked at their C-termini by two disulfide bonds; each chain is about 60–70 nm long and 2–3 nm thick. Partial digestion of fibronectin with low amounts of proteases and analysis of the fragments showed that each chain comprises several functional regions with different ligand-binding specificities (Figure 20-32a). Each region, in turn, contains multiple copies of certain sequences that can be classified into one of three types. These classifications are designated fibronectin type I, II, and III repeats, on the basis of similarities in amino acid sequence, although the sequences of any two repeats of a given type are not identical. These linked repeats give the molecule the appearance of beads on a string. The combination of different repeats composing the regions confers on fibronectin its ability to bind multiple ligands.

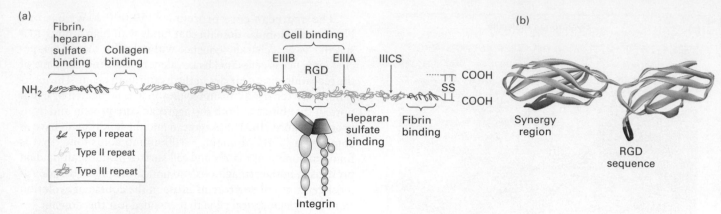

**(a)**

Fibrin, heparan sulfate binding

Collagen binding

Cell binding

EIIIB
RGD
EIIIA
IIICS

COOH
SS
COOH

NH₂

Heparan sulfate binding

Fibrin binding

Integrin

Type I repeat
Type II repeat
Type III repeat

**(b)**

Synergy region

RGD sequence

**FIGURE 20-32 Organization of fibronectin and its binding to integrin.** (a) Scale model of fibronectin is shown docked by two type III repeats to the extracellular domains of integrin. Only one of the two similar chains, which are linked by disulfide bonds near their C-termini, in the dimeric fibronectin molecule is shown. Each chain contains about 2446 amino acids and is composed of three types of repeating amino acid sequences (type I, II, or III repeats) or domains. The EIIIA, EIIIB—both type III repeats—and IIICS domain are variably spliced into the structure at locations indicated by arrows. Circulating fibronectin lacks one or both of EIIIA and EIIIB. At least five different sequences may be present in the IIICS region as a result of alternative splicing (see Figure 4-16). Each chain contains several multirepeat-containing regions, some of which contain specific binding sites for heparan sulfate, fibrin (a major constituent of blood clots), collagen, and cell-surface integrins. The integrin-binding domain is also known as the cell-binding domain. Structures of fibronectin's domains were determined from fragments of the molecule. (b) A high-resolution structure shows that the RGD binding sequence (red) extends outward in a loop from its compact type III domain on the same side of fibronectin as the synergy region (blue), which also contributes to high-affinity binding to integrins. [Adapted from D. J. Leahy et al., 1996, *Cell* **84**:161.]

One of the type III repeats in the cell-binding region of fibronectin mediates binding to certain integrins. The results of studies with synthetic peptides corresponding to parts of this repeat identified the tripeptide sequence Arg-Gly-Asp, called the *RGD sequence*, as the minimal sequence within this repeat required for recognition by those integrins. In one study, heptapeptides containing the RGD sequence or a variation of this sequence were tested for their ability to mediate the adhesion of rat kidney cells to a culture dish. The results showed that heptapeptides containing the RGD sequence mimicked intact fibronectin's ability to stimulate integrin-mediated adhesion, whereas variant heptapeptides lacking this sequence were ineffective (Figure 20-33).

A three-dimensional model of fibronectin binding to integrin based on partial structures of both fibronectin and integrin has been assembled. In a high-resolution structure of the

**EXPERIMENTAL FIGURE 20-33 A specific tripeptide sequence (RGD) in the cell-binding region of fibronectin is required for adhesion of cells.** The cell-binding region of fibronectin contains an integrin-binding hexapeptide sequence, GRGDSP in the single-letter amino acid code (see Figure 2-14). Together with an additional C-terminal cysteine (C) residue, this heptapeptide and several variants were synthesized chemically. Different concentrations of each synthetic peptide were added to polystyrene dishes that had the protein immunoglobulin G (IgG) firmly attached to their surfaces; the peptides were then chemically cross-linked to the IgG. Subsequently, cultured normal rat kidney cells were added to the dishes and incubated for 30 minutes to allow adhesion. After the nonbound cells were washed away, the relative amounts of cells that had adhered firmly were determined by staining the bound cells with a dye and measuring the intensity of the staining with a spectrophotometer. The results shown here indicate that cell adhesion increased above the background level with increasing peptide concentration for those peptides containing the RGD sequence but not for the variants lacking this sequence (modification underlined). [From M. D. Pierschbacher and E. Ruoslahti, 1984, *Proc. Nat'l. Acad. Sci. USA* **81**:5985.]

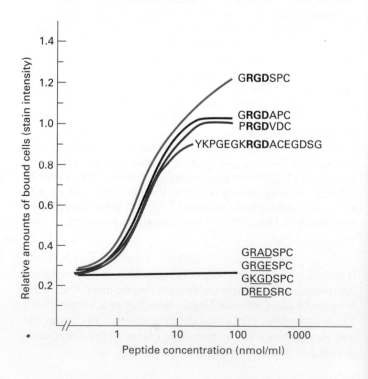

**EXPERIMENTAL FIGURE 20-34 Integrins mediate linkage between fibronectin in the extracellular matrix and the cytoskeleton.** (a) Immunofluorescent micrograph of a fixed cultured fibroblast showing colocalization of the α5β1 integrin (green) and actin-containing stress fibers (red). The cell was incubated with two types of monoclonal antibody: an integrin-specific antibody linked to a green fluorescing dye and an actin-specific antibody linked to a red fluorescing dye. Stress fibers are long bundles of actin microfilaments that radiate inward from points where the cell contacts a substratum. At the distal end of these fibers, near the plasma membrane, the coincidence of actin (red) and fibronectin-binding integrin (green) produces a yellow fluorescence. (b) Electron micrograph of the junction of fibronectin and actin fibers in a cultured fibroblast. Individual actin-containing 7-nm microfilaments, components of a stress fiber, end at the obliquely sectioned cell membrane. The microfilaments appear aligned in close proximity to the thicker, densely stained fibronectin fibrils on the outside of the cell. [Part (a) from J. Duband et al., 1988, *J. Cell Biol.* **107**:1385. Part (b) from I. J. Singer, 1979, *Cell* **16**:675; courtesy of I. J. Singer; copyright 1979, MIT.]

(a)

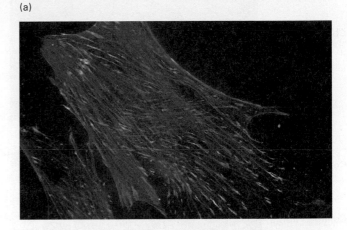

(b)

| Fibronectin fibrils | Cell exterior | Plasma membrane | Actin-containing microfilaments |

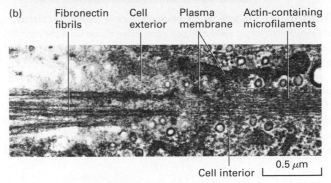

Cell interior    0.5 μm

integrin-binding fibronectin type III repeat and its neighboring type III domain, the RGD sequence is at the apex of a loop that protrudes outward from the molecule, in a position facilitating binding to integrins (Figure 20-32b). Although the RGD sequence is required for binding to several integrins, its affinity for integrins is substantially less than that of intact fibronectin or of the entire cell-binding region in fibronectin. Thus structural features near the RGD sequence in fibronectins (e.g., parts of adjacent repeats, such as the synergy region; see Figure 20-32b) and in other RGD-containing proteins enhance their binding to certain integrins. Moreover, the simple soluble dimeric forms of fibronectin produced by the liver or fibroblasts are initially in a nonfunctional conformation that binds poorly to integrins because the RGD sequence is not readily accessible. The adsorption of fibronectin to a collagen matrix or the basal lamina or, experimentally, to a plastic tissue-culture dish results in a conformational change that enhances the ability of fibronectin to bind to cells. Possibly, this conformational change increases the accessibility of the RGD sequence for integrin binding.

Microscopy and other experimental approaches (e.g., biochemical binding experiments) have demonstrated the role of integrins in cross-linking fibronectin and other ECM components to the cytoskeleton. For example, the colocalization of cytoskeletal actin filaments and integrins within cells can be visualized by fluorescence microscopy (Figure 20-34a). The binding of cell-surface integrins to fibronectin in the matrix induces the actin-cytoskeleton-dependent movement of some integrin molecules in the plane of the membrane. The ensuing mechanical tension due to the relative movement of different integrins bound to a single fibronectin dimer stretches the fibronectin. This stretching promotes self-association of fibronectins into multimeric fibrils.

The force needed to unfold and expose functional self-association sites in fibronectin is much less than that needed to disrupt fibronectin-integrin binding. Thus fibronectin molecules remain bound to integrin while cell-generated mechanical forces induce fibril formation. In effect, the integrins through adapter proteins transmit the intracellular forces generated by the actin cytoskeleton to extracellular fibronectin (inside-out signaling). Gradually, the initially formed fibronectin fibrils mature into highly stable matrix components by covalent cross-linking. In some electron micrographic images, exterior fibronectin fibrils appear to be aligned in a seemingly continuous line with bundles of actin fibers within the cell (Figure 20-34b). These observations and the results from other studies provided the first example of a molecularly well-defined adhesion receptor forming a bridge between the intracellular cytoskeleton and the extracellular matrix components—a phenomenon now known to be widespread.

## Elastic Fibers Permit Many Tissues to Undergo Repeated Stretching and Recoiling

Elastic fibers are found in the ECM of a wide variety of tissues (Figure 20-35a) that are subject to mechanical strain or deformation, such as lung expansion and contraction during breathing, the pulsatile flow of blood through blood vessels due to the heartbeat, and the stretching and contraction of skin. Elastic fibers permit the reversible elastic stretching and recoiling of these tissues.

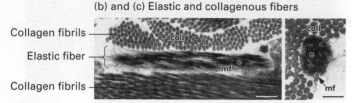

**(a) Connective tissue**

Collagen fibers

Elastic fibers

Nuclei

**(b) and (c) Elastic and collagenous fibers**

Collagen fibrils

Elastic fiber

Collagen fibrils

coll

coll

mf

e

e

mf

**FIGURE 20-35 Elastic and collagen fibers in connective tissue.**
(a) Light-microscopic view of loose connective tissue from the lung.
Elastic fibers are the thin fibers that are stained purple, collagen fibers
(bundles of collagen fibrils) are stained pink, and the nuclei of cells are
stained purple. (b) Longitudinal and (c) cross-sectional electron-
microscopic views of elastic fibers and collagen fibrils (coll) in the skin
of a mouse. The elastic fibers have a solid core of elastin (e) integrated
into and surrounded by a bundle of microfibrils (mf) Scale bars, 0.25 μm.
[Part (a) from Science Photo LIbrary; parts (b) and (c) from J. Choi et al., 2009,
*Matrix Biol.* **28**:211–220.]

The major component of an elastic fiber, which can be several hundred to several thousand nm in diameter, is an insoluble, amorphous core composed of the protein *elastin* surrounded
by a collection of 10–12 nm diameter *microfibrils* made up of
proteins fibrillin, fibulin, and associated proteins (Figure
20-35b). The core is formed by the aggregation of *tropoelastin*
molecules, elastin monomers that are covalently cross-linked
via a lysyl-oxidase-mediated process similar to that seen in collagen. Repetitive proline- and glycine-enriched hydrophobic
sequence motifs contribute to the ability of tropoelastins to self-
associate and recoil efficiently after stretching.

A variety of diseases, often involving skeletal and cardiovascular abnormalities, are consequences of mutations in the genes encoding the structural proteins of elastic
fibers or the proteins that contribute to their proper assembly.
For example, mutations in the fibrillin-1 gene cause Marfan
syndrome, whose varied symptoms can include bone overgrowth, loose joints, abnormally long extremities and face,
and cardiovascular defects due to weakness in the walls of
vessels and the aorta. There has been considerable speculation
that President Abraham Lincoln's unusually tall, elongated
body may have been a consequence of Marfan syndrome. ∎

In mammals, the synthesis of most tropoelastin only occurs
immediately before and after birth during the late fetal and
neonatal periods. Thus, most of the body's elastin must be very
durable, lasting an entire lifetime. The extraordinary stability

of elastin has been measured in a variety of ways. Pulse-chase
experiments (see Chapter 3) using radioactive amino acid administration can be used to measure the lifetime of elastin in
animals. In humans, two other methods were employed to
study the longevity of elastin and revealed that the mean lifetime of an elastin molecule in human lungs is about 70 years!
The first method takes advantage of a naturally occurring phenomenon: the slow, natural rate of conversion of L-aspartic
acid—incorporated into proteins during their synthesis—to D-
aspartic acid. Thus the age of a long-lived protein can be estimated using chemical analysis to determine the fraction of its
L-aspartic acid that has been converted over time to the D isomer, together with knowledge of the age of the tissue from
which it was isolated. The second method is a variation on the
classical pulse-chase experiments used in the laboratory. As a
consequence of nuclear weapons testing in the 1950s and
1960s, $^{14}$C was introduced into the atmosphere and hence the
food chain. This environmental $^{14}$C has been used as the radioactive "pulse" in what is essentially a pulse-chase experiment
to determine the stability of proteins of interest.

## Metalloproteases Remodel and Degrade the Extracellular Matrix

Many key physiologic processes, including tissue morphogenesis during development, control of cellular proliferation and
motility, response to injury, and even survival, require not only
the production of ECM but also its remodeling or degradation.
Because of its enormous importance as a key element in the
extracellular microenvironment of multicellular organisms, re-
modeling/degradation of the ECM must be carefully controlled.
Degradation of the ECM is often mediated by zinc-dependent
**matrix metalloproteases (MMPs)**. Given the wide array of
ECM components, it is not surprising that there are many such
metalloproteases with varying substrate specificities and sites of
expression. In many cases their names incorporate the names of
their substrates, as for the MMPs called collagenases, gelatin-
ases, elastases, and aggrecanases. Some are secreted into the
extracellular fluid, and others are closely associated with the
plasma membranes of cells, either tightly bound in a noncovalent association with the membrane or as integral membrane
proteins. Many are initially synthesized as inactive precursors
that must be specifically activated to function.

There are three major subgroups of ECM proteases based
on the enzymes' structures: MMPs (of which there are 23 in
humans), a disintegrin and metalloproteinases (ADAMs), and
ADAM with thrombospondin motifs (ADAMTSs). These proteases can degrade ECM components as well as non-ECM
components such as surface-adhesion receptors. Indeed, a key
function of ADAMs is cleaving extracellular domains from integral membrane protein substrates. One mechanism used to
control the activities of these proteases is the production of
protein inhibitors called TIMPs (tissue inhibitors of metalloproteinases) and RECK (reversion-inducing-cysteine-rich protein with kazal motifs). Some of these have their own
cell-surface receptors and functions independent of their ability
to inhibit metalloproteinases. ECM-degrading proteases are

associated with a variety of diseases, the best known of which is metastatic (spreading) cancer (see Chapter 24).

## KEY CONCEPTS of Section 20.4

### The Extracellular Matrix II: Connective Tissue

- Connective tissue, such as tendon and cartilage, differs from other solid tissues in that most of its volume is made up of extracellular matrix (ECM) rather than cells.
- The synthesis of fibrillar collagen (e.g., types I, II, and III) begins inside the cell with the chemical modification of newly made α chains and their assembly into triple-helical procollagen within the endoplasmic reticulum. After secretion, procollagen molecules are cleaved, associate laterally, and are covalently cross-linked into bundles called fibrils, which can form larger assemblies called fibers (see Figure 20-26).
- The various collagens are distinguished by the ability of their helical and nonhelical regions to associate into fibrils, to form sheets, or to cross-link other collagen types (see Table 20-4).
- Proteoglycans consist of membrane-associated or secreted core proteins covalently linked to one or more glycosaminoglycan (GAG) chains, which are linear polymers of disaccharides that are often modified by sulfation.
- Cell-surface proteoglycans such as the syndecans facilitate cell-matrix interactions and help present certain external signaling molecules to their cell-surface receptors.
- Hyaluronan, a highly hydrated GAG, is a major component of the ECM of migrating and proliferating cells. Certain cell-surface adhesion receptors bind hyaluronan to cells.
- Large proteoglycan aggregates containing a central hyaluronan molecule noncovalently bound to the core protein of proteoglycan molecules (e.g., aggrecan) contribute to the ability of the matrix to resist compressive forces (see Figure 20-31).
- Fibronectins are abundant multi-adhesive matrix proteins that play a key role in migration and cellular differentiation. They contain binding sites for integrins and ECM components (collagens, proteoglycans) and thus can attach cells to the matrix (see Figure 20-32).
- The tripeptide RGD sequence Arg-Gly-Asp, found in fibronectins and some other matrix proteins, is recognized by several integrins.
- Elastic fibers permit repeated stretching and recoiling of tissues because of their highly elastic core of cross-linked, amorphous elastin, which is surrounded by a network of microfibrils.
- The remodeling/degradation of ECM is often mediated by a large number of secreted and cell-membrane-associated zinc metalloproteases that fall into several families (MMPs, ADAMs, ADAMTSs) and whose activities are regulated by activation and by protein inhibitors (TIMPs and RECK).

## 20.5 Adhesive Interactions in Motile and Nonmotile Cells

After adhesive interactions in epithelia form during differentiation, they often are very stable and can last throughout the life span of the cells or until the epithelium undergoes further differentiation. Although such long-lasting (nonmotile) adhesion also exists in nonepithelial tissues, some nonepithelial cells must be able to crawl across or through a layer of extracellular matrix or other cells. Moreover, during development or wound healing and in certain pathologic states (e.g., cancer), epithelial cells can transform into more motile cells (the epithelial-mesenchymal transition). Changes in expression of adhesive molecules play a key role in this transformation, as they do in other biological processes involving cell movement, such as the crawling of white blood cells into tissue sites of infection. In this section, we describe various cell-surface structures that mediate transient adhesive interactions that are especially adapted for the movement of cells as well as those that mediate long-lasting adhesion. The detailed intracellular mechanisms used to generate the mechanical forces that propel cells and modify their shapes are covered in Chapters 17 and 18.

### Integrins Relay Signals between Cells and Their Three-Dimensional Environment

As already discussed, integrins connect epithelial cells to the basal lamina and, through adapter proteins, to intermediate filaments of the cytoskeleton (see Figure 20-1). That is, integrins form a bridge between the ECM and cytoskeleton; they do the same in nonepithelial cells. In nonepithelial cells, integrins in the plasma membrane also are clustered with other molecules in various adhesive structures called focal adhesions, focal contacts, focal complexes, 3-D adhesions, and fibrillar adhesions, as well as in circular adhesions called podosomes. These structures are multiprotein complexes that are collectively called the integrin adhesome. These complexes mediate mechanosensory coupling between cells and their environments, and are readily observed by fluorescence microscopy with the use of antibodies that recognize integrins or other co-clustered molecules (Figure 20-36). As with cell-matrix anchoring junctions in epithelial cells, these various adhesive structures attach nonepithelial cells to the extracellular matrix through, for example, binding to fibronectin (see Figure 20-34). They also mediate integrin association with the actin cytoskeleton and activate adhesion-dependent signals for cell growth and cell motility (see Figure 20-8).

Integrin-containing adhesive structures are dynamic (components moving in and out constantly) and contain dozens of intracellular adapter and associated proteins (more than 190 identified to date) with the potential to engage in many hundreds (more than 740) of distinct protein-protein interactions that may be subject to regulation. For example, binding sites generated by phosphorylation of integrin and its

**EXPERIMENTAL FIGURE 20-36 Integrins cluster into adhesive structures with various morphologies in nonepithelial cells.** Immunofluorescence methods were used to detect adhesive structures (green) on cultured cells. Shown here are focal adhesions (a) and 3-D adhesions (b) on the surfaces of human fibroblasts. Cells were grown directly on the flat surface of a culture dish (a) or on a three-dimensional matrix of ECM components (b). The shape, distribution, and composition of the integrin-based adhesions formed by cells vary, depending on the cells' environment. [Part (a) from B. Geiger et al., 2001, *Nature Rev. Mol. Cell Biol.* **2**:793; part (b) courtesy of K. Yamada and E. Cukierman; see E. Cukierman et al., 2001, *Science* **294**:1708–1712.]

(a) Focal adhesion

(b) 3-D adhesion

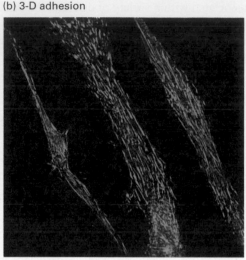

associated proteins as well as by generation of phosphorylated derivatives of phosphatidylinositol in the adjacent membrane recruit additional proteins into and can also cause release of some proteins from these multiprotein complexes. There is a tightly controlled choreography of internal signals, contributions of other signaling pathways such as those involving receptor tyrosine kinases (see Figure 20-8), and external signals (such as rigidity of the ECM) that regulate these complexes. Together they help define the precise composition and activity of the integrin multiprotein complex and the consequent influence that it has on cellular structure and activity (outside-in effect) as well as the influence of the cellular actin cytoskeleton on the ECM (inside-out effect).

Although found in many nonepithelial cells, integrin-containing adhesive structures have been studied most frequently in fibroblasts grown in cell culture on flat glass or plastic surfaces called substrata. These conditions only poorly approximate the three-dimensional ECM environment that normally surrounds such cells in vivo. When fibroblasts are cultured in three-dimensional ECM matrices derived from cells or tissues, they form adhesions to the three-dimensional ECM substratum, called 3-D adhesions. These structures differ somewhat in composition, shape, distribution, and activity from the focal or fibrillar adhesions seen in cells growing on the flat substratum typically used in cell-culture experiments (see Figure 20-36). Cultured fibroblasts with these "more natural" anchoring junctions display greater adhesion and mobility, increased rates of cell proliferation, and spindle-shaped morphologies more like those of fibroblasts in tissues than do cells cultured on flat surfaces. These observations indicate that the topological, compositional, and mechanical properties of the extracellular matrix all play a role in controlling the shape and activity of a cell. Tissue-specific differences in these matrix characteristics probably contribute to the tissue-specific properties of cells.

The importance of the 3-D environment of cells has been highlighted by cell culture studies of the morphogenesis, functioning, and stability of specialized milk-producing mammary epithelial cells and their cancerously transformed counterparts. For example, the 3-D matrix-dependent outside-in signaling mediated by integrins influences the epidermal growth-factor tyrosine-kinase-receptor signaling system and vice versa. The 3-D matrix also permits the mammary epithelial cells to generate in-vivo-like circular epithelial structures, called acini. Acini secrete the major protein constituents of milk and permit comparisons of the responses of normal and cancerous cells to potential chemotherapeutic agents. Analogous systems employing both natural and synthetic 3-D ECMs are being developed to provide more in-vivo-like conditions to study other complex tissues and organs, such as the liver.

## Regulation of Integrin-Mediated Adhesion and Signaling Controls Cell Movement

Cells can exquisitely control the strength of integrin-mediated cell-matrix interactions by regulating integrin expression levels, ligand-binding activities, or both. Such regulation is critical to the role of these interactions in cell migration and other functions involving cell movement.

**Integrin Binding** Many, if not all, integrins can exist in two conformations: a low-affinity (inactive) form and a high-affinity (active) form (Figure 20-37a). The results of structural studies and experiments investigating the binding of ligands by integrins have provided a model of the changes that take place when integrins are activated. In the inactive state, the αβ heterodimer is bent (Figure 20-37a, top; and 20-37c), the conformation of the ligand-binding site at the tip of the molecule allows only low-affinity ligand binding, and the transmembrane domains and cytoplasmic C-terminal tails of the two subunits are closely bound together. In the active state, subtle structural alterations in the conformation of the domains that form the binding site permit tighter (high-affinity) ligand binding and this is accompanied by separation of the transmembrane and cytoplasmic domains (Figure 20-37a, bottom). Activation is also accompanied by the straightening of the bent conformation into a more extended,

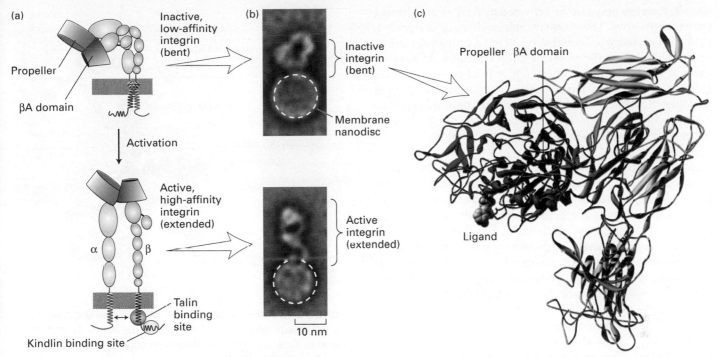

**EXPERIMENTAL FIGURE 20-37 Model for integrin activation.**
(a) Activation of integrins is thought to be due to conformational changes that include key movements near the propeller and βA domains, which increase the affinity for ligands. These are accompanied by straightening of the molecule from the inactive, low affinity, "bent" conformation (*top*) to the active, high affinity, "extended" conformation (*bottom*). Activation also involves separation (indicated by double-headed arrow, bottom) of the transmembrane and cytoplasmic domains, induced by or resulting in altered interactions with adapter proteins talin and kindlin, whose sites of binding to the cytoplasmic tail of the β chain are indicated by green and yellow ovals, respectively. (b) Single inactive (bent) integrin αIIbβ3 molecules (top panel) were incorporated into phospholipid nanodiscs (small bilayers in which the extracellular and cytoplasmic domains of the integrin are exposed to the buffer) and to some of these (bottom panel) were added the integrin-binding and activating "head" domain of the adaptor protein talin. Multiple electron microscopic images of individual discs were collected and averaged. Phospholipid nanodiscs are indicated by dashed white circles, the heights of the integrin extracellular regions that extend above the nanodiscs are indicated by brackets. (c) A molecular model of the extracellular region of αvβ3 integrin in its inactive, low-affinity ("bent") form, with the α subunit in shades of blue and the β subunit in shades of red is based on the x-ray crystal structure. The major ligand-binding sites are at the tip of the molecule, where the propeller domain of the α subunit (dark blue) and βA domain (dark red) interact. An RGD peptide ligand is shown in yellow. [Adapted from M. Arnaout et al., 2002, *Curr. Opin. Cell Biol.* **14**:641; R. O. Hynes, 2002, *Cell* **110**:673; Feng Ye et al. 2010, *J CELL BIOL.* **188**:157–73; and M. Moser et al., 2009, *Science* **324**:895–899.]

linear form in which the ligand-binding site is projected farther away from the surface of the membrane (see Figure 20-37).

These structural models provide an attractive explanation for the ability of integrins to mediate outside-in and inside-out signaling. The binding of certain ECM molecules or CAMs on other cells to the integrin's extracellular ligand-binding site would hold the integrin in the active form with separated cytoplasmic tails. Intracellular adapters could "sense" the separation of the tails and, as a result, either bind to or dissociate from the tails. The changes in these adapters could then alter the cytoskeleton and activate or inhibit intracellular signaling pathways. Conversely, changes in the metabolic/signaling state of the cells could cause intracellular adapters to bind to or dissociate from the cytoplasmic tails of integrins and thus force the tails to either separate or associate. As a consequence, the integrin would either bend (inactivate) or straighten (activate), thereby altering its interaction with the ECM or other cells. Indeed, in vitro studies of purified integrins reconstituted individually into lipid bilayer "nanodiscs" show that binding of the globular domain of the adapter protein *talin* to the cytoplasmic tail of integrin's β chain is sufficient to activate integrin, inducing an extension of the bent conformation to an extended, active form (Figure 20-37a, bottom; and 20-37b, bottom). Other studies suggest that the efficient activation of integrins in intact cells may also require the participation of another class of adapter proteins called kindlins, which bind to a distinct site on the cytoplasmic tail of integrin's β chain (see Figure 20-37 a, bottom).

Platelet function, discussed in more detail below, provides a good example of how cell-matrix interactions are modulated by controlling integrin binding activity. Platelets are cell fragments that circulate in the blood and clump together with ECM molecules to form a blood clot. In its basal state, the aIIbβ3 integrin present on the plasma membranes of platelets normally cannot bind tightly to its protein ligands (including fibrinogen and fibronectin), all of which participate in the formation of a blood clot, because it is in the inactive (bent) conformation. During clot formation, platelets bind to ECM proteins such as collagen and a large protein called von Willabrand factor through receptors that generate intracellular signals. Platelets may also be activated by ADP or the clotting enzyme thrombin. These signals induce changes in signaling pathways in the cytoplasm that result in an activating conformational change in the platelet's αIIbβ3 integrin. As a consequence, this integrin can then tightly bind to extracellular clotting proteins and participate in clot formation. People with genetic defects in the β3 integrin subunit are prone to excessive bleeding, attesting to the role of the αIIbβ3 integrin in the formation of blood clots (see Table 20-3).

**Integrin Expression** The attachment of cells to ECM components can also be modulated by altering the number of integrin molecules exposed on the cell surface. The α4β1 integrin, which is found on many hematopoietic cells, offers an example of this regulatory mechanism. For these hematopoietic cells to proliferate and differentiate, they must be attached to fibronectin synthesized by supportive (stromal) cells in the bone marrow. The α4β1 integrin on hematopoietic cells binds to a Glu-Ile-Leu-Asp-Val (EILDV) sequence in fibronectin, thereby anchoring the cells to the matrix. This integrin also binds to a sequence in a CAM called vascular CAM-1 (VCAM-1), which is present on stromal cells of the bone marrow. Thus hematopoietic cells directly contact the stromal cells as well as attach to the matrix. Late in their differentiation, hematopoietic cells decrease their expression of this integrin; the resulting reduction in the number of α4β1 integrin molecules on the cell surface is thought to allow mature blood cells to detach from the matrix and stromal cells in the bone marrow and subsequently enter the circulation.

## Connections Between the ECM and Cytoskeleton Are Defective in Muscular Dystrophy

The importance of the adhesion-receptor-mediated linkage between ECM components and the cytoskeleton is highlighted by a set of hereditary muscle-wasting diseases, collectively called muscular dystrophies. Duchenne muscular dystrophy (DMD), the most common type, is a sex-linked disorder, affecting 1 in 3300 boys, that results in cardiac or respiratory failure, usually in the late teens or early twenties. The first clue to understanding the molecular basis of this disease came from the discovery that people with DMD carry mutations in the gene encoding a protein named *dystrophin*. This very large protein was found to be a cytosolic adapter protein, binding to actin filaments and to an adhesion receptor called *dystroglycan* (Figure 20-38).

Dystroglycan is synthesized as a large glycoprotein precursor that is proteolytically cleaved into two subunits soon after it is synthesized and before it moves to the cell surface. The α subunit is a peripheral membrane protein, and the β subunit is a transmembrane protein whose extracellular domain associates with the α subunit (Figure 20-38). Multiple O-linked oligosaccharides are attached covalently to side-chain hydroxyl groups of serine and threonine residues in the α subunit. Unlike the most abundant O-linked, also called mucin-like, oligo-

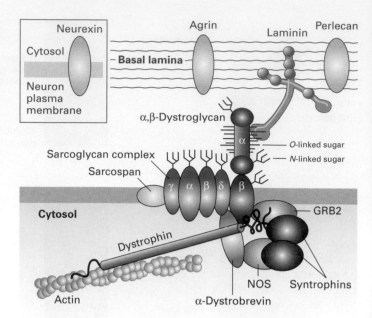

**FIGURE 20-38 Dystrophin glycoprotein complex (DGC) in skeletal muscle cells.** This schematic model shows that the DGC comprises three subcomplexes: the α,β dystroglycan subcomplex; the sarcoglycan/sarcospan subcomplex of integral membrane proteins; and the cytosolic adapter subcomplex comprising dystrophin, other adapter proteins, and signaling molecules. Through its O-linked sugars, β-dystroglycan binds to components of the basal lamina, such as laminin and perlecan, and cell-surface proteins, such as neurexin in neurons. Dystrophin—the protein defective in Duchenne muscular dystrophy—links β-dystroglycan to the actin cytoskeleton, and α-dystrobrevin links dystrophin to the sarcoglycan/sarcospan subcomplex. Nitric oxide synthase (NOS) produces nitric oxide, a gaseous signaling molecule, and GRB2 is a component of signaling pathways activated by certain cell-surface receptors (see Chapter 15). [Adapted from S. J. Winder, 2001, *Trends Biochem. Sci.* **26**:118, and D. E. Michele and K. P. Campbell, 2003, *J. Biol. Chem.* **278**(18):15457–15460.]

saccharides, in which an N-acetylgalactosamine (GalNAc) is the first sugar in the chain linked directly to the hydroxyl group of the side chain of serine or theonine (Figure 20-29b) or the linkage in proteoglycans (Figure 20-29a), many of the O-linked chains in dystroglycan are directly linked to the hydroxyl group via a mannose sugar (see Figure 20-29c).

These specialized O-linked oligosaccharides bind to various basal lamina components, including the LG domains of the multi-adhesive matrix protein laminin and the proteoglycans perlecan and agrin. The neurexins, a family of adhesion molecules expressed by neurons, also are bound via these specialized oligosaccharides, whose detailed heterogeneous structures and mechanisms of synthesis have not been fully elucidated.

The transmembrane segment of the dystroglycan β subunit associates with a complex of integral membrane proteins; its cytosolic domain binds dystrophin and other adapter proteins as well as various intracellular signaling proteins (see Figure 20-38). The resulting large, heteromeric assemblage, the *dystrophin glycoprotein complex (DGC)*, links the extracellular matrix to the cytoskeleton and signaling pathways within muscle and other types of cells. For instance, the signaling enzyme nitric oxide synthase (NOS) is associated through syntrophin with the cytosolic dystrophin subcomplex in skeletal muscle. The rise in intracellular $Ca^{2+}$ during muscle contraction activates NOS to produce nitric oxide (NO), a signaling molecule that diffuses into smooth muscle cells surrounding nearby blood vessels. NO promotes smooth muscle relaxation, leading to a local rise in the flow of blood supplying nutrients and oxygen to the skeletal muscle.

Mutations in dystrophin, other DGC components, laminin, or enzymes that add the O-linked sugars to dystroglycan can all disrupt the DGC-mediated link between the exterior and the interior of muscle cells and cause muscular dystrophies. In addition, dystroglycan mutations have been shown to greatly reduce the clustering of acetylcholine receptors on muscle cells at the neuromuscular junctions, which is also dependent on the basal lamina proteins laminin and agrin. These and possibly other effects of DGC defects apparently lead to a cumulative weakening of the mechanical stability of muscle cells as they undergo contraction and relaxation, resulting in deterioration of the cells and muscular dystrophy.

Dystroglycan provides an elegant—and medically relevant—example of the intricate networks of connectivity in cell biology. Dystroglycan was originally discovered in the context of studying DMD. However, it was later shown to be expressed in nonmuscle cells and, through its binding to laminin, to play a key role in the assembly and stability of at least some basement membranes. Thus it is essential for normal development. Additional studies led to its identification as a cell-surface receptor for the virus that causes the frequently fatal human disease Lassa fever and other related viruses, all of which bind via the same specialized O-linked sugars that mediate binding to laminin. Furthermore, dystroglycan is also the receptor on specialized cells in the nervous system—Schwann cells—to which binds the pathogenic bacterium *Mycobacterium leprae*, the organism that causes leprosy. ■

## IgCAMs Mediate Cell-Cell Adhesion in Neuronal and Other Tissues

Numerous transmembrane proteins characterized by the presence of multiple immunoglobulin domains in their extracellular regions constitute the immunoglobulin (Ig) superfamily of CAMs, or **IgCAMs**. The Ig domain is a common protein domain, containing 70–110 residues, that was first identified in antibodies, the antigen-binding immunoglobulins, but has a much older evolutionary origin in CAMs. The human, *D. melanogaster*, and *C. elegans* genomes include about 765, 150, and 64 genes, respectively, that encode proteins containing Ig domains. Immunoglobulin domains are found in a wide variety of cell-surface proteins, including T-cell receptors produced by lymphocytes and many proteins that take part in adhesive interactions. Among the IgCAMs are neural CAMs; intercellular CAMs (ICAMs), which function in the movement of leukocytes into tissues; and junction adhesion molecules (JAMs), which are present in tight junctions.

As their name implies, neural CAMs are of particular importance in neural tissues. One type, the NCAMs, primarily mediate homophilic interactions. First expressed during morphogenesis, NCAMs play an important role in the differentiation of muscle, glial, and nerve cells. Their role in cell adhesion has been directly demonstrated by the inhibition of adhesion with anti-NCAM antibodies. Numerous NCAM isoforms, encoded by a single gene, are generated by alternative mRNA splicing and by differences in glycosylation. Other neural CAMs (e.g., L1-CAM) are encoded by different genes. In humans, mutations in different parts of the L1-CAM gene cause various neuropathologies (e.g., mental retardation, congenital hydrocephalus, and spasticity).

An NCAM comprises an extracellular region with five Ig domains and two fibronectin type III domains, a single membrane-spanning segment, and a cytosolic segment that interacts with the cytoskeleton (see Figure 20-2). In contrast, the extracellular region of L1-CAM has six Ig domains and four fibronectin type III domains. As with cadherins, cis (intracellular) interactions and trans (intercellular) interactions probably play key roles in IgCAM-mediated adhesion (see Figure 20-3); however, adhesion mediated by IgCAMs is $Ca^{2+}$ independent.

The covalent attachment of multiple chains of sialic acid, a negatively charged sugar derivative, to NCAMs alters their adhesive properties. In embryonic tissues such as brain, polysialic acid constitutes as much as 25 percent of the mass of NCAMs. Possibly because of repulsion between the many negatively charged sugars in these NCAMs, cell-cell contacts are fairly transient, being made and then broken, a property necessary for the development of the nervous system. In contrast, NCAMs from adult tissues contain only one-third as much sialic acid, permitting more stable adhesions.

## Leukocyte Movement into Tissues Is Orchestrated by a Precisely Timed Sequence of Adhesive Interactions

In adult organisms, several types of white blood cells (leukocytes) participate in the defense against infection caused by bacteria and viruses or by tissue damage due to trauma or inflammation. To fight infection and clear away damaged tissue, these cells must move rapidly from the blood, where they circulate as unattached, relatively quiescent cells, into the underlying tissue at sites of infection, inflammation, or damage. We know a great deal about the movement into tissue, termed *extravasation*, of four types of leukocytes: neutrophils, which release several antibacterial proteins; monocytes, the precursors of macrophages, which can engulf and destroy foreign particles; and T and B lymphocytes, the antigen-recognizing cells of the immune system (see Chapter 23).

Extravasation requires the successive formation and breakage of cell-cell contacts between leukocytes in the blood and endothelial cells lining the vessels. Some of these contacts are mediated by **selectins**, a family of CAMs that mediate leukocyte–vascular cell interactions. A key player in these interactions is *P-selectin*, which is localized on the blood-facing surface of endothelial cells. All selectins contain a $Ca^{2+}$-dependent *lectin domain*, which is located at the distal end of the extracellular region of the molecule and recognizes

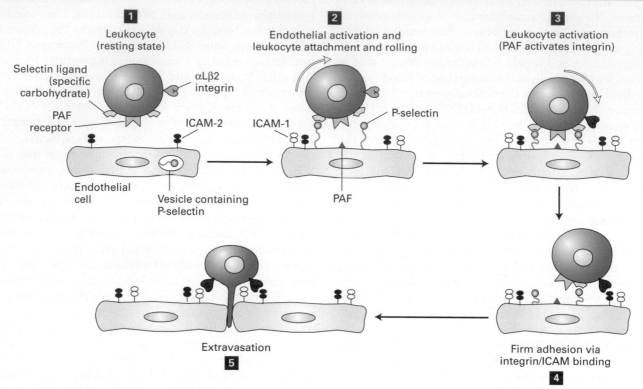

**FIGURE 20-39 Endothelium-leukocyte interactions: activation, binding, rolling, and extravasation.** Step **1**: In the absence of inflammation or infection, leukocytes and endothelial cells lining blood vessels are in a resting state. Step **2**: Inflammatory signals released only in areas of inflammation, infection, or both activate resting endothelial cells to move vesicle-sequestered selectins to the cell surface. The exposed selectins mediate weak binding of leukocytes by interacting with carbohydrate ligands on leukocytes. Blood flow forces the loosely bound leukocytes to roll along the endothelial surface of the blood vessel. Activation of the endothelium also causes synthesis of platelet-activating factor (PAF) and ICAM-1, both expressed on the cell surface. PAF and other usually secreted activators, including chemokines, then induce changes in the shapes of the leukocytes and activation of leukocyte integrins such as αLβ2, which is expressed by T lymphocytes (step **3**). The subsequent tight binding between activated integrins on leukocytes and CAMs on the endothelium (e.g., ICAM-2 and ICAM-1) results in firm adhesion (step **4**) and subsequent movement (extravasation) into the underlying tissue (step **5**). [Adapted from R. O. Hynes and A. Lander, 1992, *Cell* **68**:303.]

particular sugars in glycoproteins or glycolipids (see Figure 20-2). For example, the primary ligand for P- and E-selectins is an oligosaccharide called the *sialyl Lewis-x antigen*, a part of longer oligosaccharides present in abundance on leukocyte glycoproteins and glycolipids.

Figure 20-39 illustrates the basic sequence of cell-cell interactions leading to the extravasation of leukocytes. Various inflammatory signals released in areas of infection or inflammation first cause activation of the endothelium. P-selectin exposed on the surface of activated endothelial cells mediates the weak adhesion of passing leukocytes. Because of the force of the blood flow and the rapid "on" and "off" rates of P-selectin binding to its ligands, these "trapped" leukocytes are slowed but not stopped and literally roll along the surface of the endothelium. Among the signals that promote activation of the endothelium are **chemokines**, a group of small secreted proteins (8–12 kDa) produced by a wide variety of cells, including endothelial cells and leukocytes.

For tight adhesion to occur between activated endothelial cells and leukocytes, β2-containing integrins on the surfaces of leukocytes also must be activated indirectly by chemokines or other local activation signals such as *platelet-activating factor (PAF)*. Platelet-activating factor is unusual in that it is a phospholipid rather than a protein; it is exposed on the surface of activated endothelial cells at the same time that P-selectin is exposed. The binding of PAF or other activators to their *G protein–coupled* receptors on leukocytes leads to activation of the leukocyte integrins to their high-affinity form (see Figure 20-37). Activated integrins on leukocytes then bind to distinct IgCAMs on the surface of endothelial cells. These include ICAM-2, which is expressed constitutively, and ICAM-1. ICAM-1, whose synthesis is induced by activation, does not usually contribute substantially to leukocyte endothelial cell adhesion immediately after activation but rather participates at later times in cases of chronic inflammation. The resulting tight adhesion mediated by the $Ca^{2+}$-independent-integrin–ICAM interactions leads to the cessation of rolling and to the spreading of leukocytes on the surface of the endothelium; soon the adhered cells move between adjacent endothelial cells and into the underlying tissue. The extravasation step itself (also called *transmigration* or *diapedesis*; step **5** in Figure 20-39) requires the dissociation of otherwise stable adhesive interactions between endothelial cells that are primarily mediated by the CAM VE-cadherin. There is general agreement that the leukocyte interactions with the endothelial cells initiates outside-to-inside signaling in the endothelial cells that weakens or disrupts their VE-cadherin-mediated adhesions, permitting the paracellular movement of the leukocyte. Mechanical forces and other mechanisms may also be involved.

The selective adhesion of leukocytes to the endothelium near sites of infection or inflammation thus depends on the sequential appearance and activation of several different CAMs on the surfaces of the interacting cells. Different types of leukocytes express different integrins, though all contain the β2 subunit. Nonetheless, all leukocytes move into tissues by the same general mechanism depicted in Figure 20-39.

Many of the CAMs used to direct leukocyte adhesion are shared among different types of leukocytes and target tissues. Yet often only a particular type of leukocyte is directed to a particular tissue. How is this specificity achieved? A three-step model has been proposed to account for the cell-type specificity of such leukocyte–endothelial-cell interactions. First, endothelial activation promotes initial relatively weak, transient, and reversible binding (e.g., the interaction of selectins and their carbohydrate ligands). Without additional local activation signals, the leukocyte will quickly move on. Second, cells in the immediate vicinity of the site of infection or inflammation release or express chemical signals such as chemokines and PAF that activate only special subsets of the transiently attached leukocytes, depending on the types of chemokine receptors they express. Third, additional activation-dependent CAMs (e.g., integrins) engage their binding partners, leading to strong sustained adhesion. Only if the proper combination of CAMs, binding partners, and activation signals are engaged together with the appropriate timing at a specific site will a given leukocyte adhere strongly. Such combinatorial diversity and cross talk allows a small set of CAMs to serve diverse functions throughout the body—a good example of biological parsimony.

*Leukocyte-adhesion deficiency* is caused by a genetic defect in the synthesis of the integrin β2 subunit. People with this disorder are susceptible to repeated bacterial infections because their leukocytes cannot extravasate properly and thus fight infection within a tissue.

Some pathogenic viruses have evolved mechanisms to exploit cell-surface proteins that participate in the normal response to inflammation. For example, many of the RNA viruses that cause the common cold (rhinoviruses) bind to and enter cells through ICAM-1, and chemokine receptors can be important entry sites for human immunodeficiency virus (HIV), the cause of AIDS. Integrins appear to participate in the binding and/or internalization of a wide variety of viruses, including reoviruses (causing fever and gastroenteritis, especially in infants), adenoviruses (causing conjunctivitis, acute respiratory disease), and foot-and-mouth-disease virus (causing fever in cattle and pigs). ■

## KEY CONCEPTS of Section 20.5

### Adhesive Interactions in Motile and Nonmotile Cells

- Many cells have integrin-containing aggregates (e.g., focal adhesions, 3-D adhesions, podosomes) that physically and functionally connect cells to the extracellular matrix and facilitate inside-out and outside-in signaling.

- Via interaction with integrins, the three-dimensional structure of the ECM surrounding a cell can profoundly influence the behavior of the cell.

- Integrins exist in two conformations (bent-inactive, straight-active) that differ in the affinity for ligands and interactions with cytosolic adapter proteins (see Figure 20-37); switching between these two conformations allows regulation of integrin activity, which is important for control of cell adhesion and movements.

- Dystroglycan, an adhesion receptor, forms a large complex with dystrophin, other adapter proteins, and signaling molecules (see Figure 20-38). This complex links the actin cytoskeleton to the surrounding matrix, providing mechanical stability to muscle. Mutations in various components of this complex cause different types of muscular dystrophy.

- Neural cell-adhesion molecules, which belong to the immunoglobulin (Ig) family of CAMs, mediate $Ca^{2+}$-independent cell-cell adhesion in neural and other tissues.

- The combinatorial and sequential interaction of several types of CAMs (e.g., selectins, integrins, and ICAMs) is critical for the specific and tight adhesion of different types of leukocytes to endothelial cells in response to local signals induced by infection or inflammation (see Figure 20-39).

## 20.6 Plant Tissues

We turn now to the assembly of plant cells into tissues. The overall structural organization of plants is generally simpler than that of animals. For instance, plants have only four broad types of cells, which in mature plants form four basic classes of tissue: *dermal tissue* interacts with the environment, *vascular tissue* transports water and dissolved substances such as sugars and ions, space-filling *ground tissue* constitutes the major sites of metabolism, and *sporogenous tissue* forms the reproductive organs. Plant tissues are organized into just four main organ systems: *stems* have support and transport functions, *roots* provide anchorage and absorb and store nutrients, *leaves* are the sites of photosynthesis, and *flowers* enclose the reproductive structures. Thus at the cell, tissue, and organ levels, plants are generally less complex than most animals.

Moreover, unlike animals, plants do not replace or repair old or damaged cells or tissues; they simply grow new organs. Indeed, the developmental fate of any given plant cell is primarily based on its position in the organism rather than its lineage (see Chapter 21), whereas both are important in animals. Thus in both plants and animals a cell's direct communication with its neighbors is important. Most importantly

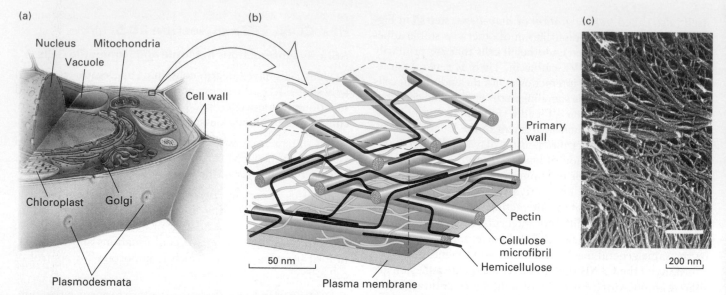

**(a)**

Nucleus   Mitochondria

Vacuole

Cell wall

Chloroplast   Golgi

Plasmodesmata

**(b)**

Primary wall

Pectin

Cellulose microfibril

Hemicellulose

50 nm

Plasma membrane

**(c)**

200 nm

**FIGURE 20-40 Structure of the plant cell wall.** (a) Overview of the organization of a typical plant cell, in which the organelle-filled cell with its plasma membrane is surrounded by a well-defined extracellular matrix called the cell wall. (b) Schematic representation of the cell wall of an onion. Cellulose and hemicellulose are arranged into at least three layers in a matrix of pectin polymers. The size of the polymers and their separations are drawn to scale. To simplify the diagram, most of the hemicellulose cross-links and other matrix constituents (e.g., extensin, lignin) are not shown. (c) Fast-freeze, deep-etch electron micrograph of the cell wall of the garden pea in which some of the pectin polysaccharides were removed by chemical treatment. The abundant thicker fibers are cellulose microfibrils, and the thinner fibers are hemicellulose cross-links (red arrowheads). [Part (b) adapted from M. McCann and K. R. Roberts, 1991, in C. Lloyd, ed., *The Cytoskeletal Basis of Plant Growth and Form,* Academic Press, p. 126, as modified in C. Somerville et al. Part (c) from T. Fujino and T. Itoh, 1998, *Plant Cell Physiol.* **39:**1315–1323.]

for this chapter and in contrast with animals, few cells in plants directly contact one another through molecules incorporated into their plasma membranes. Instead, plant cells are typically surrounded by a rigid **cell wall** that contacts the cell walls of adjacent cells (Figure 20-40a). Also in contrast with animal cells, a plant cell rarely changes its position in the organism relative to other cells. These features of plants and their organization have determined the distinctive molecular mechanisms by which plants' cells are incorporated into tissues and communicate with one another.

## The Plant Cell Wall Is a Laminate of Cellulose Fibrils in a Matrix of Glycoproteins

The plant extracellular matrix, or cell wall, which is mainly composed of polysaccharides and is ~0.2 μm thick, completely coats the outside of the plant cell's plasma membrane. This structure serves some of the same functions as those of the ECM produced by animal cells, even though the two structures are composed of entirely different macromolecules and have a different organization. About 1000 genes in the plant *Arabidopsis*, a small flowering plant also called "thale cress" (see Chapters 1 and 6), are devoted to the synthesis and functioning of its cell wall, including approximately 414 glycosyltransferase and more than 316 glycosyl hydrolase genes. Like animal-cell ECM, the plant cell wall connects cells into tissues, signals a plant cell to grow and divide, and controls the shape of plant organs. It is a dynamic structure that plays important roles in controlling the differentiation of plant cells during embryogenesis and growth and provides a barrier to protect against pathogen infection. Just as the extracellular

matrix helps define the shapes of animal cells, the cell wall defines the shapes of plant cells. When the cell wall is digested away from plant cells by hydrolytic enzymes, spherical cells enclosed by a plasma membrane are left.

Because a major function of a plant cell wall is to withstand the osmotic turgor pressure of the cell (between 14.5 and 435 pounds per square inch!), the cell wall is built for lateral strength. It is arranged into layers of **cellulose** microfibrils—bundles of 30–36 chains of long (as much as 7 μm or greater), linear, extensively hydrogen-bonded polymers of glucose in β glycosidic linkages. The cellulose microfibrils are embedded in a matrix composed of *pectin*, a polymer of D-galacturonic acid and other monosaccharides, and *hemicellulose*, a short, highly branched polymer of several five- and six-carbon monosaccharides. The mechanical strength of the cell wall depends on cross-linking of the microfibrils by hemicellulose chains (Figure 20-40b, c). The layers of microfibrils prevent the cell wall from stretching laterally. Cellulose microfibrils are synthesized on the exoplasmic face of the plasma membrane from UDP-glucose and ADP-glucose formed in the cytosol. The polymerizing enzyme, called *cellulose synthase*, moves within the plane of the plasma membrane along tracks of intracellular microtubules as cellulose is formed, providing a distinctive mechanism for intracellular/extracellular communication.

Unlike cellulose, pectin and hemicellulose are synthesized in the Golgi apparatus and transported to the cell surface, where they form an interlinked network that helps bind the walls of adjacent cells to one another and cushions them. When purified, pectin binds water and forms a gel in the presence of $Ca^{2+}$ and borate ions—hence the use of pectins in many processed foods. As much as 15 percent of the cell wall may be composed of *extensin*, a glycoprotein that contains abundant hydroxyproline and serine. Most of the hydroxyproline residues are linked to short chains of arabinose (a five-carbon monosaccharide), and the serine residues are linked to galactose. Carbohydrate accounts for about 65 percent of extensin by weight, and its protein backbone forms an extended rodlike helix with the hydroxyl or O-linked carbohydrates protruding outward. *Lignin*—a complex, insoluble polymer of phenolic residues—associates with cellulose and is a strengthening material. Like cartilage proteoglycans, lignin resists compression forces on the matrix.

The cell wall is a selective filter whose permeability is controlled largely by pectins in the wall matrix. Whereas water and ions diffuse freely across cell walls, the diffusion of large molecules, including proteins larger than 20 kDa, is limited. This limitation may account for why many plant hormones are small, water-soluble molecules, which can diffuse across the cell wall and interact with receptors in the plasma membrane of plant cells.

## Loosening of the Cell Wall Permits Plant Cell Growth

Because the cell wall surrounding a plant cell prevents it from expanding, the wall's structure must be loosened when the cell grows. The amount, type, and direction of plant-cell growth are regulated by small-molecule hormones called *auxins*. The auxin-induced weakening of the cell wall permits the expansion of the intracellular vacuole by uptake of water, leading to elongation of the cell. We can grasp the magnitude of this phenomenon by considering that, if all cells in a redwood tree were reduced to the size of a typical liver cell, the tree would have a maximum height of only 1 meter.

The cell wall undergoes its greatest changes at the **meristem** of a root or shoot tip. These sites are where cells divide and grow. Young meristematic cells are connected by thin primary cell walls, which can be loosened and stretched to allow subsequent cell elongation. After cell elongation ceases, the cell wall is generally thickened, either by the secretion of additional macromolecules into the primary wall or, more usually, by the formation of a secondary cell wall composed of several layers. Most of the cell eventually degenerates, leaving only the cell wall in mature tissues such as the xylem—the tubes that conduct salts and water from the roots through the stems to the leaves. The unique properties of wood and of plant fibers such as cotton are due to the molecular properties of the cell walls in the tissues of origin.

## Plasmodesmata Directly Connect the Cytosols of Adjacent Cells in Higher Plants

The presence of a cell wall separating cells in plants imposes barriers to cell-cell communication—and thus cell-type differentiation—not faced by animals. One distinctive mechanism used by plant cells to communicate directly is through specialized cell-cell junctions called **plasmodesmata**, which extend through the cell wall. Like gap junctions, plasmodesmata are channels that connect the cytosol of a cell with that of an adjacent cell. The diameter of the channel is about 30–60 nm, and its length can vary and be greater than 1 μm. The density of plasmodesmata varies depending on the plant and cell type, and even the smallest meristematic cells have more than 1000 interconnections with their neighbors. Although a variety of proteins and polysaccharides that are physically or functionally associated with plasmodesmata have been identified, key structural protein components of plasmodesmata and the detailed mechanisms underlying their biogenesis remain to be identified.

Molecules smaller than about 1000 Da, including a variety of metabolic and signaling compounds (ions, sugars, amino acids), generally can diffuse through plasmodesmata. However, the size of the channel through which molecules pass is highly regulated. In some circumstances, the channel is clamped shut; in others, it is dilated sufficiently to permit the passage of molecules larger than 10,000 Da. Among the factors that affect the permeability of plasmodesmata is the cytosolic $Ca^{2+}$ concentration, with an increase in cytosolic $Ca^{2+}$ reversibly inhibiting movement of molecules through these structures.

Although plasmodesmata and gap junctions resemble each other functionally with respect to forming channels for small-molecule diffusion, their structures differ dramatically in two significant ways (Figure 20-41). In plasmodesmata the plasma membranes of the adjacent plant cells merge to form a continuous channel, the *annulus*, whereas the membranes of cells at a gap junction are not continuous with each other. There are simple plasmodesmata (single pore) and complex plasmodesmata that branch into multiple channels. In addition, plasmodesmata exhibit many additional complex structural and functional characteristics. For example, they contain within the channel an extension of the endoplasmic reticulum called a *desmotubule* that passes through the annulus, which connects the cytosols of adjacent plant cells. They also have a variety of specialized proteins at the entrance of the channel and running throughout the length of the channel, including cytoskeletal, motor, and docking proteins that regulate the sizes and types of molecules that can pass through the channel. Many types of molecules spread from cell to cell through plasmodesmata, including some transcription-factor proteins, nucleic acid/protein complexes, metabolic products, and plant viruses. It appears that some of these require special chaperones to facilitate transport. Specialized kinases may also phosphorylate plasmodesmal components to regulate their activities (e.g., opening of the channels). Soluble molecules pass through the cytosolic annulus, about 3–4 nm in diameter, that lies between the plasma membrane and desmotubule, whereas membrane-bound molecules or certain proteins within the ER lumen can pass from cell to cell via the desmotubule. Plasmodesmata appear to play an especially important role in regulating the development of plant cells and tissues, as is suggested by their ability to mediate intracellular movement of transcription factors and ribonuclear protein complexes.

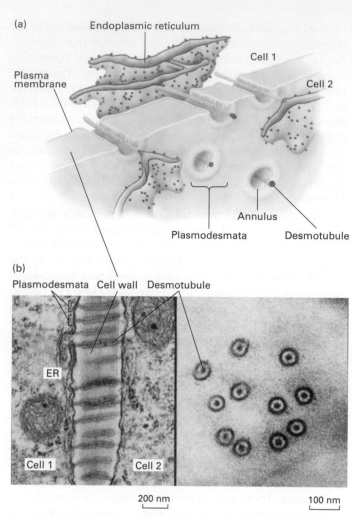

**(a)**
Endoplasmic reticulum

Plasma membrane

Cell 1

Cell 2

Annulus

Plasmodesmata

Desmotubule

**(b)**
Plasmodesmata  Cell wall  Desmotubule

ER

Cell 1

Cell 2

200 nm

100 nm

**FIGURE 20-41 Plasmodesmata.** (a) Schematic model of a plasmodesma showing the desmotubule, an extension of the endoplasmic reticulum (ER), and the annulus, a plasma-membrane-lined channel filled with cytosol that interconnects the cytosols of adjacent cells. (b) Electron micrographs of thin sections of a sugarcane leaf (brackets indicate individual plasmodesmata). (*Left*) Longitudinal view showing ER and desmotubule running through each annulus. (*Right*) Perpendicular cross-sectional views of plasmodesmata, in some of which spoke structures connecting the plasma membrane to the desmotubule can be seen. [Part (b) from K. Robinson-Beers and R. F. Evert, 1991, *Planta* **184**:307–318.]

## Only a Few Adhesive Molecules Have Been Identified in Plants

Systematic analysis of the *Arabidopsis* genome and biochemical analysis of other plant species provide no evidence for the existence of plant homologs of most animal CAMs, adhesion receptors, and ECM components. This finding is not surprising, given the dramatically different nature of cell-cell and cell-matrix/cell-wall interactions in animals and plants.

Among the adhesive-type proteins apparently unique to plants are five wall-associated kinases (WAKs) and WAK-like proteins expressed in the plasma membrane of *Arabidopsis* cells. The extracellular regions in all these proteins contain multiple epidermal growth factor (EGF) repeats, frequently found in animal cell-surface receptors, which may directly participate in binding to other molecules. Some WAKs have been shown to bind to glycine-rich proteins in the cell wall, thereby mediating membrane-wall contacts. These *Arabidopsis* proteins have a single transmembrane domain and an intracellular cytosolic tyrosine kinase domain, which may participate in signaling pathways somewhat like the receptor tyrosine kinases discussed in Chapter 16.

The results of in vitro binding assays combined with in vivo studies and analyses of plant mutants have identified several macromolecules in the ECM that are important for adhesion. For example, normal adhesion of pollen, which contains sperm cells, to the stigma or style in the female reproductive organ of the Easter lily requires a cysteine-rich protein called stigma/stylar cysteine-rich adhesin (SCA) and a specialized pectin that can bind to SCA (Figure 20-42). A small, probably ECM-embedded, ~10-kDa protein called chymocyanin works in conjunction with SCA to help direct the movement of the sperm-containing pollen tube (chemotaxis) to the ovary.

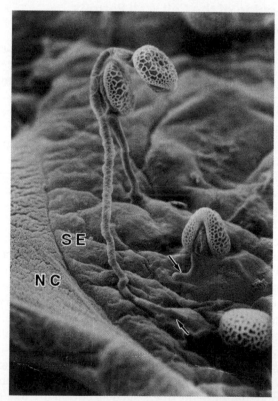

SE

NC

**EXPERIMENTAL FIGURE 20-42 An in vitro assay was used to identify molecules required for adherence of pollen tubes to the stylar matrix.** In this assay, extracellular stylar matrix collected from lily styles (SE) or an artificial matrix is dried onto nitrocellulose membranes (NC). Pollen tubes containing sperm are then added and their binding to the dried matrix is assessed. In this scanning electron micrograph, the tips of pollen tubes (arrows) can be seen binding to dried stylar matrix. This type of assay has shown that pollen adherence depends on stigma/stylar cysteine-rich adhesin (SCA) and a pectin that binds to SCA. [From G. Y. Jauh et al., 1997, *Sex Plant Reprod.* **10**:173.]

Disruption of the gene encoding glucuronyltransferase 1, a key enzyme in pectin biosynthesis, has provided a striking illustration of the importance of pectins in intercellular adhesion in plant meristems. Normally, specialized pectin molecules help hold the cells in meristems tightly together. When grown in culture as a cluster of relatively undifferentiated cells, called a callus, normal meristematic cells adhere tightly and can differentiate into chlorophyll-producing cells, giving the callus a green color. Eventually the callus will generate shoots. In contrast, mutant cells with an inactivated glucuronyltransferase 1 gene are large, associate loosely with each other, and do not differentiate normally, forming a yellow callus. The introduction of a normal glucuronyltransferase 1 gene into the mutant cells restores their ability to adhere and differentiate normally.

The paucity of plant adhesive molecules identified to date, in contrast with the many well-defined animal adhesive molecules, may be due to the technical difficulties in working with the ECM/cell wall of plants. Adhesive interactions are often likely to play different roles in plant and animal biology, at least in part because of their differences in development and physiology.

## KEY CONCEPTS of Section 20.6

### Plant Tissues

- The integration of cells into tissues in plants is fundamentally different from the assembly of animal tissues, primarily because each plant cell is surrounded by a relatively rigid cell wall.

- The plant cell wall comprises layers of cellulose microfibrils embedded within a matrix of hemicellulose, pectin, extensin, and other less abundant molecules.

- Cellulose, a large, linear glucose polymer, assembles spontaneously into microfibrils stabilized by hydrogen bonding.

- The cell wall defines the shapes of plant cells and restricts their elongation. Auxin-induced loosening of the cell wall permits elongation.

- Adjacent plant cells can communicate through plasmodesmata, junctions that allow molecules to pass through complex channels connecting the cytosols of adjacent cells (see Figure 20-41).

- Plants do not produce homologs of the common adhesive molecules found in animals. Only a few adhesive molecules unique to plants have been well documented to date.

## Perspectives for the Future

A deeper understanding of the integration of cells into tissues in complex organisms will draw on insights and techniques from virtually all subdisciplines of molecular cell biology—biochemistry, biophysics, microscopy, genetics, genomics, proteomics, and developmental biology—together with bioengineering and computer science. This area of cell biology is undergoing explosive growth.

An important set of questions for the future deals with the mechanisms by which cells detect and respond to mechanical forces on them and the extracellular matrix, as well as the influence of their three-dimensional arrangements and interactions. A related question is how this information is used to control cell and tissue structure and function. These issues involve the fields of biomechanics and mechanotransduction. Shear or other stresses can induce distinct patterns of gene expression and cell growth and can greatly alter cell metabolism and responses to extracellular stimuli. Mechanosensitive nonselective cation channels ($NSC_{MS}$), at least some of which appear to be members of the transient receptor potential, or TRP, cation channel family, are activated by the stretch of plasma membrane. They are important players in mechanotransduction, such as that involved in sensing sound in the ear, which is mediated, in part, by specialized cadherins. Most of the classes of molecules discussed in this chapter—ECM components, adhesion receptors, CAMs, intracellular adapters, and the cytoskeleton—appear to play crucial roles in mechanosensing and mechanotransduction signaling pathways. Future research should give us a far more sophisticated understanding of the roles of the three-dimensional organization of cells and ECM components and the forces acting on them under normal and pathological conditions in controlling the structures and activities of tissues. Applications of such understanding will provide new methods to explore basic cell/tissue biology and provide improved technologies for the search for novel therapies for disease.

Although junctions help play a key role in forming stable epithelial tissues and defining the shapes and functional properties of epithelia, they are not static. Remodeling in terms of replacement of older molecules with more recently synthesized molecules is ongoing, and the dynamic properties of junctions open the door to more substantial changes when necessary (the epithelial-mesenchymal transition during development, wound healing, extravasation of leukocytes, etc.). Understanding the molecular mechanisms underlying the relationship between stability and dynamic change will provide new insights into morphogenesis, maintaining tissue integrity and function, and response to (or induction of) pathology.

Numerous questions relate to intracellular signaling from CAMs and adhesion receptors. Such signaling must be integrated with other cellular signaling pathways that are activated by various external signals (e.g., growth factors) so that the cell responds appropriately and in a coordinated fashion to many different simultaneous internal and external stimuli. It appears that small GTPase proteins participate in at least some of the integrated pathways associated with signaling between cellular junctions. How are the logic circuits constructed that allow cross talk between diverse signaling pathways? How do these circuits integrate the information from these pathways? How is the combination of outside-in and inside-out signaling mediated by CAMs and adhesion receptors merged into such circuits?

We can expect ever-increasing progress in the exploration of the influence of glycobiology (the biology of oligo- and polysaccharides) on cell biology. The importance of specialized GAG sequences in controlling cellular activities, especially interactions between some growth factors and their receptors, is now clear. With the identification of the biosynthetic mechanisms by which these complex structures are generated and the development of tools to manipulate GAG structures and test their functions in cultured systems and intact animals, we can expect a dramatic increase in our understanding of the cell biology of GAGs in the next several years. There is still much to learn about the biosynthesis, structures, and functions of many other glycoconjugates, such as the O-linked sugars on dystroglycan that are essential for its binding to its ECM ligands. The new subspecialty field of *glycomics* has recently been flourishing and will contribute to our future understanding of glycobiology. Glycomics, similarly to genomics and proteomics, uses high-throughput tools, such as mass spectrometry, to perform large-scale analysis of the structures and changes in the wide range of sugar-containing molecules in cells and tissues.

A structural hallmark of CAMs, adhesion receptors, and ECM proteins is the presence of multiple domains that impart diverse functions to a single polypeptide chain. It is generally agreed that such multidomain proteins arose evolutionarily by the assembly of distinct DNA sequences encoding the distinct domains. Genes encoding multiple domains provide opportunities to generate enormous sequence and functional diversity by alternative splicing and the use of alternate promoters within a gene. Thus even though the number of independent genes in the human genome seems surprisingly small in comparison with other organisms, far more distinct protein molecules can be produced than predicted from the number of genes. Such diversity seems very well suited to the generation of proteins that take part in specifying adhesive connections in the nervous system, especially the brain. In fact, several groups of proteins expressed by neurons appear to have just such combinatorial diversity of structure. They include the protocadherins, a family of cadherins with many proteins encoded per gene (14–19 for the three genes in mammals); the neurexins, which comprise more than 1000 proteins encoded by three genes; and the Dscams, members of the IgCAM superfamily encoded by a *Drosophila* gene that has the potential to express 38,016 distinct proteins owing to alternative splicing. A continuing goal for future work will be to describe and understand the molecular basis of functional cell-cell and cell-matrix attachments—the "wiring"—in the nervous system and how that wiring ultimately permits complex neuronal control and, indeed, the intellect required to understand molecular cell biology.

## Key Terms

adapter proteins 928

adherens junction 934

adhesion receptor 927

anchoring junction 934

basal lamina 933

cadherin 935

cell-adhesion molecule (CAM) 927

cell wall 968

collagen 946

connexin 943

desmosome 935

elastin 951

epithelial-mesenchymal transition 938

epithelia 926

extracellular matrix (ECM) 927

fibrillar collagen 951

fibronectin 957

gap junction 934

glycosaminoglycan (GAG) 951

hyaluronan 956

immunoglobulin cell-adhesion molecule (IgCAM) 965

integrin 939

laminin 947

matrix metalloproteases (MMPs) 960

multi-adhesive matrix protein 946

paracellular pathway 942

plasmodesmata 969

proteoglycan 946

RGD sequence 939

selectin 965

syndecan 956

tight junction 934

## Review the Concepts

1. Describe the two phenomena that give rise to the diversity of adhesive molecules such as cadherins. What additional phenomenon gives rise to the diversity of integrins?

2. Cadherins are known to mediate homophilic interactions between cells. What is a homophilic interaction, and how can it be demonstrated experimentally for E-cadherins? What component of the extracellular environment is required for the homophilic interactions mediated by cadherins, and how can this requirement be demonstrated?

3. Together with their role in connecting the lateral membranes of adjacent epithelial cells, adherens junctions play a role in controlling cell shape. What associated intracellular structure and proteins are involved in this role?

4. What is the normal function of tight junctions? What can happen to tissues when tight junctions do not function properly?

5. Gap junctions between heart muscle cells and gap junctions between uterine myometrial smooth muscle cells form a connection that provides for rapid communication. What is this called? How is uterine myometrial smooth-muscle-cell gap-junction communication up-regulated for parturition (childbirth)?

6. What is collagen, and how is it synthesized? How do we know that collagen is required for tissue integrity?

7. Explain how changes in integrin structure mediate outside-in and inside-out signaling.

8. Compare the functions and properties of each of three types of macromolecules that are abundant in the extracellular matrix of all tissues.

9. Many proteoglycans have cell-signaling roles. Regulation of feeding behavior by syndecans in the hypothalamic region of the brain is one example. How is this regulation accomplished?

10. You have synthesized an oligopeptide containing an RGD sequence surrounded by other amino acids. What is the effect of this peptide when added to a fibroblast cell culture grown on a layer of fibronectin absorbed to the tissue culture dish? Why does this happen?

**11.** Describe the major activity and possible localization of the three major subgroups of proteins that remodel/degrade the ECM in physiological or pathological tissue remodeling. Identify a pathological condition in which these proteins play a key role.

**12.** Blood clotting is a crucial function for mammalian survival. How do the multi-adhesive properties of fibronectin lead to the recruitment of platelets to blood clots?

**13.** How do changes in molecular connections between the extracellular matrix (ECM) and cytoskeleton give rise to Duchenne muscular dystrophy?

**14.** To fight infection, leukocytes move rapidly from the blood into the tissue sites of infection. What is this process called? How are adhesion molecules involved in this process?

**15.** The structure of a plant cell wall needs to loosen to accommodate cell growth. What signaling molecule controls this process?

**16.** Compare plasmodesmata in plant cells to gap junctions in animal cells.

## Analyze the Data

Researchers have isolated two E-cadherin mutant isoforms that are hypothesized to function differently from the iso-form of the wild-type E-cadherin. An E-cadherin negative mammary carcinoma cell line was transfected with the mutant E-cadherin genes A (part a in the figure; triangles), or B (part b; triangles) or the wild-type E-cadherin gene (black circles) and compared to untransfected cells (open circles) in an aggregation assay. In this assay, cells are first dissociated by trypsin treatment and then allowed to aggregate in solution over a period of minutes. Aggregating cells from mutants A and B are presented in panels a and b respectively. To demonstrate that the observed adhesion was cadherin mediated, the cells were pretreated with a nonspecific antibody (left panel) or a function-blocking anti-E-cadherin monoclonal antibody (right panel).

**a.** Why do cells transfected with the wild-type E-cadherin gene have greater aggregation than control, untransfected cells?

**b.** From these data, what can be said about the function of mutants A and B?

**c.** Why does the addition of the anti-E-cadherin monoclonal antibody, but not the nonspecific antibody, block aggregation?

**d.** What would happen to the aggregation ability of the cells transfected with the wild-type E-cadherin gene if the assay were performed in media low in $Ca^{2+}$?

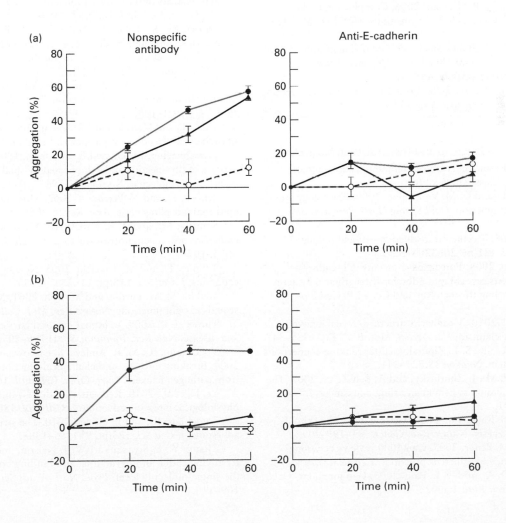

# References

## Cell-Cell and Cell-Matrix Adhesion: An Overview

Carthew, R. W. 2005. Adhesion proteins and the control of cell shape. *Curr. Opin. Genet. Dev.* **15**(4):358–363.

M. Cereijido, R. G. Contreras, and L. Shoshani. 2004. Cell adhesion, polarity, and epithelia in the dawn of metazoans. *Physiol. Rev.* **84**:1229–1262.

Gumbiner, B. M. 1996. Cell adhesion: the molecular basis of tissue architecture and morphogenesis. *Cell* **84**:345–357.

Hynes, R. O. 1999. Cell adhesion: old and new questions. *Trends Cell Biol.* **9**(12):M33–M337. Millennium issue.

Hynes, R. O. 2002. Integrins: bidirectional, allosteric signaling machines. *Cell* **110**:673–687.

Hynes, R. O. 2012. The evolution of metazoan extracellular matrix. *J. Cell Biol.* (in press)

Hynes, R. O., and Q. Zhao. 2000. The evolution of cell adhesion. *J. Cell Biol.* **150**(2):F89–F96.

Ingber, D. E. 2006. Cellular mechanotransduction: putting all the pieces together again. *FASEB J.* **20**(7):811–827.

Jamora, C., and E. Fuchs. 2002. Intercellular adhesion, signalling and the cytoskeleton. *Nature Cell Biol.* **4**(4):E101–E108.

Juliano, R. L. 2002. Signal transduction by cell adhesion receptors and the cytoskeleton: functions of integrins, cadherins, selectins, and immunoglobulin-superfamily members. *Ann. Rev. Pharmacol. Toxicol.* **42**:283–323.

Leahy, D. J. 1997. Implications of atomic resolution structures for cell adhesion. *Ann. Rev. Cell Devel. Biol.* **13**:363–393.

Thiery, J. P., and J. P. Sleeman. 2006. Complex networks orchestrate epithelial-mesenchymal transitions. *Nat. Rev. Mol. Cell Biol.* **7**:131–142.

Vogel, V. 2006. Mechanotransduction involving multimodular proteins: converting force into biochemical signals. *Annu. Rev. Biophys. Biomol. Struct.* **35**:459–488.

Vogel, V., and M. Sheetz. 2006. Local force and geometry sensing regulate cell functions. *Nat. Rev. Mol. Cell Biol.* **7**(4):265–275.

## Cell-Cell and Cell-ECM Junctions and Their Adhesion Molecules

The Cadherin Resource: http://calcium.uhnres.utoronto.ca/cadherin/pub_pages/classify/index.htm

Anderson, J. M., and C. M. Van Itallie. 2009. Physiology and function of the tight junction. *Cold Spring Harb. Perspect. Biol.* **1**(2):a002584.

Chen, C. S. 2008. Mechanotransduction—a field pulling together? *J. Cell Sci.* **121**(pt. 20):3285–3292.

Chu, Y. S., et al. 2004. Force measurements in E-cadherin-mediated cell doublets reveal rapid adhesion strengthened by actin cytoskeleton remodeling through Rac and Cdc42. *J. Cell Biol.* **167**:1183–1194.

Ciatto, C., et al. 2010. T-cadherin structures reveal a novel adhesive binding mechanism. *Nat. Struct. Mol. Biol.* **17**(3):339–347.

Clandinin, T. R., and S. L. Zipursky. 2002. Making connections in the fly visual system. *Neuron* **35**:827–841.

Conacci-Sorrell, M., J. Zhurinsky, and A. Ben-Ze'ev. 2002. The cadherin-catenin adhesion system in signaling and cancer. *J. Clin. Invest.* **109**:987–991.

Fuchs, E., and S. Raghavan. 2002. Getting under the skin of epidermal morphogenesis. *Nature Rev. Genet.* **3**(3):199–209.

Gates, J., and M. Peifer. 2005. Can 1000 reviews be wrong? Actin, alpha-catenin, and adherens junctions. *Cell* **123**(5):769–772.

Goodenough, D. A., and D. L. Paul. 2009. Gap junctions. *Cold Spring Harb. Perspect. Biol.* **1**:a002576.

Gumbiner, B. M. 2005. Regulation of cadherin-mediated adhesion in morphogenesis. *Nat. Rev. Mol. Cell Biol.* **6**:622–634.

Guttman, J. A., and B. B. Finlay. 2009. Tight junctions as targets of infectious agents. *Biochim. Biophys. Acta* **1788**(4):832–841.

Harris, T. J., and U. Tepass. 2010. Adherens junctions: from molecules to morphogenesis. *Nat. Rev. Mol. Cell Biol.* **11**(7): 502–514.

Hatzfeld, M. 2007. Plakophilins: multifunctional proteins or just regulators of desmosomal adhesion? *Biochim. Biophys. Acta* **1773**:69–77.

Hobbie, L., et al. 1987. Restoration of LDL receptor activity in mutant cells by intercellular junctional communication. *Science* **235**:69–73.

Hollande, F., A. Shulkes, and G. S. Baldwin. 2005. Signaling the junctions in gut epithelium. *Science* **2005**(277):pe13.

Hunter, A. W., R. J. Barker, C. Zhu, and R. G. Gourdie. 2005. Zonula occludens-1 alters connexin43 gap junction size and organization by influencing channel accretion. *Mol. Biol. Cell* **16**(12):5686–5698.

Jefferson, J. J., C. L. Leung, and R. K. H. Liem. 2004. Plakins: goliaths that link cell junctions and the cytoskeleton. *Nature Rev. Mol. Cell Biol.* **5**:542–553.

Laird, D. W. 2006. Life cycle of connexins in health and disease. *Biochem. J.* **394**(pt. 3):527–543.

Lee, J. M., S. Dedhar, R. Kalluri, and E. W. Thompson. 2006. The epithelial-mesenchymal transition: new insights in signaling, development, and disease *J. Cell Biol.* **172**(7):973–981.

Litjens, S. H., J. M. de Pereda, and A. Sonnenberg. 2006. Current insights into the formation and breakdown of hemidesmosomes. *Trends Cell Biol.* **16**(7):376–383.

Moser, M., K. R. Legate, R. Zent, and R. Fässler. 2009. The tail of integrins, talin, and kindlins. *Science* **324**(5929):895–899.

Müller, U. 2008. Cadherins and mechanotransduction by hair cells. *Curr. Opin. Cell Biol.* **20**(5):557–566.

Nakagawa, S., S. Maeda, and T. Tsukihara. 2011. Structural and functional studies of gap junction channels. *Curr. Opin. Struct. Biol.* **21**(1):101–108.

Oda, H., and M. Takeichi. 2011. Structural and functional diversity of cadherin at the adherens junction. *J. Cell Biol.* **193**(7):1137–1146.

Pierschbacher, M. D., and E. Ruoslahti. 1984. Cell attachment activity of fibronectin can be duplicated by small synthetic fragments of the molecule. *Nature* **309**(5963):30–33.

Schöck, F., and N. Perrimon. 2002. Molecular mechanisms of epithelial morphogenesis. *Ann. Rev. Cell Devel. Biol.* **18**:463–493.

Smutny, M., and A. S. Yap. 2010. Neighborly relations: cadherins and mechanotransduction. *J. Cell Biol.* **189**(7): 1075–1077.

Tanoue, T., and M. Takeichi. 2005. New insights into fat cadherins. *J. Cell Sci.* **118**(pt. 11):2347–2353.

Tsukita, S., M. Furuse, and M. Itoh. 2001. Multifunctional strands in tight junctions. *Nature Rev. Mol. Cell Biol.* **2**:285–293.

Turner, J. R. 2009. Intestinal mucosal barrier function in health and disease. *Nat. Rev. Immunol.* **9**(11):799–809.

Vogelmann, R., M. R. Amieva, S. Falkow, and W. J. Nelson. 2004. Breaking into the epithelial apical-junctional complex—news from pathogen hackers. *Curr. Opin. Cell Biol.* **16**(1):86–93.

Ye, F., et al.. 2010. Recreation of the terminal events in physiological integrin activation. *J. Cell Biol.* **188**(1):157–173.

Zaidel-Bar, R., and B. Geiger. 2010. The switchable integrin adhesome. *J. Cell Sci.* **123**(pt. 9):1385–1388.

Zhang, Y., S. Sivasankar, W. J. Nelson, and S. Chu. 2009. Resolving cadherin interactions and binding cooperativity at the single-molecule level. *Proc. Natl. Acad. Sci. USA* **106**(1): 109–114.

## The Extracellular Matrix I: The Basal Lamina

Boutaud, A., et al. 2000. Type IV collagen of the glomerular basement membrane: evidence that the chain specificity of network assembly is encoded by the noncollagenous NC1 domains. *J. Biol. Chem.* **275**:30716–30724.

Durbeej, M. 2010. Laminins. *Cell Tissue Res.* **339**(1):259–268.

Esko, J. D., and U. Lindahl. 2001. Molecular diversity of heparan sulfate. *J. Clin. Invest.* **108**:169–173.

Hallmann, R., et al. 2005. Expression and function of laminins in the embryonic and mature vasculature. *Physiol. Rev.* **85**:979–1000.

Hohenester, E., and J. Engel. 2002. Domain structure and organisation in extracellular matrix proteins. *Matrix Biol.* **21**(2):115–128.

Iozzo, R. V. 2005. Basement membrane proteoglycans: from cellular to ceiling. *Nature Rev. Mol. Cell Biol.* **6**(8):646–656.

Kanagawa, M., et al. 2005. Disruption of perlecan binding and matrix assembly by post-translational or genetic disruption of dystroglycan function. *FEBS Lett.* **579**(21):4792–4796.

Kruegel, J., and N. Miosge. 2010. Basement membrane components are key players in specialized extracellular matrices. *Cell Mol. Life Sci.* **67**(17):2879–2895.

Nakato, H., and K. Kimata. 2002. Heparan sulfate fine structure and specificity of proteoglycan functions. *Biochim. Biophys. Acta* **1573**:312–318.

Perrimon, N., and M. Bernfield. 2001. Cellular functions of proteoglycans: an overview. *Semin. Cell Devel. Biol.* **12**(2):65–67.

Rosenberg, R. D., et al. 1997. Heparan sulfate proteoglycans of the cardiovascular system: specific structures emerge but how is synthesis regulated? *J. Clin. Invest.* **99**:2062–2070.

Sasaki, T., R. Fassler, and E. Hohenester. 2004. Laminin: the crux of basement membrane assembly. *J. Cell Biol.* **164**(7):959–963.

## The Extracellular Matrix II: Connective Tissue

Canty, E. G., and K. E. Kadler. 2005. Procollagen trafficking, processing and fibrillogenesis. *J. Cell Sci.* **118**:1341–1353.

Comelli, E. M., et al. 2006. A focused microarray approach to functional glycomics: transcriptional regulation of the glycome. *Glycobiology* **16**(2):117–131.

Couchman, J. R. 2003. Syndecans: proteoglycan regulators of cell-surface microdomains? *Nature Rev. Mol. Cell Biol.* **4**:926–938.

Fields, G. B. 2010. Synthesis and biological applications of collagen-model triple-helical peptides. *Org. Biomol. Chem.* **8**(6):1237–1258.

Kramer, R. Z., J. Bella, B. Brodsky, and H. M. Berman. 2001. The crystal and molecular structure of a collagen-like peptide with a biologically relevant sequence. *J. Mol. Biol.* **311**:131–147.

Leitinger, B., and E. Hohenester. 2007. Mammalian collagen receptors. *Matrix Biol.* **26**(3):146–155.

Mao, J. R., and J. Bristow. 2001. The Ehlers-Danlos syndrome: on beyond collagens. *J. Clin. Invest.* **107**:1063–1069.

Orgel, J. P., T. C. Irving, A. Miller, and T. J. Wess. 2006. Microfibrillar structure of type I collagen *in situ*. *Proc. Natl. Acad. Sci. USA* **103**:9001–9005.

Sakai, T., M. Larsen, and K. Yamada. 2003. Fibronectin requirement in branching morphogenesis. *Nature* **423**:876–881.

Shaw, L. M., and B. R. Olsen. 1991. FACIT collagens: diverse molecular bridges in extracellular matrices. *Trends Biochem. Sci.* **16**(5):191–194.

Shiomi, T., V. Lemaître, J. D'Armiento, and Y. Okada. 2010. Matrix metalloproteinases, a disintegrin and metalloproteinases, and a disintegrin and metalloproteinases with thrombospondin motifs in non-neoplastic diseases. *Pathol. Int.* **60**(7):477–496.

Weiner, S., W. Traub, and H. D. Wagner. 1999. Lamellar bone: structure-function relations. *J. Struc. Biol.* **126**:241–255.

## Adhesive Interactions in Motile and Nonmotile Cells

Barresi, R., and K. P. Campbell. 2006. Dystroglycan: from biosynthesis to pathogenesis of human disease. *J. Cell Sci.* **119**(pt. 2):199–207.

Bartsch, U. 2003. Neural CAMs and their role in the development and organization of myelin sheaths. *Front. Biosci.* **8**:D477–D490.

Brummendorf, T., and V. Lemmon. 2001. Immunoglobulin superfamily receptors: cis-interactions, intracellular adapters and alternative splicing regulate adhesion. *Curr. Opin. Cell Biol.* **13**:611–618.

Cukierman, E., R. Pankov, and K. M. Yamada. 2002. Cell interactions with three-dimensional matrices. *Curr. Opin. Cell Biol.* **14**:633–639.

Even-Ram, S., and K. M. Yamada. 2005. Cell migration in 3D matrix. *Curr. Opin. Cell Biol.* **17**(5):524–532.

Geiger, B., A. Bershadsky, R. Pankov, and K. M. Yamada. 2001. Transmembrane crosstalk between the extracellular matrix and the cytoskeleton. *Nat. Rev. Mol. Cell Biol.* **2**:793–805.

Griffith, L. G., and M. A. Swartz. 2006. Capturing complex 3D tissue physiology in vitro. *Nat. Rev. Mol. Cell Biol.* **7**(3):211–224.

Lawrence, M. B., and T. A. Springer. 1991. Leukocytes roll on a selectin at physiologic flow rates: distinction from and prerequisite for adhesion through integrins. *Cell* **65**:859–873.

Nelson, C. M., and M. J. Bissell. 2005. Modeling dynamic reciprocity: engineering three-dimensional culture models of breast architecture, function, and neoplastic transformation. *Sem. Cancer Biol.* **15**(5):342–352.

Reizes, O., et al. 2001. Transgenic expression of syndecan-1 uncovers a physiological control of feeding behavior by syndecan-3. *Cell* **106**:105–116.

Rougon, G., and O. Hobert. 2003. New insights into the diversity and function of neuronal immunoglobulin superfamily molecules. *Annu. Rev. Neurosci.* **26**:207–238.

Shimaoka, M., J. Takagi, and T. A. Springer. 2002. Conformational regulation of integrin structure and function. *Ann. Rev. Biophys. Biomol. Struc.* **31**:485–516.

Somers, W. S., J. Tang, G. D. Shaw, and R. T. Camphausen. 2000. Insights into the molecular basis of leukocyte tethering and rolling revealed by structures of P- and E-selectin bound to SLe(X) and PSGL-1. *Cell* **103**:467–479.

Stein, E., and M. Tessier-Lavigne. 2001. Hierarchical organization of guidance receptors: silencing of netrin attraction by Slit through a Robo/DCC receptor complex. *Science* **291**:1928–1938.

Xiong, J. P., et al. 2001. Crystal structure of the extracellular segment of integrin aVb3. *Science* **294**:339–345.

## Plant Tissues

Bacic, A. 2006. Breaking an impasse in pectin biosynthesis. *Proc. Nat'l. Acad. Sci. USA* **103**(15):5639–5640.

Delmer, D. P., and C. H. Haigler. 2002. The regulation of metabolic flux to cellulose, a major sink for carbon in plants. *Metab. Eng.* **4**:22–28.

Iwai, H., N. Masaoka, T. Ishii, and S. Satoh. 2000. A pectin glucuronyltransferase gene is essential for intercellular attachment in the plant meristem. *Proc. Nat'l. Acad. Sci. USA* **99**:16319–16324.

Kim, S., J. Dong, and E. M. Lord. 2004. Pollen tube guidance: the role of adhesion and chemotropic molecules. *Curr. Top. Dev. Biol.* **61**:61–79.

Lee, D. K., and L. E. Sieburth. 2010. Plasmodesmata formation: poking holes in walls with ise. *Curr Biol.* **20**(11):R488–R490.

Lord, E. M., and J. C. Mollet. 2002. Plant cell adhesion: a bioassay facilitates discovery of the first pectin biosynthetic gene. *Proc. Nat'l. Acad. Sci. USA* **99**:15843–15845.

Lord, E. M., and S. D. Russell. 2002. The mechanisms of pollination and fertilization in plants. *Ann. Rev. Cell Devel. Biol.* **18**:81–105.

Lough, T. J., and W. J. Lucas. 2006. Integrative plant biology: role of phloem long-distance macromolecular trafficking. *Annu. Rev. Plant Biol.* **57**:203–232.

Lucas, W. J., B. K. Ham, and J. Y. Kim. 2009. Plasmodesmata—bridging the gap between neighboring plant cells. *Trends Cell Biol.* **19**(10):495–503.

Pennell, R. 1998. Cell walls: structures and signals. *Curr. Opin. Plant Biol.* **1**:504–510.

Roberts, A. G., and K. J. Oparka. 2003. Plasmodesmata and the control of symplastic transport. *Plant Cell Environ.* **26**:103–124.

Somerville, C., et al. 2004. Toward a systems approach to understanding plant cell walls. *Science* **306**(5705):2206–2211.

Whetten, R.W., J. J. MacKay, and R. R. Sederoff. 1998. Recent advances in understanding lignin biosynthesis. *Ann. Rev. Plant Physiol. Plant Mol. Biol.* **49**:585–609.

Zambryski, P., and K. Crawford. 2000. Plasmodesmata: gatekeepers for cell-to-cell transport of developmental signals in plants. *Ann. Rev. Cell Devel. Biol.* **16**:393–421.

# Stem Cells, Cell Asymmetry, and Cell Death

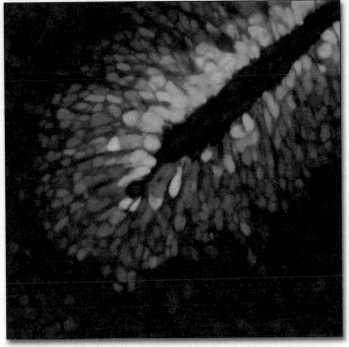

Cells being born in the developing cerebellum. All nuclei are labeled in red; the green cells are dividing and migrating into internal layers of the neural tissue. [Courtesy of Tal Raveh, Matthew Scott, and Jane Johnson.]

Many descriptions of cell division imply that the parental cell gives rise to two daughter cells that look and behave exactly like the parental cell. In other words, they imply that cell division is **symmetric** and the progeny have properties similar to the parent cell. But if this were always the case, none of the hundreds of differentiated cell types and functioning tissues present in complex multicellular organisms would ever be formed. Differences among cells can arise when two initially identical daughter cells diverge on receiving distinct developmental or environmental signals. Alternatively, the two daughter cells may differ from "birth," with each inheriting different parts of the parental cell (Figure 21-1). Daughter cells produced by such **asymmetric cell division** may differ in size, shape, and/or protein composition, or their genes may be in different states of activity or potential activity. The differences in these internal signals confer different fates on the two cells.

Striking examples of asymmetric cell division occur during early metazoan development, the focus of the first section of this chapter. The development of a new organism begins with the egg, or **oocyte**, carrying a set of chromosomes from the mother, and the **sperm**, carrying a set of chromosomes from the father. These **gametes**, or sex cells, are haploid because they have gone through meiosis (see Chapter 19). In a process called fertilization, they combine to create the initial single cell, the zygote, which has two sets of chromosomes and is therefore diploid. During embryogenesis, the zygote undergoes numerous cell divisions, both symmetric and asymmetric, ultimately giving rise to an entire organism.

Such unspecialized cells that can reproduce themselves as well as generate specific types of more specialized cells are called **stem cells**. Their name comes from the image of a plant stem, which grows upward, continuing to form more stem, while also sending off leaves and branches to the side. In this chapter we will explore various types of stem cells that differ in the variety of specialized cell types they can form. The zygote is the ultimate *totipotent* stem cell because it has the capability to generate every cell type in the body as well as the supportive placental cells that are required for embryonic development.

As we will see later in the chapter, many of the early divisions of the nematode *C. elegans* are asymmetric, and each daughter cell gives rise to a discrete set of differentiated cell

## OUTLINE

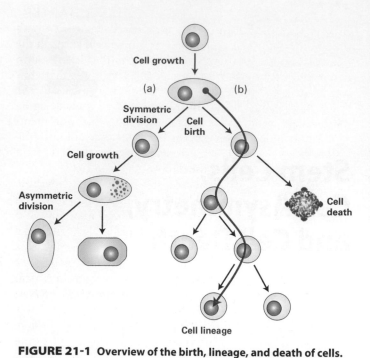

**FIGURE 21-1 Overview of the birth, lineage, and death of cells.**
Following growth, cells are "born" as the result of symmetric or asymmetric cell division. (a) The two daughter cells resulting from symmetric division are essentially identical to each other and to the parental cell. Such daughter cells subsequently can have different fates if they are exposed to different signals. The two daughter cells resulting from asymmetric division differ from birth and consequently have different fates. Asymmetric division commonly is preceded by the localization of regulatory molecules (green dots) in one part of the parent cell. (b) A series of symmetric and/or asymmetric cell divisions, called a cell lineage, gives birth to each of the specialized cell types found in a multicellular organism. The pattern of cell lineage can be under tight genetic control. Programmed cell death occurs during normal development (e.g., in the webbing that initially develops when fingers grow) and also in response to infection or toxins. A series of specific programmed events, called apoptosis, is activated in these situations.

types. In contrast, the mouse embryo passes through an eight-cell stage in which each cell can still form every tissue (both embryonic and extra-embryonic); that is, all eight cells are totipotent. At the 16-cell stage this is no longer true; some of the cells have become committed to particular differentiation paths. A group of cells called the inner cell mass (ICM) will ultimately give rise to all tissues of the embryo proper, and another set of cells will form the placental tissue. Cells such as those in the inner mass that can generate all embryonic tissues but not extra-embryonic tissues are called *pluripotent*. Cells of the ICM can be cultured in defined media, forming **embryonic stem (ES) cells.** ES cells can be grown indefinitely in culture, and each new cell remains pluripotent and can give rise to all of the tissues of an animal. We discuss the use of ES cells in uncovering the transcriptional network of gene expression underlying pluripotency and also the use of these cells in forming specific types of differentiated cells in culture for research purposes or, potentially, as "replacement parts" for outworn or diseased cells in patients.

For many years animal differentiation was thought to be unidirectional, but very recent data reveals that differentiation can be reversed experimentally; specialized, differentiated cells can be induced to revert back to an undifferentiated state. Strikingly, introducing just a small number of factors controlling the pluripotency of ES cells into multiple types of *somatic cells*, under defined conditions, can convert at least some of these somatic cells into **induced pluripotent stem (iPS) cells** that have properties seemingly indistinguishable from embryonic stem cells. As we will see in Section 21.1, iPS cells have profound utility for experimental biology and medicine.

The formation of working tissues and organs during development of multicellular organisms depends in part on specific patterns of mitotic cell division. A series of such cell divisions akin to a family tree is called a *cell lineage*. A cell lineage traces the birth order of cells as they progressively become more restricted in their developmental potential and *differentiate* into specialized cell types such as skin cells, neurons, or muscle cells (see Figure 21-1).

Many types of cells have lifetimes much less than that of the organism as a whole and so need to be constantly replaced. In mammals, for instance, cells lining the intestine and phagocytic macrophage cells live only a few days. Stem cells thus are important both during development and for replacement of outworn cells in adults. Unlike ES cells, the stem cells in the adult organism are *multipotent* stem cells: they can give rise to some types of differentiated cells found in the organism but not all of them. In the second section of this chapter we discuss several examples of multipotent stem cells, including those that give rise to germ cells, intestinal cells, and the variety of cell types found in blood.

We have already mentioned that the diversity of cells in an animal requires asymmetric cell division where the two daughter cells differ in fate. This process requires the mother cell to become asymmetric, or **polarized**, before cell division, so that cell contents are unequally distributed between the two daughters. This process of polarization is not only critical during development but also for the function of essentially all cells. For example, transporting epithelial cells, such as those that line the intestine, are polarized with their free apical surface facing the lumen to absorb nutrients and their basolateral surface contacting the extracellular matrix to transport nutrients toward the blood (see Figures 11-30 and 20-1). Other examples include cells migrating up a chemotactic gradient (see Figure 18-51) and neurons, with multiple dendrites extending from one side of the cell body that receive signals and a single axon extending from the other side that transmits signals to target cells (see Chapter 22). Thus the mechanisms that cells use to polarize are important and general aspects of their function. Not surprisingly, these mechanisms integrate elements of cell signaling pathways (see Chapters 15 and 16), cytoskeletal reorganization (see Chapters 17 and 18), and membrane trafficking (see Chapter 14). In the third section of this chapter we discuss how cells become polarized and how asymmetric cell division is critical for maintaining stem cells and their role in generating differentiated cells.

Typically we think of cell fates in terms of the differentiated cell types that are formed. A quite different cell fate, **programmed cell death**, also is absolutely crucial in the formation and maintenance of many tissues. A precise genetic regulatory system, with checks and balances, controls cell death just as other genetic programs control cell division and differentiation. In the last section of this chapter, we consider the mechanism of cell death and its regulation.

These aspects of cell biology—cell birth, the establishment of cell polarity, and programmed cell death—converge with developmental biology and are among the most important processes regulated by the signaling pathways discussed in earlier chapters.

## 21.1 Early Metazoan Development and Embryonic Stem Cells

The main focus of this section is on the first cell divisions that occur during early mammalian development and the properties of embryonic stem cells. We start with an explanation of how a single sperm is allowed to fuse with an egg, generating a zygote with a diploid genome from these two haploid germ cells.

### Fertilization Unifies the Genome

It is remarkable that a mammalian sperm is ever able to reach and penetrate the egg. For one thing, in humans each sperm is competing with more than 100 million other sperm for a single oocyte. What's more, the sperm must swim an incredible distance to reach the egg (if a sperm was the size of a person, the distance traveled would be equivalent to several miles!). And once there, the sperm must fight its way through multiple layers surrounding the egg that restrict entry. The sperm is streamlined for speed and swimming ability, but only a few dozen will reach the oocyte. The human sperm flagellum (see Chapter 18) contains about 9000 dynein motors that flex microtubules in the 50-μm axoneme.

The acrosome, found at the sperm's leading tip, is a membrane-bound compartment specialized for interaction with the oocyte. The acrosome membrane is just under the plasma membrane at the sperm head; the other side of the acrosome membrane is juxtaposed to the nuclear membrane. Inside the acrosome are soluble enzymes, including hydrolases and proteases. In proximity to the oocyte, during fertilization, the acrosome undergoes exocytosis, releasing its contents onto the surface of the oocyte. Enzymes digest the multiple egg surface layers to begin the process of sperm entry. It's a race, and the first sperm to succeed triggers a dramatic response by the oocyte that prevents *polyspermy*, the entry of other sperm that would bring in excess chromosomes.

As shown in Figure 21-2, once it reaches the egg, a sperm must first penetrate a layer of cumulus cells that surround the oocyte and then the *zona pellucida*, a gelatinous extracellular matrix composed largely of three glycoproteins called ZP1, ZP2, and ZP3. After the first sperm succeeds in fusing with the surface of the oocyte, a flux of calcium flows through the oocyte at about 5–10 μm/s, starting from the site of sperm entry. One of the effects of the calcium flux is to cause vesicles located just under the plasma membrane of the egg, the *cortical granules*, to release their contents to the outside of the plasma membrane and form a shielding fertilization membrane that blocks other sperm from entering. Finally the sperm nucleus enters the egg cytoplasm, and the egg and sperm nuclei soon fuse to create the diploid zygote nucleus.

Oocytes bring with them to the union a considerable dowry. They contain multiple circular mitochondrial DNAs whose inheritance is exclusively maternal; in mammals and many other species no sperm mitochondrial DNA enters the oocyte. Female-specific mitochondrial DNA inheritance has been used to trace maternal heritage in human history, for example following early humans from their origins in Africa. The egg cytoplasm is also packed with *maternal mRNA*: transcripts of genes essential for the earliest stages of development. There is little or no transcription during oocyte meiosis and the first embryo cleavages, so during this time the oocyte's RNA is crucial.

### Cleavage of the Mammalian Embryo Leads to the First Differentiation Events

The fertilized egg, or zygote, does not remain a single cell for long. Fertilization is quickly followed by cleavage, a series of cell divisions that take about one day each (Figure 21-3); cleavage divisions happen before the embryo is implanted in the uterus wall. Initially the cells are fairly spherical and loosely attached to each other. As demonstrated experimentally in sheep, each cell at the eight-cell stage is totipotent and has the potential to give rise to a complete animal when implanted into the uterus of a pseudopregnant animal (one treated with hormones to make her uterus responsive to embryos).

Three days after fertilization the eight-cell embryo divides again to form the 16-cell *morula* (from the Greek for "raspberry"), after which the cell affinities increase substantially and the embryo undergoes *compaction*, a process that depends in part on the surface molecule E-cadherin (see Chapter 20). The compaction process is driven by increased cell-cell adhesion that initially results in a more solid mass of cells, the *compacted morula*. In the next step some of the cell-cell adhesions locally diminish, and fluid begins to flow into an internal cavity called the *blastocoel*. Additional divisions produce a **blastocyst** stage embryo, composed of approximately 64 cells, that has separated into two cell types: **trophectoderm (TE)**, which will form extra-embryonic tissues such as the placenta, and the **inner cell mass (ICM)** (just 10–15 cells in a mouse), which gives rise to the embryo proper (Figure 21-4a). In the blastocyst, the ICM is found on one side of the blastocoel, while the TE cells form a hollow ball around the ICM and blastocoel. At this point the TE cells are in an epithelial sheet, while the ICM cells are a loose mass that can be described as **mesenchyme**.

**FIGURE 21-2 Gamete fusion during fertilization.** (a) Mammalian eggs, such as the mouse oocyte shown here, are surrounded by a ring of translucent material, the zona pellucida, which provides a binding matrix for sperm. The diameter of a mouse egg is ~70 μm, and the zona pellucida is ~6 μm thick. The polar body is a nonfunctional product of meiosis. Scale bar = 30 μm. (b) In the initial stage of fertilization, the sperm penetrates a layer of cumulus cells surrounding the egg **1** to reach the zona pellucida. Interactions between GalT, a protein on the sperm surface, and ZP3, a glycoprotein in the zona pellucida, trigger the acrosomal reaction **2**, which releases enzymes from the acrosomal vesicle. Degradation of the zona pellucida by hydrolases and proteases released during the acrosomal reaction allows the sperm to begin entering the egg **3**. Specific recognition proteins on the surfaces of egg and sperm facilitate fusion of their plasma membranes. Fusion and subsequent entry of the first sperm nucleus into the egg cytoplasm **4** and **5** trigger the release of Ca$^{2+}$ within the oocyte. Cortical granules (orange) respond to the Ca$^{2+}$ surge by fusing with the oocyte membrane and releasing enzymes that act on the zona pellucida to prevent binding of additional sperm. [Part (a) courtesy of Doug Kline; part (b) adapted from L. Wolpert et al., 2001, *Principles of Development*, 2nd ed., Oxford Press, Figure 12-22.]

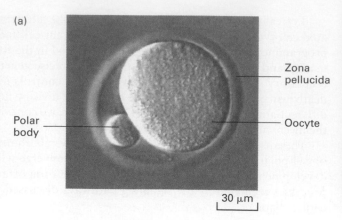

(a)

Zona pellucida

Polar body

Oocyte

30 μm

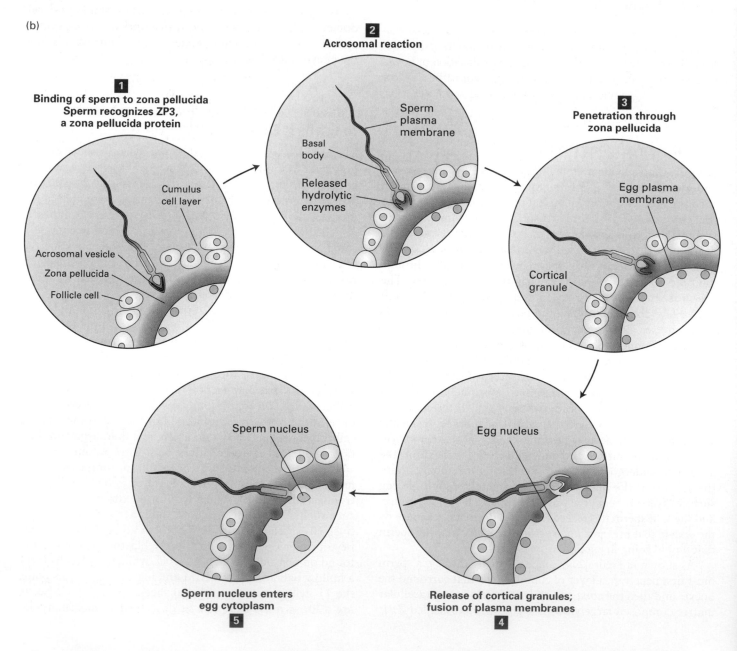

(b)

**2**
**Acrosomal reaction**

**1**
**Binding of sperm to zona pellucida**
**Sperm recognizes ZP3,**
**a zona pellucida protein**

Cumulus cell layer

Acrosomal vesicle

Zona pellucida

Follicle cell

Sperm plasma membrane

Basal body

Released hydrolytic enzymes

**3**
**Penetration through zona pellucida**

Egg plasma membrane

Cortical granule

Sperm nucleus

**Sperm nucleus enters egg cytoplasm**
**5**

Egg nucleus

**Release of cortical granules;**
**fusion of plasma membranes**
**4**

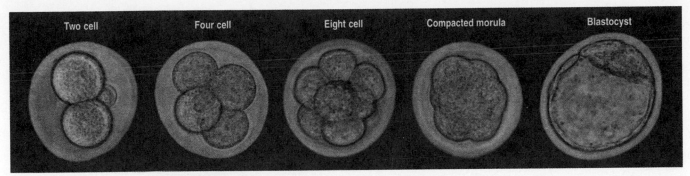

**FIGURE 21-3 Cleavage divisions in the mouse embryo.** There is little cell growth during these divisions, so that the cells become progressively smaller. See text for discussion. [Copyright Oxford University Press, T. Fleming.]

Mesenchyme, a term most commonly applied to mesoderm-derived cells, means loosely organized and loosely attached cells.

The fate of a cell in the early embryo—trophectoderm or inner cell mass—is determined by the cell's location. If an experiment is done to place a labeled cell either on the outside or the inside of a very early embryo, cells on the outside tend to form extra-embryonic tissues while cells inserted inside tend to form embryonic tissues (Figure 21-4b, c). Gene expression measurements of each stage of early development show dramatic changes in which genes are expressed. Even these very early embryos use Wnt, Notch, and TGFβ signals to regulate gene expression (see Chapter 16).

Both ICM and TE cells are stem cells: each starts its own distinct lineage and divides prolifically to produce diverse populations of cells. It is the ICM stem cells that we turn our attention to next.

## The Inner Cell Mass Is the Source of Embryonic Stem (ES) Cells

Embryonic stem (ES) cells can be isolated from the inner cell mass (ICM) of early mammalian embryos and grown indefinitely in culture when attached to a feeder cell layer (Figure 21-5a). As mentioned in the chapter introduction, cultured ES cells are pluripotent: they can differentiate into a wide range of cell types of the three primary germ layers, either in vitro or after reinsertion into a host embryo. More specifically, mouse ES cells can be injected into the blastocoel cavity of an early mouse embryo and the cell aggregate surgically transplanted into the uterus of a pseudopregnant female. The ES cells will participate in forming most if not all tissues of the resultant chimeric mice (see Figure 5-41). The injected ES cells will often give rise to functional sperm and eggs that, in turn, can generate

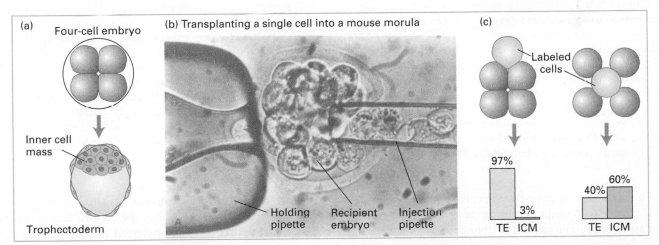

**EXPERIMENTAL FIGURE 21-4 Cell location determines cell fate in the early embryo.** (a) A four-cell embryo normally develops into a blastocyst consisting of trophectoderm (TE) cells on the outside and inner cell mass (ICM) cells inside. (b) In order to discover whether position affects the fates of cells, transplantation experiments were done with mouse embryos. First, recipient morula-stage embryos had cells removed to make room for implanted cells. Then donor morula-stage (16-cell) embryos were soaked in a dye that does not transfer between cells. Finally, labeled cells from the donor embryos were injected into inner or outer regions of the recipient embryos, as shown in the micrograph. The recipient embryo is held in place by a slight vacuum applied to the holding pipette. (c) The subsequent fates of the descendants of the transplanted labeled cells were monitored. For simplicity, four-cell recipient cells are depicted, although morula-stage embryos were used as both donors and recipients. The results, summarized in the graphs, show that outer cells overwhelmingly form trophectoderm and inner cells tend to become inner cell mass but also form considerable trophectoderm. [Parts (a) and (c) adapted from L. Wolpert et al., 2001, *Principles of Development*, 2nd ed., Oxford Press, Figure 3-12; part (b) from R. L. Gardner and J. Nichols, 1991, *Human Reprod.* **6**:25–35.]

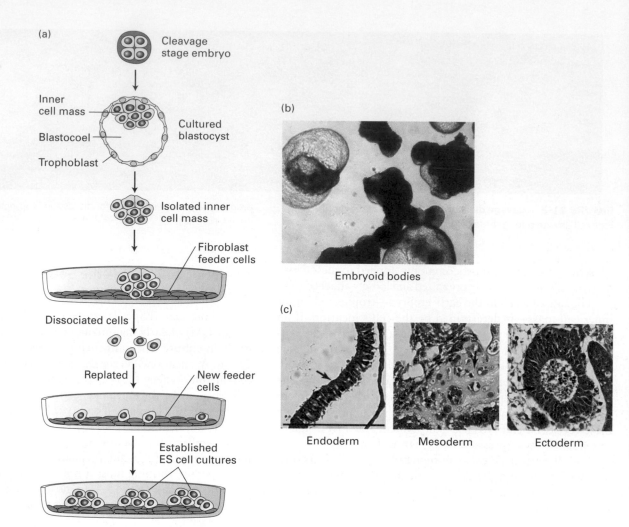

(a)

Cleavage stage embryo

Inner cell mass

Blastocoel

Trophoblast

Cultured blastocyst

Isolated inner cell mass

Fibroblast feeder cells

Dissociated cells

Replated

New feeder cells

Established ES cell cultures

(b)

Embryoid bodies

(c)

Endoderm    Mesoderm    Ectoderm

**EXPERIMENTAL FIGURE 21-5 Embryonic stem (ES) cells can be maintained in culture and can form differentiated cell types.**
(a) Human or mouse blastocysts are grown from cleavage-stage embryos produced by in vitro fertilization. The inner cell mass is separated from the surrounding extra-embryonic tissues and plated onto a layer of fibroblast cells that help to nourish the embryonic cells by providing specific protein hormones. Individual cells are replated and form colonies of ES cells, which can be maintained for many generations and can be stored frozen. ES cells can also be cultured without a fibroblast feeder layer if specific cytokines are added; leukemia inhibitory factor (LIF), for instance, supports growth of mouse ES cells by triggering

activation of the STAT3 transcription factor. (b) Embryonic stem cells allowed to differentiate in suspension culture become multicellular aggregates termed embryoid bodies (*top*). The bottom panels (c) are hematoxylin and eosin-stained sections of embryoid bodies that contain derivatives of all three germ layers that are formed from the inner cell mass during embryogenesis. Arrows in the images point to the following tissue types: (*left*) gut epithelium (endoderm), (*middle*) cartilage (mesoderm), and (*right*) neuroepithelial rosettes (ectoderm). Black bar = 100 μm. [Part (a) adapted from J. S. Odorico et al., 2001, *Stem Cells* **19**:193–204; parts (b) and (c) courtesy Drs. Lauren Surface and Laurie Boyer.]

normal living mice. In more recent experiments the host blasto-coel is treated with drugs that block mitosis so that the cells of the blastocyst become tetraploid (four copies of each chromosome), in contrast to the normal diploid ES cells that are injected into the blastocyst. In this case all of the cells in the living mice that are born after transplantation of the blastocyst aggregates derive from the donor ES cells. This is powerful evidence that single mouse ES cells indeed are pluripotent. Since these transplantation experiments cannot be done with human ES cells, formal proof that they are pluripotent is lacking.

Importantly, both human and mouse ES cells can differentiate into a wide range of cell types in culture. When placed in suspension culture, ES cells form multicellular ag-

gregates, called *embryoid bodies*, which resemble early embryos in the variety of tissues they form. When embryoid bodies are subsequently transferred to a solid surface, they grow into confluent cell sheets containing a variety of differentiated cell types, including gut epithelia, cartilage, and neural cells (Figure 21-5c). Under other conditions, ES cells have been induced to differentiate in culture into precursors for various cell types, including blood cells and pigmented epithelia; for this reason, ES cells have proved extremely useful in studies identifying the factors that commit a pluripotent cell to differentiating down a particular cell lineage.

What properties give these cells of the early embryo their remarkable plasticity? As we'll see in the next section, a variety of actors play a role: DNA methylation, transcription factors, chromatin regulators, and micro-RNAs all affect which genes become active.

## Multiple Factors Control the Pluripotency of ES Cells

During the earliest stages of embryogenesis, as the fertilized egg begins to divide, both the paternal and maternal DNA become demethylated (see the discussion of DNA methylation in Chapter 7). This happens in part because a key maintenance methyltransferase, Dnmt1, is transiently excluded from the nucleus and in part because of demethylase proteins that actively remove or "erase" these methylation marks during early development. As a result, the pattern of methylation is reset during the first few cell divisions, erasing earlier epigenetic marking of the DNA and creating a condition where cells have greater potential for diverse pathways of development. Mice engineered to lack Dnmt1 die as early embryos with drastically under-methylated DNA. ES cells prepared from such embryos are able to divide in culture, but in contrast to normal ES cells, cannot undergo in vitro differentiation.

ES cell properties are also critically dependent on the action of "master" transcription factors produced shortly after fertilization. The transcription factors Oct4, Sox2, and Nanog have essential roles in early development and are required for the specification of inner cell mass cells in the embryo as well as for specification of ES cells in culture. The expression of Oct4 and Nanog is exclusive to pluripotent cells such as the cells of the ICM and cultured ES cells. Sox2 is found in pluripotent cells, but its expression is also necessary in the multipotent neural stem cells that give rise exclusively to neuronal and glial cell types and that are discussed in the following section. Genetic studies in the mouse suggest that these regulators have distinct roles but may function in related pathways to maintain the developmental potential of pluripotent cells. For example, disruption of Oct4 or Sox2 results in the inappropriate differentiation of ICM and ES cells to trophectoderm, the cells that give rise to the placenta. However, forced expression of Oct4 in ES cells leads to a phenotype that is similar to loss of Nanog function. Thus, knowledge of the set of genes regulated by these transcription factors might reveal their essential roles during development.

The genes that are bound by these three transcription factors have been identified using chromatin immunoprecipitation experiments (see Chapter 7); each protein is found at more than a thousand chromosome locations. The target genes encode a wide variety of proteins, including the Oct4, Nanog, and Sox2 proteins themselves, forming an autoregulatory loop in which each of these transcription factors controls its own expression as well as that of the others (Figure 21-6). These transcription factors also bind to the promoter/enhancer regions of many genes encoding proteins and micro-RNAs important for the proliferation and self-renewal of ES cells.

Chromatin regulators that control gene transcription (see Chapter 7) are also important in ES cells. In *Drosophila*, Polycomb group proteins form complexes to maintain gene

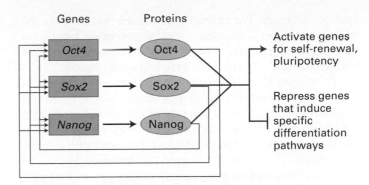

**FIGURE 21-6 Transcriptional network regulating pluripotency of ES cells.** Each of the three "master" transcription factors, Oct4, Sox2, and Nanog, binds to its own promoter as well as the promoters of the other two (black lines), forming a positive autoregulatory loop that activates transcription of each of these genes. These transcription factors also bind to the promoter/enhancer regions of many active genes encoding proteins and micro-RNAs important for the proliferation and self-renewal of ES cells (magenta lines). These factors also bind to the promoter/enhancer regions of many genes that are silenced in undifferentiated ES cells and that encode proteins and micro-RNAs essential for formation of many differentiated cell types (magenta lines). [Adapted from L. A. Boyer et al., 2006, *Curr. Opin. Genetics Dev.* **16**:455–462.]

repression states that have been previously established by DNA-binding transcription factors. Two mammalian protein complexes related to the fly Polycomb proteins, PRC1 and PRC2, are highly abundant in ES cells. Early mouse embryos lacking components of PRC2 display gastrulation defects, and ES cells lacking PRC2 functions cannot be maintained in an undifferentiated state. The PRC2 complex of proteins acts by adding methyl groups to lysine 27 of histone H3, thus altering chromatin structure to repress genes. (Note the methylation here is on an amino acid in a protein, a type of regulation distinct from the methylation of DNA.) In ES cells, PRC1 and PRC2 both silence genes whose encoded proteins or micro-RNAs (miRNAs) would otherwise induce differentiation into particular types of differentiated cells; the Polycomb proteins also maintain these genes in an epigenetic "preactivation" state such that they are poised to become activated later as part of the proper execution of specific developmental gene expression programs.

Many other regulators play important roles in controlling gene expression and maintaining pluripotency during very early development. For example, the gene encoding the miRNA let-7 is transcribed in ES cells, but the precursor RNA transcript is not cleaved to form the mature, functional miRNA. ES cells express a developmentally regulated RNA binding protein termed Lin28 that binds to the let-7 precursor RNA and prevents its cleavage. Experimental expression of mature let-7 miRNA in ES cells blocks their ability to undergo self-renewal, and thus repression of let-7 processing by Lin28 is essential for pluripotency.

The possibility of using embryonic stem cells therapeutically to restore or replace damaged tissue is fueling much research on how to induce them to differentiate into specific cell types. For example, if neurons that produce the neurotransmitter dopamine could be generated from stem cells grown in culture, it might be possible to treat people with Parkinson's disease who have lost such neurons.

By treating ES cells successively with sets of hormones, it is possible to generate cells that produce insulin. However, these cells produce only a fraction of the insulin made by a beta pancreatic islet cell, and they do not secrete insulin normally in response to changes in glucose concentration in the medium (see discussion in Chapter 16). Recently, human ES cells were triggered to differentiate into a type of islet progenitor cell termed *pancreatic endoderm*; when these cells were transplanted into mice, they generated functional beta cells that correctly processed human proinsulin into insulin (see Figure 14-24) and secreted human insulin normally in response to elevations in blood glucose. Much work is dedicated to coaxing these ES-derived endoderm cells to become normal beta cells in culture that can be used to treat human diabetic patients. However, many important questions must be answered before the feasibility of using human embryonic stem cells for therapeutic purposes can be assessed adequately. For instance, when undifferentiated ES cells are transplanted into an experimental animal, they form teratomas, a tumor that contains masses of partially differentiated cell types; thus it is essential to ensure that *all* of the ES cells used to generate an implant have indeed undergone differentiation and have lost their pluripotency. ■

Apart from their possible benefit in treating disease, ES cells have already proved invaluable for producing mouse mutants useful in studying a wide range of diseases, developmental mechanisms, behavior, and physiology. Using recombinant DNA techniques described in Chapter 5, one can eliminate or modify the function of a specific gene in ES cells (see Figure 5-40). The mutated ES cells then can be employed to produce mice with a gene knockout (see Figure 5-41). Analysis of the effects caused by deleting or modifying a gene in this way often provides clues about the normal function of the gene and its encoded protein.

## Animal Cloning Shows That Differentiation Can Be Reversed

Although different cell types may transcribe different parts of the genome, for the most part the genome is identical in all cells. Segments of the genome are rearranged and lost during development of T and B lymphocytes from hematopoietic precursors (see Chapter 23), but most somatic cells appear to have an intact genome, equivalent to that in the germ line. Evidence that at least some somatic cells have a complete and functional genome comes from the successful production of cloned animals by nuclear-transfer. In this procedure, often called somatic-cell nuclear transfer (SCNT), the nucleus of an adult somatic cell is introduced into an egg where the nucleus has been removed; the manipulated egg, which contains the diploid number of chromosomes and is equivalent to a zygote, is then implanted into a foster mother. The only source of

genetic information to guide development of the embryo is the nuclear genome of the donor somatic cell. The low efficiency of generating cloned animals by SCNT, combined with a high frequency of diseases such as obesity in the animals that are cloned, however, raises questions about how many adult somatic cells do in fact have a complete functional genome and whether those that do can be completely reprogrammed into a pluripotent undifferentiated state. Even the successes, such as the famous cloned sheep "Dolly," have some medical problems. Even if differentiated cells have a physically complete genome, clearly only parts of it are transcriptionally active (see Chapter 7). A cell could, for example, have an intact genome but be unable to properly reactivate specific genes due to inherited chromatin epigenetic states.

Further evidence that the genome of a differentiated cell can revert to having the full developmental potential characteristic of an embryonic stem cell comes from experiments where postmitotic olfactory sensory neurons were genetically marked with green fluorescence protein (GFP) and then used as donors of nuclei (Figure 21-7). When the nuclei from differentiated olfactory cells were implanted into enucleated mouse oocytes, a small fraction of them developed into blastocysts that produced GFP. The blastocysts were used to derive ES cell lines, which were then used to generate mouse embryos. These embryos, derived entirely from olfactory neuron genomes, formed healthy green-fluorescing mice. Thus, at least in some cases, the genome of a differentiated cell can be reprogrammed completely to form all tissues of a mouse.

## Somatic Cells Can Generate Induced Pluripotent Stem (iPS) Cells

Because of the inefficiency of somatic nuclear transfer, it remained unclear whether all types of somatic mammalian cells retained an intact genome and whether they could be induced to dedifferentiate into an ES-cell-like state. Shinya Yamanaka used retrovirus vectors to express a wide variety of transcription factors, singly and in combination, in cultured mouse fibroblasts. Remarkably, he found that both human and mouse cells could be reprogrammed to a pluripotent state, called an **induced pluripotent stem cell (iPS)** state, similar to that of an embryonic stem cell, following transduction with retroviruses encoding just four factors: KLF4, SOX2, OCT4, and c-MYC. Note that two of these, Sox2 and Oct4, are two of the master transcriptional regulators in ES cells discussed previously. Expression of combinations of these and other transcription factors induces iPS formation from many types of somatic cells, including antibody-producing B cells.

It was hypothesized that over time, forced expression of these genes activates expression of many cellular genes, including those encoding the Oct4, Nanog, and Sox2 pluripotency proteins that, over the course of several weeks, reprogram the somatic cells to an ES-like state. To experimentally establish this point, synthetic messenger RNAs encoding the four canonical Yamanaka transcription factors, KLF4, SOX2, OCT4, and c-MYC, were repeatedly transfected into cultured keratinocyte (skin-forming) cells. These

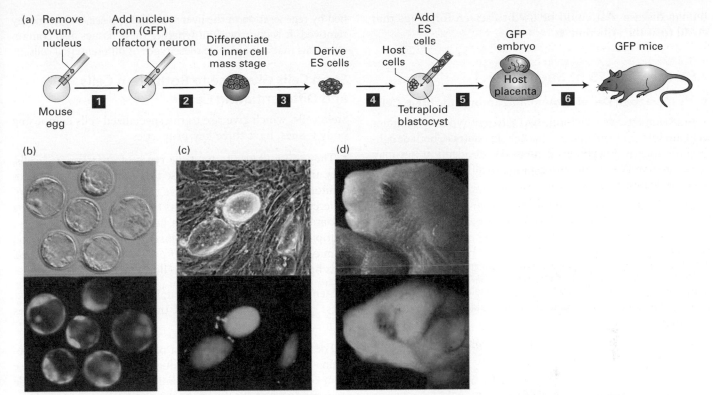

(a) Remove ovum nucleus | Add nucleus from (GFP) olfactory neuron | Differentiate to inner cell mass stage | Derive ES cells | Host cells / Add ES cells | GFP embryo | GFP mice

Mouse egg — **1** — **2** — **3** — **4** — Tetraploid blastocyst — **5** — Host placenta — **6** —

(b)  (c)  (d)

**EXPERIMENTAL FIGURE 21-7  Mice can be cloned by somatic-cell nuclear transplantation from olfactory neurons.** (a) Strategy to generate cloned ES cell lines using nuclei from olfactory sensory neurons and generation of cloned mice. **1** A nucleus from an olfactory sensory neuron isolated from a mouse that expresses green fluorescent protein (GFP) only in olfactory neurons was used to replace the nucleus of a mouse egg, and the resultant zygote was differentiated in culture until the inner cell mass stage. **2** The inner mass cells, all of which were clones of the original olfactory sensory neuron, and all of which expressed GFP, were used to generate lines of ES cells **3**, and **4** these

ES cells were injected into a tetraploid blastocyst. **5** When the blastocyst was transplanted into the uterus of a pseudopregnant mouse, the tetraploid cells from the host blastocyst could form the placenta (gray) but not the embryo proper, and **6** all of cells in the embryo proper and in the mice that were born expressed GFP. (b) Bright-field (*top*) and fluorescence images (*bottom*) of nuclear transfer blastocysts and (c) the ES cells that were isolated from the inner cell mass cells. (d) A control 12-h-old mouse (*top*) and a mouse cloned from an olfactory sensory neuron (*bottom*), all of whose cells expressed GFP. [From K. Eggan et al., 2004, *Nature* **428**:44.]

cultured cells generated normal iPS cells that had no trace of any of the exogenously added messenger RNAs, attesting to the reprogramming of keratinocytes into iPS cells by inducing expression only of normal cellular genes.

As with ES cells, single iPS cells can be experimentally introduced into a blastocyst and form all of the tissues of a mouse, including germ cells, attesting to the fact that somatic cells indeed can be reprogrammed into an embryonic state.

Because iPS cells can be derived from somatic cells of patients with difficult-to-understand diseases, they have already proved invaluable in uncovering the cellular basis of several afflictions. Consider amyotrophic lateral sclerosis (ALS), a fatal disease characterized by the degeneration of somatic motor neurons in the spinal cord, brain stem, and brain cortex. In approximately 10 percent of patients the disease is dominantly inherited (familial ALS), but in 90 percent of patients there is no apparent genetic linkage (sporadic ALS). In about 20 percent of familial ALS patients there is a point mutation in the gene encoding Cu/Zn superoxide dismutase 1 (SOD1). There is some evidence that the

mutant SOD1 protein forms aggregates that can damage cells, but whether the mutant form of SOD1 causes damage directly to the motor neurons or to the surrounding glial cells (see Figure 22-14) was not known.

In one revealing study iPS cells were derived from elderly patients with the familial form of the disease and successfully differentiated in culture to motor neurons as well as glial cells. This demonstrated the feasibility of leveraging the self-renewal of iPS cells to generate a potentially limitless supply of the cells specifically affected by ALS. In a separate study to dissect the molecular cause of ALS, motor neurons were generated from human ES or iPS cells and cultured with primary human astrocytes (a type of glial cell). About half of the motor neurons died if the glial cells expressed the mutant form of SOD1, suggesting that at least in the familial form of ALS, the defective cells are the astrocytes. Indeed, astrocytes expressing the mutant form of SOD1 secreted protein factors that were toxic to adjacent motor neurons. Determining the nature of these factors could lead to a therapy for ALS, but in any case, these experiments illustrate the value of iPS and ES cells in generating cell culture models of human disease that could be used to screen for drugs that could treat the affliction. ∎

## KEY CONCEPTS of Section 21.1

### Early Metazoan Development and Embryonic Stem Cells

• In asymmetric cell division, two different types of daughter cells are formed from one mother cell. In contrast, both daughter cells formed in symmetric divisions are identical but may have different fates if they are exposed to different external signals (see Figure 21-1).

• Specialized sperm and egg surface proteins allow the nucleus of a single mammalian sperm to enter the cytoplasm of an egg. Fusion of a haploid sperm and haploid egg nucleus generates a diploid zygote.

• Initial divisions of the mammalian embryo yield equivalent totipotent cells, but subsequent divisions yield the first differentiation event, the separation of the trophoblast epithelium from the inner cell mass.

• The inner cell mass is the source of the embryo proper as well as of embryonic stem (ES) cells.

• Cultured embryonic stem cells (ES cells) are pluripotent, capable of giving rise to all differentiated cell types of the organism with the exception of extra-embryonic tissues. They are useful in production of genetically altered mice and offer potential for therapeutic uses.

• The pluripotency of ES cells is controlled by multiple factors, including the state of DNA methylation, chromatin regulators, certain micro-RNAs, and the transcription factors Oct4, Sox2, and Nanog.

• Animal cloning establishes that cell differentiation can be reversed.

• Induced pluripotent stem (iPS) cells can be formed from somatic cells by expression of combinations of key transcription factors, including KLF4, SOX2, OCT4, and c-MYC.

## 21.2 Stem Cells and Niches in Multicellular Organisms

Many differentiated cell types are sloughed from the body or have life spans that are shorter than that of the organism. Disease and trauma also can lead to loss of differentiated cells. Since most types of differentiated cells generally do not divide, they must be replenished from nearby stem-cell populations. Postnatal (adult) animals contain stem cells for many tissues, including the blood, intestine, skin, ovaries and testes, and muscle. Even some parts of the adult brain, where little cell division normally occurs, have a population of stem cells. In muscle, stem cells are most important in healing, as relatively little cell division occurs at other times. Some other cell types, such as liver cells (hepatocytes) and insulin-producing beta pancreatic islet cells, mainly reproduce by division of already differentiated cells, as exemplified by regeneration of the liver when large pieces are surgically removed. It is controversial whether these tissues also contain stem cells that can generate these types of differentiated cells.

### Stem Cells Give Rise to Both Stem Cells and Differentiating Cells

Stem cells, which give rise to the specialized cells composing body tissues, have three key properties:

• They can give rise to multiple types of differentiated cells; they are *multipotent*. In this sense they are different from **progenitor cells,** which generally give rise to only a single type of differentiated cell. A stem cell has the capability of generating a number of different cell types but not all; that is, it is not pluripotent like an ES cell. For instance, a multipotent blood stem cell will form more of itself plus multiple types of blood cells but never a skin or a liver cell.

• Stem cells are undifferentiated; in general, they do not express proteins characteristic of the differentiated cell types formed by their descendants.

• The number of stem cells of a particular type generally remains constant over an individual's lifetime. In that sense stem cells are often said to be immortal, although no single stem cell survives for the life of the animal.

Stem cells can exhibit several patterns of cell division, outlined in Figure 21-8. A stem cell may divide symmetrically to yield two daughter stem cells identical to itself. Alternatively, a stem cell may divide asymmetrically to generate one copy of itself and one daughter stem cell that has more restricted capabilities, such as dividing for a limited time or giving rise to fewer types of progeny compared with the parental stem cell. For example, each time a stem cell in the intestine divides, one of the daughters becomes a stem cell identical to its parent while the other daughter cell divides rapidly and generates four types of differentiated intestinal cells, as we will see in greater detail shortly.

The two critical properties of stem cells that together distinguish them from all other cells are the ability to reproduce themselves indefinitely, often called *self-renewal*, and the ability to, as needed, divide asymmetrically to form one daughter stem cell identical to itself and one daughter cell of more restricted potential. Many stem-cell divisions are symmetric, producing two stem cells, but at some point some progeny need to differentiate. In this way, mitotic division of stem cells can either enlarge the population of stem cells or maintain a stem-cell population while steadily producing a stream of differentiating cells. Although some types of progenitor cells can divide symmetrically to form more of themselves, they do so only for limited periods.

### Stem Cells for Different Tissues Occupy Sustaining Niches

Stem cells need the right microenvironment. In addition to *intrinsic* regulatory signals—such as the presence of certain regulatory proteins—stem cells rely on *extrinsic* hormonal and other regulatory signals from surrounding cells to maintain

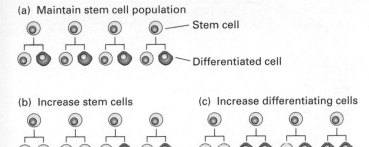

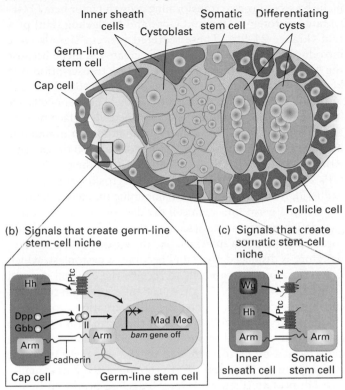

**FIGURE 21-8 Patterns of stem-cell differentiation.** Alternate patterns of stem-cell division produce more stem cells (blue) and differentiating cells (green). Stem-cell divisions meet three objectives: they must maintain the stem-cell population, sometimes they must increase the number of stem cells, and at the right time they must produce cells that go on to differentiate. (a) Stem cells can undergo asymmetric divisions, producing one stem cell and one differentiating cell. This does not increase the population of stem cells. (b) Some stem cells can divide symmetrically to increase their population, which may be useful in normal development or during recovery from injury, at the same time that others are dividing asymmetrically as in (a). (c) Cells may divide as in (b) while at the same time some stem cells produce two differentiating progeny. [From S. J. Morrison and J. Kimble, 2006, *Nature* **441**:1068–1074.]

their status as stem cells. The location where a stem-cell fate can be maintained is called a **stem-cell niche,** by analogy to an ecological niche—a location that supports the existence and competitive advantage of a particular organism. The right combination of intrinsic and extrinsic regulation, imparted by a niche, will create and sustain a population of stem cells.

In order to investigate or use stem cells, they must be found and characterized. It is often difficult to identify stem cells precisely; they are very rare and generally lack distinctive shapes and do not even divide particularly rapidly. Stem cells can be held in reserve, dividing slowly if at all, until stimulated by signals that convey the need for new cells. For example, inadequate oxygen supplies can stimulate blood stem cells to divide, and injury to the skin can stimulate regenerative cell division starting with the activation of stem cells. Some stem cells, including those that form the continuously shed epithelium of the intestine, are continuously dividing, usually at a slow rate.

## Germ-Line Stem Cells Produce Sperm and Oocytes

The *germ line* is the cell lineage that produces oocytes and sperm. It is distinct from the somatic cells that make all the other tissues but are not passed on to progeny. The germ line, like somatic-cell lineages, starts with stem cells. Stem-cell niches have been especially well defined in studies of germ-line stem cells (GSCs) in *Drosophila* and *C. elegans*. Germ-line stem cells are present in adult flies and worms, and the location of the stem cells is well known.

In the fly ovary, the niche where oocyte precursors form and begin to differentiate is located next to the tip of the *germarium* (Figure 21-9a). There are two or three germ-line stem

**FIGURE 21-9 A *Drosophila* germarium.** (a) Cross section of the germarium showing female germ-line stem cells (yellow) and some somatic stem cells (gold) in their niches and the progeny cells derived from them. The germ-line stem cells produce cytoblasts (green), which undergo four rounds of mitotic division to produce 16 interconnected cells, one of which becomes the oocyte; the somatic stem cells produce follicle cells (brown) that will make the eggshell. The cap cells (dark green) create and maintain the niche for germ-line stem cells, while the inner sheath cells (blue) produce the niche for somatic stem cells. (b) Signaling pathways that control the properties of germ-line stem cells. The signaling molecules—the TGF-β family proteins Dpp and Gbb as well as Hedgehog (Hh)—are produced by cap cells. Binding of these ligands to receptors on the surface of a germ-line stem cell, the TGF-β receptors I and II, and Ptc, respectively, results in repression of the *bam* gene by two transcription factors, Mad and Med and a co-repressor *Schnurri*. Repression of *bam* allows germ-line stem cells to renew, whereas activation of *bam* promotes differentiation. The transmembrane cell-adhesion protein E-cadherin forms the homotypic adherens junctions between germ-line stem cells and cap cells. Arm (armadillo) is the fly β-catenin and connects the cytoplasmic tails of the E-cadherin to the actin cytoskeleton; both E-cadherin and Arm are important in maintaining the stem-cell niche. (c) Signaling pathways that control the properties of somatic stem cells. The Wnt signal Wingless (Wg) is produced by the inner sheath cells and is received by the Frizzled receptor (Fz) on a somatic stem cell. Hh is similarly produced and is received by the Ptc receptor. Both receptors signal to control transcription that results in self-renewal of somatic stem cells. [Adapted from L. Li and T. Xie, 2005, *Ann. Rev. Cell Devel. Biol.* **21**:605.]

cells in this location next to a few cap cells, which create the niche by secreting two **transforming growth factor beta** (TGF-β) family proteins (Dpp and Gbb) and **Hedgehog** (**Hh**) protein (Figure 21-9b). These secreted protein signals were introduced in Chapter 16. Cap cells create the niche because the TGF-β class signals they send repress transcription of a key differentiation factor, the *Bag of marbles* (*Bam*) protein, in the neighboring germ-line stem cells. The stem cells are held in the niche by the transmembrane cell-surface protein E-cadherin, which makes adherens junctions via homotypic interactions with similar E-cadherin molecules on the cap cell. *Armadillo* (*Arm*), the fly β-catenin, connects the cytoplasmic tails of the E-cadherin to the actin cytoskeleton; like E-cadherin, Arm is important in maintaining the stem-cell niche.

When a germ-line stem cell divides, one of the resulting two daughters remains adjacent to the cap cells and is therefore maintained as a stem cell, like the mother cell. The other daughter is displaced away and becomes a cystoblast, which is too far from the cap cells to receive the cap-cell-derived signals Dpp and Gbb. As a result, Bam expression turns on, causing the germ cell to enter the differentiation program.

Separate somatic stem cells in the germarium produce follicle cells that will make the eggshell. The somatic stem cells have a niche too, created by the inner sheath cells, which produce Wingless (Wg) protein—a fly **Wnt** signal—and Hh protein (Figure 21-9c). Hedgehog produced by the cap cells may also play a role. Thus two different populations of stem cells can work in close coordination to produce different parts of an egg.

In worms, the long tubelike arms of the gonads have tips where a single cell called a distal tip cell creates a stem cell niche (Figure 21-10). The transmembrane protein Delta, produced by the distal tip cell, binds to the Notch receptor on the adjacent germ-line stem cells. The Delta/Notch signaling pathway (see Figure 16-35) in worm germ-line stem cells both represses the induction of differentiation-promoting gene products that drive entry into meiosis (i.e., germ-line differentiation) and promotes symmetric self-renewing mitotic division that creates more daughter stem cells. Meiosis is thus blocked by the Delta signal until the stem cells move beyond the range of the signal from the distal tip cell. Mutations that activate Notch in the germ-line stem cells, even in the absence of Delta signal, cause a gonadal tumor with massive numbers of extra germ-line stem cells, due to excessive mitosis and little meiosis.

The identification and characterization of *Drosophila* and *C. elegans* germ-line stem cells were important because they showed convincingly the existence of stem-cell niches and permitted experiments to identify the niche-made signals that cause cells to become and remain self-renewing stem cells. Thus a stem-cell niche is a set of cells and the signals they produce, not just a location.

Identification of specific molecules that maintain the stem-cell state in *Drosophila* and *C. elegans* brought an unexpected bonus: Some of these molecules are also used to form stem-cell niches and control stem-cell fates in mammals. For example, germ-line stem cells in the mouse testis are dependent

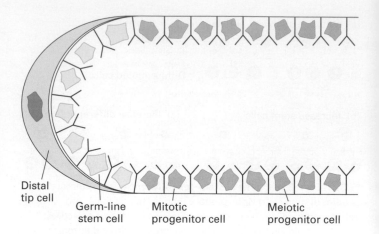

**FIGURE 21-10 Germ-line stem cells in *C. elegans*.** A cross section of the tip of the gonad shows stem cells in their niche and progeny derived from them. The single distal tip cell (green) creates and maintains the niche; it expresses Delta on its surface, which binds to the Notch receptor on the adjacent germ-line stem cells. The signals induced in these stem cells promote symmetric, self-renewing mitosis that forms two daughter germ-line stem cells. As these stem cells move outward beyond the range of the Delta signal, they continue to divide but are able to self-renew for no more than a few divisions. Eventually they become meiotic progenitor cells that form the germ cells. During these stages the cells are only partially separated by membranes ("Y" shapes) and therefore are a syncytium. [Adapted from L. Li and T. Xie, 2005, *Annu. Rev. Cell Dev. Biol.* **21**:605–631.]

on a TGF-β family signaling protein (GDNF) derived from somatic cells. Each seminiferous tubule contains a small number of germ-line stem cells that divide asymmetrically to re-create themselves and to produce spermatogonial cells. Spermatogonia proliferate and their progeny become spermatocytes, which go through the extraordinary differentiation process that builds a sperm. The niche is created by a specialized region of a Sertoli cell along with a myoid cell and a basement membrane produced by the myoid cell, though many details of the molecular signals remain to be explored.

## Intestinal Stem Cells Continuously Generate All of the Cells of the Intestinal Epithelium

The epithelium lining the small intestine is a single cell thick (see Figure 20-10) and is composed of three types of differentiated cells. The most abundant epithelial cells, the absorptive enterocytes, transport nutrients essential for survival from the intestinal lumen into the body (see Figure 11-30). The intestinal epithelium is the most rapidly self-renewing tissue in adult mammals, turning over every five days. The cells of the intestinal epithelium continuously regenerate from a stem-cell population located deep in the intestinal wall in pits called *crypts* (Figure 21-11). Pulse-chase experiments showed that intestinal stem cells produce precursor

**FIGURE 21-11 Intestinal stem cells.** Schematic drawing of an intestinal crypt showing the intestinal stem cells (yellow), their mitotic progeny (light green), and the site of terminal differentiation (dark green). The base of the crypt is the location of Paneth cells (orange), which provide a major part of the stem-cell niche and also secrete anti-microbial defense proteins called defensins that form pores in bacterial membranes, killing bacteria. Mesenchymal cells (blue) surround the base of the crypt and also secrete proteins such as Wnt that promote stem-cell fates, offsetting BMP signals that push toward differentiation. [Adapted from L. Li and T. Xie, 2005, Stem cell niche: structure and function, *Annu. Rev. Cell Dev. Biol.* **21**:605–631, and T. Sato et al., 2011, *Nature* **469**:415.]

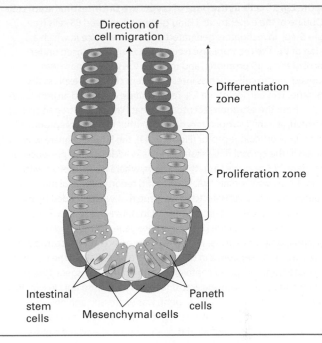

cells that proliferate and then differentiate as they ascend the sides of crypts to form the surface layer of the fingerlike gut projections called *villi*, across which intestinal absorption occurs. The time from cell birth in the crypts to the loss of cells at the tip of the villi is only about 3 to 4 days (Figure 21-12). Thus enormous numbers of cells must be produced continually to keep the epithelium intact. The production of new cells is precisely controlled: too little division would eliminate villi and lead to breakdown of the intestinal surface; too much division would create an excessively large epithelium and also might be a step toward cancer.

Experiments such as the one depicted in Figure 21-12 suggested that the intestinal stem cells were located somewhere near the bottom of the crypts, near differentiated intestinal cells called Paneth cells. But these putative stem cells have no particular morphological characteristics that reveal their remarkable abilities; which cells are the actual intestinal stem cells and which are the supportive cells that form the niche?

Prior genetic experiments had shown that Wnt signals were essential for intestinal stem cell maintenance. As evidence for the importance of Wnts, overproduction of active β-catenin (normally activated by Wnt signals; see Figure 16-30) in intestinal cells leads to excess proliferation of the intestinal epi-

thelium. Conversely, blocking the function of β-catenin by interfering with the Wnt-activated TCF transcription factor abolishes the stem cells in the intestine. Thus Wnt signaling plays a critical role in the intestinal stem cell niche as it does in the skin and other organs. Indeed, mutations that inappropriately activate Wnt signal transduction are a major contributor to the progression of colon cancer, as we will see in Chapter 24.

By analyzing a panel of genes whose expression in the intestine was induced by Wnt signaling, investigators zeroed in on Lgr5, encoding a G protein–coupled receptor, because it was expressed only in a small set of cells at the very base of the crypts. Lgr5 binds a class of secreted hormones termed R-spondins and activates intracellular signaling pathways that potentiate Wnt signaling. Lineage-tracing studies showed

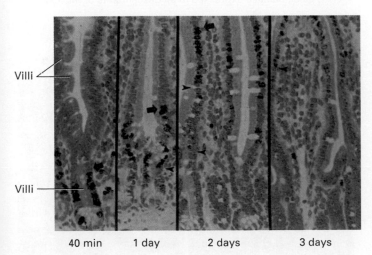

**EXPERIMENTAL FIGURE 21-12 Regeneration of the intestinal epithelium from stem cells can be demonstrated in pulse-chase experiments.** Results from a pulse-chase experiment in which radioactively labeled thymidine (the pulse) was added to a tissue culture of the intestinal epithelium. Dividing cells incorporated the labeled thymidine into their newly synthesized DNA. The labeled thymidine was washed away and replaced with nonlabeled thymidine (the chase) after a brief period; cells that divided after the chase did not become labeled. These micrographs show that 40 minutes after labeling, the entire label is in cells near the base of the crypt. At later times, the labeled cells are seen progressively farther away from their point of birth in the crypt. Cells at the top are shed. This process ensures constant replenishment of the gut epithelium with new cells. [Courtesy of C. S. Potten; from P. Kaur and C. S. Potten, 1986, *Cell Tiss. Kinet.* **19**:601.]

**EXPERIMENTAL FIGURE 21-13** Lineage-tracing studies show that the Lgr5$^+$ cells at the base of crypts are the intestinal stem cells. (a) Outline of the experiment. Using genetically altered ES cells (see Figure 5-40), investigators generated one strain of mice in which a version of the Cre recombinase (see Figure 5-42) was placed under control of the Lgr5 promoter, and thus the Cre recombinase was produced only in cells, such as intestinal stem cells, that express the Lgr5 gene. This version of the Cre recombinase contained an additional domain from the estrogen receptor (ER) that binds the estrogen analog tamoxifen; like the estrogen receptor and other nuclear receptors such as the glucocorticoid receptor (Figure 7-44), the ER-Cre chimera was retained in the cytosol unless tamoxifen was added. In the presence of tamoxifen, ER-Cre moves into the nucleus, where it can interact with *loxP* sites in the chromosomal DNA. The second reporter strain of mice contained a bacterial β-galactosidase gene that was preceded by two *loxP* sites. The blocking segment of DNA in between these *loxP* sites prevented expression of the β-galactosidase gene, and the β-galactosidase gene could only be expressed in cells where an active Cre recombinase had removed the sequence in between the two *loxP* sites. The two mice were mated, and offspring containing both marker transgenes were identified. In these mice β-galactosidase was expressed only in cells in which the Lgr5-controlled ER-Cre gene was expressed and only after addition of the estrogen analog tamoxifen to the mice. Thus only Lgr5-expressing cells—and all of their descendants—would express the β-galactosidase gene. (b) Results of the experiment. One day after adding tamoxifen to these mice, the only cells expressing β-galactosidase (measured by the blue histochemical stain) were the Lgr-expressing intestinal stem cells at the base of the crypts (*left*). Five days after tamoxifen, additional blue cells—the epithelial descendants of the intestinal stem cells—were seen migrating up the sides of the villi. Some blue stem cells remained at the bottom of the crypt. [Part (b) from N. Barker et al., 2007, *Nature* **449**:1003.]

that the descendants of these Lgr-expressing cells indeed gave rise to all of the intestinal epithelial cells (Figure 21-13). This experiment made use of gene-altered mice in which a version of the Cre recombinase (see Figure 5-42), an estrogen receptor (ER)–Cre chimera, was placed under control of the Lgr promoter; thus the ER-Cre chimera recombinase was produced only in the few putative Lgr5-expressing stem cells at the bottom of the crypts. The version of the Cre used in the study had been altered so that normally it resides inactive in the cytosol and is transferred into the nucleus only after addition of an estrogen analog (Figure 21-13a). There the Cre excises a segment of DNA, activating expression of a β-galactosidase reporter gene in these cells. Importantly, all of the descendants of these cells will also express this enzyme. Immediately after addition of the estrogen analog, the only cells expressing β-galactosidase are the stem cells in the crypts. But after a few days all of the descendant epithelial cells also expressed β-galactosidase (Figure 21-13b), showing that Lgr5 indeed is a marker of the intestinal stem cells.

In subsequent studies, single Lgr5-expressing stem cells were isolated from intestinal crypts and cultured on an extracellular matrix (see Figure 20-22) containing type IV collagen and laminin that normally underlies and supports the intestinal epithelia. These cells generated villus-like structures that

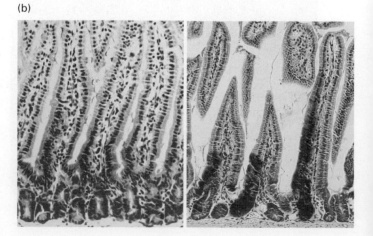

β-galactosidase expressed in all descendants of this cell

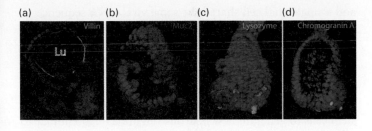

**EXPERIMENTAL FIGURE 21-14   Single Lgr5-expressing intestinal stem cells build crypt-villus structures in culture without niche cells.**  Single Lgr5-expressing cells isolated from intestinal crypts were placed in culture on a type 4 extracellular matrix (see Figure 20-22). After two weeks they formed epithelial sheets that resembled villi in structure. Staining of these organoids for specific marker proteins showed that they contained all four differentiated epithelial cell types: (a) villin, green, a marker protein for the absorptive enterocytes that are localized near the apical (luminal, Lu) surface of these cells; (b) Muc2, red, goblet cells; (c) lysozyme, green, Paneth cells; and (d) chromogranin A, green, enteroendocrine cells. The organoids were also stained with DAPI (blue) to reveal nuclei. [From T. Sato et al., 2009, *Nature* **459**:262.]

contained all four differentiated cell types found in the mature intestinal epithelium (Figure 21-14). Taken together, these experiments establish that expression of the *Lgr5* gene defines the intestinal stem cells and shows that these cells are localized at the base of the intestinal crypts interspersed between the terminally differentiated Paneth cells (see Figure 21-11).

Paneth cells produce several antibacterial proteins that protect the intestine from infections; surprising recent evidence suggests that Paneth cells also constitute a major part of the niche for the intestinal stem cells. Cultured Paneth cells produce Wnt as well as other hormones, such as EGF and a

Delta protein, that are essential for intestinal stem-cell maintenance. Co-culturing of intestinal stem cells with Paneth cells markedly improved the formation of intestinal villus-like structures, and genetic manipulations in mice that caused a reduction of Paneth cell numbers concomitantly caused a reduction in intestinal stem cells. Thus Paneth cells, which are progeny cells of the intestinal stem cells—also constitute much if not all of the niche for intestinal stem cell maintenance.

## Neural Stem Cells Form Nerve and Glial Cells in the Central Nervous System

The great interest in the formation of the nervous system and in finding better ways to prevent or treat neurodegenerative diseases has made the characterization of neural stem cells an important goal. The earliest stages of vertebrate neural development involve the rolling up of a tube of ectoderm (the cell layer that lines the outside of the embryo) that extends the length of the embryo from head to tail (Figure 21-15a).

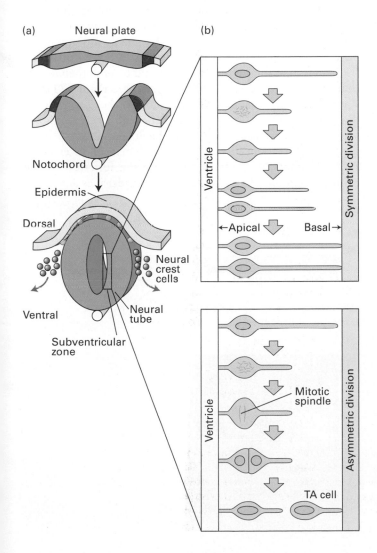

**FIGURE 21-15   Formation of the neural tube and division of neural stem cells.** (a) Early in vertebrate development a part of the ectoderm rolls up and separates from the rest of the cells. This forms the epidermis (gray) and the neural tube (blue). Near the interface between the two, neural crest cells form and then migrate to contribute to skin pigmentation, nerve formation, craniofacial skeleton, heart valves, peripheral neurons, and other structures. The notochord, a rod of mesoderm for which we are named (chordates), provides signals that affect cell fates in the neural tube. The interior of the neural tube will become a fluid-filled series of chambers called ventricles. Neural stem cells located adjacent to the ventricles, described as being in the subventricular zone (SVZ), will divide to form neurons that migrate radially outward to form the layers of the nervous system. (b) Neural stem cells in the subventricular zone can divide symmetrically (*top*) to give rise to two stem cells, side by side, both in contact with the ventricle. Alternatively stem cells can divide to produce one daughter that is a stem cell able to self-renew and another daughter, termed a transiently amplifying (TA) cell, which begins to migrate and differentiate (*bottom*). A key difference between the two patterns of division is the orientation of the mitotic spindle. [Adapted from L. R. Wolpert et al., 2001, *Principles of Development*, 2d ed., Oxford University Press.]

This *neural tube* will form the brain and spinal cord. Initially the thickness of the tube is a single layer of cells, and these cells, the embryonic neural stem cells (NSCs), will give rise to the entire central nervous system. The inside of the neural tube will form the fluid-filled compartments called *ventricles*, and the lining of the neural tube, where most cell division takes place, is called the *subventricular zone* (SVZ). The SVZ has properties of a stem-cell niche.

Labeling and tracing experiments have been done to determine how cells are born and where they go after birth. The embryonic neural stem cells that line the ventricle can divide symmetrically, producing two daughter stem cells side by side (Figure 21-15b). Alternatively they can divide asymmetrically, producing a cell that remains a stem cell and another that migrates radially outward. The migrating cells are often called *transient amplifying* (TA) *cells* because they divide rapidly multiple times to form the neural precursors called neuroblasts. Once formed, TA cells and neuroblasts migrate radially outward and form successive layers of the neural tissue in an inside-out fashion. In contrast to TA cells and neuroblasts, the stem cells remain in contact with the ventricle (see Figure 21-15b). Newly formed cells therefore traverse the layers of preexisting cells before taking up residence on the outside.

Tracing experiments with viruses have shown that a neuroblast can produce two daughters, one a neuron and one a glial cell. The experiment was done by preparing a library of retroviruses, each one containing a unique DNA sequence, so that any cell infected by a single virus would give rise to a clone of cells that all carry that particular virus's DNA sequence. In this way, all the cells that derived from a single NSC or TA precursor could be identified as a clone (Figure 21-16). The results of the experiment were surprising. First, some neurons were found to have migrated considerable distances laterally, abandoning their radial migration to the outer cortical layer. Second, in some cases, a single neuron

and a single glial cell shared the same viral DNA sequence. A neural precursor had evidently been infected and had then divided once to give rise to two quite different cell types.

For many years it was believed that no new nerve cells are formed in the adult. Most mammalian brain cells indeed stop dividing by adulthood, but some cells in the subventricular zone and in a region near the hippocampus continue to act as stem cells to generate new neurons (Figure 21-17). Similar to other types of stem cells, these neural stem cells are functionally defined by their ability to self-renew and differentiate into neural lineages, including neurons, astrocytes, and oligodendrocytes. To identify and characterize neural stem cells, cells isolated from the subventricular zone were cultured with growth factors such as FGF2 or EGF. Some of the cells survived and proliferated in an undifferentiated state; that is, they could self-renew. In the presence of other hormones these undifferentiated cells gave rise to neurons, astrocytes, or oligodendrocytes. The successful establishment of self-renewing and multipotent cells from the adult brain provides strong evidence for the presence of nerve stem-cell populations.

Some of the NSCs in the SVZ have some properties of astrocytes, such as producing glial fibrillary acidic protein (GFAP). But these NSCs can divide asymmetrically to reproduce themselves and to produce transient amplifying cells that in turn divide to form neural precursors (neuroblasts). The subventricular niche is created by mostly unknown signals from the ependymal cells that form a layer just inside the neural tube (lining the ventricle) and by endothelial cells that form blood vessels in the vicinity (see Figure 21-17). The endothelial cells, and the basal lamina they form, are in direct contact with NSCs and are believed to be essential in forming the neural stem cell niche. Each neural stem cell extends a single cilium through the ependymal cell layer to directly contact the ventricle. Though the function of the cilium is unknown, it is irresistible to consider it as a possible

**EXPERIMENTAL FIGURE 21-16 Retrovirus infection can be used to trace cell lineage.** (a) Engineered viral genome. The long terminal repeats (LTRs) are standard retroviral repeats. Viral proteins required for infection are encoded in the *gag* and in *A-env* genes. PLAP is an introduced gene for an alkaline phosphatase. Detection of this enzyme by histochemical staining is used to determine which cells carry a virus. The oligonucleotide sequence, synthesized by providing random nucleotides, is different in each virus and can be amplified by PCR, using primers for sequences that are in all viruses (blue arrows) and then sequenced. A library of more than $10^7$ distinct viruses was made. Because these viruses lack the genes required for production of new virions in infected cells, each defective virus can infect a cell only once, becomes stably integrated into the genome, and is expressed in that cell and all its descendants. (b) Tissue section showing cells infected with defective viruses. The DNA from each stained clone of cells can be extracted and amplified by PCR to determine the sequence of the infecting virus. Cells descended from the same initially infected cell will have the same oligonucleotide sequence, whereas separate infection events will give different sequences. [From J. A. Golden et al., 1995, *Proc. Nat'l. Acad. Sci. USA* **92**:5704.]

(a)

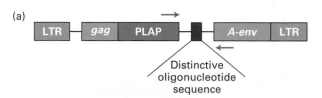

(b)

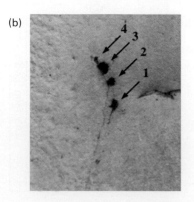

(a) Cross section of whole developing brain

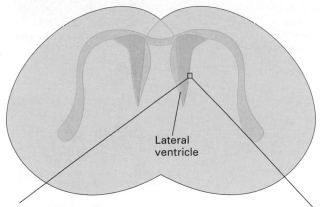

Lateral
ventricle

(b) Subventricular zone

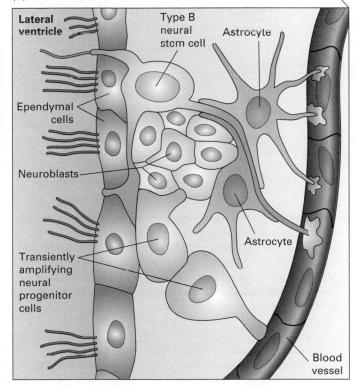

Lateral
ventricle

Type B
neural
stem cell

Astrocyte

Ependymal
cells

Neuroblasts

Astrocyte

Transiently
amplifying
neural
progenitor
cells

Blood
vessel

**FIGURE 21-17 Neural stem cell niche in the adult brain.**
(a) Cross section of the developing nervous system showing the lateral ventricle, a fluid-filled space inside the neural tube. The area just surrounding the ventricle, called the subventricular zone, is the site of neural stem cells, from which neural precursors emerge. (b) In both the embryonic and adult brain, different types of neural stem cells are located in the subventricular zone and are classified by their self-renewal and differentiation capacities, possibly representing their linear transition during neurogenesis. The type B neural stem cells express the astrocyte marker glial fibrillary acidic protein (GFAP) and are exposed to the ventricle and contact blood vessels via the apical cilia and basal ends, respectively. Type B cells have the potential to produce actively dividing (transiently amplifying) neural progenitor cells, and these subsequently generate the neuroblasts that will differentiate into neurons. [(b) Adapted from H. Suh et al., 2009, *Annu. Rev. Cell Dev. Biol.* **25**:253.]

all derive from a single type of multipotent *hematopoietic stem cell* (HSC), which gives rise to the more restricted myeloid and lymphoid progenitor stem cells that are capable of limited self-renewal (Figure 21-18). After HSCs form, numerous extracellular growth factors called **cytokines** regulate proliferation and differentiation of the precursor cells for various blood-cell lineages. Each branch of the cell-lineage tree has different cytokine regulators, allowing exquisite control of the production of specific cell types. If all blood cells are needed, for example after a bleeding injury, multiple cytokines can be produced. If only one cell type is needed, for example when a person is traveling at high altitude, **erythropoietin** is made by the kidney and stimulates the terminal proliferation and differentiation of CFU-E precursors but not other types of precursors. Erythropoietin activates several different intracellular signal transduction pathways, leading to changes in gene expression that promote formation of erythrocytes (see Figures 16-10 and 16-12). Similarly, G-CSF, a different cytokine, will stimulate proliferation and differentiation of bipotential GM-CFC macrophage/granulocyte progenitors into granulocytes, while M-CSF stimulates production of macrophages from the same progenitor cell.

The hematopoietic lineage was originally elucidated by injecting the various types of precursor cells into mice whose precursor cells had been wiped out by irradiation. By observing which blood cells were restored in these transplant experiments, researchers could infer which precursors or terminally differentiated cells (e.g., erythrocytes, monocytes) arise from a particular type of precursor.

Like other stem cells, HSCs are residents of a niche. One major cell type in the bone marrow niche is osteoblasts, bone-forming cells that are localized to the bone surface. A Delta-like ligand and proteins such as stem-cell factor (SCF), produced by niche cells, signal to Notch and SCF receptors on the surface of HSCs. Several other growth factor–receptor pairs also stimulate the HSCs, and probably other types of cells participate in forming the HSC niche.

antenna that may receive signals that would otherwise be inaccessible to the cell. The signals that create the niche are not completely characterized, but there is evidence for a blend of factors, including FGFs, BMPs, IGF, VEGF, TGFα, and BDNF. The BMPs appear to favor astrocyte differentiation over neural differentiation, one example of cell fate determination control that must remain in proper balance.

## Hematopoietic Stem Cells Form All Blood Cells

Another continuously replenished tissue is the blood, whose stem cells are located in the embryonic liver and in bone marrow in adult animals. The various types of blood cells

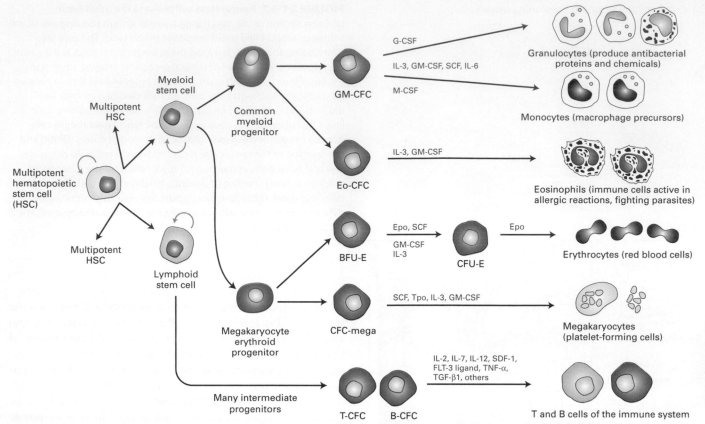

**FIGURE 21-18 Formation of blood cells from hematopoietic stem cells in the bone marrow.** Multipotent long-term repopulating hematopoietic stem cells may divide symmetrically to self-renew (blue curved arrow) or divide asymmetrically to form one daughter cell that is multipotent like the parental stem cell and another daughter with limited self-renewal capability. Ultimately this daughter generates myeloid and lymphoid stem cells; although these multipotent cells are capable of limited self-renewal, they are committed to one of the two major hematopoietic lineages. Depending on the types and amounts of cytokines present, the myeloid and lymphoid stem cells undergo rapid rounds of cell division and generate different types of progenitor cells (dark green). These progenitors are termed multipotent or unipotent in that they can give rise to several or only a single type of differentiated blood cell, respectively; they respond to one or a few specific cytokines. These progenitors are detected by their ability to form colonies containing the differentiated cell types shown at right, measured as *colony-forming cells* (*CFCs*). The colonies are detected by plating the progenitor cells in a viscous medium, formed by a polymer such as methylcellulose, such that all of the daughter differentiated cells remain localized and thus form a colony. Some of the cytokines that support this process are indicated (pink labels). GM = granulocyte-macrophage; Eo = eosinophil; E = erythrocyte; mega = megakaryocyte; T = T cell; B = B cell; CFU = colony-forming unit; CSF = colony-stimulating factor; IL = interleukin; SCF = stem-cell factor; Epo = erythropoietin; Tpo = thrombopoietin. [Adapted from M. Socolovsky et al., 1998, *Proc. Nat'l. Acad. Sci. USA* **95**:6573, and N. Noverstern et al., 2011, *Cell* **144**:296.]

Hematopoietic stem cells were detected and quantified by bone marrow transplantation experiments (Figure 21-19). The first step in these experiments was separation of the different types of hematopoietic precursors. This sorting is possible because hematopoietic stem cells and each type of progenitor produce unique combinations of cell-surface proteins that can serve as cell-type-specific markers. If bone marrow extracts are treated with fluorochrome-labeled antibodies for these markers, cells with different surface markers can be separated in a fluorescence-activated cell sorter (see Figures 9-2 and 9-3). Remarkably, such transplantation experiments revealed that a single HSC is sufficient to restore the entire blood system of an irradiated mouse. After transplantation the stem cell takes up residence in a niche in the bone marrow and divides to make more stem cells and also progenitors of the different blood lineages.

The first human bone marrow transplant was done in 1959, where a patient with end-stage (fatal) leukemia was irradiated to destroy her cancer cells. She was transfused with bone marrow cells from her identical twin sister, thus avoiding an immune response, and was in remission for three months. This pioneering beginning, which was rewarded with the Nobel Prize in Medicine in 1990, led to present-day treatments that can often lead to a complete cure of many cancers. The stem cells in the transplanted marrow can generate all types of functional blood cells, and transplants are useful in patients with certain hereditary blood diseases, including many genetic anemias (insufficient red-cell levels), genetic defects of blood cells such as the hemoglobin disorder sickle-cell anemia, and in cancer patients who have received irradiation and/or chemotherapy. Both chemotherapy and irradiation destroy the bone marrow cells as well as cancer cells. ■

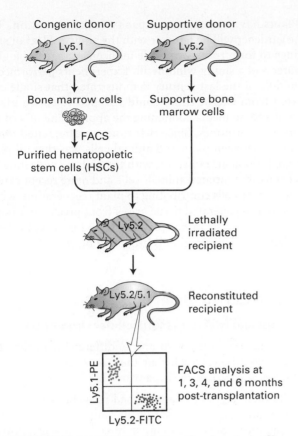

**EXPERIMENTAL FIGURE 21-19 Functional analysis of hematopoietic stem cells by bone marrow transplantation.** The two strains of mice are genetically identical except for the gene encoding a protein, termed Ly5, found on the surface of all nucleated blood cells, including all T and B lymphocytes, granulocytes, and monocytes. The proteins encoded by the two alleles of the gene, Ly5.1 and Ly5.2, can be detected by specific monoclonal antibodies. A recipient Ly5.2 mouse is lethally irradiated, then injected with stem cells purified from a Ly5.1 strain. Because the stem cells take weeks or months to produce differentiated blood cells, the recipient mouse will die unless it receives bone marrow progenitor cells from a genetically identical mouse (termed "supportive" cells) that will produce mature blood cells for the first few weeks after the transplant. At intervals after the transplant, blood or bone marrow is recovered and reacted with a blue-fluorescing monoclonal antibody to Ly5.1 and a red-fluorescing monoclonal antibody to Ly5.2. Mature blood cells that are descended from the donor stem cell are detected by FACS analysis, seen here as cells that fluoresce blue and not red. These cells can be sorted and stained with fluorescent antibodies specific for marker proteins found on different types of mature blood cells to show that a stem cell indeed is pluripotent, in that it can generate all types of lymphoid and myeloid cells. [Courtesy Dr. Chengcheng (Alec) Zhang.]

The frequency of hematopoietic stem cells is about 1 cell per $10^4$ bone marrow or embryonic liver cells. During embryonic life, stem cells often divide symmetrically, producing two daughter stem cells (see Figure 21-8); this allows the number of stem cells to increase over time and produce the large number of progenitor cells required to make all of the necessary blood cells before birth. In adult animals, hematopoietic stem cells are largely quiescent, "resting" in the $G_0$ state in the bone marrow stem cell niche. When more blood cells are needed, one or more of these cells undergoes asymmetric division, forming one stem cell like the parent and a rapidly proliferating cell that generates the progenitors illustrated in Figure 21-18.

You have probably noticed that all the molecular regulators of stem cells are familiar proteins (see Chapters 15 and 16) rather than dedicated regulators that specialize in stem-cell control. Each type of signal is used repeatedly to control cell fates and growth. These are ancient signaling systems, at least a half billion years old, for which new uses have emerged as cells, tissues, organs, and animals have evolved new variations.

## Meristems Are Niches for Stem Cells in Plants

As in their multicellular animal counterparts, the production of all tissues and organs in plants relies on small populations of stem cells. Like animal stem cells, these stem cells are defined by their ability to undergo self-renewal and also to undergo asymmetric division to generate daughter cells that produce differentiated tissues. Plant stem cells reside in specialized microenvironments, stem-cell niches where extracellular signals are produced that maintain the stem cells in a multipotent state. But since the signals that pass between stem and supportive cells are very different in these two kingdoms, stem cells and niches in plants and animals likely evolved by different pathways.

Plant stem cells are located in niches called **meristems**, which can persist for thousands of years in long-lived species such as bristlecone pines. The body axis of the plant is defined by two primary meristems that are established during embryogenesis, the *shoot apical meristem* and the *root apical meristem*. In contrast to animals, very few tissues or organs are specified during plant embryogenesis. Instead, organs such as leaves, flowers, and even germ cells are continuously generated as the plant grows and develops. The aboveground part of the plant is derived from the shoot apical meristems and the belowground part from the root apical meristems. Stem-cell identity is maintained by cellular position, rather than lineage, and intercellular signals such as hormones, mobile signaling peptides, and miRNAs.

Slowly dividing pluripotent stem cells are located at the apex of the shoot apical meristem, with more rapidly dividing multipotent "transiently amplifying" daughter cells on the periphery. Descendants of the shoot stem cells are displaced to the periphery of the meristem and are recruited to form primordia of new organs, including leaves and stems. Division ceases as cells acquire characteristics of specific cell types, and most organ growth occurs by cell expansion and elongation. New shoot stem cell niches can form in the axils of leaf primordia, which then grow to form lateral branches (Figure 21-20a). Floral meristems in turn give rise to the four floral organs—sepals, stamens, carpels, and petals—that form flowers. Unlike shoot apical meristems, floral meristems gradually become depleted as they give rise to the floral organs.

Genes required for stem-cell identity, maintenance, and cell differentiation have been defined by genetic screens for mutants exhibiting larger, smaller, or non-replenishing meristems and more recent gene expression profiling of isolated meristem cell populations. One shoot apical meristem determinant is a homeodomain transcription factor gene called

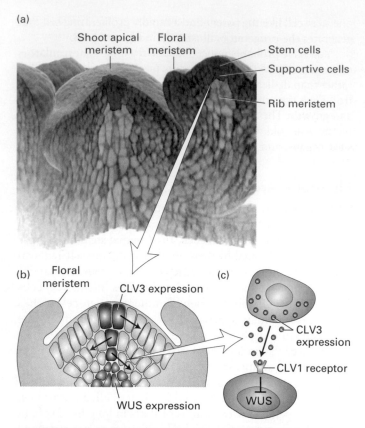

**(a)**

Shoot apical meristem

Floral meristem

Stem cells

Supportive cells

Rib meristem

**(b)** Floral meristem

CLV3 expression

WUS expression

**(c)**

CLV3 expression

CLV1 receptor

WUS

**FIGURE 21-20 Physical structure and regulatory networks in the shoot meristems of *Arabidopsis*.** The reproductive shoot apical meristem produces shoots, leaves, and more meristem. Flower production occurs when the meristem switches from leaf/shoot production to flower production, accompanied by an increase in the number of meristem cells that form floral meristems, as shown here. (a) Three-dimensional images of shoot meristems and floral meristems based on reconstruction of confocal images. Key regions of the meristems are marked in false colors: red for the stem cells, blue for the supportive cells that produce stem-cell maintenance signals, and yellow for the rib meristem. (b) Transverse section through the apex of the meristem showing CLV3 expression in stem cells (red) and WUS expression in the supportive cells (blue). The stem cells produce daughters by asymmetric division in the direction of the arrows. (c) Secreted CLV3 protein (red balls) represses WUS transcription in supportive cells; the CLV1 receptor and associated proteins are marked. [Part (a) after Robert Sablowski, 2011, *Curr. Opin. Plant Biol.* **14**:4; parts (b) and (c) after Ben Scheres, 2007, *Nat. Rev. Mol. Cell Biol.* **8**:345.]

*WUSCHEL (WUS). WUS* is required for maintenance of the stem-cell population but is expressed in the supportive cells underlying the stem cells, thus suggesting that these cells are analogous to a stem-cell niche in metazoan animals (Figure 21-20b). *WUS* promotes the expression of *CLAVATA3* (*CLV3*) in the stem cells. *CLV3* encodes a small secreted peptide that in turn binds to the CLV1 receptor on the supportive cells and negatively regulates *WUS* expression. Thus a negative feedback loop between *WUS* and *CLV3* maintains the size of the stem-cell population (Figure 21-20c). Other signals, such as the plant hormone cytokinin, are also important in regulating *WUS* expression. Root apical meristem maintenance and function is in many ways conceptually similar, although the specific genes and hormones involved are distinct.

Plants have an amazing capacity for regeneration. The home gardener will be familiar with the ability of leaf or stem cuttings to form roots with little inducement beyond a glass of water and a sunny windowsill. Experiments performed in the middle of the last century demonstrated that single cells isolated from carrot roots could regenerate entire plants when placed on media containing the appropriate mix of nutrients and hormones. Since that time, an often-cited major difference between plant and animal cells was that all plant cells are totipotent. However, with the ability to generate iPS cells from differentiated animals cells and more recent careful analyses of the cells contributing to plant regeneration, which suggest that regenerated tissue arises from preexisting populations of stem cells rather than through a process of dedifferentiation, this distinction is becoming blurred. ■

## KEY CONCEPTS of Section 21.2

### Stem Cells and Niches in Multicellular Organisms

• Stem cells are multipotent and undifferentiated; they can undergo self-renewal such that their number remains constant or increases over the organism's lifetime.

• Multipotent stem cells are formed in niches that provide signals to maintain a population of nondifferentiating stem cells. The niche must maintain stem cells without allowing their excess proliferation, and must block differentiation.

• Stem cells are prevented from differentiating by specific controls that operate in the niche. A high level of β-catenin, a component of the Wnt signaling pathway, has been implicated in preserving stem cells in the germ line and intestine by directing cells toward self-renewal division rather than differentiation states.

• Germ-line cells give rise to eggs or sperm. By definition, all other cells are somatic cells.

• In worm and fly gonads one or a few cells form the germ stem cell niche, sending signals directly to the adjacent stem cells. Daughter cells that cannot contact the niche cells undergo proliferation and differentiation into germ cells (see Figures 21-9 and 21-10).

• Populations of stem cells associated with the intestinal epithelium, brain, and many other tissues regenerate differentiated tissue cells that are damaged, sloughed, or become aged (see Figures 21-11, 21-15, and 21-18).

• Intestinal stem cells reside in the base of intestinal crypts, adjacent to Paneth cells, which form part of the niche, and are marked by expression of the Lgr5 receptor.

• Neural stem cells are found in the subventricular zone of the brain both during development and adulthood and generate both nerve and glial cells (see Figure 21-17).

• In the blood-cell lineage, different precursor types form and proliferate under the control of distinct cytokines (see Figure 21-18). This allows the body to specifically induce the replenishment of some, or all, of the necessary types.

- Plant stem cells persist for the life of the plant in the meristem. Meristem cells can give rise to a broad spectrum of cell types and structures (see Figure 21-20).

# 21.3 Mechanisms of Cell Polarity and Asymmetric Cell Division

We have discussed the importance of asymmetrical division in generating cell diversity during development and also when stem cells divide to give rise to one stem cell and one differentiated cell. What mechanisms underlie the ability of cells to become asymmetric before cell division to give rise to cells with different fates? Cell asymmetry is a concept we have met before, under the name of cell **polarity**, so let us first review what it means for a cell to be polarized.

Cell polarity—the ability of cells to organize their internal structure, resulting in changes of cell shape and regions of the plasma membrane with different protein and lipid compositions—has been introduced in several chapters. For example, we discussed how polarized epithelial cells have an apical domain with abundant microvilli separated from the basolateral domain by tight junctions (see Figures 17-1 and 20-9). To function in epithelial transport requires these cells to have different transporter proteins in the apical and basolateral membranes (Figure 11-30). As indicated earlier, not only epithelial cells but essentially all cells in the body are polarized to provide their physiological function. If a polarized cell divides and gives rise to two different daughter cells, it is said to have undergone **asymmetric cell division** (Figure 21-21a). Thus establishing cell polarity is an integral part of achieving asymmetric cell division—a cell must become polarized before dividing in order to give two different daughter cells.

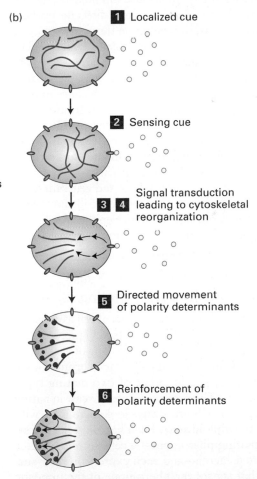

**FIGURE 21-21  General features of cell polarity and asymmetric cell division.** (a) Cell polarity requires specific determinants, including mRNAs, proteins, and lipids, to be asymmetrically localized in a cell. If the mitotic spindle is positioned so that these determinants are segregated during cell division, the two daughter cells will have different cell fate determinants. However, if the mitotic spindle is not oriented appropriately, the determinants will not be segregated properly and the daughter cells could have the same fate (not shown). (b) General hierarchy of the steps in generating a polarized cell. To know in which orientation to polarize, cells have to be exposed to a spatial cue ■. They need to have receptors or other mechanisms to sense the cue ■. After sensing the cue, signal transduction pathways ■ regulate the cytoskeleton (microtubules and/or microfilaments, depending on the system) to reorganize it in the appropriate polarized manner ■. The polarized cytoskeleton provides the framework for the transport of membrane trafficking organelles and macromolecular complexes, including fate and polarity determinants, in the cell ■. In many cases the polarity is reinforced by returning polarity determinants that have moved from the polarization site. In cases where the determinants are membrane proteins, this reinforcement cycle involves uptake by endocytosis and delivery to the polarization site ■.

## Cell Polarization and Asymmetry Before Cell Division Follow a Common Hierarchy

The polarization of a cell, whether to polarize with or without cell division, follows a general pattern, diagrammed in Figure 21-21b. In order to know in which direction to polarize, or become asymmetric, a cell must have available specific cues that provide it with spatial information (step **1**). As we will see, such cues can be localized soluble factors or signals from other cells and/or the extracellular matrix. To be receptive to these cues, cells have appropriate receptors on their surface or other machinery to sense the cues (step **2**). Once the cues have been detected, the cell has to respond appropriately by processing the incoming signal into spatial information to define the orientation of the polarity (step **3**). Generally, the next step involves the local reorganization of cytoskeletal elements, notably microfilaments and microtubules (step **4**). Now that the cell has structural asymmetry, to achieve full polarity, including molecular asymmetry, molecular motors direct the trafficking of polarity factors, that, depending on the system, may be cytoplasmic proteins and/or membrane proteins synthesized by the secretory pathway to their appropriate locations (step **5**). As we will see, in the case of many polarized cells, the polarity can often be reinforced or maintained through concentration of the polarity determinants by moving them from sites of lower concentration back to the polarized site (step **6**).

In the next section, we discuss a simple cell that shows asymmetry—a yeast cell during mating. In later sections, we turn to animal cells, where conserved polarity proteins are instrumental in interpreting polarity cues and generating cell asymmetry prior to cell division. We then describe how these same polarity proteins are used to polarize epithelial cells and finally discuss aspects of asymmetric division in stem cells.

## Polarized Membrane Traffic Allows Yeast to Grow Asymmetrically During Mating

One of the simplest and best-studied forms of cell asymmetry occurs when budding yeast cells mate (see Figure 16-23). Yeast can exist in a haploid state (a single copy of each chromosome) or diploid state (two copies of each chromosome). The haploid state can exist in either of two mating types ("sexes"), called **a** or α. The preferred state of yeast in nature is the diploid state, so **a** cells are always looking to mate with α cells to restore the diploid state. Each haploid cell type secretes a specific mating pheromone—**a** cells secrete **a** factor and α cells secrete α factor—and each express on their surface a receptor that senses the pheromone of the opposite mating type. Thus **a** cells have a receptor for α factor and α cells have a receptor for **a** factor. When you place cells of opposite mating types near each other, the receptors on each cell type bind and detect the pheromone cue of the other cell and determine its spatially highest concentration in order to know in which direction to mate. When the cells detect the opposite mating factor, two important processes occur. First, they synchronize their cell cycles by arresting at G$_0$ so that when they mate, the two haploid genomes will be at the same stage of

the cell cycle. Second, they target cell growth in the direction of the pheromone to assemble a mating projection called a "shmoo." If shmooing cells of opposite mating types touch, they will fuse at the shmoo tips and then the haploid nuclei will come together to restore the diploid state. Under conditions of plentiful nutrients, the diploid yeast cells proliferate. However, under conditions of starvation, they undergo meiosis to form stress-resistant haploid spores. When nutrients return, the spores germinate and proliferative haploid cells can grow and then mate to restore the diploid state.

By looking for mutants in yeast haploids that cannot shmoo in response to the opposite mating pheromone, researchers have identified how the asymmetric growth necessary for shmoo formation occurs (Figure 21-22). As might be anticipated, this mechanism initially involves a signal transduction pathway that establishes a polarized cytoskeleton, which in turn guides membrane traffic to the appropriate location for asymmetric growth. Activation of the mating factor receptor—a typical G protein–coupled receptor (see Figure 15-15)—results in the localized accumulation and activation of the small GTPase Cdc42 in the region of the cytoplasm closest to the pheromone source (Figure 21-22, step **1**). As we discussed in Chapter 17, small GTPases of the Rho family, of which Cdc42 is a member, regulate the assembly of the actin cytoskeleton (see Figures 17-42 and 17-43). This GTPase exists as the Cdc42-GDP inactive form and can be activated by a specific guanine nucleotide exchange factor (GEF) to the active Cdc42-GTP form. In the active form it binds and activates effector molecules. During mating, activation of the receptor leads to localized recruitment and activation of the GEF for Cdc42, thus providing localized active Cdc42-GTP. This active Ccd42-GTP leads to the local activation of a formin protein (step **2**). As we have also discussed in Chapter 17, formin proteins nucleate the assembly of polarized actin filaments, with the (+) end of the filament remaining bound to the formin (see Figure 17-13). This now provides the tracks for the transport of secretory vesicles by a myosin-V motor to the (+) ends of the filaments for localized growth and hence shmoo formation (step **3**). Notice that this mechanism requires the accumulation of polarity proteins, which includes Cdc42-GTP, to remain concentrated at the growing shmoo tip. To ensure that this polarity is maintained during shmoo growth, a directed endocytic cycle is believed to exist. In this cycle, Cdc42 that has diffused and moved away from the polarization site is internalized by endocytosis and transported back to the shmoo tip, thereby reinforcing the polarity (step **4**).

## The Par Proteins Direct Cell Asymmetry in the Nematode Embryo

The nematode worm, *Caenorhabditis elegans*, has provided a powerful model system for understanding development. The reason it was selected for study is that the animal is transparent, has a rapid life cycle, has excellent genetics, and the lineage of cells from the one-cell embryo to the adult of 959 somatic cells is invariant (Figure 21-23a, c). A critical aspect of this lineage is the first cell division, where the P0 cell—the

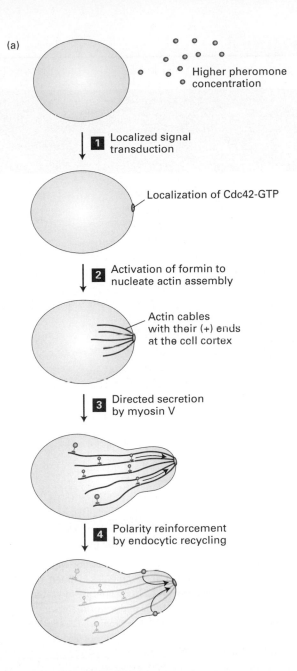

(a)

Higher pheromone concentration

**1** Localized signal transduction

Localization of Cdc42-GTP

**2** Activation of formin to nucleate actin assembly

Actin cables with their (+) ends at the cell cortex

**3** Directed secretion by myosin V

**4** Polarity reinforcement by endocytic recycling

(b)

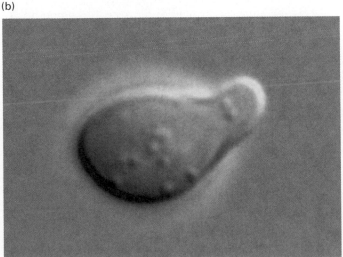

**FIGURE 21-22 Mechanism of shmoo formation in yeast.** (a) The haploid yeast cell needs to grow toward the highest concentration of mating factor, so it has a receptor on its surface that signals the cell towards the highest concentration. This signal induces the localization and activation of the small GTPase, Cdc42, to generate a higher concentration of Cdc42-GTP at this site **1**. The Cdc42-GTP locally activates a formin to nucleate and elongate actin filaments from this site **2**. Because formins bind to the (+) end of actin filaments, this end is oriented toward Cdc42-GTP and the highest concentration of the mating factor. A myosin-V motor transports secretory vesicles along the actin filament, resulting in the growth of the shmoo **3**. The polarity of the shmoo is reinforced by an endocytic cycle that constantly returns diffusing polarity factors, such as Cdc42, back along the actin filaments to the polarization site **4**. (b) DIC light-microscope image of a shmooing yeast cell. [From http://commons.wikimedia.org/wiki/File:Shmoo_yeast_S_cerevisiae.jpg.]

fertilized egg, or zygote—gives rise to the AB and P1 cells by an asymmetric cell division, and each of these two cells gives rise to different lineages. Much is known about this first asymmetric division, which is where we focus our attention.

Before the first cell division, the zygote is visibly asymmetric: cytoplasmic complexes called P granules are concentrated at the posterior end of the cell (Figure 21-23b). It turns out that during further cell divisions, these P granules always concentrate in cells that will eventually become the germ line, where they ultimately play an important role in germ-line development. As mentioned previously, the first asymmetric division of the P0 cell gives rise to the P1 cell and the larger AB cell. Following that, at the two-cell stage, the mitotic spindles are arranged at right angles to one another so that the ensuing cell divisions are also at right angles to one another (Figure 21-24a). To begin to understand how this first essential asymmetric division occurs, mutations were identified in six different genes that resulted in a symmetric first division. Since the P granules were not partitioned correctly in these mutants, the genes identified in this study were called partition defective, or *par* genes. In these mutants, P granules do not properly localize to the posterior end of the zygote, and the mitotic spindles are not oriented correctly in preparation for the second division (Figure 21-24a). A key insight came when the products of the *par* genes, namely the Par proteins, were localized. In wild-type zygotes, many of the Par proteins localize either at the cortex of the anterior half of the cell or at the cortex of the posterior half. For example, Par3 (as part of a larger complex comprising Par3, Par6, and aPKC—atypical protein kinase C) localizes anteriorly, while Par2 and Par1 localize posteriorly (Figure 21-24b). Subsequent work has shown that *antagonistic interactions* exist between these protein complexes; that is, if the Par3, Par6, aPKC complex is localized to one region, it excludes Par2 and vice versa. This is shown by the finding that the Par3, Par6, aPKC complex spreads over the whole cortex in *par2* mutants and Par2 spreads over the whole cortex in *par3* or *par6* mutants. The molecular nature of this antagonism is not fully understood, but part of it is mediated by the protein kinase aPKC phosphorylating Par2 to inhibit its ability to bind to the anterior cortex.

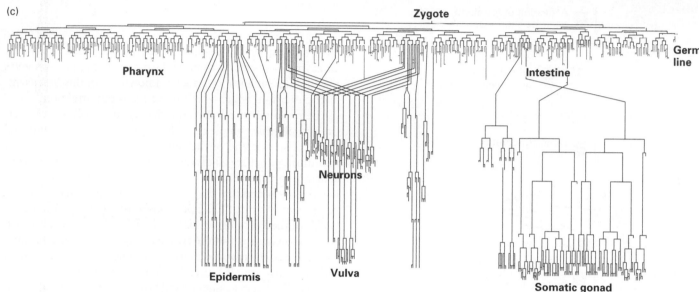

**FIGURE 21-23 Cell lineage in the nematode worm *C. elegans*.**
(a) Pattern of the first few divisions, starting with P0 (the zygote) and leading to formation of the six founder cells (yellow highlights). The first division is asymmetric, giving rise to the AB and P1 cells. The EMS cell is so named because it gives rise to most of the endoderm and mesoderm. The P4 lineage gives rise to the cells of the germ line.

(b) Micrographs of two, four and eight cell embryos with DNA stained in blue, nuclear envelope in red, and P granules in green. The P1, P2 and P3 cells, that will give rise to the germ line, are indicated. (c) The full lineage of the entire body of the worm, showing some of the tissues formed. [Part (b) from Susan Strome and Dustin Updike.]

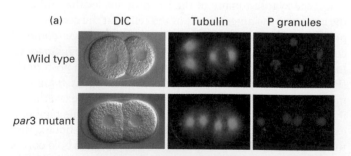

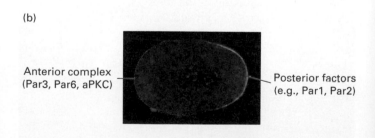

**EXPERIMENTAL FIGURE 21-24 Par proteins are asymmetrically localized in the one-cell worm embryo.** (a) DIC images of wild type and *par3* mutant embryos. Notice that in wild-type cells, the AB cell is larger than the P1 cell, whereas they are the same size in the *par3* mutant. The *par3* mutant also has a defect in spindle orientation and

P-granule (red) segregation. DNA is stained blue. (b) Complementary localization of the anterior Par complex (Par3, Par6, aPKC) (red) and posterior determinants (green) in the one-cell embryo. [Parts (a) and (b) courtesy of Diane Morton and Kenneth Kemphues.]

**FIGURE 21-25 Mechanism of segregation of the anterior Par complex in the one-cell worm embryo.** (a) Before fertilization, the cell cortex is under tension due to the activity of Rho-GEF, the exchange factor for the small GTPase, Rho. Rho-GTP activates Rho kinase, which phosphorylates the regulatory light chain of myosin II to activate it. Together with actin filaments, the active myosin II maintains tension in the cell cortex. (b) Localization of myosin II before (*top*) and after (*bottom*) fertilization. The asterisk marks the region of sperm entry. (c) Before fertilization, as the Rho-GEF is uniformly active, the cortex is under tension from the active myosin II, and the anterior Par complex (Par3/Par6/aPKC) is uniformly distributed around the cortex. On fertilization by sperm, the Rho-GEF becomes locally reduced, resulting in local deactivation of myosin II. This generates unequal tension, so the actin–myosin II contracts toward the future anterior end, moving the anterior Par complex with it. Once the anterior complex is localized, factors such as Par2 associate with the posterior cell cortex. [Part (b) from E. Munro et al., 2004, *Dev. Cell* **7:**413–424; panel (c) modified from D. St. Johnston and J. Ahringer, 2010, *Cell* **141:**757–774.]

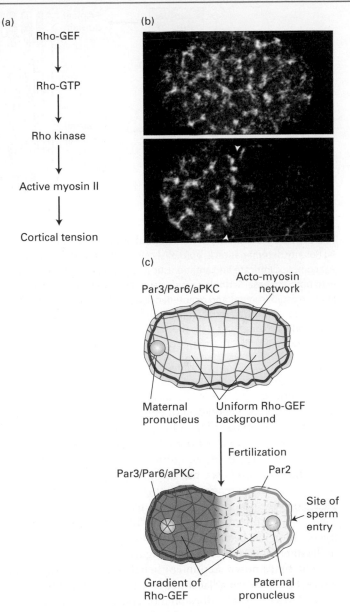

What defines the orientation of the asymmetry in the one-cell embryo? It turns out that the asymmetry is defined by the sperm entry site, which becomes the posterior end. Prior to sperm entry, the entire egg cortex is under tension provided by an actin meshwork containing active myosin II (Figure 21-25). As we discussed in Chapter 17, myosin II can form bipolar filaments that pull on actin filaments to generate tension, as is also seen in muscle and the contractile ring. Myosin II activity is regulated by a signal transduction pathway involving the small GTPase Rho (see Figure 17-42). In the unfertilized egg, Rho is maintained in its active Rho-GTP state by the uniform distribution of its activator, the nucleotide exchange factor Rho-GEF. Rho-GTP activates Rho kinase, which phosphorylates the myosin light chain of myosin II to activate it (Figure 21-25a). Sperm entry results in the local depletion of the Rho-GEF, necessary to maintain active Rho. Thus sperm entry defines the posterior region by depleting the Rho-GEF, thereby lowering active myosin II. With this local reduction in the contractile activity, the acto-myosin network contracts toward the anterior (Figure 21-25b), and as it does so, it drags (in an unknown manner) the anterior complex containing Par3, Par6, and aPKC to that end (Figure 21-25c). With the removal of the anterior complex, Par2 can now occupy the posterior cortex, and cell asymmetry is established.

It turns out that another component, Cdc42, is needed to maintain the initial asymmetry induced by the acto-myosin contraction. The active form of this GTPase, Cdc42-GTP, binds Par6 and is necessary for maintaining the complex at the anterior end, although the mechanism for this localization is not yet clear. Recent work has also implicated an endocytic reinforcement cycle, like we discussed for yeast shmoo formation, to maintain polarity. Thus the steps of responding to a cue, establishing asymmetry, and maintaining asymmetry are conserved features of both systems.

## The Par Proteins and Other Polarity Complexes Are Involved in Epithelial-Cell Polarity

In vertebrates, polarized epithelial cells use cues from adjacent cells and the extracellular matrix to orient their axis of polarization. The process of polarization is quite similar in epithelial cells of vertebrates and the fruit fly *Drosophila melanogaster*. Much of our knowledge has come from the fly system because of the ease with which mutants can be isolated and analyzed.

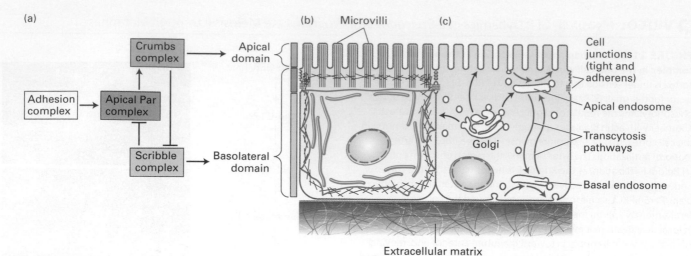

**FIGURE 21-26  Polarity establishment in epithelial cells.**
(a) Polarity determination in epithelial cells is also driven by the apical Par complex. Intricate and antagonistic interactions of the Par complex with both the basal Scribble complex and the apical Crumbs complex leads to establishment and maintenance of epithelial-cell polarity. The localization of the different complexes to membrane domains is indicated by colored bars, with the Scribble complex associating with the lateral membrane, the apical Par complex with the region at the cell junctions, and the Crumbs complex immediately apical to the Par complex. Functional epithelial polarity is maintained both by (b) a polarized cytoskeleton and (c) membrane trafficking pathways. In the biosynthetic pathway, proteins and lipids destined for the apical and basolateral domains are sorted in the Golgi complex and transported to their respective surface (red arrows). Endocytic pathways (blue arrows) regulate the abundance of proteins and lipids on each surface and sort them between surfaces by transcytosis.

Genetic screens in the fly have uncovered multiple genes necessary for the generation of epithelial-cell polarity. Analysis of the phenotypes of the mutants and proteins encoded by these genes has identified three major groups of proteins: the complex of Par3, Par6, and aPKC (in this system known as the *apical Par complex*), the Crumbs complex, and the Scribble complex. By extensive analyses of the effects of these complexes on each other when individual components are missing, a general interpretation of their contributions to epithelial-cell polarization has been achieved, although a detailed molecular understanding is still emerging (Figure 21-26a).

The first known step in epithelial-cell polarization is the interaction between adjacent cells, which in vertebrate cells is through nectin, a cell-adhesion molecule in the Ig superfamily, and a junctional protein called JAM-A. These interactions signal the cells to recruit the Par complex and to assemble adherens and tight junctions (see Figure 20-1). The Crumbs complex is recruited more apically than the Par complex, and the Scribble complex defines the basolateral surface. In the absence of the Par complex, cells cannot polarize, and it is, like in the nematode embryo, the master regulator of cell polarity. In the absence of the Scribble complex, the apical domain is greatly expanded, whereas in the absence of Crumbs, the apical domain is greatly reduced. This has led to the idea that there are antagonistic relationships between these complexes, with the apical Crumbs complex antagonizing the basolateral Scribble complex (see Figure 21-26a). Thus as is the case in the worm embryo, asymmetry is mediated by complexes working antagonistically against each other.

In a manner that is only partially understood, this arrangement of polarity proteins reorganizes the cytoskeleton, with

distinct organizations of microfilaments making up the apical and basolateral membranes. Microtubules are rather unusual in their distribution, with lateral microtubules orienting their $(-)$ ends toward the apical domain and other microtubules running perpendicular to the lateral microtubules below the microvilli and also along the base of the cell (Figure 21-26b); how these arrangements are established is not known. Membrane traffic is also polarized (Figure 21-26c). Newly made membrane proteins destined for the apical and basolateral membranes are sorted and packaged into distinct transport vesicles at the *trans*-Golgi network (TGN) and then transported to the appropriate surface. In addition, endocytic pathways from both the apical and basolateral surfaces transport missorted proteins utilizing a complex set of sorting endosomes.

In genetic screens for additional components important for epithelial-cell polarity in the fly, components of endocytic trafficking were found. For example, one such mutant affects the trafficking of the apical transmembrane protein Crumbs, so that when endocytosis is compromised, the level of Crumbs on the surface goes up and the apical domain expands. Thus epithelial polarity involves responses to spatial cues and reorganization of the cytoskeleton that provides a framework for both secretory and endocytic membrane traffic pathways for establishment and maintenance of the polarized state.

## The Planar Cell Polarity Pathway Orients Cells within an Epithelium

We have so far only discussed asymmetry in one dimension, but in many cases cells are polarized in at least two dimensions—top to bottom and front to back. Just by looking at features of

(a)

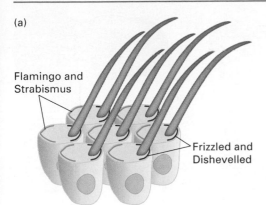

Flamingo and Strabismus

Frizzled and Dishevelled

(b)

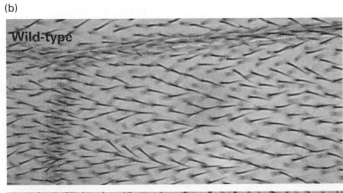

Wild-type

*Dishevelled* mutant

(c)

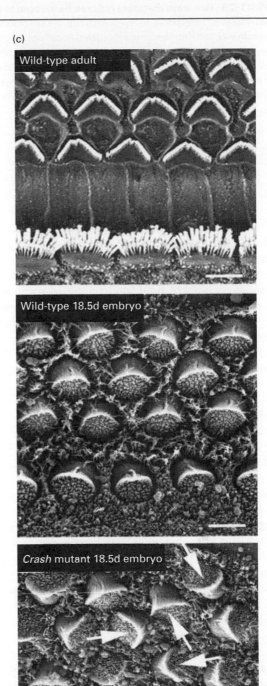

Wild-type adult

Wild-type 18.5d embryo

*Crash* mutant 18.5d embryo

**EXPERIMENTAL FIGURE 21-27 Planar-cell polarity (PCP) determines the orientation of cells.** (a) On the wing of the fly, each cell produces a hair pointing in the same direction. The directionality is determined by the asymmetric localization of components of the CPC pathway, as indicated for Frizzled, Dishevelled, Flamingo, and Strabismus, all of which are needed for orienting the hair appropriately. In (b) examples of the distribution of hairs in a wild-type fly and the *Dishevelled* mutant are shown. Notice that in the mutant, the wing cells appear normal except the hairs are not correctly oriented. (c) The sensory hair cells of the vertebrate inner ear have V-shaped arrangements of stereocilia on their surface. In the adult and 18.5-day embryo, all the cells are oriented in precisely the same way. In a mouse *Crash* mutant (the vertebrate homolog of Flamingo) defective in PCP, the cells in the 18.5-day embryo appear normal, but their relative orientations are disrupted (arrows). [Parts (a) and (b) modified from J. D. Axelrod and C. J. Tomlin, 2011, *Wiley Interdisc. Revs. Sys. Biol. Med.* **3**:588–605; part (c) from M. Fanto et al., 2004, *J. Cell Sci.* **117**:527–553.]

the animals around us, such as the scales on fish, the feathers of birds, or the hairs on your arm, it is clear that the groups of cells that give rise to these structures must be organized in a front-to-back manner in addition to a top-to-bottom manner. This type of polarity is called *planar-cell polarity* (PCP). A well-studied example from the fly is the single hair that points backward on each cell of the fly wing (Figure 21-27a). As we have seen, the fly is a system that is particularly amenable to genetic dissection. This analysis has shown that each cell responds to the

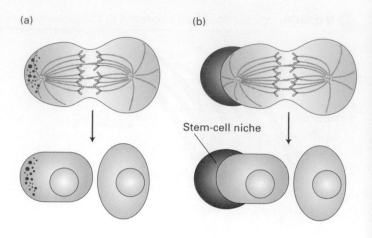

**FIGURE 21-28 Two ways that stem cells can be induced to divide asymmetrically.** (a) In response to an external cue, the cell polarizes and fate determinants (red dots) become segregated at cell division, leaving one stem cell and one differentiating cell. (b) Stem cells interacting in a stem-cell niche (red curved object) orient their mitotic spindle to give rise to a stem cell associated with the niche and a differentiating cell distant from it. [From S. J. Morrison and J. Kimble, 2006, *Nature* **441**:1068–1074.]

(a)

(b)

Stem-cell niche

planar direction of its neighbor, and components that specifically affect PCP have been identified. The overall planar polarity of an epithelium is probably determined by a gradient of some ligand. This gradient polarizes all the cells in the epithelium in the same manner, with one class of proteins, for example those encoded by the *Frizzled* and *Dishevelled* genes, on one side of each cell and a second group, for example those encoded by the *Flamingo* and *Strabismus* genes, on the other (Figure 21-27a). When components of the PCP pathway are disrupted, for example in a *Dishevelled* mutant, the epithelium is perfectly intact, but the hairs are misoriented (Figure 21-27b).

The complementary arrangement of PCP components means that the membrane protein Strabismus on the side of one cell will be adjacent to the Frizzled protein on the adjacent cell; indeed, these two proteins interact, and this interaction is probably important in coordinating PCP across an epithelium. This asymmetric distribution of PCP proteins leads, in an unknown manner, to the growth of the hair with the appropriate orientation. Although we have met Frizzled as a transmembrane receptor and Dishevelled as an adapter protein in the context of the Wnt pathway (see Figure 16-30), their role in the planar cell polarity pathway does not appear to involve Wnt but some other ligand.

Another clear example of planar-cell polarity is the sensory hair cells of the inner ear that allow you to perceive sounds. These cells have an ordered array of stereocilia arranged in a V-shaped pattern, and each cell is oriented precisely like its neighbor. In a mouse with a defect in the planar cell polarity gene *Crash* (the vertebrate homolog of fly *Flamingo*), the orderly arrangement of stereocilia within any given cell is preserved, but the relative orientations of cells to each other are disrupted (Figure 21-27c), and these types of defects can result in deafness.

## The Par Proteins Are Also Involved in Asymmetric Cell Division of Stem Cells

We have seen that stem cells often give rise to a daughter stem cell and a differentiated cell. What are the cues that set up these asymmetric cell divisions? Two types of mechanisms have been found (Figure 21-28). In one mechanism, cell fate determinants are segregated to one end of the cell before cell

division, in response to external cues. This involves the apical Par proteins, which we have already seen are instrumental in the first asymmetric division of the nematode embryo and in establishing epithelial-cell polarity. In the second mechanism, the stem cell divides with a reproducible orientation so that it remains associated with the stem-cell niche, whereas the daughter is placed away from the niche and can then differentiate. This is the situation we have already encountered in the *Drosophila* ovary, where the cap cells form a niche for the germ-line stem cells (see Figure 21-9).

A particularly well-understood example of the asymmetric division of stem cells is the formation of neurons and glial cells in the central nervous system of the fly *Drosophila* (Figure 21-29). In this system, the neuroblast stem cell arises from the neurogenic ectoderm, which is a typical epithelial layer with apical and basal surfaces. The neuroblast cell enlarges (step **1**) and moves basally into the interior of the embryo but remains in contact with the ectoderm epithelium (step **2**). It now divides asymmetrically (step **3**) to give rise to a new neuroblast stem cell and a ganglion mother cell (step **4**). The ganglion mother cell can only divide once, giving rise to two cells, either nerve and/or glial cells. The neuroblast, being a stem cell by maintaining an association with the neurogenic ectoderm niche, can divide repeatedly, giving rise to many ganglion mother cells and hence neurons and glial cells (step **5**) and so populates the central nervous system. Thus the key event is the ability of the neuroblast to divide asymmetrically (Figure 21-29b). Once again, this process involves the asymmetric accumulation of the apical Par complex—Par3, Par6, aPKC—and its positioning at the apical side of the cell (Figure 21-29c). Other factors are then positioned at the basal aspect of the cell, and the mitotic spindle is set up so that cell division segregates the polarity determinants. One of these basally localized determinants is called Miranda, a protein that associates with factors that control proliferation and differentiation (Figure 21-29c). Thus in the asymmetric division, Miranda and its associated

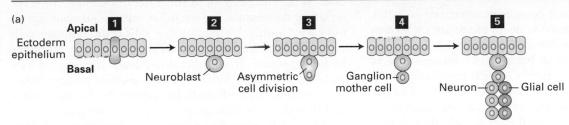

(a)

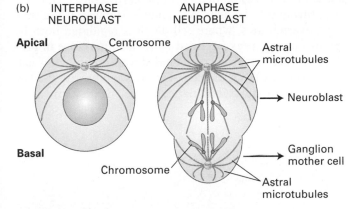

(b)

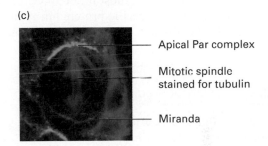

(c)

**FIGURE 21-29 Neuroblasts divide asymmetrically to generate neurons and glial cells in the central nervous system.** (a) Neuroblasts, which are stem cells, originate from the ectoderm by signals that induce them to enlarge **1**. They then move basally out of the epithelium but remain in contact with it **2**. Neuroblasts then undergo an asymmetric division **3** to give a neuroblast and a ganglion mother cell (GMC) **4**. The GMC then divides once to give two neurons or glial cells **5**. Meanwhile, the neuroblast stem cell can divide many times to give more GMCs and so populate the neural tissue. (b) The asymmetric division of the neuroblast requires the correct orientation of the mitotic spindle to give rise to a larger neuroblast and a smaller GMC cell. (c) A neuroblast at anaphase showing the segregated localization of the apical Par proteins (green) and the basal Miranda protein (red). [Part (c) from C. Cabernard and C. Q. Doe, 2009, *Dev. Cell* **17**:134–141.]

factors are segregated away from the neuroblast stem cell and into the ganglion mother cell.

## KEY CONCEPTS of Section 21.3

### Mechanisms of Cell Polarity and Asymmetric Cell Division

• Cell polarity involves the asymmetric distribution of proteins, lipids, and other macromolecules in the cell.

• Asymmetry requires cells to sense a cue, respond to it by assembling a polarized cytoskeleton, and then using this polarity to distribute factors appropriately.

• Asymmetric cell division requires cells to first become polarized, followed by division to segregate fate determinants asymmetrically.

• Mating in haploid yeast involves assembly of a mating projection (shmoo) by polarizing the cytoskeleton in the direction of highest concentration of mating pheromone and targeting secretion of cellular components for cell expansion there.

• Anterior/posterior asymmetry in the first division of the nematode worm embryo involves asymmetric F-actin/myosin-II contraction to localize the cortical anterior Par3, Par6, aPKC complex followed by the cortical association of posterior factors such as Par2.

• Apical/basal epithelial-cell polarity is also driven by the apical Par3, Par6, aPKC complex, which functions in antagonistic relationships with the apical Crumbs complex and the basal Scribble complex.

• Planar-cell polarity regulates the relative orientation of cells in a sheet.

• Asymmetric division of stem cells often involves association of the stem cell with a niche, giving rise to another stem cell and a differentiating cell.

• Asymmetric stem-cell division also involves the asymmetric distribution of the Par3, Par6, aPKC complex, which is retained in the stem cell during division, whereas fate determinants are localized away from the Par complex to end up in the differentiating cell.

## 21.4 Cell Death and Its Regulation

Programmed cell death is a counterintuitive, but essential, cell fate. During embryogenesis, death of specific cells keeps our hands from being webbed, our embryonic tails from persisting, and our brain from being filled with useless nerve connections. In fact, the majority of cells generated during brain development subsequently die. We will see in Chapter 23 how immune cells that react to normal body proteins or that produce nonfunctional antibodies are selectively killed. Many outworn muscle, epithelial, and white blood cells constantly die and need to be replaced.

Cellular interactions regulate cell death in two fundamentally different ways. First, most if not all cells in multicellular organisms require specific protein hormone signals to stay alive. In the absence of such survival signals, fre-quently referred to as **trophic factors**, cells activate a "suicide" program. Second, in some developmental contexts, including the immune system, other specific hormone signals induce a "murder" program that kills cells. Whether cells commit suicide for lack of survival signals or are murdered by killing signals from other cells, death is mediated by a common molecular pathway. In this section, we first distinguish programmed cell death from death due to tissue injury and then describe how genetic studies in the worm *C. elegans* led to elucidation of an evolutionarily conserved effector pathway that leads to cell suicide or murder. We then turn to vertebrates, where cell death is regulated both by trophic factors, as exemplified by their importance in programmed cell death in neuronal development, and cell stresses such as DNA damage. Finally we illustrate the key roles of mitochondria in initiating the cell death pathway.

---

⚡ **VIDEO:** Cells Undergoing Apoptosis

---

**FIGURE 21-30 Ultrastructural features of cell death by apoptosis.** (a) Schematic drawings illustrating the progression of morphologic changes observed in apoptotic cells. Early in apoptosis, dense chromosome condensation occurs along the nuclear periphery. The cell body also shrinks, although most organelles remain intact. Later both the nucleus and cytoplasm fragment, forming apoptotic bodies, which are phagocytosed by surrounding cells. (b) Photomicrographs comparing a normal cell (*top*) and apoptotic cell (*bottom*). Clearly visible in the latter are dense spheres of compacted chromatin as the nucleus begins to fragment. [Part (a) adapted from J. Kuby, 1997, *Immunology,* 3d ed., W. H. Freeman and Co., p. 53; part (b) from M. J. Arends and A. H. Wyllie, 1991, *Int'l. Rev. Exp. Pathol.* **32:**223.]

(a)

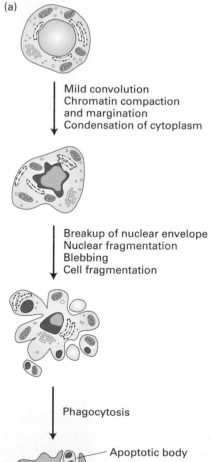

Mild convolution
Chromatin compaction
and margination
Condensation of cytoplasm

Breakup of nuclear envelope
Nuclear fragmentation
Blebbing
Cell fragmentation

Phagocytosis

Apoptotic body

Phagocytic cell

(b)

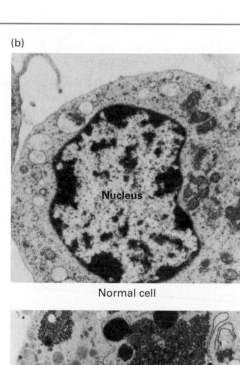

Nucleus

Normal cell

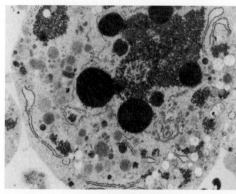

Apoptotic cell

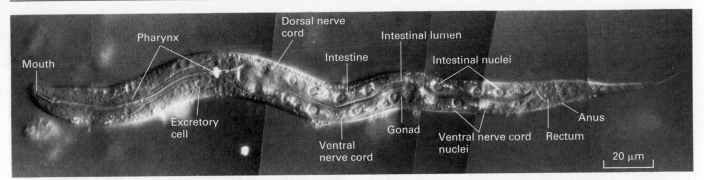

**FIGURE 21-31 Newly hatched larva of *C. elegans*.** Some of the 959 somatic-cell nuclei in the hermaphrodite form are visualized in this micrograph obtained by differential interference contrast (DIC) microscopy, sometimes called Nomarski microscopy. The most easily seen are the intestinal nuclei, which appear as round disks. [From J. E. Sulston and H. R. Horvitz, 1977, *Devel. biol.* **56**:110.]

## Programmed Cell Death Occurs Through Apoptosis

The demise of cells by programmed cell death is marked by a well-defined sequence of morphological changes, collectively referred to as **apoptosis**, a Greek word that means "dropping off" or "falling off," as leaves from a tree. Dying cells shrink, condense, and then fragment, releasing small membrane-bound apoptotic bodies, which generally are then engulfed by other cells (Figure 21-30). In dying cells, nuclei condense and the DNA is fragmented. Importantly, the intracellular constituents are not released into the extracellular milieu, where they might have deleterious effects on neighboring cells. The stereotypical changes that occur in cells during apoptosis, such as condensation of the nucleus and engulfment by surrounding cells, suggested to early scientists that this type of cell death was under the control of a strict program. This program is critical during both embryonic and adult life to maintain normal cell number and composition.

The genes involved in controlling cell death encode proteins with three distinct functions:

• "Killer" proteins are required for a cell to begin the apoptotic process.

• "Destruction" proteins do things such as digest DNA in a dying cell.

• "Engulfment" proteins are required for phagocytosis of the dying cell by another cell.

At first glance, engulfment seems to be simply an after-death cleanup process, but some evidence suggests that it is part of the final death process. For example, mutations in killer genes always prevent cells from initiating apoptosis, whereas mutations that block engulfment sometimes allow cells to persist for a while before dying. Engulfment involves the assembly of a halo of actin in the engulfing cell around the dying cell, triggered by apoptosis proteins that activate Rac, a monomeric G protein that helps regulate actin polymerization (see Figure 17-44). A signal on the surface of the dying cell also stimulates a receptor on neighboring cells, which initiates membrane changes leading to engulfment.

In contrast to apoptosis, cells that die in response to tissue damage exhibit very different morphological changes, referred to as **necrosis**. Typically, cells that undergo this process swell and burst, releasing their intracellular contents, which can damage surrounding cells and frequently causes inflammation.

## Evolutionarily Conserved Proteins Participate in the Apoptotic Pathway

The confluence of genetic studies in *C. elegans* and studies on human cancer cells suggested that an evolutionarily conserved pathway mediates apoptosis. In *C. elegans*, cell lineages are under tight genetic control and are identical in all individuals of a species. About 10 rounds of cell division or fewer create the adult worm, which is about 1 mm long and 70 μm in diameter. The adult worm has 959 somatic-cell nuclei (hermaphrodite form) or 1031 (male) (Figure 21-31). Scientists have traced the lineage of all the 947 somatic cells in *C. elegans* from the fertilized egg to the mature worm by following the development of live worms using Nomarski differential interference contrast (DIC) microscopy (see Figure 21-23c). The number of somatic cells is somewhat fewer than the number of nuclei because some cells contain multiple nuclei; that is, they are syncytial. From the zygote a series of asymmetric cell divisions produce founder cells that in turn generate all of the differentiated cells.

Of the 947 nongonadal cells generated during development of the adult hermaphrodite form, 131 undergo programmed cell death. Specific mutations have identified four genes whose encoded proteins play an essential role in controlling programmed cell death during *C. elegans* development: *ced-3*, *ced-4*, *ced-9*, and *egl-1*. In *ced-3* or *ced-4* mutants, for example, the 131 "doomed" cells survive (Figure 21-32). These mutants formed the first pieces of evidence that apoptosis was under a genetic program. The mammalian proteins that correspond most closely to the worm CED-3, CED-4, CED-9, and EGL-1 proteins are indicated in

(a)

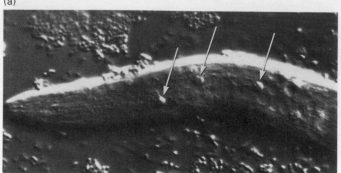

(b)

**EXPERIMENTAL FIGURE 21-32 Mutations in the *ced-3* gene block programmed cell death in *C. elegans*.** (a) Newly hatched mutant larva carry a mutation in the *ced-1* gene. Because mutations in this gene prevent engulfment of dead cells, highly refractile dead cells accumulate (arrows), facilitating their visualization. (b) Newly hatched larva with mutations in both the *ced-1* and *ced-3* genes. The absence of refractile dead cells in these double mutants indicates that no cell deaths occurred. Thus the CED-3 protein is required for programmed cell death. [From H. M. Ellis and H. R. Horvitz, 1986, *Cell* **91:**818; courtesy of Hilary Ellis.]

Figure 21-33. In discussing the worm proteins, we will include the mammalian names in parentheses to make it easier to keep the relationships clear.

The first mammalian apoptotic gene to be cloned, *bcl-2*, was isolated from human follicular lymphomas. A mutant form of this gene was created in lymphoma cells; a chromosomal rearrangement joined the protein-coding region of the *bcl-2* gene to an immunoglobulin gene enhancer. The combination results in overproduction of the Bcl-2 protein that keeps these cancer cells alive when otherwise they would become programmed to die. The human Bcl-2 protein and worm CED-9 protein are homologous; even though the two proteins are only 23 percent identical in sequence, a *bcl-2* transgene can block the extensive cell death found in *ced-9* mutant worms. Thus both proteins act as regulators that suppress the apoptotic pathway (Figure 21-33). In addition, both proteins contain a single transmembrane domain and are localized mainly to the outer mitochondrial membrane, where they serve as sensors that

**FIGURE 21-33 Evolutionary conservation of apoptosis pathways.** Similar proteins, shown in identical colors, play corresponding roles in both nematodes and mammals. (a) In nematodes, the BH3-only protein called EGL-1 binds to CED-9 on the surface of mitochondria; this interaction releases CED-4 from the CED-4/CED-9 complex. Free CED-4 then binds to and activates by autoproteolysis the caspase CED-3, which destroys cell proteins to drive apoptosis. These relationships are shown as a genetic pathway, with EGL-1 inhibiting CED-9, which in turn inhibits CED-4. Active CED-4 activates CED-3. (b) In mammals, homologs of the nematode proteins as well as many other proteins not found in worms regulate apoptosis. The Bcl-2 protein is similar to CED-9 in promoting cell survival. It does so in part by preventing activation of Apaf-1, which is similar to CED-4, and in part by other mechanisms depicted in Figure 21-38. Several types of BH3-only proteins, also detailed in Figures 21-37 and 21-38, inhibit Bcl-2 and thus allow apoptosis to proceed. Apoptotic stimuli damage mitochondria, leading to release of several proteins that stimulate cell death. In particular, cytochrome *c* released from mitochondria activates Apaf-1, which in turn activates caspase-9. This initiator caspase then activates effector caspases-3 and -7, eventually leading to cell death. See text for discussion of other mammalian proteins (SMAC/DIABLO and IAPs) that have no nematode homologs. [Adapted from S. J. Riedl and Y. Shi, 2004, *Nat. Rev. Mol. Cell Biol.* **5:**897.]

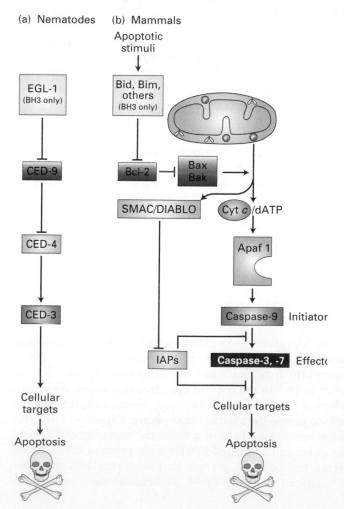

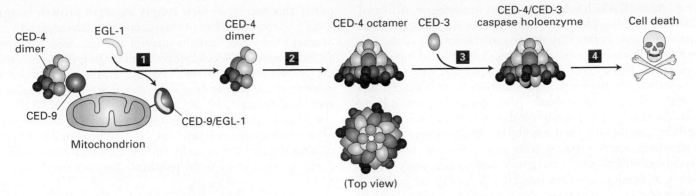

**FIGURE 21-34  Activation of CED-3 protease in *C. elegans*.** EGL-1 protein, which is produced in response to developmental signals that trigger cell death, displaces an asymmetric CED-4 dimer from its association with CED-9 on the surface of mitochondria **1**. The free CED-4 dimer combines with three others to form an octamer **2**, which binds two molecules of the CED-3 zymogen (an enzymatically inactive precursor of a protease) and triggers the conversion of the CED-3 zymogens into active CED-3 protease **3**. This effector caspase then begins to destroy cell components and thus initiate apoptosis, leading to cell death **4**. [Adapted from N. Yan et al., 2005, *Nature* **437**:831, and S. Qi et al., 2010, *Cell* **141**:446.]

control the apoptotic pathway in response to external stimuli. As we discuss next, other regulators promote apoptosis.

In the worm apoptotic pathway, CED-3 (caspase-9 in mammals) is required to destroy cell components during apoptosis. CED-4 (Apaf-1) is a protease-activating factor that causes autocleavage of (and by) the CED-3 precursor protein, creating an active CED-3 (caspase-9) protease that initiates cell death (Figures 21-33 and 21-34). Cell death does not occur in *ced-3* and *ced-4* single mutants or in *ced-9/ced-3* double mutants, whereas all cells die during embryonic life in *ced-9* mutants, so the adult form never develops. These genetic studies indicate that CED-3 and CED-4 are "killer" proteins required for cell death, that CED-9 (Bcl-2) suppresses apoptosis, and that the apoptotic pathway can be activated in all cells. Moreover, the absence of cell death in *ced-9/ced-3* double mutants suggests that CED-9 acts "upstream" of CED-3 to suppress the apoptotic pathway.

The mechanism by which CED-9 (Bcl-2) controls CED-3 (caspase-9) is now known. CED-9 protein, which is normally tethered to the outside of mitochondria, forms a complex with an asymmetric CED-4 (Apaf-1) dimer, thereby preventing activation of CED-3 by CED-4 (Figure 21-34). As a result, the cell survives. This mechanism fits with the genetics, which shows that the absence of CED-9 has no effect if CED-3 is also missing (*ced-3/ced-9* double mutants have no cell death). The three-dimensional structure of the trimeric CED-4/CED-9 complex reveals a huge contact surface between each of the two CED-4 molecules and the single CED-9 molecule; the large contact surface makes the association highly specific, but in such a way that the dissociation of the complex can be regulated.

Transcription of *egl-1*, the fourth genetically defined apoptosis regulator gene, is stimulated in *C. elegans* cells that are programmed to die, but how this is regulated is not yet clear. Newly produced EGL-1 protein binds to CED-9, alters its conformation, and catalyzes the release of CED-4 from it (Figure 21-34). Both EGL-1 and CED-9 contain a 12-amino-acid BH3 domain. Since EGL-1 lacks most of the other domains of CED-9, EGL-1 is called a *BH3-only protein*. The closest mammalian BH3-only proteins are the pro-apoptotic Bim and Bid.

Insight into how EGL-1 disrupts the CED-4/CED-9 complex comes from the molecular structure of EGL-1 (Bid/Bim) complexed with CED-9 (Bcl-2). In this complex, the BH3 domain forms the key part of the contact surface between the two proteins. CED-9 has a different conformation when bound by EGL-1 than when bound by CED-4. This finding suggests that EGL-1 binding distorts CED-9, making its interaction with CED-4 less stable. Once EGL-1 causes dissociation of the CED-4/CED-9 complex, the released CED-4 dimer joins with three other CED-4 dimers to make an octamer, which then activates CED-3 by a mechanism discussed later. Cell death soon follows (Figure 21-34).

Evidence that the steps described here are sufficient to induce apoptosis comes from experiments in which the events were reconstituted in vitro with purified proteins. CED-3, CED-4, a truncated CED-9 that lacked its transmembrane mitochondrial membrane anchor, and EGL-1 were purified, as was a CED-4/CED-9 complex. Purified CED-4 (Apaf-1) was able to accelerate the autocatalysis of purified CED-3 (caspase-9), but addition of the truncated CED-9 (Bcl-2) to the reaction mixture inhibited the autocleavage. When the CED-4/CED-9 complex was mixed with CED-3, autocleavage did *not* occur, but addition of EGL-1 to the reaction restored CED-3 autocleavage.

## Caspases Amplify the Initial Apoptotic Signal and Destroy Key Cellular Proteins

The effector proteins in the apoptotic pathway, the **caspases**, are named because they contain a key *c*ysteine residue in the catalytic site and selectively cleave proteins at sites just C-terminal to *asp*artate residues. Caspases work as homodimers, with one domain of each stabilizing the active site of the other. The principal effector caspase in *C. elegans* is CED-3, while humans have 15 different caspases. All caspases are initially made as procaspases that must be cleaved to become active. In vertebrates, initiator caspases (e.g., caspase-9) are activated by dimerization induced by binding to other types of proteins (e.g., Apaf-1), which help the initiators to aggregate. Activated initiator caspases cleave effector caspases (e.g., caspase-3) to activate them; in this way the proteolytic activity of the few activated initiator caspases becomes rapidly and hugely increased by activation of the effector caspases, leading to a massive increase in the total caspase activity level in the cell (see Figure 21-33) and cell death. Procaspases preexist in large enough numbers to accomplish the digestion of much of the cellular protein when activated by the small number of molecules that constitute the initiation signal. The various effector caspases recognize and cleave short amino acid sequences in many different target proteins. They differ in their preferred target sequences. Their specific intracellular targets include proteins of the nuclear lamina and cytoskeleton whose cleavage leads to the demise of a cell.

## Neurotrophins Promote Survival of Neurons

In mammals, but not worms, apoptosis is regulated by intracellular signals generated from many secreted and cell-surface protein hormones, as well as by many environmental stresses, such as ultraviolet irradiation and DNA damage. While the "core" apoptosis machinery in *C. elegans* is conserved in mammals, many other intracellular proteins also regulate apoptosis (see Figure 21-33, *right*).

But before plunging into these molecular details, we'll illustrate the importance of trophic factors in apoptosis by a brief analysis of the developing nervous system. When neurons grow to make connections to other neurons or to muscles, sometimes over considerable distances, more neurons grow than will eventually survive. Cell bodies of many sensory and motor neurons are located in the spinal cord and adjacent ganglia, while their long axon processes extend far outside these regions. Those that make connections prevail and survive; those that fail to connect will die.

In the early 1900s the number of neurons innervating peripheral cells was shown to depend on the size of the tissue to which they would connect, the so-called target field. For instance, removal of limb buds from the developing chick embryo leads to a reduction in the number of sensory neurons and also motor neurons innervating muscles in the bud (Figure 21-35). Conversely, grafting additional limb tissue to a limb bud leads to an increased number of neurons in corresponding regions of the spinal cord and sensory ganglia. Indeed, incremental increases in the target-field size are accompanied by commensurate incremental increases in the number of neurons innervating the target field. This relationship was found to result from the selective survival of neurons rather than changes in their differentiation or proliferation. The observation that many sensory and motor neurons die after reaching their peripheral target field suggested that these neurons compete for survival factors produced by the target tissue.

Subsequent to these early observations, scientists discovered that transplantation of a mouse sarcoma (muscle tumor) into a chick led to a marked increase in the local numbers of certain types of neurons. This finding implicated the tumor as a rich source of the presumed trophic factor. To isolate and

purify this factor, known simply as nerve growth factor (NGF), scientists used an in vitro assay in which outgrowth of neurites from sensory ganglia (nerves) was measured. Neurites are extensions of the cell cytoplasm that can grow to become the long wires of the nervous system, the **axons** and **dendrites** (see Figure 22-1). The later discovery that the submaxillary gland in the mouse also produces large quantities of NGF enabled biochemists to purify and to sequence it. A homodimer of two 118-residue polypeptides, NGF belongs to a family of structurally and functionally related trophic factors collectively referred to as **neurotrophins**. Brain-derived neuro-

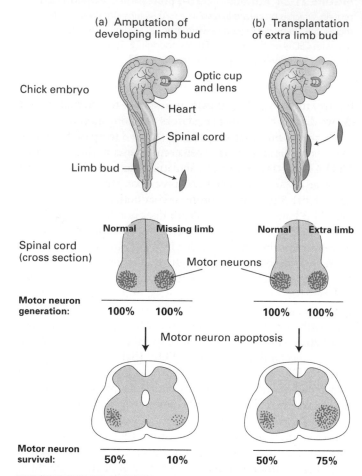

**EXPERIMENTAL FIGURE 21-35 In vertebrates the survival of motor neurons depends on the size of the muscle target field they innervate.** (a) Removal of a limb bud from one side of a chick embryo at about 2.5 days results in a marked decrease in the number of motor neurons on the affected side. In an amputated embryo (*top*), normal numbers of motor neurons are generated on both sides (*middle*). Later in development, many fewer motor neurons remain on the side of the spinal cord with the missing limb than on the normal side (*bottom*). Note that normally only about 50 percent of the motor neurons that originally are generated survive. (b) Transplantation of an extra limb bud into an early chick embryo produces the opposite effect, more motor neurons on the side with additional target tissue than on the normal side. [Adapted from D. Purves, 1988, *Body and Brain: A Trophic Theory of Neural Connections*, Harvard University Press, and E. R. Kandel, J. H. Schwartz, and T. M. Jessell, 2000, *Principles of Neural Science*, 4th ed., McGraw-Hill, p. 1054, Figure 53-11.]

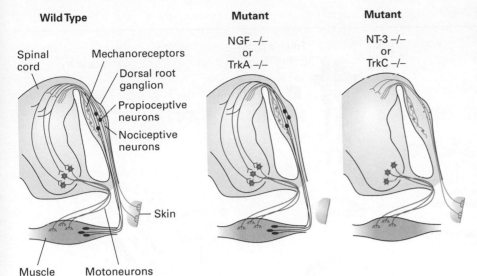

**Wild Type**

**Mutant**

NGF –/–
or
TrkA –/–

**Mutant**

NT-3 –/–
or
TrkC –/–

Spinal
cord

Mechanoreceptors

Dorsal root
ganglion

Propioceptive
neurons

Nociceptive
neurons

Skin

Muscle     Motoneurons

**EXPERIMENTAL FIGURE 21-36 Different classes of sensory neurons are lost in knockout mice lacking different trophic factors or their receptors.** In animals lacking nerve growth factor (NGF) or its receptor TrkA, small nociceptive (pain-sensing) neurons (light blue) that innervate the skin are missing. These neurons express TrkA receptor and innervate NGF-producing targets. In animals lacking either neurotrophin-3 (NT-3) or its receptor TrkC, large propioceptive neurons (red) innervating muscle spindles are missing. Muscle produces NT-3, and the propioceptive neurons express TrkC. Mechanoreceptors (orange; see Figure 22-24), another class of sensory neurons in the dorsal root ganglion, are unaffected in these mutants. [Adapted from W. D. Snider, 1994, *Cell* **77**:627.]

trophic factor (BDNF) and neurotrophin-3 (NT-3) are members of this protein family.

Neurotrophins bind to and activate a family of receptor tyrosine kinases called *Trks* (pronounced "tracks"). (The general structure of receptor tyrosine kinases and the intracellular signaling pathways they activate are covered in Chapter 16.) Each neurotrophin binds with high affinity to one Trk receptor: NGF binds to TrkA; BDNF, to TrkB; and NT-3, to TrkC. NT-3 can also bind with lower affinity to both TrkA and TrkB. These bindings—between factors and their receptors—provide a survival signal for different classes of neurons. As neurons grow from the spinal cord to the periphery, neurotrophins produced by target tissues bind to Trk receptors on the growth cones (see Figure 18-52) of the extending axons, promoting survival of neurons that successfully reach targets. In addition, neurotrophins bind to a distinct type of receptor called p75$^{NTR}$ (NTR 5 neurotrophin receptor) but with lower affinity. However, p75$^{NTR}$ forms heteromultimeric complexes with the different Trk receptors; this association increases the affinity of Trks for their ligands. Depending on the cell type, binding of NGF and BDNF to p75$^{NTR}$ in the absence of TrkA may promote cell death rather than prevent it. (The phenomenon of multiple neurotrophins interacting with multiple similar receptors is comparable to EGF-like ligands and their HER receptors, illustrated in Figure 16-7).

To critically address the role of neurotrophins in development, scientists produced mice with knockout mutations in each of the neurotrophins and their receptors. These studies revealed that different neurotrophins and their corresponding receptors are required for the survival of different classes of sensory neurons (Figure 21-36). For instance, pain-sensitive (nociceptive) neurons, which express TrkA, are selectively lost from the dorsal root ganglion of knockout mice lacking NGF or TrkA, whereas TrkB- and TrkC-expressing neurons are unaffected in such knockouts. In contrast, TrkC-expressing proprioceptive neurons, which detect the position of the limbs, are missing from the dorsal root ganglion in *TrkC* and *NT-3* mutants.

## Mitochondria Play a Central Role in Regulation of Apoptosis in Vertebrate Cells

As discussed previously, *C. elegans* CED-9 and its mammalian homolog Bcl-2 play central roles in repressing apoptosis. In nematodes, CED-9 does so by binding to and repressing the activation of CED4. In vertebrates, Bcl-2, residing in the outer mitochondrial membrane, primarily functions to maintain the low permeability of that membrane, preventing cytochrome *c* and other proteins localized to the intermembrane space (see Figure 12-16) from diffusing into the cytosol and activating apoptotic caspases.

In order to explain how Bcl-2 carries out this function and how Bcl-2 activity is regulated by trophic factors and by many environmental stimuli, we need to introduce several other important members of the *Bcl-2 family* of proteins. All members of the Bcl-2 family share a close homology in up to four characteristic regions termed the *Bcl-2 homology domains* (BH1–4 domains; Figure 21-37). Each protein has

**Pro-survival members**

BH4     BH3     BH1     BH2     TM

Bcl-2, Bcl-xL, Bcl-w, Mcl-1, A1

**Pro-apoptotic members**

Form channels in the mitochondrial outer membrane

BH3     BH1     BH2     TM

Bax, Bak, Bok

BH3-only proteins: Regulate activity of Bcl-2 and Bax/Bak proteins

BH3                     TM

Bim, Puma, Noxa, Bik, Bmf, Bad, Hrk, Bid

**FIGURE 21-37 Bcl-2 family proteins.** The Bcl-2 family is made of proteins that contain functional Bcl homology domains (BH 1-4) and are divided into three classes. Only some of the BH3-only proteins contain transmembrane (TM) domains. [After M. Giam et al., 2009, *Oncogene* **27**:S128.]

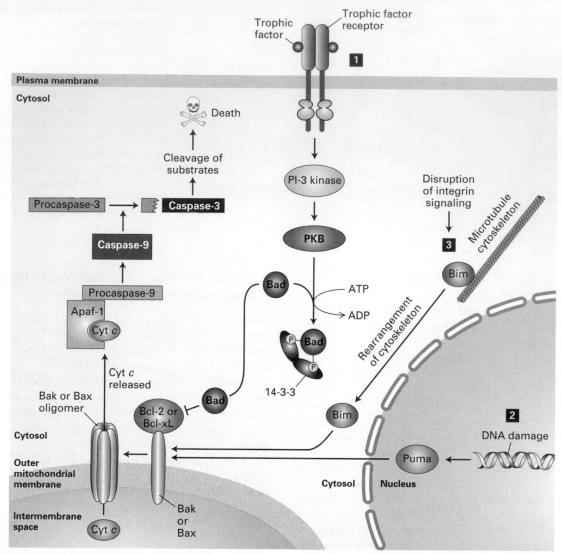

**FIGURE 21-38 Integration of multiple signaling pathways in vertebrate cells that regulate mitochondrial outer membrane permeability and apoptosis.** In healthy cells the anti-apoptotic protein Bcl-2 or its homolog Bcl-xL binds to the Bak and Bax BH3-only proteins in the outer mitochondrial membrane, blocking the ability of Bak and Bax to oligomerize and form oligomeric channels. Binding of certain BH3-only proteins, including Bim and Puma, directly to Bak and Bax causes them to form oligomeric channels in the outer mitochondrial membrane. This allows cytochrome *c* to enter the cytosol, where it binds to the adapter protein Apaf-1, promoting caspase activation that initiates the apoptotic cascade and leads to cell death. Other BH3-only proteins, including Bad, bind to Bcl-2, blocking its ability to bind to Bak and Bax and causing them to form channels in the outer mitochondrial membrane. Several stimuli trigger or repress this apoptotic pathway.

(**1**) The presence of specific trophic factors (e.g., NGF) leads to activation of the cognate receptor protein tyrosine kinase (e.g., TrkA) and activation of the PI-3 kinase PKB (protein kinase B; also called Akt) pathway (see Figure 16-26). PKB phosphorylates Bad, and phosphorylated Bad then forms a complex with the 14-3-3 protein. With Bad sequestered, it is unable to bind to Bcl-2. In the absence of trophic factors, unphosphorylated Bad binds to Bcl-2, releasing Bax and Bak and allowing them to form the oligomeric channel. (**2**) DNA damage or ultraviolet irradiation leads to induction of the BH3-only Puma protein. Puma binds to Bax and Bak, allowing them to form oligomeric channels. (**3**) Removal of a cell from its substratum disrupts integrin signaling, leading to release of the BH3-only Bim protein from the cytoskeleton. Bim also binds to Bax and Bak to promote channel formation. [After D. Ren et al., 2010, *Science* **330**:1390.]

either a pro-survival or a pro-apoptotic function. Many members of this family are single-pass transmembrane proteins and all participate in oligomeric interactions.

## The Pro-apoptotic Proteins Bax and Bak Form Pores in the Outer Mitochondrial Membrane

In vertebrate cells, Bax or Bak is required for mitochondrial damage and induction of apoptosis. These two similar pro-apoptotic proteins contain several of the BH1–4 domains (see Figure 21-37) and have a three-dimensional structure very similar to that of the pro-survival members of the family. As evidence for their role in promoting apoptosis, most mice lacking both Bax and Bak die *in utero*. Those that survive show significant developmental defects, including the persistence of interdigital webs and accumulation of extra cells in the central nervous and hematopoietic systems. Cells isolated from these mice are resistant to virtually all apoptotic stimuli. Conversely, overproduction of Bax in cultured cells induces death.

Bax and Bak reside in the outer mitochondrial membrane, normally tightly bound to Bcl-2 (Figure 21-38). When released from Bcl-2—either by being present in excess, by being displaced by binding of certain BH3-only proteins to Bcl-2, or by direct binding to other BH3-only proteins—Bax and Bak form oligomers that generate pores in the outer mitochondrial membrane. This allows release into the cytosol of mitochondrial proteins such as cytochrome *c* that, in normal healthy cells, are localized in the space between the inner and outer mitochondrial membrane. As depicted in Figure 21-33, released cytochrome *c* activates caspase-9—in part through binding to and activating Apaf-1 and in part through as yet unknown mechanisms.

As evidence for this regulatory pathway, overproduction of Bcl-2 in cultured cells blocks release of cytochrome *c* and blocks apoptosis; conversely, overproduction of Bax promotes release of cytochrome *c* into the cytosol and promotes apoptosis. Moreover, injection of cytochrome *c* into the cytosol of cells induces apoptosis. Bax and Bad oligomers, but not Bcl-2 homodimers or Bcl-2/Bax heterodimers, permit influx of ions through the outer mitochondrial membrane. It remains unclear how this ion influx triggers the release of cytochrome *c*, and it is not established whether cytochrome *c* actually moves outward through Bax/ Bak channels.

## Release of Cytochrome *c* and SMAC/DIABLO Proteins from Mitochondria Leads to Formation of the Apoptosome and Caspase Activation

The principal way cytochrome *c* in the cytosol activates apoptosis is by binding Apaf-1, the mammalian homolog of CED-4 (see Figure 21-33, *right*). In the absence of cytochrome *c*, monomeric Apaf-1 is bound to dATP. After binding cytochrome *c*, Apaf-1 cleaves its bound dATP into dADP and undergoes a dramatic assembly process into a disk-shaped heptamer, a 1.4-megadalton wheel of death called the **apoptosome** (Figure 21-39). The apoptosome serves as an activation machine for the initiator caspase, caspase-9, which is monomeric in the inactive state. Initiator caspases need to be sensitive to activation signals yet should not be activatable in an irreversible manner because accidental activation would lead to an undesirable snowball effect and rapid cell death. Significantly, caspase-9 does not require cleavage to become activated, but rather it is activated by dimerization following binding to the apoptosome. Caspase-9 then cleaves multiple molecules of effector caspases, such as caspase-3, leading to destruction of cell proteins (see Figures 21-33 and 21-38).

The three-dimensional structure of the corresponding nematode CED-4 apoptosome (see Figure 21-39c) showed how two CED-3 (caspase-9) procaspases bind adjacent to each other on the inside of the funnel-shaped octamer; these then activate each other by dimerization and proteolytic conversion, but the detailed mechanism of how this happens is unknown. The structure of the CED4 apoptosome also provided a model for the as yet unknown three-dimensional structure of the corresponding mammalian apoptosome (see Figure 21-39b, *right*).

In mammals and flies, but not worms, apoptosis is regulated by several other proteins (see Figure 21-33, *right*). A family of *i*nhibitor of *a*poptosis *p*roteins (IAPs) provides another way to restrain both initiator and effector caspases. IAPs have one or more zinc-binding domains that can bind directly to caspases and inhibit their protease activity. (Baculovirus, a type of insect virus, produces a protein that similarly binds to and inhibits caspases, thus preventing an infected cell from committing suicide which would stop a viral infection before new viruses can be made.) The inhibition of caspases by IAPs, however, creates a problem when a cell needs to undergo apoptosis. Mitochondria enter the picture once again since they are the source of a family of proteins, called SMAC/ DIABLOs, which inhibit IAPs. Assembly of Bax/Bak channels (see Figure 21-38) leads to release of SMAC/DIABLOs from mitochondria. SMAC/DIABLO then binds to IAPs in the cytosol, thereby blocking the IAPs from binding to caspases. By relieving IAP-mediated inhibition, SMAC/DIABLOs promote caspase activity and cell death.

## Trophic Factors Induce Inactivation of Bad, a Pro-apoptotic BH3-Only Protein

We saw earlier that neurotrophins such as nerve growth factor protect neurons from cell death; this is mediated by the BH3-only protein called Bad. In the absence of trophic factors, Bad is unphosphorylated and binds to Bcl-2 or the closely related anti-apoptotic protein Bcl-xL at the mitochondrial membrane (see Figure 21-38). This inhibits the ability of Bcl-2 and Bcl-xL to bind Bax and Bak, thereby allowing Bak and Bax channels to form and promoting cell death. Phosphorylated Bad, however, cannot bind to Bcl-2/Bcl-xL and is found in the cytosol complexed to the phosphoserine-binding protein 14-3-3 (see Figure 16-20).

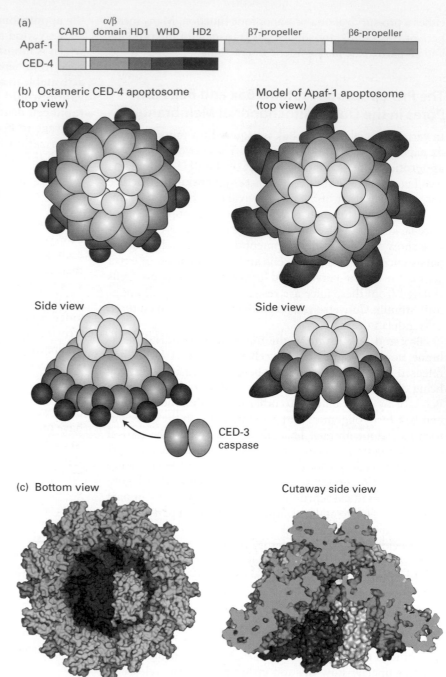

**FIGURE 21-39 Structure of the nematode apoptosome and a model for the structure of the mammalian Apaf1 apoptosome.** (a) Domains of the CED-4 protein and the corresponding mammalian Apaf1; CARD stands for N-terminal *caspase recruitment domain*. In the oligomeric apoptosome these CARD domains bind to CARD domains in the caspases. (b) Diagram of the CED4 apoptosome (*left*) and a model for the corresponding mammalian apoptosome (*right*). (c) Three-dimensional structure of the nematode octameric CED-4 apoptosome showing the binding of two CED-3 procaspases. Interaction of the apoptosome with CED-3 stimulates CED-3 dimerization, which is necessary for its activation. [From S. Qi et al., 2010, *Cell* **141**:446.]

A number of trophic factors, including NGF, induce the PI-3 kinase signaling pathway, leading to activation of protein kinase B (see Figure 16-26). Activated protein kinase B phosphorylates Bad at sites known to inhibit its pro-apoptotic activity. Moreover, a constitutively active form of protein kinase B can rescue cultured neurotrophin-deprived neurons, which otherwise would undergo apoptosis and die. These findings support the mechanism for the survival action of trophic factors depicted in Figure 21-38. In other cell types, different trophic factors may promote cell survival through post-translational modification of other components of the cell-death machinery.

## Vertebrate Apoptosis Is Regulated by BH3-Only Pro-Apoptotic Proteins That Are Activated by Environmental Stresses

Whereas worms contain a single BH3-only protein, Egl-1, mammals express at least eight, including Bad, in a cell and stress-specific manner. Their pro-apoptotic activity is regulated by diverse transcriptional and post-transcriptional mechanisms. Two, Puma and Noxa (see Figure 21-37), are transcriptionally induced by the p53 protein (see Figure 19-33); this is part of the checkpoint by which unrepaired damage to

DNA can induce apoptosis. Bim, on the other hand, is normally sequestered by the microtubule cytoskeleton by binding to a dynein light chain (see Figure 18-23). Detachment of cells from their substratum disrupts integrin signaling, rearranges the cytoskeleton, and leads to release of Bim. Both Puma and Bim likely bind directly to Bak and Bax, somehow releasing them from Bcl-2 and allowing formation of the oligomeric Bak/Bax pore and apoptosis (see Figure 21-38). Thus apoptosis of mammalian cells is regulated by a careful balance of activities of anti-apoptotic proteins such as Bcl-2 and Bcl-xL and multiple pro-apoptotic BH3-only proteins.

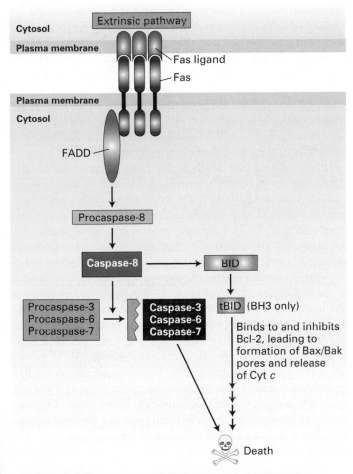

**FIGURE 21-40 Cell murder: the extrinsic apoptosis pathway.**
In this example of an extrinsic (or death-receptor-regulated) pathway found in the immune system, binding of the Fas ligand FasL (also called CD95 ligand) on the surface of one cell to the death receptor Fas on an adjacent cell leads to recruitment of the adapter protein FADD (FAS-associated death domain) and the dimerization and activation of caspase-8. Active caspase-8 then cleaves and activates caspase-3, caspase-6, and caspase-7, which then cleave vital cellular substrates and induce cell death. Cleavage of the BH3-only protein BID (BH3-interacting-domain death agonist) by caspase-8 generates a t-BID fragment that binds to Bcl-2 on the mitochondrial outer membrane, leading to release of cytochrome *c* into the cytosol and activation of the intrinsic apoptosis pathway (see Figure 21-38) as well. [Adapted from P. Bouillet and L. O'Reilly, 2009, *Nat. Rev. Immunol.* **9**:514.]

## Tumor Necrosis Factor and Related Death Signals Promote Cell Murder by Activating Caspases

Although cell death can arise as a default in the absence of survival factors, apoptosis can also be stimulated by positively acting *death signals*. For instance, tumor necrosis factor alpha (TNFα), which is released by macrophages, triggers the cell death and tissue destruction seen in certain chronic inflammatory diseases (see Chapter 23). Another important death-inducing signal, the Fas ligand, is a cell-surface protein produced by activated natural killer cells and cytotoxic T lymphocytes. This signal can trigger death of virus-infected cells, some tumor cells, and foreign graft cells.

Both TNFα and Fas ligand (also called CD95 ligand), depicted in Figure 21-40, are proteins present on the surface of one cell that bind to death receptors on an adjacent cell. These receptors have a single transmembrane domain and are activated when binding of an oligomeric ligand brings three receptor molecules into close proximity. The activated trimeric Fas receptor complex then binds a cytosolic protein called FADD (Fas-associated death domain) to the cell membrane, which then serves as an adapter to recruit and activate caspase-8, an initiator caspase. Like the other initiator caspase, caspase-9, caspase-8 is activated by dimerization following binding of two molecules to the FADD proteins recruited to an active receptor trimer. Once activated, caspase-8 activates several effector caspases and the amplification cascade begins.

Caspase-8 also cleaves the BH3-only protein *BH3-interacting-domain death agonist* (BID). The resulting t-BID fragment then binds to Bcl-2 on the mitochondrial outer membrane, leading to formation of a Bak/Bax channel, release of cytochrome *c* into the cytosol, and activation of the intrinsic apoptosis pathway (see Figure 21-38) as well.

To test the ability of the Fas receptor to induce cell death, researchers incubated cells with antibodies against the receptor. These antibodies, which bind and cross-link the cognate receptors, were found to stimulate cell death, indicating that activation of these receptors by oligomerization is sufficient to trigger apoptosis.

## KEY CONCEPTS of Section 21.4

### Cell Death and Its Regulation

- All cells require trophic factors to prevent apoptosis and thus survive. In the absence of these factors, cells commit suicide.

- Genetic studies in *C. elegans* defined an evolutionarily conserved apoptotic pathway with three major components: membrane-bound regulatory proteins, cytosolic regulatory proteins, and effector proteases called caspases in vertebrates (see Figure 21-33).

- Once activated, apoptotic proteases cleave specific intracellular substrates leading to the demise of a cell. Proteins (e.g., CED-4, Apaf-1) that bind regulatory proteins and caspases are required for caspase activation (see Figures 21-33, 21-34, and 21-39).

• Survival of motor and sensory neurons during development is mediated by neurotrophins released from target tissues that bind to Trk protein kinase receptors on the nerve growth cones (see Figure 21-36), activating an anti-apoptotic response via the PI-3 kinase pathway (see Figure 21-38).

• The Bcl-2 family contains both pro-apoptotic and anti-apoptotic proteins; most are single-pass transmembrane proteins and engage in protein-protein interactions.

• In mammals apoptosis can be triggered by oligomerization of Bax or Bak proteins in the outer mitochondrial membrane, leading to efflux of cytochrome *c* and SMAC/DIABLOs proteins into the cytosol; these then promote caspase activation and cell death.

• Bcl-2 molecules can restrain the oligomerization of Bax and Bak, inhibiting cell death.

• Pro-apoptotic BH3-only proteins (e.g., Puma, Bad) are activated by environmental stress and stimulate the oligomerization of Bax and Bak, allowing cytochrome *c* to escape into the cytosol, bind to Apaf-1, and thus activate caspases.

• Direct interactions between pro-apoptotic and anti-apoptotic proteins lead to cell death in the absence of trophic factors. Binding of extracellular trophic factors can trigger changes in these interactions, resulting in cell survival (see Figure 21-38).

• Binding of extracellular death signals, such as tumor necrosis factor and Fas ligand, to their receptors oligomerizes and activates an associated protein (FADD) that in turn triggers the caspase cascade, leading to cell murder.


## Perspectives for the Future

Cell birth, cell asymmetry, and cell death, which lie at the heart of the development, growth, and healing of an organism, are also central to disease processes, most notably cancer. Cell birth is normally carefully restricted to specific locales and times, such as the basal layer of the skin or the root meristem. Liver regenerates when there is injury, but liver cancer is prevented by restricting unnecessary growth at other times.

Some cells persist for the life of the organism, but others such as blood and intestinal cells turn over rapidly. Many cells live for a while and are then programmed to die and be replaced by others arising from a stem-cell population. Much attention is now being given to the regulation of stem cells in an effort to understand how self-renewing populations of cells are created and maintained. This has clear implications for repair of tissue: for example, to restore damaged retinas, torn cartilage, degenerating brain tissue, or failing organs. One interesting possibility is that some populations of stem cells with the potential to generate or regenerate tissue are normally eliminated by cell death during later development. If so, finding ways to selec-

tively block the death of these cells could make regeneration more likely. Could the elimination of such cells during mammalian development be the difference between an amphibian capable of limb regeneration and a mammal that is not?

ES and iPS cells will continue to provide a great deal of information about the regulatory molecules—transcription factors, chromatin- and DNA-modifying enzymes, and noncoding RNAs—and circuits that establish and maintain the pluripotent state, and that allow these cells to differentiate down specific developmental lineages. But the main interest in these cells—at least in the public's mind—is as a source of tissues to replace defective ones in many diseases. Several neurodegenerative diseases such as Alzheimer's and Parkinson's potentially could be cured *if* ES or iPS cells could be coaxed into differentiating in culture into appropriate neurons and *if* a method could be found to deliver there nerve cells to the appropriate regions of the brain. Similarly, ES and iPS cells can form normal-looking red blood cells and other types of blood cells in culture—but are these cells truly normal and completely functional, and can a bioengineering protocol be developed to make these cells in sufficient purity and numbers for transplantation into humans? Undoubtedly these problems, which are at the interface of tissue engineering and cellular and developmental biology, will eventually be solved—the question is, how soon?

Programmed cell death is the basis for the meticulous elimination of potentially harmful cells, such as autoreactive immune cells, which attack the body's own cells, or neurons that have failed to properly connect. Cell-death programs have also evolved as a defense against infection, and virus-infected cells are selectively murdered in response to death signals. Viruses, in turn, devote much of their effort to evading host defenses. Failures of programmed cell death can lead to uncontrolled cancerous growth. The proteins that prevent the death of cancer cells therefore become possible targets for drugs. As we learn in Chapter 24, many tumors contain a mixture of cells, some capable of seeding new tumors or continued uncontrolled growth and some capable only of growing in place or for a limited time. In this sense, the tumor has its own stem cells. These are now being identified and studied and are becoming vulnerable to medical intervention. One option is to manipulate the cell-death pathway by sending signals that will make cancer cells destroy themselves.


## Key Terms

apoptosis 1007
apoptosome 1013
asymmetric cell division 977
Bcl-2 family 1011
BH3-only protein 1009
caspases 1009
cell lineage 978
embryonic stem (ES) cell 978

germ line 987
induced pluripotent stem (iPS) cell 978
meristem 995
multipotent 978
planar cell polarity (PCP) 1003
pluripotent 978
polarity 997

## Review the Concepts

**1.** What two properties define a stem cell? Distinguish between a totipotent stem cell, a pluripotent stem cell, and a precursor (progenitor) cell.

**2.** Where are stem cells located in plants? Where are stem cells located in adult animals? How does the concept of stem cell differ between animal and plant systems?

**3.** In 1997, Dolly the sheep was cloned by a technique called somatic-cell nuclear transfer (or nuclear-transfer cloning). A nucleus from an adult mammary cell was transferred into an egg from which the nucleus had been removed. The egg was allowed to divide several times in culture, then the embryo was transferred to a surrogate mother who gave birth to Dolly. Dolly died in 2003 after mating and giving birth herself to viable offspring. What does the creation of Dolly tell us about the potential of nuclear material derived from a fully differentiated adult cell? Does the creation of Dolly tell us anything about the potential of an intact, fully differentiated adult cell?

**4.** Identify whether the following contain totipotent, pluripotent, or multipotent cells: (a) inner cell mass, (b) morula, (c) eight-cell embryo, (d) trophectoderm.

**5.** True or false: Differentiated somatic cells have the capacity to become reprogrammed to become other cell types. Provide one line of evidence discussed in the chapter that corroborates your response.

**6.** Explain how intestinal stem cells were first identified and then experimentally established to be multipotent stem cells.

**7.** Explain how hematopoietic stem cells were shown experimentally to be both pluripotent and capable of self-renewal.

**8.** The roundworm *C. elegans* has proved to be a valuable model organism for studies of cell birth, cell asymmetry, and cell death. What properties of *C. elegans* render it so well suited for these studies? Why is so much information from *C. elegans* experiments of use to investigators interested in mammalian development?

**9.** Asymmetric cell division often relies on cytoskeletal elements to generate or maintain the asymmetric distribution of cellular factors. In *S. cerevisiae*, what factor is localized to the bud by myosin motors? In *Drosophila* neuroblasts, what factors are localized apically by microtubules?

**10.** Discuss the role of *par* genes in generating A-P polarity in the *C. elegans* embryo.

**11.** How do studies of brain development in knockout mice support the statement that apoptosis is a default pathway in neuronal cells?

**12.** Compare and contrast cell death by apoptosis and necrosis.

**13.** Identify and list the functions of the three general classes of proteins that control cell death.

**14.** Based on your understanding of the events surrounding cell death, predict the effect(s) of the following on the ability of a cell to undergo apoptosis:

   **a.** Functional CED-9; nonfunctional CED-3

   **b.** Active Bax and cytochrome *c*; nonfunctional caspase-9

   **c.** Inactive PI-3 kinase; active Bad

**15.** TNF and Fas ligand bind cell-surface receptors to trigger cell death. Although the death signal is generated external to the cell, why do we consider the death induced by these molecules to be apoptotic rather than necrotic?

**16.** Predict the effects of the following mutations on the ability of a cell to undergo apoptosis:

   **a.** Mutation in Bad such that it cannot be phosphorylated by protein kinase B (PKB)

   **b.** Overexpression of Bcl-2

   **c.** Mutation in Bax such that it cannot form homodimers

One common characteristic of cancer cells is a loss of function in the apoptotic pathway. Which of the mutations listed above might you expect to find in some cancer cells?

**17.** How do IAPs (inhibitor of apoptosis proteins) interact with caspases to prevent apoptosis? How do mitochondrial proteins interact with IAPs to prevent inhibition of apoptosis?

## Analyze the Data

An Analyze the Data question for this chapter can be found at the *Molecular Cell Biology* website: www.whfreeman.com/lodish7e

## References

### Early Metazoan Development and Embryonic Stem Cells

Boyer, L., D. Mathur, and R. Jaenisch. 2006. Molecular control of pluripotency. *Curr. Opin. Genetics Dev.* 16:455–462.

Graf, T., and T. Enver. 2009. Forcing cells to change lineages. *Nature* 462:587–594.

Hanna, J., K. Saha, and R. Jaenisch. 2010. Pluripotency and cellular reprogramming: facts, hypotheses, unresolved issues. *Cell* 143:508–525.

Mallanna, S., and A. Rizzino. 2010. Emerging roles of microRNAs in the control of embryonic stem cells and the generation of induced pluripotent stem cells. *Dev. Biol.* 344:16–25.

Melton, C., R. Judson, and R. Blelloch. 2010. Opposing microRNA families regulate self-renewal in mouse embryonic stem cells. *Nature* 463:621.

Orkin, S., and K. Hochedlinger. 2011. Chromatin connections to pluripotency and cellular reprogramming. *Cell* 145:835–850.

Surface, L., S. Thornton, and L. Boyer. 2010. Polycomb group proteins set the stage for early lineage commitment. *Cell Stem Cell* 7:288–298.

Viswanathan, S., et al. 2008. Selective blockade of microRNA processing by Lin28. *Science* 320:97–100.

Young, R. 2011. Control of the embryonic stem cell state. *Cell* 144:940–954.

## Stem Cells and Niches in Multicellular Organisms

Birchmeier, W. 2011. Stem cells: orphan receptors find a home. *Nature* 476:287.

Copelan, E. A. 2006. Hematopoietic stem-cell transplantation. *N. Engl. J. Med.* 354:1813–1826.

Golden, J. A., S. C. Fields-Berry, and C. L. Cepko. 1995. Construction and characterization of a highly complex retroviral library for lineage analysis. *Proc. Nat'l. Acad. Sci. USA* 92:5704–5708.

He, S., D. Nakada, and S. Morrison. 2009. Mechanisms of stem cell self-renewal. *Annu. Rev. Cell Dev. Biol.* 25:377-406.

Li, L., and T. Xie. 2005. Stem cell niche: structure and function. *Annu. Rev. Cell Dev. Biol.* 21:605–631.

Mendez-Ferrer, S., et al. 2011. Mesenchymal and haematopoietic stem cells form a unique bone marrow niche. *Nature* 466:829–834.

Morrison, S. J., and J. Kimble. 2006. Asymmetric and symmetric stem-cell divisions in development and cancer. *Nature* 441:1068–1074.

Novershtern, N., et. al. 2011 Densely interconnected transcriptional circuits control cell states in human hematopoiesis *Cell* 144:296–309.

Sablowski, R. 2011. Plant stem cell niches: from signalling to execution. *Curr. Opin. Plant Biol.* 14:4–9.

Scheres, B. 2007. Stem-cell niches: nursery rhymes across kingdoms. *Nat. Rev. Mol. Cell Biol.* 8:345–354.

Suh, H., W. Deng, and P. Gage. 2009. Signaling in adult neurogenesis. *Annu. Rev. Cell Dev. Biol.* 25:253–275.

van der Flier, L. G., and H. Clevers. 2009. Stem cells, self-renewal, and differentiation in the intestinal epithelium *Annu. Rev. Physiol.* 71:241–260.

Zhang, C., and H. Lodish. 2008. Cytokine regulation of hematopoietic stem cell function. *Curr. Opin. Hematol.* 15:307–311.

## Mechanisms of Cell Polarity and Asymmetric Cell Division

Axelrod, J. D., and C. J. Tomlin. 2011. Modeling the control of planar cell polarity. *Wiley Interdisc. Revs. Sys. Biol. Med.* 3:588–605.

Cabernard C., and C. Q. Doe. 2009. Apical/basal spindle orientation is required for neuroblast homeostasis and neuronal differentiation in *Drosophila*. *Dev. Cell* 17:134–141.

Knoblich, J. A. 2008. Mechanisms of asymmetric stem cell division. *Cell* 132:583–597.

Li, R., and B. Bowerman, eds. 2010. *Symmetry Breaking in Biology*. Cold Spring Harbor Laboratory Press.

Li, R., and G. G. Gundersen. 2008. Beyond polymer polarity: how the cytoskeleton builds a polarized cell. *Nat. Rev. Mol. Cell Biol.* 9:860–873.

Mellman, I., and W. J. Nelson. 2008. Coordinated protein sorting, targeting and distribution in polarized cells. *Nat. Rev. Mol. Cell Biol.* 9:833–845.

Morrison, S. J, and J. Kimble. 2006. Asymmetric and symmetric stem-cell divisions in development and cancer. *Nature* 441:1068–1074.

Nelson, W. J. 2003. Adaption of core mechanisms to generate cell polarity. *Nature* 422:766–774.

Shivas, J. M., H. A. Morrison, D. Bilder, and A. R. Skop. 2010. Polarity and endocytosis: reciprocal regulation. *Trends in Cell Biol.* 20:445-452.

Siller, K. H., and C. Q. Doe. 2009. Spindle orientation during asymmetric cell division. *Nat. Cell Biol.* 11:365–374.

St. Johnston, D., and J. Ahringer. 2010. Cell polarity in eggs and epithelia: parallels and diversity. *Cell* 141:757–774.

Zallen, J. A. 2007. Planar polarity and tissue morphogenesis. *Cell* 129:1051–1063.

## Cell Death and Its Regulation

Adams, J. M., and S. Cory. 2007. Bcl-2-regulated apoptosis: mechanism and therapeutic potential. *Curr. Opin. Immunol.* 19:488–496.

Baehrecke, E. H. 2002. How death shapes life during development. *Nat. Rev. Mol. Cell Biol.* 3:779–787.

Bouillet, P., and L. A. O'Reilly. 2009. CD95, BIM and T cell homeostasis. *Nat. Rev. Immunol.* 9:514–519.

Giam, M., D. C. Huang, and P. Bouillet. 2008. BH3-only proteins and their roles in programmed cell death. *Oncogene* 27(suppl. 1):S128–136.

Hay, B. A., and M. Guo. 2006. Caspase-dependent cell death in *Drosophila*. *Annu. Rev. Cell Dev. Biol.* 22:623–650.

Lakhani, S. A., et al. 2006. Caspases 3 and 7: key mediators of mitochondrial events of apoptosis. *Science* 10:847–851.

Martin, S. 2010. Opening the cellular poison cabinet. *Science* 330:1330–1331.

Peter, M. 2011. Apoptosis meets necrosis. *Nature* 471:311–312.

Riedl, S. J., and G. Salvesen. 2007. The apoptosome: signalling platform of cell death *Nat. Rev. Mol. Cell Biol.* 8:405–413.

Ryoo, H. D., and E. H. Baehrecke. 2010. Distinct death mechanisms in *Drosophila* development. *Curr. Opin. Cell Biol.* 22:889–895.

Schafer, Z. T., and S. Kornbluth. 2006. The apoptosome: physiological, developmental, and pathological modes of regulation. *Devel. Cell* 10:549–561.

Teng, X., and J. Hardwick. 2010. The apoptosome at high resolution. *Cell* 141:402–404.

Yan, N., et al. 2005. Structure of the CED-4–CED-9 complex provides insights into programmed cell death in *Caenorhabditis elegans*. *Nature* 437:831–837.

# NERVE CELLS

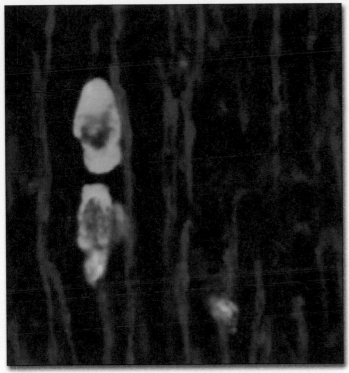

Two principal types of cells in the central nervous system: neurons (red) and glial cells (green). In this image of the developing mouse optic nerve, axons are stained with an antibody against a principal component of myelin, myelin basic protein, that surrounds the long axons. Oligodendrocytes are a type of glial cell that produce the myelin coat, and are stained with an antibody specific to the β-catenin inhibitor adenomatous polyposis coli. [From B. Emery, 2010, *Science* **330**:779.]

The nervous system regulates all aspects of bodily function and is staggering in its complexity. The 1.3-kg adult human brain—the control center that stores, computes, integrates, and transmits information—contains about $10^{11}$ nerve cells, called neurons. These neurons are interconnected by some $10^{14}$ synapses, the junction points where two or more neurons communicate. Millions of specialized neurons sense features of both the external and internal environments of organisms and transmit this information to the brain for processing and storage. Millions of other neurons regulate the contraction of muscles and the secretion of hormones. The nervous system also contains glial cells, which occupy the spaces between neurons and modulate their functions.

Despite the multiple types and shapes of neurons that are found in metazoan organisms, all nerve cells share many common properties. The structure and function of nerve cells is understood in great detail—perhaps in more detail than for any other cell type. The function of a neuron is to communicate information, which it does by two methods. *Electrical signals* process and conduct information within neurons, which are usually highly elongated cells (Figure 22-1). The electrical pulses that travel along neurons are called action potentials, and information is encoded as the frequency at which action potentials are fired. Owing to the speed of electrical transmission, neurons are champion signal transducers, much faster than cells that secrete hormones. In contrast to the electrical signals that conduct information *within* a

## OUTLINE

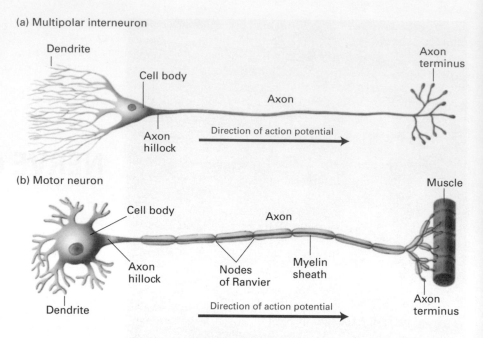

**FIGURE 22-1 Typical morphology of two types of mammalian neurons.** Action potentials arise in the axon hillock and are conducted toward the axon terminus. (a) A multipolar interneuron has profusely branched dendrites, which *receive* signals at synapses with several hundred other neurons. Small voltage changes imparted by inputs in the dendrites can sum to give rise to the more massive action potential, which starts in the hillock. A single long axon that branches laterally at its terminus *transmits* signals to other neurons. (b) A motor neuron innervating a muscle cell typically has a single long axon extending from the cell body to the effector cell. In mammalian motor neurons, an insulating sheath of myelin usually covers all parts of the axon except at the nodes of Ranvier and the axon termini. The myelin sheath is composed of cells called glia.

neuron, *chemical signals* transmit information *between* cells, utilizing processes similar to those employed by other types of signaling cells (Chapters 15 and 16).

Taken together, the electrical and chemical signaling of the nervous system allows it to detect external stimuli, integrate and process the information received, relay it to higher brain centers, and generate an appropriate response to the stimulus. For example, *sensory neurons* have specialized receptors that convert diverse types of stimuli from the environment (e.g., light, touch, sound, odorants) into electrical signals. These electrical signals are then converted into chemical signals that are passed on to other cells called *interneurons,* which convert the information back into electrical signals. Ultimately the information is transmitted to muscle-stimulating *motor neurons* or to other neurons that stimulate other types of cells, such as glands.

In this chapter we will focus on neurobiology at the cellular and molecular level. We will start by looking at the general architecture of neurons and at how they carry signals. Next, we will focus on ion flow, channel proteins, and membrane properties: how electrical pulses move rapidly along neurons. Third, we will examine communication between neurons: electrical signals traveling along a cell must be translated into a chemical pulse between cells and then back into an electrical signal in the receiving cells. In the last section we will examine neurons in several sensory tissues, including those that mediate our senses of touch, taste, and olfaction. The speed, precision, and integrative power of neural signaling enable the accurate and timely sensory perception of a swiftly changing environment.

A great deal of information about nerve cells has been gleaned from analyses of humans, mice, nematodes, and flies with mutations that affect specific functions of the nervous system. In addition, molecular cloning and structural analysis of key neuronal proteins, such as voltage-gated ion channels and receptors, have helped elucidate the cellular machinery underlying complex brain functions such as instinct, learning, memory, and emotion.

## 22.1 Neurons and Glia: Building Blocks of the Nervous System

In this section we take an initial look at the structure of neurons and how they propagate electrical and chemical signals. **Neurons** are distinguished by their elongated, asymmetric shape, by their highly localized proteins and organelles, and most of all by a set of proteins that controls the flow of ions across the plasma membrane. Because one neuron can respond to the inputs from multiple neurons, generate electrical signals, and transmit the signals to multiple neurons, a nervous system has considerable powers of signal analysis. For example, a neuron might pass on a signal only if it receives five simultaneous activating signals from input neurons. The receiving neuron measures both the total *amount* of incoming signal and whether the five signals are roughly *synchronous*. Input from one neuron to another can be either *excitatory*—combining with other signals to trigger electrical transduction in the receiving cell—or *inhibitory*, discouraging such transmission. Thus the properties and connections of individual neurons set the stage for integration and refinement of information, and the output of a nervous system is the result of its circuit properties, that is, the wiring or interconnections between neurons, and the strength of these interconnections. We will begin by looking at how signals are received and sent, and in subsequent parts of the chapter we will look at the molecular details of the machinery involved.

### Information Flows Through Neurons from Dendrites to Axons

Neurons arise from roughly spherical *neuroblast* precursors. Newly born neurons can migrate long distances before growing into dramatically elongated cells. Fully differentiated neurons take many forms, but generally share certain key features (see Figure 22-1). The nucleus is found in a rounded part of the cell called the *cell body*. Branching cell processes called **dendrites** (from the Greek for "treelike") are found at one end, and are the main structures where signals are received from other neurons via synapses. Incoming signals are also received at synapses that form on neuronal cell bodies. Neurons often have extremely long dendrites with complex branches, particularly in the central nervous system (i.e., the brain and spinal cord). This allows them to form synapses with, and receive signals from, a large number of other neurons—up to tens of thousands. Thus the converging dendritic branches allow signals from many cells to be received and integrated by a single neuron.

When a neuron is first differentiating, the end of the cell opposite the dendrites undergoes dramatic outgrowth to form a long extended arm called the **axon**, which is essentially a transmission wire. The growth of axons must be controlled so that proper connections are formed, a complex process called axon guidance that involves dynamic changes to the cytoskeleton and is discussed in Section 18.8. The diameters of axons vary from just a micrometer in certain neurons of the human brain to a millimeter in the giant fiber of the squid. Axons can be meters in length (in giraffe necks, for example), and are often partly covered with electrical insulation called the **myelin sheath** (see Figure 22-1b), which is made up of cells called glia. The insulation speeds electrical transmission and prevents short circuits. The short, branched ends of the axon at the opposite end of the neuron from the dendrites are called the *axon termini*. This is where signals are passed along to the next neuron or to another type of cell such as a muscle or hormone-secreting cell. The asymmetry of the neuron, with dendrites at one end and axon termini at the other, is indicative of the unidirectional flow of information from dendrites to axons.

## Information Moves Along Axons as Pulses of Ion Flow Called Action Potentials

Nerve cells are members of a class of *excitable cells,* which also includes muscle cells, cells in the pancreas, and some others. Like all metazoan cells, excitable cells have an inside-negative voltage or electric potential gradient across their plasma membranes, the **membrane potential** (see Chapter 11). In excitable cells this potential can suddenly become zero or even reversed, with the inside of the cell positive with respect to the outside of the plasma membrane. The membrane voltage in a typical neuron, called the *resting potential* because it is the state when no signal is in transit, is established by Na$^+$/K$^+$ ion pumps in the plasma membrane. These are the same ion pumps used by other cells to generate a resting potential. Na$^+$/K$^+$ ion pumps use energy, in the form of ATP, to move positively charged Na$^+$ ions out of the cell and K$^+$ ions inward. Subsequent movement of K$^+$ out of the cell through resting K$^+$ channels results in a net negative charge inside the cell compared with the outside. The typical resting potential of a neuron is about −60 mV.

Neurons have a language all their own. They use their unique electrical properties to send signals. The signals take the form of brief local voltage changes, from inside-negative to

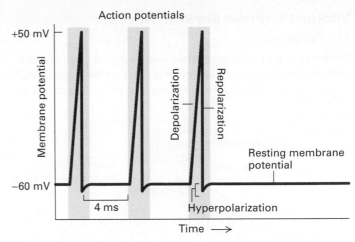

**EXPERIMENTAL FIGURE 22-2 Recording of an axonal membrane potential over time reveals the amplitude and frequency of action potentials.** An action potential is a sudden, transient depolarization of the membrane, followed by repolarization to the resting potential of about −60 mV. The axonal membrane potential can be measured with a small electrode placed into it (see Figure 11-19). This recording shows the neuron generating one action potential about every 4 milliseconds.

inside-positive, an event designated **depolarization**. A powerful surge of depolarizing voltage change, moving from one end of the neuron to the other, is called an **action potential.** "Depolarization" is somewhat of a misnomer, since the neuron suddenly goes from inside-negative to neutral to inside-positive, which could be more accurately described as depolarization followed by the opposite polarization (Figure 22-2). At the peak of an action potential, the membrane potential can be as much as +50 mV (inside-positive), a net change of about 110 mV. As we shall see in greater detail in Section 22.2, an action potential moves along the axon to the axon terminus at speeds of up to 100 meters per second. In humans, for instance, axons may be more than a meter long, yet it takes only a few milliseconds for an action potential to move along their length. Neurons can fire repeatedly after a brief recovery period, for example, every 4 milliseconds (ms), as in Figure 22-2. After the action potential passes a sector of a neuron, channel proteins and pumps restore the inside-negative resting potential (*repolarization*). The restoration process chases the action potential down the axon to the terminus, leaving the neuron ready to signal again.

Importantly, action potentials are "all or none." Once the threshold to start one is reached, a full firing occurs. The signal information is therefore carried primarily not by the intensity of the action potentials, but by the timing and frequency of them.

Some excitable cells are not neurons. Muscle contraction is triggered by motor neurons that synapse directly on excitable muscle cells (see Figure 22-1b). Insulin secretion from the Beta cells of the pancreas is triggered by neurons. In both cases the activating event involves an opening of plasma membrane channels that causes changes in the transmembrane flow of ions and in the electrical properties of the regulated cells.

## Information Flows Between Neurons via Synapses

What starts an action potential? Axon termini from one neuron are closely apposed to dendrites of another, at junctions called chemical synapses or simply **synapses** (Figure 22-3).

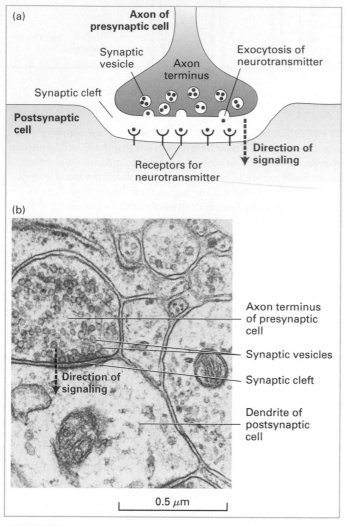

**FIGURE 22-3  A Chemical Synapse.** (a) A narrow region—the synaptic cleft—separates the plasma membranes of the presynaptic and postsynaptic cells. Arrival of action potentials in a presynaptic cell causes exocytosis at a synapse of a small number of synaptic vesicles, releasing their content of neurotransmitters (red circles). Following their diffusion across the synaptic cleft, the neurotransmitters bind to specific receptors on the plasma membrane of the postsynaptic cell. These signals either depolarize the postsynaptic membrane (making the potential inside less negative), tending to induce an action potential in the cell, or hyperpolarize the postsynaptic membrane (making the potential inside more negative), inhibiting action potential induction. (b) Electron micrograph showing a dendrite synapsing with an axon terminus filled with synaptic vesicles. In the synaptic region, the plasma membrane of the presynaptic cell is specialized for vesicle exocytosis; synaptic vesicles containing a neurotransmitter are clustered in these regions. The opposing membrane of the postsynaptic cell (in this case, a neuron) contains receptors for the neurotransmitter. [Part (b) from C. Raine et al., eds., 1981, *Basic Neurochemistry*, 3d ed., Little, Brown, p. 32.]

The axon termini of the *presynaptic cell* contain many small vesicles, termed **synaptic vesicles,** each of which is filled with a single kind of small molecule known as a **neurotransmitter.** Arrival of an action potential at a terminus triggers exocytosis of a small number of synaptic vesicles, releasing their content of neurotransmitter molecules.

Neurotransmitters diffuse across the synapse in about 0.5 ms and bind to receptors on the dendrite of the adjacent neuron. Binding of neurotransmitter triggers opening or closing of specific ion channels in the plasma membrane of *postsynaptic cell* dendrites, leading to changes in the membrane potential in this localized area of the postsynaptic cell. Generally these changes depolarize the postsynaptic membrane (making the potential less inside negative). The local depolarization, if large enough, triggers an action potential in the axon. Transmission is unidirectional, from the axon termini of the presynaptic cell to dendrites of the postsynaptic cell.

In some synapses, the effect of the neurotransmitters is to hyperpolarize and therefore lower the likelihood of an action potential in the postsynaptic cell. A single axon in the central nervous system can synapse with many neurons and induce responses in all of them simultaneously. Conversely, sometimes multiple neurons must act on the postsynaptic cell roughly synchronously to have a strong enough impact to trigger an action potential. Neuronal integration of depolarizing and hyperpolarizing signals determines the likelihood of an action potential.

Thus neurons employ a combination of extremely fast electrical transmission *along* the axon with rapid chemical communication *between* cells. Now we will look at how a chain of neurons, a circuit, can achieve a useful function.

## The Nervous System Uses Signaling Circuits Composed of Multiple Neurons

In complex multicellular animals, such as insects and mammals, neurons form signaling circuits that are constructed using three basic types of nerve cells: afferent neurons, interneurons, and efferent neurons. **Afferent neurons,** also known as sensory or receptor neurons, carry nerve impulses from receptors or sense organs *toward* the central nervous system (i.e., the brain and spinal cord). These neurons report an event that has happened, like the arrival of a flash of light or the movement of a muscle. A touch or a painful stimulus creates a sensation in the brain only after information about the stimulus travels there via afferent nerve pathways. **Efferent neurons,** also known as effector neurons, carry nerve impulses *away* from the central nervous system to generate a response. A *motor neuron,* for example, carries a signal to a muscle to stimulate its contraction (see Figure 22-1b); other effector neurons stimulate hormone secretion by endocrine cells. **Interneurons,** the largest group, relay signals from afferent to efferent neurons and to other interneurons as part of a neural pathway. An interneuron can bridge multiple neurons, allowing integration or divergence of signals and sometimes extending the reach of a signal.

In a simple type of circuit called a *reflex arc,* interneurons connect multiple sensory and motor neurons, allowing one sensory neuron to affect multiple motor neurons and one

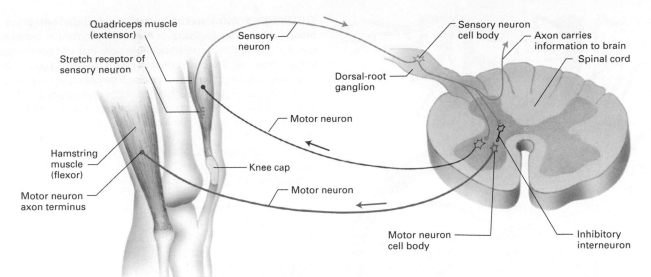

**FIGURE 22-4  The knee-jerk reflex.** A tap of the hammer stretches the quadriceps muscle, thus triggering electrical activity in the stretch receptor sensory neuron. The action potential, traveling in the direction of the top blue arrow, sends signals to the brain so we are aware of what is happening, and also to two kinds of cells in the dorsal-root ganglion that is located in the spinal cord. One cell, a motor neuron that connects back to the quadriceps (red), stimulates muscle contraction so that you kick the person who hammered your knee. The second connection activates, or "excites," an inhibitory interneuron (black). The interneuron has a damping effect, blocking activity by a flexor motor neuron (green) that would, in other circumstances, activate the hamstring muscle that opposes the quadriceps. In this way, relaxation of the hamstring is coupled to contraction of the quadriceps. This is a reflex because movement requires no conscious decision.

motor neuron to be affected by multiple sensory neurons; in this way interneurons integrate and enhance reflexes. For example, the knee-jerk reflex in humans, illustrated in Figure 22-4, involves a complex reflex arc in which one muscle is stimulated to contract while another is inhibited from contracting. The reflex also sends information to the brain to announce what happened. Such circuits allow an organism to respond to a sensory input by the coordinated action of sets of muscles that together achieve a single purpose.

These simple signaling circuits, however, do not directly explain higher-order brain functions such as reasoning, computation, and memory development. Typical neurons in the brain receive signals from up to a thousand other neurons and, in turn, can direct chemical signals to many other neurons. The output of the nervous system depends on its circuit properties—the amount of wiring, or interconnections, between neurons and the strength of these interconnections. Complex aspects of the nervous system, such as vision and consciousness, cannot be understood at the single-cell level, but only at the level of networks of nerve cells that can be studied by techniques of systems analysis. The nervous system is constantly changing; alterations in the number and nature of the interconnections between individual neurons occur, for example, in the formation of new memories.

## Glial Cells Form Myelin Sheaths and Support Neurons

For all the impressiveness of neurons, they are a minority population of cells in the human brain. **Glial cells** (also known as *neuroglia* or simply *glia*), which play many roles in the brain but do not themselves conduct electrical impulses, outnumber neurons by about 10 to 1. Of the three principal types of glia, two produce myelin sheaths—the insulation that surrounds neuronal axons (see Figure 22-1b): *oligodendrocytes* make sheaths for the central nervous system (CNS), and *Schwann cells* make them for the peripheral nervous system (PNS). (Both types of glia are discussed in more detail in Section 22.2.) *Astrocytes,* a third type of glia, provide growth factors and other signals to neurons, receive signals from neurons, and induce synapse formation between neurons.

**Astrocytes,** named for their starlike shape (Figure 22-5), constitute more than a third of the brain's mass and half of the brain's cells. Astrocytes surround many synapses and dendrites; the $Ca^{2+}$, $K^+$, $Na^+$, and $Cl^-$ channels found in astrocyte plasma membranes influence the concentration of free ions in the extracellular space, thus affecting the membrane potentials of neurons and of the astrocytes themselves. Astrocytes produce abundant extracellular matrix proteins, some of which are used as guidance cues by migrating neurons, and a host of growth factors that carry a variety of types of information to neurons. Astrocytes are joined to each other by **gap junctions,** so changes in ionic composition in a given astrocyte are communicated to adjacent astrocytes, over distances of hundreds of microns.

Some astrocytes are also critical regulators of the formation of the blood-brain barrier, the purpose of which is to control what types of molecules can travel out of the bloodstream into the brain and vice versa (see Figure 22-5). Blood vessels in the brain supply oxygen and remove $CO_2$, and deliver glucose and amino acids, with capillaries found within a few micrometers of every cell. These capillaries form the blood-brain barrier, which allows passage of oxygen and $CO_2$ across the

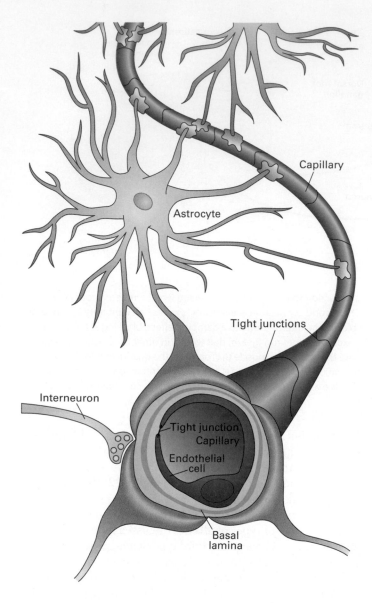

Capillary

Astrocyte

Tight junctions

Interneuron

Tight junction
Capillary

Endothelial
cell

Basal
lamina

**FIGURE 22-5 Astrocytes interact with endothelial cells at the blood-brain barrier.** Capillaries in the brain are formed by endothelial cells that are interconnected by tight junctions that are impermeable to most molecules. Transport between cells is blocked, so only small molecules that can diffuse across plasma membranes or substances specifically transported through cells can cross the barrier. Certain astrocytes surround the blood vessels, in contact with the endothelial cells, and send secreted protein signals to induce the endothelial cells to produce a selective barrier. The endothelial cells (burgundy) are ensheathed by a layer of basal lamina (orange) and contacted on the outside by astrocyte processes (tan). [From N. J. Abbott, L. Rönnbäck, and E. Hansson, 2006, *Nature Rev. Neurosci.* **7**:41–53.]

cell processes, dendrites, at one end of the cell receive chemical signals from other neurons, triggering ion flow. The electrical signal moves rapidly to axon termini at the other end of the cell (see Figure 22-1).

• A resting neuron carrying no signal has ATP-powered pumps that move ions across the plasma membrane. The outward movement of $K^+$ ions creates a net negative charge inside the cell. This voltage is called the resting potential and usually is about $-60$ mV (see Figure 22-2).

• If a stimulus causes certain ion channels to open so that certain ions can flow more freely, a strong pulse of voltage change may pass down the neuron from dendrites to axon termini. The cell goes from being $-60$ mV inside to $+50$ mV inside, relative to the extracellular fluid. This pulse is called an action potential (see Figure 22-2).

• The action potential travels down the length of the axon from the cell body to the axon termini at speeds of up to 100 meters per second.

• Neurons connect across small spaces called synapses. Since an action potential cannot jump the gap, at the axon termini of the presynaptic cell the signal is converted from electrical to chemical to stimulate the postsynaptic cell.

• Upon stimulation by an action potential, axon termini release, by exocytosis, small packets of chemicals called neurotransmitters. Neurotransmitters diffuse across the synapse and bind to receptors on the dendrites on the other side of the synapse. These receptors can induce or inhibit a new axon potential in the postsynaptic cell (see Figure 22-3).

• Neurons form circuits that usually consist of sensory neurons, interneurons, and motor neurons, as in the knee-jerk response (see Figure 22-4).

• Glial cells are ten times as abundant as neurons and serve many purposes. Oligodendrocytes and Schwann cells build the myelin insulation that coats many neurons.

• Astrocytes, another type of glial cell, wrap their processes around synapses and blood vessels and promote formation of the blood-brain barrier (see Figure 22-5). Astrocytes also secrete proteins that stimulate synapse formation.

endothelial cell wall but prevents, for example, blood-borne circulating neurotransmitters and some drugs from entering the brain. The barrier consists of a set of tight junctions (Chapter 20) that interconnect the endothelial cells that form the walls of capillaries. Surrounding astrocytes promote specialization of these endothelial cells, making them less permeable than those in capillaries found in the rest of the body.

## KEY CONCEPTS of Section 22.1

### Neurons and Glia: Building Blocks of the Nervous System

• Neurons are highly asymmetric cells composed of multiple dendrites at one end, a cell body containing the nucleus, a long axon, and axon termini.

• Neurons carry information from one end to the other using pulses of ion flow across the plasma membrane. Branched

## 22.2 Voltage-Gated Ion Channels and the Propagation of Action Potentials

In Chapter 11 we learned that an electric potential of approximately 70 mV (cytosolic face negative) exists across the plasma membrane of all cells, including resting nerve cells. This resting membrane potential is generated by outward movement of $K^+$ ions through open nongated $K^+$ channels in the plasma membrane, and is driven by the $K^+$ concentration gradient (cytosol > extracellular medium). The high cytosolic $K^+$ and low cytosolic $Na^+$ concentrations, relative to their concentrations in the extracellular medium, are generated by the plasma membrane $Na^+/K^+$ pump, which uses the energy released by hydrolysis of phosphoanhydride bonds in ATP to pump $Na^+$ outward and $K^+$ inward. The entry of $Na^+$ ions into the cytosol from the medium is thermodynamically favored, driven both by the $Na^+$ concentration gradient (extracellular medium > cytosol) and the inside-negative membrane potential (see Figure 11-25). However, most $Na^+$ channels in the plasma membrane are closed in resting cells, so little inward movement of $Na^+$ ions can occur (Figure 22-6a).

During an action potential, some of these $Na^+$ channels open, allowing inward movement of $Na^+$ ions, which depolarizes the membrane. Action potentials are propagated down the axon because a change in voltage in one part of the axon triggers the opening of channels in the next section of the axon. Such *voltage-gated channels* therefore lie at the heart of neural transmission. In this section, we first introduce some of the key properties of action potentials, which move rapidly along the axon from the cell body to the termini. We then describe how the voltage-gated channels responsible for propagating action potentials in neurons operate. In the last part of the section, we will see how the myelin sheath, produced by glial cells, increases the speed and efficiency of electrical transmission in nerve cells.

### The Magnitude of the Action Potential Is Close to $E_{Na}$ and Is Caused by $Na^+$ Influx Through Open $Na^+$ Channels

Figure 22-6b illustrates how the membrane potential will change if enough $Na^+$ channels in the plasma membrane become opened. The resulting influx of positively charged $Na^+$ ions into the cytosol will more than compensate for the efflux of $K^+$ ions through open resting $K^+$ channels. The result will be a *net* inward movement of cations, generating an excess of positive charges on the cytosolic face of the plasma membrane and a corresponding excess of negative charges on the extracellular face (owing to the $Cl^-$ ions "left behind" in the extracellular medium after influx of $Na^+$ ions). In other words, the plasma membrane becomes depolarized to such an extent that the inside face becomes positive with respect to the external face.

Recall from Chapter 11 that the equilibrium potential of an ion is the membrane potential at which there is no net flow of that ion from one side of the membrane to the other due to the balancing of two opposing forces, the ion concentration gradient and the membrane potential. At the peak of depolarization in an action potential, the magnitude of the membrane potential is very close to the $Na^+$ equilibrium potential $E_{Na}$ given by the Nernst equation (Equation 11-2), as would be expected if opening of voltage-gated $Na^+$ channels is responsible for generating action potentials. For example, the measured peak value of the action potential for the squid giant axon is +35 mV, which is close to the calculated value of $E_{Na}$ (+55 mV) based on $Na^+$ concentrations of 440 mM outside and 50 mM inside. The relationship between the magnitude of the action potential and the concentration of $Na^+$ ions inside and outside the cell has been confirmed experimentally. For instance, if the concentration of $Na^+$ ions in the solution bathing the squid axon is reduced to one-third of normal, the magnitude of the depolarization is reduced by 40 mV, nearly as predicted.

### Sequential Opening and Closing of Voltage-Gated $Na^+$ and $K^+$ Channels Generate Action Potentials

The cycle of changes in membrane potential and return to the resting value that constitutes an action potential lasts 1–2 milliseconds and can occur hundreds of times a second in a typical neuron (see Figure 22-2). These cyclical changes

(a) Resting state (cytosolic face negative)

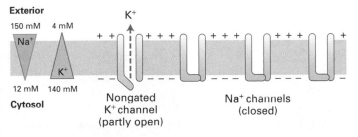

(b) Depolarized state (cytosolic face positive)

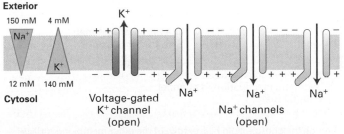

**FIGURE 22-6 Depolarization of the plasma membrane due to opening of gated $Na^+$ channels.** (a) In resting neurons, a type of nongated $K^+$ channel is open part of the time, but the more numerous gated $Na^+$ channels are closed. The movement of $K^+$ ions outward establishes the inside-negative membrane potential characteristic of most cells. (b) Opening of gated $Na^+$ channels permits an influx of sufficient $Na^+$ ions to cause a reversal of the membrane potential. In the depolarized state, voltage-gated $K^+$ channels open and subsequently repolarize the membrane. Note that the flows of ions are too small to have much effect on the overall concentration of either $Na^+$ or $K^+$ in the cytosol or exterior fluid.

in the membrane potential result from the opening and closing first of a number of *voltage-gated Na⁺ channels* (that is, channels opened by a *change* in membrane potential) in a segment of the axonal plasma membrane, and then opening and closing of *voltage-gated K⁺ channels.* The role of these channels in the generation of action potentials was elucidated in classic studies done on the giant axon of the squid, in which multiple microelectrodes can be inserted without causing damage to the integrity of the plasma membrane. However, the same basic mechanism is used by all neurons.

**Voltage-Gated Na⁺ Channels** As just discussed, voltage-gated Na⁺ channels are closed in resting neurons. A small depolarization of the membrane (as occurs when neurotransmitter stimulates a postsynaptic cell) increases the likelihood that any one channel will open; the greater the depolarization the greater the probability that a channel will open. Depolarization causes a conformational change in these channel proteins that opens a gate on the cytosolic surface of the pore, permitting Na⁺ ions to pass through the pore into the cell. Thus the greater the initial membrane depolarization, the more voltage-gated Na⁺ channels that open and the more Na⁺ ions that enter.

As Na⁺ ions flow inward through opened channels, the excess positive charges on the cytosolic face and negative charges on the exoplasmic face diffuse a short distance away from the initial site of depolarization. This *passive spread* of positive charges on the cytosolic face and negative charges on the external face depolarizes (makes the inside less negative) adjacent segments of the plasma membrane, causing opening of additional voltage-gated Na⁺ channels in these segments and an increase in Na⁺ influx. As more Na⁺ ions enter the cell, the inside of the cell membrane becomes more depolarized, causing the opening of yet more voltage-gated Na⁺ channels and even more membrane depolarization, setting into motion an explosive entry of Na⁺ ions. For a fraction of a millisecond, the permeability of this small segment of the membrane to Na⁺ becomes vastly greater than that for K⁺, and the membrane potential approaches $E_{Na}$, the equilibrium potential for a membrane permeable only to Na⁺ ions. As the membrane potential approaches $E_{Na}$, however, further net inward movement of Na⁺ ions ceases, since the concentration gradient of Na⁺ ions (outside > inside) is now offset by the inside-positive membrane potential. The action potential is, at its peak, close to the value of $E_{Na}$.

Figure 22-7 schematically depicts the critical structural features of voltage-gated Na⁺ channels and the conformational

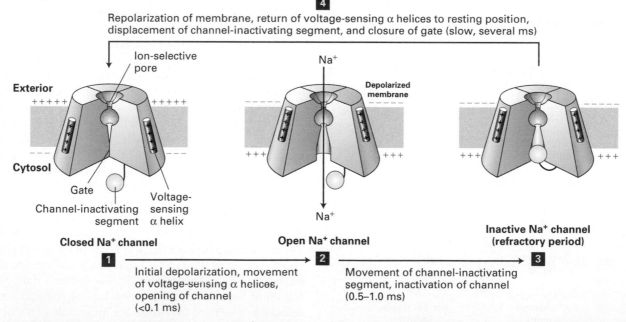

**4** Repolarization of membrane, return of voltage-sensing α helices to resting position, displacement of channel-inactivating segment, and closure of gate (slow, several ms)

Ion-selective pore

Na⁺

Exterior

+ + + + +    + + + +

Depolarized membrane

+ + + + +

Cytosol

+ + +    + + +

+ + +    + + +

Gate

Channel-inactivating segment

Voltage-sensing α helix

Na⁺

**Closed Na⁺ channel**

**1**

Initial depolarization, movement of voltage-sensing α helices, opening of channel (<0.1 ms)

**Open Na⁺ channel**

**2**

Movement of channel-inactivating segment, inactivation of channel (0.5–1.0 ms)

**Inactive Na⁺ channel (refractory period)**

**3**

**FIGURE 22-7 Operational model of the voltage-gated Na⁺ channel.** As in the K⁺ channel depicted in Figure 11-20, four transmembrane domains in the protein contribute to the central pore through which ions move. The critical components that control movement of Na⁺ ions are shown here in the cutaway views depicting three of the four transmembrane domains. In the closed, resting state, the voltage-sensing α helices, which have positively charged side chains every third residue, are attracted to the negative charges on the cytosolic side of the resting membrane. This keeps the gate segment near the cytosolic face in a "closed" position that blocks the channel, preventing entry of Na⁺ ions (step **1**). In response to a small depolarization, the voltage-sensing helices move

through the phospholipid bilayer toward the outer membrane surface, causing an immediate conformational change in the gate at the cytosolic face of the protein that opens the channel (step **2**). Within a fraction of a millisecond the channel-inactivating segment moves into the open channel, preventing passage of further ions (step **3**). Once the membrane is repolarized, the voltage-sensing helices return to the resting position, the channel-inactivating segment is displaced from the channel opening, and the gate closes; the protein reverts to the closed, resting state and can be opened again by depolarization (step **4**). [See W. A. Catterall, 2001, *Nature* **409:**988; and S. B. Long et al., 2007, *Nature* **450:**376.]

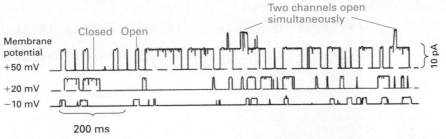

**EXPERIMENTAL FIGURE 22-8 Probability of channel opening and current flux through individual voltage-gated K⁺ channels increases with the extent of membrane depolarization.** These patch-clamp tracings were obtained from patches of neuronal plasma membrane clamped at three different potentials, +50, +20, and −10 mV. The upward deviations in the current indicate the opening of K⁺ channels and movement of K⁺ ions outward (cytosolic to exoplasmic face) across the membrane. Increasing the membrane depolarization (i.e., the clamping voltage) from −10 mV to +50 mV increases the probability a channel will open, the time it stays open, and the amount of electric current (numbers of ions) that pass through it. pA = picoamperes. [From B. Pallota et al., 1981, *Nature* **293**:471, as modified by B. Hille, 1992, *Ion Channels of Excitable Membranes*, 2d ed., Sinauer, p. 122.]

changes that cause their opening and closing. In the resting state, a segment of the protein on the cytosolic face—the *gate*—obstructs the central pore, preventing passage of ions. The channel contains four positively charged *voltage-sensing α helices*; in the resting state these helices are attracted to the inside-negative surface of the plasma membrane. A small depolarization of the membrane triggers movement of these voltage-sensing helices toward the negative charges that are building up on the exoplasmic surface, causing a conformational change in the gate that opens the channel and allows Na⁺ ion flow. After about 1 ms, further Na⁺ influx is prevented by movement of the cytosol-facing *channel-inactivating segment* into the open channel, blocking any further movement of Na⁺ ions. As long as the membrane remains depolarized, the channel-inactivating segment remains in the channel opening; during this *refractory period*, the channel is inactivated and cannot be reopened. A few milliseconds after the inside-negative resting potential is reestablished, the channel-inactivating segment swings away from the pore and the voltage-sensing α helices return to their resting position near the cytosolic surface of the membrane. Thus the channel returns to the closed resting state, once again able to be opened by depolarization. Note the important distinction between "closed" channels and those that are "inactive" as depicted in Figure 22-7.

**Voltage-Gated K⁺ Channels** The repolarization of the membrane that occurs during the refractory period is due largely to opening of voltage-gated K⁺ channels. The subsequent increased efflux of K⁺ from the cytosol removes the excess positive charges from the cytosolic face of the plasma membrane (i.e., makes it more negative), thereby restoring the inside-negative resting potential. For a brief instant, the membrane actually becomes hyperpolarized; at the peak of this **hyperpolarization**, the potential approaches $E_K$, which is more negative than the resting potential (see Figure 22-2).

Opening of the voltage-gated K⁺ channels is induced by the large depolarization of the action potential. Unlike voltage-gated Na⁺ channels, most types of voltage-gated K⁺ channels remain open as long as the membrane is depolarized, and close only when the membrane potential has returned to an inside-negative value. Because the voltage-gated K⁺ channels open slightly after the initial depolarization, at the height of the action potential, they sometimes are called *delayed K⁺ channels*. Eventually all the voltage-gated K⁺ and Na⁺ channels return to their closed resting states. The only open channels in this baseline condition are the nongated K⁺ channels that generate the resting membrane potential, which soon returns to its usual value of approximately −70 mV (see Figure 22-6a).

While the flow of Na⁺ and K⁺ ions alters membrane potential dramatically as it is depolarized, hyperpolarized, and repolarized during an action potential cycle, it is important to note that the exchange of these ions across the membrane is small compared to the overall numbers of Na⁺ and K⁺ ions in the cytosol and extracellular space. Thus, the conduction of action potentials in neurons does not *directly* require the Na⁺/K⁺ pumps that maintain their ion concentration gradients, as we will see shortly.

Voltage-gated Na⁺ channels are difficult to study using patch-clamp techniques, but the patch-clamp tracings in Figure 22-8 reveal the essential properties of voltage-gated K⁺ channels (see Figure 11-22 for a description of patch clamping). In this experiment, small segments of a neuronal plasma membrane were held clamped at different voltages, and the flux of electric charges through the patch due to flow of K⁺ ions through open K⁺ channels was measured. At the modest depolarizing voltage of −10 mV, the channels in the membrane patch open infrequently and remain open for only a few milliseconds, as judged, respectively, by the number and width of the upward blips on the tracings. Further, the ion flux through them is rather small, as measured by the electric current passing through each open channel (the height of the blips). Depolarizing the membrane further to +20 mV causes these channels to open about twice as frequently; also, more K⁺ ions move through each open channel (the height of the blips is greater) because the force driving cytosolic K⁺ ions outward is greater at a membrane

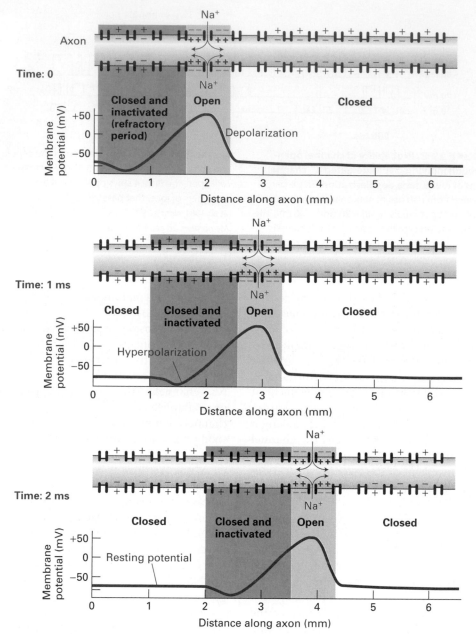

**FIGURE 22-9 Unidirectional conduction of an action potential due to transient inactivation of voltage-gated Na⁺ channels.** At time 0, an action potential (pink line) is at the 2-mm position on the axon; the Na⁺ channels at this position are open (green shading) and Na⁺ ions are flowing inward. The excess Na⁺ ions diffuse in both directions along the inside of the membrane, passively spreading the depolarization in both directions (curved pink arrows). But because the Na⁺ channels at the 1-mm position are still inactivated (red shading), they cannot yet be reopened by the small depolarization caused by passive spread; the Na⁺ channels at the "downstream" 3-mm position, in contrast, begin to open. Each region of the membrane is refractory (inactive) for a few milliseconds after an action potential has passed. Thus, the depolarization at the 2-mm site at time 0 triggers action potentials downstream only; at 1 ms an action potential is passing the 3-mm position, and at 2 ms an action potential is passing the 4-mm position.

potential of +20 mV than at −10 mV. Depolarizing the membrane further to +50 mV, the value at the peak of an action potential, causes opening of more K⁺ channels and also increases the flux of K⁺ through them. Thus, by opening during the peak of the action potential, these K⁺ channels permit the outward movement of K⁺ ions and repolarization of the membrane potential while the voltage-gated Na⁺ channels are being closed and inactivated.

More than 100 voltage-gated K⁺ channel proteins have been identified in humans and other vertebrates. As we discuss later, all these channel proteins have a similar overall structure, but they exhibit different voltage dependencies, conductivities, channel kinetics, and other functional properties. Many open only at strongly depolarizing voltages, a property required for generation of the maximal depolarization characteristic of the action potential before repolarization of the membrane begins.

## Action Potentials Are Propagated Unidirectionally Without Diminution

An action potential begins with changes that occur in a small patch of the axonal plasma membrane near the cell body. At the peak of the action potential, passive spread of the membrane depolarization is sufficient to depolarize a neighboring segment of membrane. This causes a few voltage-gated $Na^+$ channels in this region to open, thereby increasing the extent of depolarization in the region and causing an explosive opening of more $Na^+$ channels and generation of an action potential. This depolarization soon triggers opening of voltage-gated $K^+$ channels and restoration of the resting potential. The action potential thus spreads as a traveling wave away from its initial site without diminution.

As noted earlier, during the refractory period voltage-gated $Na^+$ channels are inactivated for several milliseconds. Such refractory channels cannot conduct ion movements and cannot open during this period even if the membrane is depolarized owing to passive spread. As illustrated in Figure 22-9, the inability of $Na^+$ channels to reopen during the refractory period ensures that action potentials are propagated only in one direction, from the initial axon segment where they originate to the axon termini. Because the $Na^+$ channels upstream of the location of the action potential are still inactivated they cannot be reopened by the small depolarization caused by passive spread. In contrast, the $Na^+$ channels "downstream" of the action potential begin to open.

The refractory period of the $Na^+$ channels also limits the number of action potentials that a neuron can conduct per second. This is important, since it is the frequency of action potentials that carries the information. Reopening of $Na^+$ channels upstream of an action potential (i.e., closer to the cell body) also is delayed by the membrane hyperpolarization that results from opening of voltage-gated $K^+$ channels.

## Nerve Cells Can Conduct Many Action Potentials in the Absence of ATP

The depolarization of the membrane during an action potential results from movement of just a small number of $Na^+$ ions into a neuron and does not significantly affect the intracellular $Na^+$ concentration. A typical nerve cell has about 10 voltage-gated $Na^+$ channels per square micrometer ($\mu m^2$) of plasma membrane. Since each channel passes about 5000–10,000 ions during the millisecond it is open (see Figure 11-23), a maximum of $10^5$ ions per $\mu m^2$ of plasma membrane will move inward during each action potential.

To assess the effect of this ion flux on the cytosolic $Na^+$ concentration of 10 mM (0.01 mol/L) typical of a resting axon, we focus on a segment of axon 1 micrometer ($\mu m$) long and 10 $\mu m$ in diameter. The volume of this segment is 78 $\mu m^3$, or $7.8 \times 10^{-13}$ liters, and it contains $4.7 \times 10^9$ $Na^+$ ions: $(10^{-2}$ mol/L$)$ $(7.8 \times 10^{-13}$ L$)$ $(6 \times 10^{23}$ $Na^+$/mol$)$. The surface area of this segment of the axon is 31 $\mu m^2$, and during passage of one action potential, $10^5$ $Na^+$ ions will enter per $\mu m^2$ of membrane. Thus this $Na^+$ influx increases the number of $Na^+$ ions in this segment by only one part in about 1500: $(4.7 \times 10^9) / (3.1 \times 10^6)$. Likewise, the repolarization of the membrane due to the efflux of $K^+$ ions through voltage-gated $K^+$ channels does not significantly change the intracellular $K^+$ concentration.

Because relatively few $Na^+$ and $K^+$ ions move across the plasma membrane during each action potential, the ATP-driven $Na^+/K^+$ pump that maintains the usual ion gradients plays no direct role in impulse conduction. Since the ion movements during each action potential involve only a minute fraction of the cell's $K^+$ and $Na^+$ ions, a nerve cell can fire hundreds or even thousands of times in the absence of ATP.

**All Voltage-Gated Ion Channels Have Similar Structures** Having explained how the action potential is dependent on regulated opening and closing of voltage-gated channels, we turn to a molecular dissection of these remarkable proteins. After describing the basic structure of these channels, we focus on three questions:

• How do these proteins sense changes in membrane potential?

• How is this change transduced into opening of the channel?

• What causes these channels to become inactivated shortly after opening?

The initial breakthrough in understanding voltage-gated ion channels came from analysis of fruit flies (*Drosophila melanogaster*) carrying the *shaker* mutation. These flies shake vigorously under ether anesthesia, reflecting a loss of motor control and a defect in certain motor neurons that have an abnormally prolonged action potential. Researchers suspected that the *shaker* mutation caused a defect in channel function. Cloning of the gene involved confirmed that the defective protein was a voltage-gated $K^+$ channel. The *shaker* mutation prevents the mutant channel from opening normally immediately upon depolarization. To establish that the wild-type *shaker* gene encoded a $K^+$ channel, cloned wild-type *shaker* cDNA was used as a template to produce *shaker* mRNA in a cell-free system. Expression of this mRNA in frog oocytes and patch-clamp measurements on the newly synthesized channel protein showed that its functional properties were identical with those of the voltage-gated $K^+$ channel in the neuronal membrane, demonstrating conclusively that the *shaker* gene encodes this $K^+$-channel protein.

The Shaker $K^+$ channel and most other voltage-gated $K^+$ channels that have been identified are tetrameric proteins composed of four identical subunits arranged in the membrane around a central pore. Each subunit is constructed of six membrane-spanning $\alpha$ helices, designated S1–S6, and a P segment (Figure 22-10a). The S5 and S6 helices and the P segment are structurally and functionally homologous to those in the nongated resting $K^+$ channel discussed earlier (see Figure 11-20); the S5 and S6 helices form the lining of the $K^+$ selectivity filter through which the ion travels. The S1–S4 helices form a rigid complex that functions as a voltage sensor (with positively charged side chains in S4 acting as the primary sensor). The N-terminal "ball" extending into the cytosol from S1 is the channel-inactivating segment.

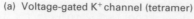

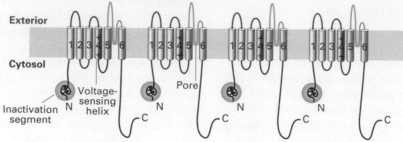

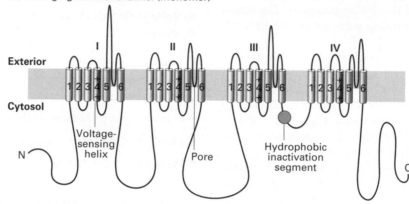

**FIGURE 22-10 Schematic depictions of the secondary structures of voltage-gated K⁺ and Na⁺ channels.** (a) Voltage-gated K⁺ channels are composed of four identical subunits, each containing 600–700 amino acids, and six membrane-spanning α helices, S1–S6. The N-terminus of each subunit, located in the cytosol and labeled N, forms a globular domain (orange ball) essential for inactivation of the open channel. The S5 and S6 helices (green) and the P segment (blue) are homologous to those in nongated resting K⁺ channels, but each subunit contains four additional transmembrane α helices. One of these, S4 (red), is the primary voltage-sensing α helix and is assisted in this role by forming a stable complex with helices S1–3. (b) Voltage-gated Na⁺ channels are monomers containing 1800–2000 amino acids organized into four transmembrane domains (I–IV) that are similar to the subunits in voltage-gated K⁺ channels. The single hydrophobic channel-inactivating segment (orange ball) is located in the cytosol between domains III and IV. Voltage-gated Ca²⁺ channels have a similar overall structure. Most voltage-gated ion channels also contain regulatory (β) subunits, which are not depicted here. [Part (a) adapted from C. Miller, 1992, *Curr. Biol.* **2**:573, and H. Larsson et al., 1996, *Neuron* **16**:387; part (b) adapted from W. A. Catterall, 2001, *Nature* **409**:988.]

Voltage-gated Na⁺ channels and Ca²⁺ channels are monomeric proteins organized into four homologous domains, I–IV (Figure 22-10b). Each of these domains is similar to a subunit of a voltage-gated K⁺ channel. However, in contrast to voltage-gated K⁺ channels, which have four channel-inactivating segments, the monomeric voltage-gated channels have a single channel-inactivating segment. Except for this minor structural difference and their varying ion permeabilities, all voltage-gated ion channels are thought to function in a similar manner and to have evolved from a monomeric ancestral channel protein that contained six transmembrane α helices. Since no molecular structure is available for a voltage-gated Ca²⁺ channel, and there is a molecular structure available only for a closed voltage-gated Na⁺ channel, our discussion focuses on voltage-gated K⁺ channels.

## Voltage-Sensing S4 α Helices Move in Response to Membrane Depolarization

Our understanding of channel-protein biochemistry is advancing rapidly owing to newly obtained crystal structures for bacterial and Shaker potassium channels and other channels.

Transmembrane proteins are notoriously difficult to produce and crystallize, posing unique challenges for the scientist. One method used to obtain crystals of these difficult membrane proteins was to surround them with bound fragments of monoclonal antibodies [F(ab)'s; Chapter 23]; in other cases they were crystallized in complexes with normal protein-binding partners. In both cases the presence of these water-soluble proteins in the complex somehow enhanced crystal formation.

The structures of the channels reveal remarkable arrangements of the voltage-sensing domains, and suggest how parts of the protein move in order to open the channel. As already noted, the K⁺-channel tetramer has a pore whose walls are formed by helices S5 and S6 (Figure 22-11a, b). Outside that core structure four arms, or "paddles," each containing helices S1–S4, protrude into the surrounding membrane and also interact with the outer sides of the S5 and S6 helices; these are the voltage sensors, and they are in minimal contact with the pore. Sensitive electrical measurements suggested that the opening of a voltage-gated Na⁺ or K⁺ channel is accompanied by the movement of 12 to 14 protein-bound positive charges from the cytosolic to the

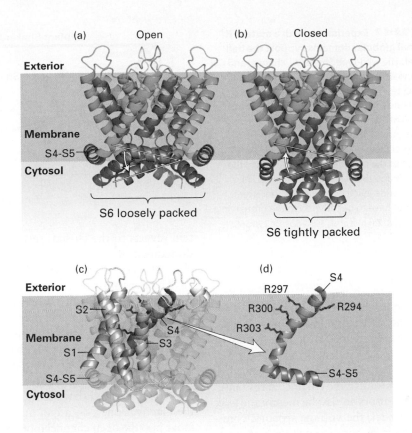

**FIGURE 22-11 Molecular structure of a voltage-sensitive K⁺ channel.** The two ribbon diagrams show models of the potassium channel in the (a) open and (b) closed states. Since the molecule is a tetramer of the same subunit, four copies of each helix are visible. The four green (S5) and blue (S6) α helices span the membrane, with the interior of the cell at the bottom and exterior at the top. Note how the helices are tightly packed at the bottom in (b), closing the channel so that the K⁺ ion cannot pass through. (Compare the distances between S5 helices as shown by the curly brackets below (a) and (b).) The S4–S5 linker (orange), located in the cytoplasm, connects the S4 helix (not shown) to the S5 helix. For clarity, helices S1 through S4 have been omitted from the model; they would normally be attached to the end of the S4–S5 linker and protrude from the molecule. (c) Three-dimensional structure of the voltage-sensing "paddles" comprising helixes S1-S4, with the four voltage-sensing arginine (R) residues in S4 (depicted in panel d). These paddles move from near the interior to the exterior of the membrane in response to depolarization. Since each one is attached to an S4–S5 linker, each linker and its attached S5 helix is moved, in turn moving S6 helices, which opens the pore. Note that the linker between S4 and S5 is pointed upward toward the exoplasmic (exterior) surface in the open channel (a), pulled upward by the outward movement of the S1-S4 paddles; in contrast the S4-S5 linker is pointing downward in the closed channel (b) when the S1-S4 paddles are nearer the cytosolic surface. (d) Three-dimensional structure of helix S4 and the S4–S5 linker. [From S. B. Long, X. Tao, E. B. Campbell, and R. MacKinnon, 2007, *Nature* **450**:376–382.]

exoplasmic surface of the membrane. The moving parts of the protein are the rigid complexes composed of helices S1–S4; S4 accounts for much of the positive charge and is the primary voltage sensor, with a positively charged lysine or arginine every third or fourth residue (Figure 21-11d). Arginines in S4 have been measured moving as much as 1.5 nm as the channel opens, which can be compared with the ~5-nm thickness of the membrane or the 1.2-nm diameter of the α helix itself.

In the resting state, the positive charges on the S1–S4 complexes (the "paddles") are attracted to the negative charges on the cytosolic face of the membrane. In the depolarized membrane, these same positive charges become attracted to the negative charges on the exoplasmic (outer) surface of the membrane, causing the S1–S4 paddles to move partly across the membrane—from the cytosolic to the exo-plasmic surface. This movement is depicted schematically in Figure 22-7 and triggers a conformational change in the protein that opens the channel.

The most unusual aspect of the voltage-sensitive channel structures is the presence of charged groups, e.g., arginines, in contact with lipid. The location of the voltage sensor helps to explain earlier experiments in which a non-voltage-sensitive channel was converted into a voltage-sensing channel by adding to it voltage-sensing domains. Such a result would seem unlikely if the voltage sensors had to be deeply embedded in the core structure.

Studies with mutant Shaker K⁺ channels support the importance of the S4 helix in voltage sensing. When one or more arginine or lysine residues in the S4 helix of the Shaker K⁺ channel were replaced with neutral or acidic residues, fewer positive charges than normal moved across the membrane in

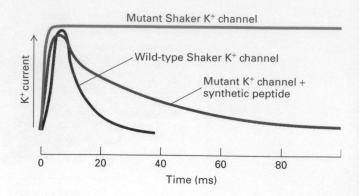

**EXPERIMENTAL FIGURE 22-12 Experiments with a mutant K⁺ channel lacking the N-terminal globular domains support the ball-and-chain inactivation model.** The wild-type Shaker K⁺ channel and a mutant form lacking the amino acids composing the N-terminal ball were expressed in *Xenopus* oocytes. The activity of the channels was monitored by the patch-clamp technique. When patches were depolarized from 0 to +30 mV, the wild-type channel opened for ~5 ms and then closed (red curve). The mutant channel opened normally, but could not close (green curve). When a chemically synthesized ball peptide was added to the cytosolic face of the patch, the mutant channel opened normally and then closed (blue curve). This demonstrated that the added peptide inactivated the channel after it opened and that the ball does not have to be tethered to the protein in order to function. [Adapted from W. N. Zagotta et al., 1990, *Science* **250**:568.]

response to a membrane depolarization, indicating that arginine and lysine residues in the S4 helix do indeed move across the membrane. The structure of the open form of a mammalian voltage-gated K⁺ channel has been contrasted with the closed structure of a different K⁺ channel. The results suggest a model for the opening and closing of the channel in response to movements of the voltage sensors across the membrane (see Figure 22-11a, b). In the model, the voltage sensors, composed of helices S1–S4, move in response to voltage and exert a torque on a linker helix that connects S4 to S5:

- In the open-channel conformation, the position of the S4–S5 linker forces the S6 helix to form a kink near the cytosolic surface (blue in Figure 22-11a) and the pore inside, near the cytosolic surface, is open. The pore's 1.2-nm diameter is sufficient to accommodate hydrated K⁺ ions.

- When the cell membrane is repolarized and the voltage sensor moves toward the cytosolic membrane surface, the S4–S5 linkers (orange in Figure 22-11b) are twisted down, toward the inside of the cell. The S6 helices are consequently straightened, squeezing the bottom of the channel closed. Thus the gate is composed of the cytosol-facing ends of the S5 and S6 helices, where the pore is narrowest.

## Movement of the Channel-Inactivating Segment into the Open Pore Blocks Ion Flow

An important characteristic of most voltage-gated channels is inactivation; that is, soon after opening they close spontaneously, forming an inactive channel that will not reopen until the membrane is repolarized. In the resting state, the globular balls at the N-termini of the four subunits in a voltage-gated K⁺ channel are free in the cytosol (Figure 22-10). Several milliseconds after the channel is opened by depolarization, one ball moves through an opening between two of the subunits and binds in a hydrophobic pocket in the pore's central cavity, blocking the flow of K⁺ ions (Figure 22-7). After a few milliseconds, the ball is displaced from the pore, and the pro-

tein reverts to the closed, resting state. The ball-and-chain domains in K⁺ channels are functionally equivalent to the channel-inactivating segment in Na⁺ channels.

The experimental results shown in Figure 22-12 demonstrate that inactivation of K⁺ channels depends on the ball domains, occurs after channel opening, and does not require the ball domains to be covalently linked to the channel protein. In other experiments, mutant K⁺ channels lacking portions of the ~40-residue chain connecting the ball to the S1 helix were expressed in frog oocytes. Patch-clamp measurements of channel activity showed that the shorter the chain, the more rapid the inactivation, as if a ball attached to a shorter chain can move into the open channel more readily. Conversely, addition of random amino acids to lengthen the normal chain slows channel inactivation.

The single channel-inactivating segment in voltage-gated Na⁺ channels contains a conserved hydrophobic motif composed of isoleucine, phenylalanine, methionine, and threonine (see Figure 22-10b). Like the longer ball-and-chain domain in K⁺ channels, this segment folds into and blocks the Na⁺-conducting pore until the membrane is repolarized (see Figure 22-7).

## Myelination Increases the Velocity of Impulse Conduction

As we have seen, action potentials can move down an unmyelinated axon without diminution at speeds up to 1 meter per second. But even such fast speeds are insufficient to permit the complex movements typical of animals. In adult humans, for instance, the cell bodies of motor neurons innervating leg muscles are located in the spinal cord, and the axons are about a meter in length. The coordinated muscle contractions required for walking, running, and similar movements would be impossible if it took 1 second for an action potential to move from the spinal cord down the axon of a motor neuron to a leg muscle. The solution is to wrap cells in insulation that increases the rate of movement of an action potential. The insulation is called a **myelin sheath** (see Figure 22-1b). The presence of a myelin sheath around an axon increases the velocity of impulse conduction to 10–100 meters per second. As a result, in a typical human motor neuron, an action potential can travel the length of a 1-meter-long axon and stimulate a muscle to contract within 0.01 seconds.

In nonmyelinated neurons, the conduction velocity of an action potential is roughly proportional to the diameter of the axon, because a thicker axon will have a greater number of ions that can diffuse. The human brain is packed with relatively small, myelinated neurons. If the neurons in the human brain were not myelinated, their axonal diameters would have to increase about 10,000-fold to achieve the same conduction velocities as myelinated neurons. Thus vertebrate brains, with their densely packed neurons, never could have evolved without myelin.

## Action Potentials "Jump" from Node to Node in Myelinated Axons

The myelin sheath surrounding an axon is formed from many glial cells. Each region of myelin formed by an individual glial cell is separated from the next region by an unmyelinated area of axonal membrane about 1 μm in length called the *node of Ranvier* (or simply, *node*; see Figure 22-1b). The axonal membrane is in direct contact with the extracellular fluid only at the nodes, and the myelin covering prevents any ion movement into or out of the axon except at the nodes. Moreover, all the voltage-gated $Na^+$ and $K^+$ channels and all the $Na^+/K^+$ pumps, which maintain the ionic gradients in the axon, are located in the nodes.

As a consequence of this localization, the inward movement of $Na^+$ ions that generates the action potential can occur only at the myelin-free nodes (Figure 22-13). The excess cytosolic positive ions generated at a node during the membrane depolarization associated with $Na^+$ movement into the cytosol as part of an action potential spread passively through the axonal cytosol to the next node with very little loss or attenuation, since they cannot cross the myelinated axonal membrane. This causes a depolarization at one node to spread rapidly to the next node and induce an action potential there, effectively permitting the action potential to jump from node to node. The transmission is called *saltatory conduction*. This phenomenon explains why the conduction velocity of myelinated neurons is about the same as that of much-larger-diameter unmyelinated neurons. For instance, a 12-μm-diameter myelinated vertebrate axon and a 600-μm-diameter unmyelinated squid axon both conduct impulses at 12 m/s.

## Two Types of Glia Produce Myelin Sheaths

Figure 22-14 shows three main types of glial cells present in the nervous system, two of which produce myelin sheaths: *oligodendrocytes* make sheaths for the central nervous system (CNS), and *Schwann cells* make them for the peripheral nervous system (PNS). Astrocytes, also shown in the figure, facilitate synapse formation and communication between neurons, and were discussed in Section 22.1. A fourth type of glia, *microglia* (not shown), constitutes a part of the CNS immune system. Microglia are not related by lineage to neurons or to other glia, and will not be discussed further.

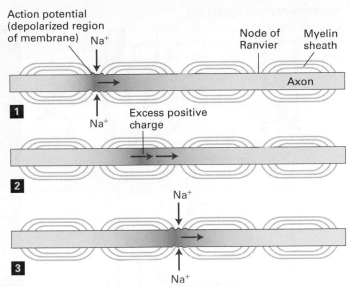

**FIGURE 22-13 Conduction of action potentials in myelinated axons.** Because the myelin layer renders the axon impermeable to ion movement across its membrane and because voltage-gated $Na^+$ channels are found only on axonal membrane at the nodes of Ranvier, the influx of $Na^+$ ions associated with an action potential can occur only at nodes. When an action potential is generated at one node (step **1**), the excess positive ions in the cytosol, which cannot move outward across the sheath, diffuse rapidly down the axon, causing sufficient depolarization at the next node (step **2**) to induce an action potential at that node (step **3**). By this mechanism the action potential jumps from node to node along the axon.

**Oligodendrocytes** Oligodendrocytes form the spiral myelin sheath around axons of the central nervous system (Figure 22-14a). Each oligodendrocyte provides myelin sheaths to segments of multiple neurons. The major protein constituents are myelin basic protein (MBP) and proteolipid protein (PLP). MBP, a peripheral membrane protein found in both the central and peripheral nervous systems (see Figure 22-15), has seven RNA splicing variants that encode different forms of the protein. It is synthesized by ribosomes located in the growing myelin sheath, an example of specific transport of mRNAs to a peripheral cell region. The localization of MBP mRNA depends on microtubules.

Damage to proteins produced by oligodendrocytes underlies a prevalent human neurological disease, multiple sclerosis (MS). MS is usually characterized by spasms and weakness in one or more limbs, bladder dysfunction, local sensory losses, and visual disturbances. This disorder—the prototype *demyelinating disease*—is caused by patchy loss of myelin in areas of the brain and spinal cord. In MS patients, conduction of action potentials by the demyelinated neurons is slowed, and the $Na^+$ channels spread outward from the nodes, lowering their nodal concentration. The cause of the disease is not known but appears to involve either the body's

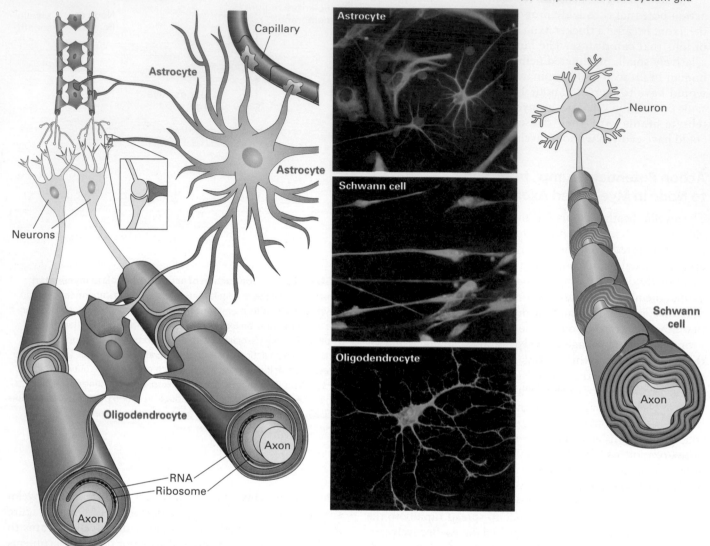

**FIGURE 22-14 Three types of glial cells.** (a) A single oligodendrocyte in the central nervous system can myelinate segments of multiple axons. Astrocytes interact with neurons but do not form myelin. (b) Each Schwann cell insulates a section of a single peripheral nervous system axon. [From B. Stevens, 2003, *Curr. Biol.* **13:**469, and adapted from D. L. Sherman and P. Brophy, 2005, *Nature Rev. Neurosci.* **6:**683–690. Photos: Courtesy of Varsha Shukla and Doug Field from NIH.]

production of auto-antibodies (antibodies that bind to normal body proteins) that react with MBP, or the secretion of proteases that destroy myelin proteins. A mouse mutant, *shiverer*, has a deletion of much of the MBP gene, leading to tremors, convulsions, and early death. Similarly, human (Pelizaeus-Merzbacher disease) and mouse (*jimpy*) mutations in the gene coding for the other major protein of CNS myelin, PLP, cause loss of oligodendrocytes and inadequate myelination. ■

**Schwann Cells** Schwann cells form myelin sheaths around peripheral nerves. A Schwann cell myelin sheath is a remarkable spiral wrap (Figure 22-14b). A long axon can have as many as several hundred Schwann cells along its length, each contributing myelin insulation to an *internode* stretch of about 1–1.5 μm of axon. For reasons that are not understood, not all axons are myelinated. Mutations in mice that eliminate Schwann cells cause the death of most neurons.

In contrast to oligodendrocytes, each Schwann cell myelinates only one axon. The sheaths are composed of about 70 percent lipid (rich in cholesterol) and 30 percent protein. In the PNS, the principal protein constituent (~80 percent) of myelin is called protein 0 ($P_0$), an integral membrane protein that has immunoglobulin (Ig) domains. MBP is also an abundant component. The extracellular Ig domains of $P_0$ bind together the surfaces of sequential wraps around the axon to compact the spiral of myelin sheath (Figure 22-15). Other proteins play this kind of role in the CNS.

 In humans, peripheral myelin, like CNS myelin, is a target of autoimmune disease, mainly involving the

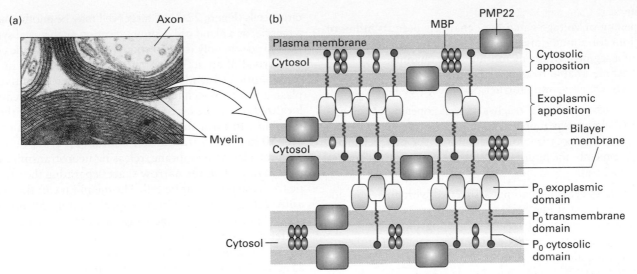

**FIGURE 22-15 Formation and structure of a myelin sheath in the peripheral nervous system.** (a) At high magnification the specialized spiral myelin membrane appears as a series of layers, or lamellae, of phospholipid bilayers wrapped around the axon. (b) Close-up view of three layers of the myelin membrane spiral. The two most abundant integral myelin membrane proteins, $P_0$ and PMP22, are produced only by Schwann cells. The exoplasmic domain of a $P_0$ protein, which has an immunoglobulin fold, associates with similar domains emanating from $P_0$ proteins in the opposite membrane surface, thereby "zippering" together the exoplasmic membrane surfaces in close apposition. These interactions are stabilized by binding of a tryptophan residue on the tip of the exoplasmic domain to lipids in the opposite membrane. Close apposition of the cytosolic faces of the membrane may result from binding of the cytosolic tail of each $P_0$ protein to phospholipids in the opposite membrane. PMP22 may also contribute to membrane compaction. Myelin basic protein (MBP), a cytosolic protein, remains between the closely apposed membranes as the cytosol is squeezed out. [Part (a) © Science VU/C. Raine/Visuals Unlimited; part (b) adapted from L. Shapiro et al., 1996, *Neuron* **17:** 435, and E. J. Arroyo and S. S. Scherer, 2000, *Histochem. Cell Biol.* **113:**1.]

formation of antibodies against $P_0$. The Guillain-Barre syndrome (GBS), also known as *acute inflammatory demyelinating polyneuropathy,* is one such disease. GBS is the most common cause of rapid-onset paralysis, occurring at a frequency of $10^{-5}$. The cause is unknown. The common inherited neurological disorder called *Charcot-Marie-Tooth disease,* which damages peripheral motor and sensory nerve function, is due to overexpression of the gene that encodes PMP22 protein, another constituent of peripheral nerve myelin. ■

Interactions between glia and neurons control the placement and spacing of myelin sheaths, and the assembly of nerve-transmission machinery at the nodes of Ranvier. Voltage-gated $Na^+$ channels and $Na^+/K^+$ pumps, for example, congregate at the nodes of Ranvier through interactions with cytoskeletal proteins. While the details of the node assembly process are not fully understood, a number of key players have been identified. In the PNS, where the process has been most studied, surface adhesion molecules in the Schwann cell membrane first interact with neuronal surface adhesion molecules. An *immunoglobulin cell-adhesion molecule* (IgCAM) in the glial membrane, called *neurofascin155,* contacts two axonal proteins, contactin and contactin-associated protein, at the edge of the node. These cell-cell contact events create boundaries at each side of the node.

The channel proteins and other molecules that will accumulate at the node are initially dispersed throughout the axons. Then axonal proteins, including two IgCAMs called *NrCAM* and *neurofascin186,* as well as ankyrin G (Chapter 17),

accumulate within the node. The two IgCAMs bind to a single transmembrane domain protein called *gliomedin* that is expressed in the glial cell. Experiments that eliminated gliomedin production showed that without it nodes do not form, so it is a key regulator. Ankyrin in the node contacts βIV spectrin, a major constituent of the cytoskeleton, thus tethering the node's protein complex to the cytoskeleton. $Na^+$ channels become associated with neurofascin186, NrCAM, and ankyrin G, firmly trapping the channel in the nodal segment of the axonal plasma membrane where it is needed. As a result of these multiple protein-protein interactions, the concentration of $Na^+$ channels is roughly a hundredfold higher in the nodal membrane of myelinated axons than in the axonal membrane of nonmyelinated neurons.

## KEY CONCEPTS of Section 22.2

### Voltage-Gated Ion Channels and the Propagation of Action Potentials

- Action potentials are sudden membrane depolarizations followed by rapid repolarization.

- An action potential results from the sequential opening and closing of voltage-gated $Na^+$ and $K^+$ channels in the plasma membrane of neurons and muscle cells (excitable cells; see Figure 22-7).

- Opening of voltage-gated $Na^+$ channels permits influx of $Na^+$ ions for about 1 ms, causing a sudden large depolarization of a segment of the membrane. The channels then close and become unable to open (refractory period) for several milliseconds, preventing further $Na^+$ flow (see Figure 22-7).

- As the action potential reaches its peak, opening of voltage-gated $K^+$ channels permits efflux of $K^+$ ions, which repolarizes and then hyperpolarizes the membrane. As these channels close, the membrane returns to its resting potential (see Figures 22-2 and 22-6).

- The excess cytosolic cations associated with an action potential generated at one point on an axon spread passively to the adjacent segment, triggering opening of voltage-gated $Na^+$ channels in the vicinity and thus propagation of the action potential along the axon.

- Because of the absolute refractory period of the voltage-gated $Na^+$ channels and the brief hyperpolarization resulting from $K^+$ efflux, the action potential is propagated in one direction only, toward the axon termini (see Figure 22-9).

- Voltage-gated $Na^+$ and $Ca^{2+}$ channels are monomeric proteins containing four domains that are structurally and functionally similar to each of the subunits in the tetrameric voltage-gated $K^+$ channels. Each domain or subunit in voltage-gated cation channels contains six transmembrane $\alpha$ helices and a nonhelical P segment that forms the ion-selectivity pore (see Figure 22-10).

- Opening of voltage-gated channels results from movement of the positively charged S1–S4 paddles toward the extracellular side of the membrane in response to a depolarization of sufficient magnitude (see Figure 22-11).

- Closing and inactivation of voltage-gated cation channels result from movement of a cytosolic "ball" segment into the open pore (see Figure 22-12a).

- Myelination, which increases the rate of impulse conduction up to a hundredfold, permits the close packing of neurons characteristic of vertebrate brains.

- In myelinated neurons, voltage-gated $Na^+$ channels are concentrated at the nodes of Ranvier. Depolarization at one node spreads rapidly with little attenuation to the next node, so that the action potential jumps from node to node (see Figure 22-13).

- Myelin sheaths are produced by glial cells that wrap themselves in spirals around neurons. Oligodendrocytes produce myelin for the CNS; Schwann cells, for the PNS (see Figure 22-14).

## 22.3 Communication at Synapses

As we have discussed, electrical pulses transmit signals along neurons, but signals are transmitted between neurons and other excitable cells mainly by chemical signals. Synapses are the junctions where *presynaptic* neurons release these chemical signals, or neurotransmitters, which then act on *postsynaptic*

target cells (Figure 22-3). A target cell may be another neuron, a muscle, or a gland cell. Communication at chemical synapses usually goes in only one direction: pre- to postsynaptic cell.

Arrival of an action potential at an axon terminus in a presynaptic cell leads to opening of voltage-sensitive $Ca^{2+}$ plasma membrane channels and an influx of $Ca^{2+}$, causing a localized rise in the cytosolic $Ca^{2+}$ concentration in the axon terminus. In turn, the rise in $Ca^{2+}$ triggers fusion of small (40–50 nm) neurotransmitter-containing synaptic vesicles with the plasma membrane, releasing neurotransmitters into the synaptic cleft, the narrow space separating the presynaptic from the postsynaptic cell. The membrane of the postsynaptic cell is located within approximately 50 nm of the presynaptic membrane, reducing the distance the neurotransmitter must diffuse.

Neurotransmitters—small, water-soluble molecules such as acetylcholine or dopamine—bind to receptors on the postsynaptic cell that, in turn, induce localized changes in the potential across its plasma membrane. If the membrane potential becomes less negative—that is, becomes depolarized—an action potential will tend to be induced in the postsynaptic cell. Such synapses are **excitatory**, and in general involve the opening of $Na^+$ channels in the postsynaptic plasma membrane. In contrast, in an **inhibitory synapse**, binding of the neurotransmitter to a receptor on the postsynaptic cell causes hyperpolarization of the plasma membrane—generation of a more inside-negative potential. Typically, hyperpolarization is the result of opening of $Cl^-$ or $K^+$ channels in the postsynaptic plasma membrane, which tends to hinder generation of an action potential.

Neurotransmitter receptors fall into two broad classes: ligand-gated ion channels, which open immediately upon neurotransmitter binding, and G protein–coupled receptors (GPCRs). Neurotransmitter binding to a GPCR induces the opening or closing of a *separate* ion-channel protein over a period of seconds to minutes. These "slow" neurotransmitter receptors were discussed in Chapter 15 along with GPCRs that bind different types of ligands and modulate the activity of cytosolic proteins other than ion channels. Here we examine the structure and operation of the excitatory *nicotinic acetylcholine receptor* found at many nerve-muscle synapses. It was the first ligand-gated ion channel to be purified, cloned, and characterized at the molecular level, and provides a paradigm for other neurotransmitter-gated ion channels.

The duration of the neurotransmitter signal depends on the amount of transmitter released by the presynaptic cell, which in turn depends on the amount of transmitter that had been stored as well as the frequency of action potentials arriving at the synapse. The duration of the signal also depends on how rapidly any unbound neurotransmitter is degraded in the synaptic cleft or transported back into the presynaptic cell. Presynaptic cell plasma membranes, as well as glia, contain transporter proteins that pump neurotransmitters across the plasma membrane back into the cell, thus keeping the extracellular concentrations of transmitter low.

In this section we focus first on how synapses form and how they control the regulated secretion of neurotransmitters

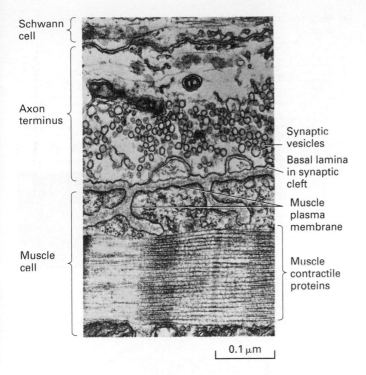

Schwann cell

Axon terminus

Muscle cell

Synaptic vesicles

Basal lamina in synaptic cleft

Muscle plasma membrane

Muscle contractile proteins

0.1 μm

**FIGURE 22-16 Synaptic vesicles in the axon terminus near the region where neurotransmitter is released.** In this longitudinal section through a neuromuscular junction, the basal lamina lies in the synaptic cleft separating the neuron from the muscle membrane, which is extensively folded. Acetylcholine receptors are concentrated in the postsynaptic muscle membrane at the top and part way down the sides of the folds in the membrane. A Schwann cell surrounds the axon terminus. [From J. E. Heuser and T. Reese, 1977, in E. R. Kandel, ed., *The Nervous System,* vol. 1, *Handbook of Physiology,* Williams and Wilkins, p. 266.]

in the context of the basic principles of vesicular trafficking outlined in Chapter 14. Next we look at the mechanisms that limit the duration of the synaptic signal, and how neurotransmitters are received and interpreted by the postsynaptic cell.

## Formation of Synapses Requires Assembly of Presynaptic and Postsynaptic Structures

Axons extend from the cell body during development, guided by signals from other cells along the way so that the axon termini will reach the correct location (see Section 18.8). As axons grow, they come into contact with their potential target cells, such as dendrites of other neurons, and often at such sites synapses form. In the CNS, synapses with presynaptic specializations occur frequently all along an axon; in contrast, motor neurons form synapses with muscle cells only at the axon termini.

Neurons cultured in isolation will not form synapses very efficiently, but when glia are added, the rate of synapse formation increases substantially. Astrocytes and Schwann cells send protein signals to neurons to stimulate the formation of synapses and then help to preserve them. One such signal is thrombospondin (TSP), a component of the extracellular matrix; mice lacking two *thrombospondin* genes have only 70 percent of the normal number of synapses in their brains. Mutual communication between neurons and the glia that surround them is frequent and complex, making the signals and information they carry an area of active research. There is also evidence that neurons form synapses on glia. While glia do not produce action potentials, they do have complex arrays of channels and ion fluxes.

At the site of a synapse, the presynaptic cell has hundreds to thousands of synaptic vesicles, some docked at the membrane

and others waiting in reserve. The release of neurotransmitter into the synaptic cleft occurs in the *active zone,* a specialized region of the plasma membrane containing a remarkable assemblage of proteins whose functions include modifying the properties of the synaptic vesicles and bringing them into position for docking and fusing with the plasma membrane. Viewed by electron microscopy, the active zone has electron-dense material and fine cytoskeletal filaments (Figure 22-16). A similarly dense region of specialized structures is seen across the synapse in the postsynaptic cell, the *postsynaptic density (PSD).* Cell-adhesion molecules that connect pre- and postsynaptic cells keep the active zone and PSD aligned. After release of synaptic vesicles in response to an action potential, the presynaptic neuron retrieves synaptic vesicle membrane proteins by *endocytosis* both within and outside the active zone.

Synapse assembly has been extensively studied at the *neuromuscular junction (NMJ)* (Figure 22-17). At these synapses **acetylcholine** is the neurotransmitter produced by motor neurons, and its receptor, AChR, is produced by the postsynaptic muscle cell. Muscle cell precursors, myoblasts, put into culture will spontaneously fuse into multinucleate myotubes that look similar to normal muscle cells. As myotubes form, AChR is produced near the center of the cell and inserted into the myotube plasma membrane, forming diffuse membrane patches (Figure 22-17a).

The formation of the neuromuscular synapse is a multistep process requiring signaling interactions between motor neurons and muscle fibers. A key player is **MuSK**, a receptor tyrosine kinase that is localized in the diffuse AChR-rich patches of the myotube plasma membrane. In ways that are not known, MuSK both induces clustering of AChRs and serves to attract the termini of growing motor neuron axons. For example, knockdown of MuSK inhibits both processes, while overexpression of MuSK in cultured muscle cells induces motor neuron growth throughout the muscle and formation of excess synapses.

Another key player is **Agrin**, a glycoprotein synthesized by developing motor neurons, transported in vesicles along axon microtubules, and secreted near the developing myotubes. Agrin binds to LRP4, a single-pass membrane-spanning protein; this stimulates an association between LRP4 and MuSK and increases MuSK kinase activity (Figure 22-17b). This leads to activation of several downstream signal transduction pathways, one of which leads to activation of Rac and Rho (see Section 17.3) and formation of clusters of

**FIGURE 22-17 Formation of the neuromuscular junction**
(a) Motor neuron-myotube interactions. Following fusion of myoblasts to form multinucleate myotubes, the nuclei synthesize acetylcholine receptor (AChR) mRNA. The nuclei near the center of each muscle fiber synthesize significantly more AChR mRNA than other nuclei. AChRs together with MuSK receptor kinases accumulate in membrane patches near the center of the cell, the prospective synaptic region of the muscle, prior to and independent of innervation; the cell is said to be "prepatterned." The motor neuron axon termini grow toward these AChR clusters and secrete the glycoprotein Agrin. Agrin, in turn, induces clustering of the AChRs (dark red) and MuSKs around the axon termini (green), forming the neuromuscular junction. (Inset) Micrograph of a synapse from a postnatal (3-week-old) mouse, viewed by staining for axons (neurofilament) and synaptic vesicles (synaptophysin), shown together in green, and AChRs, shown in red. (b) Signaling downstream of Agrin receptors. Motor axons secrete Agrin, which stabilizes postsynaptic differentiation by binding LRP4 and activating MuSK kinase activity. Phosphorylation of tyrosines in the MuSK juxtamembrane region, indicated by yellow P in circle, stimulates recruitment and tyrosine phosphorylation of Dok-7, an adaptor protein that is expressed selectively in muscle, which forms a dimer, stimulates MuSK kinase activity, and recruits the adapter protein Crk/Crk-L. Crk/Crk-L is essential to activate a Rac/Rho- and Rapsyn-dependent pathway for clustering AChRs opposite the presynaptic axon termini; this pathway involves several cytoskeletal proteins including actin and myosins. The pathway for synapse-specific transcription is less well understood but likely involves JNK kinase-dependent activation of ETS-family transcription factors that stimulate expression of multiple genes encoding synaptic proteins such as acetylcholine receptors, MuSK, LRP4, and acetylcholinesterase (AChE), the extracellular enzyme that localizes to the synaptic cleft and that degrades acetylcholine to choline and acetate. [Micrograph from Herbst, R. et al., 2002, *Development,* **129**:5449–5460.]

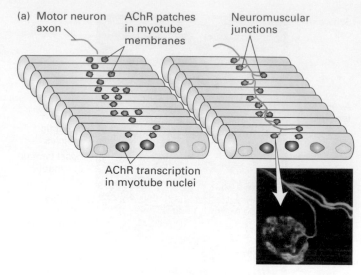

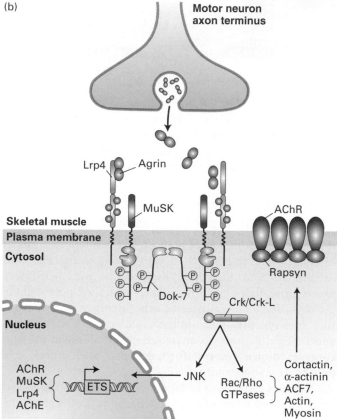

AChRs with the cytoskeletal protein rapsyn; this interaction, together with binding of other cytoskeletal proteins including actin, leads to localization of AChRs opposite the nerve termini at the neuromuscular junction. The density of acetylcholine receptors in a mature synapse reaches ~10,000–20,000/$\mu$m$^2$, while elsewhere in the plasma membrane the density is ~10/$\mu$m$^2$. Another pathway, also not well understood, leads to activation of ETS-family transcription factors and stimulation of expression of multiple genes encoding synaptic proteins such as rapsyn and AChRs.

## Neurotransmitters Are Transported into Synaptic Vesicles by H$^+$-Linked Antiport Proteins

In this section, we focus on how neurotransmitters are packaged in membrane-bound *synaptic vesicles* in the axon terminus. Numerous small molecules function as neurotransmitters at various synapses. With the exception of acetylcholine, the neurotransmitters shown in Figure 22-18 are amino acids or derivatives of amino acids. Nucleotides such as ATP and the corresponding nucleosides, which lack phosphate groups, also

function as neurotransmitters. Each neuron generally produces just one type of neurotransmitter. While the types of neurotransmitters are varied and while they operate in different parts of the nervous system, all signaling by neurotransmitters results in one of two outcomes: either the transmission of an electrical signal, or its inhibition.

All of these neurotransmitters are synthesized in the cytosol and imported into membrane-bound synaptic vesicles within axon termini, where they are stored. These vesicles are 40–50 nm in diameter, and their lumen has a low pH, generated

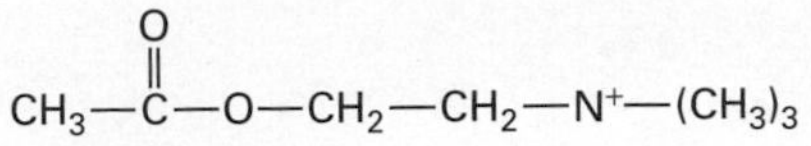

O
‖
$CH_3$—C—O—$CH_2$—$CH_2$—$N^+$—$(CH_3)_3$

**Acetylcholine**

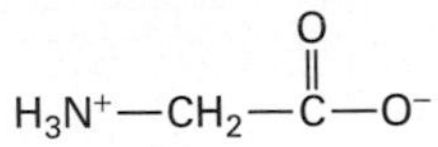

O
‖
$H_3N^+$—$CH_2$—C—$O^-$

**Glycine**

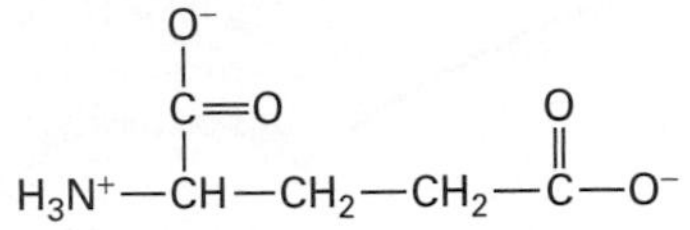

$O^-$
|
C=O    O
|      ‖
$H_3N^+$—CH—$CH_2$—$CH_2$—C—$O^-$

**Glutamate**

HO

HO —[ring]— $CH_2$—$CH_2$—$NH_3^+$

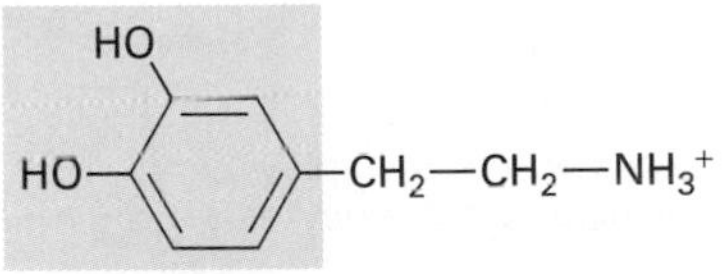

**Dopamine**
(derived from tyrosine)

HO

HO —[ring]— CH—$CH_2$—$NH_3^+$
                |
                OH

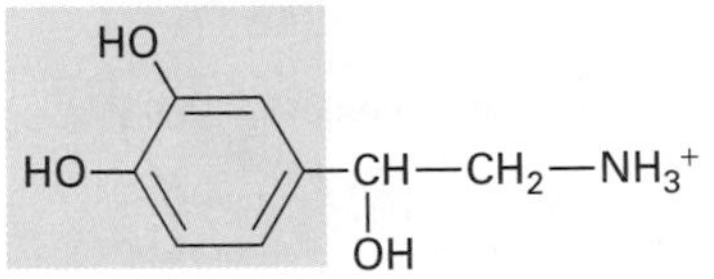

**Norepinephrine**
(derived from tyrosine)

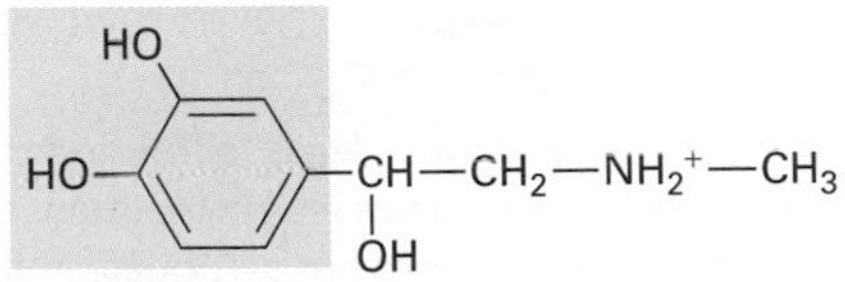

HO

HO —[ring]— CH—$CH_2$—$NH_2^+$—$CH_3$
                |
                OH

**Epinephrine**
(derived from tyrosine)

HO

[indole ring] —$CH_2$—$CH_2$—$NH_3^+$
N
H

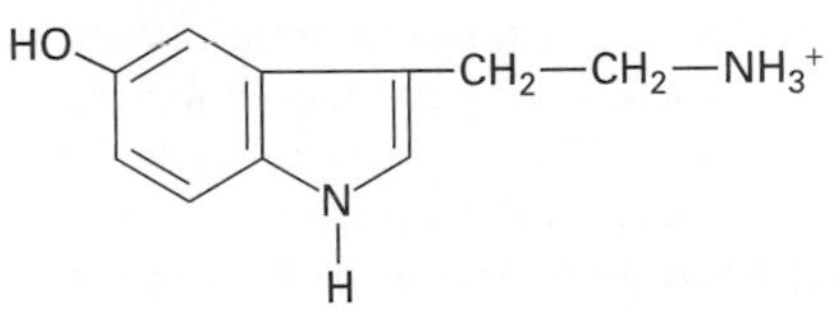

**Serotonin, or 5-hydroxytryptamine**
(derived from tryptophan)

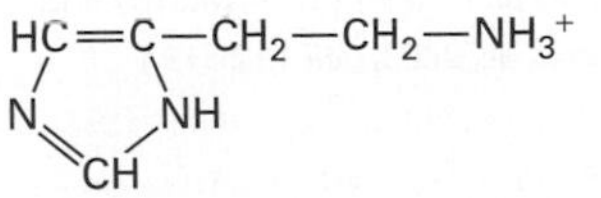

HC=C—$CH_2$—$CH_2$—$NH_3^+$
|    |
N    NH
  \ /
  CH

**Histamine**
(derived from histidine)

O
‖
$H_3N^+$—$CH_2$—$CH_2$—$CH_2$—C—$O^-$

**γ-Aminobutyric acid, or GABA**
(derived from glutamate)

**FIGURE 22-18 Structures of several small molecules that function as neurotransmitters.** Except for acetylcholine, all these are amino acids (glycine and glutamate) or derived from the indicated amino acids. The three transmitters synthesized from tyrosine, which contain the catechol moiety (blue highlight), are referred to as *catecholamines*.

by operation of a V-class proton pump in the vesicle membrane. Similar to the accumulation of metabolites in plant vacuoles (see Figure 11-29), this proton concentration gradient (vesicle lumen > cytosol) powers neurotransmitter import by ligand-specific $H^+$-linked neurotransmitter antiporters in the vesicle membrane (Figure 22-19).

For example, acetylcholine is synthesized in the cytosol from acetyl coenzyme A (acetyl CoA), an intermediate in the degradation of glucose and fatty acids, and choline in a reaction catalyzed by choline acetyltransferase:

O                                          $CH_3$
‖                                          |
$CH_3$—C—S—CoA  +  HO—$CH_2$—$CH_2$—$N^+$—$CH_3$  $\xrightarrow{\text{Choline acetyltransferase}}$
                                           |
                                           $CH_3$

**Acetyl CoA**                    **Choline**

O                              $CH_3$
‖                              |
$CH_3$—C—O—$CH_2$—$CH_2$—$N^+$—$CH_3$  + CoA—SH
                               |
                               $CH_3$

**Acetylcholine**

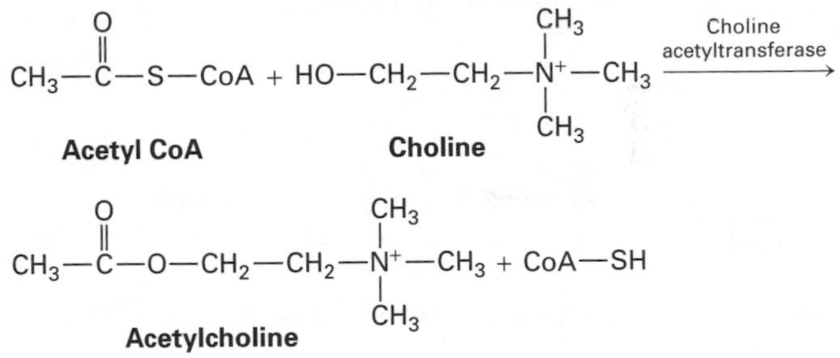

Synaptic vesicles take up and concentrate acetylcholine from the cytosol against a steep concentration gradient, using an $H^+$/acetylcholine antiporter in the vesicle membrane; as with other antiporters, the export of protons from the forming vesicle down its electrochemical gradient powers the uptake of the neurotransmitter. Curiously, the gene encoding this antiporter is contained entirely within the first intron of the gene encoding choline acetyltransferase, a mechanism conserved throughout evolution for ensuring coordinate expression of these two proteins.

Different $H^+$/neurotransmitter antiport proteins are used for import of other neurotransmitters into synaptic vesicles. For example, glutamate is imported into synaptic vesicles by proteins called *vesicular glutamate transporters (VGLUTs)*. VGLUTs are highly specific for glutamate but have rather low substrate affinity ($K_m$ = 1–3 mM). Like the acetylcholine transporter, VGLUTs are antiporters, moving glutamate into synaptic vesicles while protons move in the other direction.

## Synaptic Vesicles Loaded with Neurotransmitter Are Localized near the Plasma Membrane

A highly organized arrangement of cytoskeletal fibers in the axon terminus helps localize synaptic vesicles in the active zone. The vesicles themselves are linked together by *synapsin*, a fibrous phosphoprotein associated with the cytosolic surface of all synaptic vesicle membranes. Filaments of synapsin also radiate from the plasma membrane and bind to

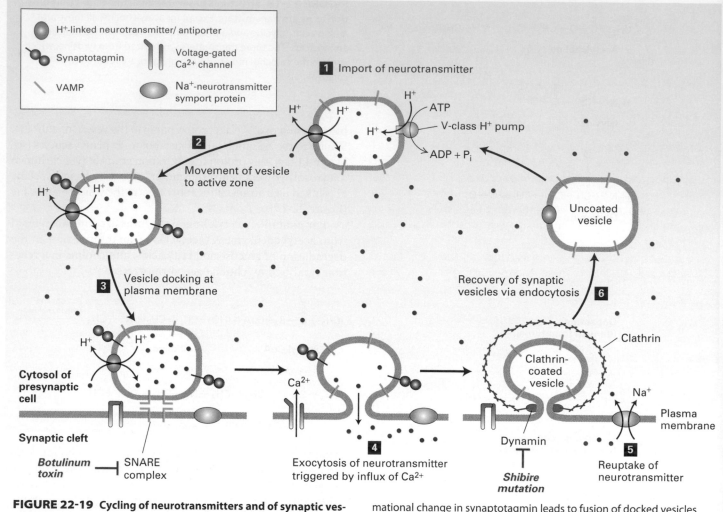

**Legend:**
- H⁺-linked neurotransmitter/antiporter
- Synaptotagmin
- Voltage-gated Ca²⁺ channel
- VAMP
- Na⁺-neurotransmitter symport protein

**1** Import of neurotransmitter

ATP
V-class H⁺ pump
ADP + Pᵢ

**2** Movement of vesicle to active zone

**3** Vesicle docking at plasma membrane

Cytosol of presynaptic cell

Synaptic cleft

**Botulinum toxin** ⊣ SNARE complex

**4** Exocytosis of neurotransmitter triggered by influx of Ca²⁺

Ca²⁺

**5** Reuptake of neurotransmitter

Na⁺
Plasma membrane

Dynamin
***Shibire mutation***

Clathrin
Clathrin-coated vesicle

**6** Recovery of synaptic vesicles via endocytosis

Uncoated vesicle

**FIGURE 22-19 Cycling of neurotransmitters and of synaptic vesicles in axon termini.** Most synaptic vesicles are formed by endocytic recycling as depicted here. The entire cycle typically takes about 60 seconds. Step **1**: The uncoated vesicles express a V-type proton pump (orange) and a single type of H⁺-neurotransmitter antiporter (blue) specific for the particular neurotransmitter, to import neurotransmitters (red dots) from the cytosol. Step **2**: Synaptic vesicles loaded with neurotransmitter move to the active zone. Step **3**: Vesicles dock at defined sites on the plasma membrane of the presynaptic cell, and the vesicle v-SNAREs called VAMP bind to the plasma membrane t-SNAREs, forming a SNARE complex. Synaptotagmin prevents membrane fusion and release of neurotransmitter. Botulinum toxin prevents exocytosis by proteolytically cleaving VAMP, the v-SNARE on vesicles. Step **4**: In response to a nerve impulse (action potential), voltage-gated Ca²⁺ channels in the plasma membrane open, allowing an influx of Ca²⁺ from the extracellular medium. The resulting Ca²⁺-induced conformational change in synaptotagmin leads to fusion of docked vesicles with the plasma membrane and release of neurotransmitters into the synaptic cleft. Synaptotagmin does not participate in the later steps of vesicle recycling or neurotransmitter import though it is still present. Step **5**: Na⁺ symporter proteins take up neurotransmitter from the synaptic cleft into the cytosol, which limits the duration of the action potential and partially recharges the cell with transmitter. Step **6**: Vesicles are recovered by endocytosis, creating uncoated vesicles, ready to be refilled and begin the cycle anew. After clathrin/AP vesicles containing v-SNARE and neurotransmitter transporter proteins bud inward and are pinched off in a dynamin-mediated process, they lose their coat proteins. Dynamin mutations such as shibire in Drosophila block the re-formation of synaptic vesicles, leading to paralysis. Unlike most neurotransmitters, acetylcholine is not recycled. [See K. Takei et al., 1996, *J. Cell Biol.* **133**:1237; V. Murthy and C. Stevens, 1998, *Nature* **392**:497; and R. Jahn et al., 2003, *Cell* **112**:519.]

vesicle-associated synapsin. These interactions probably keep synaptic vesicles close to the part of the plasma membrane facing the synapse. Indeed, synapsin knockout mice, although viable, are prone to seizures; during repetitive stimulation of many neurons in such mice, the number of synaptic vesicles that fuse with the plasma membrane is greatly reduced. Thus synapsins are thought to recruit synaptic vesicles to the active zone.

## Influx of Ca²⁺ Triggers Release of Neurotransmitters

The exocytosis of neurotransmitters from synaptic vesicles involves vesicle-targeting and fusion events similar to those that occur during the intracellular transport of secreted and plasma-membrane proteins (Chapter 14). However, two unique features critical to synapse function differ from other secretory pathways: (1) secretion is tightly coupled to arrival of an action potential at the axon terminus, and (2) synaptic vesicles are recycled locally to the axon terminus after fusion with the plasma membrane. Figure 22-19 shows the entire cycle whereby synaptic vesicles are filled with neurotransmitter, release their contents, and are recycled.

Depolarization of the plasma membrane cannot, by itself, cause synaptic vesicles to fuse with the plasma membrane. In order to trigger vesicle fusion, an action potential must be converted into a chemical signal—namely, a localized rise in the cytosolic $Ca^{2+}$ concentration. The transducers of the electrical signals are *voltage-gated $Ca^{2+}$ channels* localized to the region of the plasma membrane adjacent to the synaptic vesicles. The membrane depolarization due to arrival of an action potential opens these channels, permitting an influx of $Ca^{2+}$ ions from the extracellular medium into the region of the axon terminus near the docked synaptic vesicles. Importantly, the rise in cytosolic $Ca^{2+}$ is localized; it is also transient, as the excess $Ca^{2+}$ is rapidly pumped out of the cell by plasma membrane $Ca^{2+}$ pumps.

A simple experiment demonstrates the importance of voltage-gated $Ca^{2+}$ channels in release of neurotransmitters. A preparation of neurons in a $Ca^{2+}$-containing medium is treated with tetrodotoxin, a drug that blocks voltage-gated $Na^+$ channels and thus prevents conduction of action potentials. As expected, no neurotransmitters are secreted into the culture medium. If the axonal membrane is then artificially depolarized, by making the medium ~100 mM KCl, in the presence of extracellular $Ca^{2+}$, neurotransmitters are released from the cells because of the influx of $Ca^{2+}$ through open voltage-gated $Ca^{2+}$ channels. Indeed, patch-clamping experiments show that voltage-gated $Ca^{2+}$ channels, like voltage-gated $Na^+$ channels, open transiently upon depolarization of the membrane.

Two pools of neurotransmitter-filled synaptic vesicles are present in axon termini: those docked at the plasma membrane, which can be readily exocytosed, and the great majority that are in reserve in the active zone near the plasma membrane. Each rise in $Ca^{2+}$ generated by arrival of a single action potential triggers exocytosis of about 10 percent of the docked vesicles. Membrane proteins unique to synaptic vesicles then are specifically internalized by endocytosis, usually via the same types of clathrin-coated vesicles used to recover other plasma-membrane proteins by other types of cells. After the endocytosed vesicles lose their clathrin coat, they are rapidly refilled with neurotransmitter. The ability of many neurons to fire 50 times a second is clear evidence that the recycling of vesicle membrane proteins occurs quite rapidly. The machinery of endocytosis and exocytosis is highly conserved, and is described in more detail in Chapter 14.

## A Calcium-Binding Protein Regulates Fusion of Synaptic Vesicles with the Plasma Membrane

Fusion of synaptic vesicles with the plasma membrane of axon termini depends on **SNAREs**, the same type of proteins that mediate membrane fusion of other regulated secretory vesicles. The principal v-SNARE in synaptic vesicles (VAMP) tightly binds syntaxin and SNAP-25, the principal t-SNAREs in the plasma membrane of axon termini, to form four-helix SNARE complexes. After fusion, SNAP proteins and NSF within the axon terminus promote disassociation of VAMP from t-SNAREs, as in the fusion of secretory vesicles depicted in Figure 14-10.

Strong evidence for the role of VAMP in neurotransmitter exocytosis is provided by the mechanism of action of botulinum toxin, a bacterial protein that can cause the paralysis and death characteristic of *botulism*, a type of food poisoning. The toxin is composed of two polypeptides: One binds to motor neurons that release acetylcholine at synapses with muscle cells, facilitating entry of the other polypeptide, a protease, into the cytosol of the axon terminus. The only protein this protease cleaves is VAMP (see Figure 22-19, step **3**). After the botulinum protease enters an axon terminus, synaptic vesicles that are not already docked rapidly lose their ability to fuse with the plasma membrane because cleavage of VAMP prevents assembly of SNARE complexes. The resulting block in acetylcholine release at neuromuscular synapses causes paralysis. However, vesicles that are already docked exhibit remarkable resistance to the toxin, indicating that SNARE complexes may already be in a partially assembled, protease-resistant state when vesicles are docked on the presynaptic membrane. ■

The signal that triggers exocytosis of docked synaptic vesicles is a very localized rise in the $Ca^{2+}$ concentration in the cytosol near vesicles from 0.1 μM, characteristic of resting cells, to 1–100 μM following arrival of an action potential in stimulated cells. The speed with which synaptic vesicles fuse with the presynaptic membrane after a rise in cytosolic $Ca^{2+}$ (less than 1 ms) indicates that the fusion machinery is entirely assembled in the resting state and can rapidly undergo a conformational change leading to exocytosis of neurotransmitter (Figure 22-20). A $Ca^{2+}$-binding protein called *synaptotagmin*, located in the membrane of synaptic vesicles, is a key component of the vesicle-fusion machinery that triggers exocytosis in response to $Ca^{2+}$. A protein called complexin is thought to bind to the α-helical bundle of an assembled v-SNARE/t-SNARE complex that bridges the synaptic vesicle and plasma membranes, preventing the final fusion step. Binding of $Ca^{2+}$ to synaptotagmin relieves this inhibition, releasing complexin and allowing the fusion event to occur very rapidly. While the mechanisms by which synaptotagmin functions are debated, Figure 22-20 depicts a widely accepted model.

Several lines of evidence support a role for synaptotagmin as the $Ca^{2+}$ sensor for exocytosis of neurotransmitters. Mutant embryos of *Drosophila* and *C. elegans* that completely lack synaptotagmin fail to hatch and exhibit very reduced, uncoordinated muscle contractions. Larvae with partial loss-of-function mutations of synaptotagmin survive, but their neurons are defective in $Ca^{2+}$-stimulated vesicle exocytosis. Moreover, in mice, mutations in synaptotagmin that decrease its affinity for $Ca^{2+}$ cause a corresponding increase in the amount of cytosolic $Ca^{2+}$ needed to trigger rapid exocytosis.

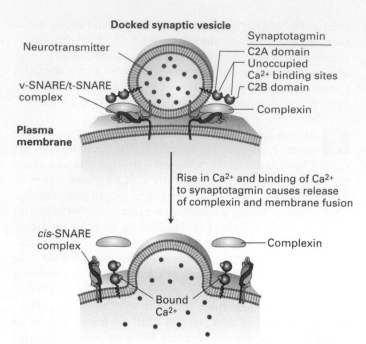

**FIGURE 22-20 Synaptotagmin-mediated fusion of synaptic vesicles with the plasma membrane.** Only a few synaptic vesicles are docked at the presynaptic plasma membrane; these are primed for fusion with the plasma membrane. The tight interconnections between the synaptic vesicle and plasma membrane are mediated in part by bundles of four α helices derived from complexes of vesicle v-SNARE and plasma membrane t-SNARE proteins (see Figure 14-10). The fusion of the two membranes is prevented by binding of complexin protein to the v-SNARE/t-SNARE complex. Synaptotagmin is composed of a short intraluminal sequence, a single transmembrane α helix that anchors it in the synaptic vesicle membrane, a linker, and two $Ca^{2+}$-binding domains termed C2A and C2B. Synaptotagmin without bound $Ca^{2+}$ may also bind to the v-SNARE/t-SNARE complex and prevent membrane fusion. A localized rise in $Ca^{2+}$ allows $Ca^{2+}$ ions to bind to synaptotagmin, altering its three-dimensional conformation. This triggers release of the complexin fusion inhibitor, binding (or altered binding) of synaptotagmin to the v-SNARE/t-SNARE complex, instantaneous membrane fusion, and release of neurotransmitters into the extracellular space.
[After T. Südhof and J. Rothman, 2009, *Science* **323**:474.]

## Fly Mutants Lacking Dynamin Cannot Recycle Synaptic Vesicles

Synaptic vesicles are formed primarily by endocytic budding from the plasma membrane of axon termini. Endocytosis usually involves clathrin-coated pits and is quite specific, in that several membrane proteins unique to the synaptic vesicles (e.g., neurotransmitter transporters) are specifically incorporated into the endocytosed vesicles and resident plasma membrane proteins (e.g., the voltage-sensitive $Ca^{2+}$ channel) remain. In this way, synaptic-vesicle membrane proteins can be reused and the recycled vesicles refilled with neurotransmitter (see Figure 22-19, step **6**).

As in the formation of other clathrin/AP-coated vesicles, pinching off of endocytosed synaptic vesicles requires the GTP-binding protein *dynamin* (see Figure 14-19). Indeed, analysis of a temperature-sensitive *Drosophila* mutant called *shibire (shi)*, which encodes the fly dynamin protein, provided early evidence for the role of dynamin in endocytosis. At the permissive temperature of 20 °C, the mutant flies are normal, but at the nonpermissive temperature of 30 °C, they are paralyzed (*shibire*, "paralyzed," in Japanese) because pinching off of clathrin-coated pits in neurons and other cells is blocked. When viewed in the electron microscope, the *shi* neurons at 30 °C show abundant clathrin-coated pits with long necks but few clathrin-coated vesicles. The appearance of nerve termini in *shi* mutants at the nonpermissive temperature is similar to that of termini from normal neurons incubated in the presence of a nonhydrolyzable analog of GTP (see Figure 14-20). Because of their inability to pinch off new synaptic vesicles, the neurons in *shi* mutants eventually become depleted of synaptic vesicles when flies are shifted to the nonpermissive temperature, leading to a cessation of synaptic signaling and to paralysis.

## Signaling at Synapses Is Terminated by Degradation or Reuptake of Neurotransmitters

Following their release from a presynaptic cell, neurotransmitters must be removed or destroyed to prevent continued stimulation of the postsynaptic cell. Signaling can be terminated by diffusion of a transmitter away from the synaptic cleft, but this is a slow process. Instead, one of two more rapid mechanisms terminates the action of neurotransmitters at most synapses.

Signaling by acetylcholine is terminated when it is hydrolyzed to acetate and choline by *acetylcholinesterase*, an enzyme localized to the synaptic cleft. Choline released in this reaction is transported back into the presynaptic axon terminus by a $Na^+$/choline symporter and used in synthesis of more acetylcholine. The operation of this transporter is similar to that of the $Na^+$-linked symporters used to transport glucose into cells against a concentration gradient (see Figure 11-26).

With the exception of acetylcholine, all the neurotransmitters shown in Figure 22-18 are removed from the synaptic cleft by transport back into the axon termini that released them. Thus these transmitters are recycled intact, as depicted in Figure 22-19 (step **5**). Transporters for GABA, norepinephrine, dopamine, and serotonin were the first to be cloned and studied. These four transport proteins are all $Na^+$-linked symporters. They are 60–70 percent identical in their amino acid sequences, and each is thought to contain 12 transmembrane α helices. As with other $Na^+$ symporters, the movement of $Na^+$ into the cell down its electrochemical gradient provides the energy for uptake of the neurotransmitter. To maintain electroneutrality, $Cl^-$ often is transported via an ion channel along with the $Na^+$ and neurotransmitter.

Neurotransmitters and their transporters are targets of a variety of powerful and sometimes devastating drugs. Cocaine binds to and inhibits the transporters for norepinephrine, serotonin, and dopamine. In particular, binding of cocaine to the dopamine transporter inhibits reuptake of dopamine, causing a higher-than-normal concentration of dopamine to remain in the synaptic cleft and prolonging the stimulation of postsynaptic neurons. Long-lasting exposure to cocaine, as occurs with habitual use, leads to down-regulation of dopamine receptors and thus altered regulation of dopaminergic signaling. It is thought that decreased dopaminergic signaling after chronic cocaine use may contribute to depressive mood disorders and sensitize important brain reward circuits to the reinforcing effects of cocaine, leading to addiction. Similarly, therapeutic agents such as the antidepressant drugs fluoxetine (Prozac) and imipramine block serotonin reuptake, and the tricyclic antidepressant desipramine blocks norepinephrine reuptake. As a result, these drugs also cause a higher-than-normal concentration of neurotransmitter to remain in the synaptic cleft and prolong the stimulation of postsynaptic neurons. Fluoxetine and similarly acting drugs such as paroxetine (Paxil) and sertraline (Zoloft) are often referred to collectively as selective serotonin reuptake inhibitors (SSRIs). ■

## Opening of Acetylcholine-Gated Cation Channels Leads to Muscle Contraction

In this section we look at how binding of neurotransmitters by receptors on postsynaptic cells leads to changes in the cells' membrane potential, using the communication between motor neurons and muscles as an example. At these synapses, called neuromuscular junctions, acetylcholine is the neurotransmitter. A single axon terminus of a frog motor neuron may contain a million or more synaptic vesicles, each containing 1000–10,000 molecules of acetylcholine; these vesicles often accumulate in rows in the active zone (see Figures 22-16 and 22-17). Such a neuron can form synapses with a single skeletal muscle cell at several hundred points.

The nicotinic acetylcholine receptor, which is expressed in muscle cells, is a *ligand-gated channel* that admits both $K^+$ and $Na^+$. These receptors are also produced in brain neurons and are important in learning and memory; acetylcholine receptor loss is observed in schizophrenia, epilepsy, drug addiction, and Alzheimer's disease. Antibodies against acetylcholine receptors constitute a major part of the autoimmune reactivity in the disease myasthenia gravis. The receptor is so named because it is bound by nicotine; it has been implicated in nicotine addiction in tobacco smokers. There are at least 14 different isoforms of the receptor, which assemble into homo- and heteropentamers with varied properties.

The effect of acetylcholine on this receptor can be determined by patch-clamping studies on isolated outside-out patches of muscle plasma membranes. Outside-out patch-clamping is a technique that measures the effects of extracellular solutes on channel receptors within the isolated patch (see Figure 11-22c). Such measurements have shown that acetylcholine causes opening of a cation channel in the receptor capable of transmitting 15,000–30,000 $Na^+$ and $K^+$ ions per millisecond. However, since the resting potential of the muscle plasma membrane is near $E_K$, the potassium equilibrium potential, opening of acetylcholine receptor channels causes little increase

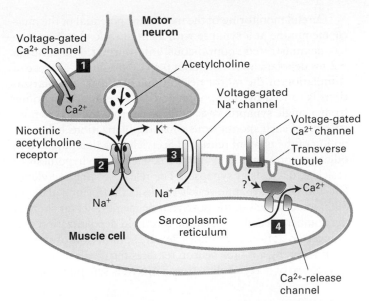

**FIGURE 22-21 Sequential activation of gated ion channels at a neuromuscular junction.** Arrival of an action potential at the terminus of a presynaptic motor neuron induces opening of voltage-gated $Ca^{2+}$ channels in the neuron (step **1**) and subsequent release of acetylcholine, which triggers opening of the ligand-gated acetylcholine receptors in the muscle plasma membrane (step **2**). The open receptor channel allows an influx of $Na^+$ and an efflux of $K^+$ from the muscle cell. The $Na^+$ influx produces a localized depolarization of the membrane, leading to opening of voltage-gated $Na^+$ channels and generation of an action potential (step **3**). When the spreading depolarization reaches transverse tubules, it is sensed by voltage-gated $Ca^{2+}$ channels in the plasma membrane. Through an unknown mechanism (indicated as ?) these channels remain closed but influence $Ca^{2+}$ channels in the sarcoplasmic reticulum membrane (a network of membrane-bound compartments in muscle), which release stored $Ca^{2+}$ into the cytosol (step **4**). The resulting rise in cytosolic $Ca^{2+}$ causes muscle contraction by mechanisms discussed in Chapter 17.

in the efflux of $K^+$ ions; $Na^+$ ions, on the other hand, flow into the muscle cell, driven by the $Na^+$ electrochemical gradient.

The simultaneous increase in permeability to $Na^+$ and $K^+$ ions following binding of acetylcholine produces a net depolarization to about −15 mV from the muscle resting potential of −85 to −90 mV. As shown in Figure 22-21, this localized depolarization of the muscle plasma membrane triggers opening of voltage-gated $Na^+$ channels, leading to generation and conduction of an action potential in the muscle cell surface membrane by the same mechanisms described previously for neurons. When the membrane depolarization reaches transverse tubules (see Figure 17-34), specialized invaginations of the plasma membrane, it acts on $Ca^{2+}$ channels in the plasma membrane apparently without causing them to open. Somehow this causes opening of adjacent $Ca^{2+}$-release channels in the sarcoplasmic reticulum membrane. The subsequent flow of stored $Ca^{2+}$ ions from the sarcoplasmic reticulum into the cytosol raises the cytosolic $Ca^{2+}$ concentration sufficiently to induce muscle contraction.

Careful monitoring of the membrane potential of the muscle membrane at a synapse with a cholinergic motor neuron has demonstrated spontaneous, intermittent, and random ~2-ms depolarizations of about 0.5–1.0 mV in the absence of stimulation of the motor neuron. Each of these depolarizations is caused by the spontaneous release of acetylcholine from a single synaptic vesicle in the neuron. Indeed, demonstration of such spontaneous small depolarizations led to the notion of the quantal release of acetylcholine (later applied to other neurotransmitters) and thereby led to the hypothesis of vesicle exocytosis at synapses. The release of one acetylcholine-containing synaptic vesicle results in the opening of about 3000 ion channels in the postsynaptic membrane, far short of the number needed to reach the threshold depolarization that induces an action potential. Clearly, stimulation of muscle contraction by a motor neuron requires the nearly simultaneous release of acetylcholine from numerous synaptic vesicles.

## All Five Subunits in the Nicotinic Acetylcholine Receptor Contribute to the Ion Channel

The acetylcholine receptor from skeletal muscle is a pentameric protein with a subunit composition of $\alpha_2\beta\gamma\delta$. These four different subunit types have considerable sequence homology with each other; on average, about 35–40 percent of the residues in any two subunits are similar. The complete receptor has five-fold symmetry, and the actual cation channel is a tapered central pore lined by homologous segments from each of the five subunits (Figure 22-22).

The channel opens when the receptor cooperatively binds two acetylcholine molecules to sites located at the interfaces of the $\alpha\delta$ and $\alpha\gamma$ subunits, as shown in Figure 22-22a. Once acetylcholine is bound to a receptor, the channel is opened within a few microseconds. Studies measuring the receptor's permeability to different small cations suggest that the open ion channel is, at its narrowest, about 0.65–0.80 nm in diameter, in agreement with estimates from electron micrographs. This would be sufficient to allow passage of both Na$^+$ and K$^+$ ions with their shell of bound water molecules.

Thus the acetylcholine receptor probably transports hydrated ions, unlike Na$^+$ and K$^+$ channels, both of which allow passage only of nonhydrated ions (see Figure 11-21).

The central ion channel is lined by five homologous transmembrane M2 $\alpha$ helices, one from each of the five subunits (see Figure 22-22c). The M2 helices are composed largely of hydrophobic or uncharged polar amino acids, but

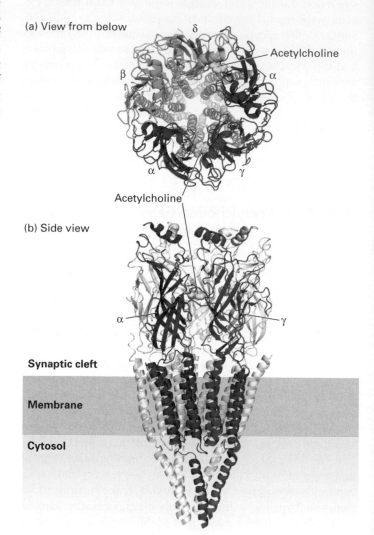

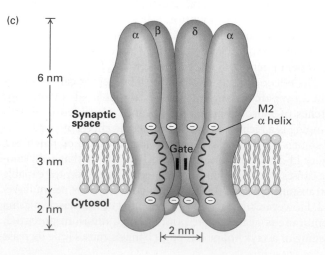

**FIGURE 22-22 Three-dimensional structure of the nicotinic acetylcholine receptor.** Three-dimensional molecular structure of the *Torpedo* nicotinic acetylcholine receptor as viewed (a) from the synaptic cleft and (b) parallel to the plane of the membrane. For clarity, only the front two subunits, $\alpha$ and $\gamma$, are highlighted in (b) (colors: $\alpha$, red; $\beta$, green; $\gamma$, blue; $\delta$, light blue). The two acetylcholine-binding sites are located about 3 nm from the membrane surface and are highlighted in yellow; only the one at the $\alpha\gamma$ interface is shown in panel b. (c) Schematic cutaway model of the pentameric receptor in the membrane. Each subunit has four membrane-spanning $\alpha$ helices, M1–M4; the M2 $\alpha$ helix (red) faces the central pore. Aspartate and glutamate side chains form two rings of negative charges, one at each end of the M2 helices, that help exclude anions from and attract cations to the channel. The gate, which is opened by binding of acetylcholine, lies within the pore. [Parts (a) and (b) from N. Unwin, 2005, *J. Mol. Biol.* **346**:967–989.]

negatively charged aspartate or glutamate residues are located at each end, near the membrane faces, and several serine or threonine residues are near the middle. Mutant acetylcholine receptors in which a single negatively charged glutamate or aspartate in one M2 helix is replaced by a positively charged lysine have been expressed in frog oocytes. Patch-clamping measurements indicate that such altered proteins can function as channels, but the number of ions that pass through during the open state is reduced. The greater the number of glutamate or aspartate residues mutated (in one or multiple M2 helices), the greater the reduction in ion conductivity. These findings suggest that aspartate and glutamate residues form a ring of negative charges on the external surface of the pore that help to screen out anions and attract $Na^+$ or $K^+$ ions as they enter the channel. A similar ring of negative charges lining the cytosolic pore surface also helps select cations for passage.

The two acetylcholine-binding sites in the extracellular domain of the receptor lie ~4–5 nm from the membrane surface (see Figure 22-22b). Binding of acetylcholine thus must trigger conformational changes in the receptor subunits that can cause channel opening at some distance from the binding sites. Receptors in isolated postsynaptic membranes can be trapped in the open or closed state by rapid freezing in liquid nitrogen. Electron microscopic images of such preparations suggest that the five M2 helices rotate relative to the vertical axis of the channel during opening and closing.

We have discussed the neuromuscular junction as an excellent example of how neurotransmitters and their receptors work. Like acetylcholine, glutamate, a principal neurotransmitter in the vertebrate brain, uses two main types of receptors. One class, termed ionotropic glutamate receptors, are ligand-gated channels that allow the flow of $K^+$, $Na^+$, and sometimes $Ca^{2+}$ in response to glutamate binding and that work along the same principles as AChR. Glutamate also binds to a second class of receptors, coupled to G proteins. Later in this chapter we will see how such G protein–coupled receptors (GPCRs) and ion channels function as receptors for odorants and tastants that activate various sensory nerve cells. To cover all of the neurotransmitter receptors, ion channels, and other signaling proteins that function in the brain would require a book much larger than this one!

## Nerve Cells Make an All-or-None Decision to Generate an Action Potential

At the neuromuscular junction, virtually every action potential in the presynaptic motor neuron triggers an action potential in the postsynaptic muscle cell that propagates along the muscle fiber. The situation at synapses between neurons, especially those in the brain, is much more complex because the postsynaptic neuron commonly receives signals from many presynaptic neurons. The neurotransmitters released from presynaptic neurons may bind to an *excitatory receptor* on the postsynaptic neuron, thereby opening a channel that admits $Na^+$ ions or both $Na^+$ and $K^+$ ions. The acetylcholine and glutamate receptors just discussed are examples of excitatory receptors, and opening of such ion channels leads to depolarization of the postsynaptic plasma membrane, promoting generation of an action potential. In contrast, binding of a neurotransmitter to an *inhibitory receptor* on the postsynaptic cell causes opening of $K^+$ or $Cl^-$ channels, leading to an efflux of additional $K^+$ ions from the cytosol or an influx of $Cl^-$ ions. In either case, the ion flow tends to hyperpolarize the plasma membrane, which inhibits generation of an action potential in the postsynaptic cell.

A single neuron can be affected simultaneously by signals received at multiple excitatory and inhibitory synapses. The neuron continuously integrates these signals and determines whether or not to generate an action potential. In this process, the various small depolarizations and hyperpolarizations generated at synapses move along the plasma membrane from the dendrites to the cell body and then to the axon hillock, where they are summed together. An action potential is generated whenever the membrane at the axon hillock becomes depolarized to a certain voltage, which can be different for different neurons, called the *threshold potential* (Figure 22-23). Thus an action potential is generated in an all-or-nothing fashion: depolarization to the threshold always leads to an action potential, whereas any depolarization that does not reach the threshold potential never induces it.

Whether a neuron generates an action potential in the axon hillock depends on the balance of the timing, amplitude, and localization of all the various inputs it receives; this signal computation differs for each type of neuron. In a sense, each neuron is a tiny analog-to-digital computer that averages all the receptor activations and electrical disturbances on its membrane (analog) and makes a decision whether or not (digital) to trigger an action potential and conduct it down the axon. An action potential will always have the same *magnitude* in any particular neuron. As we have noted, the *frequency* with which action potentials are generated in a particular neuron is the important parameter in its ability to signal other cells.

## Gap Junctions Allow Certain Neurons to Communicate Directly

*Chemical synapses* employing neurotransmitters allow one-way communication at reasonably high speed. However, sometimes signals go from cell to cell electrically, without the intervention of chemical synapses. *Electrical synapses* depend on **gap junction** channels that link two cells together (Chapter 20). The effect of gap junction connections is to perfectly coordinate the activities of joined cells. An electrical synapse is *bidirectional;* either neuron can excite the other. Electrical synapses are common in the neocortex and thalamus, for example. The key feature of electrical synapses is their speed. While it takes about 0.5–5 ms for a signal to cross a chemical synapse, transmission across an electrical synapse is almost instantaneous, on the order of a fraction of a millisecond since the cytoplasm is continuous between the cells. In addition, the presynaptic cell (the one sending the signal) does not have to reach a threshold at which it can

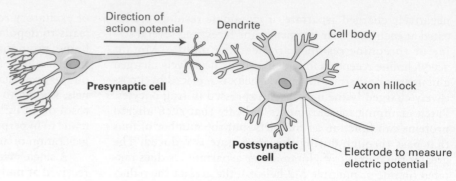

**EXPERIMENTAL FIGURE 22-23 Incoming signals must reach the threshold potential to trigger an action potential in a postsynaptic neuron.** In this example, the presynaptic neuron is generating about one action potential every 4 ms. Arrival of each action potential at the synapse causes a small change in the membrane potential at the axon hillock of the postsynaptic neuron, in this example a depolarization of ~5 mV. When multiple stimuli cause the membrane of this postsynaptic cell to become depolarized to the threshold potential, here approximately 40 mV, an action potential is induced in it.

cause an action potential in the postsynaptic cell. Instead, any electrical current continues into the next cell and causes depolarization in proportion to the current.

An electrical synapse may contain thousands of gap channels, each composed of two hemichannels, one in each apposed cell. Gap junction channels in the neuron have a structure similar to conventional gap junctions (see Figure 20-20). Each hemichannel is an assembly of six copies of the connexin protein. Since there are about 20 mammalian connexin genes, diversity in channel structure and function can arise from the different protein components. The 1.6–2.0-nm channel itself allows the diffusion of molecules up to about 1,000 Da in size and has no trouble at all accommodating ions.

## KEY CONCEPTS of Section 22.3

### Communication at Synapses

• Synapses are the junctions between a presynaptic cell and a postsynaptic cell, and consist of small gaps (see Figure 22-3).

• Communication between presynaptic and postsynaptic cells is abundant as a synapse is being formed. Cell-adhesion molecules keep the cells aligned. At the neuromuscular junction, motor neurons induce the accumulation of acetylcholine receptors in the postsynaptic muscle plasma membrane close to the forming axon terminus (see Figure 22-17).

• In presynaptic cells, low-molecular-weight neurotransmitters (e.g., acetylcholine, dopamine, epinephrine) are imported from the cytosol into synaptic vesicles by $H^+$-linked antiporters. V-class proton pumps maintain the low intravesicular pH that drives neurotransmitter import against a concentration gradient.

• Neurotransmitters (see Figure 22-18) are stored in hundreds to thousands of synaptic vesicles in the axon termini of the presynaptic cell (see Figure 22-16). When an action potential arrives there, voltage-sensitive $Ca^{2+}$ channels open and the calcium causes synaptic vesicles to fuse with the plasma membrane, releasing neurotransmitter molecules into the synapse (see Figure 22-19, step **4**).

• Neurotransmitters diffuse across the synapse and bind to receptors on the postsynaptic cell, which can be a neuron or a muscle. Chemical synapses of this sort are unidirectional (see Figure 22-3).

• Synaptic vesicles fuse with the plasma membrane using cellular machinery that is standard for exocytosis, including SNAREs, syntaxin, and SNAP proteins. Synaptotagmin protein is the calcium sensor that detects the action potential–stimulated rise in calcium that leads to the fusion (see Figure 22-20).

• Following neurotransmitter release from the presynaptic cell, vesicles are re-formed by endocytosis and recycled (see Figure 22-19, step **6**).

• Dynamin, an endocytosis protein, is critical for the formation of new synaptic vesicles, probably specifically for the "pinching off" of inbound vesicles.

• Coordinated operation of four gated ion channels at the synapse of a motor neuron and a striated muscle cell leads to release of acetylcholine from the axon terminus, depolarization of the muscle membrane, generation of an action potential, and subsequent muscle contraction (see Figure 22-21).

• The nicotinic acetylcholine receptor, a ligand-gated cation channel, contains five subunits, each of which has a transmembrane α helix (M2) that lines the channel (see Figure 22-22).

- Neurotransmitter receptors fall into two classes: ligand-gated ion channels, which permit ion passage when open, and G protein–coupled receptors, which are linked to separate ion channels.

- A postsynaptic neuron generates an action potential only when the plasma membrane at the axon hillock is depolarized to the threshold potential by the summation of small depolarizations and hyperpolarizations caused by activation of multiple neuronal receptors (see Figure 22-23).

- Electrical synapses are direct, gap junction connections between neurons. Electrical synapses, unlike chemical synapses that employ neurotransmitters, are extremely fast in signal transmission and are bidirectional.

## 22.4 Sensing the Environment: Touch, Pain, Taste, and Smell

Our bodies are constantly receiving signals from our environment—light, sound, smells, tastes, mechanical stimulation, heat, and cold. In recent years dramatic progress has been made in understanding how our senses record impressions of the outside world, and how the brain processes that information. For example, in Chapter 15 we analyzed the functions of one of the two types of photoreceptors in the human retina, the *rods*, and learned how they serve as primary recipients of visual stimulation. Rods are stimulated by weak light, like moonlight, over a range of wavelengths, while the other photoreceptors, the *cones*, mediate color vision. These photoreceptors synapse on layer upon layer of interneurons that are innervated by different combinations of photoreceptor cells. These signals are processed and interpreted by the part of the brain called the *visual cortex*, where these nerve impulses are translated into an image of the world around us.

In this section we discuss the cellular and molecular mechanisms and specialized nerve cells underlying several of our other senses: touch and pain, taste, and smell. We see how two broad classes of receptors—ion channels and G protein–coupled receptors—function in these sensing processes. As with vision, multiple interneurons connect these sensory cells with the brain, where relayed signals are converted into perceptions of the environment. For the most part, we still do not fully understand how these neural subsystems are wired. However, in the case of smell, each sensory neuron expresses a single odorant receptor, and we shall see how multiple sensory neurons that express the same receptor activate the same brain center. The connections between odorant binding and perception by the brain are thus direct and fairly well understood.

### Mechanoreceptors Are Gated Cation Channels

Our skin, especially the skin of our fingers, is highly specialized for collecting sensory information. Our whole body, in fact, has numerous **mechanosensors** embedded in its various tissues. These sensors frequently make us aware of touch, the

positions and movements of our limbs or head (proprioception), pain, and temperature, though we often go through periods when we ignore the inputs. Mammals use different sets of receptor cells to report on touch, temperature, and pain.

Many mechanosensory receptors are $Na^+$ or $Na^+/Ca^{2+}$ channels that are gated, or opened, in response to specific stimuli; activation of such receptors causes an influx of $Na^+$ or both $Na^+$ and $Ca^{2+}$ ions, leading to membrane depolarization. Examples include the stretch and touch receptors that are activated by stretching of the cell membrane; these have been identified in a wide array of cells, ranging from vertebrate muscle and epithelial cells to yeast, plants, and even bacteria.

The cloning of genes encoding touch receptors began with the isolation of mutant strains of *Caenorhabditis elegans* that were insensitive to gentle body touching. Three of the genes in which mutations were isolated—*MEC4*, *MEC6*, and *MEC10*—encode three subunits of a $Na^+$ channel in the touch-receptor cells. Studies on worms with mutations in these genes showed that these channels are necessary for transduction of a gentle body touch; biophysical studies indicated that these channels likely open directly in response to mechanical stimulation (Figure 22-24). The touch-sensitive complexes contain several other proteins essential for touch sensitivity, including subunits of novel 15-protofilament microtubules in the cytosol and specific proteins in the extracellular matrix, but precisely how they affect channel function is not yet known. Similar kinds of channels are found in bacteria and lower eukaryotes; by opening in response to membrane stretching, these chan-

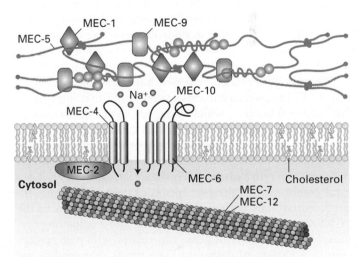

**FIGURE 22-24 A molecular model of the MEC-4 touch-receptor complex in C. *elegans*.** Mutations in any of the *MEC* genes can reduce or inactivate normal responses of the worm to a gentle body touch. The MEC-4 and MEC-10 proteins are the pore-forming subunits of the $Na^+$ channel; MEC-2 and MEC-6 are accessory subunits that enable channel activity. Mechanotransduction also requires a specialized extracellular matrix, consisting of MEC-5, a collagen isoform, and MEC-1 and MEC-9, both proteins with multiple EGF repeats. MEC-7 and MEC-12 are tubulin monomers that form novel 15-protofilament microtubules that are somehow also required for touch sensitivity. [After E. Lumpkin, K. Marshall, and A. Nelson, 2010, *J. Cell Biol.* **191**:237.]

nels may play a role in osmoregulation and the control of a constant cell volume.

A number of MEC-related molecules are expressed in mammalian neurons in the dorsal-root ganglia, and their roles in touch reception have been examined in knockout mice. For instance, disruption of the protein stomatin-like protein-3 (SLP3) causes defects in texture discrimination and loss of mechanosensitivity in a subset of mouse touch-receptor neurons.

## Pain Receptors Are Also Gated Cation Channels

Animals as diverse as snails and humans sense noxious events (the process termed nociception); pain receptors, called **nociceptors,** respond to mechanical change, heat, and certain toxic chemicals. Pain serves to alert us to events such as tissue damage that are capable of producing injury and evokes behaviors that promote tissue healing. Persistent pain in response to tissue injury is common, and many individuals suffer from chronic pain. Thus, understanding both acute and chronic pain is a major research goal, as is the development of new types of drugs to treat pain.

One of the first mammalian pain receptors to be cloned and identified was TRPV1, a $Na^+/Ca^{2+}$ channel that is found in many sensory pain neurons of the peripheral nervous system and is activated by a wide variety of exogenous and endogenous physical and chemical stimuli. The best-known activators of TRPV1 are heat greater than 43 °C, acidic pH, and capsaicin, the molecule that makes chili peppers seem hot. Activation of TRPV1 receptors leads to painful, burning sensations. Numerous TRPV1 antagonists have been developed by pharmaceutical companies as possible pain medications. However, a major side effect that has limited the utility of these drugs is that they result in an elevation in body temperature; this suggests that one "normal" function of TRPV1 is to sense and regulate body temperature, and that the drugs inhibit this function.

Another pain-related gene that has attracted wide attention is *SCN9A*, which encodes a subunit of the voltage-gated $Na^+$ channel Nav1.7, which is expressed at high density in many pain-sensitive neurons. It is not clear what Nav1.7 directly senses, but humans homozygous for nonsense mutations in *SCN9A* are totally unable to sense pain although they are otherwise normal. This suggests that Nav1.7 is a vital component in the perception of pain in humans. Indeed, individuals with certain activating mutations in the *SCN9A* gene have severe episodic pain, and a common polymorphism in the *SCN9A* gene is correlated with pain perception in a variety of diseases. Thus drugs that modify the function of the Nav1.7 channel would also be potentially useful in treating a wide array of painful conditions.

## Five Primary Tastes Are Sensed by Subsets of Cells in Each Taste Bud

We taste many chemicals, all of which are hydrophilic and nonvolatile molecules floating in saliva. All tastes are sensed on all areas of the tongue and selective cells respond preferentially to certain tastes. Like the other senses, that of taste likely evolved to increase an animal's chance of survival. Many toxic substances taste bitter or acidic, and nourishing foods are broken down into molecules that taste sweet (e.g., sugars), salty, or umami (e.g., the meaty or savory taste of monosodium glutamate and other amino acids). Animals (including humans) can never be certain exactly what enters their mouth; the sense of taste enables an animal to make a quick decision—eat it, or get rid of it. Taste is less demanding of the nervous system than olfaction, because fewer types of molecules are monitored. What is impressive is the sensitivity of taste; bitter molecules can be detected at concentrations as low as $10^{-12}$ M.

There are receptors for salty, sweet, sour, umami, and bitter tastes in all parts of the tongue. The receptors are of two different types: channel proteins for salty and sour tastes and seven-transmembrane-domain proteins (G protein–coupled receptors) for sweetness, umami, and bitterness. Specific membrane receptors that detect fatty acids are present on taste bud cells, and fatty taste may come to be recognized as a sixth basic taste quality.

Taste buds are located in bumps in the tongue called *papillae*; each bud has a pore through which fluid carries solutes inside. Each taste bud has about 50–100 taste cells (Figure 22-25a, b), which are epithelial cells but with some of the functions of neurons. Microvilli on the taste cells' apical tips bear the taste receptors, directly contacting the external environment in the oral cavity and thus experiencing wide fluctuations in food-derived molecules as well as the presence of potentially harmful compounds. Cells in the tongue and other parts of the mouth are subjected to a lot of wear and tear, and taste bud cells are continuously replaced by cell divisions in the underlying epithelium. (A taste bud cell in rats has a lifetime of 10 days.)

Reception of a taste signal causes cell depolarization that triggers action potentials; these in turn cause $Ca^{2+}$ uptake through voltage-dependent $Ca^{2+}$ channels and release of neurotransmitters (Figure 22-25c–e). Taste cells do not grow axons; instead, they signal over short distances to adjacent neurons. What is not yet understood is how the brain interprets the input from nerves downstream of all of the taste buds and tells us exactly what we are tasting.

**Bitter Taste** Bitter tastants are diverse and are detected by a diverse family of about 25–30 different G protein–coupled receptors (GPCRs) known as T2Rs. As depicted in Figure 22-25c, all of these GPCRs activate a particular $G_\alpha$ isoform, called gusducin, that is expressed only in taste cells. However, it is the released ubiquitous $G_{\beta\gamma}$ subunit of the heterotrimeric G protein that binds to and activates a specific isoform of phospholipase C, which in turn generates $IP_3$. $IP_3$ triggers $Ca^{2+}$ release from the endoplasmic reticulum (see Figure 15-36). $Ca^{2+}$ in turn binds to and opens a $Ca^{2+}$-gated $Na^+$ channel, TrpM5, leading to an influx of $Na^+$ and membrane depolarization. The combined action of elevated $Ca^{2+}$ and membrane depolarization opens the large pores of an unusual membrane channel termed Panx1, resulting in release of ATP and probably other signaling molecules into the extracellular space. ATP stimulates the nerve cells that will ultimately carry the taste information to the brain.

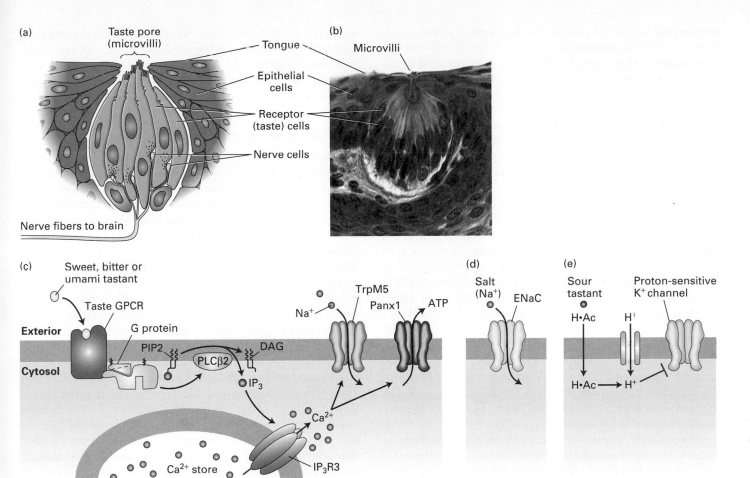

**FIGURE 22-25 The sense of taste.** Panels (a) and (b) show a mammalian taste bud and its receptors. (a) The pink cells are the taste cells. These epithelial receptor cells contact the nerve cells (yellow). The chemical signals arrive at the microvilli seen at the top. (b) Micrograph of a taste bud, showing the receptor cells. The microvilli are barely visible at the top of the taste bud, indicated by the arrows. Panels (c) through (e) show the mechanisms by which five taste qualities are recognized and transduced in taste cells. (c) Sweet, bitter, and umami ligands bind specific taste GPCRs expressed in Type II receptor cells, activating a phosphoinositide pathway that elevates cytosolic $Ca^{2+}$. $Ca^{2+}$ in turn binds to and opens a $Ca^{2+}$-gated $Na^+$ channel, TrpM5, leading to an influx of $Na^+$ and membrane depolarization. The combined action of elevated $Ca^{2+}$ and membrane depolarization opens the large pores of an unusual membrane channel termed Panx1, resulting in release of ATP and probably other signaling molecules into the extracellular space. ATP and probably these other molecules stimulate the nerve cells that will ultimately carry the information to the brain. (d) Salt is detected by direct permeation of $Na^+$ ions through membrane ion channels, including the ENaC channel, directly depolarizing the plasma membrane. (e) Organic acids like acetic acid diffuse in their protonated form (H·Ac) through the plasma membrane and dissociate into an anion and proton, acidifying the cytosol. Entry of strong acids like HCl is facilitated by a proton channel in the apical membrane of the sour-sensing cells that enables protons to reach the cytosol. Intracellular $H^+$ is believed to block a proton-sensitive $K^+$ channel (as yet unidentified) and thus depolarize the membrane. Voltage-gated $Ca^{2+}$ channels would open, leading to an elevation in cytosolic $Ca^{2+}$ that triggers exocytosis of synaptic vesicles that are not depicted. [Part (a) adapted from B. Kolb and I. Q. Whishaw, 2006, *An Introduction to Brain and Behavior*, 2d ed., Worth, p. 400; part (b) from Ed Reschke/Peter Arnold; parts (c) and (d) after N. Chaudhari and S. D. Roper, 2010, *J. Cell Biol.* **190**:285; part (e) after S. Frings, 2010, *PNAS* **107**:21955.]

Different bitter taste molecules are quite distinct in structure, which probably accounts for the need for the diverse family of T2Rs. Some T2R receptors bind only 2–4 bitter-tasting compounds, whereas others bind a wider variety of bitter compounds. The first member of the T2R family to be identified came from human genetics studies that showed an important bitterness-detection gene on chromosome 5. Mice that have five amino acid changes in the T2R protein T2R5 are unable to detect the bitter taste of cycloheximide (a protein synthesis inhibitor, see Table 9-1). Multiple T2R types

are often expressed in the same taste cell, and about 15 percent of all taste cells express T2Rs.

A dramatic gene regulation swap experiment was done to demonstrate the role of T2R proteins. Mice were engineered to express a bitter-taste receptor, a T2R protein, in cells that normally detect sweet tastants that attract mice. The mice developed a strong attraction for bitter tastes, evidently because the cells continued to send a "go and eat this" signal even though they were detecting bitter tastants. This experiment demonstrates that the specificity of taste cells is determined within the cells themselves, and that the signals they send are interpreted according to the neural connections made by that class of cells. This in turn implies a highly regulated system connecting the different classes of taste receptor cells to specific higher regions of the brain.

**Sweet and Umami Tastes** Sweet and umami tastants are detected by a GPCR family called the *T1Rs*, which are related to the T2Rs. The three mammalian T1Rs differ from one another in a small number of amino acids. The T1Rs have very large extracellular domains that comprise the taste-binding domain of the protein. In the taste-sensing glutamate receptor, the extracellular domain closes around glutamate in a way that is described as analogous to a Venus flytrap. Unlike most GPCRs, which generally function as monomers, T1Rs form homodimers and heterodimers, which is thought to increase the repertoire of molecules that can act as signals. However, the code of responses to different molecules is still under investigation. Mice lacking T1R2 or T1R3 fail to detect sugar; it is thought that the actual receptor is a heterodimer of the two. T1R3 appears to be a receptor for both sweet tastes and umami, and that is because it detects sweets when combined with T1R2 and umami when it combines with T1R1. Accordingly, taste cells express T1R1 or T1R2 but not both, as otherwise they would send an ambiguous message to the brain.

Interestingly, sweet-taste receptors are also found on the surface of certain endocrine cells in the gut; these cells also express gusducin and several other taste transduction proteins. The presence of glucose in the gut causes these cells to secrete the hormone glucagon-like peptide-1 (GLP-1), which in turn regulates appetite, and enhances insulin secretion and gut motility. Thus certain cells of the gut "taste" glucose through the same mechanisms used by taste cells of the tongue.

**Salty Taste** Salt is sensed by a member of a family of $Na^+$ channels called *ENaC channels* (Figure 22-25d). Indeed, knocking out a critical ENaC subunit in taste buds impaired salty-taste detection in mice. The influx of $Na^+$ through the channel depolarizes the taste cell, leading to neurotransmitter release. The role of ENaC channels as salt sensors is evolutionarily ancient; ENaC proteins also detect salt when expressed in insects. In *Drosophila*, taste sensors are located in multiple places including the legs, so when the fly steps on something tasty, the proboscis extends to explore it further.

**Sour Taste** Perception of sourness is due to the detection of $H^+$ ions. Many sour tastants are weak organic acids (e.g., acetic acid in vinegar), which in their protonated forms diffuse through the plasma membrane. They then dissociate into an anion and a proton, which acidifies the cytosol. Strong acids like HCl are detected by a proton channel in the apical membrane of the sour-sensing cells that enables protons to reach the cytosol. Regardless of how the intracellular $H^+$ concentration is increased, protons are believed to block

an as-yet-unidentified proton-sensitive $K^+$ channel and thus depolarize the membrane (Figure 22-25e). As with salt detection, voltage-gated $Ca^{2+}$ channels would then open, elevate cytosolic $Ca^{2+}$, and thus trigger exocytosis of neurotransmitter-filled synaptic vesicles.

## A Plethora of Receptors Detect Odors

The perception of volatile airborne chemicals imposes different demands than the perception of light, sound, touch, or taste. Light is sensed by only four rhodopsin molecules, tuned to different wavelengths. Sound is detected by mechanical effects through hairs that are tuned to different wavelengths. Touch and pain requires a small number of different gated ion channels. The sense of taste measures a small number of substances dissolved in water. In contrast to all these other senses, olfactory systems can discriminate between many hundreds of volatile molecules moving through air. Discrimination between a large number of chemicals is useful in finding food or a mate, sensing pheromones, and avoiding predators, toxins, and fires. *Olfactory receptors* work with enormous sensitivity. Male moths, for example, can detect single molecules of the signals sent drifting through the air by females. In order to cope with so many signals, the olfactory system employs a large family of olfactory receptor proteins. Humans have about 700 olfactory receptor genes, of which about half are functional (the rest are unproductive pseudogenes), a remarkably large proportion of the estimated 20,000 human genes. Mice are more efficient, with more than 1200 olfactory receptor genes, of which about 800 are functional. That means 3 percent of the mouse genome is composed of olfactory receptor genes. *Drosophila* has about 60 olfactory receptor genes. In this section we will examine how olfactory receptor genes are employed, and how the brain can recognize which odor has been sensed—the initial stages of interpretation of our chemical world. Odor molecules are called *odorants*. They have diverse chemical structures, so olfactory receptors face some of the same challenges faced by antibodies and hormone receptors—the need to bind and distinguish many variants of relatively small molecules.

Olfactory receptors are seven-transmembrane-domain proteins (Figure 22-26). In mammals, olfactory receptors are produced by cells of the nasal epithelium. These cells, called *olfactory receptor neurons (ORNs)*, transduce the chemical signal into action potentials. Each ORN extends a single dendrite to the luminal surface of the epithelium, from which immotile cilia extend to bind inhaled odorants from the air (Figure 22-27a). These olfactory sensory cilia are enriched in the odorant receptors and signal transduction proteins that mediate the initial transduction events. In *Drosophila*, ORNs have similar structures and are located in the antennae (Figure 22-27b).

In both mammals and *Drosophila* the ORNs project their axons to the next higher level of the nervous system, which in mammals is located in the olfactory bulb of the brain. The ORN axons synapse with dendrites from *mitral neurons* in mammals (called *projection neurons* in insects);

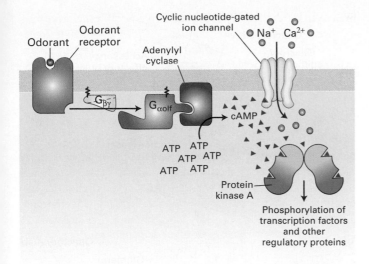

**FIGURE 22-26 Signal transduction from the olfactory GPCRs.**
Binding of an odorant to its cognate odorant receptor (OR) triggers activation of the trimeric G protein $G_{\alpha olf} \cdot G_{\beta\gamma}$, releasing the active $G_{\alpha olf} \cdot$ GTP. Activated $G_{\alpha olf} \cdot$ GTP in turn activates type III adenylyl cyclase (AC3), leading to the production of cyclic AMP (cAMP) from ATP. Molecules of cAMP bind to and open the cyclic nucleotide–gated (CNG) ion channel, leading to the influx of $Na^+$ and $Ca^{2+}$ and depolarizing the cell. cAMP also activates protein kinase A (PKA), which phosphorylates and thus regulates transcription factors and other intracellular proteins.

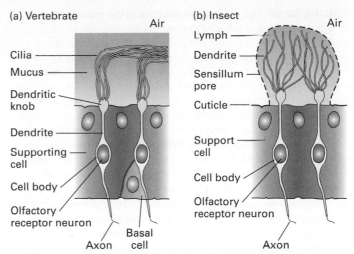

**FIGURE 22-27 Structures of olfactory receptor neurons.** Across a vast span of evolutionary distance—vertebrate and insect—olfactory receptor neurons have similar forms. (a) Vertebrate olfactory receptor neurons have one dendrite, which ends in a dendritic knob; from each dendritic knob, approximately 15 cilia extend into the nasal mucus. (b) Insect olfactory receptor neurons are morphologically similar: the bipolar neuron gives rise to a single basal axon that projects to an olfactory glomerulus in the antennal lobe. At its apical side it has a single dendritic process, from which sensory cilia extend. [Adapted from U. B. Kaupp, 2010, *Nature Rev. Neurosci.* **11**:188–200.]

these synapses occur in the clusters of synaptic structures called *glomeruli*. The mitral neurons connect to higher olfactory centers in the brain (Figure 22-28).

Humans vary markedly in their ability to detect certain odors. Some cannot detect the steroid androstenone, a compound derived from testosterone and found in human sweat. Some describe the odor as pleasant and musky, while others compare it to the smell of dirty socks. These differences are all ascribed to inactivating missense mutations in the gene encoding the single androstenone GPCR. Individuals with two copies of the wild-type allele perceive androstenone as unpleasant, whereas those possessing one or no functional alleles perceive androstenone as less unpleasant or undetectable. ■

Despite the vast number of olfactory receptors, all generate the same intracellular signals through activation of the same trimeric G protein: $G_{\alpha olf} \cdot G_{\beta\gamma}$ (see Figure 22-26). $G_{\alpha olf}$ is expressed mainly in olfactory neurons. Like $G_{\alpha s}$, the active $G_{\alpha olf} \cdot$ GTP formed after ligand binding activates an adenylyl cyclase that leads to the production of cyclic AMP (cAMP; see Figure 15-27). Two downstream signaling pathways are activated by cAMP. It binds to a site on the cytosolic face of a cyclic nucleotide–gated (CNG) $Na^+/Ca^{2+}$ channel, opening the channel and leading to an influx of $Na^+$ and $Ca^{2+}$ and local depolarization of the cell membrane. This odorant-induced depolarization in the olfactory dendrites spreads throughout the neuronal membrane, resulting in opening of voltage-gated $Na^+$ channels in the axon hillock and the generation of action potentials. Molecules of cAMP also activate protein kinase A (PKA), which phosphorylates and thus regulates transcription factors and other intracellular proteins.

## Each Olfactory Receptor Neuron Expresses a Single Type of Odorant Receptor

The key to understanding the specificity of the olfactory system is that in both mammals and insects each ORN produces only a single type of odorant receptor. Any electrical signal from that cell will convey to the brain a simple message: "my odorant is binding to my receptors." Receptors are not always completely monospecific for odorants. Some receptors can bind more than one kind of molecule, but the molecules detected are usually closely related in structure. Conversely, some odorants bind to multiple receptors.

There are about 5 million ORNs in the mouse, so on average each of the 800 or so olfactory receptor genes is active in approximately 6000 cells. There are about 2000 glomeruli (roughly 2 for each odorant receptor gene), so on average the axons from a few thousand ORNs converge on each glomerulus (see Figure 22-28). From there about 25 mitral axons per glomerulus, or a total of 50,000 mitral neurons, connect to higher brain centers. Thus the initial odorant sensing information is carried directly to higher parts of the brain without processing, a simple report of what odorant has been detected.

**FIGURE 22-28 The anatomy of olfaction in the mouse.**
(a) Schematic representation of a sagittal section through an adult mouse head. Axons of the olfactory receptor neurons (ORNs) in the main olfactory epithelium bundle to form the olfactory nerve and innervate the olfactory bulb. Each ORN of the main olfactory epithelium expresses only one odorant receptor gene. The vomeronasal organ and the accessory olfactory bulb are involved in pheromone sensing. (b) All of the olfactory receptor neurons that express a single type of receptor send their axons to the same glomerulus. In this figure each color represents the neural connections for each distinct expressed receptor. The glomeruli are located in the olfactory bulb near the brain; in the glomeruli, the ORNs synapse with *mitral neurons;* each mitral neuron has its dendrites localized to a single glomerulus and its corresponding ORNs, thus carrying information about a particular odorant to higher centers of the brain. Each glomerulus thus receives innervation from sensory neurons expressing a single odorant receptor, providing the anatomical basis of the olfactory sensory map. [After T. Komiyama and I. Luo, 2005, *Curr. Opin. Neurobiol.* **16:**67–73 and S. Demaria and J. Ngai, 2010, *J. Cell Biol.* **191:**443.]

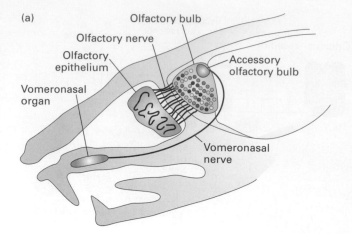

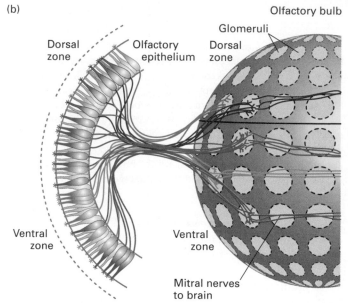

The one neuron–one receptor rule extends to *Drosophila*. Detailed studies have been done in larvae, where a simple olfactory system with only 21 ORNs uses about 10–20 olfactory receptor genes. It appears that a unique receptor is expressed in each ORN, which sends its projections to one glomerulus. ORNs can send either excitatory or inhibitory signals from their axon termini, probably in order to distinguish attractive versus repulsive odors. Similar to mammals, the axons from the ORNs end in the glomeruli, which in flies are located in the antennal lobe of the larval brain. The research in *Drosophila* began with tests of which odorants bind to which receptors (Figure 22-29a). Some odorants are detected by a single receptor, some by several, so the combinatorial pattern allows many more odorants to be distinguished than just the number of different olfactory receptors. The small total number of neurons has allowed a map to be constructed showing which odorants are detected by every glomerulus (Figure 22-29b). One striking finding was that glomeruli located near each other respond to odorants with related chemical structures, e.g., linear aliphatic compounds or aromatic compounds. The arrangement may reflect evolution of new receptors concomitant with a process of subdivision of the olfactory part of the brain.

The simple system of having each cell make only one receptor type also has some impressive difficulties to overcome: (1) Each receptor must be able to distinguish a type of odorant molecule or a set of molecules with specificity adequate to the needs of the organism. A receptor stimulated too frequently would probably not be very useful. (2) Each cell must express one and only one receptor gene product. All the other receptor genes must be turned off. At the same time, the collective efforts of all the cells in the nasal epithelium must allow the production of enough different receptors to give the animal adequate sensory versatility. It does little good to have genes for hundreds of receptors if most of them are never expressed, but it is a regulatory challenge to turn on one and only one gene in each cell and at the same time express all the receptor genes across the complete popu-

lation of cells. (3) The neuronal wiring of the olfactory system must make discrimination among odorants possible so that the brain can determine which odorants are present. Otherwise the animal might be feeling at ease and relaxed when it should be running away as fast as possible.

The solution to the first problem is the great variability of the olfactory receptor proteins, both within and between species. The solution to the second problem, the expression of a single olfactory receptor gene per cell, has been explored using transgenic mice, but the mechanism is still not understood. When an engineered olfactory receptor gene is used to produce an olfactory receptor, other genes encoding receptor proteins are turned off transcriptionally, owing to some type of feedback regulation. If an engineered olfactory receptor gene is expressed that produces a reporter protein—not an olfactory receptor protein—then other genes can still be expressed. Thus the feedback system must involve detection of the presence of a functional olfactory receptor protein.

The third problem, how the system is wired so the brain can understand which odor has been detected, has been partly answered. First, ORNs that express the same receptor send their

(a)

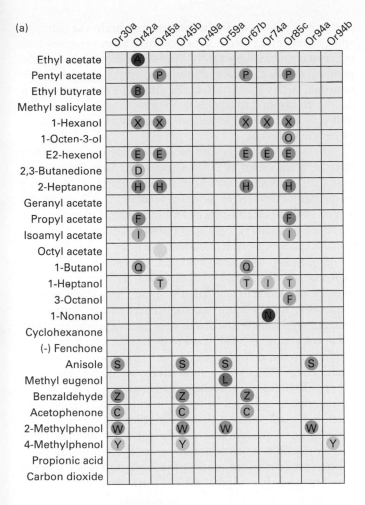

**EXPERIMENTAL FIGURE 22-29 Individual olfactory receptor types can be experimentally linked to various odorants and traced to specific glomeruli in the *Drosophila* larval olfactory system.** (a) The different olfactory receptor proteins are listed across the top, and the 27 odorants tested are shown down the left side. Colored dots indicate strong odor responses. Note that some odorants stimulate multiple receptors (e.g., pentyl acetate), while others (e.g., ethyl butyrate) act on only a single receptor. Note that many receptors, such as Or42a or Or67b, respond primarily to aliphatic compounds, whereas others, such as Or30a and Or59a, respond to aromatic compounds. (b) Spatial map of olfactory information in glomeruli of the *Drosophila* larval brain. The mapping was done by expressing a reporter gene under the control of each of the selected olfactory receptor neurons. The photograph indicates the glomeruli that receive projections from ORNs producing each of the ten indicated receptor protein types (Or42a, etc.). Also indicated are the odorants to which each receptor responds strongly. Note that with one exception (Or30a and Or45b) each glomerulus has unique sensory capacities. The exception might not be an exception if more olfactory gene expression patterns were tested. Glomeruli sensing odorants that are chemically similar tend to be situated next to one another. For example, the three glomeruli indicated by a blue solid line sense linear aliphatic compounds; those with yellow dashed lines, aromatic compounds. [Part (a) from S. A. Kreher, J. Y. Kwon, and J. R. Carlson, 2005, *Neuron* **46:**445–456. Part (b) courtesy of Jae Young Kwon, Scott Kreher, and John Carlson.]

guided to the same glomerulus destination. The full mechanism is not known, but it is clear that ORN axons respond both to their own olfactory receptor and to standard axon-guidance molecules used in other parts of the nervous system.

(b)

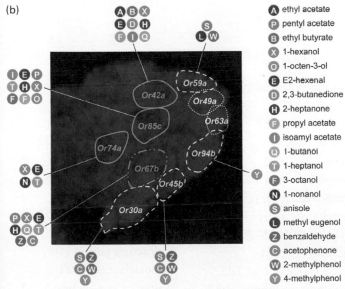

axons to the same glomerulus. Thus all cells that respond to the same odorant send processes to the same destination. In mice, a crucial clue about the patterning of the olfactory system came from the discovery that olfactory receptors play two roles in ORNs: odorant binding and, during development, axon guidance. Multiple ORN axons expressing the same receptor are

## KEY CONCEPTS of Section 23.4

### Sensing the Environment: Touch, Pain, Taste, and Smell

- Mechanoreceptors and pain receptors are gated $Na^+$ or $Na^+/Ca^{2+}$ channels.

- Touch sensitivity requires several cytoskeletal and extracellular matrix proteins as well as a gated $Na^+$ channel (see Figure 22-24).

- Five primary tastes are sensed by subsets of cells in each taste bud. Salty and sour tastes are detected by specific channel proteins, and G protein–coupled receptors detect sweetness, umami, and bitterness.

- In all cases, tastants lead to membrane depolarization and secretion of small molecules such as ATP that stimulate the adjacent neurons. Some tastant GPCRs are found in different homo- and heterodimeric combinations to detect different tastes (see Figure 22-25).

- Odorant receptors, which are seven-transmembrane G protein–coupled receptors, are encoded by a very large set of genes. Any one olfactory receptor neuron expresses one and only one olfactory receptor gene, so a signal from that cell to the brain unambiguously conveys the nature of the chemical sensed.

- ORNs that express the same receptor gene send their axons to the same glomerulus, and projection nerves (mitral neurons in mammals) carry odorant-specific information from the glomeruli to the brain (see Figures 22-27, 22-28, and 22-29).

## Perspectives for the Future

In this chapter we have provided an introduction to the remarkable properties of the nerve cells that serve as our interface with the world. The human body contains multiple types of neurons, each with its own shape, neurotransmitter, number of dendrites, length of axon, and numbers of connections with other neurons. How each of these types of cells develops in precisely the right place and makes appropriate synaptic connections with other neurons and appropriate contacts with surrounding glia remains largely a mystery. What, for instance, are the extracellular signals, transcriptional regulatory circuits, and induced or repressed proteins that tell a neuron to become myelinated or to generate a specific number of dendrites of a specific length? How does a neuron achieve its very long, polarized, branching structure? Why does one part of a neuron become a dendrite and another an axon? Why are certain key membrane proteins clustered at particular points—neurotransmitter receptors in postsynaptic densities in dendrites, $Ca^{2+}$ channels in axon termini, and $Na^+$ channels in myelinated neurons at the nodes of Ranvier? Such questions of cell shape and protein targeting also apply to other types of cells, but the morphological diversity of different types of neurons makes these particularly intriguing questions in the nervous system.

A detailed understanding of the structure and function of nerve cells will require knowledge of the three-dimensional structures of many different channels, neurotransmitter receptors, other membrane proteins, and cytoskeletal proteins. While the determination of the structure of the first voltage-gated $K^+$ channel has illuminated the mechanisms of channel gating that likely apply to other voltage-gated channels, we lack structures for any calcium channels. We do not know the structures for any of the "pain receptors" and thus we cannot rationally derive structures of putative antagonists that might be useful in pain control. We know the structures of only a few of the hundreds of G protein–coupled receptors used in the nervous system—none in the olfactory system—and thus the details of how these receptors differentiate between closely related ligands remain obscure.

From the standpoint of molecular cell biology, some of the greatest excitement has surrounded research into mechanisms of memory. In most cases memory does not depend on forming new neurons. Instead, existing neurons are modified; changes in the number and strength of synapses often underlie the establishment and endurance of memories. Current studies are directed at the molecular changes that alter synapses, both in the pre- and postsynaptic cells. For exam-ple, in many neurons in the mammalian brain the dendritic spines that emerge from dendrites and form synapses with other neurons are constantly forming and disappearing depending on the degree of stimulation of the nerve by other neurons. Such changes affect the ability of the postsynaptic neuron to respond to signals from presynaptic ones; the idea that it might relate to functional plasticity of synapses needs to be tested further.

The advances in the cell biology of the nervous system have been paralleled by extraordinary advances in exploring how neural circuitry carries out the interpretation of sensory information, analytical thought, feedback mechanisms for motor control, establishment and retrieval of memories, inheritance of instincts, regulated hormonal control, and emotional responses. Some experiments are done with noninvasive imaging technologies, observing thousands to millions of neurons and detecting global electrical activity. Others are done by observing *in vivo* a few cells at a time using inserted electrodes. This is being accomplished by improvements in imaging methods (invasive and noninvasive) combined with the development of better ways to manipulate the activities of single neurons, or of large numbers of neurons simultaneously. For instance, one can generate transgenic mice that express in specific sets of neurons an engineered $Na^+$ channel protein that is activated by light; a focused microscopic light beam will excite these cells but no others, and one can observe the consequences for the behavior of the animal. There is every reason to expect these advances to continue, an exciting prospect for understanding the brain and doing a better job of treating diseases that affect the nervous system.

## Key Terms

| | |
|---|---|
| action potential 1021 | neuromuscular junction 1037 |
| Agrin 1037 | neuron 1020 |
| astrocytes 1023 | neurotransmitters 1022 |
| axon 1021 | nociceptors 1048 |
| dendrites 1021 | node of Ranvier 1033 |
| depolarization 1021 | odorants 1050 |
| endocytosis 1037 | olfactory receptors 1050 |
| excitatory receptor 1045 | oligodendrocytes 1023 |
| glial cells 1023 | refractory period 1027 |
| glomeruli 1051 | repolarization 1021 |
| hyperpolarization 1027 | saltatory conduction 1033 |
| inhibitory receptor 1045 | Schwann cells 1023 |
| interneuron 1020 | sensory neuron 1020 |
| ligand-gated channel 1043 | synapse 1022 |
| motor neuron 1020 | synaptic vesicles 1022 |
| MuSK 1037 | voltage-gated channel 1025 |
| myelin sheath 1021 | |

## Review the Concepts

1. What is the role of glial cells in the brain and other parts of the nervous system?

2. The resting potential of a neuron is $-60$ mV inside compared with outside the cell. How is the resting potential maintained in animal cells?

3. Name the three phases of an action potential. Describe for each the underlying molecular basis and the ion involved. Why is the term *voltage-gated channel* applied to $Na^+$ channels involved in the generation of an action potential?

4. Explain how the crystal structures of potassium ion channels suggest the way in which the voltage-sensing domains interact with other parts of the proteins to open and close the ion channels. How does this structure-function relationship apply to other voltage-gated ion channels?

5. Explain why the strength of an action potential doesn't decrease as it travels down an axon.

6. Explain why the membrane potential does not continue to increase, but rather, plateaus and then decreases during the course of an action potential.

7. What does it mean to say that action potentials are "all or none"?

8. What prevents a nerve signal from traveling "backwards" toward the cell body?

9. Why is the cell unable to initiate another action potential if stimulated during the refractory period?

10. Myelination increases the velocity of action potential propagation along an axon. What is myelination? Myelination causes clustering of voltage-gated $Na^+$ channels and $Na^+/K^+$ pumps at nodes of Ranvier along the axon. Predict the consequences to action potential propagation of increasing the spacing between nodes of Ranvier by a factor of 10.

11. Describe the mechanism of action for addictive drugs such as cocaine.

12. Acetylcholine is a common neurotransmitter released at the synapse. Predict the consequences for muscle activation of decreased acetylcholine esterase activity at nerve-muscle synapses.

13. Describe the ion dynamics of the muscle-contraction process.

14. Following the arrival of an action potential in stimulated cells, synaptic vesicles rapidly fuse with the presynaptic membrane. This happens in less than 1 ms. What mechanisms allow this process to take place at such great speed?

15. Neurons, particularly those in the brain, receive multiple excitatory and inhibitory signals. What is the name of the extension of the neuron at which such signals are received? How does the neuron integrate these signals to determine whether or not to generate an action potential?

16. Explain the mechanism by which action potentials are prevented from being propagated to a postsynaptic cell if transmitted across an inhibitory synapse.

17. What is the role of dynamin in recycling synaptic vesicles? What evidence supports this?

18. Compare and contrast electrical and chemical synapses.

19. Compare the structures and functions of the receptor molecules for salty and sour taste; the taste-receptor molecules for sweetness, bitterness, and umami; and odor-receptor molecules.

## Analyze the Data

Olfaction occurs when volatile compounds bind to specific odorant receptors. In mammals, each olfactory receptor neuron in the olfactory nasal epithelium expresses a single type of odorant receptor. These odorant receptors constitute a large multigene family (>1000 members) of related proteins. Binding of odorant induces a signaling cascade that is mediated via a G protein, $G_{\alpha olf}$. Recent studies suggest that there are a small number of olfactory sensory neurons in the nasal epithelium that express members of the trace-amine–associated receptor (TAAR) family, chemoreceptors that are G protein–coupled receptors (GPCRs) but are unrelated to classical odorant receptors (see Liberles and Buck, 2006, *Nature* 442:645–650). The mouse genome encodes 15 TAAR genes while the human genome encodes 6.

a. In order to examine the expression pattern of different TAARs in the olfactory nasal epithelium, researchers localized TAAR RNA by in situ hybridization in pairwise combinations. All possible pairwise combinations of the 15 mouse TAARs were examined. A typical example of the results obtained is shown in the top set of panels in the figure below, in which TAAR6 and TAAR7 have been localized with fluorescent probes in mouse nasal epithelium. The TAAR6 probe was labeled with a green fluor, the TAAR7 probe with a red fluor. The lower set of panels shows localization of mouse odorant receptor 28 (MOR28; green), a classical odorant receptor, and TAAR6 (red). Each stained patch in the images is the staining pattern of an individual olfactory neuron. The "merge" panels show the two other images superimposed. What do these data suggest about expression patterns of the TAARs?

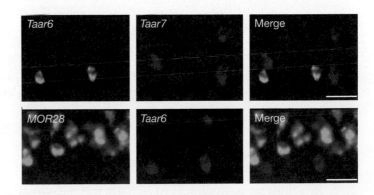

b. A number of cell lines that produce neither classical odorant receptors nor TAARs have each been transfected with the gene encoding a different TAAR. The cells have also been co-transfected with a gene encoding secreted alkaline phosphatase (SEAP) under control of a cAMP-responsive element. The cells were then exposed to various amines, as shown in the following figure, and SEAP activity in the medium was determined. The figure shows data for some representative TAARs (m = mouse, h = human). What do these data reveal about TAARs? What does the SEAP activity assay reveal about the signaling pathway utilized by chemoreception involving TAARs?

What do these data suggest may be a biological function for the TAAR5 neurons in mice? What additional studies would you conduct to support your hypothesis?

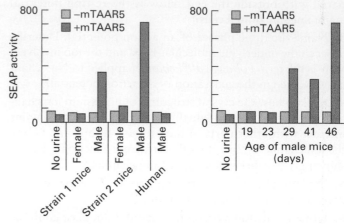

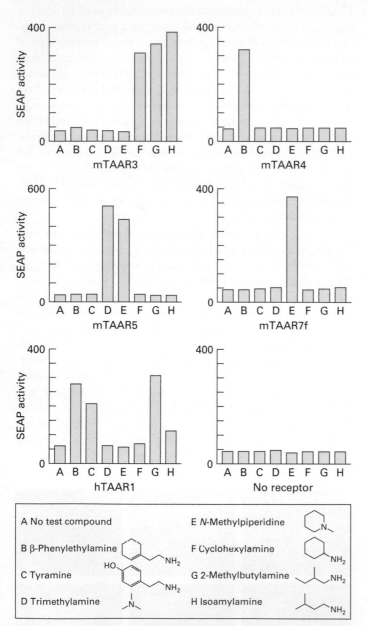

## References

### Neurons and Glia: Building Blocks of the Nervous System

Allen, N. J., and B. A. Barres. 2005. Signaling between glia and neurons: focus on synaptic plasticity. *Curr. Opin. Neurobiol.* **15**:542–548.

Bellen, H., C. Tong, and H. Tsuda. 2010. 100 years of *Drosophila* research and its impact on vertebrate neuroscience: a history lesson for the future. *Nature Rev. Neurosci.* **11**:514–522.

Freeman, M. 2010. Specification and morphogenesis of astrocytes. *Science* **330**:774–778.

Halassa, M., and P. Haydon. 2010. Integrated brain circuits: astrocytic networks modulate neuronal activity and behavior. *Ann. Rev. Physiol.* **72**:335–355.

Jan, Y. N., and L. Y. Jan. 2001. Dendrites. *Genes Dev.* **15**:2627–2641.

Jessen, K. R., and R. Mirsky. 2005. The origin and development of glial cells in peripheral nerves. *Nature Rev. Neurosci.* **6**(9):671–682.

Parker, R. J., and V. J. Auld. 2006. Roles of glia in the *Drosophila* nervous system. *Semin. Cell Dev. Biol.* **17**(1):66–77.

Sherman, D. L., and P. J. Brophy. 2005. Mechanisms of axon ensheathment and myelin growth. *Nature Rev. Neurosci.* **6**(9):683–690.

Stevens, B. 2003. Glia: much more than the neuron's side-kick. *Curr. Biol.* **13**:R469–R472.

### Voltage-Gated Ion Channels and the Propagation of Action Potentials

Catterall, W. A. 2010. Ion channel voltage sensors: structure, function, and pathophysiology. *Neuron* **67**:915–928.

Emery, B. 2010. Regulation of oligodendrocyte differentiation and myelination *Science* **330**:779–782.

Hartline, D. K., and D. R. Colman. 2007. Rapid conduction and the evolution of giant axons and myelinated fibers. *Curr. Biol.* **17**:R29–R35.

Hille, B. 2001. *Ion Channels of Excitable Membranes*, 3d ed. Sinauer Associates.

Jouhaux, E., and R. Mackinnon. 2005. Principles of selective ion transport in channels and pumps. *Science* **310**:1461–1465.

**c.** In a third set of studies, SEAP activity was measured in cells expressing mouse TAAR5 (mTAAR5) following exposure of the cells to diluted urine derived from two strains of mice or from humans, as indicated on the graphs in the next column. Mice reach puberty at about one month of age.

Long, S. B., E. B. Campbell, and R. MacKinnon. 2005. Crystal structure of a mammalian voltage-dependent *Shaker* family K⁺ channel. *Science* 309(5736):897–903.

Long, S. B., E. B. Campbell, and R. MacKinnon. 2005. Voltage sensor of Kv1.2: structural basis of electromechanical coupling. *Science* 309(5736):903–908.

Long, S. B., X. Tao, E. Campbell, and R. MacKinnon. 2007. Atomic structure of a voltage-dependent K⁺ channel in a lipid membrane-like environment. *Nature* 450:376–382.

Nave, K.-A. 2010. Myelination and support of axonal integrity by glia. *Nature* 468:244–252.

Neher, E. 1992. Ion channels for communication between and within cells. Nobel Lecture reprinted in *Neuron* 8:605–612 and *Science* 256:498–502.

Neher, E., and B. Sakmann. 1992. The patch clamp technique. *Sci. Am.* 266(3):28–35.

Payandeh, J., T. Scheuer, N. Zheng, and W. A. Catterall. 2011.The crystal structure of a voltage-gated sodium channel. *Nature* doi:10.1038/nature 10238.

## Communication at Synapses

Burden, S. J. 2011. Snapshot: neuromuscular junction. *Cell* 144:826–826 e1.

Haucke, V., E. Neher, and S. Sigrist. 2011. Protein scaffolds in the coupling of synaptic exocytosis and endocytosis. *Nature Rev. Neurosci.* 12:127–138.

Ikeda, S. R. 2001. Signal transduction. Calcium channels—link locally, act globally. *Science* 294:318–319.

Kummer, T. T., T. Misgeld, and J. R. Sanes. 2006. Assembly of the postsynaptic membrane at the neuromuscular junction: paradigm lost. *Curr. Opin. Neurobiol.* 16(1):74–82.

McCue, H., L. P. Haynes, and R. D. Burgoyne. 2010. The diversity of calcium sensor proteins in the regulation of neuronal function. *Cold Spring Harb. Perspect. Biol.* 2:a004085

Sakmann, B. 1992. Elementary steps in synaptic transmission revealed by currents through single ion channels. Nobel Lecture reprinted in *EMBO J.* 11:2002–2016 and *Science* 256:503–512.

Shen, K., and P. Scheiffele. 2010. Genetics and cell biology of building specific synaptic connectivity. *Ann. Rev. Neurosci.* 33:473–507.

Siksou, L., A. Triller, and S. Marty. 2011. Ultrastructural organization of presynaptic terminals. *Curr. Opin. Neurosci.* 21:261–268.

Südhof, T., and J. Rothman. 2009. Membrane fusion: grappling with SNARE and SM proteins. *Science* 323:474–477.

Unwin, N. 2005. Refined structure of the nicotinic acetylcholine receptor at 4Å resolution. *J. Mol. Biol.* 346:967–989.

Vrljic, M., et al. 2010. Molecular mechanism of the synaptotagmin–SNARE interaction in Ca²⁺-triggered vesicle fusion. *Nature Struct. Mol. Biol.* 17:325–331.

Wu, H., W. Xiong, and L. Mei. 2010. To build a synapse: signaling pathways in neuromuscular junction assembly. *Development* 137:1017–1033.

## Sensing the Environment: Touch, Pain, Taste, and Smell

Brochtrup, A., and T. Hummel. 2010. Olfactory map formation in the *Drosophila* brain: genetic specificity and neuronal variability. *Curr. Opin. Neurobiol.* 21:85–92.

Buck, L., and R. Axel. 1991. A novel multigene family may encode odorant receptors: a molecular basis for odor recognition. *Cell* 65:175–187.

Chaudhari, N., and S. D. Raper. 2010. The cell biology of taste. *J. Cell. Biol.* 190:285–296.

Delmas, P., J. Hao, and L. Rodat-Despoix. 2011. Molecular mechanisms of mechanotransduction in mammalian sensory neurons. *Nature Rev. Neurosci.* 12:139–153.

DeMaria, S., and J. Ngai. 2010. The cell biology of smell. *J. Cell. Biol.* 191:443–452.

Kaupp, U. B. 2010. Olfactory signaling in vertebrates and insects: differences and commonalities. *Nature Rev. Neurosci.* 11:188–200.

Lin, S. Y., and D. P. Corey. 2005. TRP channels in mechanosensation. *Curr. Opin. Neurobiol.* 15(3):350–357.

Lumpkin, E., K. Marshall, and A. Nelson. 2010. The cell biology of touch. *J. Cell. Biol.*191:237–248.

McKemy, D. D., W. M. Neuhausser, and D. Julius. 2002. Identification of a cold receptor reveals a general role for TRP channels in thermosensation. *Nature* 416:52–58.

Nelson, G., et al. 2001. Mammalian sweet taste receptors. *Cell* 106(3):381–390.

Zhang, X, and S. Firestein. 2002. The olfactory receptor gene superfamily of the mouse. *Nature Neurosci.* 5(2):124–133.

# Immunology

Dendritic cells in the skin have class II MHC molecules on their surface. Those shown here were engineered to express a class II MHC–GFP fusion protein, which fluoresces green. [Courtesy of M. Boes and H. L. Ploegh.]

Immunity is a state of protection against the harmful effects of exposure to pathogens. Host defense can take many different forms, and all successful pathogens have found ways to disarm the immune system or manipulate it to their own advantage. Host–pathogen interactions are therefore an evolutionary work in progress. This explains why we continue to be assaulted by pathogenic viruses, bacteria, and parasites. The prevalence of infectious diseases illustrates the imperfections of host defense. An immune system that could produce perfect sterilizing immunity would yield a world without pathogens, an outcome clearly at variance with life as we know it. Rather, the co-evolution of pathogens and their hosts allows pathogens, which have relatively short generation times, to continue to evolve sophisticated countermeasures, against which the host must respond by adjusting, if not improving, its defenses. Sophisticated defense comes at a price: an immune system capable of dealing with a massively diverse collection of rapidly evolving pathogens may mount an attack on the host organism's own cells and tissues, a phenomenon called *autoimmunity*.

In this chapter we deal mostly with the vertebrate immune system, with particular emphasis on those molecules, cell types, and pathways that uniquely distinguish the immune system from other types of cell and tissues. Two remarkable features that characterize the vertebrate immune system are its ability to distinguish between closely related substances (**specificity**) and its ability to recall previously experienced exposure to a foreign substance (**memory**). This is accomplished through the generation of a massively diverse set of antigen-specific receptors (**diversity**), trained on self-molecules and largely purged of self-reactive components (**tolerance**), although clearly none of these desirable goals is accomplished with absolute precision.

From a practical perspective, the powers of the immune system are exploited not only therapeutically: monoclonal antibodies are a multibillion-dollar market in the successful treatment of inflammatory conditions, autoimmunity, and cancer. The molecules that comprise the adaptive immune system—antibodies in particular—are indispensable tools for the cell biologist. Antibodies allow the visualization and isolation of the molecules they recognize, and can do so with pinpoint precision. Their ability to do so has been invaluable in the accurate description of the components that make up the cell and its organelles, and their localization, both in cells and in tissues.

## OUTLINE

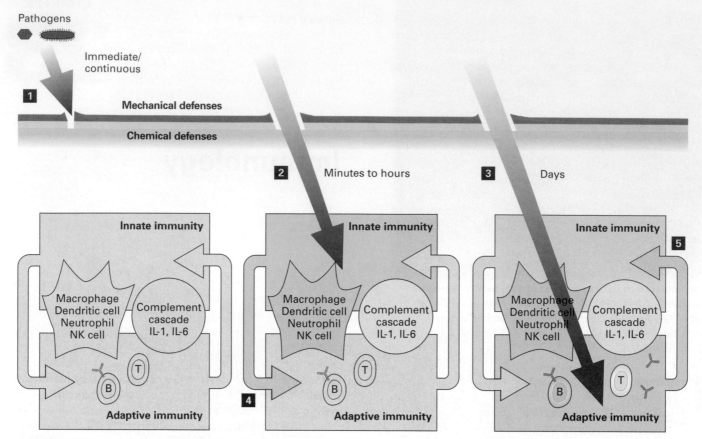

**FIGURE 23-1 The three layers of vertebrate immune defenses.**
*Left:* Mechanical defenses consist of epithelia and skin. Chemical defenses include the low pH of the gastric environment and antibacterial enzymes in tear fluid. These barriers provide continuous protection against invaders. Pathogens must physically breach these defenses (step **1**) to infect the host. *Middle:* Pathogens that have breached the mechanical and chemical defenses (step **2**) are handled by cells and molecules of the innate immune system (blue), which includes phagocytic cells (neutrophils, dendritic cells, macrophages), natural killer (NK) cells, complement proteins, and certain interleukins

(IL-1, IL-6). Innate defenses are activated within minutes to hours of infection. *Right:* Pathogens that are not cleared by the innate immune system are dealt with by the adaptive immune system (step **3**), in particular B and T lymphocytes. Full activation of adaptive immunity requires days. The products of an innate response may potentiate an ensuing adaptive response (step **4**). Likewise, the products of an adaptive immune response, including antibodies (Y-shaped icons), may enhance innate immunity (step **5**). Several cell types and secreted products straddle the fence between the innate and adaptive immune systems, and serve to connect these two layers of host defense.

Host defense comprises three layers: (1) mechanical/chemical defenses, (2) innate immunity, and (3) adaptive immunity (Figure 23-1). Mechanical and chemical defenses operate continuously. Innate immune responses, which involve cells and molecules present at all times, are rapidly activated (minutes to hours), but their ability to distinguish among many different pathogens is somewhat limited. In contrast, adaptive immune responses take several days to develop fully and are highly specific; that is, they can distinguish between closely related pathogens based on very small molecular differences in structure.

The manner in which **antigens**—any material that can evoke an immune response—are recognized and how these foreign materials are eliminated involves molecular and cell biological principles unique to the immune system. We begin this chapter with a brief sketch of the organization of the mammalian immune system, introducing the essential players of innate and adaptive immunity and describing inflammation, a local-

ized response to injury or infection that leads to the activation of immune-system cells and their recruitment to the affected site. In the next two sections, we discuss the structure and function of **antibody** (or immunoglobulin) molecules, which bind to specific molecular features on antigens, and how variability in antibody structure contributes to specific recognition of antigens. The enormous diversity of antigens that can be recognized by the immune system finds its explanation in unique rearrangements of the genetic material in B and T lymphocytes, commonly called **B cells** and **T cells,** which are the white blood cells that carry out antigen-specific recognition. These gene rearrangements not only control the specificity of antigen receptors on lymphocytes, they also determine cell-fate decisions in the course of lymphocyte development.

Although the mechanisms that give rise to antigen-specific receptors on B and T cells are very similar, the manner in which these receptors recognize antigens is very different.

The receptors on B cells can interact with intact antigens directly, but the receptors on T cells cannot. Instead, as described in Section 23.4, the receptors on T cells recognize processed (cleaved) forms of antigen, presented on the surface of target cells by glycoproteins encoded by the major histocompatibility complex (MHC). How MHC-encoded glycoproteins display these processed antigens is important for our understanding of how immune responses are initiated, not only to understand the physiology of the immune response but also for practical reasons. How can we best make antibodies that afford protection against an infectious agent? How can we raise antibodies as tools that recognize our favorite protein in a laboratory setting? Knowledge of antigen processing and presentation thus informs both **vaccine** design to protect against infectious disease and the generation of tools essential for research. MHC-encoded glycoproteins also help determine the developmental fate of T cells so that an organism's own cells and tissues (self antigens) normally do not evoke an immune response, whereas foreign antigens do. We conclude the chapter with an integrated view of the immune response to a pathogen, highlighting the collaboration between different immune-system cells that is required for an effective response.

## 23.1 Overview of Host Defenses

Because the immune system evolved to deal with pathogens, we begin our overview of host defenses by examining where typical pathogens are found and where they replicate. Then we introduce basic concepts of innate and adaptive immunity, including some of the key cellular and molecular players.

### Pathogens Enter the Body Through Different Routes and Replicate at Different Sites

Exposure to pathogens occurs via different routes. The skin itself has a surface area of ≈20 sq. ft.; the epithelial surfaces that line the airways, gastrointestinal tract, and genital tract present an even more formidable surface area of ≈4000 sq. ft. All these surfaces are continuously exposed to viruses and bacteria in the environment. Food-borne pathogens and sexually transmitted agents target the epithelia to which they are exposed. The sneeze of a flu-infected individual releases millions of virus particles in aerosolized form, ready for inhalation by the next person to be infected. Rupture of the skin, even if only by minor abrasions, or of the epithelial barriers that protect the underlying tissues, provides an easy route of entry for pathogens, which then gain access to a rich source of nutrients (for bacteria) and to the cells required for their replication (viruses).

Replication of viruses is confined strictly to the cytoplasm or nucleus of host cells, where protein synthesis and replication of the viral genetic material occur. Viruses spread to other cells either as free virus particles (virions) or by cell-to-cell spread. Many bacteria can replicate in the intercellular space, but some are specialized to invade host cells and survive there. Such intracellular bacteria reside either in membrane-delimited vesicles through which they enter cells by endocytosis or phagocytosis or in the cytoplasm if they escape from these vesicles. An effective host defense system, therefore, needs to be capable of eliminating not only cell-free viruses and free-living bacteria but also cells that harbor these pathogens.

Parasitic organisms can also cause disease. With increasingly complex life styles such as those of the protozoa that cause sleeping sickness (trypanosomes) or malaria (*Plasmodium* species), the pathogen's countermeasures also become increasingly complex. Bacteria, protozoa, and fungi—especially those that can cause disease in animals—often are called microbes.

### Leukocytes Circulate Throughout the Body and Take Up Residence in Tissues and Lymph Nodes

With the exception of erythrocytes, few cells in the course of their assigned function cover such distances as do the cells that provide immunity. The mammalian circulation serves as the necessary transport vehicle for erythrocytes, leukocytes, and platelets. Although erythrocytes never leave the circulation (their oxygen-carrying function does not require it), leukocytes (white blood cells) use the circulation exclusively for transport. The transport function of the circulation ensures delivery of lymphocytes from the sites where they are generated (bone marrow, thymus, fetal liver) to the sites where they can be activated (lymph nodes, spleen), and then to the site of infection, where they can eradicate invaders. Once lymphocytes arrive at a given location, they may leave and re-enter the circulation in the course of their tasks.

The immune system is an interconnected system of vessels, organs, and cells, divided into primary and secondary lymphoid structures (Figure 23-2). *Primary lymphoid organs*—the sites at which **lymphocytes** (the subset of leukocytes that includes B and T cells) are generated and acquire their functional properties—include the thymus, where T cells are generated, and the bone marrow, where B cells are generated. Adaptive immune responses, which require functionally competent lymphocytes, are initiated in *secondary lymphoid organs* including lymph nodes and the spleen. All of the lymphoid organs are populated by cells of hematopoietic origin (see Figure 21-18), generated in the fetal liver and throughout life in the bone marrow. The total number of lymphocytes in a young adult male is estimated to be $500 \times 10^9$, roughly 15 percent of which are found in the spleen, 40 percent in the other secondary lymphoid organs (tonsils, lymph nodes), 10 percent in the thymus, and 10 percent in the bone marrow; the remainder are circulating in the bloodstream.

Vertebrate blood vessels allow the escape of fluid from the circulation, driven by the positive arterial pressure exerted by the pumping heart. This fluid contains not only nutrients but also proteins that carry out defensive functions. To maintain homeostasis, the fluid that leaves the circulation must ultimately return and does so in the form of *lymph*, via lymphatic vessels. The total volume of lymph is up to three times the total blood volume. At their most distal ends, lymphatic vessels are open to collect the interstitial fluid that bathes the cells in tissues. The lymphatic vessels merge into larger collecting

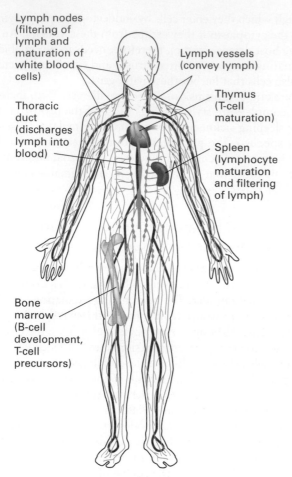

**FIGURE 23-2 The Circulatory and Lymphatic Systems.** Positive arterial pressure exerted by the pumping heart is responsible for loss of liquid from the circulation (red) into the interstitial spaces of the tissues, so that all cells of the body have access to nutrients and can dispose of waste. This interstitial fluid, whose volume is roughly three times that of all blood in the circulation, is returned to the circulation in the form of lymph, which passes through specialized anatomical structures called lymph nodes. The primary lymphoid organs, where lymphocytes are generated, are the bone marrow (B cells, T-cell precursors) and the thymus (T cells). The initiation of an immune response involves the secondary lymphoid organs (lymph nodes, spleen).

vessels, which deliver lymph to *lymph nodes*. A lymph node consists of a capsule, organized into areas that are defined by the cell types that inhabit them. Blood vessels entering a lymph node deliver B and T cells to it. The lymph that arrives in a lymph node carries cells that have encountered ("sampled") antigen, as well as soluble antigens, from the tissue drained by that particular afferent lymphatic vessel. In the lymph node, the cells and molecules required for the adaptive immune response interact, respond to the newly acquired antigenic information, and then execute the necessary effector functions to rid the body of the pathogen (Figure 23-3).

Lymph nodes can be thought of as filters in which antigenic information gathered from distal sites throughout the body is collected and displayed to the immune system in a form suitable to evoke an appropriate response. All the rele-

vant steps that lead to lymphocyte activation take place in lymphoid organs. Cells that have received proper instructions to become functionally active leave the lymph node via efferent lymphatic vessels that ultimately drain into the circulation. Such activated cells recirculate through the bloodstream and—now ready for action—may reach a location where they again leave the circulation, move into tissues, and seek out pathogenic invaders or destroy virus-infected cells.

The exit of lymphocytes and other leukocytes from the circulation, recruitment of these cells to sites of infection, processing of antigenic information, and return of immune-system cells to the circulation are all carefully regulated processes that involve specific cell-adhesion events, chemotactic cues, and the traversal of endothelial barriers, as we discuss later.

## Mechanical and Chemical Boundaries Form a First Layer of Defense Against Pathogens

As noted already, mechanical and chemical defenses form the first line of host defense against pathogens (see Figure 23-1). Mechanical defenses include the skin, epithelia, and arthropod exoskeleton, which are barriers that can be breached only by mechanical damage or through specific chemo-enzymatic attack. Chemical defenses include not only the low pH found in gastric secretions but also enzymes such as *lysozyme*, found in tear fluid, which can attack microbes directly.

The importance of mechanical defenses, which operate continuously, are immediately obvious in the case of burn victims. When the integrity of the epidermis and dermis is compromised, the rich source of nutrients in the underlying tissues is exposed, and airborne bacteria or otherwise harmless bacteria found on the skin can multiply unchecked, ultimately overwhelming the host. Viruses and bacteria have also evolved strategies to breach the integrity of these physical barriers. Enveloped viruses such as HIV, rabies virus, and influenza virus possess membrane proteins endowed with fusogenic properties. Following adhesion of a virion to the surface of the cell to be infected, direct fusion of the viral envelope with the host cell's membrane results in delivery of the viral genetic material into the host cytoplasm, where it is now available for transcription, translation, and replication (see Figures 4-47 and 4-49). Certain pathogenic bacteria (e.g., *S. aureus*) secrete collagenases that compromise the integrity of connective tissue and so facilitate entry of the bacteria.

## Innate Immunity Provides a Second Line of Defense After Mechanical and Chemical Barriers Are Crossed

The innate immune system is activated once the mechanical and chemical defenses have failed, and the presence of an invader is sensed (see Figure 23-1). The innate immune system comprises cells and molecules that are immediately available for responding to pathogens. **Phagocytes,** cells that ingest and destroy pathogens, are widespread throughout tissues and epithelia and can be recruited to sites of infection. Several soluble proteins present constitutively in the blood, or produced in response to infection or inflammation,

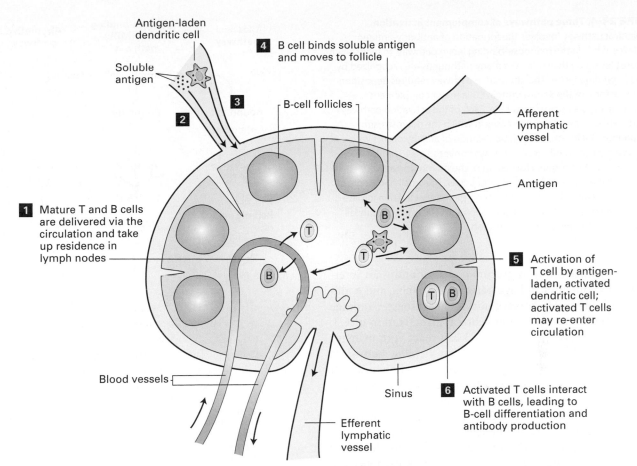

**Antigen-laden dendritic cell**

**Soluble antigen**

**2**

**3**

**4** B cell binds soluble antigen and moves to follicle

B-cell follicles

**Afferent lymphatic vessel**

**1** Mature T and B cells are delivered via the circulation and take up residence in lymph nodes

Antigen

**5** Activation of T cell by antigen-laden, activated dendritic cell; activated T cells may re-enter circulation

Blood vessels

Sinus

**6** Activated T cells interact with B cells, leading to B-cell differentiation and antibody production

Efferent lymphatic vessel

**FIGURE 23-3 Initiation of the adaptive immune response in lymph nodes.** Recognition of antigen by B and T cells (lymphocytes) located in lymph nodes initiates an adaptive immune response. Lymphocytes leave the circulation and take up residence in lymph nodes (step **1**). Lymph carries antigen in two forms—soluble antigen and antigen-laden dendritic cells; both are delivered to lymph nodes via afferent lymphatics (steps **2**, **3**). Soluble antigen is recognized by B cells (step **4**), and antigen-laden dendritic cells present antigen to T cells (step **5**). Productive interactions between T and B cells (step **6**) allow B cells to move into follicles and differentiate into plasma cells, which produce large amounts of secreted immunoglobulins (antibodies). Efferent lymphatic vessels return lymph from the lymph node to the circulation.

also contribute to innate defense. Animals that lack an adaptive immune system, such as insects, rely exclusively on innate defenses to combat infections.

**Phagocytes and Antigen-Presenting Cells** The innate immune system includes macrophages, neutrophils, and dendritic cells. All of these cells are phagocytic and come equipped with **Toll-like receptors (TLRs)**. Members of this family of cell-surface proteins detect broad patterns of pathogen-specific markers and thus are key sensors for detecting the presence of viral or bacterial invaders. Engagement of Toll-like receptors is important in eliciting effector molecules, including antimicrobial peptides. **Dendritic cells** and **macrophages** whose Toll-like receptors have detected pathogens also function as **antigen-presenting cells (APCs)** by displaying processed foreign materials to antigen-specific T cells. The structure and function of Toll-like receptors and their role in activating dendritic cells are described in detail in Section 23.6.

**Complement System** Another important component of the innate immune system is **complement,** a collection of serum proteins that can bind directly to microbial or fungal surfaces. This binding activates a proteolytic cascade that culminates in, among other things, formation of pore-forming proteins constituting the *membrane attack complex,* which is capable of permeabilizing the pathogen's protective membrane (Figure 23-4). The complement cascade is conceptually similar to the blood-clotting cascade, with amplification of the reaction at each successive stage of activation. At least three distinct pathways can activate complement. The *classical pathway* requires the presence of antibodies produced in the course of an adaptive response and bound to the surface of the microbe. How such antibodies are produced will be described below. Many microbial surfaces have physico-chemical properties, incompletely understood, that result in activation of complement via the *alternative pathway.* Finally, pathogens that contain mannose-rich cell walls activate complement through the *mannose-binding lectin pathway.* The bound lectin then triggers activation of two mannose-binding lectin-associated proteases, MASP-1 and MASP-2, which allow activation of the downstream components of the complement cascade. The three pathways converge at the activation of complement protein C3.

**FIGURE 23-4 Three pathways of complement activation.**
The classical pathway involves the formation of antibody-antigen complexes, while in the mannose-binding lectin pathway, mannose-rich structures found on the surface of many pathogens are recognized by mannose-binding lectin. The alternative pathway requires deposition of a special form of the serum protein C3, a major complement component, onto a microbial surface. Each of the activation pathways is organized as a cascade of proteases in which the downstream component is itself a protease. Amplification of activity occurs with each successive step. All three pathways converge on C3, which cleaves C5 and thus triggers formation of the membrane attack complex, leading to destruction of target cells. The small fragments of C3 and C5 generated in the course of complement activation initiate inflammation by attracting neutrophils, phagocytic cells that can kill bacteria at short range or upon ingestion.

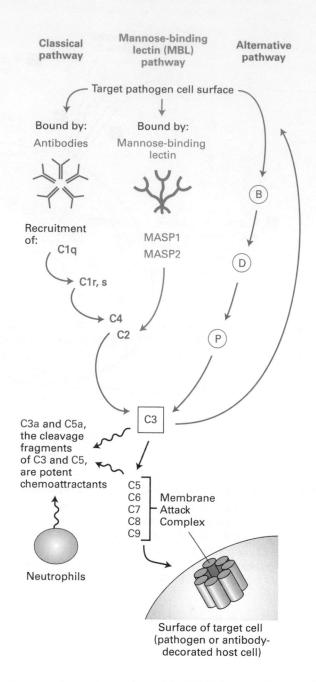

C3 is synthesized as a precursor that contains an internal, strained thioester linkage between a cysteine and a glutamate residue in close proximity. This thioester linkage becomes highly reactive upon proteolytic activation of C3. The activated thioester bond in C3 can react with primary amines or hydroxyls in close proximity, yielding a covalent bond linking C3 with a nearby protein or carbohydrate. If no such reactants are available, the thioester bond is simply hydrolyzed and renders the hydrolysis product inactive. This mode of action ensures that C3 will be covalently deposited only on antigen–antibody complexes in close proximity. Surfaces properly decorated with mannose-binding lectin or that received C3 deposits via the alternative pathway are similarly targeted. This limits the effects of complement to nearby surfaces, avoiding an inappropriate attack on cells that do not display the antigens targeted.

Regardless of the activation pathway, activated C3 unleashes the terminal components of the complement cascade, C5 through C9, culminating in formation of the membrane attack complex, which inserts itself into most biological membranes and renders them permeable. The resulting loss of electrolytes and small solutes leads to lysis and death of the target cell. Whenever complement is activated, the membrane attack complex is formed and results in death of the cell onto which this complex is deposited. The direct microbicidal effect of a fully activated complement cascade is an important protective function.

All three complement activation pathways also generate the C3a and C5a cleavage fragments, which bind to G protein–coupled receptors and function as chemoattractants for neutrophils and other cells involved in inflammation (see below). All three pathways also result in the covalent decoration of the structures targeted by complement activation with fragments of C3. Phagocytic cells make use of these C3-derived tags to recognize, ingest, and destroy the decorated particles, a process termed *opsonization*.

The complement cascade thus fulfills multiple roles in host defense: it can destroy the membranes that envelope a pathogen (bacteria, viruses); it covalently "paints" the targeted pathogen so that it may be more readily ingested by phagocytic cells capable of killing the pathogen and present-

ing its contents to cells that will initiate an adaptive immune response; and finally, the act of complement activation yields chemotactic signals to attract cells of the innate (neutrophils, macrophages, dendritic cells) and adaptive (lymphocytes) immune system to the site of infection.

**Natural Killer (NK) Cells** In addition to bacterial invaders, the innate immune system also defends against viruses. When the presence of a virus-infected cell is detected, yet other cell types of the innate immune system become active, seek out the virus-infected targets, and kill them. For instance, many virus-infected cells produce type I **interferons,** which are good at activating **natural killer (NK) cells.** Activated NK cells not only afford direct protection by eliminating the factory of new virus particles, they also secrete interferon γ

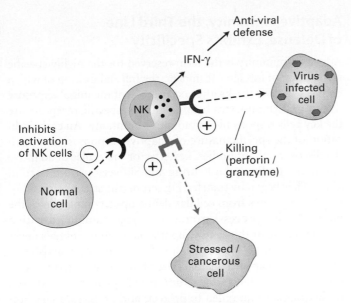

**FIGURE 23-5 Natural killer cells.** Natural killer (NK) cells are an important source of the cytokine interferon γ (IFN-γ) and can kill virus-infected and cancerous cells by means of perforins. These pore-forming proteins allow access of serine proteases called granzymes to the cytoplasm of the cell about to be killed. Granzymes also can initiate apoptosis through activation of caspases (Chapter 21). NK receptors identify infected or stressed cells and stimulate the NK cell to kill them. Other receptors identify normal cells and inhibit NK cell activation.

(IFN-γ), which is essential for orchestrating many aspects of antiviral defenses (Figure 23-5). Recognition by NK cells involves several classes of receptor, capable of delivering either stimulatory (promoting cell killing) or inhibitory signals. The interferons are classified as **cytokines,** small, secreted proteins that help regulate immune responses in a variety of ways. We will encounter other cytokines and discuss some of their receptors as the chapter progresses.

## Inflammation Is a Complex Response to Injury That Encompasses Both Innate and Adaptive Immunity

When a vascularized tissue is injured, the stereotypical response that follows is **inflammation.** Damage may be a simple paper cut or result from infection with a pathogen. Inflammation, or the inflammatory response, is characterized by four classical signs: *redness, swelling, heat,* and *pain.* These signs are caused by increased leakiness of blood vessels (vasodilation), the attraction of cells to the site of damage, and the production of soluble mediators responsible for the sensation of heat and pain. Inflammation has immediate protective value through the activation of the cell types and soluble products that together mount the innate immune response. Further, inflammation creates a local environment conducive to the initiation of the adaptive immune response. However, if it is not properly controlled, inflammation can also be a major cause of tissue damage.

Figure 23-6 depicts the key players in the inflammatory response to bacterial pathogens and the subsequent initiation

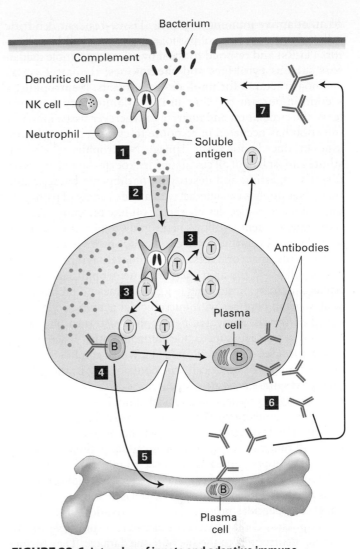

**FIGURE 23-6 Interplay of innate and adaptive immune responses against a bacterial pathogen.** Once a bacterium breaches the host's mechanical and chemical defenses, the bacterium is exposed to components of the complement cascade, as well as to cells that confer immediate protection, such as neutrophils (step **1**). Various inflammatory mediators induced by tissue damage contribute to a localized inflammatory response. Local destruction of the bacterium results in the release of bacterial antigens, which are delivered via the afferent lymphatics to the draining lymph node (step **2**). Dendritic cells acquire antigen at the site of infection, become migratory in response to microbial products, and move to the lymph node, where they activate T cells (step **3**). In the lymph node, antigen-stimulated T cells proliferate and acquire effector functions, including the ability to help B cells (step **4**), some of which may move to the bone marrow and complete their differentiation into plasma cells there (step **5**). In later stages of the immune response, activated T cells provide additional assistance to antigen-experienced B cells to yield plasma cells that secrete antigen-specific antibodies at a high rate (step **6**). Antibodies produced as a consequence of the initial exposure to bacteria act in synergy with complement to eliminate the infection (step **7**), should it persist, or afford rapid protection in the case of re-exposure to the same pathogen.

of an adaptive immune response. Tissue-resident dendritic cells sense the presence of pathogens via their Toll-like receptors (TLRs) and respond to them by releasing soluble mediators such as cytokines and **chemokines;** the latter act as chemoattractants for immune-system cells. **Neutrophils,** a second important cell type in the inflammatory response, leave the circulation and migrate to wherever tissue injury or infection has occurred in response to cytokines and chemokines produced upon tissue damage. Neutrophils, which constitute almost half of all circulating leukocytes, are phagocytic, directly ingesting and destroying pathogenic bacteria and fungi. Neutrophils can interact with a wide variety of pathogen-derived macromolecules via their Toll-like receptors. Activation of these receptors allows neutrophils to produce cytokines and chemokines; the latter can attract more leukocytes—neutrophils, macrophages, and ultimately lymphocytes (T and B cells)—to the area. Activated neutrophils can release bacteria-destroying enzymes (e.g., lysozyme and proteases) as well as small peptides with microbicidal activity, collectively called *defensins.* Activated neutrophils also turn on the enzymes that generate superoxide anion and other reactive oxygen species (see Chapter 12, p. 541), which can kill microbes at short range. Another cell type contributing to the inflammatory response is tissue-resident *mast cells.* When activated by a variety of physical or chemical stimuli, mast cells release histamine, a mediator that increases vascular permeability and thereby facilitates access to plasma proteins (e.g., complement) that can act against the invading pathogen.

A very important early response to infection or injury is activation of a variety of plasma proteases, including the proteins of the complement cascade discussed above (see Figure 23-4). The peptides produced during activation of these proteases possess chemoattractant activity, responsible for attracting neutrophils to the site of tissue damage. They further induce production of proinflammatory cytokines such as interleukin 1 and 6 (IL-1 and IL-6). The recruitment of neutrophils also depends on an increase in vascular permeability, controlled in part by lipid mediators (e.g., prostaglandins and leukotrienes) that are derived from phospholipids and fatty acids. All of these events occur rapidly, starting within minutes of injury. A failure to resolve the cause of this immediate response may result in chronic inflammation, in which cells of the adaptive immune system play an important role.

When the pathogen burden at the site of tissue damage is high, it may exceed the capacity of innate defense mechanisms to deal with them. Moreover, some pathogens have acquired, in the course of evolution, tools to disable or bypass innate immune defenses. In such situations, the adaptive immune response is required to control the infection. This adaptive response depends on specialized cells that straddle the interface between adaptive and innate immunity, including antigen-presenting cells such as macrophages and dendritic cells, which are capable of acquiring intact pathogens and of killing them upon ingestion. These antigen-presenting cells, in particular, dendritic cells, can initiate an adaptive immune response by delivering newly acquired pathogen-derived antigens to secondary lymphoid organs (see Figure 23-6).

## Adaptive Immunity, the Third Line of Defense, Exhibits Specificity

Adaptive immunity is the term reserved for the highly specific recognition of foreign substances, the full elaboration of which requires days or weeks after occurrence of the initial exposure to evolve. Lymphocytes bearing antigen-specific receptors are the key cells responsible for adaptive immunity. An early indication of the specific nature of adaptive responses came with the discovery of **antibodies,** key effector molecules of adaptive immunity, by Emil von Behring and Shibasaburo Kitasato in 1905. They began by transferring serum (the straw-colored liquid that separates from cellular debris upon completion of the blood-clotting process) from guinea pigs immunized with a sublethal dose of the deadly diphtheria toxin to animals never before exposed to the bacterium. The recipient animals were thus protected against a lethal dose of the same bacterium (Figure 23-7, *left*). Transfer of serum from animals never exposed to diphtheria toxin failed to protect, and protection was limited to the microbe that produced the diphtheria toxin. This experiment demonstrates *specificity*—that is, the ability to distinguish between two related substances, toxins, of the same class. Such specificity is a hallmark of the adaptive immune system. Even proteins that differ by a single amino acid may be distinguished by immunological means.

From these experiments, von Behring inferred the existence of corpuscles ("Antikörper"), or antibodies, as the transferable factor responsible for protection. The antibody-containing (immune) sera not only afforded protection in vivo, they also killed microbes in the test tube (Figure 23-7, *right*). Heating the immune sera to 56 °C destroyed this killing activity, but it was restored by the addition of unheated fresh serum from naive animals (i.e., animals never exposed to the microbe). This finding suggested that a second factor, now called complement, acts in synergy with antibodies to kill bacteria. We now know that von Behring's antibodies are serum proteins referred to as **immunoglobulins** and that complement is actually the series of proteases described above, which carry out pathogen destruction (see Figure 23-4). Immunoglobulins can neutralize not only bacterial toxins but also harmful agents such as viruses, by binding directly to them in a manner that prevents the virus from attaching itself to host cells. In the same vein, antibodies raised against snake venoms can be administered to the victims of snake bites to protect them from intoxication: the anti-snake venom antibodies bind to the venom, keep it from binding to its targets in the host, and in so doing neutralize it. This is called passive immunization, a procedure that can save lives by instant neutralization of a noxious substance such as a toxin. Antibodies can thus have immediate protective effects. Where today's medical advances allow the survival of individuals whose immune system is severely compromised (cancer patients receiving chemotherapy and/or radiation, transplant patients with a pharmacologically suppressed immune system, patients who suffer from HIV/AIDS, individuals with inborn deficiencies of their immune system), passive immunization can be of immediate practical importance.

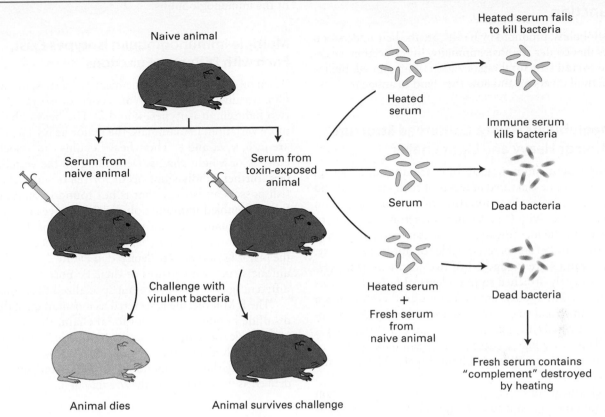

**EXPERIMENTAL FIGURE 23-7 The existence of antibody in serum from infected animals was demonstrated by von Behring and Kitasato.** Exposure of animals to a sublethal dose of diphtheria toxin (or the bacteria that produce it) elicits in their serum a substance that protects against a subsequent challenge with a lethal dose of the toxin (or the bacteria that produce it). The protective effect of this serum substance can be transferred from an animal that has been exposed to the pathogen to a naive (nonexposed) animal. When the serum recipient is subsequently exposed to a lethal dose of the bacteria, the animal survives. This effect is specific for the pathogen used to elicit the response. Serum thus contains a transferable substance (antibody) that protects against the harmful effects of a virulent pathogen. Serum harvested from these animals, said to be immune, displays bactericidal activity in vitro. Heating of immune serum destroys its bactericidal activity. Addition of fresh nonheated serum from a naive animal restores the bactericidal activity of heated immune serum. Serum thus contains another substance that complements the activity of antibodies.

## KEY CONCEPTS of Section 23.1

### Overview of Host Defenses

- Mechanical and chemical defenses provide protection against most pathogens. This protection is immediate and continuous, yet possesses little specificity. Innate and adaptive immunity provide defenses against pathogens that breach the body's mechanical/chemical boundaries (see Figure 23-1).

- The circulatory and lymphatic systems distribute the molecular and cellular players in innate and adaptive immunity throughout the body (see Figure 23-2).

- Innate immunity is mediated by the complement system (see Figure 23-4) and several types of leukocytes, the most important of which are neutrophils and other phagocytic cells such as macrophages and dendritic cells. The cells and molecules of innate immunity are deployed rapidly (minutes to hours). Molecular patterns diagnostic of the presence of pathogens can be recognized by Toll-like receptors, but the specificity of recognition is modest.

- Adaptive immunity is mediated by T and B lymphocytes. These cells require days for full activation and deployment, but they can distinguish between closely related antigens. This specificity of antigen recognition is the key distinguishing feature of adaptive immunity.

- Innate and adaptive immunity act in a mutually synergistic fashion. Inflammation, an early response to tissue injury or infection, involves a series of events that combines elements of innate and adaptive immunity (see Figure 23-6).

## 23.2 Immunoglobulins: Structure and Function

Immunoglobulins, produced by B cells, are the best-understood molecules that confer adaptive immunity. In this section we describe the overall structural organization of immunoglobulins, their structural diversity, and how they bind to antigens.

### Immunoglobulins Have a Conserved Structure Consisting of Heavy and Light Chains

Like complement, immunoglobulins are abundant serum proteins that can be classified in terms of their structural and functional properties. Fractionation of antisera, based on their functional activity (e.g., killing of microbes, binding of antigen), led to the identification of the immunoglobulins as the class of serum proteins responsible for antibody activity. Immunoglobulins are composed of two identical *heavy (H) chains,* covalently attached to two identical *light (L) chains* (Figure 23-8). The typical immunoglobulin therefore has a twofold-symmetrical structure, described as $H_2L_2$. An exception to this basic $H_2L_2$ architecture occurs in the camelids (camels, llamas, vicunas). These animals can make some immunoglobulins that are heavy-chain dimers ($H_2$) and lack light chains.

A biochemical approach was used to answer the question of how antibodies manage to distinguish between related antigens. Proteolytic enzymes were used to fragment immunoglobulins, which are rather large proteins, to identify the regions directly involved in antigen binding (see Figure 23-8). The protease papain yields monovalent fragments, called *F(ab),* that can bind a single antigen molecule, whereas the protease pepsin yields bivalent fragments, referred to as $F(ab')_2$ (F = fragment; ab = antibody). These enzymes are commonly used to convert intact immunoglobulin molecules into monovalent or bivalent reagents. Although F(ab) fragments are incapable of cross-linking antigen, $F(ab')_2$ fragments can do so, a property frequently used to cross-link and so activate surface receptors; many receptors such as the EGF receptor dimerize upon engagement of ligand, a prerequisite for full activation of downstream signaling cascades. The portion released upon papain digestion and incapable of antigen binding is called *Fc,* because of its ease of crystallization (F = fragment; c = crystallizable). This biochemical approach using proteases was followed by peptide mapping

and sequencing strategies to determine the primary structure of the immunoglobulins.

### Multiple Immunoglobulin Isotypes Exist, Each with Different Functions

Based on their distinct biochemical properties, immunoglobulins are divided into different classes, or *isotypes*. There are two light-chain isotypes, κ and λ. The heavy chains show more variation: in mammals, the major heavy-chain isotypes are μ, δ, γ, α, and ε. These heavy chains can associate with either κ or λ light chains. Depending on the vertebrate species, further subdivisions occur for the α and γ chains, and fish possess an isotype that is not found in mammals. The fully assembled immunoglobulin (Ig) derives its name from the heavy chain: μ chains yield IgM; α chains, IgA; γ chains, IgG; δ chains, IgD; and ε chains, IgE. The general structures of the major Ig isotypes are depicted in Figure 23-9. By means of unique structural features of their Fc portions, each of the different Ig isotypes carries out specialized functions.

The IgM molecule is secreted as a pentamer, stabilized by disulfide bonds and an additional chain, the J chain. In its pentameric form, IgM possesses 10 identical antigen-binding sites, which allow high-avidity interactions with surfaces that display the corresponding (cognate) antigen. Avidity is defined as the sum total of the *strength* of interactions (affinity)

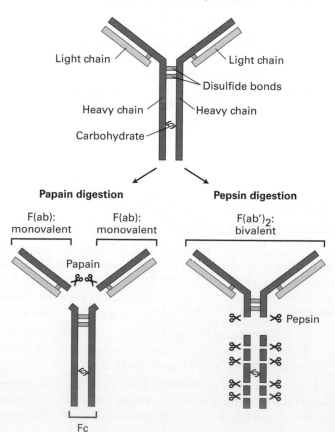

**Structure of an antibody molecule**

Light chain · Light chain · Disulfide bonds · Heavy chain · Heavy chain · Carbohydrate

**Papain digestion** · **Pepsin digestion**

F(ab): monovalent · F(ab): monovalent · $F(ab')_2$: bivalent

Papain

Pepsin

Fc

**FIGURE 23-8 The basic structure of an immunoglobulin molecule.** Antibodies are serum proteins also known as immunoglobulins. They are twofold-symmetrical structures composed of two identical heavy chains and two identical light chains. Fragmentation of antibodies with proteases yields fragments that retain antigen-binding capacity. The protease papain yields monovalent F(ab) fragments, and the protease pepsin yields bivalent $F(ab')_2$ fragments. The Fc fragment is unable to bind antigen, but this portion of the intact molecule has other functional properties.

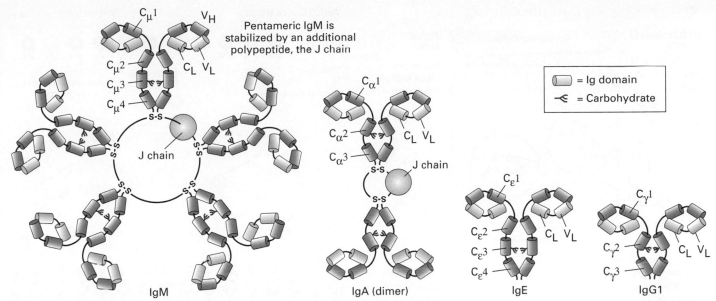

Pentameric IgM is stabilized by an additional polypeptide, the J chain

$C_\mu 1$  $V_H$
$C_\mu 2$
$C_\mu 3$  $C_L$  $V_L$
$C_\mu 4$
S-S
J chain

IgM

$C_\alpha 1$
$C_\alpha 2$  $C_L$  $V_L$
$C_\alpha 3$
S-S  J chain
S-S

IgA (dimer)

$C_\varepsilon 1$
$C_\varepsilon 2$  $C_L$  $V_L$
$C_\varepsilon 3$
$C_\varepsilon 4$

IgE

$C_\gamma 1$
$C_\gamma 2$  $C_L$  $V_L$
$C_\gamma 3$

IgG1

**FIGURE 23-9 Immunoglobulin isotypes.** The different classes of immunoglobulins, called isotypes, may be distinguished biochemically and by immunological techniques. In mouse and humans there are two light-chain isotypes ($\kappa$ and $\lambda$) and five heavy-chain isotypes ($\mu$, $\delta$, $\gamma$, $\varepsilon$, $\alpha$). Based on the identity of the heavy chain, each isotype defines a class of immunoglobulin. IgG, IgE, and IgD (not shown) are monomers with generally similar overall structures. IgM and IgA are unusual because they can occur In serum as pentamers and dimers, respectively, accompanied by an accessory subunit, the J chain, in covalent disulfide linkage. The volume-rendered depiction of the immunoglobulins highlights their modular design, with each barrel representing an individual Ig domain. Different isotypes have different functions. See Figure 23-12 for definitions of abbreviations.

of the available individual binding sites and the *number* of such binding sites. Upon deposition of IgM onto a surface that carries the antigen, the pentameric IgM molecule assumes a conformation that is highly conducive to activation of the complement cascade, an effective means of damaging the membrane onto which IgM is adsorbed and onto which complement proteins are deposited as a consequence.

The IgA molecule also interacts with the J chain, forming a dimeric structure. Dimeric IgA can bind to the polymeric IgA receptor on the basolateral side of epithelial cells, where its engagement results in receptor-mediated endocytosis. Subsequently, the IgA receptor is cleaved and dimeric IgA with the proteolytic receptor fragment (secretory piece) still attached is released from the apical side of the epithelial cell. This process, called **transcytosis,** is an effective means of delivering immunoglobulins from the basolateral side of an epithelium to the apical side (Figure 23-10a). Tear fluid and other secretions are rich in IgA and so provide protection against environmental pathogens.

The IgG isotype is important for neutralization of virus particles This isotype also helps prepare particulate antigens for acquisition by cells equipped with receptors specific for the Fc portion of IgG molecules (see below).

The immune system of the newborn is immature, and mammals transfer protective antibodies from the mother to the fetus via the mother's milk. The receptor responsible for capturing maternal IgG is the neonatal Fc receptor (FcRn), which is present on intestinal epithelial cells in rodents. By transcytosis, IgG captured on the luminal side of the newborn's intestinal tract is delivered across the gut epithelium and so makes maternal antibodies in the milk available for passive protection of the infant rodent (Figure 23-10b). In humans, FcRn is found on fetal cells that contact the maternal circulation in the placenta. Transcytosis of IgG antibodies from the maternal circulation across the placenta delivers maternal antibodies to the fetus. These maternal antibodies will protect the newborn until its own immune system is sufficiently mature to produce antibodies under its own steam. In adults, FcRn is also expressed on endothelial cells and helps control the turnover of IgG in the circulation.

As we will see in Section 23.3, the IgM and IgD isotypes are expressed as membrane bound B-cell receptors on newly generated B cells. Here the $\mu$ chains have an important role in B-cell development and activation.

## Each B Cell Produces a Unique, Clonally Distributed Immunoglobulin

The *clonal selection theory* stipulates that each lymphocyte carries an antigen-binding receptor of unique specificity. When a lymphocyte encounters the antigen for which it is specific, clonal expansion (rapid cell division) occurs and so allows an amplification of the response, culminating in clearing of the antigen (Figure 23-11). In a typical immune response, the antigen that elicits the response is of complex composition: even the simplest virus contains several distinct proteins. Many individual lymphocytes respond to a given antigen and expand in response to it, each producing its own antigen receptor of

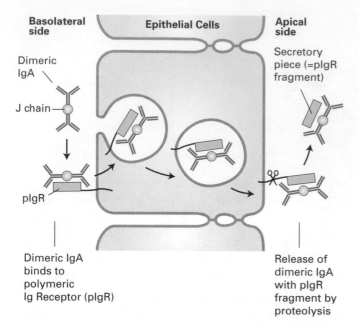

(a)

**Basolateral side** — **Epithelial Cells** — **Apical side**

Dimeric IgA

J chain

pIgR

Dimeric IgA binds to polymeric Ig Receptor (pIgR)

Secretory piece (=pIgR fragment)

Release of dimeric IgA with pIgR fragment by proteolysis

(b)

**Circulation of the neonate** — **Epithelial Cells** — **Milk in the lumen of the intestine of the neonate**

IgG

FcRn

**FIGURE 23-10 Transcytosis of IgA and IgG.** (a) IgA, found in secretions of the different mucosae, requires transport across the epithelium. IgA binds to the polymeric IgA receptor and is endocytosed. After being transported across the epithelial monolayer, a portion of the receptor is cleaved, and the IgA is released at the apical side together with a portion of the receptor, the secretory piece. (b) Suckling rodents acquire Ig from mother's milk. The newborn possesses at the apical surface of its intestinal epithelium the neonatal Fc receptor (FcRn), whose structure resembles that of class I MHC molecules (see Figure 23-21). After this receptor binds to the Fc portion of IgG, transcytosis moves the acquired IgG to the basolateral side of the epithelium. In humans, the syncytial trophoblast in the placenta expresses FcRn and so mediates acquisition of IgG from the maternal circulation and delivery to the fetus (transplacental transport).

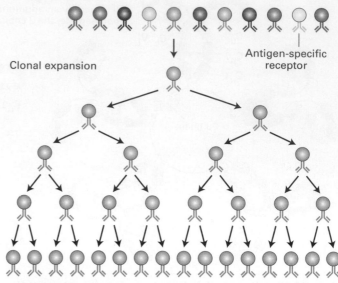

Activation of B cell

Clonal expansion

Antigen-specific receptor

**FIGURE 23-11 Clonal selection.** The clonal selection theory proposes the existence of a large set of lymphocytes, each equipped with its own unique antigen-specific receptor (indicated by different colors). The antigen that fits with the receptor carried by a particular lymphocyte allows that lymphocyte to expand clonally. From a modest number of antigen-specific cells, a large number of cells of the desired specificity (and large amounts of their secreted outputs) may be generated.

unique structure and therefore with unique binding characteristics (affinity). Because each lymphocyte is endowed with a unique receptor and clonally expands in response to antigen, this response is called *polyclonal*.

B-cell tumors, which represent malignant clonal expansions of individual lymphocytes, enabled the first molecular analysis of the processes that underlie the generation of antibody diversity. A key observation was that tumors derived from lymphocytes may produce large quantities of secreted immunoglobulins. Some of the light chains of the immunoglobulins are secreted in the urine of tumor-bearing patients. These light chains, called *Bence-Jones proteins* after their discoverers, are readily purified and afforded the first target for a protein chemical analysis.

Two key observations emerged form this work: (1) no two tumors produced light chains of the identical biochemical properties, suggesting that they were all unique in sequence; and (2) the differences in amino acid sequence that distinguish one light chain from another are not randomly distributed but occur clustered in a domain referred to as the *variable region of the light chain*, or $V_L$. This domain comprises the N-terminal $\approx 110$ amino acids. The remainder of the sequence is identical for the different light chains (provided they derive from the identical isotype, $\kappa$ or $\lambda$) and is therefore referred to as the *constant region*, or $C_L$. From the serum of tumor-bearing individuals, immunoglobulins unique to that individual patient were subsequently purified.

(a)

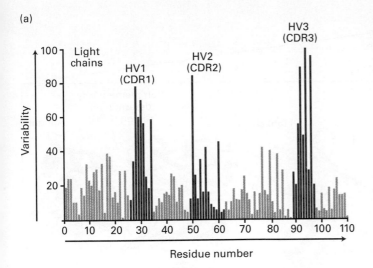

(b)

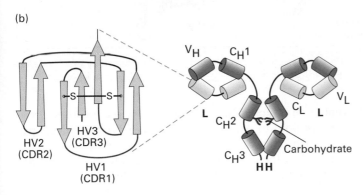

**FIGURE 23-12 Hypervariable regions and the immunoglobulin fold.** (a) Variation in amino acid variability with residue position in Ig light chains. The percentage of variable-region sequences with variant amino acids is plotted for each position in the sequence. Positions for which many different amino acid side chains are present in the data set are assigned a high variability index; those that are invariant among the sequences compared are assigned a value of 0. This analysis reveals three regions of increased variability, hypervariability (HV) regions 1, 2, and 3; these are also called complementarity-determining regions (CDRs). (b) Volume-rendered depiction of $F(ab')_2$ fragment (*right*) and ribbon diagram of a typical Ig light-chain variable domain ($V_L$) with the positions of the hypervariable regions indicated in red (*left*). The hypervariable regions are found in the loops that connect the β strands and make contact with antigen. The β strands (rendered as arrows) make up two β sheets and constitute the framework region. Note that each variable and constant domain has a characteristic three-dimensional structure, called the immunoglobulin fold. L = light chain; H = heavy chain; $V_H$ = heavy-chain variable domain; $V_L$ = light-chain variable domain; $C_H1$, $C_H2$, $C_H3$ = heavy-chain constant domains; $C_L$ = light-chain constant domain.

Sequencing of the heavy chains from these preparations revealed that the variable residues that distinguish one heavy chain from another were again concentrated in a well-demarcated domain, referred to as the *variable region of the heavy chain*, or $V_H$.

An alignment of sequences obtained from different homogeneous light-chain preparations showed a nonrandom pattern of regions of variability, revealing three *hypervariable regions*—HV1, HV2, and HV3—which are sandwiched between what are called framework regions (Figure 23-12a). (Similar alignments for the immunoglobulin heavy-chain sequences also yield hypervariable regions.) In the properly folded three-dimensional structure of immunoglobulins, these hypervariable regions are in close proximity (Figures 23-12b and 23-13) and make contact with antigen. Thus that portion of an Ig molecule containing the hypervariable regions constitutes the antigen-binding site. For this reason, hypervariable regions are also referred to as *complementarity-determining regions* (CDRs).

The difficulty of encoding in the germ line all of the information necessary to generate this enormously diverse antibody repertoire led to suggestions of unique genetic mechanisms to account for this diversity. Perhaps as many as a million different antibodies of unique specificity might be required to afford reasonable protection against the multitude of pathogens we confront in the course of a lifetime, although such estimates remain guesswork at best. Given the size of a typical antibody heavy chain and light chain (each heavy chain-light chain combination, if encoded as such, would require ≈2.5–3.5 kb, depending on the isotype), it is immediately obvious that the organism would rapidly exhaust the DNA coding capacity required to encode a set of antibody molecules of sufficient diversity to provide adequate protection against the wide array of pathogens and other foreign substances to which the organism is exposed. We shall see that, indeed, unique mechanisms are at work to create an adequately diverse set of antibodies.

## Immunoglobulin Domains Have a Characteristic Fold Composed of Two β Sheets Stabilized by a Disulfide Bond

Both the variable and constant domains of immunoglobulins fold into a compact three-dimensional structure composed exclusively of β sheets (see Figure 23-12b). A typical Ig domain contains two β sheets (one with three strands and one with four strands) held together by a disulfide bond. The residues that point inward are mostly hydrophobic and help stabilize this sandwich structure. Solvent-exposed residues show a greater frequency of polar and charged side chains. The spacing of the cysteine residues that make up the disulfide bond and a small number of strongly conserved residues characterize this evolutionarily ancient structural motif, termed the **immunoglobulin fold**. The basic immunoglobulin fold is found in numerous eukaryotic proteins that are not directly involved in antigen-specific recognition, including the Ig superfamily of cell-adhesion molecules, or IgCAMs (Chapter 20).

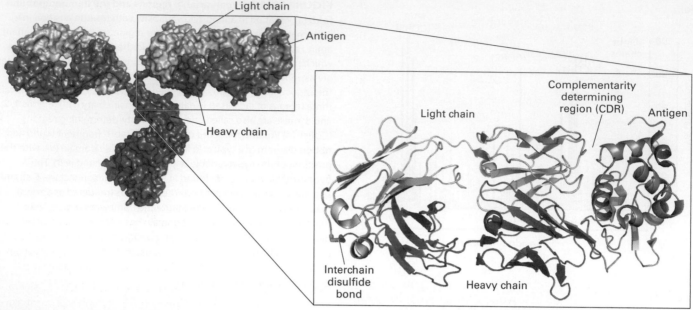

**FIGURE 23-13 Immunoglobulin structure.** This model shows the three-dimensional structure of an immunoglobulin complexed with hen egg-white lysozyme (a protein antigen) as determined by x-ray crystallography. [Based on E. A. Padlan et al., 1989 *Proc. Natl Acad. Sci USA* **86**:5938.]

The region on an antigen where it makes contact with the corresponding antibody is called an **epitope**. A protein antigen usually contains multiple epitopes, which often are exposed loops or surfaces on the protein and thus accessible to antibody molecules. Each homogeneous antibody preparation, derived from a clonal population of B cells, recognizes a single molecularly defined epitope on the corresponding antigen.

In order to solve the structure of an antibody complexed to its cognate epitope on an antigen, it is important to have a source of homogeneous immunoglobulin and the antigen in pure form. Homogeneous immunoglobulins can be obtained from B-cell tumors (malignant monoclonal expansions of immunoglobulin-secreting B cells), but in that case the antigen for which the antibody is specific is not known. The breakthrough essential for generating homogeneous antibody preparations suitable for structural analysis was the development of techniques to obtain **monoclonal antibody** produced by **hybridomas** by use of a special selection medium (see Chapter 9, pp. 402–404). The creation of immortalized cell lines that produce antibodies of defined specificity (monoclonal antibodies) has yielded essential tools for the cell biologist: monoclonal antibodies are widely used for the specific detection of macromolecules and their modifications. Monoclonal antibodies can detect proteins and their modifications (phosphorylation, nitrosylation, methylation, acetylation, etc.), complex carbohydrates, (glyco)lipids, nucleic acids, and their modifications, and have therefore found widespread use in the laboratory as well as for diagnostic purposes.

We now have detailed insights into the structure of a large number of monoclonal antibodies in complex with the antigen for which they are specific. There are no hard-and-fast rules that describe these interactions, other than that the usual rules of interaction of proteins with other (macro)molecules

apply (see Chapter 3). The CDRs make the most important contributions to the antigen-antibody interface, with a particularly prominent role for the CDR3 of the Ig heavy chain, followed by the CDR3 of the Ig light chain. How diversity is created in the CDRs will be described below.

## An Immunoglobulin's Constant Region Determines Its Functional Properties

Antibodies recognize antigen via their variable regions, but their constant regions determine many of the functional properties of antibodies. An important functional property of antibodies is their neutralizing capacity. By binding to epitopes on the surface of virus particles or bacteria, antibodies may block a productive interaction between a pathogen and receptors on host cells, thereby inhibiting (neutralizing) infection.

Antibodies attached to a virus or microbial surface can be recognized directly by cells that express receptors specific for the Fc portion of immunoglobulins. These *Fc receptors* (FcRs), which are specific for individual classes and subclasses of immunoglobulins, display considerable structural and functional heterogeneity. By means of FcR-dependent events, specialized phagocytic cells such as dendritic cells and macrophages can engage antibody-decorated particles, then ingest and destroy them in the process of opsonization. FcR-dependent events also allow some immune-system cells (e.g., monocytes and natural killer cells) to directly engage target cells that display viral or other antigens to which antibodies are attached. This engagement may induce the immune-system cells to release toxic small molecules (e.g., oxygen radicals) or the contents of cytotoxic granules, including perforins and granzymes. These proteins can attach themselves to the surface of the engaged target cell, inflict membrane damage, and so kill the target (see

Figure 23-5). This process, called *antibody-dependent cell-mediated cytotoxicity*, illustrates how cells of the innate immune system interact with, and benefit from, the products of the adaptive immune response.

Depending on the immunoglobulin isotype, antigen-antibody (immune) complexes can initiate the classical pathway of complement activation (see Figure 23-4). IgM and IgG3 are particularly good at complement activation, but all IgG classes can in principle activate complement, whereas IgA and IgE are unable to do so.

## KEY CONCEPTS of Section 23.2

### Immunoglobulins: Structure and Function

- Most immunoglobulins (antibodies) are composed of two identical heavy (H) chains and two light (L) chains, with each chain containing a variable (V) region and a constant (C) region. Proteolytic fragmentation yields monovalent F(ab) and bivalent F(ab')$_2$ fragments, which contain variable-region domains and retain antigen-binding capability (see Figure 23-8). The Fc portion contains constant-region domains and determines effector functions.

- Immunoglobulins are divided into classes based on the constant regions of the heavy chains they carry (see Figure 23-9). In mammals there are five major classes: IgM, IgD, IgG, IgA, and IgE; the corresponding heavy chains are referred to as $\mu$, $\delta$, $\gamma$, $\alpha$, and $\epsilon$. There are two major classes of light chain, $\kappa$ and $\lambda$, again characterized by the attributes of their constant regions.

- Each individual B lymphocyte expresses an immunoglobulin of unique sequence and is therefore uniquely specific for a particular antigen. Upon recognition of antigen, only a B lymphocyte that bears a receptor specific for it will be activated and expand clonally (clonal selection) (see Figure 23-11).

- The antigen specificity of antibodies is conferred by their variable domains, which contain regions of high variability, called hypervariable or complementarity-determining regions (see Figure 23-12a). These hypervariable regions are positioned at the tip of the variable domain, where they can make specific contacts with the antigen for which a particular antibody is specific.

- The repeating domains that make up immunoglobulin molecules have a characteristic three-dimensional structure, the immunoglobulin fold: it consists of two $\beta$ pleated sheets held together by a disulfide bond (see Figure 23-12b). The immunoglobulin fold is widespread in evolution and is found in many proteins other than antibodies, including an important class of cell-adhesion molecules.

- The constant regions endow antibodies with unique functional properties or effector functions, such as the capacity to bind complement, the ability to be transported across epithelia, or the ability to interact with receptors specific for the Fc portion of immunoglobulins.

## 23.3 Generation of Antibody Diversity and B-Cell Development

Pathogens have short replication times, are quite diverse in their genetic makeup, and evolve quickly, generating even more antigenic variation. An adequate defense must thus be capable of mounting an equally diverse response. Antibodies fulfill this role. The timing of the antibody response and its necessary adjustment to changes in the antigenic makeup of the pathogen in question pose unique demands on the organization and regulation of the adaptive immune system. The coding capacity of the typical vertebrate genome could not possibly encode the large number of antibodies required for adequate protection against the diversity of microbes to which the host is exposed. A unique mechanism evolved to allow not only sheer limitless variability of the antibody repertoire, but also to enable rapid adjustment of the quality of the antibodies produced, to meet the demands posed by an unfolding viral or bacterial infection. Because optimal antibody production requires assistance in the form of T-cell help, we shall see that molecular mechanisms underlying receptor diversity are fundamentally similar for B and T cells.

B cells, which are responsible for antibody production, make use of a unique mechanism by which the genetic information required for synthesis of immunoglobulin heavy and light chains is stitched together from separate DNA sequence elements, or Ig gene segments, to create a functional transcriptional unit. The act of recombination that combines Ig gene segments itself dramatically expands the variability in sequence precisely where these genetic elements are joined together. This mechanism of generating a diverse array of antibodies is fundamentally different from meiotic recombination, which occurs only in germ cells, and from alternative splicing of exons (Chapter 8). Because this recombination mechanism occurs in somatic cells but not in germ cells, it is known as *somatic gene rearrangement* or *somatic recombination*. This unusual recombination mechanism, unique to antigen receptors on B and T lymphocytes, makes it possible to specify an enormously diverse set of receptors with minimal expenditure of DNA coding space. The ability to combine at will discrete genetic elements (combinatorial diversity), in addition to the generation of yet more sequence diversity in the encoded receptors by the underlying recombination mechanisms themselves, allows adaptive immune responses against a virtually limitless array of antigens, including molecules encoded by the host.

Thus there are mechanisms at work that not only create this enormous diversity, but there are also processes that impose tolerance to curtail unwanted reactivity against "self" components; the result of such reactivity is autoimmunity. Neither mechanism is perfect: the adaptive immune system cannot generate receptors to see *all* foreign substances. Furthermore, the unavoidable price we pay for *how* we generate B- and T-cell receptors is the likelihood of self-reactive receptors (auto-immunity).

## A Functional Light-Chain Gene Requires Assembly of V and J Gene Segments

Immunoglobulin genes encoding intact immunoglobulins do not exist already assembled in the genome, ready for expression. Instead, the required gene segments are brought together and assembled in the course of B-cell development (Figure 23-14). Although the rearrangement of heavy-chain genes precedes the rearrangement of light-chain genes, we discuss light-chain genes first because of their less complex organization.

The immunoglobulin light chains are encoded by clusters of V gene segments, followed at some distance downstream by a single C segment. Each V gene segment carries its own promoter sequence and encodes the bulk of the light-chain variable region, although a small piece of the nucleotide sequence encoding the light-chain variable region is missing from the V gene segment. This missing portion is provided by one of the multiple J segments located between the V segments and the single C segment in the unrearranged κ light-chain locus (see Figure 23-14a). This J segment is a genetic element, not to be confused with the J chain, a polypeptide subunit of the pentameric IgM molecule and found also in association with IgA (see Figure 23-9). In the course of B-cell development, commitment to a particular V gene segment—a random process—results in its juxtaposition with one of the J segments, again a random choice, forming an exon encoding the entire light-chain variable region ($V_L$). The act of recombination not only generates an intact and functional light-chain gene, it also places the promoter sequence of the rearranged gene within

controlling distance of enhancer elements, located downstream of the light-chain constant-region exon, required for its transcription. Only a rearranged light-chain gene is transcribed.

**Recombination Signal Sequences** Detailed sequence analysis of the light-chain and heavy-chain loci revealed a conserved sequence element at the 3′ end of each V gene segment. This conserved element, called a *recombination signal sequence (RSS)*, is composed of heptamer and nonamer sequences separated by a 23-bp spacer. At the 5′ end of each J element, there is a similarly conserved RSS that contains a 12-bp spacer (Figure 23-15a). The 12- and 23-bp spacers separate the conserved heptamer and nonamer sequences by one and two turns of the DNA helix, respectively.

Somatic recombination is catalyzed by the RAG1 and RAG2 recombinases, which are expressed only in lymphocytes. Juxtaposition of the two gene segments to be joined is stabilized by the RAG1/RAG2 complex (Figure 23-15b). The recombinases then make a single-stranded cut at the exact boundary of each coding sequence and its adjacent RSS. Only gene segments that possess heptamer-nonamer RSSs with spacers of different lengths can engage in this type of rearrangement (the so-called 12/23-bp spacer rule). Each newly created —OH group at the site of cleavage then executes a nucleophilic attack on the complementary strand, creating a covalently closed hairpin for each of the two coding ends, and double-strand breaks at the ends of the RSSs. Protein complexes that include the Ku70 and Ku80 proteins hold this complex together, so that the ends about to be

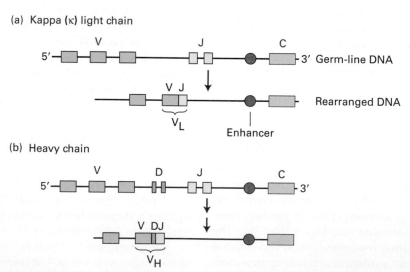

(a) Kappa (κ) light chain

(b) Heavy chain

**FIGURE 23-14 Overview of somatic gene rearrangement in immunoglobulin DNA.** The stem cells that give rise to B cells contain multiple gene segments encoding portions of immunoglobulin heavy and light chains. During development of a B cell, somatic recombination of these gene segments yields functional light-chain genes (a) and heavy-chain genes (b). Each V gene segment carries its own promoter. Rearrangement brings an enhancer close enough to the combined sequence to activate transcription. The light-chain variable region ($V_L$) is encoded by two joined gene segments, and the heavy-chain variable region ($V_H$) is encoded by three joined segments. Note that the chromosomal regions encoding immunoglobulins contain many more V, D, and J segments than shown. Also, the κ light-chain locus contains a single constant (C) segment, as shown, but the heavy-chain locus contains several distinct C segments (not shown) corresponding to the immunoglobulin isotypes.

joined remain in close proximity: double-strand breaks in chromosomes need to be repaired, and thus the ends need to be held together for resolution and repair of these breaks to proceed. The RSS ends are then covalently joined without loss or addition of nucleotides, creating a circular reaction product (deletion circle) containing the intervening DNA, which is lost from the locus altogether. The hairpin ends of the coding segments undergoing recombination then are opened and finally joined as depicted in Figure 23-15c, completing the recombination process.

The recombination mechanism just described, called *deletional joining*, occurs when the V gene segment involved has the same transcriptional orientation as the other gene segments at the light-chain locus. Some V gene segments, however, have the opposite transcriptional orientation. These are joined to J segments by a mechanism, termed *inversional joining*, in which the V segment is inverted and the intervening DNA and RSSs are not lost from the locus.

Defects in the synthesis of RAG proteins obliterates the possibility of somatic gene rearrangements. As described below, the rearrangement process is essential for B-cell development; consequently, RAG deficiency leads to the complete absence of B cells. People with defects in RAG gene function suffer from severe immunodeficiency. Targeted deletion of RAG genes in mice likewise leads to a complete defect in immunoglobulin (and T-cell receptor) gene rearrangement, resulting in a developmental block in the generation of B and T lymphocytes.

**Junctional Imprecision** In addition to the sequence variability created by the random selection of V and J gene segments to join, processing of the intermediates created in the course of recombination provides additional means for expanding the variability of immunoglobulin sequences. This additional variability is created at the junction of the segments to be joined. The opening of the hairpins at the coding ends is a key step in this process: this opening may occur symmetrically or asymmetrically (see Figure 23-15c, **4** and **5**). The protein Artemis, whose function requires the catalytic subunit of DNA-dependent protein kinase, carries out opening of the hairpins.

If opening of a hairpin is asymmetric, a short, single-stranded palindrome sequence is generated. Filling in of this overhang by DNA polymerase results in the addition of several nucleotides, called *P-nucleotides*, that were not part of the original coding region of the gene segment in question. Alternatively, the overhang may be removed by exonucleolytic attack, resulting in the removal of nucleotides from the original coding region. These possibilities apply equally to the V and the J coding regions. Symmetric opening of a hairpin retains all the original coding information. However, even if the hairpin is opened symmetrically, the ends of the DNA molecule tend to breathe, creating short single-stranded sequences, which also may be attacked exonucleolytically, resulting in removal of nucleotides.

Once the hairpins have been opened and the coding ends processed, the ends are ligated together by DNA ligase IV and

XRCC4, generating a functional light-chain gene. Inherent in the rearrangement process is *junctional imprecision* resulting in part from the addition and loss of nucleotides at the coding joints. Whenever a V and a J segment recombine, the sequence and reading frame of the VJ product cannot be predicted. Only one in three recombination reactions results in a reading frame that is compatible with light-chain synthesis.

Light-chain diversity therefore arises not only from the combinatorial usage of V and J gene segments, but also from junctional imprecision. Inspection of the three-dimensional structure of the light chain shows that the highly diverse joint, generated as a consequence of junctional imprecision, forms part of a loop—hypervariable region 3 (HV3)—that projects into the antigen-binding site and makes contact with antigen (see Figure 23-12b).

## Rearrangement of the Heavy-Chain Locus Involves V, D, and J Gene Segments

The organization of the heavy-chain locus is more complex than that of the κ light-chain locus. The heavy-chain locus contains not only a large tandem array of V gene segments (each equipped with its own promoter) and multiple J elements, but also multiple D (diversity) segments (see Figure 23-14b). Somatic recombination of a V, D, and J segment generates a rearranged sequence encoding the heavy-chain variable region ($V_H$).

At the 3' end of each V gene segment in heavy-chain DNA, there are conserved heptamer and nonamer sequences separated by spacer DNA, similar to the recombination signal sequences (RSSs) in light-chain DNA. These RSSs are also found in complementary and antiparallel configuration at the 5' end and the 3' end of each D segment (see Figure 23-15a). The J segments are similarly equipped at their 5' end with the requisite RSS. The spacer lengths in these RSSs are such that D segments can join to J segments, and V segments to already rearranged DJ segments. However, neither direct V-to-J nor D-to-D joining is allowed, in compliance with the 12/23 heptamer-nonamer rule. Heavy-chain rearrangements proceed via the same mechanisms described above for light-chain rearrangements.

In the course of B-cell development, the heavy-chain locus always rearranges first, starting with D-J rearrangement. D-J rearrangement is followed by V-DJ rearrangement. In the course of D-J and V-DJ rearrangements, terminal deoxynucleotidyltransferase (TdT) may add nucleotides to free 3' OH ends of DNA in a template-independent fashion. Up to a dozen or so nucleotides, called the *N-region*, may be added, generating additional sequence diversity at the junctions whenever D-J and V-DJ rearrangements occur (see Figure 23-15, step **7**). Only one in three rearrangements yields the proper reading frame for the rearranged VDJ sequence. If the rearrangement yields a sequence encoding a functional protein, it is called *productive*. Although the heavy-chain locus is present on two homologous chromosomes, only one productive rearrangement is permitted, as discussed below.

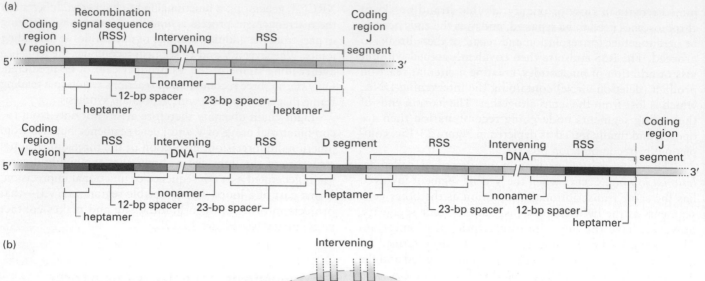

(a)

Coding
region
V region

Recombination
signal sequence
(RSS)

Intervening
DNA

RSS

Coding
region
J
segment

5′ ——— 3′

heptamer
12-bp spacer
nonamer

23-bp spacer
heptamer

Coding
region
V region

RSS

Intervening
DNA

RSS

D segment

RSS

Intervening
DNA

RSS

Coding
region
J
segment

5′ ——— 3′

heptamer
12-bp spacer
nonamer

23-bp spacer

heptamer

23-bp spacer
nonamer

12-bp spacer
heptamer

(b)

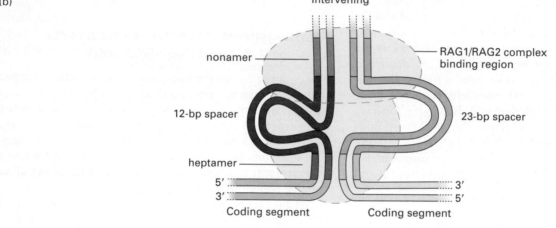

Intervening

nonamer

RAG1/RAG2 complex
binding region

12-bp spacer

23-bp spacer

heptamer

5′ ——— 3′
3′ ——— 5′

Coding segment          Coding segment

(c)

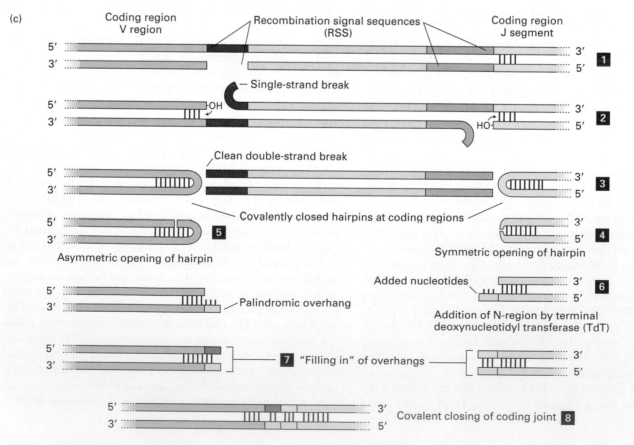

Coding region
V region

Recombination signal sequences
(RSS)

Coding region
J segment

5′          3′
3′          5′                                                                          **1**

Single-strand break

5′                                    OH                                   3′
3′                  OH          HO                                          5′        **2**

Clean double-strand break

5′                                                                              3′
3′                                                                              5′        **3**

Covalently closed hairpins at coding regions

5′          **5**                                                              3′
3′                                                                              5′        **4**

Asymmetric opening of hairpin              Symmetric opening of hairpin

Added nucleotides

5′                                                                              3′
3′ Palindromic overhang                                                         5′        **6**

Addition of N-region by terminal
deoxynucleotidyl transferase (TdT)

5′
3′        **7** "Filling in" of overhangs                                       3′
                                                                                5′

5′          3′
3′          5′        Covalent closing of coding joint **8**

**FIGURE 23-15 Mechanism of rearrangement of immunoglobulin gene segments via deletional joining.** This example depicts joining of a V segment and a J segment, as occurs at the light-chain locus (see Figure 23-14a). (a) Location of the DNA elements involved in somatic recombination of immunoglobulin gene segments at the light-chain locus (*top*) and at the heavy-chain locus (*bottom*). D segments are present in the heavy-chain, but not light-chain, locus. At the 3′ end of all V gene segments is a conserved recombination signal sequence (RSS) composed of a heptamer, a 12-bp spacer, and a nonamer. Each of the J or D segments with which a V can recombine possesses at its 5′ end a similar RSS with a 23-bp spacer. The nonamer and heptamer sequences at the 5′ end of J or D are complementary and antiparallel to those found at the 3′ end of each V when read on the same (*top*) strand. A similar arrangement exists for the RSS at the 3′ end of D and 5′ end of J at the heavy-chain locus. (b) Hypothetical model of how two coding regions to be joined may be arranged spatially, stabilized by the RAG1 and RAG2 recombinase complex. Both strands of the DNA are shown. (c) Events in joining of V and J coding regions. The germ-line DNA (**1**) is folded, bringing the segments to be joined close together, and the RAG1/RAG2 complex makes single-stranded cuts at the boundaries between the coding sequences and RSSs (**2**). The free 3′ —OH groups attack the complementary strands, creating a covalently closed hairpin at each coding end and a clean double-stranded break at each boundary with an RSS (**3**). The hairpins are opened, either symmetrically (**4**), as shown for the J segment, or asymmetrically (**5**), as shown for the V segment. Terminal deoxynucleotidyl transferase adds nucleotides in a template-independent manner to symmetrically opened hairpins (**6**, *right*), generating an overhang (yellow) of unpaired nucleotides of random sequence; asymmetric opening automatically creates a palindromic overhang (**6**, *left*). The unpaired overhangs at the ends of both the V and D coding regions are filled in by DNA polymerase (**7**) or may be excised by an exonuclease. DNA ligase IV joins the two segments generated from the V and J coding regions (**8**). Rearrangements of heavy-chain D and J segments, and V and DJ segments, occur by the same mechanism except that N-region addition does not take place. See text for additional discussion.

An enhancer located downstream of the cluster of J segments and upstream of the μ constant-region segment activates transcription from the promoter at the 5′ end of the rearranged VDJ sequence (see Figure 23-14). Splicing of the primary transcript produced from the rearranged heavy-chain gene generates a functional mRNA encoding the μ heavy chain. For both immunoglobulin heavy- and light-chain genes, somatic recombination places the promoters upstream of the V segments within functional reach of the enhancers necessary to allow transcription, so that only rearranged VJ and VDJ sequences, and not the V segments that remain in the germ-line configuration, are transcribed.

## Somatic Hypermutation Allows the Generation and Selection of Antibodies with Improved Affinities

In addition to the diversity created by somatic recombination and junctional imprecision, antigen-activated B cells can undergo *somatic hypermutation*. Upon receipt of proper additional signals, most of which are provided by T cells, expression of activation-induced deaminase (AID) is turned on. This enzyme deaminates cytosine residues to uracil. When a B cell that carries this lesion replicates, it may place an adenine on the complementary strand, thus generating a G-to-A transition (see Figure 4-35). Alternatively, the uracil may be excised by DNA glycosylase to yield an abasic site. These abasic sites, when copied, give rise to possible transitions as well as a transversion, unless the nucleotide opposite the gap is chosen to be the original G that paired with the cytosine target. Mutations thus accumulate with every successive round of B-cell division, yielding numerous mutations in the rearranged VJ and VDJ segments. Many of these mutations are deleterious, in that they reduce the affinity of the encoded antibody for antigen, but some improve the encoded antibody's affinity for antigen. B cells carrying affinity-increasing mutations have a selective advantage when they compete for the limited amount of antigen that evokes clonal selection (see Figure 23-11). The net result is generation of a B-cell population whose antibodies, as a rule, show a higher affinity for the antigen.

In the course of an immune response or upon repeated immunization, the antibody response exhibits *affinity maturation*, an increase in the average affinity of antibodies for antigen, as the result of somatic hypermutation. Antibodies produced during this phase of the immune response display affinities for antigen in the nanomolar (or better) range. For reasons that are not understood, the activity of activation-induced deaminase is focused mostly on rearranged VJ and VDJ segments, and this targeting may therefore require active transcription. The entire process of somatic hypermutation is strictly antigen-dependent and shows an absolute requirement for interactions between the B cell and certain T cells.

## B-Cell Development Requires Input from a Pre-B-Cell Receptor

As we have seen, B cells destined to make immunoglobulins must rearrange the necessary gene segments to assemble functional heavy- and light-chain genes. These rearrangements occur in a carefully ordered sequence during development of a B cell, starting with heavy-chain rearrangements. Moreover, the rearranged heavy chain is first used to build a membrane-bound receptor that executes a cell-fate decision necessary to drive further B-cell development (and antibody synthesis). Only a productive rearrangement that yields an in-frame VDJ combination can generate a complete μ heavy chain. The production of these μ chains serves as a signal to indicate to the B cell that it has successfully accomplished rearrangement, and that no further rearrangements of the heavy-chain locus on the remaining allele are required. Recall that each lymphocyte precursor starts out with two immunoglobulin locus–bearing chromosomes in the germ-line configuration. In accordance with clonal selection theory, which stipulates that each lymphocyte ought to come equipped with a single antigen-specific receptor, continued rearrangement would entail the risk of producing B cells with two different heavy chains, each with different specificity, an undesirable outcome.

Successful rearrangement of V, D, and J segments in the heavy-chain locus thus allows synthesis of a complete μ chain. B cells at this stage of development are called *pre-B cells,* as they have not yet completed assembly of a functional light-chain gene and therefore cannot engage in antigen recognition. The newly rearranged heavy-chain gene encodes the μ polypeptide, which becomes part of a signaling receptor whose expression is essential for B-cell development to proceed in orderly fashion. The μ chain made at this stage of B-cell development is a membrane-bound version.

In pre-B cells, newly made μ chains form a complex with two so-called surrogate light chains, λ5 and VpreB (Figure 23-16). The μ chain itself possesses no cytoplasmic tail and is therefore incapable of recruiting cytoplasmic components for the purpose of signal transduction. Instead, early B cells express two auxiliary transmembrane proteins, called Igα and Igβ, each of which carries in its cytoplasmic tail an immunoreceptor tyrosine-based activation motif, or ITAM. The entire complex, including μ chain, λ5, VpreB, Igα, and Igβ, constitutes the *pre-B-cell receptor (pre-BCR).* Engagement of this receptor by (unknown) suitable signals results in recruitment and activation of a Src-family tyrosine kinase, which phosphorylates tyrosine residues in the ITAMs. In their phosphorylated form, ITAMs recruit other molecules essential for signal transduction (see below). Because no functional light chains are yet part of this receptor, it is presumed to be incapable of antigen recognition.

The pre-B-cell receptor has several important functions. First, it shuts off expression of the RAG recombinases, so rearrangement of the other (allelic) heavy-chain locus cannot proceed. This phenomenon, called *allelic exclusion,* ensures that only one of the two available copies of the heavy-chain locus will be rearranged and thus expressed as a complete μ chain. Second, because of the association of the pre-B-cell receptor with Igα and Igβ, the receptor becomes a functional signal-transduction unit. Signals that emanate from the pre-BCR initiate pre-B-cell proliferation to expand the numbers

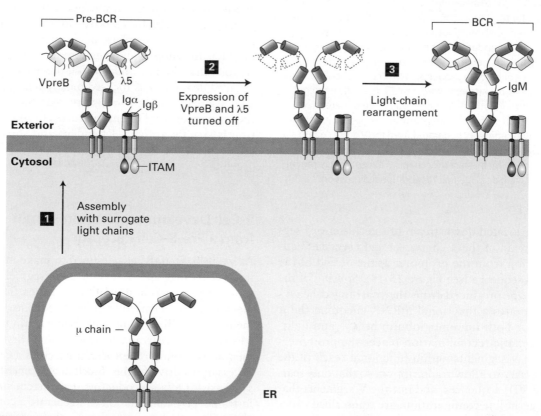

**FIGURE 23-16 Structure of the pre-B-cell receptor and its role in B-cell development.** Successful rearrangement of V, D, and J heavy-chain gene segments allows synthesis of membrane-bound μ heavy chains in the endoplasmic reticulum (ER) of a pre-B cell. At this stage, no light-chain gene rearrangement has occurred. Newly made μ chains assemble with surrogate light chains, composed of λ5 and VpreB, to yield the pre-B-cell receptor, pre-BCR (step **1**). This receptor drives proliferation of those B cells that carry it. It also suppresses rearrangement of the heavy-chain locus on the other chromosome and so mediates allelic exclusion. In the course of proliferation, the synthesis of λ5 and VpreB is shut off (step **2**), resulting in "dilution" of the available surrogate light chains and reduced expression of the pre-BCR. As a result, rearrangement of the light-chain loci can proceed (step **3**). If this rearrangement is productive, the B cell can synthesize light chains and complete assembly of the B-cell receptor (BCR), comprising a membrane-bound IgM and associated Igα and Igβ. The B cell is now responsive to antigen-specific stimulation.

of those B cells that have undergone productive D-J and V-DJ recombination.

In the course of this expansion, expression of the surrogate light-chain subunits, VpreB and λ5, subsides. The progressive dilution of VpreB and λ5 with every successive cell division allows reinitiation of expression of the RAG enzymes, which now target the κ or λ light-chain locus for recombination. A productive light-chain V-J rearrangement also shuts off rearrangement of the allelic locus (allelic exclusion). Upon completion of a successful V-J light-chain rearrangement, the B cell can make both μ heavy chains and κ or λ light chains, and assemble them into a functional B-cell receptor (BCR), which can recognize antigen (see Figure 23-16).

Once a B cell expresses a complete BCR on its cell surface, it can recognize antigen, and all subsequent steps in B-cell activation and differentiation involve antigen-specific engagement by the antigen for which that BCR is specific. The BCR not only plays a role in driving B-cell proliferation upon successful encounter with antigen, it also functions as a device for capturing and ingesting antigen, an essential step that allows the B cell to process the acquired antigen and convert it into a signal that sends out a call for assistance by T lymphocytes. This antigen-presentation function of B cells is described in later sections.

## During an Adaptive Response, B Cells Switch from Making Membrane-Bound Ig to Making Secreted Ig

As just described, the **B-cell receptor (BCR)**, a membrane-bound IgM, provides a B cell with the ability to recognize a particular antigen, an event that triggers clonal selection and proliferation of that B cell, thus increasing the number of B cells specific for the antigen (see Figure 23-11). However, key functions of immunoglobulins, such as neutralization of antigen or killing of bacteria, require that these products be released by the B cell, so that they may accumulate in the extracellular environment and act at a distance from the site where they were produced.

The choice between the synthesis of membrane-bound versus secreted immunoglobulin is made during processing of the heavy-chain primary transcript. As shown in Figure 23-17, the μ locus contains two exons (TM1 and TM2) that together encode a C-terminal domain that anchors IgM in the plasma membrane. One polyadenylation site is found upstream of these exons; a second polyadenylation site is present downstream. If the downstream poly(A) site is chosen, then further processing yields an mRNA that encodes the membrane-bound form of μ. (As described above, this choice is necessary for formation of the B-cell receptor, which includes membrane-bound IgM.) If the upstream poly(A) site is chosen, processing yields the secreted version of the μ chain. Similar arrangements are found for the other Ig constant-region gene segments, each of which can specify either a membrane-bound or a secreted heavy chain. The ability to switch between the membrane-anchored and the secreted form of immunoglobulin heavy chains by alternative use of poly-adenylation sites (*not* by alternative splicing) is so far unique to this family of gene products.

The capacity to switch from the synthesis of exclusively membrane immunoglobulin to the synthesis of secreted immunoglobulin is acquired by B cells in the course of their differentiation. Terminally differentiated B cells, called *plasma cells*, are devoted almost exclusively to the synthesis

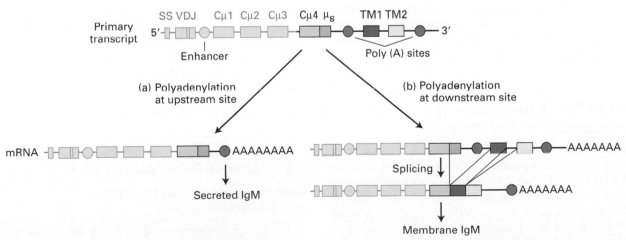

**FIGURE 23-17 Synthesis of secreted and membrane IgM.** The organization of the μ heavy-chain primary transcript is shown at the top: Cμ4 is the exon encoding the fourth μ constant-region domain; μ_s is a coding sequence unique for secreted IgM; TM1 and TM2 are exons that specify the transmembrane domain of the μ chain. Whether secreted or membrane-bound IgM is made depends on which poly(A) site is selected during processing of the primary transcript. (a) If the upstream poly(A) site is used, the resulting mRNA includes the entire Cμ4 exon and specifies the secreted form of the μ chain. (b) If the downstream poly(A) site is used, a splice donor site in the Cμ4 exon allows splicing to the transmembrane exons, yielding a mRNA that encodes the membrane-bound form of the μ chain. Similar mechanisms generate secreted and membrane-bound forms of other Ig isotypes. SS = signal sequence.

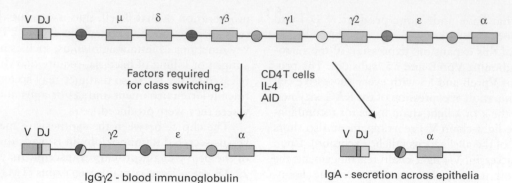

**FIGURE 23-18 Class-switch recombination in the immunoglobulin heavy-chain locus.** Class-switch recombination involves switch sites, which are repetitive sequences (colored circles) upstream of the heavy-chain constant-region genes. Recombination requires activation-induced deaminase (AID), assistance by T cells, and cytokines (e.g., IL-4) produced by certain T cells. Recombination eliminates the segment of DNA between the switch site upstream of μ exons and the constant region to which switching occurs. Class switching generates antibody molecules with the same specificity for antigen as that of the IgM-bearing B cell that mounted the original response, but with different heavy-chain constant-regions and therefore different effector functions.

of secreted antibodies (see Figure 23-6). Plasma cells synthesize and secrete several thousand antibody molecules per second. It is this ramped-up production of secreted antibodies that underlies the effectiveness of the adaptive immune response in eliminating pathogens. The protective value of antibodies is proportional to the concentration at which they are present in the circulation. Indeed, circulating antibody levels are often used as the key parameter to determine whether vaccination against a particular pathogen has been successful. The ability of plasma cells to establish adequate antibody levels is a function of their ability to secrete large amounts of immunoglobulins and so requires a massive expansion of the endoplasmic reticulum, a hallmark of plasma cells. The unfolded protein response (Chapter 13) is initiated in B cells as an essential physiological mechanism to expand the ER and prepare the differentiating B cell for its future task as a highly active secretory cell. Interference with the UPR—for example, by deletion of the XBP1 gene—abolishes the ability of B cells to turn into plasma cells.

## B Cells Can Switch the Isotype of Immunoglobulin They Make

In the immunoglobulin heavy-chain locus, the exons that encode the μ chain lie immediately downstream of the rearranged VDJ exon (Figure 23-18, *top*). This is followed by exons that specify the δ chain. Transcription of a newly rearranged immunoglobulin heavy-chain locus yields a single primary transcript that includes the μ and δ constant regions. Splicing of this large transcript determines whether a μ chain or a δ chain will be produced. Downstream of the μ/δ combination are the exons that together encode all of the other heavy-chain isotypes. Upstream of each cluster of exons (with the exception of the δ locus) encoding the different isotypes are repetitive sequences (switch sites) that are recombination-prone, presumably because of their repetitive

nature. Because each B cell necessarily starts out with surface IgM, recombination involving these sites, if it occurs, results in a *class switch* from IgM to one of the other isotypes located downstream in the array of constant-region genes (see Figure 23-18). The intervening DNA is deleted.

In the course of its differentiation, a B cell can switch sequentially. Importantly, the light chain is not affected by this process, nor is the rearranged VDJ segment with which the B cell started out on this pathway. Class-switch recombination thus generates antibodies with different constant regions but identical antigenic specificity. Each immunoglobulin isotype is characterized by its own unique constant region. As discussed previously, these constant regions determine the functional properties of the various isotypes. Class-switch recombination is absolutely dependent on the activity of activation-induced deaminase (AID) and the presence of antigen and T cells. Somatic hypermutation and class-switch recombination occur concurrently, and their combined effect allows fine tuning of the adaptive immune response with respect to the affinity of the antibodies produced and the effector functions called for.

---

**KEY CONCEPTS of Section 23.3**

### Generation of Antibody Diversity and B-Cell Development

• Functional antibody-encoding genes are generated by somatic rearrangement of multiple DNA segments at the heavy-chain and light-chain loci. These rearrangements involve V and J segments for immunoglobulin light chains, and V, D, and J segments for immunoglobulin heavy chains (see Figure 23-14).

• Rearrangement of the V and J, as well as of the V, D, and J gene segments is controlled by conserved recombination signal sequences (RSSs), composed of heptamers and nonamers separated by 12- or 23-bp spacers (see Figure 23-15). Only those segments that have spacers of different lengths can rearrange successfully.

- The molecular machinery that carries out the rearrangement process includes recombinases (RAG1 and RAG2) made only by lymphocytes and numerous other proteins that participate in nonhomologous end joining of DNA molecules in other cell types as well.

- Antibody diversity is created by the random selection of Ig gene segments to be recombined (combinatorial joining) and by the ability of the heavy and light chains produced from rearranged Ig genes to associate with many different light chains and heavy chains, respectively (combinatorial association).

- Junctional imprecision generates additional antibody diversity at the joints of the gene segments joined during somatic recombination.

- Further antibody diversity arises after B cells encounter antigen as a consequence of somatic hypermutation, which can lead to the selection and proliferation of B cells producing high-affinity antibodies, a process termed affinity maturation.

- During B-cell development, heavy-chain genes are rearranged first, leading to expression of the pre-B-cell receptor. Subsequent rearrangement of light-chain genes results in assembly of an IgM membrane-bound B-cell receptor (see Figure 23-16).

- Only one of the allelic copies of the heavy-chain locus and of the light-chain locus is rearranged (allelic exclusion), ensuring that a B cell expresses Ig with a single antigenic specificity.

- Polyadenylation of different poly(A) sites in an Ig primary transcript determines whether the membrane-bound or secreted form of an antibody is produced (see Figure 23-17).

- During an immune response, class switching allows B cells to adjust the effector functions of the immunoglobulins produced but retain their specificity for antigen (Figure 23-18).

## 23.4 The MHC and Antigen Presentation

Antibodies can recognize antigen without the involvement of any third-party molecules; the presence of antigen and antibody is sufficient for their interaction. In the course of their differentiation, B cells receive assistance from T cells, a process that will be described in some detail below. This help is antigen-specific, and the T cells responsible for providing it are *helper T cells*. Although antibodies contribute to the elimination of bacterial and viral pathogens, it is often necessary to destroy also the infected cells that serve as a source of new virus particles. This task is carried out by T cells with cytotoxic activity. These *cytotoxic T cells* and helper T cells make use of antigen-specific receptors whose genes are generated by mechanisms analogous to those used by B cells to generate immunoglobulin genes. However, T cells engage in

antigen recognition in a manner very different from B cells. The antigen-specific receptors on T cells recognize short snippets of protein antigens, presented to these receptors by membrane glycoproteins encoded by the **major histocompatibility complex (MHC)**. Various antigen-presenting cells, in the course of their normal activity, digest pathogen-derived (and self) proteins and then "post" these protein snippets (peptides) to their cell surface in a physical complex with an MHC protein. T cells can inspect these complexes, and if they detect a pathogen-derived peptide, the T cells take appropriate action, which may include killing the cell that carries the MHC-peptide complex. In this section, we describe the MHC and the proteins it encodes, and then examine how these MHC molecules are involved in antigen recognition.

### The MHC Determines the Ability of Two Unrelated Individuals of the Same Species to Accept or Reject Grafts

The major histocompatibility complex was discovered, as its name implies, as the genetic locus that controls acceptance or rejection of grafts. At a time when tissue culture had not yet been developed to the stage where tumor-derived cell lines could be propagated in the laboratory, investigators relied on serial passage in vivo of tumor tissue (transplanting a tumor from one mouse to another). It was quickly observed that a tumor that arose spontaneously in one inbred strain of mice could be propagated successfully in the strain in which it arose, but not in a genetically distinct line of mice. Genetic analysis soon showed that a single major locus was responsible for this behavior. Similarly, transplantation of healthy skin was feasible within the same strain of mice, but not when the recipient was of a genetically distinct background. Genetic analysis of transplant rejection likewise identified a single major locus that controlled acceptance or rejection, which is an immune reaction. As we now know, all vertebrates that possess an adaptive immune system have a genetic region that corresponds to the major histocompatibility complex as originally defined in the mouse.

In the mouse, the genetic region that encodes the antigens responsible for a strong graft rejection is called the *H-2 complex* (Figure 23-19a). The initial characterization of the MHC was followed by an appreciation of the genetic complexity of this region. After coarse mapping by standard genetic means (recombination within the MHC), the complete nucleotide sequence of the entire MHC was determined. The typical mammalian MHC contains dozens of genes, many encoding proteins of immunological relevance.

In humans, the discovery of the MHC relied on the characterization of antisera produced in patients who underwent multiple blood transfusions: Antigens expressed on the surface of the genetically nonidentical donor cells provoked an immune response in the recipient. The predominant target antigens recognized by these antisera are encoded by the human MHC, a genetic region also referred to as the *HLA complex* (Figure 23-19b). All vertebrate MHCs encode a highly homologous set

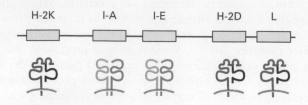

(a) Mouse MHC (H-2 complex)

H-2K   I-A   I-E   H-2D   L

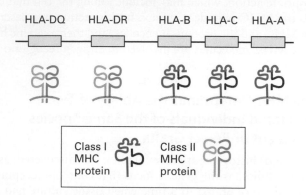

(b) Human MHC (HLA complex)

HLA-DQ   HLA-DR   HLA-B   HLA-C   HLA-A

Class I
MHC
protein

Class II
MHC
protein

**FIGURE 23-19 Organization of the major histocompatibility complex in mice and humans.** The major loci are depicted with schematic diagrams of their encoded proteins below. Class I MHC proteins are composed of an MHC-encoded single-pass transmembrane glycoprotein in noncovalent association with a small subunit, called β2-microglobulin, which is not encoded in the MHC and is not membrane bound. Class II MHC proteins consist of two nonidentical single-pass transmembrane glycoproteins, both of which are encoded by the MHC.

of proteins, although the details of organization and gene content show considerable variation between species.

The human fetus may also be considered a graft: The fetus shares only half of its genetic material with the mother, the other half being contributed by the father. Antigens encoded by this paternal contribution may differ sufficiently from their maternal counterparts to elicit an immune response in the mother. Such a response can occur because in the course of pregnancy, fetal cells that slough off into the maternal circulation can stimulate the maternal immune system to mount an antibody response against these paternal antigens. We now know that these antibodies recognize structures encoded by the human MHC. The fetus itself is spared rejection because of the specialized organization of the placenta, which prevents initiation of an immune response by the mother against fetal tissue.

## The Killing Activity of Cytotoxic T Cells Is Antigen Specific and MHC Restricted

Clearly, the function of MHC molecules is not to prevent the exchange of surgical grafts. MHC molecules play an essential role in the recognition of virus-infected cells by cytotoxic

T cells; these cells also are called cytotoxic T lymphocytes (CTLs). In virus-infected cells, MHC molecules interact with protein fragments derived from viral pathogens and display these on the cell surface where cytotoxic T cells, charged with eliminating the infection, can recognize them. How such fragments of antigen are generated and displayed for recognition by T cells will be described below. T cells that have receptors of the appropriate specificity unleash a payload of lethal molecules onto the infected target cells, destroying the target cell membrane and so killing the infected target. The destruction of these target cells can be readily measured as the release of their cytoplasmic contents when target cells physically disintegrate.

Mice that have recovered from a particular virus infection are a ready source of cytotoxic T cells that can recognize and kill target cells infected with the same virus. If T cells are obtained from a mouse that successfully cleared an infection with influenza virus, cytotoxic activity is observed against influenza-infected target cells, but not against uninfected controls (Figure 23-20). Moreover, the influenza-specific cytotoxic T cells will not kill target cells infected with a different virus, such as vesicular stomatitis virus. Cytotoxic T cells can even discriminate between closely related strains of influenza virus, and can do so with pinpoint precision: differences of a single amino acid in the viral antigen may suffice to avoid recognition and killing by cytotoxic T cells. These experiments show that cytotoxic T cells are truly antigen specific and do not simply recognize some attribute that is shared by all virus-infected cells, regardless of the identity of the virus.

In this example, it is assumed that the T cells harvested from an influenza-immune mouse are assayed on influenza-infected target cells derived from the identical strain of mouse (strain a). However, if target cells from a completely unrelated strain of mouse (strain b) are infected with the same strain of influenza and used as targets, the cytotoxic T cells from the strain a mouse are unable to kill the infected strain b target cells (see Figure 23-20b, **1** vs. **4**). It is therefore not sufficient that the antigen (an influenza-derived protein) is present; recognition by cytotoxic T cells is *restricted* by mouse strain-specific elements. Genetic mapping showed that these restricting elements are encoded by genes in the MHC. Thus cytotoxic T cells from one mouse strain immune to influenza will kill influenza-infected target cells from another mouse strain only if the two strains match at the MHC for the relevant MHC molecules. This phenomenon is therefore known as *MHC restriction*.

## T Cells with Different Functional Properties Are Guided by Two Distinct Classes of MHC Molecules

The MHC encodes two types of glycoproteins essential for immune recognition, commonly called *class I* and *class II MHC molecules.* The class I and class II MHC products are both involved in presenting antigen to T cells, but they serve two broadly distinct functions. Class I MHC products

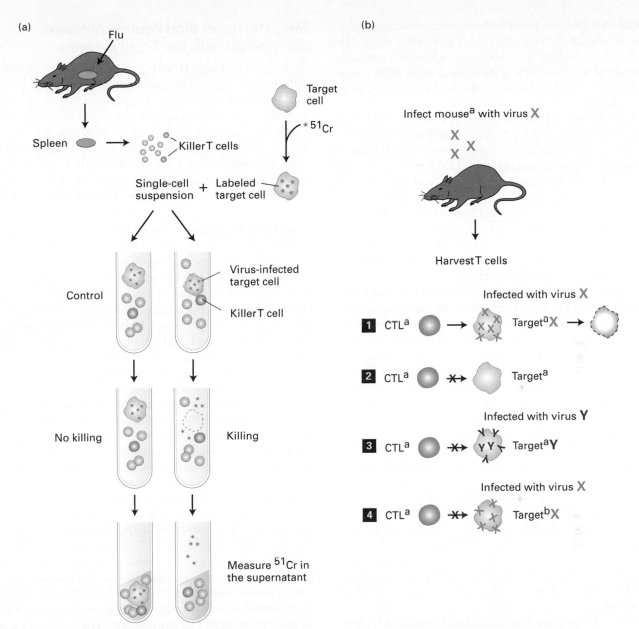

**EXPERIMENTAL FIGURE 23-20 Chromium ($^{51}$Cr) release assay allows the direct demonstration of the cytotoxicity and specificity of cytotoxic T cells in a heterogeneous population of cells.** (a) A suspension of spleen cells, containing cytotoxic (killer) T cells is prepared from mice that have been exposed to a particular virus (e.g., influenza virus) and have cleared the infection. Target cells obtained from the same strain are infected with the identical virus or left uninfected. After infection, cellular proteins are labeled nonspecifically by incubation of the target-cell suspension with $^{51}$Cr. Upon incubation of radiolabeled target cells with the suspension of T cells, killing of target cells results in release of the $^{51}$Cr-labeled proteins.

Uninfected target cells are not killed and retain their radioactive contents. Lysis of cells by cytotoxic T cells can therefore be readily detected and quantitated by measuring the radioactivity released into the supernatant. (b) Cytotoxic T lymphocytes (CTLs) harvested from mice that have been infected with virus X can be tested against various target cells to determine the specificity of CTL-mediated killing. CTLs capable of lysing virus X-infected target cells (**1**) cannot kill uninfected cells (**2**) or cells infected with a different virus, Y (**3**). When these CTLs are tested on virus X-infected targets from a strain that carries an altogether different MHC type (b), again no killing is observed (**4**). Cytotoxic T-cell activity is thus virus specific and restricted by the MHC.

mostly serve to alert cytotoxic T cells to the presence of intracellular invaders and allows the cytotoxic T cell to destroy the infected cell; this function goes hand in hand with the cell biological processes that govern the assembly and display of class I MHC products. Class II MHC products are

found on professional antigen-presenting cells: when the class II molecules on such cells present antigen, often acquired from the extracellular environment, to helper T cells, this is mostly interpreted by the immune system as a call to arms and an immediate need for an adaptive immune response that requires B-cell involvement, cytokine production, and also assistance to cytotoxic T cells. Again, the underlying cell biology that describes expression, assembly, and mode of antigen presentation by class II MHC molecules fits this functional specialization rather neatly, as we shall see below.

A comparison of the genetic maps of the mouse and human MHCs shows the presence of several class I MHC genes and several class II MHC genes, even though their arrangement shows variation between the different species (see Figure 23-19). In addition to the class I and class II MHC molecules, the MHC encodes key components of the antigen-processing and presentation machinery. Finally, the typical vertebrate MHC also encodes key components of the complement cascade. Class I and class II MHC molecules are recognized by different populations of immune system cells and therefore serve different functions.

As should be clear from the experiments outlined in Figure 23-20, cytotoxic T cells are guided in the recognition of their targets by MHC molecules. These T cells mostly use class I MHC molecules as their restriction elements, and also are characterized by the presence of the CD8 glycoprotein marker on their surface. Most, if not all, nucleated cells constitutively express class I MHC molecules and can support replication of viruses. Cytotoxic T cells recognize and kill the infected targets via the expressed class I MHC molecules that display virus-derived antigen.

As mentioned previously, B cells do not undergo final differentiation into antibody-secreting plasma cells without assistance from another subset of T cells, the *helper T cells* (or T helper cells). Helper T cells express on their surface the CD4 glycoprotein marker and use class II MHC molecules as restriction elements. The constitutive expression of class II MHC molecules is confined to so-called *professional* antigen-presenting cells, including B cells, dendritic cells, and macrophages. (Several other cell types, some in epithelia, can be induced to express class II MHC molecules, but we will not discuss these.)

The two major groups of functionally distinct T lymphocytes—cytotoxic T cells and helper T cells—can thus be distinguished based on the unique profile of membrane proteins displayed at their cell surface and by the MHC molecules they use as restriction elements:

• Cytotoxic T cells: CD8 marker; class I MHC restricted

• Helper T cells: CD4 marker; class II MHC restricted

Both CD4 and CD8 belong to the immunoglobulin (Ig) superfamily of proteins, which all include one or more Ig domains. The B-cell and T-cell receptors, polymeric IgA receptor, and many cell-adhesion molecules (Chapter 19) also belong to the Ig superfamily. The molecular basis for the strict correlation between expression of CD8 and utilization of class I MHC molecules or between expression of CD4 and utilization of class II MHC molecules as the restriction element will become evident once the structure and mode of action of MHC molecules has been described.

## MHC Molecules Bind Peptide Antigens and Interact with the T-Cell Receptor

Both class I and class II MHC molecules are highly *polymorphic;* that is, many allelic variants exist among individuals of the same species. The vertebrate immune system can respond to these allelic differences, and this ability to recognize allelic MHC variants is the underlying immunological cause for rejection of transplants that involve unrelated, genetically distinct individuals. Yet both classes of MHC molecules also are structurally similar in many respects, as are their interactions with peptides and the T-cell receptor (Figure 23-21).

**Class I MHC Molecules** Class I MHC molecules belong to the Ig superfamily and consist of two polypeptides. The larger subunit is a type I membrane glycoprotein (see Figure 13-10) encoded by the MHC. The smaller β2-microglobulin subunit is not encoded by the MHC and corresponds in structure to a free Ig domain. Originally purified from human leukocytes by digestion with papain, which releases the extracellular portion of the class I MHC molecules in intact form, these proteins are now produced by recombinant DNA technology procedures and have become important tools for the detection of antigen-specific T cells.

Implicit in the notion that MHC molecules are the targets of graft rejection is their structural variation, attributable to inherited variation (genetic polymorphism): If a recipient rejects a graft, the recipient's immune system must be capable of distinguishing unique features of the donor MHC molecules present on the graft. In fact, the genes encoded by the MHC are among the most polymorphic currently known, with over 2000 distinct allelic products identified in humans. The class I MHC molecules in humans are encoded by the HLA-A, HLA-B, and HLA-C loci (see Figure 23-19), each displaying extensive allelic variation. In the mouse, the class I MHC molecules are encoded by the H-2K and H-2D loci, each likewise with many known allelic variants. The three-dimensional structure of class I MHC molecules reveals two membrane-proximal Ig-like domains. These domains support an eight-stranded β-pleated sheet topped by two α helices. Jointly the β sheet and the helices create a cleft, closed at both ends, in which a peptide binds (see Figure 23-21a). The mode of peptide binding by a class I MHC molecule requires a peptide of rather fixed length, usually 8 to 10 amino acids, so that the ends of the peptide can be tucked into pockets that accommodate the charged amino and carboxyl groups at the termini. Further, the peptide is anchored into the peptide-binding cleft by means of a small number of amino acid side chains, each of which is accommodated by a pocket in the MHC molecule that neatly fits that particular amino acid residue (Figure 23-22a). On average, two such "specificity pockets" must be filled correctly to allow stable peptide binding. In this manner, a given MHC molecule can accommodate a large number of peptides of diverse sequence, as long as the "anchor" requirements are fulfilled.

The polymorphic residues that distinguish one allelic MHC molecule from another are mostly located in and

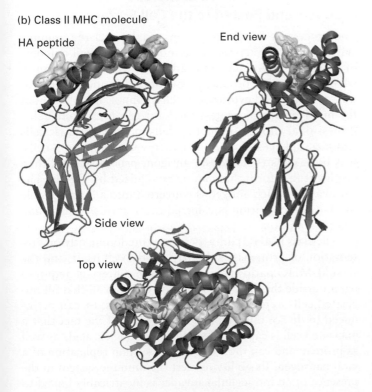

**(a) Class I MHC molecule**

HA peptide

End view

β2-microglobulin

**Class I MHC**

Side view

Top view

**(b) Class II MHC molecule**

HA peptide

End view

Side view

Top view

**FIGURE 23-21 Three-dimensional structure of class I and class II MHC molecules.** (a) Shown here is the structure of a class I MHC molecule with bound peptide as determined by x-ray crystallography. The portion of a class I MHC molecule that binds peptide consists of a β sheet composed of eight β strands and flanked by two α helices. The peptide-binding cleft is formed entirely from the MHC-encoded large subunit, which associates noncovalently with the small subunit (β2-microglobulin) encoded elsewhere. (b) Class II MHC molecules are structurally similar to class I molecules, but with several important distinctions. Both the α and β subunits of class II MHC molecules are MHC encoded and contribute to formation of the peptide-binding cleft. The peptide-binding cleft of class II MHC molecules accommodates a wider range of peptide sizes than that of class I molecules. [Part (a) based on D. N. Garboczi, 1996, *Nature* **384**:134. Part (b) based on J. Hennecke et al., 2000, *EMBO J.* **19**:5611.]

I MHC allele will therefore, as a rule, not interact with unrelated MHC molecules because of their different surface architectures (Figure 23-22b); this is the molecular basis of MHC restriction. The CD8 marker functions as a co-receptor, binding to conserved portions of the class I MHC molecules. The presence of CD8 thus "sets" the restriction specificity of any mature T cell that bears it.

**Class II MHC Molecules** The two subunits (α and β) of class II MHC molecules are both type I membrane glycoproteins and belong to the Ig superfamily. The typical mammalian MHC contains several loci that encode class II MHC molecules (see Figure 23-19). Like the large subunit of class I molecules, both the α and β subunits of class II molecules show genetic polymorphism.

The basic three-dimensional design of class II MHC molecules resembles that of class I MHC molecules: two membrane-proximal Ig-like domains support a peptide-binding portion composed of an eight-stranded β sheet and two α helices (see Figure 23-21b). For class II MHC molecules, the α and β subunits contribute equally to the construction of the peptide-binding cleft. This cleft is open at both ends and thus supports the binding of longer peptides that protrude from it. The mode of peptide binding involves pockets that accommodate specific peptide side chains, as well as contacts between side chains of the MHC molecule with main-chain atoms of the bound peptide. As for class I MHC, class II MHC polymorphisms mainly affect residues in and around the peptide-binding cleft, so that the peptide-binding specificity will usually differ among different allelic products.

A T-cell receptor that interacts with a particular class II MHC molecule will not, as a rule, interact with a different allelic molecule, not only because of the difference in the peptide-binding specificity of the allelic molecules, but also because of the polymorphisms that affect the contact residues with the T-cell receptor; as for class I MHC, this is the basis for class II MHC restricted recognition of antigen. As discussed below, class II MHC molecules evolved to present peptides generated predominantly in endosomes and lysosomes. The interactions between a peptide and class II MHC

around the peptide-binding cleft. These residues therefore determine the architecture of the peptide-binding pocket and hence the specificity of peptide binding. Further, these polymorphic residues affect the surface of the MHC molecule and hence the points of contact with the T-cell receptor. A T-cell receptor designed to interact with one particular class

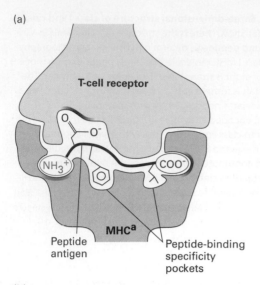

(a)

T-cell receptor

MHC<sup>a</sup>

Peptide antigen

Peptide-binding specificity pockets

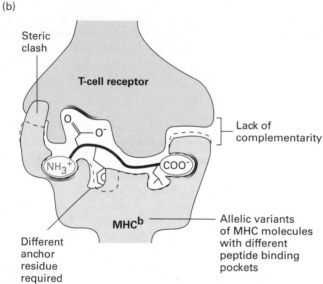

(b)

Steric clash

T-cell receptor

Lack of complementarity

MHC<sup>b</sup>

Different anchor residue required

Allelic variants of MHC molecules with different peptide binding pockets

**FIGURE 23-22 Peptide binding and MHC restriction.** (a) Peptides that bind to class I molecules are on average 8–10 residues in length, require proper accommodation of the termini, and include two or three residues that are conserved (anchor residues). Positions in class I molecules that distinguish one allele from another (polymorphic residues) occur in and around the peptide-binding cleft. The polymorphic residues in the MHC affect both the specificity of peptide binding and the interactions with T-cell receptors. Successful "recognition" of a peptide antigen–MHC complex by a T-cell receptor requires a good fit among the receptor, peptide, and MHC molecule. (b) Steric clash and a lack of complementarity between anchor residues and the MHC molecule prevent proper binding. T-cell receptors are thus restricted to specific peptide-MHC products.

molecule take place in these organelles, and class II MHC molecules are targeted specifically to those locations after their synthesis in the endoplasmic reticulum.

This targeting is accomplished by means of a chaperone called the invariant chain, a type II membrane glycoprotein (see Figure 13-10). The invariant chain (Ii) plays a key role

in the early stages of class II MHC biosynthesis by forming a trimeric structure onto which the class II MHC αβ heterodimers assemble. The final assembly product thus consists of nine polypeptides: $(\alpha\beta Ii)_3$. The interaction between Ii and the αβ heterodimer involves a stretch of Ii called the CLIP segment, which occupies the class II MHC peptide-binding cleft. Once the $(\alpha\beta Ii)_3$ complex is assembled, the complex enters the secretory pathway and is diverted to endosomes and lysosomes at the *trans*-Golgi network (see Figure 14-1). The signals responsible for this diversion are carried by the Ii cytoplasmic tail and do not obviously conform to the endosomal targeting or retrieval signals commonly found on lysosomal membrane proteins. Some of the $(\alpha\beta Ii)_3$ complexes are directed straight to the cell surface from which they may be internalized, but the vast majority end up in late endosomes.

As we saw for class I MHC molecules and their CD8 co-receptor, the CD4 co-receptor recognizes conserved features on class II MHC molecules. Any mature T cell that bears the CD4 co-receptor uses class II MHC molecules for antigen recognition.

## Antigen Presentation Is the Process by Which Protein Fragments Are Complexed with MHC Products and Posted to the Cell Surface

The process by which foreign materials enter the immune system is the key step that determines the eventual outcome of a response. A successful adaptive immune response, which includes the production of antibodies and the generation of helper and cytotoxic T cells, cannot unfold without the involvement of professional antigen-presenting cells. Professional antigen-presenting cells include dendritic cells and macrophages, both bone marrow-derived, and B cells. It is these cells that acquire antigen, process it, and then display it in a form that can be recognized by T cells. The pathway by which antigen is converted into a form suitable for T-cell recognition is referred to as *antigen processing and presentation*.

The class I MHC pathway focuses predominantly on presentation of proteins synthesized by the cell itself, and the class II MHC pathway is centered on materials acquired from outside the antigen-presenting cell. Recall that all nucleated cells express class I MHC products, or can be induced to do so; this makes sense in view of the fact that a nucleated cell is capable of synthesizing nucleic acids as well as protein, and can thus in principle sustain replication of a viral pathogen. Its ability to alert the immune system to the presence of an intracellular invader is inextricably linked to class I MHC-restricted antigen presentation. The distinction between presentation of materials synthesized by an antigen-presenting cell itself, and processing and presentation of antigen acquired from outside the cell, is by no means absolute. Together, the class I and class II pathways of antigen processing and presentation sample all of the compartments that need to be surveyed for the presence of pathogens.

Antigen processing and presentation in both the class I and II pathways may be divided into six discrete steps that are useful in comparing the two pathways: (1) acquisition of antigen, (2) tagging the antigen for destruction, (3) proteolysis, (4) delivery of peptides to MHC molecules, (5) binding of peptide to a MHC molecule, and (6) display of the peptide-loaded MHC molecule on the cell surface. In the next two sections, we describe the molecular details of each pathway.

## Class I MHC Pathway Presents Cytosolic Antigens

Figure 23-23 summarizes the six steps in the class I MHC pathway using a virus-infected cell as an example. The following discussion describes the events that occur during each step:

**1** *Acquisition of Antigen:* In the case of a virus infection, acquisition of antigen is usually synonymous with the infected state. Viruses rely on host protein synthesis to generate the building blocks for new virions. This includes the synthesis of viral cytosolic proteins as well as membrane proteins. Protein synthesis, unlike DNA replication, is an error-prone process, with a fraction of newly initiated polypeptide chains being terminated prematurely or suffering from other errors (misincorporation of amino acids, frameshifts, improper or delayed folding). These mistakes in protein synthesis affect both the host cell's own proteins and those specified by viral genomes. Such error-containing proteins must be rapidly removed so as not to clog up the cytoplasm, engage partner proteins in nonproductive interactions, or even act as dominant negative versions of a

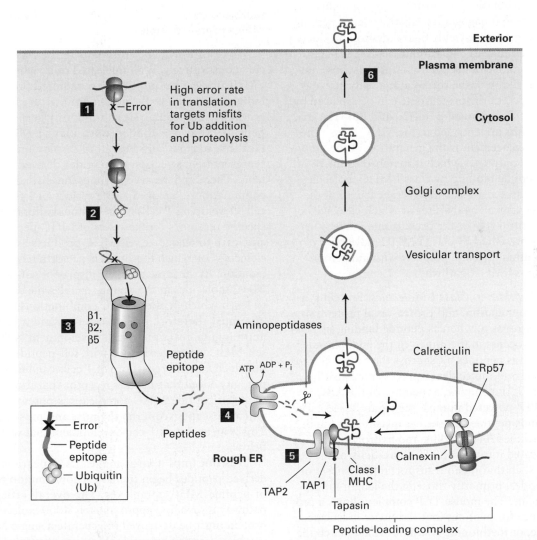

**FIGURE 23-23 Class I MHC pathway of antigen processing and presentation.** Step **1**: Acquisition of antigen is synonymous with the production of proteins with errors (premature termination, misincorporation). Step **2**: Misfolded proteins are targeted for degradation through conjugation with ubiquitin. Step **3**: Proteolysis is carried out by the proteasome. In cells exposed to interferon γ, the catalytically active β subunits of the proteasome are replaced by interferon-induced immune-specific β subunits. Step **4**: Peptides are delivered to the interior of the endoplasmic reticulum (ER) via the dimeric TAP peptide transporter. Step **5**: Peptide is loaded onto newly made class I MHC molecules within the peptide-loading complex. Step **6**: The fully assembled class I MHC–peptide complex is transported to the cell surface via the secretory pathway. See text for details.

protein. The rate of cytosolic proteolysis must be matched to the rate at which mistakes in protein synthesis and folding occur. These proteins are an important source of antigen peptides destined for presentation by class I MHC molecules. With the exception of cross-presentation (discussed below), the class I MHC pathway results in the formation of peptide-MHC complexes in which the peptides are derived from proteins synthesized by the class I MHC-bearing cell itself.

**2** *Targeting the Antigen for Destruction:* For the most part, the ubiquitin conjugation system is responsible for targeting a protein for destruction (see Chapter 3, p. 87). Ubiquitin conjugation is a covalent modification that is tightly regulated.

**3** *Proteolysis:* Ubiquitin-conjugated proteins are destroyed by proteasomal proteolysis. The proteasome is a highly processive protease that engages its substrates and, without the release of intermediates, yields final digestion products, peptides in the size range of 3–20 amino acids (see Figure 3-29). During the course of an inflammatory response and in response to interferon γ, the three catalytically active β subunits (β1, β2, β5) of the proteasome can be replaced by three immune-specific subunits: β1i, β2i, and β5i. The β1i, β2i, and β5i subunits are encoded in the MHC locus. The net result of this replacement is the generation of the *immunoproteasome,* the output of which is matched to the requirements for peptide binding by class I MHC molecules. The immunoproteasome adjusts the average length of the peptides produced as well as the sites at which cleavage occurs. Given the central role of the proteasome in the generation of peptides presented by class I MHC molecules, proteasome inhibitors interfere potently with antigen processing via the class I MHC pathway.

**4** *Delivery of Peptides to Class I Molecules:* Protein synthesis, ubiquitin conjugation, and proteasomal proteolysis all occur in the cytoplasm, whereas peptide binding by class I MHC molecules occurs in the lumen of the endoplasmic reticulum (ER). Thus peptides must cross the ER membrane to gain access to class I molecules, a process mediated by the heterodimeric TAP complex, a member of the **ABC superfamily** of ATP-powered pumps (see Figure 11-15). The TAP complex binds peptides on the cytoplasmic face and, in a cycle that includes ATP binding and hydrolysis, peptides are translocated into the ER. The specificity of the TAP complex is such that it can transport only a subset of all cytosolic peptides, primarily those in the length range of 5–10 amino acids. The mouse TAP complex shows a pronounced preference for peptides that terminate in leucine, valine, isoleucine, or methionine residues, which match the binding preference of the class I MHC molecules served by the TAP complex. The genes encoding the TAP1 and TAP2 subunits composing the TAP complex are located in the MHC locus.

**5** *Binding of Peptides to Class I Molecules:* Within the ER, newly synthesized class I MHC molecules are part of a multiprotein complex referred to as the peptide-loading complex. This complex includes two chaperones (calnexin and calreticulin) and the oxidoreductase Erp57. Another chaperone (tapasin) interacts with both the TAP complex and the class I MHC molecule about to receive peptide. The physical proximity of TAP and the class I MHC molecule is maintained by tapasin. Once peptide loading has occurred, a conformational change releases the loaded class I MHC molecule from the peptide-loading complex. This arrangement effectively ensures that only peptide-loaded molecules are displayed at the cell surface.

**6** *Display of Class I MHC–Peptide Complexes at the Cell Surface:* Once peptide loading is complete, the class I MHC–peptide complex is released from the peptide-loading complex and enters the constitutive secretory pathway (see Figure 14-1). Transfer from the Golgi to the cell surface is rapid and completes the biosynthetic pathway of a class I MHC–peptide complex.

The entire sequence of events in the class I pathway occurs constitutively in all nucleated cells, which express class I MHC molecules and the other required proteins or can be induced to do so. In the absence of a virus infection, protein synthesis and proteolysis continuously generate a stream of peptides that are loaded onto class I MHC molecules. Healthy, normal cells therefore display on their surface a representative selection of peptides derived from host proteins. There may be several thousand distinct MHC-peptide combinations displayed at the surface of a typical nucleated cell. Developing T cells in the thymus calibrate their antigen-specific receptors on these sets of MHC-peptide complexes, and learn to recognize self-MHC products as the "restriction elements" on which they must henceforth rely for antigen recognition. At the same time, the display of self-peptides by self-MHC molecules in the thymus enables the developing T cell to learn which peptide-MHC combinations are self-derived and must therefore be ignored to avoid a self-destructive autoimmune reaction. T-cell development is thus driven by self-MHC molecules loaded with self-peptides, a "template" on which a useful repertoire of T cells can be molded. Simply put, any T cell that bears a receptor that too strongly reacts with self peptide MHC complexes is potentially dangerous when it escapes from the thymus, and must be eliminated. This is an aspect of T-cell development that will be discussed below.

It is not until a virus makes its appearance that virus-derived peptides begin to make a contribution to the display of peptide-MHC complexes. The overall efficiency of this pathway is such that approximately 4000 molecules of a given protein must be destroyed to generate a single MHC-peptide complex carrying a peptide from that particular polypeptide.

An unusual mode of antigen presentation that is nonetheless crucial in the development of cytotoxic T cells is *cross-presentation.* This term refers to the acquisition by dendritic cells of apoptotic cell remnants, immune complexes, and possibly other forms of antigen by phagocytosis. By a pathway that has yet to be fully understood, these materials escape from phagosomal/endosomal compartments into the cytosol, where they are then handled according to the steps described above. Only dendritic cells are capable of cross-presentation, and so allow the loading of class I MHC molecules complexed with peptides that derive from cells other than the antigen-presenting cell itself. Autophagy (see Chapter 14) is likely to play a role in this process.

## Class II MHC Pathway Presents Antigens Delivered to the Endocytic Pathway

Although class I MHC and class II MHC molecules show a striking structural resemblance, the manner in which the two classes acquire peptide and their function in immune recognition differs greatly. Whereas the primary function of class I MHC molecules is to guide CD8-bearing cytotoxic T cells to their target cells, class II MHC molecules serve to guide CD4-bearing helper T cells to the cells with which they interact, primarily professional antigen-presenting cells. Activated CD4 T cells provide protection not only via the assistance provided to B cells in producing antibodies, but also by the complex sets of cytokines they produce, which activate phagocytic cells to clear pathogens, or help set up an inflammatory response.

As noted previously, class II MHC molecules are expressed primarily by professional antigen-presenting cells: dendritic cells and macrophages, which are phagocytic, and B cells, which are not. Hence, the class II MHC pathway of antigen processing and presentation generally occurs only in these cells. The steps in this pathway are depicted in Figure 23-24 and described below.

**1** *Acquisition of Antigen:* In the class II MHC pathway, antigen is acquired by pinocytosis, phagocytosis, or receptor-mediated endocytosis. Pinocytosis, which is rather nonspecific, involves the delivery, by a process of membrane invagination and fission, of a volume of extracellular fluid and the molecules dissolved therein. **Phagocytosis,** the ingestion of particulate materials, such as bacteria, viruses, and remnants of dead cells, involves extensive remodeling of the actin-based cytoskeleton to accommodate the incoming particle. Although phagocytosis may be initiated by specific receptor-ligand interactions, these are not always required: Even latex particles can be ingested very efficiently by macrophages. In the process of opsonization, pathogens decorated by antibodies and certain complement components are targeted to macrophages and dendritic cells, recognized by cell-surface receptors for complement components or receptors for the Fc portion of immunoglobulins, and then phagocytosed (Figure 23-25). Macrophages and dendritic cells also express several types of less selective receptors (e.g., C-type lectins, Toll-like receptors, scavenger receptors) that recognize molecular patterns on nonparticulate antigens; the bound antigen is internalized by **receptor-mediated endocytosis.** B cells, which are not phagocytic, also can acquire antigen by receptor-mediated endocytosis using their antigen-specific B-cell receptors (surface immunoglobulin) (Figure 23-26). Finally, cytosolic antigens may enter the class II MHC pathway via autophagy (see Figure 14-35). After formation of the autophagic cup, an autophagic vesicle is formed. These vesicles are of a size that can accommodate damaged organelles, and sizable quantities of cytoplasm may be encapsulated in the process. The resulting autophagosomes are destined to fuse with lysosomes, where the contents of the autophagosome then become available for digestion by lysosomal proteases.

**2** *Targeting Antigen for Destruction:* Proteolysis is required to convert intact protein antigen into peptides of a size suitable for binding to class II MHC molecules. Protein antigens are targeted for degradation by progressive unfolding, brought about by the drop in pH as proteins progress along the endocytic pathway. The pH of the extracellular environment is around pH 7.2, and that in early endosomes is between pH 6.5 and 5.5; in late endosomes and lysosomes the pH may drop to pH 4.5. ATP-powered V-class proton pumps in the endosomal and lysosomal membranes are responsible for this acidification (see Figure 11-9). Proteins that are stable at neutral pH tend to unfold when they are exposed to extremes of pH, through rupture of hydrogen bonds and destabilization of salt bridges. Further, the environment in the endosomal/lysosomal compartments is reducing, with lysosomes attaining a concentration of reducing equivalents in the millimolar range. Reduction of disulfide bonds that stabilize many extracellular proteins is the reversal of a covalent modification, catalyzed also by a thioreductase inducible by exposure to IFN-γ. The combined action of low pH and reducing environment prepares the antigens for proteolysis.

**3** *Proteolysis:* Degradation of proteins in the class II MHC pathway is carried out by a large set of lysosomal proteases, collectively referred to as cathepsins, which are either cysteine or aspartyl proteases. Other proteases, such as asparagine-specific endoprotease, also may contribute to proteolysis. A wide range of peptide fragments is produced, including some that can bind to class II MHC molecules. The lysosomal proteases operate optimally at the acidic pH within lysosomes. Consequently, agents that inhibit the activity of V-class proton pumps interfere with antigen processing, as do inhibitors of lysosomal proteases.

**4** *Encounter of Peptides with Class II Molecules:* Recall that most class II MHC molecules synthesized in the endoplasmic reticulum are directed to late endosomes. Because the peptides generated by proteolysis reside in the same topological space as the class II MHC molecules themselves, the delivery of peptide to class II MHC molecules does not require traversal of a membrane. The sole requirement, therefore, is to allow peptides and class II MHC molecules to meet. This is accomplished in the course of biosynthesis by a sorting step that is dependent on the invariant chain (Ii) and ensures delivery of the class II (αβIi)$_3$ complex to endosomal compartments.

**5** *Binding of Peptides to Class II Molecules:* The (αβIi)$_3$ complex delivered to endosomal compartments is incapable

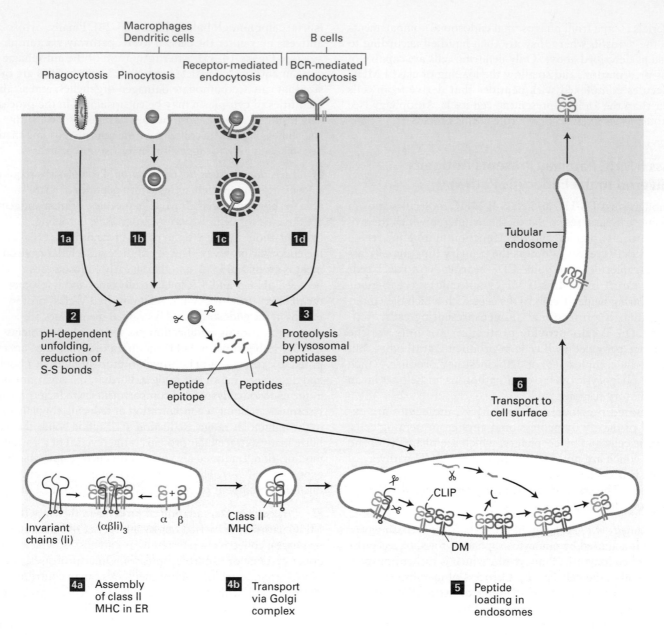

**FIGURE 23-24 Class II MHC pathway of antigen processing and presentation.** Step **1**: Particulate antigens are acquired by phagocytosis, and nonparticulate antigens by pinocytosis or endocytosis. Step **2**: Exposure of antigen to the low pH and reducing environment of endosomes and lysosomes prepares the antigen for proteolysis. Step **3**: The antigen is broken down by various proteases in endosomal and lysosomal compartments. Step **4**: Class II MHC molecules, assembled in the ER from their subunits, are delivered to endosomal/lysosomal compartments by means of signals contained in the associated invariant (Ii) chain. This delivery targets late endosomes, lysosomes, and early endosomes, ensuring that class II MHC molecules are exposed to the products of proteolytic breakdown of antigen along the entire endocytic pathway. Step **5**: Peptide loading is accomplished with the assistance of DM, a class II MHC-like chaperone protein. Step **6**: Display of peptide-loaded class II MHC molecules at the cell surface. See text for details.

of binding peptide because the peptide-binding cleft in the class II molecule is occupied by Ii. For the same reason, newly assembled class II $(\alpha\beta Ii)_3$ complexes do not compete for class I MHC-destined peptides delivered to the ER via TAP: their peptide binding site is already occupied by Ii. Recall that the ER is the site where both class I and class II MHC molecules assemble. The presence of Ii in the nascent class II MHC complex ensures that only class I MHC

molecules bind peptide in the ER. The same proteases that act on internalized antigens and degrade them into peptides act also on the $(\alpha\beta Ii)_3$ complex, resulting in removal of Ii with the exception of a small portion called the CLIP segment. Because it is firmly lodged in the class II MHC peptide-binding cleft, CLIP is resistant to proteolytic attack. The class II MHC molecules themselves are also resistant to unfolding and proteolytic attack under the conditions that

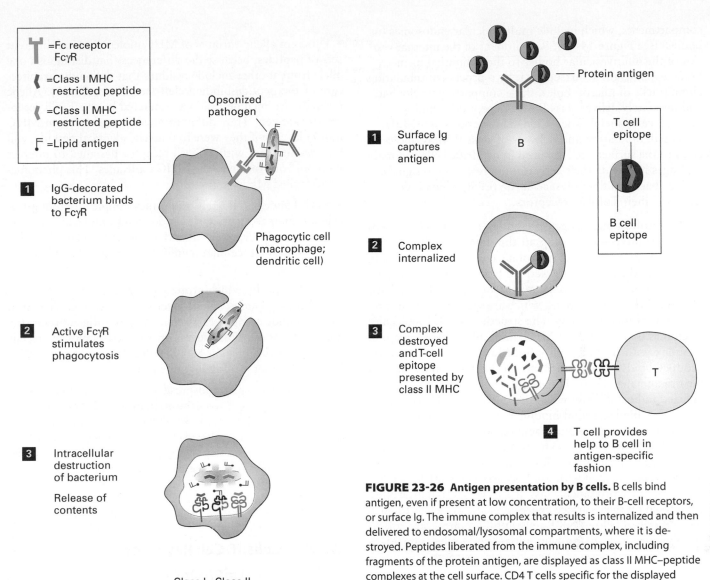

Opsonized pathogen

Phagocytic cell (macrophage; dendritic cell)

**1** IgG-decorated bacterium binds to FcγR

**2** Active FcγR stimulates phagocytosis

**3** Intracellular destruction of bacterium

Release of contents

Class I   Class II

CD1

**4** Presentation of bacterial antigens to T cells via class I cross-presentation and class II MHC

Lipid presentation via CD1

**FIGURE 23-25 Presentation of opsonized antigen by phagocytic cells.** By means of Fc receptors such as FcγR displayed on their cell surface, specialized phagocytic cells such as macrophages or dendritic cells can bind and ingest pathogens that have been decorated with antibodies (opsonization). After digestion of the phagocytosed particle (e.g., immune complex, bacterium, virus), some of the peptides produced, including fragments of the pathogen (orange), are loaded onto class II MHC molecules (green). Class II MHC–peptide complexes displayed at the surface allow activation of T cells specific for these MHC-peptide combinations. Lipid antigens are delivered to the class I MHC–like molecule CD1 (pink), whose binding site is specialized to accommodate lipids. Certain pathogen-derived peptides (purple) may be delivered to class I MHC products (blue) by means of cross-presentation. The mechanisms that underlie cross-presentation remain to be clarified.

Protein antigen

**1** Surface Ig captures antigen

B

T cell epitope

B cell epitope

**2** Complex internalized

**3** Complex destroyed and T-cell epitope presented by class II MHC

T

**4** T cell provides help to B cell in antigen-specific fashion

**FIGURE 23-26 Antigen presentation by B cells.** B cells bind antigen, even if present at low concentration, to their B-cell receptors, or surface Ig. The immune complex that results is internalized and then delivered to endosomal/lysosomal compartments, where it is destroyed. Peptides liberated from the immune complex, including fragments of the protein antigen, are displayed as class II MHC–peptide complexes at the cell surface. CD4 T cells specific for the displayed complex can now provide help to the B cell. This help is MHC restricted and antigen specific.

prevail in the endocytic pathway. The CLIP segment is removed through interaction of the αβCLIP complex with a chaperone, DM. The newly vacated peptide-binding cleft of the class II MHC molecule may now bind peptides that are abundantly present in the endocytic pathway. Although the DM protein is MHC encoded and structurally very similar to class II molecules, it does not itself bind peptides. However, newly formed class II MHC–peptide complexes are themselves susceptible to further "editing" by DM by dislodging of the peptide already bound, until the class II molecule acquires a peptide that binds so strongly that it cannot be removed by further interactions of the complex with DM. The resulting class II MHC–peptide complexes are extremely stable, with estimated half-lives in excess of 24 hours.

**6** *Display of Class II MHC–Peptide Complexes at the Cell Surface:* The newly generated class II MHC–peptide complexes are localized mostly in late endosomal compartments, which include multivesicular endosomes (or bodies) (see Figure 14-33). Recruitment of the internal vesicles of the multivesicular bodies to the delimiting membrane expands their surface area: by a process of tubulation along tracks of microtubules, these compartments elongate and ultimately deliver class II MHC–peptide complexes to the surface by membrane fusion. These events are tightly regulated: tubulation and delivery of class II MHC molecules to the surface are enhanced in dendritic cells and macrophages following their activation in response to signals, such as bacterial lipopolysaccharide (LPS), which is detected by their Toll-like receptors.

For professional antigen-presenting cells, the above steps are constitutive—happening all the time—but they can be modulated by exposure to microbial agents and cytokines. In addition to the pathways described here for class I and class II MHC products, there is a category of class I MHC-related molecules, the CD1 proteins, which are specialized in the presentation of lipid antigens. The structure of CD1 molecules resembles that of a class I MHC molecule: a heavy chain in complex with β2-microglobulin. Many species of bacteria produce lipids of a chemical structure that is not found in their mammalian hosts. These lipids can serve as antigens recognized in host defense when presented by CD1 molecules, where they bind to a lipid-binding pocket that is conceptually similar to that of the peptide-binding pocket of regular MHC molecules. Signals in the cytoplasmic tail of the CD1 heavy chain target these molecules to endolysosomal compartments, where loading with lipid occurs. The CD1-lipid complexes engage a prominent class of T cells, referred to as NKT cells based on their surface characteristics, as well as to γδ T cells described below. NKT cells fulfill an important role in cytokine production and help initiate and orchestrate adaptive immune responses via their cytokine outputs.

## KEY CONCEPTS of Section 23.4

### The MHC and Antigen Presentation

• The MHC, discovered as the genetic region responsible for acceptance or rejection of grafts, encodes two major classes of type I membrane glycoproteins, class I and class II MHC molecules. These proteins are highly polymorphic, occurring in many allelic variations in an outbred population (see Figure 23-19).

• The function of MHC products is to bind peptides and display them to antigen-specific receptors on T cells. Class I MHC molecules are found on most nucleated cells, whereas the expression of class II MHC molecules is confined largely to professional antigen-presenting cells such as dendritic cells, macrophages, and B cells.

• The organization and structure of class I and class II MHC molecules is similar and includes a region specialized for binding a wide variety of different peptides (see Figure 23-21).

• Different allelic variants of MHC molecules bind different sets of peptides, because the differences that distinguish one allele from another include residues that define the architecture of the peptide-binding cleft (see Figure 23-22). Allelic residues also include residues contacted by the T-cell receptor for antigen. Thus different allelic variants of an MHC molecule, even if they were to bind the identical peptide, will not usually react with a T-cell receptor designed to interact with only one of the allelic MHC molecules. This phenomenon is called MHC restriction.

• Class I and class II MHC molecules sample different intracellular compartments for peptide: class I molecules sample predominantly cytosolic materials, whereas class II molecules sample extracellular materials internalized by phagocytosis, pinocytosis, or receptor-mediated endocytosis.

• The process by which protein antigens are acquired, processed into peptides, and converted into surface-displayed MHC-peptide complexes is referred to as antigen processing and presentation. This process operates continuously in cells that express the relevant MHC molecules, yet can be modulated in the course of an immune response.

• Antigen processing and presentation can be divided into six discrete steps: (1) acquisition of antigen; (2) targeting the antigen for destruction; (3) proteolysis; (4) encounter of peptides with MHC molecules; (5) binding of peptide to the MHC molecule; and (6) display of the peptide-loaded MHC molecules on the cell surface (see Figures 23-25 and 23-26).

## 23.5 T Cells, T-Cell Receptors, and T-Cell Development

T lymphocytes recognize antigen through the lens of MHC molecules. The receptors entrusted with this task are structurally related to immunoglobulins. To generate these antigen-specific receptors, T cells rearrange the genes encoding the T-cell receptor (TCR) subunits by mechanisms of somatic recombination essentially identical to those used by B cells to rearrange immunoglobulin genes. The development of T cells is also strictly dependent on successful completion of the somatic gene rearrangements that yield the TCR subunits. We shall describe the subunits that mediate antigen-specific recognition, how they pair up with membrane glycoproteins essential for signal transduction, and how these complexes recognize MHC-peptide combinations.

As pointed out in the preceding section, T cells recognize antigens only together with the polymorphic MHC molecules that happen to be present in that host. In the course of T-cell development, T cells must "learn" the identity of these "self" MHC molecules and receive instructions about which MHC-peptide combinations to ignore, so as to avoid potentially catastrophic reactivity of newly generated T cells with the host's own tissues (i.e., autoimmunity). We shall describe how T cells develop, and introduce the major classes of T cells according to their function.

## The Structure of the T-Cell Receptor Resembles the F(ab) Portion of an Immunoglobulin

Much as B cells utilize the B-cell receptors to recognize antigens and then transduce signals that lead to their clonal expansion, so T cells use their **T-cell receptors.** T cells that have been activated through engagement of these antigen-specific receptors proliferate and acquire the capacity to kill antigen-bearing target cells or to secrete cytokines that will assist B cells in their differentiation. The T-cell receptor for antigen recognizes MHC molecules complexed with the appropriate peptides.

The T-cell receptor (Figure 23-27) is composed of two glycoprotein subunits, each of which is encoded by a somatically rearranged gene. The receptors are composed of either an $\alpha\beta$ pair or a $\gamma\delta$ pair. The structure of these subunits shows structural similarity to the F(ab) portion of an immunoglobulin: at the N-terminal end there is a variable domain, followed by a constant-region domain and a transmembrane segment. The cytoplasmic tails of the TCR subunits are too short to allow recruitment of cytosolic factors that assist in signal transduction. Instead, the TCR associates with the CD3 complex, a set of membrane proteins composed of $\gamma$, $\delta$, $\varepsilon$, and $\zeta$ chains. (The TCR $\gamma$ and $\delta$ subunits are not to be confused with the similarly designated subunits of the CD3 complex.) The $\varepsilon$ chain forms a noncovalent dimer with the $\gamma$ or the $\delta$ chain to yield $\delta\varepsilon$ and $\gamma\varepsilon$ complexes. The external domain of the CD3 subunits is homologous to immunoglobulin domains, and the cytoplasmic domain in each contains an ITAM domain, through which adapter molecules may be recruited upon phosphorylation of tyrosine residues in the ITAMs. The $\zeta$ chain is integrated into the CD3-TCR complex as a disulfide-bonded homodimer, and each $\zeta$ chain contains three ITAM motifs.

## TCR Genes Are Rearranged in a Manner Similar to Immunoglobulin Genes

Virtually all antigen-specific receptors generated by somatic recombination contain a subunit that is the product of V-D-J recombination (e.g., Ig heavy chain; TCR $\beta$ chain) and a

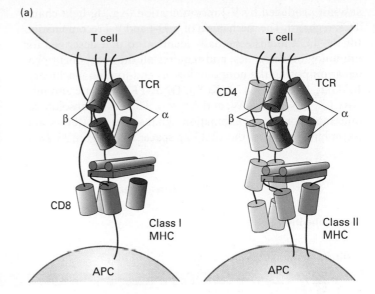

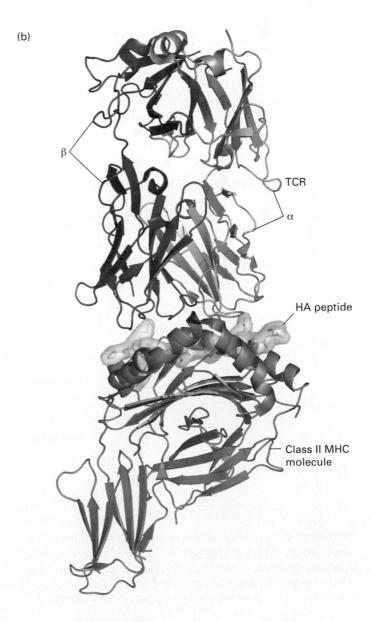

**FIGURE 23-27 Structure of the T-cell receptor and its co-receptors.** (a) The T-cell receptor (TCR) for antigen is composed of two chains, the $\alpha$ and $\beta$ subunits, which are produced by V-J and V-D-J recombination, respectively. The $\alpha\beta$ subunits must associate with the CD3 complex (see Figure 23-29) to allow the transduction of signals. The formation of a full TCR$\alpha\beta$–CD3 complex is required for surface expression. The T-cell receptor further associates with the co-receptors CD8 (light blue) or CD4 (light green), which allow interaction with conserved features of class I MHC and class II MHC molecules, respectively, on antigen-presenting cells. (b) Structure of the T-cell receptor bound to a class II MHC–peptide complex as determined by x-ray crystallography. [Part (b) based on J. Hennecke, 2000, *EMBO J.* **19**:5611.]

subunit produced by V-J recombination (e.g., Ig light chain; TCR α chain). The mechanism of V-D-J and V-J recombination for the TCR loci is essentially identical to that described for immunoglobulin genes, and requires all the component proteins composing the nonhomologous end-joining machinery: RAG-1, RAG-2, Ku70, Ku80, DNA-PK catalytic subunit, XRCC4, DNA ligase IV, and Artemis. There is an absolute requirement for the recombination signal sequences (RSSs), and recombination obeys the 12/23-bp spacer rule (Figure 23-28).

A number of noteworthy features characterize the organization and rearrangement of the TCR loci. First, the organization of the recombination signals is such that D-to-D rearrangements are allowed. Second, the enzyme terminal deoxynucleotidyltransferase (TdT) is active at the time the TCR genes rearrange, and therefore N nucleotides can be present in all rearranged TCR genes. Third, in the human and mouse, the TCR δ locus is embedded within the TCR α locus. This organization results in complete excision of the

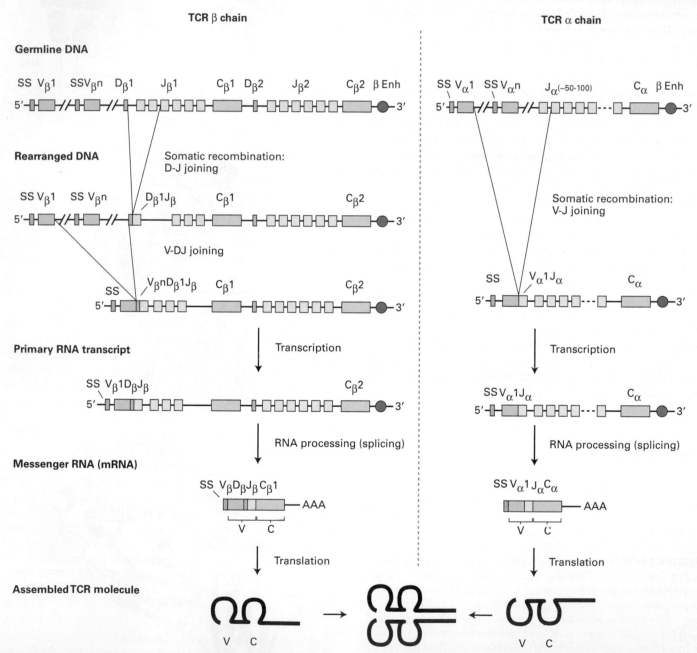

**FIGURE 23-28 Organization and recombination of TCR loci.** The organization of TCR loci is in principle similar to that of immunoglobulin loci (see Figure 23-14). *Left:* The TCR β-chain locus includes a cluster of V segments, a cluster of D segments, and several J segments, downstream of which are two constant regions. The arrangement of the recombination signals is such that not only is D-J joining allowed, but also V-DJ joining. Direct V-J joining in the TCRβ locus is not observed. *Right:* The TCR α-chain locus is composed of a cluster of V segments and a large number of J segments. SS = exon encoding signal sequence; Enh = enhancer.

interposed δ locus when TCRα rearrangement occurs, and so a choice of the TCR α locus for rearrangement precludes utilization of the δ locus, now lost by deletion. T cells that express the αβ receptor and those that express the γδ receptor are considered separate lineages with distinct functions. Among the T cells expressing γδ receptors are some capable of recognizing the CD1 molecule, which is specialized in the presentation of lipid antigens. The γδ T cells are programmed to home in on distinct anatomical sites (e.g., the epithelium lining the genital tract, the skin) and likely play a role in host defense against pathogens commonly found at these sites.

Deficiencies in the key components of the recombination apparatus, such as the RAG recombinases, preclude rearrangement of TCR genes. As we have seen for B cells, and as will be described below for T cells, development of lymphocytes is strictly dependent on the rearrangement of the antigen-receptor genes. A RAG deficiency thus prevents both B- and T-cell development.

## T-Cell Receptors Are Very Diverse, with Many of Their Variable Residues Encoded in the Junctions Between V, D, and J Gene Segments

The diversity created by somatic rearrangement of TCR genes is enormous, estimated to exceed $10^{10}$ unique receptors. Combinatorial usage of different V, D, and J gene segments makes an important contribution, as do the mechanisms of junctional imprecision and N-nucleotide addition already discussed for immunoglobulin gene rearrangements. The net result is a degree of variability in the V regions that matches that of the CDR3s of immunoglobulins (see Figure 23-12). Indeed, each of the TCR's variable regions includes three CDRs, equivalent to those in the BCR. Unlike immunoglobulin genes, however, the TCR genes do not undergo somatic hypermutation. Therefore T-cell receptors exhibit nothing equivalent to the affinity maturation of antibodies during the course of an immune response, nor is there the option of class-switch recombination or the use of alternative polyadenylation sites to create soluble and membrane-bound versions of the antigen-specific receptors.

The crystal structures of a number of T-cell receptors bound to class I MHC-peptide or class II MHC–peptide complexes have been solved. These structures show variability in how the T-cell receptor docks with the MHC-peptide complex, but the most extensive contacts in the somatically diversified CDR3 region are made with the central peptide portion of the complex, with the germ-line-encoded CDR1 and CDR2 contacting the α helices of the MHC molecules. Many of the T-cell receptors for which a structure has been solved sit down diagonally across the peptide-binding portion of the MHC-peptide complex. As a result, the T-cell receptor makes extensive contacts with the peptide cargo as well as with the α helices of the MHC molecule to which it binds. The positions at which allelic MHC molecules differ from one another frequently involve residues that directly contact the T-cell receptor, thus precluding tight binding of the "wrong" allele.

Amino acid differences that distinguish one MHC allele from another also affect the architecture of the peptide-binding cleft. Even if the MHC residues that interact directly with the T-cell receptor were shared by two MHC allelic molecules, their peptide-binding specificity is likely to differ because of amino acid differences in the peptide-binding cleft. Consequently, the TCR contact residues provided by bound peptide and essential for stable interaction with a T-cell receptor would be absent from the "wrong" MHC-peptide combination. A productive interaction with the T-cell receptor is then unlikely to occur.

## Signaling via Antigen-Specific Receptors Triggers Proliferation and Differentiation of T and B Cells

Both B-cell and T-cell receptors for antigen transduce signals by means of proteins associated with the antigen-specific portions of the receptor (i.e., Ig heavy and light chains for the BCR; α and β chains for the TCR). The cytosolic portions of the antigen-specific receptors themselves are very short, do not protrude much beyond the cytosolic leaflet of the plasma membrane, and are incapable of recruitment of downstream signaling molecules. Instead, as discussed previously, the antigen-specific receptors on T and B cells associate with auxiliary subunits that contain ITAMs (immunoreceptor tyrosine-based activation motifs). Engagement of antigen-specific receptors by ligand initiates a series of receptor-proximal events: kinase activation, modification of ITAMs, and subsequent recruitment of adapter molecules that serve as scaffolds for recruitment of yet other downstream signaling molecules.

As outlined in Figure 23-29, engagement of antigen-specific receptors activates Src-family tyrosine kinases (e.g., Lck in CD4 T cells; Lyn and Fyn in B cells). These kinases are found in close proximity to or physically associated with the antigen receptor. The active Src kinases phosphorylate the ITAMs in the antigen receptors' auxiliary subunits. In their phosphorylated forms, these ITAMs recruit and activate non–Src-family tyrosine kinases (ZAP-70 in T cells, Syk in B cells) as well as other adapter molecules. Such recruitment and activation involves phosphoinositide-specific phospholipase $C_\gamma$ and PI-3 kinases. Subsequent downstream events parallel those discussed in Chapter 16 for signaling from receptor tyrosine kinases. Antigen-specific receptors on B and T cells are perhaps best characterized as "modular" receptor tyrosine kinases, with the ligand-recognition units and kinase domains carried by separate molecules. Ultimately, signaling via antigen-specific receptors initiates transcription programs that determine the fate of the activated lymphocyte: proliferation and differentiation.

T cells depend critically on the cytokine interleukin 2 (IL-2) for clonal expansion. Following antigen stimulation of a T cell, one of the first genes to be turned on is that for IL-2. The T cell responds to its own initial burst of IL-2 and proceeds to make more IL-2, an example of autocrine stimulation and part of a positive feedback loop. An important transcription factor required for the induction of IL-2 synthesis is the NF-AT protein

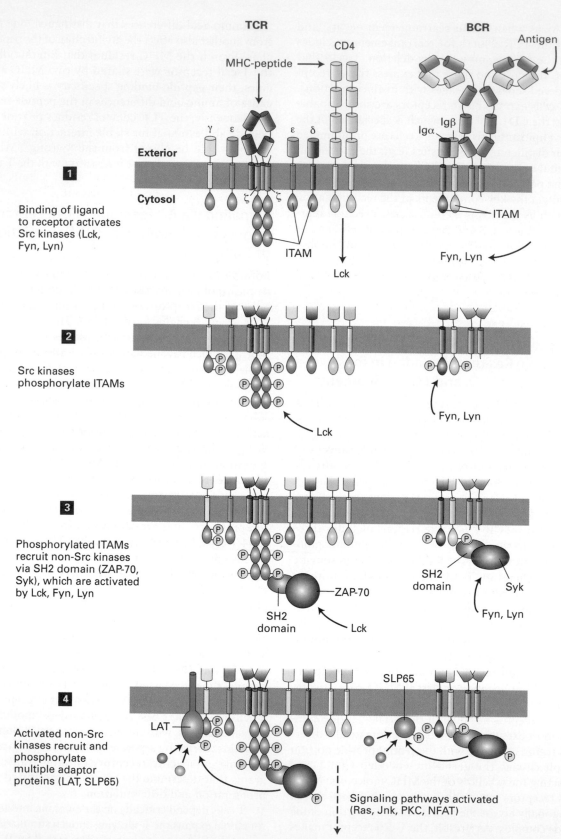

**FIGURE 23-29 Signal transduction from the T-cell receptor (TCR) and B-cell receptor (BCR).** The signal-transduction pathways used by the antigen-specific receptors of T cells (*left*) and B cells (*right*) are conceptually similar. The initial stages are depicted in this figure; downstream signaling events lead to changes in gene expression that result in proliferation and differentiation of the antigen-stimulated lymphocytes. See text for further discussion.

(nuclear factor of activated T cells). This protein is sequestered in the cytoplasm in phosphorylated form and cannot enter the nucleus unless it is dephosphorylated first. The phosphatase responsible is calcineurin, a $Ca^{2+}$-activated enzyme. The initial rise in cytosolic $Ca^{2+}$ leading to activation of calcineurin results from mobilization of ER-resident $Ca^{2+}$ stores triggered by hydrolysis of $PIP_2$ and the concomitant generation of $IP_3$ (see Figure 15-36, steps **2**–**4**).

The immunosuppressant drug cyclosporine inhibits calcineurin activity through formation of a cyclosporine-cyclophilin complex, which binds and inhibits calcineurin. If dephosphorylation of NF-AT is suppressed, NF-AT cannot enter the nucleus and participate in the up-regulation of transcription of the IL-2 gene. This precludes expansion of antigen-stimulated T cells and so leads to immunosuppression, arguably the single most important intervention that contributes to successful organ transplantation. Although the success of transplantation varies with the organ used, the availability of strong immunosuppressants such as cyclosporine has expanded enormously the possibilities of clinical transplantation. ∎

## T Cells Capable of Recognizing MHC Molecules Develop Through a Process of Positive and Negative Selection

The rearrangement of the gene segments that are assembled into a functional T-cell receptor is a stochastic event, completed on the part of the T cell without any prior knowledge of the MHC molecules with which these T-cell receptors must ultimately interact. Similar to somatic recombination of Ig heavy-chain loci in B cells, the first TCR gene segments to rearrange are the TCRβ D and J elements, followed by joining of a V segment to the newly recombined DJ. At this stage of T-cell development, productive rearrangement allows the synthesis of the TCR β chain, which is incorporated into the pre-TCR through association with the pre-T α subunit. This pre-TCR fulfills a function strictly analogous to that of the pre-BCR in B-cell development: it tells the T cell that it has successfully completed a productive rearrangement, with no need for further rearrangements on the opposing allele. It allows expansion of pre-T cells that successfully underwent rearrangement, and it imposes allelic exclusion to ensure that, as a rule, a single functional TCRβ subunit is generated for a given T cell and its descendants. RAG expression next subsides, and after the ensuing expansion phase of pre-T cells is complete, RAG expression is re-initiated to allow rearrangement of the TCRα locus, ultimately leading to the generation of T cells with a fully assembled TCR αβ receptor. Figure 23-30 illustrates the analogous steps in the development of T and B cells.

How is the newly emerging repertoire of T cells shaped so that a productive interaction with self-MHC molecules can occur? The random nature of the gene rearrangement process and the enormous variability engendered as a consequence produces a set of T-cell receptors, the vast majority of which cannot interact productively with the host MHC products. Recall that antigen processing and presentation are constitutive processes, and that in the thymus, where these selection events take place, all self-MHC molecules are necessarily occupied with peptides derived from self-proteins. These combinations of self-peptides complexed to class I and class II MHC molecules constitute the template on which the initial set of T-cell receptors must be calibrated.

T cells born with receptors that cannot engage self-MHC are useless. Such T cells will fail to perceive survival signals via the newly generated T-cell receptors and die. At the other extreme are T cells endowed with receptors that show a perfect fit for self-MHC complexed with a particular self-peptide. These T cells, if allowed to leave the thymus and take up residence in peripheral lymphoid organs, would be by definition self-reactive and could cause autoimmunity. If the number of such self-peptide MHC combinations surpasses a threshold sufficient to allow triggering of the T-cell receptor, then these T cells are instructed to die in the thymus by apoptosis, the ultimate fate of the vast majority of cells that enter and expand in the thymus. This is called negative selection and serves to purge the T-cell repertoire of overtly self-reactive T cells. Any T cell equipped with a receptor that perceives signals sufficiently strong to allow survival, but below the threshold that would impel the T cell to initiate apoptosis, receives survival signals and is positively selected. The heterogeneity of peptide-MHC complexes displayed on the surface of T cells undergoing selection makes it highly probable that the T-cell receptor interprets signals not only in a qualitative (strength, duration) manner but also in additive fashion: The summation of binding energies of the different MHC–self-peptide combinations on display helps determine the outcome of selection. This is called the *avidity model of T-cell selection.*

T cells are killed off by apoptosis only if the appropriate self-antigen is represented adequately in the thymus as MHC-peptide complexes. Proteins that are expressed in tissue-specific fashion, such as insulin in the β cells of the pancreas or the components of the myelin sheath in the nervous system, obviously do not fit this category. However, a factor called AIRE (autoimmune regulator) allows expression in a subset of epithelial cells of such tissue-specific antigens. How AIRE accomplishes this is not known, but it is widely suspected of directly regulating the transcription of the relevant genes in the thymus and at select sites in secondary lymphoid organs. Defects in AIRE lead to a failure to express these tissue-specific antigens in the thymus. In individuals who do not express AIRE, developing T cells fail to receive the full set of instructions in the thymus that lead to elimination of potentially self-reactive T cells. As a consequence, these individuals show a bewildering array of autoimmune responses, causing widespread tissue damage and disease.

TCR rearrangement occurs coincident with the gradual acquisition of the co-receptors CD4 and CD8. A key intermediate in T-cell development is a thymocyte that expresses CD4 and CD8, as well as a functional TCR-CD3 complex.

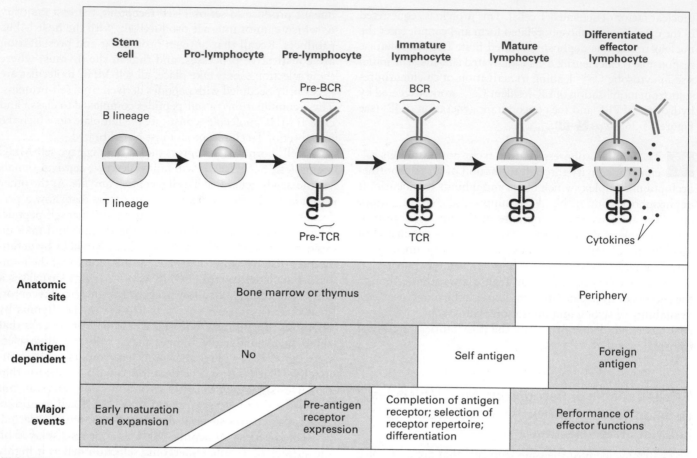

**FIGURE 23-30 Comparison of T-cell and B-cell development.**
Cell fate decisions are executed by receptors composed of either the newly rearranged μ chain (pre-BCR) or the newly rearranged β chain (pre-TCR). The pre-B- and pre-T-cell receptors serve similar functions: expansion of cells that successfully underwent rearrangement and allelic exclusion. This phase of lymphocyte development does not require antigen-specific recognition. Both the pre-BCR and pre-TCR include subunits unique to each receptor and absent from the antigen-specific receptors found on mature lymphocytes: VpreB and

λ5 (orange, green) for the pre-BCR; pre-Tα (blue) for the pre-TCR. Upon completion of the expansion phase, expression begins of the gene encoding the remaining subunit of the antigen-specific receptor: Ig light chain (light blue) for the BCR; TCR α chain (light red) for the TCR. Lymphocyte development and differentiation occur at distinct anatomical sites, and only fully assembled antigen-specific receptors (BCR, TCR) recognize antigen. Mature lymphocytes are strictly dependent on antigen recognition for their activation.

These cells are called double positive (CD4CD8+) cells and are found only as developmental intermediates in the thymus. The choice of which co-receptor (CD4 or CD8) to express determines whether a T cell will recognize class I or class II MHC molecules. How a newborn CD4CD8+ cell is instructed to become a CD8 (class I MHC restricted) T cell or CD4 (class II MHC restricted) T cell is not entirely settled.

## T Cells Require Two Types of Signal for Full Activation

All T cells require a signal via their antigen-specific receptor, the TCR, for activation, but this is not sufficient: the T cell also needs co-stimulatory signals. To perceive co-stimulatory signals, T cells carry on their surface several different receptors, of which the CD28 molecule is the best-known example. CD28 interacts with CD80 and CD86, two surface glycoproteins on the antigen-presenting cells with which the T cell interacts.

Expression of CD80 and CD86 is up-regulated when these antigen-presenting cells have themselves received the proper stimulatory signals, for example, by engagement of their Toll-like receptors (TLRs). The signals delivered via CD28 synergize with signals that emanate from the engaged T-cell receptor, all of which are required for full activation (Figure 23-31).

T cells, once activated, also express receptors that provide an attenuating or inhibitory signal upon recognition of these very same co-stimulatory molecules. The CTLA4 protein, whose expression on T cells is induced only upon activation, competes with CD28 for binding of CD80 and CD86. Because the affinity of CTLA4 for the CD80 and C86 proteins is higher than that of CD28, the inhibitory signals provided through CTLA4 will ultimately dominate the stimulatory signals that derive from engagement of CD28. Co-stimulatory molecules can thus be stimulatory or inhibitory and provide an important means of controlling the activation status and duration of a T-cell response.

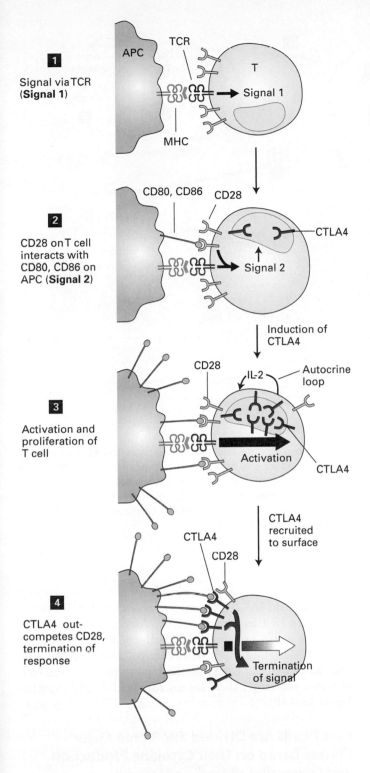

**1**

Signal via TCR
(**Signal 1**)

**2**

CD28 on T cell
interacts with
CD80, CD86 on
APC (**Signal 2**)

**3**

Activation and
proliferation of
T cell

**4**

CTLA4 out-
competes CD28,
termination of
response

**FIGURE 23-31  Signals involved in T-cell activation and termination.**
The two-signal model of T-cell activation involves recognition of an MHC-peptide complex by the T-cell receptor, which constitutes Signal 1 (step **1**), along with recognition of co-stimulatory molecules (CD80, CD86) on the surface of the antigen-presenting cell, which constitutes Signal 2 (step **2**). If costimulation is not provided, the newly engaged T cell becomes unresponsive (anergic). The provision of both Signal 1 via the T-cell receptor and Signal 2 via engagement of CD80 and CD86 by CD28 allows full activation. Full activation, in turn, leads to increased expression of CTLA4 (step **3**). After moving to the T-cell surface, CTLA4 binds CD80 and CD86, leading to inhibition of the T-cell response (step **4**). Because the affinity of CTLA4 for CD80 and CD86 is greater than that of the CD28, T-cell activation eventually is terminated.

allow a properly primed CD8 T cell to kill the target cell that bears it.

The mechanism of killing by CTLs involves two classes of proteins that act synergistically, *perforins* and *granzymes* (Figure 23-32). Perforins, which exhibit homology to the terminal components of the complement cascade composing the membrane attack complex, form pores up to 20 nm across in membranes to which they attach. The destruction of an intact permeability barrier, which leads to loss of electrolytes and other small solutes, contributes to cell death. Granzymes are delivered to and are presumed to enter the target cell, likely via the pores generated by perforin. Granzymes are serine proteases that activate effector **caspases** and so propel the target on a path of programmed cell death (apoptosis). Perforins and granzymes are packaged into cytotoxic granules, stored inside the cytotoxic T cell. Upon binding of the T-cell receptor to antigen, the cytotoxic granules and their contents are released into the cleft formed between the cytotoxic T cell and target cell. How the T cell avoids being killed upon release of granzymes and perforins into this synaptic cleft is still not known. Natural killer cells also exert cytotoxic activity and likewise rely on perforins and granzymes to kill their targets (see Figure 23-5).

## T Cells Produce an Array of Cytokines That Provide Signals to Other Immune Cells

Many lymphocytes and nonlymphoid cells in lymphoid tissue produce cytokines. These small secreted proteins instruct lymphocytes what to do by binding to specific receptors on the lymphocyte surface and initiating a transcriptional program that allows the lymphocyte to either proliferate or differentiate into an effector cell ready to exert cytotoxic activity (CD8 T cells), helper activity (CD4 T cells), or antibody-secreting activity (B cells). Cytokines that are produced by or act primarily on leukocytes are called **interleukins;** at least 32 interleukins have been recognized and molecularly characterized. Structurally related interleukins are recognized by cognate receptors with structural similarities, the interleukin-2 receptor being a particularly well-characterized example. Interleukin 2 (IL-2) was identified as T-cell growth factor and is one of the first cytokines produced when T cells

## Cytotoxic T Cells Carry the CD8 Co-receptor and Are Specialized for Killing

As we have seen, cytotoxic T cells, also called cytolytic T lymphocytes (CTLs), generally carry the CD8 glycoprotein marker and are restricted in their recognition by class I MHC molecules. They kill target cells that display the appropriate MHC-peptide combinations and do so with exquisite sensitivity: A single MHC-peptide complex suffices to

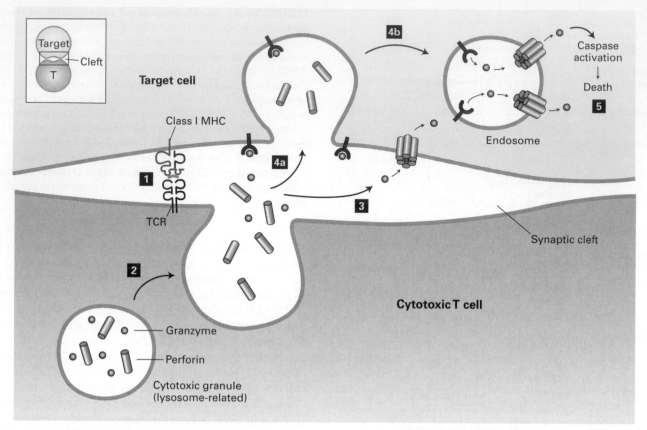

**FIGURE 23-32 Perforin- and granzyme-mediated cell killing by cytotoxic T cells.** Upon recognition of the target cell (step **1**), cytotoxic T cells form tight antigen-specific contacts with target cells. Tight contact results in the formation of a synaptic cleft, into which the contents of cytotoxic granules are released (step **2**). The content of these granules includes perforins and granzymes. Perforins form pores in the membranes onto which they adsorb, and granzymes are serine proteases that enter through the perforin pores (step **3**). Perforins are believed to act not only at the surface of the target cells but also at the surface of endosomal compartments of target cells after the perforin molecules have been internalized from the cell surface (step **4**). Once in the cytoplasm, the granzymes activate caspases, which initiate programmed cell death (step **5**).

are stimulated. IL-2 acts as an autocrine (self-acting) growth factor and drives clonal expansion of activated T cells.

Interleukin 4 (IL-4), which is produced by CD4 T cells specialized in providing help to B cells (discussed below), induces activated B cells to proliferate and to undergo class-switch recombination and somatic hypermutation. Interleukin 7 (IL-7), produced by stromal cells in the bone marrow, is essential for development of T and B cells from committed precursors. Both IL-7 and IL-15 play a role in the maintenance of T cells in the form of *memory cells*, which are antigen-experienced cells that may be called upon when re-exposure to antigen occurs. These memory T cells then rapidly proliferate and deal with the intruder. IL-2, IL-4, IL-7, IL-9, and IL-15 all rely on a common subunit for signal transduction, the common γ chain ($\gamma_c$) with α (IL-2, IL-15) and β subunits (IL-2, IL-4, IL-7, IL-9, IL-15) providing ligand specificity. Genetic defects in the $\gamma_c$ result in a near-complete failure in lymphocyte development, illustrating the importance of these cytokines not only during the effector phase of an immune response, but also in the course of lymphocyte development, where IL-7 in particular plays a key role.

The mechanism of signal transduction by cytokine receptors through the JAK/STAT pathway is described in Chapter 16 (see Figure 16-1 for quick review). Among the many genes under the control of the STAT pathway are suppressors of cytokine signaling, or SOCS proteins. These proteins are themselves induced by cytokines, bind to the activated form of JAKs, and target them for proteasomal degradation (see Figure 16-14b).

## CD4 T Cells Are Divided into Three Major Classes Based on Their Cytokine Production and Expression of Surface Markers

The major function of CD4 T cells is to provide assistance to B cells and guide their differentiation into plasma cells that secrete high-affinity antibodies. This function, carried out by helper T cells, requires both the production and secretion of cytokines, as well as direct contact between the CD4 T cell and the B cell to which it provides help.

A second type of CD4 T cell has as its major function secreting the cytokines that contribute to the establishment of an inflammatory environment. Multiple types of CD4 T cells have thus been defined based on the cytokines they produce and their functional properties. Whereas all activated T cells can produce IL-2, other cytokines are produced by particular CD4 T-cell subsets. These CD4 T cells are classified as $T_H1$ cells, characterized by the production of interferon $\gamma$ and TNF, and $T_H2$ cells, characterized by the production of IL-4 and IL-10. $T_H1$ cells, through production of IFN$\gamma$, can activate macrophages and stimulate an inflammatory response. Referred to also as *inflammatory T cells,* $T_H1$ cells nonetheless play an important role in antibody production, notably facilitating the production of complement-fixing antibodies such as IgG1 and IgG3. $T_H2$ cells, through production of IL-4, play an important role in B-cell responses that involve class-switch recombination to the IgG1 and IgE isotypes (discussed below). The combination of cytokines produced by CD4 T cells, and the interaction between the CD40 protein induced on activated T cells with CD40 ligand (CD40L) on B cells, is responsible for induction of activation-induced deaminase, and so prepares the B cell for class-switch recombination and somatic hypermutation.

Recently discovered subsets of CD4 T cells include *regulatory T cells* and $T_H17$ cells. The former can attenuate immune responses by exerting a suppressive effect on cytokine production by other T cells. These regulatory T cells restrain the activity of potentially self-reactive T cells and are important in maintaining tolerance (the absence of an immune response to self antigens). $T_H17$ cells are important in host defense against bacteria and also play a role in autoimmune disease.

## Leukocytes Move in Response to Chemotactic Cues Provided by Chemokines

Interleukins tell lymphocytes what to do by eliciting a transcriptional program that allows lymphocytes to acquire specialized effector functions. On the other hand, chemokines tell leukocytes where to go. Many cells emit chemotactic cues in the form of chemokines. When tissue damage occurs, resident fibroblasts produce a chemokine, IL-8, that attracts neutrophils to the site of damage. The regulation of lymphocyte trafficking within lymph nodes is essential for dendritic cells to attract T cells, and for T cells and B cells to meet. These trafficking steps are all controlled by chemokines.

There are approximately 40 distinct chemokines and more than a dozen chemokine receptors. One chemokine may bind to more than one receptor, and a single receptor can bind several different chemokines. This creates the possibility of generating a combinatorial code of chemotactic cues of great complexity. This code is used to allow the navigation of leukocytes from where they are generated, in the bone marrow, into the bloodstream for transport to their target destination.

Some chemokines direct lymphocytes to leave the circulation and take up residence in lymphoid organs. These migrations contribute to the population of lymphoid organs with the required sets of lymphocytes. Because these movements occur as part of normal lymphoid development, such chemokines are also referred to as *homeostatic chemokines.*

Those chemokines that serve the purpose of recruiting leukocytes to sites of inflammation and tissue damage are referred to as *inflammatory chemokines.*

Chemokine receptors are G protein–coupled receptors that function as an essential step in the regulation of cell adhesion and cell migration. Leukocytes that travel through blood vessels do so at high speed and are exposed to high hydrodynamic shear forces. For a leukocyte to traverse the endothelium and take up residence in a lymph node or seek out a site of infection in tissue, it must first slow down, a process that requires interactions of surface receptors called selectins with their ligands, which are mostly carbohydrate in nature. If chemokines are found in proximity, adsorbed to the extracellular matrix, and if the leukocyte possesses a receptor for that chemokine, activation of the chemokine receptor elicits a signal that allows integrins carried by the leukocyte to undergo a conformational change. This change results in an increase in affinity of the integrin for its ligand and causes firm arrest of the leukocyte. The leukocyte may now exit the blood vessel by a process known as *extravasation* (see Figure 20-39).

## KEY CONCEPTS of Section 23.5

### T Cells, T-Cell Receptors, and T-Cell Development

• The antigen-specific T-cell receptors are dimeric proteins consisting of $\alpha$ and $\beta$ subunits or $\gamma$ and $\delta$ subunits. T cells occur in at least two major classes based on the expression of the glycoprotein co-receptors CD4 and CD8 (see Figure 23-27).

• Cells that use class I MHC molecules as restriction elements carry CD8; those that use class II MHC molecules carry CD4. These classes of T cells are functionally distinct: CD8 T cells are cytotoxic T cells; CD4 T cells provide help to B cells and are an important source of cytokines.

• Genes encoding the TCR subunits are generated by somatic recombination of V and J segments ($\alpha$ chain) and of V, D, and J segments ($\beta$ chain); their rearrangement obeys the same rules as those defined for rearrangement of Ig genes in B cells (see Figure 23-28). Rearrangement of TCR genes occurs in the thymus and only in those cells destined to become T lymphocytes.

• A complete T-cell receptor includes the accessory CD3 complex, which is required for signal transduction. Each subunit of the CD3 complex carries in its cytoplasmic tail one or three ITAM domains; when phosphorylated, these ITAMs recruit accessory molecules involved in signal transduction (see Figure 23-29).

• In the course of T-cell development, the TCR $\beta$ locus rearranges first, encoding a functional $\beta$ subunit that is incorporated in the pre-TCR, which also includes a specialized pre-TCR subunit encoded by an unrearranged gene (see Figure 23-30). Similar to the pre-BCR, the pre-TCR mediates allelic exclusion and proliferation of those cells that successfully underwent TCR$\beta$ rearrangement.

- T cells destined to become CD8 T cells must interact with class I MHC molecules in the course of their development; those that will become CD4 T cells must interact with class II MHC molecules. Developing T cells that fail to recognize any MHC molecules at all die for lack of survival signals. T cells that interact too strongly with peptide-MHC complexes encountered during development are instructed to die (negative selection); those that have intermediate affinity for peptide-MHC complexes are allowed to mature (positive selection) and are exported from the thymus to the periphery.

- T cells are instructed where to go (cell migration) through chemotactic signals in the form of chemokines. Receptors for chemokines are G protein–coupled receptors that show some promiscuity in terms of their binding of chemokines. The complexity of the chemokine–chemokine receptor family allows precise regulation of leukocyte trafficking, both within lymphoid organs and in the periphery.

## 23.6 Collaboration of Immune-System Cells in the Adaptive Response

An effective adaptive immune response requires the presence of B cells, T cells, and antigen-presenting cells. For B cells to execute class-switch recombination and hypermutation—prerequisites for production of high-affinity antibodies—they require help from activated T cells. These T cells, in turn, can only be activated by professional antigen-presenting cells such as dendritic cells, which detect invaders using their Toll-like receptors (TLRs) and other pattern-recognition receptors, such as the C-type lectins that can recognize polysaccharides and carbohydrate determinants. The interplay between components of innate and adaptive immunity is therefore a very important aspect of adaptive immunity. This layered, interwoven nature of innate and adaptive immunity both ensures a rapid early response of immediate protective value and also primes the adaptive immune system for a specific response to any persisting invader. In this section we describe how these various elements are activated and how the relevant cell types interact.

### Toll-Like Receptors Perceive a Variety of Pathogen-Derived Macromolecular Patterns

An important part of innate defenses is the ability to immediately detect the presence of a microbial invader and respond to it. This response includes direct elimination of the invader, but it also prepares the mammalian host for a proper adaptive immune response, in particular through activation of professional antigen-presenting cells. Antigen-presenting cells are positioned throughout the epithelia (airways, gastrointestinal tract, genital tract), where contact with pathogens is most likely to occur. In the skin, a network of dendritic cells called *Langerhans cells* makes it virtually impossible for a pathogen that breaches this barrier to avoid contact with these professional antigen-presenting cells. Dendritic cells and other professional antigen-presenting cells detect the presence of bacteria and viruses through members of the Toll-like receptor (TLR) family. These proteins are named after the *Drosophila* protein Toll because of the structural and functional homology between them. *Drosophila* Toll was discovered because of its important role in dorsoventral patterning in the fruit fly, but related receptors are now recognized as capable of triggering an innate immune response in insects as well as in vertebrates.

**TLR Structure** Toll itself and all Toll-like receptors possess an extracellular domain composed of *leucine-rich repeats*. These repeats form a sickle-shaped extracellular domain believed to be involved in ligand recognition. The cytoplasmic portion of Toll-like receptors contains a domain responsible for the recruitment of adapter proteins to enable signal transduction. The signaling pathways engaged by the Toll-like receptors share many of the same components (and outcomes) as those used by receptors for the cytokine IL-1 (Figure 23-33).

The *Drosophila* Toll protein interacts with its ligand, Spaetzle, itself the product of a proteolytic conversion initiated by components of the cell wall of fungi that prey on *Drosophila*. In the fly, activation of Toll unleashes a signaling cascade that ultimately controls the transcription of genes that encode antimicrobial peptides. Although appreciated first as a cascade that controls early *Drosophila* development, it has since become clear that the Toll pathway also controls insect immunity. The activated receptor at the surface communicates with the transcriptional apparatus by means of a series of adapter proteins that activate downstream kinases interposed between the Toll-like receptor and the transcription factors that are activated by signaling from them. A key step is the ubiquitin-dependent proteasomal degradation of the Cactus protein. Its removal allows the protein Dif to enter the nucleus and initiate transcription. This pathway is highly homologous in its operation and structural composition to the NF-κB pathway in mammals (see Figure 16-34).

**Diversity of TLRs** There are approximately a dozen mammalian Toll-like receptors that can be activated by various microbial products. These receptors are expressed by a variety of cell types, but their function is crucial for the activation of dendritic cells and macrophages. Neutrophils also express Toll-like receptors. The microbial products recognized by Toll-like receptors include macromolecules found in the cell envelope of bacteria such as lipopolysaccharides (LPS), flagellins (subunits of bacterial flagella), and bacterial lipopeptides. Direct binding of at least some of these macromolecules to Toll-like receptors has been shown directly through crystallographic analysis of the relevant complexes. The presence of

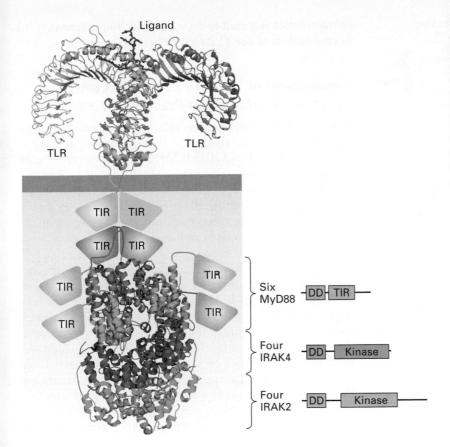

Ligand

TLR          TLR

TIR  TIR
TIR  TIR
TIR
TIR
TIR
TIR
TIR

Six MyD88 — DD–TIR
Four IRAK4 — DD–Kinase
Four IRAK2 — DD–Kinase

**FIGURE 23-33 Toll-like receptor activation.** The extracellular portions of TLRs recognize ligands of diverse chemical nature (nucleic acids, lipopolysaccharides). The cytoplasmic portions of the TLRs associate with the adapter molecule MyD88, present in six copies per complex, and recruit two types of kinase, both members of the IRAK family. These complex interactions are maintained via Toll/IL-1β receptor (TIR) homology domains and death domains (DD), as indicated in the figure. The assembled complex on the cytoplasmic side is referred to as the Myddosome. [Adapted from J. Y. Kang and J.-O. Lee, 2011, *Ann. Rev. Biochem.* **80**:917.]

distinct classes of microbial molecules is sensed by distinct receptors: for example, TLR4 for lipopolysaccharide; heterodimeric of TLR1/2 and TLR2/6 for lipopeptides; and TLR5 for flagellin. Recognition of all bacterial envelope components occurs at the cell surface.

A second set of Toll-like receptors—TLR3, TLR7, and TLR9—sense the presence of pathogen-derived nucleic acids. They do so not at the cell surface, but rather within the endosomal compartments where these receptors reside. Mammalian DNA is methylated at many CpG dinucleotides, whereas microbial DNA generally lacks this modification. TLR9 is activated by unmethylated, CpG-containing microbial DNA. Similarly, double-stranded RNA molecules present in some virus-infected cells lead to activation of TLR3. Finally, TLR7 responds to the presence of certain single-stranded RNAs. Thus the full set of mammalian TLRs allows the recognition of a variety of macromolecules that are diagnostic for the presence of bacterial, viral, and fungal pathogens.

More recently a variety of intracellular receptors for RNA and DNA have been described that recognize viral RNA and are structurally distinct form TLRs. The list of cytoplasmic receptors capable of recognizing DNA, both pathogen derived and host DNA derived, continues to grow. Several of these receptors participate in the assembly of a structure called the inflammasome (Figure 23-34), whose major function is the conversion of procaspase-1 to active caspase-1. Caspase 1 is a pro-

tease that converts proIL-1β into active IL-1β, a cytokine that elicits a strong inflammatory response. Core components of inflammasomes are proteins with leucine-rich repeats, members of the neuronal inhibitors of apoptosis (NALP) family of proteins, and the NOD proteins, so named because of the presence of nucleotide oligomerization domain. Ipaf-1, a protein related to the Apaf-1 molecule involved in apoptosis (see Chapter 21), allows the recruitment of an adaptor, ASC, to mediate complex formation with procaspase-1. This allows the conversion of procaspase-1 to active caspase-1, and conversion of proIL-1β to IL-1β. Many seemingly unrelated substances can induce assembly and activation of an inflammasome, including silica, uric acid crystals, and asbestos particles. Accordingly, inhibition of the inflammasome signaling cascade, or blocking the receptor for IL-1β, has shown therapeutic promise for a variety of inflammatory diseases.

**TLR Signal Cascade** As shown in Figure 23-33, engagement of mammalian Toll-like receptors leads to recruitment of the adapter protein MyD88, which in turn allows the binding and activation of IRAK (interleukin 1 receptor-associated kinase) proteins. After IRAK phosphorylates TNF-receptor–associated factor 6 (TRAF6), several downstream kinases come into play, leading to release of active NF-κB, a transcription factor, for translocation from the cytoplasm to the

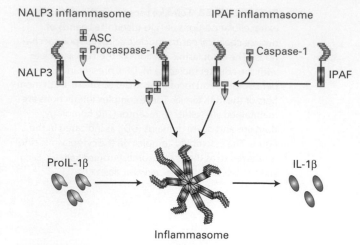

NALP3 inflammasome            IPAF inflammasome

**FIGURE 23-34 The inflammasome.** The inflammasome is a structure that senses the presence of cytoplasmic pathogen-derived nucleic acids and can be also activated by other danger signals, including particulate matter such as uric acid crystals or even asbestos. There are close to two dozen proteins that can participate in the formation of these complexes to yield inflammasomes of different composition, two of which are represented schematically. Ultimately, the fully assembled inflammasome activates caspase-1, the enzyme that converts proIL-1β into the active, cleaved form, IL-1β. NALP3 = a member of the protein family characterized by the presence of NACHT, LRR, and PYD domains; ASC = apoptosis-associated speck-like protein containing a CARD caspase recruitment domain. [Adapted from http://www.unil.ch/webdav/site/ib/shared/Tschopp/Fig1_JT.jpg.]

nucleus, where NF-κB activates various target genes (see Figure 16-35). These target genes include those encoding IL-1β and IL-6, which contribute to inflammation, as well as the genes for tumor necrosis factor (TNF) and IL-12. Expression of type I interferons, small proteins with antiviral effects, is also turned on in response to TLR signaling.

Cell responses to TLR signaling are quite diverse. For antigen-presenting cells, these responses include not only production of cytokines but also the up-regulation of co-stimulatory molecules, the surface proteins important for full activation of naive T cells. TLR signaling allows dendritic cells to migrate from where they encounter a pathogen to the draining lymph nodes, where they may interact with naive lymphocytes. Not all activated Toll-like receptors evoke the identical response. Each activated TLR controls production of a particular set of cytokines by dendritic cells. For each engaged TLR, the combination of surface proteins and the cytokine profile induced by TLR engagement thus creates a unique phenotype for the activated dendritic cell. The identity of the microbe encountered sets the pattern of the TLRs that will be activated. These, in turn, shape the differentiation pathways of activated dendritic cells in terms of the cytokines produced, the surface molecules displayed, and the chemotactic cues to which the dendritic cells respond. The mode of activation of a dendritic cell and the cytokines it produces in response create a unique microenvironment in which T cells differentiate. Here they acquire the functional characteristics required to fight the infectious agent that led to engagement of the TLRs in the first place.

## Engagement of Toll-Like Receptors Leads to Activation of Antigen-Presenting Cells

Professional antigen-presenting cells engage in continuous endocytosis, and in the absence of pathogens, they display at their surface class I and class II MHC molecules loaded with peptides derived from self-proteins. In the presence of pathogens, the Toll-like receptors on these cells are activated, inducing the antigen-presenting cells to become motile: they detach from the surrounding substratum and start to migrate in the direction of the draining lymph node, where the directional cues are provided by chemokines. An activated dendritic cell, for example, reduces its rate of antigen acquisition, up-regulates the activity of endosomal/lysosomal proteases, and increases the transfer of class II MHC–peptide complexes from the loading compartments to the cell surface. Finally, activated professional antigen-presenting cells up-regulate expression of the co-stimulatory molecules CD80 and CD86, which will allow them to activate T cells more effectively. The initial contact of a professional antigen-presenting cell with a pathogen thus results in its migration to the draining lymph node in a state that is fully capable of activating a naive T cell. Antigen is displayed in the form of peptide-MHC complexes, co-stimulatory molecules are abundantly present, and cytokines are produced that assist in setting up the proper differentiation program for the T cells to be activated.

Antigen-laden dendritic cells engage antigen-specific T cells, which respond by proliferating and differentiating. The cytokines produced in the course of this priming reaction determine whether a CD4 T cell will polarize toward an inflammatory or a helper-cell phenotype. If engagement occurs via class I MHC molecules, a CD8 T cell may develop from a precursor cytotoxic T cell to a fully active cytotoxic T cell. The activated T cells are motile and move through the lymph node, in preparation for their encounter with B cells, or, in order to contact the vasculature, enter the circulation, and leave the lymph node to execute effector functions elsewhere in the body.

## Production of High-Affinity Antibodies Requires Collaboration Between B and T Cells

To generate high-affinity antibodies, which are better at binding antigen and at neutralizing pathogens, B cells require assistance from T cells. Thus, B-cell activation requires a source of antigen to engage the B-cell receptor and the presence of activated antigen-specific T cells. Soluble antigen reaches the lymph node through the afferent lymphatics (see Figure 23-6). Bacterial growth is accompanied by the release of microbial products that can serve as antigens. If the infection is accompanied by local tissue destruction, activation of the complement cascade results in bacterial killing and concomitant release of bacterial proteins, which are also delivered via the lymphatics to the draining lymph node. Complement-modified antigens are superior in the activation of B cells through engagement of complement receptors on B cells, which serve as co-receptors for the B-cell receptor. B cells that acquire antigen via their B-cell receptors internalize the immune complex and process it for presentation via the class II MHC pathway. Antigen-experienced B cells thus convert the BCR-acquired antigen into a call for T-cell help in the form of a class II MHC–peptide complex (Figure 23-35). Note that the epitope recognized by the B-cell receptor may be quite distinct from the peptide ultimately displayed on the surface associated with a class II MHC molecule. As long as the B-cell epitope and the class II-presented peptide—a T-cell epitope—are physically linked, successful B-cell differentiation can be initiated.

The concept of linked recognition explains why there is a minimum size for molecules that can be used to successfully elicit a high-affinity antibody response. Such molecules must fulfill several criteria: they must contain the epitope seen by the B-cell receptor, undergo endocytosis and proteolysis, and a proteolyzed fragment of the molecule must bind to the allelic class II MHC molecules available in order to be presented as a class II MHC–peptide complex, which serves as a call for T-cell help. For this reason, synthetic peptides used to elicit antibodies are conjugated to carrier proteins to improve their immunogenicity, where the carrier proteins serve as the source of peptides for presentation via class II MHC products. Only through recognition of this class II MHC–peptide complex via its T-cell receptor can T cells provide the help necessary for the B cell to run its complete course of differentiation.

This concept applies equally to B cells capable of recognizing particular modifications on proteins or peptides. Antibodies that recognize the phosphorylated form of a kinase are commonly raised by immunization of experimental animals with the phosphorylated peptide in question, conjugated to a carrier protein. An appropriately specific B cell recognizes the phosphorylated site on the peptide of interest, internalizes the phosphorylated peptide and carrier, and generates a complex set of peptides by endosomal proteolysis of the carrier protein. Among these peptides there should be at least one that can bind to the class II MHC molecules carried by that B cell for a successful response to ensue. If properly displayed at the surface of the B cell, this class II MHC–peptide complex becomes the call for T-cell help, provided by CD4 T cells equipped with receptors capable of recognizing the complex of class II MHC molecule and carrier-derived peptide.

The T cell identifies, via its T-cell receptor, an antigen-experienced B cell by means of the class II MHC–peptide complexes displayed by the B cell. The B cell also displays costimulatory molecules and receptors for cytokines produced by the activated T cell (e.g., IL-4). These B cells then proliferate. Some of them differentiate into plasma cells; others are set aside and become memory B cells. The first wave of antibodies produced is always IgM. Class switching to other isotypes and somatic hypermutation (necessary for the generation and selection of high-affinity antibodies) requires the persistence of antigen or repeated exposure to antigen. In addition to cytokines, B cells require cell-cell contacts to initiate somatic hypermutation and class-switch recombina-tion. These contacts involve CD40 protein on B cells and CD40L on T cells. These proteins are members of the TNF–TNF receptor family.

## Vaccines Elicit Protective Immunity Against a Variety of Pathogens

Arguably the most important application of immunological principles is vaccines. Vaccines are materials designed to be innocuous but that can elicit an immune response for the purpose of providing protection against a challenge with the virulent version of that pathogen (Figure 23-36). It is not always known why such vaccines are as successful as they are, but in many cases, the ability to raise antibodies that can neutralize the pathogen (viruses) or that show microbicidal effects (bacteria) are good indicators of successful vaccination.

Several strategies can lead to a successful vaccine. Vaccines may be composed of live attenuated variants of more virulent pathogens. Serial passage in tissue culture or from animal to animal will often lead to *attenuation*, the molecular basis for which is not well understood. The attenuated version of the pathogen causes a mild form of the disease or causes no symptoms at all. However, by recruiting all the component parts of the adaptive immune system, such live attenuated vaccines can elicit protective levels of antibodies. These antibody levels may wane with advancing age, and repeated immunizations (booster injections) are often required to maintain full protection. Live attenuated vaccines are in use against flu, measles, mumps, and tuberculosis. In the latter case, an attenuated strain of the mycobacterium that causes the diseases is used (Bacille Calmette-Guerin; BCG). Although live attenuated poliovirus was used as a vaccine until recently, its use was discontinued because the risk of re-emergence of more virulent strains of the poliovirus outweighed the benefit. Currently, killed poliovirus is used as the vaccine of choice.

Vaccines based on the cowpox virus, a close relative of the human pathogen variola responsible for smallpox, have been used successfully to completely eradicate smallpox, the first such example of the elimination of an infectious disease. Attempts to achieve a similar feat for polio are nearing completion.

The other major type of vaccine is called a subunit vaccine. Rather than using live attenuated strains of a virulent bacterium or virus, only one of its components is used to elicit immunity. In certain cases this is sufficient to afford lasting protection against a challenge with the live, virulent source of the antigen used for vaccination. This approach has been successful in preventing infections with the hepatitis B virus. The commonly used flu vaccines are composed mostly of the envelope proteins neuraminidase and hemagglutinin (see Figure 3-10); these vaccines elicit neutralizing antibodies. For the human papillomavirus HPV 16, a serotype that causes cervical cancer, viruslike particles are generated that are composed of the virus capsid proteins but devoid of its genetic material; these viruslike particles are noninfectious, yet in many respects mimic the intact virion. The HPV vaccine now licensed for use in humans is expected to reduce the

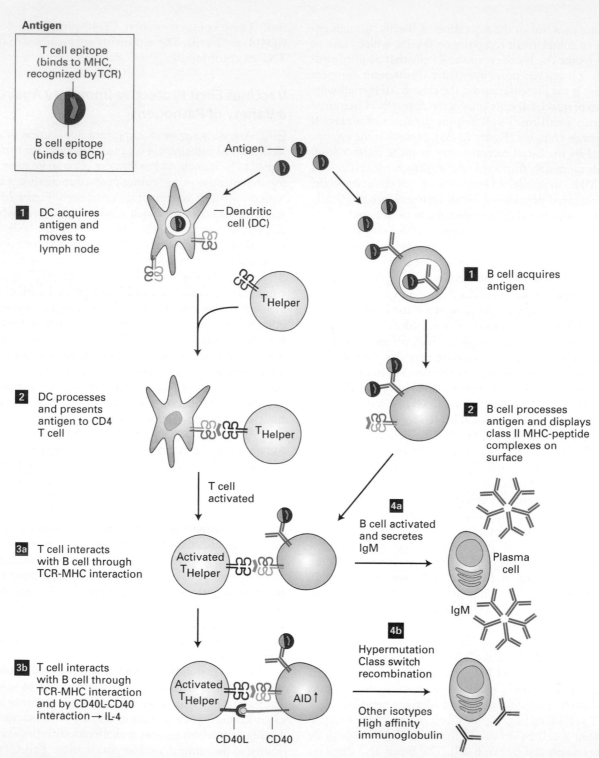

**Antigen**

T cell epitope
(binds to MHC,
recognized by TCR)

B cell epitope
(binds to BCR)

Antigen —

**1** DC acquires antigen and moves to lymph node

— Dendritic cell (DC)

T_Helper

**1** B cell acquires antigen

**2** DC processes and presents antigen to CD4 T cell

T_Helper

**2** B cell processes antigen and displays class II MHC-peptide complexes on surface

T cell activated

**3a** T cell interacts with B cell through TCR-MHC interaction

Activated T_Helper

**4a** B cell activated and secretes IgM

Plasma cell

IgM

**3b** T cell interacts with B cell through TCR-MHC interaction and by CD40L-CD40 interaction → IL-4

Activated T_Helper

AID ↑

CD40L    CD40

**4b** Hypermutation Class switch recombination

Other isotypes High affinity immunoglobulin

**FIGURE 23-35 Collaboration between T and B cells is required to initiate the production of antibodies.** *Left:* Activation of T cells by means of antigen-loaded dendritic cells (DCs). *Right:* Antigen acquisition by and subsequent activation of B cells. Step **1**: Professional antigen-presenting cells (DCs, B cells) acquire antigen. Step **2**: Antigen is internalized, processed, and presented to T cells. T-cell activation occurs when dendritic cells present antigen to T cells. Step **3a**: Activated T cells engage antigen-experienced B cells through peptide-MHC complexes displayed on the surface of the B cell. Step **3b**: T cells

that are persistently activated initiate expression of the CD40 ligand (CD40L), a prerequisite for B cells becoming fully activated and turning on the enzymatic machinery (activation-induced deaminase) to initiate class-switch recombination and somatic hypermutation. Step **4a**: A B cell that receives the appropriate instructions from CD4 helper T cells becomes an IgM-secreting plasma cell. Step **4b**: A B cell that receives signals from activated CD4 helper T cells in the form of CD40–CD40L interactions and the appropriate cytokines can class switch to other immunoglobulin isotypes and engage in somatic hypermutation.

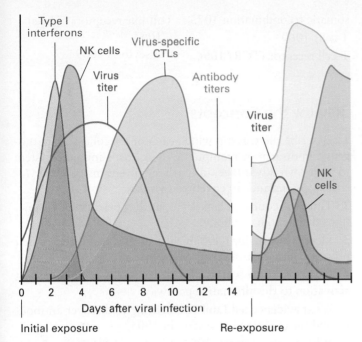

Type I interferons

NK cells

Virus-specific CTLs

Virus titer

Antibody titers

Virus titer

NK cells

| 0 | 2 | 4 | 6 | 8 | 10 | 12 | 14 |

Days after viral infection

Initial exposure                    Re-exposure

**FIGURE 23-36 Time course of a viral infection.** The initial antiviral response, seen when the numbers of infectious particles rises, includes activation of natural killer (NK) cells and production of type I interferons. These responses are part of innate defenses. The production of antibodies, as well as the activation of cytotoxic T cells (CTLs), follows, and eventually clears the infection. Re-exposure to the same virus leads to more rapid and more pronounced production of antibodies and to more rapid activation of cytotoxic T cells. A successful vaccine induces an immune response similar in some respects to that following initial exposure to a pathogen, but without causing significant symptoms of disease. If a vaccinated person subsequently is exposed to that pathogen, the adaptive immune system is primed to respond quickly and strongly.

incidence of cervical cancer in susceptible populations by perhaps as much as 80 percent, the first example of a vaccine that prevents this particular type of cancer.

From a public health perspective, cheaply produced and widely distributed vaccines are formidable tools in eradicating communicable diseases. Current efforts are aimed at producing vaccines against diseases for which no other suitable therapies are available (Ebola virus) or where socioeconomic conditions have made distribution of drugs problematic (malaria, HIV/AIDS). Through a more complete understanding of how the immune system operates, it should be possible to improve on the design of existing vaccines and extend these principles to diseases for which currently no successful vaccines are available. ∎

## KEY CONCEPTS of Section 23.6

### Collaboration of Immune-System Cells in the Adaptive Immune Response

- Antigen-presenting cells such as dendritic cells require activation by means of signals delivered to their Toll-like receptors. These receptors are broadly specific for macromolecules produced by bacteria and viruses. Engagement of Toll-like receptors activates the NF-κB signaling pathway, whose outputs include the synthesis of inflammatory cytokines (see Figure 23-33).

- Upon activation, dendritic cells become migratory and move to lymph nodes, ready for their encounter with T cells. Activation of dendritic cells also increases their display of MHC-peptide complexes and expression of co-stimulatory molecules required for initiation of a T-cell response.

- B cells require assistance of activated T cells to execute their full differentiation program to become plasma cells. Antigen-specific help is provided to B cells by activated T cells, which recognize class II MHC–peptide complexes on the surface of B cells. These B cells generate the relevant MHC-peptide complexes by internalizing antigen via BCR-mediated endocytosis, followed by antigen processing and presentation via the class II MHC pathway (see Figure 23-35).

- In addition to cytokines produced by activated T cells, B cells require cell-cell contact to initiate somatic hypermutation and class-switch recombination. This involves CD40 on B cells and CD40L on T cells.

- Important applications of the immunological concept of collaboration between T and B cells include vaccines. The most common forms of vaccines are live attenuated viruses or bacteria, which can evoke a protective immune response without causing pathology, and subunit based vaccines.

## Perspectives for the Future

Several areas of immunological research promise to have a major impact. The generation of new vaccines that are both safe and effective remains a goal of enormous practical and societal significance: HIV infection, drug-resistant tuberculosis, and malaria are three examples of major killers, each responsible for millions of deaths annually, for which no successful vaccines are currently available. Research must incorporate cell biological as well as immunological concepts to solve this unmet need. Even though some of the most successful vaccines were developed in the absence of detailed immunological knowledge, the current regulatory climate demands a more detailed understanding of the composition of a successful vaccine and why such vaccines work.

Lymphocytes are among the few cell types that can be studied as primary cells that continue to perform cell- and tissue-type specific behavior in tissue culture. This trait makes lymphocytes an attractive model in which to study details of signal transduction, to explore interactions between well-defined, different cell types under defined laboratory conditions, and to accurately measure and model the responses evoked by specific stimuli.

The ability of lymphocytes to create antigen-specific receptors of nearly limitless variability comes at a price: the generation of self-reactive receptors, which is the major contributing factor to autoimmune disease. Vertebrates have several mechanisms for keeping such self-reactive lymphocytes in check, but none of them is foolproof. We must understand how tolerance to self-antigens is generated, maintained, and ultimately broken at the onset of autoimmune disease. This understanding should help define new strategies for manipulating and controlling self-reactive lymphocytes to prevent or treat autoimmune diseases (e.g., type I diabetes, multiple sclerosis, and arthritis).

With advances in understanding of stem cells and how to use stem cells for therapy in the treatment of diseases (e.g., Parkinson's disease, muscular dystrophy, and spinal cord injuries), transplantation of heterologous stem cells (i.e., derived from an individual other than the patient) is one of several options. The immune system's ability to recognize as foreign such transplanted stem-cell derivatives is a factor that will limit the use of heterologous stem cells. Consequently, new means to suppress responses to these transplants, or to induce tolerance to them, are important goals.

More refined genetic tools continue to allow identification of additional lymphocyte subsets, often with distinct functions. Improved classification schemes will help in understanding lymphocyte function, and so open the door to selective manipulation of lymphocyte function for therapeutic gain.

Our understanding of the interplay between innate and adaptive immune responses, and the many ways in which pathogens interfere in both innate and adaptive immunity, will benefit from our increased understanding of genome sequences and genome structures, both host and pathogen. The use of genetically modified pathogenic organisms as probes for host immune function will help our understanding of basic immunology. This is a rapidly expanding area of basic cell biological research.

## Key Terms

affinity maturation 1077
antigen 1060
antigen processing and presentation 1086
autoimmunity 1059
B cells 1060
B-cell receptor (BCR) 1078
chemokines 1101
clonal selection theory 1069
complement 1063
cytokines 1065
cytotoxic T cell 1081
dendritic cells 1063
epitope 1072
Fc receptor 1072
helper T cells 1081
immunoglobulins 1066
inflammasome 1104

inflammation 1065
interleukins 1099
isotypes 1068
junctional imprecision 1075
lymphocytes 1061
macrophages 1063
major histocompatibility complex (MHC) 1081
memory cells 1100
natural killer (NK) cells 1064
neutrophils 1066
opsonization 1064
plasma cells 1079
primary lymphoid organs 1061
secondary lymphoid organs 1061

somatic recombination 1073
T cells 1060
T-cell receptor (TCR) 1096

Toll-like receptors (TLRs) 1063
transcytosis 1069

## Review the Concepts

1. Describe the ways in which each of the following pathogens can disarm or manipulate their host's immune system to their own advantage: (a) pathogenic strains of *Staphylococcus aureus* and (b) enveloped viruses.

2. Trace the course of leukocytes as they perform their functions throughout the body.

3. Identify the major mechanical and chemical defenses that protect internal tissues from microbial attack.

4. Compare/contrast the classical pathway of complement activation to the alternative pathway.

5. What evidence led Emil von Behring to discover antibodies and the complement system in 1905?

6. What is opsonization? What is the role of antibodies in this process?

7. In B cells, what mechanism ensures that only rearranged V genes are transcribed?

8. What prevents further rearrangement of immunoglobulin heavy-chain gene segments in a pre-B cell once a productive heavy-chain rearrangement has occurred?

9. How/why do antibodies undergo a class switch from producing IgM antibodies to any of the other isotypes?

10. What biochemical mechanism underlies affinity maturation of the antibody response?

11. Compare/contrast the structures of class I and class II MHC molecules. What kinds of cells express each class of MHC molecule? What are their functions?

12. Describe the six steps in antigen processing and presentation via the class I MHC restricted pathway.

13. Describe the six steps in antigen processing and presentation via the class II MHC restricted pathway.

14. What prevents self-reactive T cells from leaving the thymus?

15. Explain why T-cell–mediated autoimmune diseases are associated with particular alleles of class II MHC genes.

16. How are antigen-presenting cells and helper T cells involved in B-cell activation?

17. Outline the events in the innate and adaptive immune responses, from when a pathogen invades to clearance of the pathogen.

18. Define passive immunization and give an example.

19. How would you design a vaccine that protects against HIV infection without the possibility of infecting the patient?

20. The annual flu shot is composed of either a live attenuated virus or influenza subunits (envelope proteins neuraminidase and hemagglutinin). How does the annual flu shot protect you against infection?

**21.** Design a laboratory protocol to develop a monoclonal or polyclonal antibody against a protein of interest.

**22.** Consider a person without any functioning plasma cells. What effects would this have on the person's adaptive immune system? Innate immune system?

## Analyze the Data

To understand how ovalbumin (a protein from chicken eggs) and other foreign antigens in the cytoplasm of a cell are presented for immunosurveillance, ovalbumin can be introduced by electroporation into the cytoplasm of mouse primary B lymphocytes. The ovalbumin is cleaved, and one of its cleavage products is a peptide with the sequence SIIN-FEKL (single-letter abbreviations). When these B cells are mixed with a clonal T-cell population that specifically recognizes SIINFEKL in the context of MHC molecules, the T cells become stimulated. The graphs below show the secretion of IL-2 after these T cells were mixed with B cells that had been treated in various ways.

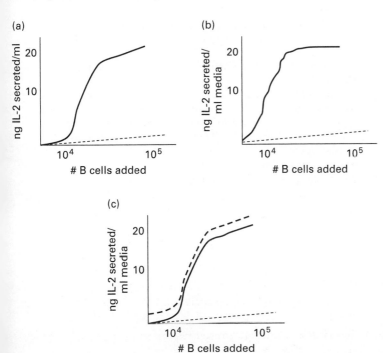

Graph A: B cells were electroporated either with ovalbumin (solid line) or with a control protein (dashed line) not recognized by this clonal T-cell population.

Graph B: B cells were first incubated with an inhibitor of lysosomal cysteine proteases (solid line) or an inhibitor of the proteasome (dashed line) and then electroporated with ovalbumin.

Graph C: B cells were exposed to a high concentration of the SIINFEKL peptide and then immediately fixed [killed]

(dotted line) with formaldehyde or incubated for 2 hours in the presence (dashed curve) or absence (solid curve) of the proteasome inhibitor and then fixed with formaldehyde.

**a.** Under what conditions is IL-2 secreted? Is it likely that the T cells or the B cells are secreting IL-2?

**b.** What information is derived from the use of the lysosomal and the proteasome inhibitors? Is the ovalbumin peptide more likely to be presented complexed with class I MHC or class II MHC molecules? What is the most likely pathway by which ovalbumin in the cytoplasm of the B cell is presented for recognition by the appropriate T cells?

**c.** Why do you think that the presence or absence of the proteasome inhibitor in experiment C had no effect on the amount of IL-2 secreted, whereas the presence of the inhibitor in experiment B shows a marked effect?

## References

### Overview of Host Defenses

Akira, S., K. Kiyoshi Takeda, and T. Kaisho. 2001. Toll-like receptors: critical proteins linking innate and acquired immunity. *Nature Immunol.* **2**:675–680.

Heyman, B. 2000. Regulation of antibody responses via antibodies, complement, and Fc receptors. *Ann. Rev. Immunol.* **18**:709–737.

Lemaitre, B., and J. Hoffmann. 2007. The host defense of *Drosophila melanogaster. Ann. Rev. Immunol.* (in press).

von Behring, E., and S. Kitasato. 1890. The mechanism of diphtheria immunity and tetanus immunity in animals. Reprinted in *Mol. Immunol.* 1991, **28**(12):1317, 1319–1320.

### Immunoglobulins: Structure and Function

Amzel, L. M., and R. J. Poljak. 1979. Three-dimensional structure of immunoglobulins. *Ann. Rev. Biochem.* **48**:961–997.

Williams, A. F., and A. N. Barclay. 1988. The immunoglobulin superfamily—domains for cell-surface recognition. *Ann. Rev. Immunol.* **6**:381–405.

### Generation of Antibody Diversity and B-Cell Development

Hozumi, N., and S. Tonegawa. 1976. Evidence for somatic rearrangement of immunoglobulin genes coding for variable and constant regions. *Proc. Nat'l Acad. Sci. USA* **73**:3628–3632.

Jung, D., et al. 2006. Mechanism and control of V(D)J recombination at the immunoglobulin heavy chain locus. *Ann. Rev. Immunol.* **24**:541–570.

Kitamura, D., et al. 1991. A B cell-deficient mouse by targeted disruption of the membrane exon of the immunoglobulin mu chain gene. *Nature* **350**:423–426.

Kitamura, D., et al. 1992. A critical role of lambda 5 protein in B cell development. *Cell* **69**:823–831.

Muramatsu, M., et al. 2000. Class switch recombination and hypermutation require activation-induced cytidine deaminase (AID), a potential RNA editing enzyme. *Cell* **102**:553–563.

Schatz, D. G., M. A. Oettinger, and D. Baltimore. 1989. The V(D)J recombination activating gene, RAG-1. *Cell* **59**:1035–1048.

### The MHC and Antigen Presentation

Bjorkman, P. J., et al. 1987. Structure of the human class I histocompatibility antigen, HLA-A2. *Nature* **329**:506–512.

Brown, J. H., et al. 1993. Three-dimensional structure of the human class II histocompatibility antigen HLA-DR1. *Nature* **364**:33–39.

Neefjes, J. J., et al. 1990. The biosynthetic pathway of MHC class II but not class I molecules intersects the endocytic route. *Cell* **61**:171–183.

Peters, P. J., et al. 1991. Segregation of MHC class II molecules from MHC class I molecules in the Golgi complex for transport to lysosomal compartments. *Nature* **349**:669–676.

Rock, K. L., et al. 1994. Inhibitors of the proteasome block the degradation of most cell proteins and the generation of peptides presented on MHC class I molecules. *Cell* **78**:761–771.

Rudolph, M. G., R. L. Stanfield, and I. A. Wilson. 2006. How TCRs bind MHCs, peptides, and coreceptors. *Ann. Rev. Immunol.* **24**:419–466.

Townsend, A. R., et al. 1984. Cytotoxic T cell recognition of the influenza nucleoprotein and hemagglutinin expressed in transfected mouse λ cells. *Cell* **39**:13–25.

Zinkernagel, R. M., and P. C. Doherty. 1974. Restriction of in vitro T cell-mediated cytotoxicity in lymphocytic choriomeningitis within a syngeneic or semiallogeneic system. *Nature* **248**:701–702.

### T Cells, T-Cell Receptors, and T-Cell Development

Dembic, Z., et al. 1986. Transfer of specificity by murine alpha and beta T-cell receptor genes. *Nature* **320**:232–238.

Kisielow, P., et al. 1988. Tolerance in T-cell-receptor transgenic mice involves deletion of nonmature CD4+8+ thymocytes. *Nature* **333**:742–746.

Lenschow, D. J., T. L. Walunas, and J. A. Bluestone. 1996. CD28/B7 system of T cell costimulation. *Ann. Rev. Immunol.* **14**:233–258.

Miller, J. F. 1961. Immunological function of the thymus. *Lancet* **30**(2):748–749.

Mombaerts, P., et al. 1992. RAG-1-deficient mice have no mature B and T lymphocytes. *Cell* **68**:869–877.

Sharpe, A. H., and A. K. Abbas. 2006. T-cell costimulation: biology, therapeutic potential, and challenges. *N. Engl. J. Med.* **355**:973–975.

Shinkai, Y., et al. 1993. Restoration of T cell development in RAG-2-deficient mice by functional TCR transgenes. *Science* **259**:822–825.

### Collaboration of Immune-System Cells in the Adaptive Immune Response

20 years of HIV science. 2003. *Nature Med.* **9**:803–843. A collection of opinion pieces on the prospects for an AIDS vaccine.

Banchereau, J. 2002. The long arm of the immune system. *Sci. Am.* **287**:52–59.

Chang, M.-H., et al. for The Taiwan Childhood Hepatoma Study Group. 1997. Universal hepatitis B vaccination in Taiwan and the incidence of hepatocellular carcinoma in children. *N. Engl. J. Med.* **336**:1855–1859.

Cytokines Online Pathfinder Encyclopedia. http://www.copewithcytokines.de.

Gross, O., et al. 2011. The inflammasome: an integrated view. *Immunol. Rev.* **243**:136–151.

Jego, G., et al. 2003. Plasmacytoid dendritic cells induce plasma cell differentiation through type I interferon and interleukin 6. *Immunity* **19**:225–234.

Kang, J. Y., and J.-O. Lee. 2011. Structural biology of the Toll-like receptor family. *Ann. Rev. Biochem.* **80**:917–941.

Koutsky, L. A., et al. 2002. A controlled trial of a human papillomavirus type 16 vaccine. *N. Engl. J. Med.* **347**:1645–1651.

Plotkin, S. A., and W. A. Orenstein. 2003. *Vaccines*, 4th ed. Saunders.

Rajewsky, K., et al. 1969. The requirement of more than one antigenic determinant for immunogenicity. *J. Exp. Med.* **129**:1131–1143.

Smith, Jane S. 1990. *Patenting the Sun: Polio and the Salk Vaccine.* Wm Morrow and Co.

Steinman, R. M., and Z. A. Cohn. 2007. Identification of a novel cell type in peripheral lymphoid organs of mice. I. Morphology, quantitation, tissue distribution. *J. Immunol.* **178**:5–25.

Steinman, R. M., and H. Hemmi. 2006. Dendritic cells: translating innate to adaptive immunity. *Curr. Top. Microbiol. Immunol.* **311**:17–58.

# Two Genes Become One: Somatic Rearrangement of Immunoglobulin Genes

N. Hozumi and S. Tonegawa, 1976, *Proc. Nat'l Acad. Sci. USA* **73**:3629

For decades, immunologists wondered how the body could generate the multitude of pathogen-fighting immunoglobulins, called antibodies, needed to ward off the vast array of different bacteria and viruses encountered in a lifetime. Clearly, these protective proteins, like all proteins, somehow were encoded in the genome. But the enormous number of different antibodies potentially produced by the immune system made it unlikely that individual immunoglobulin (Ig) genes encoded all the possible antibodies an individual might need. In studies beginning in the early 1970s, Susumu Tonegawa, a molecular biologist, laid the foundation for solving the mystery of how antibody diversity is generated.

## Background

Research on the structure of Ig molecules provided some clues about the generation of antibody diversity. First, it was shown that an Ig molecule is composed of four polypeptide chains: two identical heavy (H) chains and two identical light (L) chains. Some researchers proposed that antibody diversity resulted from different combinations of heavy and light chains. Although somewhat reducing the number of genes needed, this hypothesis still required that a large portion of the genome be devoted to Ig genes. Protein chemists then sequenced several Ig light and heavy chains. They found that the C-terminal regions of different light chains were very similar and thus were termed the constant (C) region, whereas the N-terminal regions were highly variable and thus were termed the variable (V) region. The sequences of different heavy chains exhibited a similar pattern. These findings suggested that the genome contains a small number of C genes and a much larger group of V genes.

In 1965, W. Dryer and J. Bennett proposed that two separate genes, one V gene and one C gene, encode each heavy chain and each light chain. Although this proposal seemed logical, it violated the well-documented principle that each gene encodes a single polypeptide. To avoid this objection, Dryer and Bennett suggested that a V and C gene somehow were rearranged in the genome to form a single gene, which then was transcribed and translated into a single polypeptide, either a heavy or light chain. Indirect support for this model came from DNA hybridization studies showing that only a small number of genes encoded Ig constant regions. However, until more powerful techniques for analyzing genes came on the scene, a definitive test of the novel two-gene model was not possible.

## The Experiment

Tonegawa realized that if immunoglobulin genes underwent rearrangement, then the V and C genes were most likely located at different points in the genome. The discovery of restriction endonucleases, enzymes that cleave DNA at specific sites, had allowed some bacterial genes to be mapped. However, because mammalian genomes are much more complex, similar mapping of the genes encoding V and C regions was not technically feasible. Instead, drawing on newly developed molecular biology techniques, Tonegawa devised another approach for determining whether the V and C regions were encoded by two separate genes. He reasoned that if rearrangement of the V and C genes occurs, it must happen during differentiation of Ig-secreting B cells from embryonic cells. Furthermore, if rearrangement occurs, there should be detectable differences between unrearranged germline DNA from embryonic cells and

the DNA from Ig-secreting B cells. Thus, he set out to see if such differences existed using a combination of restriction-enzyme digestion and RNA–DNA hybridization to detect the DNA fragments.

He began by isolating genomic DNA from mouse embryos and from mouse B cells. To simplify the analysis, he used a line of B-cell tumor cells, all of which produce the same type of antibody. The genomic DNA was then digested with the restriction enzyme BamHI, which recognizes a sequence that occurs relatively rarely in mammalian genomes. Thus, the DNA was broken into many large fragments. He then separated these DNA fragments by agarose gel electrophoresis, which separates biomolecules on the basis of charge and size. Since all DNA carries an overall negative charge, the fragments were separated based on their size. Next, he cut the gel into small slices and isolated the DNA from each slice. Now, Tonegawa had many fractions of DNA pieces of various sizes. He then could analyze these DNA fractions to determine if the V and C genes resided on the same fragment in both B cells and embryonic cells.

To perform this analysis, Tonegawa first isolated from the B-cell tumor cells the mRNA encoding the major type of Ig light chain, called κ. Since an RNA is complementary to one strand of the DNA from which it is transcribed, it can hybridize with this strand, forming a RNA–DNA hybrid. By radioactively labeling the entire κ mRNA, Tonegawa produced a probe for detecting which of the separated DNA fragments contained the κ-chain gene. He then isolated the 3' end of the κ mRNA and labeled it, yielding a second probe that would detect only the DNA sequences encoding the constant region of the κ chain. With these probes in hand—one specific for the combined V + C gene and one specific for C

**Embryo DNA**

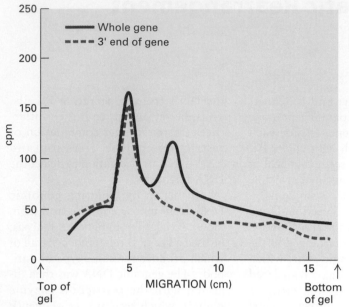

**B-Cell DNA**

**FIGURE 1 Experimental results showing that the genes encoding the variable (V) and constant (C) regions of κ light chains are rearranged during development of B cells.** These curves depict the hybridization of labeled RNA probes, specific for the entire κ gene (V + C)

and for the 3′ end that encodes the C region, to fractions of digested embryonic or B-cell DNA separated by agarose gel electrophoresis. [Adapted from N. Hozumi and S. Tonegawa, 1976, *Proc. Nat'l Acad. Sci. USA* **73**:3629.]

alone—Tonegawa was ready to compare the DNA fragments obtained from B cells and embryonic cells.

He first denatured the DNA in each of the fractions into single strands and then added one or the other labeled probe. He found that the C-specific probe hybridized to different fractions derived from embryonic and B-cell DNA (Figure 1). Even more telling, the full-length RNA probe hybridized to two *different* fractions of the embryonic DNA, suggesting that the V and C genes are not connected and that a cleavage site for BamHI lies between them. Tonegawa concluded that

during the formation of B cells, separate genes encoding the V and C regions are rearranged into a single DNA sequence encoding the entire κ light chain (Figure 2).

## Discussion

The generation of antibody diversity was a problem awaiting development of powerful molecular techniques to answer it. Tonegawa went on to clone V-region genes and prove that the rearrangement must occur somatically. These findings affected genetics as well as immunology. Whereas once it

was believed that every cell in the body contained the same genetic information, it became clear that some cells take that information and alter it to suit other purposes. In addition to somatic rearrangement, Ig genes undergo a variety of other alterations that allow the immune system to create the diverse repertoire of antibodies necessary to react to any invading organism. Our current understanding of these mechanisms rests on the foundation of Tonegawa's fundamental discovery. For this work, he received the Nobel Prize for Physiology and Medicine in 1987.

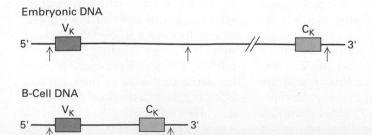

**FIGURE 2 Schematic diagram of κ light-chain DNA in embryonic cells and B cells that is consistent with Tonegawa's results.** In embryonic cells, cleavage with the BamHI restriction enzymes (red arrows) produces two different sized fragments, one containing the V gene and one containing the C gene. In B cells, the DNA is rearranged so that the V and C genes are adjacent, with no intervening cleavage site. BamHI digestion thus yields one fragment that contains both the V and C genes.

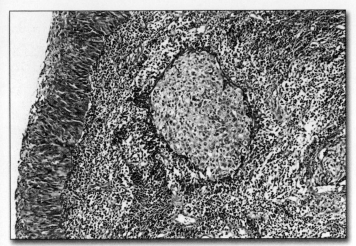

# Cancer

Nasopharyngeal carcinoma (NPC). NPC is a malignant tumor arising from the mucosal epithelium of the nasopharynx, the uppermost part of the throat. NPCs can arise due to cigarette smoking, eating nitrosamine-rich foods (such as salt-cured fish), or are the result of an Epstein-Barr Virus (EBV) infection. The section showing the NPC is stained with hematoxylin and eosin. [Biophoto Associates/Photo Researchers].

ancer causes about one-fifth of the deaths in the United States each year. Worldwide, between 100 and 350 out of 100,000 people die of cancer each year. Cancer is due to failures of the mechanisms that usually control the growth and proliferation of cells. During normal development and throughout adult life, intricate genetic control systems regulate the balance between cell birth and death in response to growth signals, growth-inhibiting signals, and death signals. Cell birth and death rates determine adult body size and the rate of growth in reaching that size. In some adult tissues, cell proliferation occurs continuously as a constant tissue-renewal strategy. Intestinal epithelial cells, for instance, live for just a few days before they die and are replaced; certain white blood cells are replaced as rapidly, and skin cells commonly survive for only 2–4 weeks before being shed. The cells in many adult tissues, however, normally do not proliferate except during healing processes. Such stable cells (e.g., hepatocytes, heart muscle cells, neurons) can remain functional for long periods or even for the entire lifetime of an organism. Cancer occurs when the mechanisms that maintain these normal growth rates malfunction to cause excess cell division.

The losses of cellular regulation that give rise to most or all cases of cancer are due to genetic damage that is often accompanied by influences of tumor-promoting chemicals, hormones, and sometimes viruses (Figure 24-1). Mutations in two broad classes of genes have been implicated in the onset of cancer: **proto-oncogenes** and **tumor-suppressor genes**. Proto-oncogenes normally promote cell growth; mutations change them into **oncogenes** whose products are excessively active in growth promotion. Oncogenic mutations usually result in either increased gene expression or production of a hyperactive product. Tumor-suppressor genes normally restrain growth, so mutations that inactivate them allow inappropriate cell division. A third, more specialized class of genes called **caretaker genes** is also often linked to cancer. Caretaker genes normally protect the integrity of the genome; when they are inactivated, cells acquire additional mutations at an increased rate—including mutations that cause the deregulation of cell growth and proliferation and lead to cancer. Many of the genes in these three classes encode proteins that help regulate cell proliferation (i.e., entry into and progression through the cell cycle) or cell death by **apoptosis**; others encode proteins that participate in repairing damaged DNA.

Cancer commonly results from mutations that arise during a lifetime's exposure to **carcinogens**, substances encountered in the environment, which include certain chemicals and

## OUTLINE

**FIGURE 24-1 Overview of changes in cells that cause cancer.**
During carcinogenesis, six fundamental cellular properties are altered, as shown here in this tumor growing within normal tissue, to give rise to the complete, most destructive cancer phenotype. Less dangerous tumors arise when only some of these changes occur. In this chapter we examine the genetic changes that result in these altered cellular properties.
[Adapted from Hanahan, D., and R. A. Weinberg, 2011, *Cell* **144**:646–674.]

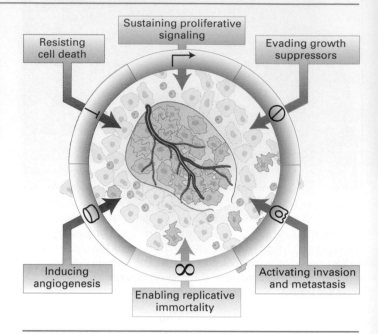

radiation. Most cancer cells have lost one or more genome maintenance and repair systems due to mutation, which may explain the large number of additional mutations they accumulate. Although DNA-repair enzymes do not directly inhibit cell proliferation, cells that have lost the ability to repair errors, gaps, or broken ends in DNA accumulate mutations in many genes, including those that are critical in controlling cell growth and proliferation. Thus loss-of-function mutations in caretaker genes such as the genes encoding DNA-repair enzymes prevent cells from correcting mutations that inactivate tumor-suppressor genes or activate oncogenes.

Cancer-causing mutations occur mostly in somatic cells, not in the germ-line cells, and somatic-cell mutations are not passed on to the next generation. However, certain inherited mutations, which are carried in the germ line, increase the probability that cancer will occur at some time. In a destructive partnership, somatic mutations can combine with inherited mutations to cause cancer.

The cancer-forming process, called *oncogenesis* or *tumorigenesis*, is an interplay between genetics and the environment. Most cancers arise after genes are altered by carcinogens or by errors in the copying and repair of genes. Even if the genetic damage occurs only in one somatic cell, division of this cell will transmit the damage to its daughter cells, giving rise to a **clone** of altered cells. Rarely, however, does mutation in a single gene lead to the onset of cancer. More typically, a series of mutations in multiple genes creates a progressively more rapidly proliferating cell type that escapes normal growth restraints, creating an opportunity for additional mutations. The cells also acquire other properties that give them an advantage, such as the ability to escape from normal epithelia and stimulate the growth of vasculature to obtain oxygen. Eventually the clone of cells grows into a **tumor**. In some cases cells from the primary tumor migrate to new sites, where they form secondary tumors, a process termed **metastasis**. Most cancer deaths are due to invasive, fast-growing metastasized tumors.

Metastasis is a complex process with many steps. It is facilitated by the tumor cells producing their own growth factors and angiogenesis factors (inducers of blood vessel growth). Motile, invasive cells are the most dangerous. Tissues that produce growth factors and readily grow new vasculature, such as bone, blood vessels, and liver, are the most vulnerable to invasion since these characteristics help to support the invaders.

Time plays an important role in cancer. Many years may be required to accumulate the multiple mutations that are required to form a tumor, so most cancers develop later in life. The requirement for multiple mutations also lowers

the frequency of cancer compared with what it would be if tumorigenesis were triggered by a single mutation. However, huge numbers of cells are, in essence, mutagenized and tested for altered growth during our lifetimes, a powerful selection in favor of these cells, which, in this case, we do not want. Cells that proliferate quickly become more abundant, undergo further genetic changes, and can become progressively more dangerous. Furthermore, cancer occurs most frequently after the age of reproduction and therefore plays a lesser role in reproductive success. So cancer is common, in part reflecting increasingly long human lives but also reflecting the lack of evolutionary selection against the disease.

In this chapter, we first introduce the properties of tumor cells and describe the multistep process of oncogenesis. Next, we consider the general types of genetic changes that lead to the unique characteristics of cancer cells and the interplay between somatic and inherited mutations. The following sections examine in detail how mutations affecting both growth-promoting and growth-inhibiting processes can result in excess cell proliferation. We conclude the chapter with a discussion of the role of carcinogens and how a breakdown in DNA-repair mechanisms due to the loss of caretaker genes can lead to oncogenesis.

## 24.1 Tumor Cells and the Onset of Cancer

Before examining in detail the genetic basis of cancer, we consider the general process of tumorigenesis and the properties of tumor cells that distinguish them from normal cells.

(a)

(b)      Tumor cells     Normal cells

**FIGURE 24-2 Gross and microscopic views of a tumor invading normal liver tissue.** (a) The gross morphology of a human liver in which a metastatic lung tumor is growing. The white protrusions on the surface of the liver are the tumor masses. (b) A light micrograph of a section of the tumor in (a) showing areas of small, dark-staining tumor cells invading a region of larger, light-staining, normal liver cells. [Courtesy of J. Braun.]

The change from a normal cell into a cancer cell commonly involves multiple steps, each one adding properties that make cells more likely to grow into a tumor. The genetic changes that underlie oncogenesis alter several fundamental properties of cells, allowing cells to evade normal growth controls and ultimately conferring the full cancer phenotype (see Figure 24-1). Cancer cells acquire a drive to proliferate that does not require an external inducing signal. They fail to sense signals that restrict cell division and continue to live when they should die. They often change their attachment to surrounding cells or the extracellular matrix, breaking loose to move away and spread the tumor. A cancer cell may, up to a point, resemble a particular type of normal, rapidly dividing cell, but the cancer cell and its progeny will be immortal. Tumors are characteristically *hypoxic* (oxygen starved), so to grow to more than a small size, tumors must obtain a blood supply. They often do so by signaling to induce the growth of blood vessels into the tumor. As cancer progresses, tumors become an abnormal organ, increasingly well adapted to growth and invasion of surrounding tissues.

Normal animal cells are often classified according to their embryonic tissue of origin, and the naming of tumors has followed suit. Malignant tumors are classified as *carcinomas* if they derive from epithelia such as endoderm (gut epithelium) or ectoderm (skin and neural epithelia) and *sarcomas* if they derive from mesoderm (muscle, blood, and connective tissue precursors). Carcinomas are by far the most common type of malignant tumor (more than 90 percent). Most tumors are solid masses, but the *leukemias*, a class of sarcomas, grow as individual cells in the blood. (The name *leukemia* is derived from the Latin for "white blood": the massive proliferation of leukemic cells can cause a patient's blood to appear milky.) The *lymphomas*, another type of malignant sarcoma, are solid tumors of lymphocytes and plasma cells. Malignant brain tumors can be derived from neural cells, and *glioblastomas* are tumors of the glial cells that constitute the major cell type in the brain.

## Metastatic Tumor Cells Are Invasive and Can Spread

Tumors arise with great frequency, especially in older individuals, but most pose little risk to their host because they are localized and of small size. We call such tumors **benign**; an example is a wart, a benign skin tumor. The cells composing benign tumors closely resemble and may function like normal cells. The cell-adhesion molecules that hold tissues together keep benign tumor cells, like normal cells, localized to the tissues where they originate. A fibrous capsule usually delineates the extent of a benign tumor and makes it an easy target for a surgeon. Benign tumors become serious medical problems only if their sheer bulk interferes with normal functions or if they secrete excess amounts of biologically active substances such as hormones. Acromegaly, the overgrowth of head, hands, and feet, for example, can occur when a benign pituitary tumor causes overproduction of growth hormone.

In contrast, cells composing a **malignant** (cancerous) tumor (Figure 24-2), usually grow and divide more rapidly than normal cells and fail to die at the normal rate. A key characteristic of malignant cells is their ability to invade nearby tissue, spreading and seeding additional tumors while the cells continue to proliferate. Some malignant tumors, such as those in the ovary or breast, remain localized and encapsulated, at least for a time. When these tumors progress, the cells invade surrounding tissues and undergo metastasis (Figure 24-3a). Most malignant cells eventually acquire the ability to metastasize. Thus the major characteristics that differentiate metastatic (or malignant) tumors from benign ones are their abilities to invade nearby tissues and spread to distant sites in the body.

Normal cells are restricted to their place in an organ or tissue by cell-cell adhesion and by physical barriers such as the *basement membrane*, which underlies layers of epithelial cells and also surrounds the endothelial cells of blood vessels (see Chapter 20). In contrast, cancer cells have acquired the ability to penetrate the basement membrane using a cell protrusion

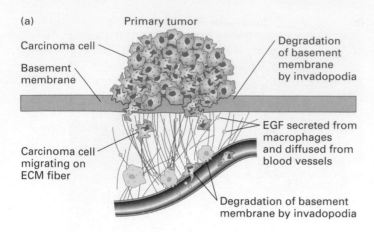

(a)

Primary tumor

Carcinoma cell

Basement membrane

Degradation of basement membrane by invadopodia

Carcinoma cell migrating on ECM fiber

EGF secreted from macrophages and diffused from blood vessels

Degradation of basement membrane by invadopodia

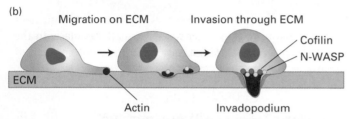

(b)

Migration on ECM

Invasion through ECM

Cofilin

N-WASP

ECM

Actin

Invadopodium

**FIGURE 24-3 Metastasis.** (a) First steps in metastasis, using breast carcinoma cells as an example. Cancer cells leave the main tumor and attack the basement membrane, using extracellular matrix (ECM) fibers to reach the blood vessels. The cancer cells can be attracted by signals such as epidermal growth factor (EGF), which can be secreted by macrophages (yellow). At the blood vessels they penetrate the layer of endothelial cells that forms the vessel walls and enter the bloodstream. (b) Carcinoma cells penetrate the extracellular matrix and blood vessel wall by extending "invadopodia," which produce matrix metalloproteases and other proteases to open up a path. [Adapted from Yamaguchi et al., 2005, *Curr. Opin. Cell Biol.* **17**:559.]

called an "invadopodium" and to migrate to distant sites in the body (Figure 24-3b). A developmental process known as the **epithelial-to-mesenchymal transition (EMT)** is thought to play a crucial role during the process of metastasis in some cancers. During normal development, the conversion of epithelial cells into mesenchymal cells is a step in the formation of some organs and tissues. An EMT requires distinct changes in patterns of gene expression and results in fundamental changes in cell morphology, such as loss of cell-cell adhesion, loss of cell polarity, and the acquisition of migratory and invasive properties. During metastasis, the EMT regulatory pathways are thought to be activated at the invasive front of tumors, producing single migratory cells. At the heart of the EMT are two transcription factors, Snail and Twist. These transcription factors promote expression of genes involved in cell migration, trigger down-regulation of cell adhesion factors such as E-cadherin, and increase the production of proteases that digest the basement membrane, thus allowing its penetration by the tumor cells. For example, many tumor cells secrete a protein (plasminogen activator) that converts the serum protein plasminogen to the active protease plasmin. Importantly, expression of many important drivers of the EMT,

such as SNAIL1 and SNAIL2, has been shown to correlate with disease relapse and patient survival in many cancers, including breast, colon, and ovarian cancer. The occurrence of the EMT indicates poor clinical outcome.

As the basement membrane disintegrates, some tumor cells will enter the bloodstream, but fewer than 1 in 10,000 cells that escape the primary tumor survive to colonize another tissue and form a secondary, metastatic tumor. Much of preventative medicine is currently focused on developing methods to identify the rare tumor cells that circulate in the bloodstream. The ability to capture the *circulating tumor cells* (CTCs) would provide not only a powerful and noninvasive tool for the early detection of cancer, but their analysis could provide insights into the nature of the disease and inform treatment.

In addition to escaping the original tumor and entering the blood, cells that will seed new tumors must then adhere to an endothelial cell lining a capillary and migrate across or through it into the underlying tissue in a process called extravasation (see Chapter 20). The multiple crossings of tissue layers that underlie malignancy require significant changes in the cells' behavior and often involve new or variant surface proteins made by malignant cells.

## Cancers Usually Originate in Proliferating Cells

In order for most oncogenic mutations to induce cancer, they must occur in dividing cells so that the mutation is passed on to many progeny cells. When such mutations occur in nondividing cells (e.g., neurons and muscle cells), they generally do not induce cancer, which is why tumors of muscle and nerve cells are rare in adults. Nonetheless, cancer does occur in tissues composed mainly of nondividing differentiated cells such as erythrocytes and most white blood cells, absorptive cells that line the small intestine, and keratinized cells that form the skin. It has therefore been suggested that the cells that initiate the tumors are not the differentiated cells themselves but rather their precursor cells. Fully differentiated cells usually do not divide. As they die or wear out, they are continually replaced by proliferation and differentiation of **stem cells**, and these cells are thought to be capable of transforming into tumor cells. Because stem cells can divide continually over the life of an organism, oncogenic mutations in their DNA can accumulate, eventually transforming them into cancer cells. Cells that have acquired these mutations have an abnormal proliferative capacity and generally cannot undergo normal processes of differentiation. In recent years it has also become clear that tumor-promoting mutations have the ability to transform a nondividing, terminally differentiated cell into a proliferating cell with precursor-like properties. Thus, in some cancers, dedifferentiation and the return of a cell to a precursor-like state could be the tumor-initiating event. Irrespective of whether tumors originate from differentiated cells that have regained the ability to proliferate or through mutation of tissue stem cells, it is now evident that in many tumors, like in normal organs, there are only certain cells with the ability to divide uncontrollably and generate new tumors; such cells are *cancer stem cells*, discussed in the next section.

## Local Environment Impacts Heterogeneous Tumor Formation by Cancer Stem Cells

Some types of tumors appear to have their own stem cells; that is, these are the only tumor cells capable of seeding a new tumor. The concept is that within a tumor, some cells will cease dividing while others can continue cancerous growth. The latter, of course, are the most dangerous and the most important to destroy with anticancer treatments. Cancer stem cells are thought to give rise to some cells with high replicative capabilities and others with more limited replicative potential. It is not yet clear how many types of tumors have stem cells that differ from the majority of the other cells in the tumor.

One way to identify cancer stem cells is to purify different classes of cells from a tumor based on differences in their surface markers, usually using a fluorescence-activated cell sorter (FACS, see Chapter 9). Transplantation tests, usually with mice, reveal which classes of cell have the ability to seed a new tumor and which do not. For example, a sample of a few hundred human brain tumor cells with an antigen called *CD133* on the surface is potent in starting new tumors in immunologically deprived mice, whereas a sample of many thousands of non-*CD133* tumor cells was unable to seed tumors. Similar findings have been made for human multiple myeloma; the majority of the cells (>95 percent) express a marker called *CD138*. The small population of cells that lacks *CD138* has considerably greater ability than the rest of the cells to start tumor growth. These findings suggest that identifying cancer stem cells and then targeting those cells specifically with drugs or antibodies may be a more effective cancer therapy than targeting bulk tumor cells. ∎

The results with cancer stem cells highlight three important points. First, tumors are not always made up of uniform cells, even if they originated from a single initiating cell. Second, a minority of the cells may be the really dangerous ones. Third, tumor cells may grow faster or slower depending on whether they find themselves in a particular environment. Just as stem cells can be kept in a dividing, nondifferentiating state by virtue of occupying a suitable niche (see Chapter 21), tumor stem cells may behave according to their surroundings. Some neighboring cells may be more conducive to tumor cell or cancer stem cell growth than others. Thus the environment of tumor cells, called the *tumor microenvironment*, can have a dramatic impact on the ability of tumor cells to grow.

The idea that the tumor microenvironment matters extends to the importance of one of the most common environments for a tumor cell: inflammatory cells. Cancers frequently arise at sites of injury or chronic infection. Immune cells migrate to sites of injury and produce growth factors there to promote healing, and the extracellular matrix is reconstructed. All of these local tissue properties may contribute to the establishment and growth of a tumor. It is estimated that up to 20 percent of cancers are linked to chronic infection. For example, persistent *Helicobacter pylori* infection is associated with gastric cancer and mucosa-associated lymphoid tissue (MALT) lymphoma. Infections with hepatitis B or C viruses increase the risk of hepatocellular carcinoma.

An inflammatory response not only can promote tumorigenesis, but tumor formation itself can trigger an inflammatory response. Certain oncogenes induce a transcriptional program that leads to the recruitment of immune cells, especially macrophages. Macrophages and other lymphocytes produce tumor-promoting cytokines and trigger an inflammatory response, providing cancer cells with additional growth factors and promoting blood vessel growth, which—as we will discuss next—is an essential aspect of tumor growth.

## Tumor Growth Requires Formation of New Blood Vessels

Tumors, whether primary or secondary, require recruitment of new blood vessels in order to grow to a large mass. In the absence of a blood supply, a tumor can grow into a mass of about $10^6$ cells, roughly a sphere 2 mm in diameter. At this point, division of cells on the outside of the tumor mass is balanced by death of those in the center from an inadequate supply of nutrients. Such tumors, unless they secrete hormones, cause few problems. However, most tumors induce the formation of new blood vessels that invade the tumor and nourish it, a process called *angiogenesis*. This complex process requires several discrete steps: degradation of the basement membrane that surrounds a nearby capillary, migration of endothelial cells lining the capillary into the tumor, division of these endothelial cells, and formation of a new basement membrane around the newly elongated capillary.

Many tumors produce growth factors that stimulate angiogenesis; other tumors somehow induce surrounding normal cells to synthesize and secrete such factors. Basic fibroblast growth factor (b-FGF), transforming growth factor α (TGFα), and vascular endothelial growth factor (VEGF), which are secreted by many tumors, all have angiogenic properties. New blood vessels nourish the growing tumor, allowing it to increase in size and thus increase the probability that additional harmful mutations will occur. The presence of an adjacent blood vessel also facilitates the process of metastasis.

In humans there are five VEGF genes and three VEGF-receptor-protein genes. VEGF expression can be induced by hypoxia, starvation of cells for oxygen that occurs when $[pO_2] < 7$ mmHg. The VEGF receptors, which are tyrosine kinases, regulate different aspects of blood vessel growth such as endothelial (blood vessel wall) cell survival and growth, cell migration, and vessel wall permeability.

The hypoxia signal is mediated by hypoxia-inducible factor (HIF-1), a transcription factor that is activated in conditions of low oxygen and then binds to and induces transcription of the VEGF gene and about 30 other genes, many of which can affect the probability of tumor growth. Among these are glycolytic enzymes such as lactate dehydrogenase; thus HIF-1 may also help tumor cells to adapt to low oxygen by turning to glycolysis rather than oxidative phosphorylation for ATP generation. HIF-1 activity is controlled in turn by an oxygen sensor composed of a prolyl hydroxylase that is

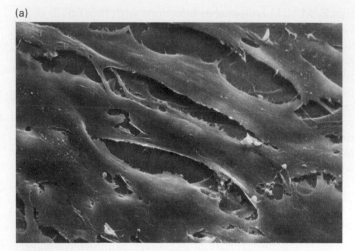 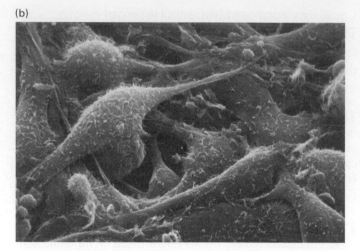

**EXPERIMENTAL FIGURE 24-4 Scanning electron micrographs reveal the organizational and morphological differences between normal and transformed 3T3 cells.** (a) Normal 3T3 cells are elongated and are aligned and closely packed in an orderly fashion. (b) 3T3 cells transformed by an oncogene encoded by Rous sarcoma virus are rounded and covered with small hairlike processes and bulbous projections. The transformed cells have lost the side-by-side organization of the normal cells and grow one atop the other. These transformed cells have many of the same properties as malignant cells. Similar changes are seen in cells transfected with DNA from human cancers containing the *ras*$^D$ oncogene. [Courtesy of L. B. Chen.]

active in normal $O_2$ levels but inactive when deprived of $O_2$. Hydroxylation of HIF-1 causes ubiquitinylation and degradation of the transcription factor, a process that is blocked when $O_2$ is low. Compounds that inhibit angiogenesis have excited much interest as potential therapeutic agents. However, their success in the clinic has thus far been limited.

## Specific Mutations Transform Cultured Cells into Tumor Cells

The morphology and growth properties of tumor cells clearly differ from those of their normal counterparts; some of these differences are also evident when cells are cultured. That mutations cause these differences was conclusively established by transfection experiments with a line of cultured mouse fibroblasts called *3T3 cells.* These cells normally grow only when attached to the plastic surface of a culture dish and are maintained at a low cell density. Because 3T3 cells stop growing when they contact other cells, they eventually form a monolayer of well-ordered cells that have stopped proliferating and are in the quiescent $G_0$ phase of the cell cycle (Figure 24-4a).

When DNA from human bladder cancer cells is transfected into cultured 3T3 cells, about one cell in a million incorporates a particular segment of the exogenous DNA that causes a phenotypic change. The progeny of the affected cell are more rounded and less adherent to one another and to the dish than are the normal surrounding cells, forming a three-dimensional cluster of cells (a focus) that can be recognized under the microscope (Figure 24-4b). Such cells, which continue to grow when the normal cells have become quiescent, have undergone oncogenic **transformation**. The transformed cells have properties similar to those of malignant tumor cells, including changes in cell morphology, ability to grow unattached to an extracellular matrix, reduced requirement for growth factors, secretion of plasminogen activator, and loss of actin microfilaments.

Figure 24-5 outlines the procedure for transforming 3T3 cells with DNA from a human bladder cancer and cloning the specific DNA segment that causes transformation. It was remarkable that a single small piece of DNA had this capability; if more than one incorporated DNA fragment had been needed to induce transformation, the experiment would have failed. Subsequent studies showed that the cloned segment included a mutant version of the cellular *ras* gene, where the glycine normally found in position 12 is replaced with a valine. This mutant was designated *ras*$^D$, where the *D* stands for "dominant." The mutation is genetically dominant because the active protein has an effect even in the presence of the other, normal *ras* allele. Normal **Ras protein**, which participates in many intracellular signal transduction pathways activated by growth factors (see Chapter 16), cycles between an inactive, "off" state with bound GDP and an active, "on" state with bound GTP. The mutated Ras$^D$ protein hydrolyzes bound GTP very slowly and therefore accumulates in the active state, sending a growth-promoting signal to the nucleus even in the absence of the hormones normally required to activate its signaling function.

The production and constitutive activation of Ras$^D$ protein is not sufficient to cause transformation of normal cells in a primary (fresh) culture of human, rat, or mouse fibroblasts. Unlike cells in a primary culture, however, cultured 3T3 cells already carry several mutations, including loss-of-function mutations in the *p19ARF* or *p53* genes, which are regulators of the cell cycle and cell survival. These mutations allow 3T3 cells to grow for an unlimited time in culture if periodically diluted and supplied with nutrients, which normal unmutated cells cannot do (see Figure 9-1b). These **immortal** 3T3 cells are transformed into full-blown tumor cells only when they produce a constitutively active Ras protein or other oncoproteins. For this reason, transfection with the *ras*$^D$ gene can transform 3T3 cells but not normal cultured primary fibroblast cells into tumor cells.

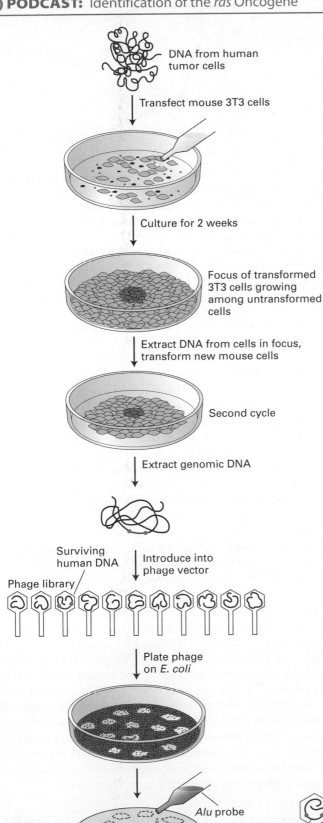

Surviving human DNA

Phage library

Introduce into phage vector

Plate phage on *E. coli*

Replica on filter paper

*Alu* probe

Oncogene

**EXPERIMENTAL FIGURE 24-5   Transformation of mouse cells with DNA from a human cancer cell permits identification and molecular cloning of the *ras^D* oncogene.** Addition of DNA from a human bladder cancer to a culture of mouse 3T3 cells causes about one cell in a million to divide abnormally and form a focus, or clone, of transformed cells. To clone the oncogene responsible for transformation, advantage is taken of the fact that most human genes have nearby repetitive DNA sequences called *Alu* sequences. DNA from the initial focus of transformed mouse cells is isolated, and the oncogene is separated from adventitious human DNA by secondary transfer to mouse cells. The adventitious DNA is human DNA that has no effect on cell transformation but just happened to end up in a cell that also contains the active oncogene. The total DNA from a secondary trans-fected mouse cell is then cloned into bacteriophage λ; only the phage that receives human DNA hybridizes with an *Alu* probe. The hybridizing phage should contain part of or all the transforming oncogene. This expected result can be proved by showing either that the phage DNA can transform cells (if the oncogene has been completely cloned) or that the cloned piece of DNA is always present in cells transformed by DNA transfer from the original bladder cancer donor cell.

A mutant *ras* gene is found in most human colon, bladder, pancreatic, and other cancers but not in normal human DNA; thus it must arise as the result of a somatic mutation in one of the tumor progenitor cells. As we will see in Section 24.2, any gene, such as *ras^D*, that encodes a protein capable of trans-forming cells in culture or contributing to cancer in animals is referred to as an **oncogene**. An oncogene arises from a normal cellular gene, a proto-oncogene, such as *ras*.

## A Multi-hit Model of Cancer Induction Is Supported by Several Lines of Evidence

As noted earlier and illustrated by the oncogenic transforma-tion of 3T3 cells, multiple mutations usually are required to convert a normal body cell into a malignant one. According to this "multi-hit" model, evolutionary (or "survival of the fittest") cancers arise by a process of clonal selection not un-like the selection of individual animals in a large population. Here is the scenario, which may or may not apply to all can-cers: A mutation in one cell, perhaps a stem cell, would give it a slight growth advantage. One of the progeny cells would then undergo a second mutation that would allow its descen-dants to grow more uncontrollably and form a small benign tumor; a third mutation in a cell within this tumor would allow it to outgrow the others and overcome constraints im-posed by the tumor microenvironment, and its progeny would form a mass of cells, each of which would have these three mutations. An additional mutation in one of these cells would allow its progeny to escape into the bloodstream and establish daughter colonies at other sites, the hallmark of metastatic cancer. This model makes two easily testable predictions.

First, all cells in a given tumor should have at least some genetic alterations in common. Systematic analysis of cells from individual human tumors supports the prediction that all the cells are derived from a single progenitor. Recall that

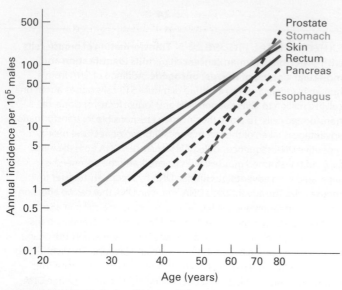

**EXPERIMENTAL FIGURE 24-6 The incidence of human cancers increases as a function of age.** The marked increase in the incidence with age is consistent with the multi-hit model of cancer induction. Note that the logarithm of annual incidence is plotted versus the logarithm of age. [From B. Vogelstein and K. Kinzler, 1993, *Trends Genet.* **9**:101.]

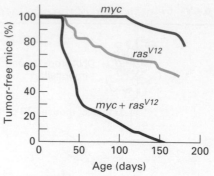

**EXPERIMENTAL FIGURE 24-7 The kinetics of tumor appearance in female mice carrying either one or two oncogenic transgenes shows the cooperative nature of multiple mutations in cancer induction.** Each of the transgenes was driven by the mouse mammary tumor virus (MMTV) breast-specific promoter. The hormonal stimulation associated with pregnancy activates the MMTV promoter and hence the overexpression of the transgenes in mammary tissue. The graph shows the time course of tumorigenesis in mice carrying either *myc* or *ras*$^{V12}$ transgenes as well as in the progeny of a cross of *myc* carriers with *ras*$^{V12}$ carriers that contain both transgenes. The results clearly demonstrate the cooperative effects of multiple mutations in cancer induction. [See E. Sinn et al., 1987, *Cell* **49**:465.]

during the fetal development of a human female each cell inactivates one of the two X chromosomes. A woman is a genetic mosaic: half her cells have one X inactivated, and the remainder have the other X inactivated. If a tumor did not arise from a single progenitor, it would be composed of a mix of cells with one or the other X inactivated. However, the opposite is observed: all the cells in a tumor taken from a woman have the same inactive X chromosome. Different tumors can be composed of cells with either the maternal or the paternal X inactive. Second, cancer incidence should increase with age because it can take decades for the required multiple mutations to occur. Assuming that the rate of mutation is roughly constant during a lifetime, then the incidence of most types of cancer would be independent of age if only one mutation were required to convert a normal cell into a malignant one. In fact, current estimates suggest 5–6 "hits," or mutations, must accumulate as the most dangerous cancer cells emerge. As the data in Figure 24-6 show, the incidence of many types of human cancer does indeed increase drastically with age.

More direct evidence that multiple mutations are required for tumor induction comes from transgenic mice. A variety of combinations of oncogenes can cooperate in causing cancer. For example, mice have been made that carry either the mutant *ras*$^{V12}$ dominant oncogene (one version of *ras*$^D$) or the c-*myc* proto-oncogene, in each case under control of a mammary-cell-specific promoter/enhancer from a retrovirus. The promoter is induced by endogenous hormone levels and tissue-specific regulators, leading to overexpression of c-*myc* or *ras*$^{V12}$ in breast tissue. Myc protein is a transcription factor that induces expression of many genes required for the transition from the G$_1$ to the S phase of the cell cycle. Heightened transcription of c-*myc* mimics previously identified oncogenic mutations that turn up c-*myc* transcription, converting the proto-oncogene into an oncogene. By itself, the c-*myc* transgene causes tumors only after

100 days and then in only a few mice; clearly only a minute fraction of the mammary cells that overproduce the Myc protein actually become malignant. Similarly, production of the mutant Ras$^{V12}$ protein alone causes tumors earlier but still slowly and with about 50 percent efficiency over 150 days. When the c-*myc* and *ras*$^{V12}$ transgenics are crossed, however, such that all mammary cells overproduce both Myc and Ras$^{V12}$, tumors arise much more rapidly and all animals succumb to cancer (Figure 24-7). Such experiments emphasize the synergistic effects of multiple oncogenes. They also suggest that the long latency of tumor formation, even in the double-transgenic mice, is due to the need to acquire still more somatic mutations.

Similar cooperative effects between oncogenes can be seen in cultured cells. For example, transfection of normal fibroblasts (*not* immortalized 3T3 fibroblasts) with either c-*myc* or activated *ras*$^D$ is not sufficient for oncogenic transformation, whereas when transfected together, the two genes cooperate to transform the cells. Deregulated levels of c-*myc* alone induce proliferation but also sensitize fibroblasts to apoptosis. Overexpression of activated *ras*$^D$ alone induces cells to go into a state where they can no longer divide, which is termed *senescence*. When the two oncogenes are expressed in the same cell, these negative cellular responses are neutralized and the cells undergo transformation. While the examples mentioned are combinations of oncogenes, it is also possible to enhance cancer rates of cultured cell transformations by combining an oncogene with the loss of a tumor-suppressor gene.

## Successive Oncogenic Mutations Can Be Traced in Colon Cancers

Studies on colon cancer provide the most compelling evidence to date for the *multi-hit model of cancer induction*. Surgeons can obtain fairly pure samples of many human

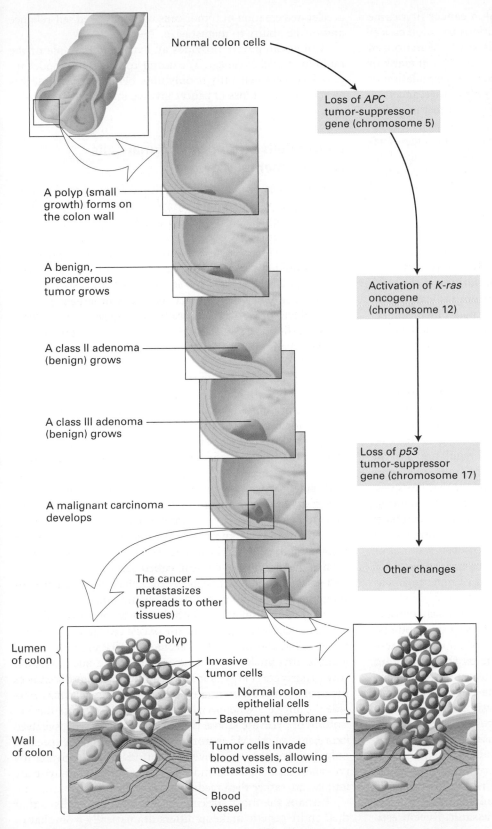

**FIGURE 24-8 The development and metastasis of human colorectal cancer and its genetic basis.** A mutation in the APC tumor-suppressor gene in a single epithelial cell causes the cell to divide, although surrounding cells do not, forming a mass of localized benign tumor cells, or a polyp. Subsequent mutations lead to expression of a constitutively active Ras protein and loss of the tumor-suppressor gene *p53*. This, together with additional yet to be identified genetic changes, generates a malignant cell. The cell continues to divide, and the progeny invade the basement membrane that surrounds the tissue. Some tumor cells spread into blood vessels that will distribute them to other sites in the body. Additional mutations permit the tumor cells to exit from the blood vessels and proliferate at distant sites; a person with such a tumor is said to have cancer. [Adapted from B. Vogelstein and K. Kinzler, 1993, *Trends Genet.* **9:**101.]

cancers, but since the tumor is observed only at one time, its exact stage of progression cannot be easily determined. An exception is colon cancer, which evolves through distinct, well-characterized morphological stages. These intermediate stages—polyps, benign adenomas, and carcinomas—can be isolated by a surgeon, allowing mutations that occur in each of the morphological stages to be identified. Numerous studies show that colon cancer arises from a series of mutations that commonly occur in a well-defined order, providing strong support for the multi-hit model (Figure 24-8).

Insight into the progression of colon cancer first came from the study of inherited predispositions to colon cancer such as familial adenomatous polyposis (FAP). Mutations in the Wnt signaling pathway have been identified in many of these syndromes, and it is now believed that deregulation of Wnt signaling results in formation of polyps (precancerous growths) on the inside of the colon wall—not just in people with inherited polyposis syndromes, but also in people afflicted with sporadic forms of colon cancer. APC (*adenomatous polyposis coli*) is a negative regulator of Wnt signaling, which promotes cell cycle entry by activating expression of the *c-myc* gene (see Chapter 16). The absence of functional APC protein thus leads to inappropriate production of Myc, and cells homozygous for *APC* mutations proliferate at a rate higher than normal and form polyps. Loss-of-function mutations in the *APC* gene are the most frequent mutations found in early stages of colon cancer. Most of the cells in a polyp contain the same one or two mutations in the *APC* gene that result in its loss or inactivation, indicating that they are clones of the cell in which the original mutation occurred. Thus *APC* is a tumor-suppressor gene, and both alleles of the *APC* gene must carry an inactivating mutation for polyps to form because cells with one wild-type *APC* gene express enough APC protein to function normally.

If one of the cells in a polyp undergoes another mutation, this time an activating mutation of the *ras* gene, its progeny divide in an even more uncontrolled fashion, forming a larger adenoma. Inactivation of the *p53* gene follows and results in the gradual loss of normal regulation and the consequent formation of a malignant carcinoma (see Figure 24-8). The *p53 protein* is a tumor suppressor that halts progression through the cell cycle in response to DNA damage. While the three "hits" listed here are certainly crucial parts of the picture, there are likely to be additional contributing genetic events. Not every colon cancer, however, acquires all the later mutations or acquires them in the order depicted in Figure 24-8. Thus different combinations of mutations may result in the same phenotype.

DNA from different human colon carcinomas generally contains mutations in all three genes—loss-of-function mutations in the tumor suppressors *APC*, and *p53* and an activating (gain-of-function) mutation in the dominant oncogene *K-ras* (one of the *ras* family of genes)—establishing that multiple mutations in the same cell are needed for the cancer to form. Some of these mutations appear to confer growth advantages at an early stage of tumor development, whereas other mutations promote the later stages, including invasion and metastasis, which are required for the malignant phenotype. The number of mutations needed for colon cancer progression may at first seem surprising, seemingly an effective barrier to tumorigenesis. Our genomes, however, are under constant assault. Recent estimates indicate that sporadically arising polyps have about 11,000 genetic alterations in each cell, though very likely only a few of these are relevant to oncogenesis. Genetic instability is a hallmark of cancer cells. This genomic instability promotes further tumor evolution, allowing for the accelerated creation of tumor cells with increased self-reliance and/or the ability to metastasize.

Colon carcinoma provides an excellent example of the multi-hit model of cancer. The degree to which this model applies to cancer generally is only now being learned, but it is clear that many types of cancer involve multiple mutations.

## Cancer Cells Differ from Normal Cells in Fundamental Ways

Cancer cells can often be distinguished from normal cells by microscopic examination. They are usually less well differentiated than normal cells or benign tumor cells. In a specific tissue, malignant cells usually exhibit the characteristics of rapidly growing cells, that is, a high nucleus-to-cytoplasm ratio, prominent nucleoli, an increased frequency of mitotic cells, and relatively little specialized structure. Normal cells stop growing when they contact other cells, eventually forming a monolayer of well-ordered cells (see Figure 24-4a). Transformed cells are less adherent, forming a three-dimensional cluster of cells (a focus) that can be recognized under the microscope (see Figure 24-4b).

Tumor cells not only differ from normal cells in their appearance, but their entire energy metabolism is rewired. Normal differentiated cells rely on mitochondrial oxidative phosphorylation to satisfy their energy needs. Cells metabolize glucose to carbon dioxide by oxidation of pyruvate through the tricarboxylic acid (TCA) cycle in the mitochondria (see Chapter 12). Only under anaerobic conditions do cells undergo anaerobic glycolysis and produce large amounts of lactate. In contrast to normal cells, most cancer cells rely on glycolysis for energy production irrespective of whether oxygen levels are high or low, producing large amounts of lactate (Figure 24-9). The use of glycolysis to produce energy even in the presence of oxygen, called aerobic glycolysis, was first discovered in cancer cells by the cell biologist Otto Warburg and is therefore called the **Warburg effect**.

The metabolism of glucose to lactate generates only 2 ATP molecules per molecule of glucose, while oxidative phosphorylation generates up to 36 molecules of ATP per molecule of glucose. It is unclear why cancer cells utilize this inefficient way to generate energy, but some of the molecular differences between normal differentiated cells and cancer cells may provide the answer. Whereas differentiated cells express the M1 isoform of pyruvate kinase (PK-M1), all cancer cells switch to expression of the M2 isoform, which is normally expressed only during embryonic development. PK-M2 is activated by tyrosine kinase signaling and results in conversion of pyruvate into lactate rather than feeding into the TCA cycle.

Perhaps the most striking feature of tumor cells is that their entire genetic makeup differs dramatically from that of normal cells. Characteristic of nearly all tumor cells is **aneuploidy**, the presence of an aberrant number of chromosomes—generally too many.

Several mechanisms have been identified that lead to aneuploidy. As discussed in Chapter 19, the DNA replication

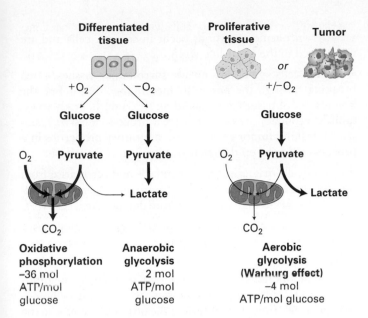

**FIGURE 24-9 Energy production in cancer cells by aerobic glycolysis.** In the presence of oxygen, nonproliferating (differentiated) cells metabolize glucose into pyruvate via glycolysis. Pyruvate is then transported into mitochondria, where it is fed into the TCA cycle. Oxygen is required as the final electron acceptor during oxidative phosphorylation. Thus when oxygen is limiting, cells metabolize pyruvate into lactate, allowing glycolysis to continue by cycling NADH back to $NAD^+$. Cancer cells and proliferating cells convert most glucose to lactate regardless of whether oxygen is present or not. The production of lactate in the presence of oxygen is called aerobic glycolysis. [Adapted from Vander Heiden et al., 2009, *Science* **324**:1029.]

checkpoint normally prevents entry into mitosis unless all chromosomes have completely replicated their DNA. The spindle assembly checkpoint prevents entry into anaphase unless all the replicated chromosomes attach properly to the spindle apparatus, and the spindle position checkpoint prevents exit from mitosis and cytokinesis if the chromosomes segregate improperly. A failure in the DNA damage and spindle assembly checkpoints causes chromosomal abnormalities and mis-segregation, respectively, leading to aneuploidy. Defects in the spindle position checkpoint cause the formation of tetraploid cells, which then through chromosome losses can also become aneuploid. The exact role of aneuploidy in tumorigenesis is being debated, but it is clear that aneuploidy causes cancer. Mutations that induce genomic instability lead to cancer in mice. Human syndromes such as mosaic variegated aneuploidy (MVA) that cause increased chromosome mis-segregation also predispose patients to certain cancers, including Wilms's tumor and rhabdomyosarcoma.

## DNA Microarray Analysis of Expression Patterns Can Reveal Subtle Differences Between Tumor Cells

Traditionally the properties of tumor and normal cells have been assessed by staining and microscopy. The prognosis for many tumors could be determined, within certain limits, from their histology. However, the appearance of cells alone has limited information content, and better ways to discern the properties of cells are desirable both to understand tumorigenesis and to arrive at meaningful and accurate decisions about prognosis and therapy.

As we have seen, genetic studies can identify the single initiating mutation or series of mutations that cause transformation of normal cells into tumor cells, as in the case of colon cancer. After these initial events, however, the cells of a tumor undergo a cascade of changes reflecting the interplay between the initiating events and signals from outside. As a result, tumor cells can become quite different, even if they arise from the same initiating mutation or mutations. Although these differences may not be recognized from the appearance of cells, they can be detected from the cells' patterns of gene expression. **DNA microarray** analysis can determine the expression of tens of thousands of genes simultaneously, permitting complex phenotypes to be defined at the molecular genetic level. (See Figures 5-29 and 5-30 for an explanation of this technique.)

The advent of DNA microarray and large-scale sequencing technology is allowing more detailed examination of tumor properties. Not surprisingly, primary tumors can often be distinguishable from metastatic tumors by the pattern of gene expression. Microarray analysis is now also routinely used to determine patient outcome and the best course of treatment for many types of cancers.

Patients affected by breast cancer have markedly different treatment responses and highly variable outcomes. Chemotherapy and hormonal therapies reduce the risk of metastases by approximately 30 percent, but 70–80 percent of these patients would have survived without the treatment. Determining which patient should receive chemotherapy to prevent metastases is thus a critical question. Microarray analyses have provided a means to make this decision. Researchers analyzed the gene expression profile of breast cancers that had not yet spread to neighboring lymph nodes (non-metastatic breast cancers). They identified 70 genes whose expression predicts the likelihood of metastasis with more than 90 percent accuracy (Figure 24-10). This "classifier" gene expression signature of poor prognosis contained genes involved in cell cycle progression, invasion, and angiogenesis, providing a biological basis for why metastasis was more likely in this patient group than in the patient group that lacked this gene expression signature. Based on these gene expression classifiers, it may now be easier to determine which patients will benefit from aggressive chemotherapies and which will not need this aggressive treatment. Similar analyses of the gene expression patterns of other tumors are likely to improve classification and diagnosis, allowing informed decisions about treatments, and also to provide insights into the properties of tumor cells. ■

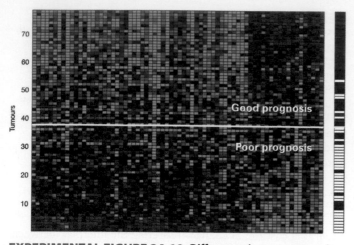

**EXPERIMENTAL FIGURE 24-10 Differences in gene-expression patterns determined by DNA microarray analysis can predict metastatic behavior of breast cancers.** Samples of mRNA were extracted from 78 breast cancer patients (age 55 and younger) that carried sporadic breast tumors that had not yet spread to neighboring lymph nodes. DNA microarray analysis of the extracted RNA determined the transcription levels of about 25,000 genes in each of the 78 experimental samples. (See Figures 5-29 and 5-30 for a description of microarray analysis.) A prognostic gene expression signature of 70 genes was then determined. First all genes were identified that were up-regulated in more than three of the 78 tumors. The correlation coefficient of the expression for each gene with disease outcome (patients who were metastasis free after five years versus patients who developed distant metastases within five years) was then calculated, and 231 genes were found to be significantly associated with disease outcome. Genes were further selected by eliminating genes with lower predictive power until a set of genes was determined whose expression patterns classified poor-prognosis patients with more than 90 percent accuracy. The classifier gene expression signature of these 70 genes is shown. In the cluster diagram, each row represents one tumor, and each vertical column contains data for a single gene of the 70-gene cluster. An intense red color indicates a high level of transcription of that gene; intense green indicates the opposite. The genes were grouped according to their similar patterns of hybridization. Patients above the yellow line have a good prognosis signature; below the yellow line the prognosis signature is poor. The metastasis status of each patient is shown to the right. White indicates patients developed distant metastases within five years after the initial diagnosis; black indicates that patients were disease free for at least five years. [From L. J. van't Veer, 2002, *Nature* **415**:530.]

## KEY CONCEPTS of Section 24.1

### Tumor Cells and the Onset of Cancer

• Cancer is a fundamental aberration in cellular behavior, touching on many aspects of molecular cell biology. Most cell types of the body can give rise to malignant tumor (cancer) cells.

• Cancer cells can multiply in the absence of at least some of the growth-promoting factors required for proliferation of normal cells and are resistant to signals that normally program cell death (apoptosis).

• Most oncogenic mutations occur in somatic cells and are not carried in the germ-line DNA.

• Cancer cells sometimes invade surrounding tissues, often breaking through the basement membranes that define the boundaries of tissues and spreading through the body to establish secondary areas of growth, a process called *metastasis*. Metastatic tumor cells acquire migratory properties in a process called the epithelial–to-mesenchymal transition.

• Cancer cells may arise from stem cells and often arise from proliferating cells. Cancer cells bear more resemblance to dividing precursor cells than to more mature differentiated cell types.

• Both primary and secondary tumors require angiogenesis, the formation of new blood vessels, in order to grow to a large mass.

• The multi-hit model, which proposes that multiple mutations are needed to cause cancer, is consistent with the genetic homogeneity of cells from a given tumor, the observed increase in the incidence of human cancers with advancing age, and the cooperative effect of oncogenic transgenes and tumor-suppressor gene mutations on tumor formation in mice.

• Colon cancer develops through distinct morphological stages that commonly are associated with mutations in specific tumor-suppressor genes and proto-oncogenes.

• Cancer cells differ from normal cells in fundamental ways. Particularly striking is the rewiring of energy metabolism toward glycolysis, a process known as the Warburg effect.

• Most human tumor cells are aneuploid, containing an abnormal number of chromosomes (usually too many). Failure of cell cycle checkpoints that normally detect unreplicated DNA, improper spindle assembly, or mis-segregation of chromosomes permits aneuploid cells to arise.

• DNA microarray analysis can identify differences in gene expression between types of tumor cells that are indistinguishable by traditional criteria and can be used to predict patient outcome.

## 24.2 The Genetic Basis of Cancer

As we have noted, mutations in three broad classes of genes—proto-oncogenes (e.g., *ras*), tumor-suppressor genes (e.g., *APC*), and caretaker genes—play key roles in cancer induction. These genes encode many kinds of proteins that help control cell growth and proliferation (Figure 24-11). Virtually all human tumors have inactivating mutations in genes whose products normally act at various cell cycle **checkpoints** to stop a cell's progress through the cell cycle if a previous step has occurred incorrectly or if DNA has been damaged. For example, most cancers have inactivating mutations in the genes coding for one or more proteins that normally restrict progression through the $G_1$ stage of the cell cycle or activating mutations in genes coding for proteins that drive the cells through the cell cycle. Likewise,

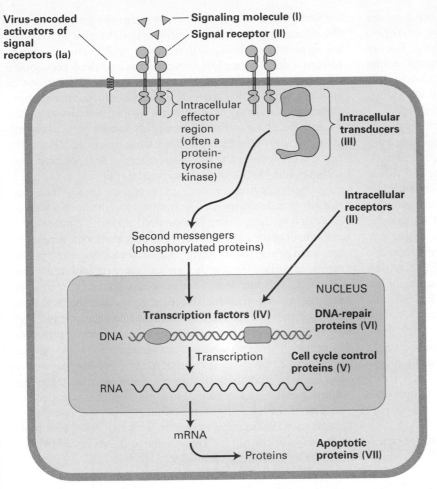

**FIGURE 24-11 Seven types of proteins that participate in controlling cell growth and proliferation.** Cancer can result from expression of mutant forms of these proteins. Mutations changing the structure or expression of proteins that normally promote cell growth generally give rise to dominantly active oncogenes. Many, but not all, extracellular signaling molecules (I), signal receptors (II), signal transduction proteins (III), and transcription factors (IV) are in this category. Cell-cycle-control proteins (V) that function to restrain cell proliferation and DNA-repair proteins (VI) are encoded by tumor-suppressor genes. Mutations in these genes act recessively, greatly increasing the probability that the mutant cells will become tumor cells or that mutations will occur in other classes. Apoptotic proteins (VII) include tumor suppressors that promote apoptosis and oncoproteins that promote cell survival. Virus-encoded proteins that activate signal receptors (Ia) also can induce cancer.

a constitutively active Ras or other activated signal transduction protein is found in several kinds of human tumors that have different origins. Thus malignancy and the intricate processes for controlling the cell cycle discussed in Chapter 19 are two faces of the same coin. In the series of events leading to growth of a tumor, oncogenes combine with tumor-suppressor mutations to give rise to the full spectrum of tumor-cell properties described in the previous section.

In this section, we consider the general types of mutations that are oncogenic and see how certain viruses can cause cancer. We also explain why some inherited mutations increase the risk for particular cancers.

## Gain-of-Function Mutations Convert Proto-oncogenes into Oncogenes

Recall that an oncogene is any gene that encodes a protein able to transform cells in culture, usually in combination with other cell alterations, or to induce cancer in animals. Of the many known oncogenes, all but a few are derived from normal cellular genes (i.e., proto-oncogenes) whose wild-type products promote cell proliferation or other features important for cancer. For example, the *ras* gene discussed previously is a proto-oncogene that encodes an intracellular signal transduction protein that

promotes controlled progression through the cell cycle; the mutant *ras$^D$* gene derived from *ras* is an oncogene whose encoded protein provides an excessive or uncontrolled growth-promoting signal. Other proto-oncogenes encode growth-promoting signaling molecules and their receptors, anti-apoptotic (cell-survival) proteins, and some transcription factors.

Conversion, or activation, of a proto-oncogene into an oncogene generally involves a *gain-of-function* mutation. At least four mechanisms can produce oncogenes from the corresponding proto-oncogenes:

1. *Point mutation* (i.e., change in a single base pair) in a proto-oncogene that results in a hyperactive or constitutively active protein product

2. *Chromosomal translocation* that fuses two genes together to produce a hybrid gene encoding a chimeric protein whose activity, unlike that of the parent proteins, often is constitutive

3. *Chromosomal translocation* that brings a growth regulatory gene under the control of a different promoter that causes inappropriate expression of the gene

4. *Amplification* (i.e., abnormal DNA replication) of a DNA segment including a proto-oncogene, so that numerous copies exist, leading to overproduction of the encoded protein

An oncogene formed by either of the first two mechanisms encodes an "oncoprotein" that differs from the normal protein encoded by the corresponding proto-oncogene. In contrast, the other two mechanisms generate oncogenes whose protein products are identical with the normal proteins; their oncogenic effect is due to production at higher-than-normal levels or in cells where they normally are not produced.

The localized amplification of DNA to produce as many as 100 copies of a given region (usually a region spanning hundreds of kilobases) is a common genetic change seen in tumors. Normally such an event would be repaired or the cell would stop cycling owing to checkpoint control, so such lesions imply a DNA-repair (caretaker) defect of some kind. This anomaly may take either of two forms: the duplicated DNA may be tandemly organized at a single site on a chromosome, or it may exist as small, independent mini-chromosome-like structures. The former case leads to a homogeneously staining region (HSR) that is visible in the light microscope at the site of the amplification; the latter case causes extra "minute" chromosomes, separate from the normal chromosomes that pepper a stained chromosomal preparation (Figure 24-12).

However they arise, the gain-of-function mutations that convert proto-oncogenes to oncogenes are genetically dominant; that is, mutation in only one of the two alleles is sufficient for induction of cancer. See Table 24-1 for a comparison of different classes of cancer-related genes.

Gene amplification may involve a small number of genes, such as the N-*myc* gene and its neighbor *DDX1*, which are amplified in neuroblastoma, or a chromosome region containing many genes. It can be difficult to determine which genes are amplified, a first step in determining which genes contribute to tumor formation. DNA microarrays or deep sequencing offer a powerful approach for finding amplified regions of chromosomes. Rather than look at gene expression, the application of microarrays or quantitative sequence analysis involves looking for abnormally abundant DNA sequences. In the microarray approach, genomic DNA from cancer cells is used to probe arrays containing fragments of genomic DNA, and spots with amplified DNA give stronger signals than control spots. In the deep-sequencing approach, the genome is broken into small fragments and sequenced. The frequency with which a particular region of the genome is sequenced provides a quantitative assessment of the copy number at a particular locus. Among the amplified genes, the strongest candidates for the relevant ones can be identified by also measuring gene expression. A breast carcinoma cell line, with four known amplified chromosome regions, was screened for amplified genes, and the expression levels of those genes were also studied on microarrays. Fifty genes were found to be amplified, but only five were also highly expressed. These five were better candidates for being new oncogenes since amplified genes that are not highly expressed are less likely to contribute to tumor growth. As microarray and whole-genome sequencing approaches become cheaper, it is possible that tumors from individual patients will routinely be analyzed in this manner to determine which oncogenic mutations underlie the tumor. Treatment could then be tailored specifically to the patient's tumor. ■

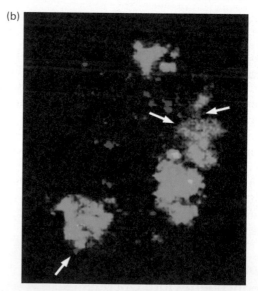

(a)

(b)

**EXPERIMENTAL FIGURE 24-12 DNA amplifications in stained chromosomes take two forms, visible under the light microscope.** (a) Homogeneously staining regions (HSRs) in a human chromosome from a neuroblastoma cell. The chromosomes are uniformly stained with a blue dye so that all can be seen. Specific DNA sequences were detected using fluorescent in situ hybridization (FISH) in which fluorescently labeled DNA clones are hybridized to denatured DNA in the chromosomes. The chromosome 4 pair is marked (red) by in situ hybridization with a large DNA clone containing the N-*myc* oncogene.

On one of the chromosome 4's an HSR is visible (green) after staining for a sequence enriched in the HSR. (b) Optical sections through nuclei from a human neuroblastoma cell that contain so-called double-minute chromosomes. The normal chromosomes are the green and blue structures; the double-minute chromosomes are the many small red dots. Arrows indicate double minutes associated with the surface or interior of the normal chromosomes. [From I. Solovei et al., 2000, *Genes Chromosomes Cancer* **29**:297–308, figs. 4 and 17.]

## TABLE 24-1 | Classes of Genes Implicated in the Onset of Cancer

| | Normal Function of Genes | Examples of Gene Products | Effect of Mutation | Genetic Properties of Mutant Gene | Origin of Mutations |
|---|---|---|---|---|---|
| Proto-oncogenes | Promote cell survival or proliferation | Anti-apoptotic proteins, components of signaling and signal transduction pathways that result in proliferation, transcription factors | Gain-of-function mutations allow unregulated cell proliferation and survival | Mutations are genetically dominant | Arise by point mutation, chromosomal translocation, amplification |
| Tumor-suppressor genes | Inhibit cell survival or proliferation | Apoptosis-promoting proteins, inhibitors of cell-cycle progression, checkpoint-control proteins that assess DNA/chromosomal damage, components of signal pathways that restrain cell proliferation | Loss-of-function mutations allow unregulated cell proliferation and survival | Mutations are genetically recessive | Arise by deletion, point mutation, methylation |
| Caretaker genes | Repair or prevent DNA damage | DNA-repair enzymes | Loss-of-function mutations allow mutations to accumulate | Mutations are genetically recessive | Arise by deletion, point mutation, methylation |

## Cancer-Causing Viruses Contain Oncogenes or Activate Cellular Proto-oncogenes

Pioneering studies by Peyton Rous beginning in 1911 led to the initial recognition that a virus could cause cancer when injected into a suitable host animal. Many years later, molecular biologists showed that Rous sarcoma virus (RSV) is a **retrovirus** whose RNA genome is reverse-transcribed into DNA, which is incorporated into the host-cell genome (see Figure 4-49). In addition to the "normal" genes present in all retroviruses, oncogenic transforming viruses such as RSV contain an oncogene: in the case of RSV, the v-*src* gene. Subsequent studies with mutant forms of RSV demonstrated that only the v-*src* gene, not the other viral genes, was required for cancer induction.

In the late 1970s, scientists were surprised to find that normal cells from chickens and other species contain a gene that is closely related to the RSV v-*src* gene. This normal cellular gene, a proto-oncogene, commonly is distinguished from the viral gene by the prefix "c" (c-*src*). RSV and other acutely transforming viruses are thought to have arisen by incorporating, or transducing, a normal cellular proto-oncogene into their genome. Subsequent mutation in the transduced gene then converted it into a dominantly acting oncogene, which can induce cell transformation even in the presence of the normal c-*src* proto-oncogene. Such viruses are called *transducing retroviruses* because their genomes contain an oncogene derived from a transduced cellular proto-oncogene. When this was first discovered, it was startling to find that these dangerous viruses were turning the animal's own genes against them.

Because its genome carries the potent v-*src* oncogene, the transducing RSV induces tumors within days. In contrast, most oncogenic retroviruses induce cancer only after a period of months or years. The genomes of these *slow-acting retroviruses*, which are weakly transforming, differ from those of transducing viruses in one crucial respect: they lack an oncogene. All slow-acting, or "long-latency," retroviruses appear to cause cancer by integrating into the host-cell DNA near a cellular proto-oncogene and activating its expression. The long terminal repeat (LTR) sequences in integrated retroviral DNA can act as an enhancer or promoter for a nearby cellular gene, thereby stimulating its transcription. For example, in the cells from tumors caused by avian leukosis virus (ALV), the retroviral DNA is inserted near the c-*myc* gene. These cells overproduce c-Myc protein; as noted earlier, overproduction of c-Myc causes abnormally rapid proliferation of cells. Slow-acting viruses act slowly for two reasons: integration near a cellular proto-oncogene (e.g., c-*myc*) is a random, rare event, and additional mutations have to occur before a full-fledged tumor becomes evident.

In natural bird and mouse populations, slow-acting retroviruses are much more common than oncogene-containing retroviruses such as Rous sarcoma virus. Thus insertional proto-oncogene activation is probably the major mechanism by which retroviruses cause cancer. Although the only retrovirus known to cause human tumors is human T-cell leukemia/lymphoma virus (HTLV), the huge investment in studying retroviruses as a model for human cancer paid off both in the discovery of cellular oncogenes and in the sophisticated understanding of retroviruses, which later accelerated progress on the HIV virus that causes AIDS.

A few DNA viruses also are oncogenic. Unlike most DNA viruses that infect animal cells (see Chapter 4), oncogenic DNA viruses integrate their viral DNA into the host-cell genome. The viral DNA contains one or more oncogenes, which permanently transform infected cells. For example, many warts and other benign tumors of epithelial cells are caused by the DNA-containing human papillomaviruses (HPV). A medically much more serious outcome of HPV infection is cervical cancer, the third most common type of cancer in women after lung and breast cancer. The Pap smear, which is used to sample the cervical tissue and screen for possible cancers, is thought to have reduced the death rate by about 70 percent. However, thousands of people still die from cervical cancer each year, and some of these deaths would have been prevented had screening been done. Fortunately, not all HPV infections lead to cancer. We will learn more about HPV oncoproteins later in the chapter.

Unlike retroviral oncogenes, which are derived from normal cellular genes and have no function for the virus except to allow their proliferation in tumors, the known oncogenes of DNA viruses are integral parts of the viral genome and are required for viral replication. As discussed later, the oncoproteins expressed from integrated viral DNA in infected cells act in various ways to stimulate cell growth and proliferation.

## Loss-of-Function Mutations in Tumor-Suppressor Genes Are Oncogenic

Tumor-suppressor genes generally encode proteins that in one way or another inhibit cell proliferation. *Loss-of-function* mutations in one or more of these "brakes" contribute to the development of many cancers. Prominent among the classes of proteins encoded by tumor-suppressor genes are these five:

1. Intracellular proteins that regulate, or inhibit, entry into the cell cycle (e.g., p16 and Rb for $G_1$)

2. Receptors or signal transducers for secreted hormones or developmental signals that inhibit cell proliferation (e.g., TGFβ and the hedgehog receptor Patched)

3. Checkpoint-control proteins that arrest the cell cycle if DNA is damaged or if the chromosome segregation apparatus is abnormal (e.g., p53)

4. Proteins that promote apoptosis

5. Enzymes that participate in DNA repair

Since generally one copy of a tumor-suppressor gene suffices to control cell proliferation, *both* alleles of a tumor-suppressor gene must be lost or inactivated in order to promote tumor development. Thus oncogenic loss-of-function mutations in tumor-suppressor genes are often genetically **recessive** (see Table 24-1). In this context "recessive" means that if there is even one working gene, producing about half the usual amount of protein product, tumor formation will be prevented. With some genes, half the amount of product is not enough, in which case the loss of just one of the two genes can lead to cancer. This kind of gene

is *haplo-insufficient*. The loss of one copy of the gene is decisive for the final phenotype, so this type of mutation is dominant. It is useful to remember, then, two processes by which cancer genes can be dominant: (1) loss of one copy of a haplo-insufficient tumor-suppressor gene, resulting in insufficient product to control growth, and (2) activation of a gene or protein that causes growth even in the presence of one normal allele, that is, a dominant oncogene (as was described in the previous section). In many cancers, tumor-suppressor genes have deletions or point mutations that prevent production of any protein or lead to production of a nonfunctional protein. Another mechanism for inactivating tumor-suppressor genes is methylation of cytosine residues in the promoter or other control elements, which inhibits their transcription. Such methylation is commonly found in nontranscribed regions of DNA (see Chapter 7).

## Inherited Mutations in Tumor-Suppressor Genes Increase Cancer Risk

Individuals with inherited mutations in tumor-suppressor genes have a hereditary predisposition for certain cancers. Such individuals generally inherit a germ-line mutation in one allele of the gene; somatic mutation of the second allele facilitates tumor progression. A classic case is retinoblastoma, which is caused by loss of function of *RB*, the first tumor-suppressor gene to be identified. As we discuss later, the protein encoded by *RB* regulates cell cycle entry.

**Hereditary versus Sporadic Retinoblastoma** Children with hereditary retinoblastoma inherit a single defective copy of the *RB* gene, sometimes seen as a small deletion on one of the two copies of chromosome 13. The children develop several retinal tumors early in life and generally in both eyes. The deletion or mutation of the normal *RB* gene on the other chromosome is an essential step in tumor formation, giving rise to a cell that produces no functional Rb protein (Figure 24-13a). Individuals with sporadic retinoblastoma, in contrast, inherit two normal *RB* alleles, each of which has undergone a loss-of-function somatic mutation in a single retinal cell (Figure 24-13b). Because losing two copies of the *RB* gene is far less likely than losing one, sporadic retinoblastoma is rare and usually affects only one eye.

If retinal tumors are removed before they become malignant, children with hereditary retinoblastoma often survive until adulthood and produce children but are at an increased risk of developing other types of tumors later in life. Because their germ cells contain one normal and one mutant *RB* allele, these individuals will, on average, pass on the mutant allele to half their children and the normal allele to the other half. Children who inherit the normal allele are normal if their other parent has two normal *RB* alleles. However, those who inherit the mutant allele have the same enhanced predisposition to develop retinal tumors as their affected parent, even though they inherit a normal *RB* allele from their other, normal parent. Thus the *tendency* to develop retinoblastoma is inherited as a dominant trait: one mutant copy is sufficient to predispose a person to develop the cancer.

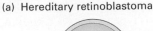
(a) Hereditary retinoblastoma

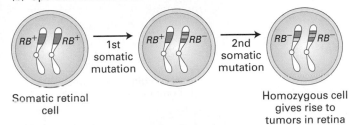

Somatic retinal cell → Somatic mutation → Homozygous cell gives rise to tumors in retina

(b) Sporadic retinoblastoma

Somatic retinal cell → 1st somatic mutation → → 2nd somatic mutation → Homozygous cell gives rise to tumors in retina

**FIGURE 24-13 Role of spontaneous somatic mutation in retinoblastoma.** This disease is marked by retinal tumors that arise from cells carrying two mutant *RB*⁻ alleles. (a) In hereditary (familial) retinoblastoma, a child inherits a normal *RB*⁺ allele from one parent and a mutant *RB*⁻ allele from the other parent. A single mutation in a heterozygous somatic retinal cell that inactivates the normal allele will produce a cell lacking any *Rb* gene function due to two mutations. (b) In sporadic retinoblastoma, a child inherits two normal *RB*⁺ alleles. Two separate somatic mutations in a particular retinal cell or its progeny are required to produce a homozygous *RB*⁻/*RB*⁻ cell.

As we will see shortly, many human tumors (not just retinal tumors) contain mutant *RB* alleles or mutations affecting other components of the same pathway; most of these arise as the result of somatic mutations. Although the hereditary retinoblastoma cases number about 100 per year in the United States, about 100,000 other cancer cases each year involve *RB* mutations acquired postconception.

**Inherited Forms of Colon and Breast Cancer** Similar hereditary predisposition for other cancers has been associated with inherited mutations in other tumor-suppressor genes. For example, individuals who inherit a germ-line mutation in one *APC* allele develop thousands of precancerous intestinal polyps (see Figure 24-8). Since there is a high probability that one or more of these polyps will progress to malignancy, such individuals have a greatly increased risk for developing colon cancer before the age of 50. Screening for polyps by colonoscopy is a good idea for people 50 or older, even when no *APC* mutation is known to be present. Likewise, women who inherit one mutant allele of *BRCA-1*, another tumor-suppressor gene, have a 60 percent probability of developing breast cancer by age 50, whereas those who inherit two normal *BRCA-1* alleles have a 2 percent probability of doing so. Heterozygous *BRCA-1* mutations also increase the lifetime risk of ovarian cancer from 2 percent to 15–40 percent. BRCA-1 protein is involved

in repairing radiation-induced DNA damage. In women with hereditary breast cancer, loss of the second *BRCA-1* allele, together with other mutations, is required for a normal breast duct cell to become malignant. However, *BRCA-1* generally is not mutated in sporadic, noninherited breast cancer.

**Loss of Heterozygosity** Clearly, then, we can inherit a propensity for cancer by receiving a damaged allele of a tumor-suppressor gene from one of our parents; that is, we are heterozygous for the mutation. That in itself typically will not cause cancer; since the remaining normal allele prevents aberrant growth, the cancer is recessive. As we have seen, subsequent loss or inactivation of the normal allele in a somatic cell, referred to as *loss of heterozygosity* (*LOH*), is usually how a cancer develops. One mechanism for LOH involves mis-segregation of the chromosomes bearing the affected tumor-suppressor gene during mitosis (Figure 24-14a). This process, also referred to as *nondisjunction*, is caused by failure of the spindle assembly checkpoint, which normally prevents a metaphase cell with an abnormal mitotic spindle from completing mitosis (see Figure 19-35). Another common mechanism for LOH is mitotic recombination between a chromatid bearing the wild-type allele and a homologous chromatid bearing a mutant allele. As illustrated in Figure 24-14b, subsequent chromosome segregation can generate a daughter cell that is homozygous for the mutant tumor-suppressor allele. A third mechanism is the deletion or mutation of the normal copy of the tumor-suppressor gene; such a deletion can encompass a large chromosomal region and need not be a precise deletion of just the tumor-suppressor gene.

Estimates vary, but hereditary cancers, that is, cancers that arise due in part to an inherited version of a gene, are thought to constitute about 10 percent of human cancers. Further work tracing contributions of human genes seems likely to increase the percentage. It is important to remember, however, that the inherited germ-line mutation alone is not sufficient to cause tumor development. In all cases, not only must the inherited normal tumor-suppressor allele be lost or inactivated, but mutations affecting other genes also are necessary for cancer to develop. Thus a person with a recessive tumor-suppressor gene mutation can be exceptionally susceptible to environmental mutagens such as radiation.

## Epigenetic Changes Can Contribute to Tumorigenesis

We have seen how mutations can undermine growth control by inactivating tumor-suppressor genes. However, these types of genes can also be silenced by repressive chromatin structures. In recent years the importance of *chromatin-remodeling complexes*, such as the SWI/SNF complex, in transcriptional control has become increasingly clear. These large and diverse multiprotein complexes have at their core an ATP-dependent helicase and often control histone modification and chromatin remodeling (see Chapter 7). SWI/SNF complexes, for example, cause changes in the positions or structures of nucleosomes, making genes accessible or inaccessible

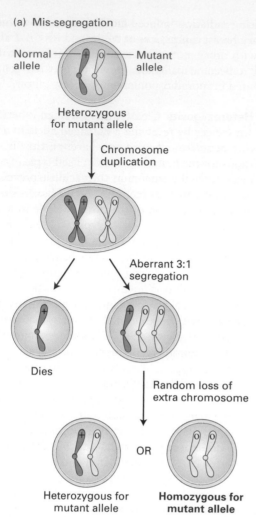

(a) Mis-segregation

Normal allele — Mutant allele

Heterozygous for mutant allele

Chromosome duplication

Aberrant 3:1 segregation

Dies

Random loss of extra chromosome

OR

Heterozygous for mutant allele

**Homozygous for mutant allele**

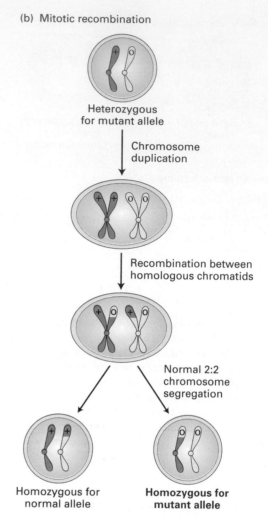

(b) Mitotic recombination

Heterozygous for mutant allele

Chromosome duplication

Recombination between homologous chromatids

Normal 2:2 chromosome segregation

Homozygous for normal allele

**Homozygous for mutant allele**

**FIGURE 24-14 Two mechanisms for loss of heterozygosity (LOH) of tumor-suppressor genes.** A cell containing one normal and one mutant allele of a tumor-suppressor gene is generally phenotypically normal. (a) If formation of the mitotic spindle is defective, then the duplicated chromosomes bearing the normal and mutant alleles may segregate in an aberrant 3:1 ratio. A daughter cell that receives three chromosomes of a type can lose one, restoring the normal 2n chromosome number. Sometimes the resultant cell will contain one normal and one mutant allele, but sometimes it will be homozygous for the mutant allele. Note that such aneuploidy (abnormal chromosome constitution) is generally damaging or lethal to relatively undifferentiated cells that have to develop into the many complex structures of an organism but can often be tolerated in clones of cells that have limited fates and duties. (b) Mitotic recombination between a chromosome with a wild-type and a mutant allele, followed by chromosome segregation, can produce a cell that contains two copies of the mutant allele.

to DNA-binding proteins that control transcription. If a gene is normally activated or repressed by SWI/SNF-mediated chromatin changes, mutations in the genes encoding the SWI or SNF proteins will cause changes in expression of the target gene. Studies with transgenic mice suggest that SWI/SNF plays a role in repressing the *E2F* genes, thereby inhibiting progression through the cell cycle. Thus loss of SWI/SNF function, just like loss of Rb function, can lead to overgrowth and perhaps cancer. Indeed, in mice, Rb protein recruits SWI/ SNF proteins to repress transcription of the *E2F* gene. Rb represses genes via this effect on E2F and by recruitment of histone deacetylases and histone methyltransferases.

Recent evidence from humans and mice has strongly implicated the *SNF5* gene in cancer. Snf5 is a core member of the SWI/SNF complex. In humans, inactivating somatic *SNF5* mu- tations causes rhabdoid tumors, which most commonly form in the kidney, and an inherited (familial) disposition to form brain and other tumors. In mice, 15–30 percent of *snf5⁻/snf5⁺* het- erozygotes develop rhabdoid tumors, and in all cases the tumor cells have lost the remaining functional allele. Since nothing was known of the mechanism involved, microarray studies were used to discover regulatory changes in the tumors. These stud- ies showed the gene-expression similarity of mouse and human tumors and that the loss of SNF5 leads to heightened expres- sion of cell cycle genes, including many regulated by E2F. With chromatin-remodeling complexes involved in so many aspects of transcriptional control, it is expected that SWI/SNF and sim- ilar complexes will be linked to many cancers. In humans, mu- tations in *Brg1*, which encodes the SWI/SNF catalytic subunit, have been found in prostate, lung, and breast tumors.

## KEY CONCEPTS of Section 24.2

### The Genetic Basis of Cancer

- Dominant gain-of-function mutations in proto-oncogenes and recessive loss-of-function mutations in tumor-suppressor genes are oncogenic.
- Among the proteins encoded by proto-oncogenes are growth-promoting signaling proteins and their receptors, signal transduction proteins, transcription factors, and apoptotic proteins (see Figure 24-11).
- An activating mutation of one of the two alleles of a proto-oncogene converts it to an oncogene. This can occur by point mutation, gene amplification, gene translocation, or overexpression.
- The first human oncogene to be identified encodes a constitutively active form of Ras, a signal transduction protein. This oncogene was isolated from a human bladder carcinoma (see Figure 24-5).
- Tumor-suppressor genes encode proteins that directly or indirectly interfere with progression through the cell cycle, such as checkpoint-control proteins that arrest the cell cycle in response to cell cycle defects, components of growth-inhibiting signaling pathways, pro-apoptotic proteins, regulators of kinase activities, and DNA-repair enzymes.
- The first tumor-suppressor gene to be recognized, *RB*, is mutated in retinoblastoma and many other tumors; some component of the RB pathway is altered in most tumors.
- Inheritance of a single mutant allele of *RB* greatly increases the probability that a specific kind of cancer will develop, as is the case for many other tumor-suppressor genes (e.g., *APC* and *BRCA-1*).
- In individuals born heterozygous for a tumor-suppressor gene, a somatic cell can undergo loss of heterozygosity (LOH) by mitotic recombination, chromosome mis-segregation, mutation, gene silencing, or deletion (see Figure 24-14).
- Mutations in caretaker genes are genetically recessive since having even one functional copy is usually sufficient to prevent serious DNA damage. However, a single copy leaves the carrier subject to possible LOH.
- Mutations affecting chromatin-modifying factors that participate in transcriptional control, such as the SWI/SNF chromatin-remodeling complex, are associated with a variety of tumors.

## 24.3 Cancer and Misregulation of Growth Regulatory Pathways

In this section, we will examine in more detail how the deregulation of growth-promoting and growth-inhibitory signaling pathways contributes to tumorigenesis. We will first discuss how the use of mouse models of human cancers has helped shape our understanding of the process of carcinogenesis. We will then examine in more detail how mutations that result in the unregulated, constitutive activity of certain proteins or in their overproduction promote cell proliferation and transformation, followed by a discussion of how loss-of-function mutations in differentiation pathways contribute to tumorigenesis. We will end this section with a description of how the molecular analysis of tumors is changing the manner in which cancer is treated. Personalized medicine—the ability to diagnose individual tumors at the molecular level and to design treatments for a patient's specific cancer—is likely to become a reality in the twenty-first century.

### Mouse Models of Human Cancer Teach Us About Disease Initiation and Progression

Genetically engineered mice have provided tremendous insights into the steps of tumor initiation and progression. For example, as shown previously in Figure 24-7, murine experiments revealed that mice overexpressing Myc develop tumors in the tissues where Myc is present at high levels, and tumor development is accelerated when Myc overproduction is combined with expression of an additional oncogene, lending support to the multi-hit model of cancer induction. The analysis of mice lacking key tumor-suppressor genes such as *RB* and *p53* also helped shape our understanding of how these genes function. However, using mouse models to study cancer is not always straightforward. Many tumor-suppressor genes serve essential functions during normal mouse development, and mice lacking both copies of these genes are not viable. The essential function of these genes early during embryogenesis precludes the study of their role in tumor progression. To circumvent this problem, researchers have begun to employ conditional "knock-in" and "knock-out" strategies that allow for the targeted activation or inactivation of a gene in a certain tissue and/or at a certain stage of development.

In the conditional mouse model, an allele of a particular oncogene or tumor suppressor gene is wild-type until activated or inactivated with exogenous chemicals or viruses in a tissue or time-specific manner. At the heart of conditional systems are the Cre and FLP recombinases. These recombinases facilitate homologous recombination between loxP and FRT sites, respectively (Figure 24-15). When the recombinases are driven by a tissue-specific promoter, recombination occurs only in the tissue that produces the recombinase. The recombinase system can be used in two ways. The recombinase target sites could flank an exon. On induction of the recombinase, the exon is lost and the gene is inactivated (Figure 24-15a). This system is especially useful to inactivate tumor suppressor genes in a tissue-specific manner. Expression of oncogenes can be controlled by introducing into the oncogene an additional exon that contains a stop codon. The oncogene with the stop-codon-containing exon is therefore not functional. However, if the additional exon is flanked by recombinase target sites, the oncogene will be produced on

**FIGURE 24-15 Conditional mouse models of cancer.** In the inactivating system (a), an exon is flanked by two loxP or FRT sites as shown. Expression of the Cre or FLP recombinase leads to homologous recombination between the two loxP and FRT sites, respectively. This leads to excision of the exon, rendering the gene nonfunctional. In the activating system (b), an additional exon with a STOP codon is introduced into the gene of interest, making the gene nonfunctional. This exon is flanked by loxP or FRT sites. When Cre or FLP recombinase is induced, the stop-codon-containing exon is recombined out and the gene of interest, such as an oncogenic form of *ras* with valine substituted for the glycine at position 12, is expressed.

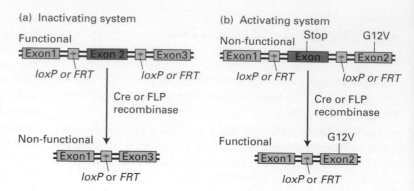

induction of the recombinase (Figure 24-15b). Using this system, researchers have examined the role of oncogenic forms of Ras in the mouse and have, using this conditional oncogenic Ras allele, created a mouse model of human lung cancer.

The development of promoters that can be regulated by exogenously added chemicals has provided an additional powerful method to control gene expression within animals. Most widely used are the Tet-On and Tet-Off systems. Each system is composed of two parts: the Tet operon promoter that regulates the expression of the gene of interest and either the transactivator tTA (in the case of Tet-Off) or the reverse transactivator rtTA (in the case of Tet-On). Both transcription factors bind to the Tet operator to induce gene expression, and both are regulatable by tetracycline (or the tetracycline analog doxycycline, more commonly used by scientists in their experiments). The difference between the two systems lies in the responses of tTA and rtTA to doxycycline binding. Doxycycline inhibits tTA from binding the promoter. Thus in the Tet-Off system, addition of doxycycline turns off transcription. In the Tet-On system, rtTA cannot bind the promoter in the absence of drug and addition of doxycycline induces transcription. Doxycycline can be administered by simply adding it to the animals' water supply. Placing the Tet transcriptional regulators under the control of tissue-specific promoters therefore allows for temporal as well as spatial control of gene expression.

Using the Tet-Off system to control Myc expression, researchers found that survival of a tumor depends on the continuous production of Myc protein. When expression of Myc was even briefly interrupted, osteogenic sarcoma cells ceased dividing and developed into mature osteocytes (Figure 24-16). It is now clear that the continuous activity of oncogenes is required for the survival of many different tumors. This dependence of a tumor on the continuous production of an oncogene has been termed **oncogene addiction** and may provide new opportunities for treatment. Specific inhibitors of oncogenes—even when applied only transiently—could lead to disease regression.

## Oncogenic Receptors Can Promote Proliferation in the Absence of External Growth Factors

Hyperactivation of a growth-inducing signaling protein due to an alteration of the protein might seem a likely mechanism of cancer, but in fact this rarely occurs. Only one such

naturally occurring oncogene, *sis*, has been discovered. The *sis* oncogene, which encodes an altered form of platelet-derived growth factor (PDGF), can aberrantly stimulate proliferation of cells that normally express the PDGF receptor. A more common event is that cancer cells begin to produce an unaltered growth factor that acts on the cell that produces it. This is called **autocrine** stimulation.

In contrast, oncogenes encoding cell-surface receptors that transduce growth-promoting signals have been associated with several types of cancer. Many of these receptors have intrinsic protein-tyrosine kinase activity in their cytosolic

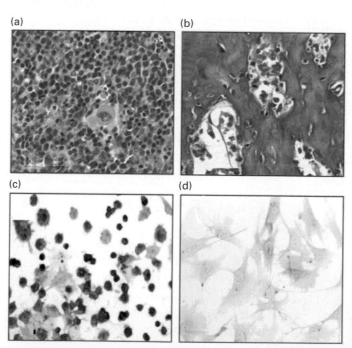

**EXPERIMENTAL FIGURE 24-16 Myc is continuously needed for tumor growth.** Transgenic mice were developed in which *myc* expression was driven by the Tet-Off system. One percent of such mice develop osteogenic sarcomas. Wild-type mice were transplanted with osteogenic sarcomas and the *myc* gene was then inactivated by treating mice with doxycyclin. This caused cessation of proliferation of osteogenic sarcomas (a) and differentiation into mature osteocytes (b). After *myc* withdrawal, cells also lost alkaline phosphatase activity, a marker for ostogenic sarcomas (c, d). Surprisingly, re-expression of Myc protein did not trigger a return to the sarcoma state. [From Meenakshi et al., 2002, *Science* **297**:102.]

domains, an activity that is quiescent until activated. Ligand binding to the external domains of these **receptor tyrosine kinases (RTKs)** leads to their dimerization and activation of their kinase activity, initiating an intracellular signaling pathway that ultimately promotes proliferation.

In some cases, a point mutation changes a normal RTK into one that dimerizes and is constitutively active even in the absence of ligand. For instance, a single point mutation converts the normal human EGF receptor 2 (Her2) into the Neu oncoprotein ("neu" for its first known role, in neuroblastoma), which is an initiator of certain mouse cancers (Figure 24-17, *left*). Similarly, human tumors called *multiple endocrine neoplasia type 2* produce a constitutively active dimeric glia-derived neurotrophic factor (GDNF) receptor that results from a point mutation in the extracellular domain. GDNF receptor and Her2 receptor are protein-tyrosine kinases, so the constitutively active forms excessively phosphorylate

their downstream target proteins. In other cases, deletion of much of the extracellular ligand-binding domain produces a constitutively active oncogenic receptor. For example, deletion of the extracellular domain of the normal EGF receptor (Figure 24-17, *right*) converts it to the dimeric ErbB oncoprotein (from erythroblastosis virus, where a viral version of the altered gene was first identified).

Mutations leading to overproduction of a normal RTK also can be oncogenic. For instance, many human breast cancers overproduce a normal Her2 receptor. As a result, the cells are stimulated to proliferate in the presence of very low concentrations of EGF and related hormones, concentrations too low to stimulate proliferation of normal cells (see Chapter 16).

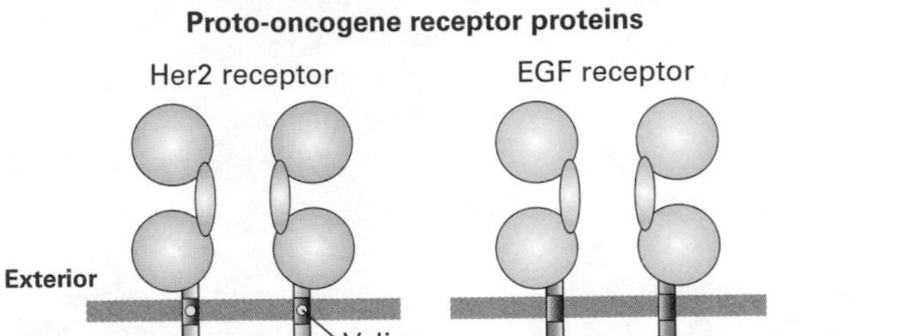

## Viral Activators of Growth-Factor Receptors Act as Oncoproteins

Viruses use their own tricks to cause cancer, presumably to increase the production of virus from infected cancer cells. For example, a retrovirus called *spleen focus-forming virus (SFFV)* induces erythroleukemia (a tumor of erythroid progenitors) in adult mice by manipulating a normal developmental signal. The proliferation, survival, and differentiation of erythroid progenitors into mature red cells absolutely require erythropoietin (Epo) and the corresponding Epo receptor (see Figure 16-8). A mutant SFFV envelope glycoprotein, termed *gp55*, is responsible for the oncogenic effect of the virus. Although gp55 cannot function as a normal retrovirus envelope protein in virus budding and infection, it has acquired the remarkable ability to bind to and activate Epo receptors in the infected cell (Figure 24-18). By inappropriately and continuously stimulating the proliferation of erythroid

**FIGURE 24-17 Effects of oncogenic mutations in proto-oncogenes that encode cell-surface receptors.** *Left:* A mutation that alters a single amino acid (valine to glutamine) in the transmembrane region of the Her2 receptor causes dimerization of the receptor, even in the absence of the normal EGF-related ligand, making the oncoprotein Neu a constitutively active kinase. *Right:* A deletion that causes loss of the extracellular ligand-binding domain in the EGF receptor leads, for unknown reasons, to constitutive activation of the kinase activity of the resulting oncoprotein ErbB.

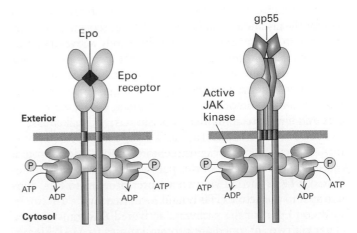

**FIGURE 24-18 Activation of the erythropoietin (Epo) receptor by the natural ligand, Epo, or a viral oncoprotein.** Binding of Epo dimerizes the receptor and induces formation of erythrocytes from erythroid progenitor cells. Normally cancers occur when progenitor cells infected by the spleen focus-forming virus produce the Epo receptor and viral gp55, both localized to the plasma membrane. The transmembrane domains of dimeric gp55 specifically bind the Epo receptor, dimerizing and activating the receptor in the absence of Epo. [See S. N. Constantinescu et al., 1999, *EMBO J.* **18**:3334.]

progenitors, gp55 induces formation of excessive numbers of erythrocytes. Malignant clones of erythroid progenitors emerge several weeks after SFFV infection as a result of further mutations in these aberrantly proliferating cells.

Another example of this phenomenon is provided by human papillomavirus (HPV), the sexually transmitted DNA virus that causes cervical cancer and genital warts. A papillomavirus protein designated *E5*, which contains only 44 amino acids, spans the plasma membrane and forms a dimer or trimer. Each E5 polypeptide can form a stable complex with one endogenous receptor for PDGF, thereby aggregating two or more PDGF receptors within the plane of the plasma membrane. This mimics hormone-mediated receptor dimerization, causing sustained receptor activation and eventually cell transformation. As we will see later, the HPV genome also codes for several other proteins that act to inhibit tumor-suppressor genes, further contributing to cell transformation. Recently a vaccine against an HPV capsid protein (L1) has been found to protect against cervical cancer caused by at least some subtypes of HPV.

## Many Oncogenes Encode Constitutively Active Signal Transduction Proteins

A large number of oncogenes are derived from proto-oncogenes whose encoded proteins aid in transducing signals from an activated receptor to a cellular target. We describe several examples of such oncogenes; each is expressed in many types of tumor cells.

**Ras Pathway Components** Among the best-studied oncogenes in this category are the $ras^D$ genes, which were the first nonviral oncogenes to be recognized. Any one of a number of changes in Ras protein can lead to its uncontrolled and therefore dominant activity. In particular, if a point mutation substitutes any amino acid for the glycine at position 12 in the Ras sequence, the normal protein is converted into a constitutively active oncoprotein. This simple mutation reduces the protein's GTPase activity, thus maintaining Ras in the active GTP-bound state. Constitutively active Ras oncoproteins are produced by many types of human tumors, including bladder, colon, mammary, skin, and lung carcinomas, neuroblastomas, and leukemias.

As we saw in Chapter 16, Ras is a key component in transducing signals from activated receptors to a cascade of protein kinases. In the first part of this pathway, a signal from an activated RTK is carried via two adapter proteins to Ras, converting it to the active GTP-bound form (see Figure 16-20). In the second part of the pathway, activated Ras transmits the signal via two intermediate protein kinases to MAP kinase. The activated MAP kinase then phosphorylates a number of transcription factors that induce synthesis of important cell-cycle and differentiation-specific proteins (see Figure 16-24). Activating Ras mutations short-circuit the first part of this pathway, making upstream activation triggered by ligand binding to the receptor unnecessary. Oncogenes encoding other altered components of the RTK/Ras/MAP kinase pathway also have been identified.

Constitutive Ras activation can also arise from a recessive loss-of-function mutation in a GTPase-activating protein (GAP). The normal GAP function is to accelerate hydrolysis of GTP and the conversion of active GTP-bound Ras to inactive GDP-bound Ras (see Figure 3-32). The loss of GAP leads to sustained Ras activation of downstream signal transduction proteins. For example, neurofibromatosis, a benign tumor of the sheath cells that surround nerves, is caused by loss of both alleles of *NF1*, which encodes a Ras GAP-type protein. Individuals with neurofibromatosis have inherited a single mutant *NF1* allele; subsequent somatic mutation in the other allele leads to formation of neurofibromas. Thus *NF1*, like *RB*, is a tumor-suppressor gene, and the tendency

(a)

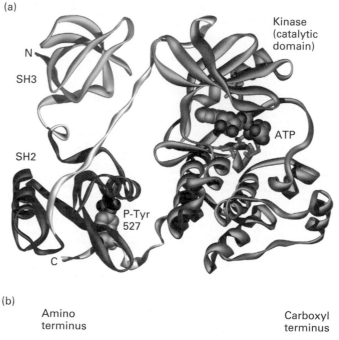

(b)

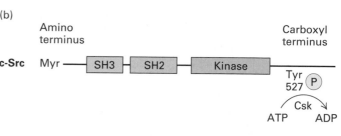

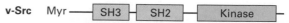

**FIGURE 24-19 Structure of Src tyrosine kinases and activation by an oncogenic mutation.** (a) Three-dimensional structure of Hck, one of several Src kinases in mammals. Phosphorylation of tyrosine 527 in the SH2 domain induces conformational strains in the SH3 and kinase domains, distorting the kinase active site so it is catalytically inactive. The kinase activity of cellular Src proteins is normally activated by removing the phosphate on tyrosine 527. (b) Domain structure of c-Src and v-Src. Phosphorylation of tyrosine 527 by Csk, another cellular tyrosine kinase, inactivates the Src kinase activity. The transforming v-Src oncoprotein encoded by Rous sarcoma virus is missing the C-terminal 18 amino acids, including tyrosine 527, and thus is constitutively active. [Part (a) from F. Sicheri et al., 1997, *Nature* **385:**602; see also T. Pawson, 1997, *Nature* **385:**582, and W. Xu et al., 1997, *Nature* **385:**595.]

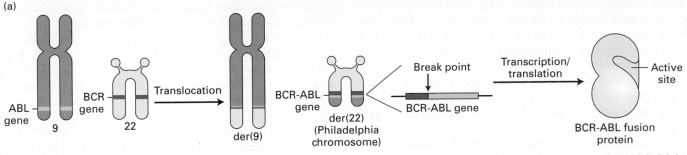

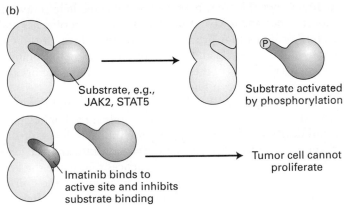

(a)

ABL gene 9    BCR gene 22 → Translocation → der(9)    BCR-ABL gene der(22) (Philadelphia chromosome)    Break point    BCR-ABL gene → Transcription/ translation → BCR-ABL fusion protein    Active site

(b)

Substrate, e.g., JAK2, STAT5 → Substrate activated by phosphorylation

Imatinib binds to active site and inhibits substrate binding → Tumor cell cannot proliferate

**FIGURE 24-20 Bcr-Abl protein kinase.** (a) Origin of the Philadelphia chromosome from a translocation of the tips of chromosomes 9 and 22 and the oncogenic fusion protein formed by the Philadelphia chromosome translocation. (b) The Bcr-Abl fusion protein is a constitutively active kinase that phosphorylates multiple signal transduction proteins. Imatinib binds to the active site of Bcr-Abl and inhibits its kinase activity. (c) Imatinib bound to the Bcr-Abl active site. [Part (c) from Nagar et al., 2002, *Cancer Research* **62**:4236.]

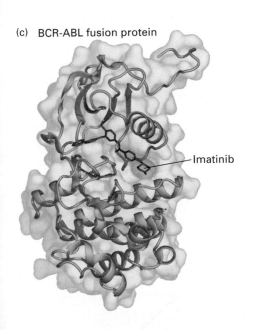

(c) BCR-ABL fusion protein

— Imatinib

to develop neurofibromatosis, like hereditary retinoblastoma, is inherited as an autosomal dominant trait.

**Src Protein Kinase** Several oncogenes encode cytosolic protein kinases that normally transduce signals in a variety of intracellular signaling pathways. Indeed, the first oncogene to be discovered, v-*src* from Rous sarcoma retrovirus, encodes a constitutively active protein-tyrosine kinase. At least eight mammalian proto-oncogenes encode a family of nonreceptor tyrosine kinases related to the v-Src protein. In addition to a catalytic domain, these kinases contain SH2 and

SH3 protein-protein interaction domains. The kinase activity of cellular Src and related proteins normally is inactivated by phosphorylation of the tyrosine residue at position 527, which is six residues from the C-terminus (Figure 24-19a). Hydrolysis of phosphotyrosine 527 by a specific phosphatase enzyme normally activates c-Src. Tyrosine 527 is often missing or altered in Src oncoproteins that have constitutive kinase activity; that is, they do not require activation by a phosphatase (Figure 24-19b).

**Abl Protein Kinase** One oncogene encoding a cytosolic nonreceptor protein kinase is generated by a chromosome translocation that fuses a part of the c-*abl* gene, which encodes a tyrosine kinase, with part of the *bcr* gene, whose function is unknown (Figure 24-20a). One consequence of the fusion is that a hybrid protein is produced, and it has odd and dangerous properties. The normal c-Abl protein promotes branching of filamentous actin and extension of cell processes, so it may function primarily to control the cytoskeleton and cell shape. The chimeric oncoproteins encoded by the *bcr-abl* oncogene form a tetramer that exhibits unregulated and continuous Abl kinase activity (Figure 24-20b). Bcr-Abl can phosphorylate and thereby activate many intracellular signal transduction proteins, and at least some of these proteins are not normal substrates of Abl. For instance, Bcr-Abl can activate JAK2 kinase and STAT5 transcription factor, which normally are activated by binding of growth factors (e.g., erythropoietin) to cell-surface receptors (see Figure 16-10). Bcr-Abl can also provide a docking site for signal transduction proteins through the Bcr part, thus potentially stimulating signal transduction.

The chromosomal translocation that forms *bcr-abl* generates the diagnostic *Philadelphia chromosome*, discovered in 1960 (see Figure 24-20a). If this translocation occurs in a hematopoietic cell in the bone marrow, the activity of the chimeric *bcr-abl* oncogene results in the initial phase of human chronic myelogenous leukemia (CML), characterized by an expansion in the number of white blood cells. A second loss-of-function mutation in a cell carrying *bcr-abl* (e.g., in *p53* or *RB*) leads to acute leukemia, which is often fatal. The CML chromosome translocation was only the first of a long series of distinctive, or "signature," chromosome translocations linked to particular forms of leukemia. Many of the gene fusions involve genes encoding transcriptional regulators, particularly transcriptional regulators of Hox genes, a group of transcription factors required for cell proliferation and differentiation during embryonic development. Each case presents an opportunity for greater understanding of the disease, earlier diagnosis, and new therapies. In the case of CML, that second step to successful therapy has already been taken.

After a painstaking search, an inhibitor of Abl kinase named imatinib (Gleevec) was identified as a possible treatment for CML in the early 1990s. Imatinib, which binds directly to the Abl kinase active site and inhibits kinase activity, is highly lethal to CML cells while sparing normal cells (Figure 24-20c). After clinical trials showing imatinib is remarkably effective in treating CML despite some side effects, it was approved by the FDA in 2001, the first cancer drug targeted to a signal transduction protein unique to tumor cells. Imatinib inhibits several other tyrosine kinases that are implicated in different cancers and has been successful in trials for treating these diseases, including forms of gastrointestinal tumors, as well. There are 90 functional tyrosine kinases encoded in the human genome, so drugs related to imatinib may be useful in controlling the activities of all these proteins. One ongoing challenge is that tumor cells can evolve to be resistant to imatinib and other such drugs, necessitating the invention of alternative drugs. ■

tion of key genes that encode growth-promoting proteins or by inhibiting transcription of growth-repressing genes.

Many nuclear proto-oncogene proteins are produced when normal cells are stimulated to grow, indicating their direct role in growth control. For example, PDGF treatment of quiescent 3T3 cells induces an approximately 50-fold increase in the production of c-Fos and c-Myc, the normal products of the *fos* and *myc* proto-oncogenes. Initially there is a transient rise of c-Fos and later a more prolonged rise of c-Myc (Figure 24-21). The levels of both proteins decline within a few hours, a regulatory effect that may, in normal cells, help to avoid cancer. c-Fos and c-Myc stimulate transcription of genes encoding proteins that promote progression through the $G_1$ phase of the cell cycle and the $G_1$-to-S transition. In tumors, the oncogenic forms of these or other transcription factors are frequently expressed at high and unregulated levels.

In normal cells, c-Fos and c-Myc mRNAs and the proteins they encode are intrinsically unstable, leading to their rapid degradation after the genes are induced. Some of the changes that turn c-*fos* from a normal gene into an oncogene involve genetic deletions of the sequences that make the Fos mRNA and protein short lived. Conversion of the c-*myc* proto-oncogene into an oncogene can occur by different mechanisms. In cells of the human tumor known as *Burkitt's lymphoma*, the c-*myc* gene is translocated to a site near the heavy-chain antibody genes, which are normally active in antibody-producing white blood cells (Figure 24-22). The c-*myc* translocation is a rare aberration of the normal DNA rearrangements that occur during maturation of antibody-producing cells. The translocated *myc* gene, now regulated by the antibody gene enhancer, is continually expressed, causing the cell to become cancerous. Localized amplification of a segment of DNA containing the *myc* gene, which

## Inappropriate Production of Nuclear Transcription Factors Can Induce Transformation

Oncogenic mutations that create oncogenes or damage tumor-suppressor genes eventually cause changes in gene expression. Experimentally this can be measured by comparing the amounts of different mRNAs produced in normal versus tumor cells. As discussed earlier, we can now measure such differences in the expression of thousands of genes with DNA microarrays (see Figure 24-10).

Since the most direct effect on gene expression is exerted by transcription factors, it is not surprising that many oncogenes encode transcription factors. Two examples are *jun* and *fos*, which initially were identified in transforming retroviruses and later found to be overexpressed in some human tumors. The c-*jun* and c-*fos* proto-oncogenes encode proteins that sometimes associate to form a heterodimeric transcription factor, called *AP1*, that binds to a sequence found in promoters and enhancers of many genes (see Figure 7-32a and Chapter 16). Both Fos and Jun also can act independently as transcription factors. They function as oncoproteins by activating transcrip-

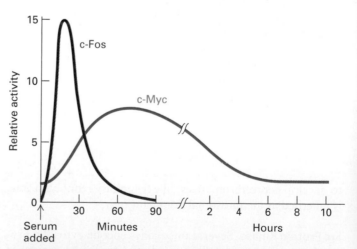

**EXPERIMENTAL FIGURE 24-21 Addition of serum to quiescent 3T3 cells yields a marked increase in the activity of two proto-oncogene products, c-Fos and c-Myc.** Serum contains factors such as platelet-derived growth factor (PDGF) that stimulate the growth of quiescent cells. One of the earliest effects of growth factors is to induce expression of c-*fos* and c-*myc*, whose encoded proteins are transcription factors. [See M. E. Greenberg and E. B. Ziff, 1984, *Nature* **311**:433.]

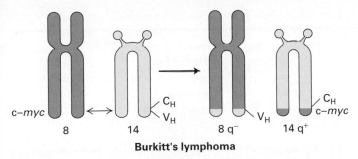

**FIGURE 24-22 Chromosomal translocation in Burkitt's lymphoma.** As a result of a translocation between chromosomes 8 and 14, the c-myc gene is placed adjacent to the gene for part of the antibody heavy chain ($C_H$), leading to overproduction of the Myc transcription factor in lymphocytes and hence their growth into a lymphoma.

occurs in several human tumors, also causes inappropriately high production of the otherwise normal Myc protein.

The *c-myc* gene encodes a basic helix-loop-helix zipper protein that acts as part of a set of interacting proteins that can dimerize in various combinations, bind to DNA, and coordinately regulate the transcription of target genes. Other members of the protein set include Mad, Max, and Mnt. Max can heterodimerize with Myc, Mad, and Mnt. Myc-Max dimers regulate genes that control proliferation, such as cyclins. Mad proteins inhibit Myc proteins, which has led to an interest in using Mad proteins, or drugs that stimulate Mad proteins, to rein in excessive Myc activity that contributes to tumor formation. Myc protein complexes affect transcription by recruiting chromatin-modifying complexes containing histone acetyltransferases (which usually stimulate transcription; see Chapter 7) to Myc target genes. Mad and Mnt work with the Sin3 co-repressor protein to bring in histone deacetylases that help to block transcription. Together, all these proteins form a regulatory network that employs protein-protein association, variations in DNA binding, and transcriptional regulation to control cell proliferation. Overproduction of Myc protein tips the scales in favor of cell division and growth.

## Aberrations in Signaling Pathways That Control Development Are Associated with Many Cancers

During normal development, secreted signals such as Hedgehog (Hh), Wnt, and TGF-β are frequently used to direct cells to particular developmental fates, which may include the property of rapid mitosis. The effects of such signals must be regulated so that growth is limited to the right time and place. Among the mechanisms available for reining in the effects of powerful developmental signals are inducible intracellular antagonists, receptor blockers, and competing signals. Mutations that prevent such restraining mechanisms from operating are likely to be oncogenic, causing inappropriate or cancerous growth.

Hh signaling, which is used repeatedly during development to control cell fates, is a good example of a signaling pathway

implicated in cancer induction. In the skin and cerebellum, one of the human Hh proteins, Sonic hedgehog, stimulates cell division by binding to and inactivating a membrane protein called Patched1 (Ptc1) (see Figure 16-33). Loss-of-function mutations in *ptc1* permit cell proliferation in the absence of an Hh signal; thus *ptc1* is a tumor-suppressor gene. People who inherit a single working copy of *ptc1* have a propensity to develop skin and brain cancer; either can occur when the remaining allele is damaged. Other people can get these diseases too if they suffer loss of both copies of the gene. Thus there are both familial (inherited) and sporadic (noninherited) cases of these diseases, just as in retinoblastoma. Mutations in other genes in the Hh signaling pathway are also associated with cancer. Some such mutations create oncogenes that turn on Hh target genes inappropriately; others are recessive mutations that affect negative regulators such as Ptc1. As is the case for a number of other tumor-suppressor genes, complete loss of Ptc1 function would lead to early fetal death since it is needed for development, so it is only the tumor cells that are homozygous *ptc1/ptc1*.

Many of the signaling pathways described in other chapters play roles in controlling embryonic development and cell proliferation in adult tissues. In recent years mutations affecting components of most of these signaling pathways have been linked to cancer. Indeed, once one gene in a developmental pathway has been linked to a type of human cancer, knowledge of the pathway gleaned from model organisms such as worms, flies, or mice allows focused investigations of the possible involvement of additional pathway genes in other cases of the cancer. For example, *APC*, a critical gene mutated on the path to colon carcinoma, is now known to be part of the Wnt signaling pathway (see Chapter 16), which led to the discovery of the involvement of β-catenin mutations in colon cancer. Mutations in tumor-suppressor developmental genes promote tumor formation in tissues where the affected gene normally helps restrain growth. Thus the tumor-suppressor gene mutations will not cause cancers in tissues where the primary role of the developmental regulator is to control cell fate—what type of cell develops—but not cell division. Mutations in developmental proto-oncogenes may induce tumor formation in tissues where an affected gene normally promotes cell proliferation or in another tissue where the gene has become aberrantly active.

Transforming growth factor β (TGF-β), despite its name, inhibits proliferation of many cell types, including most epithelial and immune-system cells. Binding of TGF-β to its receptor induces activation of cytosolic Smad transcription factors (see Figure 16-28). After translocating to the nucleus, Smads can promote expression of the gene encoding p15, an inhibitor of cyclin-dependent kinase 4 (CDK4), which causes cells to arrest in $G_1$. TGF-β signaling also promotes expression of genes encoding extracellular matrix proteins and plasminogen activator inhibitor 1 (PAI-1), which reduces the plasmin-catalyzed degradation of the matrix. Loss-of-function mutations in either TGF-β receptors or in Smads thus promote cell proliferation and probably contribute to the invasiveness and metastasis of tumor cells (Figure 24-23). Such

**FIGURE 24-23 Effect of loss of TGF-β signaling.** Binding of TGF-β, an anti-growth factor, causes activation of Smad transcription factors. In the absence of effective TGF-β signaling owing to either a receptor mutation or a SMAD mutation, cell proliferation and invasion of the surrounding extracellular matrix (ECM) increase. [See X. Hua et al., 1998, *Genes & Dev.* **12**:3084.]

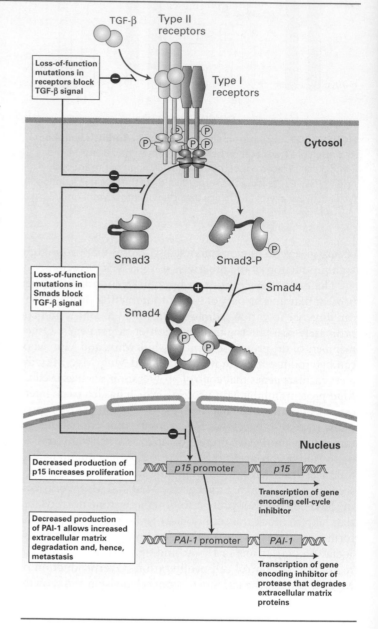

mutations have in fact been found in a variety of human cancers. For example, deletion of the *Smad4* gene occurs in many human pancreatic cancers; retinoblastoma and colon cancer cells lack functional TGF-β receptors and therefore are unresponsive to TGF-β growth inhibition.

## Molecular Cell Biology Is Changing How Cancer Is Treated

The imatinib story described earlier illustrates how genetics—the discovery of the Philadelphia chromosome and the critical oncogene it creates—together with biochemistry—the discovery of the molecular action of the Abl protein—can lead to a powerful new therapy. In general, each difference between cancer cells and normal cells provides a new opportunity to identify a specific drug or treatment that kills just the cancer cells or at least stops their uncontrolled growth. Thus knowledge of the molecular cell biology of the tumors is critical information that can be exploited by researchers to develop anticancer treatments that more precisely target just cancer cells.

⚕ Breast cancer provides a good example of how molecular cell biology techniques have affected treatments, both curative and palliative. Until the rise in the incidence of lung cancer, resulting from an increase in women smokers, breast cancer was the most deadly cancer for women, and it remains the second most frequent cause of women's cancer deaths. The cause of breast cancer is unknown, but the frequency is increased if certain mutations are carried. Breast cancers are often diagnosed during routine mammogram (x-ray) examinations. Typically, a biopsy of a 1–2-cm tissue mass is taken to check the diagnosis and is tested with antibodies to determine whether a high level of estrogen or progesterone receptor is present. These steroid receptors are capable of stimulating tumor growth and are sometimes expressed at high levels in breast cancer cells. If either receptor is present, it is exploited in the treatment. A drug called tamoxifen, which inhibits the estrogen receptor, can be used to deprive the tumor cells of a growth-stimulating hormone. The biopsy is also tested for amplification of the proto-oncogene *HER2/NEU*, which, as we have seen, encodes human EGF receptor 2. A monoclonal antibody specific for Her2 has been a strikingly successful new treatment for the subset of breast cancers that overproduce Her2. Her2 antibody injected

into the blood recognizes Her2 and causes it to be internalized, selectively killing the cancer cells without any apparent effect on normal breast (and other) cells that produce moderate amounts of Her2. Similarly, many lung cancers harbor an amplification of the EGF receptor locus. Treatment with the EGFR inhibitor erlotinib has dramatically increased life expectancy of patients with this type of lung cancer.

Breast cancer is treated with a combination of surgery, radiation therapy, and chemotherapy. The first step is surgical resection (removal) of the tumor and examination of lymph nodes for evidence of metastatic tumor, which is the major adverse prognostic factor. The subsequent treatment involves 6 weeks of radiation and 8 weeks of chemotherapy with three different types of agents. These harsh treatments are designed to kill the dividing cancer cells; however, they also cause a variety of side effects, including suppression of blood cell production, hair loss, nausea, and neuropathy. This can reduce the strength of the immune system, risking infection, and cause weakness due to poor oxygen supply. To help, patients are given the growth factor G-CSF to promote neutrophil formation (a type of white blood cell that fights bacterial and fungal infections) and erythropoietin (Epo) to stimulate red blood cell formation. Despite all this treatment, an average-risk woman (60 years old, 2-cm tumor, 1 positive lymph node) has a 30–40 percent 10-year risk of succumbing to her cancer. This risk can be reduced by 10–15 percent by hormone-blocking treatment such as tamoxifen, exploiting the molecular data that show a hormone receptor present on the cancer cells. Mortality is reduced another 5–10 percent by treatment with antibodies against the Her2/Neu oncoprotein. Thus molecular biology is having a huge impact on breast cancer victim survival rates, though still far less than one would like. ∎

To identify more genetic alterations unique to a tumor that could be exploited for new therapies, researchers now utilize RNAi technologies to identify genes that when inactivated cause tumor cells but not normal cells to die. This approach of identifying *synthetic lethal* interactions between genetic alterations, which on their own are not lethal, was first pioneered in budding yeast. With the development of genome-wide short hairpin (sh)RNAi libraries (a collection of RNAi constructs that target every gene in the human genome; see Chapter 5), this approach is now also feasible in human cells. Tumor cells and normal cells are infected with pools of shRNAi constructs, each of which harbors a unique sequence tag known as a bar code. After a period of growth the RNAi constructs can be isolated and shRNAs that were lost from the pool can be identified by deep sequencing of the remaining constructs. The shRNAs that were lost point to the target gene being essential for viability in this cell type. shRNA constructs that cause lethality in tumor cells but not normal cells suggests that a gene is essential for the survival of a tumor cell but not a normal cell. For example, this approach has been used to identify genes that when inactivated cause selective lethality in cancer cells harboring an oncogenic form of *ras*, the *K-ras* oncogene. Genes encoding proteins essential for mitotic progression such as the ubiquitin ligase APC/C (see Chapter 19) were found to be synthetic lethal with the *K-ras* allele. These proteins could provide novel targets for the development of new therapeutics for tumors that harbor oncogenic forms of *ras*.

The vision for the future of medicine is that the need for radiation, chemotherapy, and perhaps even surgery will be substantially reduced, thus reducing toxicity and collateral damage. Increased knowledge of the molecular cell biology of cancer will allow drugs to be administered that are, first, more effective and harmless and, second, tuned to the particular properties of an individual's tumor cells. In breast and lung cancer tumor treatment we can already see progress in this direction.

## KEY CONCEPTS of Section 24.3

### Cancer and the Misregulation of Growth Regulatory Pathways

- Mice in which oncogenes and tumor-suppressor genes can be expressed in a tissue-specific or temporal way teach us about how cancers arise and how they contribute to disease progression.

- Mutations or chromosomal translocations that permit RTKs for growth factors to dimerize in the absence of their normal ligands lead to constitutive receptor activity (see Figure 24-17). Such activation ultimately induces changes in gene expression that can transform cells. Overproduction of growth-factor receptors can have the same effect and lead to abnormal cell proliferation.

- Certain virus-encoded proteins can bind to and activate host-cell receptors for growth factors, thereby stimulating cell proliferation in the absence of normal signals.

- Most tumor cells produce constitutively active forms of one or more intracellular signal transduction proteins, causing growth-promoting signaling in the absence of normal growth factors.

- A single point mutation in Ras, a key transducing protein in many signaling pathways that promote cell proliferation and differentiation, reduces its GTPase activity, thereby maintaining it in an activated state.

- The activity of Src, a cytosolic signal-transducing protein-tyrosine kinase, normally is regulated by reversible phosphorylation and dephosphorylation of a tyrosine residue near the C-terminus (see Figure 24-19). The unregulated activity of Src oncoproteins that lack this tyrosine promotes abnormal proliferation of many cells.

- The Philadelphia chromosome results from a chromosomal translocation that produces the chimeric *bcr-abl* oncogene. The unregulated Abl kinase activity of the Bcr-Abl oncoprotein is responsible for its oncogenic effect. An Abl kinase inhibitor (imatinib, or Gleevec) is effective in treating chronic myelogenous leukemia (CML) and may work against a few other cancers that are driven by related kinases (see Figure 24-20).

- Inappropriate production of nuclear transcription factors such as Fos, Jun, and Myc can induce transformation. In Burkitt's lymphoma cells, *c-myc* is translocated close to an antibody gene, leading to overproduction of c-Myc (see Figure 24-22).

- Many genes that regulate normal developmental processes encode proteins that function in various signaling pathways. Their normal roles in regulating where and when growth occurs are reflected in the character of the tumors that arise when the genes are mutated.

- Loss of signaling by TGF-β, a negative growth regulator, promotes cell proliferation and development of malignancy (see Figure 24-23).

- Precise molecular analysis of early tumors is allowing the use of highly targeted drug treatments that are likely to be optimal for a particular case of a particular tumor. These refinements have allowed substantial reduction in breast cancer mortality. There are good prospects for still better targeting of treatments based on continuing increases in the understanding of tumor-cell regulation.

- The advent of molecular techniques for characterizing individual tumors is allowing the application of drugs and antibody treatments that target the properties of a particular tumor. This permits more effective treatment of individual patients and reduces the use of drugs or antibodies that will be ineffective and possibly toxic.

- Novel shRNA screening methodologies allow for the identification of genes specifically required for the survival of cancer cells, thereby facilitating the discovery of new therapeutic targets.

## 24.4 Cancer and Mutation of Cell Division and Checkpoint Regulators

The complex mechanisms regulating the eukaryotic cell cycle are prime targets for oncogenic mutations. Both positive- and negative-acting proteins precisely control the entry of cells into and their progression through the cell cycle, which consists of four main phases: $G_1$, S, $G_2$, and mitosis (see Figure 19-30). In addition, cells harbor surveillance mechanisms—known as checkpoint pathways—that ensure that cells do not enter the next phase of the cell cycle before the previous one has been completed. For example, cells that have sustained damage to their DNA normally are arrested before their DNA is replicated or in $G_2$ before chromosome segregation. This arrest allows time for the DNA damage to be repaired; alternatively, the arrested cells are directed either to commit suicide via programmed cell death or at least not to divide. The whole cell-cycle control and checkpoint systems function to prevent cells from becoming cancerous. As might be expected, mutations in this system often lead to abnormal development or contribute to cancer.

In this section we will discuss the cell cycle and checkpoint pathways that are affected in cancer. We will first describe how the pathway that controls entry into the cell cycle is mutated and misregulated in most human cancers. We will then discuss how p53 prevents tumorigenesis by helping cells to respond to DNA damage. Next we will describe how mutations in programmed cell death pathways contribute to tumorigenesis and end with a discussion of an emerging group of important oncogenes and tumor-suppressor genes: micro-RNAs.

### Mutations That Promote Unregulated Passage from $G_1$ to S Phase Are Oncogenic

Once a cell progresses past a certain point in late $G_1$, called START, it becomes irreversibly committed to entering the S phase and replicating its DNA (see Figure 19-15). D-type

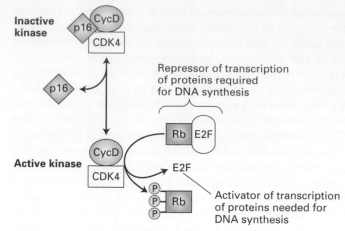

**FIGURE 24-24 START control.** Unphosphorylated Rb protein binds transcription factors collectively called *E2F* and thereby prevents E2F-mediated transcriptional activation of many genes whose products are required for DNA synthesis (e.g., DNA polymerase). The kinase activity of cyclin D–CDK4/6 phosphorylates Rb, thereby inactivating Rb and activating E2F; cyclin D–CDK4/6 activity is inhibited by p16. Overproduction of cyclin D, a positive regulator, or loss of the negative regulators p16 and Rb commonly occurs in human cancers.

cyclins, cyclin-dependent kinases (CDKs), and the Rb protein are all elements of the control system that regulates passage through START.

The pathway that controls entry into the cell cycle is estimated to be misregulated in approximately 80 percent of human cancers. At the heart of this pathway are cyclin D-CDK4/6 complexes and the transcription inhibitor Rb (Figure 24-24). The expression of D-type cyclin genes is induced by many extracellular growth factors, or **mitogens**. These cyclins assemble with their partners CDK4 or CDK6 to generate catalytically active cyclin-CDK complexes, whose kinase activity promotes progression through $G_1$. Mitogen withdrawal prior to passage through START leads to accumulation of p15 or p16. As described in Chapter 19, p15 and p16 are CDK inhibitors that bind to cyclin D–CDK4/6 complexes and inhibit their activity, thereby causing $G_1$ arrest. The transcription inhibitor Rb is controlled by cyclin D–CDK4/6 phosphorylation. Unphosphorylated Rb binds to and sequesters E2F transcription factors in the cytoplasm. E2F transcription factors stimulate transcription of genes encoding proteins required for DNA synthesis. Under normal circumstances, phosphorylation of Rb protein is initiated midway through $G_1$ by active cyclin D–CDK4/6 complexes. Rb phosphorylation is completed by cyclin E–CDK2 complexes in late $G_1$, allowing release and activation of E2F transcription factors and G-S progression. The complete phosphorylation of Rb and its disassociation from E2F irreversibly commits the cell to DNA synthesis.

Most tumors contain an oncogenic mutation that causes overproduction or loss of one of the components of the pathway that controls entry into S phase such that the cells are propelled into S phase in the absence of the proper extracellular growth signals. For example, elevated levels of cyclin D1, one of the three D-type cyclins, are found in many human cancers. In certain tumors of antibody-producing B lymphocytes, the *cyclin D1* gene is translocated such that its transcription is under control of an antibody-gene enhancer, causing elevated cyclin D1 production throughout the cell cycle irrespective of extracellular signals. (This phenomenon is analogous to the c-*myc* translocation in Burkitt's lymphoma cells discussed earlier.) That cyclin D1 can function as an oncoprotein was shown by studies with transgenic mice in which the *cyclin D1* gene was placed under control of an enhancer specific for mammary ductal cells. Initially the ductal cells underwent hyperproliferation, and eventually breast tumors developed in these transgenic mice. Amplification of the *cyclin D1* gene and concomitant overproduction of the cyclin D1 protein is common in human breast cancer; the extra cyclin D1 helps to drive cells through the cell cycle.

We have already seen that inactivating mutations in both *RB* alleles lead to childhood retinoblastoma, a relatively rare type of cancer. However, loss of *RB* gene function also is found in more common cancers that arise later in life (e.g., carcinomas of lung, breast, and bladder). These tissues, unlike retinal tissue, most likely produce other proteins (e.g., p107 and p130, both structurally related to Rb) whose function is redundant with that of Rb, and thus loss of Rb is not so critical for developing cancer in these tissues. In the retina, however, regulation of cell cycle entry appears to rely exclusively on the Rb protein, which is why patients heterozygous for the Rb gene first develop tumors in this tissue. In addition to inactivating mutations, Rb function can be eliminated by the binding of an inhibitory protein, designated E7, that is encoded by human papillomavirus (HPV), another nasty viral trick to create virus-producing tissue. At present, this is known to occur only in cervical cancer.

The proteins that function as cyclin-CDK inhibitors play an important role in regulating the cell cycle. In particular, loss-of-function mutations that prevent p16 from inhibiting cyclin D–CDK4/6 kinase activity are common in several human cancers. As Figure 24-24 makes clear, loss of p16 mimics overproduction of cyclin D1. Thus p16 normally acts as a tumor suppressor. Although the *p16* tumor-suppressor gene is deleted in some human cancers, in others the *p16* sequence is normal. In some of these latter cancers (e.g., lung cancer) the *p16* gene, or genes encoding other functionally related proteins, is inactivated by hypermethylation of its promoter region, which prevents transcription. What promotes this change in the methylation of *p16* is not known, but it prevents production of this important cell-cycle-control protein.

The locus encoding p16 is highly unusual in that it encodes no less than three tumor-suppressor genes, making it the most highly vulnerable locus in the human genome. In addition to harboring the p16-encoding gene *INK4a*, immediately upstream is the *INK4b* locus, which encodes p15, another cyclin D–CDK4/6 inhibitor (Figure 24-25). The locus codes for a key activator of the tumor suppressor p53 as well as these CDK inhibitors. p14ARF (p19ARF in the mouse) is encoded by an exon upstream of the first *INK4a* exon and shares exon 2 and exon 3 with *INK4a*. As we will see in the next section, this

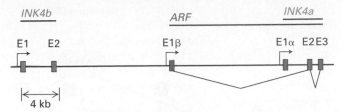

**FIGURE 24-25 The INK4b-ARF-INK4a locus.** This locus encodes three tumor-suppressor genes. Exons are designated as E. The two *INK4b* exons (orange) are located upstream of the *ARF/INK4a* locus. *ARF* (blue) is encoded by a unique E1β exon but shares exons E2 and E3 with *INK4a* (green). *INK4b* and *INK4a* encode p15 and p16 respectively. *ARF* encodes a p53 activator. [Adapted from Sherr, 2006, *Nat. Rev. Cancer* **6**:663–673.]

protein controls the stability of p53. Thus mutations in this locus would simultaneously affect the two major tumor-suppressor pathways in the cell, the Rb and p53 pathways.

## Loss of p53 Abolishes the DNA Damage Checkpoint

p53 is a central player in tumorigenesis. It is thought that most if not all human tumors either have mutations in p53 itself or proteins that regulate p53 activity. Cells with functional p53 become arrested in G₁ when exposed to DNA-damaging irradiation, whereas cells lacking functional p53 do not. Unlike other cell cycle proteins, p53 is present at very low levels in normal cells because it is extremely unstable and rapidly degraded. Mice lacking p53 are largely viable and healthy except for a predisposition to develop multiple types of tumors.

In normal mice the amount of p53 protein is heightened, a post-transcriptional response, only in stressful situations such as ultraviolet or γ irradiation, heat, and low oxygen. Irradiation by γ rays creates lesions in the DNA. Serine kinases ATM and/or ATR are recruited to these sites of damage and are activated. They then phosphorylate p53 on a serine residue in the N-terminus of the protein. This causes the protein to evade ubiquitin-mediated degradation, leading to a marked increase in its concentration (Figure 24-26). The stabilized p53 activates transcription of the gene encoding p21^CIP, which binds to and inhibits mammalian cyclin E-CDK2 and a plethora of other genes. As a result, cells with damaged DNA are arrested in G₁, allowing time for DNA repair by mechanisms discussed in Chapter 4, or the cells permanently arrest, that is, become senescent. The activity of p53 is not limited to inducing cell cycle arrest. In addition, this multipurpose tumor suppressor stimulates production of pro-apoptotic proteins (see the next section) and DNA-repair enzymes (see Figure 24-26). Senescence and apoptosis may in fact be the most important means through which p53 prevents tumor growth.

The activity of p53 normally is kept low by a protein called *Mdm2*. When Mdm2 is bound to p53, it inhibits the transcription-activating ability of p53 and at the same time, because it has E3 ubiquitin ligase activity, catalyzes the ubiquitinylation of p53, thus targeting p53 for proteasomal degradation. Phosphorylation of p53 by ATM or ATR displaces

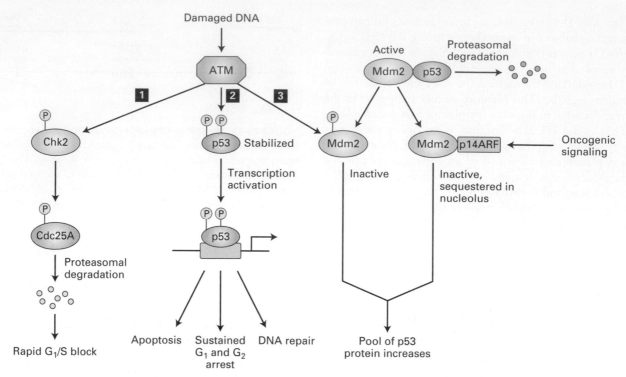

**FIGURE 24-26 G₁ arrest in response to DNA damage.** The kinase activity of ATM is activated in response to DNA damage due to various stresses (e.g., UV irradiation, heat). Activated ATM then triggers three pathways leading to arrest in $G_1$: (**1**) Chk2 is phosphorylated and, in turn, phosphorylates Cdc25A, thereby marking it for degradation and blocking its role in CDK2 activation. (**2**) In a second pathway, phosphorylation of p53 stabilizes it, permitting p53-activated expression of genes encoding proteins that cause arrest in $G_1$, promote apoptosis, or participate in DNA repair. (**3**) The third pathway is another way of controlling the pool of p53. The Mdm2 protein in its active form can form a complex with p53, inhibiting the transcription factor and causing p53 ubiquitination and subsequent degradation by the proteasome. ATM phosphorylates Mdm2 to inactivate it, causing increased stabilization of p53. In addition, Mdm2 levels are controlled by p14ARF (p19ARF in mice), which binds Mdm2 and sequesters it in the nucleolus, where it cannot access p53. The p14ARF gene is induced by high levels of mitogenic signaling that is frequently observed in cells carrying oncogenic mutations in growth factor signaling pathways. The human *Mdm2* gene is frequently amplified in sarcomas, which presumably causes excessive inactivation of p53. Similarly, p14ARF is also found mutated in some cancers.

bound Mdm2 from p53, thereby stabilizing it. Because the *Mdm2* gene is itself transcriptionally activated by p53, Mdm2 functions in an autoregulatory feedback loop with p53, perhaps normally preventing excess p53 function. The *Mdm2* gene is amplified in many sarcomas and other human tumors that contain a normal *p53* gene. Even though functional p53 is produced by such tumor cells, the elevated Mdm2 levels reduce the p53 concentration enough to abolish the p53-induced $G_1$ arrest in response to irradiation. A key regulator of the Mdm2 ubiquitin ligase is the p14ARF (p19 in mouse) protein, encoded by the multi-tumor-suppressor locus that also encodes the INK4 proteins. p14ARF binds to and sequesters Mdm2 in the nucleolus—in this manner causing stabilization of p53. Normal p14ARF levels are so low in tissues that the protein is barely detectable. This is good because it would otherwise cause p53 accumulation and hence cell cycle arrest or apoptosis. However, in response to oncogenic signaling, that is, in the presence of high levels of mitogenic signals, p14ARF transcription is induced by the E2F transcription factor (see Figure 24-26). Thus ARF is an important inhibitor of tumorigenesis since it induces p53 activation, when mitogenic

signaling reaches unphysiologically high levels, through hyperactivating mutations in the signaling pathways. For that reason, for mitogenic signaling pathways to cause uncontrolled proliferation as is seen in cancer, this p53 up-regulation must not occur. It is therefore not surprising that p53 is inactive in most human tumors, either through loss of p53 function itself or through down-regulation of positive regulators of p53 function such as p14ARF or up-regulation of negative regulators of p53 such as Mdm2.

The activity of p53 also is inhibited by a human papillomavirus (HPV) protein called *E6*. HPV encodes three proteins that contribute to its ability to induce stable transformation and mitosis in a variety of cultured cells. Two of these—E6 and E7—bind to and inhibit the p53 and Rb tumor suppressors, respectively. Acting together, E6 and E7 are sufficient to induce transformation in the absence of mutations in cell regulatory proteins. Recall that the HPV E5 protein, which causes sustained activation of the PDGF receptor, enhances proliferation of the transformed cells.

The active form of p53 is a tetramer of four identical subunits. A missense point mutation in one of the two *p53* alleles in a cell can abrogate almost all p53 activity because virtually all the oligomers will contain at least one defective subunit, and such oligomers have reduced ability to activate transcription. Oncogenic *p53* mutations thus act as **dominant negatives**, with mutations in a single allele causing a loss of function. The loss of function is incomplete, so in order to grow more rapidly, tumor cells still sometimes lose the remaining functional allele (loss of heterozygosity). As we learned in Chapter 5, dominant-negative mutations can occur in proteins whose active forms are multimeric or whose function depends on interactions with other proteins. In contrast, loss-of-function mutations in other tumor-suppressor genes (e.g., *RB*) are recessive because the encoded proteins function as monomers and mutation of a single allele has little functional consequence.

The p53 protein is a key defense mechanism against cancerous transformation. This is best illustrated by the high frequency of loss of p53 function in human cancers. Loss-of-function mutations in the *p53* gene occur in more than 50 percent of human cancers. What does it protect us against? Unlike Rb, which prevents inappropriate proliferation, p53 guards the cell from genetic changes. When the p53 $G_1$ checkpoint control does not operate properly, damaged DNA can replicate, perpetuating mutations and DNA rearrangements that are passed on to daughter cells, contributing to the likelihood of transformation into metastatic cells. At the same time, apoptosis is inhibited, contributing to evolution of transformed cells. Because of its central role in preventing tumorigenesis, researchers are intensely searching for compounds that can restore p53 function as a new way of treating a broad spectrum of human tumors.

## Apoptotic Genes Can Function as Protooncogenes or Tumor-Suppressor Genes

During normal development many cells are designated for **programmed cell death**, also known as apoptosis (see Chapter 21). Many abnormalities, including errors in mitosis, DNA damage, and an abnormal excess of cells not needed for development of a working organ, can trigger apoptosis. In some cases, cell death appears to be the default situation, with signals required to ensure cell survival. Cells can receive instructions to live and instructions to die, and a complex regulatory system integrates the various kinds of information.

If cells do not die when they should and instead keep proliferating, a tumor may form. For example, chronic lymphoblastic leukemia (CLL) occurs because cells survive when they should be dying. The cells accumulate slowly, and most are not actively dividing, but they do not die. CLL cells have chromosomal translocations that activate a gene called *bcl-2*, which we now know to be a critical blocker of apoptosis (see Figure 21-37). The resultant inappropriate overproduction of Bcl-2 protein prevents normal apoptosis and allows survival of these tumor cells. CLL tumors are therefore attributable to a failure of cell death. Another dozen or so proto-oncogenes that are normally involved in negatively regulating apoptosis have been mutated to become oncogenes. Overproduction of

their encoded proteins prevents apoptosis even when it is needed to stop cancer cells from growing.

Conversely, genes whose protein products stimulate apoptosis behave as tumor suppressors. An example is the *PTEN* gene discussed in Chapter 16. The phosphatase encoded by this gene dephosphorylates phosphatidylinositol 3,4,5-trisphosphate, a secondary messenger that functions in activation of Akt (see Figure 16-26). Cells lacking PTEN phosphatase have elevated levels of phosphatidylinositol 3,4,5-trisphosphate and active Akt, which promotes cell survival, growth, and proliferation and prevents apoptosis by several pathways. Thus PTEN acts as a pro-apoptotic tumor suppressor by decreasing the anti-apoptotic and proliferation-promoting effects of Akt.

The most common pro-apoptotic tumor-suppressor gene implicated in human cancers is *p53*. Among the genes activated by p53 are several encoding pro-apoptotic proteins such as Bax (see Figure 21-38). When most cells suffer extensive DNA damage or numerous other stresses such as hypoxia, the p53-induced expression of pro-apoptotic proteins leads to their quick demise (see Figure 24-26). While this may seem like a drastic response to DNA damage, it prevents proliferation of cells that are likely to accumulate multiple mutations. When p53 function is lost, apoptosis cannot be induced and the accumulation of mutations required for cancer to develop becomes more likely.

## Micro-RNAs Are a New Class of Oncogenic Factors

In the last few years it has become clear that noncoding RNAs, especially micro-RNAs (miRNAs), play a critical role in tumorigenesis. Generation of miRNAs typically involves the transcription of a precursor RNA that, through a number of processing steps, is trimmed down to a 20-22-nucleotide-long mature miRNA. The mature microRNAs usually bind in the 3′ untranslated region (UTR) of their target RNAs and inhibit their translation or sometimes cause degradation of the target mRNA. To date, approximately 750 miRNAs have been identified in humans and have been implicated in the regulation of as many as 30 percent of the cell's mRNAs, with fundamental roles in cell proliferation, differentiation, and apoptosis. A number of them have also been shown to function as tumor-suppressor genes or oncogenes.

The first role of miRNAs in tumorigenesis was revealed by the analysis of chromosomal region 13q14.3. This genomic region is found deleted in most cases of chronic lymphocytic leukemia (CLL), the most common leukemia in humans. Characterization of the disease causing deletion showed that two micro-RNAs, miR-15-a and miR-16-1, cause CLL. Mice mutant for both micro-RNAs develop CLL or the related cell lymphocytic leukemia. The two microRNAs appear to control cell-proliferation genes. In their absence, proliferation of B cells is increased. The let7 family of miRNAs has also been implicated in tumorigenesis. Let-7 miRNAs down-regulate the translation of Ras. Thus in the absence of the miRNA, Ras is constitutively overproduced, contributing to tumorigenesis. Let-7 miRNAs have other targets in the cell as well, and it is generally believed that each micro-RNA has multiple targets, providing ample opportunity to contribute to tumorigenesis.

Because of their function in inhibiting translation, miRNAs can function as tumor suppressor genes as well as oncogenes. miR-15-a and miR-16-1 act as tumor-suppressor genes; they normally inhibit cell proliferation, and their absence leads to cell growth. However, some miRNAs have also been found to be overexpressed in cancer, and their analysis indicates they function as oncogenes. Overexpression of miR-155 is found in many cancers and causes expansion of large B cells. Of particular interest is miR-21. This micro-RNA is overexpressed in most solid tumors, including glioblastoma, breast, lung, pancreatic, and colon tumors, and it targets several tumor-suppressor genes, among them the PTEN phosphatase. Much needs to be learned about how micro-RNAs contribute to tumorigenesis, but it is clear that through their ability to regulate many different genes, they can impact disease progression in more than one way.

## KEY CONCEPTS of Section 24.4

### Cancer and Mutation of Cell Division and Checkpoint Regulators

• Overexpression of the proto-oncogene encoding cyclin D1 or loss of the tumor-suppressor genes encoding p16 and Rb can cause inappropriate, unregulated passage through START in late $G_1$. Such abnormalities are seen in 80 percent of human tumors.

• The INK4-ARF locus represents the major tumor-suppressor locus in humans, controlling both the Rb and p53 pathways (see Figure 24-25).

• The p53 protein is a multifunctional tumor suppressor that promotes arrest in $G_1$ and $G_2$, apoptosis, and DNA repair in response to damaged DNA (see Figure 24-26).

• Loss-of-function mutations in the *p53* gene occur in more than 50 percent of human cancers. Overproduction of Mdm2, a protein that normally inhibits the activity of p53, occurs in several cancers (e.g., sarcomas) that express normal p53 protein. Thus in one way or another, the p53 stress-response pathway is inactivated to allow tumor growth.

• Human papillomavirus (HPV) encodes three oncogenic proteins: E6 (inhibits p53), E7 (inhibits Rb), and E5 (activates the PDGF receptor).

• Overproduction of anti-apoptotic proteins (e.g., Bcl-2) can lead to inappropriate cell survival and is associated with chronic lymphoblastic leukemia (CLL) and other cancers. Loss of proteins that promote apoptosis (e.g., p53 transcription factor and PTEN phosphatase) have a similar oncogenic effect.

• Micro-RNAs have been implicated in disease progression in many different types of cancer. They can function as tumor suppressors or as oncogenes.

## 24.5 Carcinogens and Caretaker Genes in Cancer

Carcinogens, which can be natural or man-made, are chemicals that cause cancer. Carcinogens cause mutations that reduce the function of tumor-suppressor genes, create oncogenes from proto-oncogenes, or damage DNA-repair systems. As our previous discussion has shown, alterations in DNA that lead to malfunction of tumor-suppressor proteins and oncoproteins are the underlying cause of most cancers. These oncogenic mutations in key growth and cell cycle regulatory genes include insertions, deletions, and base substitutions as well as chromosomal amplifications and translocations. In addition, damage to caretaker genes comprising the DNA-repair systems (see Chapter 4) leads to an increased mutation rate. Of the many mutations that accumulate, some affect cell cycle regulators, and the cells bearing these may become cancerous. Furthermore, some DNA-repair mechanisms themselves are error prone (see Figure 4-40). Those "repairs" also contribute to oncogenesis. The inability of tumor cells to maintain genomic integrity leads to formation of a heterogeneous population of malignant cells. For this reason, chemotherapy directed toward a single gene or even a group of genes is likely to be ineffective in wiping out all malignant cells. This problem adds to the interest in therapies that interfere with the blood supply to tumors, target aneuploid cells, or in other ways act on multiple types of tumor cells.

Normal dividing cells usually employ several mechanisms to prevent accumulation of detrimental mutations that could lead to cancer. One form of protection against mutation for stem cells is their relatively low rate of division, which reduces the possibility of DNA damage incurred during DNA replication and mitosis. Furthermore, the progeny of stem cells do not have the ability to divide indefinitely. After several rounds of division they exit the cell cycle, reducing the possibility of mutation-induced misregulation of cell division associated with dangerous tumors. Also, if multiple mutations are required for a tumor to grow, attract a blood supply, invade neighboring tissues, and metastasize, a low rate of replication combined with the normal low rate of mutations ($10^{-9}$) provides further shielding from cancer. However, these safeguards can be overcome if a powerful mutagen reaches the cells or if DNA repair is compromised and the mutation rate rises. When cells with stem-cell-like growth properties are mutagenized by environmental poisons and unable to efficiently repair the damage, cancer can occur.

In this section, we explore the ways in which various carcinogens act on DNA to induce cancer as well as how mutations in caretaker genes contribute to cancer by compromising the cell's ability to repair DNA damage. We end the chapter with a discussion of a special enzyme called telomerase, another guardian of the genome that protects against DNA damage, and its role in cancer cells.

## Carcinogens Induce Cancer by Damaging DNA

The ability of chemical and physical carcinogens to induce cancer is due to the DNA damage that they cause as well as the errors introduced into DNA during the cells' efforts to repair this damage. Thus carcinogens also are **mutagens**. The strongest evidence that carcinogens act as mutagens comes from the observation that cellular DNA altered by exposure of cells to carcinogens can change cultured cells, such as 3T3 cells, or cells implanted in mice into fast-growing cancer-like cells (see Figure 24-4). The mutagenic effect of carcinogens is roughly proportional to their ability to transform cells and induce cancer in animal models.

Although substances identified as chemical carcinogens have a broad range of structures with no obvious unifying features, they can be classified into two general categories. *Direct-acting carcinogens*, of which there are only a few, are mainly reactive electrophiles (compounds that seek out and react with electron-rich centers in other compounds). By chemically reacting with nitrogen and oxygen atoms in DNA, these compounds can modify bases in DNA so as to distort the normal pattern of base pairing. If these modified nucleotides are not repaired, they allow an incorrect nucleotide to be incorporated during replication. This type of carcinogen includes ethylmethane sulfonate (EMS), dimethyl sulfate (DMS), and nitrogen mustards.

In contrast, *indirect-acting carcinogens* generally are unreactive, often water-insoluble compounds that can act as potent cancer inducers only after the introduction of electrophilic centers. In animals, *cytochrome P-450 enzymes* are located in the endoplasmic reticulum of most cells and at especially high levels in liver cells. P-450 enzymes normally function to add electrophilic centers, such as OH groups, to nonpolar foreign chemicals, such as certain insecticides and therapeutic drugs, in order to solubilize them so that they can be excreted from the body. However, P-450 enzymes also can turn otherwise harmless chemicals into carcinogens. Indeed, most chemical carcinogens have little mutagenic effect until they have been modified by cellular enzymes.

## Some Carcinogens Have Been Linked to Specific Cancers

In the earliest days of cancer awareness, it became clear that at least some cancers are due to environmental poisons. For example, in 1775 it was reported that the exposure of chimney sweeps to soot caused scrotal cancer, and in 1791 the use of snuff (tobacco) was reported to be associated with nasal cancer. Environmental chemicals were originally associated with cancer through experimental studies in animals. The classic experiment is to repeatedly paint a test substance on the back of a mouse and look for development of both local and systemic tumors in the animal. Such assays led to the purification of a pure chemical carcinogen from coal tar in 1933, benzo(a)pyrene. The role of radiation in damaging chromosomes was first demonstrated in the 1920s using x-irradiated *Drosophila*. Later the ability of radiation to cause human cancer, especially leukemia, was dramatically shown by the increased rates of leukemia among survivors of the atomic bombs dropped in World War II (ionizing radiation)

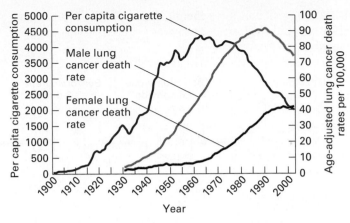

**FIGURE 24-27 Chemical carcinogenesis by tobacco smoke.** Cigarette smoking provides a clear example of a deadly form of chemical carcinogenesis. The rates of lung cancer follow the rates of smoking, with about a 30-year lag. Women began to smoke in large numbers starting in the 1960s, and starting in the 1990s, lung cancer passed breast cancer as the leading cause of women's cancer deaths. At the same time a gradual decrease in smoking rates among men starting in the 1960s has been reflected in a decrease in their lung cancer rate. [From the American Cancer Society.]

and more recently by the increase in melanoma (skin cancer) in individuals exposed to too much sunlight (UV radiation).

Although chemical carcinogens are believed to be risk factors for many human cancers, a direct linkage to specific cancers has been established only in a few cases, the most important being lung cancer and the other cancers (laryngeal, pharynx, stomach, liver, pancreas, bladder, cervix, and more) that are associated with smoking. Epidemiological studies (Figure 24-27) first indicated that cigarette smoking was the major cause of lung cancer, but the reason was unclear until the discovery that about 60 percent of human lung cancers contain inactivating mutations in the *p53* gene. The chemical *benzo(a)pyrene*, found in cigarette smoke as well as coal tar, undergoes metabolic activation in the lung (Figure 24-28) to a potent mutagen that mainly causes conversion of guanine (G) to thymine (T) bases, a transversion mutation. When applied to cultured bronchial epithelial cells, activated benzo(a) pyrene induces many mutations, including inactivating mutations at codons 175, 248, and 273 of the *p53* gene. These same positions, all within the protein's DNA-binding domain, are major mutational hot spots in human lung cancer. In fact, the nature of mutations in *p53* (and other cancer-related genes) gives clues as to the origin of the cancer. The G-to-T transversions caused by benzo(a)pyrene, for example, are present in the *p53* genes of about one-third of smokers' lung tumors. That type of mutation is relatively rare among the *p53* mutations found in other types of tumors. The carcinogen leaves its footprint. Thus there is a strong correlation between one defined chemical carcinogen in cigarette smoke and human cancer. It is likely that other chemicals in cigarette smoke induce mutations in other genes since it contains

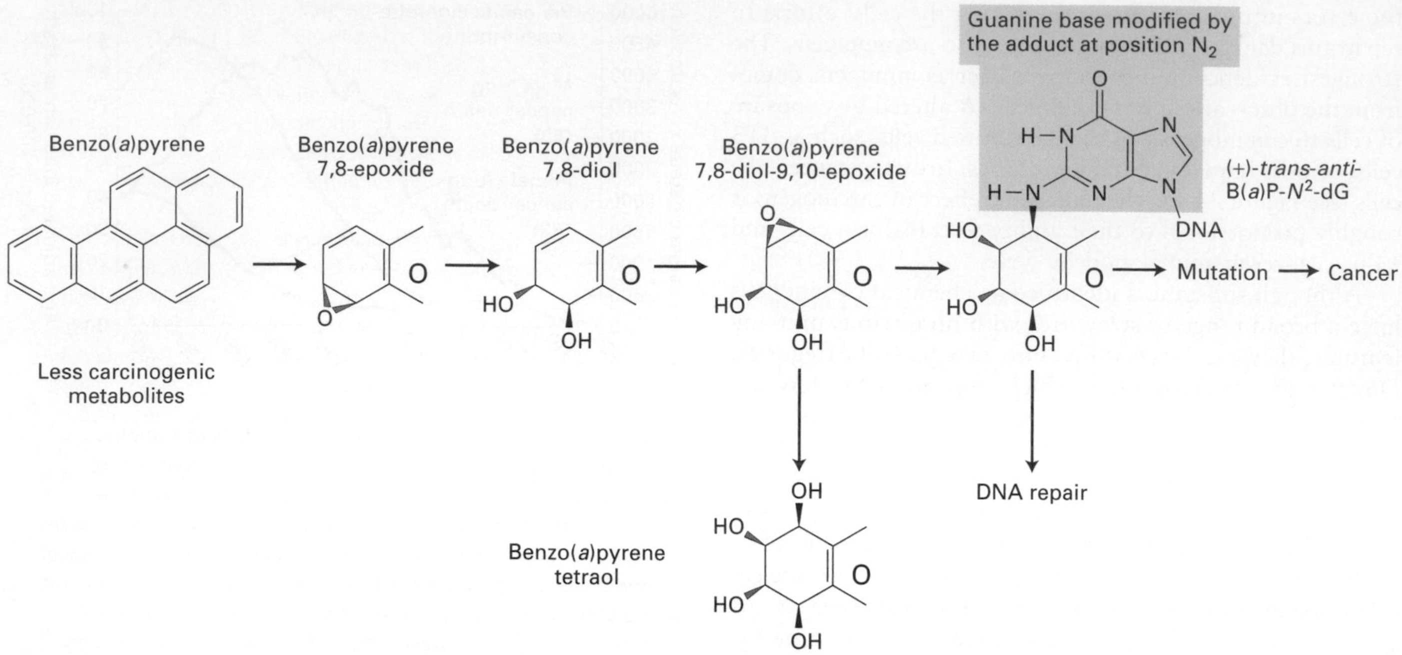

**FIGURE 24-28 Enzymatic processing of benzo(a)pyrene to a more potent mutagen and carcinogen.** Liver enzymes, particularly P-450 enzymes, modify benzo(a)pyrene in a series of reactions, producing 7,8-diol-9,10-epoxide, a highly potent mutagenic species that reacts with DNA primarily at the $N_2$ atom of a guanine base in DNA. The resulting adduct, (+)-trans-anti-B(a)P-$N^2$-dG, causes polymerase to insert an A rather than a C opposite the modified G base. Next time the DNA is replicated, a T will be inserted opposite the A, and the mutation will be complete. Horizontal arrows indicate alterations toward greater potency, while vertical arrows indicate changes in the direction of reduced toxicity. The large "O" symbol represents the rest of the multi-ring structure shown in the complete benzo(a)pyrene molecule at the left. [Adapted from E. L. Loechler, 2001, *Encyclopedia of Life Science*.]

more than 60 carcinogens. Similarly, asbestos exposure is clearly linked to mesothelioma, a type of epithelial cancer.

Lung cancer is not the only major human cancer for which a clear-cut risk factor has been identified. *Aflatoxin*, a fungal metabolite found in moldy grains, induces liver cancer (Figure 24-29a). After chemical modification by liver enzymes, aflatoxin becomes linked to G residues in DNA and induces G-to-T transversions. Aflatoxin also causes a mutation in the *p53* gene. Furthermore, cooking meat at high temperatures causes chemical reactions that form *heterocyclic amines* (HCAs), which are potent mutagens and cause colon and breast carcinomas in animal models. HCAs react with deoxyguanosine bases to form mutagenic adducts (Figure 24-29b). Exposure to other chemicals has been correlated with minor cancers. Hard evidence concerning dietary and environmental risk factors that would help us avoid other common cancers (e.g., breast, colon, and prostate cancer and leukemias) is generally lacking.

## Loss of DNA-Repair Systems Can Lead to Cancer

Even without any external carcinogen or mutagen, normal processes generate a large amount of DNA damage. The damage is due to depurination reactions, to alkylation reactions, and to the generation of reactive species such as oxygen radicals, all of which alter DNA. It has been estimated that in every cell, more than 20,000 alterations to the DNA

occur each day from just reactive oxygen species and depurination. Thus DNA repair is a crucial defense system.

The normal role of caretaker genes is to prevent or repair DNA damage. Loss of the high-fidelity DNA-repair systems that are described in Chapter 4 correlates with increased risk for cancer. For example, humans who inherit mutations in genes that encode a crucial mismatch-repair or excision-repair protein have an enormously increased probability of developing certain cancers (Table 24-2). Without proper DNA repair, people with xeroderma pigmentosum (XP) or hereditary nonpolyposis colorectal cancer (HNPCC, also known as Lynch syndrome) have a propensity to accumulate mutations in many other genes, including those that are critical in controlling cell growth and proliferation. XP causes affected people to develop skin cancer at about a thousand times the normal rate. Seven of the eight known XP genes encode components of the excision-repair machinery, and in the absence of this repair mechanism, genes that control the cell cycle or otherwise regulate cell growth and death become mutated. HNPCC genes encode components of the mismatch-repair system, and mutations in these genes are found in 20 percent of sporadic colon cancers. The cancers progress from benign polyps to full-fledged tumors much more rapidly than usual, presumably because the initial cancer cells are undergoing continuous mismatch mutagenesis without repair.

One gene frequently mutated in colon cancers because of the absence of mismatch repair encodes the type II receptor for

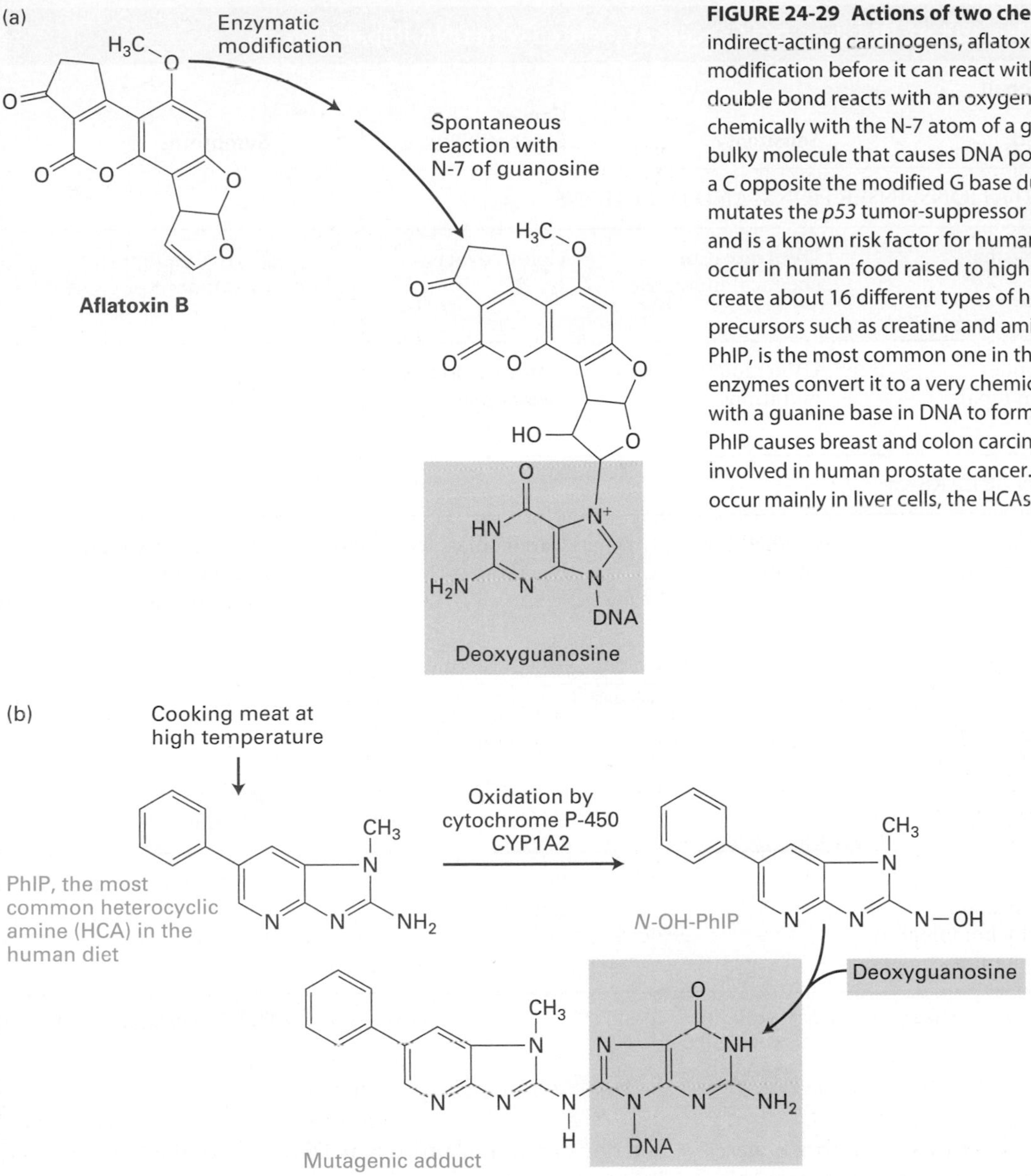

**FIGURE 24-29 Actions of two chemical carcinogens.** (a) Like all indirect-acting carcinogens, aflatoxin must undergo enzyme-catalyzed modification before it can react with DNA. In aflatoxin the colored double bond reacts with an oxygen atom, enabling it to react chemically with the N-7 atom of a guanine in DNA, forming a large bulky molecule that causes DNA polymerase to insert an A rather than a C opposite the modified G base during replication. This compound mutates the *p53* tumor-suppressor gene, causing G to T transversions, and is a known risk factor for human cancer. (b) Chemical reactions that occur in human food raised to high temperature, especially red meat, create about 16 different types of heterocyclic amines (HCAs) from precursors such as creatine and amino acids. The HCA shown here, PhIP, is the most common one in the human diet. Cytochrome P-450 enzymes convert it to a very chemically reactive form, which reacts with a guanine base in DNA to form a DNA adduct that is mutagenic. PhIP causes breast and colon carcinomas in rodents and may be involved in human prostate cancer. Although the P-450 conversions occur mainly in liver cells, the HCAs can migrate to other tissues.

TGF-β (see Figure 24-23). The gene encoding this receptor contains a sequence of 10 adenines in a row. Because of "slippage" of DNA polymerase during replication, this sequence often undergoes mutation to a sequence containing 9 or 11 adenines. If the mutation is not fixed by the mismatch-repair system, the resultant frameshift in the protein-coding sequence abolishes production of the normal receptor protein. As noted earlier, such inactivating mutations make cells resistant to growth inhibition by TGF-β, thereby contributing to the unregulated growth characteristic of these tumors. This finding attests to the importance of mismatch repair in correcting genetic damage that might otherwise lead to uncontrolled cell proliferation.

All DNA-repair mechanisms utilize a family of DNA polymerases to correct DNA damage. Nine of these polymerases, including one called *DNA polymerase β*, are capable of using templates that contain DNA adducts and other chemical mod-

ifications, even missing bases. These are called *lesion-bypass* DNA polymerases. Each member of the polymerase family has distinct capabilities to cope with particular types of DNA lesions. Presumably such polymerases are tolerated because often any repair is better than none. They are the polymerases of last resort, the ones used when more conventional and accurate polymerases are unable to perform, and they carry out a mutagenic replication process. DNA Pol β does not proofread and is overexpressed in certain tumors, perhaps because it is needed at high levels for cells to be able to divide at all in the face of a growing burden of mutations. Error-prone repair systems are thought to mediate much, if not all, of the carcinogenic effects of chemicals and radiation since it is only after the repair that a heritable mutation exists. There is growing evidence that mutations in DNA Pol β are associated with tumors. When 189 tumors were examined, 58 had mutations

**TABLE 24-2** Some Human Hereditary Diseases and Cancers Associated with DNA-Repair Defects

| Disease | DNA-Repair System Affected | Sensitivity | Cancer Susceptibility | Symptoms |
|---|---|---|---|---|
| **PREVENTION OF POINT MUTATIONS, INSERTIONS, AND DELETIONS** | | | | |
| Hereditary nonpolyposis colorectal cancer | DNA mismatch repair | UV irradiation, chemical mutagens | Colon, ovary | Early development of tumors |
| Xeroderma pigmentosum | Nucleotide excision repair | UV irradiation, point mutations | Skin carcinomas, melanomas | Skin and eye photosensitivity, keratoses |
| **REPAIR OF DOUBLE-STRAND BREAKS** | | | | |
| Bloom's syndrome | Repair of double-strand breaks by homologous recombination | Mild alkylating agents | Carcinomas, leukemias, lymphomas | Photosensitivity, facial telangiectases, chromosome alterations |
| Fanconi anemia | Repair of double-strand breaks by homologous recombination | DNA cross-linking agents, reactive oxidant chemicals | Acute myeloid leukemia, squamous-cell carcinomas | Developmental abnormalities including infertility and deformities of the skeleton; anemia |
| Hereditary breast cancer, BRCA-1 and BRCA-2 deficiency | Repair of double-strand breaks by homologous recombination | | Breast and ovarian cancer | Breast and ovarian cancer |

SOURCES: Modified from A. Kornberg and T. Baker, 1992, *DNA Replication*, 2d ed., W. H. Freeman and Company, p. 788; J. Hoeijmakers, 2001, *Nature* **411**:366; and L. Thompson and D. Schild, 2002, *Mutation Res.* **509**:49.

in the DNA Pol β gene, and most of these mutations were found neither in normal tissue from the same patient nor in the normal spectrum of mutations found in different people. Expressing two of the mutant polymerase forms in mouse cells caused them to grow with a transformed appearance, with focus formation and anchorage independence.

Double-strand breaks are particularly severe lesions because incorrect rejoining of double strands of DNA can lead to gross chromosomal rearrangements and translocations such as those that produce a hybrid gene or bring a growth regulatory gene under the control of a different promoter. Often the repair of such damage depends on using the homologous chromosome as a guide (see Figure 4-42). The B and T cells of the immune system are particularly susceptible to DNA rearrangements caused by double-strand breaks created during rearrangement of their immunoglobulin or T-cell receptor genes, explaining the frequent involvement of these gene loci in leukemias and lymphomas. *BRCA-1* and *BRCA-2*, genes implicated in human breast and ovarian cancers, encode important components of DNA-break repair systems. Cells lacking one of the BRCA

functions are unable to repair DNA where the homologous chromosome is providing the template for repair.

## Telomerase Expression Contributes to Immortalization of Cancer Cells

**Telomeres,** the physical ends of linear chromosomes, consist of tandem arrays of a short DNA sequence, TTAGGG in vertebrates. Telomeres provide the solution to the end-replication problem—the inability of DNA polymerases to completely replicate the end of a double-stranded DNA molecule (see Chapter 6). *Telomerase,* a reverse transcriptase that contains an RNA template, repeatedly adds TTAGGG repeats to chromosome ends to lengthen or maintain the 3- to 20-kb regions of repeats that decorate the ends of human chromosomes (see Figure 6-47). Embryonic cells, germ-line cells, and stem cells produce telomerase, but most human somatic cells produce only a low level of telomerase as they enter S phase. As a result of having only modest telomerase function, their telomeres shorten with each cell cycle. Extensive shortening of telomeres is recognized

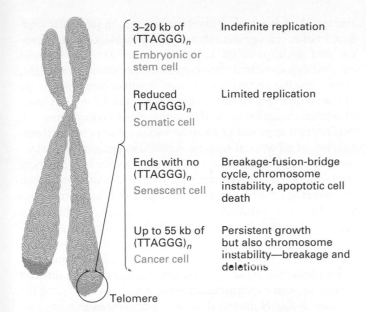

| | |
|---|---|
| 3–20 kb of (TTAGGG)$_n$<br>Embryonic or stem cell | Indefinite replication |
| Reduced (TTAGGG)$_n$<br>Somatic cell | Limited replication |
| Ends with no (TTAGGG)$_n$<br>Senescent cell | Breakage-fusion-bridge cycle, chromosome instability, apoptotic cell death |
| Up to 55 kb of (TTAGGG)$_n$<br>Cancer cell | Persistent growth but also chromosome instability—breakage and deletions |

Telomere

**FIGURE 24-30 Loss of telomeres normally limits the number of rounds of cell division.** Replication of the ends of chromosomes, the telomeres, requires a special enzyme called *telomerase*. Telomerase carries with it a short RNA template, which is used to guide the assembly of repeats of TTAGGG on the ends of chromosomes. Human embryonic cells have 8–10 kb of these repeats on each chromosome end. Because DNA polymerase requires a primer and there is none at the end of the lagging strand, it cannot replicate DNA fully at the ends (see Chapter 6). Telomerase is therefore necessary, and in its absence chromosomes shrink during each mitosis. Telomerase is present in stem cells and germ-line cells, where it is needed to keep telomeres long and allow essentially indefinite rounds of cell division. In most somatic cells, telomerase is at low or very low levels, and the length of telomeres is consequently gradually eroded. The length of the telomeres therefore provides one limit to replicative capacity. Complete loss of telomere repeats leads to triggering of DNA repair and apoptosis. Cancer cells often produce telomerase, allowing them to divide indefinitely and evade programmed death.

The prognosis for neuroblastoma, a pediatric tumor of the peripheral nervous system, can be ascertained by assessing the level of telomerase activity in tumor cells. High levels of telomerase predict a poor response to therapy, while low levels predict a good response. The level of *N-myc* protein, a transcription factor that regulates expression of telomerase, is also predictive for the outcomes of these tumors. Tumors that have amplified the *N-myc* gene to high copy number have a worse prognosis. These results show the practical importance of understanding the circuitry of telomere synthesis and its regulation. ■

Genetic approaches have further improved our understanding of the role of telomerase in cancer. The initial finding that mice homozygous for a deletion of the RNA subunit of telomerase are viable and fertile was surprising. However, after four to six generations defects began to appear in the telomerase-null mice as their very long telomeres (40–60 kb) became significantly shorter. The defects included depletion of tissues that require high rates of cell division, such as skin and intestine, and infertility.

When treated with carcinogens, telomerase-null mice develop tumors less readily than normal mice do. For example, skin papillomas induced by a combination of chemical carcinogens occur 20 times less frequently in mice lacking a functional telomerase than in normal mice, presumably because p53-triggered apoptosis is induced in response to the ever-shortening telomeres of cells that have begun to divide. However, if both telomerase and p53 are absent, there is an increased rate of epithelial tumors such as squamous-cell carcinoma, colon, and breast cancer. Mice with an *APC* mutation normally develop colon tumors, and these too are reduced if the mice lack telomerase. Some other tumors are less affected by loss of telomerase. These studies demonstrate the relevance of telomerase for unbridled cell division and make the enzyme a possible target for chemotherapy.

by the cell as a kind of DNA damage, with consequent stabilization and activation of p53 protein, leading to p53-triggered apoptosis. Figure 24-30 summarizes the effects of different numbers of telomere repeats. Complete loss of telomeres leads to end-to-end chromosome fusions and cell death.

Most tumor cells, despite their rapid proliferation rate, overcome this fate by producing telomerase. Many researchers believe that telomerase expression is essential for a tumor cell to become immortal, and specific inhibitors of telomerase have been suggested as cancer therapeutic agents. Introduction of telomerase-producing transgenes into cultured human cells that otherwise lack the enzyme can extend their life span by more than 20 doublings while maintaining telomere length. Conversely, treating human tumor cells (HeLa cells) in culture with antisense RNA against telomerase caused them to cease growth in about four weeks. Dominant-negative telomerases, such as those carrying a modified RNA template, can interfere with cancer cell growth.

## KEY CONCEPTS of Section 24.5

### Carcinogens and Caretaker Genes in Cancer

- Changes in the DNA sequence result from copying errors and the effects of various physical and chemical agents, or carcinogens. All carcinogens are mutagens; that is, they alter one or more nucleotides in DNA.

- Indirect-acting carcinogens, the most common type, must be activated before they can damage DNA. In animals, metabolic activation occurs via the cytochrome P-450 system, a pathway generally used by cells to rid themselves of noxious foreign chemicals. Direct-acting carcinogens such as EMS and DMS require no such cellular modifications in order to damage DNA.

- Benzo(*a*)pyrene, a component of cigarette smoke, causes inactivating mutations in the *p53* gene, thus contributing to the initiation of human lung tumors.

- Caretaker genes encode enzymes that repair DNA, otherwise maintain the integrity of the chromosomes, or promote the death of cells when DNA damage does occur. Mutations in caretaker genes allow the survival of cells that should die and a continuing mutagenesis of the genome that can lead to uncontrolled cell proliferation and hence cancer.

- Inherited defects in DNA-repair processes found in certain human diseases are associated with an increased susceptibility for certain cancers (see Table 24-2).

- Cancer cells, like germ cells and stem cells but unlike most differentiated cells, produce telomerase, which prevents shortening of chromosomes during DNA replication and may contribute to their immortalization. The absence of telomerase is associated with resistance to generation of certain tumors, owing to the protective p53 response.

## Perspectives for the Future

The recognition that cancer is fundamentally a genetic disease has opened enormous new opportunities for preventing and treating it. Carcinogens can now be assessed for their effects on known steps in cell cycle control. Genetic defects in the checkpoint controls for detecting damaged DNA and in the systems for repairing it can be readily recognized and used to explore the mechanisms of cancer. The multiple changes that must occur for a cell to grow into a dangerous tumor present multiple opportunities for intervention. Identifying mutated genes associated with cancer points directly toward proteins to which drugs can be targeted.

Diagnostic medicine is being transformed by our newfound ability to monitor large numbers of cell characteristics and by ever more sensitive methods to detect smaller numbers of tumor cells. The traditional methods of assessing possible tumor cells, mainly microscopy of stained cells, will be augmented or replaced by techniques for measuring the expression of tens of thousands of genes, focusing particularly on genes whose activities are identified as powerful indicators of the cell's growth properties and the patient's prognosis. Currently, DNA microarray and deep-sequencing analyses permit measurement of gene transcription and DNA copy number. In the future, techniques for systematically measuring protein content as well as proteins' modifications and localization, all important measures of cell states, will give us even more refined portraits of cells. Tumors now viewed as identical or very similar will instead be recognized as distinctly different and given appropriately different treatments. Earlier detection of tumors, based on better monitoring of cell properties, should allow more successful treatment. A focus on that particularly destructive process metastasis should be successful in identifying more of the mechanisms used by cells to migrate, attach, and invade. Manipulation of angiogenesis continues to look hopeful as a means of suffocating tumors.

The molecular cell biology of cancer provides avenues for new therapies, but prevention remains crucial and preferable to therapy. Avoidance of obvious carcinogens, in particular cigarette smoke, can significantly reduce the incidence of lung cancer and perhaps other kinds of cancer as well. Beyond minimizing exposure to carcinogens such as smoke or sunlight, certain specific approaches are now feasible. New knowledge of the involvement of human papillomavirus 16 in most cases of cervical cancer led to the development and Federal Drug Administration approval of a cancer vaccine that prevents three-quarters of all cervical cancers. Antibodies against cell-surface markers that distinguish cancer cells are a source of great hope, especially after successes with the clinical use of monoclonal antibodies against human EGF receptor 2 (Her2), a protein involved in some cases of human breast cancer. Further steps must involve medicine and science. Understanding the cell biology of cancer is a critical first step toward prevention and cure, but the next steps are hard. The success with Gleevec (imatinib) against leukemia is exceptional; many cancers remain difficult to treat and cause enormous suffering. Since *cancer* is a term for a group of highly diverse diseases, interventions that are successful for one type may not be useful for others. Despite these daunting realities, we are beginning to reap the benefits of decades of research exploring the molecular biology of the cell. We hope that many of the readers of this book will help to overcome the obstacles that remain.

## Key Terms

| | |
|---|---|
| angiogenesis 1117 | oncogene addiction 1132 |
| benign 1115 | p53 protein 1122 |
| Burkitt's lymphoma 1136 | Philadelphia |
| cancer stem cell 1116 | chromosome 1135 |
| carcinogen 1113 | proto-oncogene 1113 |
| carcinoma 1115 | Ras protein 1118 |
| caretaker gene 1113 | retinoblastoma (Rb) |
| epithelial-to-mesenchymal | protein 1128 |
| transition (EMT) 1116 | sarcoma 1115 |
| leukemia 1115 | slow-acting retrovirus 1127 |
| loss of heterozygosity | transducing retrovirus 1127 |
| (LOH) 1129 | transformation 1118 |
| malignant 1115 | tumorigenesis 1114 |
| metastasis 1114 | tumor microenvironment |
| multi-hit model of | 1117 |
| cancer induction 1120 | tumor-suppressor |
| mutagen 1145 | gene 1113 |
| oncogene 1113 | Warburg effect 1122 |

## Review the Concepts

1. What characteristics distinguish benign from malignant tumors? With respect to gene mutations, what distinguishes benign colon polyps from malignant colon carcinoma?

2. Ninety percent of cancer deaths are caused by metastatic rather than primary tumors. Define *metastasis*. Explain the

rationale for the following new cancer treatments: (a) batimastat, an inhibitor of matrix metalloproteinases and of the plasminogen activator receptor, (b) antibodies that block the function of integrins, integral membrane proteins that mediate attachment of cells to the basement membrane and extracellular matrices of various tissues, and (c) bisphosphonate, which inhibits the function of osteoclasts. Osteoclasts are cells that digest bones, for example, in remodeling during growth or healing. They can be recruited to a bone-digesting task by signals from other cells. (d) What is the importance of the EMT during metastasis?

3. Because of oxygen and nutrient requirements, cells in a tissue must reside within 100 μm of a blood vessel. Based on this information, explain why many malignant tumors often possess gain-of-function mutations in one of the following genes: *βFGF*, *TGFα*, and *VEGF*.

4. *Ras* oncogenes have been identified in human tumor cells using cell transformation assays with a line of cultured mouse fibroblasts called *3T3 cells*. How does the 3T3 cell transformation assay work? Why can transfection with a *ras* gene transform 3T3 cells but not normal cultured primary fibroblast cells?

5. Which important characteristic of tumor cells did Otto Warburg discover? How does this property contribute to the process of tumor formation?

6. What hypothesis explains the observations that incidence of human cancers increases exponentially with age? Give an example of data that confirm the hypothesis.

7. Distinguish between proto-oncogenes and tumor-suppressor genes. To become cancer promoting, do proto-oncogenes and tumor-suppressor genes undergo gain-of-function or loss-of-function mutations? Classify the following genes as proto-oncogenes or tumor-suppressor genes: *p53*, *ras*, *Bcl-2*, *jun*, *MDM2*, and *p16*.

8. Explain how DNA microarray analysis can diagnose the outcome of early stage breast cancer.

9. Despite differences in origin, cancer cells have several features in common that differentiate them from normal cells. Describe these.

10. Hereditary retinoblastoma generally affects children in both eyes, while spontaneous retinoblastoma usually occurs during adulthood only in one eye. Explain the genetic basis for the epidemiological distinction between these two forms of retinoblastoma. Explain the apparent paradox: loss-of-function mutations in tumor-suppressor genes act recessively, yet hereditary retinoblastoma is inherited as an autosomal dominant.

11. Explain the concept of loss of heterozygosity (LOH). Why do most cancer cells exhibit LOH of one or more genes? How does failure of the spindle assembly checkpoint lead to loss of heterozygosity?

12. Many malignant tumors are characterized by the activation of one or more growth-factor receptors. What is the catalytic activity associated with transmembrane growth-factor receptors such as the EGF receptor? Describe how the following events lead to activation of the relevant growth-factor receptor: (a) expression of the viral protein gp55,

(b) point mutation that converts a valine to glutamine within the transmembrane region of the Her2 receptor.

13. Describe the common signal transduction event that is perturbed by cancer-promoting mutations in the genes encoding Ras and NF-1. Why are mutations in Ras more commonly found in cancers than mutations in NF-1?

14. What is the structural distinction between the proteins encoded by c-*src* and v-*src*? How does this difference render v-*src* oncogenic?

15. Describe the mutational event that produces the *myc* oncogene in Burkitt's lymphoma. Why does the particular mechanism for generating oncogenic *myc* result in a lymphoma rather than another type of cancer? Describe another mechanism for generating oncogenic *myc*.

16. Pancreatic cancers often possess loss-of-function mutations in the gene that encodes the Smad4 protein. How does this mutation promote the loss-of-growth inhibition and highly metastatic phenotype of pancreatic tumors?

17. Why are mutations in the INK4 locus so dangerous?

18. Describe two mechanisms through which cancer-causing viruses can transform normal cells into cancer cells.

19. Explain how epigenetic changes can contribute to tumorigenesis.

20. Several strains of human papilloma virus (HPV) can cause cervical cancer. These pathogenic strains produce three proteins that contribute to host-cell transformation. What are these three viral proteins? Describe how each interacts with its target host protein.

21. Loss of p53 function occurs in the majority of human tumors. Name two ways in which loss of p53 function contributes to a malignant phenotype. Explain how benzo(*a*) pyrene can cause loss of p53 function.

22. Which human cell types possess telomerase activity? What characteristic of cancer is promoted by expression of telomerase? What concerns does this pose for medical therapies involving stem cells?

## Analyze the Data

1. Cigarette smoking is a major risk factor in the development of non-small-cell lung cancer (NSCLC), which accounts for 80 percent of all lung cancers. NSCLC is characterized by poor prognosis and resistance to chemotherapy. To understand if nicotine affects this resistance, lung cancer cell lines were treated with the chemotherapeutic drugs gemcitabine, Cisplatin, and Taxol in the presence and absence of nicotine (see Dasgupta et al., 2006, *Proc. Nat'l. Acad. Sci. USA* **103**:6332–6337).

   a. Three different NSCLC cell lines, A549, NCI-H23, and H1299, were examined by the TUNEL assay, which detects cells that are undergoing apoptosis (programmed cell death). The cells were either untreated or treated with one of the three chemotherapeutic drugs in the presence or absence of nicotine. The following data were obtained. Why are these chemotherapeutic drugs potentially useful for the treatment of lung cancer, and how does nicotine affect their potential usefulness?

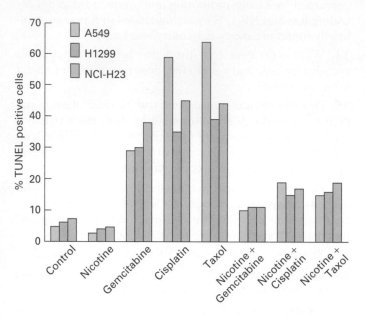

**b.** A549 cells were incubated with the drugs as in part (a) and an additional chemotherapeutic drug, camptothecin, in the presence or absence of nicotine. The cells were lysed, extracts were run on SDS gels, and then the gels were blotted and probed with antibodies against the indicated proteins. PARP is a protein that is cleaved during apoptosis. What can you deduce from these data about the effects of nicotine? What is the purpose of assessing the levels of p53, p21, and actin?

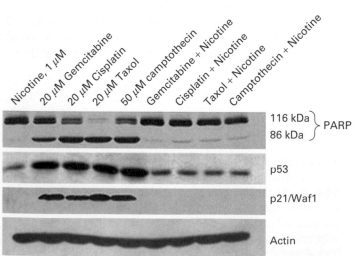

**c.** Survivin and XIAP are both members of the inhibitor-of-apoptosis (IAP) family, proteins that protect cells against apoptosis. A549 cells were treated with the chemotherapeutic drugs in the presence or absence of nicotine and presence or absence of LY294002, an inhibitor of PI-3 kinase (see Chapter 16). The levels of PARP, XIAP, survivin, and actin were assessed, as

shown in the Western blots below. The graph shows the results of a TUNEL assay in which A549 cells have been transfected with the indicated siRNAs. "None" indicates the amount of cell death with no transfected RNA. Control siRNAs 1 and 2 are irrelevant RNAs added to control for nonspecific effects of having an interfering RNA entering the cells. What do these studies suggest about the mechanism of nicotine action?

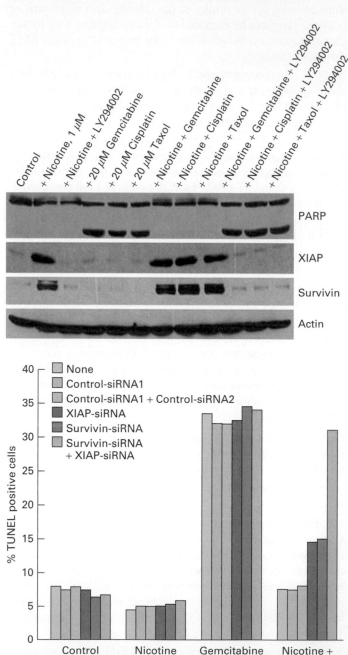

**d.** Can you provide a biological explanation for why the prognosis is poorer for patients who smoke during chemotherapy compared with those who quit smoking prior to it?

**2.** The E7 oncoprotein of the high-risk human papillomaviruses (HPV) is thought to contribute to cervical carcinogenesis at least in part by disrupting regulation of the cell cycle in cervical epithelial cells. E7 can dysregulate the cell cycle through its interaction with several cellular proteins, including the retinoblastoma suppressor protein pRb as well as the cyclin-dependent kinase inhibitor p21(Cip1).

**a.** How can transgenic mouse models be utilized to assess the impact of p21 on cell cycle dysregulation?

**b.** If the incidence of cervical disease were significantly increased in p21(−/−) mice compared with p21(+/+) mice, would this support a gain-of-function or loss-of-function role for p21 in the development of cervical cancers?

**c.** If the ability of E7 to induce cervical cancers were not significantly enhanced on the p21-null background, is this consistent with the hypothesis that the ability of E7 to inhibit p21 contributes to its carcinogenic properties?

**d.** Explain how expression of E7 mutant proteins in mouse models of cervical cancer can be utilized to assess the impact of these mutations on p21 and development of cervical carcinomas.

# References

### Introduction

Weinberg, R. A. 2006. *The Biology of Cancer.* Garland Science.

### Tumor Cells and the Onset of Cancer

Clarke, M. F., and M. Fuller. 2006. Stem cells and cancer: two faces of Eve. *Cell* 124:1111–1115.

Desgrosellier, J. S., and D. A. Cheresh. 2010. Integrins in cancer: biological implications and therapeutic opportunities. *Nature Rev. Cancer* 10(1):9–22. Review.

Egeblad, M., L. E. Littlepage, and Z. Werb. 2005. The fibroblastic coconspirator in cancer progression. *Cold Spring Harbor Symp. Quant. Biol.* 70:383–388.

Fidler, I. J. 2002. The pathogenesis of cancer metastasis: the "seed and soil" hypothesis revisited. *Nature Rev. Cancer* 3:1–6.

Folkman, J. 2006. Angiogenesis. *Annu. Rev. Med.* 57:1–18.

Grivennikov, S. I., F. R. Greten, and M. Karin. 2010. Immunity, inflammation, and cancer. *Cell* 140(6):883–899.

Hanahan, D., and R. A. Weinberg. 2011. Hallmarks of cancer: the next generation. *Cell* 144:646–674.

Huber, M. A., N. Kraut, and H. Beug. 2005. Molecular requirements for epithelial-mesenchymal transition during tumor progression. *Curr. Opin. Cell Biol.* 17:548–558.

Jain, M., et al. 2002. Sustained loss of a neoplastic phenotype by brief inactivation of MYC. *Science* 297:102–104.

Joyce, J. A., and J. W. Pollard. 2009. Microenvironmental regulation of metastasis. *Nat. Rev. Cancer* 9(4):239–252.

Kinzler, K. W., and B. Vogelstein. 1996. Lessons from hereditary colorectal cancer. *Cell* 87:159–170.

Ludwig, T. 2005. Local proteolytic activity in tumor cell invasion and metastasis. *BioEssays* 27:1181–1191.

Matsui, W., et al. 2004. Characterization of clonogenic multiple myeloma cells. *Blood* 103:2332–2336.

Maheswaran, S., and D. A. Haber. 2010. Circulating tumor cells: a window into cancer biology and metastasis. *Curr. Opin. Genet. Dev.* 20(1):96–99.

Nguyen, D. X., P. D. Bos, and J. Massagué. 2009. Metastasis: from dissemination to organ-specific colonization. *Nat. Rev. Cancer* 9(4):274–284.

Olsson, A. Y., and C. S. Cooper. 2005. The molecular basis of prostate cancer. *Brit. J. Hosp. Med. (Lond.)* 66:612–616.

Ramaswamy, S., et al. 2003. A molecular signature of metastasis in primary solid tumors. *Nature Genet.* 33:49–54.

Thiery, J. P., H. Acloque, R. Y. Huang, and M. A. Nieto. 2009. Epithelial-mesenchymal transitions in development and disease. *Cell* 139(5):871–890.

Trusolino, L., and P. M. Comoglio. 2002. Scatter-factor and semaphorin receptors: cell signalling for invasive growth. *Nature Rev. Cancer* 2:289–300.

Vander Heiden, M. G., L. C. Cantley, and C. B. Thompson. 2009. Understanding the Warburg effect: the metabolic requirements of cell proliferation. *Science* 324(5930):1029–1033.

Williams, B. R., and A. Amon. 2009. Aneuploidy: cancer's fatal flaw? *Cancer Res.* 69(13):5289–5291.

Zakarija, A., and G. Soff. 2005. Update on angiogenesis inhibitors. *Curr. Opin. Oncol.* 17:578–583.

### The Genetic Basis of Cancer

Bertucci, F., et al. 2004. Gene expression profiling of colon cancer by DNA microarrays and correlation with histoclinical parameters. *Oncogene* 23:1377–1391.

Clark, J., et al. 2002. Identification of amplified and expressed genes in breast cancer by comparative hybridization onto microarrays of randomly selected cDNA clones. *Genes Chrom. Cancer* 34:104–114.

Chen, H. Z., S. Y. Tsai, and G. Leone. 2009. Emerging roles of E2Fs in cancer: an exit from cell cycle control. *Nat. Rev. Cancer* 9(11):785–797.

Fogarty, M. P., J. D. Kessler, and R. J. Wechsler-Reya. 2005. Morphing into cancer: the role of developmental signaling pathways in brain tumor formation. *J. Neurobiol.* 64:458–475.

Grisendi, S., and P. P. Pandolfi. 2005. Two decades of cancer genetics: from specificity to pleiotropic networks. *Cold Spring Harbor Symp. Quant. Biol.* 70:83–91.

Lau, J., H. Kawahira, and M. Hebrok. 2006. Hedgehog signaling in pancreas development and disease. *Cell Mol. Life Sci.* 63:642–652.

Pollack, J. R., et al. 2002. Microarray analysis reveals a direct role of DNA copy number alteration in the transcriptional program of human breast tumors. *Proc. Nat'l. Acad. Sci. USA* 99:12963–12968.

Sasaki, T., et al. 2000. Colorectal carcinomas in mice lacking the catalytic subunit of PI(3)Kgamma. *Nature* 406:897–902.

Scambia, G., S. Lovergine, and V. Masciullo. 2006. RB family members as predictive and prognostic factors in human cancer. *Oncogene* 25:5302–5308.

Sherr, C. J., and F. McCormick. 2002. The RB and p53 pathways in cancer. *Cancer Cell* 2:103–112.

van 't Veer, L. J., et al. 2002. Gene expression profiling predicts clinical outcome of breast cancer. *Nature* 415:530–536.

West, M., et al. 2001. Predicting the clinical status of human breast cancer by using gene expression profiles. *Proc. Nat'l. Acad. Sci. USA* 98:11462–11467.

### Cancer and the Misregulation of Growth Regulatory Pathways

Bachman, K. E., and B. H. Park. 2005. Dual nature of TGF-beta signaling: tumor suppressor vs. tumor promoter. *Curr. Opin. Oncol.* 17:49–54.

Beachy, P. A., S. S. Karhadkar, and D. M. Berman. 2004. Tissue repair and stem cell renewal in carcinogenesis. *Nature* 432:324–331.

Capdeville, R., et al. 2002. Glivec (STI571, imatinib), a rationally developed, targeted anticancer drug. *Nature Rev. Drug Discov.* 1:493–502.

Downward, J. 2003. Targeting RAS signalling pathways in cancer therapy. *Nature Rev. Cancer* 3:11–22.

Gregorieff, A., and H. Clevers. 2005. Wnt signaling in the intestinal epithelium: from endoderm to cancer. *Genes Dev.* **19**:877–890.

Heyer, J., et al. 2010. Non-germline genetically engineered mouse models for translational cancer research. *Nat. Rev. Cancer* 10:470–480.

Meenakshi, J., et al. 2002. Sustained loss of a neoplastic phenotype by brief inactivation of MYC. *Science* **297**:102–104.

Rowley, J. D. 2001. Chromosome translocations: dangerous liaisons revisited. *Nat. Rev. Cancer* **1**:245–250.

Sahai, E., and C. J. Marshall. 2002. RHO-GTPases and cancer. *Nat. Rev. Cancer* **2**:133–142.

Shaulian, E., and M. Karin. 2002. AP-1 as a regulator of cell life and death. *Nature Cell Biol.* **4**:E131–E136.

Shawver, L. K., D. Slamon, and A. Ullrich. 2002. Smart drugs: tyrosine kinase inhibitors in cancer therapy. *Cancer Cell* **1**:117–123.

### Cancer and Mutation of Cell Division and Checkpoint Regulators

Bardeesy, N., et al. 2006. Both p16(Ink4a) and the p19(Arf)-p53 pathway constrain progression of pancreatic adenocarcinoma in the mouse. *Proc. Nat'l. Acad. Sci. USA* 103:5947–5952.

Burkhart, D. L., and J. Sage. 2008. Cellular mechanisms of tumour suppression by the retinoblastoma gene. *Nat. Rev. Cancer* 8(9):671–682.

Croce, C. M. 2009. Causes and consequences of microRNA dysregulation in cancer. *Nat. Rev. Genet.* 10(10):704–714.

Jiang, J., and C. Hui. 2008. Hedgehog signaling in development and cancer. *Dev. Cell.* 15(6):801–812.

Lane, D. P. 2005. Exploiting the p53 pathway for the diagnosis and therapy of human cancer. *Cold Spring Harbor Symp. Quant. Biol.* **70**:489–497.

Malumbres, M., and M. Barbacid. 2001. To cycle or not to cycle: a critical decision in cancer. *Nat. Rev. Cancer* **1**:222–231.

Massagué, J. 2008. TGFβ in cancer. *Cell* **134**:215–229.

Mooi, W. J., and D. S. Peeper. 2006. Oncogene-induced cell senescence—halting on the road to cancer. *N. Engl. J. Med.* **355**:1037–1046.

Moon, R. T., et al. 2002. The promise and perils of Wnt signaling through beta-catenin. *Science* **296**:1644–1646.

Polakis, P. 2000. Wnt signaling and cancer. *Genes Devel.* **14**:1837–1851.

Zhang, L., et al. 2000. Role of BAX in the apoptotic response to anticancer agents. *Science* **290**:989–992.

### Carcinogens and Caretaker Genes in Cancer

Bailey, S. M., and J. P. Murnane. 2006. Telomeres, chromosome instability and cancer. *Nucl. Acids Res.* **34**:2408–2417.

Batty, D., and R. Wood. 2000. Damage recognition in nucleotide excision repair of DNA. *Gene* 241:193–204.

Blackburn, E. H. 2005. Telomerase and cancer: Kirk A. Landon—AACR prize for basic cancer research lecture. *Mol. Cancer Res.* 3:477–482.

Cleaver, J. E. 2005. Cancer in xeroderma pigmentosum and related disorders of DNA repair. *Nat. Rev. Cancer* 5:564–573.

D'Andrea, A., and M. Grompe. 2003. The Fanconi anemia/BRCA pathway. *Nat. Rev. Cancer* 3:23–34.

Flores-Rozas, H., and R. Kolodner. 2000. Links between replication, recombination and genome instability in eukaryotes. *Trends Biochem. Sci.* **25**:196–200.

Friedberg, E. 2003. DNA damage and repair. *Nature* **421**:436–440.

Hoeijmakers, J. 2001. Genome maintenance mechanisms for preventing cancer. *Nature* **411**:366–374.

Jiricny, J. 2006. The multifaceted mismatch-repair system. *Nature Rev. Mol. Cell Biol.* 7:335–346.

Ju, Z., and K. L. Rudolph. 2006. Telomeres and telomerase in cancer stem cells. *Eur. J. Cancer* 42:1197–1203.

Kitagawa, R., and M. B. Kastan. 2005. The ATM-dependent DNA damage signaling pathway. *Cold Spring Harbor Symp. Quant. Biol.* 70:99–109.

Loechler, E. L. 2002. Environmental carcinogens and mutagens. In *Encyclopedia of Life Sciences*. Nature Publishing.

Muller, A., and R. Fishel. 2002. Mismatch repair and the hereditary non-polyposis colorectal cancer syndrome (HNPCC). *Cancer Invest.* 20:102–109.

O'Driscoll, M., and P. A. Jeggo. 2006. The role of double-strand break repair—insights from human genetics. *Nature Rev. Genet.* 7:45–54.

Sahin, E., and R. A. Depinho. 2010. Linking functional decline of telomeres, mitochondria and stem cells during ageing. *Nature* 464(7288):520–528.

Schärer, O. 2003. Chemistry and biology of DNA repair. *Angewandte Chemie* 42:2946–2974.

Somasundaram, K. 2002. Breast cancer gene 1 (BRCA-1): role in cell cycle regulation and DNA repair—perhaps through transcription. *J. Cell Biochem.* 88:1084–1091.

Sweasy, J. B., T. Lang, and D. DiMaio. 2006. Is base excision repair a tumor suppressor mechanism? *Cell Cycle* 5:250–259.

Thompson, L., and D. Schild. 2002. Recombinational DNA repair and human disease. *Mut. Res.* **509**:49–78.

van Gant, D., J. Hoeijmakers, and R. Kanaar. 2001. Chromosomal stability and the DNA double-stranded break connection. *Nature Rev. Genet.* **2**:196–205.

Wogan, G. N., et al. 2004. Environmental and chemical carcinogenesis. *Semin. Cancer Biol.* 14:473–486.

# GLOSSARY

Boldface terms within a definition are also defined in this glossary. Figures and tables that illustrate defined terms are noted in parentheses.

**AAA ATPase family**   A group of proteins that couple hydrolysis of ATP with large molecular movements usually associated with unfolding of protein substrates or the disassembly of multisubunit protein complexes.

**ABC superfamily**   A large group of integral membrane proteins that often function as ATP-powered **membrane transport proteins** to move diverse molecules (e.g., phospholipids, cholesterol, sugars, ions, peptides) across cellular membranes. (Figure 11-15)

**acetylcholine (ACh)**   Neurotransmitter that functions at vertebrate neuromuscular junctions and at various neuron-neuron synapses in the brain and peripheral nervous system. (Figure 22-19)

**acetyl CoA**   Small, water-soluble metabolite comprising an acetyl group linked to coenzyme A (CoA). The acetyl group is transferred to citrate in the **citric acid cycle** and is used as a carbon source in the synthesis of fatty acids, steroids, and other molecules. (Figure 12-9)

**acid**   Any compound that can donate a proton ($H^+$). The carboxyl and phosphate groups are the primary acidic groups in biological macromolecules.

**actin**   Abundant structural protein in eukaryotic cells that interacts with many other proteins. The monomeric globular form (*G-actin*) polymerizes to form actin filaments (*F-actin*). In muscle cells, F-actin interacts with **myosin** during contraction. See also **microfilament**. (Figure 17-5)

**action potential**   Rapid, transient, all-or-none electrical activity propagated in the plasma membrane of excitable cells (e.g., neurons and muscle cells) as the result of the selective opening and closing of voltage-gated $Na^+$ and $K^+$ channels. (Figures 22-2 and 22-9)

**activation domain**   A region of an activator transcription factor that will stimulate transcription when fused to a DNA-binding domain.

**activation energy**   The input of energy required to (overcome the barrier to) initiate a chemical reaction. By reducing the activation energy, an **enzyme** increases the rate of a reaction. (Figure 2-30)

**activator**   Specific **transcription factor** that stimulates transcription.

**active site**   Specific region of an enzyme that binds a **substrate** molecule(s) and promotes a chemical change in the bound substrate. (Figure 3-21)

**active transport**   Protein-mediated movement of an ion or small molecule across a membrane against its concentration gradient or electrochemical gradient driven by the coupled hydrolysis of ATP. (Figure 11-2, [1]; Table 11-1)

**adapter proteins**   Adapter proteins physically link one protein to another protein by binding to both of them. Adapter proteins directly or indirectly (via additional adapters) connect cell-adhesion molecules or adhesion receptors to elements of the cytoskeleton or to intracellular signaling proteins.

**adenosine triphosphate (ATP)**   See **ATP**.

**adenylyl cyclase**   One of several enzymes that is activated by binding of certain ligands to their cell-surface receptors and catalyzes formation of **cyclic AMP (cAMP)** from ATP; also called *adenylate cyclase*. (Figures 15-27 and 15-28)

**adhesion receptor**   Protein in the plasma membrane of animal cells that binds components of the **extracellular matrix**, thereby mediating cell-matrix adhesion. The **integrins** are major adhesion receptors. (Figure 20-1, [5])

**ADP (adenosine diphosphate)**   The product, along with inorganic phosphate, of ATP hydrolysis by ATPases.

**aerobic**   Referring to a cell, organism, or metabolic process that utilizes gaseous oxygen ($O_2$) or that can grow in the presence of $O_2$.

**aerobic oxidation**   Oxygen-requiring metabolism of sugars and fatty acids to $CO_2$ and $H_2O$ coupled to the synthesis of ATP.

**aerobic respiration**   See **aerobic oxidation**.

**afferent neurons**   Nerves that transmit signals from peripheral tissues to the central nervous system.

**agonist**   A molecule, often synthetic, that mimics the biological function of a natural molecule (e.g., a hormone).

**Agrin**   A glycoprotein synthesized by developing motor neurons that increases MuSK kinase activity in a muscle cell, facilitating development of a neuromuscular junction. (Figure 22-17)

**Akt**   A cytosolic serine/threonine kinase that is activated following binding to PI 3,4-bisphosphate and PI 3,4,5-trisphosphate; also called *protein kinase B*.

**allele**   One of two or more alternative forms of a gene. Diploid cells contain two alleles of each gene, located at the corresponding site (locus) on **homologous chromosomes**.

**allosteric**   Referring to proteins and cellular processes that are regulated by **allostery**.

**allostery**   Change in the tertiary and/or quaternary structure of a protein induced by binding of a small molecule to a specific regulatory site, causing a change in the protein's activity.

**alpha (α)   carbon atom (Cα)**   In amino acids, the central carbon atom that is bonded to four different chemical groups (except in glycine) including the **side chain**, or R group. (Figure 2-4)

**alpha (α) helix** Common protein **secondary structure** in which the linear sequence of amino acids is folded into a right-handed spiral stabilized by hydrogen bonds between carboxyl and amide groups in the backbone. (Figure 3-4)

**alternative splicing** Process by which the exons of one pre-mRNA are spliced together in different combinations, generating two or more different mature mRNAs from a single pre-mRNA. (Figure 4-16)

**amino acid** An organic compound containing at least one amino group and one carboxyl group. In the amino acids that are the **monomers** for building proteins, an amino group and carboxyl group are linked to a central carbon atom, the α carbon, to which a variable side chain is attached. (Figures 2-4 and 2-14)

**aminoacyl-tRNA** Activated form of an amino acid, used in protein synthesis, consisting of an amino acid linked via a high-energy ester bond to the 3′-hydroxyl group of a **tRNA** molecule. (Figure 4-19)

**amphipathic** Referring to a molecule or structure that has both a **hydrophobic** and a **hydrophilic** part.

**amphitelic attachment** Describes the correct attachment of chromosomes to the mitotic spindle, where sister kinetochores attach to microtubules emanating from opposite poles. (Figure 19-25)

**anaerobic** Referring to a cell, organism, or metabolic process that functions in the absence of gaseous oxygen ($O_2$).

**anaphase** Mitotic stage during which the sister **chromatids** (or duplicated homologs in meiosis I) separate and move apart (segregate) toward the spindle poles. (Figure 18-36)

**anchoring junctions** Specialized regions on the cell surface containing **cell-adhesion molecules** or **adhesion receptors**; includes *adherens junctions* and *desmosomes*, which mediate cell-cell adhesion, and *hemidesmosomes*, which mediate cell-matrix adhesion. (Figures 20-13 and 20-15)

**anaerobic respiration** Respiration in which molecules other than oxygen, such as sulfate or nitrate, are used as the final recipient of the electrons transported via the electron-transport chain.

**anaphase-promoting complex or cyclosome (APC/C)** A ubiquitin ligase that targets securin, mitotic cyclins, and other proteins for proteasomal degradation from the onset of anaphase until entry into the subsequent cell cycle.

**aneuploidy** Any deviation from the normal **diploid** number of chromosomes in which extra copies of one or more chromosomes are present or one of the normal copies is missing.

**anion** A negatively charged ion.

**antagonist** A molecule, often synthetic, that blocks the biological function of a natural molecule (e.g., hormone).

**antibody** A protein (immunoglobulin), normally produced in response to an **antigen**, that interacts with a particular site (**epitope**) on the same antigen and facilitates its clearance from the body. (Figure 3-19)

**anticodon** Sequence of three nucleotides in a tRNA that is complementary to a **codon** in an mRNA. During protein synthesis, base pairing between a codon and anticodon aligns the tRNA carrying the corresponding amino acid for addition to the growing polypeptide chain. (Figure 4-20)

**antigen** Any material (usually foreign) that elicits an immune response. For B cells, an antigen elicits formation of antibody that specifically binds the same antigen; for T cells, an antigen elicits a proliferative response, followed by production of **cytokines** or the activation of cytotoxic activity.

**antigen-presenting cell (APC)** Any cell that can digest an antigen into small peptides and display the peptides in association with class II MHC molecules on the cell surface where they can be recognized by T cells. *Professional* APCs (dendritic cells, macrophages, and B cells) constitutively express class II MHC molecules. (Figures 23-25 and 23-26)

**antiport** A type of **cotransport** in which a membrane protein (*antiporter*) transports two different molecules or ions across a cell membrane in opposite directions. See also **symport**. (Figure 11-2, [3C])

**apical** Referring to the tip (apex) of a cell, an organ, or other body structure. In the case of epithelial cells, the apical surface is exposed to the exterior of the body or to an internal open space (e.g., intestinal lumen, duct). (Figure 20-9)

**apoptosis** A genetically regulated process, occurring in specific tissues during development and disease, by which a cell destroys itself; marked by the breakdown of most cell components and a series of well-defined morphological changes; also called *programmed cel death*. See also **caspases**. (Figures 21-30, and 21-38)

**apoptosome** Large, disk-shaped heptamer of mammalian Apaf-1, a protein that assembles in response to apoptosis signals and serves as an activation machine for initiator and effector **caspases**. (Figure 21-39)

**aquaporins** A family of **membrane transport proteins** that allow water and a few other small uncharged molecules, such as glycerol, to cross biomembranes. (Figure 11-8)

**archaea** Class of **prokaryotes** that constitutes one of the three distinct evolutionary lineages of modern-day organisms; also called *archaebacteria* and *archaeans*. In some respects, archaeans are more similar to **eukaryotes** than to **bacteria** (eubacteria). (Figure 1-1)

**associated constant ($K_a$)** See **equilibrium constant**.

**aster** Structure composed of microtubules (astral fibers) that radiate outward from a **centrosome** during mitosis. (Figure 18-37)

**astrocytes** Star-shaped glial cells in the brain and spinal cord that perform many functions, including support of endothelial cells that form the blood-brain barrier, maintain extracellular ion composition, and provide nutrients to neurons.

**asymmetric carbon atom** A carbon atom bonded to four different atoms or chemical groups; also called *chiral carbon atom*. The bonds can be arranged in two different ways, producing **stereoisomers** that are mirror images of each other. (Figure 2-4)

**asymmetric cell division** Any cell division in which the two daughter cells receive the same genes but otherwise inherit different components (e.g., mRNAs, proteins) from the parental cell. (Figure 21-21a)

**ATM/ATR** Two related proteins kinases that are activated by DNA damage. Once active, they phosphorylate other proteins to initiate the cell's response to DNA damage.

**ATP (adenosine 5′-triphosphate)** A nucleotide that is the most important molecule for capturing and transferring **free energy** in cells. Hydrolysis of each of the two **phosphoanhydride bonds** in ATP releases a large amount of free energy that can be used to drive energy-requiring cellular processes. (Figure 2-31)

**ATPase** One of a large group of enzymes that catalyze hydrolysis of **ATP** to yield ADP and inorganic phosphate with release of free energy. See also **Na⁺/K⁺ ATPase** and **ATP-powered pump.**

**ATP-powered pump** Any transmembrane protein that has ATPase activity and couples hydrolysis of ATP to the active transport of an ion or small molecule across a biomembrane against its electrochemical gradient; often simply called *pump*. (Figure 11-9)

**ATP synthase** Multimeric protein complex, bound to inner mitochondrial membranes, thylakoid membranes of chloroplasts, and the bacterial plasma membrane, that catalyzes synthesis of ATP during oxidative phosphorylation and photosynthesis; also called $F_0F_1$ *complex*. (Figure 12-26a)

**ATR** See **ATM/ATR.**

**Aurora B kinase** Destabilizes faulty microtubule-kinetochore interactions by phosphorylating microtubule-binding components within the kinetochore.

**Aurora kinases** Serine/threonine kinases that play a crucial role in cell division by controlling chromatid segregation. **Aurora B kinase** destabilizes faulty microtubule-kinetochore interactions by phosphorylating microtubule-binding components within the kinetochore.

**autocrine** Referring to signaling mechanism in which a cell produces a signaling molecule (e.g., growth factor) and then binds and responds to it.

**autophagy** Literally, "eating oneself"; the process by which cytosolic proteins and organelles are delivered to the lysosome, degraded, and recycled. Autophagy involves the formation of a double-membrane vesicle called an autophagosome or autophagic vesicle. (Figure 14-35)

**autoradiography** Technique for visualizing radioactive molecules in a sample (e.g., a tissue section or electrophoretic gel) by exposing a photographic film (emulsion) or two-dimensional electronic detector to the sample. The exposed film is called an *autoradiogram* or *autoradiograph*.

**autosome** Any chromosome other than a sex chromosome.

**axon** Long process extending from the cell body of a neuron that is capable of conducting an electric impulse (**action potential**), generated at the junction with the cell body, toward its distal, branching end (the axon terminus). (Figure 22-1)

**axonal transport** Motor protein–mediated transport of organelles and vesicles along microtubules in axons of nerve cells. *Anterograde* transport occurs from cell body toward axon terminal); *retrograde* transport, from axon terminal toward cell body. (Figures 18-16 and 18-17)

**axoneme** Bundle of **microtubules** and associated proteins present in **cilia** and **flagella** and responsible for their structure and movement. (Figure 18-30)

**bacteria** Class of **prokaryotes** that constitutes one of the three distinct evolutionary lineages of modern-day organisms; also called *eubacteria*. Phylogenetically distinct from **archaea** and **eukaryotes.** (Figure 1-1)

**bacteriophage (phage)** Any virus that infects bacterial cells. Some phages are widely used as **vectors** in **DNA cloning.**

**basal** See **basolateral.**

**basal body** Structure at the base of a **cilium** or **flagellum** from which microtubules forming the **axoneme** assemble; structurally similar to a **centriole.** (Figure 18-30)

**basal lamina (pl. basal laminae)** A thin sheet-like network of extracellular-matrix components that underlies most animal epithelia and other organized groups of cells (e.g., muscle), separating them from connective tissue or other cells. (Figures 20-21 and 20-22)

**base** Any compound, often containing nitrogen, that can accept a proton (H⁺) from an acid. Also, commonly used to denote the **purines** and **pyrimidines** in DNA and RNA.

**base pair** Association of two complementary **nucleotides** in a DNA or RNA molecule stabilized by hydrogen bonding between their base components. Adenine pairs with thymine or uracil (A · T, A · U) and guanine pairs with cytosine (G · C). (Figure 4-3b)

**basic helix-loop-helix** See **helix-loop-helix, basic.**

**basolateral** Referring to the base (basal) and side (lateral) of a polarized cell, organ, or other body structure. In the case of epithelial cells, the basolateral surface abuts adjacent cells and the underlying **basal lamina.** (Figure 20-9)

**B cell** A lymphocyte that matures in the bone marrow and expresses antigen-specific receptors (membrane-bound **immunoglobulin**). After interacting with antigen, a B cell proliferates and differentiates into **antibody**-secreting *plasma cells*.

**B-cell receptor** Complex composed of an antigen-specific membrane-bound immunoglobulin molecule and associated signal-transducing Igα and Igβ chains. (Figure 23-16)

**benign** Referring to a tumor containing cells that closely resemble normal cells. Benign tumors stay in the tissue where they originate but can be harmful due to continued growth. See also **malignant.**

**beta (β)-adrenergic receptors** Seven spanning G protein–coupled receptors that bind adrenaline and related molecules, leading to activation of adenylyl cyclase.

**beta (β) sheet** A flat **secondary structure** in proteins that is created by hydrogen bonding between the backbone atoms in two different polypeptide chains or segments of a single folded chain. (Figure 3-5)

**beta (β) turn** A short U-shaped **secondary structure** in proteins. (Figure 3-6)

**bi-oriented** Indicates that the kinetochores of **sister chromatids** have attached to microtubules emanating from opposite spindle poles.

**blastocyst** Stage of mammalian embryo composed of ≈64 cells that have separated into two cell types—**trophectoderm**, which will form extra-embryonic tissues, and the **inner cell mass**, which gives rise to the embryo proper; stage that implants in the uterine wall and corresponds to the *blastula* of other animal embryos. (Figure 21-3)

**blastopore** The first opening that forms during embryogenesis of bilaterally symmetric animals, which later becomes the gut. This opening may become either the mouth or the anus.

**buffer** A solution of the acid (HA) and base ($A^-$) form of a compound that undergoes little change in pH when small quantities of strong acid or base are added at pH values near the compound's $pK_a$.

**cadherins** A family of dimeric **cell-adhesion molecules** that aggregate in adherens junctions and desmosomes and mediate $Ca^{2+}$-dependent cell-cell homophilic interactions. (Figure 20-2)

**calmodulin** A small cytosolic regulatory protein that binds four $Ca^{2+}$ ions. The $Ca^{2+}$/calmodulin complex binds to many proteins, thereby activating or inhibiting them. (Figure 3-31)

**Calvin cycle** The major metabolic pathway that fixes $CO_2$ into carbohydrates during photosynthesis; also called *carbon fixation*. It is indirectly dependent on light but can occur both in the dark and light. (Figure 12-46)

**cancer** General term denoting any of various malignant tumors, whose cells grow and divide more rapidly than normal, invade surrounding tissue, and sometimes spread (metastasize) to other sites.

**capsid** The outer proteinaceous coat of a **virus**, formed by multiple copies of one or more protein subunits and enclosing the viral nucleic acid.

**CAP site** The DNA sequence in bacteria bound by catabolite activator protein, also known as the cyclic AMP regulatory protein. (Figure 7-3)

**carbohydrate** General term for certain polyhydroxyaldehydes, polyhydroxyketones, or compounds derived from these usually having the formula $(CH_2O)_n$. Primary type of compound used for storing and supplying energy in animal cells. (Figure 2-18)

**carbon fixation** See **Calvin cycle**.

**carcinogen** Any chemical or physical agent that can cause cancer when cells or organisms are exposed to it.

**caretaker gene** Any gene whose encoded protein helps protect the integrity of the genome by participating in the repair of damaged DNA. Loss of function of a caretaker gene leads to increased mutation rates and promotes carcinogenesis.

**caspases** A class of vertebrate protein-degrading enzymes (proteases) that function in **apoptosis** and work in a cascade with each type activating the next. (Figures 21-33 and 21-38)

**catabolism** Cellular degradation of complex molecules to simpler ones usually accompanied by the release of energy. *Anabolism* is the reverse process in which energy is used to synthesize complex molecules from simpler ones.

**catalyst** A substance that increases the rate of a chemical reaction without undergoing a permanent change in its structure. Enzymes are proteins with catalytic activity, and ribozymes are RNAs that can function as catalysts. (Figure 3-20)

**cation** A positively charged ion.

**Cdc25 phosphatases** A protein phosphatase that dephosphorylates CDKs on threonine 14 and tyrosine 15, thereby activating CDKs.

**CDK-activating kinase (CAK)** Phosphorylates CDKs on a threonine residue near the active site. This phosphorylation is essential for CDK activity.

**CDK inhibitor (CKI)** Binds to cyclin-CDK complex and inhibits its activity.

**cDNA (complementary DNA)** DNA molecule copied from an mRNA molecule by **reverse transcriptase** and therefore lacking the **introns** present in the DNA of the genome.

**cell-adhesion molecules (CAMs)** Proteins in the plasma membrane of cells that bind similar proteins on other cells, thereby mediating cell-cell adhesion. Four major classes of CAMs include the **cadherins, IgCAMs, integrins,** and **selectins.** (Figures 20-1 and 20-2)

**cell-adhesion proteins** See **cell adhesion molecules (CAMs)**.

**cell cycle** Ordered sequence of events in which a eukaryotic cell duplicates its chromosomes and divides into two. The cell cycle normally consists of four phases: $G_1$ before DNA synthesis occurs; S when DNA replication occurs; $G_2$ after DNA synthesis; and M when **cell division** occurs, yielding two daughter cells. Under certain conditions, cells exit the cell cycle during $G_1$ and remain in the $G_0$ state as nondividing cells. (Figures 1-16 and 19-1)

**cell division** Separation of a cell into two daughter cells. In higher eukaryotes, it involves division of the nucleus (**mitosis**) and of the cytoplasm (**cytokinesis**); mitosis often is used to refer to both nuclear and cytoplasmic division.

**cell junctions** Specialized regions on the cell surface through which cells are joined to each other or to the extracellular matrix. (Figure 20-10; Table 20-2)

**cell line** A population of cultured cells, of plant or animal origin, that has undergone a genetic change allowing the cells to grow indefinitely. (Figure 9-1b)

**cell polarity** The ability of cells to organize their internal structure, resulting in changes of cell shape and generating regions of the plasma membrane with different protein and lipid compositions.

**cell strain** A population of cultured cells, of plant or animal origin, that has a finite life span and eventually dies, commonly after 25–50 generations. (Figure 9-1a)

**cellular respiration** See **respiration**.

**cellulose** A structural polysaccharide made of glucose units linked together by $\beta(1 \rightarrow 4)$ **glycosidic bonds.** It forms long microfibrils, which are the major component of the **cell wall** in plants.

**cell wall** A specialized, rigid extracellular matrix that lies next to the plasma membrane, protecting a cell and maintaining its shape; prominent in most fungi, plants, and prokaryotes but absent in most multicellular animals. (Figure 20-40)

**centriole** Either of two cylindrical structures within the **centrosome** of animal cells and containing nine sets of triplet microtubules; structurally similar to a **basal body**. (Figure 18-6)

**centromere** DNA sequence required for proper **segregation** of chromosomes during mitosis and meiosis; the region of mitotic chromosomes where the **kinetochore** forms and that appears constricted. (Figure 6-39)

**centrosome (cell center)** Structure located near the nucleus of animal cells that is the primary **microtubule-organizing center (MTOC)**; it contains a pair of **centrioles** embedded in a protein matrix and duplicates before mitosis, with each centrosome becoming a spindle pole. (Figures 18-6 and 18-35)

**centrosome disjunction**   Describes the process of centrosome segregation during prophase.

**channels**   Membrane proteins that transport water, ions, or small hydrophilic molecules across membranes down their concentration or electric potential gradients.

**chaperone**   Collective term for two types of proteins—*molecular chaperones* and *chaperonins*—that prevent misfolding of a target protein or actively facilitate proper folding of an incompletely folded target protein, respectively. (Figures 3-16 and 3-17)

**chaperonin**   See **chaperone**.

**checkpoint**   Any of several points in the eukaryotic **cell cycle** at which progression of a cell to the next stage can be halted until conditions are suitable.

**checkpoint pathway**   Surveillance mechanism that prevents initiation of each step in cell division until earlier steps on which it depends have been completed and mistakes that occurred during the process have been corrected.

**chemical equilibrium**   The state of a chemical reaction in which the concentration of all products and reactants is constant because the rates of the forward and reverse reactions are equal.

**chemical potential energy**   The energy stored in the bonds connecting atoms in molecules.

**chemiosmosis**   Process whereby an electrochemical proton gradient (pH plus electric potential) across a membrane is used to drive an energy-requiring process such as ATP synthesis; also called *chemiosmotic coupling*. See **proton-motive force**. (Figure 12-2)

**chemokine**   Any of numerous small, secreted proteins that function as chemotatic cues for leukocytes.

**chemotaxis**   Movement of a cell or organism toward or away from certain chemicals.

**chlorophylls**   A group of light-absorbing porphyrin pigments that are critical in **photosynthesis**. (Figure 12-33)

**chloroplast**   A specialized organelle in plant cells that is surrounded by a double membrane and contains internal chlorophyll-containing membranes (**thylakoids**) where the light-absorbing reactions of photosynthesis occur. (Figure 12-31)

**cholesterol**   A lipid containing the four-ring steroid structure with a hydroxyl group on one ring; a component of many eukaryotic membranes and the precursor of steroid hormones, bile acids, and vitamin D. (Figure 10-8c)

**chromatid**   One copy of a replicated chromosome, formed during the S phase of the cell cycle, that is joined to the other copy; also called **sister chromatid**. During mitosis, the two chromatids separate, each becoming a chromosome of one of the two daughter cells. (Figure 6-39)

**chromatin**   Complex of DNA, histones, and nonhistone proteins from which eukaryotic chromosomes are formed. Condensation of chromatin during mitosis yields the visible **metaphase** chromosomes. (Figures 6-28 and 6-30)

**chromatography, liquid**   Group of biochemical techniques for separating mixtures of molecules (e.g., different proteins) based on their mass (*gel filtration* chromatography), charge (*ion exchange* chromatography), or ability to bind specifically to other molecules (*affinity* chromatography). (Figure 3-38)

**chromosome**   In eukaryotes, the structural unit of the genetic material consisting of a single, linear double-stranded DNA molecule and associated proteins. In most prokaryotes, a single, circular double-stranded DNA molecule constitutes the bulk of the genetic material. See also **chromatin** and **karyotype**.

**cilium** (pl. **cilia**)   Short, membrane-enclosed structure extending from the surface of eukaryotic cells and containing a core bundle of **microtubules**. Cilia usually occur in groups and beat rhythmically to move a cell (e.g., single-celled organism) or to move small particles or fluid along a surface (e.g., trachea cells). See also **axoneme** and **flagellum**.

**cisterna** (pl. **cisternae**)   Flattened membrane-bounded compartment, as found in the Golgi complex and endoplasmic reticulum.

**citric acid cycle**   A set of nine coupled reactions occurring in the matrix of the **mitochondrion** in which acetyl groups are oxidized, generating $CO_2$ and reduced intermediates used to produce ATP; also called *Krebs cycle* and *tricarboxylic acid (TCA) cycle*. (Figure 12-10)

**clathrin**   A fibrous protein that with the aid of assembly proteins polymerizes into a lattice-like network at specific regions on the cytosolic side of a membrane, thereby forming a clathrin-coated pit that buds off to form a vesicle. (Figure 14-18; Table 14-1)

**cleavage**   In embryogenesis, the series of rapid cell divisions that occurs following fertilization and with little cell growth, producing progressively smaller cells; culminates in formation of the blastocyst in mammals or blastula in other animals. Also used as a synonym for the hydrolysis of molecules. (Figure 21-3)

**cleavage furrow**   Indentation in the plasma membrane that represents the initial steps in cytokinesis.

**cleavage/polyadenylation complex**   Large, multiprotein complex that catalyzes the cleavage of **pre-mRNA** at a 3′ poly(A) site and the initial addition of adenylate (A) residues to form the poly(A) tail. (Figure 8-15)

**clone**   (1) A population of genetically identical cells, viruses, or organisms descended from a common ancestor. (2) Multiple identical copies of a gene or DNA fragment generated and maintained via **DNA cloning**.

**codon**   Sequence of three nucleotides in DNA or mRNA that specifies a particular amino acid during protein synthesis; also called *triplet*. Of the 64 possible codons, three are stop codons, which do not specify amino acids and cause termination of synthesis. (Table 4-1)

**cohesin complex**   Protein complex that establishes cohesion between **sister chromatids**.

**coiled coil**   A protein **structural motif** marked by amphipathic α helical regions that can self-associate to form stable, rodlike structures in proteins; commonly found in fibrous proteins and certain transcription factors. (Figure 3-9a)

**collagen**   A triple-helical glycoprotein rich in glycine and praline that is a major component of the **extracellular matrix** and connective tissues. The numerous subtypes differ in their tissue distribution and the extracellular components and cell-surface proteins with which they associate. (Figure 20-24; Table 20-4)

**complement**   A group of constitutive serum proteins that bind directly to microbial or fungal surfaces, thereby activating a proteolytic cascade that culminates in formation of the cytolytic *membrane attack complex*. (Figure 23-4)

**complementary** (1) Referring to two nucleic acid sequences or strands that can form perfect **base pairs** with each other. (2) Describing regions on two interacting molecules (e.g., an enzyme and its substrate) that fit together in a lock-and-key fashion.

**complementary DNA (cDNA)** See **cDNA**.

**complementation** See **genetic complementation** and **functional complementation**.

**concentration gradient** In cell biology, a difference in the concentration of a substance in different regions of a cell or embryo or on different sides of a cellular membrane.

**condensin complex** Protein complex related to cohesins that compacts chromosomes and is necessary for their segregation during mitosis.

**conformation** The precise shape of a protein or other macromolecule in three dimensions resulting from the spatial location of the atoms in the molecule. (Figure 3-8)

**connexins** A family of transmembrane proteins that form **gap junctions** in vertebrates. (Figure 20-20)

**constitutive** Referring to the continuous production or activity of a cellular molecule or the continuous operation of a cellular process (e.g., constitutive secretion) that is not regulated by internal or external signals.

**contractile bundles** Bundles of **actin** and **myosin** in nonmuscle cells that function in cell adhesion (e.g., *stress fibers*) or cell movement (*contractile ring* in dividing cells).

**contractile ring** Composed of actin and myosin; located beneath the plasma membrane. During cytokinesis, its contraction pulls the membrane inward, eventually closing the connection between the two daughter cells.

**contractile vacuole** A vesicle found in many protozoans that takes up water from the cytosol and periodically discharges its contents through fusion with the plasma membrane.

**coordinately regulated** Genes whose expression are induced and repressed at the same time, as for the genes in a single bacterial operon. (Figure 4-13)

**co-oriented** Indicates that sister kinetochores attach to microtubules emanating from the *same* spindle pole rather than from opposite spindle poles.

**COPI** A class of proteins that coat transport vesicles in the **secretory pathway**. COPI-coated vesicles move proteins from the Golgi to the endoplasmic reticulum and from later to earlier Golgi cisternae. (Table 14-1)

**COPII** A class of proteins that coat transport vesicles in the **secretory pathway**. COPII-coated vesicles move proteins from the endoplasmic reticulum to the Golgi. (Table 14-1)

**cotranslational translocation** Simultaneous transport of a secretory protein into the endoplasmic reticulum as the nascent protein is still bound to the ribosome and being elongated. (Figure 13-6)

**cotransport** Protein-mediated movement of an ion or small molecule across a membrane against a concentration gradient driven by coupling to movement of a second molecule down its concentration gradient in the same (**symport**) or opposite (**antiport**) direction. (Figure 11-2, [3B, C]; Table 11-1)

**covalent bond** Stable chemical force that holds the atoms in molecules together by sharing of one or more pairs of electrons. See also **noncovalent interaction**. (Figures 2-2 and 2-6)

**CpG islands** Regions in vertebrate DNA ~100 to ~1000 bp that have an unusually high occurrence of the sequence CG. Many CpG islands function as promoters for transcription initiation, usually in both directions.

**critical cell size** Defines the size that a cell must reach before it can enter the cell cycle.

**cross-exon recognition complex** Large assembly including RNA-binding SR proteins and other components that helps delineate exons in the **pre-mRNAs** of higher eukaryotes and ensure correct **RNA splicing**. (Figure 8-13)

**crossing over** Exchange of genetic material between maternal and paternal **chromatids** during **meiosis** to produce recombined chromosomes. See also **recombination**. (Figure 5-10)

**cyclic AMP (cAMP)** A **second messenger**, produced in response to hormonal stimulation of certain G protein–coupled receptors, that activates **protein kinase A**. (Figure 15-8; Table 15-2)

**cyclic GMP (cGMP)** A **second messenger** that opens cation channels in rod cells and activates protein kinase G in vascular smooth muscle and other cells. (Figures 15-8, 15-23, and 15-37)

**cyclin** Any of several related proteins whose concentrations rise and fall during the course of the eukaryotic cell cycle. Cyclins form complexes with **cyclin-dependent kinases**, thereby activating and determining the substrate specificity of these enzymes.

**cyclin-dependent kinase (CDK)** A protein kinase that is catalytically active only when bound to a cyclin. Various cyclin-CDK complexes trigger progression through different stages of the eukaryotic cell cycle by phosphorylating specific target proteins. (Table 19-1)

**cytochromes** A group of colored, heme-containing proteins, some of which function as **electron carriers** during cellular respiration and photosynthesis. (Figure 12-14a)

**cytokine** Any of numerous small, secreted proteins (e.g., erythropoietin, G-CSF, interferons, interleukins) that bind to cell-surface receptors on blood and immune-system cells to trigger their differentiation or proliferation.

**cytokine receptor** Member of major class of cell-surface signaling receptors, including those for erythropoietin, growth hormone, interleukins, and interferons. Ligand binding leads to activation of cytosolic JAK kinases associated with the receptor, thereby initiating intracellular signaling pathways. (Figures 16-2 and 16-13)

**cytokinesis** The division of the cytoplasm following **mitosis** to generate two daughter cells, each with a nucleus and cytoplasmic organelles. (Figure 17-36)

**cytoplasm** Viscous contents of a cell that are contained within the plasma membrane but, in eukaryotic cells, outside the nucleus.

**cytoskeleton** Network of fibrous elements, consisting primarily of **microtubules**, **microfilaments**, and **intermediate filaments**, found in the cytoplasm of eukaryotic cells. The cytoskeleton provides organization and structural support for the cell and permits directed movement of organelles, chromosomes, and the cell itself. (Figures 17-1, 17-2, and 18-1)

**cytoskeletal proteins**  See **cytoskeleton**.

**cytosol**  Unstructured aqueous phase of the cytoplasm excluding organelles, membranes, and insoluble cytoskeletal components.

**cytosolic face**  The face of a cell membrane directed toward the cytosol. (Figure 10-5)

**DAG**  See **diacylglycerol**.

**dalton**  Unit of molecular mass approximately equal to the mass of a hydrogen atom ($1.66 \times 10^{-24}$ g).

**denaturation**  Drastic alteration in the **conformation** of a protein or nucleic acid due to disruption of various noncovalent interactions caused by heating or exposure to certain chemicals; usually results in loss of biological function.

**dendrite**  Process extending from the cell body of a neuron that is relatively short and typically branched and receives signals from **axons** of other neurons. (Figure 22-1)

**dendritic cells**  Phagocytic professional **antigen-presenting cells** that reside in various tissues and can detect broad patterns of pathogen markers via their **Toll-like receptors**. After internalizing antigen at a site of tissue injury or infection, they migrate to lymph nodes and initiate activation of **T cells**. (Figure 23-6)

**deoxyribonucleic acid**  See **DNA**.

**depolarization**  Decrease in the cytosolic-face negative electric potential that normally exists across the plasma membrane of a cell at rest, resulting in a less inside-negative or an inside-positive **membrane potential**.

**destruction box**  Recognition motif in APC/C substrates.

**determinant**  In the context of antibody recognition of an antigen, a region on a protein to which the antibody binds. In this context, it is synonymous with **epitope**.

**deuterosomes**  A group of bilaterally symmetric animals whose anus develops close to the blastopore and has a dorsal nerve cord. This group includes all chordates (fish, amphibians, reptiles, birds, and mammals) and echinoderms (sea stars, sea urchins).

**diacylglycerol (DAG)**  Membrane-bound **second messenger** that can be produced by cleavage of **phosphoinositides** in response to stimulation of certain cell-surface receptors. (Figures 15-8 and 15-35)

**diploid**  Referring to an organism or cell having two full sets of **homologous chromosomes** and hence two copies (**alleles**) of each gene or genetic locus. Somatic cells contain the diploid number of chromosomes ($2n$) characteristic of a species. See also **haploid**.

**dipole**  A positive charge separated in space from an equal but opposite negative charge.

**dipole moment**  A quantitative measure of the extent of charge separation, or strength, of a dipole, which for a chemical bond is the product of the partial charge on each atom and the distance between the two atoms.

**disaccharide**  A small carbohydrate (sugar) composed of two monosaccharides covalently joined by a **glycosidic bond**. (Figure 2-19)

**dissociation constant ($K_d$)**  See **equilibrium constant**.

**disulfide bond (—S—S—)**  A common covalent linkage between the sulfur atoms on two cysteine residues in different polypeptides or in different parts of the same polypeptide.

**diversity**  The entire set of antigen-specific receptors encoded by an immune system.

**DNA (deoxyribonucleic acid)**  Long linear polymer, composed of four kinds of deoxyribose **nucleotides**, that is the carrier of genetic information. See also **double helix, DNA**. (Figure 4-3)

**DNA-binding domain**  The domain of a transcription factor that binds specific, closely related DNA sequences.

**DNA cloning**  Recombinant DNA technique in which specific cDNAs or fragments of genomic DNA are inserted into a cloning **vector**, which then is incorporated into cultured host cells and maintained during growth of the host cells; also called *gene cloning*. (Figure 5-14)

**DNA damage response system**  Pathway that senses DNA damage and induces cell cycle arrest and DNA repair pathways.

**DNA library**  Collection of cloned DNA molecules consisting of fragments of the entire genome (*genomic library*) or of DNA copies of all the mRNAs produced by a cell type (*cDNA library*) inserted into a suitable cloning **vector**.

**DNA ligase**  An enzyme that links together the 3′ end of one DNA fragment with the 5′ end of another, forming a continuous strand.

**DNA microarray**  An ordered set of thousands of different nucleotide sequences arrayed on a microscope slide or other solid surface; can be used to determine patterns of gene expression in different cell types or in a particular cell type at different developmental stages or under different conditions. (Figures 5-29 and 5-30)

**DNA polymerase**  An enzyme that copies one strand of DNA (the template strand) to make the complementary strand, forming a new double-stranded DNA molecule. All DNA polymerases add deoxyribonucleotides one at a time in the 5′ → 3′ direction to the 3′ end of a short preexisting primer strand of DNA or RNA.

**DNA recombination**  The process by which two DNA molecules with similar sequences are subject to double-stranded breaks and then rejoined to generate two recombinant DNA molecules with sequences comprised of portions of each parent. (Figures 4-42 and 4-43)

**dominant**  In genetics, referring to that allele of a gene expressed in the **phenotype** of a heterozygote; the nonexpressed allele is **recessive**; also refers to the phenotype associated with a dominant allele. Mutations that produce dominant alleles generally result in a gain of function. (Figure 5-2)

**dominant-negative**  In genetics, an allele that acts in a **dominant** manner but produces an effect similar to a loss of function; generally is an allele encoding a mutant protein that blocks the function of the normal protein by binding either to it or to a protein **upstream** or **downstream** of it in a pathway.

**double helix, DNA**  The most common three-dimensional structure for cellular DNA in which the two polynucleotide strands are antiparallel and wound around each other with complementary bases hydrogen-bonded. (Figure 4-3)

**double-strand break**  Form of DNA damage where both phosphate-sugar backbones of the DNA are severed.

**downstream** (1) For a gene, the direction RNA polymerase moves during transcription, which is toward the end of the template DNA strand with a 5'-hydroxyl group. Nucleotides downstream from the +1 position (the first transcribed a nucleotide) are designated +2, +3, etc. (2) Events that occur later in a cascade of steps (e.g., signaling pathway). See also **upstream**.

**dyneins** A class of **motor proteins** that use the energy released by ATP hydrolysis to move toward the (−) end of **microtubules**. Dyneins can transport vesicles and organelles, are responsible for the movement of cilia and flagella, and play a role in chromosome movement during mitosis. (Figures 18-23 and 18-24)

**ectoderm** Outermost of the three primary cell layers of the animal embryo; gives rise to epidermal tissues, the nervous system, and external sense organs. See also **endoderm** and **mesoderm**.

**effector** Ultimate component of a signal transduction pathway that elicits a response to the transmitted signal.

**endoderm** Innermost of the three primary cell layers of the animal embryo; gives rise to the gut and most of the respiratory tract. See **ectoderm** and **mesoderm**.

**EF hand** A type of helix-loop-helix **structural motif** that occurs in many Ca$^{2+}$-binding proteins such as **calmodulin**. (Figure 3-9b)

**efferent neurons** Nerves that transmit signals from the central nervous system to peripheral tissues such as muscles and endocrine cells.

**electric potential** The energy associated with the separation of positive and negative charges. An electric potential is maintained across the plasma membrane of nearly all cells.

**electrochemical gradient** The driving force that determines the energetically favorable direction of transport of an ion (or charged molecule) across a membrane. It represents the combined influence of the ion's **concentration gradient** across the membrane and the **membrane potential**.

**electron carrier** Any molecule or atom that accepts electrons from donor molecules and transfers them to acceptor molecules in coupled **oxidation** and **reduction** reactions. (Table 12-2)

**electron transport** Flow of electrons via a series of electron carriers from reduced electron donors (e.g., NADH) to O$_2$ in the inner mitochondrial membrane, or from H$_2$O to NADP$^+$ in the thylakoid membrane of plant chloroplasts. (Figures 12-19 and 12-32)

**electron transport chain** Set of four large multiprotein complexes in the inner mitochondrial membrane plus diffusible cytochrome *c* and coenzyme Q through which electrons flow from reduced electron donors (e.g., NADH) to O$_2$. Each member of the chain contains one or more bound **electron carriers**. (Figure 12-16)

**electrophoresis** Any of several techniques for separating macromolecules based on their migration in a gel or other medium subjected to a strong electric field. (Figure 3-36)

**elongation, transcription** Addition of nucleotides to a growing polynucleotide chain, as templated by a complementary DNA coding strand. (Figure 4-11)

**elongation factor (EF)** One of a group of nonribosomal proteins required for continued **translation** of mRNA (protein synthesis) following initiation. (Figure 4-25)

**embryonic stem (ES) cells** A line of cultured cells derived from very early embryos that can differentiate into a wide range of cell types either in vitro or after reinsertion into a host embryo. (Figure 21-5)

**endergonic** Referring to reactions and processes that have a positive G and thus require an input of **free energy** in order to proceed; opposite of **exergonic**.

**endocrine** Referring to signaling mechanism in which target cells bind and respond to a **hormone** released into the blood by distant specialized secretory cells usually present in a gland (e.g., pituitary or thyroid gland).

**endocytic pathway** Cellular pathway involving. **receptor-mediated endocytocis** that internalizes extracellular materials too large to be imported by membrane transport proteins and to remove receptor proteins from the cell surface as a way to down-regulate their activity. (Figure 14-29)

**endocytosis** General term for uptake of extracellular material by invagination of the plasma membrane; includes **receptor-mediated endocytosis, phagocytosis,** and pinocytosis.

**endoplasmic reticulum (ER)** Network of interconnected membranous structures within the cytoplasm of eukaryotic cells contiguous with the outer nuclear envelope. The *rough ER*, which is associated with **ribosomes**, functions in the synthesis and processing of secreted and membrane proteins; the *smooth ER*, which lacks ribosomes, functions in lipid synthesis. (Figure 9-32)

**endosome** One of two types of membrane-bounded compartments: *early* endosomes (or endocytic vesicles), which bud off from the plasma membrane during receptor-mediated endocytosis, and **late endosomes**, which have an acidic internal pH and function in sorting of proteins to **lysosomes**. (Figures 14-1 and 14-29)

**endosymbiont** Bacterium that resides inside a eukaryotic cell in a mutually beneficial partnership. According to the endosymbiont hypothesis, both mitochondria and chloroplasts evolved from endosymbionts. (Figure 6-20)

**endothelium** The thin layer of cells that lines the interior surface of blood and lymphatic vessels.

**endothermic** Referring to reactions and processes that have a positive change in **enthalpy**, $\Delta H$, and thus must absorb heat in order to proceed; opposite of **exothermic**.

**energy charge** A measure of the fraction of total adenosine phosphates that have "high-energy" phosphoanhydride bonds, which is equal to ([ATP] + 0.5 [ADP])/([ATP] + [ADP] + [AMP]).

**enhancer** A regulatory sequence in eukaryotic DNA that may be located at a great distance from the gene it controls or even within the coding sequence. Binding of specific proteins to an enhancer modulates the rate of transcription of the associated gene. (Figure 7-22)

**enhancesome** Large nucleoprotein complex that assembles from transcription factors (activators and repressors) as they bind cooperatively to their binding sites in an **enhancer** with the assistance of DNA-bending proteins. (Figure 7-33)

**enthalpy (H)** Heat; in a chemical reaction, the enthalpy of the reactants or products is equal to their total bond energies.

**entropy (S)** A measure of the degree of disorder or randomness in a system; the higher the entropy, the greater the disorder.

**envelope** See **nuclear envelope** or **viral envelope**.

**enzyme** A protein that catalyzes a particular chemical reaction involving a specific **substrate** or small number of related substrates.

**epidermal growth factor (EGF)** One of a family of secreted signaling proteins (the *EGF family*) that is used in the development of most tissues in most or all animals. EGF signals are bound by **receptor tyrosine kinases**. Mutations in EGF signal transduction components are implicated in human cancer, including brain cancer. See **HER family**.

**epigenetic** Referring to a process that affects the expression of specific genes and is inherited by daughter cells but does not involve a change in DNA sequence.

**epinephrine** A catecholamine secreted by the adrenal gland and some neurons in response to stress; also called *adrenaline*. It functions as both a hormone and neurotransmitter, mediating "fight or flight" responses (e.g., increased blood glucose levels and heart rate).

**epithelium** (pl. **epithelia**) Sheet-like covering, composed of one or more layers of tightly adhering cells, on external and internal body surfaces. (Figure 20-9)

**epithelial-to-mesenchymal transition (EMT)** Describes a developmental program during which epithelial cells acquire the characteristics of mesenchymal cells. Cells lose adhesive properties and acquire motility.

**epitope** The part of an antigen molecule that binds to an antigen-specific receptor on B or T cells or to antibody. Large protein antigens usually possess multiple epitopes that bind to antibodies of different specificity.

**equilibrium constant ($K_{eq}$)** Ratio of forward and reverse rate constants for a reaction. For a binding reaction, A + B $\rightleftharpoons$ AB, the association constant ($K_a$) equals $K$, and the dissociation constant ($K_d$) equals $1/K$.

**erythropoietin (Epo)** A **cytokine** that triggers production of red blood cells by inducing the proliferation and differentiation of erythroid progenitor cells in the bone marrow. (Figures 16-8 and 21-23)

**E2F transcription factor complex** Transcription factor that promotes the transcription of $G_1/S$ phase cyclins and many other genes whose function is required for S phase.

**euchromatin** Less condensed portions of **chromatin** present in interphase chromosomes; includes most transcriptionally active regions. See also **heterochromatin**. (Figure 6-33a)

**eukaryotes** Class of organisms, composed of one or more cells containing a membrane-enclosed nucleus and organelles, that constitutes one of the three distinct evolutionary lineages of modern-day organisms; also called *eukarya*. Includes all organisms except **viruses** and **prokaryotes**. (Figure 1-1)

**eukaryotic translation initiation factors (eIFs)** Proteins required to initiate protein synthesis in eukaryotic cells. (Figure 4-24)

**excision-repair system, DNA** One of several mechanisms for repairing DNA damage due to spontaneous depurination or deamination or exposure to **carcinogens**. These repair systems normally operate with a high degree of fidelity and their loss is associated with increased risk for certain cancers.

**excitatory synapse** A synapse in which the neurotransmitter induces a depolarization of the postsynaptic cell, favoring generation of an action potential.

**exergonic** Referring to reactions and processes that have a negative $\Delta G$ and thus release **free energy** as they proceed; opposite of **endergonic**.

**exocytosis** Release of intracellular molecules (e.g., hormones, matrix proteins) contained within a membrane-bounded vesicle by fusion of the vesicle with the plasma membrane of a cell.

**exon** Segment of a eukaryotic gene (or of its **primary transcript**) that reaches the cytoplasm as part of a mature mRNA, rRNA, or tRNA molecule. See also **intron**.

**exon shuffling** Evolutionary process for creating new genes (i.e., new combinations of exons) from preexisting ones by recombination between introns of two separate genes or by transposition of mobile DNA elements. (Figures 6-18 and 6-19)

**exoplasmic face** The face of a cell membrane directed away from the cytosol. (Figure 10-5)

**exosome** Large exonuclease-containing complex that degrades spliced-out introns and improperly processed pre-mRNAs in the nucleus or mRNAs with shortened poly(A) tails in the cytoplasm. (Figure 8-1)

**exothermic** Referring to reactions and processes that have a negative change in **enthalpy**, $\Delta H$, and thus release heat as they proceed; opposite of **endothermic**.

**expression vector** A modified **plasmid** or virus that carries a gene or cDNA into a suitable host cell and there directs synthesis of the encoded protein; used to screen a DNA library for a gene of interest or to produce large amounts of a protein from its cloned gene (Figures 5-31 and 5-32)

**extracellular matrix (ECM)** A complex interdigitating meshwork of proteins and polysaccharides secreted by cells into the spaces between them. It provides structural support in tissues and can affect the development and biochemical functions of cells. (Table 20-1)

$F_0F_1$ **complex** See **ATP synthase**.

**facilitated transport** Protein-aided transport of an ion or small molecule across a cell membrane down its concentration gradient at a rate greater than that obtained by **simple diffusion**; also called *facilitated diffusion*. (Table 11-1)

**FAD (flavin adenine dinucleotide)** A small organic molecule that functions as an electron carrier by accepting two electrons from a donor molecule and two $H^+$ from the solution. (Figure 2-33b) **fatty acid** Any long hydrocarbon chain that has a carboxyl group at one end; a major source of energy during metabolism and a precursor for synthesis of phospholipids, triglycerides, and cholesteryl esters. (Figure 2-21; Table 2-4)

**fermentation** The conversion of some of the energy in organic molecule nutrients such as glucose into ATP via their oxidation into organic molecule "waste" products such as lactic acid or ethanol, typically involving the simultaneous cyclical reduction and oxidation of $NAD^+/NADH$.

**FG-nucleoporins** Proteins on the inner surface of the nuclear pore complex with a globular domain that forms part of the pore structure and a random coil domain of hydrophilic amino acids punctuated by short repeats rich in phenylalanine and glycine. (Figure 8-20)

**fibroblast** A common type of connective tissue cell that secretes **collagen** and other components of the **extracellular matrix;** migrates and proliferates during wound healing and in tissue culture.

**fibronectin** An abundant **multi-adhesive matrix protein** that occurs in numerous isoforms, generated by alternative splicing, in various cell types. Binds many other components of the extracellular matrix and to integrin adhesion receptors. (Figure 20-32)

**FISH** See **fluorescence in situ hybridization.**

**flagellum** (pl. flagella) Long locomotory structure (usually one per cell) extending from the surface of some eukaryotic cells (e.g.,sperm), whose whiplike bending propels the cell through a fluid medium. Bacterial flagella are smaller and much simpler structures. See also **axoneme** and **cilium.** (Figure 18-31)

**flavin adenine dinucleotide** See **FAD.**

**flippase** Protein that facilitates the movement of membrane lipids from one leaflet to the other leaflet of a phospholipid bilayer. (Figure 11-15)

**fluorescence in situ hybridization (FISH)** Any of several related techniques for detecting specific DNA or RNA sequences in cells and tissues by treating samples with fluorescent **probes** that hybridize to the sequence of interest and observing the samples by fluorescence microscopy.

**fluorescent staining** General technique for visualizing cellular components by treating cells or tissues with a fluorescent dye–labeled agent (e.g., antibody) that binds specifically to a component of interest and observing the sample by fluorescence microscopy.

**free energy (G)** A measure of the potential energy of a system, which is a function of the **enthalpy** (H) and **entropy** (S).

**functional complementation** Procedure for screening a DNA library to identify the wild-type gene that restores the function of a defective gene in a particular mutant. (Figure 5-18)

$G_0$, $G_1$, $G_2$ **phase** See **cell cycle.**

$G_1$ **CDKs** Cyclin-CDK complexes that promote entry into the cell cycle.

$G_1/S$ **phase CDKs** Cyclin-CDK complexes that promote entry into the cell cycle together with $G_1$ CDKs.

**gamete** Specialized **haploid** cell (in animals either a sperm or an egg) produced by **meiosis** of precursor **germ cells;** in sexual reproduction, union of a sperm and an egg initiates the development of a new individual.

**gap junction** Protein-lined channel connecting the cytoplasms of adjacent animal cells that allows passage of ions and small molecules between the cells. See also **plasmodesmata.** (Figure 20-20)

**gene** Physical and functional unit of heredity, which carries information from one generation to the next. In molecular terms, it is the entire DNA sequence—including **exons, introns,** and **transcription-control regions**—necessary for production of a functional polypeptide or RNA. See also **transcription unit.**

**gene conversion** A type of DNA recombination in which one DNA sequence is converted to the sequence of a second homologous DNA sequence in the same cell.

**gene control** All of the mechanisms involved in regulating **gene expression.** Most common is regulation of transcription, although mechanisms influencing the processing, stabilization, and translation of mRNAs help control expression of some genes.

**gene expression** Overall process by which the information encoded in a gene is converted into an observable **phenotype** (most commonly production of a protein).

**gene family** Set of genes that arose by duplication of a common ancestral gene and subsequent divergence due to small changes in the nucleotide sequence. (Figure 6-26)

**gene knockout** See **knockout, gene.**

**genetic code** The set of rules whereby nucleotide triplets (**codons**) in DNA or RNA specify amino acids in proteins. (Table 4-1)

**genetic complementation** Restoration of a wild-type function in diploid heterozygous cells generated from haploid cells, each of which carries a mutation in a different gene whose encoded protein is required for the same biochemical pathway. Complementation analysis can determine if recessive mutations in two mutants with the same mutant phenotype are in the same or different genes. (Figure 5-7)

**genome** Total genetic information carried by a cell or organism.

**genomics** Comparative analyses of the complete genomic sequences from different organisms and determination of global patterns of gene expression; used to assess evolutionary relations among species and to predict the number and general types of RNAs produced by an organism.

**genotype** Entire genetic constitution of an individual cell or organism, usually with emphasis on the particular alleles at one or more specific loci.

**germ cell** In sexually reproducing organisms, any cell that can potentially contribute to the formation of offspring including gametes and their immature precursors; also called *germ-line cell.* See also **somatic cell.**

**glia** Supporting cells of nervous tissue that, unlike neurons, do not conduct electrical impulses; also called *glial cells.* Of the four types, *Schwann cells* and *oligodendrocytes* produce **myelin sheaths,** *astrocytes* function in **synapse** formation, and *microglia* make **trophic factors** and serve in immune responses. (Figure 22-14)

**glucagon** A peptide hormone produced in the cells of pancreatic islets that triggers the conversion of glycogen to glucose by the liver; acts with **insulin** to control blood glucose levels.

**GLUT proteins** A family of transmembrane proteins, containing 12 membrane-spanning α helices, that transport glucose (and a few other sugars) across cell membranes down its concentration gradient. (Figure 11-5)

**glycogen** A very long, branched polysaccharide, composed exclusively of glucose units, that is the primary storage carbohydrate in animals; found primarily in liver and muscle cells.

**glycolipid** Any lipid to which a short carbohydrate chain is covalently linked; commonly found in the plasma membrane.

**glycolysis** Metabolic pathway in which sugars are degraded anaerobically to lactate or pyruvate in the cytosol with the production of ATP. (Figure 12-3)

**glycoprotein** Any protein to which one or more oligosaccharide chains are covalently linked. Most secreted proteins and many membrane proteins are glycoproteins.

**glycosaminoglycan (GAG)**   A long, linear, highly charged polymer of a repeating disaccharides in which many residues often are sulfated. GAGs are major components of the extracellular matrix, usually as components of **proteoglycans**. (Figure 20-28)

**glycosidic bond**   The covalent linkage between two monosaccharide residues formed when a carbon atom in one sugar reacts with a hydroxyl group on a second sugar with the net release of a water molecule (dehydration). (Figure 2-13)

**G protein–coupled receptor (GPCR)**   Member of a large class of cell-surface signaling receptors, including those for epinephrine, glucagon, and yeast mating factors. All GPCRs contain seven transmembrane α helices. Ligand binding leads to activation of a coupled trimeric G protein, thereby initiating intracellular signaling pathways. (Figures 15-15 and 15-17)

**Golgi complex**   Stacks of flattened, interconnected membrane-bounded compartments (cisternae) in eukaryotic cells that function in processing and sorting of proteins and lipids destined for other cellular compartments or for secretion; also called *Golgi apparatus*. (Figure 9-28)

**growth factor**   An extracellular polypeptide molecule that binds to a cell-surface receptor, triggering an intracellular signaling pathway generally leading to cell proliferation.

**GTPase superfamily**   Group of intracellular switch proteins that cycle between an inactive state with bound GDP and an active state with bound GTP. Includes the Gα subunit of **trimeric (large) G proteins**, monomeric (small) G proteins (e.g., **Ras**, Rab, Ran, and Rac), and certain **elongation factors** used in protein synthesis. (Figure 3-32)

**haploid**   Referring to an organism or cell having only one member of each pair of **homologous chromosomes** and hence only one copy (**allele**) of each gene or genetic locus. Gametes and bacterial cells are haploid. See also **diploid**.

**Hedgehog (Hh)**   A family of secreted signaling proteins that are important regulators of the development of most tissues and organs in diverse animal species. Mutations in Hh signal transduction components are implicated in human cancer and birth defects. The receptor is the Patched transmembrane protein. (Figures 16-31, 16-32 and 16-33)

**helicase**   (1) Any enzyme that moves along a DNA duplex using the energy released by ATP hydrolysis to separate (unwind) the two strands; required for DNA replication. (2) Activity of certain initiation factors that can unwind the secondary structures in mRNA during initiation of translation.

**helix-loop-helix, basic (bHLH)**   A conserved DNA-binding **structural motif**, consisting of two α helices connected by a short loop, that is found in many dimeric eukaryotic transcription factors. (Figure 7-29d)

**helix-turn-helix**   A structural motif in which two alpha helices are connected by a short stretch of connecting residues, sometimes also called a "loop." Helix-turn-helix/helix-loop-helix structural motifs can perform various functions, including binding calcium and binding DNA.

**HER family**   Group of receptors, belonging to the **receptor tyrosine kinase (RTK)** class, that bind to members of the epidermal growth factor (EGF) family of signaling molecules in humans.

Overexpression of HER2 protein is associated with some breast cancers. (Figure 16-7)

**heterochromatin**   Regions of **chromatin** that remain highly condensed and transcriptionally inactive during interphase. (Figure 6-33a)

**heterozygous**   Referring to a diploid cell or organism having two different **alleles** of a particular gene.

**hexose**   A six-carbon **monosaccharide**.

**high-energy bond**   Covalent bond that releases a large amount of energy when hydrolyzed under the usual intracellular conditions. Examples include the phosphoanhydride bonds in ATP, thioester bond in acetyl CoA, and various phosphate ester bonds.

**histone**   One of several small, highly conserved basic proteins, found in the **chromatin** of all eukaryotic cells, that associate with DNA in the **nucleosome**. (Figure 6-29)

**Holliday structure**   An intermediate in DNA recombination with four DNA strands. (Figure 4-43)

**homeodomain**   Conserved DNA-binding **structural motif** (a helix-turn-helix) found in many developmentally important transcription factors.

**homologous**   See **homologs**.

**homologous chromosome**   One of the two copies of each morphologic type of chromosome present in a **diploid** cell; also called **homolog**. Each homolog is derived from a different parent.

**homologous recombination**   See **recombination**.

**homologs**   Maternal and paternal copies of each morphologic type of chromosome present in a diploid cell; also called *homologues*.

**homology**   Similarity in characteristics (e.g., protein and nucleic acid sequences or the structure of an organ) that reflects a common evolutionary origin. Proteins or genes that exhibit homology are said to be homologous and sometimes are called homologs. In contrast, *analogy* is a similarity in structure or function that does not reflect a common evolutionary origin.

**homozygous**   Referring to a diploid cell or organism having two identical **alleles** of a particular gene.

**hormone**   Generally, any extracellular substance that induces specific responses in target cells; specifically, those signaling molecules that circulate in the blood and mediate **endocrine** signaling.

**hyaluronan**   A large, highly hydrated **glycosaminoglycan (GAG)** that is a major component of the extracellular matrix; also called *hyaluronic acid* and *hyaluronate*. It imparts stiffness and resilience as well as a lubricating quality to many types of connective tissue. (Figure 20-28a)

**hybridization, nucleic acid**   Association of two **complementary** nucleic acid strands to form double-stranded molecules, which can contain two DNA strands, two RNA strands, or one DNA and one RNA strand. Used experimentally in various ways to detect specific DNA or RNA sequences.

**hybridoma**   A **clone** of hybrid cells that are immortal and produce **monoclonal antibody**; formed by fusion of a normal antibody-producing B cell with a myeloma cell. (Figure 9-6)

**hydrocarbon**   Any compound containing only carbon and hydrogen atoms.

**hydrogen bond** A **noncovalent interaction** between an atom (commonly oxygen or nitrogen) carrying a partial negative charge and a hydrogen atom carrying a partial positive charge. Important in stabilizing the conformation of proteins and in formation of **base pairs** between nucleic acid strands. (Figure 2-8)

**hydrophilic** Interacting effectively with water. See also **polar**.

**hydrophobic** Not interacting effectively with water; in general, poorly soluble or insoluble in water. See also **nonpolar**.

**hydrophobic effect** The tendency of nonpolar molecules or parts of molecules to associate with each other in aqueous solution so as to minimize their direct interactions with water; commonly called a *hydrophobic interaction* or *bond*. (Figure 2-11)

**hyperpolarization** Increase in the magnitude of the cytosolic-face negative electric potential that normally exists across the plasma membrane of a cell at rest, resulting in a more negative **membrane potential**.

**hypertonic** Referring to an external solution whose solute concentration is high enough to cause water to move out of cells due to **osmosis**.

**hypotonic** Referring to an external solution whose solute concentration is low enough to cause water to move into cells due to **osmosis**.

**IgCAMs** A family of **cell-adhesion molecules** that contain multiple immunoglobulin (Ig) domains and mediate $Ca^{2+}$-independent cell-cell interactions. IgCAMs are produced in a variety of tissues and are components of **tight junctions**. (Figure 20-2)

**immunoblotting** Technique in which proteins separated by electrophoresis are attached to a nitrocellulose or other membrane, and specific proteins then are detected by use of labeled antibodies; also called *Western blotting*.

**immunoglobulin (Ig)** Any of the serum proteins, produced by fully differentiated **B cells**, that can function as antibodies; also occur in membrane-bound form as part of the **B-cell receptor**. Immunoglobulins are divided into five main classes (*isotypes*) that exhibit distinct functional properties. See also **antibody**. (Figures 23-8 and 23-9)

**immunoglobulin (Ig) fold** Evolutionarily ancient structural motif found in antibodies, the T-cell receptor, and numerous other eukaryotic proteins not directly involved in antigen-specific recognition; also called *Ig domain*. (Figure 23-12b)

**induced pluripotent stem (iPS) cells** A mammalian cell with properties of an embryonic stem cell that is formed from a differentiated cell type by expression of one or more transcription factors or other genes that confer pluripotency.

**inflammation** Localized response to injury or infection that leads to the activation of immune-system cells and their recruitment to the affected site; marked by the four classical signs of redness, swelling, heat, and pain. (Figure 23-6)

**inhibitory synapse** A synapse in which the neurotransmitter induces a hyperpolarization of the postsynaptic cell, inhibiting generation of an action potential.

**initiation, transcription** The process by which an RNA polymerase separates DNA strands and synthesizes the first phosphodiester bond of an RNA chain as templated by the DNA strand that enters the RNA polymerase active site. (Figure 4-11)

**initiation factor (IF)** One of a group of nonribosomal proteins that promote the proper association of ribosomes and mRNA and are required for initiation of **translation** (protein synthesis). (Figure 4-24)

**initiator** A DNA sequence that specifies transcription initiation within the sequence.

**inner cell mass (ICM)** The part of an early embryo that will form the embryo proper but not the extra-embryonic tissues, including the placenta.

**inositol 1,4,5-trisphosphate (IP₃)** Intracellular **second messenger** produced by cleavage of the membrane lipid phosphatidylinositol 4,5-bisphosphate in response to stimulation of certain cell-surface receptors. IP₃, which triggers release of $Ca^{2+}$ stored in the endoplasmic reticulum, is one of several biologically active **phosphoinositides**. (Figure 15-8; Table 15-3)

**in situ hybridization** Any technique for detecting specific DNA or RNA sequences in cells and tissues by treating samples with single-stranded RNA or DNA **probes** that hybridize to the sequence of interest. (Figure 5-28)

**insulin** A protein hormone produced in the β cells of the pancreatic islets that stimulates uptake of glucose into muscle and fat cells; acts with **glucagon** to help regulate blood glucose levels. Insulin also functions as a growth factor for many cells.

**integral membrane protein** Any protein that contains one or more hydrophobic segments embedded within the core of the **phospholipids bilayer**; also called *transmembrane protein*. (Figure 13-10)

**integrins** A large family of heterodimeric transmembrane proteins that function as adhesion receptors, promoting cell-matrix adhesion, or as cell-adhesion molecules, promoting cell-cell adhesion. (Table 20-3)

**interferons (IFNs)** Small group of cytokines that bind to cell-surface receptors on target cells inducing changes in gene expression that lead to an antiviral state or other cellular responses important in immune responses.

**interleukins (ILs)** Large group of cytokines, some released in response to inflammation, that promote proliferation and functioning of T cells and antibody-producing B cells of the immune system.

**intermediate filament** Cytoskeletal fiber (10 nm in diameter) formed by polymerization of related, but tissue-specific, subunit proteins, including **keratins, lamins,** and **neurofilaments**. (Figure 18-47; Table 18-1)

**interneurons** Nerves that receive signals from other nerve cells and that in turn transmit signals to other nerve cells.

**interphase** Long period of the cell cycle, including the $G_1$, S, and $G_2$ phases, between one M (mitotic) phase and the next. (Figures 1-16 and 19-1)

**intron** Part of a **primary transcript** (or the DNA encoding it) that is removed by splicing during RNA processing and is not included in the mature, functional mRNA, rRNA, or tRNA.

**ionic interaction** A **noncovalent interaction** between a positively charged ion (cation) and negatively charged ion (anion); commonly called *ionic bond*.

**IP₃** See **inositol 1,4,5-trisphosphate.**

**isoelectric point (pI)** The pH of a solution at which a dissolved protein or other potentially charged molecule has a net charge of zero and therefore does not move in an electric field. (Figure 3-37)

**isoform** One of several forms of the same protein whose amino acid sequences differ slightly and whose general activities are similar. Isoforms may be encoded by different genes or by a single gene whose primary transcript undergoes **alternative splicing.**

**isotonic** Referring to a solution whose solute concentration is such that it causes no net movement of water in or out of cells.

**JAK kinase** A class of protein tyrosine kinases that are bound to the cytosolic domain of cytokine receptors and are activated following cytokine binding.

**karyopherin** One of a family of nuclear transport proteins that functions as an **importin, exportin,** or occasionally both. Each karyopherin binds to a specific signal sequence in cargo proteins moving in or out of the nucleus.

**karyotype** Number, sizes, and shapes of the entire set of **metaphase** chromosomes of a eukaryotic cell. (Chapter 6 opening figure)

**keratins** A group of **intermediate filament** proteins found in epithelial cells that assemble into heteropolymeric filaments. (Figure 18-48)

**kinase** An enzyme that transfers the terminal (γ) phosphate group from ATP to a substrate. Protein kinases, which phosphorylate specific serine, threonine, or tyrosine residues, play a critical role in regulating the activity of many cellular proteins. See also **phosphatases.** (Figure 3-33)

**kinesins** A class of **motor proteins** that use energy released by ATP hydrolysis to move toward the (+) end of a **microtubule.** Kinesins can transport vesicles and organelles and play a role in chromosome movement during mitosis. (Figures 18-18 through 18-20)

**kinetic energy** Energy of movement, such as the motion of molecules.

**kinetochore** A multilayer protein structure located at or near the **centromere** of each mitotic chromosome from which microtubules extend toward the spindle poles of the cell; plays an active role in movement of chromosomes toward the poles during anaphase. (Figure 18-39)

**$K_m$** A parameter that describes the affinity of an enzyme for its substrate and equals the substrate concentration that yields the half-maximal reaction rate; also called the *Michaelis constant.* A similar parameter describes the affinity of a transport protein for the transported molecule or the affinity of a receptor for its ligand. (Figure 3-22)

**knockdown, siRNA** Technique for experimentally inhibiting translation of a specific mRNA by use of **siRNA;** useful for reducing the activity of a protein, particularly in organisms that are not amenable to classical genetic methods for isolating loss-of-function mutants.

**knockout, gene** Selective inactivation of a specific gene by replacing it with a nonfunctional (disrupted) allele in an otherwise normal organism.

**lagging strand** One of the two daughter DNA strands formed at a **replication fork** as short, discontinuous segments (Okazaki fragments), which are synthesized in the 5′ → 3′ direction and later joined. See also **leading strand.** (Figure 4-30)

**laminin** Large heterotrimeric **multi-adhesive matrix protein** that is found in all **basal lamina.** (Figure 20-23)

**lamins** A group of **intermediate filament** proteins that form a fibrous network, the **nuclear lamina,** on the inner surface of the nuclear envelope.

**late endosome** See **endosome.**

**lateral** See **basolateral.**

**lateral inhibition** Important signal-mediated developmental process that results in adjacent equivalent or near-equivalent cells assuming different fates.

**leading strand** One of the two daughter DNA strands formed at a **replication fork** by continuous synthesis in the 5′ → 3′ direction. The direction of leading-strand synthesis is the same as movement of the replication fork. See also **lagging strand.** (Figure 4-30)

**lectin** Any protein that binds tightly to specific sugars. Lectins assist in the proper folding of some glycoproteins in the endoplasmic reticulum and can be used in affinity chromatography to purify glycoproteins or as reagents to detect them in situ.

**leucine zipper** A type of coiled-coil **structural motif** composed of two α helices that form specific homo- or heterodimers; common motif in many eukaryotic transcription factors. See **coiled coil.** (Figures 7-29c and 3-9)

**ligand** Any molecule, other than an enzyme **substrate,** that binds tightly and specifically to a macromolecule, usually a protein, forming a macromolecule-ligand complex.

**linkage** In genetics, the tendency of two different loci on the same chromosome to be inherited together. The closer two loci are, the lower the frequency of **recombination** between them and the greater their linkage.

**lipid** Any organic molecule that is poorly soluble or virtually insoluble in water but is soluble in nonpolar organic solvents. Major classes include **fatty acids,** phospholipids, **steroids,** and **triglycerides.**

**lipid-anchored membrane protein** Any protein that is tethered to a cellular membrane by one or more covalently attached lipid groups, which are embedded in the phospholipids bilayer. (Figure 10-19)

**lipid raft** Microdomain in the plasma membrane that is enriched in cholesterol, sphingomyelin, and certain proteins.

**lipoprotein** Any large, water-soluble protein and lipid complex that functions in mass transfer of lipids throughout the body. See also **low-density lipoprotein (LDL).**

**liposome** Artificial spherical **phospholipid bilayer** structure with an aqueous interior that forms in vitro from phospholipids and may contain membrane proteins. (Figure 10-3c)

**long terminal repeats (LTRs)** Direct repeat sequences, containing up to 600 base pairs, that flank the coding region of integrated retroviral DNA and viral **retrotransposons.**

**low-density lipoprotein (LDL)** A class of **lipoprotein,** containing apolipoprotein B-100, that is a primary transporter of cholesterol in the form of cholesteryl esters between tissues, especially to the liver. (Figure 14-27)

**lymphocytes**   Two classes of white blood cells that can recognize foreign molecules (**antigens**) and mediate immune responses. B lymphocytes (B cells) are responsible for production of antibodies; T lymphocytes (T cells) are responsible for destroying virus and bacteria-infected cells, foreign cells, and cancer cells.

**lysis**   Destruction of a cell by rupture of the plasma membrane and release of the contents.

**lysogeny**   Phenomenon in which the DNA of a bacterial virus (bacteriophage) is incorporated into the host-cell genome and replicated along with the bacterial DNA but is not expressed. Subsequent activation leads to formation of new viral particles, eventually causing lysis of the cell.

**lysosome**   Small organelle that has an internal pH of 4–5, contains hydrolytic enzymes, and functions in degradation of materials internalized by endocytosis and of cellular components in autophagy. (Figures 9-12 and 9-32)

**M (mitotic) phase**   See **cell cycle**.

**macromolecule**   Any large, usually polymeric molecule (e.g., aprotein, nucleic acid, polysaccharide) with a molecular mass greater than a few thousand daltons.

**macrophages**   Phagocytic leukocytes that can detect broad patterns of pathogen markers via **Toll-like receptors**. They function as professional **antigen-presenting cells** and are a major source of **cytokines**.

**major histocompatibility complex (MHC)**   Set of adjacent genes that encode class I and class II **MHC molecules** and other proteins required for antigen presentation, as well as some complement proteins; called the *H-2 complex* in mice and the *HLA complex* in humans. (Figure 23-19)

**malignant**   Referring to a tumor or tumor cells that can invade surrounding normal tissue and/or undergo **metastasis**. See also **benign**.

**MAP kinase**   Any of a family of protein kinases that are activated in response to cell stimulation by many different growth factors and that mediate cellular responses by phosphorylating specific transcription factors and other target proteins. (Figures 16-21 and 16-22)

**matrix metalloproteases (MMPs)** Matrix metalloproteases (MMPs) are proteolytic enzymes that employ the metal zinc in their active sites. They operate in the extracellular space, where they cut proteins in the extracellular matrix and sometimes other proteins (e.g., some cell-surface receptors).

**maturation-promoting factor (MPF)**   Cyclin-CDK complex that has the ability to induce entry into meiosis when injected into $G_2$-resting oocytes.

**maximal velocity**   See $V_{max}$.

**mechanosensor**   Any of several types of sensory structures that are embedded in various tissues and respond to touch, the positions and movements of the limbs and head, pain, and temperature.

**mediator**   A very large multiprotein complex that forms a molecular bridge between transcriptional activators bound to an **enhancer** and to RNA polymerase II bound at a **promoter**; functions as a coactivator in stimulating transcription. (Figures 7-38 and 7-39)

**meiosis**   In eukaryotes, a special type of cell division that occurs during maturation of germ cells; comprises two successive nuclear and cellular divisions with only one round of DNA replication. Results in production of four genetically nonequivalent haploid cells (**gametes**) from an initial diploid cell. (Figure 5-3)

**membrane potential**   Electric potential difference, expressed in volts, across a membrane due to the slight excess of positive ions (cations) on one side and negative ions (anions) on the other. (Figures 11-18 and 11-19)

**membrane transport protein**   Collective term for any integral membrane protein that mediates movement of one or more specific ions or small molecules across a cellular membrane regardless of the transport mechanism. (Figure 11-2)

**memory**   The ability of an antigen-experienced immune system to respond more rapidly to a reexposure to that same antigenic stimulus.

**meristem**   Organized group of undifferentiated, dividing cells that are maintained at the tips of growing shoots and roots in plants. All the adult structures arise from meristems.

**merotelic attachment**   Indicates that a single kinetochore attaches to microtubules emanating from two opposite spindle poles.

**mesenchymal stem cell**   A class of stem cells in the bone marrow that can differentiate into fat cells, osteoblasts (bone-forming cells), and cartilage-producing cells; some may also produce muscle and other types of differentiated cells.

**mesenchyme**   Immature embryonic connective tissue, composed of loosely organized and loosely attached cells, derived from either the **mesoderm** or **ectoderm** in animals.

**mesoderm**   The middle of the three primary cell layers of the animal embryo, lying between the ectoderm and endoderm; gives rise to the notochord, connective tissue, muscle, blood, and other tissues.

**messenger RNA**   See **mRNA**.

**metaphase**   Stage of mitosis at which condensed chromosomes are aligned equidistant between the poles of the mitotic spindle but have not yet started to segregate toward the spindle poles. (Figure 18-36)

**metastasis**   Spread of cancer cells from their site of origin and establishment of areas of secondary growth.

**metazoans**   A subset of the animal kingdom that includes all multicellular animals with differentiated tissues, such as nerves and muscles.

**MHC**   See **major histocompatibility complex**.

**MHC molecules**   Glycoproteins that display peptides, derived from foreign (and self) proteins, on the surface of cells and are required for antigen presentation to **T cells**. *Class I* molecules are expressed constitutively by nearly all nucleated cells; *class II* molecules, by professional antigen-presenting cells. (Figures 23-21 and 23-22)

**micelle**   A water-soluble spherical aggregate of phospholipids or other amphipathic molecules that form spontaneously in aqueous solution. (Figure 10-3c)

**Michaelis constant**   See $K_m$.

**microfilament**   Cytoskeletal fiber ($\approx 7$ nm in diameter) that is formed by polymerization of monomeric globular (G) **actin**; also called *actin filament*. Microfilaments play an important role in muscle contraction, cytokinesis, cell movement, and other cellular functions and structures. (Figure 17-4)

**micro-RNA**   See **miRNA**.

**microtubule**   Cytoskeletal fiber ($\approx 25$ nm in diameter) that is formed by polymerization of $\alpha$, $\beta$-**tubulin** monomers and exhibits structural and functional polarity. Microtubules are important components of cilia, flagella, the mitotic spindle, and other cellular structures. (Figures 18-2 and 18-3)

**microtubule-associated protein (MAP)**   Any protein that binds to microtubules and regulates their stability. (Figures 18-13, 18-14, and 18-15)

**microtubule-organizing center**   See **MTOC**.

**microvillus (pl. microvilli)**   Small, membrane-covered projection on the surface of an animal cell containing a core of actin filaments. Numerous microvilli are present on the absorptive surface of intestinal epithelial cells, increasing the surface area for transport of nutrients. (Figures 17-4 and 20-10)

**miRNA (micro-RNA)**   Any of numerous small, endogenous cellular RNAs, 20–30 nucleotides long, that are processed from double-stranded regions of hairpin secondary structures in long precursor RNAs. A single strand of the mature miRNA associates with several proteins to form an RNA-induced silencing complex (**RISC**) that inhibits translation of a target mRNA to which the miRNA hybridizes imperfectly. Several miRNAs must hybridize to a single mRNA to inhibit its translation. See also **siRNA**. (Figures 8-25a and 8-26)

**mitochondrion (pl. mitochondria)**   Large organelle that is surrounded by two phospholipid bilayer membranes, contains DNA, and carries out **oxidative phosphorylation**, thereby producing most of the ATP in eukaryotic cells. (Figures 9-33c and 12-6)

**mitogen**   Any extracellular molecule, such as a growth factor, that promotes cell proliferation.

**mitosis**   In eukaryotic cells, the process whereby the nucleus divides, producing two genetically equivalent daughter nuclei with the diploid number of chromosomes. See also **cytokinesis** and **meiosis**. (Figure 18-36)

**mitotic CDKs**   Cyclin-CDK complexes that promote entry into and progression through mitosis.

**mitotic spindle**   A specialized temporary structure, present in eukaryotic cells during mitosis, that captures the chromosomes and then pushes and pulls them to opposite sides of the dividing cell; also called *mitotic apparatus*. (Figure 18-37)

**mobile DNA element**   See **transposable DNA element**.

**model organism**   A non-human species used to study genes, proteins, and cellular functions. Model organisms are chosen based on the expectation that discoveries made via the model organism will provide insight into other organisms.

**molecular chaperone**   See **chaperone**.

**molecular complementarity**   Lock-and-key kind of fit between the shapes, charges, hydrophobicity, and/or other physical properties of two molecules or portions thereof that allow formation of multiple **oncovalent interactions** between them at close range. (Figure 2-12)

**molecular markers, DNA-based**   DNA sequences that vary among individuals (*DNA polymorphisms*) of the same species and are useful in genetic linkage studies; includes SNPs and SSRs.

**monoclonal antibody**   Antibody produced by the progeny of a single B cell and thus a homogeneous protein that recognizes a single antigen (epitope). It can be produced experimentally by use of a **hybridoma**. (Figure 9-6)

**monomer**   Any small molecule that can be linked chemically with others of the same type to form a **polymer**. Examples include amino acids, nucleotides, and monosaccharides.

**monomeric (small) G protein**   A monomeric GTPase with a structure similar to that of the Ras protein that changes conformation when a bound GTP is hydrolyzed to GDP and phosphate. (Figure 15-7)

**monopolin complex**   Protein complex that promotes the co-orientation of **sister chromatids** during meiosis I in budding yeast.

**monosaccharide**   Any simple sugar with the formula $(CH_2O)_n$ where $n = 3$–$7$.

**monotelic attachment**   Indicates that only one of the sister kinetochore pair attached to microtubules.

**monoubiquitination**   The covalent addition of a single ubiquitin molecule to a target protein.

**motor protein**   Any member of a special class of mechanochemical enzymes that use energy from ATP hydrolysis to generate either linear or rotary motion; also called *molecular motor*. See also **dyneins, kinesins,** and **myosins**.

**mRNA (messenger RNA)**   Any RNA that specifies the order of amino acids in a protein (i.e., the primary structure). It is produced by **transcription** of DNA by RNA polymerase. In eukaryotes, the initial RNA product (primary transcript) undergoes processing to yield functional mRNA. See also **translation**. (Figure 4-15)

**mRNP-exporter**   A heterodimeric protein that binds to mRNA-containing ribonucleoprotein particles (mRNPs) and directs their export from the nucleus to cytoplasm by interacting transiently with **nucleoporins** in the nuclear pore complex. (Figure 8-22)

**MTOC (microtubule-organizing center)**   General term for any structure (e.g., centrosome, spindle pole, basal body) that organizes microtubules in cells. (Figure 18-5)

**mRNA surveillance**   The processes that lead to the degradation of a pre-mRNA or mRNA that has been improperly processed.

**multi-adhesive matrix proteins**   Group of long flexible proteins that bind to other components of the **extracellular matrix** and to cell-surface receptors, thereby cross-linking matrix components to the cell membrane. Examples include **laminin**, a major component of the basal lamina, and **fibronectin**, present in many tissues.

**multimeric**   For proteins, containing several polypeptide chains (or subunits).

**multiubiquitination**   The covalent addition of several single ubiquitin molecules, each at a distinct site, on a single target protein.

**MuSK**   A receptor tyrosine kinase localized in the myotube plasma membrane that both induces clustering of acetylcholine receptors and serves to attract the termini of growing motor neuron axons.

**mutagen** A chemical or physical agent that induces mutations.

**mutation** In genetics, a permanent, heritable change in the nucleotide sequence of a chromosome, usually in a single gene; commonly causes an alteration in the function of the gene product.

**myelin sheath** Stacked specialized cell membrane that forms an insulating layer around vertebrate **axons** and increases the speed of impulse conduction. (Figure 22-15)

**myofibril** Long, slender structures within cytoplasm of muscle cells consisting of a regular repeating array of **sarcomeres** composed of thick (**myosin**) filaments and thin (**actin**) filaments. (Figure 17-31)

**myosins** A class of **motor proteins** that have actin-stimulated ATPase activity. Myosins move along actin **microfilaments** during muscle contraction and cytokinesis and also mediate vesicle translocation. (Figure 17-22)

**NAD$^+$ (nicotinamide adenine dinucleotide)** A small organic molecule that functions as an electron carrier by accepting two electrons from a donor molecule and one H$^+$ from the solution. (Figure 2-33a)

**NADP$^+$ (nicotinamide adenine dinucleotide phosphate)** Phosphorylated form of **NAD$^+$** that is used extensively as an electron carrier in biosynthetic pathways and during photosynthesis.

**Na$^+$/K$^+$ ATPase** A P-class ATP-powered pump that couples hydrolysis of one ATP molecule to export of Na$^+$ ions and import of K$^+$ ions; is largely responsible for maintaining the normal intracellular concentrations of Na$^+$ (low) and K$^+$ (high) in animal cells; commonly called *Na$^+$/K$^+$ pump*. (Figure 11-13)

**natural killer (NK) cells** Components of the innate immune system that nonspecifically detect and kill virus-infected cells and tumor cells. (Figure 23-5)

**necrosis** Cell death resulting from tissue damage or other pathology; usually marked by swelling and bursting of cells with release of their contents. Contrast with **apoptosis**.

**negative feedback mechanism** Process where the output of a pathway inhibits its own production.

**neurofilaments (NFs)** A group of **intermediate filament** proteins, found only in neurons, that contribute to axonal structure and rate of transmission of action potentials down axons. (Figure 18-2b)

**neuron (nerve cell)** Any of the impulse-conducting cells of the nervous system. A typical neuron contains a cell body; multiple short, branched processes (**dendrites**); and one long process (**axon**). (Figure 22-1)

**neurotransmitter** Extracellular signaling molecule that is released by the presynaptic neuron at a chemical **synapse** and relays the signal to the postsynaptic cell. The response elicited by a neurotransmitter, either excitatory or inhibitory, is determined by its receptor on the postsynaptic cell. (Figures 22-18 and 22-19)

**neurotrophins** Family of structurally and functionally related

**trophic factors** that bind to receptors called Trks and are required for survival of neurons; include nerve growth factor (NGF) and brain-derived neurotrophic factor (BDNF).

**neutrophils** Phagocytic leukocytes that are attracted to sites of tissue damage and migrate into the tissue. Once activated, neutrophils secrete various chemokines, cytokines, bacteria-destroying

enzymes (e.g., lysozyme), and other products that contribute to **inflammation** and help clear invading pathogens.

**nicotinamide adenine dinucleotide** See **NAD$^+$**.

**nicotinamide adenine dinucleotide phosphate** See **NADP$^+$**.

**N-linked oligosaccharide** A branched oligosaccharide chain attached to the side-chain amino group of an asparagine residue in a glycoprotein. See also *O-linked oligosaccharide*.

**nociceptor** Mechanosensor that responds to pain associated with injury to body tissues caused by mechanical trauma, heat, electricity, or toxic chemicals.

**noncovalent interaction** Any relatively weak chemical interaction that does not involve an intimate sharing of electrons. (Figures 2-6 and 2-12)

**nonhomologous end joining (NHEJ)** A pathway that repairs double-strand breaks in DNA in which the break ends are directly ligated without the need for a homologous template.

**nonpolar** Referring to a molecule or structure that lacks any net electric charge or asymmetric distribution of positive and negative charges. Nonpolar molecules generally are less soluble in water than polar molecules and are often water insoluble.

**Northern blotting** Technique for detecting specific RNAs separated by electrophoresis by hybridization to a labeled DNA **probe**. See also **Southern blotting**. (Figure 5-27)

**nuclear body** Roughly spherical, functionally specialized region in the nucleus, containing specific proteins and RNAs; many function in the assembly of ribonucleoprotein (RNP) complexes. The most prominent type is the **nucleolus**.

**nuclear envelope** Double-membrane structure surrounding the nucleus; the outer membrane is continuous with the endoplasmic reticulum and the two membranes are perforated by **nuclear pore complexes**. (Figure 9-32)

**nuclear lamina** Fibrous network on the inner surface of the nuclear envelope composed of lamin intermediate filaments. (Figure 19-22)

**nuclear pore complex (NPC)** Large, multiprotein structure, composed largely of nucleoporins, that extends across the nuclear envelope. Ions and small molecules freely diffuse through NPCs; large proteins and ribonucleoprotein particles are selectively transported through NPCs with the aid of soluble proteins. (Figure 13-33a and b)

**nuclear receptor** Member of a class of intracellular receptors that bind lipid-soluble molecules (e.g., steroid hormones), forming ligand-receptor complexes that activate transcription; also called *steroid receptor superfamily*. (Figure 7-44d)

**nucleic acid** A polymer of **nucleotides** linked by **phosphodiester** bonds. DNA and RNA are the primary nucleic acids in cells.

**nucleic acid hybridization** See **hybridization, nucleic acid**.

**nucleocapsid** A viral **capsid** plus the enclosed nucleic acid.

**nucleolus** Large structure in the nucleus of eukaryotic cells where rRNA synthesis and processing occurs and ribosome subunits are assembled. (Figure 6-33a)

**nucleoporins** Large group of proteins that make up the **nuclear pore complex**. One class (**FG-nucleoporins**) participates in nuclear import and export.

**nucleoside** A small molecule composed of a **purine** or **pyrimidine** base linked to a pentose (either ribose or deoxyribose). (Table 2-3)

**nucleosome** Structural unit of **chromatin** consisting of a disk-shaped core of **histone** proteins around which a 147-bp segment of DNA is wrapped. (Figure 6-29)

**nucleotide** A **nucleoside** with one or more phosphate groups linked via an ester bond to the sugar moiety, generally to the 5′ carbon atom. DNA and RNA are polymers of nucleotides containing deoxyribose and ribose, respectively. (Figure 2-16 and Table 2-3)

**nucleus** Large membrane-bounded organelle in eukaryotic cells that contains DNA organized into chromosomes; synthesis and processing of RNA and ribosome assembly occur in the nucleus.

**Okazaki fragments** Short (<1000 bases), single-stranded DNA fragments that are formed during synthesis of the **lagging strand** in DNA replication and are rapidly joined by DNA ligase to form a continuous DNA strand. (Figure 4-30)

**oligopeptide** A small to medium-size linear polymer composed of amino acids connected by peptide bonds. The terms *peptide* and *oligopeptide* are often used interchangeably.

**O-linked oligosaccharide** Oligosaccharide chain that is attached to the side-chain hydroxyl group in a serine or threonine residue in a glycoprotein. See also *N*-linked oligosaccharides.

**oncogene** A gene whose product is involved either in transforming cells in culture or in inducing cancer in animals. Generally is a mutant form of a normal gene (**proto-oncogene**) for a protein involved in the control of cell growth and division. (Figure 24-11)

**oncogene addiction** Describes the observation that some cancers, despite containing numerous genetic abnormalities, depend on only a few genetic alteration to maintain their malignant phenotype. It is said that these cancers are "addicted" to certain oncogenic mutations.

**oncoprotein** A protein encoded by an **oncogene** that causes abnormal cell proliferation; may be a mutant unregulated form of a normal protein or a normal protein that is produced in excess or in the wrong time or place in an organism.

**open reading frame (ORF)** Region of sequenced DNA that is not interrupted by stop codons in one of the triplet reading frames. An ORF that begins with a start codon and extends for at least 100 codons has a high probability of encoding a protein.

**operator** Short DNA sequence in a bacterial or bacteriophage genome that binds a repressor protein and controls transcription of an adjacent gene. (Figure 7-3)

**operon** In bacterial DNA, a cluster of contiguous genes transcribed from one **promoter** that gives rise to an mRNA containing coding sequences for multiple proteins. (Figure 4-13a)

**organelle** Any membrane-limited subcellular structure found in eukaryotic cells. (Figures 1-12b, 9-32 and 9-33)

**osmosis** Net movement of water across a semipermeable membrane (permeable to water but not to solute) from a solution of lesser to one of greater solute concentration. (Figure 11-6)

**oxidation** Loss of electrons from an atom or molecule as occurs when a hydrogen atom is removed from a molecule or oxygen is added; opposite of **reduction**.

**oxidation potential** The voltage change when an atom or molecule loses an electron; a measure of the tendency of a molecule to loose an electron. For a given oxidation reaction, the oxidation potential has the same magnitude but opposite sign as the **reduction potential** for the reverse (reduction) reaction.

**oxidative phosphorylation** The phosphorylation of ADP to form ATP driven by the transfer of electrons to oxygen ($O_2$) in bacteria and mitochondria. Involves generation of a **proton-motive force** during electron transport and its subsequent use to power ATP synthesis.

**p53 protein** The product of a **tumor-suppressor gene** that plays a critical role in the arrest of cells with damaged DNA. Inactivating mutations in the *p53* gene are found in many human cancers. (Figure 24-26)

**paracrine** Referring to signaling mechanism in which a target cell responds to a signaling molecule (e.g., growth factor, neurotransmitter) that is produced by a nearby cell(s) and reaches the target by diffusion.

**patch clamping** Technique for determining ion flow through a single ion channel or across the membrane of an entire cell by use of a micropipette whose tip is applied to a small patch of the cell membrane. (Figure 11-22)

**patterning genes** Genes involved in metazoan development that determine the general organization of the animal, including the major body axes and segmentation.

**P body** Dense cytoplasmic domain, containing no ribosomes or translation factors, that functions in repression of translation and degradation of associated mRNAs; also called *cytoplasmic RNA-processing body*.

**PCR (polymerase chain reaction)** Technique for amplifying a specific DNA segment in a complex mixture by multiple cycles of DNA synthesis from short oligonucleotide primers followed by brief heat treatment to separate the complementary strands. (Figure 5-20)

**pentose** A five-carbon **monosaccharide**. The pentoses ribose and deoxyribose are present in RNA and DNA, respectively. (Figure 2-16)

**peptide** A small linear polymer composed of amino acids connected by peptide bonds. The terms *peptide* and *oligopeptide* are often used interchangeably. See also **polypeptide**.

**peptide bond** The covalent amide linkage between amino acids formed between the amino group of one amino acid and the carboxyl group of another with the net release of a water molecule (dehydration). (Figure 2-13)

**peptidoglycan** A polysaccharide chain cross-linked by peptide cross-bridges in a bacterial cell wall that confers rigidity and helps to determine the cell's shape.

**pericentriolar material** Amorphous material seen by thin-section electron microscopy surrounding the centrioles of animal cells. Pericentriolar material contains many components, including the gamma-tubulin ring complex (gamma-TuRC), which promotes nucleation of microtubule assembly. (Figure 18-6)

**peripheral membrane protein**  Any protein that associates with the cytosolic or exoplasmic face of a membrane but does not enter the hydrophobic core of the phospholipid bilayer. See also **integral membrane protein**. (Figure 10-1)

**perlecan**  A large multidomain **proteoglycan** component of the extracellular matrix (ECM) that binds to many ECM components, cell-surface molecules, and growth factors; a major component of the **basal lamina**.

**peroxisome**  Small organelle that contains enzymes for degrading fatty acids and amino acids by reactions that generate hydrogen peroxide, which is converted to water and oxygen by catalase.

**pH**  A measure of the acidity or alkalinity of a solution defined as the negative logarithm of the hydrogen ion concentration in moles per liter: $pH = -\log [H^+]$. Neutrality is equivalent to a pH of 7; values below this are acidic and those above are alkaline.

**phagocyte**  Any cell that can ingest and destroy pathogens and other particulate antigens. The primary phagocytes are neutrophils, macrophages, and dendritic cells.

**phagocytosis**  Process by which relatively large particles (e.g., bacterial cells) are internalized by certain eukaryotic cells in a process that involves extensive remodeling of the actin cytoskeleton; distinct from **receptor-mediated endocytosis**. (Figure 17-19)

**phenotype**  The detectable physical and physiological characteristics of a cell or organism determined by its **genotype**; also, the specific trait associated with a particular **allele**.

**pheromone**  A signaling molecule released by an individual that can alter the behavior or gene expression of other individuals of the same species. The yeast α and **a** mating-type factors are well-studied examples.

**phosphatase**  An enzyme that removes a phosphate group from a substrate by hydrolysis. Phosphoprotein phosphatases act with protein kinases to control the activity of many cellular proteins. (Figure 3-33)

**phosphoanhydride bond**  A type of **high-energy bond** formed between two phosphate groups, such as the γ and β phosphates and the β and α phosphates in ATP. (Figure 2-31)

**phosphodiester bond**  Chemical linkage between adjacent nucleotides in DNA and RNA; consists of two phosphoester bonds, one on the 5′ side of the phosphate and another on the 3′ side. (Figure 4-2)

**phosphoglycerides**  Amphipathic derivatives of glycerol 3-phosphate that generally consist of two hydrophobic fatty acyl chains esterified to the hydroxyl groups in glycerol and a polar head group attached to the phosphate; the most abundant lipids in biomembranes. (Figures 2-20 and 10-8a)

**phosphoinositides**  A group of membrane-bound lipids containing phosphorylated inositol derivatives; some function as **second messengers** in several signal transduction pathways. (Figures 15-35 and 16-25)

**phospholipase**  One of several enzymes that cleave various bonds in the hydrophilic end of **phospholipids**. (Figures 10-12)

**phospholipase C (PLC)**  A membrane-associated phospholipase, activated by either $G_{\alpha q}$ or $G_{\alpha o}$, that cleaves the membrane lipid phosphatidylinositol 4,5-bisphosphate to generate two second messengers, **DAG** and **IP₃**. (Figures 15-35 and 15-36a)

**phospholipid**  The major class of lipids present in biomembranes, including **phosphoglycerides** and **sphingolipids**. (Figures 10-8a, b and 2-20)

**phospholipid bilayer**  A two-layer, sheet-like structure in which the polar head groups of phospholipids are exposed to the aqueous media on either side and the nonpolar fatty acyl chains are in the center; the foundation for all biomembranes. (Figure 10- 3a, b)

**phosphorylation**  The covalent addition of a phosphate group to a molecule such as a sugar or a protein. The hydrolysis of ATP often accompanies phosphorylation, providing energy to drive the reaction and the phosphate group that is covalently added to the target molecule. Enzymes that catalyze phosphorylation are called kinases.

**photoelectron transport**  Light-driven electron transport that generates a charge separation across the **thylakoid** membrane that drives subsequent events in photosynthesis. (Figure 12-35)

**photorespiration**  A reaction pathway that competes with $CO_2$ fixation (**Calvin cycle**) by consuming ATP and generating $CO_2$, thus reducing the efficiency of photosynthesis. (Figure 12-47)

**photosynthesis**  Complex series of reactions occurring in some bacteria and in plant **chloroplasts** in which light energy is used to generate carbohydrates from $CO_2$, usually with the consumption of $H_2O$ and evolution of $O_2$.

**photosystems**  Multiprotein complexes, present in all photosynthetic organisms, that consist of light-harvesting complexes containing **chlorophylls** and a reaction center where **photoelectron transport** occurs. (Figure 12-44a)

**phragmoplast**  In plants, a temporary structure, formed during telophase, whose membranes become the plasma membranes of the daughter cells and whose contents develop into the new cell wall between them. (Figure 18-45)

**pI**  See **isoelectric point**.

**plakins**  A family of proteins that help attach **intermediate filaments** to other structures.

**plaque assay**  Technique for determining the number of infectious viral particles in a sample by culturing a diluted sample on a layer of susceptible host cells and then counting the clear areas of lysed cells (plaques) that develop. (Figure 4-45)

**plasma membrane**  The membrane surrounding a cell that separates the cell from its external environment; consists of a **phospholipids bilayer** and associated membrane lipids and proteins. (Figures 9-32, 10-1, and 10-5)

**plasmid**  Small, circular extrachromosomal DNA molecule capable of autonomous replication in a cell; commonly used as a **vector** in **DNA cloning**.

**plasmodesmata** (sing. **plasmodesma**)  Tube-like cell junctions that interconnect the cytoplasms of adjacent plant cells and are functionally analogous to **gap junctions** in animal cells. (Figure 20-41)

**point mutation**  Change of a single nucleotide in DNA, especially in a region coding for protein; can result in formation of a codon specifying a different amino acid or a stop codon. Addition or deletion of a single nucleotide will cause a shift in the **reading frame**.

**polar**  Referring to a molecule or structure with a net electric charge or asymmetric distribution of positive and negative charges. Polar molecules are usually soluble in water.

**polarity** In cell biology, the presence of functional and/or structural differences in distinct regions of a cell or cellular component.

**polarized** In cell biology, referring to any cell or subcellular structure marked by functional and structural asymmetries.

**Polo kinases** Family of protein kinases that are critical for many aspects of mitosis, such as centrosome duplication and cohesin removal from chromosomes.

**polymer** Any large molecule composed of multiple identical or similar units (**monomers**) linked by covalent bonds. (Figure 2-13)

**polymerase chain reaction** See **PCR**.

**polypeptide** Linear polymer of amino acids connected by peptide bonds, usually containing 20 or more residues. See also **protein**.

**polyribosome** A complex containing several **ribosomes**, all translating a single messenger RNA; also called *polysome*. (Figure 4-28)

**polysaccharide** Linear or branched polymer of **monosaccharides** linked by glycosidic bonds and usually containing more than 15 residues. Those with fewer than 15 residues are often called *oligosaccharides*.

**polytene chromosome** Enlarged chromosome composed of many parallel copies of itself formed by multiple cycles of DNA replication without chromosomal separation; found in the salivary glands and some other tissues of *Drosophila* and other dipteran insects. (Figure 6-43)

**polyubiquitination** The covalent addition of a chain of covalently linked ubiquitin molecules to a site on a target protein.

**polyunsaturated** Referring to a compound (e.g., fatty acid) in which two or more of the carbon-carbon bonds are double or triple bonds.

**porins** Class of trimeric transmembrane proteins through which small water-soluble molecules can cross the membrane; present in outer mitochondrial and chloroplast membranes and in the outer membrane of gram-negative bacteria. (Figure 10-18)

**positive feedback mechanism** Process where the output of a pathway promotes its own production.

**potential energy** Stored energy. In biological systems, the primary forms of potential energy are chemical bonds, concentration gradients, and electric potentials across cellular membranes.

**pre-mRNA** Precursor messenger RNA; the **primary transcript** and intermediates in RNA processing. (Figures 4-15 and 8-2)

**pre-rRNA** Large precursor ribosomal RNA that is synthesized in the nucleolus of eukaryotic cells and processed to yield three of the four RNAs present in ribosomes. (Figures 8-37 and 8-38)

**primary structure** In proteins, the linear arrangement (sequence) of amino acids within a polypeptide chain.

**primary transcript** In eukaryotes, the initial RNA product, containing **introns** and **exons**, produced by transcription of DNA. Many primary transcripts must undergo RNA processing to form the physiologically active RNA species.

**primase** A specialized RNA polymerase that synthesizes short stretches of RNA used as primers for DNA synthesis. (Figure 4-31).

**primer** A short nucleic acid sequence containing a free 3'-hydroxyl group that forms **base pairs** with a complementary template strand and functions as the starting point for addition of nucleotides to copy the template strand.

**probe** Defined RNA or DNA fragment, radioactively, fluorescently, or chemically labeled chemically labeled, that is used to detect specific nucleic acid sequences by hybridization.

**progenitor cells** A type of undifferentiated cell that, when provided with the appropriate signals, will divide and differentiate into one or a few cell types.

**programmed cell death** See **apoptosis**.

**prokaryotes** Class of organisms, including the **bacteria** (eubacteria) and **archaea,** that lack a true membrane-limited nucleus and other organelles. See also **eukaryotes**. (Figure 1-1)

**prometaphase** Second stage in mitosis, during which the nuclear envelope and nuclear lamina break down and microtubules assembled from the spindle poles "capture" chromosome pairs at specialized structures called **kinetochores**. (Figure 18-36)

**promoter** DNA sequence that determines the site of **transcription** initiation for an RNA polymerase. (Figure 4-11)

**promoter-proximal element** Any regulatory sequence in eukaryotic DNA that is located within ≈200 base pairs of the transcription start site. Transcription of many genes is controlled by multiple promoter-proximal elements. (Figure 7-22)

**prophase** Earliest stage in mitosis, during which the chromosomes condense, the duplicated centrosomes separate to become the spindle poles, and the mitotic spindle begins to form. (Figure 18-36)

**protease** Any enzyme that cleaves one or more peptide bonds in target proteins.

**proteasome** Large multifunctional protease complex in the cytosol that degrades intracellular proteins marked for destruction by attachment of multiple **ubiquitin** molecules. (Figure 3-29)

**protein** A macromolecule composed of one or more linear **polypeptide** chains and folded into a characteristic three-dimensional shape (**conformation**) in its native, biologically active state.

**protein family** Set of homologous proteins encoded by a **gene family**.

**protein domain** Distinct regions of a protein's three dimensional structure. A *functional* domain exhibits a particular activity characteristic of the protein; a *structural* domain is ≈40 or more amino acids in length, arranged in a distinct secondary or tertiary structure; a *topological* domain has a distinctive spatial relationship to the rest of the protein.

**protein kinase A (PKA)** Cytosolic enzyme that is activated by **cyclic AMP (cAMP)** and functions to phosphorylate and thus regulate the activity of numerous cellular proteins; also called *cAMP-dependent protein kinase*. (Figure 15-29)

**protein kinase B (PKB)** Cytosolic enzyme that is recruited to the plasma membrane by signal-induced **phosphoinositides** and subsequently activated; also called **Akt**. (Figure 16-26)

**protein kinase C (PKC)** Cytosolic enzyme that is recruited to the plasma membrane in response to signal-induced rise in cytosolic $Ca^{2+}$ level and is activated by membrane-bound **diacylglycerol (DAG)**. (Figure 15-36a)

**protein kinase G (PKG)**　A cytosolic protein kinase activated by cyclic GMP.

**proteoglycans**　A group of glycoproteins (e.g., perlecan and aggrecan) that contain a core protein to which is attached one or more **glycosaminoglycan (GAG)** chains. They are found in nearly all animal extracellular matrices, and some are integral membrane proteins. (Figure 20-31) **proteome** The entire complement of proteins produced by a cell.

**proteomics**　The systematic study of the amounts, modifications, interactions, localization, and functions of all or subsets of proteins at the whole-organism, tissue, cellular, and subcellular levels.

**proton-motive force**　The energy equivalent of the proton ($H^+$) concentration gradient and electric potential gradient across a membrane; used to drive ATP synthesis by **ATP synthase**, transport of molecules against their concentration gradient, and movement of bacterial flagella. (Figure 12-2)

**proto-oncogene**　A normal cellular gene that encodes a protein usually involved in regulation of cell growth or differentiation and that can be mutated into a cancer-promoting **oncogene**, either by changing the protein-coding segment or by altering its expression. (Figure 24-11)

**protostomes**　A group of bilaterally symmetric animals whose mouth develops close to the blastopore and has a ventral nerve cord. This group includes worms, insects, and mollusks.

**provirus**　The DNA of an animal virus that is integrated into a host-cell genome; during replication of the cell, the proviral DNA is replicated and appears in both daughter cells. Activation of proviral DNA leads to production and release of progeny virions.

**pseudogene**　DNA sequence that is similar to that of a functional gene but does not encode a functional product; probably arose by sequence drift of duplicated genes.

**pulse-chase**　A type of experiment in which a radioactive small molecule is added to a cell for a brief period (the pulse) and then is replaced with an excess of the unlabeled form of the same small molecule (the chase). Used to detect changes in the cellular location of a molecule or its metabolic fate over time. (Figure 3-40)

**pump**　See **ATP-powered pump.**

**purines**　A class of nitrogenous compounds containing two fused heterocyclic rings. Two purines, adenine (A) and guanine (G), are base components of nucleotides found in DNA and RNA. See also **base pair.** (Figure 2-17)

**pyrimidines**　A class of nitrogenous compounds containing one heterocyclic ring. Two pyrimidines, cytosine (C) and thymine (T), are base components of nucleotides found in DNA; in RNA, uracil (U) replaces thymine. See also **base pair.** (Figure 2-17)

**quaternary structure**　The number and relative positions of the polypeptide chains in multimeric (multisubunit) proteins. (Figure 3-10b)

**radioisotope**　Unstable form of an atom that emits radiation as it decays. Several radioisotopes are commonly used experimentally as labels in biological molecules. (Table 3-1)

**Ras protein**　A monomeric member of the **GTPase superfamily** of switch proteins that is tethered to the plasma membrane by a lipid anchor and functions in intracellular signaling pathways; activated by ligand binding to **receptor tyrosine kinases** and some other cell-surface receptors. (Figures 16-17 and 16-19)

**rate constant**　A constant that relates the concentrations of reactants to the rate of a chemical reaction.

**Rb protein**　Inhibitor of the E2F transcription factor family and thus a key regulator of cell cycle entry.

**reading frame**　The sequence of nucleotide triplets (**codons**) that runs from a specific translation start codon in an mRNA to a stop codon. Some mRNAs can be translated into different polypeptides by reading in two different reading frames. (Figure 4-18)

**receptor**　Any protein that specifically binds another molecule to mediate cell-cell signaling, adhesion, endocytosis, or other cellular process. Commonly denotes a protein located in the plasma membrane, cytosol, or nucleus that binds a specific extracellular molecule (**ligand**), which often induces a conformational change in the receptor, thereby initiating a cellular response. See also **adhesion receptor** and **nuclear receptor.** (Figures 15-1 and 16-1)

**receptor-mediated endocytosis**　Uptake of extracellular materials bound to specific cell-surface receptors by invagination of the plasma membrane to form a small membrane-bounded vesicle (early endosome). (Figure 14-29)

**receptor tyrosine kinase (RTK)**　Member of a large class of cell-surface receptors, usually with a single transmembrane domain, including those for insulin and many growth factors. Ligand binding activates tyrosine-specific protein kinase activity in the receptor's cytosolic domain, thereby initiating intracellular signaling pathways. (Figures 16-3 and 16-4)

**recessive**　In genetics, referring to that allele of a gene that is not expressed in the **phenotype** when the **dominant** allele is present; also refers to the phenotype of an individual (homozygote) carrying two recessive alleles. Mutations that produce recessive alleles generally result in a loss of the gene's function. (Figure 5-2)

**recombinant DNA**　Any DNA molecule formed in vitro by joining DNA fragments from different sources.

**recombination**　Any process in which chromosomes or DNA molecules are cleaved and the fragments are rejoined to give new combinations. Homologous recombination occurs during meiosis, giving rise to **crossing over** of homologous chromosomes. Homologous recombination and nonhomologous recombination (i.e., between chromosomes of different morphologic type) also occur during several DNA-repair mechanisms and can be carried out in vitro with purified DNA and enzymes. (Figure 5-10)

**redox reaction**　An oxidation-reduction reaction in which one or more electrons are transferred from one reactant to another.

**reduction**　Gain of electrons by an atom or molecule as occurs when a hydrogen atom is added to a molecule or oxygen is removed. The opposite of **oxidation.**

**reduction potential (E)**　The voltage change when an atom or molecule gains an electron; a measure of the tendency of a molecule to gain an electron. For a given reduction reaction, $E$ has the same magnitude but opposite sign as the **oxidation potential** for the reverse (oxidation) reaction.

**release factor (RF)**   One of two types of nonribosomal proteins that recognize stop codons in mRNA and promote release of the completed polypeptide chain, thereby terminating **translation** (protein synthesis). (Figure 4-27)

**replication fork**   Y-shaped region in double-stranded DNA at which the two strands are separated and replicated during DNA synthesis; also called *growing fork*. (Figure 4-30)

**replication origin**   Unique DNA segment present in an organism's genome at which DNA replication begins. Eukaryotic chromosomes contain multiple origins, whereas bacterial chromosomes and plasmids usually contain just one.

**reporter gene**   A gene encoding a protein that is easily assayed (e.g., β-galactosidase, luciferase). Reporter genes are used in various types of experiments to indicate activation of the promoter to which the reporter gene is linked.

**repression domain**   A region of a repressor transcription factor that will inhibit transcription when fused to a DNA-binding domain.

**repressor**   Specific **transcription factor** that inhibits transcription.

**residue**   General term for the repeating units in a polymer that remain after covalent linkage of the monomeric precursors.

**resolution**   The minimum distance between two objects that can be distinguished by an optical apparatus; also called *resolving power*.

**respiration**   The conversion of energy in nutrients into ATP via a set of reactions involving a series of protein-catalyzed, membrane-associated oxidation and reduction reactions, called an electron transport chain, that are ultimately coupled to the addition of $P_i$ to ADP (oxidative phosphorylation) to form ATP and transfer of electrons to oxygen or other inorganic electron acceptors.

**respiratory chain**   See **electron transport chain**.

**respiratory control**   Dependence of mitochondrial oxidation of NADH and $FADH_2$ on the supply of ADP and $P_i$ for ATP synthesis.

**resting $K^+$ channels**   Nongated $K^+$ ion channels in the plasma membrane that, in conjunction with the high cytosolic $K^+$ concentration produced by the **$Na^+/K^+$ ATPase**, are primarily responsible for generating the inside-negative resting **membrane potential** in animal cells.

**restriction enzyme**   Any enzyme that recognizes and cleaves a specific short sequence, the *restriction site*, in double-stranded DNA molecules; used extensively to produce **recombinant DNA** in vitro; also called *restriction endonuclease*. (Figure 5-11; Table 5-1)

**retrotransposon**   Type of eukaryotic **transposable DNA element** whose movement in the genome is mediated by an RNA intermediate and involves a reverse-transcription step. See also **transposon**. (Figure 6-8b)

**retrovirus**   Type of eukaryotic virus containing an RNA genome that replicates in cells by first making a DNA copy of the RNA.

This viral DNA is inserted into cellular chromosomal DNA, forming a **provirus**, and gives rise to further genomic RNA as well as the mRNAs for viral proteins. (Figure 4-49)

**reverse transcriptase**   Enzyme found in retroviruses that catalyzes a complex reaction in which a double-stranded DNA is synthesized from a single-stranded RNA template. (Figure 6-14)

**R group**   A portion of a molecule, or a chemical group within a larger molecule, that is covalently bonded as an appendage to the main body or core of the molecule. In an amino acid, the R group is the side chain that is attached to the alpha carbon atom and that confers the distinct characteristics of the amino acid.

**ribonucleic acid (RNA)**   See **RNA**.

**ribonucleoprotein (RNP) complex**   General term for any complex composed of proteins and RNA. Most RNA molecules are present in the cell in the form of RNPs.

**ribosome**   A large complex comprising several different **rRNA** molecules and as many as 83 proteins, organized into a large subunit and small subunit; the engine of **translation** (protein synthesis). (Figures 4-22 and 4-23)

**ribosomal RNA**   See **rRNA**.

**ribozyme**   An RNA molecule with catalytic activity. Ribozymes function in RNA splicing and protein synthesis.

**ribulose 1,5-bisphosphate carboxylase**   Enzyme located in chloroplasts that catalyzes the first reaction in the **Calvin cycle**, the addition of $CO_2$ to a five-carbon sugar (ribulose 1,5-bisphosphate) to form two molecules of 3-phosphoglycerate; also called *rubisco*. (Figure 12-45)

**RISC**   See **RNA-induced silencing complex**.

**RNA (ribonucleic acid)**   Linear, single-stranded polymer, composed of ribose **nucleotides**. mRNA, rRNA, and tRNA play different roles in protein synthesis; a variety of small RNAs play roles in controlling the stability and translation of mRNAs and in controlling chromatin structure and transcription. (Figure 4-17)

**RNA editing**   Unusual type of RNA processing in which the sequence of a pre-mRNA is altered.

**RNA-induced silencing complex (RISC)**   Large multiprotein complex associated with a short single-stranded RNA (**siRNA** or **miRNA**) that mediates degradation or translational repression of a complementary or near-complementary mRNA.

**RNA interference (RNAi)**   Functional inactivation of a specific gene by a corresponding double-stranded RNA that induces either inhibition of translation or degradation of the complementary single-stranded mRNA encoded by the gene but not that of mRNAs with a different sequence. (Figures 5-45)

**RNA polymerase**   An enzyme that copies one strand of DNA (the *template* strand) to make the **complementary** RNA strand using as substrates ribonucleoside triphosphates. (Figure 4-11)

**RNA splicing**   A process that results in removal of **introns** and joining of **exons** in pre-mRNAs. See also **spliceosome**. (Figures 8-8 and 8-9)

**rRNA (ribosomal RNA)**   Any one of several large RNA molecules that are structural and functional components of **ribosomes**.

Often designated by their sedimentation coefficient: 28S, 18S, 5.8S, and 5S rRNA in higher eukaryotes. (Figure 4-22)

**rubisco**   See **ribulose 1,5-bisphosphate carboxylase.**

**S (synthesis) phase**   See **cell cycle.**

**sarcomere**   Repeating structural unit of striated (skeletal) muscle composed of organized, overlapping thin (**actin**) filaments and thick (**myosin**) filaments and extending from one Z disk to an adjacent one; shortens during contraction. (Figures 17-31 and 17-32)

**sarcoplasmic reticulum**   Network of membranes in cytoplasm of a muscle cell that sequesters $Ca^{2+}$ ions; release of stored $Ca^{2+}$ induced by muscle stimulation triggers contraction. (Figure 17-34)

**satellite DNA**   See **simple-sequence DNA.**

**saturated**   Referring to a compound (e.g., fatty acid) in which all the carbon-carbon bonds are single bonds.

**SCF (Skp1, Cullin, F-box proteins)**   Ubiquitin-protein ligase that ubiquitinylates inhibitors of S-phase CDKs and many other proteins, marking them for degradation by proteasomes.

**second messenger**   A small intracellular molecule (e.g., cAMP, cGMP, $Ca^{2+}$, DAG, and $IP_3$) whose concentration increases (or decreases) in response to binding of an extracellular signal and that functions in **signal transduction.** (Figure 15-8)

**secondary structure**   In proteins, local folding of a polypeptide chain into regular structures including the α helix, β sheet, and β turns.

**secretory pathway**   Cellular pathway for synthesizing and sorting soluble and membrane proteins localized to the endoplasmic reticulum, Golgi, and lysosomes; plasma membrane proteins; and proteins eventually secreted from the cell. (Figure 14-1)

**segregation**   The process that distributes an equal complement of chromosomes to daughter cells during mitosis and meiosis.

**selectins**   A family of **cell-adhesion molecules** that mediate $Ca^{2+}$-dependent interactions with specific oligosaccharide moieties in glycoproteins and glycolipids on the surface of adjacent cells or in extracellular glycoproteins. (Figures 20-2 and 20-39)

**sensor**   Measures an intra- or extracellular property and converts into a signal.

**shuttle vector**   Plasmid vector capable of propagation in two different hosts. (Figure 5-17)

**side chain**   In amino acids, the variable substituent group attached to the **alpha (α) carbon atom** that largely determines the particular properties of each amino acid; also called *R group.* (Figure 2-14)

**signaling cascade**   Pathway that is activated by an intracellular or extracellular event and then transmits the signal to an **effector.**

**signaling molecule**   General term for any extracellular or intracellular molecule involved in mediating the response of a cell to its external environment or to other cells.

**signal-recognition particle (SRP)**   A cytosolic ribonucleoprotein particle that binds to the ER **signal sequence** in a nascent secretory protein and delivers the nascent chain/ribosome complex to the ER membrane, where synthesis of the protein and translocation into the ER are completed. (Figure 13-5)

**signal sequence**   A relatively short amino acid sequence within a protein that directs the protein to a specific location within the cell; also called *signal peptide* and *uptake-targeting sequence.* (Table 13-1)

**signal transduction**   Conversion of a signal from one physical or chemical form into another. In cell biology commonly refers to the sequential process initiated by binding of an extracellular signal to a receptor and culminating in one or more specific cellular responses.

**silencer**   A sequence in eukaryotic DNA that promotes formation of condensed chromatin structures in a localized region, thereby blocking access of proteins required for transcription of genes within several hundred base pairs of the silencer; also called *silencer sequence.*

**simple diffusion**   Net movement of a molecule across a membrane down its concentration gradient at a rate proportional to the gradient and the permeability of the membrane; also called *passive diffusion.*

**simple-sequence DNA**   Short, tandemly repeated sequences that are found at **centromeres** and **telomeres** as well as at other chromosomal locations and are not transcribed; also called *satellite DNA.*

**siRNA**   A small, double-stranded RNA, 21–23 nucleotides long with two single-stranded nucleotides at each end. A single strand of the siRNA associates with several proteins to form an **RNA-induced silencing complex (RISC)** that cleaves target RNAs to which the siRNA base pairs perfectly; variously called *short* or *small interfering RNA* and *small inhibitory RNA.* siRNAs can be designed to experimentally inhibit expression of specific genes. See also **miRNA.** (Figure 8-25b)

**siRNA knockdown**   A procedure that induces the degradation of a specific RNA by transfecting cells with double-stranded RNA having one strand with the same sequence as the target RNA. This double-stranded RNA is called an **siRNA** for "short-interfering RNA."

**sister chromatid resolution**   The process of untangling of the intertwined **sister chromatids** during prophase.

**sister chromatids**   The two identical DNA molecules created during DNA replication and the associated chromosomal proteins. After DNA replication, each chromosome is composed of two sister chromatids.

**Smads**   Class of transcription factors that are activated by phosphorylation following binding of members of the **transforming growth factor β (TGFβ)** family of signaling molecules to their cell-surface receptors. (Figure 16-28)

**SMC proteins**   Structural maintenance of chromosome proteins; a small family of nonhistone chromatin proteins that are critical for maintaining the morphological structure of chromosomes and their proper segregation during mitosis. Members of this family include *condensins*, which help condense chromosomes during mitosis, and *cohesins*, which link **sister chromatids** until their separation in anaphase. Bacterial SMC proteins function in the proper segregation of bacterial chromosomes to daughter cells. (Figures 6-36 and 19-27)

**SNAREs** Cytosolic and integral membrane proteins that promote fusion of vesicles with target membranes. Interaction of **v-SNAREs** on a vesicle with cognate **t-SNAREs** on a target membrane forms very stable complexes, bringing the vesicle and target membranes into close apposition. (Figure 14-10)

**snoRNA (small nucleolar RNA)** A type of small, stable RNA that functions in rRNA processing and base modification in the nucleolus.

**snRNA (small nuclear RNA)** One of several small, stable RNAs localized to the nucleus. Five snRNAs are components of the **spliceosome** and function in splicing of pre-mRNA. (Figures 8-9 and 8-11)

**somatic cell** Any plant or animal cell other than a **germ cell**.

**sorting signal** A relatively short amino acid sequence within a protein that directs the protein to particular transport vesicles as they bud from a donor membrane in the secretory or endocytic pathway. (Table 14-2)

**Southern blotting** Technique for detecting specific DNA sequences separated by electrophoresis by hybridization to a labeled nucleic acid **probe**. (Figure 5-26)

**specificity** The ability of immune cells or their products to distinguish between structurally closely related molecules.

**sperm** The male gamete—a motile haploid cell that can bind to and fuse with an egg cell, forming a zygote.

**S-phase CDKs** Cyclin-CDK complexes that promote the initiation of DNA replication.

**sphingolipid** Major group of membrane lipids, derived from sphingosine, that contain two long hydrocarbon chains and either a phosphorylated head group (sphingomyelin) or carbohydrate head group (cerebrosides, gangliosides). (Figure 10-8b)

**spindle assembly checkpoint pathway** Pathway that senses incorrect attachment of chromosomes to the mitotic spindle and induces cell cycle arrest in metaphase.

**spindle midzone** Middle portion of the mitotic spindle that plays an important role in cleavage furrow positioning in some organisms.

**spindle pole bodies** Functionally analogous structure of centrosomes in yeast.

**spindle position checkpoint pathway** Pathway that senses incorrect position of the mitotic spindle within the cell and induces cell cycle arrest in anaphase.

**spliceosome** Large ribonucleoprotein complex that assembles on a pre-mRNA and carries out **RNA splicing**. (Figure 8-11)

**SRE-binding proteins (SREBPs)** Cholesterol-dependent transcription factors, localized in the ER membrane, that are activated in response to low cellular cholesterol levels and then stimulate expression of genes encoding proteins involved in cholesterol synthesis and import as well as synthesis of other lipids. (Figure 16-37)

**starch** A very long, branched polysaccharide, composed exclusively of glucose units, that is the primary storage carbohydrate in plant cells. (Figure 12-30)

**START** Point in the cell cycle at which cells are irreversibly committed to cell division and can no longer return to the $G_1$ state.

**steady state** In cellular metabolic pathways, the condition when the rate of formation and rate of consumption of a substance are equal, so that its concentration remains constant. (Figure 2-23)

**stem cell** A self-renewing cell that can divide *symmetrically* to give rise to two daughter cells whose developmental potential is identical to that of the parental stem cell or *asymmetrically* to generate daughter cells with different developmental potentials. (Figure 21-8)

**stem-cell niche** A set of cells, extracellular matrices, and hormones that surrounds a stem cell and that maintains its stem-cell properties

**stereoisomers** Two compounds with identical molecular formulas whose atoms are linked in the same order but in different spatial arrangements. In *optical isomers*, designated D and L, the atoms bonded to an **asymmetric carbon atom** are arranged in a mirror-image fashion. *Geometric isomers* include the *cis* and *trans* forms of molecules containing a double bond.

**steroids** A group of four-ring hydrocarbons including cholesterol and related compounds. Many important hormones (e.g., estrogen and progesterone) are steroids. *Sterols* are steroids containing one or more hydroxyl groups. (Figure 10-8c)

**structural motif** A particular combination of two or more secondary structures that form a distinct three-dimensional structure that appears in multiple proteins and that often, but not always, is associated with a specific function.

**substrate** Molecule that undergoes a charge in a reaction catalyzed by an enzyme.

**substrate-level phosphorylation** Formation of ATP from ADP and $P_i$ catalyzed by cytosolic enzymes in reactions that do not depend on a proton-motive force or molecular oxygen.

**sulfhydryl group (—SH)** A substituent group present in the amino acid cysteine and other molecules consisting of a hydrogen atom covalently bonded to a sulfur atom; also called a *thiol group*.

**symmetric cell division** See **cell division**.

**symport** A type of **cotransport** in which a membrane protein (*symporter*) transports two different molecules or ions across a cell membrane in the *same* direction. See also **antiport**. (Figure 11-2, [3B])

**synapse** Specialized region between an axon terminal of a neuron and an adjacent neuron or other excitable cell (e.g., muscle cell) across which impulses are transmitted. At a *chemical* synapse, the impulse is conducted by a **neurotransmitter**; at an *electric* synapse, impulse transmission occurs via **gap junctions** connecting the pre- and postsynaptic cells. (Figure 22-3)

**synaptic vesicles** Small vesicles in axon termini that contain a neurotransmitter and that undergo exocytosis following arrival of an action potential.

**synaptonemal complex (SC)** Proteinaceous structure that mediates the association (synapsis) between homologous chromosomes during prophase of meiosis I.

**syndecans** A class of cell-surface **proteoglycans** that function in cell-matrix adhesion, interact with the cytoskeleton, and may bind external signals, thereby participating in cell-cell signaling.

**syntelic attachment** Indicates that kinetochores of a **sister chromatid** pair attach to microtubules emanating from the same pole.

**synteny** Occurrence of genes in the same order on a chromosome in two or more different species.

**TATA box** A conserved sequence in the **promoter** of many eukaryotic protein-coding genes where the transcription-initiation complex assembles. (Figure 7-14)

**T cell** A lymphocyte that matures in the thymus and expresses antigen-specific receptors that bind antigenic peptides complexed to **MHC molecules.** There are two major classes: *cytotoxic* T cells (CD8 surface marker, class I MHC restricted, kill virus-infected and tumor cells) and *helper* T cells (CD4 marker, class II MHC restricted, produce cytokines, required for activation of B cells). (Figures 23-32 and 23-35)

**T-cell receptor** A heterodimeric antigen-binding transmembrane protein containing variable and constant regions and associated with the signal-transducing multimeric CD3 complex. (Figure 23-27)

**telomere** Region at each end of a eukaryotic chromosome containing multiple tandem repeats of a short telomeric (TEL) sequence. Telomeres are required for proper chromosome **segregation** and are replicated by a special process that prevents shortening of chromosomes during DNA replication. (Figure 6-47)

**telophase** Final mitotic stage, during which the nuclear envelope re-forms around the two sets of separated chromosomes, the chromosomes decondense, and division of the cytoplasm (cytokinesis) is completed. (Figure 18-36)

**temperature-sensitive (ts) mutation** A mutation that produces a wild-type phenotype at one temperature (the permissive temperature) but a mutant phenotype at another temperature (the nonpermissive temperature). This type of mutation is especially useful in identification of genes essential for life. (Figure 5-6)

**termination, transcription** Cessation of the synthesis of an RNA chain. (Figure 4-11)

**tertiary structure** In proteins, overall three-dimensional form of a polypeptide chain, which is stabilized by multiple noncovalent interactions between side chains. (Figure 3-10a)

**thylakoids** Flattened membranous sacs in a chloroplast that can be arranged in stacks and contain the photosynthetic pigments and **photosystems.** (Figure 12-31)

**tight junction** A type of cell-cell junction between the plasma membranes of adjacent epithelial cells that prevents diffusion of macromolecules and many small molecules and ions in the spaces between cells and diffusion of membrane components between the apical and basolateral regions of the plasma membrane. (Figure 20-16)

**tolerance** The absence of an immune response to a particular antigen or set of antigens.

**Toll-like receptor (TLR)** Member of a class of cell-surface and intracellular receptors that recognize a variety of microbial products. Ligand binding initiates a signaling pathway that induces various responses depending on the cell type. (Figure 23-33)

**topogenic sequences** Segments within a protein whose sequence, number, and arrangement direct the insertion and orientation of various classes of transmembrane proteins into the endoplasmic reticulum membrane. (Figure 13-14)

**transcription** Process in which one strand of a DNA molecule is used as a template for synthesis of a **complementary** RNA by RNA polymerase. (Figures 4-10 and 4-11)

**transcription-control region** Collective term for all the DNA regulatory sequences that regulate transcription of a particular gene.

**transcription factor (TF)** General term for any protein, other than RNA polymerase, required to initiate or regulate transcription in eukaryotic cells. *General* factors, required for transcription of all genes, participate in formation of the transcription-preinitiation complex near the start site. *Specific* factors stimulate (**activators**) or inhibit (**repressors**) transcription of particular genes by binding to their regulatory sequences.

**transcription unit** A region in DNA, bounded by an initiation (start) site and termination site, that is transcribed into a single **primary transcript.**

**transcytosis** Mechanism for transporting certain substances across an epithelial sheet that combines **receptor-mediated endocytosis** and **exocytosis.** (Figures 14-25 and 23-10)

**transfection** Experimental introduction of foreign DNA into cells in culture, usually followed by expression of genes in the introduced DNA. (Figure 5-32)

**transfer RNA** See **tRNA.**

**transformation** (1) Permanent, heritable alteration in a cell resulting from the uptake and incorporation of a foreign DNA into the host-cell genome; also called *stable transfection.* (2) Conversion of a "normal" mammalian cell into a cell with cancer-like properties usually induced by treatment with a virus or other cancer-causing agent.

**transforming growth factor beta (TGFβ)** A family of secreted signaling proteins that are used in the development of most tissues in most or all animals. Members of the TGFβ family more often inhibit growth than stimulate it. Mutations in TGFβ signal transduction components are implicated in human cancer, including breast cancer. (Figures 16-27 and 16-28)

**transgene** A cloned gene that is introduced and stably incorporated into a plant or animal and is passed on to successive generations.

***trans*-Golgi network (TGN)** Complex network of membranes and vesicles that serves as a major branch point in the **secretory pathway.** Vesicles budding from this most-distal Golgi compartment carry membrane and soluble proteins to the cell surface or to lysosomes. (Figures 14-1 and 14-17)

**transition state** State of the reactants during a chemical reaction when the system is at its highest energy level; also called the **transition-state intermediate.**

**translation**  The **ribosome**-mediated assembly of a polypeptide whose amino acid sequence is specified by the nucleotide sequence in an mRNA. (Figure 4-17)

**translocon**  Multiprotein complex in the membrane of the rough endoplasmic reticulum through which a nascent secretory protein enters the ER lumen as it is being synthesized. (Figure 13-7)

**transporters**  Membrane proteins that undergo conformational changes as they move a wide variety of ions and molecules across cell membranes at a slower rate than channels. See *uniporter, symporter,* and *antiporter* in Figure 11-3.

**transport protein**  See **membrane transport protein**.

**transport vesicle**  A small membrane-bounded compartment that carries soluble and membrane "cargo" proteins in the forward or reverse direction in the **secretory pathway**. Vesicles form by budding off from the donor organelle and release their contents by fusion with the target membrane.

**transposable DNA element**  Any DNA sequence that is not present in the same chromosomal location in all individuals of a species and can move to a new position by **transposition**; also called *mobile DNA element* and *interspersed repeat.* (Table 6-1)

**transposition**  Movement of a **transposable DNA element** within the genome; occurs by a cut-and-paste mechanism or copy-and-paste mechanism depending on the type of element. (Figure 6-8)

**transposon, DNA**  A **transposable DNA element** present in prokaryotes and eukaryotes that moves in the genome by a mechanism involving DNA synthesis and transposition. See also **retrotransposon**. (Figures 6-9 and 6-10)

**triacylglycerol**  See **triglyceride**.

**triglyceride**  Major form in which fatty acids are stored and transported in animals; consists of three fatty acyl chains esterified to a glycerol molecule.

**trimeric (large) G protein**  A regulatory membrane-associated GTPase consisting of a catalytic α subunit and a β and γ subunit. When the α subunit is bound to GTP, the β and γ subunits dissociate as a heterodimer. The free α subunit and the free β-γ heterodimer can then interact with other proteins and transduce a signal across the membrane. When the α subunit hydrolyzes the GTP to GDP and phosphate, the α-GDP associates with the β-γ heterodimer, terminating signaling. (Figure 15-17)

**tRNA (transfer RNA)**  A group of small RNA molecules that function as amino acid donors during protein synthesis. Each tRNA becomes covalently linked to a particular amino acid, forming an **aminoacyl-tRNA**. (Figures 4-19 and 4-20)

**trophic factor**  Any of numerous signaling proteins required for the survival of cells in multicellular organisms; in the absence of such signals, cells often undergo "suicide" by **apoptosis**.

**trophoectoderm (TE)**  The part of an early mammalian embryo that will form the extra-embryonic tissues, including the placenta but not the embryo proper.

**t-SNAREs**  See **SNAREs**.

**tubulin**  A family of globular cytoskeletal proteins that polymerize to form the cylindrical wall of **microtubules**. (Figure 18-3)

**tumor**  A mass of cells, generally derived from a single cell, that arises due to loss of the normal regulators of cell growth; may be **benign** or **malignant**.

**tumor-suppressor gene**  Any gene whose encoded protein directly or indirectly inhibits progression through the cell cycle and in which a loss-of-function mutation is oncogenic. Inheritance of a single mutant allele of many tumor-suppressor genes (e.g., *RB, APC,* and *BRCA1*) greatly increases the risk for developing colorectal and other types of cancer. (Figures 24-8 and 24-11)

**ubiquitin**  A small protein that can be covalently linked to other intracellular proteins, thereby tagging these proteins for degradation by the **proteasome**, sorting to the lysosome, or alteration in the function of the target protein. (Figure 3-29)

**uncoupler**  Any natural substance (e.g., the protein thermogenin) or chemical agent (e.g., 2,4-dinitrophenol) that dissipates the **proton-motive force** across the inner mitochondrial membrane or thylakoid membrane of chloroplasts, thereby inhibiting ATP synthesis.

**unsaturated**  Referring to a compound (e.g., fatty acid) in which one of the carbon-carbon bonds is a double or triple bond.

**upstream**  (1) For a gene, the direction opposite to that in which RNA polymerase moves during transcription. Nucleotides upstream from the +1 position (the first transcribed nucleotide) are designated $-1$, $-2$, etc. (2) Events that occur earlier in a cascade of steps (e.g., signaling pathway). See also **downstream**.

**upstream activating sequence (UAS)**  Any protein-binding regulatory sequence in the DNA of yeast and other simple eukaryotes that is necessary for maximal gene expression; equivalent to an enhancer or promoter-proximal element in higher eukaryotes. (Figure 7-22)

**vaccine**  An innocuous preparation derived from a pathogen and designed to elicit an immune response in order to provide immunity against a future challenge by a virulent form of the same pathogen.

**van der Waals interaction**  A weak **noncovalent interaction** due to small, transient asymmetric electron distributions around atoms (dipoles). (Figure 2-10)

**vector**  In cell biology, an autonomously replicating genetic element used to carry a cDNA or fragment of genomic DNA into a host cell for the purpose of gene cloning. Commonly used vectors are bacterial plasmids and modified bacteriophage genomes. See also **expression vector** and **shuttle vector**. (Figure 5-13)

**viral envelope**  A phospholipid bilayer forming the outer covering of some viruses (e.g., influenza and rabies viruses); is derived by budding from a host-cell membrane and contains virus-encoded glycoproteins. (Figure 4-47)

**virion**  An individual viral particle.

**virus**  A small intracellular parasite, consisting of nucleic acid (RNA or DNA) enclosed in a protein coat, that can replicate only in a susceptible host cell; widely used in cell biology research. (Figure 4-44)

**$V_{max}$**  Parameter that describes the maximal velocity of an enzyme-catalyzed reaction or other process such as protein-mediated transport of molecules across a membrane. (Figures 3-22 and 11-4) **v-SNAREs** See **SNAREs**.

**Warburg effect** Named after its discoverer, Otto Warburg, describes the observation that most cancer cells predominantly produce energy by glycolysis followed by fermentation of pyruvate to lactic acid. While normal cells use this form of energy production only under the oxygen-limiting (anaerobic) condition, cancer cells metabolize glucose in this manner even in the presence of sufficient oxygen. The process is thus also referred to as *aerobic glycolysis*.

**Wee1** protein-tyrosine kinase; phosphorylates CDKs on threonine 14 and tyrosine 15 to inhibit CDK activity.

**wild type** Normal, nonmutant form of a gene, protein, cell, or organism.

**Wnt** A family of secreted signaling proteins used in the development of most tissues in most or all animals. Mutations in Wnt signal transduction components are implicated in human cancer, especially colon cancer. Receptors are Frizzled-class proteins with seven transmembrane segments. (Figure 16-30)

**x-ray crystallography** Commonly used technique for determining the three-dimensional structure of macromolecules (particularly proteins and nucleic acids) by passing x-rays through a crystal of the purified molecules and analyzing the diffraction pattern of discrete spots that results. ( Figure 3-43)

**zinc finger** Several related DNA-binding **structural motifs** composed of secondary structures folded around a zinc ion; present in numerous eukaryotic transcription factors. (Figures 3-9b and Figure 7-29a and b)

Interneurons, 1020, 1022
Interphase, 876
Interphase chromatin, transcription factors, 266
Interphase chromosomes
    by DNA amplification, 269–270
    territories, 264f, 265
Interspersed repeats. *See* Simple-sequence DNA
Intestinal stem cells
    cells at base of crypts as, 990f
    epithelial cell generation, 988–991
    illustrated, 989f
    Lgr5-expressing, 990f, 991f
    regeneration, 989f
Intracellular interactions, 928
Intraflagellar transport (IFT)
    function of, 847
    illustrated, 847f
    moving material up/down cilia and flagella, 846–847
    proteins, 756
Intronic splicing enhancers, 363
Intronic splicing silencers, 362
Introns
    defined, 127, 224
    in eukaryotes, 127, 225
    group I, 357, 389–390, 389f
    group II, 389f
    pre-mRNA (precursor mRNA), 355
    RNA splicing, 351, 352f
    self-splicing, 357, 358f
Invertebrates, experimental systems, 19–20
Inverted repeats, 236f
Ion channels, 1043–1044
    defined, 476
    elongated, 497
    G protein-coupled receptors regulation of, 693–699
    gated, 1043f
    gated cation, 1047–1048
    ion movement measurement through, 499–501
    nongated, 495–502
    novel, 501
    resting $K^+$, 497, 498f
    selectivity filter, 497–499, 499f
    sequential activation at neuromuscular junction, 1043f
    store-operated, 709
    voltage-gated ion, 1025–1036
Ion traps, 102
Ion-exchange chromatography, 97, 98f
Ionic gradients
    in biological processes, 495
    generation and maintenance, 485
Ionic interactions, 28
Ions
    $Ca^{2+}$, 486, 487
    $K^+$, 497, 498
    movement measurement, 499–501, 500f
    $Na^+$, 495, 498
    selective movement of, 495–497
    selectivity and transport in resting $K^+$ channels, 499f
    selectivity filter, 497–499, 499f
    transport proteins in accumulation of, 507, 507f
Ion-selectivity filter, 497–499, 499f
Ion-trap mass spectroscopy, 102, 103f
Iron response element-binding protein (IRE-BP), 379
Iron-sulfur clusters, 535
Isoelectric focusing (IEF), 96
Isoelectric point (pI), 96

Isoforms
    defined, 130, 933
    *Dscam*, 364
    fibronectin, 361
Isoleucine, 33, 35f

**J**
JAK kinase, 728–729, 737–738
Jun N-terminal kinases, 744
Junction adhesion molecules (JAMs), 941, 965
Junctional imprecision, 1075

**K**
$K^+$ channels
    inactivation of, 1032
    mutant, experiments, 1032f
    secondary structures of, 1030f
    Shaker, 1029
    voltage-gated, 1027–1028
    voltage-sensitive, 1031f
$K^+$ ions, 498, 499
$K^+$-channel proteins, 363–364
KDEL
    receptors, 642–643, 643f
    sorting signals, 642
Kearns-Sayre syndrome, 250
Keratinocytes, 862
Keratins, 862–863
Kinases, cyclin-dependent, 149
Kinesin-1
    anterograde transport, 834–836
    ATP cycle of movement, 837
    ATP to walk down microtubules, 837f
    heads, 839f
    heavy chains and light chains, 836–837
    as highly processive motor, 837
    movement of vesicles, 834–836
    structure of, 835f
    vesicle transport model, 836f
Kinesin-5, 852
Kinesins
    in organelle transport, 841–842
    in pole separation, 857–858
    superfamily, 836–837, 836f
Kinetic energy, 48
Kinetochore microtubules, 851
Kinetochores
    bi-oriented, 918
    co-oriented, 918
    defined, 273, 850, 876, 899
    microtubule attachment, 854
    structure of, 853f
    yeast, 853
Kleisins, 264
Knee-jerk reflex, 1023f
Knockout mice, 213–214
Kozak sequence, 140

**L**
*Lac* operon, 282–284
Lagging strand, 146, 147f
Laminin
    defined, 947
    illustrated, 948f
Lamins, 863, 898–899
Large T-antigen, 147
Late endosome, 629
Lateral inhibition, 761
LC-MS/MS
    defined, 104
    high-throughput, 108
    in protein identification in organelles, 108f
    in protein identification of complex sample, 107f

Leading edge, 776
Leading strand, 146, 147f
Leaf
    anatomy of $C_4$ plants and $C_4$ pathway, 571f
    cellular structure of, 553f
    starch, 552, 552f
Leaflets, 446
Leber's hereditary optic neuropathy, 250
Lectins, 598
Lentiviruses, 203
Lesion-bypass DNA polymerases, 1147
Leucine, 33, 35f
Leucine zippers, 66
Leucine-zipper proteins, 311
Leukemias, 1115
Leukocyte-adhesion deficiency, 967
LHCII, 565, 566f
Libraries
    cDNA, 185–186, 187f, 189f
    DNA, 185, 188
    genomic, 185, 188–191
Ligands, 78f, 724f
    binding of, 77–78
    binding site, 77
    defined, 44, 77
    low-affinity, 683f
    molecular complementarity, 676
    protein binding, 77–78, 78f
    receptor affinity for, 681–682, 684
    for receptor tyrosine kinases, 723–724
    signaling molecules as, 674
Light chains, 795
    DNA embryonic cells, 1112f
    immunoglobulins, 1068
    V and J gene segment assembly, 1074–1075
    variable region of, 1070
    V-J rearrangement, 1079
Light microscopes
    development of, 404f
    numerical aperture (NA), 405
    resolution, 403–404
Light microscopy, 404–418. *See also* Microscopy
Light reactions, 555
Light-absorbing pigments, 552–559
Light-harvesting complexes (LHCs), 557
    in cyanobacteria and plants, 558f
    defined, 555–556
    in photosynthesis efficiency, 557
Linkage disequilibrium, 209, 209f
Linkage-mapping, 207–208
Linker scanning mutations, 302, 303f
Lipid droplets, 455, 455f
Lipid rafts, 454
Lipid-anchored membrane proteins, 456
Lipids
    bilayer, 7f, 456
    binding motifs, 462
    classes in biomembranes, 448–450, 449f
    components of biomembranes, 452t
    composition, 452–454
    differential transport of, 469
    in exoplasmic/cytosolic leaflets, 453–454
    in membrane protein anchoring, 460–461, 461f
    mobility in biomembranes, 450–452
    phase transition, 451
    phosphoglycerides, 448–450
    physical properties influence, 452–453
Lipoproteins, 656
Liposomes, 445

Oligodendrocytes, 1033
Oligonucleotide probes, 188
Oligopeptides, 62
Oligosaccharides
  compartment-specific modifications, 631–632
  glycoprotein attachment, 596
  N-linked, 595–596f, 955
  O-linked, 594
  precursor, biosynthesis of, 595f
Oligosaccharyl transferase, 595
O-linked oligosaccharides
  binding to basal lamina components, 964
  defined, 594, 955
Oncogenes, 740, 1113. *See also* Cancer
  addiction, 1132
  cancer-causing viruses and, 1127–1128
  defined, 164, 1119
  mutations, effects of, 1133f
  proto-oncogenes conversion into, 1125–1126
  receptors, 1132–1133
  signal transduction proteins, 1134–1136
  successive mutations, 1120–1122
Oncogenic transformation, 1118
Oncoproteins, 751, 1133–1134
1,2-diacylglycerol (DAG), 708
  activation of protein kinase C, 709–711
  IP$_3$ pathway, 710f
  second messenger synthesis, 709f
1000 Genomes Project, 252
Oocytes
  contents, 979
  defined, 977
  germ-line stem cell production of, 987–988
  in novel channel characterization, 501, 501f
Op18/stathmin, 832–833
Open reading frames (ORFs) analysis, 253–254
Operons
  *lac*, 282–284
  *Trp*, 286, 287f, 288
Opsonization, 1064
Optical isomers. *See* Stereoisomers
Optical microscopes, 407
ORC protein, 149
Organelles. *See also specific Organelles*
  in axons, 833–834
  characterization of, 398, 424–430
  defined, 398
  DNA, 245–251
  eukaryotes, 13
  of eukaryotic cells, 424–427
  examples, 426f
  highly purified, 429
  isolation of, 424–430
  membranes, 14, 473
  parent, 634
  photosynthesis, 427
  protein composition of, 430
  protein identification in, 108f
  release, 427
  separation, 427–429, 441–442
  target, 634
  transport by microtubule motors, 841f
Organelle-specific antibodies, 429
Organisms
  descendence, 2f
  experimental, 12f
  genome sizes of, 8t
  model, 16
Organs
  development regulation in animals, 20f
  tissue organization in, 18, 18f
Origin-recognition complex (ORC), 894
Osmosis, 480
Osmotic gradient, 509

Osmotic pressure
  illustrated, 481f
  in moving water across membranes, 480–481
Osteoclasts, 510, 510f
Osteopetrosis, 510
Outer-membrane proteins, 610
Oxidation
  defined, 54
  potentials, 55
Oxidative phosphorylation, 519, 551
Oxygen-evolving complex, 562–563
Oxyntic cells. *See* Parietal cells

# P

P bodies
  defined, 372
  as translation repression sites, 376
p53
  in abolishing DNA damage checkpoint, 1141–1143
  active form, 1142–1143
  activity of, 1141
  in cancerous formation defense, 1143
  mutations as dominant negatives, 1143
Pain receptors, as gated cation channels, 1048
Palade, George, 671–672
Pancreatic endoderm, 984
Paneth cells, 989, 991
Par proteins
  in asymmetric cell division, 1004–1005
  asymmetric localization, 1000f
  in cell asymmetry, 998–1001
  in epithelial-cell polarity, 1001–1002
Paracellular pathways, 942, 942f
Paracrine signaling, 675–676
Parietal cells
  defined, 509
  in stomach acidification, 509–510f
Patch clamping
  defined, 500
  in ion movement measurement, 499–501, 500f
  in novel channel characterization, 501, 501f
  tracings, 500, 501f
Pathogens
  boundaries as defense against, 1062
  entry and replication, 1061
  vaccines and, 1105–1107
Patterning genes, 19
PC2 endoprotease, 651
P-class ion pumps, 484, 484f
PDZ domains, 941
Pentoses, 37–38
Peptide bonds
  amino acid links by, 33
  cleavage of, 92
  defined, 61
  planar, 71
  serine-protease-mediated hydrolysis of, 82f
  synthesis, 141
Peptides
  binding, 1084–1086, 1086f
  binding to class I molecules, 1088
  binding to class II molecules, 1089–1091
  defined, 62
  delivery to class I molecules, 1088
  encounter with class II molecules, 1089
  heavy sample, 104
  mass fingerprint, 104
Peptidoglycan, 39
Peptidyl-prolyl isomerases, 598
Peptidyltransferase reaction, 141
Perforins, 1099, 1100f
Pericentriolar material, 825
Peripheral membrane proteins, 456

Perlecan, 947
Permissive temperature, 176
Peroxisomal proteins
  biogenesis and division, 614f
  incorporation by different pathways, 613–615, 614f
  PTS1-directed import, 612–613, 613f
  targeting, 577, 612–615
  targeting sequence 1, 612
Peroxisomes, 427
  division of, 614–615
  oxidation of fatty acids in, 531–532, 531f
PH
  biological fluids, 45–46
  common solutions, 46f
  defined, 45
  dissociation of acids and, 47f
  enzyme activity dependence, 83f
  of late endosomes, 658–659
  maintaining with buffers, 47, 48f
Phages
  defined, 160
  temperate, 164
Phagocytes, 1062, 1063
Phagocytosis
  cytoskeleton remodel, 1089
  defined, 654, 776
  dynamics, 789f
  in recognition/removal of pathogens, 789
Phase-contrast microscopy, 405–408
Phenotype
  defined, 172
  genotype versus, 172
  mutant alleles effects on, 173f
  of mutations, 280f
Phenylalanine, 33, 35f
Pheromone-induced mating, 742f
Pheromones, 673
Phosphatases
  activation of, 677
  dephosphorylation and, 90
  in signaling pathways, 677–678
Phosphate transporter, 550
Phosphatidylinositol (PI), 708, 709f
Phosphatidylinositol-3 (PI-3)
  kinase, 745–746
  kinase pathway, 745, 747
  phosphate accumulation, 746
  phosphate generation, 746f
  PKB recruitment and activation in, 747f
Phosphoanhydride bonds, 52
Phosphodiester bonds, 33, 118
Phosphoglycerides, 448–450
  common, 41t
  defined, 40
  head groups, 41t
Phosphoinositides
  defined, 745
  signaling, 814–815, 814f
Phospholipase C (PLCs)
  activation of, 745
  defined, 708
  second messenger generation, 708–711
  SH2 domains, 745
Phospholipases
  defined, 454, 692
  specificity of, 454f
Phospholipid bilayers
  formation, 445–446, 446f
  gel and fluid forms, 450f
  lipid composition effects, 453f
  permeability of, 474f
  sealed compartment, 446–448
  structure of, 445